ELSEVIER'S DICTIONARY OF FORESTRY

ELSEVIER'S DICTIONARY OF FORESTRY

in
English, German, French and Russian

compiled by

B. DELIJSKA and P. MANOILOV
Sofia, Bulgaria

2004

ELSEVIER

Amsterdam – Boston – Heidelberg – London – New York – Oxford
Paris – San Diego – San Francisco – Singapore – Sydney – Tokyo

ELSEVIER B.V.	ELSEVIER Inc.	ELSEVIER Ltd	ELSEVIER Ltd
Sara Burgerhartstraat 25	525 B Street, Suite 1900	The Boulevard, Langford Lane	84 Theobalds Road
P.O. Box 211	San Diego, CA 92101-4495	Kidlington, Oxford OX5 1GB	London WC1X 8RR
1000 AE Amsterdam	USA	UK	UK
The Netherlands			

First edition 2004

Library of Congress Cataloging in Publication Data
A catalog record is available from the Library of Congress.

British Library Cataloguing in Publication Data
A catalogue record is available from the British Library.

ISBN: 0-444-51245-4
Transferred to digital print 2007

Printed and bound by CPI Antony Rowe, Eastbourne

PREFACE

The dictionary contains 9800 terms with about 8600 cross-references and more than 2000 tree species, used today in the broad field of forest science, management, and conservation. The terminology comprises: harvesting and forest operations, integrated forest management, agroforestry, urban forestry, wildlife and wilderness management, silviculture, forest survey and mensuration, forest ecology, forest entomology and pathology, dendrology, forest fire management, forest geomatics, forest meteorology and hydrology, forest soils and geomorphology, forest genetics and tree improvement, physico-mechanical properties of wood, wood anatomy and structure, tree physiology, wood processing etc. Many terms include definitions. Latin names of tree species and insects are also given.

The modern terms were taken from Internet sites and from current publications in journals in the respective languages.

The dictionary consists of two parts. In the first part, the *Basic Table*, the English terms are listed alphabetically and numbered consecutively. The English term is followed by its German, French and Russian equivalents. The synonyms of the English terms are also given as cross-references to the main entries, in their proper alphabetical order. The second part, *the Indexes,* contains separate alphabetical indexes of the German, French and Russian terms. The reference number(s) of every term stand(s) for the number of the English equivalent in the basic table.

The authors hope that Elsevier's Dictionary of forestry will be a valuable tool for managers, researchers, teachers, students, applied scientists, forest engineers and technicians and for everyone who takes interest in the forest science and practice.

Dr. B. Delijska
P. P. Manoilov

EXPLANATION OF SPECIAL SIGNS

1. The italics *d*, *f* and *r* in the basic table stand respectively for the German, French and Russian equivalents of the main English terms.
2. The gender of nouns is indicated as follows:

f	feminine	*fpl*	feminine plural
m	masculine	*mpl*	masculine plural
n	neuter	*npl*	neuter plural
pl	plural	*m/f*	masculine or feminine
		m/n	masculine or neuter

3. The symbol *v* designates a verb.
4. The abbreviation *adj* designates an adjective.
5. The abbreviation *adv* designates an adverb.
6. The abbreviation *lat* designates a Latin word.
7. The abbreviation (UK) means United Kingdom usage; (US) means American usage.
8. Synonyms and abbreviations are separated by semicolons.
9. Two kinds of brackets are used:

   [ ] the information can be either included or left out;

   ( ) the information does not form an integral part of the expression, but helps to clarify it.

BIBLIOGRAPHY

Bachmann P., *Glossar der Waldwachstum* (http://www.wsl.ch/forest/waldman/vorlesung/ww_gloss.ehtml)

Chokkalingam U., Jong W., *Secondary Forest: A Working Definition and Typology* (Center for International Forest Research, International Forestry Review, 3 (1) 2001, CIFOR)

Chudnoff, M., *Tropical Timbers of the World* (Agriculture Handbook No. 607, Washington, DC, US Department of Agriculture, Forest Service, p. 113, 1984)

Cumulative Glossary for Vascular Plants (http://155.187.10.12/glossary/fl-aust.html)

Diggs G.M., Lipscomb B.L., O'Kennon, R.J., *Illustrated Flora of Northcentral Texas* (Sida Botanical Miscellany No. 16, Forth Worth, Botanical Research Institute of Texas, Texas, 1999)

Dobbertin M., Schweingruber F.H., *Forest Multilingual Glossary of Dendrochronology* (http://www01.wsl.ch/glossary)

European Terminology Database (http://europa.eu.int/eurodicautom)

Ford-Robertson F.C., *Terminology of Forest Science Technology Practice and Products* (Washington, DC: Society of American Foresters, 1971)

Francis J.K., *English-Portuguese Equivalents of Forestry and Conservation Terms* (http://www.srs.fs.fed.us/pubs/gtr/gtr_so109.pdf)

Glossary of Forest Genetics (www.fgcouncil.bc.ca/docsglos.htm)

Grieg-Smith P., *Quantitative Plant Ecology* (University of California Press, Berkeley, 1983)

Helms J.A., *The Dictionary of Forestry* (Society of American Foresters, Bethesda, 1998)

Le Grand Dictionnaire Terminologique (http://www.granddictionnaire.com/btml/fra/r_motclef/index1024_1.asp)

Levitt J., *Responses of Plants to Environmental Stresses* (Volume I. Chilling, Freezing, and High Temperature Stresses, New York: Academic Press, pp. 7-18, 1980)

Lincoln R. Boxshall G., Clark P., *A Dictionary of Ecology, Evolution and Systematics* (Cambridge University Press, Cambridge, UK, 1998)

Macura P., *Elsevier's Dictionary of Botany* (Elsevier, Amsterdam, 1982)

Macura P., *Elsevier's Dictionary of Botany* (Elsevier, Amsterdam, 2002) (http://www.elsevier.com/locate/isbn/0444512292)

McPherson G., Wade E., Phillips C.B., *Glossary of Wildland Fire Management Terms* (Society of American Foresters, Bethesda, 1990)

Michael A., *Oxford Dictionary of Botany* (New York, Oxford University Press, 1992)

Mutch R.W., *Sustaining Forest Health to Benefit People, Property and Natural Resources* (proceedings of the Society of American Foresters National Convention, Richmond, VA, 1992 October 25-27)

Schuck A., Päivinen R., Hytönen T. and Pajari B., *Compilation of Forestry Terms and Definitions* (http://www.efi.fi/publications/Internal_Reports/IR_06.pdf)

Spurr S.H., Barnes B.V., *Forest Ecology* (Ronald Press, New York, 1973)

Vaucher H., *Elsevier's Dictionary of Trees and Shrubs* (Elsevier, Amsterdam, 1986) (http://www.elsevier.com/locate/isbn/0444425691)

Wenger K.F., *Forestry Handbook* (John Wiley & Sons, New York, 1984)

Биологический словарь (http://bioword.narod.ru)

Древесина. Пороки (http://ddderevo.narod.ru)

Лесная энциклопедия (http://www.forest.ru/rus/basics/glossary)

Мультитран (http://www.multitran.ru/)

Basic Table

A

* **AAC → 192**

1 **AAC apportionment**
d Jahreseinschlag-Aufteilung *f*
f répartition *f* de coupe annuelle permissible; partage *m* de coupe annuelle permissible
r долевое распределение *n* расчётной годичной лесосеки

* **Aaron's beard → 2908**

* **abaca [banana] → 5283**

2 **abandoned field; old field; old land**
(area of cleared open land no longer used for cultivation or pasture which may be in the process of reverting to forest)
d verlassener Boden *m*; leerstehendes Feld *n*
f champ *m* abandonné
r залежь *f*

3 **abandoned tree**
d verlassener Baum *m*
f arbre *m* abandonné
r поражённое дерево *n*, покинутое насекомыми

* **Abassian box → 2886**

4 **abatis; heap of trees; tree groyne; tree entanglement**
(line of defense consisting of a barrier of felled or live trees with branches)
d Baumsperre *f*
f abatis *m*
r засека *f*; завал *m* из деревьев; заграждение *n* из деревьев

5 **abaxial; abaxile; dorsal**
(facing away from the axis of an organ or organism)
d abaxial
f abaxial
r абаксиальный; направленный от оси

* **abaxial parenchyma → 9091**

* **abelmosk → 5639**

* **abelmusk → 5639**

6 **abiotic; non-biotic; nonliving** *adj*
(ref. to any factor in the environment that is non-living)
d abiotisch
f abiotique
r абиотический; неживой

7 **abiotic damage**
d abiotischer Schaden *m*
f dommage *m* abiotique
r абиотическое повреждение *n*

8 **abiotic environmental factor**
d abiotischer Umweltfaktor *m*
f facteur *m* abiotique du milieu; facteur abiotique d'ambiance
r абиотический фактор *m* окружающей среды

9 **Abney level**
(hand-held clinometer used to measure slope in percent)
d Abney-Messgerät *n*
f niveau *m* d'Abney
r ручной клинометр *m*; ручной нивелир *m* [Абнея]

10 **abnodate** *v*; **debranch** *v*; **disbranch** *v*; **knock** *v* **off; limb** *v*; **knot** *v*
(to remove the limbs from a felled tree)
d asten; ästen; ausasten; ausästen; entasten; entzweigen
f ébrancher
r обрезать сучья; обрубать сучья; очищать деревья от сучьев; отпиливать сучья

* **abnodation → 6610**

11 **aboveground; epigeous** *adj*
(growing on, or close to, the ground)
d über dem Boden; epigäisch; oberirdisch
f au-dessus du sol; épigé
r надземный

12 **aboveground biomass**
(portion of a tree, excluding the root system)
d epigäische Biomasse *f*; Biomasse über dem Boden
f biomasse *f* au-dessus du sol; biomasse épigée
r надземная биомасса *f*

* **aboveground length → 8694**

13 abruptly pinnate leaf; paripinnate leaf; even-pinnate leaf
(a pinnately-compound leaf lacking a terminal leaflet)
d paarig gefiedertes Blatt *n*
f feuille *f* paire-pennée
r парноперистый лист *m*

14 abscissed leaf; fallen leaf
d gefallendes Blatt *n*
f feuille *f* tombée
r опавший лист *m*

* **abscission of leaves → 2941**

15 abscission period
(the period of shedding of leaves, flowers, or fruits following the formation of the abscission zone)
d Entblätterungszeit *f*
f période *f* d'abscission; période d'excision
r период *m* сбрасывания

16 abscission zone
(the area of separation when a plant sheds a leaf, flower or fruit)
d Entblätterungszone *f*
f zone *f* d'abscission
r зона *f* сбрасывания

* **absent ring → 5479**

17 absolute dry wood; absolute dry timber
d absolutes Darrholz *n*
f bois *m* anhydre; bois sec absolument
r абсолютно сухая древесина *f*; абсолютно сухой лесоматериал *m*

18 absolute dry wood density; nominal density; basic density
d absolute Darrholzdichte *f*
f densité *f* nominale de bois; densité théorique de bois
r плотность *f* абсолютно сухой древесины; абсолютная плотность

19 absolute forest soil
d absoluter Waldboden *m*
f sol *m* entièrement consacré à la forestière
r абсолютная лесная почва *f*

* **absorbed moisture → 950**

20 absorbing root
r поглощающий корень *m*

21 abundance of species; species abundance
(the actual number of species per unit of area or volume)
d Artenabundanz *f*
f abondance *f* d'espèce
r распространённость *f* вида

* **abura → 4101**

* **abut → 2980**

22 abutilon; flowering maple
(Abutilon)
d Abutilon *m*; Malve *f*
f abutilon *m*
r абутилон *m*

* **acacia → 1754**

23 Acadian forest
(a transitional zone in Canada within the southeastern mixed forest defined by Rowe in 1972 as the area between the coniferous boreal region to the north and the deciduous region to the south)
d akadischer Wald *m*
f forêt *f* acadienne; forêt d'Acadie
r акадский лес *m*

24 acanthus
(a classical fleshy leaf decoration used on a wide variety of objects)
d Akanthus *m*
f feuille *f* d'acanthe; acanthe *f*
r акант *m*

25 acanthus family
(Acanthaceae)
d Akanthusgewächse *pl*
f acanthacées *pl*
r акантовые *pl*

* **acarine disease → 26**

26 acariosis; mite infestation; acarine disease; Isle of Wight disease
(infestation with itch mites)
d Akarinose *f*; Milbenbefall *m*; Milbenseuche *f*
f maladie *f* de l'île de Wight; acariose *f*
r акариоз *m*

* **acau → 7185**

f réception *f* de bois
r сдача *f* лесосеки

28 accepted chips; screened chips
(classified into several sizes by vibrated screeners)
d akzeptierte Hackspäne *mpl*
f copeaux *mpl* triés
r отсортированная щепа *f*; годная щепа

29 accessible forest
(for exploitation)
d zugänglicher Wald *m*
f forêt *f* accessible
r доступный лес *m*

30 accessible wooded area; accessible timberland; operable timberland
d betretbare Forstfläche *f*; zugängliches Waldgebiet *n*
f superficie *f* boisée accessible; région *f* forestière accessible
r доступная лесистая зона *f*; доступная лесная область *f*

31 access network
(road network used by the fire management divisions and forest contractors to fight forest fires and to log timber)
d Zugangsnetz *n*
f cloisonnement *m* d'exploitation
r сеть *f* доступа

* **accessory bud → 78**

* **accessory tree species → 7356**

32 acclimation; acclimatization
(adjustment to environmental change by an individual)
d Akklimatisation *f*; Akklimatisierung *f*
f acclimatation *f*
r акклиматизация *f*

33 accrete *v*; **coalesce** *v*
d verbinden; vereinigen; zusammenfügen
f accréter
r срастаться; обрастать

34 accreted land
(the land that emerges from the ocean or the river)
d zuwachsender Boden *m*
f terre *f* accrue
r обраставшая земля *f*

(the act of increasing or adhering by natural growth)
d Verwachsung *f*; Akkretion *f*
f accrétion *f*; adhérence *f*; conjonction *f* de croissance
r срастание *n*; обрастание *n*; зарастание *n*

* **accretion borer → 4451**

* **accretion cutting → 7472**

* **accumulating [felling] head → 5630**

* **accumulating shear → 5621**

* **accumulator basket → 8801**

* **ACE → 196**

* **acer → 5296**

36 acerate; acerous; acicular; aciculate; needle-shaped *adj*
d nadelig [zugespitzt]; nadelförmig; azikular
f aciculaire
r хвойный; игловидный; иглообразный; игольчатый

* **aceriform leaf → 2325**

37 acerous leaf; acicular leaf; needle
(long and narrow leaf of pines and related species)
d Nadel *f*
f feuille *f* aciculaire; aiguille *f*
r хвоя *f*; игольчатый лист *m*; игла *f*

38 achene; akene
(a one-seeded, dry, indehiscent fruit with the seed coat not attached to the mature ovary wall)
d Achäne *f*; Schließfrucht *f*
f akène *m*; achaine *m*; achène *m*
r семянка *f*

* **acicula → 7912**

* **acicular → 36**

* **acicular leaf → 37**

* **aciculate → 36**

39 acid brown forest soil; acid brown wooded

f sol *m* brun forestier acide
r бурая лесная кислая почва *f*

40 **acid deposition**
(precipitation and dry particulate fallout with low pH linked to other components of air pollution)
d säurige Ablagerung *f*
f dépôt *m* acide; pollution *f* acide; retombée *f* acide; précipitation *f* acide
r кислое отложение *n*

41 **acidification; turning sour; souring**
(the process of becoming acid or being converted into an acid)
d Übersäuerung *f*
f acidification *f*
r закисание *n*; закисление *n*

* **acid lime → 9348**

* **acid-loving plant → 1254**

42 **acid rain**
(rainwater that has acidity content greater than the postulated natural pH of about 5.6)
d säuriger Regen *m*
f pluie *f* acide
r кислотный дождь *m*

43 **acid-tolerant; acid-resisting**
(ref. to plants)
d säuretolerant; säurebeständig
f acidiphile
r кислотоустойчивый; кислотовыносливый

44 **acid wood; distillation wood; chemical wood**
(wood reduced to a pulp by a chemical process)
d Destillationsholz *n*; Chemieholz *n*
f bois *m* acide; bois destiné à la distillation
r химическая древесина *f*; древесина для сухой перегонки

45 **acorn**
(hard-shelled, one seeded nut of an oak, with a pointed tip and a scaly cup covering the seed base)
d Ecker *f*; Eckerich *m*; Eichel *f*
f gland *m* [de chêne]
r жёлудь *m*

46 **acorn cup**
d Eichelbecher *m*
f cupule *f*; ébauche *f* de cellule
r плюска *f* (жёлудя)

47 **acorn seeder**
(a mechanical device for sowing oak nuts)
d Eichellegeeinrichtung *f*; Eichellegemaschine *f*; Samenstreuer *m*
f semoir *m* de glands
r высевающий аппарат *m* для жёлудей; желудёвая сеялка *f*

48 **acorn weevil**
(Balaninus glandium)
d Eichelbohrer *m*; großer Eichelrüssler *m*
f balanin *m* des glands
r долгоносик *m* желудёвый

49 **across-the-grain** *adv*
(the direction at right angles to the length of the fibers and other longitudinal elements of the wood)
d quer zur Faserrichtung; quer zur Laufrichtung
f au sens transversal au fil; au sens en travers du fil; à contrefil
r поперёк волокон; поперёк годичных колец

50 **actinidia family**
(Actinidiaceae)
d Strahlengriffelgewächse *pl*
f actinidiacées *pl*
r актинидиевые *pl*

51 **actinidia vine; kiwi fruit; Chinese gooseberrry**
(Actinidia chinensis)
d [großfrüchtige] Kiwi *f*; selbstfruchtende Kiwipflanze *f*; chinesische Kiwi; chinesische Stachelbeere *f*
f actinidia *m*; kiwi *m*; groseille *f* de Chine; groseille chinoise
r актинидия *f* [китайская]; киви *n*

52 **actinomycetes**
(any of various filamentous or rod-shaped, often pathogenic microorganisms of the order Actinomycetales that are found in soil and resemble bacteria and fungi)
d Aktinomyzeten *mpl*
f actinomycètes *mpl*
r актиномицеты *mpl*

53 **actionable fire**
(generally, any fire that requires actions for suppression)

f feu *m* dangereux
r пожар *m*, возникший из-за нарушения правил пожарной безопасности

54 active cluster
(a cluster that is occupied by red-cockaded woodpeckers in a given survey year)
d aktiver Cluster *m*
f groupe *m* actif
r активная группа *f*

55 active ingredient; active substance
(that part of a herbicide or pesticide formulation which gives the substance its efficacy)
d aktive Substanz *f*; Wirkstoff *m*
f matière *f* active; substance *f* active; produit *m* actif
r активное вещество *n*

56 activity fuels
(fuels which have been directly generated or altered by management action)
d Schlagbrennholz *n*
f déchets *mpl* d'exploitation
r топлива *npl* из-за хозяйственной деятельности

57 actual cut[ting]
d tatsächtlicher Einschlag *m*
f coupe *f* actuelle
r фактическая вырубка *f*

58 actual [log] measure
(measure of logs without dull edges or slabs)
d Fertig[holz]maß *n*
f dimension *f* [de bois] finie
r размер *m* брёвен без допуска на обзол или горбыль

59 actual size
(the size of a part as practically obtained; value which may be obtained by measurement)
d tatsächtliche Größe *f*
f dimension *f* actuelle; dimension effective
r фактический размер *m*

*** acuapar → 7185**

*** acute willow → 7138**

60 adaptability
(the ability to change or be changed to fit
f adaptabilité *f*
r адаптируемость *f*; приспособляемость *f*; способность *f* к приспособлению

61 adaptive management area; AMA
(stands or forest types that require similar management practices and are grouped as one unit for the purposes of silviculture management)
d adaptierende Betriebsfläche *f*
f aire *f* de gestion adaptée
r зона *f* адаптивного управления

62 adaxial
(facing toward the axis, as the upper surface of a leaf)
d adaxial
f adaxial
r адаксиальный; направленный к оси

*** adaxial parenchyma → 9091**

*** addice → 82**

63 additional tending measures
d zusätzliche Maßnahmen *fpl*
f mesures *fpl* culturales additionnelles
r дополнительные меры *fpl* ухода

64 additive
(any substance incorporated in wood-base fiber and particle panel material)
d Additiv *n*; Zusatzstoff *m*
f produit *m* adjuvant
r вспомогательное вещество *n*

65 additives
(substances added in small amounts)
d Additive *npl*; Zusatzstoffe *mpl*
f additifs *mpl*
r добавки *fpl*

*** adherent branch → 833**

*** adhering knot → 833**

*** Adirondack spruce → 9386**

66 admixed tree species; supplementary tree species
d Beiholzart *f*; Hilfsholzart *f*; beherrschte Holzart *f*
f essence *f* forestière de mélange; essence ligneuse admixte

67 adnate; intergrown; concrescent
(growing with one side adherent to a stem)
d verwachsend
f adné; concrescent; adhérent
r сросшийся; приросший

68 adult foliage
(the mature leaves of a tree that has one type of young foliage and a distinctly different leaf shape at maturity)
d erwachsenes Laubwerk *n*
f feuillage *m* adulte
r взрослая листва *f*

69 adult leaf
d ausgewachsenes Blatt *n*
f feuille *f* adulte
r сформировавшийся лист *m*

70 adult phase
(of a tree)
d Reifungsphase *f*; Phase *f* des Reifens
f phase *f* adulte
r фаза *f* первого цветения

* **adult wood → 6964**

* **advance crop → 6527**

* **advance cultivation → 6510**

71 advance culture; preliminary culture
d Vorkultur *f*
f culture *f* anticipée
r предварительная культура *f*

72 advanced decay; intermediate decay; mature decay; typical decay
(reducing to a brown, fibrous, stringy mass which may disintegrate forming a cavity in the tree)
d fortgeschrittene Fäule *f*; Zwischenfäunis *f*
f pourriture *f* avancée; pourriture au stade intermédiaire; dégradation *f* avancée
r развившаяся гниль *f*; вторая стадия *f* гниения [древесины]; промежуточная стадия гниения

* **advance felling → 6513**

73 advance growth; volunteer growth; advance[d] regeneration; advance reproduction; preliminary planting; preregeneration
either supplement or completely obscure the trees being planted or seeded on the area)
d Vor[aus]verjüngung *f*; Voranbau *m*; Vorwuchs *m*
f croissance *f* prédominante; régénération *f* préexistante; reproduction *f* avancée; régénération préalable; rajeunissement *m* préétabli; régénération subséquente
r предварительное [естественное] возобновление *n*

74 advance of beech
d Verbuchung *f*
f envahissement *m* par le hêtre
r чрезмерное распространение *n* бука

75 advance of spruce
d Verfichtung *f*
f envahissement *m* par l'épicéa
r чрезмерное распространение *n* ели

76 advance pruning; preliminary pruning; preparatory pruning; pre-delimbing
(the pruning of bushes that have weak canes and spindly growth)
d Vorästen *n*; Vorentastung *f*
f élagage *m* avancé; pré-ébranchage *m*
r предварительная обрезка *f* [сучьев]; черновая обрезка [сучьев]

* **advance reproduction → 73**

* **advance thinning → 7472**

77 adventitious; adventive; epicormic
(originating from the trunk, such as an epicormic shoot developing from a latent meristem)
d Adventiv-; adventiv; hinzubekommend
f adventif; gourmand
r придаточный; развивающийся из спящей почки

78 adventitious bud; accessory bud
(bud from a tree bole or stem usually resulting from a branch being broken or the bole being exposed to sunlight)
d Adventivknospe *f*; Nebenknospe *f*; akzessorische Knospe *f*
f bourgeon *m* adventif; bouton *m* additionnel
r адвентивная почка *f*; придаточная почка

79 adventitious root; secondary root; root initial

d Adventivwurzel *f*; Nebenwurzel *f*
f racine *f* adventive
r придаточный корень *m*

80 adventitious shoot
(a shoot that develops outside the usual order of time and/or position)
d Adventivspross *m*
f pousse *f* adventive
r придаточный побег *m*

81 adventitious structure
(structure arising from places other than the usual, for example, roots growing from leaves or buds developing at locations other than in leaf axils or shoot apices)
d Adventivstruktur *f*
f structure *f* adventive
r придаточная структура *f*

* **adventive → 77**

* **adverse terrain → 4212**

82 adz[e]; addice
(an axlike tool with a curved blade at right angles to the handle, used for shaping wood)
d Dächsel *f*; Dechsel *f*
f tille *f*; doloire *f*; [h]erminette *f*
r тесло *n*

83 adz[e] *v*; **axe** *v*; **fell** *v* **with the axe; dub** *v*; **hew** *v*; **hack** *v*; **trim** *v*; **slab** *v*
d behauen; schnitzen
f trailler à l'hache
r тесать; обтёсывать; срезать

* **adze → 1041**

* **adzing → 4273**

* **aerating root → 86**

* **aerial cableway → 84**

* **aerial dressing → 8685**

84 aerial line; aerial ropeway; overhead rope[way]; aerial cableway; overhead cableway; skyline [cableway]
d Luftseilbahn *f*
f ligne *f* aérienne; téléphérique *m*; funiculaire *m*
r воздушная канатная дорога *f*; подвесная канатная дорога

such as balloons or helicopters)
d Luftbringen *n*
f débardage *m* aérien; hélidébardage *m*; héliportage *m*
r воздушная трелёвка *f*; лесозаготовки *fpl* с использованием летательных аппаратов

86 aerial root; aerating root; pneumatophore root; breathing root; respiratory root
(small root emitted from the stem of a plant in the open air, which, attaching themselves to the bark of trees)
d Atemwurzel *f*; Haftwurzel *f*; Luftwurzel *f*
f racine *f* aérienne
r воздушный корень *m*; дыхательный корень; пневматофор *m*

* **aerial ropeway → 84**

87 aerial seeding
(broadcast seeding of seeds or seed pellets from aircraft)
d Flugzeugsaat *f*
f ensemencement *m* aérien; semis *m* par avion
r аэросев *m*

88 aerial skidder; skyline skidder; skyline hauler; [cableway] skidder; overhead skidder; high-wire system
(a system in which a skyline is stationary and a carriage moves along it carrying logs above the ground, from the felling site to the landing)
d Luftschlepper *m*
f câblage *m* aérien
r подвесная канатная [трелёвочная] установка *f*

* **aerial skidding → 7700**

89 aerial snubbing system
(aerial system used to block the load's lengthwise movement)
d Bergabseilbahn *f*
f système *m* par câble à freiner
r подвесная канатная установка *f* для гравитационного спуска древесины; канатная установка для спуска

90 aerial survey
(a survey using aerial photographs as part of the surveying operation)
d Lufttaxation *f*
f enquête *f* par photographies aériennes; levé *m*

91 **aesthetic forest; recreation[al] forest**
d Erholungswald *m*
f forêt *f* de récréation; forêt de cure
r рекреационный лес *m*; оздоровительный лес; лес для отдыха

* **aestifruticeta → 2298**

* **aestilignosa → 8260**

* **aestisilvae → 8260**

* **aestivation → 6502**

* **A face → 3229**

92 **afara [terminalia]; limba**
(Terminalia superba)
d Limbabaum *m*
f limba *m*
r терминалия *f* пышная; лимба *f*; афара *f*

93 **afforest** *v*; **forest** *v*; **plant** *v* **trees; set** *v* **trees**
d aufforsten; bewalden; beforsten; beholzen
f boiser
r залесить; облесить; засадить лесом; сажать лес

94 **afforestation; forestation; foresting; forest planting**
(artificially establishing forest crops by planting or sowing on land that has not previously, or not recently, grown tree crops)
d Aufforstung *f*; Beforstung *f*; Waldbegründung *f*; Bewaldung *f*; Forstpflanzung *f*
f afforestation *f*; boisement *m*; création *f* de forêt
r облесение *n*; посадка *f* леса; лесопосадка *f*

95 **afforestation demands**
d Aufforstungs-Anforderungen *fpl*
f demandes *mpl* de boisement; exigences *fpl* au boisement
r лесокультурные требования *npl*

96 **afforestation of barren land; afforestation of waste land**
d Ödlandaufforstung *f*; Ödland-Bewaldung *f*
f boisement *m* de terrain nu; afforestation *f* de terrain nu
r облесение *n* пустоши

97 **afforestation of pit-heaps; covering of pit-heaps**
d Haldenaufforstung *f*

98 **afforestation of swamps**
d Mooraufforstung *f*
f boisement *m* des marais
r облесение *n* болот

* **afforestation of waste land → 96**

99 **afforestation period**
d Aufforstungsfrist *f*
f période *f* d'afforestation
r срок *m* облесения

100 **A-frame [support]**
(two wood or metal uprights mounted in the shape of the letter "A" to support lead blocks at the upper end)
d A-Ausbaugespann *n*; A-Tragwerk *n*
f support *m* en A
r промежуточная А-образная опора *f*

101 **African afzelia**
(Afzelia africana; Afzelia bipiendensis; Afzelia pachyloba)
d Doussie *f*
f doussié *m*
r дуссия *f*; афцелия *f* африканская

102 **African blackwood; blackwood; African ebony**
(Dalbergia melanoxylon)
d [afrikanisches] Ebenholz *n*
f ébène *m* d'Afrique; palissandre *m* du Sénégal
r [африканское] эбеновое дерево *n*; дальбергия *f* чёрная; дальбергия африканская

* **African boxwood → 4685**

103 **African canarium; papo canary tree**
(Canarium schweinfurthii)
d Aiele *m*
f aiélé *m*
r канариум *m* Швайнфурта

104 **African copaiba balsam tree; balsam copaiba**
(Daniellia oliveri; Daniellia klainei; Daniellia oblonga)
d Kopaivabalsam *m*; Daniellia *f*
f baume *m* de Copahu
r копайский бальзам *m*

* **African corkwood → 9042**

* **African ebony → 102**

105 **African hackberry**

f micocoulier *m* africain
r каркас *m* африканский

106 **African juniper; African pencil cedar**
(Juniperus procera)
d afrikanischer Wacholder *m*
f genévrier *m* d'Afrique; genévrier élevé
r можжевельник *m* стройный; можжевельник высокорослый

107 **African mahogany; Ivory Coast mahogany; Grand Bassam mahogany; khaya; khaja; umbaua**
(Khaya anthotheca; Khaya nyassica; Khaya senegalensis; Khaya grandifolia)
d afrikanisches Mahagoni *n*; Umbaua *m*
f acajou *m* d'Afrique; umbaua *m*; acajou blanc
r кайя *f* сенегальская; кайя крупнолистная

* **African mangrove → 6780**

108 **African nitta tree; ilomba**
(Pycnanthus angolensis)
d afrikanischer Nittabaum *m*; Ilomba *n*
f arbre *m* sacré; ilomba *m*
r паркия *f* африканская

109 **African oak; African teak**
(Oldfieldia africana; Tectona grandis)
d Teak[holz]baum *m*
f oldfieldie *f* d'Afrique
r ольдфильдия *f* африканская; тектона *f* высокая; тиковое дерево *n*; тик *m*

110 **African oil palm**
(Elaeis guineensis)
d afrikanische Ölpalme *f*; guineische Palme *f*
f élais *m* d'Afrique; élais de Guinée; palmier *m* à l'huile
r пальма *f* гвинейская масличная

* **African olive → 2712**

111 **African padauk; barwood**
(Pterocarpus soyauxii)
d roter afrikanischer Padauk *m*; afrikanisches Sandelholz *n*
f padauk *m* d'Afrique; corail *m*
r падук *m* африканский; сандал *m* Сойукса

* **African pencil cedar → 106**

112 **African prune; mueri**
(Pygeum africanum; Prunus africanum)
d afrikanischer Zwetschgenbaum *m*;
f prunier *m* d'Afrique; mueri *m*; pygeum *m*
r пиджеума *f* африканская; пигеум *m* африканский

113 **African rosewood; Rhodesian copaltree**
(Guibourtia coleosperma; Copaifera coleosperma)
d falscher Mopane *m*; Mushibi *m*
f mushibi *m*; faux bois *m* de rose
r гвибурция *f* копалоносная

* **African satinwood → 117**

114 **African scented mahogany; cedar mahogany; sapele mahogany**
(Entandrophragma cylindricum)
d Sapelli *m*
f sapelli *m* [pommelé]
r сапел[л]и *n*

115 **African sparmannia**
(Sparmannia africana)
d afrikanische Sparmannie *f*; afrikanische Zimmerlinde *f*
f sparmannie *f* d'Afrique
r спармания *f* африканская

116 **African tamarisk**
(Tamarix africana)
d afrikanische Tamariske *f*
f tamaris *m* africain
r тамариск *m* африканский

* **African teak → 109**

* **African walnut → 1087**

117 **African white mahogany; African satinwood; blonde mahogany; avodire**
(Turraeanthus africanus; Turraeanthus klainei; Bingeria africana)
d Avodire *m*
f avodiré *m*
r турраэнтус *m* [африканский]

118 **African zebrawood; zingana; zebrano**
(Microberlinia bisulcata; Microberlinia brazzavillensis)
d Zingana *n*; Zebrano *m*
f zingana *m*
r зебровое дерево *n*

119 **afrormosia; assamela; baracara**
(Pericopsis elata; Afrormosia elata; Ormosia monosperma)
d Afrormosia *f*; Kokrodua *f*

* **afterculture → 823**

120 **aftergrowth; second growth; secondary growth; regrowth; secondary thickening**
(growth traceable to the activities of a lateral cambium)
d Nachwuchs *m*; Zweitwuchs *m*
f croissance *f* secondaire
r вторичный рост *m*

121 **after-pannage**
d Nachmast *m*
f paisson *f* secondaire
r дополнительный урожай *m* плодокорма

* **after[-]planting → 823**

* **after-regeneration → 6811**

122 **after-ripening**
(biochemical or physical changes occurring in seeds, bulbs, tubers, and fruits after harvesting, when ripe in the ordinary sense)
d Nachreifen *n*
f postmaturation *f*
r послеуборочное дозревание *n*

123 **agar/block test**
(for promoting the growth of wood-destroying fungi on wooden test-blocks by nutrient agar medium)
d Agar-Klötzchenversuch *m*
f essai *m* à l'agar-agar; essai sur milieu gélosé
r испытание *n* устойчивости древесины к заражению дереворазрушающими грибами

124 **agatized wood; wood agate**
(silicified wood, most of which is chalcedony or jasper without the characteristics of agate)
d Achatholz *n*; Holzachat *m*
f bois *m* agatisé; agate *m* de bois
r деревянистый агат *m*

125 **agave family**
(Agavaceae)
d Agavengewächse *pl*
f agavacées *pl*
r агавовые *pl*

* **agba → 5731**

126 **age class**
(distinct group of trees or portion of growing stock recognized on the basis of age)
d Altersklasse *f*

127 **age[-class] distribution; order of succession**
(individuals of a population classified according to age classes or periods such as pre-reproductive or sub-adult, reproductive, and post-reproductive)
d Alterklassenlagerung *f*; Alterklassenfolge *f*
f répartition *f* des classes d'âge; suite *f* des classes d'âge
r распределение *n* по классам возраста; расположение *n* классов возраста

128 **age-class interval; age-class period**
(the number of years within the limits of a given age class)
d Altersklassenzeitraum *m*
f étendue *f* d'une classe d'âge
r интервал *m* класса возраста

129 **age-class schedule; age-class table**
d Alterklassenübersicht *f*; Alterklassentabelle *f*
f aperçu *m* des classes d'âge; table *f* des classes d'âge
r таблица *f* классов возраста

130 **age composition; age structure**
(1. an array of age classes; 2. the arrangement of age classes in a population which describes the relative strengths of the age classes)
d Alteraufbau *m*
f composition *f* d'âge
r возрастная структура *f*

131 **age determination**
(ascertaining the age of a stand or of a tree)
d Alterbestimmung *f*; Altersermittlung *f*
f détermination *f* de l'âge
r определение *n* возраста

* **age-gradation → 3893**

132 **ag[e]ing of tree; tree aging**
d Baumalterung *f*
f vieillissement *m* d'arbre
r созревание *n* дерева

133 **age of [forest] maturity; harvest age; maturity age**
(optimum harvest age of a tree from the point of view of the set objective)
d reifes Alter *n*; Haubarkeitsalter *n*
f âge *m* de maturité; âge de récolte
r возраст *m* спелости [леса]

134 **age of tree; tree age**
(in a tree born from a seed, the physiological

stemmed from an adventitious bud, only the age of the shoot can be ascertained)
d Baumalter *n*
f âge *m* d'un arbre
r возраст *m* дерева

135 age of wood; wood age
d Holzalter *n*
f âge *m* de bois
r возраст *m* древесины

136 age of [wood] cutting; cutting age; felling age; exploitable age; rotation age; removal age; final age
(age at which an even-aged stand of trees is scheduled for harvest or regeneration cutting)
d Abtriebsalter *n*; Hiebsalter *n*; wirtschaftliches Alter *n*
f âge *m* d'abattage; âge d'exploitabilité
r возраст *m* срубания [леса]; возраст рубки; возраст древостоя, равный обороту рубки; хозяйственный возраст

137 age quality of wood
d Holzdauerhaftigkeit *f*
f qualité *f* d'âge de bois
r долговечность *f* древесины

138 age ratio
(proportion of young to adults, suggesting a measure of productivity)
d Altersklassenverhältnis *n*
f proportion *f* entre des classes d'âge
r соотношение *n* классов возраста

* **agerita → 5419**

* **age structure → 130**

139 aggregate[d] retention; group retention
(retaining of trees in clumps or clusters)
d Gesamtretention *f*; gesamte Retention *f*
f rétention *f* par groupes
r удержание *n* групп; удерживание *n* в группе

* **aggregate fruit → 1839**

140 aggregate ray
(a group of small, narrow, xylem rays appearing to the unaided eye or at low magnification as a single large ray)
d Gesamtstrahl *m*
f faux rayon *m*
r сборный луч *m*; групповой луч

(tending to spread quickly)
d kräftiges Wachstum *n*
f croissance *f* agressive
r стремительный рост *m*

142 aggressive root system
d kräftiges Wurzelsystem *n*; kräftiges Wurzelnetz *n*
f système *m* de racine vigoureux
r мощная корневая система *f*

* **agricultural timber → 9570**

143 agroforestry; agrosilviculture; coppice with field crops
(collective name for land-use systems and practices in which trees and shrubs are deliberately integrated with non-woody crops and/or animals for ecological and economic purposes)
d Feldwaldbau *m*; Waldgutwirtschaft *f*; Röderwaldbetrieb *m*; Baumfeldwirtschaft *f*; Waldfeldbau *m*; Waldfeldwirtschaft *f*
f agroforesterie *f*; agrosylviculture *f*; foresterie *f* paysanne; foresterie rurale; exploitation *f* sylvo-agricole; exploitation agro-sylvicole
r переменное лесопольное хозяйство *n*; выращивание *n* леса с временным сельскохозяйственным пользованием; лесохлебное хозяйство

144 agroforestry spatial units
d Waldfeldbaueinheiten *fpl*
f unités *fpl* spatiales d'agroforesterie
r территориальные единицы *fpl* переменного лесопольного хозяйства

145 agroforestry system; agrosilvicultural system
(system for the concurrent production of agricultural crops and forest crops)
d land- und forstwirtschaftliches System *n*; Waldfeldbau-System *n*
f système *m* agroforestier; système agrosylvicole
r система *f* переменного лесопольного хозяйства

* **agrosilviculture → 143**

146 agrosilviculture method; taungya method
(a method of raising forest trees in combination with (seasonal) agricultural crops)
d Waldfeldbauverfahren *n*
f méthode *f* agrosylvicole; méthode taungya

* **aguano → 1412**

147 **ailanthus family**
(Simaroubaceae)
d Bittereschengewächse *pl*
f simar[o]ubacées *pl*
r симарубовые *pl*

148 **airborne forest fire protection**
d Luftlandewaldschutz *m* gegen Feuer
f protection *f* aérienne des forêts contre l'incendie
r авиационная охрана *f* лесов

* **air-dried lumber → 150**

149 **air-dried weight**
(weight of a substance, usually vegetation, after it has been allowed to dry to equilibrium with the atmosphere, usually without artificial heat)
d luftgetrocknete Masse *f*; Lufttrockenmasse *f*
f poids *m* sec à l'air
r воздушно-сухой весь *m*; воздушно-сухая масса *f*

150 **air-dried wood; air-seasoned wood; air-dried lumber; air-dry timber; natural-seasoned wood**
(that have been either dried by exposure to natural atmospheric conditions outdoors or in an unheated shed or dried to equilibrium with the surrounding atmosphere)
d luftgetrocknetes Schnittholz *n*; lufttrockenes Holz *n*; abgetrocknetes Nutzholz *n*
f bois *m* ressuyé; bois séché à l'aire
r воздушно-сухая древесина *f*; древесина сезонной сушки; высушенный на воздухе пиломатериал *m*; воздушно сухой лесоматериал *m*

151 **air-dried wood density**
d lufttrockene Holzdichte *f*
f densité *f* de bois séché à l'air
r плотность *f* воздушно-сухой древесины

* **air dryer → 153**

152 **air-drying; air-seasoning**
d Lufttrocknung *f*; Freilufttrocknung *f*; natürliche Holztrocknung *f*
f séchage *m* à l'air
r воздушная сушка *f*; атмосферная сушка; сушка на открытом воздухе

f assécheur *m* d'air
r воздушная сушилка *f*

* **air-dry timber → 150**

154 **air-layering; air-marcotting**
d Luftsenkervermehrung *f*
f marcottage *m* aérien
r получение *n* воздушных отводков от стебля

155 **air-pit; blind pit**
(a pit without complementary pit in an adjacent cell)
d blinde Tüpfelung *f*
f ponctuation *f* borgne; ponctuation aveugle
r межклетник *m*

156 **air pruning**
d Luftästung *f*
f élagage *m* aérien
r воздушная обрезка *f*

* **air-seasoned wood → 150**

* **air-seasoning → 152**

157 **airtanker; water bomber**
(fixed-wing aircraft fitted with tanks and equipment for dropping suppressants or retardants)
d Wasserbomber *m*
f bombardier *m* d'eau
r противопожарный самолёт *m*

* **aisle → 3564**

* **aizella → 5476**

* **akene → 38**

* **akle → 1754**

* **ako → 6414**

158 **alangium family**
(Alangiaceae)
d Alangiumgewächse *pl*
f alangiacées *pl*
r алангиевые *pl*

159 **Alaska birch**
(Betula neoalaskana)
d Alaska-Birke *f*

160 **Alaska bog willow**
(Salix fuscescens)
d braun-werdende Weide *f*
f saule *m* fauve
r ива *f* буреющая

* **Alaska larch → 250**

161 **albedo**
(measure of the solar reflectivity of a surface)
d Albedo *m*
f albédo *m*
r отражательная способность *f*; альбедо *n*

162 **albedometer**
d Albedomesser *m*
f albédomètre *m*
r альбедометр *m*

* **albur[num] → 7216**

163 **Alcock spruce**
(Picea alcockiana)
d Alcocks-Fichte *f*; zweifarbige japanische Fichte *f*
f épicéa *m* d'Alcock
r ель *f* Алькока; ель двухцветная

* **alder buckthorn → 684**

164 **alder forest; alder-wood; alder grove; alder stand**
d Erlenbruchwald *m*; Erlenbestand *m*
f au[l]naie *f*; taillis *m* d'au[l]nes
r ольховый лес *m*; ольшаник *m*; ольс *m*; олешник *m*; ольховник *m*

165 **alder swamp forest**
d Erlenbruch *m*
f au[l]naie *f* marécageuse
r болотистый ольшаник *m*

* **alder-wood → 164**

* **Aleppo oak → 3794**

* **Aleppo pine → 2220**

166 **algae** (*lat*)
(primitive plants, usually aquatic, capable of synthesizing their own food by photosynthesis)
d Algen *fpl*

* **algarroba → 1790**

* **Algerian fiber → 1717**

* **Algerian oak → 5477**

* **algerita → 5419**

167 **aliform parenchyma**
(paratracheal parenchyma with wing-like lateral extensions, as seen in cross section)
d flügelförmiges Parenchym *n*; aliformes Parenchym
f parenchyme *m* aliforme
r крыловидная паренхима *f*

168 **aligned flakeboard core; oriented flakeboard core**
d orientierter Spanplattekern *m*
f âme *m* en panneaux de grandes particules orientées; âme en panneaux de flocons orientées
r средний слой *m* [плиты] из ориентированной стружки

169 **alignment time**
(of a forest age structure)
d Ausgleichszeit *f*
f temps *m* d'alignement
r время *f* выравнивания

170 **alkalinity**
(the quality which constitutes an alkali; pH values above 7)
d Alkalinität *f*
f alcalinité *f*
r засоленность *f*

171 **alkali ratio**
d Alkaliquote *f*
f rapport *m* d'alcalin
r процент *m* активной щёлочи (от веса сухой древесины)

172 **all-aged forest; uneven-aged forest; many-aged forest**
(forest which contains trees of all ages, intermixed)
d ungleichaltriger Wald *m*
f forêt *f* de tous âges; forêt inéquienne
r разновозрастный лес *m*

173 **all-aged [forest] management; uneven-aged [forest] management**
(periodically selecting and harvesting individual trees or groups of trees from the

f gestion *f* de tous âges; gestion de forêt inéquienne
r ведение *n* разновозрастного лесного хозяйства

* **all-aged stand → 9081**

174 **all-aged structure**
(stand in which trees of most or all age classes, from seedlings to mature trees, are represented)
d ungleichaltrige Struktur *f*
f structure *f* inéquienne
r разновозрастная структура *f*

175 **Alleghany barberry**
(Barberis canadensis)
d kanadische Berberitze *f*
f vinettier *m* du Canada
r барбарис *m* канадский

* **Alleghany chink[u]apin → 1199**

176 **Alleghany hardwood forest**
(a portion of Maryland's northern hardwood forest in which black cherry, white ash, and red oak are dominant species)
d Alleghany-Laubwald *m*
f forêt *f* feuilleuse d'Alleghany
r аллеганский твёрдолиственный лес *m*

177 **Alleghany serviceberry; Alleghany shadbush; glabrous shadbush**
(Amelanchier laevis)
d Alleghany-Felsenbirne *f*; kahle Felsenbirne *f*
f amélanchier *m* glabre
r аллеганская ирга *f*

178 **Alleghany spruce beetle**
(Dendroctonus punctatus)
d Alleghany-Fichtenriesenbastkäfer *m*
f dendroctone *m* d'Alleghany
r лубоед *m* еловый аллеганский

179 **alleles; allelomorph genes; amorphous genes**
(alternative states of a gene)
d Allele *npl*; amorphe Gene *npl*
f allèles *mpl*; gènes *mpl* allélomorphiques; gènes amorphes
r аллели *mpl*; аллеломорфы *mpl*; аллельные гены *mpl*

180 **allelopathy**
(the inhibition of growth in one species of plants as a result of the release of chemical inhibitors by some plants)

result of the release of chemical inhibitors by some plants)
d Allelopathie *f*
f allélopathie *f*
r аллелопатия *f*

181 **alley**
d Traufengasse *f*; Gasse *f*
f couloir *m*; venelle *f*
r кладовая *f* на лесозаготовительном пункте

182 **alley; pocket**
(a narrow passage; especially a walk or passage in a garden or park, bordered by rows of trees or bushes)
d Stallgang *m*; Passage *f*
f allée *f*; passage *m*; couloir *m*
r узкий проход *m*; проезд *m*

183 **alley[ ]cropping; corridor planting; corridor cultivation; hedgerow intercropping**
(in which species of shrubs or trees are planted at spacing relatively closed within row and wide between rows)
d Korridorpflanzung *f*
f culture *f* en couloirs
r коридорная посадка *f*; межрядный посев *m*

184 **all-heart wood**
(lumber or other wood products that is heartwood throughout, free of sap-wood)
d vollkerniges Holz *n*
f tout bois *m* de cœur
r целиком ядровая древесина *f*; древесина с высоким содержанием ядра

185 **allied species**
d verbundene Arten *fpl*; gedopte Arten
f espèces *fpl* voisines; espèces alliées
r родственные породы *fpl*

* **alligator → 3881**

186 **alligator crusher; jaw crusher; jaw breaker; alligator squeezer**
d Backenbrecher *m*
f défibreur *m* à mâchoires; concasseur *m* à mâchoires;
broyeur *m* à mâchoires
r щековая дробилка *f*

187 **alligator juniper**

f genévrier *m* gersé
r можжевельник *m* аллигаторовой; можжевельник толстокорый

* **alligator pear → 445**

188 **alligators**
(drasting type of crazing, a pattern resembling the hide of a crocodile)
d "Alligatoren" *fpl*
f défauts *mpl* de type peau de crocodile
r дефекты *mpl* "крокодиловая кожа"

* **alligator squeezer → 186**

189 **allochthonous**
(derived from outside a system, such as leaves of terrestrial plants that fall into a stream)
d bodenfremd; allochthon
f allochtone
r аллохтонный

190 **allometric relationships**
(between two dimensions of an organism, such as root weight: shoot weight or total dry weight: stem diameter)
d allometrische Relationen *fpl*
f relations *fpl* allométriques
r аллометрические связи *fpl*

* **allotment by areas → 6842**

191 **allotment of felling areas; parcelling of crops; distribution of fellings**
d Schlageinteilung *f*
f parcellisation *f* de coupes
r распределение *n* лесосек

192 **allowable annual cut; annual allowable cut; AAC**
(amount of timber that is permitted to be cut annually from a particular area)
d zulässiger Jahreseinschlag *m*; zulässiges Soll *n*
f coupe *f* annuelle permissible; possibilité *f* annuelle de coupe; PAC
r расчётная годичная лесосека *f*; плановая годичная лесосека

193 **allowable annual harvest; annual allowable harvest**
(volume of timber or timber acreage that may be harvested from an area on a perpetual basis)
d zulässiger jährliche Erntehieb *m*; zulässige jährliche Nutzungsmenge *f*
f quantité *f* permissible annuelle d'exploitation;
r расчётная годичная лесозаготовка *f*

194 **allowable burned area**
d zulässige Brandfläche *f*; annehmbare Brandfläche
f superficie *f* brûlée admissible; surface *f* incendiée admissible; zone *f* brûlée permissible; surface brûlée tolérable
r доступная площадь *f* пожара

195 **allowable cut; prescribed cut**
(volume of timber that may be harvested during a given period to maintain sustained production)
d Abnutzungssatz *m*; Nutzungshiebssatz *m*; Hiebssatz *m*
f coupe *f* admissible; coupe permissible; coupe dirigée
r расчётная вырубка *f*; сметная вырубка

196 **allowable cut effect; ACE**
(allocation of anticipated future forest timber yields to the present allowable cut)
d Hiebssatz-Wirkung *f*
f effet *m* de coupe permissible
r эффект *m* расчётной лесосеки

197 **allowable final cut**
d Hiebssatz *m* der Abtriebsnutzung
f coupe *f* définitive admissible
r расчётная вырубка *f* главного пользования

198 **allowable intermediate cut**
d zulässige Zwischenhieb *m*
f coupe *f* intermédiaire admissible
r расчётная годичная вырубка *f* промежуточного пользования

199 **allowable sale quantity; ASQ**
(quantity of timber that may be sold from the area of suitable land covered by a plan for a period specified by the plan)
d zulässige Verkaufmenge *f*
f quantité *f* de vente admissible
r расчётное количество *n* продажи

200 **allowance for bark; bark allowance; bark deduction**
d Rindenanteil *m* am Holz; Rindenabzug *m*; Rindenberechnung *f*
f tolérance *f* de l'écorce; déduction *f* de l'écorce; défalcation *f* de l'écorce
r припуск *m* на кору; вычёт *m* коры; вычитание *n* толщины коры (в процентах от объёма ствола в коре)

201 **allowance for defect; defect deduction;**

f réfaction *f* pour anomalies; réfaction pour défaut
r учёт *m* ошибки

* **allowance for fault → 201**

202 **allowance for trim[ming]; trim allowance; trimming margin; trimming length; overlength**
(extra 2 or 3 inches left on a bucked log to allow logs with end checks, pulls, or slanting buck cuts to be trimmed to standard lumber lengths)
d Draufgabe *f*
f combe *m*; supplément *m*; marge *f* d'éboutage; surlongueur *f*
r припуск *m* на торцовку; припуск длине торцовки; качество *n* торцовки; надбавка *f*

203 **allthorn acacia**
(Acacia horrida; Acacia angustissima; Acacia karroo)
d Karroo-Akazie *f*; Kapschotendorn *m*; Weißdornakazie *f*; Prärie-Akazie *f*
f acacia *m* karroo; acacia à longues épines
r акация *f* ощетиненная; акация устрашающая

204 **alluvial forest**
(a forest on alluvial soil)
d Alluvialwald *m*
f forêt *f* alluviale
r лес *m* на аллювиальных почвах

205 **all-weather road**
d Allwetter-Straße *f*; bei jedem Wetter benutzbare Straße *f*
f route *f* [praticable par] tous les temps
r удобная дорога *f* при всякой погоде

* **alm → 7310**

* **almendro → 1003**

206 **almond-leaved pear**
(Pyrus amygdaliformis)
d mandelblättriger Birnenbaum *m*
f poirier *m* à feuilles d'amandier
r груша *f* лавролистная

* **almond[-]leaved willow → 6122**

207 **almond[-tree]; bitter almond**
(Amygdalus communis)
d Bittermandelbaum *m*; [echter] Mandelbaum *m*; echte Mandel *f*; bittere
r [горький] миндаль *m*; миндаль обыкновенный; миндаль настоящий

* **aloe → 5447**

208 **alongside delivery**
d entlange Holzlieferung *f*
f livraison *f* de bois le long du bord
r выгрузка *f* леса с борта судна на набережную

209 **along the grain; in the grain direction; in the fiber direction** *adv*
(the direction parallel with the length of the fibers and other longitudinal elements of the wood)
d faserparallel; längs zur Faser
f droit fil; dans le fil; dans le sens longitudinal des fils
r вдоль волокон

210 **alpha diversity**
(any of a variety of definitions of diversity relating to lists of species, abundance of each, and proportional abundance within a community)
d Alpha-Vielfalt *f*
f diversité *f* alpha
r альфа-разнообразие *n*

* **Alpine ash → 4054**

* **Alpine blueberry → 864**

211 **Alpine buckthorn**
(Rhamnus alpina)
d Alpen-Kreuzdorn *m*
f nerprun *m* des Alpes
r крушина *f* альпийская

212 **Alpine clematis**
(Climatis alpina)
d Alpenrebe *f*; Alpen-Waldrebe *f*; Alpen-Klematis *f*
f clématite *f* des Alpes; atragène *m* des Alpes
r клематис *m* альпийский

213 **Alpine currant**
(Ribes alpinum)
d Alpenjohannisbeere *f*
f groseillier *m* des Alpes; groseillier sauvage
r смородина *f* альпийская

214 **Alpine fir; subalpine fir**
(Abies lasiocarpa)
d Felsengebirgstanne *f*

* **Alpine gum → 4054**

215 **Alpine laburnum; Scotch laburnum**
(Laburnum alpinum)
d Alpengoldregen *m*; alpiner Bohnenbaum *m*
f cytise *m* des Alpes; aubour *m* des Alpes
r золотой дождь *m* альпийский

216 **Alpine larch; subalpine larch**
(Larix lyallii)
d Alpen-Lärche *f*
f mélèze *m* subalpin
r лиственница *f* Лайеля

217 **Alpine rocks**
d Alpensteine *mpl*
f roches *fpl* des Alpes
r гольцы *mpl* (нелесные земли; горные вершины округлой формы, поднимающиеся выше границы леса и почти лишенные лесной растительности)

* **Alpine stone pine → 8346**

* **Alpine strawberry → 9425**

* **Alpine tundra → 5594**

218 **Alpine willow**
(Salix glabra)
d kahle Weide *f*
f saule *m* glabre
r ива *f* голая

* **alstonia → 1461**

219 **alternate clear-strip system**
d Kahlschlagsmethode *f* des Streifenweisen
f système *m* par coupes en bandes alternées; système de régénération par coupes rases en bandes alternées
r система *f* сплошных чересполосных рубок

* **alternate forest crop → 4547**

220 **alternate-leaf; subopposite leaf**
(attached singly on a stem; not opposite or whorled)
d alternierendes Blatt *n*
f feuille *f* alterne
r очередной лист *m*

221 **alternate-leaf arrangement**
d alternierende Blattdisposition *f*
f arrangement *m* à feuilles alternes
r очередное листорасположение *n*

222 **alternate-leaved dogwood; green osier; pagoda dogwood**
(Cornus alternifolia)
d Hornstrauch *m*; [Pagoden-]Hartriegel *m*
f cornouiller *m* alternifolié; cornouiller à feuilles alternées
r дёрен *m* очереднолистный

223 **alternate pitting**
(type of pitting in which bordered pits are arranged with pits offset in alternate rows across the cell)
d alternierender Lochfraß *m*
f ponctuation *f* en file oblique
r очередная поровость *f*

224 **alternate-strip felling; alternate-strip cutting; leap-frog coupe**
d Springschlag *m*; Springhieb *m*
f coupe *f* alternante
r чересполосная лесосека *f*

* **alternating spiral grain → 4539**

225 **alternation of agricultural and forest crops; field and forest crops alternation**
d Land- und Holzartenwechsel *m*
f alternance *f* d'espèces forestières et agricoles
r чередование *n* лесного и сельского хозяйства

226 **alternation of tree species; change in tree species; forest crop rotation; tree species conversion; forest conversion; stand conversion**
(change from one tree species to another)
d Holzartenwechsel *m*; Umwaldung *f*; Baumartenwandel *m*; Baumartenwechsel *m*; Wechsel *m* der Holzarten
f alternance *f* d'espèces forestières; conversion *f* d'espèces arborescentes; changement *m* des essences forestières
r смена *f* древесных пород; чередование *n* древесных пород

227 **alternes**
(one of two or more plant communities that alternate with each other in space)
d vikariierende Gemeinschaften *fpl*
f communautés *fpl* vicariantes; alternes *fpl*
r одно *n* из нескольких сообществ, сменяющих друг друга на ограниченном пространстве

228 **altherbosa**
(communities with tall herbaceous plants,

f communauté *f* d'herbes hautes; communauté de macrophorbiées; communauté de mégaphorbiées
r сообщество *n* высоких трав

* **aluppe → 5244**

* **AMA → 61**

* **amabilis fir → 6025**

* **amapa-rana → 478**

* **amapa rosa → 5346**

* **amaranth → 6641**

229 **amaranth family**
(Amaranthaceae)
d Fuchsschwanzgewächse *pl*
f amarantacées *pl*
r амарантовые *pl*

* **amarela → 1008**

* **Amazon rosewood → 1006**

* **ambahu → 7519**

230 **amber forest**
d Amberwald *m*
f forêt *f* ambrée
r ископаемый лес *m*

* **amboyna → 5284**

231 **ambrosia beetle; ambrosia borer; pinhole borer; shothole borer; keyhole borer**
(Xyloterus)
d Ambrosiakäfer *m*
f scolyte *m* du bois; scolyte ambrosia
r древесинник *m*; короед *m*; калькограф *m*

232 **ambrosia beetle damage**
d Ambrosiakäfer-Schaden *m*
f dommage *m* par scolyte du bois
r повреждение *n*, вызываемое древесинником

* **ambrosia borer → 231**

* **ambuia → 1008**

* **amburana → 1417**

233 **amenity plantation; amenity planting**
(application of horticultural practices for leisure-oriented or aesthetic purposes)
d Annehmlichkeitswert *m*
f plantation *f* d'agrément; plantation récréative
r зелёное строительство *n*; озеленение *n*

234 **amenity tree**
d Annehmlichkeitsbaum *m*
f arbre *m* d'agrément
r эстетическое дерево *n*

* **ament → 1380**

235 **American alder; red alder; Oregon alder; Pacific-coast alder; western [red] alder**
(Alnus rubra; Alnus oregona)
d [nord]amerikanische Roterle *f*; Oregon-Erle *f*
f au[l]ne *m* d'Oregon; au[l]ne rouge
r ольха *f* красная

236 **American aspen; quaking aspen; trembling[-leaved] aspen; Canadian aspen**
(Populus tremuloides)
d amerikanische Zitterpappel *f*
f peuplier *m* faux-tremble [américain]
r тополь *m* осинообразный

237 **American basswood; Florida basswood; American linden; spoonwood; monkey-nut tree**
(Tilia americana; Tilia glabra)
d amerikanische Linde *f*; Schwarzlinde *f*
f tilleul *m* américain; tilleul noir
r липа *f* американская

238 **American beech**
(Fagus grandifolia)
d amerikanische Buche *f*
f hêtre *m* à grandes feuilles; hêtre américain
r бук *m* крупнолистный

239 **American bittersweet; climbing bittersweet; woody nightshade**
(Celastrus scandens)
d amerikanischer Baumwürger *m*; gemeiner Baumwürger
f célastre *m* grimpant; bourreau *m* des arbres
r бересклет *m* вьющийся; краснопузырник *m* деревогубец

240 **American black currant**
(Ribes americanum)
d amerikanische [schwarze] Johannisbeere *f*

241 **American [black] walnut; California[n] [black] walnut; Clara walnut; Claro walnut; hinds [black] walnut**
(Juglans californica)
d kalifornischer Nussbaum *m*; kalifornischer Walnussbaum *m*
f noyer *m* [noir] de Californie; noyer [noir] d'Amérique
r орех *m* [чёрный] калифорнийский

242 **American blue oak; blue oak**
(Quercus douglasii)
d amerikanische Blaueiche *f*
f chêne *m* bleu [d'Amérique]
r Дугласов дуб *m*

243 **American cedar**
(Cedrela)
d amerikanische Zeder *f*
f cèdre *m* de l'Amérique
r американский кедр *m*; цедрела *f*

244 **American cherry; [mountain] black cherry; wild black cherry; timber cherry; cabinet cherry**
(Padus serotina; Prunus serotina)
d amerikanische Spätkirshe *f*; amerikanische Traubenkirsche *f*; Vogelkirschbaum *m*; spätblühende Traubenkirsche
f merisier *m* d'Amérique; cerisier *m* tardif
r черёмуха *f* поздняя; вишня *f* американская

245 **American chestnut**
(Castanea dentata)
d amerikanische Kastanie *f*
f châtaigner *m* américain
r каштан *m* зубчатый; каштан американский

* **American coffee-tree → 1482**

* **American ebony → 1796**

246 **American elder[berry]**
(Sambucus canadensis; Sambucus mexicana)
d amerikanischer Holunder *m*; kanadischer Holunder
f sureau *m* d'Amérique; sureau du Canada; sureau blanc
r бузина *f* американская; бузина канадская

247 **American elm; Florida elm; white elm**
(Ulmus americana; Planera americana)
d amerikanische Ulme *f*; weiße Ulme; Weißulme *f*
f orme *m* d'Amérique; orme [blanc] américain
r ильм *m* американский; ильм белый

248 **American holly**
(Ilex opaca)
d amerikanische Hülse *f*; amerikanische Stechpalme *f*
f houx *m* américain
r американский падуб *m*

249 **American hornbeam; blue beech; water-beech; musclewood**
(Carpinus caroliniana)
d amerikanische Hainbuche *f*; amerikanische Weißbuche *f*; karolinische Hainbuche
f charme *m* de Caroline; charme d'Amérique
r граб *m* американский

* **American ivy → 9228**

250 **American larch; Alaska larch**
(Larix laricina)
d amerikanische Lärche *f*; kleinzapfige Lärche
f mélèze *m* laricin
r лиственница *f* американская; тамарак *m*

* **American linden → 237**

* **American mahagony → 1412**

251 **American mangrove; black mangrove; manglecito**
(Avicennia germinans)
d schwarze Mangrove *f*; Austerbaum *m*; [gemeiner] Mangelbaum *m*; Leuchterbaum *m*; Licht[er]baum *m*
f palétuvier *m* d'Amérique; manglier *m* noir; palétuvier noir; mangrove *f* noire
r чёрный мангр *m*; чёрное мангровое дерево *n*; деревокорень *m*; шандал-дерево *n*; авицения *f*

252 **American mountain[ ]ash**
(Sorbus americana)
d amerikanische Eberesche *f*
f sorbier *m* de montagnes d'Amérique
r рябина *f* американская

* **American mulberry → 6781**

* **American muskwood → 2006**

* **American nettle tree → 1777**

253 **American oil palm; oil palm**
(Elaeis oleifera)
d amerikanische Ölpalme *f*
f palmier *m* à huile américain
r американская масличная пальма *f*

254 **American plane[tree]; American plain; American sycamore; western sycamore; plane sycamore; planetree; western plane; buttonball; buttonwood**
(Platanus occidentalis)
d abendländliche Platane *f*; [nord]amerikanische Platane; Kleiderbaum *m*; Sykomore *f*; Sycamore *f*
f platane *m* occidental; platane d'Occident; platane [d'Amérique]
r американский чинар *m*; платан *m* западный; сикомор *m*

255 **American post oak; post oak**
(Quercus minor; Quercus stellata)
d Sterneiche *f*
f chêne *m* étoilé
r дуб *m* малый; дуб звездчатый

* **American red berry elder → 787**

* **American red elm → 7745**

* **American red oak → 5793**

256 **American [red] plum; August plum; goose plum**
(Prunus americana)
d amerikanische [rote] Pflaume *f*; amerikanischer Schlehdorn *m*
f prunier *m* d'Amérique; prunier américain
r слива *f* американская

257 **American silverberry; silverberry**
(Elaeagnus commutata; Elaeagnus argentea)
d amerikanischer Silberbaum *m*; Silberbeere *f*; [breitblättrige] Ölweide *f*
f chalef *m* argenté
r лох *m* серебристый

258 **American smoke-tree**
(Cotinus americanus; Cotinus obovatus; Rhus cotinoides)
d amerikanischer Rauchbaum *m*
f sumac *m* cotinus américain; fustet *m* d'Amérique
r скумпия *f* американская

259 **American storax; opossum tree; American sweet[ ]gum; satin walnut**
(Liquidambar styraciflua)
d [amerikanischer] Amberbaum *m*; Liquidambar *m*; Satinnuss-Baum *m*; Kaugummibaum *m*
f storax *m* américain; gommier *m* doux; liquidambar *m* d'Amérique; copalme *m* d'Amérique; ambrier *m*; noyer *m* satiné
r бальзам *m*; амбровое дерево

260 **American sumac[h]; staghorn sumac[h]; velvet sumac[h]**
(Rhus typhina; Rhus hirta)
d Hirschkolbensumach *m*
f sumac *m* vinaigrier; sumac de Virginie; arbre *m* aux queues de renard; bois *m* de cerf
r сумах *m* коротковолосый; сумах оленьерогий; сумах пушистый; сумах уксусный

* **American sweet[ ]gum → 259**

* **American sycamore → 254**

261 **American [white] ash; white ash; Canadian ash; Biltmore [white] ash**
(Fraxinus americana; Fraxinus alba)
d [amerikanische] Weißesche *f*; amerikanische Esche *f*
f frêne *m* blanc [d'Amérique]
r ясень *m* американский; ясень белый

* **American whitewood → 9008**

* **American woodbine creeper → 9228**

262 **American yellowwood; [Kentucky] yellowwood**
(Cladrastis lutea; Virgilia lutea)
d amerikanisches Gelbholz *n*; Gelbholzbaum *m*
f virgilier *m* [à bois jaune]; cladraste *m* à bois jaune
r виргилия *f* жёлтая; кладрастис *m* жёлтый

263 **amorpha honey locust**
(Gleditsia amorphoides)
d Gleditschie *f*; Lederhülsenbaum *m*
f févier *m* d'Argentine
r гледичия *f* аргентинская

* **amorphous genes → 179**

* **amplitude → 8673**

264 **Amur cork[ ]tree**
(Phellodendron amurensis)
d Amurkorkbaum *m*; Samtbaum *m*; mandschurischer Korkbaum *m*
f arbre *m* à liège de l'Amour; phellodendre *m* de l'Amour; arbre-liège *m* de Chine
r амурское пробковое дерево *n*; бархатное дерево;
бархат *m* амурский

265 **Amur lilac**

f lilas *m* de l'Amour
r сирень *f* амурская

266 **Amur linden**
(Tilia amurensis)
d Amurlinde *f*
f tilleul *m* de l'Amour; tilleul de Sibérie
r липа *f* амурская

267 **Amur maple**
(Acer ginnala)
d Amur-Ahorn *m*; Feuerahorn *m*
f érable *m* ginnala; érable de l'Amour; érable du fleuve amour
r клён *m* Гиннала; клён амурский; клён приречный; клён речной

268 **analysis area**
(delineated area of land subject to analyses)
d Untersuchungsgebiet *n*
f zone *f* d'analyse; zone d'enquête
r область *f* анализа

269 **analysis of annual rings; year ring analysis; tree-ring analysis**
d Jahresringenanalyse *f*
f analyse *f* de cernes annuels
r анализ *m* годичных слоёв; дендрохронологический анализ

* **anaura → 5532**

* **anchor cable → 278**

270 **anchor chain**
(heavy chain, often with spikes welded to the links, used in drag scarification)
d Ankerkette *f*; Teikette *f*; Ankertau *n*
f chaîne *f* d'ancre; chaîne de mouillage
r якорная цепь *f*

271 **anchor-chain clearing; chain-clearing; cabling**
(a method of reducing or clearing undesirable scrub by dragging through it a heavy chain between two appropriately spaced tractors)
d Verseilen *n*
f dragage *m* au câble; câblage *m*
r расчистка *f* площади от кустарника с помощью толстого троса (прикреплённого к двум тракторам)

272 **anchored tree revetment; tree spur**
d befestigter Baumsporn *m*
f arbre *m* entier disposé en épi
r фиксированная деревянная шпора *f*

d Verankerung *f*
f ancrage *m*
r якорное скрепление *n*

274 **anchoring branch**
d Astanker *m*
f branche *f* d'ancrage; crochet *m*
r анкерный сучок *m*; сучковый якорь *m*

275 **anchor log; deadman**
(a fallen tree on the shore, or a timber to which the hawser of a boom is attached)
d toter Mann *m*; Zange *f*
f palée *f* d'ancrage; bois *m* d'ancrage
r анкерное бревно *n*

276 **anchor point**
(an advantageous location from which to start constructing a fire line)
d Ankerpunkt *m*
f point *m* d'appui
r опорная площадка *f*

277 **anchor root; sinker root; deeply descending root; prop[ ]root; stilt root; brace root; supporting root**
d Ankerwurzel *f*; Festigungswurzel *f*; Unterstützungswurzel *f*
f racine *f* d'ancrage; racine d'appui; racine en échasses
r якорный корень *m*; столбовидный корень; опорный корень; поддерживающий корень

278 **anchor rope; anchor cable**
(line used to tie down a yarder to prevent tipping on a heavy pull)
d Ankerseil *n*
f corde *f* d'amarrage
r анкерный канат *m*; крепёжный канат; анкерная оттяжка *f*

279 **anchor stump**
d Ankerstumpf *m*
f souche *f* d'ancrage
r анкерный пень *m*

280 **anchor tree**
d Ankerbaum *m*
f arbre *m* d'ancrage
r анкерное дерево *n*

281 **anchor winch**
d Ankerwinde *f*
f cabestan *m*
r якорная лебёдка *f*

282 ancient forest; virgin forest; antique forest; primary forest; primordial forest; primeval forest; primal forest; original forest; untouched forest; pristine forest; wild forest; first growth; native growth; primary growth; wildwood
(area of old-growth trees that never has been harvested by humans)
d Urwald *m*; jungfräulicher Forst *m*; jungfräulicher Wald *m*; Primärwald *m*
f forêt *f* vierge; peuplement *m* originel
r девственный лес *m*; первобытный лес; первичное насаждение *n*

* **ancient pine → 6997**

* **ancillary tree species → 7356**

* **andiroba → 4039**

* **andirobeira → 4039**

* **Andrew poplar → 4766**

283 anemophilous; wind-pollinated
d windbestäubt
f anémophile
r анемофильный; ветроопыляемый

* **anemosis → 9450**

* **angiospermous forest → 2300**

* **angiospermous wood → 4124**

284 angiosperms; angiospermae (*lat*)
(the botanical name for the group of vascular flowering plants that produce seeds enclosed in an ovary; includes hardwoods, bamboos, and palms but not softwoods)
d Angiospermen *npl*; Bedecktsamer *mpl*
f angiospermes *fpl*
r покрытосемянные *pl*

285 angle-count cruising; plotless cruising; variable plot cruising; point sample cruise; point sampling
(a method for estimating standing timber stocking or volume without establishing sample plot boundaries)
d Winkelzähl-Bestandesaufnahme *f*; Punkt-Stichprobenprüfung *f*; Punkt-Stichprobenverfahren *n*
f échantillonnage *m* par placettes circulaires à rayon variable; estimation *f* [d'un peuplement] par la méthode du balayage sous angle
r пробный угловой учёт *m*

286 angledozer
d Planierraupe *f* mit schiefstellbarem Schild
f niveleur *m* à chenille avec bouclier orientable
r гусеничный трактор *m* с косым выравнителем; гусеничный трактор с поворотным выравнителем

287 angled shear
d gewinkelte Schere *f*
f cisaille *f* à lames inclinées
r изогнутые ножницы *fpl*; косоугольная саморезка *f*

* **angle grain → 4585**

288 angle of cut; cutting angle
d Schnittwinkel *m*; Schneidewinkel *m*
f angle *m* de coupe
r угол *m* среза; угол резания

289 angle planting; slit planting; T-notching; notch planting; saddle planting; wedge planting; flap planting
(prying open a cut made by a spade, mattock, or planting bar, inserting a young tree, then closing the cut on the latter by pressure)
d Klemmpflanzung *f*; Spaltpflanzung *f*; Klapp-Pflanzung *f*
f plantation *f* avec bêche; plantation en fentes; bêchage *m* en T; plantation sur ensellement; plantation faite en soulevant une motte de gazon
r посадка *f* с защемлением; отложная посадка

* **angok → 1121**

290 Angola padauk; muninga
(Pterocarpus angolensis)
d Muninga *m*; afrikanischer Padouk *m*
f muninga *m*
r птерокарпус *m* ангольский

* **angouma → 3788**

* **angsana → 5284**

291 animal community; zooc[o]enosis
(characteristic assemblage of animals associated with a forest type or ecological community)
d Zooz[o]enose *f*; Gemeinschaft *f* von Tieren

292 animal logging; animal skidding
d Zugtierbringung *f*; Frachtfuhrwesen *n*; Holztransport *m* mit Gespannen; Lasttiertransport *m*
f débardage *m* animal
r лесозаготовка *f* с применением гужевой трелёвки; гужевая трелёвка *f*

* **animal nuisible to timber → 9533**

293 anise tree
(Illicium)
d Sternanis *m*
f badianier *m*; anis *m*
r бадьян *m*; иллициум *m*

294 anise-tree family
(Illiciaceae)
d Sternanisgewächse *pl*
f illiciacées *pl*
r иллициевые *pl*; бадьяновые *pl*

* **annato tree → 4995**

* **annual allowable cut → 192**

* **annual allowable harvest → 193**

295 annual cropping
d Ernte *f*
f culture *f* annuelle
r годичный урожай *m*

296 annual cut[ting]; annual yield; year cut[ting]; year felling
d Jahreseinschlag *m*; Jahresschlag *m*; jährliche Nutzungsmenge *f*
f coupe *f* annuelle; quantité *f* annuelle d'exploitation; volume *m* exploité annuellement
r годичная лесосека *f*; годичная рубка *f*

297 annual felling removed from the forest
d jährlicher Einschlag *m* aus den Waldflächen
f coupe *f* annuelle prélevée sur la forêt
r годичная рубка *f*, снятая с леса

298 annual foliage alternation
d jährlicher Laubwechsel *m*
f alternation *f* annuelle de feuillage
r ежегодная смена *f* листвы

299 annual forest program
(the summary or aggregation of all projects that make up an integrated course of action for a given level of funding of a forest
d jährliches Forstprogramm *n*
f programme *m* forestier annuel
r годичная лесоустройственная программа *f*

300 annual growth; annual [height] increment; yearly growth
(average annual increase in the biomass of growing-stock trees of a specified area)
d jährlicher Zuwachs *m*; Jahreswuchs *m*
f croissance *f* annuelle
r годичный прирост *m*; ежегодный рост *m*

301 annual ring; year ring; growth ring; [annual] growth layer; tree ring; annulus
(layer of wood produced around a tree's stem and branches during each growing season; also called annual ring)
d Jahresring *m*; Jahrring *m*; Wuchsring *m*; Vegetationsring *m*; Zuwachsring *m*; [jährliche] Zuwachszone *f*
f cerne *m* [de croissance]; cerne annuel; anneau *m* ligneux; anneau de croissance; couche *f* d'accroissement annuelle
r годичный слой *m*; слой роста; годичное кольцо *n*; приростное кольцо; кольцо роста

302 annual ring arrangement
d Jahrringfolge *f*
f arrangement *m* de cernes annuels
r расположение *n* годичных колец

303 annual ring density
d Jahrringdichte *f*
f densité *f* de cernes annuels
r густота *f* годичных колец

304 annual ring formation
d Jahrringbildung *f*
f formation *f* de cernes annuels
r образование *n* годичных колец

* **annual ring structure → 8188**

* **annual ring width → 9398**

305 annual shoot; yearling shoot
d Jahrestrieb *m*
f rejet *m* annuel; pousse *f* annuelle
r годичный побег *m*

306 annual wildlife seed mixture
(mixture of soybean, millet, cow pea, sorghum, lespedeza, buckwheat, and other seeds from which single-season plants are grown to serve as food or protective cover for wildlife)

f mixte *m* annuel de semences sauvages; mélange *m* annuel de semences sauvages
r смесь *f* семян годовых диких растений

* **annual yield → 296**

307 annular budding; bud grafting; budding; gemmation
(specialized form of grafting, in which a scion bud and adjacent cambial tissue is slipped below the bark of the stock)
d Sprossung *f*; Okulation *f*; Okulieren *n*; Okulierung *f*
f greffage *m* d'yeux; greffage en écusson; écussonnage *m*
r прививка *f* глазком; окулирование *n*; окулировка *f*; почкование *n*; размножение *n* почкованием

308 annular saw
d Ringsäge *f*
f scie *f* annulaire
r кольцевая пила *f*

* **annulus → 301**

* **anodal [top] → 5544**

* **anonillo → 5558**

309 ant acacia
(Acacia collinsii; Acacia costaricensis)
d Ameisen-Akazie *f*
f acacia *m* de Collins
r акация *f* серповидной листвой; акация пальчатонервная

* **Antarctic beech → 6985**

310 antecedent soil water; antecedent soil moisture
(degree of wetness of the soil at the beginning of a runoff period expressed as an index or as the total inch-depth-equivalents of water stored in the soil)
d vorrangige Bodenfeuchte *f*
f humidité *f* antérieure du sol; teneur *f* en eau initiale
r предшествующая почвенная влага *f*

311 antelope [bitter]brush; bitterbrush
(Purshia tridentata; Kunzia tridentata)
d Antelopebusch *m*
f purshie *f* tridentée
r пуршия *f* трёхзубая

d Staubbeutel *m*; Anthere *f*
f anthère *m*
r пыльник *m*; пыльцевой мешок *m*

313 anthesis
(the opening of a flower bud; the exposure of stamens and stigmas to pollinating agents; the span of life of a flower)
d Anthese *f*
f anthèse *f*
r период *m* цветения; полное цветение *n*; пыление *n*

314 anthocyanosis
(presence of abnormal red-purple colo[u]ration in foliage)
d Anthocyanose *f*
f anthocyanose *f*
r антоцианоз *m*

315 anthracnose; black spot [disease]
(disease characterized by ulcer-like leaf or fruit spots and caused by fungi that produce asexual spores in the type of fruiting body called a acervulus)
d Anthracnose *f*; schwarzer Brenner *m*; Schwarzfleckenkrankheit *f*; Schwarzfleckigkeit *f*
f anthracnose *f*; moucheture *f* noire; rouille *f* noire; charbon *m*
r черная пятнистость *f*

* **anthropogenic forest → 5285**

316 anthropogenic load
(charge of human origin)
d anthropogene Ladung *f*
f charge *f* anthropogénique; charge anthropique
r антропогенная нагрузка *f*

* **anthropogenic subclimax → 2526**

* **antiaris → 6414**

* **anticlockwise spiral growth → 6943**

317 anti-erosion forests
d Erosionsschutzwalde *mpl*
f forêts *fpl* antiérosives
r противоэрозионные леса *mpl*

* **antifungal [agent] → 3771**

* **antimycotic [agent] → 3771**

318 anti-sapstain [treatment]
(a wood treatment used to prevent fungus from staining the wood)
d Saftfleckbehandlung *f*
f traitement *m* contre les taches de sève
r обработка *f* против синевы; обработка против заболонной окраски

319 anti-skid chain; anti-skid device; non-skid chain; tire chain
d Gleitschutzkette *f*
f chaîne *f* à pneu; antidérapant *m*
r цепь *f* против скольжения

* **anti-skid device → 319**

* **antlers → 401**

320 ant-tree
d Ameisenbaum *m*
f arbre *m* attaqué par des fourmis
r дерево *n*, в котором гнездятся муравьи

321 anvil blade; anvil; bed plate
(fixed steel block that provides support and resistance for the cutting blade of a single-action tree shear)
d Amboss *m*
f plaque *f* d'appui; contre-outil *m*
r опора *f* для удержания дерева; опорный рычаг *m*

322 anvil-type shear; anvil secateur; snap-cut shears; roll cut hand shears
(secateur which has a straight-edged top blade which comes down on a soft metal cutting surface)
d Amboss-Schere *f*
f sécateur *m* universel
r одноножевое срезающее устройство *n*

* **Apache pine → 2821**

* **apex → 8699**

323 aphids
(Aphididae)
d Baumläuse *fpl*; Blattläuse *fpl*
f pucerons *mpl*
r тли *fpl*

324 apical branch
d Höhenzweig *m*
f branche *f* apicale
r верхняя ветвь *f*

325 apical dominance; apical control
(control over lateral bud or shoot development exerted by plant growth regulators from a shoot apex in the current year of growth)
d Höhendominanz *f*
f dominance *f* apicale
r доминирование *n* вершины

326 apical growth; growth in height; height growth; height increment; height accretion
d Höhenzuwachs *m*; Längswachstum *n*; Höhenwuchs *m*; Höhenwachstum *n*
f incrément *m* apical; accroissement *m* en hauteur; accroissement de la hauteur
r прирост *m* в высоту; рост *m* в высоту; верхушечный рост

327 apical meristem
(growing point of the plant)
d Höhenmeristem *n*
f méristème *m* apical
r апикальная меристема *f*; верхушечная меристема

* **apical wood → 8712**

328 apotracheal parenchyma
(axial parenchyma typically independent of the pores or vessels)
d apotracheales Parenchym *n*
f parenchyme *m* apotrachéal
r апотрахеальная паренхима *f*

329 apple box
(Eucalyptus bridgesiana; Eucalyptus stuartiana)
d Bridges-Eukalyptus *m*
f gommier *m* de Bridges
r эвкалипт *m* Бриджеса

330 apple family
(Pomaceae; Malaceae)
d Pomazeen *pl*; Kernobstgehölze *npl*; Pomeen *pl*
f pomacées *pl*
r яблоневые *pl*

* **apple sawfly → 2879**

* **apple[-tree] → 1755**

331 approach grafting; veneer grafting; inarching

f greffage *m* en approche
r прививка *f* сближением; прививка черенком; аблактировка *f*

332 apricot plum
(Prunus simonii)
d Simon-Pflaume *f*; Aprikosenpflaume *f*
f prune-abricot *f*; prunier *m* abricotier
r абрикосовая слива *f*

333 apron
(a length of wood found beneath the bottom framing of a drawer, table top, chair seat etc. usually shaped and often decorated)
d Schurz *m*; Talon *m*
f allège *m*
r наличник *m*

334 apron; gutter
(on the stem for resin extraction)
d Harzablaufvorrichtung *f*; Abflussrinne *f*; Ableitungsrinne *f*; Rinne *f*
f rigole *f* d'écoulement
r [направляющий] жёлоб *m*; фартук *m*; жёлобок *m* для стока живицы [в приёмник]

335 aquasilvicultural system
(agroforestry system that combines trees with the raising of aquatic animals)
d aquasilvokulturliches System *n*; wasserforstwirtschaftliches System
f système *m* aquasylvicole
r водолесокультурная система *f*

* **Arabian juniper → 6787**

* **arabic gum → 4048**

* **arabisco → 1921**

336 aralia
(Aralia)
d Aralie *f*; Bergangelika *f*
f aralie *f*
r аралия *f*

337 aralia family
(Araliaceae)
d Araliengewächse *pl*
f araliacées *pl*
r аралиевые *pl*

* **aramatta → 8251**

(Centrolobium robustum)
d Arariba *m*
f arariba *m*
r центролобиум *m* мощный

339 araucaria family
(Araucariaceae)
d Araukariagewächse *pl*
f araucariacées *pl*
r араукариевые *pl*

* **arbor → 8797**

* **arboraceous → 9704**

340 arbor day; tree[-planting] day; forest day
d Baumpflanzungstag *m*
f jour *m* de plantation d'arbres; journée *f* de l'arbre
r день *m* древонасаждения; день посадки леса

* **arboreal → 9704**

* **arboreous → 9704**

* **arborescent → 9704**

* **arborescent aloe → 9705**

341 arborescent euphorbia; tree euphorbia; tree spurge
(Euphorbia dendroides)
d Wolfsmilch *m*
f euphorbe *f* arborescente
r молочай *m* древовидный

342 arboretum; forest garden; forest [tree] nursery; nursery garden; tree nursery; seminary
(botanical tree garden where trees are maintained for display purposes)
d Forstgarten *m*; Baumgarten *m*; Forst[baum]schule *f*; Waldbaumschule *f*; Baumschule *f*; Arboretum *n*; Kamp *m*
f arboretum *m*; pépinière *f* forestière
r лесной питомник *m*; дендрарий *m*; арборетум *m*

343 arboricide; silvicide; tree[ ]killer
(against trees and other woody growth)
d Arborizid *n*
f arboricide *m*
r арборицид *m*

344 arboriculturist; silviculturist; tree breeder; tree-grower; wood-grower; timber grower; forest tenant; forest villager; forest squatter
(cultivator licensed to live, and raise agricultural crops, on land in reserved forest)
d Baumzüchter *m*; Waldläufer *m*; Heger *m*; Oberheger *m*; Waldschaffer *m*
f arboriculteur *m*; arboricultrice *f*; pépiniériste *m/f* arboricole; producteur *m* de bois; sylviculteur *m*; tenancier *m* forestier
r воспитатель *m* деревьев; лесовод *m*; производитель *m* древесины

* **arboriform → 8876**

345 arborization
d Baumzüchtung *f*
f arborisation *f*; arborescence *f*
r древовидное разветвление *n*

* **arbor[ ]vitae → 8893**

346 arbuscle; shrubby tree; arbutus
d Bäumchen *n*; buschiger Baum *m*
f arbrisseau *m*; arbuste *m*
r древовидный кустарник *m*

347 arbuscular; arbuscle-shaped
d zwergbaumartig
f arbusculaire
r деревцеобразный

* **arbuti tree → 6020**

* **arbutus → 346**

* **arbutus-berry → 8152**

* **arc height → 4234**

* **arching → 349**

348 architectural form
(of a tree)
d architektonische Form *f*; Architekturform *f*
f forme *f* architecturale
r архитектурная форма *f*

349 arch skidding; arching
(skidding logs or trees using a mounted or trailing arch)
d Wölbung *f*; Wölben *n*
f débardage *m* à arche
r тракторная трелёвка *f* с применением арки

350 arctic bramble
(Rubus flagellaris; Rubus arcticus)
d amerikanische Taubeere *f*; Polarbrombeere *f*
f ronce *f* arctique; ronce du Nord
r княженика *f*; поленика *f*; малина *f* арктическая

* **arctic grey willow → 3926**

* **area control → 6842**

* **area determination → 3329**

* **area forester → 1988**

351 area ignition
(setting of a number of individual fires throughout an area)
d Flächenentflammung *f*
f allumage *m* de zone
r зажигание *n* по площади

352 area of isolated trees
d Fläche *f* mit einzelnen Bäumen
f surface *f* portante des arbres isolés
r участок *m* отдельных деревьев

353 area of shrub; shrubland; scrub land
(an open or closed stand of shrubs up to about 2 m tall)
d Strauchfläche *f*
f terrain *m* broussailleux; broussaille *f*
r местность *f*, покрытая кустарником; пространство *n*, покрытое кустарником; земля *f*, заросшая кустарником

* **area regulation → 6842**

* **area salvage → 7170**

354 area under canopy; area under crown cover; sheltered area; covered area
d beschirmte Fläche *f*; Überschirmungsfläche *f*
f surface *f* abritée
r площадь *f* полога; площадь под пологом

* **area under felling → 1621**

355 area under regeneration; regeneration area; planting area; compartment under regeneration
(area selected, normally in a working plan or working scheme, for regeneration generally with a specified period of time in view)

f surface *f* à régénérer; quartier *m* en régénération; parcelle *f* en régénération; surface à cultiver
r площадь *f* возобновления; участок *m* возобновления; площадь культуры

356 areca family
(Arecaceae; Palmae)
d Arecagewächse *pl*; Palmengewächse *pl*
f arécacées *pl*
r арековые *pl*

357 argyle apple; mealy stringybark [eucalyptus]; ashen-grey eucalyptus
(Eucalyptus cinerea)
d Korkenziehereukalyptus *m*; Gummi *n* des silbernen Dollars
f pomme *f* d'argyle; gomme *f* de dollar en argent; eucalyptus *m* de tire-bouchon
r аргайл *m*; эвкалипт *m* [серо]пепельный; эвкалипт серый

358 arid light forest; arid open woodland; xerophytic light forest
d arider Lichtwald *m*
f forêt *f* xérophytique; bois *m* clairsemé xérophytique
r аридное редколесье *n*; ксерофитное редколесье

* **aristate leaf → 7471**

* **Arizona ash → 9188**

359 Arizona cypress; smooth cypress
(Cupressus glabra; Cupressus arizonica)
d Arizona-Zypresse *f*
f cyprès *m* de l'Arizona
r аризонский кипарис *m*

* **Arizona ironwood → 1002**

* **Arizona pine → 9031**

360 Arizona ponderosa pine; Arizona [yellow] pine
(Pinus [ponderosa] arizonica)
d Arizonakiefer *f*
f pin *m* de l'Arizona [pondéreux]
r сосна *f* [жёлтая] аризонская

361 Arizona sycamore
(Platanus wrightii)
d Arizonaplatane *f*
f sycomore *m* de l'Arizona
r платан *m* Райта

362 Arizona walnut
(Juglans major)
d Arizona-Nussbaum *m*
f noyer *m* de l'Arizona
r аризонский орех *m*

* **Arizona white oak → 8993**

* **Arizona [yellow] pine → 360**

* **Arkansas pine → 7538**

363 Armenian oak; pontine oak; Pontic oak
(Quercus pontica)
d armenische Eiche *f*; pontische Eiche
f chêne *m* pontique
r дуб *m* понтийский

364 Armillaria root rot; spongy and butt rot
(invader - Armillaria mellea)
d Weißwurzelschwamm *m*; Wurzel- und Stammfäule *f*
f pourriture *f* blanche des racines; pourridié *m* agaric; pourriture des racines et du rejet; carie *f* racinaire et spongieuse
r белая периферическая гниль *f* корней; губчато-волокнистая корневая и напенная гниль

* **arm-saw → 3342**

* **arolla pine → 8346**

365 aromatic sumac[h]; fragrant sumac[h]; lemon sumac[h]
(Rhus aromatica; Rhus mollis)
d Gewürzsumach *m*
f sumac *m* aromatique; sumac amarante
r душистый сумах *m*

* **aromilla → 1712**

* **arrangement of plants → 6362**

* **array of stones → 6969**

366 arrest *v* in growth; stunt *v*; dwarf *v*; clog *v*
d im Wachstum aufhalten; im Wuchs aufhalten
f être arrêté dans la croissance
r останавливаться в росте

* **arriero → 5558**

* **arris → 7255**

* **arris knot → 4733**

* **arrow-shaped leaf → 4165**

367 arrow weed
(Pluchea sericea)
d Pfeilunkraut *m*
f mauvaise herbe *f* de flèche
r плюхея *f*

368 arrowwood viburnum; southern arrowwood
(Viburnum dentatum)
d gezähnter Schneeball *m*
f viorne *f* dentée
r калина *f* зубчатая

* **arrowy leaf → 4165**

369 Arroyo willow; Pacific willow
(Salix lasiolepis)
d Arroyo-Weide *f*
f saule *m* du Pacifique; saule d'Arroyo
r ива *f* Арройо

* **arson → 4436**

370 articulated growth
(rhythmic growth that results in shoot units separated by morphological discontinuities)
d Knickwachstum *n*
f incrément *m* articulé
r отчётливый рост *m*

371 articulated semi-trailer
d Sattelauflieger *m*
f semi-remorque *f* articulée
r шарнирно-сочленённый с тягачом полуприцеп *m*

372 articulated shortwood harvester
d Gelenkkurzholzerntemaschine *f*
f abatteuse-porteuse *f* de bois courts articulée
r многооперационная машина *f* с шарнирно-сочленённой рамой для заготовки сортиментов

373 articulated-steering tractor; articulated skidder
(ref. to a vehicle, hinged at the center for turning, as with a wheeled skidder)
d Knickschlepper *m*
f tracteur *m* articulé
r трелёвочный трактор *m* с рамным управлением; трелёвочный трактор с шарнирно-сочленённой рамой

374 artificial colo[u]red wood
r искусственно окрашенная древесина *f*

* **artificial forest regeneration → 6811**

375 artificial formation of stand; establishing of stand
d künstliche Bestandes[be]gründung *f*; künstliche Bestandbildung *f*
f création *f* artificielle d'un peuplement
r искусственное создание *n* насаждения; искусственное образование *n* насаждения

376 artificial pruning
d künstliches Aufasten *n*
f élagage *m* artificiel
r искусственная очистка *f* от сучьев

* **artificial reproduction → 6811**

* **artificial stand → 5285**

377 artificial thinning
d künstliche Lichtstellung *f*; künstliche Lockerung *f*
f éclaircie *f* artificielle
r искусственное изреживание *n*; искусственное прореживание *n*

378 artificial wood
d Kunstholz *n*
f bois *m* artificiel
r искусственная древесина *f*

* **artocarpe → 1011**

* **ascending of the sap → 7195**

* **asexual propagation → 9182**

* **asexual reproduction → 9182**

* **ash → 2903**

379 ash [bark] beetle
(Hylesinus)
d [bunter kleine] Eschenbastkäfer *m*
f hylésine *f* du frêne
r ясеневый лубоед *m*

* **ashen-grey eucalyptus → 357**

380 ash forest
d Eschenwald *m*
f frênaie *f*; frênière *m*
r ясеневый лес *m*

381 ash-leaved maple; ashleaf maple;

d eschenblättriger Ahorn *m*; Eschenahorn *m*
f érable *m* négundo; négundo *m*; érable aux feuilles de frêne
r клён *m* ясенелистный; клён американский

* **ash-leaved star tree → 8548**

* **Asian black birch → 2226**

382 **Asian cherry**
(Padus asiatica)
d asiatische Kirsche *f*
f merisier *m* d'Asie
r черёмуха *f* азиатская

* **Asian kaki → 4627**

383 **Asian padauk**
(Pterocarpus dalbergioides)
d Andamanen-Padauk *m*; asiatischer Padauk *m*
f padauk *m* d'Asie
r сандал *m* дальбергиевовидный; андаман *m* падук

384 **Asian white birch; broadleaf white birch**
(Betula platyphylla)
d mandschurische Birke *f*; japanische Weißbirke *f*
f bouleau *m* à larges feuilles; bouleau blanc du Japon
r берёза *f* плосколистная

* **Asiatic oak → 3937**

* **aspen → 2880**

385 **aspen blocks**
d Espenfaserholz *n*; Espenpapierholz *n*
f billots *mpl* du tremble
r осиновые балансы *mpl*

386 **aspen fallen wood**
d Espenlagerholz *n*
f bois *m* mort du tremble
r осиновый валежник *m*

387 **aspen forest**
d Espenwald *m*; Zitterpappelbestand *m*
f forêt *f* de trembles; tremblaie *f*
r осиновый лес *m*; осинник *m*

388 **aspen fuelwood**
d Espenbrennholz *n*

389 **aspen leaf beetle**
(Chrysomela crotchi brown)
d roter Espenblattkäfer *m*
f chrysomèle *f* du tremble
r осиновый листоед *m*

* **ASQ → 199**

* **assacu → 7185**

* **assamela → 119**

390 **assart; cull-land; clearing**
(woodland turned into arable land)
d Rodeland *n*
f terrain *m* défriché
r новь *f*; чищоба *f*

* **assart[ed] tree → 9141**

* **assie → 9150**

391 **assimilation of carbon dioxide; carbon dioxide assimilation**
d Kohlensäure-Assimilation *f*
f assimilation *f* chlorophyllienne
r ассимиляция *f* углекислоты

* **assistance cutting → 9320**

392 **assistant forester; associate forester**
d Hilfsförster *m*; Forstassessor *m*
f garde *m* auxiliaire
r помощник *m* лесничего

393 **associates species**
d Mischungsart *f*; Mischbaumart *f*
f essence *f* du mélange; essence associée
r сопутствующая порода *f*

394 **association**
(collection of plants with ecologically similar requirements, including one or more dominant species)
d Gemeinschaft *f*
f association *f*
r ассоциация *f*

395 **assortment hauling**
d Holzschleppen *n*
f débardage *m* d'assortiment
r вывозка *f* сортимента

* **assortment table → 3896**

396 Atlantic cedar; [blue] Atlas cedar
(Cedrus atlantica)
d atlantische Zeder *f*; Atlaszeder *f*; Silberzeder *f*
f cèdre *m* d'Algérie; cèdre de l'Atlas
r кедр *m* атласский

397 Atlantic ninebark; [common] ninebark
(Physocarpus opulifolius)
d Schneeball-Blasenspiere *f*; Virginia-Blasenspiere *f*
f physocarpe *m* nain; physocarpe à feuilles d'obier
r пузыреплодник *m* калинолистный; пузырчатник *m* калинолистный

398 Atlantic white cedar; white cedar; swamp cedar; false-cypress
(Chamaecyparis thyoides)
d Zederzypresse *f*; Kugelzypresse *f*; Weißzeder *f*
f cèdre *m* blanc [d'Amérique]; faux cyprès *m*
r кипарисовик *m* туеобразный; южный белый кедр *m*

* **Atlas cedar → 396**

399 attached stumpcutter
d angebauter Stumpfscherblock *m*
f broyeur *m* de souches attaché
r прикреплённый измельчитель *m* пней

400 attenuate *v*
(in plant morphology: growing narrower toward the base, or sometimes toward the tip)
d schwächen; mindern; dämpfen
f atténuer; réduire
r ослаблять; затухать; истощать; истощить

401 attire; horns; antlers
d Geweih *n*
f bois *m* de cerf
r рога *mpl*

* **attractant → 4509**

* **auger → 6342**

402 auger planting
(setting plants in loosened soil replaced in or brought to a dug hole using an auger)
d Bohrerpflanzung *f*
f plantation *f* à la tarière
r посадка *f* под бур

403 auger[-type] felling head; rotary cutter
f tête *f* d'abattage à fraise
r валочная фреза *f*; бурав *m* (большой); цилиндрический бур *m*

* **August plum → 256**

* **auon → 7766**

404 Australian blackwoood; blackwoood acacia
(Acacia melanoxylon)
d Schwarzholzakazie *f*
f acacia *m* à bois noir; acacia à bois dur; acacia noir d'Australie
r акация *f* [чёрная] австралийская; акация чёрнодревесная

405 Australian grass tree; grass tree; blackboy
(Xanthorroea johnsonii; Xanthorroea fulva)
d [australischer] Grasbaum *m*
f xanthorrhée *f*
r акароидин *m*; дерево *n* травяное; жёлтосмолка *f*; ксанторрея *f*

* **Australian laurel → 413**

406 Australian nothofagus; Tasmanian myrtle; Tasmanian false beech; myrtle beech; evergreen beech; silver beech
(Nothofagus cunninghamii)
d australischer Nothofagus *m*; tasmanische Scheinbuche *f*
f nothofagus *m* d'Australie; hêtre *m* austral de Tasmanie; faux hêtre de Cunningham
r нотофагус *m* Куннингама

407 Australian pine; Polynesian ironwood; horsetail tree; river-oak; she-oak
(Casuarina cunninghamiana)
d Känguru[h]baum *m*; australische Kiefer *f*; australische Flusseiche *f*
f filao *m* [à feuilles de prêle]; pin *m* australien
r казуарина *f* хвощелистная; железное дерево *n*; казуарина Куннингама

408 Australian[-]pine borer
(Chrysobothris tranquebarica)
d australischer Kiefernbohrer *m*
f agrile *f* du pin [australienne]
r златка *f* сосновая австралийская

409 Australian pink walnut; pink walnut
(Endiandra sieberi)
d rosafarbige australische Nuss *f*
f noyer *m* rose d'Australie
r криптокария *f* розовая

410 **Australian red cedar; toon**
(Toona australia; Toona ciliata; Cedrela toona)
d australische rote Zeder *f*; australisches Mahagoni *n*
f cèdre *m* rouge d'Australie
r цедрела *f* раскалывающая; цедрела австралийская; красный кедр *m* австралийский; цедрела Тоона

411 **Australian red mahogany; red mahogany; kino eucalyptus**
(Eucalyptus resinifera)
d australischer rote Mahagonibaum *m*; Harz-Eukalyptus *m*
f acajou *m* rouge australien
r эвкалипт *m* резиноносный; эвкалипт смолоносный

* **Australian silky oak → 3966**

412 **Australian umbrella-tree; [Queensland] umbrella-tree; octopus tree**
(Schefflera actinophylla; Schefflera arboricola; Brassaia actinophylla)
d Regenschirmbaum *m*; Strahlenaralie *f*; Lackblattpflanze *f*
f arbre *m* ombrelle
r шефлера *f* лучелистная

413 **Australian walnut; Queensland walnut; Australian laurel; Oriental wood; walnut bean**
(Cryptocarya palmerstonii; Endiandra palmerstonii)
d australisches Endiandra *n*
f endiandra *m* d'Australie
r криптокария *f* Пальмерстона; австралийский меданг *m*

* **Australian wattle → 5301**

414 **Australian white ash; white mountain ash**
(Eucalyptus fraxinoides)
d falsche australische Weißesche *f*
f faux frêne *m* blanc d'Australie
r эвкалипт *m* ясеневидный

415 **Austrian [black] pine; [European] black pine**
(Pinus nigra; Pinus austriaca)
d [österreichische] Schwarzkiefer *f*; Schwarzföhre *f*
f pin *m* noir [d'Autriche]
r сосна *f* чёрная [австрийская]

(Artemisia austriaca)
d österreichischer Wermut *m*
f armoise *f* d'Autriche
r полынь *f* австрийская

* **autochthonous forest → 4476**

* **autogamy → 7421**

* **autolopping → 5688**

417 **autumn-olive**
(Elaeagnus umbellata)
d Korallen-Ölweide *f*; doldige Ölweide *f*
f chalef *m* en ombrelles
r лох *m* зонтичный

* **autumn timber → 4827**

* **autumn wood → 4827**

418 **auxiliary branch; side shoot; lateral shoot**
d Geiz[trieb] *m*; Seitentrieb *m*
f pousse *f* ralentie; pousse latérale
r пазушный побег *m*; боковой побег

419 **auxiliary bud; [lateral] bud**
(a bud borne laterally on the stem in the axil of a leaf)
d Seitenknospe *f*; Sprossknospe *f*; Achselknospe *f*
f bourgeon *m* auxiliaire
r пазушная почка *f*; боковая почка; латеральная почка

420 **auxiliary standing timber; auxiliary forest standing; auxiliary trees; gap-cover; halves**
d Füllbestand *m*; Füllholz *n*
f peuplement *m* forestier de remplissage; peuplement auxiliaire
r заполняющий древостой *m*

421 **auxin**
(a phytohormone involved in many physiological processes including root and stem elongation, fruit set, cambium activity and sex determination)
d Auxin *n*
f auxine *f*
r ауксин *m*

422 **available fuel**
d tatsächlich vorhandenes brennbares Material *n*

423 **available [soil] moisture; available water** (for plants)
d nutzbare Bodenfeuchte *f*
f eau *f* disponible
r доступная вода *f* [в почве]

424 **available timber; operable timber** (timber which is available for harvest after due recognition of constraints to protect the environment and other forest uses)
d verfügbares Holz *n*
f bois *m* disponible
r доступная древесина *f*

* **available water → 423**

425 **avalanche track; avalanche forest**
d Lawinezug *m*; Bannwald *m*
f forêt *f* avalanche; couloir *m* d'avalanche
r вертикальная полоса *f* на залесенном горном склоне, лишенная древесной растительности, выбитой лавинами

* **avalanche transport → 7803**

426 **average age of stand; mean age of stand** (arithmetic mean age)
d Bestandesdurchschnittsalter *n*; Durchschnittsalter *n*
f âge *m* moyen de peuplement
r средний возраст *m* лесонасаждения

427 **average annual mortality** (average annual volume of trees 5.0 inches d.b.h. and larger that died from natural causes during the intersurvey period)
d mittlere jährliche Sterblichkeit *f*
f mortalité *f* annuelle moyenne
r средняя годичная смертность *f*

* **average annual precipitation → 5353**

428 **average annual removals** (average annual volume of trees 5.0 inches d.b.h. and larger removed from the inventory by harvesting)
d mittlere jährliche Holzernte *f*
f recette *f* moyenne annuelle
r средний годичный урожай *m* леса

429 **average annual stand depletion**
d mittlere jährliche Durchforstung *f*
f épuisement *m* annuel moyen; moyenne *f* annuelle d'extraction en éclaircie
r средний годичный объём *m* вырубки запаса насаждения

f peuplement *m* moyen de coléoptères
r средняя плотность *f* короедов

431 **average diameter of defect**
d Mitteldurchmesser *m* der Fehler
f diamètre *m* moyen de défaut
r средний диаметр *m* порока

432 **average fiber length**
d mittlere Faserlänge *f*
f longueur *f* de fibres moyenne
r средняя длина *f* волокон

* **average girth → 5440**

* **average height → 5357**

433 **average-height tree**
d Zuwachsmittelstamm *m*
f arbre *m* d'hauteur moyenne
r дерево *n* среднего роста

434 **average log; mean log; middle log; medium log**
d Bestandesmittelstamm *m*
f grume *f* moyenne; grume de moyenne longueur
r средний ствол *m* насаждения; сортимент *m* средней длины

435 **average logs to a turn; average number of tree per load**
d mittlere Stammzahl *f* pro Last
f nombre *m* moyen d'arbres par charge
r среднее число *n* брёвен в пачку; среднее число деревьев в пачке

436 **average long-term yield**
d Langzeit-Durchschnittsertrag *m*
f rendement *m* moyen à long terme
r среднемноголетний сток *m*; среднемноголетний дебит *m*

437 **average net annual growth; average net annual increment** (average annual net change in volume of trees 5.0 inches d.b.h. and larger in the absence of cutting)
d mittlerer jährlicher Nettozuwachs *m*
f accroissement *m* annuel moyen net
r чистый среднегодовой прирост *m*

* **average number of tree per load → 435**

f nombre *m* moyen d'arbres par acre
r среднее число *n* деревьев на акре

439 average total wood costs
d mittlere totale Holzwerbundskosten *pl*
f frais *mpl* d'exploitation moyens totaux
r общие средние затраты *fpl* на заготовку леса

440 average tree height; mean tree height
d mittlere Baumhöhe *f*
f hauteur *f* moyenne de l'arbre
r средняя высота *f* дерева

441 average tree length
d mittlere Baumlänge *f*
f longueur *f* moyenne de l'arbre
r средняя длина *f* хлыста

442 average tree of the stand; mean tree; type tree
d Mittelbaum *m*
f arbre *m* moyen [d'échantillonnage]
r средняя модель *f* насаждения; типичное дерево *n*

443 average volume per tree; mean tree volume
d Mittelbaumvolumen *n*
f volume *m* moyen d'arbre
r средний объём *m* дерева; средний объём хлыста

444 average yarding distance; AYD
(total yarding distance for all turns divided by total number of turns for a particular setting)
d mittlere Rückedistanz *f*
f distance *f* de roulage moyenne
r среднее расстояние *n* канатной трелёвки

445 avocado tree; alligator pear
(Persea gratissima; Persea americana)
d Avocatobaum *m*; Avocado *m*
f avocatier *m*; avocat *m*
r авокадо *n*; персея *f*; аллигаторова груша *f*

* **avodire → 117**

446 avoidable waste
(volume of timber left on the harvested area that should have been removed in accordance with the utilization standards in the cutting authority)
d anfechtbarer Holzabfall *m*; vermeidlicher Holzabraum *m*
r не неизбежные лесосечные отходы *mpl*

* **awl-shaped leaf → 8245**

447 awl-shaped needles
d wie eine Ahle geformte Nadelung *f*
f aiguilles *fpl* en forme de poinçon
r шиловидная хвоя *f*

* **awned leaf → 7471**

* **ax chop → 450**

* **axe → 83**

* **ax[e] → 453**

448 ax[e] blade
d Axtblatt *n*
f lame *f* d'hache
r лезвие *n* топора

449 ax[e] blazing; ax marking; axing
d Axtanreißen *n*; Axtanzeichen *n*
f marquage *m* à la hache
r маркировка *f* топором; затёска *f* топором; зарубка *f* топором

450 ax[e] chock; ax chop
d Axthieb *m*
f coup *m* à la hache
r удар *m* топором

451 axe eye; axe hole
d Axtöhr *n*; Auge *n* der Äxte
f œil *m* d'hache
r отверстие *n* топора; обух *m*

452 axe-face
d Axtbahn *f*
f trace *f* de frappe d'une hache; entaille *f*
r пробег *m* топора

453 axe for cutting wood; felling axe; break axe; wedge axe; split[ting] axe; splitter; cleaving axe; ax[e]; sledge axe; cleaver; whittle (US)
d Beil *n*; Fällaxt *f*; Fällhacke *f*; Forstaxt *f*; Holz[fäller]axt *f*; Holzhaueraxt *f*; Spaltaxt *f*; Spalthammer *m*; Schlägelaxt *f*
f hache *f* de bûcheron; hache de refonte; hache d'abattage; merlin *m*
r [лесорубочный] топор *m*; лесорубный топор; колун *m*; тесак *m*

* **axe-helve → 4110**

454 **axe wedge**
d Axtkeil *m*
f languette *f* de fixation d'un fer de hache
r клин *m* топора

455 **axial elements**
(a term of convenience in wood anatomy for all the cells other than those of the rays)
d axiale Elemente *npl*
f éléments *mpl* axiaux
r аксиальные элементы *mpl*

456 **axial flow combine harvester; rotary harvester**
d axiale Dreschmaschine *f*
f abatteuse-porteuse *f* à flux axial
r комбайн *m* с осевой центробежной молотилкой

457 **axial parenchyma; longitudinal parenchyma; vertical parenchyma**
(parenchyma cells derived from fusiform cambial initials)
d axiales Parenchym *n*
f parenchyme *m* axial, parenchyme longitudinal
r аксиальная паренхима *f*; продольная паренхима

* **axing → 449**

* **ax marking → 449**

* **axon hill → 1872**

* **ayan → 5732**

* **AYD → 444**

* **aye → 1090**

458 **Azores juniper; cedro; zimbro**
(Juniperus brevifolia)
d kleinblättriger Wacholder *m*
f genévrier *m* à feuilles courtes
r можжевельник *m* коротколистный

B

* **BA → 597**

* **babul [acacia] → 4048**

459 **Babylon[ian weeping] willow; witch willow; [green] weeping willow**
(Salix babylonica)
d Hängeweide *f*; Osterpalme *f*; [grüne] Trauerweide *f*; Tränenweide *f*
f saule *m* [vert] pleureur; saule de Babylone
r ива *f* плакучая; ива вавилонская

* **back[acting] shovel → 468**

460 **backbone**
(a triangular piece of wood which is placed in the apex of a V-box flume)
d Rückgrat *n*
f dos *m* de glissoire
r донный брусок *m* V-образного лотка

* **backburn → 465**

* **back[ ]country → 9404**

461 **backcountry recreation**
(near-primitive recreation experience within the natural setting of a relatively roadless area where travel is primarily by horse or foot trails or by canoe)
d Hintergrunderholung *f*
f recréation *f* d'aire naturelle; recréation d'arrière-pays
r восстановление *n* глуши; восстановление отсталых районов

462 **backcross**
(cross of a hybrid to either of its parents)
d Rückkreuzung *f*; Rückkreuzen *n*
f croisement *m* en retour; croisement en arrière; rétrocroisement *m*
r обратное скрещивание *n*

463 **backcross breeding method**
(method of improving plant characteristics in a stepwise fashion where previous gains are not lost)
d Rückkreuzung-Zuchtmethode *f*
f technique *f* d'amélioration backcross; méthode

* **backcut[ting] → 3066**

* **backdigger → 468**

* **back duster → 471**

464 **backfilling**
(to return the soil to a planting area from which it was originally dug)
d Hinterfüllung *f*
f remblai *m* de remplissage; remblayage *m*; remblaiement *m* derrière un mur de soutènement
r закладка *f* выёмки; засыпка *f* канавы

465 **backfire; backburn; counter-fire**
(blaze set in front of an advancing forest fire; an effort to check the wildfire by cutting off its fuel supply)
d Gegenbrand *m*; Gegenfeuer *n*; Rückbrennung *f*; Ausbrennen *n* durch Segenfeuer
f contrefeu *m*; retour *m* de flamme
r встречный огонь *m*; встречный пал *m*

466 **backflash**
(movement of a poison through natural root grafts, from trees that have been treated with poison, back to non-treated trees)
d Wurzelübertragung *f*
f transfert *m* de poison
r введение *n* ядовитых веществ через корни деревьев

* **background → 9236**

467 **back guy**
(line behind the spar tree, opposite the main line or skyline, which takes most of the pull in yarding logs)
d Rückanker *m*; Rückabspannung *f*
f étai-amont *m*
r оттяжной трос *m*; задняя оттяжка *f*

468 **backhoe; back[acting] shovel; pull shovel; backdigger**
d Tieflöffelbagger *m*; Löffeltiefbagger *m*
f pelle *f* rétrocaveuse; pelle-rétro *f*; rétrocaveuse *f*; fouilleuse *f*
r обратная лопата *f*; экскаватор *m* с обратной лопатой

* **backing board → 2554**

* **back line → 5411**

469 **back line**

d Fällgrenze *f*
f ligne *f* arrière
r граница *f* зоны валки; граница лесосеки, помеченная на деревьях

470 **backlog [area]; denuded [forest] area**
(inadequately stocked forestland that has been denuded)
d Rückstau *m*
f arriéré *m*
r обнаженная лесная зона *f*

471 **backpack sprayer; knapsack sprayer; knapsack duster; back duster**
(spray unit with plastic containers on a backpack frame)
d Rückenspritzgerät *n*; Rückenspritze *f*; Rückenverstäuber *m*; Motorrückensprüher *m*
f pulvérisateur *m* à dos; poudreuse *f* à dos
r ранцевый опрыскиватель *m*; спинной распылитель *m*; ранцевый опыливатель *m*

472 **back priming**
(application of paint to the back of woodwork and exterior siding to prevent moisture from getting into the wood, causing the grain to swell and the paint to peel)
d Rückankurbelung *f*; Rückvorbereiten *n*
f coloration *f* apprêtée au revers
r первое грунтование *n*; первая грунтовка *f*

473 **back rigging**
d Rückenbefestigung *f*
f fixation *f* arrière
r оснастка *f* тыловой мачты

* **back saw → 883**

474 **backsawing**
d Rückensägen *n*
f sciage *m* à doss[eret]
r пиление *n* в обратном направлении

* **backsided timber → 9079**

* **back spar → 2815**

475 **backspar trail**
(bladed or non-bladed pathway over which mobile backspar equipment travels)
d Endstützeweg *m*
f chemin *m* d'appui arrière
r путь *m* тыловой мачты

476 **backwoods**
d Hinterwälder *f*

* **bacuri → 6030**

* **bacury → 6030**

* **badana → 5558**

* **bad bevel → 2662**

* **badland[s] → 9275**

477 **baff edge; plain edge**
d Brettkante *f*
f joint *m* plat
r гладкая кромка *f* доски

* **bag → 9024**

478 **bagasse; tatajuba; amapa-rana; cow-wood**
(Bagassa guianensis)
d Bagasse *f*; zermalmter Zuckerrohrstängel *m*; Tatajuba *f*
f bagasse *f* [de cannes à sucre]; bagasse jaune; tatajuba *m*
r золотое дерево *n*; татаджуба *f*; багассе *n* Жонэ

479 **bag barker; pocket barker**
d Trommelentrindungsmaschine *f*
f écorceuse *f* à bac[s]; écorceuse pocket-barker
r бункерная установка *f* для групповой окорки

* **bag boom → 914**

* **bag boom stick → 8779**

480 **bag boom towing**
d Ringfloß *n*; Scherenflößung *f*; Schertrift *f*
f touage *m* enchaîné; radeau *m* enchaîné
r кошельный лесосплав *m*; сплав *m* кошелем

* **bahia rosewood → 1006**

* **bait → 4509**

481 **baited standing tree**
d köderter stehende Baum *m*
f arbre *m* à appât
r растущее дерево *n* с выставленной приманкой

482 **baiting; tapering**
d Ködern *n*
f appâtement *m*
r подкормка *f*; прикармливание *n* (животных в заповедниках)

f plante-appât *f*
r растение-приманка *n*

* **bakaut → 4034**

484 **Baker cypress; MacNab cypress; Modoc cypress; Siskiyou cypress**
(Cupressus bakeri; Cupressus macnabiana)
d Baker-Zypresse *f*; Macnab-Zypresse *f*
f cyprès *m* de Baker; cyprès de Macnab
r кипарис *m* Макнаба

485 **balata tree; bully tree**
(Mimusops balata)
d Balatabaum *m*
f balata *m*
r балата *f*; мимузопс *m*

* **balau → 9757**

486 **bald**
(a treeless region in forest vegetation; perhaps an area of grasses or shrubs in southern Appalachia, or a mountaintop)
d öde Fläche *f*
f chauve *f*
r оголённый район *m*

* **baldcypress → 786**

* **bale → 2038**

* **Balfour pine → 6997**

* **baling system → 7235**

487 **Balkan ash; Moraine ash**
(Fraxinus holotricha)
d Balkanesche *f*
f frêne *m* de Balkans
r ясень *m* сплошь волосистый

488 **Balkan maple**
(Acer hyrcanum)
d Balkan-Ahorn *m*
f érable *m* des Balkans
r клён *m* гирканский

489 **Balkan pine; Macedonian pine**
(Pinus peuce)
d Balkankiefer *f*; mazedonische Kiefer *f*; rumelische Kiefer
f pin *m* de Balkans; pin de Macédoine
r сосна *f* [веймутова] румелийская; сосна балканская

* **ball[ed] plant → 492**

(setting out trees with their roots left undisturbed in a dug-out clod of soil)
d Ballenpflanzung *f*
f plantation *f* en motte
r посадка *f* [сеянцев] с корнями, заделанными в ком почвы; посадка глыбой; посадка с комом земли

491 **balled roots**
(lifted seedlings with a ball of earth attached to the roots)
d Wurzeln *mpl* mit Erdballen
f racines *fpl* en mottes
r корни *mpl* с глыбой

492 **balled seedling; ball[ed] plant; balled stock; plug seedling; plug transplant**
d Pflänzling *m* mit Erdballen; Ballenpflanzling *m*
f plantule *f* en motte; plante *f* enrobée; semis *m* fiche; semis en douille
r скомканный рассад *m*; саженец *m* с глыбой; сеянец *m* с комом субстрата; саженец с комом земли

493 **balled spray**
d bewurzelter Absenker *m*
f brout *m* en racines
r отводок *m* с корнями

* **balled stock → 492**

494 **ballhoot** *v*; **ball skid** *v*
d holzrücken
f débarder de bois
r спускать брёвна со склона

* **ballhooter → 1435**

495 **ballhooting**
d Handrücken *n*
f débardage *m* à bras
r ручная трелёвка *f*

496 **ball lifter**
d Ballenausheber *m*
f extracteur *m* de mottes
r копатель *m* глыб

497 **balloon logging**
d Bringung *f* mit Ballon
f transport *m* des bois par ballon
r аэростатная трелёвка *f*; лесозаготовки *fpl* с применением аэростатов

498 **ballroom flindersia**

f flindersie *f* d'Australie
r флиндерсия *f* австралийская

* **ball-rooted seedling → 1893**

499 **ball-rooted tree**
(tree transplanted with the earth around its roots intact)
d Baum *m* mit Erdballen
f arbre *m* en mottes
r дерево *n* с комом земли

500 **ball-shaped spruce; spruce tree of spherical shape**
(Picea globosus)
d kugelförmige Fichte *f*
f épicéa *m* sphérique; épicéa en boule
r шаровидная ель *f*

501 **ball shoot; short-needled coniferous shoot**
d Bürstentrieb *m*
f broussin *m*
r щётковидный побег *m*

* **ball skid → 494**

* **balm → 9338**

* **balm cottonwood → 784**

* **balsam → 503**

* **balsam copaiba → 104**

* **balsam [fir] → 969**

502 **balsam-fir sawfly**
(Neodiprion abietis)
d Balsamtannensägewespe *f*
f diprion *m* du sapin
r пилильщик *m* пихтовый

503 **balsam[ic] resin; balsam; sweet gum; oleoresin**
d Balsamharz *n*; Harzbalsam *m*; Balsamkolophonium *n*
f baume *m*; résine *f* odorante
r бальзамическая смола *f*; бальзам *m*; живичная канифоль *f*; живица *f*

* **balsam of fir → 1294**

* **balsam of Gilead tree → 969**

* **balsam of Mecca tree → 969**

(Myroxylon balsamum; Tolifera balsamum; Myroxylon toluifera; Myroxylon pereirae)
d Tolubalsambaum *m*; Tolubalsam *m*; Tolu *m*; Perubalsambaum *m*
f baumier *m* de Tolu; tulu *m*; baumier du Pérou; arbre *m* à singe
r толуское бальзамовое дерево *n*; бальзамное дерево Толу; кабриува *f*

* **balsam poplar → 9338**

505 **balsam willow**
(Salix pyralifolia)
d Balsamweide *f*
f saule *m* à feuilles de poirier
r ива *f* бальзамовая; ива грушанколистная

506 **balsa[wood]**
(Ochroma lagopus; Ochroma boliviana)
d Balsabaum *m*; Balsa *m*
f balsa *m*
r бальза *f*; охрома *f* заячья

507 **balsa wood**
d Korkholz *n*; Balsaholz *n*
f bois *m* de balsa; bois de liège
r пробковая древесина *f*; древесина пробкового дерева

* **Baltic redwood → 1797**

* **Baltic white pine → 1805**

* **Baltic whitewood → 1805**

* **Baltimore oak → 8993**

508 **bamboo**
(any of various usually woody and/or arborescent grasses of the tropical or temperate regions belonging to the genera Bambusa, Arundinaria, Phyllostachys, Sasa, Dendrocalamus, etc., most having hard, hollow stems)
d Bambus *m*
f bambou *m*
r бамбук *m*

* **bamboo brake → 510**

509 **bamboo filament**
d Bambusfaser *f*
f fibre *m* de bambou
r волокно *n* бамбука

f forêt *f* de bambous; bambousaie *f*; bambouseraie *f*
r бамбуковый лес *m*; бамбуковая заросль *f*

511 bamboo shoots
d Bambussprossen *fpl*
f pousses *fpl* de bambou
r побеги *mpl* бамбука

* **bamboo vine → 4834**

512 banana family
(Musaceae)
d Bananengewächse *pl*
f musacées *pl*
r банановые *pl*

513 banana shrub
(Michelia figo; Michelia fuscata)
d Bananenstrausch *m*
f arbuste *m* des bananes
r михелия *f* буроватая; михелия ржавая

514 banana tree
d Bananenbaum *m*; Bananenstaude *f*
f bananier *m* [des sages]
r банановое дерево *n*

515 bandage processing; bandage treatment
d Bandagierung *f*
f procédé *m* d'imprégnation
r бандажирование *n*

* **band application → 8169**

516 band dendrometer; dendrograph
d Dendrometermessband *n*
f ruban *m* dendromètre
r ленточный дендрометр *m*

* **banded parenchyma → 5414**

517 banded tree; timber tree
d gestreifter Baum *m*
f arbre *m* rubané; arbre de haute futaie
r строевое дерево *n*

518 band girdling; band girdle
(removing a broad band of bark, from several centimeters to a meter wide)
d Bandringelung *f*
f annélation *f* totale; annélation circulaire
r ленточное кольцевание *n*

519 banding; binding
(applying a chemical or other substance to the

f cerclage *m*; pose *f* d'un lien entourant
r обвязывание *n* [стволов]; накладывание *n* ловчих поясов; обозначение *n* полосами; нанесение *n* полос; кольцевание *n*; окольцовывание *n*

* **band mill → 5049**

* **band saw → 5049**

520 band [saw]mill
d Bandsägemaschine *f*
f scierie *f* à ruban
r ленточнопильный станок *m*

521 bandsawn *adj*
d bandgesägt
f de scie à ruban
r распыленный на ленточно-пыльном станке

522 band seeding; band sowing; strip cropping; strip seeding; sowing on strips; seeding in lines; line seeding; row seeding; row sowing
(sowing of seeds in shallow furrows across a whole area on which a forest stand is to be raised)
d Streifensaat *f*; Reihensaat *f*
f ensemencement *m* par bandes; semis *m* par bandes; ensemencement en lignes; semis en lignes
r ленточный посев *m*; посев полосами; полосный посев; рядовой посев

523 band timber
d Bandholz *n*; Bindholz *n*; Reifenholz *n*
f bois *m* de bandes
r обручный лесоматериал *m*

* **Bangalay eucalyptus → 7861**

* **bangkirai → 9757**

* **bank → 8619**

* **bank[ing] ground → 5451**

* **Banksian pine → 4601**

524 bank timber bundling
d Bündeln *n* des Floßholzes im Damm
f groupage *m* de bois de flottage en bord
r береговая сплотка *f*

* **bank up → 7983**

525 banned forest

d verbannter Wald *m*; verbotener Wald
f forêt *f* interdite
r запрещённый лес *m*

526 **banyan tree; banyan fig**
(Ficus benghalensis)
d Banyanbaum *m*; bengalische Feige *f*; indischer Feigenbaum *m*
f banian *m*; figuier *m* d'Indie
r смоковница *f* бенгальская; фикус *m* бенгальский; баньян *m*

* **baobab → 5533**

* **Baptist cone → 1858**

* **baracara → 119**

* **barakaruballi → 621**

527 **Barbados cherry [malpighia]**
(Malpighia glabra; Malpighia punicifolia)
d Barbados-Kirsche *f*
f cerise *f* des Antilles; acérole *f*
r мальпигия *f* гранатоволистная

* **barbasco → 4654**

* **barberchair → 2193**

* **barberry → 1758**

528 **barberry family**
(Berberidaceae)
d Sauerdorngewächse *pl*
f berbéridacées *pl*
r барбарисовые *pl*

* **barber's chair → 2193**

529 **bardon hook; sliding [butt] hook; slide hook**
(hook used with wire rope slings for gripping trees or logs to be skidded)
d Gleithaken *m*
f crochet *m* coulissant
r чокерной замок *m*, скользящий по канату чокера

* **bare bud → 5651**

530 **bare firebreak**
d entblößter Feuerschutzstreifen *m*
f coupe-feu *m* nu
r обнажённый противопожарный барьер *m*

531 **bare[-]root; naked root; open root**
d entblößte Wurzel *f*
f racine *f* nue
r обнажённый корень *m*; оголённый корень

532 **bare-rooted planting; open-rooted planting; naked-root planting**
d Barpflanzung *f*; Pflanzung *f* mit entblößter Wurzel
f semis *m* à racines nues; plantation *f* à racines nues
r посадка *f* с обнажённой корневой системой; посадка с обнажённым корнем

533 **bare-rooted tree; bare-rooted stock**
(a tree ready for transplanting that has had the soil removed from around its roots)
d Baum *m* mit entblößter Wurzel
f arbre *m* à racines nues
r дерево *n* с обнаженным корнем

534 **bare slope**
d unbedeckter Abhang *m*
f versant *m* en friche
r безлесный [крутой] склон *m*

* **bare tree → 4874**

535 **bare wood; unfilled wood**
d unbedecktes Holz *n*
f bois *m* nu
r неотделанная древесина *f*

* **bark → 8799, 2280**

536 **barkability; peelability**
d Schälbarkeit *f*
f écorçabilité *f*
r окоркость *f*; сдирасмость *f* корки

* **bark allowance → 200**

537 **bark bar; bark strip**
d Rindenstreifen *m*
f bande *f* d'écorce
r полоска *f* коры

538 **bark basket; bark lox; basket for collecting resin**
d Harzbehälter *m*; Harzbehälterkorb *m*
f récipient *m* à résine
r сосуд *m* для живицы

539 **bark beetles; engraver beetles**
(small, cylindrical beetles of the family Scolytidae, the adult of which bores into and

d Borkenkäfer *mpl*; Buchdrucker *m*
f scolytes *mpl* de l'écorce
r короеды *mpl*

540 bark blaze; blaze; nick
d Axtschalm *m*; Schalm *m*; Anschalmen *n*; Beilschlag *m*
f signe *m* à l'écorce; flache *f* de l'écorce; carre *f*
r затёска *f* [на стволе дерева]; надрубка *f*; засечка *f*

* **bark blister[ing] → 836**

* **bark bundle → 1176**

541 bark cell
d Rindenzelle *f*
f cellule *f* de l'écorce
r клетка *f* коры

542 bark-chip mass
d Rindenspanmasse *f*
f masse *f* de copeaux à l'écorce
r древесная масса *f* с содержанием коры

543 bark-chipped streak
d Schnittrille *f*
f échauffure *f* de l'écorce
r [карро]подновка *f*

* **bark-chipper → 2281**

544 bark-chipping
d Anbringen *n* der Tropfrinne
f striation *f* de l'écorce; piquage *m*
r нанесение *n* карр узкими полосками; нанесение жёлобка

545 bark chips
d Rindenspäne *mpl*
f copeaux *mpl* à l'écorce
r щепа *f* с содержанием коры

546 bark content
(in chips)
d Rindengewicht *n*
f contenu *m* d'écorce; proportion *f* d'écorce
r содержание *n* коры

547 bark conveyor
d Rinden[zug]förderer *m*
f convoyeur *m* d'écorce
r транспортёр *m* для удаления коры; транспортёр для подачи коры

548 bark crinkle; bark flexure
r складка *f* коры

* **bark crusher → 2281**

* **bark deduction → 200**

549 bark dray
d Rindenschlitten *m*
f traîneau *m* d'écorce
r сани *pl* для вывозки коры

550 bark dust
d Rindenmehl *n*
f farine *f* d'écorce
r мука *f* из коры

551 barked; disbarked; rossed; peeled; stripped
d entrindet
f écorcé
r окорённый

* **barked → 1994**

* **bark edger → 2284**

552 barked in forest; forest-barked *adj*
d waldgeschält
f écorcé dans la forêt; écorcé forestièrement
r окорённый в лесу

553 barked log; peeled log
d waldgeschältes Schnittholz *n*; entrindetes Schnittholz
f grume *f* écorcée
r окорённое бревно *n*

554 barked strip bundles
d Bündel *fpl* von entrindete Bande
f paquets *mpl* de dosses sans écorce
r пакеты *mpl* из окорённых обрезок

555 barked wood; barked timber; disbarked wood; peeled wood; stripped wood
d entrindetes Holz *n*; Entrindungsholz *n*; abgerindetes Holz; geschältes Holz; rindenfreies Holz; Rumpenholz *n*
f bois *m* écorcé
r окорённая древесина *f*; окорённые лесоматериалы *mpl*

* **barker → 2284, 7960**

556 barker; [bark] peeler; rosser; flayer
(one who barks and smoothes the ride of a log in order that it may slide more easily; one

f écorceur *m*; ouvrier *m* écorceur
r окорщик *m*; корообдирщик *m*; рабочий *m*, занимающийся окоркой

* **bark[er] refuse → 580**

* **bark exfoliation → 3243**

557 **bark factor**
d Rindenfaktor *m*
f facteur *m* d'écorce
r фактор *m* коры

558 **bark fissure; bark wound; sore bark**
d Rindenverletzung *f*; Rindenwunde *f*
f lésion *f* de l'écorce
r рана *f* коры; травмирование *n* коры

* **bark flaking → 3243**

* **bark flexure → 548**

559 **bark formation**
d Berindung *f*
f formation *f* d'écorce
r образование *f* коры; обрастание *n* корой

560 **bark gall**
d Rindengalle *f*
f blessure *f* de l'écorce; îlot *m* d'écorce intraligneuse
r галл *m* коры

561 **bark grafting; rind grafting**
d Rindenpfropfen *n*; Pfropfen *n* hinter die Rinde
f greffage d'écorce
r прививка *f* под кору; прививка за кору

* **bark hog → 2281**

562 **bark incision**
d Rindeneinschnitt *m*
f incision *f* en écorce
r надрез *m* в коре

* **barking → 579**

563 **barking axe; peel[ing] axe**
d Schälbeil *n*; Rindenschäler *m*
f hache *f* à écorçage; écorçoir *m*
r лущильный топор *m*; корообдир *m*

564 **barking buck; barking horse**
d Entrindungsbock *m*
f chevalet *m* d'écorce
r окорочные козлы *mpl*

565 **barking depth**
d Entrindungstiefe *f*
f profondeur *f* d'écorçage
r глубина *f* окорки

566 **barking drum; debarking drum; drum [de]barker**
(large drum in which logs or billets are tumbled by mechanical rotation)
d Entrindungstrommel *f*; Trommelentrinder *m*
f écorceuse *f* à tambour
r окорочный барабан *m*; барабанная окорочная машина *f*; барабанный окорочный станок *m*

* **barking horse → 564**

* **barking in full → 1603**

* **barking iron → 7960**

567 **barking line; debarking line; debarking station; barking station**
d Entrindungslinie *f*
f poste *m* d'écorçage
r окорочная линия *f*

* **barking losses → 580**

* **barking machine → 2284**

568 **barking ring**
d Rindenring *m*
f anneau *m* d'écorçage
r окорочное кольцо *n*

* **barking saw → 4312**

* **barking spade → 8639**

* **barking spud → 8639**

* **barking station → 567**

* **barking tool → 574**

* **bark[ing] waste → 580**

* **bark knot → 833**

569 **barkless**
d ohne Rinde
f sans écorce
r без коры

* **bark lox → 538**

* **bark marker → 5102**

570 **bark meter**
d Lohemesser *m*
f couteau *m* à mesurage d'écorce
r нож *m* для измерения коры

* **bark mill → 2284**

571 **bark miner**
d Rindenbohrer *m*
f mineuse *f* de l'écorce
r пилильщик *m* коры

572 **bark mulch**
d Rindenmulch *f*
f paillis *m* d'écorce
r перегной *m* из коры

573 **bark necrosis**
d Rindennekrose *f*
f nécrose *f* de l'écorce
r некроз *m* коры

* **bark peeler → 556**

* **bark peeling → 579**

* **bark-peeling damage → 2237**

* **bark-peeling machine → 2284**

574 **bark-peeling tool; barking tool**
(a tool used to smooth the wood bark and slab to providea flat surface for gutter insertion)
d Entrindungswerkzeug *n*; Entrindungsgerät *n*
f outil *m* d'écorçage
r окорочный инструмент *m*

575 **bark plate**
d Rindenplatte *f*
f plaque *f* d'écorce
r плита *f* из коры; пласт *m* коры, снятый со ствола дерева

576 **bark pocket; bark seam**
(an opening between annual growth rings that contains bark)
d Rindentasche *f*
f poche *f* d'écorce
r карман *m* с корой; полость *f* в коре; прорость *f*

577 **bark products**
d Rindenprodukte *npl*

578 **bark ratio**
d Rindenanteil *m*
f rapport *m* d'écorce
r доля *f* коры

579 **bark removal; barking; debarking; disbarking; [bark] peeling; rossing; [bark] shaving**
d Entrinden *n*; Entrindung *f*; Entborken *n*; Schälen *n*
f écorçage *m*
r окорка *f*; снятие *n* коры; сдирание *n* коры; обдирка *f*

580 **bark residue; bark[er] refuse; bark[ing] waste; barking losses; wastage in bark; losses in bark peeling**
(bark removed from a log and also portions of wood and foreign matter such as sand, grit, or stones that may be imbedded in the bark)
d Entrindungsabfall *m*; Entrindungsabgang *m*; Schälrinde *f*; Rindenrest *m*; Entrindungsverluste *mpl*; Rindenverluste *mpl*; Schälverlust *m*
f écorce *f* restante; résidus *mpl* d'écorce; déchets *mpl* d'écorçage; pertes *fpl* à l'écorçage
r отходы *mpl* окорки; потери *fpl* при окорке; отпад *m* коры; корьё *n*

581 **bark-ringed knot**
d beringte Ast *f*
f nœud *m* encastré
r окольцованный сучок *m*

* **bark scaling → 3243**

582 **bark scar; scar**
d Rindennarbe *f*; Narbe *f*
f cicatrice *f* de l'écorce; dégât *m*; frottis *m*; frotture *f*
r рубец *m* коры; повреждение *n* коры; шрам *m*

* **bark scorch[ing] → 836**

* **bark seam → 576**

* **bark shaving → 579**

583 **bark shredder**
d Rindenschredder *m*
f broyeur *m* d'écorce
r измельчитель *m* коры

* **bark slabs → 7710**

f glissoire *f* d'écorce
r жёлоб *m* для спуска коры

585 bark store
d Rindenlager *n*
f dépôt *m* d'écorce
r склад *m* для коры

* **bark strip → 537**

* **bark stripping → 2283**

* **bark-stripping machine → 2284**

586 bark structure
d Rindenaufbau *m*
f structure *f* de l'écorce
r строение *n* коры

* **bark thickness → 8498**

587 bark[-thickness] ga[u]ge
(instrument for measuring bark thickness; this measurement is used in computations of tree growth)
d Rinden[dicken]messer *m*
f sonde *f* à l'écorce; phléomètre *m*; jauge *f* de l'écorce; micromètre *m* d'épaisseur d'écorce
r измеритель *m* толщины коры; толщинометр *m* для коры

588 barkware
d Rindenware *f*
f articles *mpl* en écorce
r корьевой товар *m*

589 bark-wood mix
d Rinde-Holz-Mischung *f*
f mixte *m* écorce-bois
r смесь *f* древесины с остатками коры

* **bark wound → 558**

590 barky; thick-barked
d grobrindig
f d'écorce grossière
r толстокорый

591 barn boss; barnman
d Pächter *m*
f chef *m* d'étable
r лицо *n*, ответственное за гужевое хозяйство

592 bar nose
d Umlenkung *f*

593 barras; scrape-resin
d Barras *m*; Bruchharz *n*; Hartharz *n*; erhärtetes Harz *n*; Scharrharz *n*
f galipot *m*; gemme *f* de grattage
r баррас *m*; твёрдая смола *f*; затвердевшая живица *f*

* **barre → 5732**

594 barrel bottle tree
(Sterculia repestris; Brachychiton rupestris)
d Glückbaum *m*; [australischer] Flaschenbaum *m*
f sterculie *f* des rochers
r стеркулия *f* утёсная

595 barrel-butted stem; churn-butted stem
(of a tree having decayed heartwood)
d Stamm *m* mit Brettwurzel
f tige *f* pattue
r закомелистый ствол *m*

* **barrel log → 4362**

* **barrel palm → 2109**

* **barrel saw → 2209**

* **barrelwood → 4362**

* **barren land → 9275**

596 barrier hedge
(a hedge planted so as to prevent runoff)
d Barriere-Hecke *f*
f haie *f* barrière
r барьерная изгородь *f*

* **barwood → 111**

597 basal area; bole area; BA
(the area in square meters of the cross-section of a tree stem, generally at breast height and inclusive of bark)
d Stammgrundfläche *f*; Grundfläche *f*; Basalfläche *f*
f surface *f* terrière [d'arbre]
r площадь *f* поверхности почвы, занимаемой стволом (дерева)

598 basal area at breast height; circular area at breast height; sectional area at breast height
(area of the cross-section of a stem at breast height)

f surface *f* circulaire à hauteur de poitrine
r площадь *f* [поперечного] сечения ствола на высоте груди

* **basal-area control → 602**

599 **basal-area factor**
(number of units of basal area per acre (or per hectare) represented by each tree)
d Grundflächen-Formfaktor *m*
f facteur *m* de surface terrière
r видовой фактор *m* поперечного сечения

600 **basal-area growth; basal-area increment**
(of a bole)
d Kreisflächenzuwachs *m*
f accroissement *m* de la surface terrière
r прирост *m* по площади сечения (ствола)

601 **basal area per hectare**
(area of the cross-section of tree stems near their base, generally at breast height and including bark, measured over 1 ha of land)
d Basalfläche *f* pro Hektar
f surface *f* terrière à l'hectare
r сумма *f* площадей поперечных сечений насаждения на гектаре

602 **basal-area regulation; basal-area control**
d Grundflächenkontrolle *f*; Kreisflächenabnutzung *f*
f contrôle *m* de la surface terrière; possibilité *f* par surface terrière
r регулирование *n* промежуточного лесопользования по площади поперечного сечения насаждения и по приросту

* **basal-area table → 8362**

603 **basal bark treatment**
(treatment for killing trees and brush in which a herbicide is applied, by sprayer or brush, to a band of bark encircling the basal portion of the stem)
d Stamm-Fuß-Behandlung *f*
f traitement *m* arboricide cortical; traitement à la base de l'arbre
r обработка *f* коры [ядохимикатами] у основания дерева; применение *n* [ядохимикатов] у основания дерева

604 **basal branch; subbranch**
d Schattenast *m*
f branche *f* à la base
r нижняя ветвь *f*

dormant buds, formed by the uniform expansion of the base of a stem)
d Wulst *m* am unteren Stamm; Stammengewulst *m*
f loupe *f* à la base [de la tige]; broussin *m* à la base
r лежащий в основе крупный нарост *m*; базальный крупный свиль *m*

* **basal shoot → 4237**

606 **basal spray[ing]; basal injection**
(treatment consisting of forcing a liquid or an encapsulated herbicide into the basal portion of a tree)
d Spray *n* am unteren Stamm
f jet *m* dirigé sur la base de la tige; injection *f* à la base de la tige
r кольцевое нанесение *n* гербицидов на кору у комля дерева

607 **base of a tree; butt-log portion of tree; butt [end]; butt of trunk; cut end; lower-end; thick end; bottom end**
(the bottom of the felled part of a tree; large end of a log)
d Baumschenkel *m*; Stammschenkel *m*; Stammende *n*; stärkeres Ende *n*; Stockabschnitt *m*; Stockbereich *m*; untere Seite *f*
f tronce *f*; fin *m* de grume; extrémité *f* de la grume; patte *f*
r комлевая часть *f* ствола [дерева]; прикорневая часть ствола; комлевой торец *m* брёвна; комлевой конец *m* брёвна; комель *m* [ствола]

608 **base population**
(the set of trees from which selections will be taken to form advanced-generation breeding and/or production populations)
d Basispopulation *f*; Grundpopulation *f*
f population *f* de base
r базовая популяция *f*

609 **base rate**
(absolute minimum price for timber)
d Basiszinssatz *m*
f taux *m* de base
r базовая расценка *f*

610 **base sale schedule**
(timber sale schedule formulated on the basis that the quantity of timber planned for sale and harvest for any future decade is equal to or greater than the planned sale and harvest

d Basisverkaufstabelle *f*
f table *f* de base de ventes
r базовая таблица *f* продажей

* **basic density → 18**

611 **basic forest management; basic silviculture**
(harvesting methods and silviculture operations including seed collecting, site preparation, artificial and natural regeneration, brushing, spacing, stand tending and etc.)
d Basiswaldbau *m*
f aménagement *m* forestier de base; sylviculture *f* de base
r базовое лесоведение *n*; базовое лесоводство *n*

612 **basic silvicultural practices**
(maintenance of the productivity of forest sites, restocking of denuded forestlands with commercial tree species)
d basiswaldbauliche Maßnahmen *fpl*
f pratiques *fpl* sylvicoles de base
r основные лесохозяйственные мероприятия *npl*

* **basic silviculture → 611**

613 **basin dam**
(a dam with a concave top, built across the mouth of a gorge to prevent an outwash fan from forming)
d Bassindamm *m*
f barrage *m* entonnoir; barrage de sommet de cône [de déjection]
r бассейновая запруда *f*

614 **basiphilous; basiphilic**
(adapted to life in alkaline soils (pH greater than 7.5))
d basiphil; kalklierend
f basiphile
r базифильный

* **basket for collecting resin → 538**

615 **basket oak; swamp chestnut oak**
(Quercus michauxii; Quercus prinus)
d Korbeiche *f*; Kastanie-Eiche *f*
f chêne *m* [châtaignier] des marais; chêne prin
r дуб *m* дубильный; дуб коровой; дуб Мишо

616 **basket planting**
d Korbpflanzung *f*

617 **basket[-]rod**
d Weidengerte *f*
f pleyon *m*; clayon *m*; osier *m*
r прутья *f*; тальник *m*

618 **basketry; basketwork; tracery**
(of tree branches and leaves)
d Korbwarenherstellung *f*
f vannerie *f*
r переплетение *n*; сплетение *n* (веток дерева)

619 **basket-willow; common osier; pollard willow; hemp willow**
(Salix viminalis)
d Korbweide *f*; Bandweide *f*; Elbweide *f*; Hanfweide *f*
f seule *m* viminal; osier *m* des vanniers; osier vert
r ива *f* корзиночная; ива прутовидная

620 **basket-willow culture**
d Korbweidenkultur *f*
f culture *f* d'osier viminal
r культура *f* корзиночной ивы

* **basketwork → 618**

621 **basralocus; barakaruballi**
(Dicorynia guianensis; Dycorynia paraensis)
d Basralocus *m*; Angelique *m*
f angélique *f* grise; angélique blanche; angélique rouge; basralocus *m*
r дикориния *f* параенская; басралокус *m*

* **basswood → 4976**

* **bast → 4502**

622 **bastard cut; rough cut**
d Grobschnitt *m*
f sciage *m* brut
r грубый распил *m*

* **bastard elm → 1777**

* **bastard grain → 7914**

* **bastard mahogany → 4039**

* **bastard sawn → 3258**

* **bastard [white] oak → 2672**

* **bastard wood → 5195**

623 **bast fiber; phloem fiber; cortex fiber**

f fibre *f* libérienne; fibre du phloème; fibre du liber
r лубяное волокно *n*; волокно флоэмы

624 bast peeling
d Entbastung *f*
f épluchage *m* de fibre libérienne
r лубяная окорка *f*; снятие *n* луба

625 bast wood
d Bastholz *n*
f bois *m* libérien
r сырьё *n* для изготовления мочала

626 batten
(a thin piece of square-cut timber used for covering panel or weatherboard edges; also for reinforcing the sides, tops or bottoms in box construction)
d Schalklatte *f*; Querstab *m*
f latte *f* [de caisserie]; barre *f* de serrage; barrette *f*; traverse *f*; tasseau *m*
r рейка *f*; планка *f*; батан *m*

627 battering; brooming; fringing
d Spalten *n*
f parement *m*
r размочаливание *n*; расщепление *n* (концов брёвен, торцов баланса)

* **bat tree → 7860**

* **Bauer eucalyptus → 857**

* **ba[u]lks → 2278**

* **bavin → 1118**

* **bavin wood → 998**

* **bay → 3936, 5451**

628 bay-cedar
(Suriana maritima)
d Bucht-Zeder *f*
f crisse *f* marine
r суриана *f* приморская

629 bay cutting; bay felling
d Buchtenhieb *m*
f coupe *f* en anse
r вырубка *f* искривлённых деревьев

* **bay felling → 629**

* **bay-gallbush → 4795**

willow
(Salix pentandra)
d Lorbeerweide *f*; Faulweide *f*
f saule-laurier *m*
r ива *f* пятитычинковая; чернотал *m*; верболоз *m*

631 bayonet-topped *adj*
(of a tree)
d bajonettspitzig
f avec pousse terminale en baïonnette
r с вершиной, направленной под углом к стволу; со штыком

632 bayrumtree
(Pimenta racemosa; Racemosus caryophyllus)
d westindischer Lorbeerölbaum *m*; westindischer Baybaum *m*
f bois *m* d'Inde; piment *m* couronné; poivre *m* de Jamaïque
r бай-дерево *n*; перец *m* гвоздичный

* **bay tree → 3936**

* **baywood → 1412**

* **BDU → 906**

* **beach pine → 5044**

633 beach plum
(Prunus maritima)
d Strandpflaume *f*; Strandpflaumebusch *m*
f prunier *m* des grèves
r слива *f* морская

634 beach rose
(Rosa rugosa)
d Heckenrose *f*; Kartoffelrose *f*
f rosier *m* rugueux
r роза *f* морщинистая

635 bead
(a small rounded, raised profile, routed along the edge of a board)
d Eckleiste *f*; Wulst *f*; Kügelchen *n*
f baguette *f*; astragale *m*
r калёвка *f*; выкружка *f*

636 bead-tree
(Abrus precatorius)
d Paternostererbse *f*
f arbre *m* à chapelet; réglisse *f* d'Amérique
r абрус *m* опасный; абрус прекаторный; абрус молитвенный

637 **beaked hazel[nut]; beaked filbert**
(Corylus cornuta; Corylus rostrata)
d Schnabelnuss *f*; Haselnuss *f* des Cornuta
f noisetier *m* à long bec
r лещина *f* рогатая

638 **beam; cant**
(log which has been sawn on two faces only)
d Stammbalken *n*; Balken *m*; Kantbalken *m*; Kantholz *n*; Eckholz *n*
f poutre *f* de bois; bille *f* dédossée; premier débit *m*
r брус *m*; тёсаное бревно *n*

* **beam tree → 8323**

639 **bean caper family; caltrop family**
(Zygophyllaceae)
d Jochblattgewächse *pl*; Doppelblattgewächse *pl*
f zygophyllacées *pl*
r парнолистниковые *pl*

* **bean goose → 3332**

* **bean-tree → 1784**

* **bean-trefoil [tree] → 1784**

* **bearberry → 1759**

* **beard → 2193**

* **bearer cable → 8745**

* **bearing cable → 8745**

640 **bearing edge**
(of a beam)
d ertragende Kante *f*; Tragkante *f*
f bord *m* d'appui
r опорный конец *m* балки; место *n* закрепления балки

641 **bearing tree; witness tree; boundary tree; lobstick**
(tree marked to identify the nearby location of a survey corner)
d Grenzenbaum *m*
f arbre *m* en production
r межевое дерево *n*; пограничное дерево

642 **bear oak; scrub oak**
(Quercus ilicifolia)
d Zwergeiche *f*
f chinquapin *m*; chêne *m* à feuilles de houx;

* **bearwood → 1544**

* **beat → 6114**

* **beaten down plant → 5040**

* **beater → 4369**

* **beating up → 823**

* **beautyberry → 3705**

* **beauty-bush → 3705**

643 **beaver tree; sweetbay [magnolia]; laurel magnolia; swamp bay; swamp magnolia**
(Magnolia virginiana; Magnolia glauca)
d Biberbaum *m*; Sumpfsassafras *m*
f arbre *m* à castor; laurier *m* des marécages; magnolia *m* de Virginie
r магнолия *f* виргинская

* **beaverwood → 1777**

644 **Bebb willow**
(Salix bebbiana)
d Bebb-Weide *f*
f saule *m* de Bebb
r ива *f* Бебба

645 **bed *v* a tree**
(to level and buffer the ground along the line on which a tree is to be felled to minimize shattering of the timber)
d Bett vorbereiten für einen Baum
f faire le lit pour la chute d'un arbre; asseoir
r подготовить ложе для дерева

* **bed [cradle] → 4840**

646 **bedding**
(a site preparation method in which special equipment is used to concentrate surface soil and forest litter into a ridge 6 to 10 inches)
d Bettzeug *n*
f billonnage *m*
r балластировка *f*

647 **bed load**
(particulates that are transported along the channel bottom in the lower layers of streamflow by rolling and bouncing)
d Bettlast *f*; Bettladung *f*
f charge *f* de fond

648 **bed planting; planting on ridges; ridge planting; timping**
(setting out young trees on a long, narrow crest of excavated soil, generally on a slice thrown up by a plough)
d in Kämmen Pflanzen *n*; Kammsaat *f*; Rabbatenpflanzung *f*
f plantation *f* sur bourrelet; plantation *f* en billons; semis *m* sur planche
r посадка *f* в гребень пласта; гребневая посадка; посадка на холмики; посадка грядками

* **bed plate → 321**

* **beeberoe → 2363**

649 **beech**
(Fagus)
d Buche *f*
f hêtre *m*
r бук *m*

650 **beech aphid; woolly beech[-leaf] aphid**
(Phyllaphis fagi)
d Buchenlaus *f*
f puceron *m* du hêtre
r тля *f* буковая

651 **beech chips**
d Buchenspäne *mpl*
f copeaux *mpl* de hêtre
r буковая щепа *f*

652 **beech family**
(Fagaceae)
d Buchengewächse *pl*
f fagacées *pl*
r буковые *pl*

653 **beech forest**
d Buchenwald *m*
f forêt *f* de hêtres
r буковый лес *m*

654 **beech grove**
d Buchenhain *m*
f hêtraie *f*
r буковая роща *f*

655 **beech leafminer**
(Lithocolletis restrictella; Phyllonorycter restrictella)
d Buchenblattminiermotte *f*
f mineuse *f* du hêtre
r пилильщик *m* минирующий буковый

656 **beech mast**
d Buch[en]mast *m*
f fainée *f*
r урожай *m* буковых орешков

657 **beech[-]nut**
d Buchecker *f*; Buchnuss *m*
f fruit *m* du hêtre; graine *f* du hêtre; faine *f*
r буковый орешек *m*

658 **beech-oil**
d Buchöl *n*
f huile *f* de faînes
r буковое масло *n*

659 **beech red heart**
d Buchenrotkern *m*
f cœur *m* rouge du hêtre
r буковое красное ядро *n*

660 **beech resin; beech tar**
d Buchenholzteer *m*; Buchenteer *m*; Buchenharz *n*
f goudron *m* de bois de hêtre; brai *m* de hêtre
r буковая смола *f*

661 **beech sapling; young beech tree; young oak tree**
d Buchenheister *m*
f haute tige *f* de hêtre
r крупный буковый саженец *m*

* **beech tar → 660**

662 **beechwood**
d Buchenholz *n*
f bois *m* de hêtre
r буковая древесина *f*

663 **beef[-]wood**
(Casuarina)
d Kasuarine *f*; Keulenbaum *m*
f casuarina *m*
r казуарина *f*

664 **beefwood family**
(Casuarinaceae)
d Kasuarinengewächse *pl*
f casuarinacées *pl*
r казуариновые *pl*

* **beetle damage → 4511**

665 **beetles**
(Coleoptera)
d Käfer *mpl*

666 bee tree
d Bienenbaum *m*
f arbre *m* à abeille; arbre aux abeilles
r дерево *n* с дуплом, заселённым пчелами

667 beetree [linden]; tile tree; white basswood
(Tilia heterophylla)
d stachelblättrige Linde *f*
f tilleul *m* blanc
r липа *f* разнолистная

* **begin hoeing → 669**

* **beginning decay → 4494**

668 beginning of the cutting
d Hiebanfang *m*
f début *m* de coupe
r первый приём *m* рубки

669 begin *v* to cut; begin *v* hoeing; lay *v* in
d anhacken; anhauen
f entamer; tousser par quintes
r подрубать; зарубать; делать затёски [на стволах]

* **beheading → 2198**

* **behen tree → 4379**

* **be[h]n tree → 4379**

* **be in leaf → 7531**

* **Belalu tree → 5526**

* **Belgian elm → 2677**

* **Belize mahogany → 1412**

* **bell heather → 3153**

* **bellied saw → 9030**

* **belly[-cut] → 1214**

* **belly palm → 2109**

670 belly plate; belly guard; [underbody] underguard
d Bauchplatte *f*
f blindage *m* [inférieur]
r защитный щит *m*; нижнее защитное ограждение *n*

671 belt-fed hog
f broyeur *m* à courroie
r рубильная машина *f* с загрузочным транспортёром

* **belt of trees → 3335**

672 belt sander; belt sanding machine
d Bandschleifmaschine *f*
f ponceuse *f* à courroie; ponceuse à bande
r ленточно-шлифовальный станок *m*

* **belt saw → 5049**

* **belt under regeneration → 6818**

673 bench axe; hand axe; hatchet
d Bankaxt *f*; Barte *f*; Bindeaxt *f*; Binderbarte *f*; Handbeil *n*
f hache *f*; cognée *f*
r плотничий топор *m*; [ручной] топор; секира *f*

674 bench[-]mark
(survey reference point, used to signify a starting point)
d Benchmark *m*; Fixpunkt *m*; Bezugsmarke *f*
f repère *f*; point *m* fixe
r репер *m*; реперный знак *m*; постоянная точка *f*

675 bench saw
d Holzsägemaschine *f*
f scie *f* d'établi
r [лесо]пильный станок *m*

* **benchwood → 4658**

* **bending → 2031**

676 bending strength
(the resistance to stress applied transversely in such a manner as to cause curvature of the member axis)
d Biegewiderstand *m*; Biegefestigkeit *f*; Knickfestigkeit *f*
f résistance *f* à la flexion
r прочность *f* на изгиб; сопротивление *n* изгибу

677 benefication
(process used to upgrade chips, making them more acceptable for pulp and paper manufacture)
d Veredelung *f*
f valorisation *f*
r обогащение *n*

* **bent growth → 2025**

678 **bentwood; pliable wood; pliable timber**
d Biegeholz *n*; Bugholz *n*; gebeugtes Holz *n*
f bois *m* courbé; bois de membrures
r гнутая древесина *f*

* **benzolive tree → 4379**

679 **bergamot tree; bergamot orange**
(Citrus aurantium ssp. bergamia)
d Bergamottenbaum *m*; Bergamotte *f*
f bergamotier *m* [à feuilles de saule]; bergamote *m*
r бергамот-дерево *m*

680 **Berlandier ash; Mexican ash**
(Fraxinus berlandieriana)
d Berlandier-Asche *f*
f cendre *m* de Berlandier; cendre mexicaine
r ясень *m* мексиканский

681 **Berlin poplar; pillar-poplar**
(Populus berolinensis)
d berlinischer Pappel *m*; Säulenpappel *m*; Lorbeer-Pappel *m*
f peuplier *m* de Berlin
r тополь *m* берлинский

682 **Bermuda cedar**
(Juniperus bermudiana)
d Bermuda-Wacholder *m*
f cèdre *m* de Bermuda
r можжевельник *m* бермудский

683 **berries**
d Beerenobst *n*; Beerenfrucht *f*
f baies *fpl*; fruits *mpl* à baies
r ягоды *fpl*

684 **berry bearing alder; alder buckthorn; glossy buckthorn**
(Frangula alnus; Rhamnus frangula)
d [gemeiner] Faulbaum *m*; Pulverbaum *m*
f bourdaine *f*; bourgène; nerprun *m* bourdaine; nerprun noir; bois *m* à poudre
r крушина *f* ольховидная; крушина ломкая; крушинник *m* ломкий

685 **berry bushes**
d Beerenkräuter *npl*
f plantes *fpl* à baies
r ягодники *mpl*

686 **best face; better face; clear face**
(of a veneer)
r лучшая пласть *f* (фанеры)

687 **best formed tree**
(tree with the straightest stems and small diameter branches)
d bestgeformter Baum *m*
f meilleur arbre *m*
r правильное дерево *n*

688 **best management practices; BMPs**
(methods, measures or practices to prevent or reduce pollutions)
d Wirtschaftlichkeit *f* der Haushaltsführung
f principes *mpl* de meilleure gestion; pratiques *fpl* de gestion d'excellence; meilleures pratiques de gestion
r наилучшие методы *mpl* управления

689 **best opening face**
d beste Anfangsseite *f*
f meilleure face *f* d'ouverture
r первый пропил *m*, обеспечивающий оптимальный выход пиломатериала

690 **beta diversity**
(any of a variety of definitions of diversity relating to differences among communities present along a gradient (e.g., soil acidity) or in the same place over time)
d Beta-Vielfalt *f*
f diversité *f* bêta
r бета-разнообразие *n*

691 **betel [nut] palm**
(Areca catechu)
d Betelnusspalme *f*; Betelpalme *f*; Katechupalme *f*; Pinangpalme *f*; Arekpalme *f*; Arecanuss *f*
f arec *m* cachou; aréquier *m*; noix *m* de bétel
r бетельная пальма *f*; арековая пальма; арека *f* катеху

692 **Bethell's process; full-cell process impregnation**
d Bethell-Verfahren *n*; Volltränkung *f*
f procédé *m* Bethell; imprégnation *f* à cellules pleines
r способ *m* пропитки древесины креозотом в вакууме

* **better face → 686**

693 **betulin**
(a resin from birch bark)
d Betulin *n*

694 biareal plot sampling
(of a forest)
d Zweiflächenstichprobe *f*
f échantillonnage *m* par placettes à deux dimensions
r выборка *f* двухразмерных пробных площадей

* **bicycle → 7698**

695 bicycle-type circular saw
d Fahrradkreissäge *f*
f scie *f* circulaire à chariot
r круглая пила *f* на тележке

696 biennial [plant]
(a plant which exists or lasts for two years; plant species that lives for 2 growing seasons)
d zweijährige Pflanze *f*
f plante *f* biennale; plante bisannuelle
r двулетник *m*; двухлетнее растение *n*

697 bifoliolate *adj*
(having just two leaflets per leaf)
d doppelblättrig
f bifeuilleté
r состоящий из двух листочков; двухлопастный

* **big-catcin willow → 7065**

* **bigcone Douglas-fir → 5588**

* **bigcone pine → 1984**

* **bigcone spruce → 1772**

698 bigfruit elm
(Ulmus macrocarpa)
d großfrüchtige Ulme *f*
f orme *m* à gros fruits
r вяз *m* крупноплодный; ильм *m* крупноплодный

699 bight; eye[-]splice; ring-splice
(loop formed by bending a rope's end back and splicing it into the line)
d Bucht *f*; Seildurchhang *m*; Seilschlaufe *f*
f épissure *f* à œil[let]
r боковой провес *m*; канатная петля *f*; тросовая петля

* **big huckleberry → 798**

* **big laurel → 7860**

* **bigleaf linden → 4801**

700 bigleaf magnolia; large-leaved [cucumber] tree; largeleaf magnolia
(Magnolia macrophylla)
d großblättrige Magnolie *f*
f magnolia *m* à grandes feuilles
r магнолия *f* крупнолистная

* **bigleaf mahogany → 1412**

701 bigleaf maple; broadleaved maple; Oregon maple
(Acer macrophyllum)
d großblättriger Ahorn *m*; Großblattahorn *m*; Oregonahorn *m*
f érable *m* à grandes feuilles; érable grandifolié
r клён *m* крупнолистный

702 bigleaf snowbell
(Styrax grandifolius)
d großblättriger Storax *m*
f aliboutier *m* à grandes feuilles
r стиракс *m* крупнолистный

* **bigleaf sumpweed → 7162**

* **bigness scale → 3981**

703 bigness scaling; full scaling
(measurement of log content in log-scale board feet, without deduction for defects)
d Vollskalierung *f*; Vollnormierung *f*
f cubage *m* brut; cubage sans réfaction
r определение *n* объёма сортиментов без скидки на дефектов

704 big sagebrush
(Artemisia tridentata)
d mexikanischer Wermut *m*
f armoise *f* commune
r полынь *f* трёхзубчатая

705 big-stick loader
(short, rotatable horizontal boom attached to a center post mounted on a pulpwood truck)
d Forstkran *m* mit horizontalem Arm
f chargeur *m* à flèche horizontale; grue *f* forestière à flèche horizontale
r погрузчик *m* в виде мачты с горизонтальной стрелой

706 bigtooth aspen; bigtooth poplar; Canadian poplar; largetoothed aspen
(Populus grandidentata)
d kanadische Pappel *f*; großzahnige Pappel

707 **bigtooth maple; canyon maple**
(Acer [saccharum] grandidentatum)
d Großzahnahornholz *n*; Canyonahornholz *n*
f érable *m* à feuilles caduques
r клён *m* крупнозубчатый

* **bigtooth poplar → 706**

* **bigtree plum → 5425**

* **bilberry → 855**

708 **bilberry pine forest**
d Blaubeer-Kiefernwald *m*
f pineraie *f* à myrtille
r сосняк-черничник *m*

709 **bilimbi tree; cucumber magnolia; cucumber-tree**
(Averrhoa bilimbi; Magnolia acuminata)
d Gurkenbaum *m*; Bilimbaum *m*; amerikanische Magnolie *f*
f bilimbi *m*; magnolia *m* d'Amérique; arbre *m* aux concombres
r магнолия *f* длиннозаострённая; магнолия остроконечная; огуречное дерево *n*; билимби *m*

710 **bilinga**
(Nauclea diderrichii; Nauclea gillettii)
d Bilinga *m*
f bilinga *m*
r науклея *f*

* **billet → 2171**

711 **billet debarker; billet debarking machine**
d Schälentringer *m*
f écorceuse *f* de billots
r окорочный станок *m* для чураков; станок для окорки чураков

712 **billet infeed conveyor**
d Schäl[zug]förderer *m*
f convoyeur *m* d'entrée de billots
r транспортёр *m* для подачи чураков

713 **billeting**
d Holzklotzschnitt *m*
f réduction *f* en billots
r распиловка *f* чураков

714 **billet outfeed conveyor**
d Schälauslaufband *m*
f convoyeur *m* de sortie de billots
r транспортёр *m* вывода чураков

* **billet saw → 1136**

* **billu → 2729**

715 **Biltmore stick; cruiser's stick; cal[l]iper stick; volume cal[l]ipers; tree scale**
(tool calibrated to measure the diameter of a tree at breast height)
d Stockkluppe *f*; Massenmessstock *m*; Kubierungskluppe *f*; Kubikkluppe *f*
f baguette *f* de Biltmore; canne-bastringue *f*; compas *m* pour mesure des volumes
r мерная линейка *f* Бильтмора; палкообразная мерная вилка *f*; шест *m* для измерения массы; мерная вилка для определения кубатуры

* **Biltmore [white] ash → 261**

* **binbinga → 1120**

716 **binder**
(a springy pole used to tighten a binding chain)
d Binder *m*; Mähbinder *m*
f tendeur *m*
r устройство *n* для удержания груза на транспортном средстве; стяжка *f*

* **binder [chain] → 5023**

* **binding → 519**

* **binding chain → 5023**

717 **binding logs**
(logs placed on the top of the chain binding a load, in order to take up the slack)
d bindende Schnittholze *npl*
f grumes *fpl* de tension
r брёвна *npl*, укладываемые поверх обвязочной цепи

* **binding wire → 5023**

718 **bintangor**
(Calophyllum)
d Bintangor *m*
f bintangor *m*
r калофиллум *m*

* **biocenose → 738**

* **biochemical systematics → 1471**

* **bioc[o]enosis → 738**

719 **biodegradables**

d Biozersetzung *f*; Biozerstörung *f*
f biodégradables *fpl*
r биоразложимые вещества *npl*

720 **biodeterioration**
d Biobeschädigung *f*
f biodétérioration *f*
r биоповреждение *n*

721 **biodiversity; biological diversity; biotic diversity**
(the diversity of plants, animals, and other living organisms in all their forms and levels of organization)
d biologische Vielfalt *f*
f diversité *f* biologique; biodiversité *f*
r биоразнообразие *n*

722 **biofuel; biomass fuel**
(product of biomass conversion; can be used directly to provide heat or electricity)
d Biokraftstoff *m*; Biobrennstoff *m*
f biocombustible *m*
r биотопливо *n*

723 **biogenesis**
d Biogenese *f*
f biogènese *f*
r биогенезис *m*

724 **biogeochemical cycle**
(circulation of chemical elements)
d biogeochemischer Kreislauf *m*
f cycle *m* biogéochimique
r круговорот *m* биохимических веществ

725 **biogeoclimatic zone**
(geographic area having similar patterns of energy flow, vegetation and soils as a result of a broadly homogenous macroclimate)
d biogeoklimatische Zone *f*
f zone *f* biogéoclimatique
r биогеоклиматическая зона *f*

* **bioindicator → 4473**

726 **bio-indicator grid**
(systematic sampling of needles/leaves in Austria carried out since 1983)
d Bioindikatornetz *n*
f réseau *m* de bio-indicateurs
r биоиндикаторная сетка *f*

* **biological diversity → 721**

727 **biological drying**
f séchage *m* biologique
r биологическая сушка *f*

728 **biological growth potential; biotic potential; reproductive potential; reproductive capacity; reproductive power**
(inherent ability of members of a population to grow in numbers within a given time and under stated environmental conditions)
d biologisches Wachstumspotential *n*; biotisches Potential *n*
f potentiel *m* de croissance biologique; potentiel biotique; capacité *f* de reproduction
r биотический потенциал *m*

729 **biological herbicide**
(naturally occurring substance or organism which kills or controls undesirable vegetation)
d biologisches Herbizid *n*; biologisches Unkrautmittel *n*
f herbicide *m* biologique
r биологический гербицид *m*

* **biological indicator → 4473**

730 **biological insecticide**
d biologisches Insektizid *n*
f insecticide *m* biologique
r биологический инсектицид *m*

* **biologic diversity → 7897**

* **biomass fuel → 722**

731 **biomass harvesting**
(harvesting of all material including limbs, tops, and unmerchantable stem and stumps, usually for energy wood)
d Holzeinschlag *m* der Biomasse
f exploitation *f* de biomasse
r заготовка *f* биомассы

732 **biome**
(major biotic community composed of all the plants and animals and smaller biotic communities, including the successional stages of an area)
d Biom *n*
f biome *m*
r биом *m*

733 **biometrics; biometry**
d Biometrie *f*; Holzmesskunde *f*
f biométrie *f*
r биометрия *f*

734 **biosphere**
(that part of the earth and atmosphere capable of supporting living organisms)
d Biosphäre *f*
f biosphère *f*
r биосфера *f*

735 **biota**
(all living organisms of an area, taken collectively)
d Biota *n*
f biote *m*
r биота *f*

736 **biotechnology**
d Gentechnik *f*
f biotechnologie *f*
r биотехнология *f*

737 **biotic**
(all of the natural living organisms in a planning area and their life processes)
d biotisch
f biotique
r биотический

738 **biotic community; bioc[o]enosis; biocenose**
(assemblage of related populations of plants and animals constituting a relatively self-sufficient ecological unit)
d Biozoenose *f*; Biocenosis *f*; Lebensgemeinschaft *f*; Artengemeinschaft *f*
f bioc[o]énose *f*; communauté *f* biotique
r биотическое сообщество *n*; биоценоз *f*; сообщество организмов

* **biotic diversity → 721**

* **biotic potential → 728**

739 **biotop[e]**
d Lebensraum *m*; Biotop *m*
f espace *m* vital; biotope *m*
r биотоп *m*

740 **biotype**
(genetically homogeneous population composed only of closely similar individuals; a genotypic race or group of organisms)
d Biotyp *m*
f biotype *m*
r биотип *m*

741 **bipinnate leaf; twice pinnate leaf; doubly pinnate leaf**
(with leaflets arranged on side branches off a
f feuille *f* bipennée
r двойноперистый лист *m*

742 **birch**
(Betula)
d Birke *f*
f bouleau *m*
r берёза *f*

743 **birch bark**
d Birkenrinde *f*
f écorce *f* de bouleau
r берёзовая кора *f*

744 **birch bark cherry; Oriental cherry**
(Prunus serrulata)
d gesägtblättrige Kirsche *f*; Tibet-Kirsche *f*
f cerisier *m* du Tibet; cerisier à feuilles en dents de scie
r вишня *f* мелкопильчатая; секура *f*

745 **birch[bark] tar**
d Birken[rinden]teer *m*; Daggert *m*
f goudron *m* de bouleau
r берёзовая смола *f*; берёзовый дёготь *m*

746 **birch blocks**
d Birkenknüppelholz *n*; Birkenrollen *fpl*
f billots *mpl* de bouleau
r берёзовые балансы *mpl*; берёзовый кругляк *m*; берёзовое коротьё *n*

747 **birch-branch; green bough**
d Maie *f*
f mai *m*
r берёзовая ветвь *f*

748 **birch charcoal**
d Birkenholzkohle *f*
f charbon *m* de bouleau
r берёзовый уголь *m*

* **birch conk → 752**

749 **birchen foliage**
d Birkenlaubwerk *n*
f feuillage *m* de bouleau
r берёзовая листва *f*

750 **birch family**
(Betulaceae)
d Birkengewächse *pl*
f bétulacées *pl*
r берёзовые *pl*

f forêt *f* de bouleau; bétulaie *f*; boulaie *f*
r берёзовый лес *m*; березняк *m*

752 **birch fungus; razor-strap fungus; birch conk**
(Polyporus betulinus; Piptoporus betulinus)
d Birkenbeule *f*
f amadou *m* de bouleau
r берёзовая губка *f*; берёзовый трутовик *m*

753 **birch grove**
d Birkenhain *m*
f bouleraie *f*; boulassière *f*
r берёзовая роща *f*

* **birch juice → 758**

754 **birchleaf miner; birch leafminer**
(Fenusa pusilla)
d Birkenblattminiermotte *f*
f petite mineuse *f* du bouleau
r пилильщик *m* минирующий берёзовый малый

* **birchleaf mountain-mahogany → 8990**

755 **birch-leaved spirea**
(Spiraea betulifolia)
d birkenblättrige Spiere *f*; birkenblättriger Spierstrauch *m*
f spirée *f* à feuilles de bouleau
r спирея *f* берёзолистная

756 **birch oil**
d Birken[rinden]öl *n*
f huile *f* de bouleau
r берёзовое масло *n*

757 **birch sap**
d Birkensaft *m*
f sève *m* de bouleau; soc *m* de bouleau
r берёзовый сок *m*

758 **birch syrup; birch juice**
d Birkensirup *m*
f sirop *m* de bouleau
r берёзовый сироп *m*

759 **birch tundra**
(all Artic and antarctic areas polewards of the taiga, and similar alpine environnements above the imberline on mountains)
d Birkentundra *f*
f taïga *f* à bouleaux
r берёзовая тундра *f*

760 **birch wax**

f cire *m* de bouleau
r берёзовый клей *m*

761 **birch weevil**
(Rhynchites betulae)
d Birken-Rüsselkäfer *m*
f charançon *m* de bouleau
r трубковёрт *m* берёзовый чёрный

762 **birchwood**
d Birkenholz *n*
f bois *m* de bouleau
r берёзовая древесина *f*

763 **bird box; nest box; bird house**
d Nistkasten *m*; Vogelkasten *m*
f nichoir *m*
r гнездовка *f*

* **bird-cherry → 2881**

764 **birddog aircraft**
(aircraft carrying the person (air attack officer) who is directing fire bombing action on a wildfire)
d Birddog-Flugzeug *m*
f aéronef *m* à coordinateur aérien
r самолёт-перехватчик *m* координатора

* **bird house → 763**

765 **bird-peck**
(a small hole or patch of distorted grain resulting from birds pecking through the growing cells in the tree)
d Schnabelhiebmahl *n*
f piquage *m*; piqûre *f*; moucheture *f*; cicatrice *f* d'un coup de bec
r птичьи поклёвки *fpl*; повреждения *npl* птицами; наклев *m* на стволе

766 **bird's eye figure; bird's eye grain**
(small localized areas in wood with the fibers indented to form small circular or elliptical figures on the tangential surface which are used for decorative purposes)
d Vogelaugenahorn-Figur *f*; Vogelaugenmaser *f*
f figure *f* à œil; mouchetures *fpl*; madrure *f* en œil de perdrix
r текстурный рисунок *m* типа "птичий глаз"; наплыв *m* "птичий глаз"; узор *m* "птичий глаз"

* **bird's eye maple → 808**

767 **bird's eye wood**
d Astlochholz *n*
f bois *m* moucheté
r древесина *f* с наплывами типа "птичий глаз"

768 **birthwort family**
(Aristolochiaceae)
d Osterluzeigewächse *pl*
f aristolochiacées *pl*
r аристолохии *fpl*; кирказоновые *pl*

* **biserrate leaf margin → 2590**

769 **Bishop['s] goutweed; dwarf elder; Bishop's elder**
(Aegopodium podagraria)
d gemeiner Giersch *m*; Geißfuß *m*
f égopode *m* podagraire; herbe *f* aux goutteux
r сныть *f* обыкновенная

770 **Bishop['s] pine; Obispo pine**
(Pinus muricata)
d Bischofskiefer *f*; Stachelkiefer *f*
f pin *m* épineux; pin de Bishop; pin de Californie
r сосна *f* мелкоигольчатая

* **bishopwood → 4645**

* **bit → 819**

* **bitch chain → 1555**

* **bitch line → 1555**

* **bite → 7232**

* **bitter almond → 207**

* **bitter angelim → 4715**

* **bitterbrush → 311**

771 **bitter buffalo[ ]berry; Canadian buffaloberry; buffalo bush; soapberry**
(Shepherdia canadensis; Lepargyrea canadensis)
d kanadischer Seifenbaum *m*
f shepherdie *f* du Canada
r шефердия *f* канадская

772 **bitter cherry**
(Prunus emarginata)
d bittere Kirsche *f*

773 **bitter nightshade**
(Solanum dulcamara)
d bittersüßer Nachtschatten *m*; Bittersüß *m*
f morelle *f* douce-amère; morelle grimpante; vigne *m* sauvage; douce-amère *f*
r паслен *m* сладко-горький

774 **bitternut [hickory]**
(Carya cordiformis; Carya amara)
d Bitternuthickory *f*; Bitternussbaum *m*
f noyer *m* amer; hickory *m* bitternut; caryer *m* cordiforme
r кария *f* сердцевидная; кария горькая

* **bitter oak → 5565**

775 **bitter orange tree; sour orange [tree]; Sevilla orange [tree]**
(Citrus aurantium)
d Pomeranzenbaum *m*; Pomeranze *f*; Bigaradie *f*; bittere Orange *f*
f bigaradier *m*; oranger *m* amer; orange *f* amère
r померанец *m* [дерево]; горький апельсин *m*; бигардия *f*

* **bitter pecan → 6130**

* **bittersweet → 7984**

* **bitter water hickory → 6130**

* **bitum → 7763**

776 **bixa family**
(Bixaceae)
d Orleangewächse *pl*
f bixacées *pl*
r биксовые *pl*

* **black alder → 2882**

* **black ant → 1762**

777 **black ash; swamp ash; water ash**
(Fraxinus nigra)
d schwarze Esche *f*; Schwarzesche *f*; Wasseresche *f*
f frêne *m* noir; frêne à feuilles de sureau
r ясень *m* чёрный; водяной ясень

778 **blackberry**
(Rubus fruticosus)
d [echte] Brombeere *f*; gemeine Brombeere; Fuchsbeere *f*
f ronce *f* des buissons; ronce des bois; ronce commune; ronce frutescente; mûrier *m* des

779 **black birch; river birch**
(Betula nigra)
d Schwarzbirke *f*
f bouleau *m* noir; bouleau de rivière
r берёза *f* чёрная; берёза вишнёвая

780 **black box**
(Eucalyptus largiflorens)
d schwarzer Eukalyptus *m*
f gummi *m* noir
r эвкалипт *m* чёрный

* **blackboy → 405**

781 **blackbutt [eucalyptus]**
(Eucalyptus pilularis)
d Blackbutt-Eukalyptus *m*; Benaroon *m*
f blackbutt *m*
r чёрный платан *m*

782 **black[cap] raspberry; thimble-berry**
(Rubus occidentalis)
d schwarze Himbeere *f*; wohlriechende Himbeere
f framboise *f*; ronce *f* occidentale
r малина *f* западная

783 **black check**
d Schwarzfleck *m*
f tache *f* noire
r чёрное пятно *n*

* **black cherry → 244**

784 **black cottonwood; balm cottonwood**
(Populus trichocarpa)
d kalifornische Balsampappel *f*; harrfrüchtige Pappel *f*
f peuplier *m* baumier californien
r тополь *m* волосистоплодный

785 **black crowberry**
(Empetrum nigrum)
d schwarze Krähenbeere *f*
f camarine *f* noire; baie *f* de cornaille; vacinet *m* noir
r шикша *f* чёрная; водяника *f*; вороника *f*

786 **black cypress; baldcypress; southern cypress; swamp cypress; Louisiana cypress**
(Taxodium distichium)
d Eibenzypresse *f*; [zweizeilige] amerikanische Sumpfzypresse *f*; Sumpftaxodie *f*; Sumpfzeder *m*
f cyprès *m* chauve; cyprès de Louisiane;
луизианский кипарис

* **black ebony [tree] → 2727**

787 **black elderberry; American red berry elder**
(Sambucus [pubens var.] melanocarpa)
d Holunderbeere *f*
f sureau *m* rouge d'Amérique; sureau arborescent
r бузина *f* красная американская

788 **black forest**
d Schwarzwald *m*
f forêt *f* noire
r чёрнолесье *n*

789 **black forest earth**
d Schwarzwaldboden *m*
f sol *m* noir de forêt
r темноцветная лесная почва *f*

790 **black-fruited| honeysuckle**
(Lonicera nigra)
d schwarze Heckenkirsche *f*; schwarzes Geißblatt *n*
f chèvrefeuille *m* à baie noire
r жимолость *f* чёрная

791 **black-green-wattle**
(Acacia mearnsii; Acacia mollissima)
d grüne Akazie *f*
f mimosa *m* vert
r акация *f* чёрная

* **black gum → 8310**

792 **blackhaw; stagberry; stagb[r]ush**
(Viburnum prunifolium)
d Schneeballbaum *m*
f viorne *f* à feuilles de prunier
r калина *f* сливолистная

793 **black hawthorn**
(Crataegus nigra; Mespilus nigra; Crataegus melanocarpa)
d Hagedorn *m*; Mehldorn *m*; Schwarzdorn *m*; schwarzfrüchtiger Weißdorn *m*; Schwarzfrucht-Weißdorn *m*; dunkler Weißdorn
f aubépine *f* noire; épine *f* écarlate
r боярышник *m* чёрный

794 **blackheaded budworm**
(Acleris variana)
d schwarzkopfiger Knospenwickler *m*

795 **blackheart; false heart**
(any of various diseases in which the central tissues blacken)
d Schwarzkern *m*; falscher Kern *m*; Falschkern *m*; dunkler Kern; fakultativer Kern; Scheinkern *m*
f faux cœur *m*; cœur olive; pseudo-cœur *m*
r внутренняя темнина *f*; ложное ядро *n*; чёрное ядро (древесины в зоне ядра)

* **blackheart sassafras → 7864**

* **black hemlock → 5588**

796 **black hills beetle; mountain pine beetle**
(Dendroctonus ponderosae; Dendroctonus montanae)
d Ponderosariesenbastkäfer *m*
f dendroctone *m* du pin ponderosa
r лубоед *m* сосны горной

797 **black-horned pine borer**
(Callidium antennatum hesperum)
d schwarz-gehörnter Kieferbohrer *m*
f callidie *f* du Pacifique
r златка *f* сосновая чёрнорогая

798 **black huckleberry; big huckleberry; thinleaf huckleberry**
(Vaccinium membranaceum; Gaylussacia baccata)
d schwarze Buckelbeere *f*
f airelle *f* à feuilles membraneuses
r гейлюссакия *f* ягодная

799 **blackjack oak**
(Quercus marilandica)
d echte Schwarzeiche *f*; Prärieeiche *f*
f chêne *m* noir d'Amérique; jaquier *m* noir; chêne du Maryland
r дуб *m* мерилендский

800 **black knot; dead knot**
d schwarzer Ast *m*; Schwarzast *m*; ausfallender Ast; Ausfallast *m*
f nœud *m* noir; nœud mort; nœud "bouchon"
r чёрный сучок *m*

* **black lancewood → 9347**

* **black lin → 7860**

801 **black line; pencilling; zone line; black zone**
(narrow brown or black lines in decayed or decaying wood)
d Zonenlinie *f*

* **black locust → 4353**

* **black mangrove → 251**

802 **black mo[u]lds fungi; green mo[u]lds fungi**
(Penicillum; Aspergillus niger)
d Schimelpilze *mpl*
f moisissure *f* noire
r пеницилл *m*; аспергилл *m*; чёрная плесень *f*

803 **black mulberry [tree]**
(Morus nigra)
d schwarzer Maulbeerbaum *m*
f mûrier *m* noir
r шелковица *f* чёрная

804 **black nightshade**
(Solanum nigrum)
d schwarzer Nachtschatten *m*
f morelle *f* noire
r паслен *m* чёрный; поздника *f*

* **black olive → 6014**

* **black pine → 415, 4601**

805 **black pine bark beetle**
(Hylastes alter)
d schwarzer Kiefernbastkäfer *m*
f hylésine *m* noir du pin
r корнежил *m* чёрный

806 **black pine leaf beetle**
(Galeruca pinicola)
d schwarzer Kiefernblattkäfer *m*
f galéruque *f* noire du pin
r листоед *m* сосновый

* **black poplar → 2884**

* **black sloe → 3260**

* **black spot [disease] → 315**

807 **black stinkwood; stinkwood [ocotea]**
(Ocotea bullata)
d Stinklorbeer *m*
f stinkwood *m*
r окотеа *f* пузырчатая

* **black spruce → 2716**

808 **black [sugar] maple; bird's eye maple; sugar maple; rock maple; hard maple**
(Acer nigrum; Acer saccharum)

f érable *m* noir; érable moucheté [d'Amérique]; érable à sucre
r клён *m* чёрный; клён твёрдый; клён птичий глаз; клён сахарный

* **black sumac[h] → 3247**

* **blackthorn → 7749**

809 **blackthorn wattle**
(Acacia mellifera)
d Schwarzdornakazie *f*
f kitir *m*
r акация *f* колючая

* **black-topped → 2272**

810 **black tupelo; tupelo-gum; pepperidge**
(Nyssa sylvatica)
d schwarzer Tupelobaum *m*; [Wald-]Tupelobaum *m*
f tupélo *m* noir; tupélo des forêts
r нисса *f* лесная

811 **black turpentine beetle**
(Dendroctonus terebrans)
d dunkler Käfer *m*
f dendroctone *m* noir de l'épinette
r лубоед *m* чёрный скипидарный

812 **black twinberry; bracted twinberry**
(Lonicera involucrata)
d behüllte Heckenkirsche *f*; Behüllte *f*
f mitchella *f*; chèvrefeuille *m* à involucre; chèvrefeuille involucré
r жимолость *f* обёрнутая; жимолость покрывальная

* **black walnut → 2715**

* **blackwattle → 2698**

* **black whortle → 5006**

813 **black willow; swamp willow**
(Salix nigra; Salix myrtilloides)
d Schwarzweide *f*; Heidelbeerweide *f*; Moorweide *f*; Sumpfweide *f*
f saule *m* noir; osier *m* noir; saule à feuilles de myrtille
r ива *f* чёрная

814 **blackwood**
(1. very dark wood of any of several blackwood trees; 2. any of several hardwood trees yielding very dark-colo[u]red)
r чёрное дерево *n*

* **blackwood → 102**

815 **black-wooded**
d schwarzholzig
f à bois noir
r чёрнодревесный

* **blackwoood acacia → 404**

* **black zone → 801**

816 **bladdernut**
(Staphylea colchica)
d kolchische Pimpernuss *f*; Klappernuss *f*
f staphylier *m*
r клекачка *f*; клокичка *f*; стафилея *f*

817 **bladdernut family**
(Staphyleaceae)
d Pimpernussgewächse *pl*
f staphylacées *pl*
r клекачковые *pl*

818 **bladder senna**
(Colutea arborescens)
d Blasenbaum *m*; [gelber] Blasenstrauch *m*
f balandier *m*
r пузырник *m* древовидный

819 **blade; cutting face; sharp end; [cutting] edge; nib; lip; bit**
d Klinge *f*; Messer *m*; Werkzeugschneide *f*; Schneide *f*; Spitze *f*
f lame *f*; couteau *m*; pointe *f*; coupant *m*; tranchant *m*
r лезвие *n*; режущая кромка *f*; остриё *n*; клинок *m*; наконечник *m*

* **bladed trail → 2934**

820 **blade of a saw; saw[ ]blade; saw web**
d Sägeblatt *n*
f lame *f* de scie
r полотно *n* пилы; пильное полотно

821 **blade setting; saw set; teeth set**
d Blatt-Schränken *n*
f affûtage *m* de lames de scie
r развод *m* зубьев пилы

* **blade with round breast → 2152**

822 **blading**
d Beschaufelung *f*

823 **blank-filling; gapping; repairing of plantation; after[-]planting; afterculture; filling[-in] planting; filling up blanks; reinforcement planting; beating up; repair planting**
(planting additional trees or shrubs and replace dead trees or shrubs)
d Lückenausbesserung *f*; Nachbesserung *f* der Bepflanzung; Ergänzung *f* der Bepflanzung
f complément *m* de plantation; complément de vides; remplacement *m* des manquants; regarni[ssage] *m*
r дополнение *n* просветов; дополнение лесных культур

* **blank in a stand → 3443**

* **blaze → 540**

824 **blaze** *v*
(to permanently mark trees, indicating those to be cut or the course of a boundary, road, or trail)
d einkerben
f griffer
r затёсывать; зарубить; клеймить

825 **blaze** *v* **a trail**
d Bahn auszeichnen
f marquer la voie; griffer la voie
r отмечать дорогу зарубками

* **blazer → 5314**

* **blazing → 5317**

826 **blazing tree; marked tree**
d ausgezeichneter Baum *m*; gestempelter Baum
f arbre *m* marqué
r помеченное дерево *n*; [от]клеймённое дерево; замаркированное дерево

* **blea → 853**

827 **bled timber**
d Harzwald *n*; gezapftes Holz *n*
f bois *m* saigné; bois expurgé
r лес *m*, отведённый в подсочку

* **bled tree → 959**

828 **bleeding; [resin] boxing; resin extraction; [resin] tapping; turpentining; dip[ping]**
(the diffusion of a colo[u]red substance such as pitch from a knot through a paint or varnish coating, resulting in an undesirable staining or
d Anzapfen *n*; Harzgewinnung *f*; Harzschrappen *n*; Harzen *n*; Schöpfen *n*
f gemmage *m*; récolte *f* de la gomme; saignée *f* des gommiers
r подсочка *f*; добыча *f* живицы; добывание *n* живицы; сбор *m* живицы

829 **blemish**
(in timber, any feature that mars its appearance without lowering its utility, e. g. blue stain in structural timber)
d Schandfleck *m*; Schande *f*; Schönheitsfehler *m*
f anomalie *f* non dépréciative
r местное обесцвечивание *n* древесины, не являющееся значительным пороком

830 **blending pin knot**
(sound knot 6.4mm (1/4 inch) or less that generally do not contains dark centers)
d verblendender Perlast *m*; verblendendes Vogelauge *n*; Stiftästchen *n*
f œil *m* de perdrix; très petit nœud *m* intermédiaire
r промежуточный игольчатый сучок *m*

* **blepharophyllous → 1580**

* **blight → 5444**

831 **blighted leaf**
d vom Brand befallenes Blatt *n*
f feuille *f* dépérie
r обесцвеченный лист *m*

832 **blind conk**
d Astbeule *f*
f renflement *m* d'une branche; bourrelet *m* d'une branche
r сучковатый нарост *m*

833 **blind knot; enclosed knot; encased knot; bark knot; occluded knot; firm knot; intergrown knot; adhering knot; adherent branch**
(a knot whose rings of annual growth are completely intergrown with those of the surrounding wood)
d eingeschlossener Ast *m*; eingewachsener Ast; Einschlussast *m*; verwachsener Ast; überwallter Ast
f nœud *m* recouvert; nœud *m* entouré d'écorce; nœud occulte; nœud adhérent
r заросший сучок *m*; плотно сросшийся со стволом сучок; сросшийся сучок; вросший сук *m*; заплывший сук

834 **blister**
d Blase *f*; Beule *f*
f boursouflure *f*; vésicule *f*; bouillon *m*
r пузырёк *m*; пузырчатое заболевание *n* растений

835 **blister figure; blister grain; landscape grain**
(a figure on smooth plain-sawn and rotary-cut surfaces that appears to consist of small, more or less widely spaced, elevated, or depressed areas of rounded contour)
d Blasefigur *f*
f figure *f* pommelée
r ноздреватый рисунок *m*; причудливая текстура *f* (из лущёной или продольно распиленной поверхности берёзы или клёна)

836 **blistering of the bark; bark blister[ing]; scorch of the bark; bark scorch[ing]**
(localized injury to bark and cambium, caused e.g. by a sudden increase in exposure of a stem or branch to high temperatures from intense sunlight)
d Rindenbrand *m*
f insolation *f* de l'écorce
r ожог *m* коры

837 **blister on a leaf**
d Blase *f* auf einem Blatt
f boursouflure *f* sur feuille
r ожог *m* листа

* **blister rust → 9376**

* **block → 7173, 8879**

* **block-bill → 1452**

838 **block cutting**
d Aushalten *n*; Nachschneiden *n*
f coupe *f* par blocs
r раскряжёвка *f* на сортименты; раскрой *m* брусков (торцованного лесоматериала)

839 **block-cutting machine**
d Blockschneidmaschine *f*
f coupeuse *f* de blocs
r станок *m* для раскроя брусков

840 **block foot**
(a cube-shaped foot, a solid block of wood, which is used generally with a square untapered leg)
d Blockfuß *m*

841 **block form**
d Blockform *f*
f forme *f* d'une grume
r форма *f* балки

842 **block handler**
d Blockanwender *m*
f chargeur *m* de billot
r подвозчик *m* баланса

* **blockhouse → 8560**

* **block measure → 5103**

843 **block mottle; mottle**
(irregular, diffuse patterns of chlorotic areas interspersed with normal green leaf tissue)
d Marmorierung *f*
f marbrure *f*
r пятнистость *f*

844 **block of beech wood**
d Buchenwaldblock *m*
f coin *m* de bois de hêtre
r буковый закреп *m*

* **block planting → 6354**

* **block strop → 4850**

* **block tender → 1454**

* **block-wood paving → 9554**

845 **blonded wood**
d Bleichholz *n*
f bois *m* blanchi
r светлая древесина *f*; отбелённая древесина

* **blonde mahogany → 117**

* **bloodtwig dogwood → 1771**

846 **bloodwood tree; campeche**
(Haematoxylon campechianum)
d Sandale *f*; Kampeche *f*
f campêche *m*
r сандаловое дерево *n*; кампешевое дерево; синий сандал *m*

* **blooming willow → 3215**

847 **blossom[ing] of a tree; tree blossom[ing]**
d Baumblüte *f*

848 **blossoming tree**
d blühender Baum *m*
f arbre *m* fleuri
r цветущее дерево *n*

849 **blowdown gap; windfall zone**
d Windbruchstelle *f*
f zone *f* de chablis; vide *m* dû aux chablis
r буреломная прогалина *f*; ветровальная прогалина

850 **blowdown [timber]; blowdowns; windthrow[n timber]; windslash; windfall [timber]; windbrake; windfall[en] trees; morefall**
(trees thrown down or with their stems broken off or other parts blown down by the wind)
d windbrüchiges Holz *n*; Bruchholz *n*; Baumbruch *m*; Sturmholz *n*; Windfallholz *n*; Windfall *m*
f bois *m* déraciné par le vent; bois de chablis; chablis *m*
r бурелом *m*; буреломная древесина *f*; ветровал *m*; ломанная древесина

851 **blowing-up of crown**
d Absprengen *n* von Baumkronen
f détachage *m* de cime par explosion
r отделение *n* кроны дерева взрывом

852 **blowing-up of stumps; stump exploding; stump blasting; uprooting by blowing-up**
d Sprengrodung *f*; Sprengroden *n*; Stocksprengung *f*
f dessouchage *m* de par explosion; défrichement *m* à l'explosif
r корчевание *n* пней взрывом; корчёвка *f* [пней] взрыванием; взрывание *n* пней

853 **blown sap; false sap; halo sap; shell rot; double sap[wood]; included sapwood; internal sapwood; blea; plethora; lunule**
(streaks or irregularly shaped areas of light-colo[u]red wood with the general appearance of normal sapwood, found embedded in the darker-colo[u]red heartwood)
d eingeschlossenes Splintholz *n*; Doppelsplint *m*; Mond[ring]fäule *f*; Mondring *m*; falscher Splint *m*
f aubier *m* inclus; aubier interne; bois *m* luné; faux aubier
r двойная заболонь *f*; внутренняя заболонь; ложная заболонь

854 **blue ash; Virginia ash**
f frêne *m* bleu [d'Amérique]; frêne anguleux
r ясень *m* четырёхгранный

* **blue Atlas cedar → 396**

* **blue beech → 249**

855 **blueberry; bilberry; whortleberry**
(Vaccinium)
d Blaubeere *f*; Waldbeere *f*; Bickbeere *f*
f airelle *f*
r черника *f*; черница *f*

856 **blue blossom [ceanothus]; California lilac**
(Ceanothus thyrsiflorus)
d Blausäckelblume *f*
f lilas *m* de Californie
r цеанотус *m* метельчатоцветный

857 **blue box; Bauer eucalyptus**
(Eucalyptus baueriana)
d Bauer-Gummi *m*
f boîte *f* bleue
r эвкалипт *m* Бауера

* **blue clematis → 4593**

* **blue damage → 860**

* **blue[d] timber → 859**

858 **blue elderberry**
(Sambucus c[a]erulea; Sambucus glauca)
d blauer Holunder *m*
f sureau *m* bleu
r бузина *f* голубая

859 **blue goods; blue[-stained] wood; blue lumber; blue[d] timber**
d angeblautes Holz *n*; verblautes Holz; Blau *n*
f bois *m* bleu
r [хвойные] пиломатериалы *mpl*, поражённые синевой; древесина *f*, поражённая синевой; засиневшая древесина

* **blue gum → 8417**

860 **blueing; blueness; blue damage**
(discolo[u]ration in the sapwood of pine)
d Verblauung *f*
f bleuissement *m*
r засинение *n*; посинение *n*; поражение *n* синевой

861 **blueing fungi; bluestain fungi**
(most common form of fungal stain occurring

f champignons *mpl* du bleu
r грибы *fpl* синевы

862 **blue Japanese oak**
(Quercus glauca; Quercus myrsinaefolia)
d japanische Blaueiche *f*
f chêne *m* glauca; chêne à feuilles de myrsine
r дуб *m* сизый; дуб японский мирзинолистный; ширакаши *m*

863 **blue-leaf birch; blue-leaved birch**
(Betula coerulea)
d Blaubirke *f*
f bouleau *m* bleu
r берёза *f* голубая

864 **blue-leaved huckleberry; Alpine blueberry; [piper] cascade blueberry**
(Vaccinium deliciosum)
d Alpenheidelbeere *f*
f airelle *f* myrtille bleue; bluet *m* du Canada
r блюэтта *f*

865 **blue-leaved stringybark**
(Eucalyptus agglomeratus)
d Blaustringybark *m*
f eucalyptus *m* bleu
r эвкалипт *m* голубой

* **blue-leaved wattle → 5936**

* **blue lumber → 859**

* **blue mallow → 7102**

866 **blue mistflower; thoroughwort; Christmas bush; Siam weed; triffid weed**
(Eupatorium odoratum; Chromolaena odorata)
d Wasserdost *m*
f herbe *f* du Laos
r посконник *m* душистый

* **blueness → 860**

* **blue oak → 242**

867 **blue paloverde**
(Parkinsonia florida; Cercidium floridum)
d Parkinsonia *f*; Paloverde-Baum *m*
f parcinsonia *m*
r паркинсония *f*

* **blue peppermint → 1052**

* **blue pine → 4308**

(Callidium antennatum antennatum)
d blauer Kieferbohrer *m*
f callidie *f* commune
r златка *f* сосновая голубая

* **bluepipe → 6292**

* **blue rot → 5050**

869 **blue sap; blue stain**
(wood defect)
d Splintbläue *f*
f bleuissement *m* de l'aubier
r синь *f* заболони; синева *f* заболони

* **blue spruce → 1726**

* **blue[-stained] wood → 859**

* **bluestain fungi → 861**

870 **bluewood [condalia]; Brazilian bluewood**
(Condalia hookeri)
d Hooker-Condalie *f*
f condalie *f* de Hooker
r кондалия *f* колючая

871 **blunt end; flush end**
(of needles)
d stumpfes Ende *n*
f pointe *f* arrondie; bout *m* franc; extrémité *f* franche
r обрубленный конец *m*

872 **blunt-leaved willow; itea**
(Salix rosmarinifolia)
d Rosmarinweide *f*
f saule *m* à feuilles de romarin
r ива *f* розмаринолистная

* **BMPs → 688**

873 **board; plank**
(a piece of lumber that is less than 38 mm in smaller dimension used for sheathing, formwork, or for further manufacture into trim and shaped products, such as siding)
d Brett *n*; Bohle *f*
f planche *f*
r доска *f*

874 **board blank**
d Bretterrohteil *m*
f platine *f* brute
r досковая заготовка *f*

f pince *f* de planche
r крюк *m* для закрепления брёвна

876 **board-foot; board foot; superfoot**
(a unit for measuring wood volume in a tree, log, or board equal to a board 1 foot long, 1 foot wide, and 1 inch thick)
d Boardfuß *m*; Brettfuß *m*
f pied-planche *m*; pied *m* de planche
r досковый фут *m*; боард-фут *m*

877 **board-foot log rule**
d Boardfußmessstab *m*
f règle *f* graduée en pied-planches; règle graduée en pieds de mesure de planches
r мерная линейка *f* определения объёмов брёвен в досковых футах

878 **board-foot log rule table; lumber scaling rule table**
d Boardfußtabelle *f*
f table *f* de cubage en pied-planches
r таблица *f* определения объёмов брёвен в досковых футах

879 **board[-foot] measure; feet board measure; board scaling**
d Brettervermessung *f*
f cubage *m* en pied-planches; cubage en pieds mesure de planche
r измерение *n* кубатуры в досковых футах

880 **boarding; [timber] planking; sheathing**
d Bretterbekleidung *f*; Bretterbeschlag *m*; Verschalung *f*; Holzabdeckung *f*; Holzbelag *m*
f planchéiage *m*; coffrage *m*; bordage *m* en bois
r обшивка *f* досками; отделка *f* деревом; обивка *f* тёсом; дощатая обшивка

881 **board [mining] slab**
d Brettschwarte *f*
f contredosse *f* [de mine]
r дощатый обапол *m*; дощатый горбыль *m*

882 **boards**
d Brettware *f*
f planches *fpl*
r доски *fpl*; тёс *m*

883 **board saw; back saw; miter saw**
(a short hand saw with very coarse teeth for cutting board for door and window frame openings)
d Längssäge *f*
f scie *f* d'encadreur; scie à dos; scie à onglet
r продольная пила *f* для распиловки на

* **board scaling → 879**

884 **board slide; plank slide; plank chute**
d Brettriese *f*
f glissoire *f* garnie de planches
r дощатый лесоспуск *m*

885 **board width**
d Brettbreite *f*
f largeur *f* de planche
r ширина *f* доски

886 **board with heart check**
d herzgespaltenes Brett *n*
f planche *f* à cadranure
r доска *f* с сердцевинной трещиной

* **boat hook → 6693**

* **bob → 5115**

887 **bobber**
d Bobber *m*
f bois *m* canard
r полностью или частично затонувшее бревно *n*

* **bobbin wood → 7926**

* **bob crotch → 5115**

888 **bobtail**
(refers to a two-axle truck)
d Stützschwanz *m*
f tracteur *m* haut-le-pied
r двухосный грузовик *m*

889 **bobtail shortwood system**
d Stützschwanzskurzholzsystem *n*
f débardage *m* en bois courts par tracteurs haut-le-pied; débardage de rondins par tracteurs haut-le-pied
r система *f* заготовки сортиментов с вывозкой их небольшими двухосными грузовиками

890 **bocage**
(a French word referring to a terrain of mixed woodland and pasture, with tortuous side-roads and lanes bounded on both sides by banks surmounted with high thick hedgerows limiting visibility)
d Hege *f*; Unzäuung *f*
f bocage *m*
r подлесок *m*; огороженный участок *m*

* **body → 8934**

* **body of timber → 3605**

891 **bodywood; stemwood; trunk timber; trunk wood; bole wood; bole billet**
(wood from the main part of a tree; the part of the bole giving quality timber)
d Stammholz *n*; Schaftholz *n*
f bois *m* de fût; bois en grume; fût *m*
r стволовая древесина *f*

* **bog bilberry → 893**

892 **bog birch; scrub birch; dwarf arctic birch**
(Betula nana; Betula glandulosa)
d Zwergbirke *f*; Alpenbirke *f*; Kriechbirke *f*; Polarbirke *f*; Morastbirke *f*
f bouleau *m* nain; bouleau glanduleux
r берёза *f* карликовая

893 **bog blueberry; bog bilberry; bog whortleberry**
(Vaccinium uliginosum)
d [echte] Moorbeere *f*; Sumpfheidelbeere *f*; Trunkelbeere *f*
f airelle *f* alpine; myrtille *f* à feuilles entières; airelle noire; airelle des marais; airelle bourbeuse
r голубика *f*; дурника *f*; пьяника *f*; гонобобель *m*

* **bog forest → 8309**

894 **bogie; buggy; bummer; janker; jinker; dolly**
(a small truck with two small wheels and a short pole, used in skidding logs)
d Bummer *m*; Bumtrip *m*; Transportkarre *f*; Dolly *n*; Rollbock *m*
f boghei *m*; bog[g]ie *m*; triqueballe *f*; éfourceau *m*
r лёгкая двухколёсная тележка *f*; трелевочные салазки *fpl*; двуколка *f*; двухколесная трелёвочная тележка

895 **bog laurel; bog kalmia**
(Kalmia polifolia; Kalmia glauca)
d Moor-Lorbeerrose *f*
f kalmia *m* à feuilles d'Andromède; calmie *f* glauque
r кальмия *f* многолистная; кальмия сизая

* **bog myrtle → 8332**

* **bog oak → 3764**

d Swamppodsol *m*
f sol *m* podzolique marécageux; sol podzolique de marais
r подзолисто-болотная почва *f*

897 **bog-rosemary [andromeda]; moorwort**
(Andromeda polifolia)
d Gränke *f*; Lavendelheide *f*; Rosmarinheide *f*; Poleigränke *f*
f romarin *m* de marais; andromède *m* [à feuilles de polium]; andromède d'Europe
r андромеда-дубровник *f*; андромеда дубровниколистная; подбел *m* дубровник; подбел многолистный

* **bog whortleberry → 893**

* **boiling-under-vacuum [process] → 5876**

* **boldo family → 5529**

* **bole → 8934**

* **bole area → 597**

* **bole billet → 891**

898 **bole form; stem form**
d Stammform *f*; Form *f* des Schaftes
f forme *f* de tronc
r форма *f* ствола

899 **bole skidding; bole removal**
d Baumstammabfuhr *f*; Stammholzrücken *n*
f débardage *m* de grumes
r трелёвка *f* хлыста

900 **bole surface**
d Stammfläche *f*
f surface *f* du tronc
r площадь *f* боковой поверхности ствола (дерева)

* **bole wood → 891**

901 **bole wood for veneer; veneer[ing] wood; wood for inlaying; peeled billet wood**
(high-quality logs of a desirable species suitable for conversion to veneer)
d Furnierholz *n*; Schäl[furnier]holz *n*
f bois *m* de tranchage; bois de déroulage
r фанерная древесина *f*; облупленная древесина

902 **bollywood; medang**

f litséa *m*
r литза *f*

* **bolster → 2038**

* **bolt → 2171**

* **bolter [saw] → 1586**

903 **bolt [length]; short log; short-cut wood; shortwood**
(short log of a length less than 120 inches suitable for manufacturing pulpwood or peeling veneer)
d Baumklotz *m*; kurzstückiges Holz *n*; Knüttel *m*; Kurzholz *n*
f bois *m* court
r бревно *n* коротномерного баланса; коротномерный сортимент *m*; коротномер *m*; коротьё *n*; короткий кряж *m*; короткий лесоматериал *m*

* **Bombay blackwood → 2728**

904 **Bombay ebony**
(Diospyros montana)
d Ceylon-Ebenholz *n*
f ébène *m* de Ceylan
r эбеновое дерево *n* из Бомбея; цейлонское эбеновое дерево

* **bond → 8115**

905 **bonding strength**
(between fibers)
d Bindekraft *f*
f cohésion *f*
r сила *f* сцепления; сила связи

906 **bone-dry unit; BDU**
(wood residue that weighs 2,400 pounds at zero percent moisture content)
d ganz-trockene Einheit *f*; ausgedörrte Einheit
f unité *f* de bois absolument sec
r совершенно сухая единица *f* древесины

907 **Bonpland willow**
(Salix bonplandiana)
d Bonpland-Weide *f*
f bois *m* de flèche; purépecha *m* en mich
r ива *f* Бонпланда

908 **bonsai tree**
(miniaturized tree trained for aesthetic effect)
d Bonsai[-Baum] *m*
f bonsaï *m*; penjing *m*
r карликовое дерево *n* в горшке

909 **bonus wood**
(wood that has been recycled)
d Bonus-Holz *n*
f bois *m* bonus
r древесина *f*, используемая для вторичной переработки

* **book matching → 3265**

* **boom → 914, 1182**

* **boomboat operator → 9569**

910 **boom chain; boom cable**
(a short chain which fastens boom sticks end to end)
d Sperrbaumkette *f*; Baumkette *f*; Ausleger *m*; Dröhnenkabel *n*
f chaîne *f* d'allingue; câble *m* d'allingue
r стрела *f*; вылет *m*

911 **boom dozer; bumpboat**
(any boat used to push or pull logs, bundles, or bags in booming ground operations)
d Schnittholzbulldozer *m*; Schwimmbulldozer *m*
f pousseur *m* de grumes
r бульдозер *m* для бревен

* **boomer → 9569**

* **boom extension → 8442**

912 **booming rafting; [timber] rafting; log cribbing**
d Bündeln *n* des Floßholzes
f groupage *m* de bois de flottage
r сплотка *f* древесины в запани; вязание *n* плотов

913 **boom loading**
d Boomladung *f*
f chargement *m* par pieu
r стреловая погрузка *f*; погрузка с помощью стрелы

* **boom man → 9569**

914 **boom of logs; log boom; raft boom; [floating] tree boom; bag boom; boom**
(floating logs tied to each other to encircle a group of logs, the whole forming a floating boom)
d Schwimmausleger *m*; Schwimmer *m*; Drifthof *m*
f élingue *f* d'arbres; gare *f*; flotteur *m* d'écartement; dalot *m* de flottage; estacade *f* [flottante]; chapelet *m*
r плавающая балансовая древесина *f*,

915 **boom post**
d Kransäule *f*
f mêt *m* de grue
r мачта *f* стрелы

* **boom reach → 6743**

* **boom stick → 8779**

* **boon → 1951**

916 **booster unit**
d Feuerlöschgeräte *npl*
f engin *m* de première intervention
r противопожарный агрегат *m*

917 **boots**
(firefighters wear all leather boots to protect their feet on the fireline and rugged terrain)
d Seilschuhe *mpl*
f bottes *f* [de protection]
r башмаки *mpl*

918 **borage**
(Borago officinalis)
d Gurkenkraut *m*; Borretsch *m*
f bourrache *m* [officinale]
r бурачник *m* [лекарственный]

919 **borage family**
(Bor[r]aginaceae)
d Borretschgewächse *pl*; Borraginazeen *pl*; Asperiofoliazeen *pl*
f bor[r]aginacées *pl*
r бурачниковые *pl*

* **border cutting → 6167**

* **bordered → 2756**

920 **bordered pit**
(a pit leading into an intercellular space between two cells, i.e., a pit without a complementary pit in an adjacent cell)
d Hofpore *f*; behöfter Tüpfel *m*; Hoftüpfel *m*
f pore *f* délignée; ponctuation *f* aréolée
r окаймлённая пора *f*

* **bordered white moth → 6262**

* **border felling → 6167**

921 **border planting; boundary planting**
(trees and/or shrubs planted at the edges of agricultural fields, for example, along field boundaries, the banks of waterways, and
f plantation *f* de bordure; plantation le long des limites
r опушечная посадка *f*

* **border protection → 6596**

* **border regeneration → 6877**

* **border seeding → 7384**

922 **border tree; edge tree; marginal tree**
d Randbaum *m*
f arbre *m* de lisière
r дерево *n* на опушке леса; опушечное дерево

923 **bore**
(the hole for the arbor in a circular saw blade)
d Bohrung *f*
f alésage *m*
r отверстие *n*; сверление *n*

924 **boreal coniferous forest**
d Nadelwald *m* in der borealen Zone; borealer Nadelwald
f forêt *f* boréale de conifères; forêt résineuse boréale
r бореальный хвойный лес *m*

925 **boreal forest; northern forest; frost-hardy forest**
(one of 3 main forest zones in the world)
d Wald *m* in der borealen Zone
f forêt *f* boréale
r бореальный лес *m*; северный лес

* **bore chipping → 4452**

926 **bore chips**
d Bohrspäne *mpl*; Hobelspäne *mpl*
f copeaux *mpl* de foreuse
r стружки *fpl* от сверления

* **bore-dust → 9728**

927 **borehole; hyphal hole**
(small breach in a cell wall created by the dissolving action of an advancing hypha, of e.g. a wood-inhabiting fungus)
d Bohrloch *n*
f pore *f* hyphale
r буровая скважина *f*

928 **borehole poisoning**
d Bohrloch-Intoxikation *f*; Vergiftungsbohren *n*
f empoisonnement *m* par forage oblique

929 **bore meal; boring powder**
d Bohrmehl *n*; Nagelspäne *mpl*
f sciure *f*; farine *f* de foreuse; son *m*
r буровая [древесная] мука *f*

* **boring → 6393**

930 **boring cut**
d Stechschnitt *n*
f trait *m* de scie plongeant
r сверлильный срез *m*

* **boring dust → 9728**

* **boring powder → 929**

931 **borrow pit**
(depression from which soil or rocks are taken to build roads, usually along side the road)
d Bodenentnahmestelle *f*; Entnahmestelle *f*
f fouille *f* d'emprunt; lieu *m* d'emprunt
r резервный карьер *m*; карьер грунта

932 **Bosisto's box; [coast-]grey box**
(Eucalyptus bosistoana; Eucalyptus bicolor)
d Bosisto-Eukalyptus *m*
f eucalyptus *m* de Bosisto
r эвкалипт *m* Босиста; эвкалипт двухцветный; серая пальма *f*

933 **bosket; bosquet; shrubbery**
(a small park or block of woodland with pathways, usually planted with small, understor[e]y trees, and often having statuary and waterworks)
d Gebüsch *n*; Gesträuch *n*; Buschwerk *n*; Strauchwerk *n*
f bosquet *m*; fruticée *f*
r кустарниковые насаждения *npl*; аллея *f*, обсаженная кустами; роща *f*; боскет *m* (аллея из формованных деревьев или кустарников)

934 **bosky; bushy; bruchy; shaggy; shrubby; shrubbed; frutescent; fruticose; dumous; dumose; dumetose**
d buschig
f touffu; buissonneux; broussailleux; frutescent; buissonnant
r кустистый; покрытый кустарником; кустарниковый

935 **Bosnian maple; Norway maple**
(Acer platanoides; Acer platanus)
d Spitzahorn *m*; Leinbaum *m*; spitzblättriger Ahorn *m*; Lehne *f*
r клён *m* остролистный; клён платановидный

936 **Bosnian pine; Heldreich pine**
(Pinus heldreichii)
d Panzerkiefer *f*; bosnische Kiefer *f*; Schlangenhaut-Kiefer *f*; balkanische Heldreichföhre *f*
f pin *m* de Bosnie; pin à écorce de sapin; pin des Balkans; pin blanchâtre; pin de Bosnie à cône rouge
r сосна *f* Гельдрейха

* **bosquet → 933**

937 **Boston ivy**
(Parthenocissus tricuspidata; Ampelopsis tricuspidata; Cissus thunbergii; Psedera tricuspidata)
d dreispitzige Jungfernrebe *f*
f vigne-vierge *f*; lierre *f* de Boston
r девичий виноград *m* [триостренный]; партеноциссус *m* триостренный

* **boswellia → 4463**

938 **botanical forest products**
(prescribed plants or fungi that occur naturally on Crown forestland)
d botanische Forstprodukte *npl*
f produits *mpl* forestiers botaniques
r ботанические леснье продукты *mpl*

939 **bo tree; pipal [tree]; pipul; peepul; sacred fig**
(Ficus religiosa)
d Pepulbaum *m* der Inder; Bobaum *m*
f figuier *m* des pagodes; arbre *m* de conseils; figuer indien; pippal *m*
r священное дерево *n*; священная смоковница *f*; фикус *m* священный; фикус религиозный (у буддистов)

940 **bottlebrush buckeye**
(Aesculus parviflora)
d Strauchkastanie *f*; Strauch-Rosskastanie *f*; Ährenparve *f*
f marronnier *m* d'Inde blanc; pavier *m* blanc
r каштан *m* конский мелкоцветковый; каштан конский мелкоцветный

* **bottle-butted → 8493**

* **bottle palm → 2109**

f arbres *mpl* en bouteille; arbres bouteilles
r бутылочные деревья *npl*

* **bottom → 4186**

942 **bottom bind**
d Hinternbinden *n*
f fixation *f* inférieure
r нижнее сжатие *n* (дерева, зависшего одним концом)

943 **bottom dog**
d Grundklammer *m*
f griffe *f* inférieure; griffe de serrage
r нижний зажим *m* для брёвен

* **bottom end → 607**

* **bottomland hardwood → 5193**

* **bottomland [riparian] forest → 5192**

944 **bottom leaf; radical leaf**
d gründliches Blatt *n*
f feuille *f* de pied
r прикорневой лист *m*

* **bottom loader → 1559**

* **bottom log → 1225**

945 **bottom sill; mudsill**
(the bed piece or bottom timber of a dam which is placed across the stream, usually resting on rocks or in the mud)
d Grundbrett *n*; Grundschwelle *f*; Grundsims *m*
f membrure *f* d'extrémité inférieure
r основное поперечное бревно *n* (сплавной плотины)

* **bottom timber landing → 4484**

* **bottom white pine → 7956**

946 **bottom wood**
d Grundholz *n*
f bois *m* de fond
r нижний ряд *m* брёвен в запани

947 **bougainvillea**
(Bougainvillea)
d Drillingsblume *f*
f bougainvillée *f*; bougainvillier *m*
r бугенвиллия *f*

* **bough → 4723**

* **boulder → 7128**

948 **boule**
(a log live edge sawn as kept together in the order of sawing)
d Blockware *fpl*
f plot *m*; boule *f*
r буль *f*

* **boultonizing → 5876**

* **Boulton process → 5876**

949 **boundary marker**
d Sichtzeichen *n* der Grenze
f balise *f* de délimitation
r ограничительный знак *m*

* **boundary planting → 921**

* **boundary tree → 641**

950 **bound water; bound moisture; combined moisture; combined water; absorbed moisture; hygroscopic moisture; imbibed moisture**
(the moisture that is intimately associated with the finer wood elements of the cell wall by adsorption and held with sufficient force to reduce the vapo[u]r pressure)
d verbindenes Wasser *n*
f eau *f* liée
r связанная влага *f*

* **bourtree → 2892**

* **bout width → 8320**

951 **bow; camber**
(a deviation from a straight line (a curve along the face of the piece of lumber) from end to end of a piece, measured at the point of greatest deviation)
d Bügel *m*; Bogen *m*
f voilement *m* longitudinal
r продольная покоробленность *f*

* **bow-saw → 3342**

952 **bow-type power-saw; power bow-saw**
d Motorbügelsäge *f*
f tronçonneuse *f* à arc
r моторная пила *f* с дугообразной шиной; лучковая моторная [цепная] пила

953 **box beam**
(a built-up beam with solid wood flanges and plywood or woodbase panel product webs)
d kubischer Tragebalken *m*
f poutre *f* en caisson
r балка *f* коробчатого сечения

954 **boxboards; caseboards; box lumber; case timber; case wood; box wood; box timber; packing-case timber; packing-case wood**
d Stabbretter *npl*; Kistenholz *n*; Emballageholz *n*; Pack[gut]holz *n*
f bois *m* de caisserie; planches *fpl* de caisse
r ящичная древесина *f*; тарная древесина; тарный лесоматериал *m*; тарный кряж *m*

955 **box cut**
(initial cut (as in coal mining) driven in a property where no open side exists; this results in a highwall on both sides of the cut)
d Box-Schnitt *m*
f entrée *f* en terre
r призматический прямой вруб *m*

956 **boxed heart; boxed pith**
(a piece of lumber in which the heart or pith is enclosed within the four sides of the piece)
d verpackter Kern *m*; eingepackter Kern
f cœur *m* inclus
r замкнутая сердцевина *f*

957 **boxed[-off]; slatting**
d gelattet
f garni de planches
r обшитый досками; дощатый

* **boxed pith → 956**

958 **boxed shake; end shake; edge shake; closed check; closed split; tight split; end check**
(a split on the end of a board)
d innerer Riss *m*; geschlossener Riss; Kantenriss *m*; Hirnholzriss *m*
f fente *f* en bout; fente refermée; fente d'arête de bois débité; fissure *f* du bord; fente non bâillante
r трещина *f*, не выходящая на пласть пиломатериала; внутренняя трещина; торцовая трещина; замкнутая трещина

959 **boxed tree; bled tree; tapped tree**
d Lachtenbaum *m*; geharzter Baum *m*
f arbre *m* flache; arbre *m* gemmé
r подсачиваемое дерево *n*; заподсоченное дерево

960 **box family**
(Buxaceae)
d Buchsbaumgewächse *pl*
f buxacées *pl*
r самшитовые *pl*

961 **box forest**
d Buchbaumwald *m*
f buxaie *f*
r самшитовый лес *m*

* **boxing → 828**

* **box lumber → 954**

962 **box piling**
d Box-Anhäufen *n*
f empilage *m* arasé-carré
r штабелёвка *f* с выравниванием торцов и расположением прокладок заподлицо с торцами

963 **box planting**
d Box-Pflanzung *f*
f plantation *f* de plantes en caissettes
r посев *m* в контейнеров для выращивания саженцев

964 **box pruning; lateral pruning; lateral cutting; side cutting**
(lateral root pruning on four sides of nursery stock in situ)
d Nebenwurzelaufasten *n*
f élagage *m* latéral [des racines]; élagage des racines latérales
r обрезка *f* боковых корней

965 **box sand myrtle**
(Leiophyllum buxifolium)
d Sandmyrte *f*; Zwerggehölz *n*
f myrte *m* de sables
r мирт *m* песчаный

* **boxthorn → 9471**

* **box timber → 954**

* **box tree → 2886**

* **box wedge → 7809**

* **box wood → 954, 2886**

966 **bracatinga**
(Mimosa scabrella bentham)
d Bracatinga *f*
f bracatinga *m*; scabrella *m* de mimosa

* **brace root → 277**

* **brachyscelreid[e] → 8121**

967 **bracken fern**
(Pteridium aquilinum)
d Adlerfarn *m*
f grande fougère *f*; fougère aigle; fougère impériale, porte-aigle *f*; fougère commune
r орляк *m* [обыкновенный]; папоротник-орляк *m*

* **bracket → 1879**

968 **bract; scale [leaf]; scale-like leaf**
(a much-reduced leaf in a flower cluster or associated with flowers; morphologically a foliar organ)
d Hochblatt *n*; Afterblatt *n*
f bractée *f*; écaille *f*
r прицветник *m*; чешуйчатый лист *m*; почечная чешуя *f*

969 **bracted balsam fir; balsam [fir]; balsam of Gilead tree; balsam of Mecca tree; Mecca myrrh tree; Gilead fir; eastern fir; var fir**
(Abies balsamea; Cammiphora opobalsamum)
d Balsamtanne *f*; Mekkabalsambaum *m*; echter Balsambaum *m*
f baumier *m* de Gilead; baumier de la Mecque; sapin *m* baumier
r пихта *f* бальзамическая; коммифора *f* бальзамная

* **bracted twinberry → 812**

* **bracteole → 6585**

* **bractlet → 6585**

* **brail → 6692**

* **brake → 1202**

* **brake block → 2263**

970 **brake rigging**
d Rückhaltetakelage *f*
f timonerie *f* de frein
r тормозной такелаж *m*

971 **brake rope**
d Rückhalteseil *n*
f corde *f* du frein
r тормозной канат *m*; тормозной трос *m*

972 **bramble**
(Rubus nessensis)
d halbaufrechte Brombeere *f*
f mûre *f* sauvage
r ежевика *f* несская

973 **bramble acacia; bramble wattle**
(Acacia victoriae)
d australischer Wüstenbaum *m*
f acacia *m* de mûre
r акация *f* терновая

* **bramble wattle → 973**

974 **branch; limb**
(a secondary woody stem growing off of the trunk or main stem of a woody plant)
d Ast *m*; Zweig *m*
f branche *f*
r ветвь *f*

975 **branch chips**
d Astspäne *mpl*
f copeaux *mpl* de branches
r щепа *f* от ветви

* **branch cluster → 1668**

* **branch coppice method → 6430**

976 **branch cutting**
d Zweigsteckling *m*
f marcotte *f* de branche
r черенок *m* от ветви

* **branch cutting → 6610**

977 **branched; branchy; limby; cladous; ramate; ramulose; ramose; ramous; ramified**
d verzweigt
f branché; ramifié; rameux
r ветвистый; суковатый; разветвлённый

978 **branched cutting strips**
d verzweigte Schlagstreifene *mpl*
f bandes *fpl* de coupe fourchues
r чередующиеся полосы *fpl* леса, разрабатываемые по обе стороны несущего каната

979 **branched knot**
d verzweigter Ast *m*; Doppelflügeast *m*
f nœud *m* groupé; nœud fourchu
r разветвлённый сучок *m*; вильчатый сучок

f racine *f* fourchue
r ветвистый корень *m*

981 brancher; knot-bumper; knotter; limber
(one who cuts and sorts the limbs from a felled trees)
d Astfänger *m*; Entaster *m*
f ébrancheur *m*
r обрубщик *m* сучьев; сучкоруб *m*; рабочий *m* на доочистке деревьев от сучьев; сучколовитель *m*; сучкоотделитель *m*

982 branch fork
d Astgabel *f*
f fourche *f* de branche
r развилина *f* сучка

* **branch formation → 4731**

983 branch-free; branchless; limbless
(of timber free from knots)
d astfrei; astlos; astrein
f sans branches; sans nœuds; ébranché; net de branches
r без ветвей; бессучковый

984 branch gathering
d Astsammeln *n*
f collection *f* de branches
r сбор *m* сучьев

985 branch hardening
d Astdürre *f*; Zweigdürre *f*
f durcissement *m* de branches
r усыхание *n* ветвей

986 branchiness; knottiness
(wood defect)
d Ästigkeit *f*; Ästungswürdigkeit *f*; Knotigkeit *f*; Verknotung *f*
f densité *f* des branches
r сучковатость *f*; суковатость *f*

987 branching; crotching; furcation; ramification
d Gabelung *f*; Abzweigung *f*; Verästelung *f*; Beastung *f*; Verzweigung *f*
f branchement *m*; embranchement *m*; ramification *f*
r [раз]ветвление *n*; ответвление *n*

988 branching trees
d Astkronenbäume *mpl*; Wipfelbäume *mpl*
f arbres *mpl* ramifiés
r ветвящиеся деревья *npl*

989 branchlessness
d Astreinheit *f*
f netteté *f* de nœuds
r бессучковатость *f*

990 branchless wood
d astfreies Holz *n*
f bois *m* sans branches
r бессучковые пиломатериалы *mpl*; бессучковая древесина *f*

991 branchlet; small branch; little branch; ramule
(a small branch or twig, generally referring to the most recent year's growth)
d Ästchen *n*; kleiner Zweig *m*
f rameau *m*; ramuscule *f*; ramille *f*
r веточка *f*; маленькая ветвь *f*

* **branch litter → 5000**

* **branch roots → 4824**

* **branch rot → 4734**

992 branch scar
d Astnarbe *f*
f cicatrice *f* de branche
r рубец *m* от сучка

* **branch shaker → 4971**

* **branch shape → 4735**

993 branch snag; branch stub; nog; knag
(broken or cut base of a branch projecting from a tree stem)
d Aststumpf *m*; Aststummel *m*
f chicot *m* de branche; moignon *m* de branche
r торчок *m* от сучка; пэнь-сломыш *m*

994 branch stem
d Seitenstamm *m*
f tige *f* latérale
r боковой ствол *m*

* **branch stub → 993**

* **branch timber → 5170**

995 branch whorl; twig burr; verticil of branches; rose
(a ring or circle of plant organs arising close together, for example, leaves from a node, or a spiral of conifer branches from a main stem)

f verticille *m*; broussin *m* de têtard
r виток *m*; завиток *m*; мутовка *f* [сучьев]; наплыв *m* на ветви; роза *f*; глазки *mpl*

* **branchwood → 5170**

* **branchy → 977**

996 **branchy crown**
d astreiche Krone *f*
f cime *f* branchue
r развесистая крона *f*

* **branding → 5317**

* **branding axe → 5314**

997 **branding equipment**
d Anzeichnengeräte *npl*
f équipement *m* de marquage
r маркировочное оборудование *n*

998 **brash; felling debris; [felling] waste; logging debris; [logging] slash[ings]; logging residues; cuttings; felling residues; bavin wood; felling refuse; chatwood**
(tree tops, branches and other woody material left on a site after logging)
d Holzabraum *m*; Afterschlag *m*; Nutzungsentgang *m*; Reiserholz *n*; Schlagabfall *m*; Schlagabraum *m*; Zweigholz *n*
f débris *m* de coupe; déchets *mpl* de coupe; résidus *mpl* de coupe
r лесосечные отходы *mpl*; обрезки *fpl*; хворост *m*; порубочные остатки *mpl*; хлам *m* от лесосеки

999 **brashing [up]; brushing through; brushing up**
(management operation in thicket-stage plantations in which dead and poorly growing lower branches are removed to a height about 2 m)
d erste Aufastung *f*; Durchreiserung *f*
f élagage *m* à hauteur d'homme
r обрезка *f* боковых ветвей и вершин деревьев

1000 **brashness**
d Unüberlegheit *f*
f fragilité *f*
r ломкость *f*; хрупкость *f*

1001 **brashness of wood**
(a condition that causes some pieces of wood abruptly without splintering at comparatively small deflections)
d Brüchigkeit *f* des Holzes; Holzsprödigkeit *f*
f fragilité *f* de bois
r хрупкость *f* древесины; ломкость *f* древесины

* **Brayley flindersia → 6665**

* **Brazilian araucaria → 1005**

* **Brazilian bluewood → 870**

1002 **Brazilian ironwood; tesota; Arizona ironwood; leopard-tree**
(Caesalpinia ferrea)
d Leopardenstrauch *m*; Leopardenbaum *m*
f pau ferro *m*
r цезальпиния *f* железная; бразильское железное дерево *n*

1003 **Brazil[ian] nut-tree; para-nut; juvia; yubia; Brazilnoot; castanheiro; almendro**
(Bertholletia excelsa)
d Paranussbaum *m*; Amazonenmandel *f*; Brasilnussbaum *m*; Jurianuss *f*
f noyer *m* de Para; noyer *m* du Brésil; châtagnier *m* du Brésil
r бертолетия *f* высокая; токота *f*; орех *m* бразильский

1004 **Brazilian pepper; Brazil peppertree**
(Schinus terebinthifolius)
d brasilianischer Pfefferbaum *m*
f baie rose *f* de Bourbon; faux poivre *m*; poivre de Bourbon; poivre rouge
r шинус *m* терепентиннолистный

1005 **Brazilian pine; Brazilian araucaria; Parana pine; curiy**
(Araucaria angustifolia; Araucaria brasiliana)
d Brasilkiefer *f*; brasilianische Araukarie *f*; Paranakiefer *f*
f sapin *m* du Brésil; pin *m* de Parana; araucaria *m* brésilien
r араукария *f* бразильская; сосна *f* Парана

1006 **Brazilian rosewood; Rio rosewood; Amazon rosewood; palissandre tree; bahia rosewood; jacaranda**
(Jacaranda brasiliana; Dalbergia nigra; Aniba parviflora)
d Brasil-Rosenholz *n*; Amazone-Rosenholz *n*; Rio-Palisander *m*; Palisanderbaum *m*; Palisanderholz *n*; Palisander *m*
f bois *m* de rose du Brésil; palissandre *m* de

r бразильское розовое дерево *n*; палисандровое дерево; палисандр *m*; дальбергия *f* чёрная; жакаринда *f* [бразильская]

1007 **Brazilian tulipwood; pinkwood; jacaranda rosa**
(Dalbergia frutescens; Dalbergia variabilis)
d [Bahia-]Rosenholz *n*; echtes Rosenholz *n*
f bois *m* de rose [véritable du Brésil]
r тюльпанное дерево *n* бразильское; тюлипвуд *m*; бразильский палисандр *m*

1008 **Brazilian walnut; [canella] imbuia; ambuia; embuya; amarela**
(Phoebe porosa; Ocotea porosa)
d Imbuia *n*
f imbuia *m*; embuia *m*
r бразильский орех *m*; имбуйя *f*

* **Brazilnoot → 1003**

* **Brazil peppertree → 1004**

1009 **Brazil trithrinax palm**
(Trithrinax brasiliensis)
d brasilianische Nadelpalme *f*; brasilianische Doppeldreizackpalme *f*
f palmier *m* du Brézil
r тритринакс *m* бразильский

1010 **Brazil wood; Pernambuco wood; pernambouc**
(Guilandia echinata; Caesalpinia echinata)
d echtes Brasilholz *n*; Pernambukholz *n*
f bois *m* de Brésil; bois-brésil *m*; pau-brasil *m*; pernambouc *m*
r бразильский лес *m*

* **breach of forest laws → 1909**

1011 **breadfruit[-tree]; artocarpe; jack[-fruit] tree; keledang; lacuch**
(Artocarpus integrifolia; Artocarpus communis; Artocarpus altilis)
d echter Brotfruchtbaum *m*; Brotbaum *m*; Keledang *m*; Jackholz *n*; Lakooch *m*; [indischer] Brotfruchtbaum *m*; Jackfruchtbaum *m*
f arbre *m* à pain; artocarpe *m*; artocarpus *m*; keledang *m*; bois *m* jack; lakuch *m*; jacquier *m*
r хлебное дерево *n* [цельнолистное]; хлебное дерево индийское

1012 **breakage of crown; crown break; crown**
f rupture *f* de la cime; cassure *f* de la cime; bris *m* de la cime
r повреждение *n* кроны; поломка *f* кроны; поломка вершины

* **break axe → 453**

1013 **breakdown of tree top; top break[age]**
d Wipfelbruch *m*; Gipfelbruch *m*
f rupture *f* d'écimage
r поломка *f* вершин

1014 **breakdown of wood tissue; wood breakdown**
d Holzabbau *m*; Holzbruch *m*; Holzzerstörung *f*; Zerfallen *n* des Holzes
f décomposition *f* de tissu ligneux
r распад *m* древесной ткани; разрушение *n* древесины

1015 **break-even-point**
d Rentabilitätsschwelle *f*; Freisetzungspunkt *m*
f seuil *m* de rentabilité; point *m* mort
r точка *f* критического объёма производства

1016 **break in a landing; rolling in a landing**
(rolling logs from the landing into a stream)
d Hinabrollen *n*
f enroulement *m*
r скатывать бревна в воду; скатка *f*; скатывание *n*

1017 **breaking-down of logs into boards; resawing operation**
d Bretterschnitt *m*; Vorschneiden *n*
f abattage *m* de grumes à planches
r распиловка *f* брёвен на пиломатериалы; распиловка брёвен на доски

1018 **break[ing-]down saw; headrig; head[ ]saw; scrag**
(the principal machine in a sawmill, used for the initial breakdown of logs by sawing along the grain)
d Hauptsäge *f*
f scie *f* de tête; scie de premier débit
r головной станок *m*; брусующий станок; брусовочный станок; станок для разделки брёвен на брусья; головная лесопильная рама *f*

* **breaking bar → 3067**

1019 **breaking length**
d Reißlänge *f*

1020 breaking of buds; budbreak[ing]; budding; sprouting; flushing
d Sprossen *n*; Ausschlagen *n*
f débourrage *m*; débourrement *m*
r распускание *n* почек; развитие *n* почек

* **break-off allowance → 9051**

* **breakthrough → 3443**

1021 breast; front portion of mo[u]ldboard
d Rumpf *m*
f versoir *m*; estomac *m*
r сердцевинная поверхность *f* доски; широкая поверхность доски

1022 breast height
(the standard height 1.3 m above ground level, at which the diameter of a standing tree is measured)
d Brusthöhe *f*
f hauteur *f* de poitrine
r высота *f* груди

1023 breast height age
(number of annual growth rings between the bark and the pith, as counted at breast height)
d Brusthöhenalter *n*
f âge *m* à hauteur de poitrine
r возраст *f* на высоте груди

* **breast height diameter → 2466**

1024 breast height girth; girth breast height
d Umfang *m* in Brusthöhe
f circonférence *f* à hauteur de poitrine
r окружность *f* на высоте груди

1025 breast work log; brow log; fender[skid]; sheer skid; glancer
(large log laid beside the track or road at a log dump or landing to prevent logs from swinging or kicking back)
d Prellholz *n*; Anschlag *m*; Abweisrechen *n*; Fender *m*
f rambarde *f* de traînage; longeron *m* de bordage
r отбойное бревно *n* [на нижней части лесоспускового волока]; бревно, укладываемое поперёк трелёвочного волока; крайнее бревно; отбойный брус *m* на спусковом волоке

* **breathing root → 86**

1026 breeder
r селекционер *m*

1027 breeding line
d Züchtungslinie *f*
f ligne *f* d'amélioration génétique; ligne de reproduction
r линия *f* скрещивания

1028 breeding of forest trees; forest tree breeding; tree breeding; forest [tree] improvement; tree improvement
(genetic study of trees to solve some specific problem or to produce a specially desired product)
d Waldbaumzüchtung *f*; Forstpflanzenzüchtung *f*; Waldbaumveredelung *f*
f amélioration *f* [génétique] des arbres forestiers
r селекция *f* и разведение *n* лесных растений; селекция древесных пород; улучшение *n* древесных пород

1029 brevifoliate
d kurzblättrig
f à feuilles courtes
r коротколистный

* **Brewer's spruce → 9330**

* **briar → 2042, 4206**

1030 bridge grafting
d Brückepropfung *f*
f greffage *m* en pont
r прививка *f* мостиком

* **bridge on rafts → 6686**

1031 bridging blue; pile discolo[u]ration
d Stapelbläue *f*
f bleuissement *m* sur pile
r синева *f* складочного леса

* **brier → 2555**

1032 brier wood; curled wood; burred timber; burled wood
d Brennreiserknüppel *mpl*; Aderholz *n*; Wirbelholz *n*
f bois *m* ronceux; bois en rondins
r древесина *f* с капами и наплывами; каповая древесина

1033 bright deals
d klare Dielen *fpl*
f madrier *m* clair sans défaut; plots *mpl* nets de

1034 **bright sapwood**
(sap-wood of natural colo[u]r or in which the stain or discolo[u]ration can be removed by surfacing to standard thickness)
d klarer Splint *m*
f aubier *m* clair (sans défauts)
r заболонь *f* натурального цвета

1035 **bright timber; unstained wood**
(wood that is free from discolo[u]rations)
d klares Nutzholz *n*
f bois *m* clair sans défauts; bois net de coloration naturelle
r пиломатериалы *mpl* натурального цвета

* **brindled beauty moth → 6147**

* **bringing cost → 7684**

1036 **Brisbane box; brush box; vinegar tree**
(Lophostemon confertus; Tristania conferta)
d Essigbaum *m*
f tristanie *f* à feuilles en bouquets
r тристания *f* свёрнутая; тристания скрученная

1037 **bristlecone fir; Santa Lucia fir**
(Abies bracteata)
d kalifornische Grannen-Tanne *f*; Santa-Lucia-Tanne *f*
f sapin *m* à bractées
r пихта *f* прелестная

* **bristlecone pine → 6997**

1038 **bristly locust; rose acacia; moss locust**
(Robinia hispida)
d Borsten-Scheinakazie *f*; Borstenrobinie *f*; rote Akazie *f*
f acacia *m* à fleurs mauves; robinier *m* hispide; robinier rose
r акация *f* мохнатая

1039 **brittleheart**
(a defect in hardwoods resulting from the presence of fibers with localized wrinkles (slip planes) that cause reduction in strength of the wood)
d zerbrechlicher Kern *m*; brüchiger Kern
f cœur *m* cassant
r хрупкая сердцевина *f*

1040 **brittle willow; crack-willow**
(Salix fragilis)
d Knackweide *f*; Bruchweide *f*; Glasweide *f*

1041 **broad axe; broad bill; squaring axe; [hollow] adze; side axe**
(axe with one face bevel[l]ed and the other flat, for hewing timber)
d Breitbeil *n*
f hache *f* demi-plate; hache à équarrir; hache d'élagage
r топор *m* с широким лезвием; плотничный топор; топор для затёски лесоматериалов

1042 **broadcast afforestation**
d Großaufforstung *f*
f afforestation *f* en plein; afforestation à la volée
r лесовосстановление *n* на больших площадей

* **broadcast application → 5990**

1043 **broadcast burn[ing]; clean burn[ing]; clear burning**
(controlled fire used as a silvicultural treatment to burn a designated area within well-defined boundaries for the purpose of reducing fuel hazards)
d Sauberbrennen *n*
f brûlage *m* extensif; brûlage général
r сплошной пал *m*; сжигание *n* начисто

* **broadcasting → 5990**

1044 **broadcast seeding; broadcast sowing; full seeding**
(sowing of seeds more or less evenly over a whole area on which a forest stand is to be raised)
d breitwürfige Saat *f*; Breitsaat *f*; Vollsaat *f*; breitverstreutes Säen *n*
f ensemencement *m* à la volée; semis *m* à volée; semis en plein; semis massif
r разбросной [по]сев *m*; посев вразброс; сплошной посев

1045 **broad-crowned; large-crowned; flat-topped**
d breitkronig
f à cime large
r ширококронный

1046 **broad drill sowing**
d Breitfurchensaat *f*
f ensemencement *m* en rangée large
r широкострочный посев *m*; широкорядный посев

f bois *m* d'arbres feuillus
r лиственный пиломатериал *m*; пиломатериал лиственных пород

* **broadleaf mahogany → 1412**

1048 **broadleaf tree; broadleaved tree; foliage tree; frondose tree**
(tree other than conifers that have (usually but not always) flat, broad leaves)
d breitblätt[e]riger Baum *m*
f arbre *m* feuillu; feuillu *m*
r широколиственное дерево *n*; лиственное дерево

* **broadleaf white birch → 384**

* **broadleaf wood → 4124**

* **broadleaved → 4800**

1049 **broadleaved euonymus; broadleaved spindle**
(Euonymus latifolius)
d breitblättriges Pfaffenhuthen *n*; Voralpenspindelbaum *m*
f fusain *m* à larges feuilles
r бересклет *m* широколистный

* **broadleaved forest → 2300**

1050 **broadleaved grasses [in oak forests]**
d breitblättrige Gräser *npl* [in Eichenwäldern]
f herbes *fpl* latifoliées [de chênaies]
r дубравное широкотравье *n*

* **broadleaved lime → 4801**

* **broadleaved maple → 701**

1051 **broadleaved ornamental tree**
d Laubzierbaum *m*
f arbre d'ornement feuillu
r лиственное декоративное дерево *n*

1052 **broadleaved peppermint; blue peppermint**
(Eucalyptus dives)
d Laubpfefferminze *f*
f menthe *f* poivrée décidue
r эвкалипт *m* широколистный; эвкалипт тимоловый; эвкалипт душистый

1053 **broadleaved roundwood**
d Laubrundholz *n*
f bois *m* rond feuillu
r круглая лиственная древесина *f*

broadleaved trees
d Laubnutzholz *n*
f bois *m* scié feuillu; bois d'œuvre feuillu
r пиломатериалы *mpl* лиственных пород

1055 **broadleaved species**
d Laubgehölz *n*; Laubholzart *f*
f espèce *f* décidue
r широколиственная порода *f*

* **broadleaved spindle → 1049**

1056 **broadleaved tea-tree; punktree; melaleuca; cajeput [tree]**
(Melaleuca linariifolia; Melaleuca quinquenervia; Melaleuca leucadendron)
d Teeölbaum *m*; Kajeputbaum *m*; Cajeputbaum *m*
f aceite *f* de Cayeput; cajeputier *m*; arbre *m* à goménol
r мелалеука *f* белодревесная

* **broadleaved tree → 1048**

* **broadleaved wood → 4124**

1057 **broadleaved woodland**
d Laubholzbodenfläche *f*
f zone *f* de forêt feuillue
r широколиственная лесная зона *f*

* **broadleaved wood of high value → 9160**

1058 **broad management class; broad management type**
(classification of timberland based on forest type and stand origin)
d extensive Bewirtschaftungsklasse *f*
f classe *f* d'exploitation extensive
r разряд *m* сплошного лесоустройства; класс *m* сплошного лесоустройства

* **broad-ringed wood → 5909**

1059 **broken chunks**
d angeschlagene Knollen *mpl*
f fragments *mpl* de bois cassés
r обломки *mpl* древесины

1060 **broken forest**
d lockerer Wald *m*; räumdiger Wald; aufgelockerter Wald; lichter Wald
f forêt *f* dégagée
r прорежённый лес *m*; разрёженный лес

f pousse *f* apicale cassée
r сломанный главный побег *m*

1062 **broken roe; broken line**
(of makore wood)
d gebrochene Textur *f*
f texture *f* en ligne cassée
r текстурный рисунок *m* в виде ломаной линии; косуля *f*; икра *f*; молоки *pl*

1063 **brominated wood**
d bromiertes Holz *n*
f bois *m* bromé
r бромированная древесина *f*

1064 **bronchial tree**
d kahler Baum *m*
f arbre *m* nu
r бронхиальное дерево *n*

1065 **bronze birch borer**
(Agrilus anxius)
d Birkensplintkäfer *m*
f agrile *m* bronzé du bouleau
r златка *f* узкотелая берёзовая бронзовая

1066 **bronzing**
(golden brown discolo[u]ration that usually appears on the lower surface of leaves and is often an advanced stage of the silvering or glazing)
d Bronzieren *n*
f bronzage *m*
r бронзирование *n*

1067 **brood bark**
d Brutrinde *f*
f écorce *f* de ponte
r кора *f*, повреждённая насекомыми

* **brood gallery → 4811**

1068 **brood tree; bug tree**
(any tree whose wood and/or bark contains large breeding populations of insects)
d Brutbaum *m*
f arbre *m* foyer; arbre de ponte
r свежозаселённое дерево *n* (насекомыми); зараженное насекомыми дерево; дерево, повреждённое насекомыми

1069 **broom**
(Cytisus)
d Bohnenbaum *m*; Geißklee *f*; Traubengeißklee *f*
f cytisaie *f*
r заросль *f* ракитника

1070 **broom bush**
d Geißklee-Gebüsch *m*
f cytisaie *f*
r заросль *f* ракитника

1071 **broomed end**
d Spaltenende *n*
f bout *m* éclaté
r размочаленный конец *m*; расщеплённый конец

* **broom hickory → 6229**

* **brooming → 627**

1072 **brooming losses**
(of log ends)
d Spaltenende-Verluste *mpl*
f pertes *fpl* par éclatement
r потери *fpl* от расщепления

1073 **broom teatree**
(Leptospermum scoparium)
d Südseemyrte *f*
f leptosperme *m* éclaté
r лептоспермум *m* метловидный

1074 **broom top**
d Besenspitze *f*
f couronne *f* en forme de balai; tête *f* éclatée
r метлообразная корона *f*

* **brow log → 1025**

1075 **brown barrel; cuttail [eucalyptus]**
(Eucalyptus fastigata)
d brauner Flascheneukalyptus *m*
f eucalyptus *m* fastigié
r равновысокий эвкалипт *m*

* **brown core → 1080**

* **brown creeper → 1807**

1076 **brown cubical rot of conifers**
(invader - Lentinus lepideus)
d kubische Braunfäule *f* der Koniferen
f carie *f* brune cubique de conifères
r бурая кубическая гниль *f* хвойных пород

1077 **brown cubical rot of timbers**
(invader - Coniophora cerebella, Coniophora puteana)
d kubische Braunfäule *f* des Holzes
f carie *f* brune cubique de bois

* **brown earth → 1079**

* **brown ebony → 4607**

1078 **brown forest podzolized soil**
d podsolierter braune Waldboden *m*
f sol *m* brun forestier podzolisé
r бурая лесная оподзоленная почва *f*

1079 **brown [forest] soil; brown earth**
(intrazonal and calcimorphic; formed on calcium-rich parent materials under deciduous forest)
d brauner Waldboden *m*
f terre *f* brune; sol *m* brun forestier; sol tempéré
r бурая [лесная] почва *f*; бурозем *m*

1080 **brown heart; brown core**
d Herzbräune *f*; Markbräune *f*; Kernhausbräune *f*
f cœur *m* brun
r бурая сердцевина *f*

1081 **brown-mottled white rot**
(invader - Pholiota adiposa)
d braunmarmorierte Weißfäule *f*
f pourriture *f* blanche marbrée
r корневая центральная гниль *f* ствола (возбудитель - гриб жирная чешуйчатка)

1082 **brown oak**
d Brauneiche *f*
f chêne *m* brun
r древесина *f* дуба с характерной коричневой окраской сердцевины

* **brown oxidation stain → 1462**

* **brown pine → 6390**

1083 **brown pocket sap rot of conifers**
(invader - Trametes americana)
d Braunsaftfäule *f* der Koniferen
f pourriture *f* brune de l'aubier par taches de conifères
r бурая ямчатая заболонная гниль *f* хвойных пород

1084 **brown rot**
(a type of wood-destroying fungus that decomposes cellulose and the associated pentosans, leaving the lignin in a more or less unaltered state)
d Braunfäule *f*; Braunfäulnis *f*
f pourriture *f* brune; moniliose *f*

1085 **brown rot of birch**
(invader - Polyporus betulinus, Piptoporous betulinus)
d Birkenbraunfäule *f*
f pourriture *f* de bouleau brune
r смешанная желтовато-бурая гниль *f* на берёзе

1086 **brown sap rot**
(invader - Trametes heteromorpha)
d Braunsaftfäule *f*
f pourriture *f* de l'aubier brune
r бурая заболонная гниль *f*

1087 **Brown's lovoa; African walnut; mansonia**
(Lovoa brownii; Lovoa klaineana; Lovoa trichilioides)
d Dibetou *m*
f dibétou *m*; noyer *m* du Gabon; noyer d'Afrique
r ловоа *f* Брауна; инкобу *m*; грецкий орех *m* из Уганды; мансония *f*

1088 **brown spire rot**
(invader - Polyporus fibrillosus)
d Braunspitzenfäule *f*
f pourriture *f* pointue brune
r бурая гниль *f* отмершей древесины

1089 **brown stain of beech; tylose formation in beech; incipient decay in beech**
d Buchenstocken *n*
f souche *f* de hêtre
r бурая гниль *f* бука

1090 **brown sterculia; aye**
(Sterculia rhinopetala)
d Lotofa *m*
f lotofa *m*
r стеркулия *f* остролепестковая

* **brown streak → 1549**

1091 **brown stringybark**
(Eucalyptus capitellata; Eucalyptus baxteri)
d brauner Stringybark *m*
f stringybark *m* brun
r эвкалипт *m* Бакстера

1092 **brown top rot**
d Spitzenbräune *f*
f pourriture *f* apicale brune
r бурая гниль *f* вершины [дерева]

1093 **brown tree pipit; tree pipit**

f pipit *m* des arbres
r лесной конек *m*

1094 **browse** *v*; **graze** *v*
d abrasen; abhüten; abweiden
f brouter; pâturer; abroutir
r обрезать побеги; удалять боковые побеги

1095 **browse; chat; slip; spray**
(buds, shoots and leaves of woody plants that serves as food for wildlife)
d Ablenker *m*; Ausschlag *m*; Lade *f*; Reis *n*; Schoss *m*
f brout *m*; stolon *m* court
r [молодой] побег *m*; росток *m*; отводок *m*

1096 **browse line; browsing level; high line; grazing line**
(uppermost limit on trees and tall shrubs to which livestock and big game browse)
d Äsungshöhe *f*
f ligne *f* d'abroutissement
r граница *f* объедания листьев и молодых побегов

* **browsing → 8806**

* **browst → 5170**

* **bruchy → 934**

1097 **bruised tree; skinned tree**
d verletzter Baum *m*; zerkleinerter Baum
f arbre *m* brisé; arbre blessé
r дерево *n*, повреждённое при трелёвке; ободранное при трелёвке дерево

* **brush → 7555**

1098 **brush barrier**
(a sediment control structure created of slash materials piled at the toe slope of a road or at the outlets of culverts, turnouts, dips, and water bars)
d Buschbarriere *f*
f barrière *f* de broussailles
r барьер *m* из кустарника

1099 **brush blade**
d Buschschneide *f*
f lame *f* débroussailleuse; lame à débroussailler
r [бульдозерное] полотно *n* для расчистки территории от зарослей кустарника

* **brush box → 1036**

* **brush-breaker plow → 1203**

* **brush buster → 7319**

1100 **brush chopper; bush hog; bush crusher**
(implement with blades mounted on a horizontal power-driven shaft, for reducing the brushes)
d Bushzerkleinerer *m*
f casseur *m* débroussailleur; coupeuse *f* à rémanents
r измельчитель *m* кустарника (тяжёлый зубчатый каток для расчистки территории от зарослей кустарника)

1101 **brush control; shrub control; scrub control**
(reduction of brush (volume, height, density, and biomass) to reduce its competition for space, moisture, light, and nutrients with plant species)
d Bürstenregelung *f*
f contrôle *m* de la croissance des arbustes
r контроль *m* кустарника

1102 **brush crew**
d Buschräumungsrotte *f*
f équipage *m* de débroussaillement; équipe *f* de débroussaillement
r бригада *f* на расчистке территории от кустарника

* **brush cutter → 7319**

* **brush cutting → 1109**

* **brush dam → 1117**

1103 **brush deflector**
(on a tractor)
d Buschablenker *m*; Buschableiter *m*
f déflecteur *m* de brousse
r устройство *n* для отклонения кустарника

* **brush digger → 1112**

* **brush disposal → 7723**

1104 **brusher**
(a person)
d Brusher *m*
f coupeur *m* de voie
r рабочий *m*, занятый на расчистке территории или дорожных трасс

1105 b[r]ush fire
(burning shrubs, brush and scrub growth, involving heavier fuel than grass)
d Buschfeuer *n*
f feu *m* de broussailles; feu de brousse
r пожар *m* кустарника

* **brush flail → 7319**

* **brush hook → 1204**

* **brushing → 1109**

1106 brushing of dry branches
d Trockenastung *f*
f ébranchage *m* de branches sèches
r обрезка *f* сухих сучьев

* **brushing through → 999**

* **brushing up → 999**

1107 brush killer
d Gestrüppvernichtungsmittel *n*
f débroussaillant *m*
r гербицид *m* для уничтожения кустарника и дёрна

1108 brushland; dumetum; brush field; bushland
(an open stand of bushes, 3–7 m high, with a canopy cover > 40%)
d Buschfläche *f*
f surface *f* couverte de buissons; brousse *f*
r территория *f*, поросшая кустарником; площадь *f* кустарника

* **brushland-breaking plough → 1203**

* **brushland harrow → 1203**

* **brushland plow → 1203**

1109 brush management; scrub clearance; brush cutting; scrub cutting; bush clearing; shrub clearing; brushing; slash thinning; slashing
(managing and manipulating stands of brush by mechanical, chemical, biological means, or by prescribed burning)
d Abbuschung *f*; Buschräumung *f*; Bürstenschnitt *m*; Abraümen *n*
f débroussaillement *m*; coupe *f* des buissons
r очистка *f* от кустарника; [вы]рубка *f* кустарника; удаление *n* кустарника; уничтожение *n* кустарника; вырубка конкурирующей растительности

(covering of branches spread on eroded land to conserve water and lessen erosion that helps establishes trees and other permanent vegetation)
d Buschmatte *f*
f paillassonnage *m* en branches
r сплетение *n* из сучьев

1111 brush *v* out
(to clear away brush from a trail, survey line, or road)
d abraümen; ausbürsten
f débroussailler
r обрезать живых и сухих ветвей; расчищать территорию от кустарника

* **brush pile → 4193**

1112 brush rake; brush digger; bush puller
(blade with teeth at the bottom, attached to a cat or skidder, used in mechanical site preparation)
d Bürstenrührstange *f*
f râteau *m* débroussailleur; débroussailleur-dessoucheur *m*; extirpateur *m*
r корчеватель *m* кустарника

1113 brush scythe; bush scythe
d Buschsense *f*
f faucille *f* forestière
r кустарниковый серп *m*

1114 brush snow fence
d Busch-Schneeschutzbepflanzung *f*; Busch-Schneeschutzhecke *f*
f haie *f* paraneige de brousse; barrière *f* pareneige de brousse; taillis *m* paraneige de brousse
r снегозащитная полоса *f* из кустарника

1115 brush stage
d Buschstadium *n*
f stade *m* de buissons
r стадия *f* кустарника

* **brushwood → 1202**

1116 brushwood; scrub remains
(the wood portions of a tree excluding the stem and roots; the wood from bushes or small branches)
d Buschholz *n*
f broussailles *fpl*; brindilles *fpl*
r древесина *f* кустарника

1117 brushwood dam; brush dam; fascine weir; fascine dike; causeway of fascines; mattress

f digue *f* fascinée; barrage *m* de correction léger mixte; barrage de brindilles; barrage de fascines
r хворостяная плотина *f*; фашинная плотина

1118 **brushwood in bundles; bundled brushwood; bundled brush check; bundle of twigs; bush of copse; bush bundle; bavin**
d Reis[ig]bündel *n*; Reisbund *n*; Reiserknüppel *m*; Schockholz *n*
f menu bois *m* de feu pour bourrées; fagot *m* de brindilles; fagot de menues branches
r пучок *m* хвороста; связка *f* хвороста; вязанка *f* хвороста

1119 **brushwood road**
d Astholzbahn *f*
f chemin *m* à broussailles
r хворостяная гать *f*; дорога *f*, выстланная хворостом

1120 **bubinga; binbinga; essingang**
(Guibourtia demeuse; Guibourtia pellegriniana; Guibourtia tessmannii)
d Bubinga *f*
f bubinga *m*
r бубинга *f*

1121 **bubinga didelotia; gombe; angok**
(Didelotia africana)
d Gombe *m*
f ekop *m* gombé
r диделотия *f* африканская

1122 **buck *v*; slash *v*; log *v***
(to cut trees into shorter lengths, such as logs or cordwood)
d abstücken; ausformen; Holz aushalten; abschalten
f tronçonner; scier un rondin
r раскряжёвывать; распиливать; разделывать

* **buckberry → 2327**

* **buckbrush ceanothus → 5660**

* **bucked wood → 1918**

1123 **bucked wood system**
d Schnittholzsystem *n*
f système *m* de débusquage
r система *f* заготовок сортиментов

1124 **bucker**
(a worker who bucks trees)
d Ablänger *m*

* **bucker → 2043**

1125 **bucker's break**
d Ablängersbruch *m*
f rupture *f* de tronçonneur
r отделение *n* горбильной части брёвен

* **bucket → 6382**

* **buck-eye → 1781**

1126 **buck-eye family; horse-chestnut family**
(Hippocastanaceae)
d Rosskastaniengewächse *pl*
f hippocastanacées *pl*
r жёлудниковые *pl*; конскокаштановые *pl*

* **bucking → 5100**

* **bucking at the diameter break → 2476**

1127 **bucking at the grade break**
d Sträuben *n* am Gradbruch
f tronçonnage *m* par la classe de qualité
r раскряжёвка *f* по предельным качественным признакам

1128 **bucking chute**
d Gerinne *f* für Querschneiden
f glissoire *f* de tronçonnage
r короткий деревянный жёлоб *m* для раскряжёвки брёвен

1129 **bucking for value**
d Sträuben *n* im Wert
f tronçonnage *m* par valeur maximale
r раскряжёвка *f* с максимальным выходом древесины по стоимости

1130 **bucking for volume**
d Sträuben *n* für Ausgabe
f tronçonnage *m* par volume maximal
r раскряжёвка *f* с максимальным выходом древесины по объёму

1131 **bucking ladder; wood ladder**
(a series of short skids laid parallel at regular intervals, used to support logs while they are being bucked)
d Holzleiter *m*
f échelle *f* de tronçonnage
r подкладки *fpl* при раскряжёвке

* **bucking mark → 8972**

1132 **bucking of log lengths**

d Querschneiden *n* des Langholzes
f tronçonnage *m* de bois long
r раскряжёвка *f* долготья; разделка *f* долготья

* **bucking point → 8972**

1133 **bucking table**
d Aushaltungstabelle *f*; Einschnitt-Tabelle *f*
f table *f* de tronçonnage
r раскряжёвочная таблица *f*; таблица для определения раскроя

1134 **bucking wedge**
d Einschnittkeil *m*
f coin *m* de tronçonnage
r клин *m* для раскряжёвки

* **buckle → 1182**

1135 **buckle guy**
(cable attached to the middle of a spar tree or steel tower to prevent its bending under strain)
d anschnallendes Halteseil *n*
f hauban *m*
r растяжка *f* для предотвращения изгиба мачты

1136 **bucksaw; billet saw; docking saw; sash saw; shortening saw; whip saw; bulh saw**
d Ablängsäge *f*; Auslängsäge *f*; Bocksäge *f*; Einschneidesäge *f*; Schlitzsäge *f*; Zapfensäge *f*
f scie *f* à bûches; scie à billettes; ébouteuse *f*; scie en long
r раскряжёвочная пила *f*

1137 **buckthorn**
(Rhamnus; Frangula)
d Kreuzdorn *m*; Faulbaum *m*
f nerprun *m*
r крушина *f*; жостер *m*; жёстер *m*

1138 **buckthorn family**
(Rhamnaceae)
d Kreuzdorngewächse *pl*
f rhamnacées *pl*
r крушиновые *pl*

1139 **buckwheat tree**
(Cliftonia monophylla)
d Buchweizenbaum *m*
f arbre *m* sarrasin
r клифтония *f* однолистная

* **bud → 419, 7944**

* **budding → 307, 1020**

1140 **budding knife**
d Okuliermesser *m*
f écussonnoir *m*
r окулировочный нож *m*

1141 **bud [eye]; eye; gem; leaf-knot; burgeon**
(dormant and unelongated stem composed of a short axis of meristem cells from which embryonic leaves, lateral buds, flower parts or all three arise)
d Knospe *f*
f bourgeon *m*
r почка *f*; глазок *m*

* **bud grafting → 307**

1142 **bud pruning; debudding; disbudding**
(removal of lateral buds from a stem to prevent them from developing into branches)
d Knospenästung *f*; Entknospungsästung *f*
f ébourgeonnage *m*; ébourgeonnement *m*
r удаление *n* почек

1143 **bud-scale**
(a modified leaf, without lamina, protecting a bud)
d Knospenschluppe *f*
f écaille *f* de bourgeon
r почечная чешуйка *f*; чешуйка почки

1144 **bud-scale scar**
d Knospenschluppe-Wunde *f*
f cicatrice *f* d'écaille de bourgeon
r рубец *m* от почечной чешуйки

1145 **budwood; graft; cleft-graft scion; scion**
(aerial plant part, often a branchlet, that is grafted onto another root-bearing plant)
d Pfropfen *n*; Edelreis *n*; Pfropfreis *n*; Impfreis *n*
f bois *m* de greffe; greffon *m*; greffe *f*; scion *m*
r черенок *m* для прививки; прививочный черенок; привой *m*; прививок *m*; привитая часть *f*

1146 **budworm**
(Choristoneura, Hedia etc.)
d Knospenwickler *m*
f tordeuse *f* des bourgeons
r листовёртка-почкоед *f*

* **buffalo bush → 771**

1147 **buffalo currant**

d Gold-Johannisbeere *f*
f gadelier *m* odorant; groseillier *m* des buffles
r смородина *f* душистая; смородина золотистая

* **buffer area → 1149**

1148 **buffer[ing] action; buffer[ing] effect**
d Pufferwirkung *f*
f effet *m* amortissant
r буферный эффект *m*

1149 **buffer strip; buffer zone; filter strip; green strip; buffer area; buffer surround; isolation strip**
(a band of forest maintained along a stream, lake, road, recreation site or different vegetative zone)
d Isolierstreifen *n*; Pufferzone *f*; Pufferstreifen *n*
f zone *f* tampon; bande *f* tampon; bande de protection; bande d'isolement; rideau *m* vert; rideau d'arbres
r буферная полоса *f*

* **buggy → 894**

1150 **bug-killed wood**
d vom Käfer getötetes Holz *n*
f bois *m* mort par attaque de coléoptères
r древесина *f*, отмершая от жуков

* **bug tree → 1068**

1151 **building [glued] plywood**
d Bau-Sperrholz *n*
f contreplaqué *m* de construction
r строительная [клеёная] фанера *f*

* **building in wood → 8570**

1152 **building pole**
d Baupfahl *m*; Baustange *f*
f perche *f* de construction
r строительная жердь *f*

1153 **building wood; building timber; construction timber; construction wood; structural timber; structural lumber; timber[-]wood**
(which has strength in relation to the anticipated structural end use, as a controlling factor in grading or selecting)
d Baunutzholz *n*; Bauholz *n*; Gerüst *n*; Rüstholz *n*; Zimmerholz *n*
f bois de construction; bois de structure

1154 **Bulgarian fir; King Boris fir**
(Abies borisii-regis)
d King-Boris-Tanne *f*
f sapin *m* de Bulgarie
r пихта *f* болгарская

* **bulh saw → 1136**

1155 **bulk density**
(measure of weight per unit of volume of a material; generally serves as an indicator of the specific gravity of wood)
d Massendichte *f*
f densité *f* apparente
r насыпная плотность *f*

1156 **bulk forest products**
d Massenforstprodukte *npl*
f produits *mpl* forestiers de masse
r основные лесоматериалы *mpl*

1157 **bulk stacking; close stacking; solid stacking; tight stacking; bulk piling; close piling; solid piling**
d Feststapelung *f*
f empilage *m* plein
r плотная штабелёвка *f*

* **bullbay → 7860**

* **bullberry → 7600**

1158 **bull block; main-line block; lead block**
(a large yarding block having a throat opening of sufficient size to allow the butt rigging to pass through)
d Haupt[leitungs]block *m*
f poulie *f* guide-support de câble tracteur
r головной блок *m* тягового каната

1159 **bull chain**
(a chain wrapped around the first log of a turn in order to check the speed)
d Bremsungskette *f*
f chaîne *f* frein
r замкнутая цепь *f* лесотранспортёра

1160 **bull donkey; grubbing winch; monkey winch; road engine; roader**
(a donkey engine mounted on a heavy sled, which is used for long-distance hauling either on the ground or on a skid road)
d Rodewinde *f*; Stumpfzieher *m*
f treuil *m* d'essouchage; treuil d'arrachage; treuil à dessoucher; treuil à déboisement

1161 bulldozer blade; KG blade
(bulldozer-mounted blade used in forestry and land-clearing operations; single spike splits and shears stumps at their base)
d Frontschild *m*
f lame *f* de bulldozer
r лопасть *f* бульдозера

1162 bulldozer piling
d Aufschichten *n*
f empilement *m* par bulldozer
r сбор *m* [порубочных остатков] бульдозером в кучи

1163 bullet planting
(setting out young trees grown in bullet-shaped rigid plastic tubes, which are injected into the ground by a spring-loaded gun, sometimes into prepared holes)
d Kugel-Pflanzung *f*
f plantation *f* en cartouche
r посадка *f* в пулеобразных трубках

* **bull hook → 1557**

* **Bull pine → 6445**

1164 bull rope; [shackle] guy rope; span rope; span wire; shackle guy cable; guy[ ]line; guy [wire]; tension cable; stiffening wire; slackpuller
(a heavy rope used to control large heavy pieces of wood)
d Abspannseil *n*; Spannseil *n*; verankertes Kabel *n*; Spanndraht *m*; Spanner *m*
f étai *m*; cordeau *m*; corde *f* de tension; câble *m* ancré; fil *m* tendeur; tendeur *m*
r растяжка *f*; оттяжка *f*; канат *m* для растяжек; натяжной канат; натяжная проволока *f*; закреплённый кабель *m*

1165 bull stick
(steel bar for punching holes under stumps for placement of dynamite or powder; tool used in splicing wire rope)
d Stierstock *m*
f bâton *m*
r палка *f*

1166 bullthorn wattle; bullthorn acacia
(Acacia cornigera; Acacia campecheana)
d Büffelhornakazie *f*; horntragende Akazie *f*
f cornigera *m*; acacia *m* épineux
r акация *f* колючая

* **bull wood → 1850**

* **bully tree → 485**

* **bummer → 894**

1167 bump
(on a pulpwood)
d Beule *f*
f broussin *m*
r неровность *f*; пузырь *m* (на поверхности фанеры)

* **bumpboat → 911**

* **bumpy log → 4721**

1168 bumpy texture; humpy texture; irregular texture
d eingebeulte Textur *f*; unregelmäßige Textur
f texture *f* bosselée; texture irrégulière
r неровная текстура *f* (в результате зарубок на растущем дереве)

* **bumpy wood → 2048**

* **bunch → 1666**

1169 bunch *v*; **bundle** *v*; **hill** *v*; **ridge** *v*
(to gather trees or logs into small piles for subsequent skidding by other equipment)
d anhäufeln; [be]häufeln; zusammenrücken
f paqueter; fagoter
r пакетировать; окучивать; вязать; укладывать в небольшие штабели

1170 bunchberry [dogwood]
(Cornus canadensis)
d Teppichhartriegel *m*
f cornouiller *m* du Canada
r дёрен *m* канадский

* **bunch [de]limber → 5323**

1171 bunching; ridging; hilling; tumping
(of timber)
d Paketierung *f*; Behäufelung *f*
f paquetage *m*; buttage *m*
r сбор *m* в кучи; окучивание *n*; пакетирование *n*

1172 bunching crew; bunch team
d Paketierungsbrigade *f*
f équipe *f* de paquetage
r бригада *f* на предварительном формировании пачек лесоматериалов (для трелёвки); звено *n* для окучивания брёвен

* **bunch planting → 6354**

* **bunch team → 1172**

1173 **bunch teamster**
d Floßbinder *m*
f collecteur *m* de paquets
r сборщик *m* брёвен; сплотщик *m*

* **bundle → 1169**

1174 **bundle bucking**
(cutting bundles or truck loads of tree-length wood or long wood into short lengths)
d Bündeleinschnitt *m*
f tronçonnage *m* par groupes
r групповая раскряжёвка *f*; раскряжёвка лесоматериалов в пакетах

* **bundled brush check → 1118**

* **bundled brushwood → 1118**

1175 **bundle limbing**
d Bündelentastung *f*
f ébranchage *m* par groupes
r групповая очистка *f* деревьев от сучьев; групповая обрезка *f* сучьев

1176 **bundle of bark; bark bundle**
d Rindenbündel *n*; Rindengebund *m*
f fagot *m* d'écorce
r пучок *m* корья; вязанка *f* коры

* **bundle of twigs → 1118**

* **bundle rigging → 7467**

1177 **bundle scars**
(tiny, somewhat circular dots within the leaf scar, caused by the breaking of the fibrovascular bundles which run through the petioles into the blades of the leaves)
d Bündelnarbe *f*
f cicatrice *f* de paquet
r рубец *m* от пучка сосудов

* **bundle wood → 9572**

1178 **bundling**
d Bündelung *f*
f javelotage *m*
r увязка *f*; увязывание *n* в пачку

* **bunk → 2038**

1179 **bunkhouse**
(the sleeping quarters of a logging crew)
d Kojehaus *n*
f colis *m*
r временный передвижной домик *m* (для лесозаготовительной бригады)

* **bunk jaw skidder → 1597**

1180 **bunk logs; bunk[ ]load; bunk tier; wing logs**
(the bottom layer of logs that rest on the bunk)
d Schemelholzschicht *f*
f charge *f* traversière; charge de fardage
r брёвна *npl*, непосредственно лежащие на балке коника; нижний ряд *m* брёвен; первый ряд погруженных на коник брёвен

* **bunya-bunya → 5566**

1181 **buoyant log**
d schwimmender Holzschnitt *m*
f grume *f* de flottage
r бревно *n*, входящее в состав бона

* **buprestid beetle → 8063**

* **burgeon → 1141**

* **burkea → 6791**

1182 **burl; woodknob; buckle; scar tissue; knob; callus; knar[l]; knog; knurl; gnarl; boom**
(a defect of wood; a large woody excrescence on trunk, branch or root stock)
d Anschwellung *f*; Aufschwellung *f*; Auswuchs *m*; Holzkropf *m*; Maserkropf *m*; Kallus *m*; Knast *m*; Locke *f*; Maserknolle *f*; Wulst *m/f*; Knorren *m*; Maserkegel *m*
f barbille *f*; bourrelet *m*; épaississement *m* callitrisoïde; callus *m*; bride *f*; bosse *f*; loupe *f*
r наплыв *m*; каллус *m*; свиль *f*; утолщение *n* на стволе дерева; вздутье *n*

* **burled wood → 1032**

* **Burma laurel → 9719**

1183 **Burma padauk**
(Pterocarpus macrocarpus)
d Burma-Padauk *m*; obstteller Padauk *m*
f arbre *m* à Kiro; padauk *m* de Burma; amboine *f*
r сандал *m* крупноплодный; бурма-падук *m*

1184 burn[ed] area; burn surface; burned-out surface; place of conflagration
(an area over which fire has recently run)
d Brandfläche *f*; Brandplatz *m*; verbrannte Fläche *f*
f surface *f* incendiée; brûlis *m*
r площадь *f*, пройденная сплошным палом; площадь пожара; гарь *f*; пожарище *n*

1185 burned chips
d Brandspäne *mpl*
f copeaux *mpl* incendiés
r обугленные частицы *fpl* щепы; подгоревшие частицы щепы

1186 burned firebreak
d ausgebrannter Schutzstreifen *m*
f coupe-feu *m* incendié
r противопожарная полоса *f*, подготовленная отжигом

1187 burned forest
d abgebrannter Wald *m*; ausgebrannter Wald
f forêt *f* incendiée
r выгоревший лес *m*; сгоревший лес

* **burned-out surface → 1184**

1188 burning block
d Brandblock *m*
f terrain *m* cerné de lignes pare-feu
r участок *m* для пуски пала; пожарный блок *m*

1189 burning edge; fire edge; firefront; fireline; head of fire
(portion of the fire upon which resources are deployed and actively engaged in suppression action)
d Feuerfront *f*; Front *f* eines Feuers; Feuerlinie *f*; Feuergrenze *f*; Feuerrand *m*
f lisière *f* d'un incendie; bordure *f* d'un incendie; front *m* d'un incendie; tête *f* d'un incendie
r кромка *f* пожара; линия *f* огня; фронт *m* пожара

1190 burning-in tool
(for trees)
d Einbrennenswerkzeug *n*
f instrument *m* de brûlage
r инструмент *m* для выжигания

1191 burning-off; burning-out; firing
d Brennen *n*
f chauffage *m*; allumage *m*
r пуск *m* пала

d Brandperiode *f*
f période *f* de feu; période favorable à l'incendie
r период *m* сжигания (порубочных остатков); пожарный период

1193 burning permit
(permit required for authorizing open burning)
d Brandgenehmung *f*
f permis *m* de feu
r билет *m* на пуск пала; разрешение *n* на пуск пала

1194 burning power; calorific power; heating power; fuel value; caloric value; calorific value; heating value
d Heizkraft *f*; Heizwert *m*
f valeur *f* calorifique
r калорийность *f*; показатель *m* теплотворной способности

1195 burning prescription
(written direction stipulating fire environment conditions, techniques, and administrative constraints necessary to achieve specified resource management objectives by using fire on a given area of land)
d Brandanordnung *f*
f ordre *m* de brûlage
r предписание *n* сжигания

* **burn scar → 3206**

* **burn surface → 1184**

1196 burnt wood; burnt timber; fire-killed wood; fire-killed timber
d Brandholz *n*
f bois *m* incendié
r горельник *m*; поражённая огнём древесина *f*

1197 burramura turpentine myrtle
(Syncarpia glomulifera; Syncarpia laurifolia)
d lorbeerblättrige Synkarpie *f*
f syncarpe *m* à feuilles de laurier
r синкарпия *f* почконосная

* **burred timber → 1032**

* **burr[ed] wood → 2048**

* **burr oak → 5566**

1198 bursera family; torchwood family
(Burseraceae)

f burséracées *pl*
r бурсеры *pl*

* **burst check → 2139**

* **bush and bog plow → 1203**

* **bush breaker → 7319**

* **bush bundle → 1118**

1199 **bush chink[u]apin; trailing chink[u]apin; [Alleghany] chink[u]apin**
(Castanea alnifolia; Castanea pumila)
d Zwergkastanie *f*; kleine Kastanie *f*
f châtaignier *m* de Virginie; châtaignier nain; chincapin *m*
r каштан *m* карликовый; карликовое каштановое дерево *n*; каштан низкорослый

* **bush clearing → 1109**

* **bush crusher → 1100**

* **bush cutter → 7319**

1200 **busheling**
(method of payment for performing piecework other than volume of timber cut, such as a car loading)
d Scheffel-Bezahlung *f*
f rémunération *f* par boisseaux
r оплата *f* по сдельным расценкам

* **busher → 8307**

1201 **bush fallow**
(an area of secondary woody growth in tropical or subtropical forests, developed between periods of field cropping)
d Waldbrache *f*
f jachère *f* forestière
r кустарниковый пар *m*

1202 **bush forest; shrub forest;scrub forest; brushwood; bushwood; stunted wood; brake; frutex**
(a dense growth of bushes)
d Ausbuschreiserholz *n*; Buschwald *m*; Strauchwald *m*; Gestrüpp *n*
f bois *m* rabougri par broussailles; forêt *f* de broussailles
r лес *m* с обилием кустарников; кустарниковый лес; [низкорослая древесно-]кустарниковая заросль *f*; кустарниковое редколесье *n*; ксерофитная

1203 **bush harrow; brushland harrow; brushland plow; brush-breaker plow; bush and bog plow; brush-breaker [plough]; brushland-breaking plough**
d Bushpflug *m*
f charrue *f* débroussailleuse
r тяжёлая кустарниковая борона *f*; кустарниковый плуг *m*

* **bush-honeysuckle → 5788**

* **bush hog → 1100**

1204 **bush knife; b[r]ush hook; rill hook; bill[hook]; machette; cutlass; panga; parang; draw hook**
d Haumesser *n*; Sense *f*; Hippe *f*; Machette *f*; Buschmesser *m*
f machette *f*; serpe *f*
r кривой нож *m* для вырубки кустарника; клювовидный нож; кусторезный нож; косарь *m*; кустарниковый нож; косяк *m*

* **bushland → 1108**

* **bushman → 9057**

* **bushman's poison → 1366**

1205 **bush nursery**
d Baumhschule *f* für Büsche; Buschbaumschule *f*
f pépinière *f* de broussailles
r рассадник *m* кустарников

1206 **bush[ ]nut; macadamia**
(Macadamia ternifolia)
d Macadamianuss *f*
f noyer *m* du Queensland; noisetier *m* d'Australie; macadamia *m*
r макадамия *f* тройчатолистная; австралийский орех *m*; киндель *m*

* **bush of copse → 1118**

* **bush palmetto → 2688**

* **bush puller → 1112**

1207 **bush red pepper**
(Capsicum frutescens)
d Cayenne-Pfeffer *m*
f piment *m* de Cayenne
r перец *m* ветвистый

* **bush scythe → 1113**

1208 **bushveld; scrub veld; veld[t]**
(African grassland that is usually nearly level, is often intermixed with scattered shrubs or trees, and is chiefly located in Eastern and Southern Africa)
d Buschveld *n*; Buschland *n*; Grasland *n*
f veld *m*
r [кустарниковый] вельд *m*

* **bushwood → 1202**

* **bushy → 934**

1209 **bushy growth; tufted growth**
d Büschelwuchs *m*
f croissance *f* en touffes
r пучкообразный рост *m*

1210 **bushy top**
(of a tree)
d Büschelspitze *f*
f sommet *m* en touffes
r пучкообразная верхушка *f*

* **bushy tree → 9472**

1211 **butcher's block**
d Metzgersblock *m*; Fleischerblock *m*
f billot *m* de boucher
r кряж *m* для рубки или колки лесоматериала

1212 **butcher's broom**
(Ruscus aculeatus)
d stechender Mäusedorn *m*
f fragon *m* [épineux]; petit houx *m*
r иглица *f* [колючая]

1213 **butcher's broom family**
(Ruscaceae)
d Traubendorngewächse *pl*
f ruscacées *pl*
r иглицевые *pl*

* **Buthan pine → 4308**

* **butt block → 1225**

* **butt carriage → 5199**

1214 **butt cut; undercut[ting]; undercut notch; cutting-down; belly[-cut]**
(deep notch cut into the base of a tree to govern the direction in which it is to fall and also to prevent the butt from splitting)
d Haukerbe *f*; Anhieb *m*
r подпил *m*; подруб *m*; подрубка *f*; подсечка *f*; комлевый рез *m*

1215 **butt cutting; butt-off; docking; butt trimming**
(squaring the end of a log or other piece of converted timber)
d Kantenkappen *n*; Abkürzen *n*; Aufstellen *n* des Holzes
f rognage *m*; finition *f*; tronçonnage *m* en dessous
r раскряжёвка *f* лесоматериала снизу; торцовка *f*; торцевание *n*; расторцовка *f*; выравнивание *n* торцов

1216 **butt damage; butt shatter**
d Erdklotzeinreißen *n*
f dommage *m* de bille de pied
r повреждение *n* комля; расщепление *n* комля; скол *m*

* **butt decay → 1233**

* **butt diameter → 2465**

1217 **butt dozing; butt digging-in**
d Erdklotzvergraben *n*
f enterrement *m* de billes de pied
r зарывание *n* комлей в почву (при волочении лесоматериалов)

* **butt [end] → 607**

1218 **buttercup family**
(Ranunculaceae)
d Hahnenfußgewächse *pl*
f renonculacées *pl*
r лютиковые *pl*

1219 **butterfly bush**
(Buddleia davidii)
d Schmetterlingsstrauch *m*; Sommerflieder *m*
f arbre *m* aux papillons; buddleia *m* du père David; lilas *m* d'été
r буддлея *f* Давида; летняя сирень *f*

1220 **butterfly family**
(Buddlejaceae)
d Schmitterlingsstrauchgewächse *pl*
f buddléiacées *pl*
r буддлеевые *pl*

1221 **butterfly tree**
(Bauhinia purpurea)
d purpurfarbener Orchideenbaum *m*; purpurfarbene Bauhinie *f*

1222 **butternut; white walnut**
(Juglans cinerea)
d Butternuss *f*; grauer Walnussbaum *m*; Graunuss *f*; Butternussbaum *m*; Ölnussbaum *m*; Ölnuss *f*
f noyer *m* à beurre; noyer gris [d'Amérique]; noyer cendré
r орех *m* серый [калифорнийский]

1223 **butter tree**
(Pentadesma butyracea Sabine)
d Butterbaum *m*
f arbre *m* à beurre
r масляное дерево *n*

* **butt hook → 1557**

1224 **butting; cutting-back**
(operation which removes only a portion of the end of a log)
d Zurückschneiden *n*
f ravalement *m*; rabattage *m*
r укорачивание *n* (длинных брёвен); откомлёвка *f*; затёска *f* конца брёвна

* **butt line → 1555**

1225 **butt log; butt length; butt block; bottom log**
(the first log above the stump)
d Baumferse *f*; Erdblock *m*; Erd[stamm]stück *m*; Erdklotz *m*
f bille *f* de pied; bille de souche
r комлевой чурак *m*; комлевое бревно *n*; толстое короткое бревно; пнёвой кряж *m*

* **butt-log portion of tree → 607**

* **butt [off] → 4666**

* **butt-off → 1215**

1226 **butt-off [block]; cull-butt; trimmed-off butt**
d Wurzelklotz *m*
f bille *f* de racine
r корневой кряж *m*

* **butt of trunk → 607**

* **buttonball → 254**

1227 **button tree; honeysuckle**
(Lonicera)
d Aalbaum *m*; Heckenkirsche *f*; Geißblatt *n*

* **buttonwood → 254**

1228 **butt pan; log bumper; butt plate; dutchman**
(a short stick placed transversely between the outer logs of a load to divert the load toward the middle and so keep any logs from falling off)
d Stoßplatte *f*; Schnittholzdämpfer *m*
f bouclier *m* [arrière]; plaque *f* d'arrêt; plaque de butée; surlongueur *f* de jonction
r прокладка *f* для предохранения брёвна; прокладка для предотвращения скатывания наружных брёвен; трелёвочный щит *m*

1229 **buttress; counterfort**
(a wall or abutment built along a stream to prevent the logs in a drive from cutting the bank or jamming)
d Strebepfeiler *m*
f barrage *m* à contreforts
r стенка *f* или береговой устой *m*, препятствующие обсыханию брёвен или образованию заторов

* **buttress → 1234**

1230 **buttressed; undulated; fluted; with eccentric trunk** *adj*
d spannrückig
f cannelé
r извилистый

1231 **buttress[ed] root**
d Brettwurzel *f*
f racine *f* cannelée
r досковидный корень *m*

* **buttressing of trunk → 8999**

1232 **butt rigging**
(system of swivels and clevises that connects the haul-back and main lines to which chokers are fastened)
d Lasthakenbefestigung *f*
f chariot-palan *m* de câble-grue
r прицепная обойма *f*

1233 **butt rot; butt decay**
(decay or rot characteristically confined to the base or lower bole of a tree)
d Stockfäule *f*; Stammendfäule *f*
f pourriture *f* du rejet
r напенная гниль *f*; комлевая гниль

* **butt shatter → 1216**

* **butt shear → 8921**

1234 **butt swell[ing]; buttress; thickening of the butt; corona**
(transition zone between stem and root at the ground line of a tree or seedling)
d Stammanlauf *m*
f empattement *m*
r закомелистость *f*

1235 **butt swelling felling**
d Fällen *n* der Stammanläufe
f abattage *m* d'empattement
r срубание *n* прикорневых наплывов

* **butt trimming → 1215**

* **butt veneer → 8219**

* **buzz saw → 1585**

* **byproducts → 9500**

C

1236 **caatinga**
(a forest composed of stunted trees and thorny bushes, found in areas of small rainfall in Brazil)
d Caatinga *m*
f caatinga *m*
r листопадный тропический лес *m*

1237 **cabbage gum**
(Eucalyptus amplifolia; Eucalyptus platyphylla)
d Kohl-Eukalyptus *m*
f gomme *f* de chou
r эвкалипт *m* с гладкой корой

1238 **cabbage palm; cabbage palmetto; cabbage tree; common palmetto**
(Sabal palmetto)
d Kohlpalme *f*; Andira *f*; Palmettopalme *f*; Sabalpalme *f*
f sabal *m*; palmetto *m*; shou *m* palmiste
r съедобная пальма *f*; пальма капустная американская; капустное дерево *n*; пальметта-сабаль *f*; сабаль *m* пальмовидный

* **cabbage palmetto → 1238**

1239 **cabbage palm terminal bud or heart**
d Endtrieb *m* der Kohlpalme
f bourgeon *m* terminal de palmier
r почка *f* финиковой пальмы

* **cabbage tree → 1238**

1240 **cabelluda**
(Eugenia tomentosa; Plinia glomerata; Myrciaria glomerata; Eugenia cabelluda)
d Cabelluda *f*
f cabelluda *m*
r евгения *f* опушённая

* **cabinet cherry → 244**

* **cabinet timber → 3776**

* **cabinet[-]wood → 3776**

1241 **cable** *v*
r трелевать лес канатными установками; трелевать лебёдкой; подтаскивать лесоматериалы с помощью каната

* **cable clip → 7699**

* **cable-crane → 7697**

* **cable-crane extraction → 1247**

1242 **cable logging machine; cable hauler; cable yarding system; cable installation**
d Seilbahnschlepper *m*
f téléphérique *m* forestier
r канатная транспортная система *f*; канатная трелёвочная установка *f*

1243 **cable profile unloaded; empty rope sag; empty rope profile**
d Leerseillinie *f*
f ligne *f* de câble de transport à vide
r линия *f* порожнего троса

1244 **cable profile when loaded; loaded rope profile**
d Vollseillinie *f*
f profil *m* en long du câble porteur
r полная тросовая линия *f*

1245 **cable road; gutter road; cableway; run[a]way; skyroad; ropeway [track]; skid[ding] trail**
(bladed or backhoe-constructed pathway where stumps are removed within the running surface)
d Anrückschneise *f*; Drahtseilbahn *f*; Kabelbahn *f*; Seilbahn *f*; Maschinenweg *m*; Drahtseilriese *f*; Skidder *m*
f voie *f* de câble; ligne *f* de débardage
r канатная дорога *f*; просека *f* для канатов; линия *f* несущего каната; линия трелёвки

* **cable [rope] → 9463**

1246 **cable skidder; winch-and-choker [equipped] skidder**
d Kabelschleifrückenschlepper *m*
f débardeur *m* à câble; débusqueur *m* à câble; tracteur *m* débusqueur
r трелёвочный трактор *m* с тросово-чокерным оборудованием; трактор с лебёдкой

* **cable transport → 1247**

1247 cable[way] logging; cable-crane extraction; cable transport; cable yarding; ropeway extraction; flying
(taking logs from the stump area to a landing using an overhead system of winch-driven cables to which logs are attached with chokers)
d Drahtseilrücken *n*; Holzabfuhr *m* mit Seilbahn; Rücken *n* mit Seilzug; Seilrücken *n*
f débardage *m* par téléphérage; débardage au câble; débardage par câble
r лесозаготовки *fpl* на базе канатных установок; канатная трелёвка *f*; тросовая трелёвка; вывозка *f* с тросом

* **cableway skidder → 88**

* **cableway skidding → 7700**

* **cable winch → 7060**

* **cable yarding system → 1242**

* **cabling → 271, 1428**

* **cabriuva → 504**

1248 cacao [tree]; cocoa [tree]; chocolate tree
(Theobroma cacao)
d Kakaobaum *m*
f cacaoyer *m*; cacaotier *m*
r шоколадное дерево *n*; какао *n*

* **cactus pear → 6533**

1249 cadaster; land register
d Flurbuch *n*
f cadastre *m*; livre *m* foncier
r книга *f* угодий

* **cajeput [tree] → 1056**

* **calabash → 1764**

1250 Calabrian pine; rect-cone Aleppo pine
(Pinus [halepensis] brutia; Pinus brutius)
d italienische Kiefer *f*
f pin *m* de Pyrénées; pin calabrais; pin de Brutie
r сосна *f* калабрийская

1251 calcicolous plant; calcicole; calciphyte
(a plant that lives in soils rich in calcium)
d Kalziphit *m*
f plante *f* calcicole

* **calcifuge [plant] → 1254**

1252 calcimorphic brown earth
d kalzimorpher Braunerdewaldboden *m*
f sol *m* brun forestier calcimorphe
r карбонатная бурая [лесная] почва *f*

1253 calciphile [plant]; calciphilous plant
(a plant that tends to be restricted to calcareous soils)
d Kalkpflanze *f*
f plante *f* calciphile
r кальцефильное растение *n*; кальцефил *m*

* **calciphilous plant → 1253**

1254 calciphobe [plant]; calcifuge [plant]; acid-loving plant
(a plant preferring alkaline soils)
d kalzifuge Pflanze *f*; kalkfeindliche Pflanze
f plante *f* calcifuge
r кальцефоб *m*; кальцефобное растение *n*

* **calciphyte → 1251**

* **calculation of [tree] volume → 8955**

* **calico tree → 1264**

* **California buckthorn → 1257**

* **California foothill pine → 7142**

* **California gall fly → 1279**

* **California lilac → 856**

* **California live oak → 1693**

* **California[n] bay → 1264**

* **California[n] bayberry → 1277**

1255 California[n] black oak; Kellog oak
(Quercus kelloggii)
d kalifornische Schwarzeiche *f*; Kellogg-Eiche *f*
f chêne *m* noir de Californie; chêne de Kellogg
r дуб *m* калифорнийский чёрный

* **California[n] [black] walnut → 241**

1256 California[n] buckeye
(Aesculus californica)
d kalifornische Rosskastanie *f*

1257 **California[n] coffeeberry; California buckthorn**
(Frangula californica; Rhamnus californica)
d kalifornischer Kreuzdorn *m*
f nerprun *m* de Californie
r крушина *f* калифорнийская

* **Californian cypress → 3891**

1258 **California[n] [fan] palm; California[n] Washington palm**
(Washingtonia filifera)
d kalifornische Fächerpalme *f*
f palmier *m* washingtonia
r вашингтония *f* нитеносная

1259 **California[n] flannelbush; flannelbush**
(Fremontodendron californicum; Fremontia californicum)
d Flanellstrauch *m*
f frémontodendron *m* de Californie; frémontie *f* à grandes fleurs
r калифорнийский скользкий вяз *m*

1260 **California[n] flatheaded borer**
(Melanophila californica)
d kalifornischer Flachkopfsbohrer *m*
f foreur *m* à tête plate de Californie
r златка *f* калифорнийская

1261 **California[n] hoptree; western hoptree**
(Ptelea crenulata)
d kalifornischer Hopfenbaum *m*
f orme *m* de Californie
r птелея *f* калифорнийская

1262 **California[n] incense-cedar**
(Calocedrus decurrens; Libocedrus decurrens)
d [kalifornische] Flusszeder *f*; Weihrauchzeder *f*
f libocèdre *m* décurrent; cèdre *m* à encens; cèdre blanc [de Californie]
r кедр *m* речной сбежистый; кедр речной благовонный; кедр калифорнийский

1263 **California[n] juniper**
(Juniperus californica)
d kalifornischer Wacholder *m*
f genévrier *m* de Californie
r можжевельник *m* калифорнийский

1264 **California[n] laurel; California[n] bay; Oregon myrtle; mountain laurel; calico tree; spice tree**
(Umbellularia californica; Kalmia latifolia)
d Oregon-Myrte *f*; Myrthenmaser *m*; kalifornischer Berglorbeer *m*
f laurier *m* de Californie; laurier des montagnes d'Amérique du Nord
r калифорнийский лавр *m*; умбеллюлярия *f* калифорнийская; кальмия *f*

1265 **California[n] lilac**
(Ceanothus arboreus)
d Säckelblumebaum *m*
f céanothe *m* en arbre
r цеанотус *m* американский

* **California[n] live oak → 1319**

* **California[n] nutmeg → 1276**

1266 **California[n] pepper[-]tree; Peruvian peppertree; peppertree**
(Schinus molle)
d peruanischer Pfefferbaum *m*; peruanischer Mastixbaum *m*
f faux poivrier *m* [commun]; poivrier d'Amérique; arbre *m* à mastic du Pérou; mondré *m*
r перечное дерево *n* калифорнийское

1267 **California[n] redbood; western Juda's tree**
(Cercis occidentalis)
d kalifornischer Judasbaum *m*
f arbre *m* de Judée de Californie
r церцис *m* западный

1268 **California[n] red fir; golden fir; western balsam fir**
(Abies magnifica)
d kalifornische Rottanne *f*; Goldtanne *f*; Prachttanne *f*; prächtige Tanne *f*
f sapin *m* rouge californien; sapin magnifique
r пихта *f* великолепная; пихта красивая

1269 **California[n] redwood; coast[al] redwood**
(Sequoia sempervirens)
d Küstensequoie *f*; immergrüne Sequoie *f*; Riesen-Mammutbaum *m*
f séquoia *m* géant de Californie; séquoia à feuilles d'if; séquoia toujours vert; séquoia côtier
r секвойя *f* вечнозелёная; калифорнийское мамонтово дерево *n*

1270 **California[n] sagebrush**
(Artemisia californica)
d kalifornischer Beifuß *m*
f armoise *f* de Californie
r полынь *f* калифорнийская

1271 **California[n] scrub oak**

f chêne *m* broussailleux de Californie; chêne buisson de Californie
r дуб *m* карликовый калифорнийский

1272 **California[n] singleleaf pinyon [pine]**
(Pinus monophylla var. californica; Pinus californiarum)
d kalifornische einblättrige Steinkiefer *f*
f pin *m* à feuilles simples de Californie
r сосна *f* [однохвойная] калифорнийская

1273 **California[n] snowbell**
(Styrax californica)
d kalifornischer Storaxbaum *m*
f aliboufier *m* de Californie
r стиракс *m* калифорнийский

1274 **California[n] sugar pine; sugar pine; true white pine; gigantic pine; Lambert's pine**
(Pinus lambertiana)
d kalifornische Zuckerkiefer *f*
f pin *m* à sucre; pin gigantesque; pin géant; pin de Lambert
r сосна *f* Ламберта; сосна сахарная [гигантская]

1275 **California[n] sycamore**
(Platanus racemosa; Platanus californica)
d traubige Platane *f*
f platane *m* de Californie
r платан *m* кистистый

1276 **California[n] torreya; California[n] nutmeg**
(Torreya californica)
d kalifornische Nusseibe *f*; Torreye-Muskatnuss *f*
f muscadier *m* de Californie; torreya *m* de Californie
r торрея *f* калифорнийская

* **California[n] Washington palm → 1258**

1277 **California[n] waxmyrtle; California[n] bayberry; Pacific waxmyrtle**
(Myrica californica)
d kalifornische Wachsmyrte *f*
f cirier *m* de Californie
r восковница *f* калифорнийская

1278 **California[n] white fir; concolor fir; white balsam; [lowland] white fir; Colorado fir**
(Abies concolor; Abies lowiana)
d kalifornische Küstentanne *f*; kalifornische Riesentanne *f*; Coloradotanne *f*; Grautanne *f*; gleichfarbene Tanne *f*; Koloradotanne *f*

1279 **California oak gall wasp; California gall fly**
(Andricus californicus)
d Eichengallwespe *f*
f cynip *m* du chêne
r галл *m* дубовой орехотворки; чернильный орешек *m*

* **California scrub oak → 7559**

* **California white oak → 9159**

1280 **calks; caulks**
(short, hobnail-like spikes in the soles and heels of boots, designed to give secure footing while walking on logs)
d Eissporne *mpl*
f calfats *mpl*
r шипы *mpl*

1281 **callery pear**
(Pyrus calleryana)
d Zierbirne *f*
f poirier *m* chanticleer; chanticleer *m*; poirier à fleurs
r груша *f* каллериана

* **cal[l]iper → 8809**

* **cal[l]iper ga[u]ge → 3653**

1282 **cal[l]iper measure**
d Kluppierung *f*; Kluppen *n*
f mesurage *m* par compas forestier
r измерение *n* кубатуры мерной вилкой

1283 **cal[l]iper rule**
d Kluppschiene *f*
f coulisseau *m* du compas forestier
r пересчётная шина *f*

* **cal[l]iper stick → 715**

* **callous growth on the edge of a scar → 2157**

* **callousing → 5862**

* **callous margin → 2157**

1284 **calluna forest type**
d Heidekrautwald *m*
f type *m* de forêt cendrée
r тип *m* леса вересковый

* **callus → 1182**

* **calorific power → 1194**

* **calorific value → 1194**

* **caltrop family → 639**

* **camalindo → 8379**

* **camber → 951**

1285 **cambial initial**
(an individual cell in the cambium)
d Kambiumzelle *f*
f initiale du cambium
r клетка *f* камбия; камбиальный инициал *m*

1286 **cambial zone; cambila region**
(a term of convenience for the layer of varying width composed of cambial initials and their undifferentiated derivatives)
d Kambiumzone *f*
f zone *f* cambiale
r камбиальная зона *f*

* **cambila region → 1286**

1287 **cambium**
(growing (generative) layer between the xylem and phloem; a lateral meristem responsible for formation of xylem and phloem)
d Holzbildungsgewebe *n*; Kambium *n*; Wachstumsschicht *f*
f assise *f* génératrice interne; cambium *m*
r камбий *m*; камбиальный слой *m*

* **cambium-shear debarker → 7074**

1288 **camel-thorn tree**
(Acacia erioloba)
d Kameldornbaum *m*
f acacia *m* ériolobe; épine-giraffe *f*
r акация *f* верблюжья

* **camp → 5071**

1289 **campeachy wood; campeche wood**
d Blauholz *n*; Allerheiligenholz *n*; Campecheholz *n*; Kampescheholz *n*; Blutholz *n*
f bois *m* de campêche
r древесина *f* кампешевого дерева

* **campeche → 846**

1290 **camp foreman; side foreman; bully**
r мастер *m* на лесозаготовках

1291 **camphor tree**
(Cinnamomum camphora; Cinnamomum officinarum)
d [echter] Kampferbaum *m*; Kampferlorbeer *m*
f camphrier *m*; arbre *m* à camphre; laurier *m* du Japon
r камфарное дерево *n*; коричник *m* камфарный; камфарный лавр *m*

* **campnosperma → 7155**

1292 **campos**
(a large grassy plain in South America, with scattered bushes and small trees)
d Campos *m*
f campos *m*
r кампос *m*; кампус *m*; кампо *m*; бразильская саванна *f* (местное название саванны в Бразилии)

1293 **camwood**
(Baphia nitida)
d Gabanholz *n*; Angolaholz *n*
f okoué *m*; attié *m*
r бафия *f* яркая

1294 **Canada balsam; Canada turpentine; balsam of fir**
(a viscous, yellowish, transparent resin obtained from the balsam fir and used as a cement for glass lenses and for mounting specimens on microscopic slides)
d kanadischer Terpentin *m*
f baume *m* du Canada; gomme *f* de sapin
r бальзам *m* канадский

* **Canada plum → 1297**

* **Canadian ash → 261**

* **Canadian aspen → 236**

* **Canadian balsam poplar → 9338**

1295 **Canadian birch tree; shrubby birch tree**
(Betula humilis)
d niedrige Birke *f*; Strauchbirke *f*
f bouleau *m* bas; bouleau frutescent; bouleau humble
r берёза *f* низкая

* **Canadian buffaloberry → 771**

1296 **Canadian hemlock; eastern hemlock; American hemlock**
(Tsuga canadensis)
d kanadische Hemlocktanne *f*; Schierlingstanne *f*
f sapin *m* du Canada; pruche *f* de l'Est; pruche du Canada; tsuga *m* [du Canada]
r тсуга *f* канадская

1297 **Canadian plum; Canada plum**
(Prunus nigra)
d Bitter-Kirsche *f*
f prunier *m* noir
r слива *f* чёрная; слива канадская

* **Canadian poplar → 706**

* **Canadian red pine → 5796**

1298 **Canadian red spruce; red spruce; yellow spruce**
(Picea rubra; Picea rubens)
d kanadische Rotfichte *f*; amerikanische Rotfichte *f*; Budsonfichte *f*; Hudsonfichte *f*
f épicéa *m* rouge du Canada; sapinette *f* rouge [du Canada]; épinette *f* rouge
r ель *f* красная

* **Canadian spruce → 9386**

1299 **Canadian yellow birch; Canadian silk-wood; Quebec birch; red birch**
(Betula alleghaniensis; Betula lutea)
d kanadische Gelbbirke *f*
f bouleau *m* jaune [canadien]
r берёза *f* жёлтая; берёза американская; берёза красная

* **Canadian yew → 9345**

* **cana-fistula → 5644**

* **Canarian oak → 5477**

1300 **canary**
(iron rod 6 feet long with a hook on one end and a handle on the other, used to pull cable or chain under bundles of pulpwood or logs for binding or hookup)
d Eisenstange *f*
f barreau *m* de fer
r железный стержень *m* для продевания троса или цепи под пачку брёвен

1301 **Canary islands date**
(Phoenix canariensis)
r финик *m* канарский

* **Canary whitewood → 9008**

* **candelabra-topped tree → 3659**

* **candelo → 2570**

1302 **candle**
(new bright green and tender growth of conifers in the spring)
d Kerze *f* [Kiefern-Trieb]
f pousse *f* en chandelle
r прямой сосновый побег *m*

* **candleberry → 7858**

1303 **candleberry [tree]; Chinese parasol [tree]; Japanese varnish-tree; phoenix tree**
(Firmiana simplex; Firmiana platanifolia; Sterculia platanifolia)
d chinesischer Sonnenschirmbaum *m*; Firnisbaum *m*; Lichtnussbaum *m*
f parasolier *m* japonais; parasol *m* chinois; arbre *m* vernis
r свечное дерево *n*

1304 **candlenut [tree]; tung[-oil] tree**
(Aleurites moluccana; Aleurites fordii; Aleurites triloba)
d Kerzenölbaum *m*; Kukuinussbaum *m*; Tungölbaum *m*; Holzölbaum *m*
f arbre *m* de t[o]ung; kemiri *m*; bancoulier *m*; bancoul *m*
r тунг *m* молукский; тунг Форда; тунговое дерево *n*

1305 **candle-shaped**
d kerzenförmig
f céréiforme
r свечевидный

1306 **candle tree; gommier**
(Parmentiera cereifera; Dacryodes excelsa)
d Kerzebaum *m*
f gommier *m* de la montagne
r катальпа *f*

1307 **cane**
(a long, often supple, woody stem)
d Stiel *m*
f canne *f*
r однолетний побег *m*; стебель *m* кустарника

1308 **cane; reed**
(Sciprus)

f roseau *m*
r камыш *m*; тростник *m*

* **canella imbuia → 1008**

* **cannon a log → 4059**

1309 **cannonball tree**
(Couroupita guianensis)
d Kanonenkugelbaum *m*
f arbre *m* aux boulets de canon; boulet *m* de canon
r пушечное дерево *n*

* **canoe birch → 6058**

* **canoe cedar → 8893**

1310 **canoewood**
d Kanuholz *n*
f bois *m* de pirogue
r древесина *f* для каноэ

1311 **canopied landscape**
(type of visual landscape to be found under an extensive, continuous crown cover of trees)
d bedeckte Landschaft *f*
f paysage *m* canopée
r ландшафт *m* под полога

* **canopy → 8810**

* **canopy closure → 1659**

* **canopy cover → 3345**

1312 **canopy [cover] class**
(any class into which crops or stands may be divided on the basis of the degree of closure)
d Schlusslockerungsklasse *f*; Beschirmungsklasse *f*
f classe *f* de couvert; classe de densité
r класс *m* сомкнутости

* **canopy density → 2393**

1313 **canopy drip**
d Beschirmungstropfen *m*
f précipitation *f* par houppier
r капельные осадки *fpl*, проникающие через полог леса

* **canopy fire → 2076**

* **canopy layer → 8150**

* **canopy opening → 4566**

1314 **canopy surface; canopy projection area**
d Schirmfläche *f*; Kronenoberfläche *f*
f surface *f* de houppier
r площадь *f* проекции полога

* **canopy thickening → 2393**

1315 **canopy tree**
d breitkroniger Baum *m*; starkkroniger Baum
f arbre *m* de dense cime
r дерево *n* с густой кроной; дерево с широкой кроной; дерево с сильно развитой кроной

* **cant → 638**

* **cant dog → 1317**

1316 **canter; chipper canter; chipping canter; chipping headrig; chipper headrig**
(a primary breakdown machine which converts logs to cants by bandsaw for further processing)
d Kanter *m*; Profilzerspaner *m*
f équarrisseuse[-déchiqueteuse] *f*; dédosseuse-déchiqueteuse *f*; coupeuse *f* pour équarrissage des grumes
r кантователь *m* (брёвен)

1317 **canthook; cant dog; rolling dog; peavie; peav[e]y; wood pick**
(stout wooden lever used in rolling logs)
d Wendehaken *m*; Sap[p]ine *f*; Sappel *m*; Kant[er]haken *m*
f sapi *m* à double croc
r кондак *m*; кантовальный крюк *m*; поворотный крюк; багровый крюк

1318 **canyon grape[vine]**
(Vitis arizonica)
d Felsenrebe *f*
f vigne *m* de canyon
r виноград *m* аризонский

1319 **canyon live oak; California[n] live oak**
(Quercus chrysolepis)
d goldschuppige Eiche *f*
f chêne *m* à écailles dorées
r дуб *m* золоточешуйчатый

* **canyon maple → 707**

* **caoba → 1412**

1320 **caoutchouc tree; Para rubber[ ]tree;**

d Hevea *f*; Parakautschukbaum *m*; Kautschukbaum *m*
f hévéa *m*; arbre *m* de Para; arbre à caoutchouc
r каучуковое дерево *n*; парагуттаперчевое дерево; гевея *f* бразильская

* **cap → 1858**

* **capability of forest → 3648**

1321 **capable-deferred forestland**
(capable forestland that has been legislatively designated or administratively designated not to be used awaiting further decision)
d fähig verzögerte Waldfläche *f*
f terrain *m* forestier d'avenir
r запланированная на будущее лесная площадь *f*

* **capable forestland → 6562**

* **capable-reserved forestland → 6563**

* **capacity class → 9774**

1322 **capacity of survival; survival rate**
(of a forest seeds)
d Überlebensrate *f*
f taux *m* de survie; probabilité *f* de survie
r сохранность *f*; выживаемость *f* (сеянцев лесных культур)

* **cape boxwood → 4685**

* **cape ebony → 7797**

1323 **cape gooseberry**
(Physalis peruviana)
d Ananaskirsche *f*; Kapstachelbeere *f*
f alkékenge *m* jaune; alkékenge du Pérou; coqueret *m* comestible; groseille *f* du cap; physalis *m*
r вишня *f* перуанская

1324 **caper family**
(Capparidaceae)
d Kaperngewächse *pl*
f capparidacées *pl*
r каперсовые *pl*

1325 **capillary water; capillary moisture**
d Kapillarwasser *n*; Porensaugwasser *n*
f eau *f* capillaire
r капиллярная вода *f*

1326 **capoeira**
r капоэйра *f*

1327 **capricorn beetles; long-horned beetles; longhorns**
(Cerambycidae)
d Holzböcke *pl*
f cornes *fpl* longues
r усачи *mpl*; дровосеки *mpl*

1328 **capsule**
(dry, thin-walled fruit containing 2 or more seeds and splitting along natural grooved lines at maturity)
d Kapsel *f*
f capsule *f*
r сухой плод *m*, раскрывающийся на несколько частей (у катальпы, ивы)

* **carapa → 4039**

1329 **carbon balance**
(difference between carbon flow into and out of component/reservoir)
d Kohlenstoffbilanz *f*
f bilan *m* du carbone
r равновесие *n* углерода

1330 **carbon cycle**
(biogeochemical cycle undergone by carbon that is utilized by organisms, later liberated upon the death and decomposition of the organisms, and returned to its original state to be reused by another organism)
d Kohlenstoffzyklus *m*
f cycle *m* du carbone
r круговорот *m* углерода; углеродный цикл *m*

* **carbon dioxide assimilation → 391**

1331 **carbonization; charring**
d Verkohlen *n*; Kohlung *f*; Kohlen *n*
f carbonisation *f*
r обугливание *n*; углежжение *n*; карбонизация *f*

1332 **carbonizing unit**
d Verkohlengerät *n*
f dispositif *m* de carbonisation
r установка *f* для переугливания древесины

1333 **carbon-nitrogen ratio; C/N ratio**
(ratio of weight of organic C to weight of total N in soil or in organic matter)
d Kohlenstoff-Stickstoffverhältnis *n*
f rapport *m* carbone-azote

* **carbon sequestration → 7452**

1334 **carbon source**
(any process or mechanism which releases a greenhouse gas, an aerosol or a precursor of a greenhouse gas into the atmosphere)
d Kohlenstoffquelle *f*
f source *f* de carbone
r источник *m* углерода

* **car camp → 9356**

1335 **carcassing timber**
(softwood used in house-building)
d Karkassenholz *n*
f bois *m* d'ossature
r лесоматериал *m* для несущих конструкций; элементы *mpl* деревянной каркасной конструкции

* **cardboard → 6059**

1336 **card of wood; trim end; log trim**
d Stammscheibe *f*; Holzscheibe *f*; Scheibe *f*; Kappscheibe *f*
f disque *m* de la tige; chute *f* de découpe
r кружок *m*, отпиленный от ствола дерева; шайба *f*; откомлёвка *f*

1337 **careful logging around regeneration**
(harvesting operation based on shelterwood cutting principles, where advanced regeneration is protected during harvesting)
d vorsichtige Abholzung *f* um Regeneration
f coupe *f* avec protection de la régénération
r тщательная заготовка *f* леса в обхвате регенерации

* **care of stands → 8456**

1338 **cargo of timber**
d Holzlast *f*
f charge *f* de bois
r груз *m* лесоматериалов

* **Caribbean [longleaf] pitch pine → 2111**

1339 **Caribbean trumpet-tree; yellow trumpet tree**
(Tabebuia aurea)
d gelber Trompetenbaum *m*; Gelbtrompetenbaum *m*
f bois *m* pissenlit
r табебуйя *f* караибская

1340 **carica family**
f caricacées *pl*
r семейство *n* папайа

1341 **carob[-tree]; locust tree**
(Ceratonia siliqua)
d Johannisbrotbaum *m*; Karobenbaum *m*
f arbre *m* à caroubes; caroubier *m*; cératonia *m*; carouge *m*
r рожковое дерево *n*

1342 **Carolina ash; pop ash**
(Fraxinus caroliniana; Fraxinus platycarpa)
d südliche Esche *f*
f frêne *m* de Caroline
r ясень *m* каролинский

1343 **Carolina basswood; Carolina linden**
(Tilia caroliniana)
d Karolinalinde *f*
f tilleul *m* de Caroline
r липа *f* каролинская

1344 **Carolina buckthorn**
(Frangula caroliniana; Rhamnus caroliniana)
d Karolina-Kreuzdorn *m*
f nerprun *m* de Caroline
r крушина *f* каролинская

* **Carolina cherry laurel → 1347**

1345 **Carolina hemlock**
(Tsuga caroliniana)
d Karolina-Hemlockstanne *f*
f tsuga *m* de Caroline
r тсуга *f* каролинская

1346 **Carolina jessamine**
(Gelsemium sempervirens)
d falscher Jasmin *m*; Carolina-Jasmin *m*
f jasmin *m* de Caroline; gelsémie *f* de Caroline
r жёлтый жасмин *m*; гельземий *m* вечнозёленый

1347 **Carolina laurelcherry; Carolina cherry laurel**
(Laurocerasus caroliniana; Prunus caroliniana)
d Karolina-Kirschlorbeer *m*
f laurier-cerise *m* de Caroline
r лавровишня *f* каролинская

* **Carolina linden → 1343**

1348 **Carolina moonseed; Carolina snailseed**
(Cocculus carolinus)
d Karolina-Kokkelstrauch *m*

1349 Carolina silverbell; shittim [wood]; opossum wood
(Halesia caroliniana; Halesia tetraptera; Mohrodendron carolinum)
d Karolina-Maiglökenbaum *m*; Galanthus *m*; Karolina-Schneeglöckchenbaum *m*
f halésia *m* de Caroline; cloche *f* de Caroline; arbre *m* perce-neige
r халезия *f* каролинская; дерево-подснежник *n*

* **Carolina snailseed → 1348**

* **Carolina willow → 1688**

1350 carpellate cone
d Stempelzapfen *m*
f cône *f* carpelle
r женская шишка *f*

1351 carpenter ant
(Camponotus)
d Rossameise *f*
f fourmi *f* charpentière
r муравей-древоточец *m*

* **carpenter barker → 1425**

* **carpenter moths → 8562**

* **carpenterworms → 8562**

1352 carpentry
d Holzbaukonstruktion *f*; Holzbauwerk *n*
f charpente *f*
r деревянное изделие *n*

* **carpentry → 9685**

* **carpentry timber → 4658**

1353 carpophore; fruit[ing] body
(a slender prolongation of the receptacle as an axis between the carpels)
d Fruchtkörper *m*; Fruchtträger *m*; Fruchthalter *m*; Karpophor *m*
f carpophore *m*
r плодовое тело *n*

* **carposome → 7928**

* **carriage building timber → 1672**

* **carriage dog → 875**

1354 carrier block
d Trägerblock *m*
f bloc *m* palette
r подкладка *f* из брёвна для сбора и транспортировки лесоматериала

* **carrier cable → 8745**

1355 carrot family
(Umbelliferae)
d Doldengewächse *pl*
f ombellifères *fpl*
r зонтичные *pl*

1356 carr [woodland]
(deciduous woodland or scrub on a permanently wet, organic soil)
d Auewald *m*
f carr *m*
r разрежённый заболоченный древостой *m*, переходящий в кустарник; территория *f* закустаренного низинного болота

1357 carrying capacity
(the maximum number of individuals of a wildlife species that an area can support during the most unfavorable time of the year)
d Belastungsfähigkeit *f*
f capacité *f* de charge
r допустимое количество *n* организмов (на единицу площади); допускаемая нагрузка *f* (рекреационных лесов)

* **carrying rope → 8745**

* **carry-up pruning → 4294**

* **cartload of wood → 8984**

1358 cart removal
(of timber)
d Karrführung *f*
f transport *m* par charrettes
r вывозка *f* трелёвочной тележкой

1359 cart-road
d Karrweg *m*
f chemin *m* utilisable par charrettes à bois
r дорога *f* для тачек

* **cartwright's timber → 1672**

* **caruto → 4605**

1360 carve *v*; **cut** *v*; **whittle** *v*
d schnitzen

* **carved veneer → 7738**

1361 **carved wood**
d geschnitztes Holz *n*
f bois *m* gravé; bois sculpté
r резьбованная древесина *f*

1362 **carver cutter**
(a tool)
d Schnitzermesser *m*; Schnitzschneidwerkzeug *n*
f sculpteur *m*
r нож *m* для резьбы

1363 **carver in wood; wood carver; wood cutter; wood engraver**
(an engraver on wood)
d Holzschnitzer *m*; Holzgraveur *m*
f ouvrier *m* sculpteur sur bois; sculpteur *m* sur bois; graveur *m* sur bois
r резчик *m* по дереву; гравёр *m* по дереву

* **carving wood → 1918**

1364 **car wood**
d Waggonladungsholz *n*
f bois *m* chargé au wagon
r древесина *f* железнодорожной доставки

* **cascade blueberry → 864**

* **cascade fir → 6025**

* **Cascade Mountain ash → 3943**

1365 **cascade Oregon grape; dwarf Oregon grape; dull Oregon grape; cascade barberry; cascade mahonia**
(Mahonia nervosa; Berberis nervosa)
d nervige Mahonie *f*
f mahonia *m* nervé
r магония *f* сетчатая

* **Cascara [buckthorn] → 1544**

1366 **casca redwater tree; redwater tree; sassy bark tree; sassy [wood]; sasswood; ordeal tree; missanda; bushman's poison**
(Erythrophleum guineense; Erythrophleum guineense)
d Tali *m*; Rotwasserbaum *m*
f tali *m*; mancône *m*; lime *m* du Gabon
r эритрофлеум *m* гвинейский

* **caseboards → 954**

(a defect in the timber caused by improper drying)
d Oberflächenverhärtung *f*
f cémentation *f*; durcissement *m* en surface
r затвердение *n* поверхности; напряжённое состояние *n*; наружное высыхание *n*

* **case timber → 954**

* **case wood → 954**

1368 **cashew apple**
(Anacardium occidentale)
d Cashewnuss *f*; Kaschubaum *m*
f anacardier *m*; acajou *m* à pommes
r анакардиум *m* западный

1369 **cashew family**
(Anacardiaceae)
d Sumachgewächse *pl*
f anacardiacées *pl*
r сумаховые *pl*

* **cask-wood → 4362**

* **cassava → 5282**

1370 **cassia bark**
d Cassia-Rinde *f*; Kassie-Rinde *f*
f écorce *f* de cassier
r кора *f* корицы; коричная корка

1371 **cassine**
(Cassine)
d Safranholzbaum *m*
f cassine *f*
r кассина *f*

1372 **castaneto; pino; podocarp[us]**
(Podocarpus guatemalensis)
d Guatemala-Podocarpus *f*
f podocarpe *m* du Guatemala
r ногоплодник *m* гватемальский

* **castanheiro → 1003**

* **casting of lumber → 9268**

1373 **castwood**
d Formholz *n*
f bois *m* coulé
r ламинированная древесностружечная плита *f* с имитационной текстурной отделкой

* **casual forest worker → 6095**

* **cataphyll → 9075**

* **cataphyllary leaf → 9075**

* **catch bark → 8775**

* **catch boom → 8136**

1374 **catch crop; substituting crop; substitution crop**
(agricultural crop introduced into and at the start of a longer-rotation forest crop, mainly to provide early financial returns)
d Ersatzgewächs *m*
f culture *f* dérobée
r почвозащитная культура *f*; ловчая культура; подсевная культура; пожнивная культура

* **catch mark → 5101**

* **catchment basin → 9280**

* **catch-water drain → 2173**

1375 **catclaw acacia**
(Acacia greggii)
d Gregg-Akazie *f*
f acacia *m* Gregg
r акация *f* Грегги

1376 **catechu tree**
(Acacia catechu)
d Catechu-Akazie *f*
f acacia *m* à cachou; cachoutier *m*
r акация *f* катехинская

1377 **caterpillar host tree**
d Raupenwirtsbaum *m*
f arbre-hôte *m* de chenilles
r дерево-хозяин *n* гусениц

* **caterpillar tractor → 7799**

1378 **catface**
(partially healed fire scar on the face of a tree; often the place where rot enters)
d Fuernarbe *f*
f tête *f* de chat
r затянувшийся пожарный рубец *m*

1379 **cathedral cut**
(a variation of the crown cut method of slicing veneers where the growth rings are exactly parallel to the slicer, producing on the face of the veneer an inverted 'V' figure)
r кафедральная рубка *f*

1380 **catkin; ament**
(compact and often drooping cluster of reduced, stalkless, and usually unisexual flowers)
d Kätzchen *n*
f chaton *m*
r серёжка *f* (на деревьях)

1381 **cat's-eye; dragon's-eye; longan**
(Dimocarpus longan)
d Longan *m*; Drachenauge *n*
f longanier *m*; longane *m*
r лонган *m*; глаза *mpl* дракона

1382 **cat's paw**
(a group of pin knots)
d Katzenpfote *f*
f gueule-de-raie *f*
r кошачья лапа *f*

* **cat spruce → 9386**

1383 **cat tree; European spindle[ ]tree; spindle[ ]tree**
(Euonymus europaeus)
d gemeiner Spindelstrauch *m*; Spindelbaum *m*; Pfaffenhütchen *n*; Pfaffenkäpplein *n*
f fusain *m* d'Europe; bois *m* carré; bonnet *m* d'évêque; bonnet de cardinal; bonnet de prêtre
r бересклет *m* [европейский]

* **cattyman → 9569**

1384 **Caucasian fir; Nordmann['s] fir; Crimean fir**
(Abies nordmanniana)
d Kaukasustanne *f*; Nordmannstanne *f*
f sapin *m* de Nordmann; sapin de Caucase
r пихта *f* кавказская; пихта Нордманна

1385 **Caucasian huckberry; Caucasian nettle tree**
(Celtis caucasica)
d kaukasischer Zürgelbaum *m*
f micocoulier *m* de Caucase
r каркас *m* кавказский

1386 **Caucasian mock orange**
(Philadelphus caucasicus)
d kaukasischer Pfeifenstrauch *m*; Zimtröschen *m*
f seringa[t] *m* de Caucase
r чубушник *m* кавказский

1387 Caucasian oak
(Quercus macranthera)
d kaukasische Eiche *f*
f chêne *m* du Caucase
r дуб *m* кавказский

1388 Caucasian wingnut
(Pterocarya fraxinifolia)
d eschenblättrige Flügelnuss *f*; kaukasische Flügelnuss
f ptérocayer *m* à feuilles de sorbier; ptérocarier à feuilles de frêne; noyer *m* du Caucase
r птерокария *f* кавказская; лапина *f* ясенелистная

1389 caudex
(the woody base of an otherwise herbaceous perennial plant)
d Stamm *m*
f caudex *m*
r ствол *m*; каудекс *m*; стержень *m*

1390 caulescent
(more or less stemmed or stem bearing, having an evident stem above ground)
d stängeltreibend; Stiel-; Stängel-; Halm-
f caulescent
r стебельный; стебельчатый

1391 cauliflory; cauliflorous
(the state of flowering from the branches or trunk)
d stammblütig
f cauliflore
r каулифлорный

1392 cauliform; stem-shaped; stemlike
(having the form of a caulis)
d stängelförmig
f cauliforme
r стеблевидный; стеблеобразный

* **caulks → 1280**

* **causeway of fascines → 1117**

1393 causing damage to forests *adj*
d forstschädlich
f causant de dommage aux forêts
r вредный для лесов

1394 caven
(Acacia caven)
d Caven *m*
f caven *m*
r акация *f* кавен; эспинильо *m*

* **Cayanne-linaloe → 3083**

1395 ceanothus
(Ceanothus)
d Säckelblume *f*
f céanothe *m*
r цеанотус *m*

* **Ceara rosewood → 4716**

* **cebil [colorado] → 6762**

1396 cedar
(Cedrus)
d Zeder *f*
f cèdre *m*
r кедр *m*; кедровое дерево *n*

1397 cedar forest; cembretum
d Zederwald *m*
f forêt *f* de cèdres; cédraie *f*
r кедровый лес *m*; кедровник *m*; кедрач *m*

1398 cedar grove; thuja light forest
d Lebensbaumlichtwald *m*
f cédrière *f*
r кедровая роща *f*

* **cedar mahogany → 114**

1399 cedar-nut
d Zirbelnuss *f*
f graine *f* de pin cembro
r кедровый орех *m*

1400 cedar of Lebanon; Lebanon cedar
(Cedrus libani; Cedrus libanotica)
d Libanonzeder *f*; Salomozeder *f*; Zeder *f*
f cèdre *m* du Liban
r кедр *m* ливанский

1401 cedar oil
d Zedernholzöl *n*
f huile *f* de cèdre
r кедровое масло *n*

* **cedar pine → 4743**

1402 cedar resin
d Zederharz *n*
f résine *f* de cèdre
r кедровая канифоль *f*

1403 cedar-tree borer
(Semanotus ligneus)
d Zederbohrer *m*

* **cedar wattle → 8268**

1404 **cedarwood**
d Zedernholz *n*; Arvenholz *n*
f bois *m* de cèdre
r древесина *f* кедра; кедровая древесина

* **cedrillo → 2006**

* **cedro → 458**

* **ceiba [yuca] → 4688**

1405 **celery [top] pine**
(Phyllcocladus)
d Blatteibe *f*; Farneibe *f*
f phylloclade *m*
r филлокладус *m*

1406 **cellular tissue**
d Zell[en]gewebe *n*
f tissue *f* cellulaire
r клеточная ткань *f*

1407 **cell wall; membrane**
d Zellwand *f*; Zellmembran *f*; Zellhaut *f*
f paroi *f* [cellulaire]; membrane *f* cellulaire
r клеточная стенка *f*; клеточная мембрана *f*; оболочка *f* клетки; клеточная оболочка

1408 **cell wall check**
(a fissure in the secondary cell wall, as in the tracheids of compression wood)
d Zellwandriss *m*
f fente *f* de la paroi
r трещина *f* клеточной стенки

* **cembra-pine → 8346**

* **cembretum → 1397**

* **center → 1951**

* **center-fed resaw → 1410**

1409 **center firing**
d Abbrennen *n* von der Mitte aus
f allumage *m* central; allumage à partir du centre; mise *f* à feu concentrique
r сжигание *n* с середины; пуск *m* пала из центра (участка)

1410 **centering resaw; resaw; center-fed resaw**
(a machine fitted with either band saws, circular, or web saws of thinner gauge than head saws)
r ребровый станок *m*

1411 **Central American cedar; Cuba cedar; Mexican cedar; Havana cedar; Honduras cedar; cigar box cedar; scented cedar**
(Cedrela odorata)
d Honduras-Zeder *f*; Zigarrenkistenzeder *f*; zentralamerikanische Zeder *f*
f cèdre *m* odorant; cèdre du Mexique; cèdre de l'Amérique centrale
r кедр *m* остиндийский; цедрела *f* душистая; душистый кедр; боссе *n*

1412 **Central American mahagony; American mahagony; Honduras mahogany; Belize mahogany; bigleaf mahogany; broadleaf mahogany; baywood; caoba; aguano**
(Swietenia macrophylla)
d Mahagonibaum *m*; Hondurasmahagoni *n*; amerikanisches Mahagoni
f acajou *m* d'Amérique; acajou américain
r свиетения *f* краснодеревная; свиетения крупнолистная; махагониевое дерево *n*; каоба *f*

* **central diameter → 2468**

* **centrally administered forest → 8041**

* **centrally managed forest → 8041**

1413 **central mean stem; stem of medium diameter**
d zentraler Mittelstamm *m*
f arbre *m* moyen central
r центральный средний ствол *m*

1414 **central rot; heart[wood] rot; pith rot**
(fungus-caused decay of a tree's heartwood)
d Herz[en]fäule *f*; Kernfäule *f*; Markfäule *f*
f pourriture *f* du cœur; carie *f* du cœur; pourriture centrale
r сердцевинная гниль *f*; гниль сердечка

1415 **central stem**
(portion of a tree between a 1-foot stump and the minimum 4-inch top diameter outside bark, or point where the central stem breaks into limbs)
d zentraler Stamm *m*
f tige *f* centrale
r центральный ствол *m*

1416 **century plant**
(Agave americana)
d amerikanische Agave *f*

1417 **cerejeira; amburana; isphingo**
(Amburana cearensis; Torresea cearensis)
d Cerejeira *m*; Amburana *m*; Isphingo *m*
f cerejeira *m*; amburana *m*; roble *m*
r сережейра *f*

1418 **certified arborist; certified arboriculturist**
(individual who is trained in the art and science of planting, caring for and maintaining individual trees)
d anerkannter Baumzüchter *m*; beglaubigter Baumzüchter
f arboriculteur *m* certifié
r кондиционный воспитатель *m* деревьев

1419 **certified pesticide applicator**
d zertifizierter Pestizidenapplikator *m*
f épandeur *m* de pesticide certifié
r апробированная машина *f* для внесения пестицидов

1420 **certified seed**
d zertifiziertes Saatgut *n*
f semence *f* homologuée; semence certifiée
r апробированные семена *npl*

* **cetico → 7519**

* **Ceylon boxwood → 4465**

1421 **Ceylon cinnamon; cinnamon tree**
(Cinnamomum zeylanicum; Cinnamomum verum; Laurus cinnamomum)
d Ceylonzimtbaum *m*; Ceylonzimt *m*; ceylonesischer Zimtstrauch *m*
f cannelier *m* de Ceylan; cannelle *f* de Ceylan
r коричное дерево *n* цейлонское; корица *f* цейлонская; коричник *m* цейлонский

* **Ceylon satinwood → 2729**

* **CFI → 1898**

* **cft → 2113**

1422 **chaff; tailings**
(lighter plant tissue fragments released in threshing)
d Abfälle *mpl* in der Baumwollverarbeitung
f débris *mpl* végétaux
r лесной подстил *m*; отходы *mpl*

* **chain bar → 4040**

* **chain barker → 1425**

1423 **chain brake; kickback brake**
d Kettenbremse *f*
f frein *m* de chaîne; frein de contrecoup
r тормоз *m* пильной цепи

* **chain-clearing → 271**

1424 **chain conveyor [for logs]; wood conveyor; chain logway**
d Ketten[zug]förderer *m*
f chaîne *f* longitudinale d'aménage de grumes; transporteur *m* à chaîne; élévateur *m* à chaîne
r цепной лесотранспортёр *m*; цепной конвейер *m*; цепной грузоподъёмник *m*

* **chain dog → 8577**

* **chain drag → 1427**

* **chain flail → 1426**

1425 **chain-flail debarker; chain barker; carpenter barker**
d Carpenter-Entringer *m*; Kettenentrinder *m*
f écorceuse *f* à chaînes [Carpenter]
r цепной окорочный станок *m*

1426 **chain-flail delimber; chain flail; flail delimber**
d Kettenflegelentastungsmaschine *f*; Flegelentastungsmaschine *f*
f ébrancheuse *f* à fléaux; fléau *m* d'ébranchage
r цепная сучкорезная машина *f*

1427 **chain for pulling; chain drag**
d Kettenschlinge *f*
f chaîne *f* de dragage
r цепной подтаскиватель *m*; тормозная цепь *f*

* **chain hook → 8577**

* **chaining → 8398**

1428 **chaining; cabling**
d Durchziehen *n* mit der Kette; Verkettung *f*; Ketten *n*; Rücken *n* mit der Kette
f passage *m* de la chaîne; débardage *m* à la chaîne; chaînage *m*
r трелёвка *f* пачек балансов, обвязанных поперёк цепью; трелёвка с цепью

* **chain logway → 1424**

1429 **chain of custody**
(process of monitoring the production and

d Bewahrungskette *f*; Verwahrungskette *f*
f continuité *f* de possession
r цепь *f* обеспечения сохранности проб

1430 **chain of the saw; cutting chain**
d Sägekette *f*
f chaîne *f* de coupe; chaîne coupante
r пильная цепь *f*

1431 **chain[ ]saw; chain ripsaw; link saw; tooth saw**
(saw which cutting elements are on an endless chain similar to a bicycle chain)
d Kettensäge *f*
f scie *f* à chaîne
r цепная пила *f*

* **chainsaw [cutting] head → 1433**

1432 **chainsaw feller-buncher; saw-head feller-buncher**
d Fäller-Bündler *m* mit Kettensäge
f abatteuse-groupeuse *f* à scie à chaîne
r валочно-пакетирующая машина *f* с пильным срезающим механизмом

1433 **chainsaw felling head; chainsaw [cutting] head; chainsaw felling attachment**
d Schlagkopf *m* mit Kettensäge
f tête *f* d'abattage à scie à chaîne
r валочная головка *f* с пилой

1434 **chainsaw operator; powersaw operator**
(a person)
d Motorsägenführer *m*; Motorsäger *m*; Sägenführer *m*
f opérateur *m* de scie à chaîne
r моторист *m* [цепной] пилы; мотопильщик *m*

* **chain sling → 8577**

1435 **chain tender; snaker; ballhooter**
(a person who rolls or slides logs down a hillside)
d Holzrücker *m*
f débardeur *m* de bois
r [рабочий-]трелёвщик *m*; трелёвщик брёвен

1436 **chain tightener; wire strainer**
d Kettenspanner *m*; Drahtspanner *m*
f tendeur *m* de chaîne; tendeur de câble; serreur *m* de chaîne
r приспособление *n* для натяжения цепи; станок *m* для натяжения цепи

f bois *m* de chaîne
r пиломатериалы *mpl* для укрепления тонких стен или перегородок

* **chain-track tractor → 7799**

1438 **chalk[bark] maple; whitebark[ed] maple**
(Acer leucoderme)
d Kreideahorn *m*
f érable *m* à blanche écorce
r клён *m* белокорый

1439 **chamaephyte; surface plant**
(a self-supporting woody plant or herbaceous evergreen perennial whose mature branch or shoot system remains perennially within 25-50 cm above ground level)
d Chamaephyt *m*; Chamäphyt *m*; Oberflächenpflanze *f*
f chamaephyte *m*; chaméphyte *m*
r хамефит *m*

1440 **chambered pith; spongy pith**
d schwammiger Kern *m*
f cœur *m* caverneux; cœur spongieux
r пористая сердцевина *f*; камерная сердцевина; губчатая сердцевина; трухлявая сердцевина

1441 **chamfer cut; splayed cut; miter cut**
d Schrägschnitt *m*
f coupe *f* chanfrein
r скошенный срез *m*; косой срез; ус *m*

* **champion tree → 6396**

* **chance → 2189**

* **change in tree species → 226**

* **changugo → 2570**

1442 **chaparral**
(a thicket of low evergreen oaks or dense tangled brushwood; dense vegetation consisting of stunted trees or bushes)
d Chaparral *n*; Tschaparral *n*
f chaparral *m*; menu taillis *m*
r чапарель *m*; колючий кустарник *m*; поросль *f* карликового дуба

* **chaparral berry → 5419**

* **chaparro → 2570**

* **charab poplar → 4468**

1443 **charcoal block**
d Holzkohlenblock *m*
f charbon *m* à souder
r древесноугольный блок *m*

1444 **charcoal burner; charcoal maker; wood burner**
(a chamber or retort, generally of metal or brick construction, for carbonizing wood to produce charcoal)
d Holzkohlenbrenner *m*; Holzkohlenvergaser *m*
f chaudière *f* à bois; four *m* à carboniser; carbonisateur *m*
r углежог *m*; древесный генератор *m*

1445 **charcoal burner; charcoal man; coaler**
(a person)
d Holzköhler *m*; Köhler *m*; Kohlenbrenner *m*
f charbonnier *m*
r угольщик *m*

* **charcoal dust → 9506**

1446 **charcoal grading; charcoal screening**
d Holzkohleneinstufung *f*
f criblage *m* de charbon de bois; classement *m* de charbon de bois
r сортировка *f* древесного угля; отсев *m* угольной мелочи

1447 **charcoal graphitization**
d Graphitisierung *f* der Holzkohlen
f graphitisation *f* de charbon de bois
r графитизация *f* древесного угля

1448 **charcoal installation**
d Holzkohleninstallation *f*
f installation *f* de carbonisation
r установка *f* для получения древесного угля

* **charcoal kiln → 1450**

* **charcoal maker → 1444**

1449 **charcoal making**
d Herkunft *f* der Holzkohle
f fabrication *f* de charbon de bois
r образование *n* древесного угля

* **charcoal man → 1445**

1450 **charcoal pile; charcoal kiln**
d Holzkohlenmeiler *m*; Meiler *m*; Meilerofen *m*
f empilement *m* de charbon; meule *f* charbonnier; meule de charbon
r древесноугольная куча *f*; костёр *m*

d Holzkohlenloch *n*
f trou *m* de charbon
r древесноугольная яма *f*

* **charcoal powder → 9506**

* **charcoal screening → 1446**

* **charcoal-wood → 9571**

1452 **charpenter's axe; block-bill**
d Zimmeraxt *f*; Zimmerbeil *n*
f hache *f* de charpentier
r топор *m* для кряжей; тесун *m*

1453 **charred wood**
d angekohltes Holz *n*
f bois *m* carbonisé
r обугливщаяся древесина *f*

* **charring → 1331**

* **charring place → 4712**

* **charring wood → 9571**

1454 **chaser; block tender**
(a member of the hauling crew on a skidroad who accompanies the turn of the logs to the landing, unhooks the grabs, and sees that they are returned to the yarding engine)
d Verfolger *m*; Jäger *m*
f chasseur *m*
r чокеровщик *m*, сопровождающий пачку брёвен (при механизированной трелёвке)

1455 **chaser-skidder [operator]**
d Ziseleur-Skidder *m*
f chasseur-débusqueur *m*
r тракторист *m*, отцепляющий брёвна от канатной установки и трелюющий их на площадку

* **chaste tree → 4961**

* **chat → 1095, 3165**

1456 **chat litter**
d Hackstreu *f*
f litière *f* hachée
r измельченная подстилка *f*

* **chatwood → 998**

* **check → 7483**

* **checking → 2003**

* **checking by shade → 8276**

1457 **check scaler; scaler; culler; screen tender**
(a person qualified to scale primary forest products and usually licensed or appointed by a government agency)
d Holzabnahmebeamter *m*; Holzaufmesser *m*; Holzvermesser *m*; Holzempfänger *m*; Scaler *m*
f mesureur *m* de bois; métreur *m* de bois
r приёмщик *m*[, производящий контрольный обмер] леса; маркировщик *m* древесины; измеритель *m* древесины

* **check stand → 4471**

1458 **check *v* the felling record**
d das Holzaufnahmebuch testen
f essayer le calepin
r подсчитывать результаты вырубки леса

1459 **check tree**
d Vergleichsbaum *m*
f arbre *m* témoin
r контрольное дерево *n*

1460 **cheeks**
d Backen *pl*; Wangen *pl*
f joue *f*
r щёки *fpl* (лезвия топора)

* **cheese block → 9561**

* **cheese tree → 4174**

1461 **cheesewood; patternwood; alstonia**
(Alstonia)
d Alstonia *f*; Emien *n*
f alstonia *m*; émien *m*
r алстония *f* конголезская

* **chemi-barking → 1463**

1462 **chemical brown-stain; kiln brown stain; brown oxidation stain; kiln stain; kiln burn**
(chemical stain that develops during kiln drying)
d chemische Braunverfärbung *f*
f tache *f* brune chimique
r химическая заболонная грибная коричневая окраска *f*; химическая коричневая окраска

1463 **chemical [de]barking; chemi-barking; chemical peeling**
d chemische Entrindung *f*
f écorçage *m* chimique
r химическая окорка *f*

1464 **chemical degradation of wood**
d chemischer Abbau *m* des Holzes
f dégradation *f* chimique de bois
r химическое разложение *n* древесины

1465 **chemical forest protection**
d chemischer Waldschutz *m*
f protection *f* chimique de la forêt
r химическая защита *f* леса

* **chemical peeling → 1463**

1466 **chemical pruning**
(application of chemicals, e.g. plant-growth regulators, to the living tree so as to kill, suppress, or inhibit lateral shoots)
d chemische Abästung *f*; chemische Ästung *f*
f élagage *m* chimique
r химический способ *m* удаления сучьев

1467 **chemical pulping**
(process in which wood fibers are separated by removing the lignin and certain other wood components through the use of chemicals)
d Holzzellstoffkochung *f*
f fabrication *f* de pâtes chimiques
r химическая варка *f* целлюлозы

1468 **chemical site preparation**
(use of herbicides to reduce competition from weeds, grasses and trees)
d chemische Vorbereitung *f* des Geländes; chemische Standortvorbereitung *f*
f préparation *f* chimique du terrain
r химическая очистка *f* поверхности почвы

1469 **chemical thinning; chemi-thinning; poison thinning**
(in which the unwanted trees are killed by treatment with herbicide)
d chemische Durchforstung *f*
f éclaircie *f* chimique
r химическое прореживание *n*

* **chemical wood → 44**

1470 **chem[i]otropism**
(the growth of a plant organ in response to an external chemical stimulus)
d Chemotropismus *m*
f chimiotropisme *m*
r хемотропизм *m*

1471 chemotaxonomy; biochemical systematics
(classification of plants on the basis of the presence and concentration of certain specific chemical compounds in them)
d biochemische Systematik *f*
f chimiotaxonomie *f*
r биохимическая систематика *f*

* **chenchen → 9125**

1472 chengal
(Balanocarpus heimii)
d Chengal *m*
f chengal *m*
r баланокарпус *m* Гейма; хенгал *m*

1473 cherry
(Padus)
d Kirsche *f*
f cerise *m*
r черёмуха *f*

1474 cherry birch; mahogany birch
(Betula lenta)
d Zuckerbirke *f*; Hainbirke *f*
f bouleau *m* flexible; bouleau acajou; bouleau merisier; merisier *m* odorant; bouleau jaune d'Amérique
r берёза *f* вишнёвая

* **cherry laurel → 2828**

1475 cherry plum; purple leaf plum; myrobalan plum
(Prunus cerasifera; Prunus divaricata; Prunus myrobalana)
d Kirschpflaume *f*; türkische Kirsche *f*; Myrobalane *f*
f prunier *m* cérasifère; prunier myrobalan; bacarinier *m*; prune *f* cerise; myrobalan *m*
r алыча *f*; дикая слива *f*; слива растопыренная; миробалан *m*; мирабель *f*; вишнеслива *f*

* **cherry stink bug → 3343**

1476 cherry[-tree]
(Cerasus)
d Kirschbaum *m*; Kirsche *f*
f cerisier *m*
r вишня *f* (дерево); вишнёвое дерево *n*

* **chest height diameter → 2466**

1477 chestnut forest
d Kastanienwald *m*

* **chestnut grove → 4005**

1478 chestnut-leaved oak
(Quercus castaneafolia)
d kastanienblättrige Eiche *f*
f chêne *m* à feuilles de châtaignier; palout *m*
r дуб *m* каштанолистный

* **chestnut oak → 5582**

* **chestnut-winged starling → 3350**

1479 chevron cuts
(modification of strip cutting where the strip is angled part way along its length)
d Risse *mpl* in einem bestimmten Winkel
f coupes *fpl* par chevrons
r переплетённые зарубки *fpl*

1480 Chickasaw plum
(Prunus angustifolia)
d Chicasawpflaume *f*
f prunier *m* chickasaw
r слива *f* узколистная

1481 chicken tree; popcorn tree; Chinese tallow[ ]tree
(Sapium sebiferum)
d chinesischer Talgbaum *m*
f arbre *m* à suif
r китайское сальное дерево *n*; восковое китайское дерево

1482 chicot; American coffee-tree; Kentucky coffee[-]tree
(Gymnocladus canadensis; Gymnocladus dioicus)
d amerikanischer Schusserbaum *m*; amerikanischer Geweihbaum *m*; Kentucky-Kaffebaum *m*; kanadischer Schusserbaum *m*
f arbre *m* à café du Kentucky; bonduc *m*; chicot *m* févier; chicot du Canada; gros févier *m*
r гимнокладус *m* двудомный; бундук *m* двудомный; бундук канадский; дерево *n* кентуккское кофейное; шикот *m*; шипот *m*

* **chief district forester → 2551**

1483 chief forester; chief forest officer; master forester; forest master; superintendent of forestry; chief forest supervisor
d Hauptförster *m*; Oberforstmeister *m*
f responsable forestier *m*; chef *m* de groupe forestier; chef de service forestier
r главный лесничий *m*

1484 **Chihuahua[n] ash**
(Fraxinus papillosa)
d Chihuahua-Esche *f*
f frêne *m* chihuahua
r ясень *m* чихуахуа

1485 **Chihuahua[n] pine; smoothleaf pine**
(Pinus leiophylla)
d mexikanische gelbe Kiefer *f*
f pin *m* jaune du Mexique
r сосна *f* Чихуахуана; сосна тонколистная

* **Chilean [incense] cedar → 6102**

1486 **Chilg[h]oza pine; Gerard pine; Nepal nut pine**
(Pinus gerardiana)
d Gerard-Kiefer *f*
f pin *m* de Gérard
r сосна *f* Жерарда

* **chim[a]era → 3820**

1487 **chimney; funnel of charcoal pile; infeed throat**
(the somewhat funnel-shaped portion of the infeed that causes the tree to move forward to the feed and downward to the anvil)
d Luftkamin *m* im Holzstapel
f cheminée *f* d'aération
r воздушная полость *f* в штабеле

* **china bark tree → 6669**

* **chinaberry → 6110**

* **Chinawood oil → 9009**

1488 **Chinese alangium**
(Alangium platanifolium)
d platanenblättriges Alangium *n*
f alangier *m* à feuilles de platane
r алангиум *m* платанолистный

1489 **Chinese angelica tree; dong quai**
(Aralia chinensis; Angelica sinensis)
d chinesische Aralie *f*; chinesischer Angelikabaum *m*; Dong Quai *m*
f angélique *f* chinoise; angélique de Chine
r аралия *f* китайская

* **Chinese arbor-vitae → 2725**

1490 **Chinese ash**
(Fraxinus chinensis; Fraxinus sinensis)
d chinesische Esche *f*

* **"Chinese beard" → 1502**

1491 **Chinese bladdernut**
(Staphylea holocarpa)
d chinesische Pimpernuss *f*
f staphylier *m* de Chine
r стафилея *f* китайская

1492 **Chinese catalpa**
(Catalpa ovata)
d kleinblütiger Trompetenbaum *m*; chinesischer Trompetenbaum
f catalpa *m* de Chine
r катальпа *f* яйцевидная; катальпа яйцевиднолистная; катальпа японская

1493 **Chinese cedrela**
(Cedrela sinensis; Cedrela serrulata; Toona sinensis)
d chinesischer Surenbaum *m*
f cédrèle *m*; acajou *m* de Chine
r тона *f* китайская; цедрела *f* китайская; азиатский кедр *m*; сиамский кедр

1494 **Chinese chestnut**
(Castanea mollissima)
d chinesische Kastanie *f*
f châtaignier *m* de Chine
r каштан *m* китайский

1495 **Chinese cinnamon**
(Cinnamomum cassia)
d chinesischer Zimtbaum *m*; Zimtkassie *f*
f cannelier *m* casse
r дерево *n* коричное китайское; корица *f* китайская; коричник *m* кассия; коричник китайский

* **Chinese cork oak → 5951**

* **Chinese date → 1783**

1496 **Chinese elm; lace[-]bark elm**
(Ulmus parvifolia)
d chinesische Ulme *f*; kleinblättrige Ulme
f orme *m* de Chine
r вяз *m* китайский; вяз мелколистный

1497 **Chinese flame tree; Chinese goldenrain tree; goldenrain tree**
(Koelreuteria bipinnata)
d chinesischer Flammenbaum *m*; chinesischer Goldregenbaum *m*
f flamboyant *m* chinois
r кельрейтия *f* дваждыперистая

1498 **Chinese flowering quince**

d chinesische Zierquitte *f*; Cathaya-Zierquitte *f*
f néflier *m* de Cathay
r хеномелес *f* китайская

* **Chinese goldenrain tree → 1497**

* **Chinese gooseberrry → 51**

1499 **Chinese hackberry**
(Celtis sinensis; Celtis bungeana)
d chinesischer Zürgelbaum *m*
f seltis *m* de Chine; micocoulier *m* de Chine
r каркас *m* китайский

1500 **Chinese hawthorn; Pekin haw**
(Crataegus pentagyna; Crataegus pinnatifida)
d fünfgriffeliger Weißdorn *m*
f aubépine *f* pinnatifide; épine *f* de Chine
r боярышник *m* пятипестичный

* **Chinese magnolia → 7225**

1501 **Chinese maple**
(Acer sinensis)
d dreiläppiger Ahorn *m*
f érable *m* de Chine
r клён *m* китайский

1502 **"Chinese moustache"; "Chinese beard"**
(a visible scar in the bark, typical for several broad leaves species)
d chinesischer Schnurrbart *m*; Chinesenbart *m*
f moustache *f* chinoise; moustache de Chine
r бровки *fpl*; китайские усы *mpl*; рубец *m* сучка на коре бука

1503 **Chinese paper birch**
(Betula alba sinensis septentrionalis; Betula albosinensis)
d rote China-Birke *f*
f bouleau *m* blanc du nord
r берёза *f* белая китайская

* **Chinese parasol [tree] → 1303**

1504 **Chinese pear; sand pear**
(Pyrus pyrifolia)
d chinesische Birne *f*; Nashi; japanische Apfelbirne *f*; Asienbirne *f*
f nashi *m*
r груша *f* китайская песочная

* **Chinese pistache → 6295**

1505 **Chinese plum yew**
(Cephalotaxus fortunei)
r тис *m* китайский

* **Chinese privet → 1799**

* **Chinese pyramid juniper → 6649**

1506 **Chinese soapberry; Kashmir soapberry**
(Sapindus mukorossi)
d Waschnuss *f*; Kashmir-Seifenbeere *f*
f sapindus *m* de Mukorossi
r сапиндус *m* Мукоросси

* **Chinese tallow[ ]tree → 1481**

1507 **Chinese tupelo**
(Nyssa sinensis)
d chinesischer Tupelobaum *m*
f tupelo *m* chinois
r нисса *f* китайская

* **Chinese wax → 8958**

1508 **Chinese white pine**
(Pinus armandii)
d Armandskiefer *f*
f pin *m* d'Armand; pin blanc de Chine; pin de David
r сосна *f* белая китайская; сосна Арманда

* **Chinese wood oil → 9009**

1509 **Chinese yellowwood**
(Cladrastis sinensis; Cladrastis amurensis)
d chinesisches Gelbholz *n*
f glycine *m* de Chine; virgilier *m* du fleuve Amour
r кладрастис *m* китайский

1510 **chinkapin oak; chinquapin oak; Mühlenberg oak; yellow [chestnut] oak**
(Quercus muehlenbergii)
d Mühlenbergseiche *f*; Gelbeiche *f*
f chêne *m* jaune; chêne à chinquapin; chêne de Mühlenberg
r бархатный дуб *m*

* **chinking board → 7521**

1511 **chip-bark segregation; chip-bark separation**
d Span-Rinde-Trennung *f*; Span-Rinde-Abscheidung *f*
f séparation *f* de copeaux et d'écorce
r разделение *n* коры и щепы

* **chipbreaker → 1534**

1512 **chip cleaner**
d Spanreiniger *m*
f dispositif *m* de nettoiement de copeaux
r устройство *n* для очистки щепы

* **chip crusher → 1534**

1513 **chip debarking**
d Hackschnitzelentrindung *f*
f écorçage *m* de copeaux
r окорка *f* щепы

* **chip disintegrator → 1534**

1514 **chip dust; wood dust; wood flour; wood meal; wood powder**
d Holzmehl *n*; Holzpulver *n*; Holzstaub *m*
f farine *f* de bois; poudre *f* de bois
r древесная пыль *f*; древесный порошок *m*; древесная мука *f*; отсев *m* щепы; мелкая фракция *f* щепы

1515 **chip formation**
d Spanbildung *f*
f formation *f* de copeau
r образование *n* стружки; стружкообразование *n*

1516 **chipfree cutting**
d spanlose Holzaufteilung *f*; spanlose Holzbearbeitung *f*
f coupe *f* sans copeaux
r безотходное резание *n*; бесстружечное резание

1517 **chip fuel; fuel chips**
d Brennspäne *mpl*; Spanbrennstoff *m*
f copeau *m* combustible
r топливная щепа *f*

1518 **chip grader**
d Schnitzelsortierer *m*
f trieur *m* de copeaux
r сортировальная машина *f* для щепы

1519 **chip groundwood; [disk] refiner groundwood**
d Spanholzschiff *m*
f copeaux *mpl* pour pâte mécanique
r древесная масса *f* из щепы

1520 **chip meter**
d Spanmesser *m*
f jauge *f* à copeaux
r счётчик *m* количества щепы

made by "chip-n-saw" machine into cants, converting part of the outside of them directly into chips without producing any sawdust)
d Chip-n-Saw-Schnittholze *npl*
f grumes *fpl* de chip-n-saw
r брёвна *npl*, идущие на обрезку на станках

1522 **chip packer**
d Hackschnitzel-Füllapparat *n*
f tasseur *m* de copeaux; bourreur *m* de copeaux
r уплотнитель *m* щепы

1523 **chip packing**
d Spanfüllung *f*
f compression *f* des copeaux
r уплотнение *n* щепы

1524 **chipped grains; ruptured grains; torn grains**
(fibers which are torn out below the finished surface, particularly around knots and curly places)
d Spaltholzfaser *fpl*; [los]gerissene Holzfaser *fpl*
f fils *mpl* brisés; fils fragmentés; grains *mpl* éclatés; veines *fpl* déchirées
r задранные волокна *npl*; разорванные волокна; расщеплённые волокна; вырванные волокна

* **chipped wood → 9508**

* **chipper → 1534, 2142**

* **chipper canter → 1316**

1525 **chipper chainsaw**
d Schnitzler-Kettensäge *f*
f scie *f* à chaîne à maillons-gouge
r пила *f* с цепью, имеющей строгающие [Г-образные] зубья

1526 **chipper discharge**
(denotes the direction chips leave the chipper housing)
d Schnitzler-Entladung *f*; Chipperentladung *f*
f décharge *f* de déchiqueteuse
r выпуск *m* щепы рубительной машины

1527 **chipper-edger**
d Besäumschnitzler *m*; Besäumfräser *m*
f déligneuse-déchiqueteuse *f*; coupeuse-déligneuse *f*
r фрезерно-обрезной станок *m*; фрезерно-пыльный станок (для выпиливания из брёвен пиломатериалов с одновременным измельчением отхода)

* **chipper headrig → 1316**

1528 **chipper infeed[er]**
(series of rollers at the front portion of the chipper where the material to be chipped enters)
d Chippereinstechbewegung *f*
f alimenteur *m* de coupeuse
r механизм *m* подачи рубительной машины

1529 **chipper[man]**
(a worker who assists tree trimmer in clearing trees, branches and brush interfering with electrical power lines. He performs all his work on the ground)
d Häcksler *m*
f conducteur *m* de coupeuse de bois
r патронщик *m*; рубщик *m* (рабочий, обслуживающий рубильную машину)

1530 **chipping**
(rewounding and extending the face, i.e. removing the solidified oleoresin, a narrow band of bark, and also of wood, so as to renew the flow of oleoresin)
d Lachtenröten *n*
f gemmage *m*; piquage *m*
r нанесение *n* подновки на карру; вздымка *f*

1531 **chipping; chopping**
(a site preparation technique in which slash, brush and logging debris are broken into smaller pieces following harvest)
d Zerkleinerung *f*; Hacken *n*
f déchiquetage *m*; mise *f* en copeaux
r дробление *n*; измельчение *n*

1532 **chipping-and-hauling operation**
d Zerkleinerungs- und Transportierungsoperation *f*
f opération *f* de déchiquetage et débardage
r операция *f* заготовки и вывозки [щепы]

* **chipping canter → 1316**

1533 **chipping crew**
d Zerspanen-Besatzung *f*
f équipe *f* de déchiquetage
r бригада *f* на заготовке щепы

* **chipping headrig → 1316**

1534 **chipping machine; chipbreaker; chipper; chopper; wood-fragmenting machine; [hammer] hog; hogger; kibbler; hog[ging] machine; chip crusher; chip disintegrator**
d Hackspänezerkleinerer *m*; Spanbrecher *m*
f déchiqueteuse *f*; broyeur *m*
r дробилка *f* [древесины]; дробилка для щепы; дезинтегратор *m*; измельчитель *m* щепы; измельчитель стружек

* **chipping operation → 3759**

* **chippings → 9508**

* **chipping slash → 9508**

* **chips → 9508**

1535 **chip screen; chip classifier**
d Hackspänesortierer *m*; Hackspänesichter *m*
f crible *m* à copeaux; tamis *m* à copeaux; classeur *m* pour copeaux
r сортировочное сито *n* для щепы

1536 **chips damage**
d Spanbruch *m*
f dégât *m* de copeaux
r повреждённость *f* частиц щепы; смятие *n* частиц щепы

1537 **chip separator**
(attachment to whole-tree chippers that separates acceptable chips from unacceptable bark, limbs, and foliage)
d Spantrenner *f*
f séparateur *m* de copeaux
r разобщитель *m* щепы

1538 **chips from forest residue**
d Holzabfällespäne *mpl*
f copeaux *mpl* de résidus forestiers
r щепа *f* из древесных отходах

1539 **chips from the stem; stem chips**
d Stammspäne *mpl*
f copeaux *mpl* de tige
r щепа *f* из ствольной части дерева; стволовая щепа

* **chip silo → 2496**

1540 **chip size**
d Spangröße *f*
f taille *f* de copeaux
r размер *m* стружки

1541 **chip streak**
d Rissabstand *m*
f strie *f* de crevasse; strie de carre
r шаг *m* подновки

1542 **chipwood**
(low grade industrial timber)
d Späneholz *n*
f panneau *m* aggloméré
r материал *m* из стружки

1543 **chir pine; longleaf Indian pine; emodi pine**
(Pinus roxburghii; Pinus longifolia)
d langblätterige Kiefer *f*
f pin *m* chir
r сосна *f* длиннохвойная; сосна Роксбурга

* **chisel → 9509**

1544 **Chittam-wood; Cascara [buckthorn]; bearwood**
(Rhamnus purshianus)
d amerikanischer Faulbaum *m*; Purgier-Faulbaum *m*
f nerprun *m* d'Amérique; nerprun de Pursh; cascara *m*
r медвежье дерево *n*; горькая кора *f*; крушина *f* американская

1545 **chlorophyll**
(green photosynthetic substance in plants which allows them to capture solar energy)
d Chlorophyll *n*
f chlorophylle *f*
r хлорофилл *m*

1546 **chlorosis**
(blanched or yellowish colo[u]ring in plants caused by nutrient or light deficiency)
d Chlorose *f*; Gelbsucht *f*
f chlorose *f*; jaunisse *f*
r хлороз *m*

1547 **chlorotic dwarf**
(an abiotic disease of Pinus strobus characterized by reduced growth, chlorosis and mottling of the needles, and premature abscission of all but current needles)
d chlorotischer Zwerg *m*
f arbre *m* nain chlorotique
r хлоротическое карликовое растение *n*; хлорозное карликовое растение

1548 **chock [block]; sill**
(a wedge or block for steadying a body and holding it motionless or for filling an unwanted space, or for blocking the movement of a wheel)
d Unterklotzung *f*; Unterlegkeil *n*; Schalkkeil *n*
f cale *f*
r поперечное бревно *n*

1549 **chocolate-brown stain; [coffee-]brown stain; fungous brown-stain; brown streak**
(a rich brown to deep chocolate-brown discolo[u]ration of the sapwood of some pines caused by a fungus)
d Braunverfärbung *f*
f coloration *f* brune fongique; tache *f* brune
r заболонная грибная коричневая окраска *f*; побурение *n*; бурая гниль *f*

1550 **chocolate family**
(Sterculiaceae)
d Sterkuliengewächse *pl*
f sterculiacées *pl*
r стеркулиевые *pl*

* **chocolate tree → 1248**

1551 **choice of tree species; tree species choice**
d Baumartenwahl *f*; Holzartenwahl *f*
f choix *m* d'espèce arborescente
r выбор *m* древесных пород

* **choice tree → 6396**

1552 **choke** *v*; **strop** *v* **up**
d drosseln; abziehen; verstopfen
f estroper
r чокеровать

1553 **chokeberry**
(Aronia)
d Apfelbeere *f*; Apfelbeerstrauch *m*
f aronia *m*; aronie *f*
r арония *f*

* **chokecherry → 1769**

1554 **choker arch skidder; choker arch unit; choker arch logging vehicle**
d Choker-Bodengewölbe *n*
f débardeur *m* à câble et arche
r трелёвочный трактор *m* с аркой и чокерами

1555 **choker [cable]; yoker; bitch line; bitch chain; butt line; hawser**
(short length of flexible wire, rope, or chain used to attach logs to a winch line or directly to a tractor)
d Drahtseilschlinge *f*; Chokerseil *n*; Choker *m*; Schleifkette *f*; Schlinge *f*; Zugtrosse *f*; Seilauge *n*
f chaîne *f* de débardage; câble *m* de traction; collier *m* à boucle; ganse *f*; lacet *m*

1556 **choker-hole**
(under tree or log)
d Stammunterhölung *f* für Kette
f saignée *f* de débusquage; creux *m* pour le pallage d'une chaîne de débardage
r углубление *n* для продевания чокера; углубление ствола для цепи

1557 **choker hook; bull hook; butt hook**
(heavy hook on the butt rigging to which chokers are attached; hook that attaches the dragline to the tackle on logs)
d Chokerhaken *m*
f crochet *m* de suspension; crochet de nœud coulant; crochet de culée
r крюк *m* для крепления чокеров; чокерный крюк; чокерный замок *m*

1558 **chokerless skidding**
d chokerloses Rücken *n*
f débusquage *m* sans lacet
r бесчокерная трелёвка *f*

1559 **chokerman; choker setter; grab setter; tong hooker; ground loader; bottom loader; send-up man; frogger; rigging slinger**
(a person in a logging operation who places the choker around the log to be hauled to the landing)
d Seilschlingemann *m*; Stammanhänger *m*; Takelage-Operator *m*
f élingueur *m*
r рабочий *m* гужевой трелёвочной бригады; чокеровщик *m*; рабочий на лесосеке, руководящий движением канатов и прицепкой груза при канатной трелёвке

1560 **choking; choker setting; hooking**
d Anhängen *n*; Chokern *n*
f accouplement *m*; étranglement *m*
r чокеровка *f*; забивка *f*; засорение *n*; скопление *n* (опилок в рубанке)

1561 **chop** *v*
(to cut into pieces)
d hacken
f hacher
r рубить; колоть; тесать; отрезать

* **chopped billet** → **2171**

1562 **chopped log**
d abgehackter Holzklotz *m*

* **chopper** → **1534, 9057**

* **chopping** → **1531**

1563 **chopping board; spring[ ]board** (*US*)
(a short board or plank, its end notched into the bole, on which the cutter stands so as to enable him to fell the tree at a level not reachable from the ground)
d Fälltisch *m*
f échafaudage *m* d'abattage; table *f* d'abattage; semelle *f* d'abattage
r доска *f*, на которой стоит лесоруб; подмостки *fpl* для валки крупных деревьев

* **chopping machine** → **9512**

* **chopping shreds** → **7553**

1564 **chord**
(straight line that joins the end points of any arc; in cable yarding, it is used to determine cable deflection)
d Sehne *f*
f corde *f*; membrure *f*
r пояс *m*; подвесная струна *f*

1565 **chord felling notch**
d Sehnenfallkerb *m*
f entaille *f* d'abattage en corde d'arc
r тангенциальный надруб *m*

* **chore** → **3740**

* **chosen tree** → **6396**

1566 **chosen tree system**
d Zukunftsstammwirtschaft *f*; Zielstammwirtschaft *f*
f système *m* de traitement sur arbres d'avenir
r хозяйство *n*, направленное на уход за лучшими стволами

* **Christmas bush** → **866**

1567 **Christmas log; Yule[tide] log**
(large log traditionally burned at Christmas)
d Julfest-Schnittholz *n*
f bûche *f* de Noël
r большое полено *n*, сжигаемое в святки

1568 **Christmas palm**
(Veitchia merillii)
d Manila-Palme *f*; Weihnachtspalme *f*

1569 Christmas tree
d Christbaum *m*; Weihnachtsbaum *m*; Weihnachtsbrauch *m*
f arbre *m* de Noël; sapin *m* de Noël
r рождественская ёлка *f*

1570 Christmas tree farming
d Weihnachtsbaumzucht *f*; Weihnachtsbaumkultur *f*
f culture *f* d'arbres de Noël
r разведение *n* рождественных ёлок

* **chrysalis → 4809**

* **chump → 2171, 3657**

1571 chunk
d Knollen *m*; Klumpen *m*; Stück *m*
f fragment *m*; morceau *m*
r комок *m*; обломка *f*

1572 chunk bucker
d Klumpenbucker *m*
f tronçonneur *m* de fragments
r рабочий *m* на подготовке лесосеки

* **chunking machine → 1574**

1573 chunk *v* up
(to clean up and pile debris after logging an area)
d Holzreste sammeln und anhäufeln
f ramasser et paqueter les débris
r собирать и окучивать лесосечные отходы

* **chunkwood chipper → 1574**

1574 chunkwood processor; chunkwood chipper; chunking machine; wood chunker
d Holzstückzerkleinerer *m*
f fragmenteur *m* [de bois cassé]
r машина *f* очистки лесосечных отходов

1575 church forest; parish forest; ecclesiastical forest
d Kirchenforst *m*; Kirchenwald *m*; Pfarrwald *m*; Pfründwald *m*
f forêt *f* d'église; forêt ecclésiastique
r церковный лес *m*

* **churn-butted → 8493**

* **churn-butted stem → 595**

1576 chute greaser; skid greaser; greaser
(one who applies skid grease to a chute)
r смазчик *m* лесоспуска

1577 chute transportation; ditch transportation
d Rutschtransport *m*
f transportation *f* par glissoire
r лесоспуск *m* по земляному лотку

* **cicatrization → 5862**

1578 cider gum; cider eucalyptus
(Eucalyptus gunnii)
d Gunnseukalyptus *m*; Mostgummibaum *m*
f eucalyptus *m* commun; gommier *m* cidre; eucalyptus de Gun
r эвкалипт *m* Ганна

* **cigar box cedar → 1411**

1579 ciliate leaf
(bearing marginal hairs; bordered with hairs)
d Cilienblatt *n*
f feuille *f* ciliée
r реснитчатый лист *m*; ресничный лист

1580 ciliate-leaved; ciliatifolious; blepharophyllous; fringe-leaved
d wimperblätt[e]rig
f ciliatifolié
r реснитчатолистный; реснитчатый

1581 Cilician fir
(Abies cilicica)
d cilicische Tanne *f*
f sapin *m* de Cilicie
r сицилианский дуб *m*

* **cinchona tree → 6669**

1582 cinnamon clethra; mountain sweet pepperbush
(Clethra acuminata)
d Scheineller *m*
f cléthre *f* blanche
r клетра *f* остроконечная

1583 cinnamon[ic] soils
d Zimtbodene *mpl*
f sols *mpl* cinnamiques; cinnamon sols
r коричневые почвы *fpl*

* **cinnamon tree → 1421**

* **Circassian walnut → 1808**

1584 circin[n]ate leaf
(curved into a circle so that the apex is nearly

f feuille *f* circinée
r свёрнутый [в трубочку] лист *m*; закрученный лист

* **circular area at breast height → 598**

* **circular bench saw → 1586**

* **circular edger → 6676**

1585 **circular saw; disk saw; rotary saw; buzz saw**
d Kreissäge *f*; Zirkularsäge *f*
f scie *f* circulaire
r круглая пила *f*; дисковая пила; циркульная пила

1586 **circular [saw]mill; radial[-arm] saw; circular bench saw; bolter [saw]**
d Sägemaschine *f* mit rollierende Einspannwagen; Auslegerkreissäge *f*; Radialarmkreissäge *f*
f scierie *f* à scie circulaire; scie *f* circulaire tous azimuts; scie radiale
r круглопильный станок *m*

1587 **circular sawn lumber**
d rundgesägtes Schnittholz *n*
f grume *f* sciée à scie circulaire
r полученный на круглопильном станке пиломатериал *m*

1588 **circular yarding method**
d Kreisrückeverfahren *n*
f méthode *f* de débardage circulaire
r круговой способ *m* канатной трелёвки

1589 **circulation of sap; sap flow; exudation of sap**
d Saftzufluss *m*; Saftausfluss *m*; Saftzufuhr *f*; Saftstrom *m*
f écoulement *m* de sève; courant *m* de la sève
r передвижение *n* сока; поток *m* сока; сокообращение *n*; сокодвижение *n*; истечение *n* сока

* **circulation period → 7077**

* **circumference at the middle → 5440**

* **CISC → 1899**

1590 **citron tree**
(Citrus medica)
d Zitronatzitronenbaum *m*
f cédratier *m*; cédrat *m*
r цитрусовое дерево *n*; цитрон *m*

1591 **citron wood**
d Zitronenholz *n*
f bois *m* de limon
r древесина *f* лимонного дерева; древесина сандаракового дерева

* **city forest → 9142**

1592 **city forester**
d Stadtoberförster *m*
f forestier *m* communal
r городской лесничий *m*

1593 **cladina forest type**
d Cladinawaldart *f*
f type *m* de forêt lichénique
r тип *m* леса лишайниковый

1594 **cladode; phyllode**
(a flat stem looking like a leaf and photosynthetic)
d Phyllode *f*
f cladode *m*; phyllode *f*
r кладодий *m*; листообразный стебель *m*

* **cladous → 977**

* **clammy azalea → 8303**

1595 **clammy locust; viscous robinia**
(Robinia viscosa; Acacia howittii)
d klebrige Robinie *f*
f robinier *m* poisseux; robinier visqueux
r акация *f* клейкая

1596 **clamp[-]bunk**
(payload bed of a forwarder, equipped with top-opening hydraulic jaws)
d Klemmschemel *m*
f pince *f* portante
r зажимный коник *m*

1597 **clamp[-bunk] skidder; bunk [grapple] skidder; tree-length transporter; bunk jaw skidder; clamp tractor**
d Klemmschemelrücker *m*
f débardeur *m* à pince portante
r трактор *m* со стрелой и коником с зажимными рычагами; трактор с коником для бесчокерной трелёвки

* **clamped tree → 9053**

* **clapboard → 7521**

* **Clara walnut → 241**

1598 clarification; lighting
d Aufhellen *n*; Umlichtung *f*; Auflichtung *f*
f clarification *f*; éclairement *m*
r осветление *n*

* **Claro walnut → 241**

* **clary → 7152**

1599 classes of protective functions of forests
d Klassen *fpl* der Waldschutzfunktionen
f classes *fpl* de fonctions protectrices de forêts
r категории *fpl* защитности лесов

1600 class of crown closure; crown closure class
(a tree classification system based on the tree's relative height, foliage density and ability to intercept light)
d Waldschlussklasse *f*
f classe *f* de fermeture [du couvert]
r класс *m* сомкнутости крон

1601 class of fire danger; fire-size class; fire [cause] class; danger class
d Klasse *f* der Brandgefahr; Waldbrandgrößenklasse *f*
f classe *f* d'incendies par surfaces
r класс *m* пожарной опасности; класс пожара; вид *m* пожара

1602 class of tree compactness
d Klasse *f* für Vollholzigkeit
f classe *f* de densité d'arbre
r класс *m* полнодревесности дерева

* **clean-barked blocks → 1609**

1603 clean barking; barking in full
d Blankschälen *n*; Weißschälen *n*
f écorçage *m* à blanc
r чистая окорка *f*; полная окорка; окорка всей поверхности

1604 clean bole; clear bole
d astreiner Schaft *m*; schaftreiner Stamm *m*; reiner Schaft
f grume *f* sans nœuds; fût *m* net [de nœuds]
r бессучковый ствол *m* ; ствол без сучьев; чистый ствол

* **clean burn[ing] → 1043**

1605 clean chips
d Blankspäne *mpl*
f copeaux *mpl* sans écorce
r стружки *fpl* без остатков коры

f coupe *f* nette et franche; tranche *f* nette; tranche sans bravure
r чистый срез *m*; ровный срез; чистый пропил *m*; гладкий пропил

1607 clean cutting; clearcut harvest; clear[-]cut[ting]; clear[-]felling; cutover; cutting-down
(removal of an entire standing crop of trees)
d Kahlabtrieb *m*; Kahlhieb *m*; Kahlschlag *m*; Totalitätshieb *m*; Abtrieb *m*
f coupe *f* à blanc; coupe rase; coupe totale
r сплошнолесосечная рубка *f*; сплошная рубка; сплошная лесосека *f*; вырубка *f*; срубка *f*

* **cleaning cutting → 7193**

1608 cleaning-up
d Feinhobeln *n*; Schlichthobeln *n*
f rabotage *m* final
r окончательное строгание *n*; чистовое строгание; прецизионное строгание (древесины)

* **clean log → 1626**

* **clean lumber → 9578**

1609 clean-peeled pulpwood; clean-barked blocks; clean-peeled blocks
d weißgeschältes Papierholz *n*; reingeschältes Papierholz; weißgeschälte Rollen *fpl*
f billots *mpl* écorcés à blanc
r чисто окорённые балансы *mpl*; окорённые набело балансы; балансы чистой окорки

* **clean stuff → 9578**

* **clean timber → 9578**

1610 clean-up crew
d Räumbrigade *f*
f équipe *f* de nettoiement
r бригада *f* на очистке [лесосек]

* **clean wood → 2351**

1611 clear *v*
d ausstocken
f défricher
r расчистить; расчищать

* **clearance → 3443, 3700**

* **clearance of branches → 6610**

1612 **clear-and-better**
d klar-und-besser
f net-et-meilleur
r высокосортный (о некоторых североамериканских породах древесины)

* **clear bole → 1604**

1613 **clear-boled; clear-shalted**
(ref. to forest or crown)
d glattschäftig
f à fût lisse
r гладкоствольный; осветлённый

1614 **clear-boled forest**
d glattschäftiger Wald *m*
f forêt *f* à fûts net
r гладкоствольный лес *m*

* **clear burning → 1043**

* **clear cut → 1606**

* **clearcut harvest → 1607**

1615 **clearcut saw**
d Freischneider *m*; Freischneidegerät *n*; Sichelsäge *f*
f scie *f* pour coupe rase
r каток-кусторез *m*

* **clear[-]cut[ting] → 1607**

* **clearcutting area → 1621**

1616 **clearcutting forest; clearfelling forest**
d Kahlschlagwald *m*
f forêt *f* de coupe totale
r лес *m* со сплошной рубкой

1617 **clearcut[ting] method; clearfelling method**
(nominal silvicultural method in which all the trees are cleared over a large area at one time in a brief period)
d Abholzungsmethode *f*; Kahlschlagsmethode *f*; Einschlagverfahren *n*; Hiebsart *f*
f méthode *f* de régénération par coupe à blanc; mode *m* de régénération par coupe totale
r метод *m* сплошной рубки

* **clear cuttings → 9578**

* **clearcut[ting] unit → 1621**

1618 **clearcutting with artificial reproduction**
d Kahlschlag *m* mit Nachverjüngung
искусственным возобновлением

1619 **clearcutting with natural reproduction**
d Kahlschlag *m* mit Naturverjüngung
f coupe *f* totale avec régénération naturelle
r сплошная рубка *f* с последующим естественным возобновлением

1620 **clearcut[ting] with reserves**
(a clearcutting method in which varying numbers of reserve trees are not harvested to attain goals other than regeneration)
d Kahlschlag *m* mit Reserve
f coupe *f* totale avec réserve; coupe à blanc avec réserves
r сплошная вырубка *f* с оставлением резервных деревьев

* **clear face → 686**

* **clear fell → 2168**

1621 **clear-felled area; clearcutting area; clearcut[ting] unit; area under felling**
d Kahlfläche *f*
f zone *f* de coupe à blanc; surface *f* d'exploitation à blanc
r площадь *f* со сплошной рубкой; лесосека *f* со сплошной рубкой

* **clear[-]felling → 1607**

* **clearfelling forest → 1616**

* **clearfelling method → 1617**

1622 **clearfelling with narrow coupes**
d Kahlschlag *m* mittels Schmalbande
f coupe *f* à blanc par bandes étroites
r сплошные рубки *fpl* узкими лесосеками

* **clearing → 390, 3443**

1623 **clearing by slash scattering**
d Reinigung *f* mittels Holzabfallstreuung
f essartage *m* par épandage de rémanents; nettoiement *m* par dispersion de rémanents
r очистка *f* [лесосек] разбрасыванием порубочных остатков

1624 **clearing of the felling site; clearance of the felling area; removal of felling area**
d Reinigung *f* des Holzschlages; Abräumung *f* des Holzschlages; Flächenräumung *f* des

f nettoiement *m* de coupe; enlèvement *m* des rémanents de coupe; nettoiement après exploitationn
r очистка *f* лесосеки; уборка *f* хлама

* **clearing timber → 7760**

1625 **clear length**
(portion of the tree between the ground and the point where the lowest limbs join the trunk)
d astreine Länge *f*
f longueur *f* sans branches
r бессучковая длина *f*

1626 **clear log; clean log**
(log without knots; first quality log)
d reines Schnittholz *n*
f grume *f* nette [de nœuds]; grume franche de nœuds
r бревно *n* без сучьев; чистое бревно

* **clear lumber → 9578**

* **clears → 9578**

* **clear-shalted → 1613**

* **clear span → 4504**

* **clear stand → 2926**

* **clear wood → 2351**

1627 **cleavability; splittability**
d Spaltbarkeit *f*; Spaltfähigkeit *f*
f aptitude *f* au clivage; fissilité *f*
r расщепляемость *f*; раскалываемость *f*; колкость *f*

1628 **cleavable timber; cleavable wood; fissile wood**
d spaltbares Holz *n*
f bois *m* clivable; bois fissile
r раскалываемый лесоматериал *m*; колкая древесина *f*

1629 **cleavage; cleaving; splitting**
(of wood)
d Absplitterung *f*; Abspaltung *f*; Splitterung *f*
f clivage *m*
r расщепление *n*; раскалывание *n*; отщепление *n*

1630 **cleavage face**
d Spaltungsebene *f*; Spaltseite *f*
сегмента колотой древесины на торце

* **cleavage strength → 6907**

1631 **cleave** *v*; **sliver** *v*
d absplittern; spalten
f cliver
r расщеплять[ся]; раскалывать[ся]

* **cleaved wood → 8563**

1632 **cleaver**
(a worker)
d Holzspalter *m*
f fendeur *m*
r дровокол *m*

* **cleaver → 453**

* **cleaver [knife] → 7960**

* **cleave saw → 5148**

* **cleaving → 1629**

* **cleaving axe → 453**

* **cleaving machine → 9512**

1633 **cleaving saw**
d Brettersäge *f*
f scie *f* de long; scie à refendre à 2 hommes
r развальная лесопильная рама *f*; распускной пильный станок *m*; продольная пила *f*

* **cleaving timber → 8563**

* **cleaving wedge → 3717**

1634 **cleft; riven; ruptured** *adj*
d Spalt-
f fendu à la dimension
r расщеплённый; расколотый

1635 **cleft grafting; wedge grafting; stock grafting**
d Spaltpfropfen *n*; Spaltpropfung *f*; Spaltveredelung *f*
f greffage *m* en fente; greffage sur fente
r прививка *f* в расщеп; прививка в щель; прививка клином

* **cleft-graft scion → 1145**

1636 **cleft leaf; lobed leaf**

f feuille *f* à lobes
r дольчатый лист *m*; раздельный лист; лопастный лист

* **cleft timber → 8563**

* **cleft wood → 8563**

1637 **clethra family**
(Clethraceae)
d Klethragewächse *pl*
f clétracées *pl*
r клетровые *pl*

1638 **climax [community]; climax association**
(the final stage of plant succession for a given environment, the vegetation being conceived as having reached a highly stable condition)
d Klimax *f*; standortgemäße Mischung *f*
f climax *m*; mélange *m* ortho-écologique
r климаксовое растительное сообщество *n*; климакс *m*; смесь *f* сообразно местопроизрастанию

1639 **climax forest; radical forest**
(a plant community dominated by trees representing the culminating stage of natural succession for that specific locality and environment)
d Klimaxforst *m*; Klimaxwald *m*; Schlusswald *m*
f forêt *f* climax; forêt climacique; forêt définitive
r коренной лес *m*; первичный лес; климакс-лес *m*

1640 **climax tree species**
d Klimax-Holzart *f*; standortgemäße Holzart *f*
f espèce *f* climax; espèce ortho-écologique
r климаксовая древесная порода *f*

1641 **climb *v* a tree; get *v* in a tree**
d Baum ersteigen; Baum erklettern
f monter d'arbre
r лезть по дереву; влезть на дерево

* **climber → 1642, 4296**

* **climbing bittersweet → 239**

1642 **climbing plant; climber; creeper; trailer**
d Kletterpflanze *f*; kriechende Pflanze *f*
f plante *f* grimpante; plante rampante
r ползучее растение *n*; лазящее растение; стелющееся растение

1643 **climbing rope**
d Kletterseil *n*
f corde *f* de sécurité
r основная верёвка *f*

1644 **climbing trees**
d Kletterbäume *mpl*
f arbres *mpl* grimpants
r стелющиеся деревья *npl*

* **clinometer → 4443**

1645 **clip *v*; lop *v*; pare *v*; prune *v***
d beschneiden
f clipser; tailler; rhabiller
r обстричь; обрезать; обрывать; откусывать

* **clipped tree → 7489**

* **clippings → 9508**

* **clockwise spiral growth → 4907**

* **clod → 7811**

* **clog → 366**

1646 **clonal seed orchard; seedling seed orchard**
d Sämlingsamenobstgarten *m*
f verger *m* à graines clonal; verger à graines de familles
r клоны *mpl* семенных деревьев, выращенных из сеянцев

1647 **clonal test; clonal trial**
(evaluation of genotypes by comparing clones in a plantation)
d klonischer Test *m*
f test *m* clonal
r испытание *n* клонов; исследование *n* клоновых признаков

1648 **clone tree**
(a tree that is reproduced by only using the cells of one tree)
d Klone *m* eines Bäumes; Klonsbaum *m*
f clone *m*
r клон *m*

1649 **cloning**
d Klone *m*
f clonage *m*
r клонинг *m*

* **close breeding → 4435**

1650 **closed canopy; congested canopy; continuous canopy; dense leaf canopy; complete canopy; full canopy; close stand; complete cover**

(in which the light cannot reach the forest floor directly)
d geschlossener Schluss *m*
f couvert *m* fermé; couverture *f* dense
r сомкнутый полог *m*; густой полог; полог крон

* **closed check → 958**

1651 **closed community**
(a plant community whose components are so completely utilizing the site as to exclude further entrants)
d geschlossene Gemeinschaft *f*
f communauté *f* fermée
r сомкнутое сообщество *n*

1652 **close[d] fire season**
(ref. to forest area)
d geschlossene Brandsaison *f*
f période *f* d'interdiction de faire du feu
r закрытый на пожароопасный сезон

* **closed forest → 3756**

1653 **closed forest area; closed wood[land]**
(forest with a minimum of 30% tree canopy cover)
d geschlossenes Waldgebiet *n*
f forêt *f* fermée
r лес *m*, закрытый для посещений; лесной заказник *m*

1654 **closed interlock system**
d geschlossenes Vorriegelungssystem *n*
f système *m* de verrouillage fermé
r замкнутая система *f* сблокированных канатов

* **closed split → 958**

* **closed wood[land] → 1653**

1655 **close grain; fine grain; fine texture; close texture**
(wood texture with small and closely spaced pores)
d feines Korn *n*; engringige Textur *f*
f fil *m* fin
r мелкослойная текстура *f*; тонкая текстура; текстура с мелкими элементами; мелкий текстурный рисунок *m*

1656 **close-grained wood; close-grown wood; fine-grown wood; fine-grained wood; narrow-pored wood**
d kleinkörniges Holz *n*; engporiges Holz; mildes Holz; feinkörniges Holz
f bois *m* de fil fin; bois à grain fin; bois à pores étroits
r мелкопористая древесина *f*; мелкозернистая древесина

1657 **close herding**
(handling a band compactly and restricting its dispersal over, and therefore use of, the range area)
d dichtes Hütten *n*
f gardiennage *m* en troupeau serré; pâturage *m* rationné
r замкнутая пастьба *f*

1658 **closely leaved**
d überblätterig; überbelaubt
f excessivement feuillu
r загущённолистный

* **closely ringed → 9470**

1659 **closeness of stand; closing of leaf canopy; canopy closure**
(the progressive reduction of space between crowns as they spread laterally, increasing canopy cover)
d Bestandesschluss *m*
f fermeture *f* du couvert
r сомкнутость *f* полога; смыкание *n* лесного полога

* **close piling → 1157**

* **close-ringed → 9470**

1660 **close-ringed wood**
(wood with narrow, inconspicuous annual rings)
d engringiges Holz *n*; feinringiges Holz
f bois *m* à cernes étroits; bois à cernes minces
r древесина *f* с узким расположением годовых колец; узкослойная древесина; мелкослойная древесина; тонкослойная древесина

* **close stack → 8552**

* **close stacking → 1157**

* **close stand → 1650**

* **close texture → 1655**

1661 **closing line**

f câble *m* ferme-benne; ligne *f* de fermeture
r закрывающий канат *m*; замыкающий канат

* **closing membrane → 6308**

* **closing of leaf canopy → 1669**

1662 clouded wood
d geadertes Holz *n*
f bois *m* tacheté
r волнистая древесина *f*; древесина [, испещрённая] с прожилками; запятнанная древесина

* **cloud forest → 5593**

1663 cloud forest belt
d Wolkenwaldstufe *f*
f étage *m* forestier au niveau de nuages
r пояс *m* лесов на уровне стояния

* **cloven billet → 7920**

* **cloven logs → 8563**

1664 clove tree
(Eugenia caryophyllata; Caryophyllus aromaticus; Syzygium aromaticum)
d Gewürznelkenbaum *m*; Gewürznelke *f*
f clou *m* de girofle; giroflier *m*; géroflier *m*
r гвоздичное дерево *n*; гвоздика *f*

1665 clump of coppice shoots
d Stockausschlagsgruppe *f*
f bouquet *m* de rejets de taillis
r группа *f* корневых отпрысков

1666 clump of trees; tree clump; clumpwood; cluster; [tree] tuft; bunch
(1. group of trees of similar ages and dimensions; 2. aggregate of stems issuing from the same root, rhizome system, or stool)
d Baum-Büschel *n*; Baumcluster *m*
f bouquet *m* d'arbres; touffe *f* d'arbres; bouquet *m*; boqueteau *m*
r пучок *m* деревьев

* **cluster-cherry → 2881**

1667 cluster development; planned unit development; PUD
(grouping structures in a housing development tightly together and using the open land thus saved for common greens and squares)
d Clusterentwicklung *f*
зон[, окружённых зелёными поясами]

* **cluster fig → 1670**

* **cluster knots → 3997**

1668 cluster of knots; knot cluster; branch cluster; dasher
d Astansammlung *f*; Holzquirl *m*; Astkranz *m*; Quirl *m*
f couronne *f* de nœuds; nœuds *mpl* en grappe
r группировка *f* сучьев

* **cluster-pine → 5309**

1669 cluster pores
d Gruppenporen *fpl*
f pores *fpl* groupées
r групповые поры *fpl*

1670 clustertree; cluster fig
(Ficus glomerata; Ficus racemosa)
d Traubenfeige *f*
f figuer *m* à grappes
r фикус *m* клубковатый

* **CMAI → 2126**

1671 cneorum family
(Cneoraceae)
d Zwergölbaumgewächse *pl*
f cnéoracées *pl*
r кнеоровые *pl*

* **C/N ratio → 1333**

1672 coach timber; coachwood; wheelwright wood; carriage building timber; cartwright's timber
d Wagnerholz *n*; Stellmacherholz *n*
f bois *m* de charronnage
r обозный лесоматериал *m*; каретный лес *m*; лесоматериал для столярных кузовных работ

* **coachwood → 1672**

1673 coal
(the most abundant fossil fuel, derived from prehistoric plant materials that were exposed to high temperatures and pressure)
d Steinkohle *f*; Kohle *f*
f charbon *m*
r уголь *m*

* **coaler → 1445**

* **coalescence → 35**

1674 **coal-off**
d Holzkohlenkahlschlag *m*
f coupe *f* totale à carbonisation
r сплошная вырубка *f* для углежжения

1675 **coarse bark**
d Grobrinde *f*
f écorce *f* grossière; écorce épaisse et rugueuse; écorce brute
r грубая кора *f*; толстая кора

1676 **coarse-fibered wood; coarse-grained wood; coarse-grown wood; coarse-textured wood**
d grobfaseriges Holz *n*; grobkörniges Holz
f bois *m* de fibres brutes; bois de fils grossiers; bois de fils bruts; bois à grain grossier
r грубоволокнистая древесина *f*; крупноволокнистая древесина

* **coarse file → 7089**

1677 **coarse fuel chips**
d Grobbrennspäne *mpl*
f copeaux *mpl* combustibles bruts
r грубая топливная щепа *f*

1678 **coarse grain; open grain; coarse texture; open texture**
d grobe Textur *f*
f grain *m* brut; grain grossier; gros grain; texture *f* brute
r широкослойная текстура *f*; грубая текстура; крупный текстурный рисунок *m*; текстура с крупными элементами

1679 **coarse-grained forest soil**
d grobkörniger Waldboden *m*
f sol *m* forestier à gros grains
r крупнослойная лесная почва *f*

* **coarse-grained wood → 1676**

1680 **coarse groundwood**
d Grobholzmasse *f*
f pâte *f* de bois brute
r грубая древесная масса *f*

* **coarse-grown wood → 1676**

1681 **coarse humus; raw humus; crude humus; mor**
(the layer of needles, leaves, and decaying matter on the forest floor)
d Auflagehumus *m*; Rohhumus *m*

1682 **coarse humus litter**
d Rohhumusstreu *f*
f couche *f* de humus brut
r грубогумусовая подстилка *f*

1683 **coarse limbing**
d grobes Entasten *n*
f ébranchage *m* brut
r грубая очистка *f* от сучьев; черновая очистка от сучьев

1684 **coarse mill residue**
(wood residue suitable for chipping such as slabs, edgings, and veneer cores)
d grobes Brackholz *n*
f rémanents *mpl* bruts d'usinage
r грубая древесина *f*, отбракованная на фабрике

1685 **coarse-porous wood; coarse-pored wood; large-pored wood**
(wood texture in which large pores are present)
d grobporiges Holz *n*
f bois *m* à pores grossiers
r крупнопористая древесина *f*

* **coarse-ringed timber → 5909**

* **coarse texture → 1678**

* **coarse-textured wood → 1676**

1686 **coarse woody debris; CWD; coarse woody residue; large woody debris; LWD**
(any large piece of woody material having a least diameter greater than 10 cm and a length greater than one meter; standing and downed dead wood in a forest)
d grobes Abfallholz *n*; grober Grobschlagabfall *m*
f débris *m* ligneux grossier; gros débris ligneux
r грубые древесные отломки *fpl*

1687 **coastal forest; littoral forest; shore forest**
d Küstenwald *m*; Uferwald *m*
f forêt *f* côtière; forêt littorale
r приморский лес *m*; прибережный лес; литоральный лес

* **coastal pine → 5309**

1688 **coastal plain willow; Carolina willow**
(Salix caroliniana; Salix wardii; Salix pitcheriana; Salix occidentalis)
d südliche Weide *f*

* **coast[al] redwood → 1269**

1689 **coastal scrub wattle; mimosa bush; sweet wattle; scented wattle**
(Acacia minuta; Acacia farnesiana; Mimosa farnesiana; Vachellia farnesiana; Acacia acicularis)
d Antillen-Akazie *f*; Kassia-Akazie *f*; süße Akazie *f*; amerikanische Akazie
f acacia *m* odorant; cassie *f*; cassier *m*; mimosa *m* de Farnèse
r акация *f* Фарнеза

1690 **coastal tropical dry forest**
d trockener Küstentropenwald *m*
f forêt *f* sèche tropicale côtière
r приморский сухой тропический лес *m*

1691 **coastal wattle; sallow wattle**
(Acacia sophorae)
d Sanddorn-Akazie *f*
f acacia *m* sophora
r акация *f* прибрежная; акация софора

* **coast ash → 4583**

1692 **coast beefwood; mountain she-oak**
(Casuarina stricta; Allocasuarina verticillata)
d Streitkolbenbaum *m*
f filao *m* dressé
r казуарина *f* аргентинская

* **coast-grey box → 932**

1693 **coast live oak; California live oak**
(Quercus agrifolia)
d kalifornische Steineiche *f*
f chêne *m* vert de Californie
r дуб *m* траволистный

* **coast rhododendron → 6023**

1694 **coast scarp forest**
d Küsteneskarpewald *m*
f forêt *f* d'escarpe côtière
r лес *m* на крутого берегового откоса

* **coast willow → 4360**

1695 **coating cover**
d beschichtende Abdeckung *f*
f couvert *m*; couverture *f*
r покровный слой *m*

1696 **coating of lichens; lichen covering**
d Flechtendecke *f*; Flechtenbehand *m*;
r лишайный покров *m*

* **cobnut [hazel] → 1778**

1697 **cocaine family**
(Erythroxylaceae)
d Rothölzergewächse *pl*
f érythroxylacées *pl*
r кокаиновые *pl*

* **cockchafer → 3502**

1698 **cockspur coral bean**
(Erythrina crista-galli; Erythrina herbacea)
d Korallenstrauch *m*
f érythrine *f* crête de coq
r эритрина *f* петушья

1699 **cockspur [hawthorn]; cockspur thorn**
(Crataegus crus-galli)
d Hahnendorn *m*; Hahnensporn-Weißdorn *m*
f aubépine *f* ergot-de-coq; aubépine de Virginie; ergot-de-coq *m*
r боярышник *m* "петуший гребень"; боярышник "петушья шпора"; боярышник шпорцевый

1700 **cocoa plum; coco-plum; icaco**
(Chrysobalanus icaco)
d Goldpflaume *f*; Ikakopflaume *f*
f icaquier *m*; icaque *f*
r икак[о] *m*; слива *f* золотая

* **cocoa [tree] → 1248**

* **cocobolo → 5426**

1701 **coco de Chile; honey palm; wine palm**
(Jubaea spectabilis; Jubaea chilensis)
d Honigpalme *f*
f cocotier *m* du Chili; jubée *f* remarquable
r медовая пальма *f*; юбея *f* видная

1702 **coconut [palm]; coconut tree; coco [palm]**
(Cocos nucifera)
d Kokospalme *f*; Schifferwanderbaum *m*
f cocotier *m*
r кокосовая пальма *f*

* **coco-plum → 1700**

1703 **cocos wood; cocowood; porcupine wood**
d Kokosholz *n*
f bois *m* de cocotier
r кокосовая древесина *f*

1704 **codominant crown class**
(forming part of the general level of the canopy and receiving full light from above)
d mitherrschende Kronenklasse *f*
f classe *f* de cimes codominantes
r класс *m* содоминирующих крон

1705 **codominant leaders**
(when two main branches of a tree are of equal strength and size, leading to the possible danger of the tree splitting)
d beherrschte Spitzentriebe *mpl*
f pousses *fpl* apicales codominantes
r кодоминантные главные побеги *mpl*; содоминантные главные побеги

1706 **codominant trees; subdominant trees**
(in stands with a closed canopy, those trees whose crowns form the general level of the canopy and receive full light from above, but comparatively little from the sides)
d mitherrschende Bäume *mpl*; beherrschte Bäume
f arbres *mpl* codominants
r кодоминантные деревья *npl*; согосподствующие деревья; сопутствующие деревья

1707 **codomination**
(trees with crowns forming the upper level of the forest canopy)
d Zusammenherrschaft *f*; Co-Herrschaft *f*
f co-dominance *f*
r кодоминантность *f*; содоминантность *f*

* **coefficient of elasticity → 5514**

1708 **coefficient of floating route filling**
d Füllungskoeffizient *m* der Flößweg
f coefficient *m* de remplissage de chemin
r коэффициент *m* заполнения лесосплавного хода

1709 **coefficient of jam compactness**
d Koeffizient *m* der Engpasskompaktheit
f coefficient *m* de densité de gorge
r коэффициент *m* полнодревесности пыжа

1710 **coefficient of raft section density**
d Kennzeichen *n* für Floßeinheitsdichte; Koeffizient *m* für Floßsektiondichte
f coefficient *m* de densité de section de flottage
r коэффициент *m* полнодревесности сплоточной единицы

1711 **coefficient of tree compactness**
f coefficient *m* de densité d'arbres
r коэффициент *m* полнодревесности

* **coffee-brown stain → 1549**

* **brown stain → 1549**

1712 **coffee-mortar; nargusta [terminalia]; aromilla**
(Terminalia amazonica; Terminalia obovata)
d Amarillo *m*; Nargusta *f*
f nargusta *m*
r терминалия *f* амазонская; терминалия обратнояйцевидная

1713 **coffee tree**
(Coffea arabica)
d Kaffee-Strauch *m*; Kaffeebaum *m*; Kaffeepflanze *f*; Bergkaffee *f*
f caféier *m*
r кофейное дерево *n*

* **coffee wood → 5426**

1714 **cohoba [tree]; yopo**
(Anadenanthera peregrina)
d Cohoba *m*; Yopo *m*
f yopo *m*
r стручковое дерево *n*; йопо *m*

1715 **cohort**
(group of individuals of the same age, recruited into a population at the same time)
d Kohorte *f*
f cohorte *f*
r когорта *f*

* **coigue → 2999**

1716 **coincident cropping**
(two or more species with similar crop duration which are grown together on the same unit of land)
d deckungsgleiche Kultur *f*
f culture *f* coïncidente
r совмещённое возделывание *n*

1717 **coir fiber; Algerian fiber**
d Kokosbastfaser *f*
f fibre *f* de cocos
r пальмовое волокно *n*; кокосовое волокно

1718 **cola tree**
(Cola acuminata)
d Kolabaum *m*

1719 cold-deciduous forest; deciduous boreal forest
d kaltabfallender Wald *m*; Laubwald *m* in der borealen Zone
f forêt *f* boréale feuillue
r зимнелистопадный лес *m*; лес холодно-умеренной области

* **cold deck → 1721**

1720 cold decking; intermediate wood stacking; intermediate wood piling
(storage the pile of logs for later transportation)
d Zwischenholzstapelung *f*
f empilage *m* de bois intermédiaire
r складирование *n* древесины на промежуточном складе; промежуточное штабелевание *n*

* **cold resistance → 3724**

1721 cold storage; cold deck; primary landing; intermediate holding site
d Zwischenlager *n*
f dépôt *m* intermédiaire
r промежуточный склад *m*; промежуточная площадка *f* с накоплением лесоматериалов

1722 cold [timber] loading
d Zwischenholzverladung *f*
f chargement *m* de bois intermédiaire
r погрузка *f* складированной древесины; погрузка древесины из запаса

* **collar → 8699**

1723 collateral bud
d kollaterale Knospe *f*
f bourgeon *m* collatéral
r коллатеральная почка *f*

1724 collected tree
(a tree taken from the wild for training as bonsai)
d gesammelter Baum *m*
f arbre *m* collectif
r коллекционированное дерево *n*

* **collecting area → 9280**

* **collecting depot → 3280**

* **collecting place → 3280**

* **collection of cones → 1863**

* **collector basket → 8621**

1725 colliery sleeper; colliery timber; pit sleeper; mine cross tie
d Grubenschwelle *f*
f traversine *f* de mine
r рудничная шпала *f*

1726 Colorado [blue] spruce; blue spruce; Parry spruce
(Picea pungens; Picea parryana)
d Blaufichte *f*; Stechfichte *f*; blaue Tanne *f*
f sapin *m* [bleu] du Colorado; épicéa *m* piquant; épicéa épineux
r ель *f* колючая; ель голубая

* **Colorado fir → 1278**

1727 Colorado pine beetle
(Dendroctonus approximatus)
d Colorado-Kiefernmarkkäfer *m*
f dendroctone *m* du pin de Colorado
r лубоед *m* сосновый колорадский

* **Colorado [pinyon] pine → 9031**

1728 colo[u]r defect
(of wood)
d Farbfehler *m*
f défaut *m* de coloration
r ненормальная окраска *f*; порок *m* окраски

1729 colo[u]red heartwood
(of a bole)
d Farbkern *m*
f cœur *m* coloré
r окрашенное ядро *n*

1730 colo[u]red wood
d angestrichenes Holz *n*; gebeiztes Holz; gefärbtes Holz; verfärbtes Holz
f bois *m* coloré
r окрашенная древесина *f*; крашеная древесина

* **colo[u]rfast wood → 2989**

* **coltie → 9696**

1731 Columbia hawthorn
(Crataegus piperi; Crataegus columbiana)
d Columbia-Hawthorn *m*
f aubépine *f* de la Colombie
r боярышник *m* колумбийский

1732 Columbian timber beetle

f scolyte *m* du bois colombien
r древесинник *m* колумбийский

* **column → 8190**

1733 **columnar cactus; tree-cactus**
(Cereus)
d Fackeldistel *f*; Kerzenkaktus *m*; Säulenkaktus *m*
f cierge *m*
r кактус *m* колонновидный; кактус цереус; цереус *m*

1734 **columnar form**
(of a bole)
d säulenartige Form *f*
f forme *f* colonnaire
r колонновидная форма *f*

1735 **columnar growth habit**
d säulenartige Wuchsform *f*
f port *m* colonnaire
r колонновидная форма *f* роста

1736 **column spruce**
d Säulenfichte *f*
f épicéa *m* colonnaire
r колонная ель *f*

1737 **Comandra blister rust; sweetfern blister rust**
(invader - Cronartium comandrae)
d Cronartium-Comandrae-Rost *m*
f rouille *f* des conifères
r ржавчина *f* сосны веймутовой и кедра сибирского

1738 **comb grain**
d Kammkorn *n*
f faux quartier *m*
r текстура *f* радиально распилённого пиломатериала

* **comb-grain[ed] wood → 6678**

* **combination beam → 1836**

1739 **combination log rule**
d kombinierte Kubiktabelle *f*
f table *f* de cubage composée
r комбинированная таблица *f* определения объёмов брёвен

1740 **combination of forest[al] and field crops; field crops and forestry in alternation**
r комбинация *f* лесных и полевых культур

* **combined moisture → 950**

* **combined water → 950**

1741 **combining sample; combined sample**
d Durchschnittsprobe *f*
f échantillon *m* composite
r смешанная выборка *f*

1742 **combustibility**
d Entflammbarkeit *f*
f combustibilité *f*
r воспламеняемость *f*; горючесть *f*

* **combustion of wood → 9568**

* **come into leaf → 7531**

1743 **comer tree**
d Kommender-Baum *m*
f arbre-arrivant *m*
r дерево-пришелец *n*

* **commerce wood → 5400**

* **commercial bole → 5393**

1744 **commercial cutting; merchantable cutting; exploitation cutting; exploitation felling; industrial harvesting**
(harvest cut that removes all merchantable timber from the area)
d Exploitationshieb *m*
f coupe *f* à blanc commercial; coupe de réalisation
r коммерческая рубка *f*; рубка леса, преследующая узкокоммерческие цели; эксплуатационная рубка (без учёта возобновления); промышленная рубка

1745 **commercial forest; industrial forest**
d Gewerbeholz *n*; Nutzwald *m*; nutzholztüchtiger Wald *m*; Wirtschaftswald *m*
f forêt *f* d'intérêt commercial
r коммерческий лес *m*; промышленный лес; лес промышленного значения; эксплуатационный лес; деловой лес

1746 **commercial forestland; commercial timberland**
(any area capable of producing 20 cubic feet of timber per acre per year that has not been

f chantier *m* forestier commercial; sol *m* d'une forêt exploitable
r лесохозяйственная почва *f*; земля *f* коммерческих лесов; земля промышленных лесов; эксплуатационная лесная земля

1747 **commercial height**
d Handelshöhe *f*
f hauteur *f* commerciale
r коммерческая высота *f*

1748 **commercially dry lumber**
d handeltrockenes Holz *n*
f bois *m* commercialement sec
r пиломатериал *m*, годный по сухости к продаже

1749 **commercial sawmill; trade sawmill**
d Handelssägewerk *n*
f scierie *f* à produits de commerce
r торговая лесопилка *f*

1750 **commercial thinning; extraction thinning**
(thinning in which harvested trees are removed from the site and used for commercial purposes)
d Handelsdurchforstung *f*
f éclaircie *f* commerciale
r промышленные рубки *fpl* ухода; коммерческие рубки ухода

* **commercial timberland → 1746**

1751 **commercial treatments**
(timber stand improvements that generate income from the sale of the trees removed)
d Handelsbehandlungen *fpl*
f traitements *mpl* commerciaux
r коммерческие обработки *fpl*

1752 **commercial tree species**
(tree species presently or prospectively suitable for industrial wood products)
d Handelsbaumarten *fpl*
f espèces *fpl* forestières commerciales
r коммерческие древесные породы *fpl*

* **commercial value → 5313**

1753 **commercial volume**
d Handelsvolumen *n*
f volume *m* commercial
r коммерческий объём *m*

* **commercial wood → 5400**

false acacia; acacia; akle
(Robinia pseudoacacia; Albizzia procera)
d [gemeine] Robinie *f*; Kameldornakazie *f*; falsche Akazie *f*; Scheinakazie *f*; weiße Robinie
f robinier *m* [commun]; [faux] acacia *m*
r робиния *f*; лжеакация *f*; белая акация *f*

* **common alder → 2882**

1755 **common apple-tree; apple[-tree]**
(Malus communis; Malus domestica)
d [gemeiner] Apfelbaum *m*; Zier-Apfel *m*; Apfel *m*
f pommier *m* [commun]
r яблоня *f* [культурная]

1756 **common apricot [tree]; abricot**
(Prunus armeniaca; Armeniaca vulgaris)
d Aprikosenbaum *m*; Aprikose *f*; Marille *f*
f abricotier *m* [d'Amérique]; abricotier commun; abricot *m* des Antilles
r абрикос *m* [обыкновенный]

1757 **common ash**
(Fraxinus excelsior)
d gewöhnliche Esche *f*; [gemeine] Esche
f frêne *m* [commun]; grand frêne; frêne d'Europe
r ясень *m* обыкновенный; ясень высокий; ясень европейский

1758 **common barberry; European barberry; barberry**
(Berberis vulgaris)
d [gemeine] Berberitze *f*; Sauerdorn *m*; gemeiner Berberitzenstrauch *m*
f épine-vinette *f*; berbéris *f* [commune]
r барбарис *m* [обыкновенный]

1759 **common bearberry; bearberry; kinnickinnick**
(Arctostaphylos uva-ursi)
d echte Bärentraube *f*; immergrüne Bärentraube
f raisin *m* d'ours; busserole *f*; arbousier *m* traînant
r толокнянка *f* аптечная; толокнянка медвежья; виноград *m* медвежий; ушко *n* медвежье; ягода *f* медвежья

1760 **common beech; European beech**
(Fagus sylvatica)
d gemeine Buche *f*
f hêtre *m* commun; hêtre d'Europe; fayard *m*; fouteau *m*

1761 **common birch; European [silver] birch; European white birch; silver birch; lady birch**
(Betula verrucosa; Betula alba; Betula pendula)
d europäische Birke *f*; Rauhbirke *f*; Warzenbirke *f*; Trauerbirke *f*; nordische Weißbirke *f*
f bouleau *m* [européen]; bouleau blanc; bouleau commun; bouleau verruqueux; bouleau en cépée
r берёза *f* бородавчатая; берёза белая; берёза провислая

1762 **common black ant; black ant**
(Lasius fuliginosus)
d Gartenameise *f*; [schwarzgraue] Wegameise *f*
f fourmi *f* noire
r муравей-древоточец *m* пахучий

* **common boxwood → 2886**

1763 **common button[ ]bush**
(Cephalanthus occidentalis)
d Knöpfchenblume *f*; Knopfblume *f*; Knopfbusch *m*
f céphalanthe *m* occidental; céphalanthe d'Occident; bois-bouton *m*
r головач *m* западный

1764 **common calabash tree; calabash**
(Crescentia cujute)
d Kalebassenbaum *m*
f calebasse *f*
r горлянковое дерево *n* обыкновенное

1765 **common camellia**
(Camellia japonica)
d Kamelie *f*
f camélia *m* [commun]
r камелия *f*

1766 **common capers**
(Capparis spinosa)
d [echter] Kapernstrauch *m*
f câprier *m* épineux
r каперсы *mpl* колючие

* **common cart-roads → 1787**

1767 **common catalog[ue] of basic materials for forest reproduction**
d gemeinsamer Katalog *m* für Ausgangsmaterial von forstlichem
pour reproduction forestière
r общий каталог *m* основных материалов для лесовозобновления

1768 **common catalpa; southern catalpa**
(Catalpa bignonioides; Bignonia catalpa)
d [amerikanischer] Trompetenbaum *m*; gemeiner Trompetenbaum
f catalpa *m* d'Amérique; catalpa [de Caroline]; arbre *m* aux haricots
r катальпа *f* бигнониевидная; катальпа сиренелистная

* **common cherry laurel → 2828**

1769 **common chokecherry; chokecherry**
(Padus virginiana; Prunus virginiana; Padus nana)
d [virginische] Traubenkirsche *f*; rote Traubenkirsche
f cerisier *m* de Virginie; cerisier noir
r черёмуха *f* виргинская

* **common coral bean → 1941**

* **common cotoneaster → 2888**

* **common cryptomeria → 4615**

1770 **common cypress; Mediterranean cypress; Italian cypress**
(Cupressus sempervirens)
d gemeine Zypresse *f*; echte Zypresse; immergrüne Zypresse; italienische Zypresse; Mittelmeerzypresse *f*
f cyprès *m* commun; cyprès pyramidal; cyprès vert de Provence; cyprès d'Italie; cyprès toujours vert
r кипарис *m* вечнозелёный

1771 **common dogwood; red dogwood; red dogtree; bloodtwig dogwood**
(Cornus sangiunea)
d [roter] Hartriegel *m*; roter Hornstrauch *m*
f cornouiller *m* rouge; cornouiller femelle; cornouiller sanguin; bois *m* punais
r дёрен *m* [кроваво-]красный; свидина *f*; глог *m*; глоговина *f*

1772 **common Douglas fir; Oregon fir; Rocky Mountains Douglas fir; [green] Douglas fir; bigcone spruce**
(Pseudotsuga taxifolia; Pseudotsuga menziesii; Pseudotsuga douglasii)

f douglasie *f*; douglas *m* [vert]; douglas bleu; sapin *m* de Douglas; sapin d'Oregon; sapin glauque
r лжетсуга *f* [тисолистная]; лжетсуга Мензиеза; Дугласова пихта *f*; дугласия *f*; орегонская сосна *f*; колумбийская сосна; жёлтая пихта; красная пихта

* **common elder → 2892**

* **common elm → 2893**

1773 **common evergreen oak; holm [oak]; holly oak; German oak**
(Quercus ilex)
d europäische Lebenseiche *f*; Grüneiche *f*; Immergrüneiche *f*; [echte] Steineiche *f*; Stechpalmen-Eiche *f*
f chêne *m* vert; chêne yeuse; yeuse *m*
r каменный дуб *m*

1774 **common fig; fig[-tree]**
(Ficus carica)
d [gewöhnlicher] Feigenbaum *m*; echter Feigenbaum; Feige *f*
f arbre *m* à carriques; [capri]figuier *m* [commun]; figuer comestible
r смоковница *f*; смоква *f*; фиговое дерево *n*; инжир *m*

* **common forest → 1812**

1775 **common gooseberry sawfly; imported currant worm**
(Nematus ribesii)
d gelbe Stachelbeerblattwespe *f*
f tenthrède *f* jaune du groseillier
r пилильщик *m*

1776 **common-grade lumber**
(lumber with obvious defects)
d Holz *n* mit Gemeinsamklasse; Holz mit ordinäre Klasse
f bois *m* de valeur ordinaire
r древесина *f* обычного качества

* **common greenbrier → 7223**

1777 **common hackberry; American nettle tree; western hackberry; bastard elm; beaverwood; hoop ash**
(Celtis occidentalis)
d [abendländischer] Zürgelbaum *m*; nordamerikanischer Zürgelbaum; Nesselbaum *m*
f micocoulier *m* [occidental]; micocoulier de

1778 **common hazel; European hazel; cobnut [hazel]; hazel[tree]; hazelnut tree**
(Corylus avellana)
d Haselnussbaum *m*; [gemeine] Hasel *f*; Hasel[nuss]strauch *m*
f noisetier *m* [commun]; coudrier *m* [pleureur]; avelinier *m*
r лещина *f* [обыкновенная]; орешник *m* лесной

1779 **common heather; sweet broom; Scotch broom; Scotch heather; heather**
(Calluna vulgaris; Erica vulgaris; Cytisus scoparius)
d [gemeines] Heidekraut *n*; Besenheide *f*; Erika *f*; Glockenheide *f*; Sommerheide *f*
f callune *f*; bruyère *f* [commune]; genêt *m* à balais; genêt commun; brande *m*
r вереск *m* [обыкновенный]; эрика *f*; ракитник *m* метельчатый

1780 **common hedgethorn; whitethorn; single-seed hawthorn**
(Crataegus monogyna; Crataegus apilfolia; Mespilus elegans)
d eingriffliger Weißdorn *m*; Einkern-Weißdorn *m*
f épine *f* blanche; aubépine *f* [à un style]; aubépine monogyne
r боярышник *m* однопестичный; боярышник одностолбчатый

* **common holly → 2899**

* **common hoptree → 6035**

* **common hornbeam → 2901**

1781 **common horse[-]chestnut; European horse[ ]chestnut; horse[-]chestnut; buck-eye**
(Aesculus hippocastanum)
d [gemeine] Rosskastanie *f*; gefülltblühende Rosskastanie
f maron[nier] *m* d'Inde [rouge]; marronnier commun
r каштан *m* конский [обыкновенный]

1782 **common jasmine; jasmine**
(Jasminum officinale)
d [echter] Jasmin *m*
f jasmin *m* [officinal]; jasmin commun
r жасмин *m* [лекарственный]

1783 **common jujube; jujube; Chinese date; dunk tree**

f jujubier *m*
r ююба *f*

* **common juniper → 3983**

1784 **common laburnum; golden[ ]chain [tree]; golden rain; bean-trefoil [tree]; bean-tree**
(Laburnum anagyroides; Cytisus laburnum; Laburnum vulgare)
d gemeiner Bohnenbaum *m*; [gemeiner] Gold[en]regen *m*
f cytise *m* [aubour]; aubour *m*; faux ébénier *m*; cytise à grappes; pluie *f* d'or; bois *m* d'arc
r золотой дождь *m*; бобовник *m* анагиролистный

1785 **common larch; European larch**
(Larix decidua; Larix europaea)
d europäische Lärche *f*; [gemeine] Lärche
f mélèze *m* [d'Europe]; mélèze commun; larix *m*
r лиственница *f* европейская; лиственница опадающая

* **common lilac → 6292**

1786 **common lime; European lime; English lime; European basswood** (*US*)
(Tilia vulgaris; Tilia europaea)
d gemeine Linde *f*; europäische Linde; Zwischenlinde *f*
f tilleul *m* d'Europe; tilleul intermédiaire
r липа *f* обыкновенная

1787 **common logging roads; common cart-roads**
d gemeinschaftliche Holzwege *mpl*
f chemins *mpl* mitoyens
r общие лесные дороги *fpl*

1788 **common lowbush blueberry; lowbush blueberry; low sweet blueberry; early-sweet blueberry; late-sweet blueberry; early-low blueberry**
(Vaccinium angustifolium)
d amerikanische Heidelbeere *f*
f airelle *f* à feuilles étroites
r голубика *f* узколистная

1789 **common mango; mango; edel**
(Mangifera indica)
d Mango[baum] *m*; Manga
f mango *m*
r манго *n* [индийское]; манговое дерево *n*

* **common maple → 8355**

algarroba
(Prosopis juliflora)
d Mesquitebaum *m*; Süßhülsenbaum *m*
f prosopis *m*; prosope *m*; mesquite *m*
r мескитовое дерево *n*; бухарник *m*; прозопис *m* серёжкоцветный; мимозка *f* [серёжкоцветная]; альгароба *f*

1791 **common mulbery; white mulberry**
(Morus alba)
d weißer Maulbeerbaum *m*
f mûrier *m* blanc
r тут *m* белый; шелковица *f* белая; тутовник *m* белый

1792 **common myrtle; myrtle [tree]**
(Myrtus communis)
d echte Myrte *f*; [gemeiner] Myrtebaum *m*; Brautmyrte *f*
f myrte *m* [commun]; nerte *m*; herbe *f* du lagui
r мирт *m* [обыкновенный]

* **common ninebark → 397**

1793 **common oak; European oak; pedunculate oak; English [brown] oak; eik**
(Quercus robur; Quercus pedunculata)
d Früheiche *f*; Stieleiche *f*; Sommereiche *f*
f chêne *m* pédonculé; chêne d'Europe; chêne commun
r дуб *m* черешчатый; дуб обыкновенный; дуб летний

* **common of firewood → 6946**

* **common osier → 619**

* **common palmetto → 1238**

1794 **common paw-paw; melon tree; papaya [tree]; papaw [tree]**
(Carica papaya)
d Pappawbaum *m*; Papau *m*; Papaya-Baum *m*; Melonenbaum *m*
f papayer *m*; arbre *m* à melon
r [карика] папайя *f*; дерево *n* дынное [обыкновенное]

1795 **common pear; pear[-tree]**
(Pyrus communis; Pirus communis)
d Holzbirnbaum *m*; Holzbirne *f*; [schweizer] Birnbaum *m*; Birne *f*; gewöhnliche Wildbirne *f*; wilder Birnbaum *m*
f poirier *m* [commun]; poirier sauvage; poirier

1796 common persimmon; persimmon; Virginian date plum; American ebony; white ebony
(Diospyros virginiana)
d Persimmon *m*; schwarze Persimone *f*; virginische Dattelpflaume *f*
f persimmon *m*; plaqueminier *m* de Virginie
r персимон *m*; хурма *f* виргинская

1797 common pine; Scot's pine; Scotch pine; Baltic redwood
(Pinus sylvestris)
d Weißkiefer *f*; gemeine Kiefer *f*; gemeine Föhre *f*; Kienbaum *m*; Kiene *f*; Tangelbaum *m*; Sandkiefer *f*; Waldkiefer *f*
f pin *m* commun; pin sylvestre; pin du Nord; pin d'écosse; pin de Riga
r сосна *f* обыкновенная; сосна лесная

1798 common pipsissewa; Prince's pine
(Chimaphila umbellata)
d Walddoldenkraut *m*; doldiges Winterlieb *n*
f chimaphile *f* ombellée
r зимолюбка *f* зонтичная; кимафила *f* зонтичная

* **common prickly-ash → 8679**

1799 common privet; tree privet; Chinese privet
(Ligustrum lucidum; Ligustrum vulgare)
d Glanzliguster *m*; Beinholz *n*; [gemeiner] Liguster *m*; Rainweide *f*; europäischer Liguster
f troène *m* [commun]; troène d'Europe
r лигуструм *m* светлый; бирючина *f* блестящая; бирючина обыкновенная

1800 common raspberry; raspberry
(Rubus idaeus)
d [gemeine] Himbeere *f*
f framboisier *m* [commun]; ronce *m* du Mont Ida
r малина *f* [обыкновенная]

* **common robinia → 1754**

1801 common sassafras; sassafras
(Sassafras albidum; Doryphora sassafras; Sassafras officinale; Sassafras varifolium)
d Sassafrasbaum *m*; Fenchelholzbaum *m*
f sassafras *m* [officinal]; laurier *m* des iroquois
r сассафрас *m* лекарственный; лавр *m* американский

d Schraubenpalme *f*
f pandanum *m*; vacquois *m*; vacoua *m*
r пандан[ус] *m*; дерево *n* винтовое

1803 common silver fir; silver fir; European silver fir
(Abies alba; Abies pectinata)
d [gemeine] Weißtanne *f*; Silbertanne *f*; Edeltanne *f*
f sapin *m* blanc; sapin pectiné; sapin argenté; sapin des Vosges
r пихта *f* серебристая; пихта белая; пихта европейская; пихта гребенчатая

* **common smoketree → 2908**

1804 common snowberry; snowberry
(Symphoricarpus albus; Symphoricarpus racemosus)
d [gemeine] Schneebeere *f*; traubige Schneebeere; Petersstrauch *m*
f symphorine *f* à fruits blancs; symphorine blanche; arbre *m* aux perles; boule *f* de cire
r снежноягодник *m* белый; снежноягодник кистистый

1805 common spruce; European spruce; Norway spruce; Baltic white pine; [Baltic] whitewood; spruce fir
(Picea excelsa; Picea abies)
d [gemeine] Fichte *f*; Rottanne *f*; Rotfichte *f*; Weißfichte *f*; Pechtanne *f*
f épicéa *m* [commun]; épicéa d'Europe; sapin *m* du nord; faux sapin; sapin rouge
r ель *f* обыкновенная; ель европейская

1806 common sweetshrub; eastern sweetshrub
(Calycanthus floridus)
d Gewürzstrauch *m*
f arbre *m* aux anémones
r каликант *m* цветущий

1807 common treecreeper; [northern] treecreeper; brown creeper
(Certhia familiaris)
d Baumläufer *m*
f grimpereau *m* des bois; grimpereau brun; grimpereau familier
r пищуха *f* обыкновенная

1808 common walnut; English walnut; European walnut; Circassian walnut; walnut
(Juglans regia)

f noyer *m* royal; noyer [commun]; noyer de France
r орех *m* [грецкий]

1809 common willow; [white] willow; European willow
(Salix alba)
d Silberweide *f*; Weißweide *f*; gelbe Trauerweide *f*
f saule *m* blanc; osier *m* blanc; osier commun; saule argenté
r ветла *f*; белолоз *m*; ива *f* серебристая; ива белая; белая верба *f*

1810 common winterberry; winterberry holly
(Ilex verticillata)
d gemeine Winterbeere *f*; amerikanische Winterbeere; Korallenhülse *f*
f houx *m* verticillé
r падуб *m* мутовчатый

1811 common witch-hazel
(Hamamelis virginiana)
d Zauberhasel *n*
f hamamélis *m* de Virginie; noisetier *m* de sorcière
r гамамелис *m* виргинский

1812 common woodland; communal forestry area; communal forest; collective forest; common forest; public forest; forest in common ownership
d Gemeinschaftswald *m*; Gemeindewald *m*; Anstaltswald *m*; Gemeindeforst *m*; Markwald *m*
f forêt *f* communale [en propriété collective]; bois *m* communal en propriété collective
r общинный лес *m*; коммунальный лес

1813 common wood[ ]louse; woodlouse; pill bug; sow bug
(Armadillidium communis)
d [gemeine] Rollassel *f*; gemeine Kugelassel *f*
f armadille *f*
r мокрица *f* [древесная]

* **common yellowwood → 5974**

1814 common yew; English yew; Irish yew; European yew; yew; God's tree
(Taxus baccata)
d Roteibe *f*; Taxbaum *m*; [gemeine] Eibe *f*; Ifenbaum *m*
f if *m* [commun]; taxis *m*
r тис *m* ягодный; тис [европейский]

1815 communal cutting area
r лесосека *f* общественного пользования

* **communal forest → 1812**

* **communal forestry → 1817**

* **communal forestry area → 1812**

* **communal forest service → 1817**

1816 community forest
d Gesellschaftswald *m*
f forêt *f* collective
r общественный лес *m*

1817 community forestry; communal forestry; communal forest service
(developed in areas marginal to agriculture, with many members of the community being landless or small-scale farmers)
d Kommunalforstverwaltung *f*
f foresterie *f* communautaire
r общинное ведение *n* лесного хозяйства; общинное лесничество *n*; коммунальное лесное управление *n*

1818 compactness of wood; wood density
d Dichte *f* des Holzes; Holzdichte *f*
f densité *f* du bois
r плотность *f* древесины

1819 compact wood
(wood over 7 cm minimum diameter at the smaller end)
d Derbholz *n*
f gros bois *m*
r крупная древесина *f*

* **compartment cutting → 1820**

* **compartment description → 2411**

1820 compartment felling; compartment cutting; cutting by area
d Flächenhieb *m*
f coupe *f* par surfaces
r площадная вырубка *f*

1821 compartment of forest; forest compartment; forest parcel; forest patch; forest block; timber compartment; wood plot; forest plot; forest enclosure
(basic territorial unit of a forest estate)
d Waldparzelle *f*; Holzeinschlagsplatz *m*; Waldstück *m*
f compartiment *m* forestier; parcelle *f* de la forêt; parcelle forestière

* **compartment under regeneration → 355**

* **compass lettuce → 6531**

1822 **compass-saw; piercing saw; scroll-saw; sweep saw; fretsaw; saber jig saw; pad saw; jig[ger] saw**
(a handsaw with a thin triangular blade for cutting curves)
d Dekupiersäge *f*; Laubsäge *f*
f scie *f* passe-partout; scie mortaiseuse; scie à découper; scie à chantourner; scie sauteuse; scie à araser
r лобзик *m*; лобзиковая пила *f*; ажурная пила; узкая пила; [узкая] ножовка *f*; ручная ножовка

1823 **compensatory planting**
(creating plantations in one area in order to replace, in part or whole, a loss of growing stock elsewhere)
d Ersatzpflanzung *f*; Ausgleichspflanzung *f*
f reboisement *m* de compensation
r компенсирующая посадка *f*; повторная посадка

1824 **competing vegetation**
(vegetation that seeks and uses the limited common resources (space, light, water, and nutrients) of a forest site needed by preferred trees for survival and growth)
d Wettbewerbsvegetation *f*
f végétation *f* compétitrice; végétation de compétition; végétation concurrentielle
r конкурентная растительность *f*

1825 **competition control**
(treatment designed to reduce the competitive effect of undesirable vegetation threatening the success of the regeneration of desirable tree species)
d Konkurrenzkontrolle *f*
f lutte *f* contre la concurrence végétale
r контроль *m* конкурентной растительности

1826 **competitively grown tree**
d durchforsteter Baum *m*
f arbre *m* en croissance concurrentielle
r дерево *n*, выросшее в густом древостое

1827 **competitor**
d Konkurrent *m*
f concurrent *m*
r конкурент *m*

* **complete canopy → 1650**

1828 **complete decay; final decay; final rot; late decay; hard rot**
(of wood)
d letzte Fäulnis *f*
f pourriture *f* finale
r конечная стадия *f* гниения; третья стадия гниения; гниль *f* в третьей стадии развития

1829 **complete leaf**
d vollständiges Blatt *n*
f feuille *f* complète
r совершенный лист *m*

* **complete stand → 3756**

* **complete tree → 3758**

1830 **complete-tree harvester; whole-tree harvester**
d Ganzbaum-Erntemaschine *f*; Vollbaum-Erntemaschine *f*
f récolteuse *f* d'arbres entiers
r лесозаготовительная машина *f* целых деревьев

1831 **complete-tree harvesting; complete-tree logging**
(harvesting of a complete tree, including the roots and crown)
d Ganzbaumernte *f*; Ganzbaumfällung *f*
f exploitation *f* intégrale
r заготовка *f* деревьев с корнями и кроной; заготовка целыми деревьями с кроной

1832 **complete-tree utilization; whole-tree utilization**
d Ganzbaumnutzung *f*; Vollbaumnutzung *f*
f usage *m* d'arbre entier
r использование *n* дерева целиком

1833 **composite [board]**
(a wood product produced by bonding wood fiber by heat and pressure)
d gemischte Holzfaserplatte *f*; zusammengesetzte Holzfaserplatte
f composite *m*
r комбинированная плита *f* (состоящая из ДСП, склеенной с ДВП)

1834 **composite family**
(Asteraceae; Compositae)
d Korbblütengewächse *pl*
f composées *fpl*
r сложноцветные *pl*

* **composite forest → 5486**

1835 composite wood products
(roundwood products manufactured into chips, wafers, strands, flakes, shavings or sawdust and then reconstituted into a variety of panel and engineered lumber products)
d zusammengesetzte Holzprodukte *npl*
f produits *mpl* du bois composites
r комбинированные древесные материалы *mpl*

* **compost → 4881**

1836 compound beam; combination beam
d kombiniertes Gebälk *n*
f poutre *f* composée
r составная балка *f*

1837 compound bud; mixed bud
d gemischte Knospe *f*
f bourgeon *m* mixte
r смешанная почка *f*

1838 compound cut
(an angled cut to both the edge and face of a board)
d Verbundschneiden *n*
f coupe *f* composée
r составное резание *n*

1839 compound fruit; aggregate fruit
d zusammengesetzte Frucht *f*; Gesamtfrucht *f*
f fruit *m* composé; fruit agrégatif
r сложный плод *m*; составной плод; сборный плод

1840 compound girder
d Holz[ver]binder *m*
f contrevent *m*; contrefiche *f*
r вязальщик *m* леса

1841 compound leaf
(leaf with more than one blade. All blades are attached to a single leaf stem)
d zusammengesetztes Blatt *n*
f feuille *f* composée
r сложный лист *m*

1842 compound middle lamella
(in wood anatomy a term of convenience for the compound layer between the secondary walls of adjacent cells, consisting of two primary cell walls and an intercellular layer of varying thickness)
d zusammengesetzte Mittellamelle *f*
f lamelle *f* moyenne composée
r составная средняя пластинка *f*

d zusammengesetzter Riss *m*
f fente *f* composée
r сложная трещина *f*

1844 compregnated wood; compreg
d comprägniertes Holz *n*
f bois *m* imprégné-densifié
r компрег *m*; древесный пластик *m*; пластифицированная древесина *f* (древесина, обработанная способом пропитки смолой под давлением)

1845 compressed fiberboard
d Holzfaserhartplatte *f*; harte Faserplatte *f*; Hartfaserplatte *f*; Hartholzplatte *f*
f carton *m* dur vulcanisé; panneau *m* de fibres [de bois] dur
r твёрдая древесноволокнистая плита *f*

* **compressed wood → 2390**

1846 compression debarking
d Kompressionentrindung *f*
f écorçage *m* par compression
r компрессионная окорка *f*

1847 compression failure
(deformation or fracture of wood fibers across the grain resulting from excessive compression along the grain)
d Kompressionsfehler *m*
f fente *f* due au refoulement des fibres
r повреждение *n* от сжатия

1848 compression parallel to the grain
d axiale Kompression *f*
f compression *f* axiale
r сжатие *n* вдоль волокон

1849 compression perpendicular to the grain
d transversale Kompression *f*
f compression *f* transversale
r сжатие *n*, перпендикулярное к направлению волокон

1850 compression wood; sway wood; hard streak; tenar; pressure wood; bull wood
(reaction wood of conifers)
d Kompressionholz *n*; Druckholz *n*
f bois *m* de compression
r креневая древесина *f*

1851 compression wood tracheid
d Druckholztracheide *f*

1852 compulsory felling
d Zwangseinschlag *m*; Zwangshieb *m*
f coupe *f* obligatoire
r принудительная рубка *f*

* **concave axe → 4390**

1853 concave saw
d Wiegensäge *f*
f scie *f* concave; lame *f* de scie circulaire concave
r вогнутая пила *f*; чашечная пила

* **concolor fir → 1278**

* **concrescent → 67**

1854 conditional clear felling; patch logging
(modification of the clearcutting system developed in patches of about 5 to 200 ha logged as single units)
d bedingte Holzaufbereitung *f*; bedingter Kahlhieb *m*
f exploitation *f* par blocs; coupe *f* à blanc conditionnelle
r лесозаготовки *fpl* с условно-сплошными [котловинными] рубками; условно-сплошные рубки *fpl*

1855 conditioning of wood
(the exposure of a wood to the influence of a prescribed atmosphere for a stipulated period of time, or until a stipulated relation is reached between material and atmosphere)
d Holzaufbereitung *f*
f conditionnement *m* de bois
r кондиционирование *n* древесины

* **conducting parenchyma → 1857**

1856 conducting root
d Leitungswurzel *f*
f racine *f* de conduction
r проводящий корень *m*

1857 conducting tissue; conducting parenchyma
(layer of cells that transports nutrients from the crown to the roots)
d Leit[ungs]gewebe *n*
f tissu *m* de conduction
r проводящая ткань *f*; проводящая паренхима *f*

* **cone → 8818**

1858 cone; cap; Baptist cone
over the end of a log to prevent its catching on stumps, roots, or other obstacles in ground skidding)
d Rückenhaube *f*; Schutzkappe *f*; Stammschlepphaube *f*
f cône *m* de débusquage
r трелёвочный конус *m*

* **cone-bearing tree → 1877**

1859 cone borer
d Zapfenbohrer *m*
f mineuse *f* des cônes
r усач *m* шишковой

* **cone collection → 1863**

* **cone crop → 2033**

1860 cone-curing shed
d Zapfentrockenschuppen *m*; Zapfentrockenspeicher *m*
f abri-séchoir *m* pour cônes; grenier *m* pour entrepôt de cônes
r сарай *m* для хранения шишек

1861 cone cutting
(useful to look for good seeds inside of a cone)
d Zapfenpflücken *n*
f cueillette *f* des cônes
r сдирание *n* шишек

1862 cone-drier
d Zapfendarre *f*
f four *m* de cônes
r шишкосушилка *f*

* **cone gatherer → 1865**

1863 cone harvesting; collection of cones; cone collection
(harvesting of cones after seed maturation but before their dispersal)
d Zapfensammlung *f*
f collection *f* de cônes
r сбор *m* шишек; заготовка *f* шишек

1864 conelet
(a cone-shaped inflorescence marked by imbricated sporophylls)
d Zäpfchen *n*
f conelet *m*
r обоеполый цветок *m* типа шишки

1865 cone rake; cone picker; cone gatherer

d Zapfenpflücker *m*
f cueilleur *m* de cônes
r шишкосборщик *m*

1866 **cone-shaped crown; spire-shaped crown**
d Spitzkrone *f*
f cime *f* en cône; cime conique; cime pointue
r конусовидная крона *f*; остроконечная крона; пирамидальная крона

1867 **cone year**
d Zapfenjahr *n*
f bonne année *f* de production de cônes; année *f* d'abondance de cônes
r урожайный год *m* шишек

1868 **confined area**
(of cutting)
d begrenzte Fläche *f*
f aire *f* limitée
r ограниченная площадь *f*

1869 **confluent parenchyma**
(coalesced aliform parenchyma forming irregular tangential or diagonal bands, as seen in cross section)
d zusammenwachsendes Parenchym *n*; konfluentes Parenchym
f parenchyme *m* confluent; parenchyme anastomosé
r конфлюэнтная паренхима *f*; сливная паренхима

* **congested canopy → 1650**

1870 **Congo senna**
(Cassia angustifolia; Senna alexandrina)
d Kassie *f*
f séné *m* [d'Alexandrie]
r кассия *f* узколистная; сена *f*

* **conical → 8401**

1871 **conical edging**
d konisches Besäumen *n*
f délignage *m* conique
r обрезка *f* [досок] на конус; обрезка кромок по сбегу

1872 **conical growth; tapering growth; axon hill; shoot of apex**
(of root)
d konischer Wuchs *m*; kegelförmiger Wuchs
f décroissance *f* forte
r конус *m* нарастания

* **coniferous board → 7828**

1873 **conifer[ous] forest; softwood forest; needle[-leaved] forest; evergreen needleleaf forest**
(forest stand or category of trees or bush that is popularly called "evergreen")
d Nadelholzwald *m*; Nadelwald *m*
f forêt *f* de résineux; forêt de conifères; forêt aciculifoliée; forêt résineuse
r хвойный лес *m*; хвойник *m*; мягколиственный лес

1874 **coniferous large-size pole; large softwood pole**
d Nadelderbstange *f*
f haute-perche *f* de résineux; perche *f* résineuse
r крупномерная хвойная жердь *f*

1875 **coniferous sawn timber**
d Nadelschnittholz *n*
f bois *m* de conifères
r пиломатериалы *mpl* хвойных пород

1876 **coniferous small-size pole**
d schwaches Nadelstangenholz *n*
f petite perche *f* de conifères
r тонкая хвойная жердь *f*

* **coniferous timber → 5708**

1877 **coniferous tree; cone-bearing tree; conifer; needle-[leaved] tree; softwood tree**
(cone-bearing, evergreen tree with needle-like, linear, or scale-like leaves)
d Nadelbaum *m*; Konifere *f*; Nadelholzbaum *m*
f conifère *m*
r хвойное дерево *n*

1878 **conifer transplant**
d Verschulnadelpflanze *f*
f transplant *m* repiqué conifère
r саженец *m* хвойных пород

1879 **conk[er]; punk; corbel; bracket**
(visible fruiting body of a wood-destroying fungus, usually indicating rot in the underlying wood)
d Pilzkonsole *f*; Konsole *f*
f faux carpophore *m*; champignon *m* en forme de console
r грибное тело *n*; консоль *m*

1880 **conk wood**
d Holz *n* mit Pilzfäule
f bois *m* à pourriture mycélienne
r поражённая грибками древесина *f*

* **connarus family → 9794**

* **connecting shackle → 2320**

* **conophorium → 5705**

1881 **conservation afforestation**
d Erhaltungsaufforstung *f*
f afforestation *f* contre l'érosion
r противоэрозионное облесение *n*; почвозащитное облесение

* **conservation of forest plantation → 8010**

1882 **conservation reserve program; CRP**
(US program designed to remove highly erodible, marginal farmland from production through a one-time cost-sharing payment to establish trees, grass, or other cover)
d Schutzreserveprogramm *n*
f programme *m* de conservation contre l'érosion
r программа *f* восстановительной консервации сильноэродируемых земель

1883 **consolidation of wooded areas; consolidation of scattered forest holdings; forest consolidation**
d Waldflurbereinigung *f*; Waldzusammenlegung *f*
f remembrement *m* forestier; remembrement de forêts
r сложение *n* лесных площадей

1884 **consolidation phase**
(often a very difficult period lasting from 1 to 5 years after tree-planting)
d Konsolidierungsphase *f*
f phase *f* de consolidation
r фаза *f* консолидации

1885 **conspicuous pin knots**
(sound knots 6.4 mm (1/4 inch) or less in diameter containing dark centers)
d auffallende Vogelaugen *fpl*; deutliche Punktaste *mpl*
f petits nœuds *mpl* apparents de bois madré
r заметные игольчатые сучки *mpl*

* **Constantinople nut → 5383**

1886 **constitution of stand; stand structure; stand composition; forest structure**
(distribution of trees in a stand, which can be described by species, vertical or horizontal spatial patterns, size of trees or tree parts, age, Waldzusammensetzung *f*; Bestandesstruktur *f*; Bestandesaufbau *m*; Waldaufbau *m*
f structure *f* de peuplement; structure de la forêt; composition *f* d'un peuplement
r состав *m* древостоя; состав насаждения; структура *f* насаждения; строение *n* леса

1887 **constructed skid trail**
(skid trail that has been constructed in advance to facilitate ground based skidding of logs)
d gebauter Ziehweg *m*
f chemin *m* de débardage construit
r конструированный [земляной] лесоспуск *m*

1888 **construction log**
d Bauschnittholz *n*
f grume *f* de construction
r строительные бревно *n*

1889 **construction of forest roads; logging road construction; forest roading; forest road construction; forest road building**
d Waldwegebau *m*; forstlicher Wegebau *m*; Waldstraßenbau *m*
f construction *f* de routes forestières; construction de chemins forestiers
r лесодорожное строительство *n*; строительство лесных дорог

* **construction timber → 1153**

* **construction wood → 1153**

1890 **consulting forester; forestry adviser; expert in forest; service forester**
(independent professional who manages forests and markets forest products for private woodland owners)
d forstlicher Berater *m*; Forstberater *m*; Forstkonsulent *m*; Begriffsumfangsforstbeamter *m*
f conseiller *m* forestier
r частный лесовод-консультант *m*; лесной советник *m*; лесовод-информатор *m*

* **consumptive water use → 2916**

1891 **contained root**
(root that does not elongate beyond the confines of the original rooting volume within a container, even when outplanted with the container removed)
d enthaltene Wurzel *f*
f racine *f* contenue

1892 container-grown; containerized
d containerisiert
f en récipient
r брикетированный

1893 container[ized] [tree] seedling; ball-rooted seedling; containerized stock
(seedling grown in small container in a controlled environment)
d im Container kultivierte Pflanze *f*; containerisierter Baumsamen *m*
f semis *m* en récipient
r брикетированный сеянец *m*

1894 container nursery
(nursery where the stock is raised individually in containers)
d containerisierte Pflanzschule *f*
f pépinière *f* [de plantes] en récipients
r питомник *m* саженцев с закрытой корневой системой

1895 container planting
(setting out of young trees (generally individually) from, or together with, receptacles containing the soil)
d containerisierte Pflanzung *f*
f plantation *f* [de plantes] en récipients
r посадка *f* с необнажённой корневой системой

* **continental Cambridge roller → 2127**

* **continual harvesting operations → 1897**

* **continuance shoot → 4237**

* **continuous canopy → 1650**

* **continuous [cover] forest → 6171**

* **continuous crops of tree → 6171**

1896 continuous-feed delimber; continuous-feed delimbing line
d Entastungsmaschine *f* mit Dauerzufürung
f ébrancheuse *f* à alimentation continue
r сучкорезная машина *f* с непрерывной подачей; сучкорезная линия *f* с непрерывной подачей

1897 continuous felling; continuous harvesting; continual harvesting operations; year-round operations
d Dauerschnitt *m*
f abattage *m* continu; fauchage *m*
круглогодовые лесозаготовки

1898 continuous forest inventory; CFI
d Dauerforsttaxation *f*
f inventaire *m* forestier permanent
r непрерывная лесная таксация *f*; таксация леса на постоянных пробных площадях (при изучении хода роста насаждений)

* **continuous forest management system → 6172**

* **continuous harvesting → 1897**

1899 continuous inventory of stand conditions; CISC
(computerized method of recording information about each individual stand of tree)
d Dauertaxation *f* der Waldzustand
f inventaire *m* continu d'état de peuplement
r непрерывная таксация *f* состояния леса

1900 continuous[-moving tree crop] harvester; swather[-harvester]
d Dauerholzerntemaschine *f*
f faucheuse[-groupeuse] *f* à action permanente
r лесозаготовительная машина *f* непрерывного действия; машина для валки "напроход"

1901 continuous pith; solid pith
d Festkern *m*
f cœur *m* solide
r сплошная сердцевина *f*; монолитная сердцевина

* **contorta pine → 5044**

1902 contorted grains
d gewundene Holzfaser *fpl*; verdrehte Holzfaser
f fils *mpl* contournés
r изогнутые волокна *npl*; скрученные волокна; текстура *f* в виде изогнутых полос

1903 contour felling; contour cut
(timber felled parallel to ground contour line)
d Konturschnitt *m*
f abattage *m* suivant les contours; abattage suivant les courbes de niveau
r валка *f* с учётом контур

1904 contour furrow; contour trench
(trench made along a contour (i.e., horizontal) line, for the purpose of checking run-off and soil loss, and conserving moisture, in a

f rigole *f* de niveau
r контурная борозда *f*; горизонтальная борозда

1905 **contour logging system**
d Kontur-Hiebssystem *n*
f système *m* de coupe suivant les courbes de niveau
r система *f* лесозаготовок с расположением волоков с учётом рельефа местности

1906 **contour planting**
d Konturpflanzung *f*
f plantation *f* en bandes de niveau
r посадка *f* по горизонталям; контурная посадка

* **contour sander → 6567**

* **contour trench → 1904**

1907 **contract haulier**
(independent truck owner or a driver working for the contractor who hauls logs from the woods to the mill)
d Vertragsabfuhrunternehmer *m*; Kontraktabfuhrunternehmer *m*
f entrepreneur *m* de transport par contrat
r водитель *m* тягача, работающий по контракту

* **contraction crack → 2422**

1908 **contractual hauling [operation]**
d Vertragsbringung *f*
f débardage *m* par contrat; débardage contractuel
r вывозка *f* леса по контракту

1909 **contravention of forest law; breach of forest laws; offence against the forest law; forest offence; rape of forest; forest trespass**
d Forstübertretung *f*; forstpolizeiliche Überschreitung *f*; Forstfrevel *m*; Waldfrevel *m*; Forstdelikt *n*
f infraction *f* forestière; délit *m* forestier
r лесное нарушение *n*; лесонарушение *n*; лесное преступление *n*

* **controlled burning → 6516**

1910 **conventional forest products**
(all commercial roundwood products except fuelwood)
d konventionelle Forstprodukte *npl*
традиционные лесные продукты

1911 **conventional ground skidding**
(any combination of rubber-tired or tracked skidding equipment)
d konventionelle Bodenseilung *f*
f débusquage *m* conventionnel
r конвенционная трелёвка *f* волоком

1912 **conventional logging**
(any combination of mechanical or hand felling and rubber-tired or tracked skidding equipment)
d konventionelle Bringung *f*
f débardage *m* conventionnel
r конвенционная трелёвка *f*

1913 **conventional wood density**
d konventionelle Holzdichte *f*
f densité *f* de bois conventionnelle
r условная плотность *f* древесины

1914 **conversion cutting**
d Umwaldungshieb *m*
f coupe *f* de conversion
r конверсионная рубка *f*

1915 **conversion of timber; wood conversion**
(transformation of natural timber into any kind of commercial product)
d Holzverarbeitung *f*; Konversion *f* von Holz
f conversion *f* de bois
r переработка *f* лесоматериалов

1916 **conversion period**
d Konversionperiode *f*
f durée *f* de la conversion
r период *m* перехода от одной системы [лесного хозяйства] к другой; ревизионный период

1917 **conversion table**
d Umrechnungstabelle *f*
f tableau *m* de conversion
r конверсионная таблица *f*

1918 **converted lumber; saw[n] goods; converted timber; carving wood; sawn timber; sawed timber; bucked wood**
(timber finished to size with a saw)
d Schnittholz *n*; besägtes Holz *n*; gesägtes Holz
f bois *m* [de construction] débité; bois scié; sciages *mpl*; bois converti
r распилённый лесоматериал *m*; пилёный лесоматериал; пиломатериал[ы] *m*[*pl*]; раскряжёванный лес *m*

1919 Cook's pine
(Araucaria cookii)
d Cook-Araukarie *f*
f araucaria *m* de Cook
r араукария *f* Кука

1920 cooperage
(containers having two round heads and a body composed of staves, such as barrels or kegs)
d Böttcherei *f*; Fassfabrikation *f*
f tonnellerie *f*
r бондарное производство *n*; бочарная продукция *f*

* **cooper's wood → 4362**

1921 copaia; futui; arabisco
(Jacaranda copaia)
d Gobaja *f*; Kopaia *f*; Karoba *f*
f gobaja *m*; boisapian *m*; faux simar[o]uba *m*
r якаранда *f* копая

1922 copaiba [copal tree]; copaiva plant
(Copaifera officinalis)
d Kopaivabaum *m*; Kopaibabaum *m*
f copayer *m*; copaïer *m* officinal; copahier *m*
r копайба *f*; копайва *f*

* **copaiva plant → 1922**

1923 copal
(a group of "hard" natural resins obtained as exudations from trees, mainly tropical species)
d Copalharz *m*; Copal *m*; Kopal *m*
f résine *f* copal; copal *m*; résine animée; manille *f*
r копал *m*

1924 copal tree; copal [blanco]
(Protium copal)
d Copalbaum *m*; Copal *m*
f arbre *m* à copal
r протиум *m* смолоносный; копал *m* смолоносный

* **Copenhagen hawthorn → 3167**

* **coping saw → 3342**

* **copper[ ]beech → 6639**

* **coppice → 2165**

1925 coppice crop
r урожай *m* послерослевого насаждения

1926 coppice cutting
d Niederwaldhieb *m*
f coupe *f* de taillis; coupe de rajeunissement; recépage *m*
r рубка *f* в порослевых древостоях

1927 coppice cutting with reserves
d Niederwaldhieb *m* mit Reserve
f recépage *m* avec réserve
r система *f* выборочной рубкой в порослевых древостоях с оставлением части молодняка и тонкомерных деревьев

1928 coppice [forest]; coppice stand; sprout forest; copse; spinney; wold; tope
(natural regeneration originating from stump sprouts, stool shoots, or root suckers)
d Ausschlagwald *m*; Forstgewächs *n*; Hackwald *m*; Reißwald *m*
f taillis *m* [simple]; basse-futaie *f*
r [молодой] порослевый лес *m*; рощица *f*; молодая поросль *f* после вырубки; низкорослый порослевый лес; порослевый древостой *m*; лесная поросль

1929 coppice forestry
(a system of regenerating a forest stand in which the cut trees produce sprouts, suckers, or shoots)
d [schlagweiser] Niederwaldbetrieb *m*; Niederwaldsystembetrieb *m*
f foresterie *f* du taillis
r низкоствольное лесосечное хозяйство *n*; [лесосечное] низкоствольное хозяйство

1930 coppice hedge
d lineares Unterholz *n*
f taillis *m* linéaire
r порослевая изгородь *f*

* **coppice-of-two-rotations system → 9033**

1931 coppice regeneration
(ability of certain hardwood species to regenerate by producing many new shoots from a cut stump)
d Verjüngung *f* durch Stockausschlag
f régénération *f* du taillis; régénération par rejets de souches
r возобновление *n* пнёвой порослью

1932 coppice selection [cut] in shelterwood
(in which only selected shoots of merchantable size are cut at each felling.

f coupe *f* par furetage
r рубка *f* в порослевом хозяйстве с оставлением части защитных деревьев; выборочная рубка в порослевых древостоях

1933 coppice shoot; sap shoot; stump sucker; parent stock; stump sprout; stool shoot
(any shoot arising from an adventitious or dormant bud near the base of a woody plant that has been cut back)
d Stockausschlag *m*; Verstockung *f*; Stockausschlagtrieb *m*
f rejet *m* de taillis; rejet de souche; brin *m* de taillis; cépée *f*
r порослевый побег *m*; поросль *f* от пня; пнёвая поросль; побег *m* от пня; пнёвой побег; корневая поросль

* **coppice stand → 1928**

1934 coppice stock; stool
(living stump capable of producing sprouts)
d Stock *m* im Niederwaldbetrieb
f souche-mère *f*
r пень *m* в низкоствольнике

1935 coppice system; sprout system; sprout method; coppicing
(in which the crop is clearcut and regenerated by stool shoots, stump sprouts, or root suckers, giving even-aged stands)
d Niederwaldsystem *n*
f taillis *m* réglé; régime *m* du taillis
r низкоствольная система *f*

* **coppice with field crops → 143**

1936 coppice with pollards
d Niederwald *m* mit Kopfholz
f taillis *m* avec têtard
r безвершинный низкоствольник *m*

* **coppice with standards → 7947**

1937 coppice with standards system; stored coppice system
d Mittelwaldbetriebsschaft *f*
f système *m* de taillis-sous-futaie; régime *m* de taillis-sous-futaie
r среднее хозяйство *n* в порослевых насаждениях с оставлением маячных деревьев

* **coppice wood → 7361**

* **copse → 1928**

1938 copsewood
d Reiserholz *n* im Niederwald
f menus bois *m* du taillis
r срубленные молодые ветви *fpl*; хворост *m* в низкоствольнике

* **coral bean → 1941**

1939 coral-bean sophora; mescal bean; Texas mountain laurel
(Sophora speciosa; Sophora secundiflora)
d Mescalbohne *m*
f mescal *m*
r софора *f* красивая; гебелия *f* красивая

1940 coralberry; Indian currant
(Symphoricarpus orbiculatus)
d Korallenbeere *f*; rote Schneebeere *f*
f groseillier *m* des Indiens; symphorine *f* à feuilles rondes
r снежноягодник *m* красноплодный; снежноягодник круглолистный; снежноягодник обыкновенный; снежноягодник округлый

1941 coral tree; common coral bean; coral bean
(Erythrina corallodendron)
d Korallenbaum *m*
f bois *m* immortel; arbre *m* corail; fève *f* de corail
r эритрина *f*; коралловое дерево *n*

* **corbel → 1879**

* **cord → 9519**

1942 cordage
d Fasertauwerk *n*; Tauwerk *n*; Takelage *f*
f cordage *m*
r снасти *pl*; такелаж *m*

1943 cordate leaf; heart-shaped leaf
(heart-shaped leaf)
d herzförmiges Blatt *n*
f feuille *f* en forme de cœur
r сердцевидный лист *m*

* **corded merchantable timber → 6230**

* **corded wood → 7982**

* **cordia wood → 4650**

1944 cordon

d Schnurbaum *m*
f cordon *m*
r кордон *m*

1945 cords per acre
d Holzstosse *mpl* pro Morgen
f cordes *fpl* par acre
r запас *m* древесины на 1 акре в кордах; объём *m* заготовленной с 1 акра древесины в кордах

1946 cords per man-hour
(quotient derived by dividing the total number of cords that a crew produces by the number of man-hours required for the production)
d Holzstosse *mpl* pro Arbeitsstunde
f cordes *fpl* par heure d'un homme
r корды *mpl* на человекочас

1947 corduroy [road]; pole-road; corduroyed trail
(logs placed transversely along a road, usually with branches intact, and covered with fill material, to "float" the road over soft subsoil)
d Balkenweg *m*; Holzgleisweg *m*; Knüppeldamm *m*; Knüppelweg *m*; Prügelweg *m*; Rollweg *m*
f chemin *m* de rondins; chemin bâchonné
r бревенчатая дорога *f*; лежневая лесовозная дорога; гать *f*

* **cordwood → 7982**

1948 core containing knots; knotty core
(central core of wood in a pruned tree that contains knots)
d astiger Kern *m*
f cœur *m* noduleux; cœur branchu
r сучковатое ядро *n*

* **core diameter → 4196**

1949 core [layer]
(of plywood)
d Mittelfurnier *n*
f âme *f*
r средний слой *m* [фанеры]; внутренний слой; серединка *f*; центральный слой фанеры

* **coreless wood → 4199**

1950 core moisture content
d Kernfeuchtketsgewicht *n*
f teneur *f* en humidité de cœur; contenu *m* en humidité
[лесоматериала]

1951 core of wood; wood core; tree kernel; heart [center]; duramen; center; [heart] pith; pith stock; medulla; marrow; boon
(the centre of a tree; the small cylinder of primary tissue of a tree stem around which the annual rings form)
d Holzkern *m*; Herz *n*; Kern *m*; Mark *n*
f moelle *f*; cœur *m*; duramen *m*
r сердцевина *f* [дерева]

1952 core particles
(of fiberboard)
d Mittel[schicht]span *m*
f particules *fpl* de moelle; plaquettes *fpl* de cœur
r стружки *fpl* среднего слоя

* **core ratio → 4200**

1953 corestock
(timber, veneer, and other wood-base panel material for use as cores in plywood)
d Furnierplatte *f*; Furnierträger *m*
f panneau *m* de [particules destiné au] placage; matériau *m* pour âmes de contreplaqués
r средний слой *m* ДСП

1954 core veneer
d Kernfurnier *n*
f placage *m* pour plis intérieurs
r шпон *m* для внутреннего слоя фанеры

* **core [wood] → 4202**

1955 coriaria family
(Coriariaceae)
d Gerberstrauchgewächse *pl*
f coriariacées *pl*
r кориариевые *pl*

* **coring borer → 4451**

* **coring instrument → 4451**

1956 cork; suber; phellem
(a tissue produced externally by the phellogen in a stem or root impermeable to water and gasses)
d Korkrinde *f*; Kork *m*; Phellem *n*
f liège *m*; suber *m*
r пробка *f*; корка *f*; феллема *f*

* **cork-bark elm → 2677**

* **cork cambium → 6201**

1957 **cork cell**
d Korkzelle *f*
f cellule *f* subéreuse
r клетка *f* пробки

1958 **cork-elm; cork-bark smooth-leaved elm**
(Ulmus suberosa)
d Korkulme *f*
f orme *m* à liège
r вяз *m* пробковый

1959 **cork-forest**
d Korkwald *m*
f forêt *f* de liège
r пробковый лес *m*

1960 **cork layer**
d Korkschicht *f*; Korkmantel *m*
f couche *f* subéreuse
r пробковый слой *m*

1961 **cork-oak; evergreen oak; cork-tree**
(Quercus suber)
d Korkeiche *f*; echter Korkbaum *m*; Pantoffelbaum *m*
f chêne-liège *m*; liège *m*; surier *m*
r дуб *m* пробковый

* **cork-pine → 2726**

1962 **cork powder**
d Korkmehl *n*
f poudre *f* de liège; poudre liégeuse
r пробковая мука *f*

* **cork-producing layer → 6201**

* **corkscrew → 5075**

1963 **corkscrew willow; Pekin willow; dragon claw willow; Hankow willow**
(Salix matsudana)
d Zickzackweide *f*; Korkenzieherweide *f*; Peking-Weide *f*
f saule *m* de Pékin; saule tortueux
r ива *f* матсудана

* **cork-tree → 1961**

* **corm → 7053**

1964 **cornel; Florida dogwood; flowering dogwood**
(Cornus florida)
cornel *m*
r дёрен *m* цветущий

1965 **Cornelian cherry [dogwood]**
(Cornus mas)
d Gelbkornel; Kornelkirschbaum *m*; Kornelkirsche *f*; [gelbe] Herlitze *f*; Tierlibaum *m*
f cornouiller *m* mâle; bois *m* de chien
r кизил *m* [мужской обыкновенный]

1966 **corner cutting**
d Eckenabschnitt *m*
f coupe *f* latérale
r боковой рез *m* (подрезание спиливаемого дерева поочерёдно с каждой стороны пропила)

* **cornering → 7578**

* **corner knot → 4733**

1967 **corner tree; pivot tree**
d Eckbaum *m*
f arbre *m* cornier; arbre pivotant
r угловое дерево *n*

* **corona → 1234**

1968 **coronal root; fibrous root**
d Faserwurzel *f*
f racine *f* fasciculée; racine fibreuse
r мочковидный корень *m*; мочковатый корень

1969 **corridor**
(a band of vegetation, usually older forest, which serves to connect distinct patches on the landscape)
d Korridor *m*
f couloir *m*
r коридор *m*

* **corridor cultivation → 183**

* **corridor planting → 183**

1970 **corridor skidding**
(logging procedure using cable yarders in which narrow clearcuts are made through a stand)
d Korridorrückung *f*
f débardage *m* par couloir
r коридорная вывозка *f*

f éclaircie *f* par couloirs
r коридорные рубки *fpl* ухода

1972 **Corsican pine**
(Pinus laricio; Pinus nigra poirentiana)
d korsische Kiefer *f*; kalabrische Kiefer
f pin *m* laricio de Corse; pin de Calabre
r сосна *f* чёрножелобчатая

1973 **cortex**
(that portion of the primary axis of a vascular plant that immediately surrounds the central cylinder; on the outside, it is enveloped by a uniseriate layer - the epidermis (stems) or the exodermis (roots))
d Kruste *f*
f cortex *m*
r корка *f*

* **cortex fiber → 623**

1974 **cortical; corticole; corticolous; epiphloedal** *adj*
(growing on the woody parts of trees and shrubs)
d wachsen in der Rinden
f cortical
r растущий на [поверхности] коры (деревьев)

1975 **cortical tissue**
d kortikales Gewebe *n*; Korkgewebe *n*; Rindengewebe *n*
f tissue *f* corticale
r ткань *f* коры; пробковая ткань *f*; корковая ткань

* **corticole → 1974**

* **corticolous → 1974**

1976 **corticolous vegetation**
d Baumrindenvegetation *f*
f végétation *f* corticole
r растительность *f* на коре деревьев

* **cost of harvesting → 8217**

* **cost of reproduction → 6350**

1977 **cost per cord**
d Kosten *pl* pro Kord
f pris *m* par corde
r себестоимость *f* на 1 корда; затраты *fpl* на 1 корд

(offered by various US state and federal agencies that pays a fixed rate or percentage of the total cost necessary to implement some forestry or agricultural practice)
d Selbstbeteiligung *f* [des Versicherten]; Selbstbehalt *m*
f participation *f* aux frais; participation aux coûts; partage *m* des frais
r распределение *n* издержек; совместное несение *n* издержек; совместное несение расходов

1979 **cost standardization in forestry**
d Aufwandsnormung *f* der Forstwirtschaft
f standardisation *f* des pertes en foresterie
r нормирование *n* затрат в лесном хозяйстве

* **cotoneaster → 5389**

1980 **co-tree**
d komplementärer Baum *m*; Ko-Baum *m*
f co-arbre *m*
r дополнительное дерево *n*; кодерево *n*

1981 **cotton root rot; Texas root rot**
(invader - Phymatotrichum omnivorum)
d Wollbaum-Wurzelschwamm *m*
f pourriture *f* des racines de coton
r техасская корневая гниль *f*

* **cottonwood → 2717**

1982 **cottonwood leaf beetle**
(Chrysomela scripta)
d roter Pappelblattkäfer *m*
f chrysomèle *f* du liard
r листоед *m* тополя; тополевый листоед

1983 **cotyledon; seminal leaf; primary leaf; seed-leaf**
(one of the first leaves to appear after germination (there may be one, two, or more); the foliar portion of the embryo as found in the seed)
d Keimblatt *n*; Keimlappen *m*; Kotyledon *n*; Samenblatt *n*
f cotylédon *m*
r семядоля *f*; семенодоля *f*

1984 **Coulter['s] pine; bigcone pine**
(Pinus coulteri)
d Coulter-Kiefer *f*
f pin *m* de Coulter
r сосна *f* Культера

1985 counterfort
(ridge of wood that develops in the angle between a lateral root and the butt of a tree, which may extend up the stem to a considerable height)
d Druckstrebe *f*
f contrefort *m*
r подпорка[-контрупор] *f*; подпора *f*; контрфорс *m*

* **counterfort → 1229**

* **counting of trees → 2837**

* **country forester → 1988**

1986 country kiln
d Landtrockner *m*
f charbonnière *f* locale
r местная сушилка *f*

1987 county forest
d Kreisforst *m*; kreiseigener Forst *m*
f forêt *f* d'arrondissement
r окружной лес *m*

1988 county forester; area forester; county forest ranger (*US*)
d Bezirkförster *m*; Revierförster *m*
f forestier *m* régional; forestier cantonal
r участковый лесничий *m*; районный лесничий

* **coupe → 5038**

* **coupe flora → 3039**

* **coupe inventory → 6831**

* **coupe inventory records → 3073**

1989 coupes on alternate strips
d Wechselstreifenhieb *m*
f coupes *fpl* par bandes alternées
r чересполосное примыкание *n* лесосек

1990 courbaril; West Indian locust
(Hymenaea courbaril; Hymenaea palustris)
d Courbaril; Jatoba; Heuschreckenbaum *m*
f courbaril *m*; jatoba *m*
r джатоба *f*; гимена *f* курбарил; курбарил *m*

* **course → 8545**

1991 cove hardwood
(timber stands in which 70 percent or more of hardwood species)
d Buchthartholz *n*; Wölbungshartholz *n*
f bois *m* feuillu profilé
r профильная твёрдая древесина *f*

1992 cover; covering
(hiding place or vegetative shelter for wildlife from predators or inclement weather)
d Deckungswert *m*
f recouvrement *m*
r покров *m*

1993 cover crop
(plants used to nurture or improve the form of a more important crop during youth by protecting it from frost, insolation, or wind)
d Schutzpflanzung *f*; Schutzbestand *m*
f culture-abri *f*; peuplement *m* d'abri
r защитное насаждение *n*; покровное насаждение

* **covered area → 354**

1994 covered with bark; under bark; barked *adj*
d berindet
f couvert d'écorce
r покрытый корой

* **covered with woods → 3394**

* **covering → 1992**

1995 covering brushwood
d Schutzbuschholz *n*
f paillis *fpl* pour recouvrir les cultures; ramilles *fpl* pour recouvrir les cultures
r покровный хворост *m*; защитный хворост

* **covering of pit-heaps → 97**

1996 cowberry; red bilberry; foxberry; fruit-ripe; lingonberry; partridgeberry; red huckleberry
(Vaccinium vitis-idaea; Mitchella repens)
d Kränsbeere *f*; Preiselbeere *f*; Steinbeere *f*; Rauschbeere *f*; rote Heidelbeere *f*
f canneberge *f* ponctuée; airelle *f* vigne d'Ida; airelle rouge; myrtille *f* à feuilles persistantes; liane *f* à squaw; airelle canche; airelle ponctuée
r брусника *f*; боровика *f*

1997 cowberry family
(Empetraceae)
d Krähenbeerengewächse *pl*

1998 cowberry pine forest
d Preiselbeer-Kiefernwald *m*
f pineraie *f* à airelle rouge
r сосняк-брусничник *m*

1999 cowberry spruce forest
d Preiselbeer-Fichtenwald *m*
f pessière *f* à airelle rouge
r ельник-брусничник *m*

2000 cow [breatnut] tree; milk tree
(Couma macrocarpa; Brosimum utile)
d Milchbaum *m*; Couma *m*
f arbre *m* au lait; couma *m*
r бросимум *m* полезный

* **cow-wood → 478**

* **coyote willow → 7183**

* **crab → 5124**

* **crabwood → 4039**

* **crack depth → 7484**

2001 crack distribution
d Rissausbreitung *f*
f distribution *f* de fentes; distribution de brisures
r распространение *n* трещин

2002 crack due to heat; heat check; heat shake
d Hitzerriss *m*
f fente *f* de retrait due à la chaleur
r трещина *f* от жары

2003 crack formation; checking; crackling; fracturing; tearing open
(of wood)
d Rissbildung *f*; Aufspringen *n*; Reißen *n*; Zerklüftung *f*
f fendillage *m*; fendillement *m*; fracturation *f*
r образование *n* трещин; трещинообразование *n*; растрескивание *n*

2004 crack length; shake length
d Risslänge *f*
f longueur *f* de crevasse
r длина *f* трещины

* **crackling → 2003**

* **crack-willow → 1040**

* **cradle → 4840**

as a result of tree uprooting and its attendant displacement of soil)
d Wiege-Hügelchen *n*
f mamelonnement *m* forestier
r седлообразный холмик *m*

2006 cramantee; American muskwood; cedrillo
(Guarea exelsa; Guarea grandifolia; Guarea trichiliodes; Huertea cubensis)
d Kramantee *f*
f cramantee *m*
r гварея *f* высокая; гуарея *f* высокая

* **cranberry forest type → 9158**

2007 crape[ ]myrtle; crepe myrtle
(Lagerstroemia indica)
d indische Lagerströmie *f*; Kräuselmyrte *f*
f largerstroemia *m* des Indes; lagerose *f*
r лагерстремия *f* индийская

* **crawler[-type] tractor → 7799**

2008 crazy drag
d verrückte Gegenkraft *f*
f petit traîneau *m* de débardage
r небольшие трелёвочные подсанки *pl*

2009 creambush [rock spirea]
(Holodiscus discolor)
d Wald-Schaumspiere *f*
f holodisque *f* décolorée
r холодискус *m* разноцветный

* **creaming → 4285**

* **cream [off] → 4272**

2010 creamy sap rot
(invader - Polyporus gilvus)
d sahnige Saftfäule *f*
f pourriture *f* de l'aubier crémeuse
r кремовая заболонная гниль *f*

* **crease → 7308**

2011 creep *v*
(ref. to plants)
d kriechen; unterlaufen
f grimper
r стлаться

* **creeper → 1642**

2012 creeping forest fire; [under]ground fire
(fire that burns in the organic material below

d kriechendes Feuer *n*; Lauffeuer *n*
f incendie *m* souterrain; incendie propageant dans la couche superficielle du sol; feu *m* de terre; feu au ras du sol
r низовой пожар *m*; ползающий огонь *m*; почвенный огонь

2013 creeping Oregon grape; creeping mahonia
(Mahonia repens; Berberis repens; Berberis sonnei)
d kriechende Mahonie *f*
f mahonia *m* rampant
r священный лиственный барбарис *m*; [скалистый] горный виноград *m*; барбарис орегонский; ползучая магония *f*

2014 creeping root
d kriechende Wurzel *f*
f racine *f* traçante
r ползучий корень *m*

2015 creeping willow
(Salix repens)
d Kriechweide *f*
f saule *m* rampant
r ива *f* ползучая

2016 creep in wood
(the deformation produced in wood by plastic flow as the result of loads acting over a period of time)
d Formbarkeit *f* des Holzes; Holzwiderling *m*
f fluage *m* de bois
r пластичность *f* древесины

* **crenate leaf margin → 2396**

2017 creosote *v* timber
d Holz mit Teeröl imprägnieren
f imperméabiliser de bois par huile de goudron
r пропитывать древесины креозотом

* **crepe myrtle → 2007**

* **crest → 9051**

* **crested swift → 8942**

* **crevice → 9507**

* **crib → 8584**

2018 cribwork; crib
d Blockbau *m*
f ouvrage *m* à claire-voie
r ряжевый переплёт *m*; сруб *m*; [костровая] крепь *f*; клеть *f*

2019 crimping; snaking; ribbing
(a more or less regular corrugation of the surface of wood caused by differential shrinkage of earlywood and latewood)
d Ondulation *f*; Wölbung *f*; Kräuselung *f*
f ondulation *f*
r волнистость *f*; ребристость *f*

2020 crinkling
(bending or twisting of foliage without breaking)
d Knittern *n*
f ridement *m*; enroulement *m*
r сморщивание *n*

2021 crinodendron family
(Elaeocarpaceae)
d Crinodendrongewächse *pl*
f éléocarpacées *pl*
r элеокарповые *pl*

2022 crippled wood; dwarfed wood
d Krüppelholz *n*
f bois *m* rabougri
r корявый лес *m*

2023 critical habitat
(a habitat of threatened or endangered species as designated by various state forest practices boards)
d kritische Umwelt *f*; kritisches Habitat *n*
f habitat *m* critique; site *m* critique
r критическое местообитание *n*

2024 critical winter forest range
(forested habitat, usually stands of mature or old-growth conifers, which provides deer and elk with resources critical to survival during severe winters)
d kritische Winterförsterei *f*
f habitat *m* forestier d'hiver critique
r критическое зимнее пастбище *n*

* **crook → 2100**

2025 crooked growth; bent growth
d Krüppelwuchs *m*; Knickwuchs *m*
f croissance *f* coudée
r искривлённый рост *m*; коленчатый рост

2026 crooked knot
d knorriger Ast *m*
f branche *f* genouillée; branche courbée
r искривлённый сучок *m*; корявый сучок

f bois *m* genouillé
r искривлённое бревно *n*

2028 **crookedness of the stems; stem crookedness**
d Krummschaftigkeit *f*
f sinuosité *f* des fûts; port *m* flexueux
r кривоствольность *f*

2029 **crooked timber; crooked wood; kneewood; knee-timber; curved timber**
d krummgewachsenes Holz *n*; Krippelholz *n*; Krummholz *n*; Knieholz *n*
f bois *m* crochu; bois coudé; bois tordu
r искривлённый лесоматериал *m*; кривоствольный лес *m*; лесоматериал с кривизной; кривулина *f*; криволесье *n*

2030 **crooked tree; curved tree**
d Krüppelbaum *m*; Krummbaum *m*
f fouet *m*; arbre *m* crochu; arbre courbé
r искривлённое дерево *n*; кривое дерево; корявое дерево

* **crooked trunk → 2154**

* **crooked wood → 2029**

2031 **crook[ing]; bending; rippling; sweep**
(tree defect characterized by a sharp bend in the main stem)
d Knick *m*; Biegung *f*; Krümmung *f*
f courbure *f*; cintrage *m*; aboutage *m*
r искривление *n*; изгиб *m*; коробление *n*; скручивание *n*; кривизна *f*

2032 **crop basal area; stand basal area; total basal area**
d Bestandesgrundfläche *f*; Bestandeskreisfläche *f*
f surface *f* terrière [unitaire] d'un peuplement
r общая сумма *f* площадей поперечного сечения деревьев на единице площади; общая площадь *f* массива; основная площадь массива

* **crop height → 4233**

* **crop of acorns → 5843**

* **crop of beech nuts → 656**

2033 **crop of cones; cone crop**
d Zapfenernte *f*; Zapfenertrag *m*
f récolte *f* de cônes
r урожай *m* [хвойных] семян; урожай шишек

2034 **crop(ping) intensity**
d Ernteintensität *f*
f intensité *f* de rentrée de la récolte
r интенсивность *f* культуры

* **cropping wood → 3732**

* **crop tree → 5569**

* **crop-tree thinning → 8512**

2035 **crossaxe; pick-axe; pit mattock**
d Kreuzhacke *f*; Kreuzhaue *f*; Kreuzaxt *f*
f pioche *f* forestière
r крестовый топор *m*

2036 **cross-band** *v*
(to place the grain of layers of wood at right angles in order to minimize shrinking and swelling)
d kreuzbandumsetzen
f contreplaquer
r скрещивать слоёв

2037 **crossband[ing]**
(a layer of veneer whose grain direction is at right angles to that of the face plies)
d Kreuzbandumsetzung *f*
f pli *m* transversal; contre[-]placage *m*
r крестообразное расположение *n* слоёв

* **crossbar → 2038**

2038 **crossbeam; crossbar; wood bunk; bunk; bale; peg beam; bolster**
(a beam on which logs rest in a trailer or a truck)
d Holm *m*; Querbaum *m*; Querträger *m*; Querbalken *m*
f arbre *m* transversal; traverse *f*
r коник *m*; поперечина *f*; поперечная балка *f*; поперечный брус *m*; траверса *f* цепного лесотранспортёра; подкладка *f* под брёвна

2039 **cross-break; transverse shake; upset**
d Querriss *m*
f fracture *f* transversale
r поперечная трещина *f*; поперечный разрыв *m*; расщепление *n* древесины поперёк волокон; крестовый меток *m*

2040 **cross core**
(of plywood)
d Quermittelfurnier *n*
f âme *f* transversale
r поперечный внутренний слой *m*; поперечная черновая рубашка *f*

* **cross-cut → 2176**

2041 **crosscut point**
d Querschneidenpunkt *m*
f point *m* de tronçonnement
r граница *f* (место на долготьё или на бревне, где оно должно быть перепилено)

2042 **cross[cut]-saw; trim saw; briar; [timber] gangsaw**
d Quersäge *f*; Bundsäge *f*
f scie *f* transversale
r поперечная пила *f*

2043 **crosscutter; bucker; log maker**
d Quersägemaschine *f*; Abkürzmaschine *f*; Ablängsägemaschine *f*; Kappmaschine *f*; Stützsäge *f*
f tronçonneuse *f*
r поперечно-пильный станок *m*; торцовочный станок; концеравнитель *m*; раскряжёвочный станок; раскряжёвочное устройство *n*

* **crosscut[ting] → 8571**

* **crosscut wood → 2051**

2044 **cross deck**
d Querbühne *f*
f rampe *f* transversale
r поперечная эстакада *f*

2045 **cross-dimensions**
(of a stack)
d Querdimensionen *fpl*
f dimensions *fpl* transversales; dimensions d'équarrissage
r поперечные размеры *mpl* (длина и ширина штабеля древесины)

2046 **cross-drain culvert**
(a culvert used to carry ditch water from one side of the road to the other)
d Querdohle *f*
f buse *f* transversale
r поперечная закрытая канава *f*

* **crosser → 6236**

2047 **crosser stain; sticker stain; stick[er] mark**
(a brown or blue stain that develops in seasoned lumber where it has been in contact with the stickers)
d Verfärbung *f* durch Lagerhölzer; Lagerholzzeichnung *f*; Lagerholzflecken *mpl*
f décoloration *f* de bois gisants; taches *fpl* apparaissantes sur de bois entreposé; tache *fpl*
от прокладок; изменение *n* цвета складочной древесины; лежневые пятна *npl*

2048 **cross-fib[e]red wood; spiral-grained wood; wavy-fib[e]red wood; wave growth timber; twisted wood; torse wood; curly-grained wood; wavy-grown wood; bumpy wood; burr[ed] wood**
(wood in which the fibers are distorted so that they have a curled appearance)
d drehwüchsiges Holz *n*; krausgewachsenes Holz; verwimmertes Holz; Spiralholz *n*; Überwallungsholz *n*;Wellenholz *n*
f bois *m* à fibres flexueuses; bois à fibres torses; bois ondulé
r косослойная древесина *f*; свилеватая древесина; извилисто-волокнистая древесина; неправильно росшая древесина

* **crossfield → 6738**

2049 **crossfield pit pair**
d Kreuzungsfeldgrubepaar *n*
f couple *m* de ponctuation par champ de croisement
r пара *f* пор поля перекрёста

2050 **crossfield pitting**
(the pitting occurring in the rectangle formed in a radial section by the walls of a ray cell and those of an axial tracheid)
d Kreuzungsfeld-Tüpfelung *f*
f ponctuation *f* par champ de croisement
r поровость *f* поля перекрёста

* **crossfire → 3246**

* **cross grain → 4585**

* **cross-grain axe → 9026**

* **cross-grained cutting → 8571**

2051 **cross-grained wood; crosscut wood; crossways wood; end-grain[ed] wood**
(wood in which the fibers deviate from a line parallel to the sides of the piece; may be either diagonal or spiral grain or a combination of the two)
d Hirnholz *n*; Querholz *n*
f bois *m* frontal; bois de bout
r торцевая древесина *f*; торец *m*; поперечина *f*

2052 **crosshaul[ing]; rolling; parbuckling**

f chargement *m* latéral
r погрузка *f* [брёвен] накатыванием; подъём *m* [брёвен] накатыванием; поперечная накатка *f*

* **crossing → 4401**

2053 crossing-the-lead
(timber felled across rough terrain or across other felled timber; this is a major cause of breakage in large timber)
d Kreuzung *f* der Leitung
f croisement *m* de guide
r пересечение *n* леса при трелёвке

* **cross-laminated wood → 3876**

2054 cross log in a stack; cross skid; roller skid; tree shoe; shoe [rake]; shoe tree
(device in the shape of a segment of a circle used to support the skyline from a spar tree)
d Bremsklotz *n*; Klotzbremse *f*; Schuhspanner *m*; Unterlage *f* für Lagerung
f sabot *m* [d'arrêt]; embauchoir *m*
r [поперечные] лежни *mpl*; башмак *m*; стойка *f* для предупреждения раскатывания брёвен; стопорное стабилизирующее устройство *n*

* **cross-pile → 6232**

2055 cross piling
d Querstapelung *f*
f empilage *m* croisé; empilage en travers
r поперечная штабелёвка *f*

2056 cross pollination; open pollination; free pollination
d Querpollination *f*
f pollinisation *f* croisée; pollinisation libre
r перекрёстное опыление *n*

2057 cross[-]ride; riding
d Querschneise *f*
f laie *f* croisée; laye *f* croisée
r поперечная просека *f*

* **cross skid → 2054**

2058 cross subsidy; cross subsidization; cross subsidizing
(for example, selling a valuable timber stand for less than market value to compensate the purchaser for having to pay for another timber stand that is worthless)
d Kreuzbeihilfe *f*
перекрёстное субсидирование

* **cross support → 4548**

2059 cross[-]trees
d Kreuz-Bäume *mpl*; Quersalinge *fpl*; Saling *fpl*
f barres *fpl* traversières
r салинг *m*

* **crossways wood → 2051**

* **crosswise cutting → 8571**

* **crotching → 987**

* **crotchline → 7742**

2060 crotch veneer
d Gabelfurnier *n*
f placage *m* ronceux
r шпон *m* из развилочной части брёвна

* **crotch [wood] → 3655**

2061 croton
(Croton)
d Krebsblume *f*
f croton *m*
r кротон *m*; каскараль *m*

2062 crowded tree; tree with one-sided crown
d eingeklemmter Stamm *m*; Stamm mit einseitiger Krone
f arbre *m* à cime écrasée; arbre à cime unilatérale; arbre très serré
r зажатий ствол *m*; ствол с однобоковой кроной

* **crowding level → 3370**

2063 crown-area index
d Kronenflächeindex *m*
f indice *m* d'aire de cime
r показатель *m* площади проекции кроны

2064 crown base; position of the crown
(point where the crown of a tree commences)
d Ansatz *m* der Krone; Kronenansatz *m*; Kronensatz *m*
f base *f* de la cime; première *f* couronne
r основание *n* кроны; место *n* прикрепления кроны

2065 crown boundary
d Kronenrand *m*

* **crown break → 1012**

2066 **crown canopy; crown cover**
(the vertical projection of the tree crown on the ground; a measurement of how much of an area is covered by the leaves in the crown of its trees)
d Kronendach *n*; Kronenschichte *f*
f couvert *m* de cime
r купол *m* кроны; верхний ярус *m* ветвей деревьев

2067 **crown chips**
d Kronenspäne *mpl*
f copeaux *mpl* de cime
r щепа *f* из вершины

2068 **crown class**
(class of tops or crowns occupying a similar position in the canopy or crown cover)
d Kronenklasse *f*
f classe *f* de cime
r класс *m* кроны

2069 **crown closure; crown contact; vinicity of crown**
(condition when the crowns of trees touch and effectively block sunlight from reaching the forest floor)
d Kronenschluss *m*
f fermeture *f* des cimes; fermeture des houppiers
r сомкнутость *f* крон; смыкание *n* крон

* **crown closure → 2094**

* **crown closure class → 1600**

2070 **crown competition**
d Kronenwettbewerb *m*
f compétition *f* des cimes; concurrence *f* des cimes
r конкуренция *f* крон; соревнование *n* крон

2071 **crown condition; crown status**
d Kronenzustand *m*
f état *m* des houppiers
r состояние *n* [развития] кроны; статут *m* кроны

* **crown contact → 2069**

* **crown correction → 2083**

* **crown cover → 2066**

* **crown damage → 1012**

2072 **crown density**
(amount, compactness or depth of foliage of a tree crown)
d Kronendichte *f*
f densité *f* du couvert d'une cime
r густота *f* кроны

* **crown depth → 2082**

* **crown [depth] ratio → 2085**

2073 **crown development**
d Kronenentwicklung *f*
f développement *m* de cime
r развитие *n* кроны

2074 **crown diameter**
d Kronendurchmesser *m*
f diamètre *m* de cime
r диаметр *m* кроны

2075 **crown/diameter ratio**
(of a standing tree, the ratio of the crown diameter in feet to the diameter breast-height over bark in inches)
d Kronendurchmesserverhältnis *n*
f rapport *m* diamètre cime/fût; rapport entre diamètre de cime et fût
r соотношение *n* диаметров кроны и ствола

* **crown edge → 2662**

2076 **crown fire; [running] crowning fire; canopy fire**
(fire that burns through the tops of living trees or brush)
d Kronenfeuer *n*; Wipfelfeuer *n*; Gipfelfeuer *n*
f feu *m* de cime[s]; incendie *m* de cimes
r верховой [беглый] пожар *m*; верховой лесной пожар; пожар крон; головная часть *f* пожара

2077 **crown forest** (*UK*)
d Hofkammerwald *m*; Kronforst *m*
f forêt *f* camérale; forêt de la couronne
r удельный лес *m* (принадлежащий королевской семье)

2078 **crown form**
d Kronenform *f*
f forme *f* de la cime
r форма *f* кроны

f formation *f* en cime
r формирование *n* кроны

2080 crown gall
d Kronenabschürfung *f*
f tumeur *f* bactérienne du collet
r корончатый галл *m*

2081 crown grafting
d Kopfveredelung *f*
f greffage *m* en couronne
r прививка *f* за кроной

2082 crown height; crown depth; crown length
d Kronenlänge *f*
f hauteur *f* de la couronne; longueur *f* de cime; longueur de houppier
r высота *f* кроны; длина *f* кроны; протяжённость *f* кроны

* **crowning fire → 2076**

2083 crowning-off; crown correction
(cutting back, in more or less systematic fashion, the crown of a tree to produce a close head of shoots)
d Kronenschnitt *m*
f têtard *m*; taille *f* de forme
r подрезка *f* кроны; обрезка *f* кроны; удаление *n* кроны

2084 crown layer
d Kronenschicht *f*
f couche *f* de cimes
r слой *m* крон

* **crown length → 2082**

2085 crown/length ratio; crown percent; crown [depth] ratio
(of a standing tree, the ratio of crown length to tree height)
d Beastungsgrad *m*
f rapport *m* cime/hauteur totale
r соотношение *n* длины кроны к высоте дерева

2086 crown map
d Kronenkarte *f*
f carte *f* des cimes
r карта *f* крон

2087 crown [of tree]; summit
(the branches and foliage of a tree)
d Baumkrone *f*; Krone *f*
f cime *f* [d'arbre]; cimier *m*; couronne *f*; faîte *f*;

* **crown percent → 2085**

2088 crown projection
d Kronenbereich *m*; Kronenprojektion[grundfläche] *f*
f projection *f* horizontale des cimes
r [горизонтальная] проекция *f* кроны

2089 crown protection systems
d Kronenschutzsysteme *npl*
f systèmes *mpl* de protection de cime
r системы *fpl* защиты крон

* **crown pruning → 2096**

2090 crown ratio
(the ratio of the leaved portion of a tree)
d Kronenprozent *n*; Kronenanteil *n*
f taux *m* de cime; pourcent *m* de cimes
r процент *m* кроны

* **crown reduction → 2096**

2091 crown-root ratio
d Kronenwurzelverhältnis *n*
f rapport *m* cime-racine
r соотношение *n* крон к корням

2092 crown rot
(invader - Phytophthora parasitica)
d Kronefäule *f*
f pourriture *f* de cime
r гниль *f* корневой шейки

* **crown rupture → 1012**

* **crown saw → 2209**

2093 crown size
d Kronenausmaß *n*; Kronengröße *f*
f taille *f* de cime
r размер *m* кроны

2094 crown spread; crown closure
(the area covered by the branches and foliage of a tree)
d Kronendeckel *m*
f couverture *f* des cime
r раскидистость *f* кроны

* **crown spring → 7940**

* **crown status → 2071**

2095 crown tending
d Kronenpflege *f*

2096 crown thinning; crown pruning; top thinning; crown reduction
(removal of trees from the dominant and codominant crown classes to favor the best trees of those same crown classes)
d Kronendurchforstung *f*; Kronenabschluss *m*; Kronenlockerung *f*
f éclaircie *f* par les cimes; taille *f* des cimes; éclaircie de mise en lumière
r разреживание *n* крон; прорубка *f* крон

* **crowntree → 5463**

2097 crown volume
d Kronenvolumen *n*
f volume *m* des cimes
r объём *m* кроны

2098 crown width
d Kronenbreite *f*
f largeur *f* de cime
r ширина *f* кроны

* **crown wood → 4284**

* **CRP → 1882**

* **crub forest → 1202**

2099 crucifer family; mustard family
(Cruciferae)
d Kreuzblütengewächse *pl*
f crucifères *pl*
r крестоцветные *pl*

2100 cruck; crook; knee
d Vorderknie *n*; Knie *n*
f coude *m*; genou *m*
r изогнутая ветка *f* дерева; имеющая ответвление ветка дерева

* **crude balsam → 6733**

* **crude humus → 1681**

* **crude resin → 6733**

* **cruiser's stick → 715**

2101 cruising axe
d kreuzende Axt *f*
f hache *f* d'échantillonnage
r лёгкий топор *m*

* **cruising crew → 8288**

* **cruising prism → 6554**

2102 cruising radius
(a loose term for the distance covered between successive points at which an individual animal is observed or traced over a specified period)
d Aktionsradius *m*
f intervalle *m* d'observation
r радиус *m* действия

2103 crumbly brown cubical rot
(invader - Fomes pinicola, Fomitopsis pinicola)
d krümelige Braunfäule *f*
f carie *f* grumeleuse brune cubique
r бурая сплошная кубическая гниль *f*

* **crumbly rot → 2433**

2104 crusher; crushing machine
d Brecher *m*; Quetscher *m*
f défibreur *m*; casseur *m*
r машина *f* для измельчения (лесосечных отходов на вырубках при подготовке к механизированным лесопосадкам); рубильная машина

2105 crushing; grinding
d Quetschen *n*; Schroten *n*
f écrasage *m*; écrasement *m*; broiement *m*; broyage *m*; râpage *m*
r измельчение *n* [древесины]

* **crushing machine → 2104**

2106 crust *v*
d verkrusten
f croûter
r заплывать; покрываться коркой

* **crutch → 3655**

* **cryptomeria → 4615**

2107 cryptopodzolic brown earth
d cryptopodsolische Braunerde *f*
f sol *m* brun podzolique
r скрытоподзолистая бурая [лесная] почва *f*

2108 crystalliferous cell
(a cell containing one or more crystals)
d kristallführende Zelle *f*
f cellule *f* cristallifère
r кристаллоносная клетка *f* (паренхимы)

2109 Cuban belly; belly palm; bottle palm; barrel palm
(Colpothrinax wrightii; Pritchardia wrightii; Crispa gastrococos)
d kubanische Bauchpalme *f*
f palmier *m* ventru
r пальма *f* барригона

2110 Cuban mahogany; Dominican mahagoni; Jamaican mahagoni; madeira
(Swietenia mahagoni; Cedrela mahagoni; Cedrus mahogani)
d echtes Mahagoni *n*
f acajou *m* de Cuba
r махагони *n* настоящее; махагониевое дерево *n* низкое; свиетения *f* махагони; акажу *n*

2111 Cuba pine; Caribbean [longleaf] pitch pine; Honduras pine; slash pine
(Pinus caribaea; Pinus elliottii)
d zentralamerikanische Kiefer *f*; karibische Kiefer
f pin *m* des Antilles; pin d'Amérique centrale
r сосна *f* Элиота; сосна караибская

* **cubby → 8822**

2112 cubical rot; cubical cracking
d würfelige Fäule *f*
f pourriture *f* cubique
r кубическая гниль *f*; кубообразная гниль; призматическая гниль

2113 cubic foot; cft
(1.728 cubic inches of wood; contains 6-10 useable board feet of wood)
d Kubikfuß *m*
f pied-cube *m*
r кубический фут *m*

2114 cubic meter of dense timber; cubic meter of trunk wood; solid cubic meter
d Kubikmeter *n*; Festmeter *n*
f mètre *m* cube solide
r кубометр *m* плотных лесоматериалов; плотный кубометр

* **cubic meter of harvested timber → 9246**

2115 cubic meter of piled wood; stacked cubic meter
d Beugholzraummeter *n*
f stère *m*; mètre *m* cube [apparent] de bois
r складочный кубометр; кубический складочный метр *m*

* **cubic meter of trunk wood → 2114**

2116 cubic meter solid volume over bark
d Vorratsfestmeter *n*
f mètres *mpl* cubes sur pied
r кубический метр *m* древесины в плотных единицах с корой

* **cubic table → 8957**

2117 cubic unit log rule
(a table of units of measurement of volume or capacity)
d Kubierungstabelle *f*; Kubierungstafel *f*; Kubiktabelle *f*
f barème *m* du cubage
r таблица *f* определения объёмов [брёвен] в кубических единицах; кубатурник *m*

* **cubing → 8955**

2118 cubing by the quarter-girth; volume determination by quarter girth
d Viertelkubierung *f*
f cubage *m* au quart déduit
r определение *n* четверти объёма

* **cucumber magnolia → 709**

* **cucumber-tree → 709**

* **cudgel → 7101**

2119 cull board
d Fehlbrett *n*
f planche *f* de rebut
r дефектная доска *f*

* **cull-butt → 1226**

* **culled forest → 7410**

* **culler → 1457**

2120 cull factor; cull percentage
d Ausschussfaktor *m*
f coefficient *m* de réfaction; pourcentage *m* de rebut
r коэффициент *m* отбраковки; процент *m* неликвидной древесины (в растущем дереве); процент объёма нетоварной части [дерева или брёвна]

2121 **cull[ing]; improvement cutting; improvement felling; tending cutting; removal of noxious trees**
(the deduction made from gross timber volume to adjust for defect)
d Verbesserungshieb *m*; Pflegehieb *m*; Durchplänterung *f*; Aushieb *m*
f coupe *f* d'amélioration; réfaction *f*; élimination *f* des plants malformés; extraction *f* de mauvais bois
r вырубка *f* малоценных деревьев; рубка *f* ухода

* **cull-land → 390**

* **cull log → 5764**

2122 **cull log chipping**
d Ausschussholzzerkleinerung *f*
f fragmentation *f* de grumes de rebut
r размельчение *n* дефектных брёвен

2123 **cull [lumber]; cull[wood]; dead culls; wood culls; rejects; offal timber; refused timber; refuse wood; off-cuts**
(trees or logs or portions thereof that are of merchantable size but are rendered unmerchantable by defects)
d Ausschlussholz *n*; Ausschlussholzware *f*; Ausschuss *m*; Nichtnutzholz *n*
f bois *m* de rebut; rebut *m*; déchet *m* de bois
r нестандартная древесина *f*; забракованная древесина; неликвидная древесина; забракованные лесоматериалы *mpl*; отбросы *mpl* древесины; неделовые древесные остатки *mpl*

* **cull percentage → 2120**

* **cull pile → 4774**

2124 **culm; stalk**
(the stem of grasses and bamboos, usually hollow except at the swollen nodes)
d Halm *m*; Culmus *m*
f chaume *m*
r стебель *m*

2125 **culmination age**
(age at which the stand, for the stated diameter limit and utilization standard, achieves its maximum average rate of volume production is maximized)
d Kulminationsalter *n*
f âge *m* culminant
r кульминационный возраст *m*

2126 **culmination of mean annual increment; CMAI**
(point where the total growth increment divided by age is at its maximum)
d mittlerer jährliche Zuwachs *m*
f culmination *f* d'accroissement moyen annuel
r кульминация *f* среднегодового прироста

2127 **cultipacker; land packer; continental Cambridge roller**
d Cambridgewalze *f*
f rouleau *m* rayonneur; rouleau Cambridge
r ребристый каток *m* (напр. для расчистки лесных площадей)

* **cultivar → 2129**

2128 **cultivate** *v*
d bebauen; anbauen
f cultiver
r выращивать; обрабатывать; возделывать

2129 **cultivated variety; cultivar**
(a clone or seedling population with enough favo[u]rable characteristics to warrant cultivation)
d Sorte *f* (in Kultur entstandene Varietät)
f cultivar *m*
r культурный сорт *m*

* **cultivation by burning over → 3176**

* **cultural costs → 6350**

2130 **cultural diversity**
d kulturelle Mannigfaltigkeit *f*
f diversité *f* culturelle
r разнообразие *n* культур

* **cultural expenses → 6350**

2131 **cultural treatments; cultural operations; cultural practices**
d Feldarbeitsgang *m*; Kulturarbeiten *fpl*
f opérations *fpl* culturales; soins *mpl* culturaux; pratiques *fpl* culturales
r агротехнический уход *m*

* **culture → 9788**

2132 **culvert**
(a metal, wooden, plastic, or concrete conduit through which surface water can flow under or across roads)
d Durchlass *m*; Dohle *f*

2133 cumaru
(Coumarouna odorata)
d Cumaru *n*
f cumaru *m*
r кумару *n*

* **cumquat → 4747**

2134 cumulative effects
(effects on biota of stress imposed by more than one mechanism)
d Kumulationswirkungen *fpl*
f effets *mpl* cumulatifs; effets par accumulation
r кумулятивное действие *n*

2135 cumulative growth curve
(of a tree)
d Summenwachstumskurve *f*
f courbe *f* de croissance cumulée
r кумулятивная кривая *f* роста; совокупная кривая роста

2136 cunit; cunet
(a unit of measure of timber volume = 100 cubic feet)
d Kunite *f*
f cunite *f*
r кьюнит *m* (единица измерения объёма плотной древесины без коры, равная 2.83 $м^3$)

2137 Cunningham araucaria
(Araucaria cunninghamii)
d Hopfenpinie *f*
f araucaria *m* de Cunningham
r араукария *f* Куннингама

2138 cunonia family
(Cunoniaceae)
d Cunoniengewächse *pl*
f cunoniales *pl*
r кунонецветные *pl*

* **cup frills → 3715**

* **cup fungi → 7144**

2139 cup gall; ring gall; burst check; ring check; ring shake; drought ring; arc shake; cup shake; round shake; ring failure
(a break in the wood along the annual growth rings; a separation of wood along the grain and parallel to the annual rings, either within or between the rings)
d Ringriss *m*; tangentialer Riss *m*; Tangentialriss *m*; Ringsbruch *m*; Ringkluft *f*;
f roulure *f*; lunure *f*
r кольцевой отлуп *m*; отлупная трещина *f*; кольцевая трещина; тангенциальная трещина; кольцеобразная трещина; ядровый отлуп; сердцевинный отлуп

2140 cupping
(a group of various operations, e. g. bark scraping, initial cutting of the face, and [re]fixing of aprons and cups, preparatory to the start of tapping)
d Anlegen *n* der Lachte
f préparation *f* du gemmage
r подготовка *f* добывания живицы

2141 cup[ping]; dish[ing]
(warping along the face of a board from edge to edge caused when one board face dries faster than the other)
d Wöllring *f*; Schüsselung *f*; Querkrümmung *f*; Anrichten *n*
f tuilage *m*; voilement *m*; déformation *f* en cuvette
r коробление *n* [доски] в поперечном направлении вследствие усушки; комбинированное коробление листовых древесных материалов; поперечное коробление

2142 cupping axe; chipper [for resin tapping]; tree facer
d Lachtenbeil *n*; Lachtenreißer *m*; Harzreißer *m*
f bridon *m*; rasclet *m*
r топор *m* для карр; топор с широкой режущей кромкой конической формы; резак *m* для карр; вздымочный хак *m*

* **cup shake → 2139**

* **curapixa → 6762**

* **curiy → 1005**

* **curl → 9185**

* **curled hard maple → 7608**

* **curled wood → 1032**

* **curling → 9187**

2143 curlleaf mountain-mahogany
(Cercocarpus ledifolius)
d kraulblätteriges Bergmahagoni *n*
f cercocarpe *m* à feuilles ondulées
r церкокарпус *m* багульниколистный

2144 curly figure; wavy figure; swirl figure; undulated figure; fluted figure
(a figure caused by irregular small undulations in the region of the knot)
d gewundener Faserverlauf *m*
f figure *f* ondulée
r свилеватая фигура *f*; извилистый пробег *m* волокон

2145 curly grain; swirl grain; wavy grain; wild grain
(a grain due to undulations in the direction of fiber alignment)
d gewellte Maserung *f*
f fil *m* sinueux; fil irrégulier; fil tourmenté; fibre *f* ondulée
r путанное волокно *n*

* **curly[-grained] growth → 9185**

* **curly-grained log → 4721**

2146 curly-grained point
d Astknorre *f*
f point *m* ondulé
r свилеватое место *n*

* **curly-grained wood → 2048**

2147 curry leaves; curry-leaf tree
(Murraya koenigii)
d Curryblätter *npl*
f feuilles *fpl* de cari; feuilles de curry; caloupilé *m*; carripoulé *m*; kaloupilé *m*
r муррая *f* Кёнига

* **curly maple → 7608**

2148 currant [bush]
(Ribes)
d Stachelbeere *f*
f groseillier *m*
r смородина *f*

* **currant rust → 9376**

2149 currant shoot borer
(Lampronia capitella)
d Rohrbohrer *m*
f perceuse *f* des pousses
r листовёртка *f* побеговая

2150 current annual increment; current annual growth
(the annual increment of wood for increment the current year)
r текущий годичный прирост *m*

* **curry-leaf tree → 2147**

* **curtidor → 6237**

* **curupay [ata] → 6762**

2151 curve cutting
d Profilschnitt *m*
f coupe *f* curviligne; chantournage *m*
r профильное резание *n*

2152 curved blade; blade with round breast
d gekrümmtes Sägeblatt *n*
f lame *f* de scie ventrue; lame de scie incurvée
r изогнутое полотно *n* пилы

2153 curved skyline
d gekrümmte Luftseilbahn *f*
f câble *m* porteur courbé
r несущий канат *m*, расположенный по кривой

2154 curved stem; crooked trunk
d gekrümmter Stamm *m*
f tige *f* courbée
r искривлённый ствол *m*

* **curved timber → 2029**

* **curved tree → 2030**

* **curve of estimated volume → 9241**

2155 curvifoliate
(ref. to leaf)
d krummbelaubt
f curvifoliacé
r изогнутозубчатый

2156 cushion
(prepared area made up of brushwood on which a tree or pieces of trees may be dropped)
d Puffer *m*
f coussin *m*; coussinet *m*
r опора *f*

2157 cushion of callus; callous margin; callous growth on the edge of a scar
d Überwallungswulst *m*; Überwallungsleiste *f*
f bourrelet *m* de recouvrement
r заплывочный желвак *m*; заплывочный нарост *m*

2158 **custard-apple family**
(An[n]onaceae)
d Anonengewächse *pl*
f an[n]onacées *pl*
r анноновые *pl*

2159 **custom[ary] measure**
(of timber)
d übliches Maß *n*
f mesure *f* coutumière
r размер *m* с допуском на обзол или горбыль

* **custom sawing → 7244**

2160 **custom string measure; full measure**
d Vollmaß *n*
f mesure *f* pleine
r измерение *n* кубатуры делением средней длины окружности на четыре

2161 **cut** *v*
d abbeilen; schneiden
f couper
r отёсывать топором

* **cut → 1360, 3037, 7232**

2162 **cut-and-pile shortwood system**
d Kurzholzhieb-und-Stapelungssystem *n*
f système *m* de débardage et empilage de bois court
r система *f* с заготовкой и окучиванием сортиментов у пня (на месте валки)

2163 **cut-and-skid crew**
d Besatzung *f* der Hieb und Rückung
f équipe *f* d'abattage et débardage
r бригада *f* на валке и трелёвке

2164 **cut-and-skid operations**
d Hieb-und-Rückungsoperationen *fpl*
f opérations *fpl* d'abattage et débardage
r обычные лесозаготовки *fpl*

2165 **cut** *v* **back; dip** *v*; **coppice** *v*; **shorten** *v*
d einstutzen
f raccourcir; ravaler
r подрезывать

* **cut back → 2175**

2166 **cutblock**
(specific area of land identified on a forest development plan, or in a licence to cut, road permit, or Christmas tree permit, within which timber is to be or has been harvested)
r расчётная лесосечная площадь *f*

2167 **cutblock adjacency requirements**
(integrated resource management requirements that specify the desired spatial relationships among cutblocks)
d Anforderungen *fpl* um Hiebflächennähe
f exigences *fpl* à la contiguïté de coupes permissibles
r потребности *fpl* близости лесосек

2168 **cut** *v* **clear; log** *v* **off; clear fell** *v*
d kahlhauen; kahlschlagen
f couper à blanc
r вырубить наголо

* **cut down → 3037**

* **cut end → 607**

* **cuticle → 4890**

2169 **cut** *v* **into; cut** *v* **open; hew** *v* **open**
d einhauen; aufhauen; aufhacken
f entamer; piocher
r взрыхлить; зарубить

* **cutlass → 1204**

* **cutleaf ground-cherry → 3984**

2170 **cut-leaved** *adj*
(a tree or shrub with leaves shaped in distinctive segments)
d schneidblätterig
f à feuilles coupées
r ясенелистный; срывной

2171 **cut log; piece of log; wood chuck; [chopped] billet; chump; bolt; stick; section**
(a short section of a tree trunk)
d Waldmurmeltier *m*; Kloben *m*; Holzblock *m*; Holzklotz *m*; Klotz *m*; Trumm *n*; Bloch *m*; Schrot *n*
f pièce *f* de bois; bûche *f*; billot *m*; billon *m*; bille *f*; billette *f*
r полено *n*; болванка *f*; чурбан *m*; чурак *m*

2172 **cut** *v* **off; hew** *v*; **rough** *v* **plane**
d abschroten; schroten; umschroten
f hacher
r отрезать; отрубать; отсекать; валить

2173 **cutoff drain; interception drain; catch-water drain; trap drain**
(a ditch constructed to intercept surface and subsurface run-off and lead it away from

d Sickergraben *m*
f fossé *f* d'assainissement
r перехватывающая дрена *f*

* **cutoff saw → 5056**

2174 **cutoff sections**
d Abschnitte *mpl*
f pièces *fpl* de bois tronçonnées
r обрезки *fpl* лесоматериалов

* **cut open → 2169**

2175 **cut *v* out; cut *v* back**
d niederhauen
f rabaisser
r стёсывать

2176 **cut out; cross-cut** *adj*
d anschnittet; abgelängt; abschaltend; Trenn-
f toisé; marqué pour la découpe
r отсечённый

* **cutover → 1607**

* **cutover area → 5066**

* **cut slope → 7753**

2177 **cut stock**
(a term of softwood stock comparable to dimension parts in hardwoods)
d anschnitteter Stamm *m*
f débit *m*
r отрезанные заготовки *fpl*

2178 **cut strip; swath**
d Arbeitsstreif *m*; Hiebszug *m*; Schlagstreifen *m*; Zeidelwald *m*
f bande *f* de coupe
r пасечная лента *f*; [лесосечная] пасека *f*; полоса *f* сплошь вырубленного леса

2179 **cut-stump treatment; stump treatment**
d Baumstumpfbehandlung *f*
f traitement *m* des souches
r обработка *f* комля

2180 **cut surface**
d Schnittfläche *f*
f surface *f* de coupe; tranche *f*
r плоскость *f* пропила

* **cuttail [eucalyptus] → 1075**

2181 **cutter**
r врубовая машина *f*; режущий станок *m*

* **cutter → 2204, 9526**

* **cutter bar → 4040**

2182 **cutter-head debarker**
d Fräskopf-Entrindungsmaschine *f*
f écorceuse *f* à réducteur
r корообдирочная машина *f* с валочной головки

2183 **cutter link**
d Schneidezahn *m*
f maillon[-gouge] *m*; chaînon *m*
r строгающая цепь *f*

* **cut-through → 3443**

2184 **cutting**
(a shoot, twig or other plant part removed from the donor plant, usually for the purpose of inducing roots to produce a steckling)
d Ableger *m*
f bouture *f*
r черенок *m*; отводок *m*

2185 **cutting**
(a rectangular portion, of specified minimum dimensions)
d Schnitt *m*
f débit-type *m*
r обрезка *f*

* **cutting → 8824**

2186 **cutting after deterioration**
d Kalamitätshieb *m*
f coupage *m* après dépérissement
r рубка *f* леса вследствие стихийных повреждений

* **cutting age → 136**

2187 **cutting along the grain**
d Adernschneiden *n*; Schnitt *m* in der Faserrichtung
f coupe *f* dans le sens du fil; coupe dans le sens axial
r резание *n* вдоль волокон; продольное резание

* **cutting angle → 288**

* **cutting area → 2189**

* **cutting authority → 6183**

* **cutting-back → 1224, 4187**

* **cutting bar → 4040**

* **cutting by area → 1820**

2188 **cutting capacity**
d haubarer Vorrat *m*
f capacité *f* de coupe
r производительность *f* резания; запас *m* отведённого в рубку насаждения

* **cutting chain → 1430**

2189 **cutting chance; logging chance; logging area; chance; tree-felling area; cutting drift; felling area; cutting area**
(area of forestland from which some or all timber has recently been cut)
d Holznutzungsfläche *f*; Nutzungsfläche *f*; Schlagfläche *f*; Fällschnittfläche *f*; Schnittfläche *f*; [abgeholzter] Aushau *m*; Abtriebsfläche *f*; Fläche *f* des Schlages; Hiebsfläche *f*
f surface *f* à exploiter; surface de coupe; aire *f* de coupe
r площадь *f* лесосеки; площадь лесного массива; территория *f* лесосеки; площадь вырубки; площадь пользования

2190 **cutting circle**
d Schnittflugkreis *m*
f cercle *m* de coupe
r окружность *f* резания; круговая траектория *f* лезвий

2191 **cutting class; felling class**
d Hiebsartsklasse *f*; Bestandsklasse *f* nach der Hiebsart
f classe *f* opérationnelle [de peuplements]
r класс *m* возраста главной рубки

2192 **cutting contract; felling contract; timber contract; logging contract**
(written, legally binding document used in the purchase of standing timber, without the acquisition of title to the land)
d Werbungsvertrag *m*; Einschlagvertrag *m*
f contrat *m* d'exploitation de bois
r контракт *m* на заготовку

2193 **cutting crest; barber's chair; barberchair; tombstone; beard; whiskers; sliver; sloven; stump pull; spaul**
(a result of faulty felling or heavy lean of the tree)
d Waldbär *m*; Waldbart *m*; Bart *m*; Waldhieb *m*; Kamm *m*; Späne *mpl* am Stock
f barbe *f*; peigne *m*; éclat *m*
r отщеп *m* [от поваленного дерева]; отщеп на пне; пнёвой козырёк *m*; козырёк, остающийся на пне

2194 **cutting cycle; felling cycle; logging cycle; cutting rotation; felling rotation; timber rotation**
(planned interval between major harvest operations in a stand)
d Hiebsumlauf *m*; Hiebsumtrieb *m*; Turnus *m*
f cycle *m* d'abattage; rotation *f* des coupes; périodicité *f* des coupes
r оборот *m* рубки

* **cutting date → 3060**

2195 **cutting depth; felling height; saw notch height; depth of cut; kerf depth**
d Fallkerbhöhe *f*; Fallkerbtiefe *f*; Schnitttiefe *f*; Schnitthöhe *f*; Arbeitshöhe *f*
f hauteur *f* à la découpe
r высота *f* пропила; высота срезания; высота спиливания; высота подпила; глубина *f* резания; высота среза; толщина *f* снимаемого за один проход слоя древесины

* **cutting direction → 2519**

* **cutting directions → 6844**

* **cutting-down → 1214, 1607**

* **cutting drift → 2189**

* **cutting edge → 819**

* **cutting face → 819, 3069**

* **cutting fee → 8195**

2196 **cutting forest**
d Schlagwald *m*; Schlägeholz *n*
f forêt *f* d'abattage
r лес *m*, отведённый под рубку

* **cutting in forest → 3564**

* **cutting in sap flow period → 3065**

* **cutting interval → 7077**

* **cutting inventory → 6831**

* **cutting licence → 6183**

2197 **cutting losses; felling losses**
d Fällungsverluste *mpl*
f pertes *fpl* d'abattage
r потери *fpl* при рубке; потери от валки; лесорубочные потери

* **cutting of hold-overs → 2969**

* **cutting of hold-over trees → 2969**

* **cutting of standards → 2969**

2198 **cutting of the top; top clipping; top riddance; decapitation; topping; pollarding; beheading**
(cutting off the top of section of a standing tree)
d Scheitelung *f*; Abkappen *n*; Abgipfen *n*; Abwipfeln *n*; Hochbesäumen *n*; Zopfung *f*; Kappung *f*
f écimage *m* [de sommets]; étêtage *m*; éhou[p]age *m*
r удаление *n* вершины; отрубка *f* вершины; декапитация *f*

2199 **cutting order; felling order**
d Hiebsordnung *f*; Schlagordnung *f*
f ordre *m* de coupes
r порядок *m* рубки; рубочный ряд *m*

* **cutting pattern → 5051**

* **cut[ting] period → 3072**

* **cutting permit → 6183**

* **cutting plan → 6332**

* **cutting point → 7668**

* **cut[ting-practice] control → 3055**

* **cutting[-practice] rules → 6844**

2200 **cut[ting] quantity; felling quantity; felled volume**
d Einschlagmenge *f*; vorgesehene Hiebsmasse *f*; vorgesehene Hiebsmenge *f*; Holzerntevolumen *n*
f volume *m* abattu; volume d'exploitation
r количество *n* вырубленной древесины; [валовой] объём *m* рубки

d Hiebsregime *n*; Hiebsplan *m*
f régime *m* d'exploitation
r организация *f* рубок

* **cutting register → 3073**

* **cutting right → 2955**

* **cutting rotation → 2194**

* **cuttings → 998**

2202 **cutting sequence; felling sequence; felling series; cutting series; succession of cuttings; succession of felling; progression of cutting**
d Hiebfortschritt *m*; Hiebsfolge *f*; Hiebsreihe *f*
f suite *f* de coupes; séquence *f* de coupage; succession *f* d'abattage; avancement *m* de la coupe
r последовательность *f* рубки

2203 **cutting system; cutting-unit method; method of cutting management**
d Hiebsführungsmethode *f*
f méthode *f* de gestion d'abattage; système *m* de gestion d'abattage
r метод *m* ведения рубки

* **cutting time → 3077**

2204 **cutting tool; cutter**
d schneidendes Werkzeug *n*
f hacheur *m*
r режущий инструмент *m*; резец *m*; нож *m*; секач *m*

* **cutting tract → 7668**

* **cutting unit → 9097**

* **cutting-unit method → 2203**

* **cutting width → 9400**

* **cutting year → 3144**

* **cutt off → 3037**

2205 **cut-to-order; cut-to-size** *adj*
(sawed to the standard panel sizes into various shapes and dimensions requested by secondary manufacturers)
d zurechtschneiden
f déligné

* **cut-to-size → 2205**

* **cut wood → 3042**

* **CWD → 1686**

2206 **cyathea family**
(Cyatheaceae)
d Farnbäumgewächse *pl*
f cyathéacées *pl*
r циатейные *pl*

2207 **cycad family; cycas family**
(Cycadaceae)
d Farnpalmen *pl*; Zykadazeen *pl*
f cycadacées *pl*
r саговниковые *pl*

2208 **cylinder chipper; drum chipper; drum chopper**
d Trommelhacker *m*; Walzenhackmaschine *f*
f déchiqueteur *m* à tambour; déchiqueteuse *f* à tambour
r барабанная рубильная машина *f*

2209 **cylinder saw; crown saw; barrel saw**
d Trommelsäge *f*; Kreisringsäge *f*; fassförmige Säge *f*
f scie *f* [en forme] cylindrique
r цилиндрическая пила *f*

2210 **cynipid gall wasps; cynipids; gall flies**
(Cynipidae)
d Gallwespen *fpl*
f cynips *mpl* [gallicoles]; cynips galligènes
r орехотворки *fpl*; галлообразующие насекомые *npl*

* **cynipids → 2210**

2211 **cypress**
(Cupressus)
d Zypresse *f*
f cyprès *m*
r кипарис *m*

2212 **cypress cone; galbulus**
d Beerenzapfen *m*
f galbule *m*
r кипарисовая шишка *f*

2213 **cypress dome**
(cypress stand with a characteristic dome-shape (tall in the center, shorter on the edges) that occurs in Stillwater swamps)
d Zypressenkuppel *f*

2214 **cypress family**
(Cypressaceae)
d Zypressengewächse *pl*
f cypressacées *pl*
r кипарисовые *pl*

2215 **cypress forest**
d Zypressenwald *m*
f cyprière *f*
r кипарисовый лес *m*; кипарисовая роща *f*

2216 **cypressoid pit**
d Zypressentüpfel *f*
f ponctuation *f* cypressoïde
r кипарисовидная поровость *f*

2217 **cypress pine**
(Callitris)
d Kallitris *f*
f callitris *m*
r каллитрис *m*

2218 **cypress pond**
(depression dominated by cypress trees and characterized by standing water and constant seepage)
d Zypressenweiher *m*
f étang *m* de cyprès
r кипарисовый пруд *m*

2219 **cypress wood**
d Zypressenholz *n*
f bois *m* de cyprès
r кипарисовая древесина *f*

2220 **Cyprus pine; Aleppo pine; kadj**
(Pinus halepensis)
d Zypernkiefer *f*; Aleppokiefer *f*; Pyrenäen-Schwarzkiefer *f*
f pin *m* de Chypre; pin d'Alep; kadj *m*
r сосна *f* аллепская; сосна ельдарская

2221 **cyrilla family; titi family**
(Cyrillaceae)
d Cyrillagewächse *pl*
f cyrillacées *pl*
r цирилловые *pl*

2222 **cytokinins; cytokines**
(a class of plant growth regulators involved in cell division, and often effective in shoot organogenesis)
d Zytokinine *npl*
f cytokinines *mpl*
r цитокинины *mpl*

D

* **d.a.b. → 2465**

2223 **dado**
(a rectangular channel cut partway into a board)
d Sockelbereich *m*; Würfel *m*
f engravure *f*; embrèvement *m*
r цоколь *m*; панель *f*

* **daedal leaf → 2317**

2224 **dagger tree; Spanish bayonet; soap tree yucca**
(Yucca elata)
d Seifenbaumyucca *f*
f arbre *m* de savon
r юкка *f* высокая

2225 **dahoon [holly]**
(Ilex cassine)
d Cassinabaum *m*; Dahoonhülse *f*; Dahoonstechpalme *f*
f houx *m* Cassine
r падуб *m* Кассине; падуб каролинский

2226 **Dahurian birch; Asian black birch**
(Betula davurica)
d mongolische Birke *f*
f bouleau *m* de Mongolie
r берёза *f* даурская

2227 **Dahurian larch**
(Larix dahurica; Larix gmelinii)
d dahurische Lärche *f*
f mélèze *m* de Dahurie
r лиственница *f* даурская

2228 **Daimyo oak**
(Quercus dentata)
d Daimyo-Eiche *f*; Kaisereiche *f*
f chêne *m* de Mongolie; chêne Daimyo
r дуб *m* зубчатый; кашива *m*

2229 **daisybush**
(Olearia)
d Olearie *f*
f oléarie *f*
r олеария *f*

water or to keep out the sea)
d Damm *m*
f barrage *m*; digue *f*
r дамба *f*; плотина *f*

2231 **dam afforestation**
d Bermenaufforstung *f*
f afforestation *f* de digues
r облесение *n* откосов, дорог, каналов или плотин

2232 **damage by deers**
d Fegeschaden *m*
f frayure *f* par cervidés
r вред *m* от оленей

* **damage by fume → 2240**

2233 **damage by game; nibbing by game; game bite; game damage**
d Wildtierschaden *m*
f dégât *m* causé par le gibier; dégât de gibier
r повреждение *n*, причиненное дичью; потрава *f* дичью

2234 **damage by insects; insect damage; insect injury**
d Insektenfraß *m*; Insekten-Fraßschaden *m*
f dégât *m* d'insectes
r повреждение *n* насекомыми; обгрызание *n* насекомыми

2235 **damage by rime**
d Reifschaden *m*
f dégât *m* dû au givre
r повреждение *n* от инея

2236 **damage by wind; wind damage; damage from storm**
d Windwurf *m*; Sturmschaden *m*
f défaut *m* par le vent; dégât *m* d'ouragan
r повреждение *n* ветром; повреждение от бури

2237 **damage caused by peeling; bark-peeling damage; injury caused by peeling; barking damage; peeling damage; scar of peeling**
d Schälschaden *m*; Schälwunde *f*
f dégât *m* d'écorçage
r повреждение *n* от окорки

* **damage done by worms → 9703**

2238 **damaged timber; damaged wood**
(timber that has been affected by injurious agents such as wind (as in the case of

f bois *m* endommagé
r повреждённая древесина *f*

2239 damaged tree
d schadhafter Baum *m*
f arbre *m* endommagé
r повреждённое дерево *n*

2240 damage due to smoke; smoke damage; damage by fume; fume damage
(injury from gases or vapo[u]rs and smoke also)
d Rauchschäden *m*
f dégât *m* par la fumée
r повреждение *n* от дыма

* **damaged wood → 2238**

* **damage from storm → 2236**

2241 damage of forest; forest damage; forest deterioration
d Waldrückgang *m*; Waldverschleuderung *f*; Waldbeschädigung *f*; Waldschäden *mpl*
f détérioration *f* de la forêt; dommage *m* de la forêt
r повреждение *n* леса; ухудшение *n* леса

2242 damage of wood; wood damage
d Holzschaden *m*
f dégât *m* de bois
r повреждение *n* древесины

* **dam for driving timber → 5039**

2243 dam[m]ar
d Dammarharz *n*
f dam[m]ar *m*; résines *fpl* dammar
r дам[м]ар *m*; дам[м]ара *f* (группа натуральных смол)

* **dam[ming] board → 4521**

* **damp rot → 9353**

* **damp wood → 3710**

2244 Damson-tree; Damson plum; wild plum
(Prunus insititia)
d Haferpflaume *f*; Haferschlehe *f*; Kriechenpflaume *f*; Krieche *f*
f prune *f* Damson; prunier *m* de Damas; prunier sauvage; damas *m*; prunéole *m*
r тернослив *m*; терносливa *f*

* **dancer → 9022**

* **danger tree → 4179**

* **danicho → 5974**

2245 danta; olborbora
(Nesogordonia papaverifera; Cisanthera papaverifera)
d Danta *n*; Akuama *n*; Kotibe *n*
f kotibe *m*; apru *m*; danta *m*; otutu *m*; ovoe *m*
r данта *f*

* **dao dragon plum → 6033**

2246 daphne; mezereon
(Daphne mezereum)
d [gemeiner] Seidelbast *m*; Kellerhals *m*
f daphné *m*; bois *m* gentil; bois *m* joli
r дафна *f*; волчник *m* [обыкновенный]

2247 dapple *adj*
(having patches of different colo[u]r, or shade of colo[u]r, as in the redwood texture)
d gefleckt
f pommelé
r пятнистый (о текстуре красного дерева)

2248 dappled shade
(method of management designed to retain 30-50% of the upper canopy cover of a wood)
d gesprenkeltes Schatten *n*
f ombre *m* pommelé
r рассеянная тень *f*

2249 dark bloodwood eucalyptus; red bloodwood
(Eucalyptus gummifera)
d rotes Bloodwood *n*
f eucalyptus *m* gummifère
r эвкалипт *m* камеденосный

2250 dark brown cubical heartwood rot
(invader - Fomes laricis, Fomitopsis officinalis)
d kubische Dunkelbraunfäule *f*
f pourriture *f* brune foncée cubique de cœur
r тёмно-бурая кубическая ядровая гниль *f*

2251 dark brown cubical rot
(invader - Polyporus sulphureus, Laetiporus sulphureus)
d kubische Braunfäule *f*
f pourriture *f* brune cubique
r бурая кубическая гниль *f*

2252 dark coniferous timber
d Dunkelnadelholz *n*

* **darkened coupe → 2253**

2253 **dark felling; darkened coupe**
d Dunkelschlag *m*; Dunkelhieb *m*
f coupe *f* sombre
r беспорядочная валка *f*; тёмная лесосека *f*

2254 **dark red seraya; dark red meranti; nemesu**
(Shorea negrosensis)
d dunkelrotes Meranti *n*
f meranti *m* rouge foncé
r сорея *f* тёмнокрасная

2255 **Darlington ash; red ash**
(Fraxinus pennsylvanica)
d amerikanische Grauesche *f*; pennsylvanische Esche *f*; Rotesche *f*
f frêne *m* rouge; frêne de Pennsylvanie
r ясень *m* пенсильванский; ясень пушистый; яс нь ланцетолистный; ясень ланцетный

* **Darlington oak → 4835**

* **Darwin blackwattle → 2698**

* **dasher → 1668**

2256 **date of inventory**
(dating of a US forest inventory based on the calendar year in which most of a inventory data were collected)
d Inventarisierungstermin *m*
f date *f* d'inventaire
r срок *m* таксации

2257 **date palm**
(Phoenix dactylifera)
d [echte] Dattelpalme *f*
f palmier *m* dattier; dattier *m*
r финиковая пальма *f*

2258 **date-plum**
(Diospyros lotus)
d Dattelpflaume *f*; Lotuspflaume *f*
f plaqueminier *m* d'Orient; faux lotier *m*; galac *m* d'Italie
r хурма *f* обыкновенная; хурма кавказская

* **David's mountain laurel → 9214**

2259 **Dawn redwood**
(Metasequoia glyptostroboides)
d chinesisches Rotholz *n*; Urwelt-Mammutbaum *m*
f métaséquoia *m* du Sichuan
r метасеквойя *f* глиптостробовидная

(cutting of trees along the edges of roads in order to reduce shading of the road surface and to speed up drying)
d Tageslichteinfall *m*
f admission *f* du jour
r просвет *m*

* **d.b.h. → 2466**

2261 **dead core**
d toter Kern *m*
f cœur *m* mort
r мёртвое ядро *n*

* **dead culls → 2123**

2262 **dead end**
d totes Streckenstück *n*; Sackgasse *f*; Sackstrecke *f*
f galerie *f* en impasse; impasse *f*; route *f* sans issue
r тупик *m*; глухой конец *m*

2263 **deadener; brake block**
(a heavy log or timber, with spikes set in the butt end, so fastened in a log slide that the logs passing under it come in contact with the spikes and have their speed retarded)
d Bremsblock *m*
f billon *m* de freinage
r тормозное бревно *n*; тормозной блок *m*

* **deadening → 6953**

2264 **dead face; dry face**
d trockene Lachte *f*
f carre *f* sèche
r сухая карра *f*

2265 **deadfallen branches**
d liegende Totäste *fpl*
f branches *fpl* mortes
r опавшие сухие сучья *mpl*; мелкий валёжник *m*

2266 **deadfall[en wood]; deadfallen timber; fallen dead[ ]wood; down[ed] timber; dry fallen wood; fallen trees**
d liegendes Trockenholz *n*; Abraum *m*; Fallholz *n*; Fegeholz *n*; anfallendes Holz *n*; Urholz *n*; Wädelholz *n*
f bois *m* sec gisant; bois de ramassage
r лежащий сухой лесоматериал *m*; валёжник *m*; валёжная древесина *f*; падалица *f*

f forêt *f* morte
r сухостойный лес *m*; погибшее насаждение *n*

2268 deadhead; sunken log; sinker
(a sunken or semi-submerged log in a waterway; a species of tree whose timber, even if air-dried, would sink in water)
d Absenker *m*; gesunkene Rundhölzer *npl*; Sinkholz *n*; Senkholz *n*
f bois *m* fondrier; bois submergé; embu *m*
r утоп *m*; топляк *m*; затонувшее дерево *n*; бревно *n*, которое не держится на плаву; затонувшие лесоматериалы *mpl*

* **dead knot → 800**

2269 dead litter
d tote Streu *f*
f litière *f* morte
r мёртвая подстилка *f*; мёртвый покров *m*

* **deadman → 275**

2270 dead pruning
d Trockenholzaufasten *n*
f élagage *m* de bois sec
r обрезка *f* сухостоя

* **dead-square log → 3755**

2271 dead [standing] tree; [standing] snag; snak
(standing dead tree more of 14 *m* from which the leaves and most of the branches have fallen)
d Leichenbaum *m*; Abständer *m*; Dürrständer *m*; stehender Trockenbaum *m*; abgestorbener Baum *m*; toter Baum; Dürrling *m*
f arbre *m* mort; arbre sec [sur pied]; chandelier *m*
r сухостойное дерево *n*; буреломное дерево; дерево, сломанное ветром; мертвяк *m*; отмершее дерево; мёртвое дерево; сушняк *m*

* **dead timber → 2274**

2272 dead-topped; black-topped
d mit abgestorbener Krone
f à cime dépérissante
r с отмирающей кроной

2273 dead-top tree; spike-top[ped] tree; staghead[ed] tree; top-kill tree; top-drying tree; dry-topped tree
d wipfeldürrer Baum *m*; Baum mit Trockenwipfel
f arbre *m* à sommet mort; arbre à sommet sec; arbre â cime desséchée
r суховершинное дерево *n*; дерево с сухой верхушкой

2274 deadwood; dead timber; skeg
(timber produced from dead standing trees)
d abgestorbenes Holz *n*; stammtrockenes Holz *n*; Totholz *n*; Steventotholz *n*; Trocknis *f*; Raftholz *n*; Skeg *n*
f bois *m* mort; bois sec sur pied; chicot *m*
r сухостой *m*; сухостойная древесина *f*; отмершая древесина; остатки *mpl* отмерших деревьев; мёртвая древесина *f*; омертвелая древесина

2275 deadwood component
(wood left in production forests for maintaining rare, or endangered living organisms)
d Dürrholzkomponente *f*
f composante *f* de bois mort
r компонент *m* сухостоя

* **de-afforest → 2524**

* **deal boards → 2278**

2276 dealer in wood; wood monger; timber dealer; timber merchant
d Holzhändler *m*; Holzkaufmann *m*
f marchand *m* du bois
r лесоторговец *m*

2277 deal frame
d Gattertrennsäge *f*; Spaltgatter *n*; Spaltgattersägemaschine *f*
f châssis *m* de madrier
r делительная лесопильная рама *f*

2278 deals; deal boards; ba[u]lks
(a large slabbed log on the headsaw, usually having one or more rounded edges, which is destined for further processing)
d Holzdielen *fpl*; Dielen *fpl*
f madriers *mpl*
r дильсы *mpl*; еловые или сосновые доски *fpl*; половица *f*; балки *fpl*

2279 death-watch beetle damage
d Klopfkäferfraß *m*; Totenuhrfraß *m*
f dégât *m* par grande vrillette; dégât par vrillette striée

2280 **debark** *v*; **bark** *v*; **decorticate** *v*; **disbark** *v*; **rind** *v*; **flay** *v*; **ross** *v*; **peel** *v*
(to remove the bark from trees or sections of trees, usually mechanical rather than manual)
d abborken; abkrusten; ablohen; Holz abrinden; entrinden; schälen
f décortiquer; écorcer
r окорять; снимать кору; сдирать кору

* **debarker → 2284**

2281 **debarker-chipper; bark-chipper; bark crusher; tan crusher; bark hog**
d Lohbrecher *m*; Lohmühle *f*
f écorceuse-déchiqueteuse *f*
r дробилка *f* для коры; кородробилка *f*; окорочно-рубильная машина *f*

* **debarking → 579**

2282 **debarking condition**
d Entrindungszustand *m*
f état *m* d'écorçage
r состояние *n* окорки

* **debarking drum → 566**

2283 **[de]barking in strips; bark stripping; stripping**
(removing the bark of a tree in narrow strips)
d Reppeln *n*
f écorçage *m* par bandes
r очистка *f* от коры полосами; окорка *f* полосами; пролыска *f*

* **debarking line → 567**

2284 **debarking machine; disbarking machine; barking machine; bark-peeling machine; peeler; bark-stripping machine; debarker; barker; bark edger; rasp[ing] machine; bark mill**
(machine for removing bark from logs or bolts)
d Rindenschälmaschine *f*; Entrindungsmaschine *f*; Entrinder *m*
f machine *f* à écorcer; écorceuse *f*
r окорочный станок *m*; короснимaтель *m*; корообдирочная установка *f*; обдирочная машина *f*; окорочная машина; корообдир *m*

* **debarking station → 567**

* **debranch → 10**

* **debranching → 6610**

2285 **debranching knife; limbing knife; pruning knife**
d Ästungsmesser *m*; Entastungsmesser *m*; Gartenmesser *m*
f couteau *m* à ébrancher; couteau d'ébranchage; couteau ébrancheur
r нож *m* для обрубки сучьев; сучкорезный нож

* **debris avalanche → 2287**

2286 **debris flow**
(mixture of soil, rock, wood debris and water which flows rapidly down steep gullies)
d Schlagabfällsstrom *m*
f flux *m* à débris
r быстрый водяной поток *m* с содержанием грунта и порубочных остатков

* **debris layer → 4999**

2287 **debris slide; debris avalanche**
d Schuttgleiten *n*
f avalanche *f* détritique; glissement *m* de débris
r сползание *n* со склона неустойчивых грунтов вместе с порубочными остатками; обломочный оползень *m*

2288 **debris-strewn forest**
d gestreuter Wald *m* mit Schlagabfäll
f forêt *f* à débris
r захламлённый лес *m* (с разбросанными сучьями, валежником и т.п.)

* **debudding → 1142**

* **decapitation → 2198**

* **decay → 2307, 6647**

* **decayed → 7212**

2289 **decayed heart[wood]; rotten heart; punky heart**
d Faulkern *m*; fauler Kern *m*
f cœur *m* pourri
r гнилая ядровая древесина *f*; гнилое ядро *n*; гнилая сердцевина *f*

2290 **decayed knot; punk knot; rotten knot; unsound knot**
(a knot which is not as hard as the surrounding wood)
d angefaulter Ast *m*; fauler Ast; Faulast *m*; verfaulter Ast

2291 decayed wood; fungous wood; spongy wood; rotten timber; rotten wood; punky wood
d rostiges Holz *n*; verschimmeltes Holz; schwammiges Holz; angefaultes Holz; Faulholz *n*; morsches Holz; verfaultes Holz; Rauchholz *n*; Schwammholz *n*
f bois *m* pourri; bois piqué
r гнилая древесина *f*; прелая древесина; заплесневелая древесина; древесина, поражённая грибной гнилью

2292 decaying wood
d verfallendes Holz *n*
f bois *m* pourrissant
r загнивающая древесина *f*

2293 decay resistance; durability against rot
d Fäulnisresistenz *f*; Fäulnisbeständigkeit *f*
f résistance *f* à la pourriture; résistance à l'altération par les champignons; imputrescibilité *f*; durabilité *f* contre la pourriture
r устойчивость *f* к загниванию; гнилоустойчивость *f*

2294 decay-resistant wood
d fäulnisbeständiges Holz *n*
f bois *m* résistant à pourriture
r устойчивая к загниванию древесина *f*

2295 decay retardant; decay retarder
d Fäulnisbremse *f*; Fäulnisverzögerer *m*
f retardant *m* contre une pourriture; retardateur *m* contre une pourriture
r замедлитель *m* гниения

* **decay retarder → 2295**

2296 decided lean tree
d starkgesenkter Baum *m*
f arbre *m* incliné fortement
r дерево *n* с ярко выраженным наклоном

2297 decidulignosa
(communities of deciduous trees and shrubs)
d Laubgemeinschaften *fpl*
f communautés *fpl* feuillues
r сообщество *n* листопадных древесных и кустарниковых пород

* **deciduous → 4877**

* **deciduous boreal forest → 1719**

2298 deciduous brush; aestifruticeta
r кустарниковые листопадные сообщества *npl*

2299 deciduous cover
(vegetation classes where 75% or more of the diagnostic vegetation is made up of tree or shrub species that shed foliage simultaneously in response to an "unfavorable" season)
d sommergrüne Bedeckung *f*
f couvert *m* décidu
r листопадный ковёр *m*

2300 deciduous forest; broadleaved forest; angiospermous forest
d laubabwerfender Wald *m*; Laubwald *m*
f forêt *f* caduque; forêt de feuillus; forêt feuillue; forêt caducifoliée; forêt décidue; forêt défeuillante
r листопадный лес *m*; [широко]лиственный лес

2301 deciduous forest biome
d Laubwald-Biom *n*
f biome *m* de la forêt décidue
r биом *m* листопадного леса

* **deciduous summer forests → 8260**

2302 deciduous [tree]
(a tree which normally loses its leaves or needles during the fall or winter)
d Laubbaum *m*
f caduque *f*; arbre *m* décidu; arbre à feuillage caduc
r листопадное дерево *n*

* **deciduous tropical forest → 5545**

* **deciduous wood → 4124**

2303 decking
(the surfacing material of a deck)
d Bohlenbelag *m*
f revêtement *m*
r настил *m*; материал *m* для настила; платформа *f*

2304 decking
d Abdeckung *f*; Deckung *f*
f platelage *m*
r выравнивание *n* комлей окучиваемых лесоматериалов

* **decking site → 5451**

f empilement *m* de grumes
r куча *f* бревен; куча лесоматериалов

2306 deck wood
d Deckholz *n*
f bois *m* de pont
r палубный кряж *m*

2307 decline *v*; **decay** *v*
d abständig werden
f dépérir
r перестоять

2308 decomposer
(organism, usually a bacterium or a fungus, that breaks down the bodies of dead plants and animals into simpler substances usable by green plants)
d Auflöser *m*; Zerleger *m*; Reduzent *m*; Zersetzer *m*
f décomposeur *m*; bioréducteur *m*; minéralisateur *m*
r редуцент *m* (организм, разлагающий органическое вещество)

2309 decomposite leaf
d mehrfach zusammengesetztes Blatt *n*
f feuille *f* décomposée
r разделённый лист *m*

2310 decomposition
(a large number of interrelated processes by which organic matter is broken down to smaller particles and to soluble forms available for plant uptake)
d Abbau *m*; Zerfall *m*; Zerlegung *f*; Zersetzung *f*
f décomposition *f*
r разложение *n*; распад *m*

* **decorticate → 2280**

2311 decreaser [plant species]
(any plant species of the original vegetation that will decrease in relative amount with continued overuse, e.g., grazing)
d abnehmende Pflanzenart *f*
f espèce *f* en évolution régressive
r вид *m* растений, выпадающий из фитоценоза после выпаса

2312 decreasing sap flow
d absteigender Saftstrom *m*
f écoulement *m* descendant de sève
r нисходящее сокодвижение *n*

d kriechend; gestreckt
f rampant
r стелющийся; ползучий

2314 decumbent crown
d liegende Krone *f*
f cime *f* décombante
r распростёртая розетка *f* листьев

2315 decurrent branching habit
d abwärts laufendes Verzweigungshabit *n*
f habit *m* décurrent de branchage
r раскидистая низбегающая форма *f* ветвления (кроны)

2316 decurrent form
(of a bole)
d abwärts laufende Form *f*
f forme *f* décurrente
r низбегающая форма *f*

* **decurrent growth habit → 2358**

* **decussate leaf arrangement → 5928**

2317 dedalous leaf; daedal leaf; patterned leaf
d ausgerandetes Blatt *n*; buchtiges Blatt; verworrenes Blatt
f feuille *f* émarginée; feuille rétuse
r выемчатый лист *m*

2318 dedusting; dust extraction
d Entstaubung *f*; Staubabführung *f*
f dépoussiérage *m*
r удаление *n* пыли; пылеотделение *n*; очистка *f* от пыли; обеспыливание *n*

* **dee → 7699**

2319 dee
(a notch cut into the side of a log near the butt to give the drag chain a better grip)
d Endeschnitt *m*
f entaille *f* de débusquage
r зарубка *f* на конце брёвна

2320 dee; connecting shackle
(D-shaped connecting link in rigging, used for high-lead cable logging)
d Drahtseilklemme *f*
f manille *f*
r прицепное кольцо *n*

2321 deep blue
d tiefe Bläue *f*

2322 deep cutting; deeping
(of balk)
d Tiefbesäumen *n*; Trennschnitt *n*
f sciage *m* par trait haut; dédoublage *m* en hauteur
r распиловка *f* на тонкие доски или рейки; раскрой *m* по толщине; роспуск *m*

2323 deep dormancy; true dormancy
d tiefer Ruhezustand *m*; tiefe Ruhephase *f*
f dormance *f* profonde
r глубокий покой *m*

2324 deep forest
d tiefer Wald *m*
f forêt *f* profonde
r большой лес *m*

* **deeping → 2322**

* **deeply descending root → 277**

* **deeply-rooted plant → 2326**

2325 deeply serrated leaf; deeply lobed leaf; aceriform leaf
(shaped like a maple leaf)
d tief eingeschnittenes Blatt *n*
f feuille *f* serrée profond
r глубоко дольчатый лист *m*; глубоко пальчатый лист; глубоко зубчатый лист

2326 deep-rooting plant; deep-rooted plant; deeply-rooted plant
d Tiefwurzler *m*
f plante *f* en racine profonde
r глубококорневое растение *n*

2327 deerberry; squaw-huckleberry; buckberry; Georgia blueberry
(Vaccinium stramineum)
d Hirschbeere *f*
f airelle *f* à longues étamines
r черника *f* сахалинская; черника волосистая; черника тычиночная

2328 deer fence
(high tensile electrified or woven wire fencing installed around an area to reduce or eliminate deer browsing of seedlings)
d Hochwildzaun *m*
f clôture *f* haute des cervidés
r высокая ограда *f* (окружающая олений вольер)

2329 deer foot; fawn-foot
(the broadened end of an axe handle, for
d Rotwildfuß *m*
f pied *m* de biche
r конец *m* топорища, уширенный для удобства удержания его рукой

2330 deer-forest
d Wald *f* für Rotwild
f forêt *f* des cervidés
r олений заповедник *m*

2331 deer glade; deer lawn
(an unplanted area in a wood specifically designed for deer population management)
d Rotwildglade
f clairière *f* des cervidés
r оленья поляна *f*

2332 deer yard
(a wintering area used by deer during severe weather)
d Wintereinstand *m*
f quartier *m* d'hivernage des cervidés
r пашня *f* оленей зимой

* **defect deduction → 201**

2333 defect inherent in wood
d innewohnender Holzmangel *m*
f défaut *m* inhérent au bois
r врождённый дефект *m* древесины; неустранимый дефект древесины

2334 defective length
d Fehlerlänge *f*
f longueur *f* de défaut
r длина *f* дефектной части ствола

2335 defective stick
d fehlerhaftes Knüppel *n*
f rondin *m* défectueux
r дефектный хлыст *m*; неделовой хлыст

* **defective tree → 7082**

2336 deferred area
(where timber harvesting or other forest development activities have been postponed for a period of time)
d Verschiebensfläche *f*
f aire *f* différée
r область *f*, запланированная на будущее

* **deferred shoot → 2848**

2337 deficit forest
(in which existing stands cannot provide enough harvest volume to maintain the

d Defizitwald *m*
f forêt *f* déficitaire
r дефицитный лес *m*

* **defoliating insects → 4858**

* **defoliation → 2941**

2338 **defoliator; defoliant; defoliating agent**
(an agent or insect that damages plants by destroying its leaves or needles)
d Laubfresser *m*
f défoliateur *m*; défoliant *m*; produit *m* défoliant; phyllophage *m*
r дефолиатор *m*; дефолиант *m*; листоед *m*

* **deforest → 2524**

2339 **deforestation; dis[af]forestation**
(the removal of trees and often the plants associated with them)
d Entwaldung *f*; Abforstung *f*; Waldabholzung *f*; Abholzung *f*
f déboisement *m*; déforestation *f*
r обезлесение *n*

2340 **deformed crown**
d bedrängte Krone *f*
f cime *f* déformée
r деформированная крона *f*

2341 **degrade deduction; degrade**
(generally, in timber and other forest produce, any process that lowers their value for any purpose)
d Verminderung *f*
f dégradation *f*
r снижение *n* качества из-за дефектов

2342 **degraded forest**
(with temporary or permanent reduction of density, structure, species composition or productivity)
d abgeflachter Wald *m*
f forêt *f* dégradée
r деградировавший лес *m*

2343 **degree of closeness**
d Beschirmungsgrad *m*; Schlussgrad *m*
f degré *m* de couverture; canopée *f*
r степень *f* сомкнутости

* **degree of mixture → 3897**

2344 **degree of stocking; percent[age] of**

f degré *m* de densité [du peuplement]; proportion *f* de surface occupée
r показатель *m* полноты насаждения; отношение *n* площади, занятой деревьями, к общей площади насаждения

* **degree of tapering → 8405**

2345 **degum** *v*
d entharzen
f dégommer
r удалять камедь; удалять смолу; обессмоливать

2346 **dehiscent** *adj*
(splitting at maturity)
d aufspringend
f déhiscent
r растрескивающийся; лопающийся

2347 **de[-]horning; severe cutting-back; vigorous cutting-back**
(severely pruning shrubs or trees, often carried to extremes of butchery)
d kräftiges Zurückschneiden *n*
f ravalement *m*
r омоложение *n* [деревьев] короткой обрезкой ветвей

2348 **delamination**
(the separation of layers in laminated wood or plywood because of failure of the adhesive)
d Delamination *f*; Schichtablösung *f*
f décollement *m* des plis
r расслаивание *n*

* **delegate eucalyptus → 4054**

* **deliberate fire → 6516**

2349 **delignification**
(removal of lignin from wood by chemical treatment)
d Delignifizierung *f*
f délignification *f*
r делигнификация *f*

2350 **delimbed non-skidded harvesting; delimbed nonslip harvesting**
d astreine Gleitschutzholzernte *f*
f exploitation *f* sans nœuds antidérapante
r обрубка *f* сучьев и трелёвка *f* хлыстов в полностью погруженном положении; заготовка *f* в полностью погруженном положении

2351 delimbed stem; clear wood; clean wood; knot-free wood
(wood formed subsequent to pruning)
d Baumschaft *m*; Langholzstamm *m*; Stammspindel *f*; astreines Holz *n*
f bois *m* sans défaut; bois sans nœuds
r очищенный от сучьев лесоматериал *m*

* **delimber → 8825**

* **delimber axe → 5167**

2352 delimber-bucker; delimber-slasher[-sorter]
d Entastungs- und Einschneidemaschine *f*
f ébrancheuse-tronçonneuse *f*
r сучкорезно-раскряжёвочная машина *f*

* **delimber-buncher → 4964**

* **delimber head → 4969**

* **delimber-topper → 4964**

* **delimbing → 6610**

* **delimbing attachment → 4969**

2353 delimbing boom; limbing boom; delimbing rail
d Entastungsarm *m*
f flèche *f* d'ébranchage
r стрела *f* сучкорезной машины

2354 delimb[ing] carriage; limbing carriage
d Entastungskarren *m*
f chariot *m* d'ébranchage
r суппорт *m* с сучкорезным ножом

2355 delimbing quality
d Entastungsqualität *f*
f qualité *f* d'ébranchage
r качество *n* очистки деревьев от сучьев

* **delimbing rail → 2353**

2356 delimb[ing] stroke; limbing stroke; limber stroke
d Entastungszug *m*
f course *f* d'ébranchage
r сучкорезный удар *m*

* **delimbing unit → 4969**

2357 deliquescent
(ramifying into fine divisions, such as
d leicht zerfließend
f déliquescent
r с разросшимися латеральными почками

2358 deliquescent growth habit; decurrent growth habit
d leicht zerfließende Wuchsform *f*
f habit *m* à croissance déliquescente
r низбегающая форма *f* роста

* **deliquescent tree → 5617**

2359 delivery of wood; timber supply; wood supply
(available timber categorized by species, end-use, and relative value)
d Holzzufuhr *f*; Holzlieferung *f*
f offre *f* de bois; fourniture *f* de bois
r доставка *f* лесоматериалов; доставка леса; обеспечение *n* лесом; снабжение *n* лесом

2360 dell
d Delle *f*
f vallée *f*; val *m*
r лесистая долина *f*; лесистая ложбина *f*; облесённая лощина *f*

2361 demanding species
(species that require large areas of favo[u]rable habitat to survive)
d abfordernde Arten *fpl*
f espèces *fpl* demandeuses
r взыскательные виды *mpl*

2362 demand of wood; timber demand; timber want
(this is a relationship between stumpage or delivered log price and the quantity of timber produced)
d Holzbedarf *m*
f besoin *m* en bois
r потребность *f* в древесине

2363 Demerara greenheart; greenheart; beeberoe
(Chlorocardium rodioei; Ocotea rodioei)
d [Demerara-]Grünharz *n*
f bois *m* de cœur vert; ébène *m* vert
r нектандра *f* Родией; оцота *f* мыльная; окотеа *f* мыльная; мыльное дерево *n*

2364 demolition wood
d Abbruchholz *n*
f bois *m* de démolition
r уничтоженная древесина *f*

2365 den; lair
(rain proof, weather-tight cavity in a tree or a space dug by animals among rocks or in soil)
d Höhle *f*; Bau *m*
f tanière *f*
r каморка *f*; берлога *f*

* **dendriform → 8876**

2366 dendriform cactus; dendritic cactus
d baumförmiger Kaktus *m*; baumartiger Kaktus
f cactier *m* dendritique
r древовидный кактус *m*

2367 dendrite
(a short fiber that conducts toward the cell body of the neuron)
d Dendrit *n*
f dendrite *f*
r древовидный отпечаток *m*; дендрит *m*

* **dendritic[al] → 8876**

* **dendritic branching habit → 2939**

* **dendritic cactus → 2366**

2368 dendritic drainage
(a river or stream tributary pattern resembling the branching of certain hardwood trees)
d baumartige Dränage
f drainage *m* dendritique
r разветвлённый дренаж *m*; дендритовый дренаж

* **dendritic form → 2939**

2369 dendrobiont
(an organism living in the forest)
d Dendrobiont *n*
f dendrobionte *m*
r дендробионт *m*

2370 dendrochemistry; forest chemistry; wood chemistry
d Forstchemie *f*; Holzchemie *f*
f chimie *f* forestière; sylvichimie *f*; chimie de bois; xylochimie *f*
r лесохимия *f*

2371 dendrochore
(a treed region)
d Dendroraum *m*
f dendrochore *f*
r дендрохора *f*; зона *f* лесов

f dendrochronologiste *m*
r дендрохронолог *m*

2373 dendrochronology
d Dendrochronologie *f*
f dendrochronologie *f*
r дендрохронология *f*

2374 dendroclimatology; tree-ring climatology
d Dendroklimatologie *f*
f dendroclimatologie *f*
r дендроклиматология *f*

2375 dendroecology
(study of ecological relationships and changes using dendrochronologically dated tree rings)
d Dendroökologie *f*
f dendroécologie *f*
r дендроэкология *f*

* **dendroenergy → 9567**

2376 dendrogeomorphology
(a subfield of dendrochronology (q v) which utilizes dated tree rings to study land forms and geomorphic processes)
d Dendrogeomorphologie *f*
f dendrogéomorphologie *f*
r дендрогеоморфология *f*

2377 dendrogram; diagrammatic tree
(a diagram showing the relationships produced by classification)
d Dendrogramm *n*
f dendrogramme *m*
r дендрограмма *f*

* **dendrograph → 516**

2378 dendrohydrology
d Dendrohydrologie *f*
f dendrohydrologie *f*
r дендрогидрология *f*

* **dendroid[al] → 8876**

2379 dendrologist
(one versed in the natural history of trees)
d Dendrologe *m*
f dendrologiste *m*
r дендролог *m*

2380 dendrology
(the study, nomenclature and identification of trees)
d Dendrologie *f*; Baumkunde *f*; Gehölzkunde *f*

2381 dendrometer
(a device for measuring the diameter of the trunk or branch of a tree)
d Dendrometer *n*; Baummesser *m*
f dendromètre *m*
r дендрометр *m*

2382 dendrometric characteristics
d dendrometrische Merkmale *npl*
f caractéristiques *fpl* dendrométriques
r дендрометрические характеристики *fpl*

2383 dendrometry
(measuring volume, growth and development of individual trees and stands and determining various products obtainable from them)
d Dendrometrie *f*
f dendrométrie *f*
r дендрометрия *f*

2384 dendrophil
d Dendrophyl *m*
f dendrophile *m*
r дендрофил *m*

2385 dendrophilous; dendrophil *adj*
(tree-loving; dwelling on or among trees)
d bäumeliebend; dendrophil
f dendrophile
r дендрофильный

2386 dendroscope
(device used to reduce the top of a large tree to a given form)
d Dendroskop *m*
f dendroscope *m*
r дендроскоп *m*

2387 dense crown
d dichte Krone *f*
f cime *f* dense
r густая крона *f*

2388 dense forest
d dichter Wald *m*
f forêt *f* dense
r густой лес *m*

* **dense leaf canopy → 1650**

2389 densely leaved; foliose *adj*
(bearing numerous leaves)
d blätterreich; dichtbelaubt; dichtblätterlich
f densifolié
r сильно облиственный; густо[об]лиственный

* **dense stand [of trees] → 3756**

* **dense timber → 8659**

* **dense wood → 8659**

2390 densified wood; compressed wood; pregwood
d Press[lagen]holz *n*; Schichtholzkunststoff *m*; Sperrfestholz *n*; Sternfestholz *n*
f bois *m* densifié; bois comprimé
r прессованная древесина *f*; древесный слоистый пластик *m*

2391 density cycle
(a completed period of rhythmic fluctuation in the density of a species, e.g. from high to low density and back to high)
d Zyklus *m* der Bestandsdichten
f cycle *m* de fluctuation de la densité
r цикл *m* плотности (насаждения)

2392 density of forest transportation network
d Waldwegenetzdichte *f*
f densité *f* de réseau routier forestier
r густота *f* лесной транспортной сети; густота сети лесных дорог

2393 density of leaf canopy; canopy density; canopy thickening
(amount of foliar cover, combining the extent of canopy closure and crown density)
d Beschirmungsdichte *f*
f densité *f* du couvert; densité d'ombrage
r густота *f* полога

* **density of stand → 3370**

2394 density of undergrowth
d Unterwuchsdichte *f*
f densité *f* de sous-bois
r густота *f* подроста

2395 density rule
(of wood by the annual rings)
d Dichteregel *f*
f règle *m* de densité
r правило *n* определения плотности

* **dentate → 7461**

2396 dentate leaf margin; serrate leaf margin; crenate leaf margin
(leaf margin that has rounded or curved

f bord *m* de feuille denté; bord de feuille crénelé
r зубчатая форма *f* края листа; пальчатая форма края листа; городчатая форма края листа

* **den tree → 4333**

* **denuded [forest] area → 470**

2397 denya; okan
(Cylicodiscus gabunensis)
d Denya *n*; Okan *n*
f okan *m*
r циликодискус *m* габунский

2398 deodar [cedar]; Himalayan cedar; Indian cedar
(Cedrus deodara)
d Himalayazeder *f*; Deodarzeder *f*; echte Zeder *f*
f cèdre *m* de l'Himalaya; cèdre déodar; déodar *m*
r кедр *m* гималайский

2399 deodar weevil
(Pissodes nemorensis)
d Deodarzeder-Rüsselkäfer *m*
f charançon *m* de déodar
r смолёвка *f* кедровая

* **department of forestry → 2552**

2400 departure
(schedule which deviates from the principle of non-declining flow by exhibiting a planned decrease in the timber sale and harvest schedule at any time in the future)
d Abzug *m*
f départ *m*
r отклонение *n*

2401 depletion allowance
(deduction from taxable income derived from a wasting asset)
d Depositengeschäft *f*
f indemnité *f* d'épuisement des réserves
r налоговая льгота *f* на доходы с невосполняемых природных ресурсов; налоговая скидка *f* на истощение недр

2402 deprecated
d missbilligt; abgelehnt
f à cœur pourri
r с гнилой сердцевиной

2403 depth gauge
d Teufenanzeiger *m*; Tiefenmesser *m*; Tiefenlehre *f*
f limiteur *m* de profondeur; guide *m* de profondeur; limiteur d'épaisseur de copeau
r ограничитель *m* глубины зарезания; останов *m*, регулирующий глубину сверления; просвет *m* между режущими кромками пильной цепи

* **depth of cut → 2195**

2404 depth of defect
d Fehlertiefe *f*
f profondeur *f* de défaut
r глубина *f* порока

2405 depth of floating route
d Tiefe *f* des Flößweges
f profondeur *f* de canal de flottage
r глубина *f* лесосплавного хода

2406 derelict stand
d verfallener Bestand *m*
f peuplement *m* abandonné
r покинутое насаждение *n*; заброшенное насаждение; запущенное насаждение

* **dereserve → 2524**

2407 deresinification; deresination
(extraction of the resins and tarry substances)
d Entharzung *f*; Entharzen *n*; Entteerung *f*
f dérésinage *m*
r обессмоливание *n*

* **derivative forest → 7346**

2408 derived savanna[h]
(savanna vegetation existing on sites formerly supporting high forest, having developed from such forest through cutting and burning and being maintained as savanna by annual or frequent grass fires)
d sekundäre Savanne *f*
f savane *f* de dégradation
r производная саванна *f*

2409 derno-podzolic soils; sod-podzolic soils; humic pods
d derno-podsolische Bodene *mpl*
f sols *mpl* podzoliques de tourbe mottière
r дерново-подзолистые почвы *fpl* (подтип подзолистых почв, формирующихся в подзоне южной тайги)

* **derrick → 5025**

2410 derrick-wood hoist; derrick crane
d Derrickkran *m*; Toppenant *m*
f mât *m* de charge de bois; grue-derrick *f*; mât-grue *m*
r кран *m* для загрузки древесного сырья; деррик[-кран] *m*; мачтовый кран

2411 description of stand; stand description; compartment description
d Bestandsbeschreibung *f*
f descriptif *m* de peuplement
r [таксационное] описание *n* насаждения

2412 desert almond
(Prunus fasciculata)
d Wüstenmandel *f*
f amande *f* de désert
r миндаль *m* пустынный

2413 desert hackberry; spiny hackberry
(Celtis pallida)
d Wüstenzürgelbaum *m*
f micocoulier *m* de désert
r каркас *m* бледный

2414 desertification
(a process by which land becomes increasingly unproductive and barren)
d Desertifikation *f*; Verwüstung *f*; Wüstenbildung *f*
f désertification *f*
r превращение *n* [степи] в пустыню; дезертификация *f*

2415 desert ironwood
(Olneya tesota)
d Wüsteneisenholz *n*
f bois *m* de fer
r железное дерево *n* пустынное

2416 desert lavender
(Hyptis emoryi)
d Wüstenlavendel *m*
f lavande *f* de désert
r пустынная лаванда *f*

* **desert Oregon grape → 3704**

2417 desert pavement
(a surface of stones and rocks remaining after finer material has been blown away)
d Wüstenpflasterboden *m*
f pavage *m* désertique
r пустынная корка *f*

2418 desert rose
f rosier *m* de désert
r роза *f* звездчатая

2419 desert shrub
d Wüstenbusch *m*
f arbuste *m* de désert
r пустынный куст *m*

* **desert sumac[h] → 5005**

* **desert thorn → 9471**

2420 desert willow
(Chilopsis linearis)
d Wüstenweide *f*
f saule *m* de désert
r губовник *m* линейный

2421 desiccated wood
d künstlich getrocknetes Holz *n*
f bois *m* séché artificiellement
r древесина *f* искусственной сушки; искусственно высушенная древесина

2422 desiccation crack; contraction crack; drought crack; season[ing] crack; seasoning stress; seasoning check; shrinkage crack; seam
(stress in timber caused by variation in shrinkage as it dries)
d Schwindriss *m*; Trockenriss *m*; Trocknungsriss *m*
f brisure *f* de séchage; fente *f* de sécheresse; fente de retrait
r трещина *f* от усушки; трещина от высыхания

2423 desiccation of wood; drying of wood; wood drying; seasoning
(drying timber to a moisture content appropriate to the conditions and purposes for which it is to be used)
d Austrocknung *f* des Holzes; Holztrocknung *f*
f séchage *m* du bois
r сушка *f* древесины

* **designated area → 9799**

2424 designated official
(a person designated by name or title to be a designated energy, mines and petroleum resources official, designated environment official, or designated forest official)
d bezeichneter Beamter *m*

2425 designated skid road; designated skid trail
(pre-planned network of skid roads or skid trails, designed to reduce soil disturbance and planned for use in subsequent forestry operations in the same area)
d vorgesehener Entstaubungsweg *m*
f chemin *m* de débardage différé
r обозначенный лесоспуск *m*

2426 designated wilderness
d bezeichnete Wildnis *f*
f zone *f* de nature protégée différée
r обозначенная девственная местность *f*; обозначенная первобытная местность

2427 desilter
d Entschlammer *m*
f dessilteur *m*
r иглоочиститель *m*

2428 desirable plant species
(species that contribute to management objectives)
d gewünschte Pflanzenarten *fpl*
f espèces *fpl* végétales privilégiées
r желательные растительные виды *mpl*

* **desired exotic species → 2430**

2429 desired future stand condition
(description of the characteristics of the future stand)
d gewünschter zukünftige Bestandeszustand *m*
f état *m* de peuplement désiré
r желанное будущее состояние *n* насаждения

2430 desired non-native species; desired exotic species
(species of plants or animals that are not indigenous to an area but which represent a significant, and usually remnant segment of a gene pool)
d gewünschte Exoten *mpl*
f espèces *fpl* exotiques désirées
r желательная неместная порода *f*; желательная инородная порода

2431 desired plant community
(plant community that produces the kind, proportion and amount of vegetation necessary for meeting or exceeding the land-use plan requirements or ecological site objectives)
r желательное растительное сообщество *n*

* **destroying polyporus → 9581**

* **destruction decay → 2433**

2432 destruction of the tropical forests
d Beispiel *n* der Tropenwaldvernichtung; Zerstörung *f* der Tropenwälder
f destruction *f* des forêts tropicales
r уничтожение *n* тропических лесов

2433 destruction rot; destruction decay; crumbly rot
d Destruktionsfäule *f*
f pourriture *f* de destruction
r трухлявость *f*

2434 destructive lumbering; rapine
d Raubwirtschaft *f*
f exploitation *f* abusive
r хищническое хозяйство *n*

* **determa → 6782**

2435 determinate growth
(pattern of development in which the apical meristem differentiates into flowers, terminating the production of additional leaves and stems)
d bestimmtes Wachstum *n*
f croissance *f* déterminée
r ограничиваемый рост *m*

* **determination of cut → 9776**

2436 determination of diameter
(of a tree stem)
d Durchmesserbestimmung *f*
f détermination *f* de diamètre
r измерение *n* диаметра; определение *n* диаметра; замер *m* толщины

2437 determination of grain inclination
d Faserneigung-Bestimmung *f*
f détermination *f* d'inclinaison du fil
r измерение *n* наклона волокон

2438 determination of standing volume; stand volume inventory; survey of the stock; stocking survey
(determination of the stocking of an area of both well-spaced and total trees; also used to generate an inventory label)
d Bestandesvorrataufnahme *f*
f détermination *f* de cubage du volume sur pied

2439 detrimental soil disturbance
(changes caused by forest practices in the physical, chemical, or biological properties of the soil)
d nachteilige Bodenstörung *f*
f détérioration *f* du sol nuisible
r повреждение *n* неустойчивого грунта

* **detritivore → 2441**

2440 detritus
(1. small pieces of dead and decomposing plants and animals; 2. detached and broken-down organic fragments of structure; 3. small organic particles such as leaves and twigs)
d Detritus *m*; Gesteinsschutt *m*; Trümmer *pl*
f détritus *m*; matériaux *mpl* détritiques; débris *m*
r наносные отложения *npl*; наносы *mpl*; детрит *m*

2441 detrivore; detritivore
(organism (plant or animal) that feeds on organic remains or organic (mostly plant) debris)
d Fäulnisfresser *m*; Detritivore *m*; Detritusfresser *m*
f détritivore *m*
r детритоядный организм *m*

2442 deutzia
(Deutzia)
d Deutzie *f*
f deutzia *f*
r дейция *f*

2443 devastated forest
(forest in which the regenerative capacity has been destroyed)
d verwüsteter Forst *m*
f forêt *f* dévastée
r истребленный лес *m*

2444 devegetation
(removing the vegetation from an area)
d Devegetation *f*
f dévégétation *f*
r устранение *n* растительности

2445 developmental stage
d Entwicklungsstufe *f*
f état *m* de développement
r фаза *f* роста и развития; стадия *f* развития

2446 developmental tendency
r тенденция *f* развития

2447 development objectives
(short-term (often 5-year) planning objectives for a specific management area)
d Entwicklungsziele *npl*
f objectifs *mpl* de développement
r цели *fpl* развития

2448 development of a stand; stand development
(growth of a stand through its various developmental stages)
d Bestandesentwicklung *f*
f développement *m* d'un peuplement; évolution *f* d'in peuplement
r развитие *n* насаждения

2449 development of branches
d Astentwicklung *f*
f développement *m* de branches
r развитие *n* ветвей

2450 deviation of log axis
d Klotzachse-Abweichung *f*; Holzscheitachse-Abweichung *f*
f déviation *f* d'axe de grume
r отклонение *n* оси брёвна

2451 Devil's-club
(Echinopanax elalum; Oplopanax horridus)
d Teufelskeule *f*
f bois *m* piquant
r эхинопанакс *m*; оплопанакс *m*; заманиха *f*

2452 Devil's-walking-stick; Hercules' club; tree aralia
(Aralia spinosa)
d Teufelskrückstock *m*; Angelikabaum *m*
f angélique *f* en arbre d'Amérique; aralia *m* épineux; angélique épineuse; bâton *m* du diable
r аралия *f* колючая; аралия шиповатая

2453 devil[-]tree
(Alstonia macrophylla)
d Dämonenbaum *m*; Weißquirlbaum *m*
f arbre *m* démon
r чёртово дерево *n*

2454 dewatering
(condition in stream channel when all the water flow occurs within the permeable streambed sediments, so no surface water is left; common in small streams with considerable accumulations of gravel)
d Entwässerung *f*

* **dewberry → 2891**

* **dewinger → 2456**

2455 **dewinging**
d Samenentflügelung *f*; Saatgutreinigung *f*
f dessalement *m* des graines
r очищение *n* семян

2456 **dewinging machine; dewinger**
d Samenflügler *m*
f désaileuse *f*
r обескрыливатель *m*

2457 **dew point; dewpoint**
d Taupunkt *m*
f point *m* de rosée
r точка *f* росы

2458 **dew willow**
(Salix rorida)
d Tauweide *f*
f saule *m* rossé
r ива *f* росистая; ива росяная; шелюга *f* сибирская

2459 **diagnostic plot sampling**
d diagnostische Grundstückabtastung *f*
f échantillonnage *m* de placettes diagnostique
r диагностическая закладка *f* пробных площадей (при лесоустройстве)

2460 **diagonal edging**
d diagonaler Dreikantenspreißel *m*
f délignage *m* diagonal
r обрезка *f* [досок] по диагонали

2461 **diagonal grain**
(a type of cross grain resulting from failure to saw parallel to the growth increments)
d Diagonalfaser *f*
f fil *m* diagonal
r диагонально наклоненное волокно *n*

2462 **diagonal-grained wood**
(wood in which the annual rings are at an angle with the axis of a piece as a result of sawing at an angle with the bark of the log)
d diagonal-gekörntes Holz *n*
f bois *m* à fil incliné
r древесина *f* с диагональным наклоном волокон

2463 **diagram log rule**
d diagrammatische Kubiktabelle *f*
f table *f* de cubage diagrammatique

* **diagrammatic tree → 2377**

2464 **dial ga[u]ge dendrometer**
d Messuhr-Dendrometer *n*
f dendromètre *m* d'indicateur à cadran
r дендрометр *m* с циферблатной шкалой

2465 **diameter above buttress; d.a.b.; diameter at butt[-end]; diameter at foot; butt diameter; diameter at stump height; d.s.h.; stump diameter; diameter at the base**
d Durchmesser *m* am stärkeren Ende; Fußdurchmesser *m*; Unterstärke *f*; Durchmesser am unteren Stammende
f diamètre *m* au-dessus d'empattements; diamètre à hauteur de souche; DHS; diamètre au gros bout
r диаметр *m* в комле[вом конце]; диаметр основания; диаметр в нижнем отрубе; нижний диаметр; диаметр [у] пня

2466 **diameter at breast height; d.b.h.; breast height diameter; chest height diameter**
(tree diameter in inches, outside bark, usually measured at 4.5 feet above ground)
d Brusthöhendurchmesser *m*
f diamètre *m* à hauteur de poitrine; DHP; diamètre à hauteur d'homme
r диаметр *m* на высоте груди; таксационный диаметр

* **diameter at butt[-end] → 2465**

2467 **diameter at felling level; diameter at ground line**
(diameter measure of a standing tree at the estimated cutting height)
d Durchmesser *m* auf Holzschlagniveau
f diamètre *m* de zone du collet; diamètre à la découpe
r диаметр *m* в месте среза

* **diameter at foot → 2465**

2468 **diameter at half tree height; half-height diameter; mean diameter; central diameter; mid-sectional diameter; mid-diameter**
d Mittelstärke *f*; Mittenstärke *f*; Mitteldurchmesser *m*
f diamètre *m* à mi-hauteur; diamètre moyen
r диаметр *m* на половине высоты дерева; срединный диаметр; толщина *f* ствола дерева на середине высоты

2469 **diameter at root collar; d.r.c.**

d Wurzelanlauf-Durchmesser *m*
f diamètre *m* d'empattement
r диаметр *m* корневой шейки

2470 **diameter at smaller end; diameter at top; top[ping] diameter; diameter at thin butt**
d Oberdurchmesser *m*; Oberstärke *f*; Zopfdurchmesser *m*
f petit diamètre *m*; diamètre au fin bout; diamètre au petit bout
r диаметр *m* в верхнем отрубе; диаметр вершинного торца; верхний диаметр *m*

* **diameter at stump height → 2465**

* **diameter at the base → 2465**

* **diameter band → 2484**

2471 **diameter class; diameter group**
(classification of trees based on tree d.b.h.)
d Durchmesserklasse *f*; Durchmesserstufe *f*; Stärkeklasse *f*
f classe *f* de diamètre
r класс *m* [толщины] диаметров; ступень *f* диаметров

2472 **diameter[-class] limit**
(maximum diameter of trees to be cut, as in a timber sales contract)
d Durchmesserklassengrenze *f*
f limite *f* de classe de diamètres
r допуск *m* диаметра; верхний или нижний предел *m* ступени толщины (ствола); отпускной диаметр *m*

* **diameter distribution → 2483**

* **diameter ga[u]ge → 3653**

* **diameter group → 2471**

* **diameter growth → 2474**

2473 **diameter-height function**
d Durchmesser-Höhe-Funktion *f*
f fonction *f* hauteur-diamètre
r зависимость *f* между диаметра и высоты

2474 **diameter increment; radial increment; radial growth; diameter growth**
(of a bole)
d Durchmesserzuwachs *m*; Radialzuwachs *m*; Dickenzuwachs *m*
f accroissement *m* en diamètre: accroissement
прирост; рост *m* в толщину

2475 **diameter inside bark; d.i.b.; diameter under bark**
(diameter of a tree or log excluding double bark thickness)
d Durchmesser *m* ohne Rinde
f diamètre *m* sans écorce
r диаметр *m* без коры

2476 **diameter-limit cut[ting]; size-limit felling; bucking at the diameter break; girth-limit cutting; felling to a diameter limit**
(removal of all merchantable trees above a specified minimum diameter)
d Dimensionsplenterung *f*; Dimensionsdurchforstung *f*
f coupe *f* au diamètre limite; coupe au diamètre minimal; abattage *m* au diamètre limite; éclaircie *f* sur dimensions
r рубка *f* от допустимого диаметра; рубка до определённого диаметра; раскряжёвка *f* по предельному диаметру; вырубка *f* крупных стволов

2477 **diameter-limit sale**
(timber sale in which all trees over a specified d.b.h. may be cut)
d Durchmesser-Begrenzungsverkauf *m*
f vente *f* à diamètre limite
r продажа *f* от допустимого диаметра

2478 **diameter of a knot**
(in the base)
d Astdurchmesser *m*
f diamètre *m* d'un nœud
r диаметр *m* сука

2479 **diameter of cutting circle**
d Schnittkreisdurchmesser *m*
f diamètre *m* de cercle de coupe
r диаметр *m* режущей части; режущий диаметр; диаметр резания

2480 **diameter outside bark; diameter over bark; d.o.b.**
(diameter of a tree or log including bark)
d Durchmesser *m* in Rinde; Durchmesser mit Rinde
f diamètre *m* sur écorce; diamètre avec écorce
r диаметр *m* с корой

2481 **diameter over stubs; d.o.s.**
(outer diameter of the knotty core in pruned trees)
d Durchmesserüberstummel *m*

2482 diameter over/under bark; girth over/under bark; over/under bark measure
(measurement of the diameter or circumference of a trunk or branch made with the bark removed/non removed)
d Durchmesser mit/ohne Rinde; Umfang *m* mit/ohne Rinde
f circonférence *f* sur/sous écorce
r окружность *f* с/без корой

2483 diameter sorting; diameter distribution
d Stärkesortierung *f*
f triage *m* par diamètre
r сортировка *f* по диаметру; распределение *n* [деревьев] по диаметру

2484 diameter tape; diameter band
(tape calibrated to measure the diameter of a tree by measuring its circumference)
d Durchmessermessband *n*; Durchmesserband *n*; Diameterbandmass *n*
f ruban *m* diamétrique; ruban de mesure du diamètre; ruban pour les diamètres
r рулетка *f* для диаметров; мерная тесьма *f* для диаметров

* **diameter under bark → 2475**

* **diamondleaf oak → 4835**

* **diamond willow → 5480**

* **d.i.b. → 2475**

* **dibble [bar] → 6344**

2485 dibble planting; dibbling; peg planting
(sowing seeds or setting out seedlings in rough holes made with a stick or peg)
d Dibbelsaat *f*; Horstsaat *f*; Häufchensaat *f*
f plantation *f* au bâton
r посадка *f* под кол

* **dichapetalum family → 8412**

2486 dicotyledon
(Dycotyledonum)
d Dikotyledone *f*; zweikeimblättrige Pflanze *f*
f dicotylédone *f*
r двудольное растение *n*

2487 dieback of bark
d Rindentod *m*
f dépérissement *m* d'écorce
r отмирание *n* коры

f dépérissement *m* de branches
r отмирание *n* ветвей

2489 dieback of coffee
(Colletotrichum coffeanum)
d Blattfallkrankheit *f* des Kaffees
f anthracnose *f* du caféier
r антракноз *m* кофейного дерева

2490 dieback wood
d abständiges Holz *n*; abgängiges Holz
f bois *m* dépérissant
r отмирающая древесина *f*

* **die-hammer → 5315**

* **die-square → 7963**

* **die-squared log → 3755**

* **die-stock → 8809**

* **difficult country → 4212**

2491 difficult site
(forest site with environmental conditions that are unfavorable for tree establishment and growth)
d schwieriger Stellungsort *m*
f station *f* difficultée
r неблагоприятное местообитание *n*

* **difficult terrain → 4212**

2492 diffuse parenchyma
(single apotracheal parenchyma strands or cells distributed irregularly among fibers, as seen in cross section)
d diffuses Parenchym *n*
f parenchyme *m* dispersé
r ветвящаяся паренхима *f*

2493 diffuse-porous wood
(wood in which the pores are of fairly uniform or only gradually changing size and distribution throughout a growth ring)
d zerstreutporiges Holz *n*
f bois *m* à pores diffus; bois à pores épars; bois à pores disséminés
r рассеянно-пористая древесина *f*; рассеянно-сосудистая [лиственная] древесина

2494 diffuse radiation
(radiant energy propagating in many different directions through a given small volume of

f rayonnement *m* diffus
r рассеянная радиация *f*

2495 **dig** *v*; **grub** *v*
d graben
f arracher
r копать; рыть

2496 **digester silo; chip silo**
d Bunker *m* für Späne; Spänebunker *m*; Hackholzbehälter *m*; Hackschnitzelsillo *n/m*
f silo *m* de copeaux
r бункер *m* для щепы

* **digger pine → 7142**

* **digging of ditches → 8962**

* **digging round a tree → 6953**

* **digitate leaf → 6037**

2497 **dilatation**
d Dilatation *f*
f dilatation *f*
r тангенциальный рост *m* ткани годичных колец древесины

2498 **dillenia family**
(Dilleniaceae)
d Dilleniengewächse *pl*
f dilléniacées *pl*
r диллениевые *pl*

2499 **dimension**
d Maß *n*; Dimension *f*
f dimension *f*
r размер *m*

2500 **dimensional classification**
(of lumber)
d Maßklassifikation *f*
f classement *m* dimensionnel
r размерностная классификация *f*

* **dimension board → 8103**

2501 **dimension bolt**
d Dimensionskurzholz *n*
f bois *m* court de dimension établie
r бревно *n* установленных размеров

2502 **dimensioned wood**
d abgemessenes Holz *n*
f bois *m* dimensionné
r измеренная древесина *f*

2503 **dimension stock; regular-sized timber**
(lumber sawn to definite dimensions)
d Dimensionsholz *n*; Einheitsholz *n*; Standardholz *n*
f bois *m* débité à dimension[s spécifiées]; bois d'échantillon
r стандартный пиломатериал *m*; пиломатериал стандартных размеров; заготовка *f* определённых размеров

2504 **dimension parts**
(hardwood stock processed to a point where the maximum waste is left at the mill, and the maximum utility is delivered to the user)
d Dimensionsteile *mpl*
f parts *mpl* de bois débité
r детали *fpl*, обработанные в размер

* **dimension stock → 2503**

* **dimension timber → 2503**

2505 **dimension timber waste**
d Brockenholz *n*
f résidus *mpl* de bois débité
r толстомерные деревянные отходы *mpl*

2506 **dimorphic branching**
(ability to grow only either vertically or horizontally)
d dimorphe Gabelung *f*; zweigestaltige Abzweigung *f*
f ramification *f* dimorphique
r диморфное разветвление *n*

* **dina → 6056**

2507 **dioecious plant**
(having male and female reproductive structures on different individuals)
d zweihäusige Pflanze *f*; dioezische Pflanze
f plante *f* dioïque
r двудомное растение *n*

2508 **dip**
(a liquid in which timber may be immersed for a short time for some specific purpose, e.g. to preserve it)
d Tauchlösung *f*
f bain *m*
r жидкость *f* для протравливания

* **dip → 828, 2165, 2514**

2509 **dip grain; plume**
(very local, slight undulations of the

d Wirbel *m*
f fil *m* dévié
r завиток *m*

2510 **diplodia twig blight of pine**
(Diplodia pinea)
d Diplodia-Triebsterben *n* der Föhre
f flétrissement *m* des pousses de pin
r диплодиоз *m* сосны

* **dipper → 6905**

* **dipping → 828**

2511 **dipping**
(immersion of seedling roots in a solution or water prior to planting)
d Tauchbeizung *f*
f désinfection *f* par immersion; trempage *m*
r отделка *f* окунанием; погружение *n* в антисептик

2512 **dipterocarp forest**
d Flügelfruchtwald *m*
f forêt *f* diptérocarpe
r диптерокарповый лес *m*

2513 **dipterocarpus family; dipterocarps**
(Dipterocarpaceae)
d Flügelfruchtgewächse *pl*
f diptérocarpacées *pl*
r диптерокарпы *mpl*

* **directed felling → 2515**

2514 **directional cut; round-up; dip; sink; stemnotch**
d Stammankerben *n*; Kerben *n*
f entaille *f* à la tige
r средняя часть *f* недопила, образуемая при боковых резах (при валке дерева)

2515 **directional felling; directed felling**
(a mechanical means to control the direction of falling timber)
d gerichtetes Fällen *n*
f abattage *m* directionnel
r направленная валка *f*

2516 **direction of fiber; direction of grain; grain direction; fiber orientation**
(in paper)
d Holzfaserrichtung *f*; Faser[ver]lauf *m*; Fiberrichtung *f*
f orientation *f* de fils; direction *f* des veines;

2517 **direction of lead**
d Rückerichtung *f*
f orientation *f* du tracé; direction *f* de guide
r направление *n* перемещения [леса при трелёвке]

2518 **direction of lean**
d Richtung *f* des Neigung
f direction *f* d'inclinaison
r направление *n* естественного наклона [дерева]

2519 **direction of the coupe; cutting direction; felling direction**
d Fällrichtung *f*; Schnittbewegung *f*; Schlagrichtung *f*
f direction *f* d'abattage; orientation *f* de coupe
r направление *n* резания; направление рубки

2520 **direct radiation**
(incoming solar radiation received, without deflection or scattering, from the sun)
d direkte Strahlung *f*
f rayonnement *m* direct
r прямая радиация *f*

* **direct sale → 7157**

2521 **direct seeding; direct sowing; seeding in situ**
(artificial systematic sowing of seeds by manual or mechanical means in an area on which a forest stand is to be raised)
d Direktsaat *f*; direkte Aussaat *f*; Frei[land]saat *f*; Freiflächensaat *f*
f ensemencement *m* direct; semis *m* direct [à la volée]; semis à la volée
r высев *m* в открытый грунт; прямой засев *m*

2522 **direct [timber] hauling; hauling without skidding**
d Direkt[forst]bringung *f*
f débardage *m* direct [de bois]
r прямая вывозка *f* [леса]; прямая доставка *f*

2523 **direct wood cost[s]**
d direkte Holzkosten *pl*
f frais *mpl* d'exploitation directs
r прямые затраты *fpl* на заготовку леса

* **dirt slide → 5116**

2524 **dis[af]forest** *v*; **deforest** *v*; **de-afforest** *v*; **dereserve** *v*; **disreserve** *v*
d entwalden
f déboiser

* **dis[af]forestation → 2339**

* **disbark → 2280**

* **disbarked → 551**

* **disbarked wood → 555**

* **disbarking → 579**

* **disbarking machine → 2284**

* **disbranch → 10**

* **disbranching → 6610**

* **disbudding → 1142**

* **disc barker → 2533**

2525 **discharge channel; discharge carrier**
d Entladungskanal *m*
f coursier *m* [d'évacuateur]
r выпускной канал *m*; спускной канал

2526 **disclimax; anthropogenic subclimax**
(a one-time climax community which has been disturbed by people or their domestic livestock, such as a deciduous forest being replaced by a cropland)
d anthropogene Subklimax *f*
f disclimax *m*; plagioclimax *m*
r дисклимакс *m*

2527 **discolo[u]ration of the sap[ ]wood**
d Entfärbung *f* des Splintes
f décoloration *f* de l'aubier
r обесцвечивание *n* заболони

2528 **discontinuous growth ring**
(a growth ring that is not present all around the stem)
d diskontinuierlicher Jahrring *m*
f cerne *m* discontinu en anneau
r прерванное годичное кольцо *n*

2529 **disc stump grinder; tree stump disposal; stump grinder; stump slicer; stump chopper**
d Scheiben-Stubbenroder *m*
f broyeur *m* de souches à disques
r дисковая машина *f* для измельчения пней

* **disengage → 6856**

* **dish[ing] → 2141**

2530 **disintegrated wood; hogged wood**
(pulpwood logs to be chipped in a hog)
d aufgelöstes Holz *n*
f bois *m* désagrégé
r измельчённая древесина *f*

2531 **disjunctive parenchyma**
(axial or radial parenchyma cells partially disjoined during the process of differentiation)
d trennendes Parenchym *n*
f parenchyme *m* disjoint
r дизъюнктивная паренхима *f*; разорванная паренхима

2532 **disjunctive tracheid**
(a tracheid partly disjoined laterally from another during differentiation; contact is maintained by means of tubular processes)
d trennende Tracheide *f*
f trachéide *f* disjointe
r дизъюнктивная сосудовидная клетка *f*

2533 **disk barker; disc barker**
d Scheibenentrinder *m*
f écorceuse *f* à disque
r дисковый окорочный станок *m*

2534 **disk chipper; disk crusher**
d Scheibenhacker *m*; Scheibenhackmaschine *f*
f coupeuse *f* à disque; concasseur *m* à disque
r дисковая рубильная машина *f*

2535 **disk ploughing; disking; harrowing**
(mechanical site preparation method of scarifying the soil by pulling an offset disk over a site)
d Scheibenpflügen *n*
f hersage *m*; labourage *m* de disque
r обработка *f* почвы дисковыми орудиями

* **disk refiner groundwood → 1519**

* **refiner groundwood → 1519**

* **disk saw → 1585**

2536 **dispersed retention**
(retaining individual trees scattered throughout a cutblock)
d zerstreutes Zurückhalten *n*
f rétention *f* dispersée

2537 **dispersed use**
(recreation activities generally associated with low-density use that takes place throughout the entire forest such as hunting, fishing, hiking, etc.)
d zerstreuter Gebrauch *m*
f utilisation *f* dispersée
r рассредоточенное использование *n*

2538 **dispersion of leaves**
d Laubverwehung *f*
f dispersion *f* de feuilles mortes sous l'effet du vent
r развевание *n* листвы

* **disreserve → 2524**

* **dissected leaf → 2549**

* **dissolvent → 8505**

* **dissolver → 8505**

2539 **distance between knots; distance between branches**
d Astabstand *m*
f écartement *m* des branches; espacement *m* entre verticilles
r расстояние *n* сучьев

2540 **distant-patrol floating method**
d Flößungsmethode *f* mittels entfernte Patrouillieren
f méthode *f* de flottage par inspection à distance
r дистанционно-патрульный способ *m* лесосплава

* **distillation wood → 44**

* **distortion of lumber → 9268**

* **distortion of wood → 9268**

* **distribution of fellings → 191**

2541 **distribution of slash**
d Verteilung *f* des Schrägstriches
f distribution *f* de déchets [de coupe]
r распределение *n* порубочных остатков

* **district forest officer → 2551**

2542 **district of forest guard; district of forest warden; forester's district; forest range district; ranger district**
d Forstwärdersbezirk *m*
район *m* лесничества

2543 **disturbed natural forest**
d gestörter natürlicher Wald *m*
f forêt *f* vierge détériorée
r нарушенный естественный лес *m*

2544 **disturbed vegetation**
d gestörte Vegetation *f*
f végétation *f* détériorée
r нарушенная растительность *f*

2545 **ditch block[ing]**
(blockage that is located directly downgrade of a cross-drain culvert or cross-ditch and designed to deflect water flow from a ditch into a cross-drain culvert)
d Blockierung *f* eines Abzugsgrabenes
f blocage *m* de fossé
r блокировка *f* канавы

* **ditcher → 8963**

* **ditching → 8962**

* **ditching system → 2614**

* **ditch transportation → 1577**

2546 **diversified slash disposal**
d Holzabfallbeseitigung *f*
f élimination *f* diversifiée des rémanents
r применение *n* нескольких методов очистки от отходов

2547 **diversion drain; diversion ditch**
(channel with a supporting ridge on the lower side constructed across a slope to intercept surface water)
d Ablenkungsgraben *n*
f fossé *m* de dérivation; fossé de ceinture; fossé de détour
r отводная канава *f*

2548 **diversion of lumber**
d Ablenkung *f* des Bauholzes
f dérivation *f* du bois
r распределение *n* досок (в процессе сортировки)

* **diversity index → 7898**

2549 **divided leaf; parted leaf; dissected leaf**
(cut or divided into numerous lobes or divisions; a deeply cut leaf, the cleft not

f feuille *f* disséquée
r рассечённый лист *m*

2550 **divi-divi**
(Caesalpinia coriaria; Caesalpinia tinctoria)
d Dividivi *n*
f divi-divi *m*
r диви-диви *n*; цезальпиния *f* красильная

2551 **divisional forest officer; district forest officer; regional forester; chief district forester**
d Bezirksforstmeister *m*
f officier *m* divisionnaire
r окружной лесничий *m*; областной лесничий

2552 **division of forestry; department of forestry; forestry department; forest service**
d Forstabteilung *f*; Forstbehörde *f*
f section *f* forestière
r лесное отделение *n*; лесное ведомство *n*; отделение лесоуправления

* **d.o.b. → 2480**

* **docking → 1215**

* **docking saw → 1136**

* **dockmackie → 5293**

2553 **dogbane family**
(Apocynaceae)
d Hundsgiftgewächse *pl*
f apocynacées *pl*
r кутровые *pl*

2554 **dog board; backing board**
(in sawing lumber on a headsaw, the last board in the log to which the carriage dogs are attached)
d Restbohle *f*
f contredosse *f*; dosse *f* de tranche; chute *f* de tranche
r зажимная доска *f*

2555 **dog-rose; wild rose; wild; brier**
(Rosa canina)
d Hundsrose *f*; Hecken-Rose *f*
f églantine *f*; églantier *m*; rosier *m* des chiens
r шиповник *m* [обыкновенный]

2556 **dogwood family; hazel family**
(Cornaceae)
Kornelkirschgewächse *pl*; Kornazeen *pl*
f cornacées *pl*
r дёренные *pl*; дёреновые *pl*; кизиловые *pl*

* **dolbeer → 8609**

* **dolly → 894**

* **domain[al] forest → 8041**

2557 **domatium (pl domatia)**
(small structures on the lower surface of a leaf in some woody dicotyledons, usually consisting of depressions, partly enclosed by leaf tissue or hairs, located in the axils of the primary veins)
d Domatium *n*
f domatie *f*
r полость *f* в растении

2558 **domesticated introduced tree species**
(introduced tree species planted outside their natural biotope, area or region)
d domestizierte eingebürgerte Baumarten *fpl*
f espèces *fpl* forestières introduites domestiquées
r акклиматизированные привнесенные древесные породы *fpl*

2559 **domestication of forest species**
d Kultivierung *f* der Baumarten
f cultivation *f* d'especes forestières
r окультуривание *n* лесных видов

2560 **domestic dry rot fungus**
(Serpula lacrymans; Merulius giganteus)
d [echter] Hausschwamm *m*
f mérule *m*; champignon *m* des maisons
r [настоящий] домовый гриб *m*

* **domestic wood → 4341**

2561 **dominance**
(an extent to which a given species predominates in a community because of its size, abundance, or coverage)
d Dominanz *f*; Vorherrschaft *f*
f dominance *f*
r доминантность *f*; доминирование *n*; преобладание *n*

2562 **dominance potential**
(relative ability of a tree or plant species to dominate a forest ecosystem)
d Dominanzpotential *n*

2563 **dominant**
(in forestry, an individual or species in the upper layers of the canopy)
d herrschend
f dominant
r господствующий; доминантный

* **dominant age → 6526**

2564 **dominant crown class**
(extending above the general level of the canopy and receiving full light from above and partly from the sides)
d herrschende Kronenklasse *f*
f classe *f* de cime dominante
r класс *m* доминирующих деревьев

2565 **dominant diameter**
d herrschender Durchmesser *m*
f diamètre *m* dominant
r доминирующий диаметр *m*

* **dominant height → 8692**

2566 **dominant species**
d vorherrschender Bestand *m*; vorherrschende Art *f*
f espèce *f* dominante
r господствующая порода *f*; преобладающая порода

* **dominant thinning → 7411**

2567 **dominant tree; guiding stick; holdover**
(tree that has escaped logging, windthrow, or fire, and occupies a dominant position in the stand)
d herrschender Baum *m*; Hauptbaum *m*
f arbre *m* dominant; survivant *m*
r господствующее дерево *n*; маячное дерево; дерево-маяк *n*

2568 **dominant tree diameter**
d Hauptbaumdurchmesser *m*
f diamètre *m* de l'arbre dominant
r диаметр *m* доминирующего дерева

* **dominated → 6011**

* **dominated crop → 8237**

* **dominated trees → 8275**

* **Dominican mahagoni → 2110**

2569 **domino felling**
(the partial cutting of multiple trees which are
d Dominoholzschlag *m*
f abattage *m* en cascade
r валка *f* подпиленных деревьев с помощью другого дерева

2570 **doncella; golden spoon; serrette; changugo; chaparro; candelo; kanoaballi**
(Byrsonima coriacea var. spicata)
d Donzella *f*
f byrsonyme *m*; moureiller *m*; suretti *m* de grand bois
r бирсонима *f* колосистая

* **dong quai → 1489**

* **donkey → 7692, 8609**

2571 **donkey logging; drum logging**
(yarding on the ground with a donkey engine, as contrasted with animal logging, or other power logging methods)
d Abholzung *f* mittels Holzladewinde
f débardage *m* au treuil
r трелёвка *f* паровой лебёдкой

2572 **dormancy; dormant state; quiescence**
(when no apparent growth or development is taking place; a form of growth regulation)
d Schlafenszeit *f*
f dormance *f*
r покой *m*; состояние *n* покоя

2573 **dormant bud; latent bud; resting bud; suppressed bud**
d Präventivknospe *f*; schlafendes Auge *n*; schlafende Knospe *f*; Zukömmling *m*
f bourgeon *m* préventif; bourgeon dormant; préventif *m*
r спящая почка *f*

2574 **dormant cutting**
d schlafender Steckling *m*
f bouture *f* dormante; bouture de rameau aoûté
r спящий черенок *m*

2575 **dormant oil; dormant spray**
(viscous solution containing pesticides and/or fungicides and applied to trees and shrubs in late fall and winter while the plant is dormant)
d ruhendes Öl *n*; Ruhendöl *n*
f huile *f* d'hiver
r латентное масло *n*

* **dormant pruning → 9462**

2576 **dormant seedling**

d schlafender Sämling *m*
f plantule *f* dormante
r спящий сеянец *m*

* **dormant spray → 2575**

* **dormant state → 2572**

* **dorsal → 5**

* **d.o.s. → 2481**

* **dote → 2578**

* **doted → 7212**

* **doted tree → 7082**

* **dote wood → 2605**

2577 **dot grid**
(transparent sheet of film (overlay) with systematically arranged dots, each dot representing a number of area units)
d Punktrasterfeld *n*
f grille *f* à points
r точечная сетка *f*

2578 **dotiness; dote**
(a slight modification of the chemical composition caused by the attack of fungus growth in the wood, generally before drying)
d Spreufleckigkeit *f*
f échauffure *f*
r пёстрая гниль *f*

* **doty → 7212**

* **doty wood → 2605**

2579 **double-bladed axe; double-bitted axe; double-billed axe**
d Bille *f*; Wendeaxt *f*
f hache *f* à double tranchant
r топор *m* с двусторонним лезвием; двусторонний топор; сдвоенный топор

* **double-bladed shear → 8921**

* **double chopper → 9026**

* **double dray → 2587**

2580 **double-frill girdling; double-hack girdling; two-hack girdling**
d Doppel-Nutringelung *f*

* **double grafting → 8691**

* **double handed crosscut saw → 9030**

* **double heart → 2583**

2581 **double-hook dog**
d Doppelthakenzange *f*
f pince *f* de crochet double
r зажим *m* для брёвен из двух крючьев

2582 **double leader**
d Doppelspitzentrieb *m*
f pousse *f* apicale double
r двойной главный побег *m*

2583 **double pith; double heart**
d Doppelkern *m*
f cœur *m* double
r двойная сердцевина *f*

* **double ring → 5616**

* **double sap[wood] → 853**

2584 **double saw**
d Doppelsäge *f*
f scie *f* double
r двусторонняя пила *f*

2585 **double-serrate[d] leaf**
d doppeleingeschnittenes Blatt *n*
f feuille *f* serrée double
r двойной пильчатой лист *m*

2586 **double-skyline system**
d Doppelluftseilsystem *n*
f système *m* de deux câbles porteurs
r подвесная канатная установка *f* с двумя несущими канатами

2587 **double-sled[ge]; twin-sled[ge]; two-sled; double dray; wag[g]on sled; jumbo**
(a type of tongueless double sled used for short-distance hauling)
d Langholzschlitten *m*; Stammholzschlitten *n*
f double traîneau *m*
r сдвоенные сани *pl*

2588 **double-tag line rigging**
d Doppeltumbaulinientakelung *f*
f câblage *m* à deux stabilisateurs
r трелёвочная оснастка *f* с закреплёнными на одной скобе двумя сборочными чокерами

f charge *f* à remorque double
r груз *m* [лесоматериалов] на двух прицепах

2590 doubly crenate leaf margin; doubly serrate leaf margin; biserrate leaf margin
(a leaf margin looking as serrated knife which scallops are themselves serrated, like waves upon waves)
d doppelgezähnter Blattrand *m*
f bord *m* de feuille serré double
r двоякогородчатая форма *f* края листа; двоякопальчатая форма края листа

* **doubly pinnate leaf → 741**

* **Douglas fir → 1772**

2591 Douglas hawthorn
(Crataegus douglasii)
d Douglas-Weißdorn *m*
f aubépine *f* de Douglas
r боярышник *m* Дугласа

* **doum [palm] → 4688**

2592 dovetail halving
d schwalbenschwanzförmiges Blatt *n*; Eckküberblattung *f* durch schwalbenschwanzförmiges Verzinken
f queue *f* d'aronde perdue
r врубка *f* на шип ласточкин хвост

* **dovetail saw → 4936**

2593 dove-tree
(Davidia involucrata)
d Taschentuchbaum *m*
f arbre *m* aux mouchoirs; arbre aux pochettes; arbre aux colombes
r давидия *f* обвёрточная

2594 dove-tree family
(Davidiaceae)
d Taubenbäumgewächse *pl*
f davidiacées *pl*
r давидиевые *pl*

2595 dowelled joint; [gib-and-]cotter joint; spline [joint]; feather; slip key
(a thin piece of wood that fits in the mating grooves cut into two pieces of wood)
d Langholzfeder *f*; Fugenbrett *n*; Verdübelung *f*
f fausse languette *f*; clé *f*; planchette *f* de joint;

* **dowel[pin] → 9553**

* **dowel saw → 6384**

* **down[ed] timber → 2266**

2596 down[ed] tree; downtree
(any tree that is lying on the ground, whether uprooted, stem-broken, or deliberately cut)
d gefällter Baum *m*
f arbre *m* abattu
r поваленное дерево *n*

* **Downey birch → 6620**

2597 downhill thinning
d Abwärtserlockerung *f*
f éclaircie *f* descendante
r рубки *fpl* ухода при трелёвке вниз по склону

2598 downhill wood skidding; downhill [wood] extraction
d Holzabreissen *n*; Holzbergung *f*; Holzsturz *m*; Wälgerung *f*
f débusquage *m* du bois descendant
r спуск *m* древесины с горы; скатывание *n* лесоматериала с горы; трелёвка *f* [древесины] вниз по склону

2599 down-the-row thinning; pre[-]thinning
(keeping smaller lateral branches)
d Auslichtung *f*; Auslichten *n*
f prédémariage *m*; distançage *m*; dégarnissage *m*
r предварительное прореживание *n*

* **downtree → 2596**

2600 down trunk
d gefällter Stumpf
f tronc *m* coupé
r срубленный ствол *m*

2601 downward chipping
d Abwärtszerkleinerung *f*
f déchiquetage *m* par le dessus
r измельчение *n* древесины при вращении режущего инструмента в вертикальной плоскости

2602 downy hawthorn
(Crataegus mollis)
d behaarter Weißdorn *m*

2603 downy serviceberry; high serviceberry; shadblow serviceberry; shadbush serviceberry
(Amelanchier arborea; Amelanchier canadensis)
d Baum-Felsenbirne *f*; kanadische Felsenbirne *f*
f amélanchier *m* arborescent; amélanchier arbre; amélanchier du Canada
r ирга *f* древовидная; ирга канадская

2604 Doyle log rule; Doyle-rule
(for determination of log volume)
d Doyle-Schnittholztabelle *f*
f table *f* de [cubage de] Doyle
r таблица *f* Дойля

* **doze → 6652**

2605 dozy wood; doty wood; dote wood; fusty wood
d stockiges Holz *n*
f bois *m* échauffé
r дряблая древесина *f*

* **drag → 5058**

* **drag cart → 5208**

2606 drag device
d Gegenkraftvorrichtung *f*
f amortisseur *m* de traînée
r цепь *f* для торможения брёвна при трелёвки вниз по склону

* **drag[ging] line → 8745**

* **drag[ging] track → 8745**

* **drag[ging] chain → 7798**

* **dragging dog → 8577**

* **dragging tractor → 7799**

2607 drag hole
(in the end of log)
d Gegenkraftbohrung *f*
f trou *m* de débusquage
r отверстие *n* для закрепления тяговой цепи

* **dragon claw willow → 1963**

* **dragon's-eye → 1381**

2608 dragon[-]tree; drago
f grand dragonnier *m*; drago *m*; montouchi *m*
r драцена *f*; драконово дерево *n*; драконник *m*

* **drag road → 5116**

* **drag rope → 8745**

2609 drag[ ]saw; reciprocating saw; slashing saw; slasher saw; turning saw
d Fuchsschwanzsäge *f*; Gegenzugsäge *f*
f tronçonneuse *f* alternative; scie *f* à refendre
r пила *f* с возвратно-поступательным движением полотна; пила с обратной резкой; пила типа "лисий хвост"

2610 drag scarification
(towing one or more rows of anchor chains, sharkfin barrels, tractor pads to break up and possibly spread slash and to loosen the forest floor and topsoil or expose mineral soil)
d Kettenskarifikation *f*
f scarifiage *m* par traînage
r цепная скарификация *f*

* **drag sled → 5115**

2611 drain
(the (annual or periodic) loss in growing stock from any cause, e.g. removal by cuttings or by mortality from fire, fungi, insects, wind etc.)
d Abfluss *m*
f décroissement *m*
r истощение *n*

2612 drainage
(the process of removal of water from soil)
d Drainage *f*; Dränage *f*
f drainage *m*
r дренаж *m*

2613 drainage ditch; infiltration ditch; outlet ditch; water furrow
d Sammelgraben *m*; Abzugsgraben *m*; Abflussgraben *m*; Abflussfurche *f*; Sauggraben *m*; Entwässerungsfurche *f*
f caniveau *m* de drainage; fossé *m* [d'écoulement]; fossé de décharge
r дренажная канава *f*; сборная канава; осушительная канава; водоотводный канал *m*; водосброс *m*

2614 drainage system; ditching system
d Entwässerungsanlage *f*
f système *m* de drainage

2615 **drain pipe; drain sump**
(pipe that helps remove excess water from a planting hole)
d Dränrohr *n*; Drän *m*; Ablaufrohr *n*; Abflussleitung *f*; Entwässerungsrohr *n*
f tuyau *m* de drainage; tube *m* de drainage
r дренажная труба *f*; дрена *f*

2616 **draw bar**
d Anhängekupplung *f*
f râtelier *m*
r сцепная тяга *f*; сцепное приспособление *n*; цепная серьга *f* (трактора); сцепной брусок *m*

2617 **drawdown**
(a process of reducing allowable annual cuts to a sustainable level)
d Vorhangherunterlassen *n*; Vorhangzuziehen *n*
f rabattement *m*
r снижение *n*; сокращение *n*; урезка *f*

* **draw hook → 1204**

* **drawing → 7995**

2618 **drawn wood**
d gezogenes Holz *n*
f bois *m* filé
r прицепная древесина *f*

2619 **draw shear**
(carrier-mounted, single-action, anvil shear used in mechanized cutting operations)
d Betragschere *f*
f cisaille *f* tractée
r цепное срезающее устройство *n*

* **dray → 5115**

* **d.r.c. → 2469**

2620 **dressed lumber; surfaced lumber; surfaced timber; shaved lumber; dressed timber; planed timber**
(timber finished to a smooth surface on one or more surfaces)
d Hobelware *f*; gehobeltes Holz *n*
f bois *m* raboté; sciage *m* raboté
r строганный пиломатериал *m*

2621 **dressed size**
(the cross-sectional dimensions of lumber after planing)
d gekleidete Größe *f*
f dimension *f* après corroyage
r размер *m* после строгания (пиломатериала)

* **dressing → 8281**

* **dressing in transit → 7244**

2622 **dried wood**
d getrocknetes Holz *n*
f bois *m* [des]séché
r высушенная древесина *f*; выдержанная древесина

2623 **drier; dry kiln; kiln**
d Dörrofen *m*; Darrofen *m*; Darre *f*
f four *m* [de sécherie]
r сушильная печь *f*; печь для сушки

* **drift → 5442**

* **driftage → 2625**

2624 **drift fence**
(obstacle fashioned from barbed wire, or woody debris that is designed to restrict or control cattle movement)
d Zaunung *f*
f barrière *f* d'interdiction
r перегон *m* для скота; скотопрогон *m*

2625 **drift floating; drift[ing]; loose floating; stream driving; driftage**
d Losflößerei *f*; Abtrift *f*; Driften *n*; Einzelstammflößerei *f*; ungebundene Flößerei *f*; Flößung *f* in Einzelstämmen; Triftflößerei *f*
f flottage *m* à bûches perdues
r молевой лесосплав *m*; сплав *m* леса россыпью

* **drifting wood → 3279**

* **drift-path → 3277**

* **driftway → 3277**

* **drift[wood] → 3279**

* **drilling → 7869**

* **drill planting → 6356**

* **drill seeding → 7869**

* **drill sowing → 7869**

2626 **drip; throughfall; net precipitation**
(all the precipitation reaching the forest floor minus the stemflow)

f précipitation *f* au sol
r осадки *mpl*, достигающие почвы; проникающие атмосферные осадки; осадки, достигающие подстилки в лесу

2627 **drip zone**
(the area around the base of a tree that lies below and within the circumference of the branches, etc., which forms its crown)
d Niederschlagszone *f*
f zone *f* de précipitation
r капельная зона *f*

2628 **drivable; floatable**
d driftbar; triftbar; floßbar
f flottable
r пригодный к сплаву; сплавной; пригодный для сплава

* **drive → 8584**

* **driver → 9569**

* **driving → 8753**

2629 **driving area; hauling area**
d Bringungsfläche *f*
f surface *f* de débardage
r площадь *f* вывозки

2630 **driving crew; floating crew**
d Flößereimannschaft *f*
f équipe *f* de flottage
r сплавная бригада *f*; бригада сплавщиков

* **driving loss → 5177**

* **driving regulation → 3276**

* **driving road → 3277**

* **driving season → 3278**

2631 **driving with dams**
d Klauswassertrift *f*
f flottage *m* par retenue d'eau
r сплав *m* по запруде

2632 **drooping crown**
d Trauerkrone *f*
f cime *f* fanaisée
r плакучая крона *f*

* **drooping juniper → 5423**

2633 **drop-crotching**
(a form of light pruning, which tends to
d Fallenlassen-Gabelung *f*
f raccourcissement *m* sur tire-sève
r обрезка *f* побегов на концах ветвей

2634 **drop-feed chipper**
d sturzgespeiste Hackmaschine *f*
f coupeuse *f* de bois à alimentation descendante
r рубильная машина *f* с вертикальной подачи сырья

2635 **drought**
(moisture unbalance when loss of water through the leaves exceeds uptake of water, predominantly by the roots)
d Dürre *f*
f sécheresse *f*
r засуха *f*

* **drought crack → 2422**

2636 **drought-deciduous forest**
d sommergrüner Wald *m*
f forêt *f* décidue à la sécheresse
r летнелистопадный лес *m*

2637 **drought hardiness**
d Dürrefestigkeit *f*; Dürreresistenz *f*
f résistance *f* à la sécheresse
r засухоустойчивость *f*

* **drought ring → 2139**

2638 **drought tree**
d verdorrter Baum *m*
f arbre *m* séchéré
r засохшее дерево *n*

2639 **drought watch**
d Dürrebewachung *f*
f surveillance *f* continue de la sécheresse
r наблюдение *n* засухи

* **droving → 8753**

* **drug → 9395**

* **drug snowbell → 3825**

* **drum chipper → 2208**

* **drum chopper → 2208**

2640 **drum chopping**
(site preparation technique in which logging debris is leveled by a bulldozer pulling a large drum filled with water)
d Trommelzerhacken *n*

* **drum [de]barker → 566**

* **drum logging → 2571**

2641 drum *v* logs
d Klötze ziehen mit Trommelwinde
f foulonner de grumes
r вытаскивать брёвна лебёдкой

* **drum saw → 5049**

* **drumstick tree → 4379**

2642 drum tender
d Trommeloperator *m*
f opérateur *m* de tambour
r оператор *m* окорочного барабана

2643 drum winch
d Trommelwinde *f*
f treuil *m* à tambour
r барабанная лебёдка *f*

* **drunken saw → 4312**

2644 drupaceous fruit; stone fruit; drupe
(fleshy fruit with a central hard core containing 1 or more seeds)
d Steinfrucht *f*
f drupe *f*; fruit *m* à noyau
r костянка *f*; косточковый плод *m*

2645 druxiness
(in wood)
d Innenfäule *f*
f pourriture *f* interne
r внутренняя гниль *f*

2646 dry barking
d trockene Entrindung *f*
f écorçage *m* sec
r сухая окорка *f*

2647 dry branch; dry knot
d Trockenast *m*; Dürrast *m*
f branche *f* sèche
r сухой сук *m*; засохший сук

2648 dry chute; dry slide; hog slide
(a prepared way, often trough-shaped, either on the surface of the ground or built up of timber or other material, down, along, and up which timber is transported, either by power or by gravity)
d Trockenriese *f*; Trockenrutsche *f*

* **dry cooperage → 7713**

* **dry face → 2264**

2649 dry facing
d Lachtentrockenheit *f*
f dessèchement *m* de la carre
r сухость *f* карры

* **dry fallen wood → 2266**

2650 dry forest
d Trockenwald *m*
f forêt *f* sèche
r сухой лес *m*; ксерофитный лес

* **drying of wood → 2423**

* **dry kiln → 2623**

* **dry knot → 2647**

2651 dryland log sorting
d Schnittholzsortierung *f* auf dem Trockenland
f triage *m* de grumes en terre
r сортировка *f* [лесоматериалов] на земле

2652 dry matter to solar radiation ratio
d Trockensubstanz/Solar[ein]strahlung-Verhältnis *n*
f rapport *m* de matière sèche par le rayonnement solaire
r соотношение *n* сухого вещества к облучённости

2653 dry matter to transpiration ratio
d Trockensubstanz/Transpiration-Verhltnis *n*
f rapport *m* de matière sèche par la transpiration
r соотношение *n* сухого вещества к испарению

2654 dry packing
(method of banding that uses a tight waterproof bandage packed with a chemical, either dry or in paste form)
d trockene Verpackung *f*
f injection *f* à sec
r сухая набивка *f*

2655 dry pruning
d Sammlerhieb *m*
f élagage *m* à sec
r обрезка *f* засохших ветвей; удаление *n* сушняка

2656 dry rot
(decay causing by fungi Merulius lacrymans

d Trockenfäule *f*
f pourriture *f* sèche
r сухая [бурая призматическая] гниль *f*

* **dry slide → 2648**

2657 **dry strength**
(of wood)
d Stärke *f* im trockenen Status; Festigkeit *f* im trockenen Zustand
f force *f* à l'état sec
r прочность *f* в сухом состоянии

* **dry-topped tree → 2273**

2658 **dry wood**
d trockenes Holz *n*; Dürrholz *n*; Trockenholz *n*
f bois *m* sec
r сухая древесина *f*

2659 **dry wood particles**
d trockene Holzpartikeln *fpl*
f plaquettes *fpl* de bois sec
r сухие древесные частицы *fpl*

2660 **dry wood weight**
d Dürrgewicht *n* des Holzes
f poids *m* de bois sec
r вес *m* сухой древесины; масса *f* сухой древесины

* **d.s.h. → 2465**

2661 **dual band-saw**
d Zweibandsägemaschine *f*
f scie *f* à ruban double
r ленточная пила *f* с двумя режущими кромками

* **dub → 83**

* **ducky willow → 6975**

2662 **dull edge; crown edge; bad bevel; rough edge; wane[y edge]**
(the presence of bark, or the lack of wood from any cause, on the edge or corner of a piece of lumber)
d Baumkante *f*; Schalkante *f*; Wahnkante *f*; Waldkante *f*; Fehlkante *f*
f flache *f*; joint *m* non déligné; bord *m* non avivé
r обзол *m*; неопиленная кромка *f*; рашпильный обрез *m*; зазубренный край *m*

2663 **dull-edged timber; rough-edged timber**
f bois *m* avivé brut
r брусья *mpl* с обзолом

* **dull Oregon grape → 1365**

* **dumetose → 934**

* **dumetum → 1108**

2664 **dummy [tree]**
(tree rigged to raise a spar tree for use in yarding)
d Dummy-Baum *m*
f arbre *m* grue
r вспомогательное дерево *n*, оснащаемое для установки трелёвочной мачты

* **dumose → 934**

* **dumous → 934**

* **dumping ground → 8664**

* **dumping place → 8664**

2665 **dumping pocket**
d Entladentasche *f*
f poche *f* de déchargement
r разгрузочный карман *m*

* **dumpsite → 9276**

* **dundathu pine → 6664**

2666 **dune afforestation**
d Dünenaufforstung *f*
f afforestation *f* de dunes
r облесение *n* дюн

2667 **dune forest**
d Dünenwald *m*
f forêt *f* dunale
r дюнный лес *m*

* **dune willow → 4360**

* **dunk tree → 1783**

* **dunnage [wood] → 8579**

2668 **duplex brush-cutter**
d Duplexallmäher *m*
f débroussailleuse *f* double
r сдвоенный каток-кусторез *m*

2669 durability of wood; strength of wood; timber strength; timber resistance
d Dauerhaltbarkeit *f* des Holzes; Festigkeit *f* des Holzes; Holzfestigkeit *f*; Resistenz *f* des Holzes
f durabilité *f* de bois; solidité *f* de bois
r прочность *f* древесины; стойкость *f* древесины

2670 durable wood
d Hartholz *n*; hartes Holz *n*
f bois *m* solide
r прочная древесина *f*

* **duramen → 1951**

2671 duramination; duraminization; pithness
(formation of duramen)
d Verkernung *f*
f duraminisation *f*
r образование *n* ядра

2672 Durand [white] oak; bastard [white] oak
(Quercus durandii; Quercus sinuata)
d Durand-Eiche *f*
f chêne *m* de Durand
r дуб *m* Дюранда

* **duranga → 6551**

2673 Durango pine
(Pinus durangensis)
d Durango-Kiefer *f*
f pin *m* de Durango
r сосна *f* Дюранго

* **durmast [oak] → 7464**

2674 duster
d Entstaubungsanlage *f*
f blutoir *m*
r пылеочиститель *m*

* **dust extraction → 2318**

2675 dusting apparatus; spraying apparatus; duster; sprayer
d Stäubegerät *n*; Bestäubungsgerät *n*; Stäubemaschine *f*
f appareil *m* à poudrer; soufflerie *f*
r опылительный аппарат *m*; опыливатель *m*

2676 dust palliatives
(chemicals or compounds applied to road surfaces to reduce dust created by traffic)
r средства *npl* для связывания пыли

* **dusty zenobia → 4350**

2677 Dutch elm; Belgian elm; cork-bark elm
(Ulmus hollandica)
d holländische Ulme *f*; belgische Ulme; Bastard-Ulme *f*
f orme *m* hollandais; orme de Hollande; orme belgique
r вяз *m* голландский

2678 Dutch elm disease
(Ceratocystis ulmi)
d Ulmenkrankheit *f*
f maladie *f* hollandaise de l'orme
r [голландская] болезнь *f* ильмов; цератоцистис *m* вязовый

* **dutchman → 1228**

2679 dutchman corner
d Dutchman-Winkel *m*
f angle *m* caché
r закрытый дорогой угол *m* осваиваемой лесосеки

2680 Dutchman [wedge]
(a block used to change direction of line lead)
d Dutchman-Keil
f coin *m* encastré
r вкладной клин *m*

2681 duty on wood; duty on timber; tariff on timber
d Holzzoll *m*
f taxe *f* prélevée sur le bois
r лесная пошлина *f*

* **dwarf → 366**

2682 dwarf almond
(Amygdalus nana)
d Zwergmandel
f amandier *m* nain
r миндаль *m* низкий

* **dwarf arctic birch → 892**

2683 dwarf blueberry; dwarf huckleberry; dwarf bilberry
(Vaccinium caespitosum)
d zwergartige Heidelbeere *f*
f airelle *f* gazonnante
r голубика *f* дернистая; голубика карликовая

d zwergartiger Brombeere *f*
f ronce *f* rabougrie
r ежевика *f* карликовая

* **dwarf bush-honeysuckle → 5788**

* **dwarf cherry → 7855**

* **dwarfed growth → 8223**

* **dwarfed wood → 2022**

* **dwarf elder → 769**

* **dwarf [fan] palm → 5384**

2685 dwarf forest
d zwergartiger Wald *m*
f forêt *f* rabougrie
r карликовый лес *m*

* **dwarf huckleberry → 2683**

* **dwarfing → 8223**

* **dwarf [mountain] pine → 5598**

2686 dwarfness; dwarfism; nanism
d Nanismus *m*
f nanisme *m*
r нанизм *m*; низкорослость *f*

2687 dwarf oak; ground-oak
(Quercus nana)
d Süntelbuche *f*
f chêne *m* nain
r карликовый дуб *m*

* **dwarf Oregon grape → 1365**

2688 dwarf palmetto; bush palmetto
(Sabal minor)
d Zwergpalmetto *m*
f petit palmier *m*; palmetto *m* nain
r карликовая пальметто *f*

* **dwarf pine → 5598**

2689 dwarf serviceberry
(Amelanchier spicata)
d ährige Felsenbirne *f*; Ährenfelsenbirne *f*
f amélanchier *m* à feuilles rondes
r ирга *f* колосистая; ирга метельчатая

* **dwarf shoot → 7548**

2690 dwarf subshrub
d Zwerghalbstrauch *m*
f sous-arbrisseau *m*
r полукустарничек *m*

* **dwarf sumac[h] → 3247**

2691 dwarf tree
(a ligneous plant which hight is less of 3 m)
d Zwergbaum *m*
f arbre *m* nain
r карликовое дерево *n*

* **dyeing wood → 2693**

2692 dyer's oak; yellow-bark oak; quercitron oak
(Quercus velutina; Quercus tinctoria)
d [amerikanische] Färbereiche *f*
f chêne *m* des teinturies; chêne quercitron; quercitron *m*
r дуб *m* [американский] красильный; дуб бархатный

2693 dyewood; dyeing wood
d Farbholz *n*
f bois *m* de teinture
r красильное дерево *n*; красильная древесина *f*

2694 dying[-off] of forests; forest death
d Waldtod *m*; Baumsterben *n*
f mort *m* des forêts
r отмирание *n* деревьев

2695 dying tree
d Kümmerer *m*; Kümmerling *m*
f arbre *m* mourant
r отмирающее дерево *n*

2696 dysgenic *adj*
(detrimental to the genetic quality of a population and future generations)
d dysgenisch
f dysgénique
r генетически опасный; дисгенический; вырождающийся; способствующий вырождению

E

2697 eared willow; round-ear willow
(Salix aurita)
d Ohrweide *f*; Öhrchenweide *f*
f saule *m* à oreillettes; petit marsault *m*; saule auriculé
r ива *f* ушастая

2698 earleaf acacia; Darwin blackwattle; blackwattle; Papuan wattle
(Acacia auriculiformis)
d Ohrblattakazie *f*
f acacia *m* auriculiforme
r ак[illegible]ция *f* загнутолистная

* **ear-leaf magnolia → 3688**

2699 early bark
d Frührinde *f*
f liber *m* initial
r ранняя кора *f*

2700 early burning
d rechtzeitiges Brennen *n*
f brûlage *m* précoce; brûlage hâtif; brûlage anticipé; incinération *f* préventive; brûlage préventif
r ранневесенний пуск *m* пала; своевременное сжигание *n*

* **early decay → 4494**

2701 early forest succession
(the early developmental stage or condition of a plant community)
d frühe Waldsukzession *f*
f succession *f* forestière initiale
r ранняя лесная сукцессия *f*

2702 early leaf
d Frühblatt *n*
f feuille *f* printanière
r ранневесенний лист *m*

* **early-low blueberry → 1788**

* **early-sweet blueberry → 1788**

2703 early thinning
d Durchläuterung *f*

2704 early[ ]wood; springwood; initial parenchyma
(the less dense, larger celled, first formed part of a growth ring)
d Früh[jahrs]holz *n*; Frühlingsholz *n*; Weitholz *n*
f bois *m* initial; bois de printemps; bois printanier
r ранняя древесина *f*; весенняя древесина

2705 earned harvest
(allows the timber manager an immediate increase in the allowable cut when the manager applies intensive management techniques that will accelerate future timber growth)
d erworbene Ernte *f*
f exploitation *f* gagnée
r заработанные лесозаготовки *fpl*

2706 earpod tree
(Enterolobium)
d Chimbuva-Baum *m*
f guanacaste *m*
r энтеролобиум *m*

* **earth → 7833**

2707 earth chute for summer use
d Sommerriesweg *m*
f glissoire *f* utilisable en été
r летний лесоспуск *m*

2708 earth chute for winter use; winter chute
d Winterriesung *f*; Winterriesweg *m*
f glissoire *f* hivernale
r зимний лесоспуск *m*; зимняя спускная дорога *f*

* **earth creep → 7835**

2709 earth[en] chute; ground slide; earth slide
(a furrow in the earth in which logs are dragged)
d Erdriese *f*
f glissoire *f* en sol naturel
r земляной [лесо]спуск *m*

* **East African antiaris → 6414**

2710 East African camphorwood
(Ocotea usambarensis)
d Stinkholz *n*
f muura *m*; camphre *m*
r окотеа *f* узамбарская; оцота *f* чёрная

d ostafrikanische Kordie *f*; Mukumari *m*
f mukumari *m*
r кордия *f* африканская

2712 **East African olive; African olive**
(Olea hochstetteri)
d Hochstetter-Olivenbaum *m*
f olivier *m* de Hochstetter
r маслина *f* Хохштетера

2713 **East African satinwood; olonvogo**
(Fagara macrophylla; Fagara tessmannii)
d Olonvogo *n*
f pigeon *m* malherbe
r фагара *f* крупнолистная; атласное дерево *n* африканское

* **East African silky oak → 3966**

2714 **East African yellowwood; podo**
(Podocarpus)
d Manio *n*; Steineibe *f*; Fußfruchtbaum *m*
f manio *m*
r ногоплодник *m*

* **eastern arborvitae → 5795**

2715 **eastern black walnut; black walnut**
(Juglans nigra)
d schwarze Walnuss *f*; Schwarznuss *f*
f noyer *m* noir
r орех *m* чёрный

2716 **eastern [Canadian] spruce; New Brunswick spruce; Quebec spruce; [swamp] black spruce; shortleaf black spruce**
(Picea mariana; Abies mariana; Abies nigra)
d [nordamerikanische] Schwarzfichte *f*
f épinette *f* noire [d'Amérique]; épicéa *m* noir [d'Amérique]; épinette à la bière
r ель *f* чёрная

2717 **eastern cottonwood; cottonwood; eastern poplar; necklace-poplar**
(Populus deltoides)
d Rosenkranzpappel *f*; virginische Pappel *f*; amerikanische Schwarzpappel *f*; kanadische Schwarzpappel
f peuplier *m* de Virginie; peuplier de la Caroline; peuplier noir d'Amérique; peuplier deltoïde
r тополь *m* дельтовидный; тополь канадский

* **eastern fir → 969**

* **eastern hemlock → 1296**

2718 **[eastern] hemlock looper**
(Lambdina fiscellaria)
d Hemlocktanne-Spanner *m*
f arpenteuse *f* de la pruche
r пяденица *f* гемлоковая

2719 **eastern hophornbeam**
(Ostrya virginiana)
d virginische Hopfenbuche *f*
f ostryer *m* de Virginie
r хмелеграб *m* виргинский

2720 **eastern larch beetle**
(Dendroctonus simplex)
d Lärchekäfer *m*
f dendroctone *m* du mélèze
r лубоед *m* лиственницы американской

* **eastern poplar → 2717**

2721 **eastern redbud**
(Cercis canadensis)
d kanadischer Judasbaum *m*; amerikanischer Judasbaum
f arbre *m* de Judée canadien; gainier *m* du Canada
r церцис *m* канадский

2722 **eastern red[ ]cedar; [Virginian] pencil cedar; red juniper**
(Juniperus virginiana; Juniperus lucayana)
d virginischer Wacholder *m*; virginische Bleistiftzeder *f*; amerikanische Bleistiftzeder
f cèdre *m* rouge; cèdre de Virginie; genévrier *m* de Virginie; genévrier rouge d'Amérique; genévrier aux crayons
r можжевельник *m* вирджинский; вирджинский карандашный кедр *m*; карандашное дерево *n*; красный кедр

2723 **eastern spruce; Oriental spruce**
(Picea orientalis)
d Sapindusfichte *f*; orientalische Fichte *f*; Kaukasusfichte *f*; morgenländische Fichte
f épicéa *m* d'Orient; sapinette *f* d'Orient
r ель *f* восточная

2724 **eastern spruce beetle; red-winged pine beetle; Sitka-spruce beetle**
(Dendroctonus piceaperda; Dendroctonus obesus)
d Sitkafichtenriesenbastkäfer *m*
f dendroctone *m* de l'épinette Sitka

* **eastern sweetshrub → 1806**

2725 **eastern thuja; Oriental thuja; Oriental arborvitae; Chinese arbor-vitae**
(Platycladus orientalis; Thuja orientalis; Biota orientalis)
d orientalischer Lebensbaum *m*; morgenländischer Lebensbaum
f thuya *m* d'Orient; thuya oriental; thuya de Chine
r восточная туя *f*; биота *f* восточная

* **eastern white cedar → 5795**

2726 **eastern white pine; [northern] white pine; Weymouth pine; cork-pine; Melancholy pine**
(Pinus strobus)
d Weimouthskiefer *f*; Weimouthsföhre; Strobe *f*
f pin *m* [du lord] Weymouth; pin baliveau; pin blanc
r сосна *f* веймутова; жёлтая сосна; квебекская сосна

2727 **East Indian ebony; true ebony; [black] ebony [tree]**
(Diospyros ebenum; Diospyros hebecarpa)
d echter Ebenholzbaum *m*; echtes Ebenholz *n*
f ébénier *m* [vrai]; ébénier d'Asie
r чёрное дерево *n*; хурма *f* эбеновая

2728 **East Indian rosewood; Indian rosewood; Bombay blackwood**
(Dalbergia latifolia; Dalbergia bariensis)
d [ost]indisches Rosenholz *n*; ostindisches Jacaranda *n*; ostindischer Palisander *m*
f palissandre *m* des Indes; palissandre d'Asie
r розовое дерево *n* индийское; дальбергия *f* широколистная

2729 **East Indian sapwood; East Indian satinwood; Ceylon satinwood; satinwood; billu**
(Chloroxylon swietenia)
d ostindisches Satinholz *n*; ostindisches Seidenholz *n*
f bois *m* satiné; citron *m* de Ceylan
r атласное дерево *n* [индийское]

2730 **East Indian walnut wood; koko wood; siris**
(Ehretia acuminata var. serrata)
d ostindische Nuss *f*; Kokko *n*; Siris *n*
f koko *m*; ehretia *m*
r эреция *f* заострённая

2731 **ebony family**
f ébénacées *pl*
r эбеновые *pl*

* **ebony [tree] → 2727**

* **eccentric peeling → 4093**

2732 **eccentric pith; eccentric heart[wood]**
d exzentrischer Kern *m*
f cœur *m* excentré
r смещенная сердцевина *f*

2733 **eccentric [stem] growth**
(wood defect)
d exzentrischer Stammwuchs *m*; Exzenter[stamm]wuchs *m*
f croissance *f* excentrique [du tronc]; développement *m* excentrique [du tronc]
r эксцентрический рост *m* ствола

* **ecclesiastical forest → 1575**

2734 **ecesis; oecesis; plant colonization**
(the process by which a plant or animal becomes established in a new habitat)
d Äquivalent *m*
f écésis *m*
r эцезис *m*; захват *m* организмом нового местообитания

2735 **eclectic thinning**
(removing an agreed proportion of trees of all sizes, species, ages and quality types throughout an area)
d eklektische Durchforstung *f*
f éclaircie *f* éclectique
r эклектичное прореживание *n*

* **ecoforestry → 2839**

* **ecological amplitude → 2740**

2736 **ecological balance; ecological stability; natural balance**
(state of dynamic equilibrium within a community of organisms in which genetic, species and ecosystem diversity remain relatively stable)
d ökologisches Gleichgewicht *n*
f équilibre *m* écologique; équilibre naturel
r экологический баланс *m*; экологическая устойчивость *f*

2737 **ecological integrity**
(quality of a natural unmanaged or managed ecosystem in which the natural ecological

f intégrité *f* écologique
r экологическая целостность *f*

2738 ecologically suitable species
(coniferous or deciduous tree species that are naturally adapted to a site's environmental conditions)
d ökologisch verwendbare Arten *fpl*
f espèce *f* écologiquement propre
r экологически подходящий вид *m*

2739 ecological niche
(the specific set of environmental/habitat conditions that permit the full development and completion of the life cycle of an organism)
d ökologische Nische *f*
f niche *f* écologique
r экологическая ниша *f*

* **ecological stability → 2736**

2740 ecological tolerance; ecological amplitude; ecological range
(range of one or more environmental conditions in which an organism or a process can function)
d ökologische Toleranz *f*
f amplitude *f* écologique
r экологическая амплитуда *f*

2741 ecology
(branch of science dealing with the relationships of organisms and their cycles and rhythms, community development, and environments)
d Ökologie *f*
f écologie *f*
r экология *f*

2742 economically operable forest stands
(for which log prices exceed harvesting costs, including profit and return to capital)
d ökonomisch funktionelle Waldbestande *mpl*
f peuplements *mpl* forestiers rentables
r рентабельные лесонасаждения *npl*

2743 economic diameter; merchantable diameter
(of trees)
d verwertbarer Mindestdurchmesser *m*; Grenzdurchmesser *m*
f diamètre *m* marchand minimal; plus petit diamètre utilisable
r рентабельный отпускной диаметр *m*; минимальный рентабельный диаметр

2744 economic forestry
d ökonomische Forstwirtschaft *f*
f foresterie *f* rentable
r рентабельное лесничество *n*

* **economic log selection → 2746**

* **economic maturity → 3146**

2745 economic rotation; financial rotation; physical rotation
d finanzielle Hiebsreife *f*
f exploitabilité *f* financière
r оборот *m* [рубки] по экономической спелости

2746 economic selection cutting; economic log selection
d wirtschaftlicher Blenderschlag *m*; wirtschaftlicher Sortimentshieb *m*
f coupe *f* économico-sélective; exploitation *f* par assortiments
r рентабельная выборочная рубка *f*; сортиментная рубка

2747 ecospecies
d Ökospezies *n*
f écoespèce *f*
r экологический вид *m*; эковид *m*

2748 ecosystem integrity
(soundness or wholeness of the processes and organisms composing the ecosystem)
d Ökosystem-Vollständigkeit *f*
f intégrité *f* de l'écosystème
r целостность *f* экосистемы

2749 ecosystem management
(use of an ecological approach to achieve productive resource management by blending social, physical, economic and biological needs and values to provide healthy ecosystems)
d Ökosystemverwaltung *f*
f gestion *f* de l'écosystème
r управление *n* экосистемы

2750 ecosystem productivity
(ability of an ecosystem to produce, grow or yield products - whether trees, shrubs or other organisms)
d Ökosystemproduktivität *f*
f productivité *f* de l'écosystème
r продуктивность *f* экосистемы

2751 ecotone
(a transition zone between two different plant

d Ökoton *m*
f écotone *m*
r экотон *m*; пограничное сообщество *n*

* **ecotype → 4079**

2752 **ecozone**
(area within which the climate conforms to a particular pattern, and hence, vegetation is generally of a particular kind)
d Ökozone *f*
f écozone *f*
r экозона *f*

2753 **ectotrophic mycorrhiza; ectomycorrhiza**
(mycorrhizal association in which the fungal hyphae form a compact mantle on the surface of the roots)
d ektotrophe Mykorrhiza *f*; Ectomycorrhiza *f*
f mycorhize *f* ectotrophe; ectomycorhize *f*
r эктотрофная микориза *f*

2754 **edaphic characteristics**
(of the soil environment (opposed to physiographic or climatic factors), which limits the existence or development of an organism or community)
d edaphische Charakteristiken *fpl*
f caractéristiques *fpl* édaphiques
r эдафические характеристики *fpl*; почвенные характеристики

* **edel → 1789**

* **edge → 819**

2755 **edge/area ratio**
(length of forest edge per cutblock area)
d Rand/Bereich-Verhältnis *n*
f rapport *m* lisière/zone
r [со]отношение *n* опушки к площадей леса

* **edge bend → 7936**

* **edge coupe → 6167**

2756 **edged; bordered**
d besäumt; vollkantig
f déligné; massicoté
r обрезной; окаймлённый; отточенный

2757 **edged timber; edged lumber**
d Beschnittholz *n*; besäumtes Schnittholz *n*; besäumtes Holz *n*

2758 **edge effect**
(habitat conditions created at or near the more-or-less well-defined boundary between ecosystems)
d Randeffekt *m*; Randwirkung *f*
f effet *m* de lisière
r влияние *n* соседнего сообщества

2759 **edge firing**
d Brennen *n* vom Rande aus
f incendie *m* par les bords
r пуск *m* пала от границ; сжигание *n* с краю (участка)

2760 **edge grain**
(referring to timber or veneer cut in a plan approximately at right angles to the annual rings)
d Spiegelschnitt *m*
f débit *m* sur maille; débité *f* sur quartier
r текстура *f* на кромке; распил *m* по кромке

* **edge-grain[ed] timber → 6678**

* **edge knot → 4733**

2761 **edge of a forest; forest edge; forest border; wooded border; forest skirt; forest margin; forest boundary; forest limit; stand margin; timber limit**
(a loose term for the limit beyond which trees cannot or do not occur)
d Wald[es]saum *m*; Waldrand *m*; Bestandesrand *m*; Waldgrenze *f*; bewaldeter Grenzstreifen *m*
f bordure *f* d'une forêt; lisière *f* de peuplement; orée *f* du bois; limite *f* boisée; limite forestière; zone *f* de combat
r опушка *f* леса; лесная кайма *f*; кромка *f* леса; контур *m* леса; край *m* насаждения

2762 **edge of cut; felling edge; felling border**
d Steilrand *m*
f front *m* de coupe
r режущее остриё *n*

2763 **edge of sample; arris of sample**
(of wood)
d Probenrand *m*; Stichprobenkante *f*
f rive *f* d'échantillon
r грань *f* образца; ребро *n* образца

* **edger → 2767, 4661**

2764 **edger operator**

f opérateur *m* de machine déligneuse
r станочник *m* обрезного станка; оператор *m* установки для обрезки досок

2765 edge rounding
d Kantenabrundung *f*
f chantournage *m* de rives
r округление *n* кромок

* **edge sander → 6567**

* **edge shake → 958**

* **edge tree → 922**

2766 edging
d Besäumung *f*
f délignage *m*
r обрезка *f* кромок; опиловка *f* кромок

* **edging and chipping cutter → 4321**

* **edgings → 5215**

2767 edging saw; edger
(sawmill machinery used to saw cants after they come off the headrig, squaring the edges and ripping the cants into lumber)
d Besäumungsmaschine *f*; Besäumanlage *f*; Besäumeinrichtung *f*; Besäumautomat *m*
f machine *f* déligneuse; déligneuse *f*
r установка *f* для обрезки; обрезное устройство *n*; обрезной станок *m*

* **education in forestry → 6565**

* **EFC → 2896**

2768 effective germinative capacity; effective germination power
(of grains)
d effektives Keimvermögen *n*
f faculté *f* germinative efficace; capacité *f* germinative efficace
r действительная всхожесть *f*

2769 effective seedling; effective stocking
(any seedling, whether natural or planted, that has survived in reasonable vigor for some arbitrary time and is so sited that it should make an effective contribution to the crop)
d wirkungsvoller Sämling *m*
f semis *m* établi; matériel *m* relatif efficace
r эффективный лесной сеянец *m*

2770 effect of undercut
r влияние *n* зарубки; напряжение *n* зарубки

2771 effects of chemicals on wood
d Effekte *mpl* der Chemikalien auf Holz
f effets *mpl* de produits chimiques sûr le bois
r воздействие *n* химикатов на древесине

* **egba → 5731**

2772 eggfruit tree
(Pouteria campechiana)
d Canistel *n*
f canistel *m*
r путерия *f* кампешевая

* **egg gallery → 4811**

2773 egg niche; egg pocket; egg notch
d Einische *f*; Eigrube *f*; Eikerbe *f*
f entaille *f* de ponte; niche *f* de ponte
r углубление *n* [в коре] для яиц [насекомых]; яйцевая ямка *f*

* **egg-shaped leaf → 5988**

* **egg tunnel → 4811**

2774 ehretia family
(Ehretiaceae)
d Ehretiagewächse *pl*
f ehrétiacées *pl*
r эрециевые *pl*

* **eik → 1793**

* **eis → 4592**

* **ekor kuda → 6784**

2775 elaeagnus willow; rosemary willow; hoary willow
(Salix elaeagnus; Salix incana)
d Grauweide *f*; Lavendelweide *f*; Uferweide *f*
f saule *m* tomenteux; saule incane; saule drapé
r ива *f* сизоватая

2776 elastic wood; flexible wood
d elastisches Holz *n*; biegsames Holz; plastisches Holz
f bois *m* élastique; bois flexible
r упругая древесина *f*; пластичная древесина

2777 Eldar pine
(Pinus eldarica)
d afghanische Kiefer *f*

* **elder → 2892**

2778 **elderberry**
(berrylike fruit of an elder used for e.g. wines and jellies)
d Holunderbeere *f*
f baie *f* de sureau
r бузинная ягода *f*

2779 **electric[-driven] saw; electric power-saw**
d elektrische Säge *f*
f scie *f* électrique
r электромоторная пила *f*

2780 **elephant tree; torote**
(Bursera microphylla)
d Weißgummibaum *m*
f arbre *m* éléphant
r бурсера *f* мелколистная

* **elevating truck → 3660**

2781 **elfin birch woodland**
d Birkenkrummholz *n*; Birkengebüschwald *n*
f forêt *f* rabougrie à bouleau
r берёзовое криволесье *n*

2782 **elfin forest; elfin wood**
(forest of small, often fantastically-shaped trees, generally covered with lichens or mosses)
d Gebüschwald *n m*
f forêt *f* de lutins
r карликовый густой лес *m*; стланик *m*

2783 **elfinwood belt**
d Krummholz- und Knieholzstufe *f*
f étage *m* des formations ligneuses rampantes et rabougries
r ярус *m* криволесий и стлаников

* **elite log → 6395**

2784 **elite seed stand**
d Elitesaatbestand *m*; Vorzugserntebestand *m*
f peuplement *m* producteur de graines élites
r элитный семенной древостой *m*; элитное семенное насаждение *n*

2785 **elite stand**
d Elitebestand *m*
f peuplement *m* d'élite
r элитное насаждение *n*

* **elite stem → 6395**

* **elkwood → 9040**

2786 **elm**
(Ulmus)
d Ulme *f*; Effe *f*; Rüster *f*
f orme *m*; ormeau *m*
r ильм *m*; берест *m*; карагач *m*; вяз *m*

2787 **elm-ash-cottonwood [forest type]**
(in which elm, ash, or cottonwood, singly or in combination, constitute a plurality of the stocking)
d Ulme-Esche-Pappel-Wald *m*
f forêt *f* de frênes, ormes et peupliers
r лес *m* из ильма, ясеня и тополи

* **elm bark beetle → 4794**

2788 **elm family**
(Ulmaceae)
d Ulmengewächse *pl*; Ulmazeen *pl*
f ulmacées *pl*
r берёстовые *pl*; ильмовые *pl*

2789 **elm forest**
d Ulmenwald *m*
f ormaie *f*
r ильмовый лес *m*; вязовая роща *f*; вязовник *m*

2790 **elm leaf miner**
(Fenusa ulmi)
d Ulmenblattminiermotte *f*
f tenthrède *f* mineuse de l'orme
r пилильщик *m* минирующий ильмовый

2791 **elm wood**
d Ulmenholz *n*
f bois *m* d'orme
r ильмовая древесина *f*

2792 **elm zelkova**
(Zelkova carpinifolia)
d kaukasische Zelkove *f*
f orme *m* du Caucase; orme à feuilles de charme; orme de Sibérie
r дзельква *f* граболистная

* **elupa → 5244**

2793 **embankment**
d Damm *m*
f endiguement *m*; endiguage *m*
r ограждение *n*

2794 **embedded grit; grit**

d eingebettetes Grit *n*
f sables *m* encastrés
r вкрапленные твёрдые частицы *fpl*

2795 embryo
(tiny, developing plant inside the seed)
d Embryo *m*; Keimling *n*; Keim *m*; Fötus *m*
f embryon *m*; germe *m*; fœtus *m*
r эмбрион *m*; зародыш *m*

* **embuya → 1008**

2796 emergence [of a plant]; plant emergence
(appearance of the plant above ground after germination in the soil)
d Aufgehen *n*
f émergence *f*
r всхожесть *f* растения; появление *n* всходов

2797 emergent [tree]
(tree whose crown at maturity projects well above the level of the highest canopy)
d Wipfelbaum *m*
f émergent *m*
r дерево *n* с хорошо разветвлённой вершиной

* **emodi pine → 1543**

* **emongi → 5731**

2798 empirical yield table
(representation of actual and forecast volumes at different stocking levels per age class)
d empirische Ertragstafel *f*
f table *f* de rendement empirique
r таблица *f* роста средних насаждений; таблица роста обыкновенных насаждений

2799 employment in forestry
d Beschäftigung *f* in der Forstwirtschaft
f emploi *m* dans la sylviculture
r занятость *f* в лесном хозяйстве

* **empress tree → 7122**

* **empty rope profile → 1243**

* **empty rope sag → 1243**

* **encased knot → 833**

* **enclosed knot → 833**

2800 enclosing
d Umzäunung *f*

* **enclosure → 5019**

* **endangered plant communities → 8531**

2801 endangered species; threatened species; red-listed species
(plant or animal population that is declining to the extent it may be extinct if it is not protected by laws or regulations)
d vom Aussterben bedrohte Art *f*
f espèce *f* menacée; espèce en danger de disparition
r вид *m*, находящийся под угрозой исчезновения; угрожаемый вид

* **end bud → 8470**

* **end check → 958**

2802 endemic species
(a species whose natural occurrence is confined to a certain region and whose distribution is relatively limited)
d einheimische Art *f*
f espèce *f* endémique
r эндемичный вид *m*; эндемик *m*

2803 end grain; end face; slash grain; SG
(the grain shown on a cross cut surface)
d Hirnfläche *f*; Stirnfläche *f*
f veinure *f* sur coupe transversale de bois; fil *m* d'extrémité
r торцевое волокно *n*

* **end-grain cutting → 2814**

* **end-grain[ed] wood → 2051**

2804 end grain turning
(turning in the end of a workpiece, which has the grain running parallel with the axis of the lathe)
d Hirnholzumlauf *m*; Querschnittumlauf *m*
f évolution *f* transversale en bois de bout
r поперечное вращение *n* [торцевого волокна]

2805 end haul[ing]
(moving excavated material from one section of the road to another or to a disposal site, during road construction or modification)
d Langrücken *n*
f débardage *m* en longueur
r длинная трелёвка *f*

2806 end hooks

d Endehaken *m*
f pince *f* d'extrémité
r торцевые крюки *mpl*

* **endless chain log haul-up → 4173**

2807 end mark
d Waldhammerzeichen *n* von Stirnseite
f empreinte *f* du marteau forestier à l'extrémité
r метка *f* [лесного молотка] на торце; клеймо *n* на торце

2808 endmast
d Endemast *m*
f mât *m* de fin
r растущее или срезанное дерево *n* в качестве промежуточной опоры

* **end moisture → 3141**

2809 endoderm[is]
(single layer of cells between the cortex and pericycle in the root)
d Endoderm *n*
f endoderme *m*
r эндодерма *f*

2810 endogenic; endogenous
(development within (endogenesis), that is, forming new tissue within the plant)
d endogenisch
f endogène
r эндогенный

2811 endogen wood
(the wood in the form of bundles or threads, irregularly distributed throughout the whole diameter, not forming annual layers, and with no distinct pith)
d Endogenholz *n*
f bois *m* endogène
r лесоматериалы *mpl* из внутренних слоёв дерева

* **endomycorrhiza → 2813**

2812 endosperm
(nutritive tissue in a seed, in angiosperms triploid and formed in the embryo sac after fertilization, in gymnosperms haploid and derived from the sterile portion of the female gametophyte)
d Endosperm *n*
f endosperme *m*
r эндосперм *m*

hyphae are present on root surfaces as individual threads that may penetrate directly into root hairs, other epidermal cells and occasionally into cortical cells)
d endotrophe Mykorrhiza *f*
f mycorhize *f* endotrophe; endomycorhize *f*
r эндотрофная микориза *f*

2814 end section; end-grain cutting
d Hirnschnitt *m*
f travers *m* du fil
r торцевой разрез *m*

* **end shake → 958**

2815 end support; tailhold; tailholt; offset tower; tail spar; back spar
(in cable logging, the anchorage at the outer end of the skyline away from the landing)
d Endstütze *f*
f support *m* terminal; appui *m* d'extrémité
r концевая опора *f*; тыловая опора

2816 end-use moisture content
d Gebrauchsfeuchte *f*
f teneur *f* en humidité exploitable
r эксплуатационная влажность *f*

2817 end wall
d Endwand *f*
f paroi *f* terminale
r торцевая стенка *f*; торцовая стена *f*

2818 energy chips
(whole-tree chips used for energy)
d Energiespäne *mpl*
f copeaux *mpl* énergétiques
r энергетическая щепа *f*

2819 energy forest plantation; energy stand
(an silvicultural system aimed at producing plants as an energy source continuously from year to year)
d Energieforstplantage *f*
f forêt *f* énergétique
r энергетическая лесная плантация *f*

2820 energy wood
(wood that has been delivered to pulp and paper mills, specifically for burning in boilers)
d Energieholz *n*
f bois *m* d'énergie; bois pour la production d'énergie
r энергетическая древесина *f*

d Engelmannskiefer *f*
f pin *m* d'Engelmann
r сосна *f* Энгельмана

2822 **Engelmann spruce; mountain spruce; silver spruce**
(Picea engelmannii)
d Engelmannsfichte *f*
f épicéa *m* d'Engelmann; épinette *f* d'Engelmann; sapin *m* de l'Arizona
r ель *f* Энгельмана

2823 **Engelmann spruce beetle**
(Dendroctonus engelmanni)
d Engelmannsfichtenkäfer *m*
f dendroctone *m* de l'épinette d'Engelmann
r лубоед *m* ели Энгельмана

2824 **Engelman spruce weevil**
(Pissodes engelmanni)
d Engelmannsfichtenrüssler *m*
f charançon *m* du pin d'Engelmann
r смолёвка *f* ели Энгельмана

2825 **engineered wood**
d technisches Holz *n*
f bois *m* de haute technologie
r высокотехнологическая древесина *f*

2826 **engineered wood products**
(composite wood product made from glued fiber, lumber and/or veneer to meet specific design criteria)
d technische Holzprodukte *npl*
f produits *mpl* du bois de haute technologie
r продукты *mpl* высокотехнологической древесины

* **engineering fabric → 3831**

* **English ash → 2903**

* **English [brown] oak → 1793**

* **English elm → 2893**

2827 **English field maple; field maple; hedge maple**
(Acer campestre)
d Feldahorn *m*; Massholder *m*; Hecken-Ahorn *m*; Masserle *f*; Kreuzbaum *m*
f érable *m* champêtre; érable commun; auzérolle *f*
r клён *m* полевой

* **English holly → 2899**

2828 **English laurel; [common] cherry laurel; spurge laurel**
(Laurocerasus officinalis; Prunus laurocerasus)
d Kirschlorbeer *m*; pontische Lorbeerkirsche *f*; Lorbeerseidelbast *m*
f laurier-cerise *m*; laurier-palme *m*; daphné *m* lauréole; lauréole *m*; daphné morillon; laurier *m* des bois
r лавровишня *f* [лекарственная]

* **English lime → 1786**

* **English plane → 2906**

* **English walnut → 1808**

* **English yew → 1814**

* **engraver beetles → 539**

2829 **enriched fallow**
(a form of agroforestry in which useful, mainly woody species are sown or planted before cultivation ceases)
d angereicherte Waldbrache *f*
f jachère *f* enrichie
r обогащённый пар *m*; улучшенный пар

2830 **enrichment plantation; enrichment planting; improvement planting**
(any planting done to improve the value of a stand, and not to establish a regular plantation)
d Verbesserungsplantage *f*
f plantation *f* d'enrichissement
r улучшающие посадки *fpl* (для увеличения доли участия ценных пород)

2831 **entire bud**
d gesamte Knospe *f*; Vollknospe *f*
f bourgeon *m* plain
r полная почка *f*

2832 **entire[-kind] leaf**
(a leaf without teeth, lobes or divisions)
d Vollblatt *n*
f feuille *f* entière
r цельный лист *m*

2833 **entire leaf margin**
(with smooth, untoothed edges)
d Vollblattrand *m*

2834 **entire tree length; full-length log; tree-length [log]; tree-length stem**
(entire tree, excluding the unmerchantable top and limbs)
d Vollholz *n*
f tronc *m* entier; rondin *m* en longueur d'arbre
r целый хлыст *m*

2835 **entire tree trunk**
d Vollholzstamm *m*
f tronc *m* d'arbre entier
r ствол *m* с корнями

2836 **entrance of slide; slide entrance**
(the opening in a splash dam through which logs pass)
d Riesmund *m*
f entrée *f* d'une glissoire; sortie *f* d'une glissoire
r горло *n* древоспуска

2837 **enumeration of trees; counting of trees; telling of trees; tree tally; stem counting**
(system of recording trees counted during a timber cruise)
d Baumauszählung *f*; Stammabzählen *n*; Bestandesaufzählung *f*
f énumération *f* d'arbres; comptage *m* d'arbres
r перечёт *m* деревьев; перечёт стволов; стволовой перечёт

2838 **enumeration [survey]; valuation survey**
(of a forest)
d Bestandesauszählung *f*
f inventaire *m* [forestier] statistique; comptage *m* du peuplement
r перечислительная таксация *f*; перечёт *m* [насаждения]

* **environmental conservancy → 2842**

2839 **environmental forestry; ecoforestry**
(forestry application of environmental principles in conducting commercial forestry on company lands)
d Umgebungsforstwirtschaft *f*; ökologischer Waldbau *m*
f foresterie *f* d'environnement; écoforesterie *f*
r природоохранное лесное хозяйство *n*; экологическое лесное хозяйство; экологическое лесничество *n*

* **environmental harvesting [of timber] → 7193**

2840 **environmental impact analysis**
(activity that considers, estimates, and natural, and social factors, the probable effects of a plan or proposed project upon local systems)
d Umweltverträglichkeitsstudium *n*
f étude *f* d'impact sur l'environnement; étude d'altéralogie
r анализ *m* влияния на окружающую среду; анализ воздействия на окружающую среду

2841 **environmentally sensitive area; ESA**
(area requiring special management attention to protect important scenic values, fish and wildlife resources, historical and cultural values, and other natural systems or processes)
d ökologisch anfälliges Gebiet *n*
f zone *f* écologiquement sensible; région *f* écologiquement sensible; zone écologiquement fragile; région écologiquement fragile
r экологически уязвимая зона *f*; район *m* с легкоуязвимой природной средой; район, уязвимый в экологическом отношении

2842 **environmental protection; environment conservation; environmental conservancy; preservation**
(action of reserving, protecting or safeguarding a portion of the natural environment from unnatural disturbance)
d Naturschutz *m*; Umweltschutz *m*; Umweltsicherung *f*
f conservation *f* de l'environnement; protection *f* de l'environnement
r охрана *f* природы; защита *f* окружающей среды; сохранение *n* природы; предохранение *n* природы

2843 **environmental rehabilitation**
(measures undertaken to remedy environmental damage done to the land)
d Umweltneugestaltung *f*
f remise *f* en état de l'environnement
r восстановление *n* окружающей среды

* **environment conservation → 2842**

2844 **ephedra**
(Ephedra)
d Meerträubel *n*; Meerträubchen *n*
f éphédra *m*; raisin *m* de mer
r хвойник *m*; эфедра *f*

2845 **ephedra family**
(Ephedraceae)
d Meerträubelgewächse *pl*

2846 ephedroid perforation plate
(a plate having a small group or circular openings (as in Ephedra))
d ephedroidische Perforationsplatte *f*
f cloison *f* perforée éphédroide; perforation *f* foraminée
r эфедроидная перфорационная пластинка *f* (в древесине)

2847 ephemeral plant
(short-lived or transitory plant)
d ephemere Pflanze *f*
f plante *f* éphémère; plante saisonnière
r однолетнее растение *n*

* **ephemeral stream → 4555**

* **epicormic → 77**

2848 epicormic branch; epicormic shoot; epicormic sprout; stem shoot; deferred shoot
(branch growing from the main stem or branch of a tree, and arising from dormant buds under the bark)
d Klebast *m*; Stammausschlag *m*; Präventivspross *m*
f pousse *f* préventive; branche *f* gourmande; rejet *m* de tige
r побег *m*, развившийся из спящей почки; стволовой побег; побег от спящей почки

2849 epidemic mass attack; mass infection; outbreak
(of injurious insects)
d Massenauftreten *n*
f attaque *f* massive; pullulation *f*
r массовое нападение *n*

* **epigeous → 11**

2850 epigeous germination
d epigäische Keimung *f*
f germination *f* épigée
r образование *n* почки на поверхности листа

2851 epinasty
(more rapid growth or elongation of the upper side of a plant organ)
d Epinastie *f*
f épinastie *f*
r эпинастия *f*

* **epiphloedal adj → 1974**

* **epiphloem → 6157**

(a plant growing on living leaves)
d Luftpflanze *f*; Aerophyt *m*; Überpflanze *f*
f plante *f* épiphylle
r растительный эктопаразит *m*; растущий на листе эктопаразит; расположенный на листьях эктопаразит

2853 epiphyte
(a plant growing on but not nourished by another plant)
d Epiphyt *m*
f épiphyte *m*
r эпифит *m*

2854 episodic growth
d episodisches Wachstum *n*
f croissance *f* épisodique
r эпизодический рост *m*

2855 epixylous; xylogenous
d xylogen; epixyl
f xylogène; épixyle
r растущий на древесине

* **equatorial rain forest → 8978**

2856 equilibrium moisture content of wood
(moisture content at which timber neither gains nor loses moisture from the surrounding atmosphere)
d Holzfeuchthegleichgewicht *n*; Gleichgewichtsfeuchte *f* des Holz
f humidité *f* du bois équilibre
r равновесная влажность *f* древесины

2857 eradication; uprooting
(eliminating the pathogen after it is established or by eliminating the plants that carry the pathogen)
d Ausrottung *f*; Wurzelausziehen *n*; Wurzelgraben *n*; Rodung *f*; Entwurzelung *f*
f éradication *f*; déracinement *m*
r выравнивание *n* с корнем; выкорчёвывание *n*; корчевание *n*

* **eradicator → 8808**

2858 erect habit
d aufrechte Wuchsform *f*
f port *m* dressé; port érigé
r прямостоячий габитус *m*

2859 eremacausis
(a gradual oxidation from exposure to air and moisture, as in the decay of old trees or of

f érémacausis *m*
r постепенное загнивание *n* под действием атмосферы

2860 Erman birch; gold birch
(Betula ermanii)
d Ermansbirke *f*; Gold-Birke *f*
f bouleau *m* d'Erman
r берёза *f* Эрмана; берёза каменная

2861 erosion
(group of processes whereby earth or rock material is worn away, loosened, or dissolved and removed from any part of the Earth's surface by wind, water, ice, or gravity)
d Erosion *f*
f érosion *f*
r эрозия *f*; выветривание *n*

* **ESA → 2841**

2862 escape
(a plant)
d Flüchtling *m*
f plante *f* subspontanée
r одичавшее экзотическое растение *n*

2863 escape-cover[t]
(vegetation dense enough to aid animals in escaping from potential enemies, including hunters; a synonym for security, security area, security cover, and hiding cover, but imprecise)
d Schutzdeckung *f*
f couvert *m* de refuge
r предохранительный настил *m*

2864 escape route; get-away route
(faller's route away of the tree)
d Laufweg *m* des Fällers
f chemin *m* d'évacuation; parcours *m* d'évacuation; itinéraire *m* d'évacuation
r пут *m* отхода (вальщика от дерева)

2865 espalier tree
d Spalierbaum *m*
f espalier *m*
r шпалера *f*

2866 essential oil; volatile oil; ethereal oil
d ätherisches Öl *n*
f huile *f* essentielle; essence *f*; huile volatile
r эфирное масло *n*; летучее масло

* **essingang → 1120**

f création *f* d'un peuplement mélangé
r образование *n* смешанного насаждения

* **establishing of stand → 375**

2868 establishment period; formation period
(time elapsing between the initiation of a new crop and its establishment)
d Begründungszeitraum *m*
f période *f* d'implantation; durée *f* de la régénération
r период *m* приживаемости

2869 estimate by eye; ocular estimate; ocular estimation
d Schätzung *f* nach Augenmaß; flüchtige Aufnahme *f*; Augenschätzung *f*; Okularschätzung *f*
f estimation *f* à vue [d'œil]
r окулярная оценка *f*; глазомерная таксация *f*; учёт *m* на глаз

* **estimating crew → 8288**

2870 estimation of forest soil; forest soil evaluation
d Waldbodenschätzung *f*
f estimation *f* du sol forestier
r оценка *f* лесной почвы

* **ethereal oil → 2866**

2871 ethnobiomedical forest reserve
d ethnobiomedizinische Waldreserve *f*
f réserve *f* forestière ethnobiomédicale
r этнобиомедицинский заповедник *m*

2872 eucalypt[us]; box
(Eucalyptus)
d Eukalyptus[baum] *m*
f eucalyptus *m*
r эвкалипт *m*

2873 eucalyptus forest
d Eukalyptuswald *m*
f eucalyptaie *f*
r эвкалиптовый лес *m*

2874 eucalypt[us] oil
d Eukalyptusöl *n*
f essence *f* d'eucalyptus
r эвкалиптовое масло *n*

2875 eucommia family
(Eucommiaceae)
d Eucommiagewächse *pl*

2876 **eucryphia family**
(Eucryphiaceae)
d Eucryphiagewächse *pl*
f eucryphiacées *pl*
r эукрифиевые *pl*

2877 **eugenic** *adj*
(favorable to the genetic quality of a population)
d eugenisch
f eugénique
r евгенический

2878 **European alder; grey alder; white alder; speckled alder**
(Alnus incana)
d Grauerle *f*; Weißerle *f*
f au[l]ne *m* blanc; au[l]ne de montagne
r ольха *f* серая; ольха белая

2879 **European apple sawfly; apple sawfly**
(Hoplocampa testudinea)
d Apfelsägewespe *f*
f hoplocampe *m* des pommes
r пилильщик *m* яблонный [плодовый]

2880 **European aspen; Swedish aspen; [Finnish] aspen; French aspen; trembling aspen**
(Populus tremula)
d Zitterpappel *f*; Zitteresche *f*; Espe *f*
f peuplier *m* tremble; tremble *m*
r осина *f*; тополь *m* дрожащий

* **European barberry → 1758**

* **European basswood → 1786**

* **European beech → 1760**

2881 **European bird-cherry; bird-cherry; pin-cherry; fire-cherry; cluster-cherry; [wild] red cherry**
(Prunus padus; Prunus racemosa)
d Ahlbaum *m*; Ahlkirsche *f*; Kitschbaum *m*; [gewöhnliche] Traubenkirsche *f*; europäische Traubenkirsche
f cerisier *m* à grappes; putier *m*; putiet *m*; bois *m* puant; merisier *m* à grappes
r [обыкновенная] черёмуха *f*; черёмуха кистевая

2882 **European [black] alder; black alder; common alder; sticky alder**
(Alnus glutinosa)
d Schwarzerle *f*; Roterle *f*; [gemeine] Erle *f*;
glutineux
r ольха *f* [чёрная]; ольха клейкая

2883 **European black currant**
(Ribes nigrum)
d schwarze Johannisbeere *f*; schwarze Ribisel *f*; Cassis *m*
f groseille *f* à grappes noires; groseille noire; cassis *m*
r смородина *f* чёрная

* **European black pine → 415**

2884 **European black poplar; black poplar; home-grown poplar**
(Populus nigra)
d Schwarzpappel *f*; Feldpapel *f*
f peuplier *m* noir; liardier *m*; liard *m*
r тополь *m* чёрный; осокорь *m*

2885 **European bladdernut**
(Staphylea pinnata)
d gefiederte Pimpernuss *f*; wilde Pimpernuss; gemeine Klappernuss *f*
f pistachier *m* sauvage; faux pistachier; staphylier *m* d'Europe
r стафилея *f* перистая; клекачка *f* пятилистная

2886 **European box[wood]; [common] boxwood; evergreen boxwood; Abassian box; box [tree]; Turkey box**
(Buxus sempervirens)
d [immergrüner] Buchs[baum] *m*; immergrüner Schneeball *m*; Buchsholz *n*
f buis *m* [commun]; bois *m* béni[t]; buis panaché; buis boule
r самшит *m* [вечнозелёный]; европейская пальма *f*; турецкая пальма

2887 **European chestnut; sweet chestnut; [Spanish] chestnut**
(Castanea vesca; Castanea sativa)
d Edelkastanie *f*; Eßkastanie *f*; Kastanie *f*; Marone *f*
f châtaignier *m*; marron *m* comestible
r каштан *m* [посевной]; каштан съедобный; каштан европейский; каштан благородный

2888 **European cotoneaster; common cotoneaster**
(Cotoneaster vulgaris; Cotoneaster integerrimus)
d gemeine Zwergmispel *f*; echte Zwergmispel
f cotonéaster *m* commun; cotonnier *m* à feuilles

2889 European cranberry; small cranberry; moorberry; fenberry
(Vaccinium oxycoccos; Oxycoccus palustris)
d Kranbeere *f*; Moosbeere *f*; Sumpfbeere *f*; Torfbeere *f*
f airelle *f* canneberge; canneberge *m*; myrtille *f* des marais
r клюква *f* [болотная]

2890 European cranberry[bush[; water elder
(Viburnum opulus)
d gemeiner Schneeball *m*; gewöhnlicher Schneeball; gefüllter Schneeball; Wasserschneeball *m*
f viorne *f* obier; viorne aquatique; boule *m* de neige; sureau *m* d'eau
r калина *f* обыкновенная; калина красная

2891 European dewberry; dewberry
(Rubus caesius)
d Acker-Brombeere *f*; Kratzbeere *f*
f ronce *f* bleuâtre
r ежевика *f* сизая

2892 European elder; [common] elder; bourtree
(Sambucus nigra)
d [schwarzer] Holunder *m*; Holler *m*; Baumholder *m*; Fliederbaum *m*; Aashornbaum *m*; Alhornbaum *m*
f sureau *m* [noir]; sureau commun; susier *m*
r бузина *f* [чёрная]

2893 European elm; English elm; common elm; field elm; nave elm; September elm; smoothleaf elm; smooth-leaved elm
(Ulmus procera; Ulmus campestris; Ulmus serotina; Planera carpinifolia; Ulmus foliaceae)
d englische Ulme *f*; Feldulme *f*; Feldrüster *m*; glattblättrige Rüster *m*; Rotrüster *m*; gemeine Ulme
f orme *m* d'Angleterre; orme commun; orme champêtre; arbre *m* à pauvre homme
r вяз *m* стройный; вяз полевой; ильм *m* поздний; берест *m* листоватый; вяз листоватый

2894 European elm bark
(Scolytus multistriatus)
d kleiner Ulmensplintkäfer *m*
f [petit] scolyte *m* européen de l'orme
r заболонник *m* вязовый; заболонник струйчатый

* **European fan palm → 5384**

2895 European fly honeysuckle; fly honeysuckle; dwarf honeysuckle; fly woodbine
(Lonicera xylosteum)
d gemeine Heckenkirsche *f*; Ahlkirsche *f*; Frauenholz *n*; Hundsbeere *f*; Hundskirsche *f*; Knochenholz *n*; rote Heckenkirsche *f*
f chèvrefeuille *m* à balais; camerisier *m* à balais; chèvrefeuille des haies
r жимолость *f* лесная; жимолость обыкновенная; жимолость пушистая; ягода *f* волчья

2896 European Forestry Commission; EFC
d europäische Forstkommission *f*
f Commission *f* européenne des forêts
r Европейская Комиссия *f* по Лесоводству

2897 European green alder; green alder
(Alnus viridis)
d Grünerle *f*; Alpenerle *f*
f au[l]ne *m* vert; au[l]ne gris d'Amérique; au[l]ne alpin; au[l]ne crispé
r ольха *f* зелёная

2898 European hackberry; Mediterranean hackberry; nettle tree; hackberry
(Celtis australis)
d südlicher Zürgelbaum *m*; europäischer Zürgelbaum; Honigbeerenbaum *m*
f micocoulier *m* de Provence; celtis *m*; fanabrégié *m*
r каменное дерево *n*; обрастница *f*; каркас *m* южный

* **European hazel → 1778**

2899 European holly; common holly; English holly; holly [tree]
(Ilex aquifolium)
d [gemeine] Hülse *f*; [gemeine] Stechpalme *f*; gewöhnliche Stechhülse *f*; Winterbeere *f*
f houx *m* [commun]
r падуб *m* [остролистный]; остролист *m*

2900 European hophornbeam; hophornbeam
(Ostrya carpinifolia)
d [europäische] Hopfenbuche *f*; gemeine Hopfenbuche
f charme *m* noir; charme-houblon *m*; ostrie *f* à feuille de charme; ostrya *m*; faux charme *m*
r хмелеграб *m* [обыкновенный]

2901 European hornbeam; common hornbeam; hornbeam; hardbeam; white beech
(Carpinus betulus)
d [gemeine] Hainbuche *f*; Weißbuche *f*; Steinbuche *f*; Hagebuche *f*
f charme *m* commun; charme [faux]; charme

* **European horse[ ]chestnut → 1781**

* **European house-borer → 4389**

* **European house longhorn → 4389**

* **European larch → 1785**

* **European lime → 1786**

2902 **European mistletoe; white mistletoe; [oak] mistletoe**
(Viscum album)
d Drudenfuß *m*; Hexenbesen *m*; Donnerbesen *m*; [Leim-]Mistel *m*
f gui *m*
r омела *f* белая

2903 **European mountainash; [English] ash; rowan [tree]**
(Sorbus aucuparia; Pyrus aucuparia)
d [nordische] Eberesche *f*; Quitschbeere *f*; Quitze *f*; Vogelbeere *f*; Vogelbeerbaum *m*; Drosselbeere *f*
f sorbier *m* [des oiseaux]; sorbier des oiseleurs
r рябина *f* [обыкновенная]; рябина сибирская

* **European oak → 1793**

2904 **European olive[-tree]; olive**
(Olea europaea; Olea sativa)
d Ölbaum *m*; Olivenbaum *m*; Olive *f*
f olivier *m* [d'Europe]
r оливковое дерево *n*; маслина *f* [европейская]; олива *f*

2905 **European pine sawfly; fox-colo[u]red sawfly; lesser pine sawfly**
(Diprion sertifer)
d rote Kiefernbuschhornblattwespe *f*; rotgelbe Kiefernbuschhornblattwespe
f diprion *m* du pin sylvestre
r пилильщик *m* рыжий сосновый

2906 **European plane; English plane; London plane[ ]tree; London sycamore tree; mapleleaf sycamore**
(Platanus hybrida; Platanus acerifolia)
d europäische Platane *f*; Hybrid-Platane *f*; ahornblättrige Platane *f*; gemeine Platane; Bastard-Platane *f*
f platane *m* commun; platane [hybride] à feuille d'érable
r платан *m* клёнолистный

2907 **European red elder; red elderberry;**
Sambucus callicarpa)
d Traubenholunder *m*; roter Holunder *m*; Bergholunder *m*; Waldholunder *m*; Hirschholunder *m*; Korallenholunder *m*
f sureau *m* rouge [d'Europe]; sureau à grappes; sureau rameux; sureau à fruits rouges; sureau pubescent; sureau de montagne
r бузина *f* красная; бузина обыкновенная

* **European [silver] birch → 1761**

* **European silver fir → 1803**

2908 **European smoketree; [common] smoketree; Venetian sumac[h]; smoke[-]tree; Aaron's beard**
(Cotinus coggygria; Psorothamnus spinosus; Dalea spinosa; Rhus cotinus)
d Perückenbaum *m*; Perückenstrauch *m*; Fisettholz *n*; Gelbholzsumach *m*
f fustet *m* [commun]; sumac *m* fustet; arbre *m* à perruque; barbe *m* de Jupiter
r скумпия *f* кожевенная; дерево *n* жёлтое; дерево физетовое; жёлтинник *m*; рай-дерево *n*; сумах *m* венецианский; сумах дубильный

* **European spindle[ ]tree → 1383**

* **European spruce → 1805**

2909 **European spruce beetle**
(Dendroctonus micans)
d Riesenbastkäfer *m*
f dendroctone *m* de l'épinette
r лубоед *m* еловой большой

2910 **European spruce sawfly**
(Gilpinia polytoma; Diprion hercyniae; Diprion polytomus)
d Fichtenbuschhornblattwespe *f*
f tenthrède *f* européenne de l'épinette; diprion *m* européen de l'épinette
r пилильщик *m* еловый общественный

* **European walnut → 1808**

* **European white birch → 1761**

* **European white elm → 7137**

2911 **European white poplar; white poplar; silver[-leaf] poplar; abele [poplar]**
(Populus alba)
d Weißpappel *f*; Silberpappel *f*
f peuplier *m* blanc [pyramidal]; peuplier

* **European willow → 1809**

* **European yew → 1814**

2912 **eutrophic; rich in nutrients**
(characteristic of waters rich in dissolved inorganic or mineral nutrient materials and exhibiting a paucity or complete lack of oxygen in the bottom waters)
d eutrophisch; nährstoffreich
f eutrophique; riche en matière nutritive
r нормально питающийся; богатый питательных веществ

2913 **eutrophication**
(enrichment of water bodies with plant nutrients, resulting in growth of algae, other plants and (eventually) animals in the water)
d Eutrophierung *f*
f eutrophisation *f*
r эвтрофикация *f*

2914 **evaporation**
(the process by which any substance is converted from a liquid state into, and carried off in, vapo[u]r)
d Verdunstung *f*
f évaporation *f*
r испарение *n*

2915 **evaporative cooling**
(absorption of heat from the atmosphere during evapotranspiration)
d Verdunstungskühlung *f*
f refroidissement *m* par évaporation d'eau
r испарительное охлаждение *n*

2916 **evapotranspiration; consumptive water use**
(the loss of water through a plant's leaves, where it evaporates)
d Gesamtverdunstung *f*; Evapotranspiration *f*; aktuelle Verdunstung *f*
f évapotranspiration *f*; consommation *f* [absolue] d'eau
r эвапотранспирация *f*

* **even-aged forest → 2920**

2917 **even-aged [forest] management**
(growing, harvesting, and reproducing timber in stands of trees of essentially the same age, at least within 10-year classes)
d gleichaltrige Waldbewirtschaftung *f*
f gestion *f* de forêt équienne
r ведение *n* одновозрастного хозяйства

proportion of the trees falls into the same age class, generally resulting in a single storey forest)
d gleichaltriger Hochwald *m*
f futaie *f* pleine
r одновозрастной высокоствольный лес *m*

2919 **even-aged silvicultural system**
(silvicultural system that is designed to regenerate and maintain an even-aged stand)
d gleichaltrige Betriebsart *f*
f régime *m* sylvicole de forêt équienne
r система *f* ведения одновозрастного хозяйства

2920 **even-aged stand; even-aged forest**
(forest composed of trees of the same age; stand of trees consisting of one or two age classes)
d gleichaltriger Bestand *m*
f forêt *f* équienne; équienne *f*
r одновозрастное насаждение *n*; одновозрастной лес *m*

2921 **even-aged yield table**
d Ertragstafel *f* des gleichaltriges Bestandes
f table *f* de production d'équienne
r таблица *f* роста одновозрастных насаждений

2922 **even[ ]flow**
(in harvest scheduling, the requirement that the harvest level in each period be equal to the harvest level in the preceding period)
d Gleichfluss *m*
f flux *m* d'exploitation homogène
r однородный поток *m*

2923 **even grain; even texture; uniform texture**
(with small difference between summer and autumn wood)
d gleichförmige Textur *f*; uniforme Textur
f texture *f* homogène
r однородная текстура *f*; текстура, равномерная по окраске и рисунку; равнослойная текстура; одноцветная текстура (весенних и летных слоёв древесины)

2924 **evening-primerose family**
(Onagraceae)
d Nachtkerzengewächse *pl*
f onagracées *pl*
r кипрейные *pl*

f longueurs *fpl* égales paires
r одинаковые долготья *npl*; [попарно] равные долготья

* **even-pinnate leaf → 13**

2926 even stand; homogeneous stand; homogeneous forest; uniform stand; pure forest; pure stand; clear stand; unmixed stand
(stand of uniform trees; stand in which at least 80% of the trees in the main crown canopy are of a single species)
d gleichförmiger Bestand *m*; Reinbestand *m*; reiner Bestand
f peuplement *m* homogène; peuplement uniforme; peuplement régulier; peuplement non mixte; peuplement pur
r однородное насаждение *n*; чистый древостой *m*; чистое насаждение; насаждение одной породы

* **even texture → 2923**

2927 everblooming acacia
(Acacia retinoides var. floribunda)
d immerblühende Akazie *f*
f mimosa *m* des fleuristes
r акация *f* ретиноидес обильноцветущая

* **evergreen beech → 406**

* **evergreen boxwood → 2886**

2928 evergreen firebreak; green firebreak; greenbreak
d Laubholzschutzstreifen *m*
f pare-feu *m* boisé; bande *f* de protection en essences feuillues
r вечнозелёный противопожарный барьер *m*; защитная полоса *f* из лиственных пород

2929 evergreen forest
d immergrüner Wald *m*
f forêt *f* toujours verte; forêt sempervirente
r вечнозелёный лес *m*

* **evergreen Himalaya oak → 5562**

* **evergreen needleleaf forest → 1873**

* **evergreen oak → 1961**

2930 evergreen rain forest
d immergrüner [tropischer] Regenwald *m*
f forêt *f* pluviale toujours verte; forêt dense

2931 evergreens
(plants that retain foliage year round)
d immergrüne Pflanzen *fpl*; Immergrüne *pl*
f plantes *fpl* à feuilles persistantes
r вечнозелёные растения *npl*

2932 evodia
(Evodia)
d Evodia *n*
f évodia *m*
r эводия *f*

2933 excavated pith; hollow pith; hollow heart
d Hohlkern *m*
f cœur *m* creux
r выемчатая сердцевина *f*; полостная сердцевина; дупло *n*

2934 excavated trail; bladed trail
(constructed trail that has a width greater than 1.5 m and a mineral soil cutbank height greater than 30 cm)
d ausgegrabene Spur *f*
f chemin *m* excavé
r выкопанная тропа *f*

* **excavator → 8963**

* **excelsior-wood → 9697**

2935 excess[ive] cutting; excessive felling; excess exploitation; overexploitation; overcut[ting]; overfelling
d Mehreinschlag *m*; Überhauung *f*; Überhieb *m*; Holzfrevel *m*; überhöhter Einschlag *m*; Raubschlag *m*; Übernutzung *f*
f surcoupe *f*; surexploitation *f*
r чрезмерная рубка *f*; хищническая рубка; переруб *m* [расчётной лесосеки]; сверхплановое лесопользование *n*

2936 excessive growth; overgrowth
d übermäßiges Wachstum *n*
f accroissement *m* excessif
r чрезмерное разрастание *n*; чрезмерный рост *m*

* **excess of growing stock → 8287**

* **excess present value → 5718**

2937 excess sapwood
d übermäßiger Splint *m*
f aubier *m* trop épais
r чрезмерная заболонь *f*

2938 excrescence; whorl trunk scar

f excroissance *f* [de bois] du tronc
r нарост *m* на стволе дерева; наплыв *m* на стволе дерева

2939 excurrent branching habit; excurrent branching form; dendritic branching habit; dendritic form
(of a crone)
d auslaufender Habitus *m*; durchlaufende Wuchsform *f*; dendritische Wuchsform
f forme *f* excurrente; forme de ramification à forte décroissance; forme dendritique; habitude *f* dendritique
r сбежистая форма *f* роста

2940 exfoliate *v*
(cast off in scales, laminae, or splinters)
d abblättern
f exfolier
r лупиться; шелушиться; сходить слоями; отслаиваться; расслаиваться

2941 exfoliation; defoliation; abscission of leaves
(stripping of leaves)
d Abblätterung *f*; Entlaubung *f*; Entblätt[er]ung *f*; Abblatten *n*; Defoliation *f*
f exfoliation *f*; défoliation *f*; défeuillaison *f*
r дефолиация *f*; обезлиствение *n*; опадение *n* почечных чешуек

2942 exogenous; exogenic; exogenetic
(ref. to plant having wood bark, and pith; a wood forming a layer between the other two, and increasing, if at all, by the animal addition of a new layer to the outside next to the bark)
d exogenisch
f exogène
r из заболонной части дерева

* **exogenous timber → 3317**

* **exogens → 3317**

* **exotics → 2944**

2943 exotic tree
d fremdländischer Baum *m*
f arbre *m* exotique
r дерево-экзот *n*

2944 exotic [tree] species; non-native species; exotics
(species introduced accidentally or intentionally to a region beyond its natural range)
essences *fpl* [forestières] étrangères
r экзоты *mpl*

2945 expected forest yield
d Walderwartungsertrag *m*
f rendement *m* forestier expectatif
r ожидаемый лесной доход *m*

2946 experimental forest; sample forest; sample stand
(designated area managed for research purposes)
d Versuchforst *m*; Probenstand *m*
f forêt *f* expérimentale; peuplement *m* d'essai
r опытный лес[ной участок] *m*; пробное насаждение *n*

* **experimental plot → 7173**

* **expert in forest → 1890**

2947 explant
(the excised fragment of a plant tissue or organ used to start a tissue culture)
d Explantat *n*
f explant *m*
r эксплантат *m*

2948 explantation
(removing (living tissue) from the natural site of growth and place in a medium for culture)
d Explantation *f*
f explantation *f*
r эксплантация *f*

* **exploitable age → 136**

2949 exploitable diameter
d schlagbarer Durchmesser *m*
f diamètre *m* d'exploitabilité
r эксплуатационный диаметр *m*

2950 exploitable diameter limit
d Grenze *f* der schlagbare Durchmesser
f limite *f* de diamètre d'exploitabilité
r минимальный эксплуатационный диаметр *m* (деревьев)

2951 exploitable forest; harvestable forest
(forest and other wooded land on which there are no legal, economic or technical restrictions on wood production)
d hiebsreifer Wald *m*
f forêt *f* exploitable
r годный для использования лес *m*;

2952 exploitable value; realization value; liquidation value; utilization value
(of a forest)
d hiebsreifer Wert *m*
f valeur *f* de consommation; valeur de réalisation immédiate
r оценка *f* древостоя, произведенная из условия вырубки всей коммерческой древесиной

* **exploitation cutting → 1744**

* **exploitation felling → 1744**

2953 exploitation map
d Nutzungskarte *f*
f carte *f* d'exploitation
r карта *f* лесозаготовок

2954 exploitation percent
d Nutzungsprozent *n*
f pourcentage *m* d'extraction
r процент *m* вырубки; показатель *m* интенсивности рубки

2955 exploitation right; logging right; cutting right
d Ausbeutungsrecht *n*
f droit *m* d'exploitation; droit de coupe
r право *n* эксплуатации; право рубить [лес]

2956 exploited forest
d ausgenutzter Wald *m*
f forêt *f* exploitée
r освоенный лес *m*

* **exploration of forests → 3549**

* **explosive wedge → 6487**

2957 exportation of timber; timber export
d Holzausfuhr *f*; Holzexport *m*
f exportation *f* de bois
r экспорт *m* леса; лесоэкспорт *m*

2958 export grade timber; export timber; export lumber
d Ausfuhrholz *n*
f bois *m* d'exportation
r экспортные лесоматериалы *mpl*; экспортируемая древесина *f*

2959 ex situ conservation
d ex situ Erhaltung *f*
f conservation *f* ex situ
r консервация *f* вне места

(which is relics of the flora and fauna or other geologic time periods and are of historic or scientific significance are considered under botanical areas or zoological areas)
d bestehend
f encore existant
r сохранившийся; дошедший до нас; ныне живущий

* **extending shoot → 5149**

2961 extension root
d Extensionswurzel *f*
f racine *f* prolongée
r ростовой корень *m*

2962 extensive clear felling
d extensiver Kahlhieb *m*
f coupe *f* à blanc extensive
r концентрированная сплошная рубка *f*

2963 extensive forest management
(protection from fire and insects; reliance on natural regeneration)
d aufwendige Forstwirtschaftsführung *f*
f aménagement *m* forestier extensif; gestion *f* forestière extensive
r экстенсивное ведение *n* лесного хозяйства; экстенсивное лесоустройство *n*

2964 exterior plywood
d Außenbauholz *n*
f contreplaqué *m* extérieur
r фанера *f* для наружной облицовки; древесина *f* для наружной облицовки

2965 external crack
(in wood)
d Außenriss *m*
f fente *f* extérieure; brisure *f* externe
r наружная трещина *f*; внешняя трещина

2966 external forest transport
d äußerer Holztransport *m*
f transportation *f* de bois éloigné
r трелёвка *f* по магистральному волоку; трелёвка на большие расстояния

2967 external yarding distance
(slope distance from the landing to the farthest point within the cutting unit boundary)
d außer Holzabfuhrsabstand *m*
f distance *f* externe de débusquage
r максимальное расстояние *n* канатной

* **extinguishment of smouldering fire → 5554**

2968 **extirpation**
(elimination of a species in part of its range)
d Exstirpation *f*; totale Vernichtung *f*
f extirpation *f*
r уничтожение *n* с корнем; прополка *f*

* **extraction distance → 9747**

2969 **extraction of old trees; cutting of hold-overs; cutting of hold-over trees; cutting of standards**
d Abtrieb *m* der Überständer; Überhälterauszug *m*
f extraction *f* d'arbres vieux; extraction de réserves
r вырубка *f* перестойных деревьев; удаление *n* перестоя; вырубка перестоя

* **extraction of timber → 8595**

* **extraction road → 7689**

* **extraction thinning → 1750**

2970 **extraction tractor**
d Zugschlepper *m*
f tracteur *m* d'extraction
r трелёвочный тягач *m*; лесовозный тягач; трактор-тягач *m*

* **extraction unit → 5089**

2971 **extraction wood**
d Extraktionsholz *n*
f bois *m* d'extraction
r экстракционная древесина *f*; древесина для экстрагирования; древесина для экстракции

2972 **extractives in the wood; wood extractives**
d Holzextraktivstoffe *mpl*; Holzinhaltsstoffe *mpl*
f substances *fpl* entrantes dans la composition du bois
r экстрактивные вещества *npl* древесины; экстракционные вещества, содержащиеся в древесину

* **extracutting → 2973**

2973 **extraordinary cutting; extracutting**
d außerordentliche Felling *f*
f coupe *f* extraordinaire

2974 **extra-period fire**
d schwerer Waldbrand *m*; Waldbrand von überdurchschnittlicher Dauer
f incendie *m* prolongé
r весьма продолжительный пожар *m*

2975 **extra-thick timber**
d überdickes Holz *n*
f bois *m* surépaissé
r особенно крупномерные строительные брёвна *npl*

2976 **extruded chipboard**
d stranggepresste Spanplatte *f*
f panneau *m* de contreplaqué extrudé
r экструзионная древесностружечная плита *f*; ДСП *f* экструзионного прессования

* **exudation of gum → 4051**

2977 **exudation of resin; resinosis; resin exudation; resin flow; resin flux; resin glut**
(an abnormal exudation of oleoresin from conifers)
d Harz[aus]fluss *m*; Harzüberfülle *f*
f résinose *f*; écoulement *m* de résine; excès *m* de résine
r [чрезмерное] смолотечение *n*; истечение *n* живицы; смолоистечение *n*

* **exudation of sap → 1589**

* **eye → 1141**

* **eye[-]splice → 699**

F

* **FAC → 3328**

* **face → 2985, 3069, 8155**

2978 **face check [and check]**
d Flächenriss *m*
f fente *f* de face
r трещина *f* на лицевой поверхности

2979 **face cord; rick**
(stove length wood with a face of 32 square feet (a stack 4 ft high and 8 ft wide) but only 16 inches deep)
d Kordel *f*; Rick *n*
f corde *f* mince; rique *m/f*
r фэйс-корд *m*; штабель *m* короткомерного баланса

2980 **face cut; slab cut; crosscut end; abut**
(the first cut made along a log on a sawbench, providing a flat surface that can be used for its further conversion; section of wood sawn and removed from a tree's base)
d Abhiebsschnitt *m*; Hirn *n*; Hirnende *n*; Hirnseite *f*; Stirnseite *f*
f coupe *f* transversale terminale; découvert *m*
r первый рез *m*; торцевой срез *m*; поперечное сечение *n* ствола

* **face hardening → 1367**

2981 **face knot**
d Seitenknoten *m*
f nœud *m* de face
r пластовый сучок *m*

2982 **face length**
d Harzrisslänge *f*; Lachtenlänge *f*
f longueur *f* de carre
r длина *f* карры; длина подновки

2983 **face log; head block**
(the log placed under the front of the skids in a skidway to raise them to the desired height; the front, bottom log on a skidway)
d Hauptblock *m*
f dernière bille *f*; bloc *m* porte-grume
r переднее нижнее бревно *n*;

2984 **face mark**
d Seitenmarke *f*
f marque *f* de face
r маркировка *f* лицевой стороны материала

* **face of lumber → 7258**

* **face ply → 2986**

* **facer → 4661**

2985 **face side; face; splash**
d Spiegel *m* [des Holzes]
f côté *m* de la face; madrure *f* du bois; maille *f*
r лицевая поверхность *f*

2986 **face veneer; outer veneer; top veneer; face ply; outer ply**
(veneer that is used for exposed surfaces in plywood)
d Außenfurnier *n*; Deckblatt *n*; Deckfurnier *n*; Vorderfurnier *n*
f placage *m* de parement; placage extérieur; placage de surface
r облицовочный шпон *m*; лицевой шпон

2987 **faciation**
(a part of a climax association which lacks some of the dominants of the normal association due to slight differences in environmental fundamentals)
d Faciation *f*
f faciation *f*
r фасциация *f*

* **facing → 7107**

* **factitious ring → 2998**

* **factor of shape → 8840**

2988 **factory lumber; Fct; factory timber**
d Geschirrholz *n*; Zeugholz *n*
f bois *m* débité industriel
r пиломатериал *m*, предназначенный для дальнейшей переработки

2989 **fadeless wood; colo[u]rfast wood**
d unverwelkliches Holz *n*
f bois *m* grand teint
r невыцветающая древесина *f*; светостойкая древесина

2990 **faggot maker**
d Wellenbinder *m*

* **faggots → 9572**

* **faggotwood → 9572**

2991 **failed patch; fail[ed] spot; failed area; failed place**
(on veneer)
d Ausfallstelle *f*; Fehlstelle *f*
f place *f* défectueuse
r фаутный участок *m*

2992 **failed plantation**
d misslungene Kultur *f*
f plantation *f* non réussie
r погибшие лесные культуры *fpl*

* **fail[ed] spot → 2991**

* **faint ring → 5479**

* **fall → 3042**

2993 **fall block**
(block in skyline systems, can be lowered to pick up loads on the ground and then raised as required for hauling them onto the landing)
d Hubrolle *f*
f poulie *f* de charge
r грузовой блок *m*; опускающийся грузоподъёмный блок; подвижный блок (трелёвочной каретки)

2994 **falldown**
(situation in which second-growth forests provide less timber than the original forests)
d Niederfallung *f*; Verfall *m*
f exploitation *f* secondaire
r невыполнение *n*; падение *n*

* **fallen dead[ ]wood → 2266**

* **fallen forest → 3042**

* **fallen leaf → 14**

* **fallen timber → 3042**

* **fallen trees → 2266**

* **fallen wood → 3042**

* **faller → 9526**

* **falling cut → 3066**

* **falling wedge → 3079**

2995 **fall of needles; needle fall; removal of needles**
d Nadelabfall *m*; Entnadelung *f*
f chute *f* naturelle des aiguilles; enlèvement *m* des aiguilles
r лишение *n* хвои; удаление *n* хвои

2996 **fall of seeds; seed fall; seed shedding**
d Samen[ab]fall *m*
f dissémination *f*; chute *f* de semence
r осыпание *n* семян

2997 **fallow land**
(land temporarily uncultivated to restrict production that is considered excessive)
d Brachland *n*
f friche *f*
r поле *n* под паром; невспаханное поле

* **fall plum → 5425**

* **false acacia → 1754**

2998 **false annual ring; false ring; factitious ring**
(one of the growth rings of a double (or multiple) ring)
d falscher Jahrring *m*; Scheinring *m*
f anneau *m* faux; faux cerne *m* [annuel]
r ложное [годичное] кольцо *n*

2999 **false beech; coigue; rauli**
(Nothofagus)
d Scheinbuche *f*; Südbuche *f*
f hêtre *m* austral
r нотофагус *m*

* **false Cape boxwood → 4685**

* **false-cypress → 398**

3000 **false fruit; spurious fruit**
d falsche Frucht *f*
f fruit *m* faux
r ложный плод *m*

* **false heart → 795**

* **false kamani → 4464**

3001 **false medullary ray**
d falscher Markstrahl *m*; Scheinmarkstrahl *m*
f faux rayon *m* médullaire
r ложный сердцевинный луч *m*

* **false ring → 2998**

d afrikanischer Kautschuk *m*
f holarrhène *f* florifère
r холларена *f* обильноцветущая

* **false sandalwood → 4322**

* **false sap → 853**

* **false sycamore [maple] → 8355**

3003 **false tamarix**
(Myricaria)
d Rispelstrausch *m*
f myricaire *f*
r мирикария *f*

3004 **false terminal bud; pseudoterminal bud**
d falsche Endknospe *f*
f bourgeon *m* terminal faux
r ложная верхушечная почка *f*; псевдоверхушечная почка

3005 **false tinder polypore**
(Fomes igniarius; Phellinus igniarius)
d Zunderschwamm *m*
f faux amadouvier *m*
r ложный трутовик *m*

3006 **false tungoiltree**
(Garcia nutans)
d falscher Tungölbaum *m*
f garcia *m* penché
r гарция *f* поникающая

* **false Weymouth pine → 6617**

3007 **family forestry**
(tested open-pollinated, polycross or full-sib families are deployed as single families or as prescribed family mixtures)
d Familienforstwirtschaft *f*
f foresterie *f* de famille
r фамильное лесничество *n*

3008 **family hedges**
(hedges used to donate cuttings, in which sibs from the same full-sib, polycross or open-pollinated family are mixed, and their cuttings are bulked)
d Familienhecken *fpl*
f haie *f* de famille
r фамильная изгородь *f*

3009 **family tree plantation**
d Stammbaumplantation *f*

3010 **fan-compound leaf**
(a compound leaf with leaflets radiating from a single point)
d Fächerfiederblatt *n*
f feuille *f* composée flabellée
r веерообразный сложный лист *m*

3011 **fan [figure]; feather**
d aufgefächerte Figur *f*; fächerförmige Figur
f figure *f* en éventail; figure flabellée
r текстурный веерообразный рисунок *m*; рисунок в виде узелков; рисунок в виде завитков

* **fanleaf palm → 5384**

* **fan palm → 5384**

3012 **fantail [butt log]**
(an overhang consisting of the fan-shaped end part)
d fächerförmige Erdblock *m*
f bille *f* de pied barrée; bille de pied voûtée
r короткое бревно *n* с веерообразным концом

3013 **fantailing**
d Gillung *f*
f voussure *f*
r раскатывание *n* в стороны свободных концов трелюемого леса

3014 **fan[-trained tree]; palmette**
d Palmette *f*
f palmette *f*
r дерево *n* с веерообразной кроной

* **farkleberry → 8926**

3015 **farm forest; farm woodland; farm woodlot**
d Farmwald *m*; Bauernwald *m*; Gutswald *m*
f forêt *f* paysanne; forêt rurale; forêt de ferme
r фермерский лес *m*; крестьянский лес

3016 **farm forestry; woodlot management**
(growing trees for timber, poles, fuelwood on farmland)
d Bauernwaldwirtschaft *f*
f traitement *m* de parcelle forestière
r крестьянское лесное хозяйство *n*

3017 **farmstead**
d Hofstatt *f*
f fermette *f*
r усадьба *f*

* **farm woodlot → 3015**

3018 fascine; hurdle
(a bundle of brushwood used to help make a road foundation on marshy or unstable ground, to form a barrier against erosion of the bank of, or to accumulate sand, silt etc.)
d Faschine *f*; Reis[ig]welle *f*
f fascine *f*; boullée *f*
r пучок *m* прутьев; фашина *f*

3019 fascine choker
d Faschine-Choker *m*
f collier *m* étrangleur de fascines
r пресс *m* для вязки фашин

* **fascine dike → 1117**

* **fascine weir → 1117**

* **fascine wood → 9572**

* **fastened tree → 9053**

* **fastening rope → 5023**

3020 fast-growing forest
d schnellwachsender Wald *m*; raschwüchsiger Wald; starkwüchsiger Wald
f forêt *f* à croissance rapide
r быстрорастущий лес *m*

3021 fast-growing tree
(tree species that matures quickly and is usually not long lived)
d schnellwüchsiger Baum *m*; starkwüchsiger Baum
f arbre *m* à croissance rapide
r быстрорастущее дерево *n*

* **fast growth → 141**

* **fastigate → 8401**

3022 fast taper
d große Abholzigkeit *f*; große Durchmesserabnahme *f*
f gros défilement *m*
r большой сбег *m*

3023 fast-taper tree
d Baum *m* mit großer Abnahme
f arbre *m* à gros défilement
r дерево *n* с большим сбегом

d Faden *m*; Brennholzklafter *f*; Klafter *f*
f brasse *f*
r сажень *f* [дров]; фатом *m*

3025 fat trees
d Alaunbäume *mpl*
f arbres *mpl* gras
r древесные породы *fpl*, запасающие в тканях жиры; древесные породы со смолистой древесиной

3026 fatwood
d geöltes Holz *n*; Ölholz *n*; Kienholz *n*
f bois *m* gras [de pin]
r древесина *f*, пропитанная маслами

* **faulty → 7212**

* **faulty tree → 7082**

* **faulty wood → 9118**

3027 faveira
(Mora gonggrijpii; Mora bucquia)
d Morabukea *m*
f morabukea *m*
r мора *f* Джонггрийпа

3028 favorable grade
(wedge used to throw a tree in the desired direction)
d günstige Neigung *f*
f pente *f* favorable
r благоприятный уклон *m*; уклон в грузовом направлении

* **fawn-foot → 2329**

* **Fct → 2988**

* **FDP → 3376**

* **FDRC → 3377**

* **feather → 2595, 3011**

* **feather-compound leaf → 6038**

* **feather crotch → 3031**

3029 feather-edged board; featherboard
d nicht parallel gesägtes Brett *n*; scharfkantiges Brett
f planche *f* débitée non parallèle

3030 feathered tree
(an young tree well furnished with branches to near ground level)
d gefiederter Baum *m*
f arbre *m* en drapeau
r перистое дерево *n*

3031 feather [grain]; feather crotch
(this type of figure results from the weaving together of wood fibers strengthening the areas around major limbs and branches)
d scharfkantige Textur *f*
f fil *m* drapé
r перистый рисунок *m* текстуры; перистая текстура *f*

3032 feather-nerved leaf
d federgenervtes Blatt *n*
f feuille *f* nervée en éventail
r перистонервный лист *m*

* **feather palm → 5384**

3033 feed chute; intake chute
d Hackbock *m*
f goulotte *f* d'alimentation
r загрузочный лоток *m*

3034 feeding root
d Futterwurzel *f*
f racine *f* d'alimentation
r питающий корень *m*

3035 feed roller; live roller
d Rolle *f*; Förderwalze *f*; Treibolle *f*
f rouleau *m* de renvoi; rouleau entraîneur; poulie *f* motrice; galet *m* moteur
r транспортный каток *m*; плавучий вал *m*

* **feet board measure → 879**

3036 feet surface measure; ft.s.m.; square-foot measure; sq.ft.m.
d Vermessung *f* im Quadratfüss
f mesurage *m* de surface par pieds carrés
r измерение *n* площади поверхности в квадратных футах

* **FEL → 3720**

* **fell → 3042**

3037 fell *v*; **cut** *v* **[down]; cutt** *v* **off**
(trees or forest)
d abholzen; ausholzen; entholzen; abtreiben; ausfällen; fällen

* **fellable → 3231**

3038 felled area
d ausgeholzte Fläche *f*; Schlagblöße *f*
f coupe *f* usée; coupe vidangée
r вырубленная площадь *f*

3039 felled-area flora; coupe flora
d Schlagflora *f*; Kahlschlagflora *f*
f flore *f* naissante à la suite d'une coupe; flore typique de coupe à blanc
r лесосечная флора *f*; флора сплошной лесосеки

3040 felled measure
d Messung *f* am Liegenden
f cubage *m* du bois gisant
r обмер *m* лежащего лесоматериала

3041 felled plantation
d entnommener Bestand *m*
f plantation *f* gisante
r вырубленное насаждение *n*

* **felled sale → 7157**

* **felled volume → 2200**

3042 felled wood; cut wood; fallen timber; fallen wood; fallen forest; fell; fall
d Baumleichen *pl*; Abholz *n*; abgefallenes Holz *n*; Baumfall *m*; geschlagenes Holz; befallendes Holz
f bois *m* gisant; abattis *m*
r срубленный лес *m*; заготовленный лес; поваленный лес; заготовленная древесина *f*

* **feller → 9526**

* **feller-buncher [shear] → 3053**

* **feller-[de]limber-buncher → 8853**

* **feller-delimber head → 3068**

* **feller[-director] → 8836**

* **feller-forwarder → 3075**

* **feller head → 3063**

3043 feller selection
(instead of foresters spot or blaze marking trees for thinning beforehand, skilled fellers work their way through stands, cutting down

f sélection *f* d'abatteur
r выбор *m* вальщика

3044 **feller shear attachment; shear [felling] head**
d Scherekopf *m*
f tête *f* d'abattage à cisaille
r ножевое срезающее устройство *n*

* **feller-skidder → 3075**

3045 **felling above buttress**
(of a tree)
d Holzeinschlag *m* über Wurzelanlauf
f abattage *m* au-dessus des empattements
r срезание *n* дерева выше области закомелистости

3046 **felling above the rootstock; tree-length felling; felling to the tree-length stage**
d Einschlag *m* über Wurzelbrut; Langholzabhieb *m*
f abattage *m* à culée blanche
r валка *f* над корневища; валка хлыстов; заготовка *f* хлыстов

* **felling age → 136**

3047 **felling and bucking**
(process of cutting down standing timber and then cutting it into specific lengths for yarding and hauling)
d Einschlag *m* und Querschneiden *n*
f abattage *m* et ébranchage *m*
r валка *f* и раскряжёвка *f*

3048 **felling-and-bucking crew**
d Ablängungsbrigade *f*
f équipe *f* d'abattage-ébranchage
r валочно-раскряжёвочная бригада *f*

* **felling [and bucking] saw → 8854**

* **felling area → 2189**

3049 **felling arm**
d Fällungsarm *m*
f bras *m* d'abattage
r лесорубный рычаг *m*

* **felling axe → 453**

3050 **felling bar; felling cylinder**
d Fällungsbarre *f*
f cylindre *m* de coupe
r лесорубная полоса *f*

d Fällungssäule *f*
f flèche *f* d'abattage
r лесорубная стрела *f*

* **felling border → 2762**

3052 **felling break; felling shake; felling fracture**
d Fällungriss *m*
f éclatement *m* d'abattage; trou *m* d'abattage; fente *f* d'abattage; fracture *f* d'abattage
r трещина *f* от валки; повреждение *n* (дерева) при валке

3053 **felling-bunching machine; feller-buncher [shear]; prebuncher**
(tractor equipped with hydraulic shears or a saw)
d Fäller-Bündler *m*; Fäll-Paketiermaschine *f*
f abatteuse-groupeuse *f*; abatteuse *f* à berceau collecteur; fardeleuse *m*
r валочно-пакетирующая машина *f* [с ножевым срезающим устройством]

* **felling by groups → 4001**

* **felling class → 2191**

3054 **felling comb**
d Bruchleiste *f*
f came *f* d'abattage
r лесорубный конёк *m*

* **felling contract → 2192**

* **felling contractor → 5072**

3055 **felling control; cut[ting-practice] control; supervision of cuttings**
(set of rules and actions that describe the allowable variation in the annual harvest rate either above or below the allowable annual cut approved by the chief forester)
d Einschlagkontrolle *f*; Einschlagüberwachung *f*
f contrôle *m* des exploitations
r контроль *m* рубки

3056 **felling cost[s]; felling value**
d Fäll[ungs]kosten *pl*; Hauungskosten *pl*; Hiebskosten *pl*; Abtriebswert *m*
f coût *m* d'abattage; valeur *f* d'abattage
r стоимость *f* [вы]рубки; затраты *fpl* на валку леса; стоимость лесосечных работ

* **felling crew → 9527**

f coin *m* pneumatique; coussin *m* d'abattage
r валочная подушка *f*

* **felling cut → 3066**

3058 felling cutter
d Baumfällschild *m*
f coupeur *m* d'abattage
r лесовалочный нож *m*

* **felling cycle → 2194**

* **felling cylinder → 3050**

3059 felling damage
d Fällungsschaden *m*
f détérioration *f* causée par l'abattage; dégâts *mpl* d'abattage; dommage *m* de coupe
r повреждение *n* от валки

3060 felling date; cutting date
d Fällungsdatum *n*; Schlagdatum *n*
f date *f* d'abattage; date de coupe
r дата *f* вырубки

* **felling debris → 998**

* **felling direction → 2519**

3061 felling distribution
d Fällungsverteilung *f*
f distribution *f* d'abattage
r порядок *m* вырубки

* **felling dog → 8829**

* **felling during the sap flow → 3065**

* **felling edge → 2762**

* **felling face → 3722**

3062 felling forestry
d schlagweise Waldwirtschaft *f*
f foresterie *f* d'abattage
r лесосечное лесное хозяйство *n*

* **felling fracture → 3052**

* **felling frame → 3684**

3063 felling head; feller head; saw-head [felling] attachment; saw attachment; saw head
d Schlagkopf *m*
f tête *f* d'abattage [à chaîne coupante]; tête

* **felling height → 2195**

3064 felling in chequer arrangement
d schachbrettförmiger Hieb *m*
f abattage *m* en quinconces
r шахматная рубка *f* леса

3065 felling in the growing season; summer felling; felling during the sap flow; cutting in sap flow period; sour felling
d Fällung *f* im Saft; Saftfällung *f*; Safthieb *m*; Sommereinschlag *m*; Sommerfällung *f*; Sommerhieb *m*
f coupe *f* d'été; abattage *m* en sève; abattage dans la période d'écoulement de sève; abattage des arbres en feuilles
r рубка *f* во время сокодвижения; рубка в соку; рубка в период движения сока; летняя рубка; рубка с оставлением поваленных деревьев с необрубленной кроной на лесосеке

3066 felling kerf; felling cut; falling cut; backcut[ting]
(the cut in a felling operation made on the opposite side from the undercut; the last of the three cuts required to fall a tree)
d Fällungsschnitt *m*; Fällungskermen *n*; Fällschnitt *m*
f trait *m* d'abattage; trait de scie à abattage
r валочный пропил *m*; задний рез *m*; вырез *m* [в дереве] со стороны, противоположной направлению валки

3067 felling lever; breaking bar
d Fäll[ungs]hebel *m*
f levier *m* d'abattage
r валочный рычаг *m*

* **felling licence → 6183**

3068 felling-limbing head; feller-delimber head; harvesting head; harvesting attachment
d Fällung-Entastungskopf *m*
f tête *f* d'abattage-ébranchage
r валочная головка *f*

* **felling losses → 2197**

* **felling machine → 8853**

3069 felling notch; cutting face; face; gullet
(an undercut in a tree preparatory to felling it)
d Fallkerbe *f*; Fallhieb *m*
f entaille *f* d'abattage; bande *f* de fracture

* **felling of burls → 3070**

3070 felling of burrs; felling of burls
d Wulstaushieb *m*; Knorrenaushieb *m*
f extraction *f* de loupes; extraction de broussins
r вырубка *f* наростов; вырубка наплывов

3071 felling of wolf trees; pruning of wolves
d Schnitt *m* in Läuterungen; Protzenaushieb *m*
f extraction *f* de loups
r обрезка *f* разросшихся деревьев

* **felling order → 2199**

* **felling pattern → 5051**

3072 felling period; logging period; felling season; logging season; cut[ting] period
d Einschlagsperiode *f*
f saison *f* d'abattage; époque *f* d'abattage; période *f* d'abattage
r период *m* рубки; лесозаготовительный сезон *m*; сезон лесозаготовок; период валки

* **felling plan → 6332**

* **felling point → 7668**

* **felling quantity → 2200**

* **felling refuse → 998**

3073 felling register; cutting register; record of cuttings; felling record; coupe inventory records; log book
d Einschlag-Aufnahmeliste *f*; Holzaufnahmebuch *n*; Holzabnahmebuch *n*
f calepin *m* de dénombrement d'abattage
r ведомость *f* лесных материалов; ведомость [сдачи-]приёмки леса

* **felling residues → 998**

* **felling returns → 9777**

* **felling rotation → 2194**

* **felling season → 3072**

* **felling sequence → 2202**

* **felling series → 2202**

* **felling shake → 3052**

3074 felling sight; sighting
d Einschlagmarkieren *n*; Hiebsignierfärbung *f*; Signierfärbung *f*; Kennfärben *n*; Signieren *n*
f repère *m* d'abattage; teinture *f* provisoire
r маркировка *f* деревьев в рубку; клеймение *n* дерева

* **felling site → 7668**

3075 felling-skidding machine; feller-forwarder; feller-skidder
d Rückemaschine *f* mit Harvester; Fäll-Rückekombine *f*; Fäll-Rückemaschine *f*
f débusqueur *m*; débusqueuse *f*
r валочно-трелёвочная машина *f*

3076 felling technique; logging technique
d Einschlagstechnik *f*
f technique *f* d'abattage
r техника *f* рубки

3077 felling time; cutting time
d Fällungszeit *f*; Einschlagzeit *f*; Hiebszeit *f*; Schlagzeit *f*; Hauungszeit *f*
f temps *m* d'abattage
r время *n* валки; время рубки

* **felling to a diameter limit → 2476**

* **felling to the full tree stage → 3761**

3078 felling to the shortwood stage; short-length logging; shortwood logging; short-log operation; log-length logging; short logging
d Kurzholzfällung *f*
f exploitation *f* en bois courts; exploitation par billes courtes
r заготовка *f* сортиментов

* **felling to the tree-length stage → 3046**

* **felling value → 3056**

* **felling waste → 998**

3079 felling wedge; falling wedge; wedge
(a wedge used to throw a tree in the desired direction)
d Holzfäll-Keilgerät *n*; Fällkeil *m*; Schrotkeil *m*
f coin *m* d'abattage
r [лесо]валочный клин *m*; клин рубки

3080 felling with delayed shedding
d Hieb *m* mit verzögerte Abführung
f abattage *m* à transportation retardée

3081 felling with the extraction of the rootstock
(felling trees and shrubs by exposing and cutting the roots)
d Auskesselung *f*; Hieb *m* mit Roden von Bäumen
f abattage *m* avec dessouchage
r валка *f* деревьев с корнями; валка с обрезкой корневой системы

3082 felloe wood
d Felgenholz *n*
f bois *m* de jante
r ступичный кряж *m*; ободная древесина *f*

* **fell with the axe → 83**

* **female [layer of] cork → 9227**

3083 female rosewood; Cayanne-linaloe
(Aniba duckei; Aniba rosaeodora)
d weibliches Rosenholz *n*
f bois *m* de rose femelle
r пальмароза *f*; розовое дерево *n*

* **female tree → 5569**

* **femel cutting → 7412**

* **femel system → 7620**

* **FEN → 3393**

* **fenberry → 2889**

* **fence → 5019**

* **fence pole → 3084**

3084 fence post; fence pole; fence stake; hedge pole; stob
d Zaunpfahl *m*; Pfahl *m*; Zaunpfosten *m*
f piquet *m* de clôture; pieu *m* de clôture
r кол *m* для ограды; кол для забора

* **fender boom → 3147**

* **fender[skid] → 1025**

* **fenestriform pit → 6287**

* **fen forest → 8309**

3085 fenwood
d Fennholz *n*
f bois *m* de forêt marécageuse
r древесина *f* болотистого леса

d Fermentationsschicht *f*; F-Horizont; F-Schicht *f*
f couche *f*; horizon *m* de fermentation; horizon F
r ферментационный слой *m*; ферментационный горизонт *m*; средний слой полуразложившегося опада

3087 fern
d Farn *m*; Farnkraut *n*
f fougère *f*
r папоротник *m*

3088 fernery; fern heath
d Farnheide *f*; Farnkrautfeld *n*; Farndickicht *n*
f fougeraie *f*
r папоротниковая заросль *f*

* **fertilization → 3090**

3089 fertilizer; manure
d Dünger *m*
f fumier *m*; engrais *m*
r удобрение *n*

3090 fertilizer application; fertilizing; fertilization
(application of chemical or organic fertilizers with the objective of increasing the unit area soil productivity)
d Düngung *f*; Begiftung *f*
f application *f* d'engrais; fertilisation *f*; fécondation *f*
r внесение *n* удобрений; внесение ядохимикатов; удобрение *n*

* **fetid buckeye → 5874**

3091 fetterbush [lyonia]
(Lyonia lucida)
d glänzende Lyonie *f*
f entrave-buisson *m*
r лиония *f* блестящая

3092 fever tree
(Acacia xanthophloea)
d Gelbfieberbaum *m*
f arbre *m* à fièvre
r акация *f* лихорадочная

* **F-horizon → 3086**

* **fiberboard → 9551**

* **fiber-building board → 9551**

f classificateur *m* de fibres
r фракционатор *m* древесной массы

3094 **fiber collapse**
(the flattening of single cells or rows of cells during the drying or pressure treatment of wood)
d Faserstauchung *f*
f effondrement *m* des fibres
r оседание *n* волокон; местное сплющивание *n* волокон

3095 **fiber crotch[ing]**
(of wood)
d Fasergabelung *f*
f fourche *f* de fibres
r разветвление *n* волокон

3096 **fibered; fibrous; fibrillar; stringy**
d faserig; faserförmig; faserartig
f fasciculé; fibreux; fibrillaire
r волокнистый

3097 **fiber length**
d Faserlänge *f*
f longueur *f* de fibres
r длина *f* волокон

* **fiber orientation → 2516**

3098 **fiber saturation point**
(the point in the seasoning or wetting of timber at which the cell cavities are free from water but cell walls are still saturated with bound water)
d Fasersättigungspunkt *m*
f point *m* de saturation de fibres
r точка *f* насыщения волокон

3099 **fiber separation**
d Fasertrennung *f*
f défibrage *m*
r разобщение *n* волокон

3100 **fiber structure**
d Faserstruktur *f*
f structure *f* de fibres
r строение *n* волокна

3101 **fiber waviness; fiber wavyness**
d Wellenfaserigkeit *f*; Unregelmäßigkeit *f* der Textur; Wimmer *m*
f gondolement *m*; ondulation *f* de fils
r свилеватость *f* [волокон]; волнообразность *f*; волнистость *f*; неоднородность *f* текстуры

* **fibreslab → 9551**

* **fibrillar → 3096**

3102 **fibril[lar] angle**
(the angle between the longitudinal axis of the cell and the direction of the fibrils in the cell wall)
d Fibrillenwinkel *m*
f angle *m* des fibrilles
r угол *m* наклона микрофибрилл

* **fibrous → 3096**

* **fibrous[-felted] board → 9551**

* **fibrous root → 1968**

3103 **fibrous structure**
d faserförmige Struktur *f*
f structure *f* fibreuse
r волокнистая структура *f*

3104 **fibrous yellow heart rot**
(invader - Echinodontium tinctorium)
d faserartige gelbe Kernfäule *f*
f pourriture *f* jaune fibreuse du cœur
r волокнистая жёлтая ядровая гниль *f*

3105 **fibro-vascular tissue; vascular tissue**
(tissue composed of woody fibers and ducts)
d Gefäßgewebe *n*
f tissu *m* vasculaire
r сосудисто-волокнистая ткань *f*

3106 **fiddleback**
(of wood)
d Flammung *f*; flammige Zeichnung *f*
f madrure *f*; moirure *f*
r узорчатость *f*

3107 **fiddle butt**
(large spruce butt log suitable for manufacture of musical instruments)
d Stamm *m* für Instrumentenholz
f grume *f* pour bois d'instruments
r инструментальный ствол *m*; комель *m*, пригодный для производства музыкальных инструментов

3108 **fiddlewood; yashnik**
(Vitex gaumeri)
d Leban *n*
f bricole-bois *m*
r прутняк *m*; витекс *m*

f afforestation *f* de champ
r полевое лесоразведение *n*

* **field and forest crops alternation → 225**

3110 field capacity
(of a soil)
d Feldkapazität *f*; Feuchtekapazität *f* des Bodens
f capacité *f* [de rétention] au champ
r полевая влагоёмкость *f*

3111 field check
d Feldkontrolle *f*
f contrôle *m* au champ
r проверка *f* на объекте; проверка в полевых условиях

* **field crops and forestry in alternation → 1740**

3112 field debarker
d mobile Entrindungsmaschine *f*
f écorceuse *f* au champ
r передвижная окорочная машина *f*

3113 field delimber; self-propelled lopper
d mobile Entastungsmaschine *f*
f ébrancheuse *f* mobile
r передвижная сучкорезная машина *f*; самоходная сучкорезная машина

* **field elm → 2893**

3114 field germination
d Feldkeimung *f*
f germination *f* au champ
r полевая всхожесть *f*

3115 field grafting
d Freilandpfropfung *f*
f greffage *m* en terrain dégagé
r прививка *f* в открытом грунте

3116 field layer; herbaceous layer; herb layer
(the vegetation at ground level in a forest or woodland; part of the understor[e]y in a forest that consists mainly of non-woody plants)
d lebende Bodendecke *f*; Krautschicht *f*
f couverture *f* vivante au-dessus du sol; strate *f* herbacée
r живой надпочвенный покров *m*; травяной ярус *m*; ярус травянистых растений; травянистый слой *m*

* **field maple → 2827**

established in or near the forest rather than near an administrative or executive headquarters)
d unständiger Kamp *m*
f pépinière *f* volante
r полевой питомник *m*

* **field planting → 5908**

3118 field-protection forest; field-safeguarding forest
d Feldschutzwald *m*
f forêt *f* de protection de champ
r полезащитный лес *m*

3119 field staff; operating employees; running trades; road service employees
d Feldpersonal *n*
f personnel *m* roulant; personnel de route; roulants *mpl*; personnel de terrain
r персонал *m* на местах; разъездной персонал

3120 field tally
(of trees)
d Feldaufnahme *f*
f inventaire *m* au champ
r полевой перечёт *m* (деревьев)

* **field woodland → 5001**

3121 field work
d Forsteinrichtungsarbeit *f*
f travail *m* aux champs
r лесоустроительные работы *fpl*

* **fig[-tree] → 1774**

* **figure → 9678**

3122 figured birch; curly birch; [wavy-]grained birch; flamy birch
d Fladerbirke *f*; Flammenbirke *f*
f bouleau *m* bariolé; bouleau flammé
r узорчатая берёза *f*

3123 figured growth
d Profilwuchs *m*
f croissance *f* figurée
r узорчатый рост *m*

3124 figured veneer
d Profilfurnier *n*
f placage *m* figuré
r секторная разделка *f* фанерного кряжа

3125 figure of density of a pile
d Umrechnungszahl *f* für Schichtholz
f coefficient *m* de cubage de bois empilé
r пересчётное число *n* для поленицы

* **figure spotting → 9187**

* **figuring → 9187**

3126 figwort family
(Scrophulariaceae)
d Braunwurzgewächse *pl*
f scrofulariacées *pl*
r норичниковые *pl*

3127 filamentous fungi
(Hyphomycetales)
d filamentöse Pilze *mpl*; Eichenwirrling *mpl*
f dédalées *fpl* du chêne; hyphomycètes *mpl*
r дубовые губки *f*; гифомицеты *mpl*

* **filbert → 4183**

* **filbert orchard → 4181**

3128 file dust
d Holzschrot *n/m*
f limature *f*
r опилки *fpl* при работе напильником

3129 filing block; filing board
d Feilenbrett *n*
f étau *m*
r подушка *f* под опиливаемое бревно

* **filled seed → 3753**

3130 filler; spackle
(a substance used to fill the holes and irregularities in planed or sanded surfaces to decrease the porosity of the surface before applying finish coatings)
d Füllstoff *m*
f mastic *m*
r заполнитель *m*

3131 filler [tree species]; muffler
(a tree or species of inferior value, retained in thinnings or cleanings in the absence of any better)
d Umfütterer *m*
f essence *f* de remplissage; manchon *m*
r древесная порода-заполнитель *f*; заполняющая порода *f*

* **fillet → 6236**

* **filling up blanks → 823**

3132 filling wedge
d Füllkeil *m*
f coin *m* d'obturation
r заполняющий клин *m*

* **filter strip → 1149**

3133 filtration zone
(area of undisturbed forest floor between roads and streams through which sediments are filtered from water that runs off roads before that water can enter streams)
d Filtrationzone *f*
f zone *f* de filtration
r зона *f* просачивания

* **final age → 136**

3134 final clearing
d abschließende Reinigung *f*
f défrichement *m* final; défrichage *m* final
r окончательная зачистка *f*

* **final crop → 3145**

3135 final crop spacing
(the density of trees prior to felling them for timber)
d Endbestandweite *f*
f écartement *m* de peuplement final
r конечный запас *m* насаждения

* **final crop tree → 5569**

3136 final cut[ting]; final felling; final harvest; principal felling; main felling (*US*); major harvest
(cutting of the mature crop of trees either as a single clearcut or a series of regeneration established)
d Abräumungshieb *m*; Abräumungsschlag *m*; Räumungshieb *m*; Abtriebnutzung *f*; Einschlag *m* der Hauptnutzung; Endnutzung *f*; Hauptnutzung *f*; Säuberungshieb *m*; Endhieb *m*; Räumungsschlag *m*
f coupe *f* définitive; coupe principale; coupe finale
r окончательная рубка *f*; главная рубка; рубка главного пользования

3137 final debarking
d Nachentrinden *n*; Nachentrindung *f*

* **final decay → 1828**

3138 **final diameter; future diameter**
d Zieldurchmesser *m*
f diamètre *m* final
r целевой диаметр *m*

3139 **final distance**
d Endabstand *m*
f distance *f* finale
r терминальное расстояние *n*

* **final felling → 3136**

* **final harvest → 3136**

* **final landing → 4484**

3140 **final location survey**
d endgültige Lageprüfung *f*
f marquage *m* définitif à piquets
r окончательная съёмка *f* трассировки

* **final m.a.i. → 3979**

* **final mean annual increment → 3979**

3141 **final moisture; end moisture**
(of wood)
d Endfeuchte *f*; Restfeuchtigkeit *f*
f teneur *f* en humidité après séchage du bois
r конечная влажность *f*; остаточная влажность

* **final rot → 1828**

* **final seed → 5338**

3142 **final stage of felling**
d abschließendes Stadium *n* des Holzschlags
f étape *f* finale de coupe
r окончательная стадия *f* рубки; окончательный приём *m* постепенной рубки

3143 **final thinning**
d Enddurchforstung *f*
f éclaircie *f* finale
r окончательные рубки *fpl* ухода; сплошные рубки

3144 **final year of growth; cutting year**
d Einschlagsjahr *n*; Schlagjahr *n*; Schlägerungsjahr *n*
r год *m* рубки

3145 **final yield; final crop; principal yield; return from final cutting**
(the trees that remain after successive thinnings and are finally felled at maturity)
d Fällungsergebnis *n*; Einschlaghöhe *f*; Hauptnutzungsetat *m*; Haubarkeitsertrag *m*; Abtriebsertrag *m*
f produit *m* principal; résultat *m* de l'abattage
r результат *m* главной рубки; доход *m* от рубки главного пользования лесом; окончательный выход *m* древесины; остающаяся часть *f* насаждения

3146 **financial maturity; maturity; economic maturity; financial exploitability**
(age of a tree, crop or stand, beyond which value-increase is insufficient to earn a specified rate of interest)
d Schlagbarkeit *f*
f âge *m* d'exploitabilité financière
r финансовая спелость *f*

* **financial rotation → 2745**

3147 **fin boom; fender boom; rudder boom; sheer boom; glancing boom**
(a boom so secured that it guides floating logs in the desired direction)
d Streichversatz *m*; Flügelschwimmer *m*; Seitenschwimmer *m*; Leitschwimmer *m*; Wehrbaum *m*; Wehrer *m*
f flotteur *m* de déviation; flotteur pilote; flotteur latéral; palplanche *f*
r реевый бон *m*; бон для направления брёвен с одной стороны реки на другую; ведущий поплавок *m*; лесонаправляющий бон; крылатый поплавок; боковой поплавок

* **fine check → 4088**

3148 **fine chips; fines**
(fine milled chips used in the production of particleboard; larger than sander dust or wood flour)
d Feinschlichtspan *m*; Schlichtspan *m*
f fines *fpl*
r тонкая чистовая стружка *f*; обрывки *fpl* волокон; очень короткие волокна *npl*; мелкая щепа *f*

* **fine cut → 7789**

3149 **fine-fibered wood; small-fibered wood**
d Faserfeinholz *n*; feinfaseriges Holz *n*

3150 **fine finish**
(coatings of paint, varnish, lacquer, wax or other material applied to high quality wood surfaces to protect and enhance their durability and appearance)
d Feinpolitur *f*
f poli *m* de qualité
r финишная обработка *f*; чистовая обработка

3151 **fine fuel chips**
d Feinbrennspäne *mpl*
f copeaux *mpl* combustibles fins
r тонкая топливная щепа *f*

* **fine grain → 1655**

* **fine-grained wood → 1656**

3152 **fine groundwood**
d Feinholzschliff *m*
f pâte *f* de bois fine
r тонкая высококачественная древесная масса *f*

* **fine-grown wood → 1656**

3153 **fine-leaved heath; heather bell; bell heather; twisted heath**
(Erica cinerea)
d graue Heide *f*; Grauheide *f*
f bruyère *f* cendrée; bréjotte *f*
r эрика *f* сизая

3154 **fineness [of wood]**
d Feinheit *f*
f finesse *f*
r мелкослойность *f* [древесины]

3155 **fine-porous; fine-pored; narrow-pored**
d engporigfeinporig
f à pores fins
r мелкопористый

3156 **fine residues**
(residues not suitable for chipping, such as sawdust, shavings, and veneer clippings)
d feine Überreste *mpl*
f rémanents *mpl* fins; résidus *mpl* fins
r мелкие остатки *mpl*

* **fine-ringed → 9470**

* **fines → 3148**

* **fine texture → 1655**

f à structure fine
r тонкотекстурный; мелкозернистый; тяжёлого механического состава

3158 **fine-toothed saw**
d Feinsäge *f*
f scie *f* à dents fines
r пила *f* с мелкими зубьями; пила для тонкого пропила

3159 **fine wood; luxury wood; precious wood; show wood**
d Edelholz *n*; Luxusholz *n*; Feinholz *n*
f bois *m* parfait; bois précieux
r древесина *f* ценных пород; ценная древесина; благородная древесина

* **fine wood shavings → 9697**

3160 **fine woodwork**
(products such as trim, panelling used for architectural woodwork to provide a high quality decorative appearance to rooms)
d feine Holzbearbeitungen *fpl*
f produits *mpl* de bois fins
r высококачественные деревянные изделия *npl*

* **fingered leaf → 6037**

3161 **finger joint; wedge joint**
(an end joint made up of several meshing wedges or fingers of wood bonded together with an adhesive)
d Keilzinkenverbindung *f*; Zinkenfuge *f*; Fingerzapfen *m*
f jointure *f* digitale; aboutage *m* digital
r шиповое соединение *n*

3162 **finger oak**
(Quercus digitata)
d Finger-Eiche *f*
f chêne *m* de doigt
r дуб *m* пальчатый

* **finger-parted leaf → 6037**

* **finger roll → 5552**

* **finger-veined leaf → 6041**

* **finish[ed] lumber → 3164**

3163 **finished size; net size; target size**

f dimension *f* prévue
r размер *m* после машинной обработки; чистовой размер; ожидаемый размер

3164 finish [wood]; finish[ed] lumber
(a collective term for upper grades of timber suitable for natural or stained finishing)
d Ausstattungsholz *n*; ausgebautes Holz *n*
f bois *m* de finition
r отделочная древесина *f*; лесоматериал *m* для столярных и внутренних отделочных работ; чистовой пиломатериал *m*

* **Finnish aspen → 2880**

3165 fir cone; chat
d Tannenzapfen *m*
f cône *m* de sapin
r пихтовая шишка *f*

3166 fire *v*; **burn** *v*
d abbrennen; ausbrennen
f échauffer; brûler
r сжигать; выжигать; жечь; обжигать; зажигать; поджигать; воспламеняться; гореть

* **fire annihilator → 3211**

* **fire beater → 3171**

* **fire belt → 3169**

3167 fireberry hawthorn; Copenhagen hawthorn; large-spurred thorn; scarlet hawthorn
(Crataegus coccinea; Crataegus intricata)
d Scharlach-Weißdorn *m*; Scharlachdorn *m*
f aubépine *f* écarlate
r боярышник *m* ярко-красный

3168 firebrand
d Brandpilz *m*
f marquage *m* au feu
r головня *f*; пожарная отметина *f*

3169 firebreak; fuelbreak; fire trace; fire belt
(strip of land maintained clear of trees and woody fuel used to stop or control the spread of fire)
d Feuerbahn *f*; Feurgestell *n*; Brandbahn *f*; Brandschneise *f*; Brandschutzstreifen *m*; Feuerschutzstreifen *m*
f coupe-feu *m*; ligne *f* d'appui; ligne pare-feu; ligne aménagée pour countrefeu
r линия *f* для встречного огня; противопожарная полоса *f*

* **fire brigade → 3183**

3170 fire camp
d Waldbrand-Lager *n*
f camp *m* principal
r базовый лагерь *m* при тушении пожара

* **fire [cause] class → 1601**

3171 fire chaser; towerman; fire[ ]fighter; fireman; firewoman; fire beater; smoke chaser
(a person observing for forest fire)
d Brandwart *m*; Feuerwehrmann *m*; Brandbekämpfungsspezialist *m*
f pompier *m*; pompière *f*; sapeur-pompier *m*; sapeuse-pompière *f*
r пожарник *m*

* **fire-cherry → 2881**

3172 fire climax; pyroclimax
(a community maintained by regular fires and therefore differing from any other climax community)
d Brandklimax *f*
f climax *m* pyrophytique; climax d'incendie; pyroclimax *m*
r послепожарный климакс *m*

3173 fire concentration
d Brandentstehenhäufigkeit *f*
f densité *f* des incendies
r частота *f* возникновения пожаров

3174 fire-cone looper; looper; measuring worm; spanworm; inch worm
(Eupithecia spermaphaga)
d Messwurm *m*; Kieferspanner *m*
f chenille *f* arpenteuse
r пяденица *f*

* **fire control → 3192**

3175 fire control planning
(planning for predicting, preventing, detecting, and influencing the spread and control of wildfires)
d Brandschutzplanung *f*
f prévision *f* de l'incendie; prévision contre l'incendie
r планирование *n* борьбы с лесным пожаром

f sartage *m*
r подсечное хозяйство *n*

3177 fire damage
d Brandschaden *m*
f dommage-incendie *m*
r ущерб *m*, причинённый пожаром; повреждение *n* от лесного пожара

3178 fire-dependent
(requiring one or more fires of varying frequency, timing, intensity, and size in order to achieve optimal conditions for population survival and/or growth)
d brandspezifisch
f dépendant d'incendie
r пожаро-зависящий

3179 fire district
d Brandgebiet *n*
f district *m* de protection contre les incendies; district de défense contre les incendies
r пожарный район *m*

3180 fire drying
d Feuertrocknung *f*
f séchage *m* par fumage à chaud
r огневая сушка *f*

3181 fire duration
(length of time that combustion occurs at a given point)
d Brandzeitdauer *f*
f durée *f* de l'incendie
r длительность *f* пожара

* **fire edge → 1189**

3182 fire exclusion
(policy of suppressing all wildland fires in an area)
d Brandverhütung *f*
f exclusion *f* de l'incendie
r недопущение *n* пожара

* **fire extinguisher → 3211**

* **fire extinguishing agent → 3211**

* **fire[ ]fighter → 3171**

3183 fire-fighting crew; fire brigade
d Löschmannschaft *f*
f équipe *f* d'incendie; équipe de lutte contre l'incendie; équipe d'extincteurs d'incendies
r бригада *f* по тушению

d Feuereinmessgerät *n*; Feuereinpeilgerät *n*
f repéreur *m* d'incendie; appareil *m* de relèvement
r прибор-пеленгатор *m* пожара

3185 fire[-free] interval; fire return interval
(time (in years) between two successive fires in a designated area)
d Feuerintervall *n*; feuerfreies Intervall *n*
f périodicité *f* des incendies
r интервал *m* возобновления пожара; периодичность *f* пожаров

* **firefront → 1189**

* **fire guard → 3191**

3186 fire[-]guard; fire stop
(strategically planned barrier, either manually or mechanically constructed, intended to stop a fire or retard its rate of spread)
d Feuerstop *m*; Feuerdamm *m*
f pare-feu *f*; barrière *f* artificielle pare-feu; digue *f* anti-incendie
r плотина *f* для задержания огня

* **fire hazard → 3205**

3187 fire hazard reduction; fuel reduction
d Verminderung *f* der Feuersgefahr
f réduction *f* du risque d'incendie; réduction du danger d'incendie
r уменьшение *n* пожарной опасности

3188 fire impact[s]
(immediately evident effect of fire on the ecosystem in terms of biophysical alterations)
d Feuerwirkung *f*
f effet *m* d'incendie
r воздействие *n* пожара

3189 fire-insurance tree
d Überbehälter *m* für den Waldbrandfall
f portegraine *m* de sauvegarde; semencier *m* conservé en prévision d'incendie
r подстрахованное семенное дерево *n*

* **fire-killed timber → 1196**

* **fire-killed wood → 1196**

* **fireline → 1189**

3190 fireline intensity
(rate of heat release per unit time per unit length of fire front)

f intensité *f* à la lisière de feu
r сила *f* пожара; интенсивность *f* линии огня

3191 fire look-out dispatcher; fire warden; fire guard
d Brandwächter *m*
f observateur-coordonnateur *m*; surveillant *m* [d'un lieu] d'incendie
r пожарник-наблюдатель *m*; пожарный наблюдатель *m*; пожарный надзиратель *m*

* **fireman → 3171**

3192 fire management; fuel management; fire control
(activities concerned with the protection of people, property and forest areas from wildfire)
d Waldbrandkontrolle *f*; Feuerleitung *f*
f asservissement *m* de l'incendie; conduite *m* de feu; défense *f* contre l'incendie
r наблюдение *n* за пожаром

3193 fire management area
(one or more parcels of land with clearly defined boundaries and with established fire management direction)
d Waldbrandbekämpfungsfläche *f*
f surface *f* organisée pour le lutte anti-incendie
r лесопожарная площадь *f*

3194 fire pack
(used to carry tools, equipment, and supplies on their backs)
d Waldbrandgepäck *n*
f paquetage *m* contre-incendie
r противопожарный набор *m*

3195 fire perimeter
d Brandumfang *m*; Brandumfanglinie *f*
f périmètre *m* de l'incendie; périmètre du foyer
r периметр *m* пожара

3196 fire-plough; fire-plow
d Waldbrandpflug *m*; Brandpflug *m*
f charrue *f* pour l'ouverture de tranchées par feu; charrue pare-feu
r противопожарный плуг *m*; плуг для тушения лесных пожаров

3197 fireproofing compound; fire-retarding agent; fire retardant
(substance that by chemical or physical action reduces flammability of combustibles)
d Feuerschutzmittel *n*; Flammschutzmittel *n*; feuerhemmender Stoff *m*
ignifugeant
r ингибитор *m* горения

3198 fireproof wood; fire-resistant wood; fireproofing wood
d feuerfestes Holz *n*
f bois *m* ignifuge
r огнестойкая древесина *f*; огнеупорная древесина

3199 fire protection cuttings
d Brandschutzfällungen *fpl*; Brandschutzhiebe *mpl*
f abattages *mpl* de protection contre l'incendie
r противопожарные разрывы *mpl*

3200 fire rake; rich tool
d Rich-Werkzeug *n*
f râteau *m* à feu
r пожарные грабли *mpl*

3201 fire regime
(patterns of fire occurrence, size, and severity - and sometimes, vegetation and fire effects as well - in a given area or ecosystem)
d Feuerregime *n*
f régime *m* des incendies
r режим *m* действия пожара

3202 fire-resistant species
(species with morphological characteristics that give it a lower probability of being injured or killed by fire)
d feuerresistente Art *f*
f espèce *f* résistante au feu
r огнеупорная порода *f*

3203 fire-resistant tree
(yellow shirts and green pants that are the trademarks of wildland firefighters)
d feuerresistenter Baum *m*
f arbre *m* résistant au feu
r огнестойкое дерево *n*

* **fire-resistant wood → 3198**

* **fire retardant → 3197**

* **fire-retarding agent → 3197**

* **fire return interval → 3185**

3204 fire ring
(of a tree)
d Feuerring *m*

3205 fire risk; fire hazard
(probability or chance of fire starting determined by the presence and activities of causative agents)
d Feuergefahr *f*; Brandgefahr *f*
f risque *m* d'incendie; danger *m* d'incendie
r опасность *f* возникновения пожара; пожарная опасность

3206 fire-scar; fire wound; burn scar
(a scar on a tree trunk left when tissues were damaged by fire)
d Feuerwunde *f*; Feuerverletzung *f*; Feuernarbe *f*; Waldbrandnarbe *f*
f cicatrice *f* d'incendie; cicatrice de feu; cicatrice de brûlure
r пожарный рубец *m*; след *m* огня; огневая рана *f*; пожарная пророст *f*; сухобокость *f*

* **fire season → 3434**

3207 fire severity
(degree to which a site has been altered or disrupted by fire)
d Feuerschwierigkeit *f*
f violence *f* d'un feu; gravité *f* d'un feu
r опасность *f* пожара

3208 fire shelter
d Feuerschutz *m*
f abri *m* contre le feu
r убежище *n* от пожара

3209 fire shovel
d Feuerschaufel *f*
f pelle *f* à feu
r кочегарная лопата *f*

* **fire-size class → 1601**

* **fire stop → 3186**

3210 fire supervision
d Feuerüberwachung *f*
f surveillance *f* du feu
r наблюдение *n* за огнем

3211 fire suppressant; fire extinguisher; fire extinguishing agent; fire annihilator
(agent directly applied to burning fuels to extinguish the flaming and smoldering or glowing stages of combustion)
d Feuerlöschmittel *n*
f produit *m* extincteur; produit d'extinction
r огнетушитель *m*

d Feuerdorn *m*
f buisson *m* ardent
r пираканта *f*

3213 fire tolerance
d Feuertoleranz *f*
f tolérance *f* d'incendie
r пожарная устойчивость *f*

3214 firetool cache
(supply of fire-fighting tools and equipment assembled in planned quantities or standard units at a strategic point for exclusive use in fire suppression)
d Löschgerätelager *n*
f dépôt *m* d'outillage contre le feu
r склад *m* тушительных орудий

3215 firetop; fireweed; willow herb; blooming willow
(Chamaenerion angustifolium; Epilobium angustifolium)
d Weidenröschen *n*
f épilobe *m* à feuille étroite; épilobe en épi
r иван-чай *m*; кипрей *m*; хаменериум *m* узколистный

3216 fire tower; lookout tower
(structures in forests which rise above the tree tops and are staffed during fire danger periods)
d Feuerbeobachtungsturm *m*; Feuerwachtturm *m*
f tour *f* ignifuge; tour à feu; tour de guet
r наблюдательная пожарная вышка *f*

* **fire trace → 3169**

3217 fire trail
(cleared area constructed around logging slash or other fire hazards in order to prevent the spread of fire to this hazardous material)
d Anmarschweg *m* zur Brandstelle
f voie *f* d'accès vers l'incendie; chemin *m* d'attaque
r путь *m* к пожарищу

* **fire warden → 3191**

3218 fire weather index; FWI
d meteorologischer Waldbrandgefahren-Index *m*
f indice *m* météorologique des incendies
r метеорологический индекс *m* пожара

* **fireweed → 3215**

* **firewood → 3740**

3219 **firewood billet**
d Brennknüppel *m*
f bûche *f* de chauffage
r топливный чурак *m*

3220 **firewood bundling machine; firewood processor**
d Feuerholz-Bündelmaschine *f*
f tronçonneuse-fendeuse *f*; scie-fendeuse *f* à bûches; fagoteuse *f*
r станок *m* для увязки дров и щепок; связочная машина *f*

3221 **firewood ends**
d Brennholzende *npl*
f bouts *mpl* de bois de feu; bouts de bois à brûler
r остатки *mpl* топливной древесины (длиною 0.46 - 1.68 м которые используются в столярных работах)

3222 **firewood harvesting**
d Brennholzeinschlag *m*
f ramassage *m* de bois de feu
r заготовка *f* дров

* **firewood processor → 3220**

3223 **fireworm**
d Blattraupe *f*
f tordeuse *f*
r гусеница *f* листовёртки

* **fire wound → 3206**

3224 **fir forest; firry forest; fir stand**
d Tannenwald *m*
f sapinière *f*
r пихтовый лес *m*; пихтовник *m*; пихтач *m*; пихтарник *m*

* **firing → 1191**

* **firm knot → 833**

3225 **firmwood**
(sound or solid wood in a log suitable for either chips or solid wood products such as lumber or veneer)
d festes Holz *n*
f bois *m* dur
r крепкая древесина *f*; твёрдая древесина

* **fir needle → 5707**

* **firry forest → 3224**

* **fir stand → 3224**

* **first attack → 4495**

3226 **first-class standard; stander**
d erstklassige Überständer *mpl*
f réserve *f* de première classe
r отборные специальные деревья *npl*, оставляемые в порослевом насаждении с одним оборотом рубки

3227 **first cleaning; first weeding; first thinning**
d erster Blenderhieb *m*; erste Blenderung *f*
f éclaircie *f* première; éclaircie initiale
r первое осветление *n*; первый приём *m* промежуточных рубок (рубки ухода)

* **first cutting → 6513**

* **first felling → 6513**

* **first growth → 282**

3228 **first pass**
(the first of two or more planned entries over a specific period of time (usually one rotation) to harvest timber)
d erster Durchlauf *m*
f premier passage *m*
r первый пропуск *m*

3229 **first-quality face; A face**
d erstklassiges Gesicht *n*
f face *f* de première qualité
r наружная пласть *f* из древесины высшего сорта

* **firsts → 9578**

* **first thinning → 3227**

* **first weeding → 3227**

* **fir-tree weevil → 4793**

* **fir wood → 7959**

* **fishpoison tree → 3284**

3230 **fish-protecting forest**
d fischgeschützender Wald *m*
f forêt *f* de protection du poisson
r рыбоохранный лес *m*

* **fissile wood → 1628**

* **fissure → 9507**

3231 **fit for felling; fellable; fit for cutting; fit for hewing; ripe for the harvest** *adj*
d geeignet zum Abhauen; abhauensreif
f exploitable d'abattage
r годный к рубке

3232 **five-leaved chastetree; negundo chastetree**
(Vitex negundo)
d chinesischer Keuschheitsbaum *m*
f troène *m* de Chine; nika *m*; lagundi *m*; vitex *m*
r прутняк *m* пятилистный; пятилистное целомудренное дерево *n*; ниргунди *m*

3233 **five-stamen tamarisk; saltcedar**
(Tamarix pentandra; Tamarix ramosissima)
d Sommertamariske *f*; kaspische Tamariske *f*
f tamarix *m* d'été
r тамариск *m* пятитычиночный

3234 **fixed-area plot sampling method**
(controlled cruise method where small plots of a fixed size are used to sample a portion of a forest area to obtain information that can be used to describe the whole area)
d Methode *f* der Festflächenstichprobenabnahme
f méthode *f* d'estimation par placettes d'échantillon fixes
r метод *m* фиксированной закладки пробных площадей

3235 **fixed-boom grapple**
d Festausleger-Greifer *m*
f grappin *m* de flèche fixe
r пачковый захват *m* на неповоротной в горизонтальной плоскости стреле (для бесчокерной трелёвки леса)

3236 **fixed-boom machine**
d Festauslegerkran *m*
f cran *m* à flèche fixe
r трелёвочная машина *f* с неповорачивающейся в горизонтальной плоскости стрелой

* **fixed pricelist of wood → 7123**

* **fixed tree → 9053**

3237 **fixed-wire logging; single-wire logging**
d Eintragung *f* mit festgelegtem Seil
канату

* **flabbiness → 6652**

* **flaccidity → 6652**

3238 **flaccid leaf**
d schlaffes Blatt *n*
f feuille *f* flasque
r увядший лист *m*

3239 **flagging**
(marking by colo[u]red plastic ribbon attached to trees or stakes to make boundaries, stakes, and other markers visible)
d Fahnenmarkierung *f*
f marquage *m* par pavillon
r обозначение *n* флагом

3240 **flag-shaped crown**
d fahnenförmige Krone *f*
f cime *f* en forme d'étendard
r флагообразная крона *f*

* **flail delimber → 1426**

3241 **flake**
(a small flat wood particle of predetermined dimensions, uniform thickness, with fiber direction in essentially in the same plane)
d Flocke *f*; Schuppe *f*; Lamelle *f*
f paillette *f*; flocon *m* de bois; feuillet *m*
r комок *m*; кусочек *m*

3242 **flakeboard**
(wood chipboard the surface of which is in the form of flakes)
d Spanplatte *f*
f panneau *m* à parement en flocons; panneau de flocons
r древесностружечная плита *f* из станочной стружки

3243 **flaking of the bark; bark exfoliation; bark flaking; bark scaling**
d Abblätterung *f* der Rinde
f détachement *m* de l'écorce en feuillets
r отслаивание *n* коры; отлуп *m* коры

* **flamboyant → 3249**

3244 **flame azalea**
(Rhododendron calendulaceum; Rhododendron luteum)
d gelbe Alpenrose *f*

3245 flame [bottle] tree; white kurrajong; lacebark
(Brachychiton discolor; Brachychiton acerifolia; Sterculia acerifolia)
d [Lacebark-]Flammenbaum *m*; weißer Kurrajong *m*
f arbre *m* ampolla; flamme *f* australienne
r стеркулия *f* клёнолистная

3246 flame figure; flamy figure; crossfire
d geflammte Textur *f*
f veinure *f* ondulée
r пламеобразная текстура *f*; пламенистый узор *m*

3247 flame-leaf sumac[h]; dwarf sumac[h]; winged suma[h]; black sumac[h]; shining sumac[h]
(Rhus copallina)
d feuerroter Essigbaum *m*
f sumac *m* à copal; sumac ailé; sumac brillant
r сумах *m* копаловый

3248 flame of the forest; palash mool
(Butea monosperma)
d Plossobaum *m*; Kinobaum *m*; Palasabaum *m*
f palas *m*
r бутия *f* односемянная; "пламя *n* леса"; палаш *m*

3249 flame[-]tree; flamboyant; peacock flower; mohur tree
(Delonix regia; Poinciana regia)
d Flamboyant *m*
f flamboyant *m*; poinciane *m*
r делоникс *m* королевский

* **flamy birch → 3122**

* **flamy figure → 3246**

* **flannelbush → 1259**

* **flap planting → 289**

3250 flared butt
d erweiterte Stammende *f*
f bille *f* épanouie; bille mandrinée
r увеличенный комель *m* дерева

3251 flat-blade chain saw
d Geradblattsäge *f*
f scie *f* à lame plate
r цепная пила *f* с прямой пильной шиной; пила с прямым полотном

f cime *f* plate
r плоская крона *f*

3253 flat cutting; flatting
(of lumber)
d Schwarteneinschnitt *m*
f sciage *m* par trait bas
r раскрой *m* по ширине

* **flat grain → 7914**

3254 flat-grain timber; flat-grain[ed] lumber; flat-sawn timber; plain[-]sawn timber; slash-grain[ed] timber; slash-sawn timber
(timber converted so that the growth rings meet the face in any part at an angle of less than 45 degrees)
d Tangentialschnittholz *n*
f bois *m* scié sur dosse; bois débité en plot
r пиломатериал *m* тангенциального распила; пиломатериал тангенциальной распиловки

3255 flat-knife chipper
d Flachmesser-Zerrhacker *m*
f déchiqueteuse *f* à couteaux plats; déchiqueteuse à lames plates
r рубильная машина *f* с плоскими ножами

3256 flatland forest; flatwood (*US*)
(a site with flat to gently-sloping topography with relatively poorly-drained, sandy soils)
d Flachlandwald *m*
f forêt *f* à plain
r лес *m* на равнине

3257 flat planting; straight planting
(planting trees directly into the ground without beds or, in some cases, without first moving logging debris)
d flache Pflanzung *f*
f plantation *f* plate
r плоскостной посев *m*; нерельефный посев

3258 flat-sawed; flat[-]sawn; plain-sawed; plain[-]sawn; bastard sawn; slash[-]sawn *adj*
(of timber converted so that the growth layers meet the face of any part at an angle less than 45 degrees)
d flachgesägt
f scié sur dosse; débité en plot
r распилённый по касательной к годовым кольцам; тангенциального распила; тангенциальной распиловки; тангенциально распилённый

* **flat-sawn timber → 3254**

3259 flat stacking
d Flachstapelung *f*
f empilage *m* à plat
r плотная укладка *f* (щепы)

* **flatting → 3253**

* **flat-topped → 1045**

* **flat-top saw → 7003**

* **flatwood → 3256**

3260 flatwoods plum; black sloe
(Prunus umbellata)
d doldenblütige Kirsche *f*
f prunier *m* ombellé
r слива *f* зонтиковидная

* **flaw → 9507**

* **flawless wood → 9578**

3261 flax family
(Linaceae)
d Leingewächse *pl*
f linacées *pl*
r льновые *pl*

* **flay → 2280**

* **flayer → 556**

* **F-layer → 3086**

3262 fleshy fruit
d fleischartige Frucht *f*
f fruit *m* charnu
r сочный плод *m*

3263 fleshy root
d fleischige Wurzel *f*
f racine *f* charnue
r мясистый корень *m*

3264 flexibility of wood
d Holzbiegsamkeit *f*
f flexibilité *f* de bois
r гибкость *f* дерева; гибкость древесины

* **flexible wood → 2776**

* **flipper → 5062**

3265 flitch matching; book matching
(sequentially saw of the lumber from the same log)
d spiegelbildliche Furnierzusammensetzung *f*
f appareillage *m* retourné; assemblage *m* au raccord
r соответствующий подбор *m* шпона

3266 flittern
d Eichenheister *m*
f haute tige *f* de chêne
r крупный дубовый саженец *m*

3267 float
(a measure of timber equal to 18 loads)
d Floß *n*
f flot *m*
r единица *f* измерения круглых лесоматериалов

* **float → 8584**

* **floatable → 2628**

* **float[ed] timber → 3279**

* **floated wood → 3279**

3268 floater [log]
d Schwimmerholz *n*
f bois *m* flottable
r лесоматериал *m*, пригодный к сплаву в свежесрубленном виде; лесоматериал, удерживаемый на плаву; бревно *n*, пригодное к сплаву

3269 float *v* timber
(to transport round timber down a waterway)
d abschwemmen; aufschwemmen; Holz [ver]flößen; Holz treiben
f flotter bois
r сплавлять лес

* **floating basin → 5105**

3270 floating basins scheme
d Bassinschema *n* der Flößerei
f schème *m* de flottage par bassin
r бассейновая схема *f* лесосплава

* **floating channel → 3277**

* **floating creek → 3277**

* **floating crew → 2630**

f distance *f* de flottage
r лесосплавная дистанция *f*

3272 **floating district**
d Flößereibezirk *m*
f district *m* de flottage
r сплавной район *m*

3273 **floating in raft sections; floating with section-rafts**
d Gestörflößerei *f*; Gestörtrift *f*
f flottage *m* de bois par radeaux fragmentés
r лесосплав *m* в сплоточных единицах; плотовой сплав *m*

3274 **floating logway; pullboat**
d Floßdurchlass *m*
f passe *f* à bois flottants
r плавающий лесотранспортёр *m*

3275 **floating mark**
d Flößereimarke *f*; Triftmarke *f*
f marque *f* de [propriété de bois destiné au] flottage
r сплавная марка *f*

* **floating period → 3278**

3276 **floating regulation; driving regulation**
d Flößereistatut *n*; Flöß[erei]ordnung *f*
f règlement *m* pour le flottage; statut *m* pour le flottage
r правила *npl* сплава

* **floating right → 6945**

3277 **floating route; floatway; float road; floating channel; floating creek; driftway; drift-path; driving road**
(a channel cleared in a swamp and used to float cypress logs from the woods to the boom at the river or mill)
d Floßweg *m*; Floßgasse *f*; Floßstraße *f*; Triftweg *m*; Triftkanal *m*; Triftpfad *m*; Triftschtraße *f*
f canal *m* de flottage; cours *m* d'eau à flottage; sentier *m* de halage; voie *f* de flottage
r сплавная трасса *f*; лесосплавной путь *m*; лесосплавный канал *m*; сплавной канал; лесосплавный ход *m*; путь молевого сплава

3278 **floating season; floating period; driving season**
d Triftperiode *f*; Floßperiode *f*
f époque *f* du flottage

* **floating tree boom → 914**

* **tree boom → 914**

* **floating with section-rafts → 3273**

3279 **floating wood; floatwood; float[ed] timber; drifting wood; floating timber; floated wood; drift[wood]; raftwood; rafted wood**
(any wood from trees or structures which has floated in water)
d Flößholz *n*; geflößtes Holz *n*; Treibholz *n*; Schwemmenholz *n*; Wasserholz *n*; Wasserware *f*; Triftholz *n*
f bois *m* flotté; bois flottant; bois échappé; bois de flottage
r сплавная древесина *f*; сплавной лесоматериал *m*; сплавляемый в плотах лес *m*; сплавной лес; плавник *m*; сплавленный лес

3280 **floating wood landing; receiving place; collecting place; collecting depot**
(of the boles in float)
d Einbindestelle *f*; Rollplatz *m*
f lieu *m* d'assemblage [de bois flottés]; place *f* d'atterrissage de bois flottés
r место *n* связывания; бревенчатый каток *m*

* **float road → 3277**

* **floatway → 3277**

* **floatwood → 3279**

3281 **flooded forest**
d bewässerter Wald *m*
f forêt *f* inondée
r затопленный лес *m*

* **flooded gum → 3550**

3282 **flooding; inundation**
(overflow, high stream flow, or inundation that comes from a river or other body of water and causes or threatens to cause property damage)
d Überschwemmung *f*; Überflutung *f*; Flut *f*
f inondation *f*; crue *f*
r затопление *n*; наводнение *n*; заливание *n*; наводок *m*

* **floodplain forest → 5192**

3283 **floor**
(a tongue and grooved piece of wood lumber

f plancher *m*; varangue *f*; radier *m*
r пол *m*

* **florecillo → 5275**

* **Florida basswood → 237**

* **Florida dogwood → 1964**

* **Florida elm → 247**

* **Florida evergreen blueberry → 7523**

3284 **Florida fishpoison tree; fishpoison tree**
(Piscidia piscipula)
d Hundholzbaum *m*
f piscidia *m*
r писцидиевое дерево *n*

3285 **Florida maple; southern sugar maple**
(Acer floridanum; Acer barbatum)
d Florida-Ahorn *m*
f érable *m* barbu
r клён *m* флоридский; клён бородатый

3286 **Florida strangler fig; strangler fig**
(Ficus aurea)
d Würgerfeige *f*
f figuier *m* blanc
r фикус *m* золотистый

3287 **Florida [swamp] privet; ink-bush**
(Forestiera segregata; Myrica segregata; Forestiera porulosa; Adelia porulosa; Adelia pinetorum)
d Florida-Liguster *m*; Florida-Hartriegel *m*
f troène *m* de Floride
r аделия *f* флоридская; форестиера *f* флоридская

3288 **Florida yew**
(Taxus floridana)
d Florida-Eibe *f*
f if *m* de Floride
r тис *m* флоридский

* **florist's willow → 3880**

* **flour berry → 8323**

* **flourishing stand → 8536**

3289 **flow chain tray sorter; planing mill sorter**
d Sortierkettenförderer *m*; Sortierkettenzug *m*
f raboteuse-trieuse *f*
r сортировочное устройство *n* потока площадка *f*

3290 **flower** *v*
d blühen
f fleurir
r цвести

3291 **flower[ing]**
d Blühen *n*
f floraison *f*
r цветение *n*

3292 **flowering ash; fragrant ash; manna[-ash]**
(Fraxinus cuspidata; Fraxinus ornus)
d Blütenesche *f*; Blumenesche *f*; Mannaesche *f*
f frêne *m* à fleurs; frêne fleuri; arbre *m* à la manne; orne *m*
r белый ясень *m*; ясень цветочный; ясень остролистный

* **flowering dogwood → 1964**

* **flowering maple → 22**

* **flowering raspberry → 6640**

* **flume → 9281**

* **fluoride injury → 6764**

3293 **flush** *v*; **germinate** *v*
d sprossen
f germer
r давать отростки; давать побеги

3294 **flush cutting**
(a pruning cut to remove a tree limb in which the cut is completely flush with the tree)
d Glattschnitt *m*
f coupe *f* à ras
r обрезка *f* заподлицо; обрубка *f* заподлицо; резание *n* в плоскости движения пилы

* **flush end → 871**

* **flushing → 1020, 7807**

* **fluted → 1230**

* **fluted figure → 2144**

3295 **fluted parting tool**
(a tool with a wedge chaped section which has a flute on the wider of the two edges)
d gerillter Einstechmeißel *m*

3296 **fluting**
(formation of flutes or channels on the roundwood surface)
d Kannelierung *f*; Profilierung *f*
f strie *f*; cannelure *f*; rupture *f* par flexion
r рифление *n*; гофрирование *n*

* **flying → 1247**

3297 **flying back of the wood; kicking up of the wood; kickback of wood**
(the sudden and dangerous jump of the butt of a falling tree as it comes down; also, the abrupt and dangerous backward movement of a chainsaw toward the operator)
d Rückschlagen *n* des Holzes
f détente *f* brusque du bois; rebond *m*; recul *m*
r обратный удар *m* древесины; отбрасывание *n* назад

3298 **flying dutchman**
(a stout snatch block through which the main line runs, and which, anchored by a cable or strap, keeps the main line pulled out of straight so as to clear obstacles)
d fliegender Holländer *m*
f flying-dutchman *m*
r открывающийся отводной блок *m* тягового каната

* **FNC → 3692**

* **fodder → 4868**

3299 **folding caliper**
d Taschenkluppe *f*
f compas *m* forestier pliant
r складная мерная вилка *f*

3300 **folding cruiser stick**
d Gelenkmaßstab *m*
f mètre *m* pliant
r складная таксационная линейка *f*

3301 **folding tower**
d Kippmast *m*
f fût *m* pliant
r складная мачта *f*

3302 **foliaceous; foliated**
d laubartig; blattartig
f foliacé
r листоватый

3303 **foliage; leafage; leafness; shoot; greenery; umbrage**
d Belaubung *f*
f feuillage *m*; système *m* foliacé; caduc *m*
r листва *f*; облиствение *n*

* **foliage bud → 4859**

* **foliage expansion → 9084**

3304 **foliage profile**
(woodland vegetation has a vertical structure. the foliage profile of a stand can be divided into layers: field, shrub and canopy layers)
d Blattwerkprofil *n*
f profil *m* de feuillage
r профиль *m* листвы

3305 **foliage spraying**
d Blattwerkbespritzung *f*
f pulvérisation *f* de feuillage
r опрыскивание *n* крон гербицидами

* **foliage tree → 1048**

3306 **foliar; foliate** *adj*
d laubig; blätterig
f foliaire
r лиственный

3307 **foliar analysis; leaf analysis**
(chemical evaluation of the status of plant nutrients or the plant-nutrient requirements of a soil by the analysis of leaves or needles)
d Blattanalyse *f*
f analyse *f* foliaire; analyse de feuille
r листовой анализ *m*; анализ листа

* **foliate → 3306, 7531**

* **foliated → 3302**

* **foliose → 2389**

* **fomes root rot → 6764**

3308 **foodchain; trophic chain**
(series of spatially associated species, having feeding relations, each of which lives (at least in part) as a predator, parasite, or absorber of the next lower down in the series)
d Nahrungskette *f*
f chaîne *f* alimentaire; chaîne trophique
r пищевая цепь *f*; трофическая связь *f*

* **food forest → 3313**

3309 food plot; food patch
(area sown or planted with grains, legumes, berries, mast-producing woody species, or other plants to provide food for wildlife)
d Wildacker *m*
f terrain *m* à viander
r питательная учётная площадь *f*

3310 fool's huckleberry; rusty menziesia
(Menziesia ferruginea)
d frostempfindliche Menziesie *f*
f menziésie *f* ferrugineuse
r менцизия *f* ржавая

3311 footage
d Gesamtmenge *f*; Gesamtlänge *f*
f estimation *f* en pieds-planches; volume *m* total en pieds-planches
r объём *m* древесины в кубических футах; общая масса *f* в кубических футах

3312 Foothill paloverde; yellow paloverde
(Parkinsonia microphylla; Cercidium microphyllum)
d kleinblättrige Parkinsonie *f*
f parcinsonia *m* à petites feuilles
r паркинсония *f* мелколистная

3313 forage forest; food forest
(growing close-planted, fast-growing tree species that can be mechanically harvested)
d Futterwald *m*
f forêt *f* de fourrage
r продовольственный лес *m*

3314 Forbes cypress; tecate cypress
(Cupressus forbesii; Cupressus guadalupensis)
d Forbes-Zypresse *f*
f cyprès *m* de Forbes
r кипарис *m* Форбеса

* **forbiddance of cutting → 6577**

3315 forced burning
d beschleunigtes Überbrennen *n*
f [éco]brûlage *m* accéléré
r ускоренное пережигание *n*

3316 foreign body
(a non-wood object in the wood)
d Fremdkörper *m*; Fremdstoff *m*
f corps *m* étranger
r инородное тело *n*

3317 foreign species timber; foreign wood;
f bois *m* exotique
r лесоматериал *m* иноземных пород; импортный лесоматериал

* **forest → 93**

3318 forest; silva lat
(complex ecological system in which trees are the dominant life form)
d Wald *m*; Forst *m*
f forêt *f*
r лес *m*

* **forest absence → 4753**

3319 forest academy
d Forstakademie *f*
f académie *f* forestière
r лесная академия *f*

* **forest access road → 3463**

3320 forest administration
d Forstadministration *f*
f administration *f* forestière
r лесная администрация *f*

3321 forest aesthetics
d Forstästhetik *f*
f esthétique *f* forestière
r лесная эстетика *f*

3322 forest affairs
d Waldgeschäfte *fpl*
f affaires *fpl* forestières
r лесное дело *n*

3323 forest agency; forest bureau; forest office
d Forestbüro *n*; Forestbürogeschäftszimmer *n*
f bureau *m* forestier
r агентство *n* по лесному хозяйству; лесная канцелярия *f*

* **forest aisle → 3564**

3324 forest[al]; sylvatic; silvan; sylvan *adj*
(of or pertaining to forests)
d waldig
f forestier
r лесной

3325 forest albedo
d Forstrückstrahlvermögen *n*
f albédo *m* forestier
r отражательная способность *f* леса

* **forestal situation → 3358**

3326 **forestal statics**
d forstliche Statik *f*
f statique *f* forestière
r лесная статика *f*

* **forest [a]melioration → 7611**

3327 **forest ants**
(Formica spp.)
d Waldameisen *fpl*
f fourmis *mpl* forestières
r лесные муравьи *mpl*

3328 **Forest Appeals Commission; FAC**
d Forst-Rekurskommission *f*
f Commission *f* de recours forestière
r лесная аппеляционная комиссия *f*

* **forest appraisal → 3411**

* **forest appraiser → 3410**

* **forest area → 3551**

3329 **forest area determination; area determination**
d Waldflächebestimmung *f*
f détermination *f* d'aire forestière
r дешифрирование *n* участков

3330 **forest area under management; forest management area; [forest] management zone**
(stands or forest types that require similar management practices and can be grouped for treatment as a management unit)
d Forstverwaltungsgebiet *n*; Betriebsfläche *f*
f zone *f* d'exploitation forestière; surface *f* de gestion
r площадь *f* [лесо]хозяйства; зона *f* лесного хозяйства

* **forest arterial road → 3463**

* **forest assessment → 3411**

* **forest assessor → 3410**

* **forest assets → 3346**

3331 **forest association; forest [living] community; forest phytoc[o]enosis**
d Waldgesellschaft *f*; forstliche Lebensgemeinschaft *f*
f association *f* forestière
r лесная ассоциация *f*; лесное [со]общество *n*; лесной фитоценоз *m*

* **forestation → 94**

* **forest backbone road → 3463**

* **forest balance → 9486**

* **forest-barked → 552**

3332 **forest bean goose; bean goose**
(Anser fabalis)
d Saatgans *f*
f oie *f* des moissons
r гуменник *m*

3333 **forest beat; forest section; forest patrol**
(a minor executive forest charge, essentially protective, commonly a subdivision of a range; the charge of a forest guard of comparable junior rank)
d Begang *m*; Belauf *m*
f triage *m* forestier
r лесной обход *m*; обход лесника

3334 **forest bed**
d Forstschicht *f*
f couche *f* forestière
r древесный слой *m* (ископаемый)

3335 **forest belt; woodland belt; wooded belt; belt of trees**
(strip of trees left standing between two clearings)
d Waldgürtel *m*; Waldstreifen *m*; Baumgürtel *m*; Forstgürtel *m*; Kulisse *f*
f ceinture *f* de forêt; ceinture forestière; coulisse *f*
r лесной пояс *m*; лесная полоса *f*; кулиса *f* (деревья, оставленные при сводке для лесовозобновления)

3336 **forest berry**
d Waldbeere *f*
f baie *f* forestière
r лесная ягода *f*

3337 **forest biological diversity; forest iodiversity**
d forstbiologische Vielfalt *f*
f diversité *f* biologique forestière
r биологическое разнообразие *n* леса

3338 **forest biological reserve**
d waldbiologisches Reservat *n*

3339 **forest biology**
d Waldbiologie *f*
f biologie *f* forestière
r лесная биология *f*; биология леса

* **forest biomass → 9490**

* **forest block → 1821**

3340 **forest bookkeeping**
d forstliche Buchführung *f*
f comptabilité *f* forestière
r лесное счетоводство *n*

* **forest border → 2761**

* **forest border cutting → 6167**

3341 **forest botany**
d Forstbotanik *f*
f botanique *f* forestière
r лесная ботаника *f*

* **forest boundary → 2761**

3342 **forest bow-saw; bow-saw; arm-saw; coping saw**
d Bogensäge *f*; Bügelsäge *f*
f scie *f* [d'abattage] à archet; scie à arc
r пила *f* с дугообразной шиной; лучковая пила

3343 **forest bug; cherry stink bug**
(Pentatoma rufipes)
d rotbeinige Baumwanze *f*
f punaise *f* à pattes rousses
r щитник *m* красноногий

* **forest bureau → 3323**

3344 **forest cadastre; forest cadastral register**
d Waldkataster *m*
f cadastre *m* forestier
r лесной кадастр *m*

3345 **forest canopy; forest cover; overhead cover; canopy cover**
(more or less continuous cover of branches and foliage formed collectively by adjacent tree crowns; the overhead branches and leaves of streamside vegetation)
d Waldschluss *m*; Bestandesschirm *m*
f manteau *m* forestier; couvert *m* forestier; couvert du peuplement
r полог *m* [леса]; лесной полог; древесный полог; полог насаждения; лесной покров

d Waldkapital *n*; forstliches Kapital *n*; Waldvermögen *n*
f capital *m* forestier; capital forestal; capital-forêt *m*
r лесной капитал *m*; лесное имущество *n*

3347 **forest cemetery**
d Waldfriedhof *m*
f cimetière *m* forestier
r лесное кладбище *n*

3348 **forest certification**
d Waldbeglaubigung *f*; Waldzertifizieren *n*
f certification *f* forestière
r лесная аттестация *f*

* **forest checking station → 3413**

3349 **forest chemicals**
(chemical substances or formulations that perform important functions in forest management)
d Forstchemikalien *fpl*
f produits *mpl* chimiques forestiers
r лесохимикаты *mpl*

* **forest chemistry → 2370**

3350 **forest chestnut-winged starling; chestnut-winged starling**
(Onychognathus fulgidus)
d Kastanienflügelstar *m*
f rufipenne *f* de forêt; étourneau *m* roupenne
r каштановокрылый скворец *m*

* **forest circle → 3382**

3351 **forest clean-up; forest cleaning**
d Waldreinigung *f*
f nettoiement *m* de forêt
r очистка *f* леса

3352 **forest clearing**
(reduction of density, with or without recovery of the product)
d Abholzen *n*
f essartage *m* des forêts
r сведение *n* лесов

3353 **forest climatology**
d Waldklimatologie *f*
f climatologie *f* forestière
r лесная климатология *f*

3354 **forest climax**
d Waldklimax *f*

3355 **forest company; logging company**
d Waldaktiengesellschaft *f*
f société *f* forestière
r лесная компания *f*; лесозаготовительная фирма *f*

* **forest compartment → 1821**

3356 **forest concession; forest privilege; timber concession**
d Forstkonzession *f*; Waldkonzession *f*
f concession *f* forestière; concession de bois
r лесная концессия *f*; лесозаготовительная концессия

3357 **forest condition inventory**
d Waldzustandsinventur *f*; Zustandserfassung *f* des Waldes
f inventaire *m* de condition forestière
r инвентаризация *f* состояния леса

3358 **forest conditions; forestal situation**
d forstliches Verhältnisse *npl*
f conditions *fpl* forestières; vie *f* de la forêt
r лесные условия *npl*

3359 **forest conservation; forest protection**
d Erhaltung *f* des Waldes; Forstschutz *m*; Waldschutz *m*; Schutz *m* des Waldes; Walderhaltung *f*
f conservation *f* de la forêt; protection *f* forestière
r охрана *f* леса; сохранение *n* леса; лесозащита *f*

3360 **forest conservation law; forest protection law**
d Walderhaltungsgesetz *n*
f loi *f* de conservation des forêts
r закон *m* об охране лесов

3361 **forest conservator**
d Waldkonservator *m*
f conservateur *m* des forêts
r служащий *m* управления охраны леса

* **forest consolidation → 1883**

3362 **forest contour map**
d Umrissforstkarte *f*
f carte *f* en courbes forestière
r лесная контурная карта *f*

* **forest conversion → 226**

* **forest country → 3551**

* **forest crop → 3605**

* **forest cropping → 9523**

* **forest crop rotation → 226**

3363 **forest crop type; timber type; stand type; forest [stand] type**
(group of forested areas or stands of similar composition which differentiates it from other such groups)
d Waldtyp *m*; Waldart *f*; Bestandesart *f*
f type *m* forestier; type de peuplement
r тип *m* леса; род *m* насаждения

* **forest cruising → 3617**

3364 **forest cultivation**
d Waldbewirtschaftung *f*
f cultivation *f* forestière
r выращивание *n* лесов

3365 **forest cultivator**
d Waldgrubber *m*
f cultivateur *m* forestier
r лесной культиватор *m*

* **forest culture → 3528**

3366 **forest cut by compartment**
d periodischer Schlagwald *m*
f coupe *f* de la forêt en compartiments
r раздельная рубка *f* леса

* **forest cutting plan → 6332**

* **forest damage → 2241**

* **forest day → 340**

* **forest death → 2694**

* **forest debris → 9707**

3367 **forest decay; forest degradation**
(a reduction in the biomass, productivity or biodiversity of a forest, and alterations in ecological functions)
d Waldverwesung *f*
f dévitalisation *f* des forêts; dégradation *f* des forêts
r вырождение *n* лесов

3368 **forest decline; forest dieback**
(progressive dying from the extremity of the

f dépérissement *m* terminal des arbres
r гибель *f* леса

* **forest deficiency → 4753**

3369 **forest defoliator**
d forstlicher Laubfresser *m*
f phyllophage *m* forestier
r лесной листоед *m*

* **forest degradation → 3367**

3370 **forest density; forest stocking; stocking; density of stand; stand density; volume density; crowding level**
(quantity of trees per unit area, usually evaluated in terms of basal area, crown cover and stocking)
d Walddichte *f*; Bewaldungsdichte *f*; Bestandesdichte *f*; Bestockungsdichte *f*
f densité *f* relative de la forêt; proportion *f* de surface occupée par la forêt; densité du peuplement
r густота *f* леса; густота древостоя; полнота *f* [лесонасаждения]; сомкнутость *f* [лесо]насаждения; полнота насаждения; полнота древостоя

* **forest depot → 8664**

3371 **forest depot camp; logging depot**
(a logging camp comprising several buildings which are to be unused for more than one year)
d Forstpunkt *m*; forstlicher Hauungsbetrieb *m*; Holzaufbereitungsstelle *f*
f quartier *m* général; dépôt *m*
r лесопункт *m*

3372 **forest design plan**
(longterm outline design plan - 20 years or more)
d Forstentwurfplan *m*; Walddesignplan *n*
f plan *m* de développement forestier
r лесоплан *m*

* **forest destroyer → 3523**

* **forest deterioration → 2241**

3373 **forest devastation; forest liquidation; forest destruction**
d Walddevastation *f*; Waldabschwendung *f*; Waldverwüstung *f*; Vernichtung *f* des Waldes; Waldvernichtung *f*
f liquidation *f* des forêts; destruction *f* de la
уничтожение *n* леса

3374 **forest development**
d Waldentwicklung *f*
f développement *m* des forêts
r развитие *n* леса

3375 **forest development plan**
(operational plan guided by the principles of integrated resource management which details the logistics of timber harvesting usually over a period of five years)
d Waldentwicklungsplan *m*
f plan *m* de développement des forêts
r план *m* лесоразработки

3376 **forest development program; FDP**
d Waldentwicklungsprogramm *n*
f programme *m* de développement forestier
r программа *f* лесопользования

3377 **Forest Development Review Committee; FDRC**
d Ausschuss *m* für die Waldentwicklungbericht
f Comité *m* pour le rapport de développement forestier
r Комиссия *f* по приёмке лесопользования

3378 **forest development roads; forest system roads**
(forest development transportation system)
d forstwirtschaftliches Wegenetz *n*
f chemins *mpl* forestiers; voirie *f* forestière
r временные лесные дороги *fpl*

* **forest devil → 8808**

* **forest dieback → 3368**

3379 **forest diseases**
d Waldkrankheiten *fpl*
f maladies *fpl* de forêts
r болезни *fpl* леса

3380 **forest dispatching service; forest inspection service**
d Aufsichtforstamt *n*; Forstinspektion[sbeamt] *f*
f inspection *f* forestière
r лесная инспекция *f*

3381 **forest distribution**
d Waldverbreitung *f*
f distribution *f* de la forêt
r распространение *n* леса

3382 **forest district; forest circle**
d Forstrevier *n*; Forstbezirk *m*; Oberförsterei *f*;

f district *m* forestier; cantonnement *m* forestier
r лесной округ *m*

3383 forest district map
d Forstrevierkarte *f*
f carte *f* de district forestier
r карта *f* лесничества

3384 forest division
d Waldabteilung *f*
f division *f* forestière
r лесное деление *n*

3385 forest draining
d Waldentwässerung *f*
f drainage *m* de la forêt
r осушение *n* леса; лесоосушение *n*

3386 forest-dried wood; wood dried in the forest
d waldtrockenes Holz *n*
f bois *m* séché à la forêt
r лесоматериал *m*, высушенный в лесу; древесина *f*, высохшая в лесу

3387 forest drill seeder
(mechanical device for sowing seed in furrowed lines)
d Forstdrill[sä]maschine *f*
f semoir *m* en ligne
r рядовая сеялка *f* для посева лесных семян

* **forest duff → 4999**

* **forest dug seedling → 9419**

3388 forest dusting; forest spraying
d Waldbestäubung *f*
f poudrage *m* de la forêt; pulvérisation *f* en forêt
r опыливание *n* леса; опыление *n* леса

3389 forest dweller
d Waldbewohner *m*
f habitant *m* forestier
r лесной обитатель *m*

3390 forest ecology
d Forstökologie *f*; Waldökologie *f*
f écologie *f* forestière
r лесная экология *f*

3391 forest economics; forest economy
(generally, that branch of forestry concerned with the forest as a productive asset subject to economic principles)
d Forstökonomie *f*
r экономика *f* леса; экономика лесного хозяйства

3392 forest ecoregion
d Wuchsgebiet *n*
f écozone *f* forestière
r лесной экорайон *m*

3393 forest ecosystem network; FEN
(planned landscape zone that serves to maintain or restore the natural connectivity within a landscape unit)
d Forstökosystemnetz *n*
f réseau *m* d'écosystèmes forestiers
r сеть *f* лесных экосистем

3394 forested; stocked; wooded; covered with woods
d bewaldet; bestockt; waldbedeckt; waldig; baumbestanden; beholzt
f boisé; couvert de bois; peuplé d'essences forestières
r покрытый лесом; облесённый; лесистый; заросший деревьями; обсаженный деревьями

* **forest[ed] area → 9542**

3395 forest[ed] ecosystem
(a spatially defined area suitable for tree growth that includes all interacting organisms and environmental components within its boundaries)
d Forstökosystem *n*
f écosystème *m* forestier
r экосистема *f* леса; лесная экосистема

3396 forested forage
d Forstfutter *n*
f fourrage *m* forestier
r лесные кормовые растения *npl*

* **forest edge → 2761**

3397 forested lowland
d bewaldetes Tiefland *n*
f plaine *f* boisée
r полесье *n*

* **forest[ed] tract → 9542**

3398 forested tract size
(area of forest within the contiguous tract containing each forest inventory and analysis sample plot)
d bewaldete Gebietgröße *f*

3399 forested wetland
(area characterized by woody vegetation taller than 20 feet where soil is at least periodically saturated or covered by water)
d bewaldetes Feuchtgebiet *n*
f terre *f* inondée forestière
r лесная заболоченная территория *f*

* **forest enclosure → 1821**

3400 forest engineer; reforestation forester
d Forstingenieur *m*
f ingénieur *m* forestier; ingénieure *f* forestière
r лесной инженер *m*; инженер лесных культур

3401 forest engineering; forest technology; forest technique
d Forstingenieurwesen *n*; Forsttechnologie *f*; Forsttechnik *f*
f génie *m* forestier; technologie *f* forestière
r лесоинженерное дело *n*; лесная технология *f*; технология ведения лесного хозяйства

3402 forest engines; logging machines; wood's equipments; forestry machinery
(machines for yarding, moving or handling logs, trees, chunks, trailers and related materials or equipment)
d Forstmaschinen *fpl*
f appareils *mpl* forestiers; engins *mpl* forestiers
r лесные машины *fpl*; машины для лесного хозяйства

3403 forest entity
d Waldwesen *n*; Waldentität *f*
f entité *f* de la forêt
r лесной объект *m*

3404 forest entomology; forest insectology; wood entomology
(studies and practices dealing with insects in relations to forests and forest products)
d Forstentomologie *f*; Forstinsektenkunde *f*; Waldinsektenkunde *f*
f entomologie *f* forestière
r лесная энтомология *f*

3405 forest environment
d Bestandesumwelt *f*
f milieu *m* forestier
r лесная среда *f*

3406 forester
d Forstmeister *m*; Förster *m*; Forstmann *m*

3407 forester house; ranger house; ranger cabin
d Forsthaus *n*
f maison *f* forestière
r лесная усадьба *f*

* **forester's district → 2542**

3408 forest estate; forest homestead; wood estate
d Waldgut *n*
f domaine *m* forestier; bienfonds *m* boisé
r лесное поместье *n*; лесное имение *n*; лесная дача *f*

* **forest estimation → 3411**

3409 forest estimation records; report of forest valuation; report of forest survey
d Taxationsmanual *n*
f procès-verbal *m* d'aménagement
r таксационная книга *f*

3410 forest estimator; forest appraiser; forest valuer; timber cruiser; forest assessor; forest looker; taxator
d Forsttaxator *m*; Forstschätzer *m*
f estimateur *m* de bois; estimateur forestier; aménagiste *m*
r лесной таксатор *m*; рабочий-измеритель *m* диаметров деревьев; лесной объездчик *m*, устанавливающий норму лесоповала

3411 forest [e]valuation; forest estimation; forest appraisal; forest assessment; forest mensuration
(the branch of forestry concerned with evaluating forest estates and their components)
d Forst[ab]schätzung *f*; Waldbewertung *f*; Waldwertschätzung *f*; Waldwertrechnung *f*
f évaluation *f* forestière; estimation *f* forestière; expertise *f* forestière
r стоимостная оценка *f* леса; лесная оценка

3412 forest expectation value; forest expectancy value; prospective value of a forest
d Walderwartungswert *m*
f valeur *f* d'avenir de la forêt
r ожидаемая стоимость *f* леса

3413 forest experimental station; forest checking station
d forstliche Versuchsanstalt *f*; forstliche Versuchsstation *f*

3414 **forest exploitation; forest use; timber exploitation; wood exploitation; forest husbandry; forest utilization**
d Waldexploitation *f*; Ausbeutung *f* des Waldes; Forst[be]nutzung *f*; Waldnutzung *f*; Holzausbeute *f*
f exploitation *f* de la forêt; mise *f* en œuvre de bois
r использование *n* лесных ресурсов; лесопользование *n*; лесоэксплуатация *f*

* **forest exploitation → 9588**

3415 **forest exploitation after deterioration**
d Kalamitätennutzung *f*
f exploitation *f* de la forêt après dépérissement
r лесопользование *n* после стихийного бедствия

3416 **forest exploitation plan**
d Schlägerungsplan *m*
f plan *m* d'exploitation de la forêt
r план *m* эксплуатации леса

* **forest exploitation size → 3637**

* **forest extraction → 9588**

3417 **forest face**
(standing timber adjacent to a clear-cut area)
d Waldseite *f*
f face *f* de la forêt
r граничный древостой *m*

3418 **forest fallowing**
d Schlagruhe *f*
f arrêt *m* d'exploitation
r покой *m* лесосеки

3419 **forest farm; tree farm**
(privately owned forest managed on a multiple use basis with timber production as an important management goal)
d Waldfarm *f*
f propriété *f* forestière de production; ferme *f* forestière
r лесная ферма *f*; принадлежащий фермеру участок *m* леса

3420 **forest farmer**
d Waldbauer *m*
f agriculteur *m* forestier
r лесной фермер *m*

3421 **forest farming**
r лесопольное хозяйство *n*

3422 **forest fauna**
d Waldfauna *f*
f faune *f* forestière
r лесная фауна *f*

* **forest feeder → 3523**

* **forest feller → 9526**

3423 **forest felling fund**
d Holzeinschlagfonds *m*
f fonds *m* d'exploitation forestière
r лесосечный фонд *m*

3424 **forest fertilization**
d Forstdüngung *f*
f fertilisation *f* forestière
r удобрение *n* лесов

3425 **forest-field crop**
d Waldfeldkultur *f*
f culture *f* sylvo-agricole
r лесопольная культура *f*

3426 **forest filter strip**
(area between a stream and construction activities that achieves sediment control by using the natural filtering capabilities of the forest floor and litter)
d Forstfilterstreifen
f bande *f* de filtre forestier
r лесная фильтрирующая полоса *f*

3427 **forest finance**
d Forstfinanzwirtschaft *f*
f finance *f* forestière
r лесная финансовая система *f*

3428 **forest fire; wood fire**
(an unplanned or unwanted natural or human-caused fire, or a prescribed fire that escapes its bounds)
d Baumfeuer *n*; Waldbrand *m*
f feu *m* de forêt; incendie *m* de forêt
r лесной пожар *m*

3429 **forest fire against the wind**
d Waldbrand *m* gegen den Wind
f feu *m* de forêt contre le vent
r лесной пожар *m* против ветра

3430 **forest-fire conservancy; forest-fire protection**

f protection *f* de forêt contre l'incendie
r охрана *f* лесов от пожаров

* **forest-fire danger → 6971**

3431 **forest-fire fighting**
(all activities concerned with controlling and extinguishing a forest fire following its detection)
d Bekämpfung *f* von Waldbränden; Waldbrandbekämpfung *f*; Feuerbekämpfung *f*; Löschen *n* eines Brandes
f lutte *f* contre les feux de forêt; extinction *f* d'un incendie forestier; suppression *f* d'un incendie forestier
r тушение *n* лесного пожара; борьба *f* с лесным пожаром

* **forest-fire hazard → 6971**

3432 **forest-fire insurance**
d Waldfeuerversicherung *f*
f assurance *f* contre l'incendie des bois et forêts
r страхование *n* леса от пожара

3433 **forest-fire prevention management**
d Waldbrandverhütungsverwaltung *f*
f gestion *f* de prévention contre les incendies de forêt
r противопожарное устройство *n* лесов

* **forest-fire protection → 3430**

3434 **forest-fire season; fire season; forest-fire risk period**
(period(s) of the year during which firs are likely to start, spread and do damage)
d Waldbrandperiode *f*; Waldbrandsaison *f*; Waldbrandzeit *f*
f saison *f* des feux forestiers; période *f* d'incendies de forêts
r лесопожарный период *m*; лесопожарный сезон *m*

3435 **forest floor; forest litter**
(1. the surface layer of a soil supporting forest vegetation; 2. the organic-matter system of the various forest soil types)
d Waldstreu *f*; Streuschicht *f*; Waldbodendecke *f*
f litière *f* forestière
r лесная подстилка *f*

3436 **forest floor disturbance**
d Waldstreustörung *f*

3437 **forest flora**
d Waldflora *f*
f flore *f* forestière
r лесная флора *f*

* **forest foreman → 5069**

3438 **forest formation**
d Waldformation *f*
f formation *f* de forêt
r лесная формация *f*; лесорастительная формация

3439 **forest forming tree species**
d forstbildende Baumart *f*
f espèce *f* créative forestière
r лесообразующая порода *f*

* **forest fragmentation → 3681**

3440 **forest fruiting tree**
d Wildobstbaum *m*
f arbre *m* fruitier forestier
r лесное плодовое дерево *n*

3441 **forest fruits**
d Waldfrüchte *fpl*
f fruits *mpl* forestiers
r лесные плоды *mpl*

* **forest fund → 8114**

3442 **forest gallery**
(an art gallery in the forest)
d Waldgalerie *f*
f galerie *f* forestière
r лесная галерея *f*

3443 **forest gap; gap in a stand; breakthrough; blank in a stand; clearing; clearance; open wood[land]; cut-through; meadow [in the wood]**
(an open grassy area surrounded by woods)
d Waldblöße *f*; Walddurchhau *m*; Waldsteig *m*; Bestandeslücke *f*; Blanke *f*; Durchbruch *m*; Kahlstelle *f*; Lichtung *f*
f terrain *m* déboisé; terrain non boisé; clairière *f*; manquant *m*; vide *m*
r [лесная] прогалина *f*; необлесённый участок *m*; редина *f*; просвет *m* в насаждении

* **forest garden → 342**

3444 **forest gene protection**
d Forstgenschutz *m*

3445 forest genesis
d Waldgenese *f*
f genèse *f* forestière
r лесной генезис *m*

3446 forest genetic resources
(the multitude of information included in all organisms within a forest)
d genetische Waldressourcen *fpl*
f ressources *fpl* génétiques de la forêt
r лесные генетические ресурсы *mpl*

3447 forest genetics
(study of heredity in forest trees)
d Waldgenetik *f*
f génétique *f* forestière
r лесная генетика *f*

* **forest glade → 8865**

3448 forest goods
d Waldwaren *pl*; Forstgut *n*
f biens *mpl* forestiers
r лесные продукты *mpl*

3449 forest grass
d Waldgras *n*
f graminée *f* forestière
r лесная трава *f*

3450 forest grassland; wooded grassland; recreation glade
(land covered with grasses and other herbaceous species, and with woody species that cover between approximately 10 to 40% of the ground)
d Waldweide *f*; Forstweide *f*; Waldgrünland *n*
f steppe-erme *f* forestière; herbage *m* forestier
r лесной луг *m*; лесная лужайка *f*

3451 forest-grassland ecotone
d Waldweide-Ekoton *m*; Grünland-Wald-Ekoton *m*
f écotone *m* forêt-pâturage
r лесолуговое пограничное сообщество *n*

3452 forest grazing; forest pasture; wood[land] pasture; woodland grazing; silvipasture; silvopasture
(integration of trees with pasture)
d Forstbeweidung *f*
f pâturage *m* en forêt; pacage *m* en foret; sylvopâturage *m*
r выпас *m* в лесу; лесная пастьба *f*

3453 forest-green *adj*
r травянисто-зелёный

3454 forest growing stock; standing stock; standing timber; standing crop; wood on the stem; stumpage; standing forest; growing timber; living residues
(the standing timber itself; the living tree component of the standing volume)
d stehendes Holz *n*; Holz auf dem Stamm; Holz auf de Stock
f bois *m* sur pied; matériel *m* sur pied
r лес *m* на корню; древостой *m*; растущий лес; на пню стоящая древесина *f*

3455 forest-grown *adj*
d forstgewachsen
f de croissance en forêt
r выросший в лесу

* **forest growth → 7610**

3456 forest guard; forest warden; forest ranger; woodranger; forest keeper; wood guard; woodward; overseer of the woods
d berittener Forstaufseher *m*; Waldhüter *m*; Waldwärter *m*; Forstschutzbeamt *m*; Schutzbeamt *m*; Unterförster *m*
f protecteur *m* en bois; garde *m* de forêt; garde-forestier *m*
r лесной сторож *m*; лесной надзиратель *m*; лесной объездчик *m*

* **forest hamlet → 3639**

3457 forest harvesting operation; [timber-]harvesting operation; logging operation; lumbering operation; wood felling practice
(any operation associated with felling and/or moving trees, logs, veneer bolts, poles, piles and other forest products from the stump to the point of delivery)
d Holzeinschlagarbeiten *fpl*; Hauungsarbeiten *fpl*; Schlagflächenarbeiten *fpl*
f abattage *m* et débardage *m*; bûcheronnage *m*
r лесозаготовительные работы *fpl*; лесосечные работы

3458 forest health
(vigo[u]r or physiological condition of a forest)
d Forstgesundheit *f*
f santé *f* de forêt
r жизнеспособность *f* леса; [санитарно-гигиеническое] состояние *n* лесов

f agents *mpl* de santé forestière
r санитарно-гигиенические вещества *npl* для оздоровления леса

3460 **forest health treatments**
(application of techniques to influence pest or beneficial organism populations, mitigate damage, or reduce the risk of future damage to forest stands)
d Forstgesundsbehandlungen *fpl*
f traitements *mpl* de santé forestière
r лесосанитарные мероприятия *npl*

3461 **forest heritage**
d Walderbe *n*
f patrimoine *m* forestier
r лесное достояние *n*

3462 **forest high school; forest university; university of forestry**
d Forsthochschule *f*; forstliche Hochschule *f*
f université *f* forestière
r высшая лесная школа *f*; лесной университет *m*

3463 **forest highway; forest access road; forest arterial road; forest backbone road; haul road**
(provide service to large land areas and usually connects with public highways or other forest arterial roads to form an integrated network of primary travel routes)
d Hauptfahrtsweg *m*; Zufahrtsweg *m*; Hauptverkehrsader *f*; Verkehrsader *f*
f grande artère *f* forestière
r магистральная лесовозная дорога *f*

* **forest history → 4314**

* **forest holding → 3543, 3568**

* **forest homestead → 3408**

3464 **forest humus**
(an accumulation of acid organic matter at the soil surface beneath a forest)
d Waldhumus *m*; Holzerde *f*
f humus *m* forestier
r лесной гумус *m*; лесной перегной *m*

* **forest husbandry → 3414**

3465 **forest hydrology**
(science that describes and analyzes the occurrence of water in forests, and its circulation)
d Waldhydrologie *f*
f hydrologie *f* forestière
r лесная гидрология *f*

* **forest hygiene → 3580**

* **forest imposts → 8422**

* **forest in common ownership → 1812**

3466 **forest industry; timber industry; wood industry; lumber industry**
d Forst[verarbeitungs]industrie *f*; Holzindustrie *f*; holzverwertende Industrie *f*; Holzwirtschaft *f*
f industrie *f* forestière; industrie du bois
r лесная промышленность *f*; лесная индустрия *f*

3467 **forest industry land**
(land owned by companies or individuals operating primary wood-using plants)
d Forstindustrieboden *m*
f terrain *m* d'industrie du bois
r земля *f* для лесной промышленности

3468 **forest industry pollution control**
d waldige und industrielle Verschmutzungskontrolle *f*
f lutte *f* contre la pollution provoquée par les industries forestières
r контроль *m* загрязнения от лесной промышленности

3469 **forest industry waste**
d Waldindustrieabfälle *mpl*
f résidus *mpl* des industries forestières
r отбросы *mpl* лесной промышленности

3470 **forest inexhaustibility**
d Unerschöpflichkeit *f* der Wälder
f abondance *f* intarissable des forêts
r неистощимость *f* леса

3471 **forest influence**
(all the modifying effects of forest cover on the environment, particularly on water supplies, soil, and micro-climate)
d Wirkungen *fpl* des Waldes
f influence *f* forestière
r влияние *n* леса; воздействие *n* леса

3472 **forest information technology; forest informatics**
d forstliche Informatik *f*
f informatique *f* forestière
r лесная информатика *f*

3473 forest in private ownership; private forest; individually owned forest
d Privatwald *m*; Privatforst *m*
f forêt *f* privée
r лес *m* частного владения; частный лес

3474 forest in ravines and gullies; gorge forest; ravine forest
d Schluchtenwald *m*
f forêt *f* de ravin
r приовражный лес *m*; байрачный лес

3475 forest insect; wood insect
d Forstinsekt *n*; Holzinsekt *n*
f insecte *m* forestier
r лесное насекомое *n*; древесное насекомое

* **forest insectology → 3404**

* **forest inspection service → 3380**

* **forest instruments → 3623**

3476 forest insurance
d Waldversicherung *f*
f assurance *f* contre les risques de la forêt
r лесное страхование *n*

3477 forest invasion
d Waldinvasion *f*; Waldeinbruch *m*
f envahissement *m* de forêt
r наступление *n* леса

3478 forest inventory; inventory of stand; timber inventory
(the process or results of field sampling (cruising) forestland to determine the quantity and quality of the timber)
d Waldinventur *f*; Holzvorratsinventur *f*; Bestandesaufnahme *f*; Bestandesinventur *f*
f inventaire *m* forestier; inventaire de bois
r учёт *m* леса; инвентаризация *f* леса; лесоинвентаризация *f*

3479 forest issues
d Waldprobleme *npl*
f problématique *f* décisionnelle forestière
r проблемы *mpl* леса

* **forest keeper → 3456**

* **forest labo[u]r → 9724**

* **forest labo[u]rer → 9654**

3480 forestland; forest stock
terre *f* forestière
r лесная земля *f*; покрытая лесом земля

3481 forest landscape
d Waldlandschaft *f*
f paysage *m* forestier
r лесной ландшафт *m*

3482 forestland[-]use plan
(plan that gathers and coordinates the direction to be followed in the overall management of a forest)
d Waldbodennutzungsplan *m*
f plan *m* d'utilisation de terrain forestier
r план *m* лесопользования

3483 forestland value
d Waldbodenwert *m*
f valeur *f* de terre forestière
r стоимость *f* лесной земли

3484 forest law; forest[ry] act
d Forstgesetz *n*; Waldgesetz *n*
f loi *f* forestière; code *m* forestier
r закон *m* о лесе; лесной закон

3485 forest legislation
d Forstgesetzgebung *f*; forstliche Gesetzgebung *f*; Forstrecht *n*
f législation *f* forestière
r лесное законодательство *n*

* **forest licence → 8605**

* **forest limit → 2761**

* **forest liquidation → 3373**

* **forest litter → 3435**

* **forest litter removal → 9153**

* **forest [living] community → 3331**

* **forest logging[-off] → 9588**

3486 forest logistics
d Forstlogistik *f*; Waldlogistik *f*
f logistique *f* forestière
r лесная логистика *f*

* **forest looker → 3410**

3487 forest-loving *adj*
d waldliebend

3488 forest management; timber management; silvicultural management; management of forests; management of woods
(application of technical forestry principles and practices and business techniques to the management of a forest)
d Forsteinrichtung *f*; Forstverwaltung *f*; Forstwirtschaftsführung *f*
f aménagement *m* forestier; gestion *f* forestière; tutelle *f* forestière
r ведение *n* лесного хозяйства; лесоводство *n*; руководство *n* лесным хозяйством

* **forest management area → 3330**

3489 forest management assistance
d Forstverwaltungshilfe *f*
f assistance *f* de gestion forestière
r консультация *f* по ведению лесного хозяйства

* **forest management by individual trees → 7657**

3490 forest management cycle
d Forstverwaltungszyklus *m*
f cycle *m* de gestion forestière
r цикл *m* ведения лесного хозяйства

3491 forest management method; mode of forest management
d Forstverwaltungsmethode *f*
f méthode *f* de gestion forestière
r метод *m* лесоустройства

3492 forest [management] plan; forest working plan; forest operational plan; plan of forest operations
(long-range plan for managing designated forest lands)
d Waldfunktionsplan *m*; forstlicher Wirtschaftsplan *m*; Forsteinrichtungsplan *m*
f plan *m* de gestion forestière; plan d'aménagement forestier
r план *m* деятельности лесничества; план [развития] лесного хозяйства; план лесонасаждений; лесоустроительный план

3493 forest management planning; forest resource planning
(planning process in forest enterprise)
d Forsteinrichtungsplanung *f*
f planification *f* de gestion forestière
r планирование *n* деятельности лесничества; планирование лесных ресурсов

d waldbauliche Ansprüche *mpl*; Forstverwaltungsansprüche *mpl*
f demandes *fpl* de gestion forestière
r требования *npl* ведения лесного хозяйства; лесоводственные требования

3495 forest management system
d Leitungsystem *n* des Waldes
f système *m* de gestion forestière
r система *f* ведения лесного хозяйства

3496 forest management unit
(area of forest land managed as a unit for fiber production and other renewable resources)
d Forstverwaltungseinheit *f*
f unité *f* d'aménagement forestier
r хозяйственная часть *f* леса

* **forest management zone → 3330**

3497 forest manager; working-plan officer
d Forsteinrichter *m*
f aménagiste *m* des forêts
r лесоустроитель *m*

* **forest mantle → 7508**

3498 forest map; stand map; map of stock
d Bewaldungskarte *f*; Waldbestandskarte *f*; Forstbestandskarte *f*; Bestockungskarte *f*
f carte *f* de forêts
r карта *f* лесов; карта лесонасаждений

3499 forest mapping
d Waldkartierung *f*
f cartographie *f* forestière
r картографирование *n* леса

* **forest margin → 2761**

3500 forest mark
d Waldzeichen *n*; Forstzeichen *n*
f marque *f* de la forêt
r лесной знак *m*

* **forest master → 1483**

3501 forest maturity
d Waldreife *f*
f maturité *f* de la forêt
r спелость *f* леса

3502 forest May beetle; May beetle; cockchafer; tree-beetle
(Melolontha melolontha)

f hanneton *m* [commun]; hanneton forestier
r майский жук *m*; хрущ *m*

* **forest mensuration → 3411**

3503 forest merchandise knowledge
d forstliche Warenkunde *f*
f science *f* de marche forestière
r лесное товароведение *n*

3504 forest meteorology
d Forstmeteorologie *f*
f météorologie *f* forestière
r лесная метеорология *f*

3505 forest microbiology
d forstliche Mikrobiologie *f*
f microbiologie *f* forestière
r лесная микробиология *f*

3506 forest microclimate
(average course or condition of the weather at a forest over a period of many years)
d Waldmikroklima *n*
f microclimat *m* forestier
r лесной микроклимат *m*

3507 forest mindedness; tree consciousness
d Waldgesinnung *f*; Waldverbundenheit *f*
f mentalité *f* forestière
r привязанность *f* к лесу

3508 forest model
d Waldmuster *n*; Waldmodell *n*
f modèle *m* de forêt
r модель *f* леса

* **forest monitoring → 5530**

3509 forest mortality
(death or destruction of forest trees as result of competition, disease, insect damage, drought, wind, fire, old age, and other factors, excluding harvesting)
d Waldsterblichkeit *f*
f mortalité *f* forestière
r смертность *f* леса

3510 forest mouse
d Waldmaus *f*; Waldwühlmaus *f*
f souris *f* forestière; mulot *m*
r лесная мышь *f*; лесная полёвка *f*

3511 forest mycology
d Forstmykologie *f*

3512 forest nature
d Waldnatur *f*
f nature *f* de la forêt; existence *f* de la forêt
r природа *f* леса

* **forest nomad → 6290**

3513 forest nomenclature
d forstliche Nomenklatur *f*
f nomenclature *f* forestière
r лесная номенклатура *f*

* **forest offence → 1909**

* **forest office → 3323**

* **forest officer → 3573**

3514 forest of small-leaved deciduous species; small-leaved forest; parvifoliate forest
d kleinblätteriger Wald *m*
f forêt *f* d'espèces à petites feuilles; forêt parvifoliée
r мелколиственный лес *m*

3515 forest of stunted trees; stand of dwarfed trees; scrub stand
d Krüppelwald *m*; Krüppelbestand *m*
f peuplement *m* rabougri
r лес *m* из корявых деревьев; корявое насаждение *n*

* **forest operation → 9724**

* **forest operational plan → 3492**

* **forest operator → 9654**

3516 forest ordinance
d Waldordonnanz *f*; Forstsatzung *f*
f ordonnance *f* forestière
r нормативный документ *n* о лесе

3517 forest outlier
(an area of forest separated from the main occurrence of its type, its presence being due generally to some local variation in ecological conditions or to past migration of vegetation associated with major climatic changes)
d Waldausreißer *m*; Waldsonderfall *m*
f relicte *f* forestière
r островной лес *m*

* **forest overseer → 3613**

3518 forest overstor[e]y; overstor[e]y; overwood; upper growth; upper stor[e]y;

d Ober[be]stand *m*; Oberholz *n*
f étage *m* dominant
r основной ярус *m* насаждения; господствующий ярус насаждения; верхний ярус насаждения

3519 forest owner; forest proprietor; timberland owner
d Waldeigentümer *m*; Lieferungsforstwirt *m*; Waldbesitzer *m*
f propriétaire *m* forestier
r лесовладелец *m*; лесной собственник *m*

* **forest ownership → 3543**

3520 forest parasites
d Forstparasiten *mpl*
f parasites *mpl* forestiers
r лесные паразиты *mpl*

* **forest parcel → 1821**

* **forest pasture → 3452**

* **forest patch → 1821**

3521 forest pathology
(science that deals with diseases of forest trees or stands, and with the deterioration of forest products by organisms)
d Forstpathologie *f*
f pathologie *f* forestière
r лесная патология *f*; патология лесных пород

* **forest patrol → 3333**

* **forest pedology → 3601**

3522 forest percent; percent of forest area; percentage of forest land
(the rate of discount used in forest valuation for computations of e.g. capital value)
d Bewaldungsprozent *n*; Waldanteil *m*
f taux *m* de placement forestier; taux de terrain forestier; pourcentage *m* de boisement
r процент *m* лесистости (степень облесённости территории, определяемая отношением площади покрытых лесом земель к общей ее площади)

3523 forest pest; forest destroyer; forest feeder
d Forstschädling *m*; Baumfeind *m*; Waldverderber *m*
f ravageur *m* forestier; animal *m* nuisible à la

3524 forest pest control
d Kampf *m* gegen Waldschädlinge
f lutte *f* contre les animaux nuisibles à la forêt
r борьба *f* с лесными вредителями

3525 forest phenological station
d forstliche phänologische Station *f*
f station *f* forestière phénologique
r лесная фенологическая станция *f*

3526 forest photogrammetry
d Forstfotogrammetrie *f*
f photogrammétrie *f* forestière
r лесная фотограмметрия *f*

* **forest phytoc[o]enosis → 3331**

3527 forest plant; hylad
d Forstpflanze *f*; Waldpflanze *f*
f plante *f* forestière; plante sylvestre
r лесное растение *n*

3528 forest plantation; timber plantation; forest culture; tree culture; arboriculture; ligniculture
(forest crop established by artificial, either by sowing or planting; woodland of definitive density of noble varieties for the production of timber)
d Forstkultur *f*; Waldkultur *f*; Holzkultur *f*
f plantation *f* forestière; forêt *f* de plantation; culture *f* forestière; arboriculture *f*; ligniculture *f*
r лесная плантация *f*; лесное насаждение *n*; лесная культура *f*; древесная культура

* **forest planting → 94**

3529 forest planting plan
d Aufforstungsplan *m*
f plan *m* de boisement
r проект *m* посадки лесных культур; план *m* облесения

3530 forest plant rearing
d Forstpflanzenahnzucht *f*
f culture *f* de plantes forestières
r разведение *n* лесных растений

* **forest plot → 1821**

3531 forest plow; forest plough; woodland plough
d Forstpflug *m*; Forststreifenpflug *m*; Waldpflug *m*

3532 **forest police**
d Forstpolizei *f*
f police *f* forestière
r лесная полиция *f*

3533 **forest policy; forest politics**
d Forstpolitik *f*
f politique *f* forestière
r лесная политика *f*

3534 **forest practices act; forest practice law; forest practices code; FPC**
(state legislation establishing minimum standards for forest practices and providing for necessary administrative procedures)
d Wald[bau]praxisgesetz *m*; Forstwirtschaftsgesetz *n*; Wald[bau]praxiscode *m*
f loi *f* relative à l'économie forestière
r закон *m* лесоводства; лесохозяйственный закон

3535 **forest preserve; preserved forest**
d Bannforst *m*; Hegeforst *m*; Hegeholz *n*; Waldschutzgebiet *n*
f forêt *f* réservée
r огражденный лес *m*

3536 **forest price; forest value**
d Waldpreis *m*; Waldwert *m*
f valeur *f* de la forêt
r [капитальная] стоимость *f* леса

* **forest privilege → 3356**

3537 **forest production; wood production; timber production**
(material that is obtained for use from a forest; includes major products such as poles and roundwood (for timber), as well as minor products such as medicinals, gums, resins, oils, fungi, honey)
d Waldproduktion *f*; Holzproduktion *f*; Walderzeugnisse *npl*; Forsterzeugnisse *npl*; Holzgegenstände *mpl*
f production *f* forestière
r лесная продукция *f*; лесопродукция *f*; изделия *npl* из древесины; лесоматериалы *mpl*

3538 **forest production zone**
d Waldproduktionszone *f*
f zone *f* de production forestière
r зона *f* лесной продукции

3539 **forest productivity; stand productivity**
r продуктивность *f* леса; производительность *f* леса

3540 **forest products future**
d Waldproduktionszukunft *f*
f futur *m* sur produits forestiers
r будущее *n* лесной продукции

3541 **forest product utilization**
d Holzausnutzung *f*
f utilisation *f* de produit forestier
r использование *n* лесоматериалов

3542 **forest profile**
(range of forest conditions that exists across the landscape, including such factors as timber species, quality, condition and age, location, elevation, topography, accessibility, and economic viability)
d Forstprofil *m*
f profil *m* de la forêt
r профиль *m* леса

3543 **forest property; forest ownership; forest holding**
d Waldeigentum *n*; Waldbesitz *m*
f propriété *f* forestière
r лесная собственность *f*

3544 **forest prophylaxy**
d Waldprophylaxe *f*
f prophylactique *f* forestière
r лесная профилактика *f*

* **forest proprietor → 3519**

* **forest protection → 3359**

* **forest protection law → 3360**

3545 **forest protection station**
d Waldschutzstation *f*
f station *f* de protection forestière
r лесозащитная станция *f*

* **forest putrid litter → 3671**

* **forest railroad → 5079**

* **forest railway → 5079**

3546 **forest range; forest station; forestry**
d Försterei *f*; Forstamt *n*
f foresterie *f*
r лесничество *n*

3547 **forest range administration**

f administration *f* de foresterie
r администрация *f* лесничества

* **forest range district → 2542**

* **forest ranger → 3456**

3548 **forest raster data set**
d Waldrasterdatenmenge *f*
f ensemble *m* de données de rastre forestier
r множество *n* данных лесного растра

* **forest reclamation → 7611**

3549 **forest reconnaissance; exploration of forests; timber reconnaissance; recce**
(a preliminary inspection or survey of a forest or range area to gain general information useful for future management)
d Waldinspektion *f*; Walderkundung *f*; Walderforschung *f*
f reconnaissance *f* forestière; inspection *f* de la forêt; étude *f* de la forêt
r исследование *n* леса; [инспекционное] обследование *n* леса; выборочная таксация *f* леса

* **forest recovery → 6821**

* **forest recreation → 6821**

3550 **forest red gum; flooded gum; horncap eucalyptus; mungurra**
(Eucalyptus tereticornis; Eucalyptus umbellata)
d Mungurra *m*; rundgehörnter Eukalyptus *m*
f eucalyptus *m* ombellé; mungurra *m*
r эвкалипт *m* вальковиднорогий

3551 **forest region; forest zone; forest area; forest country**
d Forstbereich *m*; Waldregion *f*; Waldgebiet *n*; Waldgegend *f*; Waldzone *f*
f région *f* forestière; aire *f* forestière
r лесная зона *f*; лесная область *f*; лесной район *m*; лесная местность *f*; лесистая местность

3552 **forest regions classification**
d Forstbereichsklassifizierung *f*
f classification *f* des régions forestières
r классификация *f* лесных районов

* **forest removal → 9588**

* **forest renewal → 6821**

d Forstrente *f*; Waldrente *f*
f rente *f* forestière
r лесная рента *f*; капитальный лесной доход *m*

* **forest reproduction → 6821**

3554 **forest reproduction period**
(process by which new individuals are produced from parent trees, by either sexual or asexual (vegetative) means)
d Verjüngungszeit *f*
f période *f* de reproduction
r период *m* лесовозобновления

3555 **forest reproductive material**
d Vermehrungsgut *n*
f matériel *m* forestier de reproduction
r материал *m* лесовозобновления

3556 **forest research institute**
d forstliche Forschungsanstalt *f*; Forstforschungsinstitut *n*
f institut *m* de recherche forestière
r лесной научно-исследовательский институт *m*

3557 **forest reserve; reserved forest; forest sanctuary**
(non-intervention forest which is usually not managed or exploited)
d Waldreserve *f*; Waldreservat *n*
f réserve *f* de forêt; réserve forestière; forêt *f* classée
r лесной заповедник *m*; лесной резерват *m*

3558 **forest reserve fund**
d Forstreservefonds *m*
f fonds *m* forestier de réserve
r запасный лесной фонд *m*

3559 **forest reserves; timber reserves; timber stock; real growing stock; present growing stock**
d forstliche Reserven *fpl*; Forstreserven *fpl*
f réserves *fpl* forestières
r лесные запасы *mpl*; лесные резервы *mpl*; запасы древостоя; запасы насаждения; запасы древесины на корню; запасы леса на корню

* **forest residuals → 3561**

3560 **forest residue baling**
d Presse *f* für Holzabfälle

3561 forest residues; forest residuals; wood waste; kerf waste; timber waste
(sum of wasted and unused wood in the forest, including logging residues; rough, rotten, and dead trees; and annual mortality)
d Holzabfälle *mpl*; Holzreste *mpl*
f déchets *mpl* de bois; déchets ligneux; résidus *mpl* forestiers
r отходы *mpl* древесины; древесные отходы; лесные отходы

* **forest resource planning → 3493**

3562 forest resources
(resources or values associated with forest land, including but not limited t water, wildlife, fisheries, recreation, timber, range, and heritage)
d Waldressourcen *fpl*
f ressources *fpl* forestières
r лесные ресурсы *mpl*

3563 forest resources survey
d Vermessung *f* den Waldressourcen
f enquête *f* sur les ressources forestières
r съёмка *f* лесных ресурсов

* **forest returns → 6569**

* **forest revenue → 6569**

3564 forest ride; ride; forest aisle; aisle; cutting in forest; lane; vista
(a path made for riding on horseback, especially through woodlands)
d Waldfahrt *f*; Waldschneise *f*; Schneise *f*; Gestell *n*
f laie *f* [forestière]; laye *f* forestière; ligne *f* [de parcelle]; layon *m*
r лесная квартальная сеть *f*; [лесная] просека *f*

* **forest right → 8573**

3565 forest road; forest track; wood track; wood path
(road from pit to loading dock, tipple, ramp, or preparation plant)
d Forstbahn *f*; Forstweg *m*; Forststraße *f*; Waldstraße *f*; Waldweg *m*; Waldbahn *f*
f chemin *m* forestier; route *f* forestière; piste *f* forestière
r лесная дорога *f*; дорога, проходящая по лесу; лесовозная дорожка *f*

* **forest road building → 1889**

* **forest roading → 1889**

3566 forest roadnet; forest transportation net
d forstliches Straßennetz *n*; Waldstraßennetz *n*; Waldwegenetz *n*; Waldtransportnetz *n*
f réseau *m* de pistes forestières
r сеть *f* лесовозных дорог; сеть лесных дорог; лесотранспортная сеть

* **forestry → 3546**

3567 forestry
(the management of growing timber; the art of forming or of cultivating forests)
d Forstwirtschaft *f*; Waldwirtschaft *f*
f foresterie *f*
r лесное хозяйство *n*; лесохозяйство *n*

* **forest[ry] act → 3484**

* **forestry adviser → 1890**

* **forestry department → 2552**

3568 forest[ry] enterprise; forest holding
d Forstbetrieb *m*; forstlicher Betrieb *m*; Forstbetriebswirtschaft *f*
f entreprise *f* forestière
r лесное предприятие *n*; лесохозяйственное предприятие

3569 forestry exemptions
d Forstwirtschaftsausnahmen *fpl*
f exemptions *fpl* de foresterie
r предоставление *n* льготы лесного хозяйства

3570 forestry handbook; forestry manual
d forstwirtschaftliches Handbuch *n*
f manuel *m* de foresterie
r лесохозяйственный справочник *m*; справочник лесничего

3571 forestry in harmony with nature; naturally conditioned forestry
d naturgemäße Forstwirtschaft *f*
f foresterie *f* naturelle
r природноподобное лесное хозяйство *n*

* **forestry loader → 8607**

* **forestry machinery → 3402**

* **forestry manual → 3570**

f carte *f* de foresterie
r карта *f* лесного хозяйства

3573 forest[ry] official; forest officer
(person employed who is designated be a forest officer, through name or title)
d Forstbeamter *m*
f fonctionnaire *m* forestier
r лесной чиновник *m*

3574 forestry organization
d Waldbetriebsregelung *f*
f organisation *f* forestière
r организация *f* лесного хозяйства

3575 forestry parts of green zones
d Forstwirtschaftsteilen *fpl* der Grünzonen
f partes *fpl* forestières de zones vertes
r лесопарковые части *fpl* зелёных зон

3576 forest[ry] practice; silvicultural practice
(any activity that enhances and/or recovers forest growth or harvest yield, such as site preparation, planting, thinning, fertilization, and harvesting)
d Wald[bau]praxis *f*
f pratique *f* forestière; pratique sylvicole; usages *mpl* forestiers
r практика *f* лесоводства; лесные обычаи *mpl*

3577 forestry research
d forstliche Forschung *f*
f recherche *f* sylvicole
r исследование *n* в области лесного хозяйства; лесохозяйственное исследование

3578 forest[ry] service provider
(person offering tree planting, implementation of best management practices, buyer of timber, or implementation of a silvicultural activity)
d forstwirtschaftlicher Leistungsträger *m*
f prestataire *m* de services forestiers
r поставщик *m* услуг в лесном хозяйстве

3579 forest[ry] technician; technical forester
(individual with a two-year associates degree)
d Forsttechniker *m*
f technicien *m* forestier; technicien en sylviculture
r лесной техник *m*; техник-лесовод *m*

* **forest sanctuary → 3557**

3580 forest sanitation; forest hygiene
(science which helps to prevent the occurrence of pests and to prevent their spread)
d Waldhygiene *f*
f hygiène *f* en forêt; hygiène sylvicole
r лесная санитария *f*; санитарные мероприятия *npl* в лесу

* **forest school → 7297**

3581 forest science
(science of tending woodlands and cultivation of trees)
d Forstwesen *n*; Forstkunde *f*; Waldkunde *f*
f science *f* forestière
r лесная наука *f*; научное управление *n* лесами

* **forest scientist → 6883**

* **forest section → 3333**

3582 forest sector
(that group comprising all direct stakeholders in the forest resource)
d Forstsektor *m*
f secteur *m* forestier
r лесной сектор *m*

* **forest seed → 3628**

3583 forest seed breeding
d forstliche Samenkunde *f*; Forstsamenzüchtung *f*
f sélection *f* de graines forestières
r лесное семеноведение *n*

3584 forest seed crop; forest seed yield; forest seed production
d Waldsamenernte *f*; Waldsamenertrag *m*; Waldsamenproduktion *f*
f récolte *f* des semences forestières; production *f* en graines forestières
r урожай *m* лесных семян; продукция *f* лесных семян

* **forest seed drier → 3588**

3585 forest seeder
d Waldsämaschine *f*
f semoir *m* forestier
r лесная сеялка *f*; лесная сажалка *f*

3586 forest seed establishment
d forstlicher Saatzuchtbetrieb *m*; forstlicher Samenzuchtbetrieb *m*; Forstsamenbetrieb *m*
f production *f* des graines forestières

3587 forest seed extractor
d forstlicher Samenextraktor *m*
f extracteur *m* de graines forestières
r выделитель *m* лесных семян

3588 forest seed kiln; forest seed drier
(a chamber having controlled air-flow, temperature, and relative humidity for drying wood products)
d Forstsamendarre *f*
f sécherie *f* de graines forestières
r лесная семясушилка *f*

* **forest seed production → 3584**

* **forest seed yield → 3584**

3589 forest selection method; selection method
(method of regenerating a forest stand and maintaining an uneven-aged structure by removing some trees in all size classes either singly or in small groups or strips)
d Waldauswahlverfahren *n*
f méthode *f* de jardinage
r метод *m* лесной селекции

* **forest service → 2552**

* **forest servitude → 8573**

3590 forest settlement
d Forstregelung *f*
f délimitation *f* forestière
r лесное урегулирование *n*

3591 forest site
d Waldstandort *m*
f site *m* forestier; station *f* forestière
r лесное местообитание *n*; лесорастительные условия *npl* (обозначаются буквенными и численными показателями, характеризующими плодородие (A, B, C, D) и влажность (0, 1, 2, 3, 4, 5) почвы)

3592 forest [site] capacity
d Bestandbonität *f*; Holzbonität *f*
f possibilité *f* forestière
r производительность *f* лесного насаждения; бонитет *m* лесонасаждения; бонитет леса

3593 forest site classification; site classification
(grouping of forest sites using either the composition or the productivity of the vegetation as well as soil and topographic position)
d Waldstandortklassifizierung *f*

3594 forest site factor; site factor
(ecological term referring to a physical or biological parameter used to describe and distinguish sites)
d Waldstandortsfaktor *m*
f facteur *m* de station forestière
r фактор *m* условий произрастания леса

3595 forest site mapping
d Forstkartenerkundung *f*
f cartographie *f* de sites forestiers
r картирование *n* лесорастительных условий; картирование лесов

* **forest skirt → 2761**

3596 forest sociology
d Waldsoziologie *f*
f sociologie *f* forestière
r лесная социология *f*

3597 forests of nature reserves
d Wälder in den Naturschutzgebieten *npl*; Naturschutzgebietswalde *mpl*
f forêts *fpl* de réserves naturelles
r леса *mpl* заповедников

3598 forests of sanitary protection of health resorts
d Wälder *mpl* der Kurortgesundheitsschutz; Kurortwälder *mpl*
f forêts *fpl* de protection sanitaire de stations climatiques; forêts de villégiature
r леса *mpl* санитарной охраны курортов

3599 forests of sanitary protection of water sources
d Wälder *mpl* der Wasserquellengesundheitsschutz
f forêts *fpl* de protection sanitaire de sources d'eaux
r леса *mpl* зон санитарной охраны источников водоснабжения

* **forest soil → 9547**

3600 forest soil classification
d Gütebestimmung *f* des Waldbodens
f classification *f* de sols forestiers
r классификация *f* лесных почв

* **forest soil evaluation → 2870**

3601 forest soil science; forest pedology
d Waldbodenkunde *f*

3602 forest soil structure
d Waldbodengefüge *n*
f structure *f* de sol forestier
r структура *f* лесной почвы

3603 forest soil typology
d Waldbodentypologie *f*
f typologie *f* de sols forestiers
r типология *f* лесных почв

3604 forest species
d Waldarten *fpl*
f espèces *fpl* forestières
r лесные породы *f*

* **forest spraying → 3388**

* **forest squatter → 344**

3605 forest stand; stand[ing]; forest crop; body of timber
(total number or the total weight of one or more species of plants in an area at a given time)
d Holzbestand *m*; Baumbestand *m*; Waldbestand *m*; Waldbestockung *f*; lebendes Holz *n*
f peuplement *m* forestier
r лесонасаждение *n*

* **forest [stand] type → 3363**

* **forest state fund → 3767**

* **forest station → 3546**

3606 forest statistics
d Forststatistik *f*
f statistique *f* forestière
r лесная статистика *f*

3607 forest steppe; prairie forest; wooded steppe; tree steppe
d Steppenwald *m*; Waldsteppe *f*; Baumsteppe *f*
f forêt *f* steppique; steppe *f* arborée; steppe arbustive
r степной лес *m*; лесостепь *f*

3608 forest-steppe transition
d Steppe-Wald-Übergang *m*
f transition *f* forêt-steppe
r переход *m* от леса к степи

3609 forest-steppe zones
d Steppenwaldzonen *fpl*; Waldsteppenzonen *fpl*
f zones *fpl* de steppes arborées; zones de

3610 forest stewardship
(environmentally and socially responsible use, management and development of forest resources)
d Forstverwalter[s]amt *n*
f gérance *f* forestière; intendance *f*
r рациональное использование *n* леса

3611 forest stewardship program
(cooperative, technical-assistance program designed to encourage multiple resource management on private forestland)
d Forstverwaltersprogramm *n*
f programme *m* de gérance forestière
r программа *f* рационального использования леса

* **forest stock → 3480**

* **forest stocking → 3370**

* **forest structure → 1886**

3612 forest supervision
d Forstaufsicht *f*
f supervision *f* forestière
r лесной надзор *m*

3613 forest supervisor; forest overseer
d Waldaufseher *m*; Forstaufseher *m*
f surveillant *m* forestier
r лесовод-наблюдатель *m*

3614 forest survey
d Forsterhebung *f*; Forstvermessung *f*
f arpentage *m* forestier; lever *m* forestier
r лесная съёмка *f*; межевание *n* леса

3615 forest surveying; forest survey regulation [of district]
(determination of soil conditions and topography, together with the extent, condition, composition and constitution of the forests, for such purposes as purchase or management, or as a basis for forest policies and programs)
d Waldbau *m*
f inventaire *m* forestier régional; technique *f* sylvicole
r лесоустройство *n*

* **forest survey regulation [of district] → 3615**

* **forest system roads → 3378**

f impôt *m* forestier; taxe *f* forestière; taxe de bois
r лесная такса *f*; лесной налог *m*; налог на лес

3617 forest taxation; forest cruising; timber cruising; timber cruise; forest valuation
(establishing cost basis of forest assets)
d Forsttaxation *f*; Waldtaxation *f*; Holztaxation *f*
f taxation *f* forestière
r лесная таксация *f*

* **forest team → 9527**

* **forest technique → 3401**

* **forest technology → 3401**

* **forest tenant → 344**

* **forest tending → 8456**

3618 forest tent caterpillar; tent caterpillar
(Malacosoma disstria)
d Zeltraupe *f*
f livrée *f* des forêts
r коконпряд *m* кольчатый лесной

3619 forest terminology
d Forstterminologie *f*; Forstfachausdrücke *mpl*; forstlicher Sprachgebrauch *m*
f terminologie *f* forestière
r терминология *f* по лесоводству; лесная терминология

3620 forest terminology database
d Forstterminologie-Datenbasis *f*
f base *f* de données forestières
r база *f* данных терминологии по лесоводству

3621 forest therapeutics
d Waldtherapie *f*
f thérapie *f* forestière
r лесная терапия *f*

* **forest thinners → 7760**

3622 forest throughfall
d wirksamer Holzniederschlag *m*; Holznettoniederschlag *m*
f précipitation *f* au sol forestier
r лесная капель *f*; капель, достигающая подстилки в лесу; сквозные промывочные осадки *mpl*, проникающие сквозь полог леса

d forstliche Geräte *npl*; Forstgeräte *npl*; Forstwerkzeuge *npl*
f instruments *mpl* forestiers; outils *mpl* forestiers; outillages *mpl* forestiers
r лесные инструменты *mpl*; лесные орудия *npl*

* **forest track → 3565**

3624 forest transplanting
d Waldanpflanzung *f*
f création *f* de forêts pérennes
r посадка *f* леса саженцами; лесоразведение *n* посадкой

* **forest transportation net → 3566**

3625 forest tree
d Waldbaum *m*
f arbre *m* forestier
r лесное дерево *n*

* **forest tree breeding → 1028**

* **forest [tree] improvement → 1028**

* **forest [tree] nursery → 342**

3626 forest tree planting
d Pflanzung *f* der Waldbäume; Waldbaumpflanzung *f*
f plantage *m* des arbres forestiers; plantation *f* des arbres forestiers
r посадка *f* деревьев; насаждение *n* деревьев; высадка *f* дерева в грунт

3627 forest tree population
d Waldbaumpopulation *f*
f population *f* d'arbres forestiers
r лесная популяция *f* деревьев

3628 forest tree seed; forest seed
d Forstsaatgut *n*; Forstsamengut *n*; Waldsamen *m*; Waldsämerei *f*
f semences *fpl* d'arbres forestiers; semences d'essences forestières
r лесные семена *npl*; лесосеменной материал *m*

* **forest tree species → 4714**

3629 forest tree tolerance
d Waldbaumtoleranz *f*
f tolérance *f* des arbres forestiers
r выживаемость *f* древесных пород; устойчивость *f* лесных культур

3630 forest trial
d Forstkulturprüfung *f*
f essai *m* de culture forestière
r тест *m* лесных культур

3631 forest-tundra; wooded tundra
(land areas in arctic and alpine regions devoid of large trees, varying from bare ground to various types of vegetation consisting of grasses, sedges, forbs, dwarf shrubs and trees, mosses, and lichens)
d Waldtundra *f*
f toundra *f* forestière; toundra boisée
r притундровый лес *m*; лесотундра *f*

3632 forest-type classification
d Waldtypenklassifikation *f*
f classification *f* de types forestiers
r классификация *f* типов леса

3633 forest-type labels
(symbols which are used to code information about forest types on a forest cover map, such as site, disturbance, age and height class, species, stocking)
d Waldtypetiketten *fpl*
f étiquettes *fpl* de types forestiers
r этикеты *mpl* типов леса

3634 forest-type lines
(lines on a map or aerial photo outlining forest types)
d Waldtypgrenzen *fpl*
f limites *fpl* de types forestiers
r границы *fpl* типов леса

3635 forest-type map; stand-type map; timber-type map
d Waldtypkarte *f*
f carte *f* de types forestiers
r карта *f* типов леса

3636 forest typology
d forstliche Typologie *f*; Waldtypenlehre *f*
f typologie *f* forestière
r лесная типология *f*; учение *n* о типах леса

* **forest university → 3462**

* **forest use → 3414**

3637 forest use size; forest exploitation size
d Forstnutzungsgröße *f*
f volume *m* d'utilisation de la forêt
r размер *m* лесопользования

and related uses, e.g., timber production, watershed, Christmas tree farms, recreation)
d Einteilung *f* in Waldnutzungszonen
f zonage *m* d'utilisation forestière
r зонирование *n* лесопользования

* **forest utilization → 3414**

* **forest valuation → 3617**

* **forest value → 3536**

* **forest valuer → 3410**

* **forest vegetation → 8953**

3639 forest village; forest hamlet
(a village whose lands are surrounded by government forests)
d Walddorf *n*
f village *m* forestier
r лесной посёлок *m*

* **forest villager → 344**

* **forest warden → 3456**

3640 forest wealth
d Waldreichtum *m*
f richesse *f* forestière
r лесное богатство *n*

3641 forest weed
d Forstunkraut *n*
f mauvaise herbe *f* en forêt
r лесная сорная трава *f*

3642 forestwide management requirements
d forstweite Führungsanforderungen *fpl*
f grandes exigences *fpl* de la gestion de la forêt; grands besoins *mpl* de la gestion de la forêt
r потребности *fpl* в средствах управления леса

3643 forestwide standard
(a performance criterion indicating acceptable norms, specification, or quality that actions must meet to maintain the minimum considerations for a particular resource)
d forstweiter Standard *m*
f standard *m* forestier large
r общий лесной стандарт *m*

3644 forest [wild] animals
d Tiere *npl* des Waldes; Waldtiere *npl*

3645 **forest with difficult access; inaccessible forest**
d unzugänglicher Wald *m*
f forêt *f* non accessible
r лес *m*, недоступный для освоения

* **forest with grass layer → 3917**

3646 **forest with open canopy; open forest; open stand; gappy stand**
(a forest where the canopy is not closed and a large proportion of the area does not carry trees)
d Wald *m* mit offenem Kronendach; räumdiger Stand *m*
f forêt *f* à couvert intermittent; forêt ouverte
r лес *m* с разрежённым пологом; насаждение *n* с просветами; насаждение с прогалинами

* **forest work → 9724**

* **forest worker → 9654**

* **forest working plan → 3492**

* **forest yaws → 6224**

* **forest yield → 9778**

3647 **forest yield analysis**
(study of forest yield over time using mathematical models and inventory data)
d Waldertraganalyse *f*
f analyse *f* de fertilité forestière
r анализ *m* производительности леса

3648 **forest [yield] potential; [yield] capability of forest**
d Waldpotential *n*; Waldertragsfähigkeit *f*
f potentiel *m* de la forêt; capacité *f* de production forestière
r потенциал *m* производительности леса

3649 **forest yield regulation**
(administrative and technical process which facilitates yield control, often narrowly interpreted as a process that ensures regular and sustained forest yields)
d Waldertragsregelung *f*
f règlement *m* du taux d'exploitation
r регулирование *n* лесного дохода; правила *npl* отпуска [рубки] леса; [утверждённый] порядок *m* лесопользования

f cours *m* de rentabilité forestière
r учение *n* о продуктивности лесов

* **forest zone → 3551**

3651 **forest zoology; sylvatic zoology**
d Forstzoologie *f*
f zoologie *f* forestière
r лесная зоология *f*

3652 **fork**
(that part of a tree where the main stem or the larger branches fork)
d Gabel *f*
f fourche *f*
r развилина *f*; развилка *f*

3653 **fork cal[l]iper; ca[l]liper ga[u]ge; diameter ga[u]ge**
d gegabeltes Forst-Zählwerk *n*
f compas *m* forestier fourchu
r раздвоенная лесная мерная вилка *f*

3654 **forked; furcate**
d gegabelt; gabelwüchsig
f fourchu; en fourche
r раздвоенный; растущий вилообразно; вильчатый

3655 **forked branch; crotch [wood]; crutch**
(a piece of wood taken from the fork of a tree)
d gabelförmiger Ast *m*
f branche *f* fourchue
r раздвоенная ветка *f*; костыль *f*; рассоха *f*

3656 **forked growth**
d Gabelwuchs *m*; Hochzwiesel *f*; Zwieselwuchs *m*
f développement *m* en fourche; croissance *f* fourchue
r вилообразный рост *m*; двойной побег *m*

3657 **forked stem; twin stem; chump**
(a stem defect)
d Doppelstamm *m*; Zusammenwachsen *n* zweier Bäume; Zwieselstamm *m*
f jumelle *f*; fourche *f*
r раздвоенный ствол *m*; двойной ствол

3658 **forked stem removal**
d Entzwieselung *f*
f enlèvement *m* de jumelle
r удаление *n* одного из ответвлений развилки дерева (вблизи земли)

3659 **forked tree; candelabra-topped tree**

f arbre *m* fourchu; arbre en candélabre
r дерево *n* с раздвоенным стволом; раздвоенное дерево; дерево с канделаброобразной вершиной; дерево-канделябрь *n*

3660 fork[lift] truck; fork[ ]lift; [power] lift truck; elevating truck; lift and lowering truck; lifting truck
d Gabelstapler *m*; Stapler *m*
f chariot *m* [élévateur] à fourche
r автопогрузчик *m* с вилочным захватом

3661 formation of crop; formation of stands
d Bestandsbildung *f*
f formation *f* du peuplement
r формирование *n* древостоя

* **formation of humus → 4393**

3662 formation of roundwood bundle; roundwood bundling; roundwood binding
d Rundholzstapelung *f*
f paquetage *m* de bois rond
r пакетирование *n* круглых лесоматериалов; штабелёвка *f* круглых лесоматериалов

* **formation of stands → 3661**

* **formation of wood → 4954**

* **formation period → 2868**

3663 form[ation] pruning
(early pruning of trees being grown for timber to remove unwanted (multiple) stems, leaving the straightest and most vigorous; operation consisting in improving the verticality of the trunk by preventing the development of forks or large branches)
d Formaufasten *n*
f élagage *m* à former
r обрезка *f* формирования дерева

* **formative tissue → 5402**

* **form class → 8840**

3664 form-class volume table
d Formzahlmaßentafel *f*
f table *f* de coefficients de formes [d'arbres]
r объёмная таблица *f*, учитывающая видовые числа

* **form-figure → 8840**

* **form-height → 8841**

(of a tree depending its growth)
d Formzuwachs *m*
f amélioration *f* de la forme
r видовой прирост *m*; прирост видовых высот

* **form-number → 8840**

3666 form of silvicultural system
d Waldbetriebsform *f*
f forme *f* de régime sylvicultural
r форма *f* лесохозяйства

3667 form of stand
d Bestandesform *f*
f forme *f* du peuplement
r форма *f* насаждения

3668 form point
d Kronenmittelpunkt *m*
f point *m* de forme; centre *m* de poussée
r геометрический центр *m* кроны

3669 form-point height
d Kronenmittelpunkthöhe *f*; Höhe *f* des Kronenmittelpunktes; Formpunkthöhe *f*
f coefficient *m* de centre de poussée; coefficient de point de forme
r высота *f* геометрического центра кроны

* **form-ratio → 8840**

3670 formwork timber
d Schalungsholz *n*
f bois *m* de coffrage
r дощатая опалубка *f*

3671 forna; forest putrid litter
d Förna *f*; fäulige Waldstreu *f*
f couverture *f* morte avant humification
r сгнившая подстилка *f*

* **forwarder → 3674**

3672 forwarding
(operation of moving timber products from the stump to a landing for further transport)
d Holzrücken *n* im Verladen
f débardage *m* par portage
r трелёвка *f* в погруженном положении

3673 forwarding cart
d Holzrückenkarren *m*
f chariot *m* de débardage

3674 **forwarding machine; forwarder**
(self-propelled machine, usually self-loading, that transports trees or logs by carrying them completely off the ground)
d Anhänger *m*; Forwarder *m*
f porteur *m*; dumper *m*
r машина *f* для трелёвки лесоматериалов в полностью погруженном положении; форвардер *m*

3675 **fossil fuels**
(includes coal, oil, and natural gas)
d Fossilbrennstoff *m*
f combustible *m* fossile
r органическое топливо *n*; ископаемое топливо; ископаемые виды *mpl* топлива

* **fossilized wood → 6196**

* **fossil-wood → 6196**

* **four-needled pinyon → 6082**

3676 **four-o'clock family**
(Nyctaginaceae)
d Wunderblumengewächse *pl*
f nyctaginacées *pl*
r никтагиновые *pl*

* **four-stamen tamarix → 7939**

* **four-wing sophora → 5729**

* **foxberry → 1996**

* **fox-colo[u]red sawfly → 2905**

* **foxed → 3679**

3677 **fox fire**
(phosphorescent light caused by a fungus on decaying wood)
d Fuchsfeuer *n*
f lumière *f* phosphorescente
r фосфоресцирующий свет *m* (гнилого дерева)

3678 **foxiness**
(a type of rot; the state of being foxed or discolo[u]red)
d dunkle Fäule *f*
f pourriture *f* rouge-brune; pourriture roussie
r гниль *f* красно-бурого цвета

* **foxtail → 5544**

3679 **foxy; foxed**
d schlau; fuchsartig
f moisi; piqué [par l'humidité]
r гнилой (о древесине в начальной стадии гниения)

3680 **foxy butt**
d fuchsartige Stammende *f*
f tronce *f* piquée
r гнилой комель *m*

* **FPC → 3534**

* **fracture → 9507**

* **fractured particles → 4318**

* **fracturing → 2003**

3681 **fragmentation of forests; forest fragmentation**
(breaking up by agriculture, streets, clearfelling, streams, settlements, forest roads, fencing etc. of the natural forest canopy)
d Waldfragmentierung *f*; Waldsplitterung *f*
f fragmentation *f* de forêts; morcellement *m* de forêts
r разделение *n* лесов; фрагментация *f* лесов

* **fragment screening → 6850**

* **fragrant ash → 3292**

3682 **fragrant epaulette tree**
(Pterostyrax hispida)
d Flügel-Storax *m*
f arbre *m* aux épaulettes
r птеростиракс *m* косматый; птеростиракс щетинистый

* **fragrant sumac[h] → 365**

3683 **framed lumber saw**
d umrahmte Säge *f*
f scie *f* à bûches montées
r лесная рамная пила *f*

3684 **frame of saw; saw frame; felling frame**
d Gestell *n* einer Säge
f cadre *m* d'une scie; châssis *m* d'une scie
r рама *f* пилы

* **frame saw → 5132**

3685 **frame-steer[ed] wheeled skidder**

f débusqueur *m* sur roues à châssis
r колёсный трелёвочный трактор *m* с рамным управлением

* **framework branch → 6536**

* **framework convention on global forests → 3863**

* **frameworking → 8714**

3686 **framing timber; framing [wood]**
(lumber used for structural members in a house or other building)
d Konstruktionsholz *n*; Verbandholz *n*
f bois *m* de charpente
r каркасные лесоматериалы *mpl*; конструкционный лес *m*; древесина *f* для сооружения каркаса зданий; конструкционная древесина

* **framire → 4596**

* **frankincense pine → 5029**

3687 **Franklin-tree; franklinia**
(Franklinia alatamaha; Gordonia alatamaha)
d Franklinie *f*
f franclinie *f*
r франклиния *f*

3688 **Fraser magnolia; ear-leaf magnolia; longleaf cucumber-tree**
(Magnolia fraseri)
d Fraser-Magnolie *f*
f magnolier *m* de Fraser
r магнолия *f* Фразера

3689 **Fraser['s] fir; she-balsam**
(Abies fraseri)
d Fraser-Balsamtanne *f*
f sapin *m* de Fraser
r пихта *f* Фразера

* **frass → 9728**

3690 **free-burning fire; wildfire; uncontrolled fire**
d Wildfeuer *n*; unbekämpfter Brand *m*
f incendie *m* non-combattu; feu *m* de friches; feu non maîtrisé; brûlage *m* libre
r неуправляемый пожар *m*; неуправляемый пал *m*

3691 **free canopy**
d freier Schuss *m*

3692 **free-from-noncrop competition; FNC**
(condition of a forest stand when it is established and acceptable for entry into the productive timber land base)
d freier Wettbewerb *m*
f compétition *f* de croissance libre; compétition en jachère; croissance *f* non concurrentielle
r конкуренция *f* от незасеваемых земель

3693 **free grant of timber**
d Freiholzberechtigung *f*
f droit *m* au bois de ramassage
r право *n* на бесплатный лес

3694 **free-growing; free-to-grow; open-growing**
d undurchforstet; freiwachsend
f en croissance libre
r несомкнутый; свободно растущий

3695 **free-growing assessment**
(determination for whether young trees have attained free-growing status)
d Schätzung *f* der undurchforsteten Bestände
f évaluation *f* de croissance libre
r таксация *f* свободно растущих деревьев

3696 **free-growing stand**
(stand of healthy trees of a commercially valuable species, the growth of which is not impeded by competition from plants, shrubs or other trees)
d freiwachsender Bestand *m*
f peuplement *m* en croissance libre
r свободно-растущее лесонасаждение *n*

* **free-growing tree → 5753**

3697 **free growth**
(situation in which a tree or other plant has grown with its crown more or less free from competition)
d freies Wachstum *n*
f croissance *f* libre
r свободный рост *m*

3698 **freeing of the crown; free setting of the crown; isolation of the crown**
d Kronenfreihieb *m*; Freistellung *f* der Krone; Umlichtung *f* der Krone
f dégagement *m* de la cime
r освободительная рубка *f* по кронам; освобождение *n* кроны

3699 **free moisture; free water**
(moisture found in the cell cavities of wood)
d freies Wasser *n*

* **free pollination → 2056**

* **free setting of the crown → 3698**

* **free-side bend → 7936**

3700 **free thinning; clearance**
(removal of competitive ligneous vegetation with a view to favouring growth of objective variety)
d freie Durchforstung *f*
f éclaircie *f* sélective individuelle; éclaircie libre
r рубки *fpl* простора

* **free-to-grow → 3694**

* **free water → 3699**

3701 **free wood**
d Gabholz *n*
f bois *m* libre
r лес *m*, отпускаемый бесплатно

3702 **freeze drying; lyophilization**
d Frosttrocknung *f*
f lyophilisation *f*
r сушка *f* вымораживанием

* **fre[i]jo → 4650**

* **frei jorge → 4650**

3703 **Fremont cottonwood; Fremont poplar**
(Populus fremontii)
d Fremontpappel *f*
f peuplier *m* de Frémont
r тополь *m* Фремонта

3704 **Fremont mahonia; desert Oregon grape**
(Mahonia fremontii; Berberis fremontii)
d Fremontsmahonie *f*
f mahonia *m* de Fremont
r магония *f* Фремонти

* **Fremont poplar → 3703**

* **French aspen → 2880**

3705 **French mulberry; beautyberry; beauty-bush**
(Callicarpa)
d Perlenbeerstrauch *m*; Schönbeere *f*; Schönfrucht *f*
f callicarpe *m*
r калликарпа *f*

d [französische] Tamariske *f*; gemeine Tamariske
f tamarix *m*; tamarisque *m*; tamarin *m*; tamaris *m*
r тамарикс *m* галльский; тамарикс галликский; гребенщик *m*

3707 **fresh-felled density; green density**
d Dichte *f* des frischabgehauenen Holzes
f densité *f* de bois vert
r плотность *f* свежесрубленной древесины

3708 **fresh-felled green stem**
d frischabgehauener Grünstamm *m*
f tige *f* récente
r свежесрубленный здоровый ствол *m*

3709 **freshly-cut tree; freshly-felled tree; new-fallen tree; newly fallen tree**
d frischabgehauener Baum *m*
f arbre *m* récent
r свежесрубленное дерево *n*

3710 **fresh[ly-felled] wood; greenwood; new-fallen forest; damp wood; green lumber; green [timber]; green log**
(wood which has the original moisture present in the standing tree and has not been dried below the fiber saturation point)
d Frischholz *n*; fasergesättigtes Holz *n*; frisches Holz; frisch gefälltes Holz; saftfrisches Holz; frishabgehauenes Holz; grünes Holz; grüner Wald *m*
f bois *m* récemment gisant; bois vert; bois frais; bois non ressuyé
r свежесрубленная древесина *f*; свежая древесина; сырая древесина; древесина в соку; свежая вырубка *f*; свежесрубленный лесоматериал *m*; недавно поваленный лес *m*

3711 **freshwater swamp forest**
d Süßwasser-Sumpfwald *m*
f forêt *f* de marais d'eau douce
r пресноводный болотистый лес *m*

* **fretsaw → 1822**

* **friction → 7127**

3712 **friction[al] barker**
d Reibungsentrinder *m*
f écorceuse *f* à friction
r окорочный станок *m* трением

3713 **friction debarking**

f écorçage *m* par friction
r окорка *f* трением [обрабатываемой древесины]

3714 **friction saw**
d Reibungssäge *f*
f meule *f* tronçonneuse
r фрикционная пила *f*

* **frill cuts → 3715**

3715 **frill girdles; frill cuts; cup frills**
(on the bole)
d Nutringen *mpl*
f encoches *fpl* groupées
r окольцовывающие насечки *fpl*; сгруппированные насечки

3716 **frill girdling; hack girdling; frilling; hacking**
(killing trees by inflicting a series of cuts around the bole and applying an herbicide to the wounds)
d Nutringelung *f*
f annélation *f* en encoches; incision *f* par encoches groupées; incision par encoches disjointes
r кольцевание *n* [деревьев] насечками

* **fringe-leaved → 1580**

* **fringing → 627**

* **fringing forest → 6961**

3717 **froe[-knife]; frow; cleaving wedge; splitting wedge; splitting gun**
(a long wedge-shaped blade for splitting wood into shingles etc.)
d Spaltkeil *m*
f départoir *m*
r стальной клин *m* для раскалывания брёвен

* **frogger → 1559**

* **fromager → 4688**

3718 **frond**
(the compound leaf of a fern; in mosses, a stem closely and regularly branched in one plane, resembling a fern leaf)
d Wedel *m*; Farnwedel *m*
f fronde *m*
r ваиа *f*; вайа *f*; листообразное слоевище *n*; лист *m* папоротника

* **frondesce → 7531**

d Zeit *f* der Belaubung
f feuillaison *f*; frondaison *f*
r время *n* облиствения; период *m* развития листьев

* **frondose tree → 1048**

3720 **front[-]end loader; FEL**
(wheeled or tractor loader, with a bucket or fork hinged to lifting arms, that loads or digs entirely at the front end)
d Schaufellader *m*; Frontlader *m*
f chariot *m* élévateur frontal
r фронтальной челюстной погрузчик *m*

* **front face → 3722**

3721 **frontier forests**
(large, relatively intact forest ecosystems)
d Grenzenwälder *mpl*
f forêts *fpl* frontalières
r пограничные леса *mpl*

* **front portion of mo[u]ldboard → 1021**

3722 **front [side]; felling side; front face; felling face**
d Schlagfront *f*; Schlagseite *f*
f front *m* de la coupe; face *f* d'attaque; face travaillante; face de coupe
r фронт *m* лесосеки; сторона *f* в направлении валки

* **frost callus → 3726**

3723 **frost cleft; frost crack; frost mark; frost shake; frost split**
(radial, longitudinal split in the wood of a tree, generally near the base of the bole, caused by internal stresses due to extremely cold weather)
d Frostkluft *f*; Eiskluft *f*; Frostriss *m*; Frostspalte *f*
f brisure *f* par le gel; gélivure *f*
r морозобойная трещина *f*; морозная трещина; морозобоина *f*

3724 **frost hardiness; cold resistance**
d Frostbeständigkeit *f*; Kältebeständigkeit *f*; Frosthärte *f*
f résistance *f* au froid; résistance au gel
r морозостойкость *f*; морозоустойчивость *f*; холодостойкость *f*

* **frost-hardy forest → 925**

f cœur *m* par le gel; gélivure *f* du cœur
r ложное ядро *n*, образовавшееся под воздействием мороза; морозное ядро

* **frost mark → 3723**

3726 frost rib; frost scar; frost callus
d Frostleiste *f*
f bourrelet *m* de gélivure
r морозобойный нарост *m*

3727 frost ring
d Frostring *m*
f anneau *m*; cerne *m* [traumatique] de gel; lunure *f* due au gel
r морозное кольцо *n* (потемнения камбия в годичном кольце)

* **frost scar → 3726**

* **frost shake → 3723**

* **frost split → 3723**

* **frow → 3717**

3728 frozen wood
d gefrorenes Holz *n*
f bois *m* gelé
r мёрзлая древесина *f*

3729 fructification
d Fruchtbildung *f*
f fructification *f*
r образование *n* плода

* **fruiter → 3731**

* **fruit[ing] body → 1353**

3730 fruiting spur
(on a fruit tree, a short twisted branch with rings around it, which flowers and produces fruit)
d Obstsporn *m*
f contrefort *m* de fructification
r плодоносящий побег *m*

3731 fruit[ing] tree; fruiter
d Obstbaum *m*
f arbre *m* fruitier
r плодовое дерево *n*

3732 fruiting wood; sweet wood; cropping wood
(fruit-producing wood)
d Obstbaumholz *n*

* **fruit-ripe → 1996**

3733 fruit-tree forest
d Obstbaumwald *m*
f forêt *f* d'arbres fruitiers
r лесоплодовое насаждение *n*

* **fruit-tree shaker → 8845**

* **frutescent → 934**

* **frutex → 1202**

* **fruticose → 934**

* **ft.s.m. → 3036**

3734 fuel bark
d Brennrinde *f*
f écorce *f* combustible
r топливная кора *f*; кора, используемая в качестве топлива

* **fuelbote [right] → 4388**

* **fuelbreak → 3169**

* **fuel chips → 1517**

3735 fuel continuity
(qualitative description of the distribution of fuel both horizontally and vertically)
d Kontinuität *f* des Brennmaterials
f continuité *f* du combustible
r непрерывность *f* горючего материала

3736 fuel loading
(buildup of fuels, especially easily ignited, fast-burning fuels such as pinestraw)
d Brennstoffbelastung *f*
f chargement *m* du combustible
r концентрация *f* горючего материала; запасы *mpl* горючего вещества

* **fuel management → 3192**

3737 fuel moisture [content]
(percent or fraction of oven-dry weight of fuel)
d Feuchtigkeitsgehalt *n* von Brennstoffen
f teneur *f* en eau du combustible; teneur en humidité des matières combustibles; degré *m* hygrométrique des matières combustibles
r влагосодержание *n* топлива

f besoin *m* en bois de chauffage
r потребность *f* в дровах

* **fuel reduction → 3187**

3739 **fuels**
(wildland vegetation materials that can burn)
d Brennstoffe *mpl*
f combustibles *mpl*
r топлива *npl*; горючие [материалы] *mpl*

* **fuel value → 1194**

3740 **fuelwood; firewood; splitwood; kindling wood; stovewood; chore**
(any wood intended for heating or for charcoal-making)
d Brennholz *n*; Waldbrennholz *n*; Anmacheholz *n*; Anfeuerholz *n*
f bois *m* de chauffage; bois de feu; bois à brûler; bois d'allumage; allume-feu *m*
r топливная древесина *f*; лес *m* на топливо; дровяная древесина; дрова *pl* [для растопки]

3741 **fuelwood brick**
d Brennholzklotz *m*
f briquettes *fpl* de bois
r растопочные древесные брикеты *mpl*

3742 **fuel[wood] plantation**
(setting out young trees to be hogged for burning)
d Brennholzbestand *n*
f plantation *f* énergétique
r дровяной лес *m*

3743 **fuelwood residues**
d Brennholzabschnitte *mpl*
f déchets *mpl* combustibles
r дровянные остатки *mpl*

3744 **fugacious**
d fugax
f fugace; fugitif
r быстро увядающий; быстро опадающий

3745 **full-boled; full-bodied; little tapering** *adj*
d vollstämmig; vollholzig
f de bois massif
r полнодревесный

3746 **full-boled tree; nontapering tree**
d vollhölziger Baum *m*
f arbre *m* de bois massif
r дерево *n* с полнодревесным стволом

(measuring of all trees with calipers)
d Vollkluppierung *f*; Vollkluppung *f*
f inventaire *m* intégral par compas forestier
r полное определение *n* кубатуры

* **full canopy → 1650**

* **full-cell process impregnation → 692**

3748 **full diameter**
d Volldurchmesser *m*
f diamètre *m* plain
r полный размер *m*

3749 **full drum speed**
d Volltrommelgeschwindigkeit *f*
f pleine vitesse *f* de tambour
r скорость *f* намотки каната на верхних витках

* **full-edged log → 3755**

* **full-length log → 2834**

3750 **full-length lumber**
d Ganzbaumschnittholz *n*
f bois *m* débité à pleine longueur
r пиломатериалы *mpl* по длине дерева

3751 **full mast; heavy crop**
d Vollmast *m*
f paisson *f* totale
r полный урожай *m* плодокорма

* **full measure → 2160**

3752 **fullmoon maple**
(Acer japonicum)
d japanischer Ahorn *m*
f érable *m* du Japon à feuilles pubescentes
r клён *m* японский

* **fullness of bole → 8816**

* **full scale → 3981**

* **full scaling → 703**

3753 **full seed; filled seed**
(seed showing apparently complete embryo and endosperm or megagametophyte structures, irrespective of actual viability)
d fertiger Samen *m*
f graine *f* pleine
r полновесные семена *npl*

3754 **full-sibs; sibs**
(trees with both parents in common)
d Geschwister *pl*
f pleins germains *mpl*; fratrie *f*; frères *mpl*
r родственники *mpl*; сибсы *mpl*

3755 **full-squared log; full-edged log; die-squared log; dead-square log; fully quarter[ed]-sawn timber**
d vollkantiger Holzblock *m*
f grume *f* équarrie à vive-arête; grume alignée parallèle; bois *m* scié plein sur quartier
r полнокантный пиломатериал *m*; пиломатериал точно квадратного сечения; чистообрезной пиломатериал; чисто радиальный пиломатериал

3756 **full stand; high-density stand; dense stand [of trees]; complete stand; fully-stocked stand; closed forest**
d gedrängter Bestand *m*; Vollbestand *m*; geschlossener Bestand; dichter Bestand
f peuplement *m* complètement boisé; peuplement complet; peuplement dense; peuplement serré
r высокополнотное насаждение *n*; полное насаждение; сомкнутое насаждение; густое насаждение

3757 **full standing; full stocking; normal stocking; optimum stocking; ideal stocking**
d Normalbestockung *f*; Vollbestockung *f*
f matériel *m* sur pied relatif normal; densité *f* normale d'un peuplement
r полная густота *f* древостоя; нормальная густота древостоя; сомкнутый древостой *m*

3758 **full tree; whole tree; complete tree; total tree**
(every component of the tree from the needles to the root hairs)
d kompletter Baum *m*
f arbre *m* entier
r целое дерево *n*

3759 **full-tree chipping; chipping operation**
d Komplettbaum-Zerkleinerung *f*
f mise *f* en copeau d'arbres entiers
r переработка *f* [целых] деревьев в щепу; заготовка *f* щепы

3760 **full-tree chips; whole-tree chips**
d Ganzbaumhackschnitzel *npl*; Ganzbaumhackspäne *mpl*; Ganzbaumspäne *mpl*; Vollbaumspäne *mpl*

3761 **full-tree harvesting; whole-tree harvesting; felling to the full tree stage; full-tree removal; whole-tree operation**
(tree harvesting process that includes removing the trunk, branches and in some instances the roots from a forested site)
d Ganzbaumgewinnung *f*
f exploitation *f* par arbres entiers; enlèvement *m* d'arbres entiers
r лесозаготовки *fpl* с трелёвкой целых деревьев; вывозка *f* целых деревьев [с необрубленной кроной]

* **full-tree removal → 3761**

* **full-tree skidding → 9394**

3762 **full trunk eccentricity**
d komplette Stammexzentrizität *f*
f excentricité *f* du tronc complète
r сплошная крень *f*

* **fully quarter[ed]-sawn timber → 3755**

3763 **fully-stocked area; net forest area**
d vollbestockte Waldfläche *f*
f plaine aire *f* forestière; aire forestière complètement boisé; superficie *f* forestière nette; terre *f* complètement reboisée
r полнопокрытая лесом площадь *f*

* **fully-stocked stand → 3756**

* **fume damage → 2240**

3764 **fumed oak; bog oak; pin oak; swamp oak**
(Quercus palustris)
d Mooreiche *f*; Wassereiche *f*; Sumpfeiche *f*; Spießeiche *f*
f chêne *m* de marais
r морёный дуб *m*; дуб болотный

3765 **fumigant**
d Rächermittel *n*
f fumigateur *m*
r фумигант *m*; средство *n* для окуривания

3766 **fuming**
d Räuchern *n*
f fumigation *f*
r морение *n*; окуривание *n*

* **fundamental stand → 5250**

3767 **fund of state forest; stock of state forest;**

f fonds *m* de forêt dominiale
r государственный лесной фонд *m*

* **fund of wooded forests → 8114**

3768 **fungal decay**
(of wood)
d Pilzfäule *f*
f dégradation *f* fongique; pourriture *f* mycélienne
r разложение *n*, вызываемое грибами; грибная гниль *f*

3769 **fungal disease; fungous disease; mycosis**
d Pilzerkrankung *f*; Pilzkrankheit *f*; Mycosis *f*
f maladie *f* fongique; mycose *f*
r грибное заболевание *n*; грибная болезнь *f*; микоз *m*

3770 **fungi** *pl*
(plants that lack chlorophyll and derive their nourishment directly from other organisms (parasitic fungi) or from dead organic matter (saprophytic fungi))
d Pilze *mpl*; Pilzschimmel *m*; Schwamm *m*
f champignons *mpl*
r грибы *fpl*; грибки *fpl*; губка *f*; губчатый нарост *m*; плесень *f*

3771 **fungicide; antifungal [agent]; antimycotic [agent]**
(any agent used to kill or inhibit the growth of fungi and their spores)
d Pilzvertigungsmittel *n*; pilzabtötendes Mittel *n*; Fungizid *n*; Schimmelvernichtungsmittel *n*; pilztötender Stoff *m*
f fongicide *m*; antifongique *m*
r фунгицид *m*; грибоубивающее средство *n*

3772 **fungoid growth; mildew fungus; mo[u]ld formation; mo[u]ld growth; fungous growth; fungoid mo[u]lding**
(in wood)
d Pilzwachstum *n*; Schimmelwuchs *m*
f végétation *f* mycélienne; croissance *f* de moisissure
r рост *m* грибов; рост плесени; образование *n* грибкового повреждения

3773 **fungous; punky; conky; spongy; liverish**
d schwammig
f mycélien; envahi de champignons; champignonné
r рыхлый; губчатый; грибной

* **fungous disease → 3769**

* **fungous growth → 3772**

* **fungous wood → 2291**

3774 **fungus-preserving means**
d Pilzschutzmittel *npl*
f moyennes *fpl* de préservation
r защитные средства *npl* против поражения грибами

3775 **fungus root; mycorrhiza**
(association that is probably symbiotic between the root or rhizome of a green plant and a fungus)
d Pilzwurzel *f*; Mykorrhiza *f*
f mycorrhize *f*
r грибокорень *m*; микориза *f*

* **funnel of charcoal pile → 1487**

* **furcate → 3654**

* **furcation → 987**

3776 **furniture dimension lumber; furniture stock; furniture wood; cabinet[-]wood; cabinet timber**
d Möbelholz *n*
f bois *m* d'ébénisterie; bois de mobilier; bois pour meubles
r мебельная древесина *f*; пиломатериалы *mpl* для производства мебели; мебельные лесоматериалы *mpl*

* **furrow drilling → 7869**

3777 **furrow-leaved**
d fürcheblätterig
f en sillons
r бороздчатолистный

* **furrow planting → 8964**

* **furrow sowing → 7869**

* **furry willow → 4198**

3778 **fusain**
(fossil charcoal suggesting presence of grass or forest fires)
d Faserkohle *f*

3779 fusee
(colo[u]red flares to ignite fires in burnout operations)
d Signalfeuer *n*
f feu *m* de signal
r запал *m*

3780 fusiform body; microfibril
(of wood)
d Mikrofibrille *f*
f microfibrille *f*
r микрофибрилла *f*; веретенообразное тело *n*

3781 fusiform [cambial] initial
(a cambial initial giving rise to an axial element of xylem or phloem)
d fusiformes Initial *n*; spindelförmige Kambiumzelle *f*
f initiale *f* [du cambium] fusiforme; cellule *f* [du cambium] fusiforme
r веретеновидный инициал *m* камбия

3782 fusiform crown; spindle crown
d Spindelkrone *f*
f cime *f* fusiforme; cime *f* en quenouille
r веретенообразная крона *f*

3783 fusiform parenchyma cell; intermediate fiber; substitute fiber
(an axial parenchyma cell, derived from a fusiform initial without subdivision)
d fusiforme Parenchymzelle *f*; spindelförmige Parenchymzelle
f cellule *f* fusiforme de parenchyme; fibre *f* intermédiaire; fibre cambiforme
r веретенообразная паренхимная клетка *f*

3784 fusiform ray; lenticular ray
(literally a ray that is spindle-shaped in tangential section)
d fusiformer Strahl *m*; spindelförmiger Markstrahl *m*
f rayon *m* fusiforme; rayon lenticulaire
r веретеновидный луч *m*

3785 fusiform rust
(fungus-caused disease which produces spindle-shaped swellings or cankers on the boles and branches of pines)
d spindelförmige Braunfärbung *f*
f rouille *f* fusiforme
r веретенообразная ржавчина *f*

* **fusty wood → 2605**

* **futui → 1921**

3786 future tree; Z-stem; Z-tree
d Zukunftsbaum *m*; Zukunftsstamm *m*; Z-Stamm *m*
f arbre *m* d'avenir
r дерево *n* будущего; лучшее дерево

3787 fuzzy deutzia
(Deutzia scabra)
d hohe Deutzie *f*; rauhe Deutzie
f deutzie *f* scabre; deutzie à feuilles crénelées
r дейция *f* розовая; дейция шершавая

* **fuzzy grain → 9717**

* **FWI → 3218**

G

3788 **Gaboon [mahogany]; okoume; angouma**
(Aucoumea klaineana)
d Okoume *f*; Gaboon *m*; Gabun *m*
f okoumé *m*; okumé *m*; angouma *m*
r авкумея *f* Клайна

* **gaff → 6693**

* **galbulus → 2212**

3789 **gall**
(pronounced swelling or outgrowth on a plant, usually insect or disease caused)
d Galle *f*
f galle *f*
r галл *m*

3790 **gallberry; inkberry**
(Ilex glabra)
d Tintenbeere *f*
f houx *m* glabre
r падуб *m* голый

3791 **gallery forest**
(a stretch of forest along a river in an area of otherwise open country)
d Galeriewald *m*
f forêt-galerie *f*
r галерейный лес *m*

* **gall flies → 2210**

* **gall gnats → 3793**

3792 **gallipot**
d weißes Fichtenharz *n*
f résine *f* blanche [d'épicéa]
r белая еловая смола *f*

3793 **gall midges; midges; gall gnats**
(Itonidae; Cecidomyidae)
d Gallmücken *fpl*
f cécidomyies *fpl*; mouches *fpl* des galles
r галлицы *fpl*

* **gall[-nut] → 5831**

3794 **gall oak; Aleppo oak**
(Quercus infectoria)
f chêne *m* à galles; chêne à la noix de galle; chêne d'Aleppo
r дуб *m* малоазиатский; дуб алеппский

3795 **gall rust**
d Gallrost *m*
f rouille *f* de galle; rouille tumeur
r галловая ржавчина *f*

* **game bite → 2233**

* **game damage → 2233**

3796 **game population; game stock**
d Wildtier *m*
f peuplement *m* de gibier; cheptel *m* de gibier
r количество *n* дичи; наличность *f* дичи

3797 **game species**
(any species of wildlife or fish for which hunting seasons, bag limits, and other laws and regulations have been prescribed, and which are normally harvested by hunters, trappers, poachers, and anglers)
d Wildtierart *f*
f espèce *f* de gibier
r порода *f* дичи

* **game stock → 3796**

3798 **gametophyte**
(in organisms such as trees that alternate haploid (n) and diploid (2n) phases or generations, a gametophyte is the haploid (n) gamete-producing phase or generation)
d Gametophyt *m*
f gamétophyte *m*
r гаметофит *m*

3799 **gamma diversity**
(landscape-level or regional diversity)
d Gamma-Diversität *f*
f diversité *f* gamma
r гамма-разнообразие *n*

* **gang mill → 5132**

3800 **gang of saws**
d Vollgattersäge *f*
f scie *f* à plusieurs lames; scie [multiple] à châssis
r набор *m* пил

* **gangsaw → 2042**

* **gangway → 4600**

* **gap cutting → 6106**

* **gap felling → 6106**

* **gap in a stand → 3443**

* **gapping → 823**

* **gappy stand → 3646**

3801 **gap shelterwood system**
d Lücken-Schirmschlag-System *n*; Lücken-Stellung *f*
f système *m* de coupes d'abri; système par trouées
r система *f* просветов обсеменительной рубки

3802 **gap stick**
(the pole placed across the entrance of a sorting jack to close it, when not in use)
d Abstandsstock *m*
f billot *m* de regarnissage
r бревно *n*, закрывающее сортировочный карман

3803 **gap-toothed saw**
d Äbstand-gezähnte Säge *f*
f scie *f* à dents rares
r пила *f* с редкими зубьями

3804 **garcinia; saptree**
(Garcinia)
d Garcinia *n*
f garcinia *m*
r гарциния *f*

3805 **garcinia family**
(Guttiferae)
d Johanniskrautgewächse *pl*
f guttifères *pl*
r клузиевые *pl*

* **gardening shears → 5168**

3806 **garden serviceberry; [roundleaf] serviceberry**
(Amelanchier ovalis; Amelanchier rotundifolia; Amelanchier vulgaris)
d gemeine Felsenbirne *f*; gemeine Felsenmispel *f*; rundblättrige Felsenbirne
f amélanchier *m* [sanguin]; amélanchier commun
r ирга *f* обыкновенная; ирга круглолистная; коринка *f*

remains of Mediterranean sclerophyllous woodland or maquis)
d Garrigue *n*
f garrigue *f*
r гарига *f*

* **Garry oak → 5945**

3808 **gas-fired saw; gasoline saw; gasoline-engine powered saw; petrol motor[-powered] saw**
d benzinbetriebene Motorsäge *f*; Säge *f* mit Benzinantrieb
f scie *f* à benzine
r бензиномоторная пила *f*

* **gasifyer wood → 3816**

3809 **gasoline skidder**
d Benzin-Skidder *m*
f débusqueur *m* à benzine
r трелёвочная лебёдка *f* с бензиновым двигателем

* **gate saw → 5132**

* **gean → 8330**

3810 **gear-densified wood**
d [Plast-]Presssternholz *n*
f bois *m* densifié étoilement
r древесный слоистый пластик *m* с звездообразным расположением волокон

3811 **gear[ed] drive saw**
d Zahnradgetriebesäge *f*
f scie *f* par réducteur à roue
r редукторная моторная пила *f*

* **gedu nohor → 8541**

* **geiger tree → 4799**

3812 **geissois**
(Geissois)
d Geissois *n*
f geissois *m*
r гейсоис *m*

3813 **gelatinous fiber; mucilaginous fiber**
(a fiber having a more or less unlignified inner wall with a gelatinous appearance)
d gelatinöse Faser *f*
f fibre *f* gélatineuse
r желатинозное волокно *n*

* **gemma → 4859**

* **gemmation → 307**

* **gemmule → 5474**

3814 **gene flow**
(1. exchange of genetic traits between populations by movement of individuals, gametes or spores; 2. spread of genes through crossing)
d Genfluss *m*
f flux *m* de gènes
r поток *m* генов

3815 **gene pool; gene reserve**
(total genetic information possessed by the reproductive members of a population of sexually reproducing organisms)
d Genpool *m*
f réservoir *m* de gènes
r генный фонд *m*

3816 **generator wood; gasifyer wood**
d Generatorholz *n*; Tankholz *n*
f bois *m* à gazogène
r генераторный лес *m*; генераторные дрова *pl*

* **gene reserve → 3815**

3817 **gene reserve forest**
(a parcel of native forest that is large enough to sample the natural genetic diversity)
d Genpoolwald *m*; Genfondswald *m*
f forêt *f* de réserve génétique
r лес *m* генофонда

3818 **genetically improved seedlings**
(seed or propagule that originate from a tree breeding program and that have been specifically designed to improve some attribute of seeds, seedlings, or vegetative propagules selection)
d genetischverbesserte Sämlinge *mpl*
f semis *m* génétiquement améliorée
r генетически улучшенные лесные сеянцы *mpl*

3819 **genetically modified tree; GMT**
d genetischmodifizierter Baum *m*
f arbre *m* génétiquement modifié; AGM
r генетически модифицированное дерево *n*

3820 **genetic chim[a]era; chim[a]era**
(a genetic mosaic; a tree composed of tissues
d [genetische] Chimäre *f*; Schimäre *f*
f chimère *f* [génétique]
r растение *n* с изменёнными тканями; [генетическая] химера *f*

3821 **genetic diversity**
(variation among and within species that is attributable to differences in hereditary material)
d genetische Vielfalt *f*
f diversité *f* génétique
r генетическое разнообразие *n*

3822 **genetic variation**
d genetische Variation *f*
f variation *f* génétique
r генетическая изменчивость *f*

3823 **genom[e]**
(one complete set of genes containing all the genetic information to produce an individual)
d Genom *n*
f génome *m*
r геном *m*

3824 **genotype**
(entire genetic constitution, or the sum total of genes of an organism, in contrast to the phenotype)
d Genotyp *m*
f génotype *m*
r генотип *m*

3825 **genuine storax tree; storax; styrax; [drug] snowbell; silverball**
(Styrax officinalis)
d [echter] Storaxbaum *m*; gebräuchlicher Storaxbaum
f storax *m* véritable; styrax officinal; aliboutier *m* officinal
r стиракс *m* [аптечный]; стиракс лекарственный; дерево *n* стираксовое лекарственное

3826 **genuine wood**
d Naturholz *n*
f bois *m* naturel
r натуральная древесина *f*; естественная древесина

3827 **genus**
(in wood identification, a group of species having the broader characteristics in common)
d Gattung *f*
f genre *m*
r род *m*

3828 **geometric measurement of roundwood**
d geometrische Rundholzmessung *f*
f mesurage *m* géométrique de bois rond
r геометрическое определение *n* объёма круглых лесоматериалов

3829 **geometric thinning**
d geometrische Durchforstung *f*
f éclaircie *f* géométrique
r геометрическое прореживание *n*

* **Georgia blueberry → 2327**

3830 **Georgia oak; Stone Mountain oak**
(Quercus georgiana)
d Georgia-Eiche *f*
f chêne *m* de Georgie
r дуб *m* Джорджии

* **Georgia pine → 5141**

3831 **geotextile [membrane]; engineering fabric; geofabric**
(soil reinforcement agent and filter medium, often used in road construction)
d Geotextil
f géotextile *m*; membrane *f* géotextile
r геотекстиль *m*

* **Gerard pine → 1486**

3832 **germander spirea**
(Spiraea chamaedryfolia)
d Ulmen-Spierstrauch *m*; Gamander-Spierstrauch *m*
f spirée *f* à feuilles de germandrée; spirée ormière
r спирея *f* вязолистная; таволга *f* дубравколистная

* **German oak → 1773**

3833 **germinability; germinating quality; germinative quality; sprouting ability; germinating capacity; germination capacity; germination power**
(capacity of a seed, spore, or pollen grains to germinate and develop, under given conditions)
d Keimkraft *f*; Keimfähigkeit *f*; Keimvermögen *n*
f germinabilité *f*; pouvoir *m* germinatif; viabilité *f* [germinative]; faculté *f* germinative
r всхожесть *f* семян; жизнеспособность *f*

* **germinate → 3293**

3834 **germinating energy; germinative energy; germination energy**
(percentage of seeds, spores, or pollen grains in a given sample germinating within a given period)
d Keimenergie *f*
f vigueur *f* germinative
r энергия *f* прорастания

* **germinating quality → 3833**

3835 **germination**
(breaking of the seed coat followed by the protrusion of the hypocotyl forming a root system)
d Durchwachsen *n*; Keimung *f*; Keimen *n*; Knospenbildung *f*
f germination *f*
r прорастание *n*; образование *n* почек

* **germination capacity → 3833**

* **germination energy → 3834**

3836 **germination percent[age]**
(percentage of seeds, spores, or pollen grains in a given sample that actually germinate, irrespective of time)
d Keim[ungs]prozent *n*
f taux *m* de germination
r процент *m* всхожести

* **germination power → 3833**

3837 **germination test**
(test made to determine the viability of seeds, spores, or pollen grains in a given sample)
d Keimfähigkeitsuntersuchung *f*; Keimprobe *f*; Keimversuch *m*
f essai *m* de germination; test *m* de germinabilité
r проверка *f* всхожести; испытание *n* на всхожесть

* **germinative energy → 3834**

* **germinative quality → 3833**

3838 **germplasm**
d Keimplasma *n*
f plasma *m* germinatif
r зародышевая плазма *f*

* **get-away route → 2864**

* **get in a tree → 1641**

d Geister-Eukalyptus *m*
f gommier *m*
r эвкалипт *m* папуана

* **ghost ring → 5479**

* **giant arborvitae → 8893**

3840 **giant coreopsis; giant sea dahlia**
(Coreopsis gigantea)
d Baumsonnenblume *f*; Scheinwanzenblume *f*
f coréopsis *m* géant
r ленок *m*; лептосине *m*; кореопсис *m* гигантский

* **giant eucalyptus → 3842**

3841 **giant [evergreen] chinquapin; golden chinkapin; western chinkapin**
(Chrysolepis sempervirens; Castanopsis chrysophylla; Castanea chrysophylla)
d Scheinkastanie *f*
f castanopsis *m*
r кастанопсис *m* золотистолистный

* **giant fir → 3910**

3842 **giant gum; giant eucalyptus**
(Eucalyptus regnans)
d Königseukalyptus *m*
f eucalyptus *m* ronce
r эвкалипт *m* царственный; белый горный ясень *m*

3843 **giant redbud**
(Cercis gigantea)
d Riesenjudasbaum *m*
f arbre *m* de Judée géant
r церцис *m* гигантский

* **giant sea dahlia → 3840**

3844 **giant sequoia; mammoth tree**
(Sequoia gigantea; Sequoiadendron giganteum)
d Mammutbaum *m*; Riesenbaum *m*; Riesensequoie *f*
f séquoia *m* géant; arbre *m* mammouth; wellingtonia *m* [géant]
r секвойя *f* гигантская; веллингтония *f*; секвойядендрон *m* гигантский; мамонтово[е] дерево *n*

3845 **giant trees; supercanopy trees**
d Riesenbäume *mpl*

* **gib-and-cotter joint → 2595**

* **cotter joint → 2595**

3846 **gidgee [acacia]**
(Acacia homalophylla)
d Veilchenholzakazie *f*
f yarran *m*
r акация *f* плосколистная; дерево *n* фиалковое

* **gigantic pine → 1274**

* **Gilead fir → 969**

* **ginkgo → 5245**

3847 **ginkgo family; maidenhair-tree family**
(Ginkgoaceae)
d Ginkgogewächse *pl*
f famille *f* des ginkgos; ginkgoacées *pl*
r гинкговые *pl*

3848 **gin[-]pole; standing derrick; pole derrick; guyed-mast derrick**
(a pole secured by guy ropes, to the top of which tackle for loading logs is fastened)
d Kranbaum *m*
f mât *m* de charge; mât de montage; mât de levage
r погрузочная мачта *f* с растяжками; мачта промежуточной опоры (канатной установки)

3849 **ginster**
(Genista)
d Ginster *m*
f genêt *m*
r дрок *m*

3850 **girdle**
(on the stem)
d entrindeter Messring *m*
f anneau *m* écorcé
r окорённое мерное кольцо *n*

3851 **girdle** *v*; **ringbark** *v*; **ring** *v*
(of a tree)
d beringen; ringeln
f anneler
r кольцевать; окружать; подпоясывать

3852 **girdled tree**
d geringelter Baum *m*

3853 girdle notch
d Ringelkerbe *f*; Ringelschnitt *m*
f encoche *f* circulaire; incision *f* annulaire
r кольцевой надрез *m*; кольцевание *f* [коры сучка]

3854 girdling tool
(tool removing a ring of bark for scaling)
d Ringeleisen *n*
f fer *m* à anneler
r нож *m* для кольцевания

3855 girth above buttress
d Umfang *m* über Wurzelanlauf
f circonférence *f* au-dessus des empattements
r окружность *f* у основания

* **girth at the middle → 5440**

* **girth breast height → 1024**

3856 girth increment
d Umfangszuwachs *m*
f accroissement *m* de la circonférence
r прирост *m* по окружности

* **girth-limit cutting → 2476**

3857 girth measurement
d Umfangsmessung *f*; Stärkenmessung *f*
f mesure *f* de la circonférence
r измерение *n* окружности

* **girth over/under bark → 2482**

3858 girth quotient
d Umfangsquotient *m*; Kreiskoeffizient *n*
f quotient *m* de circonférence
r коэффициент *m* окружности

* **glabrous shadbush → 177**

* **glade → 8865**

* **glancer → 1025**

* **glancing boom → 3147**

3859 gland-bearing oak
(Quercus glandulifera; Quercus serrata)
d drüsenblättrige Eiche *f*; Konara-Eiche *f*
f chêne *m* glandouilleur; chêne de la variété
r дуб *m* пильчатый; дуб железоносный; конара *m*

* **glandular ailanthus → 8892**

* **glasshouse → 3946**

* **glassy wood → 9354**

3860 gleization; gleying
(the process by which gleys are formed)
d Vergleyung *f*
f gléification *f*
r образование *n* глея в почве; глееобразование *n*; оглеение *n*

3861 gley[ed] forest grey soil
d podsolierter graue Waldboden *m*
f sol *m* gris forestier gleyifié
r глеевая серая лесная почва *f*

* **gleying → 3860**

3862 glissade skids
d Glissadehölzer *npl*
f patins *mpl* de glissade
r свежеокорённые слеги *pl*

3863 global agreement on forests; global convention on forests; framework convention on global forests
d Globalwaldvertrag *m*; Globalwaldkonvention *f*
f accord *m* global des forêts; convention *f* globale des forêts
r глобальная конвенция *f* о лесах; глобальное соглашение *n* о лесах

* **global convention on forests → 3863**

3864 global forest resources
d Globalwaldressourcen *fpl*
f ressources *fpl* des forêts globales
r ресурсы *mpl* лесах планеты

3865 global forests
d globale Wälder *mpl*; weltweite Wälder
f forêts *fpl* globales
r леса *mpl* планеты

* **global irradiance → 3866**

3866 global radiation; total radiation; total insolation; global irradiance
(total of direct solar radiation and diffuse sky radiation received by a unit horizontal surface)
d Globalstrahlung *f*

3867 **global warming**
(overall rise in the temperature of the Earth's atmosphere; a projected result of the greenhouse effect)
d Erderwärmung *f*
f réchauffement *m* planétaire
r глобальное потепление *n* [климата]

3868 **globe daisies family**
(Globul[i]ariaceae)
d Kugelblumengewächse *pl*
f globulariacées *pl*
r глобуляриевые *pl*; шаровницевые *pl*

3869 **globular crown; spherical crown**
d kugelförmige Krone *f*
f cime *f* globulaire; cime sphérique
r шаровидная крона *f*

3870 **glory bower**
(Clerodendrum)
d Lossbaum *m*; Lossstrauch *m*
f clérodendron *m*
r клеродендрон *m*

3871 **glorytree; glory-bush**
(Tibouchina)
d Schraubenbaum *m*
f tibouchine *f*
r тиб[о]ухина *f*

3872 **gloss of timber; luster**
d Glanz *m* eines Holzes
f reflet *m* du bois
r блеск *m* дерева

3873 **glossy abelia**
(Abelia grandiflora)
d großblütige Abelie *f*
f abélia *m* à grandes fleurs
r абелия *f* крупноцветная

* **glossy buckthorn → 684**

3874 **glossy oak; shingle oak** (US)
(Quercus imbricaria)
d Glanzeiche *f*; Schindeleiche *f*
f chêne *m* imbriqué; chêne à feuilles de laurier; chêne de bardeau
r дуб *m* черепитчатый

3875 **glowing combustion**
d erglühende Verbrennung *f*
f combustion *f* incandescente
r тлеющее горение *n*

3876 **glued laminated wood; glulam**
d verleimtes Holz *n*
f bois *m* lamellé-collé
r клеёный лесоматериал *m*; клеённая древесина *f*

3877 **glued wood chips**
d beleimte Holzpartikeln *fpl*
f particules *fpl* de bois gluées
r склеенные древесные частицы *fpl*; проклеенные древесные частицы; осмоленные древесные частицы

3878 **glue-laminated timber; glulam**
(structural wood product made by bonding together laminations of dimension lumber)
d verleimtes mehrschichitge Holzelement *n*
f pièce *f* de charpente lamellée-collée; élément *m* de charpente lamellé collé
r многослойный дощатоклеённой элемент *m*; деревянный клеёный дощатый элемент

* **glycyrrhiza → 4997**

* **GMT → 3819**

* **gnarl → 1182**

* **gnarled → 4737**

* **gnarled wood → 4738**

* **gnarl formation → 9187**

* **gnarly → 4737**

* **gnaw → 5730**

3879 **goat moth**
(Cossus cossus)
d Weidenbohrer *m*
f gâte-bois *m*
r древоточец *m* пахучий; древоточец ивовый

3880 **goat-willow; sallow willow; pussy willow; florist's willow; silky headed willow**
(Salix caprea; Salix discolor)
d Palmweide *f*; Salweide *f*; Bachweide *f*; Pfeifenholz *n*
f saule *m* marceau; saule marsault; saule décoloré
r ива *f* козья; бредина *f*

3881 **go-devil; alligator; sloop**
(a small sled, without a tongue, often made from the natural fork of a tree and used as an aid in skidding logs on stony or bare ground)

f traîneau *m* de débardage avec fourches; fourche *f* à débardage
r трелёвочные сани *pl* с вилами; вилообразные трелёвочные сани; волокуша *f*; короткие трелёвочные сани без дышла

* **God's tree → 1814**

* **gold birch → 2860**

* **golden apple → 4322**

* **golden[ ]chain [tree] → 1784**

* **golden chinkapin → 3841**

3882 **golden ebony; Indian ebony; Macassar ebony**
(Diospyros celebica)
d Macassar-Ebenholz *n*
f ébène *m* Macassar
r макасар *m*; хурма *f* целебесская

* **golden fir → 1268**

* **golden rain → 1784**

* **goldenrain tree → 1497**

3883 **golden-shower senna**
(Cassia fistula)
d Röhrenkassie *f*
f canéficier *m*; cassier *m* commun
r кассия *f* дудка

* **golden spoon → 2570**

3884 **golden wattle**
(Acacia pycnantha; Acacia petiolaris)
d goldene Akazie *f*
f mimosa *m* à larges feuilles; mimosa doré
r акация *f* густоцветковая

3885 **golden willow; yellow-stem white willow**
(Salix [alba] vitellina)
d gelbe Weide *f*; Dotterweide *f*
f osier *m* jaune; osier vitellin; verdoisis *m*; amarinié *m*
r ветла *f* блестяще-жёлтая

* **gombe → 1121**

* **gommier → 1306**

* **gonfolo → 5275**

(Salix gooddingii)
d Goodding-Weide *f*
f saule *m* de Goodding
r ива *f* Гудинга

3887 **goosefoot family**
(Chenopodiaceae)
d Gänsefußgewächse *pl*; Chenopodeen *pl*; Chenopodiazeen *pl*
f chénopodiacées *pl*
r лебедовые *pl*; маревые *pl*

3888 **gooseneck; shackle; yoke; scotch**
(the heavy V-shaped part of a block by which the block is attached to an object; V-shaped pair of thills joining the forward and rear sets of runners of a logging sled)
d Riesbremse *f*
f col *m* de cygne
r тормозное устройство *n* на лесоспуске; тормоз *m* лесоспуска

* **goose plum → 256**

* **goose saw → 4312**

* **gopherwood → 9767**

* **gorge forest → 3474**

3889 **gorse**
(Ulex europaeus)
d europäischer Stechginster *m*; Gaspeldorn *m*
f ajonc *m* commun; ajonc épineux; ajonc d'Europe
r улекс *m* [европейский]

* **gouge → 4328**

* **gouge blaze → 7313**

* **gouger → 7313**

3890 **governmental forestry; state[-owned] forestry**
d staatliche Forstwirtschaft *f*; Staatsforstwirtschaft *f*
f foresterie *f* dominiale
r государственное лесное хозяйство *n*

* **government forest → 8041**

3891 **Gowen cypress; Californian cypress**
(Cupressus goveniana)
d kalifornische Zypresse *f*; Gowen-Zypresse *f*

3892 grab hooks; grabs; [hook] dogs; timber hitchs; skidding hooks; logging scissors; [log skidding] grapple; skidding scissors; [skidding] tongs; timber tongs; trail dogs; timber clips
(a hook having a narrow throat, adapted to grasp any link of a chain)
d Holzzange *f*; Hebezange *f*; Rückezange *f*; Greifhaken *m*; Holzgreifer *m*; Greifer *m*
f pince *f* [de débardage]; grappin *m* [à bois]
r захватные крюки *mpl*; захваты *mpl* для брёвен; трелёвочные крюки; трелёвочные щипцы *pl*; [трелёвочный] клещевые захваты; грейфер *m*

* **grab setter → 1559**

3893 gradation of age class; age-gradation
d Altersklassenabstufung *f*; Altersabstufung *f*
f gradation *f* [des classes] d'âge
r градация *f* по классам возрасти

3894 gradation of length
d Längeabstufung *f*; Längegradation *f*
f gradation *f* par longueur
r градация *f* по длине; классификация *f* по длине

* **grade → 7750**

3895 graded-density chipboard
d Gradientdichte-Spanplatte *f*; Spanplatte *f* mit abgestufte Dichte
f panneau *m* de particules à densité progressive
r древесностружечная плита *f* различной плотности

3896 graded volume table; assortment table
d Sortimententafel *f*
f table *f* des assortiments
r сортиментная таблица *f*

* **grade in lumber → 3902**

3897 grade of mixture; degree of mixture
d Mischungsgrad *m*
f degré *m* du mélange
r степень *f* смешения

3898 grade of slenderness
(of a fiber)
d Schlankheitsgrad *m*
f degré *m* d'élancement
r отношение *n* ширины к длине

* **grade of wood → 6657**

* **grader station → 5093**

3899 grade stake; grade peg; tracing picket; ranging pole
d Absteckpfahl *m*; Absteckstange *f*; Absteckpflock *m*; Niveaupfahl *m*; Höhenpfahl *m*
f jalon *m* de nivellement; piquet *m*; broche *f* repère; fiche *f* repère
r трассировочный столбик *m*; веха *f*; вешка *f*; трассировочный колышек *m*; нивелировочный кол *m*; нивелирный кол; уровенный кол

3900 grade stamp
(a stamp placed on lumber to denote its grade)
d Qualitätstempel *m*; Stempel *m* für Qualität
f marque *f* de qualité
r штемпель *m* качества

* **gradient → 7750**

3901 grading; level[l]ing
d Planierung *f*; Planieren *n*
f planification *f*
r выравнивание *n*; планировка *f*

3902 grading of wood; sorting of wood; wood classification; grade in lumber; timber grading; culling
(classification of lumber based on quality)
d Holzsorterbildung *f*; Holzsortierung *f*; Holzklassifizierung *f*
f calibrage *m* de bois; triage *m* de bois; classification *f* du bois
r сортировка *f* лесоматериалов; сортировка леса; классификация *f* лесоматериалов

* **grading station → 5093**

3903 grading to size; sizing
d Größensortierung *f*
f calibrage *m* par les dimensions
r сортировка *f* по размерам

3904 gradual felling; successive felling
d stufenweiser Holzschlag *m*
f abattage *m* graduel
r постепенная рубка *f*

3905 graduated forester
d Diplom-Forstwirt *m*; Diplom-Forstingenieur *m*
f ingénieur *m* forestier diplômé
r дипломированный лесовод *m*

3906 graft *v*; **inoculate** *v*
d pfropfen; okulieren
f greffer; enter
r прививать; черенковать; окулировать

3907 grafted roots
d verwachsene Wurzel *fpl*
f racines *fpl* adhérentes
r сросшиеся корни *mpl*

3908 grafting
(placing a detached cutting or branch tip (scion) in close cambial contact with a rooted plant (understock) in such a manner that scion and rootstock unite)
d Propfung *f*; Pfropfen *n*
f greffage *m*
r прививка *f*; пересадка *f*

* **grafting saw → 6614**

* **grafting wax → 8958**

* **grain direction → 2516**

* **grained birch → 3122**

* **grain figure → 9678**

* **grain inclination → 4439**

* **granadillo → 5426**

* **Grand Bassam mahogany → 107**

3909 grand eucalyptus; rose gum; Tooler eucalyptus
(Eucalyptus grandis)
d Rose-Gum *m*
f rose gum *m*
r эвкалипт *m* большой

3910 grand fir; giant fir; lowland fir
(Abies grandis)
d Küstentanne *f*; Riesentanne *f*
f sapin *m* de Vancouver; sapin géant
r пихта *f* великая

* **granite gilia → 6534**

* **granite prickly phlox → 6534**

3911 granular application
(general process by which fertilizers or herbicides in the form of grains are applied to a given area)
r зернистое внесение *n*

* **grapple → 3892**

* **grapple logging vehicle → 3913**

3912 grapple[-]saw
d Greifersäge *f*
f grappin-tronçonneur *m*; grappin-scie *m*
r пила *f* на захвате; захват *m* с пилой

3913 grapple yarding system; grapple yarder; grapple skidder; grapple logging vehicle
(machine used in harvesting to bring logs into a landing)
d Zangen-Schlepper *m*
f système *m* de débusquage à pince; débardeur *m* à pince; tracteur-tireur *m* à pince; débusqueuse *f* à grappin
r канатная трелёвочная установка *f* с захватом; трелёвочный трактор *m* с [пачковым] захватом

3914 grass-cutting in the forest
d Waldgräserei *f*
f récolte *f* d'herbes en forêt
r лесной покос *m*; лесное сенокошение *n*; пользование *n* лесной травой

3915 grass family
(Gramineae)
d Gramineen *pl*
f graminées *pl*
r злаковые *pl*

3916 grass-forb [stage]
(a stage in the succession of coniferous forests that is dominated by grasses and forbs)
d Gras-Forb-Stadium *n*
f étape *f* d'herbes et phorbiés
r стадия *f* трав и полукустарников

3917 grass forest; forest with grass layer
d Graswald *m*; Wald *m* mit der Grasschicht
f forêt *f* à herbes
r лес *m* с травянистым покровом

3918 grassland-forest boundary
d Grenze *f* zwischen Grüneland und Wald
f limite *f* entre l'herbage et la forêt
r граница *f* между лесом и лугом

* **grass line → 8153**

3919 grass pine forest
d Gräser-Kiefernwald *m*

3920 **grass spruce forest**
d Gräser-Fichtenwald *m*
f pessière *f* à graminées
r травяной ельник *m*

* **grass tree → 405**

* **grating for stopping floating wood → 8136**

3921 **gravel[ed] road**
d Kiesstraße *f*; Kiesweg *m*
f chemin *m* empierré
r гравийная дорога *f*

* **gravity cable → 9464**

3922 **gravity cable logging**
(any cable system that depends on the force of gravity for downhill travel of the carriage)
d Kabelriese *f*; Seilriese *f*
f téléphérage *m* par gravité; glissoire *f* à câble
r гравитационная канатная трелёвка *f*; гравитационный проволочный лесоспуск *m*; тросный лесоспуск, работающий силой тяжести; кабельный спуск *m*; канатный лесоспуск

* **gravity wire [skidding equipment] → 9464**

3923 **gray birch; wire birch; poplar leaved birch; old-field birch**
(Betula populifolia)
d graue Birke *f*; Graubirke *f*; pappelblättrige Birke
f bouleau *m* gris; bouleau rouge; bouleau à feuilles de peuplier
r берёза *f* тополелистная

3924 **gray dogwood**
(Cornus racemosa; Cornus paniculata)
d Rispen-Hartriegel *m*
f cornouiller *m* à grappes
r дёрен *m* кистевидный

* **gray elm → 7745**

* **gray eucalyptus → 5587**

3925 **graying**
(wood defect)
d Eingrauung *f*; Vergrauen *n*
f mise *f* en gris
r посерение *n*; серая окраска *f*

3926 **grayleaf willow; arctic grey willow**
(Salix glauca)
r ива *f* серо-голубая

3927 **gray mountain holly; largeleaf holly**
(Ilex monticola)
d Berg-Stechpalme *f*
f houx *m* de montagne
r падуб *m* горный

3928 **gray oak; shin oak**
(Quercus grisea)
d graue Eiche *f*
f chêne *m* gris
r дуб *m* серый

* **gray peppermint eucalyptus → 5663**

* **gray pine → 7142**

* **graze → 1094**

3929 **grazed forest; grazing forest; wooded grazing land; grazing woodland**
(forestland that produces, at least periodically, sufficient understor[e]y vegetation that can be grazed)
d Hudewald *m*; Hutwald *m*; Weidewald *m*; Wald *m* mit Weidewirtschaft
f forêt *f* pâturable; forêt soumise au pâturage; forêt pâturée
r лес *m* с пастбищными угодьями; лес для пастбища; лес с пастьбой скота; пастбищный лес; выпасаемый лес

3930 **grazing defect; grazing damage; injury from hails; hail damage**
d Beschädigung *f* durch Vieh; Viehschaden *m*
f dégât *m* dû au bétail
r повреждение *n* от скота; повреждение пастьбой

* **grazing forest → 3929**

* **grazing line → 1096**

3931 **grazing schedule**
(sets out the class and number of livestock that can use an area described in the schedule, the dates the livestock can use the area and other prescribed information)
d Weidenplan *m*
f horaire *m* de pâturage
r план *m* пастьбы

* **grazing woodland → 3929**

* **grease bands → 8777**

3932 **greasewood**
(Sarcobatus vermiculatus)
d Greasewood *n*
f armoise *f* tridentée; larrea *m* tridenté
r жирное дерево *n*

3933 **great leadtree; tepeguaje**
(Leucaena pulverulenta)
d großes Leucaena *n*; Tepeguaje *n*
f grand[e] laucène *m/f*; laucène panaché[e]
r левкена *f* припудренная

* **great maple → 8355**

3934 **great rhododendron**
(Rhododendron maximum)
d Riesenrhododendron *m*
f rhododendron *m* géant; grand rhododendron
r рододендрон *m* крупнейший

3935 **Grecian cedar; Grecian juniper; West Asian cedar**
(Juniperus excelsa)
d griechischer Wacholder *m*; hoher Sadebaum *m*
f genévrier *m* de Grèce; genévrier d'Orient
r можжевельник *m* греческий; можжевельник высокий

* **Grecian fir → 3938**

3936 **Grecian laurel; noble laurel; [sweet] bay [tree]; true bay**
(Laurus nobilis)
d blühender Lorbeer[baum] *m*; edler Lorbeerbaum
f laurier *m* d'Apollon; laurier noble; laurier commun
r лавр *m* благородный

3937 **Grecian oak; Greek oak; Asiatic oak; Valonia[n] oak**
(Quercus graeca; Quercus macrolepis; Quercus aegilops)
d Valoneneiche *f*; Ziegenbarteiche *f*; Knopperneiche *f*; arkadische Eiche *f*
f chêne *m* de Valonia; vallonée *f*; chêne vélani; chêne à grandes écailles
r дуб *m* греческий; дуб крупночешуйчатый

3938 **Greek fir; Grecian fir**
(Abies cephalonica)
d griechische Tanne *f*; kephalonische Tanne
f sapin *m* de Grèce; sapin de Céphalonie
r пихта *f* греческая

* **green alder → 2897**

* **green ash → 2255**

3939 **greenbark ceanothus; redheart ceanothus; spiny ceanothus**
(Ceanothus spinosus)
d dornförmige Säckelblume *f*
f céanothe *m* épineux
r цеанотус *m* колючий

* **green belt → 3941**

* **green bough → 747**

* **green branch → 5013**

* **greenbreak → 2928**

* **greenbrier → 7223**

3940 **green chips**
d grüne Späne *mpl*
f copeaux *mpl* verts
r зелёная щепа *f* (получаемая при переработки целых деревьев)

3941 **green corridor; green belt; greenway; green space**
(a strip of land, usually a mile or more in width, surrounding an urban area; an area of green open space that surrounds and stretches into cities)
d Grüngürtel *m*; Grünzug *m*; Grünfläche *f*
f ceinture *f* verte; zone *f* verte
r зелёная зона *f*; лесопарковая зона; зелёный обход *m*; лесопарковый пояс *m*; пояс зелёных насаждений

* **green density → 3707**

* **green Douglas fir → 1772**

3942 **green ebony [wood]**
d grünes Ebenholz *n*
f bois *m* d'ébène vert
r древесная порода *f* из семейства эбеновых с зелёной древесиной

* **green-end production → 9116**

* **greenery → 3303**

3943 **Greene's mountain ash; western mountain ash; Sitka mountain ash; Cascade Mountain ash**

f sorbier *m* de Sitka
r рябина *f* мелкометельчатая

* **green fertilizer → 3949**

* **green firebreak → 2928**

3944 **green [grading] chain; sorting table**
(a moving belt or chain on which lumber is transported from saws to areas where workers stack the wood by certain specifications)
d Grünholzsortiertabelle *f*
f table *f* de tri[age]; table de classement
r сортировочная цепь *f* на лесозаводе

3945 **green[-]headed spruce sawfly**
(Pikonema dimmockii)
d grünköpfige Fichtenwespe *f*
f tenthrède *f* à tête verte de l'épinette
r пилильщик *m* еловой зелёноголовый

* **greenheart → 2363**

3946 **greenhouse; orchard house; pot culture house; warm house; glasshouse**
d Gewächshaus *n*
f serre *f*
r теплица *f*; оранжерея *f*

3947 **greenhouse effect**
(warming of the Earth's atmosphere caused by increasing levels of carbon dioxide and other gases in the air)
d Gewächshaus-Effekt *m*
f effet *m* de serre
r эффект *m* оранжереи

* **green log → 3710**

* **green lumber → 3710**

3948 **green mandarin; tangerine**
(Citrus reticulata)
d Mittelmeermandarine *f*; Mandarine *f*
f mandarine *f* verte
r мандарин-танжерин *m*; цитрус *m* сетчатый; мандариновая цедра *f*

3949 **green manure; plant manure; green fertilizer**
d Gründünger *m*
f engrais *m* vert; fumier *m*; lisier *m*
r зелёное удобрение *n*; сидерит *m*

* **green mastic → 5327**

3950 **green moisture content**
d Frischholzfeuchte *f*
f humidité *f* de bois vert
r влажность *f* свежесрубленной древесины

* **green mo[u]lds fungi → 802**

3951 **green oak**
d Grüneiche *f*
f chêne *m* vert
r ярко-зелёная древесина *f* дуба

* **green osier → 222**

3952 **green plum**
(Buchanania obovata)
d australische grüne Pflaume *f*
f prune *f* verte
r слива *f* зелёная

3953 **green pruning; live pruning**
(of growing tree)
d Grünastung *f*; Grünästen *n*; Stehendästen *n*; Stehendästung *f*
f élagage *m* en vert; ébranchage *m* à état vert
r обрезка *f* живых ветвей; обрезка зеленых ростков; обрубка *f* живых ветвей; отпиливание *n* живых ветвей; обрубка сучьев на стволе стоящего дерева

3954 **green rot**
d Grünfäule *f*
f pourriture *f* verte
r зелёная гниль *f*

3955 **green sawn size**
d Grünsägemaß *n*
f dimension *f* du sciage en vert
r распиловочный размер *m* свежесрубленной древесины

* **green space → 3941**

3956 **green stain**
(a green discolo[u]ration in certain hardwoods, attributed to fungus attack, e.g. Chlorosplenium aeruginosum in Quercus spp. which creates "green oak")
d grüner Fleck *m*
f tache *f* verte
r грибная зелёная окраска *f*

3957 **green state**
d Grünholzzustand *m*
f état *m* de bois vert
r сырое состояние *n* пиломатериалов; невысушенное состояние пиломатериалов

3958 green stump
d Grünstumpf *m*
f souche *f* verte
r свежий осмол *m*; неспелый осмол

* **green [timber] → 3710**

3959 green tree cut
d grüner Baumschnitt *m*
f coupe *f* d'arbres verts
r валка *f* зелёных деревьев

3960 green tree reservoir; GTR
(forested wetland inundated during the dormant period of tree growth to provide temporarily aquatic habitat, usually for waterfowl, without damaging tree survival)
d grüner Baumvorratsbehälter *m*
f réservoir *m* de réserve d'arbres verts
r хранилище *n* зелёного резерва

3961 green tree retention
(deferral of live trees of a specific species and size from harvesting, to achieve a site-specific objective)
d grünes Baumzurückhalten *n*
f rétention *f* d'arbres verts
r сохранение *n* зелёных деревьев

* **green volume → 9253**

3962 green-wattle acacia
(Acacia decurrens)
d schwarze Akazie *f*; Mimose *f*
f mimosa *m* argenté; acacia *m* "bon accueil"
r акация *f* низбегающая; акация сбегающая; мимоза *f*; ватле *n*

* **greenway → 3941**

3963 green wedge
d grüner Keil *m*
f coulée *f* de verdure; couloir *m* de verdure
r лесопарковый клин *m* (в пределах городской черты)

* **green weeping willow → 459**

3964 green weight; weight of green wood
d Frischgewicht *n*; Grüngewicht *n*
f poids *m* de bois vert
r вес *m* свежесрубленной древесины

* **greenwood → 3710**

f ombre *m* de la forêt
r лесная тень *f*

* **gregre → 6014**

3966 grevillea; East African silky oak; Australian silky oak; [red] silk-oak
(Grevillea robusta; Cardwellia sublimia)
d Grevillia *f*; afrikanische Seideneiche *f*; [rote] australische Seideeiche *f*
f grevillea *m*; grévillée *f*; chêne *m* soyeux rouge; silky-oak *m* rouge
r кружевное дерево *n*; гревиллея *f* крупная; шелковый дуб *m*; саламандр *m*

* **grey alder → 2878**

* **grey box → 932**

3967 grey-brown forest soil
d grau-brauner Waldboden *m*
f sol *m* gris-brun forestier
r серо-коричневая лесная почва *f*

3968 grey-brown sap rot
(invader - Polyporus volvatus)
d grau-braune Saftfäule *f*
f pourriture *f* de l'aubier gris-brun
r серо-бурая заболонная гниль *f*

3969 grey forest soil; grey wooded soil
d grauer Waldboden *m*
f sol *m* gris forestier
r серая лесная почва *f*

* **grey gum → 5587**

3970 grey heart
d grauer Kern *m*
f cœur *m* gris
r серое ядро *n*

3971 grey mangrove
(Avicennia marina)
d graue Mangrove *f*
f manglier *m* gris
r авиценния *f* морская

3972 grey milkwood
(Cerbera dilatata; Cerbera floribunda; Cerbera inflata; Cerbera manghas)
d Zerberusbaum *m*
f faux manguier *m*
r цербера *f* мангас

3973 grey poplar; Picardian poplar

f peuplier *m* grisaille; peuplier grisard; peuplier de Picardie
r тополь *m* седеющий

3974 **grey willow**
(Salix cinerea)
d graue Bruchweide *f*; graue Glockenheide *f*; Aschweide *f*
f saule *m* cendré; gévrine *m*
r ива *f* пепельная

* **grey wooded soil → 3969**

* **gri-gri → 6014**

3975 **grind** *v*
d schleifen; ausschleifen; geradeschleifen; schrägschleifen
f affûter; broyer
r молоть; перемалывать; перемолоть; растирать; растереть

* **grinding → 2105**

* **grinding angle → 7493**

* **grinding wood → 6061**

3976 **grip; water-bar; let**
(a ridge of packed earth or a thin pole (a pole drain) laid diagonally across the surface of a road so as to lead water)
d Wasserriegel *m*
f saignée *f*
r мелкое дренажное углубление *n*

* **grippers → 6225**

* **gripping of the log → 7315**

* **grit → 2794**

* **grit cell → 8121**

* **groove → 7308, 8155**

3977 **grooved**
d gerillt
f rainuré
r желобчатый, рифлёный

3978 **grooved timber**
r пазовая древесина *f*

3979 **gross annual increment; final mean annual increment; final m.a.i.**
(average annual volume of increment over the reference period of all trees, measured to a minimum d.b.h. of 0 cm)
d Haubarkeitsdurchschnittszuwachs *m*
f accroissement *m* moyen [annuel] final; accroissement moyen d'exploitabilité
r окончательный средний годичный прирост *m*

* **gross capacity → 3981**

* **gross increment → 8727**

3980 **gross production of wood under bark**
d Erzeugung *f* von Rohholz ohne Rinde
f production *f* de bois brut sans écorce
r грубая продукция *f* древесины без коры

3981 **gross scale; full scale; gross volume; gross capacity; bigness scale**
(measurement of log volume in which no deduction is made for defect)
d grobe Ausgabe *f*
f volume *m* brut
r грубо определённый объём *m* брёвен

3982 **ground-based systems**
(logging systems that employ ground-based equipment such as feller-bunchers, hoe chuckers, skidders, and forwarders)
d bodengestützte Systeme *npl*
f systèmes *mpl* au sol
r наземные системы *mpl*

* **ground blueberry → 7523**

* **ground cable → 3993**

3983 **ground cedar; common juniper; juniper**
Juniperus communis)
d [gemeiner] Wacholder *m*; Heidewacholder *m*; Dürrenstaude *f*; Kaddig[baum] *m*; Krammet *m*; Machandel[baum] *m*; Sachandelbaum *m*
f genévrier *m* [commun]
r можжевельник *m* [обыкновенный]; арча *f*

3984 **ground cherry; cutleaf ground-cherry; winter cherry**
(Prunus fruticosa globosa)
d Kugel-Kirsche *f*; Steppenkirsche *f*

3985 ground clearance; ground preparation; soil preparation; site preparation
(general term for removing unwanted vegetation, slashstumps, roots, and stones from a site before afforestation or reforestation)
d Baureifmachung *f* des Geländes; Standortvorbereitung *f*
f préparation *f* du site
r очистка *f* поверхности почвы

* **ground cover → 6337**

* **grounded logs salvaging → 8324**

3986 ground flora; ground vegetation; soil vegetation; surface vegetation; soil flora
(grasses, herbs and other plants growing on the forest floor)
d Bodenflora *f*
f végétation *f* du sol
r почвенная флора *f*; растительность *f* почвы

3987 groundhog line; ground skidder; haulage line; haul-in line; hauling line; inhaul line; skidder line; skidding line; sliding track; haulage rope; skidding cable
(a power skidder which skids logs on the ground)
d Lastseil *n*; Laufseil *n*; Zugseil *n*; Zugleine *f*; Rückseil *n*; Rückelinie *f*; Rückeschneise *f*; Rückegasse *f*
f câble *m* de débardage; câble de halage; layon *m* de débardage; voie *f* de rondins
r трелёвочный трос *m*; трелёвочный волок *m*

3988 ground-lead logging; ground[-line cable] logging; low-lead [cable] logging
(transporting logs from the stumps to a collecting point by dragging them along the ground with a powered cable passing through a block fastened close to ground level)
d Landrücken *n*
f téléphérage *m* au sol
r наземная трелёвка *f*

3989 ground length
(extent to which the ground around a tree is broken by ridges, gullies or swells, rock outcrops, and sharp slope changes)
d Grundlänge *f*
f longueur *f* au sol
r опорная длина *f*

3990 ground level
d Bodengleiche *f*

3991 ground-line area
d Grundlinienbereich *m*
f zone *f* du collet
r зона *f* среза

* **ground[-line cable] logging → 3988**

* **ground loader → 1559**

3992 ground meristem
d Grundmeristem *n*
f méristème *m* de base
r основная меристема *f*

* **ground-oak → 2687**

* **ground preparation → 3985**

3993 ground rope; ground cable
d tiefliegendes Zugseil *n*; Unterseil *n*
f câble *m* bas; câble inférieur
r нижний тяговый канат *m*; нижний трос *m*; низколежащий тяговой трос

* **ground skidder → 3987**

3994 ground skidding; snigging; low-line logging
(pulling logs parallel to the ground without using an arch or fairlead to raise the forward end)
d Holzrücken *n* im Schleifen; Rücken *n* durch Schleppen; Bodenseilung *f*
f débardage *m* au câble au ras du sol; traînage *m* à plat
r трелёвка *f* волоком; трелёвка тросом и лебедкой

* **ground slide → 2709**

* **ground-thinning → 8511**

* **ground vegetation → 3986**

3995 groundwater; underground water; plerotic water; phreatic water
(water below the level of the water table in the ground; water occupying the sub-surface saturated zone)
d Grundwasser *n*
f eau *f* souterraine; nappe *f* [souterraine]
r грунтовая вода *f*

3996 groundwood
d Abfallholz *n*
f rémanents *mpl* de bois

* **groundwood → 9603**

* **groundwood paper → 9634**

* **group cutting → 4001**

* **group distribution in main pocket → 9292**

* **group felling → 4001**

3997 **group knots; cluster knots**
d Gruppenknoten *mpl*
f nœuds *mpl* groupés
r групповые сучки *mpl*

3998 **group methods of measurement of volume**
d gruppenweise Methode *f* der Volumenmengen
f méthodes *fpl* de cubage
r групповые методы *mpl* измерения объёма

3999 **group of forests**
(all forests are divided into several groups)
d Waldgruppe *f*
f groupe *m* de forêts
r группа *f* лесов

4000 **group of trees**
(an aggregate of cavity trees plus a 200-foot buffer around this cluster of trees)
d Baumgruppe *f*
f groupe *m* d'arbres; touffe *f*
r [изолированная сомкнутая] группа *f* деревьев

* **group planting → 6354**

* **group regeneration → 6812**

* **group reproduction → 6812**

* **group retention → 139**

4001 **group-selection cuts; select group cutting; group-selection felling; path cutting; group cutting; group felling; felling by groups**
(removing of all merchantable timber from an area smaller than that normally recognized in timber type)
d gruppenweiser Plenterhieb *m*; gruppenweiser Schlag *m*
f jardinage *m* par bouquets; coupe *f* jardinatoire par bouquets
r группово-выборочная рубка *f*

4002 **group-selection high forest**
d Femelhochwald *m*; Plenterhochwald *m*

4003 **group-selection high oak forest**
d Eichen-Plenterwald *m*
f futaie *f* de e chênes par bouquets
r группово-выборочный дубовый высокоствольник *m*

4004 **group shelterwood cutting; group shelterwood felling**
d Gruppenschirmschlag *m*
f coupe *f* sombre par groupes
r групповая обсеменительная рубка *f*

* **grove → 5001, 9473**

4005 **grove of chestnut trees; chestnut grove**
d Kastaniengehölz *n*
f châtaigneraie *f*
r каштановая заросль *f*

* **growing of woods → 7610**

4006 **growing period; growth period; growing season; recruitment period; period of transit**
d Zuwachsperiode *f*; Einwuchszeit *f*
f période *f* d'accroissement; saison *f* de croissance
r период *m* прироста; вегетационный период; период роста

* **growing place of a tree → 4008**

4007 **growing point; vegetative point**
d Zuwachspunkt *m*
f point *m* végétatif
r точка *f* роста

* **growing season → 4006**

4008 **growing space of tree; growing place of a tree; space of growth**
d Wuchsraum *m* eines Bäumes; Standraum *m*
f aire *f* d'extension d'un arbre; zone *f* de répartition [d'un arbre]
r площадь *f* питания дерева; пространство *n*, занимаемое деревом

* **growing-stock level → 9250**

4009 **growing-stock stocking**
(absolute stocking value based only on the growing-stock trees on the plot)
d Bestandesstockung *f*; absoluter Bestandesgrad *m*
f matériel *m* absolu sur pied
r густота *f* абсолютного древостоя

* **growing-stock volume → 9250**

* **growing timber → 3454**

* **growing timber outside the forest → 5759**

4010 **growing tree; standing tree**
d stehender Baum *m*; anstehender Baum
f arbre *m* sur pied
r растущее дерево *n*; дерево на корню; оставляемое на корню дерево

* **growth → 4449**

4011 **growth and yield model**
d Wachstums- und Ertragsmodell *n*
f modèle *m* de la croissance et du rendement
r модель *f* роста и дохода

* **growth bud → 9179**

* **growth check → 4015**

4012 **growth curve**
(of a tree)
d Wachstumkurve *f*
f courbe *f* de la croissance
r кривая *f* роста

* **growth/cut ratio → 4024**

4013 **growth cycle**
d Wuchskeislauf *m*
f cycle *m* de la croissance
r цикл *m* роста; цикл развития

4014 **growth defect**
d Wuchsfehler *m*
f défaut *m* de la croissance
r дефект *m* роста; порок *m* роста

4015 **growth delay; growth retention; growth retardation; growth inhibition; stagnation of growth; growth check; growth stunting**
d Wachstumshemmung *f*; Wachstumsverzögerung *f*; Wachstumsbehinderung *f*; Wachstumsretardation *f*
f inhibition *f* de la croissance; retardement *m* de la croissance; stagnation *f* de croissance
r задержка *f* роста; замедление *n* роста; торможение *n* роста; отставание *n* в росте

4016 **growth equation**
d Wuchsgleichung *f*

4017 **growth habit; growth form; shape of growth**
d Wuchsform *f*; Habitus *m*
f port *m*; forme *f* de végétation
r форма *f* роста; характер *m* роста; биоморфа *f*

4018 **growth index**
d Zuwachsindex *m*
f indice *m* de la croissance
r показатель *m* роста

* **growth in height → 326**

* **growth inhibition → 4015**

* **growth inhibitor → 4025**

* **growth interruption → 8126**

4019 **growth in thickness**
d Dickenwachstum *n*
f croissance *f* en épaisseur
r прирост *m* в толщину

* **growth in volume → 9243**

* **growth period → 4006**

4020 **growth promoter**
d Wachstumförderer *m*
f promoteur *m* de la croissance; déclencheur *m* de la croissance; accélérateur *m* de la croissance
r стимулятор *m* роста; ускоритель *m* роста

4021 **growth-promoting** *adj*
d wachstumsfördernd
f favorisant la croissance
r ускоряющий рост; содействующий роста

4022 **growth properties**
d Wachstumseigenschaften *fpl*
f propriétés *fpl* de croissance
r характеристики *fpl* роста

* **growth rate → 6728**

4023 **growth recovery**
d Wuchserholung *f*
f restauration *f* de croissance
r восстановление *n* роста

* **growth regulator → 6340**

4024 **growth/removals ratio; growth/cut ratio**
(obtained by dividing volume of timber

d Verhältnis *n* zwischen Wachstum und Hieb
f rapport *m* croissance/abattage
r соотношение *n* объёмов прироста и рубки

4025 **growth retardant; growth inhibitor**
(slow down the growth of crop plants)
d Wuchshemmer *m*; Wuchsinhibitor *m*
f inhibiteur *m* de croissance; retardateur *m* de croissance
r замедлитель *m* роста; ингибитор *m* роста

* **growth retardation → 4015**

* **growth retention → 4015**

* **growth ring → 301**

4026 **growth-ring boundary**
(the outer limit of a growth ring)
d Jahresringengrenze *f*
f limite *f* de l'anneau de croissance; limite de cerne
r граница *f* годичного кольца (по внешнему периметру)

* **growth shake → 6675**

* **growth speed → 6728**

* **growth stunting → 4015**

* **growth table → 4457**

4027 **growth tension**
(of growing tree)
d Wuchsspannung *f*
f tension *f* de croissance
r напряжение *n* роста

* **grub → 2495, 4809**

4028 **grub *v* a tree**
d Baum ausheben; Baum ausroden; Baum entwurzeln
f décaper un arbre
r корчевать дерево; валить дерево с корнем

* **grubber → 8808**

4029 **grubbing and retention**
(removal of stumps, roots, embedded logs, organics, and unsuitable soils before or concurrently with subgrade construction)
d Wühlen *n* und Zurückhalten *n*
f essartement *m* et rétention *f*; dessouchage *m* et rétention
r корчёвка *f* и удерживание *n*

d Rodeaxt *f*; Jäthacke *f*
f hoyau *m*; binette *f*
r корчевальный топор *m*; мотыга *f* для полки

* **grubbing[-out] → 8201**

4031 **grubbings**
d Rodungen *f*
f souches *fpl* et racines *fpl*
r корчёвки *fpl*

* **grubbing-up → 8201**

* **grubbing winch → 1160**

* **grub hoe → 7812**

4032 **grubhole**
(oval, circular, or irregular hole (3/8 to 1 inch in diameter) in wood, caused by larvae and adult insects)
d Madebohrung *f*
f galerie *f* d'insecte xylophage; trou *m* d'insecte xylophage
r червоточина *f* диаметром более 3.2 мм

4033 **grub *v* out stumps; stub *v* [out]; root *v* out; stump *v* [out]**
d Stumpf roden; ausroden
f dessoucher; essoucher; abattre à culée noire
r корчевать пни

* **grumixava → 9361**

* **GTR → 3960**

4034 **guaiacol; guaia[cum]; lignum vitae; bakaut**
(Guajacum officinale; Guajacum guetamalense; Guajacum sanctum)
d [echtes] Pock[en]holz *n*; Pockholzbaum *m*; Guajakbaum *m*; Franzosenholz *n*
f gaïac *m*; guayac *m*
r гваяковое дерево *n*; гваякум *m* лечебный; бакаут *m*; гваякум священный

* **guanabana → 7856**

* **guapaque → 9190**

* **guarumo → 7519**

4035 **guatambu; pau marfim**
(Balfourodendron riedelianum)
d Guatambu
f guatambu *m*
r гуатамбу *n*

4036 **guava**
(a fruit)
d Guava *f*
f goyave *f*
r гуава *f*; гуайава *f*; гуаява *f*

4037 **guava[-tree]**
(Psidium guajava)
d Guajavaapfel *m*; Gua[ya]ve *f*
f gojavier *m*
r гуаява[-дерево] *f*

4038 **guayule**
(Parthenium argentatum)
d Guayule *f*
f guayule *m*
r гваюла *f* (каучуконосное растение)

4039 **Guiana crabwood; crabwood; bastard mahogany; andiroba; andirobeira; tangare; carapa**
(Carapa guianensis; Carapa surinamensis)
d Andiroba *n*; Krappa *n*; Karapa *n*
f carapa *m* [rouge]; andiroba *m*
r карапа *f* суринамская

* **Guiana mora → 4715**

4040 **guide bar; guide plate; cutting bar; cutter bar; sword; chain bar; chain blade**
(the stiff metal plate round whose grooved perimeter the linked cutters of a chain saw are driven)
d Schwert *n*
f guide-chaîne *f*; bras *m* coupeur
r пильный меч *m*

4041 **guide timber**
d Führerbauholz *n*
f bois *m* de guidage
r строительные леса *mpl* с направляющими рельсами для тележек

4042 **guiding log**
d führender Stamm *m*
f butée *f*
r ведущая жердь *f*

* **guiding stick → 2567**

* **gullet → 3069**

4043 **gully**
d Wasserriss *m*; Wasserrinne *f*; Gully *m*
f rigole *f*; ravine *f*

4044 **gully assessment**
(procedure for determining gully sediment and debris transport potential, and suggested management strategies)
d Bewertung *f* den Erosionsrinnen
f évaluation *f* de ravine
r оценка *f* водяных трещин

* **gully correction → 4046**

4045 **gully erosion; gullying**
d Rinnenerosion *f*
f érosion *f* en ravines; ravinement *m*
r овражная эрозия *f*; эрозия борозд

4046 **gully filling; gully correction**
d Füllung *f* den Erosionsrinnen
f remplissage *m* de ravines
r заполнение *n* оврагов

* **gullying → 4045**

4047 **gum acacia**
(Acacia senegal)
d Senegalakazie *f*; Verekbaum *m*; Dreidorn-Akazie *f*
f acacia *m* du Sénégal
r акация *f* сенегальская

4048 **gum arabic [tree]; arabic gum; babul [acacia]**
(Acacia arabica; Acacia nilotica)
d Gummiakazie *f*; Akaziengummi *n*; arabisches Gummi *n*; arabischer Gummibaum *m*; Babul *n*
f acacia *m* d'Arabie; babul *m*; acacia à gomme; gomme *f* arabique; gomme acacia
r акация *f* аравийская; акация арабская

* **gum blister → 6893**

* **gum check → 6893**

4049 **gum duct; resin duct; resin canal**
(an intercellular canal containing resin)
d Harzpore *f*; Harzgang *m*; Harzkanal *m*
f canal *m* à gommes; canal [sécréteur] résinifère
r смоляной ход *m*; смоляной канал *m*

* **gum elastic → 9716**

* **gum gall → 6893**

* **gumhar → 5255**

4050 **gum mastic**
(a resinous material obtained from the Pistacia lentiscus tree which is grown on the Island of Chios in Greece)
d Gummimastix *n*
f mastic *m*; gomme *f* mastic
r гумми *n*

* **gumming → 4051, 6895**

4051 **gummosis; gumming; exudation of gum**
(a production of gum by or in plant tissue)
d Gummifluss *m*; Gummosis *f*; Gummose *f*
f gommose *f*
r гуммоз *m*; камедетечение *n*

4052 **gum plate; resin plate**
(a resinous, septum-like structure or deposit bridging the lumen of a tracheid)
d Gummiplatte *f*
f plaque *f* résineuse
r смоляная пластинка *f* (трахеиды)

* **gum pocket → 6893**

4053 **gum [resin]**
(complex polysaccharidal substances formed by cells in reaction to wounding or infection)
d Gummi *n*; Gummistoff *m*; Gummiharz *n*
f gomme *f*; caoutchouc *m*
r камедь *f*; камедистая смола *f*; камедная смола; резина *f*

* **gum spot → 8155**

* **gum streak → 8155**

4054 **gum-top[ped] stringybark; Alpine gum; Alpine ash; delegate eucalyptus; Victorian oak; Tasmanian oak; whitetop stringybark; woollybutt; mountain white gum**
(Eucalyptus delegatensis; Eucalyptus dalrympleana)
d breitblättriger Eukalyptus *m*
f chêne *m* de Tasmanie
r эвкалипт *m* бугорчатый

4055 **gum tragacanth; kat[h]ira gum**
(a gum obtained as a dried exudate from various Asiatic and Eastern European plants of the genus Astragalus, especially Astragalus gummifer)
d Traganth *m*
f gomme *f* adragante
r трагант *m*

(any of gum bearing American or Australian trees)
d Gummibaum *m*; Kautschukbaum *m*
f arbre *m* à la gomme
r камедное дерево *n*

* **gum[-tree] → 2872**

4057 **gum[-tree] plantation**
d Gummibaumkultur *f*; Gummibaumbestand *m*
f gommeraie *f*
r плантация *f* камедных деревьев

* **gum tube → 6303**

4058 **gum vein; gum kino; kino**
(a ribbon of gum between growth rings, which may be bridged radially by wood tissue at intervals)
d Kinoharz *n*; Harzader *f*
f filon *m* de gomme; veine *f* de résine; veine résineuse; kino *m*
r смоляная жила *f*; щитовидная прожилка *f* древесины, пропитанной живицей

* **gum wood → 6901**

4059 **gun *v* a log; cannon *v* a log**
d einem Baum die Fallrichtung geben
f orienter la chute d'un arbre
r дать дереву направление валки

4060 **gungurru eucalyptus**
(Eucalyptus caesia)
d Gungurru *m*
f gungurru *m*
r эвкалипт *m* голубовато-серый

* **gun[ning] stick → 8567**

* **gurjun → 4699**

4061 **gutta-percha tree**
(Palaquium gutta)
d Guttaperchabaum *m*
f gutta-percha *f*; arbre *m* à gutta
r гуттаперчевое дерево *n*; пaляквиум *m*

4062 **guttation**
(process by which plants expel liquid water from the leaves in excess of transpiration)
d Guttation *f*; Tropfen *n*
f guttation *f*
r гуттация *f*

* **gutter → 334**

* **gutter road → 1245**

4063 **gutter trench**
(in mountain forests, a ditch dug on a slope below a fire, designed to catch rolling firebrands)
d Gossegraben *m*
f tranchée *f* collectrice; fossé *m* aval
r противопожарная канава *f*

* **guyed-mast derrick → 3848**

* **guy[ ]line → 1164**

4064 **guyline stump**
d Spannseilstumpf *m*
f tronc *m* d'étai
r пень *m* для привязки растяжек

4065 **guyline system**
d Spannseilssystem *n*
f système *m* d'étai
r система *f* погрузки с помощью захватов, подвешенных к растяжке трелёвочной мачты

4066 **guyline tractor**
d Spannseilschlepper *m*
f tracteur *m* d'étai
r трактор *m* с лебёдкой, которой используется в качестве растяжки

* **guy rope → 1164**

* **guy [wire] → 1164**

4067 **gymnomonoecious**
(presence of hermaphroditic and female flowers on the same individual)
d gynomonözisch
f gynomonoïque
r гимномоноэционный; гимномонэцичный

* **gymnospermae → 4069**

4068 **gymnosperm[ous] wood**
d Gymnospermenholz *n*
f bois *m* de gymnospermes
r голосеменная древесина *f*

4069 **gymnosperms; gymnospermae** (*lat*)
(the botanical name for the group of vascular flowering plants that produce seeds not enclosed in an ovary)
d Gymnospermen *pl*; Nacktsamer *mpl*; nacktsamige Pflanzen *fpl*

4070 **gynoduoecious**
(ref. to plant species or population in which some individuals have bisexual flowers, but some individuals have female flowers)
d gynodiözisch
f gynodioïque
r гинодиэцичный

* **gyp[p]o → 5072**

4071 **gypsy moth**
(Lymantria dispar; Porthetria dispar)
d Schwammspinner *m*
f lymantride *f* spongieuse; spongieuse *f*; bombyx *m* disparate
r непарный шелкопряд *m*; непарник *m*

* **gypsy yarder → 8609**

H

* **habillo → 7185**

4072 **habitat; natural range**
(place where an organism lives and/or the conditions of that environment including the soil, vegetation, water, and food)
d Umwelt *f*; natürliches Verbreiterungsgebiet *n*; Habitat *n*; Standort *m*; Heimat *f*
f habitat *m*; zone *f* de dispersion naturelle
r окружающая среда *f*; место *n* обитания; местообитание *n*; место распространения; область *f* естественного распространения

4073 **habitat analysis unit**
(area of land defined by geographic or administrative means and selected as the unit for evaluating the quality of habitat of a species, guild, or other taxon)
d Habitatanalyse-Einheit *f*
f unité *f* d'analyse d'habitat
r единица *f* анализа местообитания

4074 **habitat diversity**
(mix of habitats; or size variety; or numbers of layers present; or number of types; or evenness of stand sizes and shapes by type)
d Habitatdiversität *f*
f diversité *f* d'habitat
r разнообразие *n* окружающей сред

4075 **habitat effectiveness**
(percentage of available habitat that is usable by wildlife outside the hunting season)
d Wirksamkeit *f* des Lebensraumes; Lebensraumwirksamkeit *f*
f efficacité *f* d'habitat
r эффективность *f* местообитания

4076 **habitat enhancement; habitat improvement**
(any manipulation of habitat that improves its value and ability to meet specified requirements of one or more species)
d Habitatverbesserung *f*
f amélioration *f* de l'habitat
r улучшение *n* местообитания

4077 **habitat management**
(management of the forest to create environments which provide habitats (food,
d Habitatverwaltung *f*
f gestion *f* des habitats; aménagement *m* des habitats
r управление *n* местообитаний

4078 **habitat management unit; HMU**
(unit of land approximately 4,000 acres in size, the boundaries of which fall on compartment boundaries, and which contains a mix of habitat types)
d Habitatverwaltungseinheit *f*
f unité *f* d'aménagement des habitats
r единица *f* управления местообитаний

4079 **habitat type; ecotype**
(genetically distinct race or subdivision of species adapted to local habitat and climate) unique type of soil material)
d Ökotyp *m*; Habitattyp *m*
f écotype *m*
r тип *m* климаксовой растительности; экотип *m*

4080 **habit of a tree; tree habit**
d Baumgestalt *f*; Baumhabitus *m*
f port *m* d'un arbre
r внешний вид *m* дерева; габитус *m* дерева

4081 **hack**
d leichte Hacke *f*; leichte Haue *f*
f serfouette *f*
r тяпка *f* легкая; топорик *m*

* **hack → 83**

4082 **hack; incision; scotch; notch; tapping [cut]**
(unjoined cut made in the bole of a tree to permit application of a herbicide into the woody part of the stem)
d Einkerbung *f*; Kerbe *f*; Kimme *f*; Raste *f*; Vorschnitt *m*
f encoche *f*; rainette *f*
r надрез *m*; зарубка *f*; врубка *f*; надсечка *f*

4083 **hack and squirt [method]**
(method of conifer release and juvenile spacing where the bark of a tree is cut (hack) and herbicides are injected (squirt) to kill the tree)
d Zerhacken- und Spritzen-Methode *f*
f méthode *f* d'encoches et de jets de repérage
r метод *m* зарубки и инжектирования

* **hackberry → 2898**

* **hack girdling → 3716**

* **hack machine → 9512**

4084 **hack operator**
d Harzarbeiter *m*
f opérateur *m* d'entaille
r вздымщик *m*

4085 **hag**
(a small wood, or part of a wood or copse, which is markedoff or enclosed for felling, or which has been felled)
d Hag *m*
f hag *m*
r лес *m*, предназначенный для валки

4086 **haiawaballi; Sali**
(Tetragastris altissima)
d Salie *f*; Haiawaballi *n*
f cèdre *m* bagasse; encens *m* rouge; iciquier *m* cèdre
r кедр *m* багассе

* **hail damage → 3930**

4087 **hair-cap moss**
(Polytrichum communis)
d [echtes] Frauenhaarmoos *n*; gemeine Widertonmoos *n*
f polytric *m* commun
r политрихум *m* [обыкновенный]

4088 **hairline crack; micro-crack; fine check**
(wood defect)
d Haarriss *m*
f fissure *f* ténue; fente *f* ténue; gerce *f*
r тонкая трещина *f*; волосяная трещина; волосная трещина; капиллярная трещина

4089 **hairy pine borer**
(Tragosoma depsarius)
d flaumiger Kieferbohrer *m*
f longicorne *m* velu du pin
r златка *f* сосновая мохнатая

4090 **hairy poplar sawfly; poplar sawfly**
(Trichiocampus viminalis)
d flaumige Pappelsägewespe *f*
f tenthrède *f* européenne du peuplier
r пилильщик *m* тополевый

* **half-height diameter → 2468**

* **half logs → 4094**

4091 **half ring-porous** *adj*
r с полукольцевыми порами

4092 **half ripsaw**
d Halbripsäge *f*
f scie *f* demi-circulaire à refendre
r пила *f* с мелкими зубьями для продольной распиловки

4093 **half-round cutting; stay log cutting; eccentric peeling**
d Exzenterschale *f*
f déroulage *m* excentré; déroulage à cœur
r резание *n* с движением бруса по половине окружности

4094 **half rounds; half logs**
d Halbholz *n*; Halbrundholz *n*
f bois *m* demi-ronde; bois déligné en "queue de billard"
r полукруглый лесоматериал *m*; круглый лесоматериал, разрезанный вдоль

4095 **half-round sleeper; two-faced tie; pole tie**
d Halbrundschwelle *f*
f traverse *f* demi-ronde
r полукруглая шпала *f*

4096 **half-scarf; halving**
d Aufblatten *n*; Aufblattung *f*; Einblatten *n*; Blatten *n*; Überblatting *f*
f enture *f* à mi-bois
r сращивание *n* вполдерева; сплачивание *n*; врубка *f* вполдерева; врубка [брёвен или брусьев] внакладку

4097 **half-sib**
(tree with one parent in common with another)
d Halbgeschwister *pl*
f demifratrie *f*; descendance *f* uniparentale
r полусибс *m*

4098 **half-standard**
d Halbstamm *m*
f demi-tige *f*
r полуштамб *m*

4099 **half timber**
d Halb-Bohle *f*
f demi-madrier *m*
r лафет *m*; половина *f* кряжа, распилённого вдоль; полукруглое бревно *n*

4100 **half[-timber] beam**
d Halbblock *m*

4101 hallea; abura
(Mitragyna ciliata; Hallea ciliata; Mitragyna ledermannii)
d Abura *m*
f abura *m*; bahia *m*
r митрагина *f* реснитчатая

4102 halophytes
(plants that are tolerant of high levels of mineral salts in the soil)
d Salzpflanzen *fpl*; Halophyten *mpl*
f halophytes *mpl*
r галофиты *mpl*; галофитная растительность *f*

* **halo sap → 853**

* **halves → 420**

* **halving → 4096**

4103 hammer [de]barker
d Hammerschäler *m*; Hammerschälmaschine *f*
f écorceuse *f* à marteaux
r кулачковая корообдирка *f*; молотковая окорочная машина *f*

* **hammer hog → 1534**

* **hog → 1534**

4104 hammock (*US*)**; thick of the forest; thicket; bosom**
(a number of trees or shrubs growing close together; dense growth of small trees or bushes)
d Waldesdickicht *n*; Dickicht *n*; Dickung *f*
f fourré *m*
r загущённое молодое насаждение *n*; густой молодняк *m*; чаща *f*; заросль *f*; гуща *f*

* **hamper → 8712**

* **hand axe → 673**

* **hand-bag → 4112**

4105 hand barker; hand peeler
d Handentrinder *m*
f écorceur *m* à main
r окорщик *m*, работающий вручную; рабочий *m* на ручной окорке

4106 hand clipper; hand pruner
d Handschere *f*

4107 hand debarking; manual debarking
d Handentrindung *f*; Schippen *n*
f écorçage *m* manuel
r ручная окорка *f*; окорка вручную

4108 hand feller
d Handfäller *m*
f abatteur *m* à main
r вальщик *m*, осуществляющий валку деревьев вручную

4109 hand felling
d Fällung *f* mit der Hand; Handfällung *f*
f coupe *f* à main
r ручная валка *f*; валка вручную; ручная лесозаготовка *f*

4110 handle of an axe; axe-helve
d Axtstiel *m*
f manche *m* de hache
r топорище *n*

* **hand peeler → 4105**

* **hand pruner → 4106**

4111 hand[ ]saw
d Handsäge *f*
f scie *f* à main; scie égoïne; égoïne *f*
r ручная пила *f*

4112 hand-skid *v*; **hand-bag** *v*
(to move timber by hand to a point where it can be reached by horse or any other form of transport)
d rücken mit der Hand
f débarder à bras
r волочить рукой [сплавной лес к берегу]

4113 hand skidder
(one who accompanies a log as it is being dragged and places short skids beneath it)
d Handholzschleifer *m*
f débardeur *m* à bras
r рабочий *m*, подкатывающий брёвна вручную

* **hand weeding → 5287**

4114 hanger
(the ground man which scan the tree for any "hangers" that may exist)
d Aufhänger *m*
f accrocheur *m*
r установщик *m* подвесок

4115 **hanging rope**
d hängendes Seil *n*
f corde *f* accrochante
r прицепной канат *m*

* **hangover fire → 7734**

4116 **hang *v* up; lodge *v***
(in skidding, to get a load stuck in the mud or behind some obstacle)
d aufhängen
f embûcher; [s']encrouer
r повиснуть

* **hang up → 5042**

4117 **hang[-]up of logs**
d Schnittholz-Hängen *n*
f accrochage *m* de grumes
r застревание *n* баланса; образование *n* сводов

* **hang-up tree → 5041**

* **Hankow willow → 1963**

4118 **haptotropism**
(the growth (movement) of a plant organ as a response to contact or a mechanical stimulus)
d Haptotropismus *m*
f haptotropisme *m*
r гаптотропизм *m*; гаптотропия *f*

* **haras → 4686**

4119 **hardbast; hardliber**
d Hartbast *m*
f liber *m* dur; sclérenchyme *m* libérien
r твёрдый луб *m*

* **hardbeam → 2901**

* **hard beech → 8420**

* **hardboard → 9551**

4120 **hardening off**
d Verhärtung *f*
f endurcissement *m*
r затвердение *n*

4121 **hard hardwoods; hard-textured woods**
(such as sugar maple, birch, hickory, dogwood, persimmon, black locust, beech, ash, honeylocust, holly, black walnut, mulberry, and all commercial oaks)
r твёрдая широколиственная древесина *f*; древесина твёрдолиственных пород

4122 **hard hat**
d Schutzhelm *m*
f casque *f* de protection
r защитная каска *f*; защитный шлем *m*

4123 **hardiness**
(attribute of a plant that enables it to withstand environmental stresses)
d Härte *f*; Festigkeit *f*
f rusticité *f*; dureté *f*
r выносливость *f*; стойкость *f*; твёрдость *f*

* **hardleaved forest → 7301**

4124 **hardleaved wood; hardwood [lumber]; hardwood timber; broadleaved wood; [broad]leaf wood; deciduous wood; angiospermous wood**
(a conventional term for the timber of broadleaved trees)
d Laubholz *n*; Angiospermenholz *n*
f bois *m* décidu; bois de feuillus; bois feuillu
r [широко]лиственная древесина *f*

* **hardliber → 4119**

4125 **hardly combustible wood**
d schwerbrennbares Holz *n*; kaumbrennbares Holz
f bois *m* combustible difficilement
r трудно воспламеняющаяся древесина *f*

4126 **hardly processed wood**
d schwieriges Holz *n*
f bois *m* transformé difficilement
r труднообрабатываемая древесина *f*

* **hard maple → 808**

4127 **hard mast**
(fruits of oaks, hickories, pines, and beech trees that are important foods of many species of wildlife in the fall and winter)
d Hartmast *m*
f paisson *f* dure
r твёрдый плодокорм *m*

4128 **hardness of wood; wood hardness**
d Holzhärte *f*
f dureté *f* du bois
r твёрдость *f* древесины

4129 **hardpan [layer]**
(a layer of soil sufficiently clogged with clay

d Ortstein *m*; Orterde *f*
f couche *f* durcie; couche cuirasse; alios *m*
r ортштейн *m*

* **hard rot → 1828**

4130 **hard sap rot**
d Hartsaftfäule *f*
f pourriture *f* de l'aubier dure
r твёрдая заболонная гниль *f*

4131 **hard snag**
d Hartabständer *m*
f chicot *m* dur
r твёрдый сухостой *m*

* **hard streak → 1850**

* **hard-textured woods → 4121**

4132 **hardwood ash**
d Hartlaubholzasche *f*
f cendre *f* de bois dur
r древесная зола *f* твёрдолиственных пород

4133 **hardwood board**
d baumkantiges Laubholzbrett *n*; Dickte *f*
f planche *f* de feuillus
r необрезанная доска *f*; необрезная доска

4134 **hardwood cutting**
d Laubholzfällung *f*
f coupe *f* de bois feuillu
r рубка *f* широколиственных пород

4135 **hardwood cutting**
d Altholzsteckling *m*
f bouture *f* de vieux bois
r черенок *m* из старой древесины

4136 **hardwood dimension sheet; hardwood dimension lumber**
d Hartlistenholz *n*
f sciage *m* à façon de bois dur
r листовая твёрдая древесина *f*

4137 **hardwood draws**
(upland drainages containing deciduous trees and shrubs)
d Laubholz-Wasserrinnen *fpl*
f dépressions *fpl* de bois dur drainantes les prairies
r балки *fpl* широколиственной растительности

4138 **hardwooded broadleaved species; hardwoods**
broad timber groups: soft hardwoods and hard hardwoods)
d Hartlaubarten *fpl*
f feuillus *mpl*
r [твёрдо]лиственные породы *fpl*; лиственные насаждения *npl*

4139 **hardwood forest**
d Hartlaubwald *m*
f forêt *f* de bois dur
r лес *m* твёрдых пород древесины

4140 **hardwood hammock**
(an "island" of primarily hardwood trees and associated understor[e]y plants growing on an elevated, well-drained site, surrounded by vegetation characteristic of lower, wetter surroundings)
d Hartlaub-Hängematte *f*
f hamac *m* de bois dur
r гамак *m*

4141 **hardwood heart**
d Laubholzkern *m*
f cœur *m* en bois dur
r ядро *n* твёрдолиственной древесины

4142 **hardwood log**
d Hartholzbalken *m*
f grume *f* de bois dur
r чурак *m* твёрдолиственной породы

* **hardwood [lumber] → 4124**

4143 **hardwood mixture; mixture of broadleaved species**
d Hartholzbeimischung *f*
f mélange *m* d'essences feuillues
r примесь *f* широколиственной породы

4144 **hardwood pitch**
d Hartholzpech *n*
f résine *f* de bois feuillu
r пек *m* лиственной древесины

* **hardwoods → 4138**

4145 **hardwood slash**
d Laubholzreisig *n*
f brindilles *fpl* d'essences feuillues
r порубочные остатки *mpl* лиственных пород; хворост *m* лиственных пород

* **hardwood timber → 4124**

4146 **hardy-orange; trifoliate orange**
(Poncirus trifoliata)

f poncirus *m*; citronnier *m* trifolié; poncirier *m*
r понцирус *m* трёхлисточковый; лимон *m* трёхлистный

4147 hardy rubber tree
(Eucommia ulmoides)
d Guttaperchabaum *m*
f arbre *m* à gutta-percha; arbre à latex; eucommia *m*
r эукоммия *f* вязолистная; эйкоммия *f* ильмолистная

* **harewood → 8355**

4148 harm *v*
(to include any act which actually kills or injures wildlife, and emphasizes that such acts may include significant habitat modification or degradation that significantly impairs essential behavioral patterns of fish or wildlife)
d schaden; schädigen; verletzen
f nuire à; porter préjudice à
r быть вредным; вредить

* **harmful organism → 5807**

* **harrowing → 2535**

* **harsh terrain → 4212**

* **hartleaf balsam poplar → 9338**

4149 harvest *v*
(a forest)
d ernten
f récolter
r заготавливать

* **harvestable forest → 2951**

* **harvest age → 133**

4150 harvest cut[ting]; mature cutting
d Haupteinschlag *m*; Erntehieb *m*
f coupe *f* de produits au terme d'exploitabilité
r рубка *f* спелого леса

4151 harvest cutting method
(method of removing all or part of a stand of timber, either by a single cutting or a series of regeneration cuttings)
d Erntehiebsmethode *f*
f méthode *f* de coupe principale
r метод *m* рубки спелого леса

4152 harvester chain
d Erntemaschinekette *f*
f chaîne *f* de récolteuses
r система *f* лесосечных машин

4153 harvest forecast
(flow of potential timber harvests over time)
d Holzernteprognose *f*
f prognose *f* d'exploitation forestière
r прогнозирование *n* лесозаготовки

4154 harvest functions
(bucking, bunching, chipping, debarking, delimbing, felling, forwarding, loading, pilling, skidding, slashing, topping, yarding)
d Holzerntefunktionen *fpl*
f fonctions *fpl* d'exploitabilité
r операции *fpl* лесозаготовки

4155 harvest increment
(increment in the harvested part, or the above-ground biomass, of a woody perennial plant over a stated period)
d Holzernteinkrement *m*
f incrément *m* d'exploitabilité
r прирост *m* лесозаготовки

4156 harvest index
(proportion of the total plant biomass (or the above-ground biomass) produced by a crop species that forms a particular, required (harvestable) plant part)
d Holzernteindex *m*
f indice *m* d'exploitabilité
r индекс *m* лесозаготовки

* **harvesting → 9588**

* **harvesting attachment → 3068**

* **harvesting chew → 9527**

4157 harvesting equipment; logging equipment; lumberman's equipment; wood cutter's tools; wood feller's implements; logging tools; woodcutting tools
d Holzeinschlagapparatur *f*; Holzaufbereitungsmaschinen *fpl*; Holzerntgeräte *npl*; Holzhauergeräte *npl*; Holzhauerwerkzeuge *npl*; Fällungswerkzeuge *npl*
f équipement *m* d'exploitabilité; outillage *m* de bûcheron
r лесозаготовительные машины *fpl*; лесозаготовительное оборудование *n*;

* **harvesting expenses → 8217**

* **harvesting head → 3068**

4158 **harvesting method**
(mix of felling, bucking, and yarding (skidding) systems used in logging a stand of timber)
d Holzaufbereitungsmethode *f*
f méthode *f* d'exploitabilité
r метод *m* ведения лесозаготовок

4159 **harvesting of minor forest products**
d Nebennutzung *f*
f exploitation *f* de produits forestiers de cueillette
r побочное пользование *n* лесом

4160 **harvest[ing] pattern**
(spatial distribution of felled and reserve (not felled) areas across the forest landscape)
d Holzbeschaftungsmodell *n*
f modèle *m* d'exploitation
r модель *f* лесозаготовки

4161 **harvesting prescription**
(detailed plan on how, when, and where timber will be harvested from an area)
d Holzernte[an]ordnung *f*
f prescription *f* d'exploitation
r правила *npl* лесозаготовки

* **harvesting system → 5084**

4162 **harvest rate**
(rate at which timber is harvested)
d Holzaufbereitungsgeschwindigkeit *f*
f vitesse *f* d'exploitabilité
r скорость *f* лесозаготовки

4163 **harvest schedule**
(document listing the stands to be harvested year or period, usually showing types and intensities of harvests for each stand, as well as a timetable for regenerating currently non-productive areas)
d Holzerntetabelle *f*
f table *f* d'exploitations
r таблица *f* лесозаготовки

4164 **hasel alder; rough alder; hoary alder**
(Alnus rugosa)
d Bergerle *f*
f au[l]ne *m* rugueux
r ольха *f* морщинистая

(the leaf which tip is pointed and the two lower (basal) lobes are also pointed, but downward)
d spießförmiges Blatt *n*
f feuille *f* sagittale
r копьевидный лист *m*

* **hatchet → 673**

* **haulage → 8595**

* **haulage distance → 9747**

* **haulage line → 3987**

* **haulage rig → 4171**

* **haul[age] road → 7689**

* **haulage rope → 3987**

* **haulage tractor → 7799**

4166 **haulage with open rope**
d Förderung *f* mit offenem Seil
f transport *m* par câble simple
r подача *f* с открытом тросом

* **haul-back line → 5411**

4167 **haulback [tail] block; tail block**
(the block used to guide the haul-back line)
d Leerseilblock *m*; Endumlenkrolle *f*
f poulie *f* de renvoi amont; poulie de renvoi arrière
r направляющий блок *m* возвратного каната; хвостовой блок возвратного каната

* **haul-back [wire] rope → 5411**

4168 **hauler operator**
d Schlepper-Operator *m*
f opérateur *m* de débardeur
r рабочий *m*, обслуживающий канатную установку

* **hauling → 8595**

* **hauling area → 2629**

* **hauling cost[s] → 7684**

* **hauling damage → 7685**

* **haul[ing] distance → 9747**

4169 **hauling-in**
d Heranziehen *n*
f virage *m*
r подтрелёвка *f*; подтаскивание *n*

* **hauling line → 3987**

4170 **hauling period; hauling season; hauling time**
d Abfahrtszeit *f*; Bringungszeit *f*; Abfahrtsperiode *f*
f période *f* de débardage; époque *f* de vidange
r период *m* вывозки

4171 **hauling rig; haulage rig**
d Föfderungstakelage *f*
f train *m* routier
r оснастка *f* вывозки

* **hauling season → 4170**

* **hauling time → 4170**

* **hauling tractor → 7799**

* **hauling unit → 5089**

* **hauling without skidding → 2522**

* **haul-in line → 3987**

* **haul road → 3463**

4172 **haulroad feeder; spar road; [stub] spur road; log spur; secondary truck road**
d Abfuhrweg *m*; Bahnzweig *m*; Fahrweg *m*; Nebenstrecke *f*; Verbindungsweg *m*
f orifice *f* d'alimentation; voie *f* de roulage; route *f* carrossable secondaire
r лесовозный ус *m*

4173 **haul-up with [endless] jack chain; endless chain log haul-up; log chain**
(an endless spiked chain which moves logs from one point to another, usually from mill pond into the sawmill)
d Stammheber *m*; Förderkette *f*
f chaîne *f* élévatrice; chaîne monte-grume; chaîne à billes; chaîne à billots
r цепь *f* для прицепки брёвен; кондак *m* для поддержания брёвна во время раскряжёвки

* **Havana cedar → 1411**

* **having blanks → 7008**

4174 **hawthorn; May blossom; mayflower [haw]; May tree; thorn-apple tree; May bush; cheese tree**
(Crataegus oxyacantha)
d [zweigriffeliger] Weißdorn *m*; gemeiner Weißdorn
f aubépine *f* commune; aubépine à deux styles; senelle *f*; épine *f* de mai; poire *f* d'oiseau
r боярышник *m* колючий; боярышник обыкновенный; белая колючка *f*

4175 **haymaking forestland**
d Heu[machen]waldboden *m*
f terre *f* forestière de fenaison
r земельные угодья *npl* на территории лесного фонда, систематически используемые для заготовки сена

4176 **haywire**
(in cable logging)
d Verhau *m*
f câbleau *m*
r вспомогательный канат *m*

4177 **haywire operation**
(a contemptuous term for logging operation that has poor equipment)
d baufällige Operation *f*
f exploitation *f* mauvaise
r плохая лесозаготовительная операция *f*

4178 **haywire outfit**
(a poor logging equipment)
d baufällige Ausstattung *f*
f équipement *m* d'exploitation mauvais
r плохое лесозаготовительное оборудование *n*

4179 **hazardous tree; danger tree**
(a tree or any component of a tree that has sufficient structural infirmity to be identified as having a high risk of falling and causing personal or property damage)
d gefährlicher Baum *m*
f arbre *m* dangereux
r пожароопасное дерево *n*

4180 **hazardous [wood] debris**
d gefährliches Schwemmholz *n*
f matériaux *mpl* détritiques ligneux dangereux; débris *m* dangereux
r пожароопасные горючие материалы *mpl*

* **hazel alder → 7786**

* **hazel family → 2556**

4181 hazel grove; hazel coppice; filbert orchard
d Haselgebüsch *m*; Dickicht *n* aus Nussstrauchern; Haselhain *m*
f coudraie *f*; corylaie *f*
r заросль *f* лещины; орешниковая заросль; орешник *m*

4182 hazel-hen; hazel-grouse
(Bonasa bonasia)
d Haselhuhn *n*; Rothuhn *n*
f gélinotte *f* [des bois]; coq *m* des marais
r рябчик *m*

4183 hazelnut; filbert
(a fruit)
d Haselnuss *f*
f noisette *f*; aveline *f*
r лесной орех *m*

* **hazelnut tree → 1778**

* **hazel[tree] → 1778**

4184 hazel wood
d Haselholz *n*
f bois *m* de noisitier
r древесина *m* орехового дерева

* **head → 4186**

* **head block → 2983**

* **head bucker → 5069**

4185 head forest guard
d Hauptforstwärder *m*
f garde-forestier *m* principal
r лесной бригадир *m*; главный лесной сторож *m*

4186 heading; head; bottom
(any piece of lumber from which a keg or barrel head is cut)
d Bodenbretter *m*; Fassboden *m*
f fonçure *f*; fond *m*
r дно *n*; днище *n*

4187 heading[-]back; heading-in; cutting-back
(removing only a portion (of a branch or shoot), leaving another portion from which new growths can develop; cutting plants just above the ground to enable shoots to develop)
d Beschneidung *f*; Stutzen *n*
f rapprochement *m*
r омоложение *n*; укорачивание *n*

4188 head lean
(of growing tree)
d Lehne *f* nach vorne
f inclinaison *f* d'avant
r наклон *m* вперёд

* **head loader → 8695**

* **head log → 8688**

* **head of fire → 1189**

* **headrig → 1018**

* **head[ ]saw → 1018**

4189 headsaw time
(of logs)
d Hauptsägezeit *f*
f temps *m* de scie de tête
r время *n* роспуска; время развала

4190 head [spar] tree; head spar; main spar
(a tree at the landing of a skyline logging operation)
d Hauptmast *m*; Rückenmast *m*
f mât *m* de tête; cabrette *f*
r головная [естественная] мачта *f*

* **head wood cutter → 5069**

* **healing over → 5862**

4191 healthy ecosystem
(ecosystem in which structure and functions allow the maintenance of biodiversity, biotic integrity and ecological processes over time)
d gesundes Ökosystem *n*
f écosystème *m* sain
r жизнеспособная экосистема *f*

4192 healthy growth
d gesunder Wuchs *m*
f croissance *f* saine
r интенсивный рост *m*

4193 heap of brushwood; heap of branchwood; pile of brushwood; pile of branchwood; brush pile; heap of twigs
d Reisighaufen *n*
f bourrée *f*; tas *m* de brindilles
r куча *f* хвороста

* **heap of trees → 4**

f planche *f* du cœur
r сердцевинная доска *f*

* **heart [center] → 1951**

4195 **heart check** (*US*); **rift crack; heart shake; heart crack; pith check**
(a fissure originating at, and extending more or less radially from, the heart but not to the surface)
d Herzriss *m*; Kernriss *m*; Kernkluft *f*; Markriss *m*; Spaltriss *m*
f fente *f* du cœur; fente médullaire; cadranure *f*
r метиковая трещина *f*; метик *m*; сердцевинная трещина

4196 **heart diameter; core diameter**
d Kerndurchmesser *m*
f diamètre *m* de cœur
r диаметр *m* ядра

4197 **heart fiber**
d Markfaser *f*
f fibre *f* de cœur
r сердцевинное волокно *n*

* **heartleaf eucalyptus → 7605**

* **heartleaf hornbeam → 7280**

4198 **heartleaf willow; furry willow**
(Salix cordata)
d herzblättrige Weide *f*
f saule *m* à feuilles en cœur
r ива *f* сердцевиднолистная

* **heart-leaved gum → 7605**

4199 **heartless wood; coreless wood**
d kernfreies Holz *n*; Holz ohne Kern
f bois *m* sans cœur
r безъядровая древесина *f*

* **heart pith → 1951**

* **heart plank → 4194**

4200 **heart ratio; core ratio**
d Kernanteil *m*
f rapport *m* de cœur
r доля *f* ядра

4201 **heart-shaken; heart-shaky; unsound in the heart** *adj*
r трещинистый

* **heart-shaped leaf → 1943**

* **heart stain[ing] → 7994**

4202 **heart[wood]; core [wood]; true[ ]wood; inner wood**
(wood in the centre of a tree; the small cylinder of primary tissue of a tree stem around which the annual rings form)
d Kernholz *n*; kerniges Holz *n*; verkerntes Holz
f bois *m* de cœur
r ядровая древесина *f*; сердцевинная древесина; древесина центральной части ствола

4203 **heartwood decay[ing]; rotting of heart**
d Zerfall *m* des Kernholzes
f décomposition *f* de bois de cœur
r загнивание *n* ядровой древесины

* **heart[wood] rot → 1414**

4204 **heartwood/sapwood transition; sapwood/heartwood boundary**
d Kernholz/Splintholz-Übergang *m*
f transition *f* duramen-aubier
r переход *m* между сердцевинной древесины и заболони

4205 **heartwood tree species**
d Baumart *f* mit Kernholz
f espèce *f* arborescente à duramen
r ядровая древесная порода *f*

* **heat check → 2002**

4206 **heath; briar**
(Erica arborea)
d Baumheide *f*
f bruyère *f* [arborescente]; bruyère en arbre
r вереск *m* древовидный; ерика *f* древовидная

* **heather → 1779**

* **heather bell → 3153**

4207 **heath family**
(Ericaceae)
d Heidekrautgewächse *pl*
f éricacées *pl*
r вересковые *pl*

f forêt *f* à bruyère; forêt-lande *f*
r верещатниковый лес *m*; пустошный лес

* **heating power → 1194**

* **heating value → 1194**

4209 **heat kill**
(the death of foliage by the heat from a fire, without any sign of charring or browning)
d Tod *m* wegen der Hitze
f coup *m* de chaleur
r обгорание *n*

* **heat shake → 2002**

4210 **heaviness of felling; intensity of cut**
d Fällungsintensität *f*; Hiebsintensität *f*
f intensité *f* relative de la coupe
r интенсивность *f* рубки

* **heavy crop → 3751**

4211 **heavy cut**
d tiefer Schnitt *m*; Tiefschnitt *m*
f passe *f* profonde
r глубокая резка *f*

4212 **heavy going country; harsh terrain; rugged terrain; rough terrain; heavy ground; difficult country; difficult terrain; adverse terrain**
d schwieriges Gelände *n*
f terrain *m* brut
r труднопроходимая местность *f*; пересечённая местность; изрезанный рельеф *m* местности

4213 **heavy growth**
d starkes Wachstum *n*
f croissance *f* forte
r обильный рост *m*; пышный рост

4214 **heavy low thinning**
d starke Niederdurchforstung *f*
f éclaircie *f* par le bas forte
r интенсивное низовое прореживание *n*

4215 **heavy pruning**
d Schwerästung *f*
f élagage *m* fort
r восстановительная обрезка *f*; омолаживающая обрезка; сильная обрезка

* **heavy sapelle → 4744**

f gros rémanents *mpl*
r крупные порубочные остатки *mpl*; крупные отходы *mpl*

4217 **heavy thinning; strong thinning; severe thinning**
d starke Durchforstung *f*; Starkdurchforstung *f*
f éclaircie *f* forte
r сильная прорубка *f*; интенсивная проходная рубка *f*; интенсивная рубка ухода; интенсивное прореживание *n*

* **heavy timber → 4808**

4218 **heavy trees; large-sized trees**
d große Bäume *mpl*
f haute-perche *f*
r крупномерные деревья *npl*

4219 **hedge apple; horse apple; osage-orange**
(Maclura pomifera; Maclura aurantiaca)
d Osageorange *m*; apfelfrüchtiger Osagedorn *m*
f oranger *m* des osages; maclure *f* épineuse; maclure orangée
r маклюра *f* оранжевая

* **hedge euphorbia → 4470**

* **hedge lark → 9594**

* **hedge maple → 2827**

* **hedge nettle → 4222**

4220 **hedge-orchard**
(an orchard of hedged plants that produces cuttings or other tissue for vegetative propagation)
d Heckegarten *m*
f jardin *m* de haies
r питомник *m* для изгороди

* **hedge pole → 3084**

* **hedge[row] → 5019**

* **hedgerow intercropping → 183**

* **hedgerow timber → 5001**

4221 **hedgerow tree**
d Baum *m* in einer Hecke
f arbre *m* dominant dans une haie
r дерево *n* в изгороди

4222 **hedge woundwort; hedge nettle**
(Stachys silvatica)

f épiaire *m* des bois; grand épiaire; ortie *f* puante
r лесной чистец *m*

4223 **hedging**
(appearance of browse plants that have been browsed so as to appear artificially clipped)
d Bedeckung *f*
f couverture *f*
r огораживание *n*

4224 **heel; tree rest**
(the portion of old wood at the base of a cutting)
d Schneidenrücken *m*
f talon *m*
r кромка *f*

* **heeler → 5012**

* **heeling → 4226**

4225 **heel[ing] boom; tree-length boom**
(loading boom that uses tongs to heel or force one end of a log against the underside of the boom)
d Langholzausleger *m*
f flèche *f* à talon; butoir *m*
r наклоняющаяся стрела *f*; погрузочная стрела с упором

4226 **heel[ing]-in; heeling**
(storing young trees prior to planing by placing them in a trench and covering the roots or rooting portions with soil)
d Einschlagen *n*
f mise *f* en jauge; talonnage *m*
r прикопка *f*; прикапывание *n*

4227 **heel tackle; tackle**
(system of lines and blocks used to tighten the skyline)
d Heel-Tackle *n*
f ridage *m*
r полиспаст *m* для натяжения несущего каната

* **he-huckleberry → 5256**

* **height accretion → 326**

4228 **height class**
(any of the intervals into which a range of tree heights may be divided for classification and use)
d Höhenklasse *f*

4229 **height curve**
(of a trees)
d Höhenkurve *f*
f courbe *f* de hauteurs
r кривая *f* высот

4230 **height diameter**
d Höhendurchmesser *m*
f diamètre *m* à hauteur définie
r базовый диаметр *m* на высоте, составляющей определённый процент от высоты дерева

4231 **height/diameter curve**
(graphic representation of the relationship between individual tree heights and diameters used to determine tree volumes in localized areas)
d Höhe/Durchmesser-Kurbe *f*
f courbe *f* de rapport hauteur/diamètre
r кривая *f* отношения диаметра к высоте

* **height growth → 326**

* **height increment → 326**

4232 **height measuring**
(of a tree)
d Höhenmessung *f*
f mesure *f* de hauteur
r измерение *n* высоты

4233 **height of crop [stand]; stand height; crop height**
d Bestandeshöhe *f*
f hauteur *f* du peuplement
r высота *f* насаждения

4234 **height of curvature; arc height**
d Krümmungshöhe *f*; Bogenhöhe *f*
f hauteur *f* de courbure
r высота *f* прогиба кривизны [стрелы]

4235 **height of defect**
d Fehlerhöhe *f*
f hauteur *f* de défaut
r высота *f* порока

4236 **height of stump; stump height**
(distance from the ground to the top of the stump)
d Stockhöhe *f*; Stubbenhöhe *f*
f hauteur *f* de souche
r высота *f* пня

4237 **height shoot; leading shoot; basal shoot;**

f pousse *f* en hauteur
r главный побег *m*; верховой побег; основной побег; ростовой побег; стержневой побег

* **Heldreich pine → 936**

4238 **Heldreich's maple**
(Acer heldreichii)
d griechischer Ahorn *m*
f érable *m* de Heldreich
r клён *m* Гельдрайха

* **helicoidal saw → 9466**

4239 **helicopter yarding; helicopter [logging] operation**
d Holztransport *m* mit Hubschrauber; Hubschrauber-Bringung *f*
f débardage *m* par hélicoptère; transport *m* [de bois] par hélicoptère
r вертолётная трелёвка *f*; транспортировка *f* леса вертолётами

* **heli-logging → 85**

* **heliophilous tree → 7476**

* **heliotropism → 6212**

4240 **helitack crew; rappel crew**
(initial attack on wildfires involving the use of helicopters and trained crews, deployed as a complete unit)
d Hubschrauber-Löschmannschaft *f*
f équipe *f* héliportée; parachutistes *mpl* héliportés
r вертолётный десант *m*

4241 **helitanker**
d Helitanker *m*
f hélicoptère *m* porte-citerne
r вертолёт-цистерна *f*

4242 **helophyte**
(freshwater plant with perennating tissue is soil beneath the water)
d Helophyt *m*
f hélophyte *m*
r гелофит *m*

* **helper → 7359**

4243 **helper tree**
d Helferbaum *m*

4244 **hemicellulose**
(non-cellulosic polysaccharides (excluding pectins) of the intercellular layer and of the cell wall that can be extracted with aqueous alkaline solutions)
d Hemizellulose *f*
f hémicellulose *f*
r гемицеллюлоза *f*

4245 **hemicryptophyte**
d Hemicrypophyt *m*
f hémicryptophyte *m*
r гемикриптофит *m*

4246 **hemi[-]epiphyte; pseudo-liana**
(plant that begins life as an epiphyte but later sends roots into the soil)
d Hemiepiphyt *m*
f hémiépiphyte *m*
r гемиэпифит *m*

4247 **hemiepiphytic**
(remaining rooted in the ground but climbing tree trunks)
d hemiepiphyt
f hémiépiphytique
r гемиэпифитный

4248 **hemlock**
(Tsuga)
d Hemlocktanne *f*; Schierlingstanne *f*
f pruche *f*; ciguë *f*; tsuga *m*
r тсуга *f*; гемлок *m*

4249 **hemlock bark**
d Hemlockrinde *f*
f écorce *f* de tsuga
r кора *f* гемлока

4250 **hemlock sawfly**
(Neodiprion tsugae)
d Hemlock-Blattwespe *f*
f diprion *m* de la pruche
r пилильщик *m* гемлоковый

* **hemlock spruce → 5588**

4251 **hemp rope; hemp line; hemp rootwood; hemp strop**
d Hanfstrick *m*
f corde *f* de chanvre
r пеньковый канат *m*; пеньковая верёвка *f*

* **hemp tree → 4961**

* **hemp willow → 619**

4252 herbaceous vegetation
(the low-growing, non-woody plants in a forest understor[e]y, including grasses (graminoids), wildflowers, and ferns)
d Kraut *n*
f végétation *f* herbacée
r травянистая растительность *f*

* **herbaceous willow → 6228**

4253 herbage
(vegetative parts of plants)
d Gras *n*
f herbes *fpl*
r травы *fpl*

4254 herbicide; weedicide; weed killer
(chemical substances or living organisms (called bioherbicides) used to kill or control vegetation such as brush, weeds, and competing or undesirable trees)
d Herbizid *n*; Unkrautvertilgungsmittel *n*
f herbicide *m*
r гербицид *m*

4255 herbicide used for leaves
d Blattherbizid *n*
f herbicide *m* foliaire
r листный гербицид *m*

4256 herbivore
(an animal that feeds primarily on living plant substances)
d Pflanzenfresser *m*
f herbivore *m*
r растениеядное животное *n*

* **herb layer → 3116**

* **Hercules' club → 2452**

4257 herder
(one who patrols a lumber or log flume to prevent jams)
d Herdenaufseher *m*; Hirt *m*
f repousseur *m*
r рабочий *m*, предупреждающий образование заторов на лесоспуске

4258 heredity; heritability; hereditability
(that portion of the character variance due to hereditary factors as distinct from factors of environment)
d Erblichkeit *f*
f héritabilité *f*; hérédité *f*
r наследственность *f*

archaeological, paleontological, or scenic significance to the province)
d erhaltenswertes Kulturerbe *n*
f zones *fpl* de patrimoine
r зоны *fpl*, находящиеся под охраной

4260 heritage forests
(proposed name for the highly protected sites within Canada's forest land)
d Erbewalde *mpl*
f forêts *fpl* patrimoniales
r леса *mpl*, находящиеся под охраной

4261 herringbone figure; herringbone pattern
(in veneering, a hearing bone pattern is formed when successive layers of veneers are glued up so they form a mirror image)
d Fischgrätentextur *f*
f texture *f* en arêtes de poison
r укладка *f* в ёлку; ёлочный ресунок *m*

* **heterocellular ray → 4264**

4262 heteroecious
(requiring two taxonomically different hosts to be able to complete the entire life cycle, as in the rust fungi)
d heterözisch; wirtswechselnd
f hétéroïque; hétéroxène
r разнохозяйный

4263 heterogeneity
d Ungleichmäßigkeit *f*
f hétérogénéité *f*
r гетерогенность *f*; неоднородность *f*; разнородность *f*

4264 heterogeneous ray; heterocellular ray
(in the hardwoods, a ray consisting of two kinds of cells, procumbent and upright; in the softwoods, a ray consisting of ray parenchyma and ray tracheids)
d heterogener Strahl *m*
f rayon *m* ligneux hétérogène; rayon hétérocellulaire
r гетероцеллюлярный луч *m*

4265 heterogeneous ray tissue
(ray tissue in which the individual rays are composed wholly or in part of square or upright cells)
d verschiedenartige Strahlgewebe *n*
f tissu *m* des rayons hétérogènes
r ткань *f* гетероцеллюлярных лучей

4266 heterophyllous

d heterophyl
f hétérophylle
r гетерофильный; разнолистный

4267 **heterophyllous shoot**
d heterophyler Trieb *m*
f pousse *f* hétérophylle
r гетерофильный побег *m*; разнолистный побег

4268 **heteroplastic graft**
d heteroplastisches Pfropfen *n*
f greffe *f* hétéroplastique
r гетеропластический привой *m*

4269 **heterosis; hybrid vigor**
d Heterosis *f*
f hétérosis *f*; hétérose *f*; vigueur *f* hybride
r гетерозис *m*; "гибридная сила" *f*; "гибридная мощность" *f*

* **hevea → 1320**

* **hew → 83, 2172**

4270 **hewed balk**
d behauener Balken *m*
f bille *f* scié à la hache
r обтёсанный брус *m*

4271 **hewed timber; hewn timber**
(timber with or without wane, finished to size with hand tools such as an axe or adze)
d behauenes Holz n; beschlagenes Holz; bezimmertes Holz
f bois *m* scié à la hache
r тёсаный лесоматериал *m*

* **hewer → 9526**

4272 **hew *v* high-grade forest; cream *v* [off]**
d Wertholz durchhauen
f exploiter le bois de valeur; écrémer
r вырубить ценную древесину

4273 **hewing; adzing**
d Abschrubben *n*; Behauen *n*; Beschlagen *n*
f dégrossissage *m*; rabotage *m*
r утёска *f*; тёска *f*; обтёска *f*; обтёсывание *n*

* **hewn timber → 4271**

* **hew open → 2169**

* **hickory pine → 8361**

Waldbrandnarbe
f brûlure *f* interne; cicatrice *f* cachée de brûlure
r закрытая пожарная прорость *f*

* **hiemisilva → 9461**

4275 **Higan cherry**
(Prunus subhirtella)
d winterblühende Kirsche *f*; Schnee-Kirsche *f*; Higan-Kirsche *f*
f cerisier *m* à fleurs
r вишня *f* сабхиртелла

4276 **high-class forest**
d hochwertiger Wald *m*
f forêt *f* de haute classe
r высокосортный лес *m*

* **high-class timber → 4284**

* **high-climber → 4296**

4277 **high-density chipboard**
d Spanplatte *f* mit hoher Dichte
f panneau *m* de copeaux de haute densité
r тяжёлая древесностружечная плита *f*

4278 **high-density planting**
(closely planted trees (as in a woodlot) or crops)
d Pflanzung *f* mit hoher Dichte
f plantation *f* de haute densité
r высокоплотное насаждение *n*

* **high-density stand → 3756**

* **high-density timber → 8659**

4279 **high face; pulling face**
d Reißlachte *f*
f coup *m* de griffe
r резная карра *f*

4280 **high forest; highwood; high standing timber; standard forest**
(crops and stands of trees, generally of seedling origin, that normally develop a high closed canopy)
d Hochwald *m*; Baumholzbestand *m*; Hochbestand *m*; Langrohholzbestand *m*
f [haute] futaie *f*; forêt *f* dense élevée
r высокоствольный лес *m*; высокоствольное насаждение *n*; высокоствольник *m*; строевой лес

4281 **high-forest system**

d Hochwaldbetrieb *m*; Baumholzbetrieb *m*; Überhaltbetrieb *m*; Verholzzucht *f*
f régime *m* de la futaie
r высокоствольное хозяйство *n*

4282 **high forest with reserves; high forest with standards**
d Hochwald *m* mit Reserve
f futaie *f* avec surréserves; futaie réservée
r высокоствольный лес *m* с воспитанием семенных деревьев

4283 **high-forest-with-reserves system; method of high forest with reserves**
(system in which selected trees of the old crop, scattered or in groups, are retained after regeneration is completed, for the whole or a part of the next high forest rotation)
d Verholzzuchtssystem *n* mit Reserve
f système *m* de futaie avec surréserves
r высокоствольное хозяйство *n* с воспитанием семенных деревьев

* **high forest with standards → 4282**

4284 **high-grade timber; high-class timber; primary timber; select grade; [high-]quality timber; timber of superior quality; valuable wood; crown wood; uppers**
(high-quality lumber)
d hochwertiges Holz *n*; Wertholz *n*; Qualitätsholz *n*
f bois *m* de qualité; bois de [grande] valeur
r высокосортные лесоматериалы *mpl*; древесина *f* хорошего качества; лучшие сорта *mpl* брёвен

4285 **high[-]grading; creaming**
(removing the mature, high quality trees from a stand and leaving inferior species and defective trees)
d Aushieb *m* der besten Stämme
f extraction *f* des meilleurs arbres; extraction de bois de valeur; coupe *f* d'écrémage; récolte *f* d'écrémage
r вырубка *f* ценных стволов

4286 **high hazard [of forest health]**
(physical characteristics (including tree species, composition, age, and size) and biogeoclimatic factors that make a forest highly susceptible to attack by damaging agents)
d hochgefährliches Risiko *n* [für Forstgesundheit]

4287 **high herbaceous plants; tall forbs**
d Hochstauden *fpl*
f macrophorbiées *fpl*; mégaphorbiées *fpl*
r высокие травянистые растения *npl*

4288 **highleading; highlead logging; highlead skidding; highlead extraction**
(transporting logs by using a powered cable, passing through a block fastened high off the ground, to lift the front end of the logs clear of the ground while dragging them)
d Rücken *n* von hochgelegtem Kabel aus; Rückemastverfahren *n*
f traînage *m* relevé; débardage *m* au moyen d'un câble suspendu; téléphérage *m*
r полуподвесная [канатная] трелёвка *f*; трелёвка с высоко лежащего кабеля

4289 **highlead landing**
d Einseillagerplatz *m*
f dépôt *m* de traînage relevé
r погрузочная площадка *f* при канатной трелёвки полуподвесным способом

* **highlead logging → 4288**

4290 **highlead rigged balloon system**
d Ballonrückensystem *n*
f système *m* de traînage relevé par ballon
r аэростатная трелёвочная установка *f*

* **highlead skidding → 4288**

4291 **highlead system**
(logging system that uses cables rigged to a spar high above the ground so that one end of the logs can be lifted during yarding)
d hochgelegtes Drahtseilsystem *n*
f système *m* de traînage relevé
r полуподвесная канатная трелёвочная установка *f*

* **high lift loader → 6065**

* **high line → 1096**

* **highly productive timberland → 4299**

4292 **highly resinous wood**
d harzreiches Holz *n*
f bois *m* anormalement résineux
r древесина *f* с высокой смолистостью

* **high mallow → 7102**

* **high-plateau forest → 9131**

4293 **high-power cable**
d hochgespanntes Drahtseil *n*
f câble *m* de haute pouvoir
r высоконапряжённый трос *m*

* **high-pressure water debarker → 4404**

4294 **high pruning; top pruning; carry-up pruning**
(removal of live or dead branches above a man's reach)
d Hochästung *f*; Köpfen *n*
f élagage *m* en hauteur; élagage élevé; élagage des branches supérieures; écimage *m*
r обрезка *f* высоких сучьев; обрезка ветвей с помощью лестницы

4295 **high-quality hardwood**
(hardwood stand that will yield high value timber products such as face veneer, knot free lumber, furniture or specialty product stock, or flooring)
d Laubedelholz *n*
f bois *m* feuillu précieux; bois feuillu noble; essence *f* feuillue précieuse
r высококачественная лиственная порода *f*; благородная лиственная древесина *f*

* **high-quality timber → 4284**

4296 **high-rigger; rigger; [high-]climber; rig-up man; tree climber**
(a person who attaches cables & blocks to spar-trees to aid in the extraction of logs, normally two high-riggers would ascend the same tree)
d Baumsteiger *m*; Drahtseilarbeiter *m*
f ouvrier *m* grimpeur; grimpeur *m* aux arbres; ouvrier-câbleur *m*
r тросовый работник *m*

4297 **high-risk tree**
d risikoreicher Baum *m*
f arbre *m* de risque important; arbre à risque
r дерево *n*, подлежащее на срочной валке по соображениям техники безопасности

* **high serviceberry → 2603**

4298 **high-set crown**
d hochangesetzte Krone *f*
f cime *f* haute
r высокорасположенная крона *f*

4299 **high-site forestland; highly productive timberland**
r высокопродуктивная лесная земля *f*

4300 **high skyline**
d Hochluftseilbahn *f*
f câble *m* haut
r высокорасположенный несущий канат *m*

* **high standing timber → 4280**

4301 **high-stemmed tree**
d hochstämmiger Baum *m*; Hochstamm *m*; im Freien wachsender Baum
f arbre *m* de plein vent; grume *f* élancée
r высокоствольное дерево *n*

4302 **high stump**
(stump that is higher than a specified standard)
d höher Stumpf *m*
f haut tronc *m*
r высокий пень *m*

* **high thinning → 8512**

4303 **high-volume stand**
d hochvolumener Bestand *m*
f peuplement *m* à volume élevé
r насаждение *n* с большим запасом древесины

4304 **high-wheel logging**
d Rücken *n* mit hochrädrigem Rückekarren
f débardage *m* par triqueballe
r трелёвка *f* с высококолесной тележкой

* **high-wire system → 88**

* **highwood → 4280**

4305 **high yield**
(of wood)
d höher Ertrag *m*
f haut rendement *m*
r высокий выход *m*

4306 **high-yield variety**
d Hochertragsvarietät *f*
f variété *f* d'haut rendement
r высокодоходное разнообразие *n*

* **hill → 1169**

* **hill ant → 9480**

* **hilling → 1171**

* **hill toon → 4309**

4307 **Himalayan blackberry**
(Rubus discolor; Rubus procerus)
d frühe Brombeere *f*
f ronce *f* décolorée
r ежевика *f* высокая

4308 **Himalayan blue pine; Buthan pine; blue pine**
(Pinus excelsa; Pinus wallichiana)
d Tränenkiefer *f*; Himalaya-Kiefer *f*
f pin *m* [pleureur] de l'Himalaya
r сосна *f* гималайская

* **Himalayan cedar → 2398**

4309 **Himalayan cigar box cedar; Himalaya toon; hill toon**
(Cedrela serrata; Toona serrata)
d Himalaya-Surenbaum *m*
f cédrèle *m* à feuilles de châtaignier
r цедрела *f* пильчатая

4310 **Himalayan spruce; Indian spruce**
(Picea morinda)
d Morindafichte *f*; Himalayafichte *f*; indische Fichte *f*
f épicéa *m* de l'Himalaya
r ель *f* гималайская; ель индийская

* **Himalaya toon → 4309**

* **hinds [black] walnut → 241**

* **hinge → 9051**

4311 **hinged dog; hook arm**
d Wendevorrichtung *f*
f grappin *m* à articulations; grappin articulé
r поворотный рычаг *m* для удержания брёвен; поворотное устройство *n*

4312 **hinge saw; jump saw; pop-up saw; pendulum saw; swinging cross-cut saw; swing saw; goose saw; rock saw; log-cleaner saw; barking saw; drunken saw; wobble saw**
d Pendelsäge *f*; schwingende Säge *f*; Schwenksäge *f*; Schwingsäge *f*
f scie *f* à lame escamotable; scie pendulaire; scie circulaire de garde; lame *f* de scie circulaire désaxée
r маятниковая пила *f*; качающаяся пила; балансирная пила; пила с косой осью; окорочная пила

4313 **hinge wood; holding wood**
(the holding wood between the undercut and the backcut that acts as a hinge to control the direction of the fall)
d Scharnierholz *n*
f bois *m* charnière
r шарнирное бревно *n*

4314 **history of forest; forest history**
d Forstgeschichte *f*
f histoire *f* forestière
r история *f* леса

* **hitcher → 6693**

* **H-layer → 4394**

* **HMU → 4078**

* **hoarding → 6695**

* **hoary alder → 4164**

* **hoary willow → 2775**

4315 **hobblebush [viburnum]; mooseberry; hobbleberry**
(Viburnum lantanoides; Viburnum alnifolium)
d erlenblättriger Schneeball *m*
f viorne *f* à feuilles d'aulne
r калина *f* ольхолистная

4316 **hoe** *v*
d hauen
f biner
r мотыжить

4317 **hoe-chucking**
(logging system that uses an excavator or hoe to yard logs to the roadside and/or landing)
d Hacke-Aufspannen *n*
f mandrinage *m* par bineuse
r зажим *m* мотыгой

* **hoeing machine → 9512**

* **hogal → 7960**

4318 **hogged chips; fractured particles**
d Häcksel m/n; geschreddertes Holz *n*; gekörnte Holzabfälle *mpl*; Schlagspäne *mpl*
f copeaux *mpl* déchiquetés
r дроблённые отходы *mpl*; дроблёнка *f*; дроблёная щепа *f*

4319 **hogged fuel**

f copeaux *mpl* énergétiques; déchets *mpl* de bois écrasé
r измельчённая топливная древесина *f*

* **hogged wood → 2530**

* **hogger → 1534**

4320 **hog[ging]**
(reducing wood to coarse chips, for fuel or the manufacture of wood pulp and wood chipboard)
d Geschreiden *n*
f déchiquetage *m*
r дробление *n* отходов лесопыления

4321 **hogging cutter; edging and chipping cutter**
d Schneider *m*
f déchiqueteur *m* de délignage
r дробильный нож *m*

* **hog[ging] machine → 1534**

4322 **hog plum; yellow mombin; Jamaica plum; golden apple; tallow wood; tallow nut; false sandalwood**
(Ximenia americana)
d gelbe Mombinpflaume *f*; Gelbpflaume *f*
f ximénie *f* d'Amérique; prunier *m* mombine
r ксимения *f* американская

* **hog slide → 2648**

* **hoist boom → 5246**

* **hoist cylinder → 4932**

4323 **hoist[ing] rope; hoisting cable; lifting rope**
d Bergaufseilbahn *f*; Hubseil *n*
f câble *m* de levage
r подъёмный канат *m*; канат подъёмного устройства

* **hold → 9051**

* **holder of timber right → 6940**

4324 **holding boom; storage boom; receiving boom**
(a strong boom used to hold logs in storage at a sawmill)
d Bremsbaum *m*
f arbre *m* de freinage
r тормозная колода *f*

* **holding wood → 9051**

4325 **holdover**
d Überbleibsel *n*
f retardataire *m/f*
r запаздывание *n* вылёта короедов; запаздывание появления личинок насекомых

* **holdover → 2567**

* **holdover fire → 7734**

* **holdover tree → 5885**

4326 **hole digger**
d Pflanzlochmaschine *f*
f tarière *f* mécanique
r машина *f* для посадочных ямок

* **hole planting → 6355**

* **hollow adze → 1041**

4327 **hollow-backed saw**
d Hohlkreuzsäge *f*
f lame *f* de scie à dos concave
r двуручная пила *f* с искривлённой тыльной частью

4328 **hollow chisel; gouge**
(a cutting tool with a U-shaped cross-section used with the bevel rubbing)
d Hohlmeißel *m*; Hohlstemmer *m*
f gouge *f*; bédane *m* creux
r полукруглое долото *n*; полое долото; стамеска *f*

4329 **hollowed-out log**
d ausgehöhlter Stammstück *m*; gehöhlter Stammstück
f grume *f* creusée
r выдолбленное бревно *n*

* **hollow head debarker → 7074**

* **hollow heart → 2933**

4330 **hollow-horned lumber**
(timber with fissures, generally caused by drying stresses resulting from case-hardening)
d hohles Holz *n*
f bois *m* cavicorne
r лесоматериал *m* с большим количеством трещин

4331 **hollow knot**

f nœud *m* perforé
r дуплистый сучок *m*

4332 **hollowness**
d Hohlstämmigkeit *f*
f cavosité *f*
r дуплистость *f*

* **hollow pith → 2933**

* **hollow-spade → 6342**

4333 **hollow tree; den tree; cavity tree**
(a tree that typically contains numerous hollowed out cavities or holes in the bole of the tree)
d hohler Baum *m*
f arbre *m* caverneux
r дуплистое дерево *n*

4334 **hollow trunk**
d hohler Stamm *m*
f tronc *m* caverneux; cavité *f* d'un tronc causée par la pourriture
r дуплистый ствол *m*

4335 **holly family**
(Aquifoliaceae)
d Stechpalmengewächse *pl*
f aquifoliacées *pl*
r падубовые *pl*

4336 **hollyleaf cherry**
(Prunus ilicifolia)
d stechpalmenblättriger Kirschbaum *m*
f cerisier *m* à feuilles de houx
r слива *f* падуболистная

4337 **holly leaf miner**
(Phytomyza ilicis)
d Ilexminiermotte *f*
f mineuse *f* du houx
r моль-минёр *m* падубовый

4338 **hollyleaf redberry**
(Rhamnus ilicifolia)
d stechpalmenblättriger Faulbaum *m*
f nerprun *m* à feuilles de houx
r крушина *f* падуболистная

* **holly oak → 1773**

4339 **holly olive**
(Osmanthus aquifolium; Osmanthus heterophyllus)
d stachelblättrige Duftblüte *f*

* **holly [tree] → 2899**

* **holm [oak] → 1773**

4340 **holocellulose**
(the total carbohydrate fraction of wood remaining after the removal of lignin and substances extractable with solvents)
d Holozellulose *f*
f holocellulose *f*
r холоцеллюлоза *f*; голоцеллюлоза *f*

* **holt → 9051**

* **home-grown poplar → 2884**

4341 **home-grown timber; domestic wood; native wood; indigenous wood; indigenous timber**
d [ein]heimisches Holz *n*; inländisches Holz
f bois *m* indigène
r лес *m*, растущий в пределах данной страны; местная древесина *f*; древесина местных пород

4342 **home-guard type of operation**
d Heimschutz *m*
f exploitation *f* par garde
r лесозаготовки *fpl* с ежедневной доставкой рабочих из жилых центров

4343 **homeostasis**
(metabolic equilibrium actively maintained by several complex biological mechanisms that operate via the autonomic nervous system to offset disrupting changes)
d Homeostasie *f*
f homéostasie *f*
r гомеостазис *m*; оптимальное состояние *n* организма

4344 **homocellular ray; homogeneous ray**
(a ray consisting entirely of one kind of cell)
d homogener Strahl *m*
f rayon *m* homorocellulaire; rayon homogène
r гомоцеллюлярный луч *m*

* **homogeneous forest → 2926**

4345 **homogeneous ray tissue**
(ray tissue in which the individual rays are composed wholly or procumbent cells)
d homogenes Markstrahlgewebe *n*
f tissu *m* des rayons homogènes
r ткань *f* гомоцеллюлярных лучей

4346 homogeneous timber
d Homogenholz *n*
f bois *m* homogène
r однородный лесоматериал *m*

4347 homoplastic graft
(a graft between individuals of the same species or between parts of the same individual)
d homoplastisches Pfropfen *n*
f greffe *f* homoplastique
r гомопластический привой *m*

* **Honduras cedar → 1411**

* **Honduras mahogany → 1412**

* **Honduras pine → 2111**

4348 Honduras rosewood
(Dalbergia stevensonii)
d Honduras-Rosenholz *n*
f palissandre *m* du Honduras
r дальбергия *f* Стевенсона; гондурасский палисандр *m*; гондурасское розовое дерево *n*

* **honeycomb center → 5905**

* **honeycomb core → 5905**

4349 honeycombing
(splitting giving the effect of honeycomb, in lumber)
d Wabigwerden *n* [des Holzes]
f fendillement *m* au séchage; formation *f* de gerces de séchage
r образование *n* трещин усушки; образование внутренних трещин при конвективно-тепловой сушке [пиломатериалов]

* **honeycomb rot → 6313**

4350 honeycup; dusty zenobia
(Zenobia pulverulenta)
d Zenobie *f*
f zénobia *m*
r зенобия *f* припудренная

4351 honeydew drip
(honeydew precipitation reaching the forest floor)
d Honigtau-Tröpfeln *n*
f précipitation *f* [au sol] de miellat
r капанье *n* клейкой жидкости

4352 honey fungus; honey mushroom
(Armillariella mellea)
d Hallimasch *m*; honiggelber Ringling *m*
f armillaire *m*
r настоящий опёнок *m*

4353 honeylocust; sweet locust; black locust; three-thorn[ed] acacia
(Gleditsia triacanthos)
d [dreidornige] Gleditschie *f*; Schotendorn *m*; amerikanische Gleditschie
f févier *m* à trois épines; févier épineux; févier d'Amérique; arbre *m* aux escargots; gléditschie *f*; carouge *m* à miel
r гледичия *f* [обыкновенная]; трёхколючковая акация *f*

4354 honey mesquite; prairie mesquite
(Prosopis glandulosa; Prosopis chilensis)
d Mesquitebusch *m*
f légumineuse *f* mimosée épineuse
r мескитовое дерево *n* чилийское; прозопис *m* железистый; прозопис чилийский

* **honey mushroom → 4352**

* **honey palm → 1701**

* **honeysuckle → 1227**

4355 honeysuckle azalea; Pontic azalea
(Rhododendron flavum)
d pontische Azalee *f*
f rhododendron *m* pontique
r азалия *f* понтийская; рододендрон *m* понтийский

4356 honeysuckle family
(Caprifoliaceae)
d Geißblattgewächse *pl*
f caprifoliacées *pl*
r жимолостные *pl*

4357 honeytree; pagoda tree
(Sophora)
d Honigbaum *m*; Schnurbaum *m*
f arbre *m* de miel
r медоносное дерево *n*; софора *f*

4358 hook *v*
d haken; anhaken
f crocheter
r прицеплять

4359 hook
(the turned edge of the knife in a scraping

f grattoir *m*
r крюк *m*

* **hook arm → 4311**

* **hookaroon → 6225**

* **hook dogs → 3892**

* **dogs → 3892**

* **hooker → 9748**

4360 **Hooker willow; dune willow; coast willow**
(Salix hookeriana)
d Hooker-Weide *f*
f saule *m* de Hooker
r ива *f* Гукера

4361 **hook hole**
(defect resulting from manipulation of the wood)
d Hakenloch *n*
f trou *m* de griffe; trou de crochet
r дефект *m* в фанере или шпоне (образовавшийся при повреждении чурака крюком)

* **hooking → 1560**

* **hook-tender → 9748**

* **hoop ash → 1777**

4362 **hoopwood; staff wood; barrelwood; barrel log; stave log; wood for staves; stave-wood; staves; cask-wood; cooper's wood**
d Binderholz *n*; Böttcherholz *n*; Fassholz *n*; Stabholz *n*; Daubenholz *n*; Fassholzklotz *m*
f douvain *m*; merrain *m*
r бочарная древесина *f*; бондарная древесина; бондарный лесоматериал *m*; лесоматериал для выработки бочарной клёпки; клёпочный кряж *m*

4363 **hop**
(Humulus lupulus)
d Hopfen *m*
f houblon *m*
r хмель *m*

4364 **hop cone; hops**
d Blütenzapfen *m* der Hopfenpflanze; Hopfendolde *n*; Hopfenzapfen *m*
f cône *m* de houblon; houblon *m* en cône
r хмелевая шишка *f*; шишка хмеля

(Hopea)
d Merawan
f hopée *f*
r хопея *f*

* **hophornbeam → 2900**

4366 **Hoppus measure; quarter girth measure**
(a measurement in which a theoretical allowance was automatically made for wastage in conversion from round timber into a squared log)
d Hoppus-Messung *f*
f cubage *m* au quart sans déduction
r мера *f* Гоппус

* **hops → 4364**

* **hoptree → 6035**

4367 **horizontal closure of crown; horizontal canopy**
d horizontaler Kronenschluss *m*
f fermeture *f* horizontale; couvert *m* horizontal
r горизонтальная сомкнутость *f* кроны; горизонтальная сомкнутость полога

4368 **horizontal diversity**
(comparative measure of occurrence and complexity of distribution of cover types over a wide area encompassing different cover types)
d horizontale Unterschiedlichkeit *f*
f diversité *f* horizontale
r горизонтальная разность *f*

4369 **horizontal flail slasher; beater**
(a mechanism for cleaning of knots)
d Gestrüppschneider *m* mit Horizontalachse; Dreschtrommel *f*
f débroussailleuse *f* à axe horizontal; batteur *m*
r [цепной] механизм *m* ударного действия (для очистки деревьев от сучьев)

4370 **horizontal point sampling**
(method of selecting trees for measurements and of estimating stand basal area at a sample location or point sample)
d horizontales Punkt-Stichprobenverfahren *n*
f sondage *m* par points horizontaux
r таксация *f* горизонтальными точками

4371 **horizontal resistance**
d horizontale Resistenz *f*
f résistance *f* horizontale
r горизонтальное сопротивление *n*

* **hornbeam → 2901**

4372 **hornbeam forest**
d Hagebuchewald *m*; Weißbuchenwald *m*
f charmaie *f*; carpinaie *f*
r грабовый лес *m*; грабовая роща *f*; грабняк *m*

4373 **hornbeam maple; Japanese maple**
(Acer palmatum; Acer carpinifolium)
d Korallenrindenahorn *m*; japanischer Tiefrotahorn *m*; hainbuchblättriger Ahorn *m*; Fächerahorn *m*
f érable *m* japonais [à feuille laciniée]; érable à feuilles glabres
r клён *m* граболистный; клён дланевидный; клён пальмовидный

4374 **hornbeam-oak forest**
d Eichen-Hainbuchenwald *m*
f rouvraie-charmaie *f*
r дубово-грабовый лес *m*; дубняк *m* грабовый

* **horncap eucalyptus → 3550**

4375 **hornet moth**
(Aegeria apiformis)
d Hornissenglasflügler *m*
f papillon *m* frelon
r стеклянница *f* тополевая

4376 **horn knot; mule-ear knot; slash knot**
d Doppelkantenast *m*; Nagelast *f*
f cheville *f* de bois mort; nœud *m* bouchon
r роговой сучок *m*; сшивной сучок; гвоздевой сучок

* **horns → 401**

* **horse apple → 4219**

* **horse[-]chestnut → 1781**

* **horse-chestnut family → 1126**

4377 **horse dam; raft dam**
(a temporary dam made by placing large logs across a stream, in order to raise the water behind it, so as to float the rear)
d Blockdamm *m*; Balkensperre *f*
f barrage *m* de poutres
r временная запруда *f* из брёвен; плотина *f* из брёвен; ряжевая плотина

4378 **horse logging; horse skidding**
avec chevaux
r лошадиная трелёвка *f* [леса]

4379 **horseradish tree; be[h]n tree; behen tree; benzolive tree; drumstick tree**
(Moringa oleifera; Guilandina moringa; Moringa pterygosperma)
d Pferderettichbaum *m*; Behenbaum *m*; Behennussbaum *m*; Meerrettichbaum *m*
f ben *m* [oléifère]; ben ailé; moringa *f* ailée; pois *m* quénique
r моринга *f*

* **horse skidding → 4378**

* **horsetail tree → 407**

* **horse tamarind → 4851**

4380 **horse-tree; trestle tree; sawbuck; sawhorse; trestle for sawing; sawing threstle; saw[ing] jack; wood tick**
(a rack on which wood is laid for sawing by hand; especially one with X-shaped ends)
d Sägebock *m*; Längssaling *f*; Auflagebock *m*; Einschneidebock *m*; Holzbock *m*
f chevalet *m* [de sciage]; borriquete *m*
r козлы *mpl* для пилки дров; [раскряжёвочные] козлы *mpl*; козлы *pl* для распиловки [дров]; козлы для ручной разделки брёвен

4381 **host tree**
d Gastgeberbaum *m*
f arbre-hôte *m*
r дерево-хозяйн *n*

4382 **hot deck; hot skidway**
(pile of logs from which logs are hauled as soon as they are yarded; on-loading follows off-loading immediately)
d Laderampe *f*
f dépôt *m* intermédiaire de transit
r промежуточная площадка *f* без накопления лесоматериалов; погрузочная площадка с отгрузкой подтрелёванной древесины без складирования

4383 **hot loading**
d direkte Ladung *f*
f chargement *m* à chaud
r погрузка *f* [подтрелёванной древесины] без [её] складирования

4384 **hot logging**
(logging operation in which the logs are not

d direkte Abholzung *f*
f exploitation *f* sans rupture de charge
r лесозаготовки *fpl* без межоперационных запасов лесоматериалов

* **hot skidway → 4382**

4385 **hotspot**
d Vorrangstelle *f*; vordringliche Stelle *f*
f zone *f* de haute diversité biologique; point *m* chaud
r первенствующее место *n*

4386 **hot-water scarification**
d Heisswasserskarifikation *f*
f scarification *f* à eau chauffée
r скарификация *f* горячей водой

4387 **hot yarding**
d kombiniertes Fällen-Seilen *n*
f téléphérage *m* intégré
r трелёвка *f* без складирования

4388 **housebote [right]; fuelbote [right]**
(wood allowed to a tenant for repairing the house and for fuel)
d Nutzholzberechtigung *f*
f droit *m* des bois d'œuvre
r сервитут *m* на поделочный лес

4389 **house longhorn beetle; European house longhorn; European house-borer; old-house borer**
(Hylotrupes bajulus)
d Hausbock *m*
f capricorne *m* des maisons; perceur *m* du vieux bois
r чёрный домовый усач *m*

* **house longhorn beetle damage → 5134**

4390 **howel; concave axe**
(a tool used by coopers for smoothing and chamfering their work)
d Krummaxt *f*
f hache *f* à manche sinueux
r топор *m* с изогнутым по контуру дерева широким лезвием; изогнутый топор

4391 **hull**
(seed)
d Schote *f*; Hülse *f*
f parche *f* [sèche]
r шелуха *f*; скорлупа *f*; кожица *f*

4392 **Humboldt[-type] undercut**
r подпил *m* в пнёвой части

* **humic pods → 2409**

4393 **humification; formation of humus**
d Humusbildung *f*; Humifizierung *f*; Humifikation *f*
f humification *f*; formation *f* de l'humus
r гумификация *f*; процесс *m* гумусообразования

4394 **humified layer; H-layer**
d humifizierte Schicht *f*
f couche *f* humifiée; horizon *m* H
r нижний слой *m* сильно разложившегося опада

4395 **hummock[y] tundra**
d Hummock-Tundra *f*
f toundra *f* de hummock
r кочковатая тундра *f*

* **hump in a gravity yard → 4440**

* **humpy texture → 1168**

4396 **humus**
(general term for the more or less decomposed plant and animal residues in the lower organic soil layer)
d Humus *m*
f humus *m*
r гумус *m*; перегной *m*

4397 **Hungarian oak**
(Quercus frainetto; Quercus coniferta)
d ungarische Eiche *f*
f chêne *m* de Hongrie; chêne hongrois; chêne de Pannonie
r дуб *m* густой; дуб Фрайнетто

* **hung [tree] → 5041**

* **hura → 7185**

* **hurdle → 3018, 6373**

* **hurst → 7445**

4398 **hyaline layer; nucellar epidermis; perisperm**
(a nutritive tissue in an angiospermous seed, formed from the nucellus)
d hyaline Schicht *f*; Keimhülle *f*; Perispermie *f*
f périsperme *m*
r гиалиновый слой *m*

f hybride *m*
r гибрид *m*

4400 hybrid black poplar; robusta poplar
(Populus x euramericana)
d Schwarzpappel-Hybride *n*; Robustapappel *f*
f peuplier *m* noir hybride; peuplier robusta
r тополь *m* чёрный гибридный

4401 hybridization; hybridizing; crossing
d Bastardierung *f*; Hybridisierung *f*; Hybridisation *f*; Kreuzungszüchtung *f*
f hybridation *f*; croisement *m*
r гибридизация *f*; скрещивание *n*; неродственное размножение *n*

* **hybrid vigor → 4269**

4402 hydrangea family
(Hydrangeaceae)
d Hortensiengewächse *pl*
f hydrangéacées *pl*
r гортензиевые *pl*

4403 hydraulic barking
(removal of bark from round timber, such as logs, bolts or billets, by high-pressure jets of water as the pieces are mechanically rotated in a closed chamber)
d hydraulische Entrindung *f*
f écorçage *m* hydraulique
r гидравлическая окорка *f*

4404 hydraulic [de]barker; high-pressure water debarker; hydrobarker; stream[-]barker
d Wasserentrinder *m*; Wasserentrindungsmaschine *f*; hydraulischer Entrinder *m*
f écorceuse *f* hydraulique
r гидравлический окорочной станок *m*; гидравлическая окорочная машина *f*

4405 hydraulic log splitter
d hydraulischer Holzspalter *m*
f fendeuse *f* de bois hydraulique
r гидроколун *m* (для длинных брёвен)

4406 hydraulic seeding; hydroseeding
(application of seed in a water slurry that contains fertilizer, a soil binder and/or mulch)
d Anspritzsaat *f*; Nassansaat *f*; Hydrosaat *f*
f ensemencement *m* hydraulique; hydro-ensemencement *m*
r гидравлический посев *m*

* **hydraulic shovel → 4408**

tongs
d hydraulische Hebezange *f*
f pince *f* de débardage hydraulique
r гидравлический трелёвочный захват *m*

4408 hydraulic spade; hydraulic shovel
(hydraulic accessory attached to a machine used for transplanting landscape stock)
d hydraulischer Spaten *m*
f pelle *f* [hydraulique]; pelleteuse *f* hydraulique
r гидравлическая лопата *f*

* **hydraulic tongs → 4407**

4409 hydraulic tree shear
d hydraulische Baumschere *f*
f cisaille *f* [d'abattage] hydraulique
r гидравлическое ножевое устройство *n* для срезания дерева

* **hydrobarker → 4404**

* **hydrolysis of timber → 7143**

4410 hydrophyte
(a plant that is more or less restricted to moist sites)
d Hydrophyt *m*
f hydrophyte *f*
r гидрофит *m*; водное растение *n*

* **hydroseeding → 4406**

4411 hydrostatic measurement of roundwood
d hydrostatische Rundholzmessung *f*
f mesure *f* hydrostatique de bois rond
r гидростатическое определение *n* объёма круглых лесоматериалов

4412 hygiene logging
(used to prevent the spread of disease, that includes the physical and temporal separation of logging operations)
d Hygiene-Holzeinschlag *m*
f exploitation *f* d'ambiance
r санитарная лесозаготовка *f*

* **hygroscopic moisture → 950**

* **hylad → 3527**

* **hylea → 8978**

* **hylodad → 5922**

4413 **hymenopterous insects**
(Hymenoptera, an insect order including over 1,500 species of the ants, bees, wasps, and sawflies)
d Hymenopteren *pl*
f hyménoptères *mpl*; guêpes *fpl*
r перепончатокрылатые *pl*

4414 **hypha**
(a long, branching filament that collectively form the feeding structure of a fungus called the mycelium)
d Hyphe *f*
f hyphe *m*
r гифа *f*

* **hyphal hole → 927**

4415 **hypocotyl**
(the part of the stem bearing the cotyledons)
d Hypocotyl *n*
f hypocotyle *m/f*
r гипокотиль *m*; подсемядольное колено *n*

4416 **hypsometer; hypsometric altimeter**
(instrument, often a stick or other straight edge, used to measure the heights of trees on the basis of similar angles)
d Hypsometer *n*
f hypsomètre *m*
r высотомер *m*

I

* **icaco → 1700**

* **icebox → 7942**

4417 **ice-break**
d Eisbruch *m*
f bris *m* de glace
r облом *m* [сучьев] в результате ожеледи; поломка *f* от ожеледи

4418 **ice slide; [toboggan] slide**
d Eisriese *f*
f glissade *f*; glissoire *f*; toboggan *m*
r ледяной лесоспуск *m*

* **Idaho white pine → 9344**

* **ideal forest → 5781**

4419 **ideally stocked; normally stocked; optimally stocked** *adj*
(ref. to productive forest land covered with trees of merchantable species of any age)
d normalerweise bestockt
f de densité relative idéale; de densité relative normale
r идеально густой; нормально облесённый

4420 **ideal standing**
d idealer Bestand *m*
f peuplement *m* idéal
r идеальная густота *f* древостоя

* **ideal stocking → 3757**

* **identification of tree species → 8928**

4421 **idioblast; biophore; plastidule**
(a minimal element of the protoplasm capable to grow and reproduce)
d Idioblast *m*
f idioblaste *m*
r биофор *m*; идиобласт *m*

* **IFS → 4472**

4422 **ignition area**
d Entflammungsfläche *f*
f zone *f* d'allumage; surface *f* d'origine d'un
r нескольких управляемых палов; площадь воспламенения

4423 **illegal felling**
d illegaler Holzeinschlag *m*
f abattage *m* non réglementaire
r незаконная рубка *f*; безбилетная рубка

* **ilomba → 108**

* **imbibed moisture → 950**

* **imbuia → 1008**

* **immature → 4674**

4424 **immature stage**
d unreifes Stadium *n*
f stade *m* jeune
r стадия *f* неполовозрелости; неполовозрелая стадия

4425 **immature timber; immature stand; ripening stand; premature stand**
(stands of timber where the age of the leading species in a stand is less than the specified cutting age)
d unreifes Holz *n*; angehender Bestand *m*; heranreifender Bestand; heranwachsender Bestand
f bois *m* non-mûr
r неспелый древостой *m*; приспевающее насаждение *n*

* **immature wood → 4675**

4426 **imperfect heartwood**
d unvollständiges Kernholz *n*
f bois *m* de cœur imparfait
r дефектная ядровая древесина *f*

* **imperfect vessel member → 9173**

* **imported currant worm → 1775**

4427 **impregnated wood; impreg**
(improved wood in which the cell walls have been impregnated with synthetic resin so as to reduce materially its shrinking and swelling)
d imprägniertes Holz *n*; Tränkholz *n*
f impreg *m*; bois *m* imprégné
r пропитанная древесина *f*; пропитанные сортименты *mpl*

4428 **impregnating substance; impregnating agent; preservative; impregnating**

f produit *m* d'imprégnation; moyen *m* d'imprégnation
r консервирующее вещество *n*; консервант *m*; средство *n* пропитки

4429 **impregnation of wood; impregnation of timber; soaking of wood**
d Durchtränkung *f* des Holzes; Holzimprägnierung *f*; Holztränkung *f*
f imprégnation *f* du bois
r пропитка *f* древесины

4430 **impregnation under pressure**
(a process of impregnating wood products with chemicals by forcing into the structure of wood with high pressure)
d Drucktränkung *f*
f imprégnation *f* sous pression
r импрегнирование *n* под давлением

4431 **improved forest**
d vergüteter Wald *m*
f forêt *f* améliorée
r улучшенный лес *m*

4432 **improved wood; purified wood**
d vergütetes Holz *n*
f bois *m* amélioré; bois purifié
r облагороженная древесина *f*

4433 **improvement by selection**
d Pflegeeingriff zur Verbesserung
f amélioration *f* par sélection
r улучшение *n* селекцией

* **improvement cutting → 2121**

* **improvement felling → 2121**

4434 **improvement of wood; wood improvement; timber stabilization**
d Holzvergütung *f*
f amélioration *f* du bois; mise *f* en valeur du bois
r облагораживание *n* древесины

* **improvement planting → 2830**

* **inaccessible forest → 3645**

* **inarching → 331**

* **inbark → 4491**

* **inbound deck → 4485**

f reproduction *f* en consanguinité; sélection *f* consanguine; élevage *m* en consanguinité
r родственное скрещивание *n*; инбридинг *m*

4436 **incendiarism; arson**
d Waldbrandlegung *f*; Waldbrandstiftung *f*; Brandstiftung *f*
f incendie *m* prémédité
r поджог *m* [леса]

* **incendiary fire → 7466**

* **incense cedar → 6102**

* **incense juniper → 7878**

* **inch worm → 3174**

* **incidence → 8110**

4437 **incidental felling**
d zufällige Nutzung *f*; Zufallnutzung *f*
f coupe *f* de bois accidentelle
r случайная рубка *f*; бессистемная рубка

* **incipient decay in beech → 1089**

4438 **incised leaf**
(deeply notched into sharp, pointed segments)
d eingeschnittenes Blatt *n*
f feuille *f* encaissée
r надрезанный лист *m*

* **incision → 4082**

4439 **inclination of fibers; grain inclination; slope of grain**
(in lumber and other wood products, the degree of cross grain)
d Faserabweichung *f*; Faserneigung *f*
f inclinaison *f* du fil
r наклон *m* волокон

4440 **incline; hump in a gravity yard**
(a portion of a logging railroad, the grade of which is too steep for the operation of locomotives, and up or down which the log cars are raised or lowered by means of a cable and power)
d Ablaufberg *m*
f chemin *m* de fer de montagne; bosse *f* de triage
r лесовозная железная дорога *f* с лебёдочной тягой на больших уклонах

4441 **inclined cutting**

d Schräghieb *m*
f coupe *f* biaisée; coupe en tirant
r наклонное резание *n*

4442 **inclined deck**
(of timber storage)
d Ablaufbühne *f*
f plate-forme *f* inclinée
r наклонная эстакада *f*; наклонный настил *m* (лесотранспортёра)

4443 **inclinometer; clinometer**
(an instrument used to determine the slope)
d Gefällmesser *m*
f clisimètre *m*; clinomètre *m*
r клинометр *m*; склонометр *m*

* **included sapwood → 853**

4444 **income from forestry**
d Einkünfte *pl* aus Forstwirtschaft
f bénéfice *m* d'exploitation forestière
r доход *m* лесного хозяйства

* **income from thinning → 9782**

4445 **incompletely carbonized residue**
d unvollständig Rückstand *m*
f résidu *m* imparfaitement cémenté
r неполно карбонизированные отходы *mpl*; неполно обугленные остатки *fpl*

* **incomplete ring → 6087**

4446 **incomplete stand; poorly stocked stand; nonstocked stand; low-density stand; sparsely closed stand; understocked stand; light forest; thin forest; understocked forest**
d lückenhafter Bestand *m*; lichter Bestand; Lichtstand *m*; Lichtwald *m*; gelichteter Wald
f peuplement *m* clairiéré; peuplement clair; peuplement discontinu; peuplement pauvre; forêt *f* claire; clair-bois *m*
r низкополнотное насаждение *n*; низкобонитетное насаждение; разрежённый древостой *m*; редкостойный лес *m*; светлый лес; редина *f*; редколесье *n*

* **increase in quality → 6655**

4447 **increaser [plant species]**
(plant species (e.g., rangeland plants) of the original vegetation that increase in relative amount, at least for a time, under overuse)
d zunehmende Pflanzenart *f*
которого увеличивается при выпасе

4448 **increasing sap flow; upward flow of the sap**
d aufsteigender Saftstrom *m*
f écoulement *m* ascendant de sève
r восходящее сокодвижение *n*

4449 **increment; growth**
(the increase in girth, height, volume, weight or value of individual trees; the difference between the values of a variable at the end and the beginning of the measuring period)
d Zuwachs *m*
f accroissement *m*; croissance *f*; incrément *m*
r прирост *m*

4450 **incremental silviculture; intensive silviculture**
(application of cultural measures which, in addition to simply maintaining the forest cover, will allow an increase in the value or volume of the cut)
d intensiver Waldbau *m*
f sylviculture *f* supplémentaire; sylviculture intensive
r приростное лесоводство *n*; интенсивное лесоводство

4451 **increment borer; coring borer; accretion borer; coring instrument; Pressler's borer; timber tester; pole tester**
(a tool used to extract a core of wood from a tree)
d Prüfbohrer *m*; Zuwachsbohrer *m*
f tarière *f* d'accroissement; tarière de Pressler
r приростной бурав *m*

4452 **increment [boring] core; bore chipping**
(that part of the cross section of a tree extracted by an increment borer)
d Zuwachsbohrkern *m*
f carotte *f* de sondage à la tarière
r цилиндрик *m* древесины; приростно-буравое ядро *n*

4453 **increment curve**
d Zuwachskurve *f*
f courbe *f* d'accroissement
r кривая *f* прироста

* **increment due to increased light → 4944**

* **increment felling → 7472**

4454 **increment hammer**
(designed for quick check of 3/4 inch layer of

f marteau *m* d'accroissement
r приростной молоток *m*

4455 **increment in value; value increment**
d Wertzuwachs *m*
f accroissement *m* en valeur
r прирост *m* по ценности; прирост по стоимости

4456 **increment mean stem; average growth tree**
d Zuwachsmittelstamm *m*
f tige *f* moyenne par accroissement; arbre *m* d'accroissement moyen
r среднее дерево *n* по приросту; средний по приросту ствол *m*

* **increment rate → 6728**

4457 **increment table; growth table**
d Zuwachstafel *f*
f table *f* d'accroissement
r таблица *f* прироста (насаждений)

* **increment thinning → 7472**

4458 **independent crown fire; independent top fire**
d unabhängiges Gipfelfeuer *n*; unabhängiges Kronenfeuer *n*; unabhängiges Wipfelfeuer *n*
f feu *m* de cimes indépendant
r верховой устойчивый пожар *m*

4459 **indeterminate growth**
(pattern of development in which the apical meristem remains vegetative)
d unbestimmtes Wachstum *n*
f accroissement *m* non déterminé
r рост *m*, не ограничиваемый развитием верхушечной почки

* **index forest → 4471**

4460 **index of association**
(measure of the probable occurrence together of one species with another)
d Association-Index *m*
f indice *m* d'association
r индекс *m* ассоциации

4461 **index of relative abundance**
(any expression of relative abundance, e.g., a useful approximation of the total number of individuals of a species relative to time or space)
d Index *m* des relatives Reichtums
f indice *m* d'abondance relative

* **index stand → 4471**

4462 **India-charcoal trema**
(Trema orientalis)
d orientalisches Trema *n*
f micocoulier *m* d'Orient
r трема *f* восточная

4463 **India frankincense; salai tree; boswellia**
(Boswellia serrata)
d echter Weihrauchbaum *m*; Weihrauch *m*
f boswellie *f*; arbre *m* thurifère; shallaki *m*
r босвел[л]ия *f* пильчатая; ладанное дерево *n*

4464 **Indian almond [wood]; tropical almond terminalia; Malabar almond; false kamani**
(Terminalia catappa)
d Almendro *m*; Ketapang *m*; Katappenbaum *m*; Katappabaum *m*; indischer Mandelbaum *m*
f almendro *m*; talisay *m*; badam *m*
r терминалия *f* катаппа; катапан *m*

4465 **Indian boxwood; Ceylon boxwood**
(Gardenia latifolia)
d Papra *n*
f papra *m*
r гардения *f* широколистная

* **Indian cedar → 2398**

* **Indian crabwood → 6646**

* **Indian currant → 1940**

* **Indian ebony → 3882**

* **Indian fig → 6533**

* **Indian gooseberry → 5966**

4466 **Indian gum; kauri gum; kauri resin**
(the New Zealand kauri copal from both living and fossil trees of Agathis australis)
d indisches Gummi *n*
f copal *m* de kauri; gomme *f* kauri; gomme éléphantine
r смола *f* каури (новозеландского хвойного дерева)

4467 **Indian plum family**
(Flacourtiaceae)
d Flacourtiengewächse *pl*
f flacourtiacées *pl*
r флакуртиевые *pl*

4468 **Indian poplar; charab poplar**

f peuplier *m* charab
r разнолистный тополь-турангу *m*; тополь *m* разнолистный

* **Indian rosewood → 2728**

4469 **Indian rubber tree; India rubber fig; rubber-plant**
(Ficus elastica)
d Mörderfeige *f*; Gummibaum *m*
f figuier *m* [indien] élastique; figuer des Indes
r фикус *m* [каучуконосный]; фикус эластичный

* **Indian spruce → 4310**

4470 **Indian spurgetree; hedge euphorbia; sehunda**
(Euphorbia neriifolia)
d Chinakaktus *m*
f euphorbe *f* à feuilles de saule
r молочай *m* олеандролистный

* **India rubber fig → 4469**

4471 **indicating stand; check stand; indicating wood; index forest; index stand**
d Weiserbestand *m*; Vergleichsbestand *m*
f peuplement *m* témoin
r указательное насаждение *n*; сравнительное насаждение

4472 **indicative forestry strategy; IFS**
(local planning authority strategy showing the strategic impacts of afforestation according to sensitivity of different areas)
d indikative Forstwirtschaft-Strategie *f*
f stratégie *f* de foresterie indicative
r индикативная лесохозяйственная стратегия *f*

4473 **indicator organism; biological indicator; bioindicator**
(organism used to establish the presence of an environmental change through its known reaction to specific environmental phenomena)
d Bioindikator *m*
f espèce *f* indicatrice; bio-indicateur *m*; indicateur *m* biologique
r организм-индикатор *m*; определитель *m* организмов

* **indicator plot → 7173**

4474 **indicator tree**
r указательное дерево *n*

4475 **indices for forest yield**
d Waldwachstumskennzahlen *fpl*
f indicateurs *mpl* de la productivité forestière
r показатели *mpl* производительности леса

4476 **indigenous forest; native forest; autochthonous forest**
(forest where the tree species have been present in a specified location for at least X number of years)
d bodenverbessernder Wald *m*; heimischer Wald; autochthoner Wald
f forêt *f* indigène; forêt autochtone
r приуроченный к месту лес *m*

4477 **indigenous species; native species; species native to the site**
(species historically native to an area, not introduced by people)
d standortsheimische Baumarten *fpl*
f essence *f* native de la station; essence locale; espèce *f* indigène
r местная порода *f*; отечественная порода

* **indigenous timber → 4341**

* **indigenous wood → 4341**

4478 **indigo**
(Indigofera)
d Indigostrauch *m*
f indigotier *m*
r индигофера *f*

4479 **indirect wood costs**
d Gemeinholzkosten *pl*
f frais *mpl* d'exploitation indirects
r косвенные затраты *fpl* на заготовку леса

* **individual floating → 7446**

* **individually owned forest → 3473**

* **individual stand → 7650**

4480 **individual stem; single stem; single tree; isolated tree; scattered tree; whipple-tree; whiffle-tree; swingle-tree**
d Einzelstamm *m*; Einzelbaum *m*; Freilandbaum *m*; Freibaum *m*
f arbre *m* individuel; arbre isolé; palonnier *m*; polonneau *m*
r одиночное дерево *n*; отдельное дерево

* **industrial forest → 1745**

4481 **industrial forester; industry forester**
(professional employed by a wood-using industry, usually a sawmill, who purchases timber from private woodland owners)
d Gewerbeforstbeamter *m*
f abatteur *m* industriel du sciage; scieur *m*
r лесовод-производственник *m*

4482 **industrial forestry**
(large-scale, commercial tree planting for timber and other wood products)
d industrielle Forstwirtschaft *f*; kommerzielle Forstwirtschaft *f*
f foresterie *f* industrielle
r промышленное лесное хозяйство *n*

4483 **industrial fuelwood**
(wood to be used specifically by industry for production of energy)
d Nutzbrennholz *n*
f bois *m* de chauffage industriel
r поделочные дрова *pl*

* **industrial harvesting → 1744**

4484 **industrial log depot; lower depot; lower landing; bottom timber landing; final landing; logging terminal**
d Hauptlagerplatz *m*
f gare *f* d'embarquement
r нижний лесной склад *m*

* **industrial lumber → 9721**

* **industrial timber → 9721**

* **industrial wood → 9721**

* **industry chip → 6625**

* **industry forester → 4481**

* **infection → 4486**

4485 **infeed deck; inbound deck**
d Freipolter *m*
f plateforme *f* d'alimentation
r эстакада *f* для подачи брёвен; питающая площадка *f*; эстакада для подачи заготовок (на линию строгания)

* **infeed throat → 1487**

* **inferior stand → 8237**

4486 **infestation; infection**
(attack and investment by macroscopic organisms in considerable concentration, e.g. of tree crowns by budworm; timber by termites; soil or other substrates by nematodes or weeds)
d Infektion *f*; Schädlingsbefall *m*
f infestation *f*
r заражение *n*

4487 **in-field tree processing**
d Ackerlandbaumverarbeitung *f*
f traitement *m* d'arbres à la forêt
r обработка *f* деревьев в лесу

4488 **infiltration; percolation**
(flow or movement of water through the soil surface into the ground)
d Infiltration *f*; Einsickern *n*
f infiltration *f*
r пропитывание *n*; впитывание *n*; просачивание *n*; инфильтрация *f*

* **infiltration ditch → 2613**

4489 **infiltration rate; infiltration capacity**
(maximum rate at which the soil under various specified conditions can absorb falling rain or melting snow)
d Tränkreife *f*
f vitesse *f* d'infiltration
r скорость *f* фильтрации; инфильтрационная способность *f*

4490 **inga**
(Inga)
d Inga *n*
f inga *m*
r инга *f*

4491 **ingrown bark; inbark; trapped bark**
(a patch of bark, sometimes associated with a knot, that has become more or less enclosed by the growth of the tree)
d eingewachsene Rinde *f*; eingeschlossene Rinde; Rindeneinschluss *m*
f écorce *f* incluse; entre-écorce *f*; entrécorce *f*
r врастающая кора *f*; вросшая кора; тёмная пророcть *f*

4492 **ingrowth of trees; recruitment of trees**
(net volume (or number) of trees that grow large enough during a specified year to

f recrutement *m*; passage *m* à la futaie
r прирост *m* количества деревьев, достигших диаметра рубки за определённый период; прирост насаждения за счёт естественного возобновления леса

* **inhaul line → 3987**

4493 **initial afforestation**
d Erstaufforstung *f*
f afforestation *f* initiale
r первоначальное облесение *n*

* **initial attack → 4495**

4494 **initial decay; beginning decay; early decay; primary decay; marginal rot; invasive decay**
(the early stage of decay that has not proceeded far enough to soften or otherwise apparently impair the hardness of the wood)
d beginnende Fäulnis *f*; beginnender Anbruch *m*; Zerfall *m*; Zersetzung *f*
f pourriture *f* au stade initial; pourriture marginale; échauffure *f*; carie *f* marginale
r начальная стадия *f* гниения; первая стадия гниения; начинающее гниение *n*; начинающаяся прелость *f*

4495 **initial hack; first attack; initial attack**
(action taken to halt the spread or potential spread of a fire by the first fire fighting force to arrive at the fire)
d Angriffshieb *m*
f encoche *f* initiale
r первоначальный заруб *m*

4496 **initial mature inventory**
(that portion of the existing total mature forest inventory which is available for harvest)
d Anfangsreifebericht *m*
f inventaire *m* de maturité initial
r начальный перечёт *m* зрелости

* **initial parenchyma → 2704**

4497 **initial survey**
(of a forest)
d anfängliche Vermessung *f*
f arpentage *m* initial
r первичная съёмка *f*

4498 **injunction against clearing**
d Rodungsverbot *n*
f interdiction *f* de défrichement
r запрет *m* расчистки

d Forstschadinsekt *n*
f insecte *m* forestier nuisible
r лесное вредное насекомое *n*

* **injurious organism → 5807**

* **injury caused by peeling → 2237**

* **injury from hails → 3930**

* **inkberry → 3790**

* **ink-bush → 3287**

4500 **ink disease [of chestnut]**
(Phytophthora cambivora; Phytophthora cinnamoni)
d Tintenkrankheit *f* [der Kastanie]
f maladie *f* de l'encre [de marronnier]
r чернильная болезнь *f* [каштана европейского]

* **inlaid woodwork → 9590**

* **inland ceanothus → 6788**

4501 **inland forest**
d Wald *m* des Inlandes
f forêt *f* intérieure
r внутренний лес *m*

* **inlayings → 9590**

4502 **inner bark; bast; liber; sieve tissue; [secondary] phloem; inner phloem; secondary cortex; internal phloem; perimedullary phloem; intraxylary phloem**
(primary phloem internal to the primary xylem; normally, the part of the bark formed by the cambium)
d Innenrinde *f*; Rindenbast *m*; Basthaut *m*; Faserbast *m*; Phloem *n*; Bast *m*
f liber *m*; phloème *m* secondaire; phloème interne; phloème périmédullaire; filasse *f*
r внутренняя кора *f*; [внутренняя] флоэма *f*; луб *m*; лыко *n*; мочало *n*

* **inner blue → 4557**

* **inner boom → 5246**

4503 **inner gorge**
(stream reach or portion of stream that is bounded by steep hillslopes (> 40% sideslope) and terminates upslope into more

f aiguille *f* interne
r внутренняя быстрина *f*; внутренний порог *m*

* **inner phloem → 4502**

4504 **inner width; inside width; width in the clear; clear span**
d freie Weite *f*
f ouverture *f* intérieure [des fibres de bois]
r открытая даль *f*

* **inner wood → 4202**

4505 **innovation shoot; vegetative shoot; wood shoot**
(a 4 to 10inches shoot which bears only wood buds)
d vegetativer Spross *m*; Verjüngungstrieb *m*; Erneuerungsspross *m*
f pousse *f* végétative; rameau *m* à bois
r вегетативный побег *m*; обновляющий побег; подверхушечный побег

* **inoculant → 4507**

* **inoculate → 3906**

4506 **inoculation**
(deliberate introduction of material containing microorganisms into soils, nursery composts or living plants)
d Inokulation *f*; Impfung *f*; Vakzination *f*
f inoculation *f*; vaccination *f*
r инокуляция *f*; посев *m*

4507 **inoculum; inoculant; vaccine**
(culture of microorganism applied to seed to ensure symbiotic infection of the plant by the organism)
d Impfstoff *m*; Vakzin *n*; Vakzine *f*
f inoculum *m*; vaccin *m*
r прививочный материал *m*; инокулят *m*; вакцина *f*; вакцинальный биопрепарат *m*

4508 **inoperable lands**
(lands that are unsuited for timber production)
d unoperabele Boden *mpl*
f terres *fpl* inopérables
r неоперабельные земли *fpl*

4509 **insect attractant; attractant; bait**
(any substance, esp. food, used in catching birds, fish or other animals, by alluring them to a hook, snare, inclosure, or net)
d Lockmittel *n*; Lockstoff *m*; Ködergift *n*; attractif *m*; esche *f*; strouille *f*
r средство *n* приманки; приманка *f*

* **insect bands → 8777**

* **insect damage → 2234**

4510 **insect-damaged wood**
d insektenbefallenes Holz *n*
f bois *m* à dégâts d'insectes
r древесина *f*, поражённая насекомыми

* **insect ditch → 4515**

4511 **insect hole; shot hole; worm hole; worm groove; beetle damage; insect passage**
(small holes in the wood caused by insects boring; passage carved out under bark or in wood by insects feeding or laying eggs)
d Hakengang *m*; Hohlgang *m* im Holz; Fraßgang *m*; Brutgang *m*; Holzfraß *m*; Wurmfraß *m*; Wurmloch *n*; Käferfraß *m*
f galerie *f* d'insecte; piqûre *f*; galerie larvaire; trou *m* de ver
r ход *m* древоточца; ход насекомого в древесине; червоточина *f*; туннель *m*

4512 **insecticidal preservative; insecticide**
(agent or chemicals, typically used to destroy, prevent, or control insects and similar animals)
d Insekten[bekämpfungs]mittel *m*; Insektenvertilgungsmittel *m*; Insektizid *n*
f insecticide *m*
r средство *n* для уничтожения насекомых-вредителей; инсектицид *m*

* **insectifuge → 4514**

* **insect injury → 2234**

4513 **insectivores; insectivorous animals**
(animals that only eat insects)
d Insektenfresser *mpl*
f insectivores *mpl*
r насекомоядные *npl*; насекомопоедающие *npl*

* **insect passage → 4511**

4514 **insect[-]repellent; insectifuge**
d Insektenschutzmittel *n*
f insectifuge *m*
r отталкивающий *m* насекомых

4515 **insect trench; insect ditch; trap ditch; trap**

f fosse-piège *m*; fosse *f* à coléoptères
r ловчая канава *f*; ловчая яма *f*

* **insect wax → 8958**

4516 **inserted-tooth saw; insert-tooth saw; inserted-tooth sawblade**
d Sägeblatt *n* mit eingeschobenen Zähnen; Kreissägeblatt *n* mit eingeschobenen Zähnen
f lame *f* de scie circulaire à dents amovibles; lame de scie circulaire à dents rapportés
r пила *f* со вставными зубьями

* **inside face → 8602**

* **inside width → 4504**

* **insignis pine → 5548**

4517 **in situ conservation**
(the maintenance of the diversity present in and among populations of the many species used directly in agriculture, or used as sources of genes, in habitats where such diversity arose and continues to grow)
d in situ Erhaltung *f*
f conservation *f* in situ
r консервирование *n* на месте; консервирование в месте нахождения

4518 **inslope**
d Neigung *f* nach der Innenseite
f dévers *m* vers l'intérieur
r наклонение *n* во внутреннюю сторону

4519 **institution forest**
d Stiftungswald *m*; Stift[ungs]forst *m*
f forêt *f* d'établissement public; forêt de fondation
r лес *m*, завещанный учреждению

* **instruction forest → 8761**

* **instruction in forestry → 6565**

* **insufficient stocking → 9068**

* **insulating board → 4521**

4520 **insulating fiberboard; insulation fiberboard**
d Holzfaserdämmplatte *f*; Holzfaserisolierplatte *f*
f panneau *m* de bois d'isolation
r изоляционная древесно-волокнистая плита *f*

4521 **insulation board; insulating board; insulation plate; insulating plate; dam[ming] board**
(a board made from lignocellulosic fibers, usually wood or cane which are interfelted to create the principal source of bond)
d Dämmplatte *f*; Weich[faser]platte *f*
f panneau *m* [de fibres] isolant; plaque *f* isolante; panneau tendre
r изоляционный картон *m*; изоляционная плита *f*

* **insulation fiberboard → 4520**

* **intake chute → 3033**

4522 **intaken measure**
(of timber)
d Einlassmaß *n*
f mesure *f* de chargement
r погрузочный обмер *m*

* **intarsia → 9590**

4523 **integral forest yield; total forest yield**
d Integralwaldernte *f*
f rendement *m* forestier total
r полный выход *m* леса

4524 **integrated [forest] utilization**
(harvesting that converts every merchantable past of each tree according to its best end-use)
d integrierte Forstnutzung *f*
f exploitation *f* et conversion intégrées
r комплексное использование *n* заготовленного леса; интегральное пользование леса

4525 **integrated logging**
(logging operation that segregates and delivers a variety of products to mills and processors that will use them to the best advantage)
d integrierte Holzfällung *f*
f exploitation *f* intégrée
r интегрированная лесозаготовка *f*

* **integrated resource management → 5622**

4526 **integrated resource use**
(decision making process whereby all resources are identified, assessed and compared before land use or resource management decisions are made)
d integrierte Ressourcennutzung *f*

* **intensity of cut → 4210**

4527 **intensity of thinning; thinning intensity**
(measure of the combined effect of thinning weight and thinning frequency)
d Durchforstungsintensität *f*
f intensité *f* globale moyenne des éclaircies; intensité d'éclaircie
r интенсивность *f* рубок ухода

4528 **intensive forest management**
(form of forest or woodlot management in which a number of resources or roles (e.g. hydrological functions), are obtained or fulfilled at the same time over the whole area)
d intensive Forstverwaltung *f*
f aménagement *m* forestier intensif
r интенсивное ведение *n* лесного хозяйства

4529 **intensive forestry**
d intensive Forstwirtschaft *f*
f foresterie *f* intensive
r интенсивное лесное хозяйство *n*

* **intensive silviculture → 4450**

* **interbreeding → 4435**

4530 **intercellular canal; intercellular duct**
(a tubular intercellular space of indeterminate length, generally serving as a repository for resin, gum, etc.)
d Interzellulargang *m*; Interzellularraum *m*; Zwischenzellengang *m*
f canal *m* intercellulaire; canal sécréteur; méat *m* intercellulaire
r межклеточный канал *m*

4531 **intercellular cavity**
d interzelluläre Höhle *f*
f cavité *f* intercellulaire; cavité sécrétrice; poche *f* sécrétrice
r межклеточная полость *f*

* **intercellular duct → 4530**

4532 **intercellular layer; true middle lamella**
(the layer of isotropic substance between cells; this layer is largely lignin and lacks cellulose)
d interzelluläre Schicht *f*
f couche *f* intercellulaire; lamelle *f* mitoyenne; lamelle moyenne
r межклеточный слой *m*

4533 **intercellular space**
(a space between cells generally at the
d interzellulärer Raum *m*
f espace *m* intercellulaire; méat *m*
r межклеточное пространство *n*

* **interception drain → 2173**

4534 **interception loss**
(that part of the precipitation that, in the course of contact with the vegetation, evaporates and so never reaches the ground)
d Niederschlagverlust *m*
f perte *f* d'interception
r потеря *f* воды на испарение осадков, задержанных растениями

4535 **interception of precipitation**
d Beanstandung *f* der Niederschlag
f interception *f* de précipitation
r задержание *m* осадков; перехват *m* осадков

* **intercrop[ping] → 4543**

* **interculture → 4543**

* **interface → 4568**

* **intergrafting → 8691**

* **intergrown → 67**

* **intergrown knot → 833**

4536 **intergrown stem wood**
d verwachsenes Holz *n*
f bois *m* de tiges adhérentes
r древесина *f* сросшихся стволов

4537 **interim forest; transition[al] forest**
(a forest that exists or will exist until conversion to a target forest is complete)
d vorläufiger Wald *m*; Übergangswald *m*
f forêt *f* transitoire
r переходный лес *m*

4538 **interior sapstain; secret stain**
(stain of sapwood boards and dimension stock developed interiorly during seasoning, without surface evidence of its presence)
d innerer Saftfleck *m*
f tache *f* interne de l'aubier
r внутренняя заболонная окраска *f*; подслойная заболонная окраска

4539 **interlocking grain; interlocked grain; alternating spiral grain; interwoven grain; ribbon grain; ribbon figure; ribbon stripe**
(a grain where the angle of the fibers

d einander kreuzende Faser *f*; Überkreuzfaser *f*
f fil *m* entrecroisé; grain *m* rubané
r сблокированное спиральное волокно *n*

4540 **interlocking skidder; interlocking yarder**
(a device that allows the main and haul-back drums to be operated together as a single unit to maintain running line tension)
d Verriegelungsschlepper *m*
f débusqueur *m* à tambours verrouillé
r канатная трелёвочная установка *f* с лебёдкой со сблокированными барабанами

* **interlocking yarder → 4540**

4541 **intermediate branch**
d Zwischenast
f branche *f* intermédiaire
r срединная ветвь *f*

* **intermediate crop → 4547**

4542 **intermediate crown class; subdominant crown class**
(trees with crowns that extend into the canopy with dominant and codominate trees)
d Zwischenkronenklasse *f*
f classe *f* de cime intermédiaire; classe de cime subdominante
r класс *m* промежуточных крон

4543 **intermediate culture; interculture; intercrop[ping]**
(a form of multiple cropping where perennial (usually tree) crops are grown in association with other species to form a multilayered system)
d Zwischenbau *m*; Zwischenkultur *f*; Zwischenfrucht *f*
f culture *f* intermédiaire; culture intercalaire
r промежуточная культура *f*

4544 **intermediate cutting; intermediate felling; intermediate harvesting; intermediate thinning**
(any removal of trees from a stand between the time of its formation and the harvest cutting)
d Zwischenhieb *m*; Vornutzungshieb *m*; Erziehungshieb *m*
f coupe *f* intermédiaire
r воспитательная рубка *f*; предварительная рубка

* **intermediate decay → 72**

d mitteldichte Faserplatte *f*
f panneau *m* de fibres de densité moyenne
r древесноволокнистая плита *f* средней плотности

* **intermediate felling → 4544**

* **intermediate fiber → 3783**

* **intermediate harvesting → 4544**

* **intermediate holding site → 1721**

4546 **intermediate income; intermediate yield**
d Vornutzungsertrag *m*; Zwischennutzungsertrag *m*; Vorertrag *m*
f exploitation *f* intermédiaire
r предварительный доход *m*

4547 **intermediate stand; intermediate crop; alternate forest crop**
d Zwischenwald *m*
f peuplement *m* intermédiaire; forêt *f* intermédiaire
r промежуточное насаждение *n*

4548 **intermediate support [spar]; midspan support; cross support; jack line**
(spar tree or cable sling located between the head spar and tail spar to which a tree jack is attached to support a multispan skyline)
d Zwischenunterstützung *f*
f porte-grume *f* intermédiaire; support *m* intermédiaire
r промежуточная поддерживающая перекладина *f*; промежуточный поддерживающий канат *m*; опора *f*, устанавливаемая посередине пролёта; средняя опора

* **intermediate thinning → 4544**

4549 **intermediate tolerance**
(characteristics of certain tree species that allows them to survive, though not necessarily thrive, in relatively low light conditions)
d Zwischentoleranz *f*
f tolérance *f* intermédiaire
r промежуточная толерантность *f*

4550 **intermediate treatment of stand**
(any treatment in a stand during that portion of the rotation not included in the final harvest or regeneration period)
d sekundäre Klärung *f*

4551 intermediate tree; tree between overstor[e]y and understor[e]y
(which has crowns below, but still extending into, the general level of the canopy and receive a little direct light from above but none from the sides)
d zwischenständiger Baum *m*
f arbre *m* intermédiaire; arbre entre étage dominant et sous-étage
r промежуточное дерево *n*; дерево между верхним и нижним ярусами

4552 intermediate wood
(inner layers of the sapwood that are transitional between sapwood and heartwood in colo[u]r and general character)
d Übergangsholz *n*
f bois *m* intermédiaire
r промежуточная древесина *f* (внутренний слой заболони)

* **intermediate wood piling → 1720**

* **intermediate wood stacking → 1720**

* **intermediate yield → 4546**

4553 intermittent growth
(non-continuous growth without the predictable regularity of rhythmic growth)
d periodischer Wachstum *n*
f croissance *f* intermittente
r скачкообразный рост *m*; прерывистый рост

4554 intermittent planter
d periodischer Pflanzer *m*
f planteur *m* intermittent
r сезонный рабочий-сажальщик *m*

4555 intermittent stream; ephemeral stream
(a watercourse that flows in a well-defined channel only in direct response to a precipitation event. It is dry for a large part of the year)
d periodisches Gewässer *n*; zeitweilig aussetzender Wasserlauf *m*
f cours *m* d'eau temporaire; cours d'eau intermittent; cours d'eau saisonnier
r прерывающийся поток *m*; периодический поток

4556 intermittent [wood] yield
d intermittierender Schnittholzertrag *m*
f rendement *m* intermittent
r выход *m* древесины от нерегулярных рубок

d Innenbläue *f*
f bleuissement *m* interne
r внутренняя синева *f*

4558 internal forest transport
d Nahbeförderung *f*; Nahbewegung *f*; Nahtransport *m*
f transport *m* forestier interne
r внутрилесосечный транспорт *m*; [под]трелёвка *f* на короткие расстояния

* **internal phloem → 4502**

* **internal sapwood → 853**

4559 international log rule
(formula rule that allows a 1/2-inch taper for each 4 feet of length and allows for a 1/16-inch shrinkage for each inch of board thickness)
d internationale Kubiktabelle *f*
f table *f* de cubage internationale
r международная таблица *f* определения объёмов

4560 internode
(a region of the stem between two successive nodes)
d Interknoten *m*
f entre-nœud *m*
r междуузлие *n*

4561 interplant *v*
(to set young trees among existing forest growth of similar age and size, planted or natural, to bring the stand to a fully stocked condition)
d ausbessern; nachbessern; nachpflanzen
f planter entre lignes
r подсеять; посеять в междурядье

4562 interpolated cropping
(cropping sequence where two or more species are grown on the same unit of land but at least one species has a later sowing (planting) time and an earlier harvest than the other)
d einschaltene Kultur *f*
f culture *f* interpolée
r вставленная порода *f*

4563 interpretive forest site
(designated forest site and ancillary facilities to interpret, demonstrate, or facilitate the discussion of the natural environment, forest practices, and integrated resource

f site *m* forestier interprétatif
r интерпретативное лесное местообитание *n*

4564 **interruptedly pinnate leaf**
d gefiedertes Blatt *n* mit Unterbrechungen
f feuille *f* pinnée à interruptions
r прерывисто-перистый лист *m*

4565 **interruption of canopy**
d Schirmbrechung *f*
f interruption *f* de couvert
r неслитность *f* полога; изреженность *f* полога

4566 **interruption of the leaf canopy; loosening of close stand; opening of canopy; opening out of the crop; canopy opening; opening up [of a stand]**
d Schlusslockerung *f*; Lockerung *f* des Schlusses; Lichtstellung *f*
f ouverture *f* du couvert; éclaircie *f* de couvert
r изреживание *n* полога [насаждения]; изреживание сомкнутости

4567 **interseeding; intersowing; sod seeding**
(seeding into an established vegetative cover)
d Nachsaat *f*; Einsaat *f*
f ensemencement *m* sous abri; semis *m* sous abri; sursemis *m*
r посев *m* в междурядьях; посев по дернине

4568 **interspace; interface; life bar**
(the strip of bark left between two adjacent faces)
d Lebendstreifen *m*
f entrecorre *f*
r межкарровый ремень *m* (при подсочке)

* **interstitial water → 6464**

4569 **intertree distance; tree spacing; spacing between stems**
(distance between trees, which is most often regulated at the time of planting or during a harvest or thinning operation)
d Abstand *m* der Bäume; Baumabstand *m*; Stammabstand *m*
f distance *f* entre [deux] arbres
r расстояние *n* между деревьями; расстояние между стволами; размещение *n* деревьев

4570 **intertree influence**
d Beeinflussung *f* der Bäume
f influence *f* entre arbres
r взаимное влияние *n* деревьев

4571 **intervascular pitting; intervessel pitting**
(the pitting between vessel members)
d Lochfraß *m*
f ponctuation *f* intervasculaire
r межсосудистая поровость *f*

* **interwoven grain → 4539**

* **in the fiber direction → 209**

* **in the grain direction → 209**

4572 **intimacy**
(in intercropping, the degree to which different plant component species are spaced so as to be close to one another)
d Vertrautheit *f*; Innigkeit *f*
f intimité *f*
r тесная связь *f*

4573 **intolerance of shade**
(a characteristic of certain tree species that does not permit them to survive in the shade of other trees)
d Lichtbedarf *m*; Lichtbedürfnis *n*
f intolérance *f* à l'ombre; besoin *m* de lumière
r потребность *f* в свете

* **intolerant [tree] → 7476**

4574 **intolerant tree species**
(1. plant species that do not grow well in shade; 2. plant and animal species that are unable to endure significant habitat or climate change)
d Lichtbaumart *f*
f espèce *f* forestière intolérante; essence *f* de lumière
r светолюбивый древесный вид *m*

* **intraxylary phloem → 4502**

4575 **introduced pine sawfly**
(Diprion similis)
d eingebürgerte Kiefernblattwespe *f*
f diprion *m* importé du pin
r пилильщик *m* сходный сосновый

4576 **introduced species; non-indigenous species**
(nonnative species that was intentionally or unintentionally brought into an area by humans)
d eingebürgerte Arten *fpl*
f espèces *fpl* importées; essences *fpl* introduites
r вводимые породы *fpl*; интродуцированные породы

f introduction *f* d'espèces
r внесение *n* пород

* **inundation → 3282**

4578 invader
(plant species that were absent in undisturbed portions of the original vegetation and will invade under disturbance or continued overuse)
d Schädling *m*
f nuisible *m*
r вредитель *m*

* **invasive decay → 4494**

4579 inventory
(gathering of data and processing it on the quantity of physical entities (such as trees, lakes, etc.) in an area)
d Inventarisierung *f*; Inventur *f*; Zustandserfassung *f*; Durchzählung *f*
f inventaire *m*
r инвентаризация *f*; перечёт *m*; учёт *m*

* **inventory level → 9250**

* **inventory of stand → 3478**

* **inventory of volume → 5266**

4580 inverted skyline
d Rückluftseilbahn *f*
f câble *m* inverti
r обратный несущий канат *m*

4581 ipe; lapacho
(Tabebuia ipe; Tabebuia serratifolia)
d Ipebaum *m*
f ipé *m*
r табебуйя *f* ипе; ипе *n*; зелёное эбеновое дерево *n*

* **Irish yew → 1814**

4582 iroko; kambala; mvule
(Chlorophora excelsa)
d Kambala *n*; Iroko *n*
f kambala *m*; iroko *m*
r камбала *f*; ироко *n*; хлорофора *f* высокая; жёлтое дерево *n*

4583 ironbark; silvertop [ash]; coast ash
(Eucalyptus sieberii; Eucalyptus longifolia; Eucalyptus maideni)
d Eisenrindenbaum *m*
твёрдой корой; эвкалипт Майдена

4584 iron dip; raker
d Harzmesser *m*; Abziehmesser *n*; Harzspachtel *m/f*
f racloir *m* de résine; palette *f* à résine; grattoir *m*; spatule *f*
r барраскит *m*; скребок *m*; лопатка *f* для сбора живицы из приёмника; скобель *m* для смолы

* **ironwood → 6186**

* **irregular forest worker → 6095**

4585 irregular grain; oblique grain; sloping grain; angle grain; cross grain
(1. in standing trees, grain in which the fiber alignment deviates from the axis of the stem; 2. in wooden members, grain in which the fiber alignment deviates from the direction parallel to the long axis of the piece)
d Schrägfaser *f*
f fil *m* oblique; fil tranché; fibre *f* tranchée; grain *m* incliné
r волокно *n*, направленное под углом к действию силы; косое направленное волокно

4586 irregularly stocked; partially stocked *adj*
(ref. to productive forest land stocked with trees of merchantable species non-capable to utilize the complete potential of the land for growth)
d unregelmäßig bestockt
f de densité relative; de densité partielle; partiellement boisé
r неравномерно облесённый

4587 irregular shelterwood system
d irreguläres Femelschlagsystem *n*
f système *m* des coupes progressives irrégulières
r система *f* неравномерных семеннолесосечных рубок

* **irregular stand → 9081**

4588 irregular stocking
d irreguläre Bestockung *f*
f matériel *m* relatif irrégulier
r расстроенное насаждение *n*; неравномерное насаждение

* **irregular texture → 1168**

f structure *f* inéquienne irrégulière
r неравномерная разновозрастная структура *f*

* **Isle of Wight disease → 26**

4590 **isogonic growth**
d winkeltreuer Wuchs *m*
f accroissement *m* isogone
r равномерный рост *m*

* **isolated stand → 7650**

* **isolated tree → 4480**

4591 **isolating of forest**
d Waldisolation *f*
f isolement *m* de la forêt
r прочистка *f* леса

* **isolation of the crown → 3698**

* **isolation of trees → 5915**

* **isolation strip → 1149**

* **isphingo → 1417**

4592 **Italian alder; eis**
(Alnus cordata; Alnus cordifolia)
d italienische Erle *f*; klebrige Erle; herzblättrige Erle; neapolitanische Erle
f au[l]ne *m* à feuilles en cœur; grand au[l]ne; au[l]ne d'Italie
r ольха *f* сердцевидная

4593 **Italian clematis; blue clematis; single Virgin's clematis**
(Clematis viticella)
d blaue Waldrebe *f*; italiensche Waldrebe
f clématite *f* liseron; clématite petite vigne; clématite bleue
r ломонос *m* фиолетовый

* **Italian cypress → 1770**

4594 **Italian maple**
(Acer opalus)
d italienischer Ahorn *m*
f érable *m* [aux feuilles] d'obier
r итальянский клён *m*

4595 **Italian stone pine; stone pine; umbrella pine; parasol pine**
(Pinus pinea; Sciadopitys verticillata; Taxus verticillata)
pierre; pin parasol; pin à ombrelles
r пиния *f*; сосна *f* итальянская

* **itea → 872**

* **Ivory Coast mahogany → 107**

4596 **Ivory Coast terminalia; framire**
(Terminalia ivorensis)
d Framire *m*
f framiré *m*
r терминалия *f* иворенская

4597 **Ivory palm**
(Phytelephas macrocarpa)
d Elfenbeinpalme *f*; Steinnusspalme *f*
f palme *f* d'ivoire
r фителефас *m*

4598 **ivy**
(Hedera helix)
d Efeu *m*
f lierre *f*
r плющ *m* обыкновенный

4599 **ivy-leaved maple**
(Acer cissifolium)
d cissusblättriger Ahorn *m*
f érable *m* cymbalaire; érable à feuilles de frêne
r клён *m* [американский] виноградnолистный

* **ixora → 8715**

J

* **jacaranda → 1006**

* **jacaranda rosa → 1007**

* **jack[-fruit] tree → 1011**

4600 **jack-ladder; jack-slip; log jack; log-haul[-up]; gangway; log lift**
(an inclined plane up which logs are drawn, transversely or end-to-end, to the mill deck by means of an endless cable or bull chain)
d Blockförderer *m*; Blochgleitbahn *f*; Blochförderbahn *f*; Blockaufzug *m*; Jakobsleiter *m*
f convoyeur-élévateur *m*
r ковшовый элеватор *m*; нория *f*

* **jack line → 4548**

* **jack oak → 5791**

4601 **jack pine; Banksian pine; black pine; unlucky tree**
(Pinus banksiana)
d Bankskiefer *f*
f pin *m* gris [d'Amérique]; pin de Banks; pin de Hudson
r сосна *f* Банкса

4602 **jack-pine budworm**
(Choristoneura pinus pinus)
d Strauchkiefer-Knospenwickler *m*
f tordeuse *f* du pin gris
r листовёртка-почкоед *f* сосны Банкса

4603 **jack[-]pot**
(an unskilled piece of work in logging, particularly in felling, where several trees are lodged and/or crisscrossed)
d Jackpot *m*; Einsatz *m*
f gros lot *m*
r завал *m* из спиленных зависших деревьев; неправильно выполненная валка *f*

* **jack-slip → 4600**

4604 **Jaeger Joshua tree; Joshua tree; Joshua palm**
(Yucca brevifolia)
cierge *m* de l'Arizona
r юкка *f* коротколистная

4605 **jagua; caruto**
(Genipa)
d Genipa-Baum *m*
f génipa *m*
r генипа *f*

* **jajoba → 4662**

4606 **Jamaica dogwood**
(Erythrina piscipula; Piscidia erythrina)
d gemeiner Fischtänger *m*; korallenbaumartiger Fischtänger; Piscidia *n*
f bois *m* des chiens; bois iveant
r писцидия *f* ярко-красная

4607 **Jamaica ebony; cocuswood; brown ebony**
(Brya buxifolia; Brya ebenus)
d Jamaica-Kokus *m*; Jamaica-Cocus *m*; westindisches Kokus; westindischer Ebenholzbaum *m*
f cocus *m*; brya *m* faux-ébène
r брия *f* эбеновая

* **Jamaican mahagoni → 2110**

* **Jamaica plum → 4322**

* **jamb → 6475**

* **jambolan → 4646**

* **jammer → 8098**

4608 **jammer logging**
(cable logging system generally restricted to one skidding line and used for winching logs up to 300 feet from the cutting area to a log collection point)
d Rückung *f* mittels mobiler Holzlader
f débardage *m* par chargeur mobile
r трелёвка *f* мобильной погрузочной установкой

* **jam[ming] → 5096**

* **janker → 894**

4609 **Japanese alder**
(Alnus japonica)
d japanische Erle *f*
f au[l]ne *m* du Japon
r ольха *f* японская

d japanische Aralie *f*; Teufelsspazierstock *m*
f angélique *f* [en arbre] du Japon
r аралия *f* высокая

4611 Japanese anise-tree; staranise tree
(Illicium anisatum; Illicium verum)
d japanischer Sternanis *m*; Sternanispflanze *f*
f anis *m* étoilé; badianier *m* [des pagodes]; badanier du Japon; badianier sacré
r дерево *n* анисовое; иллициум *m* анисовый; иллициум священный; бадьян *m* настоящий; иллициум настоящий; сикими *n*

* **Japanese aralia → 4610**

4612 Japanese barberry
(Berberis thunbergii)
d Thunberg-Berberitze *f*
f épine-vinette *f* de Thunberg; épine-vinette du Japon; vinettier *m* de Thunberg
r барбарис *m* Тунберга

4613 Japanese beech
(Fagus crenata)
d japanische Kerbbuche *f*
f hêtre *m* à feuilles crénelées
r бук *m* с осенней окраской

4614 Japanese black pine; Thunberg pine
(Pinus thunbergii)
d japanische Schwarzkiefer *f*; Thunberg-Kiefer *f*
f pin *m* noir de Japon
r сосна *f* Тунберга

4615 Japanese cedar; sugi; common cryptomeria; cryptomeria
(Cryptomeria japonica)
d japanische Zeder *f*; japanische Zypresse *f*; Sugi *n*; japanische Sicheltanne *f*
f cryptomérie *f* du Japon; sugi *m*
r криптомерия *f* японская; японский кедр *m*

4616 Japanese chestnut
(Castanea crenata; Castanea japonica)
d japanische Kastanie *f*
f châtaignier *m* du Japon
r каштан *m* городчатый; каштан японский

4617 Japanese corktree
(Phellodendron japonicum)
d japanischer Korkbaum *m*
f philodendron *m* de Japon
r бархат *m* японский

sibirica)
d sibirische Ulme *f*
f orme *m* nain [de Sibérie]
r вяз *m* приземистый; вяз японский; вяз долинный; вяз сродный; ильмовник *m*

4619 Japanese [flowering] crab [apple]
(Malus floribunda)
d japanischer Apfel *m*; reichblütiger Zierapfel *m*
f pommier *m* japonais; pommier floribond
r яблоня *f* многоцветковая

4620 Japanese holly
(Ilex crenata)
d japanischer Zwergilex *m*; Japan-Stechpalme *f*
f houx *m* japonais
r падуб *m* городчатый

4621 Japanese hornbeam; loose-flower hornbeam
(Carpinus laxiflora; Carpinus japonica)
d rotblättrige Hainbuche *f*; japanische Hainbuche
f charme *m* du Japon
r граб *m* японский

4622 Japanese horse[ ]chestnut
(Aesculus turbinata)
d japanische Rosskastanie *f*
f marronnier *m* du Japon; marronnier de forme conique
r каштан *m* конский кубарчатый; каштан конский японский

* **Japanese lacquer tree → 4755**

4623 Japanese larch; karamatsu larch
(Larix japonica; Larix leptolepis; Larix kaempferi)
d japanische Lärche *f*
f mélèze *m* du Japon
r лиственница *f* тонкочешуйчатая; лиственница японская

* **Japanese maple → 4373**

* **Japanese mountain ash → 4742**

4624 Japanese oak; karafuto oak
(Lithocarpus glabra; Quercus grosseserrata; Quercus mongolica)
d japanische Eiche *f*; mongolische Eiche
f chêne *m* du Japon
r дуб *m* японский

4625 Japanese overgreen oak
(Quercus acuta)

f chêne *m* vert du Japon
r дуб *m* остролистный; дуб острый; агагашу *m*

4626 Japanese pagoda tree
(Sophora japonica)
d Trauerschnurbaum *m*; japanischer Pagodenbaum *m*; japanischer Perlschnurbaum *m*; japanische Sophore *f*; Tempelbaum *m*
f sophora *m* du Japon; pagode *f* japonaise
r софора *f* японская

4627 Japanese persimmon; [Asian] kaki
(Diospyros kaki)
d Kakipflaumenbaum *m*; Kakibaum *m*
f plaqueminier *m* kaki; kaki *m*; plaqueminier du Japon
r хурма *f* японская; хурма восточная

4628 Japanese persimmon fruit; kaki fruit
d Kakipflaume *f*; Kakifeige *f*
f figue *f* kaki; kaki *m*
r хурма *f* [плод]

4629 Japanese poplar
(Populus maximowiczii)
d Maximowiczspappel *f*
f peuplier *m* de Maximowicz
r тополь *m* Максимовича

4630 Japanese prickly-ash
(Zanthoxylum piperitum)
d japanischer Pfeffer *m*; Szechuanpfeffer *m*
f clavalier *m*
r японский перец *m*

4631 Japanese raspberry
(Rubus parvifolius)
d schwarze Zimthimbeere *f*
f framboise *f* noire
r малина *f* лесная; малина мелколистная

4632 Japanese red pine
(Pinus densiflorus)
d japanische Rotkiefer *f*
f pin *m* rouge du Japon
r сосна *f* густоцветная

4633 Japanese snowbell
(Viburnum plicatum)
d japanischer Schneeball *m*
f viorne *f* à feuilles plissées; viorne *f* japonaise
r калина *f* плекатум

d ostasiatische Zwergkiefer *f*
f pin *m* nain de Sibérie
r кедровый стланик *m*; сосна *f* стланиковая

4635 Japanese supplejack
(Berchemia racemosa; Berchemia magna)
d japanische Berchemie *f*
f berchénie *f* japonaise
r берхемия *f* кистистая

* **Japanese varnish-tree → 1303**

* **Japanese walnut → 7586**

4636 Japanese white pine
(Pinus parviflora)
d Mädchenkiefer *f*
f pin *m* blanc du Japon
r сосна *f* белая японская; сосна мелкоцветная

* **Japanese wineberry → 9452**

4637 Japanese wingnut; sawagurumi
(Pterocarya rhoifolia)
d japanische Flügelnuss *f*
f ptérocarier *m* du Japon
r птерокария *f* японская; лапина *f* розоволистная; савагуруми *m*

4638 Japanese wisteria
(Wisteria floribunda)
d japanische Glyzinie *f*
f glycine *m* du Japon
r вистерия *f* обильноцветущая

4639 Japanese zelkova
(Zelkova serrata)
d japanische Zelkove *f*; japanische Ulme *f*; Keakibaum *m*
f zelkova *m* du Japon; planère *f* du Japon
r дзельква *f* японская

4640 Japan quince
(Chaenomeles japonica)
d japanische Zierquitte *f*; japanische Scheinquitte *f*
f troène *m* de Japon
r айва *f* японская; хеномелес *f* японская

4641 Japan spindletree
(Euonymus japonicus)
d japanischer Spindelstrauch *m*
f fusain *m* de Japon
r бересклет *m* японский

4642 **jarrah eucalyptus**
(Eucalyptus marginata)
d Jarrah *m*
f jarrah *m*
r ярра *f*; эвкалипт *m* западноавстралийский; эвкалипт окаменелый

4643 **jarul; pyinma**
(Lagerstroemia)
d Kreppmyrte *f*; Lagerströmie *f*
f jarul *m*; banglang *m*
r лагерстремия *f*

* **jasmine → 1782**

4644 **jasmine orange**
(Murraya)
d Orangenjasmin *m*
f oranger *m* jasmin; murraya *m*
r муррея *f*; муррая *f*; кемунинг *m*

4645 **Java bishopwood; bishopwood**
(Bischofia javanica)
d malaiischer Baum *m*; Toogbaum *m*
f bois *m* l'évêque; uriam *m* d'Assam
r бишопия *f* яванская; бишофия *f* яванская; гадог *m*

4646 **Java plum; jambolan**
(Syzygium cumini; Syzygium javanicum; Eugenia jambolana)
d Javaapfelbaum *m*; Javaapfel *m*
f pomme-rose *f*; jamelongue *f*; jamelonguier *m*; faux pistachier *m*; prune *f* de Java
r евгения *f* Ямболана

* **jaw breaker → 186**

* **jaw crusher → 186**

4647 **Jeddo spruce; Jezo spruce; Yezo spruce**
(Picea jezoensis)
d Yedofichte *f*; Ajanfichte *f*
f sapin *m* du Yeso
r ель *f* аянская

4648 **Jeffrey['s] pine**
(Pinus jeffreyi)
d Jeffreyskiefer *f*
f pin *m* de Jeffrey
r сосна *f* Жеффрея

4649 **jelutong**
(Dyera costulata)
d Jelutong *m*

4650 **Jenny-wood; fre[i]jo; cordia wood; salmwood; frei jorge; laurel blanco**
(Cordia goeldiana; Cordia sagotii)
d brasilianischer Nussbaum *m*; brasilianischer Teak *m*; Freijo *m*
f freijo *m*; frei-jorge *m*
r кордия *f* Гоэлда

4651 **Jequibita rosa**
(Cariniana brasiliensis; Cariniana legalis; Cariniana pyriformis)
d Jequibita-Rosa *f*
f abarco *m*
r орех *m* американский

* **Jersey tea → 6788**

4652 **Jerusalem thorn**
(Parkinsonia aculeata)
d Jerusalemdorn *m*
f genêt *m* épineux
r паркинсония *f* колючая; паркинсония шипатая

* **Jezo spruce → 4647**

4653 **jig; joiner's cramp**
(a device used to make special cuts, guide a tool, or aid in woodworking operations)
d Einspannvorrichtung *f*; Schraubzwinge *f*
f dispositif *m* de fixation de la pièce; serre-joint *f*
r зажимное приспособление *n*; прижим *m*; струбцина *f*

* **jig[ger] saw → 1822**

* **jinker → 894**

* **joe log → 5115**

4654 **joewood; barbasco**
(Jacquinia keyensis)
d Barbasco *n*
f barbasco *m*
r атлантический ямс *m*

4655 **joggled timber**
d verschränktes Holz *n*
f bois *m* faussé
r неправильно отёсанное бревно *n*

* **John's wood → 7154**

4656 **joiner**
d Tischler *m*

* **joiner's cramp → 4653**

4657 joiner's deals
d Tischlerdielen *fpl*
f planches *fpl* de menuiserie
r столярные доски *fpl*

4658 joinery timber; joinery [wood]; benchwood; carpentry timber; small timber
(wood of small dimensions for domestic use; finished timber fixtures of buildings such as doors, windows, panelling, cupboards, etc.)
d Schreinerholz *n*; Tischlerholz *n*
f bois *m* [blanc] de menuiserie
r отборный пиломатериал *m* для столярных работ; столярный лесоматериал; поделочный лес *m*; [мелкий] поделочный пиломатериал *m*; поделочная древесина *f*

4659 joinery tools
d Tischlerwerkzeuge *npl*
f instruments *mpl* de menuiserie
r инструменты *mpl* для столярных работ

* **joinery [wood] → 4658**

4660 joining of timbers; wood joint
d Holzvereinigung *f*; Holzverbinden *n*; Holzverbindung *f*
f union *f* de bois; jointure *f* de bois; assemblage *m* de bois
r вязка *f* деревянных деталей; соединение *n* деревянных деталей

4661 jointer; facer; edger; trimming machine
(a machine for making the edges of lumber smooth, flat, and parallel)
d Abrichthobelmaschine *f*; Zusammensetzmaschine *f*
f machine *f* à jointer; dégauchisseuse *f*; dresseuse *f*; rainureuse *f*
r рубанок *m*

* **joist → 9592**

4662 jojoba; jajoba
(Simmondsia chinensis; Simmondsia californica)
d Jojoba *n*
f jojoba *m*
r жожоба *f*

* **Joshua palm → 4604**

* **Joshua tree → 4604**

(root, especially a seedling taproot, having a sharp bend greater than 90°, shaped like a J)
d J-Wurzel *f*
f racine *f* en J
r J-образный корень *m*

4664 J-shaped rod
d J-förmige Stange *f*
f tringle *m* en J
r J-образный стержень *m* для отклонения возвратного троса (канатной установки)

* **jucaro → 6014**

4665 Judas-tree
(Cercis siliquastrum)
d [gemeiner] Judasbaum *m*; Herzbaum *m*
f arbre *m* de Judée; arbre de Judas; arbre d'amour; arbre de feu; gainier *m* [commun]; bouton *m* rouge
r иудино дерево *n*; багряник *m* стручковатый

* **jujube → 1783**

* **jumbo → 2587**

4666 jump *v* butt; butt *v* [off]
(to square-cut the end of a log or other piece of converted timber)
d ein dickes Ende schneiden
f équarrir le pied d'une grume
r отрезать дефектную часть комля

4667 jumper
d Stoßbohrer *m*; Brechstange *f*
f barre *f* à mine
r прыгающий бурав *m*

4668 jumping
(moving a spar tree in an upright position to a better location at the landing)
d Springen *n*
f refoulement *m*
r перестановка *f* с места на место трелёвочной мачты в вертикальном положении

4669 jump[ing] fire; spot[ting] fire
(fire set outside the perimeter of the main fire by flying sparks or embers)
d sprungweise sich ausbreitendes Feuer *n*; Springfeuer *n*
f foyer *m* partiel; foyer secondaire; feu *m* secondaire; feu bondissant; petit foyer périphérique; foyer d'incendie sporadique; feu disséminé; foyer disséminé; foyer dispersé

* **jump saw → 4312**

4670 **jump streak**
d sprungweiser Streifen *m*
f échauffure *f* sporadique
r ступенчатая карра *f*; ребристая карра

* **Juneberry → 7462**

4671 **jungle**
(an impenetrable equatorial forest)
d Dschungel *m*
f jungle *f*
r джунгла *f*

* **juniper → 3983**

4672 **juniper forest; juniper stand**
d Wacholderwald *m*; Wacholderbestand *m*
f forêt *f* de genièvre; genévraie *f*
r можжевёловый лес *m*; арчовник *m*

4673 **junkbutt**
(badly splintered end of a felled tree that has been cut back to sound wood)
d unbrauchbarer Erdstamm *m*
f pied *m* jonque
r нарезанный комель *m*

* **junk wood → 5195**

* **jutahy → 9190**

4674 **juvenile; immature; unripe** *adj*
(that has grown past the regeneration stage but is not yet mature)
d jugendlich; unreif; juvenil
f jeune; non-mûr
r незрелый; неспелый; недоразвитый

4675 **juvenile [core] wood; immature wood; pith wood; young wood**
(inner core of xylem surrounding the pith)
d jugendliches Holz *n*; Jugendholz *n*; Jungholz *n*; Gertenholz *n*
f bois *m* de jeunesse
r ювенильная древесина *f*; недоразвитая древесина; молодая древесина с неодревесневшей тканью

4676 **juvenile foliage**
(young leaves of tree that that produces one type of foliage in new or young growth and mature leaves of a distinctly different shape)
d jugendliches Blattwerk *n*

4677 **juvenile growth**
d jugendlicher Wuchs *m*
f croissance *f* juvénile
r рост *m* перед цветением

4678 **juvenile leaf**
(leaf of young a plant or shoot from the base of an older plant)
d jugendliches Blatt *n*
f feuille *f* juvénile
r ювенильный лист *m*; распускающийся лист

4679 **juvenile phase**
(the period during the life of a tree before flowering occurs)
d jugendliche Phase *f*
f phase *f* juvénile
r фаза *f* перед цветением; фаза перед полным распусканием листьев

4680 **juvenile spacing; precommercial thinning; noncommercial thinning; non-productive thinning**
(reducing the number of trees in young stands, often carried out before the stems removed are large enough to be used or sold as a forest product)
d jugendliche Durchforstung *f*
f éclaircie *f* précommerciale; éclaircie non commerciale
r предпромышленные рубки *fpl* ухода; непромышленные рубки ухода; некоммерческие рубки ухода; прореживание *n* без сбыта продукции

4681 **juvenility**
(a phase of plant development prior to the time when the plant has the capability to flower)
d Juvenilität *f*
f juvénilité *f*
r недоразвитость *f*; отставание *n* в развитии

* **juvia → 1003**

K

4682 **kadam**
(Anthocephalus cadamba; Anthocephalus chinensis)
d Kadam *m*
f kadam *m*
r кадам *m*

* **kadj → 2220**

4683 **kahika podocarpus; kahikatea; New Zealand white pine; New Zealand dacryberry**
(Dacrycarpus dacrydioides; Podocarpus dacrydioides)
d weiße Faune *f*
f podocarpe *m* de Nouvelle-Zélande
r ногоплодник *m* дакридиевидный

* **kahikatea → 4683**

* **kaki → 4627**

* **kaki fruit → 4628**

4684 **kamala [tree]**
(Mallotus)
d Kamala *n*
f kamala *m*
r малотус *m*

4685 **kamassi [boxwood]; Knysna boxwood; African boxwood; [false] cape boxwood**
(Gonioma kamassi)
d Kamassi *m*
f kamassi *m*
r гонуама *f* камасси; камасси *m*

* **kambala → 4582**

4686 **kandis; haras**
(Garcinia cowa)
d Kandis *n*
f kandis *m*
r кандис *m*

4687 **kangaroo-thorn acacia; paradox acacia**
(Acacia paradoxa; Acacia armata)
d Känguru[h]dorn *m*; Känguru[h]-Akazie *f*
f mimosa *m* épineux

* **kanoaballi → 2570**

* **kapbaulk → 7245**

4688 **kapok[-]tree; doum [palm]; fromager; ceiba [yuca]; silk-cotton [tree]**
(Bombax pentandra; Ceiba pentandra; Eriodendron anfractuosum)
d Wollbaum *m*; Kapok[baum] *m*
f kapokier *m*; faux cotonnier *m* [soyeux]; fromager *m*
r хлопковое дерево *n*; капоковое дерево; капок *m* [настоящий пятитычиночный]; шерстяное дерево

* **kapp balk → 7245**

4689 **kapur**
(Dryobalanops aromatica; Dryobalanops lanceolata)
d Kapur *m*
f kapur *m*
r капур *m*

* **karafuto oak → 4624**

* **karamatsu larch → 4623**

4690 **Karelian birch**
(Betula rubescens var. vulgaris)
d gemaserte Birke *f*
f bouleau *m* de Norvège
r карельская берёза *f*

4691 **karri [eucalyptus]**
(Eucalyptus diversicolor)
d Karri *m*
f karri *m*
r эвкалипт *m* разноцветный

* **Kashmir soapberry → 1506**

* **kat[h]ira gum → 4055**

4692 **katsura tree**
(Cercidiphyllum japonicum)
d Kuchenbaum *m*; Lebkuchenbaum *m*; Katsurabaum *m*; Judasblattbaum *m*
f arbre *m* aux feuilles de cercis; katsura *m*; arbre [au] caramel; cercidiphylle *m* du Japon
r багряник *m* японский; круглолистник *m*

4693 **katsura-tree family**
(Cercidiphyllaceae)
d Katsuragewächse *pl*

* **kauri → 5727**

* **kauri gum → 4466**

* **kauri pine → 6664**

* **kauri resin → 4466**

* **kauta → 5532**

* **kedar → 7569**

4694 **keeping of growing stock**
d Vorratshaltung *f*
f maintien *m* du capital-bois
r содержание *n* запаса

* **kekatong → 5653**

* **keledang → 1011**

* **Kellog oak → 1255**

4695 **kempas; tualang [tree]**
(Koompassia malaccensis; Koompassia excelsa; Koompassia parvifolia)
d Kempas *n*
f kempas *m*
r кемпас *m*; компассия *f* малаккская

* **Kentucky coffee[-]tree → 1482**

* **Kentucky yellowwood → 262**

* **kerf depth → 2195**

4696 **kerf losses**
d Schnittverluste *mpl*
f pertes *fpl* au sciage
r потери *fpl* при пилении

* **kerf waste → 3561**

4697 **kermes oak**
(Quercus coccifera)
d Kermeseiche *f*
f chêne *m* kermès; kermès *m*; chêne à cochenilles
r дуб *m* кермесовый; дуб кермесоносный; дуб хермесовый

4698 **kernel fruit; pip fruit; pome [fruit]**
(fruit with fleshy outer tissue and a papery-walled inner chamber containing the seeds)
d Kernfrucht *f*

4699 **keruing; gurjun**
(Dipterocarpus gonopterus)
d Keruwing *m*; Kruwing *m*
f keruing *m*; kruen *m*
r керуинг *m*

4700 **key biotopes**
(rare and vulnerable small-sized ecosystems inside the forest area mainly in production forests, where endangered species occur)
d Schlüssel-Biotope *mpl*
f biotopes *mpl* clés
r ключевые биотопы *mpl*

4701 **key fruit; samara fruit; winged fruit**
(a dry, winged fruit which does not open at maturity)
d Flügelfrucht *f*; geflügelte Frucht *f*
f fruit *m* ailé; samare *f*
r крылатый плод *m*; крылоплод *m*; семя *n* с крылаткой; крылатое семя

* **keyhole borer → 231**

4702 **keyhole saw; lock saw**
(a saw used for cutting short-radius curves that is similar to a compass saw but usually has a narrower blade and finer teeth)
d Stichsäge *f*
f scie *f* à guichet
r пила *f* для выпиливания отверстий под замки

4703 **keying**
d Einbindung *f*
f encastrement *m*
r закрепление *n* с помощью клиньев; заклинивание *n*; закрепление шпонками

* **key lime → 9348**

4704 **key log**
(in river driving, a log which is so caught or wedged that a jam is formed and held)
d Verstopfungsstamm *m*
f grume *f* colmatante
r бревно *n*, образующее затор

* **key of wood → 9561**

4705 **keystone species**
(species that plays an important ecological role in determining the overall structure and dynamic relationships within a biotic community)
d Schlüsselart *f*

* **KG blade → 1161**

* **khaja → 107**

* **khaki gum → 9369**

* **khasya pine → 5787**

* **khaya → 107**

* **kibbler → 1534**

* **kickback brake → 1423**

4706 **kickback guard**
d Handschutz *m*
f arceau *m* de sécurité
r ограничитель *m* отбрасывания [досок] назад

* **kickback of wood → 3297**

* **kicker → 5097**

* **kicking up of the wood → 3297**

4707 **kidney-bean tree**
(Wisteria frutescens)
d amerikanische Wistarie *f*
f glycine *m* d'Amérique
r вистерия *f*; тропическая американская фасоль *f*

* **kilhig → 6645**

* **killig → 6645**

* **kill ratio → 6182**

4708 **kill strip** (*US*)
d Totstreifen *m*
f bande *f* morte
r часть *f* ствола, повреждённая насекомыми

* **kiln → 2623**

* **kiln brown stain → 1462**

* **kiln burn → 1462**

* **kiln-dried lumber → 4710**

4709 **kiln-dried saps; kiln-seasoned saps**
d ofengetrocknetes Pflanzensäfte *fpl*; künstlich getrocknete Pflanzensäfte
f sucs *mpl* végétaux séchés au four
высушенные пиломатериалы

4710 **kiln-dried wood; kiln-seasoned wood; kiln-dried lumber; oven[-]dry wood; oven[-]dry timber**
d ofengetrocknetes Holz *n*; darrtrockenes Holz
f bois *m* de construction séché au four
r высушенная в сушилке древесина *f*; древесина, высушенная печным обогревом

4711 **kilning**
(drying wood under controlled temperature and humidity)
d Darren *n*
f touraillage *m*
r сушение *n*

* **kiln-seasoned saps → 4709**

* **kiln-seasoned wood → 4710**

4712 **kiln site; place of a pile; charring place**
d Meilerstätte *f*
f place *f* à charbon de bois
r место *n* костра

* **kiln stain → 1462**

* **kindling wood → 3740**

* **kindly textured → 7822**

4713 **kind of mixture**
d Mischungsart *f*; Mischungsform *f*
f genre *m* de mélange; forme *f* de mélange
r тип *m* смешения

4714 **kind of wood; [forest] tree species; tree sort; wood sort; timber species**
d Waldbaumart *f*; Baumart *f*; Holzart *f*; Holzsorte *f*
f espèce *f* forestière; essence *f* forestière
r древесная порода *f*; сорт *m* древесины

* **King Boris fir → 1154**

* **king-nut hickory → 7507**

* **king orange → 5274**

4715 **king tree; Guiana mora; moraballi; morabukea; bitter angelim**
(Mora excelsa)
d Moraholz *n*; Mora *n*

4716 **kingwood; Ceara rosewood**
(Dalbergia cearensis)
d Königsholz *n*
f bois *m* de violette
r дальбергия *f* церенская; королевское дерево *n*; фиалковое дерево

* **kinnickinnick → 1759**

* **kino → 4058**

* **kino eucalyptus → 411**

4717 **kirikawa; marakaipo**
(Iryanthera)
d Cumala *f*
f iryanthera *m*; cumala *m*
r ириантера *f*

* **kiwi fruit → 51**

4718 **Klinki pine**
(Araucaria klinki; Araucaria hunsteinii)
d Klinki-Araukarie *f*
f pin *m* Klinki
r араукария *f* Клинки

* **knag → 993**

* **knag[gy] → 4737**

* **knaggy wood → 4738**

* **knapsack duster → 471**

* **knapsack sprayer → 471**

* **knar[l] → 1182**

* **knee → 2100**

* **knee pine → 5598**

4719 **knee root**
d Kniewurzel *f*
f racine *f* coudée
r корневой выpост *m*

* **knee-timber → 2029**

* **kneewood → 2029**

* **knife[-cut] veneer → 7738**

4720 **knife-leaf wattle**
(Acacia cultriformis)
r акация *f* ножелистная; акация культурная

* **knob → 1182, 9003**

* **knobbed → 4737**

4721 **knobby bole; curly-grained log; bumpy log**
d höckeriger Stamm *m*; knorriger Stemm
f tige *f* ondulée
r свилеватый ствол *m*; сучковатый ствол

* **knobby wood → 4738**

4722 **knobcone pine**
(Pinus attenuata)
d Höcker-Kiefer *f*
f pin *m* à cône en tubercule; pin de l'Eldorado
r сосна *f* утончённая

* **knobiness → 5742**

* **knock off → 10**

* **knog → 1182**

* **knot → 10**

4723 **knot; node; nodus; bough**
(that portion of a branch or limb that has been surrounded by subsequent growth of the stem)
d Astknoten *m*; Knoten *m*
f nœud *m*; nodus *m*
r узел *m*; сук *m*

4724 **knot area; knot zone**
d Astbereich *m*; Astfläche *f*
f zone *f* de nœuds
r зона *f* расположения сука; зона скопления сучьев

4725 **knot arrangement**
(on the tree stem)
d Aststellung *f*
f arrangement *m* de nœuds
r расположение *n* сучьев

4726 **knot base**
d Astbasis *f*
f base *f* de nœud
r основание *n* сука

* **knot borer → 4728**

4727 **knot boring**
d Astlochbohren *n*; Astbohren *n*

4728 **knot boring machine; knot borer**
d Astlochbohrer *m*; Astbohrer *m*
f perceuse *f* pour bouchonnage; machine *f* à bouchonner les nœuds
r сверловщик *m* сучков

* **knot-bumper → 981**

* **knot cluster → 1668**

4729 **knot conveyor**
d Ast[zug]förderer *m*
f convoyeur *m* de nœuds
r транспортёр *m* для сучьев

4730 **knot elevator**
d Astförderwerk *n*; Astelevator *m*
f élévateur *m* de nœuds
r подъёмник *m* для сучьев

4731 **knot formation; branch formation**
d Astbildung *f*; Zweigbildung *f*
f formation *f* de nœuds
r образование *n* сучьев; образование ветви

* **knot-free wood → 2351**

4732 **knot[ ]hole**
(opening produced when knot drops from the wood in which they were embedded)
d Astloch *n*
f trou *m* de nœud
r отверстие *n* от [выпавшего] сучка; сучковое отверстие; свищ *m*

4733 **knot on arris; arris knot; corner knot; edge knot; margin[al] knot**
d Kantenast *m*
f nœud *m* d'arête; nœud de rive
r сучок *m*, выходящий на ребро заготовки; ребровый сучок; кромочный сучок; торцовый сучок

4734 **knot rot; branch rot**
d Astfäule *f*
f pourriture *f* des branches
r гниль *f* сучьев

4735 **knot shape; branch shape**
d Astform *f*
f forme *f* de nœud
r форма *f* сучка

4736 **knot surround**
d Astumrandung *f*

* **knotted → 4737**

* **knotter → 981, 8825**

* **knottiness → 986**

* **knotting → 6610**

4737 **knotty; knotted; knag[gy]; knobbed; nodose; nodous; nodular; boughy; snaggy; snagged; gnarly; gnarled**
d knotig; ästchentragend; astreich; vielästig; krummästig
f noduleux; plein de bosses; grumeleux; noueux; à branches tordues
r сучковатый; узловатый; почковидный; шишковатый

* **knotty core → 1948**

4738 **knotty [fire]wood; knaggy wood; knotty timber; knot[ty] wood; knobby wood; gnarled wood; wood in blocks**
d knorriges Holz *n*; Knorrholz *n*; knotiges Holz; Klotzholz *n*
f bois *m* noueux; bois nouailleux; bois rustique; bois de billon
r сучковатая древесина *f*; корявая древесина; отрубки *mpl* бревен

4739 **knotty pine**
d knorrige Kiefer *f*
f pin *m* noueux
r сучковатая сосна *f*

* **knotty timber → 4738**

* **knot[ty] wood → 4738**

* **knot zone → 4724**

4740 **knuckle[-]boom [loader]; unscrambling knuckleboom; knuckleboom unscrambler**
(hydraulically operated loading boom whose mechanical action imitates the human arm)
d Scharnierboomlader *m*
f flèche *f* articulée de chargement; genouillère *f* de chargement; grue *f* de démêlage; grue d'alimentation
r шарнирный погрузчик *m*

* **knurl → 1182**

* **Knysna boxwood → 4685**

* **koko wood → 2730**

4741 **Korean ash**
(Fraxinus rhynchophylla)
d koreanische Asche *f*
f frêne *m* de Chine
r ясень *m* носолистный; ясень горный

4742 **Korean mountain ash; Japanese mountain ash**
(Sorbus commixta; Sorbus alnifolia; Sorbus discolor horticole; Sorbus reflexipetala)
d Wilfordseberesche *f*; erlenblättrige Eberesche *f*; hainbuchenblättrige Eberesche; japanische Eberesche
f alisier *m* à feuilles d'aulne
r рябина *f* смешанная

4743 **Korean [nut] pine; cedar pine; Manchurian pine**
(Pinus koraiensis)
d Mandschurie-Kiefer *f*
f pin *m* de Mandchourie
r сосна *f* кедровая маньчжурская; сосна [кедровая] корейская

4744 **kosipo; heavy sapelle; omu**
(Entandrophragma candollei)
d Kosipo *m*; Omu *m*
f ko[s]sipo *m*; omu *m*; lifuko *m*; lifaky *m*; atom-assié *m*
r энтандрофрагма *f* Кандолли; красное дерево *n* из Коссипо; косипо *n*

* **kowhai → 5729**

4745 **Koyama spruce**
(Picea koraiensis)
d Korea-Fichte *f*
f pin *m* de Corée
r ель *f* корейская

4746 **kudzu [hemp]**
(Pueraria lobata; Pueraria thunbergiana)
d Kopoubohne *f*; Kudzubohne *f*
f kudzu *m*; puéraria *m*
r пуэрария *f*

* **kummer → 8258**

4747 **kumquat; cumquat**
(Fortunella polyendra; Fortunella margarita)
d Kumquat *m*
f kumquat *m*
r кумкват *m*; кинкан *m*

L

4748 **labiate family**
(Labiatae; Lamiaceae)
d Lippenblütlengewächse *pl*
f labiées *pl*; labiatées *pl*
r губоцветные *pl*

4749 **Labrador tea ledum**
(Ledum groenlandicum)
d Grönland-Porst *m*
f lédon *m* du Groenland
r багульник *m* гренландский

* **lacebark → 3245**

* **lace[-]bark elm → 1496**

4750 **lace[-]bark pine**
(Pinus bungeana)
d Tempel-Kiefer *f*
f pin *m* Napoléon
r сосна *f* Бунге

4751 **lace bug**
(a member of the family Tingidae (Hemiptera), whose gauze-like wings and thoracic extensions give the adult insect a lacy appearance)
d Lacklaus *f*
f tigre *m*; punaise *f* réticulée
r клоп *m* азалиевый

4752 **lacewood**
(any of various Australian timber trees of the family Proteaclae (as of the genus Grevillea) having fern-like foliage and attractive mottled wood that is used in cabinetry and veneering)
d Spitzenholz *n*
f bois *m* en dentelles
r древесина *f* с кружевной текстурой

4753 **lack of forest; forest absence; forest deficiency**
d Waldmangel *m*; Waldlosigkeit *f*
f défaut *m* de forêt; manque *m* de forêt
r недостаток *m* леса; дефицит *m* леса

4754 **lack of thinning; thinning reserves**
d Durchforstungsreserven *fpl*

4755 **lacquer [varnish] tree; Japanese lacquer tree; lac tree**
(Rhus vernicifera; Schleichera oleosa)
d ostasiatischer Lackbaum *m*
f arbre *m* à laque; vernis *m* du Japon
r лаковое дерево *n*; сумах *m* лаконосный; сумах глянцевый

* **lacuch → 1011**

4756 **ladder beam; ladder pole; ladder rail**
d Leiterbaum *m*
f montant *m* d'une échelle
r лестничная жердь *f*

4757 **ladder fuels**
(fuels that provide vertical continuity between the surface fuels and crown fuels in a forest stand, thus contributing to the ease of torching and crowning)
d Feuerbrücke *f*; Feuerleiter *m*
f combustibles *mpl* étagés
r "лестничные" огни *mpl*

* **ladder pole → 4756**

* **ladder rail → 4756**

* **lady birch → 1761**

* **LAI → 4855**

* **lair → 2365**

4758 **lakeshore management area; lakeshore management zone**
(lands directly adjacent to a lake, in which forest practices standards are designed to maintain the unique combination of fish, wildlife, water)
d Küstenbetriebsfläche *f*
f zone *f* de gestion littorale; domaine *m* de gestion littoral
r береговая зона *f* управления

4759 **Lambert hazel; white filbert**
(Corylus maxima)
d Lambertshaselstrauch *m*; Lambertsnuss *f*
f noisetier *m* de Lambert; grand coudrier *m*
r лещина *f* крупная

* **Lambert's pine → 1274**

* **Lambkill kalmia → 7505**

* **lamellar root → 6378**

4760 **laminate** *v*
(to bond of two or more pieces of wood to make a single piece)
d beschichten; kaschieren; lamellieren; laminieren
f laminer
r расслаивать

4761 **laminated board; laminated sheet**
d Mehrschichtenplatte *f*
f plaque *f* stratifiée; feuille *f* laminée; panneau *m* stratifié
r ламинированная плита *f*; многослойный картон *m*

* **laminated timber → 4763**

4762 **laminated veneer lumber; LVL**
(a structural lumber product manufactured from veneers laminated so that the grain of all veneers run parallel to the axis of a member)
d Stäbchenplatte *f*
f placage *m* lamellé
r слоистый шпон *m*; многослойная фанера *f*

4763 **laminated wood; wood in layers; laminated timber**
(a built-up product made of layers or laminations of wood, all with the grain laid parallel and glued or otherwise fastened together)
d Schichtholz *n*; Lagenholz *n*; laminiertes Holz *n*
f bois *m* laminé; bois lamellé
r ламинированная древесина *f*

4764 **laminated wood bending**
d Sperrholzbiegen *n*; Schichtholzbiegen *n*
f courbage *m* de bois lamellé
r гнутьё *n* многослойной деревянной заготовки; гнутьё ламинированной древесины

4765 **Lammas shoot; Lammas growth; second[ary] shoot; late-season shoot; proleptic shoot**
(extra leader growth extension late in the growing season)
d Sommertrieb *m*; Johannistrieb *m*; Johanniswuchs *m*; Augusttrieb *m*; proleptischer Trieb *m*
f pousse *f* d'été; pousse de la Saint-Jean; pousse proleptique
r летний побег *m*; Иванов-побег *m*

4766 **lanceleaf cottonwood; Andrew poplar**
(Populus acuminata)
r ланцетолистный тополь *m*

4767 **lanceolate leaf**
(lance-shaped; broadest near the base, gradually tapering to the narrower apex)
d Lanzenblatt *n*; lanzettförmiges Blatt *n*
f feuille *f* lancéolée
r ланцетовидный лист *m*

4768 **lancewood**
(Nectandra grandis; Ocotea grandis)
d Lancewood-Baum *m*
f cèdre *m* jaune
r нектандра *f* великая

4769 **land**
(entire complex of surface and near-surface attributes of the solid portions of the surface of the Earth which are significant to people)
d Land *n*; Gelände *n*; Terrain *n*
f terre *f*; terrain *m*
r земля *f*; местность *f*

4770 **land base**
(acres of forest land that are actually available for forest management)
d Landbase *f*
f base *f* terrestre
r земная база *f*

4771 **land-based forest fire protection**
d auf der Erde gegründeter Waldbrandschutz *m*
f protection *f* de forêt contre l'incendie à terre
r наземная противопожарная охрана *f* лесов

* **landfill [site] → 9276**

* **land form → 6858**

4772 **land grader; land planer; land smoother**
d Erdhobel *m*
f niveleuse-surfaceuse *f*
r струг *m* для расчистки местности (от насаждений)

4773 **landing chaser**
(member of a logging crew who unhooks the logs at the landing and does other odd jobs)
d Landungsziseleur *m*; Landgewindestrehler *m*
f chasseur *m* à dépôt transitoire
r рабочий *m* на площадке; отцепщик *m*

* **landing gear of semi-trailer → 7435**

4774 **landing pile; cull pile**
(area of piled slash, logging residue, and stumps, created as a result of harvesting

d Abfallstapel *m*
f pile *f* de bois de rebut
r куча *f* забракованных лесоматериалов

* **landing [site] → 8664**

4775 landing spacing
d Holzlagerplatzabstand *m*
f espacement *m* entre premiers dépôts transitoires
r расстояние *n* между погрузочными площадками; расстояние между верхними складами

4776 landline location
(legal identification and accurate location of national forest property boundaries)
d Bodenleitung-Standort *m*
f localisation *f* de ligne terrestre
r местоположение *n* наземной линии

* **land packer → 2127**

* **land planer → 4772**

* **land register → 1249**

4777 landscape
(large regional units of land that are viewed as a mosaic of communities, or a unit of land with separate plant communities or ecosystems forming ecological units with distinguishable structure)
d Landschaft *f*
f paysage *m*
r ландшафт *m*; пейзаж *m*; характер *m* местности

* **landscape grain → 835**

4778 landscaping; landscape tending
(the art of placing trees, plants, and other features on a piece of land)
d Landschaftspflege *f*
f aménagement *m* paysager; aménagement paysagé; aménagement paysagiste; entretien *m* des sites
r уход *m* за ландшафтом; ландшафтная архитектура *f*; ландшафтные рубки *fpl* ухода

4779 land skidding
d Bodenrücken *n*
f débusquage *m* à terre
r трелёвка *f* по суше

* **landslip → 7835**

* **land smoother → 4772**

4780 land tenure
(holding of land and the rights that go with such holding, including all forms of holding from fee simple title to the various forms of tenancy of group ownership with allocation procedures, various kinds of lending possibilities, etc.)
d Grundbesitzverfassung *f*; Besitzverhältnisse *f*
f tenure *f* foncière
r землевладение *n*

4781 landuse classes
(commercial forest land, noncommercial forest land, reserve forest land, land area, nonforest land)
d Klassen *fpl* der Bodennutzung
f classes *fpl* d'exploitation de la terre
r классы *mpl* землепользования

4782 landuse types of forest stock; landuse types of forest fund
d Waldbodennutzungstype *mpl*
f types *mpl* d'exploitation des terres forestières
r категории *fpl* земель лесного фонда

* **lane → 3564**

* **Langbian three-leaved pine → 5787**

4783 lantana
(Lantana camara)
d Wandelröschen *n*
f lantanier *m*; lantana *m*
r лантана *f* шиповатая

4784 lapachin; yellow sucupira
(Sucupira amarella; Ferreira spectabilis)
d Lapachin *n*; gelbes Sucupira *n*
f lapachin *m*; sucupira *m* amarello
r сукупира *f* жёлтая

* **lapacho → 4581**

4785 Lapland pine
d lappländische Kiefer *f*
f pin *m* de Laponie
r сосна *f* лапландская

4786 Lapland rosebay
(Rhododendron lapponicum)
d Lappland-Alpenrose *f*
f rhododendron *m* de Laponie
r рододендрон *m* лапландский

* **lapwood → 8712**

4787 **larch**
(Larix)
d Lärche *f*; Lärchenbaum *m*
f mélèze *m*
r лиственница *f*

4788 **larch casebearer; larch leaf miner; larch-mining case moth**
(Coleophora laricella)
d Lärchenminiermotte *f*
f porte-case *m* du mélèze; coléophore *m* du mélèze; teigne *f* minière du mélèze
r чехлоноска *f* лиственничная

4789 **larch forest; larch wood; larch stand**
d Lärchenwald *m*
f méléz[e]in *m*; mélézaire *f*
r лиственничный лес *m*; лиственничник *m*; листвяг *m*

4790 **larch lappet moth**
(Tolype laricis)
d Lärchenspinner *m*
f lasiocampe *m* du mélèze
r коконопряд *m* лиственничный

* **larch leaf miner → 4788**

4791 **larch miner moth**
(Lygaeonematus erichsonii)
d Lärchenmotte *f*
f mineuse *f* du mélèze
r лиственничная моль *f*

* **larch-mining case moth → 4788**

* **larch peatmoss bog forest → 5306**

4792 **larch sawfly**
(Pristiphora erichsonii)
d große Lärchenblattwespe *f*
f mouche *f* à scie du mélèze; tenthrède *m* du mélèze
r пилильщик *m* лиственничный

* **larch stand → 4789**

* **larch wood → 4789**

4793 **large [brown] pine weevil; pine weevil; fir-tree weevil**
(Hylobius abietis)
d großer braune Nadelholzrüssler *m*; großer braune Rüsselkäfer *m*

* **large-coned Douglas-fir → 5588**

* **large-crowned → 1045**

4794 **large elm bark beetle; elm bark beetle**
(Scolytus scolytus)
d großer Ulmensplintkäfer *m*
f grand scolyte *m* de l'orme
r ильмовый заболонник-разрушитель *m*

* **large-flowered magnolia → 7860**

4795 **large gallberry; sweet gallberry; bay-gallbush**
(Ilex coriacea; Ilex lucida; Prinos lucidus)
d Glanz-Stechpalme *f*
f houx *m* clair
r падуб *m* блестящий

4796 **large knot**
(with diameter more of 3.7 cm)
d große Ast *f*
f gros nœud *m*
r крупный сучок *m*

4797 **large-knotted tree**
d Armleuchterbaum *m*
f arbre *m* de gros nœuds
r дерево *n* с крупными боковыми сучьями

4798 **large-knotted wood**
d grobästiges Holz *n*
f bois *m* à gros nœuds
r крупносучковатая древесина *f*

* **largeleaf → 4800**

4799 **largeleaf geigertree; geiger tree; orange geiger; scarlet cordia**
(Cordia sebestena)
d Scharlachkordie *f*
f mapou *m* rouge; sébestier *m*
r себестен *m*

* **largeleaf holly → 3927**

* **largeleaf magnolia → 700**

4800 **large-leaved; largeleaf; bigleaf; broadleaved**
d großblätt[e]rig
f à grandes feuilles
r крупнолистный

* **large-leaved [cucumber] tree → 700**

4801 **large-leaved lime; broadleaved lime;**

d großblättrige Linde *f*; Sommerlinde *f*
f tilleul *m* à grandes feuilles [d'Europe]
r липа *f* крупнолистная

4802 **large organic debris; LOD**
(entire trees or large pieces of trees that provide channel stability or create fish habitat diversity in a stream channel)
d großer organische Trümmer *pl*
f gros débris *m* organique
r крупные остатки *mpl* органических веществ; крупные органические остатки

4803 **large pole [forest]; thick pole; stand of high poles; high polewood**
(pole with diameter over 7 cm 1 m above butt-end)
d Derbstange *f*; starkes Stangenholz *n*
f grande perche *f*; haut perchis *m*
r крупная жердь *f*; толстый жердняк *m*

4804 **large poplar borer; poplar borer; willow borer**
(Saperda calcarata)
d großer Pappelbock *m*
f [grande] saperde *f* du peuplier; saperde chagrinée
r древоточец *m* тополевый

* **large-pored wood → 1685**

4805 **larger Mexican pine beetle**
(Dendroctonus parallelocollis)
d großer mexikanische Kiefernmarkkäfer *m*
f dendroctone *m* du pin mexicain
r лубоед *m* сосны мексиканской

4806 **large sapling**
d großer Heister *m*
f haute tige *f*
r крупный саженец *m*

4807 **large-size brushwood**
d starke Stangenreiser *npl*
f broussailles *fpl* grandes
r толстомерный хворост *m*

4808 **large[-sized] timber; thick timber; heavy timber; thick wood**
d Grobholz *n*; starkes Holz *n*; Starkholz *n*
f bois *m* fort
r крупномерный лес *m*; крупномерные лесоматериалы *mpl*; крупномерная древесина *f*; толстомерный лесоматериал; толстомер *m*

* **large softwood pole → 1874**

* **large-spurred thorn → 3167**

* **largetoothed aspen → 706**

* **large woody debris → 1686**

4809 **larva; maggot; grub; chrysalis; mite**
d Larve *f*; Made *f*
f larve *f*
r личинка *f*

* **larval burrow → 4811**

4810 **larval cradle; larval chamber; pupal cell; pupal niche**
d Puppenwiege *f*
f berceau *m* de nymphose
r кукольная колыбель *f*

4811 **larval gallery; larval mine; larval burrow; brood gallery; egg gallery; egg tunnel; mother gallery; mother tunnel; parent gallery**
(in wood)
d Larven[fraß]gang *m*; Muttergang *m*
f galerie *f* de ponte; couloir *m* de ponte
r ход *m* личинок; маточный ход

* **lashing of booms → 8578**

4812 **lasp** *v*; **whip** *v*
(the top of tree)
d auspeitschen; peitschen
f fouetter
r ухлёстывать; схлёстывать

4813 **last annual shoots**
d Endtriebe *mpl*
f pousses *fpl* récentes
r ауксибласты *mpl*; побеги *mpl* текущего года

4814 **last felling**
d letzter Hieb *m*
f abattage *m* récent
r последняя рубка *f*

4815 **late bark; late bast**
d Spätrinde *f*
f liber *m* final
r поздняя кора *f*

4816 **late burning**
d spätes Brennen *n*

* **late decay → 1828**

4817 **late forest succession**
(stage of forest succession when the majority of trees are mature)
d spätere Waldsukzession *f*; spätere Waldnachfolge *f*
f succession *f* forestière tardive
r поздняя смена *f* лесной растительности

* **latent bud → 2573**

4818 **latent period**
d Latenzstadium *n*
f période *f* latente
r скрытый период *m*; латентный период; период спячки

4819 **lateral branch; side branch**
d Seitenzweig *m*; Seitenast *m*
f branche *f* latérale
r боковая ветвь *f*; боковой сучок *m*

* **lateral bud → 419**

* **lateral cutting → 964**

4820 **lateral extension of roots**
d seitliche Wurzelerweiterung *f*
f prolongation *f* latérale des racines
r горизонтальная протяженность *f* корней

4821 **lateral hauling; side hauling**
d seitliches Heranziehen *n*
f débardage *m* latéral
r подтрелёвка *f* со стороны; подтаскивание *n* сбоку

4822 **lateral log retrieval**
d seitliche Holzstammrückgewinnung *f*
f renversement *m* latéral de grumes
r возвращение *n* [каретки] с грузом на линию трелёвки

* **lateral pruning → 964**

4823 **lateral regeneration**
d Seitenverjungung *f*
f régénération *f* latérale
r боковое возобновление *n*; возобновление от стен леса

4824 **lateral roots; laterals; branch roots; side roots**
(roots that run out from a tree horizontally just below the surface)
r боковые корни *mpl*

* **laterals → 4824**

* **lateral shoot → 418**

4825 **lateral skidding distance**
d Seitenrückenabstand *m*
f distance *f* latérale de débardage
r расстояние *n* подтаскивания лесоматериалов

* **late-season shoot → 4765**

* **late-sweet blueberry → 1788**

4826 **late thinning**
d späte Durchforstung *f*
f éclaircie *f* tardive
r позднее прореживание *n*; поздняя прорубка *f*

4827 **latewood; autumn wood; autumn timber**
(the denser wood formed during the later stages of growth of each annual ring)
d Engholz *n*; Herbstholz *n*; Spätholz *n*
f bois *m* automnal; bois final; bois tardif
r поздняя древесина *f*; осенняя древесина

4828 **latex traces; latex canals; latex ducts**
(a term used to describe the slit-like passages (as they appear in seasoned timber) running radially through the wood of certain latex-bearing trees)
d Milchröhrenspur *f*
f alvéoles *mpl* à laticifère; traces *fpl* de laticifère; laticifères *mpl*
r система *f* трубок, содержащих латекс; млечники *mpl*

4829 **latex tube**
(a laticifer enclosed in a ray)
d Milchröhre *f*; Latexröhre *f*
f laticifère *m* radial; tube *m* à latex; tube sécréteur laticifère
r латексный сосуд *m*; млечник *m*; млечный сосуд

* **lathe → 9623**

4830 **laticifer[ous] cell**
(a general term for cells containing latex)
d Milchzelle *f*
f cellule *f* laticifère
r клетка *f*, содержащая млечный сок

4831 **lauan; meranti; shorea**
(Shorea)
d Meranti *n*
f meranti *m*; méranti *m*; lauan *m*; seraya *m*; séraya *m*
r меранти *m*; лауана *f*

* **laurel → 6745**

* **laurel blanco → 4650**

* **laurel-brier → 4834**

4832 **laurel family**
(Lauraceae)
d Lorbeergewächse *pl*
f lauracées *pl*
r лавровые *pl*

4833 **laurel forest**
d Lorbeerwald *m*
f forêt *f* à laurier; laurisylve *f*
r лавровый лес *m*

4834 **laurel greenbrier; laurel-brier; bamboo vine**
(Smilax laurifolia)
d lorbeerblättrige Stechwinde *f*
f salsepareille *f* à feuilles de laurier
r сассапарель *f* лавролистная

* **laurel magnolia → 643**

4835 **laurel oak; Darlington oak; swamp laurel; diamondleaf oak**
(Quercus laurifolia; Quercus hemisphaerica)
d lorbeerblättrige Eiche *f*; Lorbeereiche *f*
f chêne *m* laurier; kalmia *m* à feuilles de laurier
r дуб *m* лавролистный

4836 **laurel sumac[h]**
(Malosma laurina; Rhus laurina)
d Lorbeersumach *m*
f sumac *m* à feuilles de laurier
r сумах *m* лавровидный

* **laurel willow → 630**

4837 **Lavalle corktree**
(Phellodendron lavallei)
d Lavalle-Korkbaum *m*
f arbre-liège *m* de Lavalle; phellodendre-liège *m* de Lavalle
r дерево *n* пробковое Лаваля

4838 **lavender bush**
r заросль *f* лаванды

4839 **Lawson [false] cypress; Oregon cedar; Port-Orford cedar**
(Chamaecyparis lawsoniana)
d Lawsons-Lebensbaumzypresse *f*; [Lawsons-]Scheinzypresse *f*
f cyprès *m* de Lawson; chamaecyparis *m* de Lawson
r кипарисовик *m* Лавсона; кедр *m* порта Орфорд

4840 **lay; bed [cradle]; [log] cradle; make bed; table**
(position on the ground where a tree will fall when severed from the stump)
d Fällungsbett *n*
f lit *m* de chute
r валочное ложе *n*

* **layered → 8130**

4841 **layered stand**
d geschichteter Bestand *m*
f peuplement *m* étagé
r ярусное насаждение *n*

4842 **layered structure; storied structure**
(forest structure characterized by the development of multiple age classes of timber)
d geschichtete Struktur *f*; schichtige Struktur
f structure *f* étagée
r слоистая структура *f*; ярусная структура

* **layering → 6582, 8147**

4843 **layering method**
d Senkvermehrungsmethode *f*
f méthode *f* de marcottage
r метод *m* размножения отводками

4844 **layer of logs in stacking**
d Rollschicht *f*
f couche *f* de rondins
r бревенчатый роликовый слой *m*

* **lay in → 669**

* **laying-in [a tree] → 7107**

4845 **laying-off; spauling**
(marking off the stem and branches of a felled tree into lengths for cross-cutting)
d Aufreißen *n* des Holzes; Stammeinreßen *n*
f marquage *m* des découpes; éclatement *m*
r отмечание *n* ствола; разметка *f* ствола и

* **laying out → 7995**

4846 **laying-up; piling; stacking; sticking**
(packing of the logs in stack)
d Stapelung *f*; Abstapeln *n*; Aufstapelung *f*; Stöckeln *n*; Zusammentragen *n*
f empilement *m*; empilage *m*; groupage *m* de bois
r укладка *f* в стопу; штабелёвка *f*; штабелирование *n*; штабелеукладка *f*; укладка в штабель

* **ld → 5022**

4847 **lead**
(the direction of the line of haul, taken from the donkey)
d Leitung *f*; Ader *f*
f itinéraire *m* de débusquage; guide *m*; tracé *m*
r конец *m* провода; рабочий канат *m*

* **lead block → 1158**

* **leader → 6536**

* **leader branch → 6536**

4848 **leading end**
(of the stem)
d Vorderflanke *f*
f extrémité *f* guidante
r передний конец *m* (трелюемого брёвна)

4849 **leading sheave**
(of a wire rope)
d Leitrolle *f*; Führungsrolle *f*
f poulie *f* guidante
r направляющий ролик *m*; ведущее колесо *n*

* **leading shoot → 4237**

* **leading tree species → 5252**

4850 **lead strap; strap; [block] strop**
(a short length of cable with an eye spliced at each end for attaching blocks to stumps, etc)
d Stropp *m*
f estrope *f*
r отрезок *m* каната (с петлями по концам для прикрепления блоков к пням)

4851 **lead[ ]tree; horse tamarind; white popinac; leucaena [tree]**
(Leucaena leucocephala; Leucaena glauca)
d Weißkopfmimose *f*

* **leaf → 7531**

4852 **leaf**
(most often green, flattened structure attached to a stem and functioning as a principal organ of photosynthesis and transpiration in most plants)
d Blatt *n*
f feuille *f*
r лист *m*

* **leafage → 3303**

* **leaf analysis → 3307**

4853 **leaf area; lamina**
d Blattfläche *f*
f superficie *f* des feuilles; limbe *m*
r листовая поверхность *f*; поверхность листа

4854 **leaf-area duration**
(leaf area index integrated over time)
d Blattflächendauer *f*
f durée *f* de limbe
r площадь *f* листовой поверхности

4855 **leaf-area index; LAI**
(a dimensionless parameter expressing the area of leaf surface over unit area of ground)
d Blattflächenindex *m*
f indice *m* [de surface] foliaire; ISF
r индекс *m* листовой поверхности; индекс площади листьев

* **leaf-area ratio → 4893**

4856 **leaf arrangement; phyllotaxy; phyllotaxis**
(the manner of arrangement of leaves on a stem)
d Blattdisposition *f*
f arrangement *m* de feuilles
r листорасположение *n*; расположение *n* листьев

4857 **leaf axil**
(the angle between a petiole and the stem)
d Blattachsel *f*
f aisselle *f* de la feuille
r влагалище *n* листа; пазуха *f* листа

* **leaf base → 4883**

4858 **leaf beetles; leaf-eating insects; defoliating insects**
(Chrysomelidae)

f insectes *mpl* phyllophages; chrysomèles *fpl*
r [жуки-]листоеды *mpl*; листоразрушающие насекомые *npl*

* **leaf blade → 4872**

* **leaf blight → 5444**

4859 leaf bud; foliage bud; gemma
(small asexual reproductive structure in e.g. liverworts and mosses that detaches from the parent and develops into a new individual; a bud which contains undeveloped leaves)
d Blattknospe *f*
f bourgeon *m* foliaire; gemme *f*
r [листовая] почка *f*; гемма *f*

4860 leaf canopy
d Blätterdach *n*; Laubdach *n*
f couvert *m* de feuilles; dais *m* de feuillage
r листовой покров *m*

4861 leaf cast
(any disease causing untimely shedding of foliage)
d Blattschüttenkrankheit *f*
f maladie *f* défoliante; rouge *m*
r сбрасывание *n* листьев из-за болезни

* **leaf cast of conifers → 6260**

4862 leaf crown
d Laubkrone *f*
f cime *f* foliaire
r лиственная крона *f*

4863 leaf curl
(a disease that causes leaves to roll up)
d Kräuselkrankheit *f* der Blätter
f enroulement *m* des bords; frisolée *f* de feuilles
r курчавость *f* листьев

4864 leaf-cutter [ant]; leaf cutting ant; parasol ant
(Atta sexdens)
d Blattschneiderameise *f*; Parasolameise *f*; Laufschneider *m*
f fourmi *m* parasol
r атта *f*; муравей *m* зонтичный; муравей-листорез *m*

* **leaf debris → 4999**

* **leaf drop → 4866**

* **leaf duff → 4999**

* **leafed → 4894**

4865 leafed crown; leafy crown
d belaubte Krone *f*
f cime *f* foliacée
r облиственная крона *f*

4866 leaf[-]fall; leaf shedding; leaf drop
d Laubfall *m*; Blattabfall *m*; Laubverlust *m*
f chute *f* [naturelle] des feuilles
r потеря *f* листьев; удаление *n* листьев; сбрасывание *n* листьев; опадение *n* листьев; обрывание *n* листьев; листопад *m*

4867 leaf fiber
d Blattfaser *f*
f fibre *f* de feuille
r волокно *n*, содержащееся в листьях

4868 leaf fodder; fodder
(leafs eaten by browsing or grazing animals)
d Laubheu *n*; Laubfutter *m*
f feuilles *fpl* sèches pour l'alimentation du bétail
r лиственный корм *m*

4869 leaf form
d Blattform *f*
f forme *f* de feuille
r форма *f* листа

4870 leaf gap
(a break in a stem's vascular tissue above the leaf trace)
d Blattlücke *f*
f fente *f* de feuille
r листовая щель *f*

4871 leafhopper
(an insect of the family Cicadellidae that sucks juices from leaves and stems, often causing permanent damage)
d Kleinzikade *f*
f cicadelle *f*
r моль-пестрянка *f*

* **leafing → 6502, 9084**

* **leaf-knot → 1141**

4872 leaf lamina; leaf blade
(flat part of a leaf, or leaflet; characteristic of deciduous trees)
d Blattspreite *f*

4873 leafless
d blattlos; kaum
f sans feuilles; aphylle
r безлистный; безлиственный

4874 leafless tree; naked tree; bare tree
d blätterloser Baum *m*
f arbre *m* sans feuilles
r безлистное дерево *n*

4875 leaflet
(one of the leaf-like subdivisions of a compound leaf)
d Blättchen *n*; Fieder *m*
f foliole *f*
r листочек *m* сложного цвета; листок *m*

* **leaf litter → 4999**

4876 leaf lobes; lobes
(the individual segments of a leaf's blade, usually rounded)
d Blattlappen *fpl*
f lobes *mpl*
r лопасти *mpl*; каверны *fpl*

4877 leaf-losing; deciduous
d laubwechselnd; abfallend
f à feuillage caduc
r листопадный

4878 leaf margin
(border of a leaf)
d Blattrand *m*
f bord *m* de feuille
r край *m* листа

4879 leaf midrib; midrib
(primary rib or vein of a leaf)
d Mittelrippe *f*
f nervure *f* médiane de feuille
r средняя жилка *f* листа; центральная жилка листа

4880 leaf miner
(the larval stage of an insect which makes galleries or burrows between the upper and lower surfaces of leaf tissue)
d Blattminiermotte *f*
f mineuse *f* des feuilles
r минирующая мушка *f*

4881 leaf mo[u]ld; compost
(a form of humus composed of decayed leaves, often used to enrich soil)
r перегной *m* из листьев; компост *m*

4882 leaf nerve; leaf vein; vein; rib
(one of the riblike vessels in the blade of the leaf)
d Blattrippe *f*; Blattader *f*; Blattnerv *m*; Rippe *f*; Flader *f*
f nervure *f*; côte *f*
r жилка *f* [листа]; листовая жилка; ребро *n*

* **leafness → 3303**

* **leaf out → 7531**

4883 leaf petiole; petiole; leaf[ ]stalk; leaf base
(stalk that connects the blade(s) to the twig)
d Blattstiel *m*; Hauptstiel *m*; Blattspindel *f*; Blattgrund *m*; Ansatzstelle *f*
f pétiole *m*; pédoncule *m*
r черешок *m* [листа]; стебелёк *m*; основание *n* листа

4884 leaf primordium
(a lateral outgrowth from the apical meristem that develops into a leaf)
d Urblatt *n*
f ébauche *f* de feuille
r листовой бугорок *m*; зачаток *m* листа

4885 leaf retaining
(applied to a plant whose leaves remain for more than one season)
d Blattbehalten *n*
f rétention *f* de feuilles
r сохранение *n* листьев

4886 leafroller moths; tube-builders
(Tortricidae)
d Wickler *mpl*; Nestwickler *mpl*
f tordeuses *fpl*; tisseuses *fpl*
r листовёртки *fpl*

4887 leaf rolling
d Nestwicklung *f*
f enroulement *m* de feuilles
r скручивание *n* листьев

4888 leaf scar
(the mark left on the twig at the point of attachment of a leafstalk when the leaf falls)
d Blattnarbe *f*
f cicatrice *f* foliaire
r листовой рубец *m*

* **leaf scorch → 7306**

4889 **leaf spot**
(any of various plant diseases that cause well-defined areas of tissue to die creating noticeable spots)
d Blattfleckenkrankheit *f*
f tacheture *f* foliaire; tache *f* foliaire; touille *f*
r пятнистость *f* листьев

* **leaf[ ]stalk → 4883**

4890 **leaf surface; cuticle**
d Blattoberfläche *f*
f surface *f* de la feuille
r листовая поверхность *f*

4891 **leaf trace**
(the strand of vascular tissue between the vascular bundle of the stem and the leaf)
d Blattspur *f*
f trace *f* foliaire
r листовой след *m*

* **leaf vein → 4882**

4892 **leaf water potential**
(total potential for water in a leaf)
d Blattwasserkraft *f*
f potentiel *m* hydraulique de feuille
r водный потенциал *m* листа

4893 **leaf weight ratio; leaf-area ratio**
(of a plant the weight of leaves compared with the total plant weight)
d Verhältnis *n* zwischen Blattfläche und Gewicht
f rapport *m* de l'aire à la masse foliaire
r отношение *n* листовой поверхности к общему весу

* **leaf wood → 4124**

4894 **leafy; leafed**
d blätterig; beblättert; belaubt; laubig
f feuillé
r облиственный; покрытый листьями; листовой

* **leafy crown → 4865**

4895 **lean**
(of a tree from a perpendicular position)
d Hängen *n*
f inclinaison *f*
r отклонение *n* (от вертикали)

4896 **leaning face**
f face *f* de suspension
r сторона *f* дерева, обращённая в направлении естественного наклона

4897 **lean tree**
d Hänger *m*; Aufhänger *m*
f arbre *m* couché
r наклоннорастущее дерево *n*; наклонное дерево

* **leap-frog coupe → 224**

4898 **leatherleaf**
(Chamaedaphne calyculata; Cassandra calyculata; Andromeda calyculata)
d Torfmyrte *f*
f cassandre *m* caliculé
r мирт *m* болотный; вереск *m* болотный; хамедафне *f* прицветничковая

4899 **leatherleaf viburnum**
(Viburnum rhytidophyllum)
d Runzeblatt-Schneeball *m*; runzelblättriger Schneeball *m*
f viorne *f* ridée; viorne à feuilles ridées
r калина *f* морщинистолистная

4900 **leatherwood**
(Dirca palustris)
d Lederholz *n*
f dirca *m* des marais; bois *m* de plomb
r дирка *f* болотная; кожевенное дерево *n* дирка

4901 **leave strip**
(a strip of timber left standing between two clear-cut areas)
d verlassener Streifen *m*; Randstreifen *m*
f rideau *m*
r оставленная семенная полоса *f*; невырубленная полоса

* **leave tree → 6887**

4902 **leavetree cut; leavetree thinning**
d Überhältershieb *m*
f coupe *f* d'arbres marqués en réserve
r рубка *f* прореживания; сплошное прореживание *n*; прореживание по всему насаждению

* **leaving of low stumps → 5197**

* **Lebanon cedar → 1400**

4903 **Lebanon oak**

f chêne *m* du Liban
r дуб *m* ливанский

4904 ledger of consignment of timber
d Holzausgabebuch *n*
f journal *m* des sorties de bois
r книга *f* расхода леса

4905 ledger of incoming timber
d Holzeinnahmerechnung *f*
f livre *m* des recettes-bois
r книга *f* лесного дохода

4906 ledum
(Ledum)
d Porst *m*
f lédon *m*
r багульник *m*

4907 left spirally growth; clockwise spiral growth
d linksdrehender Drehwuchs *m*; sonniger Drehwuchs
f croissance *f* avec torsion à gauche
r левый косослой *m*; солнечный косослой

* **legal framework → 8047**

4908 legume family
(Leguminosae; Fabaceae)
d Schmetterlingsblütengewächse *pl*
f fabacées *pl*
r бобовые *pl*

4909 lemonade sumac[h]; lemonade berry; sourberry
(Rhus integrifolia)
d ganzblättriger Essigbaum *m*
f sumac *m* à feuilles entières
r сумах *m* цельнолистный

4910 lemon-scented gum
(Eucalyptus citriodora)
d Zitronen-riechender Eukalyptus *m*
f eucalyptus *m* à odeur de citron; eucalyptus citronné
r эвкалипт *m* лимонный

* **lemon sumac[h] → 365**

4911 lemon[-tree]
(Citrus limon)
d Zitronenbaum *m*; Limonenbaum *m*; Auerzitrone *f*
f limon *m*; citronnier *m*; citrus *m* de Sicile
r лимон *m*

4912 ...
d Holzstücken *n*
f aboutage *m* de bois
r [продольное] сращивание *n* брусьев

4913 length grading; length sorting
d Längensortierung *f*
f triage *m* par longueur
r сортировка *f* по длине

4914 length measuring
d Längenmessung *f*
f mesure *f* de longueur
r измерение *n* длины

* **length sawing → 5135**

* **length sorting → 4913**

4915 length[wise] cut[ting]; longitudinal splitting; longitudinal cleavage
d Längenspaltung *f*; Längsspaltung *f*; Aufschlitzen *n*
f fendage *m*; fente *f* en longueur
r продольный разрез *m*; продольное расщепление *n* ствола дерева; продольное раскалывание *n* [при его валке]

4916 lenticel
(a specialized portion of the periderm, variously shaped, but often lenticular, consisting of loosely arranged cells that are never more than slightly suberized)
d Lentizelle *f*
f lenticelle *f*
r чечевичка *f* (поры в коре ветвей)

* **lenticular ray → 3784**

4917 leopard moth
(Zeuzera pyrina)
d Blausieb *n*
f zeuzère *f*
r древесница *f* въедливая

* **leopard-tree → 1002**

4918 lesion
(a well-marked, but limited, diseased area; a break or rupture through a tissue, especially a surface tissue)
d Verletzung *f*
f lésion *f*; blessure *f*
r поранение *n*

4919 lespedeza

f lespédéza *m*
r леспедеца *f*; леспедеза *f*

* **lesser pine sawfly → 2905**

4920 lesser pine-shoot beetle
(Myelophilus minor)
d kleiner Waldgärtner *m*
f hylésine *m* mineur
r малый лесной садовник *m*

* **let → 3976**

4921 letterwood; snakewood
(Piratinera guianensis; Brosimum guianensis)
d Schlangenholz *n*
f bois *m* d'amourette; amourette *f*; mimosa *m* d'Amazonie
r бросимум *m* Ублетта

* **leucaena [tree] → 4851**

4922 level [indicator]
d Nivellieranzeiger *m*
f indicateur *m* de niveau
r уровнемер *m*; нивелир *m*

* **level[l]ing → 3901**

4923 level[l]ing rod; level[l]ing pole; sliding staff
d Nivellierlatte *f*
f mire *f* graduée
r нивелирная рейка *f*

4924 L-F-H horizon
d L-F-H-Horizont *m*
f horizon *m* L-F-H
r лесная подстилка *f* с преобладанием в составе опада листьев, хвои, веточек

4925 liana; liane
(vinelike climbing plant that grows by supporting itself on others)
d Liane *f*
f liane *f*
r лиана *f*

* **liber → 4502**

* **liberate → 6856**

* **liberation cut[ting] → 5912**

* **liberation felling → 5912**

4926 libriform fiber
the cambial initial as infected from the length of vessel members and parenchyma strands)
d Libriformfaser *f*
f fibre *f* ligneuse libriforme; fibre ligneuse simpliciponctuée
r либриформное волокно *n*; толстостенное волокно

4927 licania
(Licania)
d Licania *n*
f licanie *f*
r ликания *f*

4928 licence to clear; permit to clear
(of a forested area)
d Rodungsbewilligung *f*; Rodungsgenehmung *f*
f licence *f* d'abattage; autorisation *f* de défrichement
r разрешение *n* на расчистку

* **licence to cut → 6183**

4929 lichen
(algae and a fungus growing in symbiotic association on solid surfaces such as rocks or tree bark)
d Flechte *f*
f lichen *m*
r лишайник *m*

* **lichen covering → 1696**

4930 lichen pine forest
d Flechten-Kiefernwald *m*
f pineraie *f* à lichens
r лишайниковый сосняк *m*; сосняк беломошник

* **licorice → 4997**

* **life bar → 4568**

4931 life of trees
d Lebensdauer *f* der Bäume
f durée *f* de vie des arbres
r продолжительность *f* жизни деревьев

* **lift and lowering truck → 3660**

4932 lift cylinder; hoist cylinder
d Hubzylinder *m*; Druckzylinder *m*
f cylindre *m* de relevage; vérin *m* de relevage
r подъёмный цилиндр *m*; цилиндр стрелы

* **lifting → 5979, 7055**

* **lifting rope → 4323**

4933 lifting the canopy; canopy lifting
(removing the lower constituents of a canopy)
d Schirmausreißen *n*
f élagage *m* de dégagement; relèvement *m* du couvert
r обрезка *f* нижних ветвей кроны; вырубка *f* нижнего яруса насаждения

4934 lifting tongs
d Stammholzhebezange *f*
f pince *f* d'élévateur à grumes
r подъёмные клещи *pl* хлыстов

* **lifting truck → 3660**

4935 lift pruning; low pruning
(removing the lower branches of trees being grown for timber)
d niedrige Ästung *f*
f élagage *m* en bas
r очистка *f* нижних сучьев

* **lift truck → 3660**

4936 light back saw; dovetail saw
d Zinkensäge *f*
f scie *f* à queues d'arondes
r шипорезная пила *f*; пила для вырезания пазов

4937 light brown cubical rot
(invader - Fomes roseus)
d hellbraune kubische Fäule *f*
f pourriture *f* brune claire cubique
r светло-бурая кубическая гниль *f*

4938 light-colo[u]red wood
d weißes Holz *n*
f bois *m* coloré clair
r древесина *f* светлых пород

4939 light crop; partial mast; scattered mast
d Sprengmast *m*
f paisson *f* dispersée
r малый урожай *m*; слабый плодокорм *m*

4940 light deficiency
d Lichtmangel *m*
f défaut *m* de lumière
r недостаток *m* света

* **light demander → 7476**

* **light-demanding tree → 7476**

* **lighter tree → 7778**

* **light felling → 5912**

* **light foliaged → 7881**

* **light forest → 4446**

4941 light forest with steppelike field layer
d Lichtwald *m* mit steppenartriger Feldschicht
f forêt *f* claire à étage herbacé steppique
r остепненный лес *m*

4942 light-framing lumber
(lumber that is 5 to 10 cm thick and 5 to 10 cm wide)
d leichtes Karkassenholz *n*
f bois *m* à charpente légère
r лёгкие каркасные лесоматериалы *mpl*

4943 light hardwoods
d leichtes Hartholz *n*
f bois *m* feuillu léger
r лиственные породы *fpl* древесины небольшой плотности

4944 light increment; increment due to increased light; open stand increment
d Lichtungszuwachs *m*
f incrément *m* de mise en lumière; incrément dû à l'éclaircie
r световой прирост *m*

* **lighting → 1598**

* **lightly wooded → 6450**

4945 lightning [damage]
(of a tree)
d Blitzschaden *m*
f dégâts *mpl* causés par la foudre
r повреждение *n* молнией; загорание *n* от молнии

4946 lightning shake; thunder shake
d Blitzrinne *f*; Blitzriss *m*
f fissure *f* interne de foudroiement
r трещина *f* молнией; трещина от удара молнии

4947 light organic solvent preservative; LOSP
(a wood preservative)
d leichtes öliges Holzschutzmittel *n*
f produit *m* léger en solvant organique
r лёгкое защитное вещество *n* на базе органических растворителей

f coupes *fpl* partielles faibles
r низкоинтенсивные выборочные рубки *fpl*

* **light pole → 7775**

4949 light red meranti
(Shorea leptoclados; Shorea leprosula; Shorea smithiana)
d hellrotes Meranti *n*
f meranti *m* rouge; séraya *m* rouge
r сорея *f* чешуйчатобороздковая; дамир мерах *m*; красный меранти *m*

4950 light slash; light wood waste
d Feinabfälle *mpl*; kleinstückige Holzabfälle *mpl*
f petits rémanents *mpl*
r мелкие порубочные остатки *mpl*; мелкие древесные отходы *mpl*

4951 light stocking
d lockere Bestockung *f*
f mini-charge *f*
r редкое молодое насаждение *n*

4952 light thinning; weak thinning
d schwache Durchforstung *f*; leichte Durchforstung
f éclaircie *f* faible; éclaircie légère
r низкоинтенсивная рубка *f* ухода; слабая прорубка *f*; легкая прорубка

* **light wood → 8660**

4953 lightwood formation
d Ölholzformation *f*; Ölholzbildung *f*
f formation *f* de bois gras
r образование *n* пнёвого осмола

* **light wood waste → 4950**

* **ligneous → 9704**

* **ligneous plant → 9714**

* **lignescent → 4955**

* **ligniculture → 3528**

4954 lignification; formation of wood; wood development; woodiness
(the process of deposition of the chemical constituent (lignin) of the woody plant cell wall)
d Holzbildung *f*; Verholzung *f*; Holzentstehung *f*
r лигнификация *f*; одревеснение *n*; образование *n* древесины

4955 lignified; lignescent
(tending to be or become ligneous or woody)
d verholzt; in Holz verwandelt
f ligneux; lignifié
r древеснеющий; лигнифицированный; содержащий лигнин; одеревенелый

4956 lignified fiber
d verholzte Faser *f*
f fibre *f* ligneuse
r лигнифицированное волокно *n*; одеревенелое волокно

* **lignified tissue → 9504**

* **ligniform → 8876**

4957 lignin
(the second most abundant component of wood. It is a cementing layer between the wood cells)
d Lignin *n*
f lignine *f*
r лигнин *m*

* **lignivorous → 9739**

4958 lignivorous insect
d holzfressendes Insekt *n*
f insecte *m* lignivore; insecte xylophage
r насекомое *n*, питающееся древесиной; деревопоедающее насекомое

* **lignocellulose → 9504**

4959 lignose
d Lignose *f*
f lignose *f*
r лигноза *f*

4960 lignotuber
(1. a woody organ arising from an accessory meristem; 2. a swollen, woody structure at the base of many eucalypts)
d Lignotuber *m*
f lignotuber *m*
r одревесневший клубень *m*

* **lignum → 9473**

* **lignum vitae → 4034**

4961 lilac chaste tree; chaste tree; hemp tree
(Vitex agnus-castus)
d Mönchspfeffer *m*; Abrahamsbaum *m*; Keuschbaum *m*
f arbre *m* au poivre; gattilier *m* [commun]; poivre *m* des moines
r прутняк *m* [обыкновенный]; витекс *m* священный; авраамово дерево *n*

4962 lily family
(Liliaceae)
d Liliengewächse *pl*
f liliacées *pl*
r лилейные *pl*

4963 lily magnolia
(Magnolia liliiflora; Magnolia discolor; Magnolia purpurea)
d Purpur-Magnolie *f*
f magnolia *m* à fleurs de lis
r магнолия *f* лилейная; магнолия лилиецветная

* **limb → 10, 974**

* **limba → 92**

* **limber → 981, 8825**

4964 limber-bunching machine; [de]limber-buncher; delimber-topper; bunch limber; lopping-bunching machine
d Astschneider-Bundwerkzeug *n*
f ébrancheuse *f* à flèche; ébrancheuse-écimeuse *f*
r сучкорезно-пакетирующая машина *f*

4965 limber pine; lumber pine; Rocky Mountains white pine
(Pinus flexilis [James])
d Nevada-Zirbeliefer *f*; biegsame Kiefer *f*
f pin *m* pliable de l'Ouest; pin blanc des Montagnes Rocheuses; pin flexible; pin souple
r сосна *f* гибкая; сосна кедровая калифорнийская

* **limber stroke → 2356**

4966 limb grader
d Astsortierer *m*; Knüpfer *m*
f trieur *m* de branches
r подборщик *m* сучьев

4967 limb growth
d Astwachstum *m*

4968 limb healing
d Astverwachsung *f*
f cicatrisation *f* de branche
r зарастание *n* сучка

* **limbing → 6610**

* **limbing boom → 2353**

* **limbing carriage → 2354**

4969 limbing head; delimber head; delimbing attachment; delimbing unit; debranching head
d Entastungskopf *f*
f tête *f* d'ébranchage; tête à ébrancher; tête ébrancheuse
r сучкорезная головка *f*; сучкорезное приспособление *n*

* **limbing knife → 2285**

* **limbing machine → 8825**

* **limbing saw → 6614**

* **limbing stroke → 2356**

* **limbless → 983**

4970 limb rust
(invader - Peridermium filamentosum)
d Astrost *m*
f rouille *f* de branches
r ржавчина *f* сучьев

4971 limb shaker; branch shaker
d Baumastschüttler *m*
f secoueur *m* de branches
r стряхиватель *m* с захватом ветвей

* **limbwood → 5170**

* **limby → 977**

4972 lime bast; Russian bast; lime bark
d Lindenrinde *f*; Lindenbast *m*
f écorce *f* de tilleul; liber *m* de tilleul; tille *f* de tilleul
r липовая кора *f*; липовое лыко *n*

4973 limed oak
d geäscherte Eiche *f*
f bois *m* de chêne chaulé
r отбелённый дуб *m*

f chênaie-tillaie *f*
r липовый дубняк *m*

4975 limestone ash
d Kalkesche *f*
f frêne *f* calciphère
r известковый ясень *m*

4976 lime[-tree]; linden; basswood
(Tilia)
d Linde *f*
f tilleul *m*
r липа *f*

4977 lime-tree blossom
d Lindenblüte *f*
f fleur *m* de tilleul
r липовый цвет *m*

4978 lime[-tree] forest; linden forest
d Lindenwald *m*
f tillaie *f*
r липовый лес *m*; липняк *m*

* **lime wood → 6196**

4979 limit for extraction
d Transportgrenze *f*
f limite *f* de transport
r ограничение *n* трелёвки

4980 limiting environmental factor
(whose presence, absence, or abundance is the main factor restricting the distribution, numbers, or condition of an organism)
d begrenzender Umweltfaktor *m*
f facteur *m* d'ambiance limitant
r ограничивающий фактор *m* окружающей среды

4981 limit of merchantability
d Marktgängigkeitsgrenze *f*
f limite *f* de qualité marchande
r предел *m* товарности (минимальный диаметр в верхнем отрубе)

4982 limit of tree growth; tree limit; tree-line; timber-line
(the height above which trees can no longer survive on a mountainside, or the northern or southern limit beyond which only stunted forms appear on a continent toward a pole)
d Baumgrenze *f*
f limite *f* des arbres; limite de végétation arborescente
r верхняя граница *f* произрастания лесов; [верхняя] граница леса; граница

* **linden → 4976**

4983 linden family
(Tiliaceae)
d Lindengewächse *pl*
f tiliacées *pl*
r липовые *pl*

* **linden forest → 4978**

4984 linden-leaved maple
(Acer distylum)
d lindenblättriger Ahorn *m*
f érable *m* á feuilles de tilleul
r клён *m* двустолбиковый

4985 linear growth
d linearer Wachstum *m*
f croissance *f* linéaire
r линейный рост *m*

4986 linear plot cruising; line-plot cruising; line-plot survey
(a sample survey employing lines of sample plots, generally laid out at regular intervals along survey lines)
d Linienstichprobeninventur *f*
f inventaire *m* par échantillonnage en lignes
r таксация *f* методом ленточных пробных площадей

4987 linear regeneration sampling
(of a forest)
d lineare Regenerationsprobennahme *f*
f échantillonnage *m* linéaire de régénération
r закладка *f* ленточных проб [для изучения возобновления]

4988 linear survey; line survey; strip survey; strip cruising; strip cruise
(in this survey, all the trees in a narrow strip are measured; the number of strips that are run through the stand depends on the width of the strip, and the number of acres that are to be measured)
d Linientaxation *f*; Linienbestandesaufnahme *f*
f inventaire *m* par échantillonnage en bandes; inventaire par sondages sur bandes rectilignes [d'orientation variées]
r таксация *f* методом линейной выборки; линейная таксация

4989 linear tree planting
(raised rows in the form of hedges or wooded strips on the sides of fields, along roads or rivers, intended for the production of wood

d lineare Waldbaumpflanzung *f*
f plantation *f* des arbres forestiers linéaire
r линейная посадка *f* деревьев

* **line camp → 7573**

* **lined-out plant → 5815**

4990 **line-intercept method**
(a method in which the data are tabulated on the basis of plants lying on a straight line cutting across the community under study)
d Methode *f* der Linienstichproben
f méthode *f* d'échantillonnage linéaire
r метод *m* закладки ленточных проб (для изучения возобновления леса)

4991 **line of leafy trees**
d Laubholzbaumreihe *f*
f rangée *f* d'arbres feuillus
r полоса *f* лиственных деревьев

* **line planting → 6356**

* **line-plot cruising → 4986**

* **line-plot survey → 4986**

* **line seeding → 522**

* **line survey → 4988**

* **line tension → 7704**

4992 **line thinning; row thinning; strip thinning; stripped-coupe felling**
(removing specified rows of trees in a plantation)
d Streifendurchforstung *f*
f éclaircie *f* en ligne; éclaircie en bandes; éclaircie en rangée; éclaircie en bordure
r прореживание *n* полосами; узколесосечные рубки *fpl* ухода; ленточные рубки ухода; полосные рубки ухода; прорубка *f* каймы

* **lingonberry → 1996**

4993 **lining out**
(transplanting seedlings or rooted cuttings in rows in a nursery bed)
d lineare Verschulung *f*
f repiquage *m* en ligne
r линейная пересадка *f*; линейная пикировка *f*

* **lining-out board → 8770**

d linearer Verschulungsbaumgarten *m*
f pépinière *f* de repiquage en ligne
r питомник *m* линейной пересадки

* **link saw → 1431**

* **lion-tailing → 6010**

* **lip → 819**

4995 **lipsticktree; annato tree**
(Bixa orellana)
d Lippenstiftbaum *m*; Annattobaum *m*; Orleansstrauch *m*
f rocouyer *m*
r бикса *f* аннатовая

* **liquidation value → 2952**

4996 **liquid displacement**
(a method of measurement of timber volume)
d Liquidverschiebung *f*
f déplacement *m* d'eau
r вытеснение *n* жидкости

* **liquid sludge → 7473**

4997 **liquorice; licorice; sweetroot; glycyrrhiza**
(a dried root of gummy texture and sweet rather astringent flavor that is the source of extracts used to mask unpleasant flavors)
d Süßholz *n*
f bois *m* de réglisse; réglisse *f*
r лакричный корень *m*; солодковый корень; лакрица *f* (высушенный корень или экстракт лакричника)

4998 **list sampling**
(of trees)
d Listenstichprobenahme *f*
f échantillonnage *m* par liste
r выборочный перечёт *m*

4999 **litterfall; leaf litter; forest duff; leaf duff; debris layer; leaf debris; L-layer**
(uppermost layer of organic debris on a forestland)
d Laubdecke *f*; Laubstreu *f*
f litière *f* de feuilles mortes; couverture *f* morte
r лиственная подстилка *f*; слой *m* лесного опада; лесной опад *m*; лиственный опад; лиственный покров *m*; сырой листовой гумус *m*

5000 **litter of branches; branch litter; twig litter**
d Aststreu *f*

* **litter raking → 6706**

* **little branch → 991**

5001 **little forest; small forest; little wood; field woodland; grove; hedgerow timber**
(a small group of trees or grove)
d Kleinwald *m*; Kleingehölz *n*; Wäldchen *n*; kleiner Wald *m*; Hain *m*; Feldgehölz *n*; Gehölz *m*; Horst *m*
f petite forêt *f*; bouquet *m* champêtre; haie *f*
r мелкий лес *m*; мелколесье *n*; роща *f*; перелесок *m*

5002 **littleleaf boxwood**
(Buxus microphylla)
d kleinblättriger Buchsbaum *m*
f buis *m* à petites feuilles
r самшит *m* мелколистный

* **littleleaf cordia → 7765**

5003 **littleleaf leadtree; littleleaf leucaena**
(Leucaena retusa)
d kleinblättriges Weißfaden *n*
f laucène *m* à petites feuilles
r левкена *f* мелколистная

* **littleleaf linden [tree] → 7767**

5004 **littleleaf mock orange**
(Philadelphus microphyllus)
d kleinblättriger Pfeifenstrauch *m*
f seringa *m* à petites feuilles
r чубушник *m* мелколистный

5005 **littleleaf sumac[h]; desert sumac[h]**
(Rhus microphylla)
d kleinblättriger Sumach *m*
f sumac *m* à petites feuilles
r сумах *m* мелколистный

5006 **little myrtlelike British bilberry; myrtle blueberry; black whortle; myrtle whortleberry; wineberry**
(Vaccinium myrtillus; Vaccinium oreophilum)
d [echte] Heidelbeere *f*; Waldheidelbeere *f*
f myrtillier *m*; myrtille *f* [à l'airelle]; brimbelle *f*; brimbellier *m*; abrétier *m*
r черника *f* миртолистная

5007 **little sound knot; pearl knot**
d gesunder Kleinast *m*
f petit nœud *m* sain
r маленький здоровый сучок *m*

5008 **littletree willow**
(Chosenia arbutifolia; Salix arbutsculoides)
d Bäumchenweide *f*
f saule-arbousier *m*
r чозения *f* крупночешуйчатая; чозения арбутилистная; чозения толкнянколистная

5009 **little walnut; Texas walnut**
(Juglans microcarpa; Juglans rupestris)
d Texas-Walnuss *f*
f noyer *m* du Texas
r орех *m* скальный; орех аризонский

* **little wood → 5001**

* **littoral forest → 1687**

5010 **live burning**
(burning of green slash progressively as it is cut)
d Verbrennen *n* von frischem Schlagabraum
f brûlage *m* immédiat; incinération *f* des rémanents frais
r сжигание *n* порубочных остатков одновременно с заготовкой леса

5011 **live deck**
d Verkehrsdeck *n*
f table *f* d'alimentation pour billes
r приёмная площадка *f* для кратковременного хранения; подвижная эстакада *f*

* **live-edged → 7963**

* **live fence → 5019**

* **live hedge → 5019**

5012 **live heel boom attachment; tree-length attachment; heeler**
d mobile Auslegervorrichtung *f*
f balancier *m* à talon
r подвижный упор *m* на стреле (для брёвна при подъёме с помощью захвата)

5013 **live knot; living knot; green branch**
d grüner Ast *m*; lebender Ast; Bruch *m*
f nœud *m* vert; branche *f* vive; branche verte
r живой сук *m*; зелёный сук; живая ветвь *f*

5014 **live load**
d Verkehrslast *f*
f charge *f* vive

5015 live oak
(Quercus virginiana)
d virginische Eiche *f*
f chêne *m* de Caroline
r дуб *m* виргинский

* **live pruning → 3953**

* **liverish → 3773**

* **live roller → 3035**

5016 live sawing; through-and-through sawing
(ripping a log by making cuts roughly parallel to the longitudinal axis, normally without turning the log)
d Auftrennen *n*; Trennschneiden *n*
f débit *m* en plot
r распил *m* насквозь

5017 live skyline; slack[ing] skyline; slackline system; slacker
(skyline that can be raised and lowered during yarding to facilitate logging)
d bewegliche Drahtseilbahn *f*; Slacker *m*
f slacker *m*; cable *m* porteur mobile
r установка *f* с опускающимся несущим канатом

5018 live tree
d lebender Baum *m*
f arbre *m* vive
r живое дерево *n*

5019 living hedge; live hedge; quickset hedge; hedge[row]; palisade; [living] fence; live fence; plant hedge; zareeba; stockade; enclosure
(bushes or shrubs or trees planted in a row and trimmed; single or double row of trees and/or bushes around fields, beside roads, rivers and canals)
d [lebender] Zaun *m*; lebende Hecke *f*
f haie *f* [vive]; clôture *f* [vive]; enceinte *f* [vive]
r [живая] изгородь *f*

* **living knot → 5013**

* **living residues → 3454**

* **living soil cover → 6337**

5020 lizard
d Rückegabel *f*
f fourche *f* de débardage
r волокуша *f* из раздвоенного дерева;

* **lizard → 5115**

* **L-layer → 4999**

5021 L-notch planting
(special form of slit planting involving two slits at right angles with the seedling placed at the apex of the L)
d L-Klemmpflanzung *f*
f plantation *f* avec fentes en L
r посадка *f* под лопату в щель L-образного профиля

5022 load; ld
(measuring unit = 1.138 m^3 round timber or 1.416 m^3 beams)
d Last *f*; Ladung *f*
f charge *f*
r лоуд *m*

5023 load binder; binder [chain]; binding chain; binding wire; toggle chain; wrapper chain; fastening rope
(chain or wire rope used to bind logs)
d Spannkette *f*; Bindedraht *m*
f câble *m* de fixation; chaîne *f* de serrage; chaîne d'attache
r обвязочный канат *m*; обвязочная цепь *f*; вязальная проволока *f*

5024 loaded distance
d Lastabstand *m*
f distance *f* de charge
r расстояние *n* трелёвки в грузовом направлении

* **loaded rope profile → 1244**

5025 loading boom; derrick
(any structure projecting from a pivot point to guide a log when lifted)
d Ladebaum *m*; Arm *m* des Laders
f pieu *m* de chargement
r погрузочная стрела *f*; грузовая стрела

* **loading deck → 5027**

5026 loading dog
(a framework supporting a deck on to which logs are hoisted from the water for loading on cars)
d Ladeklammer *m*
f clameau *m* de chargement
r опорная рама *f* погрузочной площадки (на которую брёвна выгружаются из воды для последующей отгрузки)

* **loading ledge → 5027**

* **loading platform → 5027**

5027 **loading ramp; loading site; loading deck; loading platform; loading ledge; skid platform; stillage**
(any place where round timber is assembled for further transport, commonly with a change of method)
d Ladebühne *f*; Verladeplatz *m*; Verladerampe *f*; Beladepunkt *m*; Holzverladeplatz *m*; Rahmenbühne *f*; Ladebrücke *f*; Ladesteig *m*
f rampe *f* de chargement; plate-forme *f* de chargement; quai *m* de chargement [forestier]; embarcadère *m*
r погрузочная площадка *f*; погрузочная эстакада *f*; [лесо]погрузочный пункт *m*; погрузочная платформа *f*

* **lobed leaf → 1636**

* **lobes → 4876**

5028 **lobe sinus; sinus**
(the cleft or space between two lobes of a leaf blade)
d Blattlappen-Sinus *m*; Sinus *m*
f sinus *m* [de lobe]
r полость *f* лопасти; полость; каверна *f*

5029 **loblolly pine; taeda pine; torch-pine; frankincense pine; old-field pine; rosemary pine**
(Pinus taeda)
d Loblollykiefer *f*; Fackelbaum *m*; Weihrauchkiefer *f*; Rosmarin-Fichte *f*
f pin *m* taeda; pin à l'encens; pin du Texas; pin de romarin
r сосна *f* ладанная

5030 **loblolly pine sawfly**
(Neodiprion taedae)
d Loblollykiefernblattwespe *f*
f lophyre *m* de pin taeda
r пилильщик *m* сосны ладанной

5031 **loblolly-shortleaf pine forest**
(forests in which loblolly pine, shortleaf pine, or other southern yellow pines, except longleaf or slash pine, singly or in combination, constitute a plurality of the stocking)
d Loblolly-Glattkieferwald *m*
f forêt *f* de pins taeda et pins doux

* **lobstick → 641**

5032 **local forest**
d lokaler Wald *m*
f forêt *f* locale
r лес *m* местного значения

5033 **local forest authority**
d lokale Forstdienststelle *f*
f autorité *f* d'abattage local
r местные органы *mpl* управления лесов

5034 **local forest road**
d lokale Forstbahn *f*
f chemin *m* forestier local
r лесная дорога *f* местного значения

* **locality class → 7664**

5035 **localized clear felling**
d ortsgebundener Kahlhieb *m*
f coupe *f* à blanc localisée
r локальная сплошнолесосечная рубка *f*

5036 **local resource use plan; LRUP**
(plan approved by the district manager for a portion of the provincial forest that provides area-specific resource management objectives for integrating resource use in the area)
d Nutzungsplan *m* von lokale Ressourcen
f plan *m* d'utilisation des ressources locales
r план *m* использования местных ресурсов

5037 **local volume table; regional volume table**
d lokale Massentafel *f*
f table *f* du cubage locale
r местная массовая таблица *f*

5038 **location of felling; lumber site; coupe; setting**
d Holzfällplatz *m*
f chantier *m* [forestier]; chantier d'exploitation forestière; assiette *f* de coupe; parcelle *f* de coupe; parterre *m*; trouée *f* de coupe; canton *m*; coup *m*
r лесозаготовительной участок *m*; вырубаемый участок леса

5039 **lock for log floating; logging dam; dam for driving timber; rolling dam; splash dam**
(a dam for raising the water in a shallow stream)
d Klausdamm *m*; Klausbau *m*; Klause *f*
f écluse *f* pour le flottage des bois; barrage *m* de retenue d'eau en vue du flottage

* **lock saw → 4702**

* **locustberry → 5138**

* **locust tree → 1341**

* **LOD → 4802**

* **lodge → 4116**

5040 **lodged plant; beaten down plant**
d niedergedrückte Pflanze *f*
f plante *f* écrasée
r придавленный саженец *m*

5041 **lodged tree; hang-up tree; hung [tree]**
(a tree leaning against another tree or object which prevents it from falling to the ground)
d angelehnter Baum *m*; hängengebliebener Baum; überhängender Fallbaum *m*
f arbre *m* pendant; arbre encroué
r зависшее [спиленное] дерево *n*; повиснувшее при рубке дерево; завешенное дерево

5042 **lodge[ment]; lodging; hanging; hang up**
(in felling, having a tree catch on another so that it becomes lodged)
d Hängenbleiben *n*; Einigung *f*; Häuschen *n*
f encrouage *m*; verse *f*
r зависание *n* (о дереве при валке); опирание *n* в препятствия

5043 **lodgepole cone beetle; ponderosa-pine cone beetle**
(Conophthorus ponderosae; Conophthorus contortae)
d Goldkiefernzapfenkäfer *m*
f scolyte *m* des cônes du pin lodgepole; perceur *m* de cône de pin pondéreux
r короед *m* сосны карликовой; короед, повреждающий шишки сосны жёлтой

5044 **lodgepole pine; contorta pine; shore pine; beach pine**
(Pinus contorta)
d Lodgepole-Kiefer *f*; Küstenkiefer *f*; Murray *m*
f pin *m* lodgepole; pin de Murray
r сосна *f* скрученная; сосна широкохвойная; сосна Муррея

5045 **lodgepole pine beetle**
(Dendroctonus murrayanae)
d Drehkiefernbastkäfer *m*
f dendroctone *m* du pin tordu
r лубоед *m* сосны карликовой

* **log → 1122, 8879**

* **logan → 6415**

5046 **log [and lumber] fork**
d Klauenseuche *f*
f fourche *f* combinée; fourchet *m*
r вила *f* для брёвен; вила для леса

5047 **logania family**
(Loganiaceae)
d Brechnussgewächse *pl*; Logangewächse *pl*
f loganiacées *pl*
r логаниевые *pl*

5048 **log axis; log axle**
d Klotzachse *f*; Holzscheitachse *f*; Blockachse *f*; Holzachse *f*; Stammachse *f*
f axe *m* de bille; axe de grume
r ось *f* брёвна; ось ствола дерева

5049 **log band-saw; log band mill; band mill; band saw; panel saw; drum saw; ribbon saw; strap saw; belt saw**
d Blockbandsäge *f*; Bandsäge *f*
f scie *f* à ruban [de grumes]; scie à lame-ruban; lame-ruban *f*
r ленточная пила *f* [для распиловки кряжей]

5050 **log blue; blue rot**
(a bluish or grayish discolo[u]ration of the sapwood caused by the growth of certain dark-colo[u]red fungi)
d Bläue *f*; Blaufäule *f*; Blauschaden *m*; Klotzbläue *f*; Bläuebefall *m*
f bleuissement *m* de la grume
r синева *f* древесины; синь *f*; синяя гниль *f*

* **log boat → 5115**

* **log book → 3073**

* **log boom → 914**

* **log bracket → 8577**

5051 **log breakdown pattern; felling pattern; cutting pattern**
d Einschnittmodell *n*
f modèle *m* d'abattage; schéma *m* de débit
r схема *f* разработки лесосек

* **log bumper → 1228**

* **log bunk → 5094**

5052 **log cabin; log house; log hut; Canadian**

d Blockhaus *n*; Blockhütte *f*
f maison *f* de bois; maison en rondins; hutte *f* en rondins; cabane *f* en rondins
r бревенчатый дом[ик] *m*

* **log carriage → 5199**

* **log cart → 5199, 5208**

* **logcart sulky → 5090**

5053 **log centering**
d Stammzentrierung *f*
f centrage *m* de la grume
r центрирование *n* брёвна (при окорке или лущении)

* **log chain → 4173**

* **log chipper → 9391**

* **log clamp → 5094**

* **log-cleaner saw → 4312**

* **log conversion → 5100**

5054 **log counter**
d Blockzähler *m*
f compteur *m* de grumes
r счётчик *m* брёвен

* **log cradle → 4840**

* **log cribbing → 912**

5055 **log cross-conveyor**
d Querförderband *n*
f élévateur *m* transversal de bois
r поперечный лесотранспортёр *m* для брёвен; поперечный ленточный конвейер *m*

5056 **log cutoff saw; cutoff saw; butt saw**
d Abkappsäge *f*; Kappsäge *f*; Abkürzsäge *f*
f scie *f* à tronçonner; scie de tronçonnage; tronçonneuse *f*
r торцовочная пила *f*

* **log cutter → 9526**

* **log deck → 5451**

5057 **log-deck substructure**
d Stapelplatzunterkonstruktion *f*;
r основа *f* площадки для разгрузки брёвен

* **log dog → 8577**

5058 **log drag; drag**
(load of logs or trees being skidded)
d Stammholzfuder *n*
f chargement *m* de grumes pour un véhicule
r телега *f* с хлыстами

* **log driver → 9569**

5059 **log dump; storage place on land**
d Landablage *f*
f terre *f* d'alluvion
r склад *m* [у пристани]

5060 **log elevator; log lever**
d Stammholzelevator *m*
f élévateur *m* de grumes; [convoyeur-]élévateur *m* de billes
r элеватор *m* для брёвен; бревноподъёмник *m*

5061 **log feeding; log intake**
d Blockzufuhr *f*; Stammzuführung *f*
f alimentation *f* des grumes
r подача *f* брёвна [на распиловку]

* **log fire → 8072**

5062 **log flipper; flipper; timber tosser**
d Aufladvorrichtung *f*; Ladegerät *n*; Ladevorrichtung *f*
f éjecteur *m* à bras relevables
r погрузчик *m*; устройство *n* для загрузки

* **log floater → 9569**

* **log floating → 8578**

* **log flume → 9281**

5063 **log for aircraft production**
d Flugzeugholz *n*
f bois *m* aéronautique
r древесина *f* для авиастроения; авиационный кряж *m*; авиакряж *m*

5064 **log for pole**
d Pfostenstamm *m*
f grume *f* de perches
r бревно *n* для столбов

5065 **log for ski production**
d Skiholz *n*

5066 **logged[-over] area; logged-over forest; cutover area**
(forest in which most or all of the merchantable timber has been removed)
d abgerechnete Fläche *f*; ausgeschlachtete Fläche; ausgeschlachteter Wald *m*
f aire *f* exploitée; surface *f* vidée de ses bois; forêt épuisée; forêt vidée de ses bois de valeur commerciale; bûché *m*
r расчищенная от леса область *f*; использованный лес *m*

5067 **logger**
(a person which stacks logs on ground or on truck bed, or fastens choker around log and drives tractor to skid log to loading area)
d Logger *m*
f débardeur *m*
r лесопогрузчик *m*

* **logger → 8607**

* **logger road → 7689**

* **log[ging] arch → 5090**

* **logging area → 2189**

5068 **logging blade**
(of a tractor)
d Planierschar *n*; Planierschild *m*; Polterschild *m*
f soc *m*; lame *f*
r бульдозерный отвал *m*; бульдозерный нож *m* (трелёвочного трактора)

5069 **logging boss; forest foreman; head wood cutter; saw boss; head bucker**
(supervisor or foreman of a group of fallers, normally in forest logging operations)
d Holzrottmeister *m*
f chef *m* de bûcherons
r бригадир *m* лесорубов; десятник *m* лесорубов

5070 **logging by the owner**
d Eigenhieb *m*
f exploitation *f* forestière en régie; coupe *f* en régie
r рубка *f* на свой счёт

5071 **logging camp; lumber camp; camp; shelter-box**
d Holzhauer-Camp *n*; Waldarbeitersiedlung *f*; Holzhauerhütte *f*
f cabane *f* de bûcherons; abri *m* de bûcherons;
r ... для лесорубов

* **logging chance → 2189**

* **logging company → 3355**

* **logging contract → 2192**

5072 **logging contractor; felling contractor; gyp[p]o** (*US*)
(independent logger who usually runs a small-scale logging operation)
d Fällungsunternehmer *m*; Holzeinschlagsunternehmer *m*
f exploitant *m* forestier; entrepreneur *m* d'abattage
r подрядчик *m*, ведущий лесозаготовки по контракту; предприниматель *m* рубки

* **logging costs → 8217**

* **logging crane → 8609**

* **logging crew → 9527**

* **logging cycle → 2194**

* **logging dam → 5039**

* **logging damage → 7685**

* **logging debris → 998**

* **logging depot → 3371**

5073 **logging division**
d Holzaufbereitungsabteilung *f*
f division *f* du matériel forestier
r лесозаготовительный участок *m*

5074 **logging enterprise; lumbering undertaking**
d Hauungsbetrieb *m*; Holzaufbereitungsbetrieb *m*; Holzbeschaftungsbetrieb *m*; Holz[ein]schlagsbetrieb *m*
f entreprise *m* d'exploitation [forestière]
r лесозаготовительное предприятие *n*

* **logging equipment → 4157**

* **logging industry → 5206**

5075 **logging locomotive; corkscrew**
d Holz[ein]schlagslokomotive *f*
f locomotive *f* de débardage
r лесовозный паровоз *m*

5076 logging management
d Holzaufbereitungsverwaltung *f*
f gestion *f* de débardage
r управление *n* лесозаготовками

5077 logging mat; mat
(temporary roadway constructed of hardwood lumber)
d Rückungspanvlies *n*
f tapis *m* de débardage
r лесовозный настил *m*

* **logging method → 9749**

* **logging-off → 9588**

* **logging operation → 3457**

* **logging period → 3072**

* **logging plan → 6332**

5078 logging [pole] trailer
d Holzanhänger *m*; Holzrückekarre *f*
f remorque *f* de bois
r лесовозный прицеп[-роспуск] *m*

5079 logging railroad; forest railroad; forest railway
d Waldeisenbahn *f*
f chemin *m* de fer forestier
r лесовозная железная дорога *f*; рельсовая лесовозная дорога

* **logging residues → 998**

* **logging right → 2955**

* **logging road → 7689**

* **logging road construction → 1889**

5080 logging road freight turnover
d Frachtumsatz *m* der Waldbahn
f rotation *f* de chargement de chemin forestier
r грузооборот *m* лесовозной дороги

5081 logging rule; rule for removal; order for removal
d Entfernungsregel *f*
f règlement *m* pour la vidange
r правило *n* вывозки

5082 logging safety
d Unfallschutz *m*; Unfallverhütung *f*

* **logging scissors → 3892**

* **logging season → 3072**

5083 logging setting; side
(area to be logged; a block or strip)
d Holzschlageinstellung *f*
f carreau *m* d'une coupe
r расположение *n* лесозаготовки

* **logging slash[ings] → 998**

5084 logging system; harvesting system
d Hiebssystem *n*; Holzeinschlagsystem *n*
f système *m* de coupe
r система *f* рубок; система лесозаготовок

5085 logging tax
d Hiebstaxe *f*
f impôt *m* sur l'exploitation forestière; impôt sur les opérations forestières; impôt frappant l'abattage du bois
r лесорубочная такса *f*

* **logging technique → 3076**

* **logging terminal → 4484**

5086 logging tire; logging tyre; skidder tire
d Rückenpneu *m*; Forstreifen *m*
f pneu *m* forestier
r трелёвочная шина *f*; шина трелёвочного трактора

* **logging tools → 4157**

* **logging track → 7689**

5087 logging truck-and-trailer unit; log[ging] truck; pole truck; timber truck
(vehicle used to transport logs, consisting of a cab, containing the engine and a place for the driver to sit, and a trailer on which logs are placed)
d Langholzwagen *m*; Langholztransportwagen *m*
f grumier *m*; carrosserie *f* pour [transport de] grumes
r лесовозный автомобиль[-тягач] *m*; лесовозный автопоезд *m*; [авто]лесовоз *m*; воз *m* для хлыстов

5088 logging truck loader
d Lader *m* eines Holzlastkraftwagens
f chargeur *m* de grumier
r загрузчик *m* лесовозного автопоезда

5089 logging unit; extraction unit; hauling unit
d Schreckeeinheit *f*
f unité *f* de débardage
r единица *f* вывозки

* **logging value → 8217**

* **logging vehicle → 7691**

5090 logging wheels; log[ging] arch; timber wheels; timber bob; yarding arch; [logcart] sulky; whim; tug
(any device attached to the back of a vehicle and used for raising one end of logs to facilitate movement)
d Rückebogen *m*; Sulky *n*; zweirädiger Rückekarren *m*; [zweirädiger] Rückewagen *m*
f avant-train *m* de débusquage à deux roues; arche *f* de débusquage; arche à grumes; arche à billes
r трелёвочный коник *m*; [прицепная] трелёвочная арка *f*; трелёвочная дуга *f*

* **logging winch → 7692**

5091 log gorge
d Stammschlucht *f*
f gorge *f* de grumes
r залом *m* (брёвен при лесосплаве)

* **log grab → 8577**

5092 log grading; log segregation; log survey
(classifying logs according to quality or end-use)
d Blocksortierung *f*
f triage *m* des grumes
r сортировка *f* брёвен

5093 log grading plant; grader station; grading station
d Blocksortierungsstation *f*
f poste *m* de classement des grumes
r сортировочная площадка *f* для брёвен

5094 log grapple; log clamp; log bunk
d Rundholzklammer *f*
f pince *f* de manutention; pince à grumes; esse *f*
r скоба *f* для круглого леса

* **log guide bar → 8779**

* **log guide rod → 8779**

* **log hauler → 5114**

* **log hauling → 8595**

5095 log-haul length
d Stammlänge *f* gleich dem Elevator
f longueur *f* de grumes, equale de chaîne élévatrice
r длина *m* лесоматериала, равной длине автопоезда

* **log-haul[-up] → 4600**

* **log hoist → 7692**

* **log house → 5052**

* **log hut → 5052**

* **log intake → 5061**

* **log jack → 4600**

5096 log[ ]jam[ming]; jam[ming]
(a stoppage or congestion of logs in a stream, due to an obstruction or to low water)
d Blochholzstauung *f*; Holzpropfen *n*; Fuchs *m*
f embâcle *m*; empilage *m* de billes; blocage *m* de bois de flottage
r затор *m* брёвен

5097 log kicker; kicker
d Blockabwerfer *m*; Blockabwurfvorrichtung *f*; Holzabstreifer *m*; Holzabwerfer *m*; Stammabladevorrichtung *f*
f système *m* de déplacement pour billes de bois; éjecteur *m* [de grume]
r сбрасыватель *m* [брёвен]; бревносбрасыватель *m* (с лесотранспортёра); устройство *n* для подачи кряжей

* **log-length logging → 3078**

* **log lever → 5060**

* **log lift → 4600**

5098 log lifting carriage
d Blockhebewagen *m*
f chariot *m* élévateur de bois
r тележка *f* для подъёма бревен

* **log loader → 8607**

5099 log loading
d Stammladung *f*

* **log maker → 2043**

5100 **log making; log conversion; bucking; cutting into lengths; shortening; severing**
(cutting a felled tree into specified log lengths for yarding and hauling; also, making any bucking cut on logs)
d Holzaufarbeiten *n*; Holzeinschnitt *m*; Einschneiden *n*; Holzschneiden *n*
f tronçonnage *m* [du bois]; débitage *m*
r раскряжёвка *f*; распиловка *f* [брёвен]; разделка *f* [брёвен]; разделка леса; лесопиление *n*; формовка *f*

* **logman → 9654**

5101 **log mark; bark mark; side mark; catch mark**
(a symbol chopped into the side of a log to indicate its ownership; used to identify logs)
d Holzzeichen *n*
f empreinte *f* sur grume forestier
r клеймо *n* брёвна; клеймо на бревне; марка *f* брёвна

5102 **log marker; bark marker**
(one who cuts the bark marks on logs)
d Aufreißer *m*; Messarbeiter *m*; Vermesser *m*
f marqueur *m* des grumes; marqueur des longueurs
r разметчик *m* хлыстов при раскряжёвке

5103 **log measure; log measuring; log scaling; block measure**
d Block[rund]maß *n*; Stammskalierung *f*
f mesure *f* des longueurs; mesure de volume de billes
r мера *f* для бревен; измерение *n* кубатуры; определение *n* объёма брёвна

* **log measuring → 5103**

* **log off → 2168**

* **log pass → 5116**

5104 **log passage**
d Holzstammpassage *f*
f passage *m* des grumes
r лесопровод *m*

5105 **log pond; mill pond; floating basin; rafting reservoir; storage pond**
(an area of water for storing logs, generally attached to a saw- or veneer mill, to which the logs travel via the log haul-up)
f bassin *m* de stockage; bassin à grumes; bassin de flottage; bassin de réserve; port *m* flottable; lieu *m* d'atterrissage de bois flotté
r бассейн *m* для балансов; бассейн для брёвен; бассейн для хранения леса

5106 **log-processing equipment; sawmill equipment**
d Sägewerkeinrichtung *f*; Sägemühleanlage *f*
f équipement *m* de traitement des grumes
r лесопильное оборудование *n*

5107 **log rule; log scaling rule; scaling stick; scale stick; yard stick**
(calibrated stick used to estimate wood volume in a log)
d Messstab *m*
f règle *f* graduée de cubage; ficelle *f* des marchands de bois
r мерная линейка *f* [для обмера брёвен]; градуированная линейка для измерения лесоматериалов; брёвномер *m*

* **log rule → 7112**

* **log run → 7134**

* **log running → 8578**

5108 **log salvager**
d Blockholzberger *m*
f récupérateur *m* de grumes
r заготовитель *m* повреждённых лесоматериалов

* **log [saw] carriage → 5208**

5109 **log scale**
(the lumber contents of a log or run of logs, as determined by a log rule)
d Blockholzmassenermittlung *f*
f mesure *f* de masse des billes
r мера *f* древесной массы брёвен

5110 **log scale method**
(any system of measuring round timber)
d Methode *f* der Blockholzmassenermittlung; Blockholzmassenermittlungsmethode *f*
f méthode *f* de cubage des billes
r метод *m* обмера брёвен

5111 **log scaler**
d Blockholzzähler *m*
f mesureur *m* de billes
r обмерщик *m* брёвен

5112 **log scale sale**

f vente *f* par volume de billes en pied-planches
r продажа *f* леса на корню по цене за единицу объёма фактически заготовленной древесины

* **log scale table → 7112**

* **log scaling → 5103**

* **log scaling rule → 5107**

* **log sectioning plant → 8637**

* **log segregation → 5092**

5113 **log shear**
d Holzeinschnittschere *f*
f cisaille *f* de tronçonnage de billes; cisaille à billes
r ножевое раскряжёвочное устройство *n*; раскряжёвочный нож *m*

* **logs kept under water → 9302**

5114 **log skidder; skidder; log hauler; tracked log skidder**
(self-propelled machine designed to transport trees or pans of trees by trailing or dragging)
d Bringungsschlepper *m*; Holzkraftschlepper *m*; Raupenschlepper *m*; Waldschlepper *m*
f débardeur *m*; débusqueuse *f*
r трелёвочный механизм *m*

* **log skidding grapple → 3892**

5115 **log sledge; log sled; yarding sled; drag sled; single sled; log boat; skidding sled; bob crotch; bob; joe log; travois; dray; scoot; lizard**
(a single sled used for yarding logs)
d Rückeschlitten *m*; Rückevorrichtung *f*; Holzschlitten *m*
f traîneau *m* de débusquage
r трелёвочные сани *pl*; сани для трелёвки брёвен в полупогруженном положении; подсанки *pl* для трелёвки

5116 **log slide; timber slide; timber slip; log pass; timber chute; timber[-]pass; log sluice; sluiceway; [running] slide; timber shoot; road slide; dirt slide; drag road**
(a device for moving timber past rapids and waterfalls)
d Stammriese *f*; Holzriese *f*; Holzrutsche *f*; Schleifweg *m*; Schleppweg *m*; Ziehweg *m*; Holz[ge]rinne *f*; Gerinne *f*; Riesbahn *f*;
glissoire à bois; glissoire à grumes; passe-billes *f*; lançoir *m*; passe *f* à bois
r бревноспуск *m*; эстакада *f*; [лесоспускной] лоток *m*; лесоспуск *m*; жёлоб *m*

* **logs of deciduous trees → 1047**

* **log splitter → 9512**

* **log spur → 4172**

5117 **log squaring; slab edging; slabbing**
d Kantschnitt *m*; Prismenschnitt *m*
f équarrissage *m*
r брусовка *f*; распиловка *f* с брусовкой

5118 **log stain**
d Rundholzverfärbung *f*
f tache *f* de grume
r окраска *f* брёвен

5119 **log stop**
d Blochauffangvorrichtung *f*
f système *m* d'arrêt des billes
r устройство *n* для подхватывания кряжей

5120 **log storage; wood storage; timber stockage**
d Holzlagerung *f*
f conservation *f* de bois
r хранение *n* древесины

* **log survey → 5092**

* **log trade → 8653**

5121 **log tramway; tramway**
(a light, often temporary, railroad, sometimes with wooden rails, formerly much used for transporting round timber)
d Tram[bahn] *f*
f tramway *m* forestier
r временная дорога *f*; круглолежневая лесовозная дорога

* **log trim → 1336**

5122 **log trucker**
d Holzlastwagenfahrer *m*
f entrepreneur *m* de transports forestiers
r владелец *m* лесовозного автомобиля

5123 **log trucking**
d Transport *m* mit Holz-LKW
f transport *m* de bois par grumier
r автолесовозный транспорт *m*

5124 **log turner; steam nigger; nigger; crab**
(mechanically operated hook to handle logs)
d Blochwender *m*; Blochwendevorrichtung *f*
f tourne-billes *m*; tourne-grumes *m*; renard *m* mécanique; tourne-bois *m*
r устройство *n* для поворота бревен [впереди рамной лесопильной тележке]; переворачиватель *m* брёвен

5125 **log volume**
d Schnittholzmaß *n*
f volume *m* de bois rond; volume de billes; volume de bois abattus
r общий объём *m* брёвна; объём брёвен

* **log volume table → 7112**

* **log wagon → 9256**

* **logway → 7689**

* **log yard → 8664**

* **log yield → 9780**

5126 **Lombardy poplar**
(Populus pyramidalis)
d Alleepappel *f*; Pyramidenpappel *f*
f peuplier *m* d'Italie
r тополь *m* пирамидальный; тополь итальянский

* **London plane[ ]tree → 2906**

* **London sycamore tree → 2906**

* **longan → 1381**

* **longbeak eucalyptus → 6786**

* **long beams → 5154**

5127 **long-boled; long bodied**
d hochstämmig
f élancé
r высокоствольный; длинномерный

5128 **long butt**
(section cut from the bottom log of a tree and culled because of rot or other defect)
d langer Stammabschnitt *m*
f bille *f* de pied longue
r длинная откомлёвка *f*; нижний отрезок *m* ствола

5129 **longbutt tree; thick-butted tree; tree with**
Baum mit trompetenförmigen Wurzelanlauf
f arbre *m* avec empattement renforcé; arbre en forme de pavillon
r закомелистое дерево *n*; дерево с ядровой гнилью

* **long-cut wood → 5154**

5130 **long-distance skyline**
(skyline more of 1 km)
d Fernluftseilbahn *f*
f câble *m* à distance
r воздушная канатная установка *f* большой протяжённости

5131 **long-fibered wood**
d langfaseriges Holz *n*
f bois *m* en longues fibres
r длинноволокнистая древесина *f*

5132 **long-frame saw; frame saw; sash [gangsaw]; stock gang; gang mill; sawmill; gate saw**
(a series of blades that are fixed vertically between two horizontal members; used to rip cants)
d Gattersäge[maschine] *f*; Sägegatter *n*; Rahmensäge[maschine] *f*; Vollgatter *n*
f scie *f* alternative multiple; scie [alternative] à châssis; scie à cadre
r лесопильная рама *f*; рамная пила *f*; многодисковая пила; пилорама *f*

5133 **long grain; side grain**
(of wood)
d Längsfaser *f*
f fil *m* en long
r продольное расположение *n* волокон; текстура *f* продольного распила (волокон древесины)

5134 **longhorn beetle damage; house longhorn beetle damage**
d Hausbockschaden *m*; Balkenbockschaden *m*
f dommage *m* par capricorne des maisons; défaut *m* par longicorne des charpentes
r червоточина *f* овальной формы

* **long-horned beetles → 1327**

* **longhorns → 1327**

* **longitudinal cleavage → 4915**

5135 **longitudinal cutting; length sawing; ripcut; rip sawing; [resawing] ripping**

f coupe *f* longitudinale; sciage *m* en long; débit *m* en long; dédoublage *m*; débit à traits parallèles
r продольная распиловка *f*; продольная резка *f*; продольный распил *m*; продольное распиливание *n*; распил, параллельный направлению волокон

* **longitudinal fissure → 7729**

5136 **longitudinal opening**
d Schlitzöffnung *f*
f ouverture *f* en jalousie
r продольное раскрытие *n*

* **longitudinal parenchyma → 457**

* **longitudinal splitting → 4915**

5137 **longitudinal wall**
(of wood cell)
d Längswand *f*
f membrane *f* longitudinale
r продольная стенка *f*

5138 **Long Key locustberry; locustberry**
(Byrsonima lucida; Byrsonima coriacea)
d Nancitebaum *m*
f bois *m* tan
r бирсонима *f*

* **longleaf → 5142**

* **longleaf cucumber-tree → 3688**

5139 **longleaf glorytree**
(Tibouchina longifolia)
d langblätteriger Veilchenbaum *m*
f bois *m* dents marron
r тиб[о]ухина *f* длиннолистная

* **longleaf Indian pine → 1543**

5140 **longleaf peppertree**
(Schinus longifolius)
d langblätteriger Pfefferbaum *m*
f poivrier *m* à feuilles longues
r шинус *m* длиннолистный

* **longleaf pine → 5141**

5141 **longleaf-slash pine; longleaf pine; American pitch pine; Georgia pine; southern [yellow] pine**
(Pinus palustris)
d Südkiefer *f*; Sumpfkiefer *f*; amerikanische
f pin *m* du sud; pin de marais; pitchpin *m* américain; pin de Boston; pin palustre
r сосна *f* [длиннохвойная] болотная; южная [жёлтая] сосна

5142 **long-leaved; longleaf**
d langblätterig
f à longues feuilles
r длиннолистный

* **long-length logging → 5615**

* **long-length [stem] → 5154**

5143 **long-lobed**
d langgelappt
f longue lobé
r удлинённолопастный

* **long log → 5154**

5144 **long logging; multiple-stick length operation**
d Langholzbringung *f*
f exploitation *f* en bois longs; exploitation par longues billes
r лесозаготовка *f* с трелёвкой хлыстов

* **long-log transport → 5615**

5145 **long pile; long stack**
d Langholzstapel *m*
f pile *f* de bois longs
r штабель *m* длинномерных лесоматериалов

5146 **long pocket-rot**
(invader - Hydnum abietis)
d Langlochfäule *f*
f pourriture *f* alvéolaire longue
r продолговато-ямчатая гниль *f*

* **long pole → 5154**

5147 **long run sustainable yield; LRSY**
(equal to the culmination of mean annual increment weighted by area for all productive and utilizable forest land types)
d langfristige dauerhafte Ernte *f*
f rendement *m* équilibré à long terme; production *f* équilibrée à long terme
r устойчивая крупносерийная продуктивность *f*; стабильный крупносерийный доход *m*

5148 **long saw; cleave saw; slab-saw; ripping saw; ripsaw**
(a handsaw with coarse teeth which have but

d Aufteilsäge *f*; Trennsäge *f*; Spaltsäge *f*
f scie *f* de menuisier; dédosseuse *f*
r делительная пила *f*; ребровая пила; пила для окантовки брёвен; пила для кантования брусьев

5149 **long shoot; extending shoot**
(where there is shoot dimorphism, an extended shoot contributing to the architecture of a tree)
d Langtrieb *m*
f pousse *f* allongée
r удлинённый побег *m*; длинный побег

5150 **long-span skidding**
(cable system capable of skidding logs for 3,000 feet or more)
d Großraum-Holzabfuhr *m*; weitspannender Holzabfuhr *m*
f débusquage *m* à grande portée; débusquage à longue portée
r трелёвка *f* большого пролёта; трелёвка предельного пролёта

* **long stack → 5145**

5151 **long-stemmed**
d langschäftig; langstämmig
f longiscape
r длинноствольный

* **long-tailed timber → 5154**

5152 **long-term action**
(action governed by a plan generally taking place over a period (specified) longer than 20 years from the present)
d langfristige Aktion *f*
f action *f* à long terme
r длительное действие *n*

5153 **long-term sustained yield timber capacity**
(the highest uniform wood yield that may be achieved and sustained from lands being managed for timber production under a specified management intensity and set of constraints)
d langfristige Nachhaltsertragsholzleistung *f*
f rendement *m* de bois soutenu à long terme
r устойчивая длительная производительность *f* леса

5154 **long timber; stem timber; longwood; long-cut wood; long beams; multiple-stick length; long-length [stem]; multiple-length log; long log; long-tailed timber; long pole;**
15 feet)
d langschäftiges Holz *n*; langstückiges Holz; Langholz *n*
f bois *m* long; longueur *m*
r длинномерный сортимент *m*; длинномерные лесоматериалы *mpl*; долготьё *n*; длинное бревно *n*; [древесный] хлыст *m*

5155 **longwood crew**
d Langholzbesatzung *f*; Langholzmannschaft *f*
f équipe *f* d'exploitation de bois long
r бригада *f*, заготавливающая длинномерные лесоматериалы

5156 **longwood harvester**
d Langholz-Erntemaschine *f*
f abatteuse-porteuse *f* de bois long
r машина *f* для заготовки долготья

5157 **longwood harvesting**
(timber harvesting method in which harvested trees are moved to the landing either as whole trees or as topped and limbed tree-length logs)
d Langholzernte *f*
f exploitation *f* de bois long
r заготовка *f* долготья

* **lookout tower → 3216**

* **looper → 3174**

* **loose floating → 2625**

* **loose-flower hornbeam → 4621**

5158 **loose limb; loose[-textures] knot**
(a knot that is not held firmly in place by growth or position and that cannot be relied upon to remain in place)
d unverwachsener Ast *m*; lockerer Durchfallast *m*; mürber Ast
f nœud non adhérent; nœud sautant
r несросшийся сучок *m*; выпадающий сучок; рыхлый сучок

* **loose measure → 7978**

5159 **loosened grains; unconsolidated grains**
(loosened small portions of the wood on the flat-grained surfaces of boards, usually of the tops and edges of the growth increments)
d lockerte Holzfaser *fpl*; ungeenigte Holzfaser
f fils *mpl* écailleux; fils déconsolidés; fils creux
r кольцевые трещины *fpl* между осенними и

* **loosened tree → 9309**

* **loosening of close stand → 4566**

* **loosening-out → 8515**

* **loose rock-dump → 6969**

* **loose rot → 7819**

* **loose scale → 7978**

* **loose side → 7716**

5160 **loose slash**
d loser Schlagabraum *m*
f rémanents *mpl* tendres; rémanents libres
r рыхлые порубочные остатки *mpl*

5161 **loosestrife family**
(Lythraceae)
d Weiderrichgewächse *pl*; Lythrazeen *pl*
f lythracées *pl*
r дербенниковые *pl*

5162 **loose-textured wood**
d Holz *n* mit lockere Textur
f bois *m* à texture lâche
r древесина *f* рыхлой структурой

* **loose[-textures] knot → 5158**

* **lop → 1645**

5163 **lop-and-top; top-and-lop**
d Äste *mpl* und Krone *f*
f branches *fpl* et cime *f*
r сучья *mpl* и вершина *f*[, срубаемые с поваленного дерева]; сучья с кроной

5164 **lophira family**
(Ochnaceae)
d Grätenblattgewächse *pl*
f ochnacées *pl*
r охновые *pl*

5165 **lopped tree**
d Schneidelstamm *m*; Schneitelstamm *m*
f arbre *m* à émondage
r подсечное дерево *n*

* **lopper → 8825**

* **lopping → 6610**

5166 **lopping and scattering**
ground without burning)
d Ästung *f* und Streuung *f*
f déchiquetage-épandage *m*; façonnage-épandage *m*
r приземление *n* и разбрасывание *n* (порубочных остатков)

5167 **lopping axe; pruning axe; trimming axe; delimber axe**
d Astaxt *f*; Ästeaxt *f*
f hache *f* d'ébranchage
r топор *m* для срубания сучьев; сучкорубный топор

* **lopping-bunching machine → 4964**

* **loppings → 5170**

* **lopping saw → 6614**

5168 **lopping shears; pruning shears; [tree-]pruner; gardening shears; pruning[-]scissors; secateur**
d Astschere *f*
f ébranchoir *m*; sécateur *m*
r секатор *m*; ножницы *fpl* для обрезки ветвей; ножничная ветверезка *f*

5169 **lopping tree**
d Ausästenbaum *m*
f arbre *m* ébranché
r дерево *n* с обрезанными ветвями первого порядка

5170 **lops; loppings; lopwood; branchwood; branch timber; limbwood; slashes; browst; trim**
(branches cut from a tree, generally once felled or fallen)
d Astreisig *n*; Astreiserholz *n*; Reisholz *n*; ästiges Holz *n*; Astholz *n*; Geäst[e] *n*; Gezweig *n*; Knotenholz *n*
f résidus *mpl* d'ébranchage-façonnage; rémanents *mpl* d'ébranchage-façonnage; menu-branchage *m*
r сучья *mpl*; сучняк *m*

5171 **lop[-sided] development**
(of a crown)
d einseitige Entwicklung *f*
f développement *m* unilatéral
r однобоковое развитие *n*; одностороннее развитие

* **lopwood → 5170**

d japanische Vollmispel *f*; japanische Mispel *f*; Loquate *f*
f [faux] néflier *m* du Japon; bibacier *m*; bibas *m*
r мушмула *f* японская

5173 **Lorentz red quebracho; red quebracho**
(Schinopsis lorentzii)
d roter Querbrachobaum *m*
f quebracho *m* rouge
r квербрахо *n* Лоренца

* **lorry-driver → 8983**

* **LOSP → 4947**

5174 **loss due to haulage; skidding losses**
d Bringungsverlust *m*; Kalo *m*
f pertes *fpl* au débardage
r потеря *f* при трелёвке; потеря при вывозке

* **loss due to sawing → 5178**

* **losses in bark peeling → 580**

5175 **losses in harvesting**
d Ernteverluste *mpl*
f pertes *fpl* d'exploitation; frais *mpl* d'exploitation
r лесозаготовительные затраты *fpl*

5176 **loss factors**
(reductions made to gross timber volumes to allow for decay, waste, and breakage)
d Verlustfaktoren *mpl*
f facteurs *mpl* de pertes
r факторы *mpl* потерь

* **loss from sunken timber → 5177**

5177 **loss in floating; loss from sunken timber; driving loss**
d Triftverlust *m*
f perte *f* au flottage
r потеря *f* от молевого лесосплава

5178 **loss in saw kerf; loss due to sawing**
d Schnittverlust *m*; Verschnittabschlag *m*
f perte *f* au sciage
r потеря *f* от распиловки; потеря на пропил

5179 **loss of nutrients; nutrient depletion; nutrient deficiency**
d Nährstoffverlust *m*; Nährstoffmangel *m*
f perte *f* en matières nutritives; défaut *m* en matières nutritives
r потеря *f* питательных веществ;

5180 **loss of wood substance**
d Holzsubstanzverlust *m*
f perte *f* de lignocellulose
r потеря *f* древесной массы

5181 **lot of timber; parcel of timber; portion of timber; stint**
d Holzpartie *f*; Los *n*
f partie *f* de bois; lot *m*
r партия *f* древесины; билет *m*

* **Louisiana cypress → 786**

* **lovely fir → 6025**

* **low birch → 8304**

* **low bush → 9065**

* **lowbush blueberry → 1788**

5182 **low bush tundra**
d Zwergstrauchtundra *f*
f toundra *m* à sous-arbrisseaux
r кустарничковая тундра *f*

5183 **low-density fiberboard; softboard**
d Faserplatte *f* geringer Dichte
f panneau *m* en fibres à faible densité; panneau de fibres léger
r древесноволокнистая плита *f* низкой плотности; мягкая древесноволокнистая плита

* **low-density stand → 4446**

* **low-density timber → 8660**

* **low-density wood → 8660**

5184 **low-density wood chipboard**
d Leichtspanholzplatte *f*
f panneau *m* de particules de bois à faible densité
r лёгкая древесностружечная плита *f*

* **Lowell ash → 7641**

5185 **lower crown height**
d untere Kronenhöhe *f*
f hauteur *f* à la première branche; hauteur inférieure des cimes
r нижняя высота *f* расположения кроны (от первой живой ветви, включая порослевые побеги)

f longueur *f* de cime totale
r нижняя длина *f* кроны (до первой живой ветви)

5187 **lower cut**
d unterer Schnitt *m*
f coupe *f* inférieure
r нижний рез *m* (подпила)

* **lower depot → 4484**

* **lower-end → 607**

* **lower face → 8602**

* **lower landing → 4484**

5188 **lower mountain forest**
d unterer Gebirgewald *m*; unterer Bergwald *m*
f forêt *f* montagnarde inférieure
r нижний горный лес *m*

* **lower story → 9070**

5189 **low forest**
d Niederwald *m*
f forêt *f* basse
r низколесье *n*; низкоствольник *m*

5190 **low forest cover**
d Niederwaldbedeckung *f*
f couvert *m* de taillis
r полог *m* низкоствольника

5191 **low-grade stand; low-quality stand**
(contains either a substantial number of low-quality trees, oran insufficient number or low basal area)
d geringwertiger Stamm *m*
f bas-perchis *m*
r низкопродуктивное насаждение *n*; низкокачественное лесонасаждение

* **low-grade wood → 5195**

* **low growth → 9060**

* **lowland fir → 3910**

5192 **lowland forest; bottomland [riparian] forest; floodplain forest; streamside forest**
(wooded area found within low-lying flood-prone areas)
d Auwald *m*; Auenwald *m*; Überschwemmungswald *m*
f forêt *f* de terrains en contre bas; forêt de
r [при]пойменный лес *m*; затапливаемый лес; пойма *f*

5193 **lowland hardwood; bottomland hardwood**
(stands that have at least 10 percent stocking with a forest type of oak-gum-cypress, elm-ash-cottonwood, palm, or other tropical)
d Tiefland-Laubholz *n*
f bois *m* de feuillus de bas-pays
r древесина *f* низинных твёрдолиственных пород

5194 **lowland tropical pine**
d niedrige Fichte *f*
f pin *m* de basse altitude
r низинная тропическая сосна *f*

* **lowland white fir → 1278**

* **low-lead [cable] logging → 3988**

* **low-line logging → 3994**

* **low polewood → 7775**

* **low pruning → 4935**

* **low-quality stand → 5191**

5195 **low-quality wood; low-rate timber; junk wood; bastard wood; low-grade wood; wood of inferior quality**
d geringwertiges Holz *n*; minderwertiges Holz
f bois *m* de faible qualité; bois *m* de qualité basse
r низкотоварная древесина *f*; низкосортная древесина; низкокачественная древесина; малоценная древесина

5196 **low[-stemmed] tree; bush tree**
d Niederstamm *m*
f arbre *m* à basse-tige; basse-tige *f*
r низкоствольное дерево *n*

5197 **low stumping; leaving of low stumps**
d niedrige Stumpflassen *n*
f retenue *f* de souches basses
r оставление *n* низких пней

* **low sweet blueberry → 1788**

* **low thinning → 8511**

* **LR → 7134**

* **L-root → 4663**

* **LRUP → 5036**

5198 lug tire
(tire with a cross tread and rather deep grooves)
d Greiferreifen *m*
f pneu *m* pour service hors-route
r шина *f* повышенной проходимости; шина с грунтозацепами

* **lumber business → 8653**

* **lumber camp → 5071**

5199 lumber carrier; lumber truck; timber carrier; saw-carriage; log carriage; butt carriage; straddle carrier; rackbench; timber trolley; log cart
(a traveling block used on a skyline in stem skidding)
d Holzrückekarren *m*; Holzdampfer *m*; Holzfrachtschiff *n*; Klotzwagen *m*; Blochwagen *m*
f chariot *m* de bois; chariot à billes; avant-train *m* de débardage
r телега *f* для колод; бревенная телега

5200 lumber core
(consists of glued strips of solid lumber, with one ply of crossband and one ply of face veneer on each side)
d Bauholzkern *m*
f âme *f* lattée
r серединка *f* [щита] из деревянных реек; блочно-реечное заполнение *n* [щита]

5201 lumber-core board
d Tischplatte *f* mit Bauholzkern
f contreplaqué *m* panneauté; contreplaqué à âme en bois massif
r мебельная плита *f*; столярная плита

5202 lumber drying; converted timber drying
d Schnittholztrocknung *f*
f séchage *m* de bois de sciage
r сушка *f* пиломатериалов

5203 lumber-drying kiln; timber-seasoning kiln; timber-dryer
d Schnittholztrockner *m*
f four *m* de séchage des bois; sécherie *f* de bois
r сушилка *f* для пиломатериалов

* **lumber elevator → 8581**

* **lumberer → 9654**

5204 lumber for deals; timber for deals
d Dielungsholz *n*
f bois *m* de madrier
r древесина *f* для дильс

5205 lumber ga[u]ge
d Schnittholzmeßgerät *n*
f calibre *m* de bois de sciage
r калибр *m* для измерения толщины досок

* **lumber industry → 3466**

* **lumbering → 8824**

* **lumbering costs → 8217**

5206 lumber[ing] industry; sawing industry; logging industry; sawmilling
d Holzbeschaftungsindustrie *f*; Sägeindustrie *f*
f industrie *f* [du bois] de sciage
r лесозаготовительная промышленность *f*; лесопильная промышленность

* **lumbering operation → 3457**

* **lumbering undertaking → 5074**

* **lumberjack → 9654**

5207 lumber loader; lumber unstacker
d Schnittholzlader *m*
f chargeuse *f* mécanique de bois débité
r погрузчик *m* пиломатериалов; погрузочное устройство *n* пиломатериалов

* **lumberman → 9654**

* **lumberman's equipment → 4157**

* **lumber measure → 8615**

* **lumber pine → 4965**

5208 lumber puller; log [saw] carriage; drag cart; log cart; sawmill carriage
d Blockzuführungswagen *m*; Holzrückewagen *m*; Rückenkarren *m*; Gatterwagen *m*; Rückewagen *m*; Prügelwagen *m*
f chariot *m* de bois de sciage; chariot [à grumes] de scies à cadre; chariot diviseur; chariot porte-grumes
r тележка *f* для пиломатериалов; транспортёр *m* для пиломатериалов; трелёвочная тележка; тележка для подачи

* **lumber quality → 6657**

* **lumber road → 7689**

* **lumber [saw]mill → 8637**

5209 **lumber scale**
(a volumetric measurement used in determining the amount of lumber produced)
d Schnittholzmaßstab *m*
f volume *m* en pied-planches de bois de sciage
r объём *m* деловой и дровяной древесины

* **lumber scaling rule table → 878**

5210 **lumber scanning**
d Schnittholzabtastung *f*
f échantillonnage *m* de bois de sciage
r сканирование *n* пиломатериалов

* **lumber site → 5038**

5211 **lumber stack**
d Schnittholzstapel *m*
f pile *f* de bois de sciage
r штабель *m* пиломатериалов

5212 **lumber stacking machine; lumber stacker; stapler**
(a mobile machine mounted on a wheeled or tracked chassis, equipped with a frontally mounted grapple, tusk, or forklift device for loading, unloading, stacking, or sorting of logs)
d Schnittholzabstapelanlage *f*; Abstapelautomat *m*; Abstapelgerät *n*; Abstapelvorrichtung *f*; Stapelmaschine *f*
f empileuse *f* automatique de bois de sciage; empileur *m* automatique de bois de sciage
r штабелеукладчик *m* для досок

* **lumber standardization → 8642**

5213 **lumber tally; mill tally**
(record of lumber giving the number of boards or pieces by size, grade, and species)
d Schnittholzvermessung *f*
f décompte *m* des sciages [bruts]; cubage *m* de sciages
r расчёт *m* пиломатериалов; учёт *m* пиломатериалов; обмер *m* пиломатериалов; расчёт объёма пиловочника в бревне или дереве

5214 **lumber [timber] yield; timber produce; sawing output; sawing yield; outcome of**
f rendement *m* du bois de travail; bilan *m* d'abattage
r выход *m* пиломатериала; результат *m* распиловки

* **lumber tractor → 7799**

* **lumber truck → 5199**

* **lumber unstacker → 5207**

* **lumber wagon → 9256**

5215 **lumber waste; edgings; raw waste lumber**
(waste pieces of wood cut by an edger)
d Sägewerksabfälle *mpl*; Schnittholzabfälle *mpl*
f débris *m* de sciage; résidus *m* de sciage
r отходы *mpl* лесопиления; дровяные отходы

5216 **lumber wood**
d nichtbehauenes Schnittholz *n*
f bois *m* d'œuvre non débillardé
r неотёсанные строевые лесоматериалы *mpl*

5217 **lumpsum sale**
(timber sale in which an agreed-on price for marked standing trees is set before the wood is removed)
d Pauschalverkauf *m*
f rémunération *f* forfaitaire; rémunération à prix fixe
r продажа *f* леса на корню по определённой цене за весь участок

5218 **lump wood**
d Holzstücke *mpl*
f blocs *mpl* porte-bille
r куски *mpl* дерева; обрубки *mpl*

* **lunule → 853**

* **luster → 3872**

5219 **luvisol**
(developed under forests or forest-grassland transition areas in a cool climate)
d Luvisol *m*
f luvisol *m*
r лювисоль *m*

* **luxury wood → 3159**

* **LVL → 4762**

5220 **lyctid borer; powderpost borer; powderpost beetle**
(Bostrychus capucinus; Lyctus brunneus)
d Splintholzkäfer *m*
f bostryche *m*; bostriche *m*
r короед *m* белоголовый

5221 **lymexylid damage; powderpost beetle damage**
(small holes (1/16 to 1/12 inch in diameter) filled with dry, pulverized wood, resulting from the work of beetles in seasoned and unseasoned wood)
d Holzmehlkäferschaden *m*
f dégât *m* par bostryche
r круглая червоточина *f*

* **lyophilization → 3702**

5222 **lysigenous**
(formed by a disorganization or dissolving of cells)
d lysigenisch
f lysigène
r лизогенный

5223 **lysimeter**
(an instrument for measuring the water that percolates through a certain depth of soil)
d Lysimeter *n*
f lysimètre *m*; case *f* lysimétrique
r лизиметр *m*

M

5224 **Maack cherry**
(Padus maackii)
d Amur-Kirsche *f*
f merisier *m* de Maack
r черёмуха *f* Маака

* **macadamia → 1206**

* **Macassar ebony → 3882**

5225 **macawood**
(Platymiscium)
d Trebol
f platymiscie *f*
r платимисциум *m*

5226 **Macedonian oak**
(Quercus trojana)
d makedonische Eiche *f*
f chêne *m* de Troie; chêne macédonien
r дуб *m* македонский

* **Macedonian pine → 489**

5227 **macerate** *v*; **stain** *v*
d beizen
f macérer; amaigrir; décaper
r протравить

* **machette → 1204**

5228 **machinability of wood**
d Machenschaftsfähigkeit *f* des Holzes
f façonnabilité *f* de bois
r пригодность *f* древесины к обработке

5229 **machine burn**
(dark streaks along the face of a board)
d maschinell hergestellte Brandnarbe *f*
f brûlure *f* mécanique
r ожог *m* [древесины] от перегрева инструмента

5230 **machine for chipping branches**
d Astzerkleinerungsmaschine *f*
f machine *f* à fragmenter les branches
r дробилка *f* сучьей

5231 **machine-peeled**
r окорённый машиной; окорённый на станке

* **machine planting → 5380**

5232 **machinery deck; turntable**
d Drehtisch *m*
f tourelle *f*
r поворотная платформа *f* погрузчика

* **machine weeding → 5375**

5233 **MacKenzie['s] willow**
(Salix prolixa; Salix mackenzieana)
d MacKenzie-Weide *f*
f saule *m* de Mackenzie
r ива *f* Маккензи

* **MacNab cypress → 484**

* **macro[-]element → 5235**

5234 **macrofibril**
d Makrofibrille *f*
f [macro-]fibrille *f*
r макрофибрилла *f*

5235 **macro[-]nutrient; macro[-]element; major element; major nutrient**
(chemical element (not including carbon, hydrogen, or oxygen) needed in relatively large amounts by organisms: Ca, K, Mg, N, P)
d Massennährstoff *m*; Makroelement *n*; Massenelement *n*
f macro-élément *m*
r макроэлемент *m*

5236 **macrophyte**
(any plant that can be seen with the naked, unaided eye; e.g., aquatic mosses, ferns, liverworts, rooted plants, etc.)
d Makrophyt *m*
f macrophyte *m/f*
r макрофит *m*

5237 **macropropagation**
d Makrozüchtung *f*; Makrovermehrung *f*
f macropropagation *f*
r макрораспространение *n*

* **maculata gum → 7930**

5238 **madder family**
(Rubiaceae)
d Krappgewächse *pl*; Rötegewächse *pl*
f rubiacées *pl*
r подмаренниковые *pl*

5239 **made of beechwood**
d buchisch
f fabriqué ex bois de hêtre
r сделанный из буковой древесины

* **madrone → 6020**

* **maggot → 4809**

5240 **magnolia; vaco**
(Magnolia)
d Magnolie *f*
f magnolia *m*
r магнолия *f*

5241 **magnolia family**
(Magnoliaceae)
d Magnoliengewächse *pl*
f magnoliacées *pl*
r магнолиевые *pl*

5242 **mahala mat; squaw carpet ceanothus**
(Ceanothus prostratus)
d niederlegende Säckelblume *f*
f céanothe *m* bleu
r цеанотус *m* простёртый

* **mahogany bean → 6406**

* **mahogany birch → 1474**

5243 **mahogany family**
(Meliaceae)
d Zederachgewächse *pl*
f méliacées *pl*
r мелиевые *pl*

5244 **mahua; elupa; aluppe**
(Bassia latifolia)
d Illipe *m*
f illipé *m*; bassie *f*
r бассия *f* широколистная

* **MAI → 5352**

5245 **maidenhair[-tree]; ginkgo**
(Gink[g]o biloba)
d Ginkobaum *m*; Ginkgo *n*; Mädchenhaarbaum *m*; Fächerblattbaum *m*
f ginkgo *m* [à deux lobes]; arbre *m* aux quarante écus; abricot *m* d'argent
r гинкго *m* [двухлопастный]

* **maidenhair-tree family → 3847**

* **maiden trees → 3518**

5246 **main boom; hoist boom; inner boom**
d Hauptausleger *m*
f flèche *f* principale
r основная стрела *f*

* **main branch → 6536**

* **main crop → 5250**

* **main felling → 3136**

5247 **main lateral root**
d Hauptseitenwurzel *f*
f racine *f* latérale principale
r главный боковой корень *m*

* **main-line block → 1158**

5248 **main pocket**
d Haupttasche *f*
f couloir *m* principal
r главный сортировочный коридор *m*

5249 **main ride; major ride; principal ride**
d Hauptpfad *m*; Hauptweg *m*
f laie *f* sommière
r основная просека *f*; главная просека

* **main root → 8411**

* **main spar → 4190**

5250 **main stand; fundamental stand; main crop; principal crop**
(in regular crops or stands, that portion of the growing stock retained after an intermediate cutting)
d Hauptbestand *m*; Grundbestand *m*
f peuplement *m* principal; peuplement de base
r главное насаждение *n*; основной запас *m* насаждения

5251 **main tissue layers in cross-section**
d herrschender Gewebeschichten *fpl* im Querschnitt
f couches *fpl* principales de tissue [d'arbre] en section transversale
r основные слои *mpl* ткани ствола дерева в поперечном разрезе

5252 **main tree species; principal tree species; leading tree species; chief tree species**
(species to which the silviculture of a mixed forest is primarily directed)
d Hauptbaumart *f*; Hauptholzart *f*

* **major element → 5235**

5253 **major forest production**
d Hauptwaldproduktion *f*
f production *f* forestière principale
r главная продукция *f* леса; основное лесопользование *n*

* **major harvest → 3136**

* **major nutrient → 5235**

* **major ride → 5249**

* **major root → 8411**

* **make bed → 4840**

5254 **makore**
(Dumoria heckelii; Dumoria africana)
d Makore *m*
f macore *m*
r макоре *n*

* **Malabar almond → 4464**

* **Malabar bombax → 6792**

5255 **Malay bush beech; gumhar**
(Gmelina arborea)
d asiatischer Teakbaum *m*; Gamari *m*
f gmélina *m* arborescent
r гмелина *f* древовидная; касмарья *f*

5256 **maleberry; he-huckleberry**
(Lyonia ligustrina)
d rispige Lyonie *f*
f clochette *f*
r лиония *f* бирючиновидная

5257 **male fern**
(Aspidium filixmas; Dryopteris filixmas)
d gemeiner Wurmfarn *m*; Bandwurmwurzel *f*; Geißleitere *f*; Wanzenkraut *m*
f fougère *m* male
r папоротник *m* лесной; папоротник мужской

* **male-flowered tree → 7999**

5258 **mallee [scrub]**
(any of several low-growing Australian eucalypts)
d Mallee *m*
f mallee *m*
r низкорослый эвкалипт *m*

5259 **mallow ninebark**
(Physocarpus malvaceus; Physocarpus pauciflorus)
d Oregon-Blasenspiere *f*
f physocarpe *m*
r пузыреплодник *m* мальвовый; пузырчатник *m* мальвовый

5260 **malpighia family**
(Malpighiaceae)
d Malpighiengewächse *pl*
f malpighiennes *pl*
r мальпигиевые *pl*

* **Maltese orange tree → 5274**

* **mammee sapote → 5665**

* **mammoth tree → 3844**

* **managed fire → 6516**

5261 **managed forestland**
(forest land that is being managed under a forest management plan utilizing the science of forestry)
d geleiteter Waldboden *m*
f terre *f* forestière gestionnée
r управляемая лесная земля *f*

5262 **managed harvest**
(estimated volume of timber on commercial forest land that could be cut annually for the next 10 years while improving tree stocking and bringing about a more even distribution of age classes)
d geleitete Ernte *f*
f récolte *f* gestionnée
r управляемое ведение *n* лесозаготовки

5263 **management assumptions**
(approximations of management objectives, priorities, constraints and other conditions needed to represent forest management actions in a forest planning model)
d Wirtschaftsannahmen *fpl*
f assomptions *fpl* de gestion
r ориентировочные расчёты *mpl*

* **management for rapid growth → 6724**

* **management of forests → 3488**

5264 **management of genetic resources**
d Genressourcen-Management *n*

* **management of woods → 3488**

5265 **management tables**
(tables used for commercially grown species and based on yield class)
d Wirtschaftsarttabellen *fpl*
f tables *fpl* d'espèces commerciales
r таблицы *fpl* хозяйственных видов

5266 **management-volume inventory; volume inventory; inventory of volume**
(an enumeration of such pertinent data as volume or basal area, increment and mortality of stands, made to assess silvicultural opportunities)
d Massenaufnahme *f*; Massenermittlung *f*
f inventaire *m* d'aménagement; comptage *m* de volume
r изучение *n* запаса, прироста и отпада насаждений; определение *n* запаса

* **management zone → 3330**

5267 **manchineel**
(Hippomane mancinella)
d Manzanillabaum *m*
f mancenillier *m*
r манцинелла *f*

5268 **Manchurian alder**
(Alnus hirsuta; Alnus tinctoria)
d mandschurische Erle *f*; färbe Erle
f au[l]ne *m* de Manchurian
r ольха *f* красильная

5269 **Manchurian ash; tamo**
(Fraxinus mandshurica)
d mandschurische Esche *f*
f frêne *m* de Mandchourie
r ясень *m* маньчжурский

5270 **Manchurian elm**
(Ulmus laciniata)
d geschlitzblättrige Ulme *f*
f orme *m* de Mandchourie
r вяз *m* лопастный

5271 **Manchurian maple**
(Acer mandshuricum)
d mandschurischer Ahorn *m*
f érable *m* de Mandchourie
r клён *m* маньчжурский

* **Manchurian pine → 4743**

5272 **Manchurian walnut**
f noyer *m* de Mandchourie
r орех *m* маньчжурский

5273 **mandarin[e]**
(a fruit)
d Mandarine *f*
f mandarine *f*
r мандарин *m*

5274 **mandarin[e] tree; Maltese orange tree; king orange**
(Citrus nobilis)
d Mandarinenbaum *m*; echte Mandarine *f*
f mandarinier *m*
r мандариновое дерево *n*; цитрус *m* благородный; мандарин *m* благородный

5275 **mandioqueira; florecillo; gonfolo**
(Qualea paraensis; Qualea rosea)
d Mandioqueira *n*; Berg-Gronfoeloe *n*
f qualis *m*; grignon *m* fou
r квалея *f*

* **manglecito → 251**

5276 **mangle tree; mangrove**
(Rhizophora)
d Mangrovenbaum *m*
f palétuvier *m*
r мангровое дерево *n*

5277 **mango**
(a fruit)
d Mangofrucht *f*
f mangue *f*
r манго *n*

* **mango → 1789**

5278 **mangosteen [tree]**
(Garcinia mangostana)
d Mangostanbaum *m*; Mangostane *m*
f mangoustanier *m*
r гарциния *f* мангостановая; мангостан *m*

* **mangrove → 5276**

* **mangrove community → 5280**

5279 **mangrove family**
(Rhizophoraceae)
d Mangrovebaumgewächse *pl*; Mangrovengewächse *pl*; Mangelbaumgewächse *pl*

5280 **mangrove forest; mangrove community; mangroves; tidal forest**
(open or closed stands of trees and bushes occurring in the tropics in intertidal zones, usually around the mouths of rivers, creeks and lagoons)
d Mangrovenwald *m*; Mangrovegehölze *npl*; Mangroven *fpl*
f forêt *f* de mangrove; mangroves *fpl*
r мангровый лес *m*; мангры *fpl*

5281 **mangrove swamp**
(a swampy or tidal area dominated by tropical trees or shrubs of the genera Rhizophora, Laguncularia, and Avicennia)
d Mangrovensumpf *m*
f marais *m* de mangrove
r мангровое болото *n*

5282 **manihot; cassava**
(Manihot esculenta; Manihot utilissima)
d Maniok *m*; Cassava *m*; Kassava *m*; Topioka *m*
f manioc *m* [comestible]; manioc cassave
r маниок *m* съедобный

5283 **Manila hemp; abaca [banana]**
(Musa textilis)
d Manila-Hanf *m*
f bananier *m* à fibres; abaca *m*
r банан *m* текстильный; абака *f*

5284 **Manilla padauk; narra; amboyna; angsana**
(Pterocarpus indicus)
d Manila-Padauk *m*; indisches Sandelholz *n*; Amboine *m*
f padauk *m* des Indes; loupe *f* d'Amboine
r нарра *f*; сандал *m* индийский

* **Manitoba maple → 381**

5285 **man-made forest; anthropogenic forest; artificial stand**
d künstlicher Wald *m*; Kulturwald *m*
f forêt *f* artificielle; forêt anthropique
r искусственно выращенный лес *m*; искусственный лес; искусственное лесонасаждение *n*

* **manna[-ash] → 3292**

5286 **manna gum; ribbon gum**
(Eucalyptus viminalis)
d australischer Manna-Gummibaum *m*; Ruten-Eukalyptus *m*; weißer Gummibaum *m*
f gommier *m* manna d'Australie; gommier à sucre

* **manometer → 6521**

* **mansonia → 1087**

* **manual debarking → 4107**

5287 **manual weeding; hand weeding**
d Handjäten *n*
f désherbage *m* manuel
r ручная расчистка *f* сорняков

5288 **manufactured timber**
d aufgearbeitetes Holz *n*
f bois *m* manufacturé
r промышленный лесоматериал *m*

* **manufacturing of wood → 5376**

* **manure → 3089**

* **many-aged forest → 172**

* **many-storied forest → 5633**

5289 **manzanita**
(Arctostaphylos)
d Bärentraube *f*
f busserole *f*
r толокнянка *f*

5290 **maple-beech-birch [forest type]**
(in which maple, beech, or yellow birch, singly or in combination, constitute a plurality of the stocking)
d Ahorn-Buche-Birkenwald *m*
f forêt *f* d'érable, hêtre et bouleau
r клёново-буково-берёзовый лес *m*

5291 **maple family**
(Aceraceae)
d Ahorngewächse *pl*
f acéracées *pl*
r клёновые *pl*

5292 **maple forest; maple tree wood**
d Ahornwald *m*
f érablière *f*
r клёновый лес *m*; кленовник *m*

* **mapleleaf sycamore → 2906**

5293 **mapleleaf viburnum; dockmackie**
(Viburnum acerifolium)
d ahornblättriger Schneeball *m*

5294 **maple sap; sweet water**
(clear, sweet, waterlike fluid from maple trees)
d Ahornsaft *m*
f sève *m* d'érable
r клёновый сок *m*

5295 **maple syrup**
(a liquid food derived by concentration and heat treatment of the sap of the maple tree)
d Ahornsirup *m*
f sirop *m* d'érable
r клёновый сироп *m*

5296 **maple [tree]; acer**
(Acer)
d Ahorn *m*; Ahornbaum *m*
f érable *m*
r клён *m*

* **maple tree wood → 5292**

5297 **maple wood**
d Ahornholz *n*
f bois *m* d'érable
r клёновая древесина *f*

5298 **map of crown projection**
d Kartenprojektion *f* einer Krone
f carte *f* de projection de houppiers
r карта *f* проекции кроны

* **map of stock → 3498**

5299 **map projection**
d kartografische Projektion *f*
f projection *f* cartographique
r картографическая проекция *f*

5300 **maquis**
(scrub vegetation, generally rather dense, composed of much branched, thorny and often aromatic shrubs, found in the Mediterranean region)
d Macchia *f*
f maquis *m*
r заросли *mpl* вечнозёленых кустарников; маккия *f*

* **Maracaibo boxwood → 9346**

* **marakaipo → 4717**

* **marble gall → 5831**

* **marble rot → 5573**

5301 **marblewood; mulga**
(Acacia aneura)
d australische Akazie *f*; Mulga *f*
f acacia *m* d'Australie; mulga *m*
r акация *f* безжилковая

5302 **marcescent; senescent; withering**
(withering but not falling off; marcescent leaves dry out in autumn, stay on the tree through winter and fall in the spring when new buds break)
d welkend
f marcescent
r завядающий; увядающий

* **marcotting → 6582**

5303 **marginal acres**
(areas not qualifying as standard or special components primarily because of excessive development costs, low product values, resource protection constraints, and inadequate markets)
d Grenzenmorgen *mpl*
f acres *fpl* marginales
r дополнительные участки *mpl* земли

* **marginal apotracheal parenchyma → 8475**

* **margin[al] knot → 4733**

5304 **marginal land**
(land of questionable economic capabilities for a specific purpose, probably at the financial margin, within one unit of profitability)
d grenzenmorgenes Land *n*
f terre *f* marginale
r маржинальная земля *f*; малопродуктивная земля

* **marginal regeneration → 6877**

* **marginal rot → 4494**

* **marginal seeding → 7384**

5305 **marginal thinning intensity**
(the maximum thinning intensity that can be maintained without loss in cumulative volume production)
d grenzwertige Durchforstungsintensität *f*
f intensité *f* globale des éclaircies
r предельная интенсивность *f* рубок ухода

* **marginal tree → 922**

5306 **mari; larch peatmoss bog forest**
d Torfmull-Lärchensumpfwald *m*
f mélézière *f* marécageuse sur tourbe sèche
r марь *f* (высокоствольный лиственничник на сфагновых болотах)

5307 **marine ply[-]wood**
(plywood manufactured with salt-resistant glue)
d Bootsbausperrholz *n*
f contreplaqué *m* marin
r фанера *f* для судостроения

5308 **mariola**
(Parthenium incanum)
d Mariola *f*
f mariola *m*
r мариола *f*

* **marishballi → 5532**

5309 **maritime pine; seaside pine; pinaster; cluster-pine; coastal pine**
(Pinus pinaster; Pinus maritima)
d Seestrandkiefer *f*; Seekiefer *f*; Meerkiefer *f*; Strandkiefer *f*; Sternkiefer *f*; Igelföhre *f*; Bordeauxkiefer *f*; Igelkiefer *f*; Lärchenkiefer *f*
f pin *m* maritime; pinastre *m*; pin des landes; pin côtier
r приморская сосна *f*; средиземноморская сосна

5310 **mark** *v*
d markieren; marken; auszeichnen
f marquer
r маркировать

* **marked tree → 826**

5311 **marked wood**
d angezeichnetes Holz *n*
f bois *m* marqué
r маркированная древесина *f*

* **marker → 8377**

5312 **marketing of forest products**
d Absatz *m* von Walderzeugnissen
f débouché *m* pour produits forestiers
r сбыт *m* продуктов леса

5313 **market value; sales value; commercial value; market price**
(quantitative values for resources in public demand which are traded in dollars, e.g., timber, minerals, and some types of
f valeur *f* marchande; valeur du marché; valeur commerciale; valeur vénale; prix *m* du marché
r продажная стоимость *f*

5314 **marking axe; branding axe; marking hatchet; blazer**
(axe used to stamp brands into logs)
d Beil *n* zum Auszeichnen; Schalmaxt *f*; Schalmbeil *n*
f hache *f* à flacher
r маркировочный топор[ик] *m*; топорик для затёски

5315 **marking hammer; die-hammer; paint paddle; marking cog; numbering hammer; woodman's hammer; mallet**
(a light hammer having a die for stamping letters, figures, or other distinctive devices on trees)
d Anschlaghammer *m*; Markier[ungs]hammer *m*; Numerierhammer *m*; Waldhammer *m*; Vorschlaghammer *m*
f marteau *m* forestier; marteau réceptionneur
r молоток *m* для клеймения [краской]; маркировочный молоток

* **marking hatchet → 5314**

5316 **marking of roundwood**
d Rundholzmarkierung *f*
f marquage *m* de bois rond
r маркировка *f* круглых лесоматериалов

5317 **marking of timber; timber mark[ing]; wood marking; blazing; spotting; branding**
(hammer indentation made on cut timber for identification purposes)
d Auszeichnen *n* der Schlaghölzer; Holzauszeichnung *f*; Holzmarkierung *f*; Waldzeichnung *f*
f martelage *m* de bois; marquage *m* de bois; griffage *m* de bois
r клеймение *n* лесоматериалов; маркировка *f* лесоматериалов; пометка *f* зарубками

5318 **marking of tree length**
d Längeholzmarkierung *f*
f marquage *m* de bois long
r разметка *f* хлыстов

5319 **marking rule**
(means of standardizing marking practice among individuals and for various areas of the same forest type, commonly for thinning

f règle *m* de marquage
r разметочная линейка *f*

5320 **marlberry**
(escallonoides)
d Spitzblume *f*
f ardisia *m*
r ардизия *f*

5321 **marline-spike; spike**
(a tool employed for splicing or otherwise working of a wire rope. Usually in the form of a tapered steel pin)
d Marlspiker *m*
f épissoir *m*
r шило *n* для заплётки каната

* **marmalade plum → 5665**

* **marmalade tree → 5665**

* **marquetry → 9590**

* **marri eucalyptus → 6786**

* **marrow → 1951**

* **marsh-elder → 7162**

* **marshy forest → 8309**

* **marupa → 6063**

* **mass-accretion → 9243**

5322 **massaranduba**
(Manilkara bidentata)
d Massaranduba *n*
f maçaranduba *m*
r массарандуба *f*

5323 **mass delimber; multistem delimber; multiple-tree delimber; bunch [de]limber**
d Mehrentastungsmaschine *f*; Bündel-Entaster *m*
f ébrancheuse *f* multiarbre; ébrancheuse multiple
r машина *f* для групповой обрезки сучьев

5324 **massed pitch**
(secretion of resin which has gathered where the bark has been broken)
d massiertes Pech *n*; zusammengezogenes Pech
f amas *m* d'oléorésine
r сплошной смоляной потёк *m*; сплошная слоистость *f*

* **mass infection → 2849**

5325 **mass selection**
d Massenselektion *f*; Massenauslese *f*
f sélection *f* massale; sélection de masse
r массовый отбор *m*

* **mass wasting → 7835**

5326 **mast; pannage**
(the fruit of trees such as oak, beech, sweet chestnut and also (US) the seeds of certain pines, particularly where considered as food for livestock and certain kinds of wildlife)
d Eichelmast *f*; Mast *f*
f faînée *f*; paisson *m*
r плодокорм *m*

* **master forester → 1483**

5327 **mastic tree; green mastic; real mastictree**
(Pistacia lentiscus)
d Mastixbaum *m*; Mastixstrauch *m*; Mastixpistazie *f*
f pistachier *m* lentisque; lentisque *m*; arbre *m* de mastic
r мастиковое дерево *n*; фисташка *f* мастичная

* **mast timber → 8586**

5328 **mast tree**
(a tall standing tree, trimmed and topped, and generally braced with guy lines)
d Mastbaum *m*
f arbre *m* à pilot; pylône *m*; mât *m*
r мачтовое дерево *n*; [естественная] трелёвочная мачта *f*

5329 **mast year**
(in which a tree species produces, either as an individual or a crop, an adequate amount of seed)
d Mastjahr *n*
f année *f* de glandée
r урожайный год *m*

5330 **masur birch**
(a curly-grained wood from certain birches)
d Masurbirke *f*
f bouleau *m* flammé
r древесина *f* берёзы с красивым текстурным рисунком

* **mat → 5077, 5335**

* **matched lumber → 8677**

* **match log → 5332**

5331 **match log debarking**
d Zündholzstammentrindung *f*
f écorçage *m* de billon pour allumettes
r окорка *f* спичечного кряжа

5332 **match wood; match stock; match log; match billet**
(billets suitable for peeling for match manufacture)
d Holz *n* für Holzdraht; Zündholzstamm *m*
f billon *m* pour allumettes
r древесина *f* для изготовления спичечной соломки; спичечный кряж *m*

* **match wood → 5558**

5333 **mating system**
d Paarungssystem *n*
f dispositif *m* de croisement
r система *f* разведения; система спаривания

* **matting of needles → 5706**

5334 **matting of roots**
d Verwurzelung *f*
f matage *m* des racines
r сплетение *n* корней

5335 **mattress; mat**
(in building-fiberboard manufacture, the defiberated (i.e. pulped) woody material as formed before pressing)
d Vlies *n*; Schutzmatte *f*
f mat *m*; matte *f*; tapis *m*
r ковёр *m*

* **mattress dam → 1117**

5336 **maturation gradient**
(in trees, maturation is often most advanced in the upper parts of a tree and least advanced near the base of the bole, with intermediate conditions in between)
d Reifungsgradient *m*
f gradient *m* de maturité
r градиент *m* созревания

5337 **mature** *adj*
(in even-aged management, those trees or stands that are sufficiently developed to be harvestable and that are at or near rotation age)
r спелый

* **mature cutting → 4150**

* **mature decay → 72**

5338 **mature forest; maturity stand; final seed; ripe stand; mature stand; mature timber**
(forest that has completed natural growth and development)
d haubarer Bestand *m*; reifer Bestand
f forêt *f* adulte; peuplement *m* mûr; peuplement adulte
r спелый лес *m*; спелое насаждение *n*; спелый древостой *m*

5339 **mature growth; mellowing**
d Ausreifen *n*; Reifwerden *n*; Reifen *n*
f aoûtage *m*
r созревание *n*

* **mature stand → 5338**

* **mature timber → 5338**

5340 **mature timber stage**
d Altholzstadium *n*
f stade *m* de vieux bois
r стадия *f* спелого насаждения

5341 **mature tree**
(tree that has reached the desired size or age for its intended use)
d Altbaum *m*; hiebsreifer Baum *m*
f arbre *m* adulte
r спелое дерево *n*[, предназначенное в рубку]

* **mature wood → 6964**

* **maturity → 3146**

* **maturity age → 133**

5342 **maturity class**
(trees or stands grouped according to their stage of development, from establishment to suitability for harvest)
d Ablaufklasse *f*
f classe *f* de maturité
r класс *m* спелости

* **maturity stand → 5338**

5343 **maturity stand canopy**
d Altholzschirm *m*

5344 Mauritian ebony
(Diospyros tessellaria)
d Mauritian-Ebenholz *n*
f bois *m* d'ébène noir
r хурма *f* маврикийская

5345 maximum density
(maximum allowable stand density above which stands must be spaced to a target density of well-spaced acceptable stems to achieve free-growing status)
d maximale Dichte *f*
f densité *f* maximale
r максимальная плотность *f*

* **May beetle → 3502**

* **May blossom → 4174**

* **May bush → 4174**

5346 mayflower; roble; pink trumpet-tree; rosy trumpet-tree; pink shower tree; amapa rosa; West Indian bowwood
(Tabebuia rosea; Tabebuia pentaphylla)
d Matilisguate[-Baum] *m*; [weiße] Roble *f*; amerikanische Eiche *f*
f chêne *m* de la savane; poui *m* rose; poirier-pays *m*; tabebuia *m* humifère
r майский цветок *m*

* **mayflower [haw] → 4174**

5347 May-pole; young birch tree
d Maibaum *m*
f arbre *m* de mai
r майское дерево *n*

* **May tree → 4174**

* **mazzard [cherry] → 8330**

* **MBF → 8529**

5348 Mc-Leod tool; mcleod
(combination hoe, rake and scraping tool)
d Mc-Leod-Hacke *f*; Mc-Leod-Haue *f*; Mc-Leod-Werkzeug *n*
f houe *f* Mc-Leod
r мотыга *f* Мак-Лода

* **MDF → 4545**

5349 meadow
(1. an herb-land used primarily for hay production; 2. openings in forests and
d Wiese *f*; Viehweide *f*
f pré *m*
r луг *m*; луговина *f*; поляна *f*

* **meadow [in the wood] → 3443**

5350 meadowsweet
(Spiraea latifolia)
d Spierstrauch *m*
f spirée *f*; reine-des-prés *f*
r спирея *f*; таволга *f* трёхлистная

* **mealy stringybark [eucalyptus] → 357**

5351 mean-aged forests; middle-aged forests
d Walde *mpl* von mittlerem Alter
f forêts *fpl* d'age moyen
r средневозрастные леса *mpl*

* **mean age of stand → 426**

5352 mean annual increment; MAI; mean annual accretion
(average annual increase in volume of individual trees or stands up to the specified point in time)
d jährlicher Mitteldurchschnittszuwachs *m*
f accroissement *m* moyen annuel; AMA; croît *m*; accroissement annuel moyen; AAM; accroissement moyen total
r среднегодовой прирост *m*; средний годичный прирост

5353 mean annual precipitation; mean annual rainfall; average annual precipitation
d jährlicher Mittelniederschlag *m*
f précipitation *f* moyenne annuelle
r среднегодовые осадки *fpl*

5354 mean annual yield
d jährlicher Mittelschnittholzausbeute *f*
f coupe *f* annuelle moyenne
r средняя годичная лесосека *f*

5355 mean basal area tree
d Grundflächen-Mittelstamm *m*; Grundflächenmittelstamm *m*; Kreisflächenmittelstamm *m*
f arbre *m* de surface terrière moyenne
r дерево *n* среднего ствола по поперечного сечения

* **mean diameter → 2468**

5356 mean fire [free] interval; mean fire return interval
(arithmetic average of all fire intervals

d Feuerperiodizität *f*; durchschnittliches Feuer-Intervall *n*; durchschnittliches feuerfreies Intervall *n*
f périodicité *f* moyenne des feux
r средняя периодичность *f* пожаров

* **mean fire return interval → 5356**

* **mean girdle → 5440**

* **mean girth → 5440**

5357 **mean height; average height**
d Mittelhöhe *f*
f hauteur *f* moyenne
r средняя высота *f*

5358 **mean height of predominants; predominant mean height**
d Bestandesoberhöhe *f*
f hauteur *f* moyenne d'arbres prédominants
r средняя высота *f* наиболее высоких деревьев

* **mean log → 434**

* **mean tree → 442**

* **mean tree height → 440**

5359 **mean tree piece rate**
d Mittelstammtarif *m*
f tarif *m* des arbres moyens
r тариф *m* среднего ствола

* **mean tree volume → 443**

5360 **measurable basal area**
d messbare Kreisfläche *f*
f surface *f* terrière mesurable
r измеряемая площадь *f* поперечных сечений [стволов] эксплуатационной части насаждения

5361 **measurement account**
d Messbuchung *f*
f compte *m* à la mesure
r расчёт *m* обмера; справка *f* о результатах обмера

5362 **measurement by length and mid-diameter; scaling by length and mid-diameter**
d Messung *f* nach Länge und Mittendurchmesser
f cubage *m* au moyen de la longueur et du

* **measurement list → 8376**

* **measurement mark → 8972**

5363 **measurement of growth rings**
d Jahrringmessung *f*
f mesure *f* des cernes annuels
r измерение *n* годичных слоёв; замер *m* годичных колец

5364 **measurement of timber volume; measuring of standing timber; stock taking**
d Holzmassenaufnahme *f*
f mesurage *m* de volume
r измерение *n* объёма лесоматериалов

5365 **measurement register**
(of trees)
d Kluppierungsbuch *n*
f livre *m* d'inventaire
r ведомость *f* перечёта деревьев

5366 **measurement specification; scale-book**
d Aufmaßduchliste *f*; Aufmaßliste *f*; Aufmaßbuch *n*
f spécification *f* dimensionnelle; liste *f* dimensionnelle; état *m* des mesures
r спецификация *f* размеров пиломатериалов; спецификация раскроя

5367 **measuring by top and base diameter; scaling by top and base diameter**
(of a log)
d Messung *f* nach Grund- und Zopfdurchmesser
f cubage *m* d'après le diamètre gros et au fin bout
r обмер *m* в верхнем и нижнем отрубах; обмер основного и верхнего диаметра

* **measuring chain → 8289**

* **measuring of standing timber → 5364**

* **measuring rod → 8292**

* **measuring staff → 8292**

* **measuring worm → 3174**

* **Mecca myrrh tree → 969**

5368 **mechanical damage**
d mechanische Beschädigung *f*; mechanische Verletzung *f*
f dommage *m* mécanique
r механическое повреждение *n* (зарубы, запилы, сколы, отщепи, вырывы

* **mechanical debarking → 5372**

5369 mechanical felling
d maschinelle Fällung *f*; mechanisierter Holzeinschlag *m*
f abattage *m* mécanique; abattage mécanisé
r механизированная валка *f*; механизированная рубка *f*

5370 mechanical girdling
d mechanisierte Ringelung *f*
f annélation *f* mécanique
r механизированное кольцевание *n*

5371 mechanical gum harvester
d Harzschöpfmaschine *f*
f récolteuse *f* de résine
r машина *f* для сбора живицы

* **mechanical harvesting → 5379**

5372 mechanical peeling; mechanical debarking
d mechanische Entrindung *f*
f écorçage *m* mécanique; scarifiage *m* mécanique
r механическая окорка *f*

* **mechanical planting → 5380**

* **mechanical [pulp] paper → 9634**

5373 mechanical site preparation
(any activity that involves the use of mechanical machinery to prepare a site for reforestation)
d mechanische Standortvorbereitung *f*
f préparation *f* mécanique du terrain
r механическая очистка *f* поверхности почвы

5374 mechanical thinning; systematic thinning
(removing without regard to size or quality)
d mechanische Durchforstung *f*
f éclaircie *f* mécanique; éclaircie systématique
r механический отбор *m* деревьев при рубках ухода; механизированные рубки *fpl* ухода; механическое прореживание *n*

5375 mechanical weeding; mechanized weeding; machine weeding
d mechanisches Jüten *n*; mechanische Verziehen *n*
f désherbage *m* mécanique; désherbage mécanisé
r механизированная прополка *f* сорняков

5376 mechanical wood treatment; tooling of
d mechanische Holzbearbeitung *f*
f travail *m* mécanique du bois
r механическая лесообработка *f*

5377 mechanized access and use
d mechanischer Zugriff *m* und Nutzung *f*
f accès *m* et usage *m* mécanisés
r механизированный доступ *m* и эксплуатация *f*

5378 mechanized cone harvesting
d mechanische Zapfengewinnung *f*
f collection *f* de cônes mécanisée
r механизированная заготовка *f* шишек

5379 mechanized harvesting; mechanical harvesting
(cutting with mechanized equipment, such as the carrier)
d mechanische Holzbeschaftung *f*
f exploitation *f* de forêt mécanisée
r механизированная заготовка *f*

* **mechanized logging [operation] → 6489**

5380 mechanized planting; mechanical planting; machine planting
(which opens a hole or furrow and closes it again and firms the soil about a tree seedling which is usually inserted by hand)
d mechanische Pflanzung *f*
f plantation *f* mécanisée
r механизированная посадка *f*

5381 mechanized slash disposal
d mechanische Holzabfallbeseitigung *f*
f élimination *f* des rémanents mécanisée
r механизированная очистка *f* от отходов

5382 mechanized stand tending
(any stand tending activity that involves the use of mechanical machinery to treat a stand)
d mechanische Bestandpflege *f*
f soins *mpl* culturaux au peuplement mécanisés
r механизированный уход *m* за насаждениями

* **mechanized weeding → 5375**

* **medang → 902**

* **Mediterranean brown earth → 6770**

* **Mediterranean cypress → 1770**

5383 Mediterranean hazel; tree-hazel; Turkish hazel; Constantinople nut; Turkish filbert
(Corylus colurna)
d Baumhasel *f*; türkische Hasel *f*; türkische Nuss *f*; Dicknuss *f*
f noisetier *m* méditerranéen; noisetier de Byzance; noisette *f* du Levant; noisette de Constantinople
r лещина *f* древовидная; медвежий орех *m*

5384 Mediterranean palm; feather palm; [European] fan palm; dwarf [fan] palm; fanleaf palm
(Chamaerops humilis)
d Schirmpalme *f*; Fächerpalme *f*; niedrige Zwergpalme *f*; europäische Zwergpalme; Hanfpalme *f*
f palmier *m* nain; chamérops *m*
r перистая пальма *f*; пальма веерная низкая; пальма веерная европейская; пальма карликовая; пальма хамеропс; хамеропс *m* низкий

5385 medium check
d Mittelriss *m*
f fente *f* de moyenne longueur
r трещина *f* средней длины

* **medium-density fiberboard → 4545**

5386 medium knot
(with diameter less than 3.7 cm)
d Mittelknoten *m*
f nœud *m* moyen; branche *f* de moyenne grandeur
r средний сучок *m*

* **medium log → 434**

5387 medium-sized tree
d mittelgroßer Baum *m*
f arbre *m* de dimension moyenne
r дерево *n* среднего размера

5388 medium-stocked stand
d mittelbestückter Bestand *m*
f peuplement *m* de densité moyenne
r среднеполнотное насаждение *n*

5389 medlar[-tree]; showy mespilus; cotoneaster
(Mespilus germanica; Cotoneaster conspicuus)
d [Bogen-]Zwergmispel *f*; deutsche Mispel *f*; gemeine Mispel
f néflier *m* [commun]; cotonéastre *m*;

5390 medular blemish; medullary spot; pith [ray] fleck
(an irregular strands of abnormal (often traumatic) parenchymatous tissue embedded in the wood and appearing on a longitudinal surface as a streak)
d Markfleck *m*; Zellgang *m*
f tache *f* médullaire
r сердцевинное пятно *n*; прожилки *fpl*; часть *f* сердцевинного луча

* **medulla → 1951**

* **medullary cell → 6736**

5391 medullary ray; [pith] ray
(bundles of cells which run radially between the pith to the cambium layer)
d Markstrahl *m*
f rayon *m* médullaire
r [сердцевинный] луч *m*

* **medullary spot → 5390**

5392 megagametophyte
(in a conifer, the haploid cells and tissues developing from the maternal gamete, with nutritive and other functions for the contained zygote)
d Megagametophyt *m*
f mégagamétophyte *m*; gamétophyte *m* femelle
r мегагаметофит *m*

* **melaleuca → 1056**

* **Melancholy pine → 2726**

* **melawis → 6710**

* **mellow → 7822**

* **mellowing → 5339**

* **melon tree → 1794**

* **membrane → 1407**

* **Menzies spruce → 7678**

* **meranti → 4831**

* **merawan → 4365**

* **merbau → 5476**

* **merchandiser capacities → 7726**

5393 **merchantable bole; commercial bole; merchantable log**
d Nutzholzstamm *m*; nutzholztüchtiger Stamm *m*
f tige *f* utilisable industriellement; tige marchande; bille *f* marchande
r деловой ствол *m*; товарный ствол; товарное бревно *n*

* **merchantable cutting → 1744**

* **merchantable diameter → 2743**

5394 **merchantable forest**
d Ertragswald *m*
f forêt *f* marchande
r товарный лес *m*; сортовой лес для продажи

5395 **merchantable height; salable height; merchantable length**
(point on a tree stem to which the stem is salable)
d handelsübliche Länge *f*; verkäufliche Länge
f hauteur *f* marchande; longueur *f* commerciale
r высота *f* деловой части ствола; торговая длина *f*

* **merchantable log → 5393**

5396 **merchantable sections; merchantable assortment**
(sections of the central stem of growing-stock trees that meet either pulpwood or saw-log specifications)
d handelsfähige Holzschnitte *mpl*; Nutzholzsortiment *n*
f sections *fpl* [de bois] marchandes
r товарные брёвна *npl*; деловой лесосортимент *m*

5397 **merchantable snag**
(snag that is of sufficient quality and/or volume to make it suitable for harvesting)
d Nutzdürrständer *m*
f chicot *m* marchand
r деловой сухостой *m*

* **merchantable timber → 5400**

5398 **merch[antable] top**
(smallest utilizable top)
d marktgängige Wipfel *f*
f sommet *m* marchand; sommet commercialisable
r деловой верх *m*

(tree that has attained sufficient size, quality and/or volume to make it suitable for harvesting)
d Nutzbaum *m*
f arbre *m* marchand
r товарное дерево *n*; деловое дерево; коммерчески выгодное дерево

5400 **merchantable wood; wood of commerce; commerce wood; commercial wood; merchantable timber; timber for the trade; trade timber; usable timber**
all roundwood products except fuelwood)
d Handelsholz *n*; Handelsware *f*; Waldnutzholz *n*; Nutzholz *n*
f bois *m* marchand; bois de commerce
r деловая древесина *f*; ликвидная древесина; товарная древесина; коммерческие лесоматериалы *mpl*; торговой лес *m*

5401 **merchantable [wood] volume**
(amount of sound wood in a single tree or stand that is suitable for marketing under given economic conditions)
d Nutz[holz]masse *f*
f volume *m* [de bois d'œuvre] marchand
r объём *m* деловой части [ствола]

5402 **meristem; formative tissue**
(a plant tissue composed of dividing cells and giving rise to organs)
d Fortbildungsgewebe *n*; Bildungsgewebe *n*; Meristem *n*
f méristème *m*
r образовательная ткань *f*; эмбриональная ткань; меристема *f*

5403 **meristem culture**
(apical meristem culture; a part taken (an explant) consisting only of the apical dome of tissue above the youngest leaf primordium)
d Meristemkultur *f*
f culture *f* de méristèmes
r разведение *n* меристемы

5404 **Merkus pine; Sumatran pine**
(Pinus merkusii)
d Merkus-Kiefer *f*
f pin *m* de Merkus
r сосна *f* Меркуса; тузам *m*

5405 **mersawa** (*UK*)**; bella rosa** (*US*)
(Anisoptera)
d Mersawa *m*
f ven-ven *m*

5406 meru-oak; turill
(Vitex doniana; Vitex keniensis)
d Turill *m*
f légumebois *m*
r витекс *m* кенийский

* **mescal bean → 1939**

* **mesophilic → 5408**

5407 mesophyll
(the middle and photosynthetic tissue of a leaf)
d Mesophyll *n*
f mésophile *m*
r мезофилл *m*

5408 mesophyllous; mesophilic
(adapted to life in moderately and constantly moist soils)
d mesophilig; frischezeiger
f mésophile; mésohydrique
r мезофильный

5409 mesophyllous forest; mesophytic forest
d Mesophytwald *m*; mesophiler Wald *m*
f forêt *f* mésophyte
r мезофитный лес *m*

5410 mesophyte; mesophytic plant
(a plant that requires a moderate amount of water for normal growth)
d Mesophyt *m*
f mésophyte *f*
r мезофит *m*

* **mesophytic forest → 5409**

5411 messenger; pull-back; haul-back [wire] rope; [haul-]back line; outhaul rope; return check rope; return-line
(rope used in cable logging to haul the main line and its fittings back to the point where the logs are to be attached)
d Rückholerkabel *n*; Rückholseil *n*; Umlaufseil *n*
f câble *m* [tracteur] de retour; câble de récupération
r возвратный канат *m*; возвратная верёвка *f*; возвратный трос *m*

5412 messmate [stringybark] eucalyptus
(Eucalyptus obliqua; Eucalyptus deglupta)
d Messmate-Stringybark *m*
f messmate-stringybark *m*
r эвкалипт *m* косой; эвкалипт деглюнта

d Metall-Holzklebstoff *m*
f colle *f* métal-bois
r клей *m* для склеивания металла с древесиной

5414 metatracheal parenchyma; banded parenchyma
(axial parenchyma forming concentric lines or bands, as seen in cross section)
d Bandparenchym *n*
f parenchyme *m* métatrachéal; parenchyme en couches; parenchyme concentrique
r метатрахеальная паренхима *f*

5415 metaxylem
(later-formed primary xylem, with pitted tracheary elements)
d Metaxylem *n*
f métaxylème *m*
r метаксилема *f*

* **method of cutting management → 2203**

* **method of high forest with reserves → 4283**

* **method of periods by volume → 6165**

5416 method of regeneration; regeneration method; reproduction method
d Verjüngungsmethode *f*
f méthode *f* de régénération
r метод *m* возобновления

5417 metric log measure
d metrisches Shnittholzmaß *n*
f mesure *f* des longueurs en mètres
r измерение *n* кубатуры брёвен в метрических мерах

* **Mexican ash → 680**

5418 Mexican [bald] cypress; Montezuma baldcypress
(Taxodium mucronatum)
d mexikanische Zypresse *f*; mexikanische Sumpfzypresse *f*
f cyprès *m* du Mexique; cyprès de Montezuma
r кипарис *m* болотный [мексиканский]

5419 Mexican barberry; wild currant; chaparral berry; agerita; algerita
(Mahonia trifoli[ol]ata; Berberis trifoliata)
d dreiblättrige Mahonie *f*
f mahonie *f* trifoliolée
r магония *f* трилистная

d mexikanische Eiche *f*
f chêne *m* mexicain
r дуб *m* сидячецветный мексиканский

5421 **Mexican buckeye**
(Ungnadia speciosa)
d mexikanische Kastanie *f*
f marronnier *m* mexicain
r унгнадия *f* прекрасная

* **Mexican cedar → 1411**

5422 **Mexican fir; Guatemala fir; sacred fir**
(Abies religiosa)
d Guatemalatanne *f*; Oyameltanne *f*
f sapin *m* de Guatemala
r пихта *f* религиозная

5423 **Mexican juniper; Mexican red cedar; Mexican savin; savin; drooping juniper**
(Juniperus flaccida; Juniperus gracilis)
d mexikanischer Wacholder *m*
f genévrier *m* du Mexique; genévrier flasque
r можжевельник *m* мексиканский; можжевельник провислый

* **Mexican lime → 9348**

* **Mexican mahogany → 6019**

* **Mexican manzanita → 6410**

5424 **Mexican pinyon pine; [Mexican] nut pine**
(Pinus cembroides)
d mexikanische Nusskiefer *f*; Pinyonkiefer *f*
f pin *m* pinyon
r сосна *f* кедровидная

5425 **Mexican plum; fall plum; bigtree plum**
(Prunus mexicana; Prunus lanata)
d wollige Pflaume *f*
f prunier *m* laineux
r слива *f* мексиканская

* **Mexican red cedar → 5423**

5426 **Mexican rosewood; cocobolo; partridgewood; coffee wood; granadillo**
(Dalbergia retusa; Caesalpinia granadillo; Libidibia corymbosa)
d Salamanderholz *n*; Rebhuhnholz *n*; Grenadille *f*; Partridge *f*; Cochenille *f*
f cochenille *f*; bois *m* de perdrix [des Antilles]; grenadille *f*; cocobolo *m*
r цезальпиния *f* высокая; дальбергия *f* притупленная; кокоболо *n*

* **Mexican white pine → 7867**

* **mezereon → 2246**

5427 **mezereum family**
(Thymelacaceae)
d Seidelbastgewächse *pl*
f thyméléacées *pl*
r волчниковые *pl*

5428 **michelia**
(Michelia)
d Michelie *f*
f michélie *f*
r михелия *f*

* **micro-crack → 4088**

5429 **microenvironment; microhabitat**
(small or restricted set of distinctive environmental conditions that constitute a small habitat, such as a tree stump, a dead animal, or a space between clumps of grass)
d Kleinstandort *m*; Mikrohabitat *n*; Mikroumwelt *f*
f micro-environnement *m*; microhabitat *m*; micromilieu *m*
r микросреда *f* обитания; микроместообитание *n*; микро-габитат *m*

* **microfibril → 3780**

* **micronutrients → 8740**

5430 **microorganism**
d Mikroorganismus *m*
f micro-organisme *m*
r микроорганизм *m*

5431 **micropetalous**
d kleinblütenblatterig
f à petite pétale
r мелколепестный

5432 **micropropagation**
d Mikrovermehrung *f*
f micropropagation *f*
r микрораспространение *n*

5433 **microrelief; microtopography**
(minor ground surface irregularities which, when compared from a fixed point of elevation between them, display differences in elevation of not more than 1 foot within a distance of 4 feet)
d Mikrorelief *n*

5434 microsite
(ultimate unit of the habitat; small area which exhibits localized characteristics different from the surrounding area)
d Mikrobiotop *n*
f microbiotope *m*
r микробиотоп *m*

5435 microthermophytia
d Mikrowärmepflanzen *fpl*
f microthermophytes *mpl*
r сообщество *n* бореальных лесов

* **microtopography → 5433**

* **mid-diameter → 2468**

* **middle-aged forests → 5351**

5436 middle cuts
d Mittelschnitt *m*
f seconde bille *f*
r пиломатериалы *mpl*, полученные из средней части брёвна

* **middle forest → 7947**

* **middle log → 434**

5437 middle plank
d Mittelbrett *n*
f planche *f* moyenne; planche de cœur
r срединная доска *f*

5438 middle story; midstory
(a middle canopy layer of smaller trees that occur under an overstor[e]y of trees)
d Mittelschicht *f*
f étage *m* moyen
r средний ярус *m* растительности

* **middle thinning → 5508**

5439 middlings
d Mittelware *f*
f classes *fpl* moyennes
r пиломатериалы *mpl* среднего сорта

* **midges → 3793**

5440 mid-girth; girth at the middle; mean girdle; mean girth; average girth; circumference at the middle
of a tree)
d Mittenumfang *m*
f circonférence *f* au milieu; circonférence
r ... высоты; средний охват *m*; окружность середины

* **midrib → 4879**

* **mid-sectional diameter → 2468**

* **midspan support → 4548**

* **midstory → 5438**

5441 mid[-timber] girth
(of a bole)
d Umfang *m* in halber Länge des Nützholzstammes
f circonférence *f* à mi-fût
r окружность *f* на половине высоты деловой части дерева

5442 migration; drift
(the movement of wildlife uninfluenced, at least directly, by human action)
d Migration *f*
f migration *f*
r миграция *f*

5443 migration of species
d Artenabwandern *n*
f migration *f* d'espèces
r миграция *f* видов

* **mild → 7822**

5444 mildew; smut; [leaf] blight
(any of various diseases which lead to the browning and dropping of a plant's leaves)
d Mehltau *m*; Brand *m*; Hartfäule *f*; Schütte *f*
f mildiou *m*; tache *f* des feuilles; dessèchement *m* de la feuille; dépérissement *m*
r милдью *n*; ложная мучнистая роса *f*; изгарина *f* (грибковое заболевание листьев)

* **mildew fungus → 3772**

* **mild humus → 5576**

* **milk tree → 2000**

5445 milkweed family
(Asclepiadaceae)
d Seidenpflanzengewächse *pl*
f asclépiadacées *pl*
r ластовневые *pl*

5446 milkwort family

f polygalacées *pl*
r истодовые *pl*

5447 milky mangrove; aloe
(Excoecaria agallocha; Excoecaria granulata)
d Tillai *n*; Tamil *n*; Blindbaum *m*
f palétuvier *m* aveuglant
r сальное дерево *n* Агаллоха

5448 mill
d holzverarbeitendes Werk *n*
f usine *f* de conversion du bois
r лесообрабатывающая установка *f*; лесообрабатывающая фабрика *f*

5449 mil[l]acre plot
(a sample plot of 1/1000 acre (usually 1/10 chain square) used in reproduction or vegetation surveys)
d Milacre-Parzelle *f*
f parcelle *f* à milliacre
r учётная площадь *f* 1/1000 акра

* **millboard → 6059**

5450 mill culls; shipping culls; mill debris; mill residue
of wood)
d Brackholz *n*; Sägewerkausschuss *m*; Schnittholzausschuss *m*
f rémanents *mpl* d'usinage; sciage *m* de rebut
r древесина *f*, отбракованная на фабрике; отходы *mpl* машинной обработки

5451 mill deck; pile-bottom; log deck; bank[ing] ground; rampsite; bay; raceway; decking site; stockpile
(a cleared area within a timber harvest where harvested logs are processed, piled, and loaded for transport to a sawmill or other facility)
d Stapelstelle *f*; Stapelplatz *m*; Stapelfleck *m*; Vorratslager *m*; Ganter *m*; Entladenbrücke *f*; Entladerampe *f*; Abladeplatz *m*
f chantier *m* d'empilage; premier dépôt *m* transitoire; plate-forme *f* de réception des grumes
r площадка *f* для разгрузки брёвен; разгрузочная эстакада *f*; место *n* складирования древесины; место разгрузки древесины; подштабельное место

* **milling → 9609**

5452 milling arbor
d Fräseraufnahmedorn *m*

5453 milling chipper
d Fräserkraftmühle *f*
f fraise *f* de meulage
r фрезерный стружечный станок *m*

* **mill log → 7245**

* **mill pond → 5105**

* **mill residue → 5450**

5454 mill[-]run; MR
(all the timber outturn (more particularly the merchantable outturn) of a parcel of logs or of a whole mill, without reference to grades)
d Sägewerksablauf *m*
f tout-venant *m*
r товарная продукция *f* завода

5455 mill scale
(the scale of logs made at the rafting boom or at the sawmill)
d Sägeholzvorrat *n*
f volume *m* de bois d'usinage
r объём *m* лесоматериалов, определённый на лесопильном заводе

5456 mill site
(any place in a sawmill where forest products are stored, altered, or processed)
d Werklagerhalle *f*; Fabriklagerhalle *f*
f dépôt *m* d'usine
r заводской склад *m*

* **mill tally → 5213**

5457 millwood
(timber that is used by turning mills and specialty hardwood plants)
d Werkholz *n*
f bois *m* d'usine
r заводская древесина *f*

5458 millwork
(lumber that has been remanufactured into door and window parts or decorative trim)
d Schnittware *f*; Sägeschnittware *f*; Holzfertigteile *npl*
f menuiserie *f* préfabriquée
r заготовки *fpl*, подвергшиеся первичной машинной обработке; столярные работы *fpl*

* **mimosa bush → 1689**

5459 mimosa family

f mimosacées *pl*
r мимозовые *pl*

5460 mimosa [tree]
(various mimosa trees, grown chiefly in Australia and Natal - Acacia mollissima, Acacia decurrens, Acacia dealbata, Acacia pycnantha etc)
d Mimose *f*
f mimosa *m*
r мимоза *f*

* **mine cross tie → 1725**

* **mine dump → 5464**

* **mine props → 6311**

* **mineral cycle → 5825**

* **mineral deposit → 5462**

5461 mineralized stripes
d mineralisierende Streifene *mpl*
f bandes *fpl* minéralisées
r минерализованные полосы *fpl* (полосы, очищенные от лесных горючих материалов или обработанные почвообрабатывающими орудиями либо иным способом до минерального слоя почвы)

5462 mineral stain; mineral deposit; mineral streak
(an olive to greenish-black or brown discolo[u]ration of undetermined cause in hardwoods)
d Erzlagerstätte *f*; mineralische Verfärbung *f*
f tache *f* minérale; concrétion *f* minérale
r минеральное включение *n* (в древесине в виде чёрной или коричневой полоски с зеленоватым оттенком)

* **mine tailings → 5464**

5463 mine timber; mining timber; pit timber; pit wood; crowntree
d Aufbruchholz *n*; Bergbauholz *n*; Bergholz *n*; Grubenholz *n*; Schachtholz *n*
f bois *m* de mine[s]
r шахтный лесоматериал *m*; рудничный лесоматериал; крепёжный лес *m*

5464 mine tip; mine dump; mine tailings
(for afforestation)
r шахтный отвал *m*; угольный отвал

* **mingled forest → 5486**

5465 minimum diameter; economic diameter
(of exploitable trees)
d Mindestdurchmesser *m*
f diamètre *m* minimal [d'abattage]; plus petit diamètre
r минимальный диаметр *m*

5466 minimum diameter of defect
d Fehlermindestdurchmesser *m*
f diamètre *m* minimal de défaut
r наименьшая толщина *f* вырезки порока

5467 minimum girth
(of exploitable trees)
d minimaler Stammumfang *m*
f circonférence *f* minimale de tige
r минимальная окружност *f* ствола

5468 minimum timber length
d minimale Stammlänge *f*
f longueur *f* minimale de la tige
r минимальная длина *f* деловой части ствола (до диаметра верхней части ствола)

5469 minimum utilization standard
(standard which is expressed as a maximum stump height, diameter at stump height, and top diameter and which can vary by species and timber supply area)
d minimaler Nutzungsstandard *m*
f norme *f* minimum
r норма *f* минимального использования

5470 mining slab
d Aufdeckung *f*; Balkenschlote *f*; Randschwarte *f*
f dosse *f* de mine
r рудничный горбыль *m*; обапол *m*

* **mining timber → 5463**

5471 minor forest product; side product
(such as medicinals, gums, resins, oils, fungi, honey)
d forstliches Nebenprodukt *n*
f menu produit *m*; produit forestier accessoire
r побочный продукт *m* леса

* **minor pine weevil → 7770**

5472 minus stand
d Minusbestand *m*

5473 **minus tree**
(a tree judged to be definitely inferior in some quality or qualities, i.e. poor growth habit, poor timber quality, proneness to disease or insect attack)
d Minusbaum *m*
f arbre *m* moins
r минусовое дерево *n*

5474 **minute bud; gemmule**
d Knospchen *n*
f gemmule *f*
r мелкая почка *f*; маленькая почка

5475 **miombo**
(deciduous or semi-deciduous open forest of Brachystegia and/or Isoberlinia species with a grassy root, widely distributed in Africa south of the equator)
d Miombo *n*
f miombo *m*
r миомбо *n* (ксерофильное редколесье в Восточной Африке)

5476 **mirabow; merbau; aizella**
(Intsia palembanica; Intsia bi[j]uga)
d Merbau *n*; molukkisches Eisenholz *n*
f merbau *m*
r мербау *n*

5477 **Mirbeck's oak; Canarian oak; Algerian oak**
(Quercus canariensis; Quercus mirbecki)
d kanarische Eiche *f*; Zeeneiche *f*
f chêne *m* de Mirbeck; zéen *m*
r дуб *m* Мирбека

* **missanda → 1366**

5478 **misshapen tree**
d deformierter Baum *m*; unförmiger Baum
f arbre *m* déformé
r уродливое дерево *n*; деформированное дерево

5479 **missing ring; ghost ring; faint ring; absent ring**
d fehlender Jahrring *m*
f cerne *m* absent; cerne manquant
r пропущенное годичное кольцо *n*

5480 **Missouri river willow; diamond willow**
(Salix eriocephala; Salix missourensis; Salix acutidens)
d Missouriflussweide *f*; Missouriweide *f*; Düneweide *f*; Herz-Blattweide *f*;
r ива *f* миссурийская

* **mist forest → 6700**

* **mistletoe → 2902**

5481 **mist propagation; misting**
d Sprühnebelvermehrung *f*; Nebeln *n*
f multiplication *f* sous brouillard artificiel; brumisation *f*
r укоренение *n* черенков с помощью искусственного тумана

5482 **mist spraying of foliage**
d Bestäubung *f* von Laubwerk
f poudrage *m* de la frondaison
r опыление *n* листвы

* **mite → 4809**

* **mite infestation → 26**

* **miter cut → 1441**

5483 **miter cutting**
d Gehrungsschnitt *m*
f coupe *f* à onglet
r заусовка *f*; резание *n* под углом

* **miter saw → 883**

5484 **mixed broadleaf/needleleaf forest; mixed coniferous-broadleaved forest**
d Nadel- und Laubbaummischwald *m*
f forêt *f* mixte de conifères et de feuillus
r хвойно-широколиственный лес *m*

* **mixed bud → 1837**

5485 **mixed coniferous stand; mixed coniferous forest**
d Nadelmischwald *m*
f peuplement *m* résineux mixte; résineuse *f* mélangée
r смешанный хвойный лес *m*

* **mixed cropping → 5491**

5486 **mixed forest; mingled forest; composite forest; mixed stand; multispecies stand; mixedwoods; unhomogeneous forest**
(a stand in which less than 80% of the trees in the main canopy are of a single species; includes both coniferous and deciduous trees)
d gemischter Wald *m*; Mischwald *m*;

f forêt *f* mixte; peuplement *m* mélangé; mélangé *m*; peuplement mixte; mixte *m*
r смешанный лес *m*; смешанный древостой *m*; смешанное насаждение *n*

5487 mixed grain
d Mischungsgetreide *n*; Mischungskorn *n*
f débit *m* mélangé; mélange *m* de céréales
r смешанное семя *n*

5488 mixed lengths; random lengths; random length dimension lumber
(of timber)
d gemischte Langholze *npl*
f débits *mpl* toutes longueurs
r лесоматериалы *mpl* разной длины

5489 mixed lowland hardwoods
d gemischtes Tieflandslaubholz *n*
f bois *m* de feuillus de bas-pays mixte
r древесина *f* смешанных низинных твёрдолиственных пород

5490 mixed northern hardwood
d gemischte Nordlaubholze *npl*
f bois *m* dur du nord mixte; feuillus *mpl* du nord mixtes
r смешанные северные твёрдолиственные породы *fpl*

5491 mixed planting; mixed cropping; polyculture
(simultaneous cultivation of a number of crops as opposed to stands composed of a single species)
d Mischkultur *f*
f plantation *f* mixte; culture *f* d'espèces variées; polyculture *f*
r посадка *f* с примесью сопутствующих пород; смешанная культура *f*

* **mixed seedling and sprout forest → 7947**

5492 mixed-severity fire regime
(fire regime in which fires either cause selective mortality in dominant vegetation, depending on different species' susceptibility to fire, or vary between understor[e]y and stand replacement)
d Feuerregime *n* mit gemischter Schwierigkeit
f régime *m* des incendies de gravité mixte
r пожарный режим *m* смешанной опасности

5493 mixed spruce forest; mixed spruce stand
d Mischfichtenwald *m*; Mischfichtenbestand *m*

* **mixed stand → 5486**

5494 mixed stand in groups; mixture by groups; mixture by clusters of trees
d horstweise Mischung *f*; gruppenweise Mischung
f mélange *m* par bouquets; mélange par groupes
r смесь *f* куртинами; смесь группами

* **mixed subor → 8277**

5495 mixed upland [forest]
(a forest type group consisting of pines shortleaf, pitch, loblolly (50%), and hardwoods red oak, white oak, and other hardwood (50%))
d gemischter Hügellandwald *m*
f mélange *m* montagnard
r смешанный лес *m* в холмистой местности; смешанный горный лес

* **mixedwoods → 5486**

* **mixture by clusters of trees → 5494**

* **mixture by groups → 5494**

5496 mixture by single kinds of trees; mixture by single individuals; mixture by isolated trees; tree-by-tree mixture; single-tree mixture; individual-tree mixture; stem-by-stem mixture
d Einzelmischung *f*; stammweise Mischung *f*
f mélange *m* pied à pied; mélange intime; mélange par pieds d'arbre
r подеревное смешение *n* пород; единичное смешение; единичная смесь *f*

* **mixture of broadleaved species → 4143**

5497 mixture of tree species; tree species mixing
(percentage of each recognized tree species comprising the forest type)
d Baumartenmischung *f*; Holzartenbeimischung *f*
f mélange *m* d'espèces forestières
r смешение *n* древесных пород; примесь *f* древесных пород

5498 mixture of uneven-aged trees
d stufenweise Mischung *f*
f mélange *m* par étages
r ступенчатая смесь *f*

f déchiquetteuse *f* mobile
r передвижная рубильная машина *f*

5500 mobile harvester
d mobile Erntemaschine *f*
f récolteuse *f* mobile
r передвижная лесозаготовительная машина *f*

5501 mobile logger; mobile yarder
(lightweight, two-drum yarder usually on a truck with a spar and boom)
d mobiler Holzlader *m*; Lade-Kran *m*; Mobilseilkran *m*
f chargeur *m* mobile de bois; câble-grue *m* mobile
r передвижная канатная трелёвочно-погрузочная установка *f*; передвижной лесопогрузчик *m*

* **mobile sawmill → 6469**

* **mockernut hickory → 8988**

5502 mock orange family
(Philadelphaceae)
d Pfeifenstrauchgewächse *pl*
f philadelphacées *pl*
r филадельфовые *pl*

5503 mock privet; phillyrea
(Phillyrea)
d Steinlinde *f*; Lorbeerliguster *m*
f filaria *m*
r липа *f* каменная

5504 model forest
(forest or designated area including forests and woodland for which an integrated management plan is created and implemented to achieve multiple objectives on a sustainable basis)
d Modellwald *m*
f forêt *f* modèle
r модельный лес *m*

* **model tree → 7174**

5505 model wood
d Modellholz *n*
f bois *m* modèle
r модельная древесина *f*; древесина для изготовления моделей

* **mode of forest management → 3491**

plant remains)
d Modder *m*
f moder *m*
r модер *m*; модер-гумус *m*

5507 moderately fine texture
d mittelengringige Faser *f*; gemäßigtfeine Textur *f*
f texture *f* modérément fine
r умеренно тонкая текстура *f*

5508 moderate thinning; middle thinning
d mäßige Durchforstung *f*
f éclaircie *f* modérée
r умеренная прорубка *f*; среднее прореживание *n*; умеренное прореживание; промежуточная рубка *f* средней интенсивности

5509 modified burning zone
(zone within or adjacent to a smoke-sensitive area that requires special considerations and burning techniques)
d modifizierte Brandzone *f*
f zone *f* de brûlage modifiée
r модифицированная зона *f* горения

5510 modified cut areas
(that will require modification of normal timber regeneration methods because of landscape management quality standards or filter strip criteria)
d modifizierte Schlagsfläche *f*
f aire *f* de coupe modifiée
r видоизменённая лесосека *f*

5511 modified cutting
(even-aged harvest cutting method modified to the extent that it no longer meets the definition of that even-aged harvest cutting method, but will insure the perpetuation of the desired commercial tree species)
d modifizierter Hieb *m*
f coupe *f* modifiée
r модифицированная валка *f*

5512 modified forest
(non-primary, ecologically-altered forest)
d modifizierter Wald *m*
f forêt *f* modifiée
r модифицированный лес *m*; видоизменённый лес

5513 modified wood
(wood processed by chemical treatment,

f bois *m* modifié
r модифицированная древесина *f*

* **Modoc cypress → 484**

5514 **modulus of elasticity; coefficient of elasticity**
(calculated from bending tests)
d Elastizitätsmodul *n*
f module *m* d'élasticité; module élastique; coefficient *m* d'élasticité
r модуль *m* продольной упругости; модуль упругости

5515 **modulus of rupture**
(modulus whereby the stress to which a material can be subjected before rupture, may be established)
d Biegefestigkeitsmodul *n*
f module *m* de rupture
r модуль *m* излома; модуль разрыва

* **mohur tree → 3249**

* **moist forest → 6700**

5516 **moist temperate forest**
d Regenwald *m* im temperierter Zone
f forêt *f* humide tempérée; forêt pluviale tempérée
r влажный лес *m* в умеренном климате

5517 **moist temperate mountain forest**
d Gebirgsregenwald *m* im temperierter Zone
f forêt *f* humide tempérée de montagne
r влажный горный лес *m* в умеренном климате

* **moist [tropical] mountain forest → 5593**

5518 **moisture content of wood**
(the weight of moisture contained in a piece of timber expressed as a percentage of the oven-dry weight)
d Holzwassergehalt *m*; Feuchtegehalt *m* des Holzes
f contenu *m* d'humidité du bois; teneur *f* en humidité du bois
r содержание *n* влаги в древесине; масса *f* воды в древесине

5519 **moisture distribution**
(in wood)
d Feuchtigkeitsverteilung *f*
f distribution *f* de l'humidité; répartition *f* de

5520 **moisture gradient**
(a progressive decrease (or increase) in moisture content between the core and the surface of a piece of wood)
d Feuchtigkeitsgradient *m*; Feuchtigkeitsgefälle *f*
f gradient *m* de humidité
r градиент *m* влажности

5521 **moisture of timber; wood humidity; wood moisture**
d Holzfeuchte *f*; Holzfeuchtigkeit *f*
f humidité *f* du bois
r влажность *f* древесины

5522 **moisture[-]proof; moisture-resistant**
d feuchtigkeitsbeständig; wasserundurchlässig; wasserdicht
f imperméable [à l'humidité]; résistant à l'humidité
r влагостойкий; влагоустойчивый; влагоупорный; влагонепроницаемый

5523 **moisture-resistant plywood**
d feuchtigkeitsabstoßendes Sperrholz *n*
f contreplaqué *m* résistant à l'humidité
r водостойкая фанера *f*

* **molave → 7763**

* **mold and edge sander → 6567**

5524 **moldboard plough**
d Abraumpflug *m*; Abstreichpflug *m*
f charrue *f* à versoir
r отвальный плуг *m*

5525 **mold fungi**
(Mucorales)
d Schimmelpilze *mpl*
f champignons *mpl* de moisissure
r плесневые грибы *fpl*

5526 **Molucca albizia; Belalu tree**
(Albizzia falcataria; Albizzia falcata; Falcataria moluccana; Adenanthera falcataria)
d breitlaubige Albizie *f*
f albizie *f* à graines noires
r альбиция *f* серповидная

5527 **monareal plot sampling**
d Einflächenstichproben *n*
f échantillonnage *m* par placettes à one dimension

5528 monastery forest; wood belonging to a monastery
d Klosterwald *m*; Klosterforst *m*
f forêt *f* conventuelle
r монастырский лес *m*

5529 monimia family; boldo family
(Monimiaceae)
d Monimiengewächse *pl*
f laurales *pl*
r монимиевые *pl*

5530 monitoring of forest; forest monitoring
d Forstüberwachung *f*
f surveillance *f* des forêts
r мониторинг *m* лесов

5531 monkepot tree family
(Lecythidaceae)
d Topffruchtbaumgewächse *pl*
f lécythidacées *pl*
r лецитидасовые *pl*; лецитисовые *pl*

5532 monkey apple; marishballi; kauta; anaura
(Licania macrophylla; Licania sparsipilis)
d Grigri *n*
f grigri *m*; bois *m* gris
r ликания *f* крупнолистная; спонзеуду *m*; анаура *f*

5533 monkey-bred tree; baobab
(Adansonia digitata)
d Affenbrotbaum *m*; Baobab *m*
f baobab *m*
r [африканский] баобаб *m*; адансония *f*

5534 monkey comb; tibourbou
(Apeiba aspera)
d sternförmiger gelber Affenkamm *m*; Aubletia *m*
f apeiba *m*
r апейба *f*; тибурбу *m*

* **monkey gum → 5587**

* **monkey-nut tree → 237**

* **monkey pod → 6701**

* **monkey[ ]pot → 6701**

5535 monkey puzzle tree
(Araucaria araucana)
d Schmucktanne *f*; Andentanne *f*; Araukarie *f*
f désespoir *m* des singes; araucaria *m* du Chili
r араукария *f* чилийская

5536 monocable system
d Einseilbahnsystem *n*
f transporteur *m* monocâble
r канатная установка *f* с замкнутым канатом

5537 monocarpic plant
(producing a single fruit, or having only 1 fruiting period, during the life cycle)
d einfrüchtige Pflanze *f*; einmalblühende Pflanze
f plante *f* monocarpique
r монокарпическое растение *n*; монокарпик *m*

5538 monoclimax theory
(the hypothesis that in any one environment there can be only one climax community, in contrast to the hypothesis that there can be several (polyclimax theory))
d Monoklimaxtheorie *f*
f théorie *f* de monoclimax
r теория *f* моноклимакса (единства климакса для всех сукцессий)

5539 monocormic trees
d einachsiger Baum *m*
f arbres *mpl* à un axe
r деревья *npl* с одной главной осью и боковыми ветвями

5540 monocotyledon plants; monocotyledons
(plants with one cotyledon or seed leaf)
d einsamenlappige Blütenpflanzen *fpl*; Monokotyledonen *pl*; einkeimblättrige Pflanzen *fpl*
f plantes *fpl* monocotylédones
r односеменодольные растения *npl*; однодольные *pl*

5541 monocotyledon wood
d Monokotyledonholz *n*
f bois *m* de monocotylédones
r древесина *f* однодольных растений

5542 monoculture
(cultivation of a single crop or product without using the land for other purposes)
d Monokultur *f*
f monoculture *f*
r бессменная культура *f*; бессменный посев *m*; монокультурное хозяйство *n*; монокультура *f*

5543 monoecious plants
(plants having separate male and female reproductive structures on the same

f plantes *fpl* monoïques
r однодомные растения *npl*

5544 **mononodal top; rattail [top]; anodal [top]; whorlless top; foxtail**
(of a tree)
d anormal verlängerte Baumspitze *f*; unwirtelige Baumspitze
f sommet *m* allongé anormalement; houppe *f* sans nœuds; houppe sans verticilles
r лисохвостая верхушка *f*; ненормально вытянутая верхушка

5545 **monsoon forest; tropical deciduous forest; deciduous tropical forest**
d Monsunwald *m*; Monsoonwald *m*; tropischer wechselgrüne Regenwald *m*
f forêt *f* de moussons; forêt tropicale décidue; forêt dense caducifoliée
r муссонный лес *m*; муссоновый лес

5546 **monsoon mixed forest**
d gemischter Monsunwald *m*
f forêt *f* de moussons mixte
r муссонные смешанные леса *mpl*

* **Montana larch → 9340**

* **montane forest → 5583**

5547 **Monterey cypress**
(Cupressus macrocarpa)
d Monterey-Zypresse *f*; großfrüchtige Zypresse
f cyprès *m* de Monterey; cyprès à gros fruits; cyprès de Lambert
r монтерейский кипарис *m*; кипарис крупноплодный

5548 **Monterey pine; insignis pine; radiata pine**
(Pinus radiata)
d Monterey-Föhre *f*; Montereykiefer *f*; Radiatakiefer *f*
f pin *m* de Monterey
r сосна *f* замечательная

* **Montezuma baldcypress → 5418**

5549 **Montezuma pine**
(Pinus montezumae)
d Montezumaskiefer *f*
f pin *m* de Montezuma
r сосна *f* мексиканская

5550 **Montpellier maple**
(Acer monspessulanum)
r клён *m* французский

5551 **moonseed family**
(Menispermaceae)
d Mondsamengewächse *pl*
f ménispermacées *pl*
r луносеменниковые *pl*

5552 **moonshine crotch; finger roll**
(a swirling figure found in crotch veneer)
d Mondscheingabelung *f*; Keilzinkengabelung *f*
f fourche *f* par entures multiples
r волнистый текстурный рисунок *m*; текстурный рисунок в виде завитков

* **moorberry → 2889**

* **moorwort → 897**

* **mooseberry → 4315**

5553 **mooseberry viburnum; moosewood viburnum; squashberry**
(Viburnum edule)
d essbarer Schneeball *m*
f viorne *f* comestible
r калина *f* съедобная

5554 **mopping-up a fire; extinguishment of smouldering fire**
d vollständige Ablöschung *f* und Aufräumen *n* einer Brandfläche
f extinction *f* complète et nettoiement *m* d'une surface incendiée
r полное гашение *n* и очистка *f* пожарища

5555 **mop-up time**
(elapsed time from control of a fire until organized mopping up is completed)
d Aufwischenzeit *f*
f temps *m* de nettoiement
r время *n* операции очистки защитной полосы от горючих материалов

* **mor → 1681**

* **moraballi → 4715**

* **morabukea → 4715**

* **Moraine ash → 487**

* **morefall → 850**

5556 **moringa family**

f moringacées *pl*
r моринговые *pl*

5557 morning glory family
(Convolvulaceae)
d Windengewächse *pl*
f convolvulacées *pl*
r вьюнковые *pl*

5558 morototo; match wood; badana; yagrumo macho; yarumero; anonillo; arriero
(Didymopanax morototoni; Schefflera morototonii)
d Morototo *m*
f arbre *m* de Saint Jean
r дидимопанакс *m* морототони

5559 morphologically improved seedlings
(genetically improved seedling that are grown larger than other seedling at the nursery)
d morphologisch vergütete Sämlinge *mpl*
f semis *mpl* améliorés morphologiquement
r морфологически улучшенные лесные сеянцы *mpl*

5560 Morrow honeysuckle
(Lonicera morrowii)
d Morrow-Heckenkirsche *f*
f chèvrefeuille *m* de Morrow
r жимолость *f* Моррова

5561 mortise; mortice
(a rectangular hole cut into a piece of wood to accept a tenon)
d Nute *f*; Loch *n*
f mortaise *f*
r паз *m*

5562 moru oak; evergreen Himalaya oak
(Quercus dilatata)
d Moru-Eiche *f*
f chêne *m* moru
r дуб *m* распростёртый

5563 moss
d Moos *n*
f mousse *f*
r мох *m*

* **moss capsule → 9144**

5564 moss-covering; mossy covering
d Moosdecke *f*; Moosüberzug *m*; Moosschicht *f*
f couverture *f* de mousse
r моховой покров *m*

bitter oak
(Quercus cerris; Quercus laevis)
d Zerreiche *f*; türkische Eiche *f*; Burgund-Eiche *f*; Zirneiche *f*
f chêne *m* chevelu; chêne de Bourgogne
r дуб *m* австрийский

* **moss forest → 6700**

* **moss locust → 1038**

* **mossy covering → 5564**

5566 mossy-cup oak; burr oak; bunya-bunya
(Quercus macrocarpa)
d großfrüchtige Weißeiche *f*; Kletteneiche *f*
f chêne *m* à gros glands; chêne à gros fruits
r дуб *m* австралийский; дуб крупноплодный

5567 mossy forest
d moosiger Wald *m*
f forêt *f* moussue
r альпийское криволесье *n*; моховой лес *m*

* **mossy oak → 5565**

5568 mossy swamp pine forest
d Moosbruchenkiefernwald *m*
f forêt *f* mousseuse marécageuse à pin
r заболоченный сосняк *m*

* **mother gallery → 4811**

5569 mother tree; parent tree; female tree; [final] crop tree; seed[-bearing] tree; seed bearer; seeder
(any tree whose seeds are used to produce progeny for use in genetic experimentation)
d Samen[ernte]baum *m*; Mutterbaum *m*; Samenholz *n*
f arbre *m* de peuplement final; semencier *m*; arbre porte-graines; porte-graines *m*; pied-mère *m*
r материнское дерево *n*; семенное дерево; женское дерево; семенник *m*; семеноносец *m*

5570 mother-tree archive; tree archive; tree depot; tree [gene] bank
d Samenträgerbank *m*
f collection *f* d'arbres remarquables
r плантация *f* элитных и плюсовых деревьев

* **mother tunnel → 4811**

5571 motorized access and use
(access and use by, for example, float planes, helicopters, fixed-wing aircraft, motorboats, motor bikes, all-terrain vehicles, snowmobiles, and motorized equipment)
d motorisierter Zugriff *m* und motorisierte Nutzung *f*
f accès *m* motorisé et utilisation *f* motorisée
r механизированный доступ *m* и использование *n*

5572 motor-manual felling; powersaw felling
d Fällung *f* mit motorisierter Handsäge
f abattage *m* par scie à moteur
r валка *f* ручными мотопилами

* **motor truck operator → 8983**

* **mottle → 843**

* **mottled → 6131**

5573 mottled rot; marble rot
d marmorierte Fäule *f*; Marmorfäule *f*
f pourriture *f* marbrée
r пёстрая ядровая гниль *f*; ситовая ядровая гниль

5574 mottled white trunk rot
(invader - Fomes fomentarius)
d weiße Marmorstammfäule *f*; weißer Holzstammmarmor *m*
f pourriture *f* blanche marbrée des tiges
r белая мраморовидная гниль *f* стволов

5575 mottling
(of needles)
d Scheckigwerden *n*
f bariolage *m*
r пёстрое окрашивание *n*

* **mottling → 9187**

5576 mo[u]ld; mild humus
d milder Humus *m*
f humus *m* doux
r сладкий гумус *m*; мягкий гумус

* **mo[u]lded timber → 6566**

* **mo[u]ld formation → 3772**

* **mo[u]ld growth → 3772**

* **mo[u]lding → 9613**

(a machine for making decorative mo[u]ldings and other pieces)
d Wirkmaschine *f*; Former *m*
f moulurière *f*
r фасонно-фрезерный станок *m*

5578 mounding; mound planting; step planting
(forming raised planting spots or mounds by the scooping up and inversion of a quantity of organic and mineral soil)
d Hügelpflanzung *f*
f plantation *f* sur butte; buttage *m*
r окучивание *n*; укрывание *n* землёй (саженцев после посадки)

5579 mountain alder; thinleaf alder
(Alnus tenuifolia)
d amerikanische Erle *f*
f au[l]ne *m* de Canada
r ольха *f* горная

* **mountain ash → 8482**

5580 mountain ash grove
d Ebereschengehölz *n*; Ebereschenwäldchen *n*
f sorberaie *f*
r рябинник *m*

5581 mountain birch
(Betula montana)
d Bergbirke *f*
f bouleau de montagnes
r берёза *f* горная

* **mountain black cherry → 244**

5582 mountain chestnut oak; chestnut oak; rock [chestnut] oak
(Quercus montana)
d amerikanische Kastanie-Weißcheiche *f*; Bergeiche *f*
f chêne *m* châtaignier blanc d'Amérique
r дуб *m* каштановый; каштанодуб *m*; дуб горный

* **mountain cottonwood → 5661**

* **mountain elm → 7310**

5583 mountain forest; montane forest
d Bergwald *m*; Gebirgswald *m*; Gebirgsforst *m*
f forêt *f* de montagne
r горный лес *m*

* **mountain forest land → 5586**

f podzol *m* forestier de montagne
r горно-лесная оподзоленная почва *f*

5585 **mountain forest soil**
d Gebirgeswaldboden *m*
f sol *m* forestier de montagne
r горно-лесная почва *f*

5586 **mountain forest zone; mountain forest land**
d Bergwaldgebiet *n*
f terrain *m* forestier montagnard
r горно-лесной пояс *m*

5587 **mountain grey gum; grey gum; monkey gum; spotted gum; gray eucalyptus**
(Eucalyptus cypellocarpa; Eucalyptus propinqua)
d australischer Graugummibaum *m*
f gommier *m* gris d'Australie
r эвкалипт *m* близкий

5588 **mountain hemlock; Westcoast hemlock; black hemlock; bigcone Douglas-fir; large-coned Douglas-fir; western hemlock; Pacific hemlock; hemlock spruce**
(Tsuga heterophylla; Pseudotsuga macrocarpa;Tsuga mertensiana)
d Pattons-Hemlocktanne *f*; kalifornische Hemlocktanne *f*; westamerikanische Hemlocktanne *f*; Mertens-Hemlock; Riesen-Schierlingstanne *f*
f tsuga *m* de Patton; douglas *m* de Californie; tsuga de Californie; pruche *f* subalpine; tsuga de l'ouest
r тсуга *f* западная; тсуга разнохвойная

* **mountain larch → 9340**

* **mountain laurel → 1264**

5589 **mountain maple**
(Acer spicatum)
d amerikanischer Bergahorn *m*
f érable *m* à épis; érable de montagne d'Amérique
r клён *m* кистевидный

5590 **mountain ninebark**
(Physocarpus monogynus)
d Colorado-Blasenspiere *f*
f physocarpe *m* monogyne
r однопестичный пузыреплодник *m*

* **mountain oak → 7464**

* **mountain pine → 5598**

* **mountain she-oak → 1692**

* **mountain spruce → 2822**

5591 **mountain stewartia**
(Stewartia ovata; Stewartia pentagyna)
d Bergstewartie *f*
f camélia *m* ovale
r стюартия *f* яйцевидная

* **mountain sweet pepperbush → 1582**

5592 **mountain taiga forest**
d Gebirgstaiga *f*; Gebirgstaigawald *m*
f forêt-taïga *f* de montagne; taïga *f* de montagne
r горно-таёжный лес *m*

5593 **mountain [tropical] rain forest; moist [tropical] mountain forest; cloud forest**
(a mountain forest covered by a persistent mist that creates stunted trees and abundant epiphytes)
d tropischer Gebirgsregenwald *m*; [tropischer] Nebelbergwald *m*
f forêt *f* dense humide de montagne; forêt de la zone des brouillards; forêt ombrophile montagnarde
r горный влажный [тропический] туманный лес *m*; нефелогилея *f*

5594 **mountain tundra; Alpine tundra**
d Gebirgstundra *f*
f toundra *f* alpine
r горная тундра *f*

* **mountain white gum → 4054**

* **mountain white pine → 9344**

5595 **mouracatiara**
(Mouracatiara)
d Mouracatiara *n*
f mouracatiara *m*
r муйракатиара *f*

* **movingui → 5732**

* **MR → 5454**

5596 **much-branched; strong-branched; with large branches**
d mehrästig
f à branches multiples
r сильноветвистый

5597 **mucilage cell**

d Schleimzelle *f*
f cellule *f* à mucilage
r клейкая клетка *f*

* **mucilaginous fiber → 3813**

* **mudsill → 945**

* **mueri → 112**

* **muffler → 3131**

5598 **mug[h]o pine; dwarf [mountain] pine; [Swiss] mountain pine; dwarf pine; knee pine**
(Pinus mugo; Pinus mughus; Pinus montana)
d Bergkiefer *f*; Alpenföhre *f*; Mugokiefer *f*; Latschenkiefer *f*; Höhenkiefer *f*; Legföhre *f*; Zwergkiefer *f*; Sumpfföhre *f*
f pin *m* mugo; pin de montagne; pin à crochets; pin de Briançon; pin à suiffes
r сосна *f* горная; сосна малорослая; сосна карликовая

* **Mühlenberg oak → 1510**

* **muhugu → 7609**

* **muhuhu → 7609**

* **mukumari → 2711**

* **mulberry → 8354**

5599 **mulberry family**
(Moraceae)
d Maulbeergewächse *pl*
f moracées *pl*
r тутовые *pl*

5600 **mulberry umbrella**
(Morus platanifolia)
d platanenblättriger Maulbeerbaum *m*
f mûrier *m* à feuilles de platane; mûrier kagayama
r шелковица *f* платанолистная

5601 **mulch**
(protective covering, usually organic, placed around plants to keep in moisture and prevent the growth of weeds)
d Mulch *f*
f paillis *m*; mulch *m*
r мульчирующее покрытие *n*; мульча *f*; рыхлый защитный слой *m*

forest soil, using organic residues, such as grass, straw or wood fibers to protect exposed soil and help control erosion)
d Mulchen *n*
f paillage *m*
r мульчирование *n*

5603 **mulch wood**
(typically cypress trees that are ground and sold as mulch to be used in landscaping)
d Holzmulch *f*
f mulch *m* de bois
r мульча *f* из древесных остатков

* **mule-ear knot → 4376**

* **mulga → 5301**

* **mulga eucalyptus → 6779**

5604 **mull [earth]**
(a rich soil developed under mixed forests where a suitable supply of calcium is available)
d Mull *m*
f humus à lombricides; mull *m*
r мулевый гумус; муль *m*

* **Müller's stringybark → 9765**

5605 **multiclonal mixture**
(a mixture of clones in (usually) equal proportions for use in forest plantations)
d multiklonale Mischung *f*
f mélange *m* multiclonal
r мультиклональная смесь *f*

5606 **multifunctional forest; multiple-use forest**
d Mehrzwecknutzung *f*
f forêt *f* multifonctionnelle
r многоцелевой лес *m*

5607 **multiknife chipper**
d Vielmesserhackmaschine *f*
f coupeuse *f* à couteaux multiples
r многоножевая рубильная машина *f*

5608 **multilayer chipboard**
d mehrschichtige Spanplatte *f*; Vielschichtplatte *f*
f panneau *m* multicouche; panneau contreplaqué à plis multiples; contreplaqué multiplis
r многослойная древесностружечная плита *f*

5609 **multilayer [fiber]board; multiply plywood;**

f panneau *m* de fibres multiplis; panneau de fibres à plis multiples; panneau de fibres multicouche
r многослойная древесноволокнистая плита *f*

5610 multimachine harvesting
d mehrmaschinelle Holzgewinnung *f*
f exploitation *f* forestière par machines multi-opérationelles
r лесозаготовка *f* с использованием нескольких одно- или двухоперационных машин

5611 multinodal bud
d multinodale Knospe *f*; multinodales Auge *n*
f bourgeon *m* multinodal; bourgeon polycyclique
r многоузловая почка *f*

5612 multiple-entry volume table
d Mehrfachmassentafel *f*; Mehrfachkubierungstabelle *f*
f tarif *m* de cubage en entrée multiple
r таблица *f* объёмов, измеряемых по нескольким показателям

5613 multiple forest use
(system of resource use where the forest resources in a given land unit serve more than one user)
d Mehrforstnutzung *f*
f utilisation *f* intégrée de la forêt
r многоцелевое лесопользование *n*

5614 multiple leader
d vielfacher Spitzentrieb *m*; multipeler Spitzentrieb
f pousse *f* apicale multiple; flèche *f* multiple
r сложный главный побег *m*; разветвлённый главный побег; многократный верхушечный побег

* **multiple-length log → 5154**

5615 multiple-length logging; long-length logging; long-log transport; tree-length hauling
d Langholzabfuhr *f*; Stamm[holz]abfuhr *f*; Langholztransport *m*
f transport *m* par fûts entiers; transport par troncs entiers
r заготовка *f* длинномерных сортиментов; вывозка *f* длинномерных сортиментов; вывозка хлыстов

* **multiple-purpose forestry → 5623**

5616 multiple ring; double ring
(an annual ring consisting of two (or more) growth rings)
d Scheinjahrenring *m*
f cerne *m* multiple; cerne double; couche *f* annuelle double
r двойное годичное кольцо *n*

* **multiple saw [drop] trimmer → 8971**

5617 multiple-stemmed tree; sheaf tree; deliquescent tree
d mehrfachstammter Baum *m*; Garbenbaum *m*
f arbre *m* cépée; arbre déliquescent
r многоствольное дерево *n*

* **multiple stems → 8066**

* **multiple-stick length → 5154**

* **multiple-stick length operation → 5144**

5618 multiple-strip forest management
d Vielsaumbetrieb *m*
f exploitation *f* par lisières multiples
r многокаемчатое хозяйство *n*

5619 multiple sweep
d Mehrfachkrümmung *f*
f courbure *f* multiple
r сложная кривизна *f*

* **multiple-tree accumulating head → 5630**

* **multiple-tree delimber → 5323**

5620 multiple-tree harvesting; multi-tree harvesting
d Mehrbaumernte *f*
f exploitation *f* forestière multiple; exploitation multiarbre
r групповая заготовка *f* деревьев

5621 multiple-tree shear head; multi-tree shear; multistem shear; accumulating shear
(shearhead on a feller-buncher that is capable of accumulating and holding two or more cut stems)
d Mehrbaumschere *f*
f tête *f* collectrice à cisaille; tête à cisaille à pince collectrice
r ножевой механизм *m* для группового срезания деревьев

5622 **multiple-use [forest] management; integrated resource management**
(where a forest is divided up and the various parts of it have different functions or provide different products)
d Mehrnutzungsmanagement *n*
f aménagement *m* [forestier] à objectifs intégrés
r многоцелевое ведение *n* лесного хозяйства

5623 **multiple-use forestry; multiple-purpose forestry**
(any practice of forestry that fulfils two or more objectives of management, whether products, services or other benefits)
d vielfältige Forstwirtschaft *f*
f foresterie *f* à objectifs intégrés; foresterie à objectifs multiples
r многоотраслевое лесное хозяйство *n*; многоцелевое лесное хозяйство

5624 **multiple-use sustained yield calculation; MUSYC**
(linear programming forest planning model developed by the US Forest Service)
d mehrfachverwendeter anhaltende Berechnungsertrag *m*
f calcul *m* de rendement soutenu à usage multiple
r расчёт *m* многоцелевого устойчивого дохода

* **multiplex plywood → 5609**

* **multiply plywood → 5609**

5625 **multipurpose tree**
(woody perennial that is purposefully grown to provide more than one significant contribution to the production or service functions)
d Baum *m* mit verschiedenen Verwendungen
f arbre *m* à objectifs multiples
r многоцелевое дерево *n*

5626 **multiserial ray; multiseriate ray**
(a ray consisting of several to many rows of cells in width, as viewed in the tangential section)
d vielserieller Strahl *m*
f rayon *m* multisérié
r многорядный луч *m*

5627 **multi[-]span skyline**
(skyline having one or more intermediate supports)
r многопролётная подвесная канатная установка *f*

* **multispecies stand → 5486**

5628 **multi[-]stem; with several stems**
d vielstämmig; mehrstämmig; vielschäftig
f à grand nombre de tiges; en touffe
r многоствольный

* **multistem delimber → 5323**

5629 **multistem felling**
(operation handling two or more stems at the same time)
d vielstämmiger Hieb *m*
f abattage *m* simultané de grand nombre de tiges
r валка *f* одновременно нескольких деревьев; групповая валка

5630 **multistem felling head; multiple-tree accumulating head; accumulating [felling] head; multi-tree [felling] head**
d Mehrstamm-Schlagkopf *m*
f tête *f* d'abattage [à pince] collectrice
r валочная головка *f* с накоплением срезанных деревьев

* **multistem shear → 5621**

5631 **multistorey cropping**
(multispecies crop combinations involving both annuals and perennials with an existing stand of perennials)
d Mehretagenkultur *f*; vielstufige Kultur *f*
f culture *f* pluriétagée
r многоярусная культура *f*

5632 **multistorey system**
(agroforestry system that has a number of plant components of differing stature so that several layers of canopy are formed)
d Mehretagensystem *n*
f système *m* pluriétagé
r многоярусная система *f*

5633 **multistoried forest; many-storied forest**
d Mehretagenwald *m*; Mehrschichtenwald *m*; mehrschichtiger Hochwald *m*
f forêt *f* pluriétagé; forêt à étages multiples; futaie *f* à plusieurs étages
r сложный лес *m*; многоярусный лес

5634 **multistoried stand; stairs stand**
d mehrstufiger Bestand *m*;

f peuplement *m* pluriétagé
r многоярусное насаждение *n*

5635 **multistrand log haul-up**
d mehrlitzige Blockförderbahn *f*
f chaîne *f* de billes multitoron
r многоцепной лесотранспортёр *m* для брёвен

* **multi-tree [felling] head → 5630**

* **multi-tree harvesting → 5620**

* **multi-tree shear → 5621**

* **munding → 8093**

* **mungurra → 3550**

* **municipal forest → 9142**

5636 **municipal forestry; utility forestry; urban forestry**
(cultivation and management of trees and forests for their present and potential contributions to the physiological, sociological and economic well-being of urban society)
d Gemeindeforstwirtschaft *f*
f foresterie *f* urbaine; foresterie communale
r городское лесное хозяйство *n*; парковое хозяйство; озеленение *n* городов

* **muninga → 290**

* **musanda → 9042**

* **musanga → 9042**

5637 **muscadine [grape]**
(Vitis rotundifolia)
d Muscadinerrebe *f*
f muscadine *f*
r виноград *m* круглолистный; мюскадина *f*

* **musclewood → 249**

5638 **muskeg**
(tract of partly forested peatland supporting mosses, shrubby plants, and scattered trees)
d Muskeg *n*
f fondrière *f* de mousse; muskeg *m*
r жидкая торфяная почва *f*

5639 **musk-tree; abelmusk; abelmosk; musk mallow**
d Eibisch *m*
f althéa *m*; ketmie *f* des marais
r мускусное дерево *n*; гибискус *m* мускусный

* **mustard family → 2099**

* **MUSYC → 5624**

5640 **mutation**
(sudden, heritable variation in a gene or in a chromosome structure)
d Mutation *f*
f mutation *f*
r мутация *f*

5641 **mutenye**
(Guibourtia arnoldiana)
d Mutenye *f*
f muténye *m*
r мутения *f*

5642 **mutual fire insurance**
d Feuerversicherung *f* auf Gegenseitigkeit
f assurance *f* mutuelle contre l'incendie
r взаимное страхование *n* от пожаров

5643 **mutualism**
(an interaction between the individuals of two or more species in which the growth, growth rate, an/or population size of both are increased in a reciprocally beneficial association)
d Mutualismus *m*
f mutualisme *m*
r мутуализм *m*

* **mutualist → 8358**

5644 **muzaisi; cana-fistula**
(Cassipourea verticillata)
d Muzaisi *n*
f muzaisi *m*
r кассиповрея *f*

* **mvule → 4582**

* **Myanmar laurel → 9719**

5645 **mycelium [of fungus]**
d Myzell *n*; Mizelle *f*; Myzelium *n*
f mycélium *m*
r мицелий *m*; грибница *f*

* **mycorrhiza → 3775**

5646 myrobalan family
(Combretaceae)
d Sandmandelgewächse *pl*; Langfadengewächse *pl*
f combrétacées *pl*
r комбретовые *pl*

* **myrobalan plum → 1475**

5647 myrsina family
(Myrsineceae)
d Myrsinengewächse *pl*
f myrsinacées *pl*
r мирзиновые *pl*

5648 myrsine
(Myrsine floridana)
d Florida-Myrsine *f*
f myrsine *f* de Florida
r мирзина *f* флоридская

* **myrtle beech → 406**

* **myrtle blueberry → 5006**

5649 myrtle family
(Myrtaceae)
d Myrtengewächse *pl*
f myrtacées *pl*
r миртовые *pl*

5650 myrtle forest
d Myrtenwald *m*
f forêt *f* de myrtes
r лес *m* черничный

* **myrtle [tree] → 1792**

* **myrtle whortleberry → 5006**

d schmalkroniger Baum *m*; spitzkroniger Baum; spitzgipfeliger Baum
f arbre *m* à cime étroite; arbre à cime conique
r узкокронное дерево *n*; дерево с узкой кроной; узковершинное дерево

N

5651 **naked bud; bare bud**
d kahle Knospe *f*
f bourgeon *m* nu
r голая почка *f*

* **naked root → 531**

* **naked-root planting → 532**

* **naked tree → 4874**

5652 **nakedwood; soldierwood; solider**
d Krieger *m*
f étai *m* vertical
r вертикальный брусок *m* для крепления плинтуса

5653 **Nam-Nam tree; kekatong**
(Cynometra cauliflora)
d Kekatong *m*
f kekatong *m*
r цинометра *f*

5654 **nandina; sacred bamboo**
(Nandina domestica)
d heiliger Bambus *m*; Himmelsbambus *m*; Nandine *f*
f bambou *m* sacré; bambou céleste; bambou de la Félicité
r нандина *f* домашняя

5655 **nandina family**
(Nandinaceae)
d Nandinagewächse *pl*
f nandinacées *pl*
r нандиновые *pl*

* **nanism → 2686**

5656 **nannyberry [viburnum]; sheepberry**
(Viburnum lentago)
d Schafbeere *f*; kanadischer Schneeball *m*
f viorne *f* flexible; viorne à manchettes
r калина *f* канадская

* **nargusta [terminalia] → 1712**

* **narra → 5284**

5657 **narrow-crowned tree**

5658 **narrow edge coupe**
d Schmalsaumschlag *m*
f coupe *f* étroite en lisière
r узкокаемчатая лесосека *f*

5659 **narrowleaf ash; narrow-leaved ash; southern ash**
(Fraxinus angustifolia; Fraxinus oxycarpa)
d schmalblättrige Esche *f*; Schmallblattesche *f*; spitzblättrige Esche; Edelesche *f*
f frêne *m* oxyphille; frêne à feuilles aiguës; frêne à feuilles étroites; frêne du sud
r ясень *m* остролистный

5660 **narrowleaf buckbrush; buckbrush ceanothus**
(Ceanothus cuneatus)
d burdachsche Säckelblume *f*
f céanothe *m* à feuilles aiguës
r цеанотус *m* клинолистный

5661 **narrowleaf cottonwood; mountain cottonwood**
(Populus angustifolia)
d mexikanische Pappel *f*
f peuplier *m* d'Oregon; peuplier angustifolié
r тополь *m* узколистный

* **narrow-leaved ash → 5659**

5662 **narrow-leaved everlasting pea; wood pea**
(Lathyrus latifolius)
d breitblättrige Wald-Platterbse *f*
f gesse *f* sauvage
r чина *f* многолетняя

5663 **narrow-leaved peppermint; gray peppermint eucalyptus**
(Eucalyptus radiata; Eucalyptus robertsonii)
d schmalblättrige Pfefferminze *f*
f menthe *f* poivrée
r эвкалипт *m* лучистый; эвкалипт Робертсона

5664 **narrow-leaved phillyrea**
(Phillyrea angustifolia)
d schmalblättrige Steinlinde *f*
f filaria *m* à feuilles étroites
r липа *f* каменная узколистная

* **narrow-pored → 3155**

* **narrow-pored wood → 1656**

* **narrow upright tree → 6650**

5665 **naseberry; mammee sapote; marmalade plum; sapota; sapote; marmalade tree**
(Pouteria sapota)
d Zapote *f*; Sapotillfrucht *f*
f sapote *m*; sapotillier *m*; sapotille *f*
r сапота *f*; фруктовое дерево *n*

* **national forest → 8041**

5666 **national forest management**
d Nationalforstmanagement *n*
f gestion *f* forestière nationale
r национальное ведение *n* лесного хозяйства

5667 **national forest park**
d Nationalforstpark *m*
f parc *m* national forestier
r национальный лесной парк *m*

5668 **national preserve**
d Nationalschutzgebiet *n*
f réserve *f* nationale
r национальный заказник *m*

5669 **native bacteria**
(bacteria indigenous to natural water bodies)
d native Bakterien *fpl*; heimische Bakterien
f bactéries *fpl* indigènes; bactéries natives
r эндемичные бактерии *fpl*

5670 **native elm bark beetle**
(Hylurgopinus rufipes)
d Ulme-Borkenkäfer *m*
f scolyte *m* de l'orme
r короед *m* красноногий

* **native forest → 4476**

* **native growth → 282**

* **native species → 4477**

* **native wood → 4341**

5671 **native woodland; natural woodland**
(woods mainly or entirely of locally native species)
d native Holzbodenfläche *f*; heimische Holzbodenfläche

* **natural area → 9404**

5672 **natural area [preserve]; natural reserve; [nature-]reserve; sanctuary**
(area which retains a natural or relatively natural conditions and exemplifies typical or unique vegetation and associated biotic, soil, geologic and aquatic features)
d Reservat *n*; Schongebiet *n*; Schutzgebiet *n*
f réserve *f* naturelle
r [долгосрочный] заказник *m*; заповедник *m*; резерват *m*

* **natural balance → 2736**

5673 **natural boundary**
(of a forest)
d Naturgrenze *f*
f borne *f* naturelle
r естественная граница *f*

5674 **natural camphor**
(crystalline white substance extracted from Cinnamomum camphora, with intense freshness)
d Naturkampfer *m*
f camphre *m* naturel
r природная камфара *f*; естественная камфара (из древесины и листьев камфарного дерева)

* **natural crack → 9450**

5675 **natural defect**
d Naturdefekt *m*
f dégât *m* naturel
r естественный порок *m*

5676 **natural disaster**
d Naturkatastrophe *f*
f catastrophe *f* naturelle
r природная катастрофа *f*

* **natural discolo[u]ration → 5697**

5677 **natural distribution of forests**
d natürliche Waldausbreitung *f*
f distribution *f* naturelle des forêts; répartition *f* naturelle des forêts
r естественное распространение *n* лесов

5678 **natural durability**
(of wood)
d natürliche Lebensdauer *f*

5679 natural forest
(which has evolved as sequence of natural succession but still showing anthropogenic influences or that has developed from unmanaged pastures or from fallow land)
d natürlicher Wald *m*; Naturwald *m*
f forêt *f* naturelle
r естественный лес *m*

5680 natural forest gap; open woodland of natural origin; natural open [wood]land
(land that is mostly free of trees due to the ecological conditions of the site)
d Naturwaldwiese *f*
f clairière *f* naturelle
r естественная редина *f*

5681 natural fuel
d natürlicher Brennstoff *m*
f combustible *m* naturel
r естественный [лесной] горючий материал *m*

5682 natural graft
d natürliche Pfropfrebe *f*
f greffe *f* naturelle
r естественный привой *m*

5683 naturalistic silviculture; silviculture close to nature
(a silviculture near to the nature)
d naturnaher Waldbau *m*
f sylviculture *f* naturaliste; sylviculture proche de la nature
r лесоразведение *n* близко к природе; лесное хозяйство *n* близко к природе; натуралистическое лесоразведение

5684 naturalized species
(species established as if it were a native species)
d Gastbaumart *f*
f espèce *f* naturalisée; essence *f* hôte
r натурализованный вид *m*

5685 natural lean
(of growing tree)
d natürliche Neigung *f*
f inclinaison *f* naturelle
r естественный наклон *m*

5686 natural longevity of a tree
d natürliche Lebensdauer *f* des Bäumes
f longévité *f* naturelle d'un arbre
r естественная продолжительность *f* жизни дерева

* **natural open [wood]land → 5680**

5687 natural pine
(stands that: 1. have not been artificially regenerated, 2. are classed as a pine or other softwood forest type, and 3. have at least 10% stocking)
d natürliche Kiefer *f*
f pin *m* naturel
r естественная сосна *f*

5688 natural pruning; self-pruning; autolopping; self-cleaning
(freeing of the stem of a standing tree from its branches through natural death, disintegration, and/or fall)
d natürlicher Astabfall *m*; natürliche Astreinigung *f*; Selbstreinigung *f*
f élagage *m* naturel; ébranchage *m* naturel; auto-ébranchage *m*
r естественная очистка *f* от сучьев; самоочищение *n* от сучьев; сбрасывание *n* [деревьями] ветвей; отпад *m* сучьев

* **natural range → 4072**

5689 natural range barrier
d Naturverbreitungsgebietsperre *f*
f barrière *f* d'aire naturelle
r граница *f* естественного ареала

5690 natural regeneration; natural reproduction
(trees and shrubs grown from the seed or suckers of adjacent trees or shrubs)
d natürliche Nachzucht *f*; Naturverjüngung *f*
f régénération *f* naturelle
r естественное [лесо]возобновление *n*

* **natural regrowth → 5695**

* **natural reserve → 5672**

5691 natural resource
(feature of the natural environment about what choices must be made that is of use and value in serving human needs)
d natürliche Ressource *f*
f ressource *f* naturelle
r естественный ресурс *m*

5692 natural root-grafting
d natürliches Veredeln *n* auf Wurzeln
f greffage *m* sur racine naturel
r срастание *n* корней

5693 natural rotation

d Naturrotation *f*
f rotation *f* naturelle
r естественный оборот *m*

5694 **natural rubber**
d natürlicher Kautschuk *m*
f caoutchouc *m* naturel
r естественный каучук *m*

* **natural-seasoned wood → 150**

5695 **natural seeding; selfsown seeding; self-seeding; natural regrowth; spontaneous seeding**
(dispersal by natural agents of seeds from standing trees in proximity to a regenerating area or from slash scattered over that area)
d natürliche Saat *f*; natürliche Besamung *f*; Ansamung *f*; Anwuchs *m*
f ensemencement *m* naturel; semis *m* naturel; auto-ensemencement *m*
r самосев *m*; самораспространение *n* семян; естественное обсеменение *n*; самообсеменение *n*

5696 **natural selection**
d natürliche Zuchtwahl *f*
f sélection *f* naturelle
r естественный отбор *m*

5697 **natural stain; natural discolo[u]ration**
(a discolo[u]ration on or in lumber, other than its natural colo[u]r)
d natürliche Verfärbung *f*
f décoloration *f* naturelle
r естественное обесцвечивание *n*

5698 **natural stand; unhomogeneous stand; wild stand**
(renewal of a tree crop by natural seeding, sprouting, suckering, or layering)
d natürlicher Bestand *m*
f peuplement *m* naturel
r естественное [лесо]насаждение *n*

5699 **natural thinning; self-thinning**
(tree mortality from the effect of the competition arising between trees on the same site)
d natürliche Ausscheidung *f*; natürliche Lichtstellung *f*
f éclaircie *f* naturelle
r естественное прореживание *n*; естественное изреживание *n*; самоизреживание *n* леса

f éclaircie *f* par âge naturelle
r естественное изреживание *n* [леса] из-за возраста

* **natural woodland → 5671**

* **nature-reserve → 5672**

* **naval construction wood → 8590**

5701 **naval stores**
(chemicals that can be extracted from wood)
d Harzvorräte *mpl*
f réserves *mpl* de résine; réserves résineux
r запас *m* смолы

* **nave elm → 2893**

* **necklace-poplar → 2717**

5702 **necrosis**
(due to Necria galligena: death of a cell or tissue while still part of a living organism)
d Nekrose *f*
f nécrose *f*
r нектрия *f*; некроз *m*

* **necrotic spot → 7081**

* **needle → 37**

* **needle cast → 6260**

5703 **needle cluster; needle fascicle**
d Nadelbüschel *n*
f faisceau *m* d'aiguilles
r пучок *m* хвои

* **needle fall → 2995**

* **needle fascicle → 5703**

5704 **needle foliage**
d Benadelung *f*
f frondaison *f* d'aiguilles
r охвоение *n*

5705 **needleleaf forest phytocenosis; conophorium**
d Nadelwaldgesellschaft *f*
f phytocénose *f* de forêt résineuse
r фитоценоз *m* хвойного леса

* **needle[-leaved] forest → 1873**

5706 needle litter; matting of needles
d Nadeldecke *f*; Nadelstreu *f*
f couverture *f* d'aiguilles
r хвойная подстилка *f*; хвойный опад *m*; покров *m* из хвои

5707 needle of the fir; fir needle
d Tannenadel *f*
f aiguille *f* de sapin
r пихтовая хвоинка *f*

* **needle-shaped → 36**

* **needle shedding → 6260**

5708 needle wood; softwood timber; softwoods; coniferous timber
(woods of most conifers, usually evergreen, having needles or scalelike leaves)
d Koniferen *pl*; Nadelholz *n*; Nadelware *f*
f résineux *mpl* [conifères]; bois *m* de sciage résineux
r хвойная древесина *f*; лесоматериалы *mpl* хвойных пород; хвойные [породы] *pl*

* **neemba → 6110**

* **neem [tree] → 6110**

* **negundo chastetree → 3232**

* **nemesu → 2254**

5709 Nepal alder
(Alnus nepalensis)
d Nepal-Erle *f*
f au[l]ne *m* de Népal
r ольха *f* непальская

* **Nepal nut pine → 1486**

* **nervation → 9192**

5710 nest *v*
d einbetten; nisten
f s'emboîter
r вставлять; свить [себе] гнездо; вить гнездо

* **nest box → 763**

5711 nest cavity
d Nisthöhle *f*
f trou *m* de pic
r дуплянка *f*

5712 nest planting
d Nestpflanzung *f*
f plantation *f* en nids
r гнездовой посев *m*

5713 net annual change
(increase or decrease in volume of live trees at least 5.0 inches d.b.h.; net annual change is equal to net annual growth minus average annual removals)
d Netto-Jahresveränderung *f*; Netto-Jahreswertveränderung *f*
f changement *m* net annuel
r чистое годовое изменение *n*

5714 net annual increment; net annual growth
(the gross annual increment minus natural losses)
d Netto-Jahreszuwachs *m*
f accroissement *m* net annuel
r чистый годовой прирост *m*

5715 net down procedure
(process of identifying the net land base, which is the number of hectares of forest land which actually contribute to the allowable annual cut)
d Netto-Down-Prozedur *f*
f procédure *f* régressive nette
r чистая схема *f* снижения

* **net forest area → 3763**

* **net forest yield → 5720**

5716 net increment; net growth
d Nettozuwachs *m*
f accroissement *m* net
r чистый прирост *m*; нетто-прирост *m*

5717 netleaf hackberry; paloblanco
(Celtis reticulata)
d netznerviger Zürgelbaum *m*
f micocoulier *m* réticulé
r каркас *m* сетчатый

* **net precipitation → 2626**

5718 net present value; NPV; excess present value
(stand's present worth before harvesting once costs associated with its establishment and tending have been subtracted)
d aktualisierter Wert *m*
f valeur *f* actualisée nette; VAN
r чистая приведённая стоимость *f*; рентабельность *f*

* **nettle tree → 2898**

5719 **net[t] scale; net volume**
(actual amount of merchantable wood contained in a log as opposed to the gross scale, which includes defect)
d Nettomasse *f*
f cubage *m* après réfaction; cubage corrigé; cubage net
r точно определённый объём *m* брёвен; чистая масса *f*

5720 **net yield from a forest; net forest yield**
d Netto-Waldreinertrag *m*; Netto-Waldreinrente *f*
f rendement *m* net de la forêt
r чистый доход *m* от леса

5721 **neutrophilous**
(adapted to life in neutral soils (pH from 6.5 to 7.5))
d neutrophil
f neutrophile
r нейтрофильный

* **New Brunswick spruce → 2716**

5722 **new established stand**
d neubegründeter Bestand *m*
f peuplement *m* de création nouvelle
r новосозданное насаждение *n*

* **new-fallen forest → 3710**

* **new-fallen tree → 3709**

* **new forest → 9785**

5723 **new forestry**
(forest management philosophy that attempts to retain characteristics of old-growth stands in managed stands)
d Neuforstwirtschaft *f*
f nouvelle foresterie *f*
r модерное лесное хозяйство *n*

* **newly fallen tree → 3709**

5724 **New Mexico locust**
(Robinia neomexicana)
d Neu-Mexiko-Robinie *f*
f acacia *m* de New Mexico
r акация *f* новомексиканская

5725 **new native woodland**
(new woodland especially designed and
d native Neuholzbodenfläche *f*
f superficie *f* boisée native nouvelle
r новая земля *f* естественного леса

5726 **new wood**
(a twig, branch, or stem originating in the current seasons growth)
d neues Holz *n*
f bois *m* neuf
r новая древесина *f*

* **New Zealand dacryberry → 4683**

5727 **New Zealand kauri; kauri**
(Agathis australis; Dammara australis)
d Neuseeland-Kauri *m*; Kauri *m*
f agathis *m*; kauri *m*
r агатис *m* новозеландский; даммара *f* австралийская

5728 **New Zealand pohutukawa; pohutukawa; ancient Christmas tree**
(Metrosideros excelsa)
d Pohutukawa-Baum *m*
f pohutukawa *m*
r похутукава *f*

* **New Zealand red beech → 8420**

5729 **New Zealand sophora; kowhai; four-wing sophora**
(Sophora tetraptera)
d neuseeländischer Schnurbaum *m*
f sophora *m* de la Nouvelle Zélande
r софора *f* чеырёхкрылая

* **New Zealand white pine → 4683**

* **nib → 819**

* **nibbing by game → 2233**

5730 **nibble** *v*; **gnaw** *v*
d nagen; knabbern
f ronger
r обгрызать

* **Nicaraguan [pitch] pine → 5864**

* **nick → 540**

5731 **Nigerian cedar; pink mahogany; agba; egba; emongi**
(Gossweilerodendron balsamiferum)
d Tola *m*; Agba *m*

5732 **Nigerian satinwood; ayan; barre; movingui**
(Distemonanthus benthamianus)
d Movingui *n*; afrikanisches Zitronenholz *n*; Ayan *n*
f arbre *m* des sorciers; movingui *m*
r нигерийская фагара *f*; дистемонантус *m* Бентама; мовингви *m*

* **nigger → 5124**

5733 **night logging**
d Nachtholzaufbereitung *f*
f débardage *m* nocturne
r лесозаготовки *fpl* в ночную смену; ночная трелёвка *f*

5734 **nightshade family**
(Solanaceae)
d Nachtschattengewächse *pl*
f solanacées *pl*
r пасленовые *pl*

* **nimba → 6110**

* **ninebark → 397**

* **NIPF → 5762**

5735 **nitrogen cycle**
(sequence of biochemical changes undergone by nitrogen, wherein it is used by a living organism)
d Stickstoffkreislauf *m*
f cycle *m* de l'azote
r круговорот *m* азота

5736 **nitrogen fixation**
(conversion of elemental *n* to organic combinations or to forms readily usable in biological processes)
d Stickstoffbindung *f*; Stickstoff-Fixierung *f*
f fixation *f* de l'azote
r собирание *n* азота

5737 **nitta tree; parkia**
(Parkia)
d Parkia *m*; Dourabaum *m*
f parkia *m*; parkie *f*
r паркия *f*

5738 **noble fir; red fir**
(Abies nobilis; Abies procera)
d amerikanische Edeltanne *f*; Nobilistanne *f*; pazifische Edeltanne
f sapin *m* noble [d'Amérique]
r пихта *f* благородная

5739 **noble variety**
(poorly represented variety providing timber of quality)
d edle Art *f*
f cépage *m* noble
r благородный вид *m*; благородный сорт *m*

5740 **nodding tree**
d nickender Baum *m*
f arbre *m* penché
r клонящееся дерево *n*

5741 **node**
(region of the stem where one or more leaves are attached)
d Knoten *m*
f nœud *m*
r узел *m*

* **node → 4723**

* **nodose → 4737**

5742 **nodosity; knobiness**
(the state or quality of being full of nodes)
d Verzwicktheit *f*
f densité *f* des nœuds; nodosité *f*
r узловатость *f*

* **nodous → 4737**

* **nodular → 4737**

5743 **nodular end wall**
(end wall of a parenchyma cell having a beaded appearance in sectional view)
d knotenförmiger Endwand *m*
f paroi *f* terminale noduleuse
r узловатая торцевая стенка *f*

5744 **nodulated root**
d knotenförmige Wurzel *f*
f racine *f* noduleuse
r корень *m* с клубеньками

5745 **nodulation**
(of roots)
d Knöllchenbildung *f*
f nodulation *f*
r окомкование *n*; образование *n* узелков или клубеньков

* **nodule bacteria → 7037**

* **nodus → 4723**

5746 **nomadism**
(of crops)
d Nomadentum *n*
f nomadisme *m* cultural
r кочевое возделывание *n*

* **nominal density → 18**

5747 **nominal length**
d Nominallänge *f*
f longueur *f* nominale
r номинальная длина *f*

5748 **nominal size; nominal measure**
(the rough-sawn size of a piece of lumber)
d Nominalmaß *n*; Rauhmaß *n*
f mesure *f* grossière
r простая мера *f*

* **non-biotic → 6**

5749 **non-capable forestland; non-productive forestland; unproductive forestland**
(forest land which is not capable of growing industrial crops of wood at least at the minimum biological growth potential of 20 cubic feet per acre annually)
d nichtfähiger Waldboden *m*
f terre *f* forestière non capable
r непродуктивная лесная площадь *f*

5750 **non-chargeable volume**
(all wood volume that is not included in the growth and yield projects for the selected management prescriptions used to arrive at the allowable sale quantity)
d gebührenfreier Inhalt *m*
f volume *m* de franchise
r неотносимый за счёт объём *m*

5751 **noncommercial cutting**
(cutting which does not yield a net income, usually because the trees cut are too small, are of poor quality, or are not marketable)
d nichtgewerblicher Hieb *m*; unkommerzielle Fällung *f*
f abattage *m* non commercial
r непромышленная рубка *f*

* **noncommercial thinning → 4680**

5752 **noncommercial [tree] species**
(tree species of typically small size, poor form, or inferior quality that normally do not develop into trees suitable for industrial wood products)
forestière non commerciale
r непромышленная древесная порода *f*

5753 **noncompetitively grown tree; open-grown tree; free-growing tree**
(tree with growth unimpeded by competition from plants, shrubs, or other trees and with crown that has received full light from above and from all sides throughout all or most of its life)
d undurchforsteter Baum *m*
f arbre *m* en croissance libre
r свободно-растущее дерево *n*

5754 **noncompressed fiberboard; semihardboard**
(a fiberboard or a particle board having a density ≤ 25 lb/ft³ and thickness ≥ 0.5 inch)
d Halbhartfaserplatte *f*
f panneau *m* de fibres non comprimé
r полутвёрдая древесноволокнистая плита *f*

5755 **non-declining yield**
(level of timber production planned so that the planned sale and harvest for any future decade is equal to or greater than the planned sale and harvest for the preceding period)
d nichtfallender Ertrag *m*
f rendement *m* de non déclin
r неснижающийся доход *m*

5756 **non-designated wilderness**
(areas within the provincial forest that have been zoned as wilderness through approved integrated resource management plans)
d nichtbezeichnete Wildernis *f*
f zone *f* de nature vierge non différée
r необозначенная девственная местность *f*

5757 **nonforest acres**
(areas in roads, rights-of-way for utilities, special-use areas, wildlife openings, lakes, etc.)
d Nichtholz-Acren *fpl*
f acres *fpl* non boisées
r не покрытые лесом акры *mpl*

5758 **nonforest[ed] area; nonforest land; nonstocked area**
(land that has never supported forests and land formerly forested where timber production is precluded by development or other uses)
d Nichtholzboden *m*; unbestockte Fläche *f*
f aire *f* non boisée; aire non couverte de forêt
r нелесная площадь *f*; не покрытая лесом

5759 **nonforest plantation; growing timber outside the forest; tree planting outside forest**
d Holzzucht *f* außerhalb des Waldes; Baumpflanzung *f* außerhalb des Waldes; Flurholzanbau *m*
f culture *f* d'arbres hors forêt; plantation *f* d'arbres hors forêt
r посадка *f* деревьев за пределами леса; разведение *n* леса вне лесного массива; лесоразведение *n* вне леса

5760 **nonforest use**
d Nutzung *f* außerhalb des Waldes
f utilisation *f* hors forêt
r использование *n* вне леса

5761 **nongame wildlife**
(wildlife species that are protected by state wildlife laws and can not be hunted)
d keine jagdbare Wildtiere *npl*
f faune *f* sauvage non chassable
r защищённые дикие животные *npl*

* **non-indigenous species → 4576**

5762 **nonindustrial private forestland; NIPF**
(forestland owned by a private individual, group, or corporation not involved in wood processing)
d kein industrieller wirtschaftseigene Waldboden *m*
f terre *f* forestière non industrielle
r непромышленная частная лесная земля *f*

5763 **nonlayered vegetation**
d schichtlose Vegetation *f*; außerschichtige Vegetation
f végétation *f* non stratifiée; végétation hors des strates
r неярусная растительность *f*; внеярусная растительность

* **nonlined-out plant → 9420**

* **nonliving → 6**

5764 **nonmerchantable log; unmerchantable log; cull log**
(that has not attained sufficient size, quality, and/or volume to make it suitable for harvesting)
d Nichtwirtschaftsstamm *m*
f grume *f* non marchande; grume de rebut
r неделовое бревно *n*; неделовой сортимент *m*; бракованное бревно

stand; nonmerchantable timber
d Nichtwirtschaftsbestand *m*; Nichtwirtschaftswald *m*
f peuplement *m* non marchand; bois *m* non aménagé
r негодный для продажи лесоматериал *m*; неделовой лесоматериал

* **non-native species → 2944**

5766 **nonporous wood; nonpored wood**
(wood devoid of pores or vessels)
d Holz *n* ohne Poren
f bois *m* sans vaisseaux
r непористая древесина *f*; несосудистая древесина

5767 **nonpreformed rope**
d nichtvorgeformtes drallarme Drahtseil *n*
f câble *m* non préformé
r обыкновенный раскручивающийся канат *m*

5768 **non-pressure treatment; non-pressure process**
(any treatments, e.g. diffusion, brush, where the preservative is applied to the timber without the use of more than atmospheric pressure)
d Verfahren *n* ohne Druckanwendung
f traitement *m* à la pression atmosphérique; traitement sans pression
r обработка *f* без давления

* **non-productive forestland → 5749**

* **non-productive thinning → 4680**

5769 **nonreserved forestland**
(forest land that by law or policy is available for the harvesting of forest crops)
d nichtreservierter Waldboden *m*
f terre *f* forestière non réservée
r незащищённая лесная земля *f*; лесная площадь *f*, на которой разрешены лесозаготовки

5770 **non-sclerotic**
d unverholzt
f non scléreux
r неодревесневший

5771 **nonshatter felling head**
d bruchsicherer Schlagkopf *m*
f tête *f* d'abattage à résistance au morcellement
r валочная головка *f*, не повреждающая комель срезаемого дерева

5772 **nonskyline system**
d Seilbahnsystem *n* ohne Kabel
f système *m* sans câble porteur
r канатная установка *f* без несущего каната

* **nonstocked area → 5758**

5773 **nonstocked forestland; unstocked forestland**
(timberland less than 10% stocked with growing-stock trees)
d unbestockter Waldboden *m*
f forestière *f* de densité relative nulle
r низкополнотная лесная земля *f*

* **nonstocked stand → 4446**

* **nontapering tree → 3746**

5774 **nontimber forest products**
(any commodity obtained from the forest that does not necessitate harvesting trees)
d Nebenforstprodukte *npl*
f produits *mpl* forestiers non ligneux
r лесные побочные продукты *mpl*

5775 **nontimber resources**
(resources other than timber, such as recreation, aesthetics, wildlife, fish, forage, range, water, and soils)
d holzlose Ressourcen *fpl*
f ressources *fpl* non ligneuses
r нелесные ресурсы *mpl*

5776 **nontimber resource value**
(value within the forest other than timber)
d holzloser Ressourcenwert *m*
f valeur *f* forestière non ligneuse
r стоимость *f* нелесных ресурсов

5777 **nontrail skidding**
d Rücken *n* ohne Seilbahn
f débardage *m* sans câble
r трелёвка *f* по неподготовленному волоку

5778 **non-wood products**
(goods, services and functions)
d holzlose Produkte *npl*
f ouvrages *mpl* non ligneux; produits *mpl* non dérivés du bois
r недревесная продукция *f* леса

* **Nordmann['s] fir → 1384**

5779 **normal age-class distribution**
(the ideal distribution of age classes in a forest, which would produce equal annual volume from annual or periodic fellings under the given rotation and silvicultural system)
d normale Altersklassenverteilung *f*
f répartition *f* normale par classes d'âge
r нормальное распределение *f* [насаждений] по классам возраста

5780 **normal-aged forest**
(a forest composed of a series of age classes in proportions that permit sustained yield by felling under an appropriate system)
d Normalalterswald *m*
f forêt *f* d'âge normal
r лес *m* с нормальным распределением насаждений по классам возраста

* **normal diameter → 5782**

5781 **normal forest; ideal forest**
(outdated concept, drawing on the idea of a norm or standard forest structure against which existing forest structures can be compared)
d Normalwald *m*
f forêt *f* normale
r нормальный лес *m*; идеальный лес

5782 **normal girth; normal diameter**
(of a standing tree, its estimated diameter at breast height after due allowance has been made for buttresses enveloping it there)
d Normaldurchmesser *m*
f circonférence *f* à hauteur d'homme corrigée; diamètre *m* à hauteur d'homme corrigée
r стандартный диаметр *m*; нормальный диаметр (с учётом сбега ствола)

5783 **normal increment**
(as applied to an individual stand, the increment attained under full stocking and normal health, under specified management for the locality)
d Normalzuwachs *m*
f accroissement *m* normal
r нормальный прирост *m*

* **normal kill ratio → 6182**

* **normally stocked → 4419**

5784 **normal seed stand**
(a normal stand acceptable for seed collection provided the seed bearers are carefully chosen)
d normaler Samenerntebestand *m*

5785 normal series of age classes; normal series of age gradation
d normale Alterklassenfolge *f*
f suite *f* normale des classes d'âge; série *f* normale des classes d'âge
r нормальное чередование *n* классов возраста

* **normal stocking → 3757**

5786 normal yield table
(estimated stand volume per age class at normal stocking)
d Normalertragstafel *f*
f table *f* de rendement normal
r таблица *f* роста нормальных насаждений

* **North Caroline pine → 7538**

5787 northern Burma pine; khasya pine; Benquet pine; Langbian three-leaved pine
(Pinus insularis; Pinus khasya; Pinus kesiya; Pinus langbianensis)
d Langbiankiefer *f*; Khasyakiefer *f*
f pin *m* à trois feuilles [du Langbian]; pin à trois aiguilles
r сосна *f* кхазия; сосна трёхлистная

5788 northern bush-honeysuckle; [dwarf] bush-honeysuckle
(Diervilla lonicera)
d Kapselgeißblatt *n*
f diéreville *f* chèvrefeuille
r диервил[л]а *f* жимолостная

5789 northern catalpa; catalpa
(Catalpa speciosa)
d prächtiger Trompetenbaum *m*; Hartenkatalpa *f*; Katalpa *f*
f catalpa *m* de l'ouest; bois *m* chavanon
r катальпа *f* западная; катальпа красивая; катальпа прекрасная

* **northern currant → 6769**

* **northern forest → 925**

5790 northern hardwood forest type
(an association of tree species common to the Northeastern US that includes sugar maple, red maple, yellow birch, hemlock, and American beech)
d Nordlaubholze *npl*
f feuillus *mpl* du nord
r северные твёрдолиственные породы *fpl*

5791 northern pin oak; Hill's oak [continued]
d Nordsumpfeiche *f*
f chêne *m* de marais du nord
r дуб *m* овальнолистный

5792 northern [pitch] pine; pitch pine
(Pinus rigida)
d Rigidakiefer *f*; Pech-Kiefer *f*
f pin *m* rigide; pin dur; pitchpin *m*; pin raide
r сосна *f* жёсткая

5793 northern red oak; American red oak
(Quercus borealis)
d amerikanische Roteiche *f*
f chêne *m* rouge d'Amérique
r дуб *m* красный северный; дуб красный американский

* **northern taiga open woodland → 5904**

5794 northern tree clubmoss; round-branched clubmoss; prickly tree clubmoss
(Lycopodium dendroideum)
d baumähnlicher Bärlapp *m*
f lycopode *m* dendroïde
r ликоподий *m* древовидный

* **northern treecreeper → 1807**

5795 northern white cedar; eastern white cedar; eastern arborvitae
(Thuja occidentalis)
d echter Lebensbaum *m*; abendländischer Lebensbaum
f arbre *m* de vie [de l'ouest]; thuya *m* de l'Occident; thuya occidental
r западная туя *f*

* **northern white pine → 2726**

* **Norway maple → 935**

5796 Norway pine; [Canadian] red pine
(Pinus resinosa)
d amerikanische Rotkiefer *f*; Harzföhre *f*
f pin *m* rouge d'Amérique; pin rouge du Canada
r сосна *f* смолистая

* **Norway spruce → 1805**

* **nose → 7795**

5797 nosing
(the one edge of a board worked into the form of a bull-nose)
d Kantenüberhang *m*

* **notch → 4082**

5798 **notching; notch girdling**
(removing bark and any minor buttresses by axe or saw round the base of a tree, preparatory to making the undercut)
d Abstirnen *n*; Abstirnung *f*; Vorritzen *n*
f égobelage *m*
r окольцовывание *n* дерева; подрезание *n*; прорезание *n*; кольцевание *n* V-образными насечками

* **notching spade → 6365**

* **notch planting → 289**

5799 **no till; zero till[age]; no-tillage**
(planting a crop without prior seedbed preparation into sod, crop residue, or an existing cover crop and eliminating subsequent tillage operations)
d keine Bestellung *f*; Nullackerland *n*
f préparation *f* superficielle nulle [du sol]
r нулевая обработка *f* почвы

* **not pricked-out plant → 9420**

5800 **not raised at sides chute**
d ungesattelte Riese *f*
f lançoir *m* non coffré
r лесоспуск *m* без боковых бревен

5801 **not sufficiently restocked forestland; not satisfactorily restocked forestland; not satisfactorily regenerated forestland; NSR** *adj*
(ref. to productive forest land that has been denuded and has failed partially or completely)
d hinreichend regenerierter Waldboden *m*
f terre *f* forestière insuffisamment régénérée; terre forestière incomplètement régénérée
r недостаточно восстановленная лесная земля *f*

5802 **novel species**
d Neuart *f*
f espèce *f* nouvelle
r новая порода *f*

5803 **novelty saw; planer saw**
(a hollow-ground circular saw having teeth of a special design, commonly sets of 4 peg teeth and one raker, so as to produce a particularly smooth cut)
r строгальная пила *f* с мелкими зубьями; пила для чистовой обработки

5804 **no-work zones**
(areas in which equipment and people are not allowed during forestry operations, usually for safety or ecological reasons)
d unerlaubte Zonen *fpl*
f zones *fpl* non permises
r зоны *fpl* простоя

5805 **noxious species**
(any weed so designated by the Weed Control Regulations and identified on a regional district noxious weed control list)
d Schädlingsart *f*
f essence *f* nocive
r вредный вид *m*; ядовитая порода *f*

5806 **noxious weed**
d Schadkraut *m*
f mauvaise herbe *f* nuisible
r вредный сорняк *m*

* **NPV → 5718**

* **NSR → 5801**

* **nucellar epidermis → 4398**

* **nuez encarcelada → 6130**

5807 **nuisible organism; injurious organism; pest; harmful organism**
(any organism, whether insect, pathogen, mammal, or competing vegetation, capable of causing damage to a forest crop)
d Schädlinge *n*; Schädlingsorganismus *m*
f organisme *m* nuisible; déprédateur *m*; ravageur *m*
r вредный организм *m*

* **numbering hammer → 5315**

5808 **number of species**
d Artenzahl *f*
f nombre *m* d'espèces
r число *n* пород

5809 **nun moth**
(Lymantria monacha)
d Nonne *f*
f nonne *f*
r шелкопряд-монашенка *m*

5810 nurse log
(larger and decomposing fallen log which acts as a germination substrate for tree species establishing in the understor[e]y)
d Schirmstamm *m*; Schutzstamm *m*
f grume-abri *f*
r покровное бревно *n* (питающее других деревьев)

5811 nursery bag
d Keimlingssack *m*
f sac *m* de plantules
r рассадочный мешок *m*

5812 nursery bed; transplant bed
(one of the specially prepared plots in a nursery where seed is sown or into which transplants or cuttings are put)
d Pflanzschulbeet *n*; Rabatte *f*; Verschulbeet *n*
f planche *f*; plate-bande *f*
r рассадочная грядка *f*; посадочная грядка; грядка для перешколивания

* **nursery garden → 342**

5813 nursery management
d Baumschulemanagement *n*
f gestion *f* de pépinière
r управление *n* рассадника

5814 nursery stock; snursery seedling; stock plant; transplant stock
(shrub or tree species grown in a plant nursery for planting out elsewhere)
d Verschulmaterial *n*; Verschulpflanze *f*
f matériel m de reproduction en pépinière; matériel de repiquage; plante *f* de repiquage
r сеянец *m* питомника; материал *m* для перешколивания

5815 nursery transplant; transplanted plant; transplant; schooled plant; lined-out plant; pricked-out plant
(seedling that has been replanted one or more times in a nursery to improve its size and growth potential characteristics)
d Impfreis *n*; Setzling *m*; Transplantat *m*; verschulte Pflanze *f*
f plante *f* repiquée; transplant *m*
r трансплант *m*; саженец *m*; пересаженное растение *n*; перешколенное растение

5816 nurse tree; shelter tree; nurse wood
(tree or crop of trees, either naturally occurring or introduced, used to nurture or insolation, or wind)
d Schutzbaum *m*
f arbre-abri *m*
r защитное дерево *n*; дерево, дающее защиту другому

5817 nut
(a hard, dry, single-seeded fruit which does not open at maturity)
d Nuss *f*
f noix *m*
r орех *m*

5818 nut forest
d Nusswald *n*
f forêt *f* de noyers
r ореховый лес *m*

* **nut-gall → 5831**

5819 nutlet
(one of several small, nutlike parts of a compound fruit)
d Nüsschen *n*
f nucule *f*
r орешек *m*

5820 nutmeg family
(Myristicaceae)
d Myristikazeen *pl*
f myristicacées *pl*
r мускатниковые *pl*

5821 nutmegtree; nutmeg hickory
(Carya myristiciformis)
d Muskatnussbaum *m*
f muscadier *m*; caryer *m* muscade
r кария *f* мускатная; мускатный орех *m*

5822 nut oil; walnut oil
d Walnussöl *n*; Nussöl *n*
f huile *f* de noix
r ореховое масло *n*

* **nut pine → 6082, 9031**

5823 nut plants
d Nusspflanzen *fpl*
f plantes *fpl* à noix
r орехоплодные растения *npl*

* **nutrient circulation → 5825**

5824 nutrient contents
d Nährstoffgehalt *m*

5825 **nutrient cycle; mineral cycle; nutrient circulation**
(the exchange of elements between the living and non-living components of an ecosystem)
d Nährstoffzyklus *m*; Nährstoffkreislauf *m*
f cycle *m* des substances nutritives
r круговорот *m* питательных веществ

* **nutrient deficiency → 5179**

* **nutrient depletion → 5179**

5826 **nutrients**
(the elements necessary for growth and reproduction)
d Nährstoffe *mpl*
f substances *fpl* nutritives
r питательные вещества *npl*

5827 **nut-structured gray forest soil**
d grauer nussartig-strukturierter Waldboden *m*
f sol *m* gris forestier à structure nuciforme
r серая ореховая лесная почва *f*

* **Nuttall dogwood → 9339**

5828 **Nuttall oak**
(Quercus nuttallii)
d Nuttal-Eiche *f*
f chêne *m* Nuttal
r дуб *m* нуттальский

* **Nuttall willow → 7311**

5829 **nut wood**
d Nussbaumholz *n*
f bois *m* de noix
r ореховая древесина *f*

O

5830 oak
(Quercus)
d Eiche *f*
f chêne *m*
r дуб *m*

5831 oak-apple; oak-ball; oak-gall; gall[-nut]; nut-gall; marble gall; spangle
d Gallapfel *m*; Eichapfel *m*; Eichengalle *f*
f noix *m* de galle; galle *f*; pomme *f* de chêne
r орешек *m* чернильный; орешек дубильный

* **oak association → 5836**

5832 oak-bark; tan
d Eichenrinde *f*
f écorce *f* de chêne
r дубовая кора *f*

5833 oak-bark coppice; tanbark coppice
d Schälwald *m*
f taillis *m* à écorce
r лесосека *f*, предназначенная для получения дубового корья; низкоствольник *m* на корьё

* **oak-bark forest → 5835**

5834 oak barrelwood
d Eichenbinderholz *n*
f douvain *m* de chêne
r дубовая бочарная древесина *f*

5835 oak-coppice [forest]; oak-bark forest; oak tanbark woodland
d Lohhecke *f*; Eichenschälwald *m*; Eichenniederwald *m*
f taillis *m* de chêne à écorce
r низкоствольный дубовый лес *m*; дубовая поросль *f*

* **oak eggar → 5846**

5836 oak forest; oak association; oak grove; oakery; oakwoods
d Eichenwald *m*; Eichenbestand *m*; Eichenhain *m*
f chênaie *f*; veuseraie *f*

* **oak-gall → 5831**

* **oak grove → 5836**

5837 oak-gum-cypress association
(bottomland forests consisting of 50% or more tupelo, blackgum, sweetgum, oak, and cypress, and less than 25% southern pines)
d Eiche-Eukalyptus-Zypress-Association *f*
f association *f* de chênes, eucalyptus et cyprès
r лесная ассоциация *f* из дубов, эвкалиптов и кипарисов

5838 oak-hickory [forest type]
(in which upland oaks or hickory, singly or in combination, constitute a plurality of the stocking, except where pines account for 25% to 50%, in which case the stand would be classified oak-pine)
d Eichen-Hickory-Waldtyp *m*
f type *m* de forêt chêne-hickory
r тип *m* леса из дубов и гикори

5839 oak leaf[ ]tier
(Croesia semipurpurana)
d Eichennestwickler *m*
f décophore *m* du chêne
r листовёртка *f* дубовая

5840 oak-leaved
d eichenblätt[e]rig
f à feuilles de chêne
r дуболистный

5841 oaklet; oakling
(an young oak)
d junge Eiche *f*
f chêne *m* jeune; chêneau *m*
r молодой дуб *m*; дубок *m*

5842 oak log
d Eichenstamm *m*
f grume *f* de chêne
r дубовое бревно *n*

5843 oak-mast crop; crop of acorns
d Eichelernte *f*; Eichenmast *f*
f glandée *f*
r урожай *m* жёлудей

5844 oak mildew
(Microshaera alphitoides)
d auftretende Mehltaupilze *f*
f oïdium *m* de chêne
r мучнистая роса *f* дуба

5845 oak moss
(Evernia furfuracea)
d Eichen-Moos *n*
f mousse *f* de chêne
r дубовый мох *m*

5846 oak moth; oak eggar
(Lasiocampa quercus)
d Eichenspinner *m*; Quittenvogel *m*
f bombyx *m* du chêne; bombyx du genêt
r коконопряд *m* дубовый

5847 oak-pine [forest type]
(association of tree species common to the Southeastern United States that includes loblolly pine, Virginia pine, northern red oak, and white ash)
d Eichen-Kiefernwaldtyp *m*
f type *m* de forêt chênaie-pineraie
r тип *m* леса дубово-сосновый

5848 oak sapling borer
(Goes tesselatus)
d Eichenbock *m*
f perceur *m* du chêne
r усач *m* дубовый

* **oak tanbark woodland → 5835**

5849 oak wilt
(Ceratocystis fagocearum)
d Eichwelke *f*
f flétrissement *m* du chêne
r цератоцистис *m*

5850 oak wood
d Eichenholz *n*; Räuchereiche *f*
f bois *m* de chêne
r дубовый лесоматериал *m*; древесина *f* дуба

* **oakwoods → 5836**

5851 obconic leaf
d rückkonisches Blatt *n*
f feuille *f* obconique
r обратноконусовидный лист *m*

5852 obcordate leaf
(inverted and heart-shaped; e.g., a heart-shaped leaf with the pointed end toward the stem)
d rückherzförmiges Blatt *n*
f feuille *f* obcordée
r обратносердцевидный лист *m*

* **Obispo pine → 770**

(trees grown until its maximum commercial age)
d Objektivbäume *mpl*
f arbres *mpl* objectifs; arbres d'avenir
r целевые деревья *npl*

5854 oblanceolate leaf
(reverse lance-shaped; broadest at the apex, gradually tapering to the narrower base)
d rücklanzettförmiges Blatt *n*
f feuille *f* oblancéolée
r обратноланцетовидный лист *m*

5855 obligatory selection felling
d Zwangsfemelhieb *m*; Zwangsausplenterung *f*
f coupe *f* jardinatoire obligatoire
r подневольно-выборочная рубка *f*

* **oblique grain → 4585**

5856 oblique lamination
(of wood)
d Schräglamination *f*
f lamination *f* oblique
r косая слоистость *f*

5857 oblique notching
d Schiftschnitt *m*
f égobelage *m* oblique
r поперечная вязка *f* под косым углом; врубка *f* под косым углом; косая врубка

5858 oblong leaf
(broadest at the middle, and tapering to both ends, but broader than elliptic)
d längliches Blatt *n*
f feuille *f* oblongue
r продолговатый лист *m*

5859 oboval leaf
(broadly rounded at the apex, becoming narrowed below)
d Rückovalblatt *n*
f feuille *f* obovée
r обратнояйцевидный лист *m*

* **obovate leaf → 5988**

5860 obstructing undergrowth
d versperrendes Unterholz *n*
f sous-bois *m* obstructif
r мешающий валке подрост *m*

5861 obtuse leaf
d stumpfwinkliges Blatt *n*

* **occluded knot → 833**

5862 **occlusion of wounds; healing over; cicatrization; callousing**
(process of healing of cut branch stubs by the cambium of the surrounding stem surface)
d Vernarbung *f*; Überwallung *f*
f occlusion *f*; cicatrisation *f*
r зарастание *n* рубцов; затягивание *n* карры; зарастание поверхностного повреждения; заплывание *n*

5863 **ochna**
(Ochna)
d Ochna *f*; Nagelbeere *f*
f bois *m* bouquet; arbre *m* à bouquets
r охна *f*

5864 **ocote pine; Nicaraguan [pitch] pine**
(Pinus oocarpa)
d Nicaragua-Kiefer *f*
f pin *m* du Nicaragua
r сосна *f* яйцеплодная

* **octopus tree → 412**

* **ocular estimate → 2869**

* **ocular estimation → 2869**

5865 **ocular plot-estimate; ocular-estimate-by-plot method; plot estimate method; square-foot method**
(for determining the cover and composition of range vegetation by means of plots on which the area occupied by each species is estimated by eye in square units)
d Deckungsgrad-Schätzverfahren *n*
f estimation *f* visuelle de pâturage par échantillonnage
r метод *m* глазомерного изучения (состава растительности) на учётных площадках

5866 **odd-number lengths**
(of timber)
d Schnittholze *npl* mit ungerader Länge
f débits *mpl* de longueurs impaires
r брёвна *mpl* длиной равной нечётному числу футов

5867 **odd-pinnate leaf**
(a pinnate leaf with a terminal leaflet, hence having an odd number of leaflets)
d unpaarig gefiedertes Blatt *n*
f feuille *f* pennée impaire
r непарноперистый лист *m*

* **offal timber → 2123**

* **offcut → 7712**

* **off-cuts → 2123**

* **offence against the forest law → 1909**

5868 **off-forest land**
(all lands outside forest boundary)
d Land *n* außerhalb der Wald
f terre *f* hors forêt
r земля *fpl* за границы леса

5869 **of-fibric layer**
(poorly decomposed organic soil layer)
d Oberfaserschicht *f*
f couche *f* fibrique supérieure
r верхний слой *m* неразложившегося опада мхов

* **off-road logging vehicle → 7691**

* **offset tower → 2815**

5870 **offshoot; offset**
(a shoot which derives or separates from a main stem)
d Ableger *m*
f œilleton *m*
r боковой отросток *m*

5871 **Ogeechee-lime; [sour] tupelo; Ogeechee tupelo**
(Nyssa ogeche)
d Ogeechee-Tupelobaum *m*
f miel tupélo *m*
r нисса *f* Огичи

5872 **Oglethorpe oak**
(Quercus oglethorpensis)
d Oglethorpe-Eiche *f*
f chêne *m* d'Oglethorpe
r дуб *m* Оглеторпа

5873 **oh-humic layer**
(well decomposed organic soil layer)
d Oh-Humusstreu *m*
f couche *f* humique inférieure
r нижний слой *m* сильно разложившегося опада мхов; торфянистый слой; oh-хуминовый слой

5874 **Ohio buckeye; fetid buckeye**
(Aesculus glabra)
d ranzige Kastanie *f*
f marronnier *m* américain; marronnier de

5875 oil cell
(a specialized cell of the ray or axial parenchyma containing oil, typically rounded in outline)
d Ölzelle *f*
f cellule *f* à huile
r масляная клетка *f*

* **oil palm → 253**

5876 oil seasoning; Boulton process; boultonizing; boiling-under-vacuum [process]
(the conditioning of unseasoned or partially seasoned timber for pressure treatment by heating it in a tar-oil-type preservative under vacuum and so boiling off its water)
d Boulton-Verfahren *n*
f procédé *m* Boulton
r сушка *f* в масле

* **oiltree → 9000**

* **okan → 2397**

* **okoume → 3788**

5877 okwen
(Brachystegia nigra)
d Naga *m*
f naga *m*
r брахистегия *f*

* **olborbora → 2245**

* **old field → 2**

* **old-field birch → 3923**

* **old-field pine → 5029**

5878 old-field tree
d leerstehender Baum *m*
f arbre *m* de champ
r дерево *n*, выросшее в открытое место

* **old growth → 5880**

5879 old[-]growth attributes; old[-]growth components
(structural features and other characteristics of old-growth forests)
d Altbestandattribute *npl*
f composantes *fpl* de première venue; attributs *mpl* de vieux peuplement
r характеристики *fpl* спелого леса

growth
(wooded area, usually greater than 200 years of age)
d Alt[holz]bestand *m*; Altwald *m*
f forêt *f* vieux; vieux peuplement *m*
r старый лес *m*; старое насаждение *n*

5881 oldgrowth management areas
(areas which contain, or are managed to replace, specific structural old-growth attributes and which are mapped out and treated as special management areas)
d Altholzbetriebsflächen *fpl*
f zones *fpl* de gestion de vieux peuplements
r зоны *fpl* управления спелых лесонасаждений

* **oldgrowth stand → 5880**

5882 oldgrowth stump
d Altholzstumpf *m*
f pied *m* vieux
r пень *m* многолетнего дерева

* **old-house borer → 4389**

* **old land → 2**

5883 oldman wormwood; southernwood; sagebrush
(Artemisia abrotanum)
d Eberraute *f*; Eberreis *n*; Abrandkraut *m*
f aurone *f* [mâle]; citronnelle *f*
r полынь *f* лечебная; [полын-]божье дерево *n*

5884 old standard; over-seasoned timber; overmature wood
d älteres Holz *n*; rückgängiges Holz; überständiges Holz; Überständer *m*
f bois *m* suranné
r древесина *f* перестойного дерева; перестойная древесина

5885 old tree; holdover tree; veteran
d Waldrechter *m*; überreifer Baum *m*
f arbre *m* vieux; vétéran *m*
r перестойное дерево *n*; гниющее на корню дерево

5886 old wood
(a twig, branch or stem that originated in the previous growing season or earlier)
d starkes Baumholz *n*; Altholz *n*
f vieux bois *m*
r старая древесина *f*

5887 oleander; rose bay

f laurier-rose *m*
r олеандр *m*

5888 **oleander podocarpus**
(Podocarpus nerifolius; Podocarpus bracteata)
d großblättrige Steineibe *f*
f podocarpe *m* à bractées
r ногоплодник *m* олеандролистный

* **oleaster → 9421**

5889 **oleaster family**
(Elaeagnaceae)
d Ölweidengewächse *pl*
f éléagnacées *pl*
r лоховые *pl*

* **oleoresin → 503**

* **olive → 2904**

5890 **olive ash burl**
(ash with dark wavy wood)
d Olivesche-Maser *f*
f loupe *f* de frêne olivier
r масляный ясень *m*

5891 **olive family**
(Oleaceae)
d Ölbaumgewächse *pl*
f oléacées *pl*
r маслинные *pl*

5892 **olive grove; olive plantation**
d Ölbaumpflanzung *f*; Olivenhain *m*
f oliveraie *f*; olivaie *f*; olivette *f*; olivarède *f*
r оливковая роща *f*; маслиновая роща

5893 **olive harvest**
d Olivenernte *f*
f olivaison *f*
r урожай *m* олив

* **olive plantation → 5892**

5894 **olivewood**
d Olivenholz *n*
f bois *m* d'olivier
r древесина *f* оливкового дерева

5895 **olon**
(Fagara heitzii)
d Olon *m*
f olon *m* [tendre]
r фагара *f* Хейтца

5896 **om-mesic layer**
(mesic, moderately decomposed organic soil layer)
d om-mesische Schicht *f*
f couche *f* om-mésique
r средний слой *m* полуразложившегося опада мхов

* **omu → 4744**

5897 **one-cut shelterwood method**
d einflutige Schirmschlagmethode *f*
f méthode *f* de la coupe d'abri unique
r способ *m* одноприёмной постепенной рубки

* **one-side crown → 7648**

5898 **one-side[d] growth**
(of tree stem)
d einseitiger Wuchs *m*
f croissance *f* dissymétrique
r односторонняя кривизна *f*

5899 **one-side edged lumber**
d einseitig besäumtes Holz *n*; einseitig gesägtes Holz
f bois *m* scié unilatéralement
r полуобрезной пиломатериал *m*

5900 **one-side inbark**
d einseitiger Rindeneinschluss *m*
f entre-écorce *f* superficielle
r односторонняя врастающая кора *f*

* **one-year seedling → 9756**

5901 **one-year's growth of coniferous trees**
d Jahresquirl *m*
f verticille *m* annuel
r годичная мутовка *f*

5902 **on-grade wood; sound wood**
d gesundes Holz *n*
f bois *m* sain
r здоровая древесина *f*; здоровый лесоматериал *m*; неповреждённый лесоматериал

5903 **opalized wood; wood opal**
(term applied widely to silicified wood that consists largely to wholly of opal; form of silicified wood; opal pseudomorph after wood)
d Opalholz *n*

5904 open boreal woodland; northern taiga open woodland
d borealgelichteter Nordtaigawald *m*; subarktischer Lichtwald *m*
f forêt *f* claire boréale au nord de la taïga; forêt claire subarctique
r предтундровое редколесье *n*; редкостойный северо-таёжный лес *m*

5905 open core; honeycomb core; honeycomb center
d Wabenkern *m*
f couche *f* poreuse
r пористый средний слой *m* [плиты]

5906 open defect
(a defect, natural or otherwise, which causes a hole or gap in a ply)
d augenscheinlicher Fehler *m*; Oberflächenfehler *m*
f défaut *m* ouvert; défaut apparent
r видимый порок *m*

5907 open field; open space
d Freifläche *f*; Freilage *f*
f surface *f* dégagée; surface libre
r открытое место *n*

5908 open field planting; field planting
d Freiflächenpflanzung *f*
f plantation *f* sur terrain libre
r посадка *f* в открытое место

* **open forest → 3646**

* **open forest landscape → 9310**

* **open grain → 1678**

5909 open-grained wood; broad-ringed wood; coarse-ringed timber; wide-ringed timber
(wood with wide conspicuous growth rings)
d grobringiges Holz *n*; weitringiges Holz
f bois *m* à cernes larges
r широкослойная древесина *f*; крупнослойная древесина

* **open-growing → 3694**

5910 open-growing canopy
d freiwachsende Beschirmung *f*
f houppier *m* en croissance libre
r несомкнутый полог *m*

5911 open-grown crown class
(trees with crowns that have received full particularly during early development)
d freigewachsene Kronenklasse *f*
f classe *f* de cime en croissance libre; classe de couronnes en croissance libre
r класс *m* свободно растущих крон

* **open-grown tree → 5753**

* **opening cutting → 5912**

* **opening of canopy → 4566**

5912 opening of the shelterwood; liberation cut[ting]; liberation felling; opening cutting; light felling; release cut[ting]; removal cut[ting]; disengagement cutting
d Freihieb *m*; Waldlichtung *f*; Licht[ungs]hieb *m*; Lichtschlag *m*; Lichtung *f*; Blenderhieb *m*; Blenderung *f*
f coupe *f* de dégagement; coupe d'éclaircie; coupe de libération
r вырубка *f* для освобождения; осветительная рубка *f*; рубка осветления; осветление *n*

* **opening out of the crop → 4566**

5913 opening-up
(making an area accessible, e.g. by constructing roads)
d Eröffnung *f*
f ouverture *f* de voie d'accès
r подготовка *f* доступа

* **opening up [of a stand] → 4566**

* **open knot → 9050**

5914 openness; roominess
d Freimütigkeit *f*; Offenheit *f*
f état *m* clair
r редкое стояние *n*

* **open pollination → 2056**

5915 open position of trees; isolation of trees
d Freibaumstellung *f*
f état *m* libre d'arbres; état dégagé d'arbres
r свободное стояние *n* деревьев

* **open root → 531**

* **open-rooted planting → 532**

5916 open scar
d offene Narbe *f*

5917 **open-side grapple**
d Einständer-Greifer *m*
f pince *f* de débardage à un montant
r челюстной захват *m*

* **open space → 5907**

* **open stand → 3646**

* **open stand increment → 4944**

5918 **open stand system; open wood system**
d Lichtwuchsbetrieb *m*
f méthode *f* de traitement d'une forêt favorisant de la lumière sur la frondaison
r хозяйство *n* на световой прирост

5919 **open store**
d Freilager *n*
f dépôt *m* de bois à l'air libre
r открытый склад *m*

* **open texture → 1678**

5920 **open-top van**
(van that can be loaded from above and is often covered by a tarp or mesh)
d offener Lieferwagen *m*
f [carrosserie-]fourgon *m* à toit ouvert; fourgon à caisse ouverte; fourgon à toit ouvrant
r открытый прицеп[-фургон] *m*

5921 **open vegetation; treeless vegetation**
(in habitat evaluation, all clearcuts, meadows, and other openings)
d waldlose Vegetation *f*
f végétation *f* non couverte de forêt
r редкая растительность *f*; безлесная растительность; изреженная растительность

* **open wood[land] → 3443**

* **open woodland of natural origin → 5680**

5922 **open woodland plant; hylodad**
d Pflanze *f* der offene Holzfläche
f plante *f* de terrain arboré ouvert
r растение *n* редколесья

* **open wood system → 5918**

5923 **operability line**
(line drawn on a map to differentiate between areas that are operable and those that are not)
d Bedienbarkeitslinie *f*

5924 **operable forest**
(that portion of the production forest that, under current market conditions, can be harvested at a profit)
d betriebsfähiger Wald *m*
f forêt *f* opérable
r лес *m*, пригодный к эксплуатации

* **operable timber → 424**

* **operable timberland → 30**

5925 **operating area; operating range**
(geographic sub-units of timber supply areas that have been assigned to individual major licensees for the purposes of long-term planning)
d Arbeitsbereich *m*; Betriebsbereich *m*
f zone *f* d'exploitation; plage *f* de fonctionnement; champ *m* d'action
r производственная площадь *f*

* **operating employees → 3119**

5926 **operational cruise**
(an estimate, to a specified degree of accuracy, of the volume of timber on an area to be harvested)
d betriebsbereite Bestandesaufnahme *f*
f inventaire *m* du peuplement opérationnel
r эксплуатационная таксация *f*

* **opossum tree → 259**

* **opossum wood → 1349**

5927 **opposite leaf**
(that is placed directly across from each other on the twig)
d entgegenliegendes Blatt *n*
f feuille *f* opposée
r супротивный лист *m*

5928 **opposite leaf arrangement; decussate leaf arrangement**
(a leaf arrangement which leaves grow in pairs, and each pair is positioned ninety degrees from the pair above and below it on a shoot)
d gegenständige Blattdisposition *f*
f arrangement *m* de feuilles décussé
r супротивное листорасположение *n*

5929 **opposite pitting**
(multiseriate pitting in which the pits are in

f ponctuation *f* en rangée horizontale
r супротивная поровость *f*

5930 optical cal[l]iper; optical fork
d optischer Kaliber *m*
f calibre *m* optique
r лесная оптическая мерная вилка *f*; оптическая таксационная вилка

5931 optical dendrometer
d optisches Dendrometer *n*
f dendromètre *m* optique
r оптический дендрометр *m*

* **optical fork → 5930**

5932 optimal elk cover
(a forest stand with four layers including an overstor[e]y that will intercept snow, and small openings that provide forage)
d optimale Elchdeckung *f*
f couverture *f* optimale d'élan
r оптимальный лесной покров *m*

* **optimally stocked → 4419**

5933 optimum cutting age
d Optimalabtriebsalter *n*
f âge *m* optimum d'exploitabilité
r оптимальная возраст *f* срубания

5934 optimum growing stock
d Optimalvorrat *m*
f volume *m* le plus approprié
r оптимальный запас *m* [насаждения]

* **optimum stocking → 3757**

* **orange geiger → 4799**

5935 orange tree; sweet orange
(Citrus sinensis)
d Orangenbaum *m*; Apfelsinenbaum *m*
f oranger *m* [douce]
r цитрус *m* китайский; апельсиновое дерево *n*; апельсин *m*

5936 orange wattle; blue-leaved wattle
(Acacia cyanophylla; Acacia glauca; Acacia saligna)
d Weidenblattakazie *f*; weidenblättrige Akazie *f*
f mimosa *m* eucalyptus; mimosa bleuté
r акация *f* синелистная; акация золотовенечная

5937 orangewood
r цитрусовая древесина *f*

5938 orbicular leaf
(a round leaf)
d kugelförmiges Blatt *n*
f feuille *f* orbiculaire
r округлый лист *m*

* **orchard house → 3946**

5939 orchard pruning
d Obstbäumästung *f*
f élagage *m* d'arbres fruitiers
r обрезка *f* ветвей плодовых деревьев

* **orchid tree → 6642**

* **ordeal tree → 1366**

* **order for removal → 5081**

* **order of succession → 127**

5940 ordinary forest
d ordentlicher Wald *m*
f forêt *f* [d'exploitation] ordinaire
r лес *m* с нормальным размером пользования

* **ordinary thinning → 8511**

* **Oregon alder → 235**

5941 Oregon ash
(Fraxinus oregona; Fraxinus latifolia)
d Oregon-Esche *f*
f frêne *m* d'Oregon; frêne à larges feuilles
r ясень *m* орегонский

* **Oregon cedar → 4839**

* **Oregon fir → 1772**

5942 Oregon fir sawyer
(Monochamus oregonensis)
d Douglastanne-Bockkäfer *m*
f longicorne *m* oregonais du sapin
r усач *m* чёрный орегонский

5943 Oregon grape; trailing mahonia
(Mahonia aquifolium)
d gewöhnliche Mahonie *f*; gemeine Mahonie; glanzblättrige Mahonie
f mahonie *f* à feuilles de houx; mahonia *m* commun
r магония *f* падуболистная

* **Oregon maple → 701**

* **Oregon myrtle → 1264**

5944 **Oregon vine maple; vine maple**
(Acer circinatum)
d rundblättriger Ahorn *m*; Weinblattahorn *m*
f érable *m* circiné; érable à feuilles circulaires; érable à feuilles de vigne; érable cylindrique
r клён *m* завитой

5945 **Oregon [white] oak; Garry oak**
(Quercus garryana)
d Garry-Eiche *f*
f chêne *m* de Garry; chêne garry
r дуб *m* Гарри

* **Oregon yew → 9345**

5946 **organic decay**
(of wood due to fungi and microorganisms)
d organische Fäulnis *f*
f carie *f* organique
r органическое разложение *n*

5947 **organic layer**
d organische Schicht *f*
f horizon *m* organique
r почвенный слой *m*

5948 **organic matter; organic substance**
(plant and animal material in or on the soil and in all stages of decomposition)
d organische Substanz *f*; organischer Stoff *m*
f matière *f* organique; substance *f* organique
r органическое вещество *n*

* **Oriental arborvitae → 2725**

5949 **Oriental beech**
(Fagus orientalis)
d orientalische Buche *f*
f hêtre *m* d'Orient
r бук *m* ориентальский; восточный

* **Oriental cherry → 744**

5950 **Oriental hornbeam**
(Carpinus orientalis)
d orientalische Hainbuche *f*
f charme *m* d'Orient
r граб *m* восточный; грабинник *m*

5951 **Oriental oak; Chinese cork oak**
(Quercus variabilis)
d chinesische Korkeiche *f*

5952 **Oriental planetree**
(Platanus orientalis)
d orientalische Platane *f*; morgenländische Platane
f platane *m* d'Orient
r платан *m* восточный

* **Oriental spruce → 2723**

* **Oriental thuja → 2725**

* **Oriental wood → 413**

* **oriented flakeboard core → 168**

5953 **oriented flakeboard face**
d orientierte Spanplatte-Oberseite *f*
f face *f* de panneau à flocons orientées
r наружный слой *m* плиты из ориентированной стружки

5954 **oriented strand-board; oriented structural board; OSB; oriented strand panel**
(panel products manufactured by gluing and high-temperature pressing of layers of thin wood chips)
d orientierte Strandplatte *f*
f panneau *m* de particules orientées
r плита *f* из ориентированной пряди

* **original forest → 282**

5955 **original forest vegetation**
d Urwaldvegetation *f*
f végétation *f* de la forêt vierge
r растительность *f* девственного леса

* **original growth → 6527**

* **original stand → 6527**

5956 **ormosia**
(Ormosia)
d Ormosia *f*
f bois *m* sang; ormosia *m*
r ормозия *f*

5957 **ornamental tree**
(any tree or plant that is planted for its beauty)
d Schmuckbaum *m*; Zierbaum *m*
f arbre *m* d'ornement
r декоративное дерево *n*

5958 **ornamental wood product**
d dekoratives Holzwerk *n*

* **orohylile forest → 8226**

5959 **orophyte**
(a plant inhabiting hills and mountains)
d Orophyt *m*
f orophyte *f*
r орофит *m* (растение горных поясов)

5960 **ortet**
(the original ancestor of a vegetatively propagated clone; derived from seeds)
d Ausgangpflanze *f*
f ortet *m*
r ортет *m*

5961 **orthodox seeds**
(seeds of most species of forest trees that are best stored when dried)
d altgläubiges Saatgut *n*; Orthodoxsaat *f*
f graine *m* orthodoxe; semence *f* orthodoxe
r ортодоксальный посевной материал *m*

5962 **orthotropic branching**
(of branching, stems that only grow vertically)
d orthotrope Abzweigung *f*
f branchement *m* orthotropique
r ортотропное разветвление *n*

5963 **orthotropy**
(gravitational response that produces a vertical axis)
d Orthotropie *f*
f orthotropie *f*
r ортотропия *f*

* **osage-orange → 4219**

* **OSB → 5954**

* **osier → 9439**

* **osier bed → 9431**

* **osier ground → 9431**

5964 **osier peeler**
d Weidenschälmaschine *f*
f machine *f* à peler d'osier
r станок *m* для удаления коры с веток или сучьев

* **osiery → 9436**

5965 **osmotic pressure**
d Saftdruck *m*

5966 **Otaheite gooseberry; Indian gooseberry**
(Phyllanthus acidus)
d Stachelbeerbaum *m*
f surelle *f*; cerisier *m* de Tahiti
r филлантус *m* кислый

5967 **other forestland**
(forestland other than timberland and productive reserved forest land)
d sonstiger Waldboden *m*
f autre terre *f* forestière
r иная лесная земля *f*

5968 **other removals**
(growing stock volume of trees removed from the inventory by cultural operations)
d sonstige Holzernte *f*
f autre recette *f*
r другой добив *m*

5969 **other wooded land; OWL**
(with a tree crown cover of 5-10% of trees able to reach a height of 5 m or a crown cover of more than 10% at maturity in situ and shrub or bush cover)
d sonstige Holzfläche *f*
f autre surface *f* boisée
r другая покрытая лесом земля *f*

* **outbreak → 2849**

5970 **outbreak of a forest fire**
d Ausbruch *m* eines Waldbrandes
f naissance *f* d'un incendie de forêt
r вспышка *f* лесного пожара

* **outcome of the sawing → 5214**

5971 **outcrossing**
(crossing, usually natural, to a plant of a different genotype)
d überkreuzte Befruchtung *f*
f allofécondation *f*
r ауткроссинг *m* (скрещивание особей из разных линий); непреднамеренное скрещивание *n*; случайное скрещивание

5972 **outdoor chip storage; outside chip storage**
d Freilandspanladung *f*
f dépôt-colis *m* en plein air
r открытое хранение *n* щепы

5973 **outdoor recreation; outdoor recreational activities**
d Freilufterholung *f*
f loisirs *mpl* de plein air

5974 Outeniqua yellowwood; common yellowwood; danicho
(Podocarpus falcatus; Podocarpus elongatus)
d Outeniqua-Yellowwood-Baum *m*
f bois *m* jaune
r подокарпус *m* серповидный

5975 outer bark; rhytidome; ross
(the phellem and tissues isolated by it; often enclosing pockets or cortical or phloem tissues; the phellem and the tissues isolated externally by it)
d Außenrinde *f*; Rhytidom *n*
f rhytidome *m*; écorce *f* externe
r внешний слой *m* коры

5976 outer diameter
d Außendurchmesser *m*
f diamètre *m* externe
r наружный диаметр *m*

* **outer ply → 2986**

* **outer veneer → 2986**

5977 outgrowth
(a natural consequence of development)
d Auswuchs *m*
f excroissance *f*
r перерастание *n*; разрастание *n*

* **outhaul rope → 5411**

* **outlet ditch → 2613**

* **out-of-roundness → 5984**

5978 outplant
(seedling, transplant, or cutting ready to be established on an area)
d Auspflanze *f*
f plante *f* sur le terrain; plante prête à être mise en terre
r саженец *m* перед посадки

5979 outplanting; planting-out; lifting
(loosening and removing a plant from the ground as typically practised in nurseries)
d Auspflanzen *n*; Auspflanzung *f*; Hebung *f*; Steigung *f*; Steigerung *f*
f arrachage *m*; levage *m*; hausse *f*; montée *f*
r выкопка *f*; высаживание *n*

5980 outriggers
(stabilizers that are generally found on knuckleboom loaders or mobile chippers)
d Auslegerarmbalkene *mpl*

5981 outside cal[l]iper[s]; outside gauge
d Außendimensionsmesser *m*
f compas *m* d'extérieur
r кронциркуль *m* для внешних обмеров

* **outside chip storage → 5972**

5982 outside face of timber
d Außenfläche *f* des Schnittholzes
f face *f* extérieure de grumes
r наружная пласть *f* пиломатериалов

* **outside gauge → 5981**

5983 outslope
(of a road or track surface to provide slope or camber so that water will drain from it)
d Neigung *f* nach der Außenseite
f dévers *m* vers l'extérieur
r наклонение *n* во внешнюю сторону

* **outturn → 9777**

* **outturn table → 8957**

5984 ovality; out-of-roundness
(the degree of deviation, from perfect circularity, of the cross section of the core or cladding)
d Ovalität *f*
f ovalisation *f*
r овальность *f*

5985 oval knot
d ovaler Ast *m*
f nœud *m* oval
r овальный сучок *m*

5986 ovalleaf huckleberry; ovalleaf blueberry
(Vaccinium ovalifolium)
d ovalblättrige Moorbeere *f*
f airelle *f* à feuilles ovales
r черника *f* овальнолистная

5987 ovangkol
(Guibourtia ehie)
d Amazakoue *n*
f ovengkol *m*
r гвибурция *f*

5988 ovate leaf; obovate leaf; egg-shaped leaf
(broadly rounded at the base, becoming narrowed above; broader than lanceolate)
d eiförmiges Blatt *n*
f feuille *f* ovoïdale
r яйцевидный лист *m*

* **oven[-]dry wood → 4710**

* **overage forest → 5998**

5989 **overage protective forest**
d veralteter Schutzwald *m*
f forêt *f* de protection surannée
r перестойный защитный лес *m*

5990 **overall application; broadcast application; broadcasting**
(where a fertilizer or herbicide is applied uniformly over the whole crop area, as opposed to band application)
d Großflächen-Behandlung *f*; Flächen-Behandlung *f*
f traitement *m* en plein
r сплошное внесение *n*; равномерное внесение

* **overbark volume → 8894**

5991 **overcup oak**
(Quercus lyrata)
d Leiereiche *f*
f chêne *m* à feuilles en lyre; chêne blanc aquatique
r дуб *m* лировидный

* **overcut[ting] → 2935**

* **overdense stand → 6006**

5992 **over-diameter material**
d Material *n* zum Durchmesser
f matériel *m* sur diamètre
r [лесо]материал *m* выше заданного диаметра

* **overexploitation → 2935**

* **overfelling → 2935**

* **overgrazing → 6001**

* **overgrowth → 2936**

5993 **overhang**
(the unsupported ends of long pieces of timber in a stack that is not box-end stacked)
d Überhang *m*
f empilage *m* en encorbellement
r свес *m*; выступ *m*; провес *m*; провисание *n*; неподдерживаемые концы *mpl* лесоматериалов в штабеле

* **overhead cover → 3345**

5994 **overhead hazard**
d Overhead-Risiko *n*
f risque *m* par haut
r опасность *f* для вальщика сверху

* **overhead logging → 7700**

* **overhead rope[way] → 84**

* **overhead shelter → 7511**

* **overhead skidder → 88**

* **overhead skidding → 7700**

5995 **overlanding**
(placing road construction fill over organic soil, stumps and other plant materials, corduroy or geotextiles, any of which is required to support the fill)
d Überlandung *f*
f remblayage *m*
r настил *m*

5996 **overlapping cropping**
(cropping sequence where two or more species are grown together on the same unit of land and where the sowing (planting) times do not coincide)
d Überlappungskultur *f*
f culture *f* de recouvrement
r перекрывающее возделывание *n* культур

* **overlength → 202**

5997 **overlength material**
d Überlängen *fpl*
f bois *m* de surlongueur
r [лесо]материал *m* выше заданной длины

5998 **overmature forest; overage forest; overmature stand; overmaturity stand; superannuated stand**
(forest in which, as the result of age, growth has almost entirely ceased, and decay and deterioration has accelerated)
d überalterter Wald *m*; überalterter Bestand *m*
f forêt *f* surannée; peuplement *m* suranné
r перестойный лес *m*; переспелый лес; перестойное насаждение *n*; переспелое насаждение

5999 **overmature timber stage**
d Phase *f* des überaltertes Bestands

* **overmature wood → 5884**

6000 **overmaturity**
(of a tree or stand)
d Überreife *f*
f surmaturité *f*
r перезрелость *f*

* **overmaturity stand → 5998**

6001 **overpasturing; overgrazing**
(grazing in excess of a standard)
d Weideübernutzung *f*
f surpaissance *f*; surpâturage *m*; excès *m* de pâturage
r чрезмерное пользование *n* пастьбой

6002 **overrun**
(the difference between the mill cut of merchantable lumber and the log scale)
d Mehrausbeute *f* [an Schnittholz]
f excédent *m* de rendement; surrendement *m*
r выход *m* продукции свыше расчётной; превышение *n* расчётного объёма выхода [древесины]

* **over-seasoned timber → 5884**

6003 **oversea timbers; oversea woods**
d Überseehölzer *npl*
f bois *m* d'origine d'Outre-mer
r заморские древесные породы *fpl*

* **overseer of the woods → 3456**

6004 **oversized chips**
d überdimensionale Späne *npl*
f copeaux *mpl* surdimensionnés
r нестандартная крупная щепа *f*

6005 **oversized logs**
(occasionally, log dimensions exceed the positions provided on the sample tally or Piece Scale tally forms)
d überdimensionales Rundholz *n*
f bois *m* rond surdimensionnel
r сверхкрупный круглый лес *m*

6006 **overstocked stand; overdense stand**
(situation in which trees are so closely spaced that they are competing for resources, resulting in less than full-growth potential for individual trees)
d überbestocktes Bestand *n*
f peuplement *m* trop dense; peuplement de densité relative excessive; peuplement

6007 **overstocking**
d zu hoher Bestockungsgrad *m*
f surcharge *f*
r высокая степень *f* полноты

* **overstor[e]y → 3518**

6008 **overstor[e]y removal**
(final harvest in which the cutting releases advance regeneration)
d Oberstandschlag *m*; Oberschichthieb *m*
f exploitation *f* de l'étage dominant
r рубка *f* верхнего яруса

6009 **overstor[e]y trees**
(trees that form the uppermost canopy layer in a forest of more than one story)
d Schirmbaum *m*
f arbre *m* de l'étage dominant
r дерево *n* верхнего яруса; дерево верхнего полога

* **overthinning → 8512**

6010 **overthinning of branches; lion-tailing**
(leaving only leaves on the tip of the branches resembling a lion's tail)
d Astdurchlichten *n*
f haute éclaircie *f* de branches; amenuisement *m* de branches
r верховое осветление *n* ветвей

6011 **overtopped; dominated**
(situation in which a tree cannot sufficiently extend its crown into the overstor[e]y and receive direct sunlight)
d unterdrückt; beherrscht
f supprimé; suppressé; dominé
r угнетённый; подчинённый

6012 **overtopped crown class; suppressed crown class**
(trees with crowns below the general level of the crown cover receiving no direct light either from above or from the sides)
d unterdrückte Kronnenklasse *f*
f classe *f* de cime suppressée; classe de cime dominée
r угнетённый класс *m* крон

* **overtopped trees → 8275**

6013 **overtopping**
(vegetation higher than the favored species, as in brush or deciduous species shading and

f domination *f*
r господство *n*

* **over/under bark measure → 2482**

* **overwood → 3518**

* **OWL → 5969**

6014 **oxhorn bucida; black olive; gri-gri; jucaro; ucar; gregre**
(Bucida buceras)
d schwarze Olive *f*
f bois *m* gris-gris
r маслина *f* чёрная

6015 **Ozark chinkapin**
(Castanea ozarkensis)
d Ozark-Kastanie *f*
f châtaignier *m* d'Ozark
r каштан *m* озаркийский

6016 **ozigo**
(Dacryodes buettneri)
d Ozigo *m*
f ozigo *m*
r озиго *m*

6017 **ozone needle mottle**
d Ozonsprenkelung *f* der Nadel
f marbrure *f* d'aiguille par ozone
r пятнистое повреждение *n* хвои озоном

P

6018 **pachycladous**
d dickästig
f pachybranché; pachyramifié
r имеющий толстые побеги

* **Pacific-coast alder → 235**

6019 **Pacific-coast mahogany; Mexican mahogany**
(Swietenia humilis)
d mexikanisches Mahagoni *n*
f acajou *m* de Mexique
r свиетения *f* мексиканская

* **Pacific dogwood → 9339**

* **Pacific hemlock → 5588**

6020 **Pacific madrone; madrone; arbuti tree**
(Arbutus menziesii)
d Madrona *m*
f madrona *m*; arbousier *m* d'Amérique
r мадрона *f*

6021 **Pacific ninebark**
(Physocarpus capitatus)
d Blasenspiere *f*
f bois *m* à sept écorces
r пузыреплодник *m* головчатый

6022 **Pacific poison-oak**
(Toxicodendron diversilobum)
d Gifteiche *f*
f sumac *m* irrégulièrement lobé
r токсикодендрон *m*

6023 **Pacific rhododendron; coast rhododendron**
(Rhododendron macrophyllum)
d großblättrige Alpenrose *f*
f rhododendron *m* à grandes feuilles; rosage *m* à grandes feuilles
r рододендрон *m* крупнолистный

6024 **Pacific serviceberry; saskatoon serviceberry**
(Amelanchier florida; Amelanchier alnifolia)
d blütenreiche Felsenbirne *f*
f amélanchier *m* de l'Ouest
r ирга *f* ольхолистная

6025 **Pacific silver fir; red silver fir; lovely fir; cascade fir; white fir**
(Abies amabilis; Abies pinusa)
d Purpurtanne *f*
f sapin *m* gracieux; sapin amabilis
r пихта *f* миловидная; пихта пурпурная

* **Pacific waxmyrtle → 1277**

* **Pacific willow → 369**

* **Pacific yew → 9345**

* **packing-case timber → 954**

* **packing-case wood → 954**

6026 **pack of roundwood; roundwood bundle**
d Rundholzstapel *m*; Rundholzbank *f*
f faisceau *m* de bois rond
r пакет *m* круглых лесоматериалов; штабель *m* круглых лесоматериалов

6027 **pack-sack method**
d Rucksackmethode *f*
f méthode *f* d'emballe-sac
r способ *m* лесосечных работ с заготовкой сортиментов у пня

6028 **pack-sack skidder**
d Holzrück[en]schlepper *m*; Rückenschlepper *m*; Rückenmaschine *f*
f débardeur *m* par emballe-sac
r трелёвочный трактор *m* для перевозки пачек сортиментов в полностью погруженном положении

* **pacuri → 6926**

* **pacuru → 6926**

* **pad saw → 1822**

* **pagoda dogwood → 222**

* **pagoda tree → 4357**

* **paint gun → 8884**

6029 **paint marking; paint spotting**
(of trees)
d Auszeichnung *f* mit Farbe
f marquage *m* à la peinture; marquage à la couleur
r клеймение *n* краской

* **paint paddle → 5315**

6030 **pakuri; bacury; bacuri**
(Platonia insignis)
d Gelbholz *n*; Bakuri *n*
f parcouri *m* jaune
r бакури *m*

6031 **pal[a]eozoic forest**
d Paläozoikwald *m*; paläozoisches Wald *m*
f forêt *f* paléozoïque
r палеозойский лес *m*

* **palash mool → 3248**

6032 **palatability**
(characteristic of plants or plant parts that elicit a selective grazing or browsing response by animals)
d Schmackhaftigkeit *f*
f appétissance *f*; appétibilité *f*
r поедаемость *f*

6033 **paldao; dao dragon plum**
(Dracontomelum dao)
d Paldao *n*; Dao *n*
f paldao *m*
r драконтомелум *m* дао

* **pale → 6475**

* **paleta → 9190**

* **palimira alstonia → 9363**

* **palisade → 5019**

* **palissandre tree → 1006**

6034 **pallet**
(a low wood platform on which material can be stacked to facilitate mechanical handling)
d Palette *f*
f tringle *f* de clouage
r паллета *f*; подкладка *f*; подстопное место *n*; стеллаж *m*

6035 **pallid hoptree; common hoptree; [three-leaved] hoptree**
(Ptelea trifoliata)
d dreiblättrige Lederblume *f*; gemeine Lederblume; Kleeblatt-Lederstrauch *m*; Kleestrauch *m*; Hopfenbaum *m*
f ptéléa *m* trifolié; orme *m* à trois feuilles; orme de Samarie; trèfle *m* de Virginie
r птелея *f* трёхлистная; вязовик *m* трёхлистный; кожанка *f* [трёхлистная]

f de palme
r пальмовый

6037 **palmate leaf; digitate leaf; fingered leaf; finger-parted leaf**
(divided radiately, like the fingers of a hand)
d handförmiges Blatt *n*; schwimmfüßiges Blatt
f feuille *f* palmée; feuille digitée
r лапчатый лист *m*; пальчатый лист

6038 **palmately compound leaf; pinnate[ly]-compound leaf; feather-compound leaf**
(midribs of main leaflets branching from a central midrib at several points in a feather-like pattern)
d gefingert zusammengesetztes Blatt *n*; digitates Fiederblatt *n*; fingerförmiges Fiederblatt
f feuille *f* palmaticomposée; feuille composée palmée; feuille composée pennée
r пальчатосложный лист *m*

6039 **palmately lobed leaf; palmatifid leaf**
d handförmiggelapptes Blatt *n*
f feuille *f* palmatilobée; feuille palmatifide
r дланевидно-лопастный лист *m*; пальчатолопастный лист

6040 **palmately parted leaf; palmatipartite leaf; palmatiparted leaf**
d handförmiggeteiltes Blatt *n*; handspaltiggeteiltes Blatt
f feuille *f* palmatipartite
r пальчатораздельный лист *m*; дланевиднораздельный лист

6041 **palmately veined leaf; palmatinervate leaf; palminerved leaf; finger-veined leaf**
d handförmig geädertes Blatt *n*; handnerviges Blatt
f feuille *f* palmatinervée; feuille veinée à doigt
r лист *m* с пальчатым типом жилкования; пальчатонервный лист

* **palmatifid leaf → 6039**

* **palmatinervate leaf → 6041**

* **palmatiparted leaf → 6040**

* **palmatipartite leaf → 6040**

6042 **palmatisect[ed] leaf**

f feuille *f* palmatiséquée
r дланевиднорассечённый лист *m*; дланерассечённый лист; пальчаторассечённый лист

* **palmette → 3014**

* **palm forest → 6043**

6043 **palm grove; palm forest**
d Palmenwald *m*; Palmenbestand *m*
f palmeraie *f*
r пальмовая роща *f*; пальмовник *m*; пальмовый лес *m*

* **palminerved leaf → 6041**

6044 **palm leaf**
d Palmblatt *n*
f feuille *f* de palmier
r пальмовый лист *m*

6045 **palm nut**
d Palmnuss *m*
f noix *m* de palmiste
r ядро *n* кокосового ореха; копра *f*

6046 **palm [tree]**
(any of various plants of the family Palmae, mostly tropical or subtropical monocotyledonous trees, shrubs, or vines, usually having a simple stem and a terminal crown of large pinnate or fan-shaped leaves)
d Palme *f*
f palmier *m*
r пальма *f*

6047 **Palmyra palm**
(Borassus aethiopum)
d Borassuspalme *f*
f palmier *m* rônier; rônier *m*
r пальма *f* Пальмира

* **paloblanco → 5717**

* **paludal forest → 8309**

6048 **palustrine**
(vegetated wetlands dominated by trees, shrubs, and persistent emergents)
d Palustrine *f*
f palustrine *f*
r болотистая лесная земля *f*

* **pan → 7687**

similar product, usually of standard size, such as 4x8 feet)
d Panel *n*; Platte *f*
f feuillet *m*; panneau *m*
r панель *f*

6050 **panel products**
(generally unfinished multiply products which consists of various combinations of hardwood or decorative veneer faces and inner ply materials)
d Plattenprodukte *npl*
f panneaux *mpl*
r деревянные панели *fpl*

* **panel saw → 5049**

* **panga → 1204**

* **panga panga → 9336**

6051 **panicled golden[ ]rain[ ]tree**
(Koelreuteria paniculata)
d Blasen-Esche *f*; [Rispen-]Blasenbaum *m*; chinesischer Lackbaum *m*
f savonnier *m*; arbre *m* à vernis de Chine
r кельрейтия *f* метельчатая

6052 **panicle hydrangea**
(Hydrangea paniculata)
d Rispenhortensie *f*
f hortensia *m* paniculé
r гортензия *f* метельчатая

6053 **panicle leaf**
(an arrangement of flowers consisting of several racemes)
d Rispenblatt *n*
f feuille *f* de panicule
r лист *m* метёлки

6054 **pan logging; pan skidding**
d Panabholzung *f*
f débardage *m* par souches
r трелёвка *f* волоком с использованием пэнов

6055 **pannage**
(a tax paid for the privilege of feeding swine in the woods)
d Eichellese *f*
f panage *m*; glandage *m*
r право *n* выпаса свиней в лесу

* **pannage → 5326**

6056 pao rosa; red ivorywood; dina
(Rhamnus zeyheri; Swartzia fistuloides)
d Dina *n*; Pao Rosa *f*
f pao rosa *m*; dina *m*
r шварция *f*

* **papaw [tree] → 1794**

* **papaya [tree] → 1794**

6057 paperbark maple
(Acer griseum)
d Zimtahorn *m*; grauer Ahorn *m*; Grauahorn *m*
f érable *m* à écorce de papier
r клён *m* бумажный

6058 paper birch; canoe birch
(Betula papyrifera)
d Papierbirke *f*
f bouleau *m* à papier; bouleau à canoë
r берёза *f* бумажная

6059 paper board; cardboard; millboard; pasteboard; woodpulp board
d Karton *m*; Pappe *f*
f carton *m*
r картон *m*

6060 paper mulberry; paper-tree
(Broussonetia papyrifera)
d Papiermaulbeerbaum *m*; japanischer Maulbeerbaum *m*
f broussonétie *f* à papier; mûrier *m* à papier
r бумажная шелковица *f*; брусонеция *f* бумажная; брусонетия *f* бумажная

6061 paper wood; pulpwood; grinding wood
(a soft, moist mass of wood fiber used in the manufacture of paper)
d Papierholz *n*; Zellstoffholz *n*; Zelluloseholz *n*
f bois *m* à papier; bois à pâte; bois de papeterie; bois de défibrage
r балансовая древесина *f*; балансы *mpl*

* **papo canary tree → 103**

* **Papuan wattle → 2698**

6062 paradise apple
(Malus pumila var. paradisiaca)
d Paradiesapfel *m*; Zwergapfel *m*
f pommier *m* paradis; paradis *m*
r яблоня *f* парадизка

6063 paradise tree; marupa
(Simarouba glauca)
r марупа *f*

* **paradox acacia → 4687**

6064 Paraguay tea; Para holly
(Ilex paraguensis)
d Matebaum *m*; Mate-Teestrauch *m*
f maté *m*
r падуб *m* парагвайский

6065 paralift loader; high lift loader
d Hochaufzug-Lader *m*
f chargeuse *f* à haute portée
r загрузчик *m* большой подъёмной силы

6066 parallel climbing road pattern; parallel contour road pattern
d paralleles Steigewegmodell *n*
f modèle *m* de tracé montante parallèle
r схема *f* транспортного освоения горных лесосек с одной поднимающейся вверх дорогой и примыкающими параллельными ветками по склонам

6067 parallel-cutting drum chipper
d Parallelschnitt-Walzenhackmaschine *f*
f déchiqueteuse *f* à tambour de coupe parallèle
r барабанная рубильная машина *f* параллельного резания

6068 parallel-laminated veneer
(veneer which grains have been glued parallel to one another)
d parallellaminiertes Furnier *n*
f placage *m* à plis parallèles
r параллельно-слоистый пластик *m*

6069 parallel strand lumber; PSL
(a structural wood product made by gluing together long strands of wood which have been cut from softwood veneer)
d parallel aneinandergefügtes Schnittholz *n*
f bois *m* débité à files parallélisées
r пиломатериал *m* из параллельно склеенных волокон

* **Parana pine → 1005**

* **parang → 1204**

* **para-nut → 1003**

6070 paraquat-treated wood
d paraquatbehandeltes Holz *n*
f bois *m* traité par paraquat

* **Para rubber[ ]tree → 1320**

6071 **parasite**
d Parasit *m*; Schmarotzer *m*
f parasite *m*
r паразит *m*

6072 **parasitic fungi**
d parasitische Pilze *mpl*; parasitäre Pilze
f champignons *mpl* parasitaires
r паразитические грибки *fpl*

6073 **parasitic plant**
d parasitäre Pflanze *f*
f plante *f* parasitaire
r паразитное растение *n*

* **parasol ant → 4864**

6074 **parasol leaf tree**
(Macaranga tanarius)
d malaysischer Bäum *m*
f parasol *m*
r макаранга *f*

* **parasol pine → 4595**

6075 **paratracheal parenchyma**
(axial parenchyma associated with the vessels or vascular tracheids)
d paratracheales Parenchym *n*
f parenchyme *m* paratrachéal
r паратрахеальная паренхима *f*

* **parbuckling → 2052**

* **parcelling of crops → 191**

* **parcel of timber → 5181**

6076 **pardillo**
(Cordia alliodora)
d Dominika-Rosenholz *n*; Pardillo *n*
f bois *m* de rose; laurel *m*; bois soumis
r кордия *f* аллиодора; ципре *m*

* **pare → 1645**

6077 **parenchyma strand**
(an axial series of two or more parenchyma cells derived from a single fusiform initial)
d Parenchymstrang *m*
f file *f* de cellules de parenchyme
r тяж *m* паренхимы

* **parent gallery → 4811**

d Mutterbestand *m*
f peuplement *m* servant à la récolte de graines
r материнское насаждение *n*; материнский древостой *m* (для защиты подроста)

* **parent stock → 1933**

* **parent tree → 5569**

6079 **parinarium**
(Parinarium)
d Prinarium *m*
f parinare *m*
r паринариум *m*

* **paring of the roots → 7040**

* **parings → 9508**

* **paripinnate leaf → 13**

* **parish forest → 1575**

6080 **park forest**
d Parkwald *m*
f forêt-parc *m*
r лесопарк *m*; парковый лес *m*

* **parkia → 5737**

* **parkland savanna[h] → 7229**

6081 **park-like stand**
d parkartiger Bestand *m*
f peuplement *m* de type parc
r паркоподобное насаждение *n*

6082 **Parry pinyon pine; four-needled pinyon; nut pine**
(Pinus cembroides parryana; Pinus quadrifolia)
d Nusskiefer *f*
f pinyon *m* de Parry
r сосна *f* четырёххвойная

* **Parry spruce → 1726**

* **parted leaf → 2549**

6083 **partial cutting; partial harvest**
(any cutting in which only part of the stand is harvested)
d Teilhieb *m*
f coupe *f* partielle

6084 partially intergrown knot; partially adhering knot
d halbeingewachsener Ast *m*
f nœud *m* partiellement adhérent
r частично сросшийся сучок *m*

* **partially stocked → 4586**

6085 partially stocked forestland
(where yield is significantly reduced and treatment may be considered)
d halbbestockte Waldboden *m*
f terre *f* forestière boisée partiellement
r частично облесённая лесная земля *f*

* **partial mast → 4939**

6086 partial piling
d Teilstapelung *f*
f empilage *m* partiel
r частичный сбор *m* в кучи

6087 partial ring; incomplete ring
d auskeilender Jahrring *m*
f anneau *m* [annuel] partiel; anneau discontinu; cerne *m* discontinu
r частичное годичное кольцо *n*; неполное годичное кольцо

6088 partial seeding
(seeding confined to limited areas)
d Teilsaat *f*
f semis *m* partiel; ensemencement *m* localisé
r частичный посев *m*

6089 partial slash disposal
d Teilbeseitigung *f*
f enlèvement *m* partiel de coupe
r частичное удаление *n* хлама

6090 participatory management
d teilnehmendes Management *n*
f direction *f* participative; gestion *f* participative
r совместное управление *n*

6091 particleboard; wood-particle board; wood chipboard
(panel products made from particles of wood larger than fiber)
d Spanplatte *f* [aus Holz]; Holzspanplatte *f*
f panneau *m* de particules
r древесностружечная плита *f*; ДСП

6092 particleboard building material; particleboard bind
d Bindemittel *n* für Spanplatte
древесностружечных плит

6093 part of decaying wood; part of pecky wood; part of rotten wood
d Faulholzanteil *m*
f proportion *f* de bois pourri
r доля *f* гнилой древесины

* **partridgeberry → 1996**

* **partridge rot → 6313**

* **partridgewood → 5426**

6094 parts of tree
d Baumteile *fpl*
f parties *fpl* d'un arbre
r строение *n* дерева

6095 part-time logger; casual forest worker; irregular forest worker
d unregelmäßiger Waldarbeiter *m*; Saisonarbeiter *m*; unständiger Waldarbeiter *m*; Teilzeitwaldarbeiter *m*
f boiseur *m* à temps partiel; ouvrier *m* forestier saisonnier
r частично занятый вспомогательный рабочий *m*; сезонный лесозаготовитель *m*; непостоянный лесной рабочий

6096 part-tree chips
d Teilbaumhackschnitzel *npl*; Teilbaumhackspäne *mpl*
f copeaux *mpl* en portions d'arbres
r щепа *f* из части дерева

* **parvifoliate forest → 3514**

* **pasang ijang → 7635**

6097 pass; run
(in timber harvesting, one of a planned sequence of harvesting operations designed to harvest a management unit over an extended period)
d Durchlauf *m*; Durchlass *m*; Gang *m*
f passage *m*; passe *f*
r проход *m*; пропуск *m*

6098 pass block
(light-weight block hung at the top of the spar tree and used to lift the bull block and other gear in rigging the tree)
d Durchlaufsblock *m*
f bloc *m* de passage
r блок *m* подъёмного монтажного каната

6099 passionflower family
(Passifloraceae)
d Passionsblumengewächse *pl*
f passifloracées *pl*
r страстоцветные *pl*

* **pasteboard → 6059**

6100 pasture afforestation
d Ackeraufforstung *f*
f boisement *m* de pâturage
r облесение *n* пашни

6101 pasture stand
d Hudebestand *m*; Hutebestand *m*
f peuplement *m* ouvert au pâturage
r насаждение *n* для пастьбы скота

6102 Patagonian cypress; Chilean [incense] cedar; incense cedar
(Libocedrus chilensis)
d chilenische Flusszeder *f*
f cèdre *m* du Chili; cyprès *m* du Chili
r кедр *m* речной чилийский

6103 patch; plug
(an inlay of veneer, similar to that from which the board is made, used to repair a knot-hole or other open-defect; a rough area on veneer due to oblique or twisted grain)
d gestanzter Flicken *m*; Flicken; Pfropf *m*
f pastille *f*; bouchon *m*; rugosité *f*
r заделка *f* дефектного участка; фаутный участок *m* шпона

6104 patch bark; spot bark; scorch; pocket
(a patch of bark or gum (kino) completely or partially enclosed in the wood)
d Brandfleck *m*; Brandstelle *f*
f écorçage *m* superficiel
r окорка *f* участками

6105 patch burning; spot burning
burning felling debris, grass, etc. in patches for the purpose of preparing sites for group planting or sowing)
d stellenweises Verbrennen *n*; stellenweises Brennen *n*; plätzeweises Brennen
f brûlage *m* par placettes; incinération *f* par places; brûlage localisé
r сжигание *n* площадками; сжигание *n* порубочных остатков в местах скопления; сжигание местами

6106 patch cutting; gap cutting; gap felling
(clearcut on a small area)
f coupe *f* par trouées
r котловинная рубка *f*; куртинная рубка

* **patch logging → 1854**

6107 patch planting; spot planting
(setting out young trees in small, prepared patches)
d Plätzepflanzung *f*
f plantation *f* sur placeaux
r посадка *f* в площадки биогруппами; посадка площадками

6108 patch scarifier; spot scarifier
(mechanized implement used to expose patches of mineral soil in a systematic pattern)
d Plätzenskarifikator *m*
f scarificateur *m* de placeaux
r покровосдиратель *m* площадки

* **patch seeding → 7870**

* **patch sowing → 7870**

6109 patent *adj*
(spreading, as do the branches of a tree)
d klar ersichtlich
f persistant
r проходимый

6110 paternostertree; Persian lilac; neem [tree]; margosa; neemba; nimba; chinaberry; pride of India
Melia azadirachta; Melia azedarach; Melia azedarack; Azadirachta indica)
d indischer Zedrachbaum; Paternosterbaum *m*; Nimbaum *m*; Niemaum *m*; Neem *m*
f neem *m*; nim *m*; margousier *m*; lilas *m* des Indes; arbre *m* de paternoster; acacia *m* d'Égypte
r азадирахта *f* индийская; мелия *f* ацедарах; минди *f*; маргоза *f*; ним *m*

* **path cutting → 4001**

6111 path of retreat
(from the felled tree)
d Zurückweichenweg *m*; Rückbauweg *m*
f chemin *m* de départ; piste *f* retraitante
r путь *m* отхода вальщика (от срезаемого дерева)

6112 pathogen
(microscopic organism or virus directly

f pathogène *m*; agent *m* pathogène
r патоген *m*

6113 **pathological rotation [age]**
(maximum rotation age through which a stand of trees may be grown without significant volume loss from disease)
d pathologisches Hiebsalter *n*
f âge *m* d'exploitation pathologique
r патологический оборот *m* рубки

6114 **patrol; beat**
(a minor executive forest charge, essentially protective, commonly a subdivision of a range: the charge of a forest guard of comparable junior rank)
d Patrouille *f*
f patrouille *f*
r патруль *m*

6115 **patrol man**
d Wächter *m*
f gardien *m*
r патруль *m*

* **patterned leaf → 2317**

6116 **patterned lumber**
(lumber that is shaped to a pattern or to a mo[u]lded form in addition to being surface planed)
d gemustertes Schnittholz *n*
f bois *m* façonné
r формованный пиломатериал *m*

* **patternwood → 1461**

* **paulownia → 7122**

* **pau marfim → 4035**

6117 **paving wood**
d Pfasterholz *n*
f bois *m* à pavés
r мостильная древесина *f*

6118 **pawpaw**
(Asimina)
d Papau-Baum *m*
f asiminier *m*
r асимина *f*

6119 **pay-as-cut sale; per-unit sale; unit sale**
(timber sale in which the buyer makes regular (weekly, monthly) payments based on mill receipts)
r продажа *f* по цене за единицу объёма

* **pay for hauling → 7684**

6120 **paying-in; reeling-in; warping**
(of a cable)
d Aufwickeln *n*; Aufspulen *n*
f [em]bobinage *m*
r наматывание *n*

6121 **paying-out; uncoiling; unwinding**
(of a cable)
d Abwickeln *n*; Abrollung *f*; Abrollen *n*
f déroulement *m*
r разматывание *n*

* **payung → 7563**

* **peach → 6123**

6122 **peachleaf willow; peachleaved willow; almond[-]leaved willow**
(Salix amygdaloides; Salix amygdalina; Salix triandra)
d Pfirsichblattweide *f*; Mandelweide *f*; dreimännige Weide *f*
f saule *m* à feuilles de pêcher; osier *m* à trois étamines; osier amandier; osier pâle; saule [à feuilles d']amandier
r ива *f* трёхтычинковая; ива миндалелистная; ива миндальная

6123 **peach[-]tree; peach**
(Prunus persica; Persica vulgaris)
d Pfirsichbaum *m*; rotblättriger Pfirsich *m*
f pêcher *m*
r [обыкновенный] персик *m*

6124 **peach[-]wood**
d Pfirsichholz *n*
f pêcheraie *f*
r заросли *fpl* персикового дерева

* **peacock flower → 3249**

6125 **peacock's eye**
(a type of wood texture)
d Pfauenauge *f*
f œil *m* de paon
r "павлиний глаз" *m*

* **peak-crowned tree → 5657**

6126 **peaker**
d Toppenbaum *m*
f grume *f* supérieure

6127 **pearleaf crabapple**
(Malus prinifolia)
d pflaumenblättriger Apfel *m*; Kirschenapfel *m*
f pommier *m* de Chine
r яблоня *f* китайская; яблоня сливолистная

* **pearl knot → 5007**

6128 **pear sawfly**
(Hoplocampa brevis)
d Birnensägewespe *f*
f hoplocampe du poirier
r пилильщик *m* грушевый плодовый

* **pear[-tree] → 1795**

* **pea shrub → 7564**

6129 **peaty forest humus; peat**
(generally, unconsolidated material that consists mainly of undecomposed, or only slightly decomposed, organic matter accumulated under conditions of excessive moisture)
d Waldmoorhumus *m*
f humus *m* tourbeux forestier
r лесной торфянистый гумус *m*

* **peav[e]y → 1317, 6693**

* **peavie → 1317**

6130 **pecan [hickory]; bitter pecan; [bitter] water hickory; sweet pecan; nuez encarcelada**
Carya pecan; Carya aquatica; Carya illinoensis)
d Nusshickory *f*; unechtes Hickoryholz *n*; Pekannussbaum *m*; Pekan *m*
f pacanier *m*; caryer *m*
r кария *f* пекан; пекан *m*

* **peck → 6313**

* **peckiness → 6313**

6131 **peckled; mottled**
(ref. to wood surface)
d gesprenkelt; scheckig
f tacheté; marbré
r испещрённый; с точками

6132 **pecky wood; peggy wood**
(ref. to a defect common in bald cypress)
d pilzfleckiges Holz *n*

* **pedunculate → 6193**

* **pedunculate oak → 1793**

* **peel → 2280**

6133 **peel** *v*
(to produce veneer by revolving a peeler block against a knife)
d pellen; [ab]schälen; blankschleifen; polieren; schleifen
f dérouler
r лущить; полировать; шлифовать

* **peelability → 536**

* **peeled → 551**

* **peeled billet wood → 901**

* **peeled log → 553**

6134 **peeled stick**
d Schälstück *m*
f billot *m* écorcé
r хлыст *m* с обдирами коры

6135 **peeled veneer; rotary-cut veneer**
(veneer produced by centering the entire log in a lathe and turning it against a broad cutting knife which is set into the log at a slight angle)
d Schälfurnier *n*; Rundschälfurnier *n*
f placage *m* déroulé; déroulage *m*
r окорённая фанера *f*; лущёный шпон *m*

* **peeled wood → 555**

* **peeler → 556, 2284, 9193**

* **peeler log → 9193**

* **peeling → 579**

* **peel[ing] axe → 563**

* **peeling damage → 2237**

6136 **peeling in sap flow period; sap peeling**
d Schälen *n* im Saftzeitraum; Schälen im Saftperiode
f écorçage *m* en sève
r окорка *f* в период движения сока; топорная соковая окорка

* **peeling iron → 7960**

* **peeling spade → 8639**

* **peepul → 939**

* **peg → 9561**

* **peg beam → 2038**

* **pegging [out] → 7995**

* **peggy wood → 6132**

* **peg planting → 2485**

6137 **peg root; secondary taproot**
(a secondary root from the cortical system in some [semi]parasitic plants, e. g. , Viscum, Loranthus, that grows directly downward into the tissues of the host)
d Senkwurzel *f*; sekundäre Hauptwurzel *f*
f pivot *m* secondaire; suçoir *m* secondaire
r вторичный стержневой корень *m*

6138 **peg[-tooth] saw**
(a circular or web saw having triangular-shaped teeth, i.e. peg teeth, generally used for cross-cutting as distinct from ripping)
d Lanzenzahn-Sägeblatt *n*
f lame *f* de scie à dents isocèles
r пила *f* с трёхгранными зубьями

* **pekea nut → 6188**

* **Pekin haw → 1500**

* **Pekin willow → 1963**

6139 **pellet**
(a mass of undigested hair, bones, feathers, elytra, etc. (from prey) regurgitated by a carnivorous bird or mammal)
d Gewölle *n*
f pelote *f*
r сгусток *m*

6140 **pelletizer**
d Granulator *m*
f machine *f* à fabriquer les boulettes
r дражератор *m*

6141 **pellet[ting]; seed pellet; seed dressing**
(incorporating seed in a matrix of fungicide, insecticide, repellent, colo[u]ring material or inert carrier, or any combination of these, so as to form a small ball termed a seed pellet)
d Pelletieren *n*; Samenbeizung *f*

* **pencil cedar → 2722**

* **pencilling → 801**

6142 **pencil roe; pencil line; pencil grade [texture]**
d Bleistiftfigur *f*
f texture *f* rubanée en crayon
r декоративный полосатый текстурный рисунок *m* древесины радиального распила

* **pendant → 9324**

6143 **pendant silver linden; silver pendant linden**
(Tilia petiolaris)
d hängezweigige Silberlinde *f*
f tilleul *m* argenté pleureur
r липа *f* длинночерешковая

* **pendent → 9324**

* **pendulous → 9324**

* **pendulum saw → 4312**

6144 **penetration**
d Eindringen *n*; Penetration *f*
f pénétration *f*
r проникание *n*; проникновение *n*; пенетрация *f*

* **Pensylvanian maple → 8174**

6145 **pentagonal pith**
d fünfeckiger Kern *m*
f cœur *m* pentagone
r пятиугольная сердцевина *f*; пентагональная сердцевина

* **penultimate bud → 8242**

6146 **peony family**
(Paeoniaceae)
d Pfingstrosengewächse *pl*
f péoniacées *pl*
r пионовые *pl*

6147 **peppered moth; brindled beauty moth; pepper and salt [currant] moth**
(Biston betularia)
d großer Birkenspanner *m*
f phalène *f* du bouleau
r пяденица *f* берёзовая

6148 **pepper family**

f pipéracées *pl*
r перцевые *pl*

* **pepperidge → 810**

* **peppertree → 1266**

6149 **peppervine**
(Ampelopsis arborea)
d Baumdoldenrebe *f*; Baumjungfernrebe *f*
f ampélopsis *m* arborescent
r виноградовник *m* древовидный; ампелопсис *m* древовидный

* **percentage of forest land → 3522**

6150 **percentage of non-fuel wood; percentage of timber and industrial cordwood**
d Nutzholzprozent *n*
f taux *m* de bois d'œuvre
r процент *m* поделочного леса

* **percent[age] of stocking → 2344**

* **percent of forest area → 3522**

* **perch → 6418**

* **percolation → 4488**

6151 **perennial plant; perennial**
(plant that forms annual above ground vegetation and seed structures from underground roots that persist for 3 or more years)
d mehrjährige Pflanze *f*; perennierende Pflanze
f plante *f* pérenne; plante vivace
r многолетнее растение *n*; многолетник *m*

6152 **perennial wildlife mixture**
(mixture of all or some of the following: shrub lespedeza, partridge pea, cowpea, annual lespedeza, reseeding soybeans, and other perennial plants that are beneficial to wildlife)
d mehrjährige Wildfauna *f*; perennierende Wildfauna
f mixture *f* de flore sauvage pérenne
r смесь *f* многолетних диких растений

6153 **perforated fiber**
d perforierte Faser *f*
f fibre *f* perforée; fibre-trachéide *f* ouverte; fibre-trachéide cloisonnée
r перфорированное волокно *n*

(originally imperforate) involved in the coalescence of two members of a vessel)
d Perforationsplatte *f*
f cloison *f* perforée; perforation *f*
r перфорационная пластинка *f*

6155 **perforation rim**
(the remnant of a perforation plate forming a border about a simple perforation)
d Perforationsrand *m*
f bourrelet *m*
r перфорационный обод *m*

6156 **performance-based logging**
(approval of future logging activities contingent upon a company's current practices)
d Leistungs-basierte Holzrückung *f*
f exploitation *f* du bois basée sur la performance
r лесозаготовка *f* на основе [прошлых] эксплуатационных показателей

6157 **periderm; epiphloem**
(the layers that replace the epidermis as the impermeable covering of older stems and roots, consisting of phellem (cork), phellogen and phelloderm)
d Lederkork *m*; Periderm *m*
f périderme *m*
r перидерма *f*; эпифлоэма *f*; наружный слой *m* флоэмы

* **perimedullary phloem → 4502**

6158 **periodical growth; periodical increment**
d periodischer Zuwachs *m*
f accroissement *m* périodique
r периодический прирост *m*

6159 **periodic allowable cut**
d zulässiger Periodeneinschlag *m*
f coupe *f* périodique permissible
r расчётная периодическая рубка *f*

6160 **periodic block; periodic area; regeneration block**
(part(s) of forest allocated for regeneration or other treatment during a specified period)
d Periodenfläche *f*; Periodenblock *m*
f affectation *f* [de régénération]
r периодный блок *m*; периодная площадь *f*; возобновительный блок

6161 **periodic block method of management;**

f méthode *f* d'aménagement par blocs
r метод *m* периодных блоков лесоустройства; периодно-площадной метод

6162 **periodic cut[ting]; periodic felling**
(removal of several years' accumulated AAC in one year or other period)
d Periodeneinschlag *m*
f coupe *f* périodique
r периодическая рубка *f*; периодная рубка

6163 **periodic harvesting**
d periodische Holzgewinnung *f*
f exploitation *f* forestière périodique
r периодическое лесопользование *n*; сезонные лесозаготовки *fpl*

* **periodic inventory method of management → 6161**

6164 **periodic [mean] annual increment; p.[m.]a.i.**
(the growth for any specified short period divided by the number of years in the period)
d periodischer Durchschnittszuwachs *m*
f accroissement *m* annuel [moyen] périodique; accroissement périodique moyen [annuel]; accroissement périodique annuel; APA
r периодический текущий прирост *m*; периодический средний годичный прирост

6165 **periodic method by volume; method of periods by volume**
d Massenfachwerkmethode *f*
f méthode *f* d'aménagement par volume
r периодный метод *m* определения запаса

6166 **periodic stand depletion**
d periodische Durchforstung *f*
f taux *m* d'extraction périodique en éclaircie
r периодическое изреживание *n* насаждения; периодическое промежуточное пользование *n* лесом

* **period of transit → 4006**

6167 **peripheral felling; cutting around the edges of gaps; [forest] border cutting; border felling; edge coupe**
d Rändel[ungs]hieb *m*; Randhieb *m*; Umsäumungshieb *m*; Saumschlag *m*
f coupe *f* de dégagement des lisières; coupe en lisière
r окрайная рубка *f*; окраинная рубка; рубка на границе участка

6168 **permanent access structure**
(including a road, bridge, landing, gravel pit or other similar structure, that provides access for timber harvesting)
d Dauerzugriffsstruktur *f*
f structure *f* d'accès permanent
r устойчивая структура *f* системы доступа

6169 **permanent crops**
d Dauerkulturen *fpl*
f cultures *fpl* permanentes
r плодово-ягодные плантации *fpl* (категория нелесных земель; участки, занятые ягодниками)

6170 **permanent field block method**
d Einschlagsmethode *f* der permanenter Periodenfläche
f méthode *f* d'aménagement par blocs permanents
r метод *m* постоянных периодных блоков

6171 **permanent forest; continuous [cover] forest; perpetual forest; continuous crops of tree**
d Dauerwald *m*
f forêt *f* permanente; futaie *f* permanente; futaie traitée en affectation permanente
r непрерывно-производительный лес *m*; насаждения *npl* непрерывного пользования

6172 **permanent forestry; continuous forest management system**
d Dauerwaldwirtschaft *f*
f foresterie *f* permanente
r непрерывное лесопользование *n*

6173 **permanent forest worker; permanent logger**
d ständiger Waldarbeiter *m*
f boiseur *m* permanent
r постоянный лесной рабочий *m*

6174 **permanent living firebreak**
d lebender Feuerschutzstreifen *m*
f pare-feu *m* végétal; allée *f* coupe-feu végétale
r постоянный противопожарный барьер *m* из лиственных пород

* **permanent logger → 6173**

6175 **permanent road**
(road required for long-term management of the forest)
d Dauerweg *m*
f route *f* permanente; route à viabilité

6176 **permanent sampling unit**
(within vegetation survey)
d Dauerquadrat *n*
f placeau *m* permanent
r постоянная пробная площадь *f*

6177 **permanent wilting point**
(amount of water held in the soil when plants remain wilted even though when, for a short period, the aerial parts are kept in a humid atmosphere)
d ständiger Welkepunkt *m*; Dauerwelkungspunkt *m*
f point *m* de flétrissement permanent; seuil *m* de flétrissement permanent
r устойчивая точка *f* завядания

6178 **permissible annual forest yield**
d zulässige jährliche Ergiebigkeit *f*
f rendement *m* annuel permissible de la forêt
r допустимая годичная лесосека *f*

6179 **permissible cutting**
d zulässiger Einschlag *m*
f abattage *m* permissible
r допустимая рубка *f*

6180 **permission to remove [of wood]; permit for removal [of wood]**
d Holzabgabeschein *m*; Abfuhrschein *m*; Auslieferungsschein *m*; Hauungslos *m*; Holzverabfolgungsschein *m*
f permis *m* d'enlèvement de bois; permis de vidange
r разрешение *n* на вывоз [древесины из леса или со склада]; билет *m* на вывозку леса

6181 **permitted deviation**
d zulässige Abweichung *f*
f déviation *f* permissible
r допускаемое отклонение *n* [круглых лесоматериалов]

6182 **permitted hunter-kill ratio; [normal] kill ratio**
(the proportion of (species of) a wildlife population that may be killed by hunters without diminishing subsequent yields, generally assessed annually)
d Jagd-Abschussziffer *f*; erlaubte Jagdbeute *f*
f rapport *m* de chasse [normal]; taux *m* de prélèvement permi
r допускаемый коэффициент *m* поражения

* **permit to clear → 4928**

6183 **permit to cut; cutting permit; felling permit;**
licence to cut; cutting licence; felling licence; timber permit
(legal document that authorizes the holder to harvest trees under a licence)
d Einschlagsgenehmigung *f*; Holzzettel *m*; Holzschnittgenehmung *f*
f approbation *f* de l'état d'assiette; bulletin *m* de cession de bois; permis *m* de coupe
r лесорубочный билет *m*; билет по отпуску леса; разрешение *n* рубки

* **pernambouc → 1010**

* **Pernambuco wood → 1010**

6184 **peroba**
(Aspidosperma populifolium; Aspidosperma peroba)
d Peroba-Mica *f*
f péroba *m* rose
r пероба *f*

* **perpetual forest → 6171**

6185 **Persian albizia; silk tree**
(Albizzia julibrissin)
d persische Albizie *f*; Schlafbaum *m*; Seidenakazie *f*
f arbre *m* de soie; arbre de sommeil; acacia *m* de Constantinople
r альбизия *f* ленкоранская

* **Persian lilac → 6110**

6186 **Persian parrotia; [Persian] ironwood**
(Parrotia persica)
d persische Parrotie *f*; persisches Eisenholzbaum *m*; Arga[h]nbaum *m*
f parrotie *f* de Perse; arbre *m* de fer [de Perse]
r паротия *f* персидская; [закавказское] железное дерево *n*; боккаут *m*; темир-агач *m*

* **persimmon → 1796**

6187 **persistent species**
d Dauerarten *fpl*; beständige Arten *fpl*
f espèces *fpl* persistantes
r устойчивые виды *mpl*

* **perspiration → 8321**

* **per-unit sale → 6119**

6188 **Peruvian almond; Sawari nut; pekea nut**

f caryocar *m* nucifère; noisette *f* indienne; pekeya *m*
r кариокар *m* орехоносный; орех *m* савара

* **Peruvian balsam [balm] tree → 504**

* **Peruvian peppertree → 1266**

* **pest → 5807**

6189 **pesticide**
(refer to chemicals used to destroy, prevent, or control plant or animal pests)
d Pestizid *n*
f pesticide *m*
r пестицид *m*

6190 **pesticide buffer zone**
(strip of land between the 10 m pesticide-free zone and the pesticide treatment area for preventing entry of pesticides)
d Pestizidisolierstreifen *m*; Pufferzone *f* ohne Pestiziden
f bande *f* tampon sans pesticides; bande d'isolement les pesticides
r буферная зона *f* без пестицидов

6191 **pest incidence**
(measurement of the presence and magnitude of pests within a given area)
d Pestinzidenz *f*; Schädling-Inzidenz *f*
f incidence *f* parasitaire; incidence par ravageurs
r заражение *n* вредителями

6192 **petal**
(a division of the inner set of leaf-like flower parts, usually colo[u]red)
d Kronblatt *n*; Kronenblatt *n*; Blütenblatt *n*; Blumenblatt *n*
f pétale *m*
r лепесток *m*

6193 **petiolate[d]; pedunculate** *adj*
d gestielt
f emmanché; pédonculé; pétiolé
r черенковый; черешчатый

6194 **petiolate[d] leaf**
d gestieltes Blatt *n*
f feuille *f* pétiolée
r черешковый лист *m*

* **petiole → 4883**

6195 **petrified palm wood**
d Steinpalmholz *n*
f bois *m* de palmier
r окаменевшая пальмовая древесина *f*

6196 **petrified wood; silicified wood; fossilized wood; fossil-wood; lime wood; stone wood**
silica-rich petrified wood)
d fossiles Holz *n*; Steinholz *n*
f bois *m* pétrifié; bois silicifié; bois fossilisé; stonewood *m*
r окаменелый лес *m*; ископаемая древесина *f*; окаменевшая древесина; окремнелая древесина

* **petrol motor[-powered] saw → 3808**

* **PGR → 6340**

6197 **phanerophytes**
(self-supporting woody plants or herbaceous evergreen perennials that grow taller than 25-50 cm)
d Phanerophyte *mpl*
f phanérophytes *mpl*
r фанерофиты *mpl*

6198 **phase of development**
d Entwicklungsphase *f*
f phase *f* de développement
r фаза *f* развития

* **phellem → 1956**

6199 **phellodendron**
(Phellodendron)
d Korkbaum *m*
f phellodendron *m*; arbre *m* au liège
r бархат *m*; феллодендрон *m*

6200 **phelloderm**
(a tissue that generally resembles cortical parenchyma in appearance, but which consists of the ner derivatives of the phellogen)
d Korkrindenschicht *f*; Phelloderma *n*
f phelloderme *m*; couche *f* d'écorce à liège
r слой *m* пробковой коры

6201 **phellogen; cork cambium; cork-producing layer**
(the meristem layer that produces the periderm)
d Korkkambium *n*; Phellogen *n*; Korkmutterschicht *f*
f phellogène *m*; tissu *m* phellogène; tissu de liège; couche *f* génératrice de liège

6202 phenological cycle
d phänologischer Zyklus *m*
f cycle *m* phénologique
r фенологический цикл *m*

6203 phenology
(study of timing of periodic phenomena, such as flowering, growth initiation, growth cessation)
d Phänologie *f*
f phénologie *f*
r фенология *f*

6204 phenotype
(visible characters of a plant)
d Phänotypus *m*; Phänotyp *m*; Erscheinungsbild *n*; Merkmalsbild *n*
f phénotype *m*
r фенотип *m*

* **phenotypic elite tree → 6396**

6205 phenotypic selection
(choosing trees on the basis of their phenotypic performance)
d phänotypische Selektion *f*
f sélection *f* phénotypique
r фенотипическая селекция *f*

6206 pheromon-baited insect traps
d pheromonbeköderte Insektfällen *fpl*
f pièges *mpl* à la phéromone pour insectes
r ловушки *fpl* для насекомых, наживлённые феромонами

* **phillyrea → 5503**

* **phloem → 4502**

* **phloem fiber → 623**

6207 phloem mother cell
(cell that is cut off on the outer side by the fusiform cambial initials, but which undergo further periclinal divisions before differentiating into phloem cells)
d Phloem-Mutterzelle *f*
f cellule-mère *f* du liber
r первичная клетка *f* флоэмы

6208 phloem ray
d Phloem-Strahl *m*
f rayon *m* libérien
r лубяной луч *m*; лубодревесный луч; луч флоэмы; участок *m* флоэмы между двумя сердцевинными лучами

* **phoenix tree → 1303**

6209 photodegradation of wood
d Photodegradation *f* des Holzes
f photodégradation *f* du bois
r фоторазложение *n* древесины

6210 photographical measurement of roundwood
d fotografische Rundholzmessung *f*
f mesure *f* photographique de volume de bois rond
r фотографическое определение *n* объёма круглых лесоматериалов

6211 photosynthesis
(the process by which plants convert sunlight, water and carbon dioxide into new plant tissue and oxygen)
d Fotosynthese *f*
f photosynthèse *f*
r фотосинтез *m*

6212 phototropism; heliotropism
(the growth (movement) of a plant organ as a response to the stimulus of light)
d Fototropismus *m*; Heliotropismus *m*; Fototropie *f*; Lichtwendigkeit *f*
f phototropisme *m*
r фототропизм *m*

* **phreatic water → 3995**

* **PHSP → 6506**

* **phyllode → 1594**

6213 phyllosphere
(the space immediately adjacent to, and influenced by, a plant leaf)
d Phyllosphäre *f*
f phyllosphère *f*
r филлосфера *f*

* **phyllotaxis → 4856**

* **phyllotaxy → 4856**

6214 phylum
(a taxon of high rank, the major unit of classification)
d Phylum *m*
f phylum *m*
r филюм *m*

6215 **physical maturity; physical exploitability**
(of a forest)
d physische Schlagbarkeit *f*
f exploitabilité *f* physique
r естественная спелость *f*

* **physical rotation → 2745**

6216 **physiognomic class; physiognomic group**
(defined by the relative percent canopy cover of the tree, shrub, dwarf shrub, herb, and non-vascular life form (in the uppermost strata) during the peak of the growing season)
d physiognomische Klasse *f*
f classe *f* physiognomonique
r физиогномический класс *m*

6217 **physiognomic subclass**
(determined by the predominant leaf phenology of classes defined by tree, shrub, or dwarf shrub stratum, and the average vegetation height for the herbaceous stratum)
d physiognomische Unterklasse *f*
f sous-classe *f* physiognomonique
r физиогномический подкласс *m*

6218 **physiographic class**
(measure of soil and water conditions that affect tree growth on the plot)
d physiografische Klasse *f*
f classe *f* physiographique
r физиографический класс *m*

6219 **physiographic unit**
d physiografische Einheit *f*
f unité *f* physiographique
r физиографическая единица *f*

6220 **phytocide**
(used to kill unwanted plants)
d Pflanzengift *m*; Pflanzenvernichtungsmittel *n*
f phytocide *m*
r фитоцид *m*

* **phytoc[o]enose → 6336**

* **phytoc[o]enosis → 6336**

6221 **phytohormone; plant hormone; plant growth substance**
(a broad class of naturally occurring chemicals, effective at very low concentrations for regulating plant growth and development)
d Pflanzenwuchsstoff *m*; Phytohormon *n*; Wuchsstoff *m*
r фитогормон *m*; ростовое вещество *n*

6222 **phytosanitary**
(health of plant parts)
d Pflanzenschutz *m*; Pflanzengesundheit *f*
f phytopharmaceutique *f*
r фитосанитария *f*

6223 **phytotoxicity**
(injury to plants due to exposure to a chemical)
d Phytotoxizität *f*
f phytotoxicité *f*
r фитотоксичность *f*

6224 **pian-bois; forest yaws**
d Waldgieren *n*; Holzhimbeerseuche *f*
f bouton *m* de Bahia; leishmaniose *f* forestière américaine
r древесный лейшманиоз *m*

* **Picardian poplar → 3973**

* **picaroon → 6693**

6225 **pickaroon; hookaroon; grippers**
(device with a head similar to an axe but with a point rather than a blade mounted on tie end of a handle which is used to assist in lifting and placing bolts of wood)
d Lasthaken *m*; Packhaken *m*
f sapi *m*; griffe *f* à soulever
r погрузочный крюк *m*; багор *m* с короткой деревянной ручкой

* **pick-axe → 2035**

* **pick-rear → 8324**

6226 **piece-by-piece method of measurement of volume**
d stückweise Methode *f* der Kubierung
f méthode *f* de cubage par morceaux
r поштучный метод *m* измерения объема

* **piece of log → 2171**

* **piercing saw → 1822**

6227 **piggy-back method [of yarding]**
d Huckepackverkehrsmethode *f*
f méthode *f* ferroutière; méthode piggy-back
r способ *m* трелёвки брёвен в пачке

6228 **pigmy willow; herbaceous willow**

f saule *m* herbacé
r ива *f* травянистая

* **pig-nut → 6229**

6229 **pig[nut] hickory; broom hickory; pig-nut**
(Carya glabra)
d Ferkelnuss *f*
f caryer *m* glabre
r кария *f* голая

* **pike[-]pole → 6693**

* **pile → 8619**

* **pile-bottom → 5451**

* **pile density → 7980**

* **pile discolo[u]ration → 1031**

6230 **piled merchantable timber; stacked merchantable timber; corded merchantable timber**
d Schichtnutzholz *n*
f bois *m* marchand empilé
r окученный коммерческий лесоматериал *m*; складочный поделочный лес *m*

6231 **piled slash**
d geschichteter Schlagabraum *m*
f rémanents *mpl* empilés
r порубочные остатки *mpl*, собранные в кучи

* **piled volume → 7979**

* **piled wood → 7982**

* **pile of branchwood → 4193**

* **pile of brushwood → 4193**

* **piler of lumber → 8094**

* **piler of sawn goods → 8094**

* **piles → 8588**

6232 **pile stacked crosswise; cross-pile**
d über Kreuz geschichteter Stapel *m*; Kreuzstapel *m*; Kreuzbeige *f*
f pile *f* croisée
r накрест сложенный штабель *m*; крестовый штабель

* **pile-wood → 8588**

* **piling → 4846**

6233 **piling and burning**
(piling slash after lopping, and subsequently burning the individual piles)
d Stapelung *f* und Brennen *n*
f empilage-brûlage *m* des rémanents
r сбор *m* в кучи и сжигание *n*

6234 **piling in yard**
d Stapelung *f* im Holzlager
f empilage *m* à dépôt
r штабелёвка *f* на лесном складе

6235 **piling jack; piling step**
d Ladebalken *m*
f poutre *f* de chargement
r грузовая балка *f*

6236 **piling stick; sticker; crosser; pin; stacking strip; piling strip; fillet**
(a stick or strip of wood used to separate courses in a stack and so improve air circulation)
d Hölzel *n*; Lagerholz *n*
f baguette *f*; cale *f*; épingle *f*; bois *m* d'entrepôt; latte *f* d'espacement
r прокладка *f* в штабеле [лесоматериалов]; штабельная прокладка; сложенная древесина *f*

* **pillar → 8190**

* **pillar-poplar → 681**

* **pill bug → 1813**

6237 **pilon; tapana tree; suradonni; urucurana; suradan; curtidor**
(Hyeronima alchorneoides; Hyeronima laxiflora)
d Pilon *m*; Trompillo *m*; Urukurana *m*
f pilon *m*; trompillo *m*
r хиеронима *f* альхордиевидная

6238 **pilot[ed] bit**
(a router bit fitted with a bearing above the cutter which rides on the edge of a board or template keeping the bit a fixed distance from the edge)
d Führungsmeißel *m*
f trepan *m* pilote
r направляющее долото *n*; долото лопастного типа

6239 **pilot forest; demonstration forest; type**

f forêt *f* de démonstration
r показательный лес *m*; эталонный лес

* **pin → 6236**

* **pinaster → 5309**

* **pincers → 8809**

* **pin-cherry → 2881**

6240 **pinching [out]; pinching back; stopping**
(removal or suppression of apical buds to stimulate the production of lateral branches)
d Abkneifen *n*; Pinzieren *n*
f pincement *m*; pinçage *m*
r пинцирование *n*

6241 **Pinchot['s] juniper**
(Juniperus pinchotii)
d Pinchot-Wacholder *m*
f genévrier *m* de Pinchot
r можжевельник *m* Пинчота

6242 **pin dote**
(small rotten spots on the ends of logs)
d Nadelfäule *f*
f pourriture *f* en points; échauffure *f* en points
r точечная гниль *f* (на торцах брёвна)

6243 **pineapple**
(Ananas comosus; Ananas sativus)
d Ananas *f*
f ananas *m*
r ананас *m*

6244 **pine bark**
d Strobenrinde *f*
f écorce *f* de pin [sylvestre]
r сосновая кора *f*

6245 **pine bark aphid; pine bark adelgid**
(Pineus strobi)
d Strobenrindenlaus *f*; Weymouthskiefernwollaus *f*
f puceron *m* de l'écorce du pin; chermes *m* du pin Weymouth
r хермес *m* веймутовой сосны

6246 **pine beetle**
(Blastophagus piniperda)
d Bastkäfer *m*; Kiefernmarkkäfer *m*; Waldgärtner *m*
f dendroctone *m* du pin
r лубоед-стригун *m*; лесной садовник *m*

6247 **pine bud**
r сосновая почка *f*

6248 **pine caterpillar**
(Dendrolimus punctatus)
d japanischer Kiefernspinner *m*
f chenille *f* du pin
r шелкопряд *m* сосновый крапчатый

6249 **pine chafer**
(Anomala oblivia)
d Kiefernkäfer *m*
f hanneton *m* de pin
r хрущик *m* сосновый

6250 **pine cone**
d Kiefernzapfen *m*
f cône *m* de pin; pomme *f* de pin; cocotte *f*
r сосновая шишка *f*

6251 **pine conversion**
(establishing pines on sites that were formerly occupied by hardwoods to yield trees of high quality and volume per acre)
d Kiefernkonversion *f*
f conversion *f* de pins
r смена *f* сосновых пород

6252 **pine engraver**
(Ips pini)
d Kieferkäfer *m*
f scolyte *f* du pin; rongeur *m* bostryche
r короед *m* сосновый

6253 **pine false webworm**
(Acantholyda erythrocephata)
d rotköpfige Kiefernschonungsgespinstblattwespe *f*; stahlblaue Kiefernschonungsgespinstblattwespe
f lyde *f* bleue
r пилильщик-ткач *m* красноголовый общественный

6254 **pine family**
(Pinaceae; Abietaceae)
d Kieferngewächse *pl*; Abietazeen *pl*
f pinacées *pl*
r сосновые *pl*; пихтовые *pl*

6255 **pine forest; stand of pine; pine stand; piny wood; pinery; pinetum**
d Föhrenwald *m*; Kiefernwald *m*; Rotwald *m*; Kiefernbestand *m*
f forêt *f* de pins; pineraie *f*; pinède *f*; pinière *f*; pinée *f*
r сосновый лес *m*; сосновая роща *f*;

6256 **pine gall [weevil]**
(Podapion gallicota)
d Terpentingalle *f*
f galle *f* de térébinthe
r долгоносик *m* галловый

* **pine gum → 6898**

6257 **pine hawk-moth**
(Hyloicus pinastri)
d Kiefernschwärmer *m*; Tannenpfeil *m*
f sphinx *m* à pin
r бражник *m* сосновый

* **pineland → 7953**

6258 **pine [lappet] moth**
(Dendrolimus pini)
d Kieferngluckе *f*; Kiefernspinner *m*
f lasiocampe du pin; bombyx *m* du pin
r шелкопряд *m* сосновый

6259 **pine-leaf; pine-leaved; pinifolious**
d nadelblätt[e]rig
f à feuilles de pin
r хвоелистный

6260 **pine-leaf cast; needle cast; needle shedding; leaf cast of conifers**
(any untimely shedding of foliage in needle-bearing conifers, in contrast to natural leaf-fall)
d Kiefernblatt-Schüttenkrankheit *f*
f maladie *f* de la rouille du pin
r сбрасывание *n* хвои; опадение *n* хвои; отпад *m* хвои; шютте *n*; болезнь *f* сеянцев сосны

6261 **pine-leaf miner**
(Paralechia pinifoliella)
d Kiefernblattminiermotte *f*
f mineuse *f* de feuilles du pin
r моль *f* выемчатокрылая сосновая

* **pine-leaf tube → 6273**

* **pine-leaved → 6259**

6262 **pine looper [moth]; pine moth; bordered white moth**
(Bupalus piniarius)
d [gemeiner] Kiefernspanner *m*
f arpenteuse *f* du pin; fidonie *f* du pin; phalène *f* du pin
r пяденица *f* сосновая

f oléorésine *f* de pin
r сосновый пек *m*

6264 **pine plantation**
(stands that have been artificially regenerated by planting or direct seeding and with a southern yellow pine, white pine-hemlock, or other forest type)
d Kiefernplantation *f*
f plantation *f* de pins
r сосновая плантация *f*

6265 **pine red heart rot; red ring rot**
(invader - Phellinus pini)
d Kiefernbaumschwamm *m*; Rotfäule *f* der Kiefer und Fichte
f pourriture *f* rouge du pin
r пёстро-красная гниль *f*

* **pine resin → 6898**

6266 **pine resinous stemwood**
d Harzkiefernstammholz *n*
f bois *m* de fût du pin résineux
r сосновый стволовой осмол *m*

6267 **pine-root-collar weevil**
(Hylobius radicis)
d Kiefernwurzelhalsrüssler *m*
f charançon *m* du collet de la racine du pin
r долгоносик *m* корневой сосновый

* **pinery → 6255**

6268 **pine sawfly**
(Diprion pini; Lophyrus pini)
d [gemeine] Kiefernblattwespe *f*
f lophyre *m* du pin
r пилильщик *m* обыкновенный сосновый

6269 **pine-shoot moth; shoot moth; pine-tip moth; tip moth**
(Rhyacionia buoliana)
d Kiefergallenwickler *m*
f tordeuse *f* des pousses du pin
r побеговьюн *m* [зимующий]; листовёртка *f* хвойная рыжая

* **pine spruce → 9386**

* **pine stand → 6255**

6270 **pine straw**
(fallen pine needles, primarily of longleaf and slash pines, that are raked, baled, and removed from the forest stand for sale as

f litière *f* de pins
r сосновая подстилка *f*; подстилка из сосновой хвои

6271 **pine stumpwood**
d Kiefernstockholz *n*; Kienstumpf *m*; Kiefernwurzelholz *n*
f bois *m* de souche du pin
r сосновый пнёвой осмол *m*

* **pine tar → 6898**

* **pine-tar oil → 6276**

* **pine-tip moth → 6269**

6272 **pine [tree]**
(Pinus)
d Föhre *f*; Kiefer *f*; Kien *m*
f pin *m*
r сосна *f*

6273 **pine-tube moth; pine-leaf tube**
(Argyrotaenia pinatubana; Eulia pinatubana)
d Kiefernsämlingswickler *m*
f tordeuse *f* tubicole du pin
r листовёртка *f* сосновая

* **pinetum → 6255**

6274 **pine [twist] rust**
(invader - Melampsora pinitorqua)
d Kieferndrehrost *m*
f rouille *f* [courbeuse] du pin
r ржавчина *f* сосны; сосновый вертун *m*; ржавчинное искривление *n* побегов сосны

* **pine weevil → 4793**

6275 **pine[-]wood; red deal; yellow deal**
d Kiefernholz *n*
f bois *m* de pin
r сосновая древесина *f*

6276 **pine[wood] oil; pine-tar oil**
d Kiefernharzöl *n*; Pineöl *n*; Kienöl *n*
f huile *f* [de bois] de pin
r сосновое масло *n*; хвойное масло; скипидар *m*

* **pin hole → 6461**

* **pinhole borer → 231**

* **pinifolious → 6259**

d Schindelbaum *m*
f acrocarpus *m* rose; meluang *m*
r розовый кедр *m*

6278 **pink family**
(Caryophyllaceae)
d Nelkengewächse *pl*; Karyophyllazeen *pl*
f caryophyllacées *pl*
r гвоздичные *pl*

6279 **pink ivory; umnini; umgoloty**
(Berchemia zeyheri)
d Umnini *n*; Umgoloty *n*
f ivoire *f* rose
r розовая кость *f*

* **pink mahogany → 5731**

6280 **pink muskwood; scented guarea**
(Guarea laurentii; Guarea cedrata)
d Bosse *f*
f bossé *m* foncé; bossé clair
r гварея *f* пахучая; гварея кедровая

* **pink narrowleaved ironbark [eucalyptus] → 6779**

6281 **pin knot**
(knot with diameter less than 1.2 cm)
d Nadelknoten *m*; Punktast *m*
f petit nœud *m*
r игольчатый сучок *m*; булавочный сучок

* **pink shower tree → 5346**

* **pink[ster] azalea → 6288**

* **pink trumpet-tree → 5346**

* **pink walnut → 409**

* **pinkwood → 1007**

6282 **pinnate leaf**
(a compound leaf which leaflets originate on opposite sides of a single rachis)
d gefingertes Blatt *n*
f feuille *f* pennée
r перистый лист *m*

* **pinnate[ly]-compound leaf → 6038**

6283 **pinnately-decompound leaf**
d gefiedertes gegenüberverbindete Blatt *n*

6284 **pinnatifid leaf; pinnatisect[ed] leaf; pinnately sected leaf; pinnatipartite leaf**
(cut deeply into lobes that are spaced out along the axis)
d pinnatifides Blatt *n*; fiederschnittiges Blatt
f feuille *f* pennatifide; feuille pennatilobée; feuille pennatipartite; feuille pennatiséquée
r перисторассечённый лист *m*; перистонадрезный лист; перистодольчатый лист; перистолопастный лист; перистораздельный лист

6285 **pinnatodentate leaf**
d zahngefiedertes Blatt *n*
f feuille *f* pennatidentée
r перистый лист *m* с зубчатыми долями

6286 **pinnatopectinate leaf**
d kammförmiggefiedertes Blatt *n*
f feuille *f* pennatipectinée
r перистый лист *m* с гребенчатыми долями

* **pino → 1372**

* **pin oak → 3764**

6287 **pinoid pit; fenestriform pit; windowlike pit**
(a term of convenience for the smaller types of early wood crossfield pit found in several species of Pinus)
d fensterförmige Tüpfel *f*
f ponctuation *f* pinoïde; ponctuation en fenêtre
r хвоевидная пора *f*

6288 **pinxterbloom azalea; pink[ster] azalea; purple azalea; pinxter flower; wild honeysuckle**
(Rhododendron nudiflorum; Rhododendron periclymenoides)
d nacktblütiger Rhododendron *m*
f rhododendron *m* à fleurs nues
r азалия *f* пониклая; рододендрон *m* голоцветковый

* **pinyon nut pine → 9031**

* **piny wood → 6255**

* **pioneer → 6290**

6289 **pioneer forest**
(a forest capable of invading bare sites (e.g. a newly exposed soil surface) and persisting there, until supplanted, by successional
f forêt *f* pionnière
r лес-пионер *m*

6290 **pioneer [tree species]; forest nomad; pioneer**
(tree species that has a high light requirement for seed germination and seedling establishment and is adapted to grow well in disturbed sites)
d Pionierholzart *f*; Pionierbaumart *f*
f essences *fpl* transitoires; espèce *f* pionnière
r [древесная] порода-пионер *f*; пионерная древесная порода

* **pipal [tree] → 939**

* **piped rot → 9377**

* **piper cascade blueberry → 864**

6291 **pipe rot**
d Markröhrenfäule *f*
f pourriture *f* tubulaire
r трубчатая гниль *f*

6292 **pipe tree; bluepipe; [common] lilac; Roman willow**
(Syringa vulgaris)
d [gemeiner] Flieder *m*; gewöhnlicher Fliederlilak *m*; türkischer Holunder *m*
f lilas *m* [commun]
r сирень *f* [обыкновенная]

* **pip fruit → 4698**

6293 **piptadenia**
(Piptadenia)
d Piptadenie *f*
f piptadénie *f*
r пиптадения *f*

* **pipul → 939**

6294 **pistache woodland**
d Pistazienlichtwald *m*
f bois *m* clair de pistachier
r фисташковое редколесье *n*; фисташник *m*

6295 **pistachio tree; Chinese pistache**
(Pistachia chinensis; Pistachia vera)
d Pistazienbaum *m*; echte Pistazie *f*
f pistachier *m* de Chine
r фисташка *f* обыкновенная

6296 **pistil**
(the female ovule-bearing part of a flower

f pistil *m*
r пестик *m*

6297 **pistillate-flowered tree**
(unisexual tree having only female reproductive organs)
d Stempelblütenbaum *m*
f arbre *m* pistillé
r дерево *n* с пестичными цветками; дерево с женскими цветками

* **pit → 6461**

6298 **pit aperture; pit opening**
(the opening or mouth of a pit)
d Tüpfelöffnung *f*
f orifice *f* de ponctuation
r отверстие *n* поры

6299 **pit border**
(the overarching part of the secondary cell wall)
d Tüpfelrand *m*
f aréole *f*
r окаймление *n* поры

6300 **pit canal**
(the passage from the cell lumen to the chamber of any bordered pit)
d Tüpfelkanal *m*
f canal *m* d'une ponctuation
r канал *m* поры

6301 **pit cavity**
(the entire space within a pit from the membrane to the lumen)
d Tüpfelkavität *f*
f cavité *f* d'une ponctuation
r полость *f* поры

6302 **pit chamber**
(the space between the pit membrane and the overarching pit border)
d Tüpfelkammer *f*
f chambre *m* d'une ponctuation
r поровая камера *f*

* **pith → 1951**

* **pitch blister → 6893**

* **pitch pine → 5792**

* **pitch pocket → 6893**

* **pitch seam → 6893**

* **pitch-streak → 8155**

6303 **pitch tube; resin tube; gum tube**
(tubular mass of resin that forms on the surface of bark at bark-beetle entrance holes)
d Harztrichter *m*
f bouchon *m* de résine
r смоляная трубка *f*; воронка *f* смоляного горшочка

6304 **pitch-yielding coniferous species**
d harzführende Nadelholzart *f*
f espèce *f* résineuse
r смолоносная хвойная древесная порода *f*

* **pitch-yielding tree → 6900**

6305 **pit field**
(areas on the radial walls of longitudinal coniferous tracheids bounded above and below by crassulae, containing one or more pits or devoid of pits)
d Tüpfelfeld *n*
f champ *m* de ponctuation
r поровое поле *n*

6306 **pith knot**
d Markast *f*
f nœud *m* médulleux
r гнилой сучок *m* с выпавшей сердцевиной

* **pithness → 2671**

* **pith ray → 5391**

* **pith [ray] fleck → 5390**

* **pith rot → 1414**

* **pith stock → 1951**

6307 **pith-to-cambium grafting**
d Kern-zu-Kambium-Pfropfen *n*
f greffage *m* de médulle sur cambium
r прививка *f* сердцевиной на камбий

* **pith wander → 9264**

* **pith wood → 4675**

* **pit mattock → 2035**

6308 **pit membrane; closing membrane; scalloped torus; torus**
(the part of the intercellular layer and primary cell wall that limits a pit cavity externally)
d Tüpfelschließhaut *m*
f membrane *m* de la ponctuation; torus *m* [lobé]

* **pit opening → 6298**

6309 **pit-pair**
(two complementary pits of adjacent cells)
d Tüpfelpaar *n*
f couple *f* de ponctuations
r пара *f* пор

* **pit planting → 6355**

6310 **pit-prop rotation**
d Grubenholzumtrieb *m*
f rotation *f* d'une forêt de bois de mine
r оборот *m* на рудничный лес

6311 **pit props; mine props; props**
(in mining, a round, squared or split timber that supports the roof)
d Antreibeholz *n*; Ausbauholz *n*; Pfandholz *n*; Zechnenholz *n*
f poteaux *mpl* de mines; étançon *m* de mine
r рудничная стойка *f*; рудостойка *f*; пропсы *pl*

6312 **pit sawing; whip sawing**
d Schrotsägen *n*
f sciage *m* au bois montant
r [продольная] распиловка *f* брёвен в пильной яме

* **pit sheath → 7216**

* **pit sleeper → 1725**

* **pit sowing → 6355**

6313 **pitted sap rot; pocket dry rot; honeycomb rot; partridge rot; peckiness; peck**
(characterized by finger-sized pockets of decay in the living trees of incense cedar (caused by Tyromyces amarus) and baldcypress (caused by Fomes geotropus))
d Lochfäule *f*; Wabenschwamm *m*; Taschenfäule *f*; Rebhuhnfäule *f*; Wabenfäule *f*
f carie *f* alvéolaire; pourriture par taches ou par poches
r [пёстрая] ямчатая гниль *f*; сотовая гниль

* **pit timber → 5463**

6314 **pit torus**
(a central, thicker part of a pit membrane)
d Toruskern *m*
f torus *m*
r торус *m* поры

d Klebsame *f*
f pittospore *m*; sapin *m* blanc de Tasmanie
r смолосемянник *m*

* **pit wood → 5463**

* **pivoting crane → 8343**

* **pivot tree → 1967**

* **place of a pile → 4712**

* **place of conflagration → 1184**

* **place of felling → 7668**

6316 **plagiotropic branching**
(of stems that grow only horizontally)
d plagiotrope Verzweigung *f*
f ramification *f* plagiotrope; branchement *m* plagiotrope
r плагиотропное разветвление *n*

6317 **plagiotropy**
(gravitational response that produces an oblique or horizontal axis)
d Plagiotropie *f*
f plagiotropie *f*
r плагиотропия *f*

* **plain edge → 477**

6318 **plain gum**
d Gleit-Gummi *m*
f gomme *f* lisse
r эвкалиптовая древесина *f* с простым текстурным рисунком

* **plain-sawed → 3258**

* **plain[-]sawn → 3258**

* **plain[-]sawn timber → 3254**

6319 **plains cottonwood; plains poplar**
(Populus sargentii; Populus deltoides var. occidentalis)
d Sargentspappel *f*; amerikanischer Schwarzpappel *f*
f peuplier *m* de Sargent
r тополь *m* Сарджента

* **plains poplar → 6319**

* **planed timber → 2620**

* **planer saw → 5803**

6320 **planertree; water elm**
(Planera aquatica)
d Wasserulme *f*
f orme *m* aquatique
r планера *f* водная

* **plane sole → 6323**

* **plane stock → 6323**

* **plane sycamore → 254**

* **planetree → 254**

6321 **plane tree family**
(Platanaceae)
d Platanengewächse *pl*
f platanacées *pl*
r платановые *pl*

6322 **plane[tree] grove**
d Platanenwald *m*
f platanaie *f*
r платановый лес *m*

* **planetree maple → 8355**

6323 **plane-wood; plane stock; plane sole**
d Plataneholz *n*
f rabote-bois *m*
r колодка *f* рубанка

6324 **planimeter**
(a mechanical instrument for measuring the area of a plane surface)
d Planimeter *n*
f planimètre *m*
r планиметр *m*

6325 **planing-dry wood**
d hobeltrockenes Holz *n*
f bois *m* sec à raboter; bois sec pour usinage
r искусственно высушенная древесина *f* для строгания

6326 **planing machine; planer**
(a machine to surface rough lumber)
d Tischhobelmaschine *f*; Längshobelmaschine *f*
f traîneau *m* niveleur; machine *f* à raboter; raboteuse *f*
r шлифовальный станок *m*; строгальный станок

6327 **planing mill stacker**
d Hobelwerkstapler *m*
f empileur *m* d'usine de rabotage
r укладчик *m* строганных досок

* **plank → 873**

* **plank chute → 884**

* **planking → 880**

* **plank log → 7245**

* **plank slide → 884**

6328 **plank timbers; score ply**
d Plankenholz *n*
f bois *m* de madriers
r древесина *f* для планок

6329 **planned cutting; planned felling**
d planmäßiger Einschlag *m*
f coupe *f* planifiée
r плановая рубка *f*; планомерная рубка; планируемая рубка

6330 **planned grazing system**
(system approved by the regional manager or district manager respecting the use of land for grazing and the dispersal of livestock over land)
d geplante Weidensystem *n*
f système *m* de pâturage planifié
r планированная система *f* пастьбы

* **planned ignition → 6516**

* **planned unit development → 1667**

6331 **plan of attack**
(of fire)
d Einsatzplan *m*
f plan *m* d'attaque
r план *m* тушения

6332 **plan of cuttings; logging plan; [forest] cutting plan; felling plan**
(map, along with a written plan, describing the road building, harvesting, and other related operations that are submitted for a forest officer's approval)
d Einschlagplan *m*; Hauungsplan *m*; Einschlagprogramm *n*
f règlement *m* d'exploitation
r лесосечный план *m*; план рубок

6333 **plant**
d Pflanze *f*
f plante *f*
r растение *n*

6334 **plant** *v*
d bepflanzen; pflanzen; anpflanzen
f planter; empoter
r засаживать; сажать

6335 **plantain banana**
(Musa paradisiaca)
d Banane *f*; Pisang *m*
f bananier *m*
r банан *m* десертный

6336 **plant association; plant community; phytoc[o]enosis; phytoc[o]enose**
(assemblage of plants occurring together at any point in time, thus designating no particular ecological status)
d Pflanzengemeinschaft *f*; Pflanzengesellschaft *f*; Phytozönose *f*
f association *f* végétale; communauté *f* végétale; phytocénose *f*
r растительное сообщество *n*; фитоценоз *m*

* **plant colonization → 2734**

6337 **plant cover; vegetation cover; vegetative cover; vegetable cover; ground cover; living soil cover**
(the herbaceous plants (including grasses and ferns) and the lowest shrubs occupying an area)
d Pflanzendecke *f*; Bodenbedeckung *f*; Geländebedeckung *f*
f couverture *f* végétale; tapis *m* végétal; couvert *m* végétal
r растительный покров *m*; [живой] [на]почвенный покров *m*

* **plant digging machine → 6367**

6338 **plant diversity**
(variety of food or cover for wildlife)
d Pflanzendiversität *f*
f diversité *f* de plantes
r растительное разнообразие *n*

6339 **planted forest; planted stand; stand of seedlings**
d Kulturbestand *m*; Pflanzbestand *m*
f forêt *f* plantée
r посаженный лес *m*; посадка *f*

* **plant fragments → 6371**

6340 **plant [growth] regulator; plant-growth regulator; PGR; growth regulator; plant regulating substance; plant/growth hormone**
(natural substance that regulates the enlargement, division or activation of plant cells)
d Pflanzenwachstumsregulator *m*
f régulateur *m* de croissance
r регулятор *m* роста

* **plant growth substance → 6221**

6341 **plant harvesting**
(collection of plant life including, but not limited to, bark, berries, boughs, branches, burls, cones, conks, ferns, flowers, grasses, herbs, fungi, lichens, mosses, mushrooms, roots, sedges, shrubs, sprays and twigs)
d Pflanzenernte *f*
f récolte *f* de plantes
r сбор *m* растительного урожая

* **plant hedge → 5019**

* **plant hormone → 6221**

* **planting area → 355**

6342 **planting auger; auger; planting borer; hollow-spade; semi-circular spade**
(instrument for drilling holes, used for taking soil samples)
d Pflanzenlochbohrer *m*; Pflanzenbohrer *m*; Hohlspaten *m*
f plantoir *m* à tarière; tarière *f*
r бурав *m* для посадочных ямок; сажальный бурав; полуцилиндрическая лопата *f*

6343 **planting bag**
d Pflanzentasche *f*
f poche *f* à plantes
r кузов *m* для растений

6344 **planting bar; planting stick; planting peg; dibble [bar]**
(a tool used to make holes in the ground for planting tree seedlings)
d Dibbelstock *m*; Pflock *m*; Pflanzdolch *m*; Pflanzeisen *n*; Pflanzstock *m*
f bêche *f* à planter; canne-semoir *f*; plantoir *m*
r сажальный кол *m*; колышек *m*; пикет *m*

6345 **planting board**
d Pflanzbrett *n*; Saatbrett *n*
f planche *f* à encoches
r посадочная доска *f*; сажальная доска

* **planting borer → 6342**

6346 **planting brick; planting block**
(block of loam, peat, plastic foam, etc. with a hole in the top for a seedling or transplant)
d Pflanzenbrikett *n*
f motte *f* compactée
r посадочный брикет *m*

6347 **planting by single trees**
d Einzelbaumpflanzung *f*
f plantation *f* par arbres isolés
r посадка *f* единичных деревьев

6348 **planting close to a stump**
d Abstutzung *f*; Stockachselpflanzung *f*
f plantation *f* entre empattements de racines
r посадка *f* на пазуху пня

6349 **planting crew**
d Pflanzenkolonne *f*
f équipe *f* de planteurs
r лесокультурная бригада *f*; бригада по посадке

* **plant[ing] density → 6369**

* **planting distance → 6362**

6350 **planting expenses; cultural expenses; cost of reproduction; cultural costs**
d Kulturkosten *pl*
f frais *mpl* de culture
r стоимость *f* посадки; стоимость культуры

6351 **planting gun**
(special devices of varying complexity which make holes by compression and either set or shoot a containerized seedling into the soil)
d Setzgerät *n*
f plantoir *m* à pistolet
r посадочное ружьё *n*

6352 **planting hoe**
d Pflanzenhacke *f*
f houe *f* de planteur
r мотыга *f* для посадки

6353 **planting hole; planting pit**
d Baumgrube *f*
f trou *m* d'ensemencement; trou de plantation

6354 **planting in clumps; block planting; group planting; bunch planting; multiple planting**
d Blockpflanzung *f*; Büschelpflanzung *f*; Gruppenpflanzung *f*
f plantation *f* par bouquets; plantation par groupes; plantation en blocs; plantation par touffes
r групповая посадка *f*; посадка пучками

6355 **planting in holes; pit planting; hole planting; pit sowing**
(setting out young trees in small depressions, natural or excavated, with a view to collecting and conserving moisture)
d Lochpflanzung *f*; Löcherpflanzung *f*
f plantation *f* sur buttes par trouées; plantation sur potets; plantation par trous; plantation en côté de potet
r посадка *f* в лунку; посадка в ямки; посев *m* в ямки

6356 **planting in lines; row planting; strip planting; line planting; drill planting**
(setting trees, generally in two or more parallel lines, in a long narrow area)
d Reihenpflanzung *f*; Reihenverband *m*
f plantation *f* en bandes; plantation en lignes
r рядовая посадка *f*; посадка рядами; линейная посадка

6357 **planting in step-like patches**
d Stufenpflanzung *f*; Treppenpflanzung *f*
f plantation *f* en banquettes
r ступенчатая посадка *f*

* **planting on ridges → 648**

6358 **planting on stump; stump planting; stumping**
d Stumpfpflanzung *f*
f plantage *m* des stumps
r посадка *f* на пень

* **planting-out → 5979**

* **planting peg → 6344**

* **planting period → 8665**

* **planting pit → 6353**

* **planting point → 7396**

6359 **planting pot**
d Pflanzentopf *m*

6360 **planting row**
d Pflanz[en]reihe *f*
f rang *m* de plantation
r ряд *m* лесных культур

* **planting season → 8665**

6361 **planting sleeve**
d Pflanzenröhre *f*
f tube *m* de plantation
r посадочная трубка *f*

6362 **planting space; planting distance; arrangement of plants**
d Pflanzenverband *m*; Pflanzenweite *f*
f écartement *m* des plantes
r размещение *n* саженцев

* **planting spot → 7396**

* **planting stick → 6344**

6363 **planting stock**
(part of a plant, usually consisting of the root system, together with a part of the stem, onto which is grafted a part of another plant)
d Pflanzgut *n*; Pflanzenmaterial *n*
f matériel *m* de reproduction; stock *m* de plantation
r лесокультурный посадочный материал *m*

6364 **planting under cover; underplanting; planting under shelterwood; underwood silviculture; raising an underwood; undersowing**
(planting young trees under the canopy of an existing stand)
d Pflanzung *f* unter Schirmbestand; Unterpflanzung *f*; Unterbau *m*
f plantation *f* en sous-étage; plantage *m* sous abri; sous-plantation *f*
r посадка *f* под пологом леса; посев *m* под пологом леса

6365 **planting wedge; notching spade; wedge-shaped spade**
d Keilspaten *m*; Klemmspaten *m*
f bêche *f* en coin pour plantage en fentes
r клиновидная лопата *f*; лопата для посадки в щель

* **planting with saplings → 7203**

6366 **plantlet**
d Pflänzchen *n*; Samenspross *m*

6367 **plant lifting machine; plant lifter; plant digging machine**
(machine that loosens and removes plants from the ground)
d Pflanzenausheber *m*
f arracheuse *f*
r выкопочная машина *f*; копач *m*; механическая лопата *f* для выкопки саженцев

* **plant manure → 3949**

6368 **plant percent[age]; seedling percent**
(percentage by number of seeds in a given sample that develop into seedlings at the end of a given period, generally the end of the first growing season)
d Keimlingsprozent *n*
f pourcentage *m* de réussite des semis
r процент *m* [успешной] посадки

6369 **plant population; plant[ing] density**
(the number of plants per unit area)
d Pflanzendichte *f*
f compacité *f* de plantation
r густота *f* посадки

6370 **plant production nursery**
d Pflanzschule *f*
f pépinière *f* de production
r производственный рассадник *m*

6371 **plant refuse; plant residues; vegetable refuse; plant remains; vegetable remains; plant fragments**
d Pflanzenüberreste *mpl*; Pflanzenrückstände *mpl*; Pflanzenabfälle *mpl*
f résidus *mpl* végétaux; déchets *mpl* de plantes
r растительные остатки *mpl*

* **plant regulating substance → 6340**

6372 **plant retrogression**
(process of deterioration in vegetation whereby an area becomes successively occupied by different plant communities of lower ecological order in succession)
d rückläufige Pflanzenbewegung *f*; Pflanzenrückentwicklung *f*; Pflanzenregression *f*
f rétrogression *f* des plantes; régression *f* des plantes
r ретрогрессия *f* растений

* **plant thinner → 8513**

6373 **plant tray; hurdle**

d Horde *f*
f caissette *f*
r этажерка *f*

* **plant trees → 93**

6374 **plant virus**
d Pflanzenvirus *n*
f virus *m* végétal
r растительный вирус *m*

6375 **plastic foiled chipboard; composite particle board**
d laminierte Spanplatte *f*
f panneau *m* de particules laminé
r ламинированная древесностружечная плита *f*

* **plastidule → 4421**

6376 **plateau oak**
(Quercus fusiformis)
d spindelförmige Eiche *f*
f chêne *m* à feuilles fusiformes
r дуб *m* веретенообразной; дуб веретеновидный

6377 **plate pine**
d Plattenkiefer *f*
f pin *m* à écorce en peau de serpent
r сосна *f* с широкопластинчатой корой

6378 **plate[-shaped] root; lamellar root**
d blättchenartige Wurzel *f*; lamellare Wurzel
f racine *f* lamellaire
r пластинчатый корень *m*; пластиновидный корень; листообразный корень

6379 **pleaching; intertwisting**
(a process whereby branches of woody plants are interwoven and plaited together to form an impassible hedge or very thick arbor)
d Verflechtung *f*
f entrelacement *m*
r переплетение *n* веток для изгороди

* **plenter system → 7620**

* **plenter wood → 7410**

* **plenter[wood] thinning → 7411**

* **plerotic water → 3995**

* **plethora → 853**

* **pliable wood → 678**

* **plot estimate method → 5865**

6380 **plot [identification] number**
d Flächennummer *f*; Traktnummer *f*
f numéro *m* de la placette
r номер *m* делянки

6381 **plot identification [number] of national forest inventory**
d Traktnummer *f* der nationale Waldinventur
f numéro *m* de placette de mesure de l'inventaire forestier national
r идентификация *m* делянок национальной лесной таксации

* **plotless cruising → 285**

* **plot sowing → 7870**

* **plough → 6383**

6382 **plough body; plow bottom; bucket**
d Pflugkörper *m*
f pièce o travaillante; corps *m* de la charrue
r кассета *f* для трелёвки; ковш *m*

6383 **plow** *v*; **plough** *v*
d pflügen
f labourer
r пахать; вспахивать

* **plow bottom → 6382**

* **PLS → 6636**

* **plug → 6103**

6384 **plug saw; dowel saw**
(a small cylinder saw for cutting out thin rods of wood, from whose cross-cutting short plugs and dowels can be made)
d Zylindersäge *f*
f scie *f* tubulaire à rondelles
r резец *m* для высверливания сучков; нож *m* для обрезки пробок

* **plug seedling → 492**

* **plug transplant → 492**

6385 **plumb** *v*
(to true up the vertical position of a member such as a wall, by the use of a plumb-line so

f aplomber
r определять наклон дерева

6386 **plumb end**
d Steinmeißel *m*
f poinçon *m*
r отвесная кромка *f* (стропил)

* **plume → 2509**

6387 **plume clematis; Virgin's bower**
(Clematis flammula)
d brennende Waldrebe *f*; scharfer Blazenzug *m*; scharfe Waldrebe
f clématite *f* flammette; flammule *f*; clématite brûlante; clématite odorante
r ломонос *m* жгучий

* **plume poppy → 8899**

6388 **plum [leaf] sawfly**
(Hoplocampa flava)
d [weißfüßige] Kirschblattwespe *f*; gelbbraune Pflaumensägewespe *f*
f hoplocampe *m* commun du prunier; ver cordonnier *m* des prunes
r пилильщик *m* вишневый бледноногий; пилильщик крыжовниковый бледноногий

6389 **plummet**
d Senkblei *n*; Lot *n*
f fil *m* à plomb
r лот *m* (при буксировке плотов)

6390 **plum pine; brown pine; she-oak male**
(Podocarpus elatus)
d australische bräunliche Podocarpus *f*
f podocarpe *m* brun
r подокарп *m* высокий

6391 **plum [tree]**
(Prunus domestica)
d [gemeiner] Pflaumenbaum *m*; Haus-Pflaume *f*; Zwetschenbaum *m*; Pflaume *f*
f prunier *m* [domestique]
r слива *f* [домашняя]

6392 **plum yew family**
(Cephalotaxeae)
d Kopfeibengewächse *pl*
f céphalotaxacées *pl*
r цефалотаксусовые *pl*

6393 **plunge cutting; boring**
(starting a cut in the center of a log using the
f sciage *m* en plongée; mortaisage *m*
r глубокое пиление *n*; высверливание *n*; бурение *n*

6394 **plus stand; seed-production stand; seed stand**
(any stand used as a source of seed)
d Auslesebestand *m*; Samenerntebestand *m*
f peuplement *m* plus; peuplement producteur de graines; peuplement porte-graines
r плюсовое насаждение *n*; отборное насаждение

6395 **plus stem; elite log; elite stem**
d Auslesestamm *m*; Wahlstamm *m*
f tige *f* plus; tige élite
r плюсовый ствол *m*; элитный ствол; отборный ствол

6396 **plus tree; specimen tree; [phenotypic] elite tree; superior tree; select[ed] tree; choice tree; chosen tree; champion tree**
(a tree or genotype selected on the basis of its outstanding single-copy performance, but not yet clonally tested or progeny tested)
d Plusbaum *m*; Musterbaum *m*; Auslesebaum *m*; Elitebaum *m*
f arbre *m* plus; arbre de spécimen; arbre modèle; arbre *m* sélectionné; arbre d'élite
r плюсовое дерево *n*; [фенотипическое] элитное дерево; эталонное дерево

6397 **ply**
(an individual layer in plywood)
d Lage *f*
f pli *m*
r слой *m*

* **plylog → 9193**

6398 **plywood; [wood] veneer**
(a panel made up of thin sheets of veneer which the grain direction is at right angles to each other)
d Blattholz *n*; Sperrholz *n*; Furnier *n*
f contreplaqué *m*; bois *m* de placage
r клеёная фанера *f*; слоистая древесина *f*

6399 **plywood bundling**
(an individual layer in plywood; usually a ply is a single veneer)
d Furnierbündeln *n*
f paquetage *m* de contreplaqués; groupage *m* de contreplaqués
r упаковка *f* фанеры в пачку

f liste *f* de contreplaqués
r спецификация *f* на фанеру

* **plywood log → 9193**

* **p.[m.]a.i. → 6164**

* **pneumatophore root → 86**

6401 **poacher**
d Wilddieb *m*; Wilderer *m*
f braconnier *m*
r браконьер *m*

6402 **poaching**
d Wilddieberei *f*; Wilderei *f*
f braconnage *m*
r браконьерство *n*

6403 **pochote; saqui-saqui**
(Bombacopsis quinata)
d Satinwollbaum *m*
f pochote *m*; saqui-saqui *m*
r бомбакопсис *m*

* **pocket → 182, 6104**

* **pocket barker → 479**

6404 **pocket chisel; sash chisel**
d Taschenmeißel *m*
f ciseau *m* de poche
r карманное долото *n*

* **pocket dry rot → 6313**

6405 **pockwood**
d Pockholz *n*; Guajakholz *n*
f bois *m* de gaïac
r гваяковая древесина *f*

6406 **pod mahogany; Rhodes mahogany; mahogany bean**
(Intsia; Afzelia quanzensis)
d Hülsenmahagoni *n*
f chamfuta *m*
r интсия *f*; афцелия *f* кванценская

* **podo → 2714**

6407 **podocarp family**
(Podocarpaceae)
d Steineibengewächse *pl*
f podocarpacées *pl*
r ногоплодниковые *pl*

6408 **podzolic soils**
d podsolische Boden *mpl*
f sols *mpl* podzoliques
r подзолистые почвы *fpl*; подзолы *mpl* (почвы тайги и смешанных лесов)

* **pohutukawa → 5728**

* **pointed → 7917**

6409 **pointed tip**
(of needles)
d auslaufende Spitze *f*
f extrémité *f* conique
r заострённый наконечник *m*

6410 **point-leaf manzanita; Mexican manzanita**
(Arctostaphylos pungens)
d mexikanische Bärentraube *f*
f busserole *f* mexicaine
r толокнянка *f* колючая

* **point sample cruise → 285**

* **point sampling → 285**

* **poison-dogwood → 8315**

* **poison elder[berry] → 8315**

6411 **poison girdling**
d Giftringelung *f*
f annélation *f* avec phytocide
r кольцевание *n* с внесением ядохимикатов

6412 **poison-ivy; poison-sumac[h]; poison-oak**
(Rhus toxicodendron; Rhus radicans)
d Giftsumach *m*
f arbre *m* à la puce; arbre à la gale; herbe *m* à la puce; sumac *m* grimpant; sumac empoisonné
r сумах *m* ядоносный; сумах укореняющийся; сумах канадский

* **poisonnut → 8191**

6413 **poisonous wood**
d Giftholz *n*
f bois *m* toxique
r ядовитая древесина *f*

* **poison-sumac[h] → 6412**

* **poison thinning → 1469**

6414 **poison tree; [East African] antiaris; ako**

f kirundu *m*; upas *m*
r анчара *f* африканская; анчар *m* африканский

6415 pokelogan; logan; set-back
(a bay or pocket into which logs may float during a drive)
d Logan *n*
f petite baie *f*
r временная запань *f*; карман *m*

* **poker-work → 9498**

6416 pokeweed family
(Phytolaccaceae)
d Kermesbeerengewächse *pl*
f phytolaccacées *pl*
r лаконосовые *pl*

6417 polar willow
(Salix polaris)
d Polarweide *f*
f saule *m* polaire
r ива *f* полярная

* **pole → 6475**

6418 pole; perch
(a still-young tree, from the time its lower branches begin to die, up to the time when the rate of height growth begins to slow down and crown expansion becomes marked)
d Holzstange *f*; Waldstange *f*; Rundstange *f*; Gerte *f*; Stange *f*
f perche *f*
r жердь *f*; шест *m*; средневозрастное дерево *n*

* **pole chute → 6425**

* **pole crop → 6428**

* **pole derrick → 3848**

6419 pole forest; poletimber stand; stand of poles; pole stand
(stand at least 10% stocked with live trees, of which half or more of total stocking is in poletimber and sawtimber trees, and with poletimber stocking exceeding that of sawtimber)
d Stangenwald *m*
f perchis *m*
r жердняк *m*

6420 pole pruner; pruning stick
(used to prune the upper parts of a tree or tall shrub)
d Aufastungsstange *f*
f élagueur *m* à long manche
r секатор *m* на штанге

* **pole-road → 1947**

6421 pole saw
(a saw attached to a long pole for pruning tree limbs without using a ladder)
d Stangensäge *f*
f scie *f* à long manche; scie à perches
r жердняковая пила *f*

6422 pole size
d Stangengröße *f*
f taille *f* de perche
r размер *m* жерди

6423 pole skidder
d Stangen-Schlepper *m*
f débusqueur *m* de perches
r трактор *m* для трелёвки коротья

6424 pole skids
d Stangenpolter *mpl*; Polsterhölzer *npl*; Unterlage *f*
f bachons *mpl* en perches
r подштабельные [деревянные] подкладки *fpl*; подштабельная опора *f*

6425 pole slide; pole chute
d Stangenholzriese *f*
f glissoire *f* à perches
r жердняковый лесоспуск *m*; трелёвочный волок *m* из поперечно проложенных слег

6426 pole stage
d Stangenstadium *n*
f stade *m* de perchis
r стадия *f* средневозрастного насаждения

* **pole stand → 6419**

6427 pole stock; shaft wood
d Deichselholz *n*; Deichselstange *f*
f bois *m* pour timons
r оглобельник *m*; оглобля *f*

* **pole tester → 4451**

* **pole tie → 4095**

* **poletimber stand → 6419**

* **pole wagon → 9256**

6428 **polewood; pole crop; poletimber; chockwood**
(trees that are used in making poles; typically telephone poles; trees 5 to 8.9 inches d.b.h. for softwoods and 5 to 10.9 inches for hardwoods)
d Pfostenholz *n*; Stangenholz *n*; Pfeilerholz *n*
f perchis *m*; bois *m* d'étai
r древесина *f* для столбов; столбовая древесина

* **poling → 7995**

* **pollard → 7489**

* **poll[ard] → 8702**

* **pollard[ed] tree → 7489**

6429 **pollarded wood**
d bekappenes Holz *n*
f bois *m* écimé
r древесина *f* из подстригаемых деревьев

* **pollarding → 2198**

6430 **pollard[ing] system; topping exploitation; branch coppice method**
d Kopfholzbetrieb *m*; Kopfholzwirtschaft *f*; Kopfholzzucht *f*; Schneitelbetrieb *m*
f système *m* de taillis sur têtards; exploitation *f* en têtards
r безвершинное хозяйство *n*; [подсочно-]кобловое хозяйство

6431 **pollard shoot**
d Kopfausschlag *m*; Kopfholz *n*; Kopfholzschoß *m*
f bois *m* de têtard; rejet *m* de têtard
r отпрыск *m* от безвершинника

* **pollard willow → 619**

6432 **pollen**
d Pollen *f*; Blütenstaub *m*
f pollen *m*
r пыльца *f*; цветень *f*

6433 **pollen flight**
d Pollenflug *m*
f dispersion *f* de pollens
r перенос *m* пыльцы; лёт *m* пыльцы

6434 **pollination**
r опыление *n*

* **polling → 7578**

6435 **polyareal plot sampling**
d Mehrflächenstichproben *n*
f estimation *f* par placettes d'échantillon multiples
r выборка *f* пробных площадей с множеством переменных размеров

6436 **polycarpic**
(producing flowers or spores more than once during a life cycle)
d mehrfrüchtig
f polycarpique
r поликарпический; многократно плодоносящий; многоплодный

6437 **polyclimax theory**
(the hypothesis that in any one environment there can be several climax communities is termed polyclimax theory)
d Polyklimaxtheorie *f*
f théorie *f* polyclimacique
r теория *f* поликлимакса (совокупности различных устойчивых формаций в условиях единого климата, но на разных почвах)

* **polyculture → 5491**

6438 **polygamodioecious**
(having bisexual and male flowers on some plants and bisexual and female flowers on others)
d polygamzweihäusig
f polygamodioïque
r полигамнодвудомный

6439 **polymorphism**
d Vielgestaltigkeit *f*
f polymorphisme *m*
r полиморфизм *m*

* **Polynesian ironwood → 407**

6440 **polyploid**
(having more than 2 sets of homologous chromosomes)
d Polyploid *n*
f polyploïde *m*
r полиплоид *m*

6441 **polypody**

f polypode *m*
r многоножка *f*; широколиственница *f*; полиподиум *m*

* **pome [fruit] → 4698**

6442 pomegranate family
(Punicaceae)
d Grenatapfelgewächse *pl*
f punicacées *pl*
r гранатовые *pl*

6443 pomegranate [tree]
(Punica granatum)
d Granatapfelbaum *m*; Granatapfel *m*
f grenadier *m* [d'Europe]; grenadier commun; punica *m* à fleurs
r гранатник *m*; гранатовое дерево *n*; гранат *m*

6444 pond [bald]cypress
(Taxodium ascendens)
d Pondzypresse *f*
f cyprès *m* pond; cyprès chauve pyramidal
r таксодиум *m* восходящий; таксодий *m* восходящий

6445 ponderosa pine; western yellow pine; Bull pine; bird's-eye pine
(Pinus ponderosa)
d [westliche] Gelbkiefer *f*; Schwerholzkiefer *f*; Goldkiefer *f*
f pin *m* à bois lourd; pin jaune; pin pondéreux; pin ponderosa
r сосна *f* жёлтая [западная]

6446 ponderosa-pine bark borer
(Acanthocinus spectabilis)
d Goldkiefer-Bockkäfer *m*
f perceur *m* de l'écorce de pin pondéreux
r усач *m*, повреждающий шишки сосны жёлтой

* **ponderosa-pine cone beetle → 5043**

6447 pond pine
(Pinus serotina)
d Pondkiefer *f*; Wiesenkiefer *f*
f pin *m* serotina
r сосна *f* поздняя

6448 pongame oiltree
(Derris indica)
d Derris *n*
f derris *m*
r дерис *m* индийский

* **Pontic oak → 363**

* **pontine oak → 363**

6449 poor growth; poor thrift
d schwacher Wuchs *m*; Kümmerwuchs *m*
f croissance *f* pauvre; croissance médiocre
r слабый рост *m*; плохой рост

* **poorly forested → 6450**

* **poorly stocked stand → 4446**

6450 poorly wooded; poorly forested; sparsely wooded; thinly stocked; poorly stocked; scantily stocked; lightly wooded
d schütterbewalded; licht bestockt; waldarm; waldleer
f pauvre en forêt; à peuplement clair
r бедный лесом; малолесный; покрытый лесом с низким запасом; плохо облесённый; редко облесённый; редко посаженный; редко стоящий; редкостойный

6451 poor [quality] tree
d Baum *m* mit schlechter Qualität; Baum mit niedriger Qualität
f arbre *m* de mauvaise qualité
r низкокачественное дерево *n*

* **poor thrift → 6449**

* **pop ash → 1342**

* **popcorn tree → 1481**

6452 poplar bark
d Pappelrinde *f*
f écorce *f* de peuplier
r тополевая кора *f*

* **poplar borer → 4804**

6453 poplar carpenterworm
(Acossus centerensis)
d Pappel-Zimmererwurm *m*
f charpentier *m* du peuplier
r тополевый древоточец *m*

6454 poplar forest; poplar stand; poplar grove
d Pappelbestand *m*; Pappelwald *m*
f peupleraie *f*
r тополевый лес *m*; тополевник *m*; тополевая заросль *f*

* **poplar leaved birch → 3923**

6455 **poplar nursery**
d Pappelgärtnerei *f*
f pépinière *f* de peupliers; populetum *m*
r тополевый питомник *m*

6456 **poplar plantation; [young] poplar crop; populiculture**
d Pappelkultur *f*
f culture *f* de peupliers
r тополевое насаждение *n*; тополевая культура *f*

* **poplar sawfly → 4090**

* **poplar stand → 6454**

6457 **poplar-willow forest**
d Pappel-Weidenwald *m*
f forêt *f* à peupliers et saules
r тополево-ивовый лес *m*

6458 **population density**
d Populationsdichte *f*
f densité *f* de peuplement
r плотность *f* популяции

6459 **population growth**
d Populationswachstum *m*
f croissance *f* de population
r рост *m* насаждения

6460 **population pressure**
(combined effects exerted by the individuals of a population upon the other organisms in a community and upon their physical environment)
d Populationsdruck *m*
f pression *f* de population
r нагрузка *f* популяций [на среду]

* **populiculture → 6456**

* **pop-up saw → 4312**

* **porcupine wood → 1703**

6461 **pore; pin hole; pit**
(the cross section of a vessel or of a vascular tracheid)
d Pore *f*; Tüpfel *m/n*
f pore *m*
r пора *f*; точечное отверстие *n*; неутолщённое место *n* (в оболочке клетки)

* **pored wood → 6467**

radia multiple
(a group of two more pores crowded together and flattened along the lines of contact so as to appear as subdivisions of a single pore)
d Gefäßbündel *n*
f groupe *m* de pores; file de pores; pores *mpl* accolés; chaînette *f* de pores
r группа *f* мелких сосудов; пучок *m* мелких сосудов; сосудистый пучок

6463 **pore volume; pore space; void**
(volume in the wood structure that is not occupied by wood tissue)
d Porenvolumen *n*; Hohlraumgehalt *m*; Hohlraumvolumen *n*; Porenraum *n*
f volume *m* des pores; espace *m* poreux
r объём *m* пор; пористый объём; порозное пространство *n*

6464 **pore water; interstitial water**
d Porenwasser *n*
f eau *f* poreuse; eau interstitielle
r поровая вода *f*; влага пористого пространства

6465 **porosity; porousness; poriness**
d Porigkeit *f*; Porosität *f*; Porenanteil *m*
f porosité *f*
r порозность *f*; пористость *f*; скважность *f*; ноздреватость *f*

6466 **porous fiberboard**
d poröse Faserplatte *f*
f panneau *m* de fibres [de bois] poreux
r пористая древесноволокнистая плита *f*

* **porousness → 6465**

6467 **porous wood; pored wood**
(wood with pores or vessels originating from a continuous cambium layer, featured by nearly all hardwood species)
d poriges Holz *n*; poröses Holz
f bois *m* avec vaisseaux
r пористая древесина *f*; сосудистая древесина; губчатая древесина

6468 **porrect branch**
(extended horizontally branch; stretched out branch)
d verlängerte Ast *f*
f branche *f* prolongée
r ветвь *f*, отходящая от главной оси; ветвь, направленная кнаружи от главной оси

6469 **portable sawmill; mobile sawmill**

f scierie *f* mobile; scierie transportable
r переносимая лесопильная рама *f*; передвижная лесопилка *f*

6470 **portable slasher; self-propelled slasher; roadside slasher**
d tragbares Buschhacker *m*
f tronçonneuse *f* mobile
r переносимый кусторез *m*

* **portable wood chipper → 5499**

6471 **portia tree**
(Thespesia populnea)
d Portiabaum *m*
f miro *m*; bois *m* de rose d'Océanie
r теспезия *f* обыкновенная

* **portion of timber → 5181**

* **Port-Orford cedar → 4839**

* **position of the crown → 2064**

6472 **positive selection**
d Begünstigung *f*
f sélection *f* positive
r позитивная селекция *f*

6473 **possibilities of wood use**
d Verwendungsmöglichkeiten *fpl* des Holzes
f possibilités *fpl* d'utilisation du bois
r возможности *fpl* употребления древесины

* **possibility by area → 9770**

6474 **possumhaw [viburnum]; smooth witherod**
(Viburnum nudum)
d Schneeball *m*
f alisier *m*
r калина *f* голоцветковая

* **possum oak → 9294**

* **possumwood → 7185**

6475 **post; pole; pale; stake; jamb; upright**
(a short timber up to 16 feet in length used in an upright position to support other structures for fencing)
d Holzpfahl *m*; Holzpilote *f*; Pfosten *m*
f pieu *m*; piquet *m* en bois; poteau *m*
r деревянная свая *f*; деревянный столб *m*

6476 **post-catastrophic secondary forest; post-abandonment secondary forest**
d Postkataklysmen-Sekundärwald *m*

6477 **post-extraction secondary forest**
d Posthieb-Sekundärwald *m*
f forêt *f* secondaire après abattage
r вторичный лес *m* след валки

6478 **postfire forest site**
d Postbrand-Waldstandort *m*
f site *m* forestier après d'un incendie
r лесной участок *m*, пройдённый пожаром

6479 **postfire rehabilitation**
d Postbrandneugestaltung *f*
f restauration *f* de végétation avant d'un incendie
r восстановление *n* растительности на гарях

6480 **post market**
(market of trees to be used as fence posts)
d Pfostenmarkt *m*
f marche *f* de piquets
r рынок *m* столбов

* **post oak → 255**

* **pot culture house → 3946**

6481 **potential wood yield**
(maximum yield of timber that can be obtained annually on a sustained yield basis by utilizing intensive forestry management practices)
d Holzertragvermögen *n*
f capacité *f* de production de bois
r потенциальная производительность *f* леса

6482 **pot planting**
(setting out young trees in pot-shaped receptacles having a closed or only perforated end)
d Topfpflanzung *f*
f plantation *f* en pots
r посадка *f* в горшочках; горшечная посадка

6483 **potted tree**
d Sämling *m* im Pflanztopf
f semis *m* en pot
r сеянец *m* в горшочке

* **powdered rot → 7819**

6484 **powdered wood**
d Moderholz *n*
f bois *m* en poudre
r трухлявая древесина *f*

f artificier *m*
r подрывник *m* (на корчёвке пней)

* **powderpost beetle → 5220**

* **powderpost beetle damage → 5221**

6486 **powderpost beetles**
(Bostrychidae)
d Kapuzenkäfer *mpl*
f lyctes *mpl* striés
r капюшонники *mpl*

* **powderpost borer → 5220**

6487 **powder wedge; explosive wedge**
(a device consisting of a wedge made up of two longitudinal sections that enclose a hollow chamber)
d Explosivtkeil *m*
f coin *m* explosif; coin de refend; pistolet *m* à refendre
r взрывной клин *m*; клин для раскалывания

* **power bow-saw → 952**

6488 **power [chain]saw**
(saw that is powered by a gasoline, hydraulic, or electric motor)
d Motorkettensäge *f*
f tronçonneuse *f* à chaîne mécanisée
r моторная цепная пила *f*

6489 **power[ed] logging; power skidding; power hauling; mechanized logging [operation]**
(logging in which most or all of the hand labor is replaced by machines; requires a large outlay of capital)
d Maschinenträcken *n*; motorisierte Bringung *f*; Rücken *n* mit Kraftantrieb
f débardage *m* mécanisé; débardage par traction mécanique
r механизированная канатная трелёвка *f*; механизированные лесозаготовки *fpl*

6490 **-3/2-power law of self-thinning**
(dense populations that have reached a size at which mortality occurs demonstrate a negative relationship between log mean plant weight and log stand density; this generally has a slope of -3/2)
d -3/2-Potenzgesetz *n* der Naturausscheidung
f relation *f* à la puissance -3/2 de l'éclaircie naturelle

* **power lift truck → 3660**

6491 **power[-]saw; motor[ized] saw**
d Motorsäge *f*
f scie *f* mécanisée
r моторная пила *f*

* **powersaw felling → 5572**

* **powersaw operator → 1434**

* **power skidding → 6489**

* **prairie forest → 3607**

6492 **prairie-forest soil**
d bewaldeter Prärieboden *m*
f sol *m* de prairie boisée
r лесная почва *f* прерий

* **prairie mesquite → 4354**

6493 **prairie-timber zone**
d Waldpräriezone *f*
f zone *f* de prairie arborée; zone de prairie boisée
r лесопрерийная зона *f*

6494 **prebunched wood**
d vorgebündeltes Holz *n*
f bois *m* prégroupé
r предварительное окученные деревья *npl*; предварительное уложенные в пачки деревья

* **prebuncher → 3053**

6495 **prebunching; prepiling**
(of timber)
d Vorstapelung *f*
f prébuttage *m*
r предварительное окучивание *n*; предварительное пакетирование *n*

6496 **prechilling**
(of seeds)
d Vorkühlung *f*
f réfrigération *f* préparatoire; stratification *f* froide
r предварительное охлаждение *n*

6497 **prechoke**
d Vorchoke *n*
f étranglement *m* préalable
r предварительная чокеровка *f*

6498 precipitation; rainfall
(discharge of water, in liquid or solid state, out of the atmosphere, generally upon a land or water surface)
d Niederschlag *m*
f précipitation *f*
r осадок *m*

* **precommercial thinning → 4680**

6499 precommercial treatments; precommercial operations
(forestry operations that require landowner investment, such as cleaning or weeding stands to remove trees that have little or no cash value)
d Vorhandelsbehandlungen *fpl*
f traitements *mpl* précommerciaux
r предпромышленные операции *fpl*

* **pre-delimbing → 76**

* **predominant mean height → 5358**

6500 predominant tree
d vorherrschender Baum *m*
f arbre *m* prédominant
r прегосподствующее дерево *n*; наивысшее дерево

6501 preferable species
(plant species that are preferred by animals and are grazed during first choice)
d vorzugsweise Art *f*
f espèce *f* préférable
r предпочтительный вид *m*

6502 prefoliation; vernation; aestivation; leafing
(the arrangement of the leaves within the leaf bud, as regards their folding, coiling, rolling, etc.)
d Knospenanlage *f*
f vernation *f*
r листорасположение *n* в почке; листосложение *n*; листовое почкосложение *n*

6503 preformed rope; rope without twist
d Vorformseil *n*; vorgeformtes drallarmes Drahtseil *n*
f câble *m* préformé
r нераскручивающийся канат *m*

6504 pregermination
(germination of seed, generally to the stage when the radicle is just emerging, before
f prégermination *f*; germination *f* physiologique
r предвсходовое состояние *n*

* **pregwood → 2390**

* **pre-harvesting → 6513**

6505 pre-harvest silviculture assessment; pre-harvest silviculture survey
(survey carried out on a stand prior to logging to collect specific information on the silvicultural conditions)
d vor Holzeinschlag waldbauliche Vermessung *f*
f estimation *f* sylviculturale avant la récolte
r лесохозяйственная съёмка *f* до рубки

6506 pre-harvest silviculture prescription; PHSP
(document that applies site-specific field data and develops forest management prescriptions for areas in advance of logging)
d vor Holzeinschlag waldbauliche Vorschrift *f*
f prescription *f* sylviculturale avant la récolte
r лесохозяйственное предписание *n* до рубки

* **pre-harvest silviculture survey → 6505**

6507 prehaul *v*
d vorschleppen; vorziehen
f prédébardage *m*
r трелевать лесоматериалы в полностью погруженном положении

6508 pre-holing; pre-pitting
d vorbereitende Pflanzenlochherstellung *f*
f préparation *f* des trous
r приготовление *n* посадочных ямок

6509 prejuvenility
d Vorjuvenilität *f*
f prématurité *f*
r предварительная недоразвитость *f*

6510 preliminary cultivation; advance cultivation
d Vorbau *m*
f cultivation *f* préliminaire
r подготовительная культивация *f*

* **preliminary culture → 71**

* **preliminary planting → 73**

* **preliminary pruning → 76**

6511 preload *v*

d vorspannen
f précharger
r предварительно нагружать

* **prelogging → 6513**

* **premature stand → 4425**

6512 **premium for grubbing trees**
d Rodungsprämie *f*
f prime d'arrachage; prime pour l'arrachage
r награда *f* для корчёвки деревьев

6513 **preparation cut; preparatory cut; preparatory felling; advance felling; pre-harvesting; prelogging; first cutting; first felling; preparatory cutting**
(cutting specified high-value wood products such as poles and pilings before cutting the remainder of the trees; removal of trees near the end of a rotation to open the canopy and enlarge the crowns of seed bearers)
d Vorhieb *m*; Vorbereitungshieb *m*; Vorbereitungsschlag *m*
f coupe *f* préparatoire; coupe anticipée; coupe initiale; pré-exploitation *f*
r подготовительная рубка *f*; предварительная валка *f*; зарубка *f*; вырубка *f* перед главной рубкой; наступательная рубка; начальная рубка

6514 **preparatory period; transition period**
(in cutting)
d Übergangszeitraum *m*
f période *f* de préparation; temps *m* de passage
r подготовительный период *m*; переходный период (при постепенных рубках)

* **preparatory pruning → 76**

6515 **prepared timber**
d vorbereitetes Holz *n*
f bois *m* prêt
r готовый пиломатериал *m*

* **prepiling → 6495**

* **pre-pitting → 6508**

* **preregeneration → 73**

6516 **prescribed burning; prescribed fire; controlled burning; prescription burning; planned ignition; deliberate fire; managed fire**
(wildland fire burning under specified
d kontrolliertes Abbrennen *n*; angeordnetes Brennen *n*; Brennen auf Anordnung
f brûlage *m* dirigé; brûlage contrôlé; brûlage sur ordre; brûlage sur prescription
r управляемый пал *m*; сжигание *n* по предписанию; запланированное сжигание; планированное зажигание *n*

* **prescribed cut → 195**

6517 **prescription**
(in forestry, a science-based plan for managing)
d Verordnung *f*; Anordnung *f*; Vorschrift *f*; Empfehlung *f*
f prescription *f*; recommandation *f*
r предписание *f*; рекомендация *f*

* **prescription burning → 6516**

* **present growing stock → 3559**

* **preservation → 2842**

* **preservation against blue stain → 6594**

* **preservative → 4428**

* **preserved forest → 3535**

* **preserved tree → 7892**

6518 **preset choker**
d voreinstellenes Chokerseil *n*; vorausbestimmendes Chokerseil
f collier *m* étrangleur prépositionné; collier à boucle prépositionné
r чокер *m*, заранее надетый на дерево

6519 **preskidding**
d Vorrücken *n*
f pré-débardage *m*
r окучивание *n* древесины для последующей трелёвки

6520 **pressing a tree in felling**
d Baumpressen *n* unter Fällung
f pressage *m* d'arbre au abattage
r толкание *n* дерева при валке

* **Pressler's borer → 4451**

6521 **pressure ga[u]ge; manometer**
(in kiln drying, an instrument for measuring and indicating pressure, e.g., steam pressure)
d Manometer *n*

6522 pressure on edges
d Kantendruck *m*
f pression *f* sur bouts; pression sur rives
r давление *n* на кромки

6523 pressure treatment; pressure process
d Druckverfahren *n*
f procédé *m* sous pression; procédé par pression
r обработка *f* давлением

* **pressure wood → 1850**

* **pre[-]thinning → 2599**

6524 prethinning timber stand
d undurchforsteter Bestand *m*
f peuplement *m* avant éclaircie
r насаждение *n*, не пройдённое рубками ухода

6525 pretonging
d Zangenvornehmen *n*
f élinguage *m* préliminaire
r предварительная зацепка *f* брёвен клещевым захватом

6526 prevailing age; dominant age
(of a trees in a forest)
d herrschendes Alter *n*
f âge *m* dominant
r господствующий возраст *m*

6527 previous stand; advance crop; original growth; original stand
d Vorbestand *m*
f peuplement *m* préexistant
r прежнее насаждение *n*; предыдущее насаждение

6528 price of timber party
d Holzgruppenpreis *f*
f prix *m* de bois en parties
r цена *f* лесоматериалов в партии

* **pricked-out plant → 5815**

6529 pricking-out
(transplanting seedlings that are too small to be handled by conventional methods, individually into boxes, flats, containers, etc., or into nursery beds)
d Verschulen *n*; Verschulung *f*
f repiquage *m*
r перешколивание *n*

* **prickled → 9730**

juniper
(Juniperus oxycedrus)
d Zederwacholder *m*; rotbeeriger Wacholder *m*; griechische Zeder *f*; spanische Zeder
f genévrier *m* oxycèdre; oxycèdre *m*; genévrier piquant; cade *m*
r дерево *n* карандашное; можжевельник *m* испанский; можжевельник красноватый; можжевельник красный

6531 prickly lettuce; compass lettuce
(Lactuca serriola; Lactuca scariola)
d Leberdistel *m*; Skariol; Stachellatich *m*; Stachelsalad; wilder Lattich *m*; Zaunlattich *m*
f escarole *m*; laitue *f* sauvage; escariole *f*
r латук *m* компасный; латук лесной; молокан *m* компасный; салат *m* дикий

6532 prickly parsnip
(Echinophora)
d Stacheldolde *f*
f échinophore *m*
r колючезонтичник *m*; колюченосник *m*

6533 prickly pear; cactus pear; Indian fig; tuna
(Opuntia ficus-indica)
d Kaktusfeige *f*; Stachelfeige *f*
f figue *f* de Barbarie; figue de cactus; figue d'Inde
r опунция *f* индийская; фиговое дерево *n* индийское

6534 prickly phlox; granite prickly phlox; granite gilia; rock gilia
(Leptodactylon pungens; Gilia pungens)
d Stechgilie *f*
f gilia *m* épineux
r гилия *f* колючая

* **prickly pine → 8361**

* **prickly tree clubmoss → 5794**

* **pride of India → 6110**

* **primal forest → 282**

6535 primary bark
d primäre Rinde *f*
f écorce *f* primaire
r первичная кора *f*

6536 primary branch; leader branch; leader; main branch; framework branch; scaffold branch

f branche *f* charpentière; charpentière *f*; branche de charpente; branche primaire; branche mère; mère *f*
r главная ветвь *f*; скелетная ветвь

6537 **primary cell wall**
d primäre Zellwand *f*
f paroi *m* primaire
r первичная клеточная стенка *f*

6538 **primary conversion; rough conversion**
(initial steps in the process of converting a felled or fallen tree to lumber)
d primäre Konversion *f*
f conversion *f* primaire
r первичная обработка *f*

* **primary decay → 4494**

* **primary forest → 282**

* **primary growth → 282**

6539 **primary insect**
d primäres Insekt *n*
f insecte *m* primaire
r первичное насекомое *n*

* **primary landing → 1721**

6540 **primary land unit**
d primäre Landeinheit *f*
f unité *f* de terrain primaire
r земельный выдел *m*

* **primary leaf → 1983**

6541 **primary logging road**
(road designed and maintained for a high level of use)
d primäre Rückelinie *f*
f chemin *m* de vidange primaire
r первичная гать *f*

6542 **primary needles**
d primäre Benadelung *f*
f aiguilles *fpl* primaires
r первичная хвоя *f*

6543 **primary phloem**
(first formed phloem; in stems and roots it is differentiated below the apical meristem before a definite cambium can be recognized)
d primäres Phloem *n*
f phloème *m* primaire
r первичная флоэма *f*

primary cell walls within the limits of which one or more pit-pairs usually develop)
d primordialer Tüpfel *m*; primäres Tüpfelfeld *n*
f champ *m* de ponctuation primaire
r зачаточная пора *f*; первичное поровое поле *n*

6545 **primary ray**
(in wood anatomy, a ray originating in the primary tissues and extended by cambial growth)
d primärer Strahl *m*
f rayon *m* primaire
r первичный луч *m*

6546 **primary root; radicle [root]; seminal root**
(a small structure resembling a rootlet)
d primäre Wurzel *f*; Radicule *f*
f radicule *f*; racine *f* radiculaire
r зародышевый корень *m*; первичный корень; зачаток *m* корешка; росток *m* корешка

6547 **primary succession**
(the invasion and occupation by a pioneer community of a hitherto bare area of the earth's surface)
d primäre Sukzession *f*
f succession *f* primaire
r первичная сукцессия *f*

* **primary timber → 4284**

6548 **primary transportation**
(movement of a felled tree from the stump to a landing)
d primäre Beförderung *f*; primärer Transport *m*
f transportation *f* primaire
r первичный транспорт *m*

6549 **primary wood; primary xylem**
(cluster of sieve conducting tissues formed by a primary apical meristem and surrounding the medulla)
d primäres Holz *n*; Primärholz *n*; primäres Xylem *n*
f bois *m* primaire; xylème *m* primaire; xylène *m*
r первичная древесина *f*; первичная ксилема *f*

6550 **primary wood-using plants**
(industries receiving roundwood or chips from roundwood for the manufacture of products such as veneer, pulp, and lumber)
d Primärholzverarbeitungsanlagen *fpl*
f fabriques *fpl* utilisantes un bois primaire

* **primary xylem → 6549**

6551 **primavera; white mahogany; duranga**
(Cybistax donnell-smithii; Tabebuia donnell-smithii)
d Cybistax *n*
f primavera *m*
r табебуйя *f* Донэль-Смита

6552 **prime log**
(log that is a given size and free from defects)
d erstklassiger Stammstück *m*; erster Baumstamm *m*
f grume *f* précieuse
r высококачественное бревно *n*; бревно без дефектов

6553 **prime mover**
(heavy equipment used to tow other machines such as disc trenchers for site preparation)
d Primärmaschine *f*; Primärantrieb *m*
f moteur *m* primaire
r машина *f* для первичного транспорта леса

* **primeval forest → 282**

* **primitive area → 9404**

* **primordial forest → 282**

* **primordial pit → 6544**

* **Prince's pine → 1798**

* **princess tree → 7122**

* **principal crop → 5250**

* **principal felling → 3136**

* **principal ride → 5249**

* **principal tree species → 5252**

* **principal yield → 3145**

6554 **prism angle ga[u]ge; cruising prism; wedge prism**
(optical instrument, consisting of a thin wedge of glass which establishes a fixed (critical) angle of projection in a point sample)
d Prismenwinkelmesser *m*; Schieblehre *f* mit Prismenbacken
f prisme *m*; jauge *m* angulaire
r [клиновидная] таксационная призма *f*; таксационный прицел *m*

* **private forest → 3473**

6555 **private timber holder; private forest owner**
d Privatwaldbesitzer *m*
f propriétaire *m* forestier privé
r частный лесной владелец *m*

6556 **problem forest type**
(non-merchantable forest types, including: stands of unfavo[u]rable stocking)
d problematischer Waldtyp *m*
f type *m* forestier problématique
r проблемный тип *m* леса

6557 **processionary caterpillar**
(the larva of a processionary moth)
d Prozessionsspinner *m*
f chenille *f* processionnaire; chattepelouse *f*
r походный шелкопряд *m*

6558 **procumbent habit**
d liegende Wuchsform *f*
f port *m* procombant; port *m* procumbant
r стелющийся габитус *m*

6559 **procumbent ray cell**
(a narrow ray cell with its longest axis radial)
d liegende Markstrahlzelle *f*
f cellule *f* dressée de rayon; cellule couchée de rayon
r лежачая лучевая клетка *f*

6560 **procurement forester**
(forester responsible for purchasing timber or logs for their company or business)
d Beschaffungsforstbeamter *m*
f forestier *m* responsable de l'approvisionnement
r лесничий *m*, ответственный за материально-техническое снабжение

6561 **production [timber] forest; productive [timber] forest; production stand**
(forest treated repeatedly and sometimes in order to achieve multipurpose goals)
d Erwerbswald *m*
f forêt *f* productive de bois; forêt de production
r продуктивный лес *m*

6562 **productive forestland; capable forestland**
(forest land that is capable of producing a merchantable stand within a defined period of time)
d ertragfähige Waldfläche *f*; ertragfähiger Waldboden *m*
f terre *f* forestière productive; chantier *m*

6563 **productive-reserved forestland; capable-reserved forestland**
(forest land sufficiently productive to qualify as timberland but withdrawn from timber utilization through statute or administrative regulation)
d ertragfähiger reservierter Waldboden *m*
f chantier *m* forestier productif réserve
r зарезервированная продуктивная лесная земля *f*

* **productive [timber] forest → 6561**

* **productivity class → 9774**

6564 **professional [consulting] forester; registered professional forester; RPF**
(person who performs or directs works, services, or undertakings that require specialized knowledge, training, and experience in forestry)
d berufsständischer namentliche Waldfacharbeiter *m*
f technicien *m* forestier professionnel nominatif
r зарегистрированный профессиональный лесовод *m*

6565 **professional forestry training; education in forestry; teaching in forestry; instruction in forestry; forest[al] education**
d forstliche Berufsausbildung *f*
f enseignement *m* [professionnel] forestier; apprentissage *m* forestier
r лесное [профессиональное] образование *n*; лесное обучение *n*

6566 **profiled timber; mo[u]lded timber**
d Profilholz *n*
f bois *m* profilé; bois mouluré; parement *m* profilé
r профильный пиломатериал *m*

6567 **profile sander; [mold and] edge sander; contour sander**
d Profilschleifmaschine *f*
f ponceuse *f* de chant; ponceuse de contour; ponceuse de profil
r шлифовальный станок для обработки профильных поверхностей и кромок

6568 **profitability of the forest**
d Waldrentabilität *f*
f rentabilité *f* de la forêt
r рентабельность *f* лесного хозяйства

6569 **profit in forestry; forest revenue; forest**
f revenu *m* forestier
r лесная прибыль *f*

6570 **progeny**
(the offspring of a particular plant or particular combination of one female and one male tree plant or animal)
d Nachkommenschaft *f*
f descendance *f*
r потомство *n* (растений)

6571 **progeny test; progeny trial**
(test in which the genetic constitution of an individual is evaluated from the performance of its progeny produced by some specific mating system)
d Nachkommenschaftsprüfung *f*
f test *m* de descendance
r определение *n* посевных качеств семян; проверка *f* потомства

* **progression of cutting → 2202**

* **progressive burning → 8308**

6572 **progressive clear-strip system**
(shelterwood system with clearcutting in strips that are generally not wider than the height of the adjoining trees and are generally laid out against the prevailing wind)
d Einsaumbetrieb *m*
f système *m* de régénération par coupes rases en bandes contiguës; système de traitement en lisière unique
r каёмчатое хозяйство *n*

6573 **progressive cutting; progressive felling**
d fortschreitender Hieb *m*; vorrückender Hieb
f coupe *f* de proche en proche
r поступательная рубка *f*

6574 **progressive edge**
d fortschreitender Saum *m*
f lisière *f* progressive
r прогрессирующая кайма *f*

* **progressive felling → 6573**

6575 **progress of felling**
d Hiebsfortschritt *m*
f avancement *m* des coupes
r ход *m* рубки

6576 **prohibition of clear cuttings**
d Kahlschlagverbot *n*; Kahlhiebverbot *n*

6577 prohibition of cutting; prohibition of logging; forbiddance of cutting
d Schlagverbot *n*; Hauungsverbot *n*
f interdiction *f* d'abattage
r запрет *m* вырубки

6578 project area
d Fläche *f* des Projektes
f surface *f* de projet; surface d'ouvrage
r район *m* работ

6579 prolepsis
(development of a lateral branch only after a period of dormancy as a lateral bud)
d Prolepsis *n*
f pousse *f* d'août
r пролепсис *m*

* **proleptic shoot → 4765**

6580 proliferation of buds; prolification of buds
d Knospenproliferation *f*; Knospensucht *f*; Knospung *f*; Knospenprolifikation *f*
f prolifération *f* des bourgeons
r пролиферация *f* почек

* **prop → 7270**

6581 propagation by cuttings
d Stecklingsvermehrung *f*
f reproduction *f* par boutures; bouturage *m*
r размножение *n* черенками

6582 propagation by layering; layering; marcotting
(rooting of an undetached branch, lying on or partially buried in the soil, or surrounded by moist fiber sealed in a plastic wrap)
d Vermehrung *f* durch Absenker; Senkervermehrung *f*; Ablegen *n*
f marcottage *m*; reproduction *f* par marcottes; multiplication *f* par marcottes
r размножение *n* отводками

* **propagule → 9183**

6583 proper-length [wood] material
d Eigenlängs[holz]werk *n*
f bois *m* de longueur propre
r [лесо]материал *m* заданной длины

6584 properly seasoned
d verschiffungstrocken
f correctement séché
r хорошо высушенный

f bractéole *f*
r первый лист *m*; прицветничек *m*

* **prop[ ]root → 277**

* **props → 6311**

6586 prosenchyma
(a general term for elongated cells with tapering ends)
d Prosenchym *n*
f prosenchyme *m*; sclérenchyme *m*; tissu *m* fibreux
r прозенхима *f*

* **prospective value of a forest → 3412**

6587 prospectus
(document that describes the location of a property, indicates trees marked for cutting, and states that the timber will be sold in accordance with a suitable contact)
d Prospekt *m*
f prospectus *m*; note *f* d'information
r проспект *m*; публикация *f*

* **prostrate → 2313**

6588 prostrate habit
d kriechendes Habit *n*; kriechende Wuchsform *f*
f port *m* rampant
r распростёртый габитус *m*

6589 prostrate tree form
d kriechende Baumform *f*
f forme *f* d'arbre rampante
r стелющаяся форма *f* дерева

6590 protea
(Protea)
d Schimmerbaum *m*
f protée *f*
r протея *f*

6591 protea family
(Proteaceae)
d Proteusgewächse *pl*
f protéacées *pl*
r протейные *pl*

6592 protected forest; protection forest; protective forest
(for maintaining the high biodiversity values and comply to the following general

f forêt *f* protégée; forêt de protection; forêt mise en défends; forêt mise à ban
r охраняемый лес *m*; защитный лес; заповедный лес

6593 protected wooded area
(such as provincial park, federal park, wilderness area, ecological reserve, and recreation area that has protected designations according to federal and provincial statutes)
d geschützte Forstfläche *f*
f espace *m* boisé protégé
r охраняемая залесённая местность *f*

6594 protection against blue stain; preservation against blue stain
d Bläueschutz *m*
f protection *f* contre le bleu
r предохранение *n* от синевы

* **protection forest → 6592**

6595 protection forest planting
d Schutzwaldpflanzung *f*
f plantation *f* de forêt
r посадка *f* защитных насаждений

* **protection goggles → 7150**

6596 protection of the edges of stands; border protection
d Randschutz *m*
f protection *f* de la lisière
r защита *f* опушки

6597 protective capacity of the forest
(recognized function of the forest; to protect the people, wildlife and ecosystem components from damage)
d Waldschutzkapazität *f*
f pouvoir *m* isolant de la forêt
r защитная способность *f* леса

6598 protective element
(support element providing protection to forest resources and uses)
d Schutzelement *n*
f élément *m* protecteur
r защитный элемент *m*

* **protective forest → 6592**

* **protective forest belt → 7508**

6599 protective glove; safety glove
d Handschuhe *f*

* **protective goggles → 7150**

6600 protective plants
(plants grown to protect crops, soils or land from adverse environmental factors)
d Schutzpflanzen *fpl*
f plantes *fpl* de protection
r защитные растения *f*

6601 protoplast
(the mass of protoplasm enclosed by a cell wall)
d Protoplast *m*
f protoplaste *m*
r образец *m*; оригинал *m*; первообраз *m*

6602 protoxylem
(first-formed primary xylem, with tracheary elements characterized by annular or spiral thickenings)
d Protoxylem *n*
f protoxylème *m*
r протоксилема *f*

6603 provenance; provenience
(territory of practically identical ecological conditions on which are scattered stands representing phenotypical characteristics that are usually analogous)
d Herkunft *f*; Provenienz *f*
f provenance *f*; origine *f*
r источник *m*

6604 provenance test; provenance trial; provenance check
(experiment, usually replicated, comparing trees grown from seed or cuttings collected from many parts of a species, natural range)
d Herkunftskontrolle *f*
f test *m* de provenance
r контроль *m* происхождения

* **provenience → 6603**

6605 proventitious bud
d Proventivknospe *f*
f bourgeon *m* proventif
r провентивная почка *f*

6606 provincial forest
d Provinzialwald *m*
f forêt *f* provinciale
r периферийный лес *m*; местный лес

6607 provincial forest inventory
(description of the quantity and quality of forest trees, non-wood values, etc. compiled

d Provinzialwaldinventur *f*
f inventaire *m* de forêt provinciale
r таксация *f* государственного леса

* **prune → 1645**

6608 **pruned timber**
d beschnittetes Holz *n*
f bois *m* élagué
r обессученный лесоматериал *m*

6609 **prune *v* live branches**
d wertästen; grünästen; beschnitten
f élaguer de branches en vert
r подрезать живых ветвей

* **pruner → 5168**

6610 **pruning; abnodation; branch cutting; clearance of branches; debranching; disbranching; [de]limbing; knotting; lopping; shragging; snedding; trimming[-out]; rounding out**
(removal of limbs from trees; cutting one or more branches)
d Astreinigung *f*; Ästung *f*; Astung *f*; Aufästung *f*; Entasten *n*; Aufhauen *n*; Abästung *f*; Ausästen *n*; Baumschnitt *m*
f ébranchage *m*; élagage *m*
r очистка *f* [деревьев] от сучьев; очищение от сучьев; подрезка *f* [сучьев]; обрезка *f* сучьев

* **pruning axe → 5167**

6611 **pruning chisel; pruning iron; pushing iron; pushing chisel**
d Stoßeisen *n*
f fer *m* à ébranchage
r железный долбежный шест *m*

6612 **pruning for shape; shape pruning; topiary**
(the practice of pruning full size trees or shrubs into ornamental shapes)
d Erziehungsschnitt *m*; Formschnitt *m*
f topiaire *f*; taille *f* de formation
r фигурная стрижка *f* [кустов и деревьев]; формовочная стрижка крон древесных растений; обрезка *f* с целью формирования кроны

6613 **pruning hook**
d Ästungshaken *m*
f crochet *m* d'ébranchage
r сучковый крюк *m*; кривой нож *m*

* **pruning knife → 2285**

* **pruning of wolves → 3071**

6614 **pruning saw; grafting saw; limbing saw; lopping saw**
(saw specially designed to prune standing trees)
d Ast[ungs]säge *f*; Entastungssäge *f*; Baumssäge *f*
f scie *f* à élaguer; scie à ébrancher; scie à ébranchage
r садовая пила *f*; пила для обрезки ветвей и сучьей; сучкорезка *f*

* **pruning[-]scissors → 5168**

* **pruning shears → 5168**

* **pruning stick → 6420**

6615 **pruning tree**
d Schneidelholz *n*; Schneitelholz *n*
f arbre *m* d'émonde
r дерево *n* обрезки

6616 **pruning wound**
(wound left when a branch is cut of)
d Entastungswunde *f*
f plaie *f*
r рана *f* от обрезки

* **psammophyte → 7192**

* **pseudo-liana → 4246**

6617 **pseudostrobus pine; false Weymouth pine; smoothbark Mexican pine**
(Pinus pseudostrobus)
d Pseudostrobuskiefer *f*
f pin *m* pseudostrobus
r сосна *f* ложновеймутова

* **pseudoterminal bud → 3004**

* **PSL → 6069**

6618 **pteridophytes**
(a class of flowerless plants, embracing ferns, horsetails, club mosses, quillworts, and other like plants)
d Pteridophyten *mpl*
f ptéridophytes *mpl*
r птеридофиты *mpl*

f pubérulent
r короткоопушённый; покрытый лёгким пушком; слегка опушённый

6620 **pubescent birch; Downey birch**
(Betula pubescens)
d Moorbirke *f*; Haarbirke *f*
f bouleau *m* pubescent
r берёза *f* пушистая

6621 **pubescent oak; downy oak**
(Quercus pubescens; Quercus humilis)
d weichhaarige Eiche *f*; flaumhaarige Eiche; Flaumeiche *f*; Wolleiche *f*
f chêne *m* pubescent; chêne blanc [de Provence]; chêne truffier
r пушистый дуб *m*

* **public forest → 1812**

6622 **public highway**
(highway for which public money has been spent and which is dedicated to public use by a plan deposited in the Land Titles Office for the district in which the road is situated)
d öffentliche Landstraße *f*
f chemin *m* public; voie *f* publique
r дорога *f* общественного пользования

6623 **public involvement**
(procedures for obtaining and considering the views of the general public in planning and decision-making processes)
d allgemeine Beteiligung *f*; öffentliche Betroffenheit *f*
f involvement *m* public
r привлечение *n* общественности

* **public sale of timber → 8557**

* **PUD → 1667**

* **pulai → 9363**

6624 **Pulaski tool**
(in Quebec, tool, which combines a single-bitted axe blade for chopping with a narrow blade for trenching, to clear vegetation when constructing a fireline)
d Pulaski-Werkzeug *n*
f outil *m* Pulaski
r инструмент *m* Пуласки; комбинированный топор-кирка *m*

* **pull-back → 5411**

* **puller → 8808**

* **pulling face → 4279**

* **pulling wheel → 8210**

* **pull shovel → 468**

* **pulp → 9603**

* **pulpboard → 9493**

* **pulpboard → 9551**

* **pulp bolt → 6627**

6625 **pulp chip; industry chip**
d Holzschliffspäne *mpl*
f copeau *m* de papeterie
r технологическая щепа *f*

* **pulphook → 6630**

* **pulp log → 6627**

6626 **pulp trees**
(small trees and saplings that will be ground to produce paper)
d Schleifbäume *mpl*
f arbres *mpl* à papier
r деревья *npl* для целлюлозы

* **pulpwood → 6061**

6627 **pulpwood bolt; pulp bolt; pulp[wood] stick; pulp log**
(log that does not meet the one-third merchantability standard for a saw log but contains a minimum of 50% sound wood fiber by volume)
d Schleifholzknüppel *m*; Schleifholzrundling *m*; Holzrolle *f*; Faserholzrolle *f*; Papierholzrolle *f*
f rondin *m* de bois de pâte; rondin de bois à papier; rondin de défibrage; pitoune *m*; grume *f* à papier
r балансовый кругляк *m*; бревно *n*, годное для волокнистого полуфабриката

6628 **pulpwood chipper**
d Papierholzhackmaschine *f*
f déchiqueteuse *f* de rondins de défibrage
r рубильная машина *f* для балансов

6629 **pulpwood grapple**
d Papierholz-Greifer *m*

* **pulpwood harvester → 7551**

6630 **pulpwood hook; pulphook**
d Papierholzhaken *m*
f crochet *m* de rondins à main
r крюк *m* для окучивания балансов вручную

6631 **pulpwood inventory**
d Zelluloseholzinventur *f*
f inventaire *m* de bois de trituration
r учёт *m* количества лесоматериалов (на рейде)

6632 **pulpwood logging**
d Faserholzschlag *m*
f débardage *m* de bois de trituration
r заготовка *f* балансов

* **pulp[wood] stick → 6627**

6633 **pulverized charcoal**
d pulverisierte Holzkohle *f*
f charbon *m* pulvérisé
r порошкообразный древесный уголь *m*

* **pumped tree → 5129**

6634 **pumpkin ash**
(Fraxinus tomentosa; Fraxinus profunda)
d rote amerikanische Esche *f*
f frêne *m* pubescent
r ясень *m* томентоза

* **pungent oak → 7190**

* **punk → 1879**

* **punk knot → 2290**

* **punktree → 1056**

* **punky → 3773**

* **punky heart → 2289**

6635 **punky pine**
d Schwammkiefer *f*
f pin *m* sylvestre attaqué par champignon
r сосна *f*, пораженная губок

* **punky wood → 2291**

* **pupal cell → 4810**

* **pupal niche → 4810**

6636 **pure live seed; PLS**
d reiner Lebendsamen *m*
f graine *f* pure vivante; GPV
r чистое живое семя *n*

* **pure stand → 2926**

* **purified wood → 4432**

* **purple azalea → 6288**

* **purple beech → 6639**

6637 **purplebloom maple**
(Acer pseudosieboldianum)
d koreanischer Ahorn *m*; Bergahorn *m*
f érable *m* de Corée
r клён *m* ложнозибольдовый

6638 **purpleblow maple**
(Acer truncatum)
d chinesischer Spitzahorn *m*
f érable *m* de Shantung
r клён *m* усечённый

6639 **purple European beech; copper[ ]beech; purple beech**
(Fagus sylvatica purpurea)
d Blutbuche *f*; Rotbuche *f*
f hêtre *m* à feuilles pourpres; hêtre pourpre; hêtre rouge
r бук *m* лесной тёмнопунцовый

6640 **purple-flowering raspberry; flowering raspberry**
(Rubus odoratus)
d Zimthimbeer *m*
f ronce *f* odorante; ronce framboisier du Canada
r малина *f* душистая

6641 **purpleheart [amaranth]; amaranth**
(Peltogyne paniculata; Peltogyne porphyrocardia; Peltogyne venosa)
d Amaranthbaum *m*; Amaranth *m*
f amarante *f*
r пелтогин *m* метельчатый пурпурный; амарант *m*

* **purple leaf plum → 1475**

6642 **purple orchid tree; orchid tree**
(Bauhinia variegata)
d Orchideenbaum *m*

6643 purple[ ]osier willow; purple osier; purple willow
(Salix purpurea)
d Krebsweide *f*; Purpurweide *f*
f saule *m* pourpre; osier *m* rouge; osier franc
r ива *f* пурпурная

6644 purple trumpet tree
(Tabebuia avellandae; Tabebuia impetiginosa)
d roter Lapacho *m*
f lapacho *m* pourpre; pau *m* d'Arco
r пурпурный лапачо *m*

* **purple willow → 6643**

* **pusher bar → 8905**

* **push[ing] arm → 8905**

* **pushing chisel → 6611**

* **pushing iron → 6611**

6645 push[ing] pole; killig; kilhig; sampson
(a stout pole, sometimes notched into the tree stem at one end and braced against the base of a peavey handle at the other, used to push a small tree manually in the desired direction of fall)
d Holzfällungsgabel *f*; Fällkatze *f*; Druckstange *f*
f poussoir *m*
r валочная вилка *f*; валочный шест *m*; валочная кошка *f* (для направленной валки деревьев вручную)

* **push-pull felling saw → 9030**

6646 pussur; Indian crabwood
(Carapa molluccensis)
d Njuru *n*; indisches Karapa *n*
f passur *m*; carapa *m* des Indes
r карапа *f* молюккская; нири *n*

* **pussy willow → 3880**

* **put forth leaves → 7531**

* **putlock → 7270**

* **putlog → 7270**

6647 putrefaction; decay; rotting
(the process of breakdown by bacterial action)
d Fäulnis *f*; Verrottung *f*

* **putrid → 7212**

* **putty → 9606**

* **pyinma → 4643**

6648 pyramidal habit
d pyramidale Wuchsform *f*
f port *m* pyramidal
r пирамидальный габитус *m*

6649 pyramid Chinese juniper; Chinese pyramid juniper
(Juniperus chinensis)
d chinesischer Sadebaum *m*; chinesischer Wacholder *m*
f genévrier *m* de Chine
r можжевельник *m* китайский

6650 pyramid tree; narrow upright tree
d pyramidenförmiger Baum *m*
f pyramide *f*
r дерево *n* пирамидальной формы

* **pyroclimax → 3172**

* **pyrography → 9498**

* **pyrogravure → 9498**

* **pyroligneous acid → 9688**

Q

* **Q factor → 6654**

* **QMD → 6651**

* **quadrat → 7173**

6651 **quadratic mean diameter; QMD**
(diameter of mean basal area tree)
d Mittelstammdurchmesser *m*
f diamètre *m* quadratique moyen
r средний квадратический диаметр *m*

6652 **quagginess; flaccidity; flabbiness; doze**
(wood defect)
d Schlaffheit *f*
f flaccidité *f*
r дряблость *f*; рыхлость *f*

* **quaking aspen → 236**

* **quality class → 9774**

6653 **quality control of timber**
d Holzqualitätskontrolle *f*
f contrôle *m* de la qualité du bois
r контроль *m* качества лесоматериалов

6654 **quality factor; Q factor**
(the ratio Q between numbers of trees in successive 2-inch diameter classes)
d Gütefaktor *m*; Q-Faktor *m*
f facteur *m* de qualité; facteur Q
r коэффициент *m* добротности

6655 **quality increment; increase in quality**
d Qualitätszuwachs *m*
f accroissement *m* en qualité
r качественный прирост *m*

6656 **quality of the stand; stand quality**
d Bestandsgüte *f*
f qualité *f* d'un peuplement
r качество *n* насаждения

6657 **quality of wood; wood quality; timber quality; lumber quality; grade of wood**
d Holzqualität *f*
f qualité *f* de bois

* **quality timber → 4284**

* **quantity increment → 9243**

6658 **quarter beam**
d Kreuzholz *n*; Viertelholz *n*
f poutre *f* en quart
r брус *m* из четверти брёвна

6659 **quarter cut; quarter saw[ing]; radial section; radial [sur]face; radial conversion**
(the surface exposed by both longitudinal cutting and transverse-longitudinal cutting where these run in the plane of any rays, i.e. in any plane through the pith)
d Quart[i]erschnitt *m*; Radialschnitt *m*; Riftschnitt *m*; Kreuzschnitt *m*; Radialtrennen *n*
f section *f* sur quartier; tranche *f* sur maille; coupe *f* radiale
r радиальный разрез *m*; радиальный распил *m*; радиальная распиловка *f*

6660 **quartered mast**
d vierteilter Mast *m*
f mât *m* quartier
r радиально распилённое мачтовое бревно *n*

* **quartered[-sawn] timber → 6678**

6661 **quarter girth**
(of a tree or a log)
d Viertelumfang *m*
f quart *m* de la circonférence
r четверть *f* длины окружности (дерева или брёвна)

* **quarter girth measure → 4366**

* **quartering → 6688**

* **quarter saw[ing] → 6659**

6662 **quassia**
(Quassia)
d Quassiabaum *m*
f bois *m* de quassia; quassia *m*; quassier *m*
r квассия *f*

6663 **quassia[-]wood**
(Picrasma)
d Bitterholz *n*
f picrasme *m*
r пикразма *f*; пикрасма *f*

* **Quebec birch → 1299**

* **quebracho blanco → 9378**

6664 Queensland kauri; kauri pine; dundathu pine; smoothbark kauri
(Agathis robusta; Dammara robusta)
d Kaurifichte *f*
f kauri *m* de Nouvelle Zélande; pin *m* de Kaori
r новозеландское хвойное дерево *n*

6665 Queensland maple; Brayley flindersia
(Flindersia brayleyana; Flindersia pimenteliana)
d Brayley-Flindersia *f*
f flindersie *f* de Brayley
r флиндерсия *f* Брейли

* **Queensland umbrella-tree → 412**

* **Queensland walnut → 413**

* **quercitron oak → 2692**

6666 quick-growing species
d schnellwachsende Arten *fpl*
f espèces *fpl* à croissance rapide
r быстрорастущие породы *fpl*

* **quickset hedge → 5019**

* **quiescence → 2572**

6667 quince [tree]
(Cydonia oblonga)
d Quittenbaum *m*; [gemeine] Quitte *f*
f cognassier *m*; cognier *m*; coudonnier *m*
r айва *f* [продолговатая]; айва обыкновенная

* **quincuncial planting → 6668**

6668 quincunx planting; quincuncial planting
(planting of five objects, such as trees, with one on each of four corners and one in the center)
d Fünfeckpflanzung *f*; Fünfeckverband *m*; schachbrettförmige Aussaat *f*
f plantation *f* en quinconces; plantation en pentagone
r посадка *f* в шахматном порядке; пятиконечная посадка

6669 quinine tree; cinchona tree; china bark tree
(Cinchona pubescens; Cinchona officinalis; Pinckneya pubens)
d Chinarindenbaum *m*; Chinabaum *m*; Fieberrindenbaum *m*

R

6670 **rabbit netting**
d kaninchendichtes Maschendrahtgatter *n*
f grillage *m* contre les lapins
r изгородь *f* из сеточной проволоки против кроликов

6671 **raceme**
d Traube *f*
f grappe *f*; racème *m*
r кисть *f*; гроздь *m*

* **raceway → 5451**

* **rack → 9745**

* **rackbench → 5199**

6672 **rack saw**
(a headsaw (circular or band) with a travel[l]ing table operated by rack-and-pinion, manually or by power)
d Racksäge *f*
f scie *f* avec aménage à crémaillère
r раскряжёвочная пила *f* с подвижным столом

6673 **rack-side tree**
d Baum *m* im Holzabfuhrskorridor
f arbre *m* à la fin de corridor de débusquage
r дерево *n* на краю трелёвочного коридора

6674 **rack system**
d Racksystem *n*
f système *m* de câbles
r система *f* [трелёвочных] волоков

* **radial[-arm] saw → 1586**

6675 **radial check; radial shake; ray shake; growth shake**
(a fissure more or less along the rays but not extending into the heart)
d Radialriss *m*; radialer Riss *m*; Schilferriss *m*; Strahlenriss *m*
f fente *f* radiale
r радиальная трещина *f*

* **radial conversion → 6659**

Abkürzkreissägemaschine *f*; Besäumkreisanlage *f*
f machine *f* déligneuse de scie circulaire; machine déligneuse de scie radiale
r круглопильный обрезной станок *m*; круглопильный торцовочный станок

* **radial growth → 2474**

* **radial increment → 2474**

6677 **radial intercellular canal**
(a canal extending across the grain in a radial direction, contained in a fusiform ray)
d Radialinterzellulargang *m*
f canal *m* intercellulaire radial
r радиальный межклеточный канал *m*; радиальный секреторный ход *m*

6678 **radial[ly]-sawn timber; comb-grain[ed] wood; edge-grain[ed] timber; quartered[-sawn] timber; rift[-sawn] timber; rift-grained lumber; vertical-grain[ed] timber; stripe-grained wood**
(when the rings form an angle of 45° to 90° with the wide surface of the piece)
d radialgesägtes Holz *n*; Radialholz *n*; spiegelndes Holz; spiegelgeschnittenes Schnittholz *n*; Spiegelholz *n*; kammkörniges Holz *n*
f bois *m* débité sur maille; bois radial; bois scié sur mailles; bois scié sur quartier
r пиломатериал *m* радиального распила; древесина *f* радиальной распиловки

6679 **radial parenchyma; ray parenchyma**
(parenchyma composing the rays wholly or in part)
d Radialparenchym *n*
f parenchyme *m* radial; parenchyme de rayons ligneux
r лучевая паренхима *f*

* **radial section → 6659**

* **radial shake → 6675**

6680 **radial shrinkage**
d radiales Schwingen *n*
f retrait *m* radial
r радиальная усушка *f*

* **radial [sur]face → 6659**

f gonflement *m* radial
r радиальное разбухание *n*

* **radia multiple → 6462**

* **radiata pine → 5548**

6682 **radiation**
(transfer of heat through a gas or a vacuum, other than by heating of the intervening space)
d Strahlung *f*
f rayonnement *m* [calorifique]
r радиация *f*; излучение *n*

6683 **radical bud**
d Wurzelknospe *f*
f bourgeon *m* radical
r корневая почка *f*

* **radical forest → 1639**

* **radical leaf → 944**

* **radication → 7011**

* **radicle [root] → 6546**

6684 **radicose** *adj*
d großwurzelig
f de racines grossières
r имеющий большие корни

6685 **raffia palm**
(Raffia ruffia; Raphia pedunculata)
d Raffia-Palme *f*; Bastpalme *f*
f raffia *m*
r рафия *f*

* **raft boom → 914**

6686 **raft-bridge; bridge on rafts**
d Floßbrücke *f*
f pont *m* de flottage
r плавучий мост *m*

6687 **raft bundle**
(logs bound together into a circular unit for towing)
d Floßwinde *f*
f ligature *f* d'un train de bois
r связка *f* плотов

* **raft dam → 4377**

* **rafted wood → 3279**

6688 **rafter timber; rafter wood; quartering**
(one of a series of roof support timbers that provide principal support for the roofing material)
d Sparrenholz *n*
f bois *m* à chevrons; chevron *m*
r стропильный лесоматериал *m*

* **rafting → 912**

6689 **rafting by sea; sea rafting**
d Seeflößerei *f*
f flottage *m* marin
r морской сплав *m*

6690 **rafting dogs**
(a wedge-shaped piece of metal with a ring or eye in the blunt end)
d Flößenzange *f*
f pince *f* de flottage
r штыри *mpl* с ушком

* **rafting reservoir → 5105**

* **rafting right → 6945**

6691 **raft of bundled wood**
d Bündelgestör *n*
f assemblage *m* de bois de flottage
r пучковая запань *f*

6692 **raft section; brail**
(a standard raft of logs or piling securely fastened together for long towing)
d Floßsektion *f*; Floßeinheit *f*; Gebund *n*
f section *f* de flottage
r сплоточная единица *f*; клетка *f*; сплавная секция *f*

* **raft[s]man → 9569**

6693 **raftsman's pole; pike[-]pole; hitcher; peav[e]y; picaroon; boat hook; gaff**
(a picked pole from 12 to 20 feet long, with or without a hook, used in holding boats to jams in driving and for pulling logs from brush and eddies out into the current)
d Flößerstange *f*; Flöß[erei]stange *f*; Floßhaken *m*; Greisbeil *n*; Kehrhaken *m*
f gaffe *f* [emmanchée]; croc *m* de batelier
r [лесо]сплавной багор *m*

6694 **raft with the tiers bound crosswise**
d Kreuzfloß *n*
f train *m* de flottage en croix
r крестовый плот *m*

6695 rail fence; hoarding
d Holzzaun *m*
f clôture *f* en bois
r деревянный забор *m*

6696 railroad logging; railway logging; railway removal
d Eisenbahnabfuhr *f*
f débardage *m* par voie ferrée
r лесозаготовки *fpl* с вывозкой леса по железной дороге; вывозка *f* по лесовозной железной дороге

6697 rain[drop] figure
(of wood)
d Regentropfenfigur *f*
f figure *f* à gouttes de pluie
r волнообразная текстура *f*; текстура в виде капли дождя

* **rainfall → 6498**

6698 rainfall distribution
(quantity of rain falling at different times of the year)
d Niederschlagverteilung *f*
f répartition *f* des pluies
r распределение *n* осадков

6699 rain[fall] intensity; rainfall rate
(rate at which rain is falling at any given instant, usually expressed in inches per hour)
d Niederschlagintensität *f*
f intensité *f* pluviale; pluviosité *f*
r интенсивность *f* дождевых осадков; интенсивность дождя

6700 rain[ ]forest; moist forest; moss forest; mist forest
(generally, a forest that grows in a region of heavy annual precipitation)
d Regenwald *m*; Nebelwald *m*
f forêt *f* ombrophile; forêt pluviale; forêt humide; forêt de brouillard; forêt buée
r дождевой лес *m*; влажный лес

6701 raintree; saman [tree]; monkey[ ]pot; monkey pod
(Albizzia saman; Pithecellobium saman; Samanea saman)
d [südostasiatischer] Regenbaum *m*; Saman *m*; Lecythis *m*
f arbre *m* de pluie; saman *m*; lecythis *m*
r энтеролобиум *m* саман; саман *m*

6702 raised grain
raised above the softer earlywood but not torn loose from it)
d plüschartige Faser *f*
f fil *m* pelucheux
r волнистое волокно *n*

6703 raised-grain butt wood
d plüschartiges Stammendeholz *n*
f bois *m* de fil pelucheux
r узорчатая древесина *f* комлевой или пнёвой части дерева

* **raising an underwood → 6364**

6704 raising of seedlings
d Pflanzenaufzucht *f*
f culture *f* de plantes
r возращение *n* саженцев

6705 rake *v*
d rechen; harken
f râteler; ratisser
r сгребать граблями

* **rake blade → 7042**

* **rakedozer → 8808**

* **raker → 4584**

* **raking → 7043**

6706 raking of litter; litter raking
d Streurechen *n*
f râtelage *m* de litière
r сгребание *n* подстилки

* **ramate → 977**

6707 ram[block]; rammer [log]
d Rammklotz *m*
f mouton *m*
r свайный кряж *m*

* **ramet → 9183**

6708 ramicorn branches; ramicorns
(abnormally large branches that project at sharp acute angles from the bole and are persistent)
d unregelmäßig verlängerte Äste *fpl*
f ramicornes *fpl*
r ненормально вытянутые сучья *mpl*

* **ramification → 987**

6709 ramiform
d astförmig; zweigförmig
f ramiforme
r ветвеобразный

6710 ramin; melawis
(Gonystylus bancanus; Gonystylus forbesii; Gonystylus macrophyllum)
d Ramin *m*
f ramin *m*
r рамин *m*

* **rammer [log] → 6707**

* **ramose → 977**

* **ramous → 977**

6711 ram pike
(a tree broken off by wind and with a splintered end on the portion left standing)
d Baum *m* mit Spaltenende
f arbre *m* à bout éclaté
r дерево *m* с размочаленным концом

* **rampsite → 5451**

* **ramule → 991**

* **ramulose → 977**

6712 random felled wood
d zufällig gefälltes Holz *n*
f bois *m* abattu au hasard
r отдельно лежащие поваленные деревья *npl*

6713 randomized blocks
(random placement of fixed sampling areas in which to conduct vegetation research)
d randomisierte Blocke *mpl*; Blöcke *mpl* mit zufälliger Zuteilung
f blocs *mpl* aléatoires; blocs complètement randomisés
r рандомизированные участки *mpl*

6714 randomized complete-block design
(an experimental design in which each block contains a complete replication of the treatments, which are allocated to the various units within the blocks in a randomized fashion)
d randomisierter Blockentwurfplan *m*
f plan *m* en blocs aléatoires; plan en blocs complètement randomisés
r рандомизированный план *m* блоков

* **random lengths → 5488**

6715 random road pattern
d zufälliger Wegmuster *m*
f modèle *m* de tracé aléatoire
r схема *f* освоения лесосек с хаотичным расположением веток и усок

6716 range development
(structure or excavation related to the management, for range purposes, of range land or livestock; practice, excluding grazing, that is designed to improve range conditions)
d Weideverweiterung *f*
f développement *m* de terrain de parcours
r оборудование *n* участка [лесничества]

6717 range enhancement; range improvement
(any treatment, development, or structure designed to achieve or maintain the desired plant community)
d Weideverbesserung *f*
f amélioration *f* de terrain de parcours
r улучшение *n* участка

6718 range-finder dendrometer
d Entfernungsdendrometer *n*
f télémètre-dendromètre *m*
r дальномер-дендрометр *m*

* **range improvement → 6717**

6719 range plant
d Weidepflanze *f*
f plante *f* pastorale
r пастбищное растение *n*

* **ranger cabin → 3407**

* **ranger district → 2542**

6720 range readiness
(stage of plant growth at which grazing may begin under a specific management plan without permanent damage to vegetation or soil)
d Weidebereitschaft *f*
f aptitude *f* à l'emploi de parcours
r готовность *f* участка

* **ranger house → 3407**

6721 range type
(defined area with specific physical characteristics, which differs from other areas in its ability to produce distinctive kinds and

d Weidetyp *m*
f type *m* de parcours
r тип *m* лесного участка

6722 range use plan
(operational plan that describes the range and livestock management measures that will be implemented to ensure that range resources are protected and management objectives for other identified resource values are achieved)
d Weidenutzungsplan *m*
f plan *m* d'utilisation du parcours
r план *m* использования участка

* **ranging pole → 3899**

6723 rapattack
(initial attack whereby firefighters are transported to the fire in a specially-equipped helicopter, from which they descend during a hover by means of rappelling down a rope fitted with a mechanical device to control the rate of descent)
d Schlag-Attacke *f*
f attaque *f* initiale
r ударная атака *f*

* **rape of forest → 1909**

6724 rapid forest management; management for rapid growth
d Schnellwuchsbetrieb *m*
f traitement *m* à croissance rapide
r хозяйство *n* на быстрый рост

* **rapine → 2434**

* **rappel crew → 4240**

6725 rare species
(occurring as a very few individuals or small groups at widely scattered localities over a large geographic area of what appears to be suitable habitat)
d seltene Art *f*
f espèce *f* rare
r редкая порода *f*

6726 rare woods
d seltene Holze *npl*
f bois *m* rares
r редкие породы *fpl* древесины

* **raspberry → 1800**

* **rasp[ing] machine → 2284**

6727 ...
(proportion of the watershed area allowed to be cut each year)
d Schneidegeschwindigkeit *f*
f vitesse *f* d'abattage
r скорость *f* резания

6728 rate of growth; growth rate; growth speed; increment rate
d Wachstumsgeschwindigkeit *f*
f vitesse *f* de croissance
r скорость *f* роста

6729 rate of profit in forestry
d forstlicher Zinsfuß *m*
f taux *m* de revenu forestier
r процент *m* лесной прибыли

6730 rate of spread; ROS
(speed at which a fire extends its horizontal dimensions, expressed in terms of distance per unit area of time)
d Ausdehnungsgeschwindigkeit *f*
f vitesse *f* de propagation
r скорость *f* распространения

* **rate of stocking → 8110**

* **ratline → 6732**

* **rattail [top] → 5544**

6731 rattan [palm]; sweet segde
(Calamus rotang)
d Rotan *m*
f rotin *m*; rotang *m*
r раттан *m*; ротанг *m*; индийская ротанговая пальма *f*

6732 rattling line; ratline
(a rope through which at intervals small pins are driven into the logs which are to compose a raft joint; its purpose is to hold the logs together until the boom poles can be adjusted)
d Webeleine *f*
f enfléchure *f*
r выбленочный трос *m*; линь *m*

* **rauli → 2999**

* **ravine forest → 3474**

6733 raw gum; crude balsam; raw resin; crude resin
d Rohharz *n*; Rohbalsam *m*
f gomme *f* brute; baume *m* brut; résine *f* brute

* **raw humus → 1681**

* **raw log → 9044**

6734 **raw pruning**
d grobe Ästung *f*
f élagage *m* brut
r предварительное удаление *n* сучьев

* **raw resin → 6733**

* **raw timber → 7096**

* **raw waste lumber → 5215**

* **raw[ ]wood → 7096**

* **ray → 5391**

6735 **ray [cambial] initial**
(a cambial initial giving rise to a ray cell; usually one of a group and often more or less isodiametric as seen in tangential section)
d Strahlkambiumzelle *f*
f initiale *f* de rayon
r лучевый инициал *m* камбия

6736 **ray cell; medullary cell**
d Markstrahl[en]zelle *f*
f cellule *f* de rayon
r клетка *f* сердцевинного луча

6737 **ray cell walls**
d Markstrahlszellwanden *fpl*
f parois *f* de rayon médullaire
r стенки *fpl* клеток сердцевинных лучей

6738 **ray crossing; crossfield**
(a term of convenience for the rectangle formed by the walls of a ray cell and an axial tracheid, as seen in a radial section)
d Kreuzungsfeld *n*
f champ *m* de croisement
r радиальный разрез *m* сердцевинного луча; поле *n* перекрёста (оболочек клеток лучей и осевых трахеид)

* **ray parenchyma → 6679**

* **ray shake → 6675**

6739 **ray tracheid**
(a tracheid forming part of a ray)
d Markstrahltracheide *f*
f trachéide *f* de rayon; trachéide transversale
r лучевая трахеида *f*

6740 **ray-vessel pitting**
(pitting between a ray cell and a vessel member)
d Markstrahl-Gefäß-Tüpfelung *f*
f ponctuation *f* rayon-vaisseau
r сосудисто-лучевые поры *fpl*

6741 **raze knife; scribe [knife]; scriber**
(tool for marking trees or round timber by scoring the outer surface)
d Baumreißer *m*
f rainette *f*
r нож *m* для маркировки деревьев или брёвен; резак *m* для маркировки деревьев

* **razor-strap fungus → 752**

6742 **reach**
(wood or metal structural member connecting a logging trailer to a truck tractor; distance spanned by a skyline)
d Reichweite *f*
f portée *f*
r выступающая часть *f*

6743 **reach of crane; boom reach**
d Boomweite *f*
f couverture *f* de câble
r вылет *m* стрелы; расстояние *n* стрелы

6744 **reaction wood**
(wood with distinctive anatomical and physical characteristics, formed typically in parts of leaning or crooked stems and in branches, that tends to restore the original position of the branch or stem if this has been disturbed)
d Reaktionsholz *n*
f bois *m* de réaction
r реактивная древесина *f*

* **reafforestation → 6821**

* **real growing stock → 3559**

* **realization value → 2952**

6745 **real laurel; laurel**
(Laurus)
d Lorbeer[baum] *m*; Lorbeerstrauch *m*
f laurier *m*
r лавр *m*

* **real mastictree → 5327**

6746 **rear sledge**
d Hinterschlitten *m*

6747 **reburn**
d Wiederverbrennung *f*; zweite Verbrennung *f*
f recombustion *f*
r повторный пал *m*

6748 **recalcitrant seed**
(seed with a short life span, which usually must be kept moist and warm)
d persistenter Samen *m*
f semence *f* récalcitrante
r биологически стойкое семя *n*

* **recce → 3549**

* **receiving boom → 4324**

* **receiving place → 3280**

* **reciprocating saw → 2609**

6749 **reconnaissance survey**
(preliminary visit and report usually executed rapidly and at a relatively low cost)
d Übersichtsvermessung *f*
f levé *m* de reconnaissance
r рекогносцировочная съёмка *f*

6750 **reconstituted wood**
(panel products made with strands, wafers, particles or fibers of wood)
d rekonstruiertes Holz *n*
f bois *m* reconstitué
r реконструированная древесина *f*

6751 **recording cal[l]iper**
d Registrierkluppe *f*
f compas *m* enregistreur
r регистрирующая мерная вилка *f*

* **record of cuttings → 3073**

6752 **recovery rate**
(rate of conversion of roundwood to timber products)
d Erholungsgeschwindigkeit *f*
f vitesse *f* de restauration
r скорость *f* восстановления

6753 **recovery shoot**
d Regenerationstrieb *m*
f pousse *f* de régénérescence
r регенерационный побег *m*

* **recreation[al] forest → 91**

6754 **recreation area; recreation site**
r рекреационная зона *f*; зона отдыха

6755 **recreation feature significance**
(quality, uniqueness, and availability of a recreation feature as classified in the recreation inventory)
d Erholungsobjektsignifikanz *f*
f signification *f* d'objet de récréation
r значимость *f* рекреационного объекта

6756 **recreation [features] inventory**
(identification, classification, and recording of the types and locations of biophysical recreation and cultural features, existing and potential recreation activities, feature significance and feature sensitivity)
d Erholungsinventur *f*
f inventaire *m* de récréation
r перечёт *m* рекреационных объектов

* **recreation glade → 3450**

6757 **recreation potential**
d Erholungsmöglichkeit *f*
f potentiel *m* récréatif
r рекреационный потенциал *m* (лесных ландшафтов)

6758 **recreation resource; recreation value**
(recreation feature, a scenic or wilderness feature or setting that has recreational significance or value or a recreation facility)
d Erholungsressource *f*
f ressource *f* récréative
r рекреационный ресурс *m*

* **recreation site → 6754**

6759 **recreation trail**
(trail and its ancillary facilities for recreation or to protect a recreation resource)
d Erholungspiste *f*
f train *m* de récréation
r рекреационный маршрут *m*

* **recreation value → 6758**

* **recruitment of trees → 4492**

* **recruitment period → 4006**

* **rectangular timber → 7969**

* **rect-cone Aleppo pine → 1250**

6760 **recurrent forest inventory**

f inventaire *m* forestier récurrent
r периодическая таксация *f* леса; повторная таксация леса

6761 **recurrently flushing shoot**
d regelmäßig treibender Austrieb *m*
f pousse *f* lessivée périodiquement
r побег *m*, периодически дающий отростки

* **red alder → 235**

6762 **red angico; true angico preto; curupay [ata]; cebil [colorado]; curapixa**
(Anadenanthera macrocarpa; Piptadenia macrocarpa)
d rotes Angico *n*; Kurupayra *n*
f angico *m* rouge; curupayra *m*
r пиптадения *f* крупнолистная

* **red ash → 2255**

6763 **redbay [persea]; sweet bay; silk bay**
(Persea borbonia)
d roter Lorbeerbaum *m*
f laurier-sauce *m*; laurier *m* rouge; laurier Bourbon
r лавр *m* красный; персея *f* бурбонская; персея красная

* **red beech → 8420**

6764 **red belt fungus; fir needle rust; fluoride injury; fomes root rot**
(invader - Fomitopsis pinicola, Fomes pinicola)
d Tannennadel-Rostpilz *m*
f polypore *m* marginé; amadouvier *m* des pins
r окаймлённый трутовик *m*

* **red-berried elder → 2907**

* **red-berried juniper → 6530**

* **red bilberry → 1996**

* **red birch → 1299**

* **red bloodwood → 2249**

6765 **red-brown butt rot**
(invader - Polyporus Schweinitzii, Phaeolus circinatus)
d rote-brune Stammfäule *f*
f pourriture *f* du pied brune-rouge
r красновато-бурая напенная гниль *f*

6766 **red-brown root and butt rot**
(invader - Polyporus Schweinitzii, Phaeolus circinatus)
d rote-brune Wurzel- und Stammfäule *f*
f pourriture *f* brune-rouge de la racine et du pied
r ямчатая ядровая напенная и корневая гниль *f*

6767 **red buckeye**
(Aesculus pavia)
d echte Pavie *f*; rote Pavie
f marronnier *m* de Virginie; pavier *m* rouge
r каштан *m* конский розовоцветный; каштан павия

* **red cedar → 8893**

* **red cherry → 2881**

6768 **red common oak**
(Quercus borealis maxima; Quercus rubra maxima)
d große Roteiche *f*
f gros chêne *m* rouge
r дуб *m* северный крупный

6769 **red currant; northern currant**
(Ribes rubrum; Ribes sylvestre)
d rote Johannisbeere *f*
f groseillier *m* à grappes
r смородина *f* красная

* **red deal → 6275**

6770 **reddish-brown forest soil; Mediterranean brown earth**
d rötlichbrauner Waldboden *m*; Mittelmeer-Braunerde *f*
f sol *m* brun-roux de forêt; sol brun méditerranéen
r красновато-бурая лесная почва *f*; бурая средиземноморская почва

* **red dogtree → 1771**

* **red dogwood → 1771**

* **red elderberry → 2907**

* **red fir → 5738**

6771 **red-flowered blue gum; white ironbark**
(Eucalyptus leucoxylon)
d weißer Eukalyptus *m*
f eucalyptus *m* blanc
r эвкалипт *m* белодревесный

d blutrote Johannisbeere *f*; Blutjohannisbeere *f*
f groseillier *m* à fleurs; groseillier sanguin
r смородина *f* кровяно-красная

6773 **red[-]flowering gum; scarlet eucalyptus**
(Eucalyptus ficifolia)
d Purpureukalyptus *m*; Scharlach-Eukalyptus *m*
f eucalyptus *m* à fleurs rouges; gommier *m* rouge
r эвкалипт *m* фиголистный

6774 **red[-flowering] maple; scarlet maple; swamp maple**
(Acer rubrum)
d roter Ahorn *m*; Rotahorn *m*; virginischer Ahorn; rotblütiger Ahorn
f érable *m* rouge; érable de Virginie
r клён *m* красный

* **red gum → 6786**

6775 **red-haw hawthorn**
(Crataegus sanquinea)
d Blutdorn *m*
f épine *f* rouge
r боярышник *m* кроваво-красный

6776 **red-headed pine sawfly**
(Neodiprion lecontei)
d rotköpfige Kieferblattwespe *f*
f diprion *m* de Leconte
r пилильщик *m* сосновый красноголовый

* **redheart ceanothus → 3939**

6777 **red heart rot**
(invader - Stereum sanguinolentum)
d Rotkernfäule *f*
f pourriture *f* du cœur rouge
r бурая раневая гниль *f* стволов

6778 **red heart[wood]**
d Rotkern *m*
f cœur *m* rouge
r красное ядро *n*

* **red hickory → 7482**

* **red huckleberry → 1996**

6779 **red ironbark [eucalyptus]; narrowleaved red ironbark; pink narrow-leaved ironbark [eucalyptus]; mulga eucalyptus**
(Eucalyptus sideroxylon; Eucalyptus crebra)
d roter Ironbark *m*; Eisenholzeukalyptus *m*
f ironbark *m* rouge
эвкалипт красный железнокорый

* **red ivorywood → 6056**

* **red juniper → 2722**

* **red-listed species → 2801**

* **red louro → 6782**

* **red mahogany → 411**

6780 **red mangrove [tree]; African mangrove**
(Rhizophora mucronata; Rhizophora mangle)
d afrikanische Mangrove *f*
f manglier *m* d'Afrique; palétuvier *m* rouge
r красный мангр *m*; ризофора *f* остроконечная

6781 **red mulberry; American mulberry**
(Morus rubra)
d roter Maulbeerbaum *m*
f mûrier *m* rouge
r шелковица *f* американская; тут *m* красный

6782 **red ocotea; red louro; determa; wane**
(Ocotea rubra)
d Louro vermelho *m*; Red Louro *m*; Wane *m*
f grignon *m* rouge; grignon franc
r оцота *f* красная; окотеа *f* красная; ване *n*; детерма *f*; григнон *m*

6783 **red-osier dogwood; skawbush**
(Cornus stolonifera)
d gelbästiger Hartriegel *m*
f cornouiller *m* stolonifère
r дёрен *m* американский

6784 **red pine; rimu; ekor kuda**
(Dacrydium cupressimum; Dacrydium nidulum)
d Harzeibe *f*
f dacryne *m*; dacrydion *m*
r дакридиум *m* кипарисовый; риму *m*; новозеландская красная сосна *f*

* **red pine → 5796**

6785 **red pine cone beetle**
(Conophthorus resinosae)
d Rotkiefernzapfenkäfer *m*
f scolyte *m* des cônes de pin rouge
r короед *m*, повреждающий шишки сосны смолистой

* **red quebracho → 5173**

6786 red river gum; [river] red gum; longbeak eucalyptus; marri eucalyptus
(Eucalyptus calophylla; Eucalyptus haematoxylon; Eucalyptus camaldulensis)
d roter Eukalyptus *m*; Rotgummibaum *m*
f eucalyptus *m* rouge; eucalyptus rostrata; gommier *m* de rivière
r эвкалипт *m* приречный; эвкалипт кровавокрасный; эвкалипт кроваводревесный; эвкалипт камальдульский

6787 red rock cedar; Arabian juniper; Phoenician juniper
(Juniperous phoenicia)
d rotfrüchtiger Wacholder *m*; phönizischer Sadebaum *m*
f sabine *m* maritime; genévrier *m* de Phénicie
r можжевельник *m* красноплодный

6788 red root; Jersey tea; inland ceanothus
(Ceanothus herbaceus; Ceanothus ovatus; Ceanothus intermedius)
d amerikanische Säckelblume *f*; Säckelstrauch *m*
f céanothe *m* à feuilles étroites
r цеанотус *m* яйцевиднолистный

6789 red rot
(a rot with a reddish or reddish-purple incipient stage)
d Rotfäule *f*
f pourriture *f* rouge
r красная гниль *f*

6790 red sandalwood; red sanders; sandalwood padauk
(Pterocarpus santalinus)
d roter Sandelholzbaum *m*; roter Sandelbaum *m*; rotes Sandelholz *n*
f bois *m* rouge de Kodar; santal *m* rouge
r сандал *m* глистогонный; красное сандаловое дерево *n*; птерокарпус *m* санталовый; сандал красный

6791 red seringa; burkea
(Burkea africana)
d rote Syringa *f*
f guéléba *m*
r буркея *f* африканская

6792 red silk cottontree; Malabar bombax; kondrotti
(Bombax ceiba; Bombax malabaricum)
d Semul *m*; indisches Seidenwollbaum *m*

* **red silk-oak → 3966**

* **red spruce → 1298**

* **red stain → 7215**

6793 redstem ceanothus
(Ceanothus sanguineus)
d blutrote Säckelblume *f*
f céanothe *m* sanguin
r цеанотус *m* кроваво-красный

6794 red stringybark
(Eucalyptus macrorrhyncha)
d roter Stringybark *m*
f stringybark *m* rouge
r эвкалипт *m* крупноносый; эвкалипт красный веревчатокорый

6795 red stripe
(wood defect)
d rotes Streifen *n*
f bande *f* rouge
r грибные ядровые полосы *fpl*

6796 red[-]top
d rote Krone *f*
f cime *f* rousse
r побуренные листвы *fpl* (на повреждённых деревьях)

6797 red turpentine beetle
(Dendroctonus valens)
d roter Bastkäfer *m*
f dendroctone *m* rouge de l'épinette
r лубоед *m* красный скипидарный

6798 reduced circular saw
d kugelförmige Spaltsägemaschine *f*
f scie *f* circulaire réduite
r круглопильный делительный станок *m*; круглопильный станок для продольной распиловки

6799 reduction of debris
d Aufraümen *n* der Schlagfläche; Räumen *n* des Schlagreisigs
f réduction *f* des débris
r очистка *f* [вырубок] от порубочных остатков

6800 redvein maple
(Acer rufinerve)
d rostnerviger Ahorn *m*; Schlangenhautahorn *m*
f érable *m* jaspé de gris; érable à tronc strié
r клён *m* рыжеватожилковатый

6801 red willow
(Salix rubra)
d rote Weide *f*
f saule *m* rouge; verdiau *m*
r ива *f* красная

* **red-winged pine beetle → 2724**

6802 redwood
(the soft reddish wood of either of two species of sequoia trees or either of two huge coniferous California trees)
d Rotholz *n*
f bois *m* rouge
r красное дерево *n*; красная древесина *f*

* **red[wood] ant → 9480**

* **reed → 1308**

* **reeling-in → 6120**

6803 refoliation
d sekundäre Belaubung *f*
f feuillaison *f* secondaire
r вторичное облиствение *n*

* **reforestation → 6821**

6804 reforestation backlog
(areas that need to have trees reestablished; planting, seeding, or preparing the site for natural regeneration can do this)
d Waldverjüngungszone *f*
f zone *f* de reboisement
r зона *f* для лесовозобновления

* **reforestation forester → 3400**

6805 reforestation technique
d Waldverjüngungstechnik *f*
f technique *f* de reboisement
r техника *f* лесовозобновления

6806 reforested
d wieder geforstet
f reboisé
r повторно облесённый

6807 refuse burner
(a wood waste incinerator)
d Abfallbrenner *m*
f incinérateur *m* à déchets
r печь *f* для сжигания отходов;

6808 refuse chain
d Abfallkette *f*
f chaîne *f* de déchets de coupe
r цепь *f* для сталкивания брёвен; цепь для удаления отходов

* **refuse chipper → 7721**

* **refused timber → 2123**

* **refuse grinder → 7721**

6809 refuse lac
d Kirilack *m*
f laque *f* de rebut; laque en blocs
r лак *m* из отходов

* **refuse wood → 2123**

6810 regenerating forestland; restocking forestland
d regenerierbarer Waldboden *m*
f terre *f* forestière régénérée; terre forestière en voie de repeuplement
r возобновившаяся лесная земля *f*

6811 regeneration after removal of the old growth; artificial forest regeneration; after-regeneration; artificial reproduction
(establishing a new forest by planting or direct seeding)
d Nachverjüngung *f*; künstliche Waldverjüngung *f*
f régénération *f* artificielle de forêts; post-régénération *f*
r возобновление *n* [леса] после вырубки; последующее возобновление [леса]; дополнительное возобновление; искусственное лесовозобновление *n*

* **regeneration area → 355**

* **regeneration block → 6160**

6812 regeneration by groups; group regeneration; group reproduction
d gruppenweise Verjüngung *f*; horstweise Verjüngung
f régénération *f* par groupes
r куртинное возобновление *n*; куртинный подрост *m*; групповое возобновление

6813 regeneration by strips; regeneration by strip felling
d streifenweise Verjüngung *f*

6814 **regeneration class**
d Verjüngungsklasse *f*
f classe *f* de régénération
r класс *m* возобновления

6815 **regeneration cut[ting]; reproduction cut[ting]; reproductive felling; regeneration felling; regenerative cutting; renewal felling; renovation felling**
(timber harvest designed to promote natural establishment of trees)
d Verjüngungshieb *m*; Verjüngungsschlag *m*; Erneuerungshieb *m*
f coupe *f* de régénération; coupe de rénovation
r лесовосстановительная рубка *f*; лесовозобновительная рубка

6816 **regeneration delay**
(maximum time allowed in a prescription, between the start of harvesting in the area to which the prescription applies, and the earliest date by which the prescription requires a minimum number of oacceptable well-spaced trees)
d Verjüngungsverzögern *n*
f délai *m* de régénération
r замедление *n* возобновления

* **regeneration felling → 6815**

6817 **regeneration following a catastrophe; regeneration following disaster**
d Katastrophenverjüngung *f*
f régénération *f* accidentelle
r возобновление *n* катастрофной площади

6818 **regeneration front; reproduction front; belt under regeneration**
d Verjüngungsfront *f*
f lisière *f* de régénération
r кайма *f* возобновления

6819 **regeneration initiation**
(year in which the new crop is deemed to be started at an acceptable stocking level)
d Verjüngungsinitiation *f*
f début *m* de la régénération
r начало *n* регенерации

6820 **regeneration interval; regeneration period**
(period between the seed cutting and the final cutting on a particular area under one of the shelterwood systems)
d Verjüngungszeitraum *m*; Verjüngungsperiode *f*
f durée *f* de régénération; période *f* de
возобновительный период

* **regeneration method → 5416**

6821 **regeneration of a forest; reforestation; reafforestation; forest renewal; timber regeneration; forest reproduction; forest recovery; forest recreation; restocking**
(renewing a forest after harvest whether by natural or artificial means)
d Waldverjüngung *f*; Bestandsbegründung *f*; Forstanbau *m*; Neuaufforstung *f*; Wiederaufforstung *f*; Waldverneuerung *f*; Waldwiederbegründung *f*; Forsterneuerung *f*; Wiederbestockung *f*
f régénération *f* de forêt; reboisement *m*; reforestation *f*; reconstitution *f* de forêt; repeuplement *m* de forêt; reconstitution forestière
r восстановление *n* леса; лесовосстановление *n*; лесовозобновление *n*; повторное облесение *n*; повторные лесопосадки *fpl*; облесение заново; возобновление *n* леса

6822 **regeneration of cutover stands**
d Schlagflächeverjüngung *f*
f régénération *f* des parcelles de coupe
r возобновление *n* леса на вырубках; зарастание *n* вырубок

6823 **regeneration performance assessment; RPA**
(sampling survey carried out to collect field data on the height growth, competition, and stocking of young stands)
d Wiederbestockungsperformanz-Abschätzung *f*
f estimation *f* de performance de régénération
r оценка *f* регенеративной производительности

* **regeneration period → 6820**

6824 **regeneration survey**
(inventory of the quantity and quality of regeneration over a given area)
d Wiederbestockungsaufnahme *f*
f relevé *m* de la régénération
r съёмка *f* возобновления

6825 **regeneration under a shelterwood**
d Anbau *m* unter Schirm; Schirmverjüngung *f*; Schirmbesamung *f*
f régénération *f* sous abri
r возобновление *n* леса под пологом [старых

6826 regeneration without site preparation
(the area is characterized by the absence of a manageable stand because of inadequate stocking of growing stock)
d Verjüngung *f* ohne Standortvorbereitung
f régénération *f* sans préparation de terrain
r восстановление *n* без очистки поверхности почвы

6827 regeneration with site preparation
(the area is characterized by the absence of a manageable stand because of inadequate stocking of growing stock)
d Verjüngung *f* mit Standortvorbereitung
f régénération *f* avec préparation *f* de terrain
r восстановление *n* с очистки поверхности почвы

* **regenerative cutting → 6815**

* **regional forester → 2551**

6828 regionally important species
(which are not red- or blue-listed, that require management practices that differ from standard integrated resource management guidelines in order to fulfill critical habitat needs)
d gebietsweit wichtige Arten *fpl*
f espèces *fpl* significantes régionalement
r виды *mpl* локального значения

6829 regional plan
(contains forest management alternatives based on a detailed analysis of timber supply within the region)
d Regionalplan *m*
f plan *m* régional
r региональный план *m*

* **regional volume table → 5037**

* **registered professional forester → 6564**

6830 registered selected trees
(that have documented code number and its characteristics by a responsible authority)
d eingeschriebene Auslesebäume *mpl*
f arbres *mpl* sélectionnés enregistrés
r зарегистрированные отборные деревья *npl*

* **registration of the cut → 6831**

6831 registration of the felled timber; registration of the cut; cutting inventory; coupe inventory
f inventaire *m* d'abattage; réception *f* d'une coupe
r учёт *m* рубок; учёт объёма лесозаготовок; приёмка *f* лесосеки

6832 regression of increment
d Zuwachsrückgang *m*
f diminution *f* de l'accroissement
r падение *n* прироста

* **regrowth → 120**

* **regrowth forest → 7346**

6833 regrowth[ing]
(producing a large number of shoots when a single or few stems are mechanically removed but the root system is left intact)
d Waldnachwuchs *m*
f recrutement *m* de forêt
r порослевое возобновление *n* леса

6834 regular felling; regular cutting
d planmäßiger Hieb *m*
f coupe *f* régulière; coupe réglée
r валка *f* по утверждённой схеме

6835 regularized dried timber
d kalibriertes Schnittholz *n*
f bois *m* sec calibré
r калиброванный сухой пиломатериал *m*

6836 regularized green timber
d kalibriertes Frischholz *n*
f bois *m* vert calibré
r калиброванный сырой пиломатериал *m*

6837 regularized round timber
d kalibriertes Rundholz *n*
f bois *m* rond calibré
r калиброванное оцилиндрованное бревно *n*

* **regular-sized timber → 2503**

6838 regular uneven-aged structure
(stand in which three or more distinct age classes occupy approximately equal areas and provide a balanced distribution of diameter classes)
d reguläre gleichaltrige Struktur *f*
f structure *f* inéquienne régulière
r равномерная разновозрастная структура *f*

6839 regulated acres
(commercial forest land organized for

f terres *fpl* à réglementation
r регулированные земли *fpl*

6840 regulated forest
(forest that has reached a level where the cut equals the growth)
d regulierter Wald *m*
f forêt *f* réglée
r регулированный лес *m*

6841 regulated harvest
(volume scheduled in calculations of the allowable sale quantity which is harvested from suitable commercial forest land)
d regulierte Ernte *f*
f exploitation *f* réglée
r регулированные лесозаготовки *fpl*

6842 regulation by areas; rotation by areas; allotment by areas; area regulation; area control
(controlling the annual or periodic acreage harvested from a forest, despite fluctuations in fiber-yield volumes)
d Flächenfachwerk *n*
f compartimentage *m* par contenance; compartimentage par surfaces
r периодный метод *m* по площади

6843 regulation of mixture
d Mischungsregulierung *f*
f réglage *m* du mélange
r регулирование *n* смешения

6844 regulations of cut; regulations of felling; cutting[-practice] rules; cutting directions
d Einschlagsanweisungen *fpl*; Einschlagsverfügungen *fpl*
f règlements *mpl* d'exploitation
r правила *npl* рубок [главного лесопользования]; инструкции *fpl* рубки

6845 rehabilitated secondary forest
d renovierter Sekundärwald *m*
f forêt *f* secondaire réhabilitée
r восстановлённый вторичный лес *m*

6846 rehabilitation of redundant sites; restoration of sites; site rehabilitation
(conversion of the existing unsatisfactory cover on highly productive forest sites to a cover of commercially valuable species)
d Sanierung *f* von aufgelassenen Abbaustätten
f réhabilitation *f* de sites désaffectés
r восстановление *n* резервных участков

6847 reinventory
(complete restratification of an area on recent, mid-scale aerial photographs based on extensive field work)
d repetitive Inventur *f*
f inventaire *m* secondaire
r повторный перечёт *m*

6848 reiteration
(development of shoots outside the normal expression of the architectural model of a tree)
d Wiederholung *f*
f réitération *f*
r повторение *n*

6849 reiterative complex
(the complex of meristems produced by reiteration)
d ständig wiederholender Komplex *m*
f complexe *m* réitératif
r повторяющийся комплекс *m*

6850 rejection; fragment screening; screen rejects
d Sortiererabfälle *mpl*
f refus *mpl* de classeur
r отсортированные отходы *mpl*; отходы сортирования

* **rejects → 2123**

6851 rejuvenation
(of trees)
d Verjüngung *f*
f rejuvénilisation *f*
r омоложение *n*

6852 relascope
(hand-held instrument used to measure tree height, tree basal area, stem diameter at various heights, form class, and distance)
d Spiegelrelaskop *m*; Relaskop *m*
f relascope *m* [à miroir]
r [зеркальный] реласкоп *m*

6853 relative dominance
(the number of occurrences of 1 species as a percentage of the total number of occurrences of all species)
d relative Dominanz *f*
f dominance *f* relative
r относительное доминирование *n* (процент площади сечения деревьев данного вида)

6854 relative humidity; relative moisture
(dimensionless ratio of the actual vapo[u]r

d relative Feuchtigkeit *f*; Relativfeuchtigkeit *f*
f degré *m* relatif d'humidité; humidité *f* relative
r относительная влажность *f*

* **relative moisture → 6854**

6855 **relative thinning intensity**
(periodic (annual) yield of a stand from thinnings, expressed as a percentage of its periodic annual increment)
d relative Durchforstungsintensität *f*
f intensité *f* relative d'éclaircie
r относительная интенсивность *f* рубок ухода

6856 **release** *v*; **liberate** *v*; **disengage** *v*
(to open up established trees from a situation in which they have been suppressed by thinning out undesirable trees and shrubs which have overtopped them)
d freihauen
f dégager
r освобождать вырубкой

* **release cut[ting] → 5912**

6857 **relic[t] tree**
d Reliktbaum *m*
f relique *m*
r реликтовое дерево *n*

6858 **relief; land form**
(difference in elevation between the high and low points of a land surface or the inequalities of a land surface, taken collectively)
d Relief *n*
f relief *m*
r рельеф *m*

* **reload site → 8763**

6859 **re[-]logging**
(salvaging small timber, culls, and other residuals following the main logging operation)
d Nachholzaufbereitung *f*
f ré-exploitation *f*; exploitation *f* complémentaire
r рубка *f* деревьев, оставленные после главной рубки; освоение *n* недорубов; повторные лесозаготовки *fpl*

* **remaining stand → 6889**

6860 **remediation**
(measures undertaken in respect to an area of
d Sanierung *f*; Heilerziehung *f*; Altlastensanierung *f*
f remédiation *f*; assainissement *m*; réhabilitation *f* de site
r ремедиация *f*; рекультивация *f* земель

* **remelento → 6926**

* **remnant tree → 6887**

* **removal age → 136**

* **removal cut[ting] → 5912**

6861 **removal of buttresses**
d Abbeilen *n* der Wurzelanläufe
f égobelage *m* des empattements
r обрубка *f* надземных корней перед валкой; обтёска *f* закомелистости перед валкой

* **removal of felling area → 1624**

* **removal of needles → 2995**

* **removal of noxious trees → 2121**

6862 **removal of timber; timber removal**
(merchantable volume of trees removed from the inventory by harvesting, cultural operations)
d Forstabgabe *f*; Abgabe *f* von Holz; Holzabgabe *f*; Holzverabfolgung *f*
f enlèvement *m* forestier; délivrance *f* forestière
r отпуск *m* леса (на корню); сдача *f* леса

6863 **removal period**
d Abfuhrfrist *f*
f période *f* de transport
r срок *m* вывозки (леса)

6864 **removal regeneration cutting**
(cutting a stand which has an advanced manageable understor[e]y of 1-20 year old age class reproduction)
d Verjüngungsnachhieb *m*
f coupe *f* de régénération secondaire
r окончательная лесовосстановительная рубка *f*

6865 **removing the first cork layer**
d erste Entkorkung *f*
f démasclage *m*
r удаление *n* пробки

* **renewal felling → 6815**

6866 **rengas**

f rengas *m*
r глюта *f* ренгас; ренгас *m*

* **renovation felling → 6815**

6867 **rented wooded area**
d gepachtete Forstfläche *f*
f superficie *f* boisée affermée
r арендованная залесённая местность *f*

* **repairing of plantation → 823**

* **repair planting → 823**

6868 **repand leaf margin**
(a slightly undulating leaf margin)
d wellenförmiger Blattrand *m*
f bord *m* de feuille ondulé
r волнистая форма *f* края листа

6869 **repellent; repelling agent**
d abstoßendes Mittel *n*
f produit *m* répulsif
r отталкивающее средство *n*

6870 **repetition felling**
d erneuter Einschlag *m*; Nachhiebschlag *m*
f coupe *f* répétitive
r повторная рубка *f*

* **replaced shoot → 6872**

6871 **replacement of wood**
d Holzverdrängung *f*
f refoulement *m* du bois
r вытеснение *n* древесины

6872 **replacing shoot; replaced shoot; substitute shoot**
d Ersatzspross *m*; Ersatztrieb *m*
f rejet *m* de remplacement; pousse *f* de remplacement
r новый побег *m*

6873 **replanting; transplanting**
d Umpflanzung *f*; Verpflanzung *f*
f replantation *f*
r пересадка *f*

6874 **replanting cycle**
d Umpflanzungszyklus *m*
f cycle *m* de replantation
r цикл *m* восстановления посадок

6875 **replication**
(number of times a survey or a treatment is
f réplication *f*
r повторность *f*

* **report of forest survey → 3409**

* **report of forest valuation → 3409**

6876 **reproduction by rooted cuttings**
d Fortpflanzung *f* mittels Wurzelrebe
f reproduction *f* par boutures racinées; régénération *f* par greffes racinées
r воспроизведение *n* укоренившимися черенками

* **reproduction cut[ting] → 6815**

6877 **reproduction from the side; border regeneration; marginal regeneration**
d Randverjüngung *f*
f régénération *f* à partir de la lisière
r возобновление *n* от стен леса при полосных рубках; возобновление от опушки

* **reproduction front → 6818**

* **reproduction method → 5416**

6878 **reproductive bud**
d reproduktive Knospe *f*
f bourgeon *m* reproductif
r семенная почка *f*

* **reproductive capacity → 728**

* **reproductive felling → 6815**

* **reproductive potential → 728**

* **reproductive power → 728**

6879 **resak**
(Vatica rassak; Vatica tonkinensis)
d Resak *m*
f resak *m*
r ватика *f* тонкинская; тау *m*; лантау *m*

6880 **resaw** *v*
(to saw a piece of lumber along its horizontal axis)
d krempelsägen
f refendre
r распиливать по толщине

* **resaw → 1410**

f rescié; recalibré
r распилённый по толщине; повторно распиленный

* **resawing operation → 1017**

* **resawing ripping → 5135**

* **ripping → 5135**

6882 **resawn timber**
d Trennholz *n*
f bois *m* rescié
r распилённый по толщине лесоматериал *m*

6883 **research forester; forest scientist**
d Forschungsförster *m*
f forestier-chercheur *m*
r лесовод-исследователь *m*; учёный-лесовод *m*

6884 **research natural area; scientific-purpose area**
d Forschungsnaturschutzgebiet *n*
f réserve *f* naturelle d'intérêt scientifique
r заповедник *m* научного значения

* **reserve → 5672**

6885 **reserve cutting**
(harvesting that retains live trees of a specific species and size on the area to be cut to achieve a site-specific objective)
d Hieb *m* mit Reserve
f coupe *f* avec réserves
r рубка *f* с оставлением семенных деревьев (при ведении среднего хозяйства)

* **reserved forest → 3557**

* **reserved thrifty standards method → 6888**

6886 **reserved timberland**
(timberland that has statutory or administrative restrictions prohibiting the harvest of trees)
d reservierter Holzboden *m*
f chantier *m* forestier réservé
r резервные лесные угодья *npl*; резервный лесистый участок *m*

6887 **reserved tree; remnant tree; leave tree; standard [of the second rotation]**
(tree selected to remain standing, after the rest of the stand has been felled over a younger or a new crop, for some special purpose)
f arbre *m* marqué en réserve; réserve *f*; franc-pied *m*
r резервное дерево *n*; отборное специальное дерево

* **reserves → 6889**

6888 **reserve seed-tree system; reserved thrifty standards method**
(seed tree method in which some or all of the seed trees are retained after regeneration has become established to attain goals other than regeneration)
d Starkholzzuchtsystem *n*
f méthode *f* de production de bois fort
r метод *m* разведения толстомерного леса

* **reserve sprout forest → 7947**

6889 **reserve stand; remaining stand; residual [stand]; reserves; residual trees; residuals**
(trees, pole-sized or larger, retained after the regeneration period under the clearcutting, seed tree, shelterwood, or coppice methods)
d bleibender Bestand *m*; Rückstand *m*
f peuplement *m* restant [sur pied]; arbres *mpl* réservés; anciens *mpl*; réserves *fpl*; résidu *m*
r резервное насаждение *n*; оставляемая часть *f* насаждения

6890 **residual basal area**
(live trees left standing after the completion of harvesting)
d restliche Basalfläche *f*; Restbasalfläche *f*
f surface *f* terrière résidu
r остаточная общая площадь *f* поперечных сечений [стволов]

* **residuals → 6889**

* **residual [stand] → 6889**

* **residual trees → 6889**

6891 **residual value appraisal**
(timber appraisal that subtracts logging, transportation, and manufacturing costs from the value of finished products to get an estimated stumpage price)
d Restwertabschätzung *f*
f estimation *f* de valeur résiduelle
r оценка *f* остаточной стоимости

* **residue baler → 7235**

* **resin → 8907**

* **resin bleeder → 6905**

* **resin boxing → 828**

6892 **resin boxing period**
d Harzungsdauer *f*
f période *f* de gemmage
r срок *m* подсочки

6893 **resin bubble; pitch pocket; pitch blister; gum blister; resin pocket; resin gall; resin seam; pitch seam; pitch shake; gum check; gum gall; gum pocket**
(in coniferous woods, an opening between growth rings which usually contains or has contained resin or bark or both)
d Harzbeule *f*; Harzgalle *f*; Harztasche *f*; Harzblase *f*
f poche *f* [longitudinale] de résine [dans l'écorce]; vésicule *f* de résine; poche de gomme
r смоляной кармашек *m*; смоляной желвак *m*; трещина *f*, заполненная смолой; полость *f* со смолой

* **resin canal → 4049**

* **resin collector → 6905**

6894 **resin-containing chips**
d harzige Hackspäne *mpl*
f copeaux *mpl* résineux
r смолистая щепа *f*

* **resin duct → 4049**

* **resin extraction → 828**

* **resin exudation → 2977**

* **resin flow → 2977**

* **resin flux → 2977**

* **resin gall → 6893**

* **resin gatherer → 6905**

* **resin glut → 2977**

6895 **resinification; tarring; gumming**
d Verharzung *f*; Anteeren *n*
f goudronnage *m*
r смоление *n*

6896 **resin industry**
r лесохимическая промышленность *f*

6897 **resin knife; [resin] scraper; scraping iron**
d Harzeisen *n*
f couteau *m* à résine
r резак *m* для смолы

* **resin maker → 6905**

6898 **resin of pine trees; pine gum; pine resin; pine tar**
d Föhrenharz *n*; Kiefernharz *n*; Kiefernbalsam *m*; Kieferteer *m*; Kienteer *m*; Kienharz *n*
f résine *f* de pin; goudron *m* de pin
r сосновая живица *f*; сосновая смола *f*

* **resinosis → 2977**

6899 **resinous stumpwood**
d Stockkienholz *n*; verkientes Wurzelholz *n*; harziges Wurzelholz
f bois *m* de souche anormalement résineux
r пнёвый осмол *m*

6900 **resinous tree; pitch-yielding tree**
d harzreicher Baum *m*; harzführender Baum
f arbre *m* résineux
r смолистое дерево *n*; смолоносное дерево

6901 **resin[ous] wood; gum wood**
d Alaunholz *n*; geharztes Holz *n*
f bois *m* résineux
r смолистая древесина *f*; засмолённая древесина; осмол *m*; смольё *n*

* **resin plate → 4052**

* **resin pocket → 6893**

6902 **resin-rich heartwood**
d harzreiches Kernholz *n*
f bois *m* de cœur riche en résine
r жирный осмол *m*; обогащённая смолой ядровая древесина *f*

6903 **resin ring gall**
d Ringriss *m* mit Harz
f roulure *f* de résine
r кольцевая смоляная трещина *f*

* **resin scraper → 6897, 6905**

* **resin seam → 6893**

f place *f* de résine
r участок *m* подсочки

* **resin-streak → 8155**

6905 **resin tapper; resin bleeder; resin collector; resin scraper; resin gatherer; resin maker; dipper**
(a person who collects wood resin)
d Harzsammler *m*; Harzkratzer *m*; Harzeinsammler *m*
f récolteur *m* de résine; résinier *m*; gemmeur *m*
r собиратель *m* смолы; сборщик *m* живицы

* **resin tapping → 828**

* **resin tube → 6303**

6906 **resistance**
(ability of an organism to exclude the effect of a pathogen or other damaging factor)
d Resistenz *f*; Widerstand *m*
f résistance *f*
r устойчивость *f*; невосприимчивость *f*

6907 **resistance to cleavage; resistance to fission; splitting strength; cleavage strength**
d Spaltfestigkeit *f*
f résistance *f* à la fente
r стойкость *f* против раскалывания

6908 **resistance to peeling**
d Schälwiderstand *m*
f résistance *f* à l'écorçage
r сопротивление *n* лущению

6909 **reskidding**
d Nachholzabfuhr *f*; Nachrückung *f*
f débusquage *m* répétitif
r вторичная трелёвка *f*

* **resonance bloc → 7851**

* **resonant wood → 7851**

6910 **resource-and-transportation plan**
d Ressourcen- und Transportplan *m*
f plan *m* de transportation des ressources
r проект *m* транспортного освоения [лесных] ресурсов

6911 **resource folio**
(collection of resource capability and forest inventory maps, other resource data, interpretations, and management objectives for each resource sector)
r папка *f* ресурсов

6912 **resource management zone; RMZ**
(zone of the planning area that is distinct from other zones with respect to biophysical characteristics, resource issues or resource management direction)
d Bewirtschaftungszone *f* der Ressourcen
f zone *f* de gestion des ressources
r зона *f* управления ресурсами

6913 **resource overlay**
d Ressourcenüberzug *m*
f revêtement *m* de ressources
r схема *f* сырьевых ресурсов

6914 **resource value**
(includes biological diversity, fisheries, wildlife, minerals, oil and gas, energy, water quality and quantity, recreation and tourism, natural and cultural heritage, timber, forage, wilderness and aesthetics)
d Ressourcenwert *m*
f valeur *f* des ressources
r стоимость *f* ресурсов

* **respiratory root → 86**

6915 **responsible official**
(forest service line officer with the authority and responsibility to oversee the planning process and make decisions on proposed actions)
d leitender Staatsbeamt *m*
f haut fonctionnaire *m*
r ответственный чиновник *m*

* **resting bud → 2573**

* **restocking → 6821**

* **restocking forestland → 6810**

6916 **restoration**
(return of an ecosystem or habitat to its original community structure, natural complement of species and natural functions)
d Restaurierung *f*; Restauration *f*
f restauration *f*
r восстановление *n*

* **restoration of sites → 6846**

6917 **restricted forest**
d Wald *m* mit beschränkter Nutzung
f forêt *f* à exploitation restreinte

6918 **restriction of stock**
d Bestandbeschränkung *f*
f limitation *f* de peuplement
r ограничение *n* густоты древостоя

* **result of the felling → 9777**

6919 **retention harvesting**
d Retentionsholzeinschlag *m*
f coupe *f* à rétention variable
r удерживающие лесозаготовки *fpl*

6920 **reticulate parenchyma**
(the net-like pattern formed on the cross section by rays and regularly spaced bands or lines of axial parenchyma when the bands or lines and the rays are of about the same width and distance apart)
d netzförmiges Parenchym *n*
f parenchyme *m* en réseau
r сетчатая паренхима *f*

6921 **reticulate perforation plate**
(a plate with multiple perforations having a net-like appearance (as in certain Bignoniaceae))
d netzförmige Perforationsplatte *f*
f cloison *f* perforée en réseau
r сетчатая перфорационная пластинка *f*

* **return check rope → 5411**

* **return from final cutting → 3145**

* **return-line → 5411**

* **returns from thinning → 9782**

6922 **return-trip distance**
d Rundfahrt-Distanz *f*
f distance *f* de voyage aller-retour
r расстояние *n* трелёвки в порожнем направлении

6923 **revegetation**
(reestablishing and developing a plant cover)
d Erneuerung *f* des Pflanzenwuchses; Grünverbauung *f*; Lebendverbauung *f*; biologische Verbauung *f*
f revégétation *f*; reverdissement *m*; restauration *f* de couverture végétale
r восстановление *n* растительности

6924 **reverse casehardening**
(a stressed condition in seasoned timber characterized by tension in the outer layers overconditioning)
d Rückeinsatzhärtung *f*
f cémentation *f* inverse
r напряжённое состояние *n* древесины в конечный период сушки (наружные слои находятся в сжатом состоянии, а внутренние - в растянутом)

6925 **reverse radial pattern**
d umgekehrtes Radialmodell *n*
f modèle *m* radial inverse
r обратно-радиальная схема *f* разработки делянки канатной установкой

6926 **rheedia; pacuri; pacuru; remelento**
(Rheedia)
d Rheedia *f*
f cirolier *m*; rhéédie *f*
r реедия *f*

* **rhizobium → 7037**

* **rhizogenic → 6929**

6927 **rhizome; root-stalk; understock**
(the stock or root plant onto which a shoot has been grafted to produce a new plant in bud-grafting, especially in cases of double-worked trees)
d Wurzelstock *m*; Rhizom *n*
f rhizome *m*; ensouchement *m*
r корневище *n*; ризом *m*

6928 **rhizophore**
(a leafless stem that produces roots)
d Rhizophor *m*; Wurzelträger *m*
f rhizophore *m*
r ризофор *m*

6929 **rhizophorous; rhizogenic** *adj*
(bearing roots)
d rhizophorig
f rhizophore
r образующий корни

6930 **rhizosphere**
(soil surrounding and directly influenced by plant roots)
d Rhizosphäre *f*
f rhizosphère *f*
r ризосфера *f*

* **rhizotaxis → 7011**

* **Rhodesian copaltree → 113**

* **Rhodesian teak → 9793**

* **Rhodes mahogany → 6406**

6931 **rhythmic growth**
(growth of an axis determined by a rhythm that results in periodic shoot extension alternating with dormancy)
d rhythmisches Wachstum *n*
f croissance *f* rythmique
r ритмический рост *m*

* **rhytidome → 5975**

* **rib → 4882**

6932 **ribbed blade**
d Rippenblatt *n*
f lame *f* nervurée
r рифлёное лезвие *n*

6933 **ribbed buttress**
d Rippenstrebepfeiler *m*
f empattement *m* nervuré
r ребристая закомелистость *f*

6934 **ribbed pine borer**
(Rhagium lineatum; Rhagium inquisitor)
d kleiner Kiefernzangenbock *m*
f rhagie *f* chercheuse
r рагий *m* бороздчатый

* **ribbing → 2019**

* **ribbon figure → 4539**

* **ribbon grain → 4539**

* **ribbon gum → 5286**

* **ribbon saw → 5049**

* **ribbon stripe → 4539**

6935 **ribbonwood**
d gestreiftes Holz *n*
f bois *m* rubané
r полосчатая древесина *f*

* **rich in nutrients → 2912**

6936 **richly wooded; densely wooded; richly timbered; well-wooded; thickly wooded; rich in forests; well-forested**
d waldreich; baumreich
f abondant en forêt

* **rich tool → 3200**

* **rick → 2979**

6937 **ride**
(the side of a log upon which it rests when being dragged)
d Jägersteig *m*
f layon *m*
r сторона *f* брёвна, соприкасающаяся с землёй (при трелёвке)

* **ride → 3564**

* **ridge → 1169, 7007**

6938 **ridge-and-valley road pattern**
d Tal- und Rückenwegmodell *n*
f modèle *m* de chemin en crêtes et vallées
r схема *f* транспортного освоения горных лесосек с дорогой, поднимающейся по впадине, а затем поворачивающей на гребень

* **ridge beam → 7007**

* **ridgepiece → 7007**

* **ridge planting → 648**

* **ridge pole → 7007**

* **ridging → 1171**

* **riding → 2057**

* **rift-grained lumber → 6678**

* **rift[-sawn] timber → 6678**

6939 **rig** *v*
d ausrüsten
f gréer
r оснащать

* **rigger → 4296**

* **rigging → 7467**

* **rigging slinger → 1559**

6940 **right-holder to wood; holder of timber right**
(individual, group, or company that holds a

f usager *m* au bois
r владелец *m* права на лес

6941 **right-of-way**
(strip of land over which a power line, railway line, road, etc., extends)
d Vorfahrtrecht *n*; Durchgangsrecht *n*; Wegrecht *n*; Wegebenutzungsrecht *n*
f droit *m* de passage; emprise *f* [routière]; emprise de la route
r полоса *f* отвода; трасса *f*

6942 **right-of-way clearing**
d Straßengrundstücksräumung *f*
f libération *f* d'emprise
r вырубка *f* леса на полосе отвода

6943 **right spirally growth; anticlockwise spiral growth**
d rechtsdrehender Drehwuchs *m*; widersonniger Drehwuchs
f croissance *f* avec torsion à droite
r правый косослой *m*; противосолнечный косослой

6944 **right timber**
d Berechtigungsholz *n*
f bois *m* d'usage
r сервитутная древесина *f*

6945 **right to float timber; rafting right; floating right**
d Flöß[erei]recht *n*
f droit *m* de flottage
r право *n* сплава

6946 **right to [get] firewood; common of firewood**
(an allowance to a person out of an estate to get firewood for life)
d Brennholzberechtigung *f*
f droit *m* au bois de chauffage; affouage *m*; estovoir *m*
r сервитут *m* на дрова

6947 **right to litter**
d Streuberechtigung *f*; Streunutzungsrecht *n*; Streunutzungsservitut *n*
f droit *m* de ramassage de feuilles mortes; servitude *f* de ramassage de feuilles mortes
r сервитут *m* на сбор подстилки

6948 **rig timber**
d Bohranlageholz *n*
f bois *m* d'appareil de forage
r лесоматериал *m* для буровых установок

6949 **rill erosion**
(formation of rills as a consequence of poor cultivation)
d Bächleinerosion *f*; Rillenerosion *f*
f érosion *f* par sillons
r бороздчатая эрозия *f*; бороздная эрозия

* **rill hook → 1204**

6950 **rim saw; segmental circular saw**
d kreisförmige Segmentsäge *f*
f scie *f* circulaire à segments rapportés
r дисковая сегментная пила *f*

* **rimu → 6784**

* **rind → 2280**

* **rind grafting → 561**

* **ring → 3851**

* **ringbark → 3851**

* **ring barker → 7074**

6951 **ring-barking; ring-girdling; ring-stripping; ringing**
(removing a narrow strip of bark, all around: 1. a living stem, in order to stimulate flowering or to girdle it; 2. a felled stem or a log, for under-bark diameter measurement)
d Ringeln *n*; ringförmiges Abschälen *n*
f écorçage *m* circulaire
r кольцевая окорка *f*; роторная окорка; снятие *n* коры кольцами

* **ring check → 2139**

6952 **ring eccentricity**
d Kernexzentrizität *f*
f croissance *f* excentrique du cernes
r эксцентричность *f* годичных колец; эксцентричность годичных слоёв

* **ring failure → 2139**

* **ring gall → 2139**

6953 **ring-girdling; ringing; digging round a tree; deadening**
(destruction of tissue, particularly living tissue, in a rough ring around a stem, branch or root)
d Beringelung *f*; Ringelung *f*
f annélation *f* [partielle]; annelage *m*
r кольцевое обгрызание *n*; кольцевые

* **ring-girdling → 6951**

* **ringing → 6951, 6953**

* **ring pattern → 7451**

* **ring-pored wood → 6955**

6954 **ring porosity**
d Ringporigkeit *f*
f porosité *f* en anneau; porosité en cercles concentriques
r кольцевая пористость *f*

6955 **ring-porous wood; ring-pored wood**
(wood in which the pores of one part of a growth ring are in distinct contrast in size or number (or both) with those of the other part)
d ringporiges Holz *n*
f bois *m* à pores circulaires; bois à zone poreuse
r кольцесосудистая лиственная древесина *f*; кольцепоровая древесина; кольцепористая древесина

6956 **ring rot**
(circular rot in a log; decay in a log, which follows the annual rings more or less closely)
d Ringfäule *f*
f carie *f* du bois initial
r кольцевая гниль *f*

* **ring sequence → 7451**

* **ring series → 7451**

* **ring shake → 2139**

* **ring-splice → 699**

6957 **ring[ ]spot**
(a circular area of chlorosis with a green center)
d Ringfleck *m*
f tache *f* en anneau
r кольцевое пятно *n*

* **ring-stripping → 6951**

* **ring[-type] debarker → 7074**

6958 **ring-width chronology; tree-ring chronology**
(time series of annual ring widths of one or several trees)
d Jahresringenchronologie *f*

* **Rio rosewood → 1006**

6959 **rip** *v*
(to saw along the grain)
d aufreißen; zerreißen
f refendre
r распиливать вдоль волокон

6960 **riparian afforestation**
d Uferbepflanzung *f*
f afforestation *f* ripicole; boisement *m* ripiphile; boisement rivulaire
r облесение *n* берегов

* **riparian corridor → 6963**

6961 **riparian forest; riverian forest; fringing forest; tugai**
(vegetation, with trees and shrubs, growing alongside or close to a watercourse, lake, swamp)
d Ufer[schutz]wald *m*; Wasserschutzwald *m*; Flussuferwald *m*; Tugai *m*
f forêt *f* ripicole; forêt riveraine; ripisylve *f*; tugaï *m*
r крайречный лес *m*; берегозащитный лес; водоохранный лес; тугайный лес; тугай *m*; урема *f*

* **riparian management area → 8157**

6962 **riparian reserve zone**
(that portion, if any, of the riparian management area or lakeshore management area located adjacent to a stream, wetland or lake)
d reservierte Uferzone *f*
f zone *f* réserve ripicole
r прибережная охранная зона *f*

6963 **riparian [zone]; riparian corridor**
(area adjacent to a waterway such as a river, stream, or lake)
d Uferzone *f*
f zone *f* riveraine
r зона *f*, соседняя берегам водоёмов

* **ripcut → 5135**

* **ripe for the harvest → 3231**

* **ripening stand → 4425**

* **ripe stand → 5338**

6964 **ripewood; adult wood; mature wood**
(wood which is characterized by relatively

d ausgereiftes Holz *n*; hiebreifes Holz; reifes Holz; Reifholz *n*
f bois *m* mûr; bois adulte
r спелая древесина *f*; зрелая древесина

6965 **ripewood colo[u]r**
d Altersfarbe *f* des Holzes
f couleur *f* de bois mûr
r цвет *m* спелой древесины

6966 **ripper**
(toothed blade or set of heavy tines mounted at the front or rear of a vehicle for breaking up soft rock and hard ground, and tearing out stumps and boulders)
d Längssägemaschine *f*; Brettsäge *f*; Vielblockbandsäge *f*; Langssäge *f*; Längensäge *f*
f scie *f* pour débit longitudinal; scie circulaire à refendre; dessoucheuse *f* à griffe; défonceuse *f*
r многопильный станок *m* для продольной распиловки

* **ripper → 7313**

6967 **ripping**
(mechanical penetration and shearing of range soils to depths of 3-7 cm for the purpose of breaking hardpan layers to facilitate penetration of plant roots, water, organic matter, and nutrients)
d Auflockern *n*; Auflockerung *f*; Lockerung *f*
f ripage *m*
r рыхление *n*

* **ripping saw → 5148**

6968 **ripple mark**
(fine horizontal striation visible on the tangential longitudinal surfaces of certain woods, due to the storied arrangement of the rays or of the axial elements or of both)
d horizontale Streifung *f*
f ligne *f* d'étagement
r след *m* струйчатости; волнообразный знак *m*; рифель *m*; рябь *f*

* **rippling → 2031**

6969 **riprap; array of stones; tipped stone; loose rock-dump; rock filling**
(rough stones of various sizes placed compactly or irregularly on the ground surface to prevent scouring by water or debris)
d Steinpackung *f*; Steinschüttung *f*; régulier; perré *m*
r каменная наброска *f*

* **ripsaw → 5148**

* **rip sawing → 5135**

* **rise → 8083**

6970 **rise-and-fall saw**
d Vertikalverschiebungssäge *f*
f scie *f* circulaire à niveau réglable
r круглая пила *f* с регулируемой глубиной пропила

6971 **risk of forest fire; forest-fire danger; forest-fire hazard**
(assessment of both fixed and variable factors of the fire environment, which determine the ease of ignition, rate of spread, difficulty of control, and the fire impact)
d Waldbrandgefahr *f*; Waldbrandrisiko *n*
f danger *m* d'incendie de forêt; danger de feu forestier
r опасность *f* лесного пожара

6972 **risk rating assessment**
(process of identifying the degree of risk that timber harvesting imposes on adjacent and downslope social, economic, and forest resource values)
d Gefährdungsabschätzung *f*; Risikobewertung *f*
f évaluation *f* de cote de vulnérabilité; évaluation de cote des risques; estimation *f* du risque
r оценка *f* риска

6973 **risk [tree] cutting**
(cutting to remove trees that are likely to die before the next periodic cut)
d Risikobaumhieb *m*
f coupe *f* d'arbres à risque
r рубка *f* рисковых деревьев

* **riven → 1634**

* **riven timber → 8563**

* **river birch → 779**

* **river cottonwood → 8306**

* **river driver → 9569**

* **riverian forest → 6961**

6974 **river landing**

f premier dépôt *m* transitoire à rivière
r приречный склад *m*

* **river-oak → 407**

* **river red gum → 6786**

6975 **river willow; ducky willow**
(Salix fluviatilis; Salix melanopsis)
d Flussweide *f*
f saule *m* noirâtre
r ива *f* речная

* **riving machine → 9512**

* **RMA → 8157**

* **RMZ → 6912**

* **road change → 9751**

6976 **road deactivation**
(measures taken to stabilize roads and logging trails during periods of inactivity)
d Weg-Deaktivierung *f*
f désactivation *f* de chemins
r консервация *m* путей

* **road engine → 1160**

* **roader → 1160**

6977 **roadless forest area**
d weglose Waldfläche *f*
f aire *f* forestière sans chemin
r лесная зона *f* без путей

6978 **road[ ]pattern**
d Wegschema *n*
f schème *m* de chemins
r схема *f* транспортного освоения лесосек

6979 **road permit**
(for the construction or modification of a forest road to facilitate access to timber planned for harvest)
d Weggenehmigung *f*
f permis *m* de chemins
r дорожный пропуск *m*

6980 **road prism**
(area of the ground containing the road surface, cut slope and fill slope)
d Wegprisma *n*
f prisme *m* de chemin
r призма *f* дороги

(scaling of log which is taken on the landing)
d Skalierung *f* im Weg
f cubage *m* au chemin
r обмер *m* брёвен на погрузочной площадке

* **road service employees → 3119**

6982 **roadside harvester**
d Chaussee-Erntemaschine *f*
f récolteuse *f* routière
r передвижная обрабатывающая [сучкорезно-раскряжёвочная] машина *f*

* **roadside slasher → 6470**

6983 **road[side] tree**
d Chausseebaum *m*; Straßenbaum *m*
f arbre *m* routier; arbre d'alignement routie
r придорожное дерево *n*; подорожное дерево

* **road slide → 5116**

6984 **roasted charcoal**
d geröstete Kohle *f*
f charbon *m* grillé
r муфельный древесный уголь *m*

* **robinia → 1754**

* **roble → 5346**

6985 **roble pellin; Antarctic beech**
(Nothofagus antarctica)
d antarktische Scheinbuche *f*; Roble-Pellin *n*
f roble pellin *m*; phellodendron *m* au liège de l'Amour; arbre *m* au liège de l'Amour
r антарктический нотофагус *m*

* **robusta poplar → 4400**

6986 **rock blade; stump blade**
(having scarifier teeth instead of a plain edge, for pushing large objects like tree roots and rocks off a site, leaving smaller stones, soil, etc. in place)
d Räumschild *m*; Stumpfaxt *f*
f lame de râteau; lame dessoucheuse; lame à rémanents
r зубчатое бульдозерное полотно *n*; отвал *m* [бульдозера] для корчевания пней

6987 **rock buckthorn**
(Quercus saxatilis)
d Felsen-Kreuzdorn *m*
f nerprun *m* des roches
r жестер *m* слабительный

6988 rock elm; spreading elm
(Ulmus thomasii)
d Felsenulme *f*; Flügelulme *f*
f orme *m* des roches; orme de Thomas; orme à grappes; orme liège
r вяз *m* кистевидный; вяз Томаса; вяз ветвистый

* **rock filling → 6969**

* **rock gilia → 6534**

* **rock maple → 808**

6989 rock rhododendron
(Rhododendron ferrugineum)
d braune Alpenrose *f*; Donnerrose *f*; Schneerose *f*
f rose *f* des Alpes; rhododendron *m*
r [альпийский] рододендрон *m*; рододендрон ржаволистный; рододендрон ржавый

6990 rockrose
(Cistus)
d Cistrose *f*; Zistrose *f*
f ciste *f*
r ладанник *m*

6991 rockrose family
(Cistaceae)
d Zistrosengewächse *pl*
f cistacées *pl*
r ладанниковые *pl*; цистусовые *pl*

6992 rockrose shrubbery
d Zistrose-Gestrüpp *m*
f cistaie *f*
r заросль *f* ладанников

* **rock saw → 4312**

6993 rock spray; rock spirea
(Holodiscus)
d Schaumspiere *f*
f holodisque *m*
r холодискус *m*

6994 rockwood
d Holzasbest *m*
f bois *m* de roche
r каменная древесина *f*

* **Rocky Mountains Douglas fir → 1772**

6995 Rocky Mountains juniper
(Juniperus scopulorum)
r можжевельник *m* скальный

6996 Rocky Mountains maple; soft maple
(Acer glabrum)
d Zwergahorn *m*
f érable *m* américain; érable nain de Douglas
r клён *m* голый

6997 Rocky Mountains pine; bristlecone pine; foxtail pine; ancient pine; Balfour pine
(Pinus longaeva; Pinus aristata; Pinus balfouriana)
d Grännenkiefer *f*; Borstenkiefer *f*; Fuchsschwanzkiefer *f*; Balfourskiefer *f*
f pin *m* à cône épineux des montagnes Rocheuses; pin queue de renard
r сосна *f* остевидная; сосна остистая; сосна Бальфура

* **Rocky Mountains pinyon → 9031**

* **Rocky Mountains white pine → 4965**

* **rod → 8292**

6998 rogu[e]ing
(systematic removal of individuals not desired for the perpetuation of a population)
d künstliche Zuchtwahl *f*
f élimination *f*
r искусственный отбор *m* (нежелательных деревьев в семенных древостоях)

* **rogue [tree] → 9321**

6999 roll bark
(hemlock tanbark that has not been carefully dried and hence is of inferior quality)
d zusammengerollte Lohrinde *f*
f écorce *f* à tan en rouleaux
r свёрнутое корьё *n*

* **roll cut hand shears → 322**

* **roller skid → 2054**

* **rolling → 2052**

7000 rolling chopper
d Rollzerkleinerer *m*
f rouleau *m* débroussailleur
r роликовый обрубщик *m* сучьев

* **rolling dam → 5039**

* **rolling dog → 1317**

7001 rolling-off; roll off
d Abrollen *n*
f désencrouage *m*
r поворачивание *n* [зависшего дерева с помощью рычага]

7002 rolling of stumpwood chips
d Stubbenholzspanewalze *f*
f roulage *m* de copeaux de bois gras
r вальцевание *n* смольной щепы

* **rolling stalk cutter → 7319**

7003 rolling-table saw; flat-top saw
d Tischkreissäge *f*
f scie *f* circulaire à plateau; scie à chariot libre
r роликовая пила *f*

* **roll off → 7001**

* **roll up → 8369**

7004 rollway
(a place prepared for rolling logs into a stream)
d Rollbett *n*; Überlaufstrecke *f*
f partie *f* déversante d'un barrage
r площадка *f* на крутом склоне

* **Roman willow → 6292**

7005 roof batten; roof lathe
(small timbers fixed to the top of rafters to which the roofing material is secured)
d Dachlatte *f*
f latte *f* de toit; latte voltige
r обрешетина *f*

7006 roof timbering
d Dachstuhl *m*
f boisage *m* du faîte
r деревянная крепь *f* кровли

7007 roof-tree; ridge[piece]; ridge beam; ridge pole
(the highest part of the roof at the meeting of the upper ends of the common rafters)
d Firstbalken *m*
f poutre *f* faîtière
r коньковый брус *m*; конёк *m* [крыши]

* **roominess → 5914**

7008 roomy; having blanks
d räumig; waldlos
f spacieux
r с просветами

7009 root
d Wurzel *f*
f racine *f*
r корень *m*

* **rootage → 7029**

* **root anchorage → 7029**

7010 root-and-shoot cutting; shoot cutting; stump transplant
d Wurzelsteckling *m*
f bouture *f* végétative
r вегетативный черенок *m*; побеговой черенок

7011 root arrangement; root disposal; radication; rhizotaxis
(the arrangement of the roots of plants)
d Wurzelanordnung *f*; Rhizotaxis *f*
f disposition *f* de racines; arrangement *m* de racines; rhizotaxie *f*; radication *f*
r расположение *n* корней; корнерасположение *n*

7012 root ball
(the clump of soil containing the roots of a tree, often bound in burlap)
d Wurzelballen *m*
f chevelu *m* de la racine
r корневой пучок *m*

* **root buttress → 7015**

* **root butt swelling → 7015**

7013 root-cap
d Wurzelhaube *f*; Wurzelmütze *f*; Wurzelmützchen *n*
f coiffe *f* de racine
r корневой чехлик *m*

7014 root climber
(a plant that raises itself up by means of adventitious roots, which adhere to the plants that support it)
d Wurzelschlinger *m*
f plante *f* à racines aériennes
r растение *n*, цепляющееся с помощью воздушных корней

7015 root collar; root crown; root-neck; root swelling; root butt swelling; root buttress
d Wurzelhals *m*; Wurzelanlauf *m*; Wurzelansatz *m*; Stockanschwellung *f*
f collet *m* de racine
r шейка *f* корня; корневая шейка; корневая закомелистость *f*; нарост *m* у корня;

7016 root competition
d Wurzelkonkurrenz *f*
f concurrence *f* entre racines
r межкорневая конкуренция *f*

7017 root crops
d Hackfrüchte *fpl*
f plantes *fpl* racines; plantes sarclées
r корнеплоды *mpl*

* **root crown → 7015**

* **root cutter → 7019**

7018 root cutting
d Wurzelschnittling *m*
f bouture *f* de racine
r корневой черенок *m*

7019 root cutting machine; root cutter; root pruner; root slicer
d Wurzelschneidemaschine *f*; Rübenschneider *m*; Rübenschneidemaschine *f*
f appareil *m* à trancher les racines; coupe-racines *f*
r корнерезная машина *f*; машина для обрезки корней

* **root disposal → 7011**

7020 rooted branch
d verwurzelte Ast *f*; Ableger *m*
f branche *f* racinée
r укоренившаяся ветка *f*

7021 rooted cutting
d Wurzelrebe *f*; Wurzelstange *f*
f bouture *f* racinée; greffe *f* racinée
r укоренившийся черенок *m*

* **root-end → 7046**

* **rooter → 7042**

7022 root extension
d Wurzelausdehnung *f*; Wurzelerweiterung *f*
f extension *f* de racines; prolongement *m* de racines; rallongement *m* de racines
r протяжённость *f* корней

* **root extractor → 8808**

7023 root feeders
d Wurzeltroge *mpl*
f ravageurs *mpl* de racines
r вредители *mpl* корней

d Wurzelbildung *f*
f formation *f* de racines
r корнеобразование *n*; образование *n* корней

7025 root fungus
(Fomitopsis annosa; Heterobasidion annosum)
d Wurzelpilz *m*
f champignon *m* racinaire
r корневая губка *f*

* **root gall → 7036**

7026 root graft
d Wurzelsprosse *f*
f greffe *f* racinaire
r прививка *f* в корень

7027 root growth
d Wurzelwachstum *n*
f croissance *f* de racines
r рост *m* корней

* **root habit → 7050**

7028 root hairs
(unicellular, hairlike growth from a root)
d Haarwurzeln *fpl*; Wurzelhaare *npl*
f chevelu *m* racinaire
r корневые волоски *mpl*

7029 rooting; rootage; root anchorage; root penetration
d Durchwurzelung *f*; Anwurzeln *n*; Bewurzelung *f*; Wurzelverankerung *f*
f enracinement *m*; ancrage *m* radiculaire; pénétration *f* des racines
r укоренение *n*; прикрепление *n* корнями

7030 rooting ability
d Durchwurzelungskapazität *f*
f pouvoir *m* d'enracinement
r способность *f* укоренения

7031 rooting of cutting
d Steckling-Durchwurzelung *f*
f enracinement *m* de bouture
r укоренение *n* черенка

7032 root[ing] zone; root range
(of a soil)
d Wurzelzone *f*
f horizon *m* racinaire; zone *f* radiculaire; zone de racines
r корнеобитаемый слой *m*; зона *f* [распространения] корней

7033 root length
(layer of the soil interwoven by plant roots)
d Wurzellänge *f*
f longueur *f* de racines
r длина *f* корневой части

7034 rootless
d wurzellos; ohne Wurzeln
f sans racines
r без корней; бескорешковый

7035 rootlet; rooty fiber
d Keimwurzel *f*; Wurzelfaser *f*; Würzelchen *n*
f fibre *f* radiculaire; radicelle *f*
r корешок *m*; проросток *m*; корневая мочка *f*

* **root-neck → 7015**

7036 root nodule; root gall; tubercle
d Wurzelknöllchen *n*; Tuberkel *n*
f tubercule *f*; nodule *m* racinaire
r корневой клубенёк *m*

7037 root nodule bacteria; nodule bacteria; rhizobium
(type of bacterium that has the capacity of both invading the roots of certain species of the Leguminosae and fixing atmospheric nitrogen, which is subsequently used by the host plant)
d Knöllchenbakterien *fpl*; Rhizobium *n*
f bactéries *fpl* des nodosités
r клубеньковые бактерии *fpl*

* **root out → 4033**

7038 root penetrability; root penetration capability
d Durchdringbarkeit *f* der Wurzel; Durchwurzelbarkeit *f*; Durchdringvermögen *n* der Wurzel
f pénétrabilité *f* des racines; aptitude *f* des racines à la pénétration
r способность *f* проникновения корней

* **root penetration → 7029**

7039 root-pruned
d mit verschnittener Wurzel
f à racine élaguée; rhabillé
r с обрезанным корнем

* **root pruner → 7019**

7040 root pruning; root trimming; paring of the roots
outplanting)
d Wurzelschnitt *m*; Beschneiden *n* der Wurzeln
f élagage *m* des racines; taillage *m* des racines
r подрезка *f* корней; укорачивание *n* корней; обрезка *f* корней; обрезывание *n* корней

7041 root puddling
(act or treatment of immersing, sometimes several times in close succession, the root systems of bare-root planting stock in a clay slurry with the aim of improving outplant performance)
d Wurzeleinschlämmen *n*
f pralinage *m* des racines
r замывание *n* корней

* **root-puller → 8808**

7042 root rake; rake blade; rooter
(implement, either mounted on the front of a dozer, skidder or forwarder, or trailed, having tines for collecting stumps and slash)
d Wurzelextraktor *m*
f râteau *m* déracineur
r зубчатый корчеватель *m*

7043 root raking; raking
(site preparation technique using a bulldozer or similar equipment with a blade having teeth instead of a plain edge, for pushing large, coarse woody debris and rocks)
d Wurzelrechen *n*
f râtelage *m*
r корчевание *n* бульдозером

* **root range → 7032**

7044 root rot; stump rot
(disease that destroys tree roots, often killing the tree; invaders - Armillaria, Oozonium, Thielavia etc.)
d Wurzelschwamm *m*; Wurzelfäule *f*
f pourriture *f* des racines; pourridié *m*
r корневая гниль *f*

7045 root sap
d Wurzelsaft *f*
f sève *m* de racine
r корневые выделения *npl*

7046 root shoot; stem-tuber; surculus; root-end; sprout root
(a root-bearing plant or plant part, generally a stem or root, onto which another plant part is grafted)
d Wurzelschoss *m*; Wurzelbrut *f*; Ausläufer *m*

f porte-greffe *m*; drageon *m* [racinaire]
r корневой отпрыск *m*; корневой побег *m*; волчок *m*; корневой подвой *m*

* **root slicer → 7019**

7047 root sprung tree
(one that has sustained damage to the root system due to the force of the wind acting on the tree's crown)
d Baum *m* mit gefederten Wurzeln
f arbre *m* à racines suspendues
r дерево *n* с висячими корнями

* **root-stalk → 6927**

7048 root stripping
(accidental removal of roots during lifting, handling, and planting, especially when caused by improper practices)
d Wurzelausschlachtung *f*
f dépouillement *m* des racines
r обнажение *n* корней

7049 root sucker; sucker; seeker
d Saugwurzel *f*
f drageon *m*; surgeon *m*
r сосущий корень *m*

* **root swelling → 7015**

7050 root system; root habit
d Wurzelsystem *n*; Wurzelwerk *n*; Wurzelnetz *n*
f système *m* racinaire; système radiculaire
r корневая система *f*

* **root timber → 9580**

7051 root tip
d Wurzelende *f*
f extrémité *f* de racine
r верхушка *f* корня; кончик *m* корня

7052 root-[to-]shoot ratio
(total mass or volume of the plant root system divided by the total mass or volume of the shoot system, usually on an oven-dry basis)
d Verhältnis *n* zwischen Wurzeln und Schüssen
f rapport *m* système racinaire/système foliacé
r соотношение *n* массы корней и побегов

* **root trimming → 7040**

7053 root tuber; corm
(a specialized part of a stem; a short, enlarged base of a stem where food is stored)
r корневой клубень *m*

7054 root wad
(the ball of roots, soil and rocks that remains intact when a tree, shrub, or stump is uprooted)
d Wurzelbündel *n*; Wurzelbausch *m*
f motte *f* racinaire
r корневой комок *m*

* **rootwood → 9580**

7055 root-wrenching; wrenching; lifting
(nursery operation to condition nursery stock by loosening the contact between soil and roots of seedlings in a nursery bed)
d Wurzelanheben *n*
f soulevage *m* [des racines]
r обрыв *m* корней; подрезка *f* боковых корней (сеянцев при подъёме пласта почвы в питомнике)

* **rooty fiber → 7035**

* **rope adjuster → 7058**

* **rope clamp → 7699**

* **rope clip → 7699**

7056 rope coil
d Seiltrommel *f*
f trommel *m*
r барабан *m* для намотки каната

* **rope cramp → 7699**

7057 rope guide; rope guiding
d Seil[rollen]führung *f*
f guide-câble *m*
r проведение *n* троса

7058 rope guider; rope adjuster
d Seilkürzer *m*; Seilverteiler *m*
f ajusteur *m* de câble
r канатоукладчик *m*; тросоукладчик *m*

* **rope guiding → 7057**

* **rope not carrying the load → 7714**

7059 rope of ordinary lay
d Kreuzschlagseil *n*
f câble *m* à torsions croisées
r крестовый трос *m*

* **ropeway [track] → 1245**

7060 rope winch; cable winch
d Kabelwinde *f*; Aufseilwinde *f*; Seilwinde *f*
f treuil *m* de câble
r канатная лебёдка *f*; тросовая лебёдка; лебёдка для подвесной трелёвки

* **rope without twist → 6503**

7061 ropey figure
(markings in the form of a twisted rope)
d seilartige Figur *f*
f figure *f* cordée
r волокнистая фигура *f*

7062 roping
d Längsrillen *n*; Abseilen *n*
f cordage *m*; contention *f*
r крепление *n* тросов

* **ROS → 6730**

* **rose → 995**

* **rose acacia → 1038**

* **rose bay → 5887**

7063 rosebay shrublet
d Alpenrosengesträuch *m*
f rhodoraie *f*
r заросль *f* рододендронов; рододендровник *m*

7064 rose family
(Rosaceae)
d Rosengewächse *pl*
f rosacées *pl*
r розоцветные *pl*

7065 rosegold pussy willow; big-catcin willow
(Salix gracilistyla)
d schlankgriffelige Weide *f*; Kätzchenweide *f*
f saule *m* aux chatons noirs
r ива *f* тонкостолбиковая

* **rose gum → 3909**

7066 rosemary; rosmarin
(Rosmarinus officinalis)
d Rosmarin *m*
f romarin *m*; herbe *m* aux couronnes
r розмарин *m* аптечный

* **rosemary pine → 5029**

7067 rose of Sharon; shrubby althea
(Hibiscus syriacus)
d Strauch-Eibisch *m*
f ketmie *f* de Syrie; mauve *f* en arbre
r гибискус *m* сирийский

7068 rose-tree
d Rosenstock *m*; Rosenstrauch *m*
f rosier *m*
r розовый куст *m*

* **rosmarin → 7066**

* **ross → 2280, 5975**

* **rossed → 551**

* **rosser → 556**

7069 rosser[-type] debarker
(machine that peels bark using knives)
d Entrindungsmaschine *f* mit Messer
f écorceuse *f* à couteaux rotatifs
r окорочный станок *m* с вращением брёвна

* **rossing → 579**

* **rosy trumpet-tree → 5346**

* **rot → 9529**

* **rotary barker → 7074**

* **rotary crane → 8343**

7070 rotary cutter; rotary cutting blade
d Kreisschere *f*; Kreisscherenmesser *m*
f fraise *f* d'abattage; cisaille *f* circulaire
r режущий диск *m*; роторный срезающий орган *m*; фреза *f*

* **rotary cutter attachment → 403**

* **rotary cutter [felling] head → 403**

* **rotary cutting blade → 7070**

* **rotary-cut veneer → 6135**

* **rotary harvester → 456**

* **rotary peeler → 7074**

* **rotary saw → 1585**

f débroussailleuse *f* rotative
r ротационный кусторез *m*

7072 **rotary tiller; tiller; tilling machine**
(site preparation machine using hammers, teeth, tines, or flails mounted on a horizontal drum or horizontal or vertical shaft revolving at high speed)
d Rotorkrümler *m*; Rotorackerfräse *f*; Rotationshackmaschine *f*; Rotorhacke *f*
f laboureur *m* à lames rotatives; rotobineuse *f*; cultivateur *m* rotatif
r почвенная фреза *f*

7073 **rotary-type chip screen**
d rotierender Hackspänesortierer *m*
f crible *m* à copeaux rotatif
r вращающееся барабанное сито *n* для щепы

7074 **rotating ring debarker; hollow head debarker; cambium-shear debarker; ring[-type] debarker; ring barker; rotary peeler; rotary barker**
d Lochrotorenrinder *m*; Ringrotorentrinder *m*; Rotorentrinder *m*
f écorceuse *f* à anneau rotatif; dérouleuse *f*
r роторный окорочный станок *m*; роторная корообдирка *f*

* **rotation age → 136**

7075 **rotation burning**
(prescribed burning applied at regular intervals on a specific site as a means of pest control)
d Wechselverbrennen *n*
f brûlage *m* cyclique
r циклическое сжигание *n*

* **rotation by areas → 6842**

* **rotation length → 7077**

7076 **rotation of the highest interest of forest capital**
d Umtrieb *m* des höchsten Waldreinertrages
f révolution *f* du meilleur rendement
r оборот *m* высшей лесной ренты

7077 **rotation period; rotation length; circulation period; cutting interval**
d Umtriebszeit *f*; Umlaufzeit *f*; Umschlagszeit *f*
f durée *f* de révolution; période *f* de révolution
r интервал *m* между рубками

f pourriture *f* de lésion
r раневая гниль *f*; гниль от поранения

* **rot-damaged → 7212**

* **rot of the stem → 8998**

* **rot of timber → 9529**

* **rot of wood → 9529**

* **rot speckle → 7081**

* **rotted → 7212**

7079 **rotted stick**
d Faultknüppel *m*
f billot *m* pourri
r хлыст *m* с гнилью

* **rotten → 7212**

7080 **rotten cull tree**
d fauliger Abfallbaum *m*
f arbre *m* de rebut pourri
r гнилое неделовое дерево *n*; гнилое фаутное дерево

* **rotten heart → 2289**

* **rotten knot → 2290**

7081 **rotten spot; necrotic spot; rot speckle**
d Faulstelle *f*; Faulfleck *m*
f marque *f* de pourriture; tacheture *f* de pourriture
r гнилое место *n*; [неглубокое] гнилостное пятно *n*

* **rotten timber → 2291**

7082 **rotten tree; defective tree; doted tree; faulty tree**
(tree or log or portions thereof that are of merchantable size but are rendered unmerchantable by defects; rough or rotten tree)
d morscher Baum *m*; Fehlbaum *m*
f arbre *m* de rebut
r гнилое дерево *n*; гнилушка *f*; фаутное дерево

* **rotten wood → 2291**

* **rotting → 6647**

7083 **rotting of the sapwood**
d Splintfäulnis *f*
f décomposition *f* de l'aubier
r загнивание *n* заболони

* **rough alder → 4164**

7084 **rough board**
d ungehobeltes Brett *n*; rauhes Brett; rohes Brett
f planche *f* brute; planche non sciée
r нестроганная доска *f*

* **rough conversion → 6538**

* **rough cut → 622**

7085 **rough cut wood**
(lumber that has not been dressed (surfaced) but which has been sawn, edged, and trimmed to at least show saw marks in the wood on the four longitudinal surfaces of each piece for its overall length)
d Grobschnittholz *n*
f bois *m* scié brut
r жёсткая древесина *f* в процессе резания

7086 **rough-dry lumber**
d ungehobeltes Trockenholz *n*
f bois *m* non scié sec
r нестроганный сухой пиломатериал *m*

7087 **rough-dry size**
d grobgetrocknete Größe *f*
f dimension *f* après sécheresse
r размер *m* после сушки

* **rough edge → 2662**

7088 **rough-edged board**
d grobkantiges Brett *n*
f planche *f* avivée brute; planche flacheux
r доска *f* с обзолом

* **rough-edged timber → 2663**

7089 **rough file; coarse file; rubber**
d Grobfeile *f*
f lime *f* grossière
r грубый подпилок *m*

7090 **rough green lumber**
d ungehobeltes Frischholz *n*
f bois *m* vert non scié
r нестроганный сырой пиломатериал *m*

* **rough log → 9044**

* **rough lumber → 7096**

7091 **roughly-squared wood**
d vierseitig zugerichtetes Holz *n*
f bois *m* équarri rugueux
r грубообтёсанный лесоматериал *m*

* **rough plane → 2172**

7092 **rough reduction burns**
(controlled ground fires used in pine types to reduce fire hazards caused by excessive flammable vegetation buildup)
d Verbrennungen *fpl*
f brûlures *fpl* de réduction brute
r палы *mpl*

7093 **rough-rotten-cull tree**
(live tree of commercial species that do not meet regional specifications for freedom from defect primarily because of roughness or poor form)
d grobbehauener Abfallbaum *m*
f arbre *m* abattu pourri grossièrement
r срубленное сгнившее дерево *n*

7094 **rough-sawn timber; rough-hewn timber**
(timber used in the original round form, such as in poles, posts or bridge beams)
d sägerauhes Schnittholz *n*; vorgeschnittenes Holz *n*; grobbehauenes Holz
f bois *m* équarri grossièrement
r грубоопилённая древесина *f*; грубоотёсанное бревно *n*

7095 **rough skid trail**
d Grobschienenweg *m*
f chemin *m* de débardage brute
r неустроенный трелёвочный волок *m*; неподготовленный трелёвочный волок

* **rough terrain → 4212**

7096 **rough timber; raw timber; rough lumber; rough wood; raw[ ]wood; unwrought timber; timber in the rough; wood in the rough**
(lumber that has not been dressed (surfaced) but which has been sawn, edged, and trimmed)
d unbearbeitetes Holz *n*; rauhes Holz; zugerichtetes Holz; Holzrohstoff *m*; [entrindetes] Rohholz *n*; Rohstoff *m*
f bois *m* brut
r необработанный лесоматериал *m*;

7097 rough trees
(live trees of any size that do not contain at least one merchantable 12-foot saw log, now or prospectively, because of roughness or poor form)
d Faultbäume *mpl*
f arbres *mpl* pourris
r сгнившие деревья *npl*

* **rough wood → 7096**

7098 rough wood conveyor
d Rohholzförderer *m*
f convoyeur *m* de bois brut
r конвейер *m* для неокорённого баланса

7099 rough wooded country
d gebirgiges Land *n*
f terrain *m* brut boisé
r пересечённая лесная местность *f*

* **round → 8545**

7100 round assortment
d Rundholzsortiment *n*
f assortiment *m* rond
r круглый сортимент *m*

7101 round billet; cudgel
(a short, round section of a log)
d Rundknüppel *m*; Rundstück *n*
f gourdin *m*; rondin *m*
r мелкий круглый лесоматериал *m*; мелкий кругляк *m*

* **round-branched clubmoss → 5794**

7102 round dock; blue mallow; high mallow
(Malva silvestris)
d wilde Malve *f*
f mauve *f* sauvage; meule *f* sylvestre
r лесной просвирняк *m*; мальва *f* лесная

* **round-ear willow → 2697**

* **round-edged timber → 9079**

7103 round firewood
d Brennholzknüppel *m*
f rondins *mpl* de bois de chauffage
r дровяной кругляк *m*

7104 roundheaded fir borer
(Tetropium abietis)
d rundkopfiger Fichtenkäfer *m*

7105 roundheaded pine beetle
(Dendroctonus adjunctus)
d rundkopfiger Kieferbohrer *m*
f longicorne *m* de pin
r лубоед *m* круглоголовый сосновый

* **round industrial roll → 7113**

7106 rounding down length
(in taxation)
d Abrundenlänge *f*
f longueur *f* d'arrondissement
r длина *f* округления

* **rounding out → 6610**

7107 rounding-up; facing; laying-in [a tree]
d Rundschnitt *m*
f entaille *f* circulaire
r круглая распиловка *f*; очистка *f* [основания дерева] для подготовки места подпила

7108 round knot
(a knot that is oval or circular in form)
d Rundast *m*
f nœud *m* rond
r круглый сучок *m*

7109 roundleaf dogwood
(Cornus rugosa; Cornus circinata)
d rundblättriger Hartriegel *m*
f bois *m* de calumet
r дёрен *m* морщинистый

7110 roundleaf hawthorn; eastern fireberry hawthorn
(Crataegus rotundifolia; Crataegus chrysocarpa phoenicea)
d rundblättriger Weißdorn *m*
f aubepine *f* a feuilles rondes; aubépine à fruits jaunâtres; aubépine dorée
r боярышник *m* круглолистный

* **roundleaf serviceberry → 3806**

* **round log → 7113**

* **round shake → 2139**

7111 round stick
d Rundstock *m*; runder Stock *m*
f gourdin *m*; bâton *m* rond
r дубина *f*; толстая палка *f*

* **round stock → 7113**

7112 round timber measurement table; log rule; log scale table; log volume table
(table intended to show amounts of lumber that may be sawed from logs of different sizes under various assumed conditions)
d Umwaldungstafel *f* für Holzmaßeinheiten; Rundholztabelle *f*
f table *f* de mesure de volume de billes; barème *m* de conversion
r таблица *f* определения объёмов брёвен

* **round-up → 2514**

7113 roundwood; wood in the round; round log; timber in the round; round timber; round stock; round industrial roll
(timber or fuelwood prepared in the round state)
d Rundholz *n*; Nutzrolle *f*; Nutzroller *m*; Ganzholz *n*; Prügel *m*
f bois *m* rond; gros rondin *m* d'industrie
r круглый лесоматериал *m*; [поделочный] кругляк *m*; круглый лес *m*; кругляковое полено *n*

* **roundwood binding → 3662**

* **roundwood bundle → 6026**

* **roundwood bundling → 3662**

* **roundwood cutting → 7115**

7114 roundwood grade
d Rundholzklasse *f*
f classe *f* de bois rond
r класс *m* круглого леса

7115 roundwood harvest; roundwood cutting
d Einschnitt *m* des Rundholzes; Rundholzeinschnitt *m*
f exploitation *f* de bois rond
r заготовка *f* круглых лесоматериалов; раскрой *m* круглых лесоматериалов

7116 round[wood] measure
d Rundholzmessung *f*
f mesure *f* de bois rond
r измерение *n* кубатуры круглых лесоматериалов

7117 roundwood product
(any primary product such as lumber, poles, pilings, pulp, or fuelwood, that is produced from roundwood)
d Rundholzprodukt *n*; Rundholzfabrikate *npl*

7118 roundwood waste
d Rundholzanfall *m*
f déchets *mpl* de bois rond
r отходы *mpl* круглого лесоматериала

7119 roundwood yard
d Rundholzlager *m*; Rundholzplatz *m*; Rundholzablage *f*
f dépôt *m* de grumes; dépôt de bois rond; parc *m* de grumes; parc à bois ronds
r склад *m* круглого леса

7120 rout *v*
(to cut out by gouging)
d wegjagen; aufstöbern
f détourer
r выкапывать; выкопать

* **roving agriculture → 7520**

* **rowan [tree] → 2903**

7121 row of trees; tree row
d Baumreihe *f*; Baumstreifen *n*
f rangée *f* d'arbres; ligne *f* d'arbres
r ряд *m* деревьев

* **row planting → 6356**

* **row seeding → 522**

* **row sowing → 522**

* **row thinning → 4992**

7122 royal paulownia; paulownia; empress tree; princess tree
(Paulownia imperialis; Paulownia tomentosa)
d Kaiser-Paulownie *f*; Blauglockenbaum *m*
f paulownia *m* impérial
r пауловния *f* войлочная; пауловния империалис

* **royal tree → 7892**

7123 royalty of timber; timber royalty; timber tariff; fixed pricelist of wood
d Holzpreisliste *f*
f barème *m* de prix de bois
r расценка *f* леса

* **RPA → 6823**

* **RPF → 6564**

f arbre *m* à frotture
r дерево *n*, о которое трётся крупная дичь

* **rubber → 7089**

* **rubber-plant → 4469**

7125 **rubber-tired skidder; wheel[ed] skidder; wheeled tractor**
(internal combustion wheeled vehicle specifically designed to skid logs from the forest to a transportation system)
d Radrückschlepper *m*
f tracteur *m* à pneus
r трелёвочный трактор *m* на резиновом ходу; колёсный трелёвочный трактор; колёсный тягач *m*

* **rubber[ ]tree → 1320**

7126 **rubbery wood**
d gummihaftes Holz *n*
f bois *m* souple
r [высоко]эластичная древесина *f*; резиноподобная древесина

7127 **rubbing; friction**
(of growing trees)
d Scheuern *n*
f ponçage *m*; friction *f*
r задевание *n*

7128 **rubble; boulder**
(rough stones of irregular shape and size, broken from larger masses by geological processes or by quarrying)
d Geröll *n*; Geschiebe *n*
f galet *m*; moellon *m*
r галька *f*

7129 **rub tree**
(tree used as a fender or pivot to protect the remaining stand during yarding)
d Reibenbaum *m*
f arbre *m* de friction
r растущее дерево *n*, оставленное для уменьшения отклонения канатов и каретки

7130 **ruby horsechestnut**
(Aesculus carnea)
d rotblühende Rosskastanie *f*; Purpur-Kastanie *f*
f marronnier *m* rouge; marronnier à fleurs rouges
r эскулус *m* тёмно-розовый

* **rudder boom → 3147**

(Rutaceae)
d Rautengewächse *pl*
f rutacées *pl*
r рутовые *pl*

* **rugged terrain → 4212**

* **rule for removal → 5081**

* **rumprake → 8937**

* **run → 6097**

* **run[a]way → 1245**

7132 **runner; sprout; tiller; stolon; sprig**
(a slender horizontal stem with elongated internodes)
d Ausläufer *m*; Spross *m*; Trieb *m*; Stengelchen *n*; Zweiglein *n*; Stolon *m*
f pousse *f*; rejet *m*; germe *m*; stolon *m*; rejeton *m*; talle *f*; coulant *m*
r стелющаяся ветвь *f*; [стелющийся] побег *m*; столон *m*; отпрыск *m*; ус *m*

* **running crowning fire → 2076**

7133 **running skyline**
d Laufdrahtseilbahn *f*
f câble-grue *m* à chariot autoleveur
r тягово-несущий канат *m*

* **running slide → 5116**

* **running trades → 3119**

7134 **run of logs; log run; LR**
d Holzserie *f*
f série *f* des grumes
r партия *f* брёвен

* **ruptured → 1634**

* **ruptured grains → 1524**

7135 **rush-leaf acacia**
(Acacia pinifolia; Acer juncifolia)
d kieferblättrige Akazie *f*
f acacia *m* à feuilles de pin
r акация *f* сосноволистная

* **Russian bast → 4972**

* **Russian olive → 9421**

7136 **Russian pea tree; Russian pea shrub**

f acacia *m* jaune; caraganier *m* frutescent
r карагана *f* кустарниковая; дереза *f*; чилига *f* [степная]

7137 Russian [white] elm; European white elm
(Ulmus laevis; Ulmus pedunculata; Ulmus effusa)
d Flatterulme *f*; Flatterrüster *m*
f orme *m* pédonculé; orme lisse; orme diffus
r вяз *m* гладкий

7138 Russian willow; silver willow; acute willow
(Salix acutifolia)
d kaspische Weide *f*; spitzblättrige Weide
f saule *m* caspien; saule à feuilles aiguës
r ива *f* остролистная; краснотал *m*; ива-щелюга *f*; верба *f* красная

7139 rust fungi (*lat, pl*)
(Uredinales)
d Rostpilze *mpl*
f champignons *mpl* de rouille
r ржавчинные грибы *fpl*; ржавчина *f*

* **rustic area → 9404**

7140 rusting
d Rosten *n*
f formation *f* de la rouille
r ржавление *n*

7141 rusty blackhaw; rusty nanny-berry; southern blackhaw
(Viburnum rufidulum; Viburnum rufotomentosum; Viburnum prunifolium var. ferrugineum)
d Rostschneeball *m*
f viorne *f* rouillée
r калина *f* рыжеватая

* **rusty lyonia → 8880**

* **rusty menziesia → 3310**

* **rusty staggerbush → 8880**

S

* **saber jig saw → 1822**

7142 **Sabine's pine; digger pine; gray pine; California foothill pine**
(Pinus sabiniana)
d Sabine-Föhre *f*; Sabine-Kiefer *f*
f pin *m* [de] Sabine
r сосна *f* Сабина; сосна белая калифорнийская

* **sabre-like butt → 7736**

7143 **saccharification of wood; wood saccharification; hydrolysis of timber**
(a chemical process of decomposition involving addition of water)
d Holzverzuckerung *f*
f saccharification *f* de bois
r осахаривание *n* древесины

7144 **sac fungi; cup fungi *pl***
(Ascomycetes)
d Schlauchpilze *mpl*
f ascomycètes *fpl*
r сумчатые грибы *fpl*; аскомицеты *mpl*

7145 **Sachalin cork[ ]tree**
(Phellodendron sachalinense)
d Sachalin-Korkbaum *m*
f phellodendron *m* de Sachalin
r бархат *m* сахалинский

7146 **Sachalin willow**
(Salix sachalinensis)
d Sachalin-Weide *f*
f saule *m* de Sachalin
r ива *f* сахалинская

* **sacred bamboo → 5654**

* **sacred fig → 939**

* **sacred fir → 5422**

* **saddle → 8577**

7147 **saddle grafting**
(a mode of grafting in which a deep cleft is made in the end of the scion by two sloping which is placed upon it saddlewise)
d Sattelpfropfen *n*
f greffage *m* à l'anglaise compliqué; greffage [anglais] à cheval; greffe *f* en fente anglaise
r прививка *f* седлом

* **saddle planting → 289**

7148 **saddle tree**
d Sattelbaum *m*
f bois *m* de rive d'un glissoir
r седельный арчак *m*

7149 **safety equipment**
d Schutzvorrichtung *f*
f équipement *m* de sécurité
r защитное снаряжение *n* (лесоруба)

* **safety glove → 6599**

7150 **safety goggles; protection goggles; protective goggles**
d Schutzbrille *f*
f lunettes *fpl* étanches
r защитные очки *pl* (лесоруба)

7151 **safety swede**
(lever used to tighten binders on loaded logging trucks)
d Schutzhebel *m*
f levier *m* de sécurité
r защитный рычаг *m*; страховочный рычаг

7152 **sage; clary; salvia**
(Salvia)
d Salbei *f*
f sauge *m*
r сальвия *f*

* **sagebrush → 5883**

* **sagittal leaf → 4165**

* **sagittate leaf → 4165**

7153 **sago palm; sago tree**
(Sagus laevis; Metroxylon rumphii; Phoenix lavrae; Phoenix farinifera)
d Sago[baum] *m*; Sagopalme *f*; Zapfenpalmfarn *m*; Zamie *f*
f sagou[t]ier *m*; zamie[r]
r саговая пальма *f*

7154 **Saint John's wort; John's wood**
(Hypericum perforatum)
d Johanniskraut *n*

7155 **sajo; terentang; campnosperma**
(Campnosperma)
d Campnosperma *m*
f campnosperma *m*
r кампноспермa *f*

* **salable height → 5395**

* **salai tree → 4463**

7156 **sale-area betterment**
(improvements in a forest that can be funded by K-V deposits)
d Hiebsfläche-Qualitätsverbesserung *f*
f amélioration *f* d'aire de coupe
r улучшение *n* площади лесосеки; модернизация *f* территории лесосеки

7157 **sale at the stump; stump[age] sale; felled sale; direct sale; standing sale; sale of standing timber; sale by area**
d Verkauf *m* auf dem Stamm; Holz[ver]kauf *m* auf dem Stock; Stockverkauf *m*; Verkauf von stehendem Holz; Blockverkauf *m*
f vente *f* de bois sur pied; vente après abattage
r продажа *f* леса на корню; продажа стоящего леса; продажа на месте валки

* **sale of wood → 8631**

7158 **sale[s] area**
d Verkauffläche *f*
f aire *f* de vente
r район *m* сбыта; место *n* торговли

* **sale schedule → 8634**

* **sales list → 8376**

* **sales value → 5313**

7159 **sale unit**
(timber sales arrangement in which the buyer pays for forest products removed in units)
d Verkaufeinheit *f*
f unité *f* de vente
r продажная единица *f*

7160 **sal forest**
(a forest dominated by sal (Shorea robusta))
d Salwald *m*
f forêt *f* à sal
r саловый лес *m*

* **sali → 4086**

* **sallow wattle → 1691**

* **sallow willow → 3880**

* **salmwood → 4650**

* **salt bush → 7162**

* **saltcedar → 3233**

7161 **salt chuck**
d Salzbassin *n*
f bassin *m* de stockage en eau salée
r бассейн *m* для брёвен с подсоленной водой

7162 **saltmarsh-elder; marsh-elder; salt bush; bigleaf sumpweed**
(Iva frutescens)
d Rispenkraut *m*
f ive *f* xanthifoliée; ive à feuilles de lampourde
r ива *f* кустарниковая

7163 **sal tree**
(Shorea robusta)
d Salbaum *m*
f sal *m*
r сорея *f* мощная

7164 **salt-tolerant** *adj*
d salztolerant
f tolérant au sel
r солевыносливый; солеустойчивый

7165 **salt tree**
(Halimodendron)
d Salzstrauch *m*
f caragana *m*; halimodendron *m*
r галимодендрон *n*; чамыш *m*; чингиль *m*

7166 **salvable dead trees; salvageable dead trees**
(standing or downed dead trees that were formerly growing stock and merchantable)
d tote Schadbäume *mpl*
f arbres *mpl* morts récupérables
r сухостойные повреждённые деревья *npl*

7167 **salvage**
(dead or dying trees that occur in excess of those needed for wildlife, aesthetics, or other purposes)
d geschadetes Holz *n*
f bois *m* cassé; volis *m*
r повреждённый лес *m*

7168 salvage harvest[ing of timber]; salvage cut[ting]; salvage felling; salvaging; salvage thinning
(removal of dead, damaged, or diseased trees to recover maximum value prior to deterioration)
d nachträgliches Aufarbeiten *n* schwächerer Sortimente im Schlag; Aufarbeitung *f* von Kalamitätshölzen
f coupe *f* de peuplements abandonnés; coupe résiduelle; nettoyage *m* final de menus assortiments; façonnage *m* de chablis
r рубки *fpl* сухостойных и повреждённых деревьев; рубка пострадавшего [от бедствия] леса; рубка повреждённых деревьев; заготовка *f* повреждённого леса

7169 salvage logging
(the removal, after the main logging, of the rest of the timber, with a view to supplying a different class of product)
d Schadholzhieb *m*
f exploitation *f* de récupération
r заготовка *f* низкокачественных деревьев; заготовка повреждённых деревьев

* **salvage thinning → 7168**

* **salvaging → 7168**

* **salvaging after the main drive → 8324**

7170 salvaging of timber; salvage of timber; area salvage
(salvaging dead timber within a given area)
d Totholzsammlung *f*
f sauvetage *m* de bois; récupération *f* de bois
r собирание *n* отходов; сбор *m* древесины

* **salvia → 7152**

* **saman [tree] → 6701**

* **samara fruit → 4701**

* **sample area → 7173**

7171 sample cutting; sample felling; test felling; test cutting
d Probehieb *m*; Probefällung *f*
f abattage *m* d'essai
r опытная рубка *f*; пробная рубка

* **sample forest → 2946**

7172 sample log
r базовое бревно *n*; расчётное бревно; условное бревно

7173 sample plot; indicator plot; experimental plot; sample area; test area; quadrat; block
(a small, clearly demarcated, sample area of known size on which ecological observations are made)
d Probefläche *f*; Musterfläche *f*
f surface *f* expérimentale; surface d'essai
r пробная площадь *f*; делянка *f*

* **sample stand → 2946**

* **sample stem → 8480**

7174 sample tree; model tree; test tree
d Modellbaum *m*
f arbre *m* d'échantillonnage
r модельное дерево *n* насаждения; дерево, выбранное для опыта

7175 sample tree cutting
d Modellstammfällung *f*
f coupe *f* d'arbres d'échantillonnage
r рубка *f* модельных деревьев

7176 sample tree of a given class
d Klassenprobenstamm *m*
f arbre *m* échantillon d'essai par classe
r средний ствол *m* класса

7177 sampling error
d Probenahmefehler *m*
f erreur *f* d'échantillonnage
r ошибка *f* выборки

7178 sampling unit
d Stichprobeneinheit *f*
f unité *f* d'échantillonnage
r единица *f* наблюдения; единица выборки

* **sampson → 6645**

* **sanctuary → 5672**

7179 sand afforestation
d Sandaufforstung *f*
f boisement *m* de sables
r облесение *n* песков

7180 sandal bead tree
(Adenanthera pavonina)
d indischer Korallenbaum *m*
f bois *m* de corail; crête *m* de paon; faux santal *m* rouge; papita *m*

* **sandalwood → 9382**

7181 **sandalwood family**
(Santalaceae)
d Sandelholzgewächse *pl*
f santalacées *pl*
r сандаловые *pl*

* **sandalwood padauk → 6790**

7182 **sandarac**
(Tetraclinis articulata)
d Sandarakbaum *m*; Berberthuja *f*
f sandarac *m*
r сандараковое дерево *n*; алжирская туя *f*

7183 **sandbar willow; coyote willow**
(Salix interior; Salix longifolia; Salix exigua)
d langblättrige Weide *f*
f saule *m* de l'intérieur; saule exigua; saule de coyotes; saule des bancs de sable
r ива *f* длиннолистная

7184 **sand[-binding] forest**
d sandgefesselter Wald *m*
f forêt *f* aux sables
r пескоукрепляющий лес *m*

7185 **sandbox [tree]; habillo; hura; possumwood; acau; acuapar; assacu**
(Hura crepitans)
d Possentrie *f*; Sandbüchsenbaum *m*
f assac *m*; sablier *m* blanc; arbre *m* au diable
r хура *f* скрипучая

7186 **sand cherry**
(Prunus pumila)
d kanadischer Kirschbaum *m*; Sandkirsche *f*
f cerisier *m* canadien
r вишня *f* канадская

7187 **sand-heath**
(Ceratiola)
d Hornbusch *m*
f cératiole *f*
r цератиола *f*

7188 **sand hickory**
(Carya pallida)
d Sandhickory *f*
f caryer *m* de sables
r кария *f* бледная

7189 **sanding**
(a type of woodworking)
r шлифование *n*; полирование *n*

* **San Domingo boxwood → 9346**

7190 **sandpaper oak; pungent oak**
(Quercus pungens)
d Stecheiche *f*
f chêne *m* papier de verre
r дуб *m* колючий

* **sand pear → 1504**

7191 **sand pine; scrub pine; shrub pine**
(Pinus clausa; Pinus inops)
d Drehkiefer *f*
f pin *m* tordu [latifolié]
r сосна *f* закрытая

7192 **sand plant; psammophyte**
(a plant growing in sands or sandy soils)
d Dünenpflanze *f*
f psammophyte *m*
r псаммофит *m*

7193 **sanitary felling; environmental harvesting [of timber]; sanitation cut[ting]; cleaning cutting**
(removal of dead, damaged, or susceptible trees, essentially to prevent the spread of pests or pathogens and so promote forest hygiene)
d Ausschlussaushieb *m*; Gesundheitshieb *m*; Reinigungshieb *m*
f coupe *f* d'assainissement; coupe sanitaire; coupe de nettoiement
r санитарная рубка *f*; очистительная рубка; очистная рубка

7194 **sanitation measures**
d Sanierungsmaßnahmen *fpl*
f mesures *fpl* sanitaires; mesures d'assainissement
r санитарные меры *fpl*

* **Santa Lucia fir → 1037**

* **sap → 7208, 7216**

* **sapan → 7209**

7195 **sap ascen[tion]; ascending of the sap**
d Ansteigen *n* des Saftes; Saftaufstieg *m*
f montée *f* de la sève
r восхождение *n* сока

7196 **sap burr**
d Splintauswuchs *m*

7197 **sap-clear; sap-free**
d splintfrei
f sans aubier
r без заболони; не содержащий заболони

7198 **sap displacement**
d Saftverdrängung *f*
f déplacement *m* de sève
r вытеснение *n* сока

7199 **sap-drawer**
d Saftzieher *m*
f tire-sève *m*
r крупный отпрыск *m*

* **sapele mahogany → 114**

7200 **sapel[l]e; sapelli**
(Entandrophragma)
d Bastardmahagoni *n*
f entandrophragme *m*; bois *m* rouge
r энтандрофрагма *f*; красное дерево *n*

* **sap flow → 1589**

* **sap-free → 7197**

7201 **sapium**
(Sapium)
d Klebbaum *m*; Talgbaum *m*
f bois *m* brûlant
r сапиум *m*

7202 **sapling**
(live, vigorous, and well-formed tree of commercial species, usually from 2 to 6 inches in d.b.h.)
d junger Baum *m*; Heister *m*; Schössling *m*
f élève *m*; jeune arbre *m*; brin *m*
r молодое деревцо *n*; отросток *m*; порослевый сеянец *m*

* **sapling growth → 7204**

7203 **sapling planting; planting with saplings**
d Heisterpflanzung *f*
f plantation *f* de gaules
r посадка *f* саженцев

7204 **sapling-pole stand; sapling growth; stake forest; stake wood; sapling wood thicket; young-growth stand**
d Jungbestand *m*
f gaulis *m*
r молодой древостой *m*; прутняк *m*

7205 **sapling size**
d Schösslinggröße *f*
f dimension *f* de gaule
r размер *m* отростка

7206 **sapling stage**
d Schösslingstadium *n*
f stade *m* de gaule
r стадия *f* молодняка второго класса; стадия подроста

* **sapling wood thicket → 7204**

7207 **sapodilla family**
(Sapotaceae)
d Sapotengewächse *pl*
f sapotacées *pl*
r сапотовые *pl*

7208 **sap [of wood]; wood sap; tree sap; sap; tree succus**
(the watery fluid that circulates through a tree carrying the chemical food that enables the tree to grow)
d Baumsaft *m*; Holzsaft *m*; Saft *f*
f sève *f* [arborescente]
r [древесный] сок *m*

* **sapota → 5665**

* **sapote → 5665**

7209 **sappan[wood]; sapan**
(Caesalpinia sappan)
d Sappan-Cäsalpinie *f*; Sappan-Baum *m*
f bois *m* de sappan
r саппановое дерево *n*; цезальпиния *f* саппан

7210 **sap-peeled**
d geschält im Saftperiode
f écorcé en sève
r весенней окорки; окорённый весной

* **sap peeling → 6136**

7211 **sap period**
d Saftperiode *f*
f saison *f* de montée de la sève
r сезон *m* сокодвижения; период *m* сокодвижения

7212 **saprogen[ic]; putrid; decayed; doted; doty; faulty; rotten; rotted; unsound; rot-damaged**
d saprogen; faulig; verfault; modrig; kariös
f putride; pourri; putréfié
r сапрогенный; гнилостный; вызывающий

* **saprophyte → 7214**

7213 **saprophytic fungi**
d saprophytische Pilze *mpl*
f champignons *mpl* saprophytiques
r сапрофитные грибы *fpl*

7214 **saprophytic plant; saprophyte**
(a plant, generally without green colo[u]r, which derives its food from dead, organic matter)
d Saprophyt *m*
f plante *f* saprophyte; saprophyte *f*
r сапрофит *m*

* **sap shoot → 1933**

7215 **sap[-]stain; red stain**
(any stain that predominantly affects sapwood)
d Splintfleck *m*; Saftfleck *m*; Eisenfleck *m*; roter Fleck *m*
f tache *f* [colorée] de l'aubier; tache de sève; rougissement *m*
r заболонное пятно *n*; продубина *f*; заболонная окраска *f*

* **saptree → 3804**

* **sapupira → 8251**

7216 **sapwood; sap; splintwood; pit sheath; albur[num]**
(the portion of the wood that in the living tree contains living cells and reserve materials such as starch)
d Saftholz *n*; Splintholz *n*; Splint *m*
f bois *m* d'aubier; aubier *m*
r заболонь *f*; заболонная древесина *f*

7217 **sapwood estimate**
d Splintholzabschätzung *f*
f estimation *f* d'aubier
r оценка *f* заболони

* **sapwood/heartwood boundary → 4204**

7218 **sap[wood] rot**
d Rotstreifigkeit *f*; Splintfäule *f*
f pourriture *f* d'aubier
r заболонная гниль *f*; периферическая гниль (возбудители - древесные грибы)

7219 **sapwood species**
d Splintholzart *f*

7220 **sapwood tree**
(a tree in which the wood of mature stems is of uniform colo[u]r and appears to be entirely sapwood)
d Reifholzbaum *m*; Splint[holz]baum *m*
f arbre *m* à bois parfait non distinct; arbre sans duramen
r заболоченное дерево *n*

* **saqui-saqui → 6403**

7221 **Sargent['s] cypress**
(Cupressus sargentii)
d Sargents-Zypresse *f*
f cyprès *m* de Sargent
r кипарис *m* Сарджента

7222 **Sargent['s] mountain ash**
(Sorbus sargentiana)
d Sargents-Eberesche *f*
f sorbier *m* de Sargent
r рябина *f* Сарджента

7223 **sarsaparilla; common greenbrier; greenbrier**
(Smilax officinalis)
d Stechwinde *f*
f salsepareille *f*
r сассапарель *f*

* **sash chisel → 6404**

* **sash [gangsaw] → 5132**

* **sash saw → 1136**

* **saskatoon serviceberry → 6024**

* **sassafras → 1801**

* **sasswood → 1366**

* **sassy bark tree → 1366**

* **sassy [wood] → 1366**

* **satin walnut → 259**

* **satinwood → 2729**

7224 **satisfactorily stocked**
(ref. to productive forest land that has been regenerated naturally or artificially to at least a minimum number of well-established,

f de densité relative satisfaisante; suffisamment reboisé
r достаточно восстановленный

7225 saucer magnolia; Chinese magnolia
(Magnolia soulangiana)
d Tulpen-Magnolie *f*
f magnolia *m* [hybride] de Soulange
r магнолия *f* Суланжа

7226 sausage tree
(Kigelia africana; Kigelia pinnata)
d Leberwurstbaum *m*
f arbre *m* à saucissons
r кигелия *f* африканская

7227 savanna[h]
(a grassland with scattered trees, either as individuals or clumps)
d Savanne *f*
f savane *f*
r саванна *f*

7228 savanna[h]-forest; savanna[h] woodland
(a more or less open, tropical or subtropical woodland having an undergrowth mainly of grasses, the trees being of moderate height and generally deciduous or, if evergreen, tending to have small leaves)
d Savannenwald *m*
f forêt *f* de savane
r саванный лес *m*; саванное редколесье *n*

7229 savanna[h] parkland; parkland savanna[h]
d Parklandsavanne *f*
f forêt-parc *m* à couloirs de savane
r лесопарковая саванна *f*

* **savanna[h] woodland → 7228**

* **savin → 5423**

* **sawagurumi → 4637**

7230 saw arbor; saw shaft; saw mandre; saw mandril
(the shaft and bearings that hold a power-driven saw)
d Kreissägenachse *f*
f arbre *m* d'une scie circulaire
r оправка *f* шлицевой фрезы; ось *f* круглой пилы

* **Sawari nut → 6188**

* **saw attachment → 3063**

* **saw boss → 5069**

7231 saw box; saw bow
d Sägebügel *m*
f arc *m* de la scie
r лук *m* пилы

* **sawbuck → 4380**

* **saw-carriage → 5199**

7232 saw cut; cut; bite
d Sägedurchschnitt *m*
f coupe *f* [de sciage]
r рез *m*; раскрой *m*; пропил *m*; распил *m*

7233 saw deck
d Sägerampe *f*
f plate-forme *f* de sciage
r площадка *f* для распиловки

7234 saw dog
d Sägeklammer *m*
f greffe *f* de scie
r [реактивный] упор *m* пилы

* **sawdust asthma → 9700**

7235 sawdust baling system; baling system; residue baler
d Sägemehlballensystem *n*
f presse *f* à résidus ligneux; système *m* d'emballage de sciure; ramasseuse-presse *f* à résidus
r система *f* упаковки древесных отходов

7236 sawdust burner
d Sägemehlbrenner *n*
f brûleur *m* de sciure
r топка *f* для сжигания древесных отходов

7237 sawdust compost
d Sägemehlkompost *m*
f compost *m* de sciure
r опилочный компост *m*

* **sawed timber → 1918**

7238 sawed tree
d abgesägter Stamm *m*
f arbre *m* scié
r срезанное дерево *n*

* **sawed veneer → 7261**

7239 saw filer; saw sharpener
d Sägeschärfer *m*
f dispositif *m* d'affûtage de scie
r пилоточ *m*; точильщик *m*

7240 sawflies
(Tenthredinoidea; Tenthredinidae)
d Buschhornblattwespen *fpl*; Sägewespe *f*; Blattwespen *fpl*
f tenthrèdes *fpl*; mouches *fpl* à scie
r пилильщики *mpl*

* **saw frame → 3684**

* **saw head → 3063**

* **saw-head feller-buncher → 1432**

* **saw-head [felling] attachment → 3063**

* **sawhorse → 4380**

7241 sawing
(cutting or separating with a saw)
d Sägen *n*
f sciage *m*
r пиление *n*; пилка *f*; распил *m*

7242 sawing defect
d Sägedefekt *m*
f défaut *m* de sciage
r дефект *m* распиловки

7243 sawing diameter
d Sägedurchmesser *m*
f diamètre *m* de sciage
r диаметр *m* спиливаемого дерева

7244 sawing for accounts of clients; dressing in transit; custom sawing
d Lohnschnitt *m*; Lohnsägen *n*
f sciage *m* à façon
r распиловка *f* за плату

* **sawing industry → 5206**

* **saw[ing] jack → 4380**

7245 saw[ing] log; sawlog; mill log; plank log; sawtimber [tree]; sawmill lumber; wood for sawing; kapp balk; kapbaulk
(tree at least 18 cm d.b.h. and suitable for conversion to lumber)
d Schneideholzbaum *m*; Sägeblock *m*; Sägebloch *m*; Säge[rund]holz *n*; Sägeklotz *m*; Schneideholz *n*
r пиловочное бревно *n*; пиловочный кряж *m*; капбалка *f*; капбалк *m*; пиловочник *m*

* **sawing output → 5214**

* **sawing threstle → 4380**

* **sawing yield → 5214**

7246 saw kerf; saw-slit; saw notch
(1. grooves or notches made in cutting with a saw; 2. that portion of a log, timber, or other piece of wood removed by the saw in parting the material into two pieces)
d Sägeschnitt *m*; Sägekerb *m*; Schnittfuge *f*
f trait *m* de scie; entaille *f* de sciage
r пильный надрез *m*; фуг *m* пропила

7247 saw kerf surface
d Sägeschnittfläche *f*
f surface *f* d'entaille de sciage
r поверхность *f* пропила

* **sawlog → 7245**

7248 sawlog diameter[-frequency] distribution
d Durchmesser-Häufigkeitsverteilung *f* des Sägeholzes
f distribution *f* [de fréquence] de billes par diamètre
r [частотное] распределение *n* [пиловочных] брёвен по диаметру

7249 sawlog portion
(that portion of the central stem of sawtimber trees between the stump and the sawlog top)
d Sägeholzteil *f*
f part *m* de bois de sciage
r часть *f* пиловочника

7250 sawlog scaling diameter
d Durchmesser *m* der skalieren Sägeholz
f diamètre *m* de cubage
r диаметр *m* измеряемого брёвна

7251 sawlog table
d Sägeholztabelle *f*
f table *f* des billes
r таблица *f* объёмов пиловочника

7252 sawlog top
(the point on the central stem of sawtimber trees above which a saw log can not be produced)
d Sägeholzwipfel *m*

* **saw mandre → 7230**

* **saw mandril → 7230**

* **sawmill → 5132**

* **sawmill carriage → 5208**

* **sawmill equipment → 5106**

* **sawmilling → 5206**

* **sawmill lumber → 7245**

7253 **sawmill lumberyard**
d Sägewerklagerplatz *m*
f chantier *m* de scierie
r склад *m* пиломатериалов лесопильного завода

7254 **sawmill operating cost[s]**
d Sägewerk-Bewirtschaftungskosten *pl*
f frais *mpl* d'exploitation de scierie
r эксплуатационные расходы *mpl* лесопильного цеха

* **saw[n] goods → 1918**

* **saw notch → 7246**

* **saw notch height → 2195**

* **sawn timber → 1918**

7255 **sawn timber arris; arris**
d Schnittholzkante *f*; Kante *f*
f arête *f* [de bois scié]
r ребро *n* [пиломатериала]

7256 **sawn timber depth**
d Schnittholzdicke *f*
f profondeur *f* de bois scié
r толщина *f* пиломатериала

7257 **sawn timber end**
d Schnittholzende *f*
f bout *m* de bois scié; rive *f* de bois scié
r кромка *f* пиломатериала; торец *m* пиломатериала (любая из двух противоположных более узких продольно опиленных поверхностей обрезного пиломатериала)

7258 **sawn timber face; face of lumber; lumber face**
d Fläche *f* des Schnittholzes;
r пласть *f* пиломатериала

7259 **sawn timber length**
d Blocklänge *f*; Schnittholzlänge *f*
f longueur *f* de bois scié
r длина *f* пиломатериала; длина брёвна; длина кряжа

7260 **sawn timber profiles**
d Holzprofile *npl*
f profils *mpl* de bois scié
r профили *mpl* [обработанных] пиломатериалов; профилированная древесина *f*

7261 **sawn veneer; sawed veneer**
(veneer produced by sawing)
d gesägtes Furnier *n*; Sägefurnier *n*
f placage *m* scié
r пиленая фанера *f*

7262 **saw-palmetto**
(Sabal serrulata; Serenoa repens)
d Zwerg-Sägepalme *f*
f baie *f* de sabal
r зубчатая пальметто *f*; низкорослая веерolистная пальма *f*

7263 **saw pitch**
d Kettenteilung *f*
f pas *m* de chaîne
r шаг *m* пильной цепи

* **saw set → 821**

* **saw shaft → 7230**

* **saw sharpener → 7239**

* **saw-slit → 7246**

* **saw stroke → 8182**

7264 **saw teeths**
d Sägezähne *mpl*
f dents *mpl* de scie
r пильные зубы *mpl*

7265 **sawtimber stands**
(stands at least 16.7% stocked with growing-stock trees, with half or more of this stocking in sawtimber or poletimber trees)
d Schneideholzbestand *m*
f peuplement *m* de bois de sciage
r насаждение *n* из пиловочниках

* **sawtimber volume → 9248**

7266 **sawtooth oak**
(Quercus acutissima)
d japanische Kastanieneiche *f*
f chêne *m* du Japon à feuilles de châ
r дуб *m* острейший

* **saw web → 820**

7267 **saw wedge**
d Säge[schnitt]keil *m*
f coin *m* de scie
r пильный клин *m*

7268 **sawyer**
d Sägewerkarbeiter *m*
f tronçonneur *m*; conducteur *m* de tronçonneuse
r рамщик *m* лесопильного завода

7269 **saxifrage family**
(Saxifragaceae)
d Steinbrechgewächse *pl*
f saxifragacées *pl*
r камнеломковые *pl*

* **SBFEP → 7759**

* **scaffold branch → 6536**

7270 **scaffold-pole; prop; shore; putlock; putlog**
d Rüstbaum *m*; Gerüststange *f*
f arbre *m* d'échafaudage
r стойка *f* коренных лесов; лес *m* для подмостков; жердь *f* для подмостков

7271 **scalariform parenchyma**
(the ladder-like pattern formed on the cross section by rays and regularly spaced bands or lines of axial parenchyma when the latter are distinctly narrower than the rays)
d leiterförmiges Parenchym *n*
f parenchyme *m* en échelle
r лестничная паренхима *f*

7272 **scalariform perforation plate; latticed perforation plate**
(a plate with multiple perforations elongated and parallel)
d leiterförmige Perforationsplatte *f*; scalariforme Perforationsplatte
f cloison *f* perforée scalariforme; perforation *f* en grille; perforation scalariforme
r лестничная перфорационная пластинка *f*

7273 **scalariform pitting**
d leiterförmige Tüpfelung *f*
f ponctuations *fpl* en disposition scalariforme
r лестничная поровость *f*

* **scale-book → 5366**

7274 **scale bud; scale insect**
(a member of the Homopterous super-family Coccoidea, plant-juice suckers)
d Eulecanium *m*
f lécanium *m*; coccide *f*; cochenille *f*
r чешуйчатая почка *f*

7275 **scale form**
d Abmessungstabelle *f*; Messtarif *m*
f table *f* de mesuréments
r таблица *f* замеров (число брёвен, диаметр, длина и пр.)

* **scale insect → 7274**

* **scale [leaf] → 968**

* **scale-like leaf → 968**

7276 **scale-like needles**
d schuppenförmige Benadelung *f*
f aiguilles *fpl* écailleuses
r чешуевидная хвоя *f*

7277 **scale pine**
d Schuppenkiefer *f*
f pin *m* avec écorce écailleuse
r сосна *f* с чешуйчатой корой

* **scaler → 1457**

7278 **scales**
(Coccidea)
d Schuppen *fpl*
f écailles *fpl*
r червецы *mpl*; щитовки *fpl*

* **scale stick → 5107**

* **scaling → 8955**

7279 **scaling at mid point**
(of a log)
d Mittelpunktabmessung *f*
f cubage *m* en milieu
r обмер *m* в средней части

* **scaling by length and mid-diameter → 5362**

* **scaling length of log → 8005**

* **scaling stick → 5107**

* **scalloped torus → 6308**

* **scalping → 7286**

* **scaly hornbeam → 7280**

7280 **scaly ironwood; scaly hornbeam; heartleaf hornbeam**
(Carpinus cordata)
d herzblättrige Hainbuche *f*
f charme *m* à feuilles en cœur
r граб *m* приморский; граб сердцелистный

* **scantily stocked → 6450**

7281 **scantiness of species**
d Artenarmut *f*
f pauvreté *f* en espèces
r бедность *f* видов

* **scantlings → 7969**

* **scar → 582**

7282 **scarcity of timber; timber shortage; wood hunger; wood famine; timber famine**
d Holzmangel *m*; Holznot *f*
f manque *m* de bois
r дефицит *m* древесины

7283 **scarf**
(sloping surface left on a butt log or stump as a result of the undercut)
d Fällschnittsdach *n*; Fällkerbendach *n*; Halstuch *n*; Schal *m*
f toit *m* de l'entaille d'abattage; semelle *f* du trait d'abattage
r скос *m*

7284 **scarf block**
d Schalblock *m*
f bloc *m* entaillé
r кусок *m* дерева, удаляемый из подпила

7285 **scarf joint**
(a joint made by overlapping and bolting or locking together the ends of two pieces of timber that are halved, notched, or cut away)
d Schäften *n*; Schäftfuge *f*; abgeschrägte Fuge *f*
f assemblage *m* en biseau; joint *m* en biseau; joint en sifflet; enture *f*
r косой прируб *m*

(loosening the topsoil of open areas or breaking up the forest floor to assist the germination of natural seed)
d Abziehen *n* des Bodenüberzuges; Skarifikation *f*; Ritzen *n*
f enlèvement *m* de la couverture du sol; dégazonnement *m*; scarifiage *m*; scarification *f*
r минерализация *f* поверхности почвы; сдирание *n* травяного покрова; снятие *n* дернины; скарификация *f*

* **scarifier → 7313**

* **scarifying → 7286**

* **scarlet-berried elder → 2907**

* **scarlet cordia → 4799**

* **scarlet eucalyptus → 6773**

* **scarlet hawthorn → 3167**

* **scarlet maple → 6774**

7287 **scarlet oak**
(Quercus coccinea)
d Scharlacheiche *f*
f chêne *m* écarlate
r дуб *m* американский шарлаховый

* **scar of peeling → 2237**

7288 **scar ring**
d Narbering *m*
f anneau *m* de broussin
r кольцевое повреждение *n* коры

* **scar tissue → 1182**

7289 **scar wood**
d einwüchsiges Holz *n*; vernarbendes Holz; Narbenholz *n*
f bois *m* à cicatrices
r древесина *f* с проростью

7290 **scattered knot; single knot**
d Einzelast *m*
f nœud *m* isolé; nœud unique
r одиночная ветвь *f*; одиночный сучок *m*

* **scattered mast → 4939**

7291 **scattered seed-tree system**
d Einzelüberhalt *m*

7292 **scattered species**
(used in fiberboards)
d Einzelart *f*
f espèce *f* isolée
r древесина *f* случайных пород

* **scattered stand → 7293**

7293 **scattered timber; scattered stand**
d lockerer Bestand *m*; Streubestand *m*
f bois *m* dispersé; peuplement insuffisamment boisé
r разрежённое лесонасаждение *n*

* **scattered tree → 4480**

7294 **scatter from local trees**
d Streuung *f* an Baümen am Standort
f diffusion *f* par les arbres locaux
r разброс *m* отдельных деревьев

* **scented cedar → 1411**

* **scented guarea → 6280**

* **scented wattle → 1689**

* **schefflera → 9040**

7295 **schizogenous; schizogenetic**
(formed by the separation of tissue elements owing to the splitting of the common wall between adjacent cells)
d schizogen
f schizogène
r схизогенный; шизогенетический

7296 **schizolysigenous**
(originating by the splitting of the cell wall, and developing by the breakdown of surrounding tissues)
d schizolysigenisch
f schizolysigène
r схизолизогенный; шизолизогенный

* **schooled plant → 5815**

* **school forest → 8761**

7297 **school of forestry; forest school**
d Forstschule *f*; forstliche Lehranstalt *f*
f école *f* forestière [secondaire]
r лесная школа *f*

* **sciaphilous species → 7474**

* **scion → 1145**

7298 **scion rooting**
d Pfropfreiswurzelung *f*
f affranchissement *m* du greffon
r укоренение *n* привоя

7299 **scion wood**
(shoots from which graft scions are cut)
d Edelreisholz *n*; Pfropfreisholz *n*
f scion *m* à greffon
r побеги *mpl* привоя

* **scissor-type shear → 8921**

* **scissor-type topping shear → 8947**

7300 **sclereid; sclerotic cell**
(a strengthening element that has thick, often lignified secondary walls and which commonly lacks a protoplast when mature)
d Sklereide *f*
f scléréide *f*; cellule *f* scléreuse; sclérite *m*
r склереида *f*

7301 **sclerophyl[lous] forest; hardleaved forest**
(forest characterized by the prevalence of species having leathery, generally small, leaves, and developed in a climate with relatively hot and dry seasons)
d Sklerophylwald *m*
f forêt *f* sclérophylle
r склерофильный лес *m*; жёстколистный лес; твёрдолиственный лес

* **sclerosis → 8097**

7302 **sclerotic**
d sklerotisch
f scléreux
r одревесневший

* **sclerotic cell → 7300**

7303 **sclerotic tylosis**
(a tylosis, with an exceptionally thick, laminated, lignified wall and ramiform pits)
d Stein-Thylle *f*
f thylle *m* scléreux
r одревесневший каллус *m*

* **scoot → 5115**

7304 **scoots**
(two-runner sled, without tongue or shafts,

f rebuts *mpl*
r двухполозные сани *pl* для трелёвки брёвен в погруженном положении

* **scorch → 6104**

7305 **scorch height; scorch line**
d Versengungslinie *f*
f ligne *f* rousse
r высота *f* опала (пожаром)

7306 **scorching of the leaf; leaf scorch**
(the drying and browning of leaf margins usually caused by unfavorable environmental conditions)
d Blattversengung *f*; Blattverfärbung *f*
f brûlure *f* de la feuille; grillage *m* des feuilles
r ожог *m* листьев

* **scorch line → 7305**

* **scorch of the bark → 836**

7307 **score**
d Kratzer *m*; Blockkerbe *f*
f sillon *m* linéaire de déroulage; engoujure *f*
r отметка *f*; след *m*

7308 **score** *v*; **groove** *v*; **crease** *v*
(in hewing timber, to mark with lines or with ax hacks the limits of the cut, both as to width and depth)
d rillen; ritzen; anritzen
f saigner; entailler; rainer; rainurer; marquer; réussir
r отмечать; наносить отметки

* **score ply → 6328**

7309 **scoring saw**
d Rillensäge *f*
f scie *f* à inciser; inciseur *m*
r подрезная пила *f*

* **scotch → 3888, 4082**

* **Scotch broom → 1779**

7310 **Scotch elm; mountain elm; witch elm; wych elm; alm**
(Ulmus scabra; Ulmus glabra; Ulmus montana)
d Bergulme *f*; Weißulme *f*; Bergrüster *m*; Weißrüster-Ulme *f*
f orme *m* de montagne; orme blanc; orme à

* **Scotch heather → 1779**

* **Scotch laburnum → 215**

* **Scotch pine → 1797**

* **Scot's pine → 1797**

7311 **Scouler['s] willow; Nuttall willow**
(Salix scouleriana)
d Scouler-Weide *f*
f saule *m* de Scouler
r ива *f* Неттола

* **scrag → 1018**

7312 **scrambler**
(a plant that raises itself above other vegetation in the understor[e]y with the aid of thorns or hooks)
d Verwürfler *m*; Scrambler *m*
f embrouilleur *m*; scrambleur *m*
r скрамблер *m*

* **scraper → 6897**

* **scrape-resin → 593**

* **scraping iron → 6897**

7313 **scratcher; ripper; gouge blaze; gouger; scarifier; scribing knife**
(a tool for marking trees or round timber by scoring the outer surface)
d Reißhaken *m*; Reißer *m*; Reißmesser *n*
f scarificateur *m*; griffe *f*
r покровосдиратель *m*; резак *m*

* **scree → 8378**

* **screefing → 7286**

7314 **screefing spade**
d Wühlspaten *m*
f bêche *f*
r рыхлительная лопата *f*

* **screen coupe → 8171**

* **screened chips → 28**

* **screen felling → 8171**

* **screen rejects → 6850**

7315 **screwing of the log; gripping of the log**
d Verklammerung *f* des Stammes
f fixation *f* d'une grume par griffes antifentes
r скрепление *n* ствола скобами

* **screw pine → 1802**

7316 **screw-pine family**
(Pandanaceae)
d Schraubengewächse *pl*
f pandanacées *pl*
r пандановые *pl*

7317 **scribe** *v*
(to mark for an irregular cut)
d anreißen; anzeichnen
f chantourner
r вписывать

* **scribe [knife] → 6741**

* **scriber → 6741**

* **scribing knife → 7313**

7318 **Scribner decimal rule**
(one of several log rules designed to estimate lumber yield from logs)
d Scribners Dezimalsatz *n*
f barème *m* décimal de Scribner
r десятичная таблица *f* Скрибнера

* **scrog → 9572**

* **scroll-saw → 1822**

* **scrub → 7555**

* **scrub birch → 892**

* **scrub clearance → 1109**

7319 **scrub-clearing machine; brush buster; bush breaker; b[r]ush cutter; rolling stalk cutter; scrub cutter; slash cutter; brush flail**
d Allmäher *m*
f débroussailleuse *f* [à fléaux]; coupe-broussailles *f*; équipement *m* d'essartage; défricheur *m*
r кусторез *m*; машина *f* для уничтожения кустарника и дёрна

* **scrub control → 1101**

* **scrub cutting → 1109**

d Gebüsch-Dschungel *m*
f jungle *f* arbustive
r кустарниковая джунгла *f*

* **scrub land → 353**

* **scrub oak → 642**

* **scrub pine → 7191**

7321 **scrub pulverizer; scrub slasher**
d Bruschhacker-Zerstäuber *m*
f débroussailleur-pulvériseur *m*
r кусторез-измельчитель *m*

* **scrub remains → 1116**

* **scrub stand → 3515**

7322 **scrub tundra**
d Buschtundra *f*
f toundra *f* de broussailles
r кустарниковая тундра *f*

* **scrub veld → 1208**

7323 **scrub wood**
d Kollerbusch *m*
f arbre *m* rabougri
r низкорослое дерево *n*; приземистое дерево

* **SDI → 8011**

7324 **seaboard timber**
d Seeküstenholz *n*
f bois *m* de littoral maritime
r прибережные лесоматериалы *mpl*

7325 **sea buckthorn**
(Hippophae rhamnoides)
d [gewöhnlicher] Sanddorn *m*; Haffdorn *m*
f argousier *m* [faux-nerprun]; bourdaine *f* marine; saule *m* épineux
r облепиха *f* [крушинная]

7326 **seagrape**
(Coccoloba uvifera)
d Seetraube *f*
f raisinier *m* bord-de-mer; raisin *m* marine
r кокколоба *f*

7327 **Sea Island [cotton]; tree cotton**
(Gossypium barbadense; Gossypium arboreum)
d Sea-Island-Baumwollpflanze *f*

7328 **sea-lavender family**
(Plumbaginaceae)
d Bleiwurzgewächse *pl*
f plumbaginacées *pl*
r плюмбаговые *pl*

* **seam → 2422**

* **sea rafting → 6689**

7329 **seaside alder**
(Alnus maritima)
d Küstenerle *f*
f au[l]ne *m* de bord de la mer
r ольха *f* приморская

* **seaside pine → 5309**

7330 **seasonal forest**
(forest showing well-market changes in appearance according to the season of the year)
d saisonaler Wald *m*
f forêt *f* saisonnière
r сезонный тропический лес *m*

7331 **seasonal harvesting constraints**
d saisonale Holzeinschlagsbeschränkungen *fpl*
f contraintes *fpl* d'exploitation forestière saisonnières
r ограничения *npl* во введении лесозаготовок по погодным условиям; погодные условия *npl*, влияющие на лесозаготовки

7332 **seasonal road**
d saisonaler Weg *m*
f chemin *m* saisonné
r сезонный путь *m*

* **seasoned lumber → 7335**

7333 **seasoned pine block**
d getrockneter Kiefernblock *m*
f billot *m* de pin sec; billot de pin saisonné
r сухая сосновая заготовка *f* (для рубки спичечной соломки)

7334 **seasoned stump**
d getrockneter Stumpf *m*
f souche *f* sèche; souche saisonnée
r спелый осмол *m*

7335 **seasoned wood; seasoned timber; seasoned lumber**
(timber that has been dried so that the
d abgelagertes Holz *n*
f bois *m* saisonné
r выдержанный лесоматериал *m*

* **seasoning → 2423**

* **seasoning check → 2422**

* **season[ing] crack → 2422**

7336 **seasoning pile**
d Trocknungsstapel *m*
f pile *f* de séchage
r штабель *m* высушиваемого лесоматериала

* **seasoning stress → 2422**

7337 **season *v* timber**
d Holz abhärten
f saisonner
r выдерживать лесоматериалы

7338 **seat cut**
d Sitzschnitt *m*
f coupe *f* horizontale
r горизонтальный рез *m*

* **secateur → 5168**

7339 **secondary bark**
d sekundäre Rinde *f*
f écorce *f* secondaire
r вторичная кора *f*

7340 **secondary borers**
d sekundäre Käfer *mpl*
f perceuses *fpl* de bois secondaires
r вторичные стволовые вредители *mpl*

7341 **secondary branch**
(branch formed at some distance from the main stem and borne by a primary branch)
d Ast *m* der zweiten Kategorie
f sous-charpentière *f*; branche *f* secondaire; charpentière *f* secondaire; sous-mère *f*
r ветвь *f* второго порядка

7342 **secondary bud**
d sekundäre Knospe *f*
f bourgeon *m* secondaire
r запасная пазушная почка *f*

7343 **secondary cell wall**
(that portion of the cell wall formed after the cell enlargement has been completed)
d sekundäre Zellwand *f*

7344 secondary channel
(subordinate channel in a stream reach with more than one channel; minor channel in a floodplain)
d Seitengerinne *f*
f canal *m* secondaire; cours *m* d'eau secondaire
r дополнительный канал *m*

* **secondary cortex → 4502**

7345 secondary felling
d Nachhauung *f*; Nachhieb *m*
f coupe *f* secondaire
r дополнительная рубка *f*

7346 second[ary] forest; second-growth [forest]; second-growth stand; regrowth forest; derivative forest; secondary stand
(forest growth that has developed (naturally or artificially) following the removal of the original forest)
d Sekundärwald *m*; Sekundärforst *m*
f forêt *f* de seconde venue; peuplement *m* accessoire
r вторичный лес *m*; производный лес; лесонасаждение *n*, сменившее девственный лес

7347 secondary forest garden
d Sekundärforstgarten *m*
f pépinière *f* forestière secondaire
r вторичный лесной питомник *m*

* **secondary growth → 120**

7348 secondary insect
d sekundäres Insekt *n*
f insecte *m* secondaire
r вторичное насекомое *n*

7349 secondary lateral branch
d sekundärer Seitenzweig *m*
f branche *f* latérale secondaire
r боковая ветвь *f* второго порядка

7350 secondary logging road
(road designed for relatively little use)
d sekundäres Schleifweg *m*
f chemin *m* de débardage secondaire
r дополнительная лесная дорога *f*

7351 secondary needles
d Sekundarbenadelung *f*
f aiguille *f* secondaire
r вторичная хвоя *f*

7352 secondary pests
d Sekundärschädlinge *mpl*
f ravageurs *mpl* secondaires
r вторичные древоточцы *mpl*

* **secondary phloem → 4502**

7353 secondary ray
(in wood anatomy a ray derived from the cambium (i.e. originating after the development of secondary xylem), and not extending inwards as far as the pith)
d sekundärer Strahl *m*
f rayon *m* secondaire
r вторичный луч *m*

* **secondary root → 79**

* **second[ary] shoot → 4765**

* **secondary stand → 7346**

7354 secondary succession
d sekundäre Sukzession *f*
f succession *f* secondaire
r вторичная сукцессия *f*

* **secondary taproot → 6137**

* **secondary thickening → 120**

7355 secondary timber
d untergeordnetes Holz *n*
f bois *m* secondaire
r древесина *f* второстепенного значения

7356 secondary tree species; accessory tree species; subordinate species; ancillary tree species
(species of inferior quality and/or size, and of lesser silvicultural value, associated with the principal species)
d Nebenbaumart *f*; Nebenholzart *f*
f essence *f* secondaire; essence auxiliaire
r малоценное насаждение *n*; второстепенная древесная порода *f*

* **secondary tropical [rain] forest → 8980**

* **secondary truck road → 4172**

7357 secondary xylem; secondary wood
(wood produced by the cambium)
d sekundäres Xylem *n*
f xylème *m* secondaire
r вторичная древесина *f*; древесина, образуемая камбием; вторичная ксилема *f*

7358 second-class standard

f réserve *f* de deuxième classe
r отборные специальные деревья *npl*, оставляемые в порослевом насаждении с двумя оборотами рубки

7359 **second faller; helper**
(the subordinate in a crew of two fallers)
d Helfer *m*
f aide *f*
r дополнительный рабочий *m*

* **second growth → 120**

* **second-growth [forest] → 7346**

7360 **second-growth management**
d Nachwuchspflege *f*
f petit jardinage *m*
r уход *m* за подроста

* **second-growth stand → 7346**

* **second-growth thinners → 7760**

7361 **second-growth wood; coppice wood**
d Nachwuchsholz *n*
f bois *m* de taillis
r древесина *f* порослевых пород

7362 **second length; second log**
d End-Stammstück *m*; Zopf *m*; Zopfblock *m*
f grume *f* secondaire; surbille *f*
r бревно *n*, следующее после комлевого; срединное бревно

7363 **second log diameter**
d Zopfblockdurchmesser *m*
f diamètre *m* de surbille
r диаметр *m* срединного бревна

7364 **second pass**
(next entry to harvest timber after green-up (or other recovery objective) occurs)
d sekundärer Durchlauf *m*
f passage *m* secondaire
r второй проход *m*

7365 **seconds**
d Holzware *f* zweiter Qualität; Holzware zweiter Wahl
f bois *m* de qualité secondaire; seconds *mpl*
r пиломатериалы *mpl* второго сорта

7366 **second thinning**
d sekundäre Durchforstung *f*

* **secret stain → 4538**

* **section → 2171**

* **sectional area at breast height → 598**

* **sectional measurement of volume → 9244**

7367 **sector fork**
d Sektorkaliber *m*
f calibre *m* à secteur
r секторная лесная мерная вилка *f*

7368 **sedge-grass spruce forest**
d Seggen-Gräserfichtenwald *m*
f pessière *f* à laîches et graminées
r осоково-злаковый ельник *m*

7369 **sedimentation tar**
d Absetzteer *m*
f goudron *m* de sédimentation
r отстойная смола *f*

7370 **seed**
d Samen *m*
f semence *f*
r семя *n*; семена *npl*

7371 **seed** *v*; **sow** *v*
d ansäen; aussäen; ansamen; besäen; säen
f ensemencer; semer
r сеять; посеять

7372 **seed[ ]bank**
(place in which seeds of rare plant or obsolete varieties are stored, usually vacuum-packed and under cold conditions, to prolong their viability)
d Saatbank *f*
f banque *f* de semences
r посевная банка *f*

* **seed bearer → 5569**

* **seed[-bearing] tree → 5569**

7373 **seedbed; seed plot**
(the soil prepared by natural or artificial means to promote the germination of seeds and the growth of seedlings)
d Saatbeet *n*; Saatbett *n*
f lit *m* de germination; planche *f* de semis
r посевная гряда *f*; посевное место *n*; посевное отделение *n*

7374 **seedbed density; seeding density; sowing**

f densité *f* des graines
r густота *f* посевов

7375 seed block
(generally used to describe uncut blocks of trees that are left between and around small clearcut blocks to provide seeds for natural regeneration)
d Saatblock *m*
f bloc *m* de semis
r посевной блок *m*

7376 seed collection; seed collecting; seed cropping; seed harvest
d Samengewinnung *f*; Saatgutgewinnung *f*; Saatguternte *f*; Samenernte *f*
f récolte *f* de graines
r сбор *m* семян; заготовка *f* семян; добывание *n* семян; урожай *m* семян

7377 seed collection stand
(forest stand that exhibits good characteristics of growth, form, and vigo[u]r and that is not managed for cone production)
d Samengewinnungsbestand *m*
f peuplement *m* de récolte de graines
r насаждение *n* из семенной коллекции

* **seed cropping → 7376**

7378 seed crop stand
d Erntebestand *m*
f peuplement *m* prêt pour la récolte
r ликвидная масса *f* древостоя

* **seed cut[ting] → 7399**

* **seed dressing → 6141**

* **seeder → 5569**

* **seed fall → 2996**

* **seed flask → 7381**

7379 seed furrow; seed drill; seed rill; seed[ed] strip
d Saatreihe *f*; Saatrille *f*; Saatstreifen *m*; Saatrinne *f*
f sillon *m* de semis
r посадочная борозда *f*; посевная бороздка *f*; посевная полоса *f*; посевной жёлоб *m*

7380 seed germination
d Auskeimen *n* des Samens
f germination *f* de la graine; germination des

* **seed harvest → 7376**

7381 seed horn; sowing horn; seed flask
d Saathorn *n*; Saatflasche *f*; Saatflinte *f*
f tube *m* à semence; bouteille *f* à semence; cornet *m* à semence
r рожок *m* для высева семян

7382 seeding; sowing
d Aussaat *f*
f ensemencement *m*
r сев *m*; посев *m*; высев *m*

* **seeding density → 7374**

7383 seed[ing] felling; seed[ing] cutting
d Besamlungshieb *m*; Besamlungsschlag *m*; Samenschlag *m*
f coupe *f* d'ensemencement
r обсеменительный приём *m* постепенной рубки; обсеменительная рубка *f*

7384 seeding from adjacent stand; border seeding; marginal seeding
d Randbesamung *f*; Saumbesamung *f*
f ensemencement *m* de bord; semis *m* en lisière
r обсеменение *n* [лесосек] от стен леса; опушечное обсеменение

* **seeding in lines → 522**

* **seeding in situ → 2521**

7385 seeding lath; sowing lath
(device, commonly of wood, for obtaining uniformly spaced drills in a seedbed and aiding the even distribution of hand-sown seed in them)
d Saatlatte *f*
f planche *f* semoir; latte *f* à semis
r регулятор *m* нормы высева; посевная рейка *f*

7386 seed lac
d Samenlack *m*
f laque *f* de grains
r очищенная шеллачная смола *f* в виде мелких зёрен

* **seed-leaf → 1983**

* **seedling → 7392**

* **1-0 seedling → 9756**

f baliveaux *mpl* francs de pied
r семенное возобновление *n* леса

7388 **seedling crop; seedling forest; seed-production forest**
(crops and stands of trees, generally of seedling origin, that normally develop a high closed canopy)
d Kernwald *m*; Kornwuchs *m*; Samenwald *m*
f forêt *f* issue de semence; forêt issue de graine; semis *m*
r семенной древостой *m*; семенной лес *m*

7389 **seedling evaluation**
d Samenbeschätzung *f*
f évaluation *f* d'ensemencement
r определение *n* всхожести; определение энергии прорастания

* **seedling forest → 7388**

* **seedling percent → 6368**

* **seedlings → 8917**

7390 **seedling-sapling stand; sapling-seedling stand**
(stand of trees whose average d.b.h. is between 1 and 4 inches)
d Sämlings-Schössling-Bestand *m*
f peuplement *m* de semis et de gaules
r насаждение *n* из саженцев и молодняков

* **seedling seed orchard → 1646**

7391 **seedling stage**
d Aufwuchsalter *n*; Aufwuchsstadium *n*
f stade *m* de semis
r стадия *f* молодняка первого класса; стадия молодого древостоя

* **seedling stand → 9785**

7392 **seedling tree; seedling**
(seedlings-live trees of commercial species with diameters less than 1.0 inch that are expected to survive)
d Kernbaum *m*; Sämling *m*
f arbre *m* issu de semence
r дерево *n*, выросшее из семени

7393 **seedlot**
(quantity of cones or seeds having the same species, source, quality and year of collection)
d Samengruppe *f*

* **seed origin → 7395**

* **seed pellet → 6141**

* **seed plantage → 8915**

* **seed plantation → 8915**

* **seed plot → 7373**

* **seed-production forest → 7388**

7394 **seed-production forest area**
d Waldsamentragsfläche *f*
f aire *f* de production en graines forestières
r плюсовый лесосеменной участок *m*

* **seed-production stand → 6394**

* **seed rill → 7379**

* **seed shedding → 2996**

7395 **seed source; seed origin; seed provenance**
(locality where a seed lot was collected usually defined on an eco-geographic basis by distance, elevation, precipitation, latitude, etc.)
d Saatgutherkunft *f*; Saatgutprovenienz *f*
f origine *f* des graines; issue *f* de graines; provenance *f* de graines
r происхождение *n* семян

7396 **seedspot; planting spot; planting point; sowing point**
(prepared, limited space, e.g., a small, cultivated patch, within which (tree) seeds are sown)
d Pflanzungsplatz *m*
f placeau *m* de plantation
r лесокультурное посевное место *n*; лесокультурное посадочное место

* **seed stand → 6394**

7397 **seed storage**
d Aufbewahrung *f* des Saatgutes
f conservation *f* de graines
r хранение *n* семян

7398 **seed trap**
(device for catching the seeds falling on a small area of ground, from trees or shrubs)
d Samenfalle *f*

7399 seed-tree cut[ting]; seed cut[ting]; seed-tree removal cutting; seed-tree harvest
(felling of all the trees in an area except for a few desirable individuals that provide seed for the next forest)
d Einschlag *m* unter Belassung von Überhaltern
f coupe *f* rase avec semenciers; coupe des semenciers
r семеннолесосечная рубка *f*

7400 seed-tree cutting method; shelterwood [cutting] system; seed-tree silvicultural system
(a tree removal method whereby the entire stand is removed in a final cutting except for selected single seed trees to provide the seed for reproduction)
d Einschlagmethode *f* unter Belassung von Überhältern
f mode *m* de régénération par coupe rase avec semenciers; système *m* de régénération par coupe rase avec semencier
r способ *m* сплошных рубок с оставлением семенников; система *f* семеннолесосечной рубки

* **seed-tree harvest → 7399**

* **seed-tree removal cutting → 7399**

* **seed-tree silvicultural system → 7400**

* **seeker → 7049**

7401 seep
(a constantly wet area on the forest floor where water surfaces (seeps) from an underground source)
d Sickerwasserplatz *m*
f place *f* de suintement; place d'eau filtrante
r место *n* просачивания из земли воды

* **segmental circular saw → 6950**

* **sehunda → 4470**

* **selected cutting → 7655**

* **selected felling → 7655**

* **selected-line cutting → 8168**

7402 selected red oaks
(group of several red oak species composed of cherrybark, Shumard, and northern red oaks)
d selektive Roteichenarten *fpl*

7403 selected seeds
d Zuchtsaat *f*; Zuchtsaatgut *n*
f semences *fpl* sélectionnées
r плюсовые семена *npl*

* **select[ed] tree → 6396**

7404 selected white oaks
(group of several white oak species composed of white, swamp chestnut, swamp white, chinkapin, Durand, and bur oaks)
d selektive Weißeichenarten *fpl*
f espèces *fpl* sélectionnées de chênes blancs
r выбранные породы *fpl* белого дуба

* **select grade → 4284**

* **select group cutting → 4001**

* **selection border cutting → 8168**

7405 selection coppice
d geplenterter Niederwald *m*
f taillis *m* fureté
r выборочный низкоствольник *m*

7406 selection-cut area
d Blenderschlagsfläche *f*
f aire *f* de coupe jardinatoire
r лесосека *f* с выборочной рубкой; площадь *f* с выборочной рубкой

7407 selection differential
(average phenotypic value of the selected individuals, expressed as a deviation from the population mean)
d Selektionsdifferential *n*
f différentiel *m* de sélection
r селекционный дифференциал *m*

* **selection felling → 7412**

7408 selection felling forestry
d Dunkelschlagbetrieb *m*; Femelschlagbetrieb *m*; Schirmschlagbetrieb *m*; Plenterwaldsystem *n*
f système *m* de la forêt jardinée
r выборочнолесосечное хозяйство *n*

7409 selection felling in coppice; thinning stool-shoots
d Aushieb *m* des stärkes Holzes im Niederwald; Auszugshieb *m* im Niederwald
f extraction *f* de perches du taillis; coupe *f* de taillis-sous-futaie

7410 **selection forest; selection stand; culled forest; plenter wood**
(forest in which by selective cutting has removed only certain species above a certain size)
d Femelwald *m*; Plenterwald *m*; Plenterbestand *m*; Plänterwald *m*; durchhauener Wald *m*; geplenterter Wald
f forêt *f* jardinée; peuplement *m* jardiné; futaie *f* jardinée; forêt écrémée; forêt parcourue par une coupe
r выборочный лес *m*; лес выборочной рубки; выборочный древостой *m*; насаждение *n*, в котором проведена выборочная рубка вне лесосек; выборочно прорубленный лес

* **selection logging → 7412**

* **selection method → 3589**

* **selection system → 7620**

7411 **selection thinning; selective thinning; plenter[wood] thinning; dominant thinning**
d Plenterdurchforstung *f*; Auslesedurchforstung *f*
f éclaircie *f* jardinatoire; éclaircie sélective
r выборочное прореживание *n*; селекционное изреживание *n*

7412 **selective cut[ting]; selective felling; selection logging; selection felling; femel cutting**
(removing only certain species above a certain size or value)
d Ausplenterung *f*; Blenderschlag *m*; Ausforstung *f*; Femelhieb *m*; Plenterhieb *m*; Sortenaushieb *m*; Femelschlag *m*
f coupe *f* jardinatoire
r выборочная рубка *f*; приисковая рубка

7413 **selective harvesting**
d selektive Ernte *f*; Auswahlernte *f*; Mehrstufenernte *f*; Mehrphasenernte *f*
f cueillette *f* sélective; récolte *f* décomposée
r выборочная уборка *f* [урожая]

7414 **selective [high] pruning**
d Auswahlästung *f*
f élagage *m* sélectif
r выборочная обрезка *f* сучьев или ветвей

7415 **selectively marked area**
d Fläche *f* mit selektiver Markierung

* **selective thinning → 7411**

* **self-cleaning → 5688**

* **selfcomer → 9420**

7416 **self-cutting; self-felling; self-reaping**
d Selbstwerbung *f*; Selbstwerberhieb *m*; Selbsteinschlag *m*
f exploitation *f* par l'acquéreur
r заготовка *f* на себя

* **self-grafting → 35**

7417 **self-grown**
d selbstwachsend
f de régénération naturelle
r выросший самосевом

* **selfing → 7421**

7418 **self loader**
(logging truck with a loading device, generally a knuckleboom loader, mounted behind the cab)
d Selbstlader *m*
f autochargeuse *f*
r самозагрузчик *m*

7419 **self-loading skidder**
d selbsteinladender Schlepper *m*; Selbstladeskidder *m*
f débusqueur *m* autochargeur
r самозагружающийся трелёвочный трактор *m*

7420 **self logs**
d Wertholzbloche *mpl*
f billes *fpl* de bois de valeur
r брёвна *npl* высокого качества, продаваемые поштучно; кряжи *mpl* ценной древесины

7421 **self-pollination; selfing; autogamy**
(when a female cone is pollinated with pollen from the same tree or clone)
d Selbstbestäubung *f*; Selbsten *n*
f auto-pollinisation *f*; autofécondation *f*
r самоопыление *f*

* **self-propelled lopper → 3113**

* **self-propelled slasher → 6470**

7422 **self-propelled stumpcutter**
d selbstfahrender Stumpfscherblock *m*

* **self-pruning → 5688**

* **self-reaping → 7416**

* **self-seeding → 5695**

* **selfsown plant → 9420**

* **selfsown seeding → 5695**

* **self-thinning → 5699**

7423 **selva**
(rainforest of the Amazon)
d Selva *n*
f selva *m*
r сельва *f*

* **semi-circular spade → 6342**

7424 **semideciduous forest**
d halblaubwechselnder Wald *m*
f forêt *f* à feuillage semi-caduc; forêt semi-décidu
r полулистопадный лес *m*

* **semi-dry lumber → 7528**

7425 **semi-dry tropical forest**
d halbtrockener Wald *m*
f forêt *f* dense demi-sèche
r тропический полусухой лес *m*

7426 **semi-evergreen moist broadleaf forest**
d halbimmergrüner Laubnebelwald *m*
f forêt *f* demi-sempervirante humide feuillu
r полувечнозелёный влажный широколистный лес *m*

7427 **semi-factured wood[en] products**
d Holzhalbwaren *pl*
f produits *mpl* semi-finis en bois
r лесные полутовары *mpl*

* **semihardboard → 5754**

7428 **semihardwood cutting**
(a cutting taken from a stem that has just begun to develop woody tissue, for the purpose of propagation)
d Halbhartholzhieb *m*
f coupe *f* de bois demi-dur
r рубка *f* полутвёрдой древесины

* **seminal leaf → 1983**

* **seminary → 342**

7429 **semi-natural forest**
(consists of tree species which would occur naturally on a specific site and show similarities to primary forest)
d halbnatürlicher Wald *m*
f forêt *f* demi-naturelle
r полуестественный лес *m*

7430 **semirigid core**
d halbstarrer Kern *m*
f cœur *m* semi-rigide
r рыхлая серединка *f*; рыхлое наполнение *n*

7431 **semiring porosity**
d Halbringporigkeit *f*
f porosité *f* semi-anneau
r полукольцевая пористость *f*

7432 **semiring porous wood**
d halbringporiges Holz *n*
f bois *m* à zone semi-poreuse; bois à cernes semi-poreux
r полукольцесосудистая древесина *f*

7433 **semisuspended timber skidding**
d Holzrücken *n* im Halbschleifen
f débusquage *m* de bois en état demi-suspendu
r трелёвка *f* леса в полупогруженном положении

7434 **semi-tolerant tree species**
d Halbschattbaumart *f*
f essence *f* de phénombre
r полутеневыносливая порода *f*

7435 **semi-trailer landing gear; landing gear of semi-trailer**
(dollies or portion of a trailer that holds the trailer upright when it is not being supported by the truck tractor)
d stützeinrichtunger Sattelanhänger *m*
f béquille *f* de semi-remorque; béquille de soutien
r механизм *m* подъёма опорных передних колёс полуприцепа

7436 **sen**
(Kalopanax pictus; Kalopanax ricinifolius; Acanthopanax ricinifolius)
d Baumaralie *f*
f sen *m*
r акантопанакс *m* клещевинолистный; харигира *f*

f drageonnement *m*
r появление *n* отпрысков от корней

* **send-up man → 1559**

7438 **senescence**
(deteriorative changes of an organ or whole plant preceding death)
d Seneszenz *f*
f sénescence *f*
r старение *n*; дегенерация *f*

* **senescent → 5302**

7439 **senna family**
(Caesalpiniaceae)
d Johannisbrotgewächse *pl*
f césalpiniacées *pl*
r цезальпиниевые *npl*

7440 **sensitive resource area**
(identifiable geographic unit of the forestland base that requires a specific combination of forest practices to adequately protect important resource values)
d empfindliches Ressourcengebiet *n*
f zone *f* de ressources sensibles
r зона *f* чувствительных ресурсов

7441 **sensitive slope; unstable slope**
(any slope identified as prone to mass wasting)
d instabile Böschung *f*
f pente *f* sensitive
r неустойчивый склон *m*

7442 **sensitive soils**
(forestland areas that have a moderate to very high hazard for soil compaction, erosion, displacement, mass wasting or forest floor displacement)
d empfindlicher Boden *m*
f sols *mpl* sensitifs
r чувствительные почвы *fpl*

7443 **sensitive species; vulnerable species**
(populations of these species may not be declined by their habitat or other requirements are such that they are sensitive to further disturbance)
d empfindliche Art *f*
f espèce *f* sensitive; espèce vulnérable
r чувствительная порода *f*

7444 **sensitive watershed**
(that is used for domestic purposes or that has
f bassin *m* versant sensible
r чувствительный водораздел *m*

7445 **separated forest stand; tree bunch; tree bundle; hurst**
d Baumbündel *n*
f faisceau *m* d'arbres
r колок *m*; куртинный лес *m*; куртина *f* [леса]; куртина деревьев; несколько деревьев

7446 **separate driving; individual floating**
d getrenntes Flößen *n*
f flottage *m* séparé
r разрозненный сплав *m*

7447 **separating fruit**
d trennende Frucht *f*
f fruit *m* séparé
r дробный плод *m*

7448 **separator wood**
(material that has been separated from the whole tree during the chipping process and is unacceptable for pulp and paper manufacture)
d getrenntes Holz *n*
f bois *m* séparé
r отделённая древесина *f*

* **septate fiber-tracheid → 7450**

7449 **septate parenchyma cell**
(an axial or radial parenchyma cell with one or more thin transverse walls across its lumen)
d gefächerte Parenchymzelle *f*
f cellule *f* cloisonnée de parenchyme
r перегородчатая паренхимная клетка *f*; септированная паренхимная клетка

* **septate tracheid → 8145**

7450 **septate wood fiber; septate fiber-tracheid**
(a fiber with thin transverse walls across the lumen)
d gefächerte Faser *f*
f fibre *f* ligneuse cloisonnée; fibre ligneuse septée
r септированное древесное волокно *n*

* **September elm → 2893**

7451 **sequence of rings; ring sequence; ring series; ring pattern**
d Ringenordnung *f*

7452 **sequestration of carbon; carbon sequestration**
(the uptake and storage of carbon)
d Kohlenstoff-Sequestrierung *f*
f piégeage *m* de carbone
r удаление *n* углерода

7453 **seral; successional**
(of or about the series of stages in ecological or plant succession)
d seriell; fortlaufend; [nach]folgend
f serial; de succession
r сукцессионный; серийный

7454 **seral species**
d nachfolgende Art *f*
f espèce *f* serial
r сукцессионное растение *n*

7455 **seral stage; stage of succession**
(any stage of development of an ecosystem from a disturbed, unvegetated state to a climax plant community)
d Sukzessivstadium *n*; Seralstufe *f*
f stade *m* de succession écologique; stade sériel
r стадий *m* сукцессии

7456 **Serbian spruce**
(Picea omorica)
d serbische Fichte *f*; Omorikafichte *f*
f épicéa *m* de Serbie
r ель *f* сербская

7457 **sere**
(the complete cycle of changes from the original state to the climax condition)
d Serie *f*
f série *f*; sère *f*
r смена *f* стадий развития биоценоза; ряд *m* фитоценозов; серия *f* фитоценозов (совокупность последовательно развивающихся растительных сообществ)

* **seringa → 8335**

7458 **serotinous** *adj*
(particularly applied to plant species or individuals with cones that remain on the tree without opening for one or more years)
d spätreifend
f sérotinal; tardif
r поздний; поздноцветущий; приносящий поздние плоды

7459 **serotinous cone**
d später Zapfen *m*

7460 **serrate-ciliate**
d gesägt-wimperblätt[e]rig
f serré-cilié
r пильчато-реснитчатый

7461 **serrate[d]; dentate; toothed**
d verzahnt; gezähnt; gesägt
f denté; crénelé; serré
r зубчатый; пильчатый

* **serrate leaf margin → 2396**

* **serrette → 2570**

7462 **serviceberry; Juneberry**
(Amelanchier)
d Felsenbirne *f*
f amélanchier *m*
r ирга *f*

* **serviceberry → 3806**

* **service forester → 1890**

* **service-tree → 8991**

7463 **sessile leaf**
(without a petiole (leaf stalk), that is, borne directly on the stem)
d sitzendes Blatt *n*
f feuille *f* apode; feuille sessile
r сидячий лист *m*

7464 **sessile oak; durmast [oak]; mountain oak**
(Quercus sessiliflora; Quercus petraea)
d Steineiche *f*; Darreiche *f*; Traubeneiche *f*; Wintereiche *f*; Späteiche *f*; Gebirgseiche *f*
f chêne *m* rouvre; chêne à glands sessiles; chêne sessile
r дуб *m* скальный; дуб-железняк *m*; дуб сидячецветный; дуб зимний

* **set-back → 6415**

7465 **set-back tree**
d zurückgesetzter Baum *m*
f arbre *m* en retrait
r осевшее назад дерево *n* (при валке)

7466 **set fire; incendiary fire**
d Wald-Brandstiftung *f*
f incendie *m* criminel
r пожар *m* вследствие поджога; пожар из-за пуска пала

7467 **set of lines; bundle rigging; rigging**

f timonerie *f*
r канатная оснастка *f*; тросовая оснастка; формировочный такелаж *m*

* **setting → 5038**

7468 **setting; strip**
d Trasse *f*
f bande *f* de téléphérage
r площадь *f*, осваиваемая канатной установкой с одной стоянки; лесосека *f*, осваиваемая канатной установкой с одной стоянки

7469 **setting size**
d Sägedimension *f*
f dimension *f* du sciage
r распиловочный размер *m*

7470 **settlement**
d Siedlung *f*
f établissement *m* de colons; colonisation *f*; peuplement *m*
r поселение *n*; колония *f*

* **set trees → 93**

7471 **setulose leaf; aristate leaf; awned leaf**
d Blatt *n* mit Grannen
f feuille *f* sétule; feuille barbue; feuille aristée
r ворсистый лист *m*; лист, покрытый щетинками; щетинконосный лист

* **setup → 8049**

7472 **severance felling; severance cutting; increment felling; increment thinning; accretion cutting; advance thinning**
(cleared strip cut through a stand so as to develop a wind-firm edge before making any fellings)
d Loshieb *m*
f coupe *f* de préparation de lisière; coupe de détachement
r прореживание *n* в древостое; проходная рубка *f*; охранная рубка

* **severe cutting-back → 2347**

* **severe thinning → 4217**

* **severing → 5100**

* **severity of thinning → 8519**

* **Sevilla orange [tree] → 775**

d Klärschlamm *m*
f boue *f* [d'eaux] d'égout; boue d'épuration; vase *m* de filtrage
r осадок *m* сточной воды; необработанный осадок; сырой осадок

* **SG → 2803**

* **shackle → 3888**

* **shackle guy cable → 1164**

* **shackle guy rope → 1164**

* **shadblow serviceberry → 2603**

* **shadbush serviceberry → 2603**

* **shade-bearing tree → 7478**

7474 **shade[-bearing] tree species; shade-tolerant [tree] species; tolerant species; sciaphilous species**
(tree species that will regenerate in shade so is often large seeded)
d schattenertragende Art *f*; Schattenbaumart *f*
f essence *f* d'ombre; essence sciaphile
r теневыносливая порода *f*

* **shade endurance → 7477**

7475 **shade-giving shrub**
d shattenspendender Strauch *m*
f arbuste *m* d'ombre
r затеняющий кустарник *m*

7476 **shade-intolerant tree; intolerant [tree]; light-demanding tree; light demander; heliophilous tree**
(incapable of developing and growing normally in the shade of, and in competition with, other trees)
d lichtbedürftiger Baum *m*; Lichtbaum *m*
f arbre *m* intolérant; arbre exigeant la lumière; arbre héliophile
r светолюбивое дерево *n*

* **shade-requiring tree → 7478**

7477 **shade tolerance; shade endurance**
(relative capacity of a species to become established and persist under a canopy)
d Schattenfestigkeit *f*; Schattenerträgnis *n*
f capacité *f* de support l'ombre; stabilité *f*

7478 shade-tolerant tree; tolerant tree; shade-requiring tree; shade-bearing tree
(capable of reproducing under the shade of parent trees)
d schattenertragender Baum *m*; Schattenbaum *m*
f arbre *m* d'ombre
r теневыносливое дерево *n*

* **shade-tolerant [tree] species → 7474**

7479 shade tree; shady tree
d schattiger Baum *m*; shattenspendender Baum
f arbre *m* d'ombrage; arbre obombrant
r тенистое дерево *n*; дерево, посаженное, чтобы давать тень

7480 shadetree trimming
d Abästung *f* der schattiger Baum
f élagage *m* d'arbres d'ombrage
r обрезка *f* затеняющих деревьев или сучьев

7481 shading
d Beschirmung *f*; Beschattung *f*
f ombrage *m*
r защита *f* пологом; затенение *n*

* **shady tree → 7479**

* **shaft height → 8077**

* **shaft length → 8077**

* **shaft wood → 6427**

7482 shagbark [hickory]; red hickory
(Carya ovata; Carya microcarpa)
d Hickorynussbaum *m*; Schuppenrindenhickory *f*
f caryer *m* ovale; caryer blanc; noyer *m* [blanc] d'Amérique
r гикори *m* косматый; кария *f* косматая; кария овальная; кария мелкоплодная

* **shaggy → 934**

7483 shake; split; check
(a lengthwise grain separation between growth rings, or a break through the rings caused by stresses in the standing tree or by felling and handling of the log)
d Riss *m*; Schwingung *f*; Spalt *m*
f gerçure *f*; gerce *f*; fente *f* [unilatérale]; secousse *f*
r трещина *f*

7484 shake depth
d Risstiefe *f*
f profondeur *f* de fente
r глубина *f* подновки

* **shake length → 2004**

7485 shallow beam
d Flachbalken *m*
f poutre *f*
r тонкая балка *f*

7486 shallow root; horizontal root; surface-root; superficial root
d Flachwurzel *f*; Horizontalwurzel *f*; Streichwurzel *f*; Tagwurzel *f*; Tauwurzel *f*
f racine *f* horizontale; racine superficielle
r поверхностный корень *m*; [поверхностно] стелющийся корень; горизонтальный корень

7487 shallow root system
d Flachwurzelsystem *n*
f système *m* de racines superficielles
r поверхностная корневая система *f*

7488 shallow root tree
d Flächewurzelbaum *m*
f arbre *m* à racines superficielles
r дерево *n* с поверхностными корнями

* **shamrock → 9656**

* **shanty man → 9654**

7489 shaped tree; clipped tree; pollard[ed] tree; pollard
d beschnittener Baum *m*; gekappter Baum; geköpfter Baum; Kopfholzstamm *m*; Poller *m*
f arbre *m* étêté; arbre taillé; fût *m* de têtard
r подрезанное дерево *n*; безвершинник *m*; подстриженное дерево; формованное дерево; дерево с обрезанной кроной

* **shape number → 8840**

7490 shape of a tree; tree form; tree shape
d Baumform *f*
f forme *f* d'arbre
r форма *f* дерева

* **shape of growth → 4017**

* **shape pruning → 6612**

7491 shaping
d Formgebung *f*; Modellierung *f*

7492 **sharp-edged timber**
d scharfkantiges Vollholz *n*
f bois *m* à vive arête
r острокантный брус *m*; остроконечный брус

* **sharp end → 819**

* **sharp-leafed → 7494**

7493 **sharpness angle; tooth angle; grinding angle; wedge angle**
(the angle between the cutting face of the top plate of a cutter link and the top plate itself)
d Schärfewinkel *m*
f angle *m* de bec
r угол *m* заострения; угол заточки (горизонтальной режущей кромки Г-образного звена пильной цепи)

7494 **sharp-petalled; sharp-leafed**
d scharf blätt[e]rig
f de feuille raide
r остролепестковый

* **shaved lumber → 2620**

* **shave knife → 7960**

* **shaving → 579**

* **shavings → 9508**

* **sheaf tree → 5617**

7495 **shear**
(a condition of stress or strain where parallel planes slide relative to each other; tendency of wood fibers to slide longitudinally)
d Scherbeanspruchung *f*; Scherkraft *f*; Scherspannung *f*
f cisaillement *m*
r вырыв *m*

7496 **shear boom**
d Scherbaum *m*
f grume *f* faisant partie d'une élingue de flottage
r бревно *n* для запруды; кошель *m*

7497 **shear crack**
d Scherungsriss *m*
f fente *f* par cisaillement
r трещина *f* скалывания

7498 **sheared end**
(of a tree)
r срезанный ножами конец *m*

7499 **sheared surface**
d gescherte Fläche *f*
f surface *f* cisaillée
r срезанная поверхность *f* [брёвна]; поверхность ножевого среза

7500 **shear failure**
d Scherbruch *m*
f rupture *f* par cisaillement
r разрушение *n* в результате среза; разрушение от скалывания; разрушение при срезе

7501 **shear felling; shearing**
(harvest using mechanical shears; in Christmas tree culture, to prune the branches to make dense foliage and give the tree a conical shape)
d Scherhieb *m*; Scheren *n*
f abattage *m* à la cisaille; traitement *m* à la cisaille; cisaillement *m*; cisaillage *m*
r валка *f* леса машинами с ножевым срезающим устройством; безопилочное резание *n*

* **shear [felling] head → 3044**

7502 **shear-through-thickness strength**
(of logs)
d Dichte-Scherfestigkeit *f*; Sprengkraft *f*
f résistance *f* au cisaillement par épaisseur
r прочность *f* на срез по толщине

* **sheathing → 880**

7503 **shea tree**
(Butyrospermum parkii Kotsch)
d Schibutterbaum *m*
f karité *f*
r дерево *n* ши

7504 **sheave-type grapple**
d Rillenscheibengreifer *m*
f pince *f* de débardage à poulie entaillée
r захват *m* с трособлочным управлением

* **she-balsam → 3689**

* **sheepberry → 5656**

7505 **sheep laurel; Lambkill kalmia**
(Kalmia angustifolia)
d schmalblättrige Lorbeerrose *f*

* **sheer boom → 3147**

* **sheer skid → 1025**

7506 **sheet erosion; sheet washings**
(removal of a fairly uniform layer of soil from the land surface by runoff water, without the development of conspicuous water channels)
d Flächenerosion *f*; flächenhafte Erosion *f*; flächenhafte Abtragung *f*
f érosion *f* en nappes; érosion par couches
r плоскостная эрозия *f*; поверхностная эрозия

7507 **shellbark hickory; king-nut hickory**
(Carya laciniosa)
d Königsnuss *f*
f caryer *m* lacinié; hickory *m* écailleux
r кария *f* бахромчатая

* **shell rot → 853**

* **shell shake → 8279**

* **shelly timber → 9696**

7508 **shelterbelt; protective forest belt; timberbelt; wood mantle; forest mantle**
(extended windbreak of living trees and shrubs established and maintained for the protection of farmlands over an area larger than a single farm)
d Waldschutzstreifen *m*; Waldschutzgürtel *m*; feldschützender Waldstreifen *m*; Waldmantel *m*; Sicherheitsstreifen *m*
f rideau *m* boisé de protection; bande *f* boisée de protection; rideau-abri *m*
r [поле]защитная лесная полоса *f*; лесная опушка *f*

7509 **shelterbelt establishment**
d Schutzwaldanbau *m*
f création *f* de rideau-abri
r полезащитное лесоразведение *n*

* **shelter-box → 5071**

* **sheltered area → 354**

7510 **sheltered spot seeding**
(sowing of seeds under small conical shelters of translucent or opaque, bio- or photo-degradable material)
d Schutzplätzesaat *f*
f ensemencement *m* sur placeaux sous-abri;

* **shelter tree → 5816**

7511 **shelterwood; overhead shelter**
d Schirmbestand *m*
f abri *m*
r верхний полог *m*

7512 **shelterwood compartment system**
d Schirmschlagsystem *n*
f système *m* de régénération par coupe progressive
r система *f* равномерных постепенных рубок по кварталам

7513 **shelterwood cut[ting]; shelterwood felling; shelterwood harvest; shelterwood logging**
(method of harvesting timber so that selected trees remain scattered throughout the tract to provide seeds for regeneration and shelter for seedlings)
d Schirmhieb *m*; Schirmschlag *m*
f coupe *f* d'abri
r рубка *f* осветления верхнего полога

* **shelterwood [cutting] system → 7400**

7514 **shelterwood group selection system**
d Femelschirmschlagsystem *n*; Femelschirmschlagbetrieb *m*
f régime *m* des coupes progressives d'abri par bouquets
r система *f* группово-выборочных рубок верхнего полога

* **shelterwood harvest → 7513**

* **shelterwood logging → 7513**

7515 **shelterwood method**
(regeneration method under an even-aged silvicultural system under which a portion of the mature stand is retained as a source of seed or protection)
d Schirmschlagmethode *f*
f méthode *f* de régénération d'abri par coupe progressive
r способ *m* семеннолесосечной рубки

7516 **shelterwood thinning**
d Schirmdurchforstung *f*
f éclaircie *f* d'abri
r семеннолесосечная выборочная рубка *f*

7517 **shelterwood wedge coupe; wedge**

f coupe *f* d'abri en coin; coupe d'abri en dents de scie
r клиновидновыборочная рубка *f*; клиновидная рубка верхнего полога; клинообразная рубка

7518 **shelterwood with reserves**
(a variant of the Shelterwood Method in which some or all of the shelter trees are retained, well beyond the normal period of retention, to attain goals other than regeneration)
d Schirmschlag *m* mit Reserve
f coupe *f* d'abri avec réserve
r семеннолесосечная вырубка *f* с оставлением резервных деревьев

* **she-oak → 407**

* **she-oak male → 6390**

7519 **shield-leaf pumpwood; yagrumo; cetico; ambahu; guarumo**
(Cecropia obtusifolia; Cecropia peltata)
d Guarumo *m*; Kanonenbaum *m*
f bois-canon *m*; cécropie *f* peltée
r цекропия *f* щитовидная

7520 **shifting cultivation; shifting agriculture; roving agriculture**
(primitive, unregulated method of cyclical cultivation, still widely practised in the tropics)
d Wanderfeldbau *m*; wandernde Rodungskultur *f*
f culture *f* itinérante; culture nomade
r сменная культивация *f*; сменная обработка *f*; чередующаяся обработка

7521 **shingle; wooden slat; chinking board; clapboard**
(thin rectangular piece of wood, sawn along the grain used like tiles for roofing and siding)
d Schindel *f*
f bardeau *m*
r гонт *m*

* **shingle bolt → 9575**

* **shingle oak → 3874**

* **shining sumac[h] → 3247**

7522 **shining willow**
f saule *m* brillant
r ива *f* блестящая

* **shin oak → 3928**

7523 **shiny blueberry; ground blueberry; Florida evergreen blueberry**
(Vaccinium myrsinites; Cyanococcus myrsinites; Vaccinium nitidum)
d glänzende Bickbeere *f*; Schwarzbeere *f*
f myrtille *f* moulue
r черника *f* блестящая

7524 **shipbuilding timber**
d Seebauholz *n*
f bois *m* de construction maritime
r строительный лесоматериал *m* для морских сооружений; судостроительный кряж *m*

7525 **shiplapped lumber**
(lumber rabbeted on both edges of each piece to provide a close-lapped joint)
d Faserholz *n*
f bois *m* à rive à mi-bois; bois à fibre
r пиломатериал *m*, строганный на фальц

7526 **shipping bundle**
d Versandbündel *n*
f faisceau *m* d'expédition
r пакет *m* [лесоматериалов] для отгрузки

* **shipping culls → 5450**

7527 **shipping dry** *adj*
(having a moisture content (ovendry basis) of 14% to 20%)
d trocken zur Verschiffung
f commercialement sec pour expédition
r транспортно сухой

7528 **shipping-dry lumber; semi-dry lumber**
(lumber that is partially dried to prevent stains and mold in transit)
d verladetrockenes Holz *n*; halbtrockenes Holz
f bois *m* ressuyé commercialement sec
r транспортно сухой лесоматериал *m*; пиломатериал *m* транспортной влажности; пиломатериал, частично высушенный для предотвращения засинения во время транспортировки

7529 **shipping mark**
d Verlademarke *f*
f marque *f* de chargement
r погрузочная метка *f*

* **shittim [wood] → 1349**

7530 **shive; sliver**
(in pulpwood)
d Splitter *m*
f bûchette *f*; incuit *m*
r пучок *m* волокон

* **shoe [rake] → 2054**

* **shoe tree → 2054**

* **shoot → 3303, 7944**

* **shoot bud → 9179**

* **shoot cutting → 7010**

7531 **shoot *v* leaves; foliate *v*; come *v* into leaf; leaf *v* out; put *v* forth leaves; be *v* in leaf; leaf *v*; frondesce *v***
d durchblättern
f feuillir
r покрываться листьями; выпускать листья

* **shoot moth → 6269**

* **shoot of apex → 1872**

7532 **shoot pruning**
(cutting away undesirable shoots to favor survival and growth of selected shoots)
d Schussästung *f*
f élagage *m* des rejets
r обрезка *f* побегов

* **shop lumber → 9721**

* **shore → 7270**

* **shorea → 4831**

* **shore forest → 1687**

* **shore pine → 5044**

7533 **short-boled; short bodied; short-stemmed**
d kurzschäftig
f à fût court
r короткоствольный

* **short-cut wood → 903**

7534 **short distance hauling**
d Nahbringung *f*; Nahförderung *f*

* **shorten → 2165**

* **shortening → 5100**

* **shortening saw → 1136**

7535 **short-fibered wood; short-grained wood**
d kurzfaseriges Holz *n*; anbrüchiges Holz *n*; brüchiges Holz; sprödes Holz
f bois *m* à fibres courtes
r коротковолокнистая древесина *f*; хрупкая древесина; ломкая древесина

7536 **short grain**
d Kurzfaser *f*
f fil *m* court
r короткое волокно *n*

* **short-grained wood → 7535**

7537 **shorthear**
(Calamagrostis)
d Reitgras
f gourbet *m*; calamagrostis *m*
r вейник *m*

* **shortleaf black spruce → 2716**

7538 **shortleaf pine; North Caroline pine; Arkansas pine**
(Pinus echinata; Pinus mitis)
d Glattkiefer *f*; Carolinakiefer *f*; Fichtenkiefer *f*
f pin *m* doux; pin à courtes aiguilles
r сосна *f* ежовая; сосна жёлтая

* **short-length logging → 3078**

* **short log → 903**

7539 **short logger; short-log truck**
d Holzsortiment-Kraftkarren *m*
f chariot-porteur *m* d'assortiments de bois
r сортиментовоз *m*

* **short logging → 3078**

* **short-log operation → 3078**

7540 **short-log trees**
(sawtimber-sized trees of commercial species that contain at least one merchantable 8-foot to 11-foot saw log)
d Klotzbäume *mpl*
f arbres *mpl* de bois court
r деревья *npl* для коротья

* **short-needled coniferous shoot → 501**

* **short needle yew → 9345**

7541 **shortness of grain**
d Kurzfaserigkeit *f*
f fragilité *f* de fils
r коротковолокнистость *f*

7542 **short rotation**
d Kurzumtrieb *m*; Kurzrotation *f*
f courte rotation *f*; courte révolution *f*; croissance *f* rapide
r короткий оборот *m* рубки

7543 **short-rotation bushwood**
d Buschholzbetrieb *m*
f broussailles *fpl* à rotation courte
r хозяйство *n* на кустарник

7544 **short-rotation coppice; short-rotation silviculture**
d Niederwaldbetrieb *m* mit Kurzumtrieb; Stockausschlagsbetrieb *m* mit Kurzumtrieb
f taillis *m* à courte rotation; taillis à courte révolution; taillis à croissance rapide
r рубка *f* в низкоствольном хозяйстве с коротким оборотом ротации

7545 **short-rotation forestry**
d Forstwirtschaft *f* mit Kurzumtrieb
f sylviculture *f* à courte rotation
r лесохозяйство *n* с коротким оборотом ротации

* **short-rotation silviculture → 7544**

7546 **short-rotation tree**
d schnellwachsender Baum *m*
f arbre *m* à rotation courte
r дерево *n* с коротким оборотом ротации

7547 **shorts**
(with length less of 183 mm)
d Shorts *fpl*
f courson *m*; fausse coupe *f*; petite longueur *f*; bois *m* courts
r пиломатериалы *mpl* короче стандарта

7548 **short shoot; spur shoot; dwarf shoot**
d Kurztrieb *m*
f pousse *f* courte
r укороченный побег *m*

7549 **short skid**
d Kurzschleifweg *m*

* **short-stemmed → 7533**

* **shortwood → 903**

7550 **shortwood crew**
d Kurzholzmannschaft *f*
f équipe *f* de bois courts
r бригада *f*, заготавливающая сортименты

7551 **shortwood harvester; pulpwood harvester**
d Kurzholzerserntemaschine *f*
f abatteuse-porteuse *f* de bois courts
r многооперационная машина *f* для заготовки сортиментов

* **shortwood logging → 3078**

* **shotgun → 8567**

* **shot hole → 4511**

* **shothole borer → 231**

* **show wood → 3159**

* **showy mespilus → 5389**

7552 **showy mountain ash**
(Sorbus decora)
d Labrador-Eberesche *f*; Schmuck-Eberesche *f*; amerikanische Vogelbeere *f*
f sorbier *m* des montagnes; sorbier décoratif
r рябина *f* декоративная

* **shragging → 6610**

* **shreds → 7553**

7553 **shredded chips; [chopping] shreds**
d zerkleinerte Späne *mpl*
f copeaux *mpl* fractionnés
r измельчённая щепа *f*; измельчённые стружки *fpl*; измельчённые обрезки *fpl*; тонкие древесные стружки; стружечный кряж *m*

* **shrinkage coefficient → 9653**

* **shrinkage crack → 2422**

7554 **shrinkage of wood; wood shrinkage**
(the reduction in dimension or volume which takes place in timber when the moisture content is reduced below fiber saturation point, expressed as a percentage of the original dimensions or volume)
d Desorption *f* des Holzes; Schwinden *n*

* **shrinkage value → 9653**

7555 **shrub; scrub; brush**
(perennial plant differing from a perennial herb in its persistent and woody stem(s), and less definitely from a tree in its lower stature and the general absence of a well-defined main stem)
d Busch *m*; Gesträuch *n*; Strauch *n*; Staude *f*
f buisson *m*
r кустарник *m*; куст *m*

* **shrubbed → 934**

* **shrubbery → 933**

7556 **shrubbery path**
d Gebüschweg *m*
f chemin *m* au fruticées
r дорожка *f*, обсаженная кустарником

7557 **shrub bog**
(wetland consisting of saturated organic material, such as peat and muck, and shrubs)
d Gebüsch-Sumpf *m*
f marais *m* broussailleux
r кустарниковое болото *n*

* **shrubby → 934**

* **shrubby althea → 7067**

* **shrubby birch tree → 1295**

* **shrubby tree → 346**

* **shrub clearing → 1109**

* **shrub control → 1101**

* **shrub forest → 1202**

* **shrubland → 353**

7558 **shrublet**
(a dwarf woody plant)
d Kleinbusch *m*
f petit arbrisseau *m*
r мелкий кустарник *m*; заросль *f* кустарничков

7559 **shrub live oak; turbinella oak; sonorian scrub oak; California scrub oak**
(Quercus turbinata; Quercus gamelliflora; Quercus crassilamellata)
d kalifornische Strauch-Eiche *f*

* **shrub pine → 7191**

7560 **shrub savanna[h]**
d Buschsavanne *f*
f savane *f* arbustive
r кустарниковая саванна *f*

7561 **Shumard oak**
(Quercus shumardii)
d Shumards-Eiche *f*
f chêne *m* de Shumard
r дуб *m* Шумарда

7562 **shuttle hauling**
(use of preloading trailers to reduce truck turn-around time)
d Bringung *f* im Pendelwagen
f débardage *m* à navettes
r вывозка *f* [леса] с предварительной погрузкой на прицепы

7563 **Siam rosewood; Thailand rosewood; payung; trac**
(Dalbergia cochinchinensis)
d Siam-Rosenholz *n*; Payung *n*
f palissandre *m* de Siam; bois *m* de rose du Siam; payung *m*; trac *m*
r дальбергия *f* сиамская

* **Siam weed → 866**

7564 **Siberian acacia; Siberian pea tree; pea shrub**
(Caragana arborescens)
d [gemeiner] Erbsenstrauch *m*; sibirischer Erbsenstrauch; Erbsenbaum *m*
f amarilla *m*; caraganier *m* arborescent; pois *m* de Sibérie
r карагана *f* древовидная; акация *f* жёлтая

7565 **Siberian crabapple**
(Malus baccata)
d sibirischer Apfel *m*; Beerenapfel *m*
f pommier *m* à baies
r яблоня *f* сибирская

* **Siberian elm → 4618**

7566 **Siberian fir**
(Abies sibirica; Abies pichta)
d sibirische Tanne *f*
f sapin *m* de Sibérie
r пихта *f* сибирская

7567 **Siberian larch**

f mélèze *m* de Sibérie
r лиственница *f* сибирская; лиственница Ледебура

* **Siberian pea tree → 7564**

7568 **Siberian spruce**
(Picea obovata)
d sibirische Fichte *f*
f épicéa *m* de Sibérie; sapin *m* sibérien
r ель *f* сибирская

7569 **Siberian [stone] pine; kedar**
(Pinus cembra var. sibirica)
d sibirische Zirbel-Kiefer *f*
f arolle *f* sibérienne
r сосна *f* кедровая сибирская; кедр *m* сибирский

* **sibs → 3754**

* **side → 5083**

* **side axe → 1041**

7570 **side bind**
d Seitenfestlaufen *n*
f coincement *m* latéral
r боковое сжатие *n* (дерева, подпираемого по краям и в середине); положение *n* дерева, лежащего поперёк крутого склона

7571 **side blocking**
d Seitenblockierung *f*
f blocage *m* latéral; établissement *m* de blocs à la limite [forestière]
r установка *f* блоков по границе [лесосеки]

7572 **side board**
d Seitenbrett *n*
f dernière planche *f*
r доска *f* без сердцевины; боковая доска; получистая доска

* **side branch → 4819**

7573 **side camp; line camp**
d Lager *m* an Feuerschutzstreifen
f camp *m* volant; camp avancé; camp de ligne d'arrêt
r крайний боковой лагерь *m* (при тушении пожара)

7574 **side crack**
d äußerer Riss *m*

* **side cutting → 964**

* **side foreman → 1290**

7575 **side grafting**
d Seitenpfropfen *n*
f greffage *m* latéral; greffage de côté
r прививка *f* в боковой отщеп

* **side grain → 5133**

* **side hauling → 4821**

7576 **side line**
(of growing tree)
d Neigungslinie *f*
f ligne *f* d'inclinaison
r линия *f* наклона в сторону

7577 **side lumber**
(a board from the outer portion of the log - ordinarily one produced when squaring off a log for a tie or timber)
d Seitenholz *n*
f planche *f* latérale
r пиломатериал *m* из внешней части брёвна

* **side mark → 5101**

7578 **side notching; polling; cornering**
d Wurzelanläufe *mpl*
f égobelage *m* latérale
r подрезание *n* заболони вокруг дерева; подрезание дерева с обеих сторон от подпила

* **side piece → 7712**

* **side product → 5471**

7579 **side[-pull] line**
d Seitenlinie *f*; Nebenlinie *f*
f chaîne *f* auxiliaire
r оттяжной канат *m* (для бокового перемещения передвижного несущего каната)

7580 **side ride**
d Nebenschneise *f*
f laie *f* latérale
r боковая просека *f*; второстепенная просека

* **side roots → 4824**

7581 **side shading; side shade**
d Randbeschattung *f*; Seiten[be]schatten *n*

* **side shoot → 418**

7582 **side taproot**
d Seitenhauptwurzel *f*
f racine *f* pivotante latérale
r боковой стержневой корень *m*

7583 **side winder**
(a tree fallen by another tree)
d Aufhalter *m*
f arbre *m* gisant par un autre arbre
r дерево *n*, сваленное при валке другого; дерево, на котором зависло другое спиленное дерево

7584 **Siebold Japanese ash**
(Fraxinus sieboldiana)
d chinesische Blütenesche *f*
f frêne *m* chinois à fleurs
r ясень *m* Зибольда

7585 **Siebold maple**
(Acer sieboldianum)
d Sieboldsahorn *m*
f érable *m* de Siebold
r клён *m* Зибольда

7586 **Siebold walnut; Japanese walnut**
(Juglans sieboldiana; Juglans ailanthifolia)
d japanischer Nussbaum *m*; japanische Walnuss *f*
f noyer *m* du Japon
r орех *m* Зибольда; орех айлантолистный; онигуруми *m*

7587 **Sierra alder**
(Alnus rhombifolia)
d amerikanische Weißerle *f*
f au[l]ne *m* de la Sierra
r ольха *f* ромболистная

7588 **Sierra juniper; western juniper**
(Juniperus occidentalis)
d Bennett-Wacholder *m*
f genévrier *m* de l'Occident; genévrier de Bennett
r можжевельник *m* западноамериканский

* **sieve area → 7590**

7589 **sieve cell**
(a long, slender, conducting cell of the phloem, that does not form a constituent element of a sieve tube)
d Siebzelle *f*

7590 **sieve field; sieve area**
(a depressed area in the wall of a sieve element, perforated by a sieve-like cluster of minute pores through which the protoplast is connected with that of a contiguous sieve element)
d Siebfeld *n*
f crible *m*
r ситовидное поле *n* (клеток древесины)

7591 **sieve pitting**
(an arrangement of small pits in sieve-like clusters; sievelike clustered pits in the ends of the narrow tubular processes between disjunctive cells)
d Tüpfelung *f*
f ponctuations *fpl* en tamis
r ситовидные поры *fpl*

7592 **sieve plate**
(a specialized part of the wall of a sieve tube member that has a solitary sieve area (simple sieve plate), or several closely placed sieve areas often arranged in a scalariform or reticulate manner)
d Siebplatte *f*; Lochplatte *f*
f plaque *f* criblée; cloison *f* criblée
r ситовидная пластинка *f*

* **sieve tissue → 4502**

7593 **sieve tube**
(a food-conducting tube of the phloem made up of an axial series of sieve tube members)
d Siebröhre *f*
f tube *m* criblé
r ситовидная трубка *f*

* **sighting → 3074**

7594 **Sikkim spruce**
(Picea spinulosa)
d Sikkim-Fichte *f*
f sapin *m* du Sikkim
r ель *f* шипованная; ель восточногималайская

7595 **Silesian willow**
(Salix silesiaca)
d schlesische Weide *f*
f saule *m* de Silésie
r ива *f* силезская

* **silicified wood → 6196**

* **silk bay → 6763**

7596 **silk-cotton family**
(Bombacaceae)
d Bombakaziengewächse *pl*; Wollbaumgewächse *pl*
f bombacacées *pl*
r шерстяных *pl*

* **silk-oak → 3966**

7597 **silk tassel tree**
(Garryaceae)
d Becherkätzchengewächse *pl*
f garryacées *pl*
r гарриевые *pl*

* **silk tree → 6185**

* **silky grain → 7601**

* **silky headed willow → 3880**

* **sill → 1548**

7598 **silt**
(fragments of rock or organic matter sediment most of the individual particles of which are between 0.05 and 0.002 mm in diameter)
d Schluff *m*; Silt *m*; Schlick *m*
f limon *m*
r пыль *f*; ил *m*; илистый нанос *m*

7599 **siltation; silting**
(filling-in of lakes and stream channels with soil particles, usually as a result of erosion on adjacent land)
d Verstopfung *f*; Versetzung *f*; Stockung *f*
f alluvionnement *m*; colmatage *m*; limonage *m*
r заиливание *n*; заращивание *n*; облитерация *f*; отложение *n* наносов

* **silva** *lat* **→ 3318**

* **silvan → 3324**

* **silverball → 3825**

* **silver bark → 7787**

* **silver beech → 406**

* **silverbell → 7800**

* **silverberry → 257**

* **silver birch → 1761**

7600 **silver buffaloberry; bullberry**
(Shepherdia argentea)
d Silber-Büffelbeere *f*
f shéferdie *f* argentée
r шефердия *f* серебристая

7601 **silver figure; silver grain; silky grain**
(conspicuous medullary rays in quarter-sawn lumber)
d Kernstrahlung *f*; Spiegelfaser *f*
f maille *m*; maillure *f*
r рисунок *m*, образованный блестящими сердцевинными лучами; сердцевинные лучи *mpl*; серебристая текстура *f* (распиленной древесины)

* **silver fir → 1803**

7602 **silver fir bark beetle**
(Pityokteines curvidens)
d krummzähniger Tannenborkenkäfer *m*; krummzähniger Weißtannenborkenkäfer *m*
f rongeur *m* du sapin blanc; bostryche *m* curvidente
r короед *m* пихтой серебристой

* **silver grain → 7601**

7603 **silver grain surfaced wood**
(wood with a figure produced by conspicuous rays)
d Holz *n* mit Silberfaser
f bois *m* maillé
r обтёсанная древесина *f* с сердцевинными лучами

7604 **silvergreen-wattle acacia**
(Acacia dealbata)
d australische Silberakazie *f*; echte Akazie *f*
f acacia *m* blanchâtre; mimosa *m* blanchâtre; mimosa blanchissant
r акация *f* беловатая; акация подбеленная; акация серебристая

7605 **silver gum; heart-leaved gum; heartleaf eucalyptus**
(Eucalyptus cordata)
d Silberdollar-Eukalyptus *m*
f eucalyptus *m* à feuilles cordiformes
r эвкалипт *m* сердцевидный

7606 **silver-leaf; silver blight**
(common name for the disease caused by the fungus Stereum purpureum, which gives a silvery appearance to the leaves of some of its

f plomb *m*
r леукофиллум *m*

* **silver[-leaf] poplar → 2911**

7607 **silver lime[tree]; silver linden**
(Tilia tomentosa; Tilia argentea)
d [ungarische] Silberlinde *f*
f tilleul *m* argenté
r липа *f* серебристая

7608 **silver maple; curled hard maple; curly maple**
(Acer saccharinum)
d Silberahorn *m*
f érable *m* argenté; érable blanc
r клён *m* серебристый; клён сахаристый; клён кудрявый сахарный

7609 **silver oak; muhuhu; muhugu**
(Brachylaena hutchinsii; Brachylaena merana; Brachylaena ramiflora)
d Muhuhu *m*
f hazotokana *m*; muhuhu *m*
r брахилена *f* гучинская; мухугу *m*

* **silver pendant linden → 6143**

* **silver pine → 9344**

* **silver spruce → 2822**

* **silvertop [ash] → 4583**

* **silver willow → 7138**

* **silvicide → 343**

7610 **silvics; silviculture; timber growing; growing of woods; forest growth**
(a branch of forestry that is concerned with the methods of raising and growing trees)
d Forstwissenschaft *f*; Holzzucht *f*; Holzanbau *m*; Anbau *m* in Walde; Forstzüchtung *f*; Waldzucht *f*
f sylviculture *f*; entretien *m* de la forêt; culture *f* du bois
r лесоразведение *n*; разведение *n* леса; дереворазведение *n*

7611 **silvicultural amelioration; forest [a]melioration; forest reclamation**
d Forstmelioration *f*

7612 **silvicultural behaviour**
d waldbauliches Verhalten *n*; forstliches Verhalten
f comportement *m* forestier; comportement sylvicole
r лесоводственное поведение *n*

7613 **silvicultural biomass**
(the total mass, at a given time, of living organisms of one or more species per unit forest area (species biomass) or of all the species in the community)
d waldbauliche Biomasse *f*
f biomasse *f* forestière
r лесохозяйственная биомасса *f*

7614 **silvicultural conversion**
(change from one silvicultural system to another)
d waldbauliche Konversion *f*
f conversion *f* sylvicole
r лесокультурное преобразование *n*

7615 **silvicultural decision model**
(a computer model for simulation and possibly prediction of the silvicultural behaviour)
d waldbauliches Entscheidungsmodell *n*
f modèle *m* de décision sylvicole
r модель *f* принятия лесохозяйственных решений

7616 **silvicultural efficacy**
(capacity of a herbicide indirectly to promote positive growth responses in crop trees)
d waldbauliche Wirksamkeit *f*
f efficacité *f* sylvicole
r лесохозяйственная эффективность *f*

* **silvicultural management → 3488**

* **silvicultural operation → 7623**

* **silvicultural practice → 3576**

7617 **silvicultural prescription; silviculture prescription; timber management prescription**
(statement following an examination of a compartment usually done at least once every 10 years to determine management needs of the timber)
d waldbauliche Anordnung *f*; Holzwirtschaftsvorschrift *f*

7618 silvicultural regime; silviculture regime
(series of stand tending (thinning, pruning, etc.) treatments applied after regeneration to achieve a specific stand management objective)
d waldbauliches Regime *n*
f régime *m* sylvicole; traitement *m* sylvicole
r лесохозяйственный режим *m*

7619 silvicultural rotation
(through which a species maintains satisfactory growth and reproduction on a given site)
d waldbaulicher Umtrieb *m*
f âge *m* d'exploitabilité naturelle; âge d'exploitabilité sylvicole
r оборот *m* рубки по возобновительной спелости

7620 silvicultural selection system; selection system; jardinage system; plenter system; femel system
(for removing mature timber either as single scattered individuals or in small groups at relatively short intervals, repeated indefinitely)
d Plenterbetrieb *m*; Femelbetrieb *m*
f système *m* de jardinage
r лесохозяйственная система *f* выборочной рубки

7621 silvicultural system
(planned program of treatments throughout the life of the stand to achieve stand structural objectives based on integrated resource management goals)
d waldbauliches System *n*; waldbaulicher Betrieb *m*
f système *m* sylvicole
r лесохозяйственная система *f*

7622 silvicultural thinning
d waldbauliche Durchforstung *f*
f éclaircie *f* sylvicole
r лесохозяйственная рубка *f* ухода

7623 silvicultural treatment; silviculture treatment; silvicultural operation
(any activity that enhances and/or recovers forest growth or harvest yield, such as site preparation, planting, thinning, fertilization, and harvesting)
d Holzbehandlung *f*
f traitement *m* forestier
r лесохозяйственное мероприятие *n*

* **silviculture → 7610**

* **silviculture close to nature → 5683**

* **silviculture prescription → 7617**

* **silviculture regime → 7618**

7624 silviculture survey
(sampling procedure to determine silvicultural conditions such as planting survival, free-growing status, stocking, etc., leading to management decisions)
d waldbauliche Vermessung *f*
f estimation *f* sylviculturale
r лесохозяйственная съёмка *f*

* **silviculture treatment → 7623**

* **silviculturist → 344**

7625 silvipastoral system; sylvopastoral system; silvopastoralism; sylvopastoralism
(mixed management system involving a stand of trees and undergrowth pasture on the same parcel of land; a form of agroforestry system consisting of the trees and pasture and animal components)
d Waldweidewirtschaft *f*; Forstweidesystem *n*
f système *m* sylvopastoral; sylvipastoralisme *m*
r система *f* лесного и пастбищного хозяйств; система разведения скота в лесу

* **silvipasture → 3452**

* **silvogenic soil → 9547**

* **silvopastoralism → 7625**

* **silvopasture → 3452**

7626 silvopisciculture
(growing trees as part of a fish-farming enterprise)
d Forstfischzucht *f*; Forstfischkultur *f*; Forstfischbetrieb *m*
f sylvipisciculture *f*
r рыборазведение *n* в лесу; рыбоводство *n* в лесу

* **simarouba → 9373**

7627 Simon poplar
(Populus simonii)
d chinesische Pappel *f*

7628 simple bud
d einfache Knospe *f*
f bourgeon *m* simple
r простая почка *f*

7629 simple fruit
d einfache Frucht *f*
f fruit *m* simple
r простой плод *m*

7630 simple heart shake
d einfacher Kernriss *m*; einfacher Sprung *m*
f fente *f* de cœur simple
r простая метиковая трещина *f*

7631 simple leaf
(single leaf blade with a bud at the base of the leafstem)
d einfaches Blatt *n*
f feuille *f* simple
r простой лист *m*

7632 simpleleaf chastetree
(Vitex trifolia; Vitex simplicifolia)
d indischer Wildpfeffer *m*
f vitex *m* à feuilles simples
r прутняк *m* трилистный

7633 simple perforation plate
d einfache Perforationsplatte *f*
f cloison *f* perforée simple
r простая перфорационная пластинка *f*

7634 simple sweep
d einfache Krümmung *f*
f courbure *f* simple
r простая кривизна *f*

* **simultaneous ignition → 8308**

7635 Singapore oak; pasang ijang
(Quercus conocarpa; Pasania conocarpa)
d Singapore-Eiche *f*
f chêne *m* de Singapore
r дуб *m* конусоплодный; пасанг идянг *m*

7636 single-blade [tree] shear; single-bitted axe; single-bladed axe
d Einblatt-Axt *f*
f sécateur *m* d'abattage à simple tranchant
r топор *m* с односторонним лезвием

7637 single-boled
d einläufig; einstämmig
f à tige unique
r одноствольный

7638 single-entry volume table; one-way volume table
d Einzelerfassungsmassentafel *f*
f tarif *m* de cubage à une entrée
r таблица *f* объёмов, измеряемых по одному показателю

* **single knot → 7290**

7639 single-layer [chip]board; single-layer particle board
d einschichtige Spanplatte *f*
f panneau *m* de particules homogènes
r однослойная древесностружечная плита *f*

7640 single-layered
d einschichtig
f mono-strate
r однослойный

* **single-layer particle board → 7639**

7641 singleleaf ash; Utah ash; Lowell ash
(Fraxinus anomala; Fraxinus lowellii)
d einblätt[e]rige Esche *f*
f frêne *m* à feuilles simples
r ясень *m* неправильный

7642 singleleaf pine; singleleaf pinyon
(Pinus monophylla)
d einblätt[e]rige Kiefer *f*
f pin *m* à une aiguille; pin à feuilles simples
r сосна *f* однохвойная

7643 single-line grapple
d eingleisiger Greifer *m*
f pince *f* de débardage uniligne; pince de débardage à simple voie
r одноканатный трелёвочный захват *m*

7644 single-line system; single-strop system
d Einseilsystem *n*
f système *m* uniligne
r одноканатная система *f*

7645 single-moldboard plough
(plough with one moldboard, generally right-hand, turning the whole furrow slice to one side of the furrow)
d Pflug *m* mit Einzelstreichblech; Pflug mit Einzelriester
f charrue *f* à versoir simple
r одноотвальный плуг *m*

7646 single-pole support; single-strop support
d Einschlingenunterstützung *f*

7647 single-seed fruit
d Einzelkornfrucht *f*
f fruit *m* monograine
r односемянный плод *m*

* **single-seed hawthorn → 1780**

7648 single-side crown; one-side crown
d einseitige Krone *f*
f cime *f* unilatérale
r односторонняя крона *f*

* **single sled → 5115**

7649 single-span system; single-span skyline
(skyline without intermediate support spars)
d Einfeldsystem *n*
f câble-grue *m* à franchissement single
r однопролётная канатная установка *f*

7650 single stand; individual stand; isolated stand
d Einzelbestand *m*
f peuplement *m* isolé
r одиночное насаждение *n*

* **single stem → 4480**

7651 single-stem delimber; single-tree delimber
d Einzelbaum-Entastungsmaschine *f*
f ébrancheuse *f* simple
r машина *f* для обрезки деревьев по одному

7652 single-stem total volume
d Einzelbaum-Gesamtmasse *f*
f volume *m* entier d'arbre [individuel]
r объём *m* [отдельного] дерева; объём отдельного хлыста

7653 single-storey[ed] forest; single-storied stand
d einstufiger Wald *m*; einstufiger Bestand *m*
f forêt *f* monoétagée; peuplement *m* à un seul étage
r простой древостой *m*; одноярусный древостой

7654 single-strand log haul-up
d einzellitziger Blockförderer *m*
f chaîne *f* de billes monocaténaire; élévateur *m* à une chaîne centrale
r продольный одноцепной лесотранспортёр *m*

* **single-strop support → 7646**

* **single tree → 4480**

* **single-tree delimber → 7651**

* **single-tree mixture → 5496**

7655 single-tree selection [cutting]; selected cutting; selected felling
(in which trees are removed individually, here and there, over an entire forest or stand)
d Auszugshieb *m*; Auszugshauung *f*
f coupe *f* sélective par arbre; jardinage *m* par arbre
r вырубка *f* единичных деревьев; поштучная заготовка *f* деревьев

7656 single-tree [selection] method
(method of regenerating uneven-aged stands in which individual trees of any size are removed more or less uniformly throughout the stand)
d Auszugshiebsmethode *f*
f méthode *f* de jardinage par arbre
r способ *m* поштучной заготовки деревьев

7657 single-tree [system of] management; forest management by individual trees
d Einzelstammwirtschaft *f*
f gestion *f* sélective par arbre; exploitation *f* arbre par arbre
r хозяйство *n*, направленное на уход за единичными деревьями

* **single-variable volume table → 7638**

* **single Virgin's clematis → 4593**

* **single-wire logging → 3237**

7658 singling
(of transplants)
d Verkrehlen *n*; Vereinzelung *f*
f démariage *m*
r разъединение *n*

7659 singly-choked trees
d einzeln erstickte Bäume *mpl*; einzeln gewürgte Bäume
f arbres *mpl* choqués individuellement
r деревья *npl*, зачокерованные по одному

* **sink → 2514**

* **sinker → 2268**

7660 sinking; sinkage
(of logs)
d Untersichtweite *f*; Abteufen *n*
f enfoncement *m*; fonçage *m*; immergence *f*; immersion *f*
r потопление *n*

7661 sinuate leaf margin
(imaging the outline of gentle rolling waves)
d gebuchteter Blattrand *m*
f bord *m* de feuille sinué
r выемчатая форма *f* края листа

* **sinus → 5028**

* **siphon feed gun → 8884**

* **sipo → 9150**

* **siris → 2730**

* **Siskiyou cypress → 484**

7662 site-adapted forest
(forest that consists of trees and undergrowth species and densities of those species that would naturally be found associated with the local area climate, topography, soil type, and moisture regime)
d standortgerechter Wald *m*
f forêt *f* adaptée à la station
r адаптированный к месту лес *m*; приспособленный к месту лес

* **site amelioration → 7666**

7663 site capability; site capacity
d Standort-Fähigkeit *f*; Ertragsfähigkeit *f*
f potentiel *m* de station
r возможность *f* местообитания

7664 site class; locality class
(classification of forest land in terms of inherent capacity to grow crops of industrial wood)
d Standortsgüteklasse *f*
f classe *f* de station
r класс *m* местообитания

* **site classification → 3593**

7665 site disturbance
d Standortstörung *f*
f détérioration *f* de la station
r повреждение *n* почвы и древостоя на лесосеке

7666 site improvement; site amelioration
(modifications to a given site in order to improve growing conditions for a specific species or mixture of species)
d Standortverbesserung *f*
f amélioration *f* de la station
r улучшение *n* местопроизрастания

7667 site index
d Standortbonität *f*
f indice *m* de fertilité de la station
r индекс *m* местообитания

7668 site of cutting; place of felling; [tree-]felling site; felling point; cutting tract; cutting point
d Einschlagsort *m*; Fällort *m*; Hiebsort *m*; Schnittstelle *f*
f parterre *f* de coupe; [dé]coupe *f* d'abattage; lieu *m* de la coupe; point *m* de chute; emplacement *m* de trait de scie
r место *n* рубки; место пропила; место выполнения подпила на дереве

7669 site potential tree height
(the height a dominant tree may attain given the site conditions where it occurs)
d potentielle Standortbaumhöhe *f*
f hauteur *f* potentielle d'arbres à la station
r потенциальная высота *f* деревьев местообитания

* **site preparation → 3985**

7670 site quality; site productivity
(productive capacity of a site; usually expressed as volume production of a given species per unit area or per unit of time)
d Standortsgüte *f*; Bodengüte *f*; Bodenbonität *f*
f qualité *f* de station; qualité stationnelle
r качество *n* лесорастительных условий; продуктивность *f* местообитания

* **site rehabilitation → 6846**

7671 site requirement
d standortspezifische Anforderung *f*; Standortsanspruch *m*
f besoin *m* du site [requise]; exigence *f* écologique de station
r требование *n* к местопроизрастанию

7672 site selection
d Standort[saus]wahl *f*

7673 site sensitivity
(assessment of the susceptibility of a site to soil-degrading processes, such as soil compaction, erosion, mass wasting, and forest floor displacement)
d Standortsensitivität *f*
f sensitivité *f* de la station
r чувствительность *f* местообитания

7674 site-specific *adj*
(pertaining to a specific planning unit)
d flächenbezogen
f en fonction de la surface
r зависящий от местных условий; сайт-специфичный

7675 site type
(biotic productivity unit (usually for trees) which is qualitatively defined on the basis of being relatively uniform with respect to climate, soil and natural vegetation cover)
d Standortstyp *m*
f type *m* de site; type de station
r тип *m* лесорастительных условий

7676 site utilization
(term used when indicating the proportion of a usable forest site occupied by healthy, vigorous forest crop trees at any one point in time)
d Erschließung *f* des Grundstücks
f aménagement *m* du terrain
r использование *n* местообитания

7677 Sitka alder
(Alnus sinuata)
d Sitka-Erle *f*
f au[l]ne *m* de Sitka
r ольха *f* вырезанная

* **Sitka mountain ash → 3943**

7678 Sitka spruce; Menzies spruce; tideland spruce; western spruce
(Picea sitchensis)
d Sitkafichte *f*
f sapin *m* de Sitka; épicéa *m* de Sitka
r ель *f* ситхинская

* **Sitka-spruce beetle → 2724**

7679 Sitka-spruce weevil
(Pissodes sitchensis)
d Sitkafichte-Rüsselkäfer *m*
f charançon *m* du sapin de Sitka
r смолёвка *f* ситхинской ели

(Salix sitchensis)
d Sitkaweide *f*
f saule *m* de Sitka
r ива *f* ситхинская

* **size class → 8924**

* **size-limit felling → 2476**

* **sizes of timber → 8576**

* **size speck → 8155**

* **sizing → 3903**

* **skawbush → 6783**

* **skeg → 2274**

7681 skeletal substance
(the material, mainly cellulose, comprising the framework of the cell wall)
d Gerüstsubstanz *f*
f squelette *f* cellulaire
r скелетное вещество *n*

* **skidder → 88, 5114**

7682 skidder driver
(a worker)
d Schlepperfahrer *m*
f entrepreneur *m* de débusqueur
r тракторист *m* трелёвочного трактора

* **skidder line → 3987**

7683 skidder logging
d Holzaufbereitung *f* mittels Schlepper
f débardage *m* par tracteur
r лесозаготовка *f* с применением колёсных тракторов; трелёвка *f* колёсными тракторами

* **skidder tire → 5086**

* **skidding → 8595**

* **skidding cable → 3987**

* **skid[ding] chain → 7798**

7684 skidding cost[s]; hauling cost[s]; bringing cost; pay for hauling
d Rückekosten *pl*; Rücke[r]lohn *m*
f frais *mpl* de débardage

7685 **skidding damage; hauling damage; logging damage**
d Abrückschaden *m*; Rückenschaden *m*
f dégât *m* de débardage
r повреждение *n* [деревьев] при трелёвке

* **skidding dish → 7687**

* **skid[ding] distance → 9747**

* **skidding dog → 8577**

* **skidding hooks → 3892**

* **skidding lane → 7689**

* **skidding line → 3987**

7686 **skidding log**
d Rutschenklemme *f*
f grume *f* de débardage; bille *f* de débardage
r скоба *f* с крюком для крепления брёвна к тяговой цепи

* **skidding losses → 5174**

7687 **skid[ding] pan; snigging pan; pan; skidding shoe; skidding dish**
d Rückpfanne *f*; Hemmschuh *m*; Radschuh *m*; Sperrschuh *m*
f pelle *f* de débusquage; pelle de débardage; traîne *f* à billot; traîne de débusquage
r трелёвочный пен *m*; трелёвочный поддон *m* (широкий стальной щит, поддерживающий концы брёвен при трелёвке)

7688 **skidding pattern**
d Holzabfuhrmodell *n*
f modèle *m* de débardage
r схема *f* расположения трелёвочных волоков

7689 **skid[ding] road; skidding lane; snig track; travois road; toting log; haul[age] road; logway; skidway; logging track; log truck road; logging road; logger road; lumber road; extraction road**
(road cut through the woods for skidding; any way, more or less prepared, over which logs are dragged)
d Rückeweg *m*; Rückebahn *f*; Mähnweg *m*; Hiebsweg *m*; schleppender Weg *m*
f chemin *m* de débusquage; chemin de débardage; chemin de travois; chemin de

* **skidding scissors → 3892**

* **skidding shoe → 7687**

* **skidding sled → 5115**

* **skidding tongs → 3892**

* **skid[ding] trail → 1245**

7690 **skidding trip**
d Holzabfuhrreise *f*
f voyage *m* de débardage
r грузовой рейс *m*

7691 **skidding vehicle; tracked vehicle; [off-road] logging vehicle**
d Schlepper *m*
f système *m* de débardage
r трелёвочная установка *f*

7692 **skidding winch; logging winch; log hoist; log hauler transfer; donkey**
(in logging, a portable engine mounted on a sled and equipped with a drum and cable)
d Bringungswinde *f*; Forstseilwinde *f*; Holzschleifwinde *f*; Rückenwinde *f*; Blockgleitbahn *f*; Blockförderbahn *f*; Holzrückmaschine *f*
f treuil *m* de débusquage; treuil de débardage; glissoire *f* pour billes; appareil *m* à débarder du bois
r трелёвочная лебёдка *f*; продольный лесотранспортёр *m*; конвейер *m* для кряжей; трелёвочная машина *f*

* **skid greaser → 1576**

* **skid platform → 5027**

7693 **skid poles**
(logs or poles, commonly used in pairs, on which logs are rolled)
d Gleitkufen *fpl*
f bachons *mpl*
r покаты *mpl* для закатки брёвен

7694 **skids**
([half]-round timber laid in pairs or sets to form a skidway or in skidding)
d Kufen *fpl*
f bâti *m* de longerons; longrines *fpl*
r брусья *mpl* для выстилки волока или площадки; слеги *mpl*

* **skidway → 7689**

7695 skinner
d Maschinenarbeiter *m*
f conducteur *m*
r рабочий *m* на подготовке лесоматериалов к чокеровке

7696 skipper road
(a skidding road on which poles are placed zig-zag across the road or transversely at intervals or "skips" of 4 to 6 feet)
d Skipperbahn *f*
f chemin *m* bachonné en travers; chemin bachonné en zigzag
r укреплённый трелёвочный волок *m*

* **skip-row planting → 8964**

* **skunk spruce → 9386**

* **skycar carrier → 7698**

* **skyline block → 8864**

* **skyline buggy → 7698**

7697 skyline cable-crane; cable-crane; skyline crane
(yarding system capable of moving logs laterally to a skyline as well as transporting logs either up or down a skyline to a landing)
d Kabelkran *m*; Seillaufkran *m*; Seilkran *m*
f câble-grue *m*; blondin *m*
r кабельный кран *m*; кабель-кран *m*; кабельно-подвесной кран

* **skyline [cableway] → 84**

7698 skyline carriage; skycar carrier; bicycle; skyline buggy
(wheeled device that rides back and forth on the skyline for yarding or loading)
d Kranwagen *m*; Laufwagen *m*; Laufkatze *f*
f chariot-palan *m*
r трелёвочная каретка *f* (подвесной канатной установки для трелёвки или погрузки леса)

7699 skyline clamp; rope clamp; rope cramp; rope clip; cable clip; wire clip; dee
d Seilklemme *f*; Seilschloss *m*
f pince *f* à câble; collier *m* de câbles
r зажим *m* для крепления несущего каната; канатный зажим

* **skyline crane → 7697**

* **skyline hauler → 88**

logging; skyline skidding; cableway skidding; overhead skidding; aerial skidding; skylining
(transporting logs from stumps to collecting points that uses a heavy cable stretched between high points)
d Flugbringung *f*; Holzrückung *f* mit einer Drahtseilbahn; Drahtseilholzbringung *f*; Luftrückung *f*
f traînage *m* sur câble
r подвесная трелёвка *f*; лесозаготовка *f* с подвесной канатной трелёвкой; перемещение *n* груза по несущему канату

7701 skyline logging area
d Flugbringungsfläche *f*
f aire *f* de débardage par câble
r лесосека *f*, разрабатываемая на базе подвесных канатных установок

7702 skyline range
d Luftseilbahnverbreitungsgebiet *n*
f aire *f* d'extension de câble
r протяжённость *f* несущего каната; охват *m* несущего каната

* **skyline skidder → 88**

* **skyline skidding → 7700**

7703 skyline squad
d Luftseilbahnkommando *n*
f brigade *m* de support de câble
r бригада *f*, обслуживающая канатную установку

7704 skyline tension; line tension
d Seilspannung *f*
f tension *f* de câble
r натяжение *n* несущего каната

7705 skyline tension indicator
d Seilspannungsindikator *m*
f indicateur *m* de tension de câble
r указатель *m* натяжения несущего каната; динамометр *m*

7706 skyline thinning
d Durchforstung *f* mit einer Drahtseilbahn
f éclaircie *f* par traînage sur câble
r рубки *fpl* ухода, выполняемые с помощью подвесных канатных установок

* **skyline yarding → 7700**

* **skylining → 7700**

* **slab → 83, 7712**

7707 slab barker
d Schwartenentrinder *m*
f écorceuse *f* de dosses
r окорочный барабан *m* для горбыля

* **slabbed log → 7971**

* **slabber → 7708**

* **slabbing → 5117**

7708 slab[bing] chipper; slabbing [chip] head; slabber
d Schwartenhackmaschine *f*
f tête *f* dédosseuse
r дробилка *f* горбыля

7709 slab chipping
d Schwartenzerkleinerung *f*
f déchiquetage *m* de dosse
r дробление *n* горбыля

* **slab cut → 2980**

* **slab edging → 5117**

* **slab-saw → 5148**

7710 slabs of bark; bark slabs
d Rindenplatten *fpl*
f plaques *fpl* d'écorce
r коревые плиты *fpl*

7711 slab transfer deck
d Schwartebühne *f*
f rampe *f* de transfert de dosse
r транспортёр *m* для отвода горбыля

7712 slabwood; slab; flitch [billet]; offcut; side piece
(the exterior portion of a log removed by a saw, having one flat and one curved surface)
d Holzschwarte *f*; Schwarte *f*; Heilbuttschnitte *f*
f bois *m* de dosse; dosse *f*; dosseau *m*; croûte *f*; quartelle *f*; quartelot *m*
r горбыль *m*; отщеп *m*; брус[ок] *m* неквадратного сечения

7713 slack cooperage; dry cooperage
(cooperage used as containers for dry, semidry or solid products)
d trockene Fassware *f*; trockene Böttcherei *f*
f tonnellerie *f* non étanche
r сухотарные бочки *fpl*

7714 slack[ing] rope; rope not carrying the load
d Leerseil *n*; Hängseil *n*
f câble *m* à vide; câble porteur à tension variable; câble de retour; corde *f* suspendue
r опускающийся несущий канат *m*; слабонатянутый трос *m*; порожный трос; висячий трос

7715 slack line logging; slack rope logging
d Leerseilrücken *n*
f débardage *m* par câble à vide
r трелёвка *f* установкой с опускающимся несущим канатом

* **slackline system → 5017**

* **slackpuller → 1164**

* **slack rope → 7714**

7716 slack side; loose side
d lockere Seite *f*
f face *f* distendue
r внутренняя сторона *f* ленты шпона

* **slash → 1122**

7717 slash-and-burn shifting cultivation
(cultivation by agricultural fires set to clear forests; requires a lot of land, since farmers must move their crops to new fields every few years)
d Brandrodung-Wanderfeldbau *m*
f agriculture *f* itinérante après débroussaillement et brûlis; agriculture sur brûlis
r подсечно-огневое земледелие *n*

7718 slash area
d Abraumfläche *f*
f aire *f* à rémanents; zone *f* de rémanents
r неочищенная площадь *f*

7719 slash burning
d Holzabfallverbrennung *f*; Reisigverbrennung *f*; Schlagbrennen *n*; Abraumverbrennen *n*
f incinération *f* des déchets; brûlage *m* de déchets
r сжигание *n* порубочных остатков; сжигание отходов древесины; сжигание хвороста

7720 slash-burning legislation
d Schlagbrennengesetz *n*
f législation *f* d'incinération des déchets [forestiers]

7721 **slash chopper; wood scrab grinder; [waste]wood hog; refuse chipper; refuse grinder; slasher**
d Holzabfallzerkleinerer *m*; Holzabfallfräze *f*; Abfallzerkleinerer *m*; Buschhacker *m*
f hache-bois *m*; broyeuse *f* de rémanents; broyeur *m* de déchets; hachoir *m* de déchets
r дробилка *f* для древесных отходов; дробилка для лесосечных отходов; измельчитель *m* лесосечных отходов; рубильная машина *f* для отходов; слешер *m*

* **slash clearing → 7723**

7722 **slash conk**
(Lenzites sepiaria)
d Zaunblättling *m*
f lenzite *m* des clôtures
r столбовой гриб *m*

* **slash cutter → 7319**

7723 **slash disposal; brush disposal; slash removal; slash clearing**
(treatment or handling of slash, particularly so as to reduce fire or insect hazard)
d Abfallbeseitigung *f*; Reisigräumen *n*
f élimination *f* des rémanents; enlèvement *m* des rémanents
r уборка *f* [лесосечных] отходов; уборка хвороста; очистка *f* от отходов; удаление *n* отходов; устранение *n* отбросов

7724 **slash-disposal fire**
d Abfallbeseitigungsfeuer *n*
f feu *m* d'élimination des rémanents
r огневая очистка *f* лесосек (путём сжигания порубочных остатков)

7725 **slash-disposal legislation**
d Abfallbeseitigungsgesetz *n*
f législation *f* d'élimination des rémanents
r законодательство *n* по очистке лесосек

* **slasher → 7721**

* **slasher desk → 7727**

7726 **slasher-merchandiser capacities; merchandiser capacities**
d Schneider-Leistungsvermögene *npl*
f capacités *fpl* de poste de marchandisage
r мощности *fpl* [машины] для раскряжёвки лесоматериалов

* **slasher saw → 2609**

d Holzausformungsplatz *m*; Waldausformungsplatz *m*; Ausformungsplatz *m*; Manipulationsfläche *f*
f plate-forme *f* de tronçonnage
r раскряжёвочная площадка *f*; место *n* раскряжёвки

* **slashes → 5170**

7728 **slash figure**
(wood texture where the annual rings form a more or less acute angle with the face of the board)
d Querschnittszeichnung *f*
f figure *f* de fente en bout; figure de débit en long; figure de fissure longitudinale
r текстура *f* на торце; текстура, получаемая при продольной распиловке

7729 **slash figure check; longitudinal fissure; split[ting of grain]**
d Längsriss *m*; Fasersplittung *f*
f fissure *f* longitudinale
r продольная трещина *f*

* **slash grain → 2803**

* **slash-grain[ed] timber → 3254**

* **slashing → 1109**

* **slash[ings] → 998**

* **slashing saw → 2609**

* **slash knot → 4376**

7730 **slash pile**
d Holzabfallbündel *n*
f empilement *m* des rémanents
r куча *f* порубочных остатков

* **slash pine → 2111**

* **slash removal → 7723**

* **slash sawing → 8383**

* **slash[-]sawn → 3258**

* **slash-sawn timber → 3254**

* **slash thinning → 1109**

7731 **slash utilization**

f utilisation *f* des rémanents
r использование *n* древесных отходов

7732 slat
(relatively long, narrow, thin strip of wood as used e.g. for crates, trellis work)
d Brettchen *n*; Latte *f*
f planchette *f*; latte *f*; liteau *m*
r поперечная доска *f*; [карандашная] дощечка *f*

7733 slat saw
d Kreissäge *f* für Latten
f scie *f* circulaire à lattes
r пила *f* для получения карандашной дощечки

* **slatting → 957**

* **sledge axe → 453**

* **sleeper blocks → 8591**

7734 sleeper fire; holdover fire; hangover fire
d ruher Brand *m*
f feu *m* à retardement
r спокойный пожар *m*; слабый пожар

* **sleeper wood → 8591**

7735 sleepy wood
d geröstetes Holz *n*; vertrocknetes Holz; übertrocknetes Holz
f bois *m* paresseux
r пересушенная древесина *f*; обожжённая древесина

* **sleeve planting → 9002**

7736 sleighrunner butt; sabre-like butt
d säbelförmiger Stamm *m*; säbelwüchsiger Stamm; Stamm mit Säbelwuchs
f arbre *m* à croissance en cimeterre
r саблевидный комель *m*; саблевидный ствол *m*

7737 slender-branched; with slender branches; subramose *adj*
d mit schwachen Ästen; schwachästig
f pourvu de menues branches; faiblement branchu
r со слабыми сучьями; мелкосучковатый; слабоветвистый; маловетвистый

* **sliced board → 8280**

(veneer that is cut from a block using a knife which results in individual pieces instead of a continuous sheet)
d Messerfurnier *n*
f panneau *m* de tranchage
r строганный шпон *m*; ножевая фанера *f*

* **slicer → 9623**

* **slide → 5116**

7739 slide block
d Gleitholz *n*; Gleitschuh *m*
f goulotte *f*; embranche *m*
r скользящее бревно *n*

* **slide entrance → 2836**

* **slide-ga[u]ge → 8809**

* **slide hook → 529**

* **sliding [butt] hook → 529**

7740 sliding choker
d Gleitchoker *m*
f chaîne *f* de débardage glissante
r скользящий чокер *m*

* **sliding staff → 4923**

* **sliding track → 3987**

7741 slimy rot
(invader - Stereum sulcatum)
d fadenziehende Fäule *f*
f pourriture *f* gluante
r слизистая гниль *f*

7742 sling; crotchline
(rope suspended from the end of the loading line and terminating in the end hooks)
d Anschlagseil *n*; Stropp *m*
f élingue *f*
r погрузочный строп *m*

7743 sling-wrapped bundle
d stroppeingewickeltes Bündel *n*
f faisceau *m* enrobé par élingue
r обвязанная пачка *f* лесоматериалов

* **slip → 1095**

* **slip key → 2595**

7744 slipper

f frotteur *m*
r рабочий *m*, затёсывающий концы брёвен (перед трелёвкой)

7745 **slippery elm; gray elm; [American] red elm**
(Ulmus rubra)
d glatte Ulme *f*; Rotulme *f*
f orme *m* rouge
r скользкий вяз *m*

7746 **slip road distance**
d Gleitbahnabstand *m*
f distance *f* entre glissières
r расстояние *n* между трелёвочными волокнами

* **slit planting → 289**

7747 **slitter crack**
d Trennriss *m*
f coupe *f* déchiquetée
r разрыв *m* по краю листа (при продольной резке)

* **sliver → 1631, 2193, 7530**

7748 **sloanea**
(Sloanea)
d Sloanea *m*
f quapalier *m*
r слонея *f*

7749 **sloe; blackthorn**
(Prunus spinosa)
d Schlehdorn *m*; Stechdorn *m*; Schlehenpflaume *f*; [europäischer] Schwarzdorn *m*; Schlehe *f*
f épine *m* [noir]; prunellier *m*; épinette *f*
r терн *m*; терновник *m*; слива *f* колючая

* **sloop → 3881**

7750 **slope; gradient; grade**
(the rise or fall of a ground surface, or of the line of a road, ditch or other linear construction)
d Steigung *f*; Gefälle *f*; Böschungsneigung *f*; Abhang *m*
f pente *f*; dénivellement *m*; versant *m*
r склон *m*; спуск *m*; уклон *m*; наклон *m*

7751 **slope afforestation**
d Abhangsaufforstung *f*; Abhangsbewaldung *f*
f boisement *m* sur les versants
r облесение *n* склонов

f forêt *f* sur les versants
r лес *m* на склоне

7753 **slope of cutting; slope in cut section; cut slope**
(face of an excavated bank required to lower the natural ground line to the desired road profile)
d Einschnittsböschung *f*; Böschung *f* im Abtrag
f talus *m* de déblai
r крутизна *f* откоса выёмки

* **slope of grain → 4439**

7754 **slope-protection forest**
d Abhangsschutzwald *m*
f forêt *f* de protection des talus
r склонозащитный лес *m*

* **sloping grain → 4585**

* **sloven → 2193**

7755 **slower-growing tree**
d langsam wachsender Baum *m*
f arbre *m* à croissance lente
r медленно растущее дерево *n*

7756 **slow-growing species**
d langsam wachsende Arten *fpl*
f espèces *fpl* à croissance lente
r бавнорастущие породы *fpl*

7757 **slow-growing wood**
d langsam wachsender Wald *m*
f forêt *f* à croissance lente
r медленнорастущий лес *m*

* **sluiceway → 5116**

7758 **slump**
(mass movement process in which slope failure occurs on a usually curved slip surface)
d Abkühlung *f*
f glissement *m*
r оседание *n*

* **small branch → 991**

7759 **small business forest enterprise program; SBFEP**
d Programm *n* [zur Unterstützung] von kleiner Forstunternehmen
f programme *m* [de développement] des petites entreprises forestières

* **small cranberry → 2889**

7760 **small-dimension timber; small-sized timber; clearing timber; forest thinners; second-growth thinners; smallwood**
(round timber below a standard diameter or girth but sometimes also not less than a certain minimum)
d Dünnholz *n*; schwaches Holz *n*; schwaches Sortiment *n*; Schwachholz *n*; Schwachholzrollen *fpl*
f petit bois *m*; menu bois; fin bois
r тонкомерные лесоматериалы *mpl*; тонкомер *m*; тонкомерные балансы *mpl*; тонкомерные деревья *npl*, оставленные при рубке леса; тонкомерный сортимент *m*; мелкая древесина *f*; тонкомерный лес

7761 **small[er] end; top end; top face**
(of a trunk)
d Ablass *m*; schwächeres Ende *n*; oberes Stammende *n*; Topfende *n*; Zapfende *f*
f fin *m* out; petit bout *m*
r вершинный торец *m*

7762 **smaller Mexican pine beetle**
(Dendroctonus mexicanus)
d kleiner mexikanischer Kiefernkäfer *m*
f petit dendroctone *m* du pin mexicain
r лубоед *m* мексиканский сосновый

* **small-fibered wood → 3149**

7763 **smallflower chastetree; molave; bitum**
(Vitex parviflora)
d Wildpfeffer *m*
f vitex *m* à petites fleurs
r прутняк *m* мелкоцветный

* **small forest → 5001**

7764 **small knot**
(with diameter less than 1.8 cm)
d Kleinäst *f*
f très petit nœud *m*; nœud tête d'épingle
r мелкий сучок *m*

7765 **small-leaf geigertree; littleleaf cordia**
(Cordia parvifolia)
d kleinblätterige Kordie *f*
f cordie *f* à petites feuilles
r кордия *f* мелколистная

7766 **small-leaved fig; auon**
(Ficus eugeniodes)
d kleinblätteriger Feigenbaum *m*

* **small-leaved forest → 3514**

7767 **small-leaved lime; littleleaf linden [tree]; winter linden tree**
(Tilia cordata; Tilia parvifolia; Tilia sylvestris; Tilia ulmifolia)
d kleinblättrige Steinlinde *f*; kleinblättrige Linde *f*; Winterlinde *f*
f tilleul *m* à petites feuilles
r мелколистная липа *f*; сердцевидная липа

7768 **small-leaved snowberry**
(Symphoricarpus microphyllus)
d kleinblättrige Schneebeere *f*
f symphorine *f* à petites feuilles
r снежноягодник *m* мелколистный

7769 **small-leaved stringybark; thin-leaved stringybark**
(Eucalyptus wilkinsoniana)
d kleinblättriger Stringybark *m*
f stringybark *m* à petites feuilles
r эвкалипт *m* Уилкинсона

7770 **small pine weevil; minor pine weevil**
(Pissodes notatus)
d kleiner Kiefernjungholzrüssler *m*
f petit charançon *m* du pin
r точечная смолёвка *f*

7771 **small pit-prop**
d Spitzenknüppel *m*
f rondin *m* de pointe
r тонкий рудный кругляк *m*

* **small roundwood → 7774**

7772 **small-scale forestry**
(non-industrial forestry operations)
d forstlicher Kleinbetrieb *m*
f sylviculture *f* à petite échelle
r ограниченное лесное хозяйство *n*; мелкомасштабное лесное хозяйство

7773 **small-scale mechanical harvesting system**
d mechanisiertes Kleinhiebssystem *n*
f système *m* de coupe à petite échelle
r система *f* лесозаготовок с применением лёгких лесосечных машин

7774 **small-sized round timber; small roundwood**
(normally refers to logs of 7 cm maximum top diameter)
d Klöppelholz *n*; Prügelholz *n*; Rundling *f*; Stöckerholz *n*

* **small-sized timber → 7760**

7775 small[-size] pole[wood]; low polewood; light pole; thin pole
(pole under 7 cm)
d geringes Stangenholz *n*; schwaches Stangenholz; Reiser[holz]stange *f*; Reisstange *f*
f petite perche *f*; gaule *f*
r тонкая жердь *f*; жердь-хворостина *f*; кол *m*; тонкомерное бревно *n*

7776 small stand
d Kleinbestand *m*
f petit peuplement *m*
r мелкое насаждение *n*

7777 small-stature stock
d kleinwüchsige Pflänzling *f*
f petite plantule *f*
r малорослый подвой *m*; карликовый подвой

* **small timber → 4658**

7778 small tree; lighter tree; undersized tree
d Kleinbaum *m*
f petit arbre *m*
r тонкомерное дерево *n*

7779 small-tree clearcutting; smallwood felling
d Dünnholzernte *f*; Dünnholzgewinnung *f*; Schwachholzaufarbeitung *f*
f coupe *f* de petits arbres
r сплошная рубка *f* тонкомерных деревьев; заготовка *f* тонкомерных деревьев

* **smallwood → 7760**

7780 smallwood bundle
d Dünnholzbund *n*; Dünnholzstangenbündel *n*; Stangenstoß *m*
f paquet *m* de petit bois
r пакет *m* тонкомерной древесины; пачка *f* жердняка

* **smallwood felling → 7779**

7781 smallwood form-factor
d Dünnholzformzahl *f*
f coefficient *m* de forme du petit bois
r видовое число *n* тонкомерных лесоматериалов

7782 smallwood table
d Dünnholztabelle *f*

* **smoke chaser → 3171**

* **smoke damage → 2240**

7783 smoke jumper
d Waldbandfallschirmspringer *m*
f parachutiste *m* anti-incendie
r парашютист *m* по пожару

7784 smoke management
(scheduling and conducting of a prescribed burning program under predetermined burning prescriptions and firing techniques)
d Rauch-Management *m*
f régulation *f* des émissions de fumée
r управление *n* [выделения] дыма

7785 smoke-sensitive area
(that has been identified in which smoke accumulations may cause a safety or public health hazard, or may unreasonably deny aesthetic enjoyment to the public)
d rauchempfindliches Gebiet *n*
f zone *f* sensible aux fumées
r зона *f*, чувствительная к дыму

* **smoke[-]tree → 2908**

7786 smooth alder; tag alder; hazel alder
(Alnus serrulata)
d Haselerle *f*
f au[l]ne *m* tendre
r ольха *f* мелкопильчатая

7787 smooth bark; silver bark
d Spiegelrinde *f*; Glanzrinde *f*
f écorce *f* lisse
r зеркальная кора *f*; гладкая кора

7788 smooth-barked tree species
d glattrindige Holzart *f*
f espèce *f* à écorce lisse
r гладкокорая древесная порода *f*

* **smoothbark kauri → 6664**

* **smoothbark Mexican pine → 6617**

7789 smooth cut; fine cut
d Schlichthieb *f*
f frappe *f* fine [d'une lime]; taille *f* douce
r мелкая насечка *f*

* **smooth cypress → 359**

* **smooth-edged leaf → 7790**

* **smoothleaf pine → 1485**

* **smooth-leaved elm → 2893**

7790 **smooth-margin leaf; smooth-edged leaf**
d Blatt *n* mit glättendem Rand
f feuille *f* à bord lisse
r цельнокрайний лист *m*

7791 **smooth sumac[h]**
(Rhus glabra)
d glattblätt[e]riger Sumach *m*; Scharlachsumach *m*
f sumac *m* [à bois] glabre
r сумах *m* гладкий; сумах голый

* **smooth witherod → 6474**

7792 **smo[u]ldering fire**
(combustion without flame, but usually with incandescence and moderate smoke)
d glimmendes Feuer *n*
f feu *m* couvant; feu qui couve; combustion *f* lente; feu d'allure lente
r пожар *m* в беспламенной стадии горения; тлеющий пожар

* **smut → 5444**

* **SMZ → 8157**

* **snag → 2271**

* **snagged → 4737**

7793 **snagging**
(removing or cutting away snags, on land or in water)
d Grobschliff *m*
f arasement *m* des chicots
r уборка *f* сухостойных и буреломных деревьев

* **snaggy → 4737**

7794 **snag top**
d trockener Wipfel *m*
f cime *f* sèche
r сухая вершина *f*

* **snak → 2271**

* **snaker → 1435**

* **snakewood → 4921**

* **snap-cut shears → 322**

7795 **snape; nose; snout**
(end of a log, rounded-off so as to facilitate dragging)
d abgekopptes Blockholz *n*
f nez *m*
r затёсанный конец *m* брёвна

7796 **snatch block**
(a single sheave block made so that the shell opens on one side to permit a rope to be slipped over the sheave without threading the end of it through the block)
d Umlenkrolle *f* mit Seileinlass
f poulie *f* de renvoi à gorge
r поворотный ролик *m* с пропуском троса

* **snedding → 6610**

7797 **sneezewood; cape ebony**
(Pt[a]eroxylon utile; Pteroxylon obliquum)
d Niesholzbaum *m*
f ptéroxyle *m*
r птероксилон *m* полезный

7798 **snig chain; skid[ding] chain; drag[ging] chain**
(length of chain fastened around the end of a log)
d Ziehkette *f*; Würgekette *f*; Hemmkette *f*; Rückenkette *f*
f chaîne *f* de traînage; chaîne de halage; chaîne de retenue
r цепь *f* для трелёвки брёвен; натяжная цепь; чокерная цепь; цепной чокер *m*

* **snigging → 3994**

* **snigging pan → 7687**

7799 **snigging tractor; dragging tractor; hauling tractor; haulage tractor; caterpillar tractor; crawler[-type] tractor; chain-track tractor; track-laying tractor; lumber tractor**
d Kettenschlepper *m*; Kettentraktor *m*; Rücktraktor *m*
f tracteur *m* à chenilles; tracteur sur chenilles; tracteur à chaînes
r гусеничный [трелёвочный] трактор *m*; трелёвочный трактор; лесовозный трактор

* **snig length → 9747**

* **snig track → 7689**

* **snout → 7795**

* **snowbell → 3825**

* **snowberry → 1804**

* **snow break[age] → 7801**

7800 snowdrop tree; silverbell
(Halesia)
d Schneeglöckchenbaum *m*; Maiglöckchenstrauch *m*
f halesia *m*; halésie *f*; arbre *m* aux cloches d'argent
r халезия *f*

7801 snow-fallen wood; snow break[age]; wood crushing by snow
d Schneebruch *m*
f bris *m* de neige
r снеголом *m*

7802 snow melt
d Ausaperung *f*
f fonte *f* des neiges
r подтаивание *n* снега; снеготаяние *n*

7803 snow slide; avalanche transport
(a temporary slide on a steep slope, made by dragging a large log through deep snow which is soft or thawing; when frozen solidly)
d Schneerutschbahn *f*
f glissoire *f* enneigée
r [лесо]спуск *m* по снежному лотку; снежный [лесо]спуск

7804 snub[bing] line
(a rope or cable attached to the rear bunk of a logging sled used to control he speed on steep grades; a wire rope used with a donkey for snubbing logs, or log cars)
d gleichgültig abweisendes Kabel *n*
f câble *m* à freiner
r прицепной грузосборный канат *m*; канат для спуска вагонеток с горы

* **soaking of wood → 4429**

7805 soapbark
(Quillaja saponaria)
d Seifenrindenbaum *m*
f bois *m* de Panama; panama *m*
r квилайя *f* мыльная

* **soapberry → 771**

d Sapindengewächse *pl*; Sapindazeen *pl*; Seifenbaumgewächse *pl*
f sapindacées *pl*
r сапиндовые *pl*

* **soap tree yucca → 2224**

7807 soboliferous; flushing; sprouting *adj*
(having several loosely clumped, principal stems that arise from a common underground stem system, each distinct above ground level)
d sprießend
f formant de pousses
r дающий побеги; дающий отростки; дающий поросль

7808 social forestry
(planting and tending of trees or shrubs for the well-being and betterment of local communities)
d soziale Forstwirtschaft *f*
f foresterie *f* sociale
r социальное лесное хозяйство *n*

7809 socket wedge; box wedge
(metal wedge having a socket near its head, into which a wooden peg is fitted for initially steadying the wedge while it is being driven)
d Steckkeil *m*
f coin *m* ferré
r валочный клин *m* с гнездом для деревянной ручки (для первоначальной установки в пропил)

7810 Socotra dragon[ ]tree
(Dracaena cinnabari)
d Sokotra-Drachenbaum *m*; sokotrensischer Drachenblutbaum *m*
f arbre *m* du dragon; sang-dragon *m*
r драцена *f* киноварно-красная

7811 sod; clod; sward
d Rasenplagge *f*; Plagge *f*; Rasen *m*; Sode *f*
f motte *f* [de gazon]; tourbe *f*
r дернина *f*; дёрн *m*

7812 sod cutter; grub hoe
d Abstechpflug *m*; Plaggenpflug *m*; Rasenschälpflug *m*; Rasenhacke *f*; Rasenmesser *n*; Plaggenhaue *f*; Plaggenhacke *f*
f houe *f* à gazon
r дёрнорезный плуг *m*; плуг-дёрносниматель *m*; дёрнорез *m*;

7813 sod formation; turf formation
d Berasung *f*; Verrasen *n*; Verrasung *f*
f formation *f* de tourbe
r дёрнообразование *n*; задернение *n*; одернение *n*

* **sod-podzolic soils → 2409**

* **sod seeding → 4567**

7814 soft bark
d Weichbast *m*
f liber *m* mou
r мягкий луб *m*; мягкая кора *f*

* **softboard → 5183**

7815 soft fruits
d Weichobst *n*
f fruits *mpl* mous
r мягкие плоды *mpl*

7816 soft hardwood
(soft-textured hardwood such as boxelder, red and silver maples, hackberry, loblolly-bay, sweetgum, yellow-poplar, magnolia, sweetbay, water tupelo, blackgum, sycamore, cottonwood, black cherry, willow, basswood, and elm)
d weiches Laubholz *n*
f bois *m* feuillu mou
r мягкая широколиственная древесина *f*

7817 soft heart
d weicher Kern *m*
f cœur *m* mou
r мягкая сердцевина *f*

* **soft maple → 6996**

7818 soft mast
(soft fleshy fruits potentially eaten by wildlife - persimmon, wild grapes, blackberries, blueberries, huckleberries, mulberries, plums etc.)
d Weichmast *f*
f faînée *f* tendre
r мягкий плодокорм *m*

7819 soft rot; loose rot; powdered rot
(a special type of decay developing under very wet conditions in the outer wood layers)
d Moderfäule *f*; lose Fäule *f*; weiche Fäule
f pourriture *f* molle; pourriture vaseuse; carie *f* spongieuse
r трухлявая гниль *f*

d weiche Saftfäule *f*
f pourriture *f* de l'aubier *f* molle
r мягкая заболонная гниль *f*

7821 soft snag
(snag composed primarily of wood in advanced stages of decay and deterioration, particularly in the sapwood portion)
d weicher Dürrständer *m*
f chicot *m* pourri
r гнилой сухостой *m*

7822 soft[-textured]; kindly textured; mild; mellow
d mild
f tendre
r мягкий

7823 soft[-textured] wood; softwood; soft timber
d Weichholz *n*
f bois *m* tendre
r мягкая древесина *f*; лесоматериалы *mpl* мягких пород

7824 soft white gum; virgin dip
d Jungfernharz *n*
f gemme *f* vierge
r живица *f*, собираемая в первый год подсочки; живица первого года подсочки

7825 soft white rot
(invader - Daedalea confragosa)
d milde Weißfäule *f*
f pourriture *f* blanche molle
r мягкая белая гниль *f*

7826 soft white spongy rot
(invader - Polyporus hirsutus)
d milde Weißschwammfäule *f*
f pourriture *f* blanche molle spongieuse
r мягкая белая грибная гниль *f*

* **softwood → 7823**

7827 softwood ash
d Nadelholzasche *f*; Weichholzasche *f*
f cendre *m* de bois résineux; cendre de bois mou
r древесная зола *f* хвойных пород

7828 softwood board; coniferous board
d Nadelholzbrett *n*
f planche *f* de conifère
r доска *f* из древесины хвойной породы

f bouture *f* herbacée
r черенок *m* мягколиственных пород

* **softwood forest → 1873**

7830 **softwood fuelwood**
d Nadelbrennholz *n*
f bois *m* de chauffage résineux
r хвойные дрова *pl*

* **softwoods → 5708**

7831 **softwood slash**
d Nadelholzabraum *m*
f rémanents *mpl* de bois résineux
r порубочные остатки *mpl* мягколиственных пород

7832 **softwood species**
d Weichholzart *f*
f espèce *f* de conifères
r мягкая древесная порода *f*; мягколиственная порода

* **softwood timber → 5708**

* **softwood tree → 1877**

7833 **soil; earth**
(unconsolidated mineral or organic material at the surface of the earth that is capable of supporting plant growth)
d Erdboden *m*; Boden *m*; Erde *f*
f sol *m*
r почва *f*; грунт *m*; земля *f*

7834 **soil compaction**
(decrease in volume of soil and sediment as a result of compressive stress, continued deposition above them, but also from drying, use of heavy equipment, and other causes)
d Boden[ver]dichtung *f*
f durcissement *m* du sol; tassement *m* du sol; compactage *m* du sol
r уплотнение *n* почвы

7835 **soil creep; earth creep; soil slip; landslide; landslip; mass wasting**
d Bergrutsch *m*; Erdrutsch *m*
f glissade *m* de sol naturel
r оползень *f*; обвал *m*; гравитационное перемещение *n* горных пород

7836 **soil disturbance**
(disturbance caused by a forest practice on an area covered by a silviculture prescription or
f perturbation *f* du sol
r нарушение *n* почвенного покрова; повреждение *n* почвы

7837 **soil disturbance hazard**
(assessment of the susceptibility of a soil to adverse impacts on its productive capability due to soil compaction, soil puddling, surface erosion, mineral soil displacement, mass wasting, or forest floor displacement)
d Bodenstörungsrisiko *n*
f risque *f* de perturbation du sol
r опасность *f* от повреждения почвы

* **soil flora → 3986**

7838 **soil moisture; soil water**
(water diffused in the soil immediately below the land surface)
d Bodenfeuchte *f*
f humidité *f* du sol
r почвенная влага *f*

* **soil preparation → 3985**

7839 **soil-protecting species**
d Bodenschutzarten *fpl*
f espèce *f* protectrice du sol
r почвозащитная порода *f*

* **soil slip → 7835**

* **soil vegetation → 3986**

* **soil water → 7838**

* **soldierwood → 5652**

* **soledad pine → 8717**

7840 **solid bentwood**
d gebogenes Vollholz *n*
f bois *m* courbé solide
r гнутая массивная древесина *f*

* **solid contents → 7842**

* **solid cubic meter → 2114**

* **solider → 5652**

7841 **solid glued wood**
d Brettschichtholz *n*
f bois *m* massif collé
r массивная клеёная древесина *f*

* **solid piling → 1157**

* **solid pith → 1901**

* **solid stacking → 1157**

* **solid timber → 7843**

7842 **solid volume; solid contents; solid measure**
(of wood)
d Festgehalt *m*; Festmaß *n*
f volume *m* réel
r объём *m* в плотных единицах; плотный объём; плотная кубомасса *f*

7843 **solid wood; massif wood; solid lumber; solid timber**
d Massivholz *n*; Strukturholz *n*
f bois *m* massif
r массивная древесина *f*

7844 **solid wood bending**
d Massivholzbiegen *n*
f courbage *m* de bois massif
r гнутьё *n* массивной древесины

7845 **solitary pore**
(a pore completely surrounded by other elements)
d Einzelpore *f*; solitäre Pore *f*
f pore *m* isolé
r изолированная пора *f*

7846 **somatic embryogenesis**
(process by which clones are produced by cell growth from a seed embryo)
d somatische Embryogenese *f*; somatische Embryonalentwicklung *f*
f embryogenèse *f* somatique
r соматический эмбриогенез *m*; соматическое эмбриональное развитие *n*

7847 **sombre forest**
d düsterer Wald *m*
f forêt *f* sombre
r тёмный лес *m*; угрюмый лес

7848 **sonneratia family**
(Sonneratiaceae)
d Sonnentaugewächse *pl*
f sonneratiacées *pl*
r соннератиевые *pl*

* **sonorian scrub oak → 7559**

7849 **sooty bark**
f maladie *f* de la suie; suie *f*
r почерненная кора *f* клёна

* **sorb-tree → 8991**

* **sore bark → 558**

* **sorrel tree → 7857**

7850 **sorting of roundwood**
d Rundholzsortierung *f*
f triage *m* de bois rond
r сортировка *f* круглых лесоматериалов

* **sorting of wood → 3902**

* **sorting table → 3944**

7851 **sounding timber; sounding wood; resonant wood; resonance bloc**
d Resonanzholz *n*; Gleichklangsholz *n*; Geigenholz *n*; Instrumentenholz *n*; Tonholz *n*
f bois *m* de résonance
r резонансовая древесина *f*; резонансный кряж *m*; скрипичная древесина

7852 **sound knot**
(a knot that is solid across its face, at least as hard as the surrounding wood, and shows no indication of decay)
d gesunder Ast *m*
f nœud *m* sain
r [плотный] здоровый сучок *m*

7853 **soundness of wood**
d Gesundheitszustand *m* des Holzes
f état *m* sanitaire du bois
r состояние *n* здоровья древесины

7854 **sound sapwood**
d gesundes Splintholz *n*
f aubier *m* sain
r здоровая заболонь *f*

* **sound wood → 5902**

* **sourberry → 4909**

7855 **sour cherry; dwarf cherry**
(Prunus cerasus)
d Sauerkirschenbaum *m*; Zwergkirsche *f*; Sauerkirsche *f*; Weichselkirsche *f*
f cerise *m* acide; griottier *m*; cerisier *m* nain
r вишня *f* кислая

* **sour felling → 3065**

* **sour lime → 9348**

* **sour orange [tree] → 775**

7856 **soursop; guanabana**
(Annona muricata)
d Sauersack; Sauerapfel; sauere Sobbe *f*; Stachelan[n]one *f*; stachliger Flaschenbaum *m*
f anone *f* muriquée; anone hérissée
r аннона *f* колючая; аннона игольчатая

* **sour trefoil → 9656**

* **sour tupelo → 5871**

7857 **sourwood; sorrel tree**
(Oxydendrum arboreum)
d Sauerampferbaum *m*
f andromède *m* en arbre
r андромеда *f* древовидная

* **southern arrowwood → 368**

* **southern ash → 5659**

7858 **southern bayberry; [swamp] candleberry; waxmyrtle**
(Morella cerifera; Myrica cerifera; Myrica pennsylvanica)
d Wachsmyrte *f*; Wachsbaum *m*
f cirier *m* de Pennsylvanie
r мирика *f* восконосная; восковой мирт *m*

* **southern blackhaw → 7141**

* **southern catalpa → 1768**

7859 **southern cone rust**
(invader - Cronartium strobilinum)
d Rostpilz *m*
f rouille *f* de conifères
r ржавчина *f* хвойная

* **southern cypress → 786**

7860 **southern magnolia; large-flowered magnolia; bat tree; black lin; big laurel; bullbay**
(Magnolia grandiflora)
d großblütige Magnolie *f*; immergrüne Magnolie
f magnolia *m* à grandes fleurs
r крупноцветковая магнолия *f*; крупноцветная магнолия

7861 **southern mahogany; Bangalay eucalyptus**
f eucalyptus *m* botryoïde
r эвкалипт *m* гроздевидный; эвкалипт краснодревесный клейкий

7862 **southern pine beetle**
(Dendroctonus frontalis)
d Kiefernborkenkäfer *m*; Fichtenborkenkäfer *m*
f dendroctone *m* méridional du pin
r лубоед *m* сосновый южный

7863 **southern red oak; Spanish oak**
(Quercus rubra; Quercus falcata)
d sichelblättrige Eiche *f*; Roteiche *f*
f chêne *m* rouge; chêne d'Espagne
r дуб *m* красный

7864 **southern sassafras; Tasmanian sassafras; blackheart sassafras**
(Atherosperma moschatum)
d tasmanischer Fieberbaum *m*
f sassafras *m* de Tasmanie
r аферосперма *f* мошата; сассафрас *m*

7865 **southern soapberry tree**
(Sapindus saponaria)
d Seifen[beeren]baum *m*
f arbre *m* à savonnette; savonnier *m* saponaire
r дерево *n* мыльное; сапиндус *m* мыльный; ягода *f* мыльная

* **southern sugar maple → 3285**

* **southernwood → 5883**

* **southern [yellow] pine → 5141**

7866 **southwestern pine [bark] beetle**
(Dendroctonus barberi)
d südwestlicher Kiefernbastkäfer *m*
f dendroctone *m* occidental du pin
r лубоед *m* сосновый югозападный

7867 **southwestern white pine; Mexican white pine**
(Pinus strobiformis; Pinus ayacahuite)
d weymouthsähnliche Kiefer *f*; mexikanische Weymouthskiefer *f*
f pin *m* blanc du Mexique
r сосна *f* белая юго-западная; сосна белая мексиканская

* **sow → 7371**

* **sow bug → 1813**

7868 **sowing brick**
(prepared, sometimes fertilized, block or ball of loam, peat, plastic foam, etc., into which one or more seeds are pressed)
d Einsaatballen *n*
f motte *f* à semis
r брикет *m* для посева семян

* **sowing density → 7374**

* **sowing horn → 7381**

7869 **sowing in furrows; furrow drilling; drilling; furrow sowing; sowing in trenches; drill sowing; drill seeding**
(sowing of seed in deep furrows simultaneously with disc trenching for site preparation)
d Säen *n* in Furchen; Furchensaat *f*; Rillensaat *f*; Rillensäen *n*; Drillsaat *f*; Drillen *n*
f ensemencement *m* en rangée; semis *m* en sillons
r бороздовой посев *m*; посев в борозды

7870 **sowing in patches; patch seeding; patch sowing; spot seedling; plot sowing**
(sowing of seeds within small, cultivated, or otherwise-prepared patches)
d Plattensaat *f*; Plätzesaat *f*
f ensemencement *m* sur placeaux
r посев *m* площадками

* **sowing in trenches → 7869**

* **sowing lath → 7385**

* **sowing on strips → 522**

* **sowing point → 7396**

* **space of growth → 4008**

7871 **space thinning; spacing**
d Entfernung *f*
f éclaircie *f* par espacement
r равномерное изреживание *n* насаждения

* **spacing between stems → 4569**

* **spacing block → 8093**

* **spackle → 3130**

7872 **spade budding**
d Spaten-Okulieren *n*; Spaten-Sprossung *f*

7873 **spalt**
d Splitter *m*
f noyau *m* de bardage
r обломок *m*; осколок *m*

7874 **span**
(of cableway)
d Spanne *f*
f envergure *m*; portée *f*
r пролёт *m*

* **span-dogs → 8829**

* **spangle → 5831**

* **Spanish bayonet → 2224**

7875 **Spanish box**
(Buxus balearica)
d spanischer Buchsbaum *m*
f buis *m* d'Espagne; buis de Mahon
r самшит *m* балеарский

7876 **Spanish cedar**
(Cedrela huberi; Cedrela hirsuta; Cedrela barbata; Cedrela brasiliensis)
d spanischer Zedrelenholzbaum *m*
f care *m* rouge; acajou *m* rouge
r цедрела *f* белая

* **Spanish chestnut → 2887**

* **chestnut → 2887**

7877 **Spanish fir**
(Abies pinsaro)
d spanische Tanne *f*
f sapin *m* d'Espagne; sapin d'Andalousie
r пихта *f* испанская

7878 **Spanish juniper; incense juniper**
(Juniperus thurifera)
d spanischer Wacholder *m*; Weihrauchsadebaum *m*; Weichrauchwacholder *m*
f genévrier *m* thurifère; genévrier d'Espagne; genévrier à encens; genévrier porte-encens
r можжевельник *m* ладанный

* **Spanish oak → 7863**

7879 **Spanish plane tree**
(Platanus [acerifolia] hispanica)
d spanische Platane *f*
f platane *m* d'Espagne
r платан *m* испанский

* **span wire → 1164**

* **spanworm → 3174**

* **sparkleberry → 8926**

* **spar road → 4172**

7880 **sparse foliage**
d verstreutes Laubwerk *n*
f feuillage *m* éparse
r редкая листва *f*

* **sparsely closed stand → 4446**

7881 **sparsely leaved; thin-crowned; light foliaged**
d lichtbelaubt; lichtkronig
f à feuillage clair
r редко облиственный; с изреженной кроной; с изреженным пологом

* **sparsely wooded → 6450**

7882 **sparse trees**
d verstreute Bäume *mpl*
f arbres *mpl* éparses
r разрёженные деревья *npl*

* **spar timber → 8586**

* **spatial arrangement → 7884**

7883 **spatial continuity of forest community**
d räumliche Kontinuität *f* der Waldgesellschaft
f continuité *f* spatiale de communauté forestière
r пространственная целостность *f* лесного сообщества

7884 **spatial order; spatial arrangement**
d räumliche Bestellung *f*
f agencement *m* spatial; arrangement *m* spatial
r пространственное расположение *n*

7885 **spatial variation in the forest**
d räumliche Forstvariation *f*
f variation *f* spatiale en forêt
r пространственная изменчивость *f* в лесу

7886 **spatulate leaf**
(shaped like a spatula, broadest at the tip and tapering to the base)
d spatelförmiges Blatt *n*
f feuille *f* à spatule
r лопатовидный лист *m*

* **spauling → 4845**

* **special acres → 7890**

7887 **special assortment log**
d Sondersortimentschnittholz *n*
f grume *f* d'assortiment spécial
r бревно *n* специального назначения

7888 **special cutting**
(logging activities in special areas, such as recreation areas and administrative sites, where other uses or other values override timber production values)
d Sonderhieb *m*
f coupe *f* spéciale
r специальная рубка *f*

7889 **special forest products**
(poles; posts; pilings; shakes; shingle bolts; Christmas trees; building logs; mining timbers, props, and caps; cribbing; firewood and fuel logs; hop poles; orchard props; car stakes etc.)
d Sonderforstprodukte *npl*
f produits *mpl* de forêt spéciaux
r специальные лесные продукты *mpl*

7890 **special lands; special acres**
(component of the regulated commercial forestland area that needs special timber management techniques to achieve landscape objectives)
d Sonderländer *npl*
f terres *fpl* spéciales
r земли *fpl* специального назначения

7891 **special timber**
d Spezialholz *n*
f bois *m* spécial
r древесина *f* специального назначения

7892 **special tree; preserved tree; royal tree**
d Spezialbaum *m*; klassifizierter Baum *m*
f arbre *m* spécial; arbre classé
r особо ценное дерево *n*

7893 **species**
(singular or plural term for a population or series of populations of organisms that are capable of interbreeding freely with each other but not with members of other species)
d Art *f*; Arten *fpl*
f espèce *f*; espèces *fpl*
r порода *f*; породы *fpl*; вид *m*

7894 species at risk
d Risikoarten *f*
f espèces *fpl* à risque
r виды *mpl* с риском

7895 species corresponding to the site
d standortsgerechte Baumart *f*
f essence *f* conforme à la station
r вид *m*, соответствующий лесорастительных условий

7896 species density
(total number of species per square mile)
d Artendichte *f*
f densité *f* des espèces
r густота *f* пород

7897 species diversity; biologic diversity
(number of different species occurring in some location or under some condition)
d Artendiversität *f*
f diversité *f* des espèces
r разнообразие *n* пород

7898 species diversity index; diversity index
(the ratio between the total number of species in a biotic community and some rating of the relative importance (numbers, biomass, productivity, etc.) of individual species)
d Artendiversitätsindex *m*
f indice *m* de diversité des espèces
r показатель *m* разнообразия пород

7899 species groups
d Artengruppen *fpl*
f groupes *mpl* des espèces
r группы *fpl* пород

7900 species inventory
d Arteninventarisierung *f*
f inventaire *m* d'espèces
r инвентаризация *f* пород

7901 species name card
d Namenkarte *f* der Arten
f fiche *f* nominale d'espèce
r карточка *f* с наименованием породы древесины

* **species native to the site → 4477**

7902 species richness; wealth of species
(the total number of individuals of a species in a defined area, population, or community)
r богатство *n* видов

7903 species strange to the site
d standortsfremde Baumart *f*
f essence *f* étrange à la station
r вид *m*, чужой для лесорастительных условий; интродуцированный вид

7904 specific leaf area
(area of a leaf per unit leaf weight)
d spezifische Blattfläche *f*
f aire *f* foliaire spécifique
r удельная площадь *f* листа

7905 specific leaf volume
(volume of a leaf per unit)
d spezifischer Blattvolumen *m*
f volume *m* foliaire spécifique
r удельный объём *m* листа

7906 specific nodulation of stems
d spezielle Knötchenbildung *f* auf Stengeln
f nodulation *f* spécifique de tiges
r специфическое образование *n* узелков на стволе

* **specimen tree → 6396**

7907 speckle; spot; stain
(any alteration in its natural colo[u]r of the wood substance)
d Fleck *m*
f tache *f* [colorée]; tacheture *f*
r пятно *n*; пятнышко *n*; крапинка *f*

* **speckled alder → 2878**

7908 speckled yellow rot
(invader - Trametes hispida)
d fleckige Gelbfäule *f*
f pourriture *f* jaune madrée
r пятнистая жёлтая гниль *f*; пёстрая жёлтая гниль

* **spherical crown → 3869**

7909 spicebush
(Lindera benzoin)
d Fieberstrauch *m*
f benjoin *m* odoriférant; arbre *m* à épice; laurier *m* benzoin
r линдера *f*; бензойное дерево *n*

* **spice tree → 1264**

7910 **spike[d] knot; splay knot; wing-shaped knot**
(a knot cut approximately parallel to its long axis so that the exposed section is definitely elongated)
d Flügelast *m*
f nœud *m* plat; moignon *m* pointu; nœud tranché dans le sens de la longueur [au moment de sciage]
r продолговатый [сшивной] сучок *m*; сучок, разрезанный в продольном направлении

* **spike-top[ped] tree → 2273**

* **spindle crown → 3782**

7911 **spindle palm**
(Hyophorbe verschaffeltii)
d Spindelpalme *f*
f palmier *m* fusain; palmiste *m* marron
r пальма-веретено *f*

* **spindle[ ]tree → 1383**

* **spine → 8524**

* **spinney → 1928**

7912 **spinule; acicula**
d Nadelchen *n*
f spinule *f*
r иголочка *f*; мелкая колючка *f*; шипик *m*

7913 **spiny; thorny; braky**
d dornig; schwierig
f épineux
r покрытый иглами; покрытый колючками; покрытый шипами; тернистый

* **spiny ceanothus → 3939**

* **spiny hackberry → 2413**

7914 **spiral grain; bastard grain; flat grain**
(arrangement of the fibers in a piece of timber or veneer which results from their growth in a spiral direction around the trunk of the tree)
d Fladerzeichnung *f*
f fibre *f* torse; fil *m* spiral
r спиральное волокно *n*; узорная прожилка *f* (волокон древесины)

* **spiral-grained wood → 2048**

* **spiral growth → 9021**

d Spiralsäge *f*
f scie *f* spirale
r пила *f* с косыми зубьями

7916 **spiral thickenings**
(helical ridges on the inner face of, and part of, the secondary wall)
d Schraubenverdickungen *fpl*
f épaississements *mpl* spiralés
r винтовые уплотнения *npl*

7917 **spired; cuspidate[d]; pointed**
(having a crown with several points)
d zugespitzt
f à aiguilles; cuspidé
r остроконечный; заострённый; шпилевидный; имеющий шпиль; со шпилем

* **spire-shaped crown → 1866**

* **splash → 2985**

* **splash dam → 5039**

* **splayed cut → 1441**

* **splay knot → 7910**

7918 **splice**
d Klebung *f*
f jonction *f*; épissure *f*
r сращивание *n*; заплётка *f*; сросток *m* (каната)

7919 **splice grafting**
d Kopulieren *n*; Schäften *n*
f greffage *m* à l'anglaise simple
r концевая прививка *f*; копулировка *f*

* **spline [joint] → 2595**

* **splinters → 9508**

* **splintwood → 7216**

* **split → 7483**

7920 **split billet; cloven billet**
(with a diameter more of 14 cm)
d Klobenholz *n*; Schichtbrennholz *n*; gespaltenes Holzscheit *n*
f bûche *f* refendue; bois *m* refendu
r крупные колотые дрова *pl*; дрова-коротьё *pl*; колотый сортимент *m*;

7921 **split bunk**
d Spaltschemel *m*
f traverse *f* divisée
r коник *m*, разделённый на части (для укладки подсортированных по породам или размерам лесоматериалов)

* **split logs → 8563**

* **splittability → 1627**

* **splitter → 453**

* **split timber → 8563**

* **splitting → 1629**

* **split[ting] axe → 453**

* **splitting gun → 3717**

* **split[ting of grain] → 7729**

* **splitting strength → 6907**

7922 **splitting test**
(of a tree)
d Aufteilungstest *m*
f essai *m* de fendage
r испытание *n* на раскалываемость

* **splitting wedge → 3717**

* **splitwood → 3740**

* **spongy → 3773**

* **spongy and butt rot → 364**

* **spongy pith → 1440**

* **spongy wood → 2291**

* **sponk → 8666**

7923 **spontaneous combustion**
d Eigenentzündung *f*; Selbstentzündung *f*
f combustion *f* spontanée
r самовоспламенение *n*; самовозгорание *n*

7924 **spontaneous regeneration**
d spontane Verjüngung *f*
f régénération *f* spontanée
r внезапное возобновление *n*

* **spontaneous seeding → 5695**

* **spool donkey → 8609**

7925 **spool tender; winchman**
(one who guides the cable on a spool donkey)
d Windemann *m*
f treuilliste *m*
r лебёдчик *m*

7926 **spoolwood; bobbin wood; spool bar**
(small sawed squares of wood from which spools and other small items are turned after drying)
d Spulenholz *n*
f bois *m* de bobines
r шпилевая древесина *f*; заготовки *fpl* для катушек

* **spoonwood → 237**

7927 **spore**
d Spore *f*
f spore *f*
r спора *f*

7928 **sporophore; carposome**
(a spore-bearing branch or organ: the part of the thallus of a sporophyte that develops spores)
d Feuerschwamm *m*; Sporophor *m*
f sporophore *m*
r спорофор *m*

7929 **sporophyll**
(a leaf in ferns and mosses that bears the sporangia)
d Sporophylle *f*
f sporophylle *f*
r спорофилл *m*

* **spot → 7907**

* **spot bark → 6104**

* **spot burning → 6105**

* **spot planting → 6107**

* **spot scarifier → 6108**

* **spot seedling → 7870**

7930 **spotted [brittle] gum; maculata gum**
(Eucalyptus maculata; Eucalyptus maculosa)
d gesprenkelter Eukalyptus *m*
f eucalyptus *m* tacheté; eucalyptus moucheté
r эвкалипт *m* [густо]пятнистый

7931 **spotted pine sawyer**
(Monochamus maculosus)
d gesprenkelter Kiefernbockkäfer *m*
f longicorne *m* maculé du pin
r усач *m* чёрный пятнистый

7932 **spotter**
d Aufklärer *m*
f aiguilleur *m*
r наблюдатель *m*; корректировщик *m*

* **spotting → 5317**

* **spot[ting] fire → 4669**

7933 **spot weeding**
(removing undesirable vegetation from patches)
d Punktschweißen *n*; Punktschweißung *f*
f désherbage *m* localisé; soudure *f* par points; soudage *m* par points
r расчистка *f* локальной территории

* **sprawling crown → 7935**

* **spray → 1095**

* **sprayer → 2675**

* **spraying apparatus → 2675**

* **spray[ing] gun → 8884**

7934 **spreader**
(any substance, solid or liquid, that, when added to a pesticide, herbicide, liquid fertilizer, or fire retardant, enables it to spread better over the surfaces on which it is deposited)
d Benetzungsmittel *n*; Stander *m*; Verteiler *m*; Sprühfahrzeug *m*
f dispersant *m*; écarteur *m*
r разбрасыватель *m*

7935 **spreading crown; sprawling crown**
d Ausbreitungskrone *f*
f cime *f* divariquée
r раскидистая крона *f*

* **spreading elm → 6988**

* **sprig → 7132**

7936 **spring; edge bend; free-side bend**
(longitudinal curvature of the edge of a piece of timber, not affecting the face)
r продольно-радиальное коробление *n*

* **spring[ ]board → 1563**

7937 **spring bundling**
d Frühlingsbüscheln *n*; Frühlingsschwadlegen *n*
f ballottage *m* en printemps; groupage *m* en printemps
r ранневесенняя сплотка *f*

* **springer → 9022**

7938 **spring felling; spring cutting**
d Frühlingshieb *m*
f coupe *f* en printemps
r весенняя рубка *f*

7939 **spring-flowering tamarisk; four-stamen tamarix**
(Tamarix parviflora; Tamarix tetrandra)
d viermännige Tamariske *f*; kleinblütige Tamariske
f tamaris *m* de printemps; tamaris à quatre étamines
r тамариск *m* четырёхтычинковый; тамарикс *m* крымский

7940 **spring of the crown; crown spring**
d Kronenanfang *m*
f source *f* de cime
r начало *n* кроны (зона ствола, где начинаются нижние сучья кроны)

7941 **spring-peeling season**
d Frühschälensaison *f*
f période *f* d'écorçage en printemps
r сезон *m* весенней окорки

* **springwood → 2704**

7942 **sprinkler; water box; icebox; tank**
(a large wooden tank from which water is sprinkled over logging roads during freezing weather in order to ice the surface)
d Wasserkasten *m*; Wasserbecken *n*
f caisse *f* à eau
r водоотливной ящик *m*

7943 **sprocket**
(wood harvesting tool)
d Sprossenrad *n*; Stachelrad *n*
f pignon *m* étoile
r звёздочка *f*

* **sprout → 7132**

f bourgeonner; pousser
r давать почки; почковаться

7945 **sprout chestnut forest**
d Kastanien-Niederwald *m*
f taillis *m* de châtaigniers
r каштановый низкоствольник *m*

* **sprout forest → 1928**

* **sprouting → 1020, 7807**

* **sprouting ability → 3833**

7946 **sprouting tree**
(tree that grows from the stump or sucker root of a parent tree: it is not of seed origin)
d Anflug *m*
f arbre *m* pousse; rejet *m*
r порослевое дерево *n*; дерево, дающее поросль

* **sprout method → 1935**

* **sprout root → 7046**

7947 **sprout seedling forest; middle forest; coppice with standards; reserve sprout forest; standard coppice; mixed seedling and sprout forest**
d Mittelwald *m*; Hauschicht *f*
f taillis-sous-futaie *m*
r средний лес *m*

* **sprout system → 1935**

7948 **sprout thinning**
d Niederwalddurchforstung *f*
f éclaircie *f* de bourgeons
r прореживание *n* низкоствольника

* **spruce ambrosia beetle → 8173**

7949 **spruce aphid**
(Liosomaphis abietinum; Elatobium abietinum)
d Fichtenröhrenlaus *f*
f puceron *m* de l'épinette
r тля *f* еловая трубчатая; тля пихтовая

* **spruce apple → 7952**

7950 **spruce blocks**
d Fichtenfaserholz *n*
f billots *mpl* d'épicéa

7951 **spruce budworm**
(Choristoneura fumiferana)
d Fichtenwickler *m*; nordamerikanischer Tannenknospenwickler *m*
f tordeuse *f* des bourgeons de l'épinette
r гусеница *f* листовёртки-почкоеда елового

7952 **spruce cone; spruce apple**
d Fichtenzapfen *m*
f cône *m* d'épicéa
r еловая шишка *f*

* **spruce fir → 1805**

7953 **spruce forest; spruce stand; spruce grove; pineland**
d Fichtenwald *m*
f pessière *f*
r еловый лес *m*; ельник *m*

7954 **spruce humus**
(under spruce forest)
d Mulm *m*; saurer Humus *m*
f humus *m* acide
r кислый гумус *m*

7955 **spruce needle**
d Fichtennadel *f*
f aiguille *f* d'épicéa
r еловая хвоя *f*

7956 **spruce pine; bottom white pine; Walter's pine**
(Pinus glabra)
d amerikanische Weißkiefer *f*
f pin *m* blanc américain; pin glabre
r сосна *f* голая

7957 **spruce resin**
d Fichtenharz *n*
f résine *f* d'épicéa
r еловая живица *f*; еловая смола *f*

7958 **spruce sawfly**
(Nematus abietinus)
d Fichten[nadelgespins]blattwespe *f*
f mouche *f* à scie de l'épicéa
r пилильщик *m* еловый

* **spruce stand → 7953**

7959 **spruce timber; spruce wood; fir wood**
(a timber of coniferous species)
d Fichtenware *f*; Fichtenholz *n*; Tannenholz *n*
f bois *m* d'épicéa; bois d'épinette; bois de sapin
r еловая древесина *f*; пихтовая древесина

7960 spud[der]; barking iron; barker; peeling scalpel; shave knife; peeling iron; cleaver [knife]; hogal
(tool with a narrow-shaped, curved blade used in removing bark by hand)
d Schäleisen *n*; Schälmesser *n*; Loheisen *n*; Entrindungsmesser *n*; Streifenmesser *m*; Röteisen *n*; Schaber *m*
f décortiqueur *m*; fer *m* à écorçage de grumes; fendeuse *f*
r корosниматель *m*; [окорочный] скобель *m*; короочистный нож *m*; нож для скобления коры; нож корообдирочный; нож для лущения древесины; лущильный нож

* **spunk → 8666**

7961 spurge family
(Euphorbiaceae)
d Wolfsmilchgewächse *pl*
f euphorbiacées *pl*
r молочайные *pl*

* **spurge laurel → 2828**

* **spurious fruit → 3000**

* **spur shoot → 7548**

* **sq.ft.m. → 3036**

7962 square *v*
d abkappen; kappen; quadrieren; beschlagen; kanten
f équarrir
r торцевать; подрезать торец; обтёсывать бревно со всех сторон

7963 square-cut; squared; die-square; square-edged; square-sawn; square-hewn; live-edged
d vollkantig; vierkantig; scharfkantig
f équarri; équarri-scié; équarri-dégrossi; avivé; à vive arête
r обрезанный на четыре канта

7964 square-cutting
d vollkantiger Hieb *m*
f coupe *f* d'équerre; coupe à bord droit
r обрезка *f* на четыре канта

* **squared → 7963**

7965 squared beam
d Bündel *n* im Quadrat

7966 squared end
d vollkantiges Ende *n*
f bout *m* droit; bout rectangulaire
r прямой торец *m*

* **square[d] log → 7969**

* **squared lumber → 7969**

* **squared timber → 7969**

* **squared wood → 7969**

7967 square edge
d vollkantige Kante *f*
f bord *m* droit; vive arête *f*
r прямоугольная кромка *f*; обрезная кромка

* **square-edged → 7963**

7968 square-edged board
d besäumtes Brett *n*
f planche *f* équarrie
r чистообрезная доска *f*

7969 square-edged [sawn] timber; squared timber; square-sawn timber; squared lumber; square[d] log; squared wood; squares; rectangular timber; scantlings
(in sawmilling, lumber items that have been sawed on the edges to eliminate wane or taper)
d vierkantig beschlagenes Holz *n*; Vierkantblock *m*; Balkenholz *n*; scharfkantiges Holz
f poutrage *m* équarri; bois *m* équarri [coupé]; quartier *m* [de bois]
r четырёхкантный блок *m*; четырёхкантный брус[ок] *m*; окантованный брус

7970 square [foot]; superficial foot
(= 0.929 m^3)
d Quadratfuß *m*
f pied *m* carré; pied-carré *m*; pied-planche *m* mince
r квадратный фут *m*

* **square-foot measure → 3036**

* **square-foot method → 5865**

* **square-hewn → 7963**

7971 square log; slabbed log
d vierkantiges Baumstamm *n*

* **square piling → 7973**

7972 **square ray cell**
(a ray cell approximately square as seen in radial section)
d Quadratmarkstrahl *m*
f cellule *f* carrée de rayon
r квадратная лучевая клетка *f*

* **squares → 7969**

* **square-sawn → 7963**

* **square-sawn timber → 7969**

7973 **square stacking; square piling**
d Vierkantstapelung *f*
f empilage *m* en carré
r четырёхкантная штабелёвка *f*

* **squaring axe → 1041**

* **squashberry → 5553**

* **squawbush → 9471**

* **squaw carpet ceanothus → 5242**

* **squaw-huckleberry → 2327**

7974 **stabilization of slopes**
d Abhangstabilisierung *f*
f stabilisation *f* des versants
r укрепление *n* склонов

* **stack → 8619**

* **stack base → 8664**

7975 **stack bearer; stack support**
d Stapelstütz *m*
f bois *m* équarri pour fondation de piles
r опора *f* штабеля

7976 **stack disconnector**
d Holzvereinzelner *m*; Stammvereinzelner *m*; Vereinzelungsanlage *f*
f déconnecteur *m* de piles
r разобщитель *m* пачки хлыстов; растаскиватель *m* хлыстов

* **stacked content → 7979**

7977 **stacked cubic foot**
d schichtholzer Kubikfuß *m*
f pied-cube *m* apparent; pied-cube de bois
кубический фут

7978 **stacked [cubic] measure; stack measure; loose measure; loose scale**
d Raummaß *n*; Schichtmaß *n*
f mesure *f* de bois enstéré
r измерение *n* в складочных кубических мерах; складочная мера *f*

* **stacked cubic meter → 2115**

* **stacked merchantable timber → 6230**

7979 **stacked volume; piled volume; superficial volume; stacked content**
(the total space occupied by a stack of wood, as determined by its external dimensions)
d Rauminhalt *m*; Raumgehalt *m*
f volume *m* apparent du bois empilé
r складочный объём *m*; кубатура *f*, учитываемая в складочной мере

* **stacked wood → 7982**

* **stack elevator → 8581**

* **stacking → 4846**

7980 **stack[ing] density; pile density**
(the closeness of packing of the pieces in stack, expressed as the ratio of solid volume of wood to its stacked volume)
d Stapeldichte *f*
f coefficient *m* d'enstérage
r плотность *f* укладки

7981 **stacking point**
d Stapelungspunkt *m*
f point *m* d'empilage
r место *n* окучивания лесоматериала (для трелёвки)

* **stacking strip → 6236**

7982 **stacking wood; stacked wood; stackwood; cordwood; wood in heaps; wood in piles; piled wood; corded wood**
(small diameter or low quality wood suitable for firewood, pulp, or chips)
d Beugholz *n*; Brennbündel *n*; Klafterholz *n*; Stapelholz *n*; Setzholz *n*
f bois *m* enstéré; bois d'empilage; bois en corde; bois empilé
r заштабелёванная древесина *f*; штабелированные лесоматериалы *mpl* в

* **stack measure → 7978**

* **stack support → 7975**

7983 **stack *v* [the timber]; bank *v* up**
(to pile up logs on a landing)
d aufschichten; aufstapeln; einstapeln; stapeln
f mettre [de bois] en pile
r укладывать [лесоматериалы] в штабель; штабелировать

* **stackwood → 7982**

* **staddle → 8555**

7984 **staff tree; bittersweet**
(Celastrus)
d Baumwürger *m*
f célastre *m*
r бирючина *f*

7985 **staff-tree family**
(Celastraceae)
d Baumwürgergewächse *pl*
f célastracées *pl*
r бересклетовые *pl*; краснопузырниковые *pl*

* **staff wood → 4362**

7986 **stag beetles**
(Lucanidae)
d Schröter *pl*
f lucanes *fpl*
r рогачи *mpl*

* **stagberry → 792**

* **stagb[r]ush → 792**

* **stage of succession → 7455**

7987 **staggered rows**
d auf Lücken gepflanzte Reihen *fpl*
f lignes *fpl* plantées dans les vides
r на просветах посаженные ряды *mpl*

7988 **staggered settings**
(clear-cut settings separated by uncut timber)
d verschobene Hiebsflächen *fpl*
f coupes *fpl* en échec
r лесосеки *fpl* в шахматном порядке

* **staggering → 8171**

* **staghead → 8711**

the upper crown, giving the appearance of antlers)
d Gipfeltrockenheit *f*; Zopftrocknis *f*; Wipfeltrocknis *f*; Zopftrockenheit *f*
f sécheresse *f* en cime
r суховершинность *f*

* **staghead[ed] tree → 2273**

* **staghorn sumac[h] → 260**

7990 **stagnant growth**
(due to poor site and/or excessive stocking)
d Riegelwuchs *m*
f croissance *f* stagnante; accroissement *m* bloqué
r приостановленный рост *m*; задержанный рост

7991 **stagnant stand**
d Riegelbestand *m*
f peuplement *m* stagnant
r отстающее в росте насаждение *n*

7992 **stagnated tree**
d zurückbleibender Baum *m*
f arbre *m* stagnant
r дерево *n*, отстающее в росте

* **stagnation of growth → 4015**

* **stain → 5227**

* **stain → 7907**

7993 **stained wood**
d beschmutztes Holz *n*; beflecktes Holz
f bois *m* moiré
r морёная древесина *f*; протравленная древесина

7994 **staining of heart[wood]; heart stain[ing]**
d Kernfärbung *f*
f coloration *f* du cœur
r окраска *f* ядра; окраска сердцевины

* **stairs stand → 5634**

* **stake → 6475**

* **stake forest → 7204**

* **stake wood → 7204**

7995 **staking out; pegging [out]; poling; laying out; drawing**

f jalonnage *m*; jalonnement *m*; piquetage *m*
r маркировка *f* вешками

7996 **stalactitiform rust; stalactoform rust**
(invader - Peridermium stalactiforme)
d stalaktiförmiger Rostpilz *m*
f rouille *f* stalactiforme
r сталактитовидная ржавчина *f*

* **stalk → 2124**

7997 **stalk cell**
(of pine)
d Stielzelle *f*
f cellule *f* pédonculaire
r инициалий *m* антеридиальной клетки; стебельковая клетка *f*; клетка стебелька

7998 **stamen**
(the pollen-producing organ of a flower)
d Staubblatt *n*; Staubgefäss *n*
f étamine *m*
r тычинка *f*

7999 **staminate[-flowered] tree; male-flowered tree**
(unisexual tree with only male reproductive organs)
d Staubblattbaum *m*
f arbre *m* à fleurs staminées
r дерево *n* с тычиночными цветами; дерево с мужскими цветами; мужское дерево

8000 **stand after thinning**
(trees left standing after harvesting)
d verbliebener Bestand *m*; verbleibender Bestand
f peuplement *m* après éclaircissage
r насаждение *n*, пройдённое рубками ухода

8001 **stand age**
(average age of dominant and codominant trees in the stand)
d Alter *n* des Bestandes; Bestandesalter *n*
f âge *m* du peuplement
r возраст *m* насаждения

8002 **stand age class**
(the age of the forest stand by 10 year classes)
d Bestandesalterklasse *f*
f classe *f* d'âge du peuplement
r класс *f* возраста насаждения

8003 **standard acres**
(component of the regulated commercial forestland area that needs special timber
d Standardacker *mpl*
f acres *fpl* standards
r стандартизованные акры *mpl*

* **standard coppice → 7947**

* **standard forest → 4280**

8004 **standard knot**
(with diameter less than 3.7 cm for broadleaved species and 3.1 cm for coniferous species)
d Standardknoten *m*
f nœud *m* standard
r стандартный здоровый сучок *m*

8005 **standard length of log; scaling length of log**
d Normallänge *f* von Schnittholz
f longueur *f* de grume standard
r стандартная длина *f* брёвна

* **standard [of the second rotation] → 6887**

8006 **standard-sized chips**
d standardgröße Späne *npl*
f copeaux *mpl* de taille normale
r щепа *f* стандартного размера

8007 **standard volume table**
d Standardmassentafel *f*
f table *f* de volume standard
r стандартная таблица *f* объёмов

* **stand basal area → 2032**

8008 **stand classification**
d Bestandaufgliederung *f*
f classification *f* de peuplement
r классификация *f* насаждений

* **stand composition → 1886**

8009 **stand conditions**
(descriptive measurement of a stand by the criteria of composition, health, age, size, volume, or spatial arrangement)
d Bestandesverhältnisse *npl*
f conditions *fpl* [d'accroissement] du peuplement
r таксационная характеристика *f* насаждения; условия *npl* произрастания насаждения

8010 **stand conservation; conservation of forest plantation** *r* сохранность *f* лесных культур

* **stand conversion → 226**

* **stand density → 3370**

8011 **stand density index; SDI**
(number of trees of area unit)
d Bestandesschlusszahl *f*
f indice *m* de densité du peuplement; indice de plénitude de couvert
r индекс *m* полноты насаждения; индекс густоты насаждения

8012 **stand density management diagram**
(two-dimensional graph showing the logarithmic relationship between declining mean stem frequency and increasing mean tree size)
d Bestandesdichte[management]diagramm *n*
f diagramme *m* d'aménagement de la densité de peuplement
r диаграмма *f* плотности насаждения

* **stand depletion → 8515**

* **stand description → 2411**

* **stand development → 2448**

8013 **stand dynamics**
(study of changes in forest stand structure over time, including stand behavior during and after disturbances)
d Bestandesdynamik *f*
f dynamique *f* d'un peuplement
r динамика *f* [роста] насаждения

* **stander → 3226**

8014 **stand establishment**
(process of developing a crop to the stage at which the young trees may be considered established)
d Bestandestablischment *n*
f établissement *m* d'un peuplement
r приживаемость *f* насаждения

8015 **stand form-factor; timber form-factor**
d Bestandesformzahl *f*
f coefficient *m* de forme du peuplement
r видовое число *n* насаждения

8016 **stand form height**
d Bestandesformhöhe *f*

8017 **stand growth**
d Bestandeszuwachs *m*
f croissance *f* d'un peuplement
r прирост *m* леса; прирост насаждения

* **stand height → 4233**

* **stand-height diameter → 8693**

* **stand improvement → 8643**

* **stand[ing] → 3605**

8018 **standing; uncut** *adj*
d stehend; auf dem Stock
f sur pied; sur souche
r стоячий [на корню]

* **standing crop → 3454**

* **standing crop estimation → 8290**

* **standing derrick → 3848**

* **standing forest → 3454**

* **standing line → 8043**

* **standing sale → 7157**

* **standing snag → 2271**

* **standing stock → 3454**

* **standing timber → 3454**

* **standing tree → 4010**

* **standing volume → 9250**

8019 **stand level**
(level of forest management at which a relatively homogeneous land unit can be managed under a single prescription, or set of treatments)
d Bestandesebene *f*
f niveau *m* de peuplement
r уровень *m* насаждения

8020 **stand management prescription**
(site-specific operational plan describing the nature and extend of silviculture activities planned for a free growing stand of trees)
d Waldwirtschaftsregel *f*; Waldwirtschsaftsgrundsatz *m*
f principe *m* de l'économie forestière

* **stand map → 3498**

* **stand margin → 2761**

* **stand marked for felling → 8644**

8021 **stand model**
(mathematical model that forecasts the development of a forest stand, usually in terms of mean stand attributes)
d Bestandmodell *n*
f modèle *m* de peuplement
r модель *f* насаждения

* **stand of dwarfed trees → 3515**

* **stand of high poles → 4803**

* **stand of pine → 6255**

* **stand of poles → 6419**

* **stand of seedlings → 6339**

8022 **stand of seed trees**
d Saatgutbestand *m*; Kernholzbestand *m*
f peuplement *m* d'arbres semenciers
r насаждение *n* семенных деревьев

8023 **stand of timber per acre**
d Holzvorrat *m* pro Acker
f bois *m* adulte par acre
r запас *m* древесины на акр

8024 **stand of valuable timber; valuable forests**
d Wertholzbestand *m*
f peuplement *m* en bois de valeur
r древостой *m* ценной породы; особо ценные лесные массивы *mpl*

8025 **stand origin**
(classification of forest stands describing their means of origin)
d Bestandesursprung *f*
f origine *f* de peuplement
r происхождение *n* насаждения

* **stand productivity → 3539**

* **stand quality → 6656**

8026 **stand-replacement fire regime**
(fire regime in which fires kill or top-kill aboveground parts of the dominant vegetation, changing the aboveground structure substantially)
d Feuerregime *n* zur Bestandsverjüngung
f régime *m* de feu pour la régénération du peuplement
r режим *m* пожара замены насаждения

8027 **stand replacing fire**
d Feuer *n* zur Bestandsverjüngung
f feu *m* de régénération forestière
r пуск *m* пала для лесовозобновления

* **stand series → 8032**

8028 **stand-size class**
(classification of forestland based on the diameter class distribution of live trees in the stand)
d Klasse *f* der Bestandesgroße
f classe *f* de dimension de peuplement
r класс *m* размера насаждения

8029 **stand stability tending**
d Stabilitätspflege *f*
f soins *mpl* stabilisateurs
r уход *m* за устойчивости насаждения

8030 **stand strategy**
(documented plan of stand treatments to achieve management objectives during the life of a particular stand)
d Bestandesstrategie *f*
f stratégie *f* de peuplement
r стратегия *f* [развития] насаждения

* **stand structure → 1886**

8031 **stand structure indices**
d Bestandesgefügezahlen *fpl*
f indices *mpl* pour la structure du peuplement
r показатели *mpl* состава насаждения

8032 **stand succession; stand series**
(the gradual supplanting of one stand by another)
d Bestandesfolge *f*
f série *f* de peuplement
r чередование *n* насаждения

8033 **stand table; stock table**
(a table showing the number of trees by species and diameter classes, generally per unit area of a stand)
d Bestandestafel *f*; Bestandestabelle *f*
f table *f* de peuplement; tableau *m* d'inventaire
r таблица *f* [распределения] насаждений

* **stand tending → 8456**

8034 **stand transect**

d Bestandesprobestreifen *m*
f transect *m* de peuplement
r трансекта *f* лесонасаждения

* **stand type → 3363**

* **stand-type map → 3635**

8035 stand volume class
(characterization of the forest based on either the total per-acre volume of growing-stock or the total per-acre volume of sawtimber)
d Bestandesmassenklasse *f*
f classe *f* de volumes du peuplement
r класс *m* объёмов стволов насаждений

* **stand volume inventory → 2438**

8036 stand volume table
d Bestandesmassentafel *f*
f table *f* du cubage des peuplements
r таблица *f* объёмов стволов насаждений; массовая таблица насаждения

* **stapler → 5212**

* **staranise tree → 4611**

8037 star apple
(Chrysophyllum)
d Sternapfelbaum *m*
f pommier-étoile *m*
r хризофил[лум] *m*

* **star check → 8039**

8038 starch trees
d Baumarten *fpl* mit Stärke
f espèces *fpl* forestières riches à amidon
r древесные породы *fpl*, запасающие в тканях крахмал

8039 star shake; star check
(defect in a log presenting cracks in the shape of a star in the heart)
d Sternriss *m*
f fente *f* en étoile; cœur *m* étoilé; cadranure *f* étoilée
r звёздообразная трещина *f*; сложная метиковая трещина; лучевая трещина

8040 starting point of regeneration
d Verjüngungszeitpunkt *m*
f point *m* d'initialisation du rajeunissement; échéance *f* du rajeunissement
r отправная точка *f* регенерации

forest; national forest; centrally administered forest; centrally managed forest
d Staatswald *n*; Staatsforst *m*; Bundesforst *m*; Bundeswald *m*; Nationalforst *m*
f forêt *f* nationale; forêt dominiale
r государственный лес *m*; лес государственного значения

8042 state forest inventory
d Staatswaldsinventur *f*
f inventaire *m* de forêt domaniale
r государственный учёт *m* лесов

* **state[-owned] forestry → 3890**

8043 static skyline; tight skyline; standing line
(fixed cable that does not move during logging operations)
d fixiertes Kabel *n*
f câble-grue *m* statique; câble-grue à tension fixe; câble *m* fixe
r неподвижный несущий канат *m*

8044 stationary chipper
d stationäre Hacker *m*
f déchiqueteuse *f* stationnaire
r стационарная рубильная машина *f*

8045 stationary delimber
d stationäre Entastungsmaschine *f*
f ébrancheuse *f* stationnaire
r стационарная сучкорезная машина *f*

8046 statute of woods
d Forstordnung *f*; Waldordnung *f*
f règlement *m* forestier
r лесной устав *m*

8047 statutory framework; legal framework
(where forest practices are primarily regulated by legislation)
d gesetzlicher Rahmengesetz *m*
f loi-cadre *f* légale
r законодательные рамки *fpl*

8048 stave
(piece or sawn of cleft timber of definite dimensions, forming one of the uprights in the side of a barrel, vat, etc.)
d Daube *f*
f douve *f*; dou[v]elle *f*
r материал *m* для клёпки; дуга *f*

* **stave log → 4362**

* **staves → 4362**

* **stave-wood → 4362**

8049 **stay[ing] point; setup**
(of skyline)
d Durchhaltepunkt *m*
f pointe *f* douce
r стоянка *f*

8050 **stay log**
d Spannbalken *m* zum Exzentrisch-Schälen
f barre *f* de déroulage excentré
r зажимное устройство *n* (при лущении шпона)

* **stay log cutting → 4093**

8051 **stealer of wood**
d Holzdieb *m*
f voleur *m* de bois
r похититель *m* леса

8052 **stealing of wood; theft of wood; pilferage of wood; timber trespass**
d Holzdiebstahl *m*; Holzdiebfrevel *m*
f vol *m* de bois
r похищение *n* леса

8053 **steamed groundwood**
d Braunholz *n*; Braunholzschiff *m*
f pâte *f* de rémanents
r бурая древесная масса *f*

8054 **steamed wood**
d dampfbehandeltes Holz *n*
f bois *m* étuvé
r пропаренная древесина *f*

* **steam nigger → 5124**

8055 **steam tree feller; steam sawmill**
d Dampfsägemuhle *f*; Dampfsägewerk *n*
f scierie *f* à vapeur
r валочное устройство *n* с паровым приводом; паровая лесопилка *f*

8056 **steckling**
(a rooted cutting that is sufficiently developed so that it can be or has been planted in the field)
d Steckling *m*
f planchon *m*
r высадок *m* (корень двухлетнего растения); маточный корнеплод *m*; посадочный корнеплод

(Phaenops cyanea)
d blauer Kiefernprachtkäfer *m*
f bupreste *m* bleu du pin
r синяя сосновая златка *f*

8058 **stele**
(the central core of the cylindrical axis of a vascular plant arising from the further differentiation of the plerome)
d Leitbündel *n*; Stele *f*
f stèle *f*
r стель *f*

8059 **stellate pith**
d Sternmark *m*
f moelle *f* étoilée
r звёздообразная сердцевина *f*

* **stem → 8934**

8060 **stem analysis; trunk analysis**
d Stammanalyse *f*
f analyse *f* de tige; analyse du tronc
r анализ *m* древесного ствола

8061 **stem base**
d Stammbasis *f*; Stammgrund *m*
f base *f* de grume
r основание *n* ствола

8062 **stem board**
d Stammbohle *f*; Stammbrett *n*
f planche *f* de grume
r толстая доска *f* из комлевого кряжа

8063 **stem borer; buprestid beetle**
d Stengelbohrer *m*; Prachtkäfer *m*
f foreur *m* des tiges
r стволовой древоточец *m*

8064 **stem break; stem rupture**
d Stammbruch *m*
f bris *m* de la grume
r поломка *f* ствола

8065 **stembuilder**
(a variety of woody plants used in a graft as intermediate stem piece to provide a strong trunk for standard trees, or to introduce resistance to disease or winter injury)
d Stammträger *m*
f tige *f* intermédiaire
r вставочный ствол *m*

f faisceau *m* de troncs; paquet *m* de troncs
r пачка *f* [поваленных] деревьев; куртина *f*; пакет *m* поваленных деревьев; штабель *m* хлыстов

* **stem-by-stem mixture → 5496**

* **stem chips → 1539**

8067 stem class
d Stammklasse *f*
f classe *f* de tige
r класс *m* [товарности] ствола

8068 stem classification
d Stammklassifizierung *f*
f classification *f* de tiges; classement *m* selon le diamètre du tronc
r классификация *f* стволов

* **stem contents → 9249**

* **stem counting → 2837**

* **stem crookedness → 2028**

* **stem crooking → 8325**

* **stem crotch → 8074**

* **stem crutch → 8074**

8069 stem cutting
d Stammteilung *f*
f tronçonnage *m* de tige
r раскряжёвка *f* ствола на сортименты

8070 stem cutting
d Stammsteckling *m*
f marcotte *f* de tige
r стеблевой черенок *m*

8071 stem diameter
d Holzdurchmesser *m*; Stammstärke *f*
f diamètre *m* de tige
r диаметр *m* ствола; толщина *f* ствола

8072 stem fire; log fire
d Stammfeuer *n*
f feu *m* de grume
r стволовой огонь *m*

8073 stemflow
(rainfall captured by the above-ground parts of a tree that flows down the trunk (stem) to the ground)
d [Stammabfluß *m*; Stammablauf *m*]
f [écoulement *m* de l'eau le long du tronc d'un arbre]
r стволовой сток *m*; приствольный сток атмосферных осадков

8074 stem fork; stem crutch; stem crotch
(a segment of a stem that forks)
d Stammgabel *f*
f fourche *f* de tige
r развилка *f* ствола

* **stem form → 898**

8075 stem form-factor
d Stammformzahl *f*
f coefficient *m* de forme du tronc
r видовое число *n* ствола

8076 stem leader
d Stammlode *f*
f pousse *f* apicale de tige
r главный стволовой побег *m*

8077 stem length; shaft height; shaft length
d Stammlänge *f*
f longueur *f* de grume
r длина *f* ствола; длина хлыста

8078 stemless
d stammlos; ohne Stamm
f sans tige
r бесствольный; бесстебельный

* **stemlike → 1392**

* **stemnotch → 2514**

8079 stem number
(in a surface unit)
d Stammzahl *f*
f nombre *m* de tiges
r число *n* стволов

* **stem of medium diameter → 1413**

8080 stem pitting
(a symptom of viral diseases characterized by depressions on the stem of the plant)
d Stammtüpfelung *f*
f ponctuation *f* de tige
r стволовая поровость *f*

8081 stem pruning
d Stammästung *f*
f élagage *m* de tiges
r обрезка *f* стволов

* **stem rupture → 8064**

* **stem-shaped → 1392**

* **stem shoot → 2848**

8082 **stem succulent; succulent plant**
(plant with succulent stems, without proper foliage leaves)
d Fettpflanze *f*; Sukkulente *f*
f plante *f* grasse; plante succulente; succulent *m*; plante molle
r стеблевой суккулент *m*

8083 **stem taper; taper; rise**
(decrease in thickness, generally in terms of diameter, of a tree stem or log from the base up)
d Stammdurchmesserabfall *m*; Stammabschrägung *f*
f défilement *m* de tige
r сбег *m* [ствола]; сужение *n* ствола

* **stem timber → 5154**

* **stem top → 8700**

* **stem-tuber → 7046**

* **stem volume → 9249**

* **stemwood → 891**

8084 **step cutting; step felling**
d Treppenhieb *m*
f coupe *f* à pas
r ступенчатая рубка *f*

8085 **step delimbing**
d schrittweise Entastung *f*
f ébranchage *m* par courses successive
r ступенчатая обрезка *f*

* **step felling → 8084**

8086 **steppe afforestation**
d Steppenaufforstung *f*
f afforestation *f* de steppe; afforestation steppique
r облесение *n* степи

8087 **steppe silviculture**
d Steppenwaldbau *m*
f sylviculture *f* steppique
r степное лесоразведение *n*

8088 **step-type undercut**
d schrittweiser Fallkerb *m*; Stufenfallkerb *m*
f entaille *f* d'abattage pas à pas; trait *m* de scie par en-dessous progressif
r ступенчатый подпил *m*

8089 **sterigmata** (*lat*, *pl*)
(the small woody stems to which the leaves of spruces and hemlocks are attached)
d kleine Stielchen *npl*; kurze Stielchen
f stérigmates *mpl*
r стеригмы *fpl*

8090 **stewardship forest**
(privately owned forest tract that exhibits integrated forest management to protect and enhance wildlife, timber, recreation, natural beauty, and soil and water quality)
d Verantwortungswald *m*
f forêt *f* de gestion intelligente
r лес *m* с рациональным использованием ресурсов

8091 **stewartia**
(Stewartia)
d Scheinkamelie *f*
f camélia *m*
r стюартия *f*

* **stick → 2171**

* **sticker → 6236**

* **stick[er] mark → 2047**

8092 **sticker spacing**
d Abstand *m* der Zwischenhölzer
f écartement *m* des bois de remplissage
r дистанция *f* промежуточных деревьев

* **sticker stain → 2047**

8093 **sticker [strip]; spacing block; munding**
d Zwischenholz *n*
f bois *m* de remplissage
r промежуточный лес *m*

* **sticking → 4846**

8094 **sticks stacker; piler of sawn goods; piler of lumber**
(a worker who places lumber in a pile)
d Arbeiter *m* auf dem Stapelplatz
f ouvrier *m* de chantier de bois
r укладчик *m* брёвен; рабочий *m* на складе

f éclaircie *f* par bâton
r [механическое] прореживание *n* по шаблону

* **sticky alder → 2882**

* **sticky bands → 8777**

* **sticky belts → 8777**

8096 stiff boom
d fester Schwimmer *m*
f flotteur *m* fixe
r плавун *m*

8097 stiffening; sclerosis
(of cell envelope)
d Absteifung *f*; Sklerose *f*
f sclérose *f*
r повышение *n* жёсткости; одревеснение *n*

* **stiffening wire → 1164**

8098 stiffleg; jammer
(loader with a boom that does not swing)
d Rundholzlader *m*
f chargeur *m* de grumes
r погрузчик *m* круглого леса

8099 stiffness
(capacity of a body to withstand deformation, more particularly bending, as measured by the modulus of elasticity)
d Steifheit *f*
f raideur *f*; rigidité *f*
r жёсткость *f*; прочность *f*

* **stillage → 5027**

8100 stilling basin
d Tosbecken *n*
f bassin *m* d'amortissement
r отстойник *m*; отстойный бассейн *m*

* **stilt root → 277**

* **stinger [tree] → 8528**

* **stinkwood [ocotea] → 807**

* **stint → 5181**

8101 stipule
(a leaflike outgrowth at the base of a leaf stalk)
d Nebenblatt *n*

8102 stipule scar
d Nebenblattnarbe *f*
f cicatrice *f* de stipule
r прилистниковый рубец *m*

* **stob → 3084**

* **stock → 8115, 9490**

* **stockade → 5019**

8103 stock board; dimension board
d Maßbrett *n*
f planche *f* standard
r доска *f* стандартными размерами

* **stocked → 3394**

8104 stocked forest
(land supporting tree growth)
d vorrätiger Wald *m*
f forêt *f* régénérée
r возобновившийся лес *m*

8105 stocked quadrate
(having at least one live tree seedling or regrowth)
d vorrätiges Quadrat *n*
f parcelle *f* régénérée
r возобновившийся квадрат *m*

* **stock gang → 5132**

* **stock grafting → 1635**

* **stocking → 3370**

8106 stocking class
(a numeric code representing a range of stems per hectare, sometimes estimated by crown closure on aerial photographs)
d Bestockungsklasse *f*
f classe *f* de [densité] peuplement
r класс *m* густоты насаждения

8107 stocking control
(regulation and, more particularly, the limitation of seedling populations by natural, or direct or indirect artificial factors)
d Bestockungskontrolle *f*
f surveillance *f* du matériel relatif
r управление *n* насаждением

8108 stocking guide
(reference level for the optimum proportion of

f guide *m* de stockage
r ориентир *m* на густоте насаждения

* **stocking level → 9250**

8109 stocking plan
(plan that provides objectives and strategies for land allocation and/or resource management, including regional plans, subregional plans, and local resource plans)
d Bestockungsplan *m*
f plan *m* de peuplement
r план *m* насаждения

8110 stocking rate; rate of stocking; incidence
d Bestockungsdichtekoeffizient *m*
f charge *f* moyenne de peuplement
r степень *f* плотности насаждения; степень концентрации насаждения

8111 stocking standard
(required range of healthy, well-spaced, acceptable trees)
d Bestockungsstandard *m*
f standard *m* de peuplement
r стандарт *m* насаждения

* **stocking survey → 2438**

8112 stock of land for forest planting
d Forstpflanzungsbodenfonds *m*
f fonds *m* terrestre de plantation forestière
r лесокультурный фонд *m*

8113 stock of land for hydrological melioration
d Hydromeliorationsbodenfonds *m*
f fonds *m* terrestre de hydromélioration
r гидромелиоративный фонд *m* (участки лесного фонда с избыточным увлажнением почв, предназначенные для проведения осушительных работ с целью повышения продуктивности лесов)

* **stock of state forest → 3767**

8114 stock of wooded forests; fund of wooded forests; forest fund
d Aufforstungsbestand *m*; Waldfonds *m*
f fonds *m* forestier
r лесной фонд *m*

* **stockpile → 5451**

8115 stockpile *v*; bond *v*; stock *v*
d bevorraten
f stocker; mettre en entrepôt; entreposer
r штабелировать [перекатыванием]

d Vorratholz *n* in Menge
f bois *m* stocké
r балансы *mpl* кучевого хранения

* **stock plant → 5814**

8117 stock roots
d Wildlingswurzeln *fpl*
f racines *fpl* de sauvageon
r корни *mpl* подвоя

8118 stock-scion interaction
(phenotypic effects of the rootstock on the scion)
d Wechselwirkung *f* zwischen Wurzel und Edelreis
f interaction *f* entre souche et scion
r взаимодействие *n* корня и подвоя

8119 stock-sized; cut-to-size *adj*
(into parts of given dimensions)
d stoffgeleimt
f déligné
r раскроённый в размер

* **stock taking → 5364**

* **stolon → 7132**

8120 stoma
(minute opening on the surfaces of leaves and stems through which gases (e.g., oxygen, carbon dioxide, water vapo[u]r) and some dissolved minerals pass into and out of plants)
d Spaltöffnung *f*
f stomate *m*
r устьице *n*

8121 stone cell; grit cell; brachyscelreid[e]
(an approximately isodiametric cell with a massive lignified secondary wall, which is often conspicuously laminated, and which may contain ramiform pits)
d Steinzelle *f*; Brachysklereide *f*
f cellule *f* pierreuse; brachysclérite *m*; brachyscléreide *m*
r каменистая клетка *f*

* **stone fruit → 2644**

* **Stone Mountain oak → 3830**

* **stone pine → 4595**

f sol *m* strié
r каменистые россыпи *fpl* (нелесные земли; скопления крупных камней, образовавшиеся в результате водной, ветровой эрозии, селевых потоков и речных наносов)

* **stone wood → 6196**

* **stool → 1934**

8123 stool bed
d Mutterpflanzenbeet *n*
f bâche *f* de pieds-mères
r маточная плантация *f*

* **stool shoot → 1933**

8124 stop
d Brechpunkt *m*
f point *m* d'interruption
r граница *f* (участок ствола, на котором имеется заметное уменьшение диаметра, например у толстого сука)

8125 stop-go harvester
d Erntemaschine *f* mit Halt
f abatteuse-porteuse *f* à arrêt
r лесозаготовительная машина *f* с остановкой для срезания дерева; валочная машина с остановкой для срезания дерева

* **stopping → 6240**

8126 stopping of growth; growth interruption
d Wuchsstockung *f*; Wachstumstockung *f*
f arrêt *m* de croissance
r приостановление *n* роста; остановка *f* в росте

* **storage boom → 4324**

8127 storage defect
d Lager[ungs]schaden *m*
f dégât *m* dû à l'entreposage
r повреждение *n* от хранения

* **storage place on land → 5059**

* **storage pond → 5105**

* **storax → 3825**

8128 storax family
(Styraceae)
r стираксовые *pl*

8129stored coppice cutting
(cutting in coppice in which selected stems are retained, as standards)
d Mittelwaldbetrieb *m*
f traitement *m* en taillis-sous-futaie
r рубка *f* в порослевым хозяйстве с оставлением маячных деревьев

* **stored coppice system → 1937**

8130 storied; storeyed; stratified; layered
(ref. to the axial cells and rays in wood when these are arranged in horizontal series on tangential surfaces)
d schichtig
f étagé
r ярусный

8131 storied cambium; storied elements
(cells arranged in tiers as seen on the tangential surface)
d schichtiges Kambium *n*
f cambium *m* étagé; éléments *mpl* étagés
r ярусные элементы *mpl*

8132 storied high forest; stratified high forest
(crop of trees in which the canopy can be differentiated into one or more layers)
d schichtiger Höhenbestand *m*
f futaie *f* pluriétagée
r высокоствольный древостой *m* с наличием одного или более ярусов

* **storied structure → 4842**

8133 story
(horizontal stratum or layer in a plant community)
d Etage *f*; Schicht *f*
f étage *m*
r ярус *m*

8134 story crown
d Stufenkrone *f*
f cime *f* étagée
r ярусная крона *f*

8135 stout-limbed; thick-branched
d mit starken Ästen; grobästig
f pourvu de grosses branches; à grosses branches
r с толстыми сучьями; крупносуковатый

8136 stowing grid; stowing rack; grating for stopping floating wood; terminal boom; catch boom; trap boom
d Schwimmkette *f*; Baumkette *f*; Staurechen *m*; Triftrechen *m*; Floßrechen *m*; Fangrechen *m* von Schwimmern; Triftholzbrecher *m*
f grille *f* d'arrêt de flottage; butée *f* destinée à arrêter ou à dévier le bois de flottage
r запань *f*; цепь *f* для скрепления бревен; плавучая цепь

* **stow wood → 8579**

* **straddle carrier → 5199**

8137 straight-boled timber; straight-stemmed timber; straight-grown timber
d geradstammiges Holz *n*
f bois *m* à fût rectiligne; bois à fût droit
r прямоствольный лес *m*

* **straight cut → 8141**

8138 straight-fibered tree; straight-grained tree
(in which the fibers run parallel to the axis of a piece)
d geradfaseriger Baum *m*; geradringiger Baum; schlicht gewachsener Baum
f arbre *m* à fibres droites
r дерево *n* с прямыми волокнами; прямослойное дерево

8139 straight grain
(wood texture in which the fibers and other longitudinal elements run parallel to the axis of a piece.)
d gerade Faser *f*
f fibre *f* droite; droit fil *m*
r прямослой *m*; прямое волокно *n*

* **straight-grained tree → 8138**

* **straight-grown timber → 8137**

8140 straight halving
d gerades Blatt *n*; gerade Eckūberblattung *f*
f entaille *f* droite; entaille à mi-bois
r поперечный замок *m* внакладку; врубка *f* под прямым углом вполдерева; прямая врубка; прямая накладка *f* вполдерева

8141 straight-line cutting; straight cut
d gerader Schnitt *m*; Geradschnitt *m*
f trait *m* de scie rectiligne
r прямолинейный пропил *m*; прямолинейный рез *m*

bole
d Stammgeradheit *f*; Geradschäftigkeit *f*; Zweischnürigkeit *f*; Schnürigkeit *f*
f rectitude *f* du fût
r прямоствольность *f*

* **straight planting → 3257**

* **straight-stemmed timber → 8137**

8143 straight-stemmed tree
d Geradholz *n*; geradschaftiges Holz *n*
f arbre *m* à fût rectiligne; arbre à fût droit
r прямоствольное дерево *n*

8144 straight-type undercut
d Geradfallkerb *m*
f entaille *f* rectiligne
r прямой подпил *m*

8145 strand tracheid; septate tracheid
(a tracheid of an axial series (strand) of tracheids (or of mixed tracheids and parenchyma cells), each series originating from a single cambial initial)
d gefächerte Tracheide *f*
f trachéide *f* septée; élément *m* d'une file de trachéides
r тяжевая трахеида *f*

* **strangler fig → 3286**

* **strap → 4850**

* **strap saw → 5049**

8146 strategic land[-]use planning
(planning at the regional, sub-regional and, in some cases, at the local level which results in land allocation and/or resource management direction)
d strategische Landesplanung *f*; strategische Raumplanung *f*
f planification *f* stratégique de territoire
r стратегический план *m* землепользования

* **stratificated high forest → 8132**

8147 stratification; layering; tiering
(division of a forest, or any ecosystem, into separate layers of vegetation that provide distinct niches for wildlife)
d Schichtung *f*; Stufigkeit *f*
f stratification *f*; différenciation *f*
r расслоение *n*; стратификация *f*; ярусность *f*; расположение *n* ярусами

8148 stratified density
d Stufenschluss *m*
f fermeture *f* complète à tous les étages
r ступенчатая сомкнутость *f*

8149 stratobiont
d Stratobiont *n*
f stratobionte *m*
r стратобионт *m* (животное, обитающее в лесной подстилке)

8150 stratum; canopy layer
(a distinct layer of vegetation within a forest community)
d Stratum *m*; Schicht *f*
f strate *f*; couvert *m*; ombrage *m*
r таксационный выдел *m*

* **strawberry bush → 8152**

8151 strawberry-shrub family
(Calycanthaceaea)
d Gewürzstrauchgewächse *pl*
f calycanthacées *pl*
r земляничные *pl*

8152 strawberry tree; strawberry bush; tree strawberry; arbutus-berry; fish-wood
(Arbutus unedo; Euonymus americanus)
d Erdbeerbaum *m*
f arbousier *m* [commun]; arbousier des Pyrénées; arbre *m* aux fraises; arbouse *f*
r земляничное дерево *n*

8153 straw line; grass line
(light-weight line used to change main skidding lines and tackle in cable yarding systems)
d Montageseile *f*
f câble *m* de manœuvre
r вспомогательный монтажный канат *m*

8154 streak
(defect of the wood which is not generalized in the whole member)
d Streifen *m*; Anreißen *n*
f échauffure *f* localisée; rayure *f*; tache *f*
r местная гниль *f* в виде полоски

8155 streak of resinosis; resin-streak; resin blaze; pitch-streak; gum spot; gum streak; size speck; groove; box; face
(a well-defined accumulation of resin in the form of a streak, occurring in certain conifers)
d Harzkropf *m*; Harzlach[t]e *f*; Lah[t]e *f*;
f filon *m* de résine; carre *f*
r смоляное пятно *n*; смоляная полоска *f*; засмолок *m*; карра *f*; [смоляной] потёк *m*

8156 stream-barked wood
d hydraulisch geschältes Holz *n*
f bois *m* écorcé hydrauliquement
r лесоматериалы *mpl* гидравлической окорки

* **stream[-]barker → 4404**

* **streambed → 9284**

* **stream channel → 9284**

* **stream driving → 2625**

* **streamside forest → 5192**

8157 streamside management zone; SMZ; riparian management area; RMA
(area of natural timber or vegetation protected and maintained on each side of a stream or drainage to provide habitat diversity and wildlife travel corridors and water quality)
d Uferbetriebsfläche *f*
f zone *f* de gestion riveraine
r приречная территория *f*

8158 streamside trees
d Uferbestockung *f*; Ufergehölz *n*
f boisement *m* des rives; bosquets *mpl* riverains
r прибережные деревья *npl*

8159 strength group
(group of timber species according to mechanical properties of the wood of that species)
d Stärkegruppe *f*
f groupe *m* de résistance
r группа *f* прочности

* **strength of thinning → 8519**

* **strength of wood → 2669**

* **stressed condition [of timber] → 1367**

8160 stress-graded hardwoods
d stressabgestufte Laubholzarten *fpl*
f feuillus *mpl* classés par contrainte mécanique
r лиственные породы *fpl* древесины, отсортированные по прочности

8161 striated
d gestreift

8162 strict branch
d strenger Zweig *m*
f branche *f* stricte
r прямая ветвь *f*, прижатая к основной оси

8163 stringer; main beam
(in rafts, a transverse pole to which the floating timber is tied)
d Längsbalken *m*; Längsträger *m*
f entretoise *f*
r продольная балка *m*; параллельно проложенный брус *m*

8164 stringer road
(a tram road with sawed wooden rails, used for hauling logs)
d Stringerweg *m*
f câble *m* grue à longerons
r трелёвочный волок *m* из параллельно проложенных брусьев

8165 string measure; tape measure
(of a tree)
d Bandmaß *n*
f mesure *f* à la ficelle
r обмер *m* лентой по окружности

8166 string measure under bark
d Bandmaß *n* ohne Rinde
f mesure *f* à la ficelle sans écorce
r измерение *n* лентой по окружности кубатуры лесоматериала без коры

* **stringy → 3096**

8167 stringy yellow rot
(invader - Corticium gallactinum)
d klebrige Gelbfäule *f*
f carie *f* jaune fibreuse
r волокнистая жёлтая гниль *f*

* **strip → 7468**

8168 strip-and-group cutting; strip selection [cutting]; selected-line cutting; selection border cutting
(in which groups of advance growth are freed both in the strip and closely ahead of it, along with further group cuttings to initiate regeneration)
d Femelstreifenschlag *m*; Plentersaumschlag *m*; Linienplenterung *f*; Linienplenterumschlag *m*; Saumfemelschlag *m*
f coupe *f* progressive en placettes et bandes; coupe jardinatoire par bandes
r линейная выборочная рубка *f* рубка полосами; каёмчатая выборка *f*

8169 strip application; band application
d Bandanwendung *f*; streifenweise Applikation *f*
f traitement *m* par bandes
r ленточное внесение *n*

8170 strip burning; strip firing
d streifenweises Abbrennen *n*
f écobuage *m* par bandes; incinération *f* par bandes
r сжигание *n* порубочных остатков на полосах; пуск *m* пала полосами

* **strip cropping → 522**

* **strip cruise → 4988**

* **strip cruising → 4988**

8171 strip cutting; strip felling; screen coupe; screen felling; staggering
(removal of the crop in strips in more than one operations)
d Kulissenhieb *m*; Kahlsaumschlag *m*; Gassenhieb *m*; Wechselhieb *m*; Wechselschlag *m*
f coupe *f* par bandes [de proche en proche]; coupe en coulisses; tranchée *f*; abattage *m* étagé
r сплошная полосная рубка *f*; кулисная рубка; каёмчатая рубка

8172 strip[e]
d Streifen *m*
f bande *f*
r узкая лесосека *f*; делянка *f*, отводимая одному вальщику

8173 striped ambrosia beetle; spruce ambrosia beetle
(Trypodendron lineatum)
d gemeiner Nutzholzborkenkäfer *m*; gestreifter Nutzholzborkenkäfer; liniierter Nutzholzborkenkäfer
f bostryche *m* liséré; scolyte *m* birayé
r хвойный древесинник *m*

* **stripe[d] figure → 8175**

8174 striped maple; Pensylvanian maple
(Acer pennsylvanicum)
d pennsylvanischer Ahorn *m*; Streifenahorn *m*
f érable *m* jaspé; érable de Pennsylvanie; érable

8175 **stripe grain; stripe[d] figure**
(produced by cutting timber that has an interlocked grain)
d Riegeltextur *f*; gestreifte Textur *f*; gefladerte Textur
f texture *f* rubanée; texture striée
r полосатая текстура *f*; ленточная текстура; струйчатая текстура; полосчатая текстура; слоистая текстура; текстура в виде волнистых полос

* **stripe-grained wood → 6678**

8176 **stripes of pine forests**
d Kiefernwaldreihen *fpl*
f bandes *fpl* de pineraie
r балочные леса *mpl*; байрачные леса

* **strip felling → 8171**

* **strip firing → 8170**

8177 **strip of trees**
d Baumband *n*
f bande *f* d'arbres
r полоса *f* деревьев

* **stripped → 551**

* **stripped-coupe felling → 4992**

* **stripped wood → 555**

* **stripping → 2283**

* **strip planting → 6356**

* **strip seeding → 522**

* **strip selection [cutting] → 8168**

8178 **strip shelterwood cutting**
d Saumschirmschlag *m*
f coupe *f* progressive par bandes; coupe d'abri en lisière
r полосно-постепенная рубка *f*; каёмчатая обсеменительная рубка

8179 **strip shelterwood system**
(in which regeneration cuttings are carried out on fairly wide strips)
d Saumschirmschlagssystem *n*
f mode *m* de régénération par coupes progressives en bandes
r система *f* каемчатой обсеменительной рубки

d Bandspritzung *f*
f pulvérisation *f* en bandes
r ленточное распыление *n*

* **strip survey → 4988**

* **strip thinning → 4992**

8181 **strobilaceous; strobiloid**
d zapfenförmig
f à forme strobile
r шишковидный

* **strobile → 8818**

* **strobilus → 8818**

8182 **stroke of a saw; saw stroke**
(width of the cut made by a saw blade)
d Sägezug *m*
f course *f* de scie
r пильный ход *m*

* **stroll planting → 9449**

* **strong-branched → 5596**

* **strong thinning → 4217**

* **strop → 4850**

* **strop up → 1552**

* **structural board → 9255**

8183 **structural diversity**
(degree of variety, vertically or horizontally, in the herb, shrub, and tree layers of vegetation of an area)
d strukturbedingte Unterschiedlichkeit *f*
f diversité *f* structurelle
r структурное разнообразие *n*

8184 **structural grade timber**
d Zimmerwertholz *n*; Bauwertholz *n*
f bois *m* de qualité de charpente
r ценная конструкционная древесина *f*

8185 **structural joists and planks**
d strukturelle Bohlen *fpl* und Balken *mpl*
f planches *fpl* et solives *fpl* constructives
r строительные брусья *mpl* и доски *fpl*

* **structural lumber → 1153**

8186 **structural thermal cover**

d strukturelle Wärmeschutzdeckung *f*
f couverture *f* thermique structurelle
r структурный термозащитный покров *m*

* **structural timber → 1153**

8187 structure defect
d Strukturfehler *m*
f défaut *m* de structure
r порок *m* строения

8188 structure of annual rings; annual ring structure
d Jahresringen[auf]bau *m*
f structure *f* de cernes annuels
r строение *n* годичных колец; строение годичных слоев

8189 structure of growing stock; growing-stock structure
d Bestockungsaufbau *m*
f structure *f* de la réserve de bois sur pied
r структура *f* запаса

8190 strut; pillar; column
(a structural timber resisting compressive forces along the grain)
d Strebe *f*; Stiel *m*
f poteau *m*; colonne *f*
r стойка *f*

8191 strychnine tree; poisonnut
(Strychnos nux-vomica)
d Brechnussbaum *m*; Brechnuss *f*; Curare *f*
f noix *m* vomique; vomiquier *m*; coque *f* du Levant
r кураре *n*

* **stubbing → 8201**

* **stubble ploughing → 8201**

* **stub [out] → 4033**

* **stub-puller → 8808**

* **stub spur road → 4172**

* **spur road → 4172**

* **stub-wort → 9656**

8192 stud
(slender wood structural member used as a supporting element in walls and partitions)
d Stiftbolzen *m*; Stiftschraube *f*

* **stump → 8937**

8193 stump age
(number of annual growth rings between the bark and the pith, as counted at stump height)
d Stubbenalter *n*
f âge *m* à hauteur de souche
r возраст *m* пня

* **stumpage → 3454**

8194 stumpage appraisal
d Stockwert[ab]schätzung *f*
f évaluation *f* du bois sur pied
r [материально-денежная] оценка *f* лесосек

8195 stumpage price; stumpage fee; cutting fee; timber due
(price paid for standing forest trees)
d Forstgrundpreis *m*; Stammgeld *n*; Stammpreis *m*; Preis *m* auf dem Stock
f prix *m* du bois sur pied
r цена *f* лесосек; попённая плата *f*; корневая плата

* **stump[age] sale → 7157**

8196 stumpage value; value of the stand
(the value of timber as it stands uncut in the woods)
d Bestandeswert *m*; Stockwert *m*; Kosten *pl* auf dem Stamm
f valeur *f* [des bois] sur pied; valeur du peuplement
r стоимость *f* насаждения; величина *f* лесного сырья на корню; стоимость леса на корню

* **stumpage volume → 9250**

8197 stump analysis
(analysis of a stump cross-section by counting and measuring the annual growth rings, in order to estimate the age of the tree and its past rates of radial growth)
d Stumpfanalyse *f*
f analyse *f* de souche
r определение *n* возраста дерева по пню

* **stump blade → 6986**

* **stump blasting → 852**

8198 stump chips
d Stumpfspäne *mpl*

* **stump chopper → 2529**

8199 **stump-clearing digger; stump eradicator; stump extractor; stump irradicator; stump pulling machine; [tree] stump puller; stump remover; tree stump grubber; tree stump assarting machine; stumper**
d Stubbenheber *m*; Baumstumpfrodemaschine *f*; Wurzelroder *m*; Stockrodemaschine *f*
f extracteur *m* de souches; dessoucheur *m*; dessoucheuse *f*
r корчеватель *m* пней; измельчитель *m* пней

8200 **stump crushing system; stump crusher; stump processor**
d Stumpfbrecher *m*
f fragmenteuse *f* de souches
r машина *f* для измельчения пней

* **stump diameter → 2465**

* **stumper → 8199**

* **stump eradicator → 8199**

* **stump exploding → 852**

8201 **stump extraction; stump pulling; stump grubbing; stump removal; grubbing[-out]; grubbing-up; stubble ploughing; stubbing; stumping**
(scarifying of land to break down crop stems still standing after harvest using a plough or one-way disc plough)
d Stubbenrodung *f*; Stumpfrodung *f*; Stehendrodung *f*
f extraction *f* des souches; dessouchage *m*; essartement *m*
r корчёвка *f* пней

* **stump extractor → 8199**

* **stump extractor and splitter → 8205**

8202 **stump figure**
(wood figure of a stump)
d Stumpffigur *f*
f ronce *f* de souche
r рисунок *m* пнёвой древесины

8203 **stump grapple; stump gripper**
d Stumpfgreifer *m*
f grappin *m* à souches
r пнёвый зажим *m*; зажим пней

* **stump grinder → 2529**

* **stump harvester → 8205**

8204 **stump harvesting; stump[wood] operation**
d Stockholzgewinnung *f*
f exploitation *f* de souches
r заготовка *f* пнёвого осмола; заготовка леса у пня (обрезка сучьев, раскряжёвка и штабелёвка на месте валки)

8205 **stump harvesting machine; stump harvester; stump extractor and splitter**
d Baumstumpf-Erntemaschine *f*
f extracteur-sécateur *m* de souches
r машина *f* для заготовки пней

* **stump height → 4236**

* **stumping → 6358**

* **stumping → 8201**

* **stump irradicator → 8199**

8206 **stump-jump plough; stump jumper**
d Stumpfspringer *m*
f charrue *f* sauteuse de souches
r плуг *m*, переезжающий через пни; плуг-корчеватель *m* с индивидуальными предохранителями корпусов для автоматического возвращения в рабочее положение

8207 **stump lowering**
d Stumpfverminderung *f*
f baisse *f* des souches
r срезание *n* пней (с целью уменьшения их высоты)

8208 **stump of coniferous tree**
d Koniferenstumpf *m*
f souche *f* de conifère
r пень *m* хвойного дерева

* **stump [out] → 4033**

8209 **stump plant**
(special type of cutting consisting of a short cut-back stem along with its roots, which may or may not be pruned)
d Stumpfpflanze *f*
f stump *m*
r порослевое растение *n*; вегетативное растение

* **stump planting → 6358**

* **stump pull → 2193**

* **stump pulling → 8201**

* **stump pulling machine → 8199**

8210 **stump [pulling] wheel; pulling wheel**
d Stumpfziehrad *n*
f roue *f* de dessouchage
r вытяжной ролик *m* для пней

* **stump removal → 8201**

* **stump remover → 8199**

8211 **stump-root system**
d Stumpf *m* mit Wurzelsystem
f souche *f* à racines
r пень *m* с корневой системой

* **stump rot → 7044**

8212 **stump scale [table]**
d Massentafel *f* nach Stockdurchmesser
f tarif *m* de cubage à la souche
r таблица *f* объёмов лесоматериалов по диаметру пня

8213 **stumpsite**
d Fällbereich *m*
f lieu *m* de coupe à la souche; place *f* de coupe à la souche
r участок *m* лесосеки у пня

* **stump slicer → 2529**

8214 **stump splitter**
d Baumstumpf-Verzweigter *m*; Stumpfspalter *m*
f fendeuse *f* de souches
r устройство *n* для раскалывания пней

* **stump sprout → 1933**

* **stump sucker → 1933**

8215 **stump-to-mill productivity**
d Stumpfrodung zu Holzwerksproduktivität *f*
f productivité *f* à dessouchage d'usine de conversion du bois
r производительность *f* по комплексу лесозаготовительных работ

8216 **stump-to-roadside productivity**
d Stumpfrodung zu Einkehrsproduktivität *f*
f productivité *f* à dessouchage de dépôt

8217 **stump-to-truck cost; logging costs; logging value; cost of harvesting; harvesting expenses; lumbering costs**
d Holzerntekosten *pl*; Holzverbundskosten *pl*; Verbungskosten *pl*
f pertes *fpl* d'exploitation de bois
r затраты *fpl* на заготовку [и вывозку леса]; стоимость лесозаготовок

8218 **stump trailer**
d Stumpfanhänger *m*
f remorque *f* à souches
r прицеп *m* для пней

* **stump transplant → 7010**

* **stump treatment → 2179**

8219 **stump veneer; butt veneer**
d Stumpffurnier *n*
f placage *m* de souche
r шпон *m*, полученный из пня

8220 **stumpwood**
(wood product taken from the stump portion of a severed tree)
d Erdstockholz *n*; Stockholz *n*; Stubbenholz *n*
f bois *m* de pâte empilé à la souche; bois de souche
r древесина *f* пня; пнёвая древесина; прикорневая древесина; пнёвый лесоматериал *m*

8221 **stump[wood] forwarder**
d Stubbenholz-Forwarder *m*
f porteur *m* de souches
r форвардер *m* для пней

8222 **stump[wood] forwarder-cleaner**
d Stubbenholz-Forwarder-Reiniger *m*
f porteur-nettoyeur *m* de souches
r очиститель-форвардер *m* для пней

* **stump[wood] operation → 8204**

* **stunt → 366**

8223 **stunted growth; dwarfed growth; dwarfing**
d Zwergwuchs *m*; Niedrigwüchs *m*
f croissance *f* naine; croissance rabougrie
r угнетённый рост *m*; карликовый рост

* **stunted wood → 1202**

* **styrax → 3825**

f arcossière *f*
r ольховое криволесье *n*

8225 **subalpine brown rot**
(invader - Polyporus leucospongia)
d subalpine Braunfäule *f*
f pourriture *f* brune subalpine
r бурая субальпийская гниль *f*

* **subalpine fir → 214**

8226 **subalpine forest; orohylile forest**
d subalpiner Wald *m*
f forêt *f* subalpine
r субальпийский лес *m*

* **subalpine larch → 216**

8227 **subalpine slash rot**
(invader - Stereum rugisporum)
d subalpine Schlagfäule *f*
f pourriture *f* subalpine de rémanents
r субальпийская гниль *f* мёртвой древесины

8228 **subantarctic rain [evergreen] forest**
d subantarktischer [immergrüner] Regenwald *m*
f forêt *f* subantarctique [tempérée] humide toujours verte
r влажный [вечнозелёный] субантарктический лес *m*

8229 **subarctic forest**
d subarktischer Wald *m*
f forêt *f* subarctique
r субарктический лес *m*

* **subbranch → 604**

8230 **subclimax**
(the stage in succession prior to the climax community in which further development is inhibited by the influence of some factors other than climatic factors)
d Subklimax *f*
f subclimax *m*
r предклимакс *m*

8231 **subcompartment**
(subdivision of a compartment, in which there is a difference in the tree crop that makes it suitable for different management treatment)
d Unterabteilung *f*
f sous-parcelle *f*; sous-division *f*; sous-compartiment *m*
r полуквартал *m*

* **subdominant trees → 1706**

* **subdumi → 9065**

8232 **subequatorial monsoon forest**
d subäquatorialer Monsoonwald *m*
f forêt *f* des moussons subéquatoriale
r субэкваториальный муссонный лес *m*

* **suber → 1956**

8233 **suberification; suberization**
d Verkorkung *f*
f subérification *f*; subérisation *f*
r опробковение *n*

* **subopposite leaf → 220**

8234 **subor**
d Subor *m*
f subor *m*
r суборь *f* (лес на переходных сравнительно бедных почвах)

8235 **subordinate crop; suppressed crop**
d nachrangige Kultur *f*
f culture *f* dominée
r подпологовая культура *f*

* **subordinate species → 7356**

* **subramose → 7737**

8236 **subsessile leaf**
d untersitzendes Blatt *n*
f feuille *f* sous-sessile
r лист *m* с очень слабовыраженным черешком

* **subshrub → 9065**

8237 **subsidiary crop; dominated crop; inferior stand**
d Nebenbestand *m*
f peuplement *m* secondaire
r второстепенный ярус *m* насаждения; подчинённое насаждение *n*

* **subsistence agriculture → 8238**

8238 **subsistence farming; subsistence agriculture**
d Subsistenzbetrieb *m*
f agriculture *f* de subsistance
r натуральное хозяйство *n*; малорентабельное хозяйство

* **subsoiler → 8963**

* **subsoil plough → 8963**

8239 subspecies
(of plants)
d Unterart *f*
f sous-espèce *f*
r подвид *m*

* **substitute fiber → 3783**

* **substitute shoot → 6872**

* **substituting crop → 1374**

* **substitution crop → 1374**

8240 substratum; substrate
(surface on or in which organisms live; chemical term referring to a compound or molecule upon which a reaction takes place)
d Substrat *n*; Unterlage *f*
f substrat *m*; substratum *m*
r нижний слой *m*; питательная среда *f*

8241 subsurface drainage; underground drainage; subsoil drainage
(water flow through permeable soil or rock beneath the surface of the land)
d unterirdische Entwässerung *f*; Untergrundentwässerung *f*
f drainage *m* souterrain; écoulement *m* des eaux en sous-sol
r внутрипочвенный дренаж *m*; подпочвенный дренаж

8242 subterminal bud; penultimate bud
d Subterminalknospe *f*
f bourgeon *m* avant-dernier
r боковая верхушечная почка *f*; субтерминальная почка

8243 subtropical evergreen forest
d subtropischer immergrüne Wald *m*
f forêt *f* subtropicale toujours verte
r субтропический вечнозелёный лес *m*

8244 subtropical forest
d subtropischer Wald *m*
f forêt *f* subtropicale
r субтропический лес *m*

8245 subulate leaf; awl-shaped leaf
d pfriemliches Blatt *n*

8246 succession
(gradual supplanting of one community of plants by another)
d Sukzession *f*
f succession *f*
r сукцессия *f*; последовательность *f*

* **successional → 7453**

8247 successional disking; successional mowing
(wildlife enhancement practice in which a disk harrow or rotary mower is used to knock down existing vegetation every 1 to 3 years)
d Sukzessionsmähen *n*
f fauchage *m* de succession
r последовательная косьба *f*

8248 successional stage; successional phase
(stages in the natural, ecologic progression of a stand)
d Sukzessionsstadium *m*; Sukzessionsphase *f*
f stade *m* de succession; phase *f* de succession
r стадия *f* сукцессии

* **succession of cuttings → 2202**

* **succession of felling → 2202**

8249 succession of tree species
d Holzartenfolge *f*
f succession *f* des essences
r сукцессия *f* древесных пород

* **successive felling → 3904**

8250 succulence; succulency
d Saftigkeit *f*
f succulence *f*
r суккулентность *f*; сочность *f*; мясистость *f*

* **succulent plant → 8082**

* **sucker → 7049**

* **sucking → 8571**

8251 sucupira [preta]; sapupira; aramatta
(Diplotropis purpurea; Bowdichia nitida; Bowdichia vigilioides)
d Sucupira *f*
f cœur *m* dehors
r сукупира *f*; бовдихия *f*

8252 suffruticose; suffrutescent *adj*
(very low and woody; diminutively shrubby)
d kleinbuschig

* **sugar bush → 8254**

8253 **sugar hackberry; sugarberry**
(Celtis laevigata; Celtis mississippiensis)
d Mississippi-Zürgelbaum *m*
f micocoulier *m* du Mississippi
r каркас *m* миссисипский

* **sugar maple → 808**

* **sugar pine → 1274**

8254 **sugar protea; honeyflower; sugar bush**
(Protea mellifera)
d Zuckerbusch *m*
f protée *f* mellifère
r протея *f* медоносная

8255 **sugar sumac[h]**
(Rhus ovata)
d Sumach *m*
f sumac *m* ovale
r сумах *m* сахарный

* **sugi → 4615**

8256 **suitability map**
(map that describes the current potential of a habitat to support a species)
d Eignungskarte *f*
f carte *f* d'aptitudes
r карта *f* пригодности

8257 **suitable forestland; suitable timber land**
(land that is managed for timber production)
d geeigneter Waldboden *m*
f terre *f* forestière convenante
r пригодная лесная земля *f*

8258 **sulfur tuft; kummer**
(Hypholoma fasciculare; Naematoloma fasciculare)
d grünblättriger Schwefelkopf *m*; ziegelroter Schwefelkopf
f hypholome *m* fasciculaire; hypholome en touffes
r ложные опята; ложноопёнок *m* серно-жёлтый

* **sulky → 5090**

* **Sumatran pine → 5404**

* **summer felling → 3065**

8259 **summer frost injury**
r повреждение *n* летним морозом

8260 **summergreen forests; deciduous summer forests; aestilignosa; aestisilvae**
d sommergrüne Walde *mpl*; sommergrüne Laubwalde *mpl*
f forêts *fpl* vertes d'été
r летнезелёные леса *mpl* [и кустарники]

8261 **summer leaf drop**
d Sommerblattabfall *m*
f chute *f* des feuilles d'été
r летнее сбрасывание *n* листьев

8262 **summer pruning**
d Sommerästung *f*
f élagage *m* d'été
r летняя обрезка *f* крон

8263 **summersweet clethra**
(Clethra alnifolia)
d erlenblättrige Zimterle *f*
f cléthra *f* à feuilles d'au[l]ne
r клетра *f* ольхолистная

8264 **summer[ ]wood**
d Sommerholz *n*
f bois *m* d'été
r летняя древесина *f*; летние лесоматериалы *mpl*

* **summit → 2087**

8265 **sunblister; sunburn; sunscald**
(localized injury to bark and cambium, caused e.g. by a sudden increase in exposure of a stem or branch to high temperatures from intense sunlight)
d Sonnenbrand *m*
f insolation *f*; coup *m* de soleil
r солнечный ожог *m*

8266 **sun crack**
(on the tree)
d Sonnenriss *m*
f fente *f* d'insolation
r трещина *f* от солнца

* **sunken log → 2268**

8267 **sun leaf**
d Sonnenblatt *n*
f feuille *f* solaire
r световой лист *m*

8268 **sunshine wattle; cedar wattle**
(Acacia terminalis)
d Sonnenscheinakazie *f*
f mimosa *m* terminalis
r акация *f* пограничная

* **superannuated stand → 5998**

* **supercanopy trees → 3845**

* **superficial foot → 7970**

* **superficial root → 7486**

8269 **superficial sap rot**
(invader - Shizophillum communis)
d oberflächliche Saftfäule *f*
f pourriture *f* de l'aubier superficielle
r поверхностная гниль *f*

* **superficial volume → 7979**

* **superfoot → 876**

* **superintendent of forestry → 1483**

* **superior tree → 6396**

8270 **superposed bud**
d überlagerte Knospe *f*
f pousse *f* superposée
r вышележащая почка *f*

8271 **superspecies**
d Überart *f*
f super-espèce *f*
r надвид *m*

* **supervision of cuttings → 3055**

* **supplementary tree species → 66**

8272 **supply block**
(forest land with relatively homogeneous forest characteristics, access development and management concerns)
d Versorgungsblock *m*
f bloc *m* d'approvisionnement
r питающий блок *m*

8273 **supporting pole; trainer**
(wooden support used to stabilize new plantings)
d Stützpfahl *m*
r подпорный столб *m*

* **supporting root → 277**

8274 **supporting tissue**
(of wood)
d Festigungsgewebe *n*
f tissue *f* d'appui
r опорная ткань *f*

* **suppressed bud → 2573**

* **suppressed crop → 8235**

* **suppressed crown class → 6012**

8275 **suppressed trees; overtopped trees; dominated trees**
(characterized by low growth rate and low vigo[u]r as a result of competition with overtopping trees)
d unterdrückte Bäume *mpl*
f arbres *mpl* supprimés; arbres dominés
r угнетённые деревья *npl*

8276 **suppression by shade; checking by shade**
d Schirmdruck *m*
f suppression *f* par l'ombre; effet *m* gênant de l'abri
r угнетение *n* от верхнего полога

* **suradan → 6237**

* **suradonni → 6237**

8277 **suramen; mixed subor**
d Suramen *m*; gemischter Subor *m*
f suramen *m*
r сурамень *f*; сложная суборь *f* (тип леса ельник сложный)

* **surculus → 7046**

8278 **surface blue**
d Oberflächenbläue *f*
f bleuissement *m* superficiel
r поверхностные грибные заболонные окраски *fpl*

8279 **surface check; shell shake**
(seasoning checks that develop on the surface and extend into the wood for varying distances)
d Oberflächenriss *m*
f fissure *f* superficielle; gerçure *f* en surface
r трещина *f* на пласти; пластевая трещина;

8280 surfaced board; sliced board; wrought stuff
d Hobelbrett *n*
f planche *f* rabotée
r строганная доска *f*

* **surfaced lumber → 2620**

8281 surface dressing; dressing; surfacing
(of wood smoothing down its faces or edges by planing or sanding)
d Endbearbeitung *f*; Oberflächenbehandlung *f*
f surfaçage *m*
r облицовывание *n*; сглаживание *n*

* **surfaced timber → 2620**

8282 surface fire
(fire that burns in litter and other live and dead fuels at or near the surface of the ground, mostly by flaming combustion)
d Oberflächenfeuer *n*
f feu *m* de surface
r наземный пожар *m*; напочвенный пожар; поверхностный огонь *m*

8283 surface moisture content
(of sawn timber)
d Oberflächenfeuchte *f*
f contenu *m* d'eau surfacique
r поверхностная влажность *f*; влажность поверхности

8284 surface particles
(of fiberboards)
d Außenschichtspäne *mpl*; Deck[schicht]späne *mpl*
f plaquettes *fpl* de face
r стружки *fpl* наружного слоя; стружки верхнего слоя

* **surface plant → 1439**

* **surface-root → 7486**

8285 surface soil erosion
(means for an area where a forest practice has been carried out)
d Erosion *f* der Oberbodenfläche
f érosion *f* du sol en surface
r эрозия *f* поверхностного слоя почвы

* **surface vegetation → 3986**

* **surfacing → 8281**

8286 surplus forest
(in which existing stands can provide more harvest volume than is needed to maintain the harvest at the level of long run sustained yield)
d Überschusswald *m*
f forêt *f* surplus
r избыточный лес *m*

8287 surplus of growing stock; excess of growing stock
d Übervorrat *m*
f excès *m* de réserves de bois
r избыток *m* запаса

8288 survey crew; valuation crew; estimating crew; cruising crew; cruising party
d Taxierungsmannschaft *f*; Einrichtungstrupp *m*; Schätzungsrotte *f*
f équipe *f* d'aménagement; équipe d'estimateurs; chantier *m* d'estimateurs
r лесоустроительная партия *f*; таксационная партия; бригада *f* таксаторов

8289 surveying chain; surveyor's chain; measuring chain
d Messkette *f*
f chaîne *f* d'arpenteur
r таксационная мерная лента *f*; мерная цепь *f*

8290 survey of standing crop; standing crop estimation
d Abschätzung *f* stehender Bestände
f comptage *m* du volume global
r таксация *f* насаждений; инвентаризация *f* насаждений

* **survey of the stock → 2438**

8291 surveyor growth
(merchantable volume increment on trees 5.0 inches d.b.h. and larger in the inventory at the beginning of the year and surviving to the end of the year)
d Taxierungswachstum *m*
f croissance *f* d'aménagement
r таксационный прирост *m* насаждения

* **surveyor's chain → 8289**

8292 surveyor's rod; measuring rod; measuring staff; rod
(surveying instrument made of wood and graduated in feet and tenths of a foot)
d Messlatte *f*

8293 **survival index**
(of plants)
d Überlebensindex *m*
f indice *m* de survive
r показатель *m* выживаемости

* **survival rate → 1322**

* **suspended → 9324**

* **suspended wire for gravity skidding → 9464**

8294 **sustainability**
(state or process that can be maintained indefinitely)
d Nachhaltigkeit *f*; Zukunftsfähigkeit *f*
f durabilité *f*
r устойчивость *f*

8295 **sustainable forest development**
(preservation and protection of forest)
d dauerhafte Forstentwicklung *f*; nachhaltige Forstentwicklung; tragfähige Forstentwicklung
f développement *m* durable des forêts
r устойчивое развитие *n* леса

8296 **sustainable forest management**
(management regimes applied to forestland which maintain the productive and renewal capacities as well as the genetic, species and ecological diversity of forest ecosystems)
d dauerhafter Waldbau *m*
f aménagement *m* forestier durable
r ведение *n* устойчивого лесного хозяйства; устойчивое лесоустройство *n*

* **sustainable forest production → 8299**

8297 **sustainable forestry**
(management of forested area in order to provide wood products in perpetuity, soil and watershed integrity)
d dauerhafte Forstwirtschaft *f*
f foresterie *f* durable
r устойчивое лесоводство *n*

8298 **sustainable forest use**
d nachhaltige Waldnutzung *f*
f utilisation *f* de la forêt soutenue
r постоянное пользование *n* лесом; устойчивое лесопользование *n*

8299 **sustained [forest] yield; sustained forest production; sustainable forest production system)**
d Nachhaltsholzertrag *m*
f rendement *m* soutenu de bois
r устойчивая лесная продукция *f*

8300 **sustained-yield forestry**
(yield of defined forest products of specific quality and in projected quantity that a forest can provide continuously at a given intensity of management)
d nachhaltige Forstwirtschaft *f*
f foresterie *f* à rendement soutenu; foresterie à aménagement soutenu
r лесное хозяйство *n* непрерывного дохода

8301 **Swaine jack-pine sawfly**
(Neodiprion swainei)
d Kiefernbuschhornblattwespe *f*
f diprion *m* de Swaine
r пилильщик *m* сосновый Свейна

8302 **swamp** *v*
(to clear the ground of underbrush, fallen trees, and other obstructions preparatory to constructing a logging road, opening out a gutter road, skidding with animals, or yarding with a donkey engine)
d einen Weg durch den Wald hauen
f ouvrir un chemin à travers forêt
r прорубать дорогу через лес

* **swamp ash → 777**

8303 **swamp azalea; clammy azalea; swamp-honeysuckle**
(Rhododendron viscosum)
d Sumpf-Rhododendron *m*
f azalée *f* visqueuse; azalée tardive; chèvrefeuille *m* blanc des marais
r рододендрон *m* клейкий

* **swamp bay → 643**

8304 **swamp birch; low birch**
(Betula pumila)
d amerikanische Strauch-Birke *f*
f bouleau *m* nain
r берёза *f* малорослая

8305 **swamp black gum**
(Nyssa biflora)
d zweiblütiger Tupelobaum *m*
f tupélo *m* à deux fleurs
r нисса *f* двуцветковая

* **swamp black spruce → 2716**

* **swamp cedar → 398**

* **swamp chestnut oak → 615**

8306 **swamp cottonwood; swamp poplar; river cottonwood**
(Populus heterophylla)
d Sumpfpappel *m*
f peuplier *m* des marais
r тополь *m* разнолистный; тополь болотный

* **swamp cypress → 786**

8307 **swamper; busher; gutterman**
(one who swamps)
d Reiniger *m* der Holzabfälle
f débroussailleur *m*; opérateur *m* de débroussailleuse
r рабочий *m* на расчистке лесосечных отходах

8308 **swamper burning; progressive burning; simultaneous ignition**
(burning the slash as it is piled)
d gleichzeitiges Feuerlegen *n*
f brûlage *m* à la mise en tas; brûlage progressif; allumage *m* simultané
r сжигание *n* порубочных остатков в кучах; сжигание порубочных остатков при очистке лесосек

8309 **swamp forest; fen forest; wet peatland forest; bog forest; marshy forest; paludal forest**
(an lightly treed wetland habitat, more mineral-rich than a bog due because of available groundwater; typically acidic to alkaline)
d Bruchwald *m*; Sumpfwald *m*; Moorwald *m*
f forêt *f* marécageuse; forêt sur terrain très fangeux; forêt de marais; forêt sur terrain très mouilleux
r болотистый лес *m*; заболоченный лес

8310 **swamp gum; swamp peppermint; black gum**
(Eucalyptus ovata; Eucalyptus rodwayi)
d Schwarzgummi *m*
f gommier *m* noir
r эвкалипт *m* яйцевидный

* **swamp-honeysuckle → 8303**

* **swamp laurel → 4835**

8311 **swamp logging**
r лесозаготовка *f* на слабых грунтах

* **swamp magnolia → 643**

* **swamp maple → 6774**

* **swamp oak → 3764**

8312 **swamp open woodland**
d Sumpflichtwald *m*
f bois *m* clairsemé marécageux
r болотистое редколесье *m*

* **swamp peppermint → 8310**

* **swamp poplar → 8306**

8313 **swamp spruce forest with grass cover**
d Bruch-Fichtenwald *m* mit Gräserkrautschichte
f pessière *f* marécageuse à couverture herbacée
r болотно-травяной ельник *m*

8314 **swamp-subor**
d Sumpfsubor *m*
f subor *m* marécageux
r заболоченная суборь *f*

8315 **swamp sumac[h]; poison elder[berry]; poison-dogwood**
(Rhus vernix)
d Lackbaum *m*; amerikanische Firnissumach *m*; Giftsumach *m*
f sumac *m* vénéneux; sumac lustré
r сумах *m* восковый; сумах лаковый

* **swamp tupelo → 9304**

8316 **swamp wattle; water wattle; wirilda**
(Acacia retinoides; Acacia elongata)
d Sumpfakazie *f*; harzige Akazie
f mimosa *m* [des] quatre saisons; mimosa à feuilles étroites; mimosa résineux
r акация *f* ретиноидес; акация стойкая

8317 **swamp white oak**
(Quercus bicolor)
d Weißeiche *f*
f chêne *m* bicolore
r дуб *m* дву[х]цветный

* **swamp willow → 813**

8318 **swampy-forest soil**
d sumpfiger Waldboden *m*
f sol *m* forestier marécageux
r болотно-лесная почва *f*

8319 swarm
(of insects)
d Schwarm *m*
f essaim *m*
r скопление *n*

* **swath → 2178**

* **swather[-harvester] → 1900**

8320 swath width; bout width
d Streifenbreite *f*; Bandbreite *f*
f largeur *f* de la bande
r ширина *f* пасоки

* **sway wood → 1850**

8321 sweat[ing]; perspiration
(water condensation on wood surface)
d Schweißabsonderung *f*; Schweißausbruch *m*; Schwitzen *n*; Wasseraufstieg *m*; Perspiration *f*
f ressuage *m*; remontée *f* d'eau
r выпотевание *n* (конденсация влаги с поверхности древесины)

* **Swedish aspen → 2880**

8322 Swedish mountain ash
(Sorbus intermedia; Sorbus scandica)
d schwedische Eberesche *f*; schwedische Mehlbeere *f*; Oxelbeere *f*
f sorbier *m* intermédiaire; alisier *m* de Suède
r рябина *f* промежуточная дуболистная

8323 Swedish whitebeam; whitebeam [mountain ash]; beam tree; flour berry
(Sorbus aria)
d Oxelbirne *f*; [echte] Mehlbeere *f*; Mehl[beer]baum *m*; Weißlaub *m*; Silberbaum *m*
f alisier *m* blanc; allier *m*; alouchier *m*; baie *f* de farine; sorbier *m* des Alpes
r рябина *f* ария; рябина круглолистная

* **sweep → 2031**

8324 sweeping the rear; grounded logs salvaging; pick-rear; salvaging after the main drive
d Nachtrift *f*
f nettoiement *m* et enlèvement *m* de rémanents
r дополнительный молевой лесосплав *m*

8325 sweep of the bole; sweep of stem; stem crooking
d Stammkrümmung *f*; Säbelwuchs *m*
r искривление *n* ствола; кривизна *f* ствола

* **sweep saw → 1822**

8326 sweet azalea
(Rhododendron arborescens; Azalea arborescens)
d baumartige Azalee *f*
f azalée *f* arborescente
r рододендрон *m* древовидный

* **sweet bay → 6763**

* **sweetbay [magnolia] → 643**

* **sweet bay [tree] → 3936**

8327 sweetbells
(Leucothoe)
d Leucothoe *m*
f leucothoé *f*
r леукотоэ *f*

8328 sweet birch
(Betula grossa)
d Großersbirke *f*
f bouleau *m* sucré
r берёза *f* граболистная

* **sweet broom → 1779**

8329 sweet cassava
(Manihot dulcis)
d süßer Maniok *m*
f manioc *m* doux
r маниок *m* сладкий

8330 sweet cherry; mazzard [cherry]; [wild] gean
(Prunus avium; Cerasus avium)
d Tisselkirschbaum *m*; Haferkirsche *f*; Süßkirsche *f*; Süßkirschenbaum *m*; Waldkirsche *f*; Wildkirsche *f*; Vogelkirsche *f*
f merisier *m* [des oiseaux]; cerisier *m* sauvage; cerisier à fruits doux; cerisier des bois; cerise *f* douce; guignier *m*
r вишня *f* птичья; вишня сладкая; черешка *f* птичья; черешня *f*

* **sweet chestnut → 2887**

8331 sweet crabapple
(Malus coronaria)
d Kronen-Apfel *m*
f pommier *m* coronarié; pommier odorant
r яблоня *f* венечная

8332 **sweet[-]gale; bog myrtle; wax-berry**
(Myrica gale)
d Gagel[strauch] *m*; gemeiner Gagel *m*; brabante Myrte *f*
f myrique *m* baumier; galé *m* odorant; piment *m* royal; myrte *m* de marais; myrte du Brabant
r восковник *m* обыкновенный; восковница *f* обыкновенная; восковница болотная; восковой вереск *m*; болотная мирта *f*

* **sweet gallberry → 4795**

* **sweet gum → 503**

8333 **sweetleaf**
(Symplocos)
d Rechenblume *f*; Saphirbeere *f*
f symplocos *m*; omon *m*
r симплокос *m*

* **sweet locust → 4353**

8334 **sweet mining debris; sweet waste; sweet spoils; sweet tailings**
(ref. to materials in the overburden, mining refuse, mine tailings or mining spoils that are neutral or slightly alkaline and capable of supporting certain calcium-demanding plants)
d Neutralabbruchmasse *f*
f débris *m* minier neutre; débris minier doux
r нейтральные рудничные отходы *mpl*; незасолённые рудничные отходы

8335 **sweet mock[ ]orange; seringa**
(Philadelphus coronarius; Philadelphus lewisii)
d wohlriechender Pfeifenstrauch *m*; gemeiner Pfeifenstrauch; wilder Jasmin *m*
f syringa[t] *m* à bouquets; seringa[t] *m* commun; jasmin *m* bâtard
r жасмин *m* грунтовой; жасмин дикий; жасмин ночной; жасмин садовый; чубушник *m* венечный

* **sweet orange → 5935**

* **sweet pecan → 6130**

* **sweetroot → 4997**

* **sweet segde → 6731**

8336 **sweet-spire family**
(Iteaceae)
d Rosmarienweidegewächse *pl*

* **sweet spoils → 8334**

* **sweet tailings → 8334**

* **sweet waste → 8334**

* **sweet water → 5294**

* **sweet wattle → 1689**

* **sweet willow → 630**

* **sweet wood → 3732**

* **swell-butted → 8493**

8337 **swelling**
(of sawn timber)
d Schwellen *n*; Ausbauchung *f*; Aufwölbung *f*
f gonflement *m*
r разбухание *n*; отделение *n* слоёв; отслаивание *n*; вздутье *n*; шелушение *n*

8338 **swelling value**
d Schwellenwert *m*
f valeur *f* de gonflement
r коэффициент *m* разбухания

8339 **swidden fallow secondary forest**
d Sekundärwald *m* mit Brandbrache
f forêt *f* itinérante après débroussaillement et brûlis
r вторичный лес *m* с выжиганной растительности

8340 **swifter; swiftering line; swifting tackle**
(a rope or cable placed across the end of the first tier of each raft section in order to hold the boom sticks in position)
d Schwichttakel *m*; Schwichtleine *f*
f ceinture *f*
r поворина *f*

8341 **swinging**
d schwankende Bewegung *f*; Schwanken *n*
f halage *m* [après débusquage]; balancement *m*
r поворот *m* из стороны в сторону

8342 **swinging boom grapple**
d Schwenkarm-Greifer *m*
f grappin *m* à flèche pivotante
r пачковый захват *m* на поворотной в горизонтальной плоскости стреле

8343 **swinging boom machine; swing[ing] crane; rotary crane; pivoting crane; turning**

f grue *f* à pivot; grue tournante; grue pivotante; grue à bras orientable
r трелёвочная машина *f* с поворотной стрелой

* **swinging cross-cut saw → 4312**

* **swingle-tree → 4480**

* **swing saw → 4312**

8344 **swing [skid] road**
d Schwungweg *m*
f chemin *m* de débardage pivotant
r волок *m* при вторичной трелёвке (от промежуточного склада к дороге)

8345 **swing system**
d Schwungsystem *n*
f système *m* pivotant
r канатная установка *f* для вторичной трелёвки

* **swirl figure → 2144**

* **swirl grain → 2145**

* **Swiss mountain pine → 5598**

8346 **Swiss [stone] pine; Alpine stone pine; arolla pine; cembra-pine**
(Pinus cembra)
d Zirbel-Kiefer *f*; Zirbe *f*; Arve *f*
f pin *m* cembro[t]; cembrot *m*; pin arolle
r сосна *f* кедровая европейская; кедр *m* европейский; сосна кедровая карпатская

8347 **Swiss stone pine forest**
d Arvenwald *m*
f arolaie *f*
r европейский кедровник *m*

8348 **Swiss willow**
(Salix helvetica)
d schweizerische Weide *f*
f saule *m* de Suisse
r ива *f* гельветика

8349 **switchback**
(a 180-degree direction change in a trail or road used to climb steep slopes)
d Rückschaltung *f*
f demi-tour *m*; rebroussement *m*
r смена *f* направления

8350 **swivel**
d Drehlager *n*; Drehscheibe *f*
f émerillon *m*
r шарнир *m*

8351 **swizzlestick tree**
(Quararibea turbinata)
d Espingo *m*
f bois-flot *m*
r дерево *n* суизл-стик

8352 **swollen knot**
d aufgebeulter Ast *m*
f nœud *m* vicieux recouvert
r переросший гнилой сук *m*; сук с наростом

* **sword → 4040**

8353 **sword-shaped leaf**
d säbelformiges Blatt *n*
f feuille *f* en forme d'épée
r мечевидный лист *m*

8354 **sycamine; mulberry**
(Morus)
d Maulbeerbaum *m*
f mûrier *m*
r шелковица *f*; тутовое дерево *n*; тут *m*

8355 **sycamore maple; white sycamore; sycamore [plane]; false sycamore [maple]; [English] harewood; common maple; great maple; planetree maple**
(Acer pseudoplatanus)
d [echter] Bergahorn *m*; Sykomore-Ahorn *m*; stumpfblättriger Ahorn *m*; Weißahorn *m*; Waldahorn *m*
f érable *m* sycomore; sycomore *m*; érable faux-platane; faux platane *m*
r клён *m* белый; клён ложноплатановый; явор *m* [обыкновенный]

8356 **Sydney golden wattle; Sydney acacia**
(Acacia longifolia)
d langblättrige Akazie *f*
f acacia *m* à longues feuilles
r акация *f* длиннолистная

* **sylvan → 3324**

* **sylvatic → 3324**

* **sylvatic zoology → 3651**

8357 **sylvofacies**
(a common view of a forest stand)
d Waldgesicht *n*; Waldfazies *f*

* **sylvopastoralism → 7625**

* **sylvopastoral system → 7625**

8358 **symbion[t]; symbiote; mutualist**
(organism associated with another organism to the mutual benefit of each)
d Symbiont *m*
f symbio[n]te *m*
r симбионт *m*

8359 **synecology**
d Synökologie *f*
f synécologie *f*
r синэкология *f*

8360 **synfolium**
d Synfolium *m*
f synfolium *m*
r ярус *m* в лесном фитоценозе

* **systematic thinning → 5374**

T

* **table → 4840**

* **table-land forest → 9131**

8361 **Table-Mountain pine; prickly pine; hickory pine**
(Pinus pungens)
d Tafelbergföhre *f*
f pin *m* pungens
r сосна *f* колючая

8362 **table of basal areas; basal-area table; table of sectional areas**
d Grundflächentafel *f*; Kreisflächentafel *f*
f table *f* des surfaces terrières; table des surfaces circulaires
r таблица *f* площадей поперечного сечения

8363 **table of logging**
d Forstamtstarif *m*
f table *f* de débardage
r расценки *fpl* на лесозаготовках

* **table of sectional areas → 8362**

* **tacamahac poplar → 9338**

* **tackle → 4227**

* **taeda pine → 5029**

* **tag alder → 7786**

8364 **tagging**
(of wild animals)
d Kennzeichnung *f*; Markierung *f*
f étiquetage *m*
r нанесение *n* меток; мечение *n*

8365 **tag line**
(term used in the US; smaller diameter rope used to control and direct a swinging load)
d Halteseil *n*
f câble *m* stabilisateur
r направляющий канат *m*; канат с захватом

8366 **tag line rigging**
d Halteseiltakelung *f*
f câblage *m* à stabilisateur; timonerie *f* en câble
r дополнительного чокера

8367 **taiga boreal forest**
d borealer Taigawald *m*
f forêt *f* boréale du type taïga
r таёжный бореальный лес *m*

8368 **taiga [forest]**
(northern, subarctic, coniferous forests of Europe, Asia and N. America, typically open or interspersed with bogs, which form a transition zone between the denser forests to the south and the tundra to the north)
d Taiga *f*; Taigawald *m*
f taïga *f*; forêt-taïga *f*
r тайга *f*; таёжный лес *m*

* **tail block → 4167**

8369 **tail *v* down; roll *v* up**
(to roll logs on a skidway to a point on the skids where they can be easily reached by the loading crew)
d aufrollen
f enrouler
r подкатывать брёвна по слегам (к месту погрузки)

* **tailhold → 2815**

* **tailholt → 2815**

* **tailings → 1422**

* **tail spar → 2815**

8370 **tail tractor**
d Seilschlepper *m*
f tracteur *m* à câble
r трактор *m* с канатом

8371 **tail tree**
(a tree to which the far end of the skyline is attached)
d endstützter Baum *m*
f mât *m* de queue; mât aval
r тыловая мачта *f*

8372 **tainted tree**
d verdorbener Baum *m*
f arbre *m* entaché
r подгнившее дерево *n*; подпорченное дерево

8373 **talipot [palm]**

f talipot *m*
r веерная пальма *f* талипот

* **tall forbs → 4287**

8374 **tall oil**
d Fichtenöl *n*; flüssiges Harz *n*; Tallöl *n*
f résine *f* liquide
r талловое масло *n*

* **tallow nut → 4322**

8375 **tallow tree**
(Detarium senegalense; Detarium macrocarpum)
d Mambode *n*; Bobode *n*
f boiré *m*
r дерево *n* сальное

* **tallow wood → 4322**

8376 **tally card; tally sheet; [timber] measurement list; timber list; wood list; sales list**
d Holzliste *f*
f liste *f* de bois de sciage sur dimensions préétablies
r лист *m* учёта [лесоматериалов]; список *m* лесоматериалов

8377 **tallyman; marker**
(one who records or tallies the measurements of logs as they are called by the scaler)
d Listenführer *m*
f chef *m* de comptage; pointeur *m*
r учётчик *m*

* **tally sheet → 8376**

8378 **talus; scree**
(a sloping mass of loose rocks at the base of a cliff; a large smooth mass of rock detached from its place of origin)
d Geröllhalde *f*; Schutthalde *f*; Schuttkegel *m*
f pente *f* pierreuse; éboulis *m*
r осыпь *f*

* **tamarillo → 8946**

8379 **tamarind; camalindo**
(Tamarindus indica)
d Tamarinde *f*
f tamarinier *m*
r тамаринд *m* [индийский]; индийский финик *m*

8380 **tamarind plum**
(Dialium)
d Tamarindenbaum *m*
f diale *m*
r диалиум *m*

* **tamarisk → 3706**

8381 **tamarisk brake**
d Tamariske-Dickicht *n*
f tamariçaie *f*
r заросль *f* гребенщика

8382 **tamarisk family**
(Tamaricaceae)
d Tamariskengewächse *pl*
f tamaricacées *pl*
r гребенщиковые *pl*

* **tamo → 5269**

* **tan → 5832**

* **tanbark → 8392**

* **tanbark coppice → 5833**

* **tan crusher → 2281**

* **tangare → 4039**

8383 **tangential cut; tangential section; crown cut; flat sawing; slash sawing**
(a method of slicing veneers whereby the average inclination of the growth rings to the wider face is tangential or less than 45°)
d Tangentialschnitt *m*; Fladerschnitt *m*; Herzschnitt *m*; Sehnenschnitt *m*
f coupe *f* tangentielle; sciage *m* tangentiel; sciage sur dosse
r тангенциальный распил *m*; тангенциальный разрез *m*

* **tangential expansion → 8387**

8384 **tangential grain densified wood**
d tangentialfaseres Presseschichtholz *n*
f bois *m* densifié tangentiellement
r слоистый древесный пластик *m* с тангенциальным расположением волокон

8385 **tangential grain inclination**
d Tangentialfaserneigung *f*
f inclinaison *f* du fil tangentiel
r тангенциальный наклон *m* волокон

8386 tangential shrinkage
d tangentiale Schwindung *f*
f retrait *m* tangentiel
r тангенциальная усушка *f*

8387 tangential swelling; tangential expansion
d tangentiale Quellung *f*; Tangentialschwellen *n*
f gonflement *m* tangentiel
r тангенциальное разбухание *n*

* **tangerine → 3948**

* **tangle → 8389**

8388 tangled stems
d verwickelnde Stengel *mpl*; umgebogene Stengel
f tiges *fpl* embrouillées; tiges courbées
r сплетённые стволы *mpl*

8389 tangled vegetation; tangle
d verwickelnde Vegetation *f*
f végétation *f* embrouillée
r сплетённая растительность *f*; густая заросль *f*

8390 tanglefoot
(a sticky compound placed around the base of trees, etc., to prevent insects from climbing and damaging the plant)
d Fliegenpapier *n*
f papier *m* adhésif
r липучая бумага *f*

8391 tan growth
d Gerbstoffstau *m*
f gonflement *m* tannant
r дубильный нарост *m*

* **tank → 7942**

8392 tanner's bark; ting bark; tanning bark; tanbark
(the bark of certain oak trees, especially the common oaks, used for tanning)
d Rindengerbmittel *n*; Rindengut *n*; Gerbrinde *f*; Lohrinde *f*; Gerblohe *f*
f écorce *f* riche en tan[n]in; écorce tannante
r дубильная кора *f*; дубильное корьё *n*; кора, богатая танином

8393 tannin
(astringent substance of different species of trees used in converting hide to leather)
d Tannin *n*; Gerbstoff *m*; Gerbsäure *f*
f tan[n]in *m*
r дубильное вещество *n*

8394 tanoak
(Lithocarpus densiflorus)
d Südeiche *f*
f pasanie *f*
r камнеплодник *m* густоцветковый; пазания *f* густоцветковая

8395 tan root
d Gerbwurzel *f*
f racine *f* tannante
r корень *m*, содержащий дубильные вещества

8396 tan speckle
d Lohfleck *m*
f tache *f* tannante
r дубильное пятно *n*

* **tapana tree → 6237**

8397 tape-line
d Messschnur *f*; Messleine *f*
f ficelle *f* de cubage; ruban *m* de cubage
r мерная тесьма *f*

* **tape measure → 8165**

8398 tape measurement; taping; chaining
(using a surveyor's chain or tape for linear measurements along the ground)
d Bandmessung *f*
f mesurage *m* à ruban
r измерение *n* мерной лентой

* **taper → 8083**

8399 taper coefficient
d Ausbauchungszahl *f*; Kennzeichen *n* für Abholzigkeit
f coefficient *m* de conicité
r коэффициент *m* сбега

8400 taper curve
d Abholzigkeitskurve *f*
f courbe *f* de défilement; courbe de décroissement
r кривая *f* сбежистости

8401 tapered; conical; fastigate *adj*
(having a narrow crown with ascending branches)
d spitz zugelaufen; schlank zulaufend; abholzig
f conique; effilé; fastigié; à forte décroissance
r сбежистый; суживающийся к вершине; конусообразный; конический;

8402 taper-edged timber; tapered wood
d Brechholz *n*
f bois *m* scié biaisé; bois défilé
r сбежистый лесоматериал *m*; пиломатериал, обрезанный по сбегу

8403 tapering
d Reduzieren *f*; Zuspitzung *f*
f diminution *f*; décroissance *f*
r сужение *n*; заострение *n*; придание *n* клиновидной формы

* **tapering → 482**

* **tapering growth → 1872**

8404 tapering mechanism; taper set
d Sprengvorrichtung *f*
f dispositif *m* d'alignement
r механизм *m* выравнивания брёвна; механизм установки брёвна в горизонтальное положение

8405 taperingness; degree of tapering; fall-off
(of a tree stem)
d Konizität *f*; Abholzigkeit *f*; Durchmesserabfall *m*; Durchmesserabnahme *f*
f conicité *f*
r сбежистость *f*; уменьшение *n* диаметра от комля к вершине

8406 tapering tree
d abförmiger Baum *m*; konischer Baum
f arbre *m* à forte décroissance
r сбежистое дерево *n*

8407 taper sawing
(ripping a log by making cuts parallel to the outside surface rather than roughly parallel to the longitudinal axis)
d konischer Tangentialschnitt *m*
f sciage *m* tangentiel exactement dans le fil
r пиление *n* по сбегу

8408 taper sawn
d konisch geschnitten
f scié tangentiel
r распилённый по сбегу

* **taper set → 8404**

8409 taper table
d Abholzigkeitstafel *f*
f table *f* de défilement; table de décroissement
r таблица *f* сбега

* **tapped tree → 959**

* **tapping → 828**

* **tapping [cut] → 4082**

8410 tapping intensity
d Saftschälenintensität *f*
f intensité *f* de gemmage
r интенсивность *f* подсочки; нагрузка *f* каррами (ствола)

8411 taproot; major root; main root
(main root of a tree which strikes downward with or without heavy branching)
d Hauptwurzel *f*; Herzwurzel *f*; Pfahlwurzel *f*; Stützwurzel *f*
f racine *f* pivotante; pivot *m*; pivotante *f*; racine principale
r главный корень *m*; стержневой корень; основной корень

8412 tapura family; dichapetalum family
(Dichapetalaceae; Chailletiaceae)
d Spindelbaumgewächse *pl*
f dicots *pl*
r дихапеталовые *pl*

8413 target growing stock
d Bestockungsziel *m*
f but *m* [de composition] du peuplement
r плановый запас *m* леса на корню

* **target size → 3163**

8414 target stocking standards
(number of well-spaced acceptable trees per hectare that will, in normal circumstances, produce an optimum free-growing crop)
d Zielbestockungsstandard *mpl*
f standards *mpl* de peuplement objectif
r расчётные стандарты *mpl* насаждения

* **tariff on timber → 2681**

8415 tarif[f] table [of timber]
d Tariftabelle *f*
f tarif *m* [de cubage] gradué
r таблица *f* лесной таксы (попённой платы)

* **tarring → 6895**

8416 tar-spot disease
d Runzelschorf *m*

8417 Tasmanian blue eucalyptus; blue gum
(Eucalyptus globulosus; Eucalyptus globulus; Eucalyptus gigantea)
d Eisenveilchenbaum *m*; blauer Gummibaum *m*; Blaugummibaum *m*; Fieber[heil]baum *m*
f eucalyptus *m* globuleux; gommier *m* bleu de Tasmanie; gommier bleuâtre; arbre *m* à la fièvre
r эвкалипт *m* шаровидный; эвкалипт шариковый; эвкалипт круглый

* **Tasmanian false beech → 406**

* **Tasmanian myrtle → 406**

* **Tasmanian oak → 4054**

* **Tasmanian sassafras → 7864**

* **Tasmanian whitewood → 6315**

* **tatajuba → 478**

8418 Tatarian maple
(Acer tataricum)
d tatarischer Steppenahorn *m*; tatarischer Ahorn *m*; Tatarenahorn *m*
f érable *m* de Tatarie
r клён *m* татарский; чёрноклён *m*

* **taukkyan → 9719**

* **taungya method → 146**

8419 taungya plantation
(forest crop in conjunction with a temporary agricultural crop)
d Waldfeld-Plantation *f*
f plantation *f* en taungya
r лес *m* с временным сельскохозяйственным пользованием

8420 tawhai; [New Zealand] red beech; hard beech
(Nothofagus fusca)
d Tawhai *m*; neuseeländische Rotbuche *f*
f tawhai *m*; hêtre *m* de Nouvelle Zélande
r бук *m* красный новозеландский

8421 Tawhiwhi pittosporum
(Pittosporum tenuifolium)
d schmalblättrige Klebsame *f*
f pittospore *m* de Nouvelle Zélande
r смолосемянник *m* тонколистный

8422 tax on forest
d Waldbesteuerung *f*
f impôt *m* dont on grève la forêt
r обложение *n* леса налогом

* **taxator → 3410**

8423 taxoc[o]enose; taxocene
d Taxocenosis *f*
f taxocénose *f*
r таксоценоз *m*

8424 taxodioid pit
(a crossfield pit in early wood, with a large, ovoid to circular, included aperture that is wider than the lateral space on either side between the aperture and the border)
d taxodioidische Tüpfelung *f*
f ponctuation *f* taxodioïde
r таксодиоидная поровость *f*

8425 taxodium family
(Taxodiaceae)
d Sumpfzypressengewächse *pl*
f taxodiacées *pl*
r таксодиевые *pl*

* **taxon [of vegetation] → 8427**

8426 taxonometric
d taxonometrisch
f taxonométrique; taxinométrique
r таксонометрический

8427 taxonomic[al] unit; taxon [of vegetation]
(a group or category, at any level, in a system for classifying plants or animals)
d taxonomische Einheit *f*; Taxon *m*
f taxon *m*
r таксон *m*; таксономическая единица *f*

8428 taxonomist
d Systematiker *m*; Taxonom *m*
f taxonomiste *m*; taxinomiste *m*; taxologue *m*
r таксоном *m*

8429 taxonomy
(science of classifying organisms)
d Taxonomie *f*
f taxonomie *f*; taxinomie *f*
r таксономия *f*

8430 tax on the clearing of woodland or forest
d Steuer *f* auf die Rodung von Flächen mit Holz- oder Waldbestand
f taxe *f* sur le défrichement des surfaces en

8431 taxus family
(Taxaceae)
d Eibengewächse *pl*
f taxacées *pl*
r тисовые *pl*

* **tcf → 8530**

8432 tchitola
(Oxystigma oxyphyllum; Pterogypodium oxyphyllum; Oxystigma mannii)
d Tchitola[-Baum] *m*
f tchitola *m*
r китола *f*; читола *f*

8433 teaberry; wintergreen
(Gaultheria)
d Scheinbeere *f*
f gaulthérie *f*; thé *m* des bois
r гаультерия *f*

* **teaching in forestry → 6565**

8434 tea family
(Theaceae)
d Teegewächse *pl*
f théacées *pl*
r чайные *pl*

8435 teak forest
d Teakholzbaumwald *m*; Djatiwald *m*
f teckeraie *f*
r тиковый лес *m*

8436 teakwood
d Teakholz *n*
f bois *m* de teck
r тиковая древесина *f*

* **tearing open → 2003**

8437 tea tree; tea shrub
(Melaleuca alternifolia)
d echter Teeölbaum *m*; Teebaum *m*; Teestrauch *m*; Holztee *m*
f arbre *m* à thé; melaleuca *m*
r лептоспермум *m*; чайный куст *m*; древесный чай *m*

8438 teatree oil
d Teebaumöl *n*
f huile *f* d'arbre à thé
r чайное масло *n*; масло чайного дерева

* **tecate cypress → 3314**

d technische Schlagbarkeit *f*
f exploitabilité *f* technique
r техническая спелость *f*

* **technical forester → 3579**

8440 technical rotation
d technischer Umtrieb *m*
f âge *m* d'exploitabilité technique
r оборот *m* рубки по технической спелости

8441 tee halving
d T-Aufblatten *n*
f enture *f* à mi-bois en T
r врубка *f* в Т-образном виде

* **teeth set → 821**

8442 telescopic [boom] extension; boom extension
d [teleskopische] Spritzrohrverlängerung *f*
f rallonge *f* [télescopique] de flèche; rallonge télescopique
r выдвижение *n* [телескопической] стрелы

* **telling of trees → 2837**

8443 temperate forest
(the woodland of rather mild climatic areas, composed mainly of deciduous trees)
d Wald *m* in der gemässigten Zone
f forêt *f* tempérée
r лес *m* умеренной зоны

8444 temperate forest-steppe zones
d gemäßigte Steppenwaldzonen *fpl*
f zones *fpl* de forêts steppiques tempérées
r лесостепные зоны *fpl* умеренного пояса

8445 tempered hardboard
(dense fiberboard that has been specially treated to increase its durability, strength, density, and moisture resistance)
d temperierte Holzfaserplatte *f*
f panneau *m* dur trempé; panneau de fibres extra-dur
r темперированная твёрдая древесноволокнистая плита *f*

8446 tempo of cutting
d Hiebstempo *n*
f tempo *m* d'abattage
r темп *m* рубки

8447 temporal forest
d vorübergehender Wald *m*

8448 **temporary access structure**
(access structure out a forest development plan, access management plan, logging plan, road permit or silviculture prescription)
d vorübergehende Zugriffstruktur *f*
f structure *f* d'accès transitoire
r временная структура *f* доступа

8449 **temporary bridge**
(bridge having most of its major components constructed of untreated wood)
d vorübergehende Brücke *f*
f pont *m* provisoire
r временный мост *m*; переправа *f*

8450 **temporary [land] tenure**
(non-alienated land on which the timber is alienated to private interests)
d vorübergehende Besitzverhältnisse *f*
f tenure *f* temporaire
r временное землевладение *n*

8451 **temporary mixture**
d vorübergehende Mischung *f*
f mélange *m* transitoire
r временное смешение *n*

* **tenar → 1850**

8452 **tender root**
d weiche Wurzel *f*
f racine *f* molle
r тонкий корень *m*; нежный корень

* **tending cutting → 2121**

8453 **tending of [forest] plantation**
d Kulturpflege *f*
f soins *mpl* aux plantations
r агротехнический уход *m* за лесными культурами

8454 **tending of individual trees; tree-care**
d Einzelstammpflege *f*
f soins *mpl* culturaux arbre par arbre
r уход *m* за отдельными деревьями

* **tending of new growth → 8458**

8455 **tending of sapling**
d Dickungspflege *f*
f soins *mpl* aux fourrés; soins aux gaules
r уход *m* за саженцами

stands; stand tending
(variety of forest management treatments, including spacing, fertilization, pruning, and commercial thinning)
d waldbauliche Behandlung *f*; Bestandespflege *f*; Waldpflege *f*; Bestandeserziehung *f*
f soins *mpl* culturaux au peuplement
r уход *m* за лесонасаждениями; уход за лесом

8457 **tending of the increment**
d Zuwachspflege *f*
f soins *mpl* favorisants l'accroissement
r уход *m* за приростом

8458 **tending of young stand; tending of new growth; young growth tending**
(operation comprising cleanings and thinnings)
d Jungwuchspflege *f*
f soins *mpl* culturaux de jeunesse; coupe *f* d'entretien
r уход *m* за молодняком; рубка *f* ухода в молодых насаждениях; рубка ухода в молодняках

8459 **tending path**
d Pflegegasse *f*
f piste *f* de soins culturaux
r тропа *f* ухода

8460 **tending rotation**
d Pflegeumlauf *m*; Pflegewiederkehr *f*
f périodicité *f* de soins culturaux
r оборот *m* ухода

8461 **tending unit**
d Pflegeeinheit *f*
f unité *f* d'intervention
r единица *f* ухода

8462 **tendril**
(slender, coiling, modified leaf or stem of climbing plants, such as peas, that supports the plant by attaching it to surrounding surfaces)
d Klammer *f*; Ranke *f*
f vrille *f*; crampon *m*
r усик *m*; усообразный придаток *m*

8463 **tensile densified wood**
d Plast-Pressschichtholz *n*
f bois *m* densifié de traction

8464 tensiometer
(a measuring instrument for measuring the tension in a wire or fiber or beam, the moisture content of soil or the surface tension of a liquid)
d Festigkeitsprüfer *m*
f tensiomètre *m*
r тензиометр *m*

* **tension cable → 1164**

8465 tension failure
d Zugbruch *m*
f rupture *f* par traction
r разрушение *n* при растяжении; разрыв *m* при растяжении

* **tension member → 8542**

* **tension rod → 8542**

8466 tension skidding
d Zugrücken *n*
f débardage *m* par tension
r трелёвка *f* канатной установкой с тягово-несущим канатом

8467 tension wood
(reaction wood that forms on the upper side of a leaning hardwood tree)
d Zugholz *n*
f bois *m* de tension
r тяговая древесина *f* (лиственных пород)

* **tent caterpillar → 3618**

8468 tenure management plan
(plan that relates to the management, development and use of the land to which a licence or permit is made appurtenant)
d Besitzverhältnisse-Verwaltungsplan *m*
f plan *m* de gestion de tenure
r план *m* управления землевладением

* **tepeguaje → 3933**

8469 terebinth pistache
(Pistacia terebinthus)
d Terpentinpistazie *f*
f pistachier *m* térébinthe
r теребинт *m*; терпентиновое дерево *n*; терпентинное дерево

* **terentang → 7155**

* **terminal boom → 8136**

d Endknospe *f*; Gipfelknospe *f*; Spitzenknospe *f*; Terminalknospe *f*
f bourgeon *m* terminal
r верхушечная почка *f*; вершинная почка; терминальная почка; конечная почка

8471 terminal delivery area
d Auslaufstrecke *f*
f voie *f* de réception; voie de garage
r пробег *m* лесоспуска

8472 terminal feeders
d Gipfelsprossschädlinge *mpl*
f ravageurs *mpl* des pousses
r вредители *mpl* побегов; вредители почек

8473 terminal hypertrophy
(the abnormal terminal elongation of the stem of conifers without the development of side branches)
d Fuchsschwanz *m*
f hypertrophie *f* terminale
r верхушечная гипертрофия *f* (о ненормально вытянутой верхушке дерева хвойной породы)

8474 terminal leader; terminal shoot; top shoot
d Gipfelspross *m*; Terminalspross *m*; Gipfeltrieb *m*; Längstrieb *m*
f pousse *f* terminale
r верхушечный побег *m*; вершинный побег

8475 terminal parenchyma; marginal apotracheal parenchyma
(apotracheal parenchyma cells occurring either singly or forming a more or less continuous layer of variable width at the close of a season's growth)
d terminales Parenchym *n*
f parenchyme *m* terminal
r верхушечная паренхима *f*

* **terminal shoot → 8474**

8476 termite
(Reticuliterms luciphydus Rossi)
d Termit *m*
f termite *m*
r термит *m*

8477 ternate leaf
(compound leaf divided into three, resulting in three leaves on one rachis.)
d Fingerblatt *n*

8478 terrain hazard assessment
(assessment or characterization of unstable or potentially unstable slopes on forested lands)
d Abschätzung *f* der Terraingefahren
f évaluation *f* des dangers de terrain
r оценка *f* опасности от нестабильных грунтов

* **tesota → 1002**

* **test area → 7173**

8479 test band
d Probestreifen *m*
f bande *f* d'essai
r пробная полоса *f*

* **test cutting → 7171**

* **test felling → 7171**

8480 test stem; sample stem
d Probestamm *m*
f tige *f* d'essai
r контрольное бревно *n*

* **test tree → 7174**

* **tether line → 8543**

8481 tetracentron family
(Tetracentraceae)
d Tetracentrongewächse *pl*
f tetracentracées *pl*
r тетрацентровые *pl*

8482 Texas ash; mountain ash
(Fraxinus texensis)
d Texas-Esche *f*
f frêne *m* de Texas
r ясень *m* техасский

8483 Texas buckeye
(Aesculus arguta)
d Texas-Rosskastanie *f*
f marronnier *m*
r каштан *m* конский острый

8484 Texas cedarwood
(Juniperus texana)
d Texas-Zederholz *n*
f bois *m* de cèdre du Texas
r можжевельник *m* техасский

* **Texas mountain laurel → 1939**

d Texas-Pflaume *f*
f plaqueminier *m* de Texas
r персимон *m* техасский

* **Texas root rot → 1981**

* **Texas walnut → 5009**

* **texture of grain → 9678**

* **TFL → 8835**

* **Thailand rosewood → 7563**

* **theft of wood → 8052**

8486 theoretical crosscut point
d theoretischer Querschneidenpunkt *m*
f point *m* de tronçonnement théorique
r условная метка *f*; расчётная метка (место, на котором долготьё или хлыст визуально оценено для перепиливания с целью сортировки)

8487 theory of forest management
d Waldbautheorie *f*
f théorie *f* de gestion forestière
r теория *f* лесоустройства; теория управления лесами

8488 theory of forest taxation; theory of forest valuation; theory of forest mensuration
d Waldtaxationslehre *f*
f cours *m* d'estimation forestière
r учение *n* о лесной таксации

8489 thermal conductivity of wood
(the amount of heat that will flow in 1 hr through a homogeneous material 1inch thick and 1 ft square per 1°F temperature difference between the surfaces)
d Holzwärmeleitfähigkeit *f*
f conductivité *f* thermique de bois
r теплопроводность *f* древесины

8490 thermal cover
d thermische Wärmeschutzdeckung *f*
f couverture *f* thermique
r термозащитное покрытие *n*; термозащитный покров *m*

8491 thermophilous; thermophilic
d thermophil
f thermophile
r термофильный

f thermophyte *m*
r теплолюбивое растение *n*

* **thick-barked → 590**

* **thick-branched → 8135**

8493 **thick-butted; bottle-butted; swell-butted; churn-butted**
d mit trompetenförmigem Wurzelanlauf
f à fort empattement; à empattement en forme de fond de bouteille
r с трубкообразной закомелистости; с сильной закомелистости; с бутылкообразной закомелистости

* **thick-butted tree → 5129**

* **thick end → 607**

8494 **thickening**
d Verdichtung *m*; Dichterwerden *n*
f épaississement *m*
r утолщение *n*; загущение *n*; уплотнение *n*

* **thickening of the butt → 1234**

8495 **thicket stage**
d Gebüsch-Stadium *n*; Buschwald-Stadium *n*
f stade *m* de fourré
r стадия *f* молодого леса

* **thickly wooded → 6936**

8496 **thickness**
d Dicke *f*
f épaisseur *f*
r толщина *f*

8497 **thickness ga[u]ge**
d Dickenmesser *m*
f compas *m* d'épaisseur
r измеритель *m* толщины; толщиномер *m*

8498 **thickness of bark; bark thickness**
d Rindendicke *f*; Rindenstärke *f*
f épaisseur *f* de l'écorce
r толщина *f* коры

8499 **thickness stages**
d Dickestufen *fpl*
f stades *mpl* d'épaisseur
r ступени *fpl* толщины

8500 **thickness tolerance**
d Dickentoleranz *f*; Stärkentoleranz *f*

8501 **thick planting**
d dichte Pflanzung *f*
f ensemencement *m* dense
r уплотнённый посев *m*; уплотнённая посадка *f*; загущённая посадка

* **thick pole → 4803**

8502 **thick stuff**
d dicker Balken *m*
f billot *m* épais; poutre *f* épaisse
r толстый брус *m* (толщиной более 114.3 мм)

* **thick timber → 4808**

* **thick wood → 4808**

* **thimble-berry → 782**

8503 **thin** *v*
d auslichten; durchforsten; vereinzeln; verziehen; verhacken
f éclaircir; démarier
r прореживать; прорывать

* **thin-crowned → 7881**

* **thin forest → 4446**

* **thinleaf alder → 5579**

* **thinleaf huckleberry → 798**

* **thin-leaved stringybark → 7769**

* **thinly stocked → 6450**

8504 **thinned[-out] crown; lighted crown**
d gelichtete Baumkrone *f*
f cime *f* éclairée
r изреженная крона *f* [дерева]

8505 **thinner; dissolvent; dissolver**
d Verdünnungsmittel *n*; Lösungsmittel *n*
f diluant *m*; solvant *m*
r разбавляющее вещество *n*; разбавитель *m*; разжижитель *m*

* **thinner → 8513**

8506 **thinning agents; thinnings**
d Durchschlag *m*
f produits *mpl* d'éclaircissage
r лесоматериалы *mpl*, заготовленные при

8507 thinning area
d Durchforstungsfläche *f*
f aire *f* d'éclaircie
r площадь *f* рубок ухода

8508 thinning cell
d Durchforstungszelle *f*
f plus petite surface *f* unitaire d'éclaircie; élément *m* unitaire de surface d'éclaircie
r ячейка *f* прорубки

8509 thinning cycle; thinning interval; thinning period
(time interval between thinnings in the same stand)
d Durchforstungsperiode *f*
f cycle *m* d'éclaircie
r период *m* повторяемости рубок ухода

* **thinning degree → 8519**

8510 thinning frequency
d Durchforstungsturnus *m*
f périodicité *f* des éclaircies
r повторяемость *f* рубок ухода

* **thinning from above → 8512**

* **thinning grade → 8519**

* **thinning intensity → 4527**

* **thinning interval → 8509**

8511 thinning in the lower story; low thinning; thinning from below; ordinary thinning; ground-thinning; underthinning
(removal of trees from the lower crown classes to favor those in the upper crown classes)
d Niederdurchforstung *f*; Durchforstung *f* im Beherrschten
f éclaircie *f* par le bas
r низовое прореживание *n*; рубка *f* ухода по низовому методу; прорубка *f* подчинённого яруса; низовая проходная рубка

8512 thinning in the upper story; thinning the dominant trees; thinning from above; high thinning; overthinning; crop-tree thinning
(lopping, felling, or herbicidal treatment of selected dominant trees)
d Hochdurchforstung *f*; Oberholzhieb *m*; Durchforstung *f* im Herrschenden
f éclaircie *f* par le haut
r верхнее прореживание *n*; прореживание *n* в верхнем ярусе насаждения; вырубка *f* деревьев господствующего класса

8513 thinning machine; plant thinner; thinner
d Vereinzelungsmaschine *f*; Ausdünner *m*; Verdünner *m*; Auslichter *m*
f éclaircisseuse *f*; [pré]démarieuse *f*
r прореживатель *m*

8514 thinning of seedlings
d Samendünnung *f*
f dépressage *m* des semis
r прореживание *n* посевов

8515 thinning [out]; stand depletion; loosening-out
(cutting made in an immature crop or stand primarily to accelerate diameter increment but also, by suitable selection, to improve the average form of the trees that remain)
d Durchforstung *f*; Ausbau *m*; Freihauung *f*; Lockerung *f*; Durchlichtung *f*; Vereinzelung *f*; Verlichtung *f*; Bestandes[auf]lockerung *f*
f dégagement *m*; dépressage *m*; extraction *f* du peuplement en éclaircie; épuisement *m* de peuplement; éclaircissage *m*; éclaircie *f*; démariage *m*
r прореживание *n* [насаждения]; разреживание *n*; изреживание *n*; прорывание *n*; продергивание *n*

* **thinning period → 8509**

8516 thinning regime
(term comprising the type, degree, and frequency of thinning for a given area)
d Durchforstungsregime *n*
f mode *m* d'éclaircie
r система *f* прореживания; организация *f* рубок ухода

* **thinning reserves → 4754**

* **thinnings → 8506**

* **thinning schedule → 8518**

8517 thinning series
(two or more adjacent forest plots that are thinned differently)
d Durchforstungsfolge *f*; Durchforstungsreihe *f*
f suite *f* d'éclaircies comparées
r серия *f* рубок ухода

* **thinning stool-shoots → 7409**

f table *f* d'éclaircies
r таблица *f* норм разреживания насаждений; план *m* рубок ухода

* **thinning the dominant trees → 8512**

8519 **thinning weight; thinning grade; severity of thinning; thinning degree; strength of thinning**
(degree of thinning expressed in terms of the volume removed at any one time)
d Durchforstungsgrad *m*; Durchforstungsstärke *f*
f intensité *f* du passage en éclaircie; intensité des éclaircies; dosage *m* des éclaircies; degré *m* d'intensité d'éclaircie
r вес *m* выборки при рубках ухода; объём *m* выборки при рубках ухода; степень *f* прорубки

* **thinning yield → 9782**

8520 **thin plank**
d Dünnbrett *n*
f planche *f* mince
r тонкая доска *f*; тесина *f*

* **thin pole → 7775**

8521 **thin shortwood**
d Dünnkurzholz *n*; Nutzreiser *npl*
f bois *m* court mince
r тонкомерное коротьё *n*; поделочный хворост *m*; топорник *m*

8522 **thin stem**
d Dünnstamm *m*
f tige *f* fine
r тонкомерный ствол *m*

8523 **thin [wood]**
d Dünnholzblock *m*
f billot *m* mince
r тонкий брус *m* (толщиной до 100 мм)

8524 **thorn; spine**
(a modified plant organ, especially a stem, that is stiffened and terminates in a pungent point)
d Dorn *m*
f piquant *m*; épine *f*
r шип *m*; колючка *f*

* **thorn-apple tree → 4174**

8525 **thorn bush**
r дёрн *m*

8526 **thorn forest**
d Dornwald *m*
f forêt *f* épineuse
r колючий лес *m*; колючее редколесье *n*

8527 **thornless honey locust**
(Gleditsia triacanthos var. inermis)
d dornenlose Gleditschie *f*
f févier *m* sans épines
r гледичия *f*, лишённая колючек

8528 **thorn tree; stinger [tree]**
d Dornenbaum *m*; stachliger Baum *m*
f arbre *m* épineux
r колючее дерево *n*

* **thorny → 7913**

* **thoroughwort → 866**

8529 **thousand board feet measure; MBF**
(symbol indicating 1000 board feet of wood fiber volume either in log form or after conversion into lumber)
d Brettervermessung *f* im Tausendboardfuß
f cubage *m* en milliers de pied-planches
r измерение *n* лесоматериалов в тысячах квадратных футов

8530 **thousand cubic feet; tcf**
d Tausendkubikfuß *m*
f millier *m* de pieds cubes
r тысяч кубических футов

8531 **threatened plant communities; endangered plant communities**
(ecosystems that are restricted in their distribution over a natural landscape or to a specific geographic area or a particular type of local environment)
d bedrohte Pflanzengesellschaften *fpl*
f communautés *fpl* végétales menacées
r угрожаемые растительные сообщества *npl*; растительные сообщества, находящиеся под угрозой [исчезновения]

* **threatened species → 2801**

8532 **three-edged cant**
d dreikantiger Balken *m*
f équarri *m* trilatéral
r трёхкантный брус *m*

8533 **three-flower maple**

f érable *m* à trois fleurs
r трёхцветковый клён *m*

* **three-leaved hoptree → 6035**

8534 three-line running skyline system
d Dreileitungsluftseilsystem *n*
f système *m* à trois câbles porteurs
r канатная установка *f* с тягово-несущим и вытяжным канатами

8535 three-lobed leaf
d zweigelapptes Blatt *n*
f feuille *f* trilobée
r трёхлопастный лист *m*

* **three-thorn[ed] acacia → 4353**

8536 thrifty stand; flourishing stand; thriving stand
d frohwüchsiger Bestand *m*; wuchsfreudiger Bestand
f peuplement *m* bienvenant
r дружнорастущее насаждение *n*

* **through-and-through sawing → 5016**

8537 through check; through shake
(split which crosses through a log at its diameter.)
d durchgehender Riss *m*
f fente *f* traversante; fissure *f* diamétrale
r сквозная трещина *f*

* **throughfall → 2626**

8538 throughfall sampler
d Bestandesniederschlagssammler *m*; Kronendurchlass[sammler] *m*
f collecteur *m* de précipitations
r отборщик *m* капельных проб

8539 through inbark
d durchgehender Rindeneinschluss *m*
f entre-écorce *f* transitoire
r сквозная врастающая кора *f*

* **through shake → 8537**

* **through swirl → 8789**

* **throw → 9314**

* **throw[ing] → 8824**

* **thuja → 8893**

* **Thunberg pine → 4614**

* **thunder shake → 4946**

* **thuya → 8893**

8540 t[h]ylosoid
(a proliferation of a thin-walled epithelial cell into an intercellular canal)
d Thylosoid *n*
f thylloïde *m*
r тиллоид *m*

8541 tiama; gedu nohor
(Entandrophragma angolense; Entandrophragma congoense)
d Tiama *n*
f tiama *m*
r энтандрофрагма *f* ангольская; красное дерево *n* из Тиамы

* **tibourbou → 5534**

* **tidal forest → 5280**

* **tideland spruce → 7678**

8542 tie; tension member; tension rod
(a structural member resisting tension forces along the grain)
d Zugstab *m*; gezogener Bauteil *m*; Zuganker *m*
f barre *f* tendue; membre *m* tendu; barre de tension; élément *m* tendu
r лежень *f*; перекладина *f*

8543 tie-down line; tether line
(in balloon logging)
d Anbindungskabel *n*
f câble *m* d'attache
r крепёжный канат *m* аэростата

* **tier → 8545**

8544 tie ridge
(narrow strip of ground left unexcavated so as to break the horizontal continuity of the trenching)
d Damm *m* mit Querkämmen
f billon *m* cloisonné
r нарезка *f* борозд

* **tiering → 8147**

8545 tier of logs; tier; round; course
(a single layer of stacked timber; in rafting, the group of parallel logs which are stowed in

f lit *m* [de grumes]; strate *m* de grumes; couche *f* de grumes; plan *m* de grumes
r ярус *m* брёвен; слой *m* брёвен; ряд *m* брёвен

8546 tier pile
(a multiple layer of stacked timber, generally separated from adjacent layers by stickers)
d Kastenstapel *m*
f pile *f* de lits
r штабель *m* широких досок одинаковой ширины

8547 tigertail spruce
(Picea polita)
d Rosenfichte *f*; Tigerschwanz-Fichte *f*
f épinette *f* du Japon
r ель *f* изящная; ель японская

8548 tigerwood; ash-leaved star tree; zorrowood
(Astronium fraxinifolium; Astronium gracile; Astronium graveolens)
d Goncalo Alves *m*; Urunday *m*
f astronium *m* à feuilles de frêne; bois *m* de zèbre
r астрониум *m* ясенелистный; гонкало *n*; тигровое дерево *n*

8549 tight cooperage; wet cooperage
(cooperage used as containers for liquids, semisolids, and heavy solids)
d feuchte Böttcherei *f*
f tonnellerie *f* étanche
r заливные бочки *fpl*

8550 tight knot
(that portion of a branch which is embedded in the tree trunk while the branch is alive)
d fester Ast *m*; festsitzender Ast; fester Knoten *m*
f nœud *m* fixe
r плотный сучок *m*; плотно сидящий сучок

8551 tightline system
d Drahtseilbahn *f* mit fixiertem Kabel; stationäre Luftseilbahn *f*
f système *m* de câble-grue à tension fixe
r канатная установка *f* с неподвижным несущим канатом

* **tight skyline → 8043**

* **tight split → 958**

8552 tight stack; close stack
d Feststapel *m*
f pile *f* serrée
r плотный штабель *m*; штабель *m* с плотной укладкой лесоматериалов

* **tight stacking → 1157**

8553 tile cell
(a special type of apparently empty upright ray cell of approximately the same height as the procumbent ray cells and occurring in indeterminate horizontal series usually interspersed among the procumbent cells)
d Ziegelzelle *f*
f cellule *f* palissadique [de rayon]
r трубчатая клетка *f*

* **tile tree → 667**

8554 tillage
(soil-stirring action to loosen soils and so benefit crops by supplying a suitable soil environment)
d Bodenbearbeitung *f*; Pflügung *f*
f travail *m* du sol
r обработка *f* почвы; пахота *f*; вспашка *f*

* **tiller → 7072, 7132**

8555 tiller; staddle
(erect or ascending stem that branches from the base of another or below the ground surface)
d Ausläufer *m*; Bestockungstrieb *m*; Lassreitel *m*; Reitel *m*; Lassreidel *m*; Reidel *m*; Lassreis *n*
f baliveau *m*; lais *m*
r подрост *m*; деревцо *n*, оставляемое в срубаемом лесу; отросток *m* от корня

* **tilling machine → 7072**

* **timber apron → 8654**

8556 timber assortment; wood assortment
d Holzsortiment *n*
f assortiment *m* ligneux; assortiment de bois; catégories *fpl* de bois usinés
r лесной сортимент *n*

8557 timber auction; public sale of timber
d Holzversteigerung *f*; Holzauktion *f*
f vente *f* de bois par adjudication public à la criée
r лесной аукцион *m*; лесные торги *mpl*

* **timber balance → 9486**

8558 timber base

d Holzbasis *f*
f base *f* de bois
r лесная база *f*

8559 **timber-based material; wood-base material**
d Holzwerkstoff *m*
f matériel *m* dérivé du bois
r древесный материал *m*

* **timber beast → 9654**

* **timberbelt → 7508**

* **timber bin → 8621**

8560 **timber blocking; blockhouse**
d Balkenbau *m*; Balkenwerk *n*
f fortin *m*
r шпальная клетка *f*; сруб *m* [из брёвен]; связка *f* из брусьев

* **timber bob → 5090**

8561 **timber bonds**
d Waldobligation *f*
f lettre *f* de crédit garantie par un domaine boisé
r лесная облигация *f*

8562 **timber borers; wood-destroying insects; timber-boring insects; carpenter moths; wood fretters; carpenterworms; timber worms**
(any one of numerous species of beetles whose larv[ae] bore in the wood, or beneath the bark, of trees)
d Holzbohrer *mpl*; Holzbohrkäfer *mpl*; Holzfresser *pl*; Holzzerstörer *pl*; Holzwürmer *mpl*; Nagekäfer *mpl*; Pochkäfer *mpl*
f perces-bois *mpl*; buprestes *mpl*; perceuses *fpl* de bois; vers *mpl* du bois
r дереворазрушающие насекомые *npl*; древоточцы *mpl*; древогрызы *mpl*; жуки-точильщики *mpl*

8563 **timber brought; split logs; cleaving timber; cleft timber; cloven logs; riven timber; split timber; cleft wood; cleaved wood**
d Scheitholz *n*; Spälterholz *n*; Spaltholz *n*; Scheit *n*
f bois *m* de fente
r колотый лесоматериал *m*

* **timber bunch → 9496**

8564 **timber buyer; wood buyer**
(person responsible for purchasing timber or logs for their company or business)
d Holzkäufer *m*
f acheteur *m* de bois
r покупатель *m* леса

* **timber carriage → 9395**

* **timber carrier → 5199**

8565 **timber-carrier driver**
d Hängeblockwagen *m*; Holzfrachter *m*
f conducteur *m* de triqueballe
r тележка *f* для леса; оборудование *n* для транспортировки леса

* **timber cherry → 244**

* **timber chute → 5116**

8566 **timber class**
(breakdown of inventory volume in live and salvable dead trees by tree size and tree class)
d Holzklasse *f*
f classe *f* de bois
r класс *m* лесоматериала

* **timber clips → 3892**

* **timber combine → 9516**

* **timber compartment → 1821**

8567 **timber compass; shotgun; gun[ning] stick**
(two-drum, live skyline yarding system used in uphill logging, in which the carriage moves down the skyline by gravity, is lowered to attach logs, and is then raised and pulled to the landing by the main line)
d Fällungskompass *m*; Fallrichtungsanzeiger *m*
f indicateur *m* de direction de chute
r указатель *m* направления валки; устройство *n* для определения направления падения дерева

* **timber concession → 3356**

8568 **timber connector**
(a metal ring or plate that, by being embedded in adjacent wood faces or in one wood face, acts in shear to transmit loads from one timber to another)
d Holzspreize *f*
f connecteur *m* de bois

8569 **timber conservation**
d Holzkonservierung *f*
f conservation *f* du bois
r консервирование *n* древесины

8570 **timber construction; building in wood; wooden structure**
d Holzkonstruktion *f*
f construction *f* en bois
r деревянная конструкция *f*

* **timber consumption → 9517**

* **timber contract → 2192**

8571 **timber crosscutting; crosscut[ting]; cross-grained cutting; crosswise cutting; sucking**
d Aufarbeiten *n* von Langholz; Holzablängen *n*; Querschneiden *n*; Querschnitt *m*; Zersplittern *n*
f coupe *f* transversale
r поперечный распил *m* [древесины]; поперечный [раз]рез *m*; рез при раскряжёвке

* **timber cruise → 3617**

* **timber cruiser → 3410**

* **timber cruising → 3617**

8572 **timber culvert**
d Holzdurchlass *m*; Holzrohr *n*
f ponceau *m* de bois
r деревянная [перепускная] труба *f*

* **timber cutter → 9526**

* **timber dam → 9550**

* **timber dealer → 2276**

* **timber deck → 8654**

8573 **timber deed; timber right; forest right; wood right; forest servitude**
(a legally recognized claim, possessed by a person, community or property, to a continuing share in the enjoyment of some or all of the benefits of the forest property of another)
d Holzberechtung *f*; Waldnutzungsrecht *n*; Holzrecht *n*; Holzservitut *n*
f droit *m* d'usage au bois; servitude *f* forestière; servitude de bois; afforestage *m*
r документ *m* на приобретение или

8574 **timber deficit area**
d Holzmangelgebiet *n*; Holzzuschussgebiet *n*
f zone *f* déficitaire en bois
r район *m* с недостаточным лесом

* **timber demand → 2362**

8575 **timber dependency**
(the percentage of all earnings in a county represented by timber-related earnings)
d Holzabhängigkeit *f*
f dépendance *f* de bois
r зависимость *f* леса

* **timber depot → 8664**

8576 **timber dimensions; sizes of timber**
d Holzdimensionen *fpl*
f dimensions *fpl* des bois
r размеры *mpl* древесины

8577 **timber dog; log dog; log bracket; log grab; chain dog; chain hook; chain sling; dragging dog; skidding dog; saddle**
(a hook on the end of a haul-up chain of a size to permit its being hooked into a link of the chain when the latter is looped around a log or other object)
d Klammerhaken *m*; Kettenhaken *m*; Kettensporn *m*
f clameau *m*; crampon *m*
r скоба *f* для прикрепления тяговой цепи (к бревну); зажим *m* для брёвен; скобочный крюк *m*; цепной крюк; каретка *f* бревнотаски

* **timber drag → 9395**

8578 **timber driving; lashing of booms; log running; log floating**
d [gebundene] Flößerei *f*; Flößung *f* des Holzes; Flößung *f* in Bündeln; Flößen *n*
f flottage *m* de bois [par train]; drave *f*
r сплав *m* [леса плотами]; [плотовой] лесосплав *m*

* **timber driving-out → 8595**

* **timber drug → 9395**

* **timber-dryer → 5203**

* **timber due → 8195**

8579 **timber dunnage; dunnage [wood]; stow wood**
(timber, generally of poor quality, used for

d Stauholz *n*; Abstandsholz *n*; Garnierholz *n*
f bois *m* accumulé; bois de fardage; bois d'arrimage; fardage *m*; bardis *m*
r заторный лес *m*; низкосортная древесина *f* в качестве прокладок при загрузке судна

8580 timbered
d gezimmert; zimmernd
f de bois
r бревенчатый; построенный из дерева

* **timber[ed] area → 9542**

8581 timber elevator; lumber elevator; stack elevator
d Stapel-Hebebühne *f*; Stapelvorrichtung *f*
f chariot *m* élévateur; transporteur-élévateur *m*
r подъёмный конвейер *m* для лесоматериалов

8582 timber engineering; wood engineering; wood technology
d Holztechnologie *f*
f technologie *f* de construction en bois
r технология *f* обработки древесины

* **timber estimation → 9562**

8583 timber exchange
d Holzbörse *f*; Holzhandelshof *m*
f échange *f* de bois
r лесная торговая биржа *f*; лесоторговая биржа

* **timber exploitation → 3414**

* **timber export → 2957**

* **timber extractor → 8597**

* **timber face → 8654**

* **timber famine → 7282**

8584 timber float; wood[en] float; raft; float; drive; crib
(logs fastened together to form a small raft, as in a storage pond)
d Baumflößung *f*; Holzdrift *f*; Trift *f*; Holzfloß *n*; Floß *n*; Gestör *n*
f flot *m* de bois; radeau *m*; train *m* de bois flotté; train de flottage
r лесосплав *m*; сплавной плот *m*; [лесосплавной] пучок *m* брёвен

* **timber for deals → 5204**

tree forest
d Baumholz *n*
f ancien *m*; vieille écorce *f*
r крупный лес *m*; бревенник *m*

8586 timber for masts; mast timber; spar timber
d Mast[en]holz *n*; Bemastungsholz *n*; Mittelbauholz *n*
f bois *m* à pilots; bois à poteaux
r мачтовый лес *m*; мачтовый лесоматериал *m*; древесина *f*, идущая на столбы и опоры; мачтовые брёвна *npl*; свайные брёвна; круглый строевой лес (диаметром не менее 20 см)

* **timber form-factor → 8015**

8587 timber for musical instruments
d Musikinstrumentenholz *n*
f bois *m* pour la fabrication d'instruments musicaux
r лесоматериал *m* для музыкальных инструментов

8588 timber for piling; pile-wood; piles
d Rammpfahlholz *n*; Pfahlholz *n*
f bois *m* à pieux; pieux *mpl*
r гидростроительные брёвна *npl*; гидролес *m*; свайный лес *m*

8589 timber for railway equipment
d Werkstättenhölzer *npl*
f bois *m* d'artisan; bois de voie ferrée
r древесина *f* для железной дороги

8590 timber for ship building; ship timber; naval construction wood
d Schiffsbauholz *n*
f bois *m* de marine
r корабельный лес *m*

8591 timber for sleepers; sleeper wood; sleeper blocks; timber for ties
d Schwellenholz *n*
f bois *m* à traverses
r шпальный кряж *m*; шпальный лесоматериал *m*

* **timber for the trade → 5400**

8592 timber for wag[g]on works; wag[g]on stock; wag[g]on building timber
d Waggonbauholz *n*
f bois *m* de wagons

8593 **timber-framed building**
d Balkenkonstruktion *f*; Holzbaukonstruktion *f*
f bâtiment *m* de bois
r здание *n* с деревянным каркасом

* **timber fungus → 9581**

* **timber gangsaw → 2042**

* **timber grading → 3902**

* **timber grower → 344**

* **timber growin→ 7610**

* **timber harvest → 9778**

* **timber harvesting → 9588**

8594 **timber harvesting land base**
(portion of the total area of a management unit considered to contribute to, and be available for, long-term timber supply)
d Holzerntelandbase *f*
f base *f* terrestre d'exploitation de bois
r земная база *f* для лесозаготовки

* **timber-harvesting operation → 3457**

* **harvesting operation → 3457**

8595 **timber haul[age]; timber hauling; [log] hauling; haulage; skidding; extraction of timber; transportation of wood; wood transport; timber driving-out**
(of logs from one point to another, usually from a landing to the mill or shipping point)
d Abfuhr *f* des Holzes; Holzabfuhr *f*; Ausrücken *n* des Holzes; Beförderung *f* des Holzes; Holzbringung *f*; Holzrücken *n*; Holztransport *m*
f débusquage *m*; débardage *m*; vidange *f* de bois; transportation *f* de bois; transport *m* des exploitations forestières
r трелёвка *f*; вывозка *f* лесных материалов; транспортировка *f* леса; вывозка *f* [леса]; лесотранспорт *m*

8596 **timber hauler**
d Holzaufzug *m*; Stammförderer *m*
f débusqueur *m* de bois; transporteur *m* des exploitations forestières
r лесотранспортёр *m* [для брёвен]

8597 **timber haulier; timber extractor**
r предприниматель *m* по вывозке

* **timber hauling → 8595**

8598 **timber-hauling apparatus; timber-hoist apparatus; hauling equipment**
d Holzabfuhrfahrzeug *n*; Holzfördermittel *n*; Bringungsmittel *n*
f moyen *m* de débardage
r лесотранспортный механизм *m*; лесотранспортное средство *n*; лесотаска *f* для брёвен

8599 **timber height**
(height of a tree to the point at which the stem breaks into branches)
d Holzhöhe *f*
f hauteur *m* de bois
r высота *f* [товарной части] дерева

* **timber hitchs → 3892**

* **timber-hoist apparatus → 8598**

* **timber industry → 3466**

8600 **timbering; timber supports; timberwork; wooden supports**
d Holzeinbau *m*; Holzausbau *m*; Zimmerung *f*; Verzimmerung *f*; Auszimmerung *f*; Abzimmerung *f*
f boisage *m*; étayage *m*; soutènement *m* de bois
r деревянная обшивка *f*; деревянная опалубка *f*

8601 **timber injurious to health**
d gesundheitsschädliches Holz *n*
f bois *m* nuisible
r вредная древесина *f*

8602 **timber inner face; inside face; timber lower face; lower face**
d Innenfläche *f* des Schnittholzes
f face *f* intérieure du bois
r внутренняя пласть *f* пиломатериала

* **timber in the rough → 7096**

* **timber in the round → 7113**

* **timber inventory → 3478**

8603 **timber jack; lifting jack; windlass**
d Hebewinde *f*
f cric *m* forestier; cric à tronçonner; treuil *m* de

8604 **timberland; wooded land; woodland; wooded country; timber region**
(forestland that is producing, or capable of producing, in excess of 20 cubic feet per acre per year of industrial roundwood products)
d Holzboden *m*; bewaldete Region *f*; Waldareal *n*; Waldland *n*
f région *f* boisée; parcelle *f* boisée
r лесные угодья *npl*; территория *f*, покрытая лесом; залесённая местность *f*; участок *m* строевого леса; лесистый участок

* **timberland owner → 3519**

8605 **timber lease; timber licence; forest licence**
(area-based tenures which revert to the government when merchantable timber on the area has been harvested and the land reforested)
d Holz[einschlags]lizenz *f*
f licence *f* forestière; licence d'abattage de bois
r лицензия *f* на заготовку древесины; разрешение *n* на заготовку и торговлю лесоматериалами

* **timber limit → 2761**

* **timber-line → 4982**

8606 **timberling**
d geringer Nutzholzstamm *m*
f grume *f* de bois d'œuvre
r тонкий поделочный хворост *m*

* **timber lining → 9521**

* **timber list → 8376**

8607 **timber loader; log loader; forestry loader; timber loading device; logger**
d Holzverlader *m*; Holzladevorrichtung *f*
f dispositif *m* de chargement de bois; grue *f* de manutention; grue de chargement; grue chargeuse
r устройство *n* для погрузки леса; лесопогрузчик *m*

8608 **timber loading; timber shipment**
d Aufladen *n* des Holzes; Holzladen *n*; Holzverladung *f*
f chargement *m* de bois
r погрузка *f* лесоматериалов; погрузка леса

8609 **timber loading crane; timber loading winch; loading hoist; logging crane; [spool] donkey; gypsy yarder; dolbeer**
d Holz[lade]kran *m*; Ladekran *m*; Verladewinde *f*; Waldbahnkran *m*
f treuil *m* de chargement de bois
r рельсовый лесопогрузочный кран *m*; погрузочная лебёдка *f*

* **timber loading device → 8607**

8610 **timber loading fork; timber-loading grapples**
d Ladegabel *f*; Gabellader *m*
f chargeur *m* à fourche
r вилочный лесопогрузчик *m*; вилообразный грузитель *m*; челюстной лесопогрузчик; погрузочный захват *m* для лесоматериалов

* **timber loading winch → 8609**

8611 **timber load line**
d Holzladungslinie *f*
f ligne *f* de charge pour bois en pontée
r линия *f* лесопогрузки

* **timber log → 5154**

* **timber lower face → 8602**

* **timberman → 9654**

* **timber management → 3488**

8612 **timber management compartment**
(small subdivision of forest area for purposes of orientation, administration, and silvicultural operations, and defined by permanent boundaries)
d Holzwirtschaftabteilung *f*
f compartiment *m* d'économie du bois
r управляемый лесоучасток *m*

* **timber management prescription → 7617**

8613 **timber market; wood market**
d Holzmarkt *m*
f marché *m* de bois
r лесной рынок *m*

8614 **timber market zones**
(geographic areas used as a basis for analyzing the economic effects of timber sale programs on national forests)
d Holzmarktzonen *fpl*
f zones *fpl* de marché de bois
r зоны *fpl* лесного рынка

8615 timber measure; lumber measure
d Holzmaß *n*
f mesure *f* pour le cubage des bois
r мера *f* объёма лесоматериалов

* **timber measurement list → 8376**

* **timber merchant → 2276**

8616 timber mining
d Holzmining *n*
f extraction *f* de bois sans régénération
r беспорядочное лесопользование *n*; лесопользование без восстановления лесов; эксплуатация *f* лесных богатств без восстановления

* **timber of broadleaved trees → 1054**

* **timber of superior quality → 4284**

8617 timber operability
(an economic suitability of timber for harvesting)
d Holzdurchführbarkeit *f*
f opérabilité *f* de bois
r пригодность *f* лесоматериалов к эксплуатации

* **timber[-]pass → 5116**

* **timber permit → 6183**

8618 timber physics; wood physics
d Holzphysik *f*
f physique *f* de bois
r физика *f* древесины

8619 timber pile; wood pile; pile; timber stack; wood stack; stack; bank
(an orderly arrangement of the pieces that facilitates handling, seasoning, inventory and storage)
d Holzstapel *m*; Stapel *m*; Reihe *f*; Gruppe *f*; Wall *m*
f empilage *m*
r штабель *m* лесоматериалов; копилка *f*

* **timber planking → 880**

* **timber plantation → 3528**

8620 timber plantation crop
d Holzplantationernte *f*
f plantation *f* de peuplements industriels
r деловые лесоматериалы *mpl*

d Holztasche *f*
f panier *m* collecteur
r карман-накопитель *m*

8622 timber point
d Holz[stamm]endpunkt *f*
f point *m* final de bois
r крайняя точка *f* товарной части ствола

* **timber preservation → 9627**

8623 timber-preserving means; wood-preserving means; wood preservers
d Holzschutzmittel *npl*
f moyens *mpl* de préservation du bois
r защитные средства *npl* для древесины

* **timber price → 9629**

* **timber processing → 9630**

* **timber produce → 5214**

8624 timber producing country
d Holzherstellerland *n*
f pays *m* producteur de bois
r страна-производитель *f* лесоматериалов

* **timber production → 3537**

8625 timber-productive land
(forestland that is capable of producing a merchantable stand within a reasonable length of time)
d produktiver Waldboden *m*
f terre *f* productrice de bois
r продуктивный лесоучасток *m*

8626 timber products; wood products; wood material
(roundwood products and byproducts)
d Holzwerke *npl*; Holzfabrikate *npl*
f produits *mpl* ligneux; ouvrages *mpl* en bois
r деревянные изделия *npl*

* **timber purchase → 9635**

* **timber quality → 6657**

8627 timber quantity
d Holzmenge *f*
f quantité *f* de bois
r количество *n* древесины

* **timber rafting → 912**

f guide-roues *m* en bois
r деревянные рельсы *fpl*

* **timber reconnaissance → 3549**

* **timber regeneration → 6821**

* **timber region → 8604**

* **timber removal → 6862**

* **timber reserves → 3559**

* **timber resistance → 2669**

8629 timber resources
d Holzressourcen *fpl*
f ressources *fpl* de bois
r лесоресурсы *mpl*

8630 timber resource system
(system which: 1. protect, improve, grow, and harvest timber from forestland; 2. protect, process, and utilize wood and wood-related products)
d Holzressourcensystem *n*
f système *m* de ressources forestières
r система *f* лесоресурсов

* **timber right → 8573**

* **timber rotation → 2194**

* **timber royalty → 7123**

8631 timber sale; sale of wood; wood sale
(issuance of a contract for cutting and removing designated trees)
d Holzverkauf *m*
f vente *f* de bois
r продажа *f* древесины; продажа леса

8632 timber sale licence
(defines a specific volume of timber to be harvested from a specific area)
d Holzverkauflizenz *f*
f licence *f* de vente de bois
r лицензия *f* на продажу древесины

8633 timber sale quantity
(volume of timber)
d Holzverkaufquantität *f*
f quantité *f* de bois de vente
r количество *n* продажи древесины

8634 timber sale schedule; sale schedule
by a forest plan)
d Holzverkauftabelle *f*
f table *f* de vente de bois
r каталог *m* продажей лесоматериалов

8635 timber sample; wood sample; wood specimen
d Holzmuster *n*; Holzprobe *f*
f échantillon *m* de bois
r образец *m* древесины; деревянная проба *f*

8636 timber saving
d Holzeinsparung *f*
f économie *f* de bois
r сбережение *n* древесины

8637 timber sawmill; lumber [saw]mill; log sectioning plant
d Sägewerk *n*; Sägerei *f*; Sägemühle *f*
f scierie *f* [de sciages lourds]; scierie de gros débits; usine *f* [de bois] de sciage
r лесопильный завод *m*; лесопилка *f*

8638 timber seal
(of a forest road)
d Dichtungsholz *n*
f bois *m* d'étanchéité
r деревянный слой *m*

* **timber-seasoning kiln → 5203**

8639 timber shave; peeling spade; barking spade; barking spud
d Holzschabe *f*
f vastringue *f* de bois
r окорочная лопатка *f*

* **timbershed → 9650**

* **timber shipment → 8608**

* **timber shoot → 5116**

* **timber shortage → 7282**

8640 timber-significant county
(county where the forest products industry represents a significant share of the economy)
d holzsignifikanter Bezirk *m*
f province *f* à production de bois signifiante
r район *m* с преобладающей лесодобивной промышленностью

8641 timber site index
(measure of site productivity based on tree

f indice *m* de qualité de station forestière
r показатель *m* лесного местообитания; индекс *m* лесного местообитания

* **timber sleeper → 9556**

* **timber slide → 5116**

* **timber slip → 5116**

* **timber soil → 9547**

* **timber species → 4714**

* **timber stabilization → 4434**

* **timber stack → 8619**

* **timber stand → 8585**

8642 **timber standardization; lumber standardization; wood standardization**
d Holznormung *f*
f standardisation *f* de bois; norme *f* pour le bois
r стандартизация *f* лесоматериалов; нормировка *f* древесины

8643 **timber stand improvement; TSI; stand improvement**
(all thinnings made during the life of a forest stand for the purpose of improving the composition or productivity of the stand)
d Bestandesfortschritt *m*
f amélioration *f* d'un peuplement
r улучшение *n* древостоя

8644 **timber stand marked for cutting; stand marked for felling**
d angerissener Bestand *m*; ausgezeichneter Bestand; markierter Bestand
f bois *m* marqué à la griffe; peuplement *m* griffé; peuplement martelé; peuplement flaché
r древостой *m*, отмеченный для вырубки

* **timber stand taxation → 8657**

* **timber stock → 3559**

* **timber stockage → 5120**

8645 **timber stream**
d Holzstrom *m*
f flux *m* de bois
r поток *m* лесоматериалов

* **timber strength → 2669**

8646 **timber supply analysis**
(assessment of future timber supplies over long planning horizons (more than 200 years) by using timber supply models for different scenarios identified in the planning process)
d Holzzufuhranalyse *f*
f analyse *f* d'offre de bois
r анализ *m* доставки лесоматериалов

8647 **timber supply area; TSA**
(defined by an established pattern of wood flow from management units to the primary timber-using industries)
d Holzversorgungszone *f*
f aire *f* d'offre de bois; zone *f* d'alimentation à bois
r район *m* снабжения древесиной

8648 **timber supply block; TSB**
(division of a timber supply area)
d Holzversorgungsblock *m*
f bloc *m* d'alimentation à bois
r блок *m* снабжения древесиной

8649 **timber supply model**
(analytical model (usually computer-based) that simulates the harvest and growth of collections of forest stands over several decades according to specific data and management assumptions)
d Holzversorgungsmodell *n*
f modèle *m* d'offre de bois
r модель *f* доставки лесоматериалов

* **timber supports → 8600**

8650 **timber surface**
d Holzoberfläche *f*
f surface *f* de bois
r поверхность *f* древесины

8651 **timber surplus; surplus**
d Holzüberschuss *m*
f excès *m* de bois
r избыток *m* леса; остающаяся на корню древесина *f*

* **timber tariff → 7123**

* **timber tax → 3616**

* **timber taxation → 8422**

8652 **timber tenure**
d Holzbesitz *m*

* **timber tester → 4451**

* **timber tongs → 3892**

* **timber tosser → 5062**

* **timber tract → 9542**

8653 timber trade; wood trade; log trade; lumber business
d Forstfachhandel *m*; Holzhandel *m*
f commerce *m* de bois
r торговля *f* лесом; лесоторговля *f*

* **timber tree → 517**

* **timber trespass → 8052**

* **timber trolley → 5199**

* **timber truck → 5087**

* **timber type → 3363**

* **timber-type map → 3635**

8654 timber upstream facing; timber face; timber deck; wooden protection apron; timber apron
(an upstream watertight facing of firstly naturally round timbers and later sawn timber placed over the rockfill)
d Stützboden *m* aus Holz
f masque *m* amont en bois; tapis *m* de protection en bois
r верховая грань *f* из бревен; деревянный верховой откос *m*

8655 timber user
d Holzverbraucher *m*
f utilisateur *m* de bois; consommateur *m* de bois
r потребитель *m* древесины

* **timber utilization → 9154**

* **timber volume → 9689**

8656 timber volume depletion
d Holzverknappung *f*; Walderschöpfung *f*
f épuisement *m* de volume du bois
r истощение *n* запасов леса

8657 timber volume inventory; timber stand taxation
d Vorrataufnahme *f*

* **timber wagon → 9256**

* **timber want → 2362**

* **timber ware → 9560**

* **timber waste → 3561**

8658 timber weight; weight of wood
(the weight equals the ovendry wood and the moisture it holds)
d Holzgewicht *n*
f masse *f* de bois
r масса *f* древесины

* **timber weir → 9550**

* **timber wheels → 5090**

* **timber wholesale merchant → 9392**

8659 timber with high density; high-density timber; dense timber; dense wood
d Holz *n* mit hohem Dichte; dichtes Holz
f bois *m* de haute densité
r древесина *f* высокой плотностью; плотная древесина; уплотнённая древесина

8660 timber with low density; low-density wood; light wood; low-density timber
d leichtes Holz *n*; Leichtholz *n*
f bois *m* à faible densité
r древесина *f* с небольшой плотностью; лёгкая древесина

8661 timber with low shrinkage
d Holz *n* mit niedrigem Einlaufen
f bois *m* de faible retrait
r малоусадочная древесина *f*

8662 timber with small cross-section area
d Holz *n* mit kleiner Querschnittfläche
f bois *m* de petite section transversale
r древесина *f* малой площади поперечного сечения

* **timber[-]wood → 1153**

* **timberwork → 8600**

8663 timber-work bridge
d Holzbrücke *f*
f pont *m* de charpente
r деревянный мост *m*

* **timber working machine → 9701**

8664 timber yard; wood yard; log yard; stack base; dumping place; dumping ground; forest depot; wood depot; timber depot; landing [site]
(cleared area in the woods to which logs are yarded for loading onto trucks for shipment to a processing plant)
d Holzlagerplatz *m*; Holzlagerhof *m*; Holzplatz *m*; Pollerplatz *m*; Ganterplatz *m*
f chantier *m* de bois; dépôt *m* de bois; dépôt forestier; parc *m* à bois [débités]; jetée *f*
r лесной склад *m*; лесной двор *m*; складочное место *n*; склад лесоматериалов; лесосклад *m*

8665 time for planting; planting season; planting period
d Kulturzeit *f*
f temps *m* de culture
r культурный сезон *m*

* **timping → 648**

8666 tinder; spunk; sponk; touchwood
(any readily combustible substance, that will ignite on contact with a spark)
d Zunder *m*
f amadou *m*
r сухая гнилая древесина *f*; трут *m*

8667 tinderwood; tinder fungus
(Fomes fomentarius)
d echter Zunderschwamm *m*
f amadouvier *m*; fomes *m* allume-feu
r настоящий трутовик *m*

8668 tine cultivator; tine plough; tine harrow
(plough in which the leading edge of the landside is extended forward and downward as a tine-bearing replaceable sock)
d Zinkenpflug *m*
f cultivateur *m* à éperon; charrue *f* à éperon; herse *f* à éperon
r плуг-рыхлитель *m* с черенковым ножом

* **ting bark → 8392**

8669 tipburn on leaf
d Spitzendürre *f*
f brûlure *f* au sommet du limbe; brunissement *m* des extrémités des feuilles
r солнечный ожог *m* листьев; отмирание *n* вершины листьев

* **tip moth → 6269**

* **tipping edge → 9051**

8670 tip weevil; white pine weevil
(Pissodes strobi)
d Weymouthskiefernrüsselkäfer *m*
f charançon *m* du pin blanc; charançon du pin Weymouth
r смолёвка *f* сосны веймутовой

* **tire chain → 319**

8671 tissue culture
(cultivation of plant or animal tissues in a controlled artificial environment on defined media under aseptic conditions)
d Gewebekultur *f*
f culture *f* de tissus
r [искусственно] выращенная живая ткань *f*

* **titi family → 2221**

* **T-notching → 289**

8672 tobira family
(Pittosporaceae)
d Klebsamengewächse *pl*
f pittosporacées *pl*
r питтоспоровые *pl*

* **toboggan slide → 4418**

* **slide → 4418**

* **TOF → 8891**

* **toggle chain → 5023**

8673 tolerance; amplitude
(ability of an organism or biological process to subsist under a given set of environmental conditions)
d Toleranz *f*
f tolérance *f*
r толерантность *f*; выносливость *f*

* **tolerant species → 7474**

* **tolerant tree → 7478**

* **Tolu tree → 504**

* **tombstone → 2193**

8674 tomentose rose
(Rosa tomentosa)
d Filzrose *f*

8675 ton
(a timber volume unit)
d Tonne *f*
f tonne *f*
r тонна *f*

* **tong hooker → 1559**

* **tongs → 3892**

8676 tongue and groove
d Spund *m* und Nut *f*; Feder *f* und Nut *f*
f rainure *f* et languette *f*; planches *fpl* à rainure et languette
r шпунт *m* и гребень *m*; шпунтовое соединение *n*

8677 tongue and groove lumber; matched lumber
(lumber that is edge dressed and shaped to make a close tongue and groove joint at the edges or ends, when laid edge to edge, or end to end)
d Bezugsholz *n*; angepasstes Holz *n*
f bois *m* bouveté
r доски *fpl* с пазом и гребнем

8678 tongue grafting; whip grafting
d Kopulieren *n* mit Gegenzunge
f greffage *m* par rainure
r копулировка *f* с язычком

* **Tooler eucalyptus → 3909**

* **tooling of wood → 5376**

* **toon → 410**

8679 toothache[ ]tree; common prickly-ash
(Zanthoxylum americanum; Zanthoxylum fraxineum)
d Zahnwehholz *n*
f clavalier *m* d'Amérique; frêne *m* épineux
r перец *m* американский

* **tooth angle → 7493**

* **toothed → 7461**

8680 toothed core
d gezackter Kern *m*
f cœur *m* à dents
r зубчатое ядро *n* (напр. клёна)

f feuille *f* à dents
r пильчатый лист *m*; зубчатый лист

* **tooth saw → 1431**

* **top-and-lop → 5163**

8682 top bind
(binding the upper tiers of a load of logs)
d Oberbinden *n*
f fixation *f* supérieure
r верхнее сжатие *n* (дерева, лежащего концами на возвышениях)

8683 top branch
d Gipfelzweig *m*; Hochast *m*
f branche *f* supérieure haute
r верхушечная ветвь *f*

* **top break[age] → 1013**

* **top clipping → 2198**

8684 top dog
d Oberklammer *m*
f griffe *f* supérieure
r верхний зажим *m* для брёвен

8685 top dressing; aerial dressing
(surface application of fertilizers or organic ameliorants)
d Düngerauflage *f*; Kopfdüngung *f*; Blattdüngung *f*
f préparation *f* du sol superficiel; application *f* d'engrais en surface; fumure *f* de couverture
r поверхностное внесение *n* удобрений

8686 top drying; top-kill
(in trees, death or dieback of the leader and more or less of the upper part of the crown)
d Absterben *n*; Windeldürre *f*; Gipfeldürre *f*; Wipfeldürre *f*
f descente *f* de cime; dépérissement *m* terminal
r высыхание *f* вершины; отмирание *n*; верхушечное усыхание *n*

* **top-drying tree → 2273**

* **tope → 1928**

* **top end → 7761**

8687 top [end] cut[ting]; top section
d Schnitt *m* durch die Spitzoberfläche

8688 **top [end] log; head log**
(the uppermost log cut from a bole)
d Kopfschnittholz *n*; Kopfstück *m*; Gipfelblock *m*; Wipfelstück *m*; Kopfabschnitt *m*; Kopfende *f*
f surbille *f* de tête; surbille supérieure; grume *f* de tête; surbille de découpe [marchande]; houppier *m*
r вершинник *m*; вершинное бревно *n*; верхний отруб *m*; верхний кряж *m*

8689 **top end of trunk**
d Stammkopf *m*; oberes Ende *n* des Stammes
f petit bout *m* de tige
r вершина *f* хлыста

* **top face → 7761**

8690 **top girth**
d Stammkopfumfang *m*
f circonférence *f* au sommet; circonférence au petit bout
r обхват *m* у вершины

8691 **top grafting; double grafting; intergrafting**
d Zwischenpfropfung *f*; Zwischenveredlung *f*
f surgreffage *m*; double greffage *m*
r прививка *f* в расщеплённую конечную почку; прививка в крону

* **top height → 8726**

8692 **top height; dominant height**
(average height of the hundred trees of largest diameter per hectare)
d Oberhöhe *f*; Spitzenhöhe *f*; Gipfelhöhe *f*
f hauteur *f* [de l'arbre] dominante
r доминирующая высота *f*; высота господствующих деревьев

8693 **top-height diameter; stand-height diameter**
d Mitterdurchmesser *m* der herrschenden Stammklasse
f diamètre *m* moyen des prédominants
r средний квадратический диаметр *m* наиболее толстых доминирующих деревьев

* **topiary → 6612**

* **top infected by pine blister → 8698**

* **top-kill → 8686**

* **top-kill tree → 2273**

d Länge *f* über dem Boden
f longueur *f* en crête
r высота *f* надземной части

8695 **top loader; head loader**
(when two men are engaged in loading logs on trucks or cars, one is termed head loader and the other second loader)
d Hauptlader *m*
f chargeur *m* principal
r главный грузчик *m*, укладывающий брёвна на воз лесоматериалов

8696 **top lopping**
(cutting limbs from downed tree tops so that no limbs are more than a specified length along the tree stem)
d Spitzenablösung *f*; Spitzenabsprung *f*
f élagage *m* de cime
r обрезка *f* вершин; подрезка *f* вершин; вершинная обрезка

8697 **top notch cut; upper cut**
d Oberfallkerb *m*
f entaille *f* d'abattage supérieure
r верхний рез *m* подпила

8698 **top of a foxy tree; top infected by pine blister**
d Kienzopf *m*
f cime *f* desséchée par Peridermium pini
r ветреник *m*

8699 **top of a tree; tree top; apex; collar**
d Baumspitze *f*; Baumgipfel *f*; Kopf *m*; Knoppe *f*; Wipfel *m*; Zopf *m*
f sommet *m* d'arbre; houppier *m*; houppe *f*
r вершина *f* дерева; верхняя часть *f* дерева; верхушка *f* дерева

8700 **top of bole; stem top**
(of a tree)
d Stammspitze *f*
f sommet *m* de fût; sommet de tige
r вершина *f* ствола

8701 **top of canopy**
d Schlussspitze *f*
f sommet *m* de couvert
r верх *m* полога

8702 **top *v* off; poll[ard] *v***
(of a tree)
d entwipfeln; köpfen; zopfen; abzopfen; hammeln
f écimer; étêter; éhoupper

8703 topography
(physical features of a geographic area, such as those represented on a map, taken collectively; especially, the relief and contours of the land)
d Topografie *f*
f topographie *f*
r топография *f*

8704 topophysis
(the long persistence of non-genetic, position or age effects after grafting or rooting)
d Topophysis *f*
f topophyse *m/f*
r топофиза *f*

8705 toposequence
(sequence of related soils that differ from one another, primarily because of topography as a soil formation factor)
d Toposequenz *f*
f toposéquence *f*
r топографическая серия *f* пластов [почвы]

8706 topper
(a person)
d Oberster *m*
f écimeur *m*
r рабочий *m* на обрезке вершин

* **topper knife → 8709**

* **topping → 2198**

8707 topping attachment; topping device; topping unit; topper
d Entspitzmaschine *f*; Kopfmaschine *f*
f dispositif *m* d'écimage; écimeuse *f*; décolleteuse *f*
r машина *f* для обрезки вершин

* **top[ping] diameter → 2470**

* **topping exploitation → 6430**

8708 topping height
d Entspitzenhöhe *f*
f hauteur *f* d'écimage
r высота *f* обрезки вершины

8709 topping knife; topper knife
d Köpfmesser *m*
f couteau *m* écimeur; couteau décolleteur
r нож *m* для обрезки вершин; ножевое устройство *n* для обрезки вершин

* **topping unit → 8707**

* **top pruning → 4294**

8710 top red[dening]
d Kronenrotfärbung *f*
f rougissement *m* de cime
r красное окрашивание *n* кроны

* **top riddance → 2198**

8711 top rot; staghead
d Wipfelfäule *f*
f pourriture *f* descendante; rouille *f* blanche; rouille albergine
r вершинная гниль *f*

* **tops → 8712**

* **top section → 8687**

* **top shoot → 8474**

* **top thinning → 2096**

* **top veneer → 2986**

8712 topwood; tops; apical wood; lapwood; laps; hamper
(tops left in the woods in logging)
d kronenbürtiges Holz *n*; Gipfelholz *n*; Wipfelholz *n*; Zopfholz *n*; Zopfware *f*
f bois de cime; bois terminal; bois façonné au petit bout de la grume; bois *m* provenant de la surbille
r круглый лесоматериал *m* из вершинной части деревьев; древесина *f* вершин деревьев; вершинная древесина; верхушечная древесина; вершинный лесоматериал *m*

8713 topwood harvester [assembly]; topwood harvesting equipment
d Gipfelholz-Erntemaschine *f*
f récupératrice *f* de houppiers
r заготовительный комбайн *m* вершинной древесины

8714 topworking; frameworking
(grafting of scions into the tops of trees, usually in the hope of inducing flowering by the scions)
d Bespantung *f*; Spant[en]werk *m*
f production *f* de pousses
r производство *n* прививок

8715 torch tree; ixora
(Ixora)
d Ixora *m*
f ixora *m*
r иксора *f*

8716 torchwood
(Amyris balsamifera; Amyris elemifera)
d Amyris[-Sandelholz] *n*; westindisches Sandelholz *n*
f balsamier *m* élémifère
r амирис *m*; западноиндийское сандаловое дерево *n*

* **torn grains → 1524**

* **torote → 2780**

8717 Torrey pine; soledad pine
(Pinus torreyana)
d Soledadkiefer *f*
f pin *m* de Torrey
r сосна *f* Торрея

* **torse wood → 2048**

* **torus → 6308**

8718 total age
(number of years elapsed since the germination of the seed, rooting of cuttings, or the budding of the sprout or root sucker; the average total age of the trees comprising it)
d Gesamtalter *n*
f âge *m* total
r суммарный возраст *m*; общий возраст

8719 total area
(spatial unit suggested to embrace agriculture, horticulture, forestry and agroforestry)
d Gesamtfläche *f*
f aire *f* totale
r общая площадь *f*

* **total basal area → 2032**

8720 total chance planning
(early planning over an entire development area for the best overall realization of all objectives identified by broader planning)
d Gesamthiebsflächeplannung *f*
f planification *f* de chantier totale
r общее планирование *n* лесозаготовительных участков

f coupe *f* totale par année
r общий объём *m* годичной рубки

8722 total forest area
d Gesamtwaldfläche *f*
f aire *f* forestière totale
r общая лесная площадь *f*

8723 total forest cover
d Gesamtwalddeckung *f*
f couvert *m* forestier total
r общий лесной покров *m*

8724 total forest exploitation
(including thinning and sanitation cuttings)
d Gesamtforstbenutzung *f*
f exploitation *f* forestière totale
r суммарное пользование *n* лесом

* **total forest yield → 4523**

* **total growth → 8727**

8725 total heat release
(heat released by combustion during burnout of all fuels)
d Gesamtwärmeabgabe *f*
f dégagement *m* de chaleur total
r общее количество *n* выделяемого тепла; общее тепловыделение *n*

8726 total height; top height
(height of a tree from the ground level to the top of its crown)
d Scheitelhöhe *f*
f hauteur *f* totale [d'un arbre]; hauteur de l'extrémité de la cime
r общая высота *f*; высота до верхушки

8727 total increment; total growth; gross increment
d Gesamtalterzuwachs *m*; Bruttozuwachs *m*
f accroissement *m* total; accroissement brut
r общий прирост *m*; валовой прирост (включая прирост деревьев, вырубленных при рубках ухода)

* **total insolation → 3866**

8728 total porosity
d Gesamtporosität *f*
f porosité *f* totale
r общая порозность *f*

* **total radiation → 3866**

8729 total resource plan

d Gesamtaufkommenplan *m*
f plan *m* total des ressources
r общий план *m* ресурсов

8730 total shrinkage
d Gesamtschwindung *f*; maximale Schrumpfung *f*
f retrait *m* total
r полная усушка *f*

8731 total site type
d Gesamtstandortstyp *m*
f type *m* de station total
r тип *m* биогеоценоза

8732 total swelling
d Gesamtschwellen *n*
f gonflement *m* total
r полное разбухание *n*

* **total tree → 3758**

8733 total volume of a tree; total [tree] volume
(volume of the main stem of the tree including stump and top)
d Baumholzgesamtmasse *f*; Baumholzmasse *f*; Baumholzvolumen *n*
f volume *m* total d'arbre
r общая масса *f* дерева

8734 totara [podocarpus]
(Podocarpus totara)
d Totara *m*
f totara *m*
r подокарпус *m* тотара

* **toting log → 7689**

* **touchwood → 7082**

* **touchwood → 8666**

8735 toughness
(a quality of wood which permits the material to absorb a relatively large amount of energy, to withstand repeated shocks, and to undergo considerable deformation before breaking)
d Zähigkeit *f*
f nervosité *f*; résistance *f*
r вязкость *f*

8736 towable water
d Anhängwasser *n*
f eau *f* tractable
r пригодная к сплаву вода *f*

d Bugsieren *n*
f touage *m*
r буксировка *f*

8738 tow-endless cable system
d Drahtseilsystem *n* mit geschlossene Schleppleine
f système *m* par câbles de remorque fermés
r подвесная установка *f* с несущим и двумя тяговыми замкнутыми канатами

* **towerman → 3171**

* **towing → 8737**

* **tow[ing] rope → 9007**

* **town forest → 9142**

8739 trabecula
(a cylindrical, barlike structure extending radially across the lumen of a cell from one tangential wall to the other)
d Trabekel *f*
f trabécule *f*
r трабекула *f*; перекладина *f* (клетки древесины)

* **trac → 7563**

8740 trace elements; micronutrients
d Spurenelemente *npl*
f éléments-trace *mpl*; oligo-éléments *mpl*
r микроэлементы *mpl*

* **tracery → 618**

8741 tracheary elements
(the principal water conducting elements of the xylem, mostly vessel members and tracheids)
d tracheale Elemente *npl*
f éléments *mpl* d'une fille de trachéides; éléments conducteurs
r трахеальные элементы *mpl*; сосудистые элементы

8742 tracheid
(an imperforate wood cell with bordered pits to congeneric elements)
d Leitzelle *f*; Tracheide *f*
f trachéide *f*
r трахеида *f*; проводящая клетка *f*

8743 tracing
d Trassierung *f*

* **tracing picket → 3899**

* **track cable → 8745**

* **tracked log skidder → 5114**

* **tracked vehicle → 7691**

8744 **track formation**
d Bahnkörper *m*; Unterbaukrone *f*
f palier *m* de la voie; plate-forme *f* de la voie
r строительство *n* [трелёвочных] волоков

* **track-laying tractor → 7799**

8745 **track line; track cable; carrying rope; carrier cable; bearing cable; bearer cable; traction rope; drag rope; drag[ging] line; tracking rope; drag[ging] track**
d Hängebahn *f*; Rennbahn *f*; Schwebeseilbahn *f*; Seilhängebahn *f*; Tragseil *n*
f câble *m* porteur; câble tracteur principal; câble de téléphérique
r несущий канат *m*; несущий трос *m*; тяговой канат [лебёдки]; грузовой канат; тяговой трос; тяговая верёвка *f*

8746 **tractor-drawn harvester**
d Traktorzug-Erntemaschine *f*
f récolteuse *f* au tracteur; récolteuse par traction mécanique
r прицепной лесозаготовительный комбайн *m*

8747 **tractor logging; tractor removal; tractor skidding**
(any logging method that uses a tractor as the power for transporting logs from the stumps to a collecting point, whether by dragging or carrying the logs)
d Schlepperrücken *n*; Traktorbringung *f*; Traktorenabfuhr *f*; Bringung *f* mittels Traktor
f débusquage *m* au tracteur
r тракторная трелёвка *f*; тракторная вывозка *f*; вывозка трактором

* **trade sawmill → 1749**

* **trade timber → 5400**

8748 **traditional forest**
(large, densely wooded areas that are found in remote areas with few people)
d traditioneller Wald *m*
f forêt *f* traditionnelle
r традиционный лес *m*

8749 **tragant; tragacanth [milkvetch]**
(Astragalus gummifer)
d gummigebender Tragacanth *m*; Gummitraganth *m*
f astragale *m* à gomme adragante
r астрагал *m* камеденосный

* **trail chute → 8757**

* **trail dogs → 3892**

* **trailer → 1642**

8750 **trailer chain**
d Anhängerkette *f*
f chaîne *f* de remorque
r верига *f* прицепа

8751 **trailer chipper**
d Anhänger-Hacker *m*
f découpeuse *f* à remorque
r прицепная рубильная машина *f*

8752 **trailer for round wood**
d Rundholzanhängewagen *m*; Rundholzanhänger *m*
f remorque *f* à grumes
r прицеп *m* для круглого леса

8753 **trail herding; driving; droving**
(of a group of animals)
d Viehtreiben *n* auf Routen
f déplacement *m* sur draille
r передвижение *n* стадом

8754 **trailing**
d Treiben *n*
f traînage *m* de grumes en files
r трелёвка *f* брёвен, прицеплённых друг за другом

* **trailing chink[u]apin → 1199**

8755 **trailing ends**
d angehängte Ende *npl*
f bouts *mpl* traînés
r концы *mpl* брёвен, волочащиеся по земле при трелёвке

* **trailing mahonia → 5943**

8756 **trailing shrubbery**
d angehängter Gebüsch *m*

8757 trail[ing] slide; trail chute
(a slide on which the grade is so low that animals are required to move the logs)
d Ziehbahn *f*
f glissoire *f* coulisse
r трелёвочный волок *m* по грунту [с отбойным бревном]

8758 trail of logs
d Stammschleife *f*
f traînée *f* de grumes; train *m* de grumes
r груз *m* последовательно связанных брёвен

8759 trail skidding
d Ziehwegrücken *n*
f débardage *m* par glissoire
r трелёвка *f* по подготовленному волоку

8760 trainer
(tree beneath the main canopy which by its shading and/or abrasive action hastens the natural pruning or improves the form of some other tree)
d Trainingsbaum *m*
f arbre *m* élagueur; dominé *m* élagueur
r дерево-тренажёр *m*

* **trainer → 8273**

8761 training forest; instruction forest; school forest
d Lehrforst *m*; Lehrrevier *n*; Schulwald *m*
f forêt *f* pour stagiaires
r учебный лес *m*; школьный лес

8762 training forest district; training forest station; training forest range
d Lehrforstamt *n*; Lehrforstrevier *n*
f district *m* forestier pour stagiaires
r учебное лесничество *n*

* **tramway → 5121**

8763 transfer site; reload site
d Holzumschlagpunkt *m*
f place *f* de rechargement
r перегрузочная площадка *f*; лесоперевалочный пункт *m*; пункт перегрузки

8764 transgenic [plant]
(genetically engineered plant or offspring of genetically engineered plants)
d transgene Pflanze *f*
f plante *f* transgénique
r генетически изменённое растение *n*

8765 transition area; transition zone
(of a dicotyledonous stem where the anatomy changes from that of a root to that of a stem)
d Übergangszone *f*
f zone *f* transitoire
r переходная зона *f*

* **transition period → 6514**

8766 transit raft
d Transitfloß *n*
f radeau *m* de transit
r транзитный плот *m*

8767 transpiration drying
d Transpirationstrocknung *f*
f séchage *m* par évaporation
r транспирационная сушка *f* (на лесосеке)

8768 transpiration stream
(the upward passage of water and solutes within the plant from the roots to the leaves)
d Transpirationsfluss *m*
f flux *m* de transpiration
r транспирационный поток *m*

8769 transplant *v*
d umpflanzen; verpflanzen
f repiquer
r пересаживать

* **transplant → 5815**

* **transplant bed → 5812**

* **transplanted plant → 5815**

* **transplanting → 6873**

8770 transplant[ing] board; lining-out board
(simple device having regularly spaced slots for the individual plants so as to ensure proper spacing and lining out in the new bed)
d Verschulbrett *n*
f planche *f* à repiquer; planche de repiquage
r доска *f* для перешколивания

* **transplanting machine → 8897**

8771 transplanting plough; undercutting plough; transplant undercutter
(plough used in the nursery to open trench for the roots of plants being lined out, while simultaneously backfilling it)
d Baumschulrodepflug *m*; Aushebepflug *m*; Rodepflug *m*
f charrue *f* planteuse; dégarnisseuse *f*; cale *f*

8772 **transplant lines**
(juvenile shrubs and trees planted in nurseries from seedlings or rooted cuttings, to reach acceptable size for planting elsewhere)
d Transplantatsreihen *fpl*
f lignes *fpl* de transplants
r ряды *mpl* трансплантов

8773 **transplant spacing**
d Verschulungsverband *m*
f écartement *m* des plantes repiquées
r размещение *n* перешколенных саженцев

* **transplant stock → 5814**

* **transplant undercutter → 8771**

* **transportation of wood → 8595**

* **transverse shake → 2039**

8774 **transverse sills**
d Querschwellen *fpl*
f seuils *mpl* transversaux
r брёвна *npl*, лежащие поперёк течения

* **trap bands → 8777**

8775 **trap bark; catch bark**
(of a tree)
d Fangrinde *f*; Borkenplatte *f*
f écorce *f* servante de piège
r ловчая кора *f*

8776 **trap billet; trap stick**
d Fangkloben *m*; Fangscheit *n*; Fangknüppel *m*; Fangklotz *m*
f rondin-piège *m*; quartier-piège *m*; billon-piège *m*
r ловчее полено *n*

* **trap boom → 8136**

* **trap ditch → 4515**

* **trap drain → 2173**

* **trap log → 8778**

* **trapped bark → 4491**

8777 **trap rings; trap bands; sticky belts; sticky bands; grease bands; insect bands**
d Giftringe *mpl*; Klebgürtel *mpl*; Leimringe *mpl*
f anneaux *mpl* à glu; ceintures *fpl* à glu

* **trap stick → 8776**

8778 **trap tree; trap log; trap stem**
(for forest pests)
d Fangbaum *m*
f arbre-piège *m*
r ловчее дерево *n*

* **trap trench → 4515**

8779 **trash boom; log guide bar; log guide rod; [bag] boom stick**
(floating logs tied together in rafts to be towed by boat to their destination)
d Floßring *m*; Randbohle *f*
f habillot *m*; tige *f* de guidage; estacade *f*; drôme *f*; bôme *f*
r лесонаправляющее сооружение *n*; бон *m*; оплотник *m*

8780 **trash bund**
(windrow of slash, brushwood, etc. made along the contour between patches of cultivation or planted stock in order to prevent erosion on slopes)
d Schlagabraumschwaden *m*
f andain *m* de niveau
r дамба *f* из отходов (ветровальной древесины, кустарников)

8781 **traumatic cork; wound cork**
d Wundekork *m*
f liège *m* de cicatrisation
r раневая пробка *f*

8782 **traumatic heartwood; wound heartwood**
(patches of dead sapwood that develop in the vicinity of wounds)
d Wundekern *m*
f bois *m* de cœur traumatique; cœur *m* de lésion
r травматическое ядро *n*; раневое ядро

8783 **traumatic intercellular canal; traumatic resin canal**
(a canal formed in response to injury to the living tree)
d Wundeharzkanal *m*
f canal *m* intercellulaire traumatique; canal de résine cicatriciel
r раневой смоляной канал *m*

8784 **traumatic parenchyma; wound parenchyma**
(parenchyma cells of irregular size, shape and distribution resulting from injury to the

f parenchyme *m* traumatique; parenchyme cicatriciel; parenchyme de blessure
r раневая паренхима *f*

* **traumatic resin canal → 8783**

8785 traumatic ring
d Wulstring *m*; Wundring *m*
f cerne *m* traumatique; cerne de gelée; cerne de sécheresse
r кольцевой нарост *m*; нарост в виде кольца

8786 traumatic streak on the bark
d Rindenrose *f*; Siegel *n*
f tache *f* traumatique de l'écorce
r раневое пятно *n* на коре дерева; рубец *m* от заплыва сучка

8787 traumatic tissue; wound tissue
d Wundgewebe *n*
f tissu *m* traumatique
r раневая ткань *f*

* **traumatic wood → 9734**

8788 traumatic zone
(of wood)
d traumatische Zone *f*
f zone *f* traumatique
r раневая зона *f*

* **travel[l]er's joy → 9402**

8789 traversing curl; through swirl
d durchgehender Wirbel *m*
f verticille *m* transitoire
r сквозной завиток *m*

8790 traversing knot; passing knot
d durchgehender Ast *m*
f nœud *m* transitoire; nœud transversal
r сквозной сучок *m* (сучок, выходящий на две противоположные стороны пиломатериала)

* **travois → 5115**

* **travois road → 7689**

8791 treatability
(of wood)
d Vergütbarkeit *f*
f traitabilité *f* (aptitude à subir un traitement)
r обрабатываемость *f*

* **treated timber → 8792**

(wood treated with a preservative)
d bearbeitetes Holz *n*; behandeltes Holz
f bois *m* traité
r обработанная древесина *f*; консервированная древесина; древесина, обработанная пропиткой

8793 treatment opportunity class
(identifies the physical opportunity to improve stand conditions by applying management practices)
d bearbeitungsmöglichkeitende Klasse *f*
f classe *f* d'opportunité de traitement
r класс *m* благоприятных возможностей обработки

8794 treatment prescription
(operational details required for carrying out individual silviculture activities such as site preparation and planting)
d Bearbeitungsvorschrift *f*
f prescription *f* de traitement
r предписание *n* для обработки

8795 treatment season
(season or year the planned treatment activity will be carried out)
d Bearbeitungssaison *f*
f saison *f* de traitement; période *f* de traitement
r сезон *m* обработки

8796 treatment unit
(geographic unit of productive forestland area designated in a prescription for a specific silviculture activity or series of treatments)
d Bearbeitungseinheit *f*
f unité *f* de traitement
r единица *f* обработки

8797 tree; arbor
(woody plant having a single main stem, more or less definite crown and usually attaining a height of at least 10 feet)
d Baum *m*
f arbre *m*
r дерево *n*

* **tree age → 134**

* **tree aging → 132**

* **tree aralia → 2452**

* **tree archive → 5570**

8798 tree balling machine
d baumschulige Knäuelmaschine *f*
f pelotonneuse *f* de gaules
r машина *f* для выкопки саженцев

8799 tree bark; bark
(the primary ground tissue of a stem or root between the epidermis of phellem and the vascular system)
d Baumrinde *f*; Borke *f*
f écorce *f* [d'arbre]
r [древесная] кора *f*

8800 tree-base sprayer
d Spritzgerät *n* zur Unkrautbekämpfung unter den Bäumen
f pulvérisateur *m* à désherber au pied des arbres; pulvérisateur pour désherbage sur la ligne
r опрыскиватель *m* основы дерева

8801 tree basket; accumulator basket; tree collector
d Riegelholz *n*
f berceau *m* de débardage; berceau collecteur; collecteur *m* d'arbres; panier *m* accumulateur; conteneur *m*; cheville *f* à bois
r сборщик *m* деревьев

* **tree-beetle → 3502**

8802 tree belt area
d Baumbandfläche *f*
f aire *f* de bande forestière
r площадь *f* древесного пояса

* **tree between overstor[e]y and understor[e]y → 4551**

* **tree bicycle → 8814**

8803 treebine
(Cissus)
d Klimme *f*
f cisse *f*
r циссус *m*

* **tree blossom[ing] → 847**

* **tree bole → 8934**

* **tree borer → 9016**

* **tree breeder → 344**

* **tree breeding → 1028**

d Holzzuchtplantage *f*
f plantation *f* arboricole
r плантация *f* по лесоразведению

8805 tree-breeding station
d Holzzuchtamt *n*
f station *f* de génétique forestière
r станция *f* по лесоразведению

8806 tree browsing [by game]; browsing
(feeding on the above-ground parts of trees and shrubs (buds, shoots and leaves) by livestock or wild animals)
d Wildtierverbiss *m*; Verbiss *m*; Äsung *f*
f nourriture *f* herbacée; nourriture du gibier; viandis *m*; abroutissement *m*; broûtage *m*
r объедание *n* листья и молодые побеги; ощипывание *n* листья и молодые побеги

8807 tree buckthorn; Australian red ash
(Alphitonia excelsa)
d [hohe] Alphitonia *f*
f alphitonie *f* élevée
r альфитония *f* высокая

8808 tree bulldozer; tree dozer; tree extractor; tree pusher; rakedozer; grubber; tree mover; forest devil; stub-puller; root extractor; eradicator; [root-]puller; uprooting machine; [tree] uprooter; tree lifter; tree puller-buncher; trowel
d Rodegerät *n*; Stubenrodemaschine *f*; Waldteufel *m*
f abatteur-déracineur *m*; bouteur *m* renverseur; bouldozeur-renverseur *m*; extracteur *m* d'arbres; déracineur *m*
r корчеватель *m* [деревьев]; корчевальная машина *f*; машина для выкапывания деревьев; древовал[-корчеватель] *m*

* **tree bunch → 7445**

* **tree bundle → 7445**

* **tree-by-tree mixture → 5496**

* **tree-cactus → 1733**

8809 tree cal[l]iper; cal[l]iper; pincers; die-stock; slide-ga[u]ge; tree compass
(tool to measure the diameter of a tree)
d Baumkluppe *f*; Messkluppe *f*; Kluppe *f*; Forst-Zählwerk *n*; Kaliber *m*
f pied *m* à coulisse; compas *m* bec de cigogne; bastringue *m*; compas en ciseaux; compas forestier

8810 tree canopy; canopy
d Baumschluss *m*; Baumbeschirmung *f*
f couvert *m* [d'arbre]
r полог *m* дерева; древесный покров *m*

* **tree-care → 8454**

8811 tree cavities
(hollow cavities in trees that provide resting or nesting places for wildlife)
d Baumhöhlen *fpl*
f cavités *fpl* d'arbre
r древесные полости *mpl*

* **tree cholla → 9259**

8812 tree class
(any class into which the trees forming a crop or stand may be divided for a variety of purposes)
d Baumklasse *f*
f classe *f* d'arbres
r класс *m* деревьев

8813 tree climacium moss
(Climacium dendroides)
d Bäumchenmoos *n*
f climacium *m* arborescent
r климаций *m* древовидный

* **tree climber → 4296**

8814 tree climbing device; tree bicycle
d Baumsteigegerät *n*
f appareil *m* pour montée aux arbres
r снаряд *m* для лазания

8815 tree club moss
(Lycopodium obscurum)
d verbreiteter Bärlapp *m*
f lycopode *m* foncé
r плаун *m* булавовидный; ликоподий *m*

* **tree clump → 1666**

* **tree collector → 8801**

8816 tree compactness; fullness of bole
d Vollholzigkeit *f*; Holzmassendurchschnitt *m*
f densité *f* d'arbre; volume *m* réel de bois
r полнодревесность *f*

* **tree compass → 8809**

8817 tree competition
(struggle among trees and other vegetation,
d Konkurrenz *f*; Wettbewerb *m*
f compétition *f*; concurrence *f*
r конкуренция *f*

8818 tree cone; cone; strobile; strobilus
(conical fruit of gymnospermous plants such as conifers and cycads, typically consisting of seed bearing, overlapping scales that develop pollen-bearing sacs or naked ovules or seeds)
d Nadelbaumzapfen *m*; Zapfen *m*; Strobilus *m*
f cône *m*; strobile *m*
r шишка *f*

8819 tree-conscious
d waldfreundlich
f favorable à la forêt
r благосклонный к лесу

* **tree consciousness → 3507**

* **tree cotton → 7327**

8820 tree counter
d Baumzähler *m*
f compteur *m* d'arbres
r счётчик *m* деревьев

* **treecreeper → 1807**

* **tree crop → 8844**

* **tree crop harvester → 8853**

8821 tree-crop interface
(spatial extent, above and below ground, over which some form of competition and or facilitation occurs between the tree and the crop components in an agroforestry system)
d Baum-Ernte-Interface *f*
f interface *f* entre arbres et agricultures
r взаимодействие *n* лесных и полевых культур

8822 tree cubby set; tree cubby house; cubby
d kleines Spielhaus *n*
f cabane *f* à piège
r деревянный домик *m*

* **tree cubing → 8955**

* **tree culture → 3528**

8823 tree cutter arm
d Baumschneiderarm *m*
f bras *m* sécateur de bois; bras d'abattage de

8824 tree cut[ting]; [wood] cutting; wood felling [operation]; lumbering; throw[ing]
(act of cutting down a standing tree)
d Baumfällen *n*; Holzfällen *n*; Holzrüstung *f*; Holzabhauen *n*; Holzabhieb *m*; Holzausfällung *f*; Baumschlag *m*; Holzeinschlag *m*; Hauung *f*
f abattage *m* [des arbres]; coupe *f* de bois
r [вы]рубка *f* [леса]; валка *f* [деревьев]; лесоповал *m*; отрубание *n*; срубание *n*; обрубание *n*; обрубка *f*

* **tree defect → 8830**

8825 tree delimbing machine; limbing machine; tree[- length] delimber; delimber; limber; lopper; knotter
d Astabschneider *m*; Astschneider *m*; Astholzmaschine *f*; Entastungsmaschine *f*; Baumentastmaschine *f*; Baumentastungsgerät *n*
f ébrancheuse *f*; ébrancheur *m*; machine *f* à ébrancher des arbres
r сучкорезная машина *f*; сучкорезка *f*; сучкоруб *m*

* **tree depot → 5570**

8826 tree diameter
d Baumdurchmesser *m*
f diamètre *m* d'arbre
r диаметр *m* дерева

8827 tree dimensions; tree sizes
d Baumdimensionen *fpl*; Baumgrößen *fpl*
f dimensions *fpl* des arbres
r размеры *mpl* дерева

8828 tree disease
d Baumkrankheit *f*; Baumerkrankung *f*
f maladie *f* d'arbre
r заболевание *n* дерева; болезнь *f* дерева

8829 tree dog; felling dog; span-dogs
d Baumklemme *f*
f griffe *f* d'appui
r захват *m* (для брёвен)

* **tree dozer → 8808**

* **treed pasture → 9711**

8830 tree drawback; tree defect; tree fault
(characteristics of an individual tree that
f défaut *m* d'arbre
r дефект *m* дерева; порок *m* дерева; фаут *m* дерева

8831 tree droop
d Baumauslenkung *f*
f fanaison *f* d'arbre
r свисание *n* дерева

8832 tree duck
(Dendrocygninae)
d Witwenente *f*; Witwerente *f*
f dendrocygne *m*
r древесная утка *f*

* **tree entanglement → 4**

8833 tree epiphyte
d Baumepiphyt *m*
f épiphyte *m* arboricole
r эпифит *m* на деревьях

* **tree euphorbia → 341**

* **tree exploitability → 8885**

* **tree extractor → 8808**

* **tree facer → 2142**

* **tree farm → 3419**

8834 tree farming
(any agroforestry practice that incorporates trees into farmland)
d Baumzucht *f*; Baumkultur *f*
f fermage *m* d'arbres
r разведение *n* деревьев

8835 tree farm licence; TFL
d Waldfarmlizenz *f*
f licence *f* de ferme forestière
r лицензия *f* на лесопитомник

* **tree fault → 8830**

8836 tree feller; feller[-director]
d Holzfäller *m*
f abatteuse *f* forestière
r валочная машина *f* (для направленной валки деревьев)

* **tree-felling area → 2189**

8837 **tree-fern**
(Cyathea)
d Baumfarn *m*; Farnbaum *m*
f fougère *f* arborescente; fougère en arbre
r циатея *f*; древовидный папоротник *m*

8838 **tree festival**
d Baumfest *n*
f fête *f* de l'arbre
r лесной фестиваль *m*

* **tree forest → 8585**

* **tree form → 7490**

8839 **tree formation**
d Dendritenbildung *f*
f formation *f* arborescente
r формирование *n* дерева; древесное строение *n*

8840 **tree form-class; form class; tree form-factor; form-figure; form-number; shape number; factor of shape; form-ratio**
(a ratio of the tree diameter at the top of the first log, to the diameter at breast height)
d Formklasse *f*; Formzahl *f*; Baumformzahl *f*; Formfaktor *m*
f classe *f* de forme; coefficient *m* de forme [de l'arbre]
r класс *m* формы; видовое число *n*

8841 **tree form-height; form-height**
d Formhöhe *f*
f produit *m* de la hauteur d'un arbre par le coefficient de forme
r видовая высота *f*

8842 **tree for seaside planting**
d Baum *m* für Seeküstenpflanzung
f arbre *m* pour plantations en bord de la mer
r дерево *n* для посадки у берега [моря]

8843 **tree frequency**
d Baumhäufigkeit *f*
f fréquence *f* d'arbres
r частота *f* деревьев; встречаемость *f* деревьев

* **tree-frog → 9579**

8844 **tree fruit; tree crop**
d Baumfrucht *f*
f fruit *m* d'arbre
r плод *m* дерева; плод древесных растений

(machine designed to shake a tree in order to dislodge its fruits for collection from the ground)
d Baumfruchtschüttler *m*; Baumschüttelmaschine *f*; Baumschüttler *m*
f secoueur *m* d'arbre; secoueuse *f* mécanique; agitateur *m* de fruits; marteau *m* d'arbre; vibreur *m* de troncs
r [в]стряхиватель *m* деревьев; [в]стряхиватель плодов; [в]стряхиватель плодовых или семенных деревьев

* **tree [gene] bank → 5570**

8846 **tree generation**
d Baumgeneration *f*
f génération *f* forestière
r лесное поколение *n*

8847 **tree getting**
d Baumeinstellung *f*
f pointage *m* d'arbre
r наводка *f* на дерево; захват *m* дерева

8848 **tree girth**
(the size or measurement of a tree trunk; the distance around a tree; the circumference)
d Baumumfang *m*
f circonférence *f* de l'arbre
r окружность *f* дерева; охват *m* дерева

8849 **tree grade**
(classification of sawtimber-size trees based on external characteristics as indicators of quality or value)
d Baumsorte *f*
f sorte *f* d'arbre
r сорт *m* деревьев

* **tree-grower → 344**

* **tree growth → 8858**

* **tree groyne → 4**

8850 **tree guard**
(an orchard or forest nursery product of strips of various materials that circle the trunks of young trees, protecting them from rodents and mechanical damage)
d Baumschutz *m*
f protecteur *m* pour arbre; filet *m* protecteur
r защитное ограждение *n* дерева

8851 **tree-guard sprayer**
d Baumschutzspritze *f*

* **tree habit → 4080**

8852 tree hardiness
d Baumfestigkeit *f*
f rusticité *f* d'arbre
r устойчивость *f* дерева к неблагоприятным внешним воздействиям

8853 tree harvesting machine; tree[-length] harvester; [tree crop] harvester; tree processor; felling machine; feller-[de]limber-buncher
d Baumfällmaschine *f*; Fällmaschine *f*; Forstmaschine *f*; Holzhackmaschine *f*; Erntemaschine *f*
f abatteuse-ébrancheuse *f*; abatteuse-porteuse *f*; moissonneuse *f*; récolteuse *f*
r лесозаготовительная машина *f*; валочно-сучкорезно-раскряжёвочная машина

8854 tree harvesting saw; felling [and bucking] saw
d Baumfällsäge *f*; Fällsäge *f*; Waldsäge *f*
f scie *f* à abattage; scie forestière
r пила *f* для валки леса; [лесо]валочная пила; пила-дровянка *f*; древесная пила

* **tree having a pointed top → 5657**

* **tree-hazel → 5383**

8855 tree height
(vertical distance between the extreme top of a standing tree and ground level)
d Baumhöhe *f*
f hauteur *f* de l'arbre
r высота *f* дерева

8856 tree house
(a small house perched in a tree)
d Baumhaus *n*
f maisonnette *f* montée dans un arbre
r шалаш *m* на дереве

* **tree-huckelberry → 8926**

* **tree improvement → 1028**

8857 tree inclination
d Baumneigung *f*
f inclinaison *f* d'arbre
r наклон *m* дерева

8858 tree increment; tree growth
d Baumzuwachs *m*

8859 tree increment [sample] plot
d Zuwachsschätzung-Probefläche *f*
f placette *f* d'échantillonnage d'accroissement; placette de sondage d'accroissement
r опытный участок *m* для измерения прироста [модельных деревьев]

8860 tree infected by fungi
d Schwammbaum *m*
f arbre *m* attaqué par champignons
r поврежденное грибом дерево *n*

8861 tree-injection
(deliberate introduction, by pressure or simple absorption of a chemical into the sapstream of a living tree)
d Baumimpfung *f*
f injection *f* d'arbres
r инъекция *f* химикатов в живое дерево

8862 tree-injector; tree-sprayer
(specially designed tool used to inject a solution into a living tree)
d Baumspritze *f*
f injecteur *m* d'arbre
r опрыскиватель *m* для деревьев; лесной опрыскиватель; инжектор *m*

8863 tree interplanting
(planting young trees among existing natural regeneration or previously planted trees of similar age)
d Baumnachpflanzen *n*
f plantation *f* forestière intercalaire
r дополнительная посадка *f* деревьев

8864 tree jack; skyline block
(a grooved saddle of a wood or metal rollers contained within two steel plates, attached to a tree with a strap, used as a guide for skyline, said guy or similar static line)
d Drahtseilblock *m*
f bloc *m* d'appui de câble porteur
r опорный блок *m* несущего каната

* **tree kernel → 1951**

* **tree[ ]killer → 343**

* **tree knocker → 8845**

8865 treelawn; [forest] glade
(usually narrow area between the curb and the sidewalk)
d Waldwiese *f*

8866 tree layer; tree stratum
d Baumschicht *f*
f strate *f* arborée; strate arborescente
r древесный ярус *m*; ярус древостоя

* **tree-length attachment → 5012**

* **tree-length boom → 4225**

* **tree[- length] delimber → 8825**

* **tree-length felling → 3046**

8867 tree-length forwarder
d Langholz-Forwarder *m*
f porteur *m* de troncs entiers
r форвардер *m* для долготей

8868 tree-length grapple
d Langholzgreifer *m*
f grappin *m* à troncs entiers
r захват *m* для долготей; захват для хлыстов

* **tree[-length] harvester → 8853**

8869 tree-length harvesting; tree-length logging; whole-stem logging; tree-length operation
d Volllangholzbringung *f*
f exploitation *f* par fûts entiers; exploitation par troncs entiers; exploitation en toute longueur
r лесозаготовка *f* с трелёвкой целых хлыстов

8870 tree-length harvesting method
(method of harvesting that includes felling a tree, cutting of the top and delimbing it before transport to a mill)
d Langholzbringungsmethode *f*
f méthode *f* d'exploitation par fûts entiers
r способ *m* лесозаготовок с трелёвкой хлыстов

* **tree-length hauling → 5615**

8871 tree-length loading
d Langholzladung *f*
f chargement *m* de troncs entiers
r погрузка *f* хлыстов

* **tree-length [log] → 2834**

* **tree-length logging → 8869**

* **tree-length operation → 8869**

8872 tree-length skidder operation
r лесосечные работы *fpl* с тракторной трелёвкой хлыстов

8873 tree-length skidding system
d Langholzrückensystem *n*
f système *m* de débardage en toute longueur; système de débusquage de troncs entiers; système de débardage en grande longueur
r система *f* трелёвки хлыстов

* **tree-length stem → 2834**

* **tree-length transporter → 1597**

8874 treeless; unwooded; woodless; unforested *adj*
(ref. to productive forest land that lacks trees)
d unbewaldet; waldlos; baumfrei
f sans arbres; non boisé; non couvert de forêt
r нелесной; безлесный; необлесённый; не покрытый лесом

8875 treeless period; treeless time
d waldlose Zeit *f*
f période *f* sans forêt
r безлесный период *m*

* **treeless vegetation → 5921**

* **tree lifter → 8808**

8876 treelike; tree-shape; arboriform; dendriform; dendritic[al]; dendroid[al]; ligniform *adj*
(having a form resembling a shrub or tree)
d baumartig; baumförmig; dendritisch
f dendritique; dendroïde
r древовидный; дендроидный

* **tree-like aloe → 9705**

8877 tree-like Jacob's-rod
(Asphodeline dendroides)
d baumartige Junkerlilie *f*
f bâton *m* de Jacob dendritique
r асфоделина *f* древовидная

* **tree limit → 4982**

8878 tree limonium
(Limonium arborescens)
d baumartiger Meerlavendel *m*
f lavande *f* dendritique; statice *m* arborescent
r лимониум *m* древовидный

8879 tree log; log; block
(the part of the trunk that is suitable for use as timber)
d Stammblock *m*; Holzstock *m*; Knüppelholz *n*; Holzstammsektion *f*; Stengelstück *m*
f grume *f*; billot *m*
r бревно *n*; открякёванная часть *f*; кряж *m*; отрез *m*; колода *f*

8880 tree lyonia; rusty lyonia; rusty staggerbush
(Lyonia ferruginea; Andromeda ferruginea; Andromeda rigida; Xolisma ferruginea)
d frostempfindliche Lavendelheide *f*; frostempfindliche Lyonie *f*
f andromède *f* ferrugineuse
r лиония *f* ржавая

8881 tree mallow
(Lavatera arborea)
d Baummalve *f*
f grande mauve *f*; lavatère *f* arbustive
r древовидная хатьма *f*

8882 tree marigold
(Tithonia diversifolia)
d mexikanischer Sonnenblumenstrauch *m*
f tournesol *m* mexicain; calendula *m*
r титония *f* [разнолистная]

8883 tree marking
d Anzeichnen *n* der Bäume; Auszeichnen *n* der Bäume
f marquage *m* d'arbre
r маркировка *f* дерева

8884 tree marking [paint-and-]gun; paint gun; spray[ing] gun; siphon feed gun
(low-pressure hand tool for squirting a distinctive mark of paint on trees and timber)
d Holz[farb]hammer *m*
f pistolet *m* marqueur; pistolet vaporisateur; pistolet de martelage
r маркировочный пистолет *m*; распылительный пистолет

8885 tree maturity; tree exploitability
d Baumreife *f*
f maturité *f* d'arbre
r спелость *f* дерева

8886 tree mechanism of resistance to the diseases
d Baumseuchenresistenzmechanismus *m*
f mécanisme *m* de résistance d'arbre aux maladies

8887 tree monoculture
(cultivation of a single tree species)
d Baummonokultur *f*
f monoculture *f* arborescente
r древесная монокультура *f*

8888 tree-moss
d Baumflechte *f*
f lichen *m* arboricole
r древесный лишай *m*

* **tree mover → 8808**

* **tree-nail → 9553**

8889 tree name
d Baumname *m*
f nom *m* d'arbre
r название *n* дерева

* **tree nursery → 342**

8890 tree nutrition
d Baumernährung *f*
f nutrition *f* minérale des arbres
r питание *n* деревьев

8891 tree off forest; TOF
(includes trees on arable land, land under permanent crops, permanent meadows and pastures etc.)
d Baum *m* außerhalb des Waldes
f arbre *m* dehors forêt
r дерево *n* вне леса

8892 tree-of-Heaven; [glandular] ailanthus
(Ailanthus altissima)
d drüsiger Götterbaum *m*; gemeiner Götterbaum; Götterbaum
f ailante *m* [glanduleux]; arbre *m* des dieux; arbre du ciel; faux vernis *m* du Japon
r айлант *m* высочайший

8893 tree of life; giant arborvitae; arbor[ ]vitae; thuja; thuya; [western] red cedar; canoe cedar
(Thuja plicata; Thuja gigantea)
d Riesenlebensbaum *m*; Lebensbaum *m*; Thuja *f*; Thuje *f*
f [grand] arbre *m* de vie; thuya *m* [géant]; thuya de Lobb
r туя *f* скалистая; туя гигантская; кедр *m* красный

8894 tree overbark volume; overbark volume;

f volume *m* sur écorce
r объём *m* дерева с корой

8895 tree peony
(Paeonia arborea; Paeonia suffruticosa)
d baumartige Pfingstrose *f*; strauchartige Päonie *f*; strauchartige Pfingstrose
f pivoine *f* en arbre
r пеон *m* древовидный; пеон полукустарниковый

* **tree pipit → 1093**

8896 tree plantation yielding immediate cash return
d Baumpflanzung *f* mit Immediatkassenertrag
f plantation *f* forestière à rapport financier immédiat
r лесная плантация *f* незабавным денежным доходом

* **tree[-planting] day → 340**

8897 tree-planting machine; transplanting machine; tree [trans]planter
(machine that cuts a narrow trench through the soil in which seedling roots are inserted and then held in place by closing of the trench)
d Baumpflanzmaschine *f*; Forstpflanzmaschine *f*
f planteuse *f* [d'arbres]
r лесопосадочная машина *f*

* **tree planting outside forest → 5759**

8898 tree-planting sword
d Pflanzschaufel *f*; Pflanzspaten *m*
f soc *m* à lame de rasoir
r посадочный меч *m*; сажальный меч; меч для посадки в щель

8899 tree poppy; plume poppy; bocconia
(Bocconia; Macleaya)
d Federmohn *m*
f bocconie *f*
r боккония *f*; маклейя *f*

8900 tree-portion logging; tree-section logging
d Baumrücken *n* in Portionen
f exploitation *f* par portions d'arbres entiers; exploitation d'arbres en morceaux
r поштучная трелёвка *f* дерева

* **tree privet → 1799**

f façonnage *m* d'arbre
r переработка *f* дерева

* **tree processor → 8853**

8902 tree prop; tree stake
d Baumpfahl *m*; Baumstütze *f*
f support *m* [d'arbre]; tuteur *m* d'arbre
r подпорка *f* для дерева; опора *f* для дерева; стойка *f*

8903 tree-protected shifting cultivation
d Baumschutzwanderfeldbau *m*
f culture *f* itinérante après débroussaillement et brûlis
r сменная культивация *f*, предохраняющая деревьев

8904 tree protector
d Baumschützer *m*
f corset *m* d'arbres
r предохранитель *m* деревьев

* **tree-pruner → 5168**

* **tree puller-buncher → 8808**

* **tree pusher → 8808**

8905 tree-pushing arm; push[ing] arm; tree pusher [arm]; pusher bar
d Baumaufschieber *m*; Baumeinsetzarm *m*
f bras *m* directeur; barre *f* directrice
r толкатель *m* дерева

8906 tree race
(a genetic subdivision of a tree, with more or less distinct characteristics when grown in a particular environment)
d Baumrasse *f*
f race *f* arborescente
r генетическая раса *f* одного древесного вида

* **tree reproductive success percent → 8918**

8907 tree resin; wood resin; resin; wood gum; wood pitch
(natural vegetable substance occurring in trees, especially the coniferous species)
d Baumharz *n*; Holzharz *n*; Pechharz *n*; Holzgummi *n*; Holzpech *n*
f résine *f* [de bois]; gomme *f* de bois
r древесная смола *f*; древесная камедь *f*; живица *f*; экстракционная канифоль *f*; древесный пек *m*

8908 **tree rhododendron**
(Rhododendron arboreum)
d Baum-Rhododendron *m*
f rhododendron *m* en arbre
r рододендрон *m* древесный

* **tree ring → 301**

* **tree-ring analysis → 269**

* **tree-ring chronology → 6958**

8909 **tree-ring chronology of rainfall**
d Niederschlagchronologie *f* durch Jahresringe
f chronologie *f* de précipitation par cerne de croissance
r определение *n* изменения количества осадков по годичным кольцам деревьев

8910 **tree-ring chronology of temperature**
d Temperaturchronologie *f* durch Jahresringe
f chronologie *f* de température par cernes de croissance
r определение *n* изменений температуры по годичным кольцам деревьев

* **tree-ring climatology → 2374**

8911 **tree-ring count**
d Jahrringzählung *f*
f comptage *m* des cernes annuels
r счёт *m* годичных слоев

8912 **tree-ring dating**
d Jahresringendatierung *f*
f datation *f* par cernes de croissance
r определение *n* возраста по годичным кольцам

8913 **tree-ripe[ned]** *adj*
d in Baum ausgereift
f ajouté en arbre
r дозревающий на дереве; не снимаемый до полного вызревания

* **tree row → 7121**

8914 **tree-run** *adv*
(about the fruits)
d direkt aus Baum
f directement par l'arbre
r прямо с дерева; в виде, в каком снят с дерева

* **tree sap → 7208**

* **tree scale → 715**

* **tree-section logging → 8900**

8915 **tree [seed] garden; tree seed orchard; seed plantation; seed plantage**
(multistoried agroforestry system in which a mixture of several fruit and other useful trees is cultivated)
d Baumsamenplantage *f*
f verger *m* à graines; plantation *f* issue de graines
r [лесо]семенное насаждение *n*; лесосеменная плантация *f*; лесосеменной питомник *m*

8916 **tree-seedling harvester**
d Baumsämlingerntemaschine *f*; Pflänzlingerntemaschine *f*
f récolteuse *f* de plantules
r машина *f* для выкопки сеянцев

8917 **tree seedlings; seedlings**
d Baumsämlinge *mpl*
f semis *mpl* forestiers
r всходы *mpl*; лесные сеянцы *mpl*

8918 **tree seedling [take] percent; tree reproductive success percent**
d Pflänzlinganwachsen-Prozent *n*; Fortpflanzungserfolg-Prozent *n*
f pourcentage *m* de réussite de semis forestier; pourcentage de reprise de semis forestier
r процент *m* приживаемости лесных культур (созданных посевом)

* **tree seed orchard → 8915**

8919 **tree separator**
d Forstvereinzelner *m*
f séparateur *m* d'arbres
r разобщитель *m* деревьев

8920 **tree's genus; tree's genre**
(tree taxonomic group containing one or more species)
d Baumgattung *f*
f genre *m* arborescent
r древесный род *m*

* **tree shape → 7490**

* **tree-shape → 8876**

8921 **tree shear; scissor-type shear; twin-bladed shear; double-bladed shear; butt shear**

d Baumschere *f*
f cisaille *f* d'abattage
r ножевое устройство *n* для срезания дерева

8922 tree shelter
(plastic tube that can be wrapped around the stem of hardwood seedlings to increase survival and growth)
d Baumschirm *m*
f couvert *m* [d'arbre]
r древесный щит *m*

* **tree shoe → 2054**

8923 tree shrew
(insectivorous arboreal mammal of southeast Asia that resembles a squirrel with large eyes and long sharp snout)
d Spitzmaus *f*
f musaraigne *f*
r землеройка *f*; бурозубка *f*

8924 tree size class; size class
(ranges in tree sizes representing stages in the development of a tree or stand)
d Baumgrößenklasse *f*; Größenklasse *f*
f classe *f* de dimension [d'arbre]
r класс *m* размера [дерева]

* **tree sizes → 8827**

8925 tree snake
d Waldschlange *f*
f serpent *m* forestier
r лесная змея *f*

* **tree sort → 4714**

* **tree spacing → 4569**

8926 tree sparkleberry; sparkleberry; tree-huckelberry; farkleberry
(Vaccinium arboreum)
d baumartige Heidelbeere *f*
f vaciet *m* d'arbre
r черника *f* кустарниковая

* **tree species → 4714**

* **tree species choice → 1551**

8927 tree species composition; tree species structure
d Holzartenzusammensetzung *f*
f structure *f* d'espèce arborescente
r породный древесный состав *m*

8928 tree species determination; identification of tree species; wood identification
d Holzartenbestimmung *f*
f définition *f* d'espèces arborescentes; détermination *f* des essences forestières
r определение *n* древесных пород

8929 tree species distribution
d Holzartenverteilung *f*
f distribution *f* d'espèces arborescentes
r распределение *n* древесных пород

* **tree species mixing → 5497**

* **tree species structure → 8927**

8930 trees per hour
(ref. to production)
d Bäume *mpl* pro Stunde
f arbres *mpl* par heure
r число *n* деревьев в час

* **tree-sprayer → 8862**

* **tree spur → 272**

* **tree spurge → 341**

8931 tree stage
d Baumstadium *n*
f stade *m* d'arbre
r стадия *f* дерева

* **tree stake → 8902**

8932 tree staking
(putting stakes in the soil for supporting the tree)
d Baumanpfählen *n*; Stützen *n*
f tuteurage *m* d'arbre
r опора *f* дерева вешками

* **tree stand → 8585**

8933 tree stand age
d Baumholzalter *n*; Baumbestandsalter *n*
f âge *m* d'arbre adulte
r возраст *m* крупного леса

8934 tree stem; stem; body; tree bole; bole; tree trunk; trunk
(above-ground axis of a plant; the main trunk of a tree, below the lowest branch)
d Baumstamm *m*; Holzstamm *m*; Stamm *m*; Schaft *m*
f tige *f* d'arbre; tronc *m* [d'arbre]; fût *m*

* **tree steppe → 3607**

8935 **tree stinger**
(a steel cylinder projecting beyond the face of a cutting bit that serves as a pilot or guide)
d Baumstinger *m*
f pic *m* d'abattage
r шарнирное замковое устройство *n*

* **tree stock → 9419**

* **tree stratum → 8866**

* **tree strawberry → 8152**

8936 **tree structure**
(showing the possible hierarchical organization of any object; used to depict all kinds of taxonomic knowledge)
d Baumstruktur *f*
f structure *f* arborescente
r древовидная структура *f*

8937 **tree stump; stump; rumprake**
(the base part of a tree that remains standing after the tree has been felled)
d Holzknüppel *m*; Strunk *m*; Stumpen *m*; Baumstumpf *m*; Stumpf *m*; Stummel *m*
f souche *f*; étoc *m*
r пень *m*; пенёк *m*; коротыш *m*; обрубок *m*

* **tree stump assarting machine → 8199**

* **tree stump disposal → 2529**

* **tree stump grubber → 8199**

* **stump puller → 8199**

* **tree stump puller → 8199**

* **tree succus → 7208**

8938 **tree surgeon**
(a specialist in treating damaged trees)
d Baumchirurg *m*; Baumchirurgin *f*
f chirurgien *m* arboricole; chirurgienne *f* arboricole
r рабочий *m* на подрезке веток и вершин

8939 **tree-surgery; tree work**
(care and repair of trees valued for amenity)
d Baumchirurgie *f*
f chirurgie *f* des arbres
r уход *m* за деревьями; обрезка *f* деревьев на омоложение

d Baumbesichtigung *f*; Baumaufnahme *f*
f contrôle *m* des arbres
r проверка *f* состояния деревьев

8941 **tree swallow; wood swallow**
(any one of numerous species of Old World passerine birds belonging to the genus Artamus and allied genera of the family Artamid[ae])
d Sumpfschwalbe *f*
f hirondelle *f* bicolore
r речная ласточка *f*

8942 **tree swift; crested swift**
(Hemiprocne)
d Baumsegler *m*
f martinet *m* des arbres
r древесный стриж *m*; хохлатый стриж; клехо *m*

* **tree tally → 2837**

8943 **tree temperament**
(the set of growth and development reactions shown towards its environment by a tree species during its whole life cycle)
d Baum-Temperament *n*
f tempérament *m* d'arbre
r характер *m* дерева

8944 **tree test**
d Baumtest *m*
f essai *m* d'arbre
r тест *m* дерева

* **tree-toad → 9579**

8945 **tree tobacco**
(Nicotiana glauca)
d blaugrüner Tabak *m*
f tabac *m* glauque; nicotiane *f* glauque
r табак *m* сизый; никотиана *f* сизая

8946 **tree tomato; tamarillo**
(Cyphomandra betacea)
d Tomatenbaum *m*; Baumtomate *f*
f tomate *f* en arbre; tamarille *f*; tamarillo *m*
r томатное дерево *n*

* **tree top → 8699**

8947 **tree-top cutter; scissor-type topping shear; topping shear**
d Baumspitzeschere *f*
f cisaille *f* d'écimage

8948 tree topology
d Baumtopologie *f*
f topologie *f* d'arbre
r топология *f* дерева

* **tree [trans]planter → 8897**

8949 tree trimmer
d Baumtrimmer *m*
f émondeur *m*; élagueur *m*
r рабочий *m* на подрезке деревьев

* **tree trunk → 8934**

8950 treetrunk bristle fern
(Trichomanes capillaceum)
d Haarfarn *m*
f trichomane *m*
r тонковласник *m* кривостебельный

8951 treetrunk clubmoss
(Huperzia aqualupiana; Huperzia chinensis)
d chinesische Teufelsklaue *f*
f huperzie *f* chinoise
r баранец *m* китайский

* **tree[trunk] vibrator → 8845**

* **tree tuft → 1666**

8952 tree underbark volume; underbark volume; VUB
d rindloses Stammvolumen *n*
f volume *m* [d'arbre] sous écorce; volume [d'arbre] sans écorce
r объём *m* дерева без коры

* **tree uprooter → 8808**

8953 tree vegetation; forest vegetation; woody vegetation
(forest stands or cover types consisting of a plant community made up of trees and other woody vegetation, growing more or less closely together)
d Waldvegetation *f*; Waldanbau *m*; Waldbestockung *f*
f végétation *f* arborescente
r древесная растительность *f*; лесная растительность

8954 tree viper
d Baumviper *f*
f vipère *f* d'arbre
r древесная гадюка *f*

estimate; scaling; cubing
d Baumkubierung *f*; Kubierung *f*
f détermination *f* de volume d'arbre; cubage *m*
r определение *n* объёма дерева; определение кубатуры

8956 tree volume function
(a function for measuring the volume of standing trees)
d Baumvolumenfunktion *f*
f fonction *f* de volume d'arbre
r функция *f* объёма дерева

8957 tree volume table; volume table; volumetric table; cubic table; outturn table
(showing the estimated average tree or stand volume based on given tree measurements, usually diameter and height)
d Baummassentafel *f*; Baumvorratstafel *f*
f tarif *m* d'estimation des bois sur pied; table *f* d'estimation de volume d'arbres
r таблица *f* вычисления объёмов деревьев; объёмная таблица; массовая таблица; таблица запасов

8958 tree wax; Chinese wax; insect wax; grafting wax
d Baumwachs *m*; Chinawachs *m*
f cire *m* d'arbre; cire d'insectes; mastic *m* à greffer
r древесная замазка *f*; садовый клей *m*; вар *m*

* **tree with one-sided crown → 2062**

* **tree with swollen butt → 5129**

* **tree-wool → 9697**

* **tree work → 8939**

8959 tree wound paint
d Baumwundefarbe *f*
f peinture *f* cicatrisante de bois
r раневая окраска *f* дерева

8960 tree wrapping
(the application of various materials to the trunks of trees to attempt to protect against damage by rabbits, children, wind, sunshine, etc)
d Baumdecken *n*; Baumumhüllung *f*
f capage *m*; emballage *m* d'arbre
r упаковка *f* дерева

8961 trefoil

f trèfle *m*
r трёхлистник *m*; трёхлистный клевер *m*

* **trembling aspen → 2880**

* **trembling[-leaved] aspen → 236**

* **trench digger → 8963**

* **trencher → 8963**

* **trench excavator → 8963**

8962 **trenching; ditching; digging of ditches**
(site preparation technique creating a more or less continuous furrow, with surface debris, duff, and low vegetation scattered to one side)
d Kreuz- und Querpflügen *n*; Grabenziehen *n*; Gräbenzug *m*
f creusement *m* de fossés; excavation *f* de fossés; scarifiage *m* par sillons; ruellage *m*; rigolage *m*
r рытьё *n* канав; копание *n* канав; прокладка *f* канав

8963 **trenching plough; trencher; subsoil plough; subsoiler; ditcher; [trench] excavator; trench digger**
(a plough for breaking up the compacted subsoil)
d Rajolpflug *m*; Untergrundpflug *m*; Wühlpflug *m*; Säschar *n*; Schar *n*
f trancheuse *f*; excavatrice *f*; sous-soleuse *f*; bineuse *f*
r подпочвенный плуг *m*; плуг для глубокой вспашки

8964 **trench planting; furrow planting; skip-row planting**
(setting out young trees in a shallow trench or a continuous slit)
d Einstupfen *n*; Furchenpflanzung *f*; Riefensaat *f*; Rinnensaat *f*
f plantation *f* en sillon; plantation sur sillon
r посадка *f* в [дно]борозду; посадка в щель; бороздовая посадка

* **trestle for sawing → 4380**

* **trestle tree → 4380**

8965 **trestle[-trees]**
(one of two strong bars of timber, fixed horizontally on the opposite sides of the masthead, to support the crosstrees and the frame of the top)
r устройство *n* для выгрузки леса из воды

8966 **triangulation**
(a trigonometric method of surveying; a method of determining the position of a fixed point from the angles to it from two fixed points a known distance apart)
d Triangulation *f*; Triangulierung *f*
f triangulation *f*
r триангуляция *f*

8967 **tributary system**
(of roads)
d untergeordnetes System *n*
f système *m* tributaire
r сеть *f* вспомогательных дорог

8968 **trickling water**
(water flowing on the surface of the ground without penetrating it)
d Rieselwasser *n*
f eau *f* de ruissellement
r стекающая вода *f*; просачивающаяся вода

8969 **tricuspidate leaf**
d Trikuspidalblatt *n*
f feuille *f* tricuspidienne
r трёхзубчатый лист *m*

8970 **trident maple**
(Acer buergerianum)
d Bürgersahorn *m*; Dreizahnahorn *m*
f érable *m* de Burger
r клён *m* трёхраздельный

* **triffid weed → 866**

* **trifoliate orange → 4146**

* **trim → 83, 5170**

* **trim allowance → 202**

* **trim end → 1336**

* **trimmed-off butt → 1226**

8971 **trimmer; multiple saw [drop] trimmer**
(a beam that receives the end of a header in floor framing)
d Vielblatt-Abkürz-Kreissägemaschine *f*; Vielblatt-Abkürzsäge *f*
f ébouteuse *f* multilame; ébouteuse à lames multiples; ébouteuse-tronçonneuse *f*

* **trimming axe → 5167**

* **trimming length → 202**

* **trimming machine → 4661**

* **trimming margin → 202**

8972 **trimming mark; measurement mark; bucking point; bucking mark**
(on the tree stem)
d Ablängzeichen *n*
f marque *f* de coupe; marque de tronçonnage
r метка *f* для раскряжёвки; раскряжёвочная метка (место презания ствола при раскряжёвке)

* **trimming[-out] → 6610**

* **trim saw → 2042**

* **trip → 9314**

* **trophic chain → 3308**

* **tropical almond terminalia → 4464**

* **tropical deciduous forest → 5545**

8973 **tropical dry deciduous forest**
d tropischer laubwerfende Trockenwald *m*
f forêt *f* sèche tropicale caducifoliée
r тропический сухой листопадный лес *m*

8974 **tropical dry semideciduous forest**
d tropischer halbwechselgrüne Trockenwald *m*
f forêt *f* sèche tropicale semi-décidue
r тропический полусухой листопадный лес *m*

8975 **tropical forest**
(a tropical woodland with an annual rainfall of a least 250 cm, marked by broadleaved evergreen trees forming a continuously canopy)
d Tropenwald *m*; tropischer Wald *m*
f forêt *f* tropicale
r тропический лес *m*

8976 **tropical forest ecosystem**
d Tropenwald-Ökosystem *n*
f écosystème *m* de forêt tropicale
r экосистема *f* тропического леса

8977 **tropical forestry**
d Tropenwaldwirtschaft *f*

8978 **tropical moist forest; equatorial rain forest; tropical rainforest; tropical rainfall forest; hylea**
d tropischer Regenwald *m*
f forêt *f* tropicale humide; forêt ombrophile tropicale; forêt dense équatoriale
r влажный тропический лес *m*; тропический дождевой лес; экваториальный дождевой лес; гилея *f*

8979 **tropical primeval forest**
d tropischer Urwald *m*
f forêt *f* vierge tropicale
r девственный тропический лес *m*

* **tropical rainfall forest → 8978**

* **tropical rainforest → 8978**

8980 **tropical secondary forest; secondary tropical [rain] forest**
d tropischer Sekundärwald *m*
f forêt *f* [humide] tropicale secondaire
r вторичный [влажный] тропический лес *m*

8981 **tropical wood; tropical timber**
d Tropenholz *n*
f bois *m* d'outre-mer; bois tropical
r древесина *f* тропических пород деревьев; тропическая древесина

8982 **tropophytes**
(plant that markedly changes its character, particularly its water requirements, with seasonal changes of climate, e.g. a deciduous tree)
d Tropophyte *mpl*
f tropophytes *fpl*
r тропофиты *mpl* (растения, приспособленные к сезонным изменениям климата, напр. листопадные деревья)

* **trowel → 8808**

8983 **truck driver; motor truck operator; lorry-driver**
(a teamster who skids logs with a truck)
d Lastkraftwagenfahrer *m*; Lastwagenfahrer *m*
f conducteur *m* de camion
r водитель *m* лесовоза

8984 **truckload; cartload of wood; turn**
d Holzfuder *n*; Holzfuhre *f*
f charge *f* de bois complet; voiturée *f* de bois; train *m* de bois

8985 **truck logging** (*US*); **lorry removal**
d Lastkraftwagenrückung *f*; LKW-Rückung *f*
f transport *m* par camions
r вывозка *f* леса автомобилями

* **true angico preto → 6762**

* **true bay → 3936**

8986 **true core of wood; true heart**
d wahrer Holzkern *m*
f cœur *m* vrai de bois
r настоящее ядро *n* древесины

* **true dormancy → 2323**

* **true ebony → 2727**

8987 **true firs**
(noble fir, grand fir and etc.)
d wahre Tannen *fpl*
f sapins *mpl* vrais
r настоящие пихты *fpl*

* **true heart → 8986**

8988 **true hickory; mockernut hickory; white hickory**
(Carya alba; Carya tomentosa)
d echtes Hickoryholz *n*; filzige Hickory *f*; Spottnuss *f*
f hickory *m* [véritable]
r кария *f* белая; гикори *m*

8989 **true log**
d richtiger Stamm *m*
f grume *f* droite
r правильно обработанное бревно *n*; бревно правильной формы

* **true middle lamella → 4532**

8990 **true mountain mahogany; birchleaf mountain-mahogany**
(Cercocarpus montanus; Cercocarpus betuloides)
d Bergmahagoni *m*
f acajou *m* de montagne
r горное красное дерево *n*

8991 **true service-tree; service-tree; sorb-tree**
(Sorbus domestica)
d Spierling *m*; Hauseberesche *f*
f sorbier *m* domestique; cormier *m*; corbier *m*
r рябина *f* домашняя; рябина садовая

f cubage *m* en volume réel
r измерение *n* чистого объёма

8993 **true white oak; [Arizona] white oak; Baltimore oak; stave oak**
(Quercus alba; Quercus arizonica)
d Arizona-Weißeiche *f*; amerikanische Weißeiche *f*
f chêne *m* [d'Arizona] blanc; chêne blanc d'Amérique
r дуб *m* белый; дуб аризонский

* **true white pine → 1274**

* **true[ ]wood → 4202**

8994 **trumpet creeper**
(Campsis radicans)
d amerikanische Klettertrompete *f*; Trompetenblume *f*
f bignone *m*; jasmin *m* de Virginie
r капсис *m* укореняющийся; текома *f* укореняющаяся

8995 **trumpet-creeper family**
(Bignoniaceae)
d Trompetenbaumgewächse *pl*; Bignoniengewächse *pl*; Bignoniaceen *pl*
f bignoniacées *pl*
r бигнониевые *pl*

* **trunk → 8934**

* **trunk analysis → 8060**

8996 **trunk eccentricity**
(wood defect)
d Stammexzentrizität *f*
f excentricité *f* du tronc
r эксцентричность *f* ствола; крень *f*

8997 **trunk protection**
d Stammschutz *m*
f protection *f* du tronc
r предохранение *n* ствола

8998 **trunk rot; stem rot; rot of the stem**
d Stammfäule *f*
f pourriture *f* des tiges; pourriture de la grume
r стволовая гниль *f*

* **trunk timber → 891**

8999 **trunk warping; buttressing of trunk**
d Stammrückigkeit *f*

* **trunk wood → 891**

* **TSA → 8647**

* **TSB → 8648**

* **TSI → 8643**

* **tualang [tree] → 4695**

9000 tuba-root jewelvine; oiltree
(Derris elliptica)
d Rosenwein *m*; Juwelenwein *m*
f vigne *f* grimpante
r масличное дерево *n*

* **tube-builders → 4886**

9001 tubed nursery stock
d Baumschulen *m* in Röhre
f matériel *m* de reproduction en tube
r сеянец *m* с закрытой корневой системой

9002 tube planting; sleeve planting
(setting out young trees in narrow, open-ended cylinders of various materials)
d Röhrenpflanzung *f*
f plantation *f* [de semis] en tube
r посадка *f* в тубах; посадка в трубках

9003 tuber; knob
(a greatly swollen underground shoot or stem, generally irregular in shape, which acts as a store of nutrients)
d Höcker *m*; Knolle *f*
f tuber *m*; bosse *f*; bulbe *m*
r клубень *m*; клубнелуковица *f*; луковица *f*

* **tubercle → 7036**

9004 tuberous root
d Blumenknolle *f*
f racine *f* tubéreuse
r клубневидный корень *m*

9005 tuft; tussock
(bunch of twigs or needles)
d Büschel *n*
f touffe *f*
r пучок *m*

* **tuft → 1666**

* **tufted growth → 1209**

9006 tufted trees
r неветвящиеся деревья *npl*

* **tug → 5090**

* **tugai → 6961**

9007 tugger rope; tow[ing] rope
d Schleppleine *f*; Schleppseil *n*; Schlepptau *n*; Schlepptrosse *f*; Abschleppseil *n*; Treidel *f*
f câble *m* de remorque
r буксирный канат *m*; буксирный трос *m*; буксир *m*

9008 tulip-tree; American whitewood; Canary whitewood; yellow poplar
(Liriodendron tulipifera)
d Tulpenbaum *m*; Liriodendron *m*; Kanarienholz *n*
f tulipier *m* [d'Amérique]; tulipier de Virginie; sapin *m* blanc d'Amérique; sapin blanc des Canaries
r тюльпанное дерево *n*; лириодендрон *m* тюльпанный; жёлтый тополь *m*

* **tumping → 1171**

* **tuna → 6533**

9009 tung oil; Chinawood oil; Chinese wood oil
(from tree seeds of type Aleurites)
d China-Holzöl *n*; chinesisches Öl *n*
f huile *f* de tung
r тунговое масло *n*

* **tung[-oil] tree → 1304**

* **tupelo → 5871**

* **tupelo-gum → 810**

9010 tupelo-gum family
(Nyssaceae)
d Tupelobaumgewächse *pl*
f nyssacées *pl*
r ниссовые *pl*

* **turbinella oak → 7559**

* **turf formation → 7813**

* **turill → 5406**

* **Turkey box → 2886**

* **Turkey oak → 5565**

* **Turkish hazel → 5383**

* **turn → 8984**

9011 **turnaround time**
(time it takes for a truck or tractor to be loaded and unloaded)
d Durchlaufzeit *f*; Umschaltzeit *f*
f temps *m* de rotation
r время *n* полного рейса (с грузом и порожняком); затраты *fpl* времени на рейс

* **turning crane → 8343**

* **turning saw → 2609**

* **turning sour → 41**

9012 **turning timber**
d Drechslerholz *n*
f bois *m* de tournage
r токарный лесоматериал *m*; древесина *f*, пригодная для токарных работ

* **turntable → 5232**

9013 **turpentine orchard**
d Harzbestand *m*
f atelier *m* de gemmage
r заподсоченный лес *m*; подсачиваемое насаждение *n*

* **turpentining → 828**

* **tussock → 9005**

* **twice pinnate leaf → 741**

9014 **twig beetles**
(Psoidae)
d Zweigkäfer *mpl*; Zweig[ab]stechere *mpl*
f rhynchites *mpl* coupe-bourgeons; charançons *mpl* coupe-bourgeons
r псоиды *mpl*

* **twig burr → 995**

9015 **twigging**
d Weidengerten *fpl*
f perches *fpl* d'osier
r ивовые жерди *fpl*

9016 **twig girdler; tree borer**
(Oncideres cingulata)
d Heckenprachtkäfer *m*; Schmalbauchkäfer *m*; Zickzackwurm *m*; Zweigringler *m*

* **twig litter → 5000**

9017 **twig scar**
d Knospennarbe *f*; Knospenwunde *f*
f cicatrice *f* de têtard
r побеговый рубец *m*

* **twin-bladed shear → 8921**

* **twiner → 9018**

9018 **twining plant; twiner**
d Schlingpflanze *f*; Schlinger *m*
f plante *f* volubile
r вьющееся растение *n*

9019 **twin leaders**
d Zwillingsspitzentrieb *m*
f pousse *f* apicale à double flèche
r близнецовый конечный побег *m*

* **twin-sled[ge] → 2587**

* **twin stem → 3657**

9020 **twist**
(a spiral distortion along the length of a piece of timber)
d Drehung *f*
f gauchissement *m*; torsion *f*
r закручивание *n*; скручивание *n*; извилина *f*

9021 **twisted growth; spiral growth**
(wood defect)
d Wechseldrehwuchs *m*; Drehwuchs *m*; gedrehter Faserverlauf *m*
f croissance *f* en spirale
r рост *m* по спирали; спиральный рост; косослой *m*

* **twisted heath → 3153**

* **twisted wood → 2048**

9022 **twister; dancer; springer**
(a supple pole for wedging between the frame and the loading chains of a pole wagon so as to take up the chain slack)
d Springer *m*
f sommier *m*
r гибкий кол *m* (закладываемый между рамой и грузовой цепью для её натяжения)

9023 **twistiness**
(spiral distortion along the grain)
d Drehwüchsigkeit *f*

* **twist wood → 9307**

9024 **twitch** *v*; **bag** *v*
(to skid logs or tree lengths on the ground without an antifriction device)
d reißen; zucken; zerren; schleppen
f emballer
r трелевать брёвна по земле; тянуть волоком

9025 **two-aged stand**
(stand containing two distinct age classes differing by more than 20% of the rotation age)
d zweialtriger Bestand *m*
f peuplement *m* à deux âges
r двухлетнее насаждение *n*

9026 **twobill adze; cross-grain axe; double chopper**
d Queraxt *f*; Querbeil *n*
f herminette *f*
r топор *m* с двумя лопастями; поперечный топор

9027 **two-edged cant**
d zweikantiger Balken *m*
f équarri *m* bilatéral
r двухкантный брус *m*

* **two-faced tie → 4095**

9028 **two-furrow plough**
(plough with two moldboards turning the furrow slices to the same side)
d Zweischarpflug *m*; Zweifurchenpflug *m*
f charrue *f* à deux socs [et versoirs simultanés]; charrue bisoc
r двухкорпусной плуг *m*

* **two-hack girdling → 2580**

* **two-handed saw → 9030**

* **two-leaved nut pine → 9031**

9029 **two-line grapple**
d Zweileitungsgreifer *m*
f grappin *m* à câble double
r двухканатный трелёвочный захват *m*

9030 **two-man saw; two-handed saw; double handed crosscut saw; push-pull felling saw; bellied saw**
d zweimännige Säge *f*; Zweimann[blatt]säge *f*;
r двуручная пила *f*

9031 **twoneedle pinyon; two-leaved nut pine; [pinyon] nut pine; Rocky Mountains pinyon; Colorado [pinyon] pine; Arizona pine**
(Pinus edulis)
d Koloradokiefer *f*; Arizonanusskiefer *f*; essbare Kiefer *f*
f pin *m* d'Arizona; pin comestible; pin à amandes
r сосна *f* съедобная

9032 **two-petal ash**
(Fraxinus dipetala)
d zweipetale Esche *f*
f frêne *m* à deux pétales
r ясень *m* двулепестковый

9033 **two-rotation coppice system; coppice-of-two-rotations system**
d zweihiebiger Niederwaldbetrieb *m*; Niederwaldüberhaltbetrieb *m*
f taillis *m* composé; traitement *m* en taillis avec réserves
r двухоборотная система *f* рубки порослевого древостоя; перестойное хозяйство *n* в низкоствольнике

* **two-sled → 2587**

9034 **two-stage cutting; two-stage felling**
d zweistufiger Hieb *m*
f coupe *f* en deux abattages
r двухприёмная постепенная рубка *f*

9035 **two-storied forest; two-storied stand**
(forest stand in which two height classes of considerable difference occur, the overstor[e]y and understor[e]y)
d doppelhiebiger Holzwald *m*
f peuplement *m* à deux étages
r двухъярусный лес *m*; двухъярусное насаждение *n*

9036 **two-storied high forest system**
(the two crops eventually being harvested together, or the upper one before the lower)
d doppelhiebiges Hochwaldsystem *n*
f traitement *m* en futaie à deux étages
r система *f* двухъярусного высокоствольного хозяйства

* **two-storied stand → 9035**

9037 **tylosis**
(an outgrowth from an adjacent ray or axial parenchyma cell through a pit cavity in a vessel wall, partially or completely blocking the vessel lumen)
d Thylle *f*
f thylle *m*
r тил *m*; тилла *f*

* **type forest → 6239**

* **type tree → 442**

* **typical decay → 72**

U

9038 Ubah besi
(Parastemon)
d Parastemon *m*
f parastémon *m*
r парастемон *m*

* **ucar → 6014**

* **umbaua → 107**

* **umbrage → 3303**

9039 umbrella elm; weeping elm
(Ulmus glabra horizontalis; Ulmus glabra pendula)
d Schirmulme *f*
f orme *m* parasol
r вяз *m* шершавый; вяз плакучий

9040 umbrella magnolia; schefflera; elkwood
(Magnolia tripetala; Magnolia virginiana)
d Schirmmagnolie *f*
f schefflera *m*; magnolier-parasol *m*; cucumber *m* magnolia
r магнолия *f* трёхлепестная

* **umbrella pine → 4595**

9041 umbrella thorn
(Acacia tortilis)
d Schirmakazie *f*; gewundene Akazie *f*
f acacia *m* parasol
r акация *f* зонтичная

9042 umbrella-tree; African corkwood; musanga; musanda
(Musanga cecropioides)
d Schirmbaum *m*
f parasolier *m*
r мусанг *m*; музанга *f*

* **umbrella-tree → 412**

* **umgoloty → 6279**

* **umgusi → 9793**

* **umnini → 6279**

f copeaux *mpl* non écorcés
r неокорённая щепа *f*

9044 unbarked log; raw log; rough log
d unentrindeter Stamm *m*
f grume *f* non écorcée
r неокорённое бревно *n*

9045 unbarked timber; unpeeled timber; undressed timber
d unentrindetes Holz *n*; Holz in der Rinde
f bois *m* non écorcé
r неокорённый лесоматериал *m*; неокорённая древесина *f*

9046 unclassified felling; unregulated felling
d außerordentliche Nutzung *f*; nicht klassifizierte Nutzung
f coupe *f* hors règlement; coupe non réglée
r целевая рубка *f*; неплановая рубка

9047 unclassified forest data
d nicht klassifizierte Forstdaten *pl*
f données *fpl* forestières non classées
r неклассифицированные лесные данные *pl*

9048 uncloven stacked timber
d nicht gespaltenes Holz *n*
f bois *m* non refendu
r нерасколотый лесоматериал *m*

* **uncoiling → 6121**

* **unconsolidated grains → 5159**

* **uncontrolled fire → 3690**

9049 uncontrolled forest fire
d unbekämpfter Waldbrand *m*
f incendie *m* de forêt non-combattu
r безудержно горящий лесной пожар *m*

9050 uncovered knot; open knot
d enthüllter Knoten *m*; ungedeckter Knoten; offener Ast *m*
f nœud *m* à découvert; nœud ouvert; branche *f* ouverte
r голый сучок *m*; открытый сучок

* **uncut → 8018**

* **uncut timber → 9079**

9051 uncut wood; holding wood; break-off allowance; crest; hinge; hold; holt; tipping edge
(the part of a tree left uncut until the end in

d Restbestand *m*; Vorratsanreicherung *f*
f charnière *f*
r недопил *m*; недоруб *m*

* **under bark → 1994**

* **underbark volume → 8952**

9052 **underbind**
d befestigte Baumposition *f*
f pose *f* fixe d'arbre
r защемлённое положение *n* дерева

9053 **underbind tree; fixed tree; fastened tree; clamped tree**
d befestigter Baum *m*; eingespannter Baum
f arbre *m* fixé
r защемлённое дерево *n*; зажатое дерево

* **underbody underguard → 670**

* **underguard → 670**

* **underbrush → 9065**

9054 **underburning**
d Unterbrand *m*
f brûlage *m* en sous-étage
r пал *m* под пологом леса

* **undercrop → 9070**

9055 **undercut area**
d Fallkerbfläche *f*
f zone *f* d'entaille
r площадь *f* подпила

* **undercut notch → 1214**

9056 **undercut seedling**
(nursery-grown seedling that has had its taproot cut by passing a blade or wire through the seedbed a few centimeters below ground level)
d eingeschnittener Pflänzling *m*
f plantule *f* dégagée
r подрезанный сеянец *m*

9057 **undercutter; chopper; bushman**
(a skilled woodsman who chops the undercut in trees so that they shall fall in the proper direction)
d Häcksler *m*
f boquillon *m*; bûcheur *m*
r рабочий *m*, производящий подпил

* **undercutting plough → 8771**

9058 **under-diameter material**
d unterlänges Material *n*
f matériau *m* de sous-diamètre
r [лесо]материал *m* ниже заданного диаметра

9059 **underground cutting**
d Abhieb *m* unter dem Erdboden
f exploitation *f* à culée noire
r срубка *f* под поверхностью земли

* **underground drainage → 8241**

* **underground fire → 2012**

* **ground fire → 2012**

* **underground water → 3995**

9060 **undergrowth; low growth**
(a loose term generally comprising both the herbaceous cover and the lower shrubs, and even the lowest trees, under a forest canopy)
d Unterwuchs *m*
f couverture *f* vivante
r подрост *m*

9061 **undergrowth damage; undergrowth disturbance**
d Niederdruck *m* von Jungwuchs
f perturbation *f* de sous-bois
r повреждение *n* подроста

9062 **under-length material**
d Unter-Länge-Material *n*
f matériel *m* de sous-longueur
r [лесо]материал *m* ниже заданной длины

9063 **underplant**
(young trees used for underplanting)
d Unterpflanze *f*
f plante *f* de sous-étage
r подпологовый саженец *m*

* **underplanting → 6364**

9064 **under-run**
(a loss in inventory volume, so that the amount manufactured or sold was less than was indicated by the volume of raw material)
d Unternutzung *f*

9065 **undershrub; underbrush; low bush; dwarf shrub; subshrub; subdumi**
(a plant with more or less woody branches and roots, less woody than a true shrub, but more woody than a perennial herbaceous plant)
d Unterbrusch *m*; Erdstrauch *m*; Halbstrauch *m*
f brousse *f* de sous-étage; espèce *f* sous-arbustive
r полукустарник *m*; кустарничек *m*

9066 **underside limb**
d Ast *m* auf der Unterseite
f branche *f* de surface intérieure
r сук *m*, расположенный с нижней стороны

9067 **undersized chips**
d Hackschnitzel *npl*; Hackspäne *mpl*
f copeaux *mpl* de dimension insuffisante; copeaux trop petits
r нестандартная мелкая щепа *f*

* **undersized tree → 7778**

* **undersowing → 6364**

* **understock → 6927**

* **understocked forest → 4446**

* **understocked stand → 4446**

9068 **understocking; insufficient stocking**
(stand of trees so widely spaced that, even with full growth potential realized, crown closure will not occur)
d Unterbestockung *f*
f boisement *m* insuffisant
r недостаточная сомкнутость *f*

9069 **understor[e]y fire regime**
(in which fires are generally not lethal to the dominant vegetation and do not substantially change the structure of the dominant vegetation)
d Unterholz-Feuerregime *n*
f régime *m* des incendies de sous-étage
r режим *m* пожара в нижнем ярусе

9070 **understor[e]y [of young regeneration]; lower story; underwood; undercrop**
(young trees that are growing beneath the tall mature trees in a timber stand)
d Unterstand *m*; Unterholz *n*; Niederholz *n*; Unter[holz]schicht *f*; Unterstufe *f*; Urwaldfolgebestand *m*
r второй ярус *m* насаждения; нижний ярус насаждения; подлесок *m*; подрастающий лес *m*

9071 **understor[e]y protection**
(removal of mature trees while damage to the understor[e]y is kept to a minimum)
d Unterstandschutz *m*
f protection *f* du sous-étage
r защита *f* подрастающего леса

9072 **understor[e]y trees**
(the portion of forest stand below the overstor[e]y; the shrubs and smaller trees between the forest canopy and the ground cover)
d Lasswald *m*
f arbres *mpl* de sous-étage
r деревья *npl* второго яруса; деревья нижнего яруса

* **underthinning → 8511**

9073 **underwater logging**
d Unterwasser-Holzrückung *f*
f débardage *m* engorgé
r лесозаготовка *f* на затопленных водой участках

* **underwood → 9070**

* **underwood silviculture → 6364**

9074 **undesirable tree species; weed-tree species**
(species that conflict with or do not contribute to the management objectives)
d Unholzart *f*
f essence *f* ligneuse indésirable
r сорная древесная порода *f*

9075 **undeveloped leaf; cataphyllary leaf; cataphyll**
(a leaf, or leaf-like organ, usually found below ground)
d unentwickeltes Blatt *n*
f cataphylle *f*
r кроющий лист *m*; [чешуйчатый] недоразвитый лист; катафилл *m* (почки)

9076 **undisturbed forest**
(which have been excluded from human intervention for a long time, thus allowing natural succession and natural disturbances to occur)
d nichtgestörter Wald *m*

9077 **undisturbed vegetation**
d nichtgestörte Vegetation *f*
f végétation *f* non détériorée
r ненарушенная растительность *f*

* **undressed timber → 9045**

* **undulated → 1230**

* **undulated figure → 2144**

9078 **undulated texture**
d Maser[textur] *f*
f texture *f* ondulée
r волнистая текстура *f*

9079 **unedged [sawn] timber; waney sawn timber; uncut timber; waney wood; backsided timber; round-edged timber**
(a timber sawn or hewn from the log with parallels faces but the edges left uncut, i.e. naturally rounded)
d unbesäumtes Schnittholz *n*; wahnkantiges Holz *n*; waldkantiges Holz
f bois *m* scié non avivé; bois flacheux; bois non déligné
r необрезной распиленный материал *m*; необрезной пиломатериал *m*; древесина *f* с обзолом

* **uneven-aged forest → 172**

* **uneven-aged [forest] management → 173**

9080 **uneven-aged [silvicultural] system**
(silvicultural system in which stands have an uneven-aged structure)
d ungleichaltrige Forstwirtschaft *f*
f régime *m* inéquien
r разновозрастное лесное хозяйство *n*

9081 **uneven-aged stand; unevened stand; all-aged stand; irregular stand**
(stand of trees containing three or more age classes)
d ungleichaltriger Bestand *m*; unregelmäßiger Bestand
f peuplement *m* inéquien; peuplement d'âges multiples; peuplement inégal; peuplement irrégulier
r разновозрастное насаждение *n*; неоднородное насаждение

9082 **uneven grain; uneven texture**
d ungleichartige Textur *f*
f texture *f* hétérogène
r ... слоёв древесины); текстура, неравномерная по окраске и рисунку

9083 **unfilled surface**
(of wood)
d unbesetzte Fläche *f*
f surface *f* non moulée; surface vide
r неотделанная поверхность *f*; поверхность с открытыми порами

* **unfilled wood → 535**

9084 **unfolding of leaves; foliage expansion; leafing**
d Laubausbruch *m*; Blattentfaltung *f*; Laubaustrieb *m*; Laubentfaltung *f*
f épanouissement *m* des feuilles; foliaison *f*
r распускание *n* листьев; развёртывание *n* листьев

* **unforested → 8874**

* **unhomogeneous forest → 5486**

* **unhomogeneous stand → 5698**

9085 **unifoliate leaf; unifoliolate leaf**
d Einblatt *n*
f feuille *f* unifoliaire; feuille unifoliée
r однолопастный лист *m*

9086 **uniform shelterwood [cutting]**
d gleichförmiger Shirmschlag *m*
f coupe *f* progressive uniforme
r равномерная постепенная рубка *f*

9087 **uniform shelterwood method**
(in which the canopy is opened fairly evenly throughout the regeneration area)
d Methode *f* des einheitlichen Schirmschlages
f mode *m* de régénération par coupes progressives uniformes
r способ *m* равномерных постепенных рубок

* **uniform stand → 2926**

9088 **uniform strip system**
d Breitsaumschlagbetrieb *m*
f système *m* de traitement par coupes en lisières allongées
r система *f* ширококаёмчатой рубки

* **uniform texture → 2923**

9089 **uniform volume curve**
d Einheitsmassenkurve *f*

9090 unilaterally compound pitting
(pitting in which one pit subtends two or more smaller pits in the cell adjacent)
d einseitig zusammengesetzte Tüpfelung *f*
f ponctuations *fpl* unilatéralement groupées
r односторонне комбинированная поровость *f*

9091 unilaterally paratracheal parenchyma; abaxial parenchyma; adaxial parenchyma
(paratracheal parenchyma limited to the outer (abaxial) or inner (adaxial) sides of the vessels)
d abaxiales Parenchym *n*
f parenchyme *m* abaxial
r абаксиальная паренхима *f*

9092 uninodal
(of conifers forming a single whorl of branches each year)
d einquirlig
f uninodal; monocyclique
r одномутовчатый (о типе ветвления стволов деревьев хвойных пород)

9093 uninodal bud
d einquirlige Knospe *f*
f pousse *f* uninodale
r одноузловая почка *f*

9094 uniseriate ray; uniserial ray
(a ray one cell wide as seen in tangential section; arranged in one row)
d einreihiger Strahl *m*
f rayon *m* unisérié
r однорядный луч *m*

9095 unit of chips; chip unit
(chip volume equal to 1 cord of pulpwood)
d Spaneinheit *f*
f unité *f* de copeaux
r юнит *m* щепы

9096 unit of cleared forest
d abgerechnete Waldparzelle *f*
f unité *f* de forêt défrichée
r очищенный лесной участок *m*

9097 unit of cutting; cutting unit
(under the cutting system of grading lumber, an area of face of 12 in^2)
d Hiebseinheit *f*
f débit *m* élémentaire
r рабочий блок *m* лесосеки

f base *f* de vente à unité de volume
r система *f* продажи леса на корню по цене за единицу объёма фактически заготовленной древесины

* **unit sale → 6119**

9099 unit weight
d Einheitsgewicht *m*
f poids *m* spécifique
r объёмная масса *f*

9100 universal soil loss equation; USLE
d allgemeiner Bodenverlustausgleich *m*
f équation *f* universelle d'érosion des sols
r универсальное уравнение *n* смыва почвы; универсальная формула *f* расчёта потерь почвы

* **university of forestry → 3462**

9101 unlimbed
d nicht astreinig
f non ébranché
r с необрезанными сучьями

9102 unloading point
d Entladestelle *f*
f partie *f* terminale [d'un glissoir]
r разгрузочный пункт *m*; место *n* разгрузки

* **unlucky tree → 4601**

* **unmanaged forest → 9122**

9103 unmanaged forestland
(forest land that is not subject to management under a forest management plan)
d nichtgeleiteter Waldboden *m*
f terre *f* forestière non gestionnée
r неуправляемая лесная земля *f*

* **unmerchantable log → 5764**

* **unmerchantable stand → 5765**

9104 unmerchantable tree
d Nichtwirtschaftsbaum *m*
f arbre *m* non marchand
r неделовое дерево *n*; нетоварное дерево

* **unmixed stand → 2926**

* **unpeeled chips → 9043**

9105 unproductive forest
(all forest which is not regularly managed and have an increment in volume less than 1 m^3/ha/year in the foreseeable future)
d nichtproduktiver Wald *m*
f forêt *f* non productive
r непродуктивный лес *m*

* **unproductive forestland → 5749**

9106 unproven stocking forestland
(that has had the tree cover removed or killed and where the subsequent degree of stocking has not been assessed)
d Waldboden *m* mit unbewiesene Schlussgrad
f terre *f* forestière à degré d'occupation non prouvée
r лесная земля *f* с непроверенного запаса

9107 unquantified stocked forestland
(for which the distinction of partially and fully stocked is not available)
d nichtquantifizierter bestockte Waldboden *m*
f terre *f* forestière boisée non quantifiée
r неопределённая количественно облесённая земля *f*

9108 unrecovered timber
d nicht zurückgehendes Holz *n*
f bois *m* résiduel
r остаточная древесина *f*

9109 unrecovered volume
(timber that is within the cutting specifications of the minimum utilization standards of the cutting authority and not removed from the area)
d nicht zurückgehendes Volumen *n*
f volume *m* [de bois] résiduel
r остаточный объём *m* [древостоя]

9110 unregulated acres
(commercial forestland not organized for sustained yield production of wood)
d nichtregulierte Länder *npl*
f terres *fpl* sans réglementation
r неурегулированные земли *fpl*

* **unregulated felling → 9046**

9111 unregulated forest
(commercial forest land that will not be organized for timber production under sustained-yield principles)
d nichtregulierter Wald *m*

* **unripe → 4674**

9112 unsalvaged losses
(volume of timber destroyed by natural causes)
d nichtgeborgene Verluste *mpl*
f pertes *fpl* non récupérées
r невосстановляемые затраты *fpl*

9113 unsawn
d nichtgesägt
f non scié
r нераспиленный

9114 unscrambler; unscrambling system
(a transport device regulating aligned timber disposition)
d Scharniergerät *n*
f système *m* démêleur; démêleur *m*; aligneuse *f*
r наклонный транспортёр *m*, не повреждающий поверхность досок

* **unscrambling knuckleboom → 4740**

9115 unscreened chips
d unsortierte Späne *mpl*
f copeaux *mpl* non triés
r несортированная щепа *f*

9116 unseasoned timber; green-end production; unseasoned wood
(timber in which the average moisture content exceeds 25 %)
d nichttrockenes Holz *n*
f bois *m* non séché
r невысушенная древесина *f*; невыдержанная древесина; сырые пиломатериалы *mpl*

9117 unsorted sawnwood; saw falling
d unsortiertes Schnittholz *n*
f bois *m* scié non trié
r несортированные пиломатериалы *mpl*

* **unsound → 7212**

* **unsound in the heart → 4201**

* **unsound knot → 2290**

9118 unsound wood; faulty wood
d Fehlholz *n*
f bois *m* défectueux
r фаутная древесина *f*; фаутный

9119 unspecified forest plantation
d nicht spezifizierte Forstkultur *f*
f plantation *f* forestière non spécifiée
r лесное насаждение *n* без спецификации

9120 unspecified timber-productivity
d nicht spezifizierte Holzproduktivität *f*
f productivité *f* de bois non spécifiée
r неопределённая продуктивность *f* леса

* **unstable slope → 7441**

* **unstained wood → 1035**

* **unstocked forestland → 5773**

9121 unsuitable forestland
(forest land that is not managed for timber production)
d ungeeignete Waldfläche *f*
f terre *f* forestière non appropriée
r непригодная лесная площадь *f*

9122 untended forest; unmanaged forest
d ungepflegter Wald *m*
f forêt *f* non-soignée
r неухоженный лес *m*

9123 unthinned forest ecosystem
d undurchforstetes Waldökosystem *n*
f écosystème *m* forestier non éclaircie
r экосистема *f* леса, не пройдённая рубками ухода; непрорубленная лесная экосистема

9124 untopped tree
d Baum *m* mit keiner reinigen Spitze
f arbre *m* à cime non élaguée
r дерево *n* с необрезанной вершиной; дерево с необрубленной вершиной

* **untouched forest → 282**

* **untransplanted plant → 9420**

* **unwinding → 6121**

* **unwooded → 8874**

* **unwrought timber → 7096**

9125 upas tree; chenchen
(Antiaris toxicaria)
d Upasbaum *m*
f chenchen *m*
r антиарис *m* ядовитый

f seconde taille *f* d'une lime; seconde frappe *f* d'une lime
r верхний напил *m*

9127 upgrowth
(the process or result of growing up)
d Aufwuchs *m*; Aufschlag *m*
f croissance *f* vers le haut
r произрастание *n*

9128 uphill felling
d Holzeinschlag *m* nach oben
f abattage *m* vers le haut
r валка *f* [леса] в направлении вверх по склону

9129 uphill skidding
(of logs or other loads)
d Rücken *n* nach oben
f débusquage *m* vers le haut
r трелёвка *f* вверх по склону

9130 uphill thinning
d Durchforstung *f* nach oben
f éclaircie *f* vers le haut
r рубки *fpl* ухода при трелёвке вверх по склону

9131 upland forest [area]; high-plateau forest; table-land forest
(forestland away from streams outside of riparian areas where soil and vegetative conditions are not influenced by localized streamside or wetland conditions)
d Hügellandwald *m*
f forêt *f* sur collines
r лес *m* в холмистой местности

9132 upland hardwood
(stands that have at least 10 percent stocking and classed as an oak-hickory or maple-beech-birch forest type)
d Berglaubholz *n*
f bois *m* feuillu de montagne
r широколиственный горный лес *m*

* **upper crop → 3518**

9133 upper crown height
d obere Kronenhöhe *f*
f hauteur *f* supérieure des cimes; hauteur à la première couronne
r верхняя высота *f* [расположения] кроны (от первой мутовки живых ветвей)

f longueur *f* de cime supérieure
r верхняя *f* длина кроны

* **upper cut → 8697**

9135 **upper depot; upper landing place**
d Einkehr *f*; kleiner Lagerplatz *m*
f dépôt *m* supérieur
r верхний склад *m*

* **upper growth → 3518**

9136 **upper mountain forest**
d Höchstbergwald *m*
f forêt *f* de haute montagne
r высокогорный лес *m*

9137 **upper rope**
d Oberseil *n*
f câble *m* supérieur
r верхний трос *m*

* **uppers → 4284**

9138 **upper stem portion**
(that portion of the central stem of sawtimber trees between the saw-log top and the minimum top diameter of 4 inches outside bark)
d oberer Stammanteil *m*
f section *f* de grume supérieure
r верхняя часть *f* брёвна

* **upper stor[e]y → 3518**

* **upright → 6475**

9139 **upright ray cell**
(a ray cell with its longest dimension axial)
d aufrechte Markstrahlzelle *f*
f cellule *f* de rayon dressée
r прямостоячая лучевая клетка *f*

9140 **uproot** *v*
d roden
f déraciner
r корчевать

9141 **uprooted tree; assart[ed] tree**
d entwurzelter Baum *m*
f arbre *m* déraciné
r искоренелое дерево *n*

* **uprooter → 8808**

* **uprooting → 2857**

* **uprooting machine → 8808**

* **upset → 2039**

* **upward flow of the sap → 4448**

* **upwind side → 9451**

9142 **urban forest; city forest; town forest; amenity forest; municipal forest**
(forest in an urbanized environment)
d Stadtforst *m*
f forêt *f* urbaine; forêt d'agrément
r городской лес *m*

* **urban forestry → 5636**

9143 **urgency of regeneration**
d Verjüngungsdringlichkeit *f*
f urgence *f* du rajeunissement
r неотложность *f* восстановления

9144 **urn; moss capsule**
d Moosbüchse *f*
f urne *f*
r урна *f*; урночка *f*

9145 **urn-gum**
(Eucalyptus urnigera)
d Urnen-Eukalyptus *m*
f eucalyptus *m* à urne; gommier *m* à fruits en forme d'urne
r эвкалипт *m* урновидный; эвкалипт урноносный; эвкалипт коробочконосный

* **urucurana → 6237**

* **usable timber → 5400**

9146 **used length**
(of timber)
d Gesamtschnittholzlänge *f* aus einem Stamm
f longueur *f* totale d'une grume susceptible d'être débitée en sciage; longueur de bois commercialisable
r общая длина *f* пиловочника

* **use of wood → 9154**

9147 **user-friendly forestry**
d anwenderfreundliche Waldwirtschaft *f*
f foresterie *f* amicale
r лесное хозяйство *n*, удобное для пользователя

9148 Ussurian pear
(Pyrus ussuriensis)
d Ussur-Birne *f*
f poirier *m* d'Ussur
r груша *f* уссурийская

* **Utah ash → 7641**

9149 Utah juniper
(Juniperus osteosperma; Juniperus utahensis; Sabina osteosperma)
d Utah-Johannisbeere *f*
f genévrier *m* d'Utah
r можжевельник *m* утахский

9150 utile; sipo; assie
(Entandrophragma utile)
d Sipo *m*
f sipo *m*
r энтандрофрагма *f* полезная; красное дерево *n* из Ассие; сипо *m*

* **utility forestry → 5636**

9151 utility hardwood
d anwenderfreundliche Forstwirtschaft *f*
f bois *m* feuillu dur d'utilité
r твёрдолиственная древесина *f* для конструкционных целей

9152 utilization after target diameter
d Zielstärkennutzung *f*
f exploitation *f* selon diamètre d'exploitabilité
r эксплуатация *f* согласно заданным диаметром

9153 utilization of forest litter; forest litter removal
d Bodenstreunutzung *f*
f ramassage *m* de litière
r сбор *m* лесной подстилки; пользование *n* подстилкой

9154 utilization of wood; use of wood; timber utilization; wood utilization
(dimensions and quality of timber that is actually cut and removed from an area)
d Holzverwendung *f*
f utilisation *f* de bois; emploi *m* de bois
r утилизация *f* древесины

9155 utilization percent
(of wood)
d Nutzungsprozent *n*; Ausnutzungsprozent *n*
f taux *m* de production normal
r процент *m* выхода [древесины]

(dimensions (stump height, top diameter, base diameter, and length) and quality of trees that must be cut and removed)
d Nutzungsstandard *mpl*
f standards *mpl* d'utilisation
r стандарты *mpl* использования

* **utilization value → 2952**

V

9157 **vaccinate** *v*
d impfen; vakzinieren
f vacciner
r инокулировать; делать прививку вакциной

* **vaccine → 4507**

9158 **vaccinium forest type; cranberry forest type**
d Kränbeerewaldtyp *m*
f forêt *f* de type canneberge
r тип *m* леса брусничный

* **vaco → 5240**

9159 **valley oak; California white oak**
(Quercus lobata)
d kalifornische Weißeiche *f*
f chêne *m* blanc de Californie
r дуб *m* дольчатый

* **Valonia[n] oak → 3937**

* **valuable forests → 8024**

9160 **valuable hardwood; broadleaved wood of high value**
d Edellaubholz *n*
f bois *m* de feuillus précieux; feuillu *m* noble
r ценная широколиственная древесина *f*

9161 **valuable timber tree**
d Edelbaumholz *n*
f arbre *m* à bois de valeur
r ценная поделочная порода *f* дерева; ценное строевое дерево *n*

9162 **valuable tree species**
d Wertholzart *f*; Wertbaumart *f*
f espèce *f* forestière de [grande] valeur; espèce forestière précieuse
r ценная древесная порода *f*

* **valuable wood → 4284**

* **valuation crew → 8288**

9163 **valuation of growing stock**
d Vorratsbewertung *f*

* **valuation survey → 2838**

* **value increment → 4455**

* **value of the stand → 8196**

9164 **values-at-risk; VaR**
(specific or collective set of natural resources and man-made improvements/developments that have measurable or intrinsic worth and that could or may be destroyed or otherwise altered by fire in any given area)
d Risikowert *m*; Wert *m* des Risikos
f valeurs *fpl* en risques; existences *fpl*
r стоимость *f*, подверженная риску; рисковая стоимость

9165 **vanilla tree**
(Vanilla planifolia)
d echte Vanille *f*
f vanille *f*
r ванильное дерево *n* платанолистное

* **VaR → 9164**

* **var fir → 969**

9166 **variable-area plot sampling method**
(method of timber cruising commonly used for industrial timber cruising in which sampling area (plot size) varies with tree diameter)
d Methode *f* der variable Flächenstichprobennahme
f méthode *f* d'estimation par placettes d'échantillon variables
r метод *m* переменной закладки пробных площадей

9167 **variable density yield table**
d Ertragstafel *f* mit variierbarer Dichte
f table *f* de production à densité variable
r таблица *f* роста насаждений различной полноты

* **variable plot cruising → 285**

9168 **variable plot radius**
d variabeler Grundstückradius *m*
f placette *f* à rayon variable
r переменный радиус *m* площади

9169 **variable retention**
(relatively new silvicultural system that follows nature's model by always retaining

f rétention *f* variable
r переменное выдерживание *n*; переменное задержание *n*

9170 variegation
d Panaschierung *f*; plasmatische Vererbung *f*
f variégation *f*
r вирусная мозаика *f*; пёстролистность *f*

9171 variety
d Varietät *f*
f variété *f*
r сорт *m*

* **vas → 9687**

9172 vascular plant
(plant possessing specialized conducting tissue consisting of xylem and phloem)
d Gefäßpflanze *f*
f plante *f* vasculaire
r сосудистое растение *n*

* **vascular tissue → 3105**

9173 vascular tracheid; imperfect vessel member
(an imperforate cell resembling in form and position a small vessel member)
d Gefäß-Tracheide *f*
f trachéide *f* vasculaire; élément *m* de vaisseau imparfait
r сосудистая трахеида *f*

9174 vasicentric parenchyma
(paratracheal parenchyma forming a complete sheath around a vessel, of variable width and circular or slightly oval in cross section)
d vasizentrisches Parenchym *n*
f parenchyme *m* circumvasculaire; parenchyme en manchon; parenchyme juxtavasculaire
r вазицентрическая паренхима *f*

9175 vasicentric tracheid
(a short, irregularly-formed tracheid in the immediate proximity of a vessel and not forming part of a definite axial row)
d vasizentrische Tracheide *f*
f trachéide *f* juxtavasculaire
r вазицентрическая трахеида *f*

* **vegetable cover → 6337**

* **vegetable refuse → 6371**

* **vegetable tar → 9672**

9176 vegetation
d Vegetation *f*
f végétation *f*
r растительность *f*

9177 vegetational mosaic
d Vegetationsmosaik *n*
f formations *fpl* végétales en mosaïque
r растительная мозаика *f*

* **vegetation cover → 6337**

9178 vegetative
(not flowering)
d vegetativ
f végétal
r растительный; вегетативный

9179 vegetative bud; growth bud; shoot bud; wood bud
(bud which, on opening, gives rise to a vegetative shoot)
d vegetative Knospe *f*; Holzknospe *f*
f œil *m* à bois; bourgeon *m* à bois; bouton *m* à bois
r побеговая почка *f*; древесная почка

* **vegetative cover → 6337**

9180 vegetative lot
(quantity of vegetative material or vegetative propagules having the same species, source and year of collection)
d vegetative Gruppe *f*
f partie *f* de matériel végétatif
r партия *f* вегетативного материала

9181 vegetative material
(plant parts or tissues used to produce vegetative propagules through asexual means)
d vegetatives Material *n*
f matériel *m* végétatif
r вегетативный материал *m*

* **vegetative point → 4007**

9182 vegetative propagation; asexual propagation; asexual reproduction
(reproduction by other than sexually produced seed)
d vegetative Vermehrung *f*; Fortpflanzung *f*
f multiplication *f* végétative; propagation *f* asexuée; reproduction *f* végétative

9183 **vegetative propagule; propagule; ramet**
(any spore, seed, fruit or other part of a plant or microorganism capable of producing a new plant and used as a means of dispersal)
d Brutknospe *f*; Keimknospe *f*; Propagula *n*
f ramet *m/f*; propagule *f*
r часть *f* растения для вегетативного размножения

9184 **vegetative regeneration**
d vegetative Regenerierung *f*
f régénération *f* végétative
r вегетативное возобновление *n* растений

* **vegetative shoot → 4505**

* **vein → 4882**

* **veinage → 9187**

9185 **veined growth; wavy-fibered growth; curly[-grained] growth; curl**
d Maserwuchs *m*; Schraubenwuchs *m*; Wimmerwuchs *m*; Wellenwuchs *m*
f croissance *f* madrée; croissance ondulée
r свилеватый [на]рост *m*

9186 **veined wood; figured wood**
d Maserholz *n*; gemasertes Holz *n*; maseriges Holz
f bois *m* veiné; bois madré
r древесина *f* с узорчатой текстурой; узорчатая древесина

9187 **veining; veinage; figuring; figure spotting; mottling; curling; gnarl formation**
(formation of curls or spirals)
d Maserung *f*; Maserbildung *f*
f tachage *m*; veinage *m*; roulage *m*
r образование *n* свили; развитие *n* свилеватости

* **veld[t] → 1208**

9188 **velvet ash; Arizona ash**
(Fraxinus velutina)
d Arizona-Esche *f*
f frêne *m* velours
r ясень *m* бархатный

9189 **velvet mesquite**
(Prosopis velutina)
d erstauner Mesquitebusch *m*
f prosope *m* velours
r мескитовое дерево *n* бархатное

9190 **velvet tamarind; jutahy; guapaque; paleta**
(Dialium guianense; Arouna guianensis; Dialium divaricatum)
d Palo Sangre *m*
f diale *m* de Guinée
r тамариндо *n*; диалиум *m* гвинейский

9191 **velvet tree**
(Miconia calvescens)
d Miconia *m*
f miconia *m*
r микония *f*

9192 **venation; nervation**
(the manner in which the veins are arranged in the leaf)
d Äderung *f*; Nervatur *f*
f nervation *f*
r жилкование *n*; иннервация *f*; расположение *n* жилок

* **veneer → 6398**

* **veneer bolt → 901**

9193 **veneer butt; peeler log; peeler; veneer log; veneer bolt; plywood log; plylog**
d Schälfurnierblock *m*; Furnierblock *m*; Messerblock *m*; Schälrolle *f*
f bille *f* de tranchage; grume *f* de déroulage
r фанерный кряж *m*; фанерный брус *m*; фанерное бревно *n*; облупленный кругляк *m*

9194 **veneer chipper; veneer hog**
d Furnier-Hackzerspaner *m*
f fragmenteuse *f* de placage; broyeur *m* de placage
r дробилка *f* отходов шпона

9195 **veneered chipboard**
d furnierte Spanplatte *f*
f panneau *m* de particules plaqué; panneau en agglomérés plaqué
r ДСП *f*, облицованная натуральным шпоном

* **veneer grafting → 331**

* **veneer hog → 9194**

9196 **veneering**
(facing a substrate with a thin layer of ornamental wood)
d Furnierung *f*

* **veneer[ing] wood → 901**

* **veneer lathe → 9623**

* **veneer log → 9193**

9197 veneer sheet
d Furnierschicht *f*; Furnierblatt *n*; Holzblatt *n*; Holzscheider *m*
f feuille *f* de placage
r фанерный лист *m*; фанерный слой *m*

* **Venetian sumac[h] → 2908**

9198 ventilation index; VI
(numerical value relating to the potential of the atmosphere to disperse airborne pollutants from a stationary source)
d Belüftungsindex *m*
f indice *m* d'aérage; indice de ventilation
r коэффициент *m* вентиляции

9199 vera[wood]
(Bulnesia arborea; Bulnesia sarmientoi)
d Vera *m*
f bulnésie *f* arborescente
r бульнезия *f* древовидная

9200 verdure
d Blattmasse *f*; Grünmasse *f*; Ziergrün *n*
f verdure *f*
r масса *f* листвы; зелёная листва *f*; зелёная масса; зелень *f*

9201 vermiculated
d befallen von Wurm
f attaqué par le ver
r поражённый червем

9202 vernal-autumnal pond
(a seasonal pool with surface water evident from spring thaw thru early June)
d temporärer See *m* von Herbst bis Frühjahr; temporäres Gewässer *n*
f étang *m* temporaire d'automne au printemps
r временный водоём *m*, наполняющийся водой весной и осенью

9203 vernalization
(exposure of a plant or plant organ to chilling, which subsequently brings about a development phase change)
d Tiefkühlung *f*
f vernalisation *f*
r яровизация *f*

* **vertical canopy → 9204**

9204 vertical closure [of crown]; vertical canopy
(in a crop or stand, the progressive reduction of space between crowns as they spread laterally, increasing the canopy density)
d Vertikal[kronen]schluss *m*; vertikale Kronenschluss *m*
f couvert *m* vertical au sol; aspect *m* serré d'un peuplement dans le sens vertical
r вертикальная сомкнутость *f* [кроны]; вертикальная сомкнутость полога

9205 vertical diversity
(diversity in an area that results from the complexity of the above ground structure of the vegetation)
d senkrechte Vielfalt *f*
f diversité *f* verticale
r вертикальное разнообразие *n*

* **vertical-grain[ed] timber → 6678**

* **vertical parenchyma → 457**

9206 vertical resistance
(resistance to plant pathogens conferred by specific, identified genes)
d vertikale Resistenz *f*
f résistance *f* verticale
r вертикальное сопротивление *n*

9207 vertical root
d Senkrechtwurzel *f*
f racine *f* verticale
r вертикальный корень *m*

9208 vertical structure
(of the vegetation)
d vertikale Struktur *f*
f structure *f* verticale
r вертикальная структура *f*

9209 verticillate leaf arrangement; whorl leaf [a]estivation
(arrangement along a twig or shoot in groups of three or more at each node)
d quirlige Blattdisposition *f*
f arrangement *m* de feuilles verticillé
r мутовчатое листорасположение *n*

* **verticil of branches → 995**

9210 vervain family
(Verbenaceae)
d Eisenkrautgewächse *pl*

9211 **very unstable terrain**
(terrain units classified as being in terrain class V in the coastal terrain stability classification)
d sehr rutschgefährdetes Gelände *n*; sehr instabiles Gelände; Rutschterrain *n*
f terrain *m* très instable
r весьма неустойчивая территория *f*

9212 **vessel element; vessel member; vessel segment**
(one of the cellular components of a vessel)
d faserförmiges Element *n*
f élément *m* de vaisseau
r членик *m* сосуда

9213 **vessel perforation**
(an opening from one vessel member to another)
d Gefäßdurchbrechung *f*
f perforation *f* de vaisseau
r сосудистая перфорация *f*

* **vessel segment → 9212**

9214 **vetch-leaf sophora; David's mountain laurel**
(Sophora viciifolia; Sophora davidii)
d wickenblättriger Perlschnurbaum *m*
f sophora *m* à feuilles de vesce
r софора *f* виколистная

* **veteran → 5885**

* **VI → 9198**

9215 **viability**
(population of sufficient numbers and reproductive potential to maintain its existence over time)
d Lebensfähigkeit *f*
f viabilité *f*
r жизненность *f*; живучесть *f*; жизнестойкость *f*

* **Victorian oak → 4054**

* **vigorous cutting-back → 2347**

9216 **vigorous tree**
d gutwüchsiger Baum *m*; frohwüchsiger Baum
f arbre *m* vigoureux
r жизнеспособное дерево *n*; здоровое дерево

9217 **vigo[u]r class**
(assumption of the health of a tree based on
f classe *f* de vigueur
r класс *m* мощности

9218 **village forest [area]**
(forest constituted and management for the supply of forest produce to a village)
d Dorfwald *m*
f forêt *f* villageoise
r деревенский лес *m*

9219 **village forest garden**
(larger than the homegarden, less densely planted and not so well tended)
d Dorfforstgarten *m*
f pépinière *f* forestière de village
r лесной деревенский питомник *m*

9220 **village woodlot**
d Dorfwaldstück *m*
f parcelle *f* forestière de village
r деревенский лесной участок *m*

9221 **vine family**
(Vitaceae)
d Weinrebengewächse *pl*; Rebengewächse *pl*
f vitacées *pl*
r виноградовые *pl*

* **vinegar tree → 1036**

* **vine maple → 5944**

9222 **vine [tropical rain] forest**
d [tropischer] Lianen-Regenwald *m*
f forêt *f* [dense tropicale] à lianes
r лиановый [влажный тропический] лес *m*

9223 **vineyards**
d Weinberge *f*
f vignobles *mpl*
r виноградники *mpl* (участки виноградников на территории лесного фонда)

* **vinicity of crown → 2069**

9224 **violet willow**
(Salix daphnoides)
d Seidelbastweide *f*; Schimmelweide *f*; Reifweide *f*
f saule *m* daphné
r верба *f*; шелюга *f* жёлтая; ива *f* волчниковая

9225 **violetwood**
(any of several hard purplish or reddish woods, especially kingwood)

f bois *m* d'amarante; bois violet
r древесина *f* амаранта; амарант *m*

9226 **virescence**
d Grünwerden *n*
f verdissement *m*
r зеленение *n*

9227 **virgin cork; female [layer of] cork; inferior layer of cork**
d unberührter Kork *m*
f liège *m* mâle
r девственная пробка *f*

* **virgin dip → 7824**

* **virgin forest → 282**

* **Virginia ash → 854**

9228 **Virginia creeper; American ivy; American woodbine creeper; [wild] woodbine; woody climber**
(Parthenocissus quinquefolia; Ampelopsis quinquefolia; Vitis quinquefolia)
d Jungfernrebe *f*; fünfblättrige Zaunrebe *f*; wilder Wein *m*
f vigne-vierge *f* [ordinaire]; vigne-vierge commune; parthénocisse *f* à cinq folioles
r дикий пятилистный виноград *m*

* **Virginian date plum → 1796**

* **Virginian pencil cedar → 2722**

9229 **Virginia pine**
(Pinus virginiana)
d virginische Kiefer *f*; Jersey-Kiefer *f*
f pin *m* de Virginie; pin de Jersey
r сосна *f* виргинская

9230 **Virginia-pine sawfly**
(Neodiprion pratii pratii)
d virginische Kiefernbuschhornblattwespe *f*
f diprion *m* de pin de Virginie
r пилильщик *m* виргинской сосны

* **Virgin's bower → 6387**

9231 **virgin tree**
d Urbaum *m*; Primärbaum *m*
f arbre *m* non gemmé
r девственное дерево *n*; незаподсоченное дерево

9232 **virola**
f virola *m*
r вирола *f*

9233 **viscoelastic behavior**
(of wood)
d viskoelastisches Benehmen *n*
f comportement *m* viscoélastique
r вязкоэластичное поведение *n*

* **viscous robinia → 1595**

9234 **visible crown diameter**
(diameter of a tree crown as discernible on a vertical aerial photograph)
d sichtbarer Kronendurchmesser *m*
f diamètre *m* de cime apparent; diamètre de cime visible
r различимый диаметр *m* кроны дерева

9235 **visible tree height**
(that part of the total height of a tree discernible on an aerial photograph)
d sichtbare Baumhöhe *f*
f hauteur *f* visible d'un arbre
r различимая высота *f* дерева (на аэрофотоснимке)

* **vista → 3564**

9236 **visual distance zone; background**
(area located from 3-5 miles to infinity from the viewer, characterized by perception of outline shape, landforms, and patterns of light and dark)
d Sichtentfernungszone *f*
f fond *m*; zone *f* de distance visuelle; zone de distance de visibilité
r задний план *m*

9237 **visual impact assessment**
(evaluation of the visual impact of resource development proposals on forest landscape)
d Sichtauswirkungseinschätzung *f*
f détermination *f* de l'impact visuel; évaluation *f* de l'impact visuel
r оценка *f* видимого воздействия; оценка видимого влияния

9238 **visual quality objectives; VQO**
(set of measurable goals for the management of forest visual resources)
d Sichtqualitätsziele *npl*
f objectifs *mpl* de qualité visuels
r видимые требования *npl* к качеству

9239 **vocational woods training**
d praktische Ausbildung *f* im Walde
f apprentissage *m* pratique en forêt
r практическое обучение *n* в лесу

* **void → 6463**

* **volatile oil → 2866**

9240 **volume age**
d Massenakter *n*
f âge *m* de la possibilité
r возраст *m* по массе

* **volume cal[l]ipers → 715**

9241 **volume curve; curve of estimated volume**
d Massenkurve *f*
f courbe *f* des volumes
r кривая *f* масс

* **volume density → 3370**

* **volume determination by quarter girth → 2118**

9242 **volume equation**
d Volumengleichung *f*
f équation *f* de volume
r объёмное уравнение *n*

* **volume growth → 9243**

9243 **volume increment; volume growth; growth in volume; mass-accretion; quantity increment**
d Massenzuwachs *m*
f accroissement *m* en volume
r прирост *m* по объёму; прирост по массе

* **volume inventory → 5266**

9244 **volume measurement by sections; sectional measurement of volume**
d sektionsweise Kubierung *f*
f cubage *m* par sections
r секционное определение *n* кубатуры

9245 **volume of branchwood**
d Astholzmasse *f*
f volume *m* du bois des branches
r объём *m* древесины сучьев

9246 **volume of harvested timber; cubic meter of harvested timber**
d Erntefestmeter *n*

9247 **volume of live trees**
(cubic-foot volume of sound wood in live trees at least 5.0 inches d.b.h. from a 1-foot stump to a minimum 4.0-inch top d.o.b. of the central stem)
d Grünbaumvolumen *n*
f volume *m* d'arbres verts
r объём *m* зелёных деревьев

9248 **volume of sawlog portion of sawtimber; volume of sawtimber; sawtimber volume**
(net volume in cubic-feet of the sawlog portion of sawtimber trees)
d Schneideholzvolumen *n*
f volume *m* du bois de sciage
r объём *m* пиловочника

* **volume of standing reserve → 9250**

9249 **volume of stem; stem volume; stem contents**
d Stammholzvolumen *n*; Stamminhalt *m*
f volume *m* du bois en grumes; volume des grumes
r кубатура *f* стволовой древесины; объём *m* стволовой части дерева; объём ствола

* **volume of timber → 9689**

9250 **volume of [timber-]growing stock; growing-stock volume; volume of timber stand; volume of standing reserve; stumpage volume; standing volume; growing-stock level; stocking level; inventory level**
(value or volume of a tree or group of trees as they stand uncut in the woods)
d stehender Holzvorrat *m*; Bestandesvorrat *m*; Massenvorrat *m*; stehende Holzreserve *f*
f volume *m* du matériel sur pied; réserve *f* [de bois] sur pied; niveau *m* de matériel sur pied; volume du bois sur pied
r объём *m* запаса [древостоя]; объём лесного сырья на корню

* **volume output → 9773**

* **volume production → 9773**

9251 **volume shrinkage**
d räumliches Schwinden *n*; Volumenschwindung *f*
f retrait *m* volumique
r объёмная усушка *f*

* **volume table → 8957**

9252 **volume when dry**
(of timber)
d Darrvolumen *n*; Volumen *n* in trockenem Zustand
f volume *m* à l'état sec
r объём *m* [лесоматериалов] в абсолютно сухом состоянии

9253 **volume when green; green volume**
(of wood)
d Freschmasse *f*; Freschvolumen *n*; Grünvolumen *n*
f volume *m* à l'état vert
r масса *f* свежесрубленной древесины; объём *m* древесиной зелени; объём свежесрубленных лесоматериалов

* **volume yield → 9773**

* **volunteer growth → 73**

* **VQO → 9238**

* **VUB → 8952**

* **vulnerable species → 7443**

9254 **vulnerary balsam**
d Wundebalsam *m*
f baume *m* cicatriciel
r раневой бальзам *m*

W

9255 **waferboard; structural board**
(a panel product of wafers of wood bound together by resin, heat, and pressure; can be made of species not suitable for lumber or plyboard production)
d Waffelplatte *f*
f panneau *m* de grandes particules; panneau gaufré; panneau structural
r вафельная плита *f*

* **wag[g]on building timber → 8592**

9256 **wag[g]on for long tailed timber; lumber wagon; timber wagon; log wagon; pole wagon**
d Langholzwaggon *m*
f wagon *m* pour bois long
r вагон *m* для хлыстов

* **wag[g]on sled → 2587**

* **wag[g]on stock → 8592**

9257 **wainscot [billet]; wainscot plank**
(the lower part of an interior wall when finished in a material different from that of the upper part)
d Paneellierholz *n*; Wagenschoßbohle *f*
f bois *m* à panneaux; bois refait
r панельная древесина *f*; ванчёс *m*; средняя часть *f* брёвна, остающаяся после выпиливания брусьев

9258 **wainscot oak**
d Paneellier-Eichenholz *n*
f bois *m* de chêne à panneaux; bois de chêne refait
r древесина *f* дуба радиального распила; дуб *m* для переработки на ванчёсы

* **wainscot plank → 9257**

9259 **walking-stick cholla; tree cholla**
(Opuntia imbricata)
d Cholla *f*
f cholla *m*
r опунция *f* черепитчатая

* **walnut → 1808**

9260 **walnut family**
(Junglandaceae)
d Walnuss[baum]gewächse *pl*; Juglandazeen *pl*
f juglandacées *pl*
r ореховые *pl*

9261 **walnut mountain forest**
d Walnuss-Gebirgswald *m*
f noyeraie *f* de montagne
r ореховый горный лес *m*

* **walnut oil → 5822**

9262 **walnut[-tree]**
(Juglans)
d Walnussbaum *m*; Nussbaum *m*; Walnuss *f*
f noyer *m*
r орех *m*

* **Walter's pine → 7956**

9263 **wand**
(a long, relatively thin, stem cutting, as commonly used in propagating willows)
d Streunensteckling *m*
f plançon *m*
r длинный тонкий черенок *m*

9264 **wander[ing] heart; pith wander**
(condition in which the pith deviates considerably from the growth axis and may therefore outcrop on a longitudinal cut surface of the timber)
d Umherziehenkern *m*; Streunenkern *m*
f moelle *f* divaguante
r блуждающая сердцевина *f*

9265 **wandoo eucalyptus**
(Eucalyptus wandoo)
d Wandoo *m*
f wandoo *m*
r эвкалипт *m* ванду

* **wane → 6782**

* **wane[y edge] → 2662**

9266 **waney log**
d unbesäumter Stamm *m*
f grume *f* non délignée; grume de bois flacheuse
r окорённое бревно *n*, слегка стёсанное на четыре канта

* **waney sawn timber → 9079**

* **want → 9690**

* **warm house → 3946**

* **warping → 6120**

9267 **warping drum**
d Haspelwerk *n*
f tambour *m* d'enroulement [de câble de débardage]
r мотовило *n*

9268 **warping of wood; distortion of wood; casting of lumber; distortion of lumber**
(drying defect caused by the differential shrinkage along the three axes of a piece of wood)
d Krümmen *n* des Holzes; Werfen *n* des Holzes
f distorsion *f* de bois
r коробление *n* древесины; коробление пиломатериалов; покоробленность *f* древесины

9269 **warp restraint**
(in drying lumber and other wood products the application of external loads to a pile, package, or kiln truckload to prevent or reduce warp)
d Werfenbeschränkung *f*
f restriction *f* de la cannelure
r ограничение *n* коробления

9270 **washboarding**
d Waschbrettschnitt *m*
f frisure *f*; tôle *f* ondulée
r волнистый характер *m* поверхности древесины [вследствие неравномерной усушки ранних и поздних слоев]

9271 **Washington palm**
(Washingtonia robusta)
d mexikanische Fächerpalme *f*
f palmier *m* du Mexique
r пальма *f* Вашингтона

9272 **Washoe pine; yellow pine**
(Pinus [ponderosa var.] washoensis)
d Washoe-Kiefer *f*
f pin *m* de Washoe
r сосна *f* Washoe

* **wastage in bark → 580**

* **waste → 998**

f résidus *mpl* de tronçonnage
r обрезки *fpl* концов лесоматериала

9274 **waste chips**
d Abfallhobelspan *m*; Abfallspan *m*
f copeaux *mpl* de déchets
r стружка-отходы *f* при строгании; строгальная стружка *f*

9275 **waste[ ]land; bare land; barren land; badland[s]**
(more or less barren, rough, broken land that is strongly dissected or gullied by streams)
d Ödland *n*
f terrain *m* nu; terre *f* nue; terrain en friche; mauvaise terre
r бросовые земли *fpl*; пустошь *f*; пустырь *m*; безлесье *n*

9276 **waste tips; waste site; landfill [site]; dumpsite**
d Deponie *f*
f dépotoirs *mpl*
r отвальная куча *f*; отвал *m*

9277 **waste[wood]**
d ausfallendes Holz *n*; Ausfallholz *n*
f bois *m* restant; bois résidu
r оставленная древесина *f*; брошенная древесина

9278 **wastewood chipper; wood chipper**
d Holzzerkleinerungsmaschine *f*
f découpeuse *f* de bois à rémanents
r рубильная машина *f* для древесных отходов; дробилка *f* для дров

* **wood hog → 7721**

* **wastewood hog → 7721**

* **water ash → 777**

* **water-bar → 3976**

* **water-beech → 249**

9279 **water birch; western paper birch**
(Betula occidentalis)
d Himalaja-Wasserbirke *f*
f bouleau *m* occidental; bouleau d'eau; bouleau fontinal
r берёза *f* западная

* **water bomber → 157**

9280 water catchment; water intake; watershed area; catchment basin; collecting area
(an area of land that collects and discharges water into a single main stream through a series of smaller tributaries)
d Quellgebiet *n*; Quellbezirk *m*; Einlauf-Bauwerk *n*; Wasserfassung *f*; Wasserentnahme *f*; Wassereinzugsgebiet *n*
f prise *f* d'eau; captage *m* [de l'eau]; bassin *m* versant; bassin d'alimentation d'une source
r водосборный бассейн *m*; водосбор *m*; область *f* истока

9281 water chute; wet chute; water slide; [log] flume
(a trough in which water runs, used it transporting logs, lumber or timbers)
d Holzwasserrinne *f*; Wasserriese *f*; Nassriese *f*
f glissoire *f* à l'eau pour le transport du bois; glissoire humide
r водный лесоспуск *m*; водный лоток *m*; водяной лоток; лесосплавный лоток

9282 water-conservation forest; water-regulation forest
d Gewässerschutzwald *m*; Gewässerregelungswald *m*
f forêt *f* de conservation des eaux; forêt de régularisation des eaux
r водорегулирующий лес *m*

9283 watercore; waterheart; wet heart
d Feuchtkern *m*; Nasskern *m*
f cœur *m* mouillé; cœur humide
r водослойная древесина *f* ядра; водослой *m*; водяное пятно *n*

9284 watercourse; stream channel; streambed
(open conduit either natural or artificial, which periodically or continuously contains moving water, or which forms a connecting link between two bodies of water)
d Wasserlauf *m*; fließendes Gewässer *n*; Gewässerbett *n*
f cours *m* d'eau
r канал *m*; пролив *m*

* **water elder → 2890**

* **water elm → 6320**

9285 waterfall
d Wasserfall *m*
f chute *f* d'eau; cascade *f*
r каскад *m* (уступчатая система подачи брёвен для размолёвки плота)

9286 water harvesting
(general term for any means of collecting and re-distributing runoff water)
d Wasserernte *f*
f récolte *f* de l'eau; récupération *f* de l'eau; conservation *f* des eaux de surface
r сбор *m* поверхностного стока

* **waterheart → 9283**

* **water hickory → 6130**

9287 water-holding capacity; water-withholding power
(of a soil, the amount of water that it is capable of holding, between field capacity and permanent wilting point)
d Wasserbindevermögen *m*; Wasserhaltekraft *f*
f capacité *f* de rétention en eau
r водоудерживающая способность *f*; водопоглотительная способность

* **water intake → 9280**

9288 waterjet cutting
d Schneiden *n* mit Wasserstrahl
f abattage *m* par jet d'eau
r резание *n* с помощью водяной струи

9289 waterjet debarker
d Ringstrahlentrinder *m*; Wasserstrahlentrindung *f*
f écorceuse *f* par jet d'eau
r струйно-гидравлический окорочный станок *m*

9290 waterlogged timber; waterlogged wood
d Halbsenker *m*
f bois *m* entreposé dans l'eau
r полутопляк *m*

9291 water logging
(a process or condition in which the water table reaches or rises above the ground surface, or its capillary fringe is near to the surface)
d Vernässung *f*
f engorgement *m*
r подтопление *n*

9292 water log sorting; group distribution in main pocket
(of timber)
d Schnittholzsortierung *f* auf dem Wasser

9293 watermark disease
d Wasserstreifenkrankheit *f*
f coloration *f* bactérienne
r болезнь *f*, вызывающая водослой древесины

9294 water-oak; possum oak
(Quercus nigra)
d Schwarzeiche *f*
f chêne *m* noir
r дуб *m* чёрный

* **water-regulation forest → 9282**

9295 water resources
(supply of water in a given area or basin interpreted in terms of availability of surface and underground water)
d Wasserressourcen *mpl*
f ressources *fpl* d'eau
r водяные ресурсы *mpl*

* **watershed area → 9280**

9296 watershed assessment
(present state of watersheds and the cumulative impact of proposed development on peak flows, suspended sediment, bedload, and stream channel stability within the watershed)
d Wassereinzugsabschätzung *f*
f estimation *f* des bassins versants
r оценка *f* водосборных бассейнов

9297 watershed integrity
(stable overall physical condition of the watershed (bedrock, landforms, soils, drainage ways) within which transfers of energy, matter and, especially of water occur)
d Wasserscheide-Integrität *f*
f intégrité *f* des bassins versants
r целостность *f* водосбора

9298 watershed management
(planned use of drainage basins in accordance with predetermined objectives)
d Wasserscheidenmanagement *n*
f gestion *f* des bassins versants
r управление *n* водосборными бассейнами

9299 water shoot; water sprout; water twig
d Wasserreiser *m*; Wasserreis *n*; Wasserschoss *m*
f pousse *f* humide
r водяной побег *m*

f débusquage *m* dans l'eau
r трелёвка *f* по воде; водяная трелёвка

* **water slide → 9281**

* **water sprout → 9299**

9301 water storage; wet storage
d Lagerung *f* im Wasser; Wasserlagerung *f*
f conservation *f* de bois dans l'eau
r хранение *n* в воде

9302 water-stored logs; logs kept under water
d im Wasser gelagertes Rundholz *n*; wassergelagertes Rundholz
f bois *m* rond entreposé dans l'eau
r круглый лес *m*, хранимый в воде; лежащий в воде круглый лес

9303 water timber bundling; timber bundling
d Floßbinden *n* des Holzes; Holzverband *m*
f paquetage *m* de bois dans l'eau
r сплотка *f* леса; сплотка на воде

9304 water tupelo; swamp tupelo
(Nyssa aquatica)
d Wassertupelobaum *m*
f tupélo *m* aquatique
r нисса *f* водная

* **water twig → 9299**

* **water wattle → 8316**

* **water-withholding power → 9287**

* **wave growth timber → 2048**

* **wavy-fibered growth → 9185**

* **wavy-fib[e]red wood → 2048**

* **wavy figure → 2144**

* **wavy grain → 2145**

* **wavy-grained birch → 3122**

* **wavy-grown wood → 2048**

* **wax-berry → 8332**

* **waxmyrtle → 7858**

9305 wax myrtle family

f myricacées *pl*
r мириковые *pl*; восковниковые *pl*

9306 **wax tree; wax palm**
(Toxicodendron succedaneum)
d Wachspalme *f*; Talgsumach *m*
f sumac *m* feux-vernix
r сумах *m* резиноносный

9307 **wayfaring tree; twist wood**
(Viburnum lantana)
d wolliger Schneeball *m*; flaumhaarige Schlinge *f*
f viorne *f* mancienne; viorne flexible; mancienne *f*; boule *f* de neige cotonneuse; cochère *m*
r калина *f* гордовина; калина лантана

9308 **wayside trees**
d Alleebaum *m*
f arbres *mpl* à la route
r аллейные деревья *npl*; аллея *f*; придорожные деревья

9309 **weakened tree; loosened tree**
d lockerer Baum *m*
f arbre *m* affaibli
r ослабленное дерево *n*

* **weak thinning → 4952**

9310 **weald; open forest landscape**
d offene Waldlandschaft *f*
f paysage *m* forestier ouvert
r открытый лесной ландшафт *m*

* **wealth of species → 7902**

9311 **weathering**
(a mechanical or chemical disintegration and discolo[u]ration of the surface of wood caused by exposure to light, dust and sand carried by winds and the alternate shrinking and swelling of the surface fibers brought by changes in the weather)
d Verwitterung *f*
f altération *f* atmosphérique
r воздействие *n* атмосферных условий; старение *n* в атмосферных условиях

9312 **weather-proof tree; weather-resisting tree**
d Wetterbaum *m*
f arbre *m* fouetté par le vent
r погодоустойчивое дерево *n*

9313 **web-spinning pine sawfly**
(Acantholyda hieroglyphica)
r одиночный пилильщик-ткач *m*

* **wedge → 3079**

* **wedge angle → 7493**

9314 **wedge** *v* **a tree; throw** *v*; **trip** *v*
(topple over with wedges a tree that is being felled)
d einkeilen
f coincer
r свалить дерево с помощью клина

* **wedge axe → 453**

9315 **wedge cutting**
d Keilschlag *m*
f coupe *f* en coin
r клиновидная рубка *f*; клиновой вруб *m*

* **wedge grafting → 1635**

* **wedge joint → 3161**

9316 **wedge method**
(modification of the strip shelterwood method in which cuttings begin as narrow, interior, wedge-shaped strips with the apex into the prevailing wind)
d Keilverfahren *n*
f mode *m* de régénération par coupes progressives en coin
r клиновой метод *m*

* **wedge planting → 289**

* **wedge prism → 6554**

* **wedge-shaped spade → 6365**

* **wedge shelterwood cutting → 7517**

9317 **wedge system**
d Keilschlagsystem *n*
f système *m* de régénération par coupes progressives en coin
r система *f* клиновидно-лесосечных рубок

9318 **wedging**
d Abkeilen *n*
f décolletage *m*
r расширение *n* пропила клином; расклинивание *n*

9319 **wedging cut**
d Spaltenschnitt *m*

9320 weed cutting; weeding; assistance cutting
(elimination of competing vegetation around young trees, for example, by cultivation, by the use of herbicides, or by cutting or slashing)
d Weideholzaushieb *m*; Jäten *n*; Verziehen *n*
f désherbage *m*; extraction *f* de rebut; extraction de mort-bois
r прополка *f* сорняков; расчистка *f* территории от кустарника; вырубка *f* мягких пород; вырубка сорных пород

* **weedicide → 4254**

* **weed killer → 4254**

9321 weed-tree; rogue [tree]
(any tree of a species having little or no economic value on the site in question)
d unerwünschter Baum *m*
f arbre *m* indésirable; indésirable *m*; arbre rogue; impureté *f*
r нежелательное дерево *n*

* **weed-tree species → 9074**

9322 weed wood
d Unholz *n*
f bois *m* indésirable; mort-bois *m*
r нежелательная древесина *f*

9323 weepiness
d Renkwuchs *m*
f suintement *m*
r плакучесть *f* ветвей; скручивание *n* и свисание *n* ветвей

9324 weeping; pendant; pendent; pendulous; suspended *adj*
d hängend; herabhängend; Trauer-
f pendant; pleureur
r висячий; плакучий; провислый; свисающий

* **weeping boree acacia → 9328**

9325 weeping Camperdown elm
(Ulmus glabra camperdownii)
d Weinenulme *f*
f orme de Camperdown
r ильм *m* плакучий

* **weeping elm → 9039**

9326 weeping growth habit
d Hängewuchsform *f*

9327 weeping mulberry
(Morus alba pendula)
d Hängemaulbeerbaum *m*
f mûrier *m* pleureur
r шелковица *f* плакучая

9328 weeping myall; weeping boree acacia
(Acacia pendula)
d Hänge-Akazie *f*
f acacia *m* pleureur
r акация *f* плакучая

9329 weeping purple beech
(Fagus sylvatica purpurea pendula)
d veredelte Hänge-Blutbuche *f*; Hängebuche *f*
f hêtre *m* pleureur pourpre
r бук *m* лесной плакучий

9330 weeping spruce; Brewer's spruce
(Picea breweriana)
d Hängefichte *f*; Brewersfichte *f*; Mähnenfichte *f*
f épicéa *m* pleureur; épicéa de Brewer
r ель *f* Бревера

* **weeping willow → 459**

9331 weevils
(Curculionidae)
d Rüsselkäfer *pl*
f pissodes *mpl*
r долгоносики *mpl*; слоники *mpl*

9332 weigela
(Weigela)
d Weigelie *f*; Weigela *f*
f weigela *m*
r вейгела *f*

9333 weighting tree
d Gewichtungsbaum *m*
f arbre *m* de charge
r весовое дерево *n*

9334 weight measurement to estimate volume
d Volumenabschätzung *f* mittels Gewichtmessung
f évaluation *f* de volume par barymétrie
r определение *n* объёма [лесоматериалов] взвешиванием

* **weight of green wood → 3964**

* **weight of wood → 8658**

9335 **well-stocked**
(situation in which a forest stand contains trees spaced widely enough to prevent competition yet closely enough to utilize the entire site)
d gutbestückt
f de densité normale
r нормальной густоты

* **well-wooded → 6936**

9336 **wenge; panga panga**
(Millettia laurentii)
d Wenge *f*; Panga-Panga *f*
f wengé *m*
r панга-панга *f*; венге *n*

* **West Asian cedar → 3935**

* **Westcoast hemlock → 5588**

9337 **western balsam bark beetle**
(Dryocoetes confusus; Dryocoetes abietis)
d Balsamtannenbastkäfer *m*
f scolyte *m* du sapin de l'Ouest
r [короед-]лесовик *m* пихты бальзамической

* **western balsam fir → 1268**

9338 **western balsam poplar; [Canadian] balsam poplar; hartleaf balsam poplar; tacamahac poplar; balm**
(Populus balsamifera; Populus tacamahacca)
d [kanadische] Balsampappel *f*; östliche Balsampappel
f peuplier *m* baumier [du Canada]; peuplier balsamique; peuplier balsamifère
r тополь *m* [индейский] бальзамический

* **western chinkapin → 3841**

9339 **western flowering dogwood; Pacific dogwood; Nuttall dogwood**
(Cornus nuttallii)
d [weiße] Hartriegel *f*; Nuttallsblütenhartriegel *m*
f cornouiller *m* blanc; cornouiller de Nuttall
r дёрен *m* Натела

* **western hackberry → 1777**

* **western hemlock → 5588**

* **western hoptree → 1261**

* **western Juda's tree → 1267**

9340 **western larch; Oregon larch; mountain larch; western tamarack; Montana larch**
(Larix occidentalis)
d westamerikanische Lärche *f*
f mélèze *m* de l'Ouest; mélèze occidental; mélèze d'Amérique du Nord
r лиственница *f* западная

9341 **western larch borer**
(Tetropium velutinum)
d westamerikanischer Lärchekäfer *m*
f longicorne *m* du mélèze occidental
r усач *m* лиственницы западной

* **western mountain ash → 3943**

* **western paper birch → 9279**

9342 **western pine beetle**
(Dendroctonus brevicomis)
d westamerikanischer Riesenbastkäfer *m*
f dendroctone *m* du pin de l'ouest
r лубоед *m* сосны жёлтой

* **western plane → 254**

* **western [red] alder → 235**

* **western red cedar → 8893**

9343 **western soapberry; wild china-tree**
(Sapindus drummondi)
d Drummond-Seifenbeere *f*
f savonnier *m* de Drummond
r сапиндус *m* Друммонда

* **western spruce → 7678**

* **western sycamore → 254**

* **western tamarack → 9340**

9344 **western [white] pine; mountain white pine; silver pine; Idaho white pine**
(Pinus monticola)
d westamerikanische Weymouthskiefer *f*; westliche Weymouthskiefer
f pin *m* argenté américain; pin blanc de l'Ouest
r сосна *f* горная Веймутова

* **western yellow pine → 6445**

9345 **western yew; Canadian yew; Pacific yew; Oregon yew; short needle yew**
(Taxus brevifolia; Taxus canadensis)
d kurzblättrige Eibe *f*; kanadische Eibe
f if *m* occidental; if du Canada; if à feuilles

* **West Indian bowwood → 5346**

9346 **West Indian boxwood; Maracaibo boxwood; San Domingo boxwood**
(Phyllostylon brasiliensis; Phyllostylon rhamnoides; Samaroceltis rhamnoides; Gossypiospermum praecox)
d Zapatero[-Buchsbaum] *m*; Capanema *m*; Baitoa *m*
f zapatero *m*; camasari *m*; capanema *m*
r самшит *m* [Сан Доминго]

9347 **West Indian lancewood; black lancewood; asta**
(Oxandra lanceolata)
d Lancewood *m*
f bois *m* de lance
r оксандра *f* ланцетолистная

9348 **West Indian lime; Mexican lime; acid lime; sour lime; key lime**
(Citrus aurantifolia)
d saure Limette *f*; Limettenbaum *m*; Limettenzitrone *f*
f lime *f* mexicaine; limette *f* acide; limettier *m*; lime acide; limier *m*; limonelle *f*; citron *m* vert
r лайм *m*; лимон *m* лиметта; цитрус *m* лиметта; лимон сладкий; лиметта *f*

* **West Indian locust → 1990**

9349 **West Indies satinwood**
(Zanthoxylum flavum)
d westindisches Satinholz *n*
f espénille *f*
r западноиндийское атласное дерево *n*

9350 **wet barking**
d Nassentrindung *f*
f écorçage *m* humide
r мокрая окорка *f*

* **wet chute → 9281**

* **wet cooperage → 8549**

* **wet heart → 9283**

9351 **wetland plant; hylophyte**
d Hylophyt *n*
f hylophyte *m*
r гилофит *m*

9352 **wet line**
(line established by using water with or without a wetting agent, to control wildfires
f ligne *f* traitée à l'eau; câble *m* immergé
r мокрый кабель *m*

* **wet peatland forest → 8309**

9353 **wet rot; damp rot**
d Nassfäule *f*
f pourriture *f* humide
r мокрая гниль *f*

* **wet storage → 9301**

9354 **wetwood; glassy wood**
(wood with abnormally high water content
d glasiges Holz *n*
f bois *m* humide; bois trempé
r водянистая древесина *f*

* **Weymouth pine → 2726**

* **WHA → 9413**

9355 **wheeled digger**
d Radroder *m*
f dessoucheur *m* sur roues
r колёсный корчеватель *m*

* **wheel[ed] skidder → 7125**

* **wheeled tractor → 7125**

9356 **wheel[ed-trailed] camp; car camp**
d rädriger Lager *m*
f logement *m* sur roues
r посёлок *m* с домиками-прицепами; лесопункт *m* на колесах

9357 **wheel-stamen tree family**
(Trochodendraceae)
d Radbaumgewächse *pl*
f trochodendracées *pl*
r троходендровые *pl*

* **wheelwright wood → 1672**

* **whiffle-tree → 4480**

* **whim → 5090**

* **whip → 4812**

* **whip grafting → 8678**

9358 **whip[per]**
(any slender tree that the wind causes to

f fouetteur *m*
r схлестывающее дерево *n*; дерево, схлёстывающее соседние деревья

9359 **whipping**
(of crowns)
d Peitschen *n*
f fouettage *m*
r хлестанье *n*

* **whipple-tree → 4480**

9360 **whips**
d Gerten *fpl*
f fouets *mpl*
r неделовые деревья *npl* в насаждении (лиственные деревья в хвойном насаждении)

* **whip saw → 1136**

* **whip sawing → 6312**

* **whiskers → 2193**

* **white alder → 2878**

* **white ash → 261**

9361 **white balata; grumixava**
(Micropholis melinoniana)
d Grumixava *m*
f grumixava *m*; balata *m* blanc; faux balata
r балата *f* белая

* **white balsam → 1278**

* **whitebark[ed] maple → 1438**

9362 **white[-]bark pine**
(Pinus albicaulis)
d Weißbarkenkiefer *f*
f pin *m* albicaule; pin à blanche écorce
r сосна *f* белоствольная

* **white basswood → 667**

* **whitebeam [mountain ash] → 8323**

* **white beech → 2901**

* **white box → 9369**

* **white cedar → 398**

9363 **white cheesewood; palimira alstonia; pulai**
f alstonia *m* des écoliers; pulai *m*; shaitan *m*
r алстония *f* шоларис; лами *m*; пулан *m*

9364 **white cinnamon tree, wild cinnamon**
(Canella winterana; Canella alba)
d weißer Caneel *m*; weißer Zimt *m*; Weißzimtbaum *m*
f cannelle *f* [blanche]
r корица *f* белая

9365 **white cinquefoil**
(Potentilla alba)
d weißes Fingerkraut *n*
f potentille *f* blanche
r лапчатка *f* белая

9366 **white clover**
(Trifolium repens)
d Weißklee *f*; Kriechklee *f*; weißer Wiesenklee *f*
f trèfle *m* blanc
r клевер *m* белый

9367 **white cubical heart rot**
(invader - Polyporus borealis, Abortiporus borealis)
d kubische Weißherzfäule *f*
f pourriture *f* du cœur blanche cubique
r белая ядровая кубическая гниль *f* стволов и корней ели и пихты

9368 **white cubical rot**
(invader - Polyporus adustus)
d kubische Weißfäule *f*
f pourriture *f* blanche cubique
r белая кубическая гниль *f*

* **white deal → 9389**

* **white ebony → 1796**

* **white elm → 247**

9369 **white eucalyptus; white box; white gum; khaki gum; poplar gum**
(Eucalyptus alba; Eucalyptus albens)
d Poplargummi *m*
f gommier *m* blanc
r эвкалипт *m* белый

* **white filbert → 4759**

* **white fir →1278, 6025**

9370 **white flaky rot**
(invader - Pleurotus ostreatus)
d flockige Weißfäule *f*

9371 white fleck; white spot
(for tree marking)
d Blesse *f*
f blanchis *m*
r лесная примета *f*; путь *m* [в лесу], отмеченный зарубками

* **white gum → 9369**

* **white hickory → 8988**

* **white ironbark → 6771**

* **white kurrajong → 3245**

* **white mahogany → 6551**

9372 white marble rot
d Weißstreifigfäule *f*
f pourriture *f* marbrée blanche
r белая волокнистая ядровая гниль *f*

9373 white marupa; simarouba
(Simarouba amara; Quassia amara)
d Bitterholzbaum *m*; Surinam-Bitterholz *n*; Quassie *f*; Simaroubabaum *m*; Quassiaholzbaum *m*
f bois *m* amer; simar[o]uba *m*; quassier *m* amer
r марупа *f* белая; квассия *f* горькая; дерево *n* симаруба

* **white mistletoe → 2902**

9374 white mottled rot
(invader - Fomes applanatus, Ganoderma applanatum)
d buntfleckige Weißfäule *f*
f pourriture *f* blanche bigarrée
r желтовато-белая ядровая гниль *f*

* **white mountain ash → 414**

* **white mulberry → 1791**

9375 white myrtle
(Myrtus acmeniodes)
d weißer Myrtebaum *m*
f myrte *m* blanc
r мирт *m* игольчатый

* **white oak → 8993**

* **white pine → 2726**

9376 white pine blister rust; blister rust; currant
d Säulchenrost *m*; Säulenrost *m*
f rouille *f* vésiculeuse
r серянка *f* сосны веймутовой и кедра сибирского

* **white pine weevil → 8670**

9377 white pocket-rot; piped rot
(invader - Fomes nigrolimitatus)
d Taschenweißfäule *f*
f pourriture *f* blanche par poches
r белая ямчатая гниль *f*

* **white popinac → 4851**

* **white poplar → 2911**

9378 white quebracho; quebracho blanco
(Aspidosperma quebracho-blanco)
d weißer Quebrachobaum *m*
f quebracho *m*
r квебрахо *n* белое; квебрачо *n* белое; аспидосперма *f*

9379 white-red-jack pine [forest]
(forest in which eastern white pine, red pine, or jack pine, singly or in combination, constitute a plurality of the stocking)
d Weymouth-zwerger rote Kieferwald *m*
f forêt *f* de pins Weymouth, pins blancs et pins rouges
r лес *m*, с преобладающей чёрной, смолистой и Веймутовой сосной

9380 white ring rot
(invader - Fomes robustus, Phellinus robustus)
d weiße Ringfäule *f*
f carie *f* du bois initial blanche
r жёлто-белая кольцевая гниль *f* стволов

9381 white rot
(a type of wood-destroying fungus that attacks both cellulose and lignin; the resultant mass is spongy or stringy, usually white, but may assume various shades of yellow, tan, and light brown)
d Weißfäule *f*
f pourriture *f* blanche
r белая [волокнистая] гниль *f*

9382 white sandalwood; sandalwood
(Santalum album)
d Sandel[holz]baum *m*
f santal *m* blanc

9383 **white slash rot**
(invader - Stereum hirsutum)
d Weißschlagfäule *f*
f pourriture *f* de rémanents blanche
r белая смешанная гниль *f*

9384 **white specks**
(small white pits or spots caused by a fungus in the living tree)
d weiße Flecken *fpl*
f taches *fpl* blanches
r белые пятна *npl*

9385 **white spongy rot**
(invader - Fomes annosus, Fomitopsis annosa)
d Weißschwammfäule *f*; Annosus-Wurzelschwamm *m*
f pourriture *f* mycélienne blanche; maladie *f* du rond
r белая грибная гниль *f*; пёстрая гниль корней

* **white spot → 9371**

9386 **white spruce; Canadian spruce; cat spruce; pine spruce; skunk spruce; Adirondack spruce**
(Picea canadensis; Picea glauca)
d [kanadische] Weißfichte *f*; Balsamfichte *f*; Schimmelfichte *f*
f épicéa *m* blanc; épicéa de l'Amérique du Nord; sapinette *f* blanche
r ель *f* канадская; ель белая

9387 **white stringybark [eucalyptus]**
(Eucalyptus scabra; Eucalyptus globoidea)
d Weißstringybark *m*
f eucalyptus *m* scabre
r эвкалипт *m* шероховатый

* **white sycamore → 8355**

* **whitethorn → 1780**

* **whitetop stringybark → 4054**

9388 **white trunk rot**
(invader - Fomes igniarius, Phellinus igniarius)
d Weißstammfäule *f*
f pourriture *f* blanche des tiges; pourriture blanche du tronc
r белая ядровая стволовая гниль *f* с чёрными линиями

* **white willow → 1809**

* **willow → 1809**

9389 **white[ ]wood; white deal**
(includes Engelmann spruce, all true firs, hemlocks and pines)
d Weißholz *n*
f bois *m* blanc
r белая древесина *f*

* **whitewood → 1805**

9390 **whitewood long logs**
d Fichtenlangholz *n*
f grumes *fpl* longues de bois blanc; bois *m* de sapin en grumes
r длинномерный сортимент *m* белой древесиной

* **whittle → 453, 1360**

9391 **whole-log chipper; log chipper**
d Langholzhackmaschine *f*
f déchiqueteuse *f* de bois long
r рубильная машина *f* для длинных брёвен; рубильная машина для долготей

9392 **wholesale lumberman; timber wholesale merchant**
d Holzgrößhändler *m*
f marchand *m* de bois en gros
r крупный лесоторговец *m*

9393 **whole[-section] timber**
d Voll[schnitt]holz *n*
f bois *m* entier
r цельная древесина *f*; деревянный элемент *m* сплошного сечения; крупный лесоматериал *m* (квадратный брус толщиной 30 см)

* **whole-stem logging → 8869**

* **whole tree → 3758**

* **whole-tree chips → 3760**

* **whole-tree harvester → 1830**

* **whole-tree harvesting → 3761**

9394 **whole-tree logging; full-tree skidding**
d Transport *m* ganzer Bäume; Ganzbaumverfahren *n*; Vollbaumabfuhr *f*

* **whole-tree operation → 3761**

9395 **whole-tree trailer; timber carriage; timber drag; timber drug; drug**
d Langholznachläufer *m*; Langholzanhänger *m*
f remorque *f* pour bois long; pince *f* lève-bois
r прицеп *m* для длинномерных грузов; прицеп-роспуск *m*; тягло *n*

* **whole-tree utilization → 1832**

* **whorl leaf [a]estivation → 9209**

* **whorlless top → 5544**

* **whorl trunk scar → 2938**

* **whortleberry → 855**

9396 **whortle-leaved willow**
(Salix myrsinites)
d Myrten-Weide *f*; schwarzwerdende Weide *f*
f saule *m* faux-myrte
r ива *f* миртолистная; ива чернеющая

* **wicker of willow → 9437**

* **wide-ringed timber → 5909**

9397 **widowmaker [branch]**
(a dead top or limb that has fallen and is lodged, possibly between two trees)
d abgestorbener hängender Ast *m*; abgestorbener und hängender Ast
f branche *f* morte suspendue dans la cime; branche sèche et pendante
r отмерший сук *m*

* **width in the clear → 4504**

9398 **width of annual ring; annual ring width**
d Jahresringenbreite *f*
f largeur *f* de cerne annuel
r ширина *f* годичного кольца

9399 **width of chip**
d Spanbreite *f*
f largeur *f* de copeau
r ширина *f* стружки

9400 **width of [saw] cut; width of kerf; cutting width**
d Schnittbreite *f*; Schnittweide *f*
f largueur *f* du trait de scie; largueur de coupe
r ширина *f* реза; ширина пропила; ширина резания

9401 **wild angelica**
(Angelica silvestris)
d wilder Brustwurz *m*; [Wald-]Engelwurz *m*; Waldengelwurz *m*
f angélique *f* forestière
r дудник *m* лесной

* **wild apple → 9403**

* **wild area → 9404**

* **wild black cherry → 244**

* **wild china-tree → 9343**

9402 **wild clematis; travel[l]er's joy**
(Clematis vitalba)
d [gemeine] Waldrebe *f*; gewöhnliche Waldrebe
f clématite *f* des haies; clématite des bois; clématite vigne blanche
r ломонос *m* белый

9403 **wild [crab]apple-tree; wild apple**
(Malus sylvestris)
d Holzapfelbaum *m*; Wildapfel *m*; Wildapfelbaum *m*
f pommier *m* sauvage; pommier acerbe
r яблоня *f* лесная; яблоня дикая

* **wild currant → 5419**

9404 **wilderness [area]; wild area; primitive area; rustic area; natural area; back[ ]country**
(ref. to all those portions of wildlands in which usually there are no permanent, improved, or maintained access roads or operational, fixed facilities)
d entlegene Gegend *f*; unbewohnte Gegend; Waldung *f*; Wildernis *f*
f zone *f* de nature vierge; aire *f* naturelle vierge; espace *m* naturel; randonnée *f* inculte
r дебри *pl*; девственная местность *f*; дикая местность; девственная природа *f*

* **wildfire → 3690**

* **wild forest → 282**

9405 **wild forest products; wild forest produce**
d Urwaldprodukte *npl*
f produits *mpl* forestiers de cueillette
r продукты *mpl* девственного леса

9406 **wild fruit-tree mountain forest**
d Gebirgwald *m* mit wildwachsenden Obstbäumen

* **wild gean → 8330**

* **wild grain → 2145**

* **wild honeysuckle → 6288**

9407 **wild hydrangea**
(Hydrangea arborescens)
d Waldhortensie *f*
f hortensia *m* sauvage
r гортензия *f* древовидная; аннабель *f*

9408 **wildland**
(uncultivated land other than fallow)
d Urland *n*
f friche *f*; terre *f* vierge; étendues *fpl* agrestes
r нетронутая территория *f*; нетронутый участок *m* природы; целина *f*

9409 **wildland fire**
(any nonstructural fire, other than prescribed fire, that occurs in a wildland)
d Urlandbrand *m*
f feu *m* de végétation
r пожар *m* в целине

9410 **wildland-urban interface**
(area where various structures (most notably private homes) and other human developments meet or are intermingled with forest and other vegetative fuel types)
d Interface *f* zwischen Urland und Stadt
f interface *f* entre terre vierge et urbaine
r переходная зона *f* между целиной и города

9411 **wildlife**
(raptors, threatened species, endangered species, game, and other species of vertebrates prescribed as wildlife by regulation)
d wild lebende Tiere und Pflanzen *fpl*; Wildfauna und Wildflora
f faune *f* [et flore f] sauvage; vie *f* sauvage
r дикая природа *f*; дикая флора *f* и фауна *f*; дикие животные *npl* и растения *npl*

9412 **wildlife habitat**
(all of the environmental conditions of a specific place occupied by a wildlife population)
d Tier- und Pflanzenwelthabitat *n*; Lebensraum *m* für Wildtiere und Wildflora
f habitat *m* de la vie sauvage; site *f* de la vie sauvage
r габитат *m* дикой природы

is necessary to meet the habitat requirements of one or more species of identified wildlife)
d Tier- und Pflanzenwelthabitatsfläche *f*
f aire *f* d'habitat de la vie sauvage
r площадь *f* габитата дикой природы

9414 **wildlife habitat features**
(significant mineral lick or wallow, an active nest of a bald eagle, osprey or great blue heron, or any other feature agreed to by the district manager and a designated environment official)
d Tier- und Pflanzenwelthabitatsmerkmale *npl*
f caractéristiques *fpl* de la vie sauvage
r особенности *fpl* габитата дикой природы

9415 **wildlife management**
(application of scientific and technical principles to wildlife populations and habitats to maintain such populations essentially for recreational and/or scientific purposes)
d Tier- und Pflanzenwelthabitatsmanagement *n*
f gestion *f* de la faune et flore sauvage; aménagement *m* de la vie sauvage
r управление *n* дикой природы (мероприятия, направленные на воспроизводство диких зверей и птиц)

9416 **wildlife plantings**
(agricultural crops specifically planted for wildlife in fields or small forest openings and sometimes referred to as food plots)
d Wildflora-Pflanzungen *fpl*
f plantations *fpl* sauvages
r насаждения *npl* из диких культур

9417 **wildlife species**
d Tier- und Pflanzenweltarten *fpl*
f espèces *fpl* de la faune et flore sauvage
r дикие виды *mpl*

9418 **wildlife [travel] corridor**
(forested area or other established vegetation used as travel lanes or buffer zone to connect larger stands of suitable wildlife habitat or to prevent isolation of important foraging and nesting area)
d Wildpflanzenweltkorridor *m*
f couloir *m* de la flore et la faune sauvage
r коридор *m* диких видов

9419 **wildlife tree; forest dug seedling; wild stock; tree stock**
(identified in an operational plan to provide present or future wildlife habitat)
d Wildbaum *m*

9420 wild[l]ing; selfsown plant; wild seedling; selfcomer; nonlined-out plant; untransplanted plant; not pricked-out plant
(young seedling that develops in the wild without the help of humans)
d Wildling *n*; unverschulte Pflanze *f*
f sauvageon *m*; plante *f* non repiquée
r непершколенный саженец *m*; непершколенное растение *n*; подвой *m*; дичок *m*

9421 wild olive [tree]; oleaster; Russian olive
(Elaeagnus angustifolia)
d [schmalblättrige] Ölweide *f*; schmale Ölweide; wilder Ölbaum *m*; Oleaster *m*
f olivier *m* sauvage; oléastre *m*; olivier de Bohême
r лох *m* узколистный; маслина *f* дикая

* **wild plum → 2244**

* **wild red cherry → 2881**

9422 wild red raspberry
(Rubus strigosus)
d amerikanische Himbeere *f*
f framboise *f* sauvage; framboisier *m* rouge
r малиновый куст *m*; малина *f* щетинистая; малина американская

* **wild rose → 2555**

9423 wild rosemary larch forest
d Porst-Lärchenwald *m*
f mélézin *m* à lède
r багульниковый лиственничник *m*

* **wild seedling → 9420**

9424 wild service-tree; wild serviceberry; checker-tree
(Sorbus torminalis)
d [echter] Elsbeerbaum *m*; Elsbeere *f*; Atlasbeerbaum *m*; Ruhrbirne *f*
f alisier *m* torminal; alise *f*; alisier des bois; allouchier *m*
r рябина *f* глоговина; берека *f* [лечебная]; богорошник *m*

* **wild stand → 5698**

* **wild stock → 9419**

9425 wild strawberry; wood[land] strawberry; Alpine strawberry
f fraisier *m* commun; fraisier des bois
r земляника *f* лесная

9426 wild transplant
d Wildling *m* zum Verschulen
f sauvageon *m* destiné au repiquage
r дичок *m* для перешколивания

9427 wild tree fruit
d Holzobst *n*
f fruit *m* d'arbre sauvage
r плод *m* дикорастущего дерева

* **wildwood → 282**

* **wild woodbine → 9228**

* **woodbine → 9228**

9428 willow bark
d Weidenrinde *f*
f écorce *f* de saule
r ивовая кора *f*

* **willow borer → 4804**

* **willow culture → 9431**

9429 willow family
(Salicaceae)
d Weidengewächse *pl*
f salicacées *pl*
r ивовые *pl*

9430 willow forest; willow stand; willow grove
d Weidenwald *m*; Weidenbestand *m*
f forêt *f* de saule
r ивовый лес *m*

* **willow herb → 3215**

9431 willow holt; willow plantation; willow culture; osier bed; osier ground
d Weidenheger *m*; Weidenpflanzung *f*
f plantation *f* de saules; plantation d'osier
r ивовый питомник *m*

9432 willow-leaf pear
(Pyrus salicifolia)
d weidenblättriger Birnbaum *m*; Weidenbirne *m*
f poirier *m* à feuilles de saule
r груша *f* иволистная

9433 willow-leaf spirea
(Spiraea salicifolia)
d Weidenspierstrauch *m*

9434 **willow oak**
(Quercus phellos)
d Weideneiche *f*
f chêne *m* à feuilles de saule
r дуб *m* иволистный

* **willow plantation → 9431**

9435 **willow sawfly**
(Nematus oligospilus)
d Lindenblattwespe *f*; große Weidenblattwespe *f*
f tenthrède *m* du saule
r ивовый пилильщик *m*

9436 **willow shrub formation; osiery**
d Weidengehölz *n*; Weidengebüsch *m*
f saulaie *f*; oseraie *f*; saussaie *f*
r ивняк *m*

* **willow stand → 9430**

9437 **willow stem; wicker of willow; willow twig; willow stick**
d Weidenrute *f*; Weidenzweig *m*
f brin *m* de saule; branche *f* de saule; tige *f* d'osier
r прутья *npl* для плетения; ивовые прутья; ивовая ветка *f*; ивовая лоза *f*

9438 **willow tit**
(Parus montanus)
d Mönchsmeise *f*; Weidenmeise *f*; Alpenmeise *f*
f mésange *f* boréale; mésange des saules; mésange alpestre
r гаичка *f* буроголовая

9439 **willow [tree]; osier; sallow**
(Salix)
d Weidenbaum *m*; Weide *f*
f osier *m*
r ива *f*

* **willow twig → 9437**

* **winch-and-choker [equipped] skidder → 1246**

* **winchman → 7925**

* **wind barrier → 9445**

* **windbelt → 9445**

9440 **wind-bend; wind-curved; wind-lean** *adj*
d windschief

* **windbrake → 850**

* **windbreak → 9445**

9441 **wind-curved timber**
d windschiefes Holz *n*
f bois *m* courbé par le vent
r растущий кривой лес *m*

* **wind damage → 2236**

* **windfall[en] trees → 850**

* **windfall [timber] → 850**

9442 **windfall timber treatment**
d Windwurfaufarbeitung *f*
f traitement *m* de chablis
r разработка *f* бурелома; разработка ветровала

* **windfall zone → 849**

9443 **windfirm tree**
(able to withstand strong winds)
d windfester Baum *m*
f arbre *m* stable au vent
r ветроустойчивое дерево *n*

* **windlass → 8603**

* **wind-lean → 9440**

9444 **wind-lean tree**
(condition of tree having a leaning stem, result of partial uprooting or wind action)
d windschiefer Baum *m*
f arbre *m* couché par le vent; arbre déformé par le vent
r дерево *n*, наклоненное ветром

9445 **wind mantle; windbelt; wind-protective strip; windstrip; wind barrier; windbreak**
(small-scale shelterbelt or other barrier, natural or artificial, maintained against the wind)
d Windmantel *m*; Windschutzstreifen *m*; Windschirm *m*; Windbruch *m*
f forêt *f* de protection contre le vent; rideau *m* brise-vent; brise-vent *m*
r ветроударная опушка *f* леса; лесной ветрозащитный пояс *m*; ветрозащитная полоса *f*

* **windowlike pit → 6287**

9446 **windrocking**
(movement of tree stems in the wind, which may lead to chafing of the collar and sometimes of the roots)
d Windschwingen *n*
f balancement *m* au vent
r наклон *m* дерева под воздействием ветра; качание *n* на ветру

9447 **windrow**
(slash that has been gathered into an elongated pile)
d Schwaden *m*; Schwad *m*
f andain *m*; javelle *f*
r штабель *m* древесины, уложенный вдоль дороги; вал *m* порубочных остатков

9448 **windrowing**
(of slash)
d Schwadlegen *n*
f mise *f* en andain; andainage *m*
r сбор *m* в валы; сбор в валки

9449 **windrow planting; stroll planting**
(planting between the two lanes created in windrowing)
d Pflanzung *f* zwischen Abraumwällen
f plantation *f* sur entrandain; plantation entre déchets alignés
r посадка *f* между рядами порубочных остатков; посадка между валами

9450 **wind shake; wind split; anemosis; natural crack**
(crack in a tree caused by high winds)
d Windriss *m*; Luftriss *m*
f fissure *f* par le vent; fente *f* provoquée par le vent
r ветряная трещина *f*; ветреница *f*

* **windslash → 850**

* **windstrip → 9445**

* **windthrow[n timber] → 850**

9451 **windward side; upwind side**
(upwind side of, for example, a shelterbelt)
d Luvseite *f*
f bord *m* du vent; côté *m* du vent; côté [exposé] au vent; face *f* au vent; lof *m*
r наветренная сторона *f*

* **wineberry → 5006**

* **wine palm → 1701**

(Rubus phoenicolasius)
d japanische Weinbeere *f*; rotborstige Himbeere *f*
f ronce *f* [à poils rouges]
r малина *f* пурпурноплодная

9453 **wing**
(thin, flat, dry, wing-like projection on a fruit or seed or along a twig)
d Flügel *m*
f aile *f*
r боковой лепесток *m*; крылатка *f*

9454 **winged euonymus**
(Euonymus alatus)
d Flügelspindelstrauch *m*; Korkspindel *m*
f fusain *m* ailé
r стелющийся бересклет *m*; бересклет крылатый

9455 **winged felling notch; winged undercut**
d Flügelfallkerb *m*
f arête *f* d'abattage brisée en deux pans
r двусторонний надруб *m*

* **winged fruit → 4701**

* **winged suma[h] → 3247**

* **wing logs → 1180**

9456 **wingnut**
(Pterocarya)
d Flügelnuss *f*
f noyer *m* aux ailes; noyer ailé; noyer à feuilles; ptérocaryer *m*
r птерокария *f*; лапина *f*

* **wing-shaped knot → 7910**

9457 **winter bark[-]scorch; winter sunscald**
d Winter-Rindenbrand *m*
f coup *m* de soleil hivernal
r зимний солнечный ожог *m* коры

* **winterberry holly → 1810**

* **winter cherry → 3984**

* **winter chute → 2708**

* **winter cutting → 9460**

9458 **winter [de]barking**
d Winterschälung *f*
f écorçage *m* hivernal

9459 **winter family**
(Winteraceae)
d Winterrindengewächse *pl*
f winteracées *pl*
r винтеровые *pl*

9460 **winter felling; winter cutting**
d Winterfällung *f*; Wintereinschlag *m*
f coupe *f* hivernale
r зимняя рубка *f*; валка *f* зимой

* **wintergreen → 8433**

9461 **wintergreen forest; hiemisilva**
d wintergrüner Wald *m*
f forêt *f* verte d'hiver; hiemisilve *m*
r зимнезелёный лес *m* (листопадный лес тропических и субэкваториальных широт, теряющий листву на сухое время года)

* **winter linden tree → 7767**

9462 **winter pruning; dormant pruning**
d Winteraufasten *n*; Winterschnitt *n*
f élagage *m* d'hiver; taille *f* d'hiver
r обрезка *f* в спящем состоянии; зимняя обрезка

* **winter sunscald → 9457**

* **wire birch → 3923**

9463 **wire cable; cable [rope]; wire rope**
(used for lines in yarding systems)
d Seilschlinge *f*; Seil *n*; Trosse *f*; Drahtseil *n*
f câble *m*; élingue *f*
r [проволочный] канат *m*; кабель *m* [для проволочного спуска]; трос *m*

* **wire clip → 7699**

9464 **wire gravity cable; gravity cable; gravity wire [skidding equipment]; suspended wire for gravity skidding**
d Kabel *n* für Drahtseilriese
f câble *m* pour glissoire
r проволочный канат *m* для [гравитационного] спуска лесоматериалов с гор

* **wire rope → 9463**

9465 **wire rope sling tightening**
d Drahtseilschlinge-Einspannen *n*
f serrage *m* d'élingue pour câble
r затягивание *n* канатного стропа

9466 **wire saw**
d Drahtsäge *f*
f machine *f* à scier à fil; scie *f* à fil; scie au fil hélicoïdal
r проволочная пила *f*

* **wire strainer → 1436**

* **wirilda → 8316**

* **witch elm → 7310**

9467 **witches' broom**
(an abnormal tufted growth of small branches on a tree or shrub caused by fungi or viruses)
d Hexenbesen *m*
f balai *m* de sorcière
r ведьмина метла *f*

9468 **witch-hasel family**
(Hamamelidaceae)
d Zaubernussgewächse *pl*
f hamamélidacées *pl*
r гамамелидовые *pl*; гамамалиевые *pl*

* **witch willow → 459**

* **with eccentric trunk → 1230**

9469 **witheredness**
d Welkheit *f*; Welke *f*
f état *m* fané
r вялость *f*

* **withering → 5302**

* **with large branches → 5596**

* **with several stems → 5628**

* **with slender branches → 7737**

9470 **with small annual rings; close-ringed; closely ringed; fine-ringed** *adj*
d feinumringend
f à cernes étroits
r тонкослойный

* **witness tree → 641**

* **wobble saw → 4312**

* **wold → 1928**

9471 **wolfberry; desert thorn; squawbush; boxthorn**

f lyciet *m*
r лициум *m*; лиций *m*

9472 wolf tree; bushy tree
(a tree which occupies more space in the forest than its value justifies)
d Protze *m*; Wolf *m*
f arbre *m* loup; loup *m*
r дерево-волк *n*; разросшееся дерево *n*; корявое деревцо *n*

9473 wood; lignum; grove
(lignified secondarily thickened plant tissue)
d Holz *n*
f bois *m*
r древесина *f*

* **wood acid → 9688**

9474 wood adhesive
d Holzklebstoff *m*
f colle *f* à bois
r клей *m* для древесины

* **wood agate → 124**

* **wood age → 135**

9475 wood agent; wood broker
d Holzmakler *m*
f revendeur *m* de bois
r лесной агент *m*

9476 wood alcohol; wood spirit
(a light, volatile, inflammable liquid, $CH_3.OH$, obtained by the distillation of wood)
d Holzalkohol *m*; Holzgeist *m*; Holzspiritus *m*
f alcool *m* de bois
r древесный спирт *m*

9477 wood analysis
d Holzanalyse *f*
f analyse *f* de bois
r анализ *m* древесины

9478 wood anatomy
d Anatomie *f* des Holzes; Holzanatomie *f*
f anatomie *f* de bois
r анатомия *f* древесины

9479 wood anemone
(Anemone nemorosa)
d [echtes] Buschwindröschen *n*; Anemone *f*
f anémone *f* des bois
r ветреница *f* лесная

d rote Waldameise *f*; Hügelameise *f*
f fourmi *m* rousse; fourmi rouge
r муравей *m* лесной обыкновенный; муравей лесной рыжий

9481 wood as a source of energy
d Energiegewinnung *f* aus Holz
f bois *m* comme source d'énergie
r древесина *f* как энергетический источник

9482 wood ash
d Holzasche *f*
f cendre *f* de bois
r древесная зола *f*

* **wood assortment → 8556**

9483 wood atlas
d Holzatlas *m*
f atlas *m* de bois
r атлас *m* древесины

9484 wood attacked by dry rot
d fäulebefallenes Holz *n*; Trockenfäulenholz *n*
f bois *m* à pourriture sèche
r древесина *f*, поражённая сухой гнилью; истлевшая древесина

9485 wood-attacking fungus; wood-destroying fungus
d holzbewohnender Pilz *m*; holzzerstörender Pilz
f champignon *m* lignivore
r дереворазрушающий гриб *m*

9486 wood balance; timber balance; forest balance
d Holzbilanz *f*; Forstbilanz *f*; Holzhaushalt *m*
f bilan *m* de la production ligneuse; bilan forestier
r баланс *m* древесины; лесной баланс

* **wood bale → 9496**

9487 wood-based fuel; wood fuel
d Holzbrennstoff *m*
f combustible *m* de bois
r древесное топливо *n*

* **wood-base material → 8559**

9488 wood bedstraw
(Galium silvaticum)
d Wald-Labkraut *f*
f gaillet *m* des bois
r подмаренник *m* лесной

9489 wood bending
d Holzbiegen *n*; Biegung *f* des Holzes
f courbage *m* de bois
r гнутьё *n* древесины

9490 wood biomass; forest biomass; stock
(total woody material in a forest)
d forstliche Biomasse *f*
f biomasse *f* ligneuse
r биомасса *f* древесных растений; лесная биомасса

9491 wood block
d Holzblock *m*
f pavé *m* de bois
r деревянное клише *n*; клише с гравюры на дереве; деревянная подкладка *f*

9492 wood board
d Holzplatte *f*
f plaque *f* de bois
r [древесная] плита *f*

9493 wood board; pulpboard
d Holzkarton *m*; holzhaltige Pappe *f*
f carton-bois *m*
r штамповочный картон *m*

* **wood borer → 9538**

* **wood-boring drill → 9538**

9494 wood-boring weevil damage
d Prachtkäfer-Schaden *m*
f dommage *m* par perces-bois; dommage par bupreste
r неокрашенная червоточина *f* овальной формы

* **wood breakdown → 1014**

9495 wood brick; wood briquette
d Holzbrikett *n*; Spänebrikett *n*; Holzziegel *m*
f briquette *f* [en particules] de bois; brique *f* en bois
r древесный брикет *m*; брикет из [отходов] древесины; древесностружечный брикет; деревянный брусок *m*

* **wood broker → 9475**

* **wood bud → 9179**

9496 wood bundle; timber bunch; wood bale
d Holzbund *m*; Bündel *m*; Holzpaket *n*; Langhaufen *m*
лесоматериалов; пачка *f* хлыстов

9497 wood bundling wire
d Holzbündeldraht *m*
f câble *m* d'empilement de bois
r связочная проволока *f* для стяжки пакетов лесоматериалов

* **wood bunk → 2038**

* **wood burner → 1444**

9498 woodburning; pyrography; poker-work; pyrogravure
d Holzverbrennung *f*; Holzbrandmalerei *f*
f pyrogravure *f* [sur bois]
r рисунок *m* по дереву; выжигание *n* по дереву

9499 wood burning
d Holzbrand *m*
f brûlage *m* de bois
r сжигание *n* древесины

* **wood buyer → 8564**

9500 wood byproducts; byproducts
(primary wood products (e.g. pulp chips, animal bedding, fuelwood, etc.) recycled from mill residues)
d Holznebenprodukte *npl*
f sous-produits *mpl* [de bois]
r промежуточные продукты *mpl*; побочные продукты

9501 wood carpet
(a kind of floor covering made of thin pieces of wood secured to a flexible backing, as of cloth)
d Holzteppich *m*
f tapis *m* de bois
r древесный ковёр *m*

* **wood carver → 1363**

9502 wood carving; wood cutting; wooden cut; wood engraving; xylography; woodcraft
(the act or art engraving on wood)
d Schnitzarbeit *f*; Holzaufteilung *f*; Holzschneiden *n*; Schnitzkunst *f*; Holzstich *m*
f sculpture *f* sur bois; gravure *f* sur bois
r резная работа *f*; резание *n* древесины; резьба *f* по дереву

9503 wood cell
(a slender cylindrical or prismatic cell usually tapering to a point at both ends)
d Holzzelle *f*
f cellule *f* de bois
r древесная клетка *f*; клетка древесной ткани

9504 wood cellulose; wood fiber (*US*)**; wood fibre** (*UK*)**; lignified tissue; lignocellulose**
(the carbohydrate that is the principal constituent of wood)
d Lignozellulose *f*; Zellulose *f*; Zellstoff *m*
f cellulose *f* [du bois]; lignocellulose *f*
r целлюлоза *f*; лигноцеллюлоза *f*; клетчатка *f*

9505 wood charcoal; charcoal
d Holzkohle *f*
f charbon *m* de bois
r древесный уголь *m*

9506 wood charcoal fines; charcoal dust; charcoal powder
d Holzkohlenstaub *m*; Holzkohlenklein *n*; Holzkohlenpulver *n*
f poussière *f* de charbon de bois
r древесноугольная мелочь *f*; древесноугольная пыль *f*

9507 wood check; wood chink; wood cleft; wood crack; fissure; crevice; flaw; fracture
(ruptures along the grain that develop during seasoning either because of a difference in radial and tangential shrinkage or because of uneven shrinkage of the tissue in adjacent portions of the wood)
d Holzspalte *f*; Holzriss *m*; Einriss *m*
f stagnation *f*; brisure *f*; crevasse *f*; fente *f*
r трещина *f* в древесине; раскол *m*

* **wood chemistry → 2370**

* **wood chipboard → 6091**

* **wood chipper → 9278**

9508 wood chips; chips; shavings; parings; splinters; wood scrap; chipping slash; chippings; clippings; wood particles; chipped wood
(small pieces of wood used to make pulp)
d Holzpartikeln *fpl*; Holzteilchen *npl*; Sägespäne *mpl*; Holzspäne *mpl*; Späne *mpl*; Hackholz *n*; Kleinholz *n*
f copeaux *mpl* [de bois]; plaquettes *fpl* de bois; éclats *mpl* de bois

9509 wood chisel; chisel
(a wedge-like, sharp-edged tool used for cutting or shaping timber)
d Stechbeitel *m*; Meißel *m*
f ciseau *m* à bois
r долото *n*; резец *m*

* **wood chuck → 2171**

* **wood chunker → 1574**

9510 wood clamp
d Holzklemme *f*; Holzverbindungsklemme *f*
f serre-joint *m* de bois
r прижим *m* лесоматериала

* **wood classification → 3902**

9511 wood cleaning; wood clearing
d Holzreinigung *f*
f dégagement *m* de bois
r очистка *f* древесины

9512 wood-cleaving machine; wood-splitting machine; log splitter; wood splitter; chopping machine; hack machine; hoeing machine; wood-splitting apparatus; wood-splitting attachment; cleaving machine; riving machine
d Holzspalter *m*; Holzspaltmaschine *f*; Hackmaschine *f*; Spaltmaschine *f*
f machine *f* à fendre le bois; fendeuse *f* de bois; fendeuse de bûches; hachoir *m*; dispositif *m* de fendage; dispositif d'éclatage des bûches
r дровоколка *f*; устройство *n* для продольной распиловки брёвен; устройство для раскалывания древесины; приспособление *n* для раскалывания лесоматериалов; механический колун *m*; дровокольный станок *m*

* **wood cleft → 9507**

9513 wood colo[u]r
d Holzfarbe *f*
f couleur *f* du bois
r окраска *f* древесины; цвет *m* древесины; оттенок *m* окраски древесины

9514 wood colo[u]r change
d Veränderung *f* der Holzfarbe
f changement *m* de la couleur de bois
r изменение *n* окраски древесины

9515 wood colo[u]ring
d Holzfärbung *f*

9516 **wood combine; timber combine**
d Holzkombinat *n*
f combinat *m* de l'industrie de bois
r лесокомбинат *m*

9517 **wood consumption; timber consumption**
d Gebrauch *m* des Holzes; Holzverbrauch *m*
f consommation *f* de bois
r потребление *n* лесоматериалов; лесопотребление *n*

9518 **wood container**
d Holzcontainer *m*; Holzbehälter *m*
f contenant *m* de bois
r бункер *m* для древесного сырья

* **wood conversion → 1915**

* **wood conveyor → 1424**

9519 **woodcord; cord**
(128 cubic feet of stacked roundwood containing wood and airspace, with all the pieces of similar length and lined up on approximately the same direction)
d Holzstoß *m*; Stoß *m*; Beige *f*; Kordel *f*
f corde *f* [de bois]; pile *f* de bois; staque *f*
r корд *m* [дров]; стек *m*; куча *f* дров; поленница *f* дров; вязанка *f* дров

* **wood core → 1951**

9520 **wood core plywood**
d Sperrholz-Verbundplatte *f*
f contreplaqué *m* à âme [en bois]
r фанера *f* из ядра древесины

9521 **wood covering; timber lining; wood-lagging**
d Holzdeckung *f*
f revêtement *m* en bois
r деревянный настил *m*

* **wood crack → 9507**

9522 **woodcraft**
d Weidmannskunst *f*
f aptitude *f* d'orientation en forêt
r умение *n* ориентироваться в лесу

* **woodcraft → 9502**

9523 **wood cropping; forest cropping**
d Holzernte *f*
f récolte *f* de bois
r урожай *m* леса

9524 **wood cubic meter**
(unit of measure for the volume of wood products, equivalent to the volume of a cube that measures one meter on each side)
d Forstkubukmeter *n*
f mètre *m* cube de bois
r кубометр *m* древесины на корню; кубический метр *m* древесины на корню

9525 **wood culler; wood grader; wood marker; wood sorter**
d Holzbracker *m*; Holzsortierer *m*
f classeur *m* de bois
r бракёр *m* древесины; браковщик *m* древесины; рабочий *m* на сортировке лесоматериалов

* **wood culls → 2123**

9526 **wood cutter; timber cutter; [log] cutter; hewer; [forest] feller; faller**
(a person who fells trees)
d Holzfäller *m*; Fäller *m*; Holzhacker *m*; Hauer *m*; Holzhauer *m*; Holzschläger *m*
f bûcheron *m*; abatteur *m*
r лесоруб *m*; дровосек *m*; вальщик *m* [леса]

* **wood cutter → 1363**

9527 **wood cutter's crew; felling crew; harvesting chew; logging crew; forest team**
d Holzeinschlagbrigade *f*; Holzfällerrotte *f*; Fällrotte *f*; Fällungsbrigade *f*; Holzerschaft *f*; Holzhauerrotte *f*
f équipe *f* de bûcherons; chantier *m* de bûcherons
r лесозаготовительная бригада *f*; лесосечная бригада; бригада вальщиков; бригада лесорубов

* **wood cutter's tools → 4157**

* **wood cutting → 8824**

* **wood cutting → 9502**

* **woodcutting tools → 4157**

* **wood damage → 2242**

9528 **wood damaged by fire**
d brandgeschadetes Holz *n*
f arsin *m*; bois *m* endommagé par un incendie
r древесина *f*, повреждённая пожаром

9529 **wood decay; rot of wood; rot of timber;**

f décomposition *f* du bois; pourriture *f* [du bois]
r древесная гниль *f*; гниль [древесины]

* **wood decolorization → 9536**

9530 **wood defect**
d Holzfehler *m*
f défaut *m* du bois
r дефект *m* древесины

9531 **wood deformity**
d Holzdeformität *f*
f déformité *f* du bois
r деформативность *f* древесины

9532 **wood delivery**
d Holzanfuhr *f*; Holzablage *f*
f délivrance *f* de bois
r доставка *f* древесины

* **wood density → 1818**

* **wood depot → 8664**

9533 **wood destroyer; animal nuisible to timber**
d Holzschädling *m*
f nuisible *m* du bois
r древесный вредитель *m*

9534 **wood-destroying effect**
d holzschädigende Einwirkung *f*
f effet *m* destructeur de bois
r дереворазрушающее воздействие *n*

* **wood-destroying fungus → 9485**

* **wood-destroying insects → 8562**

* **wood detail → 9548**

9535 **wood deterioration**
d Holz[be]schädigung *f*; Holzzerfall *m*
f détérioration *f* du bois
r разложение *n* древесины; распад *m* древесины

* **wood development → 4954**

9536 **wood discolo[u]ration; wood decolorization**
d Holzentfärbung *f*
f décoloration *f* du bois
r обесцвечивание *n* древесины

* **wood dormancy → 9647**

* **wood dough → 9606**

* **wood dried in the forest → 3386**

9537 **wood drier; wood dryer**
d Holzdarre *f*; Holztrockenanlage *f*
f séchoir *m* à bois
r лесосушилка *f*; сушильная установка *f* для древесины

9538 **wood drill [bit]; wood-boring drill; wood borer**
d Bohrer *m* für Holzbearbeitung; Forstnerbohrer *m*; Holzbohrer *m*
f rayon *m*; foret *m* à bois
r буравчик *m* по дереву; сверло *n* по дереву; плотничный бурав *m* (для высверливания отверстий под шурупы)

9539 **wood drink**
(a decoction or infusion of medicinal woods)
d Holzdrink *m*
f boisson *m* de bois
r напиток *m* из дерева

* **wood dryer → 9537**

* **wood drying → 2423**

* **wood dust → 1514**

* **wood dust asthma → 9700**

9540 **wood-dust filter**
d Holzmehlfilter *m*
f filtre *m* de farine de bois
r фильтр *m* для древесной пыли

9541 **wood dye**
d Holzfarbstoff *m*
f colorant *m* du bois
r древесная окраска *f*

* **wood-eating → 9739**

* **wooded → 3394**

9542 **wooded area; timber[ed] area; timber tract; woods; forest[ed] tract; forest[ed] area; wood[en] area**
(open stand of trees > 8 m high and with a canopy cover of 40% or more, usually among grasses)
d Holzbodenfläche *f*; Forstfläche *f*; bestockte Fläche *f*; bewaldete Fläche; Wald[boden]fläche *f*
f superficie *f* boisée; surface *f* boisée; surface

* **wooded belt → 3335**

9543 **wooded bog; wooded swamp**
d baumbestandener Sumpf *m*; baumbestandene Sumpfgelände *f*
f marécage *m* boisé
r лесное болото *n*

* **wooded border → 2761**

* **wooded country → 8604**

9544 **wooded firebreak**
(planted with non-palatable woody species playing the role of transhumance corridors, to be managed by the herders)
d baumbestandenes Feuerschutzgürtel *n*; bewaldete Brandschneise *f*; bewaldeter Brandschutzstreifen *m*
f coupe-feu *m* boisé
r облесённый противопожарный барьер *m*

9545 **wooded formation**
d Holzformation *f*
f formation *f* ligneuse
r лесистое образование *n*

* **wooded grassland → 3450**

* **wooded grazing land → 3929**

* **wooded land → 8604**

9546 **wooded savanna[h]; tree savanna[h]**
d bewaldete Savanne *f*
f savane *f* arborée; savane boisée
r саванна *f* с отдельно растущими деревьями

9547 **wood[ed] soil; silvogenic soil; timber soil; forest soil**
d Waldboden *m*
f sol *m* forestier; sol de forêt
r лесная почва *f*; почва под лесом

* **wooded steppe → 3607**

* **wooded swamp → 9543**

* **wooded tundra → 3631**

9548 **wood element; wood detail**
d Holzbestandteil *m*
f élément *m* de bois
r составная часть *f* древесины; деревянная деталь *f*

* **wood[en] area → 9542**

* **wooden article → 9560**

9549 **wooden box; wood case**
d Holzkiste *f*
f boîte *f* en bois; caisse *f* en bois
r деревянный ящик *m*

* **wooden cut → 9502**

9550 **wooden dam; timber dam; timber weir**
d Holzsperre *f*
f barrage *m* en bois
r деревянная плотина *f*

9551 **wood[en] fiberboard; fiberboard; fibre[-]board; fiber-building board; fibrous[-felted] board; hardboard; pulpboard; fibreslab**
(a sheet of material commonly 0.25 cm thick made from pulped wood)
d Holzfaserplatte *f*; Faserplatte *f*; Fiberbrett *n*
f panneau *m* de fibres [de bois]
r древесноволокнистая плита *f*

* **wood[en] float → 8584**

* **wood engineering → 8582**

9552 **wood engraver**
(any of several species of small beetles whose larv[ae] bore beneath the bark of trees, and excavate furrows in the wood)
d Holzborkenkäfer *m*
f scolyte *m* du cambium
r жук-топограф *m*; жук-древоточец *m*

* **wood engraver → 1363**

* **wood engraving → 9502**

9553 **wooden nail; wood dowel; dowel[pin]; tree-nail**
(a cylindrical timber rod generally without nut or thread driven into pre-drilled holes to make a joint)
d Holznagel *m*; Holzdübel *m*
f clou *m* de bois
r деревянный гвоздь *m*

9554 **wood[en] pavement; block-wood paving**
d Holzpflaster *n*; Pflasterklotz *m*
f pavage *m* de bois
r деревянная мостовая *f*

9555 wooden pin; wood[en] plug; wood slip
d Holzpfosten *m*; Holzpflock *m*
f tampon *m* de bois; bouchon *m* en bois; cheville *f* en bois; clavette *f*
r деревянная пробка *f*; деревянная заглушка *f*; деревянная втулка *f*

* **wooden protection apron → 8654**

* **wooden slat → 7521**

9556 wooden sleeper; timber sleeper
d Holzschwelle *f*
f seuil *m* en bois
r [деревянная] шпала *f*

* **wooden structure → 8570**

* **wooden supports → 8600**

9557 wooden surface
d Holzauflage *f*; Holzunterlage *f*
f surface *f* en bois
r деревянная поверхность *f*

9558 wood enterprise; wood firm
d holzbearbeitender Betrieb *m*; Holzwirtschaftsbetrieb *m*
f entreprise *f* de bois
r лесопромышленное предприятие *n*

* **wood entomology → 3404**

9559 wooden upstream face
(the face of a dam which is normally in contact with the water)
d Holzsteigfläche *f*
f face *f* amont de bois
r деревянная верховая грань *f*; деревянная напорная грань

9560 wooden ware; woodware; wooden article; timber ware
d Holzware *f*; Holzartikel *m*
f article *m* en bois; marchandise *f* en bois
r лесные товары *mpl*; деревянное оборудование *n*

9561 wooden wedge; key of wood; wood key; cheese block; wooden peg; peg
(a wedge-shaped moveable block, set up on a bunk of a wheeled or tracked vehicle, to hold the loads in place)
d Holzkeil *m*
f coin *m* de bois; clé *f* de bois
r [небольшой] деревянный клин *m*

9562 wood evaluation; timber estimation
d Holzbeurteilung *f*
f évaluation *f* du bois
r оценка *f* древесины

* **wood expert → 9659**

* **wood exploitation → 3414**

* **wood extractives → 2972**

* **wood fabrication → 5376**

9563 wood failure
d Holzdefekt *m*
f rupture *f* dans le bois
r недостаток *m* древесины

* **wood famine → 7282**

9564 wood fashioning
d Holzausformen *n*; Holzausformung *f*; Holzzurichtung *f*
f façonnage *m* du bois
r отделка *f* древесины

* **wood feller's implements → 4157**

* **wood felling [operation] → 8824**

* **wood felling practice → 3457**

* **wood fiber → 9504**

9565 wood fiber saturation
d Fasersättigung *f* des Holzes
f saturation *f* de fibres ligneuses
r насыщение *n* волокон древесины

* **wood fibre → 9504**

* **wood figure [of grain] → 9678**

9566 wood file; wood filler
d Holzfeile *f*; Raspel *f*
f lime *f* à bois
r напильник *m* по дереву; рашпиль *m*

* **wood fire → 3428**

9567 wood-fired energy; dendroenergy
d holzgefeuerte Energie *f*
f énergie *f* [dérivée] du bois
r энергия *f* от древесного топлива

f chauffage *m* au bois
r топка *f* дровами

* **wood firm → 9558**

9569 wood floater; river driver; [log] driver; boomer; rafter; raft[s]man; boom man; log floater; cattyman; boomboat operator
(one who works on a log drive)
d Flößarbeiter *m*; Holzflößer *m*; Flößer *m*; Steuermann *m*; Floßknecht *m*; Flößereiarbeiter *m*
f flotteur *m* [de bois]; compagnon *m* de flottage
r [лесо]сплавщик *m*; плотогон *m*; работник *m* по сплаву

* **wood flour → 1514**

9570 wood for agricultural requirements; agricultural timber
d Holz *n* für landwirtschaftlichen Bedarf
f bois *m* agricole
r лесоматериалы *mpl* сельскохозяйственного значения

9571 wood for charcoal; charcoal-wood; charring wood
d Köhlereiholz *n*; Kohlholz *n*; Schwelholz *n*; Verkohlungsholz *n*
f bois *m* à carboniser; charbonnette *f*
r дрова *pl* для углежжения

9572 wood for fascines; fascine wood; faggotwood; faggots; bundle wood; scrog
d Faschinenholz *n*; Bündelholz *n*
f bois *m* à fascines; rames *fpl* à fagots
r фашинный лес *m*; вязаные дрова *pl*

9573 wood for hydraulic works; wood used in water construction
d Wasserbauholz *n*
f bois *m* de construction hydraulique
r гидротехнический лес *m*

* **wood for inlaying → 901**

9574 wood for parquetry
d Parkettholz *n*
f bois *m* à parquet; lame *f* de parquet
r паркетная древесина *f*

* **wood for sawing → 7245**

9575 wood for shingles; shingle bolt
(a short split section of a log from which shingles are manufactured)
r гонтовой лес *m*

* **wood for staves → 4362**

* **wood-fragmenting machine → 1534**

9576 wood framing
(applies to the type of construction made of wood frames)
d Schalung *f* aus Holz
f coffrage *m* en bois
r деревянный каркас *m*

9577 wood-free
d holzfrei
f sans bois
r без содержания древесной массы

9578 wood free from defects; clears; clear cuttings; clear lumber; clean lumber; clean timber; clean stuff; flawless wood; firsts
(wood without knots and defects)
d ausgesuchtes Holz *n*; fehlerfreies Holz; Wahlholz *n*
f bois *m* clair de construction; bois sans défauts
r беспорочная древесина *f*; первосортные пиломатериалы *mpl*; отборные пиломатериалы; отборная древесина

* **wood fretters → 8562**

9579 wood frog; tree-frog; tree-toad
(a common North American frog (Rana sylvatica) which lives chiefly in the woods, except during the breeding season)
d Waldfrosch *m*; Baumfrosch *m*; Laubfrosch *m*
f grenouille *f* des bois; rainette *f* arboricole
r лягушка *f* древесная; древесница *f*

9580 wood from roots; rootwood; root timber
(secondary xylem of roots)
d Holz *n* der Wurzeln; Wurzelholz *n*
f bois *m* de racines
r корневая древесина *f*; древесина корней; корневой лесоматериал *m*

* **wood fuel → 9487**

9581 wood fungus; timber fungus; destroying polyporus
d Holzpilz *m*; Holzschwamm *m*; Baumschwamm *m*
f champignon *m* des murs; champignon arboricole

9582 **wood gas**
d Holzgas *n*
f gaz *m* de bois
r древесный газ *m*

9583 **wood glue**
d Holzleim *m*
f colle *f* de bois
r древесный клей *m*

* **wood grader → 9525**

* **wood grain → 9678**

* **wood-grower → 344**

* **wood guard → 3456**

9584 **wood guillotine**
d Holzquerschneider *m*
f guillotine *f* de bois
r ножевое режущее устройство *n* для дерева

* **wood gum → 8907**

9585 **wood handbook**
d Holzhandbuch *n*
f manuel *m* de bois
r справочное руководство *n* по древесине

9586 **wood handler**
d Arbeiter *m* im Holzlagerhaus
f manutentionnaire *m* de bois
r рабочий *m* на лесном складе

9587 **wood-handling equipment; wood-handling device**
d Holzbeladevorrichtung *f*
f engin *m* forestier de manutention; équipement *m* de manutention des bois
r оборудование *n* для погрузки лесоматериалов

* **wood hardness → 4128**

9588 **wood harvesting; timber harvesting; harvesting; forest extraction; forest exploitation; forest removal; [forest] logging[-off]**
(felling and long-distance transport of the growing stock (or part of it))
d Gewinnung *f* des Holzes; Holzversorgung *f*; Holzwerbung *f*; Rohholzbereitstellung *f*
f exploitation *f* forestière [d'abattage et débusquage]
r заготовка *f* леса; лесозаготовка *f*

* **wood humidity meter → 9611**

* **wood hunger → 7282**

9589 **wood hygroscopicity**
d Holzhygroskopizität *f*
f hygroscopicité *f* de bois
r гигроскопичность *f* древесины

* **wood identification → 8928**

* **wood improvement → 4434**

* **wood in blocks → 4738**

* **wood industry → 3466**

* **woodiness → 4954**

* **wood in heaps → 7982**

* **wood in layers → 4763**

9590 **wood inlaying; inlaid woodwork; marquetry; [wood] intarsia; inlayings**
d Holzmosaik *n*; Holzmosaikeinlegearbeit *f*; Intarsia *f*
f parquet-mosaïque *m*; marqueterie *f*
r интарсия *f*

* **wood in piles → 7982**

* **wood insect → 3475**

* **wood in the rough → 7096**

* **wood in the round → 7113**

9591 **wood introduction**
d Einnahmen *pl* des Holzes; Holzeinnahmen *pl*
f introduction *f* de bois
r поступление *n* древесины (с мест заготовок)

* **wood joint → 4660**

9592 **wood joist; joist**
(a piece of lumber or composite I-beam product that supports a ceiling or floor)
d Holzbalken *m*; Holzträger *m*; Tragleiste *f*; Kantel *m/n*
f solive *f*
r несущая [деревянная] балка *f*; ферма *f*; брусок *m*; несущая рейка *f*

* **wood kev → 9561**

9593 wood labeling scheme
d Holz-Bezeichnungsschema *n*
f plan *m* d'étiquetage de bois
r система *f* маркировки древесных пород и деревянных изделий

* **wood labour → 9724**

* **wood labo[u]rer → 9699**

* **wood ladder → 1131**

* **wood-lagging → 9521**

* **woodland → 8604**

* **woodland belt → 3335**

* **woodland grazing → 3452**

* **wood[land] pasture → 3452**

* **woodland plough → 3531**

* **woodland range → 9711**

* **wood[land] strawberry → 9425**

9594 wood[ ]lark; hedge lark
(Lullula arborea)
d Heidelerche *f*
f alouette *f* lulu
r лесной жаворонок *m*; юла *f*

* **wood lathe → 9623**

9595 wood lathing
d Pergel *m*
f lattis *m* de bois
r драны *f*; дранка *f*

9596 wood left partially barked
d teilgerindetes Holz *n*
f bois *m* imparfaitement écorcé à la machine
r частично окорённые лесоматериалы *mpl*

* **woodless → 8874**

9597 wood letters; wood types
d Holzbuchstaben *fpl*
f caractères *mpl* de bois
r деревянные буквы *fpl*

* **wood list → 8376**

9598 wood losses
r потери *fpl* древесины

9599 wood[ ]lot
(a small plot of trees grown for fuelwood, which can also provide small timber and poles)
d Holzparzelle *f*
f terre *f* boisée; terre à bois; lot *m* à bois
r лесное угодье *n*; лесной участок *m* фермы; огороженный забором участок

9600 woodlot licence
(allows for small-scale forestry to be practised in a described area on a sustained or perpetual yield basis)
d Holzparzellelizenz *f*
f licence *f* de lot à bois
r лицензия *f* на лесное угодье

* **woodlot management → 3016**

* **woodlouse → 1813**

9601 wood lubricant
d Holzschmiermittel *n*
f lubrifiant *m* à bois
r смазка *f* для древесины

9602 wood macrostructure
d Holzmakrostruktur *f*
f macrostructure *f* de bois
r макроструктура *f* древесины

* **woodman's hammer → 5315**

* **wood mantle → 7508**

* **wood marker → 9525**

* **wood market → 8613**

* **wood marking → 5317**

9603 wood-mass; groundwood; wood[ ]pulp; woody pulp; pulp
(soft, moist mass of wood fiber used in paper manufacture)
d Holzmasse *f*; Holzschliff *m*; Holzfaserbrei *m*; Faserbrei *m*; Holzzeug *m*; Schleifstoff *m*; Steinschliff *m*
f pâte *f* de bois mécanique; bois *m* feuillu à pâte; pâte de fibres de bois
r древесная масса *f*; [древесная] целлюлоза *f*; [древесно]волокнистая масса

f planche *f* en pâte de bois
r плита *f* из древесной массы

9605 wood massif
d Fostmassiv *n*
f massif *m* forestier
r лесной массив *m*

9606 wood mastic; wood dough; putty
d Holz-Mastix *n*; Holzteig *m*
f mastic *m* de bois; pâte *f* de bois
r мастика *f* для древесины; шпатлёвка *f* для древесины; древесная паста *f*; замазка *f*

* **wood material → 8626**

* **wood meal → 1514**

9607 wood mensuration
(volume, weight and assortment determination as interface between forest production and the quantity transformed in the industry)
d Holzmessung *f*
f mesure *f* du bois
r обмер *m* древесины

9608 wood microscopy
d Mikroskopie *f* des Holzes
f microscopie *f* de bois
r микроскопия *f* дерева

9609 wood milling; milling
d Holzfräsen *n*; Fräsen *n*
f usinage *m* [du bois à la toupie]; fraisage *m* [de bois]
r размол *m* древесины; фрезерование *n*

9610 wood mineralization
d Mineralisierung *f* des Holzes
f minéralisation *f* de bois
r минерализация *f* древесины

* **wood moisture → 5521**

9611 wood moisture meter; wood humidity meter
(a tool for determine the moisture content of wood)
d Holzfeuchtemesser *m*; Holzfeuchtemessgerät *n*
f hygromètre *m* à bois; humidimètre *m* à bois
r влагомер *m* для древесины; измеритель *m* влажности древесины

9612 wood molasse
r деревянная меласса *f*

* **wood monger → 2276**

9613 wood mo[u]lding; mo[u]lding
(a wood strip having a curved or projecting surface used for decorative purposes)
d Profilholzleiste *f*; Pressling *m*
f moulure *f* [de bois]; moulurage *m*
r декоративная накладка *f*; молдинг *m*; профилированный полосовой материал *m*

9614 wood oddments
(a motley wood assortment)
d Holzüberschuss[-Sortimenten] *mpl*
f défets *mpl* de bois
r разрозненная спецификация *f* лесотоваров

* **wood of commerce → 5400**

* **wood of inferior quality → 5195**

9615 wood oil
(a resinous oil obtained from several East Indian trees)
d Holzöl *n*
f huile *f* de bois
r древесное масло *n*

* **wood on the stem → 3454**

* **wood opal → 5903**

9616 wood panelling
(decorative treatment of walls, ceilings, doors, and furniture consisting of a series of wide, thin sheets of wood, called panels)
d Holzverkleidung *f*; Holzvertäfelung *f*
f panneautage *m* par bois; lambrissage *m* par bois
r обшивка *f* деревянными панелями; обшивка деревянными филёнками

9617 wood parenchyma; xylem parenchyma
(parenchyma occurring in the xylem)
d Holzparenchym *n*
f parenchyme *m* de bois
r древесная паренхима *f*

9618 wood parenchyma cell
d Holzparenchymzelle *f*
f cellule *f* de parenchyme
r клетка *f* древесной паренхимы

* **wood-particle board → 6091**

f longueur *f* de copeau de bois
r длина *f* древесной частицы

* **wood particles → 9508**

9620 **wood particles drying**
d Holzpartikeltrocknung *f*
f séchage *m* de copeaux de bois
r сушка *f* древесных частиц

9621 **wood-particle width**
d Holzpartikelbreite *f*
f largeur *f* de copeau de bois
r ширина *f* древесной частицы

* **wood path → 3565**

* **wood pea → 5662**

9622 **woodpeckers**
(Picidae)
d Spechte *mpl*
f pics *mpl*
r дятловые *pl*

9623 **wood-peeling machine; veneer lathe; [wood] lathe; wood-turning lathe; slicer**
(machine on which logs are peeled to yield veneer for plywood)
d Holzschälmaschine *f*; Furnierschälmaschine *f*; Rundschälmaschine *f*; Abschälmaschine *f*; Schälmaschine *f*
f tour *m* à bois; dérouleuse *f* de bois
r лущильный станок *m*

* **wood physics → 8618**

* **wood pick → 1317**

9624 **wood pigeon**
(any one of numerous species of Old World pigeons belonging to Palumbus and allied genera of the family Columbid[ae])
d Ringeltaube *f*
f pigeon *m* ramier
r вяхирь *m*; витютень *m*; голубь *m* дикий; голубь лесной

9625 **wood pigment**
d Holzpigment *n*
f pigment *m* de bois
r пигмент *m* древесины

* **wood pile → 8619**

* **wood pitch → 8907**

d Holzplastifizierung *f*
f plastification *f* de bois
r пластификация *f* древесины

* **wood plot → 1821**

* **wood powder → 1514**

9627 **wood preservation; timber preservation; wood protection**
d Holzschutz *m*
f protection *f* de bois
r защита *f* древесины

* **wood preservers → 8623**

* **wood-preserving means → 8623**

9628 **wood pretreatment**
d Holzvorbereitung *f*
f prétraitement *m* de bois
r предварительная обработка *f* древесины; подготовка *f* древесного сырья

9629 **wood price; timber price**
d Holzpreis *m*
f prix *m* de bois
r стоимость *f* лесоматериала; цена *f* леса

9630 **wood processing; woodusing; woodworking; timber processing**
d Verwertung *f* des Holzes; Holzbearbeitung *f*
f usinage *m* de bois; traitement *m* du bois; menuiserie *f*
r использование *n* древесины; обрабатывание *n* древесины; деревообработка *f*

* **wood processor → 9701**

9631 **wood procurement**
d Holzbeschaffung *f*
f approvisionnement *m* en bois
r поставка *f* леса

* **wood production → 3537**

* **wood products → 8626**

9632 **wood properties**
d Holzeingeschaften *fpl*
f propriétés *fpl* de bois
r свойства *npl* древесины

* **wood protection → 9627**

* **woodpulp board → 6059**

9633 **wood pulping**
d Holzaufschluss *m*; Kochung *f*
f production *f* de matière fibreuse
r варка *f* древесины; превращение *n* древесины в волокнистую массу

9634 **wood[pulp] paper; groundwood paper; mechanical [pulp] paper**
(paper made of wood pulp)
d holzhaltiges Papier *n*
f papier *m* de pâte mécanique
r бумага *f* из древесной массы

9635 **wood purchase; timber purchase**
d Holzeinkauf *m*
f achat *m* de bois
r покупка *f* леса

* **wood quality → 6657**

* **woodranger → 3456**

9636 **wood rasp**
d Holzreibe *f*; Holzraspel *f*
f râpe *f* à bois
r рашпиль *m* для дерева

* **wood ray → 9738**

9637 **wood-recovery plant**
d Holznutzungsanlage *f*; Holzwertstoffanlage *f*
f installation *f* de récupération de bois
r цех *m* по переработке древесины

9638 **wood refining**
d Holzraffination *f*
f raffinage *m* de bois
r рафинирование *n* древесины

9639 **wood research**
d Holzforschung *f*
f recherche *f* de bois
r исследование *n* древесины

9640 **wood reserves**
d Holzvorräte *mpl*
f réserves *fpl* de bois
r запасы *mpl* [спелой] древесины

* **wood resin → 8907**

* **wood right → 8573**

9641 **woodruff: wort**
f aspérule *f* odorante
r ясменник *m*

* **woods → 9542**

* **wood saccharification → 7143**

* **wood sale → 8631**

9642 **wood sale contract**
d Holzverkaufsvertrag *m*
f contrat *m* de vente de bois
r договор *m* на продаже леса

* **wood sample → 8635**

* **wood sap → 7208**

9643 **wood saw**
d Holzsäge *f*
f scie *f* à bois
r пила *f* для дерева

9644 **wood sawdust**
(dust or small fragments of wood made by the cutting of a saw)
d Sägemehl *n*
f sciure *f* de bois [fine]
r древесные опилки *fpl*

9645 **wood science**
d Holzkunde *f*
f science *f* de bois
r древесиноведение *n*

* **wood scrab grinder → 7721**

* **wood scrap → 9508**

9646 **wood scraper; snipper**
d Holzkratzer *m*
f gratteur *m* de bois; racleur *m* de bois
r рабочий *m*, очищающий брёвна; рабочий, обтёсывающий брёвна

* **wood's equipments → 3402**

9647 **wood sere; wood dormancy**
d Zeit *f* der Saftruhe
f repos *m* de la végétation
r время *n* покоя сокодвижения

9648 **wood shaving curls**
d Holzspanlocken *fpl*
f vrillons *mpl* à copeaux de bois
r деревянные завитые стружки *fpl*

f cisaille *f* de bois
r ножницы *pl* для дерева

9650 woodshed; timbershed
(a general raw-material-procurement territory serving a wood-using mill or wood-consuming population center)
d Holzschuppen *f*
f hangar *m* à bois; aire *f* d'approvisionnement à bois
r дровяной сарай *m*; навес *m* для хранения дров; площадь-доставчик *f* древесины

9651 wood shock hardness
d Schockhärte *f* des Holzes
f dureté *f* de choc de bois
r ударная твёрдость *f* древесины

* **wood shoot → 4505**

9652 woodshop tool; woodworking tool
d Werkzeug *n* für Holzbearbeitung; Holzbearbeitungswerkzeug *n*
f instrument *m* d'usinage de bois; outil *m* pour le travail du bois
r деревообрабатывающий инструмент *m*; инструмент для обработки древесины

* **wood shrinkage → 7554**

9653 wood shrinkage factor; shrinkage value; shrinkage coefficient
d Trocknungsstufe *f*
f coefficient *m* de retrait; valeur *f* de retrait
r коэффициент *m* усушки [древесины]

* **wood slip → 9555**

9654 wood[s]man; forest worker; lumberjack; lumberman; timberman; forest labo[u]rer; forest operator; timber beast; lumberer; shanty man; logman
(one who works in forests, performing any of a variety of jobs related to the harvesting of timber)
d Waldarbeiter *m*; Holzer *m*; Forstarbeiter *m*; Lumberjack *m*
f boiseur *m*; ouvrier *m* forestier
r лесной рабочий *m*; лесозаготовитель *m*; лесопромышленник *m*

9655 wood softening
d Erweichen *n* des Holzes
f ramollissement *m* de bois
r размягчение *n* древесины

(Oxalis acetosella)
d gemeiner Sauerklee *m*; Oxalis *m*
f pain-de-surelle *m*; oxalis *m*; oseille *f*
r кислица *f* обыкновенная

9657 wood sorrel/spruce forest
d Sauerklee/Fichtenholz *n*
f pain-de-surelle/pessière *f*
r ельник-кисличник *m*

* **wood sort → 4714**

* **wood sorter → 9525**

9658 wood sorting rule
d Holzmessanweisung *f*
f règlement *m* de cubage le bois
r инструкция *f* по обмеру леса

9659 wood specialist; wood expert
d Holzwirt *m*; Forstwirt *m*
f spécialiste *m* du bois
r специалист *m* по древесине

9660 wood species album
d Holzartalbum *n*
f album *m* d'espèces forestières
r альбом *m* древесных пород

* **wood specimen → 8635**

* **wood spirit → 9476**

* **wood splint → 9695**

* **wood splitter → 9512**

* **wood-splitting apparatus → 9512**

* **wood-splitting attachment → 9512**

* **wood-splitting machine → 9512**

* **wood stack → 8619**

9661 wood stacker
d Holzsetzer *m*; Holzstapler *m*
f empileur *m* de bois
r складчик *m* древесины

9662 wood staining
d Holzanfärben *f*; Holzverschmutzung *f*
f mordançage *m* pigmenté de bois; coloration *f* de bois
r окрашивание *n* древесины

f champignons *mpl* de coloration de bois
r деревоокрашивающие грибы *fpl*; грибы, вызывающие изменения окраски древесины

9664 wood standard
d Holzstandard *m*
f standard *m* de bois
r стандарт *m* на лесоматериалы

* **wood standardization → 8642**

9665 wood static hardness
d statische Holzhärte *f*
f dureté *f* de bois statique
r статическая твёрдость *f* древесины

* **wood storage → 5120**

9666 wood structure
d Aufbau *m* des Holzes; Holzaufbau *m*; Struktur *f* des Holzes
f structure *f* du bois
r строение *n* древесины; структура *f* древесины

9667 wood substance
d Holzsubstanz *f*
f substance *f* de bois
r древесное вещество *n*

9668 wood substance density
d Dichte *f* des Holzsubstanz
f densité *f* de lignocellulose
r плотность *f* древесинного вещества

9669 wood substitute
d Holzersatz *m*
f produit *m* de remplacement de bois
r заменитель *m* древесины

9670 wood sugar; xylose
($C_5H_{10}O_5$)
d Holzzucker *m*; Xylose *f*
f sucre *m* de bois; xylose *m*
r гидролизная глюкоза *f*; древесный сахар *m*

* **wood supply → 2359**

9671 wood survey
d Holzaufnahme *f*
f estimation *f* de bois
r таксация *f* лесоматериалов

* **wood swallow → 8941**

f goudron *m* [de bois]
r древесный дёготь *m*

9673 wood tar oil
d Holzteeröl *n*
f huile *f* de goudron de bois
r древесносмоляное масло *n*

9674 wood technique
d Holztechnik *f*
f technique *f* de bois
r лесная техника *f*

9675 wood technologist
d Holztechnologe *m*; Holztechniker *m*
f technicien *m* de bois
r техник *m* по деревообработке

* **wood technology → 8582**

9676 wood tending
(any operation carried out for the benefit of wood material)
d Holzpflege *f*
f soins *mpl* au bois
r уход *m* за древесиной

9677 wood testing
d Holzprüfung *f*
f essai *m* de bois
r испытание *n* древесины

9678 wood texture; texture of grain; wood grain; grain figure; wood figure [of grain]; figure
(the direction, size, arrangement, appearance, or quality of the fibers in wood or lumber)
d Holztextur *f*; Fasertextur *f*; Zeichnung *f* des Holzes
f texture *f* du bois; veinure *f*
r текстурный рисунок *m*; текстура *f* волокон; волокнистая текстура; узор *m* древесины; текстура древесины

* **wood tick → 4380**

* **wood ticket → 6183**

9679 wood tissue; wood[y] tissue
(general term for any long, narrow cell of wood)
d Gewebe *n* des Holzes; Holzgewebe *n*; Holzbrei *m*
f tissu *m* ligneux
r древесная ткань *f*

f toxicité *f* de bois
r ядовитость *f* древесины; токсичность *f* древесины

* **wood trachea → 9687**

* **trachea → 9687**

* **wood track → 3565**

* **wood trade → 8653**

* **wood transport → 8595**

9681 **wood-transport road**
d Holzabfuhrstraße *f*; Holzabfuhrweg *m*
f route *f* de transportation de bois
r лесовозная дорога *f*

9682 **wood turner**
(a person which shapes pieces of wood rotating in a lathe, using hand tools)
d [Holz-]Drechsler *m*
f tourneur *m* sur bois
r обораичиватель *m*

9683 **wood turning**
(an operation of wood shaping by turning)
d Holzdrehung *f*
f tournage *m* du bois
r точение *n* по дереву

* **wood-turning lathe → 9623**

9684 **wood turpentine**
d Holzterpentin *n*; Holzterpentinöl *n*
f essence *f* [de bois] de pin
r скипидар *m* из осмола

* **wood types → 9597**

* **wood used in water construction → 9573**

* **woodusing → 9630**

9685 **woodusing industry; woodworking industry; carpentry**
d Holzbearbeitungsindustrie *f*; holzbearbeitende Industrie *f*
f industrie *f* du bois ouvré
r деревообрабатывающая промышленность *f*; лесоперерабатывающая промышленность; деревообрабатывающая индустрия *f*

* **wood utilization → 9154**

d Ausnutzungsgrad *m* des Holzes
f facteur *m* d'utilisation de bois
r коэффициент *m* использования древесины

* **wood veneer → 6398**

9687 **wood vessel; [wood] trachea; vas**
(the basic units comprising wood having open ends and set one above the other so as to form continuous tubes)
d Holzgefäß *n*; Holzröhre *f*
f vaisseau *m* [de bois]; trachée *f*
r сосуд *m* древесины; древесный сосуд

9688 **wood vinegar; wood acid; pyroligneous acid**
(a complex acid liquid obtained in the dry distillation of wood, and containing large quantities of acetic acid)
d Holzessig *m*; Holzsäure *f*
f vinaigre *m* de bois
r древесный уксус *m*

9689 **wood volume; volume of timber; timber volume**
(the amount of wood in a tree or stand according to some unit of measurement)
d Holzvolumen *n*
f volume *m* de bois
r объём *m* древесины; объём лесоматериалов

9690 **wood want; want**
(the absence of wood, other than wane, from the arris or surface of a piece of timber)
d abgestumpfte Kante *f*
f épaufrure *f*
r скос *m* (дефект, из-за которого не выдерживается нужная толщина детали)

9691 **wood warbler**
(any one of numerous species of American warblers of the genus Dendroica)
d Waldlaubsänger *m*
f pouillot *m* siffleur
r пеночка-трещотка *f*

* **woodward → 3456**

* **woodware → 9560**

9692 **woodwasp damage**
d Holzwespeschaden *m*
f dégât *m* par sirex
r червоточина *f*, вызываемая жуком-рогохвостиком

d Holzwespen *fpl*
f sirex *mpl*
r рогохвосты *mpl*

* **wood waste → 3561**

9694 woodwind [instrument]
d Holzblasinstrument *m*
f instrument *m* à vent en bois
r деревянный духовой инструмент *m*

9695 wood wire; wood splint
(for matches)
d Holzdraht *m*
f fil *m* en bois; paille *f* en bois
r древесная соломка *f*; спичечная соломка

9696 wood with circular crevices inside; shelly timber; coltie
d absplintiges Holz *n*; ringschäliges Holz; kernschäliges Holz
f bois *m* roulé
r лесоматериал *m* с частичным отлупом; древесина *f* с отлупом; облупающаяся древесина; треснувший древесный ствол *m*

9697 wood wool; tree-wool; fine wood shavings; excelsior-wood
(long, curly, slender strands of wood used as an aggregate component for some particleboards and as a packing material)
d Holzwolle *f*; Waldwolle *f*
f laine *f* de bois; fibre *f* de bois; frison *m*
r древесная шерсть *f*

9698 woodwork
d Holzarbeit *f*
f travail *m* du bois
r плотничья обработка *f*; столярная обработка

9699 wood worker; wood labo[u]rer
d Holz[be]arbeiter *m*
f ouvrier *m* des bois; travailleur *m* de bois
r деревообработчик *m*

9700 wood[workers'] asthma; woodworker's lung; wood dust asthma; sawdust asthma
d Holzmehl-Asthma *n*
f pneumopathie *f* des ouvriers des bois exotiques; asthme *m* au bois; pneumonite *f* à la poussière de bois
r астма *f* из-за древесной пыли

* **woodworking → 9630**

9701 wood[-]working machine; timber working machine; wood processor
d Holzbearbeitungsmaschine *f*
f machine *f* à bois; machine-outil *m* à [travailler le] bois
r деревообрабатывающая машина *f*

9702 woodworking technique
d Holzverarbeitungstechnik *f*
f technique *f* d'usinage de bois
r метод *m* обработки древесины

* **woodworking tool → 9652**

9703 woodworm damage; damage done by worms
d Baumfraß *m*; Wurmbefall *m*
f dégât *m* de coléoptères
r повреждение *n*, вызываемое личинками древоточца

9704 woody; wooden; arborescent; arboreal; arboreous; arboraceous; ligneous *adj*
(consisting of or resembling wood; as distinguished from herbaceous)
d Holz-
f arboricole; arborescent
r древесный; деревянистый

9705 woody aloe; arborescent aloe; tree-like aloe
(Aloe arborescens; Aloe fruticosa)
d baumartige Aloe *f*; Kandelaber-Aloe *f*
f aloès *m* arborescent; corne *m* de cerf; corne de bélier
r древовидное алое *n*

* **wood yard → 8664**

9706 woody bundle; xylem bundle
d Holzbündel *m*
f faisceau *m* ligneux
r древесинный пучок *m*

* **woody climber → 9228**

9707 woody debris; forest debris
d Schlagreisig *n*
f débris *m* ligneux
r лесные горючие материалы *mpl*

9708 woody grasses
d Baumgräser *npl*
f graminées *fpl* ligneuses
r деревянистые злаки *mpl*

* **wood yield → 9778**

f à feuilles ligneuses
r древеснолистный

9710 **woody mountains**
d Waldgebirge *npl*
f montagnes *fpl* boisées
r лесистые горы *fpl*

* **woody nightshade → 239**

9711 **woody pasture-ground; treed pasture; woodland range**
(forestland (within range allotments) that produces minor amounts of forage)
d bewaldete Weide *f*; Waldlandweide *f*
f pâturage *m* boisé; pâturage arboré; pré-bois *m*
r облесённое пастбище *n*; покрытое лесом пастбище; лесное пастбище; лесной выпас *m*

9712 **woody peat**
(unconsolidated forest soil material consisting of undecomposed or slightly decomposed organic matter accumulated under conditions of excessive moisture)
d Waldmoorboden *m*; Waldtorf *m*
f sol *m* tourbeux forestier
r лесной торф *m*

9713 **woody perennial**
(plant with lignified stems that continues to grow from year to year)
d vieljährige Holzpflanze *f*
f plante *f* ligneuse pérenne
r древесное многолетнее растение *n*

9714 **woody plant; ligneous plant; xylophyte**
(with the stems and limbs containing lignin)
d holzige Pflanze *f*; Holzpflanze *f*; Holzgewächs *n*
f plante *f* ligneuse
r древесное растение *n*

* **woody pulp → 9603**

9715 **woody-rooted**
d holzwurzelig
f xylorhize
r древеснокорешковый

* **wood[y] tissue → 9679**

* **woody vegetation → 8953**

* **woolly beech[-leaf] aphid → 650**

d wolliger Kreuzdorn *m*
f nerprun *m* laineux
r бумелия *f* шерстистая

* **woollybutt → 4054**

9717 **woolly grain; fuzzy grain**
d behaarte Faser *f*
f grain *m* pelucheux
r мшистое волокно *n*; ворсистое волокно

9718 **woolly larch aphid**
(Adelges laricis)
d rote Fichtengallenlaus *f*; kleine Fichtengallenlaus *f*
f puceron *m* lanigère du mélèze
r мохнатая тля *f* лиственничная

9719 **woolly terminalia; Burma laurel; Myanmar laurel; taukkyan**
(Terminalia tomentosa; Terminalia alata; Hopea ferrea)
d Taukkyan *m*
f laurel *m* de l'Inde; sain *m*; taukkyan *m*
r терминалия *f* войлочная

9720 **woolly willow**
(Salix lanata)
d wollige Weide *f*
f saule *m* pelucheux
r ива *f* мохнатая; ива шерстистая

9721 **workable timber; industrial timber; industrial lumber; industrial wood; shop lumber**
d Industrieholz *n*
f bois *m* d'œuvre; bois de service; bois d'industrie; bois industriel
r промышленная древесина *f*; индустриальная древесина; промышленные лесоматериалы *mpl*

9722 **working block**
d Arbeitsblock *m*
f bloc *m* exploitable
r рабочий участок *m* леса

9723 **working group**
(in US, aggregate of forest stands, or forest stand and forest sites, which are grouped for the purpose of applying a common set of silvicultural treatments)
d Arbeitsgemeinschaft *f*
f section *f* d'aménagement
r хозяйство *n* (выделяется при

* **working-plan officer → 3497**

9724 **work in the forest; forest operation; forest labour; forest work; silvicultural work; wood labo[u]r**
d Waldarbeit *f*; Forstarbeit *f*
f opération *f* forestière; travail *m* forestier
r лесная работа *f*

9725 **world forest area**
d Weltwaldfläche *f*
f aire *f* forestière mondiale
r территория *f* земного шара, покрытая лесами

9726 **world forestry**
d Weltforstwirtschaft *f*
f économie *f* forestière mondiale
r мировое лесное хозяйство *n*

9727 **world lumber trade; world timber trade**
d Weltholzhandel *m*
f commerce *m* mondial de bois
r мировая лесная торговля *f*

9728 **worm-dust; bore-dust; boring dust; frass**
d Wurmmehl *n*
f farine *f* de vers
r опилки *mpl* от древоточцев; труха *f*, оставляемая насекомыми в червоточине; отходы *mpl* насекомых в червоточине

* **wormeaten → 9730**

9729 **wormeaten timber; wormeaten wood**
d wurmstichiges Holz *n*
f bois *m* vermoulu
r червоточная древесина *f*

9730 **wormed; wormy; worm-holed; wormeaten; prickled**
d wurmbefallen; wurmstichig; wurmfraßig
f piqué des vers
r червоточный

* **worm groove → 4511**

* **worm hole → 4511**

* **wort → 9641**

9731 **worthless scrubs**
d wertloses Unterholz *n*

* **wound cork → 8781**

9732 **wound gum; wound resin**
d Wundgummi *n*; Wundharz *n*
f gomme *f* de lésion; résine *f* de lésion
r раневая камедь *f*; раневая смола *f*

* **wound heartwood → 8782**

* **wound parenchyma → 8784**

* **wound rot → 7078**

9733 **wound stain**
d Wundverfärbung *f*
f décoloration *f* cicatricielle
r окрашивание *n* от ранения

* **wound tissue → 8787**

9734 **wound wood; traumatic wood**
d Wundholz *n*
f bois *m* cicatriciel; bois traumatique
r раневая древесина *f*; древесина с ненормальным строением ткани

* **wrapper chain → 5023**

* **wrenching → 7055**

* **wrought stuff → 8280**

9735 **wrought timber**
d gefertigtes Holz *n*; Bakelitholz *n*; bakelisiertes Holz
f bois *m* bakélisé
r древесина *f*, пропитанная бакелитом

* **wych elm → 7310**

X

9736 **xerophilous plants; xerophytes**
(plants capable of surviving prolonged periods of moisture deficiency)
d trockenliebende Pflanzen *fpl*; Xerophyten *mpl*
f plantes *fpl* xérophiles
r ксерофиты *mpl*; растения *npl* засушливых зон

* **xerophytic light forest → 358**

9737 **xylem**
(the tissues of the stem, branches, and roots of a woody plant lying between the pith and cambium, serving for water conduction, mechanical strength, and food storage, and characterized by the presence of tracheids or vessels)
d Holzkörper *m*; Xylem *n*
f xylème *m*
r ксилема *f*

* **xylem bundle → 9706**

* **xylem parenchyma → 9617**

9738 **xylem ray; wood ray**
(a ribbon-like aggregate of cells extending radially across the grain, so oriented that the face of the ribbon is exposed as a fleck on the quartered surface)
d Xylemstrahl *m*; Holzstrahl *m*
f rayon *m* ligneux
r ксилемный луч *m*; участок *m* ксилемы между двумя сердцевинными лучами

* **xylogenous → 2855**

* **xylography → 9502**

9739 **xylophagous; wood-eating; lignivorous**
(feeding on wood as the larvae of certain insects)
d holzfressend
f xylophage
r поедающий древесину; питающийся древесиной

* **xylophylous → 9709**

9740 **xylopia**
(Xylopia)
d Mohrenpfeffer *m*
f guile *m*
r ксилопия *f*

* **xylose → 9670**

9741 **xylotomy**
(the art of preparing sections of wood, esp. by means of a microtome, for microscopic examination)
d Xylotomie *f*
f xylotomie *f*
r ксилотомия *f*; изготовление *n* тонких древесных срезов (для микроскопического исследования строения древесины)

Y

* **yagrumo → 7519**

* **yagrumo macho → 5558**

9742 **yarder point**
d Schlepperpunkt *m*
f point *m* d'établissement de débardeur
r место *n* установки лебёдки (для привода канатной установки)

9743 **yarder rigging**
d Schleppertakelage *f*
f timonerie *f* de débardeur
r прицепное приспособление *n* для канатной трелёвки

9744 **yarder wood**
(wood brought into a yard in the form of tree lengths, logs, or bolts, to be cut into shorter lengths)
d Schnittholzware *f*
f bois *m* débardé
r складированная древесина *f*

* **yarding arch → 5090**

9745 **yarding corridor; rack**
d Holzabfuhr-Korridor *m*
f corridor *m* de débusquage; passage *m* de débusquage
r просека *f* под несущий канат; просвет *m* под несущий канат

9746 **yarding direction**
d Rückenlinie-Richtung *f*
f direction *f* de traînage sur câble
r направление *n* трелёвочного волока

9747 **yarding distance; haulage distance; haul[ing] distance; extraction distance; skid[ding] distance; snig length**
d Förderweite *f*; Rückedistanz *f*; Rückeentfernung *f*; Abfuhrsabstand *m*
f distance *f* de roulage; distance de transport; distance de débusquage
r расстояние *n* трелёвки; расстояние вывозки

9748 **yarding hook-tender; hook-tender; hooker**
turn of logs)
d Hooktender *m*; Vorarbeiter *m* am Verladeplatz
f chef *m* chargeur; chef d'équipe de chargement
r десятник *m* при складе

9749 **yarding method; logging method**
d Rückeverfahren *n*
f méthode *f* de débusquage
r способ *m* трелёвки; трелёвочный метод *m*

9750 **yarding path; yarding road**
(path followed by a turn of logs yarded by a cable method)
d Rückenlinie *f*
f ligne *f* de traînage sur câble
r линия *f* прохождения трелёвочного каната; линия лебёдочной трелёвки

9751 **yarding road change; road change**
d Rückenlinieveränderung *f*
f changement *m* de ligne de traînage
r смена *f* трелёвочного волока; перебазировка *f* несущего каната

* **yarding sled → 5115**

9752 **yarding thinning**
d Rückenlinie-Durchforstung *f*
f éclaircie *f* par ligne de traînage
r рубки *fpl* ухода с канатной трелёвкой

9753 **yard inventories**
d Lagerplatzvorräte *mpl*
f inventories *mpl* de dépôt
r запасы *mpl* склада лесоматериалов; запасы биржи лесоматериалов

9754 **yard lumber**
(a little-used term for lumber of all sizes and patterns that is intended for general building purposes having no design property requirements)
d Stapelschnittholz *n*
f bois *m* de charpente tout venant; bois de cour; bois de chantier; sciage *m* pour parc de séchage; sciage de parc
r биржевой пиломатериал *m*

9755 **yard stain**
d Lagerverfärbung *f*
f coloration *f* due à l'entreposage
r окраска *f* от хранения

* **yard stick → 5107**

* **yashnik → 3108**

* **year cut[ting] → 296**

* **year felling → 296**

9756 **yearling; one-year seedling; 1-0 seedling**
d Jährling *m*; einjähriger Sämling *m*; Lachte *f* im zweiten Jahr
f plante *f* d'un an; semis *m* de l'année; carre *f* de deuxième année
r годовой сеянец *m*; карра *f* во втором году

* **yearling shoot → 305**

* **yearly growth → 300**

* **year ring → 301**

* **year ring analysis → 269**

* **year-round operations → 1897**

9757 **yellow balau; balau; bangkirai**
(Shorea laevis; Shorea atrinervosa; Shorea glauca)
d Balau *m*
f balau *m*
r балау *m*

* **yellow-bark oak → 2692**

9758 **yellow buckeye**
(Aesculus flava; Aesculus octandra)
d gelbe Pavie *f*; gelbe Rosskastanie *f*
f marronnier *m* à fleurs jaunes
r каштан *m* конский восьмитычинковый

* **yellow [chestnut] oak → 1510**

* **yellow deal → 6275**

* **yellow earths → 9763**

9759 **yellow jasmine**
(Jasminum fruticans)
d gelber Jasmin *m*
f jasmin *m* jaune
r жасмин *m* жёлтый

9760 **yellow laminated butt rot**
(invader - Poria weirii)
d schichtete Gelbstammfäule *f*
f pourriture *f* du pied jaune laminée
r жёлтая трещиноватая напенная гниль *f*

* **yellow paloverde → 3312**

* **yellow pine → 9272**

* **yellow poplar → 9008**

9761 **yellow sap rot**
(invader - Polyporus cinnabarinus)
d Gelbsaftfäule *f*
f pourriture *f* de l'aubier jaune
r жёлтая заболонная гниль *f*

9762 **yellow slash rot**
(invader - Hymenochaete rubiginosa)
d Gelbschlagfäule *f*
f pourriture *f* jaune de rémanents
r жёлтая гниль *f* мёртвой древесины; пёстрая ядрово-заболонная гниль мёртвой древесины

9763 **yellow soils; yellow earths**
d Gelberden *fpl*
f terres *fpl* jaunes
r жёлтоземы *mpl* (почвы широколиственных лесов субтропического пояса, содержащие мало железа, развивающиеся при сильном увлажнении)

* **yellow spruce → 1298**

9764 **yellow stain**
d Gelbverfärbung *f*
f jaunissement *m* du bois
r жёлтая грибная окраска *f*

* **yellow-stem white willow → 3885**

9765 **yellow stringybark; Müller's stringybark**
(Eucalyptus triantha; Eucalyptus muellerana)
d Müller-Eukalyptus *m*
f eucalyptus *m* de Müller
r эвкалипт *m* трёхцветковый; эвкалипт Мюллера

* **yellow sucupira → 4784**

* **yellow trumpet tree → 1339**

9766 **yellow willow**
(Salix lutea)
d Gelbweide *f*
f saule *m* jaune
r ива *f* жёлтая

* **yellowwood → 262**

f bois *m* jaune
r [любая] древесина *f* жёлтого цвета

* **yew → 1814**

9768 yew grove; yew forest
d Eibenbestand *m*
f ivraie *f*
r тисовый лес *m*

9769 yew wood
d Holz *n* von Taxus baccata
f bois *m* à Taxus baccata
r древесина *f* тисового дерева

* **Yezo spruce → 4647**

* **yield assessment → 9776**

9770 yield by area; possibility by area
d Flächenhiebsatz *m*
f possibilité *f* par surfaces
r выход *m* по площади; сметная вырубка *f* по площади

9771 yield by thickness
d Dichtenhiebsatz *m*
f rendement *m* par épaisseur
r выход *m* по толщине

9772 yield by tree; yield per tree
d Baumertrag *m*
f possibilité *f* [par] pieds d'arbre; rendement *m* par arbre
r выход *m* из одного дерева

9773 yield by volume; volume yield; volume production; volume output
d Volumenleistung *f*; Massenleistung *f*; Massenertrag *m*
f possibilité *f* par volume; production *f* en volume; rendement *m* en volume
r выход *m* по объёму; доход *m* по массе; объёмная производительность *f*

* **yield calculation → 9776**

* **yield capability of forest → 3648**

9774 yield class; capacity class; productivity class; quality class
d Bonitätsklasse *f*; Güteklasse *f*; Gütegrad *m*; Bonitätsstufe *f*
f classe *f* de qualité; classe de fertilité; degré *m*
добротности

9775 yield curve
(representation of stand volume, usually as a function of stand age, in graphical or tabular form)
d Ertragskurve *f*
f courbe *f* de fertilité
r кривая *f* урожая

9776 yield determination; yield calculation; yield estimation; yield assessment; determination of cut
d Ertragsermittlung *f*; Ertragsberechnung *f*; Ertragsbestimmung *f*
f détermination *f* de la possibilité; calcul *m* de la possibilité
r расчёт *m* размера лесопользования; определение *n* годичной лесосеки

9777 yield from felling; result of the felling; felling returns; outturn
d Hiebsergebnis *n*; Hiebsertrag *m*; Bruttoholzrente *f*
f produit *m* escompté de la coupe; produit d'une coupe
r выход *m* древесины на лесосеке; результат *m* рубки; объём *m* заготовленной древесины

9778 yield of a forest; forest yield; wood yield; timber harvest
d Forstertrag *m*; Waldertrag *m*; Holzertrag *m*; Nutzholzgewinn *m*; Holzerlöse *mpl*
f rendement *m* [en matière] de la forêt; rendement *m* du bois; rentabilité forestière
r лесной доход *m*; выход *m* [деловой] древесины

9779 yield of a tree
d Ertrag *m* eines Baumes
f rendement *m* par un arbre
r выход *m* [древесины] из дерева

9780 yield of log; log yield
(lumber coming from the saw - with softwoods, it is the merchantable lumber of all grades; with hardwoods, it excludes specified lower grades)
d Stammertrag *m*; Rundholzausbeute *f*
f rendement *m* des grumes; débit *m* courant
r выход *m* пиломатериалов из брёвна

9781 yield of spruce stand
d Ertrag *m* eines Fichtenbäumes

9782 **yield of thinning; thinning yield; income from thinning; returns from thinning**
d Durchforstungsanfall *m*; Vornutzungsanfall *m*; Durchforstungsertrag *m*; Durchforstungseinnahme *f*
f produits *mpl* d'éclaircies; rendement *m* en éclaircies
r выход *m* древесины, полученный от рубок ухода; доходность *f* прорубки

* **yield per tree → 9772**

9783 **yield product**
d Nutzungsanfall *m*
f produit *m* de la coupe; quotité *f* de la coupe
r выходной продукт *m*

9784 **yield table**
(table and graphs illustrating volume per hectare of stands at a specific age)
d Ertragstafel *f*; Leistungstafel *f*
f table *f* de rendement; table de production
r таблица *f* роста насаждений

* **yoke → 3888**

* **yoker → 1555**

* **yopo → 1714**

* **young beech tree → 661**

* **young birch tree → 5347**

* **young crop → 9788**

* **young forest → 9785**

9785 **young growth; seedling stand; young forest; new forest; young stand**
d Jungwald *m*; Jungwuchs *m*; Dünnholzbestand *m*; Rohling *m*
f jeune peuplement *m*; recrû *m*
r молодняк *m*; молодое [лесное] насаждение *n*; насаждение в стадии жердняка; жердняк *m*; тонкий жердевой лес *m*

* **young-growth stand → 7204**

* **young growth tending → 8458**

9786 **young oak-bark**
d junge Eichenrinde *f*

* **young oak tree → 661**

9787 **young plant**
d Saatpflanze *f*
f jeune plante *f*
r молодое растение *n*

9788 **young plantation; young crop; culture**
d junge Kultur *f*; Kultur *f*
f jeune plantation *f*; culture *f*
r молодые лесные культуры *fpl*

* **young poplar crop → 6456**

* **poplar crop → 6456**

* **young stand → 9785**

9789 **Young' sweeping birch**
(Betula pendula youngii)
d Hängebirke *f*
f bouleau *m* pleureur; bouleau de Young
r плакучая берёза *f*

9790 **young timber stage**
d Jungholzstadium *n*
f stade *m* de bois jeune
r стадия *f* приспевающего насаждения

9791 **young tree**
d Jungbaum *m*; Jüngling *m*
f arbre *m* jeune
r молодое дерево *n*

* **young wood → 4675**

* **yubia → 1003**

9792 **yucca**
(Yucca filamentosa)
d Palmlilie *f*
f yucca *m* filamenteux
r юкка *f* нитчатая

* **Yule[tide] log → 1567**

Z

9793 **Zambezi redwood; Rhodesian teak; Rhodesian redwood; umgusi**
(Baikiaea plurijuga)
d Rhodesien-Teak *m*; Umgusi *m*
f teak *m* de Rhodésie; umgusi *m*
r родезийский тик *m*; умгузи *m*; родезийское красное дерево *n*

* **zareeba → 5019**

* **zebrano → 118**

9794 **zebrawood family; connarus family**
(Connaraceae; Rosidae)
d Rosenähnliche *pl*
f connarales *fpl*
r коннарусовые *pl*

9795 **zero-span [breaking length]**
d Nullreißlänge *f*
f mâchoire *f* jointive
r нулевой зазор *m*; нулевая зажимная длина *f*

* **zero till[age] → 5799**

9796 **zigzag edging**
d Zickzack-Randausbildung *f*; Zickzack-Bekanten *n*
f délignage *m* en chevrons; façonnage *m* des bords; débordage *m*; bordage
r обрезка *f* [досок] зигзагом

* **zimbro → 458**

* **zingana → 118**

9797 **zonal agroforestry**
(spatial planting arrangement in which the different species each system remain contiguous to some extent)
d Zonenfeldwaldbau *m*
f agroforesterie *f* locale
r зональное агролесохозяйство *n*; региональное переменное лесопольное хозяйство *n*

9798 **zonality**
d Gürtelung *f*; Zonierung *f*

* **zone line → 801**

9799 **zone of deferred development; designated area; holding zone**
(identifiable geographic unit of the forestland base that requires a specific combination of forest ractices to adequately protect important resource values)
d Gebiet *n* zurückgestellter Erschließung; Bauerwartungsland *n*
f zone *f* d'aménagement différé
r зона *f* замедленного развития

* **zooc[o]enosis → 291**

* **zorrowood → 8548**

* **Z-stem → 3786**

* **Z-tree → 3786**

9800 **zygotic forestry**
d zygotischer Wald *m*
f foresterie *f* zygotique
r зиготический лес *m*

Deutsch

Aalbaum 1227
Aashornbaum 2892
A-Ausbaugespann 100
Abästung 6610
Abästung der schattiger Baum 7480
abaxial 5
abaxiales Parenchym 9091
Abbau 2310
abbeilen 2161
Abbeilen der Wurzelanläufe 6861
Abblatten 2941
abblättern 2940
Abblätterung 2941
Abblätterung der Rinde 3243
abborken 2280
abbrennen 3166
Abbrennen von der Mitte aus 1409
Abbruchholz 2364
Abbuschung 1109
Abdeckung 2304
abendländischer Lebensbaum 5795
abendländischer Zürgelbaum 1777
abendländliche Platane 254
Abfahrtsperiode 4170
Abfahrtszeit 4170
Abfallbeseitigung 7723
Abfallbeseitigungsfeuer 7724
Abfallbeseitigungsgesetz 7725
Abfallbrenner 6807
Abfälle in der Baumwollverarbeitung 1422
abfallend 4877
Abfallhobelspan 9274
Abfallholz 3996
Abfallholzverwertung 7731
Abfallkette 6808
Abfallklötze 9273
Abfallspan 9274
Abfallstapel 4774
Abfallzerkleinerer 7721
Abfluss 2611
Abflussfurche 2613
Abflussgraben 2613
Abflussleitung 2615
Abflussrinne 334
abfordernde Arten 2361
abförmiger Baum 8406
Abforstung 2339
Abfuhr des Holzes 8595
Abfuhrfrist 6863
Abfuhrsabstand 9747
Abfuhrschein 6180
Abgabe von Holz 6862
abgängiges Holz 2490
abgebrannter Wald 1187
abgefallenes Holz 3042
abgeflachter Wald 2342
abgehackter Holzklotz 1562
abgeholzter Aushau 2189
abgekopptes Blockholz 7795
abgelagertes Holz 7335
abgelängt 2176
abgelehnt 2402
abgemessenes Holz 2502
abgerechnete Fläche 5066
abgerechnete Waldparzelle 9096
abgerindetes Holz 555
abgesägter Stamm 7238
abgeschrägte Fuge 7285
abgestorbener Baum 2271
abgestorbener hängender Ast 9397
abgestorbener und hängender Ast 9397
abgestorbenes Holz 2274
abgestumpfte Kante 9690
abgetrocknetes Nutzholz 150
Abgipfen 2198
Abhang 7750
Abhangsaufforstung 7751
Abhangsbewaldung 7751
Abhangsschutzwald 7754
Abhangsstabilisierung 7974
Abharzen 828
abhauensreif 3231
Abhiebsschnitt 2980
Abhieb unter dem Erdboden 9059
Abholz 3042
Abholzen 3352
abholzen 3037
abholzig 8401
Abholzigkeit 8405
Abholzigkeitskurve 8400
Abholzigkeitstafel 8409
Abholzung 2339
Abholzung mittels Holzladewinde 2571
Abholzungsmethode 1617
abhüten 1094
Abietazeen 6254
abiotisch 6
abiotischer Schaden 7
abiotischer Umweltfaktor 8
abkappen 7962
Abkappen 2198
Abkappsäge 5056
Abkeilen 9318
Abkneifen 6240
Abkürzen 1215
Abkürzkreissägemaschine 6676
Abkürzmaschine 2043
Abkürzsäge 5056
Abladeplatz 5451
Ablaktation 331
Ablaktieren 331
Ablänger 1124
Ablängersbruch 1125
Ablängsäge 1136
Ablängsägemaschine 2043
Ablängungsbrigade 3048
Ablängzeichen 8972
Ablass 7761
Ablaufberg 4440
Ablaufbühne 4442
Ablaufklasse 5342
Ablaufrohr 2615
Ablegen 6582
Ableger 2184, 5870, 7020
Ableitungsrinne 334
Ablenker 1095
Ablenkung des Bauholzes 2548
Ablenkungsgraben 2547
ablohen 2280
abloten 6385
Abmessungstabelle 7275
abnehmende Pflanzenart 2311
Abney-Messgerät 9
Abnutzungssatz 195
Abrahamsbaum 4961
Abrandkraut 5883
abrasen 1094
Abraum 2266
abraümen 1111
Abraümen 1109
Abraumfläche 7718
Abraumhalde 5464
Abraumpflug 5524
Abräumung des Holzschlages 1624
Abräumungshieb 3136
Abräumungsschlag 3136
Abraumverbrennen 7719
Abrichthobelmaschine 4661
Abrollen 6121, 7001
Abrollung 6121
Abrückschaden 7685
Abrundenlänge 7106
Absatz von Walderzeugnissen 5312
abschälen 6133
Abschäler 556
Abschälmaschine 9623
abschalten 1122
abschaltend 2176

Abschätzung stehender Bestände 8290
Abschleppseil 9007
abschließende Reinigung 3134
abschließendes Stadium des Holzschlags 3142
Abschnitte 2174
abschroten 2172
Abschrubben 4273
abschwemmen 3269
Abseilen 7062
Absenker 2268
Absetzteer 7369
absolute Darrholzdichte 18
absoluter Bestandesgrad 4009
absoluter Waldboden 19
absolutes Darrholz 17
absorbierende Wurzel 20
Abspaltung 1629
Abspannseil 1164
absplintiges Holz 9696
absplittern 1631
Absplitterung 1629
Absprengen von Baumkronen 851
Abstand der Bäume 4569
Abstand der Zwischenhölzer 8092
Abständer 2271
Äbstand-gezähnte Säge 3803
abständiges Holz 2490
abständig werden 2307
Abstandsholz 8579
Abstandsstock 3802
Abstapelautomat 5212
Abstapelgerät 5212
Abstapeln 4846
Abstapelvorrichtung 5212
Abstechpflug 7812
Abstecken 7995
Absteckpfahl 3899
Absteckpflock 3899
Absteckstange 3899
Absteifung 8097
absteigender Saftstrom 2312
Absterben 8686
Abstirnen 5798
Abstirnung 5798
abstoßendes Mittel 6869
Abstreichpflug 5524
abstücken 1122
Abstutzung 6348
Abteufen 7660
abtreiben 3037
Abtrieb 1607
Abtrieb der Überständer 2969
Abtriebnutzung 3136
Abtriebsalter 136
Abtriebswert 3056
Abtrift 2625
Abura 4101
Abutilon 22
Abwärtserlockerung 2597
abwärts laufende Form 2316
abwärts laufendes Verzweigungshabit 2315
Abwärtszerkleinerung 2601
abweiden 1094
Abweisrechen 1025
Abwickeln 6121
Abwipfeln 2198
Abzapfung 828
abziehen 1552
Abziehen des Bodenüberzuges 7286
Abziehmesser 4584
Abzimmerung 8600
abzopfen 8702
Abzug 2400
Abzugsgraben 2613
Abzweigung 987
Achäne 38
Achatholz 124
Achselknospe 419
Achselspross 4237
Ackeraufforstung 6100
Acker-Brombeere 2891
Ackerlandbaumverarbeitung 4487
Adaptationsfähigkeit 60
adaptierende Betriebsfläche 61
adaxial 62
Additiv 64
Additive 65
Ader 4847
Aderholz 1032
Adernschneiden 2187
Äderung 9192
Adlerfarn 967
Adventiv-; adventiv 77
Adventivknospe 78
Adventivspross 80
Adventivstruktur 81
Adventivwurzel 79
Aerophyt 2852
Affenbrotbaum 5533
afghanische Kiefer 2777
afrikanische Mangrove 6780
afrikanische Ölpalme 110
afrikanischer Kautschuk 3002
afrikanischer Nittabaum 108
afrikanischer Padouk 290
afrikanischer Pflaumenbaum 112
afrikanischer Wacholder 106
afrikanischer Zwetschgenbaum
afrikanische Seideneiche 3966
afrikanisches Mahagoni 107
afrikanische Sparmannie 115
afrikanisches Sandelholz 111
afrikanisches Zitronenholz 5732
afrikanische Tamariske 116
afrikanische Zimmerlinde 115
Afrormosia 119
Afterblatt 968
Afterschlag 998
Agar-Klötzchenversuch 123
Agavengewächse 125
Agba 5731
Ahlbaum 2881
Ahlkirsche 2881, 2895
Ahorn 5296
Ahornbaum 5296
ahornblättrige Platane 2906
ahornblättriger Schneeball 5293
Ahorn-Buche-Birkenwald 5290
Ahorngewächse 5291
Ahornholz 5297
Ahorn-Rindenbrand 7849
Ahornsaft 5294
Ahornsirup 5295
Ahornwald 5292
Ährenfelsenbirne 2689
Ährenparve 940
ährige Felsenbirne 2689
Aiele 103
Ajanfichte 4647
akadischer Wald 23
Akanthus 24
Akanthusgewächse 25
Akarinose 26
Akaziengummi 4048
Akklimatisation 32
Akklimatisierung 32
Akkretion 35
Aktinomyzeten 52
Aktionsradius 2102
aktiver Cluster 54
aktive Substanz 55
aktualisierter Wert 5718
aktuelle Verdunstung 2916
Akuama 2245
akzeptierte Hackspäne 28
akzessorische Knospe 78
Alangiumgewächse 158
Alaska-Birke 159
Alaunbäume 3025
Alaunholz 6901
Albedo 161
Albedomesser 162
Alcocks-Fichte 163
Aleppo-Eiche 3794

Alhornbaum 2892
aliformes Parenchym 167
Alkalinität 170
Alkaliquote 171
Alleebaum 9308
Alleepappel 5126
Alleghany-Felsenbirne 177
Alleghany-Fichtenriesenbastkäfer 178
Alleghany-Laubwald 176
Allele 179
Allelopathie 180
Allerheiligenholz 1289
allgemeine Beteiligung 6623
allgemeiner Bodenverlustausgleich 9100
"Alligatoren" 188
Alligator-Wacholder 187
Allmäher 7319
allochthon 189
allometrische Relationen 190
Alluvialwald 204
Allwetter-Straße 205
Almendro 4464
Alpenbirke 892
Alpenerle 2897
Alpenerlengebüsch 8224
Alpenföhre 5598
Alpengoldregen 215
Alpenheidelbeere 864
Alpenjohannisbeere 213
Alpen-Klematis 212
Alpen-Kreuzdorn 211
Alpen-Lärche 216
Alpenmeise 9438
Alpenrebe 212
Alpenrosengesträuch 7063
Alpensteine 217
Alpen-Waldrebe 212
Alpha-Vielfalt 210
Alphitonia 8807
alpiner Bohnenbaum 215
Alstonia 1461
Altbaum 5341
Altbestand 5880
Altbestandattribute 5879
Alteraufbau 130
Alterbestimmung 131
Alter des Bestandes 8001
älteres Holz 5884
Alterklassenfolge 127
Alterklassenlagerung 127
Alterklassentabelle 129
Alterklassenübersicht 129
alternierende Blattdisposition 221
alternierender Lochfraß 223
Altersermittlung 131
Altersfarbe des Holzes 6965
Altersklasse 126
Altersklassenabstufung 3893
Altersklassenverhältnis 138
Altersklassenzeitraum 128
altgläubiges Saatgut 5961
Altholz 5886
Altholzbestand 5880
Altholzbetriebsflächen 5881
Altholzschirm 5343
Altholzstadium 5340
Altholzsteckling 4135
Altholzstumpf 5882
Altlastensanierung 6860
Altwald 5880
Amaranth 6641
Amaranthbaum 6641
Amarantholz 9225
Amarillo 1712
Amazakoue 5987
Amazonenmandel 1003
Amazone-Rosenholz 1006
Amberbaum 259
Amberwald 230
Amboine 5284
Amboss 321
Amboss-Schere 322
Ambrosiakäfer 231
Ambrosiakäfer-Schaden 232
Amburana 1417
Ameisen-Akazie 309
Ameisenbaum 320
amerikanische Agave 1416
amerikanische Akazie 1689
amerikanische Blaueiche 242
amerikanische Blauesche 854
amerikanische Bleistiftzeder 2722
amerikanische Buche 238
amerikanische Eberesche 252
amerikanische Edeltanne 5738
amerikanische Eiche 5346
amerikanische Erle 5579
amerikanische Esche 261
amerikanische Färbereiche 2692
amerikanische Firnissumach 8315
amerikanische Gleditschie 4353
amerikanische Grauesche 2255
amerikanische Hainbuche 249
amerikanische Heidelbeere 1788
amerikanische Himbeere 9422
amerikanische Hülse 248
amerikanische Johannisbeere 240
amerikanische Kastanie 245
amerikanische Kastanie-Weißcheiche 5582
amerikanische Lärche 250
amerikanische Linde 237
amerikanische Magnolie 709
amerikanische Ölpalme 253
amerikanische Pflaume 256
amerikanische Platane 254
amerikanischer Amberbaum 259
amerikanischer Baumwürger 239
amerikanischer Bergahorn 5589
amerikanischer Faulbaum 1544
amerikanischer Geweihbaum 1482
amerikanischer Holunder 246
amerikanischer Judasbaum 2721
amerikanische Roteiche 5793
amerikanische rote Pflaume 256
amerikanische Roterle 235
amerikanische Rotfichte 1298
amerikanische Rotkiefer 5796
amerikanischer Rauchbaum 258
amerikanischer Schlehdorn 256
amerikanischer Schusserbaum 1482
amerikanischer Schwarzpappel 6319
amerikanischer Silberbaum 257
amerikanischer Trompetenbaum 1768
amerikanische Säckelblume 6788
amerikanische schwarze Johannisbeere 240
amerikanische Schwarzpappel 2717
amerikanisches Gelbholz 262
amerikanisches Mahagoni 1412
amerikanische Spätkirshe 244
amerikanische Stechpalme 248
amerikanische Strauch-Birke 8304
amerikanische Sumpfzypresse 786
amerikanische Taubeere 350
amerikanische Terpentinkiefer 5141
amerikanische Traubenkirsche 244
amerikanische Ulme 247
amerikanische Vogelbeere 7552
amerikanische Weißbuche 249
amerikanische Weißeiche 8993
amerikanische Weißerle 7587
amerikanische Weißesche 261
amerikanische Weißkiefer 7956
amerikanische Winterbeere 1810
amerikanische Wistarie 4707
amerikanische Zeder 243

Amur-Ahorn 267
Amurflieder 265
Amur-Kirsche 5224
Amurkorkbaum 264
Amurlinde 266
Amyris 8716
Amyris-Sandelholz 8716
Ananas 6243
Ananaskirsche 1323
Anatomie des Holzes 9478
anbauen 2128
Anbaufläche 355
Anbau in Walde 7610
Anbau unter Schirm 6825
Anbindungskabel 8543
Anbringen der Tropfrinne 544
anbrüchiges Holz 7535
Andamanen-Padauk 383
Andentanne 5535
Andira 1238
Andiroba 4039
Andrew-Pappel 4766
Anemone 9479
anerkannter Baumzüchter 1418
anfallendes Holz 2266
anfängliche Vermessung 4497
Anfangsreifebericht 4496
anfechtbarer Holzabfall 446
Anfeuerholz 3740
Anflug 7946
Anforderungen um Hiebflächennähe 2167
angebauter Stumpfscherblock 399
angeblautes Holz 859
angefaulter Ast 2290
angefaultes Holz 2291
angehängte Ende 8755
angehängter Gebüsch 8756
angehender Bestand 4425
angekohltes Holz 1453
angelehnter Baum 5041
Angelikabaum 2452
Angelique 621
angeordnetes Brennen 6516
angepasstes Holz 8677
angereicherte Waldbrache 2829
angerissener Bestand 8644
angeschlagene Knollen 1059
angestrichenes Holz 1730
angezeichnetes Holz 5311
Angiospermen 284
Angiospermenholz 4124
Angolaholz 1293
Angriffshieb 4495
anhacken 669
anhaken 4358
Anhänger 3674
Anhänger-Hacker 8751
Anhängerkette 8750
Anhängwasser 8736
anhauen 669
anhäufeln 1169
Anhieb 1214
Ankerbaum 280
Ankerkette 270
Ankerpunkt 276
Ankerseil 278
Ankerstumpf 279
Ankertau 270
Ankerwinde 281
Ankerwurzel 277
Anlachen 8155
Anlegen der Lachte 2140
Anmacheholz 3740
Anmarschweg zur Brandstelle 3217
Annattobaum 4995
annehmbare Brandfläche 194
Annehmlichkeitsbaum 234
Annehmlichkeitswert 233
Annosus-Wurzelschwamm 9385
Anonengewächse 2158
Anordnung 6517
anormal verlängerte Baumspitze 5544
anpflanzen 6334
Anreißen 8154
anreißen 7317
Anrichten 2141
anritzen 7308
Anrückschneise 1245
ansäen 7371
ansamen 7371
Ansamung 5695
Ansatz der Krone 2064
Ansatzstelle 4883
Anschalmen 540
Anschlag 1025
Anschlaghammer 5315
Anschlagseil 7742
anschnallendes Halteseil 1135
anschnittet 2176
anschnitteter Stamm 2177
Anschwellung 1182
Anspritzsaat 4406
Anstaltswald 1812
anstehender Baum 4010
Ansteigen des Saftes 7195
antarktische Scheinbuche 6985
Anteeren 6895
Antelopebusch 311
Anthere 312
Anthracnose 315
anthropogene Ladung 316
anthropogene Subklimax 2526
Antillen-Akazie 1689
Antreibeholz 6311
anwenderfreundliche Forstwirtschaft 9151
anwenderfreundliche Waldwirtschaft 9147
Anwuchs 5695
Anwurzeln 7029
Anzapfen 828
anzeichnen 7317
Anzeichnen der Bäume 8883
Anzeichnengeräte 997
Apfel 1755
Apfelbaum 1755
Apfelbeere 1553
Apfelbeerstrauch 1553
apfelfrüchtiger Osagedorn 4219
Apfelsägewespe 2879
Apfelsinenbaum 5935
apotracheales Parenchym 328
Aprikose 1756
Aprikosenbaum 1756
Aprikosenpflaume 332
aquasilvokulturliches System 335
Äquivalent 2734
arabischer Gummibaum 4048
arabisches Gummi 4048
Aralie 336
Araliengewächse 337
Arariba 338
Araukariagewächse 339
Araukarie 5535
Arbeiter auf dem Stapelplatz 8094
Arbeiter im Holzlagerhaus 9586
Arbeitsbereich 5925
Arbeitsblock 9722
Arbeitsgemeinschaft 9723
Arbeitshöhe 2195
Arbeitsstreif 2178
Arboretum 342
Arborizid 343
architektonische Form 348
Architekturform 348
Arecagewächse 356
Arecanuss 691
Arekpalme 691
Argahnbaum 6186
Arganbaum 6186
arider Lichtwald 358
Arizona-Esche 9188
Arizonakiefer 360
Arizona-Nussbaum 362
Arizonanusskiefer 9031

Arizona-Zypresse 359
arkadische Eiche 3937
Armandskiefer 1508
Arm des Laders 5025
armenische Eiche 363
Armleuchterbaum 4797
Arroyo-Weide 369
Art 7893
vom Aussterben bedrohte ~ 2801
Arten 7893
Artenabundanz 21
Artenabwandern 5443
Artenarmut 7281
Artendichte 7896
Artendiversität 7897
Artendiversitätsindex 7898
Arteneinführung 4577
Artengemeinschaft 738
Artengruppen 7899
Arteninventarisierung 7900
Artenreichtum 7902
Artenzahl 5808
Arve 8346
Arvenholz 1404
Arvenwald 8347
Aschweide 3974
asiatische Kirsche 382
asiatischer Padauk 383
asiatischer Teakbaum 5255
Asienbirne 1504
Asperiofoliazeen 919
Association-Index 4460
Ast 974
mit schwachen ~en 7737
mit starken ~en 8135
Astabschneider 8825
Astabstand 2539
Astanker 274
Astanlauf 995
Astansammlung 1668
Ast auf der Unterseite 9066
Astaxt 5167
Astbasis 4726
Astbereich 4724
Astbeule 832
Astbildung 4731
Astbohren 4727
Astbohrer 4728
Ästchen 991
ästchentragend 4737
Ast der zweiten Kategorie 7341
Astdurchlichten 6010
Astdurchmesser 2478
Astdürre 985
Ästeaxt 5167
ästen 10
Astentwicklung 2449
Äste und Krone 5163
Astfänger 981
Astfäule 4734
Astfläche 4724
Astförderer 4729
Astförderwerk 4730
Astform 4735
astförmig 6709
astfrei 983
astfreies Holz 990
Astgabel 982
Astholz 5170
Astholzbahn 1119
Astholzmaschine 8825
Astholzmasse 9245
astiger Kern 1948
ästiges Holz 5170
Ästigkeit 986
Astknorre 2146
Astknoten 4723
Astkranz 1668
Astkronenbäume 988
Astloch 4732
Astlochbohren 4727
Astlochbohrer 4728
Astlochholz 767
astlos 983
Astnarbe 992
Astquirl 995
astreich 4737
astreiche Krone 996
astrein 983
astreine Gleitschutzholzernte 2350
astreine Länge 1625
astreiner Schaft 1604
astreines Holz 2351
Astreinheit 989
Astreinigung 6610
Astreiserholz 5170
Astreisig 5170
Astrose 995
Astrost 4970
Astsäge 6614
Astsammeln 984
Astschere 5168
Astschneider 8825
Astschneider-Bundwerkzeug 4964
Astsortierer 4966
Astspäne 975
Aststellung 4725
Aststreu 5000
Aststummel 993
Astung 6610
Ästung 6610
Ästungshaken 6613
Ästungsmesser 2285
Astungssäge 6614
Ästungswürdigkeit 986
Ästung und Streuung 5166
Astverwachsung 4968
Astwachstum 4967
Astwindung 995
Astwirtel 995
Astzerkleinerungsmaschine 5230
Astzugförderer 4729
Äsung 8806
Äsungshöhe 1096
Atemwurzel 86
ätherisches Öl 2866
atlantische Zeder 396
Atlasbeerbaum 9424
Atlaszeder 396
A-Tragwerk 100
Aubletia 5534
Auenwald 5192
Auerzitrone 4911
Auewald 1356
Aufarbeiten von Langholz 8571
Aufarbeitung von Kalamitätshölzen 7168
Aufästung 6610
Aufastungsstange 6420
Aufbau des Holzes 9666
Aufbewahrung des Saatgutes 7397
Aufblatten 4096
Aufblattung 4096
Aufbruchholz 5463
Aufdeckung 5470
auffallende Vogelaugen 1885
aufforsten 93
Aufforstung 94
Aufforstungs-Anforderungen 95
Aufforstungsbestand 8114
Aufforstungsfrist 99
Aufforstungsplan 3529
aufgearbeitetes Holz 5288
aufgebeulter Ast 8352
aufgefächerte Figur 3011
Aufgehen 2796
aufgelockerter Wald 1060
aufgelöstes Holz 2530
aufhacken 2169
aufhalten
im Wachstum ~ 366
im Wuchs ~ 366
Aufhalter 7583
aufhängen 4116

Aufhauen 6610
Aufhellen 1598
Aufklärer 7932
Aufladen des Holzes 8608
Aufladen von Rundholz 2052
Aufladvorrichtung 5062
Auflagebock 4380
Auflagehumus 1681
Auflichtung 1598
Auflockern 6967
Auflockerung 6967
Auflöser 2308
Aufmaßbuch 5366
Aufmaßduchliste 5366
Aufmaßliste 5366
Aufraümen der Schlagfläche 6799
aufrechte Markstrahlzelle 9139
aufrechte Wuchsform 2858
aufreißen 6959
Aufreißen des Holzes 4845
Aufreißer 5102
Aufrollen 2052
aufrollen 8369
Aufschichten 1162
aufschichten 7983
Aufschlag 9127
Aufschlitzen 4915
Aufschwellung 1182
aufschwemmen 3269
aufseilen 1241
Aufseilwinde 7060
Aufsichtforstamt 3380
Aufspringen 2003
aufspringend 2346
Aufspulen 6120
aufstapeln 7983
Aufstapelung 4846
aufsteigender Saftstrom 4448
Aufstellen des Holzes 1215
aufstöbern 7120
Aufteilsäge 5148
Aufteilungstest 7922
Auftrennen 5016
auftretende Mehltaupilze 5844
Aufwandsnormung der Forstwirtschaft 1979
aufwendige Forstwirtschaftsführung 2963
Aufwickeln 6120
Aufwischenzeit 5555
Aufwölbung 8337
Aufwuchs 9127
Aufwuchsalter 7391
Aufwuchsstadium 7391
Auge der Äxte 451
Augenschätzung 2869
Ausaperung 7802
ausasten 10
ausästen 10
Ausästen 6610
Ausästenbaum 5169
Ausbau 8515
Ausbauchung 8337
Ausbauchungszahl 8399
Ausbauholz 6311
ausbessern 4561
Ausbeutung des Waldes 3414
Ausbeutungsrecht 2955
Ausbreitungskrone 7935
ausbrennen 3166
Ausbrennen durch Segenfeuer 465
Ausbruch eines Waldbrandes 5970
ausbürsten 1111
Ausbuschreiserholz 1202
Ausdehnungsgeschwindigkeit 6730
Ausdünner 8513
Ausfallast 800
ausfällen 3037
ausfallender Ast 800
ausfallendes Holz 9277
Ausfallholz 9277
Ausfallstelle 2991
ausformen 1122
Ausformungsplatz 7727
Ausforstung 7412
Ausfuhrholz 2958
Ausgangpflanze 5960
ausgebautes Holz 3164
ausgebrannter Schutzstreifen 1186
ausgebrannter Wald 1187
ausgedörrte Einheit 906
ausgegrabene Spur 2934
ausgehöhlter Stammstück 4329
ausgeholzte Fläche 3038
ausgenutzter Wald 2956
ausgerandetes Blatt 2317
ausgereift
in Baum ~ 8913
ausgereiftes Holz 6964
ausgeschlachtete Fläche 5066
ausgeschlachteter Wald 5066
ausgesuchtes Holz 9578
ausgewachsenes Blatt 69
ausgezeichneter Baum 826
ausgezeichneter Bestand 8644
Ausgleichspflanzung 1823
Ausgleichszeit 169
Aushalten 838
Aushebepflug 8771
Aushieb 2121
Aushieb der besten Stämme 4285
Aushieb des stärkes Holzes im Niederwald 7409
ausholzen 3037
auskeilender Jahrring 6087
Auskeimen des Samens 7380
Auskesselung 3081
Auslängsäge 1136
auslaufender Habitus 2939
auslaufende Spitze 6409
Ausläufer 7132, 8555
Ausläufer der Wurzel 7046
Auslaufstrecke 8471
Ausleger 910
Auslegerarmbalkene 5980
Auslegerkreissäge 1586
Auslesebaum 6396
Auslesebestand 6394
Auslesedurchforstung 7411
Auslesestamm 6395
auslichten 8503
Auslichten 2599
Auslichter 8513
Auslichtung 2599
Auslieferungsschein 6180
Ausnutzungsgrad des Holzes 9686
Ausnutzungsprozent 9155
auspeitschen 4812
Auspflanze 5978
Auspflanzen 5979
Auspflanzung 5979
Ausplenterung 7412
Ausreifen 5339
ausroden 4033
Ausrottung 2857
Ausrücken des Holzes 8595
ausrüsten 6939
Aussaat 7382
aussäen 7371
Ausschlag 1095
Ausschlagen 1020
ausschlagen 7944
Ausschlagwald 1928
ausschleifen 3975
Ausschlussaushieb 7193
Ausschlussholz 2123
Ausschlussholzware 2123
Ausschuss 2123
Ausschussfaktor 2120
Ausschuss für die Waldentwicklungbericht 3377
Ausschussholzzerkleinerung 2122
Außenbauholz 2964

Außenfläche des Schnittholzes 5982
Außenfurnier 2986
Außenrinde 5975
Außenriss 2965
Außenschichtspäne 8284
äußerer Holztransport 2966
äußerer Riss 7574
außer Holzabfuhrsabstand 2967
außerordentliche Felling 2973
außerordentliche Nutzung 9046
außerschichtige Vegetation 5763
Ausstattungsholz 3164
ausstocken 1611
Austerbaum 251
australische Akazie 5301
australische bräunliche Podocarpus 6390
australische Flusseiche 407
australische grüne Pflaume 3952
australische Kiefer 407
australischer Flaschenbaum 594
australischer Grasbaum 405
australischer Graugummibaum 5587
australischer Kiefernbohrer 408
australischer Manna-Gummibaum 5286
australischer Nothofagus 406
australischer rote Mahagonibaum 411
australische rote Zeder 410
australischer Teakbaum 498
australischer Wüstenbaum 973
australische Seideeiche 3966
australisches Endiandra 413
australische Silberakazie 7604
australisches Mahagoni 410
Austrocknung des Holzes 2423
Auswahlästung 7414
Auswahlernte 7413
Auswuchs 1182, 5977
auszeichnen 5310
Auszeichnen der Bäume 8883
Auszeichnen der Schlaghölzer 5317
Auszeichnung mit Farbe 6029
Auszimmerung 8600
Auszugshauung 7655
Auszugshieb 7655
Auszugshieb im Niederwald 7409
Auszugshiebsmethode 7656
autochthoner Wald 4476
Auwald 5192
Auxin 421
Avocado 445
axiale Dreschmaschine 456
axiale Elemente 455
axiale Kompression 1848
axiales Parenchym 457
Axtanreißen 449
Axtanzeichen 449
Axtbahn 452
Axtblatt 448
Axthieb 450
Axtkeil 454
Axtöhr 451
Axtschalm 540
Axtstiel 4110
Ayan 5732
azikular 36

Babul 4048
Bächleinerosion 6949
Bachweide 3880
Backen 1460
Backenbrecher 186
Bagasse 478
Bahia-Rosenholz 1007
Bahn auszeichnen 825
Bahnkörper 8744
Bahnzweig 4172
Baitoa 9346
bajonettspitzig 631
bakelisiertes Holz 9735
Bakelitholz 9735
Baker-Zypresse 484
Bakuri 6030
Balatabaum 485
Balau 9757
Balfourskiefer 6997
Balkan-Ahorn 488
Balkanesche 487
balkanische Heldreichföhre 936
Balkankiefer 489
Balken 638
Balkenbau 8560
Balkenbockschaden 5134
Balkenholz 7969
Balkenkonstruktion 8593
Balkenschlote 5470
Balkensperre 4377
Balkenweg 1947
Balkenwerk 8560
Ballenausheber 496
Ballenpflanzling 492
Ballenpflanzung 490
Ballonrückensystem 4290
Balsa 506
Balsabaum 506
Balsaholz 507
Balsambaumgewächse 1198
Balsamharz 503
Balsamkolophonium 503
Balsampappel 9338
Balsamtanne 969
Balsamtannenbastkäfer 9337
Balsamtannensägewespe 502
Balsamweide 505
Bambus 508
Bambusdickicht 510
Bambusfaser 509
Bambussprossen 511
Bambuswald 510
Banane 6335
Bananenbaum 514
Bananengewächse 512
Bananenstaude 514
Bananenstrausch 513
Bandagierung 515
Bandanwendung 8169
Bandbreite 8320
bandfütterte Zerkleinerungsmaschine 671
bandgesägt 521
Bandholz 523
Bandmaß 8165
Bandmaß ohne Rinde 8166
Bandmessung 8398
Bandparenchym 5414
Bandringelung 518
Bandsäge 5049
Bandsägemaschine 520
Bandschleifmaschine 672
Bandspritzung 8180
Bandweide 619
Bandwurmwurzel 5257
Bankaxt 673
Bankskiefer 4601
Bannforst 3535
Bannwald 425, 6592
Banyanbaum 526
Baobab 5533
Barbados-Kirsche 527
Barbasco 4654
Bärentraube 5289
Barpflanzung 532
Barras 593
Barriere-Hecke 596
Bart 2193
Barte 673
Basalfläche 597
Basalfläche pro Hektar 601
basiphil 614
Basispopulation 608
Basisverkaufstabelle 610
Basiswaldbau 611
basiswaldbauliche Maßnahmen

Basralocus 621
Bassindamm 613
Bassinschema der Flößerei 3270
Bast 4502
Bastardierung 4401
Bastardmahagoni 7200
Bastard-Platane 2906
Bastard-Ulme 2677
Bastfaser 623
Basthaut 4502
Bastholz 625
Bastkäfer 6246
Bastpalme 6685
Bau 2365
Bauchplatte 670
Bauchsäge 9030
Bauer-Gummi 857
Bauernwald 3015
Bauernwaldwirtschaft 3016
Bauerwartungsland 9799
baufällige Ausstattung 4178
baufällige Operation 4177
Bauholz 1153
Bauholzkern 5200
Baum 8797
im Freien wachsender ~ 4301
Baumabstand 4569
baumähnlicher Bärlapp 5794
Baumalter 134
Baumalterung 132
Baumanpfählen 8932
Baumaralie 7436
Baumart 4714
Baumartenmischung 5497
Baumarten mit Stärke 8038
Baumartenwahl 1551
Baumartenwandel 226
Baumartenwechsel 226
Baumartenzusammensetzung 1886
baumartig 8876
baumartige Aloe 9705
baumartige Azalee 8326
baumartige Dränage 2368
baumartige Heidelbeere 8926
baumartige Junkerlilie 8877
baumartige Pfingstrose 8895
baumartiger Meerlavendel 8878
baumartiger Kaktus 2366
Baumart mit Kernholz 4205
Baumastschüttler 4971
Baumaufnahme 8940
Baumaufschieber 8905
Baum ausheben 4028
Baumauslenkung 8831
Baum ausroden 4028
Baumband 8177
Baumbandfläche 8802
Baumbeschirmung 8810
Baumbesichtigung 8940
Baumbestand 3605
baumbestanden 3394
baumbestandener Sumpf 9543
baumbestandenes Feuerschutzgürtel 9544
baumbestandene Sumpfgelände 9543
Baumbestandsalter 8933
Baumblüte 847
Baumbruch 850
Baumbündel 7445
Baum-Büschel 1666
Bäumchen 346, 7202
Bäumchenmoos 8813
Bäumchenweide 5008
Baumchirurg 8938
Baumchirurgie 8939
Baumchirurgin 8938
Baumcluster 1666
Baumdecken 8960
Baumdimensionen 8827
Baumdoldenrebe 6149
Baumdurchmesser 8826
Baumeinsetzarm 8905
Baumeinstellung 8847
Baumeinwuchs 4492
bäumeliebend 2385
Baumentastmaschine 8825
Baumentastungsgerät 8825
Baum entwurzeln 4028
Baumepiphyt 8833
Bäume pro Stunde 8930
Baum erklettern 1641
Baumerkrankung 8828
Baumernährung 8890
Baum-Ernte-Interface 8821
Baum ersteigen 1641
Baumertrag 9772
Baumfall 3042
Baumfällen 8824
Baumfällmaschine 8853
Baumfällsäge 8854
Baumfällschild 3058
Baumfarn 8837
Baumfehler 8830
Baumfeind 3523
Baumfeldkombination 1740
Baumfeldwirtschaft 143
Baum-Felsenbirne 2603
Baumferse 1225
Baumfest 8838
Baumfestigkeit 8852
Baumflößung 8584
Baumform 7490
baumförmig 8876
baumförmiger Kaktus 2366
Baumformzahl 8840
Baumfraß 9703
baumfrei 8874
Baumfrosch 9579
Baumfrucht 8844
Baumfruchtschüttler 8845
Baum für Seeküstenpflanzung 8842
Baumgarten 342
Baumgattung 8920
Baumgeneration 8846
Baumgestalt 4080
Baumgipfel 8699
Baumgräser 9708
Baumgrenze 4982
Baumgrößen 8827
Baumgrößenklasse 8924
Baumgrube 6353
Baumgruppe 4000
Baumgürtel 3335
Baumhabitus 4080
Baumharz 8907
Baumhasel 5383
Baumhäufigkeit 8843
Baumhaus 8856
Baumheide 4206
Baumhöhe 8855
Baumhöhlen 8811
Baumholder 2892
Baumholz 8585
Baumholzalter 8933
Baumholzbestand 4280
Baumholzbetrieb 4281
Baumholzgesamtmasse 8733
Baumholzmasse 8733
Baumholzvolumen 8733
Baumhschule für Büsche 1205
Baum im Holzabfuhrskorridor 6673
Baumimpfung 8861
Baum in einer Hecke 4221
Baumjungfernrebe 6149
Baumkante 2662
baumkantige Balken 2663
baumkantiges Laubholzbrett 4133
Baumkette 910, 8136
Baumklasse 8812
Baumklemme 8829
Baumklotz 903
Baumkluppe 8809
Baumknorren 2938
Baumkrankheit 8828

Baumkultur 8834
Baumkunde 2380
Baumläufer 1807
Baumläuse 323
Baumleichen 3042
Baummalve 8881
Baummassentafel 8957
Baummesser 2381
Baum mit entblößter Wurzel 533
Baum mit Erdballen 499
Baum mit gefederten Wurzeln 7047
Baum mit großer Abnahme 3023
Baum mit keiner reinigen Spitze 9124
Baum mit niedriger Qualität 6451
Baum mit schlechter Qualität 6451
Baum mit Spaltenende 6711
Baum mit starkem Wurzelanlauf 5129
Baum mit Trockenwipfel 2273
Baum mit trompetenförmigen Wurzelanlauf 5129
Baum mit verschiedenen Verwendungen 5625
Baum mit verstarkem Wurzelanlauf 5129
Baummonokultur 8887
Baumnachpflanzen 8863
Baumname 8889
Baumneigung 8857
Baumpfahl 8902
Baumpflanzmaschine 8897
Baumpflanzung außerhalb des Waldes 5759
Baumpflanzung mit Immediatkassenertrag 8896
Baumpflanzungstag 340
Baumpieper 1093
Baumpressen unter Fällung 6520
Baumrasse 8906
baumreich 6936
Baumreife 8885
Baumreihe 7121
Baumreißer 6741
Baum-Rhododendron 8908
Baumrinde 8799
Baumrindenvegetation 1976
Baumrücken in Portionen 8900
Baumsaft 7208
Baumsamenplantage 8915
Baumsämlinge 8917
Baumsämlingerntemaschine 8916
Baumschaft 2351
Baumschenkel 607
Baumschirm 8922
Baumschlag 8824
Baumschluss 8810
Baumschneiderarm 8823
Baumschnitt 6610
Baumschule 342
Baumschulemanagement 5813
Baumschulen in Röhre 9001
baumschulige Knäuelmaschine 8798
Baumschulrodepflug 8771
Baumschüttelmaschine 8845
Baumschüttler 8845
Baumschutz 8850
Baumschützer 8904
Baumschutzspritze 8851
Baumschutzwanderfeldbau 8903
Baumschwamm 9581
Baumsegler 8942
Baumseuchenresistenz-mechanismus 8886
Baumsonnenblume 3840
Baumsorte 8849
Baumsperre 4
Baumspitze 8699
Baumspitzeschere 8947
Baumspritze 8862
Baumssäge 6614
Baumstadium 8931
Baumstamm 8934
Baumstammabfuhr 899
Baumsteigegerät 8814
Baumsteiger 4296
Baumsteppe 3607
Baumsterben 2694
Baumstinger 8935
Baumstreifen 7121
Baumstruktur 8936
Baumstumpf 8937
Baumstumpfbehandlung 2179
Baumstumpf-Erntemaschine 8205
Baumstumpfrodemaschine 8199
Baumstumpf-Verzweigter 8214
Baumstütze 8902
Baumteile 6094
Baum-Temperament 8943
Baumtest 8944
Baumtomate 8946
Baumtopologie 8948
Baumtrimmer 8949
Baumumfang 8848
Baumumhüllung 8960
Baumverarbeitung 8901
Baumviper 8954
Baumvolumenfunktion 8956
Baumvorratstafel 8957
Baumwürger 7984
Baumwürgergewächse 7985
Baumzähler 8820
Baumzucht 8834
Baumzüchter 344
Baumzüchtung 345
Baumzuwachs 8858
Baunutzholz 1153
Baupfahl 1152
Baureifmachung des Geländes 3985
Bauschnittholz 1888
Bau-Sperrholz 1151
Baustange 1152
Bauwertholz 8184
Beanstandung der Niederschlag 4535
bearbeitetes Holz 8792
Bearbeitungseinheit 8796
bearbeitungsmöglichkeitende Klasse 8793
Bearbeitungssaison 8795
Bearbeitungsvorschrift 8794
Beastung 987
Beastungsgrad 2085
bebauen 2128
Bebb-Weide 644
beblättert 4894
Becherkätzchengewächse 7597
bedeckte Landschaft 1311
Bedecktsamer 284
Bedeckung 4223
Bedienbarkeitslinie 5923
bedingte Holzaufbereitung 1854
bedingter Kahlhieb 1854
bedrängte Krone 2340
bedrohte Pflanzengesellschaften 8531
Beeinflussung der Bäume 4570
Beerenapfel 7565
Beerenfrucht 683
Beerenkräuter 685
Beerenobst 683
Beerenzapfen 2212
befallendes Holz 3042
befallen von Wurm 9201
befestigte Baumposition 9052
befestigter Baum 9053
befestigter Baumsporn 272
beflecktes Holz 7993
Beförderung des Holzes 8595
beforsten 93
Beforstung 94
Begang 3333
Begiftung 3090
Begiftungsentrindung 1463

beglaubigter Baumzüchter 1418
begrenzender Umweltfaktor 4980
begrenzte Fläche 1868
Begriffsumfangsforstbeamter 1890
Begründungszeitraum 2868
Begünstigung 6472
behaarte Faser 9717
behaarter Weißdorn 2602
behandeltes Holz 8792
behauen 83
Behauen 4273
behauener Balken 4270
behauenes Holz n 4271
behäufeln 1169
Behäufelung 1171
Behenbaum 4379
Behennussbaum 4379
beherrscht 6011
beherrschte Bäume 1706
beherrschte Holzart 66
beherrschte Spitzentriebe 1705
behöfter Tüpfel 920
beholzen 93
beholzt 3394
Behüllte 812
behüllte Heckenkirsche 812
Beige 9519
Beiholzart 66
Beil 453
Beilschlag 540
Beil zum Auszeichnen 5314
Beinholz 1799
Beispiel der Tropenwaldvernichtung 2432
beizen 5227
Bekämpfung von Waldbränden 3431
bekappenes Holz 6429
Beladepunkt 5027
Belastungsfähigkeit 1357
belaubt 4894
belaubte Krone 4865
Belaubung 3303
Belauf 3333
beleimte Holzpartikeln 3877
belgische Ulme 2677
Belüftungsindex 9198
Bemastungsholz 8586
Benadelung 5704
Benaroon 781
Benchmark 674
Benetzungsmittel 7934
bengalische Feige 526
Bennett-Wacholder 7588
Bennussgewächse 5556
bepflanzen 6334
Berasung 7813
Berberitze 1758
Berberthuja 7182
Berechtigungsholz 6944
Bergabseilbahn 89
Bergahorn 6637, 8355
Bergamotte 679
Bergamottenbaum 679
Bergangelika 336
Bergaufseilbahn 4323
Bergbauholz 5463
Bergbirke 5581
Bergeiche 5582
Bergerle 4164
Berg-Gronfoeloe 5275
Bergholunder 2907
Bergholz 5463
Bergkaffee 1713
Bergkiefer 5598
Berglaubholz 9132
Bergmahagoni 8990
Bergrüster 7310
Bergrutsch 7835
Berg-Stechpalme 3927
Bergstewartie 5591
Bergulme 7310
Bergwald 5583
Bergwaldgebiet 5586
Bergwaldspodsol 5584
berindet 1994
Berindung 559
Beringelung 6953
beringen 3851
beringte Ast 581
berittener Forstaufseher 3456
Berlandier-Asche 680
berlinischer Pappel 681
Bermenaufforstung 2231
Bermuda-Wacholder 682
berufsständischer namentlicher Waldfacharbeiter 6564
besäen 7371
besägtes Holz 1918
Besamlungshieb 7383
Besamlungsschlag 7383
Besatzung der Hieb und Rückung 2163
Besäumanlage 2767
Besäumautomat 2767
Besäumeinrichtung 2767
Besäumer 2764
Besäumfräser 1527
Besäumkreisanlage 6676
Besäumkreissägemaschine 6676
Besäumschnitzler 1527
besäumtes Holz 2757
besäumtes Schnittholz 2757
Besäumung 2766
Besäumungsmaschine 2767
Beschädigung durch Vieh 3930
Beschaffungsforstbeamter 6560
Beschäftigung in der Forstwirtschaft 2799
Beschattung 7481
Beschaufelung 822
beschichten 4760
beschichtende Abdeckung 1695
beschirmte Fläche 354
Beschirmung 7481
Beschirmungsdichte 2393
Beschirmungsgrad 2343
Beschirmungsklasse 1312
Beschirmungstropfen 1313
beschlagen 7962
Beschlagen 4273
beschlagenes Holz 4271
beschleunigtes Überbrennen 3315
beschmutztes Holz 7993
beschneiden 1645
Beschneiden der Wurzeln 7040
Beschneidung 4187
beschnitten 6609
beschnittener Baum 7489
beschnittetes Holz 6608
Beschnittholz 2757
Besenheide 1779
Besenspitze 1074
Besitzverhältnisse 4780
Besitzverhältnisse-Verwaltungsplan 8468
Bespantung 8714
Bestandaufgliederung 8008
Bestandbeschränkung 6918
Bestandbonität 3592
Bestanderhaltung 8010
Bestandesalter 8001
Bestandesalterklasse 8002
Bestandesart 3363
Bestandesaufbau 1886
Bestandesauflockerung 8515
Bestandesaufnahme 3478
Bestandesaufzählung 2837
Bestandesauszählung 2838
Bestandesdichte 3370
Bestandesdichtediagramm 8012
Bestandesdichtemanagement-diagramm 8012
Bestandesdurchschnittsalter 426
Bestandesdynamik 8013
Bestandesebene 8019
Bestandesentwicklung 2448

Bestandeserziehung 8456
Bestandesfolge 8032
Bestandesform 3667
Bestandesformhöhe 8016
Bestandesformzahl 8015
Bestandesfortschritt 8643
Bestandesgefügezahlen 8031
Bestandesgrundfläche 2032
Bestandeshöhe 4233
Bestandesinventur 3478
Bestandeskreisfläche 2032
Bestandeslockerung 8515
Bestandeslücke 3443
Bestandesmassenklasse 8035
Bestandesmassentafel 8036
Bestandesmittelstamm 434
Bestandesniederschlagssammler 8538
Bestandesoberhöhe 5358
Bestandespflege 8456
Bestandesprobestreifen 8034
Bestandesrand 2761
Bestandesschirm 3345
Bestandesschluss 1659
Bestandesschlusszahl 8011
Bestandesstockung 4009
Bestandesstrategie 8030
Bestandesstruktur 1886
Bestandestablischment 8014
Bestandestafel 8033
Bestandesumwelt 3405
Bestandesursprung 8025
Bestandesverhältnisse 8009
Bestandesvorrat 9250
Bestandesvorrataufnahme 2438
Bestandeswert 8196
Bestandeszuwachs 8017
beständige Arten 6187
Bestandmodell 8021
Bestandsbegründung 6821
Bestandsbeschreibung 2411
Bestandsbildung 3661
Bestandsgüte 6656
Bestandsklasse nach der Hiebsart 2191
Bestäubung 6434
Bestäubungsgerät 2675
Bestäubung von Laubwerk 5482
beste Anfangsseite 689
bestehend 2960
beste Seite 686
bestgeformter Baum 687
bestimmtes Wachstum 2435
bestockt 3394
bestockte Fläche 9542
Bestockungsaufbau 8189 8110
Bestockungsgrad 2344
zu hoher ~ 6007
Bestockungskarte 3498
Bestockungsklasse 8106
Bestockungskontrolle 8107
Bestockungsleiste 8108
Bestockungsplan 8109
Bestockungsstandard 8111
Bestockungstrieb 8555
Bestockungsziel 8413
Beta-Vielfalt 690
Betelnusspalme 691
Betelpalme 691
Bethell-Verfahren 692
Betragschere 2619
betretbare Forstfläche 30
Betriebsbereich 5925
betriebsbereite Bestandesaufnahme 5926
betriebsfähiger Wald 5924
Betriebsfläche 3330
Bettladung 647
Bettlast 647
Bett vorbereiten für einen Baum 645
Bettzeug 646
Betulin 693
Beugholz 7982
Beugholzraummeter 2115
Beule 834, 1167
bevorraten 8115
Bewahrungskette 1429
bewalden 93
bewaldet 3394
bewaldete Brandschneise 9544
bewaldete Fläche 9542
bewaldete Gebietgröße 3398
bewaldeter Brandschutzstreifen 9544
bewaldete Region 8604
bewaldeter Grenzstreifen 2761
bewaldeter Prärieboden 6492
bewaldete Savanne 9546
bewaldetes Feuchtgebiet 3399
bewaldetes Tiefland 3397
bewaldete Weide 9711
Bewaldung 94
Bewaldungsdichte 3370
Bewaldungskarte 3498
Bewaldungsprozent 3522
bewässerter Wald 3281
bewegliche Drahtseilbahn 5017
Bewertung den Erosionsrinnen 4044
Bewirtschaftungszone der
bewurzelter Absenker 493
Bewurzelung 7029
bezeichneter Beamter 2424
bezeichnete Wildnis 2426
bezimmertes Holz 4271
Bezirkförster 1988
Bezirksforstmeister 2551
Bezugsholz 8677
Bezugsmarke 674
Biberbaum 643
Bickbeere 855
Biegefestigkeit 676
Biegefestigkeitsmodul 5515
Biegeholz 678
Biegewiderstand 676
biegsame Kiefer 4965
biegsames Holz 2776
Biegung 2031
Biegung des Holzes 9489
Bienenbaum 666
Bigaradie 775
Bignoniaceen 8995
Bignoniengewächse 8995
Bildungsgewebe 5402
Bilimbaum 709
Bilinga 710
Bille 2579
Bindeaxt 673
Bindedraht 5023
Bindekraft 905
Bindemittel für Spanplatte 6092
bindende Schnittholze 717
Binder 716
Binderbarte 673
Binderholz 4362
Bindholz 523
Bintangor 718
Biobeschädigung 720
Biobrennstoff 722
Biocenosis 738
biochemische Systematik 1471
Biogenese 723
biogeochemischer Kreislauf 724
biogeoklimatische Zone 725
Bioindikator 4473
Bioindikatornetz 726
Biokraftstoff 722
biologische Holztrocknung 727
biologisches Herbizid 729
biologisches Insektizid 730
biologisches Unkrautmittel 729
biologisches Wachstumspotential 728
biologische Verbauung 6923
biologische Vielfalt 721
Biom 732

Biosphäre 734
Biota 735
biotisch 737
biotisches Potential 728
Biotop 739
Biotyp 740
Biozersetzung 719
Biozerstörung 719
Biozoenose 738
Birddog-Flugzeug 764
Birke 742
Birkenbeule 752
Birkenblattminiermotte 754
birkenblättriger Spierstrauch 755
birkenblättrige Spiere 755
Birkenbraunfäule 1085
Birkengebüschwald 2781
Birkengewächse 750
Birkenhain 753
Birkenholz 762
Birkenholzkohle 748
Birkenknüppelholz 746
Birkenkrummholz 2781
Birkenlaubwerk 749
Birkenöl 756
Birkenrinde 743
Birkenrindenöl 756
Birkenrindenteer 745
Birkenrollen 746
Birken-Rüsselkäfer 761
Birkensaft 757
Birkensirup 758
Birkensplintkäfer 1065
Birkenteer 745
Birkentundra 759
Birkenwachs 760
Birkenwald 751
Birnbaum 1795
Birne 1795
Birnensägewespe 6128
Bischofskiefer 770
bittere Kirsche 772
bittere Mandel 207
bittere Orange 775
Bittereschengewächse 147
Bitterholz 6663
Bitterholzbaum 9373
Bitter-Kirsche 1297
Bittermandelbaum 207
Bitternussbaum 774
Bitternuthickory 774
Bittersüß 773
bittersüßer Nachtschatten 773
Blackbutt-Eukalyptus 781
Blanke 3443
Blankschälen 1603
Blase 834
Blase auf einem Blatt 837
Blasefigur 835
Blasenbaum 818, 6051
Blasen-Esche 6051
Blasenspiere 6021
Blasenstrauch 818
Blatt 4852
vom Brand befallenes ~ 831
Blattabfall 4866
Blattachsel 4857
Blattader 4882
Blattanalyse 3307
blattartig 3302
Blattbehalten 4885
Blättchen 4875
blättchenartige Wurzel 6378
Blattdisposition 4856
Blattdüngung 8685
Blatteibe 1405
Blatten 4096
Blattentfaltung 9084
Blätterdach 4860
blätterig 3306, 4894
blätterloser Baum 4874
blätterreich 2389
Blattfallkrankheit des Kaffees 2489
Blattfaser 4867
Blattfläche 4853
Blattflächendauer 4854
Blattflächenindex 4855
Blattfleckenkrankheit 4889
Blattform 4869
blattfressende Insekte 4858
Blattgrund 4883
Blattherbizid 4255
Blattholz 6398
Blattkäfer 4858
Blattknospe 4859
Blattlappen 4876
Blattlappen-Sinus 5028
Blattläuse 323
blattlos 4873
Blattlücke 4870
Blattmasse 9200
Blattminiermotte 4880
Blatt mit glättendem Rand 7790
Blatt mit Grannen 7471
Blattnarbe 4888
Blattnerv 4882
Blattoberfläche 4890
Blattrand 4878
Blattraupe 3223
Blattrippe 4882
Blattrolle 1584
Blattschüttenkrankheit 4861
Blattspindel 4883
Blattspreite 4872
Blattspur 4891
Blattstiel 4883
Blattverfärbung 7306
Blattversengung 7306
Blattwasserkraft 4892
Blattwerk 3303
Blattwerkbespritzung 3305
Blattwerkprofil 3304
Blattwespen 7240
blattzerstörende Insekte 4858
Blau 859
Blaubeere 855
Blaubeer-Kiefernwald 708
Blaubirke 863
Bläue 5050
Bläuebefall 5050
blauer Gummibaum 8417
blauer Holunder 858
blauer Kieferbohrer 868
blauer Kiefernprachtkäfer 8057
Bläueschutz 6594
blaue Tanne 1726
blaue Waldrebe 4593
Blaufäule 5050
Blaufichte 1726
Blauglockenbaum 7122
blaugrüner Tabak 8945
Blaugummibaum 8417
Blauholz 1289
Blaupilze 861
Blausäckelblume 856
Blauschaden 5050
Blausieb 4917
Blaustringybark 865
bleibender Bestand 6889
Bleichholz 845
Bleiglanz 7606
Bleistiftfigur 6142
Bleiwurzgewächse 7328
Blenderhieb 5912
Blenderschlag 7412
Blenderschlagsfläche 7406
Blenderung 5912
Blesse 9371
Blindbaum 5447
blinde Tüpfelung 155
Blitzrinne 4946
Blitzriss 4946
Blitzschaden 4945
Bloch 2171
Blochauffangvorrichtung 5119
Blochförderbahn 4600
Blochgleitbahn 4600

Blochwender 5124
Blochwendevorrichtung 5124
Blockabwerfer 5097
Blockabwurfvorrichtung 5097
Blockachse 5048
Blockanwender 842
Blockaufzug 4600
Blockbandsäge 5049
Blockbau 2018
Blockbecken 5105
Blockdamm 4377
Blöcke mit zufälliger Zuteilung 6713
Blockförderbahn 7692
Blockförderer 4600
Blockform 841
Blockfuß 840
Blockgleitbahn 7692
Blockhaus 5052
Blockhebewagen 5098
Blockholzberger 5108
Blockholzmassenermittlung 5109
Blockholzmassenermittlungsmethode 5110
Blockholzzähler 5111
Blockhütte 5052
Blockierung eines Abzugsgrabenes 2545
Blockkerbe 7307
Blocklänge 7259
Blockmaß 5103
Blockpflanzung 6354
Blockrundmaß 5103
Blocksatz 6969
Blockschneidmaschine 839
Blocksortierung 5092
Blocksortierungsstation 5093
Blockverkauf 7157
Blockware 948
Blockzähler 5054
Blockzufuhr 5061
Blockzuführungswagen 5208
Blühen 3291
blühen 3290
blühender Baum 848
blühender Lorbeer 3936
blühender Lorbeerbaum 3936
Blumenblatt 6192
Blumenesche 3292
Blumenhartriegel 1964
Blumenknolle 9004
Blutbuche 6639
Blutdorn 6775
Blütenblatt 6192
Blütenesche 3292
blütenreiche Felsenbirne 6024 4364
Blutholz 1289
Blutjohannisbeere 6772
blutrote Johannisbeere 6772
blutrote Säckelblume 6793
Boardfuß 876
Boardfußmessstab 877
Boardfußtabelle 878
Bobaum 939
Bobber 887
Bobode 8375
Bocksäge 1136
Bocksdorn 9471
Boden 7833
über dem ~ 11
Bodenbearbeitung 8554
Bodenbedeckung 6337
Bodenbonität 7670
Bodenbretter 4186
Bodendichtung 7834
Bodenentnahmestelle 931
Bodenfeuchte 7838
Bodenflora 3986
bodenfremd 189
bodengestützte Systeme 3982
Bodengleiche 3990
Bodengüte 7670
Bodenleitung-Standort 4776
Bodenrücken 4779
Bodenschutzarten 7839
Bodenseilung 3994
Bodenstörung 7836
Bodenstörungsrisiko 7837
Bodenstreunutzung 9153
bodenverbessernder Wald 4476
Bodenverdichtung 7834
Bogen 951
Bogenhöhe 4234
Bogensäge 3342
Bogen-Zwergmispel 5389
Bohle 873
Bohlenbelag 2303
Bohnenbaum 1069
Bohranlageholz 6948
Bohrer für Holzbearbeitung 9538
Bohrerpflanzung 402
Bohrloch 927
Bohrloch-Intoxikation 928
Bohrmehl 929
Bohrspäne 926
Bohrung 923
Bombakaziengewächse 7596
Bonitätsklasse 9774
Bonitätsstufe 9774
Bonpland-Weide 907
Bonsai 908
Boomladung 913
Boomweite 6743
Bootsbausperrholz 5307
Borassuspalme 6047
Bordeauxkiefer 5309
borealer Nadelwald 924
borealer Taigawald 8367
borealgelichteter Nordtaigawald 5904
Borke 8799
Borkenkäfer 539
Borkenplatte 8775
Borraginazeen 919
Borretsch 918
Borretschgewächse 919
Borstenkiefer 6997
Borstenrobinie 1038
Borsten-Scheinakazie 1038
Böschung im Abtrag 7753
Böschungsneigung 7750
Bosisto-Eukalyptus 932
bosnische Kiefer 936
Bosse 6280
botanische Forstprodukte 938
Böttcherei 1920
Böttcherholz 4362
Boulton-Verfahren 5876
Box-Anhäufen 962
Box-Pflanzung 963
Box-Schnitt 955
brabante Myrte 8332
Bracatinga 966
Brachland 2997
Brachysklereide 8121
Brackholz 5450
Brand 5444
Brandanordnung 1195
Brandbahn 3169
Brandbekämpfungsspezialist 3171
Brandblock 1188
Brandentstehenhäufigkeit 3173
Brandfläche 1184
Brandfleck 6104
Brandgebiet 3179
Brandgefahr 3205
Brandgenehmung 1193
brandgeschadetes Holz 9528
Brandholz 1196
Brandklimax 3172
Brandkultur 3176
Brandlegung 53
Brandperiode 1192
Brandpflug 3196
Brandpilz 3168
Brandplatz 1184
Brandrodung-Wanderfeldbau

Brandschneise 3169
Brandschutzfällungen 3199
Brandschutzhiebe 3199
Brandschutzplanung 3175
Brandschutzstreifen 3169
Brandschutzstreifenintensität 3190
Brandspäne 1185
brandspezifisch 3178
Brandstelle 6104
Brandstiftung 4436
Brandumfang 3195
Brandumfanglinie 3195
Brandverhütung 3182
Brandwächter 3191
Brandwart 3171
Brandwirtschaft 3176
Brandzeitdauer 3181
brasilianische Araukarie 1005
brasilianische Doppeldreizackpalme 1009
brasilianische Nadelpalme 1009
brasilianischer Nussbaum 4650
brasilianischer Pfefferbaum 1004
brasilianischer Teak 4650
Brasilkiefer 1005
Brasilnussbaum 1003
Brasil-Rosenholz 1006
braune Alpenrose 6989
Brauneiche 1082
brauner Flascheneukalyptus 1075
brauner Stringybark 1091
brauner Waldboden 1079
Braunfäule 1084
Braunfäulnis 1084
Braunholz 8053
Braunholzschiff 8053
braunmarmorierte Weißfäule 1081
Braunsaftfäule 1086
Braunsaftfäule der Koniferen 1083
Braunspitzenfäule 1088
Braunverfärbung 1549
braun-werdende Weide 160
Braunwurzgewächse 3126
Brautmyrte 1792
Brayley-Flindersia 6665
Brecher 2104
Brechholz 8402
Brechnuss 8191
Brechnussbaum 8191
Brechnussgewächse 5047
Brechpunkt 8124
Brechstange 4667
Breitbeil 1041
breitblättrige Gräser in Eichenwäldern 1050
breitblättrige Ölweide 257
breitblättriger Baum 1048
breitblättriger Eukalyptus 4054
breitblättriges Pfaffenhuthen 1049
breitblättrige Wald-Platterbse 5662
Breitfurchensaat 1046
breitkronig 1045
breitkroniger Baum 1315
breitlaubige Albizie 5526
Breitsaat 1044
Breitsaumschlagbetrieb 9088
breitverstreutes Säen 1044
breitwürfige Saat 1044
Bremsbaum 4324
Bremsblock 2263
Bremsklotz 2054
Bremsungskette 1159
Brennbündel 7982
Brennen 1191
Brennen auf Anordnung 6516
brennende Waldrebe 6387
Brennen vom Rande aus 2759
Brennholz 3740
Brennholzabschnitte 3743
Brennholzbedarf 3738
Brennholzberechtigung 6946
Brennholzbestand 3742
Brennholzeinschlag 3222
Brennholzende 3221
Brennholzklafter 3024
Brennholzklotz 3741
Brennholzknüppel 7103
Brennknüppel 3219
Brennreiserknüppel 1032
Brennrinde 3734
Brennspäne 1517
Brennstoffbelastung 3736
Brennstoffe 3739
Brett 873
Brettbreite 885
Brettchen 7732
Bretterbekleidung 880
Bretterbeschlag 880
Bretterrohteil 874
Brettersäge 1633
Bretterschnitt 1017
Brettervermessung 879
Brettervermessung im Tausendboardfuß 8529
Brettfuß 876
Brettkante 477
Brettriese 884
Brettsäge 883
Brettware 882
Brettwurzel 1231
Brettzange 875
Brewersfichte 9330
Bridges-Eukalyptus 329
Bringung im Pendelwagen 7562
Bringung mit Ballon 497
Bringung mittels Traktor 8747
Bringungsfläche 2629
Bringungsmittel 8598
Bringungsschlepper 5114
Bringungsverlust 5174
Bringungswinde 7692
Bringungszeit 4170
Brockenholz 2505
Brombeere 778
bromiertes Holz 1063
Bronzieren 1066
Brotbaum 1011
Brotfruchtbaum 1011
Bruch 5013
Bruch-Fichtenwald mit Gräserkrautschichte 8313
Bruchharz 593
Bruchholz 850
brüchiger Kern 1039
brüchiges Holz 7535
Brüchigkeit des Holzes 1001
Bruchleiste 3054
bruchsicherer Schlagkopf 5771
Bruchwald 8309
Bruchweide 1040
Brückepropfung 1030
Bruschhacker-Zerstäuber 7321
Brusher 1104
Brusthöhe 1022
Brusthöhenalter 1023
Brusthöhendurchmesser 2466
Brusthöhenkreisfläche 598
Brutbaum 1068
Brutgang 4511
Brutknospe 9183
Brutrinde 1067
Bruttoholzrente 9777
Bruttozuwachs 8727
Bubinga 1120
Buchbaumwald 961
Buchdrucker 539
Buche 649
Buchecker 657
Buchenblattminiermotte 655
Buchengewächse 652
Buchenhain 654
Buchenharz 660
Buchenheister 661
Buchenholz 662

Buchenmast 656
Buchenrotkern 659
Buchenspäne 651
Buchenstocken 1089
Buchenteer 660
Buchenwald 653
Buchenwaldblock 844
buchisch 5239
Buchmast 656
Buchnuss 657
Buchöl 658
Buchs 2886
Buchsbaumgewächse 960
Buchsholz 2886
Bucht 699
Buchtenhieb 629
Buchthartholz 1991
buchtiges Blatt 2317
Bucht-Zeder 628
Buchweizenbaum 1139
Budsonfichte 1298
Büffelhornakazie 1166
Bügel 951
Bügelsäge 3342
Bugholz 678
Bugsieren 8737
Bummer 894
Bumtrip 894
Bündel 9496
Bündeleinschnitt 1174
Bündel-Entaster 5323
Bündelentastung 1175
Bündelgestör 6691
Bündelholz 9572
Bündel im Quadrat 7965
Bündelnarbe 1177
Bündeln des Floßholzes 912
Bündeln des Floßholzes im Damm 524
Bündelung 1178
Bündelverspannung 7467
Bündel von entrindete Bande 554
Bundesforst 8041
Bundeswald 8041
Bundsäge 2042
Bunker für Späne 2496
bunter kleine Eschenbastkäfer 379
buntfleckige Weißfäule 9374
burdachsche Säckelblume 5660
Bürgersahorn 8970
Burgund-Eiche 5565
Burma-Padauk 1183
Burseragewächse 1198
Bürstenregelung 1101
Bürstenrührstange 1112
Busch 7555
Buschableiter 1103
Buschablenker 1103
Buschbarriere 1098
Buschbaumschule 1205
Büschel 9005
Büschelpflanzung 6354
Büschelspitze 1210
Büschelwuchs 1209
Buscherbsenstrauch 7136
Buschfeuer 1105
Buschfläche 1108
Buschhacker 7721
Buschholz 1116
Buschholzbetrieb 7543
Buschhornblattwespen 7240
buschig 934
buschiger Baum 346
Buschland 1208
Buschmatte 1110
Buschmesser 1204
Buschräumung 1109
Buschräumungsrotte 1102
Buschsavanne 7560
Busch-Schneeschutzbepflanzung 1114
Busch-Schneeschutzhecke 1114
Buschschneide 1099
Buschsense 1113
Buschstadium 1115
Buschtundra 7322
Buschveld 1208
Buschwald 1202
Buschwald-Stadium 8495
Buschwerk 933
Buschwindröschen 9479
Bushpflug 1203
Bushzerkleinerer 1100
Butterbaum 1223
Butternuss 1222
Butternussbaum 1222

Caatinga 1236
Cabelluda 1240
Cajeputbaum 1056
Cambridgewalze 2127
Campecheholz 1289
Campnosperma 7155
Campos 1292
Canistel 2772
Canyonahornholz 707
Capanema 9346
Capoeira 1326
Caricagewächse 1340
Carolina-Jasmin1346
Carolinakiefer 7538
Cassava 5282
Cassia-Rinde 1370
Cassinabaum 2225
Cassis 2883
Catechu-Akazie 1376
Cathaya-Zierquitte 1498
Caven 1394
Cayenne-Pfeffer 1207
Cerejeira 1417
Ceylon-Ebenholz 904
ceylonesischer Zimtstrauch 1421
Ceylonzimt 1421
Ceylonzimtbaum 1421
Chamaephyt 1439
Chamäphyt 1439
Chaparral 1442
Charabpappel 4468
Chausseebaum 6983
Chaussee-Erntemaschine 6982
Chemieholz 44
chemische Abästung 1466
chemische Ästung 1466
chemische Braunverfärbung 1462
chemische Durchforstung 1469
chemische Entrindung 1463
chemischer Abbau des Holzes 1464
chemischer Waldschutz 1465
chemische Standortvorbereitung 1468
chemische Vorbereitung des Geländes 1468
Chemotropismus 1470
Chengal 1472
Chenopodeen 3887
Chenopodiazeen 3887
Chicasawpflaume 1480
Chihuahua-Esche 1484
chilenische Flusszeder 6102
Chimäre 3820
Chimbuva-Baum 2706
Chinabaum 6669
China-Holzöl 9009
Chinakaktus 4470
Chinarindenbaum 6669
Chinawachs 8958
Chinesenbart 1502
chinesische Aralie 1489
chinesische Birne 1504
chinesische Blütenesche 7584
chinesische Esche 1490
chinesische Kastanie 1494
chinesische Kiwi 51
chinesische Kopfeibe 1505
chinesische Korkeiche 5951
chinesische Pappel 7627

chinesischer Flammenbaum 1497
chinesischer Goldregenbaum 1497
chinesischer Keuschheitsbaum 3232
chinesischer Lackbaum 6051
chinesischer Sadebaum 6649
chinesischer Schnurrbart 1502
chinesischer Sonnenschirmbaum 1303
chinesischer Spitzahorn 6638
chinesischer Surenbaum 1493
chinesischer Talgbaum 1481
chinesischer Trompetenbaum 1492
chinesischer Tupelobaum 1507
chinesischer Wacholder 6649
chinesischer Zimtbaum 1495
chinesischer Zürgelbaum 1499
chinesisches Gelbholz 1509
chinesisches Öl 9009
chinesisches Rotholz 2259
chinesische Stachelbeere 51
chinesische Teufelsklaue 8951
chinesische Ulme 1496
chinesische Zierquitte 1498
Chip-n-Saw-Schnittholze 1521
Chippereinstechbewegung 1528
Chipperentladung 1526
Chlorophyll 1545
Chlorose 1546
chlorotischer Zwerg 1547
Choker 1555
Choker-Bodengewölbe 1554
Chokerhaken 1557
chokerloses Rücken 1558
Chokern 1560
Chokerseil 1555
Cholla 9259
Christbaum 1569
cilicische Tanne 1581
Cilienblatt 1579
cissusblättriger Ahorn 4599
Cistrose 6990
Cladinawaldart 1593
Clusterentwicklung 1667
Cochenille 5426
Co-Herrschaft 1707
Cohoba 1714
Colorado-Blasenspiere 5590
Colorado-Kiefernmarkkäfer 1727
Coloradotanne 1278
Columbia-Hawthorn 1731
comprägniertes Holz 1844
containerisiert 1892
containerisierte Pflanzschule 1894 1893
Cook-Araukarie 1919
Copal 1923, 1924
Copalbaum 1924
Copalharz 1923
Coulter-Kiefer 1984
Couma 2000
Courbaril 1990
Crinodendrongewächse 2021
Cronartium-Comandrae-Rost 1737
cryptopodsolische Braunerde 2107
Culmus 2124
Cumala 4717
Cumaru 2133
Cunoniengewächse 2138
Curare 8191
Curryblätter 2147
Cybistax 6551
Cyrillagewächse 2221

Dachlatte 7005
Dächsel 82
Dachstuhl 7006
Daggert 745
Dahoonhülse 2225
Dahoonstechpalme 2225
dahurische Lärche 2227
Daimyo-Eiche 2228
Damm 2230, 2793
Dammarharz 2243
Damm mit Querkämmen 8544
Dämmplatte 4521
Dämonenbaum 2453
dampfbehandeltes Holz 8054
dämpfen 400
Dampfsägemuhle 8055
Dampfsägewerk 8055
Daniellia 104
Danta 2245
Dao 6033
Darre 2623
Darreiche 7464
Darren 4711
Darrofen 2623
darrtrockenes Holz 4710
Darrvolumen 9252
Dattelpalme 2257
Dattelpflaume 2258
Daube 8048
Daubenholz 4362
Dauerarten 6187
Dauerforsttaxation 1898
dauerhafte Forstentwicklung 8295
dauerhafte Forstwirtschaft 8297
Dauerholzerntemaschine 1900
Dauerkulturen 6169
Dauerquadrat 6176
Dauerschnitt 1897
Dauertaxation der Waldzustand 1899
Dauerwald 6171
Dauerwaldwirtschaft 6172
Dauerweg 6175
Dauerwelkungspunkt 6177
Dauerzugriffsstruktur 6168
Dechsel 82
Deckblatt 2986
Deckfurnier 2986
Deckholz 2306
Deckschichtspäne 8284
Deckspäne 8284
Deckung 2304
deckungsgleiche Kultur 1716
Deckungsgrad-Schätzverfahren 5865
Deckungswert 1992
Defizitwald 2337
Defoliation 2941
deformierter Baum 5478
Deichselholz 6427
Deichselstange 6427
dekoratives Holzwerk 5958
Dekupiersäge 1822
Delamination 2348
Delignifizierung 2349
Delle 2360
Demerara-Grünharz 2363
Dendrit 2367
Dendritenbildung 8839
dendritisch 8876
dendritische Wuchsform 2939
Dendrobiont 2369
Dendrochronologe 2372
Dendrochronologie 2373
Dendrogeomorphologie 2376
Dendrogramm 2377
Dendrohydrologie 2378
Dendroklimatologie 2374
Dendrologe 2379
Dendrologie 2380
Dendrometer 2381
Dendrometermessband 516
Dendrometrie 2383
dendrometrische Merkmale 2382
Dendroökologie 2375
dendrophil 2385
Dendrophyl 2384
Dendroraum 2371
Dendroskop 2386
Denya 2397

Deponie 9276
Depositengeschäft 2401
Derbholz 1819
Derbstange 4803
derno-podsolische Bodene 2409
Derrickkran 2410
Derris 6448
Desertifikation 2414
Desorption des Holzes 7554
Destillationsholz 44
Destruktionsfäule 2433
Detritivore 2441
Detritus 2440
Detritusfresser 2441
deutliche Punktaste 1885
deutsche Mispel 5389
deutscher Nussbaum 1808
Deutzie 2442
Devegetation 2444
diagnostische Grundstückabtastung 2459
diagonaler Dreikantenspreißel 2460
Diagonalfaser 2461
diagonal-gekörntes Holz 2462
diagrammatische Kubiktabelle 2463
Diameterbandmass 2484
Dibbelsaat 2485
Dibbelstock 6344
Dibetou 1087
dichtbelaubt 2389
dichtblätterlich 2389
Dichte des frischabgehauenen Holzes 3707
Dichte des Holzes 1818
Dichte des Holzsubstanz 9668
dichte Krone 2387
Dichtenhiebsatz 9771
dichte Pflanzung 8501
dichter Bestand 3756
Dichteregel 2395
dichter Wald 2388
Dichterwerden 8494
Dichte-Scherfestigkeit 7502
dichtes Holz 8659
dichtes Hütten 1657
Dichtungsholz 8638
dickästig 6018
Dicke 8496
Dickenmesser 8497
Dickentoleranz 8500
Dickenwachstum 4019
Dickenzuwachs 2474
dicker Balken 8502
Dickestufen 8499
Dicknuss 5383
Dickte 4133
Dickung 4104
Dickungspflege 8455
Dielen 2278
Dielungsholz 5204
diffuses Parenchym 2492
diffuse Strahlung 2494
digitates Fiederblatt 6038
Dikotyledone 2486
Dilatation 2497
Dilleniengewächse 2498
Dimension 2499
Dimensionsdurchforstung 2476
Dimensionsholz 2503
Dimensionskurzholz 2501
Dimensionsplenterung 2476
Dimensionsteile 2504
dimorphe Gabelung 2506
Dina 6056
dioezische Pflanze 2507
Diplodia-Triebsterben der Föhre 2510
Diplom-Forstingenieur 3905
Diplom-Forstwirt 3905
direkt aus Baum 8914
Direktbringung 2522
direkte Abholzung 4384
direkte Aussaat 2521
direkte Holzkosten 2523
direkte Ladung 4383
direkte Strahlung 2520
Direktforstbringung 2522
Direktsaat 2521
diskontinuierlicher Jahrring 2528
Dita 9363
Ditabaum 9363
Dividivi 2550
Djatiwald 8435
Dogwood 1964
Dohle 2132
doldenblütige Kirsche 3260
Doldengewächse 1355
doldige Ölweide 417
doldiges Winterlieb 1798
Dolly 894
Domatium 2557
domestizierte eingebürgerte Baumarten 2558
Dominanz 2561
Dominanzpotential 2562
Dominika-Rosenholz 6076
Dominoholzschlag 2569
Dong Quai 1489
Donnerbesen 2902
Donnerrose 6989
doppelblättrig 697
doppeleingeschnittenes Blatt 2585
Doppelflügeast 979
doppelgefiedertes Blatt 741
doppelgezähnter Blattrand 2590
doppelhiebiger Holzwald 9035
doppelhiebiges Hochwaldsystem 9036
Doppelkantenast 4376
Doppelkern 2583
Doppelluftseilsystem 2586
Doppel-Nutringelung 2580
Doppelsäge 2584
Doppelspitzentrieb 2582
Doppelsplint 853
Doppelstamm 3657
doppelte Anhängerlast 2589
Doppelthakenzange 2581
Doppeltumbaulinientakelung 2588
Dorfforstgarten 9219
Dorfwald 9218
Dorfwaldstück 9220
Dorn 8524
Dornenbaum 8528
Dornengebüsch 8525
dornenlose Gleditschie 8527
dornförmige Säckelblume 3939
dornig 7913
Dornwald 8526
Dörrofen 2623
Dotterweide 3885
Douglasfichte 1772
Douglasie 1772
Douglastanne 1772
Douglastanne-Bockkäfer 5942
Douglas-Weißdorn 2591
Dourabaum 5737
Doussie 101
Doyle-Schnittholztabelle 2604
Drachenauge 1381
Drachenbaum 2608
Drahtsäge 9466
Drahtseil 9463
Drahtseilarbeiter 4296
Drahtseilbahn 1245
Drahtseilbahn mit fixiertem Kabel 8551
Drahtseilblock 8864
Drahtseilholzbringung 7700
Drahtseilklemme 2320
Drahtseilriese 1245
Drahtseilrücken 1247
Drahtseilschlinge 1555
Drahtseilschlinge-Einspannen 9465

Drahtspanner 1436
Drainage 2612
Drän 2615
Dränage 2612
Dränrohr 2615
Draufgabe 202
Drechsler 9682
Drechslerholz 9012
Drehkiefer 7191
Drehkiefernbastkäfer 5045
Drehlager 8350
Drehscheibe 8350
Drehtisch 5232
Drehung 9020
Drehwuchs 9021
drehwüchsiges Holz 2048
Drehwüchsigkeit 9023
dreiblätteriger Klee 8961
dreiblättrige Bitterorange 4146
dreiblättrige Lederblume 6035
dreiblättrige Mahonie 5419
dreiblättrige Orange 4146
dreiblättrige Zitrone 4146
dreiblütiger Ahorn 8533
Dreidorn-Akazie 4047
dreidornige Gleditschie 4353
dreikantiger Balken 8532
dreiläppiger Ahorn 1501
Dreileitungsluftseilsystem 8534
dreimännige Weide 6122
dreispitzige Jungfernrebe 937
Dreizahnahorn 8970
Dreschtrommel 4369
driftbar 2628
Driften 2625
Drifthof 914
Drillen 7869
Drillingsblume 947
Drillsaat 7869
Dröhnenkabel 910
Drosselbeere 2903
drosseln 1552
Druckholz 1850
Druckholztracheide 1851
Druckstange 6645
Druckstrebe 1985
Drucktränkung 4430
Druckverfahren 6523
Druckzylinder 4932
Drudenfuß 2902
Drummond-Seifenbeere 9343
drüsenblättrige Eiche 3859
drüsiger Götterbaum 8892
Dschungel 4671
Dummy-Baum 2664
Dünenaufforstung 2666
Düneweide 5480
Dünger 3089
Düngerauflage 8685
Düngung 3090
Dunkelhieb 2253
Dunkelnadelholz 2252
dunkelrotes Meranti 2254
Dunkelschlag 2253
Dunkelschlagbetrieb 7408
dunkle Fäule 3678
dunkler Käfer 811
dunkler Kern 795
dunkler Weißdorn 793
Dünnbrett 8520
Dünnholz 7760
Dünnholzbestand 9785
Dünnholzblock 8523
Dünnholzbund 7780
Dünnholzernte 7779
Dünnholzformzahl 7781
Dünnholzgewinnung 7779
Dünnholzstangenbündel 7780
Dünnholztabelle 7782
Dünnkurzholz 8521
Dünnstamm 8522
Duplexallmäher 2668
Durand-Eiche 2672
Durango-Kiefer 2673
durchblättern 7531
Durchbruch 3443
Durchdringbarkeit der Wurzel 7038
Durchdringvermögen der Wurzel 7038
durchforsten 8503
durchforsteter Baum 1826
Durchforstung 8515
Durchforstung im Beherrschten 8511
Durchforstung im Herrschenden 8512
Durchforstung mit einer Drahtseilbahn 7706
Durchforstung mittels Stock 8095
Durchforstung nach oben 9130
Durchforstungsanfall 9782
Durchforstungseinnahme 9782
Durchforstungsertrag 9782
Durchforstungsfläche 8507
Durchforstungsfolge 8517
Durchforstungsgrad 8519
Durchforstungsintensität 4527
Durchforstungsperiode 8509
Durchforstungsregime 8516
Durchforstungsreihe 8517
Durchforstungsreserven 4754
Durchforstungsturnus 8510
Durchforstungszelle 8508
Durchgangsrecht 6941
durchgehender Ast 8790
durchgehender Rindeneinschluss 8539
durchgehender Riss 8537
durchgehender Wirbel 8789
Durchhaltepunkt 8049
durchhauener Wald 7410
Durchlass 2132, 6097
Durchlauf 6097
durchlaufende Wuchsform 2939
Durchlaufsblock 6098
Durchlaufzeit 9011
Durchläuterung 2703
Durchlichtung 8515
Durchmesserabfall 8405
Durchmesserabnahme 8405
Durchmesser am stärkeren Ende 2465
Durchmesser am unteren Stammende 2465
Durchmesser auf Holzschlagniveau 2467
Durchmesserband 2484
Durchmesser-Begrenzungsverkauf 2477
Durchmesserbestimmung 2436
Durchmesser der skalieren Sägeholz 7250
Durchmesser-Häufigkeitsverteilung des Sägeholzes 7248
Durchmesser-Höhe-Funktion 2473
Durchmesser in Rinde 2480
Durchmesserklasse 2471
Durchmesserklassengrenze 2472
Durchmessermessband 2484
Durchmesser mit/ohne Rinde 2482
Durchmesser mit Rinde 2480
Durchmesser ohne Rinde 2475
Durchmesserstufe 2471
Durchmesserüberstummel 2481
Durchmesserzuwachs 2474
Durchplänterung 2121
Durchreiserung 999
Durchschlag 8506
durchschnittliches feuerfreies Intervall 5356
durchschnittliches Feuer-Intervall 5356
Durchschnittsalter 426
Durchschnittsprobe 1741

Durchwurzelbarkeit 7038
Durchwurzelung 7029
Durchwurzelungskapazität 7030
Durchzählung 4579
Durchziehen mit der Kette 1428
Dürrast 2647
Dürre 2635
Dürrebewachung 2639
Dürrefestigkeit 2637
Dürrenstaude 3983
Dürreresistenz 2637
Dürrgewicht des Holzes 2660
Dürrholz 2658
Dürrholzkomponente 2275
Dürrling 2271
Dürrständer 2271
düsterer Wald 7847
Dutchman-Keil 2680
Dutchman-Winkel 2679
dysgenisch 2696

Ebenholz 102
Ebenholzgewächse 2731
Eberesche 2903
Ebereschengehölz 5580
Ebereschenwäldchen 5580
Eberraute 5883
Eberreis 5883
echte Akazie 7604
echte Bärentraube 1759
echte Brombeere 778
echte Dattelpalme 2257
echte Heidelbeere 5006
echte Mandarine 5274
echte Mandel 207
echte Mehlbeere 8323
echte Moorbeere 893
echte Myrte 1792
echte Pavie 6767
echte Pechkiefer 5141
echte Pistazie 6295
echter Balsambaum 969
echter Bergahorn 8355
echter Brotfruchtbaum 1011
echter Ebenholzbaum 2727
echter Elsbeerbaum 9424
echter Feigenbaum 1774
echter Hausschwamm 2560
echter Jasmin 1782
echter Kampferbaum 1291
echter Kapernstrauch 1766
echter Korkbaum 1961
echter Lebensbaum 5795
echter Mandelbaum 207
echter Nussbaum 1808
echter Storaxbaum 3825
echter Weihrauchbaum 4463
echter Zunderschwamm 8667
echtes Brasilholz 1010
echtes Buschwindröschen 9479
echte Schwarzeiche 799
echtes Ebenholz 2727
echtes Frauenhaarmoos 4087
echtes Hickoryholz 8988
echtes Mahagoni 2110
echtes Pockenholz 4034
echtes Rosenholz 1007
echte Steineiche 1773
echte Vanille 9165
echte Zeder 2398
echte Zwergmispel 2888
echte Zypresse 1770
Eckbaum 1967
Eckenabschnitt 1966
Ecker 45
Eckerich 45
Eckholz 638
Eckleiste 635
Eckȕberblattung durch schwalbenschwanzförmiges Verzinken 2592
Ectomycorrhiza 2753
edaphische Charakteristiken 2754
Edelbaumholz 9161
Edelesche 5659
Edelholz 3159
Edelkastanie 2887
Edellaubholz 9160
Edelreis 1145
Edelreisholz 7299
Edeltanne 1803
edle Art 5739
edler Lorbeerbaum 3936
Efeu 4598
Effe 2786
Effekte der Chemikalien auf Holz 2771
effektives Keimvermögen 2768
Ehretiagewächse 2774
Eibe 1814
Eibenbestand 9768
Eibengewächse 8431
Eibenzypresse 786
Eibisch 5639
Eichapfel 5831
Eiche 5830
Eiche-Eukalyptus-Zypress-Association 5837
Eichel 45
Eichelbecher 46
Eichelbohrer 48
Eichelernte 5843
Eichellese 6055
Eichelmast 5326
Eichenbestand 5836
Eichenbinderholz 5834
eichenblätterig 5840
eichenblättrig 5840
Eichenbock 5848
Eichengalle 5831
Eichengallwespe 1279
Eichenhain 5836
Eichen-Hainbuchenwald 4374
Eichenheister 3266
Eichen-Hickory-Waldtyp 5838
Eichenholz 5850
Eichen-Kiefernwaldtyp 5847
Eichenmast 5843
Eichen-Moos 5845
Eichennestwickler 5839
Eichenniederwald 5835
Eichen-Plenterwald 4003
Eichenrinde 5832
Eichenschälwald 5835
Eichenspinner 5846
Eichenstamm 5842
Eichenwald 5836
Eichenwirrling 3127
Eichwelke 5849
eiförmiges Blatt 5988
Eigenentzündung 7923
Eigenhieb 5070
Eigenlängsholzwerk 6583
Eigenlängswerk 6583
Eignungskarte 8256
Eigrube 2773
Eikerbe 2773
einachsiger Baum 5539
einander kreuzende Faser 4539
einbetten 5710
Einbindestelle 3280
Einbindung 4703
Einblatt 9085
Einblatt-Axt 7636
Einblatten 4096
einblätterige Esche 7641
einblätterige Kiefer 7642
einblättrige Esche 7641
einblättrige Kiefer 7642
Einbrennenswerkzeug 1190
Eindringen 6144
einfache Frucht 7629
einfache Knospe 7628
einfache Krümmung 7634
einfache Perforationsplatte 7633
einfacher Kernriss 7630
einfacher Sprung 7630
einfaches Blatt 7631

einflutige Schirmschlagmethode 5897
einfrüchtige Pflanze 5537
Einfuhrholz 3317
eingebettetes Grit 2794
eingebeulte Textur 1168
eingebürgerte Arten 4576
eingebürgerte Kiefernblattwespe 4575
eingeklemmter Stamm 2062
eingepackter Kern 956
eingeschlossener Ast 833
eingeschlossene Rinde 4491
eingeschlossenes Splintholz 853
eingeschnittener Pflänzling 9056
eingeschnittenes Blatt 4438
eingeschriebene Auslesebäume 6830
eingespannter Baum 9053
eingewachsener Ast 833
eingewachsene Rinde 4491
eingleisiger Greifer 7643
Eingrauung 3925
eingriffliger Weißdorn 1780
einhauen 2169
einhäusige Pflanzen 5543
einheimische Art 2802
einheimisches Holz 4341
Einheitsgewicht 9099
Einheitsholz 2503
Einheitsmassenkurve 9089
Einigung 5042
Einische 2773
einjähriger Sämling 9756
Einkehr 9135
einkeilen 9314
einkeimblättrige Pflanzen 5540
einkerben 824
Einkerbung 4082
Einkern-Weißdorn 1780
Einkünfte aus Forstwirtschaft 4444
Einlassmaß 4522
Einlauf-Bauwerk 9280
einläufig 7637
einmalblühende Pflanze 5537
Einnahmen des Holzes 9591
einquirlig 9092
einquirlige Knospe 9093
einreihiger Strahl 9094
Einrichtungstrupp 8288
Einriss 9507
Einsaat 4567
Einsaatballen 7868
einsamenlappige Blütenpflanzen 5540
Einsaumbetrieb 6572
einschaltene Kultur 4562
einschichtig 7640
einschichtige Spanplatte 7639
Einschlagabnahme 27
Einschlag-Aufnahmeliste 3073
Einschlag der Hauptnutzung 3136
Einschlagen 4226
Einschlaghöhe 3145
Einschlagkontrolle 3055
Einschlagmarkieren 3074
Einschlagmenge 2200
Einschlagmethode unter Belassung von Überhältern 7400
Einschlagplan 6332
Einschlagprogramm 6332
Einschlagsanweisungen 6844
Einschlagsgenehmigung 6183
Einschlagsjahr 3144
Einschlagsmethode der permanenten Periodenfläche 6170
Einschlagsort 7668
Einschlagsperiode 3072
Einschlagstechnik 3076
Einschlagsverfügungen 6844
Einschlagüberwachung 3055
Einschlag über Wurzelbrut 3046
Einschlag und Querschneiden 3047
Einschlag unter Belassung von Überhaltern 7399
Einschlagverfahren 1617
Einschlagvertrag 2192
Einschlagzeit 3077
Einschlingenunterstützung 7646
Einschlussast 833
Einschneidebock 4380
Einschneiden 5100
Einschneidesäge 1136
Einschnitt des Rundholzes 7115
Einschnittkeil 1134
Einschnittmodell 5051
Einschnittsböschung 7753
Einschnittsergebnis 5214
Einschnitt-Tabelle 1133
Einseilbahnsystem 5536
Einseillagerplatz 4289
Einseilsystem 7644
einseitig besäumtes Holz 5899
einseitige Entwicklung 5171
einseitige Krone 7648
einseitiger Rindeneinschluss 5900
einseitiger Wuchs 5898
einseitig gesägtes Holz 5899
Einsickern 4488
Einspannvorrichtung 4653
einstämmig 7637
Einständer-Greifer 5917
einstapeln 7983
Einstechen 6393
einstufiger Bestand 7653
einstufiger Wald 7653
Einstupfen 8964
einstutzen 2165
Einteilung in Waldnutzungszonen 3638
Eintragung mit festgelegtem Seil 3237
einwüchsiges Holz 7289
Einwuchszeit 4006
Einzelart 7292
Einzelast 7290
Einzelbaum 4480
Einzelbaum-Entastungsmaschine 7651
Einzelbaum-Gesamtmasse 7652
Einzelbaumpflanzung 6347
Einzelbestand 7650
Einzelerfassungsmassentafel 7638
Einzelkornfrucht 7647
einzellitziger Blockförderer 7654
Einzelmischung 5496
einzeln erstickte Bäume 7659
einzeln gewürgte Bäume 7659
Einzelpore 7845
Einzelstamm 4480
Einzelstammflößerei 2625
Einzelstammpflege 8454
Einzelstammwirtschaft 7657
Einzelüberhalt 7291
Einzelverkauf 6119
Eisbruch 4417
Eisenbahnabfuhr 6696
Eisenfleck 7215
Eisenholzeukalyptus 6779
Eisenkrautgewächse 9210
Eisenrindenbaum 4583
Eisenstange 1300
Eisenveilchenbaum 8417
Eiskluft 3723
Eisriese 4418
Eissporne 1280
eklektische Durchforstung 2735
ektotrophe Mykorrhiza 2753
elastisches Holz 2776
Elastizitätsmodul 5514
Elbweide 619
elektrische Säge 2779
Elfenbeinpalme 4597
Elitebaum 6396

Elsbeerbaum 9424
Elsbeere 9424
Emballageholz 954
Embryo 2795
Emien 1461
Empfehlung 6517
empfindliche Art 7443
empfindlicher Boden 7442
empfindliches Ressourcengebiet 7440
empfindliche Wasserscheide 7444
empirische Ertragstafel 2798
Endabstand 3139
Endbearbeitung 8281
Endbestandweite 3135
Enddurchforstung 3143
Endehaken 2806
Endemast 2808
Endeschnitt 2319
Endfeuchte 3141
endgültige Lageprüfung 3140
Endhieb 3136
Endknospe 8470
Endnutzung 3136
Endoderm 2809
Endogenholz 2811
endogenisch 2810
Endosperm 2812
endotrophe Mykorrhiza 2813
End-Stammstück 7362
Endstütze 2815
Endstützeweg 475
endstützter Baum 8371
Endtrieb der Kohlpalme 1239
Endtriebe 4813
Endumlenkrolle 4167
Endwand 2817
Energieforstplantage 2819
Energiegewinnung aus Holz 9481
Energieholz 2820
Energiespäne 2818
Engelmannsfichte 2822
Engelmannsfichtenkäfer 2823
Engelmannsfichtenrüssler 2824
Engelmannskiefer 2821
Engelsüß 6441
Engelwurz 9401
Engholz 4827
englische Ulme 2893
engporiges Holz 1656
engporigfeinporig 3155
engringiges Holz 1660
engringige Textur 1655
entasten 10
Entasten 6610
Entaster 981
Entastungskopf 4969
Entastungsmaschine 8825
Entastungsmaschine mit Dauerzuführung 1896
Entastungsmesser 2285
Entastungsqualität 2355
Entastungssäge 6614
Entastungs- und Einschneidemaschine 2352
Entastungswunde 6616
Entastungszug 2356
Entbastung 624
Entblätterung 2941
Entblätterungszeit 15
Entblätterungszone 16
Entblättung 2941
entblößter Feuerschutzstreifen 530
entblößte Wurzel 531
Entborken 579
Entfärbung des Splintes 2527
Entfernung 7871
Entfernungsdendrometer 6718
Entfernungsregel 5081
Entflammbarkeit 1742
Entflammungsfläche 4422
entgegenliegendes Blatt 5927
enthaltene Wurzel 1891
entharzen 2345
Entharzen 2407
Entharzung 2407
entholzen 3037
enthüllter Knoten 9050
Entknospungsästung 1142
Entladenbrücke 5451
Entladentasche 2665
Entladerampe 5451
Entladestelle 9102
Entladungskanal 2525
entlange Holzlieferung 208
Entlaubung 2941
entlegene Gegend 9404
Entnadelung 2995
Entnahmestelle 931
entnommener Bestand 3041
Entrinden 579
entrinden 2280
Entrinder 2284
entrindet 551
entrindeter Messring 3850
entrindetes Holz 555
entrindetes Rohholz 7096
entrindetes Schnittholz 553
Entrindung 579
Entrindung durch Reibung der Hölzer 3713
Entrindungsbock 564
Entrindungsgerät 574
Entrindungsholz 555
Entrindungslinie 567
Entrindungsmaschine 2284
Entrindungsmaschine mit Messer 7069
Entrindungsmesser 7960
Entrindungstiefe 565
Entrindungstrommel 566
Entrindungsverluste 580
Entrindungswerkzeug 574
Entrindungszustand 2282
Entschlammer 2427
Entspitzenhöhe 8708
Entspitzmaschine 8707
Entstaubung 2318
Entstaubungsanlage 2674
Entteerung 2407
entwalden 2524
Entwaldung 2339
Entwässerung 2454
Entwässerungsanlage 2614
Entwässerungsfurche 2613
Entwässerungsrohr 2615
Entwicklungsphase 6198
Entwicklungsstufe 2445
Entwicklungstendenz 2446
Entwicklungsziele 2447
entwipfeln 8702
entwurzelter Baum 9141
Entwurzelung 2857
entzweigen 10
Entzwieselung 3658
ephedroidische Perforationsplatte 2846
ephemere Pflanze 2847
epigäisch 11
epigäische Biomasse 12
epigäische Keimung 2850
Epinastie 2851
Epiphyt 2853
episodisches Wachstum 2854
epixyl 2855
Erbewalde 4260
Erblichkeit 4258
Erbsenbaum 7564
Erbsenstrauch 7564
Erdbeerbaum 8152
Erdblock 1225
Erdboden 7833
Erde 7833
Erderwärmung 3867
Erdhobel 4772
Erdklotz 1225
Erdklotzeinreißen 1216

Erdrutsch 7835
Erdstammstück 1225
Erdstockholz 8220
Erdstrauch 9065
Erdstück 1225
Eremacausis 2859
Ergänzung der Bepflanzung 823
erglühende Verbrennung 3875
erhaltenswertes Kulturerbe 4259
Erhaltung
ex situ ~ 2959
in situ ~ 4517
Erhaltung des Waldes 3359
Erhaltungsaufforstung 1881
erhärtetes Harz 593
Erholungsgeschwindigkeit 6752
Erholungsinventur 6756
Erholungsmöglichkeit 6757
Erholungsobjektsignifikanz 6755
Erholungspiste 6759
Erholungsressource 6758
Erholungswald 91
Erika 1779
erlaubte Jagdbeute 6182
Erle 2882
Erlenbestand 164
erlenblättrige Eberesche 4742
erlenblättriger Schneeball 4315
erlenblättrige Zimterle 8263
Erlenbruch 165
Erlenbruchwald 164
Ermansbirke 2860
Ermittlung der tatsächlichen Masse 8992
Erneuerung des Pflanzenwuchses 6923
Erneuerungsfläche 355
Erneuerungshieb 6815
Erneuerungsspross 4505
erneuter Einschlag 6870
Ernte 295
Erntebestand 7378
Erntefestmeter 9246
Erntehieb 4150
Erntehiebsmethode 4151
Ernteintensität 2034
Erntemaschine 8853
Erntemaschinekette 4152
Erntemaschine mit Halt 8125
ernten 4149
Ernteverluste 5175
Eröffnung 5913
Erosion 2861
Erosion der Oberbodenfläche 8285
Erosionsschutzwalde 317
Ersatzpflanzung 1823
Ersatzspross 6872
Ersatztrieb 6872
Erscheinungsbild 6204
Erschließung des Grundstücks 7676
Erstaufforstung 4493
erstauner Mesquitebusch 9189
erste Aufastung 999
erste Blenderung 3227
erste Entkorkung 6865
erster Baumstamm 6552
erster Blenderhieb 3227
erster Durchlauf 3228
erstklassiger Stammstück 6552
erstklassiges Gesicht 3229
erstklassige Überständer 3226
Ertrag eines Baumes 9779
Ertrag eines Fichtenbäumes 9781
ertragende Kante 640
ertragfähiger reservierter Waldboden 6563
ertragfähiger Waldboden 6562
ertragfähige Waldfläche 6562
Ertragsberechnung 9776
Ertragsbestimmung 9776
Ertragsermittlung 9776
Ertragsfähigkeit 7663
Ertragskurve 9775
Ertragstafel 9784
Ertragstafel des gleichaltriges Bestandes 2921
Ertragstafel mit variierbarer Dichte 9167
Ertragswald 5394
erwachsenes Laubwerk 68
Erweichen des Holzes 9655
erweiterte Stammende 3250
Erwerbswald 6561
erworbene Ernte 2705
Erzeugung von Rohholz ohne Rinde 3980
Erziehungshieb 4544
Erziehungsschnitt 6612
Erzlagerstätte 5462
Esche 1757
Eschenahorn 381
Eschenbastkäfer 379
eschenblättrige Flügelnuss 1388
eschenblättriger Ahorn 381
Eschenwald 380
Espe 2880
Espenbrennholz 388
Espenlagerholz 386
Espenfaserholz 385
Espenpapierholz 385
essbare Kiefer 9031
essbarer Schneeball 5553
Essigbaum 1036
Eßkastanie 2887
Etage 8133
Etagenbestand 5634
ethnobiomedizinische Waldreserve 2871
Eucommiagewächse 2875
Eucryphiagewächse 2876
eugenisch 2877
Eukalyptus 2872
Eukalyptusbaum 2872
Eukalyptusöl 2874
Eukalyptuswald 2873
Eulecanium 7274
europäische Birke 1761
europäische Forstkommission 2896
europäische Hopfenbuche 2900
europäische Lärche 1785
europäische Lebenseiche 1773
europäische Linde 1786
europäische Platane 2906
europäischer Liguster 1799
europäischer Schwarzdorn 7749
europäischer Stechginster 3889
europäischer Zürgelbaum 2898
europäische Traubenkirsche 2881
europäische Zwergpalme 5384
Eutrophierung 2913
eutrophisch 2912
Evapotranspiration 2916
Evodia 2932
exogenisch 2942
Exoten 2944
Explantat 2947
Explantation 2948
Exploitationshieb 1744
Explosivtkeil 6487
Exstirpation 2968
Extensionswurzel 2961
extensive Bewirtschaftungsklasse 1058
extensiver Kahlhieb 2962
Extraktionsholz 2971
Exzenterschale 4093
Exzenterstammwuchs 2733
Exzenterwuchs 2733
exzentrischer Kern 2732
exzentrischer Stammwuchs 2733

Fabriklagerhalle 5456
Fächerahorn 4373
Fächerblattbaum 5245
Fächerfiederblatt 3010

Fächerpalme 5384
Faciation 2987
Fackelbaum 5029
Fackeldistel 1733
Faden 3024
fadenziehende Fäule 7741
fähig verzögerte Waldfläche 1321
fahnenförmige Krone 3240
Fahnenmarkierung 3239
fahrbares Sägegatter 6469
fahrbares Vollgatter 6469
Fahrradkreissäge 695
Fahrweg 4172
fakultativer Kern 795
Fällaxt 453
Fällbereich 8213
fällen 3037
Fällen der Stammanläufe 1235
Fallenlassen-Gabelung 2633
Fäller 9526
Fäller-Auswahl 3043
Fäller-Bündler 3053
Fäller-Bündler mit Kettensäge 1432
Fällgrenze 469
Fällhacke 453
Fällhebel 3067
Fallhieb 3069
Fallholz 2266
Fällkatze 6645
Fällkeil 3079
Fallkerbe 3069
Fällkerbendach 7283
Fallkerbfläche 9055
Fallkerbhöhe 2195
Fallkerbtiefe 2195
Fällkosten 3056
Fällmaschine 8853
Fällort 7668
Fäll-Paketiermaschine 3053
Fällrichtung 2519
einem Baum die ~ geben 4059
Fallrichtungsanzeiger 8567
Fällrotte 9527
Fäll-Rückekombine 3075
Fäll-Rückemaschine 3075
Fällsäge 8854
Fällschnitt 3066
Fällschnittfläche 2189
Fällschnittsdach 7283
Fälltisch 1563
Fällung-Entastungskopf 3068
Fällung im Saft 3065
Fällung mit der Hand 4109
Fällung mit motorisierter Handsäge 5572
Fällungsbarre 3050
Fällungsbett 4840
Fällungsbrigade 9527
Fällungsdatum 3060
Fällungsergebnis 3145
Fällungshebel 3067
Fällungsintensität 4210
Fällungskermen 3066
Fällungskompass 8567
Fällungskosten 3056
Fällungssäule 3051
Fällungsschaden 3059
Fällungsschnitt 3066
Fällungsunternehmer 5072
Fällungsverluste 2197
Fällungsverteilung 3061
Fällungswerkzeuge 4157
Fällungszeit 3077
falsche Akazie 1754
falsche australische Weißesche 414
falsche Endknospe 3004
falsche Frucht 3000
falscher Jahrring 2998
falscher Jasmin 1346
falscher Kern 795
falscher Markstrahl 3001
falscher Mopane 113
falscher Splint 853
falscher Tungölbaum 3006
Falschkern 795
Familienforstwirtschaft 3007
Familienhecken 3008
Fangbaum 8778
Fanggrube 4515
Fangkloben 8776
Fangklotz 8776
Fangknüppel 8776
Fangrechen von Schwimmern 8136
Fangrinde 8775
Fangscheit 8776
färbe Erle 5268
Färbereiche 2692
Farbfehler 1728
Farbholz 2693
Farbkern 1729
Farmwald 3015
Farn 3087
Farnbaum 8837
Farnbäumgewächse 2206
Farndickicht 3088
Farneibe 1405
Farnheide 3088
Farnkraut 3087
Farnkrautfeld 3088
Faschine 3018
Faschine-Choker 3019
Faschinendamm 1117
Faschinenholz 9572
Faserabweichung 4439
faserartig 3096
faserartige gelbe Kernfäule 3104
Faserbast 4502
Faserbrei 9603
Faserfeinholz 3149
faserförmig 3096
faserförmiges Element 9212
faserförmige Struktur 3103
Fasergabelung 3095
fasergesättigtes Holz 3710
Faserholz 7525
Faserholzrolle 6627
Faserholzschlag 6632
faserig 3096
Faserklassifizierer 3093
Faserkohle 3778
Faserlänge 3097
Faserlauf 2516
Faserneigung 4439
Faserneigung-Bestimmung 2437
faserparallel 209
Faserplatte 9551
Faserplatte geringer Dichte 5183
Faserrichtung
quer zur ~ 49
Fasersättigung des Holzes 9565
Fasersättigungspunkt 3098
Fasersplittung 7729
Faserstauchung 3094
Faserstruktur 3100
Fasertauwerk 1942
Fasertextur 9678
Fasertrennung 3099
Faserverlauf 2516
Faserwurzel 1968
Fassboden 4186
Fassfabrikation 1920
fassförmige Säge 2209
Fassholz 4362
Fassholzklotz 4362
Faulast 2290
Faulbaum 684, 1137
fäulebefallenes Holz 9484
Fäule des Holzes 9529
fauler Ast 2290
fauler Kern 2289
Faulfleck 7081
Faulholz 2291
Faulholzanteil 6093
faulig 7212
fauliger Abfallbaum 7080

Fäulnis 6647
fäulnisbeständiges Holz 2294
Fäulnisbeständigkeit 2293
Fäulnisbremse 2295
Fäulnisfresser 2441
Fäulnisresistenz 2293
Fäulnisverzögerer 2295
Faulstelle 7081
Faultbäume 7097
Faultknüppel 7079
Faulweide 630
federgenervtes Blatt 3032
Federmohn 8899
Feder und Nut 8676
Fegeholz 2266
Fegeschaden 2232
Fehlbaum 7082
Fehlbrett 2119
fehlender Jahrring 5479
Fehler-Abzug 201
fehlerfreies Holz 9578
fehlerhaftes Knüppel 2335
Fehlerhöhe 4235
Fehlerlänge 2334
Fehlermindestdurchmesser 5466
Fehlertiefe 2404
Fehlholz 9118
Fehlkante 2662
fehlkantiges Holz 2663
Fehlstelle 2991
Feige 1774
Feigenbaum 1774
Feilenbrett 3129
Feinabfälle 4950
Feinbrennspäne 3151
feine Holzbearbeitungen 3160
feines Korn 1655
feine Überreste 3156
feinfaserig 3157
feinfaseriges Holz 3149
Feinheit 3154
Feinhobeln 1608
Feinholz 3159
Feinholzschliff 3152
feinkörnig 3157
feinkörniges Holz 1656
Feinpolitur 3150
feinringiges Holz 1660
Feinsäge 3158
Feinschlichtspan 3148
feinumringend 9470
Feldahorn 2827
Feldarbeitsgang 2131
Feldaufforstung 3109
Feldaufnahme 3120
Feldgehölz 5001
Feldkapazität 3110
Feldkeimung 3114
Feldkontrolle 3111
Feldlagermeister 1290
Feldpapel 2884
Feldpersonal 3119
Feldrüster 2893
feldschützender Waldstreifen 7508
Feldschutzwald 3118
Feldulme 2893
Feldwaldbau 143
Felgenholz 3082
Felsenahorn 5550
Felsenbirne 7462
Felsengebirgstanne 214
Felsengebirgswacholder 6995
Felsen-Kreuzdorn 6987
Felsenrebe 1318
Felsenulme 6988
Femelbetrieb 7620
Femelhieb 7412
Femelhochwald 4002
Femelschirmschlagbetrieb 7514
Femelschirmschlagsystem 7514
Femelschlag 7412
Femelschlagbetrieb 7408
Femelstreifenschlag 8168
Femelwald 7410
Fenchelholzbaum 1801
Fender 1025
Fennholz 3085
fensterförmige Tüpfel 6287
Ferkelnuss 6229
Fermentationsschicht 3086
Fernluftseilbahn 5130
fertiger Samen 3753
Fertigholzmaß 58
Fertigmaß 58
Festausleger-Greifer 3235
Festauslegerkran 3236
fester Ast 8550
fester Knoten 8550
fester Schwimmer 8096
festes Holz 3225
Festgehalt 7842
Festigkeit 4123
Festigkeit des Holzes 2669
Festigkeit im trockenen Zustand 2657
Festigkeitsprüfer 8464
Festigungsgewebe 8274
Festigungswurzel 277
Festkern 1901
Festmaß 7842
Festmeter 2114
festsitzender Ast 8550
Feststapel 8552
Feststapelung 1157
Fettpflanze 8082
feuchte Böttcherei 8549
Feuchtegehalt des Holzes 5518
Feuchtekapazität des Bodens 3110
feuchtigkeitsabstoßendes Sperrholz 5523
feuchtigkeitsbeständig 5522
Feuchtigkeitsgefälle 5520
Feuchtigkeitsgehalt von Brennstoffen 3737
Feuchtigkeitsgradient 5520
Feuchtigkeitsverteilung 5519
Feuchtkern 9283
Feuerahorn 267
Feuerbahn 3169
Feuerbekämpfung 3431
Feuerbeobachtungsturm 3216
Feuerbrücke 4757
Feuerdamm 3186
Feuerdorn 3212
Feuereinmessgerät 3184
Feuereinpeilgerät 3184
feuerfestes Holz 3198
feuerfreies Intervall 3185
Feuerfront 1189
Feuergefahr 3205
Feuergrenze 1189
feuerhemmender Stoff 3197
Feuerholz-Bündelmaschine 3220
Feuerintervall 3185
Feuerleiter 4757
Feuerleitung 3192
Feuerlinie 1189
Feuerlöschgeräte 916
Feuerlöschmittel 3211
Feuernarbe 3206
Feuerperiodizität 5356
Feuerrand 1189
Feuerregime 3201
Feuerregime mit gemischter Schwierigkeit 5492
feuerresistente Art 3202
feuerresistenter Baum 3203
Feuerring 3204
feuerroter Essigbaum 3247
Feuerschaufel 3209
Feuerschutz 3208
Feuerschutzmittel 3197
Feuerschutzstreifen 3169
Feuerschutzstreifenintensität 3190
Feuerschwamm 7928
Feuerschwierigkeit 3207
Feuerstop 3186
Feuertoleranz 3213
Feuertrocknung 3180
Feuerüberwachung 3210
Feuerverletzung 3206

Feuerversicherung auf Gegenseitigkeit 5642
Feuerwachtturm 3216
Feuerwehrmann 3171
Feuerwirkung 3188
Feuerwunde 3206
Feuer zur Bestandsverjüngung 8027
Feurgestell 3169
F-Horizont 3086
Fiberbrett 9551
Fiberrichtung 2516
Fibrillenwinkel 3102
Fichte 1805
Fichtenblattwespe 7958
Fichtenborkenkäfer 7862
Fichtenbuschhornblattwespe 2910
Fichtenfaserholz 7950
Fichtenharz 7957
Fichtenholz 7959
Fichtenkiefer 7538
Fichtenlangholz 9390
Fichtennadel 7955
Fichtennadelgespinsblattwespe 7958
Fichtenöl 8374
Fichtenröhrenlaus 7949
Fichtenwald 7953
Fichtenware 7959
Fichtenwickler 7951
Fichtenzapfen 7952
Fieberbaum 8417
Fieberheilbaum 8417
Fieberrindenbaum 6669
Fieberstrauch 7909
Fieder 4875
fiederschnittiges Blatt 6284
filamentöse Pilze 3127
Filtrationzone 3133
filzige Hickory 8988
Filzrose 8674
finanzielle Hiebsreife 2745
Fingerblatt 8477
Finger-Eiche 3162
fingerförmiges Fiederblatt 6038
Fingerzapfen 3161
Firnisbaum 1303
Firstbalken 7007
fischgeschützender Wald 3230
Fischgrätentextur 4261
Fisettholz 2908
fixiertes Kabel 8043
Fixpunkt 674
Flachbalken 7485
Fläche des Projektes 6578
Fläche des Schlages 2189
Fläche mit einzelnen Bäumen 352
Fläche mit selektiver Markierung 7415
Flächen-Behandlung 5990
flächenbezogen 7674
Flächenentflammung 351
Flächenerosion 7506
Flächenfachwerk 6842
flächenhafte Abtragung 7506
flächenhafte Erosion 7506
Flächenhieb 1820
Flächenhiebsatz 9770
Flächennummer 6380
Flächenräumung des Holzschlages 1624
Flächenriss 2978
flache Pflanzung 3257
Flächewurzelbaum 7488
flachgesägt 3258
Flachlandwald 3256
Flachmesser-Zerrhacker 3255
Flachstapelung 3259
Flachwurzel 7486
Flachwurzelsystem 7487
Flacourtiengewächse 4467
Flader 4882
Fladerbirke 3122
Fladerschnitt 8383
Fladerzeichnung 7914
Flamboyant 3249
Flammenbaum 3245
Flammenbirke 3122
flammige Zeichnung 3106
Flammschutzmittel 3197
Flammung 3106
Flanellstrauch 1259
Flaschenbaum 594
Flaschenbäume 941
Flatterrüster 7137
Flatterulme 7137
Flaumeiche 6621
flaumhaarige Eiche 6621
flaumhaarige Schlinge 9307
flaumige Pappelsägewespe 4090
flaumiger Kieferbohrer 4089
Flechte 4929
Flechtenbehand 1696
Flechtendecke 1696
Flechten-Kiefernwald 4930
Flechtenschicht 1696
Fleck 7907
fleckige Gelbfäule 7908
Flegelentastungsmaschine 1426
fleischartige Frucht 3262
Fleischerblock 1211
fleischige Wurzel 3263
Fliederbaum 2892
fliegender Holländer 3298
Fliegenpapier 8390
fließendes Gewässer 9284
Flocke 3241
flockige Weißfäule 9370
Florida-Ahorn 3285
Florida-Eibe 3288
Florida-Hartriegel 3287
Florida-Liguster 3287
Florida-Myrsine 5648
Floß 3267, 8584
Flößarbeiter 9569
floßbar 2628
Floßbinden des Holzes 9303
Floßbinder 1173
Floßbrücke 6686
Floßdurchlass 3274
Floßeinheit 6692
Flößen 8578
Flößenzange 6690
Flößer 9569
Flößerei 8578
Flößereiarbeiter 9569
Flößereibezirk 3272
Flößereidistanz 3271
Flößereimannschaft 2630
Flößereimarke 3275
Flößereiordnung 3276
Flößereirecht 6945
Flößereistange 6693
Flößereistatut 3276
Flößerstange 6693
Floßgasse 3277
Floßhaken 6693
Floßholz 3279
Floßknecht 9569
Flößordnung 3276
Floßperiode 3278
Floßrechen 8136
Floßrecht 6945
Floßring 8779
Floßsektion 6692
Floßstange 6693
Floßstraße 3277
Flößung des Holzes 8578
Flößung in Bündeln 8578
Flößung in Einzelstämmen 2625
Flößungsmethode mittels entfernte Patrouillieren 2540
Floßweg 3277
Floßwinde 6687
flüchtige Aufnahme 2869
Flüchtling 2862
Flugbringung 7700
Flugbringungsfläche 7701

Flügelfallkerb 9455
flügelförmiges Parenchym 167
Flügelfrucht 4701
Flügelfruchtgewächse 2513
Flügelfruchtwald 2512
Flügelnuss 9456
Flügelschwimmer 3147
Flügelspindelstrauch 9454
Flügel-Storax 3682
Flügelulme 6988
Flugzeugholz 5063
Flugzeugsaat 87
Flurbuch 1249
Flurholzanbau 5759
Flussablage 6974
flüssiges Harz 8374
Flussuferwald 6961
Flussweide 6975
Flusszeder 1262
Flut 3282
Föfderungstakelage 4171
Föhre 6272
Föhrenharz 6898
Föhrenwald 6255
folgend 7453
Forbes-Zypresse 3314
Förderkette 4173
Förderung mit offenem Seil 4166
Förderwalze 3035
Förderweite 9747
Forestbüro 3323
Forestbürogeschäftszimmer 3323
Formaufasten 3663
Formbarkeit des Holzes 2016
Form des Schaftes 898
Former 5577
Formfaktor 8840
Formgebung 7491
Formhöhe 8841
Formholz 1373
Formklasse 8840
Formpunkthöhe 3669
Formschnitt 6612
Formzahl 8840
Formzahlmaßentafel 3664
Formzuwachs 3665
Förna 3671
Forschungsförster 6883
Forschungsnaturschutzgebiet 6884
Forst 3318
Forstabgabe 6862
Forstabschätzung 3411
Forstabteilung 2552
Forstadministration 3320
Forstakademie 3319
Forstamtstarif 8363
Forstanbau 6821
Forstarbeit 9724
Forstarbeiter 9654
Forstassessor 392
Forstästhetik 3321
Forstaufseher 3613
Forstaufsicht 3612
Forstaxt 453
Forstbahn 3565
Forstbau 1886
Forstbaumschule 342
Forstbeamter 3573
Forstbehörde 2552
Forstbenutzung 3414
Forstberater 1890
Forstbereich 3551
Forstbereichsklassifizierung 3552
Forstbestandskarte 3498
Forstbetrieb 3568
Forstbetriebswirtschaft 3568
Forstbeweidung 3452
Forstbezirk 3382
Forstbilanz 9486
forstbildende Baumart 3439
forstbiologische Vielfalt 3337
Forstbotanik 3341
Forstchemie 2370
Forstchemikalien 3349
Forstdelikt 1909
Forstdrillmaschine 3387
Forstdrillsämaschine 3387
Forstdüngung 3424
Forsteinrichter 3497
Forsteinrichtung 3488
Forsteinrichtungsarbeit 3121
Forsteinrichtungsplan 3492
Forsteinrichtungsplanung 3493
Forstentomologie 3404
Forstentwurfplan 3372
Förster 3406
Försterei 3546
Försterei-Administration 3547
Forsterhebung 3614
Forsterneuerung 6821
Forstertrag 9778
Forsterzeugnisse 3537
Forstfachausdrücke 3619
Forstfachhandel 8653
Forstfilterstreifen 3426
Forstfinanzwirtschaft 3427
Forstfischbetrieb 7626
Forstfischkultur 7626
Forstfischzucht 7626
Forstfläche 9542
Forstforschungsinstitut 3556
Forstfutter 3396
Forstgarten 342
Forstgenschutz 3444
Forstgeräte 3623
Forstgeschichte 4314
Forstgesetz 3484
Forstgesetzgebung 3485
Forstgesundheit 3458
Forstgesundsagente 3459
Forstgesundsbehandlungen 3460
Forstgewächs 1928
forstgewachsen 3455
Forstgrundpreis 8195
Forstgürtel 3335
Forstgut 3448
Forsthaus 3407
Forsthochschule 3462
Forsthydrologie 3465
Forstindustrie 3466
Forstindustrieboden 3467
Forstingenieur 3400
Forstingenieurwesen 3401
Forstinsekt 3475
Forstinsektenkunde 3404
Forstinspektion 3380
Forstinspektionsbeamt 3380
Forstkartenerkundung 3595
Forstkonsulent 1890
Forstkonzession 3356
Forstkran mit horizontalem Arm 705
Forstkubukmeter 9524
Forstkultur 3528
Forstkulturprüfung 3630
Forstkunde 3581
forstliche Berufsausbildung 6565
forstliche Biomasse 9490
forstliche Buchführung 3340
forstliche Forschung 3577
forstliche Forschungsanstalt 3556
forstliche Geräte 3623
forstliche Gesetzgebung 3485
forstliche Hochschule 3462
forstliche Informatik 3472
forstliche Lebensgemeinschaft 3331
forstliche Lehranstalt 7297
forstliche Mikrobiologie 3505
forstliche Nomenklatur 3513
forstliche phänologische Station 3525
forstlicher Berater 1890
forstlicher Betrieb 3568
forstliche Reserven 3559
forstlicher Gewinn 6569
forstlicher Hauungsbetrieb 3371

forstlicher Saatzuchtbetrieb 3586
forstlicher Samenextraktor 3587
forstlicher Samenzuchtbetrieb 3586
forstlicher Sprachgebrauch 3619
forstlicher Wegebau 1889
forstlicher Wirtschaftsplan 3492
forstlicher Zinsfuß 6729
forstliche Samenkunde 3583
forstliches Kapital 3346
forstliches Nebenprodukt 5471
forstliches Straßennetz 3566
forstliche Statik 3326
forstliches Verhalten 7612
forstliches Verhältnisse 3358
forstliche Typologie 3636
forstliche Versuchsanstalt 3413
forstliche Versuchsstation 3413
forstliche Warenkunde 3503
Forstlogistik 3486
Forstmann 3406
Forstmaschine 8853
Forstmaschinen 3402
Forstmeister 3406
Forstmelioration 7611
Forstmeteorologie 3504
Forstmykologie 3511
Forstnerbohrer 9538
Forstnutzungsgröße 3637
Forstökologie 3390
Forstökonomie 3391
Forstökosystem 3395
Forstökosystemnetz 3393
Forstordnung 8046
Forstparasiten 3520
Forstpathologie 3521
Forstpflanze 3527
Forstpflanzenahnzucht 3530
Forstpflanzenzüchtung 1028
Forstpflanzmaschine 8897
Forstpflanzung 94
Forstpflanzungsbodenfonds 8112
Forstpflug 3531
Forstpolitik 3533
Forstpolizei 3532
forstpolizeiliche Überschreitung 1909
Forstprofil 3542
Forstpunkt 3371
Forstrecht 3485
Forstregelung 3590
Forstreifen 5086
Forst-Rekurskommission 3328
Forstrente 3553
Forstreservefonds 3558
Forstreserven 3559
Forstrückstrahlvermögen 3325
Forstsaatgut 3628
Forstsamenbetrieb 3586
Forstsamendarre 3588
Forstsamengut 3628
Forstsamenzüchtung 3583
Forstsatzung 3516
Forstschadinsekt 4499
forstschädlich 1393
Forstschädling 3523
Forstschätzer 3410
Forstschätzung 3411
Forstschicht 3334
Forstschule 342, 7297
Forstschutz 3359
Forstschutzbeamt 3456
Forstseilwinde 7692
Forstsektor 3582
Forststatistik 3606
Forststeuer 3616
Forststraße 3565
Forststreifenpflug 3531
Forsttaxation 3617
Forsttaxator 3410
Forsttechnik 3401
Forsttechniker 3579
Forsttechnologie 3401
Forstterminologie 3619
Forstterminologie-Datenbasis 3620
Forstübertretung 1909
Forstüberwachung 5530
Forstunkraut 3641
Forstverarbeitungsindustrie 3466
Forstvereinzelner 8919
Forstvermessung 3614
Forstverwalteramt 3610
Forstverwaltersamt 3610
Forstverwaltersprogramm 3611
Forstverwaltung 3488
Forstverwaltungsansprüche 3494
Forstverwaltungseinheit 3496
Forstverwaltungsgebiet 3330
Forstverwaltungshilfe 3489
Forstverwaltungsmethode 3491
Forstverwaltungszyklus 3490
Forstwärdersbezirk 2542
Forstwartbezirk 3382
Forstweg 3565
Forstweide 3450
Forstweidesystem 7625
forstweite Führungsanforderungen 3642
forstweiter Standard 3643
Forstwerkzeuge 3623
Forstwesen 3581
forstwirtschaftlicher Leistungsträger 3578
forstwirtschaftliches Handbuch 3570
forstwirtschaftliches Wegenetz 3378
Forstwirtschaft mit Kurzumtrieb 7545
Forstwirtschaftsausnahmen 3569
Forstwirtschaftsführung 3488
Forstwirtschaftsgesetz 3534
Forstwirtschaftskarte 3572
Forstwirtschaftsteilen der Grünzonen 3575
Forstwissenschaft 7610
Forst-Zählwerk 8809
Forstzeichen 3500
Forstzoologie 3651
Forstzüchtung 7610
Fortbildungsgewebe 5402
fortgeschrittene Fäule 72
fortlaufend 7453
Fortpflanzung 9182
Fortpflanzung mittels Wurzelrebe 6876
Fortpflanzungserfolg-Prozent 8918
fortschreitender Hieb 6573
fortschreitender Saum 6574
Fortunes-Kopfeibe 1505
Forwarder 3674
Fossilbrennstoff 3675
fossiles Holz 6196
Fostmassiv 9605
fotografische Rundholzmessung 6210
Fotosynthese 6211
Fototropie 6212
Fototropismus 6212
Fötus 2795
Frachtfuhrwesen 292
Frachtumsatz der Waldbahn 5080
Framire 4596
Franklinie 3687
Franzosenholz 4034
französischer Ahorn 5550
französische Tamariske 3706
Fräsen 9609
Fräseraufnahmedorn 5452
Fraser-Balsamtanne 3689
Fräserkraftmühle 5453
Fraser-Magnolie 3688
Fräskopf-Entrindungsmaschine 2182
Fraßgang 4511
Frauenhaarmoos 4087

Freibaumstellung 5915
freie Durchforstung 3700
freier Schuss 3691
freier Wettbewerb 3692
freies Wachstum 3697
freies Wasser 3699
freie Weite 4504
Freifläche 5907
Freiflächenpflanzung 5908
Freiflächensaat 2521
freigewachsene Kronenklasse 5911
freihauen 6856
Freihauung 8515
Freihieb 5912
Freiholzberechtigung 3693
Freijo 4650
Freilage 5907
Freilager 5919
Freilandbaum 4480
Freilandpfropfung 3115
Freilandsaat 2521
Freilandspanladung 5972
Freilufterholung 5973
Freilufttrockner 153
Freilufttrocknung 152
Freimütigkeit 5914
Freipolter 4485
Freisaat 2521
Freischneidegerät 1615
Freischneider 1615
Freisetzungspunkt 1015
Freistellung der Krone 3698
freiwachsend 3694
freiwachsende Beschirmung 5910
freiwachsender Bestand 3696
Freizeitzone 6754
fremdes Holz 3317
Fremdkörper 3316
fremdländische Baumarten 2944
fremdländischer Baum 2943
Fremdstoff 3316
Fremontpappel 3703
Fremontsmahonie 3704
Freschmasse 9253
Freschvolumen 9253
frischabgehauener Baum 3709
frischabgehauener Grünstamm 3708
frisches Holz 3710
frischezeiger 5408
frisch gefälltes Holz 3710
Frischgewicht 3964
Frischholz 3710
frishabgehauenes Holz 3710
frohwüchsiger Baum 9216
frohwüchsiger Bestand 8536
Front eines Feuers 1189
Frontlader 3720
Frontschild 1161
Frostbeständigkeit 3724
frostempfindliche Lavendelheide 8880
frostempfindliche Lyonie 8880
frostempfindliche Menziesie 3310
Frosthärte 3724
Frostkern 3725
Frostkluft 3723
Frostleiste 3726
Frostring 3727
Frostriss 3723
Frostspalte 3723
Frosttrocknung 3702
Fruchtbildung 3729
Fruchthalter 1353
Fruchtkörper 1353
Fruchtträger 1353
Frühblatt 2702
frühe Brombeere 4307
Früheiche 1793
frühe Waldsukzession 2701
Frühholz 2704
Frühjahrsholz 2704
Frühlingsbüscheln 7937
Frühlingshieb 7938
Frühlingsholz 2704
Frühlingsschwadlegen 7937
Frührinde 2699
Frühschälensaison 7941
F-Schicht 3086
Fuchs 5096
fuchsartig 3679
fuchsartige Stammende 3680
Fuchsbeere 778
Fuchsfeuer 3677
Fuchsschwanz 8473
Fuchsschwanzgewächse 229
Fuchsschwanzkiefer 6997
Fuchsschwanzsäge 2609
Fuernarbe 1378
fugax 3744
Fugenbrett 2595
führender Stamm 4042
Führerbauholz 4041
Führungsmeißel 6238
Führungsrolle 4849
Füllbestand 420
Füllholz 420
Füllkeil 3132
Füllstoff 3130
Füllung den Erosionsrinnen 4046
Füllungskoeffizient der Floßweg 1708
fünfblättrige Zaunrebe 9228
fünfeckiger Kern 6145
Fünfeckpflanzung 6668
Fünfeckverband 6668
fünfgriffeliger Weißdorn 1500
Fungizid 3771
fürcheblätterig 3777
Furchenpflanzung 8964
Furchensaat 7869
Furnier 6398
Furnierblatt 9197
Furnierblock 9193
Furnierbündeln 6399
Furnier-Hackzerspaner 9194
Furnierholz 901
Furnierplatte 1953
Furnierschälmaschine 9623
Furnierschicht 9197
furnierte Spanplatte 9195
Furnierträger 1953
Furnierung 9196
fusiforme Parenchymzelle 3783
fusiformer Strahl 3784
fusiformes Initial 3781
Fußdurchmesser 2465
Fußfruchtbaum 2714
Futterwald 3313
Futterwurzel 3034

Gabanholz 1293
Gabel 3652
gabelförmiger Ast 3655
Gabelfurnier 2060
Gabellader 8610
Gabelrückeschlitten 3881
Gabelstapler 3660
Gabelung 987
Gabelwuchs 3656
gabelwüchsig 3654
Gabholz 3701
Gaboon 3788
Gabun 3788
Gagel 8332
Gagelstrauch 8332
Gagelstrauchgewächse 9305
Galanthus 1349
Galeriewald 3791
Gallapfel 5831
Gallapfeleiche 3794
Galle 3789
Galleiche 3794
Gallmücken 3793
Gallrost 3795
Gallwespen 2210
Gamander-Spierstrauch 3832
Gamari 5255
Gametophyt 3798
Gamma-Diversität 3799
Gang 6097
Gänsefußgewächse 3887
Ganter 5451

Ganterplatz 8664
Ganzbaumernte 1831
Ganzbaum-Erntemaschine 1830
Ganzbaumfällung 1831
Ganzbaumgewinnung 3761
Ganzbaumhackschnitzel 3760
Ganzbaumhackspäne 3760
Ganzbaumnutzung 1832
Ganzbaumschnittholz 3750
Ganzbaumspäne 3760
Ganzbaumverfahren 9394
ganzblättriger Essigbaum 4909
Ganzholz 7113
ganz-trockene Einheit 906
Garbenbaum 5617
Garcinia 3804
Garnierholz 8579
Garrigue 3807
Garry-Eiche 5945
Gartenameise 1762
Gartenmesser 2285
Gaspeldorn 3889
Gasse 181
Gassenhieb 8171
Gastbaumart 5684
Gastgeberbaum 4381
Gattersäge 5132
Gattersägemaschine 5132
Gattertrennsäge 2277
Gatterwagen 5208
Gattung 3827
geadertes Holz 1662
geäscherte Eiche 4973
Geäst 5170
Geäste 5170
gebauter Ziehweg 1887
gebeiztes Holz 1730
gebeugtes Holz 678
gebietsweit wichtige Arten 6828
Gebiet zurückgestellter Erschließung 9799
Gebirgeswaldboden 5585
gebirgiges Land 7099
Gebirgseiche 7464
Gebirgsforst 5583
Gebirgsregenwald im temperierter Zone 5517
Gebirgstaiga 5592
Gebirgstaigawald 5592
Gebirgstundra 5594
Gebirgswald 5583
Gebirgwald mit wildwachsenden Obstbäumen 9406
gebogenes Vollholz 7840
Gebrauch des Holzes 9517
gebräuchlicher Storaxbaum 3825
gebrochene Textur 1062
gebuchteter Blattrand 7661
gebührenfreier Inhalt 5750
Gebund 6692
gebundene Flößerei 8578
Gebüsch 933
Gebüsch-Dschungel 7320
Gebüsch-Stadium 8495
Gebüsch-Sumpf 7557
Gebüschwald 2782
Gebüschweg 7556
gedopte Arten 185
gedrängter Bestand 3756
gedrehter Faserverlauf 9021
geeigneter Waldboden 8257
geeignet zum Abhauen 3231
gefächerte Faser 7450
gefächerte Parenchymzelle 7449
gefächerte Tracheide 8145
Gefährdungsabschätzung 6972
gefährlicher Baum 4179
gefährliches Schwemmholz 4180
Gefälle 7750
gefallendes Blatt 14
Gefällmesser 4443
gefällter Baum 2596
gefällter Stumpf 2600
gefärbtes Holz 1730
Gefäßbündel 6462
Gefäßdurchbrechung 9213
Gefäßgewebe 3105
Gefäßpflanze 9172
Gefäß-Tracheide 9173
gefertigtes Holz 9735
gefiederte Pimpernuss 2885
gefiederter Baum 3030
gefiedertes Blatt mit Unterbrechungen 4564
gefiedertes gegenüberverbindete Blatt 6283
gefingertes Blatt 6282
gefingert zusammengesetztes Blatt 6038
gefladerte Textur 8175
geflammte Textur 3246
gefleckt 2247
geflößtes Holz 3279
geflügelte Frucht 4701
gefrorenes Holz 3728
gefülltblühende Rosskastanie 1781
gefüllter Schneeball 2890
gegabelt 3654
gegabeltes Forst-Zählwerk 3653
Gegenbrand 465
Gegenfeuer 465
gegenständige Blattdisposition 5928
gegenüberstehende Tüpfelung 5929
Gegenzugsäge 2609
geharzter Baum 959
geharztes Holz 6901
gehobeltes Holz 2620
gehöhlter Stammstück 4329
Gehölz 5001
Gehölzkunde 2380
Gehrungsschnitt 5483
Geigenholz 7851
Geißblatt 1227
Geißblattgewächse 4356
Geißfuß 769
Geißklee 1069
Geißklee-Gebüsch 1070
Geißleitere 5257
Geissois 3812
Geister-Eukalyptus 3839
Geiz 418
Geiztrieb 418
gekappter Baum 7489
gekleidete Größe 2621
geköpfter Baum 7489
gekörnte Holzabfälle 4318
gekrümmte Luftseilbahn 2153
gekrümmter Stamm 2154
gekrümmtes Sägeblatt 2152
Gelände 4769
sehr instabiles ~ 9211
sehr rutschgefährdetes ~ 9211
Geländebedeckung 6337
gelatinöse Faser 3813
gelattet 957
gelbästiger Hartriegel 6783
gelbbraune Pflaumensägewespe 6388
gelbe Alpenrose 3244
gelbe Herlitze 1965
Gelbeiche 1510
gelbe Mombinpflaume 4322
gelbe Pavie 9758
gelber Blasenstrauch 818
Gelberden 9763
gelber Jasmin 9759
gelbe Rosskastanie 9758
gelber Trompetenbaum 1339
gelbes Holz 9767
gelbes Sucupira 4784
gelbe Stachelbeerblattwespe 1775
gelbe Trauerweide 1809
gelbe Weide 3885
Gelbfieberbaum 3092
Gelbholz 6030

Gelbkiefer 6445
Gelbkornel 1965
Gelbpflaume 4322
Gelbsaftfäule 9761
Gelbschlagfäule 9762
Gelbsucht 1546
Gelbtrompetenbaum 1339
Gelbverfärbung 9764
Gelbweide 9766
geleitete Ernte 5262
geleiteter Waldboden 5261
Gelenkkurzholzerntemaschine 372
Gelenkmaßstab 3300
gelichtete Baumkrone 8504
gelichteter Wald 4446
gemaserte Birke 4690
gemasertes Holz 9186
gemäßigte Steppenwaldzonen 8444
gemäßigtfeine Textur 5507
Gemeindeforst 1812
Gemeindeforstwirtschaft 5636
Gemeindewald 1812
gemeine Berberitze 1758
gemeine Brombeere 778
gemeine Buche 1760
gemeine Eibe 1814
gemeine Erle 2882
gemeine Esche 1757
gemeine Felsenbirne 3806
gemeine Felsenmispel 3806
gemeine Fichte 1805
gemeine Föhre 1797
gemeine Hainbuche 2901
gemeine Hasel 1778
gemeine Heckenkirsche 2895
gemeine Hirnbeere 1800
gemeine Hopfenbuche 2900
gemeine Hülse 2899
gemeine Kiefer 1797
gemeine Kiefernblattwespe 6268
gemeine Klappernuss 2885
gemeine Kugelassel 1813
gemeine Lärche 1785
gemeine Lederblume 6035
gemeine Linde 1786
gemeine Mahonie 5943
gemeine Mispel 5389
gemeine Platane 2906
gemeine Quitte 6667
gemeiner Apfelbaum 1755
gemeiner Baumwürger 239
gemeiner Berberitzenstrauch 1758
gemeiner Bohnenbaum 1784
gemeiner Erbsenstrauch 7564
gemeiner Flieder 6292
gemeiner Gagel 8332
gemeiner Giersch 769
gemeiner Goldenregen 1784
gemeiner Götterbaum 8892
gemeiner Judasbaum 4665
gemeiner Kiefernspanner 6262
gemeiner Liguster 1799
gemeiner Mangelbaum 251
gemeiner Myrtebaum 1792
gemeiner Nutzholzborkenkäfer 8173
gemeine Robinie 1754
gemeiner Oleander 5887
gemeine Rollassel 1813
gemeine Rosskastanie 1781
gemeiner Pfeifenstrauch 8335
gemeiner Pflaumenbaum 6391
gemeiner Sauerklee 9656
gemeiner Schneeball 2890
gemeiner Seidelbast 2246
gemeiner Spindelstrauch 1383
gemeiner Trompetenbaum 1768
gemeiner Wacholder 3983
gemeiner Weißdorn 4174
gemeiner Wurmfarn 5257
gemeine Schneebeere 1804
gemeines Heidekraut 1779
gemeine Stechpalme 2899
gemeine Tamariske 3706
gemeine Ulme 2893
gemeine Waldrebe 9402
gemeine Walnuss 1808
gemeine Weißtanne 1803
gemeine Widertonmoos 4087
gemeine Winterbeere 1810
gemeine Zwergmispel 2888
gemeine Zypresse 1770
Gemeinholzkosten 4479
gemeinsamer Katalog für Ausgangsmaterial von forstlichem Vermehrungsgut 1767
Gemeinschaft 394
gemeinschaftliche Holzwege 1787
Gemeinschaftswald 1812
Gemeinschaft von Tieren 291
gemischte Holzfaserplatte 1833
gemischte Knospe 1837
gemischte Langholze 5488
gemischte Nordlaubholze 5490
gemischter Bestand 5486
gemischter Hügellandwald 5495
gemischter Monsunwald 5546
gemischter Subor 8277
5489
gemustertes Schnittholz 6116
Generatorholz 3816
genetische Chimäre 3820
genetische Variation 3822
genetische Vielfalt 3821
genetische Waldressourcen 3446
genetischmodifizierter Baum 3819
genetischverbesserte Sämlinge 3818
Genfluss 3814
Genfondswald 3817
Genipa-Baum 4605
Genom 3823
Genotyp 3824
Genpool 3815
Genpoolwald 3817
Genressourcen-Management 5264
Gentechnik 736
geöltes Holz 3026
geometrische Durchforstung 3829
geometrische Rundholzmessung 3828
Georgia-Eiche 3830
Geotextil 3831
gepachtete Forstfläche 6867
geplante Weidensystem 6330
geplenterter Niederwald 7405
geplenterter Wald 7410
Geradblattsäge 3251
gerade Eckîberblattung 8140
gerade Faser 8139
gerader Schnitt 8141
gerades Blatt 8140
geradeschleifen 3975
Geradfallkerb 8144
geradfaseriger Baum 8138
Geradholz 8143
geradringiger Baum 8138
geradschaftiges Holz 8143
Geradschäftigkeit 8142
Geradschnitt 8141
geradstammiges Holz 8137
Gerard-Kiefer 1486
Gerberstrauchgewächse 1955
Gerblohe 8392
Gerbrinde 8392
Gerbsäure 8393
Gerbstoff 8393
Gerbstoffstau 8391
Gerbwurzel 8395
geregelte Bodens 6839
gerichtetes Fällen 2515
gerillt 3977
gerillter Einstechmeißel 3295

geringes Stangenholz 7775
geringwertiger Stamm 5191
geringwertiges Holz 5195
Gerinne 5116
Gerinne für Querschneiden 1128
gerissene Holzfaser 1524
Geröll 7128
Geröllhalde 8378
geröstete Kohle 6984
geröstetes Holz 7735
Gerte 6418
Gerten 9360
Gertenholz 4675
Gerüst 1153
Gerüststange 7270
Gerüstsubstanz 7681
gesägt 7461
gesägtblättrige Kirsche 744
gesägtes Furnier 7261
gesägtes Holz 1918
gesägt-wimperblätterig 7460
gesägt-wimperblättrig 7460
gesammelter Baum 1724
Gesamtalter 8718
Gesamtalterzuwachs 8727
Gesamtaufkommenplan 8729
gesamte Knospe 2831
gesamte Retention 139
Gesamtfläche 8719
Gesamtforstbenutzung 8724
Gesamtfrucht 1839
Gesamthiebsflächeplannung 8720
Gesamtlänge 3311
Gesamtmenge 3311
Gesamtporosität 8728
Gesamtretention 139
Gesamtschnittholzlänge aus einem Stamm 9146
Gesamtschwellen 8732
Gesamtschwindung 8730
Gesamtstandortstyp 8731
Gesamtstrahl 140
Gesamtverdunstung 2916
Gesamtwalddeckung 8723
Gesamtwaldfläche 8722
Gesamtwärmeabgabe 8725
geschadetes Holz 7167
geschältes Holz 555
geschält im Saftperiode 7210
gescherte Fläche 7499
geschertes Ende 7498
geschichteter Bestand 4841
geschichteter Schlagabraum 6231
geschichtete Struktur 4842
Geschiebe 7128
Geschirrholz 2988
geschlossene Brandsaison 1652
geschlossene Gemeinschaft 1651
geschlossener Bestand 3756
geschlossener Riss 958
geschlossener Schluss 1650
geschlossenes Vorriegelungssystem 1654
geschlossenes Waldgebiet 1653
geschnitztes Holz 1361
geschreddertes Brennholz 4319
geschreddertes Holz 4318
Geschreiden 4320
geschützte Forstfläche 6593
geschützter Wald 6592
Geschwister 3754
Gesellschaftswald 1816
gesetzlicher Rahmengesetz 8047
gespaltenes Holzscheit 7920
gesprenkelt 6131
gesprenkelter Eukalyptus 7930
gesprenkelter Kiefernbockkäfer 7931
gesprenkeltes Schatten 2248
gestanzter Flicken 6103
Gesteinsschutt 2440
Gestell 3564
Gestell einer Säge 3684
gestempelter Baum 826
gestielt 6193
gestieltes Blatt 6194
Gestör 8584
Gestörflößerei 3273
gestörter natürlicher Wald 2543
gestörte Vegetation 2544
Gestörtrift 3273
Gesträuch 933, 7555
gestreckt 2313
gestreift 8161
gestreifter Baum 517
gestreifter Nutzholzborkenkäfer 8173
gestreiftes Holz 6935
gestreifte Textur 8175
gestreuter Wald mit Schlagabfäll 2288
Gestrüpp 1202
Gestrüppschneider mit Horizontalachse 4369
Gestrüppvernichtungsmittel 1107
gesunder Ast 7852
gesunder Kleinast 5007
gesunder Wuchs 4192
gesundes Holz 5902
gesundes Ökosystem 4191
gesundes Splintholz 7854
Gesundheitshieb 7193
7853
gesunkene Rundhölzer 2268
getrenntes Flößen 7446
getrenntes Holz 7448
getrockneter Kiefernblock 7333
getrockneter Stumpf 7334
getrocknetes Holz 2622
getuftete Bäume 9006
Gewächshaus 3946
Gewächshaus-Effekt 3947
Gewässerbett 9284
Gewässerregelungswald 9282
Gewässerschutzwald 9282
Gewebe des Holzes 9679
Gewebekultur 8671
Geweih 401
gewellte Maserung 2145
Gewerbeforstbeamter 4481
Gewerbeholz 1745
Gewichtungsbaum 9333
gewinkelte Schere 287
Gewinnung des Holzes 9588
gewöhnliche Esche 1757
gewöhnliche Mahonie 5943
gewöhnlicher Feigenbaum 1774
gewöhnlicher Fliederlilak 6292
gewöhnlicher Sanddorn 7325
gewöhnlicher Schneeball 2890
gewöhnliche Stechhülse 2899
gewöhnliche Traubenkirsche 2881
gewöhnliche Waldrebe 9402
gewöhnliche Wildbirne 1795
Gewölle 6139
gewundene Akazie 9041
gewundene Holzfaser 1902
gewundener Faserverlauf 2144
gewünschte Exoten 2430
gewünschte Pflanzenarten 2428
gewünschte Pflanzengesellschaft 2431
gewünschter zukünftige Bestandeszustand 2429
Gewürznelke 1664
Gewürznelkenbaum 1664
Gewürzstrauch 1806
Gewürzstrauchgewächse 8151
Gewürzsumach 365
gezackter Kern 8680
gezähnt 7461
gezähnter Blattrand 2396
gezähnter Schneeball 368
gezähntes Blatt 8681
gezapftes Holz 827
gezimmert 8580
gezogener Bauteil 8542

Gifteiche 6022
Giftholz 6413
Giftringe 8777
Giftringelung 6411
Giftsumach 6412, 8315
Gillung 3013
Ginkgo 5245
Ginkgogewächse 3847
Ginkobaum 5245
Ginster 3849
Gipfelblock 8688
Gipfelbruch 1013
Gipfeldürre 8686
Gipfelfeuer 2076
Gipfelhöhe 8692
Gipfelholz 8712
Gipfelholz-Erntemaschine 8713
Gipfelknospe 8470
Gipfelspross 8474
Gipfelsprossschädlinge 8472
Gipfeltrieb 8474
Gipfeltrockenheit 7989
Gipfelzweig 8683
glanzblättrige Mahonie 5943
Glanzeiche 3874
Glanz eines Holzes 3872
glänzende Bickbeere 7523
glänzende Lyonie 3091
Glanzliguster 1799
Glanzrinde 7787
Glanz-Stechpalme 4795
Glanzweide 7522
glasiges Holz 9354
Glasweide 1040
glattblätteriger Sumach 7791
glattblättriger Sumach 7791
glattblättrige Rüster 2893
glatter Schnitt 1606
glatte Ulme 7745
Glattkiefer 7538
glattrindige Holzart 7788
glattschäftig 1613
glattschäftiger Wald 1614
Glattschnitt 3294
Gleditschie 263, 4353
gleichaltrige Betriebsart 2919
gleichaltriger Bestand 2920
gleichaltriger Hochwald 2918
gleichaltrige Waldbewirtschaftung 2917
gleichfarbene Tanne 1278
Gleichfluss 2922
gleichförmige Langholze 2925
gleichförmiger Bestand 2926
gleichförmiger Shirmschlag 9086
gleichförmige Textur 2923
gleichgültig abweisendes Kabel 7804
Gleichklangsholz 7851
gleichzeitiges Feuerlegen 8308
Gleitbahnabstand 7746
Gleitchoker 7740
Gleit-Gummi 6318
Gleithaken 529
Gleitholz 7739
Gleitkufen 7693
Gleitriese 5116
Gleitschuh 7739
Gleitschutzkette 319
glimmendes Feuer 7792
Glissadehölzer 3862
globale Wälder 3865
Globalstrahlung 3866
Globalwaldkonvention 3863
Globalwaldressourcen 3864
Globalwaldvertrag 3863
Glockenheide 1779
Glückbaum 594
Gobaja 1921
Gold-Birke 2860
goldene Akazie 3884
Gold-Johannisbeere 1147
Goldkiefer 6445
Goldkiefer-Bockkäfer 6446
Goldkiefernzapfenkäfer 5043
Goldpflaume 1700
Goldregen 1784
goldschuppige Eiche 1319
Goldtanne 1268
Gombe 1121
Goncalo Alves 8548
Goodding-Weide 3886
Gossegraben 4063
Götterbaum 8892
Gowen-Zypresse 3891
graben 2495
Grabenziehen 8962
Gräbenzug 8962
Gradientdichte-Spanplatte 3895
Gramineen 3915
Granatapfel 6443
Granatapfelbaum 6443
Gränke 897
Grännenkiefer 6997
Granulator 6140
granulierte Anwendung 3911
Graphitisierung der Holzkohlen 1447
Gras 4253
Grasbaum 405
Gräser-Fichtenwald 3920
Gräser-Kiefernwald 3919
Graswald 3917
Grätenblattgewächse 5164
Grauahorn 6057
Graubirke 3923
grau-brauner Waldboden 3967
grau-braune Saftfäule 3968
graue Birke 3923
graue Bruchweide 3974
graue Eiche 3928
graue Glockenheide 3974
graue Heide 3153
graue Mangrove 3971
grauer Ahorn 6057
grauer Kern 3970
Grauerle 2878
grauer nussartig-strukturierter Waldboden 5827
grauer Waldboden 3969
grauer Walnussbaum 1222
Grauheide 3153
Graunuss 1222
Graupappel 3973
Grautanne 1278
Grauweide 2775
Greasewood 3932
Greene-Eberesche 3943
Gregg-Akazie 1375
Greifer 3892
Greiferreifen 5198
Greifersäge 3912
Greifhaken 3892
Greisbeil 6693
Grenadille 5426
Grenatapfelgewächse 6442
Grenzdurchmesser 2743
Grenze der schlagbare Durchmesser 2950
Grenzenbaum 641
Grenzenmorgen 5303
grenzenmorgenes Land 5304
Grenzenwälder 3721
Grenze zwischen Grüneland und Wald 3918
grenzwertige Durchforstungsintensität 5305
Grevillia 3966
griechischer Ahorn 4238
griechischer Wacholder 3935
griechische Tanne 3938
griechische Zeder 6530
Grigri 5532
Grobabfälle 4216
grobästig 8135
grobästiges Holz 4798
grobbehauener Abfallbaum 7093
grobbehauenes Holz 7094

grobe Ausgabe 3981
grober Grobschlagabfall 1686
grobes Abfallholz 1686
grobes Brackholz 1684
grobes Entasten 1683
grobe Textur 1678
grobfaseriges Holz 1676
Grobfeile 7089
grobgetrocknete Größe 7087
Grobholz 4808
Grobholzmasse 1680
grobkantiges Brett 7088
grobkörniger Waldboden 1679
grobkörniges Holz 1676
grobporiges Holz 1685
Grobrinde 1675
grobrindig 590
grobringiges Holz 5909
Grobschienenweg 7095
Grobschliff 7793
Grobschnitt 622
Grobschnittholz 7085
Grönland-Porst 4749
Großaufforstung 1042
Großblattahorn 701
großblätterig 4800
großblättrig 4800
großblättrige Alpenrose 6023
großblättrige Linde 4801
großblättrige Magnolie 700
großblättriger Ahorn 701
großblättriger Storax 702
großblättrige Steineibe 5888
großblütige Abelie 3873
großblütige Magnolie 7860
große Abholzigkeit 3022
große Ast 4796
große Bäume 4218
große Durchmesserabnahme 3022
große Lärchenblattwespe 4792
Größenklasse 8924
Größensortierung 3903
großer Birkenspanner 6147
großer braune Nadelholzrüssler 4793
großer braune Rüsselkäfer 4793
großer Eichelrüssler 48
großer Heister 4806
großer mexikanische Kiefernmarkkäfer 4805
großer organische Trümmer 4802
große Roteiche 6768
großer Pappelbock 4804
Großersbirke 8328
großer Ulmensplintkäfer 4794
großes Leucaena 3933
großfrüchtige Kiwi 51
großfrüchtige Ulme 698
großfrüchtige Weißeiche 5566
großfrüchtige Zypresse 5547
Großraum-Holzabfuhr 5150
großwurzelig 6684
Großzahnahornholz 707
großzahnige Pappel 706
Grubenholz 5463
Grubenholzumtrieb 6310
Grubenschwelle 1725
Grumixava 9361
grünästen 6609
Grünästen 3953
Grünastung 3953
Grünbaumvolumen 9247
grünblättriger Schwefelkopf 8258
Grundbesitzverfassung 4780
Grundbestand 5250
Grundbrett 945
Grundfläche 597
Grundflächen-Formfaktor 599
Grundflächenkontrolle 602
Grundflächenmittelstamm 5355
Grundflächen-Mittelstamm 5355
Grundflächentafel 8362
Grundholz 946
Grundklammer 943
Grundlänge 3989
gründliches Blatt 944
Grundlinienbereich 3991
Grundmeristem 3992
Grundpopulation 608
Grundschwelle 945
Grundsims 945
Gründünger 3949
Grundwasser 3995
grüne Akazie 791
grüne Douglasie 1772
Grüneiche 1773, 3951
grüner Ast 5013
grüner Baumschnitt 3959
grüner Baumvorratsbehälter 3960
grüner Fleck 3956
grüner Keil 3963
Grünerle 2897
grüner Wald 3710
grünes Baumzurückhalten 3961
grünes Ebenholz 3942
grünes Holz 3710
grüne Späne 3940
grüne Trauerweide 459
Grünfäule 3954
Grünfläche 3941
Grüngewicht 3964
Grüngürtel 3941
Grünholzsortiertabelle 3944
Grünholzzustand 3957
grünköpfige Fichtenwespe 3945
Grünland-Wald-Ekoton 3451
Grünmasse 9200
Grünsägemaß 3955
Grünstumpf 3958
Grünverbauung 6923
Grünvolumen 9253
Grünwerden 9226
Grünzug 3941
Gruppe 8619
Gruppenknoten 3997
Gruppenpflanzung 6354
Gruppenporen 1669
Gruppenschirmschlag 4004
gruppenweise Methode der Volumenmengen 3998
gruppenweise Mischung 5494
gruppenweiser Plenterhieb 4001
gruppenweiser Schlag 4001
gruppenweise Verjüngung 6812
Guajakbaum 4034
Guajakholz 6405
Guajavaapfel 4037
Guarumo 7519
Guatambu 4035
Guatemala-Podocarpus 1372
Guatemalatanne 5422
Guava 4036
Guave 4037
Guayave 4037
Guayule 4038
Guiana-Butternuss 6188
guineische Palme 110
Gully 4043
Gummi 4053
Gummiakazie 4048
Gummibaum 4469, 4056
Gummibaumbestand 4057
Gummibaumkultur 4057
Gummi des silbernen Dollars 357
Gummifluss 4051
gummigebender Tragacanth 8749
gummihaftes Holz 7126
Gummiharz 4053
Gummimastix 4050
Gummiplatte 4052
Gummistoff 4053
Gummitraganth 8749
Gummose 4051
Gummosis 4051
Gungurru 4060
Gunnseukalyptus 1578
günstige Neigung 3028

gutbestückt 9335
Gütebestimmung des Waldbodens 3600
Gütefaktor 6654
Gütegrad 9774
Güteklasse 9774
Gutswald 3015
Guttaperchabaum 4061
Guttaperchabaum 4147
Guttation 4062
gutwüchsiger Baum 9216
Gymnospermen 4069
Gymnospermenholz 4068
gynodiözisch 4070
gynomonözisch 4067

Haarbirke 6620
Haarfarn 8950
Haarriss 4088
Haarwurzeln 7028
Habitat 4072
Habitatanalyse-Einheit 4073
Habitatdiversität 4074
Habitattyp 4079
Habitatverbesserung 4076
Habitatverwaltung 4077
Habitatverwaltungseinheit 4078
Habitus 4017
Hackbock 3033
Hacke-Aufspannen 4317
hacken 1561
Hacken 1531
Hackfrüchte 7017
Hackholz 9508
Hackholzbehälter 2496
Hackmaschine 9512
Hackschnitzel 9067
Hackschnitzelentrindung 1513
Hackschnitzel-Füllapparat 1522
Hackschnitzelsillo 2496
Häcksel 4318
Häckselmaschine 2181
Häcksler 1529, 9057
Hackspäne 9067
Hackspänesichter 1535
Hackspänesortierer 1535
Hackspänezerkleinerer 1534
Hackstreu 1456
Hackwald 1928
Haferkirsche 8330
Haferpflaume 2244
Haferschlehe 2244
Haffdorn 7325
Haftwurzel 86
Hag 4085
Hagebuche 2901
Hahnendorn 1699
Hahnenfußgewächse 1218
Hahnensporn-Weißdorn 1699
Haiawaballi 4086
Hain 5001
Hainbirke 1474
hainbuchblättriger Ahorn 4373
Hainbuche 2901
hainbuchenblättrige Eberesche 4742
Haken 4359
haken 4358
Hakengang 4511
Hakenloch 4361
halbaufrechte Brombeere 972
halbbestockte Waldboden 6085
Halbblock 4100
Halb-Bohle 4099
halbeingewachsener Ast 6084
Halbgeschwister 4097
Halbhartfaserplatte 5754
Halbhartholzhieb 7428
Halbholz 4094
halbimmergrüner Laubnebelwald 7426
halblaubwechselnder Wald 7424
halbnatürlicher Wald 7429
halbringporiges Holz 7432
Halbringporigkeit 7431
halbring-porös 4091
Halbripsäge 4092
Halbrundholz 4094
Halbrundschwelle 4095
Halbschattbaumart 7434
Halbsenker 9290
Halbstamm 4098
halbstarrer Kern 7430
Halbstrauch 9065
halbtrockener Wald 7425
halbtrockenes Holz 7528
Haldenaufforstung 97
Hallimasch 4352
Halm 2124
Halm- 1390
Halophyten 4102
Halstuch 7283
Halteseil 8365
Halteseiltakelung 8366
hammeln 8702
Hammerschäler 4103
Hammerschälmaschine 4103
Handbeil 673
Handelsbaumarten 1752
Handelsbehandlungen 1751
Handelsdurchforstung 1750
handelsfähige Holzschnitte 5396
Handelssägewerk 1749
handelsübliche Länge 5395
Handelsvolumen 1753
Handelsware 5400
handeltrockenes Holz 1748
Handentrinder 4105
Handentrindung 4107
Handfäller 4108
Handfällung 4109
handförmiges Blatt 6037
handförmig geädertes Blatt 6041
handförmiggelapptes Blatt 6039
handförmiggeteiltes Blatt 6040
Handholzschleifer 4113
Handjäten 5287
handnerviges Blatt 6041
Handrücken 495
Handsäge 4111
Handschere 4106
handschnittiges Blatt 6042
Handschuhe 6599
Handschutz 4706
handspaltiggeschnittenes Blatt 6042
handspaltiggeteiltes Blatt 6040
Hanfpalme 5384
Hanfstrick 4251
Hanfweide 619
Hänge-Akazie 9328
Hängebahn 8745
Hängebirke 9789
Hängeblockwagen 8565
Hängebuche 9329
Hängefichte 9330
Hängemaulbeerbaum 9327
Hängen 4895
Hängenbleiben 5042
hängend 9324
hängendes Seil 4115
hängengebliebener Baum 5041
Hänger 4897
Hängeweide 459
Hängewuchsform 9326
hängezweigige Silberlinde 6143
Hängseil 7714
Hangwald 7752
Haptotropismus 4118
harken 6705
harrfrüchtige Pappel 784
Hartabständer 4131
Hartbast 4119
Härte 4123
harte Faserplatte 1845
Hartenkatalpa 5789
hartes Holz 2670
Hartfaserplatte 1845

Hartholz 2670
Hartholzbalken 4142
Hartholzbeimischung 4143
Hartholzpech 4144
Hartholzplatte 1845
Hartlaubarten 4138
Hartlaub-Hängematte 4140
Hartlaubholz 4121
Hartlaubholzasche 4132
Hartlaubwald 4139
Hartlistenholz 4136
Hartmast 4127
Hartriegel 222, 1771, 9339
Hartriegelgewächse 2556
Hartsaftfäule 4130
Harzablaufvorrichtung 334
Harzader 4058
Harzarbeiter 4084
Harzausfluss 2977
Harzbalsam 503
Harzbehälter 538
Harzbehälterkorb 538
Harzbestand 9013
Harzbeule 6893
Harzblase 6893
Harzeibe 6784
Harzeinsammler 6905
Harzeisen 6897
Harzen 828
Harz-Eukalyptus 411
Harzfluss 2977
Harzföhre 5796
harzführende Nadelholzart 6304
harzführender Baum 6900
Harzgalle 6893
Harzgang 4049
Harzgewinnung 828
harzige Akazie 8316
harzige Hackspäne 6894
harziges Wurzelholz 6899
Harzindustrie 6896
Harzkanal 4049
Harzkiefernstammholz 6266
Harzkratzer 6905
Harzkropf 8155
Harzlache 8155
Harzlachte 8155
Harzmesser 4584
Harzpore 4049
harzreicher Baum 6900
harzreiches Holz 4292
harzreiches Kernholz 6902
Harzreißer 2142
Harzriss 8155
Harzrisslänge 2982
Harzsammler 6905
Harzspachtel 4584
Harzstreifen 8155
Harztasche 6893
Harztrichter 6303
Harzüberfülle 2977
Harzungsdauer 6892
Harzungsfläche 6904
Harzvorräte 5701
Harzwald 827
Hasel 1778
Haselerle 7786
Haselgebüsch 4181
Haselhain 4181
Haselholz 4184
Haselhuhn 4182
Haselnuss 4183
Haselnussbaum 1778
Haselnuss des Cornuta 637
Haselnussstrauch 1778
Haspelwerk 9267
haubarer Bestand 5338
haubarer Vorrat 2188
Haubarkeitsalter 133
Haubarkeitsdurchschnittszuwachs 3979
Haubarkeitsertrag 3145
hauen 4316
einen Weg durch den Wald ~ 8302
Hauer 9526
Häufchensaat 2485
häufeln 1169
Haukerbe 1214
Haumesser 1204
Hauptausleger 5246
Hauptbaum 2567
Hauptbaumart 5252
Hauptbaumdurchmesser 2568
Hauptbestand 5250
Hauptblock 1158, 2983
Haupteinschlag 4150
Hauptleitungsblock 1158
Hauptfahrtsweg 3463
Hauptförster 1483
Hauptforstwärder 4185
Hauptholzart 5252
Hauptlader 8695
Hauptlagerplatz 4484
Hauptmast 4190
Hauptnutzung 3136
Hauptnutzungsetat 3145
Hauptpfad 5249
Hauptsäge 1018
Hauptsägezeit 4189
Hauptseitenwurzel 5247
Hauptstiel 4883
Hauptwaldproduktion 5253
Hauptweg 5249
Hauptwurzel 8411
Hausbock 4389
Hausbockschaden 5134
Häuschen 5042
Hauschicht 7947
Hauseberesche 8991
Haus-Pflaume 6391
Hausschwamm 2560
Hauung 8824
Hauungsarbeiten 3457
Hauungsbetrieb 5074
Hauungskosten 3056
Hauungslos 6180
Hauungsplan 6332
Hauungsverbot 6577
Hauungszeit 3077
Hebewinde 8603
Hebezange 3892
Hebung 5979
Heckegarten 4220
Hecken-Ahorn 2827
Heckenkirsche 1227
Heckenprachtkäfer 9016
Hecken-Rose 2555
Heckenrose 634
Heel-Tackle 4227
Hege 890
Hegeforst 3535
Hegeholz 3535
Heger 344
Heidekraut 1779
Heidekrautgewächse 4207
Heidekrautwald 1284
Heidelbeere 5006
Heidelbeerweide 813
Heidelerche 9594
Heidewacholder 3983
Heidewald 4208
Heilbuttschnitte 7712
Heilerziehung 6860
heiliger Bambus 5654
Heimat 4072
heimische Bakterien 5669
heimische Holzbodenfläche 5671
heimischer Wald 4476
heimisches Holz 4341
Heimschutz 4342
Heisswasserskarifikation 4386
Heister 7202
Heisterpflanzung 7203
Heizkraft 1194
Heizwert 1194
Helfer 7359
Helferbaum 4243

hellbraune kubische Fäule 4937
hellrotes Meranti 4949
Helophyt 4242
Hemicrypophyt 4245
Hemiepiphyt 4246
hemiepiphyt 4247
Hemizellulose 4244
Hemlock-Blattwespe 4250
Hemlockrinde 4249
Hemlocktanne 4248
Hemlocktanne-Spanner 2718
Hemmkette 7798
Hemmschuh 7687
herabhängend 9324
heranreifender Bestand 4425
heranwachsender Bestand 4425
Heranziehen 4169
Herbizid 4254
Herbstholz 4827
Herdenaufseher 4257
Herkunft 6603
Herkunft der Holzkohle 1449
Herkunftskontrolle 6604
Herlitze 1965
Herrschaft 6013
herrschend 2563
herrschende Kronenklasse 2564
herrschender Baum 2567
herrschender Durchmesser 2565
herrschender Gewebeschichten im Querschnitt 5251
herrschendes Alter 6526
Herz 1951
Herzbaum 4665
herzblättrige Erle 4592
herzblättrige Hainbuche 7280
herzblättrige Weide 4198
Herz-Blattweide 5480
Herzbräune 1080
Herzbrett 4194
Herzenfäule 1414
Herzfäule 1414
herzförmiges Blatt 1943
herzgespaltenes Brett 886
Herzriss 4195
Herzschnitt 8383
Herzwurzel 8411
heterogener Strahl 4264
heterophyl 4266
heterophyler Trieb 4267
heteroplastisches Pfropfen 4268
Heterosis 4269
heterözisch 4262
Heumachenwaldboden 4175
Heuschreckenbaum 1990
Heuwaldboden 4175
Hickorynussbaum 7482
Hiebanfang 668
Hiebfortschritt 2202
Hieb mit Reserve 6885
Hieb mit Roden von Bäumen 3081
Hieb mit verzögerte Abführung 3080
hiebreifes Holz 6964
Hiebsalter 136
Hiebsart 1617
Hiebsartsklasse 2191
Hiebseinheit 9097
Hiebsergebnis 9777
Hiebsertrag 9777
Hiebsfläche 2189
Hiebsfläche-Qualitätsverbesserung 7156
Hiebsfolge 2202
Hiebsfortschritt 6575
Hiebsführungsmethode 2203
Hiebsignierfärbung 3074
Hiebsintensität 4210
Hiebskosten 3056
Hiebsordnung 2199
Hiebsort 7668
Hiebsplan 2201
Hiebspolster 3057
Hiebsregime 2201
hiebsreifer Baum 5341
hiebsreifer Wald 2951
hiebsreifer Wert 2952
Hiebsreihe 2202
Hiebssatz 195
Hiebssatz der Abtriebsnutzung 197
Hiebssatz-Wirkung 196
Hiebssystem 5084
Hiebstaxe 5085
Hiebstempo 8446
Hiebsumlauf 2194
Hiebsumtrieb 2194
Hiebsweg 7689
Hiebszeit 3077
Hiebszug 2178
Hieb-und-Rückungsoperationen 2164
Higan-Kirsche 4275
Hilfsförster 392
Hilfsholzart 66
Himalaja-Wasserbirke 9279
Himalayafichte 4310
Himalaya-Kiefer 4308
Himalaya-Surenbaum 4309
Himalayazeder 2398
Himmelsbambus 5654
Waldboden 5801
Hinterfüllung 464
Hintergrunderholung 461
Hinternbinden 942
Hinterschlitten 6746
Hinterwälder 476
hinzubekommend 77
Hippe 1204
Hirn 2980
Hirnbeere 1800
Hirnende 2980
Hirnfläche 2803
Hirnholz 2051
Hirnholzriss 958
Hirnholzumlauf 2804
Hirnschnitt 2814
Hirnseite 2980
Hirschbeere 2327
Hirschholunder 2907
Hirschkolbensumach 260
Hirt 4257
Hitzerriss 2002
Hobelbrett 8280
Hobelspäne 926
hobeltrockenes Holz 6325
Hobelware 2620
Hobelwerkstapler 6327
hochangesetzte Krone 4298
Hochast 8683
Hochästung 4294
Hochaufzug-Lader 6065
Hochbesäumen 2198
Hochbestand 4280
Hochblatt 968
Hochdurchforstung 8512
Hochertragsvarietät 4306
hochgefährliches Risiko 4286
hochgefährliches Risiko für Forstgesundheit 4286
hochgelegtes Drahtseilsystem 4291
hochgespanntes Drahtseil 4293
Hochluftseilbahn 4300
hochproduktiver Wald 4299
Hochstamm 4301
hochstämmig 5127
hochstämmiger Baum 4301
Hochstauden 4287
Hochstaudengemeinschaften 228
Höchstbergwald 9136
Hochstetter-Olivenbaum 2712
hochvolumener Bestand 4303
Hochwald 4280
Hochwaldbetrieb 4281
Hochwald mit Reserve 4282
hochwertiger Wald 4276

Hochzwiesel 3656
Höcker 9003
höckeriger Stamm 4721
Höcker-Kiefer 4722
Hofkammerwald 2077
Hofpore 920
Hofstatt 3017
Hoftüpfel 920
hohe Alphitonia 8807
Höhe des Kronenmittelpunktes 3669
hohe Deutzie 3787
Höhe/Durchmesser-Kurbe 4231
Höhendominanz 325
Höhendurchmesser 4230
Höhenkiefer 5598
Höhenklasse 4228
Höhenkurve 4229
Höhenlinien-Graben 1904
Höhenmeristem 327
Höhenmessung 4232
Höhenpfahl 3899
Höhentrieb 4237
Höhenwachstum 326
Höhenwuchs 326
Höhenzuwachs 326
Höhenzweig 324
höher Ertrag 4305
hoher Sadebaum 3935
höher Stumpf 4302
Höhle 2365
hohler Baum 4333
hohler Knoten 4331
hohler Stamm 4334
hohles Holz 4330
Hohlgang im Holz 4511
Hohlkern 2933
Hohlkreuzsäge 4327
Hohlmeißel 4328
Hohlraumgehalt 6463
Hohlraumvolumen 6463
Hohlspaten 6342
Hohlstämmigkeit 4332
Hohlstemmer 4328
holländische Ulme 2677
Holler 2892
Holm 2038
Holozellulose 4340
Holunder 2892
Holunderbeere 787, 2778
Holz 9473
vom Käfer getötetes ~ 1150
Holz- 9704
Holzabbau 1014
Holzabdeckung 880
Holzabfallbeseitigung 2546
Holzabfällespäne 1538
Holzabfallfräze 7721
Holzabfallnutzung 7731
Holzabfallverbrennung 7719
Holzabfallverwertung 7731
Holzabfallzerkleinerer 7721
Holzabfuhr 8595
Holzabfuhrfahrzeug 8598
Holzabfuhr-Korridor 9745
Holzabfuhr mit Seilbahn 1247
Holzabfuhrmodell 7688
Holzabfuhrreise 7690
Holzabfuhrstraße 9681
Holzabfuhrweg 9681
Holzabgabe 6862
Holzabgabeschein 6180
Holzabhängigkeit 8575
Holz abhärten 7337
Holzabhauen 8824
Holzabhieb 8824
Holzablage 9532
Holzablageplatz mit Flussanschluss 6974
Holzablängen 8571
Holzabnahmebeamter 1457
Holzabnahmebuch 3073
Holzabraum 998
Holzabreissen 2598
Holz abrinden 2280
Holzabstreifer 5097
Holzabwerfer 5097
Holzachat 124
Holzachse 5048
Holzalkohol 9476
Holzalter 135
Holzanalyse 9477
Holzanatomie 9478
Holzanbau 7610
Holzanfärben 9662
Holzanfuhr 9532
Holzanhänger 5078
Holzapfelbaum 9403
Holzarbeit 9698
Holzarbeiter 9699
Holzart 4714
Holzartalbum 9660
Holzartenbeimischung 5497
Holzartenbestimmung 8928
Holzartenfolge 8249
Holzartenverteilung 8929
Holzartenwahl 1551
Holzartenwechsel 226
Holzartenzusammensetzung 8927
Holzartikel 9560
Holzasbest 6994
Holzasche 9482
Holzaufbau 9666
Holzaufbereitung 1855
Holzaufbereitung mittels Schlepper 7683
Holzaufbereitungsabteilung 5073
Holzaufbereitungsbetrieb 5074
Holzaufbereitungs-geschwindigkeit 4162
Holzaufbereitungsmaschinen 4157
Holzaufbereitungsmethode 4158
Holzaufbereitungsstelle 3371
Holzaufbereitungsverwaltung 5076
Holz auf dem Stamm 3454
Holz auf de Stock 3454
Holzauflage 9557
Holzaufmesser 1457
Holzaufnahme 9671
Holzaufnahmebuch 3073
Holzaufschluss 9633
Holzaufteilung 9502
Holzaufzug 8596
Holzauktion 8557
Holzausbau 8600
Holzausbeute 3414
Holzausfällung 8824
Holzausformen 9564
Holzausformung 9564
Holzausformungsplatz 7727
Holzausfuhr 2957
Holzausgabebuch 4904
Holz aushalten 1122
Holzausnutzung 3541
Holzauszeichnung 5317
Holzaxt 453
Holzbalken 9592
Holzbasis 8558
Holzbaukonstruktion 1352, 8593
Holzbauwerk 1352
holzbearbeitende Industrie 9685
holzbearbeitender Betrieb 9558
Holzbearbeiter 9699
Holzbearbeitung 9630
Holzbearbeitungsindustrie 9685
Holzbearbeitungsmaschine 9701
Holzbearbeitungswerkzeug 9652
Holzbedarf 2362
Holzbehälter 9518
Holzbehandlung 7623
Holzbeladevorrichtung 9587
Holzbelag 880
Holzberechtigter 6940
Holzberechtung 8573
Holzbergung 2598
Holzbeschädigung 9535

Holzbeschaftungsindustrie 5206
Holzbeschaftungsmodell 4160
Holzbesitz 8652
Holzbestand 3605
Holzbestandteil 9548
Holzbeurteilung 9562
holzbewohnender Pilz 9485
Holz-Bezeichnungsschema 9593
Holzbiegen 9489
Holzbiegsamkeit 3264
Holzbilanz 9486
Holzbildung 4954
Holzbildungsgewebe 1287
Holzbinder 1840
Holzbirnbaum 1795
Holzbirne 1795
Holzblasinstrument 9694
Holzblatt 9197
holzblätterig 9709
holzblättrig 9709
Holzblock 9491
Holzblock 2171
Holzbock 4380
Holzböcke 1327
Holzboden 8604
Holzbodenfläche 9542
Holzbohrer 8562, 9538
Holzbohrkäfer 8562
Holzbonität 3592
Holzborkenkäfer 9552
Holzbörse 8583
Holzbracker 9525
Holzbrand 9499
Holzbrandmalerei 9498
Holzbrei 9679
Holzbrennstoff 9487
Holzbrikett 9495
Holzbringung 8595
Holzbruch 1014
Holzbrücke 8663
Holzbuchstaben 9597
Holzbund 9496
Holzbündel 9706
Holzbündeldraht 9497
Holzchemie 2370
Holzcontainer 9518
Holzdampfer 5199
Holzdarre 9537
Holzdauerhaftigkeit 137
Holzdeckung 9521
Holzdefekt 9563
Holzdeformität 9531
Holz der Wurzeln 9580
Holzdichte 1818
Holzdieb 8051
Holzdiebfrevel 8052
Holzdimensionen 8576
Holzdraht 9695
Holz-Drechsler 9682
Holzdrehung 9683
Holzdrift 8584
Holzdrink 9539
Holzdübel 9553
Holzdurchführbarkeit 8617
Holzdurchlass 8572
Holzdurchmesser 8071
Holzeinbau 8600
Holzeingeschaften 9632
Holzeinkauf 9635
Holzeinnahmen 9591
Holzeinnahmerechnung 4905
Holzeinschlag 8824
Holzeinschlagapparatur 4157
Holzeinschlagarbeiten 3457
Holzeinschlagbrigade 9527
Holzeinschlag der Biomasse 731
Holzeinschlagfonds 3423
Holzeinschlag nach oben 9128
Holzeinschlagsbetrieb 5074
Holzeinschlagslizenz 8605
Holzeinschlagslokomotive 5075
Holzeinschlagsplatz 1821
Holzeinschlagsübersicht 6831
Holzeinschlagsunternehmer 5072
Holzeinschlagsystem 5084
Holzeinschlag über Wurzelanlauf 3045
Holzeinschnitt 5100
Holzeinschnittschere 5113
Holzeinsparung 8636
Hölzel 6236
Holzempfänger 1457
Holzendpunkt 8622
Holzentfärbung 9536
Holzentstehung 4954
Holzer 9654
Holzerde 3464
Holzerlöse 9778
Holzernte 9523
Holzernteanordnung 4161
Holzerntefunktionen 4154
Holzernteindex 4156
Holzernteinkrement 4155
Holzerntekosten 8217
Holzerntelandbase 8594
Holzernteordnung 4161
Holzernteprognose 4153
Holzerntetabelle 4163
Holzerntevolumen 2200
Holzerntgeräte 4157
Holzersatz 9669
Holzerschaft 9527
Holzessig 9688
Holzexport 2957
Holzextraktivstoffe 2972
Holzfabrikate 8626
Holzfällen 8824
Holzfäller 8836, 9526
Holzfälleraxt 453
Holzfällerrotte 9527
Holzfäll-Keilgerät 3079
Holzfällplatz 5038
Holzfällungsgabel 6645
Holzfarbe 9513
Holzfarbhammer 8884
Holzfarbstoff 9541
Holzfärbung 9515
Holzfaserbrei 9603
Holzfaserdämmplatte 4520
Holzfaserhartplatte 1845
Holzfaserisolierplatte 4520
Holzfaserplatte 9551
Holzfaserrichtung 2516
Holzfäule 9529
Holzfäulnis 9529
Holzfehler 9530
Holzfeile 9566
Holzfertigteile 5458
Holzfestigkeit 2669
Holzfeuchte 5521
Holzfeuchtemesser 9611
Holzfeuchtemessgerät 9611
Holzfeuchtegleichgewicht 2856
Holzfeuchtigkeit 5521
Holzfeuerung 9568
Holzfloß 8584
Holz flößen 3269
Holzflößer 9569
Holzfördermittel 8598
Holzformation 9545
Holzforschung 9639
Holzfrachter 8565
Holzfrachtschiff 5199
Holzfräsen 9609
Holzfraß 4511
holzfrei 9577
holzfressend 9739
holzfressendes Insekt 4958
Holzfresser 8562
Holzfrevel 2935
Holzfuder 8984
Holzfuhre 8984
Holz für Holzdraht 5332
Holz für landwirtschaftlichen Bedarf 9570
Holzgas 9582
Holzgefäß 9687
holzgefeuerte Energie 9567

Holzgerinne 5116
Holzgewächs 9714
Holzgewebe 9679
Holzgewicht 8658
Holzgleisweg 1947
Holzgraveur 1363
Holzgreifer 3892
Holzgrößhändler 9392
Holzgruppenpreis 6528
Holzgummi 8907
Holzhacker 9526
Holzhackmaschine 8853
Holzhalbwaren 7427
holzhaltige Pappe 9493
holzhaltiges Papier 9634
Holzhammer 8884
Holzhandbuch 9585
Holzhandel 8653
Holzhandelshof 8583
Holzhändler 2276
Holzhärte 4128
Holzharz 8907
Holzhauer 9526
Holzhaueraxt 453
Holzhauer-Camp 5071
Holzhauergeräte 4157
Holzhauerhütte 5071
Holzhauerrotte 9527
Holzhauerwerkzeuge 4157
Holzhaufen 2305
Holzhaushalt 9486
Holzherstellerland 8624
Holzhimbeerseuche 6224
Holzhöhe 8599
Holzhygroskopizität 9589
holzige Pflanze 9714
Holzimprägnierung 4429
Holz in der Rinde 9045
Holzindustrie 3466
Holzinhaltsstoffe 2972
Holzinsekt 3475
Holzkarton 9493
Holzkauf auf dem Stock 7157
Holzkäufer 8564
Holzkaufmann 2276
Holzkeil 9561
Holzkern 1951
Holzkiste 9549
Holzklasse 8566
Holzklassifizierung 3902
Holzklebstoff 9474
Holzklemme 9510
Holzklotz 2171
Holzklotzschnitt 713
Holzknospe 9179
Holzknüppel 8937
Holzkohlenbrenner 1444
Holzkohleneinstufung 1446
Holzkohleninstallation 1448
Holzkohlenkahlschlag 1674
Holzkohlenklein 9506
Holzkohlenloch 1451
Holzkohlenmeiler 1450
Holzkohlenpulver 9506
Holzkohlenstaub 9506
Holzkohlenvergaser 1444
Holzköhler 1445
Holzkombinat 9516
Holzkonservierung 8569
Holzkonstruktion 8570
Holzkörper 9737
Holzkraftschlepper 5114
Holzkran 8609
Holzkratzer 9646
Holzkropf 1182
Holzkultur 3528
Holzkunde 9645
Holzladekran 8609
Holzladen 8608
Holzladevorrichtung 8607
Holzladungslinie 8611
Holzlagerhof 8664
Holzlagerplatz 8664
Holzlagerplatzabstand 4775
Holzlagerung 5120
Holzlast 1338
Holzlastwagenfahrer 5122
Holzleim 9583
Holzleiter 1131
Holzlieferung 2359
Holzliste 8376
Holzlizenz 8605
holzlose Produkte 5778
holzlose Ressourcen 5775
holzloser Ressourcenwert 5776
Holzmakler 9475
Holzmakrostruktur 9602
Holzmangel 7282
Holzmangelgebiet 8574
Holzmarkierung 5317
Holzmarkt 8613
Holzmarktzonen 8614
Holzmaß 8615
Holzmasse 9603
Holzmassenaufnahme 5364
Holzmassendurchschnitt 8816
Holz-Mastix 9606
Holzmehl 1514
Holzmehl-Asthma 9700
Holzmehlfilter 9540
Holzmehlkäferschaden 5221
Holz-Melasse 9612
Holzmesskunde 733
Holzmessung 9607
Holzmining 8616
Holz mit Gemeinsamklasse 1776
Holz mit hohem Dichte 8659
Holz mit kleiner Querschnittfläche 8662
Holz mit lockere Textur 5162
Holz mit niedrigem Einlaufen 8661
Holz mit ordinäre Klasse 1776
Holz mit Pilzfäule 1880
Holz mit Silberfaser 7603
Holz mit Teeröl imprägnieren 2017
Holzmosaik 9590
Holzmosaikeinlegearbeit 9590
Holzmulch 5603
Holzmuster 8635
Holznagel 9553
Holznebenprodukte 9500
Holznettoniederschlag 3622
Holznormung 8642
Holznot 7282
Holznutzungsanlage 9637
Holznutzungsfläche 2189
Holzoberfläche 8650
Holzobst 9427
Holz ohne Kern 4199
Holz ohne Poren 5766
Holzöl 9615
Holzölbaum 1304
Holzpaket 9496
Holzparenchym 9617
Holzparenchymzelle 9618
Holzpartie 5181
Holzpartikelbreite 9621
Holzpartikellänge 9619
Holzpartikeln 9508
Holzpartikeltrocknung 9620
Holzparzelle 9599
Holzparzellelizenz 9600
Holzpech 8907
Holzpfahl 6475
Holzpflanze 9714
Holzpflaster 9554
Holzpflege 9676
Holzpflock 9555
Holzpfosten 9555
Holzphysik 8618
Holzpigment 9625
Holzpilote 6475
Holzpilz 9581
Holzplantationernte 8620
Holzplastifizierung 9626
Holzplatte 9492

Holzpreisliste 7123
Holzprobe 8635
Holzproduktion 3537
Holzprofile 7260
Holzpropfen 5096
Holzprüfung 9677
Holzpulver 1514
Holzqualität 6657
Holzqualitätskontrolle 6653
Holzquerschneider 9584
Holzquirl 1668
Holzraffination 9638
Holzraspel 9636
Holzrecht 8573
Holzreibe 9636
Holzreinigung 9511
Holzressourcen 8629
Holzressourcensystem 8630
Holzreste 3561
Holzreste sammeln und anhäufeln 1573
Holzriese 5116
Holzrinne 5116
Holzriss 9507
Holzrohr 8572
Holzröhre 9687
Holzrohstoff 7096
Holzrolle 6627
Holzrottmeister 5069
Holzrückekarre 5078
Holzrückekarren 5199
holzrücken 494
Holzrücken 8595
Holzrücken im Halbschleifen 7433
Holzrücken im Schleifen 3994
Holzrücken im Verladen 3672
Holzrückenkarren 3673
Holzrückenschlepper 6028
Holzrücker 1435
Holzrückewagen 5208
Holzrückmaschine 7692
Holzrückschlepper 6028
Holzrückung mit einer Drahtseilbahn 7700
Holzrüstung 8824
Holzrutsche 5116
Holzsaft 7208
Holzsäge 9643
Holzsägemaschine 675
Holzsäure 9688
Holzschabe 8639
Holzschaden 2242
holzschädigende Einwirkung 9534
Holzschädigung 9535
Holzscheibe 1336
Holzscheider 9197
Holzscheitachse 5048
Holzscheitachse-Abweichung 2450
Holzschere 9649
Holzschiene 8628
Holzschiffplatte 9604
Holzschlageinstellung 5083
Holzschläger 9526
Holzschlagsbetrieb 5074
Holzschlagslokomotive 5075
Holzschleifwinde 7692
Holzschleppen 395
Holzschliff 9603
Holzschliffspäne 6625
Holzschlitten 5115
Holzschmiermittel 9601
Holzschneiden 5100, 9502
Holzschnittgenehmung 6183
Holzschnitzer 1363
Holzschrot 3128
Holzschuppen 9650
Holzschutz 9627
Holzschutzmittel 8623
Holzschwamm 9581
Holzschwarte 7712
Holzschwelle 9556
Holzserie 7134
Holzservitut 8573
Holzsetzer 9661
holzsignifikanter Bezirk 8640
Holzsorte 4714
Holzsorterbildung 3902
Holzsortierer 9525
Holzsortierung 3902
Holzsortiment 8556
Holzsortiment-Kraftkarren 7539
Holzspalte 9507
Holzspalter 1632, 9512
Holzspaltmaschine 9512
Holzspäne 9508
Holzspanlocken 9648
Holzspanplatte 6091
Holzsperre 9550
Holzspiritus 9476
Holzspreize 8568
Holzsprödigkeit 1001
Holzstamm 8934
Holzstammendpunkt 8622
Holzstammpassage 5104
Holzstammsektion 8879
Holzstandard 9664
Holzstandortindex 8641
Holzstange 6418
Holzstapel 8619
Holzsteigfläche 9559
Holzstich 9502
Holzstock 8879
Holzstoff 9504
Holzstoß 9519
Holzstosse pro Arbeitsstunde 1946
Holzstosse pro Morgen 1945
Holzstrahl 9738
Holzstrom 8645
Holzstücke 5218
Holzstücken 4912
Holzstückzerkleinerer 1574
Holzsturz 2598
Holzsubstanz 9667
Holzsubstanzverlust 5180
Holztasche 8621
Holztaxation 3617
Holztaxe 3616
Holztechnik 9674
Holztechniker 9675
Holztechnologie 8582
Holztechnologe 9675
Holztee 8437
Holzteer 9672
Holzteeröl 9673
Holzteig 9606
Holzteilchen 9508
Holzteppich 9501
Holzterpentin 9684
Holzterpentinöl 9684
Holztextur 9678
Holztoxizität 9680
Holzträger 9592
Holztränkung 4429
Holztransport 8595
Holztransport mit Gespannen 292
Holztransport mit Hubschrauber 4239
Holz treiben 3269
Holztrockenanlage 9537
Holztrocknung 2423
Holzüberschuss 8651, 9614
Holzüberschuss-Sortimenten 9614
Holzumschlagpunkt 8763
Holzunterlage 9557
Holzverabfolgung 6862
Holzverabfolgungsschein 6180
holzverarbeitendes Werk 5448
Holzverarbeitung 1915
Holzverarbeitungstechnik 9702
Holzverband 9303
Holzverbinden 4660
Holzverbinder 1840
Holzverbindung 4660
Holzverbindungsklemme 9510

Holzverbrennung 9498
Holzverdrängung 6871
Holzvereinigung 4660
Holzvereinzelner 7976
holzverfärbende Pilze 9663
Holz verflößen 3269
Holzvergütung 4434
Holzverkauf 8631
Holzverkauf auf dem Stock 7157
Holzverkauflizenz 8632
Holzverkaufquantität 8633
Holzverkaufsvertrag 9642
Holzverkauftabelle 8634
Holzverkleidung 9616
Holzverknappung 8656
Holzverladeplatz 5027
Holzverlader 8607
Holzverladung 8608
Holzverluste 9598
Holzvermesser 1457
Holzverschmutzung 9662
Holzversorgung 9588
Holzversorgungsblock 8648
Holzversorgungsmodell 8649
Holzversorgungszone 8647
Holzversteigerung 8557
Holzvertäfelung 9616
Holzverwendung 9154
holzverwertende Industrie 3466
Holzverzuckerung 7143
Holzvolumen 9689
Holz von Taxus baccata 9769
Holzvorbereitung 9628
Holzvorräte 9640
Holzvorrat pro Acker 8023
Holzvorratsinventur 3478
Holzware 9560
Holzware zweiter Qualität 7365
Holzware zweiter Wahl 7365
Holzwärmeleitfähigkeit 8489
Holzwassergehalt 5518
Holzwasserrinne 9281
Holzverbundskosten 8217
Holzwerbung 9588
Holzwerke 8626
Holzwerkstoff 8559
Holzwertstoffanlage 9637
Holzwespen 9693
Holzwespeschaden 9692
Holzwiderling 2016
Holzwirt 9659
Holzwirtschaft 3466
Holzwirtschaftabteilung 8612
Holzwirtschaftsbetrieb 9558
Holzwirtschaftsvorschrift 7617
Holzwolle 9697
Holzzange 3892
Holzzaun 6695
Holzzeichen 5101
Holzzelle 9503
Holzzellstoffkochung 1467
Holzzerfall 9535
Holzzerkleinerungsmaschine 9278
holzzerstörender Pilz 9485
Holzzerstörer 8562
Holzzerstörung 1014
Holzzettel 6183
Holzzeug 9603
Holzziegel 9495
Holzzoll 2681
Holzzucht 7610
Holzzuchtamt 8805
Holzzucht außerhalb des Waldes 5759
Holzzuchtplantage 8804
Holzzucker 9670
Holzzufuhr 2359
Holzzufuhranalyse 8646
Holzzurichtung 9564
Holzzuschussgebiet 8574
Homeostasie 4343
homogener Strahl 4344
homogenes Markstrahlgewebe 4345
Homogenholz 4346
homoplastisches Pfropfen 4347
Hondurasmahagoni 1412
Honduras-Rosenholz 4348
Honduras-Zeder 1411
Honigbaum 4357
Honigbeerenbaum 2898
honiggelber Ringling 4352
Honigpalme 1701
Honigtau-Tröpfeln 4351
Hooker-Condalie 870
Hooker-Weide 4360
Hooktender 9748
Hopfen 4363
Hopfenbaum 6035
Hopfenbuche 2900
Hopfendolde 4364
Hopfenpinie 2137
Hopfenzapfen 4364
Hoppus-Messung 4366
Horde 6373
horizontale Resistenz 4371
horizontaler Kronenschluss 4367
horizontales Punkt-Stichprobenverfahren 4370
horizontale Streifung 6968
horizontale Unterschiedlichkeit
Hornbusch 7187
Hornissenglasflügler 4375
Hornstrauch 222
Hornstrauchgewächse 2556
horntragende Akazie 1166
Horst 5001
Horstsaat 2485
horstweise Mischung 5494
horstweise Verjüngung 6812
Hortensiengewächse 4402
Hubrolle 2993
Hubschrauber-Bringung 4239
Hubschrauber-Löschmannschaft 4240
Hubseil 4323
Hubzylinder 4932
Huckepackverkehrsmethode 6227
Hudebestand 6101
Hudewald 3929
Hudsonfichte 1298
Hügelameise 9480
Hügellandwald 9131
Hügelpflanzung 5578
Hülse 2899, 4391
Hülsenmahagoni 6406
Humboldt-Unterschnitt 4392
Humifikation 4393
humifizierte Schicht 4394
Humifizierung 4393
Hummock-Tundra 4395
Humus 4396
Humusbildung 4393
Hundholzbaum 3284
Hundsbeere 2895
Hundsgiftgewächse 2553
Hundskirsche 2895
Hundsrose 2555
Hutebestand 6101
Hutwald 3929
hyaline Schicht 4398
Hybride 4399
Hybridisation 4401
Hybridisierung 4401
Hybrid-Platane 2906
hydraulische Baumschere 4409
hydraulische Entrindung 4403
hydraulische Hebezange 4407
hydraulischer Entrinder 4404
hydraulischer Holzspalter 4405
hydraulischer Spaten 4408
hydraulisch geschältes Holz 8156
Hydromeliorationsbodenfonds 8113
Hydrophyt 4410
Hydrosaat 4406
hydrostatische Rundholzmessung

Hylophyt 9351
Hymenopteren 4413
Hyphe 4414
Hypocotyl 4415
Hypsometer 4416

idealer Bestand 4420
Idioblast 4421
Ifenbaum 1814
Igelföhre 5309
Igelkiefer 5309
Ikakopflaume 1700
Ilexminiermotte 4337
illegaler Holzeinschlag 4423
Illipe 5244
Ilomba 108
Imbuia 1008
immerblühende Akazie 2927
Immergrüne 2931
immergrüne Bärentraube 1759
Immergrüneiche 1773
immergrüne Magnolie 7860
immergrüne Pflanzen 2931
immergrüner Buchsbaum 2886
immergrüner Regenwald 2930
immergrüner Schneeball 2886
immergrüner tropischer Regenwald 2930
immergrüner Wald 2929
immergrüne Sequoie 1269
immergrüne Zypresse 1770
impfen 9157
Impfreis 1145, 5815
Impfstoff 4507
Impfung 4506
Imprägniermittel 4428
imprägniertes Holz 4427
Index des relatives Reichtums 4461
Indigostrauch 4478
indikative Forstwirtschaft-Strategie 4472
indische Fichte 4310
indische Lagerströmie 2007
indischer Brotfruchtbaum 1011
indischer Feigenbaum 526
indischer Korallenbaum 7180
indischer Mandelbaum 4464
indischer Wildpfeffer 7632
indischer Zedrachbaum 6110
indisches Gummi 4466
indisches Karapa 6646
indisches Rosenholz 2728
indisches Sandelholz 5284
indisches Seidenwollbaum 6792
Industrieholz 9721
Infiltration 4488
Inga 4490
inländisches Holz 4341
Innenbläue 4557
Innenfäule 2645
Innenfläche des Schnittholzes 8602
Innenrinde 4502
innerer Knebel 4503
innerer Riss 958
innerer Saftfleck 4538
innewohnender Holzmangel 2333
Innigkeit 4572
Inokulation 4506
insektenbefallenes Holz 4510
Insektenbekämpfungsmittel 4512
Insektenfraß 2234
Insekten-Fraßschaden 2234
Insektenfresser 4513
Insektenmittel 4512
Insektenschutzmittel 4514
Insektenvertilgungsmittel 4512
Insektizid 4512
instabile Böschung 7441
Instrumentenholz 7851
Intarsia 9590
Integralwaldernte 4523
integrierte Forstnutzung 4524
integrierte Holzfällung 4525
integrierte Ressourcennutzung 4526
intensive Forstverwaltung 4528
intensive Forstwirtschaft 4529
intensiver Waldbau 4450
Interface zwischen Urland und Stadt 9410
Interknoten 4560
intermittierender Schnittholzertrag 4556
internationale Kubiktabelle 4559
interpretativer Waldstandort 4563
interzelluläre Höhle 4531
interzellulärer Raum 4533
interzelluläre Schicht 4532
Interzellulargang 4530
Interzellularraum 4530
Inventarisierung 4579
Inventarisierungstermin 2256
Inventur 4579
Inzucht 4435
Ipebaum 4581
Iroko 4582
irreguläre Bestockung 4588
irreguläres Femelschlagsystem 4587
Isolierstreifen 1149
italienische Kiefer 1250
italienischer Ahorn 4594
italienische Steinkiefer 4595
italienische Zypresse 1770
italiensche Waldrebe 4593
Ixora 8715

Jackfruchtbaum 1011
Jackholz 1011
Jackpot 4603
Jagd-Abschussziffer 6182
Jäger 1454
Jägersteig 6937
Jahreseinschlag 296
Jahreseinschlag-Aufteilung 1
Jahresquirl 5901
Jahresring 301
Jahresringenanalyse 269
Jahresringenaufbau 8188
Jahresringenbau 8188
Jahresringenbreite 9398
Jahresringenchronologie 6958
Jahresringendatierung 8912
Jahresringengrenze 4026
Jahresschlag 296
Jahrestrieb 305
Jahreswildsaatgutmischung 306
Jahreswuchs 300
jährliche Nutzungsmenge 296
jährlicher Einschlag aus den Waldflächen 297
jährlicher Gesamthieb 8721
jährlicher Laubwechsel 298
jährlicher Mitteldurchschnittszuwachs 5352
jährlicher Mittelniederschlag 5353
jährlicher Mittelschnittholzausbeute 5354
jährlicher Zuwachs 300
jährliches Forstprogramm 299
jährliche Zuwachszone 301
Jährling 9756
Jahrring 301
Jahrringbildung 304
Jahrringdichte 303
Jahrringfolge 302
Jahrringmessung 5363
Jahrringzählung 8911
Jakobsleiter 4600
Jamaica-Cocus 4607
Jamaica-Kokus 4607
japanische Apfelbirne 1504
japanische Aralie 4610

japanische Eberesche 4742
japanische Eiche 4624
japanische Erle 4609
japanische Flügelnuss 4637
japanische Glyzinie 4638
japanische Hainbuche 4621
japanische immergrüne Eiche 4625
japanische Kastanie 4616
japanische Kastanieneiche 7266
japanische Kerbbuche 4613
japanische Lärche 4623
japanische Mispel 5172
japanischer Ahorn 3752
japanischer Apfel 4619
japanischer Kiefernspinner 6248
japanischer Korkbaum 4617
japanischer Maulbeerbaum 6060
japanischer Nussbaum 7586
japanische Rosskastanie 4622
japanische Rotkiefer 4632
japanischer Pagodenbaum 4626
japanischer Perlschnurbaum 4626
japanischer Pfeffer 4630
japanischer Schneeball 4633
japanischer Spindelstrauch 4641
japanischer Sternanis 4611
japanischer Tiefrotahorn 4373
japanischer Zwergilex 4620
japanische Scheinquitte 4640
japanische Schwarzkiefer 4614
japanische Sicheltanne 4615
japanische Sophore 4626
japanische Ulme 4639
japanische Vollmispel 5172
japanische Walnuss 7586
japanische Weinbeere 9452
japanische Weißbirke 384
japanische Zeder 4615
japanische Zelkove 4639
japanische Zierquitte 4640
japanische Zypresse 4615
Japan-Stechpalme 4620
Jarrah 4642
Jasmin 1782
Jäten 9320
Jäthacke 4030
Jatoba 1990
Javaapfel 4646
Javaapfelbaum 4646
Jeffreyskiefer 4648
Jelutong 4649
Jequibita-Rosa 4651
Jersey-Kiefer 9229
Jerusalemdorn 4652
J-förmige Stange 4664
Johannisbrotgewächse 7439
Johanniskraut 7154
Johanniskrautgewächse 3805
Johannistrieb 4765
Johanniswuchs 4765
Jojoba 4662
Josuabaum 4604
Judasbaum 4665
Judasblattbaum 4692
Jugendholz 4675
jugendlich 4674
jugendliche Durchforstung 4680
jugendliche Phase 4679
jugendlicher Wuchs 4677
jugendliches Blatt 4678
jugendliches Blattwerk 4676
jugendliches Holz 4675
Juglandazeen 9260
Jujubendorn 1783
Julfest-Schnittholz 1567
Jungbaum 9791
Jungbestand 7204
junge Eiche 5841
junge Eichenrinde 9786
junge Kultur 9788
junger Baum 7202
Jungfernharz 7824
Jungfernrebe 9228
jungfräulicher Forst 282
jungfräulicher Wald 282
Jungholz 4675
Jungholzstadium 9790
Jüngling 9791
Jungwald 9785
Jungwuchs 9785
Jungwuchspflege 8458
Jurianuss 1003
juvenil 4674
Juvenilität 4681
Juwelenwein 9000
J-Wurzel 4663

Kabelbahn 1245
Kabel für Drahtseilriese 9464
Kabelkran 7697
Kabelriese 3922
Kabelschleifrückenschlepper 1246
Kabelwinde 7060
Kadam 4682
Kaddig 3983
Kaddigbaum 3983
Käfer 665
Käferfraß 4511
Käfergraben 4515
Kaffeebaum 1713
Kahlabtrieb 1607
kahle Felsenbirne 177
kahle Knospe 5651
kahler Baum 1064
kahle Weide 218
Kahlfläche 1621
kahlhauen 2168
Kahlhieb 1607
Kahlhiebverbot 6576
Kahlsaumschlag 8171
Kahlschlag 1607
kahlschlagen 2168
Kahlschlagflora 3039
Kahlschlag mit Nachverjüngung 1618
Kahlschlag mit Naturverjüngung 1619
Kahlschlag mit Reserve 1620
Kahlschlag mittels Schmalbande 1622
Kahlschlagsmethode 1617
Kahlschlagsmethode des Streifenweisen 219
Kahlschlagverbot 6576
Kahlschlagwald 1616
Kahlstelle 3443
Kaisereiche 2228
Kaiser-Paulownie 7122
Kajeputbaum 1056
Kakaobaum 1248
Kakibaum 4627
Kakifeige 4628
Kakipflaume 4628
Kakipflaumenbaum 4627
Kaktusfeige 6533
kalabrische Kiefer 1972
Kalamitätennutzung 3415
Kalamitätshieb 2186
Kalebassenbaum 1764
Kaliber 8809
kalibriertes Frischholz 6836
kalibriertes Rundholz 6837
kalibriertes Schnittholz 6835
kalifornische Balsampappel 784
kalifornische einblättrige Steinkiefer 1272
kalifornische Fächerpalme 1258
kalifornische Flusszeder 1262
kalifornische Grannen-Tanne 1037
kalifornische Hemlocktanne 5588
kalifornische Küstentanne 1278
kalifornische Nusseibe 1276
kalifornischer Beifuß 1270
kalifornischer Berglorbeer 1264
kalifornischer Flachkopfsbohrer

kalifornische Riesentanne 1278
kalifornischer Judasbaum 1267
kalifornischer Kreuzdorn 1257
kalifornischer Nussbaum 241
kalifornische Rosskastanie 1256
kalifornische Rottanne 1268
kalifornischer Storaxbaum 1273
kalifornischer Wacholder 1263
kalifornischer Walnussbaum 241
kalifornische Schwarzeiche 1255
kalifornische Steineiche 1693
kalifornische Strauch-Eiche 7559
kalifornische Wachsmyrte 1277
kalifornische Weißeiche 9159
kalifornische Zuckerkiefer 1274
kalifornische Zwergeiche 1271
kalifornische Zypresse 3891
Kalkesche 4975
kalkfeindliche Pflanze 1254
kalklierend 614
Kalkpflanze 1253
Kallitris 2217
Kallus 1182
Kalo 5174
kaltabfallender Wald 1719
Kältebeständigkeit 3724
kalzifuge Pflanze 1254
kalzimorpher Braunerdewaldboden 1252
Kalziphit 1251
Kamala 4684
Kamassi 4685
Kambala 4582
Kambium 1287
Kambiumzelle 1285
Kambiumzone 1286
Kameldornakazie 1754
Kameldornbaum 1288
Kamelie 1765
Kamm 2193
kammförmiggefiedertes Blatt 6286
Kammkorn 1738
kammkörniges Holz 6678
Kammsaat 648
Kamp 342
Kampeche 846
Kampescheholz 1289
Kampferbaum 1291
Kampferlorbeer 1291
Kampf gegen Waldschädlinge 3524
kanadische Balsampappel 9338
kanadische Berberitze 175
kanadische Eibe 9345
kanadische Felsenbirne 2603
kanadische Pappel 706
kanadischer Holunder 246
kanadischer Judasbaum 2721
kanadischer Kirschbaum 7186
kanadische Rotfichte 1298
kanadischer Schneeball 5656
kanadischer Schusserbaum 1482
kanadischer Seifenbaum 771
kanadischer Terpentin 1294
kanadische Schwarzpappel 2717
kanadische Weißfichte 9386
Kanarienholz 9008
kanarische Dattelpalme 1301
kanarische Eiche 5477
kanarischer Drachenbaum 2608
Kandelaber-Aloe 9705
Kandelaberbaum 3659
Kandis 4686
Känguru-Akazie 4687
Kängurubaum 407
Kängurudorn 4687
Känguruh-Akazie 4687
Känguruhbaum 407
Känguruhdorn 4687
kaninchendichtes Maschendrahtgatter 6670
Kannelierung 3296
Kanonenbaum 7519
Kanonenkugelbaum 1309
Kantbalken 638
Kante 7255
Kantel 9592
kanten 7962
Kantenabrundung 2765
Kantenast 4733
Kantendruck 6522
Kantenkappen 1215
Kantenriss 958
Kantenüberhang 5797
Kanter 1316
Kanterhaken 1317
Kanthaken 1317
Kantholz 638
Kantschnitt 5117
Kanuholz 1310
Kaperngewächse 1324
Kapernstrauch 1766
Kapillarwasser 1325
Kapok 4688
Kapokbaum 4688
kappen 7962
Kappmaschine 2043
Kappsäge 5056
Kappscheibe 1336
Kappung 2198
Kapschotendorn 203
Kapstachelbeere 1323
Kapur 4689
Kapuzenkäfer 6486
Karapa 4039
karibische Kiefer 2111
kariös 7212
Karkassenholz 1335
Karoba 1921
Karobenbaum 1341
Karolina-Hemlockstanne 1345
Karolina-Kirschlorbeer 1347
Karolina-Kokkelstrauch 1348
Karolina-Kreuzdorn 1344
Karolinalinde 1343
Karolina-Maiglökenbaum 1349
Karolina-Schneeglöckchenbaum 1349
karolinische Hainbuche 249
Karpophor 1353
Karrführung 1358
Karri 4691
Karroo-Akazie 203
Karrweg 1359
Kartenprojektion einer Krone 5298
Kartoffelrose 634
kartografische Projektion 5299
Karton 6059
Karyophyllazeen 6278
kaschieren 4760
Kaschubaum 1368
Kashmir-Seifenbeere 1506
kaspische Tamariske 3233
kaspische Weide 7138
Kassava 5282
Kassia-Akazie 1689
Kassie 1870
Kassie-Rinde 1370
Kastanie 2887
Kastanie-Eiche 615
kastanienblättrige Eiche 1478
Kastanienflügelstar 3350
Kastaniengehölz 4005
Kastanien-Niederwald 7945
Kastanienwald 1477
Kastenstapel 8546
Kasuarine 663
Kasuarinengewächse 664
Katalpa 5789
Katappabaum 4464
Katappenbaum 4464
Katastrophenverjüngung 6817
Katechupalme 691
Kathedralenhieb 1379
Katsurabaum 4692
Katsuragewächse 4693

Katzenpfote 1382
Kaugummibaum 259
kaukasische Eiche 1387
kaukasische Flügelnuss 1388
kaukasischer Pfeifenstrauch 1386
kaukasischer Zürgelbaum 1385
kaukasische Zelkove 2792
Kaukasusfichte 2723
Kaukasustanne 1384
kaum 4873
kaumbrennbares Holz 4125
Kauri 5727
Kaurifichte 6664
Kautschukbaum 1320, 4056
Keakibaum 4639
kegelförmiger Wuchs 1872
Kehrhaken 6693
Keilschirmschlag 7517
Keilschlag 9315
Keilschlagsystem 9317
Keilspaten 6365
Keilverfahren 9316
Keilzinkengabelung 5552
Keilzinkenverbindung 3161
Keim 2795
Keimblatt 1983
Keimen 3835
Keimenergie 3834
Keimfähigkeit 3833
Keimfähigkeitsuntersuchung 3837
Keimhülle 4398
Keimknospe 9183
Keimkraft 3833
Keimlappen 1983
Keimling 2795
Keimlingsprozent 6368
Keimlingssack 5811
Keimplasma 3838
Keimprobe 3837
Keimprozent 3836
Keimung 3835
Keimungsprozent 3836
Keimvermögen 3833
Keimversuch 3837
Keimwurzel 7035
keine Bestellung 5799
keine jagdbare Wildtiere 5761
kein industrieller wirtschaftseigene Waldboden 5762
Kekatong 5653
Keledang 1011
Kellerhals 2246
Kellogg-Eiche 1255
Kempas 4695
Kennfärben 3074
Kennzeichen für Floßeinheitsdichte 1710
Kennzeichen für Vollholzigkeit 1711
Kennzeichnung 8364
Kentucky-Kaffebaum 1482
kephalonische Tanne 3938
Kerbe 4082
Kerben 2514
Kerbspannung 2770
Kerbwirkung 2770
Kermesbeerengewächse 6416
Kermeseiche 4697
Kern 1951
Kernanteil 4200
Kernaufschlag 7387
Kernbaum 7392
Kernbrett 4194
Kerndurchmesser 4196
Kernexzentrizität 6952
Kernfärbung 7994
Kernfäule 1414
Kernfeuchtketsgewicht 1950
kernfreies Holz 4199
Kernfrucht 4698
Kernfurnier 1954
Kernhausbräune 1080
Kernholz 4202
Kernholzbestand 8022
Kernholz/Splintholz-Übergang 4204
kerniges Holz 4202
Kernkluft 4195
Kernobstgehölze 330
Kernriss 4195
kernrissig 4201
Kernschäle 2139
kernschäliges Holz 9696
Kernstrahlung 7601
Kernwald 7388
Kernwuchs 6887
Kern-zu-Kambium-Pfropfen 6307
Keruwing 4699
Kerze 1302
Kerzebaum 1306
Kerze Kiefern-Trieb 1302
kerzenförmig 1305
Kerzenkaktus 1733
Kerzenölbaum 1304
Kesselhieb 6106
Ketapang 4464
Ketten 1428
Kettenbauholz 1437
Kettenbremse 1423
Kettenentrinder 1425
Kettenflegelentastungsmaschine
Kettenhaken 8577
Kettensäge 1431
Kettenschlepper 7799
Kettenschlinge 1427
Kettenskarifikation 2610
Kettenspanner 1436
Kettensporn 8577
Kettenteilung 7263
Kettentraktor 7799
Kettenzugförderer 1424
Keulenbaum 663
Keuschbaum 4961
Khasyakiefer 5787
Kiefer 6272
kieferblättrige Akazie 7135
Kiefergallenwickler 6269
Kieferkäfer 6252
Kiefernbalsam 6898
Kiefernbaumschwamm 6265
Kiefernbestand 6255
Kiefernblattminiermotte 6261
Kiefernblatt-Schüttenkrankheit 6260
Kiefernblattwespe 6268
Kiefernborkenkäfer 7862
Kiefernbuschhornblattwespe 8301
Kieferndrehrost 6274
Kieferngewächse 6254
Kieferngluckе 6258
Kiefernharz 6898
Kiefernharzöl 6276
Kiefernholz 6275
Kiefernkäfer 6249
Kiefernkonversion 6251
Kiefernkultur-Gespinstblattwespe 9313
Kiefernmarkkäfer 6246
Kiefernpech 6263
Kiefernplantation 6264
Kiefernsämlingswickler 6273
Kiefernschwärmer 6257
Kiefernspanner 6262
Kiefernspinner 6258
Kiefernsprosse 6247
Kiefernstockholz 6271
Kiefernstreu 6270
Kiefernteer 6898
Kiefernwald 6255
Kiefernwaldreihen 8176
Kiefernwurzelhalsrüssler 6267
Kiefernwurzelholz 6271
Kiefernzapfen 6250
Kieferspanner 3174
Kien 6272
Kienbaum 1797
Kiene 1797

Kienöl 6276
Kienpech 6263
Kienstumpf 6271
Kienteer 6898
Kienzopf 8698
Kiesstraße 3921
Kiesweg 3921
Kimme 4082
King-Boris-Tanne 1154
Kinobaum 3248
Kinoharz 4058
Kippmast 3301
Kirchenforst 1575
Kirchenwald 1575
Kirilack 6809
Kirschbaum 1476
Kirschblattwespe 6388
Kirsche 1473, 1476
Kirschenapfel 6127
Kirschlorbeer 2828
Kirschpflaume 1475
Kirundu 6414
Kistenholz 954
Kitschbaum 2881
Kiwi 51
Klafter 3024
Klafterholz 7982
Klammer 8462
Klammerhaken 8577
Klappernuss 816
Klapp-Pflanzung 289
klare Dielen 1033
klar ersichtlich 6109
klarer Splint 1034
klares Nutzholz 1035
Klärschlamm 7473
klar-und-besser 1612
Klasse der Bestandesgroße 8028
Klasse der Brandgefahr 1601
Klasse für Vollholzigkeit 1602
Klassen der Bodennutzung 4781
Klassen der Waldschutzfunktionen 1599
Klassenprobenstamm 7176
klassifizierter Baum 7892
Klauenseuche 5046
Klausbau 5039
Klausdamm 5039
Klause 5039
Klauswassertrift 2631
Klebast 2848
Klebbaum 7201
Klebgürtel 8777
klebrige Erle 2882, 4592
klebrige Gelbfäule 8167
klebrige Robinie 1595
Klebung 7918
Kleeblatt-Lederstrauch 6035
Kleestrauch 6035
Kleiderbaum 254
Kleinäst 7764
Kleinbaum 7778
Kleinbestand 7776
kleinblätterige Kordie 7765
kleinblätteriger Feigenbaum 7766
kleinblätteriger Wald 3514
kleinblättrige Linde 7767
kleinblättrige Parkinsonie 3312
kleinblättriger Buchsbaum 5002
kleinblättriger Pfeifenstrauch 5004
kleinblättriger Stringybark 7769
kleinblättriger Sumach 5005
kleinblättriger Wacholder 458
kleinblättrige Schneebeere 7768
kleinblättrige Steinlinde 7767
kleinblättriges Weißfaden 5003
kleinblättrige Ulme 1496
kleinblütenblatterig 5431
kleinblütiger Trompetenbaum 1492
kleinblütige Tamariske 7939
Kleinbusch 7558
kleinbuschig 8252
kleine Fichtengallenlaus 9718
kleine Kastanie 1199
kleiner Kiefernjungholzrüssler 7770
kleiner Kiefernzangenbock 6934
kleiner Lagerplatz 9135
kleiner mexikanischer Kiefernkäfer 7762
kleiner Ulmensplintkäfer 2894
kleiner Wald 5001
kleiner Waldgärtner 4920
kleiner Zweig 991
kleines Spielhaus 8822
kleine Stielchen 8089
Kleingehölz 5001
Kleinholz 9508
kleinkörniges Holz 1656
Kleinstandort 5429
kleinstückige Holzabfälle 4950
Kleinwald 5001
kleinwüchsige Pflänzling 7777
kleinzapfige Lärche 250
Kleinzikade 4871
Klemmpflanzung 289
Klemmschemel 1596
Klemmschemelrücker 1597
Klemmspaten 6365
Klethragewächse 1637
Kletterpflanze 1642
Kletterseil 1643
Klimax 1638
Klimaxforst 1639
Klimax-Holzart 1640
Klimaxwald 1639
Klimme 8803
Klinge 819
Klinki-Araukarie 4718
Kloben 2171
Klobenholz 7920
Klone 1649
Klone eines Bäumes 1648
klonischer Test 1647
Klonsbaum 1648
Klopfkäferfraß 2279
Klöppelholz 7774
Klosterforst 5528
Klosterwald 5528
Klotz 2171
Klotzachse 5048
Klotzachse-Abweichung 2450
Klotzbäume 7540
Klotzbläue 5050
Klotzbremse 2054
Klötze ziehen mit Trommelwinde 2641
Klotzholz 4738
Klotzsukkulenten 941
Klotzwagen 5199
Klumpen 1571
Klumpenbucker 1572
Kluppe 8809
Kluppen 1282
Kluppierung 1282
Kluppierungsbuch 5365
Kluppschiene 1283
knabbern 5730
Knackweide 1040
Knast 1182
Knick 2031
Knickfestigkeit 676
Knickschlepper 373
Knickwachstum 370
Knickwuchs 2025
Knie 2100
Knieholz 2029
Kniewurzel 4719
Knittern 2020
Knochenholz 2895
Knöllchenbakterien 7037
Knöllchenbildung 5745
Knolle 9003
Knollen 1571
Knopfblume 1763
Knopfbusch 1763

Knopperneiche 3937
Knorren 1182
Knorrenaushieb 3070
Knorrholz 4738
knorrige Kiefer 4739
knorriger Ast 2026
knorriger Stemm 4721
knorriges Holz 4738
knorriges Schnittholz 2027
Knospchen 5474
Knospe 1141
knospen 7944
Knospenanlage 6502
Knospenästung 1142
Knospenbildung 3835
Knospennarbe 9017
Knospenproliferation 6580
Knospenprolifikation 6580
Knospenschluppe 1143
Knospenschluppe-Wunde 1144
Knospensucht 6580
Knospen treiben 7944
Knospenwickler 1146
Knospenwunde 9017
Knospung 6580
Knoten 4723, 5741
knotenförmiger Endwand 5743
knotenförmige Wurzel 5744
Knotenholz 5170
knotig 4737
knotiges Holz 4738
Knotigkeit 986
Knüpfer 4966
Knüppeldamm 1947
Knüppelholz 8879
Knüppelweg 1947
Knüttel 903
Ko-Baum 1980
Kochung 9633
Köder 4509
Ködergift 4509
Ködern 482
Köderpflanze 483
köderter stehende Baum 481
Koeffizient der Engpasskompaktheit 1709
Koeffizient für Floßsektiondichte 1710
Kohle 1673
Kohlen 1331
Kohlenbrenner 1445
Kohlensäure-Assimilation 391
Kohlenstoffbilanz 1329
Kohlenstoffquelle 1334
Kohlenstoff-Sequestrierung 7452
Kohlenstoff-Stickstoffverhältnis
Köhler 1445
Köhlereiholz 9571
Kohl-Eukalyptus 1237
Kohlholz 9571
Kohlpalme 1238
Kohlung 1331
Kohorte 1715
Kojehaus 1179
Kokko 2730
Kokosbastfaser 1717
Kokosholz 1703
Kokospalme 1702
Kokrodua 119
Kolabaum 1718
kolchische Pimpernuss 816
kollaterale Knospe 1723
Kollerbusch 7323
Koloradokiefer 9031
Koloradotanne 1278
kolumbianischer Bauholzkäfer 1732
kombinierte Kubiktabelle 1739
kombiniertes Fällen-Seilen 4387
kombiniertes Gebälk 1836
Kommender-Baum 1743
kommerzielle Forstwirtschaft 4482
Kommunalforstverwaltung 1817
komplementärer Baum 1980
Komplettbaum-Zerkleinerung 3759
kompletter Baum 3758
komplette Stammexzentrizität 3762
Kompost 4881
Kompressionentrindung 1846
Kompressionholz 1850
Kompressionsfehler 1847
Konara-Eiche 3859
konfluentes Parenchym 1869
Konifere 1877
Koniferen 5708
Koniferenstumpf 8208
Königseukalyptus 3842
Königsholz 4716
Königsnuss 7507
konischer Baum 8406
konischer Tangentialschnitt 8407
konischer Wuchs 1872
konisches Besäumen 1871
konisch geschnitten 8408
Konizität 8405
Konkurrent 1827
Konkurrenz 8817
Konkurrenzkontrolle 1825
Konsole 1879
Kontinuität des Brennmaterials 3735
Kontraktabfuhrunternehmer 1907
kontrolliertes Abbrennen 6516
Kontur-Graben 1904
Kontur-Hiebssystem 1905
Konturpflanzung 1906
Konturschnitt 1903
konventionelle Bodenseilung 1911
konventionelle Bringung 1912
konventionelle Forstprodukte 1910
konventionelle Holzdichte 1913
Konvergenz-Pfropfung 331
Konversionperiode 1916
Konversion von Holz 1915
Kopaia 1921
Kopaibabaum 1922
Kopaivabalsam 104
Kopaivabaum 1922
Kopal 1923
Kopf 8699
Kopfabschnitt 8688
Kopfausschlag 6431
Kopfdüngung 8685
Kopfeibengewächse 6392
köpfen 8702
Köpfen 4294
Kopfende 8688
Kopfholz 6431
Kopfholzbetrieb 6430
Kopfholzschoß 6431
Kopfholzstamm 7489
Kopfholzwirtschaft 6430
Kopfholzzucht 6430
Kopfmaschine 8707
Köpfmesser 8709
Kopfschnittholz 8688
Kopfstück 8688
Kopfveredelung 2081
Kopoubohne 4746
Kopulieren 7919
Kopulieren mit Gegenzunge 8678
Korallenbaum 1941
korallenbaumartiger Fischtänger 4606
Korallenbeere 1940
Korallenholunder 2907
Korallenhülse 1810
Korallen-Ölweide 417
Korallenrindenahorn 4373
Korallenstrauch 1698
Korbblütengewächse 1834
Korbeiche 615
Korbpflanzung 616

Korbweidenkultur 620
Kordel 2979, 9519
Korea-Fichte 4745
koreanische Asche 4741
koreanischer Ahorn 6637
Kork 1956
Korkbaum 6199
Korkeiche 1961
Korkenziehereukalyptus 357
Korkenzieherweide 1963
Korkgewebe 1975
Korkholz 507
Korkkambium 6201
Korkmantel 1960
Korkmehl 1962
Korkmutterschicht 6201
Korkrinde 1956
Korkrindenschicht 6200
Korkschicht 1960
Korkspindel 9454
Korkulme 1958
Korkwald 1959
Korkzelle 1957
Kornazeen 2556
Kornelkirschbaum 1965
Kornelkirsche 1965
Kornelkirschgewächse 2556
Kornwuchs 7388
Korridor 1969
Korridor-Durchforstung 1971
Korridorpflanzung 183
Korridorrückung 1970
korsische Kiefer 1972
kortikales Gewebe 1975
Kosipo 4744
Kosten auf dem Stamm 8196
Kosten pro Kord 1977
Kotibe 2245
Kotyledon 1983
kräftiges Wachstum 141
kräftiges Wurzelnetz 142
kräftiges Wurzelsystem 142
kräftiges Zurückschneiden 2347
Krähenbeerengewächse 1997
Kramantee 2006
Krammet 3983
Kranbaum 3848
Kranbeere 2889
Kränbeerewaldtyp 9158
Krankheitserreger 6112
Kransäule 915
Kränsbeere 1996
Kranwagen 7698
Kranzast 995
Krappa 4039
Krappgewächse 5238
kraulblätteriges Bergmahagoni 2143
Kräuselkrankheit der Blätter 4863
Kräuselmyrte 2007
Kräuselung 2019
krausgewachsenes Holz 2048
Kraut 4252
krautige Weide 6228
Krautschicht 3116
Krebsblume 2061
Krebsweide 6643
Kreideahorn 1438
Kreisbuschhacker 7071
kreiseigener Forst 1987
Kreisfläche in Brusthöhe 598
Kreisflächenabnutzung 602
Kreisflächenmittelstamm 5355
Kreisflächentafel 8362
Kreisflächenzuwachs 600
kreisförmige Segmentsäge 6950
Kreisforst 1987
Kreiskoeffizient 3858
Kreisringsäge 2209
Kreisrückeverfahren 1588
Kreissäge 1585
Kreissägeblatt mit eingeschobenen Zähnen 4516
Kreissäge für Latten 7733
Kreissägenachse 7230
Kreisschere 7070
Kreisscherekopf 403
Kreisscherenmesser 7070
krempelgesägt 6881
krempelsägen 6880
Kreppmyrte 4643
Kreuzaxt 2035
kreuzbandumsetzen 2036
Kreuzbandumsetzung 2037
Kreuzbaum 2827
Kreuz-Bäume 2059
Kreuzbeige 6232
Kreuzbeihilfe 2058
Kreuzblumengewächse 5446
Kreuzblütengewächse 2099
Kreuzdorn 1137
Kreuzdorngewächse 1138
kreuzende Axt 2101
Kreuzfloß 6694
Kreuzhacke 2035
Kreuzhaue 2035
Kreuzholz 6658
Kreuzschlagseil 7059
Kreuzschnitt 6659
Kreuzstapel 6232
Kreuz- und Querpflügen 8962
Kreuzung der Leitung 2053
Kreuzungsfeld-Tüpfelung 2050
Kreuzungszüchtung 4401
Kriechbirke 892
Krieche 2244
kriechen 2011
kriechend 2313
kriechende Baumform 6589
kriechende Mahonie 2013
kriechende Pflanze 1642
kriechendes Feuer 2012
kriechendes Habit 6588
kriechende Wuchsform 6588
kriechende Wurzel 2014
Kriechenpflaume 2244
Kriechklee 9366
Kriechweide 2015
Krieger 5652
Krippelholz 2029
kristallführende Zelle 2108
kritisches Habitat 2023
kritische Umwelt 2023
kritische Winterförsterei 2024
Kronblatt 6192
Krone 2087
mit abgestorbener ~ 2272
Kronefäule 2092
Kronenabschluss 2096
Kronenabschürfung 2080
Kronenanfang 7940
Kronenansatz 2064
Kronenanteil 2090
Kronen-Apfel 8331
Kronenausmaß 2093
Kronenbereich 2088
Kronenbildung 2079
Kronenblatt 6192
Kronenbreite 2098
Kronenbruch 1012
kronenbürtiges Holz 8712
Kronendach 2066
Kronendeckel 2094
Kronendichte 2072
Kronendurchforstung 2096
Kronendurchlass 8538
Kronendurchlasssammler 8538
Kronendurchmesser 2074
Kronendurchmesserverhältnis 2075
Kronenentwicklung 2073
Kronenfeuer 2076
Kronenflächeindex 2063
Kronenform 2078
Kronenfreihieb 3698
Kronengröße 2093
Kronenkarte 2086
Kronenklasse 2068

Kronenmittelpunkt 3668
Kronenmittelpunkthöhe 3669
Kronenoberfläche 1314
Kronenpflege 2095
Kronenprojektion 2088
Kronenprojektiongrundfläche 2088
Kronenprozent 2090
Kronenrand 2065
Kronenrotfärbung 8710
Kronensatz 2064
Kronenschicht 2084
Kronenschichte 2066
Kronenschluss 2069
Kronenschnitt 2083
Kronenschutzsysteme 2089
Kronenspäne 2067
Kronensprengung 1012
Kronenvolumen 2097
Kronenwettbewerb 2070
Kronenwurzelverhältnis 2091
Kronenzustand 2071
Kronforst 2077
krümelige Braunfäule 2103
krummästig 4737
Krummaxt 4390
Krummbaum 2030
krummbelaubt 2155
Krümmen des Holzes 9268
krummgewachsenes Holz 2029
Krummholz 2029
Krummholz- und Knieholzstufe 2783
Krummschaftigkeit 2028
Krümmung 2031
Krümmungshöhe 4234
krummzähniger Tannenborkenkäfer 7602
krummzähniger Weißtannenborkenkäfer 7602
Krüppelbaum 2030
Krüppelbestand 3515
Krüppelholz 2022
Krüppelwald 3515
Krüppelwuchs 2025
Kruste 1973
Kruwing 4699
kubanische Bauchpalme 2109
Kubierung 8955
Kubierungskluppe 715
Kubierungstabelle 2117
Kubierungstafel 2117
Kubikfuß 2113
Kubikkluppe 715
Kubikmeter 2114
Kubiktabelle 2117
Koniferen 1076
kubische Braunfäule des Holzes 1077
kubische Dunkelbraunfäule 2250
kubischer Tragebalken 953
kubische Weißfäule 9368
kubische Weißherzfäule 9367
Kuchenbaum 4692
Kudzubohne 4746
Kufen 7694
Kugelblumengewächse 3868
Kügelchen 635
kugelförmige Fichte 500
kugelförmige Krone 3869
kugelförmiges Blatt 5938
kugelförmige Spaltsägemaschine 6798
Kugel-Kirsche 3984
Kugel-Pflanzung 1163
Kugelzypresse 398
Kukuinussbaum 1304
Kulisse 3335
Kulissenhieb 8171
Kulminationsalter 2125
Kultivierung der Baumarten 2559
Kultur 9788
Kulturarbeiten 2131
Kulturbestand 6339
kulturelle Mannigfaltigkeit 2130
Kulturkosten 6350
Kulturpflege 8453
Kulturwald 5285
Kulturzeit 8665
Kümmerer 2695
Kümmerling 2695
Kümmerwuchs 6449
Kumquat 4747
Kumulationswirkungen 2134
Kunite 2136
Kunstholz 378
künstliche Bestandbildung 375
künstliche Bestandesbegründung 375
künstliche Bestandesgründung 375
künstliche Lichtstellung 377
künstliche Lockerung 377
künstlicher Wald 5285
künstliches Aufasten 376
künstliche Waldverjüngung 6811
künstliche Zuchtwahl 6998
künstlich gefärbtes Holz 374
künstlich getrocknete Pflanzensäfte 4709
künstlich getrocknetes Holz 2421
Kurortwälder 3598
kurzblättrige Eibe 9345
kurze Stielchen 8089
Kurzfaser 7536
kurzfaseriges Holz 7535
Kurzfaserigkeit 7541
Kurzholz 903
Kurzholzeserntemaschine 7551
Kurzholzfällung 3078
Kurzholzhieb-und-Stapelungssystem 2162
Kurzholzmannschaft 7550
Kurzrotation 7542
kurzschäftig 7533
Kurzschleifweg 7549
kurzstückiges Holz 903
Kurztrieb 7548
Kurzumtrieb 7542
Küstenbetriebsfläche 4758
Küstenerle 7329
Küsteneskarpewald 1694
Küstenkiefer 5044
Küstensequoie 1269
Küstentanne 3910
Küstenwald 1687

Labrador-Eberesche 7552
Lacebark-Flammenbaum 3245
Lachte im zweiten Jahr 9756
Lachtenbaum 959
Lachtenbeil 2142
Lachtenlänge 2982
Lachtenreißer 2142
Lachtenröten 1530
Lachtentrockenheit 2649
Lackblattpflanze 412
Lacklaus 4751
Lade 1095
Ladebalken 6235
Ladebaum 5025
Ladebrücke 5027
Ladebühne 5027
Ladegabel 8610
Ladegerät 5062
Ladeklammer 5026
Ladekran 8609
Lade-Kran 5501
Laderampe 4382
Lader eines Holzlastkraftwagens 5088
Ladesteig 5027
Ladevorrichtung 5062
Ladung 5022
Lage 6397
Lagenholz 4763
Lager an Feuerschutzstreifen 7573

Lagerholzzeichnung 2047
Lagerplatzvorräte 9753
Lagerschaden 8127
Lagerströmie 4643
Lagerung im Wasser 9301
Lagerungsschaden 8127
Lagerverfärbung 9755
Lahe 8155
Lahte 8155
Lakooch 1011
Lambertshaselstrauch 4759
Lambertsnuss 4759
lamellare Wurzel 6378
Lamelle 3241
lamellieren 4760
laminieren 4760
laminiertes Holz 4763
laminierte Spanplatte 6375
Lancewood 9347
Lancewood-Baum 4768
Land 4769
Landablage 5059
Land außerhalb der Wald 5868
Landbase 4770
Landgewindestrehler 4773
Landrücken 3988
Landschaft 4777
Landschaftspflege 4778
Landtrockner 1986
land- und forstwirtschaftliches System 145
Land- und Holzartenwechsel 225
Landungsziseleur 4773
Langbiankiefer 5787
langblätterig 5142
langblätterige Kiefer 1543
langblätteriger Pfefferbaum 5140
langblätteriger Veilchenbaum 5139
langblättrige Akazie 8356
langblättrige Weide 7183
Längeabstufung 3894
Längegradation 3894
Längeholzmarkierung 5318
Längenmessung 4914
Längensäge 6966
Längensortierung 4913
Längenspaltung 4915
langer Stammabschnitt 5128
Länge über dem Boden 8694
Langfadengewächse 5646
langfaseriges Holz 5131
langfristige Aktion 5152
langfristige dauerhafte Ernte 5147
langfristige Nachhaltsertragsholzleistung
Langhaufen 9496
Langholz 5154
Langholzabfuhr 5615
Langholzabhieb 3046
Langholzanhänger 9395
Langholzausleger 4225
Langholzbesatzung 5155
Langholzbringung 5144
Langholzbringungsmethode 8870
Langholzernte 5157
Langholz-Erntemaschine 5156
Langholzfeder 2595
Langholz-Forwarder 8867
Langholzgreifer 8868
Langholzhackmaschine 9391
Langholzladung 8871
Langholzmannschaft 5155
Langholznachläufer 9395
Langholzrückensystem 8873
Langholzschlepperoperation 8872
Langholzschlitten 2587
Langholzstamm 2351
Langholzstapel 5145
Langholztransport 5615
Langholztransportwagen 5087
Langholzwagen 5087
Langholzwaggon 9256
längliches Blatt 5858
Langlochfäule 5146
Langrohholzbestand 4280
Langrücken 2805
langsam wachsende Arten 7756
langsam wachsender Baum 7755
langsam wachsender Wald 7757
Längsbalken 8163
langschäftig 5151
langschäftiges Holz 5154
Längsfaser 5133
Längshobelmaschine 6326
Längsrillen 7062
Längsriss 7729
Längssäge 883, 6966
Längssägemaschine 6966
Längssägen 5135
Längssaling 4380
Längsschnitt 5135
Längsspaltung 4915
langstämmig 5151
Längsträger 8163
Längstrieb 8474
langstückiges Holz 5154
Längswachstum 326
Längswand 5137
längs zur Faser 209
Langtrieb 5149
Langzeit-Durchschnittsertrag 436
lanzettförmiges Blatt 4767
Lapachin 4784
Lappland-Alpenrose 4786
lappländische Kiefer 4785
Lärche 1785, 4787
Lärchekäfer 2720
Lärchenbaum 4787
Lärchenkiefer 5309
Lärchenminiermotte 4788
Lärchenmotte 4791
Lärchenspinner 4790
Lärchenwald 4789
Larve 4809
Larvenfraßgang 4811
Larvengang 4811
Lassbaum 6887
Lassreidel 8555
Lassreis 8555
Lassreitel 8555
Lasswald 9072
Last 5022
Lastabstand 5024
Lasthaken 6225
Lasthakenbefestigung 1232
Lastkraftwagenfahrer 8983
Lastkraftwagenrückung 8985
Lastseil 3987
Lasttiertransport 292
Lastwagenfahrer 8983
Latenzstadium 4818
Latexröhre 4829
Latschenkiefer 5598
Latte 7732
Laub 3303
laubabwerfender Wald 2300
laubartig 3302
Laubausbruch 9084
Laubaustrieb 9084
Laubbaum 2302
Laubbuschgemeinschaften 2298
Laubdach 4860
Laubdecke 4999
Laubedelholz 4295
Laubentfaltung 9084
Lauberde 4881
Laubfall 4866
Laubfresser 2338
Laubfrosch 9579
Laubfutter 4868
Laubgehölz 1055
Laubgemeinschaften 2297
Laubheu 4868
Laubholz 4124
Laubholzart 1055
Laubholzbaumreihe 4991
Laubholzbodenfläche 1057

Laubholzreisig 4145
Laubholzschutzstreifen 2928
Laubholz-Wasserrinnen 4137
laubig 3306, 4894
Laubkrone 4862
Laubnutzholz 1054
Laubpfefferminze 1052
Laubrundholz 1053
Laubsäge 1822
Laubschnittholz 1047
Laubstammholz 1047
Laubstreu 4999
Laubverlust 4866
Laubverwehung 2538
Laubwald 2300
Laubwald-Biom 2301
Laubwald in der borealen Zone 1719
laubwechselnd 4877
Laubwerk 3303
Laubzierbaum 1051
Laufdrahtseilbahn 7133
laufender jährliche Zuwachs 2150
Lauffeuer 2012
Laufkatze 7698
Laufrichtung
quer zur ~ 49
Laufschneider 4864
Laufseil 3987
Laufwagen 7698
Laufweg des Fällers 2864
Lavalle-Korkbaum 4837
Lavendel-Gestrüpp 4838
Lavendelheide 897
Lavendelweide 2775
Lawinezug 425
Lawsons-Lebensbaumzypresse 4839
Lawsons-Scheinzypresse 4839
Leban 3108
lebende Bodendecke 3116
lebende Hecke 5019
lebender Ast 5013
lebender Baum 5018
lebender Feuerschutzstreifen 6174
lebender Zaun 5019
lebendes Holz 3605
Lebendstreifen 4568
Lebendverbauung 6923
Lebensbaum 8893
Lebensbaumlichtwald 1398
Lebensdauer der Bäume 4931
Lebensfähigkeit 9215
Lebensgemeinschaft 738
Lebenskraftsklasse 9217
Lebensraum 739
Lebensraumwirksamkeit 4075
Leberdistel 6531
Leberwurstbaum 7226
Lebkuchenbaum 4692
Lecythis 6701
Lederholz 4900
Lederhülsenbaum 263
Lederkork 6157
Leerseil 7714
Leerseilblock 4167
Leerseillinie 1243
Leerseilrücken 7715
leerstehender Baum 5878
leerstehendes Feld 2
Legföhre 5598
Lehne 935
Lehne nach vorne 4188
Lehrforst 8761
Lehrforstamt 8762
Lehrforstrevier 8762
Lehrrevier 8761
Leichenbaum 2271
leichte Durchforstung 4952
leichte Hacke 4081
leichte Haue 4081
leichtes Hartholz 4943
leichtes Holz 8660
leichtes Karkassenholz 4942
leichtes ölige Holzschutzmittel 4947
Leichtholz 8660
Leichtspanholzplatte 5184
leicht zerfließend 2357
leicht zerfließende Wuchsform 2358
Leiereiche 5991
Leim-Mistel 2902
Leimringe 8777
Leinbaum 935
Leingewächse 3261
Leistungs-basierte Holzrückung 6156
Leistungstafel 9784
Leitbündel 8058
leitender Staatsbeamt 6915
Leiterbaum 4756
leiterförmige Perforationsplatte 7272
leiterförmiges Parenchym 7271
leiterförmige Tüpfelung 7273
Leitgewebe 1857
Leitrolle 4849
Leitschwimmer 3147
Leitung 4847
Leitungsgewebe 1857
Leitungswurzel 1856
Lentizelle 4916
Leopardenbaum 1002
Leopardenstrauch 1002
Lespedezie 4919
letzte Fäulnis 1828
letzter Hieb 4814
Leuchterbaum 251
Leucothoe 8327
L-F-H-Horizont 4924
Liane 4925
Lianen-Regenwald 9222
Libanoneiche 4903
Libanonzeder 1400
Libriformfaser 4926
Licania 4927
Lichtbaum 251, 7476
Lichtbaumart 4574
Lichtbedarf 4573
Lichtbedürfnis 4573
lichtbedürftiger Baum 7476
lichtbelaubt 7881
licht bestockt 6450
Lichterbaum 251
lichter Bestand 4446
lichter Wald 1060
Lichthieb 5912
lichtkronig 7881
Lichtmangel 4940
Lichtnussbaum 1303
Lichtschlag 5912
Lichtstand 4446
Lichtstellung 4566
Lichtung 3443, 5912
Lichtungshieb 5912
Lichtungszuwachs 4944
Lichtwald 4446
Lichtwald mit steppenartriger Feldschicht 4941
Lichtwendigkeit 6212
Lichtwuchsbetrieb 5918
Lieferungsforstwirt 3519
liegende Krone 2314
liegende Markstrahlzelle 6559
liegendes Trockenholz 2266
liegende Totäste 2265
liegende Wuchsform 6558
Lignin 4957
Lignose 4959
Lignotuber 4960
Lignozellulose 9504
Liguster 1799
Liliengewächse 4962
Limbabaum 92
Limettenbaum 9348
Limettenzitrone 9348
Limonenbaum 4911

lindenblättriger Ahorn 4984
Lindenblattwespe 9435
Lindenblüte 4977
Linden-Eichenwald 4974
Lindengewächse 4983
Lindenrinde 4972
Lindenwald 4978
lineare Regenerationsprobennahme 4987
linearer Verschulungsbaumgarten 4994
linearer Wachstum 4985
lineares Unterholz 1930
lineare Verschulung 4993
lineare Waldbaumpflanzung 4989
Linienbestandesaufnahme 4988
Linienplenterumschlag 8168
Linienplenterung 8168
Linienstichprobeninventur 4986
Linientaxation 4988
liniierter Nutzholzborkenkäfer 8173
linksdrehender Drehwuchs 4907
Lippenblütlengewächse 4748
Lippenstiftbaum 4995
Liquidambar 259
Liquidverschiebung 4996
Liriodendron 9008
Listenführer 8377
Listenstichprobenahme 4998
Litsea 902
L-Klemmpflanzung 5021
LKW-Rückung 8985
Loblolly-Glattkieferwald 5031
Loblollykiefer 5029
Loblollykiefernblattwespe 5030
Loch 5561
Löcherhieb 6106
Löcherpflanzung 6355
Löcherschlag 6106
Lochfäule 6313
Lochfraß 4571
Lochpflanzung 6355
Lochplatte 7592
Lochrotorenrinder 7074
Locke 1182
lockere Bestockung 4951
lockerer Baum 9309
lockerer Bestand 7293
lockerer Durchfallast 5158
lockerer Wald 1060
lockere Seite 7716
lockerte Holzfaser 5159
Lockerung 6967, 8515
Lockerung des Schlusses 4566
Lodgepole-Kiefer 5044
Löffeltiefbagger 468
Logan 6415
Logangewächse 5047
Logger 5067
Lohbrecher 2281
Loheisen 7960
Lohemesser 570
Lohfleck 8396
Lohhecke 5835
Lohmühle 2281
Lohnsägen 7244
Lohnschnitt 7244
Lohrinde 8392
lokale Forstbahn 5034
lokale Forstdienststelle 5033
lokale Massentafel 5037
lokaler Wald 5032
Longan 1381
Loquate 5172
Lorbeer 6745
Lorbeerbaum 6745
lorbeerblättrige Eiche 4835
lorbeerblättrige Stechwinde 4834
lorbeerblättrige Synkarpie 1197
Lorbeereiche 4835
Lorbeergewächse 4832
Lorbeerliguster 5503
Lorbeer-Pappel 681
Lorbeerseidelbast 2828
Lorbeerstrauch 6745
Lorbeersumach 4836
Lorbeerwald 4833
Lorbeerweide 630
Los 5181
Löschen eines Brandes 3431
Löschgerätelager 3214
Löschmannschaft 3183
lose Fäule 7819
loser Schlagabraum 5160
Losflößerei 2625
losgerissene Holzfaser 1524
Loshieb 7472
Lossbaum 3870
Lossstrauch 3870
Lösungsmittel 8505
Lot 6389
Lotofa 1090
Lotuspflaume 2258
Louro vermelho 6782
Lückenausbesserung 823
lückenhafter Bestand 4446
Lückenhieb 6106
Lücken-Schirmschlag-System 3801
Lücken-Stellung 3801
luftgetrocknete Masse 149
luftgetrocknetes Schnittholz 150
Luftholz 9225
Luftkamin im Holzstapel 1487
Luftlandewaldschutz gegen Feuer 148
Luftpflanze 2852
Luftriss 9450
Luftrückung 7700
Luftschlepper 88
Luftseilbahn 84
Luftseilbahnkommando 7703
Luftseilbahnverbreitungsgebiet 7702
Luftsenkervermehrung 154
Lufttaxation 90
lufttrockene Holzdichte 151
lufttrockenes Holz 150
Lufttrockenmasse 149
Lufttrocknung 152
Luftwurzel 86
Lumberjack 9654
Luvisol 5219
Luvseite 9451
Luxusholz 3159
lysigenisch 5222
Lysimeter 5223
Lythrazeen 5161

Macadamianuss 1206
Macassar-Ebenholz 3882
Macchia 5300
Machandel 3983
Machandelbaum 3983
Machenschaftsfähigkeit des Holzes 5228
Machette 1204
MacKenzie-Weide 5233
Macnab-Zypresse 484
Mädchenhaarbaum 5245
Mädchenkiefer 4636
Made 4809
Madebohrung 4032
Madrona 6020
Magnolie 5240
Magnoliengewächse 5241
Mahagonibaum 1412
Mähbinder 716
Mähnenfichte 9330
Mähnweg 7689
Maibaum 5347
Maie 747
Maiglöckchenstrauch 7800
Maikäfer 3502
makedonische Eiche 5226
Makore 5254

Makrophyt 5236
Makrovermehrung 5237
Makrozüchtung 5237
malaiischer Baum 4645
malaysischer Bäum 6074
Malbaum 7124
Mallee 5258
Malpighiengewächse 5260
Malve 22
Mambode 8375
Mammutbaum 3844
Mandarine 3948, 5273
Mandarinenbaum 5274
Mandelbaum 207
mandelblättriger Birnenbaum 206
Mandelweide 6122
Mandioqueira 5275
Mandschurie-Kiefer 4743
mandschurische Birke 384
mandschurische Erle 5268
mandschurische Esche 5269
mandschurischer Ahorn 5271
mandschurischer Korkbaum 264
mandschurische Walnuss 5272
Manga 1789
Mangelbaum 251
Mangelbaumgewächse 5279
Mango 1789
Mangobaum 1789
Mangofrucht 5277
Mangostanbaum 5278
Mangostane 5278
Mangrovebaumgewächse 5279
Mangrovegehölze 5280
Mangroven 5280
Mangrovenbaum 5276
Mangrovengewächse 5279
Mangrovensumpf 5281
Mangrovenwald 5280
Manila-Hanf 5283
Manila-Padauk 5284
Manila-Palme 1568
Manio 2714
Maniok 5282
Manipulationsfläche 7727
Mannaesche 3292
Manometer 6521
Manzanillabaum 5267
Marille 1756
Mariola 5308
Mark 1951
Markast 6306
Markbräune 1080
marken 5310
Markfaser 4197
Markfäule 1414
Markierhammer 5315
markierter Bestand 8644
Markierung 8364
Markierungshammer 5315
Markierungsmaß 5319
Markriss 4195
Markröhrenfäule 6291
Markschäle 2139
Markstrahl 5391
Markstrahlenzelle 6736
Markstrahl-Gefäß-Tüpfelung 6740
Markstrahlszellwanden 6737
Markstrahltracheide 6739
Markstrahlzelle 6736
marktgängige Wipfel 5398
Marktgängigkeitsgrenze 4981
Markwald 1812
Marlspiker 5321
Marmorfäule 5573
marmorierte Fäule 5573
Marmorierung 843
Marone 2887
Marupa 6063
maschinelle Fällung 5369
maschinell geschält 5231
maschinell hergestellte Brandnarbe 5229
Maschinenarbeiter 7695
Maschinentrücken 6489
Maschinenweg 1245
Maser 9078
Maserbildung 9187
Maserholz 9186
maseriges Holz 9186
Maserkegel 1182
Maserknolle 1182
Maserkropf 1182
Masertextur 9078
Maserung 9187
Maserwuchs 9185
Maß 2499
Massaranduba 5322
Maßbrett 8103
Massenakter 9240
Massenaufnahme 5266
Massenauftreten 2849
Massenauslese 5325
Massendichte 1155
Massenelement 5235
Massenermittlung 5266
Massenertrag 9773
Massenfachwerkmethode 6165
Massenforstprodukte 1156
Massenkurve 9241
Massenleistung 9773
Massenselektion 5325
Massentafel nach Stockdurchmesser 8212
Massenvorrat 9250
Massenzuwachs 9243
Masserle 2827
Massholder 2827
massiertes Pech 5324
mäßige Durchforstung 5508
Massivholz 7843
Massivholzbiegen 7844
Maßklassifikation 2500
Mast 5326
Mastbaum 5328
Mastenholz 8586
Mastholz 8586
Mastixbaum 5327
Mastixpistazie 5327
Mastixstrauch 5327
Mastjahr 5329
Masurbirke 5330
Matebaum 6064
Material zum Durchmesser 5992
Mate-Teestrauch 6064
Matilisguate 5346
Matilisguate-Baum 5346
Maulbeerbaum 8354
Maulbeergewächse 5599
Mauritian-Ebenholz 5344
maximale Dichte 5345
maximale Schrumpfung 8730
Maximowiczspappel 4629
mazedonische Kiefer 489
Mc-Leod-Hacke 5348
Mc-Leod-Haue 5348
Mc-Leod-Werkzeug 5348
mechanische Beschädigung 5368
mechanische Bestandpflege 5382
mechanische Durchforstung 5374
mechanische Entrindung 5372
mechanische Holzabfallbeseitigung 5381
mechanische Holzbearbeitung 5376
mechanische Holzbeschaftung 5379
mechanische Pflanzung 5380
mechanischer Zugriff und Nutzung 5377
mechanisches Jüten 5375
mechanische Standortvorbereitung 5373
mechanische Verletzung 5368
mechanische Verziehen 5375
mechanische Zapfengewinnung 5378

mechanisierte Ringelung 5370
mechanisiertes Kleinhiebssystem 7773
Meerkiefer 5309
Meerrettichbaum 4379
Meerträubchen 2844
Meerträubel 2844
Meerträubelgewächse 2845
Megagametophyt 5392
Mehlbaum 8323
Mehlbeerbaum 8323
Mehlbeere 8323
Mehldorn 793
Mehltau 5444
mehrästig 5596
Mehrausbeute 6002
Mehrausbeute an Schnittholz 6002
Mehrbaumernte 5620
Mehrbaumschere 5621
Mehreinschlag 2935
Mehrentastungsmaschine 5323
Mehretagenkultur 5631
Mehretagensystem 5632
Mehretagenwald 5633
Mehrfachkrümmung 5619
Mehrfachkubierungstabelle 5612
Mehrfachmassentafel 5612
mehrfachstammter Baum 5617
mehrfachverwendeter anhaltende Berechnungsertrag 5624
mehrfach zusammengesetztes Blatt 2309
Mehrflächenstichproben 6435
Mehrforstnutzung 5613
mehrfrüchtig 6436
mehrhiebiger Bestand 5634
mehrjährige Pflanze 6151
mehrjährige Wildfauna 6152
mehrlitzige Blockförderbahn 5635
mehrmaschinelle Holzgewinnung 5610
Mehrnutzungsmanagement 5622
Mehrphasenernte 7413
Mehrschichtenbestand 5634
Mehrschichtenplatte 4761
Mehrschichtenwald 5633
mehrschichtige Faserplatte 5609
mehrschichtiger Hochwald 5633
mehrschichtige Spanplatte 5608
Mehrspannluftseilbahn 5627
mehrstämmig 5628
Mehrstamm-Schlagkopf 5630
Mehrstufenernte 7413
mehrstufiger Bestand 5634
Meilerofen 1450
Meilerstätte 4712
Meißel 9509
Mekkabalsambaum 969
Melonenbaum 1794
Meranti 4831
Merawan 4365
Merbau 5476
Meristem 5402
Meristemkultur 5403
Merkmalsbild 6204
Merkus-Kiefer 5404
Mersawa 5405
Mertens-Hemlock 5588
Mescalbohne1939
mesophiler Wald 5409
mesophilig 5408
Mesophyll 5407
Mesophyt 5410
Mesophytwald 5409
Mesquitebaum 1790
Mesquitebusch 4354
Messarbeiter 5102
messbare Kreisfläche 5360
Messbuchung 5361
Messer 819
Messerblattakazie 4720
Messerblock 9193
Messerfurnier 7738
Messkette 8289
Messkluppe 8809
Messlatte 8292
Messleine 8397
Messmate-Stringybark 5412
Messschnur 8397
Messstab 5107
Messtarif 7275
Messuhr-Dendrometer 2464
Messung am Liegenden 3040
Messung nach Grund- und Zopfdurchmesser 5367
Messung nach Länge und Mittendurchmesser 5362
Messwurm 3174
Metall-Holzklebstoff 5413
Metaxylem 5415
meteorologischer Waldbrandgefahren-Index 3218
Methode der Blockholzmassenermittlung 5110
Methode der Festflächenstichproben-abnahme 3234
Methode der Linienstichproben Flächenstichprobennahme 9166
Methode des einheitlichen Schirmschlages 9087
metrisches Shnittholzmaß 5417
Metzgersblock 1211
mexikanische Bärentraube 6410
mexikanische Eiche 5420
mexikanische Fächerpalme 9271
mexikanische gelbe Kiefer 1485
mexikanische Kastanie 5421
mexikanische Nusskiefer 5424
mexikanische Pappel 5661
mexikanischer Sonnenblumenstrauch 8882
mexikanischer Wacholder 5423
mexikanischer Wermut 704
mexikanisches Mahagoni 6019
mexikanische Sumpfzypresse 5418
mexikanische Weymouthskiefer 7867
mexikanische Zypresse 5418
Michelie 5428
Miconia 9191
Migration 5442
Mikrobiotop 5434
Mikrofibrille 3780
Mikrohabitat 5429
Mikroorganismus 5430
Mikrorelief 5433
Mikroskopie des Holzes 9608
Mikroumwelt 5429
Mikrovermehrung 5432
Mikrowärmepflanzen 5435
Milacre-Parzelle 5449
Milbenbefall 26
Milbenseuche 26
Milchbaum 2000
Milchröhre 4829
Milchröhrenspur 4828
Milchzelle 4830
mild 7822
milder Humus 5576
mildes Holz 1656
milde Weißfäule 7825
milde Weißschwammfäule 7826
Mimose 3962, 5460
Mimosengewächse 5459
mindern 400
minderwertiges Holz 5195
Mindestdurchmesser 5465
mineralische Verfärbung 5462
mineralisierende Streifene 5461
Mineralisierung des Holzes 9610
minimaler Nutzungsstandard

minimale Stammlänge 5468
Minusbaum 5473
Minusbestand 5472
Miombo 5475
Mischbaumart 393
Mischbestand 5486
Mischbestandbegründung 2867
Mischbestockung 5486
Mischfichtenbestand 5493
Mischfichtenwald 5493
Mischkultur 5491
Mischungsart 393, 4713
Mischungsform 4713
Mischungsgetreide 5487
Mischungsgrad 3897
Mischungskorn 5487
Mischungsregulierung 6843
Mischung von Wildsaatgut 306
Mischwald 5486
missbilligt 2402
Mississippi-Zürgelbaum 8253
misslungene Kultur 2992
Missouriflussweide 5480
Missouriweide 5480
Mistel 2902
mitherrschende Bäume 1706
mitherrschende Kronenklasse 1704
Mittelbauholz 8586
Mittelbaum 442
Mittelbaumvolumen 443
mittelbestückter Bestand 5388
Mittelbrett 5437
mitteldichte Faserplatte 4545
Mitteldurchmesser 2468
Mitteldurchmesser der Fehler 431
mittelengringige Faser 5507
Mittelfurnier 1949
mittelgroßer Baum 5387
Mittelhöhe 5357
Mittelknoten 5386
Mittelmeer-Braunerde 6770
Mittelmeermandarine 3948
Mittelmeerzypresse 1770
Mittelpunktabmessung 7279
Mittelrippe 4879
Mittelriss 5385
Mittelschicht 5438
Mittelschichtspan 1952
Mittelschnitt 5436
Mittelspan 1952
Mittelstammdurchmesser 6651
Mittelstammtarif 5359
Mittelstärke 2468
Mittelwald 7947
Mittelwaldbetrieb 8129
Mittenstärke 2468
Mittenumfang 5440
Mitterdurchmesser der herrschenden Stammklasse 8693
mittlere Baumhöhe 440
mittlere Baumlänge 441
mittlere Faserlänge 432
mittlere jährliche Durchforstung 429
mittlere jährliche Holzernte 428
mittlere jährliche Sterblichkeit 427
mittlere Käferpopulation 430
mittlerer jährlicher Nettozuwachs 437
mittlerer jährliche Zuwachs 2126
mittlere Rückedistanz 444
mittlere Stammzahl pro Acre 438
mittlere Stammzahl pro Last 435
mittlere totale Holzwerbundskosten 439
Mizelle 5645
Möbelholz 3776
mobile Auslegervorrichtung 5012
mobile Entastungsmaschine 3113
mobile Entrindungsmaschine 3112
mobile Erntemaschine 5500
mobile Hackmaschine 5499
mobiler Holzlader 5501
Mobilseilkran 5501
Modder 5506
Modellbaum 7174
Modellholz 5505
Modellierung 7491
Modellstamm 7172
Modellstammfällung 7175
Modellwald 5504
Moderfäule 7819
Moderholz 6484
modifizierte Brandzone 5509
modifizierter Hieb 5511
modifizierter Wald 5512
modifizierte Schlagsfläche 5510
modifiziertes Holz 5513
modrig 7212
Mohrenpfeffer 9740
molukkisches Eisenholz 5476
Mönchsmeise 9438
Mönchspfeffer 4961
Mondfäule 853
Mondring 853
Mondringfäule 853
Mondsamengewächse 5551
Mondscheingabelung 5552
Monimiengewächse 5529
Monoklimaxtheorie 5538
Monokotyledonen 5540
Monokotyledonholz 5541
Monokultur 5542
Monsoonwald 5545
Monsunwald 5545
Montageseile 8153
Monterey-Föhre 5548
Montereykiefer 5548
Monterey-Zypresse 5547
Montezumaskiefer 5549
Mooraufforstung 98
Moorbeere 893
Moorbirke 6620
Mooreiche 3764
Moor-Lorbeerrose 895
Moorwald 8309
Moorweide 813
Moos 5563
Moosbeere 2889
Moosbruchenkiefernwald 5568
Moosbüchse 9144
Moosdecke 5564
moosiger Wald 5567
Moosschicht 5564
Moosüberzug 5564
Mora 4715
Morabukea 3027
Moraholz 4715
Morastbirke 892
Mörderfeige 4469
morgenländische Fichte 2723
morgenländische Platane 5952
morgenländischer Lebensbaum 2725
Morindafichte 4310
Morototo 5558
morphologisch vergütete Sämlinge 5559
Morrow-Heckenkirsche 5560
morscher Baum 7082
morsches Holz 2291
Moru-Eiche 5562
Mostgummibaum 1578
Motorbügelsäge 952
motorisierte Bringung 6489
motorisierter Zugriff und motorisierte Nutzung 5571
Motorkettensäge 6488
Motorrückensprüher 471
Motorsäge 6491
Motorsägenführer 1434
Motorsäger 1434
Mouracatiara 5595
Movingui 5732

Mühlenbergseiche 1510
Muhuhu 7609
Mukumari 2711
Mulch 5601
Mulchen 5602
Mulga 5301
Mull 5604
Müller-Eukalyptus 9765
Mulm 7954
multiklonale Mischung 5605
multinodale Knospe 5611
multinodales Auge 5611
multipeler Spitzentrieb 5614
Mungurra 3550
Muninga 290
mürber Ast 5158
Murray 5044
Muscadinerrebe 5637
Mushibi 113
Musikinstrumentenholz 8587
Muskatnussbaum 5821
Muskeg 5638
Musterbaum 6396
Musterfläche 7173
Musterstamm 7172
Musterwald 6239
Mutation 5640
Mutenye 5641
Mutterbaum 5569
Mutterbestand 6078
Muttergang 4811
Mutterpflanzenbeet 8123
Mutualismus 5643
Muzaisi 5644
Mycosis 3769
Mykorrhiza 3775
Myristikazeen 5820
Myrobalane 1475
Myrsinengewächse 5647
Myrtebaum 1792
Myrtengewächse 5649
Myrtenwald 5650
Myrten-Weide 9396
Myrthenmaser 1264
Myzelium 5645
Myzell 5645

nachbessern 4561
Nachbesserung der Bepflanzung 823
Nachentrinden 3137
Nachentrindung 3137
nachfolgend 7453
nachfolgende Art 7454
nachhaltige Forstentwicklung 8295
Nachhaltigkeit 8294
Nachhaltsholzertrag 8299
Nachhauung 7345
Nachhieb 7345
Nachhiebschlag 6870
Nachholzabfuhr 6909
Nachholzaufbereitung 6859
Nachkommenschaft 6570
Nachkommenschaftsprüfung 6571
Nachmast 121
nachpflanzen 4561
nachrangige Kultur 8235
Nachreifen 122
Nachrückung 6909
Nachsaat 4567
Nachschneiden 838
nachteilige Bodenstörung 2439
Nachtholzaufbereitung 5733
Nachtkerzengewächse 2924
nachträgliches Aufarbeiten schwächerer Sortimente im Schlag 7168
Nachtrift 8324
Nachtschattengewächse 5734
Nachverjüngung 6811
Nachwuchs 120
Nachwuchsholz 7361
Nachwuchspflege 7360
nacktblütiger Rhododendron 6288
Nacktsamer 4069
nacktsamige Pflanzen 4069
Nadel 37
Nadelabfall 2995
Nadelbaum 1877
Nadelbaumzapfen 8818
nadelblätterig 6259
nadelblättrig 6259
Nadelbrennholz 7830
Nadelbüschel 5703
Nadelchen 7912
Nadeldecke 5706
Nadelderbstange 1874
Nadelfäule 6242
nadelförmig 36
Nadelholz 5708
Nadelholzabraum 7831
Nadelholzasche 7827
Nadelholzbaum 1877
Nadelholzbrett 7828
Nadelholz-Steckling 7829
Nadelholzwald 1873
nadelig 36
nadelig zugespitzt 36
Nadelknoten 6281
Nadelmischwald 5485
Nadelschnittholz 1875
5484
Nadelwald 1873
Nadelwaldgesellschaft 5705
Nadelwald in der borealen Zone 924
Nadelware 5708
Naga 5877
Nagekäfer 8562
Nagelast 4376
Nagelbeere 5863
Nagelspäne 929
nagen 5730
Nahbeförderung 4558
Nahbewegung 4558
Nahbringung 7534
Nahförderung 7534
Nährstoffe 5826
Nährstoffgehalt 5824
Nährstoffkreislauf 5825
Nährstoffmangel 5179
nährstoffreich 2912
Nährstoffverlust 5179
Nährstoffzyklus 5825
Nahrungskette 3308
Nahtransport 4558
Namenkarte der Arten 7901
Nancitebaum 5138
Nandinagewächse 5655
Nandine 5654
Nanismus 2686
Narbe 582
Narbenholz 7289
Narbering 7288
Nargusta 1712
Nashi 1504
Nassansaat 4406
nasse Bekämpfungslinie 9352
Nassentrindung 9350
Nassfäule 9353
Nasskern 9283
Nassriese 9281
Nationalforst 8041
Nationalforstmanagement 5666
Nationalforstpark 5667
Nationalschutzgebiet 5668
native Bakterien 5669
native Holzbodenfläche 5671
native Neuholzbodenfläche 5725
Naturdefekt 5675
naturgemäße Forstwirtschaft 3571
Naturgrenze 5673
Naturholz 3826
Naturkampfer 5674
Naturkatastrophe 5676
natürliche Astreinigung 5688

natürliche Durchforstung vom Alter 5700
natürliche Holztrocknung 152
natürliche Kiefer 5687
natürliche Lebensdauer 5678
natürliche Lebensdauer des Bäumes 5686
natürliche Lichtstellung 5699
natürliche Nachzucht 5690
natürliche Neigung 5685
natürliche Pfropfrebe 5682
natürlicher Astabfall 5688
natürlicher Bestand 5698
natürlicher Brennstoff 5681
natürliche Ressource 5691
natürlicher Kautschuk 5694
natürlicher Wald 5679
natürliche Saat 5695
natürliches Verbreiterungsgebiet 4072
natürliches Veredeln auf Wurzeln 5692
natürliche Verfärbung 5697
natürliche Waldausbreitung 5677
natürliche Zuchtwahl 5696
naturnaher Waldbau 5683
Naturrotation 5693
Naturschutz 2842
Naturschutzgebietswalde 3597
Naturverbreitungsgebietsperre 5689
Naturverjüngung 5690
Naturwald 5679
Naturwaldwiese 5680
neapolitanische Erle 4592
Nebelbergwald 5593
Nebeln 5481
Nebelwald 6700
Nebenachse 7046
Nebenbaumart 7356
Nebenbestand 8237
Nebenblatt 8101
Nebenblattnarbe 8102
Nebenforstprodukte 5774
Nebenholzart 7356
Nebenknospe 78
Nebenlinie 7579
Nebennutzung 4159
Nebenschneise 7580
Nebenstrecke 4172
Nebenwurzel 79
Nebenwurzelaufasten 964
Nebenwurzele 4824
Neem 6110
Neigung nach der Außenseite 5983
Nekrose 5702
Nelkengewächse 6278
Nepal-Erle 5709
Nervatur 9192
nervige Mahonie 1365
Nesselbaum 1777
Nestpflanzung 5712
Nestwickler 4886
Nestwicklung 4887
Netto-Down-Prozedur 5715
Netto-Jahresveränderung 5713
Netto-Jahreswertveränderung 5713
Netto-Jahreszuwachs 5714
Nettomasse 5719
Nettoniederschlag 2626
Netto-Waldreinertrag 5720
Netto-Waldreinrente 5720
Nettozuwachs 5716
netzförmige Perforationsplatte 6921
netzförmiges Parenchym 6920
netznerviger Zürgelbaum 5717
Neuart 5802
Neuaufforstung 6821
neubegründeter Bestand 5722
neues Holz 5726
Neuforstwirtschaft 5723
Neu-Mexiko-Robinie 5724
neuseeländische Rotbuche 8420
neuseeländischer Schnurbaum 5729
Neuseeland-Kauri 5727
Neutralabbruchmasse 8334
neutrophil 5721
Nevada-Zirbeliefer4965
Nicaragua-Kiefer 5864
nicht astreinig 9101
nichtbehauenes Schnittholz 5216
nichtbezeichnete Wildernis 5756
nicht entrindet Späne 9043
nichtfähiger Waldboden 5749
nichtfallender Ertrag 5755
nichtgeborgene Verluste 9112
nichtgeleiteter Waldboden 9103
nichtgesägt 9113
nicht gespaltenes Holz 9048
nichtgestörter Wald 9076
nichtgestörte Vegetation 9077
nichtgewerblicher Hieb 5751
Nichtholz-Acren 5757
Nichtholzboden 5758
nicht klassifizierte Forstdaten 9047
nicht klassifizierte Nutzung 9046
Nichtnutzholz 2123
nichtquantifizierter bestockte Waldboden 9107
nichtregulierte Länder 9110
nichtregulierter Wald 9111
nichtreservierter Waldboden 5769
nicht spezifizierte Forstkultur 9119
nicht spezifizierte Holzproduktivität 9120
nichttrockenes Holz 9116
nichtvorgeformtes drallarme Drahtseil 5767
Nichtwirtschaftsbaum 9104
Nichtwirtschaftsbestand 5765
Nichtwirtschaftsstamm 5764
Nichtwirtschaftswald 5765
nicht zurückgehendes Holz 9108
nicht zurückgehendes Volumen 9109
nickender Baum 5740
Niederdruck von Jungwuchs 9061
Niederdurchforstung 8511
Niederfallung 2994
niedergedrückte Pflanze 5040
niederhauen 2175
Niederholz 9070
niederlegende Säckelblume 5242
Niederschlag 6498
Niederschlagchronologie durch Jahresringe 8909
Niederschlagintensität 6699
Niederschlagszone 2627
Niederschlagverlust 4534
Niederschlagverteilung 6698
Niederstamm 5196
Niederwald 5189
Niederwaldbedeckung 5190
Niederwaldbetrieb 1929
Niederwaldbetrieb mit Kurzumtrieb 7544
Niederwalddurchforstung 7948
Niederwaldertrag 1925
Niederwaldhieb 1926
Niederwaldhieb mit Reserve 1927
Niederwald mit Kopfholz 1936
Niederwaldplenterhieb 1932
Niederwaldsystem 1935
Niederwaldsystembetrieb 1929
Niederwaldüberhaltbetrieb 9033
niedrige Ästung 4935
niedrige Birke 1295
niedrige Fichte 5194
niedrige Stumpflassen 5197
niedrige Zwergpalme 5384
Niedrigwüchs 8223
Niemaum 6110

nisten 5710
Nisthöhle 5711
Nistkasten 763
Niveaupfahl 3899
Nivellieranzeiger 4922
Nivellierlatte 4923
Njuru 6646
Nobilistanne 5738
Nomadentum 5746
Nominallänge 5747
Nominalmaß 5748
Nonne 5809
nordamerikanische Platane 254
nordamerikanische Roterle 235
nordamerikanischer Tannenknospenwickler 7951
nordamerikanischer Zürgelbaum 1777
nordamerikanische Schwarzfichte 2716
nordische Eberesche 2903
nordische Erle 2882
nordische Weißbirke 1761
Nordlaubholze 5790
Nordmannstanne 1384
Nordsumpfeiche 5791
Normalalterswald 5780
Normalbestockung 3757
Normaldurchmesser 5782
normale Alterklassenfolge 5785
normale Altersklassenverteilung 5779
normaler Samenerntebestand 5784
Normalertragstafel 5786
normalerweise bestockt 4419
Normallänge von Schnittholz 8005
Normalwald 5781
Normalzuwachs 5783
Novelty-Säge 5803
Nullackerland 5799
Nullreißlänge 9795
Numerierhammer 5315
Nuss 5817
Nussbaum 9262
Nussbaumholz 5829
Nüsschen 5819
Nusshickory 6130
Nusskiefer 6082
Nussöl 5822
Nusspflanzen 5823
Nusswald 5818
Nute 5561
Nutringelung 3716
Nutringen 3715
nutzbare Bodenfeuchte 423
Nutzbaum 5399
Nutzbrennholz 4483
Nutzdürrständer 5397
Nutzholz 5400
Nutzholzausbeute 3539
Nutzholzberechtigung 4388
Nutzholzgewinn 9778
Nutzholzmasse 5401
Nutzholzprozent 6150
Nutzholzsortiment 5396
Nutzholzstamm 5393
nutzholztüchtiger Stamm 5393
nutzholztüchtiger Wald 1745
Nutzmasse 5401
Nutzreiser 8521
Nutzrolle 7113
Nutzroller 7113
Nutzung außerhalb des Waldes 5760
Nutzungsanfall 9783
Nutzungsentgang 998
Nutzungsfläche 2189
Nutzungshiebssatz 195
Nutzungskarte 2953
Nutzungsplan von lokale Ressourcen 5036
Nutzungsprozent 2954, 9155
Nutzungsstandard 9156
Nutzwald 1745

Oberbestand 3518
Oberbinden 8682
Oberdurchmesser 2470
obere Kronenhöhe 9133
obere Kronenlänge 9134
oberer Stammanteil 9138
oberes Ende des Stammes 8689
oberes Stammende 7761
Oberfallkerb 8697
Oberfaserschicht 5869
Oberflächenbehandlung 8281
Oberflächenbläue 8278
Oberflächenfehler 5906
Oberflächenfeuchte 8283
Oberflächenfeuer 8282
Oberflächenpflanze 1439
Oberflächenriss 8279
Oberflächenverhärtung 1367
oberflächliche Saftfäule 8269
Oberförsterei 3382
Oberforstmeister 1483
Oberheger 344
Oberhieb 9126
Oberhöhe 8692
Oberholz 3518
Oberklammer 8684
Oberschichthieb 6008
Oberseil 9137
Oberstand 3518
Oberständer 6887
Oberstandschlag 6008
Oberstärke 2470
Oberster 8706
Objektivbäume 5853
Obstbaum 3731
Obstbäumästung 5939
Obstbaumholz 3732
Obstbaumwald 3733
Obstsporn 3730
obstteller Padauk 1183
Ochna 5863
öde Fläche 486
Ödland 9275
Ödland-Bewaldung 96
Ödlandaufforstung 96
ofengetrocknetes Holz 4710
ofengetrocknetes Pflanzensäfte 4709
offene Narbe 5916
offener Ast 9050
offener Lieferwagen 5920
offene Waldlandschaft 9310
Offenheit 5914
öffentliche Betroffenheit 6623
öffentliche Landstraße 6622
Ogeechee-Tupelobaum 5871
Oglethorpe-Eiche 5872
Oh-Humusstreu 5873
ohne Rinde 569
ohne Stamm 8078
ohne Wurzeln 7034
Ohrblattakazie 2698
Öhrchenweide 2697
Ohrweide 2697
Okan 2397
Ökologie 2741
ökologisch anfälliges Gebiet 2841
ökologische Integrität 2737
ökologische Nische 2739
ökologischer Waldbau 2839
ökologisches Gleichgewicht 2736
ökologische Toleranz 2740
ökologisch verwendbare Arten 2738
ökonomische Forstwirtschaft 2744
ökonomisch funktionelle Waldbestande 2742
Ökospezies 2747
Ökosystemproduktivität 2750
Ökosystemverwaltung 2749

Ökotyp 4079
Okoume 3788
Ökozone 2752
Okularschätzung 2869
Okulation 307
okulieren 3906
Okulieren 307
Okuliermesser 1140
Okulierung 307
Ölbaum 2904
Ölbaumgewächse 5891
Ölbaumpflanzung 5892
Oleander 5887
Olearie 2229
Oleaster 9421
Ölholz 3026
Ölholzbildung 4953
Ölholzformation 4953
Olive 2904
Olivenbaum 2904
Olivenernte 5893
Olivenhain 5892
Olivenholz 5894
Olivesche-Maser 5890
Ölnuss 1222
Ölnussbaum 1222
Olon 5895
Olonvogo 2713
Ölweide 257, 9421
Ölweidengewächse 5889
Ölzelle 5875
om-mesische Schicht 5896
Omorikafichte 7456
Omu 4744
Ondulation 2019
Opalholz 5903
Optimalabtriebsalter 5933
optimale Elchdeckung 5932
Optimalvorrat 5934
optischer Kaliber 5930
optisches Dendrometer 5931
Orangenbaum 5935
Orangenbaumholz 5937
Orangenjasmin 4644
Orchideenbaum 6642
ordentlicher Wald 5940
Oregonahorn 701
Oregon-Blasenspiere 5259
Oregon-Erle 235
Oregon-Esche 5941
Oregon-Myrte 1264
organische Fäulnis 5946
organischer Stoff 5948
organische Schicht 5947
organische Substanz 5948
orientalische Buche 5949
orientalische Platane 5952
orientalischer Lebensbaum 2725
orientalisches Trema 4462
orientierter Spanplattekern 168
orientierte Spanplatte-Oberseite 5953
orientierte Strandplatte 5954
Orleangewächse 776
Orleansstrauch 4995
Ormosia 5956
Orophyt 5959
Orterde 4129
Orthodoxsaat 5961
orthotrope Abzweigung 5962
Orthotropie 5963
ortsgebundener Kahlhieb 5035
Ortstein 4129
Osageorange 4219
ostafrikanische Kordie 2711
ostafrikanisches Antiaris 6414
ostasiatischer Lackbaum 4755
ostasiatische Zwergkiefer 4634
Osterluzeigewächse 768
Osterpalme 459
österreichischer Wermut 416
österreichische Schwarzkiefer 415
ostindische Nuss 2730
ostindischer Palisander 2728
ostindisches Jacaranda 2728
ostindisches Rosenholz 2728
ostindisches Satinholz 2729
ostindisches Seidenholz 2729
östliche Balsampappel 9338
Outeniqua-Yellowwood-Baum 5974
ovalblättrige Moorbeere 5986
ovaler Ast 5985
Ovalität 5984
Overhead-Risiko 5994
Oxalis 9656
Oxelbeere 8322
Oxelbirne 8323
Oyameltanne 5422
Ozark-Kastanie 6015
Ozigo 6016
Ozonsprenkelung der Nadel 6017

paarig gefiedertes Blatt 13
Paarungssystem 5333
Pächter 591
Packgutholz 954
Packhaken 6225
Packholz 954
Pagoden-Hartriegel 222
Paketierung 1171
Paketierungsbrigade 1172
Palasabaum 3248
Paldao 6033
Palette 6034
Palisander 1006
Palisanderbaum 1006
Palisanderholz 1006
Palmblatt 6044
Palme 6046
Palmenbestand 6043
Palmengewächse 356
Palmenwald 6043
Palmette 3014
Palmettopalme 1238
palmisch 6036
Palmlilie 9792
Palmnuss 6045
Palmweide 3880
Palo Sangre 9190
Paloverde-Baum 867
Palustrine 6048
Panabholzung 6054
Panaschierung 9170
Paneellier-Eichenholz 9258
Paneellierholz 9257
Panel 6049
Panga-Panga 9336
Pantoffelbaum 1961
Panzerkiefer 936
Pao Rosa 6056
Papau 1794
Papau-Baum 6118
Papaya-Baum 1794
Papierbirke 6058
Papierholz 6061
Papierholz-Greifer 6629
Papierholzhackmaschine 6628
Papierholzhaken 6630
Papierholzrolle 6627
Papiermaulbeerbaum 6060
Pappawbaum 1794
Pappe 6059
Pappelbestand 6454
pappelblättrige Birke 3923
Pappelgärtnerei 6455
Pappelkultur 6456
Pappelrinde 6452
Pappelwald 6454
Pappel-Weidenwald 6457
Pappel-Zimmererwurm 6453
Papra 4465
Paradiesapfel 6062
Parakautschukbaum 1320
parallel aneinandergefügtes Schnittholz 6069
paralleles Steigewegmodell 6066
parallellaminiertes Furnier 6068

Paranakiefer 1005
Paranussbaum 1003
paraquatbehandeltes Holz 6070
Parasit 6071
parasitäre Pflanze 6073
parasitäre Pilze 6072
parasitische Pilze 6072
Parasolameise 4864
Parastemon 9038
paratracheales Parenchym 6075
Pardillo 6076
Parenchymstrang 6077
parkartiger Bestand 6081
Parkettholz 9574
Parkettkiefer 5141
Parkia 5737
Parkinsonia 867
Parklandsavanne 7229
Parkwald 6080
Partridge 5426
Passage 182
Passionsblumengewächse 6099
Paternosterbaum 6110
Paternostererbse 636
pathologisches Hiebsalter 6113
Patrouille 6114
Pattons-Hemlocktanne 5588
Pauschalverkauf 5217
Payung 7563
pazifische Edeltanne 5738
Pechharz 8907
Pech-Kiefer 5792
Pechriss 8155
Pechtanne 1805
peitschen 4812
Peitschen 9359
Peitscher 9358
Pekan 6130
Pekannussbaum 6130
Peking-Weide 1963
pellen 6133
Pelletieren 6141
Pendelsäge 4312
Penetration 6144
pennsylvanische Esche 2255
pennsylvanischer Ahorn 8174
Pepulbaum der Inder 939
perennierende Pflanze 6151
perennierende Wildfauna 6152
Perforationsplatte 6154
Perforationsrand 6155
perforierte Faser 6153
Pergel 9595
Periderm 6157
Periodenblock 6160
Periodenblockeinschlagsmethode
Periodeneinschlagsmethode 6161
Periodenfläche 6160
periodische Durchforstung 6166
periodische Holzgewinnung 6163
periodischer
Durchschnittszuwachs 6164
periodischer Pflanzer 4554
periodischer Schlagwald 3366
periodischer Wachstum 4553
periodischer Zuwachs 6158
periodisches Gewässer 4555
Perispermie 4398
Perlenbeerstrauch 3705
Pernambukholz 1010
Peroba-Mica 6184
Persimmon 1796
persische Albizie 6185
persische Parrotie 6186
persisches Eisenholzbaum 6186
persistenter Samen 6748
Perspiration 8321
peruanischer Mastixbaum 1266
peruanischer Pfefferbaum 1266
Perubalsambaum 504
Perückenbaum 2908
Perückenstrauch 2908
Pestinzidenz 6191
Pestizid 6189
Pestizidisolierstreifen 6190
Peterstrauch 1804
Pfaffenhütchen 1383
Pfaffenkäpplein 1383
Pfahl 3084
Pfahlholz 8588
Pfahlwurzel 8411
Pfandholz 6311
Pfarrwald 1575
Pfasterholz 6117
Pfauenauge 6125
Pfeffergewächse 6148
Pfeifenholz 3880
Pfeifenstrauchgewächse 5502
Pfeilerholz 6428
Pfeilunkraut 367
Pferdeminze 4222
Pferderettichbaum 4379
Pferdeschlittern 4378
Pfingstrosengewächse 6146
Pfirsichbaum 6123
Pfirsichblattweide 6122
Pfirsichholz 6124
Pflanzbestand 6339
Pflanzbrett 6345
Pflänzchen 6366
Pflanzdolch 6344
Pflanze 6333
Pflanze der offene Holzfläche
5922
Pflanzeisen 6344
pflanzen 6334
Pflanzen
in Kämmen 648
Pflanzenabfälle 6371
Pflanzenaufzucht 6704
Pflanzenausheber 6367
Pflanzenbohrer 6342
Pflanzenbrikett 6346
Pflanzendecke 6337
Pflanzendichte 6369
Pflanzendiversität 6338
Pflanzenernte 6341
Pflanzenfresser 4256
Pflanzengemeinschaft 6336
Pflanzengesellschaft 6336
Pflanzengesundheit 6222
Pflanzengift 6220
Pflanzenhacke 6352
Pflanzenkolonne 6349
Pflanzenlochbohrer 6342
Pflanzenmaterial 6363
Pflanzenregression 6372
Pflanzenreihe 6360
Pflanzenröhre 6361
Pflanzenrückentwicklung 6372
Pflanzenrückstände 6371
Pflanzenschutz 6222
Pflanzentasche 6343
Pflanzentopf 6359
Pflanzenüberreste 6371
Pflanzenverband 6362
Pflanzenvernichtungsmittel 6220
Pflanzenvirus 6374
Pflanzenwachstumsregulator 6340
Pflanzenweite 6362
Pflanzenwuchsstoff 6221
Pflanzgut 6363
Pflänzlinganwachsen-Prozent
8918
Pflänzlingerntemaschine 8916
Pflänzling mit Erdballen 492
Pflanzlochmaschine 4326
Pflanzreihe 6360
Pflanzschaufel 8898
Pflanzschulbeet 5812
Pflanzschule 6370
Pflanzspaten 8898
Pflanzstock 6344
Pflanzung der Waldbäume 3626
Pflanzung mit entblößter Wurzel
532
Pflanzung mit hoher Dichte 4278
Pflanzungsplatz 7396

Pflanzung zwischen Abraumwällen 9449
Pflasterklotz 9554
Pflaume 6391
Pflaumenbaum 6391
pflaumenblättriger Apfel 6127
Pflegeeingriff zur Verbesserung 4433
Pflegeeinheit 8461
Pflegegasse 8459
Pflegehieb 2121
Pflegeumlauf 8460
Pflegewiederkehr 8460
Pflock 6344
pflügen 6383
Pflugkörper 6382
Pflug mit Einzelriester 7645
Pflug mit Einzelstreichblech 7645
Pflügung 8554
Pfosten 6475
Pfostenholz 6428
Pfostenmarkt 6480
Pfostenstamm 5064
pfriemliches Blatt 8245
Pfropf 6103
Pfropfen 1145, 3908
pfropfen 3906
Pfropfen hinter die Rinde 561
Pfropfreis 1145
Pfropfreisholz 7299
Pfropfreiswurzelung 7298
Pfründwald 1575
Phanerophyte 6197
Phänologie 6203
phänologischer Zyklus 6202
Phänotyp 6204
phänotypische Selektion 6205
Phänotypus 6204
Phase des Reifens 70
Phase des überaltertes Bestands 5999
Phellem 1956
Phelloderma 6200
Phellogen 6201
pheromonbeköderte Insektfällen 6206
Phloem 4502
Phloem-Mutterzelle 6207
Phloem-Strahl 6208
phönizischer Sadebaum 6787
Photodegradation des Holzes 6209
Phyllode 1594
Phyllosphäre 6213
Phylum 6214
physiognomische Klasse 6216
physiografische Einheit 6219
physiografische Klasse 6218
physische Schlagbarkeit 6215
Phytohormon 6221
Phytotoxizität 6223
Phytozönose 6336
Pilon 6237
pilzabtötendes Mittel 3771
Pilze 3770
Pilzerkrankung 3769
Pilzfäule 3768
pilzfleckiges Holz 6132
Pilzkonsole 1879
Pilzkrankheit 3769
Pilzschimmel 3770
Pilzschutzmittel 3774
pilztötender Stoff 3771
Pilzvertigungsmittel 3771
Pilzwachstum 3772
Pilzwurzel 3775
Pimpernussgewächse 817
Pinangpalme 691
Pinchot-Wacholder 6241
Pineöl 6276
Pinie 4595
Piniekiefer 4595
pinnatifides Blatt 6284
Pinyonkiefer 5424
Pinzieren 6240
Pionierbaumart 6290
Pionierholzart 6290
Piptadenie 6293
Pisang 6335
Piscidia 4606
Pistazienbaum 6295
Pistazienlichtwald 6294
Pistill 6296
Pitchpine 5141
Plagge 7811
Plaggenhacke 7812
Plaggenhaue 7812
Plaggenpflug 7812
plagiotrope Verzweigung 6316
Plagiotropie 6317
Planieren 3901
Planierraupe mit schiefstellbarem Schild 286
Planierschar 5068
Planierschild 5068
Planierung 3901
Planimeter 6324
Plankenholz 6328
planmäßiger Einschlag 6329
planmäßiger Hieb 6834
Plänterwald 7410
plasmatische Vererbung 9170
Plast-Presssternholz 3810
Plataneholz 6323
platanenblättriger Maulbeerbaum 5600
platanenblättriges Alangium 1488
Platanengewächse 6321
Platanenwald 6322
Platte 6049
Plattenkiefer 6377
Plattenprodukte 6050
Plattensaat 7870
Plätzenskarifikator 6108
Plätzepflanzung 6107
Plätzesaat 7870
plätzeweises Brennen 6105
Plenterbestand 7410
Plenterbetrieb 7620
Plenterdurchforstung 7411
Plenterhieb 7412
Plenterhochwald 4002
Plentersaumschlag 8168
Plenterwald 7410
Plenterwaldsystem 7408
Plossobaum 3248
Plusbaum 6396
plüschartige Faser 6702
plüschartiges Stammendeholz 6703
Pochkäfer 8562
Pockholz 4034, 6405
Pockholzbaum 4034
podsolierter braune Waldboden 1078
podsolierter graue Waldboden 3861
podsolische Boden 6408
Pohutukawa-Baum 5728
Polarbirke 892
Polarbrombeere 350
Polarweide 6417
Poleigränke 897
Polieren 7189
polieren 6133
Pollen 6432
Pollenflug 6433
Poller 7489
Pollerplatz 8664
Pollination 6434
Polsterhölzer 6424
Polterschild 5068
polygamzweihäusig 6438
Polyklimaxtheorie 6437
Polyploid 6440
Pomazeen 330
Pomeen 330
Pomeranze 775

Pondkiefer 6447
Pondzypresse 6444
pontische Azalee 4355
pontische Eiche 363
pontische Lorbeerkirsche 2828
Poplargummi 9369
Populationsdichte 6458
Populationsdruck 6460
Populationswachstum 6459
Pore 6461
Porenanteil 6465
Porenraum 6463
Porensaugwasser 1325
Porenvolumen 6463
Porenwasser 6464
poriges Holz 6467
Porigkeit 6465
poröse Faserplatte 6466
poröses Holz 6467
Porosität 6465
Porst 4906
Porst-Lärchenwald 9423
Portiabaum 6471
Possentrie 7185
Postbrandneugestaltung 6479
Postbrand-Waldstandort 6478
Posthieb-Sekundärwald 6477
Postkataklysmen-Sekundärwald 6476
potentielle Standortbaumhöhe 7669
-3/2-Potenzgesetz der Naturausscheidung 6490
prächtiger Trompetenbaum 5789
prächtige Tanne 1268
Prachtkäfer 8063
Prachtkäfer-Schaden 9494
Prachttanne 1268
praktische Ausbildung im Walde 9239
Prärie-Akazie 203
Prärieeiche 799
Präventivknospe 2573
Präventivspross 2848
Preis auf dem Stock 8195
Preiselbeere 1996
Preiselbeer-Fichtenwald 1999
Preiselbeer-Kiefernwald 1998
Prellholz 1025
Presse für Holzabfälle 3560
Pressholz 2390
Presslagenholz 2390
Pressling 9613
Presssternholz 3810
Primärantrieb 6553
Primärbaum 9231
primäre Konversion 6538
primäre Landeinheit 6540
primäre Rinde 6535
primärer Strahl 6545
primärer Transport 6548
primäre Rückelinie 6541
primärer Zweig 6536
primäres Holz 6549
primäres Insekt 6539
primäres Phloem 6543
primäres Tüpfelfeld 6544
primäre Sukzession 6547
primäres Xylem 6549
primäre Wurzel 6546
primäre Zellwand 6537
Primärholz 6549
Primärholzverarbeitungsanlagen 6550
Primärmaschine 6553
Primärwald 282
primordialer Tüpfel 6544
Prinarium 6079
Prismenschnitt 5117
Prismenwinkelmesser 6554
Privatforst 3473
Privatwald 3473
Privatwaldbesitzer 6555
Probefällung 7171
Probefläche 7173
Probehieb 7171
Probenahmefehler 7177
Probenrand 2763
Probenstand 2946
Probestamm 8480
Probestreifen 8479
problematischer Waldtyp 6556
produktiver Waldboden 8625
Profilfurnier 3124
Profilholz 6566
Profilholzleiste 9613
Profilierung 3296
Profilschleifmaschine 6567
Profilschnitt 2151
Profilwuchs 3123
Profilzerspaner 1316
Programm von kleiner Forstunternehmen 7759
Programm zur Unterstützung von kleiner Forstunternehmen 7759
Prolepsis 6579
proleptischer Trieb 4765
Propagula 9183
Propfung 3908
Prosenchym 6586
Prospekt 6587
Protoxylem 6602
Protze 9472
Protzenaushieb 3071
Provenienz 6603
Proventivknospe 6605
Provinzialwald 6606
Provinzialwaldinventur 6607
Prozessionsspinner 6557
Prüfbohrer 4451
Prügel 7113
Prügelholz 7774
Prügelwagen 5208
Prügelweg 1947
Pseudostrobuskiefer 6617
Pteridophyten 6618
Puffer 2156
Pufferstreifen 1149
Pufferwirkung 1148
Pufferzone 1149
Pufferzone ohne Pestiziden 6190
Pugeum 112
Pulai 9363
Pulaski-Werkzeug 6624
Pulverbaum 684
pulverisierte Holzkohle 6633
Punktast 6281
Punktrasterfeld 2577
Punktschweißen 7933
Punktschweißung 7933
Punkt-Stichprobenprüfung 285
Punkt-Stichprobenverfahren 285
Puppenwiege 4810
Purgier-Faulbaum 1544
Purpureukalyptus 6773
purpurfarbene Bauhinie 1221
purpurfarbener Orchideenbaum 1221
Purpurholz 9225
Purpur-Kastanie 7130
Purpur-Magnolie 4963
Purpurtanne 6025
Purpurweide 6643
pyramidale Wuchsform 6648
pyramidenförmiger Baum 6650
Pyramidenpappel 5126
Pyrenäen-Schwarzkiefer 2220

Q-Faktor 6654
Quadratfuß 7970
Quadratmarkstrahl 7972
quadrieren 7962
Qualitätsholz 4284
Qualitätstempel 3900
Qualitätszuwachs 6655
Quarterschnitt 6659
Quartierschnitt 6659

Quassie 9373
Quellbezirk 9280
Quellgebiet 9280
Queraxt 9026
Querbalken 2038
Querbaum 2038
Querbeil 9026
Querbühne 2044
Querdimensionen 2045
Querdohle 2046
Querförderband 5055
Querholz 2051
Querkrümmung 2141
Quermittelfurnier 2040
Querpollination 2056
Querriss 2039
Quersäge 2042
Quersägemaschine 2043
Quersalinge 2059
Querschneiden 8571
Querschneiden des Langholzes 1132
Querschneidenpunkt 2041
Querschneise 2057
Querschnitt 8571
Querschnittszeichnung 7728
Querschnittumlauf 2804
Querschwellen 8774
Querstab 626
Querstapelung 2055
Querträger 2038
Quetschen 2105
Quetscher 2104
Quirl 1668
quirlige Blattdisposition 9209
Quitschbeere 2903
Quitte 6667
Quittenbaum 6667
Quittenvogel 5846
Quitze 2903

Rabatte 5812
Rabbatenpflanzung 648
Rächermittel 3765
Racksäge 6672
Racksystem 6674
Radbaumgewächse 9357
Radialarmkreissäge 1586
radialer Riss 6675
radiales Schwingen 6680
radialgesägtes Holz 6678
Radialholz 6678
Radialinterzellulargang 6677
Radialparenchym 6679
Radialriss 6675
Radialschnitt 6659
Radialzuwachs 2474
Radiatakiefer 5548
Radicule 6546
rädriger Lager 9356
Radroder 9355
Radrückschlepper 7125
Radschuh 7687
Raffia-Palme 6685
Raftholz 2274
Rahmenbühne 5027
Rahmen-gesteuerter Holzschleifer 3685
Rahmenradschlepper 3685
Rahmensäge 5132
Rahmensägemaschine 5132
Rainweide 1799
Rajolpflug 8963
Ramin 6710
Rammklotz 6707
Rammpfahlholz 8588
Randbaum 922
Rand/Bereich-Verhältnis 2755
Randbesamung 7384
Randbeschattung 7581
Randbohle 8779
Randeffekt 2758
Rändelhieb 6167
Rändelungshieb 6167
Randhieb 6167
randomisierte Blocke 6713
randomisierter Blockentwurfplan 6714
Randpflanzung 921
Randschutz 6596
Randschwarte 5470
Randstreifen 4901
Randverjüngung 6877
Randwirkung 2758
Ranke 8462
ranzige Kastanie 5874
raschwüchsiger Wald 3020
Rasen 7811
Rasenhacke 7812
Rasenmesser 7812
Rasenplagge 7811
Rasenschälpflug 7812
Raspel 9566
Raste 4082
Raubschlag 2935
Raubwirtschaft 2434
rauchempfindliches Gebiet 7785
Räuchereiche 5850
Räuchern 3766
Rauchholz 2291
Rauch-Management 7784
Rauchschäden 2240
rauhes Brett 7084
rauhes Holz 7096
Rauhmaß 5748
Räumbrigade 1610
räumdiger Stand 3646
räumdiger Wald 1060
Räumen des Holzschlages 1624
Räumen des Schlagreisigs 6799
Raumgehalt 7979
räumig 7008
Rauminhalt 7979
räumliche Bestellung 7884
räumliche Forstvariation 7885
räumliche Kontinuität der Waldgesellschaft 7883
räumliches Schwinden 9251
Raummaß 7978
Räumschild 6986
Räumungshieb 3136
Räumungsschlag 3136
Raupenschlepper 5114
Raupenwirtsbaum 1377
Rauschbeere 1996
Rautengewächse 7131
Reaktionsholz 6744
Rebengewächse 9221
Rebhuhnfäule 6313
Rebhuhnholz 5426
rechen 6705
Rechenblume 8333
rechtsdrehender Drehwuchs 6943
rechtzeitiges Brennen 2700
Red Louro 6782
Reduzent 2308
Reduzieren 8403
regelmäßig treibender Austrieb 6761
Regenbaum 6701
Regenerationstrieb 6753
regenerierbarer Waldboden 6810
Regenschirmbaum 412
Regentropfenfigur 6697
Regenwald 6700
Regenwald im temperierter Zone 5516
Regieschlag 1815
Regionalplan 6829
Registrierkluppe 6751
reguläre gleichaltrige Struktur 6838
regulierte Acker 6839
regulierte Ernte 6841
regulierter Wald 6840
Reibenbaum 7129
Reibentrindung 3713
Reibungsentrinder 3712

reichblütiger Zierapfel 4619
Reichweite 6742
Reidel 8555
reif 5337
Reifen 5339
Reifenholz 523
reifer Bestand 5338
reifes Alter 133
reifes Holz 6964
Reifholz 6964
Reifholzbaum 7220
Reifschaden 2235
Reifungsgradient 5336
Reifungsphase 70
Reifweide 9224
Reifwerden 5339
Reihe 8545, 8619
auf Lücken gepflanzte ~n 7987
Reihenpflanzung 6356
Reihensaat 522
Reihenverband 6356
Reinbestand 2926
reiner Bestand 2926
reiner Lebendsamen 6636
reiner Schaft 1604
reines Schnittholz 1626
reingeschältes Papierholz 1609
Reiniger der Holzabfälle 8307
Reinigung des Holzschlages 1624
Reinigung mittels Holzabfallstreuung 1623
Reinigungshieb 7193
Reis 1095
Reisbund 1118
Reisbündel 3018
Reiserholz 998
Reiserholz im Niederwald 1938
Reiserholzstange 7775
Reiserknüppel 1118
Reiserstange 7775
Reisholz 5170
Reisigbündel 1118
Reisighaufen 4193
Reisigräumen 7723
Reisigverbrennung 7719
Reisigwelle 3018
reißen 9024
Reißen 2003
Reißer 7313
Reißhaken 7313
Reißlachte 4279
Reißlänge 1019
Reißmesser 7313
Reisstange 7775
Reißwald 1928
Reitgras 7537
rekonstruiertes Holz 6750
Relaskop 6852
relative Dominanz 6853
relative Durchforstungsintensität 6855
relative Feuchtigkeit 6854
Relativfeuchtigkeit 6854
Relief 6858
Reliktbaum 6857
Rengas 6866
Renkwuchs 9323
Rennbahn 8745
renovierter Sekundärwald 6845
Rentabilitätsschwelle 1015
repetitive Inventur 6847
Replikation 6875
Reppeln 2283
reproduktive Knospe 6878
Resak 6879
Reservat 5672
reservierter Holzboden 6886
reservierte Uferzone 6962
Resistenz 6906
Resistenz des Holzes 2669
Resonanzholz 7851
Ressourcen-Folio 6911
Ressourcenüberzug 6913
Ressourcen- und Transportplan 6910
Ressourcenwert 6914
Restauration 6916
Restaurierung 6916
Restbasalfläche 6890
Restbestand 9051
Restbohle 2554
Restfeuchtigkeit 3141
restliche Basalfläche 6890
Restwertabschätzung 6891
Retentionsholzeinschlag 6919
Revierförster 1988
Rheedia 6926
Rhizobium 7037
Rhizom 6927
Rhizophor 6928
rhizophorig 6929
Rhizosphäre 6930
Rhizotaxis 7011
Rhodesien-Teak 9793
rhythmisches Wachstum 6931
Rhytidom 5975
richtiger Stamm 8989
Richtung des Neigung 2518
Rich-Werkzeug 3200
Rick 2979
Ried 1308
Riegelholz 8801
Riegeltextur 8175
Riegelwuchs 7990
Riesbahn 5116
Riesbahnschmierer 1576
Riesbremse 3888
Rieselwasser 8968
Riesenbastkäfer 2909
Riesenbaum 3844
Riesenbäume 3845
Riesenjudasbaum 3843
Riesenlebensbaum 8893
Riesen-Mammutbaum 1269
Riesenrhododendron 3934
Riesen-Schierlingstanne 5588
Riesensequoie 3844
Riesentanne 3910
Riesmund 2836
Riftschnitt 6659
Rigidakiefer 5792
rillen 7308
Rillenerosion 6949
Rillenholz 3978
Rillensaat 7869
Rillensäen 7869
Rillensäge 7309
Rillenscheibengreifer 7504
Rinde-Holz-Mischung 589
Rindenabzug 200
Rindenanteil 578
Rindenanteil am Holz 200
Rindenaufbau 586
Rindenbast 4502
Rindenberechnung 200
Rindenbohrer 571
Rindenbrand 836
Rindenbündel 1176
Rindendicke 8498
Rindendickenmesser 587
Rindeneinschluss 4491
Rindeneinschnitt 562
Rindenfaktor 557
Rindenfalte 548
Rindenförderer 547
rindenfreies Holz 555
Rindengalle 560
Rindengebund 1176
Rindengerbmittel 8392
Rindengewebe 1975
Rindengewicht 546
Rindengut 8392
Rindenlager 585
Rindenmehl 550
Rindenmesser 587
Rindenmulch 572
Rindennarbe 582

Rindenplatte 575
Rindenplatten 7710
Rindenprodukte 577
Rindenrest 580
Rindenring 568
Rindenrose 8786
Rindenschäle 584
Rindenschäler 556, 563
Rindenschälmaschine 2284
Rindenschlitten 549
Rindenschredder 583
Rindenspäne 545
Rindenspanmasse 542
Rindenstärke 8498
Rindenstreifen 537
Rindentasche 576
Rindentod 2487
Rindenverletzung 558
Rindenverluste 580
Rindenware 588
Rindenwunde 558
Rindenzelle 541
Rindenzugförderer 547
rindloses Stammvolumen 8952
Ringelborke 2139
Ringeleisen 3854
Ringelkerbe 3853
Ringeln 6951
ringeln 3851
Ringelschnitt 3853
Ringeltaube 9624
Ringelung 6953
Ringenordnung 7451
Ringfäule 6956
Ringfleck 6957
Ringfloß 480
ringförmiges Abschälen 6951
Ringkluft 2139
ringporiges Holz 6955
Ringporigkeit 6954
Ringriss 2139
Ringriss mit Harz 6903
Ringrotorentrinder 7074
Ringsäge 308
Ringsbruch 2139
Ringschäle 2139
ringschäliges Holz 9696
Ringstrahlentrinder 9289
Rinne 334
Rinnenerosion 4045
Rinnensaat 8964
Rio-Palisander 1006
Rippe 4882
Rippenblatt 6932
Rippenstrebepfeiler 6933
Risikoarten 7894
risikoreicher Baum 4297
Risikowert 9164
Rispelstrausch 3003
Rispen-Blasenbaum 6051
Rispenblatt 6053
Rispen-Hartriegel 3924
Rispenhortensie 6052
Rispenkraut 7162
rispige Lyonie 5256
Riss 7483
Rissabstand 1541
Rissausbreitung 2001
Rissbildung 2003
Risse in einem bestimmten Winkel 1479
Risslänge 2004
Risstiefe 7484
Ritzen 7286
ritzen 7308
Robinie 1754
Roble 5346
Roble-Pellin 6985
Robustapappel 4400
Rodeaxt 4030
Rodegerät 8808
Rodeland 390
roden 9140
Rodepflug 8771
Röderwaldbetrieb 143
Rodewinde 1160
Rodung 2857
Rodungen 4031
Rodungsbewilligung 4928
Rodungsgenehmung 4928
Rodungsprämie 6512
Rodungsverbot 4498
Rohbalsam 6733
rohes Brett 7084
Rohharz 6733
Rohholz 7096
Rohholzbereitstellung 9588
Rohholzförderer 7098
Rohhumus 1681
Rohhumusstreu 1682
Rohling 9785
Rohr 1308
Rohrbohrer 2149
Röhrenkassie 3883
Röhrenpflanzung 9002
Rohstoff 7096
Rollassel 1813
Rollbett 7004
Rollbock 894
Rolle 3035
Rollplatz 3280
Rollschicht 4844
rosafarbige australische Nuss 409
Rose 995
Rose-Gum 3909
Rosenähnliche 9794
Rosenfichte 8547
Rosengewächse 7064
Rosenholz 1007
Rosenkranzpappel 2717
Rosenlorbeere 5887
Rosenstock 7068
Rosenstrauch 7068
Rosenwein 9000
Rosmarienweidegewächse 8336
Rosmarin 7066
Rosmarin-Fichte 5029
Rosmarinheide 897
Rosmarinweide 872
Rossameise 1351
Rosskastanie 1781
Rosskastaniengewächse 1126
Rosskastanienmaikäfer 3502
Rosten 7140
rostiges Holz 2291
rostnerviger Ahorn 6800
Rostpilz 7859
Rostpilze 7139
Rostschneeball 7141
Rotahorn 6774
Rotan 6731
Rotationshackmaschine 7072
rotbeeriger Wacholder 6530
rotbeinige Baumwanze 3343
rotblättrige Hainbuche 4621
rotblättriger Pfirsich 6123
rotblühende Rosskastanie 7130
rotblütiger Ahorn 6774
rotborstige Himbeere 9452
Rotbuche 6639
rote Akazie 1038
rote amerikanische Esche 6634
rote australische Seideeiche 3966
rote-brune Stammfäule 6765
rote-brune Wurzel- und Stammfäule 6766
rote China-Birke 1503
rote Fichtengallenlaus 9718
Rötegewächse 5238
rote Heckenkirsche 2895
rote Heidelbeere 1996
Roteibe 1814
Roteiche 7863
Röteisen 7960
rote Johannisbeere 6769
rote Kiefernbuschhornblattwespe 2905
rote Krone 6796

roter Ahorn 6774
roter Bastkäfer 6797
roter Espenblattkäfer 389
roter Eukalyptus 6786
roter Fleck 7215
roter Hartriegel 1771
roter Holunder 2907
roter Hornstrauch 1771
roter Ironbark 6779
roter Lapacho 6644
Roterle 2882
roter Lorbeerbaum 6763
roter Maulbeerbaum 6781
roter Pappelblattkäfer 1982
roter Querbrachobaum 5173
roter Sandelbaum 6790
roter Sandelholzbaum 6790
roter Stringybark 6794
rotes Angico 6762
rotes Bloodwood 2249
Rotesche 2255
rote Schneebeere 1940
rotes Sandelholz 6790
rotes Stinkholz 112
rotes Streifen 6795
rote Syringa 6791
rote Traubenkirsche 1769
rote Waldameise 9480
rote Weide 6801
Rotfäule 6789
Rotfäule der Kiefer und Fichte 6265
Rotfichte 1805
rotfrüchtiger Wacholder 6787
rotgelbe Kiefernbuschhornblattwespe 2905
Rotgummibaum 6786
Rotholz 6802
Rothölzergewächse 1697
Rothuhn 4182
rotierender Hackspänesortierer 7073
Rotkern 6778
Rotkernfäule 6777
Rotkiefernzapfenkäfer 6785
rotköpfige Kieferblattwespe 6776
rotköpfige Kiefernschonungsgespinstblattwespe 6253
rötlichbrauner Waldboden 6770
Rotorackerfräse 7072
Rotorentrinder 7074
Rotorhacke 7072
Rotorkrümler 7072
Rotrüster 2893
Rotulme 7745
Rotwald 6255
Rotwasserbaum 1366
Rotwildfuß 2329
Rotwildglade 2331
Rübenschneidemaschine 7019
Rübenschneider 7019
Rückabspannung 467
Rückanker 467
Rückankurbelung 472
Rückbauweg 6111
Rückbrennung 465
Rückebahn 7689
Rückebogen 5090
Rückedistanz 9747
Rückeentfernung 9747
Rückegabel 5020
Rückegasse 3987
Rückeinsatzhärtung 6924
Rückekarren 5090
Rückekosten 7684
Rückelinie 3987
Rückelohn 7684
Rückemaschine mit Harvester 3075
Rückemastverfahren 4288
Rückenbefestigung 473
Rücken durch Schleppen 3994
Rückenhaube 1858
Rückenkarren 5208
Rückenkette 7798
Rückenlinie 9750
Rückenlinie-Durchforstung 9752
Rückenlinie-Richtung 9746
Rückenlinieveränderung 9751
Rückenmaschine 6028
Rückenmast 4190
rücken mit der Hand 4112
Rücken mit der Kette 1428
Rücken mit hochrädrigem Rückekarren 4304
Rücken mit Kraftantrieb 6489
Rücken mit Seilzug 1247
Rücken nach oben 9129
Rücken ohne Seilbahn 5777
Rückenpneu 5086
Rückensägen 474
Rückenschaden 7685
Rückenschlepper 6028
Rückenschlitten 3881
Rückenspritze 471
Rückenspritzgerät 471
Rückenverstäuber 471
Rücken von hochgelegtem Kabel aus 4288
Rückenvorrichtung mit Gabeln
Rückerichtung 2517
Rückerlohn 7684
Rückeschlitten 5115
Rückeschneise 3987
Rückeverfahren 9749
Rückevorrichtung 5115
Rückewagen 5208
Rückeweg 7689
Rückezange 3892
rückgängiges Holz 5884
Rückgrat 460
Rückhalteseil 971
Rückhaltetakelage 970
rückherzförmiges Blatt 5852
Rückholerkabel 5411
Rückholseil 5411
rückkonisches Blatt 5851
Rückkreuzen 462
Rückkreuzung 462
Rückkreuzung-Zuchtmethode 463
rücklanzettförmiges Blatt 5854
rückläufige Pflanzenbewegung 6372
Rückluftseilbahn 4580
Rückovalblatt 5859
Rückpfanne 7687
Rucksackmethode 6027
Rückschaltung 8349
Rückschlagen des Holzes 3297
Rückseil 3987
Rückstand 6889
Rückstau 470
Rücktraktor 7799
Rückung mittels mobiler Holzlader 4608
Rückungspanvlies 5077
Rückvorbereiten 472
Rückzubehör 7467
ruhendes Öl 2575
Ruhendöl 2575
ruher Brand 7734
Ruhrbirne 9424
rumelische Kiefer 489
Rumpenholz 555
Rumpf 1021
Rundast 7108
rundblättrige Felsenbirne 3806
rundblättriger Ahorn 5944
rundblättriger Hartriegel 7109
rundblättriger Weißdorn 7110
runder Stock 7111
Rundfahrt-Distanz 6922
rundgehörnter Eukalyptus 3550
rundgesägtes Schnittholz 1587
Rundholz 7113
im Wasser gelagertes ~ 9302

Rundholzanhänger 8752
Rundholzanhängewagen 8752
Rundholzausbeute 9780
Rundholzbank 6026
Rundholzeinschnitt 7115
Rundholzfabrikate 7117
Rundholzklammer 5094
Rundholzklasse 7114
Rundholzlader 7119, 8098
Rundholzmarkierung 5316
Rundholzmessung 7116
Rundholzplatz 7119
Rundholzprodukt 7117
Rundholzsortierung 7850
Rundholzsortiment 7100
Rundholzstapel 6026
Rundholzstapelung 3662
Rundholztabelle 7112
Rundholzverfärbung 5118
Rundknüppel 7101
rundkopfiger Fichtenkäfer 7104
rundkopfiger Kieferbohrer 7105
Rundling 7774
Rundschälfurnier 6135
Rundschälmaschine 9623
Rundschnitt 7107
Rundstange 6418
Rundstock 7111
Rundstück 7101
Runzeblatt-Schneeball 4899
runzelblättriger Schneeball 4899
Runzelschorf 8416
Rüsselkäfer 9331
Rüstbaum 7270
Rüster 2786
Rüstholz 1153
Ruten-Eukalyptus 5286
Rutschenklemme 7686
Rutschterrain 9211
Rutschtransport 1577

Saatbank 7372
Saatbeet 7373
Saatbett 7373
Saatblock 7375
Saatbrett 6345
Saatdichte 7374
Saatflasche 7381
Saatflinte 7381
Saatgans 3332
Saatgutbestand 8022
Saatgutdichte 7374
Saatguternte 7376
Saatgutgewinnung 7376
Saatgutherkunft 7395
Saatgutprovenienz 7395
Saatlatte 7385
Saatpflanze 9787
Saatreihe 7379
Saatrille 7379
Saatrinne 7379
Saatstreifen 7379
Sabalpalme 1238
säbelförmiger Stamm 7736
säbelformiges Blatt 8353
Säbelwuchs 8325
säbelwüchsiger Stamm 7736
Sabine-Föhre 7142
Sabine-Kiefer 7142
Sachalin-Korkbaum 7145
Sachalin-Weide 7146
Sachandelbaum 3983
Säckelblume 1395
Säckelblumebaum 1265
Säckelstrauch 6788
Sackgasse 2262
Sackstrecke 2262
säen 7371
Säen in Furchen 7869
Safranholzbaum 1371
Saft 7208
Saftaufstieg 7195
Saftausfluss 1589
Saftdruck 5965
Saftfällung 3065
Saftfleck 7215
Saftfleckbehandlung 318
saftfrisches Holz 3710
Safthieb 3065
Saftholz 7216
Saftigkeit 8250
Saftperiode 7211
Saftschälen 828
Saftschälenintensität 8410
Saftstrom 1589
Saftverdrängung 7198
Saftzieher 7199
Saftzufluss 1589
Saftzufuhr 1589
Sägeblatt 820
Sägeblatt mit eingeschobenen Zähnen 4516
Sägebloch 7245
Sägeblock 7245
Sägebock 4380
Sägebügel 7231
Sägedefekt 7242
Sägedimension 7469
Sägedurchmesser 7243
Sägedurchschnitt 7232
Sägefurnier 7261
Sägegatter 5132
Sägeholzteil 7249
Sägeholzvorrat 5455
Sägeholzwipfel 7252
Sägeindustrie 5206
Sägekeil 7267
Sägekerb 7246
Sägekette 1430
Sägeklammer 7234
Sägeklotz 7245
Sägemaschine mit rollierende Einspannwagen 1586
Sägemehl 9644
Sägemehlballensystem 7235
Sägemehlbrenner 7236
Sägemehlkompost 7237
Säge mit Benzinantrieb 3808
Sägemühle 8637
Sägemühleanlage 5106
Sägen 7241
Sägenführer 1434
Sägerampe 7233
sägerauhes Schnittholz 7094
Sägerei 8637
Sägerundholz 7245
Sägeschärfer 7239
Sägeschnitt 7246
Sägeschnittfläche 7247
Sägeschnittkeil 7267
Sägeschnittware 5458
Sägespäne 9508
Sägewerk 8637
Sägewerkarbeiter 7268
Sägewerkausschuss 5450
Sägewerk-Bewirtschaftungskosten 7254
Sägewerkeinrichtung 5106
Sägewerklagerplatz 7253
Sägewerksabfälle 5215
Sägewerksablauf 5454
Sägewespe 7240
Sägezähne 7264
Sägezug 8182
Sago 7153
Sagobaum 7153
Sagopalme 7153
sahnige Saftfäule 2010
saisonale Holzeinschlags-beschränkungen 7331
saisonaler Wald 7330
saisonaler Weg 7332
Saisonarbeiter 6095
Salamanderholz 5426
Salbaum 7163
Salbei 7152
Salie 4086
Saling 2059

Salweide 3880
Salzbassin 7161
Salzpflanzen 4102
Salzstrauch 7165
salztolerant 7164
Saman 6701
Samen 7370
Samenabfall 2996
Samenbaum 5569
Samenbeizung 6141
Samenbeschätzung 7389
Samenblatt 1983
Samendünnung 8514
Samenentflügelung 2455
Samenernte 7376
Samenerntebaum 5569
Samenerntebestand 6394
Samenfall 2996
Samenfalle 7398
Samenflügler 2456
Samengewinnung 7376
Samengewinnungsbestand 7377
Samengruppe 7393
Samenholz 5569
Samenlack 7386
Samenschlag 7383
Samenspross 6366
Samenstreuer 47
Samenträgerbank 5570
Samenwald 7388
Sämling 7392
Sämling im Pflanztopf 6483
Sämlingsamenobstgarten 1646
Sämlings-Schössling-Bestand 7390
Sammelgraben 2613
Sammlerhieb 2655
Samtbaum 264
Sandale 846
Sandarakbaum 7182
Sandaufforstung 7179
Sandbüchsenbaum 7185
Sanddorn 7325
Sanddorn-Akazie 1691
Sanddüneweide 5480
Sandelbaum 9382
Sandelholzbaum 9382
Sandelholzgewächse 7181
sandgefesselter Wald 7184
Sandhickory 7188
Sandkiefer 1797
Sandkirsche 7186
Sandmandelgewächse 5646
Sandmyrte 965
Sanierung 6860
Sanierungsmaßnahmen 7194
Santa-Lucia-Tanne 1037
Sapelli 114
Saphirbeere 8333
Sapindazeen 7806
Sapindengewächse 7806
Sapindusfichte 2723
Sapine 1317
Sapotengewächse 7207
Sapotillfrucht 5665
Sappan-Baum 7209
Sappan-Cäsalpinie 7209
Sappel 1317
Sappine 1317
saprogen 7212
Saprophyt 7214
saprophytische Pilze 7213
Sargents-Eberesche 7222
Sargentspappel 6319
Sargents-Zypresse 7221
Säschar 8963
Sassafrasbaum 1801
Satinnuss-Baum 259
Satinwollbaum 6403
Sattelaufleger 371
Sattelbaum 7148
Sattelpfropfen 7147
Sauberbrennen 1043
sauberer Schnitt 1606
Säuberungshieb 3136
Sauerampferbaum 7857
Sauerapfel 7856
Sauerdorn 1758
Sauerdorngewächse 528
sauere Sobbe 7856
Sauerkirsche 7855
Sauerkirschenbaum 7855
Sauerklee/Fichtenholz 9657
Sauersack 7856
Sauggraben 2613
Saugwurzel 7049
Säulchenrost 9376
säulenartige Form 1734
säulenartige Wuchsform 1735
Säulenfichte 1736
Säulenkaktus 1733
Säulenpappel 681
Säulenrost 9376
Saumbesamung 7384
Saumfemelschlag 8168
Saumschirmschlag 8178
Saumschirmschlagssystem 8179
Saumschlag 6167
säurebeständig 43
saure Limette 9348
saurer Humus 7954
säuretolerant 43
säuriger Regen 42
Savanne 7227
Savannenwald 7228
scalariforme Perforationsplatte 7272
Scaler 1457
Schaber 7960
schachbrettförmige Aussaat 6668
schachbrettförmiger Hieb 3064
Schachtholz 5463
schaden 4148
schadhafter Baum 2239
schadhaftes Holz 2238
Schadholzhieb 7169
schädigen 4148
Schadkraut 5806
Schädling 4578
Schädlinge 5807
Schädling-Inzidenz 6191
Schädlingsart 5805
Schädlingsbefall 4486
Schädlingsorganismus 5807
Schafbeere 5656
Schaft 8934
Schäften 7285, 7919
Schäftfuge 7285
Schaftholz 891
schaftreiner Stamm 1604
Schal 7283
Schälauslaufband 714
Schälbarkeit 536
Schälbeil 563
Schalblock 7284
Schäleisen 7960
schälen 2280, 6133
Schälen 579
Schälen im Saftperiode 6136
Schälen im Saftzeitraum 6136
Schälentringer 711
Schälförderer 712
Schälfurnier 6135
Schälfurnierblock 9193
Schälfurnierholz 901
Schälholz 901
Schalkante 2662
Schalkkeil 1548
Schalklatte 626
Schalm 540
Schälmaschine 9623
Schalmaxt 5314
Schalmbeil 5314
Schälmesser 7960
Schälrinde 580
Schälrolle 9193
Schälschaden 2237
Schälstück 6134

Schälverlust 580
Schälwald 5833
Schälwiderstand 6908
Schälwunde 2237
Schälzugförderer 712
Schande 829
Schandfleck 829
Schar 8963
scharf blätterig 7494
scharf blättrig 7494
scharfer Blazenzug 6387
scharfe Waldrebe 6387
Schärfewinkel 7493
scharfkantig 7963
scharfkantiges Brett 3029
scharfkantiges Holz 7969
scharfkantiges Vollholz 7492
scharfkantige Textur 3031
Scharlachdorn 3167
Scharlacheiche 7287
Scharlach-Eukalyptus 6773
Scharlachkordie 4799
Scharlachsumach 7791
Scharlach-Weißdorn 3167
Scharnierboomlader 4740
Scharniergerät 9114
Scharnierholz 4313
Scharrharz 593
Schattenast 604
Schattenbaum 7478
Schattenbaumart 7474
schattenertragende Art 7474
schattenertragender Baum 7478
Schattenerträgnis 7477
Schattenfestigkeit 7477
schattiger Baum 7479
Schätzung der undurchforsteten Bestände 3695
Schätzung nach Augenmaß 2869
Schätzungsrotte 8288
Schaufellader 3720
Schaumspiere 6993
scheckig 6131
Scheckigwerden 5575
Scheffel-Bezahlung 1200
Scheibe 1336
Scheibenentrinder 2533
Scheibenhacker 2534
Scheibenhackmaschine 2534
Scheibenpflügen 2535
Scheiben-Stubbenroder 2529
Scheinakazie 1754
Scheinbeere 8433
Scheinbuche 2999
Scheineller 1582
Scheinjahrenring 5616
Scheinkern 795
Scheinmarkstrahl 3001
Scheinring 2998
Scheinwanzenblume 3840
Scheinzypresse 4839
Scheit 8563
Scheitelhöhe 8726
Scheitelung 2198
Scheiterholz 2305
Scheitholz 8563
Schemelholzschicht 1180
Scherbaum 7496
Scherbeanspruchung 7495
Scherbruch 7500
Scherekopf 3044
Scheren 7501
Scherenflößung 480
Scherhieb 7501
Scherkraft 7495
Scherspannung 7495
Schertrift 480
Scherungsriss 7497
Scheuern 7127
Schibutterbaum 7503
Schicht 8133, 8150
Schichtablösung 2348
Schichtbrennholz 7920
schichtete Gelbstammfäule 9760
Schichtholz 4763
Schichtholzbiegen 4764
schichtholzer Kubikfuß 7977
Schichtholzkunststoff 2390
schichtig 8130
schichtiger Höhenbestand 8132
schichtiges Kambium 8131
schichtige Struktur 4842
schichtlose Vegetation 5763
Schichtmaß 7978
Schichtnutzholz 6230
Schichtung 8147
Schieblehre mit Prismenbacken 6554
schiefe Seite 4896
Schierlingstanne 1296, 4248
Schifferwanderbaum 1702
Schiffsbauholz 8590
Schiftschnitt 5857
Schilf 1308
Schilferriss 6675
Schimäre 3820
Schimelpilze 802
Schimmelfichte 9386
Schimmelpilze 5525
Schimmelvernichtungsmittel 3771
Schimmelweide 9224
Schimmelwuchs 3772
Schindelbaum 6277
Schindeleiche 3874
Schindelholz 9575
Schippen 4107
Schirmakazie 9041
Schirmausreißen 4933
Schirmbaum 6009, 9042
Schirmbesamung 6825
Schirmbestand 7511
Schirmbrechung 4565
Schirmdruck 8276
Schirmdurchforstung 7516
Schirmfläche 1314
Schirmhieb 7513
Schirmkeilschlag 7517
Schirmmagnolie 9040
Schirmpalme 5384
Schirmschlag 7513
Schirmschlagbetrieb 7408
Schirmschlagmethode 7515
Schirmschlag mit Reserve 7518
Schirmschlagsystem 7512
Schirmstamm 5810
Schirmulme 9039
Schirmverjüngung 6825
schizogen 7295
schizolysigenisch 7296
Schlafbaum 6185
schlafende Knospe 2573
schlafender Sämling 2576
schlafender Steckling 2574
schlafendes Auge 2573
Schlafenszeit 2572
schlaffes Blatt 3238
Schlaffheit 6652
Schlagabfall 998
Schlagabfällsstrom 2286
Schlagabnahme 6831
Schlagabraum 998
Schlagabraumschwaden 8780
Schlag-Attacke 6723
Schlagaufnahme 6831
schlagbarer Durchmesser 2949
Schlagbarkeit 3146
Schlagblöße 3038
Schlagbrennen 7719
Schlagbrennengesetz 7720
Schlagbrennholz 56
Schlagdatum 3060
Schlägeholz 2196
Schlageinteilung 191
Schlägelaxt 453
Schlägerungsjahr 3144
Schlägerungsplan 3416
Schlagfläche 2189
Schlagflächenarbeiten 3457

Schlagflora 3039
Schlagfront 3722
Schlagjahr 3144
Schlagkopf 3063
Schlagkopf mit Kettensäge 1433
Schlagordnung 2199
Schlagräumung 1624
Schlagreisig 9707
Schlagrichtung 2519
Schlagruhe 3418
Schlagseite 3722
Schlagspäne 4318
Schlagstreifen 2178
Schlagverbot 6577
Schlagwald 2196
schlagweiser Niederwaldbetrieb 1929
schlagweise Waldwirtschaft 3062
Schlagzeit 3077
Schlangenhautahorn 6800
Schlangenhaut-Kiefer 936
Schlangenholz 4921
schlankgriffelige Weide 7065
Schlankheitsgrad 3898
schlank zulaufend 8401
schlau 3679
Schlauchpilze 7144
Schlehdorn 7749
Schlehe 7749
Schlehenpflaume 7749
Schleifbäume 6626
Schleife 3881
schleifen 3975, 6133
Schleifen 7189
Schleifholzknüppel 6627
Schleifholzrundling 6627
Schleifkette 1555
Schleifstoff 9603
Schleifweg 5116
Schleimzelle 5597
schleppen 9024
schleppender Weg 7689
Schlepper 7691
Schlepperfahrer 7682
Schlepper-Operator 4168
Schlepperpunkt 9742
Schlepperrücken 8747
Schleppertakelage 9743
Schleppleine 9007
Schleppseil 9007
Schlepptau 9007
Schlepptrosse 9007
Schleppweg 5116
schlesische Weide 7595
schlicht gewachsener Baum 8138
Schlichthieb 7789
Schlick 7598
Schließfrucht 38
Schlinge 1555
Schlinger 9018
Schlingpflanze 9018
Schlitzöffnung 5136
Schlitzsäge 1136
Schluchtenwald 3474
Schluff 7598
Schlüsselart 4705
Schlüssel-Biotope 4700
Schlussgrad 2343
Schlusslinie 1661
Schlusslockerung 4566
Schlusslockerungsklasse 1312
Schlussspitze 8701
Schlusswald 1639
Schmackhaftigkeit 6032
Schmalbauchkäfer 9016
schmalblättrige Esche 5659
schmalblättrige Klebsame 8421
schmalblättrige Lorbeerrose 7505
schmalblättrige Ölweide 9421
schmalblättrige Pfefferminze 5663
schmalblättrige Steinlinde 5664
schmale Ölweide 9421
schmalkroniger Baum 5657
Schmallblattesche 5659
Schmalsaumschlag 5658
Schmarotzer 6071
Schmetterlingsblütengewächse 4908
Schmetterlingsstrauch 1219
Schmitterlingsstrauchgewächse 1220
Schmuckbaum 5957
Schmuck-Eberesche 7552
Schmucktanne 5535
Schnabelhiebmahl 765
Schnabelnuss 637
Schneeball 6474
Schneeballbaum 792
Schneeball-Blasenspiere 397
Schneebeere 1804
Schneebruch 7801
Schneeglöckchenbaum 7800
Schnee-Kirsche 4275
Schneerose 6989
Schneerutschbahn 7803
Schneeweide 3926
schneidblätterig 2170
Schneide 819
Schneidegeschwindigkeit 6727
Schneideholz 7245
Schneideholzbaum 7245
Schneidelholz 6615
Schneidelstamm 5165
Schneidemaschine 2181
schneiden 2161
ein dickes Ende ~ 4666
schneidendes Werkzeug 2204
Schneiden mit Wasserstrahl 9288
Schneidenrücken 4224
Schneider 4321
Schneider-Leistungsvermögene 7726
Schneidewinkel 288
Schneidezahn 2183
Schneidmaschine 2181
Schneise 3564
Schneitelbetrieb 6430
Schneitelholz 6615
Schneitelstamm 5165
schnellwachsende Arten 6666
schnellwachsender Baum 7546
schnellwachsender Wald 3020
Schnellwuchsbetrieb 6724
schnellwüchsiger Baum 3021
Schnitt 2185
Schnittbewegung 2519
Schnittbreite 9400
Schnitt durch die Spitzoberfläche 8687
Schnittfläche 2180, 2189
Schnittflugkreis 2190
Schnittfuge 7246
Schnitthöhe 2195
Schnittholz 1918
Schnittholzabfälle 5215
Schnittholzabstapelanlage 5212
Schnittholzabtastung 5210
Schnittholzausbeute 5214
Schnittholzausschuss 5450
Schnittholzbulldozer 911
Schnittholzdämpfer 1228
Schnittholzdicke 7256
Schnittholzende 7257
Schnittholze mit ungerader Länge 5866
Schnittholzfläche 7258
Schnittholz-Hängen 4117
Schnittholzkante 7255
Schnittholzlader 5207
Schnittholzlänge 7259
Schnittholzmaß 5125
Schnittholzmaßstab 5209
Schnittholzmeßgerät 5205
Schnittholzreihe 8545
Schnittholzseite 7258
Schnittholzsortierung auf dem Trockenland 2651

Schnittholzstapel 5211
Schnittholzsystem 1123
Schnittholztrockner 5203
Schnittholztrocknung 5202
Schnittholzvermessung 5213
Schnittholzware 9744
Schnitt in der Faserrichtung 2187
Schnitt in Läuterungen 3071
Schnittkreisdurchmesser 2479
Schnittrille 543
Schnittstelle 7668
Schnitttiefe 2195
Schnittverlust 5178
Schnittverluste 4696
Schnittware 5458
Schnittweide 9400
Schnittwinkel 288
Schnitzarbeit 9502
Schnitzelsortierer 1518
schnitzen 83, 1360
Schnitzermesser 1362
Schnitzkunst 9502
Schnitzler-Entladung 1526
Schnitzler-Kettensäge 1525
Schnitzschneidwerkzeug 1362
Schnurbaum 1944, 4357
Schnürigkeit 8142
Schockhärte des Holzes 9651
Schockholz 1118
Schönbeere 3705
Schönfrucht 3705
Schongebiet 5672
Schönheitsfehler 829
Schöpfen 828
Schopfpalme 8373
Schoss 1095
Schössling 7202
Schösslinggröße 7205
Schösslingstadium 7206
Schote 4391
Schotendorn 4353
Schrägfaser 4585
Schräghieb 4441
Schräglamination 5856
schrägschleifen 3975
Schrägschnitt 1441
Schraubenbaum 3871
Schraubengewächse 7316
Schraubenpalme 1802
Schraubenverdickungen 7916
Schraubenwuchs 9185
Schraubzwinge 4653
Schreckeeinheit 5089
Schreinerholz 4658
schrittweise Entastung 8085
schrittweiser Fallkerb 8088
schroten 2172
Schröter 7986
Schrotkeil 3079
Schrotsägen 6312
Schuhspanner 2054
Schulwald 8761
Schuppe 3241
Schuppen 7278
schuppenförmige Benadelung 7276
Schuppenkiefer 7277
Schuppenrindenhickory 7482
Schurz 333
Schussästung 7532
Schüsselung 2141
Schütte 5444
schütterbewalded 6450
Schuttgleiten 2287
Schutthalde 8378
Schuttkegel 8378
Schutzbaum 5816
Schutzbeamt 3456
Schutzbestand 1993
Schutzbrille 7150
Schutzbuschholz 1995
Schutzdeckung 2863
Schutz des Waldes 3359
Schutzelement 6598
Schutzgebiet 5672
Schutzhebel 7151
Schutzhelm 4122
Schutzkappe 1858
Schutzmatte 5335
Schutzpflanzen 6600
Schutzpflanzung 1993
Schutzplätzesaat 7510
Schutzreserveprogramm 1882
Schutzstamm 5810
Schutzvorrichtung 7149
Schutzwald 6592
Schutzwaldanbau 7509
Schutzwaldpflanzung 6595
schwachästig 7737
schwache Durchforstung 4952
schwächen 400
schwächeres Ende 7761
schwacher Wuchs 6449
schwaches Holz 7760
schwaches Nadelstangenholz 1876
schwaches Sortiment 7760
schwaches Stangenholz 7775
schwache Teilhiebe 4948
schwach flaumhaarig 6619
Schwachholz 7760
Schwachholzaufarbeitung 7779
Schwaden 9447
Schwadlegen 9448
schwalbenschwanzförmiges Blatt 2592
Schwamm 3770
Schwammbaum 8860
Schwammholz 2291
schwammig 3773
schwammiger Kern 1440
schwammiges Holz 2291
Schwammkiefer 6635
Schwammspinner 4071
Schwanken 8341
schwankende Bewegung 8341
Schwarm 8319
Schwarte 7712
Schwartebühne 7711
Schwarteneinschnitt 3253
Schwartenentrinder 7707
Schwartenhackmaschine 7708
Schwartenzerkleinerung 7709
Schwarzahorn 808
Schwarzast 800
Schwarzbeere 7523
Schwarzbirke 779
Schwarzdorn 793, 7749
Schwarzdornakazie 809
schwarze Akazie 3962
schwarze Buckelbeere 798
schwarze Esche 777
schwarze Heckenkirsche 790
schwarze Himbeere 782
Schwarzeiche 9294
schwarze Johannisbeere 2883
schwarze Krähenbeere 785
schwarze Mangrove 251
schwarze Olive 6014
schwarze Persimone 1796
schwarzer Ahorn 808
schwarzer Ast 800
schwarzer Brenner 315
schwarzer Eukalyptus 780
schwarzer Holunder 2892
schwarze Ribisel 2883
schwarzer Kiefernbastkäfer 805
schwarzer Kiefernblattkäfer 806
Schwarzerle 2882
schwarzer Maulbeerbaum 803
schwarzer Nachtschatten 804
schwarzer Tupelobaum 810
Schwarzesche 777
schwarzes Geißblatt 790
schwarze Walnuss 2715
schwarze Zimthimbeere 4631
Schwarzfichte 2716
Schwarzfleck 783

Schwarzföhre 415
schwarzfrüchtiger Weißdorn 793
Schwarzfrucht-Weißdorn 793
schwarz-gehörnter Kieferbohrer 797
schwarzgraue Wegameise 1762
Schwarzgummi 8310
Schwarzholz 814
Schwarzholzakazie 404
schwarzholzig 815
Schwarzkern 795
Schwarzkiefer 415
schwarzkopfiger Knospenwickler 794
Schwarzlinde 237
Schwarznuss 2715
Schwarzpappel 2884
Schwarzpappel-Hybride 4400
Schwarzwald 788
Schwarzwaldboden 789
Schwarzweide 813
schwarzwerdende Weide 9396
Schwebeseilbahn 8745
schwedische Eberesche 8322
schwedische Mehlbeere 8322
Schweißabsonderung 8321
Schweißausbruch 8321
schweizer Birnbaum 1795
schweizerische Weide 8348
Schwelholz 9571
Schwellen 8337
Schwellenholz 8591
Schwellenwert 8338
Schwemmenholz 3279
Schwenkarm-Greifer 8342
Schwenkkran 8343
Schwenksäge 4312
Schwerästung 4215
schwerbrennbares Holz 4125
schwerer Waldbrand 2974
Schwerholzkiefer 6445
Schwert 4040
Schwichtleine 8340
Schwichttakel 8340
schwierig 7913
schwieriger Stellungsort 2491
schwieriges Gelände 4212
schwieriges Holz 4126
Schwimmausleger 914
Schwimmbulldozer 911
schwimmender Holzschnitt 1181
Schwimmer 914
Schwimmerholz 3268
schwimmfüßiges Blatt 6037
Schwimmkette 8136
Schwinden 7554
Schwingsäge 4312
Schwingung 7483
Schwitzen 8321
Schwungsystem 8345
Schwungweg 8344
Scouler-Weide 7311
Scrambler 7312
Scribners Dezimalsatz 7318
Sea-Island-Baumwollpflanze 7327
Seebauholz 7524
Seeflößerei 6689
seegrüne Weide 3926
Seekiefer 5309
Seeküstenholz 7324
Seestrandkiefer 5309
Seetraube 7326
Seggen-Gräserfichtenwald 7368
Sehne 1564
Sehnenfallkerb 1565
Sehnenschnitt 8383
Seidelbast 2246
Seidelbastgewächse 5427
Seidelbastweide 9224
Seidenakazie 6185
Seidenpflanzengewächse 5445
Seifenbaum 7865
Seifenbaumgewächse 7806
Seifenbaumyucca 2224
Seifenbeerenbaum 7865
Seifenrindenbaum 7805
Seil 9463
seilartige Figur 7061
Seilauge 1555
Seilbahn 1245
Seilbahnschlepper 1242
Seilbahnsystem ohne Kabel 5772
Seildurchhang 699
Seilführung 7057
Seilhängebahn 8745
Seilklemme 7699
Seilkran 7697
Seilkürzer 7058
Seillaufkran 7697
Seilriese 3922
Seilrollenführung 7057
Seilrücken 1247
Seilschlaufe 699
Seilschlepper 8370
Seilschlinge 9463
Seilschlingemann 1559
Seilschloss 7699
Seilschuhe 917
Seilspannung 7704
Seilspannungsindikator 7705
Seiltrommel 7056
Seilwinde 7060
Seitenast 4819
Seitenbeschatten 7581
Seitenblockierung 7571
Seitenbrett 7572
Seitenfestlaufen 7570
Seitengerinne 7344
Seitenhauptwurzel 7582
Seitenholz 7577
Seitenknospe 419
Seitenknoten 2981
Seitenkrümmung 7936
Seitenlinie 7579
Seitenmarke 2984
Seitenpfropfen 7575
Seitenrückenabstand 4825
Seitenschatten 7581
Seitenschwimmer 3147
Seitenstamm 994
Seitentrieb 418
Seitenverjungung 4823
Seitenwurzele 4824
Seitenzweig 4819
seitliche Holzstammrückgewinnung 4822
seitliches Heranziehen 4821
seitliche Wurzelerweiterung 4820
sektionsweise Kubierung 9244
Sektorkaliber 7367
Sekundarbenadelung 7351
sekundäre Belaubung 6803
sekundäre Durchforstung 7366
sekundäre Hauptwurzel 6137
sekundäre Käfer 7340
sekundäre Klärung 4550
sekundäre Knospe 7342
sekundärer Durchlauf 7364
sekundäre Rinde 7339
sekundärer Seitenzweig 7349
sekundärer Strahl 7353
sekundäre Savanne 2408
sekundäres Insekt 7348
sekundäres Schleifweg 7350
sekundäre Sukzession 7354
sekundäres Xylem 7357
sekundäre Zellwand 7343
Sekundärforst 7346
Sekundärforstgarten 7347
Sekundärschädlinge 7352
Sekundärwald 7346
Sekundärwald mit Brandbrache 8339
Selbstbehalt 1978
Selbstbestäubung 7421
Selbstbeteiligung 1978

selbsteinladender Schlepper 7419
Selbsteinschlag 7416
Selbsten 7421
Selbstentzündung 7923
selbstfahrender Stumpfscherblock 7422
selbstfruchtende Kiwipflanze 51
Selbstlader 7418
Selbstladeskidder 7419
Selbstreinigung 5688
selbstwachsend 7417
Selbstwerberhieb 7416
Selbstwerbung 7416
Selektionsdifferential 7407
selektive Ernte 7413
selektive Roteichenarten 7402
selektive Weißeichenarten 7404
seltene Art 6725
seltene Holze 6726
Selva 7423
Semul 6792
Senegalakazie 4047
Seneszenz 7438
Senkblei 6389
Senkervermehrung 6582
Senkholz 2268
senkrechte Vielfalt 9205
Senkrechtwurzel 9207
Senkvermehrungsmethode 4843
Senkwurzel 6137
Sense 1204
Seralstufe 7455
serbische Fichte 7456
Serie 7457
seriell 7453
Setzgerät 6351
Setzholz 7982
Setzling 5815
Shaitan 9363
shattenspendender Baum 7479
shattenspendender Strauch 7475
Shorts 7547
Shumards-Eiche 7561
Siam-Rosenholz 7563
sibirische Fichte 7568
sibirische Lärche 7567
sibirischer Apfel 7565
sibirischer Erbsenstrauch 7564
sibirische Tanne 7566
sibirische Ulme 4618
sibirische Zirbel-Kiefer 7569
sichelblättrige Eiche 7863
Sichelsäge 1615
Sicherheitsstreifen 7508
Sichtauswirkungseinschätzung 9237
9234
Sichtentfernungszone 9236
Sichtqualitätsziele 9238
Sichtzeichen der Grenze 949
Sickergraben 2173
Sickerwasserplatz 7401
Siebfeld 7590
Sieboldsahorn 7585
Siebplatte 7592
Siebröhre 7593
Siebzelle 7589
Siedlung 7470
Siegel 8786
Signalfeuer 3779
Signieren 3074
Signierfärbung 3074
Sikkim-Fichte 7594
Silberahorn 7608
Silberbaum 8323
Silberbeere 257
Silber-Büffelbeere 7600
Silberdollar-Eukalyptus 7605
Silberlinde 7607
Silberpappel 2911
Silbertanne 1803
Silberweide 1809
Silberzeder 396
Silt 7598
Simaroubabaum 9373
Simon-Pflaume 332
Singapore-Eiche 7635
Sinkholz 2268
Sinus 5028
Sipo 9150
Siris 2730
Sitka-Erle 7677
Sitkafichte 7678
Sitkafichtenriesenbastkäfer 2724
Sitkafichte-Rüsselkäfer 7679
Sitkaweide 7680
sitzendes Blatt 7463
Sitzschnitt 7338
Skalierung im Weg 6981
Skarifikation 7286
Skariol 6531
Skeg 2274
Skidder 1245
Skiholz 5065
Skipperbahn 7696
Sklereide 7300
Sklerophylwald 7301
Sklerose 8097
sklerotisch 7302
Slacker 5017
Slipper 7744
Sloanea 7748
Sohle 3283
Sokotra-Drachenbaum 7810
sokotrensischer Drachenblutbaum 7810
Soledadkiefer 8717
solitäre Pore 7845
Sollmaß 3163
somatische Embryogenese 7846
somatische Embryonalentwicklung 7846
Sommerästung 8262
Sommerblattabfall 8261
Sommereiche 1793
Sommereinschlag 3065
Sommerfällung 3065
Sommerflieder 1219
Sommerfrostfraß 8259
sommergrüne Bedeckung 2299
sommergrüne Laubwalde 8260
sommergrüner Wald 2636
sommergrüne Walde 8260
Sommerheide 1779
Sommerhieb 3065
Sommerholz 8264
Sommerlinde 4801
Sommerriesweg 2707
Sommertamariske 3233
Sommertrieb 4765
Sonderforstprodukte 7889
Sonderhieb 7888
Sonderländer 7890
Sondersortimentschnittholz 7887
Sonnenblatt 8267
Sonnenbrand 8265
Sonnenriss 8266
Sonnenscheinakazie 8268
Sonnentaugewächse 7848
sonniger Drehwuchs 4907
sonstige Holzernte 5968
sonstige Holzfläche 5969
sonstiger Waldboden 5967
Sorte 2129
Sortenaushieb 7412
Sortiererabfälle 6850
Sortierkettenförderer 3289
Sortierkettenzug 3289
Sortimententafel 3896
Souarinussbaum 6188
soziale Forstwirtschaft 7808
Spalierbaum 2865
Spalt- 1634, 7483
Spaltaxt 453
spaltbares Holz 1628
Spaltbarkeit 1627
Spaltblatt 1636
Spalten 627

Spaltenende-Verluste 1072
Spaltenschnitt 9319
Spälterholz 8563
Spaltfähigkeit 1627
Spaltfestigkeit 6907
Spaltgatter 2277
Spaltgattersägemaschine 2277
Spalthammer 453
Spaltholz 8563
Spaltholzfaser 1524
Spaltkeil 3717
Spaltmaschine 9512
Spaltöffnung 8120
Spaltpflanzung 289
Spaltpfropfen 1635
Spaltpropfung 1635
Spaltriss 4195
Spaltsäge 5148
Spaltschemel 7921
Spaltseite 1630
Spaltungsebene 1630
Spaltveredelung 1635
Spanbildung 1515
Spanbrecher 1534
Spanbreite 9399
Spanbrennstoff 1517
Spanbruch 1536
Späne 9508
Späne am Stock 2193
Spänebrikett 9495
Spänebunker 2496
Späneholz 1542
Spaneinheit 9095
Spanfüllung 1523
Spangröße 1540
Spanholzschiff 1519
spanische Platane 7879
spanischer Buchsbaum 7875
spanischer Wacholder 7878
spanischer Zedrelenholzbaum 7876
spanische Tanne 7877
spanische Zeder 6530
spanlose Holzaufteilung 1516
spanlose Holzbearbeitung 1516
Spanmesser 1520
Spannbalken zum Exzentrisch-Schälen 8050
Spanndraht 1164
Spanne 7874
Spanner 1164
Spannkette 5023
spannrückig 1230
Spannseil 1164
Spannseilschlepper 4066
Spannseilssystem 4065
Spanplatte aus Holz 6091
Spanplatte mit abgestufte Dichte 3895
Spanplatte mit hoher Dichte 4277
Spanreiniger 1512
Span-Rinde-Abscheidung 1511
Span-Rinde-Trennung 1511
Spantenwerk 8714
Spantrenner 1537
Spantwerk 8714
Sparrenholz 6688
spätblühende Traubenkirsche 244
späte Durchforstung 4826
Späteiche 7464
spatelförmiges Blatt 7886
Spaten-Okulieren 7872
Spaten-Sprossung 7872
spätere Waldnachfolge 4817
spätere Waldsukzession 4817
später Zapfen 7459
spätes Brennen 4816
Spätholz 4827
spätreifend 7458
Spätrinde 4815
Spechte 9622
Sperrbaumkette 910
Sperrfestholz 2390
Sperrholz 6398
Sperrholzbiegen 4764
Sperrholzliste 6400
Sperrholz-Verbundplatte 9520
Sperrschuh 7687
Spezialbaum 7892
Spezialholz 7891
spezielle Knötchenbildung auf Stengeln 7906
spezifische Blattfläche 7904
spezifischer Blattvolumen 7905
Spiegel 2985
spiegelbildliche Furnierzusammensetzung 3265
Spiegel des Holzes 2985
Spiegelfaser 7601
spiegelgeschnittenes Schnittholz 6678
Spiegelholz 6678
spiegelndes Holz 6678
Spiegelrelaskop 6852
Spiegelrinde 7787
Spiegelschnitt 2760
Spierling 8991
Spierstrauch 5350
Spießeiche 3764
spießförmiges Blatt 4165
Spindelbaum 1383
3785
spindelförmige Eiche 6376
spindelförmige Kambiumzelle 3781
spindelförmige Parenchymzelle 3783
spindelförmiger Markstrahl 3784
Spindelkrone 3782
Spindelpalme 7911
Spiralholz 2048
Spiralsäge 7915
Spitzahorn 935
spitzblättrige Eiche 4625
spitzblättrige Esche 5659
spitzblättriger Ahorn 935
spitzblättrige Weide 7138
Spitzblume 5320
Spitze 819
Spitzenablösung 8696
Spitzenabsprung 8696
Spitzenbräune 1092
Spitzendürre 8669
Spitzenhöhe 8692
Spitzenholz 4752
Spitzenknospe 8470
Spitzenknüppel 7771
Spitzentrieb 6536
spitzgipfeliger Baum 5657
Spitzkrone 1866
spitzkroniger Baum 5657
Spitzmaus 8923
spitz zugelaufen 8401
Splint 7216
Splintauswuchs 7196
Splintbaum 7220
Splintbläue 869
Splintfäule 7218
Splintfäulnis 7083
Splintfleck 7215
splintfrei 7197
Splintholz 7216
Splintholzabschätzung 7217
Splintholzart 7219
Splintholzbaum 7220
Splintholzkäfer 5220
Splitter 7530, 7873
Splitterung 1629
spontane Verjüngung 7924
Spore 7927
Sporophor 7928
Sporophylle 7929
Spottnuss 8988
Spray am unteren Stamm 606
Sprenger 6485
Sprengkraft 7502
Sprengmast 4939

Sprengvorrichtung 8404
Spreufleckigkeit 2578
sprießend 7807
Springen 4668
Springer 9022
Springfeuer 4669
Springhieb 224
Springschlag 224
Spritzgerät zur Unkrautbekämpfung unter den Bäumen 8800
Spritzrohrverlängerung 8442
sprödes Holz 7535
Spross 7132
Sprossen 1020
sprossen 3293
Sprossenrad 7943
Sprossknospe 419
Sprossung 307
Sprühfahrzeug 7934
Sprühnebelvermehrung 5481
sprungweiser Streifen 4670
sprungweise sich ausbreitendes Feuer 4669
Spulenholz 7926
Spund und Nut 8676
Spurenelemente 8740
staatliche Forstwirtschaft 3890
Staatsforst 8041
Staatsforstwirtschaft 3890
Staatswald 8041
Staatswaldfonds 3767
Staatswaldsinventur 8042
Stabbretter 954
Stäbchenplatte 4762
Stabholz 4362
Stabilitätspflege 8029
Stachelannone 7856
Stachelanone 7856
Stachelbeerbaum 5966
Stachelbeere 2148
stachelblättrige Duftblüte 4339
stachelblättrige Linde 667
Stacheldolde 6532
Stachelfeige 6533
Stachelkiefer 770
Stachellatich 6531
Stachelrad 7943
Stachelsalad 6531
stachliger Baum 8528
stachliger Flaschenbaum 7856
Stadtforst 9142
Stadtoberförster 1592
stahlblaue Kiefernschonungsgespinstblattwespe 6253
Stamabfluss 8073
Stamm 1389, 8934
Stammabfuhr 5615
Stammabladevorrichtung 5097
Stammabschrägung 8083
Stammabstand 4569
Stammabzählen 2837
Stammachse 5048
Stammanalyse 8060
Stammanhänger 1559
Stammankerben 2514
Stammanlauf 1234
Stammästung 8081
Stammausschlag 2848
Stammbalken 638
Stammbasis 8061
Stammbaumplantation 3009
Stammblock 8879
stammblütig 1391
Stammbohle 8062
Stammbrett 8062
Stammbruch 8064
Stammbündel 8066
Stammdurchmesserabfall 8083
Stammeinreßen 4845
Stammende 607
Stammendfäule 1233
Stammengewulst 605
Stammertrag 9780
Stammexzentrizität 8996
Stammfäule 8998
Stammfeuer 8072
Stammfläche 900
Stammförderer 8596
Stammform 898
Stammformzahl 8075
Stamm für Instrumentenholz 3107
Stamm-Fuß-Behandlung 603
Stammgabel 8074
Stammgeld 8195
Stammgeradheit 8142
Stammgrund 8061
Stammgrundfläche 597
Stammheber 4173
Stammholz 891
Stammholzabfuhr 5615
Stammholzelevator 5060
Stammholzfuder 5058
Stammholzhebezange 4934
Stammholzrücken 899
Stammholzschlitten 2587
Stammholzstapel 8066
Stammholzvolumen 9249
Stamminhalt 9249
Stammklasse 8067
Stammklassifizierung 8068
Stammkopfumfang 8690
Stammkrümmung 8325
Stammladung 5099
Stammlänge 8077
Stammlänge gleich dem Elevator 5095
Stammlode 8076
stammlos 8078
Stamm mit Brettwurzel 595
Stamm mit einseitiger Krone 2062
Stamm mit Säbelwuchs 7736
Stammpreis 8195
Stammriese 5116
Stammrückigkeit 8999
Stammscheibe 1336
Stammschenkel 607
Stammschleife 8758
Stammschlepphaube 1858
Stammschlucht 5091
Stammschutz 8997
Stammskalierung 5103
Stammspäne 1539
Stammspindel 2351
Stammspitze 8700
Stammstärke 8071
Stammsteckling 8070
Stammteilung 8069
Stammträger 8065
stammtrockenes Holz 2274
Stammtüpfelung 8080
Stammunterhölung für Kette 1556
Stammvereinzelner 7976
Stammvolumen mit Rinde 8894
stammweise Mischung 5496
Stammzahl 8079
Stammzentrierung 5053
Stammzuführung 5061
Standard 6887
Standardacker 8003
standardgröße Späne 8006
Standardholz 2503
Standardknoten 8004
Standardmassentafel 8007
Stander 7934
ständiger Waldarbeiter 6173
ständiger Welkepunkt 6177
ständig wiederholender Komplex 6849
Standort 4072
Standortbonität 7667
Standort-Fähigkeit 7663
standortgemäße Holzart 1640
standortgemäße Mischung 1638
standortgerechter Wald 7662
Standortsanspruch 7671

standortsfremde Baumart 7903
standortsgerechte Baumart 7895
Standortsgüte 7670
Standortsgüteklasse 7664
standortsheimische Baumarten 4477
standortspezifische Anforderung 7671
Standortstörung 7665
Standortstyp 7675
Standortverbesserung 7666
Standortvorbereitung 3985
Standortwahl 7672
Standraum 4008
Stange 6418
Stängel 1390
stängelförmig 1392
stängeltreibend 1390
Stangengröße 6422
Stangenholz 6428
Stangenholzriese 6425
Stangenpolter 6424
Stangensäge 6421
Stangen-Schlepper 6423
Stangenstadium 6426
Stangenstoß 7780
Stangenwald 6419
Stapel 8619
über Kreuz geschichteter ~ 6232
Stapelbläue 1031
Stapeldichte 7980
Stapelfleck 5451
Stapel-Hebebühne 8581
Stapelholz 7982
Stapelmaschine 5212
stapeln 7983
Stapelplatz 5451
Stapelplatzgrundbau 5057
Stapelplatzunterkonstruktion 5057
Stapelschnittholz 9754
Stapelstelle 5451
Stapelstütz 7975
Stapelung 4846
Stapelung im Holzlager 6234
Stapelungspunkt 7981
Stapelung und Brennen 6233
Stapelvorrichtung 8581
Stapler 3660
Starkdurchforstung 4217
starke Durchforstung 4217
Stärkegruppe 8159
Stärke im trockenen Status 2657
Stärkeklasse 2471
starke Niederdurchforstung 4214
Stärkenmessung 3857
starkes Baumholz 5886
starkes Holz 4808
Stärkesortierung 2483
starkes Stangenholz 4803
starke Stangenreiser 4807
starkes Wachstum 4213
starkgesenkter Baum 2296
Starkholz 4808
Starkholzzuchtsystem 6888
starkkroniger Baum 1315
starkwüchsiger Baum 3021
starkwüchsiger Wald 3020
stationäre Entastungsmaschine 8045
stationäre Hacker 8044
stationäre Luftseilbahn 8551
statische Holzhärte 9665
Staubabführung 2318
Staubbeutel 312
Staubblatt 7998
Staubblattbaum 7999
Stäubegerät 2675
Stäubemaschine 2675
Staubgefäss 7998
Staub-Palliativa 2676
Staude 7555
Stauholz 8579
Staurechen 8136
Stechbeitel 9509
Stechdorn 7749
Stecheiche 7190
stechender Mäusedorn 1212
Stechfichte 1726
Stechgilie 6534
Stechpalme 2899
stechpalmenblättriger Faulbaum 4338
stechpalmenblättriger Kirschbaum 4336
Stechpalmen-Eiche 1773
Stechpalmengewächse 4335
Stechschnitt 930
Stechwinde 7223
Steckkeil 7809
Steckling 8056
Steckling-Durchwurzelung 7031
Stecklingsvermehrung 6581
stehend 8018
Stehendästen 3953
Stehendästung 3953
stehende Holzreserve 9250
stehender Baum 4010
stehender Holzvorrat 9250
stehender Trockenbaum 2271
stehendes Holz 3454
Stehendrodung 8201
Steigung 5979, 7750
Steilrand 2762
Steinbeere 1996
Steinbrechgewächse 7269
Steinbuche 2901
Steindeckwerk 6969
Steineibe 2714
Steineibengewächse 6407
Steineiche 1773, 7464
Steinfrucht 2644
Steinholz 6196
Steinkohle 1673
Steinlinde 5503
Steinmeißel 6386
Steinnusspalme 4597
Steinpackung 6969
Steinpalmholz 6195
Steinschliff 9603
Steinschüttung 6969
Steinstreifen 8122
Stein-Thylle 7303
Steinzelle 8121
Stele 8058
stellenweises Brennen 6105
stellenweises Verbrennen 6105
Stellmacherholz 1672
Stempel 6296
Stempelblütenbaum 6297
Stempel für Qualität 3900
Stempelzapfen 1350
Stengelbohrer 8063
Stengelchen 7132
Stengelstück 8879
Steppenaufforstung 8086
Steppenkirsche 3984
Steppenwald 3607
Steppenwaldbau 8087
Steppenwaldzonen 3609
Steppe-Wald-Übergang 3608
Sterkuliengewächse 1550
Sternanis 293
Sternanisgewächse 294
Sternanispflanze 4611
Sternapfelbaum 8037
Sterneiche 255
Sternfestholz 2390
sternförmiger gelber Affenkamm 5534
Sternkiefer 5309
Sternmark 8059
Sternriss 8039
Steuer auf die Rodung von Flächen mit Holz- oder Waldbestand 8430
Steuermann 9569
Steventotholz 2274

Stichsäge 4702
Stickstoffbindung 5736
Stickstoff-Fixierung 5736
Stickstoffkreislauf 5735
Stiel 1307, 8190
Stiel- 1390
Stieleiche 1793
Stielzelle 7997
Stierstock 1165
Stiftästchen 830
Stiftbolzen 8192
Stiftforst 4519
Stiftschraube 8192
Stiftungsforst 4519
Stiftungswald 4519
Stinkholz 2710
stinkige Waldnessel 4222
Stinklorbeer 807
Stirnfläche 2803
Stirnseite 2980
Stock
 auf dem ~ 8018
Stockabschnitt 607
Stockachselpflanzung 6348
Stockanschwellung 7015
Stockausschlag 1933
Stockausschlagsbetrieb mit Kurzumtrieb 7544
Stockausschlagsgruppe 1665
Stockausschlagtrieb 1933
Stockbereich 607
Stöckeln 4846
Stöckerholz 7774
Stockfäule 1233
Stockhöhe 4236
Stockholz 8220
Stockholzgewinnung 8204
stockiges Holz 2605
Stock im Niederwaldbetrieb 1934
Stockkienholz 6899
Stockkluppe 715
Stockrodemaschine 8199
Stocksprengung 852
Stockung 7599
Stockverkauf 7157
Stockwert 8196
Stockwertabschätzung 8194
Stockwertschätzung 8194
stoffgeleimt 8119
Stolon 7132
Storaxbaum 3825
Storaxgewächse 8128
Stoß 9519
Stoßbohrer 4667
Stoßeisen 6611
Stoßplatte 1228
Strahlengriffelgewächse 50
Strahlenriss 6675
Strahlkambiumzelle 6735
Strahlung 6682
Strandkiefer 5309
Strandpflaume 633
Strandpflaumebusch 633
stranggepresste Spanplatte 2976
 bei jedem Wetter benutzbare ~ 205
Straßenbaum 6983
Straßengrundstücksräumung 6942
strategische Landesplanung 8146
strategische Raumplanung 8146
Stratobiont 8149
Stratum 8150
Sträuben am Gradbruch 1127
Sträuben für Ausgabe 1130
Sträuben im Wert 1129
Strauch 7555
strauchartige Päonie 8895
strauchartige Pfingstrose 8895
Strauchbirke 1295
Strauch-Eibisch 7067
Strauchfläche 353
Strauchkastanie 940
Strauchkiefer-Knospenwickler 4602
Strauch-Rosskastanie 940
Strauchwald 1202
Strauchwerk 933
Strebe 8190
Strebepfeiler 1229
Streichversatz 3147
Streichwurzel 7486
Streifen 8154, 8172
Streifenahorn 8174
Streifenboden 8122
Streifenbreite 8320
Streifendurchforstung 4992
Streifenmesser 7960
Streifensaat 522
streifenweise Applikation 8169
streifenweises Abbrennen 8170
streifenweise Verjüngung 6813
Streitkolbenbaum 1692
strenger Zweig 8162
stressabgestufte Laubholzarten 8160
Streuberechtigung 6947
Streubestand 7293
Streunenkern 9264
Streunensteckling 9263
Streunutzungsrecht 6947
Streunutzungsservitut 6947
Streuung an Baümen am Standort 7294
Stringerweg 8164
Strobe 2726
Strobenrinde 6244
Strobenrindenlaus 6245
Strobilus 8818
Stropp 4850, 7742
stroppeingewickeltes Bündel 7743
strukturbedingte Unterschiedlichkeit 8183
Struktur des Holzes 9666
strukturelle Bohlen und Balken 8185
strukturelle Wärmeschutzdeckung 8186
Strukturfehler 8187
Strukturholz 7843
Strunk 8937
Stubbenalter 8193
Stubbenheber 8199
Stubbenhöhe 4236
Stubbenholz 8220
Stubbenholz-Forwarder 8221
Stubbenholz-Forwarder-Reiniger 8222
Stubbenholzspanewalze 7002
Stubbenrodung 8201
Stubenrodemaschine 8808
Stück 1571
Stückverkauf 6119
stückweise Methode der Kubierung 6226
Stufenfallkerb 8088
Stufenkrone 8134
Stufenpflanzung 6357
Stufenschluss 8148
stufenweise Mischung 5498
stufenweiser Holzschlag 3904
Stufigkeit 8147
Stummel 8937
Stumpen 8937
Stumpf 8937
Stumpfanalyse 8197
Stumpfanhänger 8218
Stumpfaxt 6986
stumpfblättriger Ahorn 8355
Stumpfbrecher 8200
stumpfes Ende 871
Stumpffigur 8202
Stumpffurnier 8219
Stumpfgreifer 8203
Stumpf mit Wurzelsystem 8211
Stumpfpflanze 8209
Stumpfpflanzung 6358

Stumpfrodung zu
Einkehrsproduktivität 8216
Stumpfrodung zu
Holzwerksproduktivität 8215
Stumpfspalter 8214
Stumpfspäne 8198
Stumpfspringer 8206
Stumpfverminderung 8207
stumpfwinkliges Blatt 5861
Stumpfzieher 1160
Stumpfziehrad 8210
Sturmholz 850
Sturmschaden 2236
Sturzbett 3283
sturzgespeiste Hackmaschine
2634
Stützbock 8965
Stützboden aus Holz 8654
stützeinrichtunger Sattelanhänger
7435
Stutzen 4187
Stützen 8932
Stützpfahl 8273
Stützsäge 2043
Stützschwanz 888
Stützschwanzskurzholzsystem
889
Stützwurzel 8411
subalpine Braunfäule 8225
subalpiner Wald 8226
subalpine Schlagfäule 8227
subantarktischer immergrüner
Regenwald 8228
subantarktischer Regenwald 8228
subäquatorialer Monsoonwald
8232
subarktischer Lichtwald 5904
subarktischer Wald 8229
Subklimax 8230
Subor 8234
Subsistenzbetrieb 8238
Substrat 8240
Subterminalknospe 8242
subtropischer immergrüne Wald
8243
subtropischer Wald 8244
Sucupira 8251
Südbuche 2999
Südeiche 8394
Südkiefer 5141
südliche Esche 1342
südlicher Zürgelbaum 2898
südliche Weide 1688
südostasiatischer Regenbaum
6701
Südseemyrte 1073
Sugi 4615
Sukkulente 8082
Sukzession 8246
Sukzessionsmähen 8247
Sukzessionsphase 8248
Sukzessionsstadium 8248
Sukzessivstadium 7455
Sulky 5090
Sumach 8255
Sumachgewächse 1369
Summenwachstumskurve 2135
Sumpfakazie 8316
Sumpfbeere 2889
Sumpfeiche 3764
Sumpfföhre 5598
Sumpfheidelbeere 893
Sumpfholzrückung 8311
sumpfiger Waldboden 8318
Sumpfkiefer 5141
Sumpflichtwald 8312
Sumpfpappel 8306
Sumpf-Rhododendron 8303
Sumpfrückung 8311
Sumpfsassafras 643
Sumpfschwalbe 8941
Sumpfsubor 8314
Sumpftaxodie 786
Sumpfwald 8309
Sumpfweide 813
Sumpfzeder 786
Sumpfzypressengewächse 8425
Süntelbuche 2687
Suramen 8277
Surinam-Bitterholz 9373
süße Akazie 1689
süßer Maniok 8329
Süßholz 4997
Süßhülsenbaum 1790
Süßkirsche 8330
Süßkirschenbaum 8330
Süßwasser-Sumpfwald 3711
Swamppodsol 896
Sycamore 254
Sykomore-Ahorn 8355
Sykomore 254
Symbiont 8358
Synfolium 8360
Synökologie 8359
Systematiker 8428
Szechuanpfeffer 4630

Tafelbergföhre 8361
Tageslichteinfall 2260
Tagwurzel 7486
Taiga 8368
Taigawald 8368
Takelwerk 7467
Talgbaum 7201
Talgsumach 9306
Tali 1366
Talipotpalme 8373
Tallöl 8374
Talon 333
Tal- und Rückenwegmodell 6938
Tamarinde 8379
Tamarindenbaum 8380
Tamariske 3706
Tamariske-Dickicht 8381
Tamariskengewächse 8382
Tamil 5447
Tangelbaum 1797
tangentiale Quellung 8387
tangentialer Riss 2139
tangentiale Schwindung 8386
tangentialfaseres
Presseschichtholz 8384
Tangentialfaserneigung 8385
Tangentialriss 2139
Tangentialschnitt 8383
Tangentialschnittholz 3254
Tangentialschwellen 8387
Tankholz 3816
Tannenadel 5707
Tannenholz 7959
Tannennadel-Rostpilz 6764
Tannenpfeil 6257
Tannenwald 3224
Tannenzapfen 3165
Tannin 8393
Tariftabelle 8415
Taschenfäule 6313
Taschenkluppe 3299
Taschenmeißel 6404
Taschentuchbaum 2593
Taschenweißfäule 9377
tasmanischer Fieberbaum 7864
tasmanische Scheinbuche 406
Tatajuba 478
Tatarenahorn 8418
tatarischer Ahorn 8418
tatarischer Steppenahorn 8418
tatsächlich vorhandenes
brennbares Material 422
tatsächtliche Größe 59
tatsächtlicher Einschlag 57
Taubenbäumgewächse 2594
Tauchbeizung 2511
Tauchlösung 2508
T-Aufblatten 8441
Taukkyan 9719
Taupunkt 2457
Tausendkubikfuß 8530

Tauwurzel 7486
Tawhai 8420
Taxationsmanual 3409
Taxbaum 1814
Taxierungsmannschaft 8288
Taxierungswachstum 8291
Taxocenosis 8423
taxodioidische Tüpfelung 8424
Taxon 8427
Taxonom 8428
taxonometrisch 8426
Taxonomie 8429
taxonomische Einheit 8427
Tchitola 8432
Tchitola-Baum 8432
Teakbaum 109
Teakholz 8436
Teakholzbaum 109
Teakholzbaumwald 8435
technische Holzprodukte 2826
technischer Umtrieb 8440
technische Schlagbarkeit 8439
technisches Holz 2825
Teebaum 8437
Teebaumöl 8438
Teegewächse 8434
Teeölbaum 1056
Teer 9672
Teestrauch 8437
Teichschilf 1308
Teichsimse 1308
Teikette 270
Teilbaumhackschnitzel 6096
Teilbaumhackspäne 6096
Teilbeseitigung 6089
teilgerindetes Holz 9596
Teilhieb 6083
teilnehmendes Management 6090
Teilsaat 6088
Teilstapelung 6086
Teilzeitwaldarbeiter 6095
teleskopische Spritzrohrverlängerung 8442
Tempelbaum 4626
Tempel-Kiefer 4750
Temperaturchronologie durch Jahresringe 8910
temperierte Holzfaserplatte 8445
temporärer See von Herbst bis Frühjahr 9202
temporäres Gewässer 9202
Tepeguaje 3933
Teppichhartriegel 1170
terminales Parenchym 8475
Terminalknospe 8470
Terminalspross 8474
Terpentinpistazie 8469
Terrain 4769
testen
das Holzaufnahmebuch ~ 1458
Tetracentrongewächse 8481
Teufelskeule 2451
Teufelskrückstock 2452
Teufelsspazierstock 4610
Teufelszwirn 9471
Teufenanzeiger 2403
Texas-Esche 8482
Texas-Pflaume 8485
Texas-Rosskastanie 8483
Texas-Walnuss 5009
Texas-Zederholz 8484
theoretischer Querschneidenpunkt 8486
thermische Wärmeschutzdeckung 8490
thermophil 8491
Thermophyt 8492
Thuja 8893
Thuje 8893
Thunberg-Berberitze 4612
Thunberg-Kiefer 4614
Thylle 9037
Thylosoid 8540
Tiama 8541
Tibet-Kirsche 744
Tiefbesäumen 2322
tiefe Bläue 2321
Tiefe des Floßweges 2405
tief eingeschnittenes Blatt 2325
Tiefenlehre 2403
Tiefenmesser 2403
tiefer Ruhezustand 2323
tiefer Schnitt 4211
tiefe Ruhephase 2323
tiefer Wald 2324
Tiefkühlung 9203
Tiefland-Laubholz 5193
tiefliegendes Zugseil 3993
Tieflöffelbagger 468
Tiefschnitt 4211
Tiefwurzler 2326
Tiere des Waldes 3644
Tierlibaum 1965
Tier- und Pflanzenweltarten 9417
Tier- und Pflanzenwelthabitat 9412
Tier- und Pflanzenwelthabitatsfläche 9413
Tier- und Pflanzenwelthabitats-management 9415
9414
Tigerschwanz-Fichte 8547
Tillai 5447
Tintenbeere 3790
Tintenkrankheit 4500
Tintenkrankheit der Kastanie 4500
Tischhobelmaschine 6326
Tischkreissäge 7003
Tischler 4656
Tischlerdielen 4657
Tischlerholz 4658
Tischlerwerkzeuge 4659
Tischplatte mit Bauholzkern 5201
Tisselkirschbaum 8330
Tod wegen der Hitze 4209
Tola 5731
Toleranz 8673
Tolu 504
Tolubalsam 504
Tolubalsambaum 504
Tomatenbaum 8946
Tonholz 7851
Tonne 8675
Toogbaum 4645
Topfende 7761
Topffruchtbaumgewächse 5531
Topfpflanzung 6482
Topioka 5282
Topografie 8703
Topophysis 8704
Toposequenz 8705
Toppenant 2410
Toppenbaum 6126
Torfbeere 2889
Torfmull-Lärchensumpfwald 5306
Torfmyrte 4898
Torreye-Muskatnuss 1276
Toruskern 6314
Tosbecken 8100
totale Vernichtung 2968
Totalitätshieb 1607
Totara 8734
Totenuhrfraß 2279
toter Baum 2271
toter Kern 2261
toter Mann 275
toter Wald 2267
tote Schadbäume 7166
totes Streckenstück 2262
tote Streu 2269
Totholz 2274
Totholzsammlung 7170
Totstreifen 4708
Trabekel 8739

traditioneller Wald 8748
Traganth 4055
tragbares Buschhacker 6470
tragbares Kleinsägewerk 6469
Trägerblock 1354
tragfähige Forstentwicklung 8295
Tragkante 640
Tragleiste 9592
Tragseil 8745
Trainingsbaum 8760
Traktnummer 6380
Traktnummer der nationale Waldinventur 6381
Traktorbringung 8747
Traktorenabfuhr 8747
Traktorzug-Erntemaschine 8746
Tram 5121
Trambahn 5121
Tränenkiefer 4308
Tränenweide 459
Tränkholz 4427
Tränkreife 4489
transgene Pflanze 8764
Transitfloß 8766
Transpirationsfluss 8768
Transpirationstrocknung 8767
Transplantat 5815
Transplantatsreihen 8772
Transport ganzer Bäume 9394
Transportgrenze 4979
Transportkarre 894
Transport mit Holz-LKW 5123
transversale Kompression 1849
Trasse 7468
Trassierung 8743
Traube 6671
Traubendorngewächse 1213
Traubeneiche 7464
Trauben-Eukalyptus 7861
Traubenfeige 1670
Traubengeißklee 1069
Traubenholunder 2907
Traubenkirsche 1769, 2881
traubige Platane 1275
traubige Schneebeere 1804
Trauer- 9324
Trauerbirke 1761
Trauerkrone 2632
Trauerschnurbaum 4626
Trauerweide 459
Traufengasse 181
traumatisches Parenchym 8784
traumatische Zone 8788
Trebol 5225
treiben 7944
Treiben 8754
Treidel 9007
Trenn- 2176
Trennbandsäge 1410
trennende Frucht 7447
trennendes Parenchym 2531
trennende Tracheide 2532
Trennholz 6882
Trennriss 7747
Trennsäge 5148
Trennschneiden 5016
Trennschnitt 2322
Treppenhieb 8084
Treppenpflanzung 6357
Triangulation 8966
Triangulierung 8966
Trieb 7132
Trift 8584
triftbar 2628
Triftflößerei 2625
Triftholz 3279
Triftholzbrecher 8136
Triftkanal 3277
Triftmarke 3275
Triftperiode 3278
Triftpfad 3277
Triftrechen 8136
Triftschtraße 3277
Triftverlust 5177
Triftweg 3277
Trikuspidalblatt 8969
Trockenast 2647
Trockenastung 1106
trockene Böttcherei 7713
trockene Entrindung 2646
trockene Fassware 7713
trockene Holzpartikeln 2659
trockene Lachte 2264
trockener Küstentropenwald 1690
trockener Wipfel 7794
trockenes Holz 2658
trockene Verpackung 2654
Trockenfäule 2656
Trockenfäulenholz 9484
Trockenholz 2658
Trockenholzaufasten 2270
trockenliebende Pflanzen 9736
Trockenriese 2648
Trockenriss 2422
Trockenrutsche 2648
Trockensubstanz/Solar-einstrahlung-Verhältnis 2652
Trockensubstanz/Solarstrahlung-Verhältnis 2652
Trockensubstanz/Transpiration-Verhltnis 2653
Trockenwald 2650
Trocknungsriss 2422
Trocknungsstapel 7336
Trocknungsstufe 9653
Trog 5116
Trommelentrinder 566
Trommelentrindungsmaschine 479
Trommelhacker 2208
Trommeloperator 2642
Trommelsäge 2209
Trommelwinde 2643
Trommelzerhacken 2640
Trompetenbaum 1768
Trompetenbaumgewächse 8995
Trompetenblume 8994
Trompillo 6237
Tropenholz 8981
Tropenwald 8975
Tropenwald-Ökosystem 8976
Tropenwaldwirtschaft 8977
Tropfen 4062
tropischer Gebirgsregenwald 5593
tropischer halbwechselgrüne Trockenwald 8974
tropischer laubwerfende Trockenwald 8973
tropischer Lianen-Regenwald 9222
tropischer Nebelbergwald 5593
tropischer Regenwald 8978
tropischer Sekundärwald 8980
tropischer Urwald 8979
tropischer Wald 8975
tropischer wechselgrüne Regenwald 5545
Tropophyte 8982
Trosse 9463
Trumm 2171
Trümmer 2440
Trunkelbeere 893
Tschaparral 1442
Tuberkel 7036
Tugai 6961
Tulpenbaum 9008
Tulpen-Magnolie 7225
Tungölbaum 1304
Tupelobaum 810
Tupelobaumgewächse 9010
Tüpfel 6461
Tüpfelfarn 6441
Tüpfelfeld 6305
Tüpfelkammer 6302
Tüpfelkanal 6300
Tüpfelkavität 6301
Tüpfelöffnung 6298

Tüpfelschließhaut 6308
Tüpfelung 7591
Turill 5406
türkische Eiche 5565
türkische Galle 3794
türkische Hasel 5383
türkische Kirsche 1475
türkische Nuss 5383
türkischer Holunder 6292
Turnus 2194

überalterter Bestand 5998
überalterter Wald 5998
Überart 8271
Überbehälter für den Waldbrandfall 3189
überbelaubt 1658
überbestocktes Bestand 6006
überblätterig 1658
Überblatting 4096
Überbleibsel 4325
überdickes Holz 2975
überdimensionale Späne 6004
überdimensionales Rundholz 6005
Überflutung 3282
Übergangsholz 4552
Übergangswald 4537
Übergangszeitraum 6514
Übergangszone 8765
Überhaltbetrieb 4281
Überhälterauszug 2969
Überhältershieb 4902
Überhang 5993
überhängender Fallbaum 5041
Überhauung 2935
Überhieb 2935
überhöhter Einschlag 2935
Überkreuzfaser 4539
überkreuzte Befruchtung 5971
überlagerte Knospe 8270
Überlandung 5995
Überlängen 5997
Überlappungskultur 5996
Überlaufstrecke 7004
Überlebensindex 8293
Überlebensrate 1322
übermäßiger Splint 2937
übermäßiges Wachstum 2936
Übernutzung 2935
Überpflanze 2852
Überreife 6000
überreifer Baum 5885
Übersäuerung 41
Überschirmungsfläche 354
Überschusswald 8286
Überseehölzer 6003
Übersichtsvermessung 6749
Überständer 5884
Überständer zweiten Ranges 7358
Überständer zweiter Ordnung 7358
überständiges Holz 5884
übertrocknetes Holz 7735
Übervorrat 8287
überwallter Ast 833
Überwallung 5862
Überwallungsholz 2048
Überwallungsleiste 2157
Überwallungswulst 2157
übliches Maß 2159
Uferbepflanzung 6960
Uferbestockung 8158
Uferbetriebsfläche 8157
Ufergehölz 8158
Uferschutzwald 6961
Uferwald 1687, 6961
Uferweide 2775
Uferzone 6963
Ulmazeen 2788
Ulme 2786
Ulme-Borkenkäfer 5670
Ulme-Esche-Pappel-Wald 2787
Ulmenblattminiermotte 2790
Ulmengewächse 2788
Ulmenholz 2791
Ulmenkrankheit 2678
Ulmen-Spierstrauch 3832
Ulmenwald 2789
Umbaua 107
Umbinden 519
Umfang in Brusthöhe 1024
Umfang in halber Länge des Nützholzstammes 5441
Umfang mit/ohne Rinde 2482
Umfangsmessung 3857
Umfangsquotient 3858
Umfangszuwachs 3856
Umfang über Wurzelanlauf 3855
Umfütterer 3131
umgebogene Stengel 8388
Umgebungsforstwirtschaft 2839
umgekehrtes Radialmodell 6925
Umgoloty 6279
Umgusi 9793
Umherziehenkern 9264
Umläufe 2139
Umlaufseil 5411
Umlaufzeit 7077
Umlenkrolle mit Seileinlass 7796
Umlenkung 592
Umlichtung 1598
umpflanzen 8769
Umpflanzung 6873
Umpflanzungszyklus 6874
umrahmte Säge 3683
Umrechnungstabelle 1917
Umrechnungszahl für Schichtholz 3125
Umrissforstkarte 3362
Umsäumungshieb 6167
Umschaltzeit 9011
Umschlagszeit 7077
umschroten 2172
Umtrieb des höchsten Waldreinertrages 7076
Umtriebszeit 7077
Umwaldung 226
Umwaldungshieb 1914
Umwaldungstafel für Holzmaßeinheiten 7112
Umwelt 4072
Umweltneugestaltung 2843
Umweltschutz 2842
Umweltsicherung 2842
Umweltverträglichkeitsstudium 2840
Umzäunung 2800
unabhängiges Gipfelfeuer 4458
unabhängiges Kronenfeuer 4458
unabhängiges Wipfelfeuer 4458
unbearbeitetes Holz 7096
unbedeckter Abhang 534
unbedecktes Holz 535
unbekämpfter Brand 3690
unbekämpfter Waldbrand 9049
unberührter Kork 9227
unbesäumter Stamm 9266
unbesäumtes Schnittholz 9079
unbesetzte Fläche 9083
unbestimmtes Wachstum 4459
unbestockte Fläche 5758
unbestockter Waldboden 5773
unbewaldet 8874
unbewohnte Gegend 9404
unbrauchbarer Erdstamm 4673
undurchforstet 3694
undurchforsteter Baum 5753
undurchforsteter Bestand 6524
undurchforstetes Waldökosystem 9123
unechtes Hickoryholz 6130
unentrindeter Stamm 9044
unentrindetes Holz 9045
unentrindete Späne 9043
unentwickeltes Blatt 9075
unerlaubte Zonen 5804
Unerschöpflichkeit der Wälder

Unfallschutz 5082
Unfallverhütung 5082
unförmiger Baum 5478
ungarische Eiche 4397
ungarische Silberlinde 7607
ungebundene Flößerei 2625
ungedeckter Knoten 9050
ungeeignete Waldfläche 9121
ungeenigte Holzfaser 5159
ungehobeltes Brett 7084
ungehobeltes Frischholz 7090
ungehobeltes Trockenholz 7086
ungepflegter Wald 9122
ungesattelte Riese 5800
ungleichaltrige Forstwirtschaft 9080
ungleichaltriger Bestand 9081
ungleichaltriger Wald 172
ungleichaltrige Struktur 174
ungleichartige Textur 9082
Ungleichmäßigkeit 4263
Unholz 9322
Unholzart 9074
uniforme Textur 2923
unkommerzielle Fällung 5751
unkommerzielle Holzart 5752
Unkrautvertilgungsmittel 4254
unoperabele Boden 4508
unpaarig gefiedertes Blatt 5867
unregelmäßig bestockt 4586
unregelmäßiger Bestand 9081
unregelmäßiger Waldarbeiter 6095
unregelmäßige Textur 1168
unregelmäßige ungleichaltrige Struktur 4589
Unregelmäßigkeit der Textur 3101
unregelmäßig verlängerte Äste 6708
unreif 4674
unreifes Holz 4425
unreifes Stadium 4424
unsichtbare Waldbrandnarbe 4274
unsortierte Späne 9115
unsortiertes Schnittholz 9117
unständiger Kamp 3117
unständiger Waldarbeiter 6095
Unterabteilung 8231
Unterart 8239
Unterbau 6364
Unterbaukrone 8744
Unterbestockung 9068
Unterbrand 9054
Unterbrusch 9065
unterdrückt 6011
untere Kronenhöhe 5185
untere Kronenlänge 5186
unterer Bergwald 5188
unterer Gebirgewald 5188
unterer Schnitt 5187
untere Seite 607
Unterförster 3456
untergeordnetes Holz 7355
untergeordnetes System 8967
Untergrundentwässerung 8241
Untergrundpflug 8963
Unterholz 9070
Unterholz-Feuerregime 9069
Unterholzschicht 9070
unterirdische Entwässerung 8241
Unterklotzung 1548
Unterlage 6424
Unterlage 8240
Unterlage für Lagerung 2054
Unter-Länge-Material 9062
unterlänges Material 9058
unterlaufen 2011
Unterlegkeil 1548
Unternutzung 9064
Unterpflanze 9063
Unterpflanzung 6364
Unterschicht 9070
Unterseil 3993
Untersichtweite 7660
untersitzendes Blatt 8236
Unterstand 9070
Unterstandschutz 9071
Unterstärke 2465
Unterstufe 9070
Unterstützungswurzel 277
Untersuchungsgebiet 268
Unterwasser-Holzrückung 9073
Unterwuchs 9060
Unterwuchsdichte 2394
Unüberlegheit 1000
unverholzt 5770
unverschulte Pflanze 9420
unverwachsener Ast 5158
unverwelkliches Holz 2989
unvollständiges Kernholz 4426
unvollständig Rückstand 4445
unwirtelige Baumspitze 5544
Unzäuung 890
unzugänglicher Wald 3645
Upasbaum 9125
Urbaum 9231
Urblatt 4884
Urholz 2266
Urland 9408
Urlandbrand 9409
Urnen-Eukalyptus 9145
Urwald 282
Urwaldfolgebestand 9070
Urwaldprodukte 9405
Urwaldvegetation 5955
Urwelt-Mammutbaum 2259
Ussur-Birne 9148
Utah-Johannisbeere 9149

Vakzin 4507
Vakzination 4506
Vakzine 4507
vakzinieren 9157
Valoneneiche 3937
variabeler Grundstückradius 9168
Varietät 9171
vasizentrisches Parenchym 9174
vasizentrische Tracheide 9175
Vegetation 9176
Vegetationsmosaik 9177
Vegetationsring 301
vegetativ 9178
vegetative Gruppe 9180
vegetative Knospe 9179
vegetative Regenerierung 9184
vegetativer Spross 4505
vegetatives Material 9181
vegetative Vermehrung 9182
Veilchenholzakazie 3846
Vera 9199
veralteter Schutzwald 5989
veränderliche Retention 9169
Veränderung der Holzfarbe 9514
verankertes Kabel 1164
Verankerung 273
Verantwortungswald 8090
Verästelung 987
Verbandholz 3686
verbannter Wald 525
Verbesserungshieb 2121
Verbesserungsplantage 2830
verbinden 33
verbindenes Wasser 950
Verbindungsweg 4172
Verbiss 8806
verblautes Holz 859
Verblauung 860
verbleibender Bestand 8000
verblendender Perlast 830
verblendendes Vogelauge 830
verbliebener Bestand 8000
verbotener Wald 525
verbrannte Fläche 1184
verbreiteter Bärlapp 8815
Verbrennen von frischem Schlagabraum 5010
Verbrennungen 7092

Verbundschneiden 1838
Verbungskosten 8217
verdeckte Waldbrandnarbe 4274
Verdichtung 8494
verdorbener Baum 8372
verdorrter Baum 2638
verdrehte Holzfaser 1902
Verdübelung 2595
Verdünner 8513
Verdünnungsmittel 8505
Verdunstung 2914
Verdunstungskühlung 2915
veredelte Hänge-Blutbuche 9329
Veredelung 677
vereinigen 33
vereinzeln 8503
Vereinzelung 7658, 8515
Vereinzelungsanlage 7976
Vereinzelungsmaschine 8513
Verekbaum 4047
Verfahren ohne Druckanwendung 5768
Verfall 2994
verfallendes Holz 2292
verfallener Bestand 2406
verfärbtes Holz 1730
Verfärbung durch Lagerhölzer 2047
verfault 7212
verfaulter Ast 2290
verfaultes Holz 2291
Verfichtung 75
Verflechtung 6379
Verfolger 1454
verfügbares Holz 424
Vergiftungsbohren 928
Vergleichsbaum 1459
Vergleichsbestand 4471
Vergleyung 3860
Vergrauen 3925
Vergütbarkeit 8791
vergüteter Wald 4431
vergütetes Holz 4432
verhacken 8503
Verhältnis zwischen Blattfläche und Gewicht 4893
Verhältnis zwischen Wachstum und Hieb 4024
Verhältnis zwischen Wurzeln und Schüssen 7052
Verhärtung 4120
Verharzung 6895
Verhau 4176
verholzt 4955
verholzte Faser 4956
Verholzung 4954
4283
Verjüngung 6851
Verjüngung durch Stockausschlag 1931
Verjüngung mit Standortvorbereitung 6827
Verjüngung ohne Standortvorbereitung 6826
Verjüngungsabteilung 355
Verjüngungsdringlichkeit 9143
Verjüngungsfläche 355
Verjüngungsfront 6818
Verjüngungshieb 6815
Verjüngungsinitiation 6819
Verjüngungsklasse 6814
Verjüngungsmethode 5416
Verjüngungsnachhieb 6864
Verjüngungsperiode 6820
Verjüngungsschlag 6815
Verjüngungstrieb 4505
Verjüngungsverzögern 6816
Verjüngungszeit 3554
Verjüngungszeitpunkt 8040
Verjüngungszeitraum 6820
Verkauf auf dem Stamm 7157
Verkaufeinheit 7159
Verkauffläche 7158
verkäufliche Länge 5395
Verkauf pro Blockholzmasse 5112
Verkaufswert 5313
Verkauf von stehendem Holz 7157
Verkehrsader 3463
Verkehrsdeck 5011
Verkehrslast 5014
Verkehrswert 5313
verkerntes Holz 4202
Verkernung 2671
Verkettung 1428
verkientes Wurzelholz 6899
Verklammerung des Stammes 7315
Verknotung 986
Verkohlen 1331
Verkohlengerät 1332
Verkohlungsholz 9571
Verkorkung 8233
Verkrehlen 7658
verkrusten 2106
Verlademarke 7529
Verladeplatz 5027
Verladerampe 5027
verladetrockenes Holz 7528
Verladewinde 8609
verlängerte Ast 6468
verlassener Streifen 4901
verleimtes Holz 3876
verleimtes mehrschichitge Holzelement 3878
verletzen 4148
verletzter Baum 1097
Verletzung 4918
Verlichtung 8515
Verlustfaktoren 5176
Vermehrung durch Absenker 6582
Vermehrungsgut 3555
vermeidlicher Holzabraum 446
Vermesser 5102
Vermessung
vor Holzeinschlag
waldbauliche ~ 6505
Vermessung den Waldressourcen 3563
Vermessung im Quadratfüss 3036
Verminderung 2341
Verminderung der Feuersgefahr 3187
vernarbendes Holz 7289
Vernarbung 5862
Vernässung 9291
Vernichtung des Waldes 3373
Verordnung 6517
verpackter Kern 956
verpflanzen 8769
Verpflanzung 6873
Verrasen 7813
Verrasung 7813
Verriegelungsschlepper 4540
Verrottung 6647
verrückte Gegenkraft 2008
Versandbündel 7526
Verschalung 880
Verschiebensfläche 2336
verschiedenartige Strahlgewebe 4265
verschiffungstrocken 6584
verschimmeltes Holz 2291
Verschnittabschlag 5178
verschobene Hiebsflächen 7988
verschränktes Holz 4655
Verschulbeet 5812
Verschulbrett 8770
Verschulen 6529
Verschulmaterial 5814
Verschulnadelpflanze 1878
Verschulpflanze 5814
verschulte Pflanze 5815
Verschulung 6529
Verschulungsverband 8773
Verseilen 271

Versorgungsblock 8272
versperrendes Unterholz 5860
Verstockung 1933
verstopfen 1552
Verstopfung 7599
Verstopfungsstamm 4704
verstreute Bäume 7882
verstreutes Laubwerk 7880
Versuchforst 2946
Verteiler 7934
Verteilung des Schrägstriches 2541
vertikale Kronenschluss 9204
vertikale Resistenz 9206
vertikale Struktur 9208
Vertikalkronenschluss 9204
Vertikalschluss 9204
Vertikalverschiebungssäge 6970
Vertragsabfuhrunternehmer 1907
Vertragsbringung 1908
Vertrautheit 4572
vertrocknetes Holz 7735
verwachsend 67
verwachsener Ast 833
verwachsenes Holz 4536
verwachsene Wurzel 3907
Verwachsung 35
Verwahrungskette 1429
Verwaltung der ungleichaltrige Forst 173
verwandelt
in Holz ~ 4955
Verwandtschaftszucht 4435
Verwendungsmöglichkeiten des Holzes 6473
verwertbarer Mindestdurchmesser 2743
Verwertung des Holzes 9630
verwickelnde Stengel 8388
verwickelnde Vegetation 8389
verwimmertes Holz 2048
Verwitterung 9311
verworrenes Blatt 2317
Verwürfler 7312
verwurzelte Ast 7020
Verwurzelung 5334
verwüsteter Forst 2443
Verwüstung 2414
verzahnt 7461
Verziehen 9320
verziehen 8503
Verzimmerung 8600
verzweigt 977
verzweigter Ast 979
verzweigte Schlagstreifene 978
Verzweigung 987
Viehtreiben auf Routen 8753
Viehweide 5349
vielästig 4737
Vielblatt-Abkürz-Kreissägemaschine 8971
Vielblatt-Abkürzsäge 8971
Vielblockbandsäge 6966
vielfacher Spitzentrieb 5614
vielfältige Forstwirtschaft 5623
Vielgestaltigkeit 6439
vieljährige Holzpflanze 9713
Vielmesserhackmaschine 5607
Vielsaumbetrieb 5618
vielschäftig 5628
Vielschichtplatte 5608
vielserieller Strahl 5626
vielstämmig 5628
vielstämmiger Hieb 5629
vielstufige Kultur 5631
Vierkantblock 7969
vierkantig 7963
vierkantig beschlagenes Holz 7969
vierkantiges Baumstamm 7971
Vierkantstapelung 7973
viermännige Tamariske 7939
vierseitig zugerichtetes Holz 7091
vierteilter Mast 6660
Viertelholz 6658
Viertelkubierung 2118
Viertelumfang 6661
vikariierende Gemeinschaften 227
Violettholz 9225
Virginia-Blasenspiere 397
virginische Bleistiftzeder 2722
virginische Dattelpflaume 1796
virginische Eiche 5015
virginische Hopfenbuche 2719
virginische Kiefer 9229
virginische Kiefernbuschhornblattwespe 9230
virginische Pappel 2717
virginischer Ahorn 6774
virginischer Wacholder 2722
virginische Traubenkirsche 1769
Virola 9232
viskoelastisches Benehmen 9233
Vlies 5335
Vogelaugenahorn 808
Vogelaugenahorn-Figur 766
Vogelaugenmaser 766
Vogelbeerbaum 2903
Vogelbeere 2903
Vogelkasten 763
Vogelkirschbaum 244
Vollbaum-Erntemaschine 1830
Vollbaumnutzung 1832
Vollbaumspäne 3760
Vollbestand 3756
Vollbestandsfaktor 2344
vollbestockte Waldfläche 3763
Vollbestockung 3757
Vollblatt 2832
Vollblattrand 2833
Volldurchmesser 3748
Vollgatter 5132
Vollgattersäge 3800
Vollholz 9393
vollholzig 3745
vollhölziger Baum 3746
Vollholzigkeit 8816
Vollholzigkeitsfaktor 1711
Vollholzstamm 2835
vollkantig 2756, 7963
vollkantige Kante 7967
vollkantiger Hieb 7964
vollkantiger Holzblock 3755
vollkantiges Ende 7966
vollkerniges Holz 184
Vollkluppierung 3747
Vollkluppung 3747
Vollknospe 2831
Volllagholz 2834
Volllangholzbringung 8869
Vollmaß 2160
Vollmast 3751
Vollnormierung 703
Vollsaat 1044
Vollschnittholz 9393
Vollseillinie 1244
Vollskalierung 703
vollstämmig 3745
vollständige Ablöschung und Aufräumen einer Brandfläche 5554
vollständiges Blatt 1829
Volltränkung 692
Volltrommelgeschwindigkeit 3749
Volumenabschätzung mittels Gewichtmessung 9334
Volumeneinheit-Kaufgrundlage 9098
Volumengleichung 9242
Volumen in trockenem Zustand 9252
Volumenleistung 9773
Volumenschwindung 9251
Voralpenspindelbaum 1049
Voranbau 73
Vorarbeiter am Verladeplatz

vorausbestimmendes Chokerseil 6518
Vorausverjüngung 73
Vorbau 6510
vorbereitende Pflanzenlochherstellung 6508
vorbereitetes Holz 6515
Vorbereitungshieb 6513
Vorbereitungsschlag 6513
Vorbestand 6527
Vorblatt 6585
Vorchoke 6497
Vorderflanke 4848
Vorderfurnier 2986
Vorderknie 2100
vordringliche Stelle 4385
voreinstellenes Chokerseil 6518
Vorentastung 76
Vorertrag 4546
Vorfahrtrecht 6941
Vorformseil 6503
vorgebündeltes Holz 6494
vorgeformtes drallarmes Drahtseil 6503
vorgeschnittenes Holz 7094
vorgesehene Hiebsmasse 2200
vorgesehene Hiebsmenge 2200
vorgesehener Entstaubungsweg 2425
Vorhandelsbehandlungen 6499
Vorhangherunterlassen 2617
Vorhangzuziehen 2617
Vorherrschaft 2561
vorherrschende Art 2566
vorherrschender Baum 6500
vorherrschender Bestand 2566
Vorhieb 6513
Vorjuvenilität 6509
Vorkeimung 6504
Vorkühlung 6496
Vorkultur 71
vorläufiger Wald 4537
Vornutzungsanfall 9782
Vornutzungsertrag 4546
Vornutzungshieb 4544
vorrangige Bodenfeuchte 310
Vorrangstelle 4385
Vorrataufnahme 8657
Vorratholz in Menge 8116
vorrätiger Wald 8104
vorrätiges Quadrat 8105
Vorratsanreicherung 9051
Vorratsaufnahme 2438
Vorratsbewertung 9163
Vorratsfestmeter 2116
Vorratshaltung 4694
Vorrücken 6519
vorrückender Hieb 6573
Vorschlaghammer 5315
vorschleppen 6507
Vorschneiden 1017
Vorschnitt 4082
Vorschrift 6517
vor Holzeinschlag
waldbauliche ~ 6506
vorsichtige Abholzung um Regeneration 1337
vorspannen 6511
Vorstapelung 6495
vorübergehende Besitzverhältnisse 8450
vorübergehende Brücke 8449
vorübergehende Mischung 8451
vorübergehender Wald 8447
vorübergehende Zugriffstruktur 8448
Vorverjüngung 73
Vorwald 6289
Vorwuchs 73
vorziehen 6507
Vorzugserntebestand 2784
vorzugsweise Art 6501

Wabenfäule 6313
Wabenkern 5905
Wabenschwamm 6313
Wabigwerden 4349
Wabigwerden des Holzes 4349
Wacholder 3983
Wacholderbestand 4672
Wacholderwald 4672
Wachsbaum 7858
wachsen in der Rinden 1974
Wachsmyrte 7858
Wachspalme 9306
Wachstumförderer 4020
Wachstumkurve 4012
Wachstumsbehinderung 4015
Wachstumseigenschaften 4022
wachstumsfördernd 4021
Wachstumsgeschwindigkeit 6728
Wachstumshemmung 4015
Wachstumsretardation 4015
Wachstumsschicht 1287
Wachstumstockung 8126
Wachstums- und Ertragsmodell 4011
Wachstumsverzögerung 4015
Wächter 6115
Wädelholz 2266
Waffelplatte 9255
Wagenschoßbohle 9257
Wagnerholz 1672
Wahlholz 9578
Wahlstamm 6395
Wahnkante 2662
wahnkantige Balken 2663
wahnkantiges Holz 9079
wahrer Holzkern 8986
wahre Tannen 8987
Wald 3318
Waldabholzung 2339
Waldabschwendung 3373
Waldabteilung 3384
Waldahorn 8355
Waldaktiengesellschaft 3355
Waldameisen 3327
Waldanbau 8953
Waldanpflanzung 3624
Waldanteil 3522
Waldarbeit 9724
Waldarbeiter 9654
Waldarbeitersiedlung 5071
Waldareal 8604
waldarm 6450
Waldart 3363
Waldarten 3604
Waldaufbau 1886
Waldaufseher 3613
Waldausformungsplatz 7727
Waldausreißer 3517
Waldauswahlverfahren 3589
Waldbahn 3565
Waldbahnkran 8609
Waldbandfallschirmspringer 7783
Waldbär 2193
Waldbart 2193
Waldbau 3615
Waldbauer 3420
waldbauliche Anordnung 7617
waldbauliche Ansprüche 3494
waldbauliche Behandlung 8456
waldbauliche Biomasse 7613
waldbauliche Durchforstung 7622
waldbauliche Konversion 7614
waldbaulicher Betrieb 7621
waldbaulicher Umtrieb 7619
waldbauliches Entscheidungsmodell 7615
waldbauliches Regime 7618
waldbauliches System 7621
waldbauliches Verhalten 7612
waldbauliche Vermessung 7624
waldbauliche Wirksamkeit 7616
Waldbaum 3625
Waldbaumart 4714
Waldbaumpflanzung 3626
Waldbaumpopulation 3627

Waldbaumveredelung 1028
Waldbaumzüchtung 1028
Waldbaupraxis 3576
Waldbaupraxiscode 3534
Waldbaupraxisgesetz 3534
Waldbautheorie 8487
waldbedeckt 3394
Waldbeere 855, 3336
Waldbeglaubigung 3348
Waldbegründung 94
Waldbeschädigung 2241
Waldbesitz 3543
Waldbesitzer 3519
Waldbestand 3605
Waldbestandskarte 3498
Waldbestäubung 3388
Waldbesteuerung 8422
Waldbestockung 3605, 8953
Waldbetriebsform 3666
Waldbetriebsregelung 3574
Waldbewertung 3411
Waldbewirtschaftung 3364
Waldbewohner 3389
Waldbezirk 3382
Waldbiologie 3339
waldbiologisches Reservat 3338
Waldblöße 3443
Waldboden 3480
Waldboden 9547
Waldbodendecke 3435
Waldbodenfläche 9542
Waldbodengefüge 3602
Waldbodenkunde 3601
Waldboden mit unbewiesene Schlussgrad 9106
Waldbodennutzungsplan 3482
Waldbodennutzungstype 4782
Waldbodenschätzung 2870
Waldbodentypologie 3603
Waldbodenwert 3483
Waldbrache 1201
Waldbrand 3428
zu bekämpfender ~ 53
Waldbrandbekämpfung 3431
Waldbrandbekämpfungsfläche 3193
Waldbrandgefahr 6971
Waldbrand gegen den Wind 3429
Waldbrandgepäck 3194
Waldbrandgrößenklasse 1601
Waldbrandkontrolle 3192
Waldbrand-Lager 3170
Waldbrandlegung 4436
Waldbrandnarbe 3206
Waldbrandperiode 3434
Waldbrandpflug 3196
Waldbrandschutz 3430
auf der Erde gegründeter ~ 4771
Waldbrandschutzverhütung 3430
Wald-Brandstiftung 7466
Waldbrandstiftung 4436
Waldbrandverhütungsverwaltung 3433
Waldbrand von überdurchschnittlicher Dauer 2974
Waldbrandzeit 3434
Waldbrennholz 3740
Wäldchen 5001
Walddesignplan 3372
Wald des Inlandes 4501
Walddevastation 3373
Walddichte 3370
Walddistrikt 3382
Walddoldenkraut 1798
Walddorf 3639
Walddurchhau 3443
Waldeigentum 3543
Waldeigentümer 3519
Waldeinbruch 3477
Waldeisenbahn 5079
Wald-Engelwurz 9401
Waldengelwurz 9401
Waldentität 3403
Waldentwässerung 3385
Waldentwicklung 3374
Waldentwicklungsplan 3375
Waldentwicklungsprogramm 3376
Walderbe 3461
Walderdbeere 9425
Walderde 3480
Wälder der Kurortgesundheitsschutz 3598
Wälder der Wasserquellengesundheitsschutz 3599
Walderforschung 3549
Walderhaltung 3359
Walderhaltungsgesetz 3360
Wälder in den Naturschutzgebieten 3597
Walderkundung 3549
Walderschöpfung 8656
Waldertrag 9778
Waldertraganalyse 3647
Waldertragsfähigkeit 3648
Waldertragskunde 3650
Waldertragsregelung 3649
Walderwartungsertrag 2945
Walderwartungswert 3412
Waldessaum 2761
Waldesschatten 3965
Walde von mittlerem Alter 5351
Waldexploitation 3414
Waldfahrt 3564
Waldfarm 3419
Waldfarmlizenz 8835
Waldfauna 3422
Waldfazies 8357
Waldfeldbau 143
Waldfeldbaueinheiten 144
Waldfeldbau-System 145
Waldfeldbauverfahren 146
Waldfeldfarm 3421
Waldfeldkultur 3425
Waldfeld-Plantation 8419
Waldfeldwirtschaft 143
Waldfeuerversicherung 3432
Waldfläche 9542
Waldflächebestimmung 3329
Waldflora 3437
Waldflurbereinigung 1883
Waldfonds 8114
Waldformation 3438
Waldfragmentierung 3681
waldfreundlich 8819
Waldfrevel 1909
Waldfriedhof 3347
Waldfrosch 9579
Waldfrüchte 3441
Waldfunktionsplan 3492
Wald für Rotwild 2330
Waldgalerie 3442
Waldgärtner 6246
Waldgebiet 3551
Waldgebirge 9710
Waldgegend 3551
Waldgenese 3445
Waldgenetik 3447
Waldgeschäfte 3322
waldgeschält 552
waldgeschältes Schnittholz 553
Waldgesellschaft 3331
Waldgesetz 3484
Waldgesicht 8357
Waldgesinnung 3507
Waldgieren 6224
Waldgras 3449
Waldgräserei 3914
Waldgrenze 2761
Waldgrubber 3365
waldgrün 3453
Waldgrünland 3450
Waldgruppe 3999
Waldgürtel 3335
Waldgut 3408

Waldhammerzeichen von Stirnseite 2807
Waldheidelbeere 5006
Waldhieb 2193
Waldholunder 2907
Waldhortensie 9407
Waldhumus 3464
Waldhüter 3456
Waldhygiene 3580
waldig 3324, 3394
waldige und industrielle Verschmutzungskontrolle 3468
Wald in der borealen Zone 925
Wald in der gemässigten Zone 8443
Waldindustrieabfälle 3469
Waldinsektenkunde 3404
Waldinspektion 3549
Waldinvasion 3477
Waldinventur 3478
Waldisolation 4591
Waldkante 2662
waldkantiges Holz 9079
Waldkapital 3346
Waldkartierung 3499
Waldkataster 3344
Waldkiefer 1797
Waldkirsche 8330
Waldklimatologie 3353
Waldklimax 3354
Waldkonservator 3361
Waldkonzession 3356
Waldkrankheiten 3379
Waldkultur 3528
Waldkunde 3581
Wald-Labkraut 9488
Waldland 8604
Waldlandschaft 3481
Waldlandweide 9711
Waldlaubsänger 9691
Waldläufer 344
waldleer 6450
Waldlichtung 5912
waldliebend 3487
Waldlogistik 3486
waldlos 7008, 8874
waldlose Zeit 8875
Waldlosigkeit 4753
waldlose Vegetation 5921
Waldmaikäfer 3502
Waldmangel 4753
Waldmantel 7508
Waldmaus 3510
Waldmeister 9641
Waldmikroklima 3506
Wald mit der Grasschicht 3917
Wald mit offenem Kronendach 3646
Wald mit Weidewirtschaft 3929
Waldmodell 3508
Waldmoorboden 9712
Waldmoorhumus 6129
Waldmurmeltier 2171
Waldmuster 3508
Waldnachwuchs 6833
Waldnatur 3512
Waldnutzholz 5400
Waldnutzung 3414
Waldnutzungsrecht 8573
Waldobligation 8561
Waldökologie 3390
Waldordnung 8046
Waldordonnanz 3516
Waldparzelle 1821
Waldpflanze 3527
Waldpflege 8456
Waldpflug 3531
Waldpotential 3648
Waldpräriezone 6493
Waldpraxis 3576
Waldpraxiscode 3534
Waldpraxisgesetz 3534
Waldpreis 3536
Waldprobleme 3479
Waldproduktion 3537
Waldproduktionszone 3538
Waldproduktionszukunft 3540
Waldprophylaxe 3544
Waldrand 2761
Waldrasterdatenmenge 3548
Waldrebe 9402
Waldrechter 5885
Waldregion 3551
waldreich 6936
Waldreichtum 3640
Waldreife 3501
Waldreinigung 3351
Waldrentabilität 6568
Waldrente 3553
Waldreservat 3557
Waldreserve 3557
Waldressourcen 3562
Waldrückgang 2241
Waldsäge 8854
Waldsämaschine 3585
Waldsamen 3628
Waldsamenernte 3584
Waldsamenertrag 3584
Waldsamenproduktion 3584
Waldsamentragsfläche 7394
Waldsämerei 3628
Waldschaffer 344
Wald-Schaumspiere 2009
Waldschlange 8925
Waldschlepper 5114
Waldschluss 3345
Waldschlussklasse 1600
Waldschneise 3564
Waldschutz 3359
Waldschutzgebiet 3535
Waldschutzgürtel 7508
Waldschutzkapazität 6597
Waldschutzstation 3545
Waldschutzstreifen 7508
Waldseite 3417
Waldsonderfall 3517
Waldsoziologie 3596
Waldsplitterung 3681
Waldstandort 3591
Waldstandortklassifizierung 3593
Waldstandortsfaktor 3594
Waldstange 6418
Waldsteig 3443
Waldsteppe 3607
Waldsteppenzonen 3609
Waldsterben 3368
Waldsterblichkeit 3509
Waldstraße 3565
Waldstraßenbau 1889
Waldstraßennetz 3566
Waldstreifen 3335
Waldstreu 3435
Waldstreustörung 3436
Waldstück 1821
Waldtaxation 3617
Waldtaxationslehre 8488
Waldtaxe 3616
Waldteufel 8808
Waldtherapie 3621
Waldtiere 3644
Waldtod 2694
Waldtorf 9712
Waldtransportnetz 3566
waldtrockenes Holz 3386
Waldtundra 3631
Wald-Tupelobaum 810
Waldtyp 3363
Waldtypenklassifikation 3632
Waldtypenlehre 3636
Waldtypetiketten 3633
Waldtypgrenzen 3634
Waldtypkarte 3635
Waldung 9404
Waldvegetation 8953
Waldverbreitung 3381
Waldverbundenheit 3507
Waldverderber 3523

Waldverjüngungszone 6804
Waldvermögen 3346
Waldverneuerung 6821
Waldvernichtung 3373
Waldverschleuderung 2241
Waldversicherung 3476
Waldverwesung 3367
Waldverwüstung 3373
Waldwachstumskennzahlen 4475
Waldwaren 3448
Waldwärter 3456
Waldweg 3565
Waldwegebau 1889
Waldwegenetz 3566
Waldwegenetzdichte 2392
Waldweide 3450
Waldweide-Ekoton 3451
Waldweidewirtschaft 7625
Waldwert 3536
Waldwertrechnung 3411
Waldwertschätzung 3411
Waldwesen 3403
Waldwiederbegründung 6821
Waldwiese 8865
Waldwirtschaft 3567
Waldwirtschaftsregel 8020
Waldwirtschsaftsgrundsatz 8020
Waldwolle 9697
Waldwühlmaus 3510
Waldzeichen 3500
Waldzeichnung 5317
Waldzertifizieren 3348
Waldziest 4222
Waldzone 3551
Waldzucht 7610
Waldzugsäge 9030
Waldzusammenlegung 1883
Waldzusammensetzung 1886
Waldzustandsinventur 3357
Wälgerung 2598
Wall 8619
Walnuss 9262
Walnuss 1808
Walnussbaum 9262
Walnussbaumgewächse 9260
Walnuss-Gebirgswald 9261
Walnussgewächse 9260
Walnussöl 5822
Walzenhackmaschine 2208
Wandelröschen 4783
Wanderfeldbau 7520
wandernde Rodungskultur 7520
Wandoo 9265
Wane 6782
Wangen 1460
Wanzenkraut 5257
Waschnuss 1506
Washoe-Kiefer 9272
Wasseraufstieg 8321
Wasserbauholz 9573
Wasserbecken 7942
Wasserbindevermögen 9287
Wasserbomber 157
wasserdicht 5522
Wasserdost 866
Wassereiche 3764
Wassereinzugsabschätzung 9296
Wassereinzugsgebiet 9280
Wasserentnahme 9280
Wasserentrinder 4404
Wasserentrindungsmaschine 4404
Wasserernte 9286
Wasseresche 777
Wasserfall 9285
Wasserfassung 9280
wasserforstwirtschaftliches
System 335
Wassergarten 5105
wassergelagertes Rundholz 9302
Wasserhaltekraft 9287
Wasserholz 3279
Wasserkasten 7942
Wasserlager 5105
Wasserlagerplatz 5105
Wasserlagerung 9301
Wasserlauf 9284
Wasserreis 9299
Wasserreiser 9299
Wasserressourcen 9295
Wasserriegel 3976
Wasserriese 9281
Wasserrinne 4043
Wasserriss 4043
Wasserrücken 9300
Wasserscheide-Integrität 9297
Wasserscheidenmanagement 9298
Wasserschneeball 2890
Wasserschoss 9299
Wasserschutzwald 6961
Wasserstrahlentrindung 9289
Wasserstreifenkrankheit 9293
Wassertupelobaum 9304
Wasserulme 6320
wasserundurchlässig 5522
Wasserware 3279
Webeleine 6732
Wechsel der Holzarten 226
Wechseldrehwuchs 9021
Wechselhieb 8171
Wechselschlag 8171
Wechselstreifenhieb 1989
Wechselverbrennen 7075
Wedel 3718
Wegameise 1762
Weg-Deaktivierung 6976
Wegebenutzungsrecht 6941
Weggenehmigung 6979
wegjagen 7120
weglose Waldfläche 6977
Wegprisma 6980
Wegrecht 6941
Wegschema 6978
Wehrbaum 3147
Wehrer 3147
weibliches Rosenholz 3083
Weichbast 7814
weiche Fäule 7819
weicher Dürrständer 7821
weicher Kern 7817
weiche Saftfäule 7820
weiches Laubholz 7816
weiche Wurzel 8452
Weichfaserplatte 4521
weichhaarige Eiche 6621
Weichholz 7823
Weichholzart 7832
Weichholzasche 7827
Weichmast 7818
Weichobst 7815
Weichplatte 4521
Weichrauchwacholder 7878
Weichselkirsche 7855
Weide 9439
Weidebereitschaft 6720
Weideholzaushieb 9320
Weidenbaum 9439
Weidenbestand 9430
Weidenbirne 9432
Weidenblattakazie 5936
weidenblättrige Akazie 5936
weidenblättriger Birnbaum 9432
Weidenbohrer 3879
Weideneiche 9434
Weidengebüsch 9436
Weidengehölz 9436
Weidengerte 617
Weidengerten 9015
Weidengewächse 9429
Weidenheger 9431
Weidenmeise 9438
Weidenpflanzung 9431
Weidenplan 3931
Weidenrinde 9428
Weidenröschen 3215
Weidenrute 9437
Weidenschälmaschine 5964
Weidenspierstrauch 9433
Weidenutzungsplan 6722

Weidepflanze 6719
Weiderrichgewächse 5161
Weidetyp 6721
Weideübernutzung 6001
Weideverbesserung 6717
Weideverweiterung 6716
Weidewald 3929
Weidmannskunst 9522
Weigela 9332
Weigelie 9332
Weihnachtsbaum 1569
Weihnachtsbaumkultur 1570
Weihnachtsbaumzucht 1570
Weihnachtsbrauch 1569
Weihnachtspalme 1568
Weihrauch 4463
Weihrauchkiefer 5029
Weihrauchsadebaum 7878
Weihrauchzeder 1262
Weimouthsföhre 2726
Weimouthskiefer 2726
Weinberge 9223
Weinblattahorn 5944
Weinenulme 9325
Weinrebengewächse 9221
Weiserbaum 4474
Weiserbestand 4471
Weiserstamm 4474
Weißahorn 8355
Weißbarkenkiefer 9362
Weißbuche 2901
Weißbuchenwald 4372
Weißdorn 4174
Weißdornakazie 203
weiße Faune 4683
weiße Flecken 9384
weiße Hartriegel 9339
Weißeiche 8317
weiße Marmorstammfäule 5574
weißer Caneel 9364
weißer Eukalyptus 6771
weißer Gummibaum 5286
weißer Holzstammmarmor 5574
weiße Ringfäule 9380
weißerKurrajong 3245
Weißerle 2878
weißer Maulbeerbaum 1791
weißer Myrtebaum 9375
weiße Robinie 1754
weiße Roble 5346
weißer Quebrachobaum 9378
weißer Wiesenklee 9366
weißer Zimt 9364
weißer Zürgelbaum 105
Weißesche 261
weißes Fichtenharz 3792
weiße Ulme 247
Weißfäule 9381
Weißfichte 1805, 9386
weißfüßige Kirschblattwespe 6388
weißgeschälte Rollen 1609
weißgeschältes Papierholz 1609
Weißgummibaum 2780
Weißholz 9389
Weißkiefer 1797
Weißklee 9366
Weißkopfmimose 4851
Weißlaub 8323
Weißpappel 2911
Weißquirlbaum 2453
Weißrüster-Ulme 7310
Weißschälen 1603
Weißschlagfäule 9383
Weißschwammfäule 9385
Weißstammfäule 9388
Weißstreifigfäule 9372
Weißstringybark 9387
Weißtanne 1803
Weißulme 247, 7310
Weißweide 1809
Weißwurzelschwamm 364
Weißzeder 398
Weißzimtbaum 9364
Weitholz 2704
weitringiges Holz 5909
weitspannender Holzabfuhr 5150
Welke 9469
welkend 5302
Welkheit 9469
Wellenbinder 2990
Wellenfaserigkeit 3101
wellenförmiger Blattrand 6868
Wellenholz 2048
Wellenwuchs 9185
Weltforstwirtschaft 9726
Weltholzhandel 9727
Weltwaldfläche 9725
weltweite Wälder 3865
Wendeaxt 2579
Wendehaken 1317
Wendevorrichtung 4311
Wenge 9336
Werbungsvertrag 2192
Werfenbeschränkung 9269
Werfen des Holzes 9268
Werkholz 5457
Werklagerhalle 5456
Werkstättenhölzer 8589
Werkzeug für Holzbearbeitung 9652
Werkzeugschneide 819
Wert des Risikos 9164
Wertholz 4284
Wertholzart 9162
Wertholzbestand 8024
Wertholzbloche 7420
Wertholz durchhauen 4272
wertloses Unterholz 9731
Wertzuwachs 4455
westamerikanische Hemlocktanne 5588
westamerikanische Lärche 9340
westamerikanischer Lärchekäfer 9341
westamerikanischer Riesenbastkäfer 9342
westamerikanische Weymouthskiefer 9344
westindischer Baybaum 632
westindischer Ebenholzbaum 4607
westindischer Lorbeerölbaum 632
westindisches Kokus 4607
westindisches Sandelholz 8716
westindisches Satinholz 9349
westliche Gelbkiefer 6445
westliche Weymouthskiefer 9344
Wettbewerb 8817
Wettbewerbsvegetation 1824
Wetterbaum 9312
weymouthsähnliche Kiefer 7867
Weymouthskiefernrüsselkäfer 8670
Weymouthskiefernwollaus 6245
Weymouth-zwerger rote Kieferwald 9379
wickenblättriger Perlschnurbaum 9214
Wickler 4886
widersonniger Drehwuchs 6943
Widerstand 6906
Wiederaufforstung 6821
Wiederbestockung 6821
Wiederbestockungsaufnahme 6824
Wiederbestockungsperformanz-Abschätzung 6823
wieder geforstet 6806
Wiederholung 6848
Wiederverbrennung 6747
wie eine Ahle geformte Nadelung 447
Wiege-Hügelchen 2005
Wiegensäge 1853
Wiese 5349
Wiesenkiefer 6447
Wildacker 3309

Wildbaum 9419
Wilddieb 6401
Wilddieberei 6402
wilde Malve 7102
wilde Pimpernuss 2885
wilder Birnbaum 1795
wilder Brustwurz 9401
Wilderei 6402
Wilderer 6401
wilder Jasmin 8335
wilder Lattich 6531
Wildernis 9404
wilder Ölbaum 9421
wilder Wein 9228
Wildfauna und Wildflora 9411
Wildfeuer 3690
Wildflora-Pflanzungen 9416
Wildkirsche 8330
wild lebende Tiere und Pflanzen 9411
Wildling 9420
Wildlingswurzeln 8117
Wildling zum Verschulen 9426
Wildobstbaum 3440
Wildpfeffer 7763
Wildpflanzenweltkorridor 9418
Wildtier 3796
Wildtierart 3797
Wildtierschaden 2233
Wildtierverbiss 8806
Wilfordseberesche 4742
Wimmer 3101
Wimmerwuchs 9185
wimperblätterig 1580
wimperblättrig 1580
windbestäubt 283
Windbruch 9445
windbrüchiges Holz 850
Windbruchstelle 849
Windeldürre 8686
Windemann 7925
Windengewächse 5557
Windfall 850
Windfallholz 850
windfester Baum 9443
Windmantel 9445
Windriss 9450
windschief 9440
windschiefer Baum 9444
windschiefes Holz 9441
Windschirm 9445
Windschutzstreifen 9445
Windschwingen 9446
Windwurf 2236
Windwurfaufarbeitung 9442
winkeltreuer Wuchs 4590
Winteraufasten 9462
Winterbeere 2899
winterblühende Kirsche 4275
Wintereiche 7464
Wintereinschlag 9460
Wintereinstand 2332
Winterfällung 9460
wintergrüner Wald 9461
Winterlinde 7767
Winterriesung 2708
Winterriesweg 2708
Winter-Rindenbrand 9457
Winterrindengewächse 9459
Winterschälung 9458
Winterschnitt 9462
Wipfel 8699
Wipfelbaum 2797
Wipfelbäume 988
Wipfelbruch 1013
Wipfeldürre 8686
wipfeldürrer Baum 2273
Wipfelfäule 8711
Wipfelfeuer 2076
Wipfelholz 8712
Wipfelstück 8688
Wipfeltrocknis 7989
Wirbel 2509
Wirbelholz 1032
Wirkmaschine 5577
wirksamer Holzniederschlag 3622
wirksamer Niederschlag 2626
Wirksamkeit des Lebensraumes 4075
Wirkstoff 55
Wirkungen des Waldes 3471
wirkungsvoller Sämling 2769
wirtschaftlicher Blenderschlag 2746
wirtschaftlicher Sortimentshieb 2746
wirtschaftliches Alter 136
Wirtschaftlichkeit der Haushaltsführung 688
Wirtschaftsannahmen 5263
Wirtschaftsarttabellen 5265
Wirtschaftswald 1745
Wirtschaftswaldboden 1746
wirtswechselnd 4262
Witwenente 8832
Witwerente 8832
wohlriechende Himbeere 782
wohlriechender Pfeifenstrauch 8335
Wölben 349
Wölbung 349, 2019
Wölbungshartholz 1991
Wolfsmilchgewächse 7961
Wolkenwaldstufe 1663
Wollbaum 4688
Wollbaumgewächse 7596
Wollbaum-Wurzelschwamm 1981
Wolleiche 6621
wollige Pflaume 5425
wolliger Kreuzdorn 9716
wolliger Schneeball 9307
wollige Weide 9720
Wöllring 2141
Wuchserholung 4023
Wuchsfehler 4014
Wuchsform 4017
wuchsfreudiger Bestand 8536
Wuchsgebiet 3392
Wuchsgleichung 4016
Wuchshemmer 4025
Wuchsinhibitor 4025
Wuchskeislauf 4013
Wuchsraum eines Bäumes 4008
Wuchsring 301
Wuchsspannung 4027
Wuchsstockung 8126
Wuchsstoff 6221
Wühlen und Zurückhalten 4029
Wühlpflug 8963
Wühlspaten 7314
Wulst 635, 1182
Wulst am unteren Stamm 605
Wulstaushieb 3070
Wulstring 8785
Wundebalsam 9254
Wundeharzkanal 8783
Wundekern 8782
Wundekork 8781
Wunderblumengewächse 3676
Wundfäule 7078
Wundgewebe 8787
Wundgummi 9732
Wundharz 9732
Wundholz 9734
Wundring 8785
Wundverfärbung 9733
Würfel 2223
würfelige Fäule 2112
Würgekette 7798
Würgerfeige 3286
Wurmbefall 9703
wurmbefallen 9730
Wurmfraß 4511
wurmfraßig 9730
Wurmloch 4511
Wurmmehl 9728
wurmstichig 9730
wurmstichiges Holz 9729

mit verschnittener ~ 7039
Wurzelanheben 7055
Wurzelanlauf 7015
mit trompetenförmigem ~ 8493
Wurzelanlauf-Durchmesser 2469
Wurzelanläufe 7578
Wurzelanordnung 7011
Wurzelansatz 7015
Wurzelausdehnung 7022
Wurzelausschlachtung 7048
Wurzelausschlag 7046
Wurzelausziehen 2857
Wurzelballen 7012
Wurzelbausch 7054
Wurzelbildung 7024
Wurzelbrut 7046
Wurzelbündel 7054
Würzelchen 7035
Wurzeleinschlämmen 7041
Wurzelende 7051
Wurzelerweiterung 7022
Wurzelextraktor 7042
Wurzelfaser 7035
Wurzelfäule 7044
Wurzelgraben 2857
Wurzelhaare 7028
Wurzelhals 7015
Wurzelhaube 7013
Wurzelholz 9580
Wurzelklotz 1226
Wurzelknöllchen 7036
Wurzelknolle 7053
Wurzelknospe 6683
Wurzelkonkurrenz 7016
Wurzellänge 7033
wurzellos 7034
Wurzelmützchen 7013
Wurzelmütze 7013
Wurzelnausschlagen 7437
Wurzelnetz 7050
Wurzeln mit Erdballen 491
Wurzelpilz 7025
Wurzelrebe 7021
Wurzelrechen 7043
Wurzelroder 8199
Wurzelsaft 7045
Wurzelschlinger 7014
Wurzelschneidemaschine 7019
Wurzelschnitt 7040
Wurzelschnittling 7018
Wurzelschoss 7046
Wurzelschwamm 7044
Wurzelsprosse 7026
Wurzelstange 7021
Wurzelsteckling 7010
Wurzelträger 6928
Wurzeltroge 7023
Wurzelübertragung 466
Wurzel- und Stammfäule 364
Wurzelverankerung 7029
Wurzelwachstum 7027
Wurzelwerk 7050
Wurzelzone 7032
Wüstenbildung 2414
Wüstenbusch 2419
Wüsteneisenholz 2415
Wüstenlavendel 2416
Wüstenmandel 2412
Wüstenpflasterboden 2417
Wüstenrose 2418
Wüstenweide 2420
Wüstenzürgelbaum 2413

Xerophyten 9736
Xylem 9737
Xylemstrahl 9738
xylogen 2855
Xylose 9670
Xylotomie 9741

Yedofichte 4647
Yopo 1714

Zähigkeit 8735
zahngefiedertes Blatt 6285
Zahnradgetriebesäge 3811
Zahnwehholz 8679
Zamie 7153
Zange 275
Zangen-Schlepper 3913
Zangenvornehmen 6525
Zapatero 9346
Zapatero-Buchsbaum 9346
Zäpfchen 1864
Zapfen 8818
Zapfenbohrer 1859
Zapfendarre 1862
Zapfende 7761
Zapfenernte 2033
Zapfenertrag 2033
zapfenförmig 8181
Zapfenjahr 1867
Zapfenpalmfarn 7153
Zapfenpflücken 1861
Zapfenpflücker 1865
Zapfensäge 1136
Zapfensammlung 1863
Zapfentrockenschuppen 1860
Zapfentrockenspeicher 1860
Zapote 5665
Zauberhasel 1811
Zaunblättling 7722
Zaunlattich 6531
Zaunpfahl 3084
Zaunpfosten 3084
Zaunung 2624
Zebrano 118
Zechnenholz 6311
Zeder 1396, 1400
Zederachgewächse 5243
Zederbohrer 1403
Zederharz 1402
Zedernholz 1404
Zedernholzöl 1401
Zederwacholder 6530
Zederwald 1397
Zederzypresse 398
Zeeneiche 5477
Zeichnung des Holzes 9678
Zeidelwald 2178
Zeit der Belaubung 3719
Zeit der Saftruhe 9647
zeitweilig aussetzender Wasserlauf 4555
Zellengewebe 1406
Zellgang 5390
Zellgewebe 1406
Zellhaut 1407
Zellmembran 1407
Zellstoff 9504
Zellstoffholz 6061
Zellulose 9504
Zelluloseholz 6061
Zelluloseholzinventur 6631
Zellwand 1407
Zellwandriss 1408
Zeltraupe 3618
Zenobie 4350
zentralamerikanische Kiefer 2111
zentralamerikanische Zeder 1411
zentraler Mittelstamm 1413
zentraler Stamm 1415
Zerberusbaum 3972
zerbrechlicher Kern 1039
Zerfall 2310, 4494
Zerfall des Kernholzes 4203
Zerfallen des Holzes 1014
zergliedertes Blatt 2549
Zerhacken- und Spritzen-Methode 4083
zerkleinerter Baum 1097
zerkleinerte Späne 7553
Zerkleinerung 1531
Zerkleinerungs- und Transportierungsoperation 1532
Zerklüftung 2003

zermalmter Zuckerrohrstängel 478
Zerreiche 5565
zerreißen 6959
zerren 9024
Zersetzer 2308
Zersetzung 2310
Zersetzung 4494
Zerspanen-Besatzung 1533
Zersplittern 8571
Zerstörung der Tropenwälder 2432
zerstreuter Gebrauch 2537
zerstreutes Zurückhalten 2536
zerstreutporiges Holz 2493
zertifizierter Pestizidenapplikator 1419
zertifiziertes Saatgut 1420
Zeugholz 2988
Zickzack-Bekanten 9796
Zickzack-Randausbildung 9796
Zickzackweide 1963
Zickzackwurm 9016
ziegelroter Schwefelkopf 8258
Ziegelzelle 8553
Ziegenbarteiche 3937
Ziehbahn 8757
Ziehkette 7798
Ziehweg 5116
Ziehwegrücken 8759
Zielbestockungsstandard 8414
Zieldurchmesser 3138
Zielstammwirtschaft 1566
Zielstärkennutzung 9152
Zier-Apfel 1755
Zierbaum 5957
Zierbirne 1281
Ziergrün 9200
Zigarrenkistenzeder 1411
Zimmeraxt 1452
Zimmerbeil 1452
Zimmerholz 1153
zimmernd 8580
Zimmerung 8600
Zimmerwertholz 8184
Zimtahorn 6057
Zimtbodene 1583
Zimthimbeer 6640
Zimtkassie 1495
Zimtröschen 1386
Zingana 118
Zinkenfuge 3161
Zinkenpflug 8668
Zinkensäge 4936
Zirbe 8346
Zirbel-Kiefer 8346
Zirneiche 5565
Ziseleur-Skidder 1455
Zistrose 6990
Zistrose-Gestrüpp 6992
Zistrosengewächse 6991
Zitronatzitronenbaum 1590
Zitronenbaum 4911
Zitronenholz 1591
Zitronen-riechender Eukalyptus 4910
Zitteresche 2880
Zitterpappel 2880
Zitterpappelbestand 387
Zonenfeldwaldbau 9797
Zonenlinie 801
Zonierung 9798
Zoozenose 291
Zoozoenose 291
Zopf 7362, 8699
Zopfblock 7362
Zopfblockdurchmesser 7363
Zopfdurchmesser 2470
zopfen 8702
Zopfholz 8712
Zopftrockenheit 7989
Zopftrocknis 7989
Zopfung 2198
Zopfware 8712
Z-Stamm 3786
Züchter 1026
Zuchtsaat 7403
Zuchtsaatgut 7403
Züchtungslinie 1027
zucken 9024
Zuckerahorn 808
Zuckerbirke 1474
Zuckerbusch 8254
Zufahrtsweg 3463
zufällige Nutzung 4437
zufälliger Wegmuster 6715
zufällig gefälltes Holz 6712
Zufallnutzung 4437
zufriedenstellend regeneriert 7224
zugänglicher Wald 29
zugängliches Waldgebiet 30
Zugangsnetz 31
Zuganker 8542
Zugbruch 8465
zugerichtetes Holz 7096
zugespitzt 7917
Zughaken 4359
Zugholz 8467
Zugleine 3987
Zugrücken 8466
Zugschlepper 2970
Zugseil 3987
Zugtrosse 1555
Zukömmling 2573
Zukunftsbaum 3786
Zukunftsfähigkeit 8294
Zukunftsstamm 3786
Zukunftsstammwirtschaft 1566
zulässige Abweichung 6181
zulässige Brandfläche 194
zulässige Hiebfläche 2166
zulässige jährliche Ergiebigkeit 6178
zulässige jährliche Nutzungsmenge 193
zulässiger Einschlag 6179
zulässiger Jahreseinschlag 192
zulässiger jährliche Erntehieb 193
zulässiger Periodeneinschlag 6159
zulässiges Soll 192
zulässige Verkaufmenge 199
zulässige Zwischenhieb 198
Zunder 8666
Zunderschwamm 3005
Zündholzstamm 5332
Zündholzstammentrindung 5331
zunehmende Pflanzenart 4447
zurechtschneiden 2205
Zürgelbaum 1777
zurückbleibender Baum 7992
zurückgesetzter Baum 7465
zurückgreifende Forstinventur 6760
zurückgreifende Waldinventur 6760
Zurückschneiden 1224
Zurückweichenweg 6111
zusammenfügen 33
zusammengerollte Lohrinde 6999
zusammengesetzte Frucht 1839
zusammengesetzte Holzfaserplatte 1833
zusammengesetzte Holzprodukte 1835
zusammengesetzte Mittellamelle 1842
zusammengesetzter Riss 1843
zusammengesetztes Blatt 1841
zusammengezogenes Pech 5324
Zusammenherrschaft 1707
zusammenrücken 1169
Zusammensetzmaschine 4661
Zusammentragen 4846
zusammenwachsendes Parenchym 1869
Zusammenwachsen zweier Bäume 3657

Zusatzstoff 64
Zusatzstoffe 65
Zuspitzung 8403
Zustandserfassung 4579
Zustandserfassung des Waldes 3357
Zuwachs 4449
Zuwachsbohrer 4451
Zuwachsbohrkern 4452
zuwachsender Boden 34
Zuwachshammer 4454
Zuwachsindex 4018
Zuwachskurve 4453
Zuwachsmittelstamm 4456
Zuwachsperiode 4006
Zuwachspflege 8457
Zuwachspunkt 4007
Zuwachsring 301
Zuwachsrückgang 6832
Zuwachsschätzung-Probefläche 8859
Zuwachstafel 4457
Zuwachszone 301
Zwangsausplenterung 5855
Zwangseinschlag 1852
Zwangsfemelhieb 5855
Zwangshieb 1852
zweialtriger Bestand 9025
Zweibandsägemaschine 2661
zweiblütiger Tupelobaum 8305
zweifarbige japanische Fichte 163
Zweiflächenstichprobe 694
Zweifurchenpflug 9028
Zweig 974
Zweigabstechere 9014
Zweigbildung 4731
Zweigdürre 985
zweigelapptes Blatt 8535
zweigestaltige Abzweigung 2506
zweigförmig 6709
Zweigholz 998
Zweigkäfer 9014
Zweiglein 7132
zweigriffeliger Weißdorn 4174
Zweigringler 9016
Zweigstechere 9014
Zweigsteckling 976
Zweigtod 2488
Zweigwurzel 980
zweihäusige Pflanze 2507
zweihiebiger Niederwaldbetrieb 9033
zweijährige Pflanze 696
zweikantiger Balken 9027
zweikeimblättrige Pflanze 2486
Zweileitungsgreifer 9029
Zweimannsäge 9030
zweipetale Esche 9032
zweirädiger Rückekarren 5090
zweirädiger Rückewagen 5090
Zweischarpflug 9028
Zweischnürigkeit 8142
zweispuriger Schlitten 7304
zweistufiger Hieb 9034
zweite Verbrennung 6747
Zweitwuchs 120
zweizeilige amerikanische Sumpfzypresse 786
Zwergahorn 6996
Zwergapfel 6062
zwergartige Heidelbeere 2683
zwergartiger Brombeere 2684
zwergartiger Wald 2685
Zwergbaum 2691
zwergbaumartig 347
Zwergbirke 892
Zwergeiche 642
Zwerggehölz 965
Zwerghalbstrauch 2690
Zwergkastanie 1199
Zwergkiefer 5598
Zwergkirsche 7855
Zwergmandel 2682
Zwergmispel 5389
Zwergölbaumgewächse 1671
Zwergpalmetto 2688
Zwerg-Sägepalme 7262
Zwergstrauchtundra 5182
Zwergwuchs 8223
Zwetschenbaum 6391
Zwiesel 3659
Zwieselbaum 3659
Zwieselstamm 3657
Zwieselwuchs 3656
Zwille 3659
Zwillingsspitzentrieb 9019
Zwischenast 4541
Zwischenbau 4543
Zwischenfäunis 72
Zwischenfrucht 4543
Zwischenhieb 4544
Zwischenholz 8093
Zwischenholzstapelung 1720
Zwischenholzverladung 1722
Zwischenkronenklasse 4542
Zwischenkultur 4543
Zwischenlager 1721
Zwischenlinde 1786
Zwischennutzungsertrag 4546
Zwischenpfropfung 8691
zwischenständiger Baum 4551
Zwischentoleranz 4549
Zwischenwald 4547
Zwischenzellengang 4530
zygotischer Wald 9800
Zykadazeen 2207
Zyklus der Bestandsdichten 2391
Zylindersäge 6384
Zypernkiefer 2220
Zypresse 2211
Zypressengewächse 2214
Zypressenholz 2219
Zypressenkuppel 2213
Zypressentüpfel 2216
Zypressenwald 2215
Zypressenweiher 2218
Zytokinine 2222

Français

AAM 5352
abaca 5283
abarco 4651
abatis 4
abattage 8824
abattage à culée blanche 3046
abattage à la cisaille 7501
abattage à transportation retardée 3080
abattage au-dessus des empattements 3045
abattage au diamètre limite 2476
abattage avec dessouchage 3081
abattage continu 1897
abattage dans la période d'écoulement de sève 3065
abattage de grumes à planches 1017
abattage d'empattement 1235
abattage des arbres 8824
abattage des arbres en feuilles 3065
abattage d'essai 7171
abattage directionnel 2515
abattage en cascade 2569
abattage en quinconces 3064
abattage en sève 3065
abattage étagé 8171
abattage et débardage 3457
abattage et ébranchage 3047
abattage graduel 3904
abattage mécanique 5369
abattage mécanisé 5369
abattage non commercial 5751
abattage non réglementaire 4423
abattage par jet d'eau 9288
abattage par scie à moteur 5572
abattage permissible 6179
abattage récent 4814
abattages de protection contre l'incendie 3199
abattage simultané de grand nombre de tiges 5629
abattage suivant les contours 1903
abattage suivant les courbes de niveau 1903
abattage vers le haut 9128
abatteur 9526
abatteur à main 4108
abatteur-déracineur 8808
abatteur industriel du sciage 4481
abatteuse à berceau collecteur 3053
abatteuse-ébrancheuse 8853
abatteuse forestière 8836
abatteuse-groupeuse 3053
abatteuse-porteuse 8853
abatteuse-porteuse à arrêt 8125
abatteuse-porteuse à flux axial 456
abatteuse-porteuse de bois courts 7551
abatteuse-porteuse de bois courts articulée 372
abatteuse-porteuse de bois long 5156
abattis 3042
abattre 3037
abattre à culée noire 4033
abaxial 5
abélia à grandes fleurs 3873
abiotique 6
abondance d'espèce 21
abondance intarissable des forêts 3470
abondant en forêt 6936
abord de nœud 4736
aboutage 2031
aboutage de bois 4912
aboutage digital 3161
abrétier 5006
abri 7511
abri contre le feu 3208
abricot d'argent 5245
abricot des Antilles 1756
abricotier 1756
abricotier commun 1756
abricotier d'Amérique 1756
abri de bûcherons 5071
abri-séchoir pour cônes 1860
abroutir 1094
abroutissement 8806
abura 4101
abutilon 22
acacia 1754
acacia à bois dur 404
acacia à bois noir 404
acacia à cachou 1376
acacia à feuilles de pin 7135
acacia à fleurs mauves 1038
acacia à gomme 4048
acacia à longues épines 203
acacia à longues feuilles 8356
acacia auriculiforme 2698
acacia blanchâtre 7604
acacia "bon accueil" 3962
acacia d'Arabie 4048
acacia d'Australie 5301
acacia de Collins 309
acacia de Constantinople 6185
acacia d'Égypte 6110
acacia de mûre 973
acacia épineux 1166
acacia ériolobe 1288
acacia Gregg 1375
acacia jaune 7136
acacia karroo 203
acacia noir 3962
acacia noir d'Australie 404
acacia odorant 1689
acacia parasol 9041
acacia pleureur 9328
acacia sophora 1691
académie forestière 3319
acajou américain 1412
acajou à pommes 1368
acajou blanc 107
acajou d'Afrique 107
acajou d'Amérique 1412
acajou de Chine 1493
acajou de Cuba 2110
acajou de Mexique 6019
acajou de montagne 8990
acajou rouge 7876
acajou rouge australien 411
acanthacées 25
acanthe 24
acariose 26
accélérateur de la croissance 4020
accès et usage mécanisés 5377
accès motorisé et utilisation motorisée 5571
acclimatation 32
accord global des forêts 3863
accouplement 1560
accréter 33
accrétion 35
accrochage de grumes 4117
accrocheur 4114
accroissement 4449
accroissement annuel du diamètre 2474
accroissement annuel moyen 5352
accroissement annuel moyen net 437
accroissement annuel moyen périodique 6164
accroissement annuel périodique 6164
accroissement bloqué 7990
accroissement brut 8727
accroissement d'arbre 8858
accroissement de branche 4967
accroissement de la circonférence 3856
accroissement de la hauteur 326
accroissement de la surface terrière 600

accroissement en hauteur 326
accroissement en qualité 6655
accroissement en valeur 4455
accroissement en volume 9243
accroissement excessif 2936
accroissement isogone 4590
accroissement moyen annuel 5352
accroissement moyen annuel final 3979
accroissement moyen d'exploitabilité 3979
accroissement moyen final 3979
accroissement moyen total 5352
accroissement net 5716
accroissement net annuel 5714
accroissement non déterminé 4459
accroissement normal 5783
accroissement périodique 6158
accroissement périodique annuel 6164
accroissement périodique moyen 6164
accroissement périodique moyen annuel 6164
accroissement total 8727
aceite de Cayeput 1056
acéracées 5291
acérole 527
achaine 38
achat de bois 9635
achène 38
acheteur de bois 8564
aciculaire 36
acidification 41
acidiphile 43
acres marginales 5303
acres non boisées 5757
acres standards 8003
acrocarpus rose 6277
actinidia 51
actinidiacées 50
actinomycètes 52
action à long terme 5152
adaptabilité 60
adaxial 62
additifs 65
adhérence 35
adhérent 67
administration de foresterie 3547
administration forestière 3320
admission du jour 2260
adné 67
adventif 77
aéronef à coordinateur aérien 764
affaires forestières 3322
afforestage 8573
afforestation 94
afforestation à la volée 1042
afforestation contre l'érosion 1881
afforestation de champ 3109
afforestation de digues 2231
afforestation de dunes 2666
afforestation de steppe 8086
afforestation de terrain nu 96
afforestation en plein 1042
afforestation initiale 4493
afforestation ripicole 6960
afforestation steppique 8086
affouage 6946
affranchissement du greffon 7298
affûtage de lames de scie 821
affûter 3975
afrormosia 119
agate de bois 124
agathis 5727
agavacées 125
agave américain 1416
agba 5731
âge à hauteur de poitrine 1023
âge à hauteur de souche 8193
âge culminant 2125
âge d'abattage 136
âge d'arbre adulte 8933
âge de bois 135
âge de la possibilité 9240
âge de maturité 133
âge de récolte 133
âge d'exploitabilité 136
âge d'exploitabilité financière 3146
âge d'exploitabilité naturelle 7619
âge d'exploitabilité sylvicole 7619
âge d'exploitabilité technique 8440
âge d'exploitation pathologique 6113
âge dominant 6526
âge d'un arbre 134
âge du peuplement 8001
âge moyen de peuplement 426
agencement spatial 7884
agent ignifuge 3197
agent pathogène 6112
agents de santé forestière 3459
âge optimum d'exploitabilité 5933
âge total 8718
agitateur de fruits 8845
AGM 3819
agriculteur forestier 3420
agriculture de subsistance 8238
agriculture itinérante après
agriculture sur brûlis 7717
agrile bronzé du bouleau 1065
agrile du pin 408
agrile du pin australienne 408
agroforesterie 143
agroforesterie locale 9797
agrosylviculture 143
aide 7359
aiélé 103
aiguille 37
à ~es 7917
aiguille d'épicéa 7955
aiguille de sapin 5707
aiguille interne 4503
aiguilles écailleuses 7276
aiguille secondaire 7351
aiguilles en forme de poinçon 447
aiguilles primaires 6542
aiguilleur 7932
ailante 8892
ailante glanduleux 8892
aile 9453
aimant les bois 3487
aire à rémanents 7718
aire d'approvisionnement à bois 9650
aire de bande forestière 8802
aire d'éclaircie 8507
aire de coupe 2189
aire de coupe jardinatoire 7406
aire de coupe modifiée 5510
aire de débardage par câble 7701
aire de gestion adaptée 61
aire de production en graines forestières 7394
aire de récréation 6754
aire de vente 7158
aire d'extension de câble 7702
aire d'extension d'un arbre 4008
aire d'habitat de la vie sauvage 9413
aire différée 2336
aire d'offre de bois 8647
aire exploitée 5066
aire foliaire spécifique 7904
aire forestière 3551
aire forestière complètement boisé 3763
aire forestière mondiale 9725
aire forestière sans chemin 6977
aire forestière totale 8722
aire limitée 1868
airelle 855
airelle à feuilles étroites 1788
airelle à feuilles membraneuses 798

airelle alpine 893
airelle bourbeuse 893
airelle canche 1996
airelle canneberge 2889
airelle des marais 893
airelle gazonnante 2683
airelle myrtille bleue 864
airelle noire 893
airelle ponctuée 1996
airelle rouge 1996
airelle vigne d'Ida 1996
aire marquée sélectivement 7415
aire naturelle vierge 9404
aire non boisée 5758
aire non couverte de forêt 5758
aire totale 8719
aisselle de la feuille 4857
ajonc commun 3889
ajonc d'Europe 3889
ajonc épineux 3889
ajouté en arbre 8913
ajusteur de câble 7058
akène 38
alangiacées 158
alangier à feuilles de platane 1488
albédo 161
albédo forestier 3325
albédomètre 162
albizie à graines noires 5526
album d'espèces forestières 9660
alcalinité 170
alcool de bois 9476
alésage 923
algues 166
aliboufier de Californie 1273
aliboutier à grandes feuilles 702
aliboutier officinal 3825
aligneuse 9114
alimentation des grumes 5061
alimenteur de coupeuse 1528
alios 4129
alise 9424
alisier 6474
alisier à feuilles d'aulne 4742
alisier blanc 8323
alisier des bois 9424
alisier de Suède 8322
alisier torminal 9424
alkékenge du Pérou 1323
alkékenge jaune 1323
allée 182
allée coupe-feu végétale 6174
allège 333
allèles 179
allélopathie 180
allier 8323
allouchier 9424
allumage 1191
allumage à partir du centre 1409
allumage central 1409
allumage de zone 351
allumage simultané 8308
allume-feu 3740
alluvionnement 7599
almendro 4464
aloès arborescent 9705
alouchier 8323
alouette lulu 9594
alphitonie élevée 8807
alstonia 1461
alstonia des écoliers 9363
altération atmosphérique 9311
alternance d'espèces forestières 226
alternance d'espèces forestières et agricoles 225
alternation annuelle de feuillage 298
alternes 227
althéa 5639
alvéoles à laticifère 4828
AMA 5352
amadou 8666
amadou de bouleau 752
amadouvier 8667
amadouvier des pins 6764
amaigrir 5227
amande amère 207
amande de désert 2412
amandier 207
amandier commun 207
amandier nain 2682
amarantacées 229
amarante 6641
amarilla 7564
amarinié 3885
amas d'oléorésine 5324
amboine 1183
ambrier 259
amburana 1417
âme 1949
âme en panneaux de flocons orientées 168
âme en panneaux de grandes particules orientées 168
amélanchier 3806, 7462
amélanchier à feuilles rondes 2689
amélanchier arborescent 2603
amélanchier arbre 2603
amélanchier commun 3806
amélanchier de l'Ouest 6024
amélanchier sanguin 3806
âme lattée 5200
amélioration d'aire de coupe 7156
amélioration de la forme 3665
amélioration de la station 7666
amélioration de l'habitat 4076
amélioration des arbres forestiers 1028
amélioration de terrain de parcours 6717
amélioration du bois 4434
amélioration d'un peuplement 8643
amélioration génétique des arbres forestiers 1028
amélioration par sélection 4433
amélioration sylvicole 7611
aménagement à objectifs intégrés 5622
aménagement de la vie sauvage 9415
aménagement des habitats 4077
aménagement du terrain 7676
aménagement forestier 3488
aménagement forestier à objectifs intégrés 5622
aménagement forestier de base 611
aménagement forestier durable 8296
aménagement forestier extensif 2963
aménagement forestier intensif 4528
aménagement paysagé 4778
aménagement paysager 4778
aménagement paysagiste 4778
aménagiste 3410
aménagiste des forêts 3497
amenuisement de branches 6010
âme transversale 2040
amortisseur de traînée 2606
amourette 4921
ampélopsis arborescent 6149
amplitude écologique 2740
anacardiacées 1369
anacardier 1368
analyse de bois 9477
analyse de cernes annuels 269
analyse de fertilité forestière 3647
analyse de feuille 3307
analyse de souche 8197
analyse de tige 8060
analyse d'offre de bois 8646
analyse du tronc 8060
analyse foliaire 3307

ancien 8585
anciens 6889
ancrage 273
ancrage radiculaire 7029
andain 9447
andainage 9448
andain de niveau 8780
andiroba 4039
andromède 897
andromède à feuilles de polium 897
andromède d'Europe 897
andromède en arbre 7857
andromède ferrugineuse 8880
anémone des bois 9479
anémophile 283
angélique blanche 621
angélique chinoise 1489
angélique de Chine 1489
angélique du Japon 4610
angélique en arbre d'Amérique 2452
angélique en arbre du Japon 4610
angélique épineuse 2452
angélique forestière 9401
angélique grise 621
angélique rouge 621
angico rouge 6762
angiospermes 284
angle caché 2679
angle de bec 7493
angle de coupe 288
angle des fibrilles 3102
angouma 3788
animal nuisible à la forêt 3523
animaux sauvages forestiers 3644
anis 293
anis étoilé 4611
anneau 3727
anneau annuel partiel 6087
anneau de broussin 7288
anneau d'écorçage 568
anneau de croissance 301
anneau discontinu 6087
anneau écorcé 3850
anneau faux 2998
anneau ligneux 301
anneau partiel 6087
anneaux à glu 8777
année d'abattage 3144
année d'abondance de cônes 1867
année de coupe 3144
année de glandée 5329
année d'exploitation 3144
annelage 6953
annélation 6953
annélation double par encoches groupées 2580
annélation en encoches 3716
annélation mécanique 5370
annélation partielle 6953
annélation totale 518
anneler 3851
anneleur des rameaux 9016
annonacées 2158
anomalie non dépréciative 829
anonacées 2158
anone hérissée 7856
anone muriquée 7856
anthère 312
anthèse 313
anthocyanose 314
anthracnose 315
anthracnose du caféier 2489
antidérapant 319
antifongique 3771
aoûtage 5339
APA 6164
apeiba 5534
aperçu des classes d'âge 129
aphylle 4873
aplomber 6385
apocynacées 2553
appareil à débarder du bois 7692
appareil à poudrer 2675
appareil à trancher les racines 7019
appareil de relèvement 3184
appareillage retourné 3265
appareil pour montée aux arbres 8814
appareils forestiers 3402
appât 4509
appâtement 482
appétibilité 6032
appétissance 6032
application d'engrais 3090
application d'engrais en surface 8685
application sous forme granulaire 3911
apprentissage forestier 6565
apprentissage pratique en forêt 9239
approbation de l'état d'assiette 6183
approvisionnement en bois 9631
appui d'estrope unique 7646
appui d'extrémité 2815
apru 2245
aptitude à l'emploi de parcours 6720
pénétration 7038
aptitude d'orientation en forêt 9522
aquifoliacées 4335
araliacées 337
aralia épineux 2452
aralie 336
arariba 338
arasement des chicots 7793
araucaria brésilien 1005
araucariacées 339
araucaria de Cook 1919
araucaria de Cunningham 2137
araucaria du Chili 5535
arborescence 345
arborescent 9704
arboretum 342
arboricide 343
arboricole 9704
arboriculteur 344
arboriculteur certifié 1418
arboricultrice 344
arboriculture 3528
arborisation 345
arbouse 8152
arbousier 8152
arbousier commun 8152
arbousier d'Amérique 6020
arbousier des Pyrénées 8152
arbousier traînant 1759
arbre 8797
sans ~s 8874
arbre à abeille 666
arbre à anneau écorcé 3852
arbre à appât 481
arbre abandonné 3
arbre à basse-tige 5196
arbre abattu 2596
arbre abattu pourri grossièrement 7093
arbre à beurre 1223
arbre à bois de valeur 9161
arbre à bois parfait non distinct 7220
arbre à bouquets 5863
arbre à bout éclaté 6711
arbre-abri 5816
arbre à café du Kentucky 1482
arbre à camphre 1291
arbre à caoutchouc 1320
arbre à caroubes 1341
arbre à carriques 1774
arbre à castor 643
arbre à chapelet 636
arbre à cime conique 5657
arbre à cime desséchée 2273

arbre à cime non élaguée 9124
arbre à cime unilatérale 2062
arbre à copal 1924
arbre à croissance en cimeterre 7736
arbre à croissance lente 7755
arbre à croissance rapide 3021
arbre adulte 5341
arbre à émondage 5165
arbre à épice 7909
arbre à feuillage caduc 2302
arbre affaibli 9309
arbre à fibres droites 8138
arbre à fièvre 3092
arbre à fleurs staminées 7999
arbre à forte décroissance 8406
arbre à frotture 7124
arbre à fût droit 8143
arbre à fût rectiligne 8143
arbre à goménol 1056
arbre à gros défilement 3023
arbre à gutta 4061
arbre à gutta-percha 4147
arbre à Kiro 1183
arbre à la fièvre 8417
arbre à la fin de corridor de débusquage 6673
arbre à la gale 6412
arbre à la gomme 4056
arbre à la manne 3292
arbre à la puce 6412
arbre à laque 4755
arbre à latex 4147
arbre à liège de l'Amour 264
arbre à mastic du Pérou 1266
arbre à melon 1794
arbre ampolla 3245
arbre à objectifs multiples 5625
arbre à orchidées 1221
arbre à pain 1011
arbre à pauvre homme 2893
arbre à perruque 2908
arbre à pilot 5328
arbre à racines nues 533
arbre à racines superficielles 7488
arbre à racines suspendues 7047
arbre à risque 4297
arbre à rotation courte 7546
arbre-arrivant 1743
arbre à saucissons 7226
arbre à savonnette 7865
arbre à singe 504
arbre à sommet mort 2273
arbre à sommet sec 2273
arbre à suif 1481
8860
arbre attaqué par des fourmis 320
arbre au caramel 4692
arbre au diable 7185
arbre au lait 2000
arbre au liège 6199
arbre au liège de l'Amour 6985
arbre au poivre 4961
arbre aux abeilles 666
arbre aux anémones 1806
arbre aux boulets de canon 1309
arbre aux cloches d'argent 7800
arbre aux colombes 2593
arbre aux concombres 709
arbre aux épaulettes 3682
arbre aux escargots 4353
arbre aux feuilles de cercis 4692
arbre aux fraises 8152
arbre aux haricots 1768
arbre auxiliaire 4243
arbre aux mouchoirs 2593
arbre aux papillons 1219
arbre aux perles 1804
arbre aux pochettes 2593
arbre aux quarante écus 5245
arbre aux queues de renard 260
arbre avec empattement renforcé 5129
arbre à vernis de Chine 6051
arbre blessé 1097
arbre brisé 1097
arbre caramel 4692
arbre caverneux 4333
arbre cépée 5617
arbre classé 7892
arbre collectif 1724
arbre corail 1941
arbre cornier 1967
arbre couché 4897
arbre couché par le vent 9444
arbre courbé 2030
arbre crochu 2030
arbre d'accroissement moyen 4456
arbre d'agrément 234
arbre d'alignement routie 6983
arbre d'amour 4665
arbre d'ancrage 280
arbre dangereux 4179
arbre d'avenir 3786
arbre de bois massif 3746
arbre d'échafaudage 7270
arbre de champ 5878
arbre d'échantillonnage 7174
arbre de charge 9333
arbre décidu 2302
arbre de dimension moyenne 5387
arbre de fer 6186
arbre de fer de Perse 6186
arbre de feu 4665
arbre déformé 5478
arbre déformé par le vent 9444
arbre de freinage 4324
arbre de friction 7129
arbre de gros nœuds 4797
arbre de gutta-percha 4061
arbre de haute futaie 517
arbre dehors forêt 8891
arbre de Josué 4604
arbre de Judas 4665
arbre de Judée 4665
arbre de Judée canadien 2721
arbre de Judée de Californie 1267
arbre de Judée géant 3843
arbre de l'étage dominant 6009
arbre déliquescent 5617
arbre de lisière 922
arbre d'élite 6396
arbre de mai 5347
arbre de mastic 5327
arbre de mauvaise qualité 6451
arbre de miel 4357
arbre démon 2453
arbre d'émonde 6615
arbre de Noël 1569
arbre de Para 1320
arbre de paternoster 6110
arbre de peuplement final 5569
arbre de plein vent 4301
arbre de pluie 6701
arbre de ponte 1068
arbre déraciné 9141
arbre de rebut 7082
arbre de rebut pourri 7080
arbre de risque important 4297
arbre de Saint Jean 5558
arbre de savon 2224
arbre de sciage 7245
arbre des dieux 8892
arbre de soie 6185
arbre de sommeil 6185
arbre de spécimen 6396
arbre des sorciers 5732
arbre de surface terrière moyenne 5355
arbre de toung 1304
arbre de tung 1304
arbre de vie 5795, 8893
arbre de vie de l'ouest 5795
arbre d'hauteur moyenne 433
arbre d'ombrage 7479

arbre dominant dans une haie 4221
arbre d'ornement 5957
arbre d'ornement feuillu 1051
arbre du ciel 8892
arbre du dragon 7810
arbre d'une scie circulaire 7230
arbre ébranché 5169
arbre échantillon d'essai par classe 7176
arbre élagueur 8760
arbre éléphant 2780
arbre en candélabre 3659
arbre en croissance concurrentielle 1826
arbre en croissance libre 5753
arbre encroué 5041
arbre endommagé 2239
arbre en drapeau 3030
arbre en forme de pavillon 5129
arbre en mottes 499
arbre en production 641
arbre en retrait 7465
arbre entaché 8372
arbre entier 3758
arbre entier disposé en épi 272
arbre entre étage dominant et sous-étage 4551
arbre épineux 8528
arbre étêté 7489
arbre exigeant la lumière 7476
arbre exotique 2943
arbre feuillu 1048
arbre fixé 9053
arbre flache 959
arbre fleuri 848
arbre forestier 3625
arbre fouetté par le vent 9312
arbre fourchu 3659
arbre foyer 1068
arbre fruitier 3731
arbre fruitier forestier 3440
arbre gemmé 959
arbre génétiquement modifié 3819
arbre gisant par un autre arbre 7583
arbre grue 2664
arbre héliophile 7476
arbre-hôte 4381
arbre-hôte de chenilles 1377
arbre incliné fortement 2296
arbre indésirable 9321
arbre indicateur 4474
arbre individuel 4480
arbre intermédiaire 4551
arbre intolérant 7476
arbre jeune 9791
arbre-liège de Chine 264
arbre-liège de Lavalle 4837
arbre loup 9472
arbre mammouth 3844
arbre marchand 5399
arbre marqué 826
arbre marqué en réserve 6887
arbre modèle 6396
arbre moins 5473
arbre mort 2271
arbre mourant 2695
arbre moyen 442
arbre moyen central 1413
arbre moyen d'échantillonnage 442
arbre nain 2691
arbre nain chlorotique 1547
arbre non gemmé 9231
arbre non marchand 9104
arbre nu 1064
arbre obombrant 7479
arbre ombrelle 412
arbre orchidée 6642
arbre penché 5740
arbre pendant 5041
arbre perce-neige 1349
arbre-piège 8778
arbre pistillé 6297
arbre pivotant 1967
arbre plus 6396
arbre porte-fraise 5452
arbre porte-graines 5569
arbre pour plantations en bord de la mer 8842
arbre pousse 7946
arbre prédominant 6500
arbre rabougri 7323
arbre récent 3709
arbre résineux 6900
arbre résistant au feu 3203
arbre rogue 9321
arbre routier 6983
arbre rubané 517
arbre sacré 108
arbres à la route 9308
arbre sans duramen 7220
arbre sans feuilles 4874
arbres à papier 6626
arbre sarrasin 1139
arbres à un axe 5539
arbre sauvage 9419
arbres bouteilles 941
arbres cespiteux 9006
arbres choqués individuellement 7659
arbres d'avenir 5853
arbres de bois court 7540
arbres de sous-étage 9072
arbres dominés 8275
arbre sec 2271
arbre sécheré 2638
arbre sec sur pied 2271
arbre sélectionné 6396
arbres en bouteille 941
arbres éparses 7882
arbres géants 3845
arbres gras 3025
arbres grimpants 1644
arbres morts récupérables 7166
arbres objectifs 5853
arbres par heure 8930
arbre spécial 7892
arbres pourris 7097
arbres ramifiés 988
arbres réservés 6889
arbres sélectionnés enregistrés 6830
arbres supprimés 8275
arbre stable au vent 9443
arbre stagnant 7992
arbres tuftés 9006
arbre sur pied 4010
arbre taillé 7489
arbre témoin 1459
arbre thurifère 4463
arbre-tige 2777
arbre transversal 2038
arbre très serré 2062
arbre vernis 1303
arbre vieux 5885
arbre vigoureux 9216
arbre vive 5018
arbrisseau 346
arbrisseaux sans valeur 9731
arbusculaire 347
arbuste 346
arbuste de désert 2419
arbuste des bananes 513
arbuste d'ombre 7475
arc de la scie 7231
arceau de sécurité 4706
arche à billes 5090
arche à grumes 5090
arche de débusquage 5090
arcossière 8224
ardisia 5320
arécacées 356
arec cachou 691
aréole 6299
aréquier 691
arête 7255

arête de bois scié 7255
argousier 7325
argousier faux-nerprun 7325
aristolochiacées 768
armadille 1813
armillaire 4352
armoise commune 704
armoise d'Autriche 416
armoise de Californie 1270
armoise tridentée 3932
arolaie 8347
arolle sibérienne 7569
aronia 1553
aronie 1553
arpentage forestier 3614
arpentage initial 4497
arpenteuse de la pruche 2718
arpenteuse du pin 6262
arrachage 5979
arracher 2495
arracheuse 6367
arrangement à feuilles alternes 221
arrangement de cernes annuels 302
arrangement de feuilles 4856
arrangement de feuilles décussé 5928
arrangement de feuilles verticillé 9209
arrangement de nœuds 4725
arrangement de racines 7011
arrangement spatial 7884
arrêt de croissance 8126
arrêt d'exploitation 3418
arriéré 470
arrondi 5797
arsin 9528
article en bois 9560
articles en écorce 588
artificier 6485
artocarpe 1011
artocarpus 1011
asclépiadacées 5445
ascomycètes 7144
asiminier 6118
aspect serré d'un peuplement dans le sens vertical 9204
aspérule odorante 9641
assac 7185
assainissement 6860
assamela 119
assécheur d'air 153
assemblage au raccord 3265
assemblage de bois 4660
assemblage de bois de flottage
asseoir 645
asservissement de l'incendie 3192
assiette de coupe 5038
assimilation chlorophyllienne 391
assise génératrice interne 1287
assistance de gestion forestière 3489
association 394
association de chênes, eucalyptus et cyprès 5837
association forestière 3331
association végétale 6336
assomptions de gestion 5263
assortiment de bois 8556
assortiment ligneux 8556
assortiment rond 7100
assurance contre les risques de la forêt 3476
assurance contre l'incendie des bois et forêts 3432
assurance mutuelle contre l'incendie 5642
asthme au bois 9700
astragale 635
astragale à gomme adragante 8749
astronium à feuilles de frêne 8548
atelier de gemmage 9013
atlas de bois 9483
atom-assié 4744
atragène des Alpes 212
attaque initiale 6723
attaque massive 2849
attaqué par le ver 9201
atténuer 400
attié 1293
attractif 4509
attributs de vieux peuplement 5879
aubage 822
aubépine 1780
aubépine à deux styles 4174
aubepine a feuilles rondes 7110
aubépine à fruits jaunâtres 7110
aubépine à un style 1780
aubépine commune 4174
aubépine de Douglas 2591
aubépine de la Colombie 1731
aubépine de Virginie 1699
aubépine dorée 7110
aubépine duveteuse 2602
aubépine écarlate 3167
aubépine ergot-de-coq 1699
aubépine monogyne 1780
aubépine noire 793
aubépine pinnatifide 1500
aubier clair 1034
aubier inclus 853
aubier interne 853
aubier sain 7854
aubier trop épais 2937
aubour 1784
aubour des Alpes 215
aulnaie 164
aulnaie marécageuse 165
aulne à feuilles en cœur 4592
aulne alpin 2897
aulne blanc 2878
aulne commun 2882
aulne crispé 2897
aulne de bord de la mer 7329
aulne de Canada 5579
aulne de la Sierra 7587
aulne de Manchurian 5268
aulne de montagne 2878
aulne de Népal 5709
aulne de Sitka 7677
aulne d'Italie 4592
aulne d'Oregon 235
aulne du Japon 4609
aulne glutineux 2882
aulne gris d'Amérique 2897
aulne noir 2882
aulne rouge 235
aulne rugueux 4164
aulne tendre7786
aulne vert 2897
aunaie 164
aunaie marécageuse 165
aune à feuilles en cœur 4592
aune alpin 2897
aune blanc 2878
aune commun 2882
aune crispé 2897
aune de bord de la mer 7329
aune de Canada 5579
aune de la Sierra 7587
aune de Manchurian 5268
aune de montagne 2878
aune de Népal 5709
aune de Sitka 7677
aune d'Italie 4592
aune d'Oregon 235
aune du Japon 4609
aune glutineux 2882
aune gris d'Amérique 2897
aune noir 2882
aune rouge 235
aune tendre 7786
aune rugueux 4164
aune vert 2897
aurone 5883

auto-ébranchage 5688
auto-ensemencement 5695
autofécondation 7421
auto-pollinisation 7421
autorisation de défrichement 4928
autorité d'abattage local 5033
autre recette 5968
autre surface boisée 5969
autre terre forestière 5967
auxine 421
auzérolle 2827
avalanche détritique 2287
avancement de la coupe 2202
avancement des coupes 6575
avant-train de débardage 5199
avant-train de débusquage à deux roues 5090
avec pousse terminale en baïonnette 631
aveline 4183
avelinier 1778
avelinier 1778
avivé 7963
avocat 445
avocatier 445
avodiré 117
axe de bille 5048
axe de grume 5048
azalée arborescente 8326
azalée jaune 3244
azalée tardive 8303
azalée visqueuse 8303

babul 4048
bacarinier 1475
bâche de pieds-mères 8123
bachons 7693
bachons en perches 6424
bactéries des nodosités 7037
bactéries indigènes 5669
bactéries natives 5669
badam 4464
badanier du Japon 4611
badianier 293, 4611
badianier des pagodes 4611
badianier sacré 4611
bagasse 478
bagasse de cannes à sucre 478
bagasse jaune 478
baguette 635, 6236
baguette de Biltmore 715
bahia 4101
baie de cornaille 785
baie de farine 8323
baie de sabal 7262
baie rose de Bourbon 1004
baies 683
bain 2508
baisse des souches 8207
balai de sorcière 9467
balancement 8341
balancement au vent 9446
balancier à talon 5012
balandier 818
balanin des glands 48
balata 485
balata blanc 9361
balau 9757
balise de délimitation 949
baliveau 8555
baliveaux francs de pied 7387
balle de bois 9496
ballottage en printemps 7937
balsa 506
balsamier élémifère 8716
bambou 508
bambou céleste 5654
bambou de la Félicité 5654
bambou sacré 5654
bambousaie 510
bambouseraie 510
bananier 6335
bananier 514
bananier à fibres 5283
bananier des sages 514
bancoul 1304
bancoulier 1304
bande 8172
bande boisée de protection 7508
bande d'arbres 8177
bande d'écorce 537
bande de coupe 2178
bande de filtre forestier 3426
bande de fracture 3069
bande de protection 1149
bande de protection en essences feuillues 2928
bande d'essai 8479
bande de téléphérage 7468
bande d'isolement 1149
bande d'isolement les pesticides 6190
bande morte 4708
bande rouge 6795
bandes de coupe fourchues 978
bandes de pineraie 8176
bandes minéralisées 5461
bande tampon 1149
bande tampon sans pesticides 6190
banglang 4643
baobab 5533
barbasco 4654
barbe 2193
barbe de Jupiter 2908
barbille 1182
bardeau 7521
bardis 8579
barème décimal de Scribner 7318
barème de conversion 7112
barème de prix de bois 7123
barème du cubage 2117
bariolage 5575
barrage 2230
barrage à contreforts 1229
barrage de brindilles 1117
barrage de correction léger mixte 1117
barrage de fascines 1117
barrage de poutres 4377
barrage de retenue d'eau en vue du flottage 5039
barrage de sommet de cône 613
barrage de sommet de cône de déjection 613
barrage en bois 9550
barrage entonnoir 613
barre à mine 4667
barreau de fer 1300
barre de déroulage excentré 8050
barre de serrage 626
barre de tension 8542
barre directrice 8905
barres traversières 2059
barre tendue 8542
barrette 626
barrière artificielle pare-feu 3186
barrière d'aire naturelle 5689
barrière de broussailles 1098
barrière d'interdiction 2624
barrière pareneige de brousse 1114
base de bois 8558
base de données forestières 3620
base de grume 8061
base de la cime 2064
base de nœud 4726
base de vente à unité de volume 9098
base terrestre 4770
base terrestre d'exploitation de bois 8594
basiphile 614
bas-perchis 5191
basralocus 621
basse-futaie 1928
basse-tige 5196

bassin d'alimentation d'une source 9280
bassin d'amortissement 8100
bassin de flottage 5105
bassin de réserve 5105
bassin de stockage 5105
bassin de stockage en eau salée 7161
bassin versant 9280
bassin versant sensible 7444
bastringue 8809
bâti de longerons 7694
bâtiment de bois 8593
bâton 1165
bâton de Jacob dendritique 8877
bâton du diable 2452
bâton rond 7111
batteur 4369
baume 503
baume brut 6733
baume cicatriciel 9254
baume de Copahu 104
baume du Canada 1294
baumier de Gilead 969
baumier de la Mecque 969
baumier de Tolu 504
baumier du Pérou 504
bêchage en T 289
bêche 7314
bêche à planter 6344
bêche en coin pour plantage en fentes 6365
bédane creux 4328
ben 4379
ben ailé 4379
bénéfice d'exploitation forestière 4444
benjoin odoriférant 7909
ben oléifère 4379
béquille de semi-remorque 7435
béquille de soutien 7435
berbéridacées 528
berbéris 1758
berbéris commune 1758
berceau collecteur 8801
berceau de débardage 8801
berceau de nymphose 4810
berchénie japonaise 4635
bergamote 679
bergamotier 679
bergamotier à feuilles de saule 679
besoin de lumière 4573
besoin du site 7671
besoin du site requise 7671
besoin en bois 2362
bétulaie 751
bétuline 693
bibacier 5172
bibas 5172
bienfonds boisé 3408
biens forestiers 3448
bifeuilleté 697
bigaradier 775
bignone 8994
bignoniacées 8995
bilan d'abattage 5214
bilan de la production ligneuse 9486
bilan du carbone 1329
bilan forestier 9486
bilimbi 709
bilinga 710
bille 2171, 7245
bille à sciage 7245
bille de débardage 7686
bille dédossée 638
bille de pied 1225
bille de pied barrée 3012
bille de pied longue 5128
bille de pied voûtée 3012
bille de racine 1226
bille de souche 1225
bille de tranchage 9193
bille épanouie 3250
bille mandrinée 3250
bille marchande 5393
bille scié à la hache 4270
billes de bois de valeur 7420
billette 2171
billon 2171
billon cloisonné 8544
billon de freinage 2263
billonnage 646
billon-piège 8776
billon pour allumettes 5332
billot 8879
billot 2171
billot de boucher 1211
billot de pin saisonné 7333
billot de pin sec 7333
billot de regarnissage 3802
billot écorcé 6134
billot épais 8502
billot mince 8523
billot pourri 7079
billots de bouleau 746
billots d'épicéa 7950
billots du tremble 385
billots écorcés à blanc 1609
biner 4316
binette 4030
biocénose 738
biocénose forestière 3331
biocoénose 738
biocombustible 722
biodégradables 719
biodétérioration 720
biodiversité 721
biogènese 723
bio-indicateur 4473
biologie forestière 3339
biomasse au-dessus du sol 12
biomasse épigée 12
biomasse forestière 7613
biomasse ligneuse 9490
biome 732
biome de la forêt décidue 2301
biométrie 733
bioréducteur 2308
biosphère 734
biote 735
biotechnologie 736
biotique 737
biotope 739
biotopes clés 4700
biotype 740
bixacées 776
blackbutt 781
blanchis 9371
blessure 4918
blessure de l'écorce 560
bleuissement 860
bleuissement d'arbres 847
bleuissement de la grume 5050
bleuissement de l'aubier 869
bleuissement foncé 2321
bleuissement interne 4557
bleuissement superficiel 8278
bleuissement sur pile 1031
blindage 670
blindage inférieur 670
blocage de bois de flottage 5096
blocage de fossé 2545
blocage latéral 7571
bloc d'alimentation à bois 8648
bloc d'approvisionnement 8272
bloc d'appui de câble porteur 8864
bloc de passage 6098
bloc de semis 7375
bloc entaillé 7284
bloc exploitable 9722
bloc palette 1354
bloc porte-grume 2983
blocs aléatoires 6713
blocs complètement randomisés 6713

bluet du Canada 864
blutoir 2674
bobinage 6120
bocage 890
bocconie 8899
boggie 894
boghei 894
bogie 894
boiré 8375
bois 9473
de ~ 8580
sans ~ 9577
bois à bardeaux 9575
bois abattu au hasard 6712
bois à brûler 3740
bois à carboniser 9571
bois accumulé 8579
bois à cernes étroits 1660
bois à cernes larges 5909
bois à cernes minces 1660
bois à cernes semi-poreux 7432
bois à charpente légère 4942
bois à chevrons 6688
bois à cicatrices 7289
bois acide 44
bois à dégâts d'insectes 4510
bois adulte 6964
bois adulte par acre 8023
bois aéronautique 5063
bois à fagots 7774
bois à faible densité 8660
bois à fascines 9572
bois à fibre 7525
bois à fibres courtes 7535
bois à fibres fines 3149
bois à fibres flexueuses 2048
bois à fibres torses 2048
bois à fil incliné 2462
bois à fût droit 8137
bois à fût rectiligne 8137
bois agatisé 124
bois à gazogène 3816
boisage 8600
boisage du faîte 7006
bois à grain fin 1656
bois à grain grossier 1676
bois agricole 9570
bois à gros nœuds 4798
bois amélioré 4432
bois amer 9373
bois anhydre 17
bois anormalement résineux 4292
bois à panneaux 9257
bois à papier 6061
bois à parquet 9574
bois à pâte 6061
bois à pieux 8588
bois à pilots 8586
bois à pores circulaires 6955
bois à pores diffus 2493
bois à pores disséminés 2493
bois à pores épars 2493
bois à pores étroits 1656
bois à pores grossiers 1685
bois à poteaux 8586
bois à poudre 684
bois à pourriture mycélienne 1880
bois à pourriture sèche 9484
bois à rive à mi-bois 7525
bois artificiel 378
bois à sept écorces 6021
bois à Taxus baccata 9769
bois à texture lâche 5162
bois à traverses 8591
bois automnal 4827
bois avec vaisseaux 6467
bois avivé 2757
bois à vive arête 7492
bois avivé brut 2663
bois à zone poreuse 6955
bois à zone semi-poreuse 7432
bois bakélisé 9735
bois béni 2886
bois bénit 2886
bois blanc 9389
bois blanc de menuiserie 4658
bois blanchi 845
bois bleu 859
bois bonus 909
bois bouquet 5863
bois-bouton 1763
bois bouveté 8677
bois-brésil 1010
bois bromé 1063
bois brûlant 7201
bois brut 7096
bois canard 887
bois-canon 7519
bois carbonisé 1453
bois carré 1383
bois cassé 7167
bois cavicorne 4330
bois chargé au wagon 1364
bois charnière 4313
bois chavanon 5789
bois cicatriciel 9734
bois clair de construction 9578
bois clair de pistachier 6294
bois clair sans défauts 1035
bois clairsemé marécageux 8312
bois clairsemé xérophytique 358
bois clivable 1628
bois coloré clair 4938
bois combustible difficilement 4125
bois commercialement sec 1748
bois comme source d'énergie 9481
bois communal en propriété collective 1812
bois comprimé 2390
bois converti 1918
bois coudé 2029
bois coulé 1373
bois courbé 678
bois courbé par le vent 9441
bois courbé solide 7840
bois court 903
bois court de dimension établie 2501
bois court mince 8521
bois courts 7547
bois crochu 2029
bois d'allumage 3740
bois d'amarante 9225
bois d'amourette 4921
bois d'ancrage 275
bois d'appareil de forage 6948
bois d'arbre fruitier 3732
bois d'arbres feuillus 1047
bois d'arc 1784
bois d'arrimage 8579
bois d'artisan 8589
bois d'aubier 7216
bois de balsa 507
bois de bandes 523
bois débardé 9744
bois d'ébène noir 5344
bois d'ébène vert 3942
bois d'ébénisterie 3776
bois de billon 4738
bois débité 1918
bois débité à dimension 2503
bois débité à dimensions spécifiées 2503
bois débité à files parallélisées 6069
bois débité à pleine longueur 3750
bois débité en plot 3254
bois débité industriel 2988
bois débité sur maille 6678
bois de bobines 7926
bois de bouleau 762
bois de bout 2051
bois de Brésil 1010
bois de caisserie 954
bois de calumet 7109

bois de cèdre du Texas 8484
bois de cerf 260, 401
bois de chablis 850
bois de chaîne 1437
bois de chantier 9754
bois d'échantillon 2503
bois de charpente 3686
bois de charpente tout venant 9754
bois de charronnage 1672
bois de chauffage 3740
bois de chauffage de tremble 388
bois de chauffage industriel 4483
bois de chauffage résineux 7830
bois de chêne 5850
bois de chêne à panneaux 9258
bois de chêne chaulé 4973
bois de chêne refait 9258
bois de chien 1965
bois décidu 4124
bois de cime 8712
bois de cocotier 1703
bois de cœur 4202
bois de cœur imparfait 4426
bois de cœur riche en résine 6902
bois de cœur traumatique 8782
bois de cœur vert 2363
bois de coffrage 3670
bois de colombage 8192
bois de commerce 5400
bois de compression 1850
bois de conifères 1875
bois de construction 1153
bois de construction débité 1918
bois de construction hydraulique 9573
bois de construction maritime 7524
bois de construction séché au four 4710
bois de corail 7180
bois de cour 9754
bois de cyprès 2219
bois de défibrage 6061
bois de démolition 2364
bois de déroulage 901
bois de dosse 7712
bois de faible qualité 5195
bois de faible retrait 8661
bois de fardage 8579
bois défectueux 9118
bois de fente 8563
bois de fer 2415
bois de feu 3740
bois de feuillus 4124
bois de feuillus de bas-pays 5193
bois de feuillus précieux 9160
bois de fibres brutes 1676
bois défilé 8402
bois de fil fin 1656
bois de fil pelucheux 6703
bois de fils bruts 1676
bois de fils grossiers 1676
bois de finition 3164
bois de flèche 907
bois de flottage 3279
bois de fond 946
bois de forêt marécageuse 3085
bois de fructification 3732
bois de fût 891
bois de fût du pin résineux 6266
bois de gaïac 6405
bois de grande valeur 4284
bois de greffe 1145
bois de guidage 4041
bois de gymnospermes 4068
bois de haute densité 8659
bois de haute technologie 2825
bois de hêtre 662
bois de jante 3082
bois de jeunesse 4675
bois de lance 9347
bois de liège 507
bois délígné en "queue de billard" 4094
bois de limon 1591
bois de littoral maritime 7324
bois de longueur propre 6583
bois de madrier 5204
bois de madriers 6328
bois de marine 8590
bois de membrures 678
bois de menuiserie 4658
bois de mine 5463
bois de mines 5463
bois demi-ronde 4094
bois de mobilier 3776
bois de monocotylédones 5541
bois d'empilage 7982
bois d'énergie 2820
bois de noisitier 4184
bois de noix 5829
bois densifié 2390
bois densifié de traction 8463
bois densifié étoilement 3810
bois densifié tangentiellement 8384
bois d'entrepôt 6236
bois dents marron 5139
bois de palmier 6195
bois de Panama 7805
bois de papeterie 6061
bois de perdrix 5426
bois de perdrix des Antilles 5426
bois dépérissant 2490
bois de petite section transversale 8662
bois d'épicéa 7959
bois de pin 6275
bois d'épinette 7959
bois de pirogue 1310
bois de placage 6398
bois de plomb 4900
bois de pont 2306
bois de printemps 2704
bois de qualité 4284
bois de qualité basse 5195
bois de qualité de charpente 8184
bois de qualité secondaire 7365
bois de quassia 6662
bois d'érable 5297
bois déraciné par le vent 850
bois de racines 9580
bois de ramassage 2266
bois de réaction 6744
bois de rebut 2123
bois de réglisse 4997
bois de remplissage 8093
bois de résonance 7851
bois de rive d'un glissoir 7148
bois de roche 6994
bois de rose 1007, 6076
bois de rose d'Océanie 6471
bois de rose du Brésil 1006
bois de rose du Siam 7563
bois de rose femelle 3083
bois de rose véritable du Brésil 1007
bois désagrégé 2530
bois de sapin 7959
bois de sapin en grumes 9390
bois de sappan 7209
bois des chiens 4606
bois de sciage 7245
bois de sciage résineux 5708
bois de service 9721
bois de souche 8220
bois de souche anormalement résineux 6899
bois de souche du pin 6271
bois desséché 2622
bois destiné à la distillation 44
bois de structure 1153
bois de surlongueur 5997
bois d'étai 6428
bois de taillis 7361
bois d'étanchéité 8638
bois d'été 8264

bois de tension 8467
bois de têtard 6431
bois de tiges adhérentes 4536
bois de tournage 9012
bois de tranchage 901
bois de valeur 4284
bois de valeur ordinaire 1776
bois de violette 4716
bois de voie ferrée 8589
bois de wagons 8592
bois d'exportation 2958
bois d'extraction 2971
bois de zèbre 8548
bois dimensionné 2502
bois d'Inde 632
bois d'industrie 9721
bois dispersé 7293
bois disponible 424
bois d'œuvre 9721
bois d'œuvre feuillu 1054
bois d'œuvre non débillardé 5216
bois d'olivier 5894
bois d'oranger 5937
bois d'origine d'Outre-mer 6003
bois d'orme 2791
bois d'ossature 1335
bois d'outre-mer 8981
bois dur 3225
bois dur du nord mixte 5490
bois d'usage 6944
bois d'usine 5457
boisé 3394
bois échappé 3279
bois échauffé 2605
bois écimé 6429
bois écorcé 555
bois écorcé hydrauliquement 8156
bois élagué 6608
bois élastique 2776
boisement 94
boisement de pâturage 6100
boisement de pentes minières 97
boisement de sables 7179
boisement des marais 98
boisement des rives 8158
boisement de terrain nu 96
boisement insuffisant 9068
boisement ripiphile 6960
boisement rivulaire 6960
boisement sur les versants 7751
bois empilé 7982
bois en corde 7982
bois en dentelles 4752
bois endogène 2811
bois endommagé 2238
bois endommagé par un incendie
bois en longues fibres 5131
bois en poudre 6484
bois en rondins 1032
bois enstéré 7982
bois entier 9393
bois entreposé dans l'eau 9290
bois équarri 7969
bois équarri coupé 7969
bois équarri grossièrement 7094
bois équarri pour fondation de piles 7975
bois équarri rugueux 7091
boiser 93
bois étuvé 8054
boiseur 9654
boiseur à temps partiel 6095
boiseur permanent 6173
bois exotique 3317
bois expurgé 827
bois façonné 6116
bois façonné au petit bout de la grume 8712
bois faussé 4655
bois feuillu 4124
bois feuillu à pâte 9603
bois feuillu de montagne 9132
bois feuillu dur 4121
bois feuillu dur d'utilité 9151
bois feuillu léger 4943
bois feuillu mou 7816
bois feuillu noble 4295
bois feuillu précieux 4295
bois feuillu profilé 1991
bois filé 2618
bois final 4827
bois fissile 1628
bois flacheux 9079
bois flexible 2776
bois-flot 8351
bois flottable 3268
bois flottant 3279
bois flotté 3279
bois fondrier 2268
bois fort 4808
bois fossilisé 6196
bois frais 3710
bois frontal 2051
bois gelé 3728
bois genouillé 2027
bois gentil 2246
bois gisant 3042
bois grand teint 2989
bois gras 3026
bois gras de pin 3026
bois gravé 1361
bois gris 5532
bois humide 9354
bois ignifuge 3198
bois immortel 1941
bois imparfaitement écorcé à la machine 9596
bois imprégné 4427
bois imprégné-densifié 1844
bois incendié 1196
bois inconnu 1777
bois indésirable 9322
bois indigène 4341
bois industriel 9721
bois initial 2704
bois intermédiaire 4552
bois iveant 4606
bois jack 1011
bois jaune 5974, 9767
bois joli 2246
bois lamellé 4763
bois lamellé-collé 3876
bois laminé 4763
bois l'évêque 4645
bois libérien 625
bois libre 3701
bois long 5154
bois luné 853
bois madré 9186
bois maillé 7603
bois manufacturé 5288
bois marchand 5400
bois marchand empilé 6230
bois marqué 5311
bois marqué à la griffe 8644
bois massicoté 2757
bois massif 7843
de ~ 3745
bois massif collé 7841
bois modèle 5505
bois modifié 5513
bois moiré 7993
bois mort 2274
bois mort du tremble 386
bois mort par attaque de coléoptères 1150
bois moucheté 767
bois mouluré 6566
bois mûr 6964
bois mycélien 6132
bois naturel 3826
bois net de coloration naturelle 1035
bois neuf 5726
bois noir 814
à ~ 815
bois non aménagé 5765
bois non déligné 9079

bois non refendu 9048
bois non ressuyé 3710
bois non scié sec 7086
bois non séché 9116
bois nouailleux 4738
bois noueux 4738
bois nu 535
bois nuisible 8601
bois ondulé 2048
bois opalisé 5903
bois paresseux 7735
bois parfait 3159
bois pétrifié 6196
bois piquant 2451
bois piqué 2291
bois pissenlit 1339
bois pour la fabrication d'instruments musicaux 8587
bois pour la production d'énergie 2820
bois pour meubles 3776
bois pourri 2291
bois pourrissant 2292
bois pour timons 6427
bois précieux 3159
bois prégroupé 6494
bois prêt 6515
bois primaire 6549
bois printanier 2704
bois profilé 6566
bois provenant de la surbille 8712
bois puant 2881
bois punais 1771
bois purifié 4432
bois raboté 2620
bois rabougri 2022
bois rabougri par broussailles 1202
bois radial 6678
bois rainuré 3978
bois rares 6726
bois récemment gisant 3710
bois reconstitué 6750
bois refait 9257
bois refendu 7920
bois rescié 6882
bois résidu 9277
bois résiduel 9108
bois résineux 6901
bois résineux sombre 2252
bois résistant à pourriture 2294
bois ressuyé 150
bois ressuyé commercialement sec 7528
bois restant 9277
bois rond calibré 6837
bois rond entreposé dans l'eau 9302
bois rond feuillu 1053
bois rond surdimensionnel 6005
bois rouge 7200
bois rouge 6802
bois rouge de Kodar 6790
bois roulé 9696
bois rubané 6935
bois rustique 4738
bois saigné 827
bois sain 5902
bois saisonné 7335
bois sang 5956
bois sans branches 990
bois sans cœur 4199
bois sans défaut 2351
bois sans défauts 9578
bois sans nœuds 2351
bois sans vaisseaux 5766
bois satiné 2729
bois scié 1918
bois scié à la hache 4271
bois scié avivé 2757
bois scié biaisé 8402
bois scié brut 7085
bois scié feuillu 1054
bois scié non avivé 9079
bois scié non trié 9117
bois scié plein sur quartier 3755
bois scié sur dosse 3254
bois scié sur mailles 6678
bois scié sur quartier 6678
bois scié unilatéralement 5899
bois sculpté 1361
bois sec 2658
bois sec absolument 17
bois sec à raboter 6325
bois sec calibré 6835
bois sec gisant 2266
bois séché 2622
bois séché à la forêt 3386
bois séché à l'aire 150
bois séché artificiellement 2421
bois secondaire 7355
bois sec pour usinage 6325
bois sec sur pied 2274
bois séparé 7448
bois silicifié 6196
bois solide 2670
boisson de bois 9539
bois soumis 6076
bois souple 7126
bois spécial 7891
bois stocké 8116
bois surépaissé 2975
bois sur pied 3454
bois tacheté 1662
bois tan 5138
bois tardif 4827
bois tendre 7823
bois terminal 8712
bois tordu 2029
bois toxique 6413
bois traité 8792
bois traité par paraquat 6070
bois transformé difficilement 4126
bois traumatique 9734
bois trempé 9354
bois tropical 8981
bois veiné 9186
bois vermoulu 9729
bois vert 3710
bois vert calibré 6836
bois vert non scié 7090
bois violet 9225
boîte bleue 857
boîte en bois 9549
bombacacées 7596
bombardier d'eau 157
bombax 6792
bombyx disparate 4071
bombyx du chêne 5846
bombyx du genêt 5846
bombyx du pin 6258
bôme 8779
bonduc 1482
bonne année de production de cônes 1867
bonnet de cardinal 1383
bonnet de prêtre 1383
bonnet d'évêque 1383
bonsaï 908
boqueteau 1666
boquillon 9057
boraginacées 919
bordage 9796
bordage en bois 880
bord d'appui 640
bord de feuille 4878
bord de feuille crénelé 2396
bord de feuille denté 2396
bord de feuille entier 2833
bord de feuille ondulé 6868
bord de feuille serré double 2590
bord de feuille sinué 7661
bord droit 7967
bord du vent 9451
bord non avivé 2662
bordure d'une forêt 2761

borraginacées 919
borriquete 4380
bosquet 933
bosquets riverains 8158
bosse 1182, 9003
bossé clair 6280
bosse de triage 4440
bossé foncé 6280
bostriche 5220
bostryche 5220
bostryche curvidente 7602
bostryche liséré 8173
boswellie 4463
botanique forestière 3341
bottes 917
bottes de protection 917
bouchon 6103
bouchon de résine 6303
bouchon en bois 9555
bouchonnage des noeuds 4727
bouclier 1228
bouclier arrière 1228
boue d'eaux d'égout 7473
boue d'égout 7473
boue d'épuration 7473
bougainvillée 947
bougainvillier 947
bouillon 834
boulaie 751
boulassière 753
bouldozeur-renverseur 8808
boule 948
bouleau 742, 1761
bouleau acajou 1474
bouleau à canoë 6058
bouleau à feuilles de peuplier 3923
bouleau à larges feuilles 384
bouleau à papier 6058
bouleau bariolé 3122
bouleau bas 1295
bouleau blanc 1761
bouleau blanc du Japon 384
bouleau blanc du nord 1503
bouleau bleu 863
bouleau commun 1761
bouleau d'Alaska 159
bouleau d'eau 9279
bouleau de Mongolie 2226
bouleau de montagnes 5581
bouleau de Norvège 4690
bouleau de rivière 779
bouleau d'Erman 2860
bouleau de Young 9789
bouleau en cépée 1761
bouleau européen 1761
bouleau fontinal 9279
bouleau frutescent 1295
bouleau glanduleux 892
bouleau gris 3923
bouleau humble 1295
bouleau jaune 1299
bouleau jaune canadien 1299
bouleau jaune d'Amérique 1474
bouleau merisier 1474
bouleau nain 892, 8304
bouleau noir 779
bouleau occidental 9279
bouleau pleureur 9789
bouleau pubescent 6620
bouleau rouge 3923
bouleau sucré 8328
bouleau verruqueux 1761
boule de cire 1804
boule de neige 2890
boule de neige cotonneuse 9307
bouleraie 753
boulet de canon 1309
boullée 3018
bouquet 1666
bouquet champêtre 5001
bouquet d'arbres 1666
bouquet de rejets de taillis 1665
bourdaine 684
bourdaine marine 7325
bourgène 684
bourgeon 1141
bourgeon à bois 9179
bourgeon adventif 78
bourgeon auxiliaire 419
bourgeon avant-dernier 8242
bourgeon collatéral 1723
bourgeon de pin 6247
bourgeon de sapin 6247
bourgeon dormant 2573
bourgeon foliaire 4859
bourgeon mixte 1837
bourgeon multinodal 5611
bourgeonner 7944
bourgeon nu 5651
bourgeon plain 2831
bourgeon polycyclique 5611
bourgeon préventif 2573
bourgeon proventif 6605
bourgeon radical 6683
bourgeon reproductif 6878
bourgeon secondaire 7342
bourgeon simple 7628
bourgeon terminal 8470
bourgeon terminal de palmier 1239
bourgeon terminal faux 3004
bourrée 4193
bourrelet 1182, 6155
bourrelet de gélivure 3726
bourrelet de recouvrement 2157
bourrelet d'une branche 832
bourreur de copeaux 1522
boursouflure 834
boursouflure sur feuille 837
bout de bois scié 7257
bout droit 7966
bout éclaté 1071
bouteille à semence 7381
bouteur renverseur 8808
bout franc 871
bouton à bois 9179
bouton additionnel 78
bouton de Bahia 6224
bouton rouge 4665
bout rectangulaire 7966
bouts de bois à brûler 3221
bouts de bois de feu 3221
bouts traînés 8755
bouturage 6581
bouture 2184
bouture de racine 7018
bouture de rameau aoûté 2574
bouture de vieux bois 4135
bouture dormante 2574
bouture herbacée 7829
bouture racinée 7021
bouture végétative 7010
bracatinga 966
brachyscléreide 8121
brachysclérite 8121
braconnage 6402
braconnier 6401
bractée 968
bractéole 6585
brai de hêtre 660
branche 974
    à ~s multiples 5596
    à ~s tordues 4737
    à grosses ~s 8135
    sans ~s 983
branché 977
branche à la base 604
branche apicale 324
branche charpentière 6536
branche courbée 2026
branche d'ancrage 274
branche de charpente 6536
branche de moyenne grandeur 5386
branche de saule 9437
branche de surface intérieure 9066

branche gourmande 2848
branche intermédiaire 4541
branche latérale 4819
branche latérale secondaire 7349
branchement 987
branchement orthotropique 5962
branchement plagiotrope 6316
branche mère 6536
branche morte suspendue dans la cime 9397
branche ouverte 9050
branche primaire 6536
branche prolongée 6468
branche racinée 7020
branche sèche 2647
branche sèche et pendante 9397
branche secondaire 7341
branches et cime 5163
branches mortes 2265
branche stricte 8162
branche supérieure haute 8683
branche verte 5013
branche vive 5013
brande 1779
bras coupeur 4040
bras d'abattage 3049
bras d'abattage de bois 8823
bras directeur 8905
bras-porte-couteaux 8823
brasse 3024
bras sécateur de bois 8823
bréjotte 3153
bricole-bois 3108
bride 1182
bridon 2142
brigade de support de câble 7703
brimbelle 5006
brimbellier 5006
brin 7202
brin de saule 9437
brin de taillis 1933
brindilles 1116
brindilles d'essences feuillues 4145
brique en bois 9495
briquette de bois 9495
briquette en particules de bois 9495
briquettes de bois 3741
bris de glace 4417
bris de la cime 1012
bris de la grume 8064
bris de neige 7801
brise-vent 9445
brisure 9507
brisure de séchage 2422
brisure par le gel 3723
broche porte-fraise 5452
broche repère 3899
broiement 2105
bronzage 1066
broussaille 353
broussailles 1116
broussailles à rotation courte 7543
broussailles grandes 4807
broussailleux 934
brousse 1108
brousse de sous-étage 9065
broussin 501, 1167
broussin à la base 605
broussin d'aubier 7196
broussin de têtard 995
broussonétie à papier 6060
brout 1095
broûtage 8806
brout en racines 493
brouter 1094
broyage 2105
broyer 3975
broyer de souche à remorque automotrice 7422
broyeur 1534
broyeur à courroie 671
broyeur à mâchoires 186
broyeur d'écorce 583
broyeur de déchets 7721
broyeur de placage 9194
broyeur de souches à disques 2529
broyeur de souches attaché 399
broyeuse de rémanents 7721
brûlage accéléré 3315
brûlage à la mise en tas 8308
brûlage anticipé 2700
brûlage contrôlé 6516
brûlage cyclique 7075
brûlage de bois 9499
brûlage de déchets 7719
brûlage dirigé 6516
brûlage en sous-étage 9054
brûlage extensif 1043
brûlage général 1043
brûlage hâtif 2700
brûlage immédiat 5010
brûlage libre 3690
brûlage localisé 6105
brûlage par placettes 6105
brûlage précoce 2700
brûlage préventif 2700
brûlage progressif 8308
brûlage sur ordre 6516
brûlage sur prescription 6516
brûleur de sciure 7236
brûlis 1184
brûlure au sommet du limbe 8669
brûlure de la feuille 7306
brûlure interne 4274
brûlure mécanique 5229
brûlures de réduction brute 7092
brumisation 5481
brunissement des extrémités des feuilles 8669
bruyère 1779, 4206
bruyère arborescente 4206
bruyère cendrée 3153
bruyère commune 1779
bruyère en arbre 4206
brya faux-ébène 4607
bubinga 1120
bûche 2171
bûché 5066
bûche de chauffage 3219
bûche de Noël 1567
bûche refendue 7920
bûcheron 9526
bûcheronnage 3457
bûchette 7530
bûcheur 9057
buddléiacées 1220
buddleia du père David 1219
buis 2886
buis à petites feuilles 5002
buis boule 2886
buis commun 2886
buis de Mahon 7875
buis d'Espagne 7875
buis panaché 2886
buisson 7555
buisson ardent 3212
buissonnant 934
buissonneux 934
bulbe 9003
bulletin de cession de bois 6183
bulnésie arborescente 9199
bupreste bleu du pin 8057
bupreste du cèdre 1403
bupreste du thuja 1403
buprestes 8562
bureau forestier 3323
burséracées 1198
buse 2132
buse transversale 2046
busserole 1759, 5289
busserole mexicaine 6410
but de composition du peuplement 8413
but du peuplement 8413
butée 4042

butoir 4225
buttage 1171, 5578
buxacées 960
buxaie 961
byrsonyme 2570

caatinga 1236
cabane à piège 8822
cabane de bûcherons 5071
cabane en rondins 5052
cabelluda 1240
cabestan 281
câblage 271
câblage à deux stabilisateurs 2588
câblage aérien 88
câblage à stabilisateur 8366
câble 9463
câble à distance 5130
câble à freiner 7804
câble ancré 1164
câble à torsions croisées 7059
câbleau 4176
câble à vide 7714
câble bas 3993
câble d'allingue 910
câble d'attache 8543
câble de débardage 3987
câble de fixation 5023
câble de halage 3987
câble de haute pouvoir 4293
câble de levage 4323
câble de manœuvre 8153
câble d'empilement de bois 9497
câble de récupération 5411
câble de remorque 9007
câble de retour 7714, 5411
câble de téléphérique 8745
câble de traction 1555
câble ferme-benne 1661
câble fixe 8043
câble-grue 7697
câble-grue à chariot autoleveur 7133
câble-grue à franchissement multiple 5627
câble-grue à franchissement single 7649
câble grue à longerons 8164
câble-grue à tension fixe 8043
câble-grue mobile 5501
câble-grue statique 8043
câble haut 4300
câble immergé 9352
câble inférieur 3993
câble inverti 4580
câble non préformé 5767
7714
câble porteur courbé 2153
cable porteur mobile 5017
câble pour glissoire 9464
câble préformé 6503
câbler 1241
câble stabilisateur 8365
câble supérieur 9137
câble tracteur de retour 5411
câble tracteur principal 8745
cabrette 4190
cacaotier 1248
cacaoyer 1248
cachoutier 1376
cactier dendritique 2366
cactus de Barbarie 4604
cadastre 1249
cadastre forestier 3344
cade 6530
cadranure 8039
cadranure étoilée 8039
cadre d'une scie 3684
caduc 3303
caduque 2302
caféier 1713
caisse à eau 7942
caisse en bois 9549
caissette 6373
cajeputier 1056
calamagrostis 7537
calcul de la possibilité 9776
calcul de rendement soutenu à usage multiple 5624
cale 1548, 6236, 8771
calebasse 1764
calendula 8882
calepin de dénombrement d'abattage 3073
calfats 1280
calibrage de bois 3902
calibrage par les dimensions 3903
calibre à secteur 7367
calibre de bois de sciage 5205
calibre optique 5930
callicarpe 3705
callidie commune 868
callidie du Pacifique 797
callitris 2217
callune 1779
callus 1182
calmie glauque 895
caloupilé 2147
calycanthacées 8151
camarine noire 785
camasari 9346
cambrure 7936
came d'abattage 3054
camélia 1765, 8091
camélia commun 1765
camélia ovale 5591
camerisier à balais 2895
camp avancé 7573
camp de ligne d'arrêt 7573
campêche 846
camphre 2710
camphre naturel 5674
camphrier 1291
campnosperma 7155
campos 1292
camp principal 3170
camp volant 7573
canal à gommes 4049
canal de flottage 3277
canal de résine cicatriciel 8783
canal d'une ponctuation 6300
canal intercellulaire 4530
canal intercellulaire radial 6677
canal intercellulaire traumatique 8783
canal résinifère 4049
canal secondaire 7344
canal sécréteur 4530
canal sécréteur résinifère 4049
canéficier 3883
canistel 2772
caniveau de drainage 2613
canne 1307
canne-bastringue 715
canneberge 2889
canneberge ponctuée 1996
cannelé 1230
cannelier casse 1495
cannelier de Ceylan 1421
cannelle 9364
cannelle blanche 9364
cannelle de Ceylan 1421
cannelure 3296
cannelure du tronc 8999
cannelure mâle 2595
canne-semoir 6344
canopée 2343
canton 5038
cantonnement forestier 3382
caoutchouc 4053
caoutchouc naturel 5694
capacité au champ 3110
capacité de charge 1357
capacité de coupe 2188
capacité de production de bois 6481
capacité de production forestière

capacité de rétention au champ 3110
capacité de rétention en eau 9287
capacité de support l'ombre 7477
capacité germinative efficace 2768
capacités de poste de marchandisage 7726
capage 8960
capanema 9346
capital forestal 3346
capital forestier 3346
capital-forêt 3346
capoeira 1326
capparidacées 1324
capricorne des maisons 4389
câprier épineux 1766
caprifiguier commun 1774
caprifoliacées 4356
capsule 1328
captage 9280
captage de l'eau 9280
caractères de bois 9597
caractéristiques de la vie sauvage 9414
caractéristiques dendrométriques 2382
caractéristiques édaphiques 2754
caragana 7165
caraganier arborescent 7564
caraganier frutescent 7136
carapa 4039
carapa des Indes 6646
carapa rouge 4039
carbonisateur 1444
carbonisation 1331
care rouge 7876
caricacées 1340
carie alvéolaire 6313
carie brune cubique de bois 1077
carie brune cubique de conifères 1076
carie du bois initial 6956
carie du bois initial blanche 9380
carie du cœur 1414
carie grumeleuse brune cubique 2103
carie jaune fibreuse 8167
carie marginale 4494
carie organique 5946
carie racinaire et spongieuse 364
carie spongieuse 7819
carotte de sondage à la tarière 4452
caroubier 1341
carouge 1341
carpophore 1353
carr 1356
carre 540, 8155
carreau d'une coupe 5083
carre de deuxième année 9756
carre sèche 2264
carripoulé 2147
carrosserie-fourgon à toit ouvert 5920
carrosserie pour grumes 5087
carrosserie pour transport de grumes 5087
carte d'aptitudes 8256
carte de district forestier 3383
carte de foresterie 3572
carte de forêts 3498
carte de projection de houppiers 5298
carte des cimes 2086
carte de types forestiers 3635
carte d'exploitation 2953
carte en courbes forestière 3362
cartographie de sites forestiers 3595
cartographie forestière 3499
carton 6059
carton-bois 9493
carton dur vulcanisé 1845
caryer 6130
caryer blanc 7482
caryer cordiforme 774
caryer de sables 7188
caryer glabre 6229
caryer lacinié 7507
caryer muscade 5821
caryer ovale 7482
caryocar nucifère 6188
caryophyllacées 6278
cascade 9285
cascara 1544
case lysimétrique 5223
casque de protection 4122
cassandre caliculé 4898
casseur 2104
casseur débroussailleur 1100
cassie 1689
cassier 1689
cassier commun 3883
cassine 1371
cassis 2883
cassure de la cime 1012
castanopsis 3841
casuarina 663
casuarinacées 664
catalogue commun des matériaux de base pour reproduction
catalpa d'Amérique 1768
catalpa de Caroline 1768
catalpa de Chine 1492
catalpa de l'ouest 5789
cataphylle 9075
catastrophe naturelle 5676
catégories de bois usinés 8556
caudex 1389
caulescent 1390
cauliflore 1391
cauliforme 1392
causant de dommage aux forêts 1393
caven 1394
cavité d'une ponctuation 6301
cavité d'un tronc causée par la pourriture 4334
cavité intercellulaire 4531
cavités d'arbre 8811
cavité sécrétrice 4531
cavosité 4332
céanothe 1395
céanothe à feuilles aiguës 5660
céanothe à feuilles étroites 6788
céanothe bleu 5242
céanothe en arbre 1265
céanothe épineux 3939
céanothe sanguin 6793
cécidomyies 3793
cécropie peltée 7519
cédraie 1397
cédrat 1590
cédratier 1590
cèdre 1396
cèdre à encens 1262
cèdre bagasse 4086
cèdre blanc 398, 1262
cèdre blanc d'Amérique 398
cèdre blanc de Californie 1262
cèdre d'Algérie 396
cèdre de Bermuda 682
cèdre de l'Amérique 243
cèdre de l'Amérique centrale 1411
cèdre de l'Atlas 396
cèdre de l'Himalaya 2398
cèdre déodar 2398
cèdre de Virginie 2722
cèdre du Chili 6102
cèdre du Liban 1400
cèdre du Mexique 1411
cèdre jaune 4768
cédrèle 1493
cédrèle à feuilles de châtaignier 4309
cèdre odorant 1411
cèdre rouge 2722

ceinture 8340
ceinture de forêt 3335
ceinture forestière 3335
ceintures à glu 8777
ceinture verte 3941
célastracées 7985
célastre 7984
célastre grimpant 239
cellule à huile 5875
cellule à mucilage 5597
cellule carrée de rayon 7972
cellule cloisonnée de parenchyme 7449
cellule couchée de rayon 6559
cellule criblée 7589
cellule cristallifère 2108
cellule de bois 9503
cellule de l'écorce 541
cellule de parenchyme 9618
cellule de rayon 6736
cellule de rayon dressée 9139
cellule dressée de rayon 6559
cellule du cambium fusiforme 3781
cellule fusiforme 3781
cellule fusiforme de parenchyme 3783
cellule laticifère 4830
cellule-mère du liber 6207
cellule palissadique 8553
cellule palissadique de rayon 8553
cellule pédonculaire 7997
cellule pierreuse 8121
cellule scléreuse 7300
cellule subéreuse 1957
cellulose 9504
cellulose du bois 9504
celtis 2898
cembrot 8346
cémentation 1367
cémentation inverse 6924
cendre de Berlandier 680
cendre de bois 9482
cendre de bois dur 4132
cendre de bois mou 7827
cendre de bois résineux 7827
cendre mexicaine 680
centrage de la grume 5053
centre de poussée 3668
cépage noble 5739
cépée 1933
céphalanthe d'Occident 1763
céphalanthe occidental 1763
céphalotaxacées 6392
cératiole 7187
cercidiphylle du Japon 4692
cerclage 519
cercle de coupe 2190
cercocarpe à feuilles ondulées 2143
céréiforme 1305
cerejeira 1417
cerise 1473
cerise acide 7855
cerise des Antilles 527
cerise douce 8330
cerisier 1476
cerisier à feuilles de houx 4336
cerisier à feuilles en dents de scie 744
cerisier à fleurs 4275
cerisier à fruits doux 8330
cerisier à grappes 2881
cerisier amer 772
cerisier buissonnant 3984
cerisier canadien 7186
cerisier des bois 8330
cerisier de Tahiti 5966
cerisier de Virginie 1769
cerisier du Tibet 744
cerisier nain 7855
cerisier noir 1769
cerisier sauvage 8330
cerisier tardif 244
cerne 301
  à ~s étroits 9470
cerne absent 5479
cerne annuel 301
cerne de croissance 301
cerne de gel 3727
cerne de gelée 8785
cerne de sécheresse 8785
cerne d'incendie 3204
cerne discontinu 6087
cerne discontinu en anneau 2528
cerne double 5616
cerne manquant 5479
cerne multiple 5616
cerne noir 801
cerne traumatique 8785
cerne traumatique de gel 3727
cerne traumatique d'incendie 3204
certification forestière 3348
césalpiniacées 7439
cœur
  à ~ crevassé 4201
  à ~ pourri 2402
chablis 850
chaînage 1428
chaîne à billes 4173
chaîne à pneu 319
chaîne auxiliaire 7579
chaîne coupante 1430
chaîne d'allingue 910
chaîne d'ancre 270
chaîne d'arpenteur 8289
chaîne d'attache 5023
chaîne de billes monocaténaire 7654
chaîne de billes multitoron 5635
chaîne de coupe 1430
chaîne de débardage 1555
chaîne de débardage glissante 7740
chaîne de déchets de coupe 6808
chaîne de dragage 1427
chaîne de halage 7798
chaîne de mouillage 270
chaîne de récolteuses 4152
chaîne de remorque 8750
chaîne de retenue 7798
chaîne de serrage 5023
chaîne de traînage 7798
chaîne élévatrice 4173
chaîne frein 1159
chaîne longitudinale d'aménage de grumes 1424
chaîne monte-grume 4173
chaîne trophique 3308
chaînette de pores 6462
chaînon 2183
chalef argenté 257
chalef en ombrelles 417
chamaecyparis de Lawson 4839
chamaephyte 1439
chambre d'une ponctuation 6302
chaméphyte 1439
chamérops 5384
chamfuta 6406
champ abandonné 2
champ d'action 5925
champ de croisement 6738
champ de ponctuation 6305
champ de ponctuation primaire 6544
champignon arboricole 9581
champignon des maisons 2560
champignon des murs 9581
champignon en forme de console 1879
champignon lignivore 9485
champignonné 3773
champignon racinaire 7025
champignons 3770
champignons de coloration de bois 9663

champignons du bleu 861
champignons parasitaires 6072
champignons saprophytiques 7213
chandelier 2271
changement de la couleur de bois 9514
changement de ligne de traînage 9751
changement des essences forestières 226
changement net annuel 5713
chanticleer 1281
chantier 5038
chantier de bois 8664
chantier de bûcherons 9527
chantier d'empilage 5451
chantier de scierie 7253
chantier d'estimateurs 8288
chantier d'exploitation forestière 5038
chantier d'exploitation permissible 2166
chantier forestier 5038
chantier forestier commercial 1746
chantier forestier productif 6562
chantier forestier productif réserve 6563
chantier forestier réservé 6886
chantournage 2151
chantournage de rives 2765
chantourner 7317
chaparral 1442
chapelet 914
charançon de bouleau 761
charançon de déodar 2399
charançon du collet de la racine du pin 6267
charançon du pin blanc 8670
charançon du pin d'Engelmann 2824
charançon du pin Weymouth 8670
charançon du sapin de Sitka 7679
charançons coupe-bourgeons 9014
charbon 315, 1673
charbon à souder 1443
charbon de bois 9505
charbon de bouleau 748
charbon grillé 6984
charbonnette 9571
charbonnier 1445
charbonnière locale 1986
charbon pulvérisé 6633
charge anthropogénique 316
charge à remorque double 2589
charge de bois 1338
charge de bois complet 8984
charge de fardage 1180
charge de fond 647
chargement à chaud 4383
chargement de bois 8608
chargement de bois intermédiaire 1722
chargement de grumes pour un véhicule 5058
chargement des grumes 5099
chargement de troncs entiers 8871
chargement du combustible 3736
chargement latéral 2052
chargement par pieu 913
charge moyenne de peuplement 8110
charge traversière 1180
chargeur à flèche horizontale 705
chargeur à fourche 8610
chargeur de billot 842
chargeur de grumes 8098
chargeur de grumier 5088
chargeur mobile de bois 5501
chargeur principal 8695
chargeuse à haute portée 6065
chargeuse mécanique de bois débité 5207
charge vive 5014
chariot à billes 5199
chariot à fourche 3660
chariot à grumes de scies à cadre 5208
chariot de bois 5199
chariot de bois de sciage 5208
chariot d'ébranchage 2354
chariot de débardage 3673
chariot de scies à cadre 5208
chariot diviseur 5208
chariot élévateur 8581
chariot élévateur à fourche 3660
chariot élévateur de bois 5098
chariot élévateur frontal 3720
chariot-palan 7698
chariot-palan de câble-grue 1232
chariot porte-grumes 5208
chariot-porteur d'assortiments de bois 7539
charmaie 4372
charme 2901
charme à feuilles en cœur 7280
charme blanc 2901
charme bouleau 2901
charme commun 2901
charme d'Orient 5950
charme du Japon 4621
charme faux 2901
charme-houblon 2900
charme noir 2900
charnière 9051
charpente 1352
charpentier du peuplier 6453
charpentière 6536
charpentière secondaire 7341
charrue à deux socs 9028
charrue à deux socs et versoirs simultanés 9028
charrue à éperon 8668
charrue à versoir 5524
charrue à versoir simple 7645
charrue bisoc 9028
charrue débroussailleuse 1203
charrue forestière 3531
charrue pare-feu 3196
charrue planteuse 8771
charrue pour l'ouverture de tranchées par feu 3196
charrue sauteuse de souches 8206
chasseur 1454
chasseur à dépôt transitoire 4773
chasseur-débusqueur 1455
châssis de madrier 2277
châssis d'une scie 3684
châtagnier du Brésil 1003
châtaigneraie 4005
châtaigner américain 245
châtaignier 2887
châtaignier de Chine 1494
châtaignier de Virginie 1199
châtaignier d'Ozark 6015
châtaignier du Japon 4616
châtaignier nain 1199
chaton 1380
chattepelouse 6557
chaudière à bois 1444
chauffage 1191
chauffage au bois 9568
chaume 2124
chauve 486
chef chargeur 9748
chef de bûcherons 5069
chef de campement 1290
chef de comptage 8377
chef de groupe forestier 1483
chef d'équipe de chargement 9748
chef de service forestier 1483
chef d'étable 591
chemin à broussailles 1119
chemin au fruticées 7556
chemin bâchonné 1947

chemin d'appui arrière 475
chemin d'attaque 3217
chemin de débardage 7689
chemin de débardage brute 7095
chemin de débardage construit 1887
chemin de débardage différé 2425
chemin de débardage pivotant 8344
chemin de débardage secondaire 7350
chemin de débusquage 7689
chemin de départ 6111
chemin de fer de montagne 4440
chemin de fer forestier 5079
chemin de rondins 1947
chemin de travois 7689
chemin d'évacuation 2864
chemin de vidange de bois 7689
chemin de vidange primaire 6541
cheminée d'aération 1487
chemin empierré 3921
chemin excavé 2934
chemin forestier 3565
chemin forestier local 5034
chemin public 6622
chemin saisonné 7332
chemins forestiers 3378
chemins mitoyens 1787
chemin utilisable par charrettes à bois 1359
chênaie 5836
chênaie-tillaie 4974
chenchen 9125
chêne 5830
chêne à chinquapin 1510
chêne à cochenilles 4697
chêne à écailles dorées 1319
chêne à feuilles de châtaignier 1478
chêne à feuilles de houx 642
chêne à feuilles de laurier 3874
chêne à feuilles de myrsine 862
chêne à feuilles de saule 9434
chêne à feuilles en lyre 5991
chêne à feuilles fusiformes 6376
chêne à galles 3794
chêne à glands sessiles 7464
chêne à grandes écailles 3937
chêne à gros fruits 5566
chêne à gros glands 5566
chêne à la noix de galle 3794
chêneau 5841
chêne bicolore 8317
chêne blanc 6621, 8993
chêne blanc aquatique 5991
chêne blanc de Provence 6621
chêne bleu 242
chêne bleu d'Amérique 242
chêne broussailleux de Californie 1271
chêne brun 1082
chêne buisson 642
chêne buisson de Californie 1271
chêne châtaignier blanc d'Amérique 5582
chêne châtaignier des marais 615
chêne chevelu 5565
chêne commun 1793
chêne Daimyo 2228
chêne d'Aleppo 3794
chêne d'Arizona blanc 8993
chêne de bardeau 3874
chêne de Bourgogne 5565
chêne de Caroline 5015
chêne de doigt 3162
chêne de Durand 2672
chêne de Garry 5945
chêne de Georgie 3830
chêne de Hongrie 4397
chêne de Kellogg 1255
chêne de la savane 5346
chêne de la variété 3859
chêne de marais 3764
chêne de marais du nord 5791
chêne de Mirbeck 5477
chêne de Mongolie 2228
chêne de Mühlenberg 1510
chêne de Pannonie 4397
chêne de Shumard 7561
chêne de Singapore 7635
chêne des marais 615
chêne d'Espagne 7863
chêne des teinturies 2692
chêne de Tasmanie 4054
chêne de Troie 5226
chêne d'Europe 1793
chêne de Valonia 3937
chêne d'Oglethorpe 5872
chêne du Caucase 1387
chêne du Japon 4624
chêne du Japon à feuilles de châ 7266
chêne du Liban 4903
chêne du Maryland 799
chêne écarlate 7287
chêne étoilé 255
chêne garry 5945
chêne glandouilleur 3859
chêne glauca 862
chêne gris 3928
chêne hongrois 4397
chêne jeune 5841
chêne kermès 4697
chêne laurier 4835
chêne-liège 1961
chêne liège de Chine 5951
chêne macédonien 5226
chêne mexicain 5420
chêne moru 5562
chêne nain 2687
chêne noir 9294
chêne noir d'Amérique 799
chêne noir de Californie 1255
chêne Nuttal 5828
chêne papier de verre 7190
chêne pédonculé 1793
chêne pontique 363
chêne prin 615
chêne pubescent 6621
chêne quercitron 2692
chêne rouge 7863
chêne rouge d'Amérique 5793
chêne rouvre 7464
chêne sessile 7464
chêne soyeux rouge 3966
chêne truffier 6621
chêne vélani 3937
chêne vert 1773, 3951
chêne vert de Californie 1693
chêne vert du Japon 4625
chêne yeuse 1773
chengal 1472
chenille arpenteuse 3174
chenille du pin 6248
chenille processionnaire 6557
chénopodiacées 3887
cheptel de gibier 3796
chermes du pin Weymouth 6245
chevalet 4380, 8965
chevalet d'écorce 564
chevalet de sciage 4380
chevelu de la racine 7012
chevelu racinaire 7028
cheville à bois 8801
cheville de bois mort 4376
cheville en bois 9555
chèvrefeuille 1227
chèvrefeuille à baie noire 790
chèvrefeuille à balais 2895
chèvrefeuille à involucre 812
chèvrefeuille blanc des marais 8303
chèvrefeuille de Morrow 5560
chèvrefeuille des haies 2895
chèvrefeuille involucré 812
chevron 6688
chicot 2274

chicot dur 4131
chicot févier 1482
chicot marchand 5397
chicot pourri 7821
chimaphile ombellée 1798
chimère 3820
chimère génétique 3820
chimie de bois 2370
chimie forestière 2370
chimiotaxonomie 1471
chimiotropisme 1470
chincapin 1199
chinquapin 642
chirurgie des arbres 8939
chirurgien arboricole 8938
chirurgienne arboricole 8938
chlorophylle 1545
chlorose 1546
choix d'espèce arborescente 1551
choix du site 7672
cholla 9259
chronologie de précipitation par cerne de croissance 8909
chronologie de température par cernes de croissance 8910
chronologie par cernes annuels 6958
chrysomèle du liard 1982
chrysomèle du tremble 389
chrysomèles 4858
chute d'eau 9285
chute de découpe 1336
chute de semence 2996
chute des feuilles 4866
chute des feuilles d'été 8261
chute de tranche 2554
chute naturelle des aiguilles 2995
chute naturelle des feuilles 4866
cicadelle 4871
cicatrice cachée de brûlure 4274
cicatrice de branche 992
cicatrice de brûlure 3206
cicatrice d'écaille de bourgeon 1144
cicatrice de feu 3206
cicatrice de l'écorce 582
cicatrice de paquet 1177
cicatrice de stipule 8102
cicatrice de têtard 9017
cicatrice d'incendie 3206
cicatrice d'un coup de bec 765
cicatrice foliaire 4888
cicatrice ouverte 5916
cicatrisation 5862
cicatrisation de branche 4968
cierge 1733
ciliatifolié 1580
cime 2087
à ~ dépérissante 2272
à ~ large 1045
cime branchue 996
cime conique 1866
cime d'arbre 2087
cime décombante 2314
cime déformée 2340
cime dense 2387
cime desséchée par Peridermium pini 8698
cime divariquée 7935
cime éclairée 8504
cime en cône 1866
cime en forme d'étendard 3240
cime en quenouille 3782
cime étagée 8134
cime fanaisée 2632
cime foliacée 4865
cime foliaire 4862
cime fusiforme 3782
cime globulaire 3869
cime haute 4298
cime plate 3252
cime pointue 1866
cime rousse 6796
cime sèche 7794
cime sphérique 3869
cimetière forestier 3347
cime unilatérale 7648
cimier 2087
cinchona 6669
cinnamon sols 1583
cintrage 2031
circonférence à hauteur de poitrine 1024
circonférence à hauteur d'homme corrigée 5782
circonférence à mi-fût 5441
circonférence au-dessus des empattements 3855
circonférence au milieu 5440
circonférence au petit bout 8690
circonférence au sommet 8690
circonférence de l'arbre 8848
circonférence minimale de tige 5467
circonférence moyenne 5440
circonférence sur/sous écorce 2482
circonscription de garde-forestier 2542
cire d'arbre 8958
cire de bouleau 760
cire d'insectes 8958
cirolier 6926
cisaillage 7501
cisaille à billes 5113
cisaille à lames inclinées 287
cisaille circulaire 7070
cisaille d'abattage 8921
cisaille d'abattage hydraulique 4409
cisaille de bois 9649
cisaille d'écimage 8947
cisaille de tronçonnage de billes 5113
cisaille hydraulique 4409
cisaillement 7495, 7501
cisaille tractée 2619
ciseau à bois 9509
ciseau à couper 3295
ciseau de poche 6404
cisse 8803
cistacées 6991
cistaie 6992
ciste 6990
citron de Ceylan 2729
citronnelle 5883
citronnier 4911
citronnier trifolié 4146
citron vert 9348
citrus de Sicile 4911
cladode 1594
cladraste à bois jaune 262
clair-bois 4446
clairière 3443
clairière des cervidés 2331
clairière forestière 8865
clairière naturelle 5680
clameau 8577
clameau de chargement 5026
clarification 1598
classe d'âge 126
classe d'âge du peuplement 8002
classe d'arbres 8812
classe de bois 8566
classe de bois rond 7114
classe de cime 2068
classe de cime dominante 2564
classe de cime dominée 6012
classe de cime en croissance libre 5911
classe de cime intermédiaire 4542
classe de cimes codominantes 1704
classe de cime subdominante 4542
classe de cime suppressée 6012
classe de couronnes en croissance

classe de densité 1312
classe de densité d'arbre 1602
classe de densité peuplement 8106
classe de diamètre 2471
classe de dimension 8924
classe de dimension d'arbre 8924
classe de dimension de peuplement 8028
classe de fermeture 1600
classe de fermeture du couvert 1600
classe de fertilité 9774
classe de forme 8840
classe de hauteurs 4228
classe de maturité 5342
classe de peuplement 8106
classe de qualité 9774
classe de régénération 6814
classe de station 7664
classe de tige 8067
classe de vigueur 9217
classe de volumes du peuplement 8035
classe d'exploitation extensive 1058
classe d'incendies par surfaces 1601
classe d'opportunité de traitement 8793
classement de charbon de bois 1446
classement dimensionnel 2500
classement selon le diamètre du tronc 8068
classe opérationnelle 2191
classe opérationnelle de peuplements 2191
classe physiognomonique 6216
classe physiographique 6218
classes de fonctions protectrices de forêts 1599
classes d'exploitation de la terre 4781
classes moyennes 5439
classeur de bois 9525
classeur pour copeaux 1535
classificateur de fibres 3093
classification de la station forestière 3593
classification de peuplement 8008
classification de sols forestiers 3600
classification des régions forestières 3552
classification de tiges 8068
classification du bois 3902
clavalier 4630
clavalier d'Amérique 8679
clavette 9555
clayon 617
clé 2595
clé de bois 9561
clématite bleue 4593
clématite brûlante 6387
clématite des Alpes 212
clématite des bois 9402
clématite des haies 9402
clématite flammette 6387
clématite liseron 4593
clématite odorante 6387
clématite petite vigne 4593
clématite vigne blanche 9402
clérodendron 3870
cléthra à feuilles d'aulne 8263
cléthra à feuilles d'aune 8263
cléthre blanche 1582
clétracées 1637
climacium arborescent 8813
climatologie forestière 3353
climax 1638
climax d'incendie 3172
climax forestier 3354
climax pyrophytique 3172
clinomètre 4443
clipser 1645
clisimètre 4443
clivage 1629
cliver 1631
cloche de Caroline 1349
clochette 5256
cloison criblée 7592
cloisonnement d'exploitation 31
cloison perforée 6154
cloison perforée en réseau 6921
cloison perforée éphédroide 2846
cloison perforée scalariforme 7272
cloison perforée simple 7633
clonage 1649
clone 1648
clôture 5019
clôture en bois 6695
clôture haute des cervidés 2328
clôture vive 5019
clou de bois 9553
clou de girofle 1664
cnéoracées 1671
co-arbre 1980
coccide 7274
cochenille 5426, 7274
cochère 9307
cocotier du Chili 1701
cocotte 6250
cocus 4607
code forestier 3484
co-dominance 1707
coefficient de centre de poussée 3669
coefficient de conicité 8399
coefficient de cubage de bois empilé 3125
coefficient de densité d'arbres 1711
coefficient de densité de gorge 1709
coefficient de densité de section de flottage 1710
coefficient de forme 8840
coefficient de forme de l'arbre 8840
coefficient de forme du petit bois 7781
coefficient de forme du peuplement 8015
coefficient de forme du tronc 8075
coefficient de hauteur de peuplement 8016
coefficient d'élasticité 5514
coefficient d'enstérage 7980
coefficient de point de forme 3669
coefficient de réfaction 2120
coefficient de remplissage de chemin 1708
coefficient de retrait 9653
cœur 1951
cœur à dents 8680
cœur branchu 1948
cœur brun 1080
cœur cassant 1039
cœur caverneux 1440
cœur coloré 1729
cœur creux 2933
cœur dehors 4784
cœur dehors 8251
cœur de lésion 8782
cœur double 2583
cœur en bois dur 4141
cœur étoilé 8039
cœur excentré 2732
cœur gris 3970
cœur humide 9283
cœur inclus 956
cœur mort 2261
cœur mou 7817
cœur mouillé 9283

cœur par le gel 3725
cœur pentagone 6145
cœur pourri 2289
cœur rouge 6778
cœur rouge du hêtre 659
cœur semi-rigide 7430
cœur solide 1901
cœur spongieux 1440
cœur vrai de bois 8986
coffrage 880
coffrage en bois 9576
cognassier 6667
cognée 673
cognier 6667
cohésion 905
cohorte 1715
coiffe de racine 7013
coincement latéral 7570
coincer 9314
coin d'abattage 3079
coin de bois 9561
coin de bois de hêtre 844
coin de refend 6487
coin de scie 7267
coin de tronçonnage 1134
coin d'obturation 3132
coin encastré 2680
coin explosif 6487
coin ferré 7809
coin pneumatique 3057
colatier 1718
col de cygne 3888
coléophore du mélèze 4788
colis 1179
colle à bois 9474
collecteur d'arbres 8801
collecteur de paquets 1173
collecteur de précipitations 8538
collection d'arbres remarquables 5570
collection de branches 984
collection de cônes 1863
collection de cônes mécanisée 5378
colle de bois 9583
colle métal-bois 5413
collet de racine 7015
collier à boucle 1555
collier à boucle prépositionné 6518
collier de câbles 7699
collier étrangleur de fascines 3019
collier étrangleur prépositionné 6518
colmatage 7599
colonisation 7470
coloration apprêtée au revers 472
coloration bactérienne 9293
coloration brune fongique 1549
coloration de bois 9662
coloration du cœur 7994
coloration due à l'entreposage 9755
coloriage du bois 9515
combe 202
combinaison de cultures sylvo-agricoles 1740
combinat de l'industrie de bois 9516
combrétacées 5646
combustibilité 1742
combustible de bois 9487
combustible fossile 3675
combustible immédiat 422
combustible naturel 5681
combustibles 3739
combustibles étagés 4757
combustion incandescente 3875
combustion lente 7792
combustion spontanée 7923
Comité pour le rapport de développement forestier 3377
commerce de bois 8653
commerce mondial de bois 9727
commercialement sec pour expédition 7527
Commission de recours forestière 3328
Commission européenne des forêts 2896
communauté animale 291
communauté biotique 738
communauté de macrophorbiées 228
communauté de mégaphorbiées 228
communauté d'herbes hautes 228
communauté fermée 1651
communauté forestière 3331
communautés des broussailles feuillues 2298
communautés feuillues 2297
communautés végétales menacées 8531
communautés vicariantes 227
communauté végétale 6336
communauté végétale désirée 2431
compacité de plantation 6369
compactage du sol 7834
compagnon de flottage 9569
compartimentage par contenance 6842
compartiment d'économie du bois 8612
compartiment forestier 1821
compas bec de cigogne 8809
compas d'épaisseur 8497
compas d'extérieur 5981
compas en ciseaux 8809
compas enregistreur 6751
compas forestier 8809
compas forestier fourchu 3653
compas forestier pliant 3299
compas pour mesure des volumes 715
compétition 8817
compétition de croissance libre 3692
compétition des cimes 2070
compétition en jachère 3692
complément de plantation 823
complément de vides 823
complexe réitératif 6849
comportement forestier 7612
comportement sylvicole 7612
comportement viscoélastique 9233
composante de bois mort 2275
composantes de première venue 5879
composées 1834
composite 1833
composition d'âge 130
composition d'un peuplement 1886
compost 4881
compost de sciure 7237
compression axiale 1848
compression des copeaux 1523
compression transversale 1849
comptabilité forestière 3340
comptage d'arbres 2837
comptage des cernes annuels 8911
comptage de volume 5266
comptage du peuplement 2838
comptage du volume global 8290
compte à la mesure 5361
compteur d'arbres 8820
compteur de grumes 5054
concasseur à disque 2534
concasseur à mâchoires 186
concession de bois 3356
concession forestière 3356
concrescent 67
concrétion minérale 5462
concurrence 8817

concurrent 1827
condalie de Hooker 870
conditionnement de bois 1855
conditions d'accroissement du peuplement 8009
conditions du peuplement 8009
conditions forestières 3358
conducteur 7695
conducteur de camion 8983
conducteur de coupeuse de bois 1529
conducteur de triqueballe 8565
conducteur de tronçonneuse 7268
conductivité thermique de bois 8489
conduite de feu 3192
cône 8818
cône carpelle 1350
cône de débusquage 1858
cône de houblon 4364
cône d'épicéa 7952
cône de pin 6250
cône de sapin 3165
conelet 1864
cône sérotinale 7459
conicité 8405
conifère 1877
conique 8401
conjonction de croissance 35
connarales 9794
connecteur de bois 8568
conseiller forestier 1890
conservateur des forêts 3361
conservation de bois 5120
conservation de bois dans l'eau 9301
conservation de graines 7397
conservation de la forêt 3359
conservation de l'environnement 2842
conservation de plantation forestière 8010
conservation des eaux de surface 9286
conservation du bois 8569
conservation ex situ 2959
conservation in situ 4517
consommateur de bois 8655
consommation absolue d'eau 2916
consommation d'eau 2916
consommation de bois 9517
construction de chemins forestiers 1889
construction de routes forestières 1889
construction en bois 8570
contention 7062
contenu d'eau surfacique 8283
contenu d'écorce 546
contenu d'humidité du bois 5518
contenu en humidité 1950
continuité de possession 1429
continuité du combustible 3735
continuité spatiale de communauté forestière 7883
contraintes d'exploitation forestière saisonnières 7331
contrat de vente de bois 9642
contrat d'exploitation de bois 2192
contredosse 881, 2554
contredosse de mine 881
contrefeu 465
contrefiche 1840
contrefil
à contrefil 49
contrefort 1985
contrefort de fructification 3730
contre-outil 321
contre-placage 2037
contreplacage 2037
contreplaqué 6398
contreplaqué à âme 9520
contreplaqué à âme en bois 9520
contreplaqué à âme en bois massif 5201
contreplaqué de construction 1151
contreplaqué extérieur 2964
contreplaqué marin 5307
contreplaqué multiplis 5608
contreplaqué panneauté 5201
contreplaquer 2036
contreplaqué résistant à l'humidité 5523
contrevent 1840
contrôle au champ 3111
contrôle de la croissance des arbustes 1101
contrôle de la qualité du bois 6653
contrôle de la surface terrière 602
contrôle des arbres 8940
contrôle des exploitations 3055
convention globale des forêts 3863
conversion de bois 1915
conversion de pins 6251
conversion d'espèces arborescentes 226
conversion primaire 6538
conversion sylvicole 7614
convolvulacées 5557
convoyeur de nœuds 4729
convoyeur d'entrée de billots 712
convoyeur de sortie de billots 714
convoyeur-élévateur 4600
convoyeur-élévateur de billes 5060
copahier 1922
copaïer officinal 1922
copal 1923
copal de kauri 4466
copalme d'Amérique 259
copayer 1922
copeau combustible 1517
copeau de papeterie 6625
copeaux 9508
copeaux à l'écorce 545
copeaux combustibles bruts 1677
copeaux combustibles fins 3151
copeaux de bois 9508
copeaux de branches 975
copeaux déchiquetés 4318
copeaux de cime 2067
copeaux de déchets 9274
copeaux de dimension insuffisante 9067
copeaux de foreuse 926
copeaux de hêtre 651
copeaux de l'arbre entier 3760
copeaux de résidus forestiers 1538
copeaux de souche 8198
copeaux de taille normale 8006
copeaux de tige 1539
copeaux énergétiques 2818, 4319
copeaux en portions d'arbres 6096
copeaux fractionnés 7553
copeaux incendiés 1185
copeaux non écorcés 9043
copeaux non triés 9115
copeaux pour pâte mécanique 1519
copeaux résineux 6894
copeaux sans écorce 1605
copeaux surdimensionnés 6004
copeaux triés 28
copeaux trop petits 9067
copeaux verts 3940
coq des marais 4182
coque du Levant 8191
coqueret comestible 1323
corail 111
corbier 8991
cordage 1942, 7062
corde 1564, 9519
corde accrochante 4115
cordeau 1164

corde de chanvre 4251
corde de sécurité 1643
corde de tension 1164
corde du frein 971
corde mince 2979
cordes par acre 1945
cordes par heure d'un homme 1946
corde suspendue 7714
cordie à petites feuilles 7765
cordon 1944
coréopsis géant 3840
coriariacées 1955
cormier 8991
cormus 7053
cornacées 2556
corne de bélier 9705
corne de cerf 9705
cornel 1964
cornes longues 1327
cornet à semence 7381
cornigera 1166
cornouiller à feuilles alternées 222
cornouiller à fleurs 1964
cornouiller à grappes 3924
cornouiller alternifolié 222
cornouiller blanc 9339
cornouiller de Floride 1964
cornouiller de Nuttall 9339
cornouiller du Canada 1170
cornouiller femelle 1771
cornouiller mâle 1965
cornouiller rouge 1771
cornouiller sanguin 1771
cornouiller stolonifère 6783
corps de la charrue 6382
corps étranger 3316
correctement séché 6584
corridor de débusquage 9745
corset d'arbres 8904
cortex 1973
cortical 1974
corylaie 4181
côte 4882
côté au vent 9451
côté de la face 2985
côté du vent 9451
côté exposé au vent 9451
coton de Sea Island 7327
cotonéaster commun 2888
cotonéastre 5389
cotonnier 5389
cotonnier à feuilles entières 2888
cotylédon 1983
couche 3086
couche d'accroissement annuelle 301
couche de cimes 2084
couche d'écorce à liège 6200
couche de grumes 8545
couche de humus brut 1682
couche de lichens 1696
couche de rondins 4844
couche durcie 4129
couche fibrique supérieure 5869
couche forestière 3334
couche génératrice de liège 6201
couche humifiée 4394
couche humique inférieure 5873
couche intercellulaire 4532
couche om-mésique 5896
couche poreuse 5905
couches principales de tissue d'arbre en section transversale 5251
couches principales de tissue en section transversale 5251
couche subéreuse 1960
coude 2100
coudonnier 6667
coudraie 4181
coudrier 1778
coudrier pleureur 1778
coulant 7132
coulée de verdure 3963
couleur de bois mûr 6965
couleur du bois 9513
coulisse 3335
coulisseau du compas forestier 1283
couloir 181, 182, 1969
couloir d'avalanche 425
couloir de la flore et la faune sauvage 9418
couloir de ponte 4811
couloir de verdure 3963
couloir principal 5248
couma 2000
coup 5038
coupage après dépérissement 2186
coup à la hache 450
coupant 819
coup de chaleur 4209
coup de griffe 4279
coup de soleil 8265
coup de soleil hivernal 9457
coupe à blanc 1607
coupe à blanc avec réserves 1620
coupe à blanc commercial 1744
coupe à blanc conditionnelle
coupe à blanc localisée 5035
coupe à blanc par bandes étroites 1622
coupe à bord droit 7964
coupe actuelle 57
coupe admissible 195
coupe alternante 224
coupe à main 4109
coupe annuelle 296
coupe annuelle moyenne 5354
coupe annuelle permissible 192
coupe annuelle prélevée sur la forêt 297
coupe anticipée 6513
coupe à onglet 5483
coupe à pas 8084
coupe à ras 3294
coupe à rétention variable 6919
coupe au diamètre limite 2476
coupe au diamètre minimal 2476
coupe avec protection de la régénération 1337
coupe avec réserves 6885
coupe biaisée 4441
coupe-broussailles 7319
coupe chanfrein 1441
coupe composée 1838
coupe curviligne 2151
coupe d'abattage 7668
coupe d'abri 7513
coupe d'abri avec réserve 7518
coupe d'abri en coin 7517
coupe d'abri en dents de scie 7517
coupe d'abri en lisière 8178
coupe d'amélioration 2121
coupe dans le sens axial 2187
coupe dans le sens du fil 2187
coupe d'arbres à risque 6973
coupe d'arbres d'échantillonnage 7175
coupe d'arbres marqués en réserve 4902
coupe d'arbres verts 3959
coupe d'assainissement 7193
coupe de bois 8824
coupe de bois accidentelle 4437
coupe de bois demi-dur 7428
coupe de bois feuillu 4134
coupe déchiquetée 7747
coupe d'éclaircie 5912
coupe de conversion 1914
coupe d'écrémage 4285
coupe de dégagement 5912
coupe de dégagement des lisières 6167
coupe de détachement 7472

coupe de la forêt en compartiments 3366
coupe de libération 5912
coupe de nettoiement 7193
coupe d'ensemencement 7383
coupe d'entretien 8458
coupe de petits arbres 7779
coupe de peuplements abandonnés 7168
coupe de préparation de lisière 7472
coupe de proche en proche 6573
coupe de produits au terme d'exploitabilité 4150
coupe d'équerre 7964
coupe de rajeunissement 1926
coupe de réalisation 1744
coupe de régénération 6815
coupe de régénération secondaire 6864
coupe de rénovation 6815
coupe des buissons 1109
coupe de sciage 7232
coupe des semenciers 7399
coupe de taillis 1926
coupe de taillis-sous-futaie 7409
coupe d'été 3065
coupe dirigée 195
coupe économico-sélective 2746
coupe en anse 629
coupe en cathédrale 1379
coupe en coin 9315
coupe en coulisses 8171
coupe en deux abattages 9034
coupe en lisière 6167
coupe en printemps 7938
coupe en régie 5070
coupe en tirant 4441
coupe étroite en lisière 5658
coupe extraordinaire 2973
coupe-feu 3169
coupe-feu boisé 9544
coupe-feu incendié 1186
coupe-feu nu 530
coupe finale 3136
coupe hivernale 9460
coupe horizontale 7338
coupe hors règlement 9046
coupe inférieure 5187
coupe initiale 6513
coupe intermédiaire 4544
coupe intermédiaire admissible 198
coupe jardinatoire 7412
coupe jardinatoire obligatoire 5855
coupe jardinatoire par bouquets 4001
coupe latérale 1966
coupe longitudinale 5135
coupe modifiée 5511
coupe nette et franche 1606
coupe non réglée 9046
coupe obligatoire 1852
coupe par bandes 8171
coupe par bandes de proche en proche 8171
coupe par blocs 838
coupe par furetage 1932
coupe par surfaces 1820
coupe partielle 6083
coupe par trouées 6106
coupe périodique 6162
coupe périodique permissible 6159
coupe permissible 195
coupe planifiée 6329
coupe préparatoire 6513
coupe principale 3136
coupe progressive en placettes et bandes 8168
coupe progressive par bandes 8178
coupe progressive uniforme 9086
couper 2161, 3037
couper à blanc 2168
coupe-racines 7019
coupe radiale 6659
coupe rase 1607
coupe rase avec semenciers 7399
coupe réglée 6834
coupe régulière 6834
coupe répétitive 6870
coupe résiduelle 7168
coupe sanitaire 7193
coupe sans copeaux 1516
coupe secondaire 7345
coupe sélective par arbre 7655
coupes en échec 7988
coupe sombre 2253
coupe sombre par groupes 4004
coupes par bandes alternées 1989
coupes par chevrons 1479
coupes partielles faibles 4948
coupe spéciale 7888
coupe tangentielle 8383
coupe totale 1607
coupe totale à carbonisation 1674
coupe totale avec post-régénération 1618
coupe totale avec régénération naturelle 1619
coupe transversale 8571
coupe transversale terminale 2980
coupeur d'abattage 3058
coupeur de voie 1104
coupeuse 2181
coupeuse à couteaux multiples 5607
coupeuse à disque 2534
coupeuse à rémanents 1100
coupeuse de blocs 839
coupeuse de bois à alimentation descendante 2634
coupeuse-déligneuse 1527
coupe usée 3038
coupeuse pour équarrissage des grumes 1316
coupe vidangée 3038
couple de ponctuation par champ de croisement 2049
couple de ponctuations 6309
courant de la sève 1589
courbage de bois 9489
courbage de bois lamellé 4764
courbage de bois massif 7844
courbaril 1990
courbe d'accroissement 4453
courbe de croissance cumulée 2135
courbe de décroissement 8400
courbe de défilement 8400
courbe de fertilité 9775
courbe de hauteurs 4229
courbe de la croissance 4012
courbe de rapport hauteur/diamètre 4231
courbe des volumes 9241
courbe des volumes unitaires 9089
courbé par le vent 9440
courbure 2031
courbure de tronc 8325
courbure de tige 8325
courbure multiple 5619
courbure simple 7634
couronne 2087
couronne de nœuds 1668
couronne en forme de balai 1074
cours d'eau 9284
cours d'eau à flottage 3277
cours d'eau intermittent 4555
cours d'eau saisonnier 4555
cours d'eau secondaire 7344
cours d'eau temporaire 4555
cours de rentabilité forestière 3650
cours d'estimation forestière 8488

coursier 2525
coursier d'évacuateur 2525
courson 7547
courte révolution 7542
courte rotation 7542
coussin 2156
coussin d'abattage 3057
coussinet 2156
coût d'abattage 3056
couteau 819
couteau à ébrancher 2285
couteau à mesurage d'écorce 570
couteau à résine 6897
couteau d'ébranchage 2285
couteau décolleteur 8709
couteau ébrancheur 2285
couteau écimeur 8709
couvert 1695, 8150, 8810, 8922
couvert d'arbre 8810, 8922
couvert de bois 3394
couvert décidu 2299
couvert de cime 2066
couvert d'écorce 1994
couvert de feuilles 4860
couvert de refuge 2863
couvert de taillis 5190
couvert du peuplement 3345
couvert fermé 1650
couvert forestier 3345
couvert forestier total 8723
couvert horizontal 4367
couvert libre 3691
couverture 1695, 4223
couverture d'aiguilles 5706
couverture de câble 6743
couverture de mousse 5564
couverture dense 1650
couverture de peuplement adulte 5343
couverture des cime 2094
couverture morte 4999
couverture morte avant humification 3671
couverture optimale d'élan 5932
couverture thermique 8490
couverture thermique structurelle 8186
couverture végétale 6337
couverture vivante 9060
couverture vivante au-dessus du sol 3116
couvert végétal 6337
couvert vertical au sol 9204
cramantee 2006
crampon 8462, 8577
cran à flèche fixe 3236
création de forêt 94
création de forêts pérennes 3624
création de rideau-abri 7509
création d'un peuplement mélangé 2867
crénelé 7461
crête de paon 7180
creusement de fossés 8962
creux pour le pallage d'une chaîne de débardage 1556
crevasse 9507
criblage de charbon de bois 1446
crible 7590
crible à copeaux 1535
crible à copeaux rotatif 7073
cric à tronçonner 8603
cric forestier 8603
crisse marine 628
croc de batelier 6693
crochet 274
crochet coulissant 529
crochet d'ébranchage 6613
crochet de culée 1557
crochet de nœud coulant 1557
crochet de rondins à main 6630
crochet de suspension 1557
crocheter 4358
croisement 4401
croisement de guide 2053
croisement en arrière 462
croisement en retour 462
croissance 4449
de ~ en forêt 3455
croissance agressive 141
croissance annuelle 300
croissance annuelle récente 2150
croissance avec torsion à droite 6943
croissance avec torsion à gauche 4907
croissance coudée 2025
croissance d'aménagement 8291
croissance de moisissure 3772
croissance de population 6459
croissance de racines 7027
croissance déterminée 2435
croissance dissymétrique 5898
croissance d'un peuplement 8017
croissance en épaisseur 4019
croissance en spirale 9021
croissance en touffes 1209
croissance épisodique 2854
croissance excentrique 2733
croissance excentrique du cernes 6952
croissance excentrique du tronc
croissance forte 4213
croissance fourchue 3656
croissance intermittente 4553
croissance juvénile 4677
croissance libre 3697
en ~ 3694
croissance linéaire 4985
croissance madrée 9185
croissance médiocre 6449
croissance naine 8223
croissance non concurrentielle 3692
croissance ondulée 9185
croissance pauvre 6449
croissance prédominante 73
croissance rabougrie 8223
croissance rapide 7542
croissance rythmique 6931
croissance saine 4192
croissance secondaire 120
croissance stagnante 7990
croissance vers le haut 9127
croît 5352
croton 2061
croûte 7712
croûter 2106
crucifères 2099
crue 3282
cryptomérie du Japon 4615
cubage 8955
cubage après réfaction 5719
cubage au chemin 6981
cubage au moyen de la longueur et du diamètre au milieu d'une grume 5362
cubage au quart déduit 2118
cubage au quart sans déduction 4366
cubage brut 703
cubage corrigé 5719
cubage d'après le diamètre gros et au fin bout 5367
cubage de sciages 5213
cubage du bois gisant 3040
cubage en milieu 7279
cubage en milliers de pied-planches 8529
cubage en pied-planches 879
cubage en pieds mesure de planche 879
cubage en volume réel 8992
cubage net 5719
cubage par sections 9244
cubage sans réfaction 703
cucumber magnolia 9040
cueillette des cônes 1861

culmination d'accroissement moyen annuel 2126
cultivar 2129
cultivateur à éperon 8668
cultivateur forestier 3365
cultivateur rotatif 7072
cultivation d'especes forestières 2559
cultivation forestière 3364
cultivation préliminaire 6510
cultiver 2128
culture 9788
culture-abri 1993
culture annuelle 295
culture anticipée 71
culture coïncidente 1716
culture d'arbres de Noël 1570
culture d'arbres hors forêt 5759
culture de méristèmes 5403
culture de peupliers 6456
culture de plantes 6704
culture de plantes forestières 3530
culture de recouvrement 5996
culture dérobée 1374
culture d'espèces variées 5491
culture de tissus 8671
culture dominée 8235
culture d'osier viminal 620
culture du bois 7610
culture en couloirs 183
culture forestière 3528
culture intercalaire 4543
culture intermédiaire 4543
culture interpolée 4562
culture itinérante 7520
culture itinérante après débroussaillement et brûlis 8903
culture nomade 7520
culture pluriétagée 5631
cultures permanentes 6169
culture sylvo-agricole 3425
cumala 4717
cumaru 2133
cunite 2136
cunoniales 2138
cupule 46
curupayra 6762
curvifoliacé 2155
cuspidé 7917
cyathéacées 2206
cycadacées 2207
cycle biogéochimique 724
cycle d'abattage 2194
cycle d'éclaircie 8509
cycle de fluctuation de la densité
cycle de la croissance 4013
cycle de l'azote 5735
cycle de replantation 6874
cycle des substances nutritives 5825
cycle du carbone 1330
cycle phénologique 6202
cylindre de coupe 3050
cylindre de relevage 4932
cynip du chêne 1279
cynips 2210
cynips gallicoles 2210
cynips galligènes 2210
cyprès 2211
cyprès à gros fruits 5547
cyprès chauve 786
cyprès chauve pyramidal 6444
cyprès commun 1770
cyprès de Baker 484
cyprès de Californie 3891
cyprès de Forbes 3314
cyprès de Gowen 3891
cyprès de Lambert 5547
cyprès de l'Arizona 359
cyprès de Lawson 4839
cyprès de Louisiane 786
cyprès de Macnab 484
cyprès de Monterey 5547
cyprès de Montezuma 5418
cyprès de Sargent 7221
cyprès d'Italie 1770
cyprès du Chili 6102
cyprès du Mexique 5418
cyprès faux 786
cyprès pond 6444
cyprès pyramidal 1770
cypressacées 2214
cyprès toujours vert 1770
cyprès vert de Provence 1770
cyprière 2215
cyrillacées 2221
cytisaie 1070
cytise 1069, 1784
cytise à grappes 1784
cytise aubour 1784
cytise des Alpes 215
cytokinines 2222

dacrydion 6784
dacryne 6784
dais de feuillage 4860
dalot de flottage 914
damar 2243
damas 2244
dammar 2243
danger de feu forestier 6971
danta 2245
daphné 2246
daphné lauréole 2828
daphné morillon 2828
datation par cernes de croissance 8912
date d'abattage 3060
date de coupe 3060
date d'inventaire 2256
dattier 2257
davidiacées 2594
débardage 8595
débardage à arche 349
débardage à bras 495
débardage à courte distance 7534
débardage aérien 85
débardage à la chaîne 1428
débardage à navettes 7562
débardage animal 292
débardage au câble 1247
débardage au câble au ras du sol 3994
débardage au marais 8311
débardage au moyen d'un câble suspendu 4288
débardage au treuil 2571
débardage avec chevaux 4378
débardage contractuel 1908
débardage conventionnel 1912
débardage d'assortiment 395
débardage de bois de trituration 6632
débardage de grumes 899
débardage de rondins par tracteurs haut-le-pied 889
débardage direct 2522
débardage direct de bois 2522
débardage en bois courts par tracteurs haut-le-pied 889
débardage engorgé 9073
débardage en grande longueur 9394
débardage en longueur 2805
débardage forestier au marais 8311
débardage latéral 4821
débardage mécanisé 6489
débardage nocturne 5733
débardage par câble 1247
débardage par câble à vide 7715
débardage par chargeur mobile 4608
débardage par contrat 1908
débardage par couloir 1970
débardage par glissoire 8759
débardage par hélicoptère 4239

débardage par téléphérage 1247
débardage par tension 8466
débardage par tracteur 7683
débardage par traction mécanique 6489
débardage par triqueballe 4304
débardage par voie ferrée 6696
débardage sans câble 5777
débarder à bras 4112
débarder de bois 494
débardeur 5067, 5114
débardeur à bras 4113
débardeur à câble 1246
débardeur à câble et arche 1554
débardeur à pince 3913
débardeur à pince portante 1597
débardeur de bois 1435
débardeur par emballe-sac 6028
débit 2177
débitage 5100
débit à traits parallèles 5135
débit courant 9780
débité en plot 3258
débit élémentaire 9097
débit en long 5135
débit en plot 5016
débité sur quartier 2760
débit mélangé 5487
débits de longueurs impaires 5866
débits toutes longueurs 5488
débit sur maille 2760
débit-type 2185
déboisement 2339
déboiser 2524
débordage 9796
débouché pour produits forestiers 5312
débourrage 1020
débourrement 1020
débris 2440
débris dangereux 4180
débris de coupe 998
débris de sciage 5215
débris ligneux 9707
débris ligneux grossier 1686
débris minier doux 8334
débris minier neutre 8334
débris végétaux 1422
débroussaillant 1107
débroussaillement 1109
débroussailler 1111
débroussailleur 8307
débroussailleur-dessoucheur 1112
débroussailleur-pulvériseur 7321
débroussailleuse 7319
débroussailleuse à axe horizontal parallèle 6067
débroussailleuse double 2668
débroussailleuse rotative 7071
débusquage 8595
débusquage à grande portée 5150
débusquage à longue portée 5150
débusquage à terre 4779
débusquage au tracteur 8747
débusquage conventionnel 1911
débusquage dans l'eau 9300
débusquage de bois en état demi-suspendu 7433
débusquage de bois long par tracteurs 8872
débusquage du bois descendant 2598
débusquage répétitif 6909
débusquage sans lacet 1558
débusquage vers le haut 9129
débusqueur 3075
débusqueur à benzine 3809
débusqueur à câble 1246
débusqueur à tambours verrouillé 4540
débusqueur autochargeur 7419
débusqueur de bois 8596
débusqueur de perches 6423
débusqueur sur roues à châssis 3685
débusqueuse 3075, 5114
débusqueuse à grappin 3913
début de coupe 668
début de la régénération 6819
décaper 5227
décaper un arbre 4028
décharge de déchiqueteuse 1526
déchet de bois 2123
déchets combustibles 3743
déchets de bois 3561
déchets de bois écrasé 4319
déchets de bois rond 7118
déchets d'écorçage 580
déchets de coupe 998
déchets de coupe évitables 446
déchets de plantes 6371
déchets d'exploitation 56
déchets ligneux 3561
déchiquetage 1531, 4320
déchiquetage de dosse 7709
déchiquetage-épandage 5166
déchiquetage par le dessus 2601
déchiqueteur à tambour 2208
déchiqueteur de délignage 4321
déchiqueteuse 1534
déchiqueteuse à couteaux plats 3255
déchiqueteuse à lames plates 3255
déchiqueteuse de bois long 9391
déchiqueteuse de rondins de défibrage 6628
déchiqueteuse mobile 5499
déchiqueteuse stationnaire 8044
déclencheur de la croissance 4020
décollement des plis 2348
décolletage 9318
décolleteuse 8707
décoloration cicatricielle 9733
décoloration de bois gisants 2047
décoloration de l'aubier 2527
décoloration du bois 9536
décoloration naturelle 5697
décomposeur 2308
décomposition 2310
décomposition de bois de cœur 4203
décomposition de l'aubier 7083
décomposition de tissu ligneux 1014
décomposition du bois 9529
décompte des sciages 5213
décompte des sciages bruts 5213
déconnecteur de piles 7976
décophore du chêne 5839
décortiqué par machine 5231
décortiquer 2280
décortiqueur 7960
découpe d'abattage 7668
découpeuse à remorque 8751
découpeuse de bois à rémanents 9278
découvert 2980
décroissance 8403
à forte ~ 8401
décroissance forte 1872
décroissement 2611
dédalées du chêne 3127
dédosseuse 5148
dédosseuse-déchiqueteuse 1316
dédoublage 5135
dédoublage en hauteur 2322
dédoubleuse 1410
déduction de l'écorce 200
défalcation de l'écorce 200
défaut apparent 5906
défaut d'arbre 8830
défaut de coloration 1728
défaut de forêt 4753
défaut de la croissance 4014
défaut de lumière 4940
défaut de sciage 7242
défaut de structure 8187
défaut du bois 9530

défaut inhérent au bois 2333
défaut ouvert 5906
défaut par le vent 2236
défaut par longicorne des charpentes 5134
défauts de type peau de crocodile 188
défense contre l'incendie 3192
défets de bois 9614
défeuillaison 2941
défibrage 3099
défibreur 2104
défibreur à mâchoires 186
défilement de tige 8083
définition d'espèces arborescentes 8928
déflecteur de brousse 1103
défoliant 2338
défoliateur 2338
défoliation 2941
défonceuse 6966
déforestation 2339
déformation en cuvette 2141
déformité du bois 9531
défrichage final 3134
défrichement à l'explosif 852
défrichement final 3134
défricher 1611
défricheur 7319
dégagement 8515
dégagement anticipé 2703
dégagement de bois 9511
dégagement de chaleur total 8725
dégagement de la cime 3698
dégager 6856
dégarnissage 2599
dégarnisseuse 8771
dégât 582
dégât causé par le gibier 2233
dégât de bois 2242
dégât de coléoptères 9703
dégât de copeaux 1536
dégât d'écorçage 2237
dégât de débardage 7685
dégât de gibier 2233
dégât d'insectes 2234
dégât d'ouragan 2236
dégât dû à l'entreposage 8127
dégât dû au bétail 3930
dégât dû au givre 2235
dégât naturel 5675
dégât par bostryche 5221
dégât par grande vrillette 2279
dégât par la fumée 2240
dégât par sirex 9692
dégât par vrillette striée 2279
dégauchisseuse 4661
dégazonnement 7286
dégommer 2345
dégradation 2341
dégradation avancée 72
dégradation chimique de bois 1464
dégradation des forêts 3367
dégradation fongique 3768
degré de couverture 2343
degré de densité 2344
degré de densité du peuplement 2344
degré d'élancement 3898
degré de qualité 9774
degré d'intensité d'éclaircie 8519
degré du mélange 3897
degré hygrométrique des matières combustibles 3737
degré relatif d'humidité 6854
dégrossissage 4273
déhiscent 2346
délai de régénération 6816
délignage 2766
délignage conique 1871
délignage diagonal 2460
délignage en chevrons 9796
déligné 2205, 2756, 8119
déligneuse 2767
déligneuse-déchiqueteuse 1527
délignification 2349
délimitation forestière 3590
déliquescent 2357
délit forestier 1909
délivrance de bois 9532
délivrance forestière 6862
demandes de boisement 95
demandes de gestion forestière 3494
démariage 7658
démariage 8515
démarier 8503
démarieuse 8513
démasclage 6865
démêleur 9114
demifratrie 4097
demi-madrier 4099
demi-tige 4098
demi-tour 8349
dendrite 2367
dendritique 8876
dendrobionte 2369
dendrochore 2371
dendrochronologie 2373
dendrochronologiste 2372
dendroclimatologie 2374
dendroctone de l'épinette d'Engelmann 2823
dendroctone de l'épinette Sitka 2724
dendroctone du mélèze 2720
dendroctone du pin 6246
dendroctone du pin de Colorado 1727
dendroctone du pin de l'ouest 9342
dendroctone du pin mexicain 4805
dendroctone du pin ponderosa 796
dendroctone du pin tordu 5045
dendroctone méridional du pin 7862
dendroctone noir de l'épinette 811
dendroctone occidental du pin 7866
dendroctone rouge de l'épinette 6797
dendrocygne 8832
dendroécologie 2375
dendrogéomorphologie 2376
dendrogramme 2377
dendrohydrologie 2378
dendroïde 8876
dendrologie 2380
dendrologiste 2379
dendromètre 2381
dendromètre d'indicateur à cadran 2464
dendromètre optique 5931
dendrométrie 2383
dendrophile 2384, 2385
dendroscope 2386
dénivellement 7750
densifolié 2389
densité
de ~ normale 9335
de ~ partielle 4586
de ~ relative 4586
de ~ relative idéale 4419
de ~ relative normale 4419
de ~ relative satisfaisante 7224
densité apparente 1155
densité d'arbre 8816
densité de bois conventionnelle 1913
densité de bois séché à l'air 151
densité de bois vert 3707
densité de cernes annuels 303
densité de lignocellulose 9668
densité de peuplement 6458
densité de réseau routier forestier

densité des espèces 7896
densité des graines 7374
densité des incendies 3173
densité des nœuds 5742
densité de sous-bois 2394
densité d'ombrage 2393
densité du bois 1818
densité du couvert 2393
densité du couvert d'une cime 2072
densité du peuplement 3370
densité maximale 5345
densité nominale de bois 18
densité normale d'un peuplement 3757
densité relative de la forêt 3370
densité théorique de bois 18
denté 7461
dents de scie 7264
déodar 2398
départ 2400
départoir 3717
dépendance de bois 8575
dépendant d'incendie 3178
dépérir 2307
dépérissement 5444
dépérissement de branches 2488
dépérissement d'écorce 2487
dépérissement de la forêt 3373
dépérissement terminal 8686
dépérissement terminal des arbres 3368
déplacement d'eau 4996
déplacement de sève 7198
déplacement sur draille 8753
de pores demi-circulaires 4091
dépôt 3371
dépôt acide 40
dépôt-colis en plein air 5972
dépôt de bois 8664
dépôt de bois à l'air libre 5919
dépôt de bois rond 7119
dépôt d'écorce 585
dépôt de grumes 7119
dépôt de traînage relevé 4289
dépôt d'outillage contre le feu 3214
dépôt d'usine 5456
dépôt forestier 8664
dépôt intermédiaire 1721
dépôt intermédiaire de transit 4382
dépotoirs 9276
dépôt supérieur 9135
dépouillement des racines 7048
dépoussiérage 2318
dépressage des semis 8514
dépressions de bois dur drainantes les prairies 4137
déracinement 2857
déraciner 9140
déracineur 8808
dérésinage 2407
dérivation du bois 2548
dernière bille 2983
dernière planche 7572
déroulage 6135
déroulage à cœur 4093
déroulage excentré 4093
déroulement 6121
dérouler 6133
dérouleuse 7074
dérouleuse de bois 9623
derris 6448
désactivation de chemins 6976
désaileuse 2456
descendance 6570
descendance uniparentale 4097
descente de cime 8686
descriptif de peuplement 2411
désencrouage 7001
désertification 2414
désespoir des singes 5535
désherbage 9320
désherbage localisé 7933
désherbage manuel 5287
désherbage mécanique 5375
désherbage mécanisé 5375
déshydratation 2454
désinfection par immersion 2511
dessalement des graines 2455
dessèchement de la carre 2649
dessèchement de la feuille 5444
dessilteur 2427
dessouchage 8201
dessouchage de par explosion 852
dessouchage et rétention 4029
dessoucher 4033
dessoucheur 8199
dessoucheur sur roues 9355
dessoucheuse 8199
dessoucheuse à griffe 6966
destruction de la forêt 3373
destruction des forêts tropicales 2432
détachage de cime par explosion 851
détachement de l'écorce en feuillets 3243
détente brusque du bois 3297
détérioration causée par l'abattage 3059
détérioration du bois 9535
détérioration du sol nuisible 2439
détermination d'aire forestière 3329
détermination de cubage du volume sur pied 2438
détermination de diamètre 2436
détermination de l'âge 131
détermination de la possibilité 9776
détermination de l'impact visuel 9237
détermination des essences forestières 8928
détermination de volume d'arbre 8955
détermination d'inclinaison du fil 2437
détourer 7120
détritivore 2441
détritus 2440
deutzia 2442
deutzie à feuilles crénelées 3787
deutzie scabre 3787
dévégétation 2444
développement de branches 2449
développement de cime 2073
développement des forêts 3374
développement de terrain de parcours 6716
développement d'un peuplement 2448
développement durable des forêts 8295
développement en fourche 3656
développement excentrique 2733
développement excentrique du tronc 2733
développement groupé 1667
développement unilatéral 5171
dévers vers l'extérieur 5983
dévers vers l'intérieur 4518
déviation d'axe de grume 2450
déviation permissible 6181
dévitalisation des forêts 3367
DHP 2466
DHS 2465
diagramme d'aménagement de la densité de peuplement 8012
diale 8380
diale de Guinée 9190
diamètre à hauteur définie 4230
diamètre à hauteur de poitrine 2466
diamètre à hauteur de souche 2465

diamètre à hauteur d'homme corrigée 5782
diamètre à la découpe 2467
diamètre à mi-hauteur 2468
diamètre au-dessus d'empattements 2465
diamètre au fin bout 2470
diamètre au gros bout 2465
diamètre au petit bout 2470
diamètre avec écorce 2480
diamètre d'arbre 8826
diamètre de cercle de coupe 2479
diamètre de cime 2074
diamètre de cime apparent 9234
diamètre de cime visible 9234
diamètre de cœur 4196
diamètre de cubage 7250
diamètre de l'arbre dominant 2568
diamètre d'empattement 2469
diamètre de sciage 7243
diamètre de surbille 7363
diamètre de tige 8071
diamètre d'exploitabilité 2949
diamètre de zone du collet 2467
diamètre dominant 2565
diamètre du cœur noueux 2481
diamètre d'un nœud 2478
diamètre externe 5976
diamètre final 3138
diamètre marchand minimal 2743
diamètre minimal 5465
diamètre minimal d'abattage 5465
diamètre minimal de défaut 5466
diamètre moyen 2468
diamètre moyen de défaut 431
diamètre moyen des prédominants 8693
diamètre plain 3748
diamètre quadratique moyen 6651
diamètre sans écorce 2475
diamètre sur écorce 2480
dibétou 1087
dicots 8412
dicotylédone 2486
diéreville chèvrefeuille 5788
différenciation 8147
différentiel de sélection 7407
diffusion par les arbres locaux 7294
digue 2230
digue anti-incendie 3186
digue fascinée 1117
dilatation 2497
dilléniacées 2498
diluant 8505
dimension 2499
dimension après sécheresse 7087
dimension de bois finie 58
dimension de gaule 7205
dimension de parcelle boisée 3398
dimension du sciage 7469
dimension du sciage en vert 3955
dimension effective 59
dimension finie 58
dimension prévue 3163
dimensions d'équarrissage 2045
dimensions des arbres 8827
dimensions des bois 8576
dimensions transversales 2045
diminution 8403
diminution de l'accroissement 6832
dina 6056
diprion de la pruche 4250
diprion de Leconte 6776
diprion de pin de Virginie 9230
diprion de Swaine 8301
diprion du pin sylvestre 2905
diprion du sapin 502
diprion européen de l'épinette 2910
diprion importé du pin 4575
diptérocarpacées 2513
dirca des marais 4900
directement par l'arbre 8914
direction d'abattage 2519
direction de guide 2517
direction des fibres 2516
direction des veines 2516
direction de traînage sur câble 9746
direction d'inclinaison 2518
direction participative 6090
disclimax 2526
dispersant 7934
dispersion de feuilles mortes sous l'effet du vent 2538
dispersion de pollens 6433
dispositif d'affûtage de scie 7239
dispositif d'alignement 8404
dispositif de carbonisation 1332
dispositif de chargement de bois 8607
dispositif d'écimage 8707
dispositif d'éclatage des bûches 9512
dispositif de croisement 5333
dispositif de fendage 9512
dispositif de fixation de la pièce 4653
dispositif de nettoiement de
disque de la tige 1336
dissémination 2996
distançage 2599
distance de charge 5024
distance de débusquage 9747
distance de flottage 3271
distance de roulage 9747
distance de roulage moyenne 444
distance de transport 9747
distance de voyage aller-retour 6922
distance entre arbres 4569
distance entre deux arbres 4569
distance entre glissières 7746
distance externe de débusquage 2967
distance finale 3139
distance latérale de débardage 4825
distorsion de bois 9268
distribution d'abattage 3061
distribution de billes par diamètre 7248
distribution de brisures 2001
distribution de déchets 2541
distribution de déchets de coupe 2541
distribution de fentes 2001
distribution de fréquence de billes par diamètre 7248
distribution de la forêt 3381
distribution de l'humidité 5519
distribution d'espèces arborescentes 8929
distribution naturelle des forêts 5677
district de défense contre les incendies 3179
district de flottage 3272
district de protection contre les incendies 3179
district forestier 3382
district forestier pour stagiaires 8762
diversité alpha 210
diversité bêta 690
diversité biologique 721
diversité biologique forestière 3337
diversité culturelle 2130
diversité de plantes 6338
diversité des espèces 7897
diversité d'habitat 4074
diversité gamma 3799
diversité génétique 3821
diversité horizontale 4368

divi-divi 2550
division du matériel forestier 5073
division forestière 3384
doloire 82
domaine de gestion littoral 4758
domaine forestier 3408
domatie 2557
dominance 2561
dominance apicale 325
dominance relative 6853
dominant 2563
domination 6013
dominé 6011
dominé élagueur 8760
dommage abiotique 7
dommage de bille de pied 1216
dommage de coupe 3059
dommage de la forêt 2241
dommage-incendie 3177
dommage mécanique 5368
dommage par bupreste 9494
dommage par capricorne des maisons 5134
dommage par perces-bois 9494
dommage par scolyte du bois 232
données forestières non classées 9047
dormance 2572
dormance profonde 2323
dosage des éclaircies 8519
dos de glissoire 460
dosse 7712
dosseau 7712
dosse de mine 5470
dosse de tranche 2554
double greffage 8691
double traîneau 2587
douce-amère 773
douelle 8048
douglas 1772
douglas bleu 1772
douglas de Californie 5588
douglasie 1772
douglas vert 1772
doussié 101
douvain 4362
douvain de chêne 5834
douve 8048
douvelle 8048
dragage au câble 271
drageon 7046, 7049
drageonnement 7437
drageon racinaire 7046
drago 2608
drainage 2612
drainage souterrain 8241
drave 8578
dresseuse 4661
droit au bois de chauffage 6946
droit au bois de ramassage 3693
droit de coupe 2955
droit de flottage 6945
droit de passage 6941
droit de ramassage de feuilles mortes 6947
droit des bois d'œuvre 4388
droit d'exploitation 2955
droit d'usage au bois 8573
droit fil 209, 8139
drôme 8779
drupe 2644
dumper 3674
durabilité 8294
durabilité contre la pourriture 2293
durabilité de bois 2669
durabilité naturelle 5678
duramen 1951
duraminisation 2671
durcissement 4954
durcissement de branches 985
durcissement du sol 7834
durcissement en surface 1367
durée de la conversion 1916
durée de la régénération 2868
durée de limbe 4854
durée de l'incendie 3181
durée de régénération 6820
durée de révolution 7077
durée de vie des arbres 4931
dureté 4123
dureté de bois statique 9665
dureté de choc de bois 9651
dureté du bois 4128
dynamique d'un peuplement 8013
dysgénique 2696

eau capillaire 1325
eau de ruissellement 8968
eau disponible 423
eau interstitielle 6464
eau liée 950
eau poreuse 6464
eau souterraine 3995
eau tractable 8736
ébauche de cellule 46
ébauche de feuille 4884
ébénacées 2731
ébène d'Afrique 102
ébène de Ceylan 904
ébénier 2727
ébénier d'Asie 2727
ébénier vrai 2727
ébéniste 4656
éboulis 8378
ébourgeonnage 1142
ébourgeonnement 1142
ébouteuse 1136
ébouteuse à lames multiples 8971
ébouteuse multilame 8971
ébouteuse-tronçonneuse 8971
ébranchage 6610
ébranchage à état vert 3953
ébranchage brut 1683
ébranchage de branches sèches 1106
ébranchage naturel 5688
ébranchage par courses successive 8085
ébranchage par groupes 1175
ébranché 983
ébrancher 10
ébrancheur 981, 8825
ébrancheuse 8825
ébrancheuse à alimentation continue 1896
ébrancheuse à fléaux 1426
ébrancheuse à flèche 4964
ébrancheuse-écimeuse 4964
ébrancheuse mobile 3113
ébrancheuse multiarbre 5323
ébrancheuse multiple 5323
ébrancheuse simple 7651
ébrancheuse stationnaire 8045
ébrancheuse-tronçonneuse 2352
ébranchoir 5168
écaille 968
écaille de bourgeon 1143
écailles 7278
écartement de peuplement final 3135
écartement des bois de remplissage 8092
écartement des branches 2539
écartement des plantes 6362
écartement des plantes repiquées 8773
écarteur 7934
écésis 2734
échafaudage d'abattage 1563
échange de bois 8583
échantillon composite 1741
échantillon de bois 8635
échantillonnage de bois de sciage 5210

échantillonnage linéaire de régénération 4987
échantillonnage par liste 4998
échantillonnage par placettes à deux dimensions 694
échantillonnage par placettes à one dimension 5527
échantillonnage par placettes circulaires à rayon variable 285
échantillonnage par points 285
échauffer 3166
échauffure 2578, 4494
échauffure de l'écorce 543
échauffure en points 6242
échauffure localisée 8154
échauffure sporadique 4670
échéance du rajeunissement 8040
échelle de tronçonnage 1131
échinophore 6532
écimage 2198, 4294
écimage [e sommets 2198
écimer 8702
écimeur 8706
écimeuse 8707
éclaircie 8515
éclaircie artificielle 377
éclaircie chimique 1469
éclaircie commerciale 1750
éclaircie d'abri 7516
éclaircie de bourgeons 7948
éclaircie de couvert 4566
éclaircie de mise en lumière 2096
éclaircie descendante 2597
éclaircie éclectique 2735
éclaircie en bandes 4992
éclaircie en bordure 4992
éclaircie en ligne 4992
éclaircie en rangée 4992
éclaircie faible 4952
éclaircie finale 3143
éclaircie forte 4217
éclaircie géométrique 3829
éclaircie initiale 3227
éclaircie jardinatoire 7411
éclaircie légère 4952
éclaircie libre 3700
éclaircie mécanique 5374
éclaircie modérée 5508
éclaircie naturelle 5699
éclaircie non commerciale 4680
éclaircie par âge naturelle 5700
éclaircie par bâton 8095
éclaircie par couloirs 1971
éclaircie par espacement 7871
éclaircie par le haut 8512
éclaircie par les cimes 2096
éclaircie par ligne de traînage 9752
éclaircie par traînage sur câble 7706
éclaircie précommerciale 4680
éclaircie première 3227
éclaircie secondaire 7366
éclaircie sélective 7411
éclaircie sélective individuelle 3700
éclaircie sur dimensions 2476
éclaircie sylvicole 7622
éclaircie systématique 5374
éclaircie tardive 4826
éclaircie vers le haut 9130
éclaircir 8503
éclaircissage 8515
éclaircisseuse 8513
éclairement 1598
éclat 2193
éclatement 4845
éclatement d'abattage 3052
éclats de bois 9508
écluse pour le flottage des bois 5039
écobrûlage accéléré 3315
écobuage par bandes 8170
écoespèce 2747
écoforesterie 2839
école forestière 7297
école forestière secondaire 7297
écologie 2741
écologie forestière 3390
économie de bois 8636
économie de la foresterie 3391
économie forestière 3391
économie forestière mondiale 9726
écorçabilité 536
écorçage 579
écorçage à blanc 1603
écorçage chimique 1463
écorçage circulaire 6951
écorçage de billon pour allumettes 5331
écorçage de copeaux 1513
écorçage en sève 6136
écorçage final 3137
écorçage hivernal 9458
écorçage humide 9350
écorçage hydraulique 4403
écorçage manuel 4107
écorçage mécanique 5372
écorçage par bandes 2283
écorçage sec 2646
écorçage superficiel 6104
écorce 8799
sans ~ 569
écorcé 551
écorce à tan en rouleaux 6999
écorce brute 1675
écorce combustible 3734
écorcé dans la forêt 552
écorce d'arbre 8799
écorce de bouleau 743
écorce de cassier 1370
écorce de chêne 5832
écorce de peuplier 6452
écorce de pin 6244
écorce de pin sylvestre 6244
écorce de ponte 1067
écorce de saule 9428
écorce de taillis de chêne 9786
écorce de tilleul 4972
écorce de tsuga 4249
écorcé en sève 7210
écorce épaisse et rugueuse 1675
écorce externe 5975
écorcé forestièrement 552
écorce grossière 1675
d'~ 590
écorce incluse 4491
écorce lisse 7787
écorce primaire 6535
écorcer 2280
écorce restante 580
écorce riche en tanin 8392
écorce riche en tannin 8392
écorce secondaire 7339
écorce servante de piège 8775
écorce tannante 8392
écorceur 556
écorceur à main 4105
écorceuse 2284
écorceuse à anneau rotatif 7074
écorceuse à bac 479
écorceuse à bacs 479
écorceuse à chaînes 1425
écorceuse à chaînes Carpenter 1425
écorceuse à couteaux rotatifs 7069
écorceuse à disque 2533
écorceuse à friction 3712
écorceuse à marteaux 4103
écorceuse à réducteur 2182
écorceuse à tambour 566
écorceuse au champ 3112
écorceuse de billots 711
écorceuse-déchiqueteuse 2281

écorceuse par jet d'eau 9289
écorceuse pocket-barker 479
écorçoir 563
écosystème de forêt tropicale 8976
écosystème forestier 3395
écosystème forestier non éclaircie 9123
écosystème sain 4191
écotone 2751
écotone forêt-pâturage 3451
écotype 4079
écoulement ascendant de sève 4448
écoulement de l'eau de pluie 8073
écoulement de l'eau de pluie le long du fût d'un arbre 8073
écoulement de résine 2977
écoulement descendant de sève 2312
écoulement des eaux en sous-sol 8241
écoulement de sève 1589
écozone 2752
écozone forestière 3392
écrasage 2105
écrasement 2105
écrémer 4272
ectomycorhize 2753
écussonnage 307
écussonnoir 1140
effet amortissant 1148
effet de coupe permissible 196
effet de l'entaille d'abattage 2770
effet de lisière 2758
effet de serre 3947
effet destructeur de bois 9534
effet d'incendie 3188
effet gênant de l'abri 8276
effets cumulatifs 2134
effets de produits chimiques sûr le bois 2771
effets par accumulation 2134
efficacité d'habitat 4075
efficacité sylvicole 7616
effilé 8401
effondrement des fibres 3094
éfourceau 894
églantier 2555
églantine 2555
égobelage 5798
égobelage des empattements 6861
égobelage latérale 7578
égobelage oblique 5857
éhouage 2198
éhoupage 2198
éhoupper 8702
ehretia 2730
ehrétiacées 2774
éjecteur 5097
éjecteur à bras relevables 5062
éjecteur de grume 5097
ekop gombé 1121
élagage 6610
élagage aérien 156
élagage à former 3663
élagage à hauteur d'homme 999
élagage artificiel 376
élagage à sec 2655
élagage avancé 76
élagage brut 6734
élagage chimique 1466
élagage d'arbres d'ombrage 7480
élagage d'arbres fruitiers 5939
élagage de bois sec 2270
élagage de cime 8696
élagage de dégagement 4933
élagage des branches supérieures 4294
élagage des racines 7040
élagage des racines latérales 964
élagage des rejets 7532
élagage d'été 8262
élagage de tiges 8081
élagage d'hiver 9462
élagage élevé 4294
élagage en bas 4935
élagage en hauteur 4294
élagage en vert 3953
élagage fort 4215
élagage latéral 964
élagage latéral des racines 964
élagage naturel 5688
élagage sélectif 7414
élaguer de branches en vert 6609
élagueur 8949
élagueur à long manche 6420
élais d'Afrique 110
élais de Guinée 110
élancé 5127
éléagnacées 5889
élément de bois 9548
élément de charpente lamellé-collé 3878
élément de vaisseau 9212
élément de vaisseau imparfait 9173
élément d'une file de trachéides 8145
élément protecteur 6598
éléments d'une fille de trachéides 8741
éléments étagés 8131
éléments-trace 8740
élément tendu 8542
élément unitaire de surface d'éclaircie 8508
éléocarpacées 2021
élevage en consanguinité 4435
élévateur à chaîne 1424
élévateur à une chaîne centrale 7654
élévateur de billes 5060
élévateur de grumes 5060
élévateur de nœuds 4730
élévateur transversal de bois 5055
élève 7202
élimination 6998
élimination des plants malformés 2121
élimination des rémanents 7723
élimination des rémanents mécanisée 5381
élimination diversifiée des rémanents 2546
élinguage préliminaire 6525
élingue 7742, 9463
élingue d'arbres 914
élingueur 1559
élongis 8965
embâcle 5096
emballage d'arbre 8960
emballer 9024
embarcadère 5027
embauchoir 2054
embobinage 6120
emboîter
s'~ 5710
embranche 7739
embranchement 987
embrèvement 2223
embrouilleur 7312
embryogenèse somatique 7846
embryon 2795
embu 2268
embûcher 4116
embuia 1008
émergence 2796
émergent 2797
émerillon 8350
émien 1461
emmanché 6193
émondeur 8949
empattement 1234
à ~ en forme de fond de bouteille 8493

empétracées 1997
empierrement régulier 6969
empilage 4846, 8619
empilage à dépôt 6234
empilage à plat 3259
empilage arasé-carré 962
empilage-brûlage des rémanents 6233
empilage croisé 2055
empilage de billes 5096
empilage de bois intermédiaire 1720
empilage en carré 7973
empilage en encorbellement 5993
empilage en travers 2055
empilage partiel 6086
empilage plein 1157
empilement 4846
empilement de charbon 1450
empilement de grumes 2305
empilement des rémanents 7730
empilement par bulldozer 1162
empileur automatique de bois de sciage 5212
empileur de bois 9661
empileur d'usine de rabotage 6327
empileuse automatique de bois de sciage 5212
emplacement de trait de scie 7668
emploi dans la sylviculture 2799
emploi de bois 9154
empoisonnement par forage oblique 928
empoter 6334
empreinte du marteau forestier à l'extrémité 2807
empreinte sur grume forestier 5101
emprise 6941
emprise de la route 6941
emprise routière 6941
encastrement 4703
enceinte 5019
enceinte vive 5019
encens rouge 4086
enclosure 2800
encoche 4082
encoche circulaire 3853
encoche du trait d'abattage 9319
encoche initiale 4495
encoches groupées 3715
encore existant 2960
encrouage 5042
encrouer
s'~ 4116
endiandra d'Australie 413
endoderme 2809
endogène 2810
endomycorhize 2813
endosperme 2812
endurcissement 4120
énergie dérivée du bois 9567
énergie du bois 9567
enfléchure 6732
enfoncement 7660
engin de première intervention 916
engin forestier de manutention 9587
engins forestiers 3402
engorgement 9291
engoujure 7307
engrais 3089
engrais vert 3949
engravure 2223
enlèvement d'arbres entiers 3761
enlèvement de jumelle 3658
enlèvement de la couverture du sol 7286
enlèvement des aiguilles 2995
enlèvement des rémanents 7723
enlèvement des rémanents de coupe 1624
enlèvement forestier 6862
enlèvement partiel de coupe 6089
enquête par photographies aériennes 90
enquête sur les ressources forestières 3563
enracinement 7029
enracinement de bouture 7031
enrobage 6141
enrochement 6969
enrochement réglé 6969
enroulement 1016, 2020
enroulement de feuilles 4887
enroulement des bords 4863
enrouler 8369
enseignement 6565
enseignement professionnel forestier 6565
ensemble de données de rastre forestier 3548
ensemencement 7382
ensemencement aérien 87
ensemencement à la volée 1044
ensemencement de bord 7384
ensemencement dense 8501
ensemencement direct 2521
ensemencement en lignes 522
ensemencement en rangée 7869
ensemencement en rangée large
ensemencement localisé 6088
ensemencement naturel 5695
ensemencement par bandes 522
ensemencement sous abri 4567
ensemencement sur placeaux 7870
ensemencement sur placeaux abrité 7510
ensemencement sur placeaux sous-abri 7510
ensemencer 7371
ensouchement 6927
entaille 452
entaille à la tige 2514
entaille à mi-bois 8140
entaille circulaire 7107
entaille d'abattage 3069
entaille d'abattage en corde d'arc 1565
entaille d'abattage pas à pas 8088
entaille d'abattage supérieure 8697
entaille de débusquage 2319
entaille de Humboldt 4392
entaille de ponte 2773
entaille de sciage 7246
entaille droite 8140
entailler 7308
entaille rectiligne 8144
entamer 669, 2169
entandrophragme 7200
enter 3906
enterrement de billes de pied 1217
entité de la forêt 3403
entomologie forestière 3404
entourage 2800
entrave-buisson 3091
entrécorce 4491
entrecorre 4568
entre-écorce 4491
entre-écorce superficielle 5900
entre-écorce transitoire 8539
entrée d'une glissoire 2836
entrée en terre 955
entrelacement 6379
entre-nœud 4560
entreposer 8115
entrepreneur d'abattage 5072
entrepreneur de débusqueur 7682
entrepreneur de transport de bois 8597
entrepreneur de transport par contrat 1907
entrepreneur de transports forestiers 5122

entreprise d'exploitation forestière 5074
entreprise forestière 3568
entretien de cimes 2095
entretien de la forêt 7610
entretien des sites 4778
entretoise 8163
enture 7285
enture à mi-bois 4096
enture à mi-bois en T 8441
énumération d'arbres 2837
envahi de champignons 3773
envahissement de forêt 3477
envahissement par le hêtre 74
envahissement par l'épicéa 75
envergure 7874
épaisseur 8496
épaisseur de l'écorce 8498
épaississement 8494
épaississement callitrisoïde 1182
épaississements spiralés 7916
épandeur de pesticide certifié 1419
épanouissement des feuilles 9084
épaufrure 9690
éphédra 2844
éphédracées 2845
épiaire des bois 4222
épicéa 1805
épicéa blanc 9386
épicéa colonnaire 1736
épicéa commun 1805
épicéa d'Alcock 163
épicéa de Brewer 9330
épicéa de l'Amérique du Nord 9386
épicéa de l'Himalaya 4310
épicéa d'Engelmann 2822
épicéa de Serbie 7456
épicéa de Sibérie 7568
épicéa de Sitka 7678
épicéa d'Europe 1805
épicéa d'Orient 2723
épicéa en boule 500
épicéa épineux 1726
épicéa noir 2716
épicéa noir d'Amérique 2716
épicéa piquant 1726
épicéa pleureur 9330
épicéa rouge du Canada 1298
épicéa sphérique 500
épigé 11
épilobe à feuille étroite 3215
épilobe en épi 3215
épinastie 2851
épine 7749, 8524, 8525
épine de mai 4174
épine écarlate 793
épine-giraffe 1288
épine noir 7749
épine rouge 6775
épinette 7749, 7949
épinette à la bière 2716
épinette d'Engelmann 2822
épinette du Japon 8547
épinette noire 2716
épinette noire d'Amérique 2716
épinette rouge 1298
épineux 7913
épine-vinette 1758
épine-vinette de Thunberg 4612
épine-vinette du Japon 4612
épingle 6236
épiphyte 2853
épiphyte arboricole 8833
épissoir 5321
épissure 7918
épissure à œil 699
épissure à œillet 699
épixyle 2855
épluchage de fibre libérienne 624
époque d'abattage 3072
époque de vidange 4170
époque du flottage 3278
épuisement annuel moyen 429
épuisement de peuplement 8515
épuisement de volume du bois 8656
équarri 7963
équarri bilatéral 9027
équarri-dégrossi 7963
équarrir 7962
équarrir le pied d'une grume 4666
équarri-scié 7963
équarrissage 5117
équarrisseuse 1316
équarrisseuse-déchiqueteuse 1316
équarri trilatéral 8532
équation de la croissance 4016
équation de volume 9242
équation universelle d'érosion des sols 9100
équienne 2920
équilibre écologique 2736
équilibre naturel 2736
équipage de débroussaillement 1102
équipe d'abattage-ébranchage 3048
équipe d'abattage et débardage 2163
équipe d'aménagement 8288
équipe de débroussaillement 1102
équipe de déchiquetage 1533
équipe de flottage 2630
équipe de lutte contre l'incendie 3183
équipe de nettoiement 1610
équipe de paquetage 1172
équipe de planteurs 6349
équipe d'estimateurs 8288
équipe d'exploitation de bois long 5155
équipe d'extincteurs d'incendies 3183
équipe d'incendie 3183
équipe héliportée 4240
équipement de manutention des bois 9587
équipement de marquage 997
équipement de sécurité 7149
équipement d'essartage 7319
équipement de traitement des grumes 5106
équipement d'exploitabilité 4157
équipement d'exploitation mauvais 4178
érable 5296
érable à blanche écorce 1438
érable à écorce de papier 6057
érable à épis 5589
érable à feuilles caduques 707
érable à feuilles circulaires 5944
érable à feuilles de frêne 4599
érable à feuilles de platane 935
érable á feuilles de tilleul 4984
érable à feuilles de vigne 5944
érable à feuilles glabres 4373
érable à grandes feuilles 701
érable américain 6996
érable argenté 7608
érable à sucre 808
érable à trois fleurs 8533
érable à tronc strié 6800
érable aux feuilles de frêne 381
érable aux feuilles d'obier 4594
érable barbu 3285
érable blanc 7608
érable champêtre 2827
érable circiné 5944
érable commun 2827
érable cylindrique 5944
érable cymbalaire 4599
érable d'Amérique 8174
érable de Burger 8970
érable de Chine 1501
érable de Corée 6637
érable de Heldreich 4238

érable de montagne d'Amérique 5589
érable de Montpellier 5550
érable de Norvège 935
érable de Pennsylvanie 8174
érable des Balkans 488
érable de Shantung 6638
érable de Siebold 7585
érable de Tatarie 8418
érable de Virginie 6774
érable d'obier 4594
érable du fleuve amour 267
érable du Japon à feuilles pubescentes 3752
érable faux-platane 8355
érable faux-sycomore 935
érable ginnala 267
érable grandifolié 701
érable japonais 4373
érable japonais à feuille laciniée 4373
érable jaspé 8174
érable jaspé de gris 6800
érable moucheté 808
érable moucheté d'Amérique 808
érable nain de Douglas 6996
érable négundo 381
érable noir 808
érable plane 935
érable rouge 6774
érable sycomore 8355
érablière 5292
éradication 2857
érémacausis 2859
ergot-de-coq 1699
éricacées 4207
erminette 82
érosion 2861
érosion du sol en surface 8285
érosion en nappes 7506
érosion en ravines 4045
érosion par couches 7506
érosion par sillons 6949
erreur d'échantillonnage 7177
érythrine crête de coq 1698
érythroxylacées 1697
escariole 6531
escarole 6531
escarre 8416
esche 4509
espace boisé protégé 6593
espace intercellulaire 4533
espacement entre premiers dépôts transitoires 4775
espacement entre verticilles 2539
espace naturel 9404
espalier 2865
espèce 7893
espèce à écorce lisse 7788
espèce arborescente à duramen 4205
espèce aubier 7219
espèce clé 4705
espèce climax 1640
espèce créative forestière 3439
espèce décidue 1055
espèce de conifères 7832
espèce de gibier 3797
espèce dominante 2566
espèce écologiquement propre 2738
espèce en danger de disparition 2801
espèce endémique 2802
espèce en évolution progressive 4447
espèce en évolution régressive 2311
espèce forestière 4714
espèce forestière de grande valeur 9162
espèce forestière de valeur 9162
espèce forestière intolérante 4574
espèce forestière non commerciale 5752
espèce forestière précieuse 9162
espèce indicatrice 4473
espèce indigène 4477
espèce isolée 7292
espèce menacée 2801
espèce naturalisée 5684
espèce nouvelle 5802
espèce ortho-écologique 1640
espèce pionnière 6290
espèce préférable 6501
espèce principale 5252
espèce protectrice du sol 7839
espèce rare 6725
espèce résineuse 6304
espèce résistante au feu 3202
espèces *7893*
espèces à croissance lente 7756
espèces à croissance rapide 6666
espèces alliées 185
espèces à risque 7894
espèces de la faune et flore sauvage 9417
espèces demandeuses 2361
espèce sensitive 7443
espèce serial 7454
espèces exotiques 2944
espèces exotiques désirées 2430 1752
espèces forestières exotiques 2944
espèces forestières introduites domestiquées 2558
espèces forestières riches à amidon 8038
espèces importées 4576
espèce sous-arbustive 9065
espèces persistantes 6187
espèces sélectionnées de chêne rouge 7402
espèces sélectionnées de chênes blancs 7404
espèces significantes régionalement 6828
espèces végétales privilégiées 2428
espèces voisines 185
espèce vulnérable 7443
espénille 9349
essai à l'agar-agar 123
essai d'arbre 8944
essai de bois 9677
essai de culture forestière 3630
essai de fendage 7922
essai de germination 3837
essaim 8319
essai sur milieu gélosé 123
essartage des forêts 3352
essartage par épandage de rémanents 1623
essartement 8201
essartement et rétention 4029
essayer le calepin 1458
esse 5094
essence 2866
essence associée 393
essence auxiliaire 7356
essence conforme à la station 7895
essence de bois de pin 9684
essence de lumière 4574
essence de phénombre 7434
essence de pin 9684
essence de remplissage 3131
essence d'eucalyptus 2874
essence d'ombre 7474
essence du mélange 393
essence étrange à la station 7903
essence feuillue précieuse 4295
essence forestière 4714
essence forestière de mélange 66
essence forestière non commerciale 5752
essence hôte 5684

essence locale 4477
essence native de la station 4477
essence nocive 5805
essence principale 5252
essence sciaphile 7474
essence secondaire 7356
essences étrangères 2944
essences forestières étrangères 2944
essences introduites 4576
essences transitoires 6290
essoucher 4033
estacade 914, 8779
estacade flottante 914
esthétique forestière 3321
estimateur de bois 3410
estimateur forestier 3410
estimation à vue 2869
estimation à vue d'œil 2869
estimation d'aubier 7217
estimation de bois 9671
estimation de performance de régénération 6823
estimation des bassins versants 9296
estimation de valeur résiduelle 6891
estimation du capital bois 9163
estimation d'un peuplement par la méthode du balayage sous angle constant 285
estimation du risque 6972
estimation du sol forestier 2870
estimation en pieds-planches 3311
estimation forestière 3411
estimation par la méthode du balayage sous angle constant 285
estimation par placettes d'échantillon multiples 6435
estimation sylviculturale 7624
estimation sylviculturale avant la récolte 6505
estimation visuelle de pâturage par échantillonnage 5865
estomac 1021
estovoir 6946
estrope 4850
estroper 1552
établissement de blocs à la limite 7571
établissement de blocs à la limite forestière 7571
établissement de colons 7470
établissement des graines 8014
étage 8133
étagé 8130
étage des formations ligneuses rampantes et rabougries 2783
étage dominant 3518
étage forestier au niveau de nuages 1663
étage moyen 5438
étai 1164
étai-amont 467
étai vertical 5652
étamine 7998
étançon de mine 6311
étang de cyprès 2218
étang temporaire d'automne au printemps 9202
étape de peuplement suranné 5999
étape d'herbes et phorbiés 3916
étape finale de coupe 3142
état clair 5914
état de bois vert 3957
état d'écorçage 2282
état de développement 2445
état dégagé d'arbres 5915
état de peuplement désiré 2429
état des houppiers 2071
état des mesures 5366
état fané 9469
état libre d'arbres 5915
état sanitaire du bois 7853
étau 3129
étayage 8600
étendue d'une classe d'âge 128
étendues agrestes 9408
étêtage 2198
étêter 8702
étiquetage 8364
étiquettes de types forestiers 3633
étoc 8937
étourneau roupenne 3350
étranglement 1560
étranglement préalable 6497
être arrêté dans la croissance 366
étude d'altéralogie 2840
étude de la forêt 3549
étude d'impact sur l'environnement 2840
eucalyptaie 2873
eucalyptus 2872
eucalyptus à feuilles cordiformes 7605
eucalyptus à fleurs rouges 6773
eucalyptus à odeur de citron 4910
eucalyptus à urne 9145
eucalyptus botryoïde 7861
eucalyptus citronné 4910
eucalyptus commun 1578
eucalyptus de Bosisto 932
eucalyptus de Gun 1578
eucalyptus de Müller 9765
eucalyptus de tire-bouchon 357
eucalyptus fastigié 1075
eucalyptus globuleux 8417
eucalyptus gummifère 2249
eucalyptus moucheté 7930
eucalyptus ombellé 3550
eucalyptus ronce 3842
eucalyptus rostrata 6786
eucalyptus rouge 6786
eucalyptus scabre 9387
eucalyptus tacheté 7930
eucommia 4147
eucommiacées 2875
eucryphiacées 2876
eugénique 2877
euphorbe à feuilles de saule 4470
euphorbe arborescente 341
euphorbiacées 7961
eutrophique 2912
eutrophisation 2913
évaluation de cote des risques 6972
évaluation de cote de vulnérabilité 6972
évaluation de croissance libre 3695
évaluation de l'impact visuel 9237
évaluation d'ensemencement 7389
évaluation de ravine 4044
évaluation des dangers de terrain 8478
évaluation de volume par barymétrie 9334
évaluation du bois 9562
évaluation du bois sur pied 8194
évaluation forestière 3411
évaporation 2914
évapotranspiration 2916
évodia 2932
évolution d'in peuplement 2448
évolution transversale en bois de bout 2804
excavation de fossés 8962
excavatrice 8963
excédent de rendement 6002
excentricité du tronc 8996
excentricité du tronc complète 3762
excès de bois 8651
excès de pâturage 6001

excessivement feuillu 1658
exclusion de l'incendie 3182
excroissance 5977
excroissance de bois du tronc 2938
excroissance du tronc 2938
exemptions de foresterie 3569
exfoliation 2941
exfolier 2940
exigence écologique de station 7671
exigences à la contiguïté de coupes permissibles 2167
exigences au boisement 95
existence de la forêt 3512
existences 9164
exogène 2942
expertise forestière 3411
explant 2947
explantation 2948
exploitabilité financière 2745
exploitabilité physique 6215
exploitabilité technique 8439
exploitable d'abattage 3231
exploitant forestier 5072
exploitation abusive 2434
exploitation à culée noire 9059
exploitation agro-sylvicole 143
exploitation arbre par arbre 7657
exploitation avec chevaux 4378
exploitation complémentaire 6859
exploitation d'ambiance 4412
exploitation d'arbres en morceaux 8900
exploitation de biomasse 731
exploitation de bois long 5157
exploitation de bois rond 7115
exploitation de forêt mécanisée 5379
exploitation de la forêt 3414
exploitation de la forêt après dépérissement 3415
exploitation de l'étage dominant 6008
exploitation de produits forestiers de cueillette 4159
exploitation de récupération 7169
exploitation de souches 8204
exploitation du bois basée sur la performance 6156
exploitation en bois courts 3078
exploitation en bois longs 5144
exploitation en têtards 6430
exploitation en toute longueur 8869
exploitation forestière 9588
exploitation forestière d'abattage et débusquage 9588
exploitation forestière en régie 5070
exploitation forestière multiple 5620
exploitation forestière par machines multi-opérationelles 5610
exploitation forestière périodique 6163
exploitation forestière totale 8724
exploitation gagnée 2705
exploitation intégrale 1831
exploitation intégrée 4525
exploitation intermédiaire 4546
exploitation mauvaise 4177
exploitation multiarbre 5620
exploitation par arbres entiers 3761
exploitation par assortiments 2746
exploitation par billes courtes 3078
exploitation par blocs 1854
exploitation par fûts entiers 8869
exploitation par garde 4342
exploitation par l'acquéreur 7416
exploitation par lisières multiples 5618
exploitation par longues billes 5144
exploitation par portions d'arbres entiers 8900
exploitation par troncs entiers 8869
exploitation réglée 6841
exploitation sans nœuds antidérapante 2350
exploitation sans rupture de charge 4384
exploitation secondaire 2994
exploitation selon diamètre d'exploitabilité 9152
exploitation sylvo-agricole 143
exploiter le bois de valeur 4272
exportation de bois 2957
extension de racines 7022
extinction complète et nettoiement d'une surface incendiée 5554
extinction d'un incendie forestier 3431
extirpateur 1112
extirpation 2968
extracteur d'arbres 8808
extracteur de mottes 496
extracteur de souches 8199
extracteur-sécateur de souches 8205
extraction d'arbres vieux 2969
extraction de bois de valeur 4285
extraction de bois sans régénération 8616
extraction de broussins 3070
extraction de loupes 3070
extraction de loups 3071
extraction de mauvais bois 2121
extraction de mort-bois 9320
extraction de perches du taillis 7409
extraction de rebut 9320
extraction de réserves 2969
extraction des meilleurs arbres 4285
extraction des souches 8201
extraction du peuplement en éclaircie 8515
extrémité cisaillée 7498
extrémité conique 6409
extrémité de la grume 607
extrémité de racine 7051
extrémité franche 871
extrémité guidante 4848

fabacées 4908
fabrication de charbon de bois 1449
fabrication de pâtes chimiques 1467
fabriques utilisantes un bois primaire 6550
fabriqué ex bois de hêtre 5239
face amont de bois 9559
face au vent 9451
face d'attaque 3722
face de bois 7258
face de bois scié 7258
face de coupe 3722
face de la forêt 3417
face de panneau à flocons orientées 5953
face de première qualité 3229
face de suspension 4896
face distendue 7716
face extérieure de grumes 5982
face intérieure du bois 8602
face la plus belle 686
face meilleure 686
face nette 686
face travaillante 3722
faciation 2987

façonnage d'arbre 8901
façonnage de chablis 7168
façonnage des bords 9796
façonnage du bois 9564
façonnage-épandage 5166
facteur abiotique d'ambiance 8
facteur abiotique du milieu 8
facteur d'ambiance limitant 4980
facteur d'écorce 557
facteur de qualité 6654
facteur de station forestière 3594
facteur de surface terrière 599
facteur d'utilisation de bois 9686
facteur Q 6654
facteurs de pertes 5176
faculté germinative 3833
faculté germinative efficace 2768
fagacées 652
fagot de brindilles 1118
fagot d'écorce 1176
fagot de menues branches 1118
fagoter 1169
fagoteur 2990
fagoteuse 3220
faiblement branchu 7737
faine 657
fainée 656, 5326
faînée tendre 7818
faire le lit pour la chute d'un arbre 645
faisceau d'aiguilles 5703
faisceau d'arbres 7445
faisceau de bois rond 6026
faisceau de troncs 8066
faisceau d'expédition 7526
faisceau enrobé par élingue 7743
faisceau ligneux 9706
faîte 2087
famille des ginkgos 3847
fanabrégié 2898
fanaison d'arbre 8831
fardage 8579
fardeleuse 3053
farine de bois 1514
farine d'écorce 550
farine de foreuse 929
farine de vers 9728
fasciculé 3096
fascine 3018
fastigié 8401
fauchage 1897
fauchage de succession 8247
faucheuse à action permanente 1900
faucheuse-groupeuse à action permanente 1900
faune forestière 3422
faune sauvage 9411
faune sauvage non chassable 5761
fausse coupe 7547
fausse languette 2595
faux acacia 1754
faux amadouvier 3005
faux aubier 853
faux balata 9361
faux bois de rose 113
faux carpophore 1879
faux cerne 2998
faux cerne annuel 2998
faux charme 2900
faux cœur 795
faux cotonnier 4688
faux cotonnier soyeux 4688
faux cyprès 398
faux ébénier 1784
faux frêne blanc d'Australie 414
faux hêtre de Cunningham 406
faux-houx 4339
faux lotier 2258
faux manguier 3972
faux néflier du Japon 5172
faux pistachier 2885
faux pistachier 4646
faux platane 8355
faux poivre 1004
faux poivrier 1266
faux poivrier commun 1266
faux quartier 1738
faux rayon 140
faux rayon médullaire 3001
faux santal rouge 7180
faux sapin 1805
faux simarouba 1921
faux simaruba 1921
faux vernis du Japon 8892
favorable à la forêt 8819
favorisant la croissance 4021
fayard 1760
fécondation 3090
fendage 4915
fendeur 1632
fendeuse 7960
fendeuse de bois 9512
fendeuse de bois hydraulique 4405
fendeuse de bûches 9512
fendeuse de souches 8214
fendillage 2003
fendillement 2003
fendillement au séchage 4349
fendu à la dimension 1634
fente 7483, 9507
fente d'arête de bois débité 958
fente de cœur simple 7630
fente de face 2978
fente de feuille 4870
fente de la paroi 1408
fente de moyenne longueur 5385
fente de retrait 2422
fente de retrait due à la chaleur 2002
fente de sécheresse 2422
fente d'insolation 8266
fente du cœur 4195
fente due au refoulement des fibres 1847
fente en bout 958
fente en étoile 8039
fente en longueur 4915
fente extérieure 2965
fente médullaire 4195
fente non bâillante 958
fente par cisaillement 7497
fente provoquée par le vent 9450
fente radiale 6675
fente refermée 958
fente ténue 4088
fente traversante 8537
fente unilatérale 7483
fer à anneler 3854
fer à ébranchage 6611
fer à écorçage de grumes 7960
fermage d'arbres 8834
fermage forestier 3421
ferme forestière 3419
fermette 3017
fermeture complète à tous les étages 8148
fermeture des cimes 2069
fermeture des houppiers 2069
fermeture du couvert 1659
fermeture horizontale 4367
fertilisation 3090
fertilisation forestière 3424
fête de l'arbre 8838
feu à retardement 7734
feu au ras du sol 2012
feu bondissant 4669
feu couvant 7792
feu d'allure lente 7792
feu dangereux 53
feu de broussailles 1105
feu de brousse 1105
feu de cime 2076
feu de cimes 2076
feu de cimes indépendant 4458
feu de forêt 3428
feu de forêt contre le vent 3429

feu d'élimination des rémanents 7724
feu de régénération forestière 8027
feu de signal 3779
feu de surface 8282
feu de terre 2012
feu de végétation 9409
feu disséminé 4669
feuillage 3303
à ~ caduc 4877
à ~ clair 7881
feuillage adulte 68
feuillage de bouleau 749
feuillage éparse 7880
feuillage juvénile 4676
feuillaison 3719
feuillaison secondaire 6803
feuille 4852
à grandes ~s 4800
à longues ~s 5142
à ~s coupées 2170
à ~s courtes 1029
à ~s de chêne 5840
à ~s de pin 6259
à ~s ligneuses 9709
de ~ raide 7494
sans ~s 4873
feuillé 4894
feuille à bord lisse 7790
feuille acérée 8245
feuille aciculaire 37
feuille à dents 8681
feuille adulte 69
feuille à lobes 1636
feuille alterne 220
feuille apode 7463
feuille aristée 7471
feuille à spatule 7886
feuille barbue 7471
feuille bipennée 741
feuille ciliée 1579
feuille circinée 1584
feuille complète 1829
feuille composée 1841
feuille composée flabellée 3010
feuille composée palmée 6038
feuille composée pennée 6038
feuille d'acanthe 24
feuille décomposée 2309
feuille de palmier 6044
feuille de panicule 6053
feuille dépérie 831
feuille de pied 944
feuille de placage 9197
feuille digitée 6037
feuille encaissée 4438
feuille en forme de cœur 1943
feuille en forme d'épée 8353
feuille entière 2832
feuille flasque 3238
feuille juvénile 4678
feuille laminée 4761
feuille lancéolée 4767
feuille nervée en éventail 3032
feuille obconique 5851
feuille obcordée 5852
feuille oblancéolée 5854
feuille oblongue 5858
feuille obovée 5859
feuille obtuse 5861
feuille opposée 5927
feuille orbiculaire 5938
feuille ovoïdale 5988
feuille paire-pennée 13
feuille palmaticomposée 6038
feuille palmatifide 6039
feuille palmatilobée 6039
feuille palmatinervée 6041
feuille palmatipartite 6040
feuille palmatiséquée 6042
feuille palmée 6037
feuille pennatidentée 6285
feuille pennatifide 6284
feuille pennatilobée 6284
feuille pennatipartite 6284
feuille pennatipectinée 6286
feuille pennatiséquée 6284
feuille pennée 6282
feuille pennée anticompound 6283
feuille pennée impaire 5867
feuille pétiolée 6194
feuille pinnée à interruptions 4564
feuille printanière 2702
feuille rétuse 2317
feuille sagittale 4165
feuilles de cari 2147
feuilles de curry 2147
feuille serrée double 2585
feuille serrée profond 2325
feuille sessile 7463
feuille sétule 7471
feuille simple 7631
feuille solaire 8267
feuille sous-sessile 8236
feuilles sèches pour l'alimentation du bétail 4868
feuillet 3241, 6049
feuille ternée 8477
feuille tombée 14
feuille unifoliaire 9085
feuille unifoliée 9085
feuille veinée à doigt 6041
feuillir 7531
feuillu 1048
feuillu noble 9160
feuillus 4138
feuillus classés par contrainte mécanique 8160
feuillus du nord 5790
feuillus du nord mixtes 5490
feu non maîtrisé 3690
feu qui couve 7792
feu secondaire 4669
fève de corail 1941
févier à trois épines 4353
févier d'Amérique 4353
févier d'Argentine 263
févier épineux 4353
févier sans épines 8527
fibre cambiforme 3783
fibre de bambou 509
fibre de bois 9697
fibre de cocos 1717
fibre de cœur 4197
fibre de feuille 4867
fibre droite 8139
fibre du liber 623
fibre du phloème 623
fibre gélatineuse 3813
fibre intermédiaire 3783
fibre libérienne 623
fibre ligneuse 4956
fibre ligneuse cloisonnée 7450
fibre ligneuse libriforme 4926
fibre ligneuse septée 7450
fibre ligneuse simpliciponctuée 4926
fibre ondulée 2145
fibre perforée 6153
fibre radiculaire 7035
fibre torse 7914
fibre-trachéide cloisonnée 6153
fibre-trachéide ouverte 6153
fibre tranchée 4585
fibreux 3096
fibrillaire 3096
fibrille 5234
ficelle de cubage 8397
ficelle des marchands de bois 5107
fiche nominale d'espèce 7901
fiche repère 3899
fidonie du pin 6262
figue de Barbarie 6533
figue de cactus 6533

figuer à grappes 1670
figuer comestible 1774
figuer des Indes 4469
figuer indien 939
figuier 1774
figuier à petites feuilles 7766
figuier blanc 3286
figuier des pagodes 939
figuier d'Indie 526
figuier élastique 4469
figuier indien élastique 4469
figure à gouttes de pluie 6697
figure à œil 766
figure cordée 7061
figure de débit en long 7728
figure de fente en bout 7728
figure de fissure longitudinale 7728
figure en éventail 3011
figure flabellée 3011
figure ondulée 2144
figure pommelée 835
fil
dans le ~ 209
dans le sens longitudinal des ~s 209
filao 407
filao à feuilles de prêle 407
filao dressé 1692
fil à plomb 6389
filaria 5503
filaria à feuilles étroites 5664
filasse 4502
fil court 7536
fil dévié 2509
fil d'extrémité 2803
fil diagonal 2461
fil drapé 3031
file de cellules de parenchyme 6077
file de pores 6462
fil en bois 9695
fil en long 5133
fil entrecroisé 4539
filet protecteur 8850
fil fin 1655
fil irrégulier 2145
fil oblique 4585
filon de gomme 4058
filon de résine 8155
fil pelucheux 6702
fils brisés 1524
fils contournés 1902
fils creux 5159
fils déconsolidés 5159
fils écailleux 5159
fil spiral 7914
fil tendeur 1164
fil tourmenté 2145
fil tranché 4585
filtre de farine de bois 9540
finance forestière 3427
fin bois 7760
fin de grume 607
fines 3148
finesse 3154
finition 1215
fin out 7761
fissilité 1627
fissure diamétrale 8537
fissure du bord 958
fissure interne de foudroiement 4946
fissure longitudinale 7729
fissure par le vent 9450
fissure superficielle 8279
fissure ténue 4088
fixation arrière 473
fixation de l'azote 5736
fixation d'une grume par griffes antifentes 7315
fixation inférieure 942
fixation supérieure 8682
flaccidité 6652
flache 2662
flache de l'écorce 540
flacourtiacées 4467
flamboyant 3249
flamboyant chinois 1497
flamme australienne 3245
flammule 6387
fléau d'ébranchage 1426
flèche articulée de chargement 4740
flèche à talon 4225
flèche d'abattage 3051
flèche d'ébranchage 2353
flèche multiple 5614
flèche principale 5246
flétrissement des pousses de pin 2510
flétrissement du chêne 5849
fleur de tilleul 4977
fleurir 3290
flexibilité de bois 3264
flexure d'écorce 548
flindersie d'Australie 498
flindersie de Brayley 6665
flocon de bois 3241
floraison 3291
flore forestière 3437
flore naissante à la suite d'une 3039
flot 3267
flot de bois 8584
flottable 2628
flottage à bûches perdues 2625
flottage de bois 8578
flottage de bois par radeaux fragmentés 3273
flottage de bois par train 8578
flottage marin 6689
flottage par retenue d'eau 2631
flottage séparé 7446
flotter bois 3269
flotteur 9569
flotteur de bois 9569
flotteur d'écartement 914
flotteur de déviation 3147
flotteur fixe 8096
flotteur latéral 3147
flotteur pilote 3147
fluage de bois 2016
flux à débris 2286
flux de bois 8645
flux de gènes 3814
flux de transpiration 8768
flux d'exploitation homogène 2922
flying-dutchman 3298
fœtus 2795
foliacé 3302
foliaire 3306
foliaison 9084
folio des ressources 6911
foliole 4875
fomes allume-feu 8667
fonçage 7660
fonction de volume d'arbre 8956
fonction hauteur-diamètre 2473
fonctionnaire forestier 3573
fonctions d'exploitabilité 4154
fonçure 4186
fond 4186, 9236
fondation de chantier d'empilage 5057
fondrière de mousse 5638
fonds de forêt dominiale 3767
fonds d'exploitation forestière 3423
fonds forestier 8114
fonds forestier de réserve 3558
fonds terrestre de hydromélioration 8113
fonds terrestre de plantation forestière 8112
fongicide 3771
fonte des neiges 7802

foresterie à aménagement soutenu 8300
foresterie amicale 9147
foresterie à objectifs intégrés 5623
foresterie à objectifs multiples 5623
foresterie à rendement soutenu 8300
foresterie communale 5636
foresterie communautaire 1817
foresterie d'abattage 3062
foresterie de famille 3007
foresterie d'environnement 2839
foresterie dominiale 3890
foresterie durable 8297
foresterie du taillis 1929
foresterie industrielle 4482
foresterie intensive 4529
foresterie naturelle 3571
foresterie paysanne 143
foresterie permanente 6172
foresterie rentable 2744
foresterie rurale 143
foresterie sociale 7808
foresterie tropicale 8977
foresterie urbaine 5636
foresterie zygotique 9800
forestier 3324, 3406
forestier cantonal 1988
forestier-chercheur 6883
forestier communal 1592
forestière de densité relative nulle 5773
forestier régional 1988
forestier responsable de l'approvisionnement 6560
forêt 3318
foret à bois 9538
forêt à bruyère 4208
forêt acadienne 23
forêt accessible 29
forêt aciculifoliée 1873
forêt à couvert intermittent 3646
forêt à croissance lente 7757
forêt à croissance rapide 3020
forêt adaptée à la station 7662
forêt à débris 2288
forêt adulte 5338
forêt à étages multiples 5633
forêt à exploitation restreinte 6917
forêt à feuillage semi-caduc 7424
forêt à herbes 3917
forêt à laurier 4833
forêt à lianes 9222
forêt améliorée 4431
forêt anthropique 5285
forêt à peupliers et saules 6457
forêt à plain 3256
forêt artificielle 5285
forêt à sal 7160
forêt autochtone 4476
forêt aux sables 7184
forêt avalanche 425
forêt basse 5189
forêt boréale 925
forêt boréale de conifères 924
forêt boréale du type taïga 8367
forêt boréale feuillue 1719
forêt buée 6700
forêt caducifoliée 2300
forêt caduque 2300
forêt camérale 2077
forêt châtaignier 1477
forêt claire 4446
forêt claire à étage herbacé steppique 4941
forêt claire boréale au nord de la taïga 5904
forêt claire subarctique 5904
forêt classée 3557
forêt climacique 1639
forêt climax 1639
forêt collective 1816
forêt communale 1812
forêt communale en propriété collective 1812
forêt conventuelle 5528
forêt côtière 1687
forêt d'abattage 2196
forêt d'Acadie 23
forêt d'âge normal 5780
forêt d'agrément 9142
forêt d'arbres fruitiers 3733
forêt d'arrondissement 1987
forêt de bambous 510
forêt de bas-pays 5192
forêt de bois dur 4139
forêt de bouleau 751
forêt de brouillard 6700
forêt de broussailles 1202
forêt de cèdres 1397
forêt décidue 2300
forêt décidue à la sécheresse 2636
forêt de conifères 1873
forêt de conservation des eaux 9282
forêt de coupe totale 1616
forêt de cure 91
forêt de démonstration 6239
forêt de ferme 3015
forêt déficitaire 2337
forêt définitive 1639
forêt de fondation 4519
forêt de fourrage 3313
forêt de frênes, ormes et peupliers 2787
forêt dégagée 1060
forêt de genièvre 4672
forêt de gestion intelligente 8090
forêt d'église 1575
forêt dégradée 2342
forêt de haute classe 4276
forêt de haute montagne 9136
forêt de hêtres 653
forêt de la couronne 2077
forêt de la zone des brouillards 5593
forêt de liège 1959
forêt de lutins 2782
forêt de mangrove 5280
forêt de marais 8309
forêt de marais d'eau douce 3711
forêt demi-naturelle 7429
forêt demi-sempervirante humide feuillu 7426
forêt de montagne 5583
forêt de montagne à arbres fruitiers sauvages 9406
forêt de moussons 5545
forêt de moussons mixte 5546
forêt de myrtes 5650
forêt de noyers 5818
forêt dense 2388
forêt dense caducifoliée 5545
forêt dense demi-sèche 7425
forêt dense élevée 4280
forêt dense équatoriale 8978
forêt dense humide de montagne 5593
forêt dense humide toujours verte 2930
forêt dense tropicale à lianes 9222
forêt de pins 6255
forêt de pins taeda et pins doux 5031
forêt de pins Weymouth, pins blancs et pins rouges 9379
forêt de plantation 3528
forêt de production 6561
forêt de protection 6592
forêt de protection contre le vent 9445
forêt de protection de champ 3118
forêt de protection des talus 7754
forêt de protection du poisson

forêt d'érable, hêtre et bouleau 5290
forêt de ravin 3474
forêt de récréation 91
forêt de régularisation des eaux 9282
forêt de réserve génétique 3817
forêt de résineux 1873
forêt de saule 9430
forêt de savane 7228
forêt d'escarpe côtière 1694
forêt des cervidés 2330
forêt de seconde venue 7346
forêt des moussons subéquatoriale 8232
forêt d'espèces à petites feuilles 3514
forêt d'établissement public 4519
forêt de terrains en contre bas 5192
forêt de tous âges 172
forêt de trembles 387
forêt de type canneberge 9158
forêt dévastée 2443
forêt d'exploitation ordinaire 5940
forêt de zone d'inondation 5192
forêt d'intérêt commercial 1745
forêt diptérocarpe 2512
forêt dominiale 8041
forêt dunale 2667
forêt ecclésiastique 1575
forêt écrémée 7410
forêt énergétique 2819
forêt épineuse 8526
forêt épuisée 5066
forêt équienne 2920
forêt expérimentale 2946
forêt exploitable 2951
forêt exploitée 2956
forêt fermée 1653
forêt feuilleuse d'Alleghany 176
forêt feuillue 2300
forêt-galerie 3791
forêt humide 6700
forêt humide tempérée 5516
forêt humide tempérée de montagne 5517
forêt humide tropicale secondaire 8980
forêt incendiée 1187
forêt indigène 4476
forêt inéquienne 172
forêt inondable 5192
forêt inondée 3281
forêt interdite 525
forêt intérieure 4501
forêt issue de semence 7388
forêt itinérante après débroussaillement et brûlis 8339
forêt jardinée 7410
forêt-lande 4208
forêt littorale 1687
forêt locale 5032
forêt marchande 5394
forêt marécageuse 8309
forêt mésophyte 5409
forêt mise à ban 6592
forêt mise en défends 6592
forêt mixte 5486
forêt mixte de conifères et de feuillus 5484
forêt modèle 5504
forêt modifiée 5512
forêt monoétagée 7653
forêt montagnarde inférieure 5188
forêt morte 2267
forêt mousseuse marécageuse à pin 5568
forêt moussue 5567
forêt multifonctionnelle 5606
forêt nationale 8041
forêt naturelle 5679
forêt noire 788
forêt non accessible 3645
forêt non détériorée 9076
forêt non productive 9105
forêt non-soignée 9122
forêt normale 5781
forêt ombrophile 6700
forêt ombrophile montagnarde 5593
forêt ombrophile tropicale 8978
forêt opérable 5924
forêt ordinaire 5940
forêt ouverte 3646
forêt paléozoïque 6031
forêt-parc 6080
forêt-parc à couloirs de savane 7229
forêt parcourue par une coupe 7410
forêt parvifoliée 3514
forêt pâturable 3929
forêt pâturée 3929
forêt paysanne 3015
forêt permanente 6171
forêt pionnière 6289
forêt plantée 6339
forêt pluriétagé 5633
forêt pluviale 6700
forêt pluviale tempérée 5516
forêt privée 3473
forêt productive de bois 6561
forêt profonde 2324
forêt protégée 6592
forêt provinciale 6606
forêt rabougrie 2685
forêt rabougrie à bouleau 2781
forêt régénérée 8104
forêt réglée 6840
forêt réservée 3535
forêt résineuse 1873
forêt résineuse boréale 924
forêt ripicole 6961
forêt riveraine 6961
forêt rurale 3015
forêt saisonnière 7330
forêt sans réglementation 9111
forêts antiérosives 317
forêt sclérophylle 7301
forêts d'age moyen 5351
forêts de protection sanitaire de sources d'eaux 3599
forêts de protection sanitaire de stations climatiques 3598
forêts de réserves naturelles 3597
forêts de villégiature 3598
forêt sèche 2650
forêt sèche tropicale caducifoliée 8973
forêt sèche tropicale côtière 1690
forêt sèche tropicale semi-décidue 8974
forêt secondaire après abattage 6477
forêt secondaire postcatastrophique 6476
forêt secondaire réhabilitée 6845
forêt semi-décidu 7424
forêt sempervirente 2929
forêts frontalières 3721
forêts globales 3865
forêt sombre 7847
forêt soumise au pâturage 3929
forêts patrimoniales 4260
forêt steppique 3607
forêt subalpine 8226
forêt subantarctique humide toujours verte 8228
forêt subantarctique tempérée humide toujours verte 8228
forêt subarctique 8229
forêt subtropicale 8244
forêt subtropicale toujours verte 8243
forêt surannée 5998
forêt sur collines 9131

forêt sur terrain très fangeux 8309
forêt sur terrain très mouilleux 8309
forêts vertes d'été 8260
forêt-taïga 8368
forêt-taïga de montagne 5592
forêt tempérée 8443
forêt temporelle 8447
forêt toujours verte 2929
forêt traditionnelle 8748
forêt transitoire 4537
forêt tropicale 8975
forêt tropicale décidue 5545
forêt tropicale humide 8978
forêt tropicale secondaire 8980
forêt urbaine 9142
forêt verte d'hiver 9461
forêt vidée de ses bois de valeur commerciale 5066
forêt vierge 282
forêt vierge détériorée 2543
forêt vierge tropicale 8979
forêt vieux 5880
forêt villageoise 9218
forêt xérophytique 358
foreur à tête plate de Californie 1260
foreur des tiges 8063
formant de pousses 7807
formation arborescente 8839
formation de bois gras 4953
formation de cernes annuels 304
formation de copeau 1515
formation d'écorce 559
formation de forêt 3438
formation de gerces de séchage 4349
formation de la rouille 7140
formation de l'humus 4393
formation de nœuds 4731
formation de racines 7024
formation de tourbe 7813
formation du bois 4954
formation du peuplement 3661
formation en cime 2079
formation ligneuse 9545
formations végétales en mosaïque 9177
forme
à ~ strobile 8181
forme architecturale 348
forme colonnaire 1734
forme d'arbre 7490
forme d'arbre rampante 6589
forme décurrente 2316
forme de feuille 4869
forme dendritique 2939
forme de nœud 4735
forme de ramification à forte décroissance 2939
forme de régime sylvicultural 3666
forme de tronc 898
forme de végétation 4017
forme d'une grume 841
forme du peuplement 3667
forme excurrente 2939
fortin 8560
fossé 2613
fosse à coléoptères 4515
fossé aval 4063
fossé d'assainissement 2173
fossé de ceinture 2547
fossé d'écoulement 2613
fossé de décharge 2613
fossé de dérivation 2547
fossé de détour 2547
fosse-piège 4515
fouet 2030
fouets 9360
fouettage 9359
fouetter 4812
fouetteur 9358
fougeraie 3088
fougère 3087
fougère aigle 967
fougère arborescente 8837
fougère commune 967
fougère en arbre 8837
fougère impériale, porte-aigle 967
fougère male 5257
fouille d'emprunt 931
fouilleuse 468
foulonner de grumes 2641
four 2623
four à carboniser 1444
fourche 3652, 3657
en ~ 3654
fourche à débardage 3881
fourche combinée 5046
fourche de branche 982
fourche de débardage 5020
fourche de fibres 3095
fourche de tige 8074
fourche par entures multiples 5552
fourchet 5046
fourchu 3654
four de cônes 1862
four de séchage des bois 5203
four de sécherie 2623
fourgon à toit ouvrant 5920
fourmi charpentière 1351
fourmi noire 1762
fourmi parasol 4864
fourmi rouge 9480
fourmi rousse 9480
fourmis forestières 3327
fourniture de bois 2359
fourrage forestier 3396
fourré 4104
fouteau 1760
foyer d'incendie sporadique 4669
foyer dispersé 4669
foyer disséminé 4669
foyer partiel 4669
foyer secondaire 4669
fracturation 2003
fracture d'abattage 3052
fracture transversale 2039
fragilité 1000
fragilité de bois 1001
fragilité de fils 7541
fragment 1571
fragmentation de forêts 3681
fragmentation de grumes de rebut 2122
fragmenteur 1574
fragmenteur de bois cassé 1574
fragmenteuse de placage 9194
fragmenteuse de souches 8200
fragments de bois cassés 1059
fragon 1212
fragon épineux 1212
fraisage 9609
fraisage de bois 9609
frais de culture 6350
frais de débardage 7684
frais d'exploitation 5175
frais d'exploitation de scierie 7254
frais d'exploitation directs 2523
frais d'exploitation indirects 4479
frais d'exploitation moyens totaux 439
fraise d'abattage 7070
fraise de meulage 5453
fraisier commun 9425
fraisier des bois 9425
framboise 782
framboise noire 4631
framboise sauvage 9422
framboisier 1800
framboisier commun 1800
framboisier rouge 9422
framiré 4596
franclinie 3687

frappe fine d'une lime 7789
fratrie 3754
frayure par cervidés 2232
freijo 4650
frei-jorge 4650
frein de chaîne 1423
frein de contrecoup 1423
frémontie à grandes fleurs 1259
frémontodendron de Californie
1259
frênaie 380
frêne 1757
frêne à deux pétales 9032
frêne à feuilles aiguës 5659
frêne à feuilles de sureau 777
frêne à feuilles étroites 5659
frêne à feuilles simples 7641
frêne à fleurs 3292
frêne à larges feuilles 5941
frêne anguleux 854
frêne blanc 261
frêne blanc d'Amérique 261
frêne bleu 854
frêne bleu d'Amérique 854
frêne calciphère 4975
frêne chihuahua 1484
frêne chinois 1490
frêne chinois à fleurs 7584
frêne commun 1757
frêne de Balkans 487
frêne de Caroline 1342
frêne de Chine 4741
frêne de Mandchourie 5269
frêne de Pennsylvanie 2255
frêne de Texas 8482
frêne d'Europe 1757
frêne d'Oregon 5941
frêne du sud 5659
frêne épineux 8679
frêne fleuri 3292
frêne noir 777
frêne oxyphille 5659
frêne pubescent 6634
frêne rouge 2255
frêne velours 9188
frênière 380
fréquence d'arbres 8843
frères 3754
friche 2997, 9408
friction 7127
frisolée de feuilles 4863
frison 9697
frisure 9270
fromager 4688, 6792
frondaison 3719
frondaison d'aiguilles 5704
front de la coupe 3722
front d'un incendie 1189
frotteur 7744
frottis 582
frotture 582
fructification 3729
fruit agrégatif 1839
fruit ailé 4701
fruit à noyau 2644
fruit à pépins 4698
fruit charnu 3262
fruit composé 1839
fruit d'arbre 8844
fruit d'arbre sauvage 9427
fruit du hêtre 657
fruit faux 3000
fruit monograine 7647
fruits à baies 683
fruit séparé 7447
fruits forestiers 3441
fruit simple 7629
fruits mous 7815
frutescent 934
fruticée 933
fruticée traînée 8756
fugace 3744
fugitif 3744
fumier 3089, 3949
fumigateur 3765
fumigation 3766
fumure de couverture 8685
funiculaire 84
fusain 3778
fusain ailé 9454
fusain à larges feuilles 1049
fusain de Japon 4641
fusain d'Europe 1383
fustet 2908
fustet commun 2908
fustet d'Amérique 258
fût 891, 8934
à ~ court 7533
à ~ lisse 1613
futaie 4280
futaie à plusieurs étages 5633
futaie avec surréserves 4282
futaie de e chênes par bouquets
4003
futaie jardinée 7410
futaie par bouquets 4002
futaie permanente 6171
futaie pleine 2918
futaie pluriétagée 8132
futaie réservée 4282
futaie traitée en affectation
permanente 6171
fût net de nœuds 1604
fût pliant 3301
futur sur produits forestiers 3540

gadelier odorant 1147
gaffe 6693
gaffe emmanchée 6693
gaïac 4034
gaillet des bois 9488
gainier 4665
gainier commun 4665
gainier du Canada 2721
galac d'Italie 2258
galbule 2212
galé odorant 8332
galerie de ponte 4811
galerie d'insecte 4511
galerie d'insecte xylophage 4032
galerie en impasse 2262
galerie forestière 3442
galerie larvaire 4511
galéruque noire du pin 806
galet 7128
galet moteur 3035
galipot 593
galle 3789, 5831
galle de térébinthe 6256
gamétophyte 3798
gamétophyte femelle 5392
ganse 1555
gants protecteurs 6599
garcia penché 3006
garcinia 3804
garde auxiliaire 392
garde de forêt 3456
garde-forestier 3456
garde-forestier principal 4185
gardien 6115
gardiennage en troupeau serré
1657
gare 914
gare d'embarquement 4484
garni de planches 957
garrigue 3807
garryacées 7597
gâte-bois 3879
gattilier 4961
gattilier commun 4961
gauchissement 9020, 9023
gaule 7775
gaulis 7204
gaulthérie 8433
gaz de bois 9582
geissois 3812
gélinotte 4182
gélinotte des bois 4182

gelsémie de Caroline 1346
gelure estivale 8259
gemmage 828, 1530
gemme 4859
gemme de grattage 593
gemmeur 6905
gemme vierge 7824
gemmule 5474
génération forestière 8846
gènes allélomorphiques 179
gènes amorphes 179
genèse forestière 3445
genêt 3849
genêt à balais 1779
genêt commun 1779
genêt épineux 4652
génétique forestière 3447
genévraie 4672
genévrier 3983
genévrier à encens 7878
genévrier à feuilles courtes 458
genévrier aux crayons 2722
genévrier commun 3983
genévrier d'Afrique 106
genévrier de Bennett 7588
genévrier de Californie 1263
genévrier de Chine 6649
genévrier de Grèce 3935
genévrier de l'Occident 7588
genévrier de Phénicie 6787
genévrier de Pinchot 6241
genévrier des Montagnes Rocheuses 6995
genévrier d'Espagne 7878
genévrier de Virginie 2722
genévrier d'Orient 3935
genévrier du Mexique 5423
genévrier d'Utah 9149
genévrier élevé 106
genévrier flasque 5423
genévrier gersé 187
genévrier oxycèdre 6530
genévrier piquant 6530
genévrier porte-encens 7878
genévrier rouge d'Amérique 2722
genévrier thurifère 7878
génie forestier 3401
génipa 4605
génome 3823
génotype 3824
genou 2100
genouillère de chargement 4740
genre 3827
genre arborescent 8920
genre de mélange 4713
géotextile 3831
gerçure 7483
gerçure en surface 8279
germe 2795, 7132
germer 3293
germinabilité 3833
germination 3835
germination au champ 3114
germination de la graine 7380
germination des semences 7380
germination épigée 2850
germination physiologique 6504
géroflier 1664
gesse sauvage 5662
gestion de débardage 5076
gestion de forêt équienne 2917
gestion de forêt inéquienne 173
gestion de la faune et flore sauvage 9415
gestion de l'écosystème 2749
gestion de pépinière 5813
gestion de prévention contre les incendies de forêt 3433
gestion des bassins versants 9298
gestion des habitats 4077
gestion des ressources génétiques 5264
gestion de tous âges 173
gestion forestière 3488
gestion forestière extensive 2963
gestion forestière nationale 5666
gestion participative 6090
gestion sélective par arbre 7657
gévrine 3974
gilia épineux 6534
ginkgo 5245
ginkgoacées 3847
ginkgo à deux lobes 5245
giroflier 1664
gland 45
glandage 6055
gland de chêne 45
glandée 5843
gléditschie 4353
gléification 3860
glissade 4418
glissade de sol naturel 7835
glissement 7758
glissement de débris 2287
glissoire 4418
glissoire à bois 5116
glissoire à câble 3922
glissoire à grumes 5116
glissoire à l'eau pour le transport du bois 9281
glissoire à perches 6425
glissoire coulisse 8757
glissoire de tronçonnage 1128
glissoire enneigée 7803
glissoire en sol naturel 2709
glissoire garnie de planches 884
glissoire hivernale 2708
glissoire humide 9281
glissoire pour billes 7692
glissoire pour le transport des grumes 5116
glissoire sèche 2648
glissoire utilisable en été 2707
globulariacées 3868
glycine d'Amérique 4707
glycine de Chine 1509
glycine du Japon 4638
gmélina arborescent 5255
gobaja 1921
gojavier 4037
gomme 4053
gomme acacia 4048
gomme adragante 4055
gomme arabique 4048
gomme brute 6733
gomme de bois 8907
gomme de chou 1237
gomme de dollar en argent 357
gomme de lésion 9732
gomme de sapin 1294
gomme éléphantine 4466
gomme kauri 4466
gomme lisse 6318
gomme mastic 4050
gommeraie 4057
gommier 3839
gommier à fruits en forme d'urne 9145
gommier à sucre 5286
gommier blanc 9369
gommier bleuâtre 8417
gommier bleu de Tasmanie 8417
gommier cidre 1578
gommier de Bridges 329
gommier de la montagne 1306
gommier de rivière 6786
gommier doux 259
gommier gris d'Australie 5587
gommier manna d'Australie 5286
gommier noir 8310
gommier rouge 6773
gommose 4051
gondolement 3101
gonflement 8337
gonflement radial 6681
gonflement tangentiel 8387
gonflement tannant 8391
gonflement total 8732

goudron de bois 9672
goudron de bois de hêtre 660
goudron de bouleau 745
goudron de pin 6898
goudron de sédimentation 7369
goudronnage 6895
gouge 4328
goulotte 7739
goulotte d'alimentation 3033
gourbet 7537
gourdin 7101, 7111
gourmand 77
goyave 4036
GPV 6636
gradation d'âge 3893
gradation des classes d'âge 3893
gradation par longueur 3894
gradient de humidité 5520
gradient de maturité 5336
grain brut 1678
graine de pin cembro 1399
graine du hêtre 657
graine orthodoxe 5961
graine pleine 3753
graine pure vivante 6636
grain grossier 1678
grain incliné 4585
grain rubané 4539
grains éclatés 1524
graisseur de glissoire 1576
graisseur de lançoir 1576
graminée forestière 3449
graminées 3915
graminées ligneuses 9708
grand arbre de vie 8893
grand aulne 4592
grand aune 4592
grand charançon du pin 4793
grand coudrier 4759
grand dragonnier 2608
grande artère forestière 3463
grande fougère 967
grande laucène 3933
grande mauve 8881
grande perche 4803
grand épiaire 4222
grande saperde du peuplier 4804
grandes exigences de la gestion de la forêt 3642
grand frêne 1757
grand laucène 3933
grand rhododendron 3934
grands besoins de la gestion de la forêt 3642
grand scolyte de l'orme 4794
graphitisation de charbon de bois
grappin 3892
grappin à articulations 4311
grappin à bois 3892
grappin à bois courts 6629
grappin à câble double 9029
grappin à flèche pivotante 8342
grappin articulé 4311
grappin à souches 8203
grappin à troncs entiers 8868
grappin de flèche fixe 3235
grappin-scie 3912
grappin-tronçonneur 3912
gratteur de bois 9646
grattoir 4359, 4584
graveur sur bois 1363
gravité d'un feu 3207
gravure sur bois 9502
gréer 6939
greffage 3908
greffage à cheval 7147
greffage à l'anglaise compliqué 7147
greffage à l'anglaise simple 7919
greffage anglais à cheval 7147
greffage d'écorce 561
greffage de côté 7575
greffage de médulle sur cambium 6307
greffage d'yeux 307
greffage en approche 331
greffage en couronne 2081
greffage en écusson 307
greffage en fente 1635
greffage en pont 1030
greffage en terrain dégagé 3115
greffage latéral 7575
greffage par bêche 7872
greffage par rainure 8678
greffage sur fente 1635
greffage sur racine naturel 5692
greffe 1145
greffe de scie 7234
greffe en fente anglaise 7147
greffe hétéroplastique 4268
greffe homoplastique 4347
greffe naturelle 5682
greffer 3906
greffe racinaire 7026
greffe racinée 7021
greffon 1145
grenadier 6443
grenadier commun 6443
grenadier d'Europe 6443
grenadille 5426
grenier pour entrepôt de cônes 1860
grévillée 3966
griffage de bois 5317
griffe 7313
griffe à soulever 6225
griffe d'appui 8829
griffe de serrage 943
griffe inférieure 943
griffer 824
griffer la voie 825
griffe supérieure 8684
grignon fou 5275
grignon franc 6782
grignon rouge 6782
grigri 5532
grillage contre les lapins 6670
grillage des feuilles 7306
grille à points 2577
grille d'arrêt de flottage 8136
grimper 2011
grimpereau brun 1807
grimpereau des bois 1807
grimpereau familier 1807
grimpeur aux arbres 4296
griottier 7855
gros bois 1819
gros chêne rouge 6768
gros débris ligneux 1686
gros débris organique 4802
gros défilement 3022
groseille à grappes noires 2883
groseille chinoise 51
groseille de Chine 51
groseille du cap 1323
groseille noire 2883
groseillier 2148
groseillier à fleurs 6772
groseillier à grappes 6769
groseillier des Alpes 213
groseillier des buffles 1147
groseillier des Indiens 1940
groseillier noir américain 240
groseillier sanguin 6772
groseillier sauvage 213
gros févier 1482
gros grain 1678
gros lot 4603
gros nœud 4796
gros rémanents 4216
gros rondin d'industrie 7113
groupage de bois 4846
groupage de bois de flottage 912
groupage de bois de flottage en bord 524
groupage de contreplaqués 6399
groupage en printemps 7937
groupe actif 54

groupe de pores 6462
groupe de résistance 8159
groupes des espèces 7899
grue à bras orientable 8343
grue à pivot 8343
grue chargeuse 8607
grue d'alimentation 4740
grue de chargement 8607
grue de démêlage 4740
grue de manutention 8607
grue-derrick 2410
grue forestière à flèche horizontale 705
grue pivotante 8343
grue tournante 8343
grume 8879
grume-abri 5810
grume alignée parallèle 3755
grume à papier 6627
grume à sciage 7245
grume colmatante 4704
grume creusée 4329
grume d'assortiment spécial 7887
grume de bois d'œuvre 8606
grume de bois dur 4142
grume de bois flacheuse 9266
grume d'échantillonnage 7172
grume de chêne 5842
grume de construction 1888
grume de débardage 7686
grume de déroulage 9193
grume de flottage 1181
grume de moyenne longueur 434
grume de perches 5064
grume de rebut 5764
grume de ski 5065
grume de tête 8688
grume droite 8989
grume écorcée 553
grume élancée 4301
grume équarrie 7971
grume équarrie à vive-arête 3755
grume faisant partie d'une élingue de flottage 7496
grume franche de nœuds 1626
grume hachée 1562
grumeleux 4737
grume moyenne 434
grume nette 1626
grume nette de nœuds 1626
grume non délignée 9266
grume non écorcée 9044
grume non marchande 5764
grume pour bois d'instruments 3107
grume précieuse 6552
grumes de chip-n-saw 1521
grumes de tension 717
grume secondaire 7362
grumes longues de bois blanc 9390
grume supérieure 6126
grumier 5087
grumixava 9361
guanacaste 2706
guatambu 4035
guayac 4034
guayule 4038
guéléba 6791
guêpes 4413
gueule-de-raie 1382
gui 2902
guide 4847
guide-câble 7057
guide-chaîne 4040
guide de profondeur 2403
guide de stockage 8108
guide-roues en bois 8628
guignier 8330
guile 9740
guillotine de bois 9584
gummi noir 780
gungurru 4060
gutta-percha 4061
guttation 4062
guttifères 3805
gymnospermes 4069
gynodioïque 4070
gynomonoïque 4067

habillot 8779
habit à croissance déliquescente 2358
habit à croissance pleureur 9326
habitant forestier 3389
habitat 4072
habitat critique 2023
habitat de la vie sauvage 9412
habitat forestier d'hiver critique 2024
habit décurrent de branchage 2315
habitude dendritique 2939
hachage par tambour 2640
hache 673
hache à double tranchant 2579
hache à écorçage 563
hache à équarrir 1041
hache à flacher 5314
hache à manche sinueux 4390
hache-bois 7721
hache d'abattage 453
hache d'échantillonnage 2101
hache de charpentier 1452
hache d'élagage 1041
hache demi-plate 1041
hache de refonte 453
hacher 1561, 2172
hacheur 2204
hacheuse 2181
hachoir 9512
hachoir de déchets 7721
hag 4085
haie 5001, 5019
haie barrière 596
haie de famille 3008
haie paraneige de brousse 1114
haie vive 5019
halage 8341
halage après débusquage 8341
halesia 7800
halésia de Caroline 1349
halésie 7800
halimodendron 7165
halophytes 4102
hamac de bois dur 4140
hamamélidacées 9468
hamamélis de Virginie 1811
hangar à bois 9650
hanneton 3502
hanneton commun 3502
hanneton de pin 6249
hanneton forestier 3502
haptotropisme 4118
hauban 1135
hausse 5979
haute éclaircie de branches 6010
haute futaie 4280
haute-perche 4218
haute-perche de résineux 1874
haute tige 4806
haute tige de chêne 3266
haute tige de hêtre 661
hauteur à la découpe 2195
hauteur à la première branche 5185
hauteur à la première couronne 9133
hauteur commerciale 1747
hauteur de bois 8599
hauteur d'écimage 8708
hauteur de courbure 4234
hauteur de défaut 4235
hauteur de la couronne 2082
hauteur de l'arbre 8855
hauteur de l'arbre dominante 8692
hauteur de l'extrémité de la cime 8726

hauteur dominante 8692
hauteur du peuplement 4233
hauteur inférieure des cimes 5185
hauteur marchande 5395
hauteur moyenne 5357
hauteur moyenne d'arbres prédominants 5358
hauteur moyenne de l'arbre 440
hauteur potentielle d'arbres à la station 7669
hauteur supérieure des cimes 9133
hauteur totale 8726
hauteur totale d'un arbre 8726
hauteur visible d'un arbre 9235
haut fonctionnaire 6915
haut perchis 4803
haut rendement 4305
haut tronc 4302
hazotokana 7609
hélicoptère porte-citerne 4241
hélidébardage 85
héliportage 85
hélophyte 4242
hémicellulose 4244
hémicryptophyte 4245
hémiépiphyte 4246
hémiépiphytique 4247
herbage forestier 3450
herbe à la puce 6412
herbe aux couronnes 7066
herbe aux goutteux 769
herbe du lagui 1792
herbe du Laos 866
herbes 4253
herbes latifoliées 1050
herbes latifoliées de chênaies 1050
herbicide 4254
herbicide biologique 729
herbicide foliaire 4255
herbivore 4256
hérédité 4258
héritabilité 4258
herminette 82, 9026
hersage 2535
herse à éperon 8668
hétérogénéité 4263
hétéroïque 4262
hétérophylle 4266
hétérose 4269
hétérosis 4269
hétéroxène 4262
hêtraie 654
hêtre 649
hêtre à feuilles crénelées 4613
hêtre américain 238
hêtre austral 2999
hêtre austral de Tasmanie 406
hêtre commun 1760
hêtre de Nouvelle Zélande 8420
hêtre d'Europe 1760
hêtre d'Orient 5949
hêtre pleureur pourpre 9329
hêtre pourpre 6639
hêtre rouge 6639
hévéa 1320
hickory 8988
hickory bitternut 774
hickory écailleux 7507
hickory véritable 8988
hiemisilve 9461
hippocastanacées 1126
hirondelle bicolore 8941
histoire forestière 4314
holarrhène florifère 3002
holocellulose 4340
holodisque 6993
holodisque décolorée 2009
homéostasie 4343
hopée 4365
hoplocampe commun du prunier 6388
hoplocampe des pommes 2879
hoplocampe du poirier 6128
horaire de pâturage 3931
horizon 3086
horizon de fermentation 3086
horizon H 4394
horizon L-F-H 4924
horizon organique 5947
horizon racinaire 7032
hormone végétale 6221
hortensia paniculé 6052
hortensia sauvage 9407
houblon 4363
houblon en cône 4364
houe à gazon 7812
houe de planteur 6352
houe Mc-Leod 5348
houppe 8699
houppe sans nœuds 5544
houppe sans verticilles 5544
houppier 8688, 8699
houppier en croissance libre 5910
houx 2899
houx américain 248
houx Cassine 2225
houx clair 4795
houx commun 2899
houx de montagne 3927
houx glabre 3790
hoyau 4030
huile d'arbre à thé 8438
huile de bois 9615
huile de bois de pin 6276
huile de bouleau 756
huile de cèdre 1401
huile de faînes 658
huile de goudron de bois 9673
huile de noix 5822
huile de pin 6276
huile de tung 9009
huile d'hiver 2575
huile essentielle 2866
huile volatile 2866
humidimètre à bois 9611
humidité antérieure du sol 310
humidité de bois vert 3950
humidité du bois 5521
humidité du bois équilibre 2856
humidité du sol 7838
humidité libre 3699
humidité relative 6854
humification 4393
humus 4396
humus acide 7954
humus à lombricides 5604
humus brut 1681
humus doux 5576
humus forestier 3464
humus tourbeux forestier 6129
huperzie chinoise 8951
hutte de bûcherons 5071
hutte en rondins 5052
hybridation 4401
hybride 4399
hydrangéacées 4402
hydro-ensemencement 4406
hydrologie forestière 3465
hydrophyte 4410
hygiène en forêt 3580
hygiène sylvicole 3580
hygromètre à bois 9611
hygroscopicité de bois 9589
hylésine du frêne 379
hylésine mineur 4920
hylésine noir du pin 805
hylobe du pin 4793
hylophyte 9351
hyménoptères 4413
hypertrophie terminale 8473
hyphe 4414
hypholome en touffes 8258
hypholome fasciculaire 8258
hyphomycètes 3127
hypocotyle 4415
hypsomètre 4416

icaquier 1700
iciquier cèdre 4086
idioblaste 4421
if 1814
if à feuilles courtes 9345
if à prunes 1505
if commun 1814
if de Floride 3288
if du Canada 9345
if occidental 9345
illiciacées 294
illipé 5244
ilomba 108
îlot d'écorce intraligneuse 560
imbuia 1008
immergence 7660
immersion 7660
impasse 2262
imperméabiliser de bois par huile de goudron 2017
imperméable 5522
imperméable à l'humidité 5522
impôt dont on grève la forêt 8422
impôt forestier 3616
impôt frappant l'abattage du bois 5085
impôt sur les opérations forestières 5085
impôt sur l'exploitation forestière 5085
impreg 4427
imprégnation à cellules pleines 692
imprégnation du bois 4429
imprégnation sous pression 4430
impureté 9321
imputrescibilité 2293
incendie criminel 7466
incendie de cimes 2076
incendie de forêt 3428
incendie de forêt non-combattu 9049
incendie non-combattu 3690
incendie par les bords 2759
incendie prémédité 4436
incendie prolongé 2974
incendie propageant dans la couche superficielle du sol 2012
incendie souterrain 2012
incidence parasitaire 6191
incidence par ravageurs 6191
incinérateur à déchets 6807
incinération des déchets 7719
incinération des rémanents frais 5010
incinération préventive 2700
inciseur 7309
incision annulaire 3853
incision en écorce 562
incision par encoches disjointes 3716
incision par encoches groupées 3716
inclinaison 4895
inclinaison d'arbre 8857
inclinaison d'avant 4188
inclinaison du fil 4439
inclinaison du fil tangentiel 8385
inclinaison naturelle 5685
incrément 4449
incrément apical 326
incrément articulé 370
incrément de mise en lumière 4944
incrément d'exploitabilité 4155
incrément dû à l'éclaircie 4944
incuit 7530
indemnité d'épuisement des réserves 2401
indésirable 9321
indicateur biologique 4473
indicateur de direction de chute 8567
indicateur de niveau 4922
indicateur de tension de câble 7705
indicateurs de la productivité forestière 4475
indice d'abondance relative 4461
indice d'aérage 9198
indice d'aire de cime 2063
indice d'association 4460
indice de densité du peuplement 8011
indice de diversité des espèces 7898
indice de fertilité de la station 7667
indice de la croissance 4018
indice de plénitude de couvert 8011
indice de qualité de station forestière 8641
indice de surface foliaire 4855
indice de survive 8293
indice de ventilation 9198
indice d'exploitabilité 4156
indice foliaire 4855
indice météorologique des incendies 3218
indices pour la structure du
industrie de résine 6896
industrie de sciage 5206
industrie du bois 3466
industrie du bois de sciage 5206
industrie du bois ouvré 9685
industrie forestière 3466
infestation 4486
infiltration 4488
influence entre arbres 4570
influence forestière 3471
informatique forestière 3472
infraction forestière 1909
inga 4490
ingénieure forestière 3400
ingénieur forestier 3400
ingénieur forestier diplômé 3905
inhibiteur de croissance 4025
inhibition de la croissance 4015
initiale de rayon 6735
initiale du cambium 1285
initiale du cambium fusiforme 3781
initiale fusiforme 3781
injecteur d'arbre 8862
injection à la base de la tige 606
injection à sec 2654
injection d'arbres 8861
inoculation 4506
inoculum 4507
inondation 3282
insecte forestier 3475
insecte forestier nuisible 4499
insecte lignivore 4958
insecte primaire 6539
insecte secondaire 7348
insectes phyllophages 4858
insecte xylophage 4958
insecticide 4512
insecticide-appât 4509
insecticide biologique 730
insectifuge 4514
insectivores 4513
insolation 8265
insolation de l'écorce 836
inspection de la forêt 3549
inspection forestière 3380
installation de carbonisation 1448
installation de récupération de bois 9637
institut de recherche forestière 3556
instrument à vent en bois 9694
instrument de brûlage 1190
instrument d'usinage de bois 9652
instruments de menuiserie 4659

intégrité des bassins versants 9297
intégrité écologique 2737
intendance 3610
intensité à la lisière de feu 3190
intensité d'éclaircie 4527
intensité de gemmage 8410
intensité de rentrée de la récolte 2034
intensité des éclaircies 8519
intensité du passage en éclaircie 8519
intensité globale des éclaircies 5305
intensité globale moyenne des éclaircies 4527
intensité pluviale 6699
intensité relative d'éclaircie 6855
intensité relative de la coupe 4210
interaction entre souche et scion 8118
interception de précipitation 4535
interdiction d'abattage 6577
interdiction de coupe à blanc 6576
interdiction de défrichement 4498
interface entre arbres et agricultures 8821
interface entre terre vierge et urbaine 9410
interfinancement 2058
interruption de couvert 4565
intervalle d'observation 2102
intimité 4572
intolérance à l'ombre 4573
introduction de bois 9591
introduction d'espèces 4577
inventaire 4579
inventaire au champ 3120
inventaire continu d'état de peuplement 1899
inventaire d'abattage 6831
inventaire d'aménagement 5266
inventaire de bois 3478
inventaire de bois de trituration 6631
inventaire de condition forestière 3357
inventaire de forêt domaniale 8042
inventaire de forêt provinciale 6607
inventaire de maturité initial 4496
inventaire de récréation 6756
inventaire d'espèces 7900
inventaire de volume du bois 8657
inventaire forestier 3478
inventaire forestier permanent 1898
inventaire forestier récurrent 6760
inventaire forestier régional 3615
inventaire forestier statistique 2838
inventaire intégral par compas forestier 3747
inventaire par échantillonnage en bandes 4988
inventaire par échantillonnage en lignes 4986
inventaire par sondages sur bandes rectilignes 4988
inventaire par sondages sur bandes rectilignes d'orientation variées 4988
inventaire secondaire 6847
inventaire statistique 2838
inventories de dépôt 9753
involvement public 6623
ipé 4581
iroko 4582
ironbark à feuilles longues 4583
ironbark rouge 6779
iryanthera 4717
ISF 4855
isolement de la forêt 4591
issue de graines 7395
itéacées 8336
itinéraire de débusquage 4847
itinéraire d'évacuation 2864
ive à feuilles de lampourde 7162
ive xanthifoliée 7162
ivoire rose 6279
ivraie 9768
ixora 8715

jacaranda de Brasil 1006
jachère enrichie 2829
jachère forestière 1201
jacquier 1011
jalon de nivellement 3899
jalonnage 7995
jalonnement 7995
jambe de force 8273
jamelongue 4646
jamelonguier 4646
jaquier noir 799
jardinage par arbre 7655
jardinage par bouquets 4001
jardin de haies 4220
jarrah 4642
jarul 4643
jasmin 1782
jasmin de Caroline 1346
jasmin de Virginie 8994
jasmin jaune 9759
jasmin officinal 1782
jatoba 1990
jauge à copeaux 1520
jauge angulaire 6554
jauge de l'écorce 587
jaunisse 1546
jaunissement du bois 9764
javelle 9447
javelotage 1178
jelutong 4649
jet dirigé sur la base de la tige 606
jetée 8664
jeune 4674
jeune arbre 7202
jeune peuplement 9785
jeune plantation 9788
jeune plante 9787
joint en biseau 7285
joint en sifflet 7285
joint non déligné 2662
joint plat 477
jointure de bois 4660
jointure digitale 3161
jojoba 4662
jonction 7918
joue 1460
jour de plantation d'arbres 340
journal des sorties de bois 4904
journée de l'arbre 340
jubée remarquable 1701
juglandacées 9260
jujubier 1783
jumelle 3657
jungle 4671
jungle arbustive 7320
juvénilité 4681

kadam 4682
kadj 2220
kaki 4627, 4628
kalmia à feuilles d'Andromède 895
kalmia à feuilles de laurier 4835
kalmia à feuilles étroites 7505
kalmie à larges feuilles 1264
kaloupilé 2147
kamala 4684
kamassi 4685
kambala 4582
kandis 4686
kapokier 4688
kapur 4689
karité 7503

kauri 5727
kauri de Nouvelle Zélande 6664
kekatong 5653
keledang 1011
kemiri 1304
kempas 4695
kermès 4697
keruing 4699
ketmie des marais 5639
ketmie de Syrie 7067
kino 4058
kirundu 6414
kitir 809
kiwi 51
koko 2730
kokrodua 119
kolatier 1718
kosipo 4744
kossipo 4744
kotibe 2245
kruen 4699
kudzu 4746
kumquat 4747

labiatées 4748
labiées 4748
labourage de disque 2535
labourer 6383
laboureur à lames rotatives 7072
lacet 1555
lagerose 2007
lagundi 3232
laie 3564
laie croisée 2057
laie forestière 3564
laie latérale 7580
laie sommière 5249
laine de bois 9697
lais 8555
laitue sauvage 6531
lakuch 1011
lambrissage par bois 9616
lame 819, 5068
lame à débroussailler 1099
lame à rémanents 6986
lame débroussailleuse 1099
lame de bulldozer 1161
lame de parquet 9574
lame de râteau 6986
lame de scie 820
lame de scie à dents isocèles 6138
lame de scie à dos concave 4327
lame de scie circulaire à dents amovibles 4516
lame de scie circulaire à dents 1853
lame de scie circulaire désaxée 4312
lame de scie incurvée 2152
lame de scie ventrue 2152
lame dessoucheuse 6986
lame d'hache 448
lamelle de feuille 4872
lamelle mitoyenne 4532
lamelle moyenne 4532
lamelle moyenne composée 1842
lame nervurée 6932
lame-ruban 5049
lamination oblique 5856
laminer 4760
lançoir 5116
lançoir non coffré 5800
lançoir sec 2648
languette de fixation d'un fer de hache 454
lantana 4783
lantanier 4783
lapachin 4784
lapacho pourpre 6644
laque de grains 7386
laque de rebut 6809
laque en blocs 6809
largerstroemia des Indes 2007
largeur de cerne annuel 9398
largeur de cime 2098
largeur de copeau 9399
largeur de copeau de bois 9621
largeur de la bande 8320
largeur de planche 885
largueur de coupe 9400
largueur du trait de scie 9400
larix 1785
larrea tridenté 3932
larve 4809
lasiocampe du mélèze 4790
lasiocampe du pin 6258
laticifère radial 4829
laticifères 4828
latte 626, 7732
latte à semis 7385
latte d'arpenteur 8292
latte de caisserie 626
latte d'espacement 6236
latte de toit 7005
latte voltige 7005
lattis de bois 9595
lauan 4831
laucène à petites feuilles 5003
laucène glauque 4851
laucène panaché 3933
laucène panachée 3933
laurel 6076
laurel de l'Inde 9719
lauréole 2828
laurier 6745
laurier benzoin 7909
laurier Bourbon 6763
laurier-cerise 2828
laurier-cerise de Caroline 1347
laurier commun 3936
laurier d'Apollon 3936
laurier de montagne 1264
laurier des bois 2828
laurier des iroquois 1801
laurier des marécages 643
laurier du Japon 1291
laurier noble 3936
laurier-palme 2828
laurier-rose 5887
laurier rouge 6763
laurier-sauce 6763
laurisylve 4833
lavande 4838
lavande de désert 2416
lavande dendritique 8878
lavatère arbustive 8881
laye croisée 2057
laye forestière 3564
layon 3564, 6937
layon de débardage 3987
lécanium 7274
lécythidacées 5531
lecythis 6701
lédon 4906
lédon du Groenland 4749
législation d'élimination des rémanents 7725
législation d'incinération des déchets 7720
législation d'incinération des déchets forestiers 7720
législation forestière 3485
légumebois 5406
légumineuse mimosée épineuse 4354
leishmaniose forestière américaine 6224
lenticelle 4916
lentisque 5327
lenzite des clôtures 7722
leptosperme éclaté 1073
lésion 4918
lésion de l'écorce 558
lespédéza 4919
lettre de crédit garantie par un domaine boisé 8561
leucaena glauque 4851

levé aérien 90
levé de reconnaissance 6749
lever forestier 3614
levier d'abattage 3067
levier de sécurité 7151
liane 4925
liane à squaw 1996
liant de panneau de particules 6092
liard 2884
liardier 2884
liber 4502
libération d'emprise 6942
liber de tilleul 4972
liber dur 4119
liber final 4815
liber initial 2699
liber mou 7814
libocèdre décurrent 1262
licanie 4927
licence d'abattage 4928
licence d'abattage de bois 8605
licence de ferme forestière 8835
licence de lot à bois 9600
licence de vente de bois 8632
licence forestière 8605
lichen 4929
lichen arboricole 8888
liège 1956, 1961
liège de cicatrisation 8781
liège mâle 9227
lierre 4598
lierre de Boston 937
lieu d'assemblage 3280
lieu d'assemblage de bois flottés 3280
lieu d'atterrissage de bois flotté 5105
lieu de coupe à la souche 8213
lieu de la coupe 7668
lieu d'emprunt 931
lifaky 4744
lifuko 4744
ligature d'un train de bois 6687
ligne 3564
ligne aérienne 84
ligne aménagée pour countrefeu 3169
ligne arrière 469
ligne d'abroutissement 1096
ligne d'amélioration génétique 1027
ligne d'appui 3169
ligne d'arbres 7121
ligne de câble de transport à vide 1243
ligne de débardage 1245
ligne de fermeture 1661
ligne de parcelle 3564
ligne de reproduction 1027
ligne d'étagement 6968
ligne de traînage sur câble 9750
ligne d'inclinaison 7576
ligne pare-feu 3169
ligne rousse 7305
lignes de transplants 8772
lignes plantées dans les vides 7987
ligne traitée à l'eau 9352
ligneux 4955
ligniculture 3528
lignification 4954
lignifié 4955
lignine 4957
lignocellulose 9504
lignose 4959
lignotuber 4960
lilas 6292
lilas commun 6292
lilas de Californie 856
lilas de l'Amour 265
lilas des Indes 6110
lilas d'été 1219
liliacées 4962
limature 3128
limba 92
limbe 4853
limbe de feuille 4872
lime à bois 9566
lime acide 9348
lime du Gabon 1366
lime grossière 7089
lime mexicaine 9348
limette acide 9348
limettier 9348
limier 9348
limitation de peuplement 6918
limite boisée 2761
limite de cerne 4026
limite de cime 2065
limite de classe de diamètres 2472
limite de diamètre d'exploitabilité 2950
limite de l'anneau de croissance 4026
limite de qualité marchande 4981
limite des arbres 4982
limite de transport 4979
limite de végétation arborescente 4982
limite entre l'herbage et la forêt 3918
limiteur d'épaisseur de copeau 2403
limiteur de profondeur 2403
limon 4911, 7598
limonage 7599
limonelle 9348
linacées 3261
line d'exploitation 5923
liquidambar d'Amérique 259
liquidation des forêts 3373
lisier 3949
lisière de peuplement 2761
lisière de régénération 6818
lisière d'un incendie 1189
lisière progressive 6574
liste de bois de sciage sur dimensions préétablies 8376
liste de contreplaqués 6400
liste dimensionnelle 5366
lit 8545
lit de chute 4840
lit de germination 7373
lit de grumes 8545
liteau 7732
litière de feuilles mortes 4999
litière de menues branches 5000
litière de pins 6270
litière forestière 3435
litière hachée 1456
litière morte 2269
litséa 902
livraison de bois le long du bord 208
livre des recettes-bois 4905
livre d'inventaire 5365
livrée des forêts 3618
livre foncier 1249
lobes 4876
localisation de ligne terrestre 4776
locomotive de débardage 5075
lof 9451
loganiacées 5047
logement sur roues 9356
logistique forestière 3486
loi-cadre légale 8047
loi de conservation des forêts 3360
loi forestière 3484
loi relative à l'économie forestière 3534
loisirs de plein air 5973
longane 1381
longanier 1381
longeron de bordage 1025
longévité naturelle d'un arbre

longicorne de sapin 7104
longicorne du mélèze occidental 9341
longicorne maculé du pin 7931
longicorne oregonais du sapin 5942
longicorne velu du pin 4089
longiscape 5151
longrines 7694
longue lobé 5143
longueur 5154
longueur au sol 3989
longueur commerciale 5395
longueur d'arrondissement 7106
longueur de bois commercialisable 9146
longueur de bois scié 7259
longueur de carre 2982
longueur de cime 2082
longueur de cime supérieure 9134
longueur de cime totale 5186
longueur de copeau de bois 9619
longueur de crevasse 2004
longueur de défaut 2334
longueur de fibres 3097
longueur de fibres moyenne 432
longueur de grume 8077
longueur de grumes, equale de chaîne élévatrice 5095
longueur de grume standard 8005
longueur de houppier 2082
longueur de racines 7033
longueur de rupture 1019
longueur en crête 8694
longueur minimale de la tige 5468
longueur moyenne de l'arbre 441
longueur nominale 5747
longueur sans branches 1625
longueurs égales paires 2925
longueur totale d'une grume susceptible d'être débitée en sciage 9146
lophyre de pin taeda 5030
lophyre du pin 6268
lot 5181
lot à bois 9599
lot de semences 7393
lotofa 1090
loup 9472
loupe 1182
loupe à la base 605
loupe à la base de la tige 605
loupe d'Amboine 5284
loupe de frêne olivier 5890
lubrifiant à bois 9601
lucanes 7986
lunure 2139
lunure due au gel 3727
lutte contre la concurrence végétale 1825
lutte contre la pollution provoquée par les industries forestières 3468
lutte contre les animaux nuisibles à la forêt 3524
lutte contre les feux de forêt 3431
luvisol 5219
lyciet 9471
lycopode dendroïde 5794
lycopode foncé 8815
lyctes striés 6486
lyde bleue 6253
lyde champêtre 9313
lymantride spongieuse 4071
lyophilisation 3702
lysigène 5222
lysimètre 5223
lythracées 5161

macadamia 1206
maçaranduba 5322
macérer 5227
machette 1204
machine à bois 9701
machine à bouchonner les nœuds 4728
machine à couper 2181
machine à ébrancher des arbres 8825
machine à écorcer 2284
machine à fabriquer les boulettes 6140
machine à fendre le bois 9512
machine à fragmenter les branches 5230
machine à jointer 4661
machine à peler d'osier 5964
machine à raboter 6326
machine à scier à fil 9466
machine déligneuse 2767
machine déligneuse de scie circulaire 6676
machine déligneuse de scie radiale 6676
machine-outil à bois 9701
machine-outil à travailler le bois 9701
mâchoire jointive 9795
maclure épineuse 4219
maclure orangée 4219
macore 5254
macro-élément 5235
macrophyte 5236
macropropagation 5237
macrostructure de bois 9602
madrier clair sans défaut 1033
madriers 2278
madrona 6020
madrure 3106
madrure du bois 2985
madrure en œil de perdrix 766
magnolia 5240
magnolia à fleurs de lis 4963
magnolia à grandes feuilles 700
magnolia à grandes fleurs 7860
magnoliacées 5241
magnolia d'Amérique 709
magnolia de Soulange 7225
magnolia de Virginie 643
magnolia hybride de Soulange 7225
magnolier de Fraser 3688
magnolier-parasol 9040
mahonia commun 5943
mahonia de Fremont 3704
mahonia nervé 1365
mahonia rampant 2013
mahonie à feuilles de houx 5943
mahonie trifoliolée 5419
mahot rouge 4715
mai 747
maille 2985, 7601
maillon 2183
maillon-gouge 2183
maillure 7601
maintien du capital-bois 4694
maison de bois 5052
maison en rondins 5052
maison forestière 3407
maisonnette montée dans un arbre 8856
maladie d'arbre 8828
maladie défoliante 4861
maladie de la rouille du pin 6260
maladie de la suie 7849
maladie de l'encre 4500
maladie de l'encre de marronnier 4500
maladie de l'île de Wight 26
maladie du rond 9385
maladie fongique 3769
maladie hollandaise de l'orme 2678
maladies de forêts 3379
mallee 5258
malpighiennes 5260
mamelonnement forestier 2005
mancenillier 5267

mancienne 9307
mancône 1366
mandarine 5273
mandarine verte 3948
mandarinier 5274
mandrinage par bineuse 4317
manglier d'Afrique 6780
manglier gris 3971
manglier noir 251
mango 1789
mangoustanier 5278
mangrove noire 251
mangroves 5280
mangue 5277
manille 1923, 2320
manio 2714
manioc 5282
manioc cassave 5282
manioc comestible 5282
manioc doux 8329
manomètre 6521
manquant 3443
manque de bois 7282
manque de forêt 4753
manteau forestier 3345
manuel de bois 9585
manuel de foresterie 3570
manutentionnaire de bois 9586
mapou rouge 4799
maquis 5300
marais broussailleux 7557
marais de mangrove 5281
marbré 6131
marbrure 843
marbrure d'aiguille par ozone 6017
marcescent 5302
marchand de bois en gros 9392
marchand du bois 2276
marchandise en bois 9560
marché de bois 8613
marche de piquets 6480
marcottage 6582
marcottage aérien 154
marcotte de branche 976
marcotte de tige 8070
marécage boisé 9543
marge d'éboutage 202
margousier 6110
mariola 5308
maron d'Inde 1781
maronnier d'Inde rouge 1781
marquage à la couleur 6029
marquage à la hache 449
marquage à la peinture 6029
marquage au feu 3168
marquage de bois long 5318
marquage de bois rond 5316
marquage définitif à piquets 3140
marquage des découpes 4845
marquage par pavillon 3239
marque de chargement 7529
marque de coupe 8972
marque de face 2984
marque de flottage 3275
marque de la forêt 3500
marque de pourriture 7081
marque de propriété de bois destiné au flottage 3275
marque de qualité 3900
marque de tronçonnage 8972
marqué pour la découpe 2176
marquer 5310, 7308
marquer la voie 825
marqueterie 9590
marqueur des grumes 5102
marqueur des longueurs 5102
marron comestible 2887
marronnier 8483
marronnier à fleurs jaunes 9758
marronnier à fleurs rouges 7130
marronnier américain 5874
marronnier commun 1781
marronnier de Californie 1256
marronnier de forme conique 4622
marronnier de l'Ohio 5874
marronnier de Virginie 6767
marronnier d'Inde blanc 940
marronnier du Japon 4622
marronnier fétide 5874
marronnier mexicain 5421
marronnier rouge 7130
marteau d'accroissement 4454
marteau d'arbre 8845
marteau forestier 5315
marteau réceptionneur 5315
martelage de bois 5317
martinet des arbres 8942
marupa 6063
masque amont en bois 8654
masse de bois 8658
masse de copeaux à l'écorce 542
massicoté 2756
massif forestier 9605
mastic 3130, 4050
mastic à greffer 8958
mastic de bois 9606
mat 5335
mât 5328
matage des racines 5334
mât aval 8371
mât de fin 2808
mât de levage 3848
mât de montage 3848
mât de queue 8371
mât de tête 4190
maté 6064
matériau de sous-diamètre 9058
matériau pour âmes de contreplaqués 1953
matériaux détritiques 2440
matériaux détritiques ligneux dangereux 4180
matériel absolu sur pied 4009
matériel de repiquage 5814
matériel de reproduction 6363
matériel de reproduction en pépinière 5814
matériel de reproduction en tube 9001
matériel dérivé du bois 8559
matériel de sous-longueur 9062
matériel forestier de reproduction 3555
matériel relatif efficace 2769
matériel relatif irrégulier 4588
matériel sur diamètre 5992
matériel sur pied 3454
matériel sur pied relatif normal 3757
matériel végétatif 9181
mât-grue 2410
matière active 55
matière de végétation 6221
matière organique 5948
mât quartier 6660
matte 5335
maturité d'arbre 8885
maturité de la forêt 3501
mauvaise herbe de flèche 367
mauvaise herbe en forêt 3641
mauvaise herbe nuisible 5806
mauvaise terre 9275
mauve en arbre 7067
mauve sauvage 7102
méat 4533
méat intercellulaire 4530
mécanisme de résistance d'arbre aux maladies 8886
mégagamétophyte 5392
mégaphorbiées 4287
meilleur arbre 687
meilleure face d'ouverture 689
meilleures pratiques de gestion 688
melaleuca 8437
mélangé 5486

mélange de céréales 5487
mélange d'espèces forestières 5497
mélange d'essences feuillues 4143
mélange intime 5496
mélange montagnard 5495
mélange multiclonal 5605
mélange ortho-écologique 1638
mélange par bouquets 5494
mélange par étages 5498
mélange par groupes 5494
mélange par pieds d'arbre 5496
mélange pied à pied 5496
mélange transitoire 8451
mélasse de bois 9612
mélézaire 4789
mélèze 1785, 4787
mélèze commun 1785
mélèze d'Amérique du Nord 9340
mélèze de Dahurie 2227
mélèze de l'Ouest 9340
mélèze de Sibérie 7567
mélèze d'Europe 1785
mélèze du Japon 4623
mélézein 4789
mélèze laricin 250
mélèze occidental 9340
mélèze subalpin 216
mélézière marécageuse sur tourbe sèche 5306
mélézin 4789
mélézin à lède 9423
méliacées 5243
meluang 6277
membrane cellulaire 1407
membrane de la ponctuation 6308
membrane géotextile 3831
membrane longitudinale 5137
membre tendu 8542
membrure 1564
membrure d'extrémité inférieure 945
ménispermacées 5551
ménisperme de Caroline 1348
mentalité forestière 3507
menthe poivrée 5663
menthe poivrée décidue 1052
menu bois 7760
menu bois de feu pour bourrées 1118
menu-branchage 5170
menuiserie 9630
menuiserie préfabriquée 5458
menuisier 4656
menu produit 5471
menus bois du taillis 1938
meranti 4831
méranti 4831
meranti rouge 4949
meranti rouge foncé 2254
merbau 5476
mère 6536
merisier 8330
merisier à grappes 2881
merisier d'Amérique 244
merisier d'Asie 382
merisier de Maack 5224
merisier des oiseaux 8330
merisier odorant 1474
méristème 5402
méristème apical 327
méristème de base 3992
merlin 453
merrain 4362
mérule 2560
mésange alpestre 9438
mésange boréale 9438
mésange des saules 9438
mescal 1939
mésohydrique 5408
mésophile 5407, 5408
mésophyte 5410
mesquite 1790
messmate-stringybark 5412
mesurage à ruban 8398
mesurage de surface par pieds carrés 3036
mesurage de volume 5364
mesurage géométrique de bois rond 3828
mesurage par compas forestier 1282
mesure à la ficelle 8165
mesure à la ficelle sans écorce 8166
mesure coutumière 2159
mesure de bois enstéré 7978
mesure de bois rond 7116
mesure de chargement 4522
mesure de hauteur 4232
mesure de la circonférence 3857
mesure de longueur 4914
mesure de masse des billes 5109
mesure des cernes annuels 5363
mesure des longueurs 5103
mesure des longueurs en mètres 5417
mesure de volume de billes 5103
mesure du bois 9607
mesure grossière 5748
mesure hydrostatique de bois rond 4411
mesure pleine 2160
mesure pour le cubage des bois 8615
mesures culturales additionnelles 63
mesures d'assainissement 7194
mesures sanitaires 7194
mesureur de billes 5111
mesureur de bois 1457
métaséquoia du Sichuan 2259
métaxylème 5415
mêt de grue 915
météorologie forestière 3504
méthode agrosylvicole 146
méthode d'aménagement par blocs 6161
méthode d'aménagement par blocs permanents 6170
méthode d'aménagement par volume 6165
méthode d'échantillonnage linéaire 4990
méthode de coupe principale 4151
méthode de cubage des billes 5110
méthode de cubage par morceaux 6226
méthode de débardage circulaire 1588
méthode de débusquage 9749
méthode de flottage par inspection à distance 2540
méthode de gestion d'abattage 2203
méthode de gestion forestière 3491
méthode de jardinage 3589
méthode de jardinage par arbre 7656
méthode de la coupe d'abri unique 5897
méthode de marcottage 4843
méthode d'emballe-sac 6027
méthode d'encoches et de jets de repérage 4083
méthode de production de bois fort 6888
méthode de régénération 5416
méthode de régénération d'abri par coupe progressive 7515
méthode de régénération par coupe à blanc 1617
méthode d'estimation par placettes d'échantillon fixes 3234
méthode d'estimation par

méthode de traitement d'une forêt favorisant de la lumière sur la frondaison 5918
méthode d'exploitabilité 4158
méthode d'exploitation par fûts entiers 8870
méthode de rétrocroisement 463
méthode ferroutière 6227
méthode piggy-back 6227
méthodes de cubage 3998
méthode taungya 146
mètre cube apparent de bois empilé 2115
mètre cube de bois 9524
mètre cube de bois empilé 2115
mètre cube solide 2114
mètre pliant 3300
mètres cubes sur pied 2116
métreur de bois 1457
mettre de bois en pile 7983
mettre en entrepôt 8115
mettre en pile 7983
meule charbonnier 1450
meule de charbon 1450
meule sylvestre 7102
meule tronçonneuse 3714
michélie 5428
micocoulier 1777
micocoulier africain 105
micocoulier de Caucase 1385
micocoulier de Chine 1499
micocoulier de désert 2413
micocoulier de Provence 2898
micocoulier de Virginie 1777
micocoulier d'Orient 4462
micocoulier du Mississippi 8253
micocoulier occidental 1777
micocoulier réticulé 5717
miconia 9191
microbiologie forestière 3505
microbiotope 5434
microclimat forestier 3506
micro-environnement 5429
microfibrille 3780
microhabitat 5429
micromètre d'épaisseur d'écorce 587
micromilieu 5429
micro-organisme 5430
micropropagation 5432
microrelief 5433
microscopie de bois 9608
microthermophytes 5435
miel tupélo 5871
migration 5442
migration d'espèces 5443
millepertuis 7154
millepertuis commun 7154
millier de pieds cubes 8530
mimosa 5460
mimosa à feuilles étroites 8316
mimosa à larges feuilles 3884
mimosa argenté 3962
mimosa blanchâtre 7604
mimosa blanchissant 7604
mimosa bleuté 5936
mimosacées 5459
mimosa couteau 4720
mimosa d'Amazonie 4921
mimosa de Farnèse 1689
mimosa des fleuristes 2927
mimosa des quatre saisons 8316
mimosa doré 3884
mimosa épineux 4687
mimosa eucalyptus 5936
mimosa quatre saisons 8316
mimosa résineux 8316
mimosa terminalis 8268
mimosa vert 791
minéralisateur 2308
minéralisation de bois 9610
mineuse de feuilles du pin 6261
mineuse de l'écorce 571
mineuse des cônes 1859
mineuse des feuilles 4880
mineuse du hêtre 655
mineuse du houx 4337
mineuse du mélèze 4791
mini-charge 4951
miombo 5475
mire 8292
mire graduée 4923
miro 6471
mise à feu concentrique 1409
mise en andain 9448
mise en copeau d'arbres entiers 3759
mise en copeaux 1531
mise en gris 3925
mise en jauge 4226
mise en œuvre de bois 3414
mise en valeur du bois 4434
mitchella 812
mixte 5486
mixte annuel de semences sauvages 306
mixte écorce-bois 589
mixture de flore sauvage pérenne 6152
mode d'éclaircie 8516
mode de régénération par coupe rase avec semenciers 7400
mode de régénération par coupes progressives en coin 9316
mode de régénération par coupes progressives uniformes 9087
mode de régénération par coupe totale 1617
modèle d'abattage 5051
modèle de chemin en crêtes et vallées 6938
modèle de débardage 7688
modèle de décision sylvicole 7615
modèle de forêt 3508
modèle de la croissance et du rendement 4011
modèle de peuplement 8021
modèle de tracé aléatoire 6715
modèle de tracé montante parallèle 6066
modèle d'exploitation 4160
modèle d'offre de bois 8649
modèle radial inverse 6925
moder 5506
module d'élasticité 5514
module de rupture 5515
module élastique 5514
moelle 1951
moelle divaguante 9264
moelle étoilée 8059
moellon 7128
moignon de branche 993
moignon pointu 7910
moirure 3106
moisi 3679
moisissure noire 802
moissonneuse 8853
mondré 1266
moniliose 1084
monoculture 5542
monoculture arborescente 8887
monocyclique 9092
mono-strate 7640
montagnes boisées 9710
montant d'une échelle 4756
montée 5979
montée de la sève 7195
monter d'arbre 1641
montouchi 2608
mor 1681
mora 4715
morabukea 3027
moracées 5599
morceau 1571
morcellement de forêts 3681
mordançage pigmenté de bois 9662

morelle noire 804
moringa ailée 4379
moringacées 5556
mortaisage 6393
mortaise 5561
mortalité annuelle moyenne 427
mortalité forestière 3509
mort-bois 9322
mort des forêts 2694
moteur primaire 6553
motte 7811
motte à semis 7868
motte compactée 6346
motte de gazon 7811
motte racinaire 7054
mouche à scie de l'épicéa 7958
mouche à scie du mélèze 4792
mouches à scie 7240
mouches des galles 3793
moucheture 765
moucheture noire 315
mouchetures 766
mouillefer 4339
moulurage 9613
moulure 9613
moulure de bois 9613
moulurière 5577
mouracatiara 5595
moureiller 2570
mousse 5563
mousse de chêne 5845
moustache chinoise 1502
moustache de Chine 1502
mouton 6707
movingui 5732
moyen de débardage 8598
moyen d'imprégnation 4428
moyenne annuelle d'extraction en éclaircie 429
moyennes de préservation 3774
moyens de préservation du bois 8623
mueri 112
muhuhu 7609
mukumari 2711
mulch 5601
mulch de bois 5603
mulga 5301
mull 5604
mulot 3510
multiplication par marcottes 6582
multiplication sous brouillard artificiel 5481
multiplication végétative 9182
mungurra 3550
muninga 290
mûrier 8354
mûrier à feuilles de platane 5600
mûrier à papier 6060
mûrier blanc 1791
mûrier des haies 778
mûrier kagayama 5600
mûrier noir 803
mûrier pleureur 9327
mûrier rouge 6781
murraya 4644
musacées 512
musaraigne 8923
muscadier 5821
muscadier de Californie 1276
muscadine 5637
mushibi 113
muskeg 5638
mutation 5640
muténye 5641
mutualisme 5643
muura 2710
muzaisi 5644
mycélien 3773
mycélium 5645
mycologie forestière 3511
mycorhize ectotrophe 2753
mycorhize endotrophe 2813
mycorrhize 3775
mycose 3769
myricacées 9305
myricaire 3003
myrique baumier 8332
myristicacées 5820
myrobalan 1475
myrsinacées 5647
myrsine de Florida 5648
myrtacées 5649
myrte 1792
myrte blanc 9375
myrte commun 1792
myrte d'Amérique du Nord 1264
myrte de marais 8332
myrte de sables 965
myrte d'Oregon 1264
myrte du Brabant 8332
myrtille 5006
myrtille à feuilles entières 893
myrtille à feuilles persistantes 1996
myrtille à l'airelle 5006
myrtille des marais 2889
myrtille moulue 7523

naga 5877
naissance d'un incendie de forêt 5970
nappe 3995
nappe souterraine 3995
nargusta 1712
nashi 1504
nature de la forêt 3512
ncrouer 4116
nécrose 5702
nécrose de l'écorce 573
neem 6110
néflier 5389
néflier commun 5389
néflier de Cathay 1498
néflier du Japon 5172
négundo 381
nerprun 1137
nerprun à feuilles de houx 4338
nerprun bourdaine 684
nerprun d'Amérique 1544
nerprun de Californie 1257
nerprun de Caroline 1344
nerprun de Pursh 1544
nerprun des Alpes 211
nerprun des roches 6987
nerprun laineux 9716
nerprun noir 684
nerte 1792
nervation 9192
nervosité 8735
nervure 4882
nervure médiane de feuille 4879
net de branches 983
net-et-meilleur 1612
netteté de nœuds 989
nettoiement après exploitationn 1624
nettoiement de coupe 1624
nettoiement de forêt 3351
nettoiement et enlèvement de rémanents 8324
nettoiement par dispersion de rémanents 1623
nettoyage final de menus assortiments 7168
neutrophile 5721
nez 7795
nez du guide 592
niche de ponte 2773
niche écologique 2739
nichoir 763
nicotiane glauque 8945
nika 3232
nim 6110
niveau d'Abney 9
niveau de matériel sur pied 9250
niveau de peuplement 8019
niveau du sol 3990

niveleuse-surfaceuse 4772
nodosité 5742
nodulation 5745
nodulation spécifique de tiges 7906
nodule racinaire 7036
noduleux 4737
nodus 4723
nœud 4723, 5741
sans ~s 983
nœud "bouchon" 800
nœud à découvert 9050
nœud adhérent 833
nœud bouchon 4376
nœud d'arête 4733
nœud de face 2981
nœud de rive 4733
nœud encastré 581
nœud entouré d'écorce 833
nœud fixe 8550
nœud fourchu 979
nœud groupé 979
nœud isolé 7290
nœud médulleux 6306
nœud mort 800
nœud moyen 5386
nœud noir 800
nœud non adhérent 5158
nœud occulte 833
nœud ouvert 9050
nœud oval 5985
nœud partiellement adhérent 6084
nœud perforé 4331
nœud plat 7910
nœud pourri 2290
nœud recouvert 833
nœud rond 7108
nœud sain 7852
nœud sautant 5158
nœuds en grappe 1668
nœuds groupés 3997
nœud standard 8004
nœud tête d'épingle 7764
nœud tranché dans le sens de la longueur 7910
nœud tranché dans le sens de la longueur au moment de sciage 7910
nœud transitoire 8790
nœud transversal 8790
nœud unique 7290
nœud vert 5013
nœud vicieux recouvert 8352
noisetier 1778
noisetier à long bec 637
noisetier commun 1778
noisetier de Lambert 4759
noisetier de sorcière 1811
noisetier méditerranéen 5383
noisette 4183
noisette de Constantinople 5383
noisette du Levant 5383
noisette indienne 6188
noix 5817
noix de bétel 691
noix de cola 1718
noix de galle 5831
noix de palmiste 6045
noix vomique 8191
nomadisme cultural 5746
nombre d'espèces 5808
nombre de tiges 8079
à grand ~ 5628
nombre moyen d'arbres par acre 438
nombre moyen d'arbres par charge 435
nom d'arbre 8889
nomenclature forestière 3513
non boisé 8874
non couvert de forêt 8874
non ébranché 9101
non-mûr 4674
nonne 5809
non scié 9113
non scléreux 5770
norme minimum 5469
norme pour le bois 8642
note d'information 6587
nothofagus d'Australie 406
noueux 4737
nourriture du gibier 8806
nourriture herbacée 8806
nouvelle foresterie 5723
noyau de bardage 7873
noyer 1808, 9262
noyer à beurre 1222
noyer à feuilles 9456
noyeraie de montagne 9261
noyer ailé 9456
noyer amer 774
noyer aux ailes 9456
noyer blanc d'Amérique 7482
noyer cendré 1222
noyer commun 1808
noyer d'Afrique 1087
noyer d'Amérique 241, 7482
noyer de Californie 241
noyer de France 1808
noyer de l'Arizona 362
noyer de Mandchourie 5272
noyer de Para 1003
noyer du Gabon 1087
noyer du Japon 7586
noyer du Queensland 1206
noyer du Texas 5009
noyer gris 1222
noyer gris d'Amérique 1222
noyer noir 2715
noyer noir d'Amérique 241
noyer noir de Californie 241
noyer rose d'Australie 409
noyer royal 1808
noyer satiné 259
nucule 5819
nuire à 4148
nuisible 4578
nuisible du bois 9533
numéro de la placette 6380
numéro de placette de mesure de l'inventaire forestier national 6381
nutrition minérale des arbres 8890
nyctaginacées 3676
nyssacées 9010

objectifs de développement 2447
objectifs de qualité visuels 9238
observateur-coordonnateur 3191
obtenteur 1026
occlusion 5862
ochnacées 5164
œil à bois 9179
œil de paon 6125
œil de perdrix 830
œil d'hache 451
œilleton 5870
officier divisionnaire 2551
offre de bois 2359
oïdium de chêne 5844
oie des moissons 3332
okan 2397
okoué 1293
okoumé 3788
okumé 3788
oldfieldie d'Afrique 109
oléacées 5891
oléarie 2229
oléastre 9421
oléorésine de pin 6263
oligo-éléments 8740
olivaie 5892
olivaison 5893
olivarède 5892
oliveraie 5892
olivette 5892
olivier 2904
olivier de Bohême 9421

olivier sauvage 9421
olon 5895
olon tendre 5895
ombellifères 1355
ombrage 7481, 8150
ombrage latéral 7581
ombre de la forêt 3965
ombre pommelé 2248
omon 8333
omu 4744
onagracées 2924
ondulation 2019
ondulation de fils 3101
opale de bois 5903
opale xyloïde 5903
opérabilité de bois 8617
opérateur de débardeur 4168
opérateur de débroussailleuse 8307
opérateur de machine déligneuse 2764
opérateur d'entaille 4084
opérateur de scie à chaîne 1434
opérateur de tambour 2642
opération de déchiquetage et débardage 1532
opération forestière 9724
opérations culturales 2131
opérations d'abattage et débardage 2164
orange amère 775
oranger 5935
oranger amer 775
oranger des osages 4219
oranger douce 5935
oranger jasmin 4644
orchidée du pauvre 6642
ordonnance forestière 3516
ordre de brûlage 1195
ordre de coupes 2199
orée du bois 2761
organisation forestière 3574
organisme nuisible 5807
orientation de coupe 2519
orientation de fils 2516
orientation du tracé 2517
orienter la chute d'un arbre 4059
orifice d'alimentation 4172
orifice de ponctuation 6298
origine 6603
origine de peuplement 8025
origine des graines 7395
ormaie 2789
orme 2786
orme à feuilles de charme 2792
orme à feuilles larges 7310
orme à liège 1958
orme américain 247
orme aquatique 6320
orme à trois feuilles 6035
ormeau 2786
orme bâtard 1777
orme belgique 2677
orme blanc 7310
orme blanc américain 247
orme champêtre 2893
orme commun 2893
orme d'Amérique 247
orme d'Angleterre 2893
orme de Californie 1261
orme de Camperdown 9325
orme de Chine 1496
orme de Hollande 2677
orme de Mandchourie 5270
orme de montagne 7310
orme de Samarie 6035
orme de Sibérie 2792
orme des roches 6988
orme de Thomas 6988
orme diffus 7137
orme du Caucase 2792
orme hollandais 2677
orme liège 6988
orme lisse 7137
orme nain 4618
orme nain de Sibérie 4618
orme parasol 9039
orme pédonculé 7137
orme rouge 7745
ormosia 5956
orne 3292
orophyte 5959
ortet 5960
orthotropie 5963
ortie puante 4222
oseille 9656
oseraie 9436
osier 617, 9439
osier amandier 6122
osier à trois étamines 6122
osier blanc 1809
osier commun 1809
osier des vanniers 619
osier franc 6643
osier jaune 3885
osier noir 813
osier pâle 6122
osier rouge 6643
osier vert 619
osier vitellin 3885
ostrie à feuille de charme 2900
ostrya 2900
outil d'écorçage 574
outillage de bûcheron 4157
outillages forestiers 3623
outil pour le travail du bois 9652
outil Pulaski 6624
outils forestiers 3623
ouverture de voie d'accès 5913
ouverture du couvert 4566
ouverture en jalousie 5136
ouverture intérieure 4504
ouverture intérieure des fibres de bois 4504
ouvrage à claire-voie 2018
ouvrage en bois ornemental 5958
ouvrages en bois 8626
ouvrages non ligneux 5778
ouvrier-câbleur 4296
ouvrier de chantier de bois 8094
ouvrier des bois 9699
ouvrier écorceur 556
ouvrier forestier 9654
ouvrier forestier saisonnier 6095
ouvrier grimpeur 4296
ouvrier sculpteur sur bois 1363
ouvrir un chemin à travers forêt 8302
ovalisation 5984
ovengkol 5987
ovoe 2245
oxalis 9656
oxycèdre 6530
ozigo 6016

PAC 192
pacage en foret 3452
pacanier 6130
pachybranché 6018
pachyramifié 6018
padauk d'Afrique 111
padauk d'Asie 383
padauk de Burma 1183
padauk des Indes 5284
pagode japonaise 4626
paillage 5602
paillassonnage en branches 1110
paille en bois 9695
paillette 3241
paillis 5601
paillis d'écorce 572
paillis pour recouvrir les cultures 1995
pain-de-surelle 9656
pain-de-surelle/pessière 9657
paisson 5326
paisson dispersée 4939
paisson dure 4127

palas 3248
paldao 6033
palée d'ancrage 275
palette à résine 4584
palétuvier 5276
palétuvier aveuglant 5447
palétuvier d'Amérique 251
palétuvier noir 251
palétuvier rouge 6780
palier de la voie 8744
palissandre d'Asie 2728
palissandre de Rio 1006
palissandre de Siam 7563
palissandre des Indes 2728
palissandre du Honduras 4348
palissandre du Sénégal 102
palissandre noir 1006
palissandre noir de Brésil 1006
palliatifs de poussière 2676
palme
de ~ 6036
palme d'ivoire 4597
palmeraie 6043
palmette 3014
palmetto 1238
palmetto nain 2688
palmier 6046
palmier à huile américain 253
palmier à l'huile 110
palmier dattier 2257
palmier de Noël 1568
palmier des Canaries 1301
palmier du Brézil 1009
palmier du Mexique 9271
palmier fusain 7911
palmier nain 5384
palmier rônier 6047
palmier ventru 2109
palmier washingtonia 1258
palmiste marron 7911
palonnier 4480
palout 1478
palplanche 3147
palustrine 6048
panage 6055
panama 7805
pandanacées 7316
pandanum 1802
panier accumulateur 8801
panier collecteur 8621
panneau 6049
panneau aggloméré 1542
panneau à parement en flocons 3242
panneau contreplaqué à plis multiples 5608
2976
panneau de copeaux de haute densité 4277
panneau de fibres 9551
panneau de fibres à plis multiples 5609
panneau de fibres de bois 9551
panneau de fibres de bois dur 1845
panneau de fibres de bois poreux 6466
panneau de fibres de densité moyenne 4545
panneau de fibres dur 1845
panneau de fibres extra-dur 8445
panneau de fibres isolant 4521
panneau de fibres léger 5183
panneau de fibres multicouche 5609
panneau de fibres multiplis 5609
panneau de fibres non comprimé 5754
panneau de fibres poreux 6466
panneau de flocons 3242
panneau de grandes particules 9255
panneau de particules 6091
panneau de particules à densité progressive 3895
panneau de particules de bois à faible densité 5184
panneau de particules destiné au placage 1953
panneau de particules homogènes 7639
panneau de particules laminé 6375
panneau de particules orientées 5954
panneau de particules plaqué 9195
panneau de placage 1953
panneau de tranchage 7738
panneau dur trempé 8445
panneau en agglomérés plaqué 9195
panneau en fibres à faible densité 5183
panneau gaufré 9255
panneau isolant 4521
panneau multicouche 5608
panneau stratifié 4761
panneau structural 9255
panneautage par bois 9616
panneau tendre 4521
panneaux 6050
papier adhésif 8390
papier de pâte mécanique 9634
papillon frelon 4375
papita 7180
papra 4465
paquetage 1171
paquetage contre-incendie 3194
paquetage de bois dans l'eau 9303
paquetage de bois rond 3662
paquetage de contreplaqués 6399
paquet de bois 9496
paquet de petit bois 7780
paquet de troncs 8066
paqueter 1169
paquets de dosses sans écorce 554
parachutiste anti-incendie 7783
parachutistes héliportés 4240
paradis 6062
parasite 6071
parasites forestiers 3520
parasol 6074
parasol chinois 1303
parasolier 9042
parasolier japonais 1303
parastémon 9038
parc à bois 8664
parc à bois débités 8664
parc à bois ronds 7119
parc de grumes 7119
parcelle à milliacre 5449
parcelle boisée 8604
parcelle de coupe 5038
parcelle de la forêt 1821
parcelle en régénération 355
parcelle forestière 1821
parcelle forestière de village 9220
parcelle régénérée 8105
parcellisation de coupes 191
parche 4391
parche sèche 4391
parcinsonia 867
parcinsonia à petites feuilles 3312
parc national forestier 5667
parcouri jaune 6030
parcours d'évacuation 2864
pare-feu 3186
pare-feu boisé 2928
pare-feu végétal 6174
parement 627
parement profilé 6566
parenchyme abaxial 9091
parenchyme aliforme 167
parenchyme anastomosé 1869
parenchyme apotrachéal 328
parenchyme axial 457
parenchyme cicatriciel 8784

parenchyme concentrique 5414
parenchyme confluent 1869
parenchyme de blessure 8784
parenchyme de bois 9617
parenchyme de rayons ligneux 6679
parenchyme disjoint 2531
parenchyme dispersé 2492
parenchyme en couches 5414
parenchyme en échelle 7271
parenchyme en manchon 9174
parenchyme en réseau 6920
parenchyme juxtavasculaire 9174
parenchyme longitudinal 457
parenchyme métatrachéal 5414
parenchyme paratrachéal 6075
parenchyme radial 6679
parenchyme terminal 8475
parenchyme traumatique 8784
parinare 6079
parkia 5737
parkie 5737
paroi 1407
paroi cellulaire 1407
paroi primaire 6537
parois de rayon médullaire 6737
paroi secondaire 7343
paroi terminale 2817
paroi terminale noduleuse 5743
parquet-mosaïque 9590
parrotie de Perse 6186
partage de coupe annuelle permissible 1
partage des frais 1978
part de bois de sciage 7249
parterre 5038
parterre de coupe 7668
partes forestières de zones vertes 3575
parthénocisse à cinq folioles 9228
participation aux coûts 1978
participation aux frais 1978
particules de bois gluées 3877
particules de moelle 1952
partie de bois 5181
partie de matériel végétatif 9180
partie déversante d'un barrage 7004
partiellement boisé 4586
parties d'un arbre 6094
partie terminale 9102
partie terminale d'un glissoir 9102
parts de bois débité 2504
pasanie 8394
pas de chaîne 7263
passage 182, 6097
passage de la chaîne 1428
passage des grumes 5104
passage secondaire 7364
passe 6097
passe à bois 5116
passe à bois flottants 3274
passe-billes 5116
passe profonde 4211
passifloracées 6099
passur 6646
pastille 6103
pâte de bois 9606
pâte de bois brute 1680
pâte de bois fine 3152
pâte de bois mécanique 9603
pâte de fibres de bois 9603
pâte de rémanents 8053
pathogène 6112
pathologie forestière 3521
patins de glissade 3862
patrimoine forestier 3461
patrouille 6114
patte 607
pâturage arboré 9711
pâturage boisé 9711
pâturage en forêt 3452
pâturage rationné 1657
pâturer 1094
pau-brasil 1010
pau d'Arco 6644
pau ferro 1002
paulownia impérial 7122
pauvre en forêt 6450
pauvreté en espèces 7281
pavage de bois 9554
pavage désertique 2417
pavé de bois 9491
pavier blanc 940
pavier glabre 5874
pavier rouge 6767
paysage 4777
paysage canopée 1311
paysage forestier 3481
paysage forestier ouvert 9310
pays producteur de bois 8624
payung 7563
pêcher 6123
pêcheraie 6124
pédoncule 4883
pédonculé 6193
peigne 2193
peinture cicatrisante de bois 8959
pekeya 6188
pelle 4408
pelle à feu 3209
pelle de débardage 7687
pelle-rétro 468
pelle rétrocaveuse 468
pelleteuse hydraulique 4408
pelote 6139
pelotonneuse de gaules 8798
pendant 9324
pénétrabilité des racines 7038
pénétration 6144
pénétration des racines 7029
penjing 908
pente 7750
pente favorable 3028
pente pierreuse 8378
pente sensitive 7441
péoniacées 6146
pépinière de broussailles 1205
pépinière de peupliers 6455
pépinière de plantes en récipients 1894
pépinière de production 6370
pépinière de repiquage en ligne 4994
pépinière en récipients 1894
pépinière forestière 342
pépinière forestière de village 9219
pépinière forestière secondaire 7347
pépinière volante 3117
pépiniériste arboricole 344
perces-bois 8562
perceur de cône de pin pondéreux 5043
perceur de l'écorce de pin pondéreux 6446
perceur du chêne 5848
perceur du vieux bois 4389
perceuse des pousses 2149
perceuse pour bouchonnage 4728
perceuses de bois 8562
perceuses de bois secondaires 7340
perche 6418
perche de construction 1152
perche résineuse 1874
perches d'osier 9015
perchis 6419, 6428
perforation 6154
perforation de vaisseau 9213
perforation en grille 7272
perforation foraminée 2846
perforation scalariforme 7272
périderme 6157
périmètre de l'incendie 3195
périmètre du foyer 3195
période d'abattage 3072

période d'afforestation 99
période d'écorçage en printemps 7941
période de débardage 4170
période de feu 1192
période de gemmage 6892
période de préparation 6514
période de régénération 6820
période de reproduction 3554
période de révolution 7077
période de traitement 8795
période de transport 6863
période d'excision 15
période d'implantation 2868
période d'incendies de forêts 3434
période d'interdiction de faire du feu 1652
période favorable à l'incendie 1192
période latente 4818
période sans forêt 8875
périodicité des coupes 2194
périodicité des éclaircies 8510
périodicité des incendies 3185
périodicité de soins culturaux 8460
périodicité moyenne des feux 5356
périsperme 4398
permis de chemins 6979
permis de coupe 6183
permis de feu 1193
permis d'enlèvement de bois 6180
permis de vidange 6180
pernambouc 1010
péroba rose 6184
perré 6969
persimmon 1796
persistant 6109
personnel de route 3119
personnel de terrain 3119
personnel roulant 3119
perte au flottage 5177
perte au sciage 5178
perte d'interception 4534
perte de lignocellulose 5180
perte en matières nutritives 5179
pertes à l'écorçage 580
pertes au débardage 5174
pertes au sciage 4696
pertes d'abattage 2197
pertes de bois 9598
pertes d'exploitation 5175
pertes d'exploitation de bois 8217
pertes non récupérées 9112
pertes par éclatement 1072
perturbation de sous-bois 9061
perturbation du sol 7836
pessière 7953
pessière à airelle rouge 1999
pessière à graminées 3920
pessière à laîches et graminées 7368
pessière marécageuse à couverture herbacée 8313
pessière mixte 5493
pesticide 6189
pétale 6192
à petite ~ 5431
pétiole 4883
pétiolé 6193
petit arbre 7778
petit arbrisseau 7558
petit bois 7760
petit bois rond 7774
petit bout 7761
petit bout de tige 8689
petit charançon du pin 7770
petit dendroctone du pin mexicain 7762
petit diamètre 2470
plus ~ 5465
plus ~ utilisable 2743
petite baie 6415
petite forêt 5001
petite longueur 7547
petite mineuse du bouleau 754
petite perche 7775
petite perche de conifères 1876
petite plantule 7777
petit foyer périphérique 4669
petit houx 1212
petit jardinage 7360
petit marsault 2697
petit nœud 6281
très ~ 7764
très ~ intermédiaire 830
petit nœud sain 5007
petit palmier 2688
petit peuplement 7776
petit scolyte européen de l'orme 2894
petits nœuds apparents de bois madré 1885
petits rémanents 4950
petit traîneau de débardage 2008
peuplé d'essences forestières 3394
peuplement 7470
peuplement abandonné 2406
peuplement accessoire 7346
peuplement à deux âges 9025
peuplement à deux étages 9035 8000
peuplement à un seul étage 7653
peuplement auxiliaire 420
peuplement avant éclaircie 6524
peuplement à volume élevé 4303
peuplement bienvenant 8536
peuplement clair 4446
à ~ 6450
peuplement clairiéré 4446
peuplement complet 3756
peuplement complètement boisé 3756
peuplement d'abri 1993
peuplement d'âges multiples 9081
peuplement d'arbres semenciers 8022
peuplement de base 5250
peuplement de bois de sciage 7265
peuplement de création nouvelle 5722
peuplement de densité moyenne 5388
peuplement de densité relative excessive 6006
peuplement de fourré 1925
peuplement de gibier 3796
peuplement d'élite 2785
peuplement dense 3756
peuplement de récolte de graines 7377
peuplement de semis et de gaules 7390
peuplement d'essai 2946
peuplement de type parc 6081
peuplement discontinu 4446
peuplement en bois de valeur 8024
peuplement en croissance libre 3696
peuplement étagé 4841
peuplement excessivement reboisé 6006
peuplement flaché 8644
peuplement forestier 3605
peuplement forestier de remplissage 420
peuplement griffé 8644
peuplement homogène 2926
peuplement idéal 4420
peuplement inégal 9081
peuplement inéquien 9081
peuplement insuffisamment boisé 7293
peuplement intermédiaire 4547
peuplement irrégulier 9081

peuplement martelé 8644
peuplement mélangé 5486
peuplement mixte 5486
peuplement moins 5472
peuplement moyen de coléoptères 430
peuplement mûr 5338
peuplement naturel 5698
peuplement non marchand 5765
peuplement non mixte 2926
peuplement originel 282
peuplement ouvert au pâturage 6101
peuplement pauvre 4446
peuplement pluriétagé 5634
peuplement plus 6394
peuplement porte-graines 6394
peuplement préexistant 6527
peuplement prêt pour la récolte 7378
peuplement principal 5250
peuplement producteur de graines 6394
peuplement producteur de graines élites 2784
peuplement producteur de graines normal 5784
peuplement pur 2926
peuplement rabougri 3515
peuplement régulier 2926
peuplement résineux mixte 5485
peuplement restant 6889
peuplement restant sur pied 6889
peuplement secondaire 8237
peuplement serré 3756
peuplement servant à la récolte de graines 6078
peuplements forestiers rentables 2742
peuplement stagnant 7991
peuplement suranné 5998
peuplement témoin 4471
peuplement trop dense 6006
peuplement uniforme 2926
peupleraie 6454
peuplier à grands dents 706
peuplier angustifolié 5661
peuplier argenté 2911
peuplier balsamifère 9338
peuplier balsamique 9338
peuplier baumier 9338
peuplier baumier californien 784
peuplier baumier du Canada 9338
peuplier blanc 2911
peuplier blanc pyramidal 2911
peuplier charab 4468
peuplier de la Caroline 2717
peuplier deltoïde 2717
peuplier de Maximowicz 4629
peuplier de Picardie 3973
peuplier de Sargent 6319
peuplier de Simon 7627
peuplier des marais 8306
peuplier de Virginie 2717
peuplier d'Hollande 2911
peuplier d'Italie 5126
peuplier d'Oregon 5661
peuplier faux-tremble 236
peuplier faux-tremble américain 236
peuplier grisaille 3973
peuplier grisard 3973
peuplier lancéolé 4766
peuplier noir 2884
peuplier noir d'Amérique 2717
peuplier noir hybride 4400
peuplier robusta 4400
peuplier tremble 2880
phalène du bouleau 6147
phalène du pin 6262
phanérophytes 6197
phase adulte 70
phase de consolidation 1884
phase de développement 6198
phase de succession 8248
phase juvénile 4679
phellodendre de l'Amour 264
phellodendre-liège de Lavalle 4837
phellodendron 6199
phellodendron au liège de l'Amour 6985
phellodendron de Sachalin 7145
phelloderme 6200
phellogène 6201
phénologie 6203
phénotype 6204
philadelphacées 5502
philodendron de Japon 4617
phléomètre 587
phloème interne 4502
phloème périmédullaire 4502
phloème primaire 6543
phloème secondaire 4502
photodégradation du bois 6209
photogrammétrie forestière 3526
photosynthèse 6211
phototropisme 6212
phylloclade 1405
phyllode 1594
phyllophage 2338
phyllophage forestier 3369
physalis 1323
physique de bois 8618
physocarpe 5259
physocarpe à feuilles d'obier 397
physocarpe monogyne 5590
physocarpe nain 397
phytocénose 6336
phytocénose de forêt résineuse 5705
phytocide 6220
phytohormone 6221
phytolaccacées 6416
phytopharmaceutique 6222
phytotoxicité 6223
pic d'abattage 8935
picrasme 6663
pics 9622
pièce de bois 2171
pièce de charpente lamellée-collée 3878
pièce o travaillante 6382
pièces de bois tronçonnées 2174
pied
sur ~ 8018
pied à coulisse 8809
pied bloc 840
pied-carré 7970
pied carré 7970
pied-cube 2113
pied-cube apparent 7977
pied-cube de bois empilé 7977
pied de biche 2329
pied de planche 876
pied jonque 4673
pied-mère 5569
pied-planche 876
pied-planche mince 7970
pied vieux 5882
piégeage de carbone 7452
piège à semences 7398
pièges à la phéromone pour insectes 6206
pieu 6475
pieu de chargement 5025
pieu de clôture 3084
pieux 8588
pigeon malherbe 2713
pigeon ramier 9624
pigment de bois 9625
pignon 4595
pignon étoile 7943
pile croisée 6232
pile de bois 9519
pile de bois de rebut 4774
pile de bois de sciage 5211
pile de bois longs 5145

pile pleine 8552
pilon 6237
piment couronné 632
piment de Cayenne 1207
piment royal 8332
pin 6272
pin à amandes 9031
pin à blanche écorce 9362
pin à bois lourd 6445
pinacées 6254
pin à cône en tubercule 4722
pin à cône épineux des montagnes Rocheuses 6997
pin à courtes aiguilles 7538
pin à crochets 5598
pin à écorce de sapin 936
pin à écorce en peau de serpent 6377
pin à feuilles simples 7642
pin à feuilles simples de Californie 1272
pin albicaule 9362
pin à l'encens 5029
pin à ombrelles 4595
pin argenté américain 9344
pin arolle 8346
pinastre 5309
pin à sucre 1274
pin à suiffes 5598
pin à trois feuilles 5787
pin à trois feuilles du Langbian 5787
pin à une aiguille 7642
pin australien 407
pin avec écorce écailleuse 7277
pin baliveau 2726
pin blanc 2726
pin blanc américain 7956
pin blanc de Chine 1508
pin blanc de l'Ouest 9344
pin blanc des Montagnes Rocheuses 4965
pin blanc du Japon 4636
pin blanc du Mexique 7867
pin blanchâtre 936
pinçage 6240
pin calabrais 1250
pince 3892
pince à câble 7699
pince à grumes 5094
pince de crochet double 2581
pince de débardage 3892
pince de débardage à poulie entaillée 7504
pince de débardage à simple voie 7643
pince de débardage hydraulique 4407
pince de débardage uniligne 7643
pince de flottage 6690
pince d'élévateur à grumes 4934
pince de manutention 5094
pince de planche 875
pince d'extrémité 2806
pince lève-bois 9395
pin cembro 8346
pin cembrot 8346
pincement 6240
pince portante 1596
pin chir 1543
pin comestible 9031
pin commun 1797
pin côtier 5309
pin d'Alep 2220
pin d'Amérique centrale 2111
pin d'Arizona 9031
pin d'Armand 1508
pin de Balkans 489
pin de Banks 4601
pin de basse altitude 5194
pin de Bishop 770
pin de Bosnie 936
pin de Bosnie à cône rouge 936
pin de Boston 5141
pin de Briançon 5598
pin de Brutie 1250
pin de Calabre 1972
pin de Californie 770
pin de Chypre 2220
pin de Corée 4745
pin d'écosse 1797
pin de Coulter 1984
pin de David 1508
pin de Durango 2673
pin de Gérard 1486
pin de Hudson 4601
pin de Jeffrey 4648
pin de Jersey 9229
pin de Kaori 6664
pin de Lambert 1274
pin de Laponie 4785
pin de l'Arizona 360
pin de l'Arizona pondéreux 360
pin de l'Eldorado 4722
pin de l'Himalaya 4308
pin de Macédoine 489
pin de Mandchourie 4743
pin de marais 5141
pin de Merkus 5404
pin de montagne 5598
pin de Monterey 5548
pin de Montezuma 5549
pin de Parana 1005
pin de pierre 4595
pin de Pyrénées 1250
pin de Riga 1797
pin de romarin 5029
pin de Sabine 7142
pin des Antilles 2111
pin des Balkans 936
pin des landes 5309
pin de Torrey 8717
pin de Virginie 9229
pin de Washoe 9272
pin doux 7538
pin du lord Weymouth 2726
pin du Nicaragua 5864
pin du Nord 1797
pin dur 5792
pin du sud 5141
pin du Texas 5029
pinède 6255
pinée 6255
pin épineux 770
pineraie 6255
pineraie à airelle rouge 1998
pineraie à graminées 3919
pineraie à lichens 4930
pineraie à myrtille 708
pin flexible 4965
pin géant 1274
pin gigantesque 1274
pin glabre 7956
pin gris 4601
pin gris d'Amérique 4601
pinière 6255
pin jaune 6445
pin jaune du Mexique 1485
pin Klinki 4718
pin laricio de Corse 1972
pin lodgepole 5044
pin maritime 5309
pin mugo 5598
pin nain de Sibérie 4634
pin Napoléon 4750
pin naturel 5687
pin noir 415
pin noir d'Autriche 415
pin noir de Japon 4614
pin noueux 4739
pin palustre 5141
pin parasol 4595
pin pignon 4595
pin pinier 4595
pin pinyon 5424
pin pleureur de l'Himalaya 4308
pin pliable de l'Ouest 4965
pin pondéreux 6445

pin pungens 8361
pin queue de renard 6997
pin raide 5792
pin rigide 5792
pin rouge d'Amérique 5796
pin rouge du Canada 5796
pin rouge du Japon 4632
pin Sabine 7142
pin serotina 6447
pin souple 4965
pin sylvestre 1797
pin sylvestre attaqué par champignon 6635
pin taeda 5029
pin tordu 7191
pin tordu latifolié 7191
pin Weymouth 2726
pinyon de Parry 6082
pioche forestière 2035
piocher 2169
pipéracées 6148
pipit des arbres 1093
pippal 939
piptadénie 6293
piquage 544, 765, 1530
piquant 8524
piqué 3679
piqué des vers 9730
piqué par l'humidité 3679
piquet 3899
piquetage 7995
piquet de clôture 3084
piquet en bois 6475
piqûre 4511
piqûre 765
piscidia 3284
pissodes 9331
pistachier de Chine 6295
pistachier lentisque 5327
pistachier sauvage 2885
pistachier térébinthe 8469
piste de débardage 7689
piste de soins culturaux 8459
piste forestière 3565
piste retraitante 6111
pistil 6296
pistolet à refendre 6487
pistolet de martelage 8884
pistolet marqueur 8884
pistolet vaporisateur 8884
pitchpin 5792
pitchpin américain 5141
pitoune 6627
pittosporacées 8672
pittospore 6315
pittospore de Nouvelle Zélande
pivot 8411
pivotante 8411
pivot secondaire 6137
placage 9196
placage à plis parallèles 6068
placage de parement 2986
placage déroulé 6135
placage de souche 8219
placage de surface 2986
placage extérieur 2986
placage figuré 3124
placage lamellé 4762
placage pour plis intérieurs 1954
placage ronceux 2060
placage scié 7261
place à charbon de bois 4712
placeau de plantation 7396
placeau permanent 6176
place d'atterrissage de bois flottés 3280
place d'eau filtrante 7401
place de coupe à la souche 8213
place défectueuse 2991
place de rechargement 8763
place de résine 6904
place de suintement 7401
placette à rayon variable 9168
placette d'échantillonnage d'accroissement 8859
placette de sondage d'accroissement 8859
plage de fonctionnement 5925
plagioclimax 2526
plagiotropie 6317
plaie 6616
plaine aire forestière 3763
plaine boisée 3397
planche 873, 5812
planche à cadranure 886
planche à encoches 6345
planche à repiquer 8770
planche avivée brute 7088
planche brute 7084
planche débitée non parallèle 3029
planche de cœur 5437
planche de conifère 7828
planche de feuillus 4133
planche de grume 8062
planche de rebut 2119
planche de repiquage 8770
planche de semis 7373
planche du cœur 4194
planche en pâte de bois 9604
planche équarrie 7968
planche flacheux 7088
planche mince 8520
planche moyenne 5437
planche non sciée 7084
plancher 3283
planche rabotée 8280
planches 882
planches à rainure et languette 8676
planches de caisse 954
planches de menuiserie 4657
planche semoir 7385
planches et solives constructives 8185
planche standard 8103
planchette 7732
planchette de joint 2595
planchon 8056
plançon 9263
plan d'aménagement forestier 3492
plan d'attaque 6331
plan de boisement 3529
plan de clivage 1630
plan de développement des forêts 3375
plan de développement forestier 3372
plan de gestion de tenure 8468
plan de gestion forestière 3492
plan de grumes 8545
plan de peuplement 8109
plan d'étiquetage de bois 9593
plan de transportation des ressources 6910
plan d'exploitation de la forêt 3416
plan d'utilisation des ressources locales 5036
plan d'utilisation de terrain forestier 3482
plan d'utilisation du parcours 6722
plan en blocs aléatoires 6714
plan en blocs complètement randomisés 6714
planère du Japon 4639
planification 3901
planification de chantier totale 8720
planification de gestion forestière 3493
planification stratégique de territoire 8146
planimètre 6324
plan régional 6829
plantage des arbres forestiers

plantage sous abri 6364
plantation à la tarière 402
plantation à racines nues 532
plantation arboricole 8804
plantation au bâton 2485
plantation avec bêche 289
plantation avec fentes en L 5021
plantation d'agrément 233
plantation d'arbres hors forêt 5759
plantation de bordure 921
plantation de forêt 6595
plantation de gaules 7203
plantation de haute densité 4278
plantation d'enrichissement 2830
plantation de peuplements industriels 8620
plantation de pins 6264
plantation de plantes en caissettes 963
plantation de plantes en récipients 1895
plantation des arbres forestiers 3626
plantation des arbres forestiers linéaire 4989
plantation de saules 9431
plantation de semis en tube 9002
plantation d'osier 9431
plantation en bandes 6356
plantation en bandes de niveau 1906
plantation en banquettes 6357
plantation en billons 648
plantation en blocs 6354
plantation en cartouche 1163
plantation en côté de potet 6355
plantation énergétique 3742
plantation en fentes 289
plantation en lignes 6356
plantation en motte 490
plantation en nids 5712
plantation en panier 616
plantation en pentagone 6668
plantation en pots 6482
plantation en quinconces 6668
plantation en récipients 1895
plantation en sillon 8964
plantation en sous-étage 6364
plantation en taungya 8419
plantation entre déchets alignés 9449
plantation entre empattements de racines 6348
plantation en tube 9002
plantation faite en soulevant une motte de gazon 289
financier immédiat 8896
plantation forestière familiale 3009
plantation forestière intercalaire 8863
plantation forestière non spécifiée 9119
plantation gisante 3041
plantation issue de graines 8915
plantation le long des limites 921
plantation mécanisée 5380
plantation mixte 5491
plantation non réussie 2992
plantation par arbres isolés 6347
plantation par bouquets 6354
plantation par groupes 6354
plantation par touffes 6354
plantation par trous 6355
plantation plate 3257
plantation récréative 233
plantations sauvages 9416
plantation sur bourrelet 648
plantation sur butte 5578
plantation sur buttes par trouées 6355
plantation sur ensellement 289
plantation sur entrandain 9449
plantation sur placeaux 6107
plantation sur potets 6355
plantation sur sillon 8964
plantation sur terrain libre 5908
plante 6333
plante-appât 483
plante à racines aériennes 7014
plante biennale 696
plante bisannuelle 696
plante calcicole 1251
plante calcifuge 1254
plante calciphile 1253
plante de repiquage 5814
plante de sous-étage 9063
plante de terrain arboré ouvert 5922
plante dioïque 2507
plante d'un an 9756
plante écrasée 5040
plante en racine profonde 2326
plante enrobée 492
plante éphémère 2847
plante épiphylle 2852
plante forestière 3527
plante grasse 8082
plante grimpante 1642
plante ligneuse 9714
plante ligneuse pérenne 9713
plante molle 8082
plante parasitaire 6073
plante pastorale 6719
plante pérenne 6151
plante prête à être mise en terre 5978
planter 6334
plante rampante 1642
planter entre lignes 4561
plante repiquée 5815
plantes à baies 685
plantes à feuilles persistantes 2931
plante saisonnière 2847
plantes à noix 5823
plante saprophyte 7214
plantes de protection 6600
plantes monocotylédones 5540
plantes monoïques 5543
plantes racines 7017
plantes sarclées 7017
plante subspontanée 2862
plante succulente 8082
plante sur le terrain 5978
plantes xérophiles 9736
plante sylvestre 3527
plante transgénique 8764
planteur intermittent 4554
planteuse 8897
planteuse d'arbres 8897
plante vasculaire 9172
plante vivace 6151
plante volubile 9018
plantoir 6344
plantoir à pistolet 6351
plantoir à tarière 6342
plan total des ressources 8729
plantule 6366
plantule dégagée 9056
plantule dormante 2576
plantule en motte 492
plaque criblée 7592
plaque d'appui 321
plaque d'arrêt 1228
plaque de bois 9492
plaque de butée 1228
plaque d'écorce 575
plaque isolante 4521
plaqueminier de Texas 8485
plaqueminier de Virginie 1796
plaqueminier d'Orient 2258
plaqueminier du Japon 4627
plaqueminier kaki 4627
plaque résineuse 4052
plaques d'écorce 7710
plaque stratifiée 4761
plaquettes de bois 9508

plaquettes de face 8284
plasma germinatif 3838
plastification de bois 9626
platanacées 6321
platanaie 6322
platane 254
platane à feuille d'érable 2906
platane commun 2906
platane d'Amérique 254
platane de Californie 1275
platane d'Espagne 7879
platane d'Occident 254
platane d'Orient 5952
platane hybride à feuille d'érable 2906
platane occidental 254
plate-bande 5812
plateforme d'alimentation 4485
plate-forme de chargement 5027
plate-forme de la voie 8744
plate-forme de réception des grumes 5451
plate-forme de sciage 7233
plate-forme de tronçonnage 7727
plate-forme inclinée 4442
platelage 2304
platine brute 874
platymiscie 5225
plein de bosses 4737
pleine vitesse de tambour 3749
pleins germains 3754
pleureur 9324
pleyon 617
pli 6397
pli transversal 2037
plomb 7606
plot 948
plots nets de coloration naturelle 1033
pluie acide 42
pluie d'or 1784
plumbaginacées 7328
pluviosité 6699
pneu forestier 5086
pneumonite à la poussière de bois 9700
pneumopathie des ouvriers des bois exotiques 9700
pneu pour service hors-route 5198
poche à plantes 6343
poche d'écorce 576
poche de déchargement 2665
poche de gomme 6893
poche de résine 6893
poche longitudinale de résine
pochote 6403
podocarpacées 6407
podocarpe à bractées 5888
podocarpe brun 6390
podocarpe de Nouvelle-Zélande 4683
podocarpe du Guatemala 1372
podzol forestier de montagne 5584
pohutukawa 5728
poids de bois sec 2660
poids de bois vert 3964
poids sec à l'air 149
poids spécifique 9099
poinciane 3249
poinçon 6386
pointage d'arbre 8847
point chaud 4385
point d'appui 276
point de chute 7668
point de flétrissement permanent 6177
point de forme 3668
point d'empilage 7981
point de rosée 2457
point de saturation de fibres 3098
point d'établissement de débardeur 9742
point de tronçonnement 2041
point de tronçonnement théorique 8486
point d'initialisation du rajeunissement 8040
point d'interruption 8124
pointe 819
pointe arrondie 871
pointe douce 8049
pointeur 8377
point final de bois 8622
point fixe 674
point mort 1015
point ondulé 2146
point végétatif 4007
poire d'oiseau 4174
poirier 1795
poirier à feuilles d'amandier 206
poirier à feuilles de saule 9432
poirier à fleurs 1281
poirier chanticleer 1281
poirier commun 1795
poirier d'Ussur 9148
poirier-pays 5346
poirier sauvage 1795
poirier véritable 1795
pois de Sibérie 7564
pois quénique 4379
poivre des moines 4961
poivre rouge 1004
poivrier à feuilles longues 5140
poivrier d'Amérique 1266
police forestière 3532
poli de qualité 3150
politique forestière 3533
pollen 6432
pollinisation 6434
pollinisation croisée 2056
pollinisation libre 2056
pollution acide 40
polonneau 4480
polycarpique 6436
polyculture 5491
polygalacées 5446
polygamodioïque 6438
polymorphisme 6439
polyploïde 6440
polypode 6441
polypore marginé 6764
polytric commun 4087
pomacées 330
pomme d'argyle 357
pomme de chêne 5831
pomme de pin 6250
pommelé 2247
pomme-rose 4646
pommier 1755
pommier à baies 7565
pommier acerbe 9403
pommier commun 1755
pommier coronarié 8331
pommier de Chine 6127
pommier-étoile 8037
pommier floribond 4619
pommier japonais 4619
pommier odorant 8331
pommier paradis 6062
pommier sauvage 9403
pompier 3171
pompière 3171
ponçage 7127, 7189
ponceau 2132
ponceau de bois 8572
ponceuse à bande 672
ponceuse à courroie 672
ponceuse de chant 6567
ponceuse de contour 6567
ponceuse de profil 6567
poncirier 4146
poncirus 4146
ponctuation aréolée 920
ponctuation aveugle 155
ponctuation borgne 155
ponctuation cypressoïde 2216

ponctuation en file oblique 223
ponctuation en rangée horizontale 5929
ponctuation intervasculaire 4571
ponctuation par champ de croisement 2050
ponctuation pinoïde 6287
ponctuation rayon-vaisseau 6740
ponctuations en disposition scalariforme 7273
ponctuations en tamis 7591
ponctuations unilatéralement groupées 9090
ponctuation taxodioïde 8424
pont de charpente 8663
pont de flottage 6686
pont provisoire 8449
population d'arbres forestiers 3627
population de base 608
populetum 6455
pore 6461
à ~s fins 3155
pore délignée 920
pore hyphale 927
pore isolé 7845
pores accolés 6462
pores groupées 1669
porosité 6465
porosité en anneau 6954
porosité en cercles concentriques 6954
porosité semi-anneau 7431
porosité totale 8728
port 4017
port colonnaire 1735
port dressé 2858
port d'un arbre 4080
porte-case du mélèze 4788
portée 6742, 7874
portegraine de sauvegarde 3189
porte-graines 5569
porte-greffe 7046
porte-grume intermédiaire 4548
port érigé 2858
porter préjudice à 4148
porteur 3674
porteur de souches 8221
porteur de troncs entiers 8867
porteur-nettoyeur de souches 8222
port flexueux 2028
port flottable 5105
port procombant 6558
port procumbant 6558
port pyramidal 6648
pose fixe d'arbre 9052
possibilité annuelle de coupe 192
possibilité forestière 3592
possibilité par pieds d'arbre 9772
possibilité par surfaces 9770
possibilité par surface terrière 602
possibilité par volume 9773
possibilité pieds d'arbre 9772
possibilités d'utilisation du bois 6473
poste de classement des grumes 5093
poste d'écorçage 567
postmaturation 122
post-régénération 6811
pot à plantation 6359
poteau 6475, 8190
poteau de soutien 8273
poteaux de mines 6311
potentiel biotique 728
potentiel de croissance biologique 728
potentiel de dominance 2562
potentiel de la forêt 3648
potentiel de station 7663
potentiel hydraulique de feuille 4892
potentiel récréatif 6757
potentille blanche 9365
poudrage de la forêt 3388
poudrage de la frondaison 5482
poudre de bois 1514
poudre de liège 1962
poudre liégeuse 1962
poudreuse à dos 471
pouillot siffleur 9691
poui rose 5346
poulie de charge 2993
poulie de renvoi à gorge 7796
poulie de renvoi amont 4167
poulie de renvoi arrière 4167
poulie guidante 4849
poulie guide-support de câble tracteur 1158
poulie motrice 3035
pourcentage de boisement 3522
pourcentage de rebut 2120
pourcentage de reprise de semis forestier 8918
pourcentage de réussite de semis forestier 8918
pourcentage de réussite des semis 6368
pourcentage d'extraction 2954
pourcent de cimes 2090
pourri 7212
pourriture 9529
pourriture alvéolaire longue 5146
pourriture apicale brune 1092
pourriture au stade initial 4494
pourriture au stade intermédiaire 72
pourriture avancée 72
pourriture blanche 9381
pourriture blanche bigarrée 9374
pourriture blanche cubique 9368
pourriture blanche des racines 364
pourriture blanche des tiges 9388
pourriture blanche du tronc 9388
pourriture blanche efflorescente 9370
pourriture blanche marbrée 1081
pourriture blanche marbrée des tiges 5574
pourriture blanche molle 7825
pourriture blanche molle spongieuse 7826
pourriture blanche par poches 9377
pourriture brune 1084
pourriture brune claire cubique 4937
pourriture brune cubique 2251
pourriture brune de l'aubier par taches de conifères 1083
pourriture brune foncée cubique de cœur 2250
pourriture brune-rouge de la racine et du pied 6766
pourriture brune subalpine 8225
pourriture centrale 1414
pourriture cubique 2112
pourriture d'aubier 7218
pourriture de bouleau brune 1085
pourriture de cime 2092
pourriture de destruction 2433
pourriture de la grume 8998
pourriture de l'aubier brune 1086
pourriture de l'aubier crémeuse 2010
pourriture de l'aubier dure 4130
pourriture de l'aubier gris-brun 3968
pourriture de l'aubier jaune 9761
pourriture de l'aubier molle 7820
pourriture de l'aubier superficielle 8269
pourriture de lésion 7078
pourriture de rémanents blanche 9383
pourriture des branches 4734

pourriture des racines de coton 1981
pourriture des racines et du rejet 364
pourriture des tiges 8998
pourriture du bois 9529
pourriture du cœur 1414
pourriture du cœur blanche cubique 9367
pourriture du cœur rouge 6777
pourriture du pied brune-rouge 6765
pourriture du pied jaune laminée 9760
pourriture du rejet 1233
pourriture en points 6242
pourriture finale 1828
pourriture gluante 7741
pourriture humide 9353
pourriture interne 2645
pourriture jaune de rémanents 9762
pourriture jaune fibreuse du cœur 3104
pourriture jaune madrée 7908
pourriture marbrée 5573
pourriture marbrée blanche 9372
pourriture marginale 4494
pourriture molle 7819
pourriture mycélienne 3768
pourriture mycélienne blanche 9385
pourriture par taches ou par poches 6313
pourriture pointue brune 1088
pourriture rouge 6789
pourriture rouge-brune 3678
pourriture rouge du pin 6265
pourriture roussie 3678
pourriture sèche 2656
pourriture subalpine de rémanents 8227
pourriture tubulaire 6291
pourriture vaseuse 7819
pourriture verte 3954
pourvu de grosses branches 8135
pourvu de menues branches 7737
pousse 7132
pousse adventive 80
pousse allongée 5149
pousse annuelle 305
pousse apicale à double flèche 9019
pousse apicale cassée 1061
pousse apicale de tige 8076
pousse apicale double 2582
pousse d'août 6579
pousse de la Saint-Jean 4765
pousse de régénérescence 6753
pousse de remplacement 6872
pousse d'été 4765
pousse en chandelle 1302
pousse en hauteur 4237
pousse hétérophylle 4267
pousse humide 9299
pousse latérale 418
pousse lessivée périodiquement 6761
pousse préventive 2848
pousse proleptique 4765
pousser 7944
pousse ralentie 418
pousses apicales codominantes 1705
pousses de bambou 511
pousses récentes 4813
pousse superposée 8270
pousse terminale 8474
pousse uninodale 9093
pousse végétative 4505
pousseur de grumes 911
poussière de charbon de bois 9506
poussoir 6645
poutrage équarri 7969
poutre 7485
poutre à demi-bois 4100
poutre avivée 7965
poutre composée 1836
poutre de bois 638
poutre de chargement 6235
poutre en caisson 953
poutre en quart 6658
poutre épaisse 8502
poutre faîtière 7007
pouvoir d'enracinement 7030
pouvoir germinatif 3833
pouvoir isolant de la forêt 6597
pralinage des racines 7041
pratique forestière 3576
pratiques culturales 2131
pratiques de gestion d'excellence 688
pratiques sylvicoles de base 612
pratique sylvicole 3576
pré 5349
pré-bois 9711
prébuttage 6495
précharger 6511
précipitation 6498
précipitation acide 40
précipitation au sol 2626
précipitation au sol de miellat
précipitation de miellat 4351
précipitation moyenne annuelle 5353
précipitation par houppier 1313
prédébardage 6507
pré-débardage 6519
prédémariage 2599
prédémarieuse 8513
pré-ébranchage 76
pré-exploitation 6513
prégermination 6504
prématurité 6509
premier débit 638
premier dépôt transitoire 5451
premier dépôt transitoire à rivière 6974
première couronne 2064
premier passage 3228
préparation chimique du terrain 1468
préparation des trous 6508
préparation du gemmage 2140
préparation du site 3985
préparation du sol superficiel 8685
préparation mécanique du terrain 5373
préparation superficielle nulle 5799
préparation superficielle nulle du sol 5799
prescription 6517
prescription d'économie du bois 7617
prescription de traitement 8794
prescription d'exploitation 4161
prescription sylviculturale avant la récolte 6506
pressage d'arbre au abattage 6520
pressage des résidus 3560
presse à résidus ligneux 7235
pression de population 6460
pression de sève 5965
pression sur bouts 6522
pression sur rives 6522
prestataire de services forestiers 3578
prétraitement de bois 9628
préventif 2573
prévision contre l'incendie 3175
prévision de l'incendie 3175
primavera 6551
prime d'arrachage 6512
prime pour l'arrachage 6512
principe de l'économie forestière 8020

prise d'eau 9280
prisme 6554
prisme de chemin 6980
pris par corde 1977
prix de bois 9629
prix de bois en parties 6528
prix du bois sur pied 8195
prix du marché 5313
probabilité de survie 1322
problématique décisionnelle forestière 3479
procédé Bethell 692
procédé Boulton 5876
procédé d'imprégnation 515
procédé par pression 6523
procédé sous pression 6523
procédure régressive nette 5715
procès-verbal d'aménagement 3409
producteur de bois 344
production de bois brut sans écorce 3980
production de matière fibreuse 9633
production de pousses 8714
production en graines forestières 3584
production en volume 9773
production équilibrée à long terme 5147
production forestière 3537
production forestière principale 5253
productivité à dessouchage de dépôt 8216
productivité à dessouchage d'usine de conversion du bois 8215
productivité de bois non spécifiée 9120
productivité de l'écosystème 2750
productivité forestière 3539
produit actif 55
produit adjuvant 64
produit de bois rond 7117
produit défoliant 2338
produit de la coupe 9783
produit de la hauteur d'un arbre par le coefficient de forme 8841
produit de remplacement de bois 9669
produit d'extinction 3211
produit d'imprégnation 4428
produit d'une coupe 9777
produit escompté de la coupe
produit forestier accessoire 5471
produit ignifugeant 3197
produit léger en solvant organique 4947
produit principal 3145
produit répulsif 6869
produits chimiques forestiers 3349
produits de bois fins 3160
produits d'éclaircies 9782
produits d'éclaircissage 8506
produits d'écorce 577
produits de forêt spéciaux 7889
produits du bois composites 1835
produits du bois de haute technologie 2826
produits forestiers botaniques 938
produits forestiers conventionnels 1910
produits forestiers de cueillette 9405
produits forestiers de masse 1156
produits forestiers non ligneux 5774
produits ligneux 8626
produits non dérivés du bois 5778
produits semi-finis en bois 7427
profil de feuillage 3304
profil de la forêt 3542
profil en long du câble porteur 1244
profils de bois scié 7260
profondeur de bois scié 7256
profondeur de canal de flottage 2405
profondeur d'écorçage 565
profondeur de défaut 2404
profondeur de fente 7484
prognose d'exploitation forestière 4153
programme de conservation contre l'érosion 1882
programme de développement des petites entreprises forestières 7759
programme de développement forestier 3376
programme de gérance forestière 3611
programme des petites entreprises forestières 7759
programme forestier annuel 299
projection cartographique 5299
projection horizontale des cimes 2088
prolifération des bourgeons 6580
prolongement de racines 7022
promoteur de la croissance 4020
propagation asexuée 9182
propagule 9183
prophylactique forestière 3544
proportion de bois pourri 6093
proportion d'écorce 546
proportion de surface occupée 2344
proportion de surface occupée par la forêt 3370
proportion entre des classes d'âge 138
propriétaire forestier 3519
propriétaire forestier privé 6555
propriété forestière 3543
propriété forestière de production 3419
propriétés de bois 9632
propriétés de croissance 4022
prosenchyme 6586
prosope 1790
prosope velours 9189
prosopis 1790
prospectus 6587
protéacées 6591
protecteur en bois 3456
protecteur pour arbre 8850
protection aérienne des forêts contre l'incendie 148
protection chimique de la forêt 1465
protection contre le bleu 6594
protection de bois 9627
protection de fonds génétique forestier 3444
protection de forêt contre l'incendie 3430
protection de forêt contre l'incendie à terre 4771
protection de la lisière 6596
protection de l'environnement 2842
protection du sous-étage 9071
protection du tronc 8997
protection forestière 3359
protée 6590
protée mellifère 8254
protoplaste 6601
protoxylème 6602
provenance 6603
provenance de graines 7395
province à production de bois signifiante 8640
pruche 4248
pruche de l'Est 1296

prune-abricot 332
prune cerise 1475
prune Damson 2244
prune de Java 4646
prunellier 7749
prunéole 2244
prune verte 3952
prunier 6391
prunier abricotier 332
prunier américain 256
prunier cérasifère 1475
prunier cerise 772
prunier chickasaw 1480
prunier d'Afrique 112
prunier d'Amérique 256
prunier de Damas 2244
prunier des grèves 633
prunier domestique 6391
prunier laineux 5425
prunier mombine 4322
prunier myrobalan 1475
prunier noir 1297
prunier ombellé 3260
prunier sauvage 2244
psammophyte 7192
pseudo-cœur 795
ptéléa trifolié 6035
ptéridophytes 6618
ptérocarier à feuilles de frêne 1388
ptérocarier du Japon 4637
ptérocaryer 9456
ptérocayer à feuilles de sorbier 1388
ptéroxyle 7797
pubérulent 6619
puceron de l'écorce du pin 6245
puceron de l'épinette 7949
puceron du hêtre 650
puceron lanigère du mélèze 9718
pucerons 323
puéraria 4746
pulai 9363
pullulation 2849
pulvérisateur à désherber au pied des arbres 8800
pulvérisateur à dos 471
pulvérisateur à écran de protection d'arbre 8851
pulvérisateur pour désherbage sur la ligne 8800
pulvérisation de feuillage 3305
pulvérisation en bandes 8180
pulvérisation en forêt 3388
punaise à pattes rousses 3343
punaise réticulée 4751
purépecha en mich 907
purshie tridentée 311
putier 2881
putiet 2881
putréfaction 6647
putréfié 7212
putride 7212
pygeum 112
pylône 5328
pyramide 6650
pyroclimax 3172
pyrogravure 9498
pyrogravure sur bois 9498

quai de chargement 5027
quai de chargement forestier 5027
qualis 5275
qualité d'âge de bois 137
qualité de bois 6657
qualité d'ébranchage 2355
qualité de station 7670
qualité d'un peuplement 6656
qualité stationnelle 7670
quantité annuelle d'exploitation 296
quantité de bois 8627
quantité de bois de vente 8633
quantité de vente admissible 199
quantité permissible annuelle d'exploitation 193
quapalier 7748
quart de la circonférence 6661
quartelle 7712
quartelot 7712
quartier 7969
quartier de bois 7969
quartier d'hivernage des cervidés 2332
quartier en régénération 355
quartier général 3371
quartier-piège 8776
quassia 6662
quassier 6662
quassier amer 9373
quebracho 9378
quebracho rouge 5173
quercitron 2692
queue d'aronde perdue 2592
quinquina 6669
quotient de circonférence 3858
quotité de la coupe 9783

rabaisser 2175
rabattage 1224
rabattement 2617
rabotage 4273
raboteuse 6326
raboteuse-trieuse 3289
raccourcir 2165
raccourcissement sur tire-sève 2633
race arborescente 8906
racème 6671
racine 7009
  à ~ élaguée 7039
  de ~s grossières 6684
  sans ~s 7034
racine absorbante 20
racine adventive 79
racine aérienne 86
racine cannelée 1231
racine charnue 3263
racine contenue 1891
racine coudée 4719
racine d'alimentation 3034
racine d'ancrage 277
racine d'appui 277
racine de conduction 1856
racine en échasses 277
racine en J 4663
racine fasciculée 1968
racine fibreuse 1968
racine fourchue 980
racine horizontale 7486
racine lamellaire 6378
racine latérale principale 5247
racine molle 8452
racine noduleuse 5744
racine nue 531
racine pivotante 8411
racine pivotante latérale 7582
racine principale 8411
racine prolongée 2961
racine radiculaire 6546
racines adhérentes 3907
racines de sauvageon 8117
racines en mottes 491
racines latérales 4824
racine superficielle 7486
racine tannante 8395
racine traçante 2014
racine tubéreuse 9004
racine verticale 9207
racleur de bois 9646
racloir de résine 4584
radeau 8584
radeau de transit 8766
radeau enchaîné 480
radication 7011
radicelle 7035
radicule 6546
radier 3283

raffinage de bois 9638
raideur 8099
rainer 7308
rainette 4082, 6741
rainette arboricole 9579
rainuré 3977
rainure et languette 8676
rainurer 7308
rainureuse 4661
raisin de mer 2844
raisin d'ours 1759
raisinier bord-de-mer 7326
raisin marine 7326
rajeunissement préétabli 73
rallonge de flèche 8442
rallongement de racines 7022
rallonge télescopique 8442
rallonge télescopique de flèche 8442
ramassage de bois de feu 3222
ramassage de litière 9153
ramasser et paqueter les débris 1573
ramasseuse-presse à résidus 7235
rambarde de traînage 1025
rameau 991
rameau à bois 4505
rames à fagots 9572
ramet 9183
rameux 977
ramicornes 6708
ramification 987
ramification dimorphique 2506
ramification plagiotrope 6316
ramifié 977
ramiforme 6709
ramille 991
ramilles pour recouvrir les cultures 1995
ramin 6710
ramollissement de bois 9655
rampant 2313
rampe de chargement 5027
rampe de transfert de dosse 7711
rampe transversale 2044
ramuscule 991
randonnée inculte 9404
rang de plantation 6360
rangée d'arbres 7121
rangée d'arbres feuillus 4991
râpage 2105
râpe à bois 9636
rapport carbone-azote 1333
rapport cime/hauteur totale 2085
rapport cime-racine 2091
rapport de chasse 6182
rapport de chasse normal 6182
rapport de cœur 4200
rapport d'écorce 578
rapport de l'aire à la masse foliaire 4893
rapport de matière sèche par la transpiration 2653
rapport de matière sèche par le rayonnement solaire 2652
rapport diamètre cime/fût 2075
rapport entre diamètre de cime et fût 2075
rapport lisière/zone 2755
rapport système racinaire/système foliacé 7052
rapprochement 4187
rasclet 2142
râteau à feu 3200
râteau débroussailleur 1112
râteau déracineur 7042
râtelage 7043
râtelage de litière 6706
râteler 6705
râtelier 2616
ratisser 6705
ravageur 5807
ravageur forestier 3523
ravageurs de racines 7023
ravageurs des pousses 8472
ravageurs secondaires 7352
ravalement 1224, 2347
ravaler 2165
ravine 4043
ravinement 4045
rayon 9538
rayon fusiforme 3784
rayon hétérocellulaire 4264
rayon homogène 4344
rayon homorocellulaire 4344
rayon lenticulaire 3784
rayon libérien 6208
rayon ligneux 9738
rayon ligneux hétérogène 4264
rayon médullaire 5391
rayon multisérié 5626
rayonnement 6682
rayonnement calorifique 6682
rayonnement diffus 2494
rayonnement direct 2520
rayonnement global 3866
rayon primaire 6545
rayon secondaire 7353
rayon unisérié 9094
rayure 8154
reboisé 6806
1823
rebond 3297
rebroussement 8349
rebut 2123
rebuts 7304
recalibré 6881
recépage 1926
recépage avec réserve 1927
réception de bois 27
réception d'une coupe 6831
recette moyenne annuelle 428
réchauffement planétaire 3867
recherche de bois 9639
recherche sylvicole 3577
récipient
en ~ 1892
récipient à résine 538
récolte de bois 9523
récolte décomposée 7413
récolte de cônes 2033
récolte d'écrémage 4285
récolte de graines 7376
récolte de la gomme 828
récolte de l'eau 9286
récolte de plantes 6341
récolte des semences forestières 3584
récolte d'herbes en forêt 3914
récolte gestionnée 5262
récolter 4149
récolteur de résine 6905
récolteuse 8853
récolteuse au tracteur 8746
récolteuse d'arbres entiers 1830
récolteuse de plantules 8916
récolteuse de résine 5371
récolteuse mobile 5500
récolteuse par traction mécanique 8746
récolteuse routière 6982
recombustion 6747
recommandation 6517
reconnaissance forestière 3549
reconstitution de forêt 6821
reconstitution forestière 6821
recouvrement 1992
recréation d'aire naturelle 461
recréation d'arrière-pays 461
recrû 9785
recrutement 4492
recrutement de forêt 6833
rectitude du fût 8142
recul 3297
récupérateur de grumes 5108
récupération de bois 7170
récupération de l'eau 9286

réduction du danger d'incendie 3187
réduction du risque d'incendie 3187
réduction en billots 713
réduire 400
ré-exploitation 6859
réfaction 2121
réfaction pour anomalies 201
réfaction pour défaut 201
refendeuse 1410
refendre 6880, 6959
reflet du bois 3872
reforestation 6821
refoulement 4668
refoulement du bois 6871
réfrigération préparatoire 6496
refroidissement par évaporation d'eau 2915
refus de classeur 6850
regarni 823
regarnissage 823
régénération accidentelle 6817
régénération à partir de la lisière 6877
régénération artificielle de forêts 6811
régénération avec préparation de terrain 6827
régénération de forêt 6821
régénération des parcelles de coupe 6822
régénération du taillis 1931
régénération latérale 4823
régénération naturelle 5690
de ~ 7417
régénération par bandes 6813
régénération par greffes racinées 6876
régénération par groupes 6812
régénération par rejets de souches 1931
régénération préalable 73
régénération préexistante 73
régénération sans préparation de terrain 6826
régénération sous abri 6825
régénération spontanée 7924
régénération subséquente 73
régénération végétative 9184
régime de la futaie 4281
régime des coupes progressives d'abri par bouquets 7514
régime des incendies 3201
régime des incendies de gravité mixte 5492
régime des incendies pour remplacement de peuplement 8026
régime de taillis-sous-futaie 1937
régime d'exploitation 2201
régime du taillis 1935
régime inéquien 9080
régime sylvicole 7618
régime sylvicole de forêt équienne 2919
région boisée 8604
région écologiquement fragile 2841
région écologiquement sensible 2841
région forestière 3551
région forestière accessible 30
région forestière inexploitée 476
réglage du mélange 6843
règle de densité 2395
règle de marquage 5319
règle graduée de cubage 5107
règle graduée en pied-planches 877
règle graduée en pieds de mesure de planches 877
règlement de cubage le bois 9658
règlement d'exploitation 6332
règlement du taux d'exploitation 3649
règlement forestier 8046
règlement pour la vidange 5081
règlement pour le flottage 3276
règlements d'exploitation 6844
réglisse 4997
réglisse d'Amérique 636
régression des plantes 6372
régulateur de croissance 6340
régulation des émissions de fumée 7784
réhabilitation de site 6860
réhabilitation de sites désaffectés 6846
reine-des-prés 5350
réitération 6848
rejet 7132, 7946
rejet annuel 305
rejet de remplacement 6872
rejet de souche 1933
rejet de taillis 1933
rejet de têtard 6431
rejet de tige 2848
rejeton 7132
rejuvénilisation 6851
relascope 6852
l'éclaircie naturelle 6490
relations allométriques 190
relevé de la régénération 6824
relèvement du couvert 4933
relicte forestière 3517
relief 6858
relique 6857
rémanents bruts d'usinage 1684
rémanents de bois 3996
rémanents de bois résineux 7831
rémanents d'ébranchage-façonnage 5170
rémanents d'usinage 5450
rémanents empilés 6231
rémanents fins 3156
rémanents libres 5160
rémanents tendres 5160
remblai de remplissage 464
remblaiement derrière un mur de soutènement 464
remblayage 464, 5995
remédiation 6860
remembrement de forêts 1883
remembrement forestier 1883
remise en état de l'environnement 2843
remontée d'eau 8321
remorque à grumes 8752
remorque à souches 8218
remorque de bois 5078
remorque pour bois long 9395
remplacement des manquants 823
remplissage de ravines 4046
rémunération à prix fixe 5217
rémunération forfaitaire 5217
rémunération par boisseaux 1200
renard mécanique 5124
rendement annuel permissible de la forêt 6178
rendement de bois soutenu à long terme 5153
rendement de la forêt 9778
rendement de non déclin 5755
rendement de peuplement résineux 9781
rendement des grumes 9780
rendement du bois 9778
rendement du bois de travail 5214
rendement en éclaircies 9782
rendement en matière de la forêt 9778
rendement en volume 9773
rendement équilibré à long terme 5147
rendement forestier expectatif 2945

rendement moyen à long terme 436
rendement net de la forêt 5720
rendement par arbre 9772
rendement par épaisseur 9771
rendement par un arbre 9779
rendement soutenu de bois 8299
renflement d'une branche 832
rengas 6866
renonculacées 1218
rentabilité de la forêt 6568
rentabilité forestière 9778
rente forestière 3553
renversement latéral de grumes 4822
répartition de coupe annuelle permissible 1
répartition de l'humidité 5519
répartition des classes d'âge 127
répartition des pluies 6698
répartition naturelle des forêts 5677
répartition normale par classes d'âge 5779
repère 674
repère d'abattage 3074
repéreur d'incendie 3184
repeuplement de forêt 6821
repiquage 6529
repiquage en ligne 4993
repiquer 8769
replantation 6873
réplication 6875
repos de la végétation 9647
repousseur 4257
représentant habilité 2424
reproduction avancée 73
reproduction en consanguinité 4435
reproduction par boutures 6581
reproduction par boutures racinées 6876
reproduction par marcottes 6582
reproduction végétative 9182
resak 6879
rescié 6881
réseau de bio-indicateurs 726
réseau d'écosystèmes forestiers 3393
réseau de pistes forestières 3566
réserve 6887
réserve biologique forestière 3338
réserve de bois sur pied 9250
réserve de deuxième classe 7358
réserve de forêt 3557
réserve de première classe 3226
ethnobiomédicale 2871
réserve isolée 7291
réserve nationale 5668
réserve naturelle 5672
réserve naturelle d'intérêt scientifique 6884
réserves 6889
réserves de bois 9640
réserves d'éclaircies 4754
réserves de résine 5701
réserves forestières 3559
réserves résineux 5701
réserve sur pied 9250
réservoir de gènes 3815
réservoir de réserve d'arbres verts 3960
résidu 6889
résidu imparfaitement cémenté 4445
résidus de bois débité 2505
résidus de bois évitables 446
résidus d'ébranchage-façonnage 5170
résidus d'écorce 580
résidus de coupe 998
résidus de sciage 5215
résidus des industries forestières 3469
résidus de tronçonnage 9273
résidus fins 3156
résidus forestiers 3561
résidus végétaux 6371
résine 8907
résine animée 1923
résine blanche 3792
résine blanche d'épicéa 3792
résine brute 6733
résine copal 1923
résine de bois 8907
résine de bois feuillu 4144
résine de cèdre 1402
résine de lésion 9732
résine d'épicéa 7957
résine de pin 6898
résine liquide 8374
résine odorante 503
résines dammar 2243
résineuse mélangée 5485
résineux 5708
résineux conifères 5708
résinier 6905
résinose 2977
résistance 6906, 8735
résistance à la fente 6907
résistance à la flexion 676
résistance à l'altération par les
résistance à la sécheresse 2637
résistance à l'écorçage 6908
résistance au cisaillement par épaisseur 7502
résistance au froid 3724
résistance au gel 3724
résistance horizontale 4371
résistance verticale 9206
résistant à l'humidité 5522
responsable forestier 1483
ressource naturelle 5691
ressource récréative 6758
ressources d'eau 9295
ressources de bois 8629
ressources des forêts globales 3864
ressources forestières 3562
ressources génétiques de la forêt 3446
ressources non ligneuses 5775
ressuage 8321
restauration 6916
restauration de couverture végétale 6923
restauration de croissance 4023
restauration de végétation avant d'un incendie 6479
restriction de la cannelure 9269
résultat de l'abattage 3145
retardant contre une pourriture 2295
retardant incendie 3197
retardataire 4325
retardateur contre une pourriture 2295
retardateur de croissance 4025
retardateur de flammes 3197
retardement de la croissance 4015
rétention d'arbres verts 3961
rétention de feuilles 4885
rétention dispersée 2536
rétention par groupes 139
rétention variable 9169
retenue de souches basses 5197
retombée acide 40
retour de flamme 465
retrait de bois 7554
retrait radial 6680
retrait tangentiel 8386
retrait total 8730
retrait volumique 9251
rétrocaveuse 468
rétrocroisement 462
rétrogression des plantes 6372
réussir 7308
revégétation 6923

reverdissement 6923
revêtement 2303
revêtement de ressources 6913
revêtement en bois 9521
révolution du meilleur rendement 7076
rhabillé 7039
rhabiller 1645
rhagie chercheuse 6934
rhamnacées 1138
rhéédie 6926
rhizome 6927
rhizophoracées 5279
rhizophore 6928, 6929
rhizosphère 6930
rhizotaxie 7011
rhododendron 6989
rhododendron à fleurs nues 6288
rhododendron à grandes feuilles 6023
rhododendron de Laponie 4786
rhododendron en arbre 8908
rhododendron géant 3934
rhododendron jaune 3244
rhododendron pontique 4355
rhodoraie 7063
rhynchites coupe-bourgeons 9014
rhytidome 5975
riche en matière nutritive 2912
richesse en espèces 7902
richesse forestière 3640
ridage 4227
rideau 4901
rideau-abri 7508
rideau boisé de protection 7508
rideau brise-vent 9445
rideau d'arbres 1149
rideau vert 1149
ridement 2020
rigidité 8099
rigolage 8962
rigole 4043
rigole d'écoulement 334
rigole de niveau 1904
ripage 6967
ripisylve 6961
rique 2979
risque de perturbation du sol 7837
risque d'incendie 3205
risque important 4286
risque important de santé forestière 4286
risque par haut 5994
rive de bois scié 7257
rive d'échantillon 2763
robinier 1754
robinier poisseux 1595
robinier rose 1038
robinier visqueux 1595
roble 1417
roble pellin 6985
roches des Alpes 217
rocouyer 4995
rognage 1215
romarin 7066
romarin de marais 897
ronce 9452
ronce à poils rouges 9452
ronce arctique 350
ronce bleuâtre 2891
ronce commune 778
ronce décolorée 4307
ronce des bois 778
ronce des buissons 778
ronce de souche 8202
ronce du Mont Ida 1800
ronce du Nord 350
ronce framboisier du Canada 6640
ronce frutescente 778
ronce occidentale 782
ronce odorante 6640
ronce rabougrie 2684
rondin 7101
rondin de bois à papier 6627
rondin de bois de pâte 6627
rondin de défibrage 6627
rondin défectueux 2335
rondin de pointe 7771
rondin en longueur d'arbre 2834
rondin-piège 8776
rondins de bois de chauffage 7103
ronger 5730
rongeur bostryche 6252
rongeur du sapin blanc 7602
rônier 6047
rosacées 7064
rosage à grandes feuilles 6023
roseau 1308
rose des Alpes 6989
rose gum 3909
rosier 7068
rosier de désert 2418
rosier des chiens 2555
rosier rugueux 634
rosier tomenteux 8674
rotang 6731
rotation de chargement de chemin forestier 5080
rotation des coupes 2194
rotation d'une forêt de bois de mine 6310
rotobineuse 7072
roue de dessouchage 8210
rouge 4861
rougissement 7215
rougissement de cime 8710
rouille albergine 8711
rouille blanche 8711
rouille courbeuse du pin 6274
rouille de branches 4970
rouille de conifères 7859
rouille de galle 3795
rouille des conifères 1737
rouille du pin 6274
rouille fusiforme 3785
rouille noire 315
rouille stalactiforme 7996
rouille tumeur 3795
rouille vésiculeuse 9376
roulage 9187
roulage de copeaux de bois gras 7002
roulants 3119
rouleau Cambridge 2127
rouleau débroussailleur 7000
rouleau de renvoi 3035
rouleau entraîneur 3035
rouleau rayonneur 2127
roulure 2139
roulure de résine 6903
route à viabilité permanente 6175
route carrossable secondaire 4172
route de transportation de bois 9681
route forestière 3565
route permanente 6175
route praticable par tous les temps 205
route sans issue 2262
route tous les temps 205
rouvraie-charmaie 4374
ruban de cubage 8397
ruban de mesure du diamètre 2484
ruban dendromètre 516
ruban diamétrique 2484
ruban pour les diamètres 2484
rubiacées 5238
ruellage 8962
rufipenne de forêt 3350
rugosité 6103
rupture dans le bois 9563
rupture d'écimage 1013
rupture de la cime 1012
rupture de tronçonneur 1125
rupture par cisaillement 7500
rupture par flexion 3296

rusticité 4123
rusticité d'arbre 8852
rutacées 7131

sabal 1238
sabine maritime 6787
sables encastrés 2794
sablier blanc 7185
sabot 2054
sabot d'arrêt 2054
sabre 8325
saccharification de bois 7143
sac de plantules 5811
sagouier 7153
sagoutier 7153
saignée 3976
saignée de débusquage 1556
saignée des gommiers 828
saigner 7308
sain 9719
saison d'abattage 3072
saison de croissance 4006
saison de montée de la sève 7211
saison des feux forestiers 3434
saison de traitement 8795
saisonner 7337
sal 7163
salicacées 9429
salsepareille 7223
salsepareille à feuilles de laurier 4834
saman 6701
samare 4701
sandarac 7182
sang-dragon 7810
santalacées 7181
santal blanc 9382
santal rouge 6790
santé de forêt 3458
sapelli 114
sapelli pommelé 114
saperde chagrinée 4804
saperde du peuplier 4804
sapeur-pompier 3171
sapeuse-pompière 3171
sapi 6225
sapi à double croc 1317
sapin à bractées 1037
sapin à liège 214
sapin amabilis 6025
sapin argenté 1803
sapin baumier 969
sapin blanc 1803
sapin blanc d'Amérique 9008
sapin blanc des Canaries 9008
sapin blanc de Tasmanie 6315
sapin concolor 1278
sapindacées 7806
sapin d'Andalousie 7877
sapin de Bulgarie 1154
sapin de Caucase 1384
sapin de Céphalonie 3938
sapin de Cilicie 1581
sapin de Douglas 1772
sapin de Fraser 3689
sapin de Grèce 3938
sapin de Guatemala 5422
sapin de l'Arizona 2822
sapin de Noël 1569
sapin de Nordmann 1384
sapin de Sibérie 7566
sapin de Sitka 7678
sapin d'Espagne 7877
sapin des Vosges 1803
sapin de Vancouver 3910
sapin d'Oregon 1772
sapin du Brésil 1005
sapin du Canada 1296
sapin du Colorado 1278, 1726
sapin du nord 1805
sapindus de Mukorossi 1506
sapin du Sikkim 7594
sapin du Yeso 4647
sapinette 7949
sapinette blanche 9386
sapinette d'Orient 2723
sapinette rouge 1298
sapinette rouge du Canada 1298
sapin géant 3910
sapin glauque 1772
sapin gracieux 6025
sapinière 3224
sapin magnifique 1268
sapin noble 5738
sapin noble d'Amérique 5738
sapin pectiné 1803
sapin rouge 1805
sapin rouge californien 1268
sapin sibérien 7568
sapin subalpin 214
sapins vrais 8987
sapotacées 7207
sapote 5665
sapotille 5665
sapotillier 5665
saprophyte 7214
saqui-saqui 6403
sartage 3176
sassafras 1801
sassafras de Tasmanie 7864
sassafras officinal 1801
saturation de fibres ligneuses
saulaie 9436
saule à feuilles aiguës 7138
saule à feuilles d'amandier 6122
saule à feuilles de myrtille 813
saule à feuilles de pêcher 6122
saule à feuilles de poirier 505
saule à feuilles de romarin 872
saule à feuilles en cœur 4198
saule amandier 6122
saule à oreillettes 2697
saule-arbousier 5008
saule argenté 1809
saule auriculé 2697
saule aux chatons noirs 7065
saule blanc 1809
saule brillant 7522
saule caspien 7138
saule cassant 1040
saule cendré 3974
saule daphné 9224
saule d'Arroyo 369
saule de Babylone 459
saule de Bebb 644
saule décoloré 3880
saule de coyotes 7183
saule de désert 2420
saule de diamant 5480
saule de Goodding 3886
saule de Hooker 4360
saule de l'intérieur 7183
saule de Mackenzie 5233
saule de Pékin 1963
saule de Sachalin 7146
saule des bancs de sable 7183
saule de Scouler 7311
saule de Silésie 7595
saule de Sitka 7680
saule des plaines costales 1688
saule de Suisse 8348
saule drapé 2775
saule du Pacifique 369
saule épineux 7325
saule exigua 7183
saule fauve 160
saule faux-myrte 9396
saule fragile 1040
saule glabre 218
saule glauque 3926
saule herbacé 6228
saule incane 2775
saule jaune 9766
saule-laurier 630
saule marceau 3880
saule marsault 3880
saule noir 813
saule noirâtre 6975

saule polaire 6417
saule pourpre 6643
saule rampant 2015
saule rossé 2458
saule rouge 6801
saule tomenteux 2775
saule tortueux 1963
saule vert-de-mer 3926
saule vert pleureur 459
saussaie 9436
sauvageon 9420
sauvageon destiné au repiquage 9426
sauvetage de bois 7170
savane 7227
savane arborée 9546
savane arbustive 7560
savane boisée 9546
savane de dégradation 2408
savonnier 6051
savonnier de Drummond 9343
savonnier saponaire 7865
saxifragacées 7269
scabrella de mimosa Bentham 966
scarifiage 7286
scarifiage mécanique 5372
scarifiage par sillons 8962
scarifiage par traînage 2610
scarificateur 7313
scarificateur de placeaux 6108
scarification 7286
scarification à eau chauffée 4386
schefflera 9040
schéma de débit 5051
schème de chemins 6978
schème de flottage par bassin 3270
schizogène 7295
schizolysigène 7296
sciage 7241
sciage à doss 474
sciage à dosseret 474
sciage à façon 7244
sciage à façon de bois dur 4136
sciage au bois montant 6312
sciage brut 622
sciage de parc 9754
sciage de rebut 5450
sciage en long 5135
sciage en plongée 6393
sciage par trait bas 3253
sciage par trait haut 2322
sciage pour parc de séchage 9754
sciage raboté 2620
sciages 1918
sciage sur dosse 8383
le fil 8407
scie à abattage 8854
scie à araser 1822
scie à arc 3342
scie à archet 3342
scie à benzine 3808
scie à billettes 1136
scie à bois 9643
scie à bûches 1136
scie à bûches montées 3683
scie à cadre 5132
scie à chaîne 1431
scie à chaîne à maillons-gouge 1525
scie à chantourner 1822
scie à chariot libre 7003
scie à châssis 3800
scie à châssis 5132
scie à découper 1822
scie à dents fines 3158
scie à dents rares 3803
scie à deux hommes 9030
scie à dos 883
scie à ébranchage 6614
scie à ébrancher 6614
scie à élaguer 6614
scie à fil 9466
scie à guichet 4702
scie à inciser 7309
scie à lame escamotable 4312
scie à lame plate 3251
scie à lame-ruban 5049
scie à long manche 6421
scie alternative à châssis 5132
scie alternative multiple 5132
scie à main 4111
scie annulaire 308
scie à onglet 883
scie à perches 6421
scie à plusieurs lames 3800
scie à queues d'arondes 4936
scie à refendre 2609
scie à refendre à 2 hommes 1633
scie à ruban 5049
de ~ 521
scie à ruban de grumes 5049
scie à ruban double 2661
scie à tronçonner 5056
scie au fil hélicoïdal 9466
scie avec aménage à crémaillère 6672
scie circulaire 1585
scie circulaire à chariot 695
scie circulaire à lattes 7733
scie circulaire à niveau réglable 6970
scie circulaire à segments rapportés 6950
scie circulaire de garde 4312
scie circulaire novelty 5803
scie circulaire réduite 6798
scie circulaire tous azimuts 1586
scie concave 1853
scie cylindrique 2209
scie d'abattage à archet 3342
scie de long 1633
scie de menuisier 5148
scie demi-circulaire à refendre 4092
scie d'encadreur 883
scie de premier débit 1018
scie d'établi 675
scie de tête 1018
scie de tronçonnage 5056
scie double 2584
scie égoïne 4111
scie électrique 2779
scie en forme cylindrique 2209
scie en long 1136
scie-fendeuse à bûches 3220
scie forestière 8854
scie mécanisée 6491
scie mortaiseuse 1822
scie multiple à châssis 3800
science de bois 9645
science de marche forestière 3503
science de sol forestier 3601
science forestière 3581
scie par réducteur à roue 3811
scie passe-partout 1822
scie pendulaire 4312
scie pour coupe rase 1615
scie pour débit longitudinal 6966
scie radiale 1586
scierie 8637
scierie à produits de commerce 1749
scierie à ruban 520
scierie à scie circulaire 1586
scierie à vapeur 8055
scierie de gros débits 8637
scierie de sciages lourds 8637
scierie mobile 6469
scierie transportable 6469
scier un rondin 1122
scie sauteuse 1822
scie spirale 7915
scié sur dosse 3258
scié tangentiel 8408
scie transversale 2042
scie tubulaire à rondelles 6384
scieur 4481

sciure 929
sciure de bois 9644
sciure de bois fine 9644
scléréide 7300
sclérenchyme 6586
sclérenchyme libérien 4119
scléreux 7302
sclérite 7300
sclérose 8097
scolyte ambrosia 231
scolyte birayé 8173
scolyte de l'orme 5670
scolyte des cônes de pin rouge 6785
scolyte des cônes du pin lodgepole 5043
scolyte du bois 231
scolyte du bois colombien 1732
scolyte du cambium 9552
scolyte du pin 6252
scolyte du sapin de l'Ouest 9337
scolyte européen de l'orme 2894
scolytes 665
scolytes de l'écorce 539
scrambleur 7312
scrofulariacées 3126
sculpter sur bois 1360
sculpteur 1362
sculpteur sur bois 1363, 9502
sébestier 4799
sécateur 5168
sécateur à main 4106
sécateur d'abattage à simple tranchant 7636
sécateur universel 322
séchage à l'air 152
séchage biologique 727
séchage de bois de sciage 5202
séchage de copeaux de bois 9620
séchage du bois 2423
séchage par évaporation 8767
séchage par fumage à chaud 3180
sécheresse 2635
sécheresse en cime 7989
sécherie de bois 5203
sécherie de graines forestières 3588
séchoir à bois 9537
seconde bille 5436
seconde frappe d'une lime 9126
seconde taille d'une lime 9126
seconds 7365
secoueur d'arbre 8845
secoueur de branches 4971
secoueuse mécanique 8845
secousse 7483
section de flottage 6692
section de grume supérieure 9138
section forestière 2552
sections de bois marchandes 5396
sections marchandes 5396
section sur quartier 6659
sécurité de coupe 5082
sélection consanguine 4435
sélection d'abatteur 3043
sélection de graines forestières 3583
sélection de masse 5325
sélection massale 5325
sélection naturelle 5696
sélectionneur 1026
sélection phénotypique 6205
sélection positive 6472
seltis de Chine 1499
selva 7423
semelle d'abattage 1563
semelle du trait d'abattage 7283
semence 7370
semence certifiée 1420
semence homologuée 1420
semence orthodoxe 5961
semence récalcitrante 6748
semences d'arbres forestiers 3628
semences d'essences forestières 3628
semences sélectionnées 7403
semencier 5569
semencier conservé en prévision d'incendie 3189
semer 7371
semi-remorque articulée 371
semis 7388
semis à la volée 2521
semis améliorés morphologiquement 5559
semis à racines nues 532
semis à volée 1044
semis de l'année 9756
semis direct 2521
semis direct à la volée 2521
semis en douille 492
semis en lignes 522
semis en lisière 7384
semis en plein 1044
semis en pot 6483
semis en récipient 1893
semis en sillons 7869
semis établi 2769
semis fiche 492
semis forestiers 8917
semis génétiquement améliorée 3818
semis par avion 87
semis par bandes 522
semis partiel 6088
semis sous abri 4567
semis sur planche 648
semoir de glands 47
semoir en ligne 3387
semoir forestier 3585
sen 7436
séné 1870
séné d'Alexandrie 1870
senelle 4174
sénescence 7438
sensitivité de la station 7673
sentier de halage 3277
séparateur d'arbres 8919
séparateur de copeaux 1537
séparation de copeaux et d'écorce 1511
séquence de cernes 7451
séquence de coupage 2202
séquoia à feuilles d'if 1269
séquoia côtier 1269
séquoia géant 3844
séquoia géant de Californie 1269
séquoia toujours vert 1269
seraya 4831
séraya 4831
séraya rouge 4949
sère 7457
serfouette 4081
serial 7453
série 7457
série de peuplement 8032
série des grumes 7134
série normale des classes d'âge 5785
seringa à petites feuilles 5004
seringa commun 8335
seringa de Caucase 1386
seringat commun 8335
seringat de Caucase 1386
sérotinal 7458
serpe 1204
serpent forestier 8925
serrage d'élingue pour câble 9465
serre 3946
serré 7461
serré-cilié 7460
serre-joint 4653
serre-joint de bois 9510
serreur de chaîne 1436
servitude de bois 8573
servitude de ramassage de feuilles mortes 6947
servitude forestière 8573

seuil de rentabilité 1015
seuil en bois 9556
seuils transversaux 8774
seule viminal 619
sève 7208
sève arborescente 7208
sève de bouleau 757
sève d'érable 5294
sève de racine 7045
shaitan 9363
shallaki 4463
shéferdie argentée 7600
shepherdie du Canada 771
shou palmiste 1238
sifflet d'abattage 1214
signe à l'écorce 540
signification d'objet de récréation 6755
silky-oak rouge 3966
en ~s 3777
sillon de semis 7379
sillon linéaire de déroulage 7307
silo de copeaux 2496
simarouba 9373
simaroubacées 147
simaruba 9373
simarubacées 147
sinuosité des fûts 2028
sinus 5028
sinus de lobe 5028
sipo 9150
sirex 9693
sirop de bouleau 758
sirop d'érable 5295
site critique 2023
site de la vie sauvage 9412
site forestier 3591
site forestier après d'un incendie 6478
site forestier interprétatif 4563
slacker 5017
soc 5068
soc à lame de rasoir 8898
soc de bouleau 757
société forestière 3355
société végétale forestière 3331
sociologie forestière 3596
soins au bois 9676
soins aux fourrés 8455
soins aux gaules 8455
soins aux plantations 8453
soins culturaux 2131
soins culturaux arbre par arbre 8454
soins culturaux au peuplement 8456
soins culturaux de jeunesse 8458
soins favorisants l'accroissement 8457
soins stabilisateurs 8029
sol 7833
au-dessus du ~ 11
solanacées 5734
sol brun forestier 1079
sol brun forestier acide 39
sol brun forestier calcimorphe 1252
sol brun forestier podzolisé 1078
sol brun méditerranéen 6770
sol brun podzolique 2107
sol brun-roux de forêt 6770
sol de forêt 9547
sol de prairie boisée 6492
sol d'une forêt exploitable 1746
sol entièrement consacré à la forestière 19
sol forestier 9547
sol forestier à gros grains 1679
sol forestier de montagne 5585
sol forestier marécageux 8318
sol gris-brun forestier 3967
sol gris forestier 3969
sol gris forestier à structure nuciforme 5827
sol gris forestier gleyifié 3861
solidité de bois 2669
solive 9592
sol noir de forêt 789
sol podzolique de marais 896
sol podzolique marécageux 896
sols cinnamiques 1583
sols podzoliques 6408
sols podzoliques de tourbe mottière 2409
sols sensitifs 7442
sol strié 8122
sol tempéré 1079
sol tourbeux forestier 9712
solvant 8505
sommet allongé anormalement 5544
sommet commercialisable 5398
sommet d'arbre 8699
sommet de bille 7252
sommet de couvert 8701
sommet de fût 8700
sommet de tige 8700
sommet en touffes 1210
sommet marchand 5398
sommier 9022
son 929
sondage par points horizontaux
sonneratiacées 7848
sophora à feuilles de vesce 9214
sophora de la Nouvelle Zélande 5729
sophora du Japon 4626
sorberaie 5580
sorbier 2903
sorbier décoratif 7552
sorbier de montagnes d'Amérique 252
sorbier des Alpes 8323
sorbier de Sargent 7222
sorbier de Sitka 3943
sorbier des montagnes 7552
sorbier des oiseaux 2903
sorbier des oiseleurs 2903
sorbier domestique 8991
sorbier intermédiaire 8322
sorte d'arbre 8849
sortie d'une glissoire 2836
souche 8937
sur ~ 8018
souche à racines 8211
souche d'ancrage 279
souche de conifère 8208
souche de hêtre 1089
souche-mère 1934
souche saisonnée 7334
souche sèche 7334
souches et racines 4031
souche verte 3958
soudage par points 7933
soudure par points 7933
soufflerie 2675
soulevage 7055
soulevage des racines 7055
source de carbone 1334
source de cime 7940
souris forestière 3510
sous-arbrisseau 2690
sous-bois 9070
sous-bois obstructif 5860
sous-charpentière 7341
sous-classe physiognomonique 6217
sous-compartiment 8231
sous-division 8231
sous-espèce 8239
sous-étage 9070
sous-frutescent 8252
sous-mère 7341
sous-parcelle 8231
sous-plantation 6364
sous-produits 9500
sous-produits de bois 9500
sous-soleuse 8963

spacieux 7008
sparmannie d'Afrique 115
spatule 4584
spécialiste du bois 9659
spécification dimensionnelle 5366
sphinx à pin 6257
spinule 7912
spirée 5350
spirée à feuille de saule 9433
spirée à feuilles de bouleau 755
spirée à feuilles de germandrée 3832
spirée ormière 3832
spongieuse 4071
spore 7927
sporophore 7928
sporophylle 7929
squelette cellulaire 7681
stabilisateurs 5980
stabilisation des versants 7974
stabilité ombreuse 7477
stade d'arbre 8931
stade de bois jeune 9790
stade de buissons 1115
stade de fourré 8495
stade de gaule 7206
stade de perchis 6426
stade de semis 7391
stade de succession 8248
stade de succession écologique 7455
stade de vieux bois 5340
stade jeune 4424
stades d'épaisseur 8499
stade sériel 7455
stagnation 9507
stagnation de croissance 4015
standard de bois 9664
standard de peuplement 8111
standard forestier large 3643
standardisation de bois 8642
standardisation des pertes en foresterie 1979
standards de peuplement objectif 8414
standards d'utilisation 9156
staphylacées 817
staphylier 816
staphylier de Chine 1491
staphylier d'Europe 2885
staque 9519
statice arborescent 8878
station de génétique forestière 8805
station de protection forestière 3545
station forestière 3591
station forestière phénologique 3525
statique forestière 3326
statistique forestière 3606
statut pour le flottage 3276
stèle 8058
steppe arborée 3607
steppe arbustive 3607
steppe-erme forestière 3450
sterculiacées 1550
sterculie des rochers 594
stère 2115
stérigmates 8089
stinkwood 807
stipule 8101
stock de plantation 6363
stocker 8115
stolon 7132
stolon court 1095
stomate 8120
stonewood 6196
storax américain 259
storax véritable 3825
strate 8150
strate arborée 8866
strate arborescente 8866
strate de grumes 8545
stratégie de foresterie indicative 4472
stratégie de peuplement 8030
strate herbacée 3116
strate inférieure 9070
stratification 8147
stratification froide 6496
stratobionte 8149
striation de l'écorce 544
strie 3296
strié 8161
strie de carre 1541
strie de crevasse 1541
stringybark à petites feuilles 7769
stringybark brun 1091
stringybark rouge 6794
strobile 8818
strouille 4509
structure
à ~ fine 3157
structure adventive 81
structure arborescente 8936
structure d'accès permanent 6168
structure d'accès transitoire 8448
structure de cernes annuels 8188
structure de fibres 3100
structure de la forêt 1886
structure de la réserve de bois sur
structure de peuplement 1886
structure de sol forestier 3602
structure d'espèce arborescente 8927
structure du bois 9666
structure étagée 4842
structure fibreuse 3103
structure inéquienne 174
structure inéquienne irrégulière 4589
structure inéquienne régulière 6838
structure verticale 9208
stump 8209
styracées 8128
styrax officinal 3825
subclimax 8230
suber 1956
subérification 8233
subérisation 8233
subor 8234
subor marécageux 8314
substance active 55
substance attractive 4509
substance de bois 9667
substance de croissance 6221
substance organique 5948
substances entrantes dans la composition du bois 2972
substances nutritives 5826
substrat 8240
substratum 8240
succession 8246
de ~ 7453
succession d'abattage 2202
succession des essences 8249
succession forestière initiale 2701
succession forestière tardive 4817
succession primaire 6547
succession secondaire 7354
succulence 8250
succulent 8082
suçoir secondaire 6137
sucre de bois 9670
sucs végétaux séchés au four 4709
sucupira amarello 4784
suffisamment reboisé 7224
sugi 4615
suie 7849
suintement 9323
suite d'éclaircies comparées 8517
suite de coupes 2202
suite des classes d'âge 127
suite normale des classes d'âge 5785
sumac à bois glabre 7791

sumac à feuilles entières 4909
sumac ailé 3247
sumac amarante 365
sumac à petites feuilles 5005
sumac aromatique 365
sumac brillant 3247
sumac cotinus américain 258
sumac de Virginie 260
sumac empoisonné 6412
sumac feux-vernix 9306
sumac fustet 2908
sumac glabre 7791
sumac grimpant 6412
sumac irrégulièrement lobé 6022
sumac lustré 8315
sumac ovale 8255
sumac vénéneux 8315
sumac vinaigrier 260
super-espèce 8271
superficie boisée 9542
superficie boisée accessible 30
superficie boisée affermée 6867
superficie boisée native 5671
superficie boisée native nouvelle 5725
superficie brûlée admissible 194
superficie des feuilles 4852
superficie forestière nette 3763
supervision forestière 3612
supplément 202
support 8902
support d'arbre 8902
support en A 100
support intermédiaire 4548
support terminal 2815
suppressé 6011
suppression d'un incendie forestier 3431
suppression par l'ombre 8276
supprimé 6011
suramen 8277
surbille 7362
surbille de découpe 8688
surbille de découpe marchande 8688
surbille de tête 8688
surbille supérieure 8688
surcharge 6007
surcoupe 2935
sureau 2892
sureau à fruits rouges 2907
sureau à grappes 2907
sureau arborescent 787
sureau blanc 246
sureau bleu 858
sureau commun 2892
sureau de montagne 2907
sureau du Canada 246
sureau noir 2892
sureau pubescent 2907
sureau rameux 2907
sureau rouge 2907
sureau rouge d'Amérique 787
sureau rouge d'Europe 2907
surelle 5966
suretti de grand bois 2570
surexploitation 2935
surfaçage 8281
surface
  en fonction de la ~ 7674
  plus petite ~ unitaire d'éclaircie 8508
surface abritée 354
surface à cultiver 355
surface à exploiter 2189
surface à régénérer 355
surface boisée 9542
surface brûlée tolérable 194
surface circulaire à hauteur de poitrine 598
surface cisaillée 7499
surface couverte de buissons 1108
surface de bois 8650
surface de coupe 2180, 2189
surface de coupe communale 1815
surface de débardage 2629
surface dégagée 5907
surface de gestion 3330
surface de houppier 1314
surface de la feuille 4890
surface d'entaille de sciage 7247
surface de projet 6578
surface de sol boisé 9542
surface d'essai 7173
surface d'exploitation à blanc 1621
surface d'origine d'un incendie 4422
surface d'ouvrage 6578
surface du tronc 900
surface en bois 9557
surface expérimentale 7173
surface forestière 9542
surface incendiée 1184
surface incendiée admissible 194
surface libre 5907
surface non moulée 9083
surface organisée pour le lutte anti-incendie 3193
surface portante des arbres isolés 352
surface terrière d'arbre 597
surface terrière d'un peuplement 2032
surface terrière mesurable 5360
surface terrière résidu 6890
surface terrière unitaire d'un peuplement 2032
surface vide 9083
surface vidée de ses bois 5066
surgeon 7049
surgreffage 8691
surier 1961
surlongueur 202
surlongueur de jonction 1228
surmaturité 6000
surpaissance 6001
surpâturage 6001
surrendement 6002
sursemis 4567
surveillance continue de la sécheresse 2639
surveillance des forêts 5530
surveillance du feu 3210
surveillance du matériel relatif 8107
surveillant d'incendie 3191
surveillant d'un lieu d'incendie 3191
surveillant forestier 3613
survivant 2567
susier 2892
sycomore 8355
sycomore de l'Arizona 361
sylvichimie 2370
sylviculteur 344
sylviculture 7610
sylviculture à courte rotation 7545
sylviculture à petite échelle 7772
sylviculture de base 611
sylviculture intensive 4450
sylviculture naturaliste 5683
sylviculture proche de la nature 5683
sylviculture steppique 8087
sylviculture supplémentaire 4450
sylvipastoralisme 7625
sylvipisciculture 7626
sylvofaciès 8357
sylvopâturage 3452
symbionte 8358
symbiote 8358
symphorine à feuilles rondes 1940
symphorine à fruits blancs 1804
symphorine à petites feuilles 7768
symphorine blanche 1804
symplocos 8333

synfolium 8360
syringa à bouquets 8335
syringat à bouquets 8335
système agroforestier 145
système agrosylvicole 145
système aquasylvicole 335
système à trois câbles porteurs 8534
système d'arrêt des billes 5119
système de câble-grue à tension fixe 8551
système de câbles 6674
système de coupe 5084
système de coupe à petite échelle 7773
système de coupes d'abri 3801
système de coupe suivant les courbes de niveau 1905
système de débardage 7691
système de débardage en grande longueur 8873
système de débardage en toute longueur 8873
système de débardage et empilage de bois court 2162
système de débusquage 1123
système de débusquage à pince 3913
système de débusquage de troncs entiers 8873
système de déplacement pour billes de bois 5097
système de deux câbles porteurs 2586
système de drainage 2614
système de futaie avec surréserves 4283
système de gestion d'abattage 2203
système de gestion forestière 3495
système de jardinage 7620
système de la forêt jardinée 7408
système d'emballage de sciure 7235
système démêleur 9114
système de pâturage planifié 6330
système de racines superficielles 7487
système de racine vigoureux 142
système de régénération par coupe progressive 7512
système de régénération par coupe rase avec semencier 7400
système de régénération par coupes rases en bandes alternées 219
système de régénération par coupes rases en bandes contiguës 6572
système de ressources forestières 8630
système des coupes progressives irrégulières 4587
système d'étai 4065
système de taillis-sous-futaie 1937
système de taillis sur têtards 6430
système de traînage relevé 4291
système de traînage relevé par ballon 4290
système de traitement en lisière unique 6572
système de traitement par coupes en lisières allongées 9088
système de traitement sur arbres d'avenir 1566
système de verrouillage fermé 1654
système foliacé 3303
système par câble à freiner 89
système par câbles de remorque fermés 8738
système par coupes en bandes alternées 219
système par trouées 3801
système pivotant 8345
système pluriétagé 5632
système racinaire 7050
système radiculaire 7050
système sans câble porteur 5772
systèmes au sol 3982
systèmes de protection de cime 2089
système sylvicole 7621
système sylvopastoral 7625
système tributaire 8967
système uniligne 7644

tabac glauque 8945
tabebuia humifère 5346
tableau de conversion 1917
tableau d'inventaire 8033
table d'abattage 1563
table d'accroissement 4457
table d'alimentation pour billes 5011
table de base de ventes 610
table d'éclaircies 8518
table de classement 3944
table de coefficients de formes d'arbres 3664
table de cubage composée 1739
table de cubage de Doyle 2604
table de cubage diagrammatique 2463
table de cubage en pied-planches 878
table de cubage internationale 4559
table de débardage 8363
table de décroissement 8409
table de défilement 8409
table de Doyle 2604
table de mesure de volume de billes 7112
table de mesuréments 7275
table de petit bois 7782
table de peuplement 8033
table de production 9784
table de production à densité variable 9167
table de production d'équienne 2921
table de rendement 9784
table de rendement empirique 2798
table de rendement normal 5786
table des assortiments 3896
table des billes 7251
table des classes d'âge 129
table des surfaces circulaires 8362
table des surfaces terrières 8362
table d'estimation de volume d'arbres 8957
table de tri 3944
table de triage 3944
table de tronçonnage 1133
table de vente de bois 8634
table de volume standard 8007
table d'exploitations 4163
table du cubage des peuplements 8036
table du cubage locale 5037
tables d'espèces commerciales 5265
tachage 9187
tache 7907, 8154
tache brune 1549
tache brune chimique 1462
tache colorée 7907
tache colorée de l'aubier 7215
tache de baguette 2047
tache de grume 5118
tache de l'aubier 7215
tache de sève 7215

tache foliaire 4889
tache interne de l'aubier 4538
tache médullaire 5390
tache minérale 5462
tache noire 783
taches apparaissantes sur de bois entreposé 2047
taches blanches 9384
tache tannante 8396
tacheté 6131
tache traumatique de l'écorce 8786
tacheture 7907
tacheture de pourriture 7081
tacheture foliaire 4889
tache verte 3956
taïga 8368
taïga à bouleaux 759
taïga de montagne 5592
taillage des racines 7040
taille de cime 2093
taille de copeaux 1540
taille de formation 6612
taille de forme 2083
taille de perche 6422
taille des cimes 2096
taille d'hiver 9462
taille douce 7789
tailler 1645
taillis 1928
taillis à courte révolution 7544
taillis à courte rotation 7544
taillis à croissance rapide 7544
taillis à écorce 5833
taillis avec têtard 1936
taillis composé 9033
taillis d'aulnes 164
taillis d'aunes 164
taillis de châtaigniers 7945
taillis de chêne à écorce 5835
taillis fureté 7405
taillis linéaire 1930
taillis paraneige de brousse 1114
taillis réglé 1935
taillis simple 1928
taillis-sous-futaie 7947
tali 1366
talipot 8373
talisay 4464
talle 7132
talon 4224
talonnage 4226
talus de déblai 7753
tamaricacées 8382
tamariçaie 8381
tamarille 8946
tamarinier 8379
tamaris 3706
tamaris africain 116
tamaris à quatre étamines 7939
tamaris de printemps 7939
tamarisque 3706
tamarix 3706
tamarix d'été 3233
tambour d'enroulement 9267
tambour d'enroulement de câble de débardage 9267
tamis à copeaux 1535
tampon de bois 9555
tanière 2365
tanin 8393
tannin 8393
tapis 5335
tapis de bois 9501
tapis de débardage 5077
tapis de protection en bois 8654
tapis végétal 6337
tardif 7458
tarière 6342
tarière d'accroissement 4451
tarière de Pressler 4451
tarière mécanique 4326
tarif de cubage à la souche 8212
tarif de cubage à une entrée 7638
tarif de cubage en entrée multiple 5612
tarif de cubage gradué 8415
tarif des arbres moyens 5359
tarif d'estimation des bois sur pied 8957
tarif gradué 8415
tas de brindilles 4193
tasseau 626
tassement du sol 7834
tasseur de copeaux 1522
tatajuba 478
taukkyan 9719
taux de base 609
taux de bois d'œuvre 6150
taux de cime 2090
taux de germination 3836
taux de placement forestier 3522
taux de prélèvement permi 6182
taux de production normal 9155
taux de revenu forestier 6729
taux de survie 1322
taux de terrain forestier 3522
taux d'extraction périodique en éclaircie 6166
tawhai 8420
taxacées 8431
taxation forestière 3617
taxe prélevée sur le bois 2681
taxe sur le défrichement des surfaces en nature de bois ou de forêts 8430
taxinométrique 8426
taxinomie 8429
taxinomiste 8428
taxis 1814
taxocénose 8423
taxodiacées 8425
taxodier chauve 786
taxologue 8428
taxon 8427
taxonométrique 8426
taxonomie 8429
taxonomiste 8428
tchitola 8432
teak de Rhodésie 9793
technicien de bois 9675
technicien en sylviculture 3579
technicien forestier 3579
technicien forestier professionnel nominatif 6564
technique d'abattage 3076
technique d'amélioration backcross 463
technique de bois 9674
technique de reboisement 6805
technique d'usinage de bois 9702
technique sylvicole 3615
technologie de construction en bois 8582
technologie forestière 3401
teckeraie 8435
teigne minière du mélèze 4788
teinture provisoire 3074
télémètre-dendromètre 6718
téléphérage 4288
téléphérage au sol 3988
téléphérage intégré 4387
téléphérage par gravité 3922
téléphérage par gravité à câble fixe 3237
téléphérique 84
téléphérique forestier 1242
tempérament d'arbre 8943
tempo d'abattage 8446
temps d'abattage 3077
temps d'alignement 169
temps de culture 8665
temps de nettoiement 5555
temps de passage 6514
temps de rotation 9011
temps de scie de tête 4189
tenancier forestier 344
tendance évolutive 2446

tendeur de chaîne 1436
tendre 7822
teneur en eau du combustible 3737
teneur en eau initiale 310
teneur en éléments 5824
teneur en humidité après séchage du bois 3141
teneur en humidité de cœur 1950
teneur en humidité des matières combustibles 3737
teneur en humidité du bois 5518
teneur en humidité exploitable 2816
tensiomètre 8464
tension de câble 7704
tension de croissance 4027
tenthrède à tête verte de l'épinette 3945
tenthrède du mélèze 4792
tenthrède du saule 9435
tenthrède européenne de l'épinette 2910
tenthrède européenne du peuplier 4090
tenthrède jaune du groseillier 1775
tenthrède mineuse de l'orme 2790
tenthrèdes 7240
tenure de bois 8652
tenure foncière 4780
tenure temporaire 8450
terminologie forestière 3619
termite 8476
terrain 4769
terrain à viander 3309
terrain broussailleux 353
terrain brut 4212
terrain brut boisé 7099
terrain cerné de lignes pare-feu 1188
terrain couvert de forêt 3480
terrain déboisé 3443
terrain défriché 390
terrain d'industrie du bois 3467
terrain en friche 9275
terrain forestier 3480
terrain forestier d'avenir 1321
terrain forestier montagnard 5586
terrain non boisé 3443
terrain nu 9275
terrain très instable 9211
terre 4769
terre à bois 9599
terre accrue 34
terre brune 1079
terre complètement reboisée 3763
terre d'alluvion 5059
terre forestière 3480
terre forestière à degré d'occupation non prouvée 9106
terre forestière boisée non quantifiée 9107
terre forestière boisée partiellement 6085
terre forestière convenante 8257
terre forestière de fenaison 4175
terre forestière de haute productivité 4299
terre forestière en voie de repeuplement 6810
terre forestière gestionnée 5261
terre forestière incomplètement régénérée 5801
terre forestière insuffisamment régénérée 5801
terre forestière non appropriée 9121
terre forestière non capable 5749
terre forestière non gestionnée 9103
terre forestière non industrielle 5762
terre forestière non réservée 5769
terre forestière productive 6562
terre forestière régénérée 6810
terre hors forêt 5868
terre inondée forestière 3399
terre marginale 5304
terre nue 9275
terre productrice de bois 8625
terres à réglementation 6839
terres inopérables 4508
terres jaunes 9763
terres sans réglementation 9110
terres spéciales 7890
terre vierge 9408
terril 5464
test clonal 1647
test de descendance 6571
test de germinabilité 3837
test de provenance 6604
têtard 2083
tête 2087
tête abatteuse 3063
tête à cisaille à pince collectrice 5621
tête à ébrancher 4969
tête collectrice à cisaille 5621
tête d'abattage 3063
tête d'abattage à cisaille 3044
tête d'abattage à fraise 403
tête d'abattage à pince collectrice 5630
tête d'abattage à résistance au morcellement 5771
tête d'abattage à scie à chaîne 1433
tête d'abattage collectrice 5630
tête d'abattage-ébranchage 3068
tête d'ébranchage 4969
tête de chat 1378
tête dédosseuse 7708
tête d'un incendie 1189
tête ébrancheuse 4969
tête éclatée 1074
tetracentracées 8481
texture bosselée 1168
texture brute 1678
texture du bois 9678
texture en arêtes de poison 4261
texture en ligne cassée 1062
texture hétérogène 9082
texture homogène 2923
texture irrégulière 1168
texture modérément fine 5507
texture ondulée 9078
texture rubanée 8175
texture rubanée en crayon 6142
texture striée 8175
théacées 8434
thé des bois 8433
théorie de gestion forestière 8487
théorie de monoclimax 5538
théorie polyclimacique 6437
thérapie forestière 3621
thermophile 8491
thermophyte 8492
thuya 8893
thuya de Chine 2725
thuya de Lobb 8893
thuya de l'Occident 5795
thuya d'Orient 2725
thuya géant 8893
thuya occidental 5795
thuya oriental 2725
thylle 9037
thylle scléreux 7303
thylloïde 8540
thyméléacées 5427
tiama 8541
tibouchine 3871
tige
à ~ unique 7637
sans ~ 8078
tige centrale 1415

tige de guidage 8779
tige d'essai 8480
tige d'osier 9437
tige élite 6395
tige fine 8522
tige intermédiaire 8065
tige latérale 994
tige marchande 5393
tige moyenne par accroissement 4456
tige ondulée 4721
tige pattue 595
tige plus 6395
tige récente 3708
tiges courbées 8388
tiges embrouillées 8388
tige utilisable industriellement 5393
tigre 4751
tiliacées 4983
tillaie 4978
tille 82
tille de tilleul 4972
tilleul 4976
tilleul à grandes feuilles 4801
tilleul à grandes feuilles d'Europe 4801
tilleul américain 237
tilleul à petites feuilles 7767
tilleul argenté 7607
tilleul argenté pleureur 6143
tilleul blanc 667
tilleul de Caroline 1343
tilleul de l'Amour 266
tilleul de Sibérie 266
tilleul d'Europe 1786
tilleul intermédiaire 1786
tilleul noir 237
timonerie 7467
timonerie de débardeur 9743
timonerie de frein 970
timonerie en câble d'attache 8366
tire-sève 7199
tisseuses 4886
tissu de conduction 1857
tissu de liège 6201
tissu des rayons hétérogènes 4265
tissu des rayons homogènes 4345
tissue cellulaire 1406
tissue corticale 1975
tissue d'appui 8274
tissu fibreux 6586
tissu ligneux 9679
tissu phellogène 6201
tissu traumatique 8787
toisé 2176
toit de l'entaille d'abattage 7283
tola 5731
tola branca 5731
tôle ondulée 9270
tolérance 8673
tolérance de l'écorce 200
tolérance d'épaisseur 8500
tolérance des arbres forestiers 3629
tolérance d'incendie 3213
tolérance intermédiaire 4549
tolérant au sel 7164
tomate en arbre 8946
tonne 8675
tonnellerie 1920
tonnellerie étanche 8549
tonnellerie non étanche 7713
topiaire 6612
topographie 8703
topologie d'arbre 8948
topophyse 8704
toposéquence 8705
tordeuse 3223
tordeuse à tête noire de l'épinette 794
tordeuse des bourgeons 1146
tordeuse des bourgeons de l'épinette 7951
tordeuse des pousses du pin 6269
tordeuse du pin gris 4602
tordeuses 4886
tordeuse tubicole du pin 6273
torreya de Californie 1276
torsion 9020
torus 6308, 6314
torus lobé 6308
totara 8734
touage 8737
touage enchaîné 480
touffe 4000, 9005
en ~ 5628
touffe d'arbres 1666
touffu 934
touille 4889
toundra alpine 5594
toundra à sous-arbrisseaux 5182
toundra boisée 3631
toundra de broussailles 7322
toundra de hummock 4395
toundra forestière 3631
tour à bois 9623
tour à feu 3216
touraillage 4711
tourbe 7811
tour ignifuge 3216
tournage du bois 9683
tourne-billes 5124
tourne-grumes 5124
tournesol mexicain 8882
tourneur sur bois 9682
tousser par quintes 669
tout bois de cœur 184
tout-venant 5454
toxicité de bois 9680
trabécule 8739
trac 7563
traçage 8743
tracé 4847
trace de frappe d'une hache 452
trace foliaire 4891
traces de laticifère 4828
trachée 9687
trachéide 8742
tracheïde de bois de compression 1851
trachéide de rayon 6739
trachéide disjointe 2532
trachéide juxtavasculaire 9175
trachéide septée 8145
trachéide transversale 6739
trachéide vasculaire 9173
tracteur à câble 8370
tracteur à chaînes 7799
tracteur à chenilles 7799
tracteur à pneus 7125
tracteur articulé 373
tracteur débusqueur 1246
tracteur d'étai 4066
tracteur d'extraction 2970
tracteur haut-le-pied 888
tracteur sur chenilles 7799
tracteur-tireur à pince 3913
trailler à l'hache 83
traînage à plat 3994
traînage de grumes en files 8754
traînage relevé 4288
traînage sur câble 7700
train de bois 8984
train de bois flotté 8584
train de flottage 8584
train de flottage en croix 6694
train de grumes 8758
train de récréation 6759
traîne à billot 7687
traîneau arrière de débardage 6746
traîneau d'écorce 549
traîneau de débardage avec fourches 3881
traîneau de débusquage 5115

traînée de grumes 8758
train routier 4171
traitabilité 8791
trait d'abattage 3066
trait de scie 7246
trait de scie à abattage 3066
trait de scie au petit bout 8687
trait de scie par en dessous 1214
trait de scie par en-dessous progressif 8088
trait de scie plongeant 930
trait de scie rectiligne 8141
traitement à croissance rapide 6724
traitement à la base de l'arbre 603
traitement à la cisaille 7501
traitement à la pression atmosphérique 5768
traitement arboricide cortical 603
traitement contre les taches de sève 318
traitement d'arbres à la forêt 4487
traitement de chablis 9442
traitement de parcelle forestière 3016
traitement des souches 2179
traitement du bois 9630
traitement en futaie à deux étages 9036
traitement en plein 5990
traitement en taillis avec réserves 9033
traitement en taillis-sous-futaie 8129
traitement forestier 7623
traitement intermédiaire 4550
traitement par bandes 8169
traitement sans pression 5768
traitements commerciaux 1751
traitements de santé forestière 3460
traitements précommerciaux 6499
traitement sylvicole 7618
tramway forestier 5121
tranchant 819
tranche 2180
tranchée 8171
tranchée collectrice 4063
tranche nette 1606
tranche sans bravure 1606
tranche sur maille 6659
trancheuse 8963
transect de peuplement 8034
transfert de poison 466
transition duramen-aubier 4204
transition forêt-steppe 3608
transportation de bois 8595
transportation de bois éloigné 2966
transportation par glissoire 1577
transportation primaire 6548
transport de bois par grumier 5123
transport de bois par hélicoptère 4239
transport des bois par ballon 497
transport des exploitations forestières 8595
transporteur à chaîne 1424
transporteur des exploitations forestières 8596
transporteur-élévateur 8581
transporteur monocâble 5536
transport forestier interne 4558
transport par câble simple 4166
transport par camions 8985
transport par charrettes 1358
transport par fûts entiers 5615
transport par hélicoptère 4239
transport par troncs entiers 5615
transversal au fil
au sens ~ 49
travail aux champs 3121
travail du bois 9698
travail du sol 8554
travail forestier 9724
travailleur de bois 9699
travail mécanique du bois 5376
travers du fil 2814
au sens en ~ 49
traverse 626, 2038
traverse demi-ronde 4095
traverse divisée 7921
traversine de mine 1725
trèfle 8961
trèfle blanc 9366
trèfle de Virginie 6035
tremblaie 387
tremble 2880
trempage 2511
trepan pilote 6238
treuil à déboisement 1160
treuil à dessoucher 1160
treuil à tambour 2643
treuil d'arrachage 1160
treuil de câble 7060
treuil de chargement de bois 8609
treuil de débardage 7692
treuil de débusquage 7692
treuil de levage 8603
treuil d'essouchage 1160
treuilliste 7925
triage de grumes en eau 9292
triage de grumes en terre 2651
triage des grumes 5092
triage forestier 3333
triage par diamètre 2483
triage par longueur 4913
triangulation 8966
trichomane 8950
trieur de branches 4966
trieur de copeaux 1518
tringle de clouage 6034
tringle en J 4664
triqueballe 894
tristanie à feuilles en bouquets 1036
trochodendracées 9357
troène 1799
troène commun 1799
troène de Chine 3232
troène de Floride 3287
troène de Japon 4640
troène d'Europe 1799
trommel 7056
trompillo 6237
tronc 8934
tronc caverneux 4334
tronc coupé 2600
tronc d'arbre 8934
tronc d'arbre entier 2835
tronc d'étai 4064
tronce 607
tronc entier 2834
tronce piquée 3680
tronçonnage 5100
tronçonnage de bois long 1132
tronçonnage de tige 8069
tronçonnage du bois 5100
tronçonnage en dessous 1215
tronçonnage par groupes 1174
tronçonnage par la classe de qualité 1127
tronçonnage par valeur maximale 1129
tronçonnage par volume maximal 1130
tronçonner 1122
tronçonneur 1124, 7268
tronçonneur de fragments 1572
tronçonneuse 2043, 5056
tronçonneuse à arc 952
tronçonneuse à chaîne mécanisée 6488
tronçonneuse alternative 2609
tronçonneuse-fendeuse 3220
tronçonneuse mobile 6470
tropophytes 8982

trou de crochet 4361
trou de débusquage 2607
trou de griffe 4361
trou de nœud 4732
trou d'ensemencement 6353
trou de pic 5711
trou de plantation 6353
trou de ver 4511
trou d'insecte xylophage 4032
trouée de coupe 5038
tsuga 1296, 4248
tsuga de Californie 5588
tsuga de Caroline 1345
tsuga de l'ouest 5588
tsuga de Patton 5588
tsuga du Canada 1296
tube à latex 4829
tube à semence 7381
tube criblé 7593
tube de drainage 2615
tube de plantation 6361
tuber 9003
tubercule 7036
tube sécréteur laticifère 4829
tugaï 6961
tuilage 2141
tulipier 9008
tulipier d'Amérique 9008
tulipier de Virginie 9008
tulu 504
tumeur bactérienne du collet 2080
tupélo à deux fleurs 8305
tupélo aquatique 9304
tupelo chinois 1507
tupélo des forêts 810
tupélo noir 810
turbinella 7559
tutelle forestière 3488
tuteur 8273
tuteurage d'arbre 8932
tuteur d'arbre 8902
tuyau de drainage 2615
type de forêt cendrée 1284
type de forêt chênaie-pineraie 5847
type de forêt chêne-hickory 5838
type de forêt lichénique 1593
type de parcours 6721
type de peuplement 3363
type de site 7675
type de station 7675
type de station total 8731
type forestier 3363
type forestier problématique 6556
types d'exploitation des terres forestières 4782

ulmacées 2788
umbaua 107
umgusi 9793
uninodal 9092
union de bois 4660
unité d'aménagement des habitats 4078
unité d'aménagement forestier 3496
unité d'analyse d'habitat 4073
unité de bois absolument sec 906
unité d'échantillonnage 7178
unité de copeaux 9095
unité de débardage 5089
unité de forêt défrichée 9096
unité de terrain primaire 6540
unité de traitement 8796
unité de vente 7159
unité d'intervention 8461
unité physiographique 6219
unités spatiales d'agroforesterie 144
université forestière 3462
upas 6414
urgence du rajeunissement 9143
uriam d'Assam 4645
urne 9144
usage d'arbre entier 1832
usager au bois 6940
usages forestiers 3576
usinage 9609
usinage de bois 9630
usinage du bois à la toupie 9609
usine de bois de sciage 8637
usine de conversion du bois 5448
usine de sciage 8637
utilisateur de bois 8655
utilisation de bois 9154
utilisation de la forêt soutenue 8298
utilisation de produit forestier 3541
utilisation de ressource intégrée 4526
utilisation des rémanents 7731
utilisation dispersée 2537
utilisation hors forêt 5760
utilisation intégrée de la forêt 5613

vaccin 4507
vaccination 4506
vacciner 9157
vaciet d'arbre 8926
vacinet noir 785
vaisseau 9687
vaisseau de bois 9687
val 2360
valeur actualisée nette 5718
valeur calorifique 1194
valeur commerciale 5313
valeur d'abattage 3056
valeur d'avenir de la forêt 3412
valeur de consommation 2952
valeur de gonflement 8338
valeur de la forêt 3536
valeur de réalisation immédiate 2952
valeur de retrait 9653
valeur des bois sur pied 8196
valeur des ressources 6914
valeur de terre forestière 3483
valeur du marché 5313
valeur du peuplement 8196
valeur forestière non ligneuse 5776
valeur marchande 5313
valeurs en risques 9164
valeur sur pied 8196
valeur vénale 5313
vallée 2360
vallonée 3937
valorisation 677
VAN 5718
vanille 9165
vannerie 618
varangue 3283
variation génétique 3822
variation spatiale en forêt 7885
variégation 9170
variété 9171
variété d'haut rendement 4306
vase de filtrage 7473
vastringue de bois 8639
végétal 9178
végétation 9176
végétation arborescente 8953
végétation compétitrice 1824
végétation concurrentielle 1824
végétation corticole 1976
végétation de compétition 1824
végétation de la forêt vierge 5955
végétation de sous-étage 9070
végétation détériorée 2544
végétation du sol 3986
végétation embrouillée 8389
végétation herbacée 4252
végétation hors des strates 5763
végétation mycélienne 3772
végétation non couverte de forêt 5921

veinage 9187
veine de résine 4058
veine résineuse 4058
veines déchirées 1524
veinure 9678
veinure ondulée 3246
veinure sur coupe transversale de bois 2803
veld 1208
venelle 181
vente à diamètre limite 2477
vente après abattage 7157
vente de bois 8631
vente de bois par adjudication public à la criée 8557
vente de bois sur pied 7157
vente par unité de volume 6119
vente par volume de billes en pied-planches 5112
ven-ven 5405
verbénacées 9210
ver cordonnier des prunes 6388
verdiau 6801
verdissement 9226
verdoisis 3885
verdure 9200
verger à graines 8915
verger à graines clonal 1646
verger à graines de familles 1646
vérin de relevage 4932
vernalisation 9203
vernation 6502
vernis du Japon 4755
versant 7750
versant en friche 534
vers du bois 8562
verse 5042
versoir 1021
vert-forestier 3453
verticille 995
verticille annuel 5901
verticille transitoire 8789
vésicule 834
vésicule de résine 6893
vétéran 5885
viabilité 9215
viabilité 3833
viabilité germinative 3833
viandis 8806
vibreur de troncs 8845
vidange de bois 8595
vide 3443
vide dû aux chablis 849
vie de la forêt 3358
vieille écorce 8585
vieillissement d'arbre 132
vieux peuplement 5880
vigne de canyon 1318
vigne grimpante 9000
vigne sauvage 773
vigne-vierge 937, 9228
vigne-vierge commune 9228
vigne-vierge ordinaire 9228
vignobles 9223
vigueur germinative 3834
vigueur hybride 4269
village forestier 3639
vinaigre de bois 9688
vinettier de Thunberg 4612
vinettier du Canada 175
violence d'un feu 3207
viorne à feuilles d'aulne 4315
viorne à feuilles de prunier 792
viorne à feuilles d'érable 5293
viorne à feuilles plissées 4633
viorne à feuilles ridées 4899
viorne à manchettes 5656
viorne aquatique 2890
viorne comestible 5553
viorne dentée 368
viorne flexible 9307
viorne flexible 5656
viorne japonaise 4633
viorne mancienne 9307
viorne obier 2890
viorne ridée 4899
viorne rouillée 7141
vipère d'arbre 8954
virage 4169
virgilier 262
virgilier à bois jaune 262
virgilier du fleuve Amour 1509
virola 9232
virus végétal 6374
vitacées 9221
vitesse d'abattage 6727
vitesse de croissance 6728
vitesse de propagation 6730
vitesse de restauration 6752
vitesse d'exploitabilité 4162
vitesse d'infiltration 4489
vitex 3232
vitex à feuilles simples 7632
vitex à petites fleurs 7763
vive-arête 7967
à ~ 7963
voie d'accès vers l'incendie 3217
voie de câble 1245
voie de flottage 3277
voie de garage 8471
voie de réception 8471
voie de rondins 3987
voilement 2141
voilement longitudinal 951
voilement longitudinal de rive 7936
voirie forestière 3378
voiturée de bois 8984
vol de bois 8052
voleur de bois 8051
volis 7167
volume abattu 2200
volume à l'état sec 9252
volume à l'état vert 9253
volume apparent du bois empilé 7979
volume brut 3981
volume commercial 1753
volume d'arbre sans écorce 8952
volume d'arbre sous écorce 8952
volume d'arbres verts 9247
volume de billes 5125
volume de bois 9689
volume de bois abattus 5125
volume de bois d'œuvre marchand 5401
volume de bois d'usinage 5455
volume de bois résiduel 9109
volume de bois rond 5125
volume de franchise 5750
volume des cimes 2097
volume des grumes 9249
volume des pores 6463
volume d'exploitation 2200
volume du bois des branches 9245
volume du bois de sciage 9248
volume du bois en grumes 9249
volume du bois sur pied 9250
volume du matériel sur pied 9250
volume d'utilisation de la forêt 3637
volume en pied-planches de bois de sciage 5209
volume entier d'arbre 7652
volume entier d'arbre individuel 7652
volume exploité annuellement 296
volume foliaire spécifique 7905
volume le plus approprié 5934
volume marchand 5401
volume moyen d'arbre 443
volume permissible d'exploitation annuellement 193
volume réel 7842
volume réel de bois 8816
volume réel de récolte après façonnage 9246

volume sous écorce 8952
volume sur écorce 8894
volume total d'arbre 8733
volume total en pieds-planches 3311
vomiquier 8191
voussure 3013
voûte de cyprès 2213
voyage de débardage 7690
vrille 8462
vrillons à copeaux de bois 9648

wagon pour bois long 9256
wandoo 9265
weigela 9332
wellingtonia 3844
wellingtonia géant 3844
wengé 9336
winteracées 9459

xanthorrhée 405
ximénie d'Amérique 4322
xylème 9737
xylème primaire 6549
xylème secondaire 7357
xylène 6549
xylochimie 2370
xylogène 2855
xylophage 9739
xylorhize 9715
xylose 9670
xylotomie 9741

yarran 3846
yeuse 1773
yeuseraie 5836
yopo 1714
yucca filamenteux 9792

zamie 7153
zamier 7153
zapatero 9346
zéen 5477
zelkova du Japon 4639
zénobia 4350
zeuzère 4917
zingana 118
zonage d'utilisation forestière 3638
zonalité 9798
zone biogéoclimatique 725
zone brûlée permissible 194
zone cambiale 1286
zone d'abscission 16
zone d'alimentation à bois 8647
zone d'allumage 4422
zone de brûlage modifiée 5509
zone de chablis 849
zone de combat 2761
zone de coupe à blanc 1621
zone de dispersion naturelle 4072
zone de distance de visibilité 9236
zone de distance visuelle 9236
zone déficitaire en bois 8574
zone de filtration 3133
zone de forêt feuillue 1057
zone de gestion des ressources 6912
zone de gestion littorale 4758
zone de gestion riveraine 8157
zone de haute diversité biologique 4385
zone de nature protégée différée 2426
zone de nature vierge 9404
zone de nature vierge non différée 5756
zone de nœuds 4724
zone d'enquête 268
zone d'entaille 9055
zone de prairie arborée 6493
zone de prairie boisée 6493
zone de précipitation 2627
zone de production forestière 3538
zone de racines 7032
zone de reboisement 6804
zone de rémanents 7718
zone de répartition 4008
zone de répartition d'un arbre 4008
zone de ressources sensibles 7440
zone d'exploitation 5925
zone d'exploitation forestière 3330
zone du collet 3991
zone écologiquement fragile 2841
zone écologiquement sensible 2841
zone radiculaire 7032
zone réserve ripicole 6962
zone riveraine 6963
zones de forêts steppiques tempérées 8444
zones de forêt-steppe 3609
zones de gestion de vieux peuplements 5881
zones de marché de bois 8614
zones de patrimoine 4259
zones de steppes arborées 3609
zone sensible aux fumées 7785
zones non permises 5804
zone traumatique 8788
zone verte 3941
zoocénose 291
zoologie forestière 3651
zygophyllacées 639

Русский

абака 5283
абаксиальная паренхима 9091
абаксиальный 5
абелия крупноцветная 3873
абиотический 6
абиотический фактор окружающей среды 8
абиотическое повреждение 7
аблактировка 331
абрикос 1756
абрикос обыкновенный 1756
абрикосовая слива 332
абрус молитвенный 636
абрус опасный 636
абрус прекаторный 636
абсолютная лесная почва 19
абсолютная плотность 18
абсолютно сухая древесина 17
абсолютно сухой лесоматериал 17
абутилон 22
авиакряж 5063
авиационная охрана лесов 148
авиационный кряж 5063
авицения 251
авиценния морская 3971
авкумея Клайна 3788
авокадо 445
авраамово дерево 4961
австралийский меданг 413
австралийский орех 1206
автолесовоз 5087
автолесовозный транспорт 5123
автопогрузчик с вилочным захватом 3660
агава американская 1416
агавовые 125
агагашу 4625
агатис новозеландский 5727
агентство по лесному хозяйству 3323
агролесомелиорация 7611
агротехнический уход 2131
агротехнический уход за лесными культурами 8453
адаксиальный 62
адансония 5533
адаптированный к месту лес 7662
адаптируемость 60
адвентивная почка 78
аделия флоридская 3287
аденантера павлинняя 7180
администрация лесничества 3547
ажурная пила 1822
азалия пониклая 6288
азалия понтийская 4355
азиатский кедр 1493
айва 6667
айва обыкновенная 6667
айва продолговатая 6667
айва японская 4640
айлант высочайший 8892
акадский лес 23
акажу 2110
акант 24
акантовые 25
акантопанакс клещевинолистный 7436
акариоз 26
акароидин 405
акация арабская 4048
акация австралийская 404
акация аравийская 4048
акация безжилковая 5301
акация беловатая 7604
акация верблюжья 1288
акация вооруженная 4687
акация Грегги 1375
акация густоцветковая 3884
акация длиннолистная 8356
акация жёлтая 7564
акация жестколистная 4687
акация загнутолистная 2698
акация золотовенечная 5936
акация зонтичная 9041
акация кавен 1394
акация катехинская 1376
акация клейкая 1595
акация колючая 1166
акация культурная 4720
акация лихорадочная 3092
акация медоносная 809
акация мохнатая 1038
акация низбегающая 3962
акация новомексиканская 5724
акация ножелистная 4720
акация ощетиненная 203
акация пальчатонервная 309
акация плакучая 9328
акация плосколистная 3846
акация пограничная 8268
акация подбеленная 7604
акация прибрежная 1691
акация ретиноидес 8316
акация ретиноидес обильноцветущая 2927
акация сбегающая 3962
акация сенегальская 4047
акация серебристая 7604
акация серповидной листвой
акация сосноволистная 7135
акация софора 1691
акация стойкая 8316
акация терновая 973
акация устрашающая 203
акация Фарнеза 1689
акация чёрная 791
акация чёрная австралийская 404
акация чёрнодревесная 404
акклиматизация 32
акклиматизированные привнесенные древесные породы 2558
аксиальная паренхима 457
аксиальные элементы 455
активная группа 54
активное вещество 55
актинидиевые 50
актинидия 51
актинидия китайская 51
актиномицеты 52
алангиевые 158
алангиум платанолистный 1488
алжирская туя 7182
аллеганская ирга 177
аллеганский твёрдолиственный лес 176
аллейные деревья 9308
аллели 179
аллеломорфы 179
аллелопатия 180
аллельные гены 179
аллея 9308
аллея, обсаженная кустами 933
аллигаторова груша 445
аллометрические связи 190
аллохтонный 189
алстония конголезская 1461
алстония шоларис 9363
алыча 1475
альбедо 161
альбедометр 162
альбизия ленкоранская 6185
альбиция серповидная 5526
альбом древесных пород 9660
альгароба 1790
альпийский рододендрон 6989
альпийское криволесье 5567
альфа-разнообразие 210
альфитония высокая 8807
амарант 6641, 9225
амарантовые 229
амбровое дерево 259
амендола 966
американская берёза 159

американский кедр 243
американский падуб 248
американский стиракс 259
американский усач 9016
американский чинар 254
амирис 8716
ампелопсис древовидный 6149
амурское пробковое дерево 264
анакардиум западный 1368
анализ влияния на окружающую среду 2840
анализ воздействия на окружающую среду 2840
анализ годичных слоёв 269
анализ доставки лесоматериалов 8646
анализ древесины 9477
анализ древесного ствола 8060
анализ листа 3307
анализ производительности леса 3647
ананас 6243
анатомия древесины 9478
анаура 5532
андаман падук 383
андромеда древовидная 7857
андромеда-дубровник 897
андромеда дубровниколистная 897
анемофильный 283
анизоптера 5405
анкерная оттяжка 278
анкерное бревно 275
анкерное дерево 280
анкерный канат 278
анкерный пень 279
анкерный сучок 274
аннабель 9407
аннона игольчатая 7856
аннона колючая 7856
анноновые 2158
антарктический нотофагус 6985
антиарис ядовитый 9125
антоцианоз 314
антракноз кофейного дерева 2489
антропогенная нагрузка 316
анчара африканская 6414
анчар африканский 6414
апейба 5534
апельсин 5935
апельсиновое дерево 5935
апикальная меристема 327
апотрахеальная паренхима 328
апробированная машина для внесения пестицидов 1419
аралия 336
аралия высокая 4610
аралия китайская 1489
аралия колючая 2452
аралия шиповатая 2452
араукариевые 339
араукария бразильская 1005
араукария Клинки 4718
араукария Кука 1919
араукария Куннингама 2137
араукария чилийская 5535
арборетум 342
арборицид 343
аргайл 357
ардизия 5320
арека катеху 691
арековая пальма 691
арековые 356
арендованная залесённая местность 6867
аридное редколесье 358
аризонский кипарис 359
аризонский орех 362
аристолохии 768
арония 1553
архитектурная форма 348
арча 3983
арчовник 4672
асимина 6118
аскомицеты 7144
аспергилл 802
аспидосперма 9378
ассимиляция углекислоты 391
ассоциация 394
астма из-за древесной пыли 9700
астрагал камеденосный 8749
астрониум ясенелистный 8548
асфоделина древовидная 8877
атлантический ямс 4654
атлас древесины 9483
атласное дерево 2729
атласное дерево африканское 2713
атласное дерево индийское 2729
атмосферная сушка 152
атта 4864
ауксибласты 4813
ауксин 421
ауткроссинг 5971
афара 92
аферосперма мошата 7864
африканский баобаб 5533
африканское эбеновое дерево 102
афцелия кванценская 6406
аэросев 87
аэростатная трелёвка 497
аэростатная трелёвочная установка 4290
аэротаксация 90

бразильское железное дерево 1002
брия эбеновая 4607
бавнорастущие породы 7756
багассе Жонэ 478
багор с короткой деревянной ручкой 6225
багровый крюк 1317
багряниковые 4693
багряник стручковатый 4665
багряник японский 4692
багульник 4906
багульник гренландский 4749
багульниковый лиственничник 9423
бадьян 293
бадьян настоящий 4611
бадьяновые 294
база данных терминологии по лесоводству 3620
базальный крупный свиль 605
базифильный 614
базовая популяция 608
базовая расценка 609
базовая таблица продажей 610
базовое бревно 7172
базовое лесоведение 611
базовое лесоводство 611
базовый диаметр на высоте, составляющей определённый процент от высоты дерева 4230
базовый лагерь при тушении пожара 3170
бай-дерево 632
байрачные леса 8176
байрачный лес 3474
бакаут 4034
бакури 6030
баланокарпус Гейма 1472
баланс древесины 9486
балансирная пила 4312
балансовая древесина 6061
балансовый кругляк 6627
балансы 6061
балансы кучевого хранения 8116
балансы чистой окорки 1609
балата 485

балка коробчатого сечения 953
балки 2278
балки широколиственной растительности 4137
балластировка 646
балочные леса 8176
бальза 506
бальзам 259, 503
бальзамическая смола 503
бальзам канадский 1294
бальзамное дерево Толу 504
бамбук 508
бамбуковая заросль 510
бамбуковый лес 510
банан десертный 6335
банановое дерево 514
банановые 512
банан текстильный 5283
бандажирование 515
баньян 526
баобаб 5533
барабан для намотки каната 7056
барабанная лебёдка 2643
барабанная окорочная машина 566
барабанная рубильная машина 2208
барабанная рубильная машина параллельного резания 6067
барабанный окорочный станок 566
баранец китайский 8951
барбарис 1758
барбарис канадский 175
барбарис обыкновенный 1758
барбарисовые 528
барбарис орегонский 2013
барбарис Тунберга 4612
баррас 593
барраскит 4584
бархат 6199
бархат амурский 264
бархатное дерево 264
бархатный дуб 1510
бархат сахалинский 7145
бархат японский 4617
барьер из кустарника 1098
барьерная изгородь 596
басралокус 621
бассейн для балансов 5105
бассейн для брёвен 5105
бассейн для брёвен с подсоленной водой 7161
бассейн для хранения леса 5105
бассейновая запруда 613
бассия широколистная 5244
батан 626
баугиния пёстрая 6642
баугиния пурпурная 1221
бафия яркая 1293
башмак 2054
башмаки 917
бедность видов 7281
бедный лесом 6450
безбилетная рубка 4423
безвершинник 7489
безвершинное хозяйство 6430
безвершинный низкоствольник 1936
безлесная земля 5758
безлесная растительность 5921
безлесный 8874
безлесный крутой склон 534
безлесный период 8875
безлесный склон 534
безлесье 9275
безлиственный 4873
безлистное дерево 4874
безлистный 4873
безопилочное резание 7501
безотходное резание 1516
безудержно горящий лесной пожар 9049
безъядровая древесина 4199
белая акация 1754
белая верба 1809
белая волокнистая гниль 9381
белая волокнистая ядровая гниль 9372
белая гниль 9381
белая грибная гниль 9385
белая древесина 9389
белая еловая смола 3792
белая колючка 4174
белая кубическая гниль 9368
белая мраморовидная гниль стволов 5574
белая периферическая гниль корней 364
белая смешанная гниль 9383
белая ядровая кубическая гниль стволов и корней ели и пихты 9367
белая ядровая стволовая гниль с чёрными линиями 9388
белая ямчатая гниль 9377
белое сандаловое дерево 9382
белое сандальное дерево 9382
белолоз 1809
белые пятна 9384
белый горный ясень 3842
бензиномоторная пила 3808
бензойное дерево 7909
бергамот-дерево 679
береговая зона управления 4758
береговая сплотка 524
берегозащитный лес 6961
берёза 742
берёза американская 1299
берёза белая 1761
берёза белая китайская 1503
берёза бородавчатая 1761
берёза бумажная 6058
берёза вишнёвая 779, 1474
берёза голубая 863
берёза горная 5581
берёза граболистная 8328
берёза даурская 2226
берёза жёлтая 1299
берёза западная 9279
берёза каменная 2860
берёза карликовая 892
берёза красная 1299
берёза малорослая 8304
берёза низкая 1295
бёреза плосколистная 384
берёза провислая 1761
берёза пушистая 6620
берёза тополелистная 3923
берёза чёрная 779
берёза Эрмана 2860
березняк 751
берёзовая ветвь 747
берёзовая губка 752
берёзовая древесина 762
берёзовая кора 743
берёзовая листва 749
берёзовая роща 753
берёзовая смола 745
берёзовая тундра 759
берёзовое коротьё 746
берёзовое криволесье 2781
берёзовое масло 756
берёзовые 750
берёзовые балансы 746
берёзовый дёготь 745
берёзовый клей 760
берёзовый кругляк 746
берёзовый лес 751
берёзовый сироп 758
берёзовый сок 757
берёзовый трутовик 752
берёзовый уголь 748
берека 9424
берека лечебная 9424
бересклет 1383
бересклет вьющийся 239

бересклетовые 7985
бересклет широколистный 1049
бересклет японский 4641
берест 2786
берест листоватый 2893
берёстовые 2788
берлога 2365
бертолетия высокая 1003
берхемия кистистая 4635
бескорешковый 7034
бесполовое размножение 9182
беспорочная древесина 9578
беспорядочная валка 2253
беспорядочное лесопользование 8616
бессистемная рубка 4437
бессменная культура 5542
бессменный посев 5542
бесствольный 8078
бесстебельный 8078
бесстружечное резание 1516
бессучковатость 989
бессучковая длина 1625
бессучковая древесина 990
бессучковые пиломатериалы 990
бессучковый 983
бессучковый ствол 1604
бесчокерная трелёвка 1558
бета-разнообразие 690
бетельная пальма 691
бетулин 693
бигардия 775
бигнониевые 8995
бикса аннатовая 4995
биксовые 776
билет 5181
билет на вывозку леса 6180
билет на пуск пала 1193
билет по отпуску леса 6183
билимби 709
биогенезис 723
биогеоклиматическая зона 725
биоиндикаторная сетка 726
биологическая сушка 727
биологический гербицид 729
биологический инсектицид 730
биологически стойкое семя 6748
биологическое разнообразие леса 3337
биология леса 3339
биом 732
биомасса древесных растений 9490
биометрия 733
биоповреждение 720
биоразложимые вещества 719
биоразнообразие 721
биосфера 734
биота 735
биота восточная 2725
биотехнология 736
биотип 740
биотический 737
биотический потенциал 728
биотическое сообщество 738
биотоп 739
биотопливо 722
биофор 4421
биохимическая систематика 1471
биоценоз 738
биржевой пиломатериал 9754
бирсонима 5138
бирсонима колосистая 2570
бирючина 7984
бирючина блестящая 1799
бирючина обыкновенная 1799
бишопия яванская 4645
бишофия яванская 4645
благоприятный уклон 3028
благородная древесина 3159
благородная лиственная древесина 4295
благородный вид 5739
благородный сорт 5739
благосклонный к лесу 8819
блеск дерева 3872
близкий транспорт 7534
близнецовый конечный побег 9019
блокировка канавы 2545
блок подъёмного монтажного каната 6098
блок снабжения древесиной 8648
блочно-реечное заполнение 5200
блочно-реечное заполнение щита 5200
блочный фут 840
блуждающая сердцевина 9264
блюэтта 864
боард-фут 876
бобовник анагиролистный 1784
бобовые 4908
бовдихия 8251
богатство видов 7902
богатый лесом 6936
богатый питательных веществ 2912
боккаут 6186
боккония 8899
боковая верхушечная почка 8242
боковая ветвь 4819
боковая ветвь второго порядка 7349
боковая доска 7572
боковая почка 419
боковая просека 7580
боковая трещина 7574
боковое возобновление 4823
боковое затенение 7581
боковое сжатие 7570
боковой лепесток 9453
боковой отросток 5870
боковой побег 418
боковой поплавок 3147
боковой провес 699
боковой рез 1966
боковой ствол 994
боковой стержневой корень 7582
боковой сучок 4819
боковые корни 4824
болванка 2171
болезни леса 3379
болезнь, вызывающая водослой древесины 9293
болезнь дерева 8828
болезнь ильмов 2678
болезнь сеянцев сосны 6260
болотистая лесная земля 6048
болотистое редколесье 8312
болотистый лес 8309
болотистый ольшаник 165
болотная мирта 8332
болотно-лесная почва 8318
болотно-травяной ельник 8313
большое полено, сжигаемое в святки 1567
большой лес 2324
большой сбег 3022
бомбакопсис 6403
бомбакс сейба 6792
бон 8779
бондарная древесина 4362
бондарное производство 1920
бондарный лесоматериал 4362
бон для направления брёвен с одной стороны реки на другую 3147
бонитет леса 3592
бонитет лесонасаждения 3592
бор 6255
бореальный лес 925

бороздная эрозия 6949
бороздовая посадка 8964
бороздовой посев 7869
бороздчатая эрозия 6949
бороздчатолистный 3777
борьба с лесными вредителями 3524
борьба с лесным пожаром 3431
босвелия пильчатая 4463
босвеллия пильчатая 4463
боскет 933
боссе 1411
ботанические леснье продукты 938
бочарная древесина 4362
бочарная продукция 1920
боярышник "петуший гребень" 1699
боярышник "петушья шпора" 1699
боярышник Дугласа 2591
боярышник колумбийский 1731
боярышник колючий 4174
боярышник кроваво-красный 6775
боярышник круглолистный 7110
боярышник мягкий 2602
боярышник обыкновенный 4174
боярышник однопестичный 1780
боярышник одностолбчатый 1780
боярышник пятипестичный 1500
боярышник чёрный 793
боярышник шпорцевый 1699
боярышник ярко-красный 3167
бражник сосновый 6257
бразильская саванна 1292
бразильский лес 1010
бразильский орех 1008
бразильский палисандр 1007
бразильское розовое дерево 1006
бракёр древесины 9525
бракованное бревно 5764
бракованный лес 9118
браковщик древесины 9525
браконьер 6401
браконьерство 6402
брахилена гучинская 7609
брахистегия 5877
бревенная телега 5199
бревенник 8585
бревенчатый дом 5052
бревенчатый домик 5052
бревенчатый каток 3280
бревенчатый роликовый слой 4844
брёвна высокого качества, продаваемые поштучно 7420
брёвна длиной равной нечётному числу футов 5866
брёвна, идущие на обрезку на станках 1521
брёвна, лежащие поперёк течения 8774
брёвна, непосредственно лежащие на балке коника 1180
брёвна, укладываемые поверх обвязочной цепи 717
бревно 8879
бревно без дефектов 6552
бревно без сучьев 1626
бревно, входящее в состав бона 1181
бревно, годное для волокнистого полуфабриката 6627
бревно для запруды 7496
бревно для столбов 5064
бревно, закрывающее сортировочный карман 3802
бревно короткомерного баланса 903
бревно, которое не держится на плаву 2268
брёвномер 5107
бревно, образующее затор 4704
бревноподъёмник 5060
бревно правильной формы 8989
бревно, пригодное к сплаву 3268
бревносбрасыватель 5097
бревно, следующее после комлевого 7362
бревно специального назначения 7887
бревноспуск 5116
бревно, укладываемое поперёк трелёвочного волока 1025
бревно установленных размеров 2501
бредина 3880
бригада вальщиков 9527
бригада, заготавливающая длинномерные лесоматериалы 5155
бригада лесорубов 9527
бригада на валке и трелёвке 2163
бригада на заготовке щепы 1533
бригада на очистке 1610
бригада на очистке лесосек 1610
бригада на предварительном формировании пачек лесоматериалов 1172
бригада на расчистке территории от кустарника 1102
бригада, обслуживающая канатную установку 7703
бригада по посадке 6349
бригада по тушению 3183
бригада сплавщиков 2630
бригада таксаторов 8288
бригадир лесорубов 5069
брикет для посева семян 7868
брикет из древесины 9495
брикет из отходов древесины 9495
брикетированный 1892
брикетированный сеянец 1893
бровки 1502
бромированная древесина 1063
бронзирование 1066
бронхиальное дерево 1064
бросимум полезный 2000
бросимум Ублетта 4921
бросовые земли 9275
брошенная древесина 9277
брус 638
брус для удержания брёвен при накатке 2983
брус из половины брёвна 4100
брус из четверти брёвна 6658
брус неквадратного сечения 7712
брусника 1996
брусовка 5117
брусовочный станок 1018
брусок 9592
брусок неквадратного сечения 7712
брусонетия бумажная 6060
брусонеция бумажная 6060
брусующий станок 1018
брусья для выстилки волока или площадки 7694
брусья с обзолом 2663
бубинга 1120
бугенвиллия 947

будущее лесной продукции 3540
бузина 2892
бузина американская 246
бузина голубая 858
бузина канадская 246
бузина красная 2907
бузина красная американская 787
бузина обыкновенная 2907
бузина чёрная 2892
бузинная ягода 2778
бук 649
бук европейский 1760
бук красный новозеландский 8420
бук крупнолистный 238
бук лесной 1760
бук лесной плакучий 9329
бук лесной тёмнопунцовый 6639
бук обыкновенный 1760
буковая древесина 662
буковая роща 654
буковая смола 660
буковая щепа 651
буковое красное ядро 659
буковое масло 658
буковые 652
буковый закреп 844
буковый лес 653
буковый орешек 657
бук ориентальский 5949
буксир 9007
буксирный канат 9007
буксирный трос 9007
буксировка 8737
бук с осенней окраской 4613
булавочный сучок 6281
буль 948
бульдозер для бревен 911
бульдозерное полотно для расчистки территории от зарослей кустарника 1099
бульдозерный нож 5068
бульдозерный отвал 5068
бульнезия древовидная 9199
бумага из древесной массы 9634
бумажная шелковица 6060
бумелия шерстистая 9716
бундук двудомный 1482
бундук канадский 1482
бункер для древесного сырья 9518
бункер для щепы 2496
бурав 403
бурав для посадочных ямок 6342
буравчик по дереву 9538
бурачник 918
бурачник лекарственный 918
бурачниковые 919
бурая гниль 1549
бурая гниль бука 1089
бурая гниль вершины 1092
бурая гниль вершины дерева 1092
бурая гниль отмершей древесины 1088
бурая древесная масса 8053
бурая заболонная гниль 1086
бурая кубическая гниль 2251
бурая кубическая гниль древесины, хранящейся на складах 1077
бурая кубическая гниль хвойных пород 1076
бурая лесная кислая почва 39
бурая лесная оподзоленная почва 1078
бурая лесная почва 1079
бурая почва 1079
бурая раневая гниль стволов 6777
бурая сердцевина 1080
бурая сплошная кубическая гниль 2103
бурая средиземноморская почва 6770
бурая субальпийская гниль 8225
бурая ядровая гниль 1084
бурая ямчатая заболонная гниль хвойных пород 1083
бурелом 850
буреломная древесина 850
буреломная прогалина 849
буреломное дерево 2271
бурение 6393
буркея африканская 6791
бурма-падук 1183
буровая древесная мука 929
буровая мука 929
буровая скважина 927
бурозем 1079
бурозубка 8923
бурсера мелколистная 2780
бурсеры 1198
бутия односемянная 3248
бутылочные деревья 941
буферная зона без пестицидов
буферный эффект 1148
бухарник 1790
быстро опадающий 3744
быстрорастущее дерево 3021
быстрорастущие породы 6666
быстрорастущий лес 3020
быстро увядающий 3744
быстрый водяной поток с содержанием грунта и порубочных остатков 2286
быть вредным 4148

вагон для хлыстов 9256
вагонный лесоматериал 8592
вазицентрическая паренхима 9174
вазицентрическая трахеида 9175
ваиа 3718
вайа 3718
вакцина 4507
вакцинальный биопрепарат 4507
валёжная древесина 2266
валёжник 2266
валить 2172, 3037
валить дерево с корнем 4028
валка 8824
валка "напроход" 1897
валка в направлении вверх по склону 9128
валка вручную 4109
валка деревьев 8824
валка деревьев с корнями 3081
валка зелёных деревьев 3959
валка зимой 9460
валка и раскряжёвка 3047
валка леса в направлении вверх по склону 9128
валка леса машинами с ножевым срезающим устройством 7501
валка над корневища 3046
валка одновременно нескольких деревьев 5629
валка подпиленных деревьев с помощью другого дерева 2569
валка по утверждённой схеме 6834
валка ручными мотопилами 5572
валка с обрезкой корневой системы 3081
валка с учётом контур 1903
валка хлыстов 3046

валочная вилка 6645
валочная головка 3068
валочная головка, не повреждающая комель срезаемого дерева 5771
валочная головка с накоплением срезанных деревьев 5630
валочная головка с пилой 1433
валочная кошка 6645
валочная машина 8836
валочная машина с остановкой для срезания дерева 8125
валочная пила 8854
валочная подушка 3057
валочная фреза 403
валочное ложе 4840
валочное устройство с паровым приводом 8055
валочно-пакетирующая машина 3053
валочно-пакетирующая машина с ножевым срезающим устройством 3053
валочно-пакетирующая машина с пильным срезающим механизмом 1432
валочно-раскряжёвочная бригада 3048
валочно-сучкорезно-раскряжёвочная машина 8853
валочно-трелёвочная машина 3075
валочный клин 3079
валочный клин с гнездом для деревянной ручки 7809
валочный пропил 3066
валочный рычаг 3067
валочный шест 6645
вал порубочных остатков 9447
вальцевание смольной щепы 7002
вальщик 9526
вальщик леса 9526
вальщик, осуществляющий валку деревьев вручную 4108
ване 6782
ванильное дерево платанолистное 9165
ванчёс 9257
вар 8958
варка древесины 9633
ватика тонкинская 6879
ватле 3962
введение ядовитых веществ через корни деревьев 466
вводимые породы 4576
вегетативное возобновление растений 9184
вегетативное размножение 9182
вегетативное растение 8209
вегетативный 9178
вегетативный материал 9181
вегетативный побег 4505
вегетативный черенок 7010
вегетационный период 4006
ведение лесного хозяйства 3488
ведение одновозрастного хозяйства 2917
ведение разновозрастного лесного хозяйства 173
ведение устойчивого лесного хозяйства 8296
ведомость лесных материалов 3073
ведомость перечёта деревьев 5365
ведомость приёмки леса 3073
ведомость сдачи-приёмки леса 3073
ведущая жердь 4042
ведущее колесо 4849
ведущий поплавок 3147
ведьмина метла 9467
веерная пальма талипот 8373
веерообразный сложный лист 3010
вейгела 9332
вейник 7537
величина лесного сырья на корню 8196
веллингтония 3844
вельд 1208
венге 9336
верба 9224
верба красная 7138
вербеновые 9210
верболоз 630
вереск 1779
вереск болотный 4898
вереск древовидный 4206
вереск обыкновенный 1779
вересковые 4207
веретеновидный инициал камбия 3781
веретеновидный луч 3784
веретенообразная крона 3782
веретенообразная паренхимная клетка 3783
веретенообразная ржавчина
верещатниковый лес 4208
верига прицепа 8750
вертикальная полоса на залесенном горном склоне, лишенная
древесной растительности, выбитой лавинами 425
вертикальная сомкнутость 9204
вертикальная сомкнутость кроны 9204
вертикальная сомкнутость полога 9204
вертикальная структура 9208
вертикальное разнообразие 9205
вертикальное сопротивление 9206
вертикальный брусок для крепления плинтуса 5652
вертикальный корень 9207
вертолётная трелёвка 4239
вертолётный десант 4240
вертолёт-цистерна 4241
верхнее бревно воза 6126
верхнее дерево воза 6126
верхнее сжатие 8682
верхний диаметр 2470
верхний зажим для брёвен 8684
верхний или нижний предел ступени толщины 2472
верхний кряж 8688
верхний напил 9126
верхний отруб 8688
верхний полог 7511
верхний рез подпила 8697
верхний склад 9135
верхний слой неразложившегося опада мхов 5869
верхний трос 9137
верхний ярус ветвей деревьев 2066
верхний ярус насаждения 3518
верхняя ветвь 324
верхняя высота кроны 9133
верхняя высота расположения кроны 9133
верхняя граница леса 4982
верхняя граница произрастания лесов 4982
верхняя длина кроны 9134
верхняя часть брёвна 9138
верхняя часть дерева 8699
верхняя часть пиловочника 7252
верховая грань из бревен 8654

верховое прореживание 8512
верховой беглый пожар 2076
верховой лесной пожар 2076
верховой побег 4237
верховой пожар 2076
верховой устойчивый пожар 4458
верх полога 8701
верхушечная ветвь 8683
верхушечная гипертрофия 8473
верхушечная древесина 8712
верхушечная меристема 327
верхушечная паренхима 8475
верхушечная почка 8470
верхушечное усыхание 8686
верхушечный побег 8474
верхушечный рост 326
верхушка дерева 8699
верхушка корня 7051
вершина
с ~ой, направленной под углом к стволу 631
вершина дерева 8699
вершина ствола 8700
вершина хлыста 8689
вершинная гниль 8711
вершинная древесина 8712
вершинная обрезка 8696
вершинная почка 8470
вершинник 8688
вершинное бревно 8688
вершинный лесоматериал 8712
вершинный побег 8474
вершинный торец 7761
вес выборки при рубках ухода 8519
весенней окорки 7210
весенняя древесина 2704
весенняя рубка 7938
весовое дерево 9333
вес свежесрубленной древесины 3964
вес сухой древесины 2660
весьма неустойчивая территория 9211
весьма продолжительный пожар 2974
ветвеобразный 6709
ветвистый 977
ветвистый корень 980
ветвление 987
ветвь 974
без ~ей 983
ветвь второго порядка 7341
ветвь, направленная кнаружи от главной оси 6468
ветвящаяся паренхима 2492
ветвящиеся деревья 988
ветла 1809
ветла блестяще-жёлтая 3885
веточка 991
ветреник 8698
ветреница 9450
ветреница лесная 9479
ветровал 850
ветровальная прогалина 849
ветрозащитная полоса 9445
ветроопыляемый 283
ветроударная опушка леса 9445
ветроустойчивое дерево 9443
ветряная трещина 9450
веха 3899
вечнозелёные растения 2931
вечнозелёный дождевой лес 2930
вечнозелёный лес 2929
вечнозелёный противопожарный барьер 2928
вешка 3899
взаимное влияние деревьев 4570
взаимное страхование от пожаров 5642
взаимодействие корня и подвоя 8118
взаимодействие лесных и полевых культур 8821
вздутье 1182, 8337
вздымка 1530
вздымочный хак 2142
вздымщик 4084
взрослая листва 68
взрывание пней 852
взрывной клин 6487
взрыхлить 2169
взыскательные виды 2361
вид 7893
в ~е, в каком снят с дерева 8914
видимые требования к качеству 9238
видимый порок 5906
вид, находящийся под угрозой исчезновения 2801
видовая высота 8841
видовая высота насаждения 8016
видовое число 8840
видовое число насаждения 8015
видовое число ствола 8075
видовое число тонкомерных
видовой фактор поперечного сечения 599
видоизменённая лесосека 5510
видоизменённый лес 5512
вид пожара 1601
вид растений, выпадающий из фитоценоза после выпаса 2311
вид растений, обилие представителей которого увеличивается при выпасе 4447
вид, соответствующий лесорастительных условий 7895
вид, чужой для лесорастительных условий 7903
виды локального значения 6828
виды с риском 7894
вила для брёвен 5046
вила для леса 5046
вилообразные трелёвочные сани 3881
вилообразный грузитель 8610
вилообразный рост 3656
вилочный лесопогрузчик 8610
вильчатый 3654
вильчатый сучок 979
виноград аризонский 1318
виноград круглолистный 5637
виноград медвежий 1759
виноградники 9223
виноградовник древовидный 6149
виноградовые 9221
винтеровые 9459
винтовые уплотнения 7916
виргилия жёлтая 262
вирджинский карандашный кедр 2722
вирола 9232
вирусная мозаика 9170
вистерия 4707
вистерия обильноцветущая 4638
висячий 9324
висячий трос 7714
витекс 3108
витекс кенийский 5406
витекс священный 4961
виток 995
вить гнездо 5710
витютень 9624
вишнёвое дерево 1476
вишнеслива 1475

вишня канадская 7186
вишня кислая 7855
вишня мелкопильчатая 744
вишня перуанская 1323
вишня птичья 8330
вишня сабхиртелла 4275
вишня сладкая 8330
вкладной клин 2680
вкрапленные твёрдые частицы 2794
влагалище листа 4857
влага пористого пространства 6464
влагомер для древесины 9611
влагонепроницаемый 5522
влагосодержание топлива 3737
влагостойкий 5522
влагоупорный 5522
влагоустойчивый 5522
владелец лесовозного автомобиля 5122
владелец права на лес 6940
влажность в центре поперечного сечения 1950
влажность в центре поперечного сечения лесоматериала 1950
влажность древесины 5521
влажность поверхности 8283
влажность свежесрубленной древесины 3950
влажный вечнозелёный субантарктический лес 8228
влажный горный лес в умеренном климате 5517
влажный лес 6700
влажный лес в умеренном климате 5516
влажный субантарктический лес 8228
влажный тропический лес 8978
влезть на дерево 1641
влекомые наносы 647
влияние зарубки 2770
влияние леса 3471
влияние соседнего сообщества 2758
внезапное возобновление 7924
внеочередная рубка 2973
внесение пород 4577
внесение удобрений 3090
внесение ядохимикатов 3090
внешний вид дерева 4080
внешний слой коры 5975
внешняя трещина 2965
внеярусная растительность
внутренний порог 4503
внутренний слой 1949
внутренняя быстрина 4503
внутренняя влажность 1950
внутренняя гниль 2645
внутренняя заболонная окраска 4538
внутренняя заболонь 853
внутренняя кора 4502
внутренняя пласть пиломатериала 8602
внутренняя синева 4557
внутренняя сторона ленты шпона 7716
внутренняя темнина 795
внутренняя трещина 958
внутренняя флоэма 4502
внутрилесосечный транспорт 4558
внутриотраслевое субсидирование 2058
внутрипочвенный дренаж 8241
вогнутая пила 1853
водитель лесовоза 8983
водитель тягача, работающий по контракту 1907
водное растение 4410
водный лесоспуск 9281
водный лоток 9281
водный потенциал листа 4892
водолесокультурная система 335
водоотводный канал 2613
водоотливной ящик 7942
водоохранный лес 6961
водопоглотительная способность 9287
водорегулирующий лес 9282
водоросли 166
водосбор 9280
водосборный бассейн 9280
водосброс 2613
водослой 9283
водослойная древесина ядра 9283
водостойкая фанера 5523
водоудерживающая способность 9287
водяная борозда 4043
водяная трелёвка 9300
водяная трещина 4043
водяника 785
водяниковые 1997
водянистая древесина 9354
водяное пятно 9283
водяной жёлоб 4043
водяной ясень 777
водяные ресурсы 9295
возвратная верёвка 5411
возвратный канат 5411
возвратный трос 5411
возвращение каретки с грузом на линию трелёвки 4822
возвращение с грузом на линию трелёвки 4822
воздействие атмосферных условий 9311
воздействие леса 3471
воздействие пожара 3188
воздействие химикатов на древесине 2771
возделывать 2128
воз для хлыстов 5087
воздушная канатная дорога 84
воздушная канатная установка большой протяжённости 5130
воздушная обрезка 156
воздушная полость в штабеле 1487
воздушная сушилка 153
воздушная сушка 152
воздушная таксация 90
воздушная трелёвка 85
воздушно-сухая древесина 150
воздушно-сухая масса 149
воздушно-сухой весь 149
воздушно сухой лесоматериал 150
воздушный корень 86
воз лесоматериалов, перевозимых за один рейс 8984
возможности употребления древесины 6473
возможность местообитания 7663
возобновившаяся лесная земля 6810
возобновившийся квадрат 8105
возобновившийся лес 8104
возобновительный блок 6160
возобновительный период 6820
возобновление катастрофной площади 6817
возобновление леса 6821
возобновление леса на вырубках 6822
возобновление леса под пологом 6825
возобновление леса под

возобновление леса после вырубки 6811
возобновление от опушки 6877
возобновление от стен леса 4823
возобновление от стен леса при полосных рубках 6877
возобновление пнёвой порослью 1931
возобновление после вырубки 6811
возраст дерева 134
возраст древесины 135
возраст древостоя, равный обороту рубки 136
возраст крупного леса 8933
возраст на высоте груди 1023
возраст насаждения 8001
возрастная структура 130
возраст пня 8193
возраст по массе 9240
возраст рубки 136
возраст спелости 133
возраст спелости леса 133
возраст срубания 136
возраст срубания леса 136
возращение саженцев 6704
волнистая древесина 1662
волнистая текстура 9078
волнистая форма края листа 6868
волнистое волокно 6702
волнистость 2019, 3101
волнистый текстурный рисунок 5552
волнистый характер поверхности древесины 9270
волнистый характер поверхности древесины вследствие неравномерной усушки ранних и поздних слоев 9270
волнообразная текстура 6697
волнообразность 3101
волнообразный знак 6968
волокнистая жёлтая гниль 8167
волокнистая жёлтая ядровая гниль 3104
волокнистая масса 9603
волокнистая структура 3103
волокнистая текстура 9678
волокнистая фигура 7061
волокнистый 3096
волокно
вдоль ~он 209
углом к действию силы 4585
волокно, содержащееся в листьях 4867
волокно флоэмы 623
волок при вторичной трелёвке 8344
волокуша 3881
волокуша из раздвоенного дерева 5020
волосная трещина 4088
волосяная трещина 4088
волочить рукой 4112
волочить рукой сплавной лес к берегу 4112
волчник 2246
волчник обыкновенный 2246
волчниковые 5427
волчок 7046
вороника 785
воронка смоляного горшочка 6303
ворсистое волокно 9717
ворсистый лист 7471
восковник обыкновенный 8332
восковниковые 9305
восковница болотная 8332
восковница калифорнийская 1277
восковница обыкновенная 8332
восковое китайское дерево 1481
восковой вереск 8332
восковой мирт 7858
воспитатель деревьев 344
воспитательная рубка 4544
воспламеняемость 1742
воспламеняться 3166
воспроизведение укоренившимися черенками 6876
восстановительная обрезка 4215
восстановление 6916
восстановление без очистки поверхности почвы 6826
восстановление глуши 461
восстановление леса 6821
восстановление окружающей среды 2843
восстановление отсталых районов 461
восстановление растительности 6923
восстановление растительности на гарях 6479
восстановление резервных участков 6846
поверхности почвы 6827
восстановлённый вторичный лес 6845
восточная туя 2725
восточный 5949
восходящее сокодвижение 4448
восхождение сока 7195
вписывать 7317
впитывание 4488
враг леса 3523
врастающая кора 4491
вращающееся барабанное сито для щепы 7073
вредители корней 7023
вредители побегов 8472
вредители почек 8472
вредитель 4578
вредить 4148
вредная древесина 8601
вредный вид 5805
вредный для лесов 1393
вредный организм 5807
вредный сорняк 5806
вред от оленей 2232
временная дорога 5121
временная запань 6415
временная запруда из брёвен 4377
временная структура доступа 8448
временное землевладение 8450
временное смешение 8451
временные лесные дороги 3378
временный водоём, наполняющийся водой весной и осенью 9202
временный лес 8447
временный мост 8449
временный передвижной домик 1179
время валки 3077
время выравнивания 169
время облиствения 3719
время операции очистки защитной полосы от горючих материалов 5555
время покоя сокодвижения 9647
время полного рейса 9011
время развала 4189
время роспуска 4189
время рубки 3077
врождённый дефект древесины 2333
вросшая кора 4491

врубка брёвен или брусьев внакладку 4096
врубка внакладку 4096
врубка вполдерева 4096
врубка в Т-образном виде 8441
врубка на шип ласточкин хвост 2592
врубка под косым углом 5857
врубка под прямым углом вполдерева 8140
врубовая машина 2181
вспахивать 6383
вспашка 8554
вспомогательное вещество 64
вспомогательное дерево 4243
вспомогательное дерево, оснащаемое для установки трелёвочной мачты 2664
вспомогательный канат 4176
вспомогательный монтажный канат 8153
вспышка лесного пожара 5970
вставленная порода 4562
вставлять 5710
вставочный ствол 8065
встречаемость деревьев 8843
встречный огонь 465
встречный пал 465
встряхиватель деревьев 8845
встряхиватель плодов 8845
встряхиватель плодовых или семенных деревьев 8845
всходы 8917
всхожесть растения 2796
всхожесть семян 3833
вторая стадия гниения 72
вторая стадия гниения древесины 72
вторичная древесина 7357
вторичная клеточная стенка 7343
вторичная кора 7339
вторичная ксилема 7357
вторичная сукцессия 7354
вторичная трелёвка 6909
вторичная хвоя 7351
вторичное насекомое 7348
вторичное облиствение 6803
вторичные древоточцы 7352
вторичные стволовые вредители 7340
вторичный влажный тропический лес 8980
вторичный лес 7346
вторичный лесной питомник 7347
вторичный лес след валки 6477
вторичный луч 7353
вторичный очаг пожара 4669
вторичный рост 120
вторичный стержневой корень 6137
вторичный тропический лес 8980
второй приём промежуточных рубок 7366
второй проход 7364
второй ярус насаждения 9070
второстепенная древесная порода 7356
второстепенная просека 7580
второстепенный ярус насаждения 8237
выбленочный трос 6732
выбор вальщика 3043
выбор древесных пород 1551
выборка двухразмерных пробных площадей 694
выборка пробных площадей с множеством переменных размеров 6435
выборка пробных площадей с одним постоянным размером 5527
выбор местообитания 7672
выборочная обрезка сучьев или ветвей 7414
выборочная рубка 7412
выборочная рубка в порослевых древостоях 1932
выборочная рубка полосами 8168
выборочная таксация леса 3549
выборочная уборка 7413
выборочная уборка урожая 7413
выборочное прореживание 7411
выборочнолесосечное хозяйство 7408
выборочно прорубленный лес 7410
выборочный древостой 7410
выборочный лес 7410
выборочный низкоствольник 7405
выборочный перечёт 4998
выбранные породы белого дуба 7404
выбранные породы красного дуба 7402
выветривание 2861
вывозка 8595
сортиментов 5615
вывозка леса 8595
вывозка леса автомобилями 8985
вывозка леса по контракту 1908
вывозка леса с предварительной погрузкой на прицепы 7562
вывозка лесных материалов 8595
вывозка по лесовозной железной дороге 6696
вывозка сортимента 395
вывозка с предварительной погрузкой на прицепы 7562
вывозка с тросом 1247
вывозка трактором 8747
вывозка трелёвочной тележкой 1358
вывозка хлыстов 5615
вывозка целых деревьев 3761
вывозка целых деревьев с необрубленной кроной 3761
выгоревший лес 1187
выгрузка леса с борта судна на набережную 208
выдвижение стрелы 8442
выдвижение телескопической стрелы 8442
выделитель лесных семян 3587
выдержанная древесина 2622
выдержанный лесоматериал 7335
выдерживать лесоматериалы 7337
выдолбленное бревно 4329
выемчатая сердцевина 2933
выемчатая форма края листа 7661
выемчатый лист 2317
выживаемость 1322
выживаемость древесных пород 3629
выжигание по дереву 9498
выжигать 3166
вызывающий гниение 7212
выкапывать 7120
выкопанная тропа 2934
выкопать 7120
выкопка 5979
выкопочная машина 6367
выкопочный плуг 8771
выкорчёвывание 2857
выкружка 635
вылет 910
вылет стрелы 6743
вымывание питательных

выпадающий сучок 5158
выпасаемый лес 3929
выпас в лесу 3452
выпиливать 1360
выпотевание 8321
выпускать листья 7531
выпускной канал 2525
выпуск щепы рубительной машины 1526
выравнивание 3901
выравнивание комлей окучиваемых лесоматериалов 2304
выравнивание с корнем 2857
выравнивание торцов 1215
выращенная живая ткань 8671
выращивание леса с временным сельскохозяйственным пользованием 143
выращивание лесов 3364
выращивать 2128
вырванные волокна 1524
вырезать 1360
вырез в дереве со стороны, противоположной направлению валки 3066
вырез со стороны, противоположной направлению валки 3066
вырождающийся 2696
вырождение лесов 3367
выросший в лесу 3455
выросший самосевом 7417
вырубаемый участок леса 5038
вырубать 3037
вырубить лес 2524
вырубить наголо 2168
вырубить ценную древесину 4272
вырубка 1607
вырубка деревьев господствующего класса 8512
вырубка для освобождения 5912
вырубка единичных деревьев 7655
вырубка жердняка из низкоствольника 7409
вырубка искривлённых деревьев 629
вырубка конкурирующей растительности 1109
вырубка крупных стволов 2476
вырубка кустарника 1109
вырубка леса 8824
вырубка малоценных деревьев 2121
вырубка мягких пород 9320
вырубка наплывов 3070
вырубка наростов 3070
вырубка нижнего яруса насаждения 4933
вырубка перед главной рубкой 6513
вырубка перестойных деревьев 2969
вырубка перестоя 2969
вырубка сорных пород 9320
вырубка ценных стволов 4285
вырубленная площадь 3038
вырубленное насаждение 3041
вырыв 7495
высадка дерева в грунт 3626
высадок 8056
высаживание 5979
высверливание 6393
высверливание сучков 4727
высев 7382
высевающий аппарат для жёлудей 47
высев в открытый грунт 2521
высокая ограда 2328
высокая степень опасности 4286
высокая степень полноты 6007
высокие травянистые растения 4287
высокий выход 4305
высокий пень 4302
высокогорный лес 9136
высокодоходное разнообразие 4306
высококачественная лиственная порода 4295
высококачественное бревно 6552
высококачественные деревянные изделия 3160
высоконапряжённый трос 4293
высокоплотное насаждение 4278
высокополнотное насаждение 3756
высокопродуктивная лесная земля 4299
высокорасположенная крона 4298
высокорасположенный несущий канат 4300
высокосортные лесоматериалы 4284
высокоствольник 4280
высокоствольное дерево 4301
высокоствольное насаждение 4280
высокоствольное хозяйство 4281
высокоствольное хозяйство с воспитанием семенных деревьев 4283
высокоствольный 5127
высокоствольный древостой с наличием одного или более ярусов 8132
высокоствольный лес 4280
высокоствольный лес с воспитанием семенных деревьев 4282
высокотехнологическая древесина 2825
высокоэластичная древесина 7126
высота геометрического центра кроны 3669
высота господствующих деревьев 8692
высота груди 1022
высота деловой части ствола 5395
высота дерева 8599, 8855
высота до верхушки 8726
высота кроны 2082
высота надземной части 8694
высота насаждения 4233
высота обрезки вершины 8708
высота опала 7305
высота пня 4236
высота подпила 2195
высота порока 4235
высота прогиба кривизны 4234
высота прогиба кривизны стрелы 4234
высота пропила 2195
высота спиливания 2195
высота среза 2195
высота срезания 2195
высота товарной части дерева 8599
высотомер 4416
выступ 5993
выступающая часть 6742
высушенная в сушилке древесина 4710
высушенная древесина 2622
высушенные в сушилке пиломатериалы 4709
высушенный на воздухе

высыхание вершины 8686
вытаскивать брёвна лебёдкой 2641
вытеснение древесины 6871
вытеснение жидкости 4996
вытеснение сока 7198
вытяжной ролик для пней 8210
выход деловой древесины 9778
выход древесины 9778
выход древесины из дерева 9779
выход древесины на лесосеке 9777
выход древесины от нерегулярных рубок 4556
выход древесины, полученный от рубок ухода 9782
выход из дерева 9779
выход из одного дерева 9772
выходной продукт 9783
выход пиломатериала 5214
выход пиломатериалов из брёвна 9780
выход по объёму 9773
выход по площади 9770
выход по толщине 9771
выход продукции свыше расчётной 6002
вычёт коры 200
вычитание толщины коры 200
вышележащая почка 8270
вьюнковые 5557
вьющееся растение 9018
вяз 2786
вязальная проволока 5023
вязальщик леса 1840
вязание плотов 912
вязанка дров 9519
вязанка коры 1176
вязанка хвороста 1118
вязаные дрова 9572
вязать 1169
вяз ветвистый 6988
вяз гладкий 7137
вяз голландский 2677
вяз горный 7310
вяз долинный 4618
вязка деревянных деталей 4660
вяз кистевидный 6988
вяз китайский 1496
вязкость 8735
вязкоэластичное поведение 9233
вяз крупноплодный 698
вяз листоватый 2893
вяз лопастный 5270
вязовик трёхлистный 6035
вязовник 2789
вяз плакучий 9039
вяз полевой 2893
вяз приземистый 4618
вяз пробковый 1958
вяз сродный 4618
вяз стройный 2893
вяз Томаса 6988
вяз шершавый 9039
вяз японский 4618
вялость 9469
вяхирь 9624

григнон 6782
габитат дикой природы 9412
габитус дерева 4080
гадог 4645
гаичка буроголовая 9438
галерейный лес 3791
галимодендрон 7165
галл 3789
галл дубовой орехотворки 1279
галлицы 3793
галл коры 560
галловая ржавчина 3795
галлообразующие насекомые 2210
галофитная растительность 4102
галофиты 4102
галька 7128
гамак 4140
гамамалиевые 9468
гамамелидовые 9468
гамамелис виргинский 1811
гаметофит 3798
гамма-разнообразие 3799
гаптотропизм 4118
гаптотропия 4118
гардения широколистная 4465
гарига 3807
гарриевые 7597
гарциния 3804
гарциния мангостановая 5278
гарция поникающая 3006
гарь 1184
гать 1947
гаультерия 8433
гварея высокая 2006
гварея кедровая 6280
гварея пахучая 6280
гваюла 4038
гваяковая древесина 6405
гваяковое дерево 4034
гваякум лечебный 4034
гвибурция копалоносная 113
гвоздевой сучок 4376
гвоздика 1664
гвоздичное дерево 1664
гвоздичные 6278
гебелия красивая 1939
гевея бразильская 1320
гейлюссакия ягодная 798
гейсоис 3812
гелофит 4242
гельземий вечнозёленый 1346
гемикриптофит 4245
гемицеллюлоза 4244
гемиэпифит 4246
гемиэпифитный 4247
гемлок 4248
гемма 4859
генераторные дрова 3816
генераторный лес 3816
генетическая изменчивость 3822
генетическая раса одного древесного вида 8906
генетическая химера 3820
генетически изменённое растение 8764
генетически модифицированное дерево 3819
генетически опасный 2696
генетически улучшенные лесные сеянцы 3818
генетическое разнообразие 3821
генипа 4605
генный фонд 3815
геном 3823
генотип 3824
геометрический центр кроны 3668
геометрическое определение объёма круглых лесоматериалов 3828
геометрическое прореживание 3829
геотекстиль 3831
гербицид 4254
гербицид для уничтожения кустарника и дёрна 1107
гетерогенность 4263
гетерозис 4269
гетеропластический привой 4268
гетерофильный 4266
гетерофильный побег 4267
гетероцеллюлярный луч 4264

гибискус мускусный 5639
гибискус сирийский 7067
гибкий кол 9022
гибкость дерева 3264
гибкость древесины 3264
гибрид 4399
гибридизация 4401
"гибридная мощность" 4269
"гибридная сила" 4269
гигроскопичность древесины 9589
гидравлическая лопата 4408
гидравлическая окорка 4403
гидравлическая окорочная машина 4404
гидравлический окорочной станок 4404
гидравлический посев 4406
гидравлический трелёвочный захват 4407
гидравлическое ножевое устройство для срезания дерева 4409
гидроколун 4405
гидролес 8588
гидролизная глюкоза 9670
гидромелиоративный фонд 8113
гидростатическое определение объёма круглых лесоматериалов 4411
гидростроительные брёвна 8588
гидротехнический лес 9573
гидрофит 4410
гикори 8988
гикори косматый 7482
гилея 8978
гилия колючая 6534
гилофит 9351
гимена курбарил 1990
гимнокладус двудомный 1482
гимномоноэционный 4067
гимномонэцичный 4067
гинкго 5245
гинкговые 3847
гинкго двухлопастный 5245
гинодиэцичный 4070
гипокотиль 4415
гифа 4414
гифомицеты 3127
главная ветвь 6536
главная древесная порода 5252
главная продукция леса 5253
главная просека 5249
главная рубка 3136
главное насаждение 5250
укладывающий брёвна на воз лесоматериалов 8695
главный корень 8411
главный лесничий 1483
главный лесной сторож 4185
главный побег 4237
главный сортировочный коридор 5248
главный стволовой побег 8076
гладкая кора 7787
гладкая кромка доски 477
гладкий пропил 1606
гладкокорая древесная порода 7788
гладкоствольный 1613
гладкоствольный лес 1614
глаза дракона 1381
глазки 995
глазок 1141
глазомерная таксация 2869
гледичия 4353
гледичия аргентинская 263
гледичия, лишённая колючек 8527
гледичия обыкновенная 4353
глеевая серая лесная почва 3861
глееобразование 3860
глобальная конвенция о лесах 3863
глобальная радиация 3866
глобальное потепление 3867
глобальное потепление климата 3867
глобальное соглашение о лесах 3863
глобуляриевые 3868
глог 1771
глоговина 1771
глубина лесосплавного хода 2405
глубина окорки 565
глубина подновки 7484
глубина порока 2404
глубина резания 2195
глубокая резка 4211
глубокие заболонные грибные окраски 2321
глубокий покой 2323
глубоко дольчатый лист 2325
глубокое пиление 6393
глубоко зубчатый лист 2325
глубококорневое растение 2326
глубоко пальчатый лист 2325
глухой конец 2262
глюта ренгас 6866
гмелина древовидная 5255
гниение 6647
гнилая сердцевина 2289
с гнилой сердцевиной 2402
гнилая древесина 2291
гнилая ядровая древесина 2289
гнилое дерево 7082
гнилое место 7081
гнилое неделовое дерево 7080
гнилое фаутное дерево 7080
гнилое ядро 2289
гнилой 3679
гнилой комель 3680
гнилой сухостой 7821
гнилой сучок 2290
гнилой сучок с выпавшей сердцевиной 6306
гнилостное пятно 7081
гнилостный 7212
гнилоустойчивость 2293
гнилушка 7082
гниль 9529
гниль в третьей стадии развития 1828
гниль древесины 9529
гниль корневой шейки 2092
гниль красно-бурого цвета 3678
гниль от поранения 7078
гниль сердечка 1414
гниль сучьев 4734
гниющее на корню дерево 5885
гнутая древесина 678
гнутая массивная древесина 7840
гнутьё древесины 9489
гнутьё ламинированной древесины 4764
гнутьё массивной древесины 7844
гнутьё многослойной деревянной заготовки 4764
годичная лесосека 296
годичная лесоустройственная программа 299
годичная мутовка 5901
годичная рубка 296
годичная рубка, снятая с леса 297
годичное кольцо 301
годичный побег 305
годичный прирост 300
годичный слой 301
годичный урожай 295
годная щепа 28
годный для использования лес 2951
годный к рубке 3231

год рубки 3144
голая почка 5651
голландская болезнь ильмов 2678
головач западный 1763
головная естественная мачта 4190
головная лесопильная рама 1018
головная мачта 4190
головная часть пожара 2076
головной блок тягового каната 1158
головной станок 1018
головня 3168
голосеменная древесина 4068
голосеменные 4069
голосеменные растения 4069
голоцеллюлоза 4340
голубика 893
голубика дернистая 2683
голубика карликовая 2683
голубика узколистная 1788
голубь дикий 9624
голубь лесной 9624
голый сучок 9050
гольцы 217
гомеостазис 4343
гомопластический привой 4347
гомоцеллюлярный луч 4344
гондурасский палисандр 4348
гондурасское розовое дерево 4348
гонкало 8548
гонобобель 893
гонт 7521
гонтовой лес 9575
гонуама камасси 4685
горбыль 7712
горельник 1196
гореть 3166
горизонтальная борозда 1904
горизонтальная проекция кроны 2088
горизонтальная протяженность корней 4820
горизонтальная разность 4368
горизонтальная сомкнутость кроны 4367
горизонтальная сомкнутость полога 4367
горизонтальное сопротивление 4371
горизонтальный корень 7486
горизонтальный рез 7338
горло древоспуска 2836
горная тундра 5594
горное красное дерево 8990
горно-лесная оподзоленная почва 5584
горно-лесная почва 5585
горно-лесной пояс 5586
горно-таёжный лес 5592
горный виноград 2013
горный влажный тропический туманный лес 5593
горный влажный туманный лес 5593
горный лес 5583
горный плодовый лес 9406
городское лесное хозяйство 5636
городской лес 9142
городской лесничий 1592
городчатая форма края листа 2396
гортензиевые 4402
гортензия древовидная 9407
гортензия метельчатая 6052
горшечная посадка 6482
горькая вишня 772
горькая кора 1544
горький апельсин 775
горький миндаль 207
горючесть 1742
горючие 3739
горючие материалы 3739
господство 6013
господствующая порода 2566
господствующее дерево 2567
господствующий 2563
господствующий возраст 6526
господствующий ярус насаждения 3518
госсвелеродендрон бальзамный 5731
государственное лесное хозяйство 3890
государственный лес 8041
государственный лесной фонд 3767
государственный учёт лесов 8042
готовность участка 6720
готовый пиломатериал 6515
гофрирование 3296
граб 2901
граб американский 249
граб восточный 5950
граб европейский 2901
грабинник 5950
грабняк 4372
грабовый лес 4372
граб приморский 7280
граб сердцелистный 7280
граб японский 4621
гравёр по дереву 1363
гравийная дорога 3921
гравитационная канатная трелёвка 3922
гравитационное перемещение горных пород 7835
гравитационный проволочный лесоспуск 3922
градация по длине 3894
градация по классам возрасти 3893
градиент влажности 5520
градиент созревания 5336
градуированная линейка для измерения лесоматериалов 5107
гранат 6443
гранатник 6443
гранатовое дерево 6443
гранатовые 6442
граница 2041, 8124
граница годичного кольца 4026
граница естественного ареала 5689
граница зоны валки 469
граница кроны 2065
граница леса 4982
граница лесосеки, помеченная на деревьях 469
граница между лесом и лугом 3918
граница объедания листьев и молодых побегов 1096
граница распространения дерева 4982
граница распространения леса 4982
граница эксплуатации 5923
границы типов леса 3634
граничный древостой 3417
грань образца 2763
графитизация древесного угля 1447
гребенщик 3706
гребенщиковые 8382
гребневая посадка 648
гревиллея крупная 3966
грейфер 3892
грецкий орех из Уганды 1087
грибки 3770
грибная болезнь 3769
грибная гниль 3768

грибное заболевание 3769
грибное тело 1879
грибной 3773
грибные ядровые полосы 6795
грибокорень 3775
грибопятнистая древесина 6132
грибоубивающее средство 3771
грибы 3770
грибы, вызывающие изменения окраски древесины 9663
грибы синевы 861
гроздь 6671
грубая древесина, отбракованная на фабрике 1684
грубая древесная масса 1680
грубая кора 1675
грубая очистка от сучьев 1683
грубая продукция древесины без коры 3980
грубая текстура 1678
грубая топливная щепа 1677
грубоволокнистая древесина 1676
грубогумусовая подстилка 1682
грубой гумус 1681
грубообтёсанный лесоматериал 7091
грубоопилённая древесина 7094
грубо определённый объём брёвен 3981
грубоотёсанное бревно 7094
грубые древесные отломки 1686
грубый подпилок 7089
грубый распил 622
груз лесоматериалов 1338
груз лесоматериалов на двух прицепах 2589
груз на двух прицепах 2589
грузовая балка 6235
грузовая стрела 5025
грузовой блок 2993
грузовой канат 8745
грузовой рейс 7690
грузооборот лесовозной дороги 5080
груз последовательно связанных брёвен 8758
грунт 7833
грунтовая вода 3995
группа деревьев 4000
группа корневых отпрысков 1665
группа лесов 3999
группа мелких сосудов 6462
групповая валка 5629
групповая заготовка деревьев 5620
групповая обрезка сучьев 1175
групповая обсеменительная рубка 4004
групповая очистка деревьев от сучьев 1175
групповая посадка 6354
групповая раскряжёвка 1174
группово-выборочная рубка 4001
группово-выборочный высокоствольник 4002
группово-выборочный дубовый высокоствольник 4003
групповое возобновление 6812
групповой луч 140
групповые методы измерения объёма 3998
групповые поры 1669
групповые сучки 3997
группы пород 7899
груша 1795
груша иволистная 9432
груша каллериана 1281
груша китайская песочная 1504
груша лавролистная 206
груша обыкновенная 1795
груша уссурийская 9148
грядка для перешколивания 5812
гуава 4036
гуайава 4036
гуарея высокая 2006
гуатамбу 4035
гуаява 4036, 4037
гуаява-дерево 4037
губка 3770
губовник линейный 2420
губоцветные 4748
губчатая древесина 6467
губчатая сердцевина 1440
губчато-волокнистая корневая и напенная гниль 364
губчатый 3773
губчатый нарост 3770
гужевая трелёвка 292
гуменник 3332
гумификация 4393
гумми 4050
гуммоз 4051
гумус 4396
гусеница листовёртки 3223
гусеница листовёртки-почкоеда елового 7951
выравнителем 286
гусеничный трактор с поворотным выравнителем 286
гусеничный трелёвочный трактор 7799
густая заросль 8389
густая крона 2387
густое насаждение 3756
густо заросший лесом 6936
густой лес 2388
густой молодняк 4104
густой полог 1650
густолиственный 2389
густооблиственный 2389
густота абсолютного древостоя 4009
густота годичных колец 303
густота древостоя 3370
густота кроны 2072
густота леса 3370
густота лесной транспортной сети 2392
густота подроста 2394
густота полога 2393
густота пород 7896
густота посадки 6369
густота посевов 7374
густота сети лесных дорог 2392
гуттаперчевое дерево 4061
гуттация 4062
гуща 4104

давать отростки 3293
давать побеги 3293
давать почки 7944
давидиевые 2594
давидия обвёрточная 2593
давление на кромки 6522
дакридиум кипарисовый 6784
дальбергия африканская 102
дальбергия притупленная 5426
дальбергия Стевенсона 4348
дальбергия чёрная 102
дальбергия сиамская 7563
дальбергия церенская 4716
дальбергия чёрная 1006
дальбергия широколистная 2728
дальномер-дендрометр 6718
дамар 2243
дамара 2243
дамба 2230
дамба из отходов 8780
дамир мерах 4949
даммар 2243

данта 2245
дата вырубки 3060
дать дереву направление валки 4059
дафна 2246
дающий отростки 7807
дающий побеги 7807
дающий поросль 7807
движущийся груз 5014
двойная заболонь 853
двойная сердцевина 2583
двойная серия окольцовывающих насечек 2580
двойное годичное кольцо 5616
двойной главный побег 2582
двойной пильчатой лист 2585
двойной побег 3656
двойной ствол 3657
двоиноперистый лист 741
двоякогородчатая форма края листа 2590
двоякопальчатая форма края листа 2590
двоякоперистосложный лист 6283
двудольное растение 2486
двудомное растение 2507
двуколка 894
двулетник 696
двуручная пила 9030
двуручная пила с искривлённой тыльной частью 4327
двусторонний надруб 9455
двусторонний топор 2579
двусторонняя пила 2584
двухканатный трелёвочный захват 9029
двухкантный брус 9027
двухколесная трелёвочная тележка 894
двухкорпусной плуг 9028
двухлетнее насаждение 9025
двухлетнее растение 696
двухлопастный 697
двухоборотная система рубки порослевого древостоя 9033
двухосный грузовик 888
двухполозные сани для трелёвки брёвен в погруженном положении 7304
двухприёмная постепенная рубка 9034
двухъярусное насаждение 9035
двухъярусный лес 9035
девичий виноград триостренный 937
девственная местность 9404
девственная природа 9404
девственная пробка 9227
девственное дерево 9231
девственный лес 282
девственный тропический лес 8979
дегенерация 7438
деградировавший лес 2342
дезертификация 2414
дезинтегратор 1534
действительная всхожесть 2768
дейция 2442
дейция розовая 3787
дейция шершавая 3787
декапитация 2198
декоративная накладка 9613
декоративное дерево 5957
декоративное деревянное изделие 5958
декоративный полосатый текстурный рисунок древесины радиального распила 6142
делать затёски 669
делать затёски на стволах 669
делать прививку вакциной 9157
делигнификация 2349
делительная лесопильная рама 2277
делительная пила 5148
деловая древесина 5400
деловое дерево 5399
деловой верх 5398
деловой лес 1745
деловой лесосортимент 5396
деловой ствол 5393
деловой сухостой 5397
деловые лесоматериалы 8620
делоникс королевский 3249
делянка 7173
делянка, отводимая одному вальщику 8172
дендрарий 342
дендрит 2367
дендритовый дренаж 2368
дендробионт 2369
дендрогеоморфология 2376
дендрогидрология 2378
дендрограмма 2377
дендроидный 8876
дендроклиматология 2374
дендролог 2379
дендрология 2380
характеристики 2382
дендрометрия 2383
дендрометр с циферблатной шкалой 2464
дендроскоп 2386
дендрофил 2384
дендрофильный 2385
дендрохора 2371
дендрохронолог 2372
дендрохронологический анализ 269
дендрохронология 2373
дендроэкология 2375
день древонасаждения 340
день посадки леса 340
дербенниковые 5161
деревенский лес 9218
деревенский лесной участок 9220
дерево 8797
дерево анисовое 4611
дерево будущего 3786
дерево верхнего полога 6009
дерево верхнего яруса 6009
дерево в изгороди 4221
дерево винтовое 1802
дерево, в котором гнездятся муравьи 320
дерево вне леса 8891
дерево-волк 9472
дерево, выбранное для опыта 7174
дерево, выросшее в густом древостое 1826
дерево, выросшее в открытое место 5878
дерево, выросшее из семени 7392
дерево, дающее защиту другому 5816
дерево, дающее поросль 7946
дерево для посадки у берега 8842
дерево для посадки у берега моря 8842
дерево дынное 1794
дерево дынное обыкновенное 1794
дерево жёлтое 2908
дерево каменное 1777
дерево-канделябрь 3659
дерево карандашное 6530
дерево кентуккское кофейное 1482
дерево коралловое 7180
деревокорень 251

дерево-маяк 2567
дерево между верхним и нижним ярусами 4551
дерево мыльное 7865
дерево, наклоненное ветром 9444
дерево на корню 4010
дерево, на котором зависло другое спиленное дерево 7583
дерево на краю трелёвочного коридора 6673
дерево на опушке леса 922
деревообрабатывающая индустрия 9685
деревообрабатывающая машина 9701
деревообрабатывающая промышленность 9685
деревообрабатывающий инструмент 9652
деревообработка 9630
деревообработчик 9699
дерево обрезки 6615
дерево, о которое трётся крупная дичь 7124
деревоокрашивающие грибы 9663
дерево, отстающее в росте 7992
дерево пирамидальной формы 6650
дерево, повреждённое насекомыми 1068
дерево, повреждённое при трелёвке 1097
дерево, подлежащее на срочной валке по соображениям техники безопасности 4297
дерево-подснежник 1349
деревопоедающее насекомое 4958
дерево, посаженное, чтобы давать тень 7479
дерево-пришелец 1743
дерево пробковое Лаваля 4837
дереворазведение 7610
дереворазрушающее воздействие 9534
дереворазрушающие насекомые 8562
дереворазрушающий гриб 9485
дерево сальное 8375
дерево с большим сбегом 3023
дерево, сваленное при валке другого 7583
дерево с веерообразной кроной 7047
дерево с густой кроной 1315
дерево с дуплом, заселённым пчелами 666
дерево с женскими цветками 6297
дерево симаруба 9373
дерево с канделаброобразной вершиной 3659
дерево с комом земли 499
дерево с коротким оборотом ротации 7546
дерево с крупными боковыми сучьями 4797
дерево, сломанное ветром 2271
дерево с мужскими цветами 7999
дерево с необрезанной вершиной 9124
дерево с необрубленной вершиной 9124
дерево с обнаженным корнем 533
дерево с обрезанной кроной 7489
дерево с обрезанными ветвями первого порядка 5169
дерево с пестичными цветками 6297
дерево с поверхностными корнями 7488
дерево с полнодревесным стволом 3746
дерево с прямыми волокнами 8138
дерево с раздвоенным стволом 3659
дерево с размочаленным концом 6711
дерево среднего размера 5387
дерево среднего роста 433
дерево среднего ствола по поперечного сечения 5355
дерево с сильно развитой кроной 1315
дерево с сухой верхушкой 2273
дерево стираксовое лекарственное 3825
дерево с тычиночными цветами 7999
дерево с узкой кроной 5657
дерево суизл-стик 8351
дерево, схлёстывающее соседние деревья 9358
дерево с хорошо разветвлённой вершиной 2797
дерево с ярко выраженным наклоном 2296
дерево травяное 405
дерево-тренажёр 8760
дерево фиалковое 3846
дерево физетовое 2908
дерево-хозяин гусениц 1377
дерево-хозяйн 4381
дерево ши 7503
дерево-экзот 2943
деревцеобразный 347
деревцо, оставляемое в срубаемом лесу 8555
деревья второго яруса 9072
деревья-гиганты 3845
деревья для коротья 7540
деревья для целлюлозы 6626
деревья, зачокерованные по одному 7659
деревья нижнего яруса 9072
деревья с одной главной осью и боковыми ветвями 5539
деревянистые злаки 9708
деревянистый 9704
деревянистый агат 124
деревянная верховая грань 9559
деревянная втулка 9555
деревянная деталь 9548
деревянная заглушка 9555
деревянная конструкция 8570
деревянная крепь кровли 7006
деревянная меласса 9612
деревянная мерная рейка 8292
деревянная мостовая 9554
деревянная напорная грань 9559
деревянная обшивка 8600
деревянная опалубка 8600
деревянная перепускная труба 8572
деревянная плотина 9550
деревянная поверхность 9557
деревянная подкладка 9491
деревянная проба 8635
деревянная пробка 9555
деревянная распорка 8568
деревянная свая 6475
деревянная труба 8572
деревянная шпала 9556
деревянное изделие 1352
деревянное клише 9491
деревянное оборудование 9560
деревянные буквы 9597
деревянные завитые стружки 9648
деревянные изделия 8626

деревянный брусок 9495
деревянный верховой откос 8654
деревянный гвоздь 9553
деревянный домик 8822
деревянный духовой инструмент 9694
деревянный забор 6695
деревянный каркас 9576
деревянный клеёный дощатый элемент 3878
деревянный клин 9561
деревянный мост 8663
деревянный настил 9521
деревянный слой 8638
деревянный столб 6475
деревянный элемент сплошного сечения 9393
деревянный ящик 9549
дереза 7136
дёрен американский 6783
дёрен канадский 1170
дёрен кистевидный 3924
дёрен красный 1771
дёрен кроваво-красный 1771
дёрен морщинистый 7109
дёрен Натела 9339
дёренные 2556
дёреновые 2556
дёрен очереднолистный 222
дёрен цветущий 1964
дерис индийский 6448
дёрн 7811, 8525
дернина 7811
дерново-подзолистые почвы 2409
дёрнообразование 7813
дёрнорез 7812
дёрнорезный плуг 7812
дёрноснимaтель 7812
деррик 2410
деррик-кран 2410
десятичная таблица Скрибнера 7318
десятник лесорубов 5069
десятник при складе 9748
детали, обработанные в размер 2504
детерма 6782
детрит 2440
детритоядный организм 2441
дефект в фанере или шпоне 4361
дефект дерева 8830
дефект древесины 9530
дефектная доска 2119
дефектный лесоматериал 9118
дефектный хлыст 2335
дефект распиловки 7242
дефект роста 4014
дефекты "крокодиловая кожа" 188
дефицит древесины 7282
дефицит леса 4753
дефицитный лес 2337
дефолиант 2338
дефолиатор 2338
дефолиация 2941
деформативность древесины 9531
деформированная крона 2340
деформированное дерево 5478
дешифрирование участков 3329
джатоба 1990
джунгла 4671
дзелькваграболистная 2792
дзельква японская 4639
диагностическая закладка пробных площадей 2459
диагонально наклоненное волокно 2461
диаграмма плотности насаждения 8012
диаграммная таблица определения объёмов брёвен 2463
диалиум 8380
диалиум гвинейский 9190
диаметр без коры 2475
диаметр в верхнем отрубе 2470
диаметр вершинного торца 2470
диаметр в комле 2465
диаметр в комлевом конце 2465
диаметр в месте среза 2467
диаметр в нижнем отрубе 2465
диаметр дерева 8826
диаметр доминирующего дерева 2568
диаметр измеряемого брёвна 7250
диаметр корневой шейки 2469
диаметр кроны 2074
диаметр на высоте груди 2466
диаметр на половине высоты дерева 2468
диаметр основания 2465
диаметр пенька 2481
диаметр пня 2465
диаметр режущей части 2479
диаметр резания 2479
диаметр с корой 2480
диаметр срединного бревна 7363
диаметр ствола 8071
диаметр сука 2478
диаметр у пня 2465
диаметр ядра 4196
диви-диви 2550
диделотия африканская 1121
дидимопанакс морототони 5558
диервила жимолостная 5788
диервилла жимолостная 5788
дизъюнктивная паренхима 2531
дизъюнктивная сосудовидная клетка 2532
дикая местность 9404
дикая природа 9411
дикая слива 1475
дикая флора и фауна 9411
дикие виды 9417
дикие животные и растения 9411
дикий пятилистный виноград 9228
дикориния параенская 621
диллениевые 2498
дильсы 2278
диморфное разветвление 2506
динамика насаждения 8013
динамика роста насаждения 8013
динамометр 7705
диплодиоз сосны 2510
дипломированный лесовод 3905
диптерокарповый лес 2512
диптерокарпы 2513
дира 4649
дирка болотная 4900
дисгенический 2696
дисклимакс 2526
дисковая машина для измельчения пней 2529
дисковая пила 1585
дисковая рубильная машина 2534
дисковая сегментная пила 6950
дисковый окорочный станок 2533
дистанционно-патрульный способ лесосплава 2540
дистанция промежуточных деревьев 8092
дистемонантус Бентама 5732
дихапеталовые 8412
дичок 9420
дичок для перешколивания 9426

дланевиднораздельный лист 6040
дланевиднорассечённый лист 6042
дланерассечённый лист 6042
длина брёвна 7259
длина волокон 3097
длина дефектной части ствола 2334
длина древесной частицы 9619
длина карры 2982
длина корневой части 7033
длина кроны 2082
длина кряжа 7259
длина лесоматериала, равной длине автопоезда 5095
длина округления 7106
длина пиломатериала 7259
длина подновки 2982
длина ствола 8077
длина трещины 2004
длина хлыста 8077
длинная откомлёвка 5128
длинная трелёвка 2805
длинноволокнистая древесина 5131
длинное бревно 5154
длиннолистный 5142
длинномерные лесоматериалы 5154
длинномерный 5127
длинномерный сортимент 5154
длинномерный сортимент белой древесиной 9390
длинноствольный 5151
длинный побег 5149
длинный тонкий черенок 9263
длительное действие 5152
длительность пожара 3181
днище 4186
дно 4186
добавки 65
добывание живицы 828
добывание семян 7376
добыча живицы 828
договор на продаже леса 9642
договорные лесные продукты 1910
дождевой лес 6700
дозревающий на дереве 8913
документ на приобретение или владением леса 8573
долговечность древесины 137
долговременный путь 6175
долгоносик большой сосновый 4793
долгоносики 9331
долгоносик корневой сосновый 6267
долгосрочный заказник 5672
долготьё 5154
долевое распределение расчётной годичной лесосеки 1
долото 9509
долото лопастного типа 6238
дольчатый лист 1636
доля гнилой древесины 6093
доля коры 578
доля ядра 4200
доминантность 2561
доминантный 2563
доминирование 2561
доминирование вершины 325
доминирующая высота 8692
доминирующий диаметр 2565
домовый гриб 2560
донный брусок V-образного лотка 460
доокорка 3137
дополнение лесных культур 823
дополнение просветов 823
дополнительная лесная дорога 7350
дополнительная посадка деревьев 8863
дополнительная рубка 7345
дополнительное возобновление 6811
дополнительное дерево 1980
дополнительные меры ухода 63
дополнительные участки земли 5303
дополнительный канал 7344
дополнительный молевой лесосплав 8324
дополнительный рабочий 7359
дополнительный урожай плодокорма 121
допускаемая нагрузка 1357
допускаемое отклонение 6181
допускаемое отклонение круглых лесоматериалов 6181
допускаемый коэффициент поражения 6182
допуск диаметра 2472
допуск на толщину 8500
допустимая годичная лесосека 6178
допустимая рубка 6179
допустимое количество 1119
дорога для тачек 1359
дорога общественного пользования 6622
дорога, проходящая по лесу 3565
дорожка, обсаженная кустарником 7556
дорожный пропуск 6979
доска 873
доска без сердцевины 7572
доска для перешколивания 8770
доска из древесины хвойной породы 7828
доска клинообразного сечения 3029
доска, на которой стоит лесоруб 1563
доска с непараллельными плоскостями распила 3029
доска с обзолом 7088
доска с сердцевинной трещиной 886
доска стандартными размерами 8103
доски 882
доски с пазом и гребнем 8677
досковая заготовка 874
досковидный корень 1231
досковый фут 876
доставка древесины 9532
доставка леса 2359
доставка лесоматериалов 2359
достаточно восстановленный 7224
доступная вода 423
доступная вода в почве 423
доступная древесина 424
доступная лесистая зона 30
доступная лесная область 30
доступная площадь пожара 194
доступный лес 29
доход лесного хозяйства 4444
доходность прорубки 9782
доход от рубки главного пользования лесом 3145
доход по массе 9773
дошедший до нас 2960
дощатая обшивка 880
дощатая опалубка 3670
дощатый 957
дощатый горбыль 881
дощатый лесоспуск 884
дощатый обапол 881
дощечка 7732
драгирование семян 6141

драконово дерево 2608
драконтомелум дао 6033
дранка 9595
дрань 9595
драцена 2608
драцена киноварно-красная 7810
древесина 9473
на пню стоящая ~ 3454
древесина амаранта 9225
древесина берёзы с красивым текстурным рисунком 5330
древесина болотистого леса 3085
древесина вершин деревьев 8712
древесина в соку 3710
древесина второстепенного значения 7355
древесина высокой плотностью 8659
древесина, высохшая в лесу 3386
древесина, высушенная печным обогревом 4710
древесина для авиастроения 5063
древесина для дильс 5204
древесина для железной дороги 8589
древесина для изготовления моделей 5505
древесина для изготовления спичечной соломки 5332
древесина для каноэ 1310
древесина для наружной облицовки 2964
древесина для планок 6328
древесина для сооружения каркаса зданий 3686
древесина для столбов 6428
древесина для сухой перегонки 44
древесина для экстрагирования 2971
древесина для экстракции 2971
древесина дуба 5850
древесина дуба радиального распила 9258
древесина дуба с характерной коричневой окраской сердцевины 1082
древесина железнодорожной доставки 1364
древесина жёлтого цвета 9767
древесина, идущая на столбы и деревьев 3732
древесина из подстригаемых деревьев 6429
древесина искусственной сушки 2421
древесина, испещрённая с прожилками 1662
древесина, используемая для вторичной переработки 909
древесина как энергетический источник 9481
древесина кампешевого дерева 1289
древесина кедра 1404
древесина корней 9580
древесина кустарника 1116
древесина лимонного дерева 1591
древесина малой площади поперечного сечения 8662
древесина местных пород 4341
древесина низинных твёрдолиственных пород 5193
древесина, обработанная пропиткой 8792
древесина, образуемая камбием 7357
древесина обычного качества 1776
древесина однодольных растений 5541
древесина оливкового дерева 5894
древесина орехового дерева 4184
древесина, отбракованная на фабрике 5450
древесина, отмершая от жуков 1150
древесина перестойного дерева 5884
древесина пня 8220
древесина, повреждённая пожаром 9528
древесина, поражённая грибной гнилью 2291
древесина, поражённая насекомыми 4510
древесина, поражённая синевой 859
древесина, поражённая сухой гнилью 9484
древесина порослевых пород 7361
древесина, пригодная для 507
древесина, пропитанная бакелитом 9735
древесина, пропитанная маслами 3026
древесина радиальной распиловки 6678
древесина рыхлой структурой 5162
древесина сандаракового дерева 1591
древесина светлых пород 4938
древесина с высоким содержанием ядра 184
древесина с высокой смолистостью 4292
древесина с диагональным наклоном волокон 2462
древесина сезонной сушки 150
древесина с капами и наплывами 1032
древесина с кружевной текстурой 4752
древесина случайных пород 7292
древесина смешанных низинных твёрдолиственных пород 5489
древесина с наплывами типа "птичий глаз" 767
древесина с небольшой плотностью 8660
древесина с ненормальным строением ткани 9734
древесина с обзолом 9079
древесина с отлупом 9696
древесина специального назначения 7891
древесина с прожилками 1662
древесина с проростью 7289
древесина сросшихся стволов 4536
древесина с узким расположением годовых колец 1660
древесина с узорчатой текстурой 9186
древесина твёрдолиственных пород 4121
древесина тисового дерева 9769
древесина тропических пород деревьев 8981
древесина хорошего качества 4284
древесина ценных пород 3159

древесинник 231
древесинник колумбийский 1732
древесинный пучок 9706
древесиноведение 9645
древесная гадюка 8954
древесная гниль 9529
древесная грибка 9581
древесная губка 9581
древесная замазка 8958
древесная зола 9482
древесная зола твёрдолиственных пород 4132
древесная зола хвойных пород 7827
древесная камедь 8907
древесная клетка 9503
древесная кора 8799
древесная культура 3528
древесная масса 9603
без содержания древесной массы 9577
древесная масса из щепы 1519
древесная масса с содержанием коры 542
древесная монокультура 8887
древесная мука 1514
древесная окраска 9541
древесная паренхима 9617
древесная паста 9606
древесная пила 8854
древесная плита 9492
древесная порода 4714
древесная порода-заполнитель 3131
древесная порода из семейства эбеновых с зелёной древесиной 3942
древесная порода-пионер 6290
древесная почка 9179
древесная пыль 1514
древесная растительность 8953
древесная смола 8907
древесная соломка 9695
древесная ткань 9679
древесная утка 8832
древесная целлюлоза 9603
древесная шерсть 9697
древесная щепа 9508
древеснеющий 4955
древесница 9579
древесница въедливая 4917
древесноволокнистая масса 9603
древесноволокнистая плита низкой плотности 5183
древесноволокнистая плита средней плотности 4545
древесное вещество 9667
древесное масло 9615
древесное многолетнее растение 9713
древесное насекомое 3475
древесное растение 9714
древесное строение 8839
древесное сырьё 7096
древесное топливо 9487
древеснокорешковый 9715
древеснолистный 9709
древесносмоляное масло 9673
древесностружечная плита 6091
древесностружечная плита из станочной стружки 3242
древесностружечная плита различной плотности 3895
древесностружечный брикет 9495
древесноугольная куча 1450
древесноугольная мелочь 9506
древесноугольная пыль 9506
древесноугольная яма 1451
древесноугольный блок 1443
древесные опилки 9644
древесные отходы 3561
древесные полости 8811
древесные породы, запасающие в тканях жиры 3025
древесные породы, запасающие в тканях крахмал 8038
древесные породы со смолистой древесиной 3025
древесные стружки 9508
древесные частицы 9508
древесный 9704
древесный брикет 9495
древесный вредитель 9533
древесный газ 9582
древесный генератор 1444
древесный гриб 9581
древесный дёготь 9672
древесный клей 9583
древесный ковёр 9501
древесный лейшманиоз 6224
древесный лишай 8888
древесный материал 8559
древесный пек 8907
древесный пластик 1844
древесный покров 8810
древесный полог 3345
древесный порошок 1514
древесный род 8920
2390
древесный слоистый пластик с звездообразным расположением волокон 3810
древесный слой 3334
древесный сок 7208
древесный сосуд 9687
древесный спирт 9476
древесный ствол 8934
древесный стриж 8942
древесный уголь 9505
древесный уксус 9688
древесный хлыст 5154
древесный чай 8437
древесный щит 8922
древесный ярус 8866
древовал 8808
древовал-корчеватель 8808
древовидная структура 8936
древовидная хатьма 8881
древовидное алое 9705
древовидное разветвление 345
древовидный 8876
древовидный кактус 2366
древовидный кустарник 346
древовидный отпечаток 2367
древовидный папоротник 8837
древогрызы 8562
древостой 3454
древостой, отмеченный для вырубки 8644
древостой ценной породы 8024
древоточец ивовый 3879
древоточец пахучий 3879
древоточец тополевый 4804
древоточцы 8562
дрена 2615
дренаж 2612
дренажная канава 2613
дренажная сеть 2614
дренажная труба 2615
дробилка 1534
дробилка горбыля 7708
дробилка для древесных отходов 7721
дробилка для дров 9278
дробилка для коры 2281
дробилка для лесосечных отходов 7721
дробилка для щепы 1534
дробилка древесины 1534
дробилка отходов шпона 9194
дробилка сучьей 5230
дробильный нож 4321
дроблёная щепа 4318

дробление отходов лесопыления 4320
дроблёнка 4318
дроблённые отходы 4318
дробный плод 7447
дрова 3740
дрова в кордах 7982
дрова для растопки 3740
дрова для углежжения 9571
дрова-коротьё 7920
дровокол 1632
дровоколка 9512
дровокольный станок 9512
дровосек 9526
дровосек еловый круглоголовый 7104
дровосеки 1327
дровяная древесина 3740
дровянные остатки 3743
дровяной кругляк 7103
дровяной лес 3742
дровяной сарай 9650
дровяные отходы 5215
дрок 3849
другая покрытая лесом земля 5969
другой добив 5968
дружнорастущее насаждение 8536
дряблая древесина 2605
дряблость 6652
ДСП 6091
ДСП, облицованная натуральным шпоном 9195
ДСП экструзионного прессования 2976
дуб 5830
дуб австралийский 5566
дуб австрийский 5565
дуб алеппский 3794
дуб американский красильный 2692
дуб американский шарлаховый 7287
дуб аризонский 8993
дуб бархатный 2692
дуб белый 8993
дуб болотный 3764
дуб веретеновидный 6376
дуб веретенообразной 6376
дуб виргинский 5015
дуб Гарри 5945
дуб горный 5582
дуб греческий 3937
дуб густой 4397
дуб двухцветный 8317
дуб для переработки на ванчёсы 9258
дуб дольчатый 9159
дуб дубильный 615
дуб Дюранда 2672
дуб-железняк 7464
дуб железоносный 3859
дуб звездчатый 255
дуб зимний 7464
дуб золоточешуйчатый 1319
дуб зубчатый 2228
дуб иволистный 9434
дуб изменчивый 5951
дубильная кора 8392
дубильное вещество 8393
дубильное корьё 8392
дубильное пятно 8396
дубильный нарост 8391
дубина 7111
дуб кавказский 1387
дуб калифорнийский чёрный 1255
дуб карликовый 642
дуб карликовый калифорнийский 1271
дуб каштановый 5582
дуб каштанолистный 1478
дуб кеглевидный 7559
дуб кермесовый 4697
дуб кермесоносный 4697
дуб колючий 7190
дуб конусоплодный 7635
дуб коровой 615
дуб красильный 2692
дуб красный 7863
дуб красный американский 5793
дуб красный северный 5793
дуб крупноплодный 5566
дуб крупночешуйчатый 3937
дуб лавролистный 4835
дуб летний 1793
дуб ливанский 4903
дуб лировидный 5991
дуб македонский 5226
дуб малоазиатский 3794
дуб малый 255
дуб мерилендский 799
дуб Мирбека 5477
дуб Мишо 615
дуб нуттальский 5828
дубняк 5836
дубняк грабовый 4374
дуб обыкновенный 1793
дуб овальнолистный 5791
дубовая бочарная древесина
дубовая поросль 5835
дубовая роща 5836
дубово-грабовый лес 4374
дубовое бревно 5842
дубовые губки 3127
дубовый лес 5836
дубовый лесоматериал 5850
дубовый мох 5845
дуб Оглеторпа 5872
дубок 5841
дуболистный 5840
дуб острейший 7266
дуб остролистный 4625
дуб острый 4625
дуб пальчатый 3162
дуб пильчатый 3859
дуб понтийский 363
дуб пробковый 1961
дубрава 5836
дубравное широкотравье 1050
дуб распростёртый 5562
дуб северный крупный 6768
дуб серый 3928
дуб сидячецветный 7464
дуб сидячецветный мексиканский 5420
дуб сизый 862
дуб скальный 7464
дуб траволистный 1693
дуб Фрайнетто 4397
дуб хермесовый 4697
дуб черепитчатый 3874
дуб черешчатый 1793
дуб чёрный 9294
дуб Шумарда 7561
дуб японский 4624
дуб японский мирзинолистный 862
дуга 8048
дугласия 1772
Дугласова пихта 1772
Дугласов дуб 242
дудник лесной 9401
дуплистое дерево 4333
дуплистость 4332
дуплистый ствол 4334
дуплистый сучок 4331
дупло 2933
дуплянка 5711
дурника 893
дуссия 101
душистый кедр 1411
душистый сумах 365
дыхательный корень 86
дюнный лес 2667
дятловые 9622

евгения опушённая 1240
евгения Ямболана 4646
Европейская Комиссия по Лесоводству 2896
европейская пальма 2886
европейский кедровник 8347
единая кривая масс 9089
единица анализа местообитания 4073
единица выборки 7178
единица вывозки 5089
единица измерения круглых лесоматериалов 3267
единица наблюдения 7178
единица обработки 8796
единица управления местообитаний 4078
единица ухода 8461
единичная вырубка в низкоствольнике 7409
единичная смесь 5496
единичное смешение 5496
ежевика 778
ежевика высокая 4307
ежевика карликовая 2684
ежевика лесная 778
ежевика несская 972
ежевика сизая 2891
ежегодная смена листвы 298
ежегодный рост 300
еловая балансовая древесина 7950
еловая древесина 7959
еловая живица 7957
еловая смола 7957
еловая хвоя 7955
еловая шишка 7952
еловые балансы 7950
еловые или сосновые доски 2278
еловый лес 7953
ёлочный ресунок 4261
ель Алькока 163
ель аянская 4647
ель белая 9386
ель Бревера 9330
ель восточная 2723
ель восточногималайская 7594
ель гималайская 4310
ель голубая 1726
ель двухцветная 163
ель европейская 1805
ель изящная 8547
ель индийская 4310
ель канадская 9386
ель колючая 1726
ельник 7953
ельник-брусничник 1999
ельник-кисличник 9657
ель обыкновенная 1805
ель сербская 7456
ель сибирская 7568
ель ситхинская 7678
ель чёрная 2716
ель шипованная 7594
ель Энгельмана 2822
ель японская 8547
ерика древовидная 4206
естественная граница 5673
естественная древесина 3826
естественная камфара 5674
естественная очистка от сучьев 5688
естественная продолжительность жизни дерева 5686
естественная редина 5680
естественная сосна 5687
естественная спелость 6215
естественная трелёвочная мачта 5328
естественное возобновление 5690
естественное изреживание 5699
естественное изреживание из-за возраста 5700
естественное изреживание леса из-за возраста 5700
естественное лесовозобновление 5690
естественное лесонасаждение 5698
естественное насаждение 5698
естественное обесцвечивание 5697
естественное обсеменение 5695
естественное прореживание 5699
естественное распространение лесов 5677
естественный горючий материал 5681
естественный каучук 5694
естественный лес 5679
естественный лесной горючий материал 5681
естественный наклон 5685
естественный оборот 5693
естественный отбор 5696
естественный порок 5675
естественный привой 5682
естественный ресурс 5691
жакаринда бразильская 1006
жасмин 1782
жасмин грунтовой 8335
жасмин дикий 8335
жасмин жёлтый 9759
жасмин лекарственный 1782
жасмин ночной 8335
жасмин садовый 8335
желанное будущее состояние насаждения 2429
желательная инородная порода 2430
желательная неместная порода 2430
желательное растительное сообщество 2431
желательные растительные виды 2428
желатинозное волокно 3813
железное дерево 407, 6186
железное дерево пустынное 2415
железный долбежный шест 6611
железный стержень для продевания троса или цепи под пачку брёвен 1300
жёлоб 334, 5116
жёлоб для спуска коры 584
жёлобок для стока живицы 334
жёлобок для стока живицы в приёмник 334
желобчатый, рифлёный 3977
жёлтая гниль мёртвой древесины 9762
жёлтая грибная окраска 9764
жёлтая заболонная гниль 9761
жёлтая пихта 1772
жёлтая сосна 2726
жёлтая трещиноватая напенная гниль 9760
жёлтинник 2908
жёлто-белая кольцевая гниль стволов 9380
желтовато-белая ядровая гниль 9374
жёлтое дерево 4582
жёлтоземы 9763
жёлтосмолка 405
жёлтый жасмин 1346
жёлтый тополь 9008
желудёвая сеялка 47
жёлудниковые 1126
жёлудь 45
женская шишка 1350
женское дерево 5569

заготавливать 4149
заготовительный комбайн вершинной древесины 8713
заготовитель повреждённых лесоматериалов 5108
заготовка балансов 6632
заготовка биомассы 731
заготовка в полностью погруженном положении 2350
заготовка деревьев с корнями и кроной 1831
заготовка длинномерных сортиментов 5615
заготовка долготья 5157
заготовка дров 3222
заготовка круглых лесоматериалов 7115
заготовка леса 9588
заготовка леса у пня 8204
заготовка на себя 7416
заготовка низкокачественных деревьев 7169
заготовка определённых размеров 2503
заготовка пнёвого осмола 8204
заготовка повреждённого леса 7168
заготовка повреждённых деревьев 7169
заготовка семян 7376
заготовка сортиментов 3078
заготовка тонкомерных деревьев 7779
заготовка хлыстов 3046
заготовка целыми деревьями с кроной 1831
заготовка шишек 1863
заготовка щепы 3759
заготовки для катушек 7926
заготовки, подвергшиеся первичной машинной обработке 5458
заготовленная древесина 3042
заготовленный лес 3042
заграждение из деревьев 4
загрузочный лоток 3033
загрузчик большой подъёмной силы 6065
загрузчик лесовозного автопоезда 5088
загущение 8494
загущённая посадка 8501
загущённое молодое насаждение 4104
загущённолистный 1658
заделка дефектного участка 6103
задержание осадков 4535
задержанный рост 7990
задержка роста 4015
задернение 7813
задние салазки 6746
задний план 9236
задний рез 3066
задняя оттяжка 467
задранные волокна 1524
зажатий ствол 2062
зажатое дерево 9053
зажигание по площади 351
зажигать 3166
зажим для брёвен 8577
зажим для брёвен из двух крючьев 2581
зажим для крепления несущего каната 7699
зажим мотыгой 4317
зажимная доска 2554
зажимное приспособление 4653
зажимное устройство 8050
зажимный коник 1596
зажим пней 8203
зазубренный край 2662
заиливание 7599
закавказское железное дерево 6186
заказник 5672
закисание 41
закисление 41
закладка выёмки 464
закладка ленточных проб 4987
закладка ленточных проб для изучения возобновления 4987
заклинивание 4703
закомелистое дерево 5129
закомелистость 1234
с бутылкообразной ~и 8493
с сильной ~и 8493
с трубкообразной ~и 8493
закомелистый ствол 595
закон лесоводства 3534
закон об охране лесов 3360
законодательные рамки 8047
законодательство по огневой очистке лесосек 7720
законодательство по очистке лесосек 7725
закон о лесе 3484
закрепление с помощью клиньев 4703
закрепление шпонками 4703
закручивание 9020
закрывающий канат 1661
закрытая канава 2132
закрытая пожарная прорость 4274
закрытый дорогой угол осваиваемой лесосеки 2679
закрытый на пожароопасный сезон 1652
залежь 2
залесённая местность 8604
залесённая площадь 9542
залесить 93
заливание 3282
заливные бочки 8549
залом 5091
замазка 9606
заманиха 2451
замаркированное дерево 826
замедление возобновления 6816
замедление роста 4015
замедлитель гниения 2295
замедлитель роста 4025
заменитель древесины 9669
замер годичных колец 5363
замер толщины 2436
заметные игольчатые сучки 1885
замкнутая пастьба 1657
замкнутая сердцевина 956
замкнутая система сблокированных канатов 1654
замкнутая трещина 958
замкнутая цепь лесотранспортёра 1159
заморские древесные породы 6003
замывание корней 7041
замыкающая плёнка поры 6308
замыкающий канат 1661
занятость в лесном хозяйстве 2799
заострение 8403
заострённый 7917
заострённый наконечник 6409
западная туя 5795
западноиндийское атласное дерево 9349
западноиндийское сандаловое дерево 8716
запаздывание вылёта короедов 4325
запаздывание появления

запань 8136
запас древесины на 1 акре в кордах 1945
запас древесины на акр 8023
запасная пазушная почка 7342
запасный лесной фонд 3558
запас отведённого в рубку насаждения 2188
запас смолы 5701
запасы биржи лесоматериалов 9753
запасы горючего вещества 3736
запасы древесины 9640
запасы древесины на корню 3559
запасы древостоя 3559
запасы леса на корню 3559
запасы насаждения 3559
запасы склада лесоматериалов 9753
запасы спелой древесины 9640
запланированная на будущее лесная площадь 1321
запланированное сжигание 6516
заплесневелая древесина 2291
заплётка 7918
заплывание 5862
заплывать 2106
заплывочный желвак 2157
заплывочный нарост 2157
заплывший сук 833
заповедник 5672
заповедник научного значения 6884
заповедный лес 6592
заподсоченное дерево 959
заподсоченный лес 9013
запоздалое сжигание 4816
заполнение оврагов 4046
заполнитель 3130
заполняющая порода 3131
заполняющий древостой 420
заполняющий клин 3132
запрет вырубки 6577
запрет расчистки 4498
запрет сплошной рубки 6576
запрещённый лес 525
запруда 5039
запущенное насаждение 2406
запятнанная древесина 1662
заработанные лесозаготовки 2705
заражение 4486
заражение вредителями 6191
зараженное насекомыми дерево
зарастание вырубок 6822
зарастание поверхностного повреждения 5862
зарастание рубцов 5862
зарастание сучка 4968
заращивание 7599
зарегистрированные отборные деревья 6830
зарегистрированный профессиональный лесовод 6564
зарезервированная продуктивная лесная земля 6563
зародыш 2795
зародышевая плазма 3838
зародышевый корень 6546
заросли вечнозёленых кустарников 5300
заросли персикового дерева 6124
заросль 4104
заросль гребенщика 8381
заросль кустарничков 7558
заросль лаванды 4838
заросль ладанников 6992
заросль лещины 4181
заросль ракитника 1070
заросль рододендронов 7063
заросший деревьями 3394
заросший сучок 833
зарубать 669
зарубить 824, 2169
зарубка 4082, 6513
зарубка на дереве, предназначенном к валке 3069
зарубка на конце брёвна 2319
зарубка топором 449
зарывание комлей в почву 1217
засадить лесом 93
засаживать 6334
засека 4
засечка 540
засиневшая древесина 859
засинение 860
засмолённая древесина 6901
засмолок 8155
засоленность 170
засорение 1560
засохшее дерево 2638
засохший сук 2647
застревание баланса 4117
засуха 2635
засухоустойчивость 2637
засыпка канавы 464
затвердение 4120
затвердение поверхности 1367
затенение 7481
затеняющий кустарник 7475
затёсанный конец брёвна 7795
затёска 540
затёска конца брёвна 1224
затёска на стволе дерева 540
затёска топором 449
затёсывать 824
затонувшее дерево 2268
затонувшие лесоматериалы 2268
затопление 3282
затопленный лес 3281
затор брёвен 5096
заторный лес 8579
затраты времени на рейс 9011
затраты на валку леса 3056
затраты на заготовку 8217
затраты на заготовку и вывозку леса 8217
затраты на 1 корд 1977
затраты на трелёвку 7684
затухать 400
затягивание канатного стропа 9465
затягивание карры 5862
затянувшийся пожарный рубец 1378
заусовка 5483
захват 8829
захват дерева 8847
захват для долготей 8868
захват для хлыстов 8868
захватные крюки 3892
захват организмом нового местообитания 2734
захват с пилой 3912
захват с трособлочным управлением 7504
захваты для брёвен 3892
захламлённый лес 2288
зачаток корешка 6546
зачаток листа 4884
зачаточная пора 6544
заштабелёванная древесина 7982
защемлённое дерево 9053
защемлённое положение дерева 9052
защита древесины 9627
защита окружающей среды 2842
защита опушки 6596
защита подрастающего леса

защитная каска 4122
защитная лесная полоса 7508
защитная полоса из лиственных пород 2928
защитная способность леса 6597
защитное дерево 5816
защитное насаждение 1993
защитное ограждение дерева 8850
защитное снаряжение 7149
защитные очки 7150
защитные перчатки 6599
защитные растения 6600
защитные средства для древесины 8623
защитные средства против поражения грибами 3774
защитный лес 6592
защитный рычаг 7151
защитный хворост 1995
защитный шлем 4122
защитный щит 670
защитный элемент 6598
защищённые дикие животные 5761
защищённый посев площадками 7510
звёздообразная сердцевина 8059
звёздообразная трещина 8039
звёздочка 7943
звено для окучивания брёвен 1172
зверобой 7154
здание с деревянным каркасом 8593
здоровая древесина 5902
здоровая заболонь 7854
здоровое дерево 9216
здоровый лесоматериал 5902
здоровый сучок 7852
зебровое дерево 118
зелёная гниль 3954
зелёная зона 3941
зелёная листва 9200
зелёная масса 9200
зелёная щепа 3940
зеленение 9226
зелёное строительство 233
зелёное удобрение 3949
зелёное эбеновое дерево 4581
зелёный обход 3941
зелёный сук 5013
зелень 9200
земельные угодья на используемые для заготовки сена 175
земельный выдел 6540
землевладение 4780
землеройка 8923
земли специального назначения 7890
земля 4769, 7833
земля для лесной промышленности 3467
земля естественного леса 5671
земля за границы леса 5868
земля, заросшая кустарником 353
земля коммерческих лесов 1746
земляника лесная 9425
земляничное дерево 8152
земляничные 8151
земляной лесоспуск 2709
земляной спуск 2709
земля промышленных лесов 1746
земная база 4770
земная база для лесозаготовки 8594
зенобия припудренная 4350
зеркальная кора 7787
зеркальный реласкоп 6852
зернистое внесение 3911
зиготический лес 9800
зимнезелёный лес 9461
зимнелистопадный лес 1719
зимний лесоспуск 2708
зимний солнечный ожог коры 9457
зимняя обрезка 9462
зимняя окорка 9458
зимняя рубка 9460
зимняя спускная дорога 2708
зимолюбка зонтичная 1798
злаковые 3915
златка калифорнийская 1260
златка сосновая австралийская 408
златка сосновая голубая 868
златка сосновая мохнатая 4089
златка сосновая чёрнорогая 797
златка узкотелая берёзовая бронзовая 1065
значимость рекреационного объекта 6755
золотое дерево 478
золотой дождь 1784
золотой дождь альпийский 215
зона адаптивного управления 61
зона замедленного развития 9799
зона корней 7032
зона лесного хозяйства 3330
зона лесной продукции 3538
зона лесов 2371
зональное агролесохозяйство 9797
зональность 9798
зона отдыха 6754
зона просачивания 3133
зона расположения сука 4724
зона распространения корней 7032
зона сбрасывания 16
зона скопления сучьев 4724
зона, соседняя берегам водоёмов 6963
зона среза 3991
зона управления ресурсами 6912
зона, чувствительная к дыму 7785
зона чувствительных ресурсов 7440
зонирование лесопользования 3638
зонтичные 1355
зоны лесного рынка 8614
зоны, находящиеся под охраной 4259
зоны простоя 5804
зоны управления спелых лесонасаждений 5881
зрелая древесина 6964
зубчатая пальметто 7262
зубчатая форма края листа 2396
зубчатое бульдозерное полотно 6986
зубчатое ядро 8680
зубчатый 7461
зубчатый корчеватель 7042
зубчатый лист 8681

ива 9439
ива Арройо 369
ива бальзамовая 505
ива Бебба 644
ива белая 1809
ива блестящая 7522
ива Бонпланда 907
ива буреющая 160
ива вавилонская 459
ива волчниковая 9224
ива гельветика 8348
ива голая 218

ива Гукера 4360
ива длиннолистная 7183
ива жёлтая 9766
ива каролинская 1688
ива козья 3880
ива корзиночная 619
ива красная 6801
ива кустарниковая 7162
ива ломкая 1040
ива Маккензи 5233
ива матсудана 1963
ива миндалелистная 6122
ива миндальная 6122
ива миртолистная 9396
ива миссурийская 5480
ива мохнатая 9720
ива Неттола 7311
Иванов-побег 4765
иван-чай 3215
ива остролистная 7138
ива пепельная 3974
ива плакучая 459
ива ползучая 2015
ива полярная 6417
ива прутовидная 619
ива пурпурная 6643
ива пятитычинковая 630
ива речная 6975
ива розмаринолистная 872
ива росистая 2458
ива росяная 2458
ива сахалинская 7146
ива сердцевиднолистная 4198
ива серебристая 1809
ива серо-голубая 3926
ива сизоватая 2775
ива силезская 7595
ива ситхинская 7680
ива тонкостолбиковая 7065
ива травянистая 6228
ива трёхтычинковая 6122
ива ушастая 2697
ива хрупкая 1040
ива чёрная 813
ива чернеющая 9396
ива шерстистая 9720
ива-щелюга 7138
ивняк 9436
ивовая ветка 9437
ивовая кора 9428
ивовая лоза 9437
ивовые 9429
ивовые жерди 9015
ивовые прутья 9437
ивовый лес 9430
ивовый пилильщик 9435
иглица 1212
иглица колючая 1212
иглицевые 1213
игловидный 36
иглообразный 36
иглоочиститель 2427
иголочка 7912
игольчатый 36
игольчатый лист 37
игольчатый сучок 6281
идеальная густота древостоя 4420
идеально густой 4419
идеальный лес 5781
идентификация делянок национальной лесной таксации 6381
идиобласт 4421
избыток запаса 8287
избыток леса 8651
избыточный лес 8286
известковый ясень 4975
извилина 9020
извилистая затёска конца брёвна 5797
извилисто-волокнистая древесина 2048
извилистость ствола 8999
извилистый 1230
извилистый пробег волокон 2144
изгарина 5444
изгиб 2031
изгородь 5019
изгородь из сеточной проволоки против кроликов 6670
изготовление тонких древесных срезов 9741
изделия из древесины 3537
излучение 6682
измельчение 1531, 2105
измельчение древесины 2105
измельчение древесины при вращении режущего инструмента в вертикальной плоскости 2601
измельчение лесосечных отходов 2640
измельчённая древесина 2530
измельченная подстилка 1456
измельчённая топливная древесина 4319
измельчённая щепа 7553
измельчённые обрезки 7553
измельчённые стружки 7553
измельчитель лесосечных отходов 7721
измельчитель пней 8199
измельчитель стружек 1534
измельчитель щепы 1534
изменение окраски древесины 9514
изменение цвета складочной древесины 2047
измерение в складочных кубических мерах 7978
измерение высоты 4232
измерение годичных слоёв 5363
измерение диаметра 2436
измерение длины 4914
измерение кубатуры 5103
измерение кубатуры брёвен в метрических мерах 5417
измерение кубатуры в досковых футах 879
измерение кубатуры делением средней длины окружности на четыре 2160
измерение кубатуры круглых лесоматериалов 7116
измерение кубатуры мерной вилкой 1282
измерение лентой по окружности кубатуры лесоматериала без коры 8166
измерение лесоматериалов в тысячах квадратных футов 8529
измерение мерной лентой 8398
измерение наклона волокон 2437
измерение объёма лесоматериалов 5364
измерение окружности 3857
измерение площади поверхности в квадратных футах 3036
измерение чистого объёма 8992
измеренная древесина 2502
измеритель влажности древесины 9611
измеритель древесины 1457
измеритель толщины 8497
измеритель толщины коры 587
измеряемая площадь поперечных сечений стволов эксплуатационной части насаждения 5360
измеряемая площадь поперечных сечений

изогнутая ветка дерева 2100
изогнутое полотно пилы 2152
изогнутозубчатый 2155
изогнутые волокна 1902
изогнутые ножницы 287
изогнутый топор 4390
изолированная пора 7845
изолированная сомкнутая группа деревьев 4000
изоляционная древесно-волокнистая плита 4520
изоляционная плита 4521
изоляционный картон 4521
изреженная крона 8504
с изреженной кроной 7881
изреженная крона дерева 8504
изреженная растительность 5921
изреженность полога 4565
изреженный полог
с изреженным пологом 7881
изреживание 8515
изреживание полога 4566
изреживание полога насаждения 4566
изреживание сомкнутости 4566
изрезанный рельеф местности 4212
изучение запаса, прироста и отпада насаждений 5266
изымать лесную площадь из пользования 2524
икак 1700
икако 1700
икра 1062
иксора 8715
ил 7598
илистый нанос 7598
иллициевые 294
иллициум 293
иллициум анисовый 4611
иллициум настоящий 4611
иллициум священный 4611
ильм 2786
ильм американский 247
ильм белый 247
ильм горный 7310
ильм крупноплодный 698
ильмовая древесина 2791
ильмовник 4618
ильмовые 2788
ильмовый заболонник-разрушитель 4794
ильмовый лес 2789
ильм плакучий 9325
ильм поздний 2893
дерева 2100
имеющей строгающие зубья 1525
имеющий большие корни 6684
имеющий быстрый сбег диаметра к вершине 8401
имеющий толстые побеги 6018
имеющий шпиль 7917
импортный лесоматериал 3317
импрегнирование под давлением 4430
иная лесная земля 5967
инбридинг 4435
инвентаризация 4579
инвентаризация леса 3478
инвентаризация насаждений 8290
инвентаризация пород 7900
инвентаризация состояния леса 3357
инга 4490
ингибитор горения 3197
ингибитор роста 4025
индекс ассоциации 4460
индекс густоты насаждения 8011
индекс лесного местообитания 8641
индекс лесозаготовки 4156
индекс листовой поверхности 4855
индекс местообитания 7667
индекс относительной распространённости 4461
индекс площади листьев 4855
индекс полноты насаждения 8011
индигофера 4478
индийская ротанговая пальма 6731
индийский финик 8379
индикативная лесохозяйственная стратегия 4472
индустриальная древесина 9721
инжектор 8862
инженер лесных культур 3400
инжир 1774
инициалий антеридиальной клетки 7997
инкобу 1087
иннервация 9192
инокулировать 9157
инокулят 4507
инокуляция 4506
инородное тело 3316
леса 3549
инструкции рубки 6844
инструкция по обмеру леса 9658
инструментальный ствол 3107
инструмент для выжигания 1190
инструмент для обработки древесины 9652
инструмент Пуласки 6624
инструменты для столярных работ 4659
интарсия 9590
интегральное пользование леса 4524
интегрированная лесозаготовка 4525
интенсивная проходная рубка 4217
интенсивная рубка ухода 4217
интенсивное ведение лесного хозяйства 4528
интенсивное лесное хозяйство 4529
интенсивное лесоводство 4450
интенсивное низовое прореживание 4214
интенсивное прореживание 4217
интенсивность дождевых осадков 6699
интенсивность дождя 6699
интенсивность культуры 2034
интенсивность линии огня 3190
интенсивность подсочки 8410
интенсивность рубки 4210
интенсивность рубок ухода 4527
интенсивный рост 4192
интервал возобновления пожара 3185
интервал класса возраста 128
интервал между рубками 7077
интерпретативное лесное местообитание 4563
интродуцированные породы 4576
интродуцированный вид 7903
интсия 6406
инфильтрационная способность 4489
инфильтрация 4488
инъекция химикатов в живое дерево 8861
ипе 4581
ирга 7462

ирга колосистая 2689
ирга круглолистная 3806
ирга метельчатая 2689
ирга обыкновенная 3806
ирга ольхолистная 6024
ириантера 4717
ироко 4582
ископаемая древесина 6196
ископаемое топливо 3675
ископаемые виды топлива 3675
ископаемый лес 230
искоренелое дерево 9141
искривление 2031
искривление ствола 8325
искривлённое бревно 2027
искривлённое дерево 2030
искривлённый лесоматериал 2029
искривлённый рост 2025
искривлённый ствол 2154
искривлённый сучок 2026
искусственная древесина 378
искусственная очистка от сучьев 376
искусственно выращенная живая ткань 8671
искусственно выращенный лес 5285
искусственно высушенная древесина 2421
искусственно высушенная древесина для строгания 6325
искусственно высушенные пиломатериалы 4709
искусственное изреживание 377
искусственное лесовозобновление 6811
искусственное лесонасаждение 5285
искусственное образование насаждения 375
искусственное прореживание 377
искусственное создание насаждения 375
искусственно окрашенная древесина 374
искусственный лес 5285
искусственный отбор 6998
испарение 2914
испарительное охлаждение 2915
испещрённый 6131
использование вне леса 5760
использование дерева целиком
использование древесных отходов 7731
использование лесных ресурсов 3414
использование лесоматериалов 3541
использование местообитания 7676
использованный лес 5066
испытание древесины 9677
испытание клонов 1647
испытание на всхожесть 3837
испытание на раскалываемость 7922
испытание устойчивости древесины к заражению дереворазрушающими грибами 123
исследование в области лесного хозяйства 3577
исследование древесины 9639
исследование клоновых признаков 1647
исследование леса 3549
истечение живицы 2977
истечение сока 1589
истлевшая древесина 9484
истодовые 5446
история леса 4314
источник 6603
источник углерода 1334
истощать 400
истощение 2611
истощение запасов леса 8656
истощить 400
истребление леса 3373
истребленный лес 2443
итальянский клён 4594
итеевые 8336
иудино дерево 4665

йопо 1714

кабель 9463
кабель для проволочного спуска 9463
кабель-кран 7697
кабельно-подвесной кран 7697
кабельный кран 7697
кабельный спуск 3922
кабриува 504
каверна 5028
каверны 4876
кадам 4682
каёмчатая выборка 8168
каёмчатая обсеменительная
каёмчатое хозяйство 6572
казуарина 663
казуарина аргентинская 1692
казуарина Куннингама 407
казуарина хвощелистная 407
казуариновые 664
кайма возобновления 6818
кайя крупнолистная 107
кайя сенегальская 107
какао 1248
кактус колонновидный 1733
кактус цереус 1733
калёвка 635
калибр 8809
калибр для измерения толщины досок 5205
калиброванное оцилиндрованное бревно 6837
калиброванный сухой пиломатериал 6835
калиброванный сырой пиломатериал 6836
каликант цветущий 1806
калина голоцветковая 6474
калина гордовина 9307
калина зубчатая 368
калина канадская 5656
калина клёнолистная 5293
калина красная 2890
калина лантана 9307
калина морщинистолистная 4899
калина обыкновенная 2890
калина ольхолистная 4315
калина плекатум 4633
калина рыжеватая 7141
калина сливолистная 792
калина съедобная 5553
калифорнийский лавр 1264
калифорнийский скользкий вяз 1259
калифорнийское мамонтово дерево 1269
калликарпа 3705
каллитрис 2217
каллус 1182
калорийность 1194
калофиллум 718
калькограф 231
кальмия 1264
кальмия многолистная 895
кальмия сизая 895
кальмия узколистная 7505
кальцефил 1253
кальцефильное растение 1253

кальцефобное растение 1254
камасси 4685
камбала 4582
камбиальная зона 1286
камбиальный инициал 1285
камбиальный слой 1287
камбий 1287
камбий пробковый 6201
камедетечение 4051
камедистая смола 4053
камедная смола 4053
камедное дерево 4056
камедь 4053
камелия 1765
каменистая клетка 8121
каменистые россыпи 8122
каменная древесина 6994
каменная наброска 6969
каменное дерево 2898
каменный дуб 1773
камерная сердцевина 1440
камнеломковые 7269
камнеплодник густоцветковый 8394
каморка 2365
кампешевое дерево 846
кампноспермa 7155
кампо 1292
кампос 1292
кампус 1292
камфарное дерево 1291
камфарный лавр 1291
камыш 1308
канал 9284
канал поры 6300
канариум Швайнфурта 103
канат 1555, 9463
канат для растяжек 1164
канат для спуска вагонеток с горы 7804
канатная дорога 1245
канатная лебёдка 7060
канатная оснастка 7467
канатная петля 699
канатная транспортная система 1242
канатная трелёвка 1247
канатная трелёвочная установка 1242
канатная трелёвочная установка с захватом 3913
канатная трелёвочная установка с лебёдкой со сблокированными барабанами 4540
канатная установка без вторичной трелёвки 8345
канатная установка для спуска 89
канатная установка с замкнутым канатом 5536
канатная установка с неподвижным несущим канатом 8551
канатная установка с тягово-несущим и вытяжным канатами 8534
канатный зажим 7699
канатный лесоспуск 3922
канатоукладчик 7058
канат подъёмного устройства 4323
канат с захватом 8365
кандис 4686
кантовальный крюк 1317
кантователь 1316
каоба 1412
капанье клейкой жидкости 4351
капбалк 7245
капбалка 7245
капель, достигающая подстилки в лесу 3622
капельная зона 2627
капельные осадки, проникающие через полог леса 1313
каперсовые 1324
каперсы колючие 1766
капиллярная вода 1325
капиллярная трещина 4088
капитальная стоимость леса 3536
капитальный лесной доход 3553
каповая древесина 1032
капок 4688
капок настоящий пятитычиночный 4688
капоковое дерево 4688
капоэйра 1326
капсис укореняющийся 8994
капур 4689
капустное дерево 1238
капюшонники 6486
карагана древовидная 7564
карагана кустарниковая 7136
карагач 2786
карандашная дощечка 7732
карандашное дерево 2722
карапа молюккская 6646
карапа суринамская 4039
карбонатная бурая лесная почва
карбонизация 1331
карельская берёза 4690
каретка бревнотаски 8577
каретный лес 1672
карика папайя 1794
кариокар орехоносный 6188
кария бахромчатая 7507
кария белая 8988
кария бледная 7188
кария голая 6229
кария горькая 774
кария косматая 7482
кария мелкоплодная 7482
кария мускатная 5821
кария овальная 7482
кария пекан 6130
кария сердцевидная 774
каркас африканский 105
каркас бледный 2413
каркас западный 1777
каркас кавказский 1385
каркас китайский 1499
каркас миссисипский 8253
каркасные лесоматериалы 3686
каркас сетчатый 5717
каркас южный 2898
карликовая пальметто 2688
карликовое дерево 2691
карликовое дерево в горшке 908
карликовое каштановое дерево 1199
карликовый густой лес 2782
карликовый дуб 2687
карликовый лес 2685
карликовый подвой 7777
карликовый рост 8223
карман 6415
карман-накопитель 8621
карманное долото 6404
карман с корой 576
карра 8155
карра во втором году 9756
карроподновка 543
карта крон 2086
карта лесничества 3383
карта лесного хозяйства 3572
карта лесов 3498
карта лесозаготовок 2953
карта лесонасаждений 3498
карта пригодности 8256
карта проекции кроны 5298
карта типов леса 3635
картирование лесов 3595
картирование лесорастительных условий

картографическая проекция 5299
картон 6059
карточка с наименованием породы древесины 7901
карьер грунта 931
каскад 9285
каскараль 2061
касмарья 5255
кассета для трелёвки 6382
кассина 1371
кассиоврея 5644
кассия дудка 3883
кассия узколистная 1870
кастанопсис золотистолистный 3841
каталог продажей лесоматериалов 8634
катальпа 1306
катальпа бигнониевидная 1768
катальпа западная 5789
катальпа красивая 5789
катальпа прекрасная 5789
катальпа сиренелистная 1768
катальпа яйцевидная 1492
катальпа яйцевиднолистная 1492
катальпа японская 1492
катапан 4464
катафилл 9075
категории защитности лесов 1599
категории земель лесного фонда 4782
каток-кусторез 1615
каудекс 1389
каулифлорный 1391
каучуковое дерево 1320
кафедральная рубка 1379
качание на ветру 9446
качающаяся пила 4312
качественный прирост 6655
качество древесины 6657
качество лесоматериалов 6657
качество лесорастительных условий 7670
качество насаждения 6656
качество очистки деревьев от сучьев 2355
качество торцовки 202
кашива 2228
каштан 2887
каштан американский 245
каштан благородный 2887
каштан городчатый 4616
каштан европейский 2887
каштан китайский 1494
каштан конский 1781
каштан конский восьмитычинковый 9758
каштан конский голый 5874
каштан конский калифорнийский 1256
каштан конский кубарчатый 4622
каштан конский мелкоцветковый 940
каштан конский мелкоцветный 940
каштан конский обыкновенный 1781
каштан конский острый 8483
каштан конский розовоцветный 6767
каштан конский японский 4622
каштан низкорослый 1199
каштановая заросль 4005
каштановая роща 1477
каштановокрылый скворец 3350
каштановый лес 1477
каштановый низкоствольник 7945
каштанодуб 5582
каштан озаркийский 6015
каштан павия 6767
каштан посевной 2887
каштан съедобный 2887
каштан японский 4616
квадратная лучевая клетка 7972
квадратный фут 7970
квалея 5275
квассия 6662
квассия горькая 9373
квебекская сосна 2726
квебрахо белое 9378
квебрачо белое 9378
кверебрахо Лоренца 5173
квилайя мыльная 7805
кедр 1396
кедр атласский 396
кедрач 1397
кедр багассе 4086
кедр гималайский 2398
кедр европейский 8346
кедр калифорнийский 1262
кедр красный 8893
кедр ливанский 1400
кедровая древесина 1404
кедровая канифоль 1402
кедровая роща 1398
кедровник 1397
кедровый лес 1397
кедровый орех 1399
кедровый стланик 4634
кедр остиндийский 1411
кедр порта Орфорд 4839
кедр речной благовонный 1262
кедр речной сбежистый 1262
кедр речной чилийский 6102
кедр сибирский 7569
кельрейтия дваждыперистая 1497
кельрейтия метельчатая 6051
кемпас 4695
кемунинг 4644
керуинг 4699
киви 51
кигелия африканская 7226
кизил 1965
кизил мужской обыкновенный 1965
кизиловые 2556
кизильник 5389
кизильник обыкновенный 2888
кимафила зонтичная 1798
киндель 1206
кинкан 4747
кипарис 2211
кипарис болотный 786, 5418
кипарис болотный мексиканский 5418
кипарис вечнозелёный 1770
кипарис Говена 3891
кипарис крупноплодный 5547
кипарис Макнаба 484
кипарисовая древесина 2219
кипарисовая роща 2215
кипарисовая шишка 2212
кипарисовидная поровость 2216
кипарисовик Лавсона 4839
кипарисовик туеобразный 398
кипарисовые 2214
кипарисовый купол 2213
кипарисовый лес 2215
кипарисовый пруд 2218
кипарис Сарджента 7221
кипарис Форбеса 3314
кипрей 3215
кипрейные 2924
кирказоновые 768
кислица обыкновенная 9656
кислое отложение 40
кислотный дождь 42
кислотовыносливый 43
кислотоустойчивый 43
кислый гумус 7954
кисть 6671

китола 8432
кладовая на лесозаготовительном пункте 181
кладодий 1594
кладрастис жёлтый 262
кладрастис китайский 1509
класс благоприятных возможностей обработки 8793
класс бонитета 9774
класс бонитета леса 9774
класс возобновления 6814
класс возраста 126
класс возраста главной рубки 2191
класс возраста насаждения 8002
класс деревьев 8812
класс густоты насаждения 8106
класс диаметров 2471
класс добротности 9774
класс доминирующих деревьев 2564
классификация лесных местообитаний 3593
классификация лесных почв 3600
классификация лесных районов 3552
классификация лесоматериалов 3902
классификация насаждений 8008
классификация по длине 3894
классификация стволов 8068
классификация типов леса 3632
класс кроны 2068
класс круглого леса 7114
класс лесоматериала 8566
класс местообитания 7664
класс мощности 9217
класс объёмов стволов насаждений 8035
класс пожара 1601
класс пожарной опасности 1601
класс полнодревесности дерева 1602
класс промежуточных крон 4542
класс размера 8924
класс размера дерева 8924
класс размера насаждения 8028
класс свободно растущих крон 5911
класс содоминирующих крон 1704
класс спелости 5342
класс сплошного лесоустройства 1058
класс ствола 8067
класс товарности ствола 8067
класс толщины диаметров 2471
класс формы 8840
классы землепользования 4781
клевер белый 9366
клеевые кольца 8777
клеёная фанера 6398
клеённая древесина 3876
клеёный лесоматериал 3876
клей для древесины 9474
клей для склеивания металла с древесиной 5413
клейкая клетка 5597
клеймение дерева 3074
клеймение краской 6029
клеймение лесоматериалов 5317
клеймённое дерево 826
клеймить 824
клеймо брёвна 5101
клеймо на бревне 5101
клеймо на торце 2807
клекачка 816
клекачка пятилистная 2885
клекачковые 817
клематис альпийский 212
клён 5296
клён американский 381
клён американский виноградполистный 4599
клён амурский 267
клён белокорый 1438
клён белый 8355
клён бородатый 3285
клён бумажный 6057
клён виноградполистный 4599
клён Гельдрайха 4238
клён Гиннала 267
клён гирканский 488
клён голый 6996
клён граболистный 4373
клён двустолбиковый 4984
клён дланевидный 4373
клён завитой 5944
клён Зибольда 7585
клён кистевидный 5589
клён китайский 1501
клён красный 6774
клён крупнозубчатый 707
клён крупнолистный 701
клён кудрявый сахарный 7608
клён ложнозибольдовый 6637
клёновая древесина 5297
кленовник 5292
клёново-буково-берёзовый лес 5290
клёновые 5291
клёновый лес 5292
клёновый сироп 5295
клёновый сок 5294
клён остролистный 935
клён пальмовидный 4373
клён пенсильванский 8174
клён платановидный 935
клён полевой 2827
клён приречный 267
клён птичий глаз 808
клён речной 267
клён рыжеватожилковатый 6800
клён сахаристый 7608
клён сахарный 808
клён серебристый 7608
клён татарский 8418
клён твёрдый 808
клён трёхраздельный 8970
клён усечённый 6638
клён флоридский 3285
клён французский 5550
клён чёрный 808
клён японский 3752
клён ясенелистный 381
клёпочный кряж 4362
клеродендрон 3870
клетка 6692
клетка древесной паренхимы 9618
клетка древесной ткани 9503
клетка камбия 1285
клетка коры 541
клетка пробки 1957
клетка сердцевинного луча 6736
клетка, содержащая млечный сок 4830
клетка стебелька 7997
клеточная мембрана 1407
клеточная оболочка 1407
клеточная стенка 1407
клеточная ткань 1406
клетра ольхолистная 8263
клетра остроконечная 1582
клетровые 1637
клетчатка 9504
клеть 2018
клехо 8942
клещевые захваты 3892
климакс 1638

климаксовая древесная порода 1640
климаксовое растительное сообщество 1638
климаций древовидный 8813
клин для раскалывания 6487
клин для раскряжёвки 1134
клиновидная лопата 6365
клиновидная рубка 9315
клиновидная рубка верхнего полога 7517
клиновидная таксационная призма 6554
клиновидновыборочная рубка 7517
клиновой вруб 9315
клиновой метод 9316
клиновый вруб 9319
клинок 819
клинометр 4443
клинообразная рубка 7517
клин рубки 3079
клин топора 454
клифтония однолистная 1139
клише с гравюры на дереве 9491
клокичка 816
клон 1648
клонинг 1649
клоны семенных деревьев, выращенных из сеянцев 1646
клонящееся дерево 5740
клоп азалиевый 4751
клубень 9003
клубеньковые бактерии 7037
клубневидный корень 9004
клубнелуковица 9003
клузиевые 3805
клювовидный нож 1204
клюква 2889
клюква болотная 2889
ключевой вид 4705
ключевые биотопы 4700
кнеоровые 1671
книга лесного дохода 4905
книга расхода леса 4904
книга угодий 1249
княженика 350
королевское дерево 4716
кобловое хозяйство 6430
ковёр 5335
ковш 6382
ковшовый элеватор 4600
когорта 1715
кодерево 1980
1705
кодоминантные деревья 1706
кожанка 6035
кожанка трёхлистная 6035
кожевенное дерево дирка 4900
кожица 4391
козлы 4380
козлы для пилки дров 4380
козлы для распиловки 4380
козлы для распиловки дров 4380
козлы для ручной разделки брёвен 4380
козырёк, остающийся на пне 2193
кокаиновые 1697
кокколоба 7326
коккулюс каролинский 1348
кокоболо 5426
коконопряд дубовый 5846
коконопряд лиственничный 4790
кокосовая древесина 1703
кокосовая пальма 1702
кокосовое волокно 1717
кол 7775
кола 1718
кол для забора 3084
кол для ограды 3084
коленчатый рост 2025
колёсный корчеватель 9355
колёсный трелёвочный трактор 7125
колёсный трелёвочный трактор с рамным управлением 3685
колёсный тягач 7125
количество вырубленной древесины 2200
количество дичи 3796
количество древесины 8627
количество продажи древесины 8633
колкая древесина 1628
колкость 1627
коллатеральная почка 1723
коллекционированное дерево 1724
коллекция семян 7393
колода 8879
колодка рубанка 6323
колок 7445
колония 7470
колонная ель 1736
колонновидная форма 1734
колонновидная форма роста
колотый лесоматериал 8563
колотый сортимент 7920
колоть 1561
колумбийская сосна 1772
колун 453
колышек 6344
кольцевание 519, 3853
кольцевание V-образными насечками 5798
кольцевание деревьев насечками 3716
кольцевание коры сучка 3853
кольцевание насечками 3716
кольцевание с внесением ядохимикатов 6411
кольцевать 3851
кольцевая гниль 6956
кольцевая окорка 6951
кольцевая пила 308
кольцевая пористость 6954
кольцевая смоляная трещина 6903
кольцевая трещина 2139
кольцевое нанесение гербицидов на кору у комля дерева 606
кольцевое обгрызание 6953
кольцевое повреждение коры 7288
кольцевое пятно 6957
кольцевой надрез 3853
кольцевой нарост 8785
кольцевой отлуп 2139
кольцевые повреждения 6953
кольцевые трещины между осенними и весенними слоями древесины 5159
кольцеобразная трещина 2139
кольцепористая древесина 6955
кольцепоровая древесина 6955
кольцесосудистая лиственная древесина 6955
кольцованное дерево 3852
кольцо роста 301
колючее дерево 8528
колючее редколесье 8526
колючезонтичник 6532
колюченосник 6532
колючий кустарник 1442
колючий лес 8526
колючка 8524
комбайн с осевой центробежной молотилкой 456
комбинация лесных и полевых

комбинированная таблица определения объёмов брёвен 1739
комбинированное коробление листовых древесных материалов 2141
комбинированные древесные материалы 1835
комбинированный топор-кирка 6624
комбретовые 5646
комель 607
комель, пригодный для производства музыкальных инструментов 3107
комель ствола 607
Комиссия по приёмке лесопользования 3377
комлевая гниль 1233
комлевая часть ствола 607
комлевая часть ствола дерева 607
комлевое бревно 1225
комлевой конец брёвна 607
комлевой торец брёвна 607
комлевой чурак 1225
комлевый рез 1214
коммерческая высота 1747
коммерческая рубка 1744
коммерчески выгодное дерево 5399
коммерческие древесные породы 1752
коммерческие лесоматериалы 5400
коммерческие обработки 1751
коммерческие рубки ухода 1750
коммерческий лес 1745
коммерческий объём 1753
коммифора бальзамная 969
коммунальное лесное управление 1817
коммунальный лес 1812
комок 1571, 3241
компассия малаккская 4695
компенсирующая посадка 1823
комплексное использование заготовленного леса 4524
комплексное пользование ресурсов 4526
компонент сухостоя 2275
компост 4881
компрег 1844
компрессионная окорка 1846
конара 3859
конвейер для кряжей 7692
конвенционная трелёвка 1912
конвенционная трелёвка волоком 1911
конверсионная рубка 1914
конверсионная таблица 1917
кондак 1317
кондак для поддержания брёвна во время раскряжёвки 4173
кондалия колючая 870
кондиционирование древесины 1855
кондиционный воспитатель деревьев 1418
конёк 7007
конёк крыши 7007
конец провода 4847
конец топорища, уширенный для удобства удержания его рукой 2329
конечная влажность 3141
конечная почка 8470
конечная стадия гниения 1828
конечный запас насаждения 3135
коник 2038
коник, разделённый на части 7921
конический 8401
конкурент 1827
конкурентная растительность 1824
конкуренция 8817
конкуренция крон 2070
конкуренция от незасеваемых земель 3692
коннарусовые 9794
коноприд кольчатый лесной 3618
консервант 4428
консервация вне места 2959
консервация путей 6976
консервирование в месте нахождения 4517
консервирование древесины 8569
консервирование на месте 4517
консервированная древесина 8792
консервирующее вещество 4428
конскокаштановые 1126
консоль 1879
конструированный земляной лесоспуск 1887
конструированный лесоспуск 1887
конструкционная древесина
конструкционный пиломатериал 1153
консультация по ведению лесного хозяйства 3489
контракт на заготовку 2192
контроль загрязнения от лесной промышленности 3468
контроль качества лесоматериалов 6653
контроль конкурентной растительности 1825
контроль кустарника 1101
контрольное бревно 8480
контрольное дерево 1459
контроль происхождения 6604
контроль рубки 3055
контрфорс 1985
контур леса 2761
контурная борозда 1904
контурная посадка 1906
конус нарастания 1872
конусовидная крона 1866
конусообразный 8401
конфлюэнтная паренхима 1869
концевая опора 2815
концевая прививка 7919
концентрация горючего материала 3736
концентрированная сплошная рубка 2962
концеравнитель 2043
концы брёвен, волочащиеся по земле при трелёвке 8755
кончик корня 7051
коньковый брус 7007
копайба 1922
копайва 1922
копайский бальзам 104
копал 1923
копал смолоносный 1924
копание канав 8962
копатель глыб 496
копать 2495
копач 6367
копилка 8619
копра 6045
копулировка 7919
копулировка с язычком 8678
копьевидный лист 4165
кора 8799
без ~ы 569
корабельный лес 8590
кора, богатая танином 8392
кора гемлока 4249
кора, используемая в качестве топлива 3734

кора, повреждённая насекомыми 1067
корд 9519
корд дров 9519
кордия аллиодора 6076
кордия африканская 2711
кордия Гоэлда 4650
кордия мелколистная 7765
кордон 1944
корды на человекочас 1946
коревые плиты 7710
корен не выходящий из границ контейнера 1891
коренной лес 1639
корень 7009
J-образный ~ 4663
без ~ей 7034
корень с клубеньками 5744
корень, содержащий дубильные вещества 8395
кореопсис гигантский 3840
корешок 7035
кориариевые 1955
коридор 1969
коридор диких видов 9418
коридорная вывозка 1970
коридорная посадка 183
коридорные рубки ухода 1971
коринка 3806
корица белая 9364
корица китайская 1495
корица цейлонская 1421
коричная корка 1370
коричневая гниль 1084
коричневые почвы 1583
коричник камфарный 1291
коричник кассия 1495
коричник китайский 1495
коричник цейлонский 1421
коричное дерево цейлонское 1421
корка 1956, 1973
корковая ткань 1975
корневая гниль 7044
корневая губка 7025
корневая древесина 9580
корневая закомелистость 7015
корневая лапа 7015
корневая мочка 7035
корневая плата 8195
корневая поросль 1933
корневая почка 6683
корневая система 7050
корневая центральная гниль ствола 1081
корневая шейка 7015
корневой клубенёк 7036
корневой клубень 7053
корневой комок 7054
корневой кряж 1226
корневой лесоматериал 9580
корневой наплыв 7015
корневой отпрыск 7046
корневой побег 7046
корневой подвой 7046
корневой пучок 7012
корневой черенок 7018
корневой чехлик 7013
корневые волоски 7028
корневые выделения 7045
корнежил чёрный 805
корнеобитаемый слой 7032
корнеобразование 7024
корнеплоды 7017
корнерасположение 7011
корнерезная машина 7019
корни подвоя 8117
корни с глыбой 491
коробление 2031
коробление в поперечном направлении вследствие усушки 2141
коробление доски в поперечном направлении вследствие усушки 2141
коробление древесины 9268
коробление пиломатериалов 9268
кородробилка 2281
короед 231
короед белоголовый 5220
короед красноногий 5670
короед-лесовик пихты бальзамической 9337
короед пихтой серебристой 7602
короед, повреждающий шишки сосны жёлтой 5043
короед, повреждающий шишки сосны смолистой 6785
короед сосновый 6252
короед сосны карликовой 5043
короеды 539
корончатый галл 2080
корообдир 563, 2284
корообдирочная машина с валочной головки 2182
корообдирочная установка 2284
корообдирщик 556
короочистный нож 7960
коросниматель 2284, 7960
короткие трелёвочные сани без
короткий деревянный жёлоб для раскряжёвки брёвен 1128
короткий кряж 903
короткий лесоматериал 903
короткий оборот рубки 7542
коротковолокнистая древесина 7535
коротковолокнистость 7541
короткое бревно с веерообразным концом 3012
короткое волокно 7536
коротколистный 1029
короткомер 903
короткомерный сортимент 903
короткоопушённый 6619
короткоствольный 7533
коротыш 8937
коротьё 903
корректировщик 7932
корчевальная лебёдка 1160
корчевальная машина 8808
корчевальный топор 4030
корчевание 2857
корчевание бульдозером 7043
корчевание пней взрывом 852
корчеватель 8808
корчеватель деревьев 8808
корчеватель кустарника 1112
корчеватель пней 8199
корчевать 9140
корчевать дерево 4028
корчевать пни 4033
корчёвка взрыванием 852
корчёвка и удерживание 4029
корчёвка пней 8201
корчёвка пней взрыванием 852
корчёвки 4031
корьё 580
корьевой товар 588
корявая древесина 4738
корявое дерево 2030
корявое деревцо 9472
корявое насаждение 3515
корявый древостой 1202
корявый лес 2022
корявый сучок 2026
косарь 1204
косая врубка 5857
косая слоистость 5856
косвенные затраты на заготовку леса 4479
косипо 4744
косое направленное волокно 4585
косой прируб 7285

косослойная древесина 2048
косоугольная саморезка 287
костёр 1450
косточковый плод 2644
костровая крепь 2018
костыль 3655
костянка 2644
косуля 1062
косяк 1204
котловинная рубка 6106
кофейное дерево 1713
кочевое возделывание 5746
кочегарная лопата 3209
кочковатая тундра 4395
кошачья лапа 1382
кошель 7496
кошельный лесосплав 480
коэффициент вентиляции 9198
коэффициент добротности 6654
коэффициент заполнения лесосплавного хода 1708
коэффициент использования древесины 9686
коэффициент окружности 3858
коэффициент отбраковки 2120
коэффициент полнодревесности 1711
коэффициент полнодревесности пыжа 1709
коэффициент полнодревесности сплоточной единицы 1710
коэффициент разбухания 8338
коэффициент сбега 8399
коэффициент усушки 9653
коэффициент усушки древесины 9653
край листа 4878
край насаждения 2761
крайнее бревно 1025
крайний боковой лагерь 7573
крайняя точка товарной части ствола 8622
крайречный лес 6961
кран для загрузки древесного сырья 2410
крапинка 7907
красильная древесина 2693
красильное дерево 2693
красная гниль 6789
красная древесина 6802
красная пихта 1772
красновато-бурая лесная почва 6770
красновато-бурая напенная
красное дерево из Ассие 9150
красное дерево из Коссипо 4744
красное дерево из Тиамы 8541
красное окрашивание кроны 8710
красное сандаловое дерево 6790
красное ядро 6778
краснолесье 6255
краснопузырник деревогубец 239
краснопузырниковые 7985
краснотал 7138
красный кедр 2722
красный кедр австралийский 410
красный мангр 6780
красный меранти 4949
крашеная древесина 1730
крашение древесины 9515
кремовая заболонная гниль 2010
креневая древесина 1850
крень 8996
крепёжный канат 278
крепёжный канат аэростата 8543
крепёжный лес 5463
крепкая древесина 3225
крепление тросов 7062
крепь 2018
крестовый меток 2039
крестовый плот 6694
крестовый топор 2035
крестовый трос 7059
крестовый штабель 6232
крестообразное расположение слоёв 2037
крестоцветные 2099
крестьянский лес 3015
крестьянское лесное хозяйство 3016
кривая высот 4229
кривая масс 9241
кривая отношения диаметра к высоте 4231
кривая прироста 4453
кривая роста 4012
кривая сбежистости 8400
кривая урожая 9775
кривизна 2031
кривизна ствола 8325
кривое дерево 2030
кривой нож 6613
кривой нож для вырубки кустарника 1204
криволесье 2029
кривулина 2029
криптокария Пальмерстона 413
криптокария розовая 409
криптомерия японская 4615
кристаллоносная клетка 2108
критическое зимнее пастбище 2024
критическое местообитание 2023
кромка 4224
кромка леса 2761
кромка пиломатериала 7257
кромка пожара 1189
кромочный сучок 4733
крона 2087
крона дерева 2087
крон-циркуль 8809
кронциркуль для внешних обмеров 5981
кротон 2061
кроющий лист 9075
круглая лиственная древесина 1053
круглая пила 1585
круглая пила на тележке 695
круглая пила с регулируемой глубиной пропила 6970
круглая распиловка 7107
круглая червоточина 5221
круглогодовые лесозаготовки 1897
круглолежневая лесовозная дорога 5121
круглолистник 4692
круглопильный делительный станок 6798
круглопильный обрезной станок 6676
круглопильный станок 1586
круглопильный станок для продольной распиловки 6798
круглопильный торцовочный станок 6676
круглый лес 7113
круглый лесоматериал 7113
круглый лесоматериал из вершинной части деревьев 8712
круглый лесоматериал, разрезанный вдоль 4094
круглый лес, хранимый в воде 9302
круглый сортимент 7100
круглый строевой лес 8586
круглый сучок 7108

круговая траектория лезвий 2190
круговой способ канатной трелёвки 1588
круговорот азота 5735
круговорот биохимических веществ 724
круговорот питательных веществ 5825
круговорот углерода 1330
кружевное дерево 3966
кружок, отпиленный от ствола дерева 1336
крупная древесина 1819
крупная жердь 4803
крупноволокнистая древесина 1676
крупное полено 7920
крупнолистный 4800
крупномерная древесина 4808
крупномерная хвойная жердь 1874
крупномерные деревья 4218
крупномерные лесоматериалы 4808
крупномерный лес 4808
крупнопористая древесина 1685
крупнослойная древесина 5909
крупнослойная лесная почва 1679
крупносуковатый 8135
крупносучковатая древесина 4798
крупноцветковая магнолия 7860
крупноцветная магнолия 7860
крупные колотые дрова 7920
крупные органические остатки 4802
крупные остатки органических веществ 4802
крупные отходы 4216
крупные порубочные остатки 4216
крупный буковый саженец 661
крупный дубовый саженец 3266
крупный лес 8585
крупный лесоматериал 9393
крупный лесоторговец 9392
крупный отпрыск 7199
крупный саженец 4806
крупный сучок 4796
крупный текстурный рисунок 1678
крутизна откоса выёмки 7753
крушина 1137
крушина калифорнийская 1257
крушина каролинская 1344
крушина ломкая 684
крушина ольховидная 684
крушина падуболистная 4338
крушинник ломкий 684
крушиновые 1138
крылатка 9453
крылатое семя 4701
крылатый плод 4701
крылатый поплавок 3147
крыловидная паренхима 167
крылоплод 4701
крюк 4359
крюк для закрепления брёвна 875
крюк для крепления чокеров 1557
крюк для окучивания балансов вручную 6630
кряж 8879
кряж для рубки или колки лесоматериала 1211
кряжи ценной древесины 7420
ксанторрея 405
ксерофитная древесно-кустарниковая ассоциация 1202
ксерофитное редколесье 358
ксерофитный лес 2650
ксерофиты 9736
ксилема 9737
ксилемный луч 9738
ксилопия 9740
ксилотомия 9741
ксимения американская 4322
курбарил 1990
кубатура стволовой древесины 9249
кубатура, учитываемая в складочной мере 7979
кубатурник 2117
кубическая гниль 2112
кубический метр древесины в плотных единицах с корой 2116
кубический метр древесины на корню 9524
кубический складочный метр 2115
кубический фут 2113
кубический фут поленицы 7977
кубометр древесины на корню 9524
кубометр плотных лесоматериалов 2114
кубообразная гниль 2112
кузов для растений 6343
кукольная колыбель 4810
кулачковая корообдирка 4103
кулиса 3335
кулисная рубка 8171
кульверт 2132
кульминационный возраст 2125
кульминация среднегодового прироста 2126
культура корзиночной ивы 620
культурный сезон 8665
культурный сорт 2129
куманика 778
кумару 2133
кумкват 4747
кумулятивная кривая роста 2135
кумулятивное действие 2134
кунонецветные 2138
купол кроны 2066
кураре 8191
куртина 7445, 8066
куртина деревьев 7445
куртина леса 7445
куртинная рубка 6106
куртинное возобновление 6812
куртинный лес 7445
куртинный подрост 6812
курчавость листьев 4863
куски дерева 5218
кусок дерева, удаляемый из подпила 7284
кусочек 3241
куст 7555
кустарник 7555
кустарниковая джунгла 7320
кустарниковая заросль 1202
кустарниковая саванна 7560
кустарниковая тундра 7322
кустарниковое болото 7557
кустарниковое редколесье 1202
кустарниковые листопадные сообщества 2298
кустарниковые насаждения 933
кустарниковый 934
кустарниковый вельд 1208
кустарниковый лес 1202
кустарниковый нож 1204
кустарниковый пар 1201
кустарниковый плуг 1203
кустарниковый серп 1113
кустарничек 9065
кустарничковая тундра 5182
кустистый 934
кусторез 7319

кутровые 2553
куча бревен 2305
куча дров 9519
куча забракованных лесоматериалов 4774
куча лесоматериалов 2305
куча порубочных остатков 7730
куча хвороста 4193
кьюнит 2136

лавровые 4832
лавр 6745
лавр американский 1801
лавр благородный 3936
лавр красный 6763
лавровишня 2828
лавровишня каролинская 1347
лавровишня лекарственная 2828
лавровый лес 4833
лагерстремия 4643
лагерстремия индийская 2007
ладанник 6990
ладанниковые 6991
ладанное дерево 4463
лазящее растение 1642
лайм 9348
лак из отходов 6809
лаковое дерево 4755
лаконосовые 6416
лакрица 4997
лакричный корень 4997
лами 9363
ламинированная древесина 4763
ламинированная древесностружечная плита 6375
ламинированная древесностружечная плита с имитационной текстурной отделкой 1373
ламинированная плита 4761
ландшафт 4777
ландшафтная архитектура 4778
ландшафтные рубки ухода 4778
ландшафт под полога 1311
лантана шиповатая 4783
лантау 6879
ланцетовидный лист 4767
ланцетолистный тополь 4766
лапа с режущим элементом для дерева 8823
лапина 9456
лапина розоволистная 4637
лапина ясенелистная 1388
ластовневые 5445
латексный сосуд 4829
латентное масло 2575
латентный период 4818
латеральная почка 419
с разросшимися латеральными почками 2357
латук компасный 6531
латук лесной 6531
лауана 4831
лафет 4099
лебёдка для корчёвки пней 1160
лебёдка для подвесной трелёвки 7060
лебедовые 3887
лебёдчик 7925
левкена мелколистная 5003
левкена припудренная 3933
левкена сизая 4851
левый косослой 4907
лёгкая двухколёсная тележка 894
лёгкая древесина 8660
лёгкая древесностружечная плита 5184
легкая прорубка 4952
лёгкие каркасные лесоматериалы 4942
лёгкий топор 2101
лёгкое защитное вещество на базе органических растворителей 4947
ледяной лесоспуск 4418
лежачая лучевая клетка 6559
лежащий в воде круглый лес 9302
лежащий в основе крупный нарост 605
лежащий сухой лесоматериал 2266
лежень 8542
лежневая лесовозная дорога 1947
лежневые пятна 2047
лежни 2054
лезвие 819
лезвие топора 448
лезть по дереву 1641
ленок 3840
ленточная пила 5049
ленточная пила для распиловки кряжей 5049
ленточная пила с двумя режущими кромками 2661
ленточная текстура 8175
ленточное внесение 8169
ленточнопильный станок 520
ленточно-шлифовальный станок 672
ленточные рубки ухода 4992
ленточный дендрометр 516
ленточный посев 522
лепесток 6192
лептосине 3840
лептоспермум 8437
лептоспермум метловидный 1073
лес 3318
леса заповедников 3597
леса зон санитарной охраны источников водоснабжения 3599
леса, находящиеся под охраной 4260
леса планеты 3865
леса санитарной охраны курортов 3598
лес в холмистой местности 9131
лес выборочной рубки 7410
лес генофонда 3817
лес государственного значения 8041
лес для отдыха 91
лес для пастбища 3929
лес для подмостков 7270
лес, завещанный учреждению 4519
лес, закрытый для посещений 1653
лес из ильма, ясеня и тополи 2787
лес из корявых деревьев 3515
лесистая долина 2360
лесистая ложбина 2360
лесистая местность 3551
лесистое образование 9545
лесистые горы 9710
лесистый 3394
лесистый участок 8604
лес местного значения 5032
лес на аллювиальных почвах 204
лес на корню 3454
лес на крутого берегового откоса 1694
лес на равнине 3256
лес на склоне 7752
лес на топливо 3740
лесная администрация 3320
лесная академия 3319
лесная апелляционная

лесная ассоциация из дубов, эвкалиптов и кипарисов 5837
лесная аттестация 3348
лесная база 8558
лесная биология 3339
лесная биомасса 9490
лесная ботаника 3341
лесная галерея 3442
лесная генетика 3447
лесная гидрология 3465
лесная глушь 476
лесная дача 3408
лесная делянка 1821
лесная дорога 3565
лесная дорога местного значения 5034
лесная заболоченная территория 3399
лесная земля 3480
лесная земля с непроверенного запаса 9106
лесная змея 8925
лесная зона 3551
лесная зона без путей 6977
лесная зоология 3651
лесная индустрия 3466
лесная инспекция 3380
лесная информатика 3472
лесная кайма 2761
лесная канцелярия 3323
лесная капель 3622
лесная квартальная сеть 3564
лесная климатология 3353
лесная компания 3355
лесная контурная карта 3362
лесная концессия 3356
лесная культура 3528
лесная логистика 3486
лесная лужайка 3450
лесная мерная вилка 8809
лесная местность 3551
лесная метеорология 3504
лесная микология 3511
лесная микробиология 3505
лесная мышь 3510
лесная наука 3581
лесная номенклатура 3513
лесная область 3551
лесная облигация 8561
лесная оптическая мерная вилка 5930
лесная опушка 7508
лесная опытная станция 3413
лесная оценка 3411
лесная пастьба 3452
лесная плантация незабавным денежным доходом 8896
лесная площадь 9542
лесная площадь, на которой разрешены лесозаготовки 5769
лесная подстилка 3435
лесная подстилка с преобладанием в составе опада листьев, хвои, веточек 4924
лесная полёвка 3510
лесная политика 3533
лесная полиция 3532
лесная полоса 3335
лесная поляна 8865
лесная полянка 8865
лесная популяция деревьев 3627
лесная поросль 1928
лесная почва 9547
лесная почва прерий 6492
лесная пошлина 2681
лесная прибыль 6569
лесная примета 9371
лесная прогалина 3443
лесная продукция 3537
лесная промышленность 3466
лесная просека 3564
лесная профилактика 3544
лесная работа 9724
лесная рамная пила 3683
лесная растительность 8953
лесная рента 3553
лесная сажалка 3585
лесная санитария 3580
лесная семясушилка 3588
лесная сеялка 3585
лесная собственность 3543
лесная сорная трава 3641
лесная социология 3596
лесная среда 3405
лесная статика 3326
лесная статистика 3606
лесная съёмка 3614
лесная такса 3616
лесная таксация 3617
лесная тень 3965
лесная терапия 3621
лесная терминология 3619
лесная техника 9674
лесная технология 3401
лесная типология 3636
лесная торговая биржа 8583
лесная трава 3449
лесная усадьба 3407 3525
лесная ферма 3419
лесная фильтрирующая полоса 3426
лесная финансовая система 3427
лесная флора 3437
лесная формация 3438
лесная фотограмметрия 3526
лесная школа 7297
лесная щепа 3760
лесная экология 3390
лесная экосистема 3395
лесная энтомология 3404
лесная эстетика 3321
лесная ягода 3336
лес, недоступный для освоения 3645
лесничество 3546
лесничий 3406
лесничий, ответственный за материально-техническое снабжение 6560
лесное богатство 3640
лесное болото 9543
лесное ведомство 2552
лесное вредное насекомое 4499
лесное деление 3384
лесное дело 3322
лесное дерево 3625
лесное достояние 3461
лесное законодательство 3485
лесное имение 3408
лесное имущество 3346
лесное кладбище 3347
лесное местообитание 3591
лесное нарушение 1909
лесное насаждение 3528
лесное насаждение без спецификации 9119
лесное насекомое 3475
лесное образование 6565
лесное обучение 6565
лесное общество 3331
лесное отделение 2552
лесное пастбище 9711
лесное плодовое дерево 3440
лесное поколение 8846
лесное поместье 3408
лесное почвоведение 3601
лесное право 8573
лесное предприятие 3568
лесное преступление 1909
лесное профессиональное образование 6565
лесное растение 3527

лесное сообщество 3331
лесное страхование 3476
лесное счетоводство 3340
лесное товароведение 3503
лесное угодье 9599
лесное урегулирование 3590
лесное хозяйство 3567
лесное хозяйство близко к природе 5683
лесное хозяйство непрерывного дохода 8300
лесное хозяйство по выращиванию сеянцев 3586
лесное хозяйство, удобное для пользователя 9147
лесной 3324
лесной агент 9475
лесной аукцион 8557
лесной баланс 9486
лесной биологический резерват 3338
лесной бригадир 4185
лесной ветрозащитный пояс 9445
лесной вредитель 3523
лесной выпас 9711
лесной генезис 3445
лесной гумус 3464
лесной двор 8664
лесной деревенский питомник 9219
лесной дичок 9419
лесной доход 9778
лесной жаворонок 9594
лесной заказник 1653
лесной закон 3484
лесной заповедник 3557
лесной знак 3500
лесной инженер 3400
лесной кадастр 3344
лесной капитал 3346
лесной квартал 1821
лесной компас 8809
лесной конек 1093
лесной культиватор 3365
лесной ландшафт 3481
лесной листоед 3369
лесной луг 3450
лесной массив 9605
лесной микроклимат 3506
лесной надзиратель 3456
лесной надзор 3612
лесной налог 3616
лесной научно-исследовательский институт 3556
лесной объездчик 3456
лесной объездчик, устанавливающий норму лесоповала 3410
лесной объект 3403
лесной округ 3382
лесной опад 4999
лесной опрыскиватель 8862
лесной опрыскиватель с защитой 8851
лесной орех 4183
лесной перегной 3464
лесной питомник 342
лесной плуг 3531
лесной подстил 1422
лесной пожар 3428
лесной пожар против ветра 3429
лесной покос 3914
лесной покров 3345
лесной полог 3345
лесной посёлок 3639
лесной пояс 3335
лесной просвирняк 7102
лесной рабочий 9654
лесной район 3551
лесной резерват 3557
лесной рынок 8613
лесной садовник 6246
лесной сектор 3582
лесной сервитут 8573
лесной склад 8664
лесной собственник 3519
лесной советник 1890
лесной сортимент 8556
лесной сторож 3456
лесной таксатор 3410
лесной техник 3579
лесной торф 9712
лесной торфянистый гумус 6129
лесной университет 3462
лесной устав 8046
лесной участок 1821
лесной участок, пройдённый пожаром 6478
лесной участок фермы 9599
лесной фермер 3420
лесной фестиваль 8838
лесной фитоценоз 3331
лесной фонд 8114
лесной чиновник 3573
лесной чистец 4222
лесной экорайон 3392
лесные генетические ресурсы 3446
лесные животные 3644
лесные запасы 3559
лесные звери 3644
лесные инструменты 3623
лесные кормовые растения 3396
лесные машины 3402
лесные муравьи 3327
лесные обычаи 3576
лесные орудия 3623
лесные отходы 3561
лесные паразиты 3520
лесные плоды 3441
лесные побочные продукты 5774
лесные полутовары 7427
лесные породы 3604
лесные продукты 3448
лесные резервы 3559
лесные ресурсы 3562
лесные семена 3628
лесные сеянцы 8917
лесные товары 9560
лесные торги 8557
лесные угодья 8604
лесные условия 3358
лесовалочная головка 3063
лесовалочная пила 8854
лесовалочный клин 3079
лесовалочный нож 3058
лесовик пихты бальзамической 9337
лесовладелец 3519
лесовладение 8652
лесовод 344
лесовод-информатор 1890
лесовод-исследователь 6883
лесовод-наблюдатель 3613
лесовод-производственник 4481
лесоводственное поведение 7612
лесоводственные требования 3494
лесоводство 3488
лесовоз 5087
лесовозная дорога 9681
лесовозная дорожка 3565
лесовозная железная дорога 5079
лесовозная железная дорога с лебёдочной тягой на больших уклонах 4440
лесовозный автомобиль 5087
лесовозный автомобиль-тягач 5087
лесовозный автопоезд 5087
лесовозный настил 5077

лесовозный прицеп-роспуск 5078
лесовозный трактор 7799
лесовозный тягач 2970
лесовозный ус 4172
лесовозобновительная рубка 6815
лесовозобновление 6821
лесовосстановительная рубка 6815
лесовосстановление 6821
лесовосстановление на больших площадей 1042
лесодорожное строительство 1889
лесозаготовитель 9654
лесозаготовительная бригада 9527
лесозаготовительная концессия 3356
лесозаготовительная машина 8853
лесозаготовительная машина непрерывного действия 1900
лесозаготовительная машина с остановкой для срезания дерева 8125
лесозаготовительная машина целых деревьев 1830
лесозаготовительная промышленность 5206
лесозаготовительная фирма 3355
лесозаготовительное оборудование 4157
лесозаготовительное предприятие 5074
лесозаготовительной участок 5038
лесозаготовительные затраты 5175
лесозаготовительные машины 4157
лесозаготовительные работы 3457
лесозаготовительный сезон 3072
лесозаготовительный участок 5073
лесозаготовка 9588
лесозаготовка на затопленных водой участках 9073
лесозаготовка на основе прошлых эксплуатационных показателей 6156
лесозаготовка на основе сообщество 3451
лесозаготовка на слабых грунтах 8311
лесозаготовка с использованием нескольких одно- или двухоперационных машин 5610
лесозаготовка с подвесной канатной трелёвкой 7700
лесозаготовка с применением гужевой трелёвки 292
лесозаготовка с применением колёсных тракторов 7683
лесозаготовка с трелёвкой хлыстов 5144
лесозаготовка с трелёвкой целых хлыстов 8869
лесозаготовки без межоперационных запасов лесоматериалов 4384
лесозаготовки в ночную смену 5733
лесозаготовки на базе канатных установок 1247
лесозаготовки с вывозкой леса по железной дороге 6696
лесозаготовки с ежедневной доставкой рабочих из жилых центров 4342
лесозаготовки с использованием летательных аппаратов 85
лесозаготовки с применением аэростатов 497
лесозаготовки с трелёвкой целых деревьев 3761
лесозаготовки с условно-сплошными котловинными рубками 1854
лесозаготовки с условно-сплошными рубками 1854
лесозащита 3359
лесозащитная станция 3545
лесоинвентаризация 3478
лесоинженерное дело 3401
лесокомбинат 9516
лесокультурная бригада 6349
лесокультурное посадочное место 7396
лесокультурное посевное место 7396
лесокультурное преобразование 7614
лесокультурные требования 95
лесокультурный посадочный материал 6363
лесолюбивый 3487
лесоматериал, высушенный в лесу 3386
лесоматериал выше заданного диаметра 5992
лесоматериал выше заданной длины 5997
лесоматериал для буровых установок 6948
лесоматериал для выработки бочарной клёпки 4362
лесоматериал для музыкальных инструментов 8587
лесоматериал для несущих конструкций 1335
лесоматериал для столярных и внутренних отделочных работ 3164
лесоматериал для столярных кузовных работ 1672
лесоматериал заданной длины 6583
лесоматериал иноземных пород 3317
лесоматериал ниже заданного диаметра 9058
лесоматериал ниже заданной длины 9062
лесоматериал, пригодный к сплаву в свежесрубленном виде 3268
лесоматериал с большим количеством трещин 4330
лесоматериал с кривизной 2029
лесоматериал с частичным отлупом 9696
лесоматериал, удерживаемый на плаву 3268
лесоматериалы 3537
лесоматериалы гидравлической окорки 8156
лесоматериалы, заготовленные при рубках ухода 8506
лесоматериалы из внутренних слоёв дерева 2811
лесоматериалы мягких пород 7823
лесоматериалы от рубок ухода 8506
лесоматериалы разной длины 5488
лесоматериалы сельскохозяйственного значения 9570
лесоматериалы хвойных пород

лесонаправляющее сооружение 8779
лесонаправляющий бон 3147
лесонарушение 1909
лесонасаждение 3605
лесонасаждение, сменившее девственный лес 7346
лесообрабатывающая установка 5448
лесообрабатывающая фабрика 5448
лесообразующая порода 3439
лесоосушение 3385
лесопарк 6080
лесопарковая зона 3941
лесопарковая саванна 7229
лесопарковые части зелёных зон 3575
лесопарковый клин 3963
лесопарковый пояс 3941
лесоперевалочный пункт 8763
лесоперерабатывающая промышленность 9685
лесопиление 5100
лесопилка 8637
лесопильная промышленность 5206
лесопильная рама 5132
лесопильное оборудование 5106
лесопильный завод 8637
лесопильный станок 675
лесоплан 3372
лесоплодовое насаждение 3733
лесоповал 8824
лесопогрузочный пункт 5027
лесопогрузчик 5067, 8607
лесопожарная площадь 3193
лесопожарный период 3434
лесопожарный сезон 3434
лесопользование 3414
лесопользование без восстановления лесов 8616
лесопользование после стихийного бедствия 3415
лесопольная культура 3425
лесопольное хозяйство 3421
лесопосадка 94
лесопосадочная машина 8897
лесопотребление 9517
лесопрерийная зона 6493
лесопровод 5104
лесопродукция 3537
лесопромышленник 9654
лесопромышленное предприятие 9558
лесоразведение 7610
лесоразведение близко к природе 5683
лесоразведение вне леса 5759
лесоразведение посадкой 3624
лесорастительная формация 3438
лесорастительные условия 3591
лесоресурсы 8629
лесоруб 9526
лесорубная полоса 3050
лесорубная стрела 3051
лесорубный конёк 3054
лесорубный рычаг 3049
лесорубный топор 453
лесорубочная такса 5085
лесорубочные инструменты 4157
лесорубочные потери 2197
лесорубочный билет 6183
лесорубочный топор 453
лесосанитарные мероприятия 3460
лесосека общественного пользования 1815
лесосека, осваиваемая канатной установкой с одной стоянки 7468
лесосека, предназначенная для получения дубового корья 5833
лесосека, разрабатываемая на базе подвесных канатных установок 7701
лесосека с выборочной рубкой 7406
лесосека со сплошной рубкой 1621
лесосеки в шахматном порядке 7988
лесосеменная плантация 8915
лесосеменное насаждение 8915
лесосеменное хозяйство 3586
лесосеменной материал 3628
лесосеменной питомник 8915
лесосечная бригада 9527
лесосечная пасека 2178
лесосечная флора 3039
лесосечное лесное хозяйство 3062
лесосечное низкоствольное хозяйство 1929
лесосечные отходы 998
не неизбежные ~ 446
лесосечные работы 3457
лесосечные работы с
лесосечный план 6332
лесосечный фонд 3423
лесосклад 8664
лесосплав 8578, 8584
лесосплав в сплоточных единицах 3273
лесосплавная дистанция 3271
лесосплавная плотина 5039
лесосплавной багор 6693
лесосплавной путь 3277
лесосплавной пучок брёвен 8584
лесосплавный канал 3277
лесосплавный лоток 9281
лесосплавный ход 3277
лесосплавщик 9569
лесоспуск 5116
лесоспуск без боковых бревен 5800
лесоспускной лоток 5116
лесоспуск по земляному лотку 1577
лесоспуск по снежному лотку 7803
лесостепи 3609
лесостепные зоны 3609
лесостепные зоны умеренного пояса 8444
лесостепь 3607
лесосушилка 9537
лесотаска для брёвен 8598
лес, отведённый в подсочку 827
лес, отведённый под рубку 2196
лесоторговая биржа 8583
лесоторговец 2276
лесоторговля 8653
лес, отпускаемый бесплатно 3701
лесотранспорт 8595
лесотранспортёр 8596
лесотранспортёр для брёвен 8596
лесотранспортная сеть 3566
лесотранспортное средство 8598
лесотранспортный механизм 8598
лесотундра 3631
лесоустроитель 3497
лесоустроительная партия 8288
лесоустроительные работы 3121
лесоустроительный план 3492
лесоустройство 3615
лесоучасток 1821
лесохимикаты 3349

лесохимия 2370
лесохлебное хозяйство 143
лесохозяйственная биомасса 7613
лесохозяйственная почва 1746
лесохозяйственная рекомендация 7617
лесохозяйственная рубка ухода 7622
лесохозяйственная система 7621
лесохозяйственная система выборочной рубки 7620
лесохозяйственная съёмка 7624
лесохозяйственная съёмка до рубки 6505
лесохозяйственная эффективность 7616
лесохозяйственное исследование 3577
лесохозяйственное мероприятие 7623
лесохозяйственное правило 8020
лесохозяйственное предписание до рубки 6506
лесохозяйственное предприятие 3568
лесохозяйственный закон 3534
лесохозяйственный принцип 8020
лесохозяйственный режим 7618
лесохозяйственный справочник 3570
лесохозяйство 3567
лесохозяйство с коротким оборотом ротации 7545
лесоэксплуатация 3414
лесоэкспорт 2957
леспедеза 4919
леспедеца 4919
лес-пионер 6289
лес, предназначенный для валки 4085
лес преобладающей сосны ладанной и сосны ежовой 5031
лес, пригодный к эксплуатации 5924
лес промышленного значения 1745
лес, растущий в пределах данной страны 4341
лес с временным сельскохозяйственным пользованием 8419
лес с нормальным распределением насаждений по классам возраста 5780
лес с обилием кустарников 1202
лес с ограниченным размером пользования 6917
лес со сплошной рубкой 1616
лес с пастбищными угодьями 3929
лес с пастьбой скота 3929
лес, с преобладающей чёрной, смолистой и Веймутовой сосной 9379
лес с разрежённым пологом 3646
лес с рациональным использованием ресурсов 8090
лес с травянистым покровом 3917
лес твёрдых пород древесины 4139
лестничная жердь 4756
лестничная паренхима 7271
лестничная перфорационная пластинка 7272
лестничная поровость 7273
"лестничные" огни 4757
лес умеренной зоны 8443
лес холодно-умеренной области 1719
лес частного владения 3473
лес черничный 5650
летнее сбрасывание листьев 8261
летнезелёные леса 8260
летнезелёные леса и кустарники 8260
летнелистопадный лес 2636
летние лесоматериалы 8264
летний лесоспуск 2707
летний побег 4765
летняя древесина 8264
летняя обрезка крон 8262
летняя рубка 3065
летняя сирень 1219
лёт пыльцы 6433
летучее масло 2866
леукотоэ 8327
леукофиллум 7606
лецитидасовые 5531
лецитисовые 5531
лещина 1778
лещина древовидная 5383
лещина крупная 4759
лжеакация 1754
лжетсуга 1772
лжетсуга Мензиеза 1772
лжетсуга тисолистная 1772
лиана 4925
лиановый влажный тропический лес 9222
лиановый лес 9222
либриформное волокно 4926
лигнин 4957
лигнификация 4954
лигнифицированное волокно 4956
лигнифицированный 4955
лигноза 4959
лигноцеллюлоза 9504
лигуструм светлый 1799
лизиметр 5223
лизогенный 5222
ликания 4927
ликания крупнолистная 5532
ликвидамбар смолоносный 259
ликвидная древесина 5400
ликвидная масса древостоя 7378
ликвидный кубометр 9246
ликоподий 8815
ликоподий древовидный 5794
лилейные 4962
лимба 92
лиметта 9348
лимон 4911
лимониум древовидный 8878
лимон лиметта 9348
лимон сладкий 9348
лимон трёхлистный 4146
линдера 7909
линейная выборочная рубка каёмчато-выборочная лесосечная рубка 8168
линейная пересадка 4993
линейная пикировка 4993
линейная посадка 6356
линейная посадка деревьев 4989
линейная таксация 4988
линейный рост 4985
линия для встречного огня 3169
линия лебёдочной трелёвки 9750
линия лесопогрузки 8611
линия наклона в сторону 7576
линия несущего каната 1245
линия огня 1189
линия порожнего троса 1243
линия прохождения

линия трелёвки 1245
линь 6732
лиония бирючиновидная 5256
лиония блестящая 3091
лиония ржавая 8880
липа 4976
липа американская 237
липа амурская 266
липа длинночерешковая 6143
липа каменная 5503
липа каменная узколистная 5664
липа каролинская 1343
липа крупнолистная 4801
липа обыкновенная 1786
липа разнолистная 667
липа серебристая 7607
липняк 4978
липовая кора 4972
липовое лыко 4972
липовые 4983
липовый дубняк 4974
липовый лес 4978
липовый цвет 4977
липучая бумага 8390
лириодендрон тюльпанный 9008
лисохвостая верхушка 5544
лист 4852
листва 3303
лиственная древесина 4124
лиственная крона 4862
лиственная подстилка 4999
лиственница 4787
лиственница американская 250
лиственница даурская 2227
лиственница европейская 1785
лиственница западная 9340
лиственница Лайеля 216
лиственница Ледебура 7567
лиственница опадающая 1785
лиственница сибирская 7567
лиственница тонкочешуйчатая 4623
лиственница японская 4623
лиственничная моль 4791
лиственничник 4789
лиственничный лес 4789
лиственное декоративное дерево 1051
лиственное дерево 1048
лиственные насаждения 4138
лиственные породы 4138
лиственные породы древесины небольшой плотности 4943
лиственные породы древесины,
лиственный 3306
лиственный корм 4868
лиственный лес 2300
лиственный опад 4999
лиственный пиломатериал 1047
лиственный покров 4999
листвяг 4789
лист метёлки 6053
листный гербицид 4255
листоватый 3302
листовая жилка 4882
листовая пластинка 4872
листовая поверхность 4890
листовая поверхность 4853
листовая почка 4859
листовая твёрдая древесина 4136
листовая щель 4870
листовёртка дубовая 5839
листовёртка побеговая 2149
листовёртка-почкоед 1146
листовёртка-почкоед сосны Банкса 4602
листовёртка-почкоед чёрноголовая 794
листовёртка сосновая 6273
листовёртка хвойная рыжая 6269
листовёртки 4886
листовое почкосложение 6502
листовой 4894
листовой анализ 3307
листовой бугорок 4884
листовой покров 4860
листовой рубец 4888
листовой след 4891
листоед 2338
листоед сосновый 806
листоед тополя 1982
листоеды 4858
листок 4875
листообразное слоевище 3718
листообразный корень 6378
листообразный стебель 1594
листопад 4866
листопадное дерево 2302
листопадный 4877
листопадный ковёр 2299
листопадный лес 2300
листопадный тропический лес 1236
листоразрушающие насекомые 4858
листорасположение 4856
листорасположение в почке 6502
лист папоротника 3718
лист, покрытый щетинками 7471
лист с очень слабовыраженным черешком 8236
лист с пальчатым типом жилкования 6041
лист учёта 8376
лист учёта лесоматериалов 8376
литза 902
литоральный лес 1687
лицевая поверхность 2985
лицевой шпон 2986
лицензия на заготовку древесины 8605
лицензия на лесное угодье 9600
лицензия на лесопитомник 8835
лицензия на продажу древесины 8632
лиций 9471
лициум 9471
лицо, ответственное за гужевое хозяйство 591
личинка 4809
лишайник 4929
лишайниковый сосняк 4930
лишайный покров 1696
лишение хвои 2995
лобзик 1822
лобзиковая пила 1822
ловоа Брауна 1087
ловушки для насекомых, наживлённые феромонами 6206
ловчая канава 4515
ловчая кора 8775
ловчая культура 1374
ловчая яма 4515
ловчее дерево 8778
ловчее полено 8776
ловчие кольца 8777
логаниевые 5047
ложная верхушечная почка 3004
ложная заболонь 853
ложная мучнистая роса 5444
ложное годичное кольцо 2998
ложное кольцо 2998
ложное ядро 795
ложное ядро, образовавшееся под воздействием мороза 3725
ложноопёнок серно-жёлтый 8258
ложные опята 8258
ложный плод 3000

ложный трутовик 3005
локальная сплошнолесосечная рубка 5035
ломанная древесина 850
ломкая древесина 7535
ломкость 1000
ломкость древесины 1001
ломонос белый 9402
ломонос жгучий 6387
ломонос фиолетовый 4593
лонган 1381
лопасти 4876
лопастный лист 1636
лопасть бульдозера 1161
лопата для посадки в щель 6365
лопатка для сбора живицы из приёмника 4584
лопатовидный лист 7886
лопающийся 2346
лот 6389
лоток 5116
лоуд 5022
лох зонтичный 417
лоховые 5889
лох серебристый 257
лох узколистный 9421
лошадиная трелёвка 4378
лошадиная трелёвка леса 4378
луб 4502
лубодревесный луч 6208
лубоед ели восточной 2724
лубоед ели ситхинской 2724
лубоед ели Энгельмана 2823
лубоед еловой большой 2909
лубоед еловый аллеганский 178
лубоед красный скипидарный 6797
лубоед круглоголовый сосновый 7105
лубоед лиственницы американской 2720
лубоед мексиканский сосновый 7762
лубоед сосновый колорадский 1727
лубоед сосновый югозападный 7866
лубоед сосновый южный 7862
лубоед сосны горной 796
лубоед сосны жёлтой 9342
лубоед сосны карликовой 5045
лубоед сосны мексиканской 4805
лубоед-стригун 6246
лубоед чёрный скипидарный 811
лубяной луч 6208
луг 5349
луговина 5349
луизианский кипарис 786
луковица 9003
лук пилы 7231
луносеменниковые 5551
лупиться 2940
луч 5391
лучевая паренхима 6679
лучевая трахеида 6739
лучевая трещина 8039
лучевый инициал камбия 6735
лучковая моторная пила 952
лучковая моторная цепная пила 952
лучковая пила 3342
луч флоэмы 6208
лучшая пласть 686
лучшее дерево 3786
лучшие сорта брёвен 4284
лущёный шпон 6135
лущильный нож 7960
лущильный станок 9623
лущильный топор 563
лущить 6133
лыжный кряж 5065
лыко 4502
льновые 3261
любая древесина жёлтого цвета 9767
любисоль 5219
лютиковые 1218
лягушка древесная 9579

магистральная лесовозная дорога 3463
магнолиевые 5241
магнолия 5240
магнолия виргинская 643
магнолия длиннозаострённая 709
магнолия крупнолистная 700
магнолия лилейная 4963
магнолия лилиецветная 4963
магнолия остроконечная 709
магнолия Суланжа 7225
магнолия трёхлепестная 9040
магнолия Фразера 3688
магония падуболистная 5943
магония сетчатая 1365
магония трилистная 5419
магония Фремонти 3704
мадрона 6020
майский жук 3502
майский цветок 5346
1206
макаранга 6074
макасар 3882
маккия 5300
маклейя 8899
маклюра оранжевая 4219
макоре 5254
макрораспространение 5237
макроструктура древесины 9602
макрофибрилла 5234
макрофит 5236
макроэлемент 5235
максимальная плотность 5345
максимальное расстояние канатной трелёвки 2967
маленькая ветвь 991
маленькая почка 5474
маленький здоровый сучок 5007
малина 1800
малина американская 9422
малина арктическая 350
малина душистая 6640
малина западная 782
малина лесная 4631
малина мелколистная 4631
малина обыкновенная 1800
малина пурпуроплодная 9452
малина щетинистая 9422
маловетвистый 7737
малолесный 6450
малопродуктивная земля 5304
малорентабельное хозяйство 8238
малорослый подвой 7777
малотус 4684
малоусадочная древесина 8661
малоценная древесина 5195
малоценное насаждение 7356
малый лесной садовник 4920
малый урожай 4939
мальва лесная 7102
мальпигиевые 5260
мальпигия гранатоволистная 527
мамонтово дерево 3844
мамонтовое дерево 3844
манго 1789, 5277
манговое дерево 1789
манго индийское 1789
мангостан 5278
мангровое болото 5281
мангровое дерево 5276
мангровые 5279
мангровый лес 5280

мандарин благородный 5274
мандариновая цедра 3948
мандариновое дерево 5274
мандарин-танжерин 3948
маниок сладкий 8329
маниок съедобный 5282
манометр 6521
мансония 1087
манцинелла 5267
маргоза 6110
маревые 3887
маржинальная земля 5304
мариола 5308
марка брёвна 5101
маркированная древесина 5311
маркировать 5310
маркировка вешками 7995
маркировка дерева 8883
маркировка деревьев в рубку 3074
маркировка круглых лесоматериалов 5316
маркировка лесоматериалов 5317
маркировка лицевой стороны материала 2984
маркировка топором 449
маркировочное оборудование 997
маркировочный молоток 5315
маркировочный пистолет 8884
маркировочный топор 5314
маркировочный топорик 5314
маркировщик древесины 1457
марупа 6063
марупа белая 9373
марь 5306
маслина 2904
маслина дикая 9421
маслина европейская 2904
маслина Хохштетера 2712
маслина чёрная 6014
маслинные 5891
маслиновая роща 5892
масличное дерево 9000
масло чайного дерева 8438
масляная клетка 5875
масляное дерево 1223
масляный ясень 5890
масса воды в древесине 5518
масса древесины 8658
масса листвы 9200
массарандуба 5322
масса свежесрубленной древесины 9253
масса сухой древесины 2660 7841
массовая таблица 8957
массовая таблица насаждения 8036
массовое нападение 2849
массовый отбор 5325
мастер на лесозаготовках 1290
мастика для древесины 9606
мастиковое дерево 5327
материал выше заданного диаметра 5992
материал выше заданной длины 5997
материал для клёпки 8048
материал для настила 2303
материал для перешколивания 5814
материал заданной длины 6583
материал из стружки 1542
материал лесовозобновления 3555
материал ниже заданного диаметра 9058
материал ниже заданной длины 9062
материально-денежная оценка лесосек 8194
материнский древостой 6078
материнский пробковой слой 6201
материнское дерево 5569
материнское насаждение 6078
маточная плантация 8123
маточный корнеплод 8056
маточный ход 4811
махагониевое дерево 1412
махагониевое дерево низкое 2110
махагони настоящее 2110
мачта промежуточной опоры 3848
мачта стрелы 915
мачтовое дерево 5328
мачтовые брёвна 8586
мачтовый кран 2410
мачтовый лес 8586
мачтовый лесоматериал 8586
машина для валки "напроход" 1900
машина для выкапывания деревьев 8808
машина для выкопки саженцев 8798
машина для выкопки сеянцев 8916
машина для групповой обрезки 5156
машина для заготовки пней 8205
машина для измельчения 2104
машина для измельчения пней 8200
машина для обрезки вершин 8707
машина для обрезки деревьев по одному 7651
машина для обрезки корней 7019
машина для первичного транспорта леса 6553
машина для посадочных ямок 4326
машина для сбора живицы 5371
машина для трелёвки лесоматериалов в полностью погруженном положении 3674
машина для уничтожения кустарника и дёрна 7319
машина очистки лесосечных отходов 1574
машины для лесного хозяйства 3402
маятниковая пила 4312
маячное дерево 2567
мебельная древесина 3776
мебельная плита 5201
мебельные лесоматериалы 3776
мегагаметофит 5392
медвежий орех 5383
медвежье дерево 1544
медленно растущее дерево 7755
медленнорастущий лес 7757
медовая пальма 1701
медоносное дерево 4357
международная таблица определения объёмов 4559
междуузлие 4560
межевание леса 3614
межевое дерево 641
межкарровый ремень 4568
межклетник 155
межклеточная полость 4531
межклеточное пространство 4533
межклеточный канал 4530
межклеточный слой 4532
межкорневая конкуренция 7016
межрядный посев 183
межсосудистая поровость 4571
мезофилл 5407
мезофильный 5408

мелалеука белодревесная 1056
мелиевые 5243
мелия ацедарах 6110
мелкая древесина 7760
мелкая колючка 7912
мелкая насечка 7789
мелкая почка 5474
мелкая фракция щепы 1514
мелкая щепа 3148
мелкие древесные отходы 4950
мелкие остатки 3156
мелкие порубочные остатки 4950
мелкий валёжник 2265
мелкий круглый лесоматериал 7101
мелкий кругляк 7101
мелкий кустарник 7558
мелкий лес 5001
мелкий поделочный пиломатериал 4658
мелкий сучок 7764
мелкий текстурный рисунок 1655
мелкое дренажное углубление 3976
мелкое насаждение 7776
мелкозернистая древесина 1656
мелкозернистый 3157
мелколепестный 5431
мелколесье 5001
мелколиственный лес 3514
мелколистная липа 7767
мелкомасштабное лесное хозяйство 7772
мелкопористая древесина 1656
мелкопористый 3155
мелкослойная древесина 1660
мелкослойная текстура 1655
мелкослойность 3154
мелкослойность древесины 3154
мелкосучковатый 7737
менцизия ржавая 3310
мера Гоппус 4366
мера для бревен 5103
мера древесной массы брёвен 5109
меранти 4831
мера объёма лесоматериалов 8615
мербау 5476
мёрзлая древесина 3728
меристема 5402
мерная вилка для определения
мерная линейка Бильтмора 715
мерная линейка для обмера брёвен 5107
мерная линейка определения объёмов брёвен в досковых футах 877
мерная рейка 8292
мерная тесьма 8397
мерная тесьма для диаметров 2484
мерная цепь 8289
мерсава 5405
мёртвая древесина 2274
мёртвая подстилка 2269
мёртвое дерево 2271
мёртвое ядро 2261
мёртвый покров 2269
мертвяк 2271
мескитовое дерево 1790
мескитовое дерево бархатное 9189
мескитовое дерево чилийское 4354
местная гниль в виде полоски 8154
местная древесина 4341
местная массовая таблица 5037
местная порода 4477
местная сушилка 1986
местное обесцвечивание древесины, не являющееся значительным пороком 829
местное сплющивание волокон 3094
местность 4769
местность, покрытая кустарником 353
местные органы управления лесов 5033
местный лес 6606
место выполнения подпила на дереве 7668
место закрепления балки 640
место костра 4712
местообитание 4072
место обитания 4072
место окучивания лесоматериала 7981
местоположение наземной линии 4776
место прикрепления кроны 2064
место пропила 7668
место просачивания из земли воды 7401
место разгрузки 9102
место раскряжёвки 7727
место распространения 4072
место рубки 7668
место связывания 3280
место складирования древесины 5451
место торговли 7158
место установки лебёдки 9742
метаксилема 5415
метасеквойя глиптостробовидная 2259
метатрахеальная паренхима 5414
метеорологический индекс пожара 3218
метик 4195
метиковая трещина 4195
метка для раскряжёвки 8972
метка лесного молотка на торце 2807
метка на торце 2807
метлообразная корона 1074
метод ведения лесозаготовок 4158
метод ведения рубки 2203
метод возобновления 5416
метод глазомерного изучения на учётных площадках 5865
метод закладки ленточных проб 4990
метод зарубки и инжектирования 4083
метод лесной селекции 3589
метод лесоустройства 3491
метод обмера брёвен 5110
метод обработки древесины 9702
метод обратного скрещивания 463
метод переменной закладки пробных площадей 9166
метод периодных блоков лесоустройства 6161
метод постоянных периодных блоков 6170
метод разведения толстомерного леса 6888
метод размножения отводками 4843
метод рубки спелого леса 4151
метод сплошной рубки 1617
метод фиксированной закладки пробных площадей 3234
механизированная валка 5369
механизированная заготовка 5379

механизированная канатная трелёвка 6489
механизированная очистка от отходов 5381
механизированная посадка 5380
механизированная прополка сорняков 5375
механизированная рубка 5369
механизированное кольцевание 5370
механизированные лесозаготовки 6489
механизированные рубки ухода 5374
механизированный доступ и использование 5571
механизированный доступ и эксплуатация 5377
механизированный уход за насаждениями 5382
механизм выравнивания брёвна 8404
механизм подачи рубительной машины 1528
механизм подъёма опорных передних колёс полуприцепа 7435
механизм ударного действия 4369
механизм установки брёвна в горизонтальное положение 8404
механизм устойчивости дерева к заболеваниям 8886
механическая лесообработка 5376
механическая лопата для выкопки саженцев 6367
механическая окорка 5372
механическая очистка поверхности почвы 5373
механический колун 9512
механический отбор деревьев при рубках ухода 5374
механическое повреждение 5368
механическое прореживание 5374
механическое прореживание по шаблону 8095
меч для посадки в щель 8898
мечевидный лист 8353
мечение 8364
мешающий валке подрост 5860
миграция 5442
миграция видов 5443
микориза 3775
микробиотоп 5434
микро-габитат 5429
микроместообитание 5429
микроорганизм 5430
микрораспространение 5432
микрорельеф 5433
микроскопия дерева 9608
микросреда обитания 5429
микрофибрилла 3780
микроэлементы 8740
милдью 5444
мимоза 3962, 5460
мимозка 1790
мимозка серёжкоцветная 1790
мимозовые 5459
мимузопс 485
миндаль 207
миндаль настоящий 207
миндаль низкий 2682
миндаль обыкновенный 207
миндаль пустынный 2412
минди 6110
минерализация древесины 9610
минерализация поверхности почвы 7286
минерализованные полосы 5461
минеральное включение 5462
минимальная длина деловой части ствола 5468
минимальная окружност ствола 5467
минимальный диаметр 5465
минимальный рентабельный диаметр 2743
минимальный эксплуатационный диаметр 2950
минирующая мушка 4880
минусовое дерево 5473
минусовое насаждение 5472
миомбо 5475
мирабель 1475
мирзина флоридская 5648
мирзиновые 5647
мирика восконосная 7858
мирикария 3003
мириковые 9305
миробалан 1475
мировая лесная торговля 9727
мировое лесное хозяйство 9726
мирт 1792
мирт болотный 4898
мирт игольчатый 9375
мирт обыкновенный 1792
миртовые 5649
михелия 5428
михелия буроватая 513
михелия ржавая 513
мицелий 5645
млечник 4829
млечники 4828
млечный сосуд 4829
многодисковая пила 5132
многокаемчатое хозяйство 5618
многократно плодоносящий 6436
многократный верхушечный побег 5614
многолесной 6936
многолетнее растение 6151
многолетник 6151
многоножевая рубильная машина 5607
многоножка 6441
многооперационная машина для заготовки сортиментов 7551
многооперационная машина с шарнирно-сочленённой рамой для заготовки сортиментов 372
многоотраслевое лесное хозяйство 5623
многопильный станок для поперечного раскроя пиломатериалов 8971
многопильный станок для продольной распиловки 6966
многоплодный 6436
многопролётная подвесная канатная установка 5627
многорядный луч 5626
многослойная древесноволокнистая плита 5609
многослойная древесностружечная плита 5608
многослойная фанера 4762
многослойный дощатоклеённой элемент 3878
многослойный картон 4761
многостволовый 5628
многоствольное дерево 5617
многоузловая почка 5611
многоцелевое ведение лесного хозяйства 5622
многоцелевое дерево 5625
многоцелевое лесное хозяйство 5623

многоцелевой лес 5606
многоцепной лесотранспортёр для брёвен 5635
многоярусная культура 5631
многоярусная система 5632
многоярусное насаждение 5634
многоярусный лес 5633
множество данных лесного растра 3548
мовингви 5732
модель доставки лесоматериалов 8649
модель леса 3508
модель лесозаготовки 4160
модель насаждения 8021
модельная древесина 5505
модельное дерево насаждения 7174
модельный лес 5504
модель принятия лесохозяйственных решений 7615
модель роста и дохода 4011
модер 5506
модер-гумус 5506
модернизация территории лесосеки 7156
модерное лесное хозяйство 5723
модифицированная валка 5511
модифицированная древесина 5513
модифицированная зона горения 5509
модифицированный лес 5512
модуль излома 5515
модуль продольной упругости 5514
модуль разрыва 5515
модуль упругости 5514
можжевёловый лес 4672
можжевельник 3983
можжевельник аллигаторовой 187
можжевельник бермудский 682
можжевельник вирджинский 2722
можжевельник обыкновенный 3983
можжевельник высокорослый 106
можжевельник греческий 3935
можжевельник западноамериканский 7588
можжевельник испанский 6530
можжевельник коротколистный 458
можжевельник красноватый 6530
можжевельник красноплодный 6787
можжевельник красный 6530
можжевельник ладанный 7878
можжевельник мексиканский 5423
можжевельник высокий 3935
можжевельник Пинчота 6241
можжевельник провислый 5423
можжевельник скальный 6995
можжевельник стройный 106
можжевельник техасский 8484
можжевельник толстокорый 187
можжевельник утахский 9149
мокрая гниль 9353
мокрая окорка 9350
мокрица 1813
мокрица древесная 1813
мокрый кабель 9352
молдинг 9613
молевой лесосплав 2625
молодая древесина с неодревесневшей тканью 4675
молодая дубовая кора 9786
молодая листва 4676
молодая поросль после вырубки 1928
молодняк 9785
молодое дерево 9791
молодое деревцо 7202
молодое лесное насаждение 9785
молодое насаждение 9785
молодое растение 9787
молодой древостой 7204
молодой дуб 5841
молодой побег 1095
молодой порослевый лес 1928
молодые лесные культуры 9788
молокан компасный 6531
молоки 1062
молотковая окорочная машина 4103
молоток для клеймения 5315
молоток для клеймения краской 5315
молоть 3975
молочай древовидный 341
молочайные 7961
моль выемчатокрылая сосновая 6261
моль-минёр падубовый 4337
моль-пестрянка 4871
монастырский лес 5528
монимиевые 5529
мониторинг лесов 5530
монокарпик 5537
монокарпическое растение 5537
монокультура 5542
монокультурное хозяйство 5542
монолитная сердцевина 1901
монтерейский кипарис 5547
мора высокая 4715
мора Джонггрийпа 3027
морёная древесина 7993
морение 3766
морёный дуб 3764
моринга 4379
моринговые 5556
морозная трещина 3723
морозное кольцо 3727
морозное ядро 3725
морозобоина 3723
морозобойная трещина 3723
морозобойный нарост 3726
морозостойкость 3724
морозоустойчивость 3724
морской сплав 6689
морфологически улучшенные лесные сеянцы 5559
морщинистый желвак 8416
мостильная древесина 6117
мотовило 9267
мотопильщик 1434
моторист пилы 1434
моторист цепной пилы 1434
моторная пила 6491
моторная пила с дугообразной шиной 952
моторная цепная пила 6488
мотыга для полки 4030
мотыга для посадки 6352
мотыга для снимания дёрнины 7812
мотыга Мак-Лода 5348
мотыжить 4316
мох 5563
мохнатая тля лиственничная 9718
моховой лес 5567
моховой покров 5564
мочало 4502
мочковатый корень 1968
мочковидный корень 1968
мощная корневая система 142

мощности машины для раскряжёвки лесоматериалов 7726
мужское дерево 7999
музанга 9042
муйракатиара 5595
мука из коры 550
мулевый гумус 5604
муль 5604
мультиклональная смесь 5605
мульча 5601
мульча из древесных остатков 5603
мульчирование 5602
мульчирующее покрытие 5601
муравей-древоточец 1351
муравей-древоточец пахучий 1762
муравей зонтичный 4864
муравей лесной обыкновенный 9480
муравей лесной рыжий 9480
муравей-листорез 4864
муррая 4644
муррая Кёнига 2147
муррея 4644
мусанг 9042
мускатниковые 5820
мускатный орех 5821
мускусное дерево 5639
мусоросжигательная печь 6807
муссонные смешанные леса 5546
муссонный лес 5545
муссоновый лес 5545
мутация 5640
мутения 5641
мутовка 995
мутовка сучьев 995
мутовчатое листорасположение 9209
мутуализм 5643
муфельный древесный уголь 6984
мухугу 7609
мучнистая роса дуба 5844
мушмула 5389
мушмула японская 5172
мшистое волокно 9717
мыльное дерево 2363
мюскадина 5637
мягкая белая гниль 7825
мягкая белая грибная гниль 7826
мягкая древесина 7823
плита 5183
мягкая заболонная гниль 7820
мягкая кора 7814
мягкая сердцевина 7817
мягкая широколиственная древесина 7816
мягкие плоды 7815
мягкий 7822
мягкий гумус 5576
мягкий луб 7814
мягкий плодокорм 7818
мягколиственная порода 7832
мягколиственный лес 1873
мясистость 8250
мясистый корень 3263

наблюдатель 7932
наблюдательная пожарная вышка 3216
наблюдение за огнем 3210
наблюдение за пожаром 3192
наблюдение засухи 2639
набор пил 3800
навес для хранения дров 9650
наветренная сторона 9451
наводка на дерево 8847
наводнение 3282
наводок 3282
награда для корчёвки деревьев 6512
нагрузка каррами 8410
нагрузка на рейс 8984
нагрузка популяций 6460
нагрузка популяций на среду 6460
надбавка 202
надвид 8271
надземная биомасса 12
надземный 11
надрез 4082
надрезанный лист 4438
надрез в коре 562
надрубка 540
надсечка 4082
название дерева 8889
наземная противопожарная охрана лесов 4771
наземная трелёвка 3988
наземные системы 3982
наземный пожар 8282
назначенный чиновник 2424
наивысшее дерево 6500
наилучшие методы управления 688
наименьшая толщина вырезки порока 5466
наклев на стволе 765
наклон 7750
наклон волокон 4439
наклон вперёд 4188
наклон дерева 8857
наклон дерева под воздействием ветра 9446
наклонение во внешнюю сторону 5983
наклонение во внутреннюю сторону 4518
наклоненный ветром 9440
наклонная эстакада 4442
наклонное дерево 4897
наклонное резание 4441
наклоннорастущее дерево 4897
наклонность 9023
наклонный настил 4442
наклонный транспортёр, не повреждающий поверхность досок 9114
наклоняющаяся стрела 4225
наконечник 819
накрест сложенный штабель 6232
наличник 333
наличность дичи 3796
налог на лес 3616
налоговая льгота на доходы с невосполняемых природных ресурсов 2401
налоговая скидка на истощение недр 2401
наматывание 6120
нандина домашняя 5654
нандиновые 5655
нанесение жёлобка 544
нанесение карр узкими полосками 544
нанесение меток 8364
нанесение подновки на карру 1530
нанесение полос 519
нанизм 2686
наносить отметки 7308
наносные отложения 2440
наносы 2440
напенная гниль 1233
напильник по дереву 9566
напиток из дерева 9539
наплыв 1182
наплыв на ветви 995
наплыв на стволе дерева 2938
наплыв "птичий глаз" 766
напочвенный пожар 8282
направление волокон 2516

направление естественного наклона 2518
направление естественного наклона дерева 2518
направление перемещения 2517
направление перемещения леса при трелёвке 2517
направление резания 2519
направление рубки 2519
направление трелёвочного волока 9746
направленная валка 2515
направленный к оси 62
направленный от оси 5
направляющее долото 6238
направляющий блок возвратного каната 4167
направляющий жёлоб 334
направляющий канат 8365
направляющий ролик 4849
напряжение зарубки 2770
напряжение роста 4027
напряжённое состояние 1367
напряжённое состояние древесины в конечный период сушки 6924
нарезанный комель 4673
нарезка борозд 8544
нарост в виде кольца 8785
нарост на стволе дерева 2938
нарост у корня 7015
нарра 5284
наружная пласть из древесины высшего сорта 3229
наружная пласть пиломатериалов 5982
наружная трещина 2965
наружное высыхание 1367
наружный диаметр 5976
наружный слой плиты из ориентированной стружки 5953
наружный слой флоэмы 6157
нарушение почвенного покрова 7836
нарушенная растительность 2544
нарушенный естественный лес 2543
насаждение, в котором проведена выборочная рубка вне лесосек 7410
насаждение в стадии жердняка 9785
насаждение деревьев 3626
насаждение для пастьбы скота 7265
насаждение из саженцев и молодняков 7390
насаждение из семенной коллекции 7377
насаждение, не пройдённое рубками ухода 6524
насаждение одной породы 2926
насаждение, пройдённое рубками ухода 8000
насаждение с большим запасом древесины 4303
насаждение семенных деревьев 8022
насаждение с прогалинами 3646
насаждение с просветами 3646
насаждения из диких культур 9416
насаждения непрерывного пользования 6171
насекомое, питающееся древесиной 4958
насекомопоедающие 4513
насекомоядные 4513
наследственность 4258
настил 2303, 5995
настоящее ядро древесины 8986
настоящие пихты 8987
настоящий домовый гриб 2560
настоящий опёнок 4352
настоящий трутовик 8667
наступательная рубка 6513
наступление леса 3477
насыпная плотность 1155
насыщение волокон древесины 9565
натурализованный вид 5684
натуралистическое лесоразведение 5683
натуральная древесина 3826
натуральное хозяйство 8238
натяжение несущего каната 7704
натяжная проволока 1164
натяжная цепь 7798
натяжной канат 1164
науклея 710
научное управление лесами 3581
национальное ведение лесного хозяйства 5666
национальный заказник 5668
национальный лесной парк 5667
начало кроны 7940
начало регенерации 6819
начальный перечёт зрелости 4496
начинающаяся прелость 4494
начинающее гниение 4494
неблагоприятное местообитание 2491
небольшие трелёвочные подсанки 2008
небольшой деревянный клин 9561
неветвящиеся деревья 9006
невосприимчивость 6906
невосстановляемые затраты 9112
невспаханное поле 2997
невыдержанная древесина 9116
невыполнение 2994
невырубленная полоса 4901
невысушенная древесина 9116
невысушенное состояние пиломатериалов 3957
невыцветающая древесина 2989
неглубокое гнилостное пятно 7081
негодный для продажи лесоматериал 5765
недавно поваленный лес 3710
неделовое бревно 5764
неделовое дерево 9104
неделовой лесоматериал 5765
неделовой сортимент 5764
неделовой хлыст 2335
неделовые деревья в насаждении 9360
неделовые древесные остатки древесины 2123
недопил 9051
недопущение пожара 3182
недоразвитая древесина 4675
недоразвитость 4681
недоразвитый 4674
недоразвитый лист 9075
недоруб 9051
недостаток древесины 9563
недостаток леса 4753
недостаток питательных веществ 5179
недостаток света 4940
недостаточная сомкнутость 9068
недостаточно восстановленная лесная земля 5801
недревесная продукция леса 5778
нежелательная древесина 9322
нежелательное дерево 9321

незаконная рубка 4423
незаподсоченное дерево 9231
незасолённые рудничные отходы 8334
незащищённая лесная земля 5769
незрелый 4674
неистощимость леса 3470
нейтральные рудничные отходы 8334
нейтрофильный 5721
неклассифицированные лесные данные 9047
некоммерческие рубки ухода 4680
некроз 5702
некроз коры 573
нектандра великая 4768
нектандра Родией 2363
нектрия 5702
нелесная земля 5758
нелесная площадь 5758
нелесной 8874
нелесные ресурсы 5775
неликвидная древесина 2123
ненарушенная растительность 9077
ненарушенный лес 9076
ненормальная окраска 1728
ненормально вытянутая верхушка 5544
ненормально вытянутые сучья 6708
необлесённый 8874
необлесённый участок 3443
необозначенная девственная местность 5756
необработанный лесоматериал 7096
необработанный осадок 7473
необрезанная доска 4133
необрезанные сучья
с необрезанными сучьями 9101
необрезная доска 4133
необрезной пиломатериал 9079
необрезной распиленный материал 9079
неодинаково окрашенная текстура 9082
неоднородная текстура 9082
неоднородное насаждение 9081
неоднородность 4263
неоднородность текстуры 3101
неодревесневший 5770
неокорённая древесина 9045
неокорённый лесоматериал 9045
неокрашенная червоточина овальной формы 9494
неоперабельные земли 4508
неопиленная кромка 2662
неопределённая количественно облесённая земля 9107
неопределённая продуктивность леса 9120
неотделанная древесина 535
неотделанная поверхность 9083
неотёсанные строевые лесоматериалы 5216
неотложность восстановления 9143
неотносимый за счёт объём 5750
неочищенная живица 6733
неочищенная площадь 7718
непарник 4071
непарноперистый лист 5867
непарный шелкопряд 4071
неперешколенное растение 9420
неперешколенный саженец 9420
неплановая рубка 9046
неповреждённый лесоматериал 5902
неподвижный несущий канат 8043
неподготовленный трелёвочный волок 7095
неподдерживаемые концы лесоматериалов в штабеле 5993
не покрытая лесом земля 5758
не покрытая лесом площадь 5758
не покрытые лесом акры 5757
не покрытый лесом 8874
неполнодревесный 8401
неполное годичное кольцо 6087
неполно карбонизированные отходы 4445
неполно обугленные остатки 4445
неполовозрелая стадия 4424
непористая древесина 5766
непостоянный лесной рабочий 6095
неправильно выполненная валка 4603
неправильно отёсанное бревно 4655
непреднамеренное скрещивание 5971
непрерывная валка 1897
непрерывная лесная таксация 1898
непрерывная таксация состояния леса 1899
непрерывное лесопользование 6172
непрерывно-производительный лес 6171
непрерывность горючего материала 3735
непрерывные лесозаготовки 1897
непригодная лесная площадь 9121
непродуктивная лесная площадь 5749
непродуктивный лес 9105
непромышленная древесная порода 5752
непромышленная рубка 5751
непромышленная частная лесная земля 5762
непромышленные рубки ухода 4680
непрорубленная лесная экосистема 9123
неравномерная разновозрастная структура 4589
неравномерное насаждение 4588
неравномерно облесённый 4586
нерасколотый лесоматериал 9048
нераскручивающийся канат 6503
нераспиленный 9113
нерельефный посев 3257
неровная текстура 1168
неровность 1167
неродственное размножение 4401
несколько деревьев 7445
неслитность полога 4565
неснижающийся доход 5755
не снимаемый до полного вызревания 8913
несомкнутый 3694
несомкнутый полог 5910
несортированная щепа 9115
несортированные пиломатериалы 9117
несосудистая древесина 5766
неспелый 4674

несплошные рубки 6083
несросшийся сучок 5158
нестандартная древесина 2123
нестандартная крупная щепа 6004
нестандартная мелкая щепа 9067
нестроганная доска 7084
нестроганный пиломатериал 7096
нестроганный сухой пиломатериал 7086
нестроганный сырой пиломатериал 7090
несущая балка 9592
несущая деревянная балка 9592
несущая рейка 9592
несущий канат 8745
несущий канат, расположенный по кривой 2153
несущий трос 8745
нетоварное дерево 9104
нетронутая территория 9408
нетронутый участок природы 9408
нетто-прирост 5716
неуправляемая лесная земля 9103
неуправляемый пал 3690
неуправляемый пожар 3690
неурегулированные земли 9110
неурегулированный лес 9111
неустойчивый склон 7441
неустранимый дефект древесины 2333
неустроенный трелёвочный волок 7095
неутолщённое место 6461
неухоженный лес 9122
нефелогилея 5593
неярусная растительность 5763
нивелир 4922
нивелирная рейка 4923
нивелирный кол 3899
нивелировочный кол 3899
нигерийская фагара 5732
нижнее защитное ограждение 670
нижнее сжатие 942
нижний горный лес 5188
нижний диаметр 2465
нижний зажим для брёвен 943
нижний лесной склад 4484
нижний отрезок ствола 5128
нижний рез 5187
нижний ряд брёвен 1180
нижний слой 8240
нижний слой сильно разложившегося опада 4394
нижний слой сильно разложившегося опада мхов 5873
нижний трос 3993
нижний тяговый канат 3993
нижний ярус насаждения 9070
нижняя ветвь 604
нижняя высота расположения кроны 5185
нижняя длина кроны 5186
низбегающая форма 2316
низбегающая форма роста 2358
низинная тропическая сосна 5194
низкобонитетное насаждение 4446
низкоинтенсивная рубка ухода 4952
низкоинтенсивные выборочные рубки 4948
низкокачественная древесина 5195
низкокачественное дерево 6451
низкокачественное лесонасаждение 5191
низколежащий тяговой трос 3993
низколесье 5189
низкополнотная лесная земля 5773
низкополнотное насаждение 4446
низкопродуктивное насаждение 5191
низкорослая вееролистная пальма 7262
низкорослая древесно-кустарниковая заросль 1202
низкорослое дерево 7323
низкорослость 2686
низкорослый порослевый лес 1928
низкорослый эвкалипт 5258
низкосортная древесина 5195
низкосортная древесина в качестве прокладок при загрузке судна 8579
низкоствольная система 1935
низкоствольник 5189
низкоствольник на корьё 5833
низкоствольное дерево 5196
низкоствольное лесосечное хозяйство 1929 5835
низкотоварная древесина 5195
низовая проходная рубка 8511
низовое прореживание 8511
низовой пожар 2012
никотиана сизая 8945
никтагиновые 3676
ним 6110
ниргунди 3232
нири 6646
нисса водная 9304
нисса двуцветковая 8305
нисса китайская 1507
нисса лесная 810
нисса Огичи 5871
ниссовые 9010
нисходящее сокодвижение 2312
новая древесина 5726
новая земля естественного леса 5725
новая порода 5802
новозеландская красная сосна 6784
новозеландское хвойное дерево 6664
новосозданное насаждение 5722
новый побег 6872
новь 390
ногоплодник 2714
ногоплодник гватемальский 1372
ногоплодник дакридиевидный 4683
ногоплодниковые 6407
ногоплодник олеандролистный 5888
нож 2204
нож для измерения коры 570
нож для кольцевания 3854
нож для лущения древесины 7960
нож для маркировки деревьев или брёвен 6741
нож для обрезки вершин 8709
нож для обрезки пробок 6384
нож для обрубки сучьев 2285
нож для резьбы 1362
нож для скобления коры 7960
ножевая фанера 7738
ножевое раскряжёвочное устройство 5113
ножевое режущее устройство для дерева 9584
ножевое срезающее устройство 3044

ножевое устройство для обрезки вершин 8709
ножевое устройство для срезания дерева 8921
ножевой механизм для группового срезания деревьев 5621
нож корообдирочный 7960
ножницы для дерева 9649
ножницы для обрезки ветвей 5168
ножничная ветверезка 5168
ножовка 1822
ноздреватость 6465
ноздреватый рисунок 835
номер делянки 6380
номинальная длина 5747
норичниковые 3126
нория 4600
нормальная густота древостоя 3757
нормальное распределение насаждений по классам возраста 5779
нормальное распределение по классам возраста 5779
нормальное чередование классов возраста 5785
нормальной густоты 9335
нормально облесённый 4419
нормально питающийся 2912
нормальный диаметр 5782
нормальный лес 5781
нормальный прирост 5783
нормальный семенной древостой 5784
норма минимального использования 5469
нормативный документ о лесе 3516
нормирование затрат в лесном хозяйстве 1979
нормировка древесины 8642
носок пильной шины 592
нотофагус 2999
нотофагус Куннингама 406
ночная трелёвка 5733
нулевая зажимная длина 9795
нулевая обработка почвы 5799
нулевой зазор 9795
ныне живущий 2960

обапол 5470
обвал 7835
обвязанная пачка лесоматериалов 7743
обвязывание 519
обвязывание стволов 519
обгорание 4209
обгрызание насекомыми 2234
обгрызать 5730
обдирка 579
обдирочная машина 2284
обезвершинивать 8702
обезвоживание 2454
обезлесение 2339
обезлесить 2524
обезлиствение 2941
обескрыливатель 2456
обеспечение лесом 2359
обеспыливание 2318
обессмоливание 2407
обессмоливать 2345
обессученный лесоматериал 6608
обесцвеченный лист 831
обесцвечивание древесины 9536
обесцвечивание заболони 2527
обжигать 3166
обзол 2662
обивка тёсом 880
обильный рост 4213
облагораживание древесины 4434
облагороженная древесина 4432
областной лесничий 2551
область анализа 268
область естественного распространения 4072
область, запланированная на будущее 2336
область истока 9280
облепиха 7325
облепиха крушинная 7325
облесение 94
облесение берегов 6960
облесение болот 98
облесение дюн 2666
облесение заново 6821
облесение откосов, дорог, каналов или плотин 2231
облесение пашни 6100
облесение песков 7179
облесение пустоши 96
облесение рудных разработок 97
облесение склонов 7751
облесение степи 8086
облесённая лощина 2360
облесённое пастбище 9711
барьер 9544
облесить 93
облиствение 3303
облиственная крона 4865
облиственный 4894
облитерация 7599
облицовочный шпон 2986
облицовывание 8281
обложение леса налогом 8422
облом в результате ожеледи 4417
обломка 1571
обломки, влекомые потоком по ложу 647
обломки древесины 1059
обломок 7873
обломочный оползень 2287
облом сучьев в результате ожеледи 4417
облупающаяся древесина 9696
облупленная древесина 901
облупленный кругляк 9193
обмер брёвен на погрузочной площадке 6981
обмер в верхнем и нижнем отрубах 5367
обмер в средней части 7279
обмер древесины 9607
обмер лежащего лесоматериала 3040
обмер лентой по окружности 8165
обмер основного и верхнего диаметра 5367
обмер пиломатериалов 5213
обмер по длине и среднему диаметру 5362
обмерщик брёвен 5111
обнажение корней 7048
обнаженная лесная зона 470
обнажённый корень 531
обнажённый противопожарный барьер 530
обновляющий побег 4505
обогащение 677
обогащённая смолой ядровая древесина 6902
обогащённый пар 2829
ободная древесина 3082
ободранное при трелёвке дерево 1097
обоеполый цветок типа шишки 1864
обожжённая древесина 7735
обозначение полосами 519
обозначение флагом 3239

обозначенная первобытная местность 2426
обозначенный лесоспуск 2425
обозный лесоматериал 1672
обойный гвоздик с двумя ножками 8192
оболочка клетки 1407
оборачиватель 9682
оборот высшей лесной ренты 7076
оборот на рудничный лес 6310
оборот по экономической спелости 2745
оборот рубки 2194
оборот рубки по возобновительной спелости 7619
оборот рубки по технической спелости 8440
оборот рубки по экономической спелости 2745
оборот ухода 8460
оборудование для погрузки лесоматериалов 9587
оборудование для транспортировки леса 8565
оборудование участка 6716
оборудование участка лесничества 6716
обрабатываемость 8791
обрабатывание древесины 9630
обрабатывать 2128
обработанная древесина 8792
обработанный в соответствии с размерами 2205
обработка без давления 5768
обработка давлением 6523
обработка деревьев в лесу 4487
обработка комля 2179
обработка коры у основания дерева 603
обработка коры ядохимикатами у основания дерева 603
обработка почвы 8554
обработка почвы дисковыми орудиями 2535
обработка против заболонной окраски 318
обработка против синевы 318
образец 6601
образец древесины 8635
образование ветви 4731
образование внутренних трещин при конвективно-тепловой сушке 4349
образование внутренних пиломатериалов 4349
образование глея в почве 3860
образование годичных колец 304
образование грибкового повреждения 3772
образование древесины 4954
образование древесного угля 1449
образование корней 7024
образование коры 559
образование плода 3729
образование пнёвого осмола 4953
образование почек 3835
образование почки на поверхности листа 2850
образование свили 9187
образование сводов 4117
образование смешанного насаждения 2867
образование стружки 1515
образование сучьев 4731
образование трещин 2003
образование трещин усушки 4349
образование узелков или клубеньков 5745
образование ядра 2671
образовательная ткань 5402
образующий корни 6929
обраставшая земля 34
обрастание 35
обрастание корой 559
обрастать 33
обрастница 2898
обратная лопата 468
обратное скрещивание 462
обратноконусовидный лист 5851
обратноланцетовидный лист 5854
обратно-радиальная схема разработки делянки канатной установкой 6925
обратносердцевидный лист 5852
обратнояйцевидный лист 5859
обратный несущий канат 4580
обратный удар древесины 3297
обрезанный корень
с обрезанным ~ем 7039
обрезанный на четыре канта 7963
обрезать 1645
обрезать живых и сухих ветвей
обрезать сучья 10
обрезка 2185
обрезка боковых ветвей и вершин деревьев 999
обрезка боковых корней 964
обрезка вершин 8696
обрезка ветвей плодовых деревьев 5939
обрезка ветвей с помощью лестницы 4294
обрезка в спящем состоянии 9462
обрезка высоких сучьев 4294
обрезка деревьев на омоложение 8939
обрезка досок зигзагом 9796
обрезка досок на конус 1871
обрезка досок по диагонали 2460
обрезка живых ветвей 3953
обрезка заподлицо 3294
обрезка засохших ветвей 2655
обрезка затеняющих деревьев или сучьев 7480
обрезка зеленых ростков 3953
обрезка зигзагом 9796
обрезка корней 7040
обрезка кромок 2766
обрезка кромок по сбегу 1871
обрезка кроны 2083
обрезка на конус 1871
обрезка на четыре канта 7964
обрезка нижних ветвей кроны 4933
обрезка побегов 7532
обрезка побегов на концах ветвей 2633
обрезка по диагонали 2460
обрезка разросшихся деревьев 3071
обрезка стволов 8081
обрезка сухих сучьев 1106
обрезка сухостоя 2270
обрезка сучьев 6610
обрезка с целью формирования кроны 6612
обрезка формирования дерева 3663
обрезки 998
обрезки концов лесоматериала 9273
обрезки лесоматериалов 2174
обрезная кромка 7967
обрезное устройство 2767
обрезной 2756
обрезной пиломатериал 2757

обрешетина 7005
обрубание 8824
обрубать сучья 10
обрубка 8824
обрубка живых ветвей 3953
обрубка заподлицо 3294
обрубка надземных корней перед валкой 6861
обрубка сучьев и трелёвка хлыстов в полностью погруженном положении 2350
обрубка сучьев на стволе стоящего дерева 3953
обрубки 5218
обрубленный конец 871
обрубок 8937
обрубщик сучьев 981
обручный лесоматериал 523
обрывание листьев 4866
обрывать 1645
обрывки волокон 3148
обрыв корней 7055
обсаженный деревьями 3394
обсеменение лесосек от стен леса 7384
обсеменение от верхнего полога 6825
обсеменение от стен леса 7384
обсеменительная рубка 7383
обсеменительный приём постепенной рубки 7383
обследование леса 3549
обстричь 1645
обтёсанная древесина с сердцевинными лучами 7603
обтёсанный брус 4270
обтёска 4273
обтёска закомелистости перед валкой 6861
обтёсывание 4273
обтёсывать 83
обтёсывать бревно со всех сторон 7962
обугленные частицы щепы 1185
обугливание 1331
обуглившаяся древесина 1453
обух 451
обхват у вершины 8690
обход лесника 3333
обшивка деревянными панелями 9616
обшивка деревянными филёнками 9616
обшивка досками 880
общая длина пиловочника 9146
общая лесная площадь 8722
общая масса в кубических футах 3311
общая масса дерева 8733
общая площадь 8719
общая площадь массива 2032
общая порозность 8728
общая сумма площадей поперечного сечения деревьев на единице площади 2032
общее количество выделяемого тепла 8725
общее планирование лесозаготовительных участков 8720
общее тепловыделение 8725
общественный лес 1816
общие лесные дороги 1787
общие средние затраты на заготовку леса 439
общий возраст 8718
общий каталог основных материалов для лесовозобновления 1767
общий лесной покров 8723
общий лесной стандарт 3643
общий облик 8357
общий объём брёвна 5125
общий объём годичной рубки 8721
общий план ресурсов 8729
общий прирост 8727
общинное ведение лесного хозяйства 1817
общинное лесничество 1817
общинный лес 1812
объедание листья и молодые побеги 8806
объём брёвен 5125
объём в абсолютно сухом состоянии 9252
объём в плотных единицах 7842
объём выборки при рубках ухода 8519
объём деловой и дровяной древесины 5209
объём деловой части 5401
объём деловой части ствола 5401
объём дерева 7652
объём дерева без коры 8952
объём дерева с корой 8894
объём древесиной зелени 9253
объём древесины 9689
объём древесины сучьев 9245
объём заготовленной древесины 9777
объём заготовленной с 1 акра древесины в кордах 1945
объём запаса 9250
объём запаса древостоя 9250
объём зелёных деревьев 9247
объём кроны 2097
объём лесного сырья на корню 9250
объём лесоматериалов 9689
объём лесоматериалов в абсолютно сухом состоянии 9252
объём лесоматериалов, определённый на лесопильном заводе 5455
объёмная масса 9099
объёмная производительность 9773
объёмная таблица 8957
объёмная таблица, учитывающая видовые числа 3664
объёмная усушка 9251
объёмное уравнение 9242
объём отдельного дерева 7652
объём отдельного хлыста 7652
объём пиловочника 9248
объём пор 6463
объём рубки 2200
объём свежесрубленных лесоматериалов 9253
объём ствола 9249
объём стволовой части дерева 9249
обыкновенная черёмуха 2881
обыкновенный персик 6123
обыкновенный раскручивающийся канат 5767
обычные лесозаготовки 2164
овальность 5984
овальный сучок 5985
овраг 4043
овражная эрозия 4045
оглеение 3860
оглобельник 6427
оглобля 6427
огневая очистка лесосек 7724
огневая рана 3206
огневая сушка 3180
огнестойкая древесина 3198
огнестойкое дерево 3203
огнетушитель 3211

оголённый корень 531
оголённый район 486
огораживание 2800, 4223
огороженный забором участок 9599
огороженный участок 890
ограждение 2793
огражденный лес 3535
ограничение густоты древостоя 6918
ограничение коробления 9269
ограничение трелёвки 4979
ограничения во введении лесозаготовок по погодным условиям 7331
ограниченная площадь 1868
ограниченное лесное хозяйство 7772
ограничиваемый рост 2435
ограничивающий фактор окружающей среды 4980
ограничитель глубины зарезания 2403
ограничительный знак 949
ограничитель отбрасывания досок назад 4706
ограничитель отбрасывания назад 4706
огуречное дерево 709
одеревенелое волокно 4956
одеревенелый 4955
одернение 7813
одинаковые долготья 2925
одиночная ветвь 7290
одиночное дерево 4480
одиночное насаждение 7650
одиночный перестой 7291
одиночный пилильщик-ткач 9313
одиночный сучок 7290
одичавшее экзотическое растение 2862
однобоковое развитие 5171
одновозрастное насаждение 2920
одновозрастной высокоствольный лес 2918
одновозрастной лес 2920
однодольные 5540
однодомные растения 5543
одно из нескольких сообществ, сменяющих друг друга на ограниченном пространстве 227
одноканатная система 7644
одноканатный трелёвочный
однолетний побег 1307
однолопастный лист 9085
одномутовчатый 9092
односножевое срезающее устройство 322
одноосная трелёвочная тележка 3673
одноотвальный плуг 7645
однопестичный пузыреплодник 5590
однопролётная канатная установка 7649
однородная текстура 2923
однородное насаждение 2926
однородный лесоматериал 4346
однородный поток 2922
однорядный луч 9094
односеменодольные растения 5540
односемянный плод 7647
однослойная древесностружечная плита 7639
однослойный 7640
одноствольный 7637
одностороннее развитие 5171
односторонне комбинированная поровость 9090
односторонняя врастающая кора 5900
односторонняя кривизна 5898
односторонняя крона 7648
одноузловая почка 9093
одноцветная текстура 2923
одноярусный древостой 7653
одревесневший 7302
одревесневший каллус 7303
одревесневший клубень 4960
одревеснение 4954, 8097
ожидаемая стоимость леса 3412
ожидаемый лесной доход 2945
ожидаемый размер 3163
ожог древесины от перегрева инструмента 5229
ожог коры 836
ожог листа 837
ожог листьев 7306
ожог от перегрева инструмента 5229
оздоровительный лес 91
озеленение 233
озеленение городов 5636
озиго 6016
окаймление поры 6299
окаймлённая пора 920
окаймлённый 2756
окаменевшая пальмовая древесина 6195
окаменелое дерево 5903
окаменелый лес 6196
окантованный брус 7969
окольцованный сучок 581
окольцовывание 519
окольцовывание дерева 5798
окольцовывающие насечки 3715
окомкование 5745
окончательная зачистка 3134
окончательная лесовосстановительная рубка 6864
окончательная рубка 3136
окончательная стадия рубки 3142
окончательная съёмка трассировки 3140
окончательное строгание 1608
окончательные рубки ухода 3143
окончательный выход древесины 3145
окончательный приём постепенной рубки 3142
окончательный средний годичный прирост 3979
окорённая древесина 555
окорённая фанера 6135
окорённое бревно 553
окорённое бревно, слегка стёсанное на четыре канта 9266
окорённое мерное кольцо 3850
окорённые лесоматериалы 555
окорённые набело балансы 1609
окорённый 551
окорённый весной 7210
окорённый в лесу 552
окорённый машиной 5231
окорённый на станке 5231
окорка 579
окорка в период движения сока 6136
окорка вручную 4107
окорка всей поверхности 1603
окорка мёрзлой древесины 9458
окорка полосами 2283
окорка спичечного кряжа 5331
окорка трением 3713
окорка трением обрабатываемой древесины 3713

окоркость 536
окорочная линия 567
окорочная лопатка 8639
окорочная машина 2284
окорочная пила 4312
окорочное кольцо 568
окорочно-рубильная машина 2281
окорочные козлы 564
окорочный барабан 566
окорочный барабан для горбыля 7707
окорочный инструмент 574
окорочный скобель 7960
окорочный станок 2284
окорочный станок для чураков 711
окорочный станок с вращением брёвна 7069
окорочный станок трением 3712
окорщик 556
окорщик, работающий вручную 4105
окорять 2280
окотеа красная 6782
окотеа мыльная 2363
окотеа пузырчатая 807
окотеа узамбарская 2710
окраинная рубка 6167
окрайная рубка 6167
окраска брёвен 5118
окраска древесины 9513
окраска древесины от прокладок 2047
окраска от хранения 9755
окраска сердцевины 7994
окраска ядра 7994
окрашенная древесина 1730
окрашенное ядро 1729
окрашивание древесины 9662
окрашивание от ранения 9733
окремнелая древесина 6196
округление кромок 2765
округлый лист 5938
окружать 3851
окружающая среда 4072
окружной лес 1987
окружной лесничий 2551
окружность дерева 8848
окружность на высоте груди 1024
окружность на половине высоты деловой части дерева 5441
окружность резания 2190
окружность ствола дерева на середине высоты 5440
окружность у основания 3855
оксандра ланцетолистная 9347
окулирование 307
окулировать 3906
окулировка 307
окулировка с древесиной 7872
окулировочный нож 1140
окультуривание лесных видов 2559
окулярная оценка 2869
окуривание 3766
окученный коммерческий лесоматериал 6230
окучивание 1171, 5578
окучивание древесины для последующей трелёвки 6519
окучивать 1169
олеандр 5887
олеария 2229
олений заповедник 2330
оленья поляна 2331
олешник 164
олива 2904
оливковая роща 5892
оливковое дерево 2904
ольдфильдия африканская 109
ольс 164
ольха 2882
ольха белая 2878
ольха вырезанная 7677
ольха горная 5579
ольха зелёная 2897
ольха клейкая 2882
ольха красильная 5268
ольха красная 235
ольха мелкопильчатая 7786
ольха морщинистая 4164
ольха непальская 5709
ольха приморская 7329
ольха ромболистная 7587
ольха серая 2878
ольха сердцевидная 4592
ольха чёрная 2882
ольха японская 4609
ольховник 164
ольховое криволесье 8224
ольховый лес 164
ольшаник 164
омела белая 2902
омертвелая древесина 2274
омолаживающая обрезка 4215
омоложение 4187, 6851
омоложение деревьев короткой обрезкой ветвей 2347
онигуруми 7586
опавшие сухие сучья 2265
опавший лист 14
опадение листьев 4866
опадение почечных чешуек 2941
опадение хвои 6260
опасность возникновения пожара 3205
опасность для вальщика сверху 5994
опасность лесного пожара 6971
опасность от повреждения почвы 7837
опасность пожара 3207
оператор окорочного барабана 2642
оператор установки для обрезки досок 2764
операции лесозаготовки 4154
операция заготовки и вывозки 1532
операция заготовки и вывозки щепы 1532
опилки от древоточцев 9728
опилки при работе напильником 3128
опиловка кромок 2766
опилочный компост 7237
опирание в препятствия 5042
описание насаждения 2411
оплата по сдельным расценкам 1200
оплопанакс 2451
оплотник 8779
оползень 7835
опора 2156
опора дерева вешками 8932
опора для дерева 8902
опора для удержания дерева 321
опора, устанавливаемая посередине пролёта 4548
опора штабеля 7975
опорная длина 3989
опорная площадка 276
опорная рама погрузочной площадки 5026
опорная ткань 8274
опорный блок несущего каната 8864
опорный конец балки 640
опорный корень 277
опорный рычаг 321
оправка шлицевой фрезы 7230
определение возраста 131

определение возраста по годичным кольцам 8912
определение всхожести 7389
определение годичной лесосеки 9776
определение диаметра 2436
определение древесных пород 8928
определение запаса 5266
определение запаса древесины на корню 2438
определение запаса насаждения 2438
определение изменений температуры по годичным кольцам деревьев 8910
определение изменения количества осадков по годичным кольцам деревьев 8909
определение кубатуры 8955
определение объёма брёвна 5103
определение объёма взвешиванием 9334
определение объёма дерева 8955
определение объёма лесоматериалов взвешиванием 9334
определение объёма сортиментов без скидки на дефектов 703
определение посевных качеств семян 6571
определение четверти объёма 2118
определение энергии прорастания 7389
определитель организмов 4473
определять наклон дерева 6385
опробковение 8233
опрыскивание крон гербицидами 3305
опрыскиватель для деревьев 8862
опрыскиватель основы дерева 8800
оптимальная возраст срубания 5933
оптимальное состояние организма 4343
оптимальный запас 5934
оптимальный запас насаждения 5934
оптическая таксационная вилка 5930
оптический дендрометр 5931
опунция индийская 6533
опунция черепитчатая 9259
опускающийся грузоподъёмный блок 2993
опускающийся несущий канат 7714
опустошение леса 3373
опушечная посадка 921
опушечное дерево 922
опушечное обсеменение 7384
опушка леса 2761
опыление 6434
опыление леса 3388
опыление листвы 5482
опыливание леса 3388
опыливатель 2675
опылительный аппарат 2675
опытная рубка 7171
опытный лес 2946
опытный лесной участок 2946
опытный участок для измерения прироста 8859
опытный участок для измерения прироста модельных деревьев 8859
оранжерея 3946
организация лесного хозяйства 3574
организация рубок 2201
организация рубок ухода 8516
организм-индикатор 4473
органическое вещество 5948
органическое разложение 5946
органическое топливо 3675
орегонская сосна 1772
орех 1808, 5817, 9262
орех айлантолистный 7586
орех американский 4651
орех аризонский 5009
орех бразильский 1003
орех грецкий 1808
орех Зибольда 7586
орех калифорнийский 241
орех маньчжурский 5272
ореховая древесина 5829
ореховое масло 5822
ореховые 9260
ореховый горный лес 9261
ореховый лес 5818
орехоплодные растения 5823
орехотворки 2210
орех савара 6188
орех серый 1222
орех скальный 5009
орех чёрный калифорнийский 241
орех чёрный 2715
орешек 5819
орешек дубильный 5831
орешек чернильный 5831
орешник 4181
орешник лесной 1778
орешниковая заросль 4181
оригинал 6601
ориентир на густоте насаждения 8108
ориентировочные расчёты 5263
орляк 967
орляк обыкновенный 967
ормозия 5956
орофит 5959
ортет 5960
ортодоксальный посевной материал 5961
ортотропия 5963
ортотропное разветвление 5962
ортштейн 4129
осадки, достигающие подстилки в лесу 2626
осадки, достигающие почвы 2626
осадок 6498
осадок сточной воды 7473
осахаривание древесины 7143
осветительная рубка 5912
осветление 1598, 5912
осветлённый 1613
освободительная рубка по кронам 3698
освобождать вырубкой 6856
освобождение кроны 3698
освоение недорубов 6859
освоенный лес 2956
осевшее назад дерево 7465
оседание 7758
оседание волокон 3094
осенняя древесина 4827
осина 2880
осинник 387
осиновые балансы 385
осиновые дрова 388
осиновый валежник 386
осиновый лес 387
осиновый листоед 389
осколок 7873
ослабленное дерево 9309
ослабленные разъединённые волокна 5159
ослаблять 400

осмоленные древесные частицы 3877
оснастка вывозки 4171
оснастка тыловой мачты 473
оснащать 6939
основание кроны 2064
основание листа 4883
основание ствола 8061
основание сука 4726
основа площадки для разгрузки брёвен 5057
основная верёвка 1643
основная меристема 3992
основная площадь массива 2032
основная просека 5249
основная стрела 5246
основное лесопользование 5253
основное поперечное бревно 945
основной запас насаждения 5250
основной корень 8411
основной побег 4237
основной ярус насаждения 3518
основные лесоматериалы 1156
основные лесохозяйственные мероприятия 612
основные слои ткани ствола дерева в поперечном разрезе 5251
особенно крупномерные строительные брёвна 2975
особенности габитата дикой природы 9414
особо ценное дерево 7892
особо ценные лесные массивы 8024
осоково-злаковый ельник 7368
осокорь 2884
оставление низких пней 5197
оставленная древесина 9277
оставленная семенная полоса 4901
оставляемая часть насаждения 6889
оставляемое на корню дерево 4010
останавливаться в росте 366
остановка в росте 8126
останов, регулирующий глубину сверления 2403
остатки отмерших деревьев 2274
остатки топливной древесины 3221
остаточная влажность 3141
поперечных сечений 6890
остаточная общая площадь поперечных сечений стволов 6890
остаточный объём 9109
остаточный объём древостоя 9109
остающаяся на корню древесина 8651
остающаяся часть насаждения 3145
остепненный лес 4941
остриё 819
островной лес 3517
острокантный брус 7492
остроконечная крона 1866
остроконечный 7917
остроконечный брус 7492
остролепестковый 7494
остролист 2899
осушение 2454
осушение леса 3385
осушительная канава 2613
осушительная установка 2614
осыпание семян 2996
осыпь 8378
ось брёвна 5048
ось круглой пилы 7230
ось ствола дерева 5048
отбелённая древесина 845
отбелённый дуб 4973
отбойное бревно 1025
отбойное бревно на нижней части лесоспускового волока 1025
отбойный брус на спусковом волоке 1025
отборная древесина 9578
отборное насаждение 6394
отборное специальное дерево 6887
отборные пиломатериалы 9578
отборные специальные деревья, оставляемые в порослевом насаждении с двумя оборотами рубки 7358
отборные специальные деревья, оставляемые в порослевом насаждении с одним оборотом рубки 3226
отборный пиломатериал для столярных работ 4658
отборный ствол 6395
отборщик капельных проб 8538
отбрасывание назад 3297
отбросы древесины 2123
отвал 9276
отвал бульдозера для корчевания пней 6986
отвал для корчевания пней 6986
отвальная куча 9276
отвальный плуг 5524
отверстие 923
отверстие для закрепления тяговой цепи 2607
отверстие от выпавшего сучка 4732
отверстие от сучка 4732
отверстие поры 6298
отверстие топора 451
отвесная кромка 6386
ответвление 987
ответственный чиновник 6915
отводная канава 2547
отводок 1095, 2184
отводок с корнями 493
отделение горбильной части брёвен 1125
отделение кроны дерева взрывом 851
отделение лесоуправления 2552
отделение слоёв 8337
отделённая древесина 7448
отделка деревом 880
отделка древесины 9564
отделка окунанием 2511
отделочная древесина 3164
отдельное дерево 4480
отдельно лежащие поваленные деревья 6712
отделят вершину 8702
отдых на природе 5973
отдых на свежем воздухе 5973
отёсывать топором 2161
отечественная порода 4477
отклеймённое дерево 826
отклонение 2400, 4895
отклонение оси брёвна 2450
откомлёвка 1224, 1336
открывающийся отводной блок тягового каната 3298
открытая даль 4504
открытая пророcть 5916
открытое место 5907
открытое хранение щепы 5972
открытый лесной ландшафт 9310
открытый прицеп 5920
открытый прицеп-фургон 5920
открытый склад 5919
открытый сучок 9050
откряжёванная часть 8879

отложная посадка 289
отлуп коры 3243
отлупная трещина 2139
отмирающая крона
с отмирающей кроной 2272
отмершая древесина 2274
отмершее дерево 2271
отмерший сук 9397
отметка 7307
отмечание ствола 4845
отмечать 7308
отмечать дорогу зарубками 825
отмирание 8686
отмирание вершины листьев 8669
отмирание ветвей 2488
отмирание деревьев 2694
отмирание коры 2487
отмирающая древесина 2490
отмирающее дерево 2695
относительная влажность 6854
относительная интенсивность рубок ухода 6855
относительное доминирование 6853
отношение листовой поверхности к общему весу 4893
отношение опушки к площадей леса 2755
отношение площади, занятой деревьями, к общей площади насаждения 2344
отношение углерода к азоту 1333
отношение ширины к длине 3898
отпад коры 580
отпад сучьев 5688
отпад хвои 6260
отпиливание живых ветвей 3953
отпиливать сучья 10
отправная точка регенерации 8040
отпрыск 7132
отпрыск от безвершинника 6431
отпуск леса 6862
отпускной диаметр 2472
отравление деревьев через буревые скважины 928
отравление через буревые скважины 928
отражательная способность 161
отражательная способность
отрезанные заготовки 2177
отрезать 1561, 2172
отрезать дефектную часть комля 4666
отрезок каната 4850
отросток 7202
отросток от корня 8555
отрубание 8824
отрубать 2172
отрубка вершины 2198
отрубки бревен 4738
отсев угольной мелочи 1446
отсев щепы 1514
отсекать 2172
отсечённый 2176
отслаивание 8337
отслаивание коры 3243
отслаиваться 2940
отсортированная щепа 28
отсортированные отходы 6850
отставание в развитии 4681
отставание в росте 4015
отстающее в росте насаждение 7991
отстойная смола 7369
отстойник 8100
отстойный бассейн 8100
отталкивающее средство 6869
отталкивающий насекомых 4514
оттенение опушки 7581
оттенок окраски древесины 9513
отточенный 2756
оттяжка 1164
оттяжной канат 7579
оттяжной трос 467
отходы 1422
отходы древесины 3561
отходы круглого лесоматериала 7118
отходы лесопиления 5215
отходы машинной обработки 5450
отходы насекомых в червоточине 9728
отходы окорки 580
отходы сортирования 6850
отцепщик 4773
отчётливый рост 370
отщеп 2193, 7712
отщепление 1629
отщеп на пне 2193
отщеп от поваленного дерева 2193
охват дерева 8848
охна 5863
охновые 5164
охрана леса 3359
охрана лесного генофонда 3444
охрана лесов от пожаров 3430
охрана природы 2842
охранная рубка 7472
охраняемая залесённая местность 6593
охраняемый лес 6592
охрома заячья 506
оценка видимого влияния 9237
оценка видимого воздействия 9237
оценка водосборных бассейнов 9296
оценка водяных трещин 4044
оценка древесины 9562
оценка древостоя, произведенная из условия вырубки всей коммерческой древесиной 2952
оценка заболони 7217
оценка запаса 9163
оценка лесной почвы 2870
оценка лесосек 8194
оценка опасности от нестабильных грунтов 8478
оценка остаточной стоимости 6891
оценка регенеративной производительности 6823
оценка риска 6972
оцота красная 6782
оцота мыльная 2363
оцота чёрная 2710
очень короткие волокна 3148
очередная поровость 223
очередное листорасположение 221
очередной лист 220
очистительная рубка 7193
очиститель-форвардер для пней 8222
очистка вырубок от порубочных остатков 6799
очистка деревьев от сучьев 6610
очистка для подготовки места подпила 7107
очистка древесины 9511
очистка леса 3351
очистка лесосеки 1624
очистка лесосек разбрасыванием

очистка основания дерева для подготовки места подпила 7107
очистка от коры полосами 2283
очистка от кустарника 1109
очистка от отходов 7723
очистка от порубочных остатков 6799
очистка от пыли 2318
очистка от сучьев 6610
очистка поверхности почвы 3985
очистка разбрасыванием порубочных остатков 1623
очистная рубка 7193
очищать деревья от сучьев 10
очищение от сучьев 6610
очищение семян 2455
очищенная шеллачная смола в виде мелких зёрен 7386
очищенный лесной участок 9096
очищенный от сучьев лесоматериал 2351
ошибка выборки 7177
ощипывание листья и молодые побеги 8806
паротия персидская 6186

"павлиний глаз" 6125
падалица 2266
падение 2994
падение прироста 6832
падуб 2899
падуб блестящий 4795
падуб голый 3790
падуб горный 3927
падуб городчатый 4620
падуб каролинский 2225
падуб Кассине 2225
падуб мутовчатый 1810
падубовые 4335
падуб остролистный 2899
падуб парагвайский 6064
падук африканский 111
паз 5561
пазания густоцветковая 8394
пазовая древесина 3978
пазуха листа 4857
пазушная почка 419
пазушный побег 418
пакет для отгрузки 7526
пакетирование 1171
пакетирование круглых лесоматериалов 3662
пакетировать 1169
пакет лесоматериалов 9496
пакет лесоматериалов для отгрузки 7526
пакет поваленных деревьев 8066
пакет тонкомерной древесины 7780
пакеты из окорённых обрезок 554
палаш 3248
палеозойский лес 6031
палисандр 1006
палисандровое дерево 1006
палка 1165
палкообразная мерная вилка 715
паллета 6034
пал под пологом леса 9054
палубный кряж 2306
палы 7092
пальма 6046
пальма барригона 2109
пальма Вашингтона 9271
пальма веерная европейская 5384
пальма веерная низкая 5384
пальма-веретено 7911
пальма гвинейская масличная 110
пальма капустная американская 1238
пальма карликовая 5384
пальма Манила 1568
пальма Пальмира 6047
пальмароза 3083
пальма хамеропс 5384
пальметта-сабаль 1238
пальмовая роща 6043
пальмовник 6043
пальмовое волокно 1717
пальмовый 6036
пальмовый лес 6043
пальмовый лист 6044
пальчатая форма края листа 2396
пальчатолопастный лист 6039
пальчатонервный лист 6041
пальчатораздельный лист 6040
пальчаторассечённый лист 6042
пальчатосложный лист 6038
пальчатый лист 6037
паляквиум 4061
панга-панга 9336
пандан 1802
пандановые 7316
панданус 1802
папайя 1794
папка ресурсов 6911
папоротник 3087
папоротник лесной 5257
папоротник мужской 5257
папоротниковая заросль 3088
папоротник-орляк 967
парагуттаперчевое дерево 1320
паразит 6071
паразитические грибки 6072
паразитное растение 6073
параллельно проложенный брус 8163
параллельно-слоистый пластик 6068
пара пор 6309
пара пор поля перекрёста 2049
парастемон 9038
паратрахеальная паренхима 6075
парашютист по пожару 7783
паринариум 6079
паркетная древесина 9574
паркинсония 867
паркинсония колючая 4652
паркинсония мелколистная 3312
паркинсония шипатая 4652
паркия 5737
паркия африканская 108
парковое хозяйство 5636
парковый лес 6080
паркоподобное насаждение 6081
парнолистниковые 639
парноперистый лист 13
паровая лесопилка 8055
партеноциссус триостренный 937
партия брёвен 7134
партия вегетативного материала 9180
партия древесины 5181
партия семян 7393
пасанг идянг 7635
пасека 2178
пасечная лента 2178
пасленовые 5734
паслен сладко-горький 773
паслен чёрный 804
пастбищное растение 6719
пастбищный лес 3929
патоген 6112
патологический оборот рубки 6113
патология лесных пород 3521

пауловния войлочная 7122
пауловния империалис 7122
пахать 6383
пахота 8554
пачка деревьев 8066
пачка жердняка 7780
пачка поваленных деревьев 8066
пачка хлыстов 9496
пачковый захват на неповоротной в горизонтальной плоскости стреле 3235
пачковый захват на поворотной в горизонтальной плоскости стреле 8342
пашня оленей зимой 2332
пейзаж 4777
пекан 6130
пек лиственной древесины 4144
пелтогин метельчатый пурпурный 6641
пенёк 8937
пенетрация 6144
пеницилл 802
пеночка-трещотка 9691
пентагональная сердцевина 6145
пень 8937
пень в низкоствольнике 1934
пень для привязки растяжек 4064
пеньковая верёвка 4251
пеньковый канат 4251
пень многолетнего дерева 5882
пень с корневой системой 8211
пень хвойного дерева 8208
пеон древовидный 8895
пеон полукустарниковый 8895
первая грунтовка 472
первая стадия гниения 4494
первенствующее место 4385
первичная гать 6541
первичная древесина 6549
первичная клетка флоэмы 6207
первичная клеточная стенка 6537
первичная кора 6535
первичная ксилема 6549
первичная обработка 6538
первичная сукцессия 6547
первичная съёмка 4497
первичная флоэма 6543
первичная хвоя 6542
первичное насаждение 282
первичное насекомое 6539
первичный лес 1639
первичный луч 6545
первичный транспорт 6548
первобытный лес 282
первое грунтование 472
первое осветление 3227
первоначальное облесение 4493
первоначальный заруб 4495
первообраз 6601
первосортные пиломатериалы 9578
первый лист 6585
первый приём промежуточных рубок 3227
первый приём рубки 668
первый пропил, обеспечивающий оптимальный выход пиломатериала 689
первый пропуск 3228
первый рез 2980
первый ряд погруженных на коник брёвен 1180
перебазировка несущего каната 9751
переворачиватель брёвен 5124
перегной 4396
перегной из коры 572
перегной из листьев 4881
перегон для скота 2624
перегородчатая паренхимная клетка 7449
перегрузочная площадка 8763
передвижение сока 1589
передвижение стадом 8753
передвижная канатная трелёвочно-погрузочная установка 5501
передвижная лесозаготовительная машина 5500
передвижная лесопилка 6469
передвижная обрабатывающая машина 6982
передвижная обрабатывающая сучкорезно-раскряжёвочная машина 6982
передвижная окорочная машина 3112
передвижная рубильная машина 5499
передвижная сучкорезная машина 3113
передвижной лесопогрузчик 5501
переднее нижнее бревно 2983
перекладина 8542, 8739
перекрёстное опыление 2056
перекрёстное субсидирование 2058
перекрывающее возделывание культур 5996
перелесок 5001
перемалывать 3975
переменная нагрузка 5014
переменное выдерживание 9169
переменное задержание 9169
переменное лесопольное хозяйство 143
переменный радиус площади 9168
перемещение груза по несущему канату 7700
перемолоть 3975
переносимая лесопильная рама 6469
переносимый кусторез 6470
перенос пыльцы 6433
переплетение 618
переплетение веток для изгороди 6379
переплетённая посадка 616
переплетённые зарубки 1479
перепончатокрылатые 4413
переправа 8449
переработка дерева 8901
переработка деревьев в щепу 3759
переработка лесоматериалов 1915
переработка целых деревьев в щепу 3759
перерастание 5977
переросший гнилой сук 8352
переруб 2935
переруб расчётной лесосеки 2935
пересадка 3908, 6873
пересаженное растение 5815
пересаживать 8769
пересечение леса при трелёвке 2053
пересечённая лесная местность 7099
пересечённая местность 4212
переспелое насаждение 5998
переспелый лес 5998
перестановка с места на место трелёвочной мачты в вертикальном положении 4668
перестойная древесина 5884

перестойное хозяйство в низкоствольнике 9033
перестойный защитный лес 5989
перестойный лес 5998
перестоять 2307
пересушенная древесина 7735
пересчётная шина 1283
пересчётное число для поленицы 3125
перехват осадков 4535
перехватывающая дрена 2173
переход между сердцевинной древесины и заболони 4204
переходная зона 8765
переходная зона между целиной и города 9410
переходный лес 4537
переходный период 6514
переход от леса к степи 3608
перец американский 8679
перец ветвистый 1207
перец гвоздичный 632
перечёт 2838, 4579
перечёт деревьев 2837
перечёт насаждения 2838
перечёт рекреационных объектов 6756
перечёт стволов 2837
перечислительная таксация 2838
перечное дерево калифорнийское 1266
перешколенное растение 5815
перешколивание 6529
перидерма 6157
периметр пожара 3195
период валки 3072
период вывозки 4170
период естественного возобновления 6820
периодическая рубка 6162
периодическая таксация леса 6760
периодический средний годичный прирост 6164
периодический поток 4555
периодический прирост 6158
периодический текущий прирост 6164
периодическое изреживание насаждения 6166
периодическое лесопользование 6163
периодическое промежуточное пользование лесом 6166
периодная площадь 6160
периодная рубка 6162
периодно-площадной метод 6161
периодный блок 6160
периодный метод определения запаса 6165
периодный метод по площади 6842
период перехода от одной системы к другой 1916
период перехода от одной системы лесного хозяйства к другой 1916
период повторяемости рубок ухода 8509
период приживаемости 2868
период прироста 4006
период развития листьев 3719
период роста 4006
период рубки 3072
период сбрасывания 15
период сжигания 1192
период сокодвижения 7211
период спячки 4818
период цветения 313
перистая пальма 5384
перистая текстура 3031
перистодольчатый лист 6284
перистое дерево 3030
перистолопастный лист 6284
перистонадрезный лист 6284
перистонервный лист 3032
перистораздельный лист 6284
перисторассечённый лист 6284
перистый лист 6282
перистый лист с гребенчатыми долями 6286
перистый лист с зубчатыми долями 6285
перистый рисунок текстуры 3031
периферийный лес 6606
периферическая гниль 7218
пероба 6184
персея 445
персея бурбонская 6763
персея красная 6763
персик 6123
персимон 1796
персимон техасский 8485
персонал на местах 3119
перфорационная пластинка 6154
перфорационный обод 6155
перфорированное волокно 6153
пестик 6296
пестицид 6189
пёстрая гниль 2578
пёстрая гниль корней 9385
пёстрая жёлтая гниль 7908
пёстрая ядровая гниль 5573
пёстрая ядрово-заболонная гниль мёртвой древесины 9762
пёстрая ямчатая гниль 6313
пёстрое окрашивание 5575
пёстро-красная гниль 6265
пёстролистность 9170
петля цепи 1555
печь для сжигания отходов 6807
печь для сжигания твёрдых отходов 6807
печь для сушки 2623
пигеум африканский 112
пигмент древесины 9625
пиджеума африканская 112
пикет 6344
пикразма 6663
пикрасма 6663
пила для валки леса 8854
пила для выпиливания отверстий под замки 4702
пила для вырезания пазов 4936
пила для дерева 9643
пила для кантования брусьев 5148
пила для обрезки ветвей и сучьей 6614
пила для окантовки брёвен 5148
пила для получения карандашной дощечки 7733
пила для резки гипсокартонных листов 883
пила для тонкого пропила 3158
пила для чистовой обработки 5803
пила-дровянка 8854
пила на захвате 3912
пила с возвратно-поступательным движением полотна 2609
пила с дугообразной шиной 3342
пила с косой осью 4312
пила с косыми зубьями 7915
пила с мелкими зубьями 3158
пила с мелкими зубьями для продольной распиловки 4092
пила с обратной резкой 2609

пила с прямым полотном 3251
пила с редкими зубьями 3803
пила с трёхгранными зубьями 6138
пила с цепью, имеющей строгающие Г-образные зубья 1525
пила типа "лисий хвост" 2609
пиленая фанера 7261
пиление 7241
пиление в обратном направлении 474
пиление по сбегу 8407
пилёный лесоматериал 1918
пилильщик 1775
пилильщик виргинской сосны 9230
пилильщик вишневый бледноногий 6388
пилильщик гемлоковый 4250
пилильщик грушевый плодовый 6128
пилильщик еловой зелёноголовый 3945
пилильщик еловый 7958
пилильщик еловый общественный 2910
пилильщики 7240
пилильщик коры 571
пилильщик крыжовниковый бледноногий 6388
пилильщик лиственничный 4792
пилильщик минирующий берёзовый малый 754
пилильщик минирующий буковый 655
пилильщик минирующий ильмовый 2790
пилильщик обыкновенный сосновый 6268
пилильщик пихтовый 502
пилильщик рыжий сосновый 2905
пилильщик сосновый красноголовый 6776
пилильщик сосновый Свейна 8301
пилильщик сосны ладанной 5030
пилильщик сходный сосновый 4575
пилильщик-ткач красноголовый общественный 6253
пилильщик тополевый 4090
пилильщик яблонный 2879
пилка 7241
пиловочник 7245
пиловочное бревно 7245
пиловочный кряж 7245
пиломатериал 1918
пиломатериал, годный по сухости к продаже 1748
пиломатериал для вагоностроения 8592
пиломатериал из внешней части брёвна 7577
пиломатериал из параллельно склеенных волокон 6069
пиломатериал лиственных пород 1047
пиломатериал, обрезанный по сбегу 8402
пиломатериал, предназначенный для дальнейшей переработки 2988
пиломатериал радиального распила 6678
пиломатериал стандартных размеров 2503
пиломатериал, строганный на фальц 7525
пиломатериал тангенциального распила 3254
пиломатериал тангенциальной распиловки 3254
пиломатериал точно квадратного сечения 3755
пиломатериал транспортной влажности 7528
пиломатериал, частично высушенный для предотвращения засинения во время транспортировки 7528
пиломатериалы 1918
пиломатериалы второго сорта 7365
пиломатериалы для производства мебели 3776
пиломатериалы для укрепления тонких стен или перегородок 1437
пиломатериалы короче стандарта 7547
пиломатериалы лиственных пород 1054
пиломатериалы натурального цвета 1035
пиломатериалы по длине дерева 3750
пиломатериалы, поражённые синевой 859
пиломатериалы среднего сорта 5439
пиломатериалы хвойных пород 1875
пилорама 5132
пилоточ 7239
пильная цепь 1430
пильное полотно 820
пильные зубы 7264
пильный клин 7267
пильный меч 4040
пильный надрез 7246
пильный станок 675
пильный ход 8182
пильчато-реснитчатый 7460
пильчатый 7461
пильчатый лист 8681
пиния 4595
пинцирование 6240
пионерная древесная порода 6290
пионовые 6146
пиптадения 6293
пиптадения крупнолистная 6762
пираканта 3212
пирамидальная крона 1866
пирамидальный габитус 6648
писцидиевое дерево 3284
писцидия ярко-красная 4606
питание деревьев 8890
питательная среда 8240
питательная учётная площадь 3309
питательные вещества 5826
питающая площадка 4485
питающий блок 8272
питающий корень 3034
питающийся древесиной 9739
питомник для изгороди 4220
питомник линейной пересадки 4994
питомник саженцев с закрытой корневой системой 1894
питтоспоровые 8672
пихта бальзамическая 969
пихта белая 1803
пихта благородная 5738
пихта болгарская 1154
пихта великая 3910
пихта великолепная 1268
пихта гребенчатая 1803
пихта греческая 3938
пихта европейская 1803

пихта красивая 1268
пихта миловидная 6025
пихта Нордманна 1384
пихта одноцветная 1278
пихта прелестная 1037
пихта пурпурная 6025
пихта религиозная 5422
пихтарник 3224
пихта серебристая 1803
пихта сибирская 7566
пихта субальпийская 214
пихта Фразера 3689
пихтач 3224
пихтовая древесина 7959
пихтовая хвоинка 5707
пихтовая шишка 3165
пихтовник 3224
пихтовые 6254
пихтовый лес 3224
пищевая цепь 3308
пищуха обыкновенная 1807
плавающая балансовая древесина, окружённая брёвнами бона 914
плавающий бон 914
плавающий лесотранспортёр 3274
плавник 3279
плавун 8096
плавучая цепь 8136
плавучий вал 3035
плавучий мост 6686
плагиотропия 6317
плагиотропное разветвление 6316
плакучая берёза 9789
плакучая крона 2632
плакучая форма роста 9326
плакучесть ветвей 9323
плакучий 9324
пламенистый узор 3246
пламеобразная текстура 3246
"пламя леса" 3248
план деятельности лесничества 3492
планера водная 6320
планиметр 6324
планирование борьбы с лесным пожаром 3175
планирование деятельности лесничества 3493
планирование лесных ресурсов 3493
планированная система пастьбы 6330
планированное зажигание 6516
план использования местных ресурсов 5036
план использования участка 6722
планка 626
план лесного хозяйства 3492
план лесонасаждений 3492
план лесопользования 3482
план лесоразработки 3375
план насаждения 8109
план облесения 3529
плановая годичная лесосека 192
плановая рубка 6329
плановый запас леса на корню 8413
планомерная рубка 6329
план пастьбы 3931
план развития лесного хозяйства 3492
план рубок 6332
план рубок ухода 8518
плантация камедных деревьев 4057
плантация по лесоразведению 8804
плантация элитных и плюсовых деревьев 5570
план тушения 6331
план управления землевладением 8468
план эксплуатации леса 3416
пластевая трещина 8279
пластинка листа 4872
пластиновидный корень 6378
пластинчатый корень 6378
пластификация древесины 9626
пластифицированная древесина 1844
пластичная древесина 2776
пластичность древесины 2016
пласт коры, снятый со ствола дерева 575
пластовый сучок 2981
пласть пиломатериала 7258
платан восточный 5952
платан западный 254
платан испанский 7879
платан кистистый 1275
платан клёнолистный 2906
платановые 6321
платановый лес 6322
платан Райта 361
платимисциум 5225
платформа 2303
плаун булавовидный 8815
плесневые грибы 5525
плита 9492
плита из древесной массы 9604
плита из коры 575
плита из ориентированной пряди 5954
плод дерева 8844
плод дикорастущего дерева 9427
плод древесных растений 8844
плодовое дерево 3731
плодовое тело 1353
плодово-ягодные плантации 6169
плодокорм 5326
плодоносящий побег 3730
плоская крона 3252
плоскостная эрозия 7506
плоскостной посев 3257
плоскость пропила 2180
плотина 2230
плотина для задержания огня 3186
плотина для сплавки брёвен 5039
плотина из брёвен 4377
плотная древесина 8659
плотная кубомасса 7842
плотная укладка 3259
плотная штабелёвка 1157
плотничий топор 673
плотничный бурав 9538
плотничный топор 1041
плотничья обработка 9698
плотно сидящий сучок 8550
плотно сросшийся со стволом сучок 833
плотность абсолютно сухой древесины 18
плотность воздушно-сухой древесины 151
плотность древесинного вещества 9668
плотность древесины 1818
плотность популяции 6458
плотность свежесрубленной древесины 3707
плотность укладки 7980
плотный здоровый сучок 7852
плотный кубометр 2114
плотный объём 7842
плотный сучок 8550
плотный штабель 8552
плотовой лесосплав 8578
плотовой сплав 3273
плотогон 9569

плохое лесозаготовительное оборудование 4178
плохой рост 6449
плохо облесённый 6450
площадка для разгрузки брёвен 5451
площадка для распиловки 7233
площадка на крутом склоне 7004
площадная вырубка 1820
площадь боковой поверхности ствола 900
площадь возобновления 355
площадь воспламенения 4422
площадь вывозки 2629
площадь вырубки 2189
площадь габитата дикой природы 9413
площадь-доставчик древесины 9650
площадь древесного пояса 8802
площадь культуры 355
площадь кустарника 1108
площадь лесного массива 2189
площадь лесосеки 2189
площадь лесохозяйства 3330
площадь листовой поверхности 4854
площадь одновременного пуска нескольких управляемых палов 4422
площадь, осваиваемая канатной установкой с одной стоянки 7468
площадь питания дерева 4008
площадь поверхности почвы, занимаемой стволом 597
площадь подпила 9055
площадь под пологом 354
площадь пожара 1184
площадь полога 354
площадь пользования 2189
площадь поперечного сечения ствола на высоте груди 598
площадь проекции полога 1314
площадь, пройденная сплошным палом 1184
площадь рубок ухода 8507
площадь с выборочной маркировкой 7415
площадь с выборочной рубкой 7406
площадь сечения ствола на высоте груди 598
площадь со сплошной рубкой 1621
плуг для выкопки саженцев 8771
плуг для глубокой вспашки 8963
плуг для тушения лесных пожаров 3196
плуг-корчеватель с индивидуальными предохранителями корпусов для автоматического возвращения в рабочее положение 8206
плуг, переезжающий через пни 8206
плуг-рыхлитель с черенковым ножом 8668
плюмбаговые 7328
плюска 46
плюсовое дерево 6396
плюсовое насаждение 6394
плюсовые семена 7403
плюсовый лесосеменной участок 7394
плюсовый ствол 6395
плюхея 367
плющ обыкновенный 4598
пнёвая древесина 8220
пнёвая поросль 1933
пнёвая щепа 8198
пневматофор 86
пнёвой козырёк 2193
пнёвой кряж 1225
пнёвой побег 1933
пнёвый зажим 8203
пнёвый лесоматериал 8220
пнёвый осмол 6899
побег 1095, 6366, 7132
побеги бамбука 511
побеги привоя 7299
побеги текущего года 4813
побеговая почка 9179
побеговой черенок 7010
побеговый рубец 9017
побеговьюн 6269
побеговьюн зимующий 6269
побег от пня 1933
побег от спящей почки 2848
побег, периодически дающий отростки 6761
побег, развившийся из спящей почки 2848
побочное пользование лесом 4159
побочные продукты 9500
побочный продукт леса 5471
побурение 1549
поваленный лес 3042
поверхностная влажность 8283
поверхностная гниль 8269
поверхностная корневая система 7487
поверхностная трещина 8279
поверхностная эрозия 7506
поверхностное внесение удобрений 8685
поверхностно стелющийся корень 7486
поверхностные грибные заболонные окраски 8278
поверхностный корень 7486
поверхностный огонь 8282
поверхность древесины 8650
поверхность листа 4853
поверхность ножевого среза 7499
поверхность пропила 7247
поверхность с открытыми порами 9083
повиснувшее при рубке дерево 5041
повиснуть 4116
поворачивание 7001
поворачивание зависшего дерева с помощью рычага 7001
поворина 8340
поворот из стороны в сторону 8341
поворотная платформа погрузчика 5232
поворотное устройство 4311
поворотный крюк 1317
поворотный ролик с пропуском троса 7796
поворотный рычаг для удержания брёвен 4311
повреждение ветром 2236
повреждение, вызываемое древесинником 232
повреждение, вызываемое личинками древоточца 9703
повреждение деревьев при трелёвке 7685
повреждение древесины 2242
повреждение комля 1216
повреждение коры 582
повреждение кроны 1012
повреждение леса 2241
повреждение лесной подстилки 3436
повреждение летним морозом 8259

повреждение неустойчивого грунта 2439
повреждение от бури 2236
повреждение от валки 3059
повреждение от дыма 2240
повреждение от инея 2235
повреждение от лесного пожара 3177
повреждение от окорки 2237
повреждение от сжатия 1847
повреждение от скота 3930
повреждение от хранения 8127
повреждение пастьбой 3930
повреждение подроста 9061
повреждение почвы 7836
повреждение почвы и древостоя на лесосеке 7665
повреждение при валке 3052
повреждение при трелёвке 7685
повреждение, причиненное дичью 2233
повреждения птицами 765
повреждённая древесина 2238
поврежденное грибом дерево 8860
повреждённое дерево 2239
повреждённость частиц щепы 1536
повреждённый лес 7167
повторение 6848
повторная посадка 1823
повторная рубка 6870
повторная таксация леса 6760
повторное облесение 6821
повторно облесённый 6806
повторно распиленный 6881
повторность 6875
повторные лесозаготовки 6859
повторные лесопосадки 6821
повторный пал 6747
повторный перечёт 6847
повторяемость рубок ухода 8510
повторяющийся комплекс 6849
повышение жёсткости 8097
погибшее насаждение 2267
погибшие лесные культуры 2992
поглощающий корень 20
погодные условия, влияющие на лесозаготовки 7331
погодоустойчивое дерево 9312
пограничное дерево 641
пограничное сообщество 2751
пограничные леса 3721
погружение в антисептик 2511
погрузка брёвен накатыванием 2052
погрузка древесины из запаса 1722
погрузка леса 8608
погрузка лесоматериалов 8608
погрузка накатыванием 2052
погрузка подтрелёванной древесины без её складирования 4383
погрузка складированной древесины 1722
погрузка сортиментов 5099
погрузка с помощью стрелы 913
погрузка хлыстов 8871
погрузочная лебёдка 8609
погрузочная мачта с растяжками 3848
погрузочная метка 7529
погрузочная платформа 5027
погрузочная площадка 5027
погрузочная площадка при канатной трелёвки полуподвесным способом 4289
погрузочная площадка с отгрузкой подтрелёванной древесины без складирования 4382
погрузочная стрела 5025
погрузочная стрела с упором 4225
погрузочная эстакада 5027
погрузочное устройство пиломатериалов 5207
погрузочный захват для лесоматериалов 8610
погрузочный крюк 6225
погрузочный обмер 4522
погрузочный пункт 5027
погрузочный строп 7742
погрузчик 5062
погрузчик в виде мачты с горизонтальной стрелой 705
погрузчик круглого леса 8098
погрузчик пиломатериалов 5207
подача брёвна 5061
подача брёвна на распиловку 5061
подача с открытом тросом 4166
подбел дубровник 897
подбел многолистный 897
подборщик сучьев 4966
подвесная канатная трелёвочная установка 88
подвесная канатная установка 88
подвесная канатная установка для гравитационного спуска древесины 89
подвесная канатная установка с двумя несущими канатами 2586
подвесная струна 1564
подвесная трелёвка 7700
подвесная установка с несущим и двумя тяговыми замкнутыми канатами 8738
подвид 8239
подвижная эстакада 5011
подвижный блок 2993
подвижный упор на стреле 5012
подвозная дорога 7689
подвозчик баланса 842
подвой 9420
подгнившее дерево 8372
подгоревшие частицы щепы 1185
подготовительная культивация 6510
подготовительная рубка 6513
подготовительный период 6514
подготовить ложе для дерева 645
подготовка добывания живицы 2140
подготовка доступа 5913
подготовка древесного сырья 9628
поддерживающий корень 277
поделочная древесина 4658
поделочные дрова 4483
поделочный кругляк 7113
поделочный лес 4658
поделочный пиломатериал 4658
поделочный хворост 8521
подеревное смешение пород 5496
поджигать 3166
поджог 4436
поджог леса 4436
подзолисто-болотная почва 896
подзолистые почвы 6408
подзолы 6408
подкатывать брёвна по слегам 8369
подкладка 6034
подкладка из брёвна для сбора и транспортировки

подкладки при раскряжёвке 1131
подкормка 482
подлесок 890, 9070
подмаренник лесной 9488
подмаренниковые 5238
подмостки для валки крупных деревьев 1563
подневольно-выборочная рубка 5855
подновка 543
подокарп высокий 6390
подокарпус серповидный 5974
подокарпус тотара 8734
подорожное дерево 6983
подпил 1214
подпил в пнёвой части 4392
подпологовая культура 8235
подпологовый саженец 9063
подпора 1985
подпорка 1985
подпорка для дерева 8902
подпорка-контрупор 1985
подпорный столб 8273
подпорченное дерево 8372
подпочвенный дренаж 8241
подпочвенный плуг 8963
подпоясывать 3851
подрастающий лес 9070
подрезание 5798
подрезание дерева с обеих сторон от подпила 7578
подрезание заболони вокруг дерева 7578
подрезанное дерево 7489
подрезанный сеянец 9056
подрезать живых ветвей 6609
подрезать торец 7962
подрезка 6610
подрезка боковых корней 7055
подрезка вершин 8696
подрезка корней 7040
подрезка кроны 2083
подрезка сучьев 6610
подрезная пила 7309
подрезывать 2165
подрост 8555, 9060
подруб 1214
подрубать 669
подрубка 1214
подрывник 6485
подрядчик, ведущий лесозаготовки по контракту 5072
подсанки для трелёвки 5115
подсачиваемое дерево 959
подсевная культура 1374
подсемядольное колено 4415
подсечка 1214
подсечное дерево 5165
подсечное хозяйство 3176
подсечно-огневое земледелие 7717
подсеять 4561
подслойная заболонная окраска 4538
подсочка 828
подсочно-кобловое хозяйство 6430
подстилка из сосновой хвои 6270
подстилка от ветвей 5000
подстопное место 6034
подстрахованное семенное дерево 3189
подстриженное дерево 7489
подсчитывать результаты вырубки леса 1458
подтаивание снега 7802
подтаскивание 4169
подтаскивание сбоку 4821
подтаскивать лесоматериалы с помощью каната 1241
подтопление 9291
подтрелёвка 4169
подтрелёвка на короткие расстояния 4558
подтрелёвка со стороны 4821
подушка под опиливаемое бревно 3129
подчинённая древесная порода 66
подчинённое насаждение 8237
подчинённый 6011
подштабельная опора 6424
подштабельное место 5451
подштабельные деревянные подкладки 6424
подштабельные подкладки 6424
подъём брёвен накатыванием 2052
подъём накатыванием 2052
подъёмник для сучьев 4730
подъёмные клещи хлыстов 4934
подъёмный канат 4323
подъёмный конвейер для лесоматериалов 8581
подъёмный цилиндр 4932
поедаемость 6032
поедающий древесину 9739
пожар в беспламенной стадии горения 7792
пожарной безопасности 53
пожар вследствие поджога 7466
пожар в целине 9409
пожар из-за пуска пала 7466
пожарище 1184
пожар крон 2076
пожар кустарника 1105
пожарная опасность 3205
пожарная отметина 3168
пожарная пропость 3206
пожарная устойчивость 3213
пожарник 3171
пожарник-наблюдатель 3191
пожарное кольцо 3204
пожарные грабли 3200
пожарный блок 1188
пожарный наблюдатель 3191
пожарный надзиратель 3191
пожарный период 1192
пожарный район 3179
пожарный режим смешанной опасности 5492
пожарный рубец 3206
пожаро-зависящий 3178
пожароопасное дерево 4179
пожароопасные горючие материалы 4180
пожароопасные материалы 422
пожнивная культура 1374
позднее прореживание 4826
поздний 7458
поздника 804
поздноцветущий 7458
поздняя древесина 4827
поздняя кора 4815
поздняя прорубка 4826
поздняя смена лесной растительности 4817
поздняя шишка 7459
позитивная селекция 6472
пойма 5192
пойменный лес 5192
показатели производительности леса 4475
показатели состава насаждения 8031
показатель выживаемости 8293
показатель интенсивности рубки 2954
показатель лесного местообитания 8641
показательный лес 6239
показатель площади проекции кроны 2063
показатель полноты насаждения 2344

показатель роста 4018
показатель теплотворной способности 1194
покаты для закатки брёвен 7693
покинутое насаждение 2406
покой 2572
покой лесосеки 3418
покоробленность древесины 9268
покров 1992
покров из хвои 5706
покровное бревно 5810
покровное насаждение 1993
покровный слой 1695
покровный хворост 1995
покровосдиратель 7313
покровосдиратель площадки 6108
покрываться коркой 2106
покрываться листьями 7531
покрытая лесом земля 3480
покрытое лесом пастбище 9711
покрытосемянные 284
покрытый иглами 7913
покрытый колючками 7913
покрытый корой 1994
покрытый кустарником 934
покрытый лёгким пушком 6619
покрытый лесом 3394
покрытый лесом с низким запасом 6450
покрытый листьями 4894
покрытый шипами 7913
покупатель леса 8564
покупка леса 9635
пол 3283
полевая влагоёмкость 3110
полевая всхожесть 3114
полевое лесоразведение 3109
полевой перечёт 3120
полевой питомник 3117
полезащитная лесная полоса 7508
полезащитное лесоразведение 7509
полезащитный лес 3118
поленика 350
поленница дров 9519
полено 2171
поле перекрёста 6738
поле под паром 2997
полесье 3397
ползающий огонь 2012
ползучая магония 2013
ползучее растение 1642
ползучий 2313
поликарпический 6436
полиморфизм 6439
полиплоид 6440
полиподиум 6441
полирование 7189
полировать 6133
полиспаст для натяжения несущего каната 4227
политрихум 4087
политрихум обыкновенный 4087
полная густота древостоя 3757
полная окорка 1603
полная почка 2831
полная тросовая линия 1244
полная усушка 8730
полновесные семена 3753
полнодревесность 8816
полнодревесный 3745
полное гашение и очистка пожарища 5554
полное насаждение 3756
полное определение кубатуры 3747
полное разбухание 8732
полное цветение 313
полнокантный пиломатериал 3755
полнопокрытая лесом площадь 3763
полностью или частично затонувшее бревно 887
полнота 3370
полнота древостоя 3370
полнота лесонасаждения 3370
полнота насаждения 3370
полный выход леса 4523
полный размер 3748
полный урожай плодокорма 3751
половина кряжа, распилённого вдоль 4099
половица 2278
полог 3345
полог дерева 8810
полог крон 1650
полог леса 3345
полог насаждения 3345
полог низкоствольника 5190
полог спелого насаждения 5343
полое долото 4328
положение дерева, лежащего поперёк крутого склона 7570
поломка вершин 1013
поломка вершины 1012
поломка ствола 8064
полоса деревьев 8177
полоса лиственных деревьев 4991
полоса отвода 6941
полоса сплошь вырубленного леса 2178
полосатая текстура 8175
полоска коры 537
полосная выборочная рубка 8168
полосное возобновление 6813
полосно-постепенная рубка 8178
полосные рубки ухода 4992
полосный посев 522
полостная сердцевина 2933
полость 5028
полость в коре 576
полость в растении 2557
полость лопасти 5028
полость поры 6301
полость со смолой 6893
полосчатая древесина 6935
полосчатая текстура 8175
полотно для расчистки территории от зарослей кустарника 1099
полотно пилы 820
полувечнозелёный влажный широколистный лес 7426
полуестественный лес 7429
полуквартал 8231
полукольцевая пора с ~ми 4091
полукольцевая пористость 7431
полукольцесосудистая древесина 7432
полукруглая шпала 4095
полукруглое бревно 4099
полукруглое долото 4328
полукруглый лесоматериал 4094
полукустарник 9065
полукустарниковый 8252
полукустарничек 2690
полулистопадный лес 7424
полуобрезной пиломатериал 5899
полуподвесная канатная трелёвка 4288
полуподвесная канатная трелёвочная установка 4291
полуподвесная трелёвка 4288
полусибс 4097
полутвёрдая

полутеневыносливая порода 7434
полутопляк 9290
полуцилиндрическая лопата 6342
получение воздушных отводков от стебля 154
полученный на круглопильном станке пиломатериал 1587
получистая доска 7572
полуштамб 4098
полын-божье дерево 5883
полынь австрийская 416
полынь калифорнийская 1270
полынь лечебная 5883
полынь трёхзубчатая 704
пользование лесной травой 3914
пользование подстилкой 9153
поляна 5349
померанец 775
померанец дерево 775
пометка зарубками 5317
помеченное дерево 826
помощник лесничего 392
пониженное использование 9064
понцирус трёхлисточковый 4146
попарно равные долготья 2925
попённая плата 8195
поперёк волокон 49
поперёк годичных колец 49
поперечина 2038, 2051
поперечная балка 2038
поперечная вязка под косым углом 5857
поперечная доска 7732
поперечная закрытая канава 2046
поперечная накатка 2052
поперечная пила 2042
поперечная просека 2057
поперечная трещина 2039
поперечная черновая рубашка 2040
поперечная штабелёвка 2055
поперечная эстакада 2044
поперечное бревно 1548
поперечное вращение 2804
поперечное вращение торцевого волокна 2804
поперечное коробление 2141
поперечное сечение ствола 2980
поперечно-пильный станок
поперечные размеры 2045
поперечный брус 2038
поперечный внутренний слой 2040
поперечный замок внакладку 8140
поперечный ленточный конвейер 5055
поперечный лесотранспортёр для брёвен 5055
поперечный разрез 8571
поперечный разрыв 2039
поперечный распил 8571
поперечный распил древесины 8571
поперечный рез 8571
поперечный топор 9026
поплавок 914
пора 6461
поражение синевой 860
поражённая грибками древесина 1880
поражённая огнём древесина 1196
поражённое дерево, покинутое насекомыми 3
поражённый гнилью 7212
поражённый червем 9201
поранение 4918
пористая древесина 6467
пористая древесноволокнистая плита 6466
пористая сердцевина 1440
пористость 6465
пористый объём 6463
пористый средний слой 5905
пористый средний слой плиты 5905
поровая вода 6464
поровая камера 6302
поровое поле 6305
поровость поля перекрёста 2050
порода 7893
порода дичи 3797
порода-пионер 6290
породный древесный состав 8927
породы 7893
порожный трос 7714
порозное пространство 6463
порозность 6465
порок дерева 8830
порок окраски 1728
порок роста 4014
порок строения 8187
порослевая изгородь 1930
порослевое дерево 7946
порослевое растение 8209
порослевый древостой 1928
порослевый лес 1928
порослевый побег 1933
порослевый сеянец 7202
поросль карликового дуба 1442
поросль от пня 1933
порошкообразный древесный уголь 6633
порубочные остатки 998
порубочные остатки вместе с валежником и ветровалом 3996
порубочные остатки лиственных пород 4145
порубочные остатки мягколиственных пород 7831
порубочные остатки, собранные в кучи 6231
порядок вырубки 3061
порядок лесопользования 3649
порядок рубки 2199
посадка 6339
посадка в борозду 8964
посадка в горшочках 6482
посадка в гребень пласта 648
посадка в дно борозду 8964
посадка в лунку 6355
посадка в открытое место 5908
посадка в площадки биогруппами 6107
посадка в пулеобразных трубках 1163
посадка в трубках 9002
посадка в тубах 9002
посадка в шахматном порядке 6668
посадка в щель 8964
посадка в ямки 6355
посадка глыбой 490
посадка грядками 648
посадка деревьев 3626
посадка деревьев за пределами леса 5759
посадка единичных деревьев 6347
посадка защитных насаждений 6595
посадка леса 94
посадка леса саженцами 3624
посадка между валами 9449
посадка между рядами порубочных остатков 9449
посадка на пазуху пня 6348

посадка площадками 6107
посадка по горизонталям 1906
посадка под бур 402
посадка под кол 2485
посадка под лопату в щель L-образного профиля 5021
посадка под пологом леса 6364
посадка пучками 6354
посадка рядами 6356
посадка саженцев 7203
посадка сеянцев с корнями, заделанными в ком почвы 490
посадка с защемлением 289
посадка с комом земли 490
посадка с корнями, заделанными в ком почвы 490
посадка с необнажённой корневой системой 1895
посадка с обнажённой корневой системой 532
посадка с обнажённым корнем 532
посадка с примесью сопутствующих пород 5491
посадочная борозда 7379
посадочная грядка 5812
посадочная доска 6345
посадочная лунка 6353
посадочная трубка 6361
посадочное ружьё 6351
посадочный брикет 6346
посадочный горшочек 6359
посадочный корнеплод 8056
посадочный меч 8898
посаженный лес 6339
посев 4506, 7382
посев в борозды 7869
посев в контейнеров для выращивания саженцев 963
посев в междурядьях 4567
посев вразброс 1044
посев в ямки 6355
посевная банка 7372
посевная бороздка 7379
посевная гряда 7373
посевная полоса 7379
посевная рейка 7385
посевное место 7373
посевное отделение 7373
посевной блок 7375
посевной жёлоб 7379
посев площадками 7870
посев по дернине 4567
посев под пологом леса 6364
посёлок лесозаготовителей 5071
посёлок с домиками-прицепами 9356
посерение 3925
посеять 7371
посеять в междурядье 4561
посинение 860
посконник душистый 866
последняя рубка 4814
последовательная косьба 8247
последовательность 8246
последовательность годичных кольцов 7451
последовательность рубки 2202
последующее возобновление 6811
последующее возобновление леса 6811
послепожарный климакс 3172
послеуборочное дозревание 122
поставка леса 9631
поставщик услуг в лесном хозяйстве 3578
постепенная рубка 3904
постепенное загнивание под действием атмосферы 2859
постоянная пробная площадь 6176
постоянная точка 674
постоянное пользование лесом 8298
постоянный лесной рабочий 6173
постоянный противопожарный барьер из лиственных пород 6174
построенный из дерева 8580
поступательная рубка 6573
поступление древесины 9591
потёк 8155
потенциал доминирования 2562
потенциал производительности леса 3648
потенциальная высота деревьев местообитания 7669
потенциальная производительность леса 6481
потери древесины 9598
потери от валки 2197
потери от расщепления 1072
потери при окорке 580
потери при пилении 4696
потери при рубке 2197
потеря воды на испарение
потеря древесной массы 5180
потеря листьев 4866
потеря на пропил 5178
потеря от молевого лесосплава 5177
потеря от распиловки 5178
потеря питательных веществ 5179
потеря при вывозке 5174
потеря при трелёвке 5174
поток генов 3814
поток лесоматериалов 8645
поток сока 1589
потомство 6570
потопление 7660
потрава дичью 2233
потребитель древесины 8655
потребление лесоматериалов 9517
потребности близости лесосек 2167
потребности в средствах управления леса 3642
потребность в древесине 2362
потребность в дровах 3738
потребность в свете 4573
похититель леса 8051
похищение леса 8052
походный шелкопряд 6557
похутукава 5728
почва 7833
почва под лесом 9547
почвенная влага 7838
почвенная флора 3986
почвенная фреза 7072
почвенные характеристики 2754
почвенный огонь 2012
почвенный покров 6337
почвенный слой 5947
почвозащитная культура 1374
почвозащитная порода 7839
почвозащитное облесение 1881
почерненная кора клёна 7849
почечная чешуйка 1143
почечная чешуя 968
почка 1141, 4859
почка финиковой пальмы 1239
почкование 307
почковаться 7944
почковидный 4737
поштучная заготовка деревьев 7655
поштучная трелёвка дерева 8900
поштучный метод измерения

появление отпрысков от корней 7437
пояс 1564
пояс зелёных насаждений 3941
пояс лесов на уровне стояния 1663
поясность 9798
правила лесозаготовки 4161
правила отпуска леса 3649
правила отпуска рубки леса 3649
правила рубок 6844
правила рубок главного лесопользования 6844
правила сплава 3276
правило вывозки 5081
правило определения плотности 2395
правильное дерево 687
правильно обработанное бревно 8989
право выпаса свиней в лесу 6055
право на бесплатный лес 3693
право рубить 2955
право рубить лес 2955
право сплава 6945
право эксплуатации 2955
правый кособлой 6943
практика лесоводства 3576
практическое обучение в лесу 9239
превращение в пустыню 2414
превращение древесины в волокнистую массу 9633
превращение степи в пустыню 2414
превышение расчётного объёма выхода 6002
превышение расчётного объёма выхода древесины 6002
прегосподствующее дерево 6500
предварительная валка 6513
предварительная зацепка брёвен клещевым захватом 6525
предварительная культура 71
предварительная недоразвитость 6509
предварительная обработка древесины 9628
предварительная обрезка 76
предварительная обрезка сучьев 76
предварительная рубка 4544
предварительное возобновление 73
предварительное естественное возобновление 73
предварительное окученные деревья 6494
предварительное окучивание 6495
предварительное охлаждение 6496
предварительное пакетирование 6495
предварительное прореживание 2599
предварительное удаление сучьев 6734
предварительное уложенные в пачки деревья 6494
предварительно нагружать 6511
предварительный доход 4546
предвсходовое состояние 6504
предел товарности 4981
предельная интенсивность рубок ухода 5305
предклимакс 8230
предоставление льготы лесного хозяйства 3569
предохранение от синевы 6594
предохранение природы 2842
предохранение ствола 8997
предохранитель деревьев 8904
предохранительный брус 2983
предохранительный настил 2863
предписание 6517
предписание для обработки 8794
предписание сжигания 1195
предпочтительный вид 6501
предприниматель по вывозке 8597
предприниматель рубки 5072
предприятия для первичной обработки древесины 6550
предпромышленные операции 6499
предпромышленные рубки ухода 4680
предтундровое редколесье 5904
предшествующая почвенная влага 310
предыдущее насаждение 6527
прежнее насаждение 6527
прелая древесина 2291
прерванное годичное кольцо 2528
прерывающийся поток 4555
прерывисто-перистый лист 4564
прерывистый рост 4553
пресноводный болотистый лес 3711
пресс для вязки фашин 3019
прессованная древесина 2390
прецизионное строгание 1608
прибережная охранная зона 6962
прибережные деревья 8158
прибережные лесоматериалы 7324
прибережный лес 1687
прибор-пеленгатор пожара 3184
прививать 3906
прививка 3908
прививка в боковой отщеп 7575
прививка в корень 7026
прививка в крону 8691
прививка в открытом грунте 3115
прививка в расщеп 1635
прививка в расщеплённую конечную почку 8691
прививка в щель 1635
прививка глазком 307
прививка за кору 561
прививка за кроной 2081
прививка клином 1635
прививка мостиком 1030
прививка под кору 561
прививка сближением 331
прививка седлом 7147
прививка сердцевиной на камбий 6307
прививка черенком 331
прививок 1145
прививочный материал 4507
прививочный черенок 1145
привитая часть 1145
привлечение общественности 6623
привой 1145
привязанность к лесу 3507
пригодная к сплаву вода 8736
пригодная лесная земля 8257
пригодность древесины к обработке 5228
пригодность лесоматериалов к эксплуатации 8617
пригодный для сплава 2628
пригодный к сплаву 2628

приготовление посадочных ямок 6508
придавленный саженец 5040
придание клиновидной формы 8403
придаточная почка 78
придаточная структура 81
придаточный 77
придаточный корень 79
придаточный побег 80
придорожное дерево 6983
придорожные деревья 9308
приёмка лесосеки 6831
приёмная площадка для кратковременного хранения 5011
приёмщик 1457
приёмщик, производящий контрольный обмер леса 1457
приживаемость насаждения 8014
прижим 4653
прижим лесоматериала 9510
приземистое дерево 7323
приземление и разбрасывание 5166
призма дороги 6980
призматическая гниль 2112
призматический прямой вруб 955
приисковая рубка 7412
прикапывание 4226
прикармливание 482
прикопка 4226
прикорневая древесина 8220
прикорневая часть ствола 607
прикорневой кап 7015
прикорневой лист 944
прикрепление корнями 7029
прикреплённый измельчитель пней 399
прилистник 8101
прилистниковый рубец 8102
приманка 4509
применение нескольких методов очистки от отходов 2546
применение у основания дерева 603
применение ядохимикатов у основания дерева 603
примесь древесных пород 5497
примесь широколиственной породы 4143
приморская сосна 5309
лес 1690
принадлежащий фермеру участок леса 3419
приносящий поздние плоды 7458
принудительная рубка 1852
приовражный лес 3474
приостановление роста 8126
приостановленный рост 7990
припойменный лес 5192
припуск длине торцовки 202
припуск на кору 200
припуск на торцовку 202
приречная территория 8157
приречный склад 6974
природа леса 3512
природная камфара 5674
природная катастрофа 5676
природная стойкость 5678
природноподобное лесное хозяйство 3571
природоохранное лесное хозяйство 2839
прирост 4449
прирост в высоту 326
прирост видовых высот 3665
прирост в толщину 4019
прирост дерева 8858
прирост количества деревьев, достигших диаметра рубки за определённый период 4492
прирост леса 8017
прирост лесозаготовки 4155
прирост насаждения 8017
прирост насаждения за счёт естественного возобновления леса 4492
приростно-буравое ядро 4452
приростное кольцо 301
приростное лесоводство 4450
приростной бурав 4451
приростной молоток 4454
прирост по диаметру 2474
прирост по массе 9243
прирост по объёму 9243
прирост по окружности 3856
прирост по площади сечения 600
прирост по стоимости 4455
прирост по ценности 4455
приросший 67
приспевающее насаждение 4425
приспособление для натяжения цепи 1436
лесоматериалов 9512
приспособленный к месту лес 7662
приспособляемость 60
приствольный сток атмосферных осадков 8073
притундровый лес 3631
приуроченный к месту лес 4476
прицветник 968
прицветничек 6585
прицеп для длинномерных грузов 9395
прицеп для круглого леса 8752
прицеп для пней 8218
прицеплять 4358
прицепная древесина 2618
прицепная обойма 1232
прицепная рубильная машина 8751
прицепная трелёвочная арка 5090
прицепное кольцо 2320
прицепное приспособление для канатной трелёвки 9743
прицепной грузосборный канат 7804
прицепной канат 4115
прицепной лесозаготовительный комбайн 8746
прицеп-роспуск 9395
причудливая текстура 835
пробег лесоспуска 8471
пробег топора 452
пробка 1956
пробковая древесина 507
пробковая мука 1962
пробковая ткань 1975
пробковый лес 1959
пробковый слой 1960
проблемный тип леса 6556
проблемы леса 3479
пробная площадь 7173
пробная полоса 8479
пробная рубка 7171
пробное насаждение 2946
пробный угловой учёт 285
проведение троса 7057
провентивная почка 6605
проверка в полевых условиях 3111
проверка всхожести 3837
проверка на объекте 3111
проверка потомства 6571
проверка состояния деревьев 8940

провислый 9324
проводящая клетка 8742
проводящая паренхима 1857
проводящая ткань 1857
проводящий корень 1856
проволочная пила 9466
проволочный канат 9463
проволочный канат для гравитационного спуска лесоматериалов с гор 9464
проволочный канат для спуска лесоматериалов с гор 9464
прогалина 3443
прогнивший 7212
прогнозирование лесозаготовки 4153
программа восстановительной консервации сильноэродируемых земель 1882
программа лесопользования 3376
программа мелких лесных предприятий 7759
программа развития мелких лесных предприятий 7759
программа рационального использования леса 3611
прогрессирующая кайма 6574
продажа древесины 8631
продажа леса 8631
продажа леса на корню 7157
продажа леса на корню по определённой цене за весь участок 5217
продажа леса на корню по цене за единицу объёма фактически заготовленной древесины 5112
продажа на месте валки 7157
продажа от допустимого диаметра 2477
продажа по цене за единицу объёма 6119
продажа стоящего леса 7157
продажная единица 7159
продажная стоимость 5313
продергивание 8515
продовольственный лес 3313
продолговато-ямчатая гниль 5146
продолговатый лист 5858
продолговатый сучок 7910
продолговатый сшивной сучок 7910
продолжительность жизни
продольная паренхима 457
продольная пила 1633
продольная пила для распиловки на доски 883
продольная покоробленность 951
продольная распиловка 5135
продольная распиловка брёвен в пильной яме 6312
продольная резка 5135
продольная стенка 5137
продольная трещина 7729
продольное раскалывание 4915
продольное раскалывание при его валке 4915
продольное раскрытие 5136
продольное распиливание 5135
продольное расположение волокон 5133
продольное расщепление ствола дерева 4915
продольное резание 2187
продольное сращивание брусьев 4912
продольно-радиальное коробление 7936
продольный лесотранспортёр 7692
продольный одноцепной лесотранспортёр 7654
продольный разрез 4915
продольный распил 5135
продубина 7215
продуктивная лесная земля 6562
продуктивность леса 3539
продуктивность местообитания 7670
продуктивность экосистемы 2750
продуктивный лес 6561
продуктивный лесоучасток 8625
продукт из круглого лесоматериала 7117
продукты высокотехнологической древесины 2826
продукты девственного леса 9405
продукты переработки коры 577
продукция лесных семян 3584
проезд 182
проект транспортного освоения лесных ресурсов 6910
проект транспортного освоения ресурсов 6910
проекция кроны 2088
прожилки 5390
прозенхима 6586
прозопис железистый 4354
прозопис серёжкоцветный 1790
прозопис чилийский 4354
производитель древесины 344
производительность елового насаждения 9781
производительность леса 3539
производительность лесного насаждения 3592
производительность по комплексу лесозаготовительных работ 8215
производительность по комплексу лесосечных работ 8216
производительность резания 2188
производная саванна 2408
производный лес 7346
производственная площадь 5925
производственный рассадник 6370
производство прививок 8714
произрастание 9127
происхождение насаждения 8025
происхождение семян 7395
прокладка в штабеле 6236
прокладка в штабеле лесоматериалов 6236
прокладка для предотвращения скатывания наружных брёвен 1228
прокладка для предохранения брёвна 1228
прокладка канав 8962
проклеенные древесные частицы 3877
пролепсис 6579
пролёт 7874
пролив 9284
пролиферация почек 6580
пролыска 2283
промежуточная А-образная опора 100
промежуточная древесина

промежуточная обработка насаждения 4550
промежуточная одноканатная опора 7646
промежуточная площадка без накопления лесоматериалов 4382
промежуточная площадка с накоплением лесоматериалов 1721
промежуточная поддерживающая перекладина 4548
промежуточная рубка средней интенсивности 5508
промежуточная стадия гниения 72
промежуточная толерантность 4549
промежуточное дерево 4551
промежуточное насаждение 4547
промежуточное штабелевание 1720
промежуточные продукты 9500
промежуточный игольчатый сучок 830
промежуточный лес 8093
промежуточный поддерживающий канат 4548
промежуточный склад 1721
промышленная древесина 9721
промышленная рубка 1744
промышленное лесное хозяйство 4482
промышленные лесоматериалы 9721
промышленные рубки ухода 1750
промышленный лес 1745
промышленный лесоматериал 5288
проникание 6144
проникающие атмосферные осадки 2626
проникновение 6144
пропаренная древесина 8054
пропил 7232
пропитанная древесина 4427
пропитанные сортименты 4427
пропитка древесины 4429
пропитывание 4488
пропитывать древесины креозотом 2017
прополка 2968
пропуск 6097
пропущенное годичное кольцо 5479
прорастание 3835
проращивание семян 7380
прорежённый лес 1060
прореживание 8515
прореживание без сбыта продукции 4680
прореживание в древостое 7472
прореживание насаждения 8515
прореживание низкоствольника 7948
прореживание по всему насаждению 4902
прореживание полосами 4992
прореживание посевов 8514
прореживание по шаблону 8095
прореживатель 8513
прореживать 8503
прорезание 5798
проросток 6366, 7035
прорость 576
прорубать дорогу через лес 8302
прорубка господствующего яруса 8512
прорубка каймы 4992
прорубка крон 2096
прорубка подчинённого яруса 8511
прорывание 8515
прорывать 8503
просачивание 4488
просачивающаяся вода 8968
просвет 2260
с ~ами 7008
просвет в насаждении 3443
просвет между режущими кромками пильной цепи 2403
просвет под несущий канат 9745
просека 3564
просека для канатов 1245
просека под несущий канат 9745
проспект 6587
простая кривизна 7634
простая мера 5748
простая метиковая трещина 7630
простая перфорационная пластинка 7633
простая почка 7628
простой древостой 7653
пространственная изменчивость в лесу 7885
пространственная целостность лесного сообщества 7883
пространственное расположение 7884
пространство, занимаемое деревом 4008
пространство, покрытое кустарником 353
протейные 6591
протея 6590
протея медоносная 8254
противопожарная канава 4063
противопожарная полоса 3169
противопожарная полоса, подготовленная отжигом 1186
противопожарная просека 3169
противопожарное устройство лесов 3433
противопожарные разрывы 3199
противопожарный агрегат 916
противопожарный барьер 3169
противопожарный набор 3194
противопожарный плуг 3196
противопожарный самолёт 157
противосолнечный косослой 6943
противоэрозионное облесение 1881
противоэрозионные леса 317
протиум смолоносный 1924
протоксилема 6602
протравить 5227
протравленная древесина 7993
протяжённость корней 7022
протяжённость кроны 2082
протяжённость несущего каната 7702
профили обработанных пиломатериалов 7260
профили пиломатериалов 7260
профилирование 822
профилированная древесина 7260
профилированный полосовой материал 9613
профиль леса 3542
профиль листвы 3304
профильная твёрдая древесина 1991
профильное резание 2151
профильный пиломатериал 6566

проходная рубка 7472
процент активной щёлочи 171
процент всхожести 3836
процент вырубки 2954
процент выхода 9155
процент выхода древесины 9155
процент кроны 2090
процент лесистости 3522
процент лесной прибыли 6729
процент неликвидной древесины 2120
процент объёма нетоварной части 2120
процент объёма нетоварной части дерева или брёвна 2120
процент поделочного леса 6150
процент посадки 6368
процент приживаемости лесных культур 8918
процент успешной посадки 6368
процесс гумусообразования 4393
прочистка леса 4591
прочистки 2703
прочная древесина 2670
прочность 8099
прочность в сухом состоянии 2657
прочность древесины 2669
прочность на изгиб 676
прочность на срез по толщине 7502
прутняк 3108, 4961, 7204
прутняк мелкоцветный 7763
прутняк обыкновенный 4961
прутняк пятилистный 3232
прутняк трилистный 7632
прутья 617
прутья для плетения 9437
прыгающий бурав 4667
прыжками распространяющийся пожар 4669
прямая ветвь, прижатая к основной оси 8162
прямая врубка 8140
прямая вывозка 2522
прямая вывозка леса 2522
прямая доставка 2522
прямая накладка вполдерева 8140
прямая радиация 2520
прямое волокно 8139
прямой сосновый побег 1302
прямой торец 7966
прямолинейный пропил 8141
прямолинейный рез 8141
прямо с дерева 8914
прямослой 8139
прямослойное дерево 8138
прямоствольное дерево 8143
прямоствольность 8142
прямоствольный лес 8137
прямостоячая лучевая клетка 9139
прямостоячий габитус 2858
прямоугольная балка 7965
прямоугольная кромка 7967
прямые затраты на заготовку леса 2523
псаммофит 7192
псевдоверхушечная почка 3004
псоиды 9014
птелея калифорнийская 1261
птелея трёхлистная 6035
птеридофиты 6618
птерокария 9456
птерокария кавказская 1388
птерокария японская 4637
птерокарпус ангольский 290
птерокарпус санталовый 6790
птероксилон полезный 7797
птеростиракс косматый 3682
птеростиракс щетинистый 3682
птичьи поклёвки 765
публикация 6587
пузырёк 834
пузыреплодник головчатый 6021
пузыреплодник калинолистный 397
пузыреплодник мальвовый 5259
пузырник древовидный 818
пузырчатник калинолистный 397
пузырчатник мальвовый 5259
пузырчатое заболевание растений 834
пузырь 1167
пулан 9363
пункт перегрузки 8763
пурпурный лапачо 6644
пуршия трёхзубая 311
пуск пала 1191
пуск пала для лесовозобновления 8027
пуск пала из центра 1409
пустошный лес 4208
пустошь 9275
пустынная корка 2417
пустынная лаванда 2416
пустынный куст 2419
пустырь 9275
путанное волокно 2145
путерия кампешевая 2772
путь в лесу, отмеченный зарубками 9371
путь к пожарищу 3217
путь молевого сплава 3277
путь, отмеченный зарубками 9371
путь отхода 2864
путь отхода вальщика 6111
путь тыловой мачты 475
пучковая запань 6691
пучковязатель 2990
пучкообразная верхушка 1210
пучкообразный рост 1209
пучок 9005
пучок брёвен 8584
пучок волокон 7530
пучок деревьев 1666
пучок корья 1176
пучок лесоматериалов 9496
пучок мелких сосудов 6462
пучок прутьев 3018
пучок хвои 5703
пучок хвороста 1118
пушечное дерево 1309
пушистый дуб 6621
пуэрария 4746
пыление 313
пылеотделение 2318
пылеочиститель 2674
пыль 7598
пыльник 312
пыльца 6432
пыльцевой мешок 312
пышный рост 4213
пьяника 893
пэнь-сломыш 993
пяденица 3174
пяденица берёзовая 6147
пяденица гемлоковая 2718
пяденица сосновая 6262
пятиконечная посадка 6668
пятилистное целомудренное дерево 3232
пятиугольная сердцевина 6145
пятнистая жёлтая гниль 7908
пятнистое повреждение хвои озоном 6017
пятнистость 843

пятно 7907
пятнышко 7907

работник по сплаву 9569
рабочий блок лесосеки 9097
рабочий гужевой трелёвочной бригады 1559
рабочий, занимающийся окоркой 556
рабочий, занятый на расчистке территории или дорожных трасс 1104
рабочий, затёсывающий концы брёвен 7744
рабочий-измеритель диаметров деревьев 3410
рабочий канат 4847
рабочий на доочистке деревьев от сучьев 981
рабочий на лесном складе 9586
рабочий на лесосеке, руководящий движением канатов и прицепкой груза при канатной трелёвке 1559
рабочий на обрезке вершин 8706
рабочий на площадке 4773
рабочий на подготовке лесоматериалов к чокеровке 7695
рабочий на подготовке лесосеки 1572
рабочий на подрезке веток и вершин 8938
рабочий на подрезке деревьев 8949
рабочий на расчистке лесосечных отходах 8307
рабочий на ручной окорке 4105
рабочий на складе 8094
рабочий на сортировке лесоматериалов 9525
рабочий, обслуживающий канатную установку 4168
рабочий, обтёсывающий брёвна 9646
рабочий, очищающий брёвна 9646
рабочий, подкатывающий брёвна вручную 4113
рабочий, предупреждающий образование заторов на лесоспуске 4257
рабочий, производящий подпил 9057
рабочий-трелёвщик 1435
равновесная влажность древесины 2856
равновысокий эвкалипт 1075
равномерная постепенная рубка 9086
равномерная разновозрастная структура 6838
равномерное внесение 5990
равномерное изреживание насаждения 7871
равномерный рост 4590
равнослойная текстура 2923
равные долготья 2925
рагий бороздчатый 6934
радиальная распиловка 6659
радиальная трещина 6675
радиальная усушка 6680
радиальное разбухание 6681
радиально распилённое мачтовое бревно 6660
радиальный межклеточный канал 6677
радиальный прирост 2474
радиальный разрез 6659
радиальный разрез сердцевинного луча 6738
радиальный распил 6659
радиальный секреторный ход 6677
радиация 6682
радиус действия 2102
разбавитель 8505
разбавляющее вещество 8505
разбрасыватель 7934
разбросной посев 1044
разбросной сев 1044
разброс отдельных деревьев 7294
разбухание 8337
развальная лесопильная рама 1633
развевание листвы 2538
разведение деревьев 8834
разведение леса 7610
разведение леса вне лесного массива 5759
разведение лесных растений 3530
разведение меристемы 5403
разведение рождественских ёлок 1570
развёртывание листьев 9084
развесистая крона 996
разветвление 987
разветвление волокон 3095
разветвлённый 977
разветвлённый дренаж 2368
разветвлённый сучок 979
развивающийся из спящей почки 77
развившаяся гниль 72
развилина 3652
развилина сучка 982
развилка 3652
развилка ствола 8074
развитие ветвей 2449
развитие кроны 2073
развитие леса 3374
развитие насаждения 2448
развитие почек 1020
развитие свилеватости 9187
развод зубьев пилы 821
разгрузочная эстакада 5451
разгрузочный карман 2665
разгрузочный пункт 9102
раздвоенная ветка 3655
раздвоенная лесная мерная вилка 3653
раздвоенное дерево 3659
раздвоенный 3654
раздвоенный ствол 3657
разделение коры и щепы 1511
разделение лесов 3681
разделённый лист 2309
разделка 5100
разделка брёвен 5100
разделка долготья 1132
разделка леса 5100
разделывать 1122
раздельная рубка леса 3366
раздельный лист 1636
разжижитель 8505
различимая высота дерева 9235
различимый диаметр кроны дерева 9234
разложение 2310
разложение, вызываемое грибами 3768
разложение древесины 9535
разматывание 6121
размельчение дефектных брёвен 2122
размер 2499
размер брёвен без допуска на обзол или горбыль 58
размер жерди 6422
размер кроны 2093
размер лесонасаждения 3398
размер лесопользования 3637
размерностная классификация 2500
размер отростка 7205

размер после строгания 2621
размер после сушки 7087
размер с допуском на обзол или горбыль 2159
размер стружки 1540
размеры дерева 8827
размеры древесины 8576
разметка ствола и веток срубленного дерева 4845
разметка хлыстов 5318
разметочная линейка 5319
разметчик хлыстов при раскряжёвке 5102
размещение деревьев 4569
размещение перешколенных саженцев 8773
размещение саженцев 6362
размножение отводками 6582
размножение почкованием 307
размножение черенками 6581
размол древесины 9609
размочаленный конец 1071
размочаливание 627
размягчение древесины 9655
разновозрастная структура 174
разновозрастное лесное хозяйство 9080
разновозрастное насаждение 9081
разновозрастный лес 172
разнолистный 4266
разнолистный побег 4267
разнолистный тополь-турангу 4468
разнообразие культур 2130
разнообразие окружающей сред 4074
разнообразие пород 7897
разнородность 4263
разнохозяйный 4262
разобщение волокон 3099
разобщитель деревьев 8919
разобщитель пачки хлыстов 7976
разобщитель щепы 1537
разорванная паренхима 2531
разорванные волокна 1524
разработка бурелома 9442
разработка ветровала 9442
разрастание 5977
разрежённое лесонасаждение 7293
разрежённые деревья 7882
разрежённый древостой 4446
разрежённый заболоченный древостой, переходящий в
разреживание 8515
разреживание крон 2096
разрез у вершины 8687
разрешение на вывоз 6180
разрешение на вывоз древесины из леса или со склада 6180
разрешение на заготовку и торговлю лесоматериалами 8605
разрешение на пуск пала 1193
разрешение на расчистку 4928
разрешение рубки 6183
разрозненная спецификация лесотоваров 9614
разрозненный сплав 7446
разросшееся дерево 9472
разрушение в результате среза 7500
разрушение древесины 1014
разрушение от скалывания 7500
разрушение при растяжении 8465
разрушение при срезе 7500
разрывная длина 1019
разрыв по краю листа 7747
разрыв при растяжении 8465
разряд высот 4228
разряд сплошного лесоустройства 1058
разъединение 7658
разъездной персонал 3119
рай-дерево 2908
район лесничества 2542
районный лесничий 1988
район работ 6578
район сбыта 7158
район с легкоуязвимой природной средой 2841
район снабжения древесиной 8647
район с недостаточным лесом 8574
район с преобладающей лесодобивной промышленностью 8640
район, уязвимый в экологическом отношении 2841
ракита 1040
ракитник 1069
ракитник метельчатый 1779
рама пилы 3684
рамин 6710
рамная пила 5132
рамная тележка 5208
рана коры 558
рана от обрезки 6616
рандомизированные участки 6713
рандомизированный план блоков 6714
раневая гниль 7078
раневая древесина 9734
раневая зона 8788
раневая камедь 9732
раневая окраска дерева 8959
раневая паренхима 8784
раневая пробка 8781
раневая смола 9732
раневая ткань 8787
раневое пятно на коре дерева 8786
раневое ядро 8782
раневой бальзам 9254
раневой смоляной канал 8783
ранневесенний лист 2702
ранневесенний пуск пала 2700
ранневесенняя сплотка 7937
ранние рубки ухода 2703
ранняя древесина 2704
ранняя кора 2699
ранняя лесная сукцессия 2701
ранцевый опрыскиватель 471
ранцевый опыливатель 471
раскалываемость 1627
раскалываемый лесоматериал 1628
раскалывание 1629
раскалывать 1631
раскалываться 1631
раскатывание в стороны свободных концов трелюемого леса 3013
раскидистая крона 7935
раскидистая низбегающая форма ветвления 2315
раскидистость кроны 2094
расклинивание 9318
раскол 9507
расколотое бревно 7920
расколотый 1634
раскроённый в размер 8119
раскроенный по определённой схеме 2205
раскрой 7232
раскрой брусков 838
раскрой круглых лесоматериалов 7115
раскрой по толщине 2322
раскрой по ширине 3253
раскряжёванный лес 1918

раскряжёвка лесоматериала снизу 1215
раскряжёвка лесоматериалов в пакетах 1174
раскряжёвка на сортименты 838
раскряжёвка по предельному диаметру 2476
раскряжёвка по предельным качественным признакам 1127
раскряжёвка с максимальным выходом древесины по объёму 1130
раскряжёвка с максимальным выходом древесины по стоимости 1129
раскряжёвка ствола на сортименты 8069
раскряжёвочная метка 8972
раскряжёвочная пила 1136
раскряжёвочная пила с подвижным столом 6672
раскряжёвочная площадка 7727
раскряжёвочная таблица 1133
раскряжёвочное устройство 2043
раскряжёвочные козлы 4380
раскряжёвочный нож 5113
раскряжёвочный станок 2043
раскряжёвщик 1124
раскряжёвывать 1122
распад 2310
распад древесины 9535
распад древесной ткани 1014
распил 7232, 7241
распилённый лесоматериал 1918
распилённый по касательной к годовым кольцам 3258
распилённый по сбегу 8408
распилённый по толщине 6881
распилённый по толщине лесоматериал 6882
распиливать 1122
распиливать вдоль волокон 6959
распиливать по толщине 6880
распил насквозь 5016
распиловка 5100
распиловка брёвен 5100
распиловка брёвен в пильной яме 6312
распиловка брёвен на доски 1017
распиловка брёвен на
распиловка на тонкие доски или рейки 2322
распиловка с брусовкой 5117
распиловка чураков 713
распиловочный размер 7469
распиловочный размер свежесрубленной древесины 3955
распил, параллельный направлению волокон 5135
распил по кромке 2760
расположение годичных колец 302
расположение жилок 9192
расположение классов возраста 127
расположение корней 7011
расположение лесозаготовки 5083
расположение листьев 4856
расположение сучьев 4725
расположение ярусами 8147
расположенный на листьях эктопаразит 2852
распорная поперечина 8568
распределение брёвен по диаметру 7248
распределение влаги 5519
распределение деревьев по диаметру 2483
распределение досок 2548
распределение древесных пород 8929
распределение издержек 1978
распределение лесосек 191
распределение осадков 6698
распределение по диаметру 2483
распределение по классам возраста 127
распределение порубочных остатков 2541
распростёртая розетка листьев 2314
распростёртый габитус 6588
распространение леса 3381
распространение трещин 2001
распространённость вида 21
распускание листьев 9084
распускание почек 1020
распускающийся лист 4678
распускной пильный станок 1633
распыленный на ленточно-пыльном станке 521
распылительный пистолет
рассадник кустарников 1205
рассадочная грядка 5812
рассадочный мешок 5811
рассечённый лист 2549
рассеянная радиация 2494
рассеянная тень 2248
рассеянное выдерживание 2536
рассеянно-пористая древесина 2493
рассеянно-сосудистая древесина 2493
рассеянно-сосудистая лиственная древесина 2493
расслаивание 2348
расслаивать 4760
расслаиваться 2940
расслоение 8147
рассоха 3655
рассредоточенное задержание 2536
рассредоточенное использование 2537
расстояние вывозки 9747
расстояние между верхними складами 4775
расстояние между деревьями 4569
расстояние между погрузочными площадками 4775
расстояние между стволами 4569
расстояние между трелёвочными волокнами 7746
расстояние от погрузочной площадки до дальнего конца лесосеки 2967
расстояние подтаскивания лесоматериалов 4825
расстояние стрелы 6743
расстояние сучьев 2539
расстояние трелёвки 9747
расстояние трелёвки в грузовом направлении 5024
расстояние трелёвки в порожнем направлении 6922
расстроенное насаждение 4588
растаскиватель хлыстов 7976
растение 6333
растение, обитающее на известковых почвах 1251
растение-приманка 483
растение редколесья 5922

растение, цепляющееся с помощью воздушных корней 7014
растениеядное животное 4256
растения засушливых зон 9736
растереть 3975
растирать 3975
растительная мозаика 9177
растительное разнообразие 6338
растительное сообщество 6336
растительность 9176
растительность девственного леса 5955
растительность на коре деревьев 1976
растительность почвы 3986
растительные остатки 6371
растительные сообщества, находящиеся под угрозой 8531
растительные сообщества, находящиеся под угрозой исчезновения 8531
растительный 9178
растительный вирус 6374
растительный покров 6337
растительный эктопаразит 2852
растопочные древесные брикеты 3741
расторцовка 1215
растрескивание 2003
растрескивающийся 2346
растущее дерево 4010
растущее дерево, оставленное для уменьшения отклонения канатов и каретки 7129
растущее дерево с выставленной приманкой 481
растущее или срезанное дерево в качестве промежуточной опоры 2808
растущий вилообразно 3654
растущий кривой лес 9441
растущий лес 3454
растущий на древесине 2855
растущий на коры 1974
растущий на листе эктопаразит 2852
растущий на поверхности коры 1974
растяжка 1164
растяжка для предотвращения изгиба мачты 1135
расценка леса 7123
расчёт многоцелевого устойчивого дохода 5624
расчётная вырубка 195
расчётная вырубка главного пользования 197
расчётная годичная вырубка промежуточного пользования 198
расчётная годичная лесозаготовка 193
расчётная годичная лесосека 192
расчётная лесосечная площадь 2166
расчётная метка 8486
расчётная периодическая рубка 6159
расчётное бревно 7172
расчётное количество продажи 199
расчётные стандарты насаждения 8414
расчёт обмера 5361
расчёт объёма пиловочника в бревне или дереве 5213
расчёт пиломатериалов 5213
расчёт размера лесопользования 9776
расчистить 1611
расчистка локальной территории 7933
расчистка площади от кустарника с помощью толстого троса 271
расчистка территории от кустарника 9320
расчищать 1611
расчищать территорию от кустарника 1111
расчищенная от леса область 5066
расширение пропила клином 9318
расщепление 627, 1629
расщепление древесины поперёк волокон 2039
расщепление комля 1216
расщеплённые волокна 1524
расщеплённый 1634
расщеплённый конец 1071
расщепляемость 1627
расщеплять 1631
расщепляться 1631
раттан 6731
рафинирование древесины 9638
рафия 6685
рашпиль 9566
рашпиль для дерева 9636
рашпильный обрез 2662
реактивная древесина 6744
реактивный упор пилы 7234
ребристая закомелистость 6933
ребристая карра 4670
ребристость 2019
ребристый каток 2127
ребро 4882, 7255
ребровая пила 5148
ребровый станок 1410
ребровый сучок 4733
ребро образца 2763
ребро пиломатериала 7255
ревизионный период 1916
регенерационный побег 6753
региональное переменное лесопольное хозяйство 9797
региональный план 6829
регистрирующая мерная вилка 6751
регулирование лесного дохода 3649
регулирование промежуточного лесопользования по площади поперечного сечения насаждения и по приросту 602
регулирование смешения 6843
регулированные земли 6839
регулированные лесозаготовки 6841
регулированный лес 6840
регулятор нормы высева 7385
регулятор роста 6340
редина 3443, 4446
редкая листва 7880
редкая порода 6725
редкая растительность 5921
редкие породы древесины 6726
редкое молодое насаждение 4951
редкое стояние 5914
редколесье 4446
редко облесённый 6450
редко облиственный 7881
редко посаженный 6450
редкостойный 6450
редкостойный лес 4446
редкостойный северо-таёжный лес 5904
редко стоящий 6450
редукторная моторная пила 3811
редуцент 2308

режим действия пожара 3201
режим пожара в нижнем ярусе 9069
режим пожара замены насаждения 8026
режущая кромка 819
режущее остриё 2762
режущий диаметр 2479
режущий диск 7070
режущий инструмент 2204
режущий станок 2181
рез 7232
резак 7313
резак для карр 2142
резак для маркировки деревьев 6741
резак для смолы 6897
резание вдоль волокон 2187
резание в плоскости движения пилы 3294
резание древесины 9502
резание под углом 5483
резание с движением бруса по половине окружности 4093
резание с помощью водяной струи 9288
резать 1360
резерват 5672
резервное дерево 6887
резервное насаждение 6889
резервные лесные угодья 6886
резервный карьер 931
резервный лесистый участок 6886
резервы леса на корню 3559
резервы прорубки 4754
резец 2204, 9509
резец для высверливания сучков 6384
резина 4053
резиноподобная древесина 7126
резная карра 4279
резная работа 9502
резонансный кряж 7851
резонансовая древесина 7851
рез при раскряжёвке 8571
результат главной рубки 3145
результат распиловки 5214
результат рубки 9777
резчик по дереву 1363
резьба по дереву 9502
резьбованная древесина 1361
рейка 626
рекогнисцировочная съёмка 6749
рекомендация 6517
рекреационная зона 6754
рекреационный лес 91
рекреационный маршрут 6759
рекреационный потенциал 6757
рекреационный ресурс 6758
рекультивация земель 6860
реласкоп 6852
реликтовое дерево 6857
рельеф 6858
рельсовая лесовозная дорога 5079
рельсовый лесопогрузочный кран 8609
ремедиация 6860
ренгас 6866
рентабельная выборочная рубка 2746
рентабельное лесничество 2744
рентабельность 5718
рентабельность лесного хозяйства 6568
рентабельные лесонасаждения 2742
рентабельный отпускной диаметр 2743
репер 674
реперный знак 674
реснитчатолистный 1580
реснитчатый 1580
реснитчатый лист 1579
ресничный лист 1579
ресурсы лесах планеты 3864
ретрогрессия растений 6372
речная ласточка 8941
ржавление 7140
ржавчина 7139
ржавчина сосны 6274
ржавчина сосны веймутовой и кедра сибирского 1737
ржавчина сучьев 4970
ржавчина хвойная 7859
ржавчинное искривление побегов сосны 6274
ржавчинные грибы 7139
ризом 6927
ризосфера 6930
ризофор 6928
ризофора остроконечная 6780
ризофоровые 5279
риму 6784
рисковая стоимость 9164
рисунок в виде завитков 3011
рисунок в виде узелков 3011
рисунок, образованный блестящими сердцевинными лучами 7601
рисунок по дереву 9498
ритмический рост 6931
рифель 6968
рифление 3296
рифлёное лезвие 6932
рифлёный резец 3295
робиния 1754
ровный срез 1606
рога 401
рогачи 7986
роговой сучок 4376
рогохвосты 9693
род 3827
родезийский тик 9793
родезийское красное дерево 9793
род насаждения 3363
рододендровник 7063
рододендрон 6989
рододендрон голоцветковый 6288
рододендрон древесный 8908
рододендрон древовидный 8326
рододендрон клейкий 8303
рододендрон крупнейший 3934
рододендрон крупнолистный 6023
рододендрон лапландский 4786
рододендрон понтийский 4355
рододендрон ржаволистный 6989
рододендрон ржавый 6989
родственники 3754
родственное скрещивание 4435
родственные породы 185
рождественская ёлка 1569
рождественская пальма 1568
рожковое дерево 1341
рожок для высева семян 7381
роза 995
роза войлочная 8674
роза звездчатая 2418
роза морщинистая 634
розмарин аптечный 7066
розовая кость 6279
розовое дерево 3083
розовое дерево индийское 2728
розовый кедр 6277
розовый куст 7068
розоцветные 7064
роликовая пила 7003
роликовый обрубщик сучьев 7000
роспуск 2322
рост в высоту 326
рост ветви 4967

рост корней 7027
рост насаждения 6459
рост, не ограничиваемый развитием верхушечной почки 4459
ростовое вещество 6221
ростовой корень 2961
ростовой побег 4237
росток 1095, 6366
росток корешка 6546
рост перед цветением 4677
рост плесени 3772
рост по спирали 9021
рост сука 4967
ротанг 6731
ротационный кусторез 7071
роторная корообдирка 7074
роторная окорка 6951
роторный окорочный станок 7074
роторный срезающий орган 7070
роща 933, 5001
рощица 1928
рубанок 4661
рубец коры 582
рубец от заплыва сучка 8786
рубец от почечной чешуйки 1144
рубец от пучка сосудов 1177
рубец от сучка 992
рубец сучка на коре бука 1502
рубильная машина 2104
рубильная машина для балансов 6628
рубильная машина для длинных брёвен 9391
рубильная машина для долготей 9391
рубильная машина для древесных отходов 9278
рубильная машина для отходов 7721
рубильная машина с вертикальной подачи сырья 2634
рубильная машина с загрузочным транспортёром 671
рубильная машина с плоскими ножами 3255
рубить 1561, 3037
рубка 8824
рубка в верхнем ярусе насаждения 8512
рубка верхнего яруса 6008
оборотом ротации 7544
рубка во время сокодвижения 3065
рубка в период движения сока 3065
рубка в порослевом хозяйстве с оставлением части защитных деревьев 1932
рубка в порослевым хозяйстве с оставлением маячных деревьев 8129
рубка в порослевых древостоях 1926
рубка в соку 3065
рубка главного пользования 3136
рубка деревьев, оставленные после главной рубки 6859
рубка до определённого диаметра 2476
рубка за пределами отведённой лесосеки 2973
рубка кустарника 1109
рубка леса вследствие стихийных повреждений 2186
рубка леса, преследующая узкокоммерческие цели 1744
рубка модельных деревьев 7175
рубка на границе участка 6167
рубка на свой счёт 5070
рубка осветления 5912
рубка осветления верхнего полога 7513
рубка от допустимого диаметра 2476
рубка повреждённых деревьев 7168
рубка полутвёрдой древесины 7428
рубка пострадавшего леса 7168
рубка пострадавшего от бедствия леса 7168
рубка прореживания 4902
рубка рисковых деревьев 6973
рубка с оставлением поваленных деревьев с необрубленной кроной на лесосеке 3065
рубка с оставлением семенных деревьев 6885
рубка спелого леса 4150
рубка ухода 2121
рубка ухода в молодняках 8458
рубка ухода в молодых
методу 8511
рубка широколиственных пород 4134
рубки простора 3700
рубки с оставлением поваленных деревьев на лесосеке 3080
рубки сухостойных и повреждённых деревьев 7168
рубки ухода, выполняемые с помощью подвесных канатных установок 7706
рубки ухода при трелёвке вверх по склону 9130
рубки ухода при трелёвке вниз по склону 2597
рубки ухода с канатной трелёвкой 9752
рубочный ряд 2199
рубщик 1529
рудничная стойка 6311
рудничная шпала 1725
рудничный горбыль 5470
рудничный лесоматериал 5463
рудостойка 6311
руководство лесным хозяйством 3488
рулетка для диаметров 2484
рутовые 7131
ручная валка 4109
ручная лесозаготовка 4109
ручная ножовка 1822
ручная окорка 4107
ручная пила 4111
ручная расчистка сорняков 5287
ручная трелёвка 495
ручной клинометр 9
ручной нивелир 9
ручной нивелир Абнея 9
ручной секатор 4106
ручной топор 673
рыбоводство в лесу 7626
рыбоохранный лес 3230
рыборазведение в лесу 7626
рынок столбов 6480
рыть 2495
рытьё канав 8962
рыхлая серединка 7430
рыхление 6967
рыхлительная лопата 7314
рыхлое наполнение 7430
рыхлость 6652
рыхлые порубочные остатки 5160

рыхлый сучок 5158
рычаг для подъёма брёвен 8603
рябина 2903
рябина американская 252
рябина ария 8323
рябина глоговина 9424
рябина декоративная 7552
рябина домашняя 8991
рябина круглолистная 8323
рябина мелкометельчатая 3943
рябина обыкновенная 2903
рябина промежуточная дуболистная 8322
рябина садовая 8991
рябина Сарджента 7222
рябина сибирская 2903
рябина смешанная 4742
рябинник 5580
рябчик 4182
рябь 6968
ряд
на просветах посаженные ~ы 7987
ряд брёвен 8545
ряд деревьев 7121
ряд лесных культур 6360
рядовая посадка 6356
рядовая сеялка для посева лесных семян 3387
рядовой посев 522
ряд фитоценозов 7457
ряды трансплантов 8772
ряжевая плотина 4377
ряжевый переплёт 2018

сабаль пальмовидный 1238
саблевидный комель 7736
саблевидный ствол 7736
савагуруми 4637
саванна 7227
саванна с отдельно растущими деревьями 9546
саванное редколесье 7228
саванный лес 7228
саговая пальма 7153
саговниковые 2207
садовая пила 6614
садовый клей 8958
сажальная доска 6345
сажальный бурав 6342
сажальный кол 6344
сажальный меч 8898
сажать 6334
сажать лес 93
саженец 5815
саженец перед посадки 5978
саженец хвойных пород 1878
сажень 3024
сажень дров 3024
сайт-специфичный 7674
саламандр 3966
салат дикий 6531
салинг 2059
саловый лес 7160
сальвия 7152
сальное дерево Агаллоха 5447
саман 6701
самовозгорание 7923
самовоспламенение 7923
самозагружающийся трелёвочный трактор 7419
самозагрузчик 7418
самоизреживание леса 5699
самолёт-перехватчик координатора 764
самообсеменение 5695
самоопыление 7421
самоочищение от сучьев 5688
самораспространение семян 5695
самосев 5695
самоходная сучкорезная машина 3113
самоходный измельчитель пней 7422
самшит 2886, 9346
самшит балеарский 7875
самшит вечнозелёный 2886
самшит мелколистный 5002
самшитовые 960
самшитовый лес 961
самшит Сан Доминго 9346
сандал глистогонный 6790
сандал дальбергиевовидный 383
сандал индийский 5284
сандал красный 6790
сандал крупноплодный 1183
сандаловое дерево 846
сандаловые 7181
сандал Сойукса 111
сандараковое дерево 7182
сани для вывозки коры 549
сани для трелёвки брёвен в полупогруженном положении 5115
санитарная лесозаготовка 4412
санитарная рубка 7193
санитарно-гигиенические вещества для оздоровления леса 3459
санитарно-гигиеническое 3580
санитарные меры 7194
сапели 114
сапелли 114
сапиндовые 7806
сапиндус Друммонда 9343
сапиндус Мукоросси 1506
сапиндус мыльный 7865
сапиум 7201
сапота 5665
сапотовые 7207
саппановое дерево 7209
сапрогенный 7212
сапрофит 7214
сапрофитные грибы 7213
сарай для хранения шишек 1860
сассапарель 7223
сассапарель лавролистная 4834
сассафрас 7864
сассафрас лекарственный 1801
сбег 8083
сбег ствола 8083
сбежистая форма роста 2939
сбежистое дерево 8406
сбежистость 8405
сбежистый 8401
сбежистый лесоматериал 8402
сбережение древесины 8636
сблокированное спиральное волокно 4539
сбор бульдозером в кучи 1162
сбор в валки 9448
сбор в валы 9448
сбор в кучи 1171
сбор в кучи и сжигание 6233
сбор древесины 7170
сбор живицы 828
сбор лесной подстилки 9153
сборная канава 2613
сборный луч 140
сборный плод 1839
сбор поверхностного стока 9286
сбор порубочных остатков бульдозером в кучи 1162
сбор растительного урожая 6341
сбор семян 7376
сбор сучьев 984
сбор шишек 1863
сборщик брёвен 1173
сборщик деревьев 8801
сборщик живицы 6905
сбрасывание ветвей 5688
сбрасывание деревьями ветвей

сбрасывание листьев из-за болезни 4861
сбрасывание хвои 6260
сбрасыватель 5097
сбрасыватель брёвен 5097
сбыт продуктов леса 5312
свайные брёвна 8586
свайный кряж 6707
свайный лес 8588
свалить дерево с помощью клина 9314
сведение лесов 3352
свежая вырубка 3710
свежая древесина 3710
свежеокорённые слеги 3862
свежесрубленная древесина 3710
свежесрубленное дерево 3709
свежесрубленный здоровый ствол 3708
свежесрубленный лесоматериал 3710
свежий осмол 3958
свежозаселённое дерево 1068
сверление 923
сверлильный срез 930
сверловщик сучков 4728
сверло по дереву 9538
свёрнутое корьё 6999
свёрнутый в трубочку лист 1584
свёрнутый лист 1584
сверхкрупный круглый лес 6005
сверхплановое лесопользование 2935
свес 5993
светлая древесина 845
светло-бурая кубическая гниль 4937
светло-жёлтая смешанная гниль 9370
светлые доски без синевы 1033
светлый лес 4446
световой лист 8267
световой прирост 4944
светолюбивое дерево 7476
светолюбивый древесный вид 4574
светостойкая древесина 2989
свечевидный 1305
свечное дерево 1303
свидина 1771
свиетения краснодеревная 1412
свиетения крупнолистная 1412
свиетения махагони 2110
свилеватая фигура 2144
свилеватое место 2146
свилеватость 3101
свилеватость волокон 3101
свилеватый нарост 9185
свилеватый рост 9185
свилеватый ствол 4721
свиль 1182
свисание дерева 8831
свисающий 9324
свить гнездо 5710
свить себе гнездо 5710
свищ 4732
свободная влага 3699
свободное стояние деревьев 5915
свободно-растущее дерево 5753
свободно-растущее лесонасаждение 3696
свободно растущий 3694
свободный покров 3691
свободный рост 3697
своевременное сжигание 2700
свойства древесины 9632
связанная влага 950
связка из брусьев 8560
связка плотов 6687
связка хвороста 1118
связочная машина 3220
связочная проволока для стяжки пакетов лесоматериалов 9497
связующее древесностружечных плит 6092
связующее средство древесностружечных плит 6092
священная смоковница 939
священное дерево 939
священный лиственный барбарис 2013
сглаживание 8281
сгнившая подстилка 3671
сгнившие деревья 7097
сгоревший лес 1187
сгребание подстилки 6706
сгребать граблями 6705
сгруппированные насечки 3715
сгусток 6139
сдача леса 6862
сдача лесосеки 27
сдвоенные сани 2587
сдвоенный каток-кусторез 2668
сдвоенный топор 2579
сделанный из буковой
сдирание коры 579
сдирание травяного покрова 7286
сдирание шишек 1861
сдирать кору 2280
себестен 4799
себестоимость на 1 корда 1977
себестоимость трелёвки 7684
сев 7382
северные твёрдолиственные породы 5790
северный лес 925
седельный арчак 7148
седлообразный холмик 2005
сезон весенней окорки 7941
сезон лесозаготовок 3072
сезон лесосплава 3278
сезонные лесозаготовки 6163
сезонный лесозаготовитель 6095
сезонный путь 7332
сезонный рабочий-сажальщик 4554
сезонный тропический лес 7330
сезон обработки 8795
сезон сокодвижения 7211
секатор 5168
секатор на штанге 6420
секач 2204
секвойя вечнозелёная 1269
секвойя гигантская 3844
секвойядендрон гигантский 3844
секира 673
секторная лесная мерная вилка 7367
секторная разделка фанерного кряжа 3124
секура 744
секционное определение кубатуры 9244
селекционер 1026
селекционное изреживание 7411
селекционный дифференциал 7407
селекция древесных пород 1028
селекция и разведение лесных растений 1028
сельва 7423
семейство папайа 1340
семена 7370
семенная почка 6878
семенник 5569
семенное возобновление леса 7387

семенной древостой 7388
семенной лес 7388
семеннолесосечная выборочная рубка 7516
семеннолесосечная вырубка с оставлением резервных деревьев 7518
семеннолесосечная рубка 7399
семенодоля 1983
семеномер 7398
семеноносец 5569
семечковый плод 4698
семя 7370
семядоля 1983
семянка 38
семя с крылаткой 4701
сена 1870
септированная паренхимная клетка 7449
септированное древесное волокно 7450
серая лесная почва 3969
серая окраска 3925
серая ореховая лесная почва 5827
серая пальма 932
сервитут на дрова 6946
сервитут на лес 8573
сервитут на поделочный лес 4388
сервитут на сбор подстилки 6947
сервитутная древесина 6944
сердцевидная липа 7767
сердцевидный лист 1943
сердцевина 1951
сердцевина дерева 1951
сердцевинная гниль 1414
сердцевинная доска 4194
сердцевинная древесина 4202
сердцевинная поверхность доски 1021
сердцевинная трещина 4195
сердцевинное волокно 4197
сердцевинное пятно 5390
сердцевинные лучи 7601
сердцевинный луч 5391
сердцевинный отлуп 2139
серебристая текстура 7601
серединка 1949
серединка из деревянных реек 5200
серединка щита из деревянных реек 5200
сережейра 1417
серёжка 1380
серия фитоценозов 7457
серо-бурая заболонная гниль 3968
серое ядро 3970
серо-коричневая лесная почва 3967
серянка сосны веймутовой и кедра сибирского 9376
сетчатая паренхима 6920
сетчатая перфорационная пластинка 6921
сеть вспомогательных дорог 8967
сеть доступа 31
сеть лесных дорог 3566
сеть лесных экосистем 3393
сеть лесовозных дорог 3566
сеянец 6366
сеянец в горшочке 6483
сеянец питомника 5814
сеянец с закрытой корневой системой 9001
сеянец с комом субстрата 492
сеять 7371
сжатие вдоль волокон 1848
сжатие, перпендикулярное к направлению волокон 1849
сжигание древесины 9499
сжигание местами 6105
сжигание начисто 1043
сжигание отходов древесины 7719
сжигание площадками 6105
сжигание по предписанию 6516
сжигание порубочных остатков 7719
сжигание порубочных остатков в кучах 8308
сжигание порубочных остатков в местах скопления 6105
сжигание порубочных остатков на полосах 8170
сжигание порубочных остатков одновременно с заготовкой леса 5010
сжигание порубочных остатков при очистке лесосек 8308
сжигание с краю 2759
сжигание с середины 1409
сжигание хвороста 7719
сжигать 3166
сиамский кедр 1493
сибсы 3754
сидерит 3949
сидячий лист 7463
сила пожара 3190
сила связи 905
сила сцепления 905
сильная обрезка 4215
сильная прорубка 4217
сильноветвистый 5596
сильно облиственный 2389
симарубовые 147
симбионт 8358
симплокос 8333
синева древесины 5050
синева заболони 869
синева складочного леса 1031
синий сандал 846
синкарпия почконосная 1197
синь 5050
синь заболони 869
синэкология 8359
синяя гниль 5050
синяя сосновая златка 8057
сипо 9150
сирень 6292
сирень амурская 265
сирень обыкновенная 6292
система ведения лесного хозяйства 3495
система ведения одновозрастного хозяйства 2919
система волоков 6674
система выборочной рубкой в порослевых древостоях с оставлением части молодняка и тонкомерных деревьев 1927
система группово-выборочных рубок верхнего полога 7514
система двухъярусного высокоствольного хозяйства 9036
система заготовки сортиментов с вывозкой их небольшими двухосными грузовиками 889
система заготовок сортиментов 1123
система каемчатой обсеменительной рубки 8179
система клиновидно-лесосечных рубок 9317
система лесного и пастбищного хозяйств 7625
система лесозаготовок 5084
система лесозаготовок с

система лесозаготовок с расположением волоков с учётом рельефа местности 1905
система лесоресурсов 8630
система лесосечных машин 4152
система маркировки древесных пород и деревянных изделий 9593
система неравномерных семеннолесосечных рубок 4587
система переменного лесопольного хозяйства 145
система погрузки с помощью захватов, подвешенных к растяжке трелёвочной мачты 4065
система продажи леса на корню по цене за единицу объёма фактически заготовленной древесины 9098
система прореживания 8516
система просветов обсеменительной рубки 3801
система равномерных постепенных рубок по кварталам 7512
система разведения 5333
система разведения скота в лесу 7625
система рубок 5084
система семеннолесосечной рубки 7400
система с заготовкой и окучиванием сортиментов у пня 2162
система спаривания 5333
система сплошных чересполосных рубок 219
система трелёвки хлыстов 8873
система трелёвочных волоков 6674
система трубок, содержащих латекс 4828
система упаковки древесных отходов 7235
система ширококаёмчатой рубки 9088
системы защиты крон 2089
ситовая ядровая гниль 5573
ситовидная клетка 7589
ситовидная пластинка 7592
ситовидная трубка 7593
сицилианский дуб 1581
скалистый горный виноград 2013
сканирование пиломатериалов 5210
скарификация 7286
скарификация горячей водой 4386
скатка 1016
скатывание 1016
скатывание лесоматериала с горы 2598
скатывать бревна в воду 1016
скачкообразный рост 4553
скважность 6465
сквозная врастающая кора 8539
сквозная трещина 8537
сквозной завиток 8789
сквозной сучок 8790
сквозные промывочные осадки, проникающие сквозь полог леса 3622
скелетная ветвь 6536
скелетное вещество 7681
скипидар 6276
скипидар из осмола 9684
склад 5059
склад для коры 585
складирование древесины на промежуточном складе 1720
складированная древесина 9744
складка коры 548
склад круглого леса 7119
склад лесоматериалов 8664
складная мачта 3301
складная мерная вилка 3299
складная таксационная линейка 3300
складочная мера 7978
складочное место 8664
складочный кубический фут 7977
складочный кубометр 2115
складочный лес 7982
складочный объём 7979
складочный поделочный лес 6230
склад пиломатериалов лесопильного завода 7253
склад тушительных орудий 3214
склад у пристани 5059
складчик древесины 9661
склеенные древесные частицы 3877
склереида 7300
склонозащитный лес 7754
склонометр 4443
скоба для круглого леса 5094
скоба для прикрепления тяговой цепи 8577
скоба с крюком для крепления брёвна к тяговой цепи 7686
скобель 7960
скобель для смолы 4584
скобочный крюк 8577
скобы для балансовой древесины 6629
скол 1216
скользкий вяз 7745
скользящее бревно 7739
скользящий чокер 7740
скомканный рассад 492
скопление 1560, 8319
скорлупа 4391
скорость восстановления 6752
скорость лесозаготовки 4162
скорость намотки каната на верхних витках 3749
скорость распространения 6730
скорость резания 6727
скорость роста 6728
скорость фильтрации 4489
скос 7283, 9690
скотопрогон 2624
скошенный срез 1441
скребок 4584
скремблер 7312
скрепление ствола скобами 7315
скрещивание 4401
скрещивать слоёв 2036
скрипичная древесина 7851
скрученные волокна 1902
скручивание 2031, 9020
скручивание и свисание ветвей 9323
скручивание листьев 4887
скрытоподзолистая бурая лесная почва 2107
скрытоподзолистая бурая почва 2107
скрытый период 4818
скумпия американская 258
скумпия кожевенная 2908
слабая прорубка 4952
слабоветвистый 7737
слабонатянутый трос 7714
слабые сучья
со слабыми сучьями 7737
слабый плодокорм 4939
слабый пожар 7734

сланик 8756
слеги 7694
слегка опушённый 6619
след 7307
следкатастрофический вторичный лес 6476
след огня 3206
след от прокладок 2047
след струйчатости 6968
след сучка 4736
слешер 7721
слива 6391
слива американская 256
слива домашняя 6391
слива зелёная 3952
слива золотая 1700
слива зонтиковидная 3260
слива канадская 1297
слива колючая 7749
слива кустарниковая 3984
слива мексиканская 5425
слива морская 633
слива падуболистная 4336
слива растопыренная 1475
слива узколистная 1480
слива чёрная 1297
сливная паренхима 1869
слизистая гниль 7741
сложение лесных площадей 1883
сложенная древесина 6236
сложная кривизна 5619
сложная метиковая трещина 8039
сложная суборь 8277
сложная трещина 1843
сложноцветные 1834
сложный главный побег 5614
сложный ельник 5493
сложный лес 5633
сложный лист 1841
сложный плод 1839
слоистая древесина 6398
слоистая структура 4842
слоистая текстура 8175
слоистый древесный пластик с параллельным расположением волокон 8463
слоистый древесный пластик с тангенциальным расположением волокон 8384
слоистый шпон 4762
слой 6397
слой брёвен 8545
слой пробковой коры 6200
слой роста 301
сломанный главный побег 1061
слонея 7748
слоники 9331
служащий управления охраны леса 3361
случайная рубка 4437
случайное скрещивание 5971
смазка для древесины 9601
смазчик лесоспуска 1576
смена древесных пород 226
смена направления 8349
смена сосновых пород 6251
смена стадий развития биоценоза 7457
смена трелёвочного волока 9751
сменная культивация 7520
сменная культивация, предохраняющая деревьев 8903
сменная обработка 7520
смертность леса 3509
смесь группами 5494
смесь древесины с остатками коры 589
смесь куртинами 5494
смесь многолетних диких растений 6152
смесь семян годовых диких растений 306
смесь сообразно местопроизрастанию 1638
сметная вырубка 195
сметная вырубка по площади 9770
смешанная выборка 1741
смешанная желтовато-бурая гниль на берёзе 1085
смешанная культура 5491
смешанная почка 1837
смешанное насаждение 5486
смешанное семя 5487
смешанные северные твёрдолиственные породы 5490
смешанный горный лес 5495
смешанный древостой 5486
смешанный лес 5486
смешанный лес в холмистой местности 5495
смешанный хвойный лес 5485
смешение древесных пород 5497
смещенная сердцевина 2732
смоковница бенгальская 526
смола каури 4466
смолёвка ели Энгельмана 2824
смолёвка кедровая 2399
смолёвка ситхинской ели 7679
смолёвка сосны веймутовой 8670
смоление 6895
смолистая древесина 6901
смолистая щепа 6894
смолистое дерево 6900
смолоистечение 2977
смолоносная хвойная древесная порода 6304
смолоносное дерево 6900
смолосемянник 6315
смолосемянник тонколистный 8421
смолотечение 2977
смольё 6901
смоляная жила 4058
смоляная пластинка 4052
смоляная полоска 8155
смоляная трубка 6303
смоляное пятно 8155
смоляной желвак 6893
смоляной канал 4049
смоляной кармашек 6893
смоляной потёк 8155
смоляной ход 4049
смородина 2148
смородина альпийская 213
смородина американская 240
смородина душистая 1147
смородина золотистая 1147
смородина красная 6769
смородина кровяно-красная 6772
смородина чёрная 2883
сморщивание 2020
смыкание крон 2069
смыкание лесного полога 1659
смятие частиц щепы 1536
снабжение лесом 2359
снаряд для лазания 8814
снасти 1942
снегозащитная полоса из кустарника 1114
снеголом 7801
снеготаяние 7802
снежноягодник белый 1804
снежноягодник кистистый 1804
снежноягодник красноплодный 1940
снежноягодник круглолистный 1940

снежноягодник обыкновенный 1940
снежноягодник округлый 1940
снежный лесоспуск 7803
снежный спуск 7803
снижение 2617
снижение качества из-за дефектов 2341
снимать кору 2280
сныть обыкновенная 769
снятие дернины 7286
снятие коры 579
снятие коры кольцами 6951
снятие луба 624
собирание азота 5736
собирание отходов 7170
собиратель смолы 6905
собирать и окучивать лесосечные отходы 1573
совершенно сухая единица древесины 906
совершенный лист 1829
совместное несение издержек 1978
совместное несение расходов 1978
совместное управление 6090
совмещённое возделывание 1716
совокупная кривая роста 2135
согосподствующие деревья 1706
содействующий роста 4021
содержание влаги в древесине 5518
содержание запаса 4694
содержание коры 546
содержание питательных веществ 5824
содержащий лигнин 4955
содоминантность 1707
содоминантные главные побеги 1705
соединение деревянных деталей 4660
соединение на шпонке 2595
созревание 5339
созревание дерева 132
сок 7208
соковое давление 5965
сокодвижение 1589
сокообращение 1589
сокращение 2617
солевыносливый 7164
солеустойчивый 7164
солнечный косослой 4907
солодковый корень 4997
соматический эмбриогенез 7846
соматическое эмбриональное развитие 7846
сомкнутая трещина 8279
сомкнутое насаждение 3756
сомкнутое сообщество 1651
сомкнутость крон 2069
сомкнутость лесонасаждения 3370
сомкнутость насаждения 3370
сомкнутость полога 1659
сомкнутый древостой 3757
сомкнутый полог 1650
соннератиевые 7848
сообщество бореальных лесов 5435
сообщество высоких трав 228
сообщество листопадных древесных и кустарниковых пород 2297
сообщество организмов 738
соответствующий подбор шпона 3265
соотношение диаметров кроны и ствола 2075
соотношение длины кроны к высоте дерева 2085
соотношение классов возраста 138
соотношение крон к корням 2091
соотношение массы корней и побегов 7052
соотношение объёмов прироста и рубки 4024
соотношение опушки к площадей леса 2755
соотношение содержания углерода и азота 1333
соотношение сухого вещества к испарению 2653
соотношение сухого вещества к облучённости 2652
сопротивление изгибу 676
сопротивление лущению 6908
сопутствующая древесная порода 66
сопутствующая порода 393
сопутствующие деревья 1706
соревнование крон 2070
сорея мощная 7163
сорея тёмнокрасная 2254
сорея чешуйчатобороздковая 4949
сорная древесная порода 9074
сорт древесины 4714
сортиментная рубка 2746
сортиментная таблица 3896
сортиментовоз 7539
сортимент средней длины 434
сортировальная машина для щепы 1518
сортировка брёвен 5092
сортировка древесного угля 1446
сортировка круглых лесоматериалов 7850
сортировка леса 3902
сортировка лесоматериалов 3902
сортировка лесоматериалов на воде 9292
сортировка лесоматериалов на земле 2651
сортировка на воде 9292
сортировка на земле 2651
сортировка по диаметру 2483
сортировка по длине 4913
сортировка по размерам 3903
сортировочная площадка для брёвен 5093
сортировочная цепь на лесозаводе 3944
сортировочное сито для щепы 1535
сортировочное устройство потока пиломатериалов с цепными транспортёрами 3289
сортовой лес для продажи 5394
сосна 6272
сосна Washoe 9272
сосна аллепская 2220
сосна аризонская 360
сосна Арманда 1508
сосна балканская 489
сосна Бальфура 6997
сосна Банкса 4601
сосна белая калифорнийская 7142
сосна белая китайская 1508
сосна белая мексиканская 7867
сосна белая юго-западная 7867
сосна белая японская 4636
сосна белоствольная 9362
сосна болотная 5141
сосна Бунге 4750
сосна веймутова 2726
сосна веймутова румелийская 489
сосна виргинская 9229

сосна гималайская 4308
сосна голая 7956
сосна горная 5598
сосна горная Веймутова 9344
сосна густоцветная 4632
сосна длиннохвойная 1543
сосна длиннохвойная болотная 5141
сосна Дюранго 2673
сосна ежовая 7538
сосна ельдарская 2220
сосна жёлтая 6445, 7538
сосна жёлтая аризонская 360
сосна жёлтая западная 6445
сосна Жерарда 1486
сосна жёсткая 5792
сосна Жеффрея 4648
сосна закрытая 7191
сосна замечательная 5548
сосна итальянская 4595
сосна калабрийская 1250
сосна калифорнийская 1272
сосна караибская 2111
сосна карликовая 5598
сосна кедровая европейская 8346
сосна кедровая калифорнийская 4965
сосна кедровая карпатская 8346
сосна кедровая корейская 4743
сосна кедровая маньчжурская 4743
сосна кедровая сибирская 7569
сосна кедровидная 5424
сосна колючая 8361
сосна корейская 4743
сосна Культера 1984
сосна кхазия 5787
сосна ладанная 5029
сосна Ламберта 1274
сосна лапландская 4785
сосна лесная 1797
сосна ложновеймутова 6617
сосна малорослая 5598
сосна мексиканская 5549
сосна мелкоигольчатая 770
сосна мелкоцветная 4636
сосна Меркуса 5404
сосна Муррея 5044
сосна обыкновенная 1797
сосна однохвойная 7642
сосна однохвойная калифорнийская 1272
сосна остевидная 6997
сосна остистая 6997
сосна Парана 1005
сосна Роксбурга 1543
сосна румелийская 489
сосна Сабина 7142
сосна сахарная 1274
сосна сахарная гигантская 1274
сосна скрученная 5044
сосна смолистая 5796
сосна стланиковая 4634
сосна с чешуйчатой корой 7277
сосна с широкопластинчатой корой 6377
сосна съедобная 9031
сосна тонколистная 1485
сосна Торрея 8717
сосна трёхлистная 5787
сосна Тунберга 4614
сосна утончённая 4722
сосна чёрная 415
сосна чёрная австрийская 415
сосна чёрножелобчатая 1972
сосна четырёххвойная 6082
сосна Чихуахуана 1485
сосна широкохвойная 5044
сосна Элиота 2111
сосна эльдарская 2777
сосна Энгельмана 2821
сосна яйцеплодная 5864
сосновая древесина 6275
сосновая живица 6898
сосновая кора 6244
сосновая плантация 6264
сосновая подстилка 6270
сосновая почка 6247
сосновая роща 6255
сосновая смола 6898
сосновая шишка 6250
сосновое масло 6276
сосновое насаждение 6255
сосновые 6254
сосновый вертун 6274
сосновый лес 6255
сосновый пек 6263
сосновый пнёвой осмол 6271
сосновый стволовой осмол 6266
сосняк 6255
сосняк беломошник 4930
сосняк-брусничник 1998
сосняк-черничник 708
состав древостоя 1886
состав насаждения 1886
составная балка 1836
составная средняя пластинка 1842
составная часть древесины 9548
составное резание 1838
составной плод 1839
состояние кроны 2071
состояние лесов 3458
состояние окорки 2282
состояние покоя 2572
состояние развития кроны 2071
состоящий из двух листочков 697
сосуд для живицы 538
сосуд древесины 9687
сосудистая древесина 6467
сосудистая перфорация 9213
сосудистая трахеида 9173
сосудисто-волокнистая ткань 3105
сосудистое растение 9172
сосудисто-лучевые поры 6740
сосудистые элементы 8741
сосудистый пучок 6462
сосущий корень 7049
сотовая гниль 6313
софора 4357
софора виколистная 9214
софора красивая 1939
софора чеырёхкрылая 5729
софора японская 4626
сохранение зелёных деревьев 3961
сохранение леса 3359
сохранение листьев 4885
сохранение природы 2842
сохранившийся 2960
сохранность 1322
сохранность лесных культур 8010
социальное лесное хозяйство 7808
сочность 8250
сочный плод 3262
спармания африканская 115
спелая древесина 6964
спелое дерево 5341
спелое дерево, предназначенное в рубку 5341
спелое насаждение 5338
спелость дерева 8885
спелость леса 3501
спелый 5337
спелый древостой 5338
спелый лес 5338
спелый осмол 7334
специалист по древесине 9659
специальная рубка 7888
специальные лесные продукты 7889
спецификация на фанеру 6400
спецификация размеров

специфическое образование узелков на стволе 7906
спинной распылитель 471
спиральное волокно 7914
спиральный рост 9021
спирея 5350
спирея берёзолистная 755
спирея вязолистная 3832
список лесоматериалов 8376
спичечная соломка 9695
спичечный кряж 5332
сплав 8578
сплав кошелем 480
сплавленный лес 3279
сплав леса плотами 8578
сплав леса россыпью 2625
сплавляемый в плотах лес 3279
сплавлять лес 3269
сплавная бригада 2630
сплавная древесина 3279
сплавная марка 3275
сплавная секция 6692
сплавная трасса 3277
сплавной 2628
сплавной багор 6693
сплавной канал 3277
сплавной лес 3279
сплавной лесоматериал 3279
сплавной период 3278
сплавной плот 8584
сплавной район 3272
сплавной сезон 3278
сплав по запруде 2631
сплавщик 9569
сплачивание 4096
сплетение 618
сплетение из сучьев 1110
сплетение корней 5334
сплетённая растительность 8389
сплетённые стволы 8388
сплотка древесины в запани 912
сплотка леса 9303
сплотка на воде 9303
сплоточная единица 6692
сплотщик 1173
сплошная вырубка для углежжения 1674
сплошная вырубка с оставлением резервных деревьев 1620
сплошная крень 3762
сплошная лесосека 1607
сплошная полосная рубка 8171
сплошная рубка 1607
сплошная рубка с 6227
сплошная рубка с последующим искусственным возобновлением 1618
сплошная рубка тонкомерных деревьев 7779
сплошная сердцевина 1901
сплошная слоистость 5324
сплошное внесение 5990
сплошное прореживание 4902
сплошной пал 1043
сплошной посев 1044
сплошной смоляной потёк 5324
сплошнолесосечная рубка 1607
сплошные рубки 3143
сплошные рубки узкими лесосеками 1622
спокойный пожар 7734
сползание со склона неустойчивых грунтов вместе с порубочными остатками 2287
спонзеуду 5532
спора 7927
спорофилл 7929
спорофор 7928
способ выращивания леса с временным сельскохозяйственным пользованием 146
способ лесозаготовок с трелёвкой хлыстов 8870
способ лесосечных работ с заготовкой сортиментов у пня 6027
способность к приспособлению 60
способность проникновения корней 7038
способность укоренения 7030
способ однеприёмной постепенной рубки 5897
способ поштучной заготовки деревьев 7656
способ пропитки древесины креозотом в вакууме 692
способ равномерных постепенных рубок 9087
способ семеннолесосечной рубки 7515
способ сплошных рубок с оставлением семенников 7400
способствующий вырождению 2696
справка о результатах обмера 5361
справочник лесничего 3570
справочное руководство по древесине 9585
спуск 7750
спускать брёвна со склона 494
спуск древесины с горы 2598
спуск леса по неподвижному несущему канату 3237
спускной канал 2525
спуск по неподвижному несущему канату 3237
спуск по снежному лотку 7803
спящая почка 2573
спящий сеянец 2576
спящий черенок 2574
сравнительное насаждение 4471
срастание 35
срастание корней 5692
срастаться 33
сращивание 7918
сращивание брусьев 4912
сращивание вполдерева 4096
средиземноморская сосна 5309
срединная ветвь 4541
срединная доска 5437
срединное бревно 7362
срединный диаметр 2468
средневозрастное дерево 6418
средневозрастные леса 5351
среднегодовой прирост 5352
среднегодовые осадки 5353
среднее дерево по приросту 4456
среднее прореживание 5508
среднее расстояние канатной трелёвки 444
среднее хозяйство в порослевых насаждениях с оставлением маячных деревьев 1937
среднее число брёвен в пачку 435
среднее число деревьев в пачке 435
среднее число деревьев на акре 438
среднемноголетний дебит 436
среднемноголетний сток 436
среднеполнотное насаждение 5388
средний возраст лесонасаждения 426
средний годичный объём

средний годичный прирост 5352
средний годичный урожай леса 428
средний диаметр порока 431
средний квадратический диаметр 6651
средний квадратический диаметр наиболее толстых доминирующих деревьев 8693
средний лес 7947
средний объём дерева 443
средний объём хлыста 443
средний охват 5440
средний по приросту ствол 4456
средний слой 1949
средний слой ДСП 1953
средний слой из ориентированной стружки 168
средний слой плиты из ориентированной стружки 168
средний слой полуразложившегося опада 3086
средний слой полуразложившегося опада мхов 5896
средний слой фанеры 1949
средний ствол класса 7176
средний ствол насаждения 434
средний сучок 5386
средний ярус растительности 5438
средняя высота 5357
средняя высота дерева 440
средняя высота наиболее высоких деревьев 5358
средняя годичная лесосека 5354
средняя годичная смертность 427
средняя длина волокон 432
средняя длина хлыста 441
средняя жилка листа 4879
средняя модель насаждения 442
средняя опора 4548
средняя периодичность пожаров 5356
средняя плотность короедов 430
средняя часть брёвна, остающаяся после выпиливания брусьев 9257
резах 2514
средства для связывания пыли 2676
средство для окуривания 3765
средство для уничтожения насекомых-вредителей 4512
средство приманки 4509
средство пропитки 4428
срезание дерева выше области закомелистости 3045
срезание пней 8207
срезанная поверхность 7499
срезанная поверхность брёвна 7499
срезанное дерево 7238
срезанные кустарники 9731
срезанный ножами конец 7498
срезать 83
срезать вершину 8702
срок вывозки 6863
срок облесения 99
срок подсочки 6892
срок таксации 2256
сросток 7918
сросшиеся корни 3907
сросшийся 67
сросшийся сучок 833
сруб 2018, 8560
срубание 8824
срубание прикорневых наплывов 1235
срубать 3037
сруб из брёвен 8560
срубка 1607
срубка под поверхностью земли 9059
срубленное сгнившее дерево 7093
срубленные молодые ветви 1938
срубленный лес 3042
срубленный ствол 2600
срывной 2170
стабильный крупносерийный доход 5147
стадий сукцессии 7455
стадия дерева 8931
стадия кустарника 1115
стадия молодняка второго класса 7206
стадия молодняка первого класса 7391
стадия молодого древостоя 7391
стадия молодого леса 8495
стадия неполовозрелости 4424
стадия подроста 7206
стадия приспевающего насаждения 9790
стадия развития 2445
стадия спелого насаждения 5340
стадия средневозрастного насаждения 6426
стадия сукцессии 8248
стадия трав и полукустарников 3916
сталактитовидная ржавчина 7996
стальной клин для раскалывания брёвен 3717
стамеска 4328
стандартизация лесоматериалов 8642
стандартизованные акры 8003
стандарт на лесоматериалы 9664
стандарт насаждения 8111
стандартная длина брёвна 8005
стандартная таблица объёмов 8007
стандартный диаметр 5782
стандартный здоровый сучок 8004
стандартный пиломатериал 2503
стандарты использования 9156
станок для натяжения цепи 1436
станок для окорки чураков 711
станок для разделки брёвен на брусья 1018
станок для раскроя брусков 839
станок для увязки дров и щепок 3220
станок для удаления коры с веток или сучьев 5964
станочник обрезного станка 2764
станция по лесоразведению 8805
старая древесина 5886
старение 7438
старение в атмосферных условиях 9311
старое насаждение 5880
старый лес 5880
статическая твёрдость древесины 9665
статут кроны 2071
стафилея 816

стационарная рубильная машина 8044
стационарная сучкорезная машина 8045
ствол 1389
ствол без сучьев 1604
ствол дерева 8934
стволовая гниль 8998
стволовая древесина 891
стволовая поровость 8080
стволовая щепа 1539
стволовой древоточец 8063
стволовой огонь 8072
стволовой перечёт 2837
стволовой побег 2848
стволовой сток 8073
ствол с корнями 2835
ствол с однобоковой кроной 2062
стебелёк 4883
стебель 2124, 8934
стебельковая клетка 7997
стебель кустарника 1307
стебельный 1390
стебельчатый 1390
стеблевидный 1392
стеблевой суккулент 8082
стеблевой черенок 8070
стеблеобразный 1392
стек 9519
стекающая вода 8968
стеклянница тополевая 4375
стеллаж 6034
стель 8058
стелющаяся ветвь 7132
стелющаяся форма дерева 6589
стелющееся растение 1642
стелющиеся деревья 1644
стелющийся 2313
стелющийся бересклет 9454
стелющийся габитус 6558
стелющийся корень 7486
стелющийся побег 7132
стенка или береговой устой, препятствующие обсыханию брёвен или образованию заторов 1229
стенки клеток сердцевинных лучей 6737
степень концентрации насаждения 8110
степень плотности насаждения 8110
степень прорубки 8519
степень смешения 3897
степной лес 3607
стержень 1389
J-образный ~ для отклонения возвратного троса 4664
стержневой корень 8411
стержневой побег 4237
стеригмы 8089
стеркулиевые 1550
стеркулия клёнолистная 3245
стеркулия остролепестковая 1090
стеркулия утёсная 594
стёсывать 2175
стимулятор роста 4020
стиракс 3825
стиракс аптечный 3825
стиракс калифорнийский 1273
стиракс крупнолистный 702
стиракс лекарственный 3825
стираксовое дерево 259
стираксовые 8128
стланик 2782, 8756
стлаться 2011
стоимостная оценка леса 3411
стоимость вырубки 3056
стоимость культуры 6350
стоимость леса 3536
стоимость леса на корню 8196
стоимость лесной земли 3483
стоимость лесозаготовок 8217
стоимость лесоматериала 9629
стоимость лесосечных работ 3056
стоимость насаждения 8196
стоимость нелесных ресурсов 5776
стоимость, подверженная риску 9164
стоимость посадки 6350
стоимость ресурсов 6914
стоимость рубки 3056
стойка 8190, 8902
стойка для предупреждения раскатывания брёвен 2054
стойка коренных лесов 7270
стойкость 4123
стойкость древесины 2669
стойкость против раскалывания 6907
столбовая древесина 6428
столбовидный корень 277
столбовой гриб 7722
столон 7132
столяр 4656
столярная обработка 9698
столярные работы 5458
столярный лесоматериал 4658
стопорное стабилизирующее устройство 2054
сторона брёвна, соприкасающаяся с землёй 6937
сторона в направлении валки 3722
сторона дерева, обращённая в направлении естественного наклона 4896
стоянка 8049
стоячий 8018
стоячий на корню 8018
страна-производитель лесоматериалов 8624
страстоцветные 6099
стратегический план землепользования 8146
стратегия насаждения 8030
стратегия развития насаждения 8030
стратификация 8147
стратобионт 8149
страхование леса от пожара 3432
страховочный рычаг 7151
стрела 910
стрела сучкорезной машины 2353
стреловая погрузка 913
стремительный рост 141
строгальная пила с мелкими зубьями 5803
строгальная стружка 9274
строгальный станок 6326
строганная доска 8280
строганный пиломатериал 2620
строганный шпон 7738
строгающая цепь 2183
строевое дерево 517
строевой лес 4280
строение волокна 3100
строение годичных колец 8188
строение годичных слоев 8188
строение дерева 6094
строение древесины 9666
строение коры 586
строение леса 1886
строительная жердь 1152
строительная клеёная фанера 1151
строительная фанера 1151
строительные бревно 1888

строительные леса с направляющими рельсами для тележек 4041
строительные лесоматериалы 1153
строительный лесоматериал для морских сооружений 7524
строительство волоков 8744
строительство лесных дорог 1889
строительство сосредоточенных жилых зон, окружённых зелёными поясами 1667
строительство сосредоточенных зон 1667
строительство трелёвочных волоков 8744
стропильный лесоматериал 6688
струбцина 4653
струг для расчистки местности 4772
стружечный кряж 7553
стружка-отходы при строгании 9274
стружки 9508
стружки без остатков коры 1605
стружки верхнего слоя 8284
стружки наружного слоя 8284
стружки от сверления 926
стружки среднего слоя 1952
стружкообразование 1515
струйно-гидравлический окорочный станок 9289
струйчатая текстура 8175
структура древесины 9666
структура запаса 8189
структура лесной почвы 3602
структура насаждения 1886
структурное разнообразие 8183
структурный термозащитный покров 8186
стручковое дерево 1714
стряхиватель деревьев 8845
стряхиватель плодов 8845
стряхиватель плодовых или семенных деревьев 8845
стряхиватель с захватом ветвей 4971
ступени толщины 8499
ступенчатая карра 4670
ступенчатая обрезка 8085
ступенчатая посадка 6357
ступенчатая рубка 8084
ступенчатый подпил 8088
ступень диаметров 2471
ступичный кряж 3082
стюартия 8091
стюартия яйцевидная 5591
стяжка 716
субальпийская гниль мёртвой древесины 8227
субальпийский лес 8226
субарктический лес 8229
суборь 8234
субтерминальная почка 8242
субтропический вечнозелёный лес 8243
субтропический лес 8244
субэкваториальный муссонный лес 8232
судостроительный кряж 7524
сужение 8403
сужение ствола 8083
суживающийся к вершине 8401
сук 4723
суккулентность 8250
суковатость 986
суковатый 977
сук, расположенный с нижней стороны 9066
сук с наростом 8352
сукупира 8251
сукупира жёлтая 4784
сукцессионное растение 7454
сукцессионный 7453
сукцессия 8246
сукцессия древесных пород 8249
сумах венецианский 2908
сумах восковый 8315
сумах гладкий 7791
сумах глянцевый 4755
сумах голый 7791
сумах дубильный 2908
сумах канадский 6412
сумах копаловый 3247
сумах коротковолосый 260
сумах лавровидный 4836
сумах лаковый 8315
сумах лаконосный 4755
сумах мелколистный 5005
сумаховые 1369
сумах оленьерогий 260
сумах пушистый 260
сумах резиноносный 9306
сумах сахарный 8255
сумах укореняющийся 6412
сумах уксусный 260
сумах цельнолистный 4909
сечений насаждения на гектаре 601
суммарное пользование лесом 8724
суммарный возраст 8718
сумчатые грибы 7144
суппорт с сучкорезным ножом 2354
супротивная поровость 5929
супротивное листорасположение 5928
супротивный лист 5927
сурамень 8277
суриана приморская 628
сухая бурая призматическая гниль 2656
сухая вершина 7794
сухая гнилая древесина 8666
сухая гниль 2656
сухая древесина 2658
сухая карра 2264
сухая набивка 2654
сухая окорка 2646
сухая сосновая заготовка 7333
сухие древесные частицы 2659
сухобокость 3206
суховершинное дерево 2273
суховершинность 7989
сухой лес 2650
сухой лесоспуск 2648
сухой лоток 2648
сухой плод, раскрывающийся на несколько частей 1328
сухой сук 2647
сухостой 2274
сухостойная древесина 2274
сухостойное дерево 2271
сухостойные повреждённые деревья 7166
сухостойный лес 2267
сухость карры 2649
сухотарные бочки 7713
сучковатая древесина 4738
сучковатая сосна 4739
сучковатое ядро 1948
сучковатость 986
сучковатый 4737
сучковатый нарост 832
сучковатый ствол 4721
сучковое отверстие 4732
сучковый крюк 6613
сучковый якорь 274
сучколовитель 981
сучкоотделитель 981
сучкорезка 6614, 8825
сучкорезная головка 4969

сучкорезная машина 8825
сучкорезная машина с непрерывной подачей 1896
сучкорезное приспособление 4969
сучкорезно-пакетирующая машина 4964
сучкорезно-раскряжёвочная машина 2352
сучкорезный нож 2285
сучкорезный удар 2356
сучкоруб 981, 8825
сучкорубный топор 5167
сучняк 5170
сучок, выходящий на ребро заготовки 4733
сучок, разрезанный в продольном направлении 7910
сучья 5170
сучья и вершина 5163
сучья и вершина, срубаемые с поваленного дерева 5163
сучья с кроной 5163
сушение 4711
сушилка для пиломатериалов 5203
сушильная печь 2623
сушильная установка для древесины 9537
сушка в масле 5876
сушка вымораживанием 3702
сушка древесины 2423
сушка древесных частиц 9620
сушка на открытом воздухе 152
сушка пиломатериалов 5202
сушняк 2271
сформировавшийся лист 69
схема освоения лесосек с хаотичным расположением веток и усок 6715
схема разработки лесосек 5051
схема расположения трелёвочных волоков 7688
схема сырьевых ресурсов 6913
схема транспортного освоения горных лесосек с дорогой, поднимающейся по впадине, а затем поворачивающей на гребень 6938
схема транспортного освоения горных лесосек с одной поднимающейся вверх дорогой и примыкающими параллельными схема транспортного освоения
схизолизогенный 7296
схлёстывать 4812
схлестывающее дерево 9358
сходить слоями 2940
сцепная тяга 2616
сцепное приспособление 2616
сцепной брусок 2616
счёт годичных слоев 8911
счётчик брёвен 5054
счётчик деревьев 8820
счётчик количества щепы 1520
сшивной сучок 4376
съедобная пальма 1238
съёмка возобновления 6824
съёмка лесных ресурсов 3563
сырая древесина 3710
сырая живица 6733
сырая смола 6733
сырое состояние пиломатериалов 3957
сырой гумус 1681
сырой листовой гумус 4999
сырой осадок 7473
сырые пиломатериалы 9116
сырьё для изготовления мочала 625

табак сизый 8945
табачный сучок 2290
табебуйя Донэль-Смита 6551
табебуйя ипе 4581
табебуйя караибская 1339
таблица вычисления объёмов деревьев 8957
таблица для определения раскроя 1133
таблица Дойля 2604
таблица замеров 7275
таблица запасов 8957
таблица классов возраста 129
таблица лесной таксы 8415
таблица лесозаготовки 4163
таблица насаждений 8033
таблица норм разреживания насаждений 8518
таблица объёмов, измеряемых по нескольким показателям 5612
таблица объёмов, измеряемых по одному показателю 7638
таблица объёмов лесоматериалов по диаметру пня 8212
таблица объёмов пиловочника 7251
таблица объёмов стволов лесоматериалов 7782
таблица определения объёмов брёвен 7112
таблица определения объёмов брёвен в досковых футах 878
таблица определения объёмов брёвен в кубических единицах 2117
таблица определения объёмов в кубических единицах 2117
таблица площадей поперечного сечения 8362
таблица прироста 4457
таблица распределения насаждений 8033
таблица роста насаждений 9784
таблица роста насаждений различной полноты 9167
таблица роста нормальных насаждений 5786
таблица роста обыкновенных насаждений 2798
таблица роста одновозрастных насаждений 2921
таблица роста средних насаждений 2798
таблица сбега 8409
таблицы хозяйственных видов 5265
таволга дубравколистная 3832
таволга иволистная 9433
таволга трёхлистная 5350
таёжный бореальный лес 8367
таёжный лес 8368
тайга 8368
такелаж 1942
такса для очистки лесной земли или леса 8430
таксационная книга 3409
таксационная мерная лента 8289
таксационная партия 8288
таксационная призма 6554
таксационная характеристика насаждения 8009
таксационное описание насаждения 2411
таксационный выдел 8150
таксационный диаметр 2466
таксационный прирост насаждения 8291
таксационный прицел 6554
таксация горизонтальными точками 4370

таксация древесного запаса 8657
таксация леса на постоянных пробных площадях 1898
таксация лесоматериалов 9671
таксация методом ленточных пробных площадей 4986
таксация методом линейной выборки 4988
таксация методом угловых проб 285
таксация насаждений 8290
таксация свободно растущих деревьев 3695
таксодиевые 8425
таксодий восходящий 6444
таксодиоидная поровость 8424
таксодиум восходящий 6444
таксон 8427
таксоном 8428
таксонометрический 8426
таксономическая единица 8427
таксономия 8429
таксоценоз 8423
талловое масло 8374
тальник 617
тамарак 250
тамарикс галликский 3706
тамарикс галльский 3706
тамарикс крымский 7939
тамаринд 8379
тамаринд индийский 8379
тамариндо 9190
тамариск африканский 116
тамариск пятитычиночный 3233
тамариск четырёхтычинковый 7939
тангенциальная трещина 2139
тангенциальная усушка 8386
тангенциального распила 3258
тангенциальное разбухание 8387
тангенциальной распиловки 3258
тангенциально распилённый 3258
тангенциальный надруб 1565
тангенциальный наклон волокон 8385
тангенциальный разрез 8383
тангенциальный распил 8383
тангенциальный рост ткани годичных колец древесины 2497
тариф среднего ствола 5359
тарный лесоматериал 954
татаджуба 478
тау 6879
твёрдая древесина 3225
твёрдая древесноволокнистая плита 1845
твёрдая заболонная гниль 4130
твёрдая смола 593
твёрдая широколиственная древесина 4121
твёрдолиственная древесина для конструкционных целей 9151
твёрдолиственные породы 4138
твёрдолиственный лес 7301
твёрдость 4123
твёрдость древесины 4128
твёрдый луб 4119
твёрдый плодокорм 4127
твёрдый сухостой 4131
текома укореняющаяся 8994
текстура в виде волнистых полос 8175
текстура в виде изогнутых полос 1902
текстура в виде капли дождя 6697
текстура волокон 9678
текстура древесины 9678
текстура на кромке 2760
текстура на торце 7728
текстура, неравномерная по окраске и рисунку 9082
текстура, получаемая при продольной распиловке 7728
текстура продольного распила 5133
текстура, равномерная по окраске и рисунку 2923
текстура радиально распилённого пиломатериала 1738
текстура с крупными элементами 1678
текстура с мелкими элементами 1655
текстурный веерообразный рисунок 3011
текстурный рисунок 9678
текстурный рисунок в виде завитков 5552
текстурный рисунок в виде ломаной линии 1062
текстурный рисунок типа "птичий глаз" 766
2150
телега для колод 5199
телега с хлыстами 5058
тележка для леса 8565
тележка для пиломатериалов 5208
тележка для подачи брёвен 5208
тележка для подъёма бревен 5098
тележка лесопильной рамы 5208
темир-агач 6186
тёмная лесосека 2253
тёмная поросль 4491
тёмно-бурая кубическая ядровая гниль 2250
тёмнохвойный лес 2252
темноцветная лесная почва 789
тёмный лес 7847
темперированная твёрдая древесноволокнистая плита 8445
темп рубки 8446
тенам 5405
тенденция развития 2446
теневыносливая порода 7474
теневыносливое дерево 7478
теневыносливость 7477
тензиометр 8464
тенистое дерево 7479
теория лесоустройства 8487
теория моноклимакса 5538
теория поликлимакса 6437
теория управления лесами 8487
теплица 3946
теплолюбивое растение 8492
теплопроводность древесины 8489
теребинт 8469
терминалия амазонская 1712
терминалия войлочная 9719
терминалия иворенская 4596
терминалия катаппа 4464
терминалия обратнояйцевидная 1712
терминалия пышная 92
терминальная почка 8470
терминальное расстояние 3139
терминология по лесоводству 3619
термит 8476
термозащитное покрытие 8490
термозащитный покров 8490
термофильный 8491
терн 7749

терносли 2244
тернослива 2244
терпентинное дерево 8469
терпентиновое дерево 8469
территориальные единицы переменного лесопольного хозяйства 144
территория закустаренного низинного болота 1356
территория земного шара, покрытая лесами 9725
территория лесосеки 2189
территория, покрытая лесом 8604
территория, поросшая кустарником 1108
тёс 882
тесак 453
тёсаное бревно 638
тёсаный лесоматериал 4271
тесать 83, 1561
тесина 8520
тёска 4273
тесло 82
тесная связь 4572
теспезия обыкновенная 6471
тест дерева 8944
тест лесных культур 3630
тесун 1452
тетрацентровые 8481
техасская корневая гниль 1981
техника безопасности на лесозаготовках 5082
техника лесовозобновления 6805
техника рубки 3076
техник-лесовод 3579
техник по деревообработке 9675
техническая спелость 8439
технологическая щепа 6625
технология ведения лесного хозяйства 3401
технология обработки древесины 8582
тибоухина 3871
тибоухина длиннолистная 5139
тибурбу 5534
тибухина 3871
тибухина длиннолистная 5139
тигровое дерево 8548
тик 109
тиковая древесина 8436
тиковое дерево 109
тиковый лес 8435
тиллоид 8540
тип биогеоценоза 8731
типичное дерево 442
тип климаксовой растительности 4079
тип леса 3363
тип леса брусничный 9158
тип леса вересковый 1284
тип леса дубово-сосновый 5847
тип леса из дубов и гикори 5838
тип леса лишайниковый 1593
тип лесного участка 6721
тип лесорастительных условий 7675
типология лесных почв 3603
тип смешения 4713
тис 1814
тис европейский 1814
тис китайский 1505
тис коротколистный 9345
тисовые 8431
тисовый лес 9768
тис флоридский 3288
тис ягодный 1814
титония 8882
титония разнолистная 8882
ткань гетероцеллюлярных лучей 4265
ткань гомоцеллюлярных лучей 4345
ткань коры 1975
тлеющее горение 3875
тлеющий пожар 7792
тли 323
тля буковая 650
тля еловая трубчатая 7949
тля пихтовая 7949
товарная древесина 5400
товарная продукция завода 5454
товарное бревно 5393
товарное дерево 5399
товарные брёвна 5396
товарный лес 5394
товарный ствол 5393
токарный лесоматериал 9012
токота 1003
токсикодендрон 6022
токсические пояса 8777
токсичность древесины 9680
толерантность 8673
толкание дерева при валке 6520
толкатель дерева 8905
толокнянка 5289
толокнянка аптечная 1759
толстая доска из комлевого кряжа 8062
толстая кора 1675
толстая палка 7111
толстое короткое бревно 1225
толстокорый 590
толстомер 4808
толстомерные деревянные отходы 2505
толстомерный лесоматериал 4808
толстомерный хворост 4807
толстостенное волокно 4926
толстый брус 8502
толстый жердняк 4803
толстый сук
с толстыми сучьями 8135
толуское бальзамовое дерево 504
толщина 8496
толщина коры 8498
толщина пиломатериала 7256
толщина снимаемого за один проход слоя древесины 2195
толщина ствола 8071
толщина ствола дерева на середине высоты 2468
толщиномер 8497
толщинометр для коры 587
томатное дерево 8946
тона китайская 1493
тонкая балка 7485
тонкая высококачественная древесная масса 3152
тонкая доска 8520
тонкая жердь 7775
тонкая текстура 1655
тонкая топливная щепа 3151
тонкая трещина 4088
тонкая хвойная жердь 1876
тонкая чистовая стружка 3148
тонкие древесные стружки 7553
тонкий брус 8523
тонкий жердевой лес 9785
тонкий корень 8452
тонкий поделочный хворост 8606
тонкий рудный кругляк 7771
тонковласник кривостебельный 8950
тонковолокнистая древесина 3149
тонкомер 7760
тонкомерное бревно 7775
тонкомерное дерево 7778

тонкомерные деревья, оставленные при рубке леса 7760
тонкомерные лесоматериалы 7760
тонкомерный кругляк 7774
тонкомерный лес 7760
тонкомерный сортимент 7760
тонкомерный ствол 8522
тонкослойная древесина 1660
тонкослойный 9470
тонкотекстурный 3157
тонна 8675
топка для сжигания древесных отходов 7236
топка дровами 9568
топлива 3739
топлива из-за хозяйственной деятельности 56
топливная древесина 3740
топливная кора 3734
топливная щепа 1517
топливный чурак 3219
топляк 2268
топографическая серия пластов 8705
топографическая серия пластов почвы 8705
топография 8703
тополевая заросль 6454
тополевая кора 6452
тополевая культура 6456
тополевник 6454
тополевое насаждение 6456
тополево-ивовый лес 6457
тополевый древоточец 6453
тополевый лес 6454
тополевый листоед 1982
тополевый питомник 6455
топология дерева 8948
тополь разнолистный 8306
тополь бальзамический 9338
тополь белый 2911
тополь берлинский 681
тополь болотный 8306
тополь волосистоплодный 784
тополь дельтовидный 2717
тополь дрожащий 2880
тополь индейский бальзамический 9338
тополь итальянский 5126
тополь канадский 2717
тополь крупнозубчатый 706
тополь Максимовича 4629
тополь осинообразный 236
тополь пирамидальный 5126
тополь седеющий 3973
тополь серебристый 2911
тополь Симона 7627
тополь узколистный 5661
тополь Фремонта 3703
тополь чёрный 2884
тополь чёрный гибридный 4400
топор 453, 673
топор для затёски лесоматериалов 1041
топор для карр 2142
топор для кряжей 1452
топор для срубания сучьев 5167
топорик 4081
топорик для затёски 5314
топорище 4110
топорная соковая окорка 6136
топорник 8521
топор с двумя лопастями 9026
топор с двусторонним лезвием 2579
топор с изогнутым по контуру дерева широким лезвием 4390
топор с односторонним лезвием 7636
топор с широким лезвием 1041
топор с широкой режущей кромкой конической формы 2142
топофиза 8704
торговая длина 5395
торговая лесопилка 1749
торговля лесом 8653
торговой лес 5400
торец 2051
торец колотой древесины 1630
торец пиломатериала 7257
торможение роста 4015
тормоз лесоспуска 3888
тормозная колода 4324
тормозная цепь 1427
тормозное бревно 2263
тормозное устройство на лесоспуске 3888
тормозной блок 2263
тормозной канат 971
тормозной такелаж 970
тормозной трос 971
тормоз пильной цепи 1423
торрея калифорнийская 1276
торус 6308
торус поры 6314
торфянистый слой 5873
торцевание 1215
торцевать 7962
торцевое волокно 2803
торцевой разрез 2814
торцевой срез 2980
торцевые крюки 2806
торцовая стена 2817
торцовая трещина 958
торцовка 1215
торцовочная пила 5056
торцовочный станок 2043
торцовый сучок 4733
торчок от сучка 993
точение по дереву 9683
точечная гниль 6242
точечная сетка 2577
точечная смолёвка 7770
точечное отверстие 6461
точильщик 7239
точка
с ~ми 6131
точка критического объёма производства 1015
точка насыщения волокон 3098
точка роста 4007
точка росы 2457
точно определённый объём брёвен 5719
трабекула 8739
траверса цепного лесотранспортёра 2038
травматическое ядро 8782
травмирование коры 558
травы 4253
травянистая растительность 4252
травянисто-зелёный 3453
травянистый слой 3116
травяной ельник 3920
травяной сосняк 3919
травяной ярус 3116
трагант 4055
традиционные лесные продукты 1910
традиционный лес 8748
трактор для трелёвки коротья 6423
тракторист, отцепляющий брёвна от канатной установки и трелюющий их на площадку 1455
тракторист трелёвочного трактора 7682
тракторная вывозка 8747
тракторная трелёвка 8747
тракторная трелёвка с применением арки 349
трактор с канатом 8370

трактор с лебёдкой 1246
трактор с лебёдкой, которой используется в качестве растяжки 4066
трактор со стрелой и коником с зажимными рычагами 1597
трактор-тягач 2970
транзитный плот 8766
трансекта лесонасаждения 8034
транспирационная сушка 8767
транспирационный поток 8768
трансплант 5815
транспортёр вывода чураков 714
транспортёр для отвода горбыля 7711
транспортёр для пиломатериалов 5208
транспортёр для подачи коры 547
транспортёр для подачи чураков 712
транспортёр для сучьев 4729
транспортёр для удаления коры 547
транспортировка леса 8595
транспортировка леса вертолётами 4239
транспортно сухой 7527
транспортно сухой лесоматериал 7528
транспортный каток 3035
трасса 6941
трассировка 8743
трассировочный колышек 3899
трассировочный столбик 3899
трахеальные элементы 8741
трахеида 8742
трахеида креневой древесины 1851
требование к местопроизрастанию 7671
требования ведения лесного хозяйства 3494
трелевать брёвна по земле 9024
трелевать лебёдкой 1241
трелевать лес канатными установками 1241
трелевать лесоматериалы в полностью погруженном положении 6507
трелёвка 8595
трелёвка без складирования 4387
трелёвка большого пролёта друг за другом 8754
трелёвка вверх по склону 9129
трелёвка вниз по склону 2598
трелёвка волоком 3994
трелёвка волоком с использованием пэнов 6054
трелёвка в погруженном положении 3672
трелёвка деревьев 9394
трелёвка древесины вниз по склону 2598
трелёвка канатной установкой с тягово-несущим канатом 8466
трелёвка колёсными тракторами 7683
трелёвка леса в полупогруженном положении 7433
трелёвка мобильной погрузочной установкой 4608
трелёвка на большие расстояния 2966
трелёвка на короткие расстояния 4558
трелёвка паровой лебёдкой 2571
трелёвка пачек балансов, обвязанных поперёк цепью 1428
трелёвка по воде 9300
трелёвка по магистральному волоку 2966
трелёвка по неподготовленному волоку 5777
трелёвка по подготовленному волоку 8759
трелёвка по суше 4779
трелёвка предельного пролёта 5150
трелёвка с высококолесной тележкой 4304
трелёвка с высоко лежащего кабеля 4288
трелёвка с цепью 1428
трелёвка тросом и лебедкой 3994
трелёвка установкой с опускающимся несущим канатом 7715
трелёвка хлыста 899
трелёвочная арка 5090
трелёвочная дорога 7689
трелёвочная дуга 5090
трелёвочная каретка 7698
бензиновым двигателем 3809
трелёвочная мачта 5328
трелёвочная машина 7692
трелёвочная машина с неповорачивающейся в горизонтальной плоскости стрелой 3236
трелёвочная машина с поворотной стрелой 8343
трелёвочная оснастка с закреплёнными на одной скобе двумя сборочными чокерами 2588
трелёвочная оснастка с применением дополнительного чокера 8366
трелёвочная тележка 5208
трелёвочная установка 7691
трелёвочная шина 5086
трелёвочные крюки 3892
трелёвочные подсанки из раздвоенного дерева 5020
трелевочные салазки 894
трелёвочные сани 5115
трелёвочные сани с вилами 3881
трелёвочные щипцы 3892
трелёвочный волок 3987
трелёвочный волок из параллельно проложенных брусьев 8164
трелёвочный волок из поперечно проложенных слег 6425
трелёвочный волок по грунту 8757
трелёвочный волок по грунту с отбойным бревном 8757
трелёвочный клещевые захваты 3892
трелёвочный коник 5090
трелёвочный конус 1858
трелёвочный метод 9749
трелёвочный механизм 5114
трелёвочный пен 7687
трелёвочный поддон 7687
трелёвочный трактор 7799
трелёвочный трактор для перевозки пачек сортиментов в полностью погруженном положении 6028
трелёвочный трактор на резиновом ходу 7125

трелёвочный трактор с захватом 3913
трелёвочный трактор с пачковым захватом 3913
трелёвочный трактор с рамным управлением 373
трелёвочный трактор с тросово-чокерным оборудованием 1246
трелёвочный трактор с шарнирно-сочленённой рамой 373
трелёвочный трос 3987
трелёвочный тягач 2970
трелёвочный щит 1228
трелёвщик 1435
трелёвщик брёвен 1435
трема восточная 4462
треснувший древесный ствол 9696
третья стадия гниения 1828
трёхзубчатый лист 8969
трёхкантный брус 8532
трёхколючковая акация 4353
трёхлистник 8961
трёхлистный клевер 8961
трёхлопастный лист 8535
трёхцветковый клён 8533
трещина 7483
трещина в древесине 9507
трещина, заполненная смолой 6893
трещина клеточной стенки 1408
трещина молнией 4946
трещина на лицевой поверхности 2978
трещина на пласти 8279
трещина, не выходящая на пласть пиломатериала 958
трещина от валки 3052
трещина от высыхания 2422
трещина от жары 2002
трещина от солнца 8266
трещина от удара молнии 4946
трещина от усушки 2422
трещина скалывания 7497
трещина средней длины 5385
трещинистый 4201
трещиноватая ядровая гниль 1084
трещинообразование 2003
триангуляция 8966
тристания свёрнутая 1036
тристания скрученная 1036
тритринакс бразильский 1009
тройчатый лист 8477
фасоль 4707
тропическая древесина 8981
тропический дождевой лес 8978
тропический лес 8975
тропический полусухой лес 7425
тропический полусухой листопадный лес 8974
тропический сухой листопадный лес 8973
тропическое лесное хозяйство 8977
тропофиты 8982
трос 9463
тросный лесоспуск, работающий силой тяжести 3922
тросовая лебёдка 7060
тросовая оснастка 7467
тросовая петля 699
тросовая трелёвка 1247
тросовый работник 4296
тросоукладчик 7058
тростник 1308
трофическая связь 3308
троходендровые 9357
трубковёрт берёзовый чёрный 761
трубчатая гниль 6291
трубчатая клетка 8553
трудно воспламеняющаяся древесина 4125
труднообрабатываемая древесина 4126
труднопроходимая местность 4212
трут 8666
труха, оставляемая насекомыми в червоточине 9728
трухлявая гниль 7819
трухлявая древесина 6484
трухлявая сердцевина 1440
трухлявость 2433
тсуга 4248
тсуга западная 5588
тсуга канадская 1296
тсуга каролинская 1345
тсуга разнохвойная 5588
тугай 6961
тугайный лес 6961
тузам 5404
тунг молукский 1304
тунговое дерево 1304
тунговое масло 9009
тунг Форда 1304
туннель 4511
турецкая пальма 2886
турраэнтус 117
турраэнтус африканский 117
тут 8354
тут белый 1791
тут красный 6781
тутовник белый 1791
тутовое дерево 8354
тутовые 5599
тушение лесного пожара 3431
туя гигантская 8893
туя скалистая 8893
тщательная заготовка леса в обхвате регенерации 1337
тыловая мачта 8371
тыловая опора 2815
тысяч кубических футов 8530
тычинка 7998
тюлипвуд 1007
тюльпанное дерево 9008
тюльпанное дерево бразильское 1007
тягло 9395
тяговая верёвка 8745
тяговая древесина 8467
тяговой канат 8745
тяговой канат лебёдки 8745
тяговой трос 8745
тягово-несущий канат 7133
тяжевая трахеида 8145
тяжёлая древесностружечная плита 4277
тяжёлая кустарниковая борона 1203
тяжёлого механического состава 3157
тяж паренхимы 6077
тянуть волоком 9024
тяпка легкая 4081

убежище от пожара 3208
уборка лесосечных отходов 7723
уборка отходов 7723
уборка сухостойных и буреломных деревьев 7793
уборка хвороста 7723
уборка хлама 1624
увеличенный комель дерева 3250
увядающий 5302
увядший лист 3238
увязка 1178
увязывание в пачку 1178
углежжение 1331
углежог 1444

углубление в коре для яиц насекомых 2773
углубление для продевания чокера 1556
углубление для яиц 2773
углубление ствола для цепи 1556
угнетение от верхнего полога 8276
угнетённые деревья 8275
угнетённый 6011
угнетённый класс крон 6012
угнетённый рост 8223
угол заострения 7493
угол заточки 7493
угол наклона микрофибрилл 3102
угол резания 288
угол среза 288
уголь 1673
угольный отвал 5464
угольщик 1445
угрожаемые растительные сообщества 8531
угрожаемый вид 2801
угрюмый лес 7847
удаление вершины 2198
удаление кроны 2083
удаление кустарника 1109
удаление листьев 4866
удаление одного из ответвлений развилки дерева 3658
удаление отходов 7723
удаление перестоя 2969
удаление почек 1142
удаление пробки 6865
удаление пыли 2318
удаление сушняка 2655
удаление углерода 7452
удаление хвои 2995
удалять боковые побеги 1094
удалять камедь 2345
удалять смолу 2345
ударная атака 6723
ударная твёрдость древесины 9651
удар топором 450
удельная площадь листа 7904
удельный лес 2077
удельный объём листа 7905
удержание групп 139
удерживание в группе 139
удерживающие лесозаготовки 6919
удлинённолопастный 5143
погоде 205
удобрение 3089, 3090
удобрение лесов 3424
узел 4723, 5741
узкая лесосека 8172
узкая ножовка 1822
узкая пила 1822
узкий проход 182
узковершинное дерево 5657
узкокаемчатая лесосека 5658
узкокронное дерево 5657
узколесосечные рубки ухода 4992
узкослойная древесина 1660
узловатая торцевая стенка 5743
узловатость 5742
узловатый 4737
узор древесины 9678
узорная прожилка 7914
узор "птичий глаз" 766
узорчатая берёза 3122
узорчатая древесина 9186
узорчатая древесина комлевой или пнёвой части дерева 6703
узорчатость 3106
узорчатый рост 3123
указатель направления валки 8567
указатель натяжения несущего каната 7705
указательное дерево 4474
указательное насаждение 4471
укладка в ёлку 4261
укладка в стопу 4846
укладка в штабель 4846
укладчик брёвен 8094
укладчик строганных досок 6327
укладывать в небольшие штабели 1169
укладывать в штабель 7983
укладывать лесоматериалы в штабель 7983
уклон 7750
уклон в грузовом направлении 3028
укорачивание 1224, 4187
укорачивание корней 7040
укоренение 7029
укоренение привоя 7298
укоренение черенка 7031
укоренение черенков с помощью искусственного тумана 5481
укороченный побег 7548
укосина 5980
укрепление склонов 7974
укреплённый трелёвочный волок 7696
укрывание землёй 5578
улекс 3889
улекс европейский 3889
улучшающие посадки 2830
улучшение древесных пород 1028
улучшение древостоя 8643
улучшение местообитания 4076
улучшение местопроизрастания 7666
улучшение площади лесосеки 7156
улучшение селекцией 4433
улучшение участка 6717
улучшенный лес 4431
улучшенный пар 2829
умбеллюлярия калифорнийская 1264
умгузи 9793
умение ориентироваться в лесу 9522
уменьшение диаметра от комля к вершине 8405
уменьшение пожарной опасности 3187
умеренная прорубка 5508
умеренное прореживание 5508
умеренно тонкая текстура 5507
унгнадия прекрасная 5421
универсальная формула расчёта потерь почвы 9100
универсальное уравнение смыва почвы 9100
уничтожение кустарника 1109
уничтожение леса 3373
уничтожение с корнем 2968
уничтожение тропических лесов 2432
уничтоженная древесина 2364
упаковка дерева 8960
упаковка древесных отходах 3560
упаковка фанеры в пачку 6399
уплотнение 8494
уплотнение почвы 7834
уплотнение щепы 1523
уплотнённая древесина 8659
уплотнённая посадка 8501
уплотнённый посев 8501
уплотнитель щепы 1522
упор пилы 7234

управление выделения дыма 7784
управление генетическими ресурсами 5264
управление дикой природы 9415
управление дыма 7784
управление лесозаготовками 5076
управление местообитаний 4077
управление насаждением 8107
управление рассадника 5813
управление экосистемы 2749
управляемая лесная земля 5261
управляемое ведение лесозаготовки 5262
управляемый лесоучасток 8612
управляемый пал 6516
упругая древесина 2776
уравнение хода роста насаждений 4016
урезка 2617
урема 6961
урна 9144
урночка 9144
уровенный кол 3899
уровень земли 3990
уровень насаждения 8019
уровень поверхности земли 3990
уровнемер 4922
уродливое дерево 5478
урожай буковых орешков 656
урожай жёлудей 5843
урожай леса 9523
урожай лесных семян 3584
урожайный год 5329
урожайный год шишек 1867
урожай олив 5893
урожай послерослевого насаждения 1925
урожай семян 2033, 7376
урожай хвойных семян 2033
урожай шишек 2033
ус 1441, 7132
усадьба 3017
усач дубовый 5848
усачи 1327
усач кедровый 1403
усач лиственницы западной 9341
усач, повреждающий шишки сосны жёлтой 6446
усач чёрный орегонский 5942
усик 8462
ускоренное пережигание 3315
ускоритель роста 4020
ускоряющий рост 4021
условия произрастания насаждения 8009
условная метка 8486
условная плотность древесины 1913
условное бревно 7172
условно-сплошные рубки 1854
усообразный придаток 8462
установка блоков по границе 7571
установка блоков по границе лесосеки 7571
установка для обрезки 2767
установка для переугливания древесины 1332
установка для получения древесного угля 1448
установка с опускающимся несущим канатом 5017
установщик подвесок 4114
устойчивая длительная производительность леса 5153
устойчивая к загниванию древесина 2294
устойчивая крупносерийная продуктивность 5147
устойчивая лесная продукция 8299
устойчивая структура системы доступа 6168
устойчивая точка завядания 6177
устойчивое лесоводство 8297
устойчивое лесопользование 8298
устойчивое лесоустройство 8296
устойчивое развитие леса 8295
устойчивость 6906, 8294
устойчивость дерева к неблагоприятным внешним воздействиям 8852
устойчивость к загниванию 2293
устойчивость лесных культур 3629
устойчивые виды 6187
устранение отбросов 7723
устранение растительности 2444
устройство для выгрузки леса
устройство для определения направления падения дерева 8567
устройство для отклонения кустарника 1103
устройство для очистки щепы 1512
устройство для поворота бревен 5124
устройство для поворота бревен впереди рамной лесопильной тележке 5124
устройство для погрузки леса 8607
устройство для подачи кряжей 5097
устройство для подхватывания кряжей 5119
устройство для продольной распиловки брёвен 9512
устройство для раскалывания древесины 9512
устройство для раскалывания пней 8214
устройство для удержания груза на транспортном средстве 716
устьице 8120
усушка древесины 7554
усыхание ветвей 985
утверждённый порядок лесопользования 3649
утёска 4273
утилизация древесины 9154
утолщение 8494
утолщение на стволе дерева 1182
утоп 2268
ухлёстывать 4812
уход за деревьями 8939
уход за древесиной 9676
уход за кроной 2095
уход за ландшафтом 4778
уход за лесом 8456
уход за лесонасаждениями 8456
уход за молодняком 8458
уход за отдельными деревьями 8454
уход за подроста 7360
уход за приростом 8457
уход за саженцами 8455
уход за устойчивости насаждения 8029
ухудшение леса 2241
участковый лесничий 1988

участок ксилемы между двумя сердцевинными лучами 9738
участок леса 1821
участок лесника 2542
участок лесничества 2542
участок лесосеки у пня 8213
участок отдельных деревьев 352
участок подсочки 6904
участок строевого леса 8604
участок флоэмы между двумя сердцевинными лучами 6208
учебное лесничество 8762
учебный лес 8761
учение о лесной таксации 8488
учение о продуктивности лесов 3650
учение о типах леса 3636
учёный-лесовод 6883
учёт 4579
учёт количества лесоматериалов 6631
учёт леса 3478
учёт на глаз 2869
учётная площадь 1/1000 акра 5449
учёт объёма лесозаготовок 6831
учёт ошибки 201
учёт пиломатериалов 5213
учёт рубок 6831
учётчик 8377
ушко медвежье 1759
ущерб, причинённый пожаром 3177

фагара крупнолистная 2713
фагара Хейтца 5895
фаза консолидации 1884
фаза первого цветения 70
фаза перед полным распусканием листьев 4679
фаза перед цветением 4679
фаза развития 6198
фаза роста и развития 2445
фактическая вырубка 57
фактический размер 59
фактор коры 557
фактор условий произрастания леса 3594
факторы потерь 5176
фамильная изгородь 3008
фамильная лесная плантация 3009
фамильное лесничество 3007
фанера для судостроения 5307
фанера из ядра древесины 9520
фанерная древесина 901
фанерное бревно 9193
фанерный брус 9193
фанерный кряж 9193
фанерный лист 9197
фанерный слой 9197
фанерование 9196
фанерофиты 6197
фартук 334
фасонирование 7491
фасонно-фрезерный станок 5577
фасциация 2987
фатом 3024
фаут дерева 8830
фаутная древесина 9118
фаутное дерево 7082
фаутный 7212
фаутный лесоматериал 9118
фаутный участок 2991
фаутный участок шпона 6103
фашина 3018
фашинная плотина 1117
фашинный лес 9572
феллема 1956
феллодендрон 6199
фенологический цикл 6202
фенология 6203
фенотип 6204
фенотипическая селекция 6205
фенотипическое элитное дерево 6396
ферма 9592
ферментационный горизонт 3086
ферментационный слой 3086
фермерский лес 3015
фиалковое дерево 4716
фиговое дерево 1774
фиговое дерево индийское 6533
фигурная стрижка 6612
фигурная стрижка кустов и деревьев 6612
физика древесины 8618
физиогномический класс 6216
физиогномический подкласс 6217
физиографическая единица 6219
физиографический класс 6218
фиксированная деревянная шпора 272
фикус 4469
фикус бенгальский 526
фикус клубковатый 1670
фикус мелколистный 7766
фикус религиозный 939
фикус священный 939
фикус эластичный 4469
филадельфовые 5502
филлантус кислый 5966
филлокладус 1405
филлосфера 6213
фильтр для древесной пыли 9540
филюм 6214
финансовая спелость 3146
финик канарский 1301
финиковая пальма 2257
финишная обработка 3150
фисташка мастичная 5327
фисташка обыкновенная 6295
фисташковое редколесье 6294
фисташник 6294
фителефас 4597
фитогормон 6221
фитосанитария 6222
фитотоксичность 6223
фитоценоз 6336
фитоценоз хвойного леса 5705
фитоцид 6220
флагообразная крона 3240
флакуртиевые 4467
флиндерсия австралийская 498
флиндерсия Брейли 6665
флора сплошной лесосеки 3039
флоэма 4502
форвардер 3674
форвардер для долготей 8867
форвардер для пней 8221
форестиера флоридская 3287
форма балки 841
форма дерева 7490
форма кроны 2078
форма лесохозяйства 3666
форма листа 4869
форма насаждения 3667
форма роста 4017
форма ствола 898
форма сучка 4735
формирование 7491
формирование дерева 8839
формирование древостоя 3661
формирование кроны 2079
формировочный такелаж 7467
формованное дерево 7489
формованный пиломатериал 6116
формовка 5100
формовочная стрижка крон

фотографическое определение объёма круглых лесоматериалов 6210
фоторазложение древесины 6209
фотосинтез 6211
фототропизм 6212
фрагментация лесов 3681
фракционатор древесной массы 3093
франклиния 3687
фреза 7070
фрезерная оправка 5452
фрезерно-обрезной станок 1527
фрезерно-пыльный станок 1527
фрезерный стружечный станок 5453
фрезерование 9609
фрикционная пила 3714
фронтальной челюстной погрузчик 3720
фронт лесосеки 3722
фронт пожара 1189
фруктовое дерево 5665
фуг пропила 7246
фумигант 3765
фунгицид 3771
функция объёма дерева 8956
фэйс-корд 2979
фюзен 3778

халезия 7800
халезия каролинская 1349
хамедафне прицветничковая 4898
хаменериум узколистный 3215
хамеропс низкий 5384
хамефит 1439
характер дерева 8943
характеристики роста 4022
характеристики спелого леса 5879
характер местности 4777
характер роста 4017
харигира 7436
хвоевидная пора 6287
хвоелистный 6259
хвойная древесина 5708
хвойная подстилка 5706
хвойник 1873, 2844
хвойное дерево 1877
хвойное масло 6276
хвойно-широколиственный лес 5484
хвойные 5708
поражённые синевой 859
хвойные породы 5708
хвойный 36
хвойный древесинник 8173
хвойный лес 1873
хвойный опад 5706
хворост 998
хворост в низкоствольнике 1938
хворост лиственных пород 4145
хворостяная гать 1119
хворостяная плотина 1117
хвостовой блок возвратного каната 4167
хвоя 37
хемотропизм 1470
хенгал 1472
хеномелес китайская 1498
хеномелес японская 4640
хермес веймутовой сосны 6245
хиеронима альхордиевидная 6237
химера 3820
химическая варка целлюлозы 1467
химическая древесина 44
химическая заболонная грибная коричневая окраска 1462
химическая защита леса 1465
химическая коричневая окраска 1462
химическая окорка 1463
химическая очистка поверхности почвы 1468
химический способ удаления сучьев 1466
химическое прореживание 1469
химическое разложение древесины 1464
хинное дерево 6669
хинона 6669
хищническая рубка 2935
хищническое хозяйство 2434
хлам от лесосеки 998
хлебное дерево 1011
хлебное дерево индийское 1011
хлебное дерево цельнолистное 1011
хлестанье 9359
хлопковое дерево 4688
хлопчатник барбадосский 7327
хлороз 1546
хлорозное карликовое растение 1547
хлоротическое карликовое растение 1547
хлыст 5154
хлыст с гнилью 7079
хлыст с обдирами коры 6134
хмелевая шишка 4364
хмелеграб 2900
хмелеграб виргинский 2719
хмелеграб обыкновенный 2900
хмель 4363
ход древоточца 4511
ход личинок 4811
ход насекомого в древесине 4511
ход рубки 6575
хозяйственная часть леса 3496
хозяйственный возраст 136
хозяйство 9723
хозяйство на быстрый рост 6724
хозяйство на кустарник 7543
хозяйство, направленное на уход за единичными деревьями 7657
хозяйство, направленное на уход за лучшими стволами 1566
хозяйство на световой прирост 5918
холларена обильноцветущая 3002
холодискус 6993
холодискус разноцветный 2009
холодостойкость 3724
холоцеллюлоза 4340
хопея 4365
хорда сегмента колотой древесины на торце 1630
хорошо высушенный 6584
хохлатый стриж 8942
хранение в воде 9301
хранение древесины 5120
хранение семян 7397
хранилище зелёного резерва 3960
хризофил 8037
хризофиллум 8037
хронология годичных колец 6958
хронология по ширине годичных колец 6958
хрупкая древесина 7535
хрупкая сердцевина 1039
хрупкость 1000
хрупкость древесины 1001
хрущ 3502
хрущик сосновый 6249
хуминовый слой

хурма 4628
хурма виргинская 1796
хурма восточная 4627
хурма кавказская 2258
хурма маврикийская 5344
хурма обыкновенная 2258
хурма плод 4628
хурма целебесская 3882
хурма эбеновая 2727
хурма японская 4627

цвести 3290
цвет древесины 9513
цветение 3291
цветение деревьев 847
цветень 6432
цвет спелой древесины 6965
цветущее дерево 848
цеанотус 1395
цеанотус американский 1265
цеанотус клинолистный 5660
цеанотус колючий 3939
цеанотус кроваво-красный 6793
цеанотус метельчатоцветный 856
цеанотус простёртый 5242
цеанотус яйцевиднолистный 6788
цедрела 243
цедрела австралийская 410
цедрела белая 7876
цедрела душистая 1411
цедрела китайская 1493
цедрела пильчатая 4309
цедрела раскалывающая 410
цедрела Тоона 410
цезальпиниевые 7439
цезальпиния высокая 5426
цезальпиния железная 1002
цезальпиния красильная 2550
цезальпиния саппан 7209
цейлонское эбеновое дерево 904
цекропия щитовидная 7519
целевая рубка 9046
целевой диаметр 3138
целевые деревья 5853
целиком ядровая древесина 184
целина 9408
цели развития 2447
целлюлоза 9504, 9603
целое дерево 3758
целостность водосбора 9297
целостность экосистемы 2748
целый хлыст 2834
цельная древесина 9393
листа 2833
цельный лист 2832
цена леса 9629
цена лесоматериалов в партии 6528
цена лесосек 8195
ценная древесина 3159
ценная древесная порода 9162
ценная конструкционная древесина 8184
ценная поделочная порода дерева 9161
ценная широколиственная древесина 9160
ценное строевое дерево 9161
центральная жилка листа 4879
центральный слой фанеры 1949
центральный средний ствол 1413
центральный ствол 1415
центрирование брёвна 5053
центролобиум мощный 338
цепная пила 1431
цепная пила с прямой пильной шиной 3251
цепная серьга 2616
цепная скарификация 2610
цепная сортировочная площадка 3289
цепная сучкорезная машина 1426
цепное срезающее устройство 2619
цепной грузоподъёмник 1424
цепной конвейер 1424
цепной крюк 8577
цепной лесотранспортёр 1424
цепной механизм ударного действия 4369
цепной окорочный станок 1425
цепной подтаскиватель 1427
цепной чокер 7798
цепь для прицепки брёвен 4173
цепь для скрепления бревен 8136
цепь для сталкивания брёвен 6808
цепь для торможения брёвна при трелёвки вниз по склону 2606
цепь для трелёвки брёвен 7798
цепь для удаления отходов 6808
цепь обеспечения сохранности проб 1429
цепь против скольжения 319
цератоцистис вязовый 2678
цербера мангас 3972
цереус 1733
церковный лес 1575
церкокарпус багульниколистный 2143
церцис гигантский 3843
церцис западный 1267
церцис канадский 2721
цефалотаксусовые 6392
цех по переработке древесины 9637
циатейные 2206
циатея 8837
цикл ведения лесного хозяйства 3490
цикл восстановления посадок 6874
циклическое сжигание 7075
цикл плотности 2391
цикл развития 4013
цикл роста 4013
циликодискус габунский 2397
цилиндрик древесины 4452
цилиндрическая пила 2209
цилиндрический бур 403
цилиндр стрелы 4932
цинометра 5653
ципре 6076
цирилловые 2221
циркульная пила 1585
циссус 8803
цистусовые 6991
цитокинины 2222
цитрон 1590
цитрус благородный 5274
цитрус китайский 5935
цитрус лиметта 9348
цитрусовая древесина 5937
цитрусовое дерево 1590
цитрус сетчатый 3948
цоколь 2223

чайное масло 8438
чайные 8434
чайный куст 8437
чамыш 7165
чапарель 1442
частичная вырубка 6083
частичное годичное кольцо 6087
частичное удаление хлама 6089
частично занятый вспомогательный рабочий 6095

частично окорённые лесоматериалы 9596
частично сросшийся сучок 6084
частичный посев 6088
частичный сбор в кучи 6086
частный лес 3473
частный лесной владелец 6555
частный лесовод-консультант 1890
частота возникновения пожаров 3173
частота деревьев 8843
частотное распределение пиловочных брёвен по диаметру 7248
часть пиловочника 7249
часть растения для вегетативного размножения 9183
часть сердцевинного луча 5390
часть ствола, повреждённая насекомыми 4708
чашечная пила 1853
чаща 4104
челюстной захват 5917
челюстной лесопогрузчик 8610
червецы 7278
червоточина 4511
червоточина, вызываемая жуком-рогохвостиком 9692
червоточина, вызываемая жуком-точильщиком 2279
червоточина диаметром более 3,2 мм 4032
червоточина овальной формы 5134
червоточная древесина 9729
червоточный 9730
чередование древесных пород 226
чередование лесного и сельского хозяйства 225
чередование насаждения 8032
чередующаяся обработка 7520
чередующиеся полосы леса, разрабатываемые по обе стороны несущего каната 978
черёмуха 1473, 2881
черёмуха азиатская 382
черёмуха виргинская 1769
черёмуха кистевая 2881
черёмуха Маака 5224
черёмуха поздняя 244
черенковать 3906
черенок для прививки 1145
черенок из старой древесины 4135
черенок мягколиственных пород 7829
черенок от ветви 976
чересполосная лесосека 224
чересполосное примыкание лесосек 1989
черешка птичья 8330
черешковый лист 6194
черешня 8330
черешок 4883
черешок листа 4883
черешчатый 6193
чёрная зона 801
чёрная линия 801
чёрная плесень 802
черная пятнистость 315
черника 855
черника блестящая 7523
черника волосистая 2327
черника кустарниковая 8926
черника миртолистная 5006
черника овальнолистная 5986
черника сахалинская 2327
черника тычиночная 2327
чернильная болезнь 4500
чернильная болезнь каштана европейского 4500
чернильный орешек 1279
черница 855
черновая обрезка 76
черновая обрезка сучьев 76
черновая очистка от сучьев 1683
чёрнодревесный 815
чёрное дерево 814, 2727
чёрное мангровое дерево 251
чёрное пятно 783
чёрное ядро 795
чёрноклён 8418
чёрнолесье 788
чернотал 630
чёрный домовый усач 4389
чёрный мангр 251
чёрный платан 781
чёрный сучок 800
чёртово дерево 2453
четверть длины окружности 6661
четырёхкантная штабелёвка 7973
четырёхкантное бревно 7971
четырёхкантный блок 7969
четырёхкантный брус 7969
чечевичка 4916
чешуевидная хвоя 7276
чешуйка почки 1143
чешуйчатая почка 7274
чешуйчатый лист 968
чешуйчатый недоразвитый лист 9075
чилига 7136
чилига степная 7136
чина многолетняя 5662
чингиль 7165
число деревьев в час 8930
число пород 5808
число стволов 8079
чистая масса 5719
чистая окорка 1603
чистая приведённая стоимость 5718
чистая схема снижения 5715
чистовая обработка 3150
чистовая окорка 3137
чистовое строгание 1608
чистовой пиломатериал 3164
чистовой размер 3163
чистое бревно 1626
чистое годовое изменение 5713
чистое живое семя 6636
чистое насаждение 2926
чистообрезная доска 7968
чистообрезной пиломатериал 3755
чисто окорённые балансы 1609
чисто радиальный пиломатериал 3755
чистый годовой прирост 5714
чистый доход от леса 5720
чистый древостой 2926
чистый прирост 5716
чистый пропил 1606
чистый среднегодовой прирост 437
чистый срез 1606
чистый ствол 1604
читола 8432
чищоба 390
членик сосуда 9212
чозения арбутилистная 5008
чозения крупночешуйчатая 5008
чозения толкнянколистная 5008
чокер 1555
чокер, заранее надетый на дерево 6518
чокерная цепь 7798
чокерной замок, скользящий по канату чокера 529

чокерный трос 1555
чокеровать 1552
чокеровка 1560
чокеровщик 1559
чокеровщик, сопровождающий пачку брёвен 1454
чрезмерная заболонь 2937
чрезмерная рубка 2935
чрезмерное пользование пастьбой 6001
чрезмерное разрастание 2936
чрезмерное распространение бука 74
чрезмерное распространение ели 75
чрезмерное смолотечение 2977
чрезмерный рост 2936
чубушник венечный 8335
чубушник кавказский 1386
чубушник мелколистный 5004
чувствительная порода 7443
чувствительность местообитания 7673
чувствительные почвы 7442
чувствительный водораздел 7444
чурак 2171
чурак твёрдолиственной породы 4142
чурбан 2171

шаг пильной цепи 7263
шаг подновки 1541
шайба 1336
шалаш для лесорубов 5071
шалаш на дереве 8856
шандал-дерево 251
шарнир 8350
шарнирное бревно 4313
шарнирное замковое устройство 8935
шарнирно-сочленённый с тягачом полуприцеп 371
шарнирный погрузчик 4740
шаровидная ель 500
шаровидная крона 3869
шаровницевые 3868
шахматная рубка леса 3064
шахтный лесоматериал 5463
шахтный отвал 5464
шварция 6056
шейка корня 7015
шелковица 8354
шелковица американская 6781
шелковица белая 1791
шелковица плакучая 9327
шелковица чёрная 803
шелковый дуб 3966
шелкопряд-монашенка 5809
шелкопряд сосновый 6258
шелкопряд сосновый крапчатый 6248
шелуха 4391
шелушение 8337
шелушиться 2940
шелюга жёлтая 9224
шелюга сибирская 2458
шерстяное дерево 4688
шерстяных 7596
шест 6418
шест для измерения массы 715
шефердия канадская 771
шефердия серебристая 7600
шефлера лучелистная 412
ширакаши 862
шизогенетический 7295
шизолизогенный 7296
шикот 1482
шикша чёрная 785
шиловидная хвоя 447
шиловидный лист 8245
шило для заплётки каната 5321
шина повышенной проходимости 5198
шина с грунтозацепами 5198
шина трелёвочного трактора 5086
шинус длиннолистный 5140
шинус терепентиннолистный 1004
шип 8524
шипик 7912
шиповник 2555
шиповник обыкновенный 2555
шиповое соединение 3161
шипорезная пила 4936
шипот 1482
шипы 1280
ширина годичного кольца 9398
ширина доски 885
ширина древесной частицы 9621
ширина кроны 2098
ширина пасоки 8320
ширина пропила 9400
ширина реза 9400
ширина резания 9400
ширина стружки 9399
широкая поверхность доски 1021
ширококронный 1045
широколиственная древесина 1057
широколиственная порода 1055
широколиственница 6441
широколиственное дерево 1048
широколиственный горный лес 9132
широколиственный лес 2300
широкорядный посев 1046
широкослойная древесина 5909
широкослойная текстура 1678
широкострочный посев 1046
шишка 8818
шишка хмеля 4364
шишковатый 4737
шишковидный 8181
шишкосборщик 1865
шишкосушилка 1862
школьный лес 8761
шлифовальный станок 6326
шлифовальный станок для обработки профильных поверхностей и кромок 6567
шлифование 7189
шлифовать 6133
шоколадное дерево 1248
шпала 9556
шпалера 2865
шпальная клетка 8560
шпальный кряж 8591
шпальный лесоматериал 8591
шпатлёвка для древесины 9606
шпил
со ~ем 7917
шпилевая древесина 7926
шпилевидный 7917
шпон для внутреннего слоя фанеры 1954
шпон из развилочной части брёвна 2060
шпон, полученный из пня 8219
шпунт и гребень 8676
шпунтовое соединение 8676
шрам 582
штабелёвка 4846
штабелёвка круглых лесоматериалов 3662
штабелёвка на лесном складе 6234
штабелёвка с выравниванием торцов и расположением прокладок заподлицо с торцами 962
штабелеукладка 4846
штабелеукладчик для досок

штабелированные
лесоматериалы в кордах
7982
штабелировать 7983
штабелировать 8115
штабелировать перекатыванием
8115
штабель высушиваемого
лесоматериала 7336
штабель длинномерных
лесоматериалов 5145
штабель древесины, уложенный
вдоль дороги 9447
штабель короткомерного
баланса 2979
штабель круглых
лесоматериалов 6026
штабель лесоматериалов 8619
штабельная прокладка 6236
штабель пиломатериалов 5211
штабель с плотной укладкой
лесоматериалов 8552
штабель хлыстов 8066
штабель широких досок
одинаковой ширины 8546
штамповочный картон 9493
штемпель качества 3900
штык
со ~ом 631
штыри с ушком 6690
шютте 6260
щёки 1460
щековая дробилка 186
щепа 9508
щепа из вершины 2067
щепа из древесных отходах
1538
щепа из ствольной части дерева
1539
щепа из целых деревьев 3760
щепа из части дерева 6096
щепа от ветви 975
щепа с содержанием коры 545
щепа стандартного размера
8006
щепки 9508
щетинконосный лист 7471
щётковидный побег 501
щитник красноногий 3343
щитовидная прожилка
древесины, пропитанной
живицей 4058
щитовки 7278

эбеновое дерево 102
эбеновое дерево из Бомбея 904
эвкалипт 2872
эвкалипт Бакстера 1091
эвкалипт Бауера 857
эвкалипт белодревесный 6771
эвкалипт белый 9369
эвкалипт близкий 5587
эвкалипт большой 3909
эвкалипт Босиста 932
эвкалипт Бриджеса 329
эвкалипт бугорчатый 4054
эвкалипт вальковиднорогий
3550
эвкалипт ванду 9265
эвкалипт Ганна 1578
эвкалипт голубовато-серый
4060
эвкалипт голубой 865
эвкалипт гроздевидный 7861
эвкалипт густопятнистый 7930
эвкалипт двухцветный 932
эвкалипт деглюнта 5412
эвкалипт длиннолистный 4583
эвкалипт душистый 1052
эвкалипт железистый 6779
эвкалипт железнодревесный
6779
эвкалипт западноавстралийский
4642
эвкалипт камальдульский 6786
эвкалипт камеденосный 2249
эвкалипт кистевидный 6779
эвкалипт коробочконосный
9145
эвкалипт косой 5412
эвкалипт краснодревесный
клейкий 7861
эвкалипт красный
веревчатокорый 6794
эвкалипт красный
железнокорый 6779
эвкалипт кроваводревесный
6786
эвкалипт кровавокрасный
6786
эвкалипт круглый 8417
эвкалипт крупноносый 6794
эвкалипт лимонный 4910
эвкалипт лучистый 5663
эвкалипт Майдена 4583
эвкалипт Мюллера 9765
эвкалиптовая древесина с
простым текстурным
рисунком 6318
эвкалиптовое масло 2874
эвкалиптовый лес 2873
эвкалипт окаменелый 4642
эвкалипт приречный 6786
эвкалипт прутовидный 5286
эвкалипт прутьевидный 5286
эвкалипт пятнистый 7930
эвкалипт разноцветный 4691
эвкалипт резиноносный 411
эвкалипт Робертсона 5663
эвкалипт с гладкой корой 1237
эвкалипт сердцевидный 7605
эвкалипт серопепельный 357
эвкалипт серый 357
эвкалипт смолоносный 411
эвкалипт с твёрдой корой 4583
эвкалипт тимоловый 1052
эвкалипт трёхцветковый 9765
эвкалипт Уилкинсона 7769
эвкалипт урновидный 9145
эвкалипт урноносный 9145
эвкалипт фиголистный 6773
эвкалипт царственный 3842
эвкалипт чёрный 780
эвкалипт шариковый 8417
эвкалипт шаровидный 8417
эвкалипт шероховатый 9387
эвкалипт широколистный 1052
эвкалипт яйцевидный 8310
эвкалипт ясеневидный 414
эвкоммиевые 2875
эводия 2932
эвтрофикация 2913
эдафические характеристики
2754
эйкоммия ильмолистная 4147
экваториальный дождевой лес
8978
экзоты 2944
эклектичное прореживание
2735
эковид 2747
экозона 2752
экологическая амплитуда 2740
экологическая ниша 2739
экологическая устойчивость
2736
экологическая целостность
2737
экологический баланс 2736
экологический вид 2747
экологически подходящий вид
2738
экологически уязвимая зона
2841
экологическое лесничество
2839
экологическое лесное хозяйство
2839

экономика лесного хозяйства 3391
экосистема леса 3395
экосистема леса, не пройдённая рубками ухода 9123
экосистема тропического леса 8976
экотип 4079
экотон 2751
экскаватор с обратной лопатой 468
эксплантат 2947
эксплантация 2948
эксплуатационная влажность 2816
эксплуатационная лесная земля 1746
эксплуатационная рубка 1744
эксплуатационная таксация 5926
эксплуатационные расходы лесопильного цеха 7254
эксплуатационный диаметр 2949
эксплуатационный лес 1745
эксплуатация лесных богатств без восстановления 8616
эксплуатация согласно заданным диаметром 9152
экспортируемая древесина 2958
экспорт леса 2957
экспортные лесоматериалы 2958
экстенсивное ведение лесного хозяйства 2963
экстенсивное лесоустройство 2963
экстрактивные вещества древесины 2972
экстракционная древесина 2971
экстракционная канифоль 8907
экстракционные вещества, содержащиеся в древесину 2972
экструзионная древесностружечная плита 2976
эксцентрический рост ствола 2733
эксцентричность годичных колец 6952
эксцентричность годичных слоёв 6952
эксцентричность ствола 8996
эктотрофная микориза 2753
эластичная древесина 7126
элементы деревянной каркасной конструкции 1335
элеокарповые 2021
элитное дерево 6396
элитное насаждение 2785
элитное семенное насаждение 2784
элитный семенной древостой 2784
элитный ствол 6395
эмбрион 2795
эмбриональная ткань 5402
эндемик 2802
эндемичные бактерии 5669
эндемичный вид 2802
эндогенный 2810
эндодерма 2809
эндосперм 2812
эндотрофная микориза 2813
энергетическая древесина 2820
энергетическая лесная плантация 2819
энергетическая щепа 2818
энергия от древесного топлива 9567
энергия прорастания 3834
энтандрофрагма 7200
энтандрофрагма ангольская 8541
энтандрофрагма Кандолли 4744
энтандрофрагма полезная 9150
энтенам 5405
энтеролобиум 2706
энтеролобиум саман 6701
эпизодический рост 2854
эпинастия 2851
эпифит 2853
эпифит на деревьях 8833
эпифлоэма 6157
эрециевые 2774
эреция заострённая 2730
эрика 1779
эрика сизая 3153
эритрина 1941
эритрина петушья 1698
эритрофлеум гвинейский 1366
эрозия 2861
эрозия борозд 4045
эрозия поверхностного слоя почвы 8285
эскулус тёмно-розовый 7130
эспинильо 1394
эстакада 5116
эстакада для подачи брёвен 4485
эстакада для подачи заготовок
этажерка 6373
эталонное дерево 6396
эталонный лес 6239
этикеты типов леса 3633
этнобиомедицинский заповедник 2871
эукоммия вязолистная 4147
эукрифиевые 2876
эфедра 2844
эфедровые 2845
эфедроидная перфорационная пластинка 2846
эфирное масло 2866
эффективность местообитания 4075
эффективный лесной сеянец 2769
эффект оранжереи 3947
эффект расчётной лесосеки 196
эхинопанакс 2451
эцезис 2734

юбея видная 1701
ювенильная древесина 4675
ювенильный лист 4678
южная жёлтая сосна 5141
южная сосна 5141
южный белый кедр 398
южный кипарис 786
юкка высокая 2224
юкка коротколистная 4604
юкка нитчатая 9792
юла 9594
юнит щепы 9095
ююба 1783

яблоневые 330
яблоня 1755
яблоня венечная 8331
яблоня дикая 9403
яблоня китайская 6127
яблоня культурная 1755
яблоня лесная 9403
яблоня многоцветковая 4619
яблоня парадизка 6062
яблоня сибирская 7565
яблоня сливолистная 6127
явор 8355
явор обыкновенный 8355
ягода волчья 2895
ягода медвежья 1759
ягода мыльная 7865
ягодники 685
ягоды 683
ядовитая древесина 6413
ядовитая порода 5805

ядровая древесная порода 4205
ядровый отлуп 2139
ядро кокосового ореха 6045
ядро твёрдолиственной древесины 4141
яйцевая ямка 2773
яйцевидный лист 5988
якаранда копая 1921
якорная лебёдка 281
якорная цепь 270
якорное скрепление 273
якорный корень 277
яма для посадки дерева 6353
ямчатая гниль 6313
ямчатая ядровая напенная и корневая гниль 6766
японский кедр 4615
японский перец 4630
ярко-зелёная древесина дуба 3951
яровизация 9203
ярра 4642
ярус 8133
ярус брёвен 8545
ярус в лесном фитоценозе 8360
ярус древостоя 8866
ярус криволесий и стлаников 2783
ярусная крона 8134
ярусная структура 4842
ярусное насаждение 4841
ярусность 8147
ярусные элементы 8131
ярусный 8130
ярус травянистых растений 3116
ясеневый лес 380
ясеневый лубоед 379
ясенелистный 2170
ясень американский 261
ясень бархатный 9188
ясень белый 261
ясень высокий 1757
ясень горный 4741
ясень двулепестковый 9032
ясень европейский 1757
ясень Зибольда 7584
ясень каролинский 1342
ясень китайский 1490
ясень ланцетный 2255
ясень ланцетолистный 2255
ясень маньчжурский 5269
ясень мексиканский 680
ясень неправильный 7641
ясень носолистный 4741
ясень обыкновенный 1757
ясень остролистный 5659
ясень пенсильванский 2255
ясень пушистый 2255
ясень сплошь волосистый 487
ясень техасский 8482
ясень томентоза 6634
ясень цветочный 3292
ясень чёрный 777
ясень четырёхгранный 854
ясень чихуахуа 1484
ясменник 9641
ячейка прорубки 8508
ящичная древесина 954